MATHEMATICS
for
JEE Main and Advanced

Volume 2

SAS Kirmani
Chairman
Erose Educational Infotech Private Limited
New Delhi

Kumar Supravin
Senior Consultant
Haryana Water Resources Authority
Chandigarh

Nishu Kumar Govil
Professor, IIT JEE Mathematics
Physics Wallah
Noida

MATHEMATICS FOR JEE MAIN AND ADVANCED – VOLUME 2

UNIVERSITIES PRESS (INDIA) PRIVATE LIMITED

Registered office
3-6-747/1/A & 3-6-754/1, Himayatnagar, Hyderabad 500 029, Telangana, India
info@universitiespress.com; www.universitiespress.com

Distributed by
Orient Blackswan Private Limited

Registered office
3-6-752 Himayatnagar, Hyderabad 500 029, Telangana, India

Other offices
Bengaluru / Chennai / Guwahati / Hyderabad
Kolkata / Mumbai / New Delhi / Noida / Patna

First published 2026

502279

ISBN: 978-93-49750-04-3

Typeset in Times NR MT Std 10 *by*
SRS Technologies, Puducherry

Printed in India by
B.B Press, Tronica City, Ghaziabad, U.P. 201 103

Published by
Universities Press (India) Private Limited
3-6-747/1/A & 3-6-754/1, Himayatnagar, Hyderabad 500 029, Telangana, India

Preface

The journey of mastering Mathematics for competitive examinations such as JEE Main and Advanced demands a unique blend of conceptual clarity, analytical thinking, and rigorous practice. With this vision, we are pleased to present Volume 1 and Volume 2 of our Mathematics series, meticulously crafted to serve as a comprehensive resource for students preparing for both school-level Board Examinations (Classes 11 and 12) and competitive entrances alike.

Volume 1 consists of 13 chapters, aligned with the Class 11 syllabus, while Volume 2 includes 12 chapters, mapped to the Class 12 curriculum. Together, these volumes provide full coverage of the Mathematics syllabus prescribed by most education boards, as well as the latest pattern and expectations of the JEE Mains and Advanced examinations.

Each chapter in both volumes is structured with the following pedagogical elements:

- **Theory with Illustrations:** Concepts are introduced in a clear, student-friendly manner and reinforced with well-chosen illustrative examples to enhance comprehension.
- **Solved Examples:** A wide array of fully solved examples bridge the gap between theory and problem-solving techniques.
- **Recapitulation and Formulae:** A quick recap section aids in last-minute revision and consolidation of important results and formulae.
- **Worksheets and Quizzes:** Carefully designed to make students more reflexive and responsive, these exercises help to sharpen their application skills in a time-bound environment.

The Exercise Section in each chapter is a hallmark of this series, containing a diverse set of problems categorized and graded in accordance with the latest trends seen in JEE Main and JEE Advanced. The structure ensures that students develop both speed and accuracy, two crucial parameters for success in competitive exams.

To assist students further, Hints and Complete Solutions have been provided for all questions, enabling self-assessment and independent learning.

Our goal through this book is not merely to prepare students for an examination but to ignite a love for mathematics, nurturing confidence and clarity that goes beyond academic boundaries.

We extend our sincere gratitude to all those who have contributed to this endeavour specially Mr Tariq Habib, Former Head of the Department of Mathematics, GD Goenka Public School, Delhi, and Former Principal, Jamia University School, Delhi, for helping in the final corrections. We are thankful to Mr Akshay and Md Anas for their transcription and the PATS India team for their proof-reading and editing .

We are indebted to our family members, colleagues and friends who helped us to bring this book to fruition. Finally, special thanks and recognition goes to Mr Thomas Mathew Rajesh, Mr Kalol Das and Mr Ramesh M R of University Press (India) Private Limited for their diligence and attention to detail.

We welcome feedback and suggestions from teachers, students, and parents to improve this work further in subsequent editions.

With best wishes for your mathematical journey ahead,

S A S Kirmani
Kumar Supravin
Nishu Kumar Govil

About the Authors

SAS Kirmani is the Founder of Erose Educational Infotech Private Limited, a pan-India Educational and Research Company headquartered in New Delhi. He founded Erose in 1992 and is the chairman of the organisation. Under his leadership and vision, Erose has successfully guided and trained aspirants for competitive examinations, IIT-JEE, Medical and IAS. Kirmani articulated, designed and implemented the Global Education and Training model to prepare students for various entrance examinations, which has become a huge success. Under the model, he teaches students to focus on answering the maximum number of questions in optimal time by using the concepts for both straightforward application and lateral thinking.

An electrical engineer from HBTI Kanpur (renamed as HBTU Kanpur in 2016), Kirmani is on the advisory board of many schools and educational institutions to provide academic and administrative support. He was a member of the Draft Formulation Committee of Computer Literacy Awareness Program (CLASS) jointly sponsored by NCERT and Ministry of Education in the year 1984. In his initial years, he worked as senior physics faculty at Delhi Public School, R K Puram, New Delhi.

Kumar Supravin obtained his B.Tech. degree and a Ph.D. in Civil Engineering from IIT Delhi. With a wealth of experience gained over the years in teaching students the techniques of preparing for IIT JEE, he shares his subject knowledge with the aspirants at Erose Infotech to help them face the exams with confidence. His philosophy centres on fostering a deep understanding of fundamental concepts rather than relying solely on rote memorization. A former member of the IAS and IFS cadre, he is currently Senior Consultant with the Haryana Water Resources Authority, Chandigarh.

Nishu Kumar Govil is an engineering graduate from Aligarh Muslim University. His childhood flair for mathematics later transformed into a passion for teaching the subject. His unique pedagogy and out-of-the-box techniques for problem-solving have helped a large number of students clear JEE for over the past two decades. He has been associated with some of the premier coaching institutes in India, including Erose and FIITJEE. As a senior faculty member at Erose, he has also served internationally, teaching at the Indian Embassy School in Dammam and Delhi Public School, Riyadh, Saudi Arabia. He also taught as senior faculty member of Erose for the Edusat Punjab Project and was part of academia that assisted in educational programmes on Toppers TV. He also served as a Consulting Editor with TV18 Educational Channel, shaping content for a wider student audience. He currently teaches mathematics for IIT JEE aspirants at Physics Wallah, Noida.

Contents

CHAPTER 1

Relations and Functions

Objectives

In this chapter on Relations and Functions, we shall study the following topics:

1. Types of relation
2. Definition of function
3. The difference between relations and functions
4. Types of functions
5. Graph of functions.

RELATIONS

Types of Relation

Reflexive: A relation is called **Reflexive** if $\forall a \in A$, aRa or $(a, a) \in R$.

Symmetric: A relation R on a set A is defined as a symmetric relation if $(a, b) \in R \Rightarrow (b, a) \in R$ That is, $aRb \Rightarrow bRa$, where $a, b \in A$.

Anti-symmetric: A relation is called anti-symmetric if whenever aRb and bRa then $a = b$.

Transitive: A relation is called transitive, if whenever aRb and bRc exist then a is related to c. In other words, if $(a, b), (b, c) \in R$ then (a, c) should also belong to R.

ILLUSTRATION 1.1

If A = {1, 2, 3, 4}, define relations on A which have properties of being:

(a) reflexive, transitive but not symmetric.

(b) symmetric but neither reflexive nor transitive.

(c) reflexive, symmetric and transitive.

Solution

Given that, A = {1, 2, 3}.

(i) Let R_1 = **{(1, 1), (1, 2), (1, 3), (2, 3), (2, 2), (1, 3), (3, 3)}**

R_1 **is reflexive as (1, 1), (2, 2) and (3, 3) lie is** R_1.

R_1 **is transitive as (1, 2)** $\in R_1$, **(2, 3)** $\in R_1$ $\Rightarrow (1, 3) \in R_1$

Now, $(1, 2) \in R_1 \Rightarrow (2, 1) \notin R_1$.

(ii) Let R_2 = **{(1, 2), (2, 1)}**

Now, $(1, 2) \in R_2$, $(2, 1) \in R_2$.

So, it is symmetric, and clearly R_2 **is not reflexive**, as $(1, 1) \notin R_2$.

Also, R_2 **is not transitive as (1, 2)** $\in R_2$, $(2, 1) \in R_2$ **but (1, 1)** $\notin R_2$.

(iii) Let R_3 = **{(1, 1), (1, 2), (1, 3), (2, 1), (2, 2), (2, 3), (3, 1), (3, 2), (3, 3)}**

R_3 **is reflexive as (1, 1) (2, 2)** and **(3, 3)** $\in R_1$.

R_3 **is symmetric as (1, 2), (1, 3), (2, 3)** $\in R_1$

$\Rightarrow$ $(2, 1), (3, 1), (3, 2) \in R_1$.

Therefore, R_3 is reflexive, symmetric and transitive.

If a relation is reflexive, symmetric and transitive then it is called **Equivalence relation**.

ILLUSTRATION 1.2

Let n be a fixed positive integer. Define a relation R in Z as follows: $\forall a, b \in Z$, aRb if and only if $a - b$ is divisible by n. Show that R is an equivalence relation.

Solution

Given $\forall a, b \in Z$, aRb if and only if $a - b$ is divisible by n.

Now, for $aRa \Rightarrow (a - a)$ is divisible by n, which is true for any integer a as '0' is divisible by n.

Thus, R is reflexive.

Now, aRb

So, $(a - b)$ is divisible by n.

$\Rightarrow$ $-(b - a)$ is divisible by n.

$\Rightarrow$ $(b-a)$ is divisible by n.

$\Rightarrow$ bRa

Thus, R is symmetric.

Let aRb and bRc.

Then, $(a-b)$ is divisible by n and $(b-c)$ is divisible by n.

So, $(a-b)+(b-c)$ is divisible by n.

$\Rightarrow$ $(a-c)$ is divisible by n.

$\Rightarrow$ aRc

Thus, R is transitive.

So, R is an equivalence relation.

A relation R that is Reflexive, Antisymmetric and Transitive is called an **Order** or **Partial order.**

Inverse Relation

Let $R \subseteq A \times B$ be a relation from A to B.

The inverse relation of R (denoted by R^{-1}) is a relation from B to A defined as

$$R^{-1} = \{(b, a) : (a, b) \in R\}$$

If $(a, b) \in R$, then $(b, a) \in R^{-1}, \forall a \in A, b \in B$.

Domain of R^{-1} = Range of R

Range of R^{-1} = Domain of R

$$(R^{-1})^{-1} = R$$

ILLUSTRATION 1.3

A relation R is defined on the set Z of integers as follows:

$$(x, y) \in R \Leftrightarrow x^2 + y^2 = 25$$

Express R and R^{-1} as the sets of ordered pairs and hence find their respective domains.

Solution

We have,

$$(x, y) \in R \Leftrightarrow x^2 + y^2 = 25 \Leftrightarrow y = \pm\sqrt{25 - x^2}$$

Let $x = 0 \Rightarrow y = \pm 5$

$$(0, 5) \in R \text{ and } (0, -5) \in R$$

$$x = \pm 3 \Rightarrow y = \sqrt{25-9} = \pm 4$$

$$(3, 4) \in R, (-3, 4) \in R, (3, -4) \in R \text{ and } (-3, -4) \in R$$

$$x = \pm 4 \Rightarrow y = \sqrt{25-16} = \pm 3$$

$$(4, 3) \in R, (-4, 3) \in R, (4, -3) \in R \text{ and } (-4, -3) \in R$$

$$x = \pm 5 \Rightarrow y = \sqrt{25-25} = 0$$

$$(5, 0) \in R \text{ and } (-5, 0) \in R$$

We also find that for any other integral value of x, y is not an integer.

$$R = \{(0,5), (0,-5), (3,4), (-3,4), (3,-4), (-3,-4), (4,3), (-4,3), (4,-3), (-4,-3), (5,0), (-5,0)\}$$

$$R^{-1} = \{(5,0), (-5,0), (4,3), (4,-3), (-4,3), (-4,-3), (3,4), (3,-4), (-3,4), (-3,-4), (0,5), (0,-5)\}$$

Clearly, domain $(R) = \{0, 3, -3, 4, -4, 5, -5\}$ = domain (R^{-1})

Identity Relation

The identity relation on a set A is the set of ordered pairs belonging to $A \times A$ and is denoted by I_A.

$$I_A = \{(a, a) : a \in A\}$$

i.e., every element of A is related to only itself.

R is an identity relation if $(a, b) \in R$ iff $a = b$, $a \in A, b \in A$.

$$I_A^{-1} = I_A$$

Domain of I_A = Range of I_A = A

Universal Relation

If A be a set and R is the set $A \times A$, then R is called universal relation of A.

If $R = A \times A$, then R is universal relation in A.

Void Relation

ϕ is called the empty or void relation if $\phi \subseteq A \times A$.

ILLUSTRATION 1.4

ϕ (Null set) for A = {1, 2}

(i) It is not reflexive as $1 \not R 1$ or $\not R 2$. (Alternatively, we can write $(1, 1) \notin \phi$ or $(2, 2) \notin \phi$)

(ii) It is symmetric as we cannot find a condition where aRb but $b \not R a$.

Misconception 1

(i) A common mistake is to consider this relation as non-symmetric because of non-availability of aRb for all $a, b \in A$. It should always be remembered that a violation of the condition that aRb but $b \not R a$ can only prove that the relation is not symmetric)

(ii) It is transitive as we cannot find three elements a, b and c such that aRb, bRc but $a \not R c$.

ILLUSTRATION 1.5

Let R be a relation on Integers (Z)

$R = \{(x, y): xRy \text{ if } x - y \text{ is divisible by } 3\}$

(i) It is reflexive as $\forall\, a \in R$, $a - a = 0$, therefore divisible by 3.

(ii) It is symmetric as whenever aRb

$\Rightarrow \quad a - b = 3I \Rightarrow b - a = -3I = 3(-I)$,

therefore, $\quad bRa$.

where I is an Integer.

(iii) It is transitive as whenever aRb and bRc

$\Rightarrow a - b = 3k_1,\ b - c = 3k_2$

where $k_1, k_2 \in I$

(i.e., k_1, k_2 are integers)

$\Rightarrow \quad a - c = a - b + b - c = 3(k_1 + k_2)$

$\Rightarrow \quad a - c$ is divisible by 3.

Therefore, Relation R is an **Equivalence Relation.** Moreover, three subsets of Z are thus formed.

$$Z_1 = \{\ldots -3, 0, 3, 6 \ldots\}$$

$$Z_2 = \{\ldots -2, 1, 4, 7 \ldots\}$$

$$Z_3 = \{\ldots -1, 2, 5, 8 \ldots\}$$

These subsets are called **Equivalence Classes** because:

(a) In each of Z_1, Z_2 and Z_3, all the elements are related to each other (in individual sets.)

(b) No element of one set is related to any element of another set.

(c) Z_1, Z_2, Z_3 are disjoint sets, and union of these three is equal to set of integers Z.

ILLUSTRATION 1.6

Two points A and B in a plane are related if OA = OB, where O is fixed point. This relation is

(a) partial order relation.

(b) equivalence relation.

(c) reflexive but not symmetric.

(d) reflexive but not transitive.

Solution

Obviously, the relation is reflexive.

Also, for any two points A and B in the plane OA = OB $\Rightarrow$ OB = OA. Therefore, the relation is a symmetric relation. Also, for any three points A, B and C in the plane

$$OA = OB \text{ and } OB = OC \Rightarrow OA = OC$$

$\therefore$ The relation is transitive relation.

$\therefore$ The relation is equivalence relation.

ILLUSTRATION 1.7

In the set $N \times N$, the relation R is defined by $(a, b) R(c, d) \Leftrightarrow ad = bc$. Then R is

(a) partial order relation.

(b) equivalence relation.

(c) reflexive and transitive but not symmetric.

(d) symmetric and transitive but not reflexive.

Solution

The relation is defined on $N \times N$. Since the elements of $N \times N$ are ordered pairs, the relation on $N \times N$ relates two ordered pairs. Now,

(i) $\forall a, b \in N\ ab = ba$

$\therefore \quad (a, b)R(a, b) \forall (a, b) \in N \times N$.

Thus, R is reflexive

(ii) Let $(a, b)R(c, d) \Leftrightarrow ad = bc\ \forall (a, b)$ and $(c, d) \in N \times N$

$\Rightarrow \quad bc = ad \Rightarrow cb = da$

$\Rightarrow \quad (c, d)R(a, b)$.

(iii) Again, let $(a, b)R(c, d)$ and $(c, d)\ R\ (e, f)$

$\Rightarrow \quad ad = bc$ and $cf = de$

$\Rightarrow \quad adcf = bcde \Rightarrow af = de \Rightarrow (a, b)R(e, f)$

Thus, R is transitive.

Thus, R is equivalence relation.

WORKSHEET – 1

1. Let $A = \{1, 2, 3, \ldots, 15, 16\}$. Let R be the equivalence relation on $A \times A$ defined by $(a, b)R(c, d)$ if and only if $ad = bc$. Find the equivalence class $[(1, 3)]$.

2. Show that the relation R in the set $A = \{1, 2, 3, 4, 5\}$ is given by

$R = \{(a, b) : |a - b| \text{ is even}\}$, is an equivalence relation. Show that all the elements of $\{1, 3, 5\}$ are related to each other and all the elements of {2, 4} are related to each other, but no element of {1, 3, 5} is related to any element of {2, 4}.

3. Show that the relation R in the set A of points in a plane given by R = {(P, Q): Distance of the point P from the origin is same as the distance of the point Q from the origin}, is an equivalence relation. Further, show that the set of all points related to a point P ≠ (0, 0) is the circle passing through P with origin as centre.

4. Show that the relation R defined in the set A of all triangles as R = {(T_1, T_2): T_1 is similar to T_2}, is equivalence relation. Consider three right-angle triangles T_1 with sides 3, 4, 5; T_2 with sides 5, 12, 13; and T_3 with sides 6, 8, 10. Which triangles among T_1, T_2 and T_3 are related?

5. Show that the relation R defined in the set A of all polygons as $R = \{(P_1, P_2): P_1 \text{ and } P_2$ have same number of sides$\}$, is an equivalence relation. What is the set of all elements in A related to the right-angle triangle T with sides 3, 4 and 5?

6. Let A = {*a*, *b*, *c*} and the relation R be defined on A as follows:
   R = {(*a*, *a*), (*b*, *c*), (*a*, *b*)}.
   Then, write the minimum number of ordered pairs to be added in R to make R reflexive and transitive.

7. Let R be relation defined on the set of natural number N as follows:
   R = {(*x*, *y*): $x \in N$, $y \in N$, $2x + y = 41$}. Find the domain and range of the relation R. Also verify whether R is reflexive, symmetric and transitive.

8. Each of the following defines a relation on N:
   (i) *x* is greater than *y*; $x, y \in N$
   (ii) $x + y = 10$; $x, y \in N$
   (iii) *xy* is square of an integer; $x, y \in N$
   (iv) $x + 4y = 10$; $x, y \in N$.
   Determine which of the above relations are reflexive, symmetric and transitive.

FUNCTIONS

Definition of Function

Function can be easily defined with the help of the concept of mapping. Let A and B be two non-empty sets. A function *f* of A into B (or from A to B), written as $f: A \to B$, assigns to each element of set A, one and only one element of set B. For example, when we study circles, if we take area as *y* and radius as *x*, we have $y = \pi x^2$, which is a rule (correspondence) that tells us to calculate a unique (single) output value of *y* for each possible input value of the radius *x*.

Note:

(i) $f: A \to B$ is a function if each element *x* in A has a unique image (corresponding element) $f(x)$ in B.

(ii) $f: A \to B$ is not a function if there is an element in A which does not have an *f*-image in B.

(iii) $f: A \to B$ is not a function if there is an element in A which has more one *f*-image in B (functions can't be multi-valued.)

ILLUSTRATION 1.8

Is $g = \{(1,1), (2,3), (3,5), (4,7)\}$ a function? If *g* is described by $g(x) = \alpha x + \beta$, then what value should be assigned to α and β?

Solution

Given, $g = \{(1,1), (2,3), (3,5), (4,7)\}$

We will examine all the three tests given in the note above.

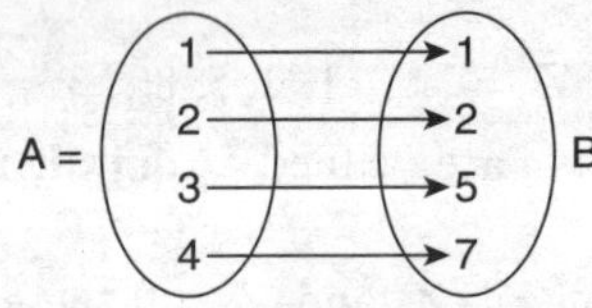

(i) Each element in A has unique image in B.

(ii) Each element of A has image in B (i.e., no element in A is left out.)

(iii) No element in A is related to two or more elements in B.

It is seen that every element of the domain has a unique image. So, *g* is a function.

Now, also given that $g(x) = \alpha x + \beta$

So, we have

$$g(1) = \alpha(1) + \beta = 1$$

$$\alpha + \beta = 1 \quad \text{...(i)}$$

and $g(2) = \alpha(2) + \beta = 3$

$$2\alpha + \beta = 3 \quad \text{...(ii)}$$

Solving (i) and (ii), we have

$$\alpha = 2 \text{ and } \beta = -1$$

Therefore, $g(x) = 2x - 1$

ILLUSTRATION 1.9

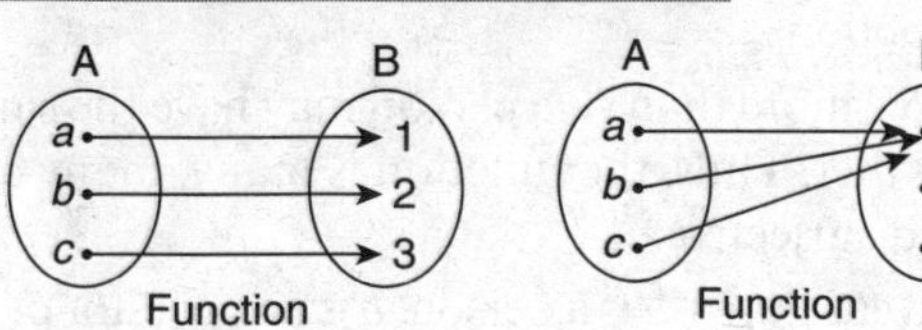

Function Function

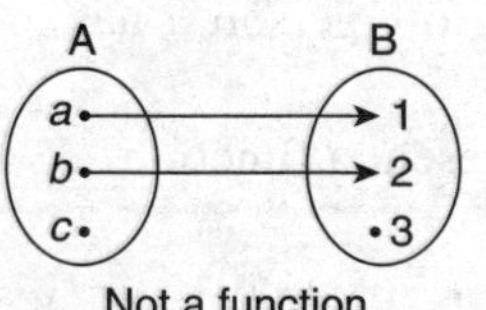

Not a function

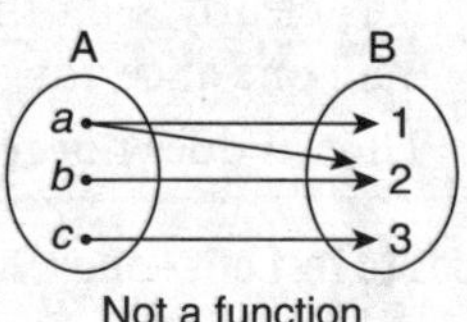

Not a function

Discussion

(i) Let f: R+ $\rightarrow$ R where $y = \pm\sqrt{x}$. This can't be considered a function as each $x \in R^+$ would have two images namely $\pm\sqrt{x}$. Thus, it would not be a function.

(ii) Let f : R $\rightarrow$ R where $y = x^3$. Here, for each $x \in R$ we would have a unique value of y in set R (as cube of any two distinct real numbers is distinct). Hence, it would represent a function.

Let f: X $\rightarrow$ Y be a function. In general, set X and Y could be any arbitrary non-empty sets. But at this level we would confine ourselves only to real-valued functions.

Sets 'X' is called domain of the function 'f'. Set 'Y' is called the co-domain of the function 'f'. The set of images of different elements of set X is called the range of the function 'f'. It is obvious that range could be subset of co-domain as we may have a few elements in the co-domain which are not the images of any element of the set X. Range is also called domain of variation.

Domain of function 'f' is normally represented as Domain (f). Range is represented as Range (f).

ILLUSTRATION 1.10

Find domain of the following functions:

(i) $y = \sqrt{x-1}$ **(ii)** $y = \sqrt{x^2 - 3x + 2}$

(iii) $y = \dfrac{1}{\sqrt{x^2 - 3x + 2}}$ **(iv)** $y = \sqrt{\log_{10} x}$

Solution

(i) For $y \in$ R, $x - 1 \geq 0$

$\Rightarrow$ $x \geq 1$

Domain $x \in [1, \infty)$

(ii) $y = \sqrt{(x-1)(x-2)}$

for $y \in$ R, $(x - 1)(x - 2) \geq 0$

$x \in (-\infty, 1] \cup [2, \infty)$

(iii) $y = \dfrac{1}{\sqrt{(x-1)(x-2)}}$

$(x - 1)(x - 2) > 0$

$x \in (-\infty, 1) \cup (2, \infty)$

(iv) $y = \sqrt{\log_{10} x}$

$\log_{10} x \geq 0 \Rightarrow x \geq 1$

$x \in [1, \infty)$

Note:

For finding the **range** there is no specific way. We have to go into the graphical details of the function to ascertain the range.

ILLUSTRATION 1.11

Find the range of the following functions:

(i) $y = \dfrac{1}{x^2 + 1}$ **(ii)** $y = 2 + \sin x$

(iii) $y = e^{-|x|}$

Solution

(i) Since $x^2 + 1 \geq 1$

Therefore $\dfrac{1}{x^2+1}$ will be less than or equal to 1.

Also $\dfrac{1}{x^2+1}$ will never be zero. $y = \dfrac{1}{x^2+1}$ $y \in (0, 1]$

(ii) Since $-1 \leq \sin x \leq 1$

$1 \leq 2 + \sin x \leq 3$

Range of $f(x) = 2 + \sin x$ is [1, 3]

(iii) $y = e^{-|x|}$

When $x \geq 0$ $y = e^{-x}$

When $x < 0$ $y = e^{x}$

The curve of $y = f(x)$ is varies from 1 to 0.

Hence, range of $f(x)$ is [1, 0).

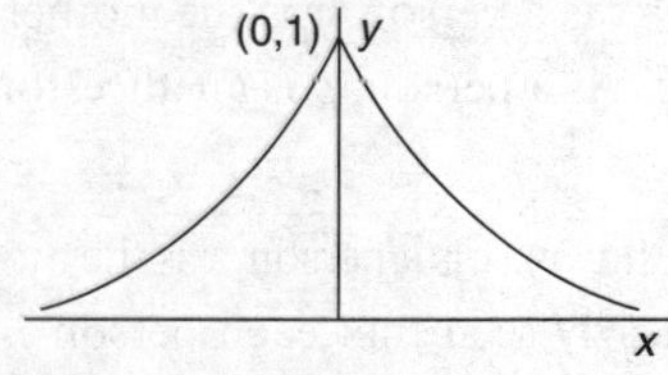

The Difference Between Relations and Functions

A relation R from a set A to a set B is a subset of A × B, the Cartesian product of A and B.

A function is a special case of a relation. A relation f from a set A to a set B is said to be a function if every element of set A has one and only one image in set B. Therefore, a function f from A to B is a subset of A × B such that (i) every element of A is necessarily the first element of some ordered pair (ii) no two ordered pairs have the same first element.

Every function is a relation but every relation is not necessarily a function.

Types of Function

One–One (Injective) Function

A function $f(x)$ is one-to-one on a domain D if $f(x_1) \neq f(x_2)$ whenever $x_1 \neq x_2$ or $f(x_1) = f(x_2)$ whenever $x_1 = x_2$.

Two different values of x do not have the same value of y.

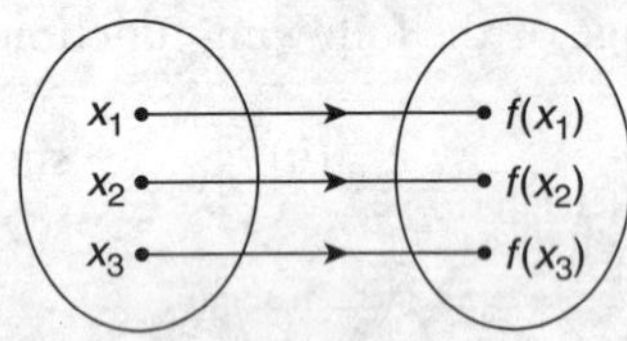

ILLUSTRATION 1.12

(i) $f(x) = 2x + 3$ where $x \in R$

$$\therefore f(x_1) = f(x_2);\ x_1, x_2 \in R$$
$$x_1 = x_2$$

∴ The function is one-to-one.

(ii) $f(x) = \sqrt{x}$ is one-to-one on any domain of non-negative numbers because $\sqrt{x_1} \neq x_2$ whenever $x_1 \neq x_2$.

(iii) $g(x) = \cos x$ is not one–one because for many different values of x we get the same value of $g(x)$

Are the following set of ordered pairs functions? If so, examine whether the mapping is injective or surjective.

(i) $\{(x, y)$: x is a person, y is the mother of $x\}$.

(ii) $\{(a, b)$: a is a person, b is an ancestor of $a\}$.

Solution

(i) Given, $\{(x, y)$: x is a person, y is the mother of $x\}$

It is clearly seen that each person x has only one biological mother.

Hence, the above set of ordered pairs make a function.

Now more than one person may have the same mother. Thus, the function is many–many one and surjective.

(ii) Given, $\{(a, b)$: a is a person, b is an ancestor of $a\}$

It is clearly seen that any person a has more than one ancestor.

Thus, it does not represent a function.

Methods to test one–one

(a) Analytically: If $x_1, x_2 \in A$, then

$$f(x_1) = f(x_2) \Rightarrow x_1 = x_2$$

or equivalently,

$$x_1 \neq x_2 \Rightarrow f(x_1) \neq f(x_2)$$

This is also a reminder of the standard concept A ⇒ B can be written as

$$\text{Not B} \Rightarrow \text{Not A}$$

(b) Graphically: If any line parallel to the x-axis cuts the graph of the function at most at one point, then the function is one–one.

(c) Monotonically: If any function is entirely increasing or decreasing in the whole domain, then $f(x)$ is one–one.

> **Tip**
> To find one–one or injective mapping, we mainly use Graphical or Monotonic method.

Number of one–one mappings

Let $f: A \to B$ be a mapping such that A and B are finite sets having m and n elements respectively, $(n > m)$.

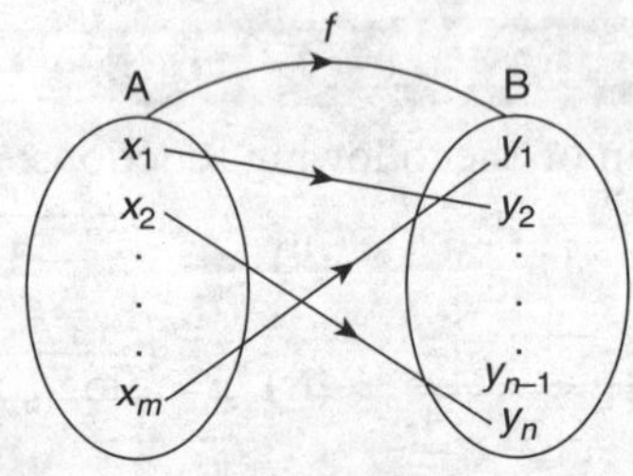

$$= \begin{cases} {}^nP_m & n \geq m \\ 0 & n < m \end{cases} = \begin{cases} {}^nC_m \cdot m! & n \geq m \\ 0 & n < m \end{cases}$$

> **Tip**
> All functions that are either strictly increasing or decreasing in their domain are one–one functions.

ILLUSTRATION 1.13

Prove that the mapping $f : A \to B$ given by $f(x) = 2x + 5$ is one-one.

Solution

Let us consider set A = {−1, 2, 4} and then B = {3, 9, 13}, i.e., two different values of x do not have same values of $f(x)$.

Hence, $f(x)$ is one-one.

I. Analytical Method

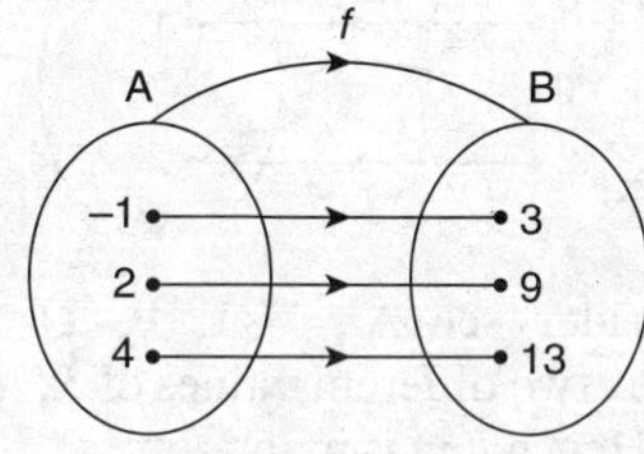

Let $x_1, x_2 \in A$

then $f(x_1) = f(x_2) \Rightarrow 2x_1 + 5 = 2x_2 + 5$

$\Rightarrow \quad 2x_1 = 2x_2 \Rightarrow x_1 = x_2$

or if $x_1 \neq x_2$

$\Rightarrow \quad 2x_1 \neq 2x_2$

$\Rightarrow \quad 2x_1 + 5 \neq 2x_2 + 5$

$\Rightarrow \quad f(x_1) \neq f(x_2)$

Hence, $f(x)$ is one–one.

II. Graphical Method

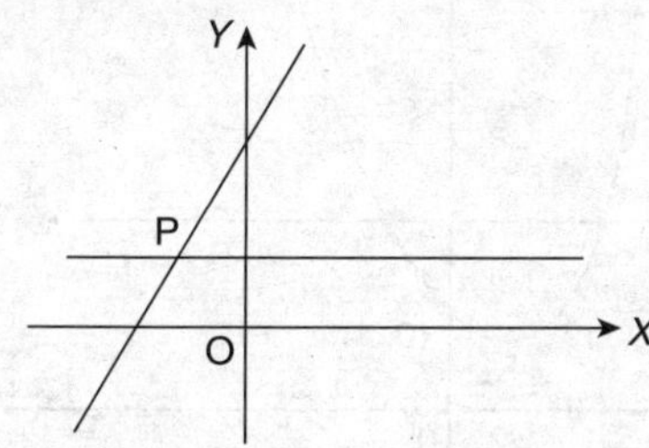

First draw the graph of $y = 2x + 5$

Since any line parallel to x-axis cuts the graph at one point P hence $f(x)$ is one–one.

III. Monotonically

Since $f(x) = 2x + 5$

$\therefore \quad f'(x) = 2 > 0$

$\Rightarrow \quad f(x)$ is strictly increasing function.

Hence, $f(x)$ is one–one.

Note: This method can be better understood after learning increasing–decreasing functions in Chapter 6 on application of derivatives.

ILLUSTRATION 1.14

Let $f(x) = ax^3 + bx^2 + cx + d\sin x$, then find the condition that $f(x)$ is always one–one function.

Solution

Given $f(x) = ax^3 + bx^2 + cx + d\sin x$

Now for $f(x)$ to be one–one, either f is entirely increasing or entirely decreasing.

$$\therefore \quad f'(x) = 3ax^2 + 2bx + c + d\cos x \geq 3ax^2 + 2bx + c - |d|$$

For $f'(x) > 0\ \forall\, x \in R$

$$a > 0 \text{ and } 4b^2 - 4.\ 3a(c - |d|) < 0$$

$$\Rightarrow \quad b^2 < 3a(c - |d|)$$

and

$$f'(x) = 3ax^2 + 2bx + c + d\cos x \leq 3ax^2 + 2bx + c + |d|$$

for $f'(x) < 0\ \forall\, x \in R$

$$a < 0 \text{ and } 4b^2 - 12a(c + |d|) < 0$$

$$\therefore \quad b^2 < 3a(c + |d|)$$

$$b^2 < 3a(c - |d|)$$

Many–One Function

If there exist at least two distinct elements in domain whose images are same then the function is known as Many–one. Then $f(x_1) \neq f(x_2)$ but $x_1 \neq x_2$

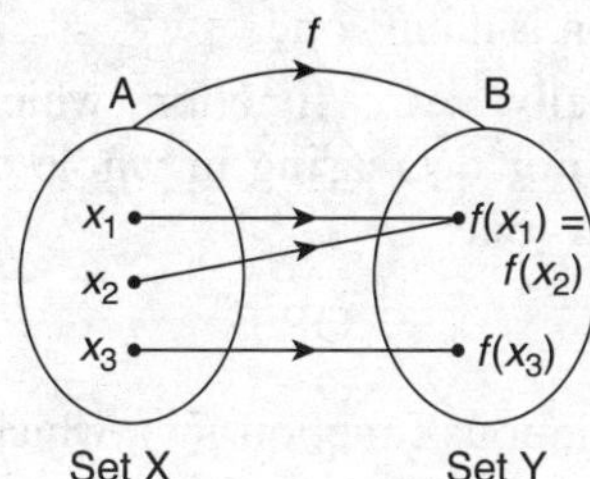

ILLUSTRATION 1.15

If f : R → R is defined by $f(x) = -|x|$, then f is ______.

(a) one–one

(b) many–one

(c) neither one-one nor many-one

(d) None of these.

Solution

$$f(1) = -1$$
$$f(-1) = -1.$$

Hence, f is many–one.

Also, by figure it is clear that f is not strictly decreasing or increasing. Hence, many–one.

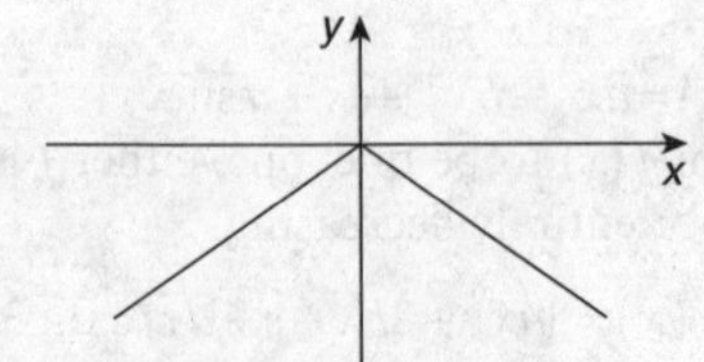

ILLUSTRATION 1.16

If $f: X \rightarrow [1, \infty)$ be a function defined as $f(x) = 1 + 3x^3$. Find the super set of all the sets X such that $f(x)$ is one–one.

Solution

Note that

$$f(x) \geq 1 \Rightarrow 1 + 3x^3 \geq 1 \Rightarrow x^3 \geq 0 \; x \in [0, \infty)$$

Moreover, for $x_1, x_2 \in [0, \infty)$, $x_1 \neq x_2$

$$\Rightarrow \quad 1 + 3x_1^3 \neq 1 + 3x_2^2 \Rightarrow f(x_1) \neq f(x_2)$$

Thus $F: X \rightarrow [1, \infty)$ is one-one for $x \in [0, \infty)$

Methods to Test Many–One

(a) **Analytically:** If $x_1, x_2 \in A$, then $f(x_1) = f(x_2) \Rightarrow x_1 \neq x_2$

(b) **Graphically:** If any line parallel to x-axis cuts the graph of the function at least at two points, then the function is many–one.

(c) **Monotonically:** Any function which is neither increasing nor decreasing in whole domain, then $f(x)$ is many–one.

OR

Any continuous function $f(x)$, which has at least one local maxima or local minima, is many–one.

Number of Many–One Mappings

Let $f: A \rightarrow B$ be a mapping such that A and B are finite sets having m and n elements respectively (where $n > m$.)

Let $A = \{x_1, x_2, x_3, \ldots x_m\}$ and $B = \{y_1, y_2, y_3, \ldots y_n\}$

$\because$ Total number of mappings from A to B = n^m and number of one–one mappings from A to B = ${}^nP_m$

Hence, number of many–one mapping = $n^m - {}^nP_m$ and if $n < m$, then all the n^m mappings are many-one mappings.

ILLUSTRATION 1.17

Prove that the mapping $f: A \rightarrow B$ given by $f(x) = x^2 + x + 1 \; \forall \; x \in R$ is many–one.

Solution

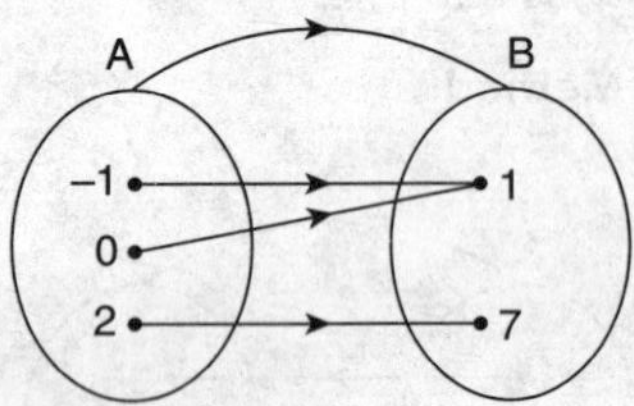

Let us consider set A = {1, 0, 2} and then B = {1, 7}, i.e., two different values of x, $f(x)$ has the same value. Hence $f(x)$ is many–one.

I. Analytical Method

Let $x_1, x_2 \in A$

Then $f(x_1) = f(x_2) \Rightarrow x_1^2 + x_1 + 1 = x_2^2 + x_2 + 1$

$$\Rightarrow x_1^2 - x_2^2 + x_1 - x_2 = 0$$
$$\Rightarrow (x_1 - x_2)(x_1 + x_2 + 1) = 0$$
$$\Rightarrow x_1 \neq x_2$$
$$\{\because \text{If } x_1 = -1, x_2 = 0\}$$

Hence, $f(x)$ is many–one.

II. Graphical Method

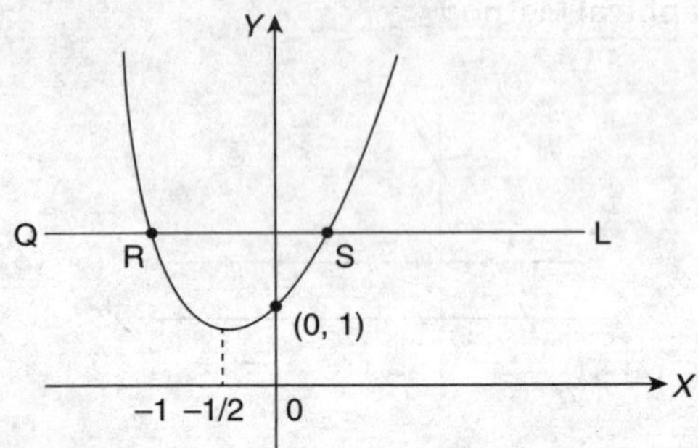

Let $y = f(x) = x^2 + x + 1$

$$\Rightarrow \quad y = \left(x + \frac{1}{2}\right)^2 + \frac{3}{4} \text{ or } \left(x + \frac{1}{2}\right)^2 = \left(y - \frac{3}{4}\right)$$

which is a parabola whose vertex is at $\left(-\frac{1}{2}, \frac{3}{4}\right)$.

It cuts the y-axis at (0, 1) and neither cuts nor touches the x-axis.

The graph of $y = x^2 + x + 1$ is as shown in the figure.

Since any line QL parallel to x-axis cuts the graph at two points Rand S, $f(x)$ is many–one.

III. Monotonically

Since $f(x)=x^2+x+1$

$$\therefore f'(x)=2x+1 \Rightarrow f'(x)>0 \text{ if } x>-1/2$$

$$\text{and } f'(x)<0 \text{ if } x<-1/2$$

Therefore, $f(x)$ is neither increasing nor decreasing and continuous everywhere, having minimum value $\frac{3}{4}$ at $x=\frac{-1}{2}$.

Hence, $f(x)$ is many–one.

Onto and Into Functions

[ONTO functions are also known as surjective]

Let us consider a function f: X → Y, where X is domain and Y is co-domain. If each element in the co-domain 'Y' has at least one pre-image in the domain X that is, for every $x \in$ X such that $f(x) = y$, then f is ONTO. Or in simple words, if range is exactly the co-domain, then f will be onto.

On the other hand, if there exists at least one element in the co-domain Y which is not an image of any element in the domain X, then f is INTO or in simple words, if range is proper subset of the co-domain, then f will be into.

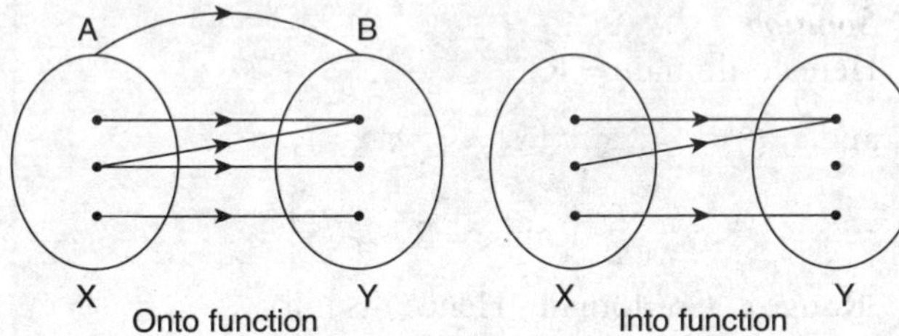

Onto function   Into function

e.g. $f: R \to R$ where $f(x) = \sin x$ is INTO

$f: R \to [-1, 1]$ where $f(x) = \sin x$ is ONTO

$f: R \to [-2, 2]$ $f(x) = \sin x$ is INTO

Note: A function which is both one–one and ONTO is also known as Bijective function.

Method to Test Onto Mapping

Let f : A → B be a mapping, let y be an arbitrary element in B; then $y = f(x)$, where $x \in$ A. Then express x in terms of y. Now if $x \in$ A $\forall y \in$ B, then f is onto.

Number of Onto Mappings

Let f : A → B be a mapping such that A and B are finite sets having m and n elements respectively. Then the number of onto (surjective) mappings from A to B

$$=\begin{cases} \sum_{r=1}^{n}(-1)^{n-r}\cdot {}^nC_r\cdot r^m, & \text{if } 1\le n\le m \\ 0 & \text{if } n>m \end{cases}$$

ILLUSTRATION 1.18

Find the number of surjections from A = {1, 2, 3, …, n}, $n \ge 2$ to B = {a, b, c}.

Solution

Method 1: Number of surjections from A to B is

$$=\sum_{r=1}^{3}(-1)^{3-r}\cdot {}^3C_r\cdot r^n$$

$$=(-1)^2\cdot {}^3C_1\cdot(1)^n+(-1)^1\cdot {}^3C_2\cdot 2^n+(-1)^0\cdot {}^3C_3\cdot 3^n$$

$$=3-3\cdot 2^n+3^n=3^n-3(2^n-1)$$

Method 2: The number of surjections from A to B is the coefficient of x^n in $n!(e^x-1)^3$.

or coefficient of x^n in $n!(e^{3x}-{}^3C_1e^{2x}+{}^3C_2e^x-1)$

or coefficient of x^n in $n!(e^{3x}-3e^{2x}+3e^x-1)$ is

$$n!\left(\frac{3^n}{n!}-3\cdot\frac{2^n}{n!}+3\cdot\frac{1^n}{n!}\right)=3^n-3(2^n-1)$$

Method 3: The total number of mappings from A to B is n^m, i.e., 3^n, of which three functions $f_1(x)=a$, $f_2(x)=b$ and $f_3(x)=c\ \forall x\in A$ are no surjections.

∴ No. of not surjections

$$={}^3C_1 2^n-{}^3C_2(1)^n=3(2^n-1)$$

Hence, no. of surjections from A to B is $3^n-3(2^n-1)$.

ILLUSTRATION 1.19

If the function $f: R \to A$ is given by $f(x)=\frac{x^2}{x^2+1}$ is surjection, then find A.

Solution

Since $y = f(x)$ is onto,

$$\therefore \quad R_f=C_f=A$$

We have $y = f(x) = \dfrac{x^2}{x^2+1}$

$\Rightarrow \quad x^2y + y = x^2$

$\Rightarrow \quad y = x^2(1-y)$

$\Rightarrow \quad x^2 = \dfrac{y}{1-y}$

$\Rightarrow \quad x = \pm\sqrt{\left(\dfrac{y}{1-y}\right)}$ exists if

$\Rightarrow \quad \dfrac{y}{1-y} \geq 0 \Rightarrow \dfrac{y}{y-1} \leq 0$

$\therefore \quad 0 \leq y < 1 \Rightarrow y \in [0,1]$

$\therefore \quad R_f = [0,1]$

Hence, A = [0, 1]

ILLUSTRATION 1.20

Let the function f: R → R be defined by $f(x) = \cos x$, $\forall\, x \in$ R. Show that f is neither one–one nor onto.

Solution

We have,

$$f: \text{R} \to \text{R}, f(x) = \cos x$$

Now,

$$f(x_1) = f(x_2)$$

$$\cos x_1 = \cos x_2$$

$$x_1 = 2n\pi \pm x_2,\ n \in \text{Z}$$

It is seen that the above equation has infinite solutions for x_1 and x_2.

Hence, $f(x)$ is a many–one function.

Also, the range of cos x is [−1, 1], which is subset of the given co-domain R.

Therefore, the given function is not onto.

Method to Test Into Mapping

Let $f: \text{A} \to \text{B}$ be a mapping, let y be an arbitrary element in B and $y = f(x)$. Express x in terms of y. Now if $x \notin \text{A} \,\forall\, y \in \text{B}$, then f is into.

Number of Into Mappings

Let $f: \text{A} \to \text{B}$ be a mapping such that A and B are finite sets having m and n clements respectively. Then number of into mapping from A to B

= Total mapping – Number of onto mappings

$$= \begin{cases} \displaystyle\sum_{r=1}^{n-1} (-1)^{r-1} \cdot {}^nC_r \cdot (n-r)^m & if\ n \leq m \\ n^m & n > m \end{cases}$$

Note: Total number of constant functions from A to B are n.

ILLUSTRATION 1.21

If $f: \text{R} \to \text{R}$ be a function such that $f(x) = \sin x$, then show that f is into and if $f: \text{R} \to [-1, 1]$, then f is onto.

Solution

$\because \quad -1 \leq \sin x \leq 1$

$\because \quad \text{R}_f = [-1,1]$

For first part

$$\text{R}_f \subset \text{C}_f \Rightarrow f \text{ is into}$$

and for second part $\text{R}_f = \text{C}_f \Rightarrow f$ is onto.

ILLUSTRATION 1.22

If $f: \text{R} \to \text{R}$ be a function defined as

$$f(x) = \begin{cases} x|x|-9 & x \in Q \\ x|x|-\sqrt{5} & x \notin Q \end{cases}, \text{ show that } f \text{ is an into.}$$

Solution

Here co-domain = R

and $\quad f(x) \neq \sqrt{5}\ \forall\, x \in \text{R}$

$\therefore \quad \text{Range} = \text{R} \sim \{\sqrt{5}\}$

Range ⊂ Co-domain. Hence f is into.

ILLUSTRATION 1.23

If functions f: A → B and g: B → A satisfy $gof = \text{I}_\text{A}$, then show that f is one–one and g is onto.

Solution

Given,

$$f: \text{A} \to \text{B and } g: \text{B} \to \text{A satisfy } gof = \text{I}_\text{A}$$

It is clearly seen that function g is inverse of f.

So, f has to be one–one and onto.

Hence, g is also one–one and onto.

From the above discussion, we can categorize the functions into the following four types:

1. **One–One Onto (Bijective) Mapping:** A mapping $f : A \to B$ is said to be one–one onto or bijective if and only if
   (i) It is one–one, i.e., $f(x_1) = f(x_2) \Rightarrow x_1 = x_2 \; \forall x_1, x_2 \in A$.
   (ii) It is onto, i.e., $\forall$, $y \in B$, there exists $x \in A$ such that $f(x) = y$.

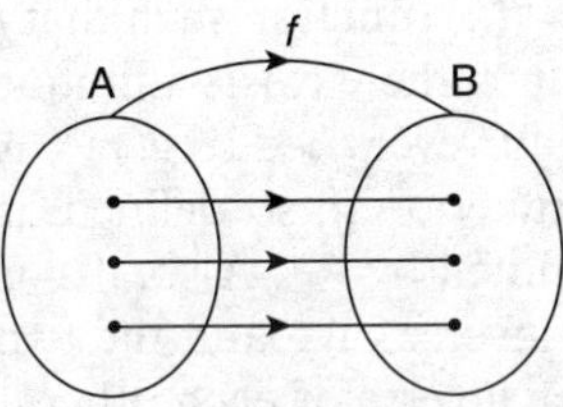

$f : A \to B$ is one–one onto

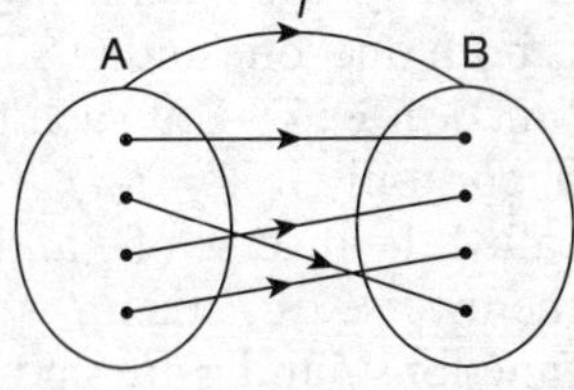

$f : A \to B$ is one–one onto

2. **One–One Into Mapping:** A mapping $f : A \to B$ is said to be one–one into if and only if
   (i) It is one–one, i.e., $f(x_1) = f(x_2) \Rightarrow x_1 = x_2 \; \forall x_1, x_2 \in A$.
   (ii) It is into i.e., $\forall \in B$ there exists $x \in A$.

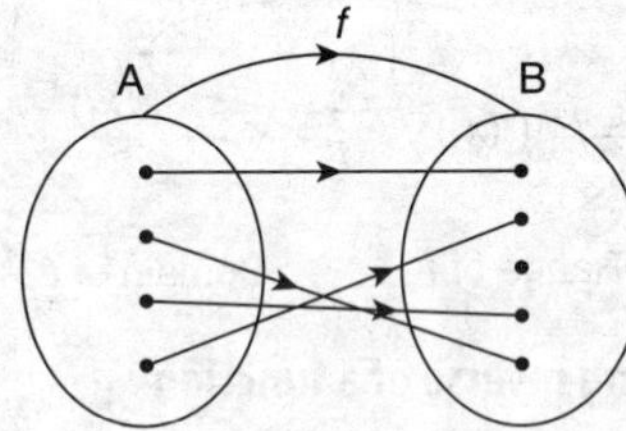

$f : A \to B$ is one–one into

3. **Many–One Onto Mapping:** A mapping $f : A \to B$ is said to be many–one onto if and only if
   (i) It is many–one, i.e., $x_1 \neq x_2 \Rightarrow f(x_1) = f(x_2) \; \forall \; x_1, x_2 \in A$.
   (ii) It is onto, i.e., $\forall \; y \in B$ there exists $x \in A$ such that $f(x) = y$.

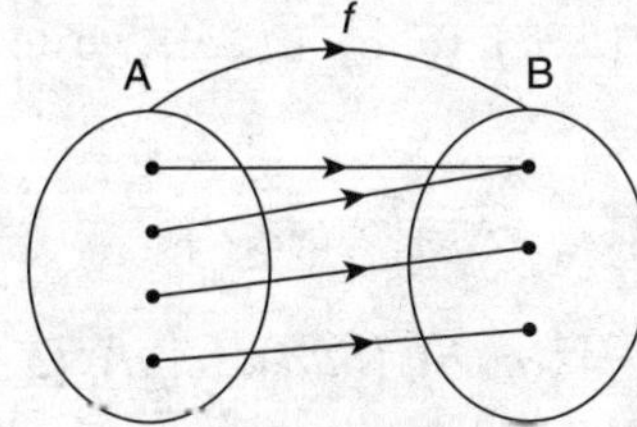

$f : A \to B$ is many–one onto

4. **Many–One Into Mapping:** A mapping $f : A \to B$ is said to be many–one into if and only if
   (i) It is many–one, i.e., $x_1 \neq x_2 \Rightarrow f(x_1) = f(x_2) \; \forall \; x_1, x_2 \in A$.
   (ii) It is into, i.e., $\forall \; y \in B$ there exists $x \in A$.

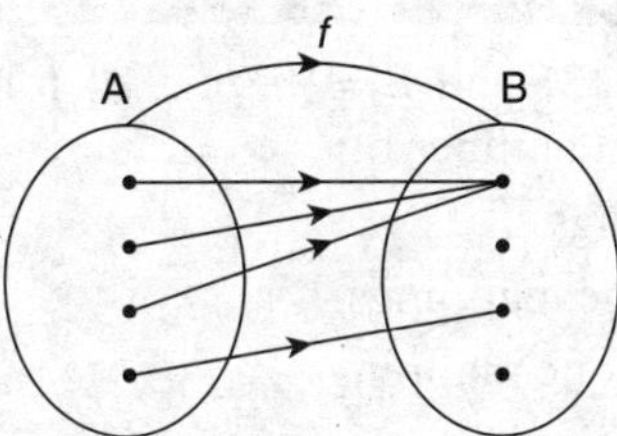

$f : A \to B$ is many–one into

The number of functions for various mappings are given below:

Let $f : A \to B$ be a mapping such that A and B are finite sets having m and n elements respectively. Then,

Description of mappings	Number of functions (Mappings)
1. Total number of mappings from A to B	n^m
2. Total number of one–one mapping from A to B	$\begin{cases} {}^nP_m & n \geq m \\ 0 & n < m \end{cases}$
3. Total number of many-one mapping from A to B	$\begin{cases} n^m - {}^nP_m & n \geq m \\ n^m & n < m \end{cases}$
4. Total number of onto (surjective) mappings from A to B	$\begin{cases} \sum_{r=0}^{n-1} (-1)^r \cdot {}^nC_r \cdot (n-r)^m, & n < m \\ n! & n = m \\ 0 & n > m \end{cases}$
5. Total number of one–one (bijective) mappings from A to B	$n!$, $n = m$
6. Total number of into mappings from A to B	$\begin{cases} \sum_{r=0}^{n-1} (-1)^{r-1} \cdot {}^nC_r \cdot (n-r)^m, & n < m \\ n^m & n > m \end{cases}$
7. Total number of constant mappings from A to B	n

ILLUSTRATION 1.24

If A = {1, 2, 3} and B = {a, b, c, d} for $f: A \to B$, find the total number of:

(i) mappings
(ii) one–one mappings
(iii) many–one mappings
(iv) onto mappings
(v) into mappings
(vi) constant mappings

Solution

Here $m = 3$ and $n = 4$ $(n > m)$

(i) Total number of mappings $= n^m = 4^3 = 64$

(ii) Total number of one–one mappings $= {}^nP_m = {}^4P_3 = 4 \cdot 3 \cdot 2 = 24$

(iii) Total number of many–one mappings $= n^m - {}^nP_m = 64 - 24 = 40$

(iv) Total number of onto mappings = 0 (Here $n > m$)

(v) Total number of into mappings $= n^m = 4^3 = 64$ (Here $n > m$)

(vi) Total number of constant mappings $= n = 4$

ILLUSTRATION 1.25

If A = {1, 2, 3, 4} and B = {a, b, c} for $f: A \to B$, find the total number of:

(i) mappings
(ii) one-one mappings
(iii) many-one mappings
(iv) onto mappings
(v) into mappings
(vi) constant mappings

Solution

Here $m = 4$ and $n = 3$ $(n < m)$

(i) Total number of mappings $= n^m = 3^4 = 81$

(ii) Total number of one–one mappings = 0 $(\because n < m)$

(iii) Total number of many–one mappings $= n^m = 3^4 = 81$ $(\because n < m)$

(iv) Total number of onto mappings

$$= \sum_{r=0}^{2} (-1)^r \cdot {}^3C_r \cdot (3-r)^4$$

$$= {}^3C_0 (3-0)^4 - {}^3C_1 \cdot (3-1)^4 + {}^3C_2 \cdot (3-2)^4$$

$$= 81 - 48 + 3 = 36$$

(v) Total number of into mappings

$$= \sum_{r=1}^{2} (-1)^{r-1} \cdot {}^3C_r \cdot (3-r)^4$$

$$= {}^3C_1 \cdot (3-1)^4 - {}^3C_2 \cdot (3-2)^4 = 48 - 3 = 45$$

(vi) Total number of constant mappings $= n = 3$

Inverse Function

If $f: X \to Y$ be defined by $y = f(x)$ function such that f is both one–one and ONTO, then there exists a unique function $g: Y \to X$ such that for every $y \in Y$, $g(y) = x$ if and only if $y = f(x)$. The function g, so defined, is called the inverse of f and denoted by f^{-1}. Also, if g is the inverse of f, then f is the inverse of g and the two functions f and g are said to be inverse of each other.

The necessary condition for a function to have an inverse is that the function must be one–one and ONTO (i.e., Bijective). Whenever an inverse function is defined, the range of the original function becomes the domain of the inverse function

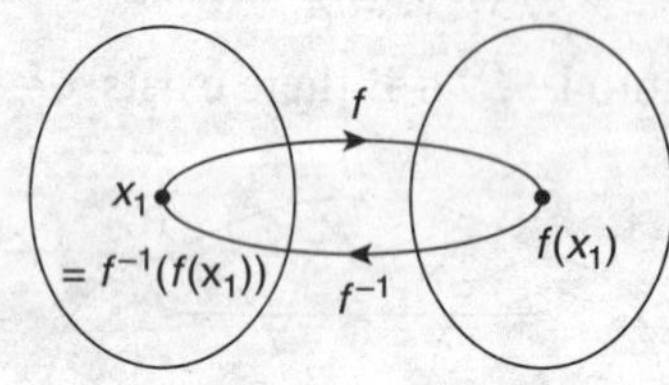

Method of finding inverse of a function

Step 1: Solve the equation $y = f(x)$ for x in terms of y.
Step 2: Interchange x and y. The resulting formula will be $y = f^{-1}(x)$.
Tip: To draw the graph of f^{-1}, reflect the graph of $f(x)$ about $y = x$ line.

ILLUSTRATION 1.26

If $A = \{a, b, c, d\}$ and the function $f = \{(a, b), (b, d), (c, a), (d, c)\}$, write f^{-1}.

Solution

$$A = \{a, b, c, d\}$$

and $$f = \{(a, b), (b, d), (c, a), (d, c)\}$$

So,

$$f^{-1} = \{(b, a), (d, b), (a, c), (c, d)\}$$

ILLUSTRATION 1.27

Let $f: R \to R$ be the function defined by $f(x) = 2x - 3$, $\forall x \in R$. Write f^{-1}.

Solution

The given function is

$$f(x) = 2x - 3, \forall x \in R$$

Let $y = 2x - 3$

$$x = (y + 3)/2$$

Thus,

$$f^{-1}(x) = (x + 3)/2$$

ILLUSTRATION 1.28

(i) Inverse of $y = \frac{2x+1}{x-1}$

Step 1: $xy - y = 2x + 1$

$\Rightarrow$ $x(y - 2) = y - 1$

$$x = \frac{y-1}{y-2}$$

Step 2: $y = f^{-1}(x) = \frac{x-1}{x-2}$

Inverse of $f(x) = \frac{x-1}{x-2}$

ILLUSTRATION 1.29

Find the inverse of the function $f(x) = \ln(x^2 + 3x + 1); x \in [1, 3]$. Assume it to be an onto function.

Solution

Given $f(x) = \ln(x^2 + 3x + 1)$

$\therefore$ $f'(x) = \frac{2x+3}{(x^2+3x+1)} > 0 \forall x \in [1,3]$

which is a strictly increasing function. Thus $f(x)$ is injective, given that $f(x)$ is onto.

Hence the given function $f(x)$ is invertible.

Now let $y = f(x) = \ln(x^2 + 3x + 1)$.

Then $x - f^{-1}(y)$ ...(i)

and $y = \ln(x^2 + 3x + 1)$

$$e^y = x^2 + 3x + 1$$

$$x^2 + 3x + 1 - e^y = 0$$

$\therefore$ $x = \frac{-3 \pm \sqrt{9 - 4 \cdot 1 \cdot (1 - e^y)}}{2} = \frac{-3 \pm \sqrt{(5 + 4e^y)}}{2}$

$= \frac{-3 + \sqrt{(5 + 4e^y)}}{2}$ $(\because x \in [1,3])$ ...(ii)

From (i) and (ii), we get $f^{-1}(y) = \frac{-3 + \sqrt{(5 + 4e^y)}}{2}$

Hence $f^{-1}(x) = \frac{-3 + \sqrt{(5 + 4e^x)}}{2}$

ILLUSTRATION 1.30

If the function $f: [1, \infty) \to [1, \infty)$ is defined by $f(x) = 2^{x(x-1)}$. Is $f(x)$ invertible? If so, find its inverse.

Solution

Given $f(x) = 2^{x(x-1)}$

$\therefore$ $f'(x) = 2^{x(x-1)} \ln 2 \cdot (2x - 1) > 0 \forall x \in [1, \infty)$

which is strictly increasing function.

Thus $f(x)$ is injective.

Also, for $x \to 1$, $f(x) \to 1$

and for $x \to \infty$, $f(x) \to \infty$

Hence, $R_f = [1, \infty) = C_f$

So, $f(x)$ is onto

$\Rightarrow$ $f(x)$ is both one–one and onto.

$\therefore$ $f(x)$ is invertible.

To find f^{-1}

Let f^{-1} be the inverse function of f, then by the rule of identity

$$fof^{-1}(x) = x$$

$\Rightarrow$ $f(f^{-1}(x)) = x$

$\Rightarrow$ $2^{f^{-1}(x)(f^{-1}(x)-1)} = x$ $(\because f(x) = 2^{x(x-1)})$

$\Rightarrow$ $f^1(x)(f^{-1})(x) - 1) = \log_2 x$

$\Rightarrow$ $(f^{-1}(x))^2 - f^{-1}(x) - \log_2 x = 0$

$\therefore$ $f^{-1}(x) - \frac{1 \pm \sqrt{1 + 4\log_2 x}}{2}$

as $f^{-1}(x) \in [1, \infty)$

$\therefore$ $f^{-1}(x) = \frac{1}{2}\left(1 + \sqrt{(1 + 4\log_2 x)}\right)$

ILLUSTRATION 1.31

Using the definition, prove that the function f: A→ B is invertible if and only if f is both one–one and onto.

Solution

Let f: A → B be many–one function.

Let $f(a) = p$ and $f(b) = p$

So, for inverse function we will have $f^{-1}(p) = a$ and $f^{-1}(p) = b$

Thus, in this case inverse function is not defined as we have two images, a and b, for one pre-image p.

But for f to be invertible, it must be one–one.

Now, let f: A → B is not onto function.

Let B = $\{p, q, r\}$ and range of f be $\{p, q\}$.

Here image r has not any pre-image, which will have no image in set A.

And for f to be invertible, it must be onto.

Thus, f is invertible if and only if f is both one–one and onto.

A function f = X → Y is invertible iff f is a bijective function.

Composite Functions

Let us consider two functions

$$f : X \to Y_1 \text{ and } g : Y_1 \to Y$$

We define function $h : X \to Y$ such that $h(x) = g(f(x))$. To obtain $h(x)$, we first take the f-image of an element $x \in X$ so that $f(x) \in Y$, which is the domain of $g(x)$. Then take g-image of $f(x)$, i.e., $g(f(x))$ which would be an element of Y.

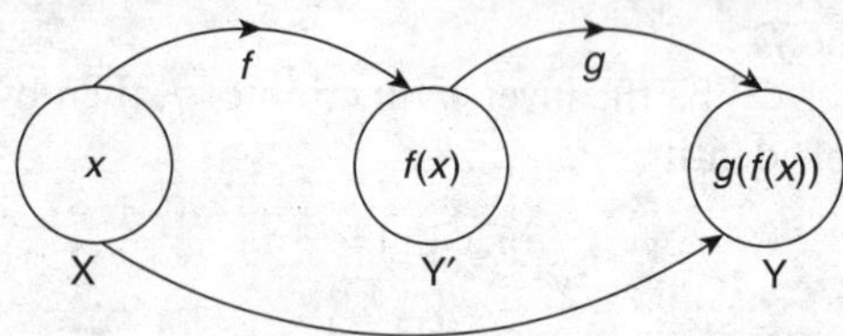

The function h defined above is called a composite of f and g and is denoted by gof. Thus $(gof)x = g(f(x))$

Similarly, $(fog)x = f(g(x))$

In general, $fog \neq gof$

Properties of Composition of Functions

1. The composition of function is not commutative i.e., $fog \neq gof$
2. The composition of function is associative, i.e., if f, g, h are three functions such that $(fog)oh$ and $fo(goh)$ exists, then $(fog)oh = fo(goh)$

(3) Let $f : A \to B$, $g : B \to C$ be two functions, then
(i) f and g are injective $\Rightarrow gof$ is injective
(ii) f and g are surjective $\Rightarrow gof$ is surjective
(iii) f and g are bijective $\Rightarrow gof$ is bijective

(4) Let $f : A \to B$, $g : B \to C$ be two functions, then
(i) $gof : A \to C$ is injective $\Rightarrow f : A \to B$ is injective
(ii) $gof : A \to C$ is surjective $\Rightarrow g : B \to C$ is surjective
(iii) $gof : A \to C$ is injective and $g : B \to C$ is surjective $\Rightarrow g : B \to C$ is injective
(iv) $gof : A \to C$ is surjective and $g : B \to C$ is injective $\Rightarrow f : A \to B$ is surjective

(5) An injective mapping from a finite set to itself is bijective.

(6) The composition of any function with the identify function is the function itself.
i.e., $f : A \to B$, then $foI_A = I_B of = f$
where I_A and I_B are the identity functions of A and B respectively.

ILLUSTRATION 1.32

If $f : R \to R$ and $g : R \to R$ be two mappings such that $f(x) = \sin x$ and $g(x) = x^2$, then

(a) show that $fog \neq gof$,

(b) find the values of $(fog)\sqrt{\pi}/2$ and $(gof)\pi/3$.

Solution

(a) Let $x \in R$

$$\therefore (fog)x = f\{g(x)\} = f(x^2) \quad \{\because g(x) = x^2\}$$

$$= \sin x^2 \{\because f(x) = \sin x\} \quad \text{...(i)}$$

And $(gof)x = g\{f(x)\}$

$$= g\{\sin x\} \; (\because f(x) = \text{s in } x)$$

$$= \sin^2 x \quad \text{...(ii)}$$

From (i) and (ii), $(fog)x \neq (gof)x \forall x \in R$

Hence, $fog \neq gof$

(b) From (i), $(fog)x = \sin x^2$

$$\therefore \quad (fog)\frac{\sqrt{\pi}}{2} = \sin\frac{\pi}{4} = \frac{1}{\sqrt{2}}$$

and from (ii), $(gof)x = \sin^2 x$

$$(gof)\frac{\pi}{3} = \sin^2\frac{\pi}{3} = \left(\sqrt{3}/2\right)^2 = 3/4$$

ILLUSTRATION 1.33

Let $f(x) = 1 + x^2$ and $f(g(x)) = 1 + x^2 - 2x^3 + x^4$, then find $g(x)$.

Solution

$$f(g(x)) = 1 + x^2 - 2x^3 + x^4 = 1 + x^2(1 - 2x + x^2)$$

$$= 1 + x^2(1-x)^2 = 1 + \{x(1-x)\}^2$$

$$= 1 + |x(1-x)|^2$$

$$\therefore \quad g(x) = |x(1-x)| \quad (\because f(x) = 1 + x^2)$$

ILLUSTRATION 1.34

If the mappings f and g are given by

$f = \{(1, 2), (3, 5), (4, 1)\}$ and $g = \{(2, 3), (5, 1), (1, 3)\}$

then write down the pairs in the mappings fog and gof.

Solution

$\because \quad f = \{(1, 2), (3, 5), (4, 1)\}$

$\therefore \quad f(1) = 2, f(3) = 5, f(4) = 1 \quad \ldots(1)$

and $\quad g = \{(2, 3), (5, 1), (1, 3)\}$

$\therefore \quad g(2) = 3, g(5) = 1, g(1) = 3 \quad \ldots(2)$

Again $D_f = \{1, 3, 4\}, R_f = \{2, 5, 1\}$ and

$D_g = \{2, 5, 1\}, R_g = \{1, 3\}$

Then gof mapping is defined in the following way:

$\therefore \quad (gof)(1) = g\{f(1)\} = g(2) = 3$

$(gof)(3) = g\{f(3)\} = g(5) = 1$

and $\quad (gof)(4) = g\{f(4)\} = g(1) = 3$

Hence, $gof = \{(1,3),(3,1),(4,3)\}$

Now since range of $g \subset$ domain of f

$\therefore \quad fog$ is defined.

The fog mapping is defined in the following way:

$\therefore \quad (fog)(2) = f\{g(2)\} = f(3) = 5$

$(fog)(5) = f\{g(5)\} = f(1) = 2$ and

$(fog(1) = f\{g(1)\} = f(3) = 5$

Hence, $fog = \{(2,5),(5,2),(1,5)\}$

ILLUSTRATION 1.35

If $f(x) = \begin{cases} 2+x & \text{if} \quad x \geq 0 \\ 2-x & \text{if} \quad x < 0 \end{cases}$, then find $(fof)(x)$.

Solution

Given

$$f(x) = \begin{cases} 2+x, & \text{if} \quad x \geq 0 \\ 2-x & \text{if} \quad x < 0 \end{cases} = 2 + |x|$$

$$\therefore \quad (fof)(x) = f(f(x)) = f(2+|x|)$$

$$= 2 + |2 + |x|| = 2 + 2 + |x| = 4 + |x|$$

$$= \begin{cases} 4+x, & \text{if} \quad x \geq 0 \\ 4-x, & \text{if} \quad x < 0 \end{cases}$$

ILLUSTRATION 1.36

Functions f, g: R $\to$ R are defined, respectively, by $f(x) = x^2$ **+ 3*x* + 1,** $g(x) = 2x - 3$, find

(i) fog **(ii)** gof **(iii)** fof **(iv)** gog

Solution

Given, $f(x) = x^2 + 3x + 1$, $g(x) = 2x - 3$

(i) $fog = f(g(x))$

$= f(2x - 3)$

$= (2x - 3)^2 + 3(2x - 3) + 1$

$= 4x^2 + 9 - 12x + 6x - 9 + 1$

$= 4x^2 - 6x + 1$

(ii) $gof = g(f(x))$

$= g(x^2 + 3x + 1)$

$= 2(x^2 + 3x + 1) - 3$

$= 2x^2 + 6x - 1$

(iii) $fof = f(f(x))$

$= f(x^2 + 3x + 1)$

$= (x^2 + 3x + 1)^2 + 3(x^2 + 3x + 1) + 1$

$= x^4 + 9x^2 + 1 + 6x^3 + 6x + 2x^2 + 3x^2 + 9x + 3 + 1$

$= x^4 + 6x^3 + 14x^2 + 15x + 5$

(iv) $gog = g(g(x))$

$= g(2x - 3)$

$= 2(2x - 3) - 3$

$= 4x - 6 - 3$

$= 4x - 9$

Even and Odd Functions

If $f: X \to Y$ is a real-valued function such that for all $x \in D$.

If $f(-x) = f(x)$ for every $x \in D$ then f will be an even function.

and if $f(-x) = -f(x)$ for every $x \in D$ then f will be an odd function.

A function can be even or odd or neither.

> **Tip**
>
> Even functions are symmetric about y-axis while odd functions are symmetric in diagonally opposite quadrants.
>
> e.g. $\sin x$, $\tan x$, x^3 are odd functions.
>
> $|x|$, $\cos x$, $\sec x$, x^2 are even functions.
>
> e^x, $\log_e x$ are neither even nor odd.

ILLUSTRATION 1.37

Which of the following functions are even, odd, neither?

(i) $f(x) = x^2 \sin x$

(ii) $f(x) = \sqrt{1+x+x^2} - \sqrt{1-x+x^2}$

(iii) $f(x) = \log(x+\sqrt{1+x^2})$

(iv) $f(x) = \sin x - \cos x$

Solution

(i) $f(-x) = -x^2 \sin x$

$f(-x) = -f(x)$; hence odd.

(ii) $f(-x) = \sqrt{1-x+x^2} - \sqrt{1+x+x^2}$

$f(-x) = -\left[\sqrt{1+x+x^2} - \sqrt{1-x+x^2}\right]$

$f(-x) = -f(x)$; hence odd.

(iii) $f(-x) = \log(-x+\sqrt{1+x^2})$

$$= \log\left[\frac{(\sqrt{1+x^2}-x)(\sqrt{1+x^2}+x)}{\sqrt{1+x^2}+x}\right]$$

$$= \log\left[\frac{1}{\sqrt{1+x^2}+x}\right]$$

$f(-x) = -\log[x+\sqrt{1+x^2}]$

$f(-x) = -f(x)$; hence odd.

(iv) $f(x) = \sin x - \cos x$

$f(-x) = -\sin x - \cos x$

Hence, $f(x)$ is neither even nor odd.

Even Function

A function $y = f(x)$ is said to be an even function if $f(-x) = f(x) \; \forall \, x \in D_f$.

e.g. $f(x) = \cos 2x$ is even because

$$f(-x) = \cos(-2x) = \cos 2x = f(x)$$

The graph of an even function $y = f(x)$ is symmetrical about the y-axis, i.e., if point (x, y) lies on the graph, then $(-x, y)$ will also lie on the graph or we say that if any point lies on the graph, then its image about the y-axis also lies on the graph. A few examples are given below:

(i)

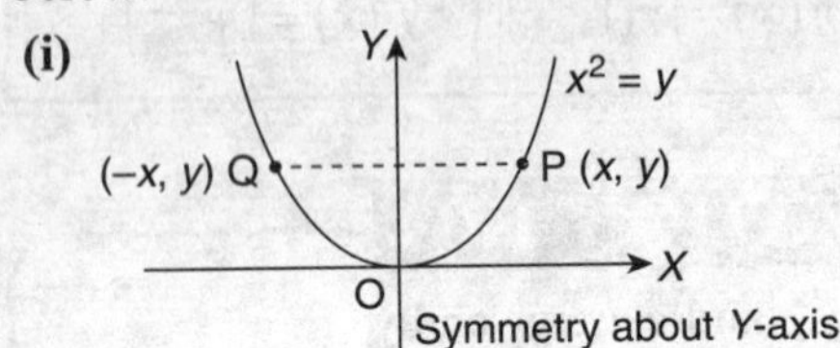

Symmetry about Y-axis

(ii)

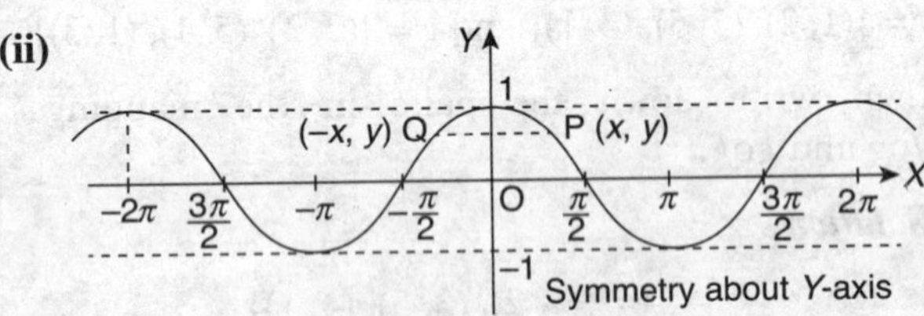

Symmetry about Y-axis

(iii)

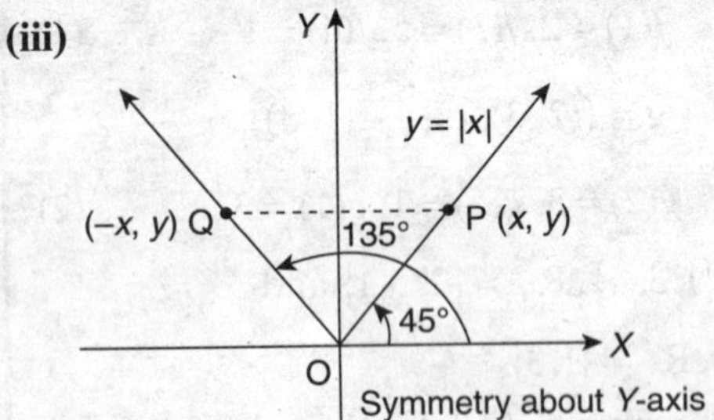

Symmetry about Y-axis

Odd Function

A function $y = f(x)$ is said to be an odd function if $f(-x) = -f(x) \; \forall \, x \in D_f$

e.g., $f(x) = x^3 + x$ is odd because $f(-x) = -x^3 - x = -f(x)$

The graph of an odd function $y = f(x)$ is symmetrical in opposite quadrants, i.e., if point (x, y) lies on the graph, then $(-x, -y)$ will also lies on the graph or we say that if any point lies on the graph, then its image about origin also lies on the graph. A few examples are given below:

(i)

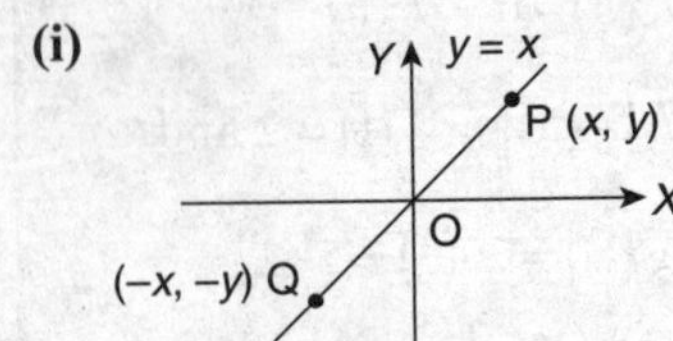

(ii)

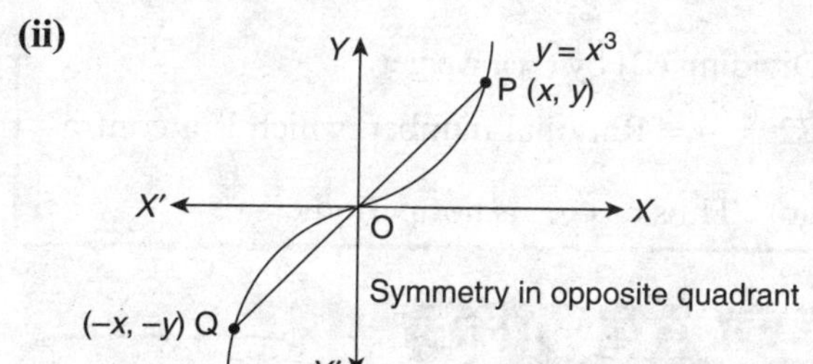

(iii)

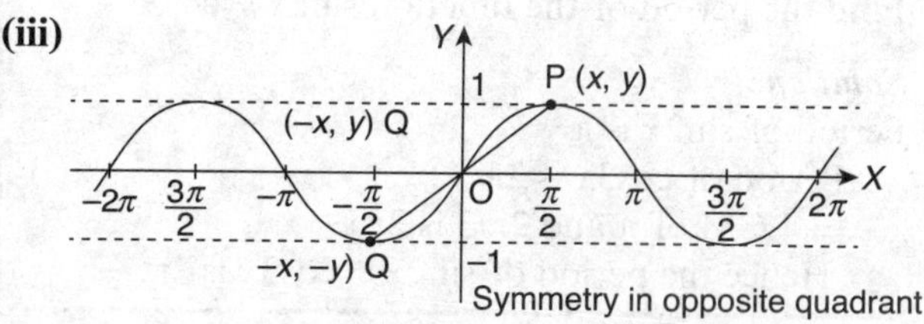

Important Points Regarding Even and Odd Functions

1. It is not essential that every function is even or odd. It is possible to have some functions which are neither even nor odd functions.
2. Inverse of an even function is not defined.
3. $f(x) = 0$ is the only function which is defined, on the entire number line, as even and odd at the same time.
4. Every function can be expressed as the sum of an even and an odd function.

   i.e., $f(x)=\frac{1}{2}\{f(x)+f(-x)\}+\frac{1}{2}\{f(x)-f(-x)\}$

   $=\{\text{Even function}\}+\{\text{Odd function}\}$
5. If $f(x)-f(-x)=0$, then $f(x)$ is an even function and if $f(x)+f(-x)=0$, then $f(x)$ is an odd function.
6. Every even function $y=f(x)$ is many–one $\forall\, x\in D_f$.
7. The first derivative of an even function is an odd function and vice-versa.

The table of composition of even-odd function is worth noting:

f(x)	g(x)	f(x) + g(x)	f(x) − g(x)	f(x) g(x)	$\frac{f(x)}{g(x)}$	$(fog)x$
Even	Even	Even	Even	Even	Even	Even
Even	Odd	Neither even nor odd	Neither even nor odd	Odd	Odd	Even
Odd	Even	Neither even nor odd	Neither even nor odd	Odd	Odd	Even
Odd	Odd	Odd	Odd	Even	Even	Odd

ILLUSTRATION 1.38

If $f(t)=\frac{t}{e^t-1}+\frac{t}{2}+1$, prove that $f(t)$ is an even function.

Solution

$$f(t)=\frac{t}{e^t-1}+\frac{t}{2}+1 \qquad \text{...(i)}$$

$$\therefore \quad f(-t)=\frac{-t}{e^{-t}-1}-\frac{t}{2}+1$$

$$=\frac{te^t}{e^t-1}-\frac{t}{2}+1 \qquad \text{(ii)}$$

Subtract (ii) from (i), we get

$$f(t)-f(-t)=\frac{t}{(e^t-1)}(1-e^t)+t=-t+t=0$$

$$\Rightarrow \quad f(-t)=f(t).$$

Hence $f(t)$ is an even function.

ILLUSTRATION 1.39

Show that $f(x)$, given by $f(x+y)=f(x)+f(y)\forall\ x,y\in R$, is an odd function.

Solution

Given $\quad f(x+y)=f(x)+f(y) \qquad \text{...(i)}$

Replace y by $-x$, we have

$$f(0)=f(x)+f(-x) \qquad \text{...(ii)}$$

Now put $x = y = 0$ in (i), we have

$$f(0)=f(0)+f(0)\Rightarrow f(0)=0$$

Then from (ii), we get

$$0=f(x)+f(-x)\Rightarrow f(-x)=-f(x)$$

Hence, $f(x)$ is an odd function

ILLUSTRATION 1.40

Express ae^x as the sum of an even and an odd function.

Solution

Let $f(x)=ae^x$, then $f(-x)=ae^{-x}$

Now $f(x)=\frac{1}{2}\{f(x)+f(-x)\}+\frac{1}{2}\{f(x)-f(-x)\}$

$=\frac{1}{2}\left(ae^{x}+ae^{-x}\right)+\frac{1}{2}\left(ae^{x}-ae^{-x}\right)$

$=(\text{Even function})+(\text{Odd function})$

Periodic Functions

A function $f : X \to Y$ is said to be a periodic function if there exists a positive real number p such that $f(x + p) = f(x)$ for all $x \in X$.

sin x, cos x, sec x, cosec x are periodic functions with period 2π.

tan x, cot x are periodic with period π.

If $f(x)$ is periodic with period p, then $f(ax + b)$ where $a, b \in R$ $(a \neq 0)$ is also periodic with period $\frac{p}{|a|}$

ILLUSTRATION 1.41

Find the period of the following, if exists:

(i) sin(2x + 3)

(ii) tan(9x + 19)

Solution

(i) sin x has a period of 2π

sin(2x + 3) has a period of $\frac{2\pi}{2}$, i.e., π

(ii) tan(9x + 19) will have a period of $\frac{\pi}{9}$.

ILLUSTRATION 1.42

Show that function $\cos\sqrt{x}$ is not periodic.

Solution

If possible, let there exist $p \neq 0$ such that $\cos\sqrt{x+p}=\cos\sqrt{x}$ for all $x \geq 0$.

(We are assuming the $\cos\sqrt{x}$ is periodic with period p)

Putting $x = 0$, we get

$$\cos\sqrt{p}=\cos 0=1$$

Or $\sqrt{p}=2k\pi$,

where k is a non-zero integer ...(A)

Again, putting $x = p$, we get

$$\cos\sqrt{2p}=\cos\sqrt{p}=\cos 0=1$$

Or $\sqrt{2p}=2m\pi$,

where m is a non-zero integer ...(B)

Dividing (B) by (A), we get

$\sqrt{2}=\frac{m}{k}$ = Rational number, which is a contradiction. Thus cos $\sqrt{x}$ is not periodic.

ILLUSTRATION 1.43

Find the period of the function sin2x + cos3x

Solution

Period of sin2x is π

Period of cos3x is 2π/3

⇒ LCM of π and 2π/3 is 2π

Hence the period of sin2x + cos3x is 2π.

Graph of Functions

1. $y = f(x) = \sin x$

Domain → (–∞, ∞), Range → [–1, 1]

Period → 2π; Nature → odd;

Interval in which the inverse can be obtained → [–π/2, π/2]

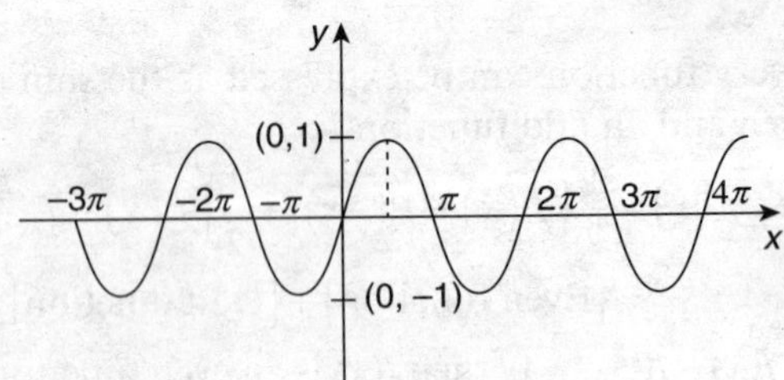

2. $y = f(x) = \cos x$

Domain → (–∞, ∞); Range → [–1, 1];

Period → 2π; Nature → even;

Interval in which the inverse can be obtained → [0, π]

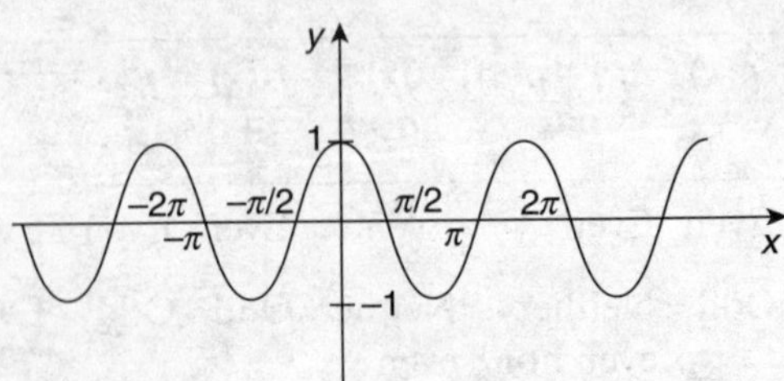

3. $y = f(x) = \tan x$

Domain → $R \sim (2n+1)\ \pi/2,\ n \in I$;

Range → (–∞, ∞)

Period → π; Nature → odd;

Interval in which the inverse can be obtained → (–π/2, π/2)

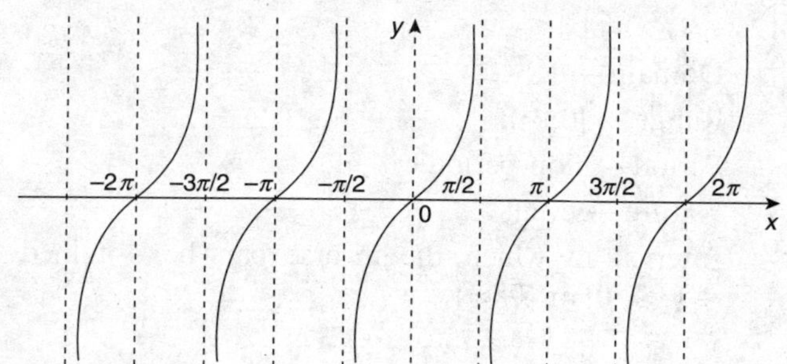

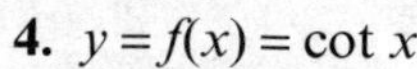

4. $y = f(x) = \cot x$

Domain $\rightarrow R \sim n\pi,\ n \in I$;

Range $\rightarrow (-\infty, \infty)$

Period $\rightarrow \pi$;

Nature $\rightarrow$ odd;

Interval in which the inverse can be obtained $\rightarrow (0, \pi)$

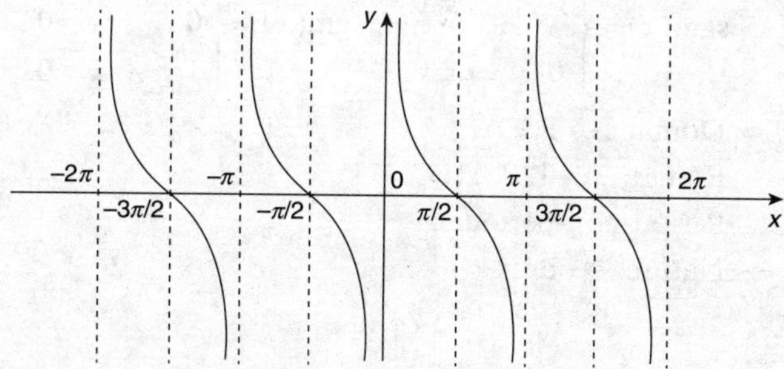

5. $y = f(x) = \sec x$

Domain $\rightarrow R \sim (2n+1)\,\pi/2,\ n \in I$;

Range $\rightarrow (-\infty, -1] \cup [1, \infty)$

Period $\rightarrow 2\pi$;

Nature $\rightarrow$ even;

Interval in which the inverse can be obtained $\rightarrow \left[0, \dfrac{\pi}{2}\right) \cup \left(\dfrac{\pi}{2}, \pi\right]$

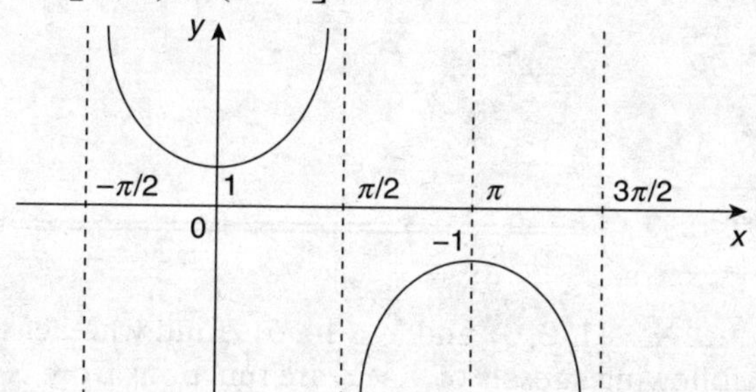

6. $y = f(x) = \operatorname{cosec} x$

Domain $\rightarrow R \sim n\pi,\ n \in I$;

Range $\rightarrow (-\infty, -1] \cup [1, \infty)$

Period $\rightarrow 2\pi$;

Nature $\rightarrow$ odd;

Interval in which the inverse can be obtained $\rightarrow [-\pi/2, 0) \cup (0, \pi/2]$

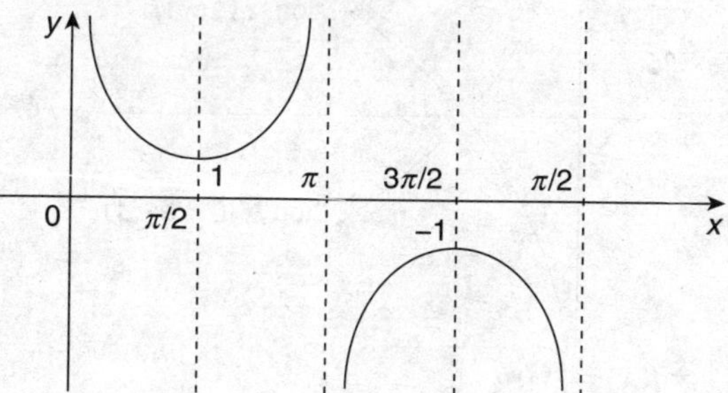

7. $y = f(x) = a^x,\ a > 1$

Domain R;

Range $\rightarrow (0, \infty)$;

Period $\rightarrow$ non-periodic;

Nature $\rightarrow$ neither odd nor even

Interval in which the inverse can be obtained $\rightarrow (-\infty, \infty)$

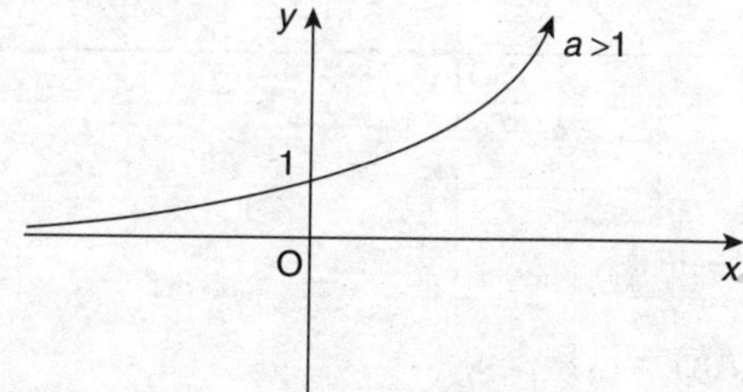

8. $y = f(x) = a^x,\ 0 < a < 1$

Domain $\rightarrow R$;

Range $\rightarrow (0, \infty)$

Period $\rightarrow$ non-periodic;

Nature $\rightarrow$ neither odd nor even;

Interval in which the inverse can be obtained $\rightarrow (-\infty, \infty)$

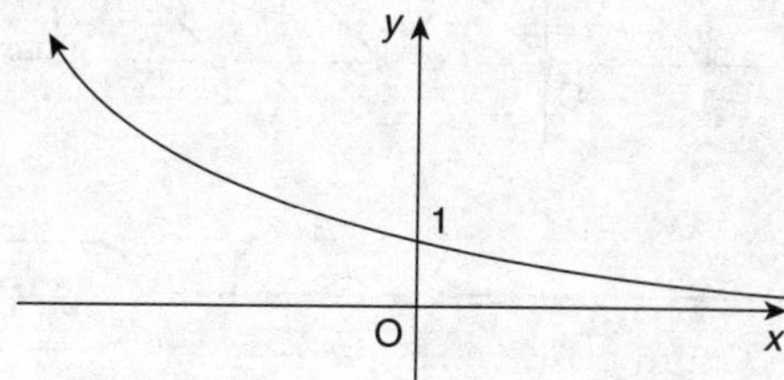

9. $y = \log_a x,\ a > 0$ and $\neq 1$

Domain $\rightarrow (0, \infty)$;

Range $\rightarrow (-\infty, \infty)$;

Period $\rightarrow$ non-periodic;

Nature $\rightarrow$ neither odd nor even

Interval in which the inverse can be obtained $\rightarrow (0, \infty)$

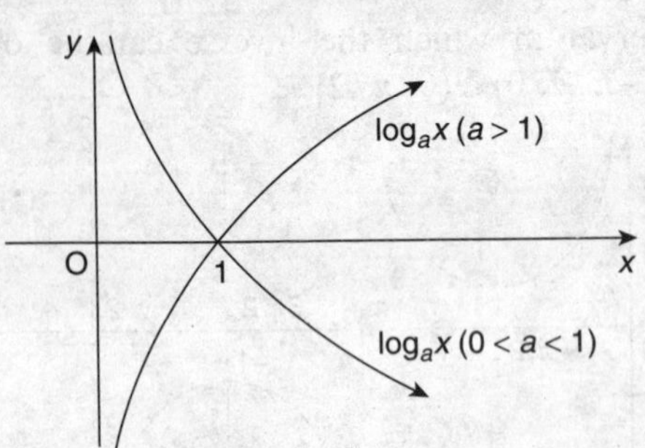

10. $y = f(x) = [x]$

Domain → R;

Range → I

Period → non periodic;

Nature → neither even nor odd

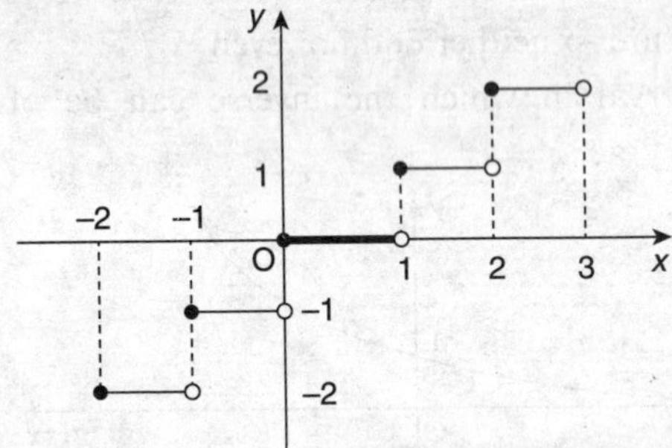

11. $y = f(x) = \{x\}$

Domain → R;

Range → [0, 1);

Period → 1;

Nature → neither even nor odd

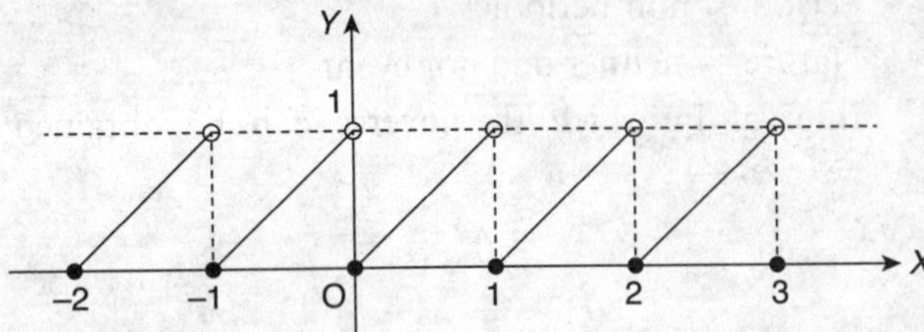

12. $y = f(x) = |x|$

Domain → R;

Range → [0, ∞);

Period → non-periodic;

Nature → even;

Interval in which the inverse can be obtained → (–∞, 0] or [0, ∞)

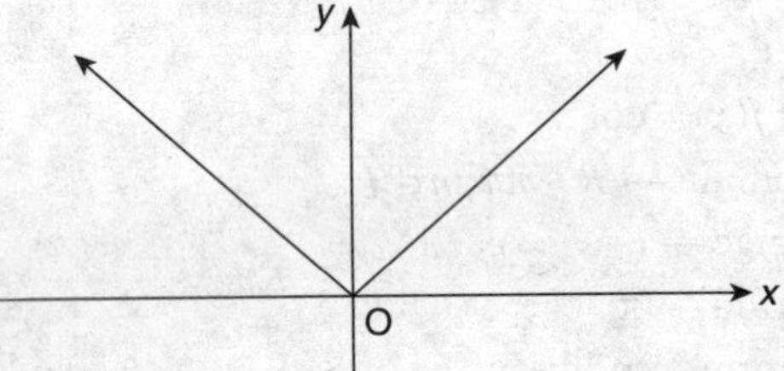

13. $y = f(x) = \text{sgm}\,(x)$

$$\text{sgm}(x) = \begin{cases} \dfrac{|x|}{x}, & x \neq 0 \\ 0, & x = 0 \end{cases} \quad \text{or } \text{sgm}(x) = \begin{cases} -1, & x < 0 \\ 0, & x = 0 \\ 1, & x > 0 \end{cases}$$

Domain → R;

Range → {–1, 0, 1};

Period-non-periodic;

Nature → odd;

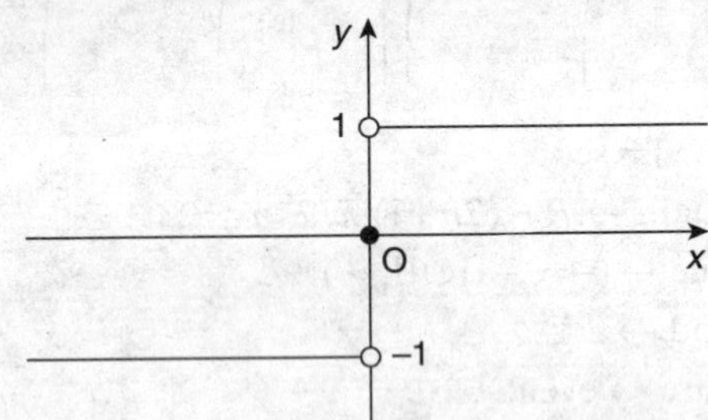

WORKSHEET – 2

1. Let D be the domain of the real-valued function f defined by $f(x) = \sqrt{(25 - x^2)}$. Then, write D.
2. Let f, g: R → R be defined by $f(x) = 2x + 1$ and $g(x) = x^2 - 2$, ∀ $x \in$ R, respectively. Then, find gof.
3. If f: R → R is defined by $f(x) = x^2 - 3x + 2$, write $f(f(x))$.
4. If the mappings f and g are given by f = {(1, 2), (3, 5), (4, 1)} and g = {(2, 3), (5, 1), (1, 3)}, write fog.
5. Let X = {1, 2, 3} and Y = {4, 5}. Find whether the following subsets of X×Y are functions from X to Y or not.

(i) f = {(1, 4), (1, 5), (2, 4), (3, 5)}

(ii) g = {(1, 4), (2, 4), (3, 4)}

(iii) h = {(1, 4), (2, 5), (3, 5)}

(iv) k = {(1, 4), (2, 5)}.

6. Let f: R → R be the function defined by $f(x) = 1/(2 - \cos x)$ ∀ $x \in$ R. Then, find the range of f.

7. Given A = {2, 3, 4}, B = {2, 5, 6, 7}. Construct an example of each of the following:

(a) An injective mapping from A to B

(b) A mapping from A to B which is not injective

(c) A mapping from B to A.

8. Let A = R – {3}, B = R – {1}. Let f: A → B be defined by $f(x)=\left(\frac{x-2}{x-3}\right)$ $\forall\, x \in$ A. Then show that f is bijective.

SOLVED EXAMPLES

Example 1. Show that the function is odd function $f(x)=4\sqrt{1-x^3}-4\sqrt{1+x^3}$

Solution

The condition for odd function is $f(-x) = -f(x)$

$$f(-x)=4\sqrt{1-(-x)^3}-4\sqrt{1+(-x)^3}$$

$$=4\sqrt{1+x^3}-4\sqrt{1-x^3}$$

$$=-4(\sqrt{1-x^3}-4\sqrt{1+x^3})=-f(x)$$

Hence $f(x)$ is an odd function.

Example 2. Show that the function is an even function: $f(x) = e^x + e^{-x}$

Solution

$$f(-x)=e^{-x}+e^{-(-x)}$$

$$=e^{-x}+e^{x}=f(x)$$

$$\therefore \quad f(-x)=+f(x)$$

Hence $f(x)$ is an even function.

Example 3. Let $f: R \to R$ defined by $f(x) = 3x + 6$ show that f is one–one.

Solution

Given $f(x) = 3x + 6$ and $x \in$ R.

I Method: To test f is one–one

$$f(x_1)=f(x_2) \Rightarrow x_1=x_2$$

$$3x_1+6=3x_2+6 \Rightarrow 3x_1=3x_2$$

$$\therefore \quad x_1=x_2,$$

$\therefore$ f is one–one

II Method: To test f is one–one.

$$f'(x)=6>0$$

i.e., it increases strictly in R.

$\therefore$ $f(x)$ is one–one.

Example 4. Let $f: R \to R$ defined by $f(x)=\frac{x^2}{1+x^2}$.

Prove that f is neither one–one nor ONTO.

Solution

First Method: To test f is one–one.

Because f is even function in its domain.

Hence it is many–one.

Second Method: To test if f is onto.

Let us find the range of f.

Let $\frac{x^2}{1+x^2}=y \Rightarrow x^2=y+yx^2$

Now x is defined if $\frac{y}{1-y}\ge 0$

$\Rightarrow 0 \le y < 1$, i.e., $y \in [0, 1)$

Range of f is $[0, 1) \ne R$

Co-domain of f. Hence 'f' is not onto.

$\therefore f$ is neither one-one nor onto.

Example 5. Find the form of composite function (fog) where

If $f(x)=-1+|x-1|, -1\le x\le 3$

$g(x)=2-|x+1|, -2\le x\le 2$

Solution

$fog(x) = f(g(x))$

$= f[2 - |x + 1|]; -2 \le x \le 2$

$= f(t)$; let $t = 2 - |x + 1|$

$= -1 + |t - 1|;$

where $-1 \le t \le 3$ and $-2 \le x \le 2$

$= -1 + |2 - |x + 1| - 1|;$

where $-1 \le 2-|x+1| \le 3$

and $-2 \le x \le 2$ ...(i)

Now solve $-1\le 2-|x+1|\le 3$

$\Rightarrow \quad -3\le -|x+1|\le 1$

$\Rightarrow \quad 3\ge |x+1|\ge -1$

$\Rightarrow \quad |x+1|\ge -1$ and $|x+1|\le 3$

$\Rightarrow \quad x \in R$ and $-3 \le x+1 \le 3$

$\Rightarrow \quad x \in R$ and $-4 \le x \le 2$

$\Rightarrow \quad -4 \le x \le 2$

$\therefore \quad$ and $-2 \le x \le 2$ ...(ii)

$fog(x)$ is defined as the intersection of

$\therefore \quad fog(x) = -1 + |1 - x + 1|; -2 \le x \le 2$

Example 6. Find the domain of the following:

(i) $\sqrt{x^2 - 5x + 6}$

(ii) $\sqrt{1-2x} + 3\sin^{-1}\frac{3x-1}{2}$

(iii) $\frac{1}{\log_{10}(1-x)} - x + 2$

(iv) $(1-3x)^{1/3} + 3\cos^{-1}\frac{2x-1}{3} + 3\tan^{-1}x$

(v) $6\sqrt{4^x + 8^{(2/3)(x-2)} - 52 - 2^{2(x-1)}}$

Solution

(i) $f(x)$ is defined.

$$x^2 - 5x + 6 \ge 0$$

$$x^2 - 5x + 6 = 0$$

$$(x-3)(x-2) = 0$$

Either $(x-3) = 0$ or $(x-2) = 0$

$\therefore \quad x = 2, 3$

Domain $= R = (-\infty, 2] \cup [3, \infty)$

(ii) $\sqrt{1-2x}$ takes real values $1 - 2x \ge 0$, i.e., $x \le 1/2$ and for $\sin^{-1}\frac{3x-1}{2}$, it lies in the interval $-1 \le \frac{3x-1}{2} \le 1$

$$-1 \le 3x \le 3 \Rightarrow \frac{-1}{3} \le x \le 1$$

Domain $= (-\infty, 1/2] \cap \left[\frac{-1}{3}, 1\right] = \left[\frac{-1}{3}, \frac{1}{2}\right]$

(iv) $\log_{10}(1-x)$ is defined if $1 - x > 0$ and $\frac{1}{\log_{10}(1-x)}$ is defined for $x \in (-\infty, 1)$

Also, $\sqrt{x+2}$ if $x + 2 \ge 0 \Rightarrow x \ge -2$

Domain $= [-2, 1) \sim \{0\}$

(v) Domain of $(1-3x)$ is R and $3^{\tan^{-1}x}$ is also R.

The domain of $3\cos^{-1}\frac{2x-1}{3} \Rightarrow -1 \le \frac{2x-1}{3} \le 1$

$$-1 \le x \le 2$$

$\therefore \quad$ Domain of f is $[-1, 2]$.

(vi) $f(x)$ is defined.

$$4^x + 8^{(2/3)(x-2)} - 52 - 2^{2(x-1)} \ge 0$$

$$2^{2x} + 2^{2(x-2)} - 2^{2(x-1)} \ge 52$$

$$2^{2x}[1 + 2^{-4} - 2^{-2}] = 52$$

$$2^{2x}\left[\frac{13}{16}\right] \ge 52 \Rightarrow 2^{2x} \ge 64 = 2^6$$

Hence $x \ge 3$, so domain of f is $[3, \infty)$.

Example 7. Find the range of the following functions:

(a) $\frac{x^2}{1+x^2}$ **(b)** $\log_e(3x^2 - 4x + 5)$

(c) $\frac{x^2 - x}{x^2 + 2x}$ **(d)** $\frac{x^2}{x^4 + 1}$

(e) $\sin x + \sin 2x$

Solution

(a) y is defined for every $x \in$ R. Hence, domain is R.

$\therefore \quad y = \frac{x^2}{1+x^2}$, we have

$$x^2y + y = x^2 \Rightarrow x^2 = \frac{y}{1-y}$$

$\therefore \quad x = \pm\sqrt{\frac{y}{1-x}}$ is defined, $\frac{y}{1-y} \ge 0$

$\Rightarrow \quad y \in [0, 1)$

(b) y is defined $(3x^2 - 4x + 5) > 0$

The discriminant is $(-4)^2 - 4 \times 3 \times 5 = -44 < 0$

The coefficient of x^2 is +ve.

Hence $(3x^2 - 4x + 5) > 0$ (for every $x \in$ R.)

$\therefore \quad$ Domain is R

Now, $\quad y = \log_e(3x^2 - 4x + 5)$

$$3x^2 - 4x + 5 = e^y$$

$$3x^2 - 4x + (5 - e^y) = 0$$

Since x is real, $D \ge 0$

i.e., $4^2 - 4.3(5 - e^y) \ge 0$

$$12e^y \ge 44 \Rightarrow e^y \ge \frac{11}{3}$$

Hence range is $\left[\log\left(\frac{11}{3}\right), \infty\right)$

(c) Domain is $x \in R - \{-2, 0\}$

Now $y = \frac{x-1}{x+2}$ where $x \ne 0$

$\Rightarrow \quad xy + 2y = x - 1$

$$x(y-1) = -(1 + 2y)$$

$$x = \frac{1+2y}{1-y}$$

$\therefore$ x is real if $y - 1 \neq 0 \Rightarrow y \neq 1$

Also, because $y = \dfrac{x-1}{x+2}$ for $x \neq 0$.

Hence, we have to exclude the value of y obtain when $x \to 0$

$$\lim_{x \to 0} \frac{x-1}{x+2} = -\frac{1}{2}$$

$\therefore$ Range is $y \in R - \left\{\dfrac{-1}{2}, 1\right\}$

(d) Clearly, $f(x) \geq 0$ for every $x \in R$

More over $(x^2 - 1)^2 = x^4 - 2x^2 + 1 \geq 0$ for every $x \in R$

So $$\frac{x^2}{x^4+1} \leq \frac{1}{2}$$

Hence range of f is [0, ½]

(d) The given function is the sum of two sines. Therefore, the minimum and maximum values could be –2 and +2, provided they are attainable. We easily note that sin and sin2x cannot become –1 simultaneously. Therefore, the range of the given function will be a proper subset of [–1, 1]. We will have to resort to calculus techniques.

Example 8. Let C be the set of complex numbers. Prove that the mapping f: C→R given by $f(z) = |z|$, $\forall\, z \in$ C, is neither one-one nor onto.

Solution

Given, f: C → R such that $f(z) = |z|$, $\forall\, z \in$ C

$$z = 6 + 8i$$

Then,

$$f(6 + 8i) = |6 + 8i| = \sqrt{(6^2 + 8^2)} = \sqrt{100} = 10$$

and, for $z = 6 - 8i$

$$f(6 - 8i) = |6 - 8i| = \sqrt{(6^2 + 8^2)} = \sqrt{100} = 10$$

Hence, $f(z)$ is many–one.

Also, $|z| \geq 0$, $\forall\, z \in$ C

But the co-domain given is 'R'

Therefore, $f(z)$ is not onto.

Example 9. Give an example of a mapping

(i) which is one-one but not onto.

(ii) which is not one-one but onto.

(iii) which is neither one-one nor onto.

Solution

(i) Let f: N → N, be a mapping defined by $f(x) = x^2$

For $f(x_1) = f(x_2)$

Then, $x_1^2 = x_2^2$

$x_1 = x_2$ (Since $x_1 + x_2 = 0$ is not possible)

Further, f is not onto, as for $1 \in$ N, there does not exist any x in N such that $f(x) = 2x + 1$.

(ii) Let f: R → [0, ∞), be a mapping defined by $f(x) = |x|$

Then, we see that $f(x)$ is not one–one as $f(2) = f(-2)$.

But $|x| \geq 0$, so range is [0, ∞].

Therefore, $f(x)$ is onto.

(iii) Let f: R → R, be a mapping defined by $f(x) = x^2$

Then clearly $f(x)$ is not one–one as $f(1) = f(-1)$.

Also range of $f(x)$ is [0, ∞).

Therefore, $f(x)$ is neither one–one nor onto.

Example 10. Let A = [–1, 1]. Then, discuss whether the following functions defined on A are one–one, onto or bijective:

(i) $f(x) = x/2$ (ii) $g(x) = |x|$

(iii) $h(x) = x|x|$ (iv) $k(x) = x^2$

Solution

Given, A = [–1, 1]

(i) f: [–1, 1] → [–1, 1], $f(x) = x/2$

Let $f(x_1) = f(x_2)$

$$x_1/2 = x_2$$

So, $f(x)$ is one–one.

Also, $x \in [-1, 1]$

$$x/2 = f(x) = [-1/2, 1/2]$$

Hence, the range is a subset of co-domain 'A'

So, $f(x)$ is not onto.

Therefore, $f(x)$ is not bijective.

(ii) $g(x) = |x|$

Let $g(x_1) = g(x_2)$

$$|x_1| = |x_2|$$

$$x_1 = \pm x_2$$

So, $g(x)$ is not one–one

Also, $g(x) = |x| \geq 0$, for all real x

Hence, the range is [0, 1], which is subset of co-domain 'A'

So, $f(x)$ is not onto

Therefore, $f(x)$ is not bijective.

(iii) $h(x) = x|x|$

Let $h(x_1) = h(x_2)$

$$x_1|x_1| = x_2|x_2|$$

If $x_1, x_2 > 0$

$$x_1^2 = x_2^2$$
$$x_1^2 - x_2^2 = 0$$
$$(x_1 - x_2)(x_1 + x_2) = 0$$
$$x_1 = x_2 \text{ (as } x_1 + x_2 \neq 0)$$

Similarly, for $x_1, x_2 < 0$, we have $x_1 = x_2$.
We see that for x_1 and x_2 of opposite sign, $x_1 \neq x_2$.
Hence, $f(x)$ is one–one.
For $x \in [0, 1], f(x) = x^2 \in [0, 1]$
For $x < 0, f(x) = -x^2 \in [-1, 0)$
Hence, the range is $[-1, 1]$.
So, $h(x)$ is onto.
Therefore, $h(x)$ is bijective.

(iv) $k(x) = x^2$
Let $k(x_1) = k(x_2)$

$$x_1^2 = x_2^2$$
$$x_1 = \pm x_2$$

Therefore, $k(x)$ is not one−one.

Example 11. Let A = {1, 2, 3, ..., 9} and R be the relation in A × A defined by (*a*, *b*) R (*c*, *d*) if $a + d = b + c$ for (*a*, *b*), (*c*, *d*) in A × A. Prove that R is an equivalence relation and also obtain the equivalent class [(2, 5)].

Solution
Given, A = {1, 2, 3, ... 9} and (*a*, *b*) R (*c*, *d*) if $a + d = b + c$ for $(a, b), (c, d) \in A \times A$.
Let (*a*, *b*)R(*a*, *b*)
So, $a + b = b + a$, $\forall\, a, b \in A$, which is true for any $a, b \in A$.
Therefore, R is reflexive.
Let (*a*, *b*)R(*c*, *d*)
Then,

$$a + d = b + c \Rightarrow c + b = d + a$$
$$(c, d)R(a, b)$$

Therefore, R is symmetric.
Let (*a*, *b*)R(*c*, *d*) and (*c*, *d*)R(*e*, *f*)

$$a + d = b + c \text{ and } c + f = d + e$$
$$a + d = b + c \text{ and } d + e = c + f$$
$$(a + d) - (d + e = (b + c) - (c + f)$$
$$a - e = b - f$$
$$a + f = b + e$$
$$(a, b)R(e, f)$$

So, R is transitive.
Therefore, R is an equivalence relation.
and [(2, 5) = (1, 4), (2, 5), (3, 6), (4, 7), (5, 8), (6, 9)] is the equivalent class under relation R.

RECAPITULATION (FORMULAE, THEOREM, IMPORTANT RESULTS)

1. Universal relation is the relation R in A given by $R = A \times A$.
2. Empty relation is the relation R in A given by $R = \phi \subset A \times A$.
3. Reflexive relation R in A is a relation with $(a, a) \in R\ \forall a \in X$.
4. Symmetric relation R in A is a relation satisfying $(a, b) \in R$ implies $(b, a) \in R$.
5. Transitive relation R in A is a relation satisfying $(a, b) \in R$ and $(b, c) \in R$ implies that $(a, c) \in R$.
6. Equivalence relation R in A is a relation, which is reflexive, symmetric and transitive.
7. Equivalence class [a] containing $a \in A$ for an equivalence relation R in A is the subset of A containing all elements *b* related to *a*.
8. A function $f : X \to Y$ is defined as one–one, also called injective function, if the images of distinct elements of X under *f* are distinct, i.e., if no elements in X have the same image in Y, i.e., $x_1, x_2 \in X, f(x_1) = f(x_2) \Rightarrow x_1 = x_2$ or $x_1 \neq x_2 \Rightarrow f(x_1) \neq f(x_2)$.
9. To test whether a function is one–one (injective) consider $x_1, x_2 \in X$ and assume that $f(x_1) = f(x_2)$ always gives $x_1 = x_2$, then the function $f : X \to Y$ is one–one, otherwise *f* is not one-one.
10. If there are at least two elements in the domain whose images are the same, the function is known as many–one.
    Let $f : X \to Y$, if two or more than two elements in X have the same image in Y, then *f* is said to be many-one.
11. Note that a function will either be one–one or many–one.
12. **Into Function:** A function $f : X \to Y$ is called into function, if there exists at least one element in Y which has no pre-image in X.

13. A function $f : X \to Y$ is defined to be onto, which is also called surjective function, if every element of Y is the image of some element of X under f, i.e., for every $y \in Y$ there exists an element $x \in X$ such that $f(x) = y$.
14. Note that a function will either be into or onto.
15. **One–One Into Function:** A function $f : X \to Y$ is said to be one–one into function if different elements in X have different f images in Y and there exists an element in Y having no pre-image in A.
16. **One–One Onto Function (Bijection):** A function $f : X \to Y$ is said to be one–one onto if different dements in X have different f images in Y and there is no element in Y having no pre-image in X.
17. A function which is both one–one and onto is called bijective function.

QUIZ

1. Is the relation aRb iff a divides b; $a, b \in N$ is reflexive?
2. What do you mean by partial order relation?
3. What is an Equivalence relation?
4. What is the relation between Domain and Range of an identity relation on a set A (IA)?
5. Under what condition two functions $f : A \to B$ and $g : C \to D$ are identical?
6. What is the basic difference between a relation and function?
7. The statement "If f is invertible, then f must be one–one and onto" is true. Is the converse of this statement also true?
8. Consider $f : \{4, 5, 6\} \to \{a, b, c\}$ given by $f(4) = a, f(5) = b, f(6) = c$ what is the type of this function.
9. Find f^{-1} for the above function.
10. f and g are two functions. Is the composition fog is same as gof?

JEE MAIN

Objective Questions

One Option Correct

1. Let $g(x) = 1 + x - [x]$ and $f(x) = \begin{cases} -1, & x < 0 \\ 0, & x = 0 \\ 1, & x > 0 \end{cases}$, then for all x, $f[g(x)]$ is equal to ________.

   (a) x (b) 1
   (c) $f(x)$ (d) $g(x)$

2. If $f(x) = \cos(\log x)$, then $f(x).f(y) - \frac{1}{2}\left[f\left(\frac{x}{y}\right) + f(xy)\right]$ has the values ______.

   (a) –1 (b) $\frac{1}{2}$
   (c) –2 (d) None of these

3. Let function f: R → R be defined by $f(x) = 2x + \sin x$, for $x \in$ R. Then f is

   (a) one to one and onto.
   (b) one-to-one but not onto.
   (c) onto but not one-to-one.
   (d) neither one-to-one nor onto.

4. If f: R → R and is defined by $f(x) = \frac{1}{2 - \cos 3x}$ for each $x \in$ R, then the range of f is ______.

   (a) [1/3, 1] (b) [1/3, 1]
   (c) [1, 2] (d) [1, 2]

5. If f : → R and g : R → are defined by $f(x) = x - [x]$ and $g(x) = [x]$ for $x \in$ R, where $[x]$ is the greatest integer not exceeding x, then for every $x \in$ R, $f(g(x))$ is equal to ______.

   (a) x (b) 0
   (c) $f(x)$ (d) $g(x)$

6. If $f : R \to R$ is defined by $f(x) = [2x] - 2[x]$ for $x \in$ R, where $[x]$ is the greatest integer not exceeding x, then the range of f is ______.

   (a) $\{x \in R : 0 \le x \le 1\}$ (b) $\{0,1\}$
   (c) $\{x \in R : x > 0\}$ (d) $\{x \in R : x \le 0\}$

7. If $f : R \to R$ is defined by

$$f(x)\begin{cases} x+4 & \text{for} \quad x<-4 \\ 3x+2 & \text{for} \quad -4 \le x < 4 \\ x-4 & \text{for} \quad x \ge 4 \end{cases}$$

Then the correct matching of List I from List II is

List I	List II
(A) $f(-5)+f(-4)$	(P) 14
(B) $f(\lvert f(-8)\rvert)$	(Q) 4
(C) $f(f(-7)+f(3))$	(R) –11
(D) $(f(f(f(0))))=1$	(S) –1
	(T) 1
	(U) 0

	A	B	C	D
(a)	(R)	(U)	(Q)	(T)
(b)	(R)	(S)	(Q)	(T)
(c)	(S)	(R)	(Q)	(P)
(d)	(R)	(U)	(T)	(Q)

8. The function $f : C \to C$ defined by $f(x)=\dfrac{ax+b}{cx+d}$ for $x \in C$ where $bd \neq 0$ reduces to a constant function, if ______.
(a) $a = c$ (b) $b = d$
(c) $ad = bc$ (d) $ab = cd$

9. If N denotes the set of all positive integers and if $f : N \to N$ is defined by $f(n) =$ the sum of positive divisors of n then, $f(2^k.3)$, where k is a positive integer, is ______.
(a) $2^{k+1}-1$ (b) $2(2^{k+1}-1)$
(c) $3(2^{k+1}-1)$ (d) $4(2^{k+1}-1)$

10. If $f : N \to Z$ is defined by

$$f(n)=\begin{cases} 2 & \text{if} \quad n=3k, k \in Z \\ 10 & \text{if} \quad n=3k+1, k \in Z \\ 0 & \text{if} \quad n=3k+2, k \in Z \end{cases}$$

then $\{n \in N : f(n) > 2\} =$ ______.
(a) {3, 6, 4} (b) {1, 4, 7}
(c) {4, 7} (d) {7}

11. The function $f : R \to R$ is defined by $f(x) = 3^{-x}$. Observe the following statements:
I. f is one–one
II. f is onto
III. f is a decreasing function
Out of these, the true statement are ______.
(a) Only I, II (b) Only II, III
(c) Only I, III (d) I, II, III

12. Let $A = \{x \in R, x \neq 0, -4 \le x \le 4\}$ and $f : A \to R$ defined by $f(x)=\dfrac{|x|}{x}$ for $x \in A$. Then the range of f is ______.
(a) $\{1, -1\}$ (b) $\{x : 0 \le x \le 1\}$
(c) 1 (d) $\{x : -4 \le x \le 0\}$

13. Let R be the real line. Consider the following subsets of the plane R × RS
$S = \{(x, y) : y = x + 1 \text{ and } 0 < x < 2\}$
$T = \{x, y) : x - y \text{ is an integer}\}$
Which of the following is true?
(a) Both S and T are equivalence relations on R.
(b) S is an equivalence relation on R but T is not.
(c) T is an equivalence relation on R but S is not.
(d) Neither S nor T is an equivalence relation on R.

14. Let W denote the words in the English dictionary. Define the relation, R is:
(a) reflexive, symmetric and not transitive
(b) reflexive, symmetric and transitive
(c) reflexive, not symmetric and transitive
(d) not reflexive, symmetric and transitive.

15. If f: R → R satisfies $f(x+y)=f(x)+f(y)$, for all
$x, y \in R$ and $f(1)=7$, then $\sum_{r=1}^{n} f(r)$ is ______.
(a) $\dfrac{7n}{2}$ (b) $\dfrac{7(n+1)}{2}$
(c) $7n(n+1)$ (d) $\dfrac{7n(n+1)}{2}$

16. Suppose $f(x)=(x+1)^2$ for $x \ge -1$. If $g(x)$ is the function whose graph is the reflection of the graph of $f(x)$ with respect to the line $y = x$, then $g(x)$ equals ______.
(a) $-\sqrt{x}-,1\ x \ge 0$ (b) $\dfrac{1}{(x+1)^2}, x > -1$
(c) $\sqrt{x+1}\ , x \ge -1$ (d) $\sqrt{x}-1, x \ge 0$

17. Let function $f : R \to R$ be defined by $f(x) = 2x + \sin x$ for $x \in R$. Then f is
(a) one–one and onto.
(b) one–one but NOT onto.
(c) onto NOT one–one.
(d) neither one–one nor onto.

18. $f(x)=\dfrac{1}{x^2}$, is
(a) symmetric about x-axis.
(b) symmetric about y-axis.
(c) symmetric about x- and y-axes.
(d) None of these.

19. If $f(x)=\log\left(x+\sqrt{x^2+1}\right)$, then $f(x)$ is
(a) even function. (b) odd function.
(c) periodic function. (d) None of these.

20. Which of the following is one-one function ______.
(a) e^x (b) e^2
(c) $\sin x$ (d) None of these.

21. The relation of "congruence modulo" is
(a) reflexive only.
(b) symmetric only.
(c) transitive only.
(d) an equivalence relation.

22. If f is any function, then $\frac{1}{2}[f(x)+f(-x)]$ is always
(a) odd function.
(b) even function.
(c) neither even nor odd function.
(d) one-one function.

23. Let R be a relation on the set N of natural number defined by $n\text{R}m$ If n divides m. Then R is
(a) Reflexive and Symmetric.
(b) Transitive and Symmetric.
(c) Equivalence.
(d) Reflexive, Transitive but Not Symmetric.

24. Set A has 3 elements and the set B has 4 elements. Then the number of injective mappings that can defined A to B is ______.
(a) 144 (b) 12
(c) 24 (d) 64

25. Let f : R → R be defined by $f(x)=x^2+1$. Then, pre-images of 17 and –3, respectively, are ______.
(a) $\phi,\{4,-4\}$ (b) $\{3,-3\},\phi$
(c) $\{4,-4\},\phi$ (d) $\{4,-4,\{2,-2\}$

26. Let $f(x)$ be defined for all $x>0$ and be continuous. Let $f(x)$ satisfy $f\left(\frac{x}{y}\right)=f(x)-f(y)$ for all x, y and $f(e)=1$. Then
(a) $f(x)$ is bounded.
(b) $f\left(\frac{1}{x}\right)\to 0$ as $x\to 0$
(c) $xf(x)\to 1$ as $x\to 0$
(d) $f(x)=\log x$

27. If X and Y are two non-empty sets where $f:\text{X}\to\text{Y}$, is function is defined such that $f(\text{C})=\{f(x):x\in\text{C}\}$ for $\text{C}\subseteq\text{X}$ and $f^{-1}(\text{D})=\{x:f(x)\in\text{D}\}$ for $\text{D}\subseteq\text{Y}$ for any $\text{A}\subseteq\text{Y}$ and $\text{B}\subseteq\text{Y}$, then
(a) $f^{-1}\{f(\text{A})\}=\text{A}$
(b) $f^{-1}\{a(\text{A})\}=A$, only if $f(\text{X})=\text{Y}$
(c) $f\{f^{-1}(B)\}=B$, only if $B\subseteq f(x)$
(d) $f\{f^{-1}(B)\}=B$

28. If $f(x)=\sin x+\cos x$, $g(x)=x^2-1$, then $g\{f(x)\}$ is invertible in the domain ______.
(a) $\left[0,\frac{\pi}{2}\right]$ (b) $\left[-\frac{\pi}{4},\frac{\pi}{4}\right]$
(c) $\left[-\frac{\pi}{2},\frac{\pi}{2}\right]$ (d) $[0,\pi]$

29. Suppose $f(x)=(x+1)^2$ for $x\geq -1$. If $g(x)$ is the function whose graph is reflection of the graph of $f(x)$ with respect to the line $y=x$, then $g(x)$ equals ______.
(a) $-\sqrt{x}-1, x\geq 0$ (b) $\frac{1}{(x+1)^2}, x>-1$
(c) $\sqrt{x+1}, x\geq -1$ (d) $\sqrt{x}-1, x\geq 0$

30. If $f(x)=3x-5$, then $f^{-1}(x)$
(a) is given by $\frac{1}{3x-5}$.
(b) is given by $\frac{x+5}{3}$.
(c) does not exist because f is not one–one.
(d) does not exist because f is not onto.

31. Which of the following functions is periodic?
(a) $f(x)=x-[x]$, where $[x]$ denotes the greatest integer less than or equal to the real number x
(b) $f(x)=\sin(1/x)$ for $x\neq 0$, $f(0)=0$
(c) $f(x)=x\cos x$
(d) None of the above.

32. In order that a relation R defined on a non-empty set A is an equivalence relation, it is sufficient if R
(a) is reflexive.
(b) is symmetric.
(c) is transitive.
(d) possesses all the above three properties.

33. With reference to a universal set, the inclusion of a subset in another is a relation, which is
(a) Symmetric only. (b) Equivalence only.
(c) Reflexive only. (d) None of these.

34. Let A = {1, 2, 3}. Then number of equivalence relation containing (1, 2) is ______.
(a) 1 (b) 2
(c) 3 (d) 4

35. Given the relation $R = \{(1,2),(2,3)\}$ in the set $A = \{1,2,3\}$. Then the minimum number of ordered pairs, which when added to R, makes it an equivalence relation is ______.
(a) 5 (b) 6
(c) 7 (d) 8

36. For real x, let $f(x) = x^3 + 5x + 1$. Then
(a) f is one–one but not onto in R.
(b) f is onto R but not one–one in R.
(c) f is one–one and onto in R.
(d) f is neither one–one nor onto in R.

37. Let $f : N \to Y$ be a function defined as $f(x) = 4x + 3$ where $Y = \{y \in N : y = 4x + 3 \text{ for some } x \in N\}$. Show that f is invertible and its inverse is ______.
(a) $g(y) = 4 + \frac{y+3}{4}$ (b) $g(y) = \frac{y+3}{4}$
(c) $g(y) = \frac{y-3}{4}$ (d) $g(y) = \frac{3y+4}{3}$

38. If $\phi(x)$ is the inverse of the function $f(x)$ and $f'(x) = \frac{1}{1+x^5}$ then $\frac{d}{dx}\phi(x)$ is ______.
(a) $\frac{1}{1+\{\phi(x)\}^5}$ (b) $\frac{1}{1+\{f(x)\}^5}$
(c) $1+\{\phi(x)\}^5$ (d) $1+f(x)$

39. The value of $(A \cup B \cup C) \cap (A \cap B^c \cap C^c) \cap C^c$ is
(a) $B \cap C^c$ (b) $B^c \cap C^c$
(c) $B \cap C$ (d) $A \cap B \cap C$

40. The domain of the function $f(x) = \frac{\cos^{-1} x}{[x]}$ is ______.
(a) $[-1, 0) \cup \{1\}$ (b) $[-1, 1]$
(c) $[-1, 1]$ (d) None of these.

41. A set A has 3 elements and another set B has 6 elements. Then ______.
(a) $3 \le n(A \cup B) \le 6$ (b) $3 \le n(A \cup B) \le 9$
(c) $6 \le n(A \cup B) \le 9$ (d) $0 \le n(A \cup B) \le 9$

42. If $f(x) = \frac{x}{x-1}$, then $\underbrace{(fofof\ldots of)(x)}_{19\,\text{times}}$ is equal to ______.
(a) $\frac{x}{x-1}$ (b) $\left(\frac{x}{x-1}\right)^{19}$
(c) $\frac{19x}{x-1}$ (d) x

43. If A = {x, y} then the power set of A is ______.
(a) $\{x^y, y^x\}$
(b) $\{\phi, x, y\}$
(c) $\{\phi, \{x\}, \{2y\}\}$
(d) $\{\phi, \{\{x\}, \{y\}, \{x, y\}\}$

44. Which of the following is a true statement?
(a) $\{a\} \in \{a, b, c\}$ (b) $\{a\} \subseteq \{a, b, c\}$
(c) $\phi \in \{a, b, c\}$ (d) None of these.

45. The number of reflexive relations of a set with four elements is equal to ______.
(a) 2^{16} (b) 2^{12}
(c) 2^8 (d) 2^4

46. If A, B, C is 3 non-empty sets, then $(A \cap B) \cap (B \cap C) \cap (C \cap A)$ is equal to ______.
(a) $A \cap B \cap C$ (b) $A \cup B \cup C$
(c) ϕ (d) None of these.

47. If sets A and B are defined as
$A = \{(x, y), y = e^x, x \in R\}$
$B = \{(x, y), y = x, x \in R\}$
Then, ______.
(a) $B \subset A$ (b) $A \subset B$
(c) $A \cap B = \phi$ (d) $A \cup B = A$

48. Let A = {x : x is a multiple of 3} and B = {x : x is a multiple of 5}. Then $A \cap B$ is given by ______.
(a) {3, 6, 9, ...} (b) {5, 10, 15, 20, ...}
(c) {15, 30, 45, ...} (d) None of these.

49. If $f(x+y, x-y) = xy$, then the arithmetic mean of $f(x, y)$ and $f(y, x)$ is ______.
(a) x (b) y
(c) 0 (d) None of these.

50. The domain of the function $f(x) = \log_e(x - [x])$ is ______.
(a) R (b) R – Z
(c) $(0, +\infty)$ (d) Z

51. Let $A = R - 3\{3\}, B = R - \{1\}$. Let $f : A \to B$ be defined by $f(x) = \frac{x-2}{x-3}$. Then

(a) f is bijective.
(b) f is one–one but not onto.
(c) f is onto but not one–one.
(d) None of these.

52. Which of the following functions is inverse of itself?

(a) $f(x) = \frac{1-x}{1+x}$ (b) $f(x) = 3^{\log x}$
(c) $f(x) = 3^{x(x+1)}$ (d) None of these.

53. If $A = \{1,2,3\}$ and $B = \{3,8\}$, then $(A \cup B) \times (A \cap B)$ is ______.

(a) {(3, 1), (3, 3), (3, 8)}
(b) {(1, 3), (2, 3), (3, 3), (8, 3)}
(c) {(1, 2), (2, 2), (3, 3), (8, 8)}
(d) {(8, 3), (8, 2), (8, 1), (8, 8)}

54. The number of onto mappings from the set $A = \{1, 2, \ldots, 100\}$ to set $B = \{1, 2\}$ is ______.

(a) $2^{100} - 2$ (b) 2^{100}
(c) $2^{99} - 2$ (d) 2^{99}

55. Suppose $A_1, A_2, \ldots, A_{30}$ are thirty sets each having 5 elements and $B_1, B_2, \ldots, B_n$ are n sets each with 3 elements, let $\bigcup_{i=1}^{30} A_i = \bigcup_{j=1}^{n} B_j = S$ and each element of S belongs to exactly 10 of the $A_i's$ and exactly 9 of the $B_i's$. Then n is equal to ______.

(a) 115 (b) 83
(c) 45 (d) None of these.

56. If $f : R \to c$ is defined by $f(x) = e^{21x}$ for $x \in R$, then f is (where c, denotes set of all complex numbers) ______.

(a) one–one
(b) onto
(c) one-one and onto
(d) neither one–one nor onto

Questions from Previous Years

57. Let R_1 and R_2 be two relations defined as follows:
$R_1 = \{(a,b) \in R^2 : a^2 + b^2 \in Q\}$ and
$R_2 = \{(a,b) \in R^2 : a^2 + b^2 \notin Q\}$ where Q is the set of all rational numbers. Then:

(a) Neither R_1 nor R_2 is transitive.
(b) R_2 is transitive but R_1 is not transitive.
(c) R_1 is transitive but R_2 is not transitive.
(d) R_1 and R_2 are both transitive.

58. If $R = \{(x, y) : y \in Z, x^2 + 3y^2 \le 8\}$ is a relation on the set integer, then the domain of R^{-1} is ______.

(a) $\{-2, -1, 1, 2\}$ (b) $\{0, 1\}$
(c) $\{-2, -1, 0, 1, 2\}$ (d) $\{-1, 0, 1,\}$

59. For a suitable chosen real constant a, let a function $f : R - \{a\} \to R$ be defined by $f(x) = \frac{a-x}{a+x}$. Further suppose that for any real number $x \ne -a$ and $f(x) \ne -a$, $(fof)(x) = x$. Then $f\left(-\frac{1}{2}\right)$ is equal to ______.

(a) $\frac{1}{3}$ (b) $-\frac{1}{3}$
(c) -3 (d) 3

60. The inverse function $f(x) = \frac{8^{2x} - 8^{-2x}}{8^{2x} + 8^{-2x}}, x \in (-1, 1)$ is

(a) $\frac{1}{4}(\log_8 e)\log_e\left(\frac{1-x}{1+x}\right)$
(b) $\frac{1}{4}\log_e\left(\frac{1-x}{1+x}\right)$
(c) $\frac{1}{4}(\log_8 e)\log_e\left(\frac{1+x}{1-x}\right)$
(d) $\frac{1}{4}\log_e\left(\frac{1+x}{1-x}\right)$

61. For $x, \in \left(0, \frac{3}{2}\right)$, let $f(x) = \sqrt{x}$, $g(x) = \tan x$ and $h(x) = \frac{1-x^2}{1+x^2}$. If $\phi(x) = ((hof)og)(x)$, then $\phi\left(\frac{\pi}{3}\right)$ is equal to ______.

(a) $\tan\frac{\pi}{12}$ (b) $\tan\frac{11\pi}{12}$
(c) $\tan\frac{7\pi}{12}$ (d) $\tan\frac{5\pi}{12}$

62. Let $f(x) = x^2, x \in R$. For any $A \subseteq R$, define $g(A) = \{x \in R : f(x) \in A\}$. If $S = [0, 4]$, then which one of the following statements is not true?

(a) $f(g(S)) = S$ (b) $g(f(S)) \neq S$
(c) $g(f(S)) = g(S)$ (d) $f(g(S)) \neq f(S)$

63. Let $\sum_{k=1}^{10}(a+k) = 16(2^{10}-1)$, where the function f satisfies $f(x+y) = f(x)f(y)$ for all natural numbers x, y and $f(1) = 2$. Then, the natural number a is ______.
(a) 2 (b) 4
(c) 3 (d) 16

64. If $f(x) = \log_e\left(\frac{1-x}{1+x}\right), |x| < 1$, then $f\left(\frac{2x}{1+x^2}\right)$ is equal to ______.
(a) $2f(x)$ (b) $2f(x^2)$
(c) $(f(x))^2$ (d) $-2f(x)$

65. For $x \in R - \{0, 1\}$, let $f_1(x) = \frac{1}{x}$, $f_2(x) = 1-x$ and $f_3(x) = \frac{1}{1-x}$ be three given functions. If a function, $J(x)$ satisfies $(f_2 oJo f_1)(x) = f_3(x)$, then $J(x)$ is equal to ______.
(a) $f_2(x)$ (b) $f_3(x)$
(c) $f_1(x)$ (d) $\frac{1}{x}f_3(x)$

66. If the function $f: R - \{1, -1\} \to A$ defined by $f(x) = \frac{x^2}{1-x^2}$, is surjective, then A is equal to ______.
(a) $R - \{-1\}$ (b) $[0, \infty)$
(c) $R - [-1, 0)$ (d) $R - (-1, 0)$

67. Let a function $f: (0, \infty) \to (0, \infty)$ be defined by $f(x) = \left|1 - \frac{1}{x}\right|$. Then, f is
(a) injective only.
(b) both injective as well as surjective.
(c) not injective but it is surjective.
(d) neither injective nor surjective.

68. Let N be the set of natural numbers and two functions f and g be defined as $f, g: N \to N$ such that

$$f(n) = \begin{cases} \frac{n+1}{2}; & \text{if } n \text{ is odd} \\ \frac{n}{2}; & \text{if } n \text{ is even} \end{cases}$$

and $g(n) = n - (-1)^n$. Then, fog is
(a) one–one but not onto.
(b) onto but not one–one.
(c) both one–one and onto.
(d) neither one–one nor onto.

69. Let $A = \{x \in R : x \text{ is not a positive integer}\}$. Define a function $f: A \to R$ as $f(x) = \frac{2x}{x-1}$, then f is
(a) injective but not surjective.
(b) not injective.
(c) surjective but not injective.
(d) neither injective nor surjective.

70. The function $f: R \to \left[-\frac{1}{2}, \frac{1}{2}\right]$ defined as $f(x) = \frac{x}{1+x^2}$ is
(a) invertible.
(b) injective but not surjective.
(c) surjective but not injective.
(d) neither injective nor surjective.

71. The function $f: [0, 3] \to [1, 29]$, defined by $f(x) = 2x^3 - 15x^2 + 36x + 1$, is
(a) one–one and onto.
(b) onto but not one–one.
(c) one–one but not onto.
(d) neither one–one nor onto.

72. Let $a, b, c \in R$. If $f(x) = ax^2 + bx + c$ be such that $a + b + c = 3$ and $f(x+y) = f(x) + f(y) + xy, \forall\, x, y \in R$, then $\sum_{n=1}^{10} f(n)$ is equal to ______.
(a) 330 (b) 165
(c) 190 (d) 255

73. Let $f(x) = x^2$ and $g(x) = \sin x$ for all $x \in R$. Then, the set of all x satisfying $(fogogof)(x) = (gogof)(x)$, where $(fog)(x) = f(g(x))$, is ______.
(a) $\pm\sqrt{n\pi}, n \in \{0, 1, 2, \ldots\}$
(b) $\pm\sqrt{n\pi}, n \in \{1, 2 \ldots\}$
(c) $\pi/2 + 2n\pi, n \in \{\ldots, -2, -1, 0, 1, 2, \ldots\}$
(d) $2n\pi, n \in \{\ldots, -2, -1, 0, 1, 2, \ldots\}$

Integer and Numerical Value Type

1. If

$$f(x) = \sin^2 x + \sin^2\left(x + \frac{\pi}{3}\right) + \cos x \cos\left(x + \frac{\pi}{3}\right)$$

and $g\left(\frac{5}{4}\right) = 1$, then $(gof)(x) =$ ______.

2. If $f(x) = (a - x^n)^{1/n}$, where $a > 0$ and n is a positive integer, then $f[f(x)] =$ ______.

3. The function
$f(x)=\frac{x^3}{3}+(m-1)x^2+(m+5)x+11$ is invertible only for k integer values of m. Then k must be ______.

4. There are exactly two distinct linear functions, _____ and _____, which map $\{-1,1\}$ onto $\{0,2\}$.

5. A function $f:\text{IR}\to\text{IR}$, where IR, is the set of real numbers, is defined by $f(x)=\frac{\alpha x^2+6x-8}{\alpha+6x-8x^2}$. Find the interval of values of α for which f is onto. Are the functions one-to-one for $\alpha=3$? Justify your answer.

6. Let A and B be two sets each with a finite number of elements. Assume that there is an injective mapping from A to B and that there is an injective mapping from B to A. Prove that there is a bijective mapping from A to B.

Assertion and Reason

(a) Both Statement 1 and Statement 2 are True and Statement 2 is correct explanation of Statement 1.
(b) Both Statement 1 and Statement 2 are True and Statement 2 is not correct explanation of Statement 1.
(c) Statement 1 is True, Statement 2 is False.
(d) Statement 1 is False, Statement 2 is True.

1. Consider the function $f(n)=\sin\frac{1}{n}$ where $n\in N$.

Statement 1: The function cannot become 0 because

Statement 2: $n\to\infty \sin\frac{1}{n}\to 0$.

2. Consider the function $f(x)=\cot(\sin x), g(x)=\cot x$.
Statement 1: Domain of $f(x)$ and $g(x)$ are same
because
Statement 2: $\cot(\sin x=cox)$ has infinite solutions.

3. Let f be a function from A to B, g is a function from B to C and h is a function from C to D then
Statement 1: $((ho(gof(x))=((hog)of(x))$
because
Statement 2: composition of function is an associative operation.

4. Consider the function: $f(x)=x^{1/x}\ (x>0)$
Let $g(x)$ be inverse of $f(x)$.
Statement 1: $g'(x)=\frac{1}{x(1+\log g(x))}$ because
Statement 2: $g(f(x))=x$

5. **Statement 1:** Let $f(x)$ be a function whose derivative is $\frac{1}{1+x^5}$.
If $g(x)$ be inverse of $f(x)$ then $g'(x)=\frac{1}{1+(g(x))^5}$ because
Statement 2: $g(f(x))=f(g(x))=x$

JEE ADVANCED

Objective Questions

One Option Correct

1. Let $f(x)=\frac{\alpha x}{x+1}, x\neq -1$. Then, for what value of ± is $f[f(x)]=x$?
(a) $\sqrt{2}$ (b) $-\sqrt{2}$
(c) 1 (d) −1

2. If $g\{f(x)\}=|\sin x|$ and $f\{g(x)\}=(\sin\sqrt{x})^2$, then
(a) $f(x)=\sin^2 x, g(x)=\sqrt{x}$
(b) $f(x)=\sin x, g(x)=|x|$
(c) $f(x)=x^2, g(x)=\sin\sqrt{x}$
(d) f and g cannot be determined.

3. Let $f(x)=|x-1|$. Then, ______.
(a) $f(x^2)=\{f(x)\}^2$
(b) $f(x+y)=f(x)+f(y)$
(c) $f(|x|)=|f(x)|$
(d) None of the above.

4. If (o) is a binary operation such that $aob=ab+1$ where $a, b\in Z$, then the given function is
(a) commutative and associative
(b) commutative but not associative

(c) associative but not commutative
(d) neither associative nor commutative.

5. Let
$R = \{(3,3),(6,6),(9,9),(12,12),(6,12),(3,9),(3,1)\}$
be the relation on the set A = {3, 6, 9, 12}. The relation is
(a) reflexive only.
(b) reflexive and transitive only.
(c) reflexive and symmetric only.
(d) an equivalence relation.

6. A = {1, 2, 3}. Then the number of relations containing (1, 2) and (2, 3), which are reflexive and transitive but not symmetric is ______.
(a) 1 (b) 2
(c) 3 (d) 4

7. Let R be the real line. Consider the following subsets of the plane $R \times R$
$S = \{(x, y) : y = x+1 \text{ and } 0 < x < 2\}$
$T = \{(x, y) : x - y \text{ is an integer}\}$
Which of the following is true?
(a) T is an equivalence relation on R, but S is not.
(b) Neither S nor T is an equivalence relation on R.
(c) Both S and T equivalence relation on R.
(d) S is an equivalence relation on R.

8. Let $f : R \to R$ be defined as $f(x) = x^4$
(a) f is one–one onto.
(b) f is many–one onto.
(c) f is one–one but not onto.
(d) f is neither one–one nor onto.

9. If $f : R \to R$ be defined as $f(x) = 3x$
(a) f is one–one onto.
(b) f is many–one onto.
(c) f is one–one but not onto.
(d) f is neither one–one nor onto.

10. A function f from the set of natural numbers to integers
$$f(n) = \begin{cases} \dfrac{n-1}{2}, & \text{when } n \text{ is odd} \\ -\dfrac{n}{2}, & \text{when } n \text{ is even} \end{cases} \text{ is}$$
(a) is one–one but not onto.
(b) onto but not one–one.
(c) one–one and onto both.
(d) neither one–one nor onto.

11. A set contains 2n + 1 element. The number of subsets of this set containing more than n elements is equal to ______.
(a) $2^n - 1$ (b) 2^n
(c) $2^n + 1$ (d) 2^{2n}

12. If the function $f : R \to R$ is defined by $f(x) = |x|(x - \sin x)$, then which of the following statements is TRUE?
(a) f is one–one, but NOT onto.
(b) f is onto, but NOT one–one.
(c) f is BOTH one–one and onto.
(d) f is NEITHER one–one NOR onto.

One or More Than One Option Correct

1. If $f(x) = x^3 + x^2 f'(1) + xf''(2) + f'''(3) \forall x \in R$, then $f(x)$ is
(a) one–one and onto.
(b) one–one and into.
(c) many–one and onto.
(d) non-invertible.

2. If $g(x) = \left(4\cos^4 x - 2\cos 2x - \frac{1}{2}\cos 4x - x^7\right)^{\frac{1}{7}}$ then ________.
(a) $g(g(101)) = 100$ (b) $g(g(101)) = 101$
(c) $g(g)(0) = 0$ (d) $g(g(-1)) = -1$

3. If $f : R \to R$, $f(x) = e^{-|x|} - e^x$ is a given function, then which of the following are correct:
(a) f is many–one into function
(b) f is many on onto function
(c) range of f is $[0, \infty]$
(d) range of f is $[-\infty, 0]$.

4. If $f(x) = \cos[\pi^2]x + \cos[-\pi^2]x$, where $[x]$ stands for the greatest integer function, then ______.
(a) $f(\pi/2) = -1$ (b) $f(\pi) = 1$
(c) $f(-\pi) = 0$ (d) $f(\pi/4) = 1$

5. Let $g(x)$ be a function defined on $[-1, 1]$. If the area of the equilateral triangle with two of its vertices at (0, 0) and $[x, g(x)]$ is $\frac{\sqrt{3}}{4}$, then the function $g(x)$ is ______.
(a) $g(x) = \pm\sqrt{1-x^2}$ (b) $g(x) = \sqrt{1-x^2}$
(c) $g(x) = -\sqrt{1-x^2}$ (d) $g(x) = \sqrt{1+x^2}$

6. If $y = f(x) = \dfrac{x+2}{x-1}$, then ______.
(a) $x = f(y)$
(b) $f(1) = 3$
(c) y increases with x for $x < 1$
(d) f is a rational function of x

7. Which of the following functions is not injective?
(a) $f(x) = |x+1|, x \in [-1, 0]$
(b) $f(x) = x + 1/x, x \in (0, \infty)$
(c) $f(x) = x^2 + 4x - 5$
(d) $f(x) = e^{-x}, x \in [0, \infty)$

8. Let $f : R \to R$ be defined by $f(x) = [x]$ and $g(x) = \dfrac{3-2x}{4}$ then
(a) f is neither one–one nor onto.
(b) g is one–one but f is not one–one.
(c) f is one–one and g is onto.
(d) neither f nor g is onto.

9. If $f : R \to R$ is a real valued function given by, $f(x) = |x^2 - 4|x| + 12$ then, $f(x) = a$ has
(a) two distinct real roots if $a > 3$.
(b) four distinct real roots if $3 > a > 1$.
(c) can have at most 8 real roots.
(d) have some of real-valued functions given.

10. Let $f(x)$ and $g(x)$ be two real valued functions given by, $f(x) = -\ln x$ and $g(x) = e^{-x}$.
Let $h(x) = f(x) - x$ and $m(x) = g(x) - x$. Furthermore, let the number of solutions of $h(x) = 0$ and $m(x) = 0$ be a and b, then
(a) $a \neq b$
(b) $a = b$
(c) $a = 1$ and $b = 1$
(d) None of these.

11. Let $f(x)$ be invertible function and let $f^{-1}(x)$ be its inverse. Let equation $\{f^{-1}(x)\} = f^{-1}(x)$ has two real roots α and β (within domain of $f(x)$), then
(a) $f(x) = x$ also have same two real roots.
(b) $f^{-1}(x) = x$ also have same two real roots.
(c) $f(x) = f^{-1}(x) = x$ also have same two real roots.
(d) area formed by $(0, 0)$ $(\alpha, f(\alpha)$ and $(\beta, f(\beta))$ is 1 unit.

Questions from Previous Years

12. Let $f(x) = \sin\left[\dfrac{\pi}{6}\sin\left(\dfrac{\pi}{2}\sin x\right)\right]$ for all $x \in R$ and $g(x) = \dfrac{\pi}{2}\sin x$ for all $x \in R$. Let $(fog)(x)$ denote $f\{g(x)\}$ and $(gof)(x)$ denote $g\{f(x)\}$. Then, which of the following is/are true?
(a) Range of f is $\left[-\dfrac{1}{2}, \dfrac{1}{2}\right]$
(b) Range of fog is $\left[-\dfrac{1}{2}, \dfrac{1}{2}\right]$
(c) $\lim\limits_{x \to 0} \dfrac{f(x)}{g(x)} = \dfrac{\pi}{6}$
(d) There is an $x \in R$ such that $(gof)(x) = 1$

13. Let $f : (0, 1) \to R$ be defined by $f(x) = \dfrac{b-x}{1-bx}$, where b is a constant such that $0 < b < 1$. Then,
(a) f is not invertible on (0, 1)
(b) $f \neq f^{-1}$ on (0, 1) and $f'(b) = \dfrac{1}{f'(0)}$
(c) $f = f^{-1}$ on (0, 1) and $f'(b) = \dfrac{1}{f'(0)}$
(d) f^{-1} is differentiable on (0, 1).

Matrix Type

1. Match the Column I with Column II.

Column I	Column II
(A) Function $f : \left[0, \dfrac{\pi}{3}\right] \to [0,1]$ defined by $f(x) = \sqrt{\sin x}$ is	(P) one-to-one function
(B) Function $f : (1, \infty) \to (1, \infty)$ defined by $f(x) = \dfrac{x+3}{x-1}$ is	(Q) many–one function
(C) Function $f : \left[-\dfrac{\pi}{2}, \dfrac{4\pi}{3}\right] \to [-1,1]$ defined by $f(x) = \sin x$ is	(R) into function
(D) Function $f : (2, \infty) \to [8, \infty)$ defined by $f(x) = \dfrac{x^2}{x-2}$ is	(S) onto function

2. Match the following (Take f as R to R):

Column I	Column II
(A) one–one into	(P) $\sin x$
(B) many–one into	(Q) x^3+x+1
(C) many–one onto	(R) x^3-x+1
(D) one–one onto	(S) $\frac{2x+1}{3x+4}$

3. Match the following functions with their inverses:

(A) $f(x)=x^2-x+1 \quad \left(-\infty,\frac{1}{2}\right]\to\left[\frac{3}{4},\infty\right)$	(P) $f^{-1}(x)=\frac{-1-\sqrt{4x-3}}{2}$
(B) $f(x)=x^2-x+1 \quad \left[\frac{1}{2},\infty\right)\to\left[\frac{3}{4},\infty\right)$	(Q) $f^{-1}(x)=\frac{1+\sqrt{4x-3}}{2}$
(C) $f(x)=x^2+x+1 [1,\infty)\to[3,\infty)$	(R) $f^{-1}(x)=\frac{\left(1-\sqrt{4x-3}\right)}{2}$
(D) $f(x)=x^2+x+1 \quad (-\infty,-1]\to[1,\infty)$	(S) $f^{-1}(x)=\frac{-1+\sqrt{4x-3}}{2}$

4. Let $f_1:\mathrm{R}\to\mathrm{R}$, $f_2:[0,\infty]\to\mathrm{R}$, $f_3:\mathrm{R}\to\mathrm{R}$ and $f_4:\mathrm{R}\to[0,\infty)$ be defined by

$$f_1(x)=\begin{cases}|x|, & \text{if } x<0\\ e^x, & \text{if } x\ge 0\end{cases};$$

$$f_2(x)=x^2; f_3(x)=\begin{cases}\sin x, & \text{if } x<0\\ x, & \text{if } x\ge 0\end{cases}$$

$$\text{and } f_4(x)=\begin{cases}f_2[f_1(x)], & \text{if } x<0\\ f_2[f_1(x)]-1, & \text{if } x\ge 0\end{cases}$$

Column I	Column II
(A) f_4 is	(P) onto but not one–one
(B) f_3 is	(Q) neither continuous nor one–one
(C) $f_2 o f_1$ is	(R) differentiable but not one–one
(D) f_2 is	(S) continuous and one–one

Integer and Numerical Value Type

1. If the functional relation $f(x)+f\left(\frac{1}{1-x}\right)=\frac{2(1-2x)}{x(1-x)}$ is satisfying by the function $f(x)=\frac{x+1}{\lambda(x-1)}$, then find value of λ.

2. The least period of the function $\cos(\cos x)+\sin(\cos x)\sin 4x$ is $k\frac{\pi}{2}$, then value of k must be ______.

3. If $f(x+y)=f(x)+f(y)$ for all x, y and $f(1)\frac{1}{2007}$, then value of (2007) must be ______.

4. If $[x]$ denotes greatest integer $\le x$, the function $[\tan x]$, $\left(0<x<\frac{5\pi}{12}\right)$ takes exactly k values exactly. The numerical quality k should be ______.

5. Let S be the set of points (x, y) given by $\mathrm{S}=\{(x,y);\ x^2+y^2-10x+16=0\}$ and if $f:\mathrm{S}\to\mathrm{R}$ be given by $f(x,y)=\frac{y}{x}$. If range of f is $\left[-\frac{3}{k},\frac{3}{k}\right]$ where $k>0$, then k must be ______.

6. Let $S=\begin{Bmatrix}(x, y), x, y\in R\\ \{x^2+y^2-6x-8y+21=0\}\end{Bmatrix}$

Let if $f:S\to R$ be a mapping given by $f(x,y)=\sqrt{x^2+y^2}$,

Then range of f is a finite interval of length ______.

7. There are exactly f integers for which the function $f(x) = {}^{16-x}C_{2x-1} + {}^{20-3x}C_{4x-5}$ is defined. k must be _______.

8. The function $f(x) = \dfrac{x+5}{\sqrt{x^2+1}}$ takes exactly k integer values, then k must be _______.

9. The fundamental period of the function $\dfrac{\sin x + \sin 3x + \sin 5x + \sin 7x}{\cos x + \cos 3x + \cos 5x + \cos 7x}$ is $\dfrac{\pi}{k}$, then k must be _______.

10. The range of the function $\dfrac{x+m}{x^2+1}$, $(m \in R)$ contains the interval $[0,1]$. If $m > \dfrac{3}{k}$, then k must be _______.

11. If $0 \le x \le 100$, then the total number of integer values taken by the function $[x]+[2x]+[3x]+[4x]+[5x]$ is _______.

Comprehension Passage

Comprehension 1

Let $f(x) = \begin{cases} 2x+a & : x \ge -1 \\ bx^2+3 & : x < -1 \end{cases}$ and

$g(x) = \begin{cases} x+4 & : 0 \le x \le 4 \\ -3x-2 & : -2 < x < 0 \end{cases}$

1. $g(f(x))$ is not defined if _______.
(a) $a \in (6,\infty), b \in (5,\infty)$ (b) $a \in (4,6), b \in (5,\infty)$
(c) $a \in (6,\infty), b \in (0,1)$ (d) $a \in (4,6), b \in (1,5)$

2. If domain of $g(f(x))$ is $[-1, 2]$, then
(a) $a = 1, b > 5$ (b) $a = 2, b > 7$
(c) $a = 2, b > 10$ (d) $a = 0, b \in R$

3. If $a = 2$ and $b = 3$ then range of $g(f(x))$ is _______.
(a) $(-2, 8]$ (b) $(0, 8]$
(c) $[4, 8]$ (d) $[-1, 8]$

Comprehension 2

Let $f: R \to R$ is a function satisfying $f(2-x) = f(2+x)$ and $f(20-x) = f(x)$, $\forall x \in R$. For this function f answers the following.

1. If $f(0) = 5$, then minimum possible number of values of x satisfying $f(x) = 5$, for $x \in [0, 170]$, is _______.
(a) 21 (b) 12
(c) 11 (d) 22

2. Graph of $y = f(x)$ is
(a) symmetrical about $x = 18$.
(b) symmetrical about $x = 5$.
(c) symmetrical about $x = 8$.
(d) symmetrical about $x = 20$.

3. If $f(2) \ne f(6)$, then
(a) fundamental period of $f(x)$ is 1.
(b) fundamental period of $f(x)$ may be 1.
(c) period of $f(x)$ can't be 1.
(d) fundamental period of $f(x)$ is 8.

Comprehension 3

A function f from a set X to Y is called onto, if for every $y \in$ Y, there exists $x \in$ X such that $f(x) = y$. Unless the contrary is specified, a real function is onto if it takes all real values, otherwise it is called into function.

Thus, if X and Y are finite sets, then f cannot be onto if Y contains more element than X.

1. The polynomial function $a_0x^n + a_1x^{n-1} + a_2x^{n-2} + \ldots\ldots + a_n = 0$, where $a_0 \ne 0$, is onto
(a) for all positive integers n.
(b) for all even positive integers n.
(c) for all odd positive integers n.
(d) for no positive integer.

2. The function $f(x) = \dfrac{x^2+2x+c}{x^2+4x+3c}$ is onto, if
(a) $0 < c << 2$ (b) $0 < c < 4$
(c) $-\dfrac{1}{2} < c < \dfrac{1}{2}$ (d) $0 < c < 1$

3. Which of the following is not true?
(a) A one–one function from the set $\{a, b, c\}$ to $\{\alpha, \beta, \gamma\}$ is onto also.
(b) An onto function from an infinite set to a finite set cannot be one–one.
(c) An onto function is always invertible.
(d) The functions $\tan x$ and $\cot x$ are onto.

Comprehension 4

Functions in mathematics may satisfy some functional relations.

For example, the functions $f(x)=3x$ satisfies $f(x+y)=f(x)+f(y)$ and $f(x)=\log x$ satisfies $f(xy)=f(x)+f(y)$. From the given functional relations, we can determine several things about the functions. At times, the functional is completely determined if it is continuous and differentiable.

1. If $f(x + y) = f(x) + f(y)$ for all x, y then $f\left(-\dfrac{3}{5}\right)$ must be equal to _______.

(a) $f(1)$ (b) $\frac{3}{5}f(1)$

(c) $-\frac{3}{5}$ (d) $-\frac{3}{5}f(1)$

2. If $f(x) + f(y) = f\left(x\sqrt{1-y^2}+y\sqrt{1-x^2}\right)$, then ______.

(a) $f\left(4x^2+3x\right)+3f(x)=0$

(b) $f\left(3x-4x^3\right)+3f(x)=0$

(c) $f\left(4x^3+3x\right)-3f(x)=0$

(d) $f\left(4x^3-3x\right)+3f(x)=0$

3. If $f(x)+f\left(\frac{1}{1-x}\right)=\frac{2(1-2x)}{x(1-x)}$ for all $x \neq 0$ and $x \neq 1$, then $f(x)$

(a) cannot be determined.

(b) must be $\frac{1+x}{1-x}$.

(c) must be $\frac{x+1}{x-1}$.

(d) should be identically zero.

Comprehension 5

Let $f(x)$ is a function. We define $f^2(x)=f(f(x))$ $f^3(x)=f(f(f(x)))$ and so on.

In certain cases, $f^n(x)$ can be proved by induction. We may also say that $f^{-(n)}(x)$ is the inverse function, f^{-1} being applied n times.

1. Let $f(x) = 3 + 4x$
Then $f^3(x)$ must be equal to ______.

(a) $27 + 64x$ (b) $64 + 27x$

(c) $63 + 64x$ (d) $64 + 63x$

2. For the same $f(x)$, $f^n(x)$ must be ______.

(a) $3^n + 4^n x$ (b) $4^n + 3^n x$

(c) $(3+4x)^n$ (d) $\left(4^n-1\right)+4^n x$

3. $f^{-n}(x)$ must be equal to ______.

(a) $\left(4^{-n}-1\right)+4^{-n}x$ (b) $\frac{1}{4^n-1+4^n x}$

(c) $\frac{1}{4^n-1}+\frac{1}{4^n}x$ (d) $\frac{1}{4^n-1}+\frac{1}{4^n x}$

ANSWER KEY

JEE Main

Objective Questions

One Option Correct

1. (b) **2.** (d) **3.** (a) **4.** (b) **5.** (b) **6.** (b) **7.** (a) **8.** (c) **9.** (c)
10. (b) **11.** (c) **12.** (a) **13.** (c) **14.** (a) **15.** (d) **16.** (d) **17.** (a) **18.** (b)
19. (b) **20.** (a) **21.** (d) **22.** (b) **23.** (d) **24.** (c) **25.** (c) **26.** (d) **27.** (c)
28. (b) **29.** (d) **30.** (b) **31.** (c) **32.** (d) **33.** (c) **34.** (b) **35.** (c) **36.** (c)
37. (c) **38.** (c) **39.** (b) **40.** (a) **41.** (a) **42.** (a) **43.** (d) **44.** (b) **45.** (b)
46. (a) **47.** (c) **48.** (c) **49.** (c) **50.** (a) **51.** (a) **52.** (a) **53.** (b) **54.** (a)
55. (c) **56.** (d) **57.** (a) **58.** (d) **59.** (d) **60.** (c) **61.** (b) **62.** (c) **63.** (c)
64. (a) **65.** (b) **66.** (c) **67.** (d) **68.** (b) **69.** (a) **70.** (c) **71.** (b) **72.** (a)
73. (b)

Integer and Numerical Value Type

1. (–1) **2.** (x) **3.** (6) **4.** (x + 1) and (–x + 1)

Assertion and Reason

1. (b) **2.** (c) **3.** (a) **4.** (a) **5.** (d)

JEE Advanced

Objective Questions

One Option Correct

1. (d) **2.** (a) **3.** (d) **4.** (b) **5.** (a) **6.** (d) **7.** (a) **8.** (d) **9.** (a)
10. (c) **11.** (d) **12.** (c)

One or More Than One Option Correct

1. (c, d) **2.** (b, c, d) **3.** (a, d) **4.** (a, c) **5.** (b, c) **6.** (a, d) **7.** (b, c) **8.** (a, b)
9. (a, b, c, d) **10.** (b, c) **11.** (a, b, c) **12.** (a, b, c) **13.** (b)

Matrix Type

1. A → P, R; B → P, S; C → Q, S; D → Q, S
2. A → S; B → P; C → R; D → Q
3. A → R; B → Q; C → S; D → P
4. A → P; B → R; C → Q; D → S

Integer and Numerical Value Type

1. $\left(\dfrac{x+1}{x-1}\right)$ **2.** (1) **3.** (1) **4.** (4) **5.** (4) **6.** (4) **7.** (2) **8.** (6) **9.** (4)
10. (4) **11.** (100)

Comprehension Passage

Comprehension 1

1. (a) **2.** (a) **3.** (c)

Comprehension 2

1. (a) **2.** (a) **3.** (c)

Comprehension 3

1. (c) **2.** (d) **3.** (c)

Comprehension 4

1. (d) **2.** (d) **3.** (c)

Comprehension 5

1. (c) **2.** (d) **3.** (a)

HINTS AND EXPLANATIONS

Worksheet – 1

1. Equivalence class [(1, 3)]

$= \{(a,b) \in A \times A : (a,b) R (1,3)\}$

$= \{(a,b) \in A \times A : a(3) = b(1)\}$

$= \left\{(a,b) \in A \times A : \dfrac{a}{b} = \dfrac{1}{3}\right\}$

Hence, [(1, 3)] = {(1, 3), (2, 6), (3, 9), (4, 12), (5, 15)}.

2. Let $a \in A$
So, $|a - a| = 0$ (which is an even number).
Therefore, R is reflexive.
Let $(a,b) \in R$
Now, $|a - b|$ is even,
Hence $|a - b|$ and $|b - a|$ are both even.
Therefore, $(b, a) \in R$
Therefore, R is symmetric.
Let $(a, b) \in R$ and $(b, c) \in R$.
$\Rightarrow |a - b|$ is even and $|b - c|$ is even.
$\Rightarrow (a - b)$ is even and $(b - c)$ is even.
$\Rightarrow (a - c) = (a - b) + (b - c)$ ($\Rightarrow$ is even
$\Rightarrow |a - c|$ is even.
$\Rightarrow (a, c) \in R$
Therefore, R is transitive.
Therefore, the given relation R is an equivalence relation.
All the elements of set {1, 3, 5} are all odd. Hence, the modulus of the difference of any two elements will be an even number. So, all the elements of this set are related to each other.
All elements of {2, 4} are even while all the element of {1, 3, 5} are odd; so no element of {1, 3, 5} can be related to any element of {2, 4}. Therefore, the absolute value of the difference between the two elements (from each of these two subsets) will not be an even value.

3. The given relation is R = {(P, Q): Distance of P from the origin is the same as the distance of Q from the origin}
Since, (P, P) ∈ R, the relation R is reflexive.
Let (P, Q) ∈ R. The distance of P from the origin is the same as the distance of Q from the origin. Similarly, the distance of Q from the origin will be the same as the distance of P from the origin. So, (Q, P) ∈ R.
Therefore, R is symmetric.
Let (P, Q) (Q, S) ∈ R
The distance of P from the origin is the same as the distance of Q from the origin and the distance of Q from the origin is the same as the distance of S from the origin. So, the distance of S from the origin will be same as the distance of P from the origin. So, (P, S) ∈ R.
Therefore, R is transitive.
Therefore, the relation R is an equivalence relation.
The set of points related to P ≠ (0, 0) will be those points whose distance from the origin is same as the distance of P from the origin and will form a circle with the centre as origin and this circle passes through P.

4. The given relation is) R = $\{(T_1, T_2) : T_1$ is similar to $T_2\}$.
The relation R is reflexive since every triangle is similar to itself.
If $(T_1, T_2) \in$ R then T_1 is similar to T_2.
$\Rightarrow T_2$ is similar to T_1.
$\Rightarrow (T_1, T_2) \in R$
Therefore, R is symmetric.
Let $(T_1, T_2), (T_2, T_3) \in R$.
$\Rightarrow T_1$ is similar to T_2 and T_2 is similar to T_3.
$\Rightarrow T_1$ is similar to T_3.
$\Rightarrow (T_1, T_3) \in R$
Therefore, R is transitive.
Therefore, the relation R is an equivalence relation.
$$\frac{3}{6}=\frac{4}{8}=\frac{5}{10}\left(=\frac{1}{2}\right)$$
Since the corresponding sides of triangles T_1 and T_3 are in the same ratio, triangle T_1 is similar to triangle T_3.
Hence, T_1 is related to T_3.

5. $R=\{(P_1,P_2): P_1$ and P_2 have same number of sides$\}$.
Since $(P_1,P_2)\in R$ as same polygon has same number of sides.
The relation R is reflexive.
Let $(P_1,P_2)\in R$.
$\Rightarrow P_1$ and P_2 have same number of sides.
$\Rightarrow P_2$ and P_1 have same number of sides.
$\Rightarrow (P_2, P_1) \in R$
Therefore, R is symmetric.
Let $(P_1,P_2),(P_2,P_3)\in R$.
$\Rightarrow P_1$ and P_2 have same number of sides.
$\Rightarrow P_2$ and P_3 have same number of sides.
$\Rightarrow P_1$ and P_3 have same number of sides.
$\Rightarrow (P_1, P_3)\in R$
Therefore, R is transitive.
Therefore, the relation R is an equivalence relation.
The elements in A related to right-angled triangle (T) with sides 3, 4 and 5 are the polygons having 3 sides.

6. Given relation, R = $\{(a, a), (b, c), (a, b)\}$
To make R as reflexive, we should add (b, b) and (c, c) to R and to make R as transitive we should add (a, c) to R.
Hence, the minimum number of ordered pairs to be added are (b, b), (c, c) and (a, c), i.e., 3.

7. Given function: R = $\{(x, y): x \in N, y \in N, 2x + y = 41\}$.
So, the domain = {1, 2, 3, ..., 20} $[y \in N]$
To find the range, we have
R = {(1, 39), (2, 37), (3, 35), ..., (19, 3), (20, 1)}
Thus, range of the function = {1, 3, 5, ..., 39}
R is not reflexive as (2, 2) ∉ R as $2 \times 2 + 2 \neq 41$
Also, R is not symmetric as (1, 39) ∈ R but (39, 1) ∉ R
Further R is not transitive as (11, 19) ∉ R, (19, 3) ∉ R; but (11, 3) ∉ R.
Therefore, R is neither reflexive nor symmetric and nor transitive.

8. (i) Given, $x > y$; $x, y \in N$.
If $(x, x) \in R$, then $x > x$, which is not true for any $x \in N$.
Thus, R is not reflexive.
Let $(x, y) \in R$
$\Rightarrow xRy$
$\Rightarrow x > y$
So, $y > x$ is not true for any $x, y \in N$
Hence, R is not symmetric.
Let xRy and yRz
$\Rightarrow x > y$ and $y > z$
$\Rightarrow x > z$
$\Rightarrow xRz$
Hence, R is transitive.

(ii) $x + y = 10;\ x, y \in N$

$R = \{(x, y); x + y = 10, x, y \in N\}$

$R = \{(1,9),(2,8),(3,7),(4,6),(5,5),(6,4),(7,3),(8,2),(9,1)\}$

$(1, 1) \notin R$

So, R is not reflexive.

$(x, y) \in R \Rightarrow (y, x) \in R$

Therefore, R is symmetric.

Now $(1, 9) \in R$, $(9, 1) \in R$, but $(1, 1) \notin R$

Therefore, R is not transitive.

(iii) Given, xy is the square of an integer $x, y \in N$.

$R = \{(x, y) : xy \text{ is a square of an integer } x, y \in N\}$

$(x, x) \in R,\ \forall\ x \in N$

As x^2 is square of an integer for any $x \in N$,

Thus, R is reflexive.

If $(x, y) \in R \Rightarrow (y, x) \in R$

So, R is symmetric.

Now, if xy is square of an integer and yz is square of an integer,

then, let $xy = m^2$ and $yz = n^2$ for some $m, n \in Z$

$x = m^2/y$ and $z = x^2/y$

$xz = m^2n^2/y^2$, which is square of an integer.

Thus, R is transitive.

(iv) $x + 4y = 10;\ x, y \in N$

$R = \{(x, y) : x + 4y = 10; x, y \in N\}$

$R = \{(2, 2), (6, 1)\}$

$(1, 1) \notin R$

Hence, R is not symmetric.

$(x, y) \in R \Rightarrow x + 4y = 10$

and $(y, z) \in R \Rightarrow y + 4z = 10$

$\Rightarrow x - 16z = -30$

$\Rightarrow (x, z) \notin R$

Therefore, R is not transitive.

Worksheet – 2

1. Given, $f(x) = \sqrt{(25 - x^2)}$

The function is defined if $25 - x^2 \geq 0$

Therefore, $x^2 \leq 25$

$-5 \leq x \leq 5$, then domain of the given function is $[-5, 5]$.

2. Given,

$f(x) = 2x + 1$ and $g(x) = x^2 - 2,\ \forall\ x \in R$

Thus, $gof = g(f(x))$

$= g(2x + 1)$

$= (2x + 1)^2 - 2$

$= 4x^2 + 4x - 1$

3. Given, $f(x) = x^2 - 3x + 2$

Then, we have

$f(f(x)) = f(x^2 - 3x + 2)$

$= (x^2 - 3x + 2)^2 - 3(x^2 - 3x + 2) + 2,$

$= x^4 + 9x^2 + 4 - 6x^3 + 4x^2 - 12x - 3x^2 + 9x - 6 + 2$

$= x^4 - 6x^3 + 10x^2 - 3x$

Therefore, $f(f(x)) = x^4 - 6x^3 + 10x^2 - 3x$

4. $f = \{(1, 2), (3, 5), (4, 1)\}$ and $g = \{(2, 3), (5, 1), (1, 3)\}$

Now,

$fog\ (2) = f(g(2)) = f(3) = 5$

$fog\ (5) = f(g(5)) = f(1) = 2$

$fog\ (1) = f(g(1)) = f(3) = 5$

Therefore, $fog = \{(2, 5), (5, 2), (1, 5)\}$

5. Given, $X = \{1, 2, 3\}$ and $Y = \{4, 5\}$

So,

$X \times Y = \{(1,4),(1,5),(2,4),(2,5),(3,4),(3,5)\}$

(i) $f = \{(1,4),(1,5),(2,4),(3,5)\}$

f is not a function as $f(1) = 4$ and $f(1) = 5$

Hence, pre-image of '1' is not unique.

(ii) $g = \{(1, 4), (2, 4), (3, 4)\}$

We see that g is a function in which each element of the domain has unique image.

(iii) $h = \{(1,4), (2, 5), (3, 5)\}$

We see that h is a function as each pre-image with a unique image.

And, function h is many–one as $h(2) = h(3) = 5$

(iv) $k = \{(1, 4), (2, 5)\}$

Function k is not a function as '3' has not any image under the mapping.

6. $f(x) = 1/(2 - \cos x)\ \forall\ x \in R$

Let $y = 1/(2 - \cos x)$

$2y - y\cos x = 1$

$\cos x = (2y - 1)/y$

$\cos x = 2 - 1/y$

Hence, $-1 \leq \cos x \leq 1$

So,

$-1 \leq 2 - 1/y \leq 1$

$-3 \leq -1/y \leq -1$

$1 \leq -1/y \leq 3$

$1/3 \leq y \leq 1$

Thus, the range of the given function is [1/3, 1].

7. Given, $A = \{2, 3, 4\}$, $B = \{2, 5, 6, 7\}$

(i) Let f: $A \to B$ denote a mapping.

$f = \{(x, y): y = x + 3\}$ or

$f = \{(2, 5), (3, 6), (4, 7)\}$, which is an injective mapping.

(ii) Let g: $A \to B$ denote a mapping such that $g = \{(2, 2), (3, 2), (4, 5)\}$, which is not an injective mapping.

(iii) Let h: B $\to$ A denote a mapping such that $h = \{(2, 2), (5, 3), (6, 4), (7, 4)\}$, which is one of the mapping from B to A.

8. $A = R - \{3\}, B = R - \{1\}$ and

f: $A \to B$ be defined by $f(x) = x - 2/x - 3 \forall x \in A$

Hence,

$f(x) = (x - 3 + 1)/(x - 3) = 1 + 1/(x - 3)$

Let $f(x_1) = f(x_2)$

$$1 + \frac{1}{x_1 - 3} = 1 + \frac{1}{x_2 - 3}$$

Solve the above to get $x_1 = x_2$

So, $f(x)$ is an injective function.

Now let $y = (x - 2)/(x - 3)$

$x - 2 = xy - 3y$

$x(1 - y) = 2 - 3y$

$x = (3y - 2)/(y - 1)$

$y \in R - \{1\} = B$

Thus, $f(x)$ is onto or subjective.

Therefore, $f(x)$ is a bijective function.

Quiz

1. Yes

Since, $\frac{a}{a} = 1 \Rightarrow$ every number divides itself, hence R is reflexive.

2. R is a partial order relation if it is reflexive, transitive; but at the same time, it is antisymmetric.

3. A relation R on a non-empty set A is called an equivalence relation if and only if it is Reflexive, Symmetric as well as transitive.

4. Domain of IA = Range of I = A

5. The two functions are identical if and only if

(i) Domain of f = Co-domain of g

(ii) Co-domain of f = Co-domain of g

(iii) $f(x) = g(x) \forall x \in$ domain

6. A relation from a set A to a set B is a subset of $A \times B$, the Cartesian product $(A \times B)$, whereas a function is a special case of relation. A relation from a set A to a set B is said to be a function for every element of set A if it has one and only one image in set B.

7. Yes

8. Since the distinct elements 4, 5, 6 of $\{4, 5, 6\}$ have different images a, b, c in $\{a, b, c\}$. So, f is one–one.

Also, range $\{a, b, c\}$ = codomain $\{a, b, c\}$

Therefore f is one–one and onto (Bijective)

9. $f = \{(4, a), (5, b), (6, c)\}$

The function is one–one onto, therefore it is invertible.

And $f^{-1} = \{(a, 4), (b, 5), (c, 6)\}$

10. Not necessarily same.

JEE Main

Objective Questions

One Option Correct

1. (b) $g(x) = 1 + x - [x]$ is greater than 1 since $x - [x] > 0$

$f[g(x)] = 1$, since $f(x) = 1$ for all $x > 0$

2. (d) Given, $f(x)\cos(\log x)$

$$\therefore f(x).f(y) - \frac{1}{2}\left[f\left(\frac{x}{y}\right) + f(xy)\right]$$

$$= \cos(\log x).\cos(\log y)$$
$$-\frac{1}{2}[\cos(\log x - \log y) + \cos(\log x + \log y)]$$

$$= \cos(\log x).\cos(\log y)$$
$$-\frac{1}{2}(2\cos(\log x).\cos(\log y)]$$

$$= \cos(\log x).\cos(\log y)$$
$$-\cos(\log x).\cos(\log y) = 0$$

3. (a) Let $f(x) = 2x + \sin x - \beta$

where β is any real number; then $f(x)$ is continuous,

$f(-\infty) = -\infty, f(\infty) = \infty \Rightarrow f(x) = 0$ for some x

$\Rightarrow 2x + \sin x = \beta$ for some x.

$\Rightarrow f$ is onto

Again, $f'(x) = 2 + \cos x > 0$ for all x.

$\Rightarrow f$ is one–one

4. (b) $2 - \cos 3x \in [1, 3]$

$$\Rightarrow \frac{1}{2 - \cos 3x} \in \left[\frac{1}{3}, 1\right]$$

5. (b) $f(g(x)) = [x] - [[x]] = [x] - [x] = 0$

6. (b)

x	$f(x)$
$\left[0, \frac{1}{2}\right)$	0
$\left[\frac{1}{2}, 1\right)$	1
$\left[1, \frac{3}{2}\right)$	0
$\left[\frac{3}{2}, 2\right)$	1

Thus $f(x)$ will take only two integer values, 0 and 1 only.

7. (a)

$f(-5) = -1,\ f(-4) = 3(-4) + 2 = 10$

$\Rightarrow f(-5) + f(-4) = -11 \quad \Rightarrow (A) \rightarrow (iii)$

$f(-8) = -8 + 4 = -4 \quad \Rightarrow |f(-8)| = 4$

Now $f(4) = 0 \quad \Rightarrow (B) \rightarrow (vi)$

$\Rightarrow$ Choice (b) and (c) are ruled out.

Now $f(-7) + f(3) = -3 + 11 = 8$

and $f(8) = 4 \quad \Rightarrow (C) \rightarrow (ii)$

$\Rightarrow (d)$ is ruled out.

$\Rightarrow (a)$ is correct.

8. (c) If $ad = bc$ then $d = \frac{bc}{a}$

$$\Rightarrow \quad \frac{ax+b}{cx+d} = \frac{ax+b}{cx+\frac{bc}{a}} = \frac{a}{c}$$

9. (c) The sum of positive divisors of 3.2^k is given by

$$f(3.2^k) = (1 + 2 + 2^2 + ...2^k)(1+3)$$

$$= \frac{(2^{k+1} - 1)}{2-1}.4 = 4(2^{k+1} - 1)$$

10. (b) $f(n) > 2$ is possible if n is a natural number of the type $3k + 1$

Since 1, 4, 7 are the numbers of the type $3k + 1$.

11. (c) I is correct since $\frac{d}{dx}3^{-x} > 0$ for all x.

II is not correct since 3^{-x} takes only positive values.

III is again correct.

Since $\frac{d}{dx}3^{-x} = 3^{-x}(-\log 3)$

12. (a) $f(x)$ is either -1 or 1. (It is -1 for $x \in [-4, 0]$ and 1 for [0.4])

$\Rightarrow$ Range is $\{1, -1\}$

13. (c) S is not reflexive since $y = x + 1$ is never true.

$\Rightarrow$ Choices (a) and (b) are wrong.

T is reflexive since $x - x$ is an integer.

T is symmetric since if $x - y$ is an integer, then $y - x$ is also an integer.

T is transitive since if $x - y$ is an integer, $y - z$ is an integer; then $x - z$ is integer as $x - z = x - y + y - z$

$\Rightarrow$ (c) is correct.

14. (a) R is reflexive since x and x have at least one word common.

R is symmetric, obviously.

If words x and y have at least one word common, y and z have at least one word common, then x and z may not have at least one word common.

15. (d) It is easily shown that $f(n) = n\,f(1)$

$$\Rightarrow \sum f(r) = f(1) \times \frac{n(n+1)}{2} = \frac{7n(n+1)}{2}$$

$\Rightarrow$ (d) is correct.

16. (d) $g(x)$ is inverse of $f(x)$ $x + 1 = \pm\sqrt{y}$

$\Rightarrow x + 1 = \pm\sqrt{y} \qquad (\because x > -1)$

$\Rightarrow f^{-1}(x)$ is $-1 + \sqrt{x}$

$\Rightarrow$ (d) is correct.

17. (a) $f'(x) = 2\cos x > 0$ all x

$\Rightarrow f$ is one−one

Moreover $f(x)$ takes all the values from $-\infty$ to $+\infty$ and as it is continuous, it takes all the values between $-\infty$ and ∞.

18. (b) Since $f(-x) = f(x)$

$f(x)$ is symmetric about y-axis.

(b) is correct.

19. (b) $f(-x) = \log(-x + \sqrt{x^2 + 1})$

$$= \log \frac{(-x + \sqrt{x^2+1})(\sqrt{x^2+1} + x)}{(\sqrt{x^2+1} + x)}$$

$$= \log \frac{1}{\sqrt{x^2+1} + x} = -f(x)$$

20. (a) Only e^x is one−one.

21. (d) Any "Congruence Modulo" is an equivalence relation; therefore, option (d) is correct.

22. (b) If $g(x)=\dfrac{f(x)+f(-x)}{2}$

Then $g(-x)=\dfrac{f(-x)+f(x)}{2}=g(x)$

$\Rightarrow g(x)$ is even.

$\Rightarrow$ (b) is correct.

23. (d) Since n divides n, $\forall n \in \mathrm{N}$, R is reflexive. R is not symmetric since for 3, 6 $\in$ N, 3R6 $\neq$ 6R3. R is transitive since for n, m, r. Whenever n/m and $m/r \Rightarrow n/r$, i.e., if n divides m and m divides r, then n will divide r.

24. (c) The total number of injective mappings from the set containing 3 elements into the set containing 4 elements is ${}^4P_3=4!=24$

25. (c) $f^{-1}(17)=x \Rightarrow f(x)=17$ or $x^2+1=17$

$\Rightarrow x=\pm 4$ or $f^{-1}(17)=\{4,-4\}$

and for

$f^{-1}(-3)=x \Rightarrow f(x)=-3$

$\Rightarrow x^2+1=-3 \Rightarrow x^2=-4$

and hence $f^{-1}(-3)=\phi$.

26. (d) A well-known function satisfying

$f\left(\dfrac{x}{y}\right)=f(x)-f(y)$ is $\log x$

Whence, choices (a), (b) (c) are ruled out. Since $\log x$ is neither bounded, nor does $\log\dfrac{1}{x}\to 0$ as $x\to 0$ and also $x \log x \to 0$ as $x \to 0$

Thus choice (d) must be correct

27. (c) Since, only (c) satisfy given definition

i.e., $f\{f^{-1}(\mathrm{B})\}=\mathrm{B}$

Only, if $B \subseteq f(x)$

28. (b) By definition of composition of function.

$g(f(x))=(\sin x+\cos x)^2-1$, is invertible (i.e., bijective)

$\Rightarrow f\{f(x)\}=\sin 2x$ is bijective

We know, $\sin x$ is bijective only when $x\in\left[-\dfrac{\pi}{2},\dfrac{\pi}{2}\right]$

Thus, $g\{f(x)\}$ is bijective, if $-\dfrac{\pi}{2}\le 2x\le\dfrac{\pi}{2}$

$\Rightarrow -\dfrac{\pi}{4}\le x\le\dfrac{\pi}{4}$

29. (d) It is only to find the inverse.

Let $y=(x)=(x+1)^2$ for $x\ge -1$

$\pm\sqrt{y}=x+1, x\ge -1$

$\Rightarrow \sqrt{y}=x+1 \Rightarrow y\ge 0, x+1\ge 0$

$\Rightarrow x=\sqrt{y}-1 \Rightarrow f^{-1}(y)=\sqrt{y}-1$

$\Rightarrow f^{-1}(x)=\sqrt{x}-1 \Rightarrow x\ge 0$

30. (b) Given, $f(x)=3x-5$

Let $y=f(x)=3x-5 \quad \Rightarrow y+5=3x$

$\Rightarrow x=\dfrac{y+5}{3}\Rightarrow f^{-1}(y)=\dfrac{y+5}{3}\Rightarrow f^{-1}(x)=\dfrac{x+5}{3}$

31. (c) Clearly, $f(x)=x-[x]=\{x\}$ which has period 1.

And $\sin\dfrac{1}{x}$, $x\cos x$ are non-periodic functions.

32. (d) A relation R in a set A is said to be an equivalence relation if R is reflexive, symmetric and transitive

Hence, (d) is correct.

33. (c) As every set is a subset of itself, so the relation, the inclusion of a subset in another, is reflexive.

Again, 'A is a subset of B does not imply B is a subset of A'; therefore, the relation, the inclusion of subset in another, is not symmetric. As the relation, the inclusion of a subset in another is not symmetric, so the relation cannot be an equivalence relation.

Hence, (c) is the correct answer.

34. (b) Given A = {1, 2, 3}.

$$\mathrm{A}\times\mathrm{A}=\{(1,1),(1,2),(1,3),(2,1),(2,2),(2,3),(3,1),(3,2),(3,3)\}$$

When 1 is related to 3,

Equivalence relationship containing (1, 2) is

$$\mathrm{R}_1=\{(1,1),(1,2),(1,3),(2,1),(2,2),(2,3),(3,1),(3,2),(3,3)\}$$

When 1 is related to 3,

Equivalence relationship containing (1, 2) is

$\mathrm{R}_2=\{(1,1),(1,2),(2,1),(2,2),(3,3)\}$

Hence, the number of equivalence relationship containing (1, 2) is 2. (b) is the correct answer.

35. (c) Set A = {1, 2, 3} and $\mathrm{R}=\{(1,2),(2,3)\}$

R is reflexive if $\{(1,2),(2,3),(3,3)\}\in\mathrm{R}$,

Again, (1, 2) $\in$ R, and (2, 3) $\in$ R,

$\therefore$ R is symmetric if (2, 1) $\in$ R, (3, 2) $\in$ R,

Now,

$R=\{(1,1),(2,2),(3,3),(1,2),(2,1),(2,3),(3,2)\}$

$\therefore$ R is transitive if $(1, 2) \in R$, and $(2, 3) \in R \Rightarrow (1, 3) \in R$,

Also, R is symmetric if (3, 1) is also added.

Hence, R is an equivalence relation if the number of ordered pairs to be added is 7.

$\therefore$ (c) is the correct answer.

36. (c) $f(x)=x^3+5x+1 \Rightarrow f'(x)=3x^2+5$

Now as $x^2>0$ for all $x \neq 0$, so $f'(x)=3x^2+5>0 \forall x \in R$

Thus $f(x)$ is strictly increasing function on R.

Therefore, $f(x)$ is one–one.

Again, as $f(x)=x^3+5x+1$ is a polynomial function, it is continuous in R.

f is continuous at every point of its domain R.

It takes all values in $(-\infty,\infty)$. Therefore, $f(x)$ is onto. Hence, $f(x)$ is one–one and onto.

$\therefore$ (c) is the correct answer.

37. (c) Solving x in terms of y we get

$x=\dfrac{y-3}{4}$

or $g(y)=\dfrac{y-3}{4}$

38. (c) $g(f(x))=x=f(g(x))$

$\Rightarrow g'(f(x))f'(x)=1$

$\Rightarrow g'(f(x))=\dfrac{1}{f'(x)}=1+x^5$

$\Rightarrow g'(f(x))=1+(g(x))^5$

$\Rightarrow g'(x)=1+(g(x))^5$

39. (b) Let us denote each individual (adjoin portion) in the figure as x, y, z, u, v, w, t, s

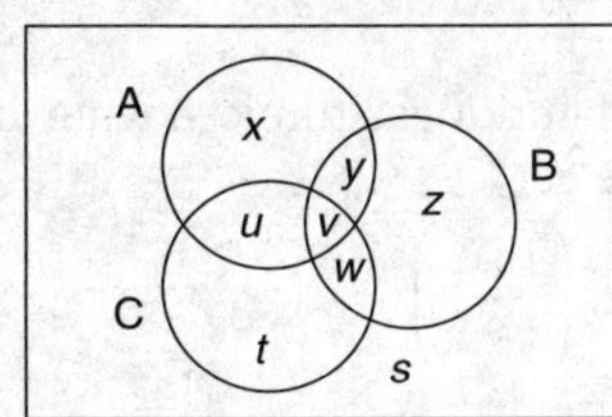

$A \cup B \cup C = x+y+u+v+w+z+t$

$B^c = x+u+t+s$

$C^c = x+y+z+s$

$B^c \cap C^c = x+s$

$\Rightarrow (B^c \cap C^c) \cap A = x$

$\Rightarrow (B^c \cap C^c \cap A) \cap C^c = x$

$\Rightarrow (A \cup B \cup C) \cap [B^c \cap C^c \cap A^c \cap C^c] = x$

40. (a) $\cos^{-1}x$ is defined if $x \in [-1,1]$

but on [0, 1), $[x] = 0$

Also, at $x = 1$, $[x] = 1$

$\Rightarrow$ Domain of $\dfrac{\cos^{-1}x}{[x]}$ is $[-1,0)\cup\{1\}$

41. (a) $n(A \cup B)$ cannot exceed $n(A)+n(B)$ and cannot be less than $n(A)$.

42. (a) $f(x)=\dfrac{x}{x-1}$

$\Rightarrow f(f(x))=\dfrac{\frac{x}{x-1}}{\frac{x}{x-1}-1}=x$

$\Rightarrow f(f(f(x)))=\dfrac{x}{x-1}$ and so on.

$\Rightarrow (fofof\ldots)(19\,\text{times})=\dfrac{x}{x-1}$

43. (d) Power set is set of all subsets.

44. (b) (a) is incorrect since $\{a\}$ is a set, not an element.

(c) is incorrect since ϕ is a set.

45. (b) If S is a set with four elements 1, 2, 3, 4 then $S \times S$ will contain 16 ordered pairs. Any reflexive relation on S will contain four ordered pairs (1, 1), (2, 2), (3, 3), (4, 4), essentially.

The remaining 12 ordered pairs may or may not be there.

$\Rightarrow 2^{12}$ reflexive relations are possible.

46. (a) (a) is correct and follows from the Venn diagram.

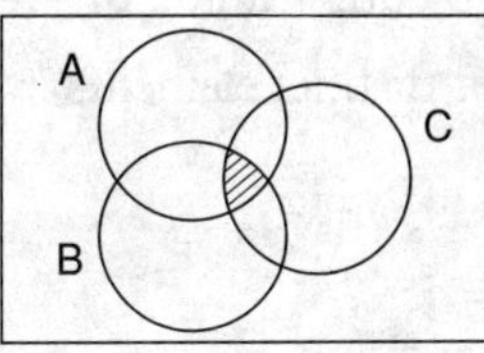

47. (c) $e^x > x$ for all x

$\Rightarrow A \cap B = \phi$

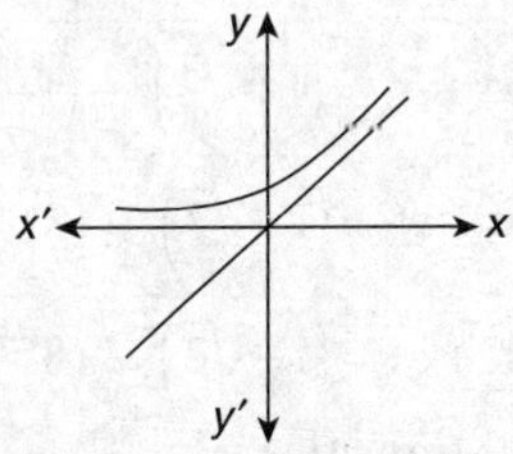

48. (c) $A\cap B$ will be a set containing multiple of 15.

49. (c) Put $x+y=u, \quad x-y=v$

Then $f(u,v)=\frac{u+v}{2}\cdot\frac{u-v}{2}$

$\Rightarrow f(x,y)=\frac{1}{4}(x^2-y^2)$

$\Rightarrow f(y,x)=\frac{1}{4}(y^2-x^2)$

$\Rightarrow$ Arithmetic mean of $f(x,y)$ and $f(y,x)$ is zero.

50. (a) $x-[x]$ is always non-negative.
$\Rightarrow f(x)$ is defined for all x.

51. (a) $f(x)$ is one–one and onto.

52. (a) For (a) $y=\frac{1-x}{1+x}\Rightarrow x=\frac{1-y}{1+y}$

$\Rightarrow f^{-1}(x)$ is $\frac{1-x}{1+x}$

53. (b) $A\cup B=\{1,2,3,8\}$ $A\cap B=\{3\}$
$\Rightarrow (A\cup B)\times(A\cap B)=\{(1,3),(2,3),(3,3),(8,3)\}$

54. (a) Total functions $=2^{100}$
There will be two functions which will not be onto (all elements of set A to 1 or all elements of set A to 2.)
$\Rightarrow$ Total functions $=2^{100}-2$

55. (c) $S=A_1\cup A_2\cup\ldots\cup A_3=B_1\cup B_2\cup\ldots\cup B_n$
Since each A_i contains 5 elements, number of elements in S (with repetitions)
$=5\times 30=150$
But each element of S appears exactly 10 times.
$\Rightarrow$ Number of distinct elements in $S=\frac{150}{10}=15$
Similarly, by other union of B_i's.
Number of distinct elements in $S=\frac{3n}{9}$, we must have $\frac{3n}{9}$
$=15\Rightarrow n=45$

56. (d) $f(x)$ is not one-one since $f(\pi)=f(2\pi)$
$f(x)$ is not onto since negative complex numbers are not f images of any reals.

57. (a) For R_1 let $a=1+\sqrt{2}, b=1-\sqrt{2}, c=8^{1/4}$

$aR_1b\Rightarrow a^2+b^2=(1+\sqrt{2})^2+(1-\sqrt{2})^2=6\in Q$

$bR_1c\Rightarrow b^2+c^2=(1-\sqrt{2})^2+(8^{1/4})^2=3\in Q$

$aR_1c\Rightarrow a^2+c^2=(1+\sqrt{2})^2+(8^{1/4})^2=3+4\sqrt{2}\notin Q$

$\therefore R_1$ is not transitive

For R_2 let $a=1+\sqrt{2}, b=\sqrt{2}, c=1-\sqrt{2}$

$aR_2b\Rightarrow a^2+b^2=(1+\sqrt{2})^2+(\sqrt{2})^2=5+2\sqrt{2}\notin Q$

$bR_2c\Rightarrow b^2+c^2=(\sqrt{2})^2+(1-\sqrt{2})^2=5-2\sqrt{2}\notin Q$

$bR_2c\Rightarrow b^2+c^2=(\sqrt{2})^2+(1-\sqrt{2})^2=5-2\sqrt{2}\notin Q$

$\therefore R_2$ is not transitive

58. (d) Since, $R=\{(x,y): y\in Z, x^2+3y^2\le 8\}$
$\therefore R=\{(1,1),(2,1),(1,-1),(0,1),(1,0)\}$
$\Rightarrow D_{R^{-1}}=\{-1,0,1\}$

59. (d) For a given function $f:R-\{a\}\to R$ defined by

$f(x)=\frac{a-x}{a+x}\because (fof)(x)=x$

$\Rightarrow f\left(\frac{a-x}{a+x}\right)=x\Rightarrow\frac{a-\frac{a-x}{a+x}}{a+\frac{a-x}{a+x}}=x$

$\Rightarrow\frac{a^2+ax-a+x}{a^2+ax+a-x}=x$

$\Rightarrow a^2+ax-a+x=a^2x+ax^2+ax-x^2$

$\Rightarrow a(a-1)=(a^2-1)x+x^2(a-1)$

$\Rightarrow (a-1)\left[x^2+(a+1)x-a\right]=0\Rightarrow a=1$

So, $f(x)=\frac{1-x}{1+x}\quad\therefore f\left(-\frac{1}{2}\right)=\frac{1+(1/2)}{1-(1/2)}=3$

60. (c) Given function

$f(x)=\frac{8^{2x}-8^{-2x}}{8^{2x}+8^{-2x}}, x\in(-1,1)=\frac{8^{4x}-1}{8^{4x}+1}=y\text{ (let)}$

On applying componendo and dividendo law, we get

$8^{4x}=\frac{1+y}{1-y}$

On applying logarithmic having base '8' both sides, we get

$4x=\log_8\left(\frac{1+y}{1-y}\right)$

$\Rightarrow x=\frac{1}{4}\log_8\left(\frac{1+y}{1-y}\right)=\frac{1}{4}(\log_8 e)\log_e\left(\frac{1+y}{1-y}\right)$

{by base changes property of logarithms: $\log_a b=\log_a e.\log_e b$}

By interchanging the variables x and y, we get the inverse function of $f(x)$ and it is

$f^{-1}(x)=\frac{1}{4}(\log_8 e)\log_e\left(\frac{1+x}{1-x}\right)$.

61. (b) Given, for $x \in \left(0, \frac{3}{2}\right)$, functions

$$f(x) = \sqrt{x} \quad \ldots(i)$$

$$g(x) = \tan x \quad \ldots(ii)$$

$$\text{and } h(x) = \frac{1-x^2}{1+x^2} \quad \ldots(iii)$$

Also given, $\phi(x) = ((hof)og)(x) = (hof)(g(x))$

$$= h(f(g(x))) = h(f(\tan x))$$

$$= h(\sqrt{\tan x}) = \frac{1-(\sqrt{\tan x})^2}{1+(\sqrt{\tan x})^2}$$

$$= \frac{1-\tan x}{1+\tan x} = \tan\left(\frac{\pi}{4} - x\right)$$

Now, $\phi\left(\frac{\pi}{3}\right) = \tan\left(\frac{\pi}{4} - \frac{\pi}{3}\right)$

$$= \tan\left(\frac{3\pi - 4\pi}{12}\right) = \tan\left(-\frac{\pi}{12}\right)$$

$$= -\tan\left(\frac{\pi}{12}\right) = \tan\left(\pi - \frac{\pi}{12}\right) = \tan\left(\frac{11\pi}{12}\right)$$

62. (c) Given, functions $f(x) = x^2, x \in R$

And $g(A) = \{x \in R : f(x) \in A\}; A \subseteq R$

Now, for $S = [0, 4]$

$$g(S) = \{x \in R : f(x) \in S = [0,4]\}$$

$$= \{x \in R : x^2 \in [0,4]\}$$

$$= \{x \in R : x \in [-2,2]\}$$

$\Rightarrow g(S) = [-2,2]$

So, $f(g(S)) = [0,4] = S$

Now, $f(S) = \{x^2 : x \in S = [0,4]\} = [0,16]$

And $g(f(S)) = \{x \in R : f(x) \in f(S) = [0,16]\}$

$$= \{x \in R : f(x) \in [0,16]\}$$

$$= \{x \in R : x^2 \in [0,16]\}$$

$$= \{x \in R : x \in [-4,4]\} = [-4,4]$$

From above, it is clear that $g(f(S)) = g(S)$.

63. (c) Given $f(x+y) = f(x)f(y)$

Let $f(x) = \lambda^x$ [where $\lambda > 0$]

$\because f(1) = 2$ [given]

$\therefore \lambda = 2$

So, $\sum_{k=1}^{10}(a+k) = \sum_{k=1}^{10}\lambda^{(a+k)} = \lambda^a\left(\sum_{k=1}^{10}\lambda^k\right)$

$$= 2^a\left[2^1 + 2^2 + 2^3 + \ldots + 2^{10}\right]$$

$$= 2^a\left[\frac{2(2^{10}-1)}{2-1}\right]$$

[by using formula of sum of n terms of a GP having first term a and common ratio r,

$$S_n = \frac{a(r^n - 1)}{r-1}, \text{ where } r > 1\Bigg]$$

$$\Rightarrow 2^{a+1}(2^{10}-1) = 16(2^{10}-1) \quad \text{(given)}$$

$$\Rightarrow 2^{a+1} = 16 = 2^4 \Rightarrow a+1 = 4 \Rightarrow a = 3$$

64. (a) Given, $f(x) = \log_e\left(\frac{1-x}{1+x}\right), |x| < 1$, then

$$f\left(\frac{2x}{1+x^2}\right) = \log_e\left(\frac{1-\frac{2x}{1+x^2}}{1+\frac{2x}{1+x^2}}\right) \quad \left[\because \left|\frac{2x}{1+x^2}\right| < 1\right]$$

$$= \log_e\left(\frac{\frac{1+x^2-2x}{1+x^2}}{\frac{1+x^2+2x}{1+x^2}}\right) = \log_e\left(\frac{(1-x)^2}{(1+x)^2}\right)$$

$$= \log_e\left(\frac{1-x}{1+x}\right)^2$$

$$= 2\log_e\left(\frac{1-x}{1+x}\right) \quad [\because \log_e |A|^m = m\log_e |A|]$$

$$= 2f(x) \quad \left[\because f(x) = \log_e\left(\frac{1-x}{1+x}\right)\right]$$

65. (b) $f_1(x) = \frac{1}{x}$, $f_2(x) = 1-x$ and $f_3(x) = \frac{1}{1-x}$

Also, we have $(f_2 o J o f_1)(x) = f_3(x)$

$$\Rightarrow f_2((J o f_1)(x)) = f_3(x)$$

$$\Rightarrow f_2(J(f_1(x))) = f_3(x)$$

$$\Rightarrow 1 - J(f_1(x)) = \frac{1}{1-x}$$

$$\left[\because f_2(x) = 1-x \text{ and } f_3(x) = \frac{1}{1-x}\right]$$

$$\Rightarrow 1 - J\left(\frac{1}{x}\right) = \frac{1}{1-x} \quad \left[\because f_1(x) = \frac{1}{x}\right]$$

$$\Rightarrow J\left(\frac{1}{x}\right) = 1 - \frac{1}{1-x}$$

$$= \frac{1-x-1}{1-x} = \frac{-x}{1-x}$$

Now, put $\frac{1}{x} = X$, then

$$J(X) = \frac{\frac{-1}{X}}{1-\frac{1}{X}} \qquad \left[\because x = \frac{1}{X}\right]$$

$$= \frac{-1}{X-1} = \frac{1}{1-X}$$

$\Rightarrow J(X) = f_3(X)$ or $J(x) = f_3(x)$

66. (c) Given function $f : \text{R} - \{1, -1\} \to \text{A}$ defined as

$f(x) = \frac{x^2}{1-x^2} = y$ (let)

$\Rightarrow x^2 = y(1-x^2) \qquad [\because x^2 \neq 1]$

$\Rightarrow x^2(1+y) = y \Rightarrow x^2 = \frac{y}{1+y}$ [provided $y \neq -1$]

$\because x^2 \geq 0 \Rightarrow \frac{y}{1+y} \geq 0 \Rightarrow y \in (-\infty, -1) \cup [0, \infty)$

Since, for surjective function, range of f = codomain

$\therefore$ Set A should be $\text{R} - [-1, 0)$.

67. (d) We have,

$$f(x) = \left|1 - \frac{1}{x}\right| = \begin{cases} -\frac{(x-1)}{x}, & \text{if } 0 < x \leq 1 \\ \frac{x-1}{x}, & \text{if } x > 1 \end{cases}$$

$$= \begin{cases} \frac{1}{x} - 1, & \text{if } 0 < x \leq 1 \\ 1 - \frac{1}{x}, & \text{if } x > 1 \end{cases}$$

Now, let us draw the graph of $y = f(x)$.

Note that when $x \to 0$, then $f(x) \to \infty$, when $x = 1$, then $f(x) = 0$, and when $x \to \infty$, then $f(x) \to 1$.

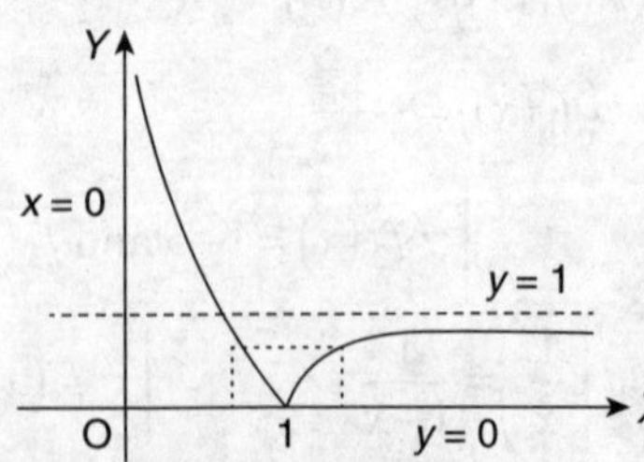

Clearly, $f(x)$ is not injective because if $f(x) < !$, then f is many–one, as shown in figure.

Also, $f(x)$ is not surjective because range of $f(x)$ is $[0, \infty]$ and but in the problem, the co-domain is $(0, \infty)$, which is wrong.

$\therefore f(x)$ is neither injective nor surjective.

68. (b) Given, $f(n) = \begin{cases} \frac{n+1}{2}; & \text{if } n \text{ is odd} \\ \frac{n}{2}; & \text{if } n \text{ is even} \end{cases}$

and $g(n) = n - (-1)^n = \begin{cases} n+1, & \text{if } n \text{ is odd} \\ n-1, & \text{if } n \text{ is even} \end{cases}$

Now, $f(g(n)) = \begin{cases} f(n+1), & \text{if } n \text{ is odd} \\ f(n-1), & \text{if } n \text{ is even} \end{cases}$

$$= \begin{cases} \frac{n+1}{2}, & \text{if } n \text{ is odd} \\ \frac{n-1+1}{2} = \frac{n}{2}, & \text{if } n \text{ is even} \end{cases}$$

$$= f(x)$$

[if n is odd, then $(n+1)$ is even and if n is even, then $(n-1)$ is odd]

Clearly, the function is not one–one as $f(2) = f(1) = 1$

but it is an onto function.

[$\because$ If $m \in \text{N}$ (codomain) is odd, then $2m \in \text{N}$ (domain) such that $f(2m) = m$, and if $m \in \text{N}$ (codomain) is even, then

$2m - 1 \in \text{N}$ (domain) such that $f(2m-1) = m$]

$\therefore$ The function is onto but not one–one.

69. (a) We have a function $f : \text{A} \to \text{R}$ defined as, $f(x) = \frac{2x}{x-1}$

One-one Let $x_1, x_2 \in \text{A}$ such that

$f(x_1) = f(x_2) \Rightarrow \frac{2x_1}{x_1 - 1} = \frac{2x_2}{x_2 - 1}$

$\Rightarrow 2x_1x_2 - 2x_1 = 2x_1x_2 - 2x_2 \Rightarrow x_1 = x_2$

Thus, $f(x_1) = f(x_2)$ has only solution, $x_1 = x_2$

onto

Let $x = 2$, then $f(2) = \frac{2 \times 2}{2-1} = 4$

But $x = 2$ is not in the domain, and $f(x)$ is one–one function

$\therefore f(x)$ can never be 4.

Similarly, $f(x)$ cannot take many values.

Hence, $f(x)$ is into (not surjective)

$\therefore f(x)$ is injective but not surjective.

70. (c) We have, $f(x)=\dfrac{x}{1+x^2}$

$\therefore f\left(\dfrac{1}{x}\right)=\dfrac{\dfrac{1}{x}}{1+\dfrac{1}{x^2}}=\dfrac{x}{1+x^2}=f(x)$

$\therefore f\left(\dfrac{1}{2}\right)=f(2)$ or $f\left(\dfrac{1}{3}\right)=f(3)$ and so on.

So, $f(x)$ is a many–one function.

Again, let $y=f(x)\Rightarrow y=\dfrac{x}{1+x^2}$

$\Rightarrow\ y+x^2y=x\Rightarrow yx^2-x+y=0$

As $x\in R$

$\therefore (-1)^2-4(y)(y)\geq 0\Rightarrow 1-4y^2\geq 0\Rightarrow y\in\left[\dfrac{-1}{2},\dfrac{1}{2}\right]$

$\therefore$ Range = Co-domain = $\left[\dfrac{-1}{2},\dfrac{1}{2}\right]$

So, $f(x)$ is surjective.

71. (b) Description of situation to find range in given domain $[a,b]$, put $f'(x)=0$ and find $x=\alpha_1,\alpha_2,\ldots\alpha_n\in[a,b]$ Now, find $\{f(a),f(\pm_1),f(\pm_2),\ldots,f(\pm_n),f(b)\}$.

Its greatest and least values give the range.

Now, $f:[0,3]\to[1,29]$

$f(x)=2x^3-15x^2+36x+1$

$\therefore f'(x)=6x^2-30x+36=6(x^2-5x+6)$

$=6(x-2)(x-3$

+ − + (sign scheme on number line with points 2, 3)

For given domain [0, 3], $f(x)$ is increasing as well as decreasing $\Rightarrow$ many–one.

Now, put $f'(x)=0\Rightarrow x=2,3$

Thus, for range $f(0)=1, f(2)=29, f(3)=28$

$\Rightarrow$ Range $\in[1,29]$

$\therefore$ Onto but not one–one.

72. (a) We have, $f(x)=ax^2+bx+c$

Now, $f(x+y)=f(x)+f(y)+xy$

Put $y=0\Rightarrow f(x)=f(x)+f(0)+0$

$\Rightarrow f(0)=0$

$\Rightarrow c=0$

Again, put $y=-x$

$\therefore f(0)=f(x)+f(-x)-x^2$

$\Rightarrow 0=ax^2+bx+ax^2-bx-x^2$

$\Rightarrow 2ax^2-x^2=0$

$\Rightarrow a=\dfrac{1}{2}$

Also, $a+b+c=3$

$\Rightarrow\dfrac{1}{2}+b+0=3\Rightarrow b=\dfrac{5}{2}$

$\therefore f(x)=\dfrac{x^2+5x}{2}$

Now, $f(n)=\dfrac{n^2+5n}{2}=\dfrac{1}{2}n^2+\dfrac{5}{2}n$

$\therefore\ \sum_{n=1}^{10}f(n)=\dfrac{1}{2}\sum_{n=1}^{10}n^2+\dfrac{5}{2}\sum_{n=1}^{10}n$

$=\dfrac{1}{2}.\dfrac{10\times11\times21}{6}+\dfrac{5}{2}\times\dfrac{10\times11}{2}$

$=\dfrac{385}{2}+\dfrac{375}{2}=\dfrac{660}{2}=330$

73. (b) $(gof)(x)=\sin x^2$

$go(gof)(x)=\sin(\sin x^2)$

$(fogogof)(x)=(\sin(\sin x^2))^2$

Again, $(gof)(x)=\sin x^2$

$(gogof)(x)=\sin(\sin x^2)$

Given, $(fogogof)(x)=(gogof)(x)$

$\Rightarrow(\sin(\sin x^2))^2=\sin(\sin x^2)$

$\Rightarrow\sin(\sin x^2)\{\sin(\sin x^2)-1\}=0$

$\Rightarrow\sin(\sin x^2)=0$ or $\sin(\sin x^2)=1$

$\Rightarrow\sin x^2=0$ or $\sin x^2=\dfrac{\pi}{2}$

$\therefore\ x^2=n\pi$

[$\sin x^2=\dfrac{\pi}{2}$ is not possible as $-1\leq\sin\theta\leq1$]

$x=\pm\sqrt{n\pi}$

Integer and Numerical Value Type

1. $f(x)=\sin^2x+\sin^2\left(x+\dfrac{\pi}{3}\right)+\cos x\cos\left(x+\dfrac{\pi}{3}\right)$

$\Rightarrow\ f(x)=\sin^2x+\left(\sin x\cos\dfrac{\pi}{3}+\cos x\sin\dfrac{\pi}{3}\right)^2$

$+\cos x\cos(x+\pi/3)$

$\Rightarrow f(x)=\sin^2x+\left(\dfrac{\sin x.1}{2}+\dfrac{\cos x.\sqrt{3}}{2}\right)^2$

$+\cos x\left(\cos x\cos\dfrac{\pi}{3}-\sin x\sin\dfrac{\pi}{3}\right)$

$$\Rightarrow f(x)=\sin^2 x+\frac{\sin^2 x}{4}+\frac{3\cos^2 x}{4}$$
$$+\frac{2.\sqrt{3}}{4}\sin x\cos x+\frac{\cos^2 x}{2}-\cos x\sin x.\frac{\sqrt{3}}{2}$$
$$=\sin^2 x+\frac{\sin^2 x}{4}+\frac{3\cos^2 x}{4}+\frac{\cos^2 x}{2}$$
$$=\frac{5}{4}\sin^2 x+\frac{5}{4}\cos^2 x=\frac{5}{4}$$

and $gof(x)=g\{f(x)\}=g(5/4)=1$

2. Given, $f(x)=(a-x^n)^{1/n}$

$\Rightarrow f[f(x)]=[a-\{(a-x^n)^{1/n}\}^n]^{1/n}=(x^n)^{1/n}=x$

$\therefore f[f(x)]=x$

3. $f'(x)=x^2+2(m-1)x+(m+5)$

If $f(x)$ is invertible it should be one–one.

$\Rightarrow f'(x)$ must be ≥ 0

$\Rightarrow 4(m-1)^2-4(m+5)\leq 0$

$\Rightarrow m\varepsilon[-1,4]$

The integers contained in this interval are –1, 0, 1, 2, 3 and 4

$\Rightarrow k=6$

4. Let $y=ax+b$ and $y=cx+d$ be two linear functions.

When $x=-1$, $y=0$ and $x=1, y=2$, then

$0=-a+b$ and $a+b=2\Rightarrow a=b=1$

$\therefore y=x+1$ ...(i)

Again, when $x=-1$, $y=2$ and $x=1, y=0$, then

$-c+d=2$ and $c+d=0$

$\Rightarrow d=1$ and $c=-1\Rightarrow y=-x+1$ ...(ii)

Hence, the two linear functions are $y=x+1$ and $y=-x+1$.

5. Let $y=\dfrac{\alpha x^2+6x-8}{\alpha+6x-8x^2}$

$\Rightarrow \alpha y+6xy-8x^2y=\alpha x^2+6x-8$

$\Rightarrow -\alpha x^2-8x^2y+6xy-6x+\alpha y+8=0$

$\Rightarrow \alpha x^2+8x^2y-6xy+6x-\alpha y-8=0$

$\Rightarrow x^2(\alpha+8y)+6x(1-y)-(8+\alpha y)=0$

Since x is real,

$\Rightarrow B^2-4AC\geq 0$

$\Rightarrow 36(1-y)^2+4(\alpha+8y)(8+\alpha y)\geq 0$

$\Rightarrow 9(1-2y+y^2)+[8\alpha+(64+\alpha^2(y+8\alpha y^2]\geq 0$

$\Rightarrow y^2(9+8\alpha)+y(46+\alpha^2)+9+8\alpha\geq 0$ ...(i)

$\Rightarrow A>0, D\leq 0, \Rightarrow 9+8\alpha>0$

and $(46+\alpha^2)^2-4(9+8\alpha)^2\leq 0$

$\Rightarrow \alpha>-9/8$ and

$[46+\alpha^2-2(9+8\alpha)][46+\alpha^2+2(9+8\alpha)]\leq 0$

$\Rightarrow \alpha>-9/8$

and $[(\alpha-2)(\alpha-14)](\alpha+8)^2\leq 0\Rightarrow \alpha>-9/8$

and $(\alpha-2)(\alpha-14)\leq 0$ $[\because (\alpha+8)^2\geq 0]$

$\Rightarrow \alpha>-9/8$ and $2\leq\alpha\leq 14\Rightarrow 2\leq\alpha\leq 14$

Thus, $f(x)=\dfrac{\alpha x^2+6x-8}{\alpha+6x-8x^2}$ will be onto, if

$$2\leq\frac{\alpha x^2+6x-8}{\alpha+6x-8x^2}\leq 14$$

Again, when $\alpha=3$

$f(x)=\dfrac{3x^2+6x-8}{3+6x-8x^2}$, in this case $f(x)=0$

$\Rightarrow 3x^2+6x-8=0$

$$\Rightarrow x=\frac{-6\pm\sqrt{36+96}}{6}$$
$$=\frac{-6\pm\sqrt{132}}{6}=\frac{1}{3}\left(-3\pm\sqrt{33}\right)$$

This shows that

$$f\left[\frac{1}{3}\left(-3+\sqrt{33}\right)\right]=f\left[\frac{1}{3}\left(-3-\sqrt{33}\right)\right]=0$$

Therefore, f is not one-to-one.

6. Since there is an injective mapping from A to B, each element of A has unique image in B. Similarly, there is also an injective mapping from B to A, each element of B has a unique image in A. Or in other words, there is one-to-one onto mapping from A to B. Thus, there is bijective mapping from A to B.

Assertion and Reason

1. (b) Statement 1 is true since $\sin\frac{1}{n}=0\Rightarrow\frac{1}{n}=K\pi$, which is always irrational.

Statement 2 is correct but it does not explain Statement 1.

2. (c) Statement 1 is correct since $\cot(\sin x)$ is defined if $\sin x\neq n\pi$, or $\cot(\sin x)$ is defined if $\sin x\neq 0(|\sin x|\leq 1)$, or $x\neq n\pi$ which is also the domain of $g(x)$.

Statement 2 is not correct since $\cot(\sin x)=\cot x$

$\Rightarrow \sin x=n\pi+x$

which is not possible for any n since $|\sin x|\leq 1$ and $[n\pi]\geq\pi(n=0, x=0$ is a possibility, but for $x=0$, both $f(x)$ and $g(x)$ are not defined.

3. (a) By well-known theorem, Statement 1 is true and follows by Statement 2.

4. (a) Statement 2 is obviously true and proves Statement 1.

Indeed $g(f(x)) = x \Rightarrow g'(f(x))f'(x) = 1$

$\Rightarrow g'(f(x)) = \dfrac{1}{f'(x)} = \dfrac{1}{x^x(1+\log x)}$

On replacing x by $g(x)$, we get

$g'(f(x)) = \dfrac{1}{g(g)^{g(x)}(1+\log g(x))}$

$\Rightarrow g'(x) = \dfrac{1}{x(1+\log g(x))}$

$(\because f(g(x)) = x$ and $(g(x))^{g(x)} = x)$

5. (d) $g(f(x)) = x \Rightarrow g'(f(x))f'(x) = 1$

$\Rightarrow g'(f(x)) = \dfrac{1}{f'(x)} = 1 + x^5$

$\Rightarrow g'(f(x)) = 1 + (g(x))^5$

$\Rightarrow$ Statement 1 is incorrect and Statement 2 is true.

JEE Advanced

Objective Questions

One Option Correct

1. (d) Given, $f(x) = \dfrac{\alpha x}{x+1}$

$f[f(x)] = f\left(\dfrac{\alpha x}{x+1}\right) = \dfrac{\alpha\left(\dfrac{\alpha x}{x+1}\right)}{\dfrac{\alpha x}{x+1}+1}$

$= \dfrac{\dfrac{\alpha^2 x}{x+1}}{\dfrac{\alpha x+(x+1)}{x+1}} = \dfrac{\alpha^2 x}{(\alpha+1)x+1} = x$ [given] ...(i)

$\Rightarrow \alpha^2 x = (\alpha+1)x^2 + x$

$\Rightarrow x[\alpha^2 - (\alpha+1)x - 1] = 0$

$\Rightarrow x(\alpha+1)(\alpha-1-x) = 0$

$\Rightarrow \alpha - 1 = 0$ and $\alpha + 1 = 0$

$\Rightarrow \alpha = -1$

But $\alpha = 1$ does not satisfy the Eq. (i)

2. (a) Let $f(x) = \sin^2 x$ and $g(x) = \sqrt{x}$

Now, $fog(x) = f[g(x)] = f(\sqrt{x}) = \sin^2\sqrt{x}$

and

$gof(x) = g[f(x)] = g(\sin^2 x) = \sqrt{\sin^2 x} = |\sin x|$

Again, let $f(x) = \sin x$, $g(x) = |x|$

$fog(x) = f[g(x)] = f(|x|)$

$= \sin|x| \neq (\sin\sqrt{x})^2$

When $f(x) = x^2$, $g(x) = \sin\sqrt{x}$

$fog(x) = f[g(x)] = f(\sin\sqrt{x}) = (\sin\sqrt{x})^2$

and $(gof)(x) = g[f(x)] = g(x^2) = \sin\sqrt{x^2}$

$= \sin|x| \neq |\sin x|$

3. (d) Given, $f(x) = |x-1|$

$\therefore f(x^2) = |x^2 - 1|$

and $\{f(x)\}^2 = (x-1)^2$

$\Rightarrow f(x^2) \neq (f(x))^2$, hence (a) is false.

Also, $f(x+y) = |x+y-1|$

and $f(x) = |x-1|$

$f(y) = |y-1|$

$\Rightarrow f(x+y) \neq f(x) + f(y)$, hence (b) is false.

$f(|x|) = ||x| - 1|$

and $|f(x)| = ||x-1|| = |x-1|$

$\therefore f(|x|) \neq |f(x)|$, hence (c) is false.

4. (b) o is commutative.

Since $boa = ba + 1 = ab + 1 = aob$

Now, $(aob)oc = (ab+1)oc = (ab+1)c + 1$

$= abc + c + 1$

and $ao(boc) = ao(bc+1) = a(bc+1) + 1$

$= abc + a + 1$

$\Rightarrow (aob)oc \neq ao(boc)$

$\Rightarrow$ o is non-associative.

$\Rightarrow$ (b) is correct.

5. (a)

$R = \{(3,3), (6,6), (9,9), (12,12), (6,12), (3,9), (3,1)\}$

$A = \{3, 6, 9, 12\}$

$\forall a \in a, (a, a) \in R$

$3 \in A, (3,3) \in R$ $6 \in A, (6,6) \in R$

$9 \in A, (9,9) \in R$ $12 \in A, (12,12) \in R$

$\therefore$ R is reflexive

Again, $(6, 12) \in R$, but $(6, 12) \notin R$

$\Rightarrow$ R is not symmetric

Also $(3, 6) \in R$, $(6, 12) \in R$, but $(3, 12) \notin R$

$\Rightarrow$ Hence the relation is reflexive only.

$\therefore$ (a) is the correct answer.

6. (d) The smallest reflexive relation on set A containing (1, 2) and (2, 3) is

$R = \{(1,1),(2,2),(3,3),(1,2),(2,3)\}$

Since, (1, 2) $\in$ R, and (2, 3) $\in$ R, but (1, 3) $\notin$ R. So, R is not transitive. To make it transitive, we have to include (1, 3) in R. Including (1, 3) in R, we get

$R_1 = \{(1,1),(2,2),(3,3),(1,2),(2,3),(1,3)\}$

This is reflexive and transitive but not symmetric.

Now, if we add the pair (2, 1) to R_1 to get R_2, then the relating R_2 is still reflexive and transitive but not symmetric. Similarly, by adding (3, 2) and (3, 1) respectively to R_1, we get

$R_3 = \{(1,1),(2,2),(3,3),(1,2),(2,3),(3,2)\}$

$R_4 = \{(1,1),(2,2),(3,3),(1,2),(2,3),(3,1)\}$

These relations are reflexive and transitive but not symmetric

We observe that out of ordered pairs (2, 1), (3, 2) and (3, 1), if we add any two ordered pairs at a time to R_1, then to maintain the relation as transitive, we will be forced to add the remaining third pair and in this process the relation will become symmetric also, which is not required. Hence, the total number of reflexive, but not symmetric relations containing (1, 2) and (2, 3) is four.

So, (d) is the correct answer.

7. (a) We see that (1, 1) $\notin$ S, so is not reflexive

Again, (1, 2) $\in$ S but (2, 1) $\notin$ S. Therefore, S is not symmetric, and hence, S is not an equivalence relation.

Now as $x - x = 0 \in Z$, so T is a reflexive relation.

Again if $x - y \in Z$, then $y - x \in Z$. Therefore, T is a symmetric relation.

If $x - y = I_1$, $y - z = I_2$, then $x - z - I_1 + I_2 \in Z$

So, T is a transitive relation. Since the relation T is reflexive, symmetric and transitive, therefore T is an equivalence relation on R but S is not.

$\therefore$ (a) is the correct answer.

8. (d) $f: R \to R$ be defined as $f(x) = x^4$

$f(-1) = (-1)^4 = 1$

$f(1) = 1^4 = 1$

$\therefore$ The images of two different elements –1 and 1 of domain R are the same element of range R.

$\therefore f$ is not one–one.

Again, no negative number of the co-domain of f is the image of any element of domain of f.

$\therefore f$ of not onto

So, f is neither one–one nor onto

Hence, (d) is the correct answer.

9. (a) Let $f: R \to R$ be defined as $f(x) = 3x$

If we take two non-zero real numbers x_1, x_2, then their f images are $f(x_1)$ and $f(x_2)$, respectively.

$f(x_1) = 3x_1, f(x_2) = 3x_2$

$f(x_1) = f(x_2) \Rightarrow 3x_1 = 3x_2 \Rightarrow x_1 = x_2$

i.e., no element in the range is the image of more than one element in the domain.

$\therefore f$ is one–one.

Let $\in R$, (co-domain), $\exists\, x = \dfrac{y}{3} \in$ R (domain)

Such that $f(x) = f\left(\dfrac{y}{3}\right) = 3.\dfrac{y}{3} = y$.

$\therefore f$ is onto.

Hence, f is one–one onto.

10. (c) Here $f: N \to Z$, where N is the set of natural numbers and Z is the set of integers.

When n is odd, $f(n) = \dfrac{n-1}{2}$. When n is even

$f(n) = -\dfrac{n}{2}$

Now $f(1) = \dfrac{1-1}{2} = 0$ $f(2) = \dfrac{-2}{2} = -1$

$f(3) = \dfrac{3-1}{2} = -1$ $f(4) = \dfrac{-4}{2} = -2$

$f(5) = \dfrac{5-1}{2} = 2$ $f(6) = \dfrac{-6}{2} = -3$ etc

We see that different elements in N have different images in Z and every element of Z is the image of any element of N under f.

Hence, f is one–one and onto both.

$\therefore$ (c) is the correct answer.

11. (d) The number of given subsets

${}^{2n+1}C_{n+1} + {}^{2n+1}C_{n+2} + \ldots + {}^{2n+1}C_{2n+1} = x$ (say)

Now,

$${}^{2n+1}C_0 + {}^{2n+1}C_1 + {}^{2n+1}C_2 + \ldots + {}^{2n+1}C_n + {}^{2n+1}C_{n+1} + \ldots + {}^{2n+1}C_{2n+1} = 2^{2n+1}$$

$$= {}^{2n+1}C_{2n+1} + {}^{2n+1}C_{2n} + \ldots + {}^{2n+1}C_{n+1} + x$$

$$= 2^{2n+1} \left(\because {}^{2n+1}C_0 = {}^{2n+1}C_{2n+1}, {}^{2n+1}C_1 = {}^{2n+1}C_{2n}\right)$$

$\Rightarrow x + x = 2^{2n+1}$

$\Rightarrow x = 2^n$

12. (c) The given function, $f: R \to R$ is

$f(x) = |x|(x - \sin x)$ ...(i)

The function f is an odd and continuous function and as $\lim\limits_{x \to \infty} f(x) = \infty$ and $\lim\limits_{x \to -\infty} f(x) = -\infty$, so range is R.

Therefore, f is an onto function.

$$f(x)=\begin{cases} x(x-\sin x), & x\ge 0 \\ -x(x-\sin x), & x<0 \end{cases}$$

$$\therefore f'(x)=\begin{cases} 2x-\sin x-x\cos x, & x>0 \\ -2x+\sin x+x\cos x, & x<0 \end{cases}$$

$$\begin{cases} (x-\sin x)+x(1-\cos x), & x>0 \\ (-x+\sin x)-x(1-\cos x), & x<0 \end{cases}$$

For $x>0$, $x-\sin x>0$ and $x(1-\cos x)>0$

$\therefore f'(x)>0\,\forall\, x\in(0,\infty)$

$\Rightarrow f$ is strictly increasing function, $\forall\, x\in(0,\infty)$.

Similarly, for $x\langle 0,-x+\sin x\rangle 0$

and $(-x)(1-\cos x)>0$,

therefore, $f'(x)>0\,\forall\, x\in(-\infty,0)$

$\Rightarrow f$ is strictly increasing function, $\forall\, x\in(0,\infty)$

Therefore f is a strictly increasing function for $x\in \mathrm{R}$ and it implies that f is one–one function.

One or More Than One Option Correct

1. (c, d)

Let $f(x)=x^3+ax^2+bx+c$

$\Rightarrow f'(x)=3x^2+2ax+b$

$f''(x)=6x+2a$

$f'''(x)=6$

$\therefore c=6,\ b=12+2a$ and $a=3+2a+b$

2. (b, c, d)

$$g(x)=\left[4\cos^4 x-2\left(2\cos^2 x-1\right)-\frac{1}{2}\left(2\cos^2 2x-1\right)-x^7\right]^{1/7}$$

$$\Rightarrow g(x)=\left(\frac{3}{2}-x^7\right)^{\frac{1}{7}}$$

$g(g(x))=x$

3. (a, d)

$$f(x)=\begin{cases} 0 & x<0 \\ e^{-x}-e^{x} & x\ge 0 \end{cases}$$

Range $=(-\infty,0)$

$\therefore$ many–one into

4. (a, c)

Since, $f(x)=\cos\left[\pi^2\right]x+\cos\left[-\pi^2\right]x$

$\Rightarrow f(x)=\cos(9)x+\cos(-10)x$

[using $\left[\pi^2\right]=9$ and $\left[-\pi^2\right]=-10$]

$$\therefore f\left(\frac{\pi}{2}\right)=\cos\frac{9\pi}{2}+\cos 5\pi=-1$$

$$f(\pi)=\cos 9\pi+\cos 10\pi=-1+1=0$$

$$f(-\pi)=\cos 9\pi+\cos 10\pi=-1+1=0$$

$$f\left(\frac{\pi}{4}\right)=\cos\frac{9\pi}{4}+\cos\frac{10\pi}{4}=\frac{1}{\sqrt{2}}+0=\frac{1}{\sqrt{2}}$$

Hence, (a) and (c) are correct options.

5. (b, c)

Since, are of equilateral triangle $=\frac{\sqrt{3}}{4}(\mathrm{BC})^2$

$$\Rightarrow \frac{\sqrt{3}}{4}=\frac{\sqrt{3}}{4}.\left[x^2+g^2(x)\right]\Rightarrow g^2(x)=1-x^2$$

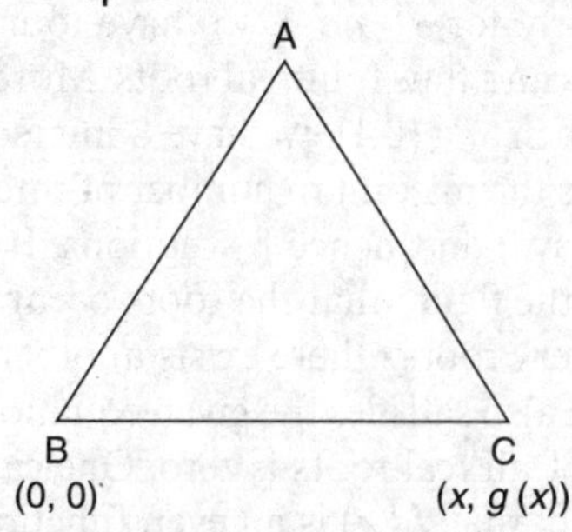

$\Rightarrow g(x)=\sqrt{1-x^2}$ or $-\sqrt{1-x^2}$

Hence, (b) and (c) are the correct options.

6. (a, d)

Given, $y=f(x)=\dfrac{x+2}{x-1}$

$\Rightarrow yx-y=x+2 \Rightarrow x(y-1)=y+2$

$\Rightarrow x=\dfrac{y+2}{y-1}\Rightarrow x=f(y)$

Here, $f(1)$ does not exist, so domain $\in \mathrm{R}-\{1\}$

$$\frac{dy}{dx}=\frac{(x-1).1-(x+2).1}{(x-1)^2}=-\frac{3}{(x-1)^2}$$

$\Rightarrow f(x)$ is decreasing for all $x\in\mathrm{R}-\{1\}$.

Also, f is a rational function of x.

Hence, (a) and (d) are correct options.

7. (b, c)

(a) Since $x\in[-1,0]$, the function $|x+1|$ is one–one.

$\Rightarrow$ Choice (a) is not possible

(b) The function given in choice (b) has a maxima at $x=1$. Hence, it cannot be one–one.

$\Rightarrow$ Choice (b) is correct.

(c) The function given in choice (c) is not one–one since it has minima at $x=-2$

$\Rightarrow$ Choice (c) is also correct.

(d) Option (d) cannot be correct since e^{-x} is essentially injective or one–one

8. (a, b)
$f(x)=[x]$ is obviously not one–one and as it takes only integer values, it cannot be onto. The function $g(x)$ is a polynomial one–one and onto.
$\Rightarrow$ (a) and (b) are correct choices.
(c) is wrong since $[x]$ cannot be one–one.
(d) is also wrong since $\frac{3-2x}{4}$ is onto.

9. (a, b, c, d)
Draw $y=f(x)$ and $y=a$.
Number of intersection points gives number of solutions.
It is clear for $a>3$, we have two intersections and similarly for $3>a>1$, we have four intersection points and thus four real roots. Moreover, we can see when, $a\in(0,1)$, we have 8 intersection points (this is the maximum number of intersections we can have) and hence 8 solutions. It is also clear from the figure that the roots occur in pairs, i.e., for every root α there exists another number $-\alpha$ which also satisfies the given equation. Hence the sum of all real roots is zero. One can see it from the fact that $f(x)$ is an even function and hence for every real root α, $-\alpha$ is also the root. Thus, all four options are correct.

10. (b, c)
First we can see that $h(x)=f(x)-x=0$
$\Rightarrow f(x)=x$ equation will have only one root.
$\because$ We have only one intersection point and hence only one solution is possible.

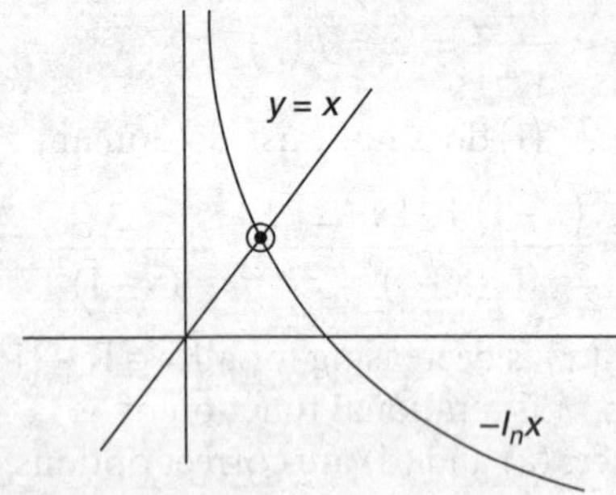

Similarly, for $m(x)=0$
$\Rightarrow g(x)=x$, we have only one solution.
$\therefore a=1$ and $b=1$

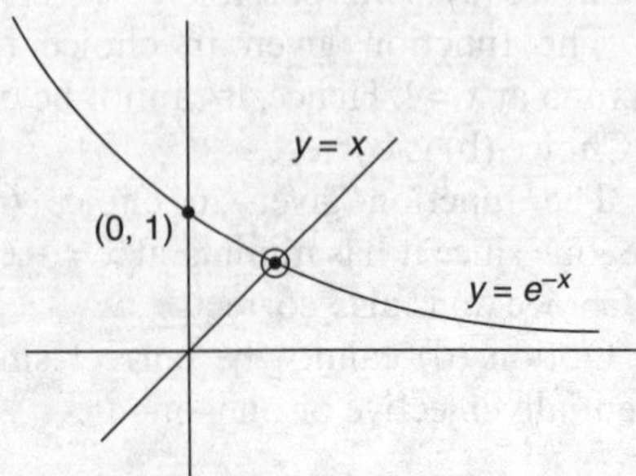

Moreover, one can see that $f(x)$ and $g(x)$ are inverse of each other. Thus $y=f(x)$ and $y=g(x)=f^{-1}(x)$ will intersect only on $y=x$ line. Thus, $a=b$ (in general).

11. (a, b, c)
We know that $y=f(x)$ and $y=f^{-1}(x)$ intersect at $y=x$ line.
Or in other words, reflection of $y=f(x)$ in $y=x$ line gives $y=f^{-1}(x)$. Thus

$$f\left(f^{-1}(x)\right)=f^{-1}(x)$$

$$x=f^{-1}(x) \quad \text{(i)}$$

and $f(x)=x$ (ii)

and $f(x)=f^{-1}(x)=x$ (iii)

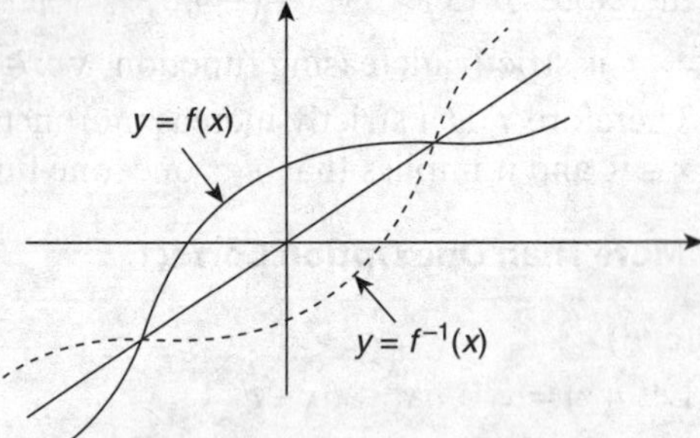

All have same roots in (a) and (b). Moreover, (0, 0) $(\alpha, f(\alpha))$ and $(\beta, f(\beta))$ are collinear and hence area of triangle formed by them is 0.

12. (a), (b), (c)

$$f(x)=\sin\left[\frac{\pi}{6}\sin\left(\frac{\pi}{2}\sin x\right)\right], x\in \mathrm{R}$$

$$=\sin\left(\frac{\pi}{6}\sin\theta\right), \theta\in\left[-\frac{\pi}{2},\frac{\pi}{2}\right]$$

where $\theta=\frac{\pi}{2}\sin x$

$$=\sin\alpha, \alpha\in\left[-\frac{\pi}{6},\frac{\pi}{6}\right], \text{where } \alpha=\frac{\pi}{6}\sin\theta$$

$$\therefore f(x)\in\left[-\frac{1}{2},\frac{1}{2}\right]$$

Hence, range of $f(x)\in\left[-\frac{1}{2},\frac{1}{2}\right]$

So, option (a) is correct.

(b) $f\{g(x)\}=f(t), t\in\left[-\frac{\pi}{2},\frac{\pi}{2}\right]$

$$\Rightarrow f(t)\in\left[-\frac{1}{2},\frac{1}{2}\right]$$

$\therefore$ Option (b) is correct.

(c) $\lim_{x\to 0}\frac{f(x)}{g(x)}=\lim_{x\to 0}\frac{\sin\left[\frac{\pi}{6}\sin\left(\frac{\pi}{2}\sin x\right)\right]}{\frac{\pi}{2}(\sin x)}$

$=\lim_{x\to 0}\frac{\sin\left[\frac{\pi}{6}\sin\left(\frac{\pi}{2}\sin x\right)\right]\frac{\pi}{6}\sin\left(\frac{\pi}{2}\sin x\right)}{\frac{\pi}{6}\sin\left(\frac{\pi}{2}\sin x\right)}\cdot\frac{\frac{\pi}{6}\sin\left(\frac{\pi}{2}\sin x\right)}{\left(\frac{\pi}{2}\sin x\right)}$

$=1\times\frac{\pi}{6}\times 1=\frac{\pi}{6}$

$\therefore$ Option (c) is correct.

(d) $g\{f(x)\}=1$

$\Rightarrow \frac{\pi}{2}\sin\{f(x)\}=1$

$\Rightarrow \sin\{f(x)\}=\frac{2}{\pi}$ ...(i)

But $f(x)\in\left[-\frac{1}{2},\frac{1}{2}\right]\subset\left[-\frac{\pi}{6},\frac{\pi}{6}\right]$

$\therefore \sin\{f(x)\}\in\left[-\frac{1}{2},\frac{1}{2}\right]$ ...(ii)

$\Rightarrow \sin\{f(x)\}\neq\frac{2}{\pi},$ [from Eqs (i) and (ii)]

i.e., No solution.

Option (d) is not correct.

13. (b) Here, $f(x)=\frac{b-x}{1-bx}$, where $0<b<1, 0<x<1$

For a function to be invertible, it should be one–one onto.

$\therefore$ Check range:

Let $f(x)=y \Rightarrow y=\frac{b-x}{1-bx}$

$\Rightarrow y-bxy=b-x \Rightarrow x(1-by)=b-y$

$\Rightarrow x=\frac{b-y}{1-by}$, where $0<x<1$

$0<\frac{b-y}{1-by}\left\langle 1\Rightarrow \frac{b-y}{1-by}\right\rangle 0$ and $\frac{b-y}{1-by}<1$

$\rightarrow y\langle b \text{ or } y\rangle\frac{1}{b}$ ...(i)

$\frac{(b-1)(y+1)}{1-by}<0-1<y<\frac{1}{b}$ ...(ii)

From Eqs (i) and (ii), we get

$y\in\left(-1,\frac{1}{b}\right)\subset$ Co-domain

Matrix Type

1. A → P, R; B → P, S; C → Q, S; D → Q, S

(A) $f'(x)=\frac{1}{2\sqrt{\sin x}}\cos x$

$f'(x)$ is positive if $x\in\left[0,\frac{\pi}{3}\right]$.

f is a one-to-one function.

Since $0\le x\le\frac{\pi}{3}$

$0\le\sin x\le\frac{\sqrt{3}}{2}$

$0\le\sqrt{\sin x}\le\sqrt{\frac{\sqrt{3}}{2}}<1$

f is an into function.

(B) $f(x)=\frac{x+3}{x-1}$

$f'(x)=\frac{(x-1)1-(x+3).1}{(x-1)^2}$

$f'(x)=\frac{-4}{(x-1)^2}$

$f'(x)<0$ Hence $f(x)$ is one-to-one.

Since $x>1$

$\therefore$ Range of $y=\frac{x+3}{x-1}$ is $(1,\infty)$

f is onto function.

(C) $-\frac{\pi}{2}\le x\le\frac{4\pi}{3}$

$f(x)=\sin x$

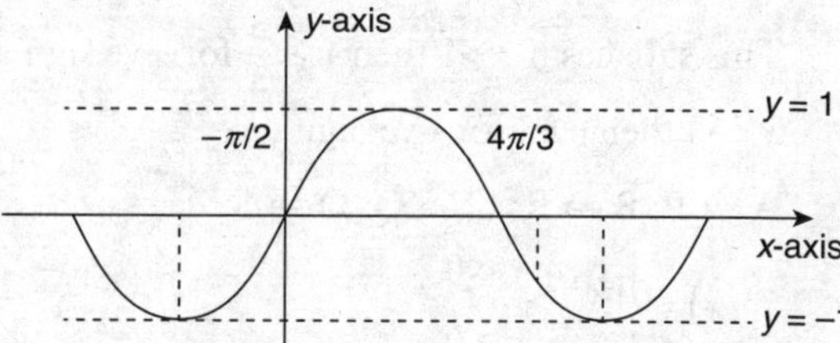

From graph, $f(x)$ is many-one and onto.

(D) $f(x)=\frac{x^2}{x-2}$

$f'(x)=\frac{(x-2).2x-x^2}{(x-2)^2}$

$f'(x)=\frac{x^2-4x}{(x-2)^2}$

$\therefore f'(x)<0$ if $2<x<4$

$f'(x)>0$ if $x>4$

$f(x)$ is many–one.

$f(4) = 8$ (is the least value of $f(x)$)
$\therefore$ range $= [8, \infty)$
$\therefore f(x)$ is onto.

2. A $\rightarrow$ S; B $\rightarrow$ P; C $\rightarrow$ R; D $\rightarrow$ Q

(A) Since $\frac{2x+1}{3x+4}$ will not attain the value $\frac{2}{3}$ but is one–one.

(B) Since $\sin x$ is neither one–one nor does it take all real values.

(C) $x^3 - x + 1$ is not one–one $\left(\frac{dy}{dx} = 0 \text{ at } x = \pm\frac{1}{\sqrt{3}}\right)$ but being a cubic polynomial, it is certainly onto.

(D) $x^3 + x + 1$ is certainly onto and is one–one since $\frac{dy}{dx} > 0$ for all x.

3. A $\rightarrow$ R; B $\rightarrow$ Q; C $\rightarrow$ S; D $\rightarrow$ P

$y = x^2 - x + 1 \Rightarrow x^2 - x + 1 - y = 0$

$$\Rightarrow x = \frac{1 \pm \sqrt{4y-3}}{2}$$

If $x < \frac{1}{2}$ then –ve sign is taken.

If $x > \frac{1}{2}$ then +ve sign is taken.

$\Rightarrow$ Again $y = x^2 + x + 1$

$\Rightarrow x^2 + x + 1 - y = 0$

$$x = \frac{-1 \pm \sqrt{4y-3}}{2}$$

This satisfies if $x>1$ then $y \geq \frac{3}{4}$ for +ve sign and $x<-1$ then $y \geq 1$ for –ve sign.

4. A $\rightarrow$ P; B $\rightarrow$ R; C $\rightarrow$ Q; D $\rightarrow$ S

$$f_1(x) = \begin{cases} |x|, & x < 0 \\ e^x, & x \geq 0 \end{cases}$$

$f_2(x) = x^2, x \geq 0$

$$f_3(x) = \begin{cases} \sin x, & x < 0 \\ x, & x \geq 0 \end{cases}$$

$$f_4(x) = \begin{cases} f_2[f_1(x)], & x < 0 \\ f_2[f_1(x)] - 1, & x \geq 0 \end{cases}$$

$$\text{Now, } f_2(f_1(x)) = \begin{cases} x^2, & x < 0 \\ e^{2x}, & x \geq 0 \end{cases}$$

$$\Rightarrow f_4 = \begin{cases} x^2, & x < 0 \\ e^{2x} - 1, & x \geq 0 \end{cases}$$

As $f_4(x)$ is continuous, $f_4'(x) = \begin{cases} 2x, & x < 0 \\ 2e^{2x}, & x > 0 \end{cases}$

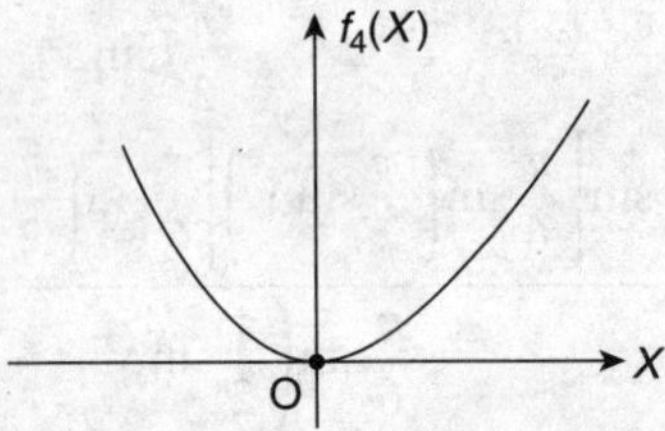

Graph for $f_4(x)$

$f_4'(0)$ is not defined. Its range is $[0, \infty)$.

Thus, range = co-domain = $[0, \infty)$; thus f_4 is onto.

Also, horizontal line (drawn parallel to X-axis) meets the curve more than once; thus function is not one–one.

Integer and Numerical Value Type

1. $f(x) + f\left(\frac{1}{1-x}\right) = \frac{2}{x} - \frac{2}{1-x}$

Put $x = \frac{1}{1-t}$ $\quad 1 - x = 1 - \frac{1}{1-t} = -\frac{t}{1-t}$

$$\therefore f\left(\frac{1}{1-t}\right) + f\left(\frac{t-1}{t}\right) = 2(1-t) - \frac{2(1-t)}{-t}$$

$$\therefore f\left(\frac{1}{1-x}\right) + f\left(\frac{x-1}{x}\right) = 2(1-x) + 2\left(\frac{1-x}{x}\right)$$

Put $\frac{x-1}{x} = t$

$x - 1 = xt \qquad x = \frac{1}{1-t}$

$1 - x = 1 - \frac{1}{1-t} = \frac{-t}{1-t}$

$\frac{1}{1-x} = \frac{t-1}{t}$

$$\therefore f\left(\frac{t-1}{t}\right) + f(t) = \frac{2t}{t-1} + \left(\frac{2t}{t-1}\right)\left(\frac{1-t}{1}\right)$$

$$\therefore f\left(\frac{x-1}{x}\right) + f(x) = \frac{2x}{x-1} - 2x$$

$$\therefore 2f(x) = \frac{2}{x} - \frac{2}{1-x} - 2(1-x) - \frac{2(1-x)}{x} + \frac{2x}{x-1} - 2x$$

$$= \frac{2}{x} - \frac{2}{1-x} - 2 + 2x - \frac{2}{x} + 2 + \frac{2x}{x-1} - 2x$$

$$= \frac{2x}{x-1} + \frac{2}{x-1} = 2\left(\frac{x+1}{x-1}\right)$$

$$\therefore f(x) = \frac{x+1}{x-1}$$

2. k must be 1 since the least period is $\frac{\pi}{2}$.

3. We can easily get $f(n) = nf(1)$, where n is a natural number, where
$f(2007) = 2007f(1) = 1$

4. $\tan\frac{\pi}{4} = 1, \tan\frac{\pi}{3} = \sqrt{3} \tan\frac{5\pi}{12} = 2 + \sqrt{3} = 3.73$

$\Rightarrow$ There must exist two values between $\frac{\pi}{3}$ and $\frac{\pi}{12}$ such that $\tan x = 2, \tan x = 3$

These $[\tan x]$ take values 0, 1, 2, 3.
$\Rightarrow k = 4$

5. S is a circle shown in the figure.

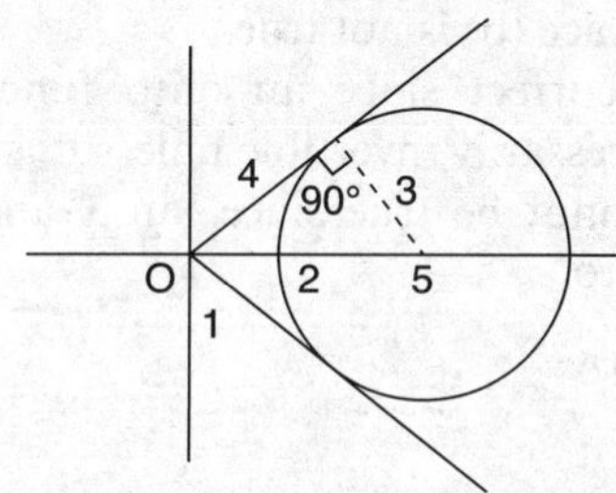

Indeed $f(x, y)$ is the slope of the line joining (0, 0) and (x, y). The minimum and maximum slopes are $-\frac{3}{4}$ and $+\frac{3}{4}$.
$\Rightarrow k = 4$

6. $f(x, y)$ is distance of (x, y) from (0, 0). The minimum and maximum distances of the circle $x^2 + y^2 - 6x - 8y + 21 = 0$ from origin are 3 and 7.

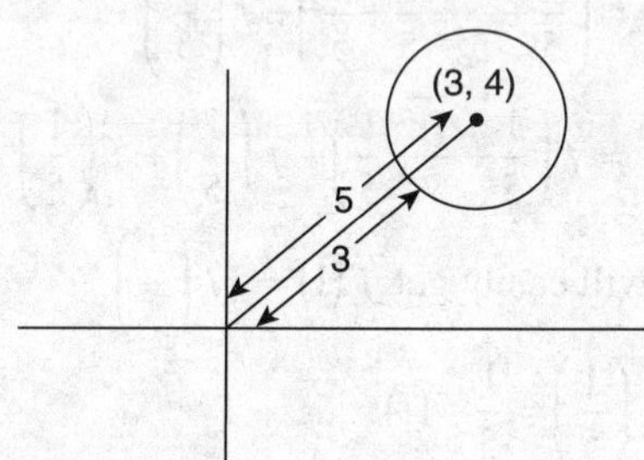

$\Rightarrow$ Range $f = [3,7]$
$\Rightarrow k = 4$

7. We must have $16 - x > 0; 2x - 1 \ge 0$
$20 - 3x > 0, 4x - 5 \ge 0$
We must also have $16 - x \ge 2x - 1$ and $30 - 3x \ge 4x - 5$.
The intersection of all inequations is {2, 3}.
$\Rightarrow k = 2$

8. $f(x) = \frac{x+5}{\sqrt{x^2+1}}\ f'(x) = \frac{1-5x}{(x^2+1)^{3/2}}$

$\Rightarrow \frac{1}{5}$ is a point of maxima.

$\lim_{x\to-\infty} f(x) = -1, f\left(\frac{1}{5}\right) = \sqrt{26}$

$\lim_{x\to+\infty} f(x) = 1$

$\Rightarrow$ Range of $f(x)$ is $(-1, \sqrt{26})$.

The integrals contained in this interval are 0, 1, 2, 3, 4, 5.
$\Rightarrow f(x)$ takes six integer values.
$\Rightarrow k = 6$

9. The function is reducible to $\tan 4x$ by trigonometric transformations.
$\Rightarrow$ Period $= \frac{\pi}{4} \Rightarrow k = 4$

10. $y = \frac{x+m}{x^2+1}$ ...(i)

$yx^2 - x + (y - m) = 0$ ...(ii)

We must have $D \ge 0$ for $y \in [0,1]$
$\Rightarrow 1 - 4y(y - m) \ge 0$ for $y \in [0,1]$
$\Rightarrow f(y) = 4y^2 - 4my - 1 \le 0$ for $y \in [0,1]$
This will be possible if $f(0) < 0, f(1) < 0$
$\Rightarrow -1 < 0, 3 - 4m < 0$
$\Rightarrow m > \frac{3}{4} \Rightarrow k = 4$

11. Let us observe the function in [0, 1) only.
Since $f(x)$ changes $\frac{1}{5}, \frac{2}{5}, \frac{3}{5}, \frac{4}{5}; \frac{1}{4}, \frac{2}{4}, \frac{3}{4}; \frac{1}{3}, \frac{2}{3}; \frac{1}{2}$
(already covered), the function $f(x)$ takes 10 integer values.
Total values taken by the function
$= 10 \times 100 + 1 = 1001$

Comprehension Passage

Comprehension 1

1. (a)

$$g(f(x)) = \begin{cases} 2x + a + 4 & ; \quad 0 \le 2x + a \le 4 \quad x \ge -1 \\ -3(2x + a) - 2 & ; \quad -2 < 2x + a < 0 \quad x \ge -1 \\ bx^2 + 3 + 4 & ; \quad 0 \le bx^2 + 3 \le 4 \quad x < -1 \\ -3(bx^2 + 3) - 2 & ; \quad -2 < bx^2 + 3 < 0 \quad x < -1 \end{cases}$$

$g(f(x))$ is defined in the following:
For $x \ge -1$
$2x + a \ge -2 + a \ \Rightarrow -2 + a \le 2x + a \le 4\ a \le 6$
Other part
$2x + a \ge -2 + a \ \Rightarrow -2 + a < 0 \ \Rightarrow a < 2$

For $x < -1 \Rightarrow x^2 > 1$
$b > 0 \Rightarrow bx^2 + 3 > b + 3 \quad b + 3 < bx^2 + 3 \le 4$
$b < 1 \Rightarrow 0 < b < 1$
Do the other part by yourself.
It is obtained as $b \in (-3, 0)$
Similarly, b can be in $(-5, 0)$
Thus, if a, b are in option (a) $g(f(x))$ is not defined.

2. (a) $x \in [-1, 2]$
$\Rightarrow -1 \le x \le 2$
$\Rightarrow -2 \le 2x \le 4$
$\Rightarrow -2 + a \le 2x + a \le 4 + a$
$\Rightarrow -2 + a \le -2$ and $4 + a \ge 4$, i.e., $a = 0$
b can be take any real value.

3. (c) If $a = 2, b = 3$
$$f(x) = \begin{cases} 2x + 2 & : \quad x \ge -1 \\ 3x^2 + 3 & : \quad x < -1 \end{cases}$$
Range of $f(x)$ is $[0, \infty)$.

Comprehension 2

1. (a) Since period of $f(x)$ is $2(10 - 2) = 16$
$f(0) = f(16) = f(32) = \ldots. = f(160) = 5$
$\therefore$ There are at least 11 values of x for which $f(x) = 5$
$f(0) = f(4) = f(16)$
Due to symmetry in one period length $f(x) = 5$ has solution other then 0, 16, 32.
$\therefore$ minimum possible number of values of x is $10 + 11 = 21$.

2. (a) Obvious by definition

3. (c) If 1 is a period, then $f(x) = f(x + 1), \forall x \in R$
$\Rightarrow f(2) = f(3) = f(4) = f(5) = f(6)$
which contradicts the given hypothesis that $f(2) \ne f(6)$.
$\therefore$ 1 cannot be period of $f(x)$.

Comprehension 3

1. (c) If the degree is odd then
$f(-\infty) = -\infty, f(\infty) = \infty$
$\Rightarrow f(x)$ will attain all values; hence a polynomial of odd degree is essentially onto.

2. (d) Let $y = \dfrac{x^2 + 2x + c}{x^2 + 4x + 3c}$ ...(i)

Then $(y - 1)x^2 + x(4y - 2) + 3cy - c = 0$ ...(ii)

If range of (i) is R, then roots of (ii) must be real for all y.

$\Rightarrow (4y - 2)^2 - 4(y - 1)(3cy - c) \ge 0$ for all y.
$\Rightarrow (4 - 3c)y^2 - 4(1 - c)y + 1 - c \ge 0$ for all y. ...(iii)

Now (iii) is true for all y if
$4 - 3c > 0$ and $16(1 - c)^2 - 4(4 - 3c)(1 - c) \le 0$
or $c < \dfrac{4}{3}$ and $4(1 - c)[4(1 - c) - (4 - 3c)] \le 0$
We easily get $0 < c < 1$.

3. (c)
(a) Since the statements written in choice (a) are true, it is not the correct choice.
(b) Again, an onto function from an infinite set to a finite set cannot be one–one. Therefore, choice (b) is not true.
(c) is correct since an onto function is not necessarily invertible unless it is one–one.
(d) cannot be true since tan x and cot x are onto.

Comprehension 4

1. (d)
On putting $x = y = 0$, we get $f(0) = f(0) + f(0)$
$\Rightarrow f(0) = 0$
Again, on putting $y = -x$, we get
$f(0) = f(x) + f(-x)$ ...(*)
$\Rightarrow f(-x) = -f(x)$
Now
$$f(1) = f\left(\frac{1}{5} + \frac{1}{5} + \frac{1}{5} + \frac{1}{5} + \frac{1}{5}\right)$$
$$= \left(\frac{1}{5} + \frac{1}{5} + \frac{1}{5} + \frac{1}{5}\right) + f\left(\frac{1}{5}\right)$$
$$= f\left(\frac{1}{5} + \frac{1}{5} + \frac{1}{5}\right) + f\left(\frac{1}{5}\right) + f\left(\frac{1}{5}\right) \text{ and so on.}$$
We will easily get $f(1) = 5f\left(\dfrac{1}{5}\right)$
$$\Rightarrow f\left(\frac{1}{5}\right) = \frac{1}{5}f(1)$$
Now, $f\left(\dfrac{3}{5}\right) = f\left(\dfrac{1}{5}\right) + f\left(\dfrac{1}{5}\right) + f\left(\dfrac{1}{5}\right)$
$$= 3f\left(\frac{1}{5}\right) = \frac{3}{5}f(1)$$
Now, $f\left(-\dfrac{3}{5}\right) = -f\left(\dfrac{3}{5}\right)$ from (*)
whence choice (d) follows.

2. (d) It is clear that $f(0) = 0$ and f is odd (replace y by $-x$.)
If we replace y by x, then

$$2f(x) = f\left(2x\sqrt{1-x^2}\right)$$

Again, if we replace y by $2x\sqrt{1-x^2}$, then

$$f(x)+f\left(2x\sqrt{1-x^2}\right) = f\left(3x-4x^3\right)$$
$$= -f\left(4x^3-3x\right)$$
$$\Rightarrow 3f(x) = -f\left(4x^3-3x\right)$$
$$\Rightarrow f\left(4x^3-3x\right)+3f(x) = 0$$

3. (c) We can form three equations in variables

$$f(x), f\left(\frac{1}{1-x}\right) \text{and } f\left(\frac{x-1}{x}\right)$$

Recall that

If $a+b=\gamma, \quad b+c=\alpha, \quad c+a=\beta$, then

$a+b+c = \frac{1}{2}(\alpha+\beta+\gamma)$ and

$a = \frac{1}{2}(\alpha+\beta+\gamma)-\alpha = \frac{\beta+\gamma-\alpha}{2}$, etc.

Now the first equation is

$f(x)+f\left(\frac{1}{1-x}\right) = \frac{2(1-2x)}{x(1-x)}$ i.e., $a+b=\gamma$

Replace x by $\frac{1}{1-x}$ and $\frac{x-1}{x}$ in succession and form two more equations.

Alternative solution:

Proceeding with the choices, (a), (b), (d) are ruled out and (c) fits into the given functional equation.

Comprehension 5

Let $f(x)$ is a function. We define

$$f^2(x) = f(f(x))$$

$f^3(x) = f(f(f(x)))$ and so on.

In certain cases, $f^n(x)$ can be proved by induction. We may also say that $f^{-(n)}(x)$ is the inverse function f^{-1} being applied n times.

1. (c) $f(f(x)) = 3+4(3+4x) = 15+16x$

$$f^3(x) = f\left(f^2(x)\right) = 3+4(15+16x)$$
$$= 63 + 64x.$$

2. (d) (a), (b), (c) for $n = 1$ for and (d) for $n = 2$ turns out to be true. Let us prove it by induction.

If $f^n(x) = \left(4^n-1\right)+4^n x$ for some n, then

$$f^{n+1}(x) = f\left(f^n(x)\right)$$
$$= 3+4\left[\left(4^n-1\right)+4^n x\right]$$
$$= \left(4^{n+1}-1\right)+4^{n+1}x$$

$\Rightarrow$ Result is true for $n + 1$.

3. (a) $y = 3+4x$

$$\Rightarrow x = \frac{y-3}{4}$$
$$\Rightarrow f^{-1}(x) = \frac{x-3}{4}$$

Putting $n = 1$ in choice (a), we get

$$f^{-1}(x) = \left(\frac{1}{4}-1\right)+\frac{1}{4}x = \frac{x-3}{4}$$

All other choices are ruled out for $n = 1$

$\Rightarrow$ (a) is the correct answer.

CHAPTER 2

Inverse Trigonometric Functions

Objectives

In this chapter on Inverse Trigonometric Functions, we shall study the following topics:

1. Definition, Range, Domain and Principal value of trigonometric function
2. Graphs of inverse trigonometric functions
3. Properties of inverse trigonometric function.

Introduction

We are already familiar with Inverse Functions from our study of the previous chapter. The inverse function denoted by f^{-1} exists if the function f (f is the function whose inverse is f^{-1}) is one-one and onto.

A function which is not one-one and onto is not invertible. The trigonometric functions are not one-one and onto over their natural domains and ranges. Therefore, they are invertible. But if we restrict the domains and ranges of trigonometric functions their inverse exists.

In this chapter, we shall study inverse trigonometric functions in limit range, their properties and behavior through graphical representation.

Quick Review of Inverse Function

Let us have a function $f: X \to Y$ such that $f(x) = y$ is one-one and onto. We can define a unique function $g: Y - X$ such that $g(y) = x$ where $x \in X; y \in Y$. Therefore, the domain of g = range of f and the range of g = domain of f. The function g is inverse of f and denoted by f^{-1}.

The function g is also one-one onto. We can conclude the discussion with the following result:

(i) $g = f^{-1}; f = g^{-1}$

(ii) $g^{-1} = (f^{-1})^{-1} = f$

(iii) $(f^{1}of)(x) = f^{-1}(f(x)) = f^{-1}(y) = x$

(iv) $(fof^{-1})(y) = f(f^{-1}(y)) = f(x) = y$

Earlier, we stated the trigonometric functions are invertible in limited domain and ranges. Let us see what this means with an example of sin functions. The domain of sin function is the set of all real values and range in the closed interval $[-1, 1]$. Restricting its domain to $\left[-\frac{\pi}{2}, \frac{\pi}{2}\right]$, the function becomes invertible.

But this is not the only interval in which $\sin^{-1}$ is invertible. There are many other intervals in which sin function is invertible, which can be better visualized by the sin graph.

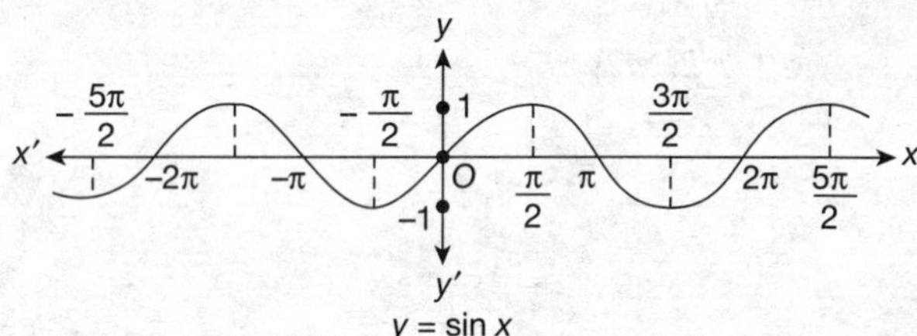

y = sin x

The other intervals are ...., $\left[-\frac{3\pi}{2}, \frac{\pi}{2}\right]$, $\left[\frac{\pi}{2}, \frac{3\pi}{2}\right]$, $\left[\frac{3\pi}{2}, \frac{5\pi}{2}\right]$,...

The Inverse of sin Function

The inverse of sin function is denoted by $\sin^{-1}$ and read as sin inverse of x or arc sin of x.

Obviously, the domain of $\sin^{-1} x$ becomes $[-1, 1]$ and range could be any of the intervals as stated above. In another way corresponding to each such interval, we get a branch of the function $\sin^{-1}$. The branch with range $\left[-\frac{\pi}{2}, \frac{\pi}{2}\right]$ is called principal value branch of $\sin^{-1}$ function.

$$\sin^{-1}[-1, 1] \to \left[-\frac{\pi}{2}, \frac{\pi}{2}\right]$$

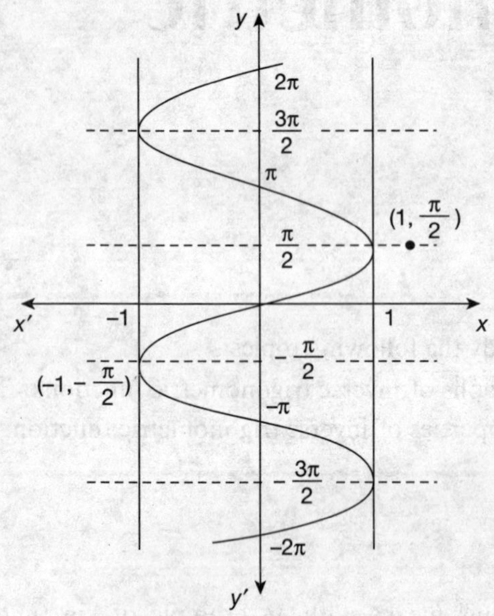

The graph of $y = \sin^{-1} x$

Note: $\sin^{-1} x$ is inverse of sin function do not misunderstand it as $\frac{1}{\sin x}$.

$$(\sin x)^{-1} = \frac{1}{\sin x}$$

Below are the domain (values of x) and principal values (values of y) of all the inverse trigonometric functions and their graph.

(i) $y = \sin^{-1} x$

$x \in [-1, 1]$

$y \in [-\pi/2, \pi/2]$

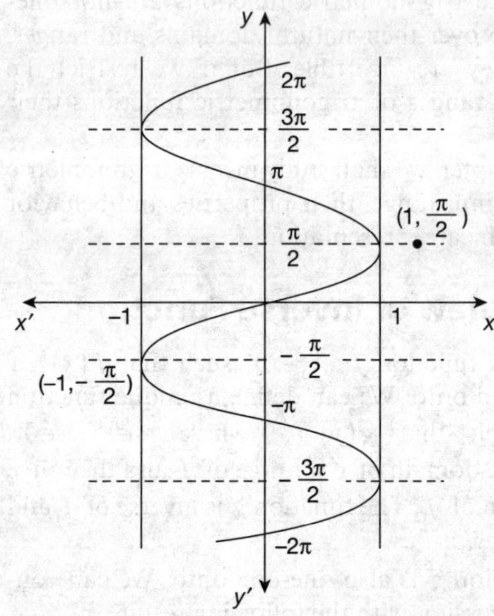

(*Continued*)

(ii) $y = \cos^{-1} x$

$x \in [-1, 1]$

$y \in [0, \pi]$

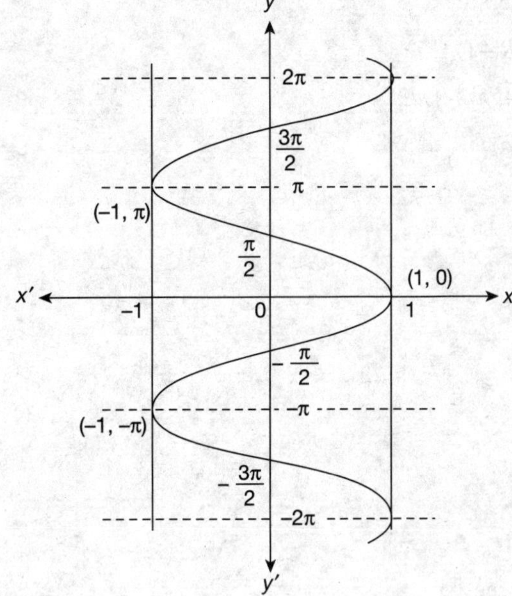

(iii) $y = \tan^{-1} x$

$x \in R$

$y \in (-\pi/2, \pi/2)$

(iv) $y = \cot^{-1} x$

$x \in R$

$y \in (0, \pi)$

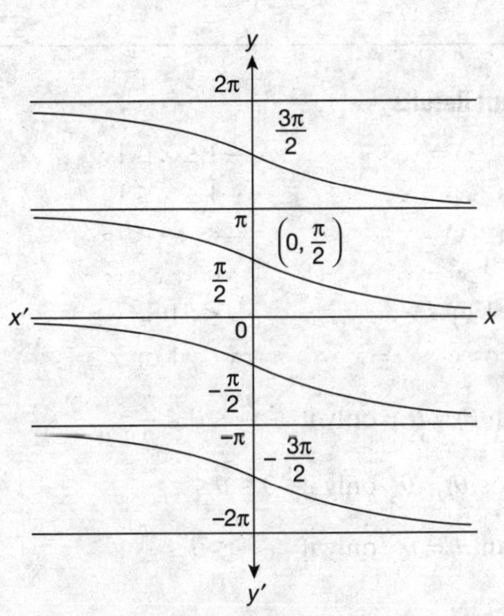

(*Continued*)

(v) $y = \sec^{-1} x$

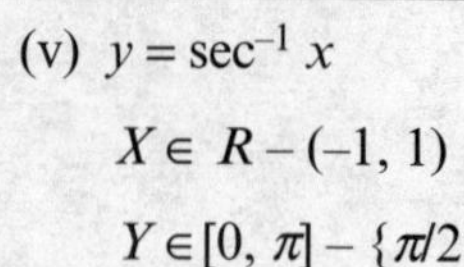

$X \in R - (-1, 1)$

$Y \in [0, \pi] - \{\pi/2\}$

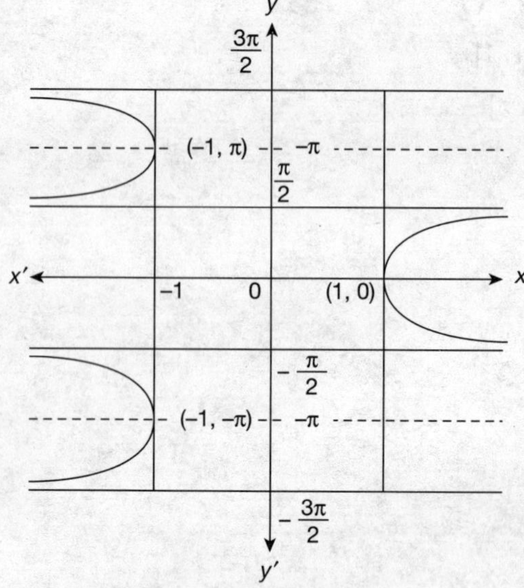

(vi) $y = \text{cosec}^{-1} x$

$x \in R - (-1, 1)$

$y \in (-\pi/2, \pi/2) - \{0\}$

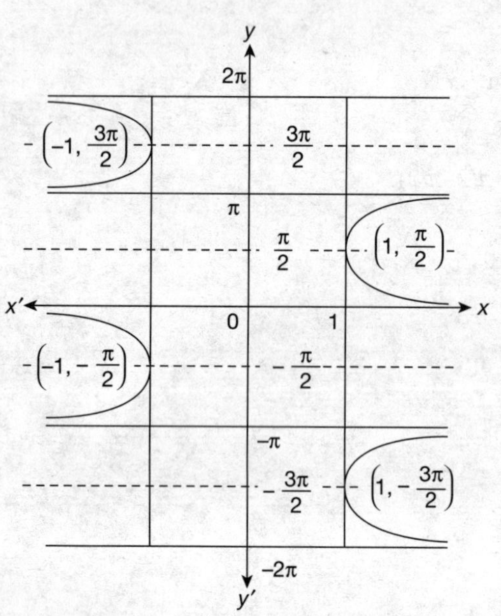

Some Important Results

(i) $\sin(\sin^{-1} x) = x \qquad -1 \le x \le 1$

$\cos(\cos^{-1} x) = x \qquad -1 \le x \le 1$

$\tan(\tan^{-1} x) = x \qquad -\infty < x < \infty$

$\cot(\cot^{-1} x) = x \qquad -\infty < x < \infty$

$\sec(\sec^{-1} x) = x \qquad x \le -1 \text{ or } x \ge 1$

$\text{cosec}(\text{cosec}^{-1} x) = x \qquad x \le -1 \text{ or } x \ge 1$

(ii) $\sin^{-1}(\sin\theta) = \theta$ only if $-\dfrac{\pi}{2} \le \theta \le \dfrac{\pi}{2}$

$\cos^{-1}(\cos\theta) = \theta$ only if $0 \le \theta \le \pi$

$\tan^{-1}(\tan\theta) = \theta$ only if $-\dfrac{\pi}{2} \le \theta \le \dfrac{\pi}{2}$

$\cot^{-1}(\cot\theta) = \theta$ only if $0 < \theta < \pi$

$\sec^{-1}(\sec\theta) = \theta$ only if $0 \le \theta < \dfrac{\pi}{2}$ or $\dfrac{\pi}{2} < \theta \le \pi$

$\text{cosec}^{-1}(\text{cosec}\,\theta) = \theta$ only if $-\dfrac{\pi}{2} \le \theta < 0$ or $0 < \theta \le \dfrac{\pi}{2}$

(iii) $\sin^{-1}(-x) = -\sin^{-1} x \qquad \forall\, x \in [-1, 1]$

$\cos^{-1}(-x) = \pi - \cos^{-1} x \qquad \forall\, x \in [-1, 1]$

$\tan^{-1}(-x) = -\tan^{-1} x \qquad \forall\, x \in R$

$\cot^{-1}(-x) = \pi - \cot^{-1} x \qquad \forall\, x \in R$

$\sec^{-1}(-x) = \pi - \sec^{-1} x \qquad \forall\, x \in (-\infty, -1] \cup [1, \infty)$

$\text{cosec}^{-1}(-x) = -\text{cosec}^{-1} x \qquad \forall\, x \in (-\infty, -1] \cup [1, \infty)$

(iv) $\sin^{-1} x = \text{cosec}^{-1}\left(\frac{1}{x}\right) \quad -1 \le x < 0 \text{ or } 0 < x \le 1$

$\cos^{-1} x = \sec^{-1}\left(\frac{1}{x}\right) \quad -1 \le x < 0 \text{ or } 0 < x \le 1$

$\tan^{-1} x = \cot^{-1}\left(\frac{1}{x}\right), x > 0$

(v) $\sin^{-1} x + \cos^{-1} x = \pi/2 \quad \forall x \in [-1, 1]$

$\tan^{-1} x + \cot^{-1} x = \pi/2 \quad x \in R$

$\sec^{-1} x + \text{cosec}^{-1} x = \pi/2 \quad -\infty < x \le -1 \text{ or } 1 \le x < \infty$

(vi) $\tan^{-1} x + \tan^{-1} y = \tan^{-1}\left(\frac{x+y}{1-xy}\right);$ if $xy < 1$

$\tan^{-1} x - \tan^{-1} y = \tan^{-1}\left(\frac{x-y}{1+xy}\right),$ if $xy > -1$

$\sin^{-1} x + \sin^{-1} y = \sin^{-1}\left[x\sqrt{1-y^2} + y\sqrt{1-x^2}\right],$

if $-1 \le x, y \le 1$ and $x^2 + y^2 \le 1$

or if $xy < 0$ and $x^2 + y^2 > 1$

$\sin^{-1} x - \sin^{-1} y = \sin^{-1}\left[x\sqrt{1-y^2} - y\sqrt{1-x^2}\right],$

if $-1 \le x, y \le 1$ and $x^2 + y^2 \le 1$

or if $xy > 0$ and $x^2 + y^2 > 1$

$\cos^{-1} x + \cos^{-1} y = \cos^{-1}\left[xy - \sqrt{1-x^2}\sqrt{1-y^2}\right],$

if $-1 \le x, y \le 1$ and $x + y \ge 0$

$\cos^{-1} x - \cos^{-1} y = \cos^{-1}\left[xy + \sqrt{1-x^2}\sqrt{1-y^2}\right],$

if $x \le y$ and $-1 \le x, y \le 1$

ILLUSTRATION 2.1

Evaluate the following:

(a) $\tan^{-1}(-1)$ **(b)** $\tan^{-1}\left(\sin\left(\frac{-\pi}{2}\right)\right)$

(c) $\sin^{-1}\left(-\frac{\sqrt{3}}{2}\right)$ **(d)** $\cos^{-1}\left(\cos\frac{13\pi}{6}\right)$

(e) $\tan\left(\tan^{-1}(-4)\right)$

Solution

(a) $\tan\theta = -1$

$\Rightarrow \tan^{-1}(-1) = -\frac{\pi}{4} \quad \left\{-\frac{\pi}{4} \in \text{range of } \tan^{-1} x\right\}$

(b) $\tan^{-1}\left(\sin\left(\frac{-\pi}{2}\right)\right) = \tan^{-1}\left(-\sin\left(\frac{\pi}{2}\right)\right)$

$= \tan^{-1}(-1) = -\frac{\pi}{4}$

(c) $\sin\left(-\frac{\pi}{3}\right) = \frac{-\sqrt{3}}{2} \Rightarrow \sin^{-1}\left(-\frac{\sqrt{3}}{2}\right) = \frac{-\pi}{3}$

$\left\{\frac{-\pi}{3} \in \text{range of } \sin^{-1} x\right\}$

(d) $\cos^{-1}\left(\cos\frac{13\pi}{6}\right) = \cos^{-1}\left(\cos\left(2\pi + \frac{\pi}{6}\right)\right)$

$= \cos^{-1}\left(\cos\frac{\pi}{6}\right)$

(e) $\tan\left(\tan^{-1} x\right) = x, \forall x \in R, \tan\left(\tan^{-1}(-4)\right) = -4$

ILLUSTRATION 2.2

Find the value of $\sin\left[2\cot^{-1}\left(\frac{-5}{12}\right)\right]$

Solution

Let $\cot^{-1}\left(\frac{-5}{12}\right) = y$. Then $\cot y = \frac{-5}{12}$.

Now $\sin\left[2\cot^{-1}\left(\frac{-5}{12}\right)\right] = \sin 2y$

$= 2\sin y \cos y = 2\left(\frac{12}{13}\right)\left(\frac{-5}{13}\right)$

$\left[\cot y < 0, \text{ so } y \in \left(\frac{\pi}{2}, \pi\right)\right]$

$= \frac{-120}{169}$

ILLUSTRATION 2.3

Sketch the graph for

(i) $y = \sin^{-1}(\sin x)$ **(ii)** $y = \cos^{-1}(\cos x)$

(iii) $y = \tan^{-1}(\tan x)$

Solution

(i) We know that $y = \sin^{-1}(\sin x)$ is periodic with period 2π.

$\therefore$ to draw this graph, we should draw the graph for one interval of length 2π and repeat for the entire values of x.

As we know, $\sin^{-1}(\sin x) =$

$$\begin{cases} x; & -\dfrac{\pi}{2} \le x \le \dfrac{\pi}{2} \\ (\pi - x); & -\dfrac{\pi}{2} \le \pi - x < \dfrac{\pi}{2}\left(\text{i.e., } \dfrac{\pi}{2} \le x \le \dfrac{3\pi}{2}\right) \end{cases}$$

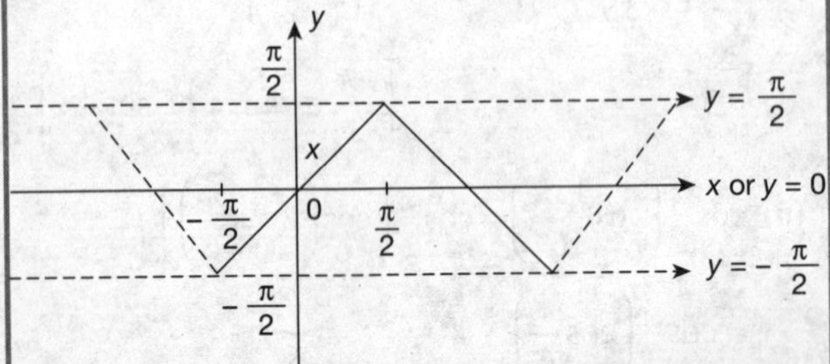

Thus, the graph for $y = \sin^{-1}(\sin x)$, is a straight line up and a straight line down with slopes 1 and –1 respectively lying between $\left[-\dfrac{\pi}{2}, \dfrac{\pi}{2}\right]$.

(ii) $y = \cos^{-1}(\cos x)$ is periodic with period 2π.

$\therefore$ To draw this graph, we should draw the graph for one interval of length 2π and repeat for entire values of x of length 2π.

As we know,

$$\cos^{-1}(\cos x) = \begin{cases} x; & 0 \le x \le \pi \\ 2\pi - x; & 0 \le 2\pi - x \le \pi \end{cases}$$

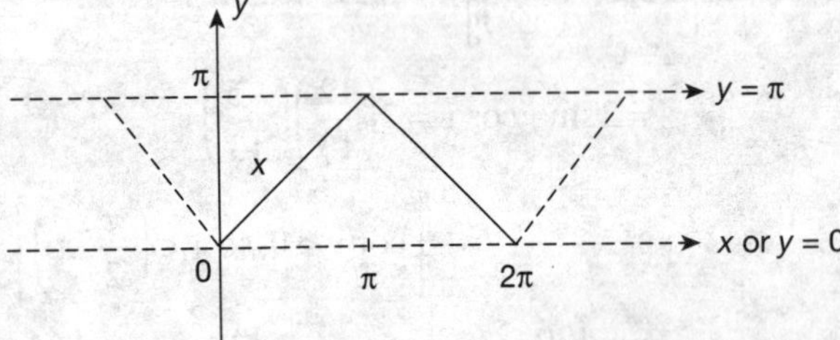

(iii) $y = \tan^{-1}(\tan x)$ is periodic with period π.

$\therefore$ to draw this graph, we should draw the graph for one interval of length π and repeat for entire values of x.

As we know; $\tan^{-1}(\tan x) = \left\{x; -\dfrac{\pi}{2} < x < \dfrac{\pi}{2}\right\}$

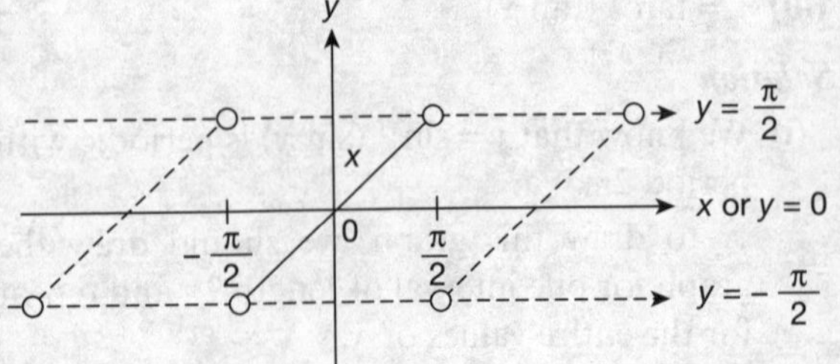

ILLUSTRATION 2.4

Sketch the graph for

(i) $y = \sin(\sin^{-1} x)$ **(ii)** $y = \cos(\cos^{-1} x)$

(iii) $y = \tan(\tan^{-1} x)$

Solution

(i) Sketch for $y = \sin(\sin^{-1} x)$

We know, domain, $x \in [-1, 1]$

and range $y = x \Rightarrow y \in [-1, 1]$

Hence, we should sketch $y = \sin(\sin^{-1} x)$ only when $x \in [-1, 1]$ and $y = x$.

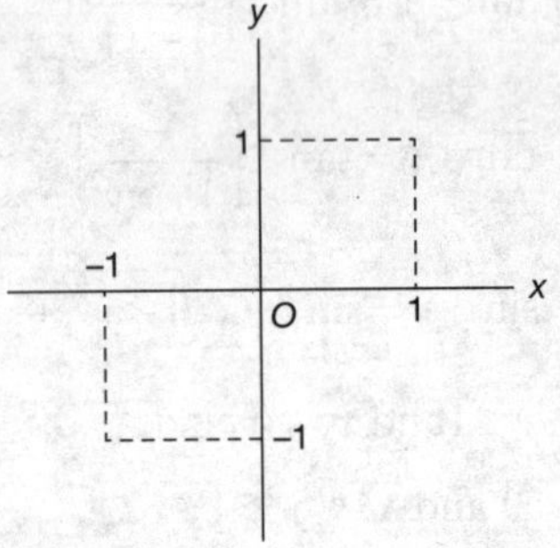

(ii) Sketch for the curve $y = \cos(\cos^{-1} x)$

We know, domain, $x \in [-1, 1]$

and range $y = x \Rightarrow y \in [-1, 1]$

Hence, we should sketch $y = \cos(\cos^{-1} x) = x$ only when $x \in [-1, 1]$. So, its graph could be plotted as shown in figure.

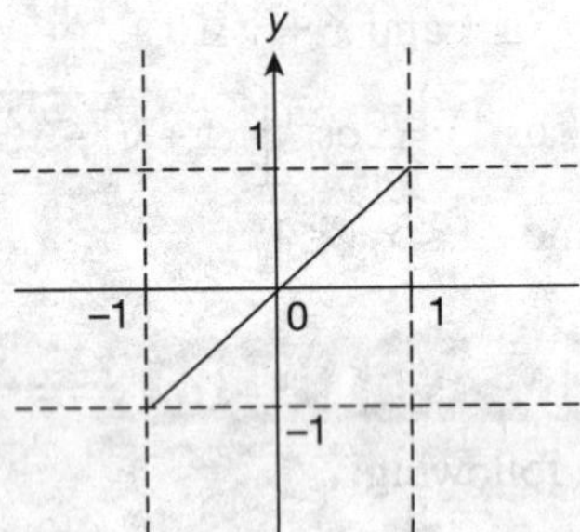

(iii) Sketch for the curve $y = \tan(\tan^{-1} x)$

We know,

Domain, $x \in R$ (i.e., $-\infty < x < \infty$)

and Range $y = x \Rightarrow y \in R$.

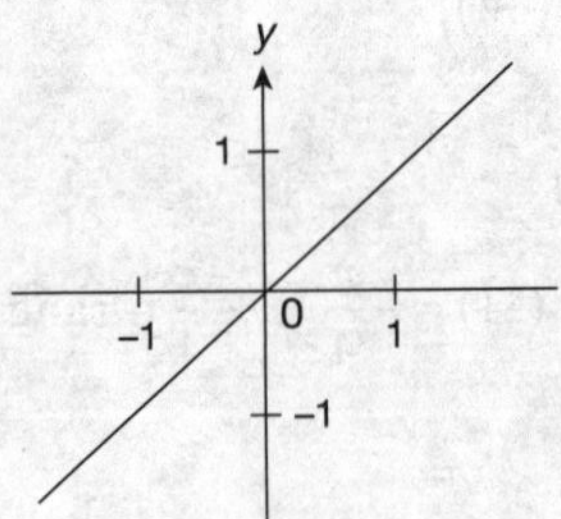

ILLUSTRATION 2.5

Find the value of the following:

(i) $\sin\left[\frac{\pi}{3}-\sin^{-1}\left(-\frac{1}{2}\right)\right]$

(ii) $\sin\left[\tan^{-1}(-\sqrt{3})+\cos^{-1}\left(-\frac{\sqrt{3}}{2}\right)\right]$

Solution

(i) $\sin\left[\frac{\pi}{3}-\sin^{-1}\left(-\frac{1}{2}\right)\right]$

$$= \sin\left[\frac{\pi}{3}-\left(-\frac{\pi}{6}\right)\right]$$

$$= \sin\left(\frac{\pi}{3}+\frac{\pi}{6}\right) = \sin\frac{\pi}{2} = 1$$

(ii) $\sin\left[\tan^{-1}(-\sqrt{3})+\cos^{-1}\left(-\frac{\sqrt{3}}{2}\right)\right]$

$$= \sin\left[\frac{-\pi}{3}+\frac{5\pi}{6}\right] = \sin\frac{\pi}{2} = 1$$

ILLUSTRATION 2.6

Express $\sin^{-1}\left(2x\sqrt{1-x^2}\right)$ in terms of $\sin^{-1} x$.

Solution

Let $\quad \sin^{-1} x = \theta$

$\Rightarrow \quad x = \sin\theta \quad \text{and} \quad -\frac{\pi}{2} \le \theta \le \frac{\pi}{2}$

i.e., $\quad -\pi \le 2\theta \le \pi$

The given expression may be written as

$$\sin^{-1}\left(2x\sqrt{1-x^2}\right) = \sin^{-1}\sin 2\theta$$

$$= 2\theta$$

$$= 2\sin^{-1} x$$

Hence,

$$\sin^{-1}\left(2x\sqrt{1-x^2}\right) = 2\sin^{-1} x$$

ILLUSTRATION 2.7

Find all positive integral solution of the equation $\tan^{-1} x + \cot^{-1} y = \tan^{-1} 3$.

Solution

We have to find values of x and y, which are positive as well as integer.

$$\tan^{-1} x + \cot^{-1} y = \tan^{-1} 3$$

$$\tan^{-1} x + \tan^{-1}\frac{1}{y} = \tan^{-1} 3 \qquad (y > 0)$$

or $$\tan^{-1}\left(\frac{x+\frac{1}{y}}{1-\frac{x}{y}}\right) = \tan^{-1} 3$$

$$\tan\tan^{-1}\left(\frac{x+\frac{1}{y}}{1-\frac{x}{y}}\right) = \tan\tan^{-1} 3$$

$\Rightarrow$ $$\frac{x+\frac{1}{y}}{1-\frac{x}{y}} = 3$$

or $$3x + 1 = y(3 - x)$$

$$y = \frac{3x+1}{3-x}$$

When x is positive, numerator is (+) ve. For y to be positive, denominator must be (+) ve

i.e., $\quad 3 - x > 0 \quad \text{or} \quad x < 3 \;\Rightarrow\; x = 1, 2$

for $\quad x = 1 \;\Rightarrow\; y = 2$

$\quad x = 2 \;\Rightarrow\; y = 7$

WORKSHEET

1. Prove that $\tan^{-1} n + \cot^{-1}(n+1) = \tan^{-1}\left(n^2+n+1\right)$
2. Prove that
$\tan^{-1}\left(\frac{a-b}{1+ab}\right)+\tan^{-1}\left(\frac{b-c}{1+bc}\right)=\tan^{-1}a-\tan^{-1}c.$
3. The number of real solutions of
$\tan^{-1}\sqrt{x(x+1)}+\sin^{-1}\sqrt{x^2+x+1}=\frac{\pi}{2}$ is:
(a) zero (b) one (c) two (d) infinite
4. Prove that $\tan^{-1}x+\cot^{-1}x=\tan^{-1}x\left(x^2+x+1\right)$
5. Prove that
$\tan^{-1}\left(\frac{1}{4}\right)+\tan^{-1}\left(\frac{2}{9}\right)=\frac{1}{2}\cos^{-1}\left(\frac{3}{5}\right)=\sin^{-1}\left(\frac{1}{\sqrt{5}}\right)$
6. If $\cos^{-1}\frac{x}{a}+\cos^{-1}\frac{y}{b}=\alpha$, prove that
$\frac{x^2}{a^2}-\frac{2xy}{ab}\cos\alpha+\frac{x^2}{a^2}=\sin^2\alpha$
7. If $\sin^{-1}\left(\frac{3}{x}\right)+\sin^{-1}\left(\frac{4}{x}\right)=\frac{\pi}{2}$, then find the value of x
8. $\tan^{-1}\frac{x}{y}-\tan^{-1}\frac{x-y}{x+y}$ is equal to (where $x>y>0$)
(a) $\frac{-\pi}{4}$ (b) $\frac{\pi}{4}$ (c) $\frac{3\pi}{4}$ (d) None.
9. $\cos^{-1}\frac{2}{\sqrt{5}}+\tan^{-1}\left(\frac{1}{3}\right)$ is equal to
(a) $\tan^{-1}\frac{1}{7}$ (b) $\tan^{-1}\frac{2}{3\sqrt{5}}$ (c) $\frac{\pi}{2}$ (d) $\frac{\pi}{4}$
10. If $\theta=\cos^{-1}\left(\frac{4}{5}\right)+\tan^{-1}\left(\frac{2}{3}\right)$ then,
(a) $\cot\theta=\frac{17}{6}$ (b) $\cos\theta=\frac{16}{5\sqrt{13}}$
(c) $\tan\theta=\frac{17}{6}$ (d) $\sin\theta=\frac{18}{5\sqrt{3}}$

SOLVED EXAMPLES

Example 1. Find the value of $\cos^{-1}(\cos 12)-\sin^{-1}(\sin 12)$

Solution

Let $P=\cos^{-1}(\cos 12)-\sin^{-1}(\sin 12)$

$=\cos^{-1}(\cos(4\pi-12))-\sin^{-1}(\sin(12-4\pi))$

$=4\pi-12-(12-4\pi)$

$=8\pi-24$

Example 2. Prove: $\cos^{-1}(-x)=\pi-\cos^{-1}x\ \forall\ x\in[-1,1]$

Solution

Since $x\in[-1,1]$

$\Rightarrow$ $-x\in[-1,1]$

Let, $\cos^{-1}(-x)=\theta$, so $\theta\in[0,\pi]$

$\Rightarrow$ $-x=\cos\theta$

$x=-\cos\theta=\cos(\pi-\theta)$

$\Rightarrow$ $\cos^{-1}x=\cos^{-1}\cos(\pi-\theta)$

Since $\pi-\theta\in[0,\pi]$

$\Rightarrow$ $\cos^{-1}\cos(\pi-\theta)=\pi-\theta$

$\Rightarrow$ $\cos^{-1}x=\pi-\theta$

$\cos^{-1}x>\frac{\pi}{2}-\cos^{-1}x \quad \left(\sin^{-1}x+\cos^{-1}x=\frac{\pi}{2}\right)$

$2\cos^{-1}x>\frac{\pi}{2}$

$\cos^{-1}x>\frac{\pi}{4}$

$\Rightarrow$ $\cos^{-1}(-x)=\pi-\cos^{-1}x$

Example 3. The number of solutions of the equation $\sin\left(\frac{\pi x}{2\sqrt{3}}\right)=x^2-2\sqrt{3}x+4$ is

(a) 0 **(b)** 2 **(c)** More than 2 **(d)** 1

Solution

We know that $-1\le\sin\left(\frac{\pi x}{2\sqrt{3}}\right)\le 1$.

Therefore, we must have

$-1\le x^2-2\sqrt{3}x+4\le 1$

$\Rightarrow$ $-1\le(x-\sqrt{3})^2+1\le 1$

$-2\le(x-\sqrt{3})^2\le 0$

but square of a real number cannot be negative; therefore, we must have $(x-\sqrt{3})^2 = 0$

$$\Rightarrow \quad x = \sqrt{3}$$

Now, $x = \sqrt{3}$ also satisfies the equation

$\therefore$ Ans. (d)

Example 4. The equation $\sin^4 x + \cos^4 x + \sin 2x + \alpha = 0$ has solution for

(a) $-\frac{1}{2} \le \alpha \le \frac{1}{2}$ **(b)** $-3 \le \alpha \le 1$

(c) $-\frac{3}{2} \le \alpha \le \frac{1}{2}$ **(d)** $-1 \le \alpha \le 1$

Solution

Let $\sin 2x = y$. Then the given equation becomes

$$y^2 - 2y - 2(1+\alpha) = 0, \text{ where } -1 \le y \le 1$$

for real y, $D \ge 0$.

$$\Rightarrow \quad 3+2\alpha \ge 0$$

$$\Rightarrow \quad \alpha \ge -\frac{3}{2}$$

Also $\quad -1 \le y \le 1 \Rightarrow -1 \le 1-\sqrt{3+2\alpha} \le 1$

$$\Rightarrow \quad 3+2\alpha \le 4 \quad \Rightarrow \alpha \le \frac{1}{2}.$$

Thus, $\quad -\frac{3}{2} \le \alpha \le \frac{1}{2}$

$\therefore$ Ans. (c)

Example 5. If $\sin^{-1}\left(x - \frac{x^2}{2} + \frac{x^3}{4} \ldots\right) + \cos^{-1}\left(x^2 - \frac{x^4}{2} + \frac{x^6}{4} \ldots\right) = \frac{\pi}{2}$, for $0 < |x| < \sqrt{2}$ then x equals

(a) $\frac{1}{2}$ **(b)** 1 **(c)** $-\frac{1}{2}$ **(d)** -1

Solution

Given $\quad \sin^{-1} A + \cos^{-1} B = \frac{\pi}{2}$

But $\quad \sin^{-1} A + \cos^{-1} A = \frac{\pi}{2}$

$$\Rightarrow \quad A = B$$

$$\Rightarrow \quad x - \frac{x^2}{2} + \frac{x^3}{4} - \ldots = x^2 - \frac{x^4}{2} + \frac{x^6}{4} \ldots$$

$$\Rightarrow \quad \frac{x}{1-\left(-\frac{x}{2}\right)} = \frac{x^2}{1-\left(-\frac{x^2}{2}\right)}$$

$$x(2+x^2) = x^2(2+x)$$

$$= 2x(x-1) = 0$$

$$x = 0, 1$$

$\therefore$ Ans. (b)

Example 6. If $\cos^{-1} x > \sin^{-1} x$, then

(a) $0 \le x < \frac{1}{\sqrt{2}}$ **(b)** $-1 < x < 0$

(c) $x < 0$ **(d)** $-1 \le x < \frac{1}{\sqrt{2}}$

Solution

We know $\cos^{-1} x$, $\sin^{-1} x$ are defined when $-1 \le x \le 1$

since $\quad \cos^{-1} x > \sin^{-1} x$

$$\Rightarrow \quad 2\cos^{-1} x > \cos^{-1} x + \sin^{-1} x = \pi/2$$

$$\Rightarrow \quad \cos^{-1} x > \frac{\pi}{4}$$

$$\cos(\cos^{-1} x) < \cos\left(\frac{\pi}{4}\right)$$

($\because \cos^{-1} x$ is decreasing function)

or $\quad x < \frac{1}{\sqrt{2}}$

Hence $\quad -1 \le x < \frac{1}{\sqrt{2}}$

$\therefore$ Ans. (d)

Example 7. $\lim_{n\to\infty} \sum_{r=1}^{n} \tan^{-1}\left(\frac{1}{2r^2}\right)$ is equal to

(a) $\frac{\pi}{4}$ (b) $\frac{\pi^2}{4}$ (c) $\frac{\pi}{2}$ (d) $\frac{\pi^2}{16}$

Solution

$$\tan^{-1}\frac{1}{2r^2} = \tan^{-1}\frac{2}{4r^2} = \tan^{-1}\frac{2}{1+4r^2-1}$$

$$= \tan^{-1}\frac{(2r+1)-(2r-1)}{1+(2r+1)(2r-1)}$$

$$= \tan^{-1}(2r+1) - \tan^{-1}(2r-1) \quad \ldots\text{(i)}$$

Putting $r = 1, 2, 3, \ldots n$ in equation (i) and adding, we get

$$\sum_{r=1}^{n} \tan^{-1}\left(\frac{1}{2r^2}\right) = \tan^{-1}(2r+1) - \tan^{-1}(2r-1)$$

$$\lim_{n\to\infty} S_n = \tan^{-1}(2n+1) - \tan^{-1} 1$$

$$\therefore \quad S_n = \tan^{-1}\infty - \tan^{-1} 1$$

$$= \frac{\pi}{2} - \frac{\pi}{4} = \frac{\pi}{4}$$

$\therefore \quad$ Ans. (a)

Example 8. If $\sin^{-1}x+\sin^{-1}y=\frac{2\pi}{3}$ and $\cos^{-1}x-\cos^{-1}y=\frac{\pi}{3}$, then $(x, y) =$

(a) (1, 1) **(b)** $\left(\frac{1}{2},1\right)$ **(c)** $\left(1,\frac{1}{2}\right)$ **(d)** $\left(\frac{1}{3},\frac{1}{3}\right)$

Solution

The second relation is rewritten as

$$\left(\frac{\pi}{2}-\sin^{-1}x\right)-\left(\frac{\pi}{2}-\sin^{-1}y\right)=\frac{\pi}{3}$$

or $$\sin^{-1}y-\sin^{-1}x=\frac{\pi}{3}$$

$\Rightarrow$ $$\sin^{-1}y=90° \quad \text{and} \quad \sin^{-1}x=30°$$

$\therefore$ Ans. (b)

Example 9. If $\tan^{-1}a+\tan^{-1}b=\sin^{-1}1-\tan^{-1}c$ then

(a) $a+b+c=ab+bc+ca$ **(b)** $\frac{1}{a}+\frac{1}{b}+\frac{1}{c}=\frac{1}{abc}$

(c) $a+b+c=abc$ **(d)** $ab+bc+ca=abc$

Solution

Given relation $\tan^{-1}a+\tan^{-1}b+\tan^{-1}c=\sin^{-1}1$

$\Rightarrow$ $$\tan^{-1}\left(\frac{a+b+c-abc}{1-ab-bc-ca}\right)$$

$\Rightarrow$ $$1-ab-bc-ca=0$$

$\Rightarrow$ $$\frac{1}{abc}=\frac{1}{a}+\frac{1}{b}+\frac{1}{c}$$

$\therefore$ Ans: (b)

Example 10. If a and b are the roots of $\sin^{-1}\left(x^2-3x+\frac{5}{2}\right)=\frac{\pi}{6}$, Then the value of a^2+b^2 is

(a) 1 **(b)** 5 **(c)** 4 **(d)** 9

Solution

$$x^2-3x+\frac{5}{2}=\frac{1}{2}$$

$$x=1, 2$$

$$a=1, b=2$$

$$a^2+b^2=1^2+2^2=5$$

$\therefore$ Ans. (b)

Example 11. If $\frac{\pi}{2}\le x\le\frac{3\pi}{2}$, then $\sin^{-1}(\sin^{-1}x)$, is equal to

(a) $\pi+x$ **(b)** x **(c)** $\pi-x$ **(d)** $-x$

Solution

We have, $\frac{\pi}{2}\le x\le\frac{3\pi}{2}$

$\Rightarrow$ $$-\frac{\pi}{2}\le x-\pi\le\frac{\pi}{2}$$

$\Rightarrow$ $$-\frac{\pi}{2}\le \pi-x\le\frac{\pi}{2}$$

$\Rightarrow$ $$\sin^{-1}\{\sin(\pi-x)\}=\pi-x$$

$\therefore$ Ans. (c)

Example 12. If $x^2+y^2+z^2=r^2$, then

$$\tan^{-1}\left(\frac{xy}{zr}\right)+\tan^{-1}\left(\frac{yz}{xr}\right)+\tan^{-1}\left(\frac{xz}{yr}\right)=?$$

(a) $\frac{\pi}{2}$ **(b)** 0 **(c)** π **(d)** None of these.

Solution

$$\tan^{-1}\left(\frac{xy}{zr}\right)+\tan^{-1}\left(\frac{yz}{xr}\right)+\tan^{-1}\left(\frac{zx}{yr}\right)$$

$$=\tan^{-1}\left[\frac{\frac{xy}{zr}+\frac{yzxr}{xr}+\frac{xz}{yr}-\frac{xyz}{r^3}}{1-\left(\frac{x^2+y^2+z^2}{r^2}\right)}\right]=\tan^{-1}\infty=\frac{\pi}{2}$$

Alternatively, $x=y=z=\frac{1}{\sqrt{3}}$ and $r=1$

Then, $$3\tan^{-1}\left(\frac{1}{\sqrt{3}}\right)=\frac{\pi}{2}$$

$\therefore$ Ans. (a)

Example 13. The greatest and least value of $(\sin^{-1}x)^3+(\cos^{-1}x)^3$ are:

(a) $\frac{7\pi^3}{8},\frac{\pi^3}{32}$ **(b)** $-\frac{\pi}{2},\frac{\pi}{2}$

(c) $\frac{\pi^3}{8},\frac{\pi^3}{8}$ **(d)** None of these.

Solution

$$\left(\sin^{-1}x\right)^3+\left(\cos^{-1}\right)^3$$

$$=\left(\sin^{-1}x+\cos^{-1}x\right)^3-3\sin^{-1}\cos^{-1}x\left(\sin^{-1}x+\cos^{-1}x\right)$$

$$=\frac{\pi^3}{8}-\frac{3\pi}{2}\sin^{-1}x\left(\frac{\pi}{2}-\sin^{-1}x\right)$$

Take $\sin^{-1}x=t$

$$=\frac{\pi^3}{8}-\frac{3\pi}{2}t\left(\frac{\pi}{2}-t\right)$$

$$=\frac{\pi^3}{8}+\frac{3\pi}{2}\left[t^2-2\times t\times\frac{\pi}{4}+\frac{\pi^2}{16}-\frac{\pi^2}{16}\right]$$

$$= \frac{\pi^3}{8} + \frac{3\pi}{2}\left[\left(t - \frac{\pi}{4}\right)^2\right] - \frac{3\pi^3}{32} \Rightarrow \frac{\pi^3}{32} + \frac{3\pi}{2}\left(\sin^{-1}x - \frac{\pi}{4}\right)^2.$$

The least value is $\frac{\pi^3}{32}$.

the greatest value is

$$= \frac{\pi^3}{32} + \frac{9\pi^2}{16} \times \frac{3\pi}{2} = \frac{7\pi^3}{8}.$$

Example 14. If x_1, x_2, x_3, x_4, are roots of the equation $x^4 - x^3 \sin 2\beta + x^2 \cos 2\beta - x \cos \beta - \sin \beta = 0$, then $\tan^{-1} x_1 + \tan^{-1} x_2 + \tan^{-1} x_3 + \tan^{-1} x_4 =$

(a) $\pi - \beta$ **(b)** $-\beta$ **(c)** $\frac{\pi}{2} - \beta$ **(d)** β

Solution

We have,

$$\sum x_1 = \sin 2\beta$$
$$\sum x_1x_2 = \cos 2\beta$$
$$\sum x_1x_2x_3 = \cos\beta$$
$$\sum x_1x_2x_3x_4 = -\sin\beta$$

$\therefore$ $\tan^{-1} x_1 + \tan^{-1} x_2 + \tan^{-1} x_3 + \tan^{-1} x_4$

$$= \tan^{-1}\left(\frac{\sin 2\beta - \cos\beta}{1 - \cos 2\beta \sin\beta}\right)$$

$$= \tan^{-1}\left(\frac{(2\sin\beta - 1)\cos\beta}{\sin\beta(2\sin\beta - 1)}\right)$$

$$= \tan^{-1}(\cos\beta) = \tan^{-1}\left[\tan\left(\frac{\pi}{2} - \beta\right)\right]$$

$$= \frac{\pi}{2} - \beta$$

$\therefore$ Ans. (c)

Example 15. The number of real solutions of $\tan^{-1}\sqrt{x(x+1)} + \sin^{-1}\sqrt{x^2 + x + 1} = \frac{\pi}{2}$ is

(a) 0 **(b)** 1 **(c)** 2 **(d)** ∞

Solution

$$\tan^{-1}\sqrt{x(x+1)} = \cos^{-1}\left(\frac{1}{\sqrt{x^2 + x + 1}}\right)$$

$$\tan^{-1}\left(\sqrt{x^2 + x + 1}\right) + \cos^{-1}\left(\frac{1}{\sqrt{x^2 + x + 1}}\right) = \frac{\pi}{2}$$

$$\left[\because \sin^{-1} x + \cos^{-1} x = \frac{\pi}{2}\right]$$

$$x^2 + x + 1 = \frac{1}{\sqrt{x^2 + x + 1}}$$
$$x^2 + x + 1 = 1$$
$$x(+x + 1) = 0$$
$$x = 0, -1$$

$\therefore$ Ans. (c)

Example 16. If $\tan^{-1}\frac{a}{x} + \tan^{-1}\frac{b}{x}\tan^{-1} + \tan^{-1}\frac{c}{x} +$ $\tan^{-1}\frac{d}{x} = \frac{\pi}{2}$, Then $x^4 - x^2\sum ab + abcd$, is equal to

(a) –1 **(b)** 1 **(c)** 0 **(d)** 2

Solution

$$\tan^{-1}\left(\frac{(a+b)x}{x^2 - ab}\right) = \frac{\pi}{2} - \tan^{-1}\left(\frac{(c+d)x}{x^2 - cd}\right)$$

$$\tan^{-1}\left(\frac{(a+b)x}{x^2 - ab}\right) = \cos^{-1}\left(\frac{(c+d)x}{x^2 - cd}\right)$$

$$\left(x^2 - ab\right)\left(x^2 - cd\right) = (a+b)\cdot(c+d)^2$$

$$x^4 - x^2\sum ab + abcd = 0$$

$\therefore$ Ans. (c)

Example 17. If $\sec^{-1}\sqrt{1+x^2} + \operatorname{cosec}^{-1}\frac{\sqrt{1+y^2}}{y} +$ $\cot^{-1}\left(\frac{1}{z}\right) = \pi$ then $x + y + z$ is equal to

(a) 0 **(b)** xyz **(c)** $2xyz$ **(d)** $3xyz$

Solution

Let $x = \tan\theta,\ y = \tan\phi$

Now, $\tan^{-1} x + \tan^{-1} y + \tan^{-1} z = \pi$

$$\tan^{-1}\left[\frac{x + y + z - xyz}{1 - (xy + yz + zx)}\right] = \pi$$

$$x + y + z = xyz$$

$\therefore$ Ans. (b)

RECAPITULATION (FORMULAE, THEOREM, IMPORTANT RESULTS)

1. The value of an inverse trigonometric function which lies in its principal value branch is called the principal value of that inverse trigonometric function.

2. The domains and ranges (principal value branches) of inverse trigonometric functions.

Functions	Domain	Range (Principal Value Branches)
$y=\sin^{-1}x$	$[-1,1]$	$\left[-\frac{\pi}{2},\frac{\pi}{2}\right]$
$y=\cos^{-1}x$	$[-1,1]$	$[0,\pi]$
$y=\operatorname{cosec}^{-1}x$	$R-(-1,1)$	$\left[-\frac{\pi}{2},\frac{\pi}{2}\right],\{0\}$
$y=\sec^{-1}x$	$R-(-1,1)$	$[0,\pi]-\left\{\frac{\pi}{2}\right\}$
$y=\tan^{-1}x$	R	$\left(-\frac{\pi}{2},\frac{\pi}{2}\right)$
$y=\cot^{-1}x$	R	$(0,\pi)$

3. For suitable values of domain, we have
   (i) $y=\sin^{-1}x \Rightarrow x=\sin y$
   (ii) $x=\sin y \Rightarrow y=\sin^{-1}x$
   (iii) $\sin(\sin^{-1}x)=x$
   (iv) $\sin^{-1}(\sin x)=x$
   (v) $\sin^{-1}\frac{1}{x}=\operatorname{cosec}^{-1}x$
   (vi) $\cos^{-1}(-x)=\pi-\cos^{-1}x$
   (vii) $\cos^{-1}\frac{1}{x}=\sec^{-1}x$
   (viii) $\cot^{-1}x=\pi+\tan^{-1}\left(\frac{1}{x}\right);\ x<0$
   (ix) $\cot^{-1}(-x)=\pi-\cot^{-1}x$
   (x) $\tan^{-1}\frac{1}{x}=\cot^{-1}x$
   (xi) $\sec^{-1}(-x)=\pi-\sec^{-1}x$
   (xii) $\sin^{-1}(-x)=-\sin^{-1}x$
   (xiii) $\tan^{-1}(-x)=-\tan^{-1}x$
   (xiv) $\tan^{-1}x+\cot^{-1}x=\frac{\pi}{2}$
   (xv) $\operatorname{cosec}^{-1}(-x)=-\operatorname{cosec}^{-1}x$
   (xvi) $\sin^{-1}x+\cos^{-1}x=\frac{\pi}{2}$
   (xvii) $\operatorname{cosec}^{-1}x+\sec^{-1}x=\frac{\pi}{2}$
   (xviii) $\tan^{-1}x+\tan^{-1}y=\tan^{-1}\frac{x+y}{1-xy}$
   (xix) $2\tan^{-1}x=\tan^{-1}\frac{2x}{1-x^2}$
   (xx) $\tan^{-1}x-\tan^{-1}y=\tan^{-1}\frac{x-y}{1+xy}$
   (xxi) $2\tan^{-1}x=\sin^{-1}\frac{2x}{1+x^2}=\cos^{-1}\frac{1-x^2}{1+x^2}$

QUIZ

1. If trigonometric functions are not one-one onto then how can we find the inverse of a trigonometric function?
2. What is the domain and range of $\cos^{-1}$?
3. Does the principal value of inverse trigonometric functions lie in all the quadrants?
4. What do you mean by principal values of an inverse trigonometric function?
5. What do you mean by principal value branch?
6. What is the difference between $\sin^{-1}$ and Sin^{-1}? (In the first case s is in small letter and in the second case S is in capital letter.)
7. What is the values of $\operatorname{cosec}^{-1}(\operatorname{cosec}\theta)$?
8. Is the expression $\sin^{-1}x+\cos^{-1}x=\operatorname{cosec}^{-1}x+\sec^{-1}x$ is true?
9. What is the principal value of $\tan^{-1}1+\sin^{-1}\left(-\frac{1}{2}\right)$?
10. Find the value of $\sin^{-1}\left(\sin\frac{3\pi}{5}\right)$.

JEE MAIN

Objective Questions

One Option Correct

1. $\cos\left[\frac{\pi}{3}-\cos^{-1}\left(-\frac{1}{2}\right)\right]$ is equal to

(a) $-\frac{1}{2}$ (b) $\frac{1}{2}$

(c) $\frac{\sqrt{3}}{2}$ (d) $-\frac{\sqrt{3}}{2}$

2. $\sin^{-1}x+\sin^{-1}\left(\frac{1}{x}\right)+\cos^{-1}x+\cos^{-1}\left(\frac{1}{x}\right)$ is equal to

(a) $\frac{\pi}{2}$ (b) 0

(c) π (d) None of these.

3. $\cos^{-1}\cos\left(\frac{7\pi}{6}\right)+\tan^{-1}\tan\left(\frac{3\pi}{4}\right)+\sin^{-1}\sin\left(-\frac{2\pi}{3}\right)$ is equal to

(a) $\frac{\pi}{2}$ (b) $\frac{\pi}{4}$

(c) $\frac{\pi}{6}$ (d) $\frac{5\pi}{6}$

4. If $\sin^{-1}x+\cot^{-1}\left(\frac{1}{2}\right)=\frac{\pi}{2}$, then x is equal to

(a) $\frac{2}{\sqrt{5}}$ (b) $\frac{1}{\sqrt{5}}$

(c) 0 (d) $\frac{\sqrt{2}}{\sqrt{5}}$

5. Which of the following is true?

(a) $\sin(\cos^{-1}x)=\cos(\sin^{-1}x)$

(b) $\cos(\tan^{-1}x)=\tan(\sec^{-1}x)$

(c) $\sec(\tan^{-1}x)=\tan(\sec^{-1}x)$

(d) All the above.

6. If $\sin^{-1}x+\sin^{-1}y=\frac{2\pi}{3}$, then $\cos^{-1}x+\cos^{-1}y=$

(a) π (b) $\frac{2\pi}{3}$

(c) $\frac{\pi}{3}$ (d) $\frac{\pi}{6}$

7. Two angles of a triangle are $\cot^{-1}2$ and $\cot^{-1}3$. Then the third angle is ______.

(a) $\frac{3\pi}{4}$ (b) $\frac{\pi}{6}$

(c) $\frac{\pi}{4}$ (d) $\frac{\pi}{3}$

8. The number of real solutions to $\tan^{-1}\sqrt{x(x+1)}+\sin^{-1}\sqrt{x^2+x+1}=\frac{\pi}{2}$, is______.

(a) two (b) zero

(c) one (d) infinite

9. If $0<\sin^{-1}x<1$ and $1+\sin^{-1}x+(\sin^{-1}x)^2+\ldots+\infty=2$ then x is equal to ______.

(a) $\frac{\pi}{6}$ (b) $\frac{\pi}{3}$

(c) $\frac{\pi}{4}$ (d) $\frac{\pi}{12}$

10. If $\cos^{-1}\frac{x}{2}+\cos^{-1}\frac{y}{3}=\theta$, then $9x^2-12xy\cos\theta+4y^2$ is equal to ______.

(a) $36\cos^2\theta$ (b) 36

(c) $36\sin^2\theta$ (d) $-36\sin^2\theta$

11. If $\tan^{-1}3+\tan^{-1}y=\tan^{-1}8$, then y is equal to ______.

(a) 5 (b) $\frac{5}{14}$

(c) $\frac{1}{5}$ (d) $\frac{14}{5}$

12. If $A=\tan^{-1}\left(\frac{x\sqrt{3}}{2k-x}\right)$ and $B=\tan^{-1}\left(\frac{2x-k}{k\sqrt{3}}\right)$ then the value of $A-B$ is ______.

(a) 60° (b) 30°

(c) 0° (d) 45°

13. The value of $\cot\left(\operatorname{cosec}^{-1}\frac{5}{3}+\tan^{-1}\frac{2}{3}\right)$ is

(a) $\frac{3}{17}$ (b) $\frac{4}{17}$

(c) $\frac{5}{17}$ (d) $\frac{6}{17}$

14. If $\cos^{-1}x-\cos^{-1}\frac{y}{2}=\alpha$, then $4x^2-4xy\cos\alpha+y^2$ is equal to ______.

(a) $4\sin^2\alpha$ (b) $-4\sin^2\alpha$
(c) $2\sin^2\alpha$ (d) 4

15. $\tan\left(2\cos^{-1}\frac{3}{5}\right)=$ ______.

(a) $\frac{7}{25}$ (b) $\frac{24}{25}$
(c) $-\frac{24}{7}$ (d) $\frac{8}{3}$

16. If $\cot^{-1}\sqrt{\cos\theta}-\tan^{-1}\sqrt{\cos\theta}=x$, then $\sin x=$

(a) $\frac{1-\cos\theta}{1+\cos\theta}$ (b) $\frac{1+\cos\theta}{1-\cos\theta}$
(c) $\frac{2\sqrt{\cos\theta}}{1+\cos\theta}$ (d) $\frac{2\sqrt{\cos\theta}}{1-\cos\theta}$

17. If $\sin^{-1}\left(x-\frac{x^2}{2}+\frac{x^3}{4}-\ldots\right)+$
$\cos^{-1}\left(x^2-\frac{x^4}{2}+\frac{x^6}{4}-\ldots\right)=\frac{\pi}{2}$ for $0<|x|<\sqrt{2}$,
then x equals ______.

(a) $\frac{1}{2}$ (b) 1
(c) $-\frac{1}{2}$ (d) -1

18. Let $\cos(2\tan^{-1}x)=\frac{1}{2}$, then the value of x is ______.

(a) $\sqrt{3}$ (b) $\frac{1}{\sqrt{3}}$
(c) $1-\sqrt{3}$ (d) $1-\frac{1}{\sqrt{3}}$

19. If $\sqrt{3}+i=(a+ib)(c+id)$, then
$\tan^{-1}\left(\frac{b}{a}\right)+\tan^{-1}\left(\frac{d}{c}\right)$ has the value ______.

(a) $\frac{\pi}{3}+2n\pi,\ n\in I$ (b) $n\pi+\frac{\pi}{6},\ n\in I$
(c) $n\pi-\frac{\pi}{3},\ n\in I$ (d) $2n\pi-\frac{\pi}{3},\ n\in I$

20. The Solution set of the equation
$\sin^{-1}x=2\tan^{-1}x$, is ______.

(a) $\{1, 2\}$ (b) $\{-1, 2\}$
(c) $\{-1, 1, 0\}$ (d) $\left\{1, \frac{1}{2}, 0\right\}$

21. If $\sin^{-1}\frac{2a}{1+a^2}-\cos^{-1}\frac{1-b^2}{1+b^2}=\tan^{-1}\frac{2x}{1-x^2}$, then the value of x is ______.

(a) a (b) b
(c) $\frac{a+b}{1-ab}$ (d) $\frac{a-b}{1+ab}$

22. The equation $\sin^{-1}x-\cos^{-1}x=\cos^{-1}\left(\frac{\sqrt{3}}{2}\right)$ has

(a) no solution.
(b) unique solution.
(c) infinite number of solutions.
(d) None of these.

23. If $4\sin^{-1}x+\cos^{-1}x=\pi$, then x is equal to ____.

(a) $\sqrt{\frac{9}{4}}$ (b) $\frac{1}{2}$
(c) $\frac{\sqrt{3}}{2}$ (d) $\frac{1}{\sqrt{2}}$

24. $\tan^{-1}\left(\frac{1}{2}\right)+\tan^{-1}\left(\frac{1}{5}\right)+\tan^{-1}\left(\frac{1}{8}\right)$ is equal to ______.

(a) $\frac{\pi}{3}$ (b) $\frac{\pi}{6}$
(c) $\frac{\pi}{4}$ (d) $\frac{\pi}{2}$

25. $2\tan^{-1}\left(\frac{1}{5}\right)+\sec^{-1}\left(\frac{5\sqrt{2}}{7}\right)+2\tan^{-1}\left(\frac{1}{8}\right)$ is equal to ______.

(a) $\frac{\pi}{2}$ (b) $\frac{\pi}{3}$
(c) $\frac{\pi}{4}$ (d) $\frac{\pi}{6}$

26. $3\sin^{-1}\left(\frac{2x}{1+x^2}\right)-4\cos^{-1}\left(\frac{1-x^2}{1+x^2}\right)+$
$2\tan^{-1}\left(\frac{2x}{1-x^2}\right)=\frac{\pi}{3}$. Then x is equal to ______.

(a) $\frac{1}{\sqrt{3}}$ (b) $\sqrt{3}$
(c) 1 (d) ∞

27. The value of $\sin^{-1}(\sin 10°)$ is ______.

(a) 10 (b) $3\pi-10$
(c) $10-3\pi$ (d) None

28. $\tan^{-1}\left(\frac{c_1x-y}{c_1y+x}\right)+\tan^{-1}\left(\frac{c_2-c_1}{1+c_2c_1}\right)+$ $\tan^{-1}\left(\frac{c_3-c_2}{1+c_3c_2}\right)+\ldots+\tan^{-1}\left(\frac{1}{c_r}\right)$ is equal to ______.

(a) $\tan^{-1}\left(\frac{y}{x}\right)$ (b) $\tan^{-1}\left(\frac{x}{y}\right)$

(c) $\tan\left(\frac{x}{y}\right)$ (d) None

29. The value of $2\tan^{-1}\frac{1}{3}+\tan^{-1}\frac{1}{7}$ is ______.

(a) $\frac{\pi}{4}$ (b) $\frac{\pi}{2}$

(c) π (d) None.

30. $\cos^{-1}\frac{1}{2}+2\sin^{-1}\frac{1}{2}$ is equal to ______.

(a) $\frac{\pi}{4}$ (b) $\frac{\pi}{6}$

(c) $\frac{\pi}{3}$ (d) $\frac{2\pi}{3}$

31. The value of $\cos^{-1}\left(\cos\frac{5\pi}{3}\right)+\sin^{-1}\left(\cos\frac{5\pi}{3}\right)$ is ______.

(a) $\pi/2$ (b) $5\pi/3$

(c) $10\pi/3$ (d) 0

32. $2\tan^{-1}(\cos x)=\tan^{-1}(\operatorname{cosec}^2 x)$, then $x=$ ______.

(a) $\frac{\pi}{2}$ (b) π

(c) $\frac{\pi}{6}$ (d) $\frac{\pi}{3}$

33. If $\cos^{-1}p+\cos^{-1}q+\cos^{-1}r=\pi$, then $p^2+q^2+r^2+2pqr$ is equal to _____.

(a) 3 (b) 1

(c) 2 (d) –1

34. If $4\sin^{-1}x+\cos^{-1}x=\pi$, then x is equal to ______.

(a) 0 (b) $\frac{1}{2}$

(c) $-\frac{1}{2}$ (d) 1

35. The principal value of $\sin^{-1}\left(\sin\frac{5\pi}{3}\right)$ is ______.

(a) $\frac{-5\pi}{3}$ (b) $\frac{5\pi}{3}$

(c) $\frac{-\pi}{3}$ (d) $\frac{4\pi}{3}$

36. The value of $\tan\left[\cos^{-1}\left(\frac{4}{5}\right)+\tan^{-1}\left(\frac{2}{3}\right)\right]$ is ______.

(a) $\frac{6}{17}$ (b) $\frac{7}{16}$

(c) $\frac{16}{7}$ (d) None

37. $\sin^{-1}\frac{4}{5}+2\tan^{-1}\frac{1}{3}$ is equal to ______.

(a) $\frac{\pi}{3}$ (b) $\frac{\pi}{4}$

(c) $\frac{\pi}{2}$ (d) 0

38. If $\sin^{-1}x+\sin^{-1}(1-x)=\cos^{-1}x$, then x to ______.

(a) $\{1, 0\}$ (b) $\{-1, 1\}$

(c) $\left\{0,\frac{1}{2}\right\}$ (d) $\{2, 0\}$

39. $\cos\left[\cos^{-1}\left(-\frac{1}{7}\right)+\sin^{-1}\left(-\frac{1}{7}\right)\right]$ is equal to _____.

(a) $-\frac{1}{3}$ (b) 0

(c) $\frac{1}{3}$ (d) $\frac{4}{9}$

40. If $\sin^{-1}x-\cos^{-1}x=\frac{\pi}{6}$, then x is equal to _____.

(a) $\frac{1}{2}$ (b) $\frac{\sqrt{3}}{2}$

(c) $-\frac{1}{2}$ (d) $-\frac{\sqrt{3}}{2}$

41. The number of real solutions of $\tan^{-1}\sqrt{x(x+1)}+\sin^{-1}\sqrt{x^2+x+1}=\pi/2$ is ______.

(a) zero (b) one

(c) two (d) infinite

42. If $\tan^{-1}\left(\frac{a}{x}\right)+\tan^{-1}\left(\frac{b}{x}\right)=\frac{\pi}{2}$, then x is equal to _____.

(a) $\sqrt{ab}$ (b) $\sqrt{2ab}$

(c) $2ab$ (d) ab

43. The value of $\sin\left(\sin^{-1}\frac{1}{3}+\sec^{-1}3\right)+\cos\left(\tan^{-1}\frac{1}{2}+\tan^{-1}2\right)$ is ______.

(a) 1 (b) 2

(c) 3 (d) 4

44. The value of $\cot^{-1} 9 + \text{cosec}^{-1} \frac{\sqrt{41}}{4}$ is given by ______.

(a) 0 (b) $\frac{\pi}{4}$

(c) $\tan^{-1} 2$ (d) $\frac{\pi}{2}$

45. $\sin\left[\frac{\pi}{2} - \sin^{-1}\left(-\frac{\sqrt{3}}{2}\right)\right]$ is equal to ______.

(a) $\frac{\sqrt{3}}{2}$ (b) $-\frac{1}{2}$

(c) $\frac{1}{2}$ (d) $-\frac{\sqrt{3}}{2}$

46. $\cos^{-1} \cos\left(-\frac{17\pi}{4}\right)$ is equal to ______.

(a) $\frac{\pi}{4}$ (b) $\frac{17\pi}{4}$

(c) $\frac{3\pi}{4}$ (d) $\frac{5\pi}{4}$

47. $\cot\left(\frac{\pi}{4} - 2\cot^{-1} 3\right)$ is equal to ______.

(a) 10 (b) 7
(c) 2 (d) 6

48. $\sin\left(\sin^{-1}\frac{1}{3} + \sec^{-1} 3\right) + \cot\left(\tan^{-1}\frac{1}{2} + \tan^{-1} 2\right)$ is equal to ______.

(a) 4 (b) 2
(c) 0 (d) 1

49. If $a > b > 0$, then the value of

$\tan^{-1}\left(\frac{a}{b}\right) + \tan^{-1}\left(\frac{a+b}{a-b}\right)$ depends on ______.

(a) a
(b) b
(c) Both a and b
(d) Neither a nor b.

50. $\sec^2\left(\tan^{-1} 2\right) + \text{cosec}^2\left(\cot^{-1} 3\right)$ is equal to ____.

(a) 15 (b) 13
(c) 6 (d) 5

51. If $\left(\tan^{-1} x\right)^2 - \left(\cot^{-1} x\right)^2 = \left(\frac{\pi}{2\sqrt{3}}\right)^2$ then x is equal to ______.

(a) $\frac{1}{\sqrt{3}}$ (b) $\sqrt{3}$

(c) 1 (d) 3

52. If $A = \sin\left(\cot^{-1} x\right)$ and $x = 9$, then $\frac{1}{A^2}$ is equal to ______.

(a) 81 (b) 82
(c) 80 (d) 9

53. If $A = \sec^{-1}\left(\frac{x+1}{x-1}\right) + \sin^{-1}\left(\frac{x-1}{x+1}\right)$ and

$B = \text{cosec}^{-1}\left(\frac{2x+3}{3x+2}\right) + \cos^{-1}\left(\frac{3x+2}{2x+3}\right)$, then ____.

(a) $A = B = \frac{\pi}{2}$ (b) $A + B = \frac{\pi}{2}$

(c) $A > B$ (d) $A < B$

54. $\cos^{-1} x$ is equal to ______.

(a) $2\sin^{-1}\sqrt{\frac{1+x}{2}}$ (b) $2\cos^{-1}\sqrt{\frac{1-x}{2}}$

(c) $2\cos^{-1}\sqrt{\frac{1+x}{2}}$ (d) None of these.

55. If $\sin^{-1}\left(\frac{3}{5}\right) + \cos^{-1}\left(\frac{12}{13}\right) = \sin^{-1} c$, then c is equal to ______.

(a) $\frac{24}{65}$ (b) $\frac{16}{65}$

(c) $\frac{56}{65}$ (d) $\frac{65}{56}$

56. If $\sin^{-1}\left(x - \frac{x^2}{2} + \frac{x^3}{4} - \ldots\right) +$

$\cos^{-1}\left(x^2 - \frac{x^4}{2} + \frac{x^6}{4} - \ldots\right) = \frac{\pi}{2}$ for $0 < |x| < \sqrt{2}$,

then x equals ______.

(a) 1/2 (b) 1
(c) –1/2 (d) –1

57. $\tan\left(\tan^{-1} x + \tan^{-1} y + \tan^{-1} z\right) -$

$\cot\left(\cot^{-1} x + \cot^{-1} y + \cot^{-1} z\right)$ is equal to ____.

(a) 0 (b) 1

(c) $\frac{3\pi}{2}$ (d) $2(x+y+z)$

58. $\cos^{-1}\left(\frac{15}{17}\right) + 2\tan^{-1}\left(\frac{1}{5}\right)$ is equal to ______.

(a) $\frac{\pi}{4}$ (b) $\frac{\pi}{2}$

(c) $\cos^{-1}\left(\frac{140}{221}\right)$ (d) $\cos^{-1}\left(\frac{171}{221}\right)$

59. If we consider only the principal values of the inverse trigonometric functions, then the value of $\tan\left(\cos^{-1}\frac{1}{5\sqrt{2}}-\sin^{-1}\frac{4}{\sqrt{17}}\right)$ is ______.

(a) $\frac{\sqrt{29}}{3}$ (b) $\frac{29}{3}$

(c) $\frac{\sqrt{3}}{29}$ (d) $\frac{3}{29}$

60. $\sin\left(\frac{1}{2}\cos^{-1}\frac{4}{5}\right)$ is equal to ______.

(a) $\frac{1}{10}$ (b) $-\frac{1}{\sqrt{10}}$

(c) $\frac{1}{\sqrt{10}}$ (d) $-\frac{1}{10}$

61. The value of $\cos^{-1}\sqrt{\frac{2}{3}}-\cos^{-1}\left(\frac{\sqrt{6}+1}{2\sqrt{3}}\right)$ is equal to ______.

(a) $\frac{\pi}{6}$ (b) $\frac{\pi}{3}$

(c) $\frac{\pi}{2}$ (d) $\frac{\pi}{4}$

62. If $A=\tan^{-1}x$, then the value of $\sin 2A$ is ______.

(a) $\frac{2x}{1+x^2}$ (b) $\frac{2x}{1-x^2}$

(c) $\frac{2x}{\sqrt{1-x^2}}$ (d) None of these.

63. If $\cos\left(2\sin^{-1}x\right)=\frac{1}{9}$, then x is equal to ______.

(a) $\frac{2}{3},-\frac{2}{3}$

(b) only $\frac{2}{3}$

(c) only $-\frac{2}{3}$

(d) neither $\frac{2}{3}$, not $-\frac{2}{3}$

64. If we consider only the principal value of the inverse trigonometric functions, the value of $\tan\left[\cos^{-1}\frac{1}{5\sqrt{2}}-\sin^{-1}\frac{4}{\sqrt{17}}\right]$ is ______.

(a) $\frac{\sqrt{29}}{3}$ (b) $\frac{3}{29}$

(c) $\frac{29}{3}$ (d) $\frac{\sqrt{3}}{29}$

65. If $\tan^{-1}(2A)+\tan^{-1}(3A)=\frac{\pi}{4}$, then A is equal to ______.

(a) $\frac{1}{3}$ (b) $\frac{1}{6}$

(c) $\frac{1}{2}$ (d) $\frac{1}{10}$

66. $\cot^{-1}\left(\frac{xy+1}{x-y}\right)+\cot^{-1}\left(\frac{yz+1}{y-z}\right)+\cot^{-1}\left(\frac{zx+1}{z-x}\right)$ is equal to ______.

(a) $\frac{\pi}{2}$ (b) π

(c) 0 (d) 1

67. $\cot\left(\cos^{-1}\frac{7}{25}\right)$ is equal to ______.

(a) $\frac{24}{25}$ (b) $\frac{7}{25}$

(c) $\frac{7}{24}$ (d) $\frac{25}{7}$

68. If $\tan(x+y)=33$ and $y=\tan^{-1}3$, then x will be ______.

(a) $\tan^{-1}(0.3)$ (b) $\tan^{-1}\left(\frac{1}{18}\right)$

(c) 0.3 (d) $\tan^{-1}(1.3)$

69. $\tan\left[\frac{1}{2}\tan^{-1}\frac{12}{5}\right]$ is equal to ______.

(a) $\frac{2}{3}$ (b) $\frac{11}{3}$

(c) $\frac{5}{3}$ (d) $\frac{3}{11}$

70. If x takes negative permissible value, then $\sin^{-1}x=$ ______.

(a) $\pi-\cos^{-1}\sqrt{1-x^2}$ (b) $\cos^{-1}\sqrt{x^2-1}$

(c) $\cos^{-1}\sqrt{1-x^2}$ (d) $-\cos^{-1}\sqrt{1-x^2}$

71. $\cos^{-1}\left(\cos\frac{5\pi}{3}\right)+\sin^{-1}\left(\cos\frac{5\pi}{3}\right)$ is equal to ______.

(a) $\frac{\pi}{2}$ (b) $\frac{5\pi}{3}$

(c) $\frac{10\pi}{3}$ (d) 0

72. The value of $\tan^{-1}1+\tan^{-1}2+\tan^{-1}3$ is ______.

(a) $-\pi$ (b) 1

(c) 0 (d) π

73. The value of $\sin^{-1}\left[\cos\left(\frac{33\pi}{5}\right)\right]$ is ______.

(a) $-\frac{\pi}{10}$ (b) $\frac{7\pi}{5}$

(c) $\frac{3\pi}{5}$ (d) $\frac{\pi}{10}$

74. $\cos\left[\cos^{-1}\left(-\frac{1}{7}\right)+\sin^{-1}\left(-\frac{1}{7}\right)\right]$ is equal to ______.

(a) 0 (b) $\frac{1}{3}$

(c) $\frac{4}{9}$ (d) $-\frac{1}{3}$

75. $\tan\left(2\sin^{-1}\frac{5}{13}\right)$ is equal to ______.

(a) $\frac{51}{78}$ (b) $\frac{60}{39}$

(c) $\frac{120}{119}$ (d) $\frac{100}{119}$

76. The value of $\tan\left[\frac{1}{2}\cos^{-1}\left(\frac{\sqrt{5}}{3}\right)\right]$ is ______.

(a) $\frac{1}{2}(3-\sqrt{5})$ (b) $\frac{3+\sqrt{5}}{2}$

(c) $3+\sqrt{5}$ (d) None of these.

77. The value of $\cot^{-1}\left[\frac{\sqrt{1-\sin x}+\sqrt{1+\sin x}}{\sqrt{1-\sin x}-\sqrt{1+\sin x}}\right]$ is ______. $(0<x<2\pi)$

(a) $2\pi-x$ (b) $2\pi-\frac{x}{2}$

(c) $\pi-\frac{x}{2}$ (d) $\frac{x}{2}$

78. $\sin^{-1}\frac{3}{5}+\tan^{-1}\frac{1}{7}$, is equal to ______.

(a) $\cos^{-1}\left(\frac{4}{5}\right)$ (b) π

(c) $\frac{\pi}{2}$ (d) $\frac{\pi}{4}$

79. If $A=\cos^2\left[\tan^{-1}\left(\sin\left(\cot^{-1}3\right)\right)\right]$, then $\frac{11}{10}A$ is equal to ______.

(a) 41 (b) 9

(c) 1 (d) 3

80. If $\sin^{-1}x+\sin^{-1}y+\sin^{-1}z=\frac{3\pi}{2}$, then value of $x^{100}+y^{100}+z^{100}-\frac{9}{x^{101}+y^{101}+z^{101}}$ is ______.

(a) 3 (b) 2

(c) 0 (d) 1

81. If $\cos^{-1}x+\cos^{-1}y+\cos^{-1}z=\pi$, then $x^2+y^2+z^2+2xyz$ is equal to ______.

(a) 0 (b) 1

(c) 5 (d) −1

82. If $\tan^{-1}x=\sin^{-1}\frac{3}{\sqrt{10}}$, then x is equal to ______.

(a) $\frac{1}{3}$ (b) $-\frac{1}{3}$

(c) 3 (d) −3

83. If $\tan\left(\cos^{-1}x\right)=\sin\left(\cot^{-1}\frac{1}{2}\right)$, then x is equal to ______.

(a) $\frac{3}{\sqrt{5}}$ (b) $\frac{\sqrt{5}}{3}$

(c) $\frac{2}{\sqrt{5}}$ (d) $\frac{1}{\sqrt{5}}$

84. If $\tan^{-1}2=2\tan^{-1}x$, then $\left(x^2+x-1\right)$ is equal to ______.

(a) 1 (b) 2

(c) 0 (d) None

85. If $0<x<1$, then $\sqrt{1+x^2}\left[\left\{x\cos\left(\cot^{-1}x\right)+\sin\left(\cot^{-1}\right)\right\}^2-1\right]^{\frac{1}{2}}$ is ______.

(a) $\frac{x}{\sqrt{1+x^2}}$ (b) x

(c) $x\sqrt{1+x^2}$ (d) $\sqrt{1+x^2}$

86. The value of x for which $\sin\left(\cot^{-1}(1+x)\right)=\cos\left(\tan^{-1}x\right)$ is ______.

(a) 1/2 (b) 1

(c) 0 (d) −1/2

Questions from Previous Years

87. A possible value of $\tan\left(\frac{1}{4}\sin^{-1}\frac{\sqrt{63}}{8}\right)$ is:

(a) $\frac{1}{\sqrt{7}}$ (b) $2\sqrt{2}-1$

(c) $\sqrt{7}-1$ (d) $\frac{1}{2\sqrt{2}}$

88. If $\left(\sin^{-1} x\right)^2 - \left(\cos^{-1} x\right)^2 = a;\ 0 < x < 1,\ a \neq 0$ then the value of $2x^2 - 1$ is _____.

(a) $\cos\left(\frac{4a}{\pi}\right)$ (b) $\sin\left(\frac{2a}{\pi}\right)$

(c) $\sin\left(\frac{2a}{\pi}\right)$ (d) $\sin\left(\frac{4a}{\pi}\right)$

89. The number of real roots of the equation $\tan^{-1}\sqrt{x(x+1)} + \sin^{-1}\sqrt{x^2+x+1} = \frac{\pi}{4}$ is _____.

(a) 1 (b) 2
(c) 4 (d) 0

90. The number of solutions of the equation $\sin^{-1}\left[x^2+\frac{1}{3}\right] + \cos^{-1}\left[x^2-\frac{2}{3}\right] = x^2$, for $x \in [-1,1]$ and $[x]$ denotes the greatest integer less than or equal to x, is _____.

(a) infinite (b) 2
(c) 4 (d) 0

91. If $\frac{\sin^{-1} x}{a} = \frac{\cos^{-1} x}{b} = \frac{\tan^{-1} y}{c};\ 0 < x < 1$, then the value of $\cos\left(\frac{\pi c}{a+b}\right)$ is _____.

(a) $1 - y^2$ (b) $\frac{1-y^2}{y\sqrt{y}}$

(c) $\frac{1-y^2}{1+y^2}$ (d) $\frac{1-y^2}{2y}$

92. If $0 < a, b, 1$ and $\tan^{-1} a + \tan^{-1} b = \frac{\pi}{4}$, then the value of $(a+b) - \left(\frac{a^2+b^2}{2}\right) + \left(\frac{a^3+b^3}{3}\right) - \left(\frac{a^4+b4}{4}\right) + \ldots$ is _____.

(a) $e^2 - 1$ (b) $\log_e\left(\frac{e}{2}\right)$

(c) e (d) $\log_e 2$

93. $2\pi - \left(\sin^{-1}\frac{4}{5} + \sin^{-1}\frac{5}{13} + \sin^{-1}\frac{16}{65}\right)$ is equal to _____.

(a) $\frac{\pi}{2}$ (b) $\frac{5\pi}{2}$

(c) $\frac{3\pi}{2}$ (d) $\frac{7\pi}{2}$

94. If S is the sum of the first 10 terms of the series $\tan^{-1}\left(\frac{1}{3}\right) + \tan^{-1}\left(\frac{1}{7}\right) + \tan^{-1}\left(\frac{1}{13}\right) + \tan^{-1}\left(\frac{1}{21}\right) + \ldots$, then $\tan(S)$ is equal to:

(a) $\frac{5}{6}$ (b) $\frac{5}{11}$

(c) $-\frac{6}{5}$ (d) $\frac{10}{11}$

95. The value of $\sin^{-1}\left(\frac{12}{63}\right) - \sin^{-1}\left(\frac{3}{5}\right)$ is equal to:

(a) $\pi - \sin^{-1}\left(\frac{63}{65}\right)$ (b) $\frac{\pi}{2} - \sin^{-1}\left(\frac{56}{65}\right)$

(c) $\frac{\pi}{2} - \cos^{-1}\left(\frac{9}{65}\right)$ (d) $\pi - \cos^{-1}\left(\frac{33}{65}\right)$

96. If $\cos^{-1} x - \cos^{-1}\frac{y}{2} = \alpha$, where $-1 \le x \le 1$, $-2 \le y \le 2$, $x \le \frac{y}{2}$, then for all x, y, $4x^2 - 4xy\cos\alpha + y^2$ is equal to:

(a) $4\sin^2\alpha$
(b) $2\sin^2\alpha$
(c) $4\sin^2\alpha - 2x^2y^2$
(d) $4\cos^2\alpha + 2x^2y^2$

97. Considering only the principal values of inverse functions, the set

$$A = \left\{x \ge 0 : \tan^{-1}(2x) + \tan^{-1}(3x) = \frac{\pi}{4}\right\}$$

(a) contains two elements.
(b) contains more than two elements.
(c) is a singleton.
(d) is an empty set.

98. The value of $\cot\left(\sum_{n=1}^{19}\cot^{-1}\left(1+\sum_{p=1}^{n}2p\right)\right)$ is:

(a) $\frac{21}{19}$ (b) $\frac{19}{21}$

(c) $\frac{22}{23}$ (d) $\frac{23}{22}$

99. If $\alpha = \cos^{-1}\left(\frac{3}{5}\right)$, $\beta = \tan^{-1}\left(\frac{1}{3}\right)$, where $0 < \alpha, \beta < \frac{\pi}{2}$, then $\alpha - \beta$ is equal to:

(a) $\tan^{-1}\left(\frac{9}{5\sqrt{10}}\right)$ (b) $\cos^{-1}\left(\frac{9}{5\sqrt{10}}\right)$

(c) $\tan^{-1}\left(\frac{9}{14}\right)$ (d) $\sin^{-1}\left(\frac{9}{5\sqrt{10}}\right)$

100. If $\cos^{-1}\left(\frac{2}{3x}\right)+\cos^{-1}\left(\frac{3}{4x}\right)=\frac{\pi}{2}\left(x>\frac{3}{4}\right)$, then x is equal to:

(a) $\frac{\sqrt{145}}{12}$ (b) $\frac{\sqrt{145}}{10}$

(c) $\frac{\sqrt{146}}{12}$ (d) $\frac{\sqrt{145}}{11}$

101. The value of

$\tan^{-1}\left[\frac{\sqrt{1+x^2}+\sqrt{1-x^2}}{\sqrt{1+x^2}-\sqrt{1-x^2}}\right]$, $|x|<\frac{1}{2}, x\neq 0$, is equal to ______.

(a) $\frac{\pi}{4}+\frac{1}{2}\cos^{-1}x^2$ (b) $\frac{\pi}{4}+\cos^{-1}x^2$

(c) $\frac{\pi}{4}-\frac{1}{2}\cos^{-1}x^2$ (d) $\frac{\pi}{4}-\cos^{-1}x^2$

102. Let $\tan^{-1}y=\tan^{-1}x+\tan^{-1}\left(\frac{2x}{1-x^2}\right)$, where $|x|<\frac{1}{\sqrt{3}}$. Then a value of y is:

(a) $\frac{3x-x^3}{1+3x^2}$ (b) $\frac{3x+x^3}{1+3x^2}$

(c) $\frac{3x-x^3}{1-3x^2}$ (d) $\frac{3x+x^3}{1-3x^2}$

103. If $f(x)=2\tan^{-1}x+\sin^{-1}\left(\frac{2x}{1+x^2}\right), x>1$, then $f(5)$ is equal to:

(a) $\tan^{-1}\left(\frac{65}{156}\right)$ (b) $\frac{\pi}{2}$

(c) π (d) $4\tan^{-1}(5)$

104. A value of x satisfying the equation $\sin\left[\cot^{-1}(1+x)\right]=\cos\left[\tan^{-1}x\right]$, is ______.

(a) $-\frac{1}{2}$ (b) -1

(c) 0 (d) $\frac{1}{2}$

105. The principal value of $\tan^{-1}\left(\cot\frac{43\pi}{4}\right)$ is:

(a) $-\frac{3\pi}{4}$ (b) $\frac{3\pi}{4}$

(c) $-\frac{\pi}{4}$ (d) $\frac{\pi}{4}$

106. The number of solutions of the equation $\sin^{-1}x=2\tan^{-1}x$ (in principal values) is:

(a) 1 (b) 4

(c) 2 (d) 3

107. If x, y, z are in AP and $\tan^{-1}x, \tan^{-1}y$ and $\tan^{-1}z$ are also in AP, then

(a) $x=y=z$ (b) $2x=3y=6z$

(c) $6x=3y=2z$ (d) $6x=4y=3z$

108. Let $x\in(0,1)$. The set of all x such that $\sin^{-1}x>\cos^{-1}x$, is the interval:

(a) $\left(\frac{1}{2},\frac{1}{\sqrt{2}}\right)$ (b) $\left(\frac{1}{\sqrt{2}},1\right)$

(c) (0, 1) (d) $\left(0,\frac{\sqrt{3}}{2}\right)$

Integer and Numerical Value Type

1. The vector of $\cot\left(\operatorname{cosec}^{-1}\frac{5}{3}+\tan^{-1}\frac{2}{3}\right)$ is equal to ______.

2. The greater of the two angles $A=2\tan^{-1}\left(2\sqrt{2}-1\right)$ and $B=3\sin^{-1}(1/3)+\sin^{-1}(3/5)$ is______.

3. The numerical value of $\tan\left\{2\tan^{-1}\left(\frac{1}{5}\right)-\frac{\pi}{4}\right\}$ is equal to__________.

4. Let a, b, c be positive real numbers. Let

$\theta=\tan^{-1}\sqrt{\frac{a(a+b+c)}{bc}}+\tan^{-1}\sqrt{\frac{b(a+b+c)}{ca}}+\tan^{-1}\sqrt{\frac{c(a+b+c)}{ab}}$. Then $\tan\theta=$________

5. Find the value of: $\cos\left(2\cos^{-1}x+\sin^{-1}x\right)$ at $x=\frac{1}{5}$, where $0\leq\cos^{-1}x\leq\pi$ and $-\pi/2\leq\sin^{-1}x\leq\pi/2$.

Assertion and Reason

Each of these questions contains two statements – Statement 1 (Assertion) and Statement 2 (Reason). Each of these questions has four alternative answers.

Only one of them is the correct answer. Select the correct answer.

(a) Both Statement 1 and Statement 2 are True and Statement 2 is the correct explanation for Statement 1.

(b) Both Statement 1 and Statement 2 are True but Statement 2 is not the correct explanation for Statement 1.

(c) Statement 1 is True, Statement 2 is False.

(d) Statement 1 is False, Statement 2 is True.

1. **Statement 1:** $\text{cosec}^{-1}\left(\frac{1}{2}+\frac{1}{\sqrt{2}}\right) > \sec^{-1}\left(\frac{1}{2}+\frac{1}{\sqrt{2}}\right)$

Statement 2: $\text{cosec}^{-1}x > \sec^{-1}x$ if $1 \le x < \sqrt{2}$

2. Let $f(x) = \sin^{-1}\left(\frac{2x}{1+x^2}\right)$

Statement 1: $f'(2) = -\frac{2}{5}$

Statement 2:

$\sin^{-1}\left(\frac{2x}{1+x^2}\right) = \pi - 2\tan^{-11} x, \quad \forall x > 1.$

3. **Statement 1:** $\sin^{-1}\left(\frac{1}{\sqrt{e}}\right) > \tan^{-1}\left(\frac{1}{\sqrt{\pi}}\right)$

Statement 2: $\sin^{-1} x > \tan^{-1} y$ for $x > y, \forall x, y \in (0,1)$

4. **Statement 1:** The equation $\left(\sin^{-1} x\right)^3 + \left(\cos^{-1} x\right)^3 - a\pi^3 = 0$ has a solution for all $a \ge \frac{1}{32}$.

Statement 2: For any $x \in R, \sin^{-1} x + \cos^{-1} x = \frac{\pi}{2}$ and $0 \le \left(\sin^{-1} x - \frac{\pi}{4}\right)^2 \le \frac{9\pi^2}{16}$.

JEE ADVANCED

Objective Questions

One Option Correct

1. If $\tan^{-1} a + \tan^{-1} b = \sin^{-1} 1 - \tan^{-1} c$ then:

(a) $a+b+c = ab+bc+ca$

(b) $\frac{1}{a}+\frac{1}{b}+\frac{1}{c} = \frac{1}{abc}$

(c) $a+b+c = abc$

(d) $ab+bc+ca = abc$

2. If $\sin^{-1} x = \frac{\pi}{5}$, for some $x \in (-1, 1)$, then the value of $\cos^{-1} x$ is ______.

(a) $\frac{9\pi}{10}$ (b) $\frac{7\pi}{10}$

(c) $\frac{5\pi}{10}$ (d) $\frac{3\pi}{10}$

3. If a and b are the root of $\sin^{-1}\left(x^2 - 3x + \frac{5}{2}\right) = \frac{\pi}{6}$, then the value of $a^2 + b^2$ is ______.

(a) 1 (b) 5

(c) 4 (d) 9

4. If $3 \sin^{-1} x = \pi + \sin^{-1}(3x - 4x^3)$, then ______.

(a) $\frac{1}{2} \le |x| \le 1$ (b) $|x| \le 1$

(c) $|x| \le \frac{1}{2}$ (d) None of these.

5. If α and β are two values of x which satisfy the equation $\sin^{-1} x + \sin^{-1} (1 - x) = \cos^{-1} x$, then ______.

(a) $\alpha + \beta = 1$ (b) $\alpha + \beta = 0$

(c) $\alpha\beta = 0$ (d) $\alpha\beta = \frac{1}{2}$

6. The greatest and least value of $(\sin^{-1} x)^3 + (\cos^{-1} x)^3$ are ______.

(a) $\frac{7\pi^3}{8}, \frac{\pi^3}{32}$ (b) $-\frac{\pi}{2}, \frac{\pi}{2}$

(c) $\frac{\pi^3}{8}, \frac{\pi^3}{8}$ (d) None of these.

7. If x_1, x_2, x_3, x_4, are roots of the equation $x^4 - x^3 \sin 2\beta + x^2 \cos 2\beta - x \cos\beta - \sin \beta = 0$, then $\tan^{-1} x_1 + \tan^{-1} x_2 + \tan^{-1} x_3 + \tan^{-1} x_4 =$ ______.

(a) $\pi - \beta$ (b) $-\beta$

(c) $\frac{\pi}{2} - \beta$ (d) β

8. If $A = \sin^{-1}\left(-\frac{\sqrt{2}}{2}\right) + \cos^{-1}\left(-\frac{1}{2}\right)$ and $B = \tan^{-1}(-\sqrt{3}) - \cot^{-1}\left(-\frac{\sqrt{3}}{3}\right)$, then:

(a) $A+B = -\frac{\pi}{12}$ (b) $A-B=0$
(c) $A+B=-\frac{7\pi}{12}$ (d) $A+B=\frac{\pi}{2}$

9. The number of real solutions of $\tan^{-1}\sqrt{x(x+1)} + \sin^{-1}\sqrt{x^2+x+1} = \frac{\pi}{2}$ is ______.

(a) 0 (b) 1
(c) 2 (d) ∞

10. If $\tan^{-1}\frac{a}{x} + \tan^{-1}\frac{b}{x}\tan^{-1} + \tan^{-1}\frac{c}{x} + \tan^{-1}\frac{d}{x} = \frac{\pi}{2}$,
Then $x^4 - x^2\sum ab + abcd$, is equal to ______.

(a) –1 (b) 1
(c) 0 (d) 2

11. $\cos^{-1}\left[\frac{\sqrt{1-\sin x}+\sqrt{1+\sin x}}{\sqrt{1-\sin x}-\sqrt{1+\sin x}}\right]$ is equal to ______.

(a) $\frac{x}{2}$ (b) $\pi - x$
(c) $2\pi - x$ (d) $\pi - \frac{x}{2}$

12. The value of $\tan^{-1}\left(\frac{1}{3}\right) + \tan^{-1}\left(\frac{1}{7}\right) + \tan^{-1}\left(\frac{1}{13}\right) + \ldots$ is ______ $\left(\text{where } T_n = \tan^{-1}\left(\frac{1}{n^2+n+1}\right)\right)$

(a) $\tan^{-1}\left(\frac{n+2}{n}\right)$ (b) $\tan^{-1}\left(\frac{n+2}{2n}\right)$
(c) $\tan^{-1}\left(\frac{n}{n+2}\right)$ (d) $\tan^{-1}\left(\frac{2n}{n+2}\right)$

13. A root of the equation $17x^2 + 17x\tan\left[2\tan^{-1}\left(\frac{1}{5}\right) - \frac{\pi}{4}\right] - 10 = 0$ is ______.

(a) 1 (b) –1
(c) $\frac{10}{17}$ (d) $\frac{17}{10}$

14. If $\cos^{-1} x = \alpha$ $(0 < x < 1)$ and $\sin^{-1}\left(2x\sqrt{1-x^2}\right) + \sec^{-1}\left(\frac{1}{2x^2-1}\right) = \frac{2\pi}{3}$ then the value of $\tan^{-1} 2x$ is ______.

(a) $\frac{\pi}{6}$ (b) $\frac{\pi}{3}$
(c) $\frac{\pi}{4}$ (d) $\frac{\pi}{2}$

15. If $\sec^{-1}\sqrt{1+x^2} + \text{cosec}^{-1}\frac{\sqrt{1+y^2}}{y} + \cot^{-1}\left(\frac{1}{z}\right) = \pi$ then $x+y+z$ is equal to ______.

(a) 0 (b) xyz
(c) $2xyz$ (d) $3xyz$

16. $\sum_{r=1}^{n} \sin^{-1}\left(\frac{\sqrt{r}-\sqrt{r-1}}{\sqrt{r(r+1)}}\right)$ is equal to ______.

(a) $\tan^{-1}(\sqrt{n}) - \frac{\pi}{4}$ (b) $\tan^{-1}(\sqrt{n}+1) - \frac{\pi}{4}$
(c) $\tan^{-1}(\sqrt{n})$ (d) $\tan^{-1}(\sqrt{n}+1)$

17. If $S = \sum_{r=1}^{10} \tan^{-1}\left(\frac{2r}{2+r^2+r^4}\right)$, then 55 cot S is equal to ______.

(a) 55 (b) 56
(c) $\frac{1}{56}$ (d) $\frac{1}{55}$

18. The value of $\cot\left(\sum_{n=1}^{23} \cot^{-1}\left(1 + \sum_{k=1}^{n} 2k\right)\right)$ is ______.

(a) $\frac{23}{25}$ (b) $\frac{25}{23}$
(c) $\frac{23}{24}$ (d) $\frac{24}{23}$

One or More Than One Option Correct

1. Let $f:(-\infty, -1) \to A$ be a function defined by $f(x) = \tan^{-1}\left(\frac{2x}{1-x^2}\right)$. If (x) is a bijective function, then set A is not ______.

(a) $\left[0, \frac{\pi}{2}\right)$ (b) $\left(0, \frac{\pi}{2}\right)$
(c) $\left(\frac{\pi}{4}, \frac{\pi}{2}\right)$ (d) $\left(-\frac{\pi}{2}, \frac{\pi}{4}\right)$

2. If $f(x) = \cos^{-1}\left(\frac{5\cos x - 12\sin x}{13}\right)$, then ______.

(a) $f'(\pi/3) = 1$ (b) $f'(\pi/6) = 0$
(c) $f(0) = \tan^{-1}\frac{12}{5}$
(d) Range of $f(x)$ is $\left[0, \frac{\pi}{2}\right]$

3. If $2^{\frac{2\pi}{\sin^{-1}x}} - 2(a+2)2^{\frac{\pi}{\sin^{-1}x}} + 8a < 0$ for at least one real x, then
(a) $\frac{1}{8} \le a < 2$
(b) $a < 2$
(c) $a \in (5, \infty)$
(d) $a \in \left(-\infty, \frac{1}{8}\right) \cup (2, \infty)$

4. If $\sum_{n=1}^{\infty} \cot^{-1}\left(\frac{n^2}{8}\right) = \frac{a}{b}\pi$, where $\frac{a}{b}$ is a rational number in its lowest form, then the correct option(s) is/are:
(a) $a - b = 3$ (b) $a + b = 11$
(c) $a + b = 10$ (d) $a - b = 4$

5. Values of x for which $\sin^{-1}\left(x^2 - \frac{x^4}{3} + \frac{x^6}{9} - \ldots\right) + \cos^{-1}\left(x^4 - \frac{x^8}{3} + \frac{x^{12}}{9} - \ldots\right) = \frac{\pi}{2}$, where $0 \le |x| \le \sqrt[4]{3}$ is/are ______.
(a) 0 (b) 1
(c) –1 (d) 2

6. If α is a real number for which $f(x) = \ln \cos^{-1} x$ is defined, then a possible value of $[\alpha]$ is ______.
(a) 0 (b) 1
(c) –1 (d) –2

7. Which of the following is a rational number?
(a) $\sin\left(\tan^{-1} 3 + \tan^{-1}\frac{1}{3}\right)$
(b) $\cos\left(\frac{\pi}{2} - \sin^{-1}\frac{3}{4}\right)$
(c) $\log_2\left(\sin\left(\frac{1}{4}\sin^{-1}\frac{\sqrt{63}}{8}\right)\right)$
(d) $\tan\left(\frac{1}{2}\cos^{-1}\frac{\sqrt{5}}{3}\right)$

8. The principal values of $\sin^{-1}\left(\sin\frac{2\pi}{3}\right)$ is:
(a) $-\frac{2\pi}{3}$ (b) $\frac{2\pi}{3}$
(c) $\frac{4\pi}{3}$ (d) None of these.

Questions from Previous Years

9. For any positive integer n, let $S_n : (0, \infty) \to \mathbb{R}$ be defined by $s_n(x) = \sum_{k=1}^{n} \cot^{-1}\left(\frac{1 + k(k+1)x^2}{x}\right)$ where for any $x \in \mathbb{R}$, $\cot^{-1} x \in (0, \pi)$ and $\tan^{-1}(x) \in \left(-\frac{\pi}{2}, \frac{\pi}{2}\right)$. Then which of the following statements is (are) TRUE?
(a) $S_{10}(x) = \frac{\pi}{2} - \tan^{-1}\left(\frac{1+11x^2}{10x}\right)$ for all $x > 0$.
(b) $\lim_{n\to\infty} \cot(S_n(x)) = x$, for all $x > 0$.
(c) The equation $S_3(x) = \frac{\pi}{4}$ has a root in $(0, \infty)$.
(d) $\tan(S_n(x)) \le \frac{1}{2}$, for all $n \ge 1$ and $x > 0$.

10. For non-negative integers n, let
$$f(n) = \frac{\sum_{k=0}^{n} \sin\left(\frac{k+1}{n+2}\pi\right)\sin\left(\frac{k+2}{n+2}\pi\right)}{\sum_{k=0}^{n} \sin^2\left(\frac{k+1}{n+2}\pi\right)}.$$
Assuming $\cos^{-1} x$ takes values in $[0, \pi]$, which of the following options is/are correct?
(a) $\lim_{n\to\infty} f(n) = \frac{1}{2}$
(b) $f(4) = \frac{\sqrt{3}}{2}$
(c) If $\alpha = \tan(\cos^{-1} f(6))$, then $\alpha^2 + 2\alpha - 1 = 0$
(d) $\sin(7\cos^{-1} f(5)) = 0$

11. If $\alpha = 3\sin^{-1}\left(\frac{6}{11}\right)$ and $\beta = \cos^{-1}\left(\frac{4}{9}\right)$, where the inverse trigonometric functions take only the principal values, then the correct option(s) is (are) ______.
(a) $\cos\beta > 0$
(b) $\sin\beta > 0$
(c) $\cos(\alpha + \beta) > 0$
(d) $\cos\alpha < 0$

Matrix Type

1. [.] represents greatest integer function in parts (A), (B) and (C)

Column I	Column II
(A) If $f(x)=\sin^{-1}x$ and $\lim_{x\to\frac{1}{2}^+} f(3x-4x^3)=a-3\lim_{x\to\frac{1}{2}^+} f(x)$, then $[a]=$	(P) 2
(B) If $f(x)=\tan^{-1}g(x)$ where $g(x)=\dfrac{3x-x^3}{1-3x^2}$ and $\lim_{h\to 0}\dfrac{f(a+3h)-f(a)}{3h}=\dfrac{3}{1+a^2}$, when $\dfrac{-1}{\sqrt{3}}<a>\dfrac{1}{\sqrt{3}}$, then find $\left[\lim_{h\to 0}\dfrac{f\left(\frac{1}{2}+6h\right)-f\left(\frac{1}{2}\right)}{6h}\right]$.	(Q) 3
(C) If $\cos^{-1}(4x^3-3x)=a+b\cos^{-1}x$ for $-1<x<\dfrac{-1}{2}$, then $[a+b+2]=$	(R) 4
(D) If $(x)=\cos^{-1}(4x^3-3x)$ and $\lim_{x\to\frac{1}{2}^+} f'(x)=a$ and $\lim_{x\to\frac{1}{2}^-} f'(x)=b$, then $a+b-3=$	(S) –2
	(T) –3

2. Let (x, y) be such that $\sin^{-1}(ax)+\cos^{-1}(y)+\cos^{-1}(bxy)=\dfrac{\pi}{2}$.

Match the statement in Column I with statement in Column II and indicate your answer by darkening the appropriate bubbles in the 4×4 matrix given in the ORS.

Column I	Column II
(A) If $a=1$ and $b=0$, then (x, y)	(P) lies on the circle $x^2+y^2=1$
(B) If $a=1$and $b=1$, then (x, y)	(Q) lies on $(x^2-1)(y^2-1)=0$
(C) If $a=1$ and $b=2$, then (x, y)	(R) lies on $y=x$
(D) If $a=2$ and $b=2$, then (x, y)	(S) lies on $(4x^2-1)(y^2-1)=0$

3. Match the following:

Column I	Column II
(A) $\sum_{i=1}^{\infty}\tan^{-1}\left(\dfrac{1}{2i^2}\right)=t$, then $\tan t=$	(P) 1
(B) Sides a, b, c of a triangle ABC are in AP and $\cos\theta_1=\dfrac{a}{b+c}$, $\cos\theta_2=\dfrac{b}{a+c}$, $\cos\theta_3=\dfrac{c}{a+b}$, then $\tan^2\left(\dfrac{\theta_1}{2}\right)+\tan^2\left(\dfrac{\theta_3}{2}\right)=$	(Q) $\dfrac{\sqrt{5}}{3}$
(C) A line is perpendicular to $x+2y+2z=0$ and passes through (0, 1, 0). The perpendicular distance of this line from the origin is	(R) 3

4. Match the Column I with Column II.

Column I	Column II
(A) Difference of greatest and least value of $\sqrt{2}(\sin 2x - \cos 2x)$ is	(P) $\frac{\pi}{4}$
(B) Difference of greatest and least value of $x^2 - 4x + 3$, $x \in [1, 3]$, is	(Q) $\frac{\pi}{6}$
(C) Greatest value of $\tan^{-1}\frac{1-x}{1+x}$, $x \in [0, 1]$, is	(R) 4
(D) Difference of greatest and least value of $\cos^{-1} x^2$, $x \in \left[-\frac{1}{\sqrt{2}}, \frac{1}{\sqrt{2}}\right]$, is	(S) 1
	(T) $\frac{\pi}{3}$

5. Match the Column I with Column II.

Column I	Column II
(A) Value of x satisfying $\sin^{-1} x - \cos^{-1} x = \cos^{-1}\left(\frac{\sqrt{3}}{2}\right)$ is	(P) < 0
(B) Value of $\sec^2(\tan^{-1}2) + \text{cosec}^2(\cot^{-1}3)$ is	(Q) < 1
(C) Greatest value of x, satisfying $\sin^{-1} x - \cos^{-1} x = \sin^{-1}(3x - 2)$ is	(R) ≥ 1
(D) Value of $\sin^{-1}(\sin 5)$ is	(S) > 0
	(T) > 2

Question from Previous Year

6. Match Column I with Column II.

Column I	Column II
(A) $\left(\frac{1}{y^2}\left(\frac{\cos(\tan^{-1} y) + y\sin(\tan^{-1} y)}{\cot(\sin^{-1} y) + \tan(\sin^{-1} y)}\right)^2 + y^4\right)^{1/2}$ takes value	(P) $\frac{1}{2}\sqrt{\frac{5}{3}}$
(B) If $\cos x + \cos y + \cos z = 0 = \sin x + \sin y + \sin z$ then Possible value of $\cos\frac{x-y}{2}$ is	(Q) $\sqrt{2}$
(C) If $\cos\left(\frac{\pi}{4} - x\right)\cos 2x + \sin x \sin 2 \sec x = \cos x \sin 2x \sec x + \cos\left(\frac{\pi}{4} + x\right)\cos 2x$ then possible Value of $\sec x$ is	(R) $\frac{1}{2}$
(D) If $\cot\left(\sin^{-1}\sqrt{1-x^2}\right) = \sin\left(\tan^{-1}\left(x\sqrt{6}\right)\right)$, $x \neq 0$ Then possible value of x is	(S) 1

Integer and Numerical Value Type

1. $\tan^{-1}\left[\dfrac{3\sin 2\alpha}{5+3\cos 2\alpha}\right]+\tan^{-1}\left[\dfrac{\tan\alpha}{4}\right]=\lambda\alpha,$ then find the value of λ, where $-\dfrac{\pi}{2}<\alpha<\dfrac{\pi}{2}$.

2. Find the number of solutions of the equation $\left(\tan^{-1}x\right)^2+\left(\cot^{-1}x\right)^2=\dfrac{5\pi^2}{8}$.

3. If $\dfrac{\pi}{2}\le x\le\dfrac{3\pi}{2}$, then $\sin^{-1}(\sin^{-1}x)$, is equal to __.

4. $\sin^{-1}\left[\cos\left(\sin^{-1}\dfrac{\sqrt{2}-\sqrt{3}}{4}+\cos^{-1}\dfrac{\sqrt{12}}{4}+\sec^{-1}\sqrt{2}\right)\right]$ is equal to _____.

5. If $x^2+y^2+z^2=r^2$, then $\tan^{-1}\left(\dfrac{xy}{zr}\right)+\tan^{-1}\left(\dfrac{yz}{xr}\right)+\tan^{-1}\left(\dfrac{xz}{yr}\right)=?$

Questions from Previous Years

6. The value of $\sec^{-1}\left(\dfrac{1}{4}\sum_{k=0}^{10}\sec\left(\dfrac{7\pi}{12}+\dfrac{k\pi}{2}\right)\sec\left(\dfrac{7\pi}{12}+\dfrac{(k+1)\pi}{2}\right)\right)$ in the interval $\left[-\dfrac{\pi}{4},\dfrac{3\pi}{4}\right]$ equals ______.

7. The number of real solutions of the equation $\sin^{-1}\left(\sum_{i=1}^{\infty}x^{i+1}-x\sum_{i-l}^{\infty}\left(\dfrac{x}{2}\right)^i\right)=\dfrac{\pi}{2}-\cos^{-1}\left(\sum_{i=1}^{\infty}\left(-\dfrac{x}{2}\right)^i-\sum_{i=1}^{\infty}(-x)^i\right)$ lying in the interval $\left(-\dfrac{1}{2},\dfrac{1}{2}\right)$ is_____.
(Here, the inverse trigonometric functions $\sin^{-1}x$ and $\cos^{-1}x$ assume values in $\left[-\dfrac{\pi}{2},\dfrac{\pi}{2}\right]$ and $[0,\pi]$ respectively.)

Comprehension Passage

Comprehension 1

It is given that $A=\left(\tan^{-1}x\right)^3+\left(\cot^{-1}x\right)^3$ where $x>0$ and $B=\left(\cos^{-1}t\right)^2+\left(\sin^{-1}t\right)^2$ where $t\in\left[0,\dfrac{1}{\sqrt{2}}\right]$ and $\sin^{-1}x+\cos^{-1}x=\dfrac{\pi}{2}$ for $-1\le x\le 1$ and $\tan^{-1}x+\cot^{-1}x=\dfrac{\pi}{2}$ for all $x\in R$.

1. The interval in which A lies is _____.
(a) $\left[\dfrac{\pi^3}{7},\dfrac{\pi^3}{2}\right)$ (b) $\left[\dfrac{\pi^3}{32},\dfrac{\pi^3}{8}\right)$
(c) $\left(\dfrac{\pi^3}{40},\dfrac{\pi^3}{10}\right)$ (d) None of these.

2. The maximum value of B is ______.
(a) $\dfrac{\pi^2}{8}$ (b) $\dfrac{\pi^2}{16}$
(c) $\dfrac{\pi^2}{4}$ (d) None of these.

3. If least value of A is λ and maximum value of B is μ, then $\cot^{-1}\cot\left(\dfrac{\lambda-\mu\pi}{\mu}\right)=$ _____.
(a) $\dfrac{\pi}{8}$ (b) $-\dfrac{\pi}{8}$
(c) $\dfrac{7\pi}{8}$ (d) $-\dfrac{7\pi}{8}$

Comprehension 2

Given that

$$\tan^{-1}\left(\frac{2x}{1-x^2}\right)=\begin{cases}2\tan^{-1}x & |x|<1\\ -\pi+2\tan^{-1}x\ , & x>1\\ \pi+2\tan^{-1}x\ , & x<-1\end{cases}$$

$$\sin^{-1}\frac{2x}{1+x^2}=\begin{cases}2\tan^{-1}x & \text{is } |x|\le 1\\ \pi-2\tan^{-1}x & \text{is } x>1\\ -\left(\pi+2\tan^{-1}x\right) & \text{is } x<-1\end{cases}$$

and $\sin^{-1}x+\cos^{-1}x=\dfrac{\pi}{2}$ for $-1\le x\le 1$.

1. If $\sin^{-1}\left(\dfrac{4x}{x^2+4}\right)+2\tan^{-1}\left(-\dfrac{x}{2}\right)$ is independent from x then:
(a) $x\in[-3,4]$ (b) $x\in[-2,2]$
(c) $x\in[-1,1]$ (d) $x\in[1,\infty]$

2. If $\cos^{-1}\dfrac{6x}{1+9x^2}=-\dfrac{\pi}{2}+2\tan^{-1}3x$, then $x\in$
(a) $\left(\dfrac{1}{3},\infty\right)$ (b) $(-1,\infty)$
(c) $(-\infty,-1)$ (d) None of these.

3. If $(x-1)\left(x^2+1\right)>0$, then $\sin\left(\dfrac{1}{2}\tan^{-1}\dfrac{2x}{1-x^2}-\tan^{-1}x\right)=$
(a) 1 (b) $\dfrac{1}{\sqrt{2}}$
(c) -1 (d) None of these.

ANSWER KEY

JEE Main

Objective Questions

One Option Correct

1. (b)	**2.** (c)	**3.** (b)	**4.** (b)	**5.** (a)	**6.** (c)	**7.** (a)	**8.** (a)	**9.** (a)
10. (c)	**11.** (c)	**12.** (b)	**13.** (d)	**14.** (a)	**15.** (c)	**16.** (a)	**17.** (b)	**18.** (b)
19. (b)	**20.** (c)	**21.** (d)	**22.** (b)	**23.** (b)	**24.** (c)	**25.** (c)	**26.** (a)	**27.** (b)
28. (b)	**29.** (a)	**30.** (d)	**31.** (a)	**32.** (d)	**33.** (b)	**34.** (b)	**35.** (c)	**36.** (d)
37. (c)	**38.** (c)	**39.** (b)	**40.** (b)	**41.** (c)	**42.** (a)	**43.** (a)	**44.** (b)	**45.** (c)
46. (a)	**47.** (b)	**48.** (d)	**49.** (d)	**50.** (a)	**51.** (b)	**52.** (b)	**53.** (a)	**54.** (c)
55. (c)	**56.** (b)	**57.** (a)	**58.** (c)	**59.** (d)	**60.** (c)	**61.** (a)	**62.** (a)	**63.** (a)
64. (b)	**65.** (b)	**66.** (c)	**67.** (c)	**68.** (a)	**69.** (a)	**70.** (d)	**71.** (a)	**72.** (d)
73. (a)	**74.** (a)	**75.** (c)	**76.** (a)	**77.** (c)	**78.** (d)	**79.** (c)	**80.** (c)	**81.** (b)
82. (c)	**83.** (b)	**84.** (c)	**85.** (c)	**86.** (d)	**87.** (a)	**88.** (b)	**89.** (d)	**90.** (d)
91. (c)	**92.** (d)	**93.** (c)	**94.** (a)	**95.** (b)	**96.** (a)	**97.** (c)	**98.** (a)	**99.** (d)
100. (a)	**101.** (a)	**102.** (c)	**103.** (c)	**104.** (a)	**105.** (c)	**106.** (a)	**107.** (a)	**108.** (b)

Integer and Numerical Value Type

1. $\left(\frac{6}{17}\right)$ **2.** $(A > B)$ **3.** $\left(\frac{-7}{17}\right)$ **4.** (0) **5.** $\frac{-2\sqrt{6}}{5}$

Assertion and Reason

1. (a) **2.** (a) **3.** (a) **4.** (a)

JEE Advanced

Objective Questions

One Option Correct

1. (b)	**2.** (d)	**3.** (b)	**4.** (a)	**5.** (c)	**6.** (a)	**7.** (c)	**8.** (c)	**9.** (c)
10. (c)	**11.** (d)	**12.** (c)	**13.** (a)	**14.** (b)	**15.** (b)	**16.** (c)	**17.** (b)	**18.** (b)

One or More Than One Option Correct

1. (a, c, d)	**2.** (a, c)	**3.** (c, d)	**4.** (a, b)
5. (a, b, c)	**6.** (a, c)	**7.** (a, b, c)	**8.** (d)
9. (a, b)	**10.** (b, c, d)	**11.** (b, c, d)	

Matrix Type

1. $A \to Q; B \to P; C \to S; D \to T$

2. $A \to P; B \to Q; C \to P; D \to S$

3. $A \to P; B \to R; C \to Q$

4. $A \to R; B \to S; C \to P; D \to Q$

5. $A \to Q, S; B \to R, S, T; C \to R, S; D \to P, Q$

6. $A \to S; B \to R; C \to Q; D \to P$

Integer and Numerical Value Type

1. (α) **2.** (–1) **3.** $(\pi - x)$ **4.** (0) **5.** $\left(\frac{\pi}{2}\right)$ **6.** (0) **7.** (2)

Comprehension Passage

Comprehension 1

1. (b) **2.** (c) **3.** (a)

Comprehension 2

1. (b) **2.** (a) **3.** (c)

HINTS AND EXPLANATIONS

Worksheet

1. $\tan^{-1} n + \cot^{-1}(n+1)$

$= \tan^{-1} n + \tan^{-1} \dfrac{1}{n+1} = \tan^{-1}\left(\dfrac{n+\dfrac{1}{n+1}}{1-n\dfrac{1}{n+1}}\right)$

$= \tan^{-1}\left(\dfrac{n^2+n+1}{n+1-n}\right) = \tan^{-1}\left(\dfrac{n^2+n+1}{1}\right)$

$= \tan^{-1}\left(n^2+n+1\right)$

2. $\text{LHS} = \tan^{-1}\left(\dfrac{a-b}{1+ab}\right) + \tan^{-1}\left(\dfrac{b-c}{1+bc}\right)$

$= \left(\tan^{-1} a - \tan^{-1} b\right) + \left(\tan^{-1} b - \tan^{-1} c\right)$

$\left[\because \tan^{-1} x - \tan^{-1} y = \tan^{-1}\left(\dfrac{x-y}{1+xy}\right)\right]$

$= \tan^{-1} a - \tan^{-1} b + \tan^{-1} b - \tan^{-1} c$

$= \tan^{-1} a - \tan^{-1} c = \text{RHS}$. Hence proved

3. (c) Let $\tan^{-1}\sqrt{x(x+1)} = \alpha$

$\Rightarrow \tan\alpha = \dfrac{\sqrt{x(x+1)}}{1}$

$\therefore \cos\alpha = \dfrac{1}{\sqrt{x^2+x+1}} \Rightarrow \alpha = \cos^{-1}\dfrac{1}{\sqrt{x^2+x+1}}$

So, $\tan^{-1}\sqrt{x(x+1)} = \cos^{-1}\dfrac{1}{\sqrt{x^2+x+1}}$

We have $\tan^{-1}\sqrt{x(x+1)} + \sin^{-1}\sqrt{x^2+x+1} = \dfrac{\pi}{2}$

$\Rightarrow \cos^{-1}\dfrac{1}{\sqrt{x^2+x+1}} = \dfrac{\pi}{2} - \sin^{-1}\sqrt{x^2+x+1}$

$\Rightarrow \cos^{-1}\dfrac{1}{\sqrt{x^2+x+1}} = \cos^{-1}\sqrt{x^2+x+1}$

$\Rightarrow \dfrac{1}{\sqrt{x^2+x+1}} = \sqrt{x^2+x+1}$

$\Rightarrow x^2+x+1 = 1 \Rightarrow x(x+1) = 0; x = 0, -1$

Hence, (c) is the correct answer.

4. $\text{LHS} = \tan^{-1}x + \tan^{-1}\dfrac{1}{x+1}$

$= \tan^{-1}\left(\dfrac{x+\dfrac{1}{x+1}}{1-x\times\dfrac{1}{x+1}}\right) = \tan^{-1}x\left(\dfrac{x^2+x+1}{x+1-x}\right)$

$= \tan^{-1}x\left(\dfrac{x^2+x+1}{1}\right) = \tan^{-1}\left(x^2+x+1\right) = \text{RHS}$

5. $\text{LHS} = \tan^{-1}\dfrac{1}{4} + \tan^{-1}\dfrac{2}{9}$

$= \tan^{-1}\left(\dfrac{\dfrac{1}{4}+\dfrac{2}{9}}{1-\dfrac{1}{4}\times\dfrac{2}{9}}\right) = \tan^{-1}\dfrac{1}{2}$

6. We have $\cos^{-1}\dfrac{x}{a} = \alpha - \cos^{-1}\dfrac{y}{b}$

$\Rightarrow \dfrac{x}{a} = \cos\left(\alpha - \cos^{-1}\dfrac{y}{b}\right)$

$\Rightarrow \dfrac{x}{a} = \cos\alpha\cos\left(\cos^{-1}\dfrac{y}{b}\right) + \sin\alpha\sin\left(\cos^{-1}\dfrac{y}{b}\right)$

$\Rightarrow \dfrac{x}{a} = \cos\alpha\cos\left(\cos^{-1}\dfrac{y}{b}\right) + \sin\alpha\sin\left(\sin^{-1}\sqrt{1-\dfrac{y^2}{b^2}}\right)$

$\Rightarrow \dfrac{x}{a} = (\cos\alpha)\dfrac{y}{b} + \sin\alpha\sqrt{\left(1-\dfrac{y^2}{b^2}\right)}$

$\therefore \left(\dfrac{x}{a} - \left(\cos\alpha.\dfrac{y}{b}\right)\right) = \sin\alpha\sqrt{\left(1-\dfrac{y^2}{b^2}\right)}$

Squaring both sides, we get

$\dfrac{x^2}{a^2} + \dfrac{y^2}{b^2}\cos^2\alpha - \dfrac{2xy}{ab}\cos\alpha = \sin^2\alpha\left(1-\dfrac{y^2}{b^2}\right)$

$\Rightarrow \dfrac{x^2}{a^2} - \dfrac{2xy}{ab}\cos\alpha + \dfrac{y^2}{b^2}\left(\cos^2\alpha + \sin^2\alpha\right) = \sin^2\alpha$

$\Rightarrow \dfrac{x^2}{a^2} - \dfrac{2xy}{ab}\cos\alpha + \dfrac{y^2}{b^2} = \sin^2\alpha$

7. If $\sin^{-1}\left(\dfrac{3}{x}\right) + \sin^{-1}\left(\dfrac{4}{x}\right) = \dfrac{\pi}{2}$

$\sin^{-1}\left(\dfrac{3}{x}\right) = \dfrac{\pi}{2} - \sin^{-1}\left(\dfrac{4}{x}\right)$

$\dfrac{3}{x} = \sin\left\{\dfrac{\pi}{2} - \sin^{-1}\left(\dfrac{4}{x}\right)\right\}$

$\dfrac{3}{x} = \cos\left\{\sin^{-1}\left(\dfrac{4}{x}\right)\right\}$

$\dfrac{3}{x} = \cos\left\{\cos^{-1}\dfrac{\sqrt{x^2-16}}{x}\right\}$

$$\frac{3}{x} = \frac{\sqrt{x^2-16}}{x}$$
$$3 = \sqrt{x^2-16}$$
$$x = 5.$$

8. (b) $\tan^{-1}\left(\frac{x}{y}\right) - \tan^{-1}\left(\frac{x-y}{x+y}\right)$

$$= \tan^{-1}\left[\frac{\frac{x}{y} - \frac{x-y}{x+y}}{1+\left(\frac{x}{y}\right)\left(\frac{x-y}{x+y}\right)}\right]$$
$$= \tan^{-1}\left[\frac{x^2+xy-xy+y^2}{xy+y^2+x^2-xy}\right]$$
$$= \tan^{-1}\left[\frac{x^2+y^2}{x^2+y^2}\right] = \tan^{-1}(1) = \frac{\pi}{4}.$$

9. (d) $\cos^{-1}\frac{2}{\sqrt{5}} + \tan^{-1}\left(\frac{1}{3}\right)$

$$= \tan^{-1}\left(\frac{1}{2}\right) + \tan^{-1}\left(\frac{1}{3}\right)$$
$$= \tan^{-1}(1)$$
$$= \tan^{-1}\left(\tan\frac{\pi}{4}\right) = \frac{\pi}{4}$$

10. (c) Let $\cos^{-1}\left(\frac{4}{5}\right) = \alpha$

$$\because \cos\alpha = \frac{4}{5},\ \sin\alpha = \frac{3}{5},\ \tan\alpha = \frac{3}{4}$$
$$\tan^{-1}\left(\frac{3}{2}\right) = \beta$$
$$\because \sin\beta = \frac{2}{\sqrt{13}},\ \cos\beta = \frac{3}{\sqrt{13}}$$
$$\theta = \alpha + \beta$$
$$\because \tan\theta = \frac{17}{6}$$

Quiz

1. Trigonometric functions, if restricted to some domain and range, become invertible.
2. Domain $[-1, 1]$;
   Range $[-\pi, 0], [0, \pi], [\pi, 2\pi], \ldots$
3. No, it never lies in the third quadrant.
4. The smallest numerical value, either positive or negative, is called the principal value of the inverse trigonometric function.
5. The position of the graph of an inverse trigonometric function in which principal value of the trigonometric function lies is called principal value branch of the function.
6. $\sin^{-1} x$ denotes all angles where sin is x and $\sin^{-1} x$ denotes the principal value.
7. θ for $-\frac{\pi}{2} \le \theta\ \frac{\pi}{2}, \theta \ne 0$
8. Yes

   Let $\sin^{-1}x = \theta \Rightarrow x = \sin\theta = \cos\left(\frac{\pi}{2} - \theta\right)$
   $$\cos^{-1}x = \frac{\pi}{2} - \theta = \frac{\pi}{2} - \sin^{-1}x$$
   $$\Rightarrow \sin^{-1}x + \cos^{-1}x = \frac{\pi}{2}$$
   Similarly, we can show that $\tan^{-1} + \cot^{-1}x = \frac{\pi}{2}$.
9. Let $\tan^{-1}1 = \theta$
   $$\tan\theta = 1 \Rightarrow \theta = \frac{\pi}{4}$$
   let $\sin^{-1}\left(-\frac{1}{2}\right) = \theta$ then $\sin\theta = -\frac{1}{2}$

   or $\theta = -\frac{\pi}{6}$
   $$\tan^{-1}1 + \sin^{-1}\left(-\frac{1}{2}\right) = \theta + \theta$$
   $$= \frac{\pi}{4} - \frac{\pi}{6} = \frac{\pi}{12}$$
10. $\sin^{-1}\left(\sin\frac{3\pi}{5}\right) = \frac{3\pi}{5}$

    but $\frac{3\pi}{5} \notin \left[-\frac{\pi}{2}, \frac{\pi}{2}\right]$
    $$\sin^{-1}\left(\sin\frac{3\pi}{5}\right) = \sin^{-1}\left(\pi - \frac{2\pi}{5}\right)$$
    $$= \sin^{-1}\left(\sin\frac{2\pi}{5}\right) = \frac{2\pi}{5}$$
    $\frac{2\pi}{5}$ lies in principal value branch of $\sin^{-1}$.

JEE Main

Objective Questions

One Option Correct

1. (b) $\cos^{-1}(-x) = \pi - \cos^{-1} x$
   $$\cos\left[\frac{\pi}{3} - \left\{\pi - \cos^{-1}\left(\frac{1}{2}\right)\right\}\right]$$
   $$= \cos\left[\frac{\pi}{3} - \pi + \cos^{-1}\left(\frac{1}{2}\right)\right]$$

$$=\cos\left[\frac{\pi}{3}-\pi+\frac{\pi}{3}\right]$$

$$=\cos\left(-\frac{\pi}{3}\right)=\frac{1}{2}$$

2. (c) We know that,

$$\sin^{-1}\left(\frac{1}{x}\right)=\text{cosec}^{-1}x,\ \cos^{-1}\left(\frac{1}{x}\right)=\sec^{-1}x$$

$\because$ Now, $(\sin^{-1}x+\cos^{-1}x)+(\sec^{-1}x+\text{cos}ec^{-1}x)$

$$=\frac{\pi}{2}+\frac{\pi}{2}=\pi$$

3. (b) $\cos^{-1}\cos\left(\pi+\frac{\pi}{6}\right)+\tan^{-1}\tan\left(\pi-\frac{\pi}{4}\right)$

$$+\sin^{-1}\left[-\sin\frac{2\pi}{3}\right]$$

$$=\cos^{-1}\left\{-\cos\frac{\pi}{6}\right\}+\tan^{-1}\left\{-\tan\frac{\pi}{4}\right\}-\sin^{-1}\sin\frac{2\pi}{3}$$

$$=\pi-\cos^{-1}\left(\cos\frac{\pi}{6}\right)-\tan^{-1}\left(\tan\frac{\pi}{4}\right)$$

$$-\sin^{-1}\sin\left(\pi-\frac{\pi}{3}\right)$$

$$=\pi-\frac{\pi}{6}-\frac{\pi}{4}-\frac{\pi}{3}=\frac{\pi}{4}$$

4. (b)

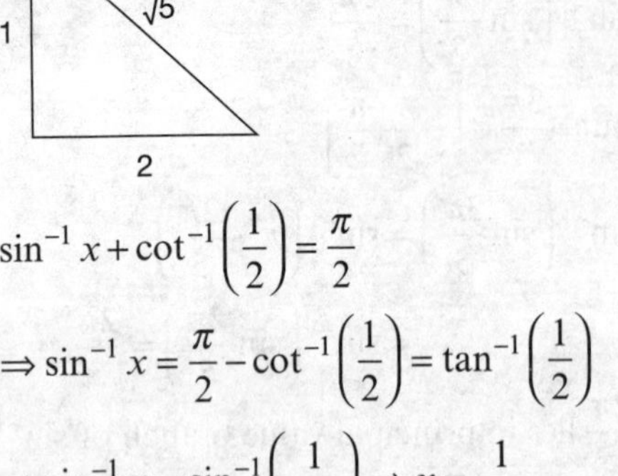

$$\sin^{-1}x+\cot^{-1}\left(\frac{1}{2}\right)=\frac{\pi}{2}$$

$$\Rightarrow\sin^{-1}x=\frac{\pi}{2}-\cot^{-1}\left(\frac{1}{2}\right)=\tan^{-1}\left(\frac{1}{2}\right)$$

$$\Rightarrow\sin^{-1}x=\sin^{-1}\left(\frac{1}{\sqrt{5}}\right)\Rightarrow x=\frac{1}{\sqrt{5}}$$

5. (a) $\sin(\cos^{-1}x)=\cos(\sin^{-1}x)$ (By conversion property)

6. (c) $\sin^{-1}x+\sin^{-1}y=\frac{2\pi}{3}$

$$\Rightarrow\frac{\pi}{2}-\cos^{-1}x+\frac{\pi}{2}-\cos^{-1}y=\frac{2\pi}{3}$$

$$\Rightarrow\cos^{-1}x+\cos^{-1}y=\pi-\frac{2\pi}{3}=\frac{\pi}{3}$$

7. (a) The sum of the two given angles is

$$\cot^{-1}2+\cot^{-1}3=\tan^{-1}\left(\frac{1}{2}\right)+\tan^{-1}\left(\frac{1}{3}\right)$$

$$\Rightarrow\tan^{-1}(1)=\frac{\pi}{4}$$

So, the third angle is $\pi-\frac{\pi}{4}=\frac{3\pi}{4}$

8. (a) We have,

$$\tan^{-1}\sqrt{x(x+1)}+\sin^{-1}\sqrt{x^2+x+1}=\frac{\pi}{2}$$

$$\Rightarrow\cos^{-1}\frac{1}{\sqrt{(x^2+x)^2+1}}+\sin^{-1}\sqrt{(x^2+x+1)}=\frac{\pi}{2}$$

$$\Rightarrow\cos^{-1}\frac{1}{\sqrt{(x^2+x)^2+1}}=\frac{\pi}{2}-\sin^{-1}\sqrt{(x^2+x+1)}$$

$$\Rightarrow\cos^{-1}\frac{1}{\sqrt{(x^2+x)^2+1}}=\cos^{-1}\sqrt{(x^2+x+1)}$$

$$\Rightarrow\frac{1}{\sqrt{(x^2+x)^2+1}}=\sqrt{(x^2+x+1)}$$

$$\Rightarrow 1=(x^2+x+1)\left[(x^2+x)^2+\right]$$

$$\Rightarrow(x^2+x)^3(x^2+x)^2+(x^2+x)+1=1$$

$$\Rightarrow(x^2+x)\left\{(x^2+x)^2+(x^2+x)+1\right\}=0$$

$$\Rightarrow x^2+x=0$$

$$\Rightarrow x=0,-1$$

9. (a) $\frac{1}{1-\sin^{-1}x}=2$ (Infinite series of GP)

$$1-\sin^{-1}x=\frac{1}{2}$$

$$\sin^{-1}x=\frac{1}{2}$$

$$x=\frac{\pi}{6}$$

10. (c) From given relation

$$\cos^{-1}\left(\frac{x}{2}\cdot\frac{y}{3}-\sqrt{1-\frac{x^4}{4}}\sqrt{1-\frac{y^2}{9}}=\theta\right)$$

$$\Rightarrow\left(\frac{xy}{6}-\cos\theta\right)^2=\left(1-\frac{x^2}{4}\right)\left(1-\frac{y^2}{9}\right)$$

$$\Rightarrow 9x^2+4y^2-12xy\cos\theta=36-36\cos^2\theta$$
$$=36\sin^2\theta$$

11. (c) $\tan^{-1}3+\tan^{-1}y=\tan^{-1}8$

$\Rightarrow \tan^{-1}y=\tan^{-1}8-\tan^{-1}3$

$\Rightarrow \tan^{-1}y=\tan^{-1}\left(\frac{8-3}{1+8.3}\right)$

$\Rightarrow \tan^{-1}y=\tan^{-1}\left(\frac{5}{25}\right)=\tan^{-1}\left(\frac{1}{5}\right)$

$\Rightarrow y=\frac{1}{5}$

12. (b) We have,

$$\tan(A-B)=\frac{\tan A-\tan B}{1+\tan A\tan B}$$

$$=\frac{\frac{x\sqrt{3}}{2k-x}-\frac{2x-k}{k\sqrt{3}}}{1+\frac{x\sqrt{3}}{2k-x}\cdot\frac{2x-k}{k\sqrt{3}}}=\frac{1}{\sqrt{3}}$$

$A-B=30°$

13. (d)

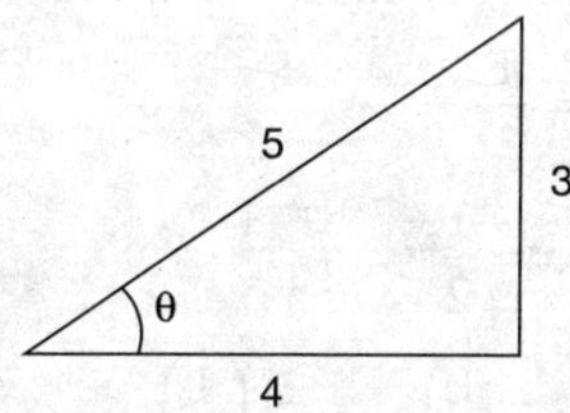

Let $\text{cosec}^{-1}\frac{5}{3}=\theta$

$\therefore\ \tan^{-1}\frac{3}{4}=\theta$

Hence, $\tan^{-1}\frac{3}{4}+\tan^{-1}\frac{2}{3}$

$$=\tan^{-1}\frac{\frac{3}{4}+\frac{2}{3}}{1-\frac{3}{4}\times\frac{2}{3}}=\tan^{-1}\frac{17}{6}$$

$\therefore\ \cot\left(\tan^{-1}\frac{17}{6}\right)=\cot\left(\cot^{-1}\frac{6}{17}\right)=\frac{6}{17}$

14. (a) $\cos^{-1}x-\cos^{-1}\frac{y}{2}=\alpha$

$$\cos^{-1}\left[\frac{xy}{2}-\sqrt{(1-x^2).\left(1-\frac{y^2}{4}\right)}\right]=\alpha$$

$$\frac{xy}{2}-\sqrt{(1-x^2).\left(1-\frac{y^2}{4}\right)}=\cos\alpha$$

$$\left(\frac{xy}{2}-\cos\alpha\right)^2=(1-x^2).\left(1-\frac{y^2}{4}\right)$$

$$\frac{x^2y^2}{4}+\cos^2\alpha-xy\cos\alpha=1+\frac{x^2y^2}{4}-x^2-\frac{y^2}{4}$$

$$-xy\cos\alpha+x^2+\frac{y^2}{4}=1-\cos^2\alpha$$

$$4x^2+y^2-4xy\cos\alpha=4\sin^2\alpha$$

15. (c) $\tan\left(2\cos^{-1}\frac{3}{5}\right)=\tan\theta$

Let $2\cos^{-1}\frac{3}{5}=\theta$

$\cos\frac{\theta}{2}=\frac{3}{5}$

$\because 2\cos^2\frac{\theta}{2}-1=\cos\theta$

$\therefore \cos\theta=-\frac{7}{25}$

$\tan\theta=-\frac{24}{7}$

16. (a)

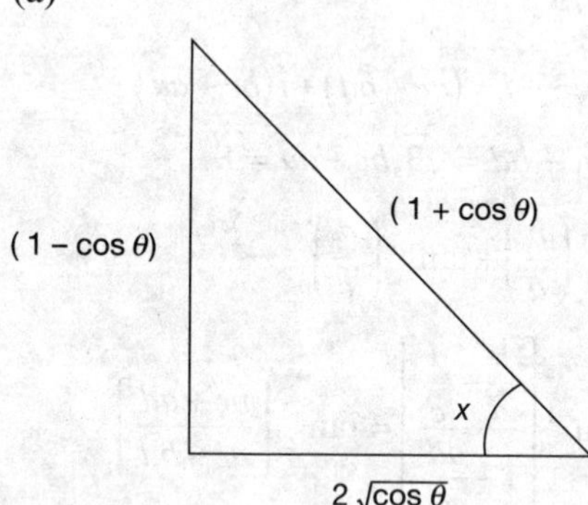

$$\cot^{-1}\sqrt{\cos\theta}-\tan^{-1}\sqrt{\cos\theta}=x$$

$$\tan^{-1}\left(\frac{1}{\sqrt{\cos\theta}}\right)-\tan^{-1}\sqrt{\cos\theta}=x$$

$$\tan^{-1}\left[\frac{\frac{1}{\sqrt{\cos\theta}}-\sqrt{\cos\theta}}{1+\left(\frac{1}{\sqrt{\cos\theta}}\right)\left(\sqrt{\cos\theta}\right)}\right]=x$$

$$\frac{1-\cos\theta}{2\sqrt{\cos\theta}}=\tan x$$

Then, $\sin x=\frac{1-\cos\theta}{1+\cos\theta}$

i.e., $\sin x=\tan^2\left(\frac{\theta}{2}\right)$

17. (b) We know that

$$\sin^{-1}x+\cos^{-1}x=\frac{\pi}{2}$$

$$\left(x-\frac{x^3}{3}+\frac{x^5}{5}-\ldots\infty\right)=\left(1-\frac{x^2}{3}+\frac{x^4}{5}-\ldots\right)$$

Put,
Both sides are equal.

18. (b) $\cos\left(2\tan^{-1}x\right)=\frac{1}{2}$

$$\cos\left[\cos^{-1}\left(\frac{1-x^2}{1+x^2}\right)\right]=\frac{1}{2}$$

$$\frac{1-x^2}{1+x^2}=\frac{1}{2} \quad \text{[Use componendo and dividendo]}$$

$$\frac{2}{-2x^2}=\frac{1+2}{1-2}=\frac{3}{-1}$$

$$x^2=\frac{1}{3}$$

$$x=\frac{1}{\sqrt{3}}$$

19. (b) $\sqrt{3}+i=(ac-bd)+i(bc+ad)$

$\because\ ac-bd=\sqrt{3}, bc+ad=1$

$$\tan^{-1}\left(\frac{b}{a}\right)+\tan^{-1}\left(\frac{d}{c}\right)$$

$$=\tan^{-1}\left[\frac{\frac{b}{a}+\frac{d}{c}}{1-\frac{bd}{ac}}\right]=\tan^{-1}\left[\frac{bc+ad}{ac-bd}\right]$$

$$=\tan^{-1}\left[\frac{1}{\sqrt{3}}\right]=\frac{\pi}{6}$$

So, general solution is $n\pi+\frac{\pi}{6}$.

20. (c) $\sin^{-1}x=2\tan^{-1}x$

$$\sin^{-1}(x)=\sin^{-1}\left(\frac{2x}{1+x^2}\right)$$

$$x=\frac{2x}{1+x^2}$$

$$x\left(1+x^2\right)-2x=0$$

$$x\left[1+x^2-2\right]=0$$

$$x\left(x^2-1\right)=0$$

$$x=0 \text{ or } x^2=1$$

$$x=0,1,-1.$$

21. (d) $2\tan^{-1}a-2\tan^{-1}b=2\tan^{-1}x$

$$\tan^{-1}\left(\frac{a-b}{1+ab}\right)=\tan^{-1}x$$

$$x=\frac{a-b}{1+ab}$$

22. (b) $\sin^{-1}x-\cos^{-1}x=\cos^{-1}\left(\frac{\sqrt{3}}{2}\right)$

$$\sin^{-1}x-\cos^{-1}x=\frac{\pi}{6} \qquad \text{(i)}$$

We know,

$$\sin^{-1}x+\cos^{-1}x=\frac{\pi}{2} \qquad \text{(ii)}$$

By (i) and (ii)

$$\sin^{-1}x=\frac{\pi}{3} \quad \text{and} \quad \cos^{-1}x=\frac{\pi}{6}$$

$$x=\frac{\sqrt{3}}{2} \qquad x=\frac{\sqrt{3}}{2}$$

23. (b) $3\sin^{-1}x+(\sin^{-1}x+\cos^{-1}x)=\pi$

$$\Rightarrow 3\sin^{-1}x=\frac{\pi}{2}$$

$$\Rightarrow \sin^{-1}x=\frac{\pi}{6} \because x=\frac{1}{2}$$

24. (c) $\left(\tan^{-1}\frac{1}{2}+\tan^{-1}\frac{1}{5}\right)+\tan^{-1}\left(\frac{1}{8}\right)$

$$=\tan^{-1}\left(\frac{\frac{1}{2}+\frac{1}{5}}{1-\frac{1}{10}}\right)+\tan^{-1}\left(\frac{1}{8}\right)$$

$$=\tan^{-1}\left(\frac{7}{9}\right)+\tan^{-1}\left(\frac{1}{8}\right)=\tan^{-1}\left(\frac{65}{65}\right)$$

$$=\tan^{-1}(1)=\frac{\pi}{4}$$

25. (c) We have

$$2\left[\tan^{-1}\frac{1}{5}+\tan^{-1}\frac{1}{8}\right]+\sec^{-1}\left(\frac{5\sqrt{2}}{7}\right)$$

$$=2\tan^{-1}\left(\frac{13}{39}\right)+\tan^{-1}\left(\frac{1}{7}\right)$$

$$=\tan^{-1}\left(\frac{3}{4}\right)+\tan^{-1}\left(\frac{1}{7}\right)$$

$$\left[\because 2\tan^{-1}x=\tan^{-1}\left(\frac{2x}{1-x^2}\right)\right]$$

$$=\tan^{-1}(1)=\frac{\pi}{4}$$

26. (a) $3(2\tan^{-1}x)-4\left(2\tan^{-1}x\right)+2(2\tan^{-1}x)=\frac{\pi}{3}$

$2\tan^{-1}x=\frac{\pi}{3}$

$\tan^{-1}x=\frac{\pi}{6}\qquad \because x=\tan\frac{\pi}{6}=\frac{1}{\sqrt{3}}$

27. (b) $\sin^{-1}\sin\theta=\theta$

$\frac{-\pi}{2}\le\theta\le\frac{\pi}{2}$

But $\sin^{-1}\sin 10\ne 10$

$\sin^{-1}\sin(3\pi-10)=\sin^{-1}\sin(\pi-10)$

$=\sin^{-1}\sin 10$

28. (b) We have,

$$\tan^{-1}\left[\frac{\frac{x}{y}-\frac{1}{c_1}}{1+\left(\frac{x}{y}\right)\left(\frac{1}{c_1}\right)}\right]+\tan^{-1}\left[\frac{\frac{1}{c_1}-\frac{1}{c_2}}{1+\frac{1}{c_1c_2}}\right]+\ldots+\tan^{-1}\left(\frac{1}{c_n}\right)$$

$$=\tan^{-1}\left(\frac{x}{y}\right)-\tan^{-1}\left(\frac{1}{c_1}\right)+\tan^{-1}\left(\frac{1}{c_1}\right)-\tan^{-1}\left(\frac{1}{c_2}\right)+\ldots$$

$$-\tan^{-1}\left(\frac{1}{c_n}\right)+\tan^{-1}\left(\frac{1}{c_n}\right)$$

$$=\tan^{-1}\left(\frac{x}{y}\right)$$

29. (a) $\tan^{-1}\left(\frac{2\left(\frac{1}{3}\right)}{1-\left(\frac{1}{3}\right)^2}\right)+\tan^{-1}\left(\frac{1}{7}\right)$

$=\tan^{-1}\left(\frac{\frac{2}{3}}{1-\frac{1}{9}}\right)+\tan^{-1}\left(\frac{1}{7}\right)$

$=\tan^{-1}\left(\frac{3}{4}\right)+\tan^{-1}\left(\frac{1}{7}\right)$

$=\tan^{-1}(1)=\frac{\pi}{4}$

30. (d) $\cos^{-1}\left(\frac{1}{2}\right)+\sin^{-1}\left(\frac{1}{2}\right)+\sin^{-1}\left(\frac{1}{2}\right)$

$=\frac{\pi}{2}+\sin^{-1}\left(\frac{1}{2}\right)$

$=\frac{\pi}{2}+\frac{\pi}{6}=\frac{2\pi}{3}\quad\left[\therefore\ \sin^{-1}x+\cos^{-1}x=\frac{\pi}{2}\right]$

31. (a) $\cos^{-1}\cos\left(\frac{5\pi}{3}\right)+\sin^{-1}\left(\cos\frac{5\pi}{3}\right)$

$=\frac{5\pi}{3}+\frac{\pi}{2}\cos^{-1}\left(\cos\frac{5\pi}{3}\right)=\frac{\pi}{2}.$

32. (d) $\tan^{-1}\left(\frac{2\cos x}{1-\cos x}\right)=\tan^{-1}\left(\operatorname{cosec}^2x\right)$

$\frac{2\cos x}{\sin^2 x}=\frac{1}{\sin^2 x}$

$2\cos x=1$

$\cos x=\frac{1}{2},x=\frac{\pi}{3}.$

33. (b) Let $\cos^{-1}p=\cos^{-1}q=\cos^{-1}r=\frac{\pi}{3}$

$\because\ p=q=r=\frac{1}{2}$

$p^2+q^2+r^2+2pqr=1.$

34. (b) Put $x=\frac{1}{2}$

$=4\sin^{-1}\left(\frac{1}{2}\right)+\cos^{-1}\left(\frac{1}{2}\right)$

$=4\left(\frac{\pi}{6}\right)+\frac{\pi}{3}=\frac{2\pi}{3}+\frac{\pi}{3}=\pi.$

35. (c) $\sin^{-1}\sin\left(\frac{5\pi}{3}\right)=\sin^{-1}\sin\left(2\pi-\frac{\pi}{3}\right)$

$=\sin^{-1}\left(-\sin\frac{\pi}{3}\right)$

$=-\sin^{-1}\sin\frac{\pi}{3}=\frac{-\pi}{3}.$

36. (d) Let $\cos^{-1}\frac{4}{5}=\theta\Rightarrow\cos\theta=\frac{4}{5}$

Now $\sin\theta=\sqrt{1-\cos^2\theta}=\sqrt{1-\frac{16}{25}}=\frac{3}{5}$

$\therefore\tan\theta=\frac{\sin\theta}{\cos\theta}=\frac{3}{4}\Rightarrow\theta=\tan^{-1}\frac{3}{4}$

$\therefore\ \tan\left[\cos^{-1}\left(\frac{4}{5}\right)+\tan^{-1}\left(\frac{2}{3}\right)\right]$

$=\tan\left[\tan^{-1}\frac{3}{4}+\tan^{-1}\frac{2}{3}\right]$

$=\tan\left[\tan^{-1}\left(\frac{\frac{3}{4}+\frac{2}{3}}{\frac{1-3}{4}+\frac{2}{3}}\right)\right]=\frac{17}{12}\times\frac{12}{6}=\frac{17}{6}$

37. (c) $\sin^{-1}\frac{4}{5}+2\tan^{-1}\left(\frac{1}{3}\right)$

$=\sin^{-1}\left(\frac{4}{5}\right)+\cos^{-1}\left(\frac{4}{5}\right)$

$$\left[\because\ 2\tan^{-1}x=\cos^{-1}\left(\frac{1-x^2}{1+x^2}\right)\right]$$

$=\frac{\pi}{2}\qquad\left(\because\ \sin^{-1}x+\cos^{-1}x=\frac{\pi}{2}\right)$

38. (c) $\sin\left[\sin^{-1}x+\sin^{-1}(1-x)\right]=\sin\left(\cos^{-1}x\right)$

$x\sqrt{1-\left(1-x^2\right)}+\left(\sqrt{1-x^2}\right)(1-x)=\sqrt{1-x^2}$

After simplification

$x=0$ or $x=\frac{1}{2}$

$x\in\left\{0,\frac{1}{2}\right\}$

39. (b) We know that

$\cos^{-1}x-\sin^{-1}x=\frac{\pi}{2}$

$\because\ \cos\left(\frac{\pi}{2}\right)=0$

40. (b)

$\sin^{-1}x+\cos^{-1}x=\frac{\pi}{2}$ (By property)

$\sin^{-1}x-\cos^{-1}x=\frac{\pi}{6}$ (Given)

$2\sin^{-1}x=\frac{2\pi}{3}$

$\sin^{-1}x=\frac{\pi}{3}$

$x=\sin\frac{\pi}{3}=\frac{\sqrt{3}}{2}$

41. (c) $\tan^{-1}\sqrt{x(x+1)}=\frac{\pi}{2}-\sin^{-1}\sqrt{x^2+x+1}$

$\Rightarrow\tan^{-1}\sqrt{x(x+1)}=\cos^{-1}\sqrt{x^2+x+1}$

$\Rightarrow\cos^{-1}\frac{1}{\sqrt{x^2+x+1}}=\cos^{-1}\sqrt{x^2+x+1}$

$\Rightarrow x^2+x+1=1\Rightarrow x(x+1)=0$

$\therefore\ x=0,-1$ are the only two real solutions.

42. (a) $\tan^{-1}\left(\frac{a}{x}\right)+\tan^{-1}\left(\frac{b}{x}\right)=\frac{\pi}{2}$

$\tan^{-1}\left[\frac{\frac{a}{x}+\frac{b}{x}}{1-\frac{ab}{x^2}}\right]=\frac{\pi}{2}$

$\frac{(a+b)x}{x^2-ab}=\frac{1}{0}$

$x^2=ab$

$x=\sqrt{ab}$

43. (a)

$\sin\left(\sin^{-1}\frac{1}{3}+\cos^{-1}\frac{1}{3}\right)+\cos\left(\tan^{-1}\frac{1}{2}+\cot^{-1}\frac{1}{2}\right)$

$\because\left(\begin{aligned}\sin^{-1}x+\cos^{-1}x=\frac{\pi}{2}\\ \cot^{-1}x+\tan^{-1}x=\frac{\pi}{2}\end{aligned}\right)$

$\because\ \sin\frac{\pi}{2}+\cos\frac{\pi}{2}=1.$

44. (b) $\cot^{-1}(9)+\operatorname{cosec}^{-1}\left(\frac{\sqrt{41}}{4}\right)$

$=\tan^{-1}\left(\frac{1}{9}\right)+\tan^{-1}\left(\frac{4}{5}\right)$

$=\tan^{-1}\left[\frac{\frac{1}{9}+\frac{4}{5}}{1-\frac{4}{45}}\right]$

$=\tan^{-1}\left[\frac{5+36}{41}\right]$

$=\tan^{-1}(1)$

$=\frac{\pi}{4}$

45. (c) $\sin\left(\frac{\pi}{2}+\frac{\pi}{3}\right)=\sin\frac{5\pi}{6}=\frac{1}{2}$

46. (a) We have $\cos^{-1}\cos\left(\frac{17\pi}{4}\right)$

$\cos^{-1}\cos\left(4\pi+\frac{\pi}{4}\right)=\cos^{-1}\cos\frac{\pi}{4}=\frac{\pi}{4}$

47. (b) $\cot\left[\tan^{-1}(1)-2\tan^{-1}\left(\frac{1}{3}\right)\right]$

$=\cot\left[\tan^{-1}(1)-\tan^{-1}\left(\frac{3}{4}\right)\right]$

$=\cot\left[\tan^{-1}\left(\frac{1}{7}\right)\right]$

$=\cot\left(\cot^{-1}7\right)$

$=7$

48. (d)

$$\sin\left(\sin^{-1}\frac{1}{3}+\cos^{-1}\frac{1}{3}\right)+\cos\left(\tan^{-1}2+\cot^{-1}2\right)$$

$$=\sin\left(\frac{\pi}{2}\right)+\cos\left(\frac{\pi}{2}\right)=1\left[\because \sin^{-1}x+\cos^{-1}x=\frac{\pi}{2}\right]$$

49. (d) $\tan^{-1}\left[\dfrac{\dfrac{a}{b}+\left(\dfrac{a+b}{a-b}\right)}{1-\left(\dfrac{a}{b}\right)\left(\dfrac{a+b}{a-b}\right)}\right]=\tan^{-1}(-1)$

50. (a) $\sec^2\left(\tan^{-1}2\right)+\text{cosec}^2\left(\cot^{-1}3\right)$

$$=\sec^2\left(\sec^{-1}\sqrt{5}\right)+\text{cosec}^2\left(\text{cosec}^{-1}\sqrt{10}\right)$$

$$=5+10=15$$

51. (b) $\left(\tan^{-1}x+\cot^{-1}x\right)\cdot\left(\tan^{-1}x-\cot^{-1}x\right)=\dfrac{\pi^2}{12}$

$$\tan^{-1}x-\cot^{-1}x=\frac{\pi}{6}$$

$$\tan^{-1}x-\left(\frac{\pi}{2}-\tan^{-1}x\right)=\frac{\pi}{6}$$

then, $x=\sqrt{3}$

52. (b) $A=\sin\left(\sin^{-1}\dfrac{1}{\sqrt{1+x^2}}\right)=\dfrac{1}{\sqrt{1+x^2}}$

$$\frac{1}{A^2}=1+x^2$$

$$=1+9^2=82$$

53. (a) $A=\cos^{-1}\left(\dfrac{x-1}{x+1}\right)+\sin^{-1}\left(\dfrac{x-1}{x+1}\right)=\dfrac{\pi}{2}$

$$B=\sin^{-1}\left(\frac{3x+2}{2x+3}\right)+\cos^{-1}\left(\frac{3x+2}{2x+3}\right)=\frac{\pi}{2}$$

$$A=B=\frac{\pi}{2}$$

54. (c) Let $\cos^{-1}x=y$

So, $x=\cos y$

$$\sqrt{\frac{1+x}{2}}=\sqrt{\frac{1+\cos y}{2}}=\cos\left(\frac{y}{2}\right)$$

$$\because 2\cos^{-1}\sqrt{\frac{1+x}{2}}=2\cos^{-1}\left(\cos\frac{y}{2}\right)$$

$$=2\left(\frac{y}{2}\right)=y=\cos^{-1}x$$

55. (c) $\sin^{-1}\left(\dfrac{3}{5}\right)+\sin^{-1}\left(\dfrac{5}{13}\right)=\sin^{-1}c$

$$\sin^{-1}\left(\frac{56}{65}\right)=\sin^{-1}c \qquad c=\frac{56}{65}$$

56. (b)

$$\sin^{-1}\left(x-\frac{x^2}{2}+\frac{x^3}{4}\ldots\right)+\cos^{-1}\left(x^2-\frac{x^4}{2}+\frac{x^6}{4}\ldots\right)=\frac{\pi}{2}$$

$$\Rightarrow\cos^{-1}\left(x^2-\frac{x^4}{2}+\frac{x^6}{4}\ldots\right)=\frac{\pi}{2}-\sin^{-1}\left(x-\frac{x^2}{2}+\frac{x^3}{4}\ldots\right)$$

$$\Rightarrow\cos^{-1}\left(x^2-\frac{x^4}{2}+\frac{x^6}{4}\ldots\right)=\cos^{-1}\left(x^2-\frac{x^2}{2}+\frac{x^3}{4}+\ldots\right)$$

$$\Rightarrow x^2-\frac{x^4}{2}+\frac{x^6}{4}\ldots=x^2-\frac{x^2}{2}+\frac{x^3}{4}+\ldots$$

On both sides we have GP of infinite terms.

$$\therefore\quad \frac{x^2}{1-\left(\frac{-x^2}{2}\right)}=\frac{x}{1-\left(-\frac{x}{2}\right)}\Rightarrow\frac{2x^2}{2+x^2}=\frac{2x}{2+x}$$

$$\Rightarrow 2x+x^3=2x^2+x^3\Rightarrow x(x-1)=0$$

$$\Rightarrow x=0,1 \text{ but } 0<|x|<\sqrt{2}\Rightarrow x=1$$

57. (a) $\cot\left[\cot^{-1}x+\cot^{-1}y+\cot^{-1}z\right]$ can be written as

$$\cot\left[\left(\frac{\pi}{2}-\tan^{-1}x\right)+\left(\frac{\pi}{2}-\tan^{-1}y\right)+\left(\frac{\pi}{2}-\tan^{-1}z\right)\right]$$

By the given expression

$$\tan\left(\tan^{-1}x+\tan^{-1}y+\tan^{-1}z\right)$$

$$\left[\frac{3\pi}{2}-\left(\tan^{-1}x+\tan^{-1}y+\tan^{-1}z\right)\right]=0$$

$$\left[\because \cot\left(\frac{3\pi}{2}-\theta\right)=\tan\theta\right]$$

58. (c) $\cos^{-1}\left(\dfrac{15}{17}\right)+\cos^{-1}\left(\dfrac{1-\left(\dfrac{1}{5}\right)^2}{1+\left(\dfrac{1}{5}\right)^2}\right)$

$$=\cos^{-1}\left(\frac{15}{17}\right)+\cos^{-1}\left(\frac{12}{13}\right)$$

$$=\cos^{-1}\left[\frac{15}{17}\times\frac{12}{13}-\frac{8}{17}\times\frac{5}{13}\right]$$

$$=\cos^{-1}\left(\frac{140}{221}\right)$$

59. (d) Let $\cos^{-1}\dfrac{1}{5\sqrt{2}}=\alpha\Rightarrow\cos\alpha=\dfrac{1}{5\sqrt{2}}$

Now, $\sin\alpha=\sqrt{1-\cos^2\alpha}=\sqrt{1-\dfrac{1}{50}}=\dfrac{7}{5\sqrt{2}}$

$\therefore \tan\alpha = \dfrac{\sin\alpha}{\cos\alpha} = 7 \Rightarrow \alpha = \tan^{-1} 7$

$\Rightarrow \cos^{-1}\dfrac{1}{5\sqrt{2}} = \tan^{-1} 7$

Also suppose $\sin^{-1}\dfrac{4}{\sqrt{17}} = \beta \Rightarrow \sin\beta = \dfrac{4}{\sqrt{17}}$

Now $\cos\beta = \sqrt{1-\sin^2\beta} = \dfrac{1}{\sqrt{17}}$

$\therefore \tan\beta = \dfrac{\sin\beta}{\cos\beta} = 4$

$\Rightarrow \beta = \tan^{-1} 4 \Rightarrow \sin^{-1}\dfrac{4}{\sqrt{17}} = \tan^{-1} 4$

$\therefore \tan\left[\cos^{-1}\dfrac{1}{5\sqrt{2}} - \sin^{-1}\dfrac{4}{\sqrt{17}}\right]$

$= \tan\left[\tan^{-1} 7 - \tan^{-1} 4\right]$

$= \tan\left(\tan^{-1}\dfrac{7-4}{1+7\times 4}\right)$

$= \tan\left(\tan^{-1}\left(\dfrac{3}{29}\right)\right) = \dfrac{3}{29}$

60. (c) Exp. $= \sin\left(\dfrac{1}{2}\times 2\theta\right)$, where $2\theta = \cos^{-1}\dfrac{4}{5}$

But $\cos 2\theta = \dfrac{4}{5}$

$\Rightarrow \sin\theta = \sqrt{\dfrac{1-\cos 2\theta}{2}} = \sqrt{\dfrac{1-4/5}{2}} = \dfrac{1}{\sqrt{10}}$

61. (a) $\cos^{-1}\left(\sqrt{\dfrac{2}{3}}\right) - \cos^{-1}\left(\dfrac{\sqrt{6}+1}{2\sqrt{3}}\right)$

$= \tan^{-1}\left(\dfrac{1}{\sqrt{2}}\right) - \tan^{-1}\left(\dfrac{\sqrt{3}-\sqrt{2}}{1+\sqrt{6}}\right)$

$= \tan^{-1}\left[\dfrac{\dfrac{1}{\sqrt{2}} - \dfrac{\sqrt{3}-\sqrt{2}}{1+\sqrt{6}}}{1+\left(\dfrac{\sqrt{3}-\sqrt{2}}{\sqrt{2}\left(1+\sqrt{6}\right)}\right)}\right]$

$= \tan^{-1}\left[\dfrac{1+\sqrt{6}-\sqrt{6}+2}{\sqrt{2}+2\sqrt{3}+\sqrt{3}-\sqrt{2}}\right]$

$= \tan^{-1}\left(\dfrac{1}{\sqrt{3}}\right) = \dfrac{\pi}{6}$

62. (a) $A = \tan^{-1} x$

$\Rightarrow x = \tan A$

Now, $\sin 2A = \dfrac{2\tan A}{1+\tan^2 A} = \dfrac{2x}{1+x^2}$

63. (a) Let $2\sin^{-1} x = \theta$,

then, $\sin\dfrac{\theta}{2} = x$,

Now, $\cos\left(2\sin^{-1} x\right) = \dfrac{1}{9}$

$\Rightarrow 1 - 2\sin^2\dfrac{\theta}{2} = \dfrac{1}{9}$

$\Rightarrow 1 - 2x^2 = \dfrac{1}{9}$

$\Rightarrow x = \pm\dfrac{2}{3}$

64. (b) $\cos^{-1}\dfrac{1}{5\sqrt{2}} = \tan^{-1}(7)$

$\cos^{-1}\dfrac{4}{\sqrt{17}} = \tan^{-1}(4)$

Now,

$\tan\left[\tan^{-1}(7) - \tan^{-1}(4)\right] = \tan\left[\tan^{-1}\left(\dfrac{7-4}{1+28}\right)\right]$

$= \dfrac{3}{29}$

65. (b) From the given equation

$\dfrac{2A+3A}{1-6A^2} = 1$

$\Rightarrow 6A^2 + 5A - 1 = 0$

$\Rightarrow A = \dfrac{1}{6}, -1$

$\therefore A = \dfrac{1}{6}$

66. (c)

$\cot^{-1}\left(\dfrac{xy+1}{x-y}\right) + \cot^{-1}\left(\dfrac{yz+1}{y-z}\right) + \cot^{-1}\left(\dfrac{zx+1}{z-x}\right)$

$= \left(\cot^{-1} x - \cot^{-1} y\right) + \left(\cot^{-1} y - \cot^{-1} z\right)$

$+ \left(\cot^{-1} z - \cot^{-1} x\right) = 0$

67. (c) $\cot\left(\cos^{-1}\dfrac{7}{25}\right)$

$\Rightarrow \cot\left[\cot^{-1}\left(\dfrac{\dfrac{7}{25}}{\sqrt{1-\left(\dfrac{7}{25}\right)^2}}\right)\right]$

$\Rightarrow \cot\left[\cot^{1}\left(\dfrac{7}{24}\right)\right]$

$\Rightarrow \dfrac{7}{24}$

68. (a) $\tan(x+y)=33$

$\Rightarrow x+y=\tan^{-1}33$

$\Rightarrow x+\tan^{-1}3=\tan^{-1}33$

$\Rightarrow x=\tan^{-1}33-\tan^{-1}3$

$\Rightarrow x=\pi+\tan^{-1}\left(\dfrac{33-3}{1+33.3}\right)$

$\Rightarrow x=\pi+\tan^{-1}\left(\dfrac{3}{10}\right)$

$\Rightarrow \tan x=\tan\pi+\tan[\tan^{-1}(0.3)]$

$\Rightarrow \quad \tan x=0+0.3$

$\Rightarrow \quad x=\tan^{-1}(0.3)$

69. (a) Let $\dfrac{1}{2}\tan^{-1}\dfrac{12}{5}=\alpha$

then, $\tan 2\theta=\dfrac{12}{5}$

$\Rightarrow \dfrac{2\tan\alpha}{1-\tan^2\alpha}=\dfrac{12}{5}$

$\Rightarrow 6\tan^2\alpha+5\tan\alpha-6=0$

$\Rightarrow \tan\alpha=\dfrac{2}{3}$

70. (d) Let $\sin^{-1}x=y$

Then, $x=\sin y$

Since, $-1\le x\le 0$,

therefore, $-\dfrac{\pi}{2}\le\sin^{-1}x\le 0$

$\Rightarrow -\dfrac{\pi}{2}\le y\le 0$

we have, $\cos y=\sqrt{1-\sin^2 y}$

$\Rightarrow \cos y=\sqrt{1-x^2}$

for $0\le y\le\pi$ ... (i)

Now, $-\dfrac{\pi}{2}\le y\le 0$

$\Rightarrow \dfrac{\pi}{2}\ge -y\ge 0$

$\Rightarrow \cos(-y)=\sqrt{1-x^2}$ [from (i)]

$\Rightarrow -y=\cos^{-1}\sqrt{1-x^2}$

$\Rightarrow y=-\cos^{-1}\sqrt{1-x^2}$

71. (a) $\cos^{-1}\left(\cos\dfrac{5\pi}{3}\right)+\sin^{-1}\left(\cos\dfrac{5\pi}{3}\right)$

$=\dfrac{5\pi}{3}+\sin^{-1}\left\{\sin\left(\dfrac{\pi}{2}-\dfrac{5\pi}{3}\right)\right\}$

$=\dfrac{5\pi}{3}+\dfrac{\pi}{2}-\dfrac{5\pi}{3}=\dfrac{\pi}{2}$

72. (d) $\tan^{-1}1+\tan^{-1}2+\tan^{-1}3$

$=\tan^{-1}1+\pi+\tan\left(-\dfrac{5}{5}\right)$

$\left[\because\ \tan^{-1}x+\tan^{-1}y=\pi+\tan^{-1}\left(\dfrac{x+y}{1-xy}\right) \quad \therefore\ xy>1\right]$

$=\tan^{-1}1+\pi+\tan^{-1}(-1)=\pi$

73. (a) We have, $\sin^{-1}\left[\cos\left(6\pi+\dfrac{3\pi}{5}\right)\right]$

$=\sin^{-1}\left[\cos\left(\dfrac{3\pi}{5}\right)\right]=\sin^{-1}\left[\sin\left(\dfrac{\pi}{2}-\dfrac{3\pi}{5}\right)\right]$

$=\sin^{-1}\left[\sin\left(-\dfrac{\pi}{10}\right)\right]=-\dfrac{\pi}{10}$

74. (a) $\cos\left[\cos^{-1}\left(-\dfrac{1}{7}\right)+\sin^{-1}\left(-\dfrac{1}{7}\right)\right]$

$\because \cos^{-1}\left(-\dfrac{1}{7}\right)+\sin^{-1}\left(-\dfrac{1}{7}\right)=\dfrac{\pi}{2}$

$\therefore \cos\left[\dfrac{\pi}{2}\right]=0$

75. (c) $\tan\left(2\sin^{-1}\dfrac{5}{13}\right)$

Exp. $=\tan\left[2\tan^{-1}\left(\dfrac{5}{12}\right)\right]$

$=\tan\left[\tan^{-1}\dfrac{120}{119}\right]=\dfrac{120}{119}$

76. (a) Let $\cos^{-1}\left(\dfrac{\sqrt{5}}{3}\right)=\alpha$,

then $\cos\alpha=\dfrac{\sqrt{5}}{3}$

where, $0<\alpha<\dfrac{\pi}{2}$

We have to find

$\tan\dfrac{\alpha}{2}=\sqrt{\dfrac{1-\cos\alpha}{1+\cos\alpha}}=\sqrt{\dfrac{1-\dfrac{\sqrt{5}}{3}}{1+\dfrac{\sqrt{5}}{3}}}$

$=\sqrt{\dfrac{3-\sqrt{5}}{3+\sqrt{5}}}=\sqrt{\dfrac{3-\sqrt{5}}{9-5}}$

$=\dfrac{1}{2}(3-\sqrt{5})$

77. (c) Since $1 \pm \sin x = \left(\cos\frac{x}{2} \pm \sin\frac{x}{2}\right)^2$

$$\cos^{-1}\left\{\frac{\sqrt{1-\sin x}+\sqrt{1+\sin x}}{\sqrt{1-\sin x}-\sqrt{1+\sin x}}\right\}$$

$$= \cot^{-1}\left\{\frac{\left[\left(\cos\frac{x}{2}-\sin\frac{x}{2}\right)+\left(\cos\frac{x}{2}+\sin\frac{x}{2}\right)\right]}{\left[\left(\cos\frac{x}{2}-\sin\frac{x}{2}\right)+\left(\cos\frac{x}{2}+\sin\frac{x}{2}\right)\right]}\right\}$$

$$= \cot^{-1}\left\{-\cot\frac{x}{2}\right\} = \cot^{-1}\left\{\cot\left(\pi-\frac{x}{2}\right)\right\}$$

$$= \left(\pi - \frac{x}{2}\right)$$

78. (d) $\sin^{-1}\frac{3}{5}+\tan^{-1}\frac{1}{7}$

$$= \tan^{-1}\frac{3}{4}+\tan^{-1}\frac{1}{7} \Rightarrow \tan^{-1}\left(\frac{\frac{3}{4}+\frac{1}{7}}{1-\frac{3}{4}\frac{1}{7}}\right)$$

$$= \tan^{-1}(1) \qquad \Rightarrow \tan^{-1}\left(\tan\frac{\pi}{4}\right)$$

$$= \frac{\pi}{4}$$

79. (c) $A = \cos^2\left[\tan^{-1}\left\{\sin\left(\sin^{-1}\frac{1}{\sqrt{10}}\right)\right\}\right]$

$$= \cos^2\left[\tan^{-1}\frac{1}{\sqrt{10}}\right]$$

$$= \cos^2\left[\cos^{-1}\sqrt{\frac{10}{11}}\right]$$

$$A = \frac{10}{11}$$

Then, $\frac{11}{10}A$ is equal to 1

80. (c) We know that, $\left|\sin^{-1}x\right| \le \frac{\pi}{2}$, therefore,

$$\sin^{-1}x+\sin^{-1}y+\sin^{-1}z = \frac{3\pi}{2}$$

$$\Rightarrow \sin^{-1}x = \sin^{-1}y = \sin^{-1}z = \frac{\pi}{2}$$

$$\Rightarrow x = y = z = \sin\frac{\pi}{2} = 1$$

$$\therefore x^{100}+y^{100}+z^{100}-\frac{9}{x^{101}+y^{101}+z^{101}}$$

$$= 3-\frac{9}{3} = 0$$

81. (b)

Method-1: $\cos^{-1}x+\cos^{-1}y = \pi-\cos^{-1}z$

$$xy-\left(\sqrt{1-x^2}\right)\left(\sqrt{1-y^2}\right) = -z$$

$$(xy+z)^2 = \left(1-x^2\right)\left(1-y^2\right)$$

$$x^2+y^2+z^2+2xy = 1$$

Method-2: Put $x = \frac{1}{2} = y = z$

Then, $x^2+y^2+z^2+2xy = 3\left(\frac{1}{2}\right)^2+2\left(\frac{1}{2}\right)^3$

$$= 3\left(\frac{1}{4}\right)+2\left(\frac{1}{8}\right)$$

$$= \frac{3}{4}+\frac{1}{4} = 1$$

82. (c) $\tan^{-1}x = \sin^{-1}\frac{3}{\sqrt{10}}$

$$\Rightarrow \tan^{-1}x = \tan^{-1}\left[\frac{\frac{3}{\sqrt{10}}}{\sqrt{1-\left(\frac{3}{\sqrt{10}}\right)^2}}\right]$$

$$= \tan^{-1}\left[\frac{\frac{3}{\sqrt{10}}}{\sqrt{\frac{10-9}{10}}}\right]$$

$$\Rightarrow \tan^{-1}x = \tan^{-1}\left[\frac{\frac{3}{\sqrt{10}}}{\frac{1}{\sqrt{10}}}\right]$$

$$\Rightarrow \tan^{-1}x = \tan^{-1}(3)$$

$$\Rightarrow x = 3$$

83. (b) $\tan\left(\cos^{-1}x\right) = \sin\left(\cot^{-1}\frac{1}{2}\right)$

$$\Rightarrow \tan\left[\tan^{-1}\frac{\sqrt{1-x^2}}{x}\right] = \sin\left[\sin^{-1}2\sqrt{5}\right]$$

$$\Rightarrow \frac{\sqrt{1-x^2}}{x} = \frac{2}{\sqrt{5}}$$

$$\Rightarrow x = \frac{\sqrt{5}}{3}$$

84. (c) $\because\ 2=\tan 2\theta$

where $x=\tan\theta$

$\tan 2\theta=\dfrac{2x}{1-x^2}=2$

$x=1-x^2$

$x^2+x-1=0$

85. (c) $\sqrt{1+x^2}\left[\left\{x\cos\left(\cot^{-1}x\right)+\sin\left(\cot^{-1}\right)\right\}^2-1\right]^{\frac{1}{2}}$

$$=\sqrt{1+x^2}\left[\left\{x\cos\left(\cos^{-1}\left(\frac{x}{\sqrt{1+x^2}}\right)\right)\right.\right.$$

$$\left.\left.\left(+\sin^{-1}\left(\frac{1}{\sqrt{1+x^2}}\right)\right)\right\}^2-1\right]^{\frac{1}{2}}$$

$$=\sqrt{1+x^2}\left[\left\{x.\frac{x}{\sqrt{1+x^2}}+\frac{1}{\sqrt{1+x^2}}\right\}^2-1\right]^{\frac{1}{2}}$$

$$=\sqrt{1+x^2}\left[\left(\sqrt{1+x^2}\right)^2-1\right]^{\frac{1}{2}}=x\sqrt{1+x^2}$$

86. (d) $\sin\left(\cot^{-1}(1+x)\right)=\cos\left(\tan^{-1}x\right)$

$$\Rightarrow\sin\left[\sin^{-1}\left(\frac{1}{\sqrt{1+(1+x)^2}}\right)\right]$$

$$=\cos\left[\cos^{-1}\left(\frac{1}{\sqrt{1+x^2}}\right)\right]$$

$$\Rightarrow\frac{1}{\sqrt{1+(1+x)^2}}=\frac{1}{\sqrt{1+x^2}}$$

$\Rightarrow 1+1+2x+x^2=1+x^2$

$\Rightarrow 2x+1=0$

$\Rightarrow x=-\dfrac{1}{2}$

87. (a) Let $\dfrac{1}{4}\sin^{-1}\dfrac{\sqrt{63}}{8}=0$

$\sin 4\theta=\dfrac{\sqrt{63}}{8}$

$\cos 4\theta=\dfrac{1}{8}$ $\quad\left[\because\cos\theta=\sqrt{1-\sin^2\theta}\right]$

$\Rightarrow 2\cos^2 2\theta-1=\dfrac{1}{8}$ $\quad\left[\because\cos 2\theta=2\cos^2\theta-1\right]$

$\Rightarrow 2\cos^2 2\theta=\dfrac{9}{16}\Rightarrow\cos 2\theta=\dfrac{3}{4}$

$\Rightarrow 2\cos^2\theta-1=\dfrac{3}{4}\Rightarrow\cos^2\theta=\dfrac{7}{8}$

$\Rightarrow\cos\theta=\dfrac{\sqrt{7}}{2\sqrt{2}}$

$\therefore\tan\theta=\dfrac{1}{\sqrt{7}}$ $\quad\left[\because\ \sin\theta=\sqrt{\sec^2\theta-1}\right]$

88. (b) It is given that, $a=\left(\sin^{-1}x\right)^2-\left(\cos^{-1}x\right)^2$

$=\left(\sin^{-1}x+\cos^{-1}x\right)\left(\sin^{-1}x-\cos^{-1}x\right)$

$=\dfrac{\pi}{2}\left(\dfrac{\pi}{2}-\cos^{-1}x-\cos^{-1}x\right)$

$\Rightarrow 2\cos^{-1}x=\dfrac{\pi}{2}-\dfrac{2a}{\pi}$

$\Rightarrow\ \cos^{-1}\left(2x^2-1\right)=\dfrac{\pi}{2}-\dfrac{2a}{\pi}$

$\left[\because\ 2\cos^{-1}x=\cos^{-1}\left(2x^2-1\right)\right]$

$\Rightarrow 2x^2-1=\cos\left(\dfrac{\pi}{2}-\dfrac{2a}{\pi}\right)$

89. (d) $\tan^{-1}\sqrt{x(x+1)}+\sin^{-1}\sqrt{x^2+x+1}=\dfrac{\pi}{4}$

For the equation to be defined,

$x_2+x\geq 0$

$\Rightarrow x^2+x+1\leq 1$

$\therefore$ Only possibility that the equation is defined

$x^2+x=0\Rightarrow x=0;x=-1$

So, none of these values satisfy the given equation.

Hence, number of real roots = 0

90. (d) We have, $\sin^{-1}\left[x^2+\dfrac{1}{3}\right]+\cos^{-1}\left[x^2-\dfrac{2}{3}\right]=x^2$

$-1\leq x\leq 1\Rightarrow 0\leq x^2\leq 1$

$\Rightarrow\dfrac{1}{3}\leq x^2+\dfrac{1}{3}\leq\dfrac{4}{3}$

So, $\left[x^2+\dfrac{1}{3}\right]=0$ or 1

Hence, LHS is always equal to π

Hence, $x^2=p$ has no solution in $[-1,1]$

91. (c) Let $\frac{\sin^{-1}x}{a}=\frac{\cos^{-1}x}{b}=\frac{\tan^{-1}y}{c}=\lambda$

$\therefore \sin^{-1}x=a\lambda, \cos^{-1}=b\lambda$ and $\tan^{-1}y=c\lambda$

We know that

$\sin^{-1}x+\cos^{-1}=\frac{\pi}{2}\Rightarrow(a+b)=\lambda=\frac{\pi}{2}$

$\therefore \lambda=\frac{\pi}{2(a+b)}$

Now, $\tan^{-1}y=\frac{\pi c}{2(a+b)}$

$\therefore \cos\left(\frac{\pi c}{a+b}\right)=\cos\left(2\tan^{-1}y\right)$

$=\cos\left(\cos^{-1}\left(\frac{1-y^2}{1+y^2}\right)\right)=\frac{1-y^2}{1+y^2}$

92. (d) $\tan^{-1}a+\tan^{-1}b=\tan^{-1}\left(\frac{a+b}{1-ab}\right)=\frac{\pi}{4}$

$\Rightarrow \frac{a+b}{1-ab}=1\Rightarrow a+b+ab=1$

$\Rightarrow (1+a)(1+b)=2$ (i)

Given that

$\left(a-\frac{a^2}{2}+\frac{a^3}{3}+..\right)+\left(b-\frac{b^2}{2}+\frac{b^3}{3}+..\right)$

$=\log_e(1+a)+\log_e(1+b)$

$=\log_e(1+a)(1+b)=\log_e 2$ [from (i)]

93. (c) $2\pi-\left(\sin^{-1}\frac{4}{5}+\sin^{-1}\frac{5}{13}+\sin^{-1}\frac{16}{65}\right)$

$=2\pi-\left(\tan^{-1}\frac{4}{3}+\tan^{-1}\frac{5}{12}+\tan^{-1}\frac{16}{63}\right)$

$\left[\because \sin^{-1}\frac{4}{5}=\tan^{-1}\frac{4}{3}\right]$

$=2\pi-\left\{\tan^{-1}\left(\frac{\frac{4}{3}+\frac{5}{12}}{1-\frac{4}{3}.\frac{5}{12}}\right)+\tan^{-1}\frac{16}{63}\right\}$

$=2\pi-\left(\tan^{-1}\frac{63}{16}+\tan^{-1}\frac{16}{63}\right)$

$=2\pi-\left(\tan^{-1}\frac{63}{16}+\cot^{-1}\frac{63}{16}\right)$

$=2\pi-\frac{\pi}{2}=\frac{3\pi}{2}$

94. (a) $S=\tan^{-1}\left(\frac{1}{3}\right)+\tan^{-1}\left(\frac{1}{7}\right)+\tan^{-1}\left(\frac{1}{13}\right)+...$

up to 10 times

$=\tan^{-1}\left(\frac{2-1}{1+2.1}\right)+\tan^{-1}\left(\frac{3-2}{1+3.1}\right)$

$+\tan^{-1}\left(\frac{4-3}{1+3.4}\right)+...+\tan^{-1}\left(\frac{11-10}{1+11.10}\right)$

$=\left(\tan^{-1}2-\tan^{-1}1\right)+\left(\tan^{-1}3-\tan^{-1}2\right)$

$+\left(\tan^{-1}4-\tan^{-1}3\right)+....+\left(\tan^{-1}11-\tan^{-1}10\right)$

$\tan^{-1}11-\tan^{-1}1=\tan^{-1}\left(\frac{11-1}{1+11.1}\right)=\tan^{-1}\left(\frac{5}{6}\right)$

$\therefore \quad \tan(S)=\frac{5}{6}$

95. (b) $-\sin^{-1}\left(\frac{3}{5}\right)+\sin^{-1}\left(\frac{12}{13}\right)$

$=-\sin^{-1}\left(\frac{3}{5}\times\frac{5}{13}-\frac{12}{13}\times\frac{4}{5}\right)$

$\left(\because \; xy\,e''0 \text{ and } x^2+y^2d''1\right)$

$\left[\because \sin^{-1}x-\sin^{-1}y=\sin^{-1}\left\{x\sqrt{1-y^2}-y\sqrt{1-x^2}\right\}\right]$

$=\sin^{-1}\left(\frac{-33}{65}\right)=\sin^{-1}\left(\frac{33}{65}\right)$

$=\cos^{-1}\left(\frac{56}{65}\right)=\frac{\pi}{2}-\sin^{-1}\left(\frac{56}{65}\right)$

96. (a) Given, $\cos^{-1}x-\cos^{-1}\frac{y}{2}=\alpha$

$\Rightarrow \cos^{-1}\left(\frac{xy}{2}+\sqrt{1-x^2}\sqrt{1-\frac{y^2}{4}}\right)=\alpha$

$\Rightarrow \frac{xy}{2}+\frac{\sqrt{1-x^2}\sqrt{4-y^2}}{2}=\cos\theta$

$\Rightarrow xy+\sqrt{1-x^2}\sqrt{4-y^2}=2\cos\alpha$

$\Rightarrow (xy-2\cos\alpha)^2=\left(1-x^2\right)\left(4-y^2\right)$

$\Rightarrow x^2y^2+4\cos^2\alpha-4xy\cos\alpha=4-y^2-4x^2+x^2y^2$

$\Rightarrow 4x^2-4xy\cos\alpha+y^2=4\sin^2\alpha$

97. (c) Consider, $\tan^{-1}(2x)+\tan^{-1}(3x)=\frac{\pi}{4}$

$\Rightarrow \tan^{-1}\left(\frac{5x}{1-6x^2}\right)=\frac{\pi}{4}$

$\Rightarrow \dfrac{5x}{1-6x^2} = 1 \Rightarrow 5x = 1-6x^2$

$\Rightarrow 6x^2 + 5x - 1 = 0$

$\Rightarrow (6x-1)(x+1) = 0$

$\Rightarrow x = \dfrac{1}{6}$, (as $x \ge 0$)

Therefore, A is a singleton set.

98. (a) $\cot\left(\sum_{n=1}^{19}\cot^{-1}\left(1+\sum_{p=1}^{n}2p\right)\right)$

$= \cot\left(\sum_{n=1}^{19}\cot^{-1}(1+n(n+1))\right)$

$= \cot\left(\sum_{n=1}^{19}\tan^{-1}\left(\dfrac{(n+1)-n}{1+(n+1)n}\right)\right)$

$\left[\cot^{-1}x = \tan^{-1}\left(\dfrac{1}{x}\right) : \text{for } x > 0\right]$

$= \cot\left(\sum_{n=1}^{19}\tan^{-1}(n+1) - \tan^{-1}n\right)$

$= \cot\left(\tan^{-1}20 - \tan^{-1}1\right)$

$= \cot\left(\tan^{-1}\left(\dfrac{20-1}{1+20\times 1}\right)\right)$

$= \cot\left(\tan^{-1}\left(\dfrac{19}{21}\right)\right) = \cot\cot^{-1}\left(\dfrac{21}{19}\right) = \dfrac{21}{19}$

99. (d) $\because \cos\alpha = \dfrac{3}{5}$, then $\sin\alpha = \dfrac{4}{5}$

$\Rightarrow \tan\alpha = \dfrac{4}{3}$ and $\tan\beta = \dfrac{1}{3}$

$\because \tan(\alpha-\beta) = \dfrac{\tan\alpha - \tan\beta}{1+\tan\alpha.\tan\beta}$

$= \dfrac{\frac{4}{3}-\frac{1}{3}}{1+\frac{4}{9}} = \dfrac{1}{\frac{13}{9}} = \dfrac{9}{13}$

$\therefore \alpha - \beta = \tan^{-1}\left(\dfrac{9}{13}\right) = \sin^{-1}\left(\dfrac{9}{5\sqrt{10}}\right)$

$= \cos^{-1}\left(\dfrac{13}{5\sqrt{10}}\right)$

100. (a) $\cos^{-1}\left(\dfrac{2}{3x}\right) + \cos^{-1}\left(\dfrac{3}{4x}\right) = \dfrac{\pi}{2}; \left(x > \dfrac{3}{4}\right)$

$\Rightarrow \cos^{-1}\left(\dfrac{2}{3x}\right) = \dfrac{\pi}{2} - \cos^{-1}\left(\dfrac{3}{4x}\right)$

$\Rightarrow \cos^{-1}\left(\dfrac{2}{3x}\right) = \sin^{-1}\left(\dfrac{3}{4x}\right)$

$\left[\because \sin^{-1}x + \cos^{-1}x = \dfrac{\pi}{2}\right]$

Put $\sin^{-1}\left(\dfrac{3}{4x}\right) = \theta \Rightarrow \sin\theta = \dfrac{3}{4x}$

$\Rightarrow \cos\theta = \sqrt{1-\sin^2\theta} = \sqrt{1-\dfrac{9}{16x^2}}$

$\Rightarrow \theta = \cos^{-1}\left(\dfrac{\sqrt{16x^2-9}}{4x}\right)$

$\therefore \cos^{-1}\left(\dfrac{2}{3x}\right) = \cos^{-1}\left(\dfrac{\sqrt{16x^2-9}}{4x}\right)$

$\Rightarrow \dfrac{2}{3x} = \dfrac{\sqrt{16x^2-9}}{4x}$

$\Rightarrow x^2 = \dfrac{64+81}{9\times 16}$

$\Rightarrow x = \pm\sqrt{\dfrac{145}{144}}$

$\Rightarrow x = \dfrac{\sqrt{145}}{12}\left(\because x > \dfrac{3}{4}\right)$

101. (a) Let $x^2 = \cos 2\theta; \;\Rightarrow \theta = \dfrac{1}{2}\cos^{-1}x^2$

$\Rightarrow \tan^{-1}\left[\dfrac{\sqrt{1+x^2}}{\sqrt{1+x^2}} - \dfrac{\sqrt{1-x^2}}{\sqrt{1-x^2}}\right]$

$= \left[\dfrac{\sqrt{1+\cos 2\theta}+\sqrt{1-\cos 2\theta}}{\sqrt{1+\cos 2\theta}-\sqrt{1-\cos 2\theta}}\right]$

$\Rightarrow \tan^{-1}\left[\dfrac{1+\tan\theta}{1-\tan\theta}\right] = \tan^{-1}\left[\tan\left(\dfrac{\pi}{4}+\theta\right)\right]$

$= \dfrac{\pi}{4} + \dfrac{1}{2}\cos^{-1}x^2$

102. (c) Given that, $\tan^{-1}y = \tan^{-1}x\left[\dfrac{2x}{1-x^2}\right]$

$= \tan^{-1}x + 2\tan^{-1}x = 3\tan^{-1}x$

$\tan^{-1}y = \tan^{-1}\left[\dfrac{3x-x^3}{1-3x^2}\right]$

$\Rightarrow y - \dfrac{3x-x^3}{1-3x^2}$

103. (c) $f(x) = 2\tan^{-1}x + \sin^{-1}\left(\dfrac{2x}{1+x^2}\right)$

$\Rightarrow f(x) = 2\tan^{-1}x + \pi - 2\tan^{-1}x$

$\Rightarrow f(x)=\pi$

$\Rightarrow f(5)=\pi$

104. (a) $\sin\left[\cot^{-1}(1+x)\right]=\cos\left(\tan^{-1}x\right)$

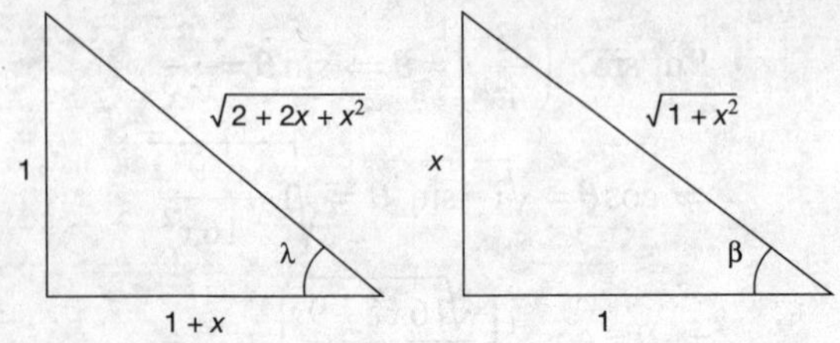

Let $\cot\lambda=1+x$

$\tan\beta=x$

$\Rightarrow \sin\lambda=\cos\beta$

$\Rightarrow \dfrac{1}{\sqrt{x^2+2x+2}}=\dfrac{1}{1\sqrt{1+x^2}}$

$\Rightarrow x^2+2x+2=x^2+1$

$\Rightarrow x=\dfrac{-1}{2}$

105. (c) $\tan^{-1}\left(\cot\dfrac{43\pi}{4}\right)=\tan^{-1}\left[\cot\left(10\pi+\dfrac{3\pi}{4}\right)\right]$

$=\tan^{-1}\left[\cot\dfrac{3\pi}{4}\right]$ $\quad[\because \cot(2n\pi+\theta)=\cot\theta]$

$=\tan^{-1}\left[\tan\left(\dfrac{\pi}{2}-\dfrac{3\pi}{4}\right)\right]$

$=\dfrac{\pi}{2}-\dfrac{3\pi}{4}=\dfrac{2\pi-3\pi}{4}=\dfrac{-\pi}{4}$

106. (a) Given equation is

$\sin^{-1}x=2\tan^{-1}x$

Now, this equation has only one solution

$\therefore$ LHS $=\sin^{-1}1=\dfrac{\pi}{2}$

and RHS $=2\tan^{-1}1=2\times\dfrac{\pi}{4}=\dfrac{\pi}{2}$

also, $x=1$ gives angle value as $\dfrac{\pi}{4}$ and $\dfrac{5\pi}{4}$

$\dfrac{5\pi}{4}$ is outside the principal value.

107. (a) Since, x, y, z are in AP

$\therefore 2y=x+z$

Also, we have

$2\tan^{-1}y=\tan^{-1}x+\tan^{-1}(z)$

$\Rightarrow \tan^{-1}\left(\dfrac{2y}{1-y^2}\right)=\tan^{-1}\left(\dfrac{x+z}{1-xz}\right)$

$\Rightarrow \dfrac{x+z}{1-y^2}=\dfrac{x+z}{1-xz}(\because 2y=x+z)$

$\Rightarrow y^2=xz$ or $x+z=0\Rightarrow x=y=z=0$

108. (b) Given $\sin^{-1}x>\cos^{-1}x$ where $x\in(0,1)$

$\Rightarrow \sin^{-1}x>\dfrac{\pi}{2}-\sin^{-1}x$

$\Rightarrow 2\sin^{-1}x>\dfrac{\pi}{2}\Rightarrow \sin^{-1}x>\dfrac{\pi}{4}$

$\Rightarrow x>\sin\dfrac{\pi}{4}\Rightarrow x>\dfrac{1}{\sqrt{2}}$

Maximum value of $\sin^{-1}x$ is $\dfrac{\pi}{2}$

So, maximum value of x is 1. So, $x\in\left(\dfrac{1}{\sqrt{2}},1\right)$.

Integer and Numerical Value Type

1. $\operatorname{cosec}^{-1}\dfrac{5}{3}=\tan^{-1}\dfrac{3}{4}$

$\cot\left(\tan^{-1}\dfrac{3}{4}+\tan^{-1}\dfrac{2}{3}\right)=\cot\left(\tan^{-1}\dfrac{\left(\dfrac{3}{4}+\dfrac{2}{3}\right)}{1-\dfrac{3}{4}\times\dfrac{2}{3}}\right)$

$=\cot\left(\tan^{-1}\dfrac{17}{6}\right)$

$=\cot\left(\cot^{-1}\dfrac{6}{17}\right)=\dfrac{6}{17}$

2. $A=2\tan^{-1}\left(2\sqrt{2}-1\right)=2\tan^{-1}(2\times1.414-1)$

$=2\tan^{-1}(1.828)>2\tan^{-1}\sqrt{3}=\dfrac{2\pi}{3}$

$\therefore A>\dfrac{2\pi}{3}$ $\quad\ldots$(i)

Now, $B=3\sin^{-1}\left(\dfrac{1}{3}\right)+\sin^{-1}\left(\dfrac{3}{5}\right)$

$=\sin^{-1}\left[3\times\dfrac{1}{3}-4\times\dfrac{1}{27}\right]+\sin^{-1}\left(\dfrac{3}{5}\right)$

$=\sin^{-1}\dfrac{23}{27}+\sin^{-1}(0.6)$

$=\sin^{-1}(0.852)+\sin^{-1}(0.6)<\sin^{-1}\left(\dfrac{\sqrt{3}}{2}\right)$

$+\sin^{-1}\left(\dfrac{\sqrt{3}}{2}\right)=\dfrac{2\pi}{3}$

$\therefore\ B < \dfrac{2\pi}{3}$ ...(ii)

From (i) and (ii), A > B

3. $\tan\left(2\tan^1\dfrac{1}{5}-\dfrac{\pi}{4}\right)$

$$= \tan\left[\tan^1\left(\frac{\frac{2}{5}}{1-\left(\frac{1}{5}\right)^2}\right)-\tan^1(1)\right]$$

$$= \tan\left[\tan^1\left(\frac{5}{12}\right)-\tan^1(1)\right] = \tan\left[\tan^1\left(\frac{\frac{5}{12}-1}{1+\frac{5}{12}}\right)\right]$$

$$= \tan\left(\tan^1\left(\frac{-7}{17}\right)\right) = \frac{-7}{17}$$

4. Let $a+b+c=u$, then

$$\theta = \tan^1\sqrt{\frac{au}{bc}}+\tan^1\sqrt{\frac{bu}{ca}}+\tan^1\sqrt{\frac{cu}{ab}}$$

$$\because\ \sqrt{\frac{au}{bc}}\times\sqrt{\frac{bu}{ca}} = \frac{u}{c} = \frac{a+b+c}{c} > 1$$

[$\because$ a,b,c are positive real numbers]

$$\therefore\ \theta = \pi+\tan^1\left[\frac{\sqrt{\frac{au}{ca}}+\sqrt{\frac{bu}{ca}}}{1-\sqrt{\frac{au}{ca}}\sqrt{\frac{bu}{ca}}}\right]+\tan^{-1}\sqrt{\frac{cu}{ab}}$$

$[\because \tan^{-1}x+\tan^{-1}y$

$= \pi+\tan^{-1}\left(\dfrac{x+y}{1-xy}\right)$ *when* $xy>1]$

$$\theta = \pi+\tan^{-1}\left[\frac{\frac{a+b}{\sqrt{abc}}\sqrt{u}}{1-\frac{u}{c}}\right]+\tan^{-1}\sqrt{\frac{cu}{ab}}$$

$$\left[\because\ \sqrt{\frac{au}{bc}}\times\sqrt{\frac{bu}{ca}} = \frac{u}{c}\right]$$

$$\theta = \pi+\tan^{-1}\left[\frac{(u-c)\sqrt{u}}{\sqrt{abc}}\times\frac{c}{-(u-c)}\right]+\tan^{-1}\sqrt{\frac{cu}{ab}}$$

$$\theta = \pi-\tan^{-1}\sqrt{\frac{uc}{ab}}+\tan^{-1}\sqrt{\frac{cu}{ab}} = \pi$$

$[\because\ \tan^{-1}(-x) = -\tan^{-1}x]$

$\therefore\ \tan\theta = \tan\pi = 0$

5. $\cos\left(2\cos^{-1}x+\sin^{-1}x\right)$

$= \cos\left(\cos^{-1}x+\cos^{-1}x+\sin^{-1}x\right)$

$= \cos\left(\cos^{-1}x+\dfrac{\pi}{2}\right)$ $\quad\left\{\because \cos^{-1}x+\sin^{-1}x = \dfrac{\pi}{2}\right\}$

$= -\sin\left(\cos^{-1}x\right) = -\sqrt{1-\cos^2\left(\cos^{-1}x\right)}$

$= -\sqrt{1-[\cos\left(\cos^{-1}x\right)]^2} = -\sqrt{1-x^2}$

At $x = \dfrac{1}{5}$, $\cos\left(2\cos^{-1}x+\sin^{-1}x\right) = -\sqrt{1-x^2}$

$= -\sqrt{1-1/25} = \dfrac{-2\sqrt{6}}{5}$

Assertion and Reason

1. (a) $\operatorname{cosec}^{-1}x > \sec^{-1}x$

$\operatorname{cosec}^{-1}x > \dfrac{\pi}{2}-\operatorname{cosec}^{-1}x$

$\operatorname{cosec}^{-1}x > \dfrac{\pi}{4}$

$1 \le x < \sqrt{2}$ and $\left(\dfrac{1}{2}+\dfrac{1}{\sqrt{2}}\right) \in [1,\sqrt{2})$

Statement 2 is true and explains Statement 1.

2. (a) $f(x) = \sin^{-1}\left(\dfrac{2x}{1+x^2}\right) = \pi-2\tan^{-1}x,\ x \ge 1$

$f'(x) = -\dfrac{2x}{1+x^2} \quad \Rightarrow \quad f'(2) = -\dfrac{2}{5}$

Statement 1 is True, Statement 2 is True; Statement 2 is the correct explanation for Statement 1.

3. (a) $\sin^{-1}x = \tan^{-1}\dfrac{x}{\sqrt{1-x^2}} > \tan^{-1}x > \tan^{-1}y$

$\left\{\because x>y, \dfrac{x}{\sqrt{1-x^2}} > x\right\}$

$\therefore$ Statement 2 is true.

$e < \pi$

$\dfrac{1}{\sqrt{e}} > \dfrac{1}{\sqrt{\pi}}$

by Statement 2

$\sin^{-1}\left(\dfrac{1}{\sqrt{e}}\right) > \tan^{-1}\left(\dfrac{1}{\sqrt{e}}\right) > \tan^{-1}\left(\dfrac{1}{\sqrt{\pi}}\right)$

Statement 1 is true.

4. (a) $\sin^{-1}x \in \left[-\frac{\pi}{2}, \frac{\pi}{2}\right]$

$\Rightarrow -\frac{3\pi}{4} \le \left(\sin^{-1}x - \frac{\pi}{4}\right) \le \frac{\pi}{4}$

$0 \le \left(\sin^{-1}x - \frac{\pi}{4}\right)^2 \le \frac{9}{16}\pi^2$ (i)

Statement 2 is true.

$(\sin^{-1}x)^3 + (\cos^{-1}x)^3 = a\pi^3$

$\Rightarrow \left(\sin^{-1}x + \cos^{-1}x\right)$

$\left[\left(\sin^{-1}x + \cos^{-1}x\right)^2 - 3\sin^{-1}x\cos^{-1}x\right] = a\pi^3$

$\Rightarrow \frac{\pi^2}{4} - 3\sin^{-1}x\cos^{-1}x = 2a\pi^2$

$\Rightarrow \sin^{-1}x\left(\frac{\pi}{2} - \sin^{-1}x\right) = \frac{\pi^2}{12}(1-8a)$

$\Rightarrow \sin^{-1}x\left(\frac{\pi}{2} - \sin^{-1}x\right) = \frac{\pi^2}{12}(1-8a)$

$\Rightarrow \sin^{-1}x\left(\frac{\pi}{2} - \sin^{-1}x\right) = \frac{\pi^2}{48}(32a-1)$

Putting these values in equation (i)

$0 \le \frac{\pi^2}{48}(32a-1) \le \frac{9}{16}\pi^2$

$\Rightarrow 0 \le 32a - 1 \le 27$

$\frac{1}{32} \le a \le \frac{7}{8}$

JEE Advanced

Objective Questions

One Option Correct

1. (b) Given relation

$\tan^{-1}a + \tan^{-1}b + \tan^{-1}c = \sin^{-1}1$

$\Rightarrow \tan^{-1}\left(\frac{a+b+c-abc}{1-ab-bc-ca}\right)$

$\Rightarrow 1 - ab - bc - ca = 0$

$\Rightarrow \frac{1}{abc} = \frac{1}{a} + \frac{1}{b} + \frac{1}{c}$

2. (d) $\sin^{-1}x + \cos^{-1}x = \frac{\pi}{2}$

$\Rightarrow \cos^{-1}x = \frac{\pi}{2} - \sin^{-1}x = \frac{\pi}{2} - \frac{\pi}{5} = \frac{3\pi}{10}$

3. (b) $x^2 - 3x + \frac{5}{2} = \frac{1}{2}$

$x - 1, 2$

$a = 1, b = 2$

$a^2 + b^2 = 1^2 + 2^2 = 5$

4. (a) $3\sin^{-1}x = \sin^{-1}\left(3x - 4x^3\right) \forall x \in \left[-\frac{1}{2}, \frac{1}{2}\right]$

$3\sin^{-1}x = \pi + \sin^{-1}\left(3x - 4x^3\right)$

$\forall x \in \left[-1, \frac{1}{2}\right] \cup \left[\frac{1}{2}, 1\right]$

5. (c) Let $\sin^{-1}x = \theta$

$x = \sin x = \theta = \cos\left(\frac{\pi}{2} - \theta\right)$

We have,

$2\sin^{-1}x + \sin^{-1}(1-x) = \cos^{-1}x + \sin^{-1}x$

$\sin^{-1}(1-x) = \frac{\pi}{2} - 2\theta$

$1 - x = \cos 2\theta = 2\cos^2\theta - 1$

$1 - x = 1 - 2x^2$

$x = 0$ or $\frac{1}{2}$

$a = 0, \beta = \frac{1}{2}$

$a\beta = 0, a + \beta = \frac{1}{2}$

6. (a) $\left(\sin^{-1}x\right)^3 + \left(\cos^{-1}\right)^3$

$= \left(\sin^{-1}x + \cos^{-1}x\right)^3 - 3$

$\sin^{-1}\cos^{-1}x\left(\sin^{-1}x + \cos^{-1}x\right)$

$= \frac{\pi^3}{8} - 3\left(\sin^{-1}x\cos^{-1}x\right)\frac{\pi}{2}$

$= \frac{\pi^3}{8} - \frac{3\pi}{2}\sin^{-1}x\left(\frac{\pi}{2} - \sin^{-1}x\right)$

$= \frac{\pi^3}{8} - \frac{3\pi 2}{8}\sin^{-1}x + \frac{3\pi}{2}\left(\sin^{-1}x\right)^2$

$= \frac{\pi^3}{8} + \frac{3\pi}{2}\left[\left(\sin^{-1}x\right)^2 - \frac{3\pi}{2}\sin^{-1}x\right]$

$= \frac{\pi^3}{8} + \frac{3\pi}{2}\left[\left(\sin^{-1}x - \frac{\pi}{4}\right)^2\right] - \frac{3\pi^3}{32}$

$= \frac{\pi^3}{32} + \frac{3\pi}{2}\left(\sin^{-1}x - \frac{\pi}{4}\right)^2$

the least value is $\frac{\pi^3}{32}$.

the greatest value is

$$= \frac{\pi^3}{32} + \frac{9\pi^2}{16} \times \frac{3\pi}{2} = \frac{7\pi^3}{8}.$$

7. (c) We have,

$$\sum x_1 = \sin 2\beta$$
$$\sum x_1 x_2 = \cos 2\beta$$
$$\sum x_1 x_2 x_3 = \cos\beta$$
$$\sum x_1 x_2 x_3 x_4 = -\sin\beta$$
$$\therefore\ \tan^{-1} x_1 + \tan^{-1} x_2 + \tan^{-1} x_3 + \tan^{-1} x_4$$
$$= \tan^{-1}\left(\frac{\sin 2\beta - \cos\beta}{1 - \cos 2\beta \sin\beta}\right)$$
$$= \tan^{-1}\left(\frac{(2\sin\beta - 1)\cos\beta}{\sin\beta(2\sin\beta - 1)}\right)$$
$$= \tan^{-1}(\cos\beta) = \tan^{-1}\left[\tan\left(\frac{\pi}{2} - \beta\right)\right]$$
$$= \frac{\pi}{2} - \beta$$

8. (c) $A = -\frac{\pi}{4} + \frac{2\pi}{3}$

$$B = -\frac{\pi}{4} + \frac{2\pi}{3}$$
$$A + B = -\frac{7\pi}{12}$$

9. (c) $\tan^{-1}\sqrt{x(x+1)} = \cos^{-1}\left(\frac{1}{\sqrt{x^2+x+1}}\right)$

$$\tan^{-1}\left(\sqrt{x^2+x+1}\right) + \cos^{-1}\left(\frac{1}{\sqrt{x^2+x+1}}\right) = \frac{\pi}{2}$$
$$\left[\therefore\ \sin^{-1} x + \cos^{-1} x = \frac{\pi}{2}\right]$$
$$x^2 + x + 1 = \frac{1}{\sqrt{x^2+x+1}}$$
$$x^2 + x + 1 = 1$$
$$x(+x+1) = 0$$
$$x = 0, -1$$

10. (c) $\tan^{-1}\left(\frac{(a+b)x}{x^2 - ab}\right) = \frac{\pi}{2} - \tan^{-1}\left(\frac{(c+d)x}{x^2 - cd}\right)$

$$\tan^{-1}\left(\frac{(a+b)x}{x^2 - ab}\right) = \cos^{-1}\left(\frac{(c+d)x}{x^2 - cd}\right)$$
$$(x^2 - ab)(x^2 - cd) = (a+b).(c+d)^2$$
$$x^4 - x^2\sum ab + abcd = 0$$

11. (d) After rationalization, we get

$$\cot^{-1}\left[\frac{1+\cos x}{-\sin x}\right] = \cot^{-1}\left[\frac{2\cos^2\frac{x}{2}}{2\sin\left(\frac{x}{2}\right).\cos\left(\frac{x}{2}\right)}\right]$$
$$= \cot^{-1}\left(-\cot\frac{x}{2}\right)$$
$$= \cot^{-1}\left[\cot\left(\pi - \frac{x}{2}\right)\right]$$
$$= \pi - \frac{x}{2}$$

12. (c) $T_n = \tan^{-1}\left(\frac{1}{n^2+n+1}\right)$

$$T_n = \tan^{-1}\left(\frac{(r+1) - r}{1 + r(r+1)}\right)$$
$$= \tan^{-1}(r+1) - \tan^{-1}(r)$$
$$S_n = \sum_{r=1}^{n} T_r$$
$$= \sum_{r=1}^{n}\left[\tan^{-1}(r+1) - \tan^{-1}(r)\right]$$
$$S_n = \tan^{-1}(n+1) - \tan^{-1}(n)$$
$$= \tan^{-1}\left(\frac{n}{n+2}\right)$$

13. (a) $2\tan^{-1}\left(\frac{1}{5}\right) = \frac{5}{12}$

$$17x^2 - 17x\tan\left(\frac{\pi}{4} - 2\tan^{-1}\left(\frac{1}{5}\right)\right) - 10 = 0$$
$$17x^2 - 17x - 10 = 0$$
$$x = 1$$
$$x = -\frac{10}{17}$$

14. (b) $\cos^{-1} x = \alpha$

$$\therefore\ x = \cos\alpha$$
$$\sin^{-1}(\sin 2\alpha) + \sec^{-1}(\sec 2\alpha) = \frac{2\pi}{3}$$

$2\alpha + 2\alpha = \frac{2\pi}{3}$ or $\alpha = \frac{\pi}{6}$

$x = \cos\frac{\pi}{6} = \frac{\sqrt{3}}{2}$

$\because\ \tan^{-1}(2x) = \tan^{-1}(\sqrt{3}) = \frac{\pi}{3}$

15. (b) Let $x = \tan\theta, \quad y = \tan\phi$

Now, $\tan^{-1}x + \tan^{-1}y + \tan^{-1}z = \pi$

$$\tan^{-1}\left[\frac{x+y+z-xyz}{1-(xy+yz+zx)}\right] = \pi$$

$x + y + z = xyz$

16. (c) $\sin^{-1}\left(\frac{\sqrt{r}-\sqrt{r-1}}{\sqrt{r(r+1)}}\right) = \tan^{-1}\left(\frac{\sqrt{r}-\sqrt{r-1}}{1+\sqrt{r}\sqrt{(r-1)}}\right)$

$= \tan^{-1}(\sqrt{r}) - \tan^{-1}(\sqrt{r-1})$

Now, $\sum_{r=1}^{n}\left(\tan^{-1}\sqrt{r} - \tan^{-1}\sqrt{r-1}\right) = \tan^{-1}\sqrt{n}$

17. (b) Since, $\tan^{-1}\left(\frac{2r}{2+r^2+r^4}\right)$

$$\tan^{-1}\left[\frac{(1+r+r^2)-(1-r+r^2)}{1+(1+r+r^2).(1-r+r^2)}\right]$$

$\tan^{-1}(1+r+r^2) - \tan^{-1}(1-r+r^2)$

$$S = \sum_{r=1}^{10}\left[\tan^{-1}(1+r+r^2) - \tan^{-1}(1-r+r^2)\right]$$

$S = \tan^{-1}(1+10+10^2) - \tan^{-1}(1)$

$$\tan^{-1}\left(\frac{10+10^2}{2+10+10^2}\right) = \tan^{-1}\left(\frac{55}{56}\right)$$

$\tan S = \frac{55}{56}$

$\because\ \cot S = \frac{56}{55}$

Then $55 \cot S = 56$.

18. (b) $\cot^{-1}\left(1+\sum_{k=1}^{n}2k\right) = \cot^{-1}[1+n(n+1)]$

$$= \tan^{-1}\left[\frac{(n+1)-n}{1+(n+1)n}\right] = \tan^{-1}(n+1) - \tan^{-1}n$$

$$\therefore \sum_{n=1}^{23}\left[\tan^{-1}(n+1) - \tan^{-1}n\right] = \tan^{-1}24 - \tan^{-1}1$$

$$= \tan^{-1}\frac{23}{25}$$

$$\therefore \cot\left[\sum_{n=1}^{23}\cot^{-1}\left(1+\sum_{k=1}^{n}2k\right)\right] = \cot\left[\tan^{-1}\frac{23}{25}\right] = \frac{25}{23}$$

One or More Than One Option Correct

1. (a, c, d)

Let $\tan^{-1}x = \theta$, where $\theta \in \left(-\frac{\pi}{2}, -\frac{\pi}{4}\right)$

$f(x) = \tan^{-1}(\tan 2\theta) = 2\theta + \pi + 2\tan^{-1}x$

2. (a, c)

$$y = \cos^{-1}\left(\frac{5}{13}.\cos x - \sqrt{1-\left(\frac{5}{13}\right)^2}.\sqrt{1-\cos^2 x}\right)$$

$$\Rightarrow y = \cos^{-1}\left(\frac{5}{13}\right) + \cos^{-1}(\cos x) = \cos^{-1}\frac{5}{13} + x$$

3. (c, d) Let

$$2^{\frac{\pi}{\sin^{-1}x}} = t \in \left(0, \frac{1}{4}\right] \cup [4, \infty)$$

$\Rightarrow (t-2a)(t-4) < 0$

4. (a, b)

$$t_n = \tan^{-1}\left(\frac{8}{n^2}\right) = \tan^{-1}\left(\frac{2}{1+\frac{n^2}{4}-1}\right)$$

$$\Rightarrow t_n = \tan^{-1}\left(\frac{n}{2}+1\right) - \tan^{-1}\left(\frac{n}{2}-1\right)$$

$$\Rightarrow \sum_{n=1}^{\infty} t_n = \frac{7\pi}{4}$$

5. (a, b, c)

$$\sin^{-1}\left(\frac{x^2}{1+\frac{x^2}{3}}\right) = \sin^{-1}\left(\frac{x^4}{1+\frac{x^4}{3}}\right)$$

$$\Rightarrow \frac{x^2}{3+x^2} = \frac{x^4}{3+x^4}$$

$\Rightarrow x = 0, 1, -1$

6. (a, c) Domain of $f(x)=\ell n\cos^{-1}x$

is $x\in[-1,1)$

$\therefore\ [\alpha]=-1\text{ or }0$

7. (a, b, c)

(a) $\sin\left(\tan^{-1}3+\tan^{-1}\frac{1}{3}\right)=\sin\frac{\pi}{2}=1$

(b) $\cos\left(\frac{\pi}{2}-\sin^{-1}\frac{3}{4}\right)=\cos\left(\cos^{-1}\frac{3}{4}\right)=\frac{3}{4}$

(c) $\sin\left(\frac{1}{4}\sin^{-1}\frac{\sqrt{63}}{8}\right)$

Let $\sin^{-1}\frac{\sqrt{63}}{8}=\theta$

So $\sin\theta=\frac{\sqrt{63}}{8}$ if $\cos\theta=\frac{1}{8}$

we have $\cos\frac{\theta}{2}=\sqrt{\frac{1+\cos\theta}{2}}=\frac{3}{4}$

$$\sin\frac{\theta}{4}=\sqrt{\frac{1-\cos\frac{\theta}{2}}{2}}=\frac{1}{2\sqrt{2}}$$

Now $\log_2\sin\left(\frac{1}{4}\sin^{-1}\frac{\sqrt{63}}{8}\right)=\log_2\frac{1}{2\sqrt{2}}=-\frac{3}{2}$

(d) $\cos^{-1}\frac{\sqrt{5}}{3}=\theta$

$\cos\theta=\frac{\sqrt{5}}{3}\quad\therefore\tan\frac{\theta}{2}=\frac{3-\sqrt{5}}{2}$ which is irrational

8. (d) The principal value of $\sin^{-1}\left(\sin\frac{2\pi}{3}\right)$

$$=\sin^{-1}\left(\sin\left(\pi-\frac{\pi}{3}\right)\right)$$

$$=\sin^{-1}\left(\sin\frac{\pi}{3}\right)=\frac{\pi}{3}$$

9. (a, b) Given that

$$S_n(x)=\sum_{k=1}^{n}\cot^{-1}\left(\frac{1+k(k+1)x^2}{x}\right)$$

$$=\sum_{k=1}^{n}\tan^{-1}\left(\frac{x}{1+kx(kx+x)}\right)$$

$$=\sum_{k=1}^{n}\tan^{-1}\left(\frac{(kx+x)-(kx)}{1+(kx+x)(kx)}\right)$$

$$\Rightarrow S_n(x)=\tan^{-1}(nx+x)-\tan^{-1}x$$

$$=\tan^{-1}\left(\frac{n\pi}{1+(n+1)x^2}\right)$$

(a) $S_{10}(x)=\tan^{-1}\frac{10x}{1+11x^2}$

$$=\frac{\pi}{2}-\tan^{-1}\left(\frac{1+11x^2}{10x}\right)(x>0)$$

(Option (a) is correct)

(b) $\lim\limits_{n\to\infty}\cot(S_n(x))$

$$=\lim_{n\to\infty}\cot^{-1}\left(\frac{1+(n+1)x^2}{nx}\right)$$

$$=\lim_{n\to\infty}\frac{\frac{1}{n}+\left(1+\frac{1}{n}\right)x^2}{x}=x(x>0)$$

(Option (b) is correct)

(c) $S_3(x)=\tan^{-1}\frac{3x}{1+4x^2}=\frac{\pi}{4}$

$\Rightarrow 4x^2-3x+1=0\Rightarrow x\notin\mathbb{R}\quad$ [$\because D$ is negative]

(Option (c) is incorrect)

(d) For $x=1$

$\tan(S_n(x))=\frac{n}{n+2}\ge\frac{1}{2}$

For $n\ge 3$

(Option (d) is incorrect)

10. (b, c, d)

$$f(n)=\frac{\sum_{k=0}^{n}2\sin\left(\frac{k+1}{n+2}\pi\right)\sin\left(\frac{k+2}{n+2}\pi\right)}{\sum_{k=0}^{n}2\sin^2\left(\frac{k+1}{n+2}\right)\pi}$$

where n is non negative integer

$$=\frac{\sum_{k=0}^{n}\left[\cos\left(\frac{\pi}{n+2}\right)-\cos\frac{(2k+3)\pi}{n+2}\right]}{\sum_{k=0}^{n}\left[1-\cos\frac{2(k+1)\pi}{n+2}\right]}$$

$$=\frac{(n+1)\cos\left(\frac{\pi}{n+2}\right)-\left[\cos\frac{3\pi}{n+2}+\cos\frac{5\pi}{n+2}+\ldots+\cos\frac{(2n+3)\pi}{n+2}\right]}{n+1-\left[\cos\frac{2\pi}{n+2}+\cos\frac{4\pi}{n+2}+\ldots+\cos\frac{2(n+1)\pi}{n+2}\right]}$$

$$=\frac{(n+1)\cos\left(\frac{\pi}{n+2}\right)-\frac{\frac{\sin(n+1)\pi}{n+2}}{\sin\left(\frac{\pi}{n+2}\right)}.\cos\frac{(2n+6)\pi}{2(n+2)}}{n+1-\frac{\frac{\sin(n+1)\pi}{n+2}}{\sin\left(\frac{\pi}{n+2}\right)}.\cos\frac{(2n+4)\pi}{2(n+2)}}$$

$$=\frac{(n+1)\cos\frac{\pi}{n+2}+\cos\frac{\pi}{n+2}}{n+1+1}-\frac{(n+2)\cos\left(\frac{\pi}{n+2}\right)}{n+2}$$

$\therefore f(n)=\cos\left(\frac{\pi}{n+2}\right)$

$\lim_{n\to\infty} f(n)=\lim_{n\to\infty}\cos\left(\frac{\pi}{n+2}\right)=1$

$\therefore$ Option (a) is incorrect.

$f(4)=\cos\left(\frac{\pi}{4+2}\right)=\cos\frac{\pi}{6}=\frac{\sqrt{3}}{2}$

$\therefore$ Option (b) is correct.

If $\alpha=\tan^{-1}\left(\cos^{-1} f(6)\right)$

$=\tan\left(\cos^{-1}\left(\cos\frac{\pi}{8}\right)\right)=\tan\frac{\pi}{8}$

Now $\tan\frac{\pi}{4}=1\Rightarrow\frac{2\tan\frac{\pi}{8}}{1-\tan^2\frac{\pi}{8}}=1$

$\Rightarrow\frac{2\alpha}{1-\alpha^2}=1\Rightarrow\alpha^2+2\alpha-1=0$

$\therefore$ Option (c) is correct.

$\sin\left(7\cos^{-1} f(5)\right)=\sin\left(7\cos^{-1}\left(\cos\frac{\pi}{7}\right)\right)$

$=\sin\left(7\times\frac{\pi}{7}\right)$

$=\sin\pi=0$

$\therefore$ Option (d) is correct.

11. (b, c, d)

$\alpha=3\sin^{-1}\frac{6}{11}>3\sin^{-1}\frac{1}{2}=\frac{\pi}{2}\Rightarrow\alpha>\frac{\pi}{2}$

$\therefore \cos\alpha<0$

$\beta=3\cos^{-1}\frac{4}{9}>3\cos^{-1}\frac{1}{2}=\pi\Rightarrow\beta>\pi$

$\therefore \cos\beta<0$ and $\sin\beta<0$

Now $\alpha+\beta>\frac{3\pi}{2}$ $\qquad\therefore \cos(\alpha+\beta)>0$

Matrix Type

1. A → Q; B → P; C → S; D → T

(A) $\sin^{-1}\left(3x-4x^3\right)=\pi-3\sin^{-1}x$ if $\frac{1}{2}<x<1$

$\therefore \lim_{x\to\frac{1}{2}^+} f\left(3x-4x^3\right)=\lim_{x\to\frac{1}{2}^+}\left(\pi-3\sin^{-1}x\right)$

$=\pi-3\lim_{x\to\frac{1}{2}^+}\sin^{-1}x$

$\therefore a=\pi$

$\therefore [a]=3$

(B) $f(x)=\tan^{-1}\left(\frac{3x-x^3}{1-3x^2}\right)=3\tan^{-1}x$, when $-\frac{1}{\sqrt{3}}<x<\frac{1}{\sqrt{3}}$

If $-\frac{1}{\sqrt{3}}<a<\frac{1}{\sqrt{3}}$, then

$\lim_{h\to0}\frac{f(a+3h)-f(a)}{3h}=\frac{3}{1+a^2}\Rightarrow f'(a)=\frac{3}{1+a^2}$

$=\lim_{h\to0}\frac{f\left(\frac{1}{2}+6h\right)-f\left(\frac{1}{2}\right)}{6h}=f'\left(\frac{1}{2}\right)=\frac{12}{5}$

$\therefore$ Required value = 2

(C) $\cos^{-1}\left(4x^3-3x\right)=\cos^{-1}(\cos3\theta)=3\theta-2\pi$ $\left\{\because \frac{2\pi}{3}<\theta<\pi\right.$

$=-2\pi+3\cos^{-1}x$

$\therefore\ [a+b+2]=[-2\pi+3+2]=-2$

(D) $f(x)=\cos^{-1}\left(4x^3-3x\right)$

$$=\cos^{-1}(\cos3\theta)=\begin{cases}3\theta, & 0<\theta<\frac{\pi}{3}\\ 2\pi-3\theta, & \frac{\pi}{3}<\theta<\frac{\pi}{2}\end{cases}$$

$$=\begin{cases}3\cos^{-1}x, & \frac{1}{2}<x<1\\ 2\pi-3\cos^{-1}x, & 0<x<\frac{1}{2}\end{cases}$$

$$\therefore\ f'(x)=\begin{cases}\frac{-3}{\sqrt{1-x^2}}, & \frac{1}{2}<x<1\\ \frac{3}{\sqrt{1-x^2}}, & 0<x<\frac{1}{2}\end{cases}$$

$a = \lim_{x \to \frac{1}{2}^+} f'(x) = -2\sqrt{3}$

$b = \lim_{x \to \frac{1}{2}^-} f'(x) = 2\sqrt{3}$

$\therefore \ a + b - 3 = -3$

2. A → P; B → Q; C → P; D → S

$\sin^{-1}(ax) + \cos^{-1} y + \cos^{-1}(bxy) = \frac{\pi}{2}$

$\Rightarrow \cos^{-1} y + \cos^{-1}(bxy) = \frac{\pi}{2} - \sin^{-1}(ax) = \cos^{-1}(ax)$

Let $\cos^{-1} y = \alpha, \cos^{-1}(bxy) = \beta, \cos^{-1}(ax) = \gamma$

Then $y = \cos\alpha, bxy = \cos\beta, ax = \cos\gamma$

$\therefore$ We get $\alpha + \beta = \gamma$ and $\cos\beta = bxy$

$\Rightarrow \cos(\gamma - \alpha) = \cos\beta = bxy$

$\Rightarrow \cos\gamma\cos\alpha + \sin\gamma\sin\alpha = bxy$

$\Rightarrow axy + \sin\gamma\sin\alpha = bxy$

$\Rightarrow (a-b)xy = -\sin\alpha\sin\gamma$

$\Rightarrow (a-b)^2 x^2y^2 = sin^2\alpha\sin^2\gamma$

$= (1 - \cos^2\alpha)(1 - \cos^2\gamma)$

$\Rightarrow (a-b)^2 x^2y^2 = (1-y^2)(1-a^2x^2)$ (i)

(A) For $a = 1$, $b = 0$, equation (i) reduces to

$x^2y^2 = (1-x^2)(1-y^2) \ \Rightarrow x^2 + y^2 = 1$

(B) For $a = 1$, $b = 1$, equation (i) becomes

$(1-x^2)(1-y^2) = 0 \ \Rightarrow (x^2-1)(y^2-1) = 0$

(C) For $a = 1$, $b = 2$, equation (i) reduces to

$x^2y^2 = (1-x^2)(1-y^2) \ \Rightarrow x^2 + y^2 = 1$

(D) For $a = 2$, $b = 2$ equation (i) reduces to

$0 = (1-4x^2)(1-y^2) \ \Rightarrow (4x^2-1)(y^2-1) = 0$

3. A → P; B → R; C → Q

(A) $t = \sum_{i=1}^{\infty} \tan^{-1}\left(\frac{1}{2i^2}\right)$

$= \sum_{i=1}^{\infty} \tan^{-1}\left[\frac{(2i+1)-(2i-1)}{1+4i^2-1}\right]$

$t = \left(\frac{1}{2i^2}\right) = \sum_{i=1}^{\infty}\left[\frac{-(2i-1)}{1+4i^2-1}\right]$

$= \sum_{i=1}^{\infty} \tan^{-1}\left[\tan^{-1}(2i+1) - \tan^{-1}(2i-1)\right]$

$= \tan^{-1} 3 - \tan^{-1} 1 + \tan^{-1} 5 - \tan^{-1} 3 + \ldots$

$+ \tan^{-1}(2n+1) - \tan^{-1}(2n-1) + \ldots.\infty$

$= \lim_{n\to\infty}\left[\tan^{-1}(2n+1) - \tan^{-1} 1\right]$

$= \lim_{n\to\infty} \tan^{-1}\left[\frac{2n}{1+(2n+1)}\right] = \lim_{n\to\infty} \tan^{-1}\left[\frac{1}{1+\frac{1}{n}}\right]$

$= \tan^{-1}(1) = \frac{\pi}{2} \Rightarrow \tan t = 1.$ (A) → (p)

(B) $\because$ a, b, c are in AP $\Rightarrow 2b = a + c$

Now $\cos\theta_1 = \frac{a}{b+c} \Rightarrow \frac{1-\tan^2\frac{\theta_1}{2}}{1+\tan^2\frac{\theta_1}{2}} = \frac{2}{b+c}$

$\Rightarrow \tan^2\frac{\theta_1}{2} = \frac{b+c-a}{b+c+a}$

Similarly, $\tan^2\frac{\theta_3}{2} = \frac{a+b-c}{a+b+c}$

$\Rightarrow \tan^2\frac{\theta_1}{2} + \tan^2\frac{\theta_3}{2} = \frac{2b}{a+b+c}$

$= \frac{2b}{3b} = \frac{2}{3}$, (B) → (r)

(C) Equation of line through (0, 1, 0) and perpendicular to $x + 2y + 2z = 0$ is $\frac{x}{1} = \frac{y-1}{2} = \frac{z}{2} = \lambda$

For some value of λ, the foot of perpendicular from origin to line is $(\lambda, 2\lambda+1, 2\lambda)$
Direction ratios of this $\perp$ from origin are $\lambda, 2\lambda+1, 2\lambda$

$\therefore 1.\lambda + 2(2\lambda+1) + 2.2\lambda = 0 \Rightarrow \lambda = -\frac{2}{9}$

$\therefore$ Foot of perpendicular is $\left(-\frac{2}{9}, \frac{5}{9}, -\frac{4}{9}\right)$

Hence required distance

$= \sqrt{\frac{4}{81} + \frac{25}{81} + \frac{16}{81}} = \sqrt{\frac{45}{81}} = \frac{\sqrt{5}}{3}$(C) → (q)

4. A → R; B → S; C → P; D → Q

(A) The difference = 2 – (–2) = 4

(B) Let $f(x) = x^2 - 4x + 3$

$f'(x) = 2x - 4 = 0 \ \Rightarrow x = 2$

$f(1) = 0, f(2) = -1, f(3) = 0$

$\therefore$ |greatest value – least value| = 1

(C) $\tan^{-1}\dfrac{1-x}{1+x}=\tan^{-1}1-\tan^{-1}x$

$\therefore$ greatest value $=\dfrac{\pi}{4}$

(D) $\therefore$ greatest value $=\dfrac{\pi}{2}$, least value $=\dfrac{\pi}{3}$

$\therefore$ difference $=\dfrac{\pi}{6}$

5. A → Q, S; B → R, S, T; C → R, S; D → P, Q

(A) Given $\sin^{-1}x-\cos^{-1}x=\dfrac{\pi}{6}$

Also, $\sin^{-1}x+\cos^{-1}x=\dfrac{\pi}{2}$

Solving $x=\dfrac{\sqrt{3}}{2}$

(B) $\sec^2\left(\tan^{-1}2\right)+\operatorname{cosec}^2\left(\cot^{-1}3\right)$

$=1+\left(\tan\left(\tan^{-1}2\right)\right)^2+1+\left(\cot\left(\cot^{-1}3\right)\right)^2$

$=15$

(C) Given equation is

$\dfrac{\pi}{2}-2\cos^{-1}x=\sin^{-1}(3x-2)$

Or $3x-2=\cos\left(2\cos^{-1}x\right)=2\cos^2\left(\cos^{-1}x\right)-1$

$=2x^2-1$

$\Rightarrow 2x^2-3x+1=0\Rightarrow x=1$ or $\dfrac{1}{2}$

(D) $\sin 5=\sin(5-2\pi)$

$\Rightarrow\ \sin^{-1}(\sin 5)=\sin^{-1}(\sin(5-2\pi))$

$=5-2\pi$

6. A → S; B → R; C → Q; D → P

(A)

$$=\left[\frac{1}{y^2}\left(\frac{\cos\left(\tan^{-1}y\right)+y\sin\left(\tan^{-1}y\right)}{\cot\left(\sin^{-1}y\right)+\tan\left(\sin^{-1}y\right)}\right)^2+y^4\right]^{\frac{1}{2}}$$

$$=\left[\frac{1}{y^2}\left(\frac{\cos\left(\cos^{-1}\dfrac{1}{\sqrt{1+y^2}}\right)+y\sin\left(\sin^{-1}\dfrac{y}{\sqrt{1+y^2}}\right)}{\cot\left(\cot^{-1}\dfrac{\sqrt{1-y^2}}{y}\right)+\tan\left(\tan^{-1}\dfrac{y}{\sqrt{1-y^2}}\right)}\right)^2+y^4\right]^{\frac{1}{2}}$$

$$=\left[\frac{1}{y^2}\left(\frac{\dfrac{\sqrt{1+y^2}}{1}}{\dfrac{1}{y\left(\sqrt{1-y^2}\right)}}\right)^2+y^4\right]^{\frac{1}{2}}$$

$=\left(1-y^4+y^4\right)^{\frac{1}{2}}=1$ $\qquad\therefore$ (A → S)

(B) $\cos x+\cos y=-\cos z$ (i)

And $\sin x+\sin y=-\sin z$ (ii)

On squaring (i) and (ii) then adding, we get

$(\cos x+\cos y)^2+(\sin x+\sin y)^2=\cos^2 z+\sin^2 z$

$\Rightarrow 2+2\cos(x-y)=1$

$\Rightarrow 4\cos^2\dfrac{x-y}{2}=1\Rightarrow\cos\dfrac{x-y}{2}=\pm\dfrac{1}{2}$

$\therefore$ (B → R)

(C) $\cos\left(\dfrac{\pi}{4}-x\right)\cos 2x+\sin x\sin 2x\sec x$

$=\cos x\sin 2x\sec x+\cos\left(\dfrac{\pi}{4}+x\right)\cos 2x$

$\Rightarrow\cos 2x\left[\cos\left(\dfrac{\pi}{4}-x\right)-\cos\left(\dfrac{\pi}{4}+x\right)\right]$

$=\sin 2x\sec x(\cos x-\sin x)$

$\Rightarrow=2\sin x\dfrac{\pi}{4}\sin x\cos 2x=2\sin x(\cos x-\sin x)$

$\Rightarrow 2\sin x\left[\dfrac{1}{\sqrt{2}}\left(\cos^2 x-\sin^2 x\right)-(\cos x-\sin x)\right]=0$

$\Rightarrow 2\sin x(\cos x-\sin x)\left(\dfrac{\cos x-\sin x}{\sqrt{2}}-1\right)=0$

$\Rightarrow\sin x=0$ or $\tan x=1$ or $\cos\left(x-\dfrac{\pi}{4}\right)=1$

$\Rightarrow x=0$ or $\dfrac{x}{4}\Rightarrow\sec x=1$ or $\sqrt{2}$

$\therefore$ (C → Q)

(D) $\cos\left(\sin^{-1}\sqrt{1-x^2}\right)=\sin\left(\tan^{-1}x\sqrt{6}\right)$

$\Rightarrow\dfrac{x}{\sqrt{1-x^2}}=\dfrac{x\sqrt{6}}{\sqrt{1+6x^2}}\Rightarrow x=\dfrac{1}{2}\sqrt{\dfrac{5}{3}}$

$\therefore$ (D → P)

Hence, A → S; B → R; C → Q; D → P.

Integer and Numerical Value Type

1. $\tan^{-1}\left[\frac{3\sin 2\alpha}{5+3\cos 2\alpha}\right]+\tan^{-1}\left[\frac{\tan\alpha}{4}\right]$

$= \tan^{-1}\left(\frac{6\tan\alpha}{8+2\tan^2\alpha}\right)+\tan^{-1}\left(\frac{\tan\alpha}{4}\right)$

$= \tan^{-1}\left(\frac{\frac{3\tan\alpha}{4+\tan^2\alpha}+\frac{\tan\alpha}{4}}{1-\frac{3\tan^2\alpha}{16+4\tan^2\alpha}}\right)\left\{\because \frac{3\tan^2\alpha}{16+4\tan^2\alpha}<1\right\}$

$= \tan^{-1}(\tan\alpha)=\alpha$

2. $(\tan^{-1}x)^2+(\cot^{-1}x)^2$

$=(\tan^{-1}x+\cot^{-1}x)^2-2\tan^{-1}x\left(\frac{\pi}{2}-\tan^{-1}x\right)$

$=\frac{\pi^2}{4}-\pi\tan^{-1}x+2(\tan^{-1}x)^2=\frac{5\pi^2}{8}$

$\therefore\ \tan^{-1}x=\frac{2\pi}{3},-\frac{\pi}{4}$

$\tan^{-1}x=-\frac{\pi}{4}$ $\quad\left\{\tan^{-1}x\neq\frac{2\pi}{3}\right\}$

$\therefore x=-1$ is the solution

3. We have, $\frac{\pi}{2}\le x\le\frac{3\pi}{2}$

$\Rightarrow -\frac{\pi}{2}\le x-\pi\le\frac{\pi}{2}\Rightarrow -\frac{\pi}{2}\le\pi-x\le\frac{\pi}{2}$

$\Rightarrow \sin^{-1}\{\sin(\pi-x)\}=\pi-x$

4. Since,

$\sin^{-1}\frac{\sqrt{2-\sqrt{3}}}{2}=\sin^{-1}\frac{\sqrt{4-2\sqrt{3}}}{2\sqrt{2}}\sin^{-1}\left(\frac{\sqrt{3}-}{2\sqrt{2}}\right)=\frac{\pi}{12}$

$\cos^{-1}\frac{\sqrt{12}}{4}=\cos^{-1}\frac{\sqrt{3}}{2}=\frac{\pi}{6},\ \cos^{-1}\sqrt{2}=\frac{\pi}{4}$

$\text{Exp.}=\sin^{-1}\left[\cot\left(\frac{\pi}{12}+\frac{\pi}{6}+\frac{\pi}{4}\right)\right]$

$=\sin^{-1}\left(\cos\frac{\pi}{2}\right)=\sin^{-1}(0)=0$

5. $\tan^{-1}\left(\frac{xy}{zr}\right)+\tan^{-1}\left(\frac{yz}{xr}\right)+\tan^{-1}\left(\frac{zx}{yr}\right)$

$=\tan^{-1}\left[\frac{\frac{xy}{zr}+\frac{yzxr}{xr}+\frac{xz}{yr}-\frac{xyz}{r^3}}{1-\left(\frac{x^2+y^2+z^2}{r^2}\right)}\right]=\tan^{-1}\infty=\frac{\pi}{2}$

Alternatively, $x=y=z=\frac{1}{\sqrt{3}}$ and $r=1$

then, $3\tan^{-1}\left(\frac{1}{\sqrt{3}}\right)=\frac{\pi}{2}$

6. $\sec^{-1}\left(\frac{1}{4}\sum_{k=0}^{10}\sec\left(\frac{7\pi}{12}+\frac{k\pi}{2}\right)\sec\left(\frac{7\pi}{12}+\frac{(k+1)\pi}{2}\right)\right)$

$=\sec^{-1}\left[\frac{1}{2}\sum_{k=0}^{10}\frac{1}{\left(\frac{7\pi}{12}+\frac{k\pi}{2}\right)\cos\left(\frac{7\pi}{12}+\frac{k\pi}{2}+\frac{\pi}{2}\right)}\right]$

$=\sec^{-1}\left[\frac{1}{2}\sum_{k=0}^{10}\frac{-1}{\sin\left(\frac{7\pi}{6}+k\pi\right)}\right]$

$=\sec^{-1}\left[\frac{1}{2}\sum_{k=0}^{10}\frac{-1}{\sin\left((k+1)\pi+\frac{\pi}{6}\right)}\right]$

If k is an even integer, then

$\sin\left((k+1)\pi+\frac{\pi}{6}=-\sin\frac{\pi}{6}=\frac{-1}{2}\right)$

If k is an odd integer, then $\sin\left((k+1)\pi+\frac{\pi}{6}\right)$

$\sin\frac{\pi}{6}=\frac{1}{2}$

$\therefore\ \sum_{k=0}^{9}\sin\left((k+1)\pi+\frac{\pi}{6}\right)=0$

Hence $\sec^{-1}\left[\frac{1}{2}\sum_{k=0}^{10}\frac{-1}{\sin(k+1)\pi+\frac{\pi}{6}}\right]$

$\sec^{-1}\left[\frac{1}{2}\left(\frac{-1}{-\frac{1}{2}}\right)\right]=\sec^{-1}(1)=0$

7. $\sin^{-1}\left(\sum_{i=1}^{\infty}x^{i+1}-\sum_{i=1}^{\infty}\left(\frac{x}{2}\right)^i\right)$

$=\frac{\pi}{2}\ \cos^{-1}\left(\sum_{i=1}^{\infty}\left(-\frac{x}{2}\right)^i-\sum_{i=1}^{\infty}(-x)^2\right)$

$\sin^{-1}\left(\frac{x^2}{1-x}-x.\frac{\frac{x}{2}}{1-\frac{x}{2}}\right)=\sin^{-1}\left(\frac{-\frac{x}{2}}{1+\frac{x}{2}}-\frac{-x}{1+x}\right)$

[$\because$ sum of infinite terms of a GP $=\dfrac{a}{1-r}$, if $|r|-1$]

$$\Rightarrow \frac{x^2}{1-x}-\frac{x^2}{2-x}=\frac{-x}{2+x}+\frac{x}{1+x}$$

$$\Rightarrow \frac{x^2}{1-x}-\frac{x}{1+x}+\frac{x}{2+x}-\frac{x^2}{2-x}=0$$

$$\Rightarrow \frac{x\left(x^2+2x-1\right)}{1-x^2}+\frac{x\left(2-3x-x^2\right)}{4-x^2}=0$$

$$\Rightarrow x\left[x^3+2x^2+5x-2\right]=0$$

$$\Rightarrow x=0 \text{ or } x^3+2x^2+5x-2=0=p(x) \quad \text{(say)}$$

We observe that $p(0)<0$ and $p\left(\frac{1}{2}\right)>0$

$\therefore$ One root of $p(x)=0$ lies in $\left(0,\frac{1}{2}\right)$

Thus, two solutions lie between $-\frac{1}{2}$ and $\frac{1}{2}$.

Comprehension Passage

Comprehension 1

1. (b) $A=\left(\tan^{-1}x\right)^3+\left(\cot^{-1}x\right)^3$

$$A=\left(\tan^{-1}x+\cot^{-1}x\right)^3$$
$$-3\tan^{-1}x\cot^{-1}x\left(\tan^{-1}x+\cot^{-1}x\right)$$

$$\Rightarrow A=\left(\frac{\pi}{2}\right)^3-3\tan^{-1}x\cot^{-1}x\frac{\pi}{2}$$

$$\Rightarrow A=\frac{\pi^3}{8}-\frac{3\pi}{2}\tan^{-1}x\left(\frac{\pi}{2}-\tan^{-1}x\right)$$

$$\Rightarrow A=\frac{\pi^3}{32}+\frac{3\pi}{2}\left(\tan^{-1}x-\frac{\pi}{4}\right)^2 \text{ ;as } x>0;$$

$$\frac{\pi^3}{32}\le A<\frac{\pi^3}{8}$$

2. (c) $B=\left(\sin^{-1}t\right)^2+\left(\cos^{-1}t\right)^2$

$$B=\left(\sin^{-1}t+\cos^{-1}t\right)^2-2\sin^{-1}t\cos^{-1}t$$

$$B=\frac{\pi^2}{4}-2\sin^{-1}t\left(\frac{\pi}{2}-\sin^{-1}t\right)$$

$$B=\frac{\pi^2}{8}+2\left(\sin^{-1}t-\frac{\pi}{4}\right)^2$$

$$B^{\max}=\frac{\pi^2}{8}+2.\frac{\pi^2}{16}=\frac{\pi^2}{4}$$

3. (a) $\lambda=\dfrac{\pi^3}{32}$ $\qquad \mu=\dfrac{\pi^2}{4}$

$$\frac{\lambda}{\mu}=\frac{\pi}{8}$$

$$\frac{\lambda-\mu\pi}{\mu}=\frac{\pi}{8}-\pi=\frac{-7\pi}{8}$$

$$\cot^{-1}\cot\left(\frac{\lambda-\mu\pi}{\mu}\right)=\cot^{-1}\cot\left(-\frac{7\pi}{8}\right)=\frac{\pi}{8}$$

Comprehension 2

1. (b) $\sin^{-1}\left(\dfrac{4x}{x^2+4}\right)+2\tan^{-1}\left(-\dfrac{x}{2}\right)$

$$=\sin^{-1}\left(\frac{x.\frac{x}{2}}{\left(\frac{x}{2}\right)^2+1}\right)-2\tan^{-1}\frac{x}{2}$$

$$=2\tan^{-1}\frac{x}{2}-2\tan^{-1}\frac{x}{2}=0$$

Here, $\left|\frac{x}{2}\right|\le 1$

$|x|\le 2$

$\Rightarrow -2\le x\le 2$

2. (a) $\cos^{-1}\dfrac{6x}{1+9x^2}=-\dfrac{\pi}{2}+2\tan^{-1}3x$

$$\Rightarrow \frac{\pi}{2}-\sin^{-1}\frac{6x}{1+9x^2}=-\frac{\pi}{2}+2$$

$$\Rightarrow \sin^{-1}\frac{6x}{1+9x^2}=\pi-2\tan^{-1}3x$$

$$\Rightarrow \sin^{-1}\frac{2.3x}{1+(3x)^2}=\pi-2\tan^{-1}3x$$

Above is true when $3x>1$

$$\Rightarrow x>\frac{1}{3};\ x\in\left(\frac{1}{3},\infty\right)$$

3. (c) $(x-1)\left(x^2+1\right)>0$

$\Rightarrow x>1$

$$\therefore \sin\left[\frac{1}{2}\tan^{-1}\left(\frac{2x}{1-x^2}\right)-\tan^{-1}x\right]$$

$$=\sin\left[\frac{1}{2}\left(-\pi+2\tan^{-1}x\right)-\tan^{-1}x\right]$$

$$=\sin\left(-\frac{\pi}{2}\right)=-1$$

CHAPTER

3 Matrices

Objectives

In this chapter on Matrices, we shall study the following topics:

1. Matrices and their types
2. Transpose of a matrix
3. Symmetric and skew matrix
4. Hermitian and skew hermitian matrix
5. Nilpotent matrix
6. Matrix method of solving a non-homogeneous system of linear equations
7. Properties of symmetric and skew-symmetric matrices
8. Solution of homogeneous system of linear equation
9. Operations on matrices
10. Adjoint of a square matrix.

Matrices and Their Types

Matrices are one of the most powerful tools in mathematics. The evolution of the concept of matrices is the result of an attempt to obtain compact and simple methods of solving systems of linear equations. Matrix notation and operations are used in electronic spreadsheet programs for personal computer, which in turn is used in different areas of business and science like budgeting, sales projection, cost estimation, analyzing the results of an experiment, etc. Also, many physical operations such as magnification, rotation and reflection through a plane can be represented mathematically by matrices. Matrices are also used in cryptography, genetics, economics, sociology, modern psychology and industrial management.

A collection of mn numbers written by means of m rows and n columns is called an $m \times n$ matrix. The matrix is written as

$$A = \begin{bmatrix} a_{11} & a_{12} & a_{13} & \cdots & a_{1n} \\ a_{21} & a_{22} & a_{23} & \cdots & a_{2n} \\ a_{31} & a_{32} & a_{33} & \cdots & a_{3n} \end{bmatrix} = (a_{ij})_{m\times n}$$

where a_{ij} denotes the element of i-th row and j-th column or (i, j) th element. The order of the matrix A is $m \times n$. If $n = 1$, then $m \times 1$ matrix it is called a **column matrix**.

Thus $\begin{pmatrix} 5 \\ 4 \\ 3 \end{pmatrix}$ is a 3×1 column matrix. Again, if $m = 1$, then $1 \times n$ matrix is called a **row matrix**.

Horizontal Matrix

An $m \times n$ matrix is called a horizontal matrix if $m < n$.

e.g., $$A = \begin{bmatrix} 1 & 0 & 2 \\ 3 & 4 & 3 \end{bmatrix}$$

Vertical Matrix

An $m \times n$ matrix is called a vertical matrix if $m > n$.

e.g., $$A = \begin{bmatrix} 1 & 0 \\ 2 & 4 \\ 3 & 5 \end{bmatrix}$$

Null Matrix or Zero Matrix

An $(m \times n)$ matrix in which all the elements are zero is called a zero matrix or null matrix of the type $m \times n$ and is denoted by $O_{m\times n}$ or O.

Thus, the matrices $\begin{bmatrix} 0 & 0 \\ 0 & 0 \\ 0 & 0 \end{bmatrix}, \begin{bmatrix} 0 & 0 & 0 \\ 0 & 0 & 0 \\ 0 & 0 & 0 \end{bmatrix}, \begin{bmatrix} 0 & 0 & 0 & 0 \\ 0 & 0 & 0 & 0 \end{bmatrix}$

are zero or null matrices of type 3×2, 3×3 and 2×4 respectively.

Unit Matrix or Identity Matrix

If in a scalar matrix the constant number on the diagonal is 1 then it is called unit matrix or identity matrix. For example, $\begin{bmatrix} 1 & 0 \\ 0 & 1 \end{bmatrix}$ and $\begin{bmatrix} 1 & 0 & 0 \\ 0 & 1 & 0 \\ 0 & 0 & 1 \end{bmatrix}$ are 2×2

and 3 × 3 identity matrices, which are denoted by I_2 and I_3 respectively. An identity matrix of $n \times n$ can also be defined as

$$a_{ij} = \begin{cases} 1 & \text{if } i = j \\ 0 & \text{if } i \neq j \end{cases} \quad 1 \le i, j \le n$$

Square Matrix

A matrix in which number of rows and number of columns are equal is called a square matrix.

Various Types of Square Matrix

Diagonal Matrix

A square matrix in which all the non-diagonal element are zero is called a diagonal matrix. For example, $\begin{bmatrix} 3 & 0 & 0 \\ 0 & 5 & 0 \\ 0 & 0 & -7 \end{bmatrix}$ is a 3×3 diagonal matrix.

Scalar Matrix

If, in a diagonal matrix, all the diagonal elements are equal to a constant then it is called scalar matrix. For example,

$$\begin{bmatrix} 3 & 0 & 0 \\ 0 & 3 & 0 \\ 0 & 0 & 3 \end{bmatrix} \text{ is a scalar matrix.}$$

Trace of a Matrix

The trace of a square matrix is defined as sum of the diagonal elements. If A is a square matrix of order $n \times n$, then

$$\text{Trace A} = \sum_{i=1}^{n} a_{ii}$$

We denote Trace A by trA.

Determinant of a Square Matrix

If A is a square matrix, then its determinant is denoted by |A| or det A. If det A ≠ 0, the matrix is said to be non-singular; otherwise, it is called singular.

Triangular Matrix

$$\text{Let A be a square matrix A} = \begin{bmatrix} a_{11} & a_{12} & a_{13} & \cdots & a_{1n} \\ a_{21} & a_{22} & a_{23} & \cdots & a_{2n} \\ a_{31} & a_{32} & a_{33} & \cdots & a_{3n} \\ \cdots & \cdots & \cdots & \cdots & \cdots \\ a_{n1} & a_{n2} & a_{n3} & \cdots & a_{nn} \end{bmatrix}$$

Any element of A is a_{ij} and for $i = j$ we get the diagonal elements. All those elements which are above the principal diagonal are of the type $a_{12}, a_{13}, a_{23}, a_{34}, a_{3n} \ldots$ or in general a_{ij} where $i < j$ and all those elements which are below the principal diagonal are of the type $a_{21}, a_{31}, a_{n3}, \ldots$ so on or in general a_{ij} where $i > j$.

Upper Triangular Matrix

A square matrix (a_{ij}) is called an upper triangular matrix if $a_{ij} = 0$ when $i > j$. In other words, all elements below the principal diagonal are zero.

Lower Triangular Matrix

A square matrix (a_{ij}) is called lower triangular matrix if $a_{ij} = 0$ when $i < j$.

In other words, all the elements above the principal diagonal should be zero.

$$\text{for} \quad \text{e.g.,} \begin{bmatrix} 3 & -2 & 4 \\ 0 & 2 & -3 \\ 0 & 0 & 7 \end{bmatrix} \text{ and } \begin{bmatrix} 2 & 0 & 0 \\ 3 & 4 & 0 \\ 2 & 8 & 6 \end{bmatrix}$$

are upper and lower triangular matrices respectively.

Equality of Matrices

Two matrices $A = [a_{ij}]_{m \times n}$, $B = [b_{ij}]_{m \times n}$ are said to be equal and written as A = B if and only if

(a) they are of the same order and

(b) each element of A is equal to the corresponding element of B.

$$\text{Let } A = \begin{bmatrix} 1 & 2 & 0 \\ 3 & 2 & 4 \end{bmatrix}, B = \begin{bmatrix} 1 & 2 & 0 \\ 3 & 2 & 4 \end{bmatrix}$$

Here A and B are equal matrices and we write A = B

ILLUSTRATION 3.1

Find the value of x, y, z and w which satisfy the matrix equation $\begin{bmatrix} x+3 & 2y+x \\ z-1 & 4w-8 \end{bmatrix} = \begin{bmatrix} -x-1 & 0 \\ 3 & 2w \end{bmatrix}$

Solution

As the two matrices are equal and are of order 2×2,

$$x + 3 = -x - 1 \quad \ldots(1)$$

$$2y + x = 0 \quad \ldots(2)$$

$$z - 1 = 3 \quad \ldots(3)$$

$$4w - 8 = 2w \quad \ldots(4)$$

on solving the above equations, we get

$$x = -2, y = 1, z = 4, w = 4$$

Transpose of a Matrix

If A is an $m \times n$ matrix, then transpose of A written as A′ or A^T is defined as an $n \times m$ matrix whose (i, j) th element is replaced by (j, i)th element of A, i.e., A′ is obtained by interchanging rows and columns of A. It can be easily verified that

(a) $(A + B)' = A' + B'$ provided A + B is defined.
(b) $(kA)' = kA'$ k is a real or complex number.
(c) $(A')' = A$
(d) $(AB)' = B'A'$ provided AB is defined.

Remark: For square matrices of the same order, the conditions given on order are not required.

ILLUSTRATION 3.2

If $A = \begin{bmatrix} 1 & 2 \\ 3 & 0 \end{bmatrix}$ and $B = \begin{bmatrix} 3 & 4 \\ 1 & 6 \end{bmatrix}$ then $(AB)^T$ equals

(a) $\begin{bmatrix} 5 & 16 \\ 9 & 16 \end{bmatrix}$ **(b)** $\begin{bmatrix} 5 & 9 \\ 16 & 12 \end{bmatrix}$

(c) $\begin{bmatrix} 5 & 9 \\ 4 & 3 \end{bmatrix}$ **(d)** None of these.

Solution

$$AB = \begin{bmatrix} 3+2 & 4+12 \\ 9+0 & 12+0 \end{bmatrix} = \begin{bmatrix} 5 & 16 \\ 9 & 12 \end{bmatrix}$$

$$\therefore \quad (AB)^T = \begin{bmatrix} 5 & 9 \\ 16 & 12 \end{bmatrix}$$

ILLUSTRATION 3.3

If $A = \begin{bmatrix} 2 & -1 \\ -7 & 4 \end{bmatrix}$ and $B = \begin{bmatrix} 4 & 1 \\ 7 & 2 \end{bmatrix}$ then $B^T A^T$ is equal to

(a) $\begin{bmatrix} 1 & 0 \\ 0 & 1 \end{bmatrix}$ **(b)** $\begin{bmatrix} 1 & 1 \\ 1 & 1 \end{bmatrix}$ **(c)** $\begin{bmatrix} 0 & 1 \\ 1 & 0 \end{bmatrix}$ **(d)** $\begin{bmatrix} 1 & 0 \\ 0 & 0 \end{bmatrix}$

Solution

$$B^T A^T = \begin{bmatrix} 4 & 7 \\ 1 & 2 \end{bmatrix} \begin{bmatrix} 2 & -7 \\ -1 & 4 \end{bmatrix}$$

$$= \begin{bmatrix} 8-7 & -28+28 \\ 2-2 & -7+8 \end{bmatrix} - \begin{bmatrix} 1 & 0 \\ 0 & 1 \end{bmatrix}$$

Symmetric and Skew Matrices

A square matrix A is said to be symmetric if A = A′. A square matrix is said to be skew-symmetric if A = –A′.

ILLUSTRATION 3.4

Express the matrix A as the sum of a symmetric and a skew-symmetric matrix where $A = \begin{bmatrix} 4 & 2 & -3 \\ 1 & 3 & -6 \\ -5 & 0 & -7 \end{bmatrix}$

Solution

$$A' = \begin{bmatrix} 4 & 1 & -5 \\ 2 & 3 & 0 \\ -3 & -6 & -7 \end{bmatrix}$$

Then $$A + A' = \begin{bmatrix} 8 & 3 & -8 \\ 3 & 6 & -6 \\ -8 & -6 & -14 \end{bmatrix}$$

and $$A - A' = \begin{bmatrix} 0 & 1 & 2 \\ -1 & 0 & -6 \\ -2 & 6 & 0 \end{bmatrix}$$

$$\therefore \quad A = \frac{1}{2}(A + A') + \frac{1}{2}(A - A')$$

$$= \begin{bmatrix} 4 & 1.5 & -4 \\ 1.5 & 3 & -3 \\ -4 & -3 & -7 \end{bmatrix} + \begin{bmatrix} 0 & 0.5 & 1 \\ -0.5 & 0 & -3 \\ -1 & 3 & 0 \end{bmatrix} = \begin{bmatrix} 4 & 2 & -3 \\ 1 & 3 & -6 \\ -5 & 0 & -7 \end{bmatrix}$$

Hermitian and Skew-Hermitian Matrix

A square matrix $A = a$ is said to be a Hermitian matrix $A^\theta = A$ and is said to be skew Hermitian if $A^\theta = -A$.

Remark

(a) The diagonal elements of Hermitian Matrix must be all real

(b) The diagonal elements of a skew-Hermitian matrix should be zero

Orthogonal Matrix

A square matrix A is said to be orthogonal if $AA' = A'A = I$

Idempotent Matrix

A square matrix A is called idempotent if $A^2 = A$

Involuntary Matrix

A matrix such that $A^2 = I$ is called involuntary matrix.

Nilpotent Matrix

A square matrix A is called a nilpotent matrix if there exists a positive integer k such that $A^k = 0$. If k is the least positive integer such that $A^k = 0$, then k is called the index of the nilpotence of matrix A.

ILLUSTRATION 3.5

$$A = \begin{bmatrix} \cos\alpha & \sin\alpha \\ \sin\alpha & \cos\alpha \end{bmatrix}$$

Solution

A is orthogonal because $A^{-1} = \begin{bmatrix} \cos\alpha & \sin\alpha \\ -\sin\alpha & \cos\alpha \end{bmatrix} = A^T$

In fact, every unit matrix is orthogonal. Determinant of orthogonal matrix is −1 or 1.

ILLUSTRATION 3.6

$\begin{bmatrix} \frac{1}{2} & \frac{1}{2} \\ \frac{1}{2} & \frac{1}{2} \end{bmatrix}$ is an idempotent matrix, because

$$A^2 = \begin{bmatrix} \frac{1}{4}+\frac{1}{4} & \frac{1}{4}+\frac{1}{4} \\ \frac{1}{4}+\frac{1}{4} & \frac{1}{4}+\frac{1}{4} \end{bmatrix} = \begin{bmatrix} \frac{1}{2} & \frac{1}{2} \\ \frac{1}{2} & \frac{1}{2} \end{bmatrix} = A$$

Also, $A = \begin{bmatrix} 1 & 0 \\ 0 & 0 \end{bmatrix}$ and $\begin{bmatrix} 0 & 0 \\ 0 & 1 \end{bmatrix}$ are idempotent matrices because $A^2 = A$ and $B^2 = B$.

In fact, every unit matrix is idempotent.

ILLUSTRATION 3.7

$A = \begin{bmatrix} 1 & 0 \\ 0 & 1 \end{bmatrix}$ is an involuntary matrix because

$$A^2 = \begin{bmatrix} 1 & 0 \\ 0 & 1 \end{bmatrix} = I$$

In fact, every unit matrix is involuntary.

ILLUSTRATION 3.8

$A = \begin{bmatrix} 0 & 0 \\ 1 & 0 \end{bmatrix}$ is a nilpotent matrix because

$A^2 = \begin{bmatrix} 0 & 0 \\ 0 & 0 \end{bmatrix}$, (Here P = 2)

Determinant of every nilpotent matrix is 0.

Matrix Method of Solving a Non-homogeneous System of Linear Equations

Consider the system of linear equations:

$$a_1x + b_1y + c_1z = d_1$$
$$a_2x + b_2y + c_2z = d_2$$
$$a_3x + b_3y + c_3z = d_3 \quad \ldots\text{(i)}$$

If $A = \begin{bmatrix} a_1 & b_1 & c_1 \\ a_2 & b_2 & c_2 \\ a_3 & c_3 & c_3 \end{bmatrix}, X = \begin{bmatrix} x \\ y \\ z \end{bmatrix}$ and $B = \begin{bmatrix} d_1 \\ d_2 \\ d_3 \end{bmatrix}$

The above system can be written as AX = B

(i) If A is a non-singular matrix, then the system of equations given by AX = B has a unique solutions given by $X = A^{-1}$.

(ii) If A is a singular matrix, and (Adj A) D = 0, then the system of equations AX = D is consistent and has infinitely many solutions.

(iii) If A is a singular matrix, and (Adj A) D ≠ 0, then the system of equation AX = D is inconsistent or has no solution.

Properties of Symmetric and Skew-symmetric Matrices

(i) If A is a square matrix, then $A + A^T$, AA^T, $A^T A$ are symmetric matrices while $A - A^T$ is a skew-symmetric matrix.

(ii) If A is a symmetric Matrix, then −A, KA, A^T, A^n, A^{-1}, $B^T AB$ are also symmetric matrices where $n \in N$, $K \in R$ and B is a square matrix of order that of A.

(iii) If A is a skew-symmetric matrix, then

(a) A^{2n} is a symmetric matrix for $n \in N$.

(b) A^{2n+1} is a skew-symmetric matrices for $n \in N$.

(c) kA is also skew-symmetric matrix where $k \in R$.

(d) $B^T AB$ is also skew-symmetric matrix where B is a square matrix of order that of A.

(iv) If A, B are two symmetric matrices, then

(a) $A \pm B$, $AB + BA$ are also symmetric matrices.

(b) $AB - BA$ is a skew-symmetric matrix.

(c) AB is a symmetric matrix when $AB = BA$.

(v) If A, B are two skew-symmetric matrices, then

(a) $A \pm B$, $AB - BA$ are skew-symmetric matrices.

(b) $AB + BA$ is a symmetric matrix.

(vi) If A is a skew-symmetric matrix and C is a column matrix, then $C^T AC$ is a zero matrix.

(vii) Every square matrix A can uniquely be expressed as sum of a symmetric and skew-symmetric matrix, i.e., $A = \left[\frac{1}{2}\left(A + A^T\right)\right] + \left[\frac{1}{2}\left(A - A^T\right)\right]$

ILLUSTRATION 3.9

If $A = \begin{bmatrix} -1 & 7 \\ 2 & 3 \end{bmatrix}$, skew-symmetric part of A is

(a) $\begin{bmatrix} -1 & \frac{9}{2} \\ -\frac{9}{2} & 3 \end{bmatrix}$ **(b)** $\begin{bmatrix} 0 & -\frac{5}{2} \\ \frac{5}{2} & 0 \end{bmatrix}$

(c) $\begin{bmatrix} -1 & -\frac{9}{2} \\ \frac{9}{2} & 3 \end{bmatrix}$ **(d)** $\begin{bmatrix} 0 & \frac{5}{2} \\ -\frac{5}{2} & 0 \end{bmatrix}$

Solution

Let $A = B + C$, where $B = \frac{1}{2}\left(A + A^T\right)$ and $C = \frac{1}{2}\left(A - A^T\right)$ are respectively symmetric and skew-symmetric parts of A.

Now

$$C = \frac{1}{2}\left\{\begin{bmatrix} -1 & 7 \\ 2 & 3 \end{bmatrix} - \begin{bmatrix} -1 & 2 \\ 7 & 3 \end{bmatrix}\right\} = \frac{1}{2}\begin{bmatrix} 0 & 5 \\ -5 & 0 \end{bmatrix} = \begin{bmatrix} 0 & \frac{5}{2} \\ -\frac{5}{2} & 0 \end{bmatrix}$$

ILLUSTRATION 3.10

If $A = \begin{bmatrix} 1 & 2 \\ 3 & 0 \end{bmatrix}$, $B = \begin{bmatrix} 3 & 4 \\ 1 & 6 \end{bmatrix}$ then which of the following statements is true?

(a) $AB = BA$ **(b)** $A^2 = B$

(c) $(AB)^T = \begin{bmatrix} 5 & 9 \\ 16 & 12 \end{bmatrix}$ **(d)** None of these.

Solution

We have $(AB)_{11} = 1.3 + 2.1 = 5$

$(BA)_{11} = 3.1 + 4.3 = 15$

$\therefore \quad AB \neq BA$

Again $\left(A^2\right)_{11} = 1.1 + 2.3$

Also, $(AB)^T = B^T A^T = \begin{bmatrix} 3 & 1 \\ 4 & 6 \end{bmatrix}\begin{bmatrix} 1 & 3 \\ 2 & 0 \end{bmatrix}$

$= \begin{bmatrix} 3+2 & 9+0 \\ 4+12 & 12+0 \end{bmatrix} = \begin{bmatrix} 5 & 9 \\ 16 & 12 \end{bmatrix}$ is correct.

Solution of Homogeneous System of Linear Equations

If d_1, d_2, d_3 are the zeroes of a system of linear equations, it is said to be homogeneous. Let $AX = 0$ be a homogeneous system of n linear equations. If A is non-singular, then the system of equations will have a unique solution, i.e., trivial solution and if A is singular then the system of equation will have infinitely many solutions.

ILLUSTRATION 3.11

Write down in matrix form, the system of equations $2x - y + 3z = 9$, $x + y + z = 6$, $x - y + z = 2$ and solve it.

Solution

The given system of equations can be written in matrix form as $AX = B$

where $A = \begin{bmatrix} 2 & -1 & 3 \\ 1 & 1 & 1 \\ 1 & -1 & 1 \end{bmatrix}$, $X = \begin{bmatrix} x \\ y \\ z \end{bmatrix}$, $B = \begin{bmatrix} 9 \\ 6 \\ 2 \end{bmatrix}$

We have $|A| = 2 \neq 0$, i.e., A is non-singular and therefore A^{-1} exists.

$$A^{-1} = \frac{\text{adj}\,A}{|A|} = \begin{bmatrix} -1 & 1 & 2 \\ 0 & \frac{1}{2} & -\frac{1}{2} \\ 1 & -\frac{1}{2} & -\frac{3}{2} \end{bmatrix}$$

$$\therefore \quad X = A^{-1}B = \begin{bmatrix} -1 & 1 & 2 \\ 0 & \frac{1}{2} & -\frac{1}{2} \\ 1 & -\frac{1}{2} & -\frac{3}{2} \end{bmatrix} \begin{bmatrix} 9 \\ 6 \\ 2 \end{bmatrix}$$

$$\Rightarrow \quad X = \begin{bmatrix} x \\ y \\ z \end{bmatrix} = \begin{bmatrix} 1 \\ 2 \\ 3 \end{bmatrix}, \text{ i.e., } (x, y, z) \equiv (1, 2, 3)$$

ILLUSTRATION 3.12

Solve the system of equations $x + 3y - 2z = 0$, $2x - y + 4z = 0$, $x - 11y + 14z = 0$.

Solution

The given system of equations in the matrix form are written as below:

$$\begin{bmatrix} 1 & 3 & -2 \\ 2 & -1 & 4 \\ 1 & -11 & 14 \end{bmatrix} \begin{bmatrix} x \\ y \\ z \end{bmatrix} = \begin{bmatrix} 0 \\ 0 \\ 0 \end{bmatrix}$$

$$AX = 0$$

$$|A| = 1(-14+44) - 3(28-4) - 2(-22+1)$$

$$= 30 - 72 + 42 = 0 \qquad \ldots\text{(i)}$$

and therefore, the system has infinitely many solutions.

Taking the first two equations,

$$x + 3y = 2z \qquad \ldots\text{(ii)}$$

$$2x - y = -4z \qquad \ldots\text{(iii)}$$

and solving in terms of z, we get

$$x = \frac{-10z}{7}, \; y = \frac{8}{7}z \qquad \ldots\text{(iv)}$$

Put these values of x and y in third equation $x - 11 + 14z = 0$, its value comes to be equal to zero.

Now if $z = 7k$, then $x = -10k$, $y = 8k$

where k is any arbitrary constant, then the required solution is

$$x = -10k, \; y = 8k \text{ and } z = 7k$$

Operations on Matrices

Addition of Two Matrices

If A and B are two matrices of the same order then the addition of A and B (written as A + B) is obtained by adding the corresponding elements of A and B.

Note that

$$A + B = B + A, \; A + 0 = A$$

Properties

(i) Matrix addition is commutative.

If A and B be two $m \times n$ matrices, then A + B = B + A.

(ii) Matrix addition is associative.

If A, B and C be three matrices each of order $m \times n$, then

$(A + B) + C = A + (B + C)$

(iii) Existence of Additive Identity

If O be the $m \times n$ matrix, each of whose elements is zero, then A + O = A = O + A for every $m \times n$ matrix A. O is called Additive Identity.

(iv) Existence of Additive Inverse

Let $A = [a_{ij}]_{m\times n}$. Then the negative of the matrix A is defined as the matrix $[-a_{ij}]_{m\times n}$ and is denoted by –A.

$A + (-A) = 0$

e.g., If $A = \begin{bmatrix} 1 & -2 & 0 \\ 3 & 2 & -5 \end{bmatrix}$ then $-A = \begin{bmatrix} -1 & 2 & 0 \\ -3 & -2 & 5 \end{bmatrix}$

(v) Cancellation laws hold good for addition:

If A, B, C be any three $m - n$ matrices, then

$A + B = A + C \Rightarrow B = C$

(vi) $k(A + B) = kA + kB$ where k is scalar.

(vii) $\text{tr}(A \pm B) = \text{tr}(A) \pm \text{tr}(B)$

Subtraction of One Matrix from Another

Let A and B be two $m \times n$ matrices. Then the difference between A and B is denoted by A – B and is defined by A – B = A + (–B)

A – B will also be an $m \times n$ matrix. In order to find A – B, the elements of B must be subtracted from the corresponding elements of A.

e.g., $A = \begin{bmatrix} 1 & 2 & 3 \\ 3 & -1 & 0 \end{bmatrix}, B = \begin{bmatrix} 2 & -1 & 4 \\ 5 & 0 & 6 \end{bmatrix}$

then

$$A - B = \begin{bmatrix} 1-2 & 2+1 & 3-4 \\ 3-5 & -1-0 & 0-6 \end{bmatrix} = \begin{bmatrix} -1 & 3 & -1 \\ -2 & -1 & -6 \end{bmatrix}$$

ILLUSTRATION 3.13

If $A = \begin{bmatrix} 1 & 4 \\ 3 & 2 \\ 2 & 5 \end{bmatrix}, B = \begin{bmatrix} -1 & -2 \\ 0 & 5 \\ 3 & 1 \end{bmatrix}$ find the matrix D such that A + B – D = 0, i.e., zero matrix.

Solution

$$D = \begin{bmatrix} 0 & 2 \\ 3 & 7 \\ 5 & 6 \end{bmatrix}$$

ILLUSTRATION 3.14

If X and Y two matrices such that $X - Y = \begin{bmatrix} 3 & 2 \\ -1 & 0 \end{bmatrix}$ and $X + Y = \begin{bmatrix} 1 & -2 \\ 3 & 4 \end{bmatrix}$ then Y matrix is ______.

(a) $\begin{bmatrix} 2 & 0 \\ 1 & 2 \end{bmatrix}$ **(b)** $\begin{bmatrix} -1 & -2 \\ 3 & 4 \end{bmatrix}$

(c) $\begin{bmatrix} -1 & -2 \\ 2 & 2 \end{bmatrix}$ **(d)** None of these.

Solution

Given that

$$X - Y = \begin{bmatrix} 3 & 2 \\ -1 & 0 \end{bmatrix} \quad \ldots(1)$$

and

$$X + Y = \begin{bmatrix} 1 & -2 \\ 3 & 4 \end{bmatrix} \quad \ldots(2)$$

Subtracting (2) from (1)

$$-2Y = \begin{bmatrix} 3 & 2 \\ -1 & 0 \end{bmatrix} - \begin{bmatrix} 1 & -2 \\ 3 & 4 \end{bmatrix}$$

$$(-2)Y = \begin{bmatrix} 3-1 & 2-(2) \\ -1-3 & 0-4 \end{bmatrix} = \begin{bmatrix} 2 & 4 \\ -4 & -4 \end{bmatrix}$$

$$\Rightarrow \quad Y = -\frac{1}{2}\begin{bmatrix} 2 & 4 \\ -4 & -4 \end{bmatrix} = \begin{bmatrix} -1 & -2 \\ 2 & 2 \end{bmatrix}$$

ILLUSTRATION 3.15

If $\begin{bmatrix} 1 & 0 \\ 3 & -4 \end{bmatrix} + \begin{bmatrix} a & 1 \\ -1 & b \end{bmatrix} = \begin{bmatrix} 2 & 1 \\ 2 & -2 \end{bmatrix}$, then the values of a, b are ______.

(a) 1, –2 **(b)** –1, 2 **(c)** –1, –2 **(d)** 1, 2

Solution

Here $\begin{bmatrix} 1+a & 0+1 \\ 3-1 & -4+b \end{bmatrix} = \begin{bmatrix} 2 & 1 \\ 2 & -2 \end{bmatrix}$

$\Rightarrow \quad 1+a = 2$ and $-4+b = -2$

$\Rightarrow \quad a = 1, b = 2$

Multiplication with a Scalar

If A is an $m \times n$ matrix and k is a real or complex number then the matrix kA is defined as a matrix, every element of which is k times the corresponding element of A.

ILLUSTRATION 3.16

If $A = \begin{bmatrix} 5 & 3 & 2 \\ 2 & -1 & 0 \end{bmatrix}$ then $3A = \begin{bmatrix} 15 & 9 & 6 \\ 6 & -3 & 0 \end{bmatrix}$

ILLUSTRATION 3.17

If $X = \begin{bmatrix} 1 & a \\ 0 & 1 \end{bmatrix}$ and $3X - \begin{bmatrix} 2 & 3 \\ 0 & 2 \end{bmatrix} = \begin{bmatrix} 1 & 3 \\ 0 & 1 \end{bmatrix}$ then a is equal to ______.

(a) 1 **(b)** 2 **(c)** 0 **(d)** –2

Solution

$$3X = \begin{bmatrix} 3 & 3a \\ 0 & 3 \end{bmatrix}$$

$$\Rightarrow \quad LHS = \begin{bmatrix} 3-2 & 3a-3 \\ 0-0 & 3-2 \end{bmatrix} = \begin{bmatrix} 1 & 3a-3 \\ 0 & 1 \end{bmatrix}$$

Now by equality of two matrices, we have

$$3a - 3 = 3 \Rightarrow a = 2$$

ILLUSTRATION 3.18

If $A=\begin{bmatrix}\cos^2\theta & \cos\theta\sin\theta\\ \cos\theta\sin\theta & \sin^2\theta\end{bmatrix}$ and $B=\begin{bmatrix}\cos^2\phi & \cos\phi\sin\phi\\ \cos\phi\sin\phi & \sin^2\phi\end{bmatrix}$, and $AB=0$, then $\theta-\phi$ is equal to ______.

(a) 0 **(b)** even multiple of $\left(\frac{\pi}{2}\right)$ **(c)** odd multiple of $\left(\frac{\pi}{2}\right)$ **(d)** odd multiple of π

Solution

Here

$$AB=\begin{bmatrix}\cos^2\theta\cos^2\phi+\cos\theta\sin\theta\cos\phi\sin\phi & \cos^2\theta\cos\phi\sin\phi+\cos\theta\sin\theta\sin^2\phi\\ \cos\theta\sin\theta\cos^2\phi+\sin^2\theta\cos\phi\sin\phi & \cos\theta\sin\theta\cos\phi\sin\phi+\sin^2\theta\sin^2\phi\end{bmatrix}$$

$$=\begin{bmatrix}\cos\theta\cos\phi\cos(\theta-\phi) & \cos\theta\sin\phi\cos(\theta-\phi)\\ \sin\theta\cos\phi\cos(\theta-\phi) & \sin\theta\sin\phi\cos(\theta-\phi)\end{bmatrix}=\begin{bmatrix}0&0\\0&0\end{bmatrix}, \text{ if } \cos(\theta-\phi)=0$$

Now $\cos(\theta-\phi)=0\Rightarrow\theta-\phi$, is an odd multiple of $\frac{\pi}{2}$

Properties of Scalar Multiplication

If A, B are matrices of the same order and λ, μ are any two scalars, then ______.

(i) $\lambda(A+B)=\lambda A+\lambda B$

(ii) $(\lambda+\mu)A=\lambda A+\mu A$

(iii) $\lambda(\mu A)=(\lambda\mu A)=\mu(\lambda A)$

(iv) $(-\lambda A)=-(\lambda A)=\lambda(-A)$

(v) $\text{tr}(kA)=k\,\text{tr}(A)$

Multiplication of Two Matrices

If A is an $m\times n$ matrix and B is an $n\times p$ matrix, then the product AB of A and B is defined as an $m\times p$ matrix whose (*i*, *j*)th element is obtained by finding sum and product of elements of *i*th row in left matrix A and elements of *j*th column of right matrix B.

ILLUSTRATION 3.19

If $A=\begin{bmatrix}5&3&2\\2&4&0\end{bmatrix}_{2\times3}$, $B=\begin{bmatrix}-5&2&7\\4&3&-1\\2&4&8\end{bmatrix}_{3\times3}$

then AB is a 2×3 matrix

i.e., $AB=\begin{bmatrix}c_{11}&c_{12}&c_{13}\\c_{21}&c_{22}&c_{23}\end{bmatrix}$

Now,

c_{11} = Sum of product of element of 1st row in A and 1st column in B

$$=5\times(-5)+3(4)+2(2)=-9$$

c_{12} = Sum of product of element of 1st row in A and second column in B

$$=5\times2+3\times3+24\times4=27$$

c_{13} = Sum of product of element of 1st row in A and third column in B

$$=5\times7+3\times(-1)+2\times8=48$$

c_{21} = Sum of product of elements of second row in A and first column in B

$$=2\times(-5)+4\times4+0\times2=16$$

Similarly, $c_{22}=2\times2+4\times3+0\times4=16$,

$$c_{23}=2\times7+4\times(-1)+0\times8=10$$

Thus $AB=\begin{bmatrix}-9&27&48\\16&16&10\end{bmatrix}$

ILLUSTRATION 3.20

If $A=\begin{bmatrix}0&1&2\\1&2&3\\2&3&4\end{bmatrix}$ and $B=\begin{bmatrix}1&-2\\-1&0\\2&-1\end{bmatrix}$, then find AB.

Is BA defined?

Solution

Matrix A is of the order 3×3 and matrix B is of the order 3×2. Since the number of columns of A = number of rows of B (each being = 3).

$\therefore$ Product AB is defined.

$$AB = \begin{bmatrix} 0 & 1 & 2 \\ 1 & 2 & 3 \\ 2 & 3 & 4 \end{bmatrix} \begin{bmatrix} 1 & -2 \\ -1 & 0 \\ 2 & -1 \end{bmatrix}$$

$$= \begin{bmatrix} 0.1+1.-1+2.2 & 0.-2+1.0+2.-1 \\ 1.1+2.-1+3.2 & 1.-2+2.0+3.-1 \\ 2.1+3.-1+4.2 & 2.-2+3.0+4.-1 \end{bmatrix}$$

$$= \begin{bmatrix} 3 & -2 \\ 5 & -5 \\ 7 & -8 \end{bmatrix}$$

Again, since number of columns of B ≠ number of rows of A.

∴ Product BA is not possible.

ILLUSTRATION 3.21

If, $\begin{bmatrix} 1 & 2 \\ 2 & 1 \end{bmatrix} \begin{bmatrix} x \\ y \end{bmatrix} = \begin{bmatrix} 5 \\ 4 \end{bmatrix}$ then:

(a) $x = 2, y = 1$ **(b)** $x = 1, y = 2$
(c) $x = 3, y = 2$ **(d)** $x = 2, y = 3$

Solution

The given matrix equation can be written as

$$\begin{bmatrix} x+2y \\ 2x+y \end{bmatrix} = \begin{bmatrix} 5 \\ 4 \end{bmatrix}$$

$\Rightarrow$ $x + 2y = 5$ and $2x + y = 4$

$\Rightarrow$ $x = 1, y = 2$

ILLUSTRATION 3.22

If $A = \begin{bmatrix} 2 & -1 \\ -1 & 2 \end{bmatrix}$ and $A^2 - 4A - nI = 0$, then n is equal to

(a) 3 **(b)** −3 **(c)** 1/3 **(d)** −1/3

Solution

$$A^2 = \begin{bmatrix} 5 & -4 \\ -4 & 5 \end{bmatrix}, 4A = \begin{bmatrix} 8 & -4 \\ -4 & 8 \end{bmatrix}, nI = \begin{bmatrix} n & 0 \\ 0 & n \end{bmatrix}$$

$\Rightarrow$ $A^2 - 4A - nI$

$$= \begin{bmatrix} 5-8-n & -4+4-0 \\ -4+4-0 & 5-8-n \end{bmatrix}$$

$$= \begin{bmatrix} -3-n & 0 \\ 0 & -3-n \end{bmatrix}$$

∴ $A^2 - 4A - nI = 0$

$\Rightarrow$ $\begin{bmatrix} -3-n & 0 \\ 0 & -3-n \end{bmatrix} = \begin{bmatrix} 0 & 0 \\ 0 & 0 \end{bmatrix}$

$\Rightarrow$ $-3 - n = 0$

$\Rightarrow$ $n = -3$

Remarks

(a) If AB is defined, BA may not be defined. If BA is also defined then AB may not be equal to BA. Thus in general AB ≠ BA.

(b) A(BC) = (AB)(C),
provided all products are defined.

(c) A(B + C) = AB + AC,
provided all products and additions are defined.

(d) If all matrices A, B, C, ... are square matrices of the same order 3 × 3 then all product say AB, BA, (AB)(C) etc. and all integral power of any of them say A^2, A^3 etc. are well defined (A^2 = A.A, $A^3 = A^2A$).

Properties of Matrix Multiplication

(i) If AB = 0 (null matrix), it does not necessarily imply that A or B is a null matrix.

For instance, if $A = \begin{bmatrix} 1 & 1 \\ 1 & 1 \end{bmatrix}$ and $B = \begin{bmatrix} 1 & -1 \\ -1 & 1 \end{bmatrix}$,

then the product AB is a null matrix, although neither A nor B is a null matrix.

(ii) Multiplication of matrices is not always commutative

AB may or may not be equal to BA

(iii) Multiplication of matrices is associative

(AB)C = A(BC)

If A, B, C are $m \times n$, $n \times p$ and $p \times q$ matrices respectively.

(iv) Multiplication of matrices is distributive w.r.t. addition of matrices:

If A is an $m \times n$ matrix and B and C are both $n \times p$ matrices, then

A(B + C) = AB + AC

(v) Multiplication of Matrix by Unit matrix, I

AI = IA = A if conformability for multiplication is satisfied.

(vi) Multiplication of Matrix by Null Matrix:

OA = AO = O where O null matrix

(vii) Whenever AB and BA both exist and are matrices of same order, it is not necessary that AB = BA e.g., $A=\begin{bmatrix}1 & 0\\0 & -1\end{bmatrix}, B=\begin{bmatrix}0 & 1\\1 & 0\end{bmatrix}$

then $AB=\begin{bmatrix}1 & 0\\0 & -1\end{bmatrix}\begin{bmatrix}0 & 1\\1 & 0\end{bmatrix}=\begin{bmatrix}0 & 1\\-1 & 0\end{bmatrix}$

$BA=\begin{bmatrix}0 & 1\\1 & 0\end{bmatrix}\begin{bmatrix}1 & 0\\0 & -1\end{bmatrix}=\begin{bmatrix}0 & -1\\1 & 0\end{bmatrix}$

Thus $AB \neq BA$

(viii) If $AB = AC \not\Rightarrow B = C$
(Cancellation Law is not applicable)

(ix) If $AB = 0$, it does not mean that $A = 0$ or $B = 0$. Again, product of two non- zero matrix may be zero matrix.

(x) $tr(AB) = tr(BA)$

Remarks

(i) The multiplication of two diagonal matrices is again a diagonal matrix.

(ii) The multiplication of two triangular matrices is again a triangular matrix.

(iii) The multiplication of two scalar matrices is also a scalar matrix.

(iv) If A and B are two matrices of the same order, then

(a) $(A+B)^2 = A^2+B^2+AB+BA$

(b) $(A-B)^2 = A^2+B^2-AB-BA$

(c) $(A-B)(A+B) = A^2-B^2+AB-BA$

(d) $(A+B)(A-B) = A^2-B^2-AB+BA$

(e) $A(-B) = (-A)B = -(AB)$

Conjugate of a Matrix

If A is square matrix whose entries are complex numbers, if we replace every element of A by its conjugate, we get a matrix called conjugate of the matrix A. This matrix is denoted by $\bar{A}$.

Note that:

(i) $\left(\bar{\bar{A}}\right) = A$

(ii) $\overline{(A+B)} = \bar{A}+\bar{B}$

(iii) $\overline{(kA)} = k\bar{A}$, k being any real number

(iv) $\overline{(AB)} = \bar{A}\bar{B}$

Transposed Conjugate of a Matrix

If A is square matrix, then $\bar{A}$ is its conjugate. If we take transpose of $\bar{A}$, we get a matrix called transposed conjugate of A. We denote this matrix by A^θ note that

(i) $(A^\theta)^\theta = A$

(ii) $(A+B)^\theta = A^\theta + B^\theta$

(iii) $(kA)^\theta = kA^\theta$, k being any complex number

(iv) $(AB)^\theta = B^\theta A^\theta$

Adjoint of a Square Matrix

If A is square matrix, then adjoint of A (written as adj A) is obtained by following method.

Step 1: Replace every element of A by its cofactor. The matrix thus obtained is called cofactor matrix.

Step 2: Take the transpose of cofactor matrix.

For 2×2 square matrices, adjoint can be easily found as follows:

If $A=\begin{bmatrix}a & b\\c & d\end{bmatrix}$ then, adj $A=\begin{bmatrix}d & -b\\-c & a\end{bmatrix}$

(i.e., interchanging the diagonal elements and changing the sign of remaining elements.)

The following properties of adjoint are not easy to verify in general but can be verified for numerical cases (A and B are square matrices of the same order).

(i) (Adj A)′ = Adj A′

(ii) Adj (AB) = (Adj B) (Adj A)

(iii) A(Adj A) = (Adj A) (A) = (det A) I

(iv) det (Adj A) = $(\det A)^2$

(v) Adj (Adj A) = $(\det A)^{n-2}$ A

ILLUSTRATION 3.23

Find the adjoint of the matrix $A=\begin{bmatrix}1 & 1 & 1\\1 & 2 & -3\\2 & -1 & 3\end{bmatrix}$

Solution

First we have to find cofactors, $C_{11}, C_{12}, \ldots$ etc.

$$C_{11}=(-1)^{1+1}\begin{vmatrix}2 & -3\\-1 & 3\end{vmatrix}=3$$

$$C_{12}=(-1)^{1+2}\begin{vmatrix}1 & -3\\2 & 3\end{vmatrix}=-9$$

$$C_{13}=(-1)^{1+3}\begin{vmatrix}1 & 2\\2 & -1\end{vmatrix}=-5$$

Similarly, we can find all cofactors,

$C_{21}, C_{22}, C_{23}, C_{31}, C_{32}, C_{33}$

$$\text{Adj}A=\begin{bmatrix}C_{11} & C_{21} & C_{31}\\C_{12} & C_{22} & C_{32}\\C_{13} & C_{23} & C_{33}\end{bmatrix}=\begin{bmatrix}3 & -4 & -5\\-9 & 1 & 4\\-5 & 3 & 1\end{bmatrix}$$

ILLUSTRATION 3.24

If $A = \begin{bmatrix} 1 & 3 & 5 \\ 3 & 5 & 1 \\ 5 & 1 & 3 \end{bmatrix}$, then adj A is equal to:

(a) $\begin{bmatrix} 14 & -4 & -22 \\ -4 & -22 & 14 \\ -22 & 14 & -4 \end{bmatrix}$ **(b)** $\begin{bmatrix} -14 & 4 & 22 \\ 4 & 22 & -14 \\ 22 & -14 & 4 \end{bmatrix}$

(c) $\begin{bmatrix} 14 & 4 & -22 \\ 4 & -22 & -14 \\ -22 & -14 & -4 \end{bmatrix}$ **(d)** None of these.

Solution

$$\text{adj } A = \begin{bmatrix} 14 & -4 & -22 \\ -4 & -22 & 14 \\ -22 & 14 & -4 \end{bmatrix}^T = \begin{bmatrix} 14 & -4 & -22 \\ -4 & -22 & 14 \\ -22 & 14 & -4 \end{bmatrix}$$

ILLUSTRATION 3.25

If $A = \begin{matrix} 2 & 0 & 0 \\ 2 & 2 & 0 \\ 2 & 2 & 2 \end{matrix}$, then adj (adj A) is equal to

(a) $8\begin{bmatrix} 1 & 0 & 0 \\ 1 & 1 & 0 \\ 1 & 1 & 1 \end{bmatrix}$ **(b)** $16\begin{bmatrix} 1 & 0 & 0 \\ 1 & 1 & 0 \\ 1 & 1 & 1 \end{bmatrix}$

(c) $64\begin{bmatrix} 1 & 0 & 0 \\ 1 & 1 & 0 \\ 1 & 1 & 1 \end{bmatrix}$ **(d)** None of these.

Solution

$$|A| = \begin{vmatrix} 2 & 0 & 0 \\ 2 & 2 & 0 \\ 2 & 2 & 2 \end{vmatrix} = 8$$

Now

$$\text{adj (adj A)} = |A|^{3-2} A = 8\begin{bmatrix} 2 & 0 & 0 \\ 2 & 2 & 0 \\ 2 & 2 & 2 \end{bmatrix} = 16\begin{bmatrix} 1 & 0 & 0 \\ 1 & 1 & 0 \\ 1 & 1 & 1 \end{bmatrix}$$

Inverse of a Matrix

If A is a square matrix of order $n \times n$ such that det A or $|A| \neq 0$ then A is called non-singular. For every non-singular matrix A, there exists another non-singular matrix B such that AB = BA = I where I is the identity matrix of the same order. We write B = A^{-1}. Since A(Adj A) = (Adj A) (A) = |A| I.

We have $$A^{-1} = \frac{1}{|A|}(\text{Adj A})$$

The following results on inverse are important for invertible matrices.

(i) $(A^{-1})' = (A')^{-1}$

(ii) $\left(A^{-1}\right)^k = \left(A^k\right)^{-1}$ for any positive integer k

(iii) $(AB)^{-1} = B^{-1}A^{-1}$ (reversal laws)

$(ABC)^{-1} = C^{-1}B^{-1}A^{-1}$

ILLUSTRATION 3.26

If $A = \begin{bmatrix} 1 & -1 & 1 \\ 2 & -1 & 0 \\ 1 & 0 & 0 \end{bmatrix}$, find A^2 and show that $A^2 = A^{-1}$

Solution

$$A^2 = A \cdot A = \begin{bmatrix} 1 & -1 & 1 \\ 2 & -1 & 0 \\ 1 & 0 & 0 \end{bmatrix}\begin{bmatrix} 1 & -1 & 1 \\ 2 & -1 & 0 \\ 1 & 0 & 0 \end{bmatrix}$$

$$A^2 = \begin{bmatrix} 0 & 0 & 1 \\ 0 & -1 & 2 \\ 1 & -1 & 1 \end{bmatrix} \quad \ldots\text{(i)}$$

$$|A| = 1 \quad \ldots\text{(ii)}$$

$$C_{11} = (-1)^{1+1}\begin{vmatrix} -1 & 0 \\ 0 & 0 \end{vmatrix} = 0$$

$$C_{12} = (-1)^{1+2}\begin{vmatrix} -2 & 0 \\ 1 & 0 \end{vmatrix} = 0$$

Similarly, we can find $C_{13}, C_{21}, C_{22}, C_{23}, C_{31}, C_{32}, C_{33}$

$$\text{Adj. } A = \begin{bmatrix} 0 & 0 & 1 \\ 0 & -1 & 2 \\ 1 & -1 & 1 \end{bmatrix}$$

$$A^{-1} = \frac{\text{Adj. A}}{|A|} = \begin{bmatrix} 0 & 0 & 1 \\ 0 & -1 & 2 \\ 1 & -1 & 1 \end{bmatrix} \quad \ldots\text{(iii)}$$

So, from (i) and (ii), $A^2 = A^{-1}$

ILLUSTRATION 3.27

Inverse matrix of $\begin{bmatrix} 2 & -3 \\ -4 & 2 \end{bmatrix}$ is

(a) $-\frac{1}{8}\begin{bmatrix} 2 & 3 \\ 4 & 2 \end{bmatrix}$ **(b)** $-\frac{1}{8}\begin{bmatrix} 2 & 4 \\ 3 & 2 \end{bmatrix}$

(c) $\frac{1}{8}\begin{bmatrix} 2 & 3 \\ 4 & 2 \end{bmatrix}$ **(d)** $\begin{bmatrix} 2 & 3 \\ 4 & 2 \end{bmatrix}$

Solution

Let the given matrix is A, then

$$|A| = -8 \text{ and adj } A = \begin{bmatrix} 2 & 4 \\ 3 & 2 \end{bmatrix}^T = \begin{bmatrix} 2 & 3 \\ 4 & 2 \end{bmatrix}$$

$$\therefore \quad A^{-1} = \frac{1}{|A|}\text{adj } A = -\frac{1}{8}\begin{bmatrix} 2 & 3 \\ 4 & 2 \end{bmatrix}$$

ILLUSTRATION 3.28

If $A = \begin{bmatrix} 0 & -1 & 2 \\ 2 & -2 & 0 \end{bmatrix}$, $B = \begin{bmatrix} 0 & 1 \\ 1 & 0 \\ 1 & 1 \end{bmatrix}$ and $M = AB$, then M^{-1} is equal to

(a) $\begin{bmatrix} 2 & -2 \\ 2 & 1 \end{bmatrix}$ **(b)** $\begin{bmatrix} 1/3 & 1/3 \\ -1/3 & 1/6 \end{bmatrix}$

(c) $\begin{bmatrix} 1/3 & -1/3 \\ 1/3 & 1/6 \end{bmatrix}$ **(d)** $\begin{bmatrix} 1/3 & -1/3 \\ -1/3 & 1/6 \end{bmatrix}$

Solution

$$M = \begin{bmatrix} 0 & -1 & 2 \\ 2 & -2 & 0 \end{bmatrix}\begin{bmatrix} 0 & 1 \\ 1 & 0 \\ 1 & 1 \end{bmatrix} = \begin{bmatrix} 1 & 2 \\ -2 & 2 \end{bmatrix}$$

$$|M| = 6, \text{adj} M = \begin{bmatrix} 2 & -2 \\ 2 & 1 \end{bmatrix}$$

$$\therefore \quad M^{-1} = \frac{1}{6}\begin{bmatrix} 2 & -2 \\ 2 & 1 \end{bmatrix} = \begin{bmatrix} \frac{1}{3} & -\frac{1}{3} \\ \frac{1}{3} & \frac{1}{6} \end{bmatrix}$$

ILLUSTRATION 3.29

$$\begin{bmatrix} 1 & -\tan\frac{\theta}{2} \\ \tan\frac{\theta}{2} & 1 \end{bmatrix}\begin{bmatrix} 1 & \tan\frac{\theta}{2} \\ -\tan\frac{\theta}{2} & 1 \end{bmatrix}^{-1}$$

is equal to

(a) $\begin{bmatrix} \sin\theta & -\cos\theta \\ \cos\theta & \sin\theta \end{bmatrix}$ **(b)** $\begin{bmatrix} \cos\theta & \sin\theta \\ -\sin\theta & \cos\theta \end{bmatrix}$

(c) $\begin{bmatrix} \cos\theta & -\sin\theta \\ \sin\theta & \cos\theta \end{bmatrix}$ **(d)** None of these.

Solution

$$\begin{bmatrix} 1 & \tan\frac{\theta}{2} \\ -\tan\frac{\theta}{2} & 1 \end{bmatrix}^{-1} = \frac{1}{\sec^2\frac{\theta}{2}}\begin{bmatrix} 1 & -\tan\frac{\theta}{2} \\ \tan\frac{\theta}{2} & 1 \end{bmatrix}$$

$\therefore$ Product

$$= \frac{1}{\sec^2\frac{\theta}{2}}\begin{bmatrix} 1 & -\tan\frac{\theta}{2} \\ \tan\frac{\theta}{2} & 1 \end{bmatrix}\begin{bmatrix} 1 & -\tan\frac{\theta}{2} \\ \tan\frac{\theta}{2} & 1 \end{bmatrix}$$

$$= \frac{1}{\sec^2\frac{\theta}{2}}\begin{bmatrix} 1-\tan^2\frac{\theta}{2} & -2\tan\frac{\theta}{2} \\ 2\tan\frac{\theta}{2} & 1-\tan^2\frac{\theta}{2} \end{bmatrix}$$

$$= \begin{bmatrix} \cos^2\frac{\theta}{2}-\sin^2\frac{\theta}{2} & -2\sin\frac{\theta}{2}\cos\frac{\theta}{2} \\ 2\sin\frac{\theta}{2}\cos\frac{\theta}{2} & \cos^2\frac{\theta}{2}-\sin^2\frac{\theta}{2} \end{bmatrix}$$

$$= \begin{bmatrix} \cos\theta & -\sin\theta \\ \sin\theta & \cos\theta \end{bmatrix}$$

WORKSHEET

1. If $A = \begin{bmatrix} 1 & -3 & 2 \\ 2 & k & 5 \\ 4 & 2 & 1 \end{bmatrix}$ is a singular matrix, then k is equal to

(a) −1 (b) 8 (c) 4 (d) −8

2. If $A = \begin{pmatrix} 2 & -1 \\ -7 & 4 \end{pmatrix}$ and $B = \begin{pmatrix} 4 & 1 \\ 7 & 2 \end{pmatrix}$, then which statement is true?

(a) $AA^T = I$ (b) $BB^T = I$

(c) $AB \neq BA$ (d) $(AB)^T = I$

3. If $A = \begin{bmatrix} 1 & 2 & 3 \\ 5 & 0 & 4 \\ 2 & 6 & 7 \end{bmatrix}$ then adj A is equal to ______.

(a) $\begin{bmatrix} -24 & 4 & 8 \\ 4 & 1 & 2 \\ 8 & 11 & -11 \end{bmatrix}$ (b) $\begin{bmatrix} -24 & 4 & 8 \\ 4 & 1 & 11 \\ 30 & -2 & -10 \end{bmatrix}$

(c) $\begin{bmatrix} -24 & 4 & 8 \\ -27 & 1 & 11 \\ 30 & -2 & -10 \end{bmatrix}$ (d) None of these.

4. If $k \begin{bmatrix} -1 & 2 & 2 \\ 2 & -1 & 2 \\ 2 & 2 & -1 \end{bmatrix}$ is an orthogonal matrix, then k is equal to ______.

(a) 1 (b) 1/2 (c) 1/3 (d) None of these.

5. If $M(\alpha) = \begin{bmatrix} \cos\alpha & -\sin\alpha & 0 \\ \sin\alpha & \cos\alpha & 0 \\ 0 & 0 & 1 \end{bmatrix}$

$M(\beta) = \begin{bmatrix} \cos\beta & 0 & \sin\beta \\ 0 & 1 & 0 \\ \sin\beta & 0 & \cos\beta \end{bmatrix}$ then $\left[M(\alpha)M(\beta)\right]^{-1}$ is equals to ______.

(a) $M(\beta)M(\alpha)$ (b) $M(-\alpha)M(-\beta)$

(c) $M(-\beta)M(-\alpha)$ (d) $-M(\beta)M(\alpha)$

6. If $A = \begin{bmatrix} 4 & 1 \\ 2 & 3 \end{bmatrix}$, then |A| is equal to ______.

(a) 12 (b) −10 (c) 10 (d) 5

7. If $A = \begin{bmatrix} 1 & 2 & 3 \\ 2 & 3 & 1 \\ 3 & 1 & 2 \end{bmatrix}$ then adj (adj A) =

(a) $\begin{bmatrix} -18 & 36 & -54 \\ 36 & -54 & 18 \\ -54 & 18 & -36 \end{bmatrix}$ (b) $-\begin{bmatrix} 18 & 36 & 54 \\ 36 & 54 & 18 \\ 54 & 18 & 36 \end{bmatrix}$

(c) $18\begin{bmatrix} 1 & 2 & 3 \\ 2 & 3 & 1 \\ 3 & 1 & 2 \end{bmatrix}$ (d) None of these.

8. If $A = \begin{bmatrix} 1 & 0 \\ 1 & 1 \end{bmatrix}$ then A^{-n} is equal to ______.

(a) $\begin{bmatrix} 1 & 0 \\ n & 1 \end{bmatrix}$ (b) $\begin{bmatrix} 1 & 0 \\ -n & -1 \end{bmatrix}$

(c) $\begin{bmatrix} 1 & 0 \\ -n & 1 \end{bmatrix}$ (d) None of these.

9. Prove that if A, B be two non-singular matrices of the same order, then AB is also non-singular and $(AB)^{-1} = B^{-1}A^{-1}$ (reversal law of inverse).

10. Show that

$$\begin{bmatrix} \cos\theta & -\sin\theta \\ \sin\theta & \cos\theta \end{bmatrix} = \begin{bmatrix} 1 & -\tan\frac{1}{2}\theta \\ \tan\frac{1}{2}\theta & 1 \end{bmatrix}\begin{bmatrix} 1 & \tan\frac{1}{2}\theta \\ -\tan\frac{1}{2}\theta & 1 \end{bmatrix}^{-1}$$

SOLVED EXAMPLES

Example 1. If $A = \begin{bmatrix} 1 & 1 \\ 0 & 1 \end{bmatrix}$, Show that $A^n = \begin{bmatrix} 1 & n \\ 0 & 1 \end{bmatrix}$

Solution

We have

$$A^2 = \begin{bmatrix} 1 & 1 \\ 0 & 1 \end{bmatrix}\begin{bmatrix} 1 & 1 \\ 0 & 1 \end{bmatrix} = \begin{bmatrix} 1+0 & 1+1 \\ 0+0 & 0+1 \end{bmatrix} = \begin{bmatrix} 1 & 2 \\ 0 & 1 \end{bmatrix} \quad ...(1)$$

$$A^3 = A^2A = \begin{bmatrix} 1 & 2 \\ 0 & 1 \end{bmatrix}\begin{bmatrix} 1 & 1 \\ 0 & 1 \end{bmatrix}$$

$$= \begin{bmatrix} 1.1+2.0 & 1.1+2.1 \\ 0.1+1.0 & 0.1+1.1 \end{bmatrix} = \begin{bmatrix} 1 & 3 \\ 0 & 1 \end{bmatrix} \quad ...(2)$$

From (1) and (2), we observe that the required result is true for $n = 2, 3$. To prove the required result, we shall use the method of induction. We shall assume the result to be true for an index n, (where n is a +ve integer) and by actual multiplication, we shall show that it holds for the next index $n + 1$.

Thus, let $A^n = \begin{bmatrix} 1 & n \\ 0 & 1 \end{bmatrix}$; then

$$A^{n+1} = A^nA = \begin{bmatrix} 1 & n \\ 0 & 1 \end{bmatrix}\begin{bmatrix} 1 & 1 \\ 0 & 1 \end{bmatrix}$$

$$= \begin{bmatrix} 1.1+n.0 & 1.1+n.1 \\ 0.1+1.0 & 0.1+1.1 \end{bmatrix} = \begin{bmatrix} 1 & n+1 \\ 0 & 1 \end{bmatrix}$$

This shows that if the result is true for the index n, it is also true for the next index $n + 1$. But we have already seen that it is true for $n = 2, 3$, etc. Hence by induction, the required result follows.

Example 2. Prove that the product of matrices

$$A = \begin{bmatrix} \cos^2\theta & \cos\theta\sin\theta \\ \cos\theta\sin\theta & \sin^2\theta \end{bmatrix}, \qquad B = \begin{bmatrix} \cos^2\phi & \cos\phi\sin\phi \\ \cos\phi\sin\phi & \sin^2\phi \end{bmatrix}$$

is zero when θ and ϕ differ by an odd multiple $\pi/2$.

Solution

Here, we have AB

$$= \begin{bmatrix} \cos^2\theta & \cos\theta\sin\theta \\ \cos\theta\sin\theta & \sin^2\theta \end{bmatrix} \begin{bmatrix} \cos^2\phi & \cos\phi\sin\phi \\ \cos\phi\sin\phi & \sin^2\phi \end{bmatrix}$$

$$\begin{bmatrix} \cos^2\theta\cos^2\phi + \cos\theta\sin\theta\cos\phi\sin\phi & \cos^2\theta\cos\phi\sin\phi + \cos\theta\sin\theta\sin^2\phi \\ \cos\theta\sin\theta\cos^2\phi + \sin^2\theta\cos\phi\sin\phi & \cos\theta\sin\theta\cos\phi\sin\phi + \sin^2\theta\sin^2\phi \end{bmatrix}$$

$$\begin{bmatrix} \cos\theta\cos\phi(\cos\theta\cos\phi + \sin\theta\sin\phi) & \cos\theta\sin\phi(\cos\theta\cos\phi + \sin\theta\sin\phi) \\ \sin\theta\cos\phi(\cos\theta\cos\phi + \sin\theta\sin\phi) & \sin\theta\sin\phi(\cos\theta\cos\phi + \sin\theta\sin\phi) \end{bmatrix}$$

$$= \begin{bmatrix} \cos\theta\cos\phi\cos(\theta\sim\phi) & \cos\theta\sin\phi\cos(\theta\sim\phi) \\ \sin\theta\cos\phi\cos(\theta\sim\phi) & \sin\theta\sin\phi\cos(\theta\sim\phi) \end{bmatrix}$$

Now, where $\theta \sim \phi =$ on odd multiple of $\pi/2$, $\cos(\theta \sim \phi) = 0$

Hence
$$AB = \begin{bmatrix} (\cos\theta\cos\phi).0 & (\cos\theta\sin\phi).0 \\ (\sin\theta\cos\phi).0 & (\sin\theta\sin\phi).0 \end{bmatrix}$$

$$= \begin{bmatrix} 0 & 0 \\ 0 & 0 \end{bmatrix}$$

Example 3. The inverse of a non-singular matrix is unique.

Solution

If possible, let B and C be two inverses of the same matrix A. Then by definition,

$$AB = BA = I \quad ...(1)$$

$$AC = CA = I \quad ...(2)$$

and from (1), we have

$$C(AB) = (I) = C$$

Also, from (2), we have

$$(CA)B = IB = B$$

Hence, C = B, i.e., the inverse is unique.

Example 4. Show that the matrix $A = \begin{bmatrix} \cos\theta & \sin\theta \\ -\sin\theta & \cos\theta \end{bmatrix}$ is orthogonal.

Solution

We have

$$AA' = \begin{bmatrix} \cos\theta & \sin\theta \\ -\sin\theta & \cos\theta \end{bmatrix} \begin{bmatrix} \cos\theta & -\sin\theta \\ \sin\theta & \cos\theta \end{bmatrix}$$

$$= \begin{bmatrix} \cos^2\theta + \sin^2\theta & -\sin\theta\cos\theta + \cos\theta\sin\theta \\ -\sin\theta\cos\theta + \cos\theta\sin\theta & \sin^2\theta + \cos^2\theta \end{bmatrix}$$

$$= \begin{bmatrix} 1 & 0 \\ 0 & 1 \end{bmatrix} = I$$

Hence A is orthogonal.

Example 5. Show that every square matrix can be uniquely expressed as the sum of a symmetric and a skew-symmetric matrix.

Solution

We have,

$$A = \frac{1}{2}A + \frac{1}{2}A' + \frac{1}{2}A - \frac{1}{2}A'$$

$$= \frac{1}{2}(A + A') + \frac{1}{2}(A - A') \qquad \ldots(1)$$

Now, $(A + A')' = A' + (A')' = A' + A = A + A'$

and $(A - A')' = A' - (A')' = A' - A = -(A - A')$

Therefore A + A′ is symmetric and A − A′ is skew-symmetric.

Thus, from (1) it is clear that

$$A = P + Q \qquad \ldots(2)$$

where $P = \frac{1}{2}(A + A')$ is symmetric, $Q = (A - A')$ is skew-symmetric, hence, we have established that any square matrix can be expressed as the sum of a symmetric and a skew-symmetric matrix.

To show that the representation is unique, let there be another way of writing A in the form

$$A = R + S \qquad \ldots(3)$$

where R is symmetric and S is skew-symmetric

Then $\qquad A' = R' + S' = R - S \qquad \ldots(4)$

From (3) and (4), $A + A' = 2R$ and $A - A' = 2S$

$$\text{Hence } R = \frac{1}{2}(A + A') = P \text{ and } S = \frac{1}{2}(A - A') = Q,$$

⇒ There is only one way of expressing a square matrix as the sum of a symmetric and a skew-symmetric matrix.

Example 6. If P is a square non-singular matrix whose determinant is unity, show that if $\text{Adj } P^{-1} = (\text{Adj } P)^{-1}$.

Hence show that if Adj B = A, |Q| = 1, then $\text{Adj } Q^{-1}BP^{-1} = PAQ$.

Solution

Note the following results:

(i) $A(\text{Adj } A) = (\text{Adj } A)A = |A|\, I_n$ (where |A| is det A)

(ii) $|AB| = |A|\,|B|$

(iii) $\text{Adj }(AB) = (\text{Adj } B)(\text{Adj } A)$

(iv) $(AB)^{-1} = B^{-1}A^{-1}$

Put $A = P^{-1}$ in (i) ⇒ $PP^{-1}\text{ Adj } P^{-1} = P(\text{Adj } P^{-1})P^{-1}$

$$\Rightarrow \qquad \text{Adj } P^{-1} = P(\text{Adj } P^{-1})P^{-1} = P\,|P^{-1}|\,I_n = P$$

$$(\because |P^{-1}| = 1 \text{ as } |PP^{-1}| = |P|\,|P^{-1}| = |I_n|)$$

Thus, $\text{Adj } P^{-1} = P.$ (**)

But $\qquad P \text{ Adj } P = |P|\, I_n = I_n \Rightarrow (\text{Adj } P)^{-1} = P$

Whence it follows from (**) that $\text{Adj } P^{-1} = (\text{Adj } P)^{-1}$.

The second result easily follows from (iii) and (iv).

$$\text{Adj } Q^{-1}BP^{-1} = \text{Adj }((Q^{-1}B)P^{-1})$$

$$= (\text{Adj } P^{-1})\text{ Adj } Q^{-1}B = (\text{Adj } P^{-1})(\text{Adj } B)(\text{Adj } Q^{-1})$$

$$= (\text{Adj } P)^{-1} A (\text{Adj } Q)^{-1} = PAQ$$

$(\because$ Adj B = A given and $\text{Adj } P^{-1} = P$, $\text{Adj } Q^{-1} = Q$ by (**))

Example 7. If $A = \begin{bmatrix} 1 & 0 & 3 \\ 2 & 1 & 1 \\ 0 & 0 & 2 \end{bmatrix}$, then |adj (adj A)| is equal to ______.

Solution

$$|A| = \begin{bmatrix} 1 & 0 & 3 \\ 2 & 1 & 1 \\ 0 & 0 & 2 \end{bmatrix} = 2$$

$$\therefore \qquad |\text{adj}(\text{adj } A)| = |A|^{(n-1)^2} = |A|^{2^2} \qquad [\because \text{Here } n = 3]$$

$$= 2^4 = 16$$

Example 8. If $I=\begin{bmatrix}1 & 0\\ 0 & 1\end{bmatrix}, J=\begin{bmatrix}0 & 1\\ -1 & 0\end{bmatrix}$ and $B=\begin{bmatrix}\cos\theta & \sin\theta\\ -\sin\theta & \cos\theta\end{bmatrix}$, then B equals ______.

(a) $I\cos\theta+J\sin\theta$ (b) $I\cos\theta-J\sin\theta$
(c) $I\sin\theta+J\cos\theta$ (d) $-I\cos\theta+J\sin\theta$

Solution

Here $\begin{bmatrix}\cos\theta & \sin\theta\\ -\sin\theta & \cos\theta\end{bmatrix}=\begin{bmatrix}\cos\theta & 0\\ 0 & \cos\theta\end{bmatrix}+\begin{bmatrix}0 & \sin\theta\\ -\sin\theta & 0\end{bmatrix}$

$=\cos\theta\begin{bmatrix}1 & 0\\ 0 & 1\end{bmatrix}+\sin\theta\begin{bmatrix}0 & 1\\ -1 & 0\end{bmatrix}$

$=I\cos\theta+J\sin\theta$

RECAPITULATION (FORMULAE, THEOREM, IMPORTANT RESULTS)

1. A matrix is an ordered rectangular array of numbers or functions.
2. A matrix having m rows and n columns is called a matrix of order $m\times n$.
3. $\left[a_{ij}\right]_{m\times 1}$ is a column matrix.
4. $\left[a_{ij}\right]_{1\times n}$ is a row matrix.
5. An $m\times n$ matrix is a square matrix if $m=n$.
6. $A=\left[a_{ij}\right]_{m\times n}$ is a diagonal matrix if $a_{ij}=0$, when $i\neq j$.
7. $A=\left[a_{ij}\right]_{n\times n}$ is a scalar matrix if $a_{ij}=0$, when $i\neq j$, $a_{ij}=k$, (k is some constant), when $i=j$.
8. $A=\left[a_{ij}\right]_{n\times n}$ is an identity matrix, if $a_{ij}=1$, when $i=j$, $a_{ij}=0$, when $i\neq j$.
9. A zero matrix has all its elements as zero.
10. $A=\left[a_{ij}\right]=\left[b_{ij}\right]=B$ if (i) A and B are of same order, (ii) $a_{ij}=b_{ij}$ for all possible values of i and j.
11. $kA=k\left[a_{ij}\right]_{m\times n}=\left[k\left(a_{ij}\right)\right]_{m\times n}$
12. $-A=(-1)A$
13. $A-B=A+(-1)B$
14. $A+B=B+A$
15. $(A+B)+C=A+(B+C)$, where A, B and C are of same order.
16. $k(A+B)=kA+kB$, where A and B are of same order, k is constant.
17. $(k+l)A=kA+lA$, where k and l are constant.
18. If $A=\left[a_{ij}\right]_{m\times n}$ and $B=\left[b_{jk}\right]_{n\times p}$, then $AB=C=\left[c_{jk}\right]_{n\times p}$, where $c_{ik}=\sum_{j=1}^{n}a_{ij}b_{jk}$
19. (i) A (BC) = (AB) C,
    (ii) A (B + C) = AB + AC,
    (iii) (A + B) C = AC + BC
20. If $A=\left[a_{ij}\right]_{m\times n}$, then A' or $A^{T}=\left[a_{ij}\right]_{n\times m}$
21. (i) $(A')'=A$,
    (ii) $(kA)'=kA'$,
    (iii) $(A+B)'=A'+B'$,
    (iv) $(AB)'=B'A'$
22. A is a symmetric matrix if $A'=A$.
23. A is a skew-symmetric matrix if $A'=-A$.
24. Any square matrix can be represented as the sum of a symmetric and a skew-symmetric matrix.
25. Elementary operations of a matrix are as follows:
    (i) $R_i\leftrightarrow R_j$ or $C_i\leftrightarrow C_j$
    (ii) $R_i\to kR_j$ or $C_i\to kC_j$
    (iii) $R_i\to R_i+kR_j$ or $C_i\to C_i+kC_j$
26. If A and B are two square matrices such that AB = BA = I, then B is the inverse matrix of A and is denoted by A^{-1} and A is the inverse of B.
27. Inverse of a square matrix, if it exists, is unique.

QUIZ

1. What is a matrix?
2. What is a square matrix?
3. If a matrix has 12 elements all together, then what are different possible orders it can have.
4. What is a scalar matrix?
5. Is matrix product commutative?
6. Is matrix multiplication associative?
7. What do you mean by trace of a matrix?
8. What is transpose of a matrix?
9. What is an upper triangular matrix?
10. What is a singular matrix?

JEE MAIN

Objective Questions

One Option Correct

1. Let A be a square matrix all of whose entries are integers. Then which one of the following is true?
(a) If det A $\neq \pm 1$, then A^{-1} exists and all its entries are non-integers.
(b) If det A $= \pm 1$, then A^{-1} exists and all its entries are integers.
(c) If det A $= \pm 1$, then A^{-1} need not exist.
(d) If det A $= \pm 1$, then A^{-1} exits but all its entries are not necessarily integers.

2. Let $A = \begin{bmatrix} 5 & 5\alpha & \alpha \\ 0 & \alpha & 5\alpha \\ 0 & 0 & 5 \end{bmatrix}$. If $|A^2| = 25$, then $|\alpha|$ equal to ______.
(a) $\frac{1}{5}$ (b) 5^2
(c) 1 (d) 5

3. If $A^2 - A + I = 0$, then inverse of A is ______.
(a) $I + A$ (b) $A - I$
(c) $I - A$ (d) A

4. If $A = \begin{bmatrix} 1 & 0 \\ 1 & 1 \end{bmatrix}$ and $I = \begin{bmatrix} 1 & 0 \\ 0 & 1 \end{bmatrix}$, then which one of the following holds for all $n \geq 1$ (by the principal of mathematical induction)?
(a) $A^n - nA + (n-1)I$
(b) $A^n = nA - (n-1)I$
(c) $A^n = 2^{n-1}A + (n-1)I$
(d) $A^n = 2^{n-1}A - (n-1)I$

5. Let $A = \begin{bmatrix} 0 & 0 & -1 \\ 0 & -1 & 0 \\ -1 & 0 & 0 \end{bmatrix}$. The only correct statement about the matrix A is
(a) A^{-1} does not exist (b) A is zero matrix
(c) $A = (-I)\, I_3$ (d) $A^2 = I$

6. Let $A = \begin{bmatrix} 1 & -1 & 1 \\ 2 & 1 & -3 \\ 1 & 1 & 1 \end{bmatrix}$ and $B\,(10) = \begin{bmatrix} 4 & 2 & 2 \\ -5 & 0 & \alpha \\ 1 & -2 & 3 \end{bmatrix}$ if B is the inverse of matrix A, then α is ______.
(a) −1 (b) 5
(c) 2 (d) −2

7. If $A = \begin{bmatrix} a & 0 & 0 \\ 0 & b & 0 \\ 0 & 0 & c \end{bmatrix}$, then A^{-1} is
(a) $\begin{bmatrix} \frac{1}{a} & 0 & 0 \\ 0 & \frac{1}{b} & 0 \\ 0 & 0 & \frac{1}{c} \end{bmatrix}$
(b) $\begin{bmatrix} -\frac{1}{a} & 0 & 0 \\ 0 & -\frac{1}{b} & 0 \\ 0 & 0 & \frac{1}{c} \end{bmatrix}$
(c) $\begin{bmatrix} a & 0 & 0 \\ 0 & b & 0 \\ 0 & 0 & \frac{1}{c} \end{bmatrix}$
(d) None of these.

8. Let $A = \begin{bmatrix} 1 & 2 \\ -5 & 1 \end{bmatrix}$ and $A^{-1} = xA + yI$, then the values of x and y are:

(a) $x = -\frac{1}{11}, y = \frac{2}{11}$ (b) $x = -\frac{1}{1}, y = -\frac{2}{11}$

(c) $x = \frac{1}{11}, y = \frac{2}{11}$ (d) $x = \frac{1}{11}, y = -\frac{2}{11}$

9. If $A = \begin{bmatrix} 1 & x \\ x^2 & 4y \end{bmatrix}$ and $B = \begin{bmatrix} -3 & 1 \\ 1 & 0 \end{bmatrix}$, adj A + B = $\begin{bmatrix} 1 & 0 \\ 0 & 1 \end{bmatrix}$, then the values of x and y are ______.

(a) 1, 1 (b) ±1, 1
(c) 1, 0 (d) None of these.

10. If A is a singular matrix, then A adj (A) is a
(a) Scalar matrix
(b) Zero matrix
(c) Identity matrix
(d) Orthogonal matrix

11. If A is 3×4 matrix and B is a matrix such that A′B and BA′ are both defined, then order of B is
(a) 3×4 (b) 3×3
(c) 4×4 (d) 4×3

12. If $A = \begin{bmatrix} 0 & 1 \\ 0 & 0 \end{bmatrix}$, I is the unit matrix of order 2 and a, b are arbitrary constants, then $(aI + bA)^2$ is equal to _____.

(a) $a^2I - abA$ (b) $a^2I + 2abA$

(c) $a^2I + b^2A$ (d) None of these.

13. If $\begin{bmatrix} 1 & 1 & 1 \\ 1 & -2 & -2 \\ 1 & 3 & 1 \end{bmatrix}\begin{bmatrix} x \\ y \\ z \end{bmatrix} = \begin{bmatrix} 0 \\ 3 \\ 4 \end{bmatrix}$, then $\begin{bmatrix} x \\ y \\ z \end{bmatrix}$ is equal to:

(a) $\begin{bmatrix} 0 \\ 1 \\ 1 \end{bmatrix}$ (b) $\begin{bmatrix} 1 \\ 2 \\ -3 \end{bmatrix}$

(c) $\begin{bmatrix} 5 \\ -2 \\ 1 \end{bmatrix}$ (d) $\begin{bmatrix} 1 \\ -2 \\ 3 \end{bmatrix}$

14. If $A = \begin{bmatrix} 6 & 8 & 5 \\ 4 & 2 & 3 \\ 9 & 7 & 1 \end{bmatrix}$ is the sum of symmetric matrix B and skew-symmetric matrix C, then B is:

(a) $\begin{bmatrix} 6 & 6 & 7 \\ 6 & 2 & 5 \\ 7 & 5 & 1 \end{bmatrix}$ (b) $\begin{bmatrix} 0 & 2 & -2 \\ -2 & 5 & -2 \\ 2 & 2 & 0 \end{bmatrix}$

(c) $\begin{bmatrix} 6 & 6 & 7 \\ -6 & 2 & -5 \\ -7 & 5 & 1 \end{bmatrix}$ (d) $\begin{bmatrix} 0 & 6 & -2 \\ -2 & 5 & -2 \\ -2 & -2 & 0 \end{bmatrix}$

15. If A and B are 2×2 matrices, then which of the following is true?
(a) $(A+B)^2 = A^2 + B^2 + 2AB$
(b) $(A-B)^2 = A^2 + B^2 - 2AB$
(c) $(A-B)(A+B) = A^2 + AB - AB - B^2$
(d) $(A-B)(A+B) = A^2 - B^2$

16. If $A = \begin{bmatrix} 0 & 0 & 1 \\ 0 & 1 & 0 \\ 1 & 0 & 0 \end{bmatrix}$, then A^{-1} is:

(a) –A (b) A
(c) 1 (d) None of these.

17. Let A, B and C be $n \times n$ matrices. Which one of the following is a correct statement?
(a) If AB = AC, then B = C.
(b) If $A^3 + 2A^2 + 3A + 5I = 0$; then A is invertible.
(c) If $A^2 = 0$, then A = 0.
(d) None of these.

18. If $A = \begin{bmatrix} 4 & 2 \\ 3 & 4 \end{bmatrix}$, $|\text{adj}\,A|$ is equal to ______.

(a) 16 (b) 10
(c) 6 (d) None of these.

19. Matrix $A = \begin{bmatrix} 1 & 0 & -k \\ 2 & 1 & 3 \\ k & 0 & 1 \end{bmatrix}$ is invertible for ______.

(a) $k = 1$ (b) $k = -1$
(c) $k = 0$ (d) all real k

20. If $A = \begin{bmatrix} 0 & 3 \\ 2 & 0 \end{bmatrix}$ and $A^{-1} = \lambda(A)$, then $\lambda =$ ______.

(a) $-\frac{1}{6}$ (b) $\frac{1}{3}$

(c) $-\frac{1}{3}$ (d) $\frac{1}{6}$

21. If $A = \begin{bmatrix} 1 & 1 \\ 1 & 1 \end{bmatrix}$ then $A^{100} =$ _____.

(a) $2^{100}A$ (b) $2^{99}A$
(c) $2^{101}A$ (d) None of these.

22. If $A = \begin{bmatrix} -2 & 6 \\ -5 & 7 \end{bmatrix}$, then adj A = ______.

(a) $\begin{bmatrix} 7 & -6 \\ 5 & -2 \end{bmatrix}$ (b) $\begin{bmatrix} 2 & -6 \\ 5 & -7 \end{bmatrix}$

(c) $\begin{bmatrix} 7 & -5 \\ 6 & -2 \end{bmatrix}$ (d) None of these.

23. If $A = \begin{bmatrix} i & -i \\ -i & i \end{bmatrix}$ and $B = \begin{bmatrix} 1 & -1 \\ -1 & 1 \end{bmatrix}$, then A^8 equals ______.

(a) 4 B (b) 128 B
(c) −128 B (d) −64 B

24. Inverse of the matrix $\begin{bmatrix} \cos 2\theta & -\sin 2\theta \\ \sin 2\theta & \cos 2\theta \end{bmatrix}$ is ____.

(a) $\begin{bmatrix} \cos 2\theta & -\sin 2\theta \\ \sin 2\theta & \cos 2\theta \end{bmatrix}$ (b) $\begin{bmatrix} \cos 2\theta & \sin 2\theta \\ \sin 2\theta & -\cos 2\theta \end{bmatrix}$

(c) $\begin{bmatrix} \cos 2\theta & \sin 2\theta \\ \sin 2\theta & \cos 2\theta \end{bmatrix}$ (d) $\begin{bmatrix} \cos 2\theta & \sin 2\theta \\ -\sin 2\theta & \cos 2\theta \end{bmatrix}$

25. What must be the matrix X if

$2X + \begin{bmatrix} 1 & 2 \\ 3 & 4 \end{bmatrix} = \begin{bmatrix} 3 & 8 \\ 7 & 2 \end{bmatrix}$?

(a) $\begin{bmatrix} 1 & 3 \\ 2 & -1 \end{bmatrix}$ (b) $\begin{bmatrix} 1 & -3 \\ 2 & -1 \end{bmatrix}$

(c) $\begin{bmatrix} 2 & 6 \\ 4 & -2 \end{bmatrix}$ (d) $\begin{bmatrix} 2 & -6 \\ 4 & -2 \end{bmatrix}$

26. If A is a skew-symmetric matrix of order n and C is a column matrix of order $n \times 1$, then $C^T AC$ is

(a) an identity matrix of order n.
(b) an identity matrix of order 1.
(c) a zero matrix of order 1.
(d) None of these.

27. If a square matrix A is such that $AA^T = I = A^TA$, then |A| is equal to ______.

(a) 0 (b) ±1
(c) ±2 (d) None of these.

28. If $\begin{bmatrix} x+y+z \\ x+y \\ y+z \end{bmatrix} = \begin{bmatrix} 9 \\ 5 \\ 7 \end{bmatrix}$, then the value of (x, y, z) is ______.

(a) (4, 3. 2) (b) (3, 2, 4)
(c) (2, 3, 4) (d) None of these.

29. If A is $n \times n$ matrix, then adj (adj A) =

(a) $|A|^{n-1} A$ (b) $|A|^{n-2} A$
(c) $|A|^n n$ (d) None of these.

30. $\cos\theta \begin{bmatrix} \cos\theta & \sin\theta \\ -\sin\theta & \cos\theta \end{bmatrix} + \sin\theta \begin{bmatrix} \sin\theta & -\cos\theta \\ \cos\theta & \sin\theta \end{bmatrix}$ is equal to ______.

(a) $\begin{bmatrix} 0 & 0 \\ 0 & 0 \end{bmatrix}$ (b) $\begin{bmatrix} 1 & 0 \\ 0 & 0 \end{bmatrix}$

(c) $\begin{bmatrix} 0 & 1 \\ 1 & 0 \end{bmatrix}$ (d) $\begin{bmatrix} 1 & 0 \\ 0 & 1 \end{bmatrix}$

31. If $\begin{bmatrix} 3 & 1 \\ 4 & 1 \end{bmatrix} X = \begin{bmatrix} 5 & -1 \\ 2 & 3 \end{bmatrix}$, then X is equal to ______.

(a) $\begin{bmatrix} -3 & 4 \\ 14 & -13 \end{bmatrix}$ (b) $\begin{bmatrix} 3 & -4 \\ -14 & 13 \end{bmatrix}$

(c) $\begin{bmatrix} 3 & 4 \\ 14 & 13 \end{bmatrix}$ (d) $\begin{bmatrix} -3 & 4 \\ -14 & 13 \end{bmatrix}$

32. If matrix $A = \begin{vmatrix} 1 & 0 & -1 \\ 3 & 4 & 5 \\ 0 & 6 & 7 \end{vmatrix}$ and its inverse is denoted by $A^{-1} = \begin{vmatrix} a_{11} & a_{12} & a_{13} \\ a_{21} & a_{22} & a_{23} \\ a_{31} & a_{32} & a_{33} \end{vmatrix}$, then the value of a_{23} is equal to ______.

(a) $\frac{21}{20}$ (b) $\frac{1}{5}$

(c) $-\frac{2}{5}$ (d) $\frac{2}{5}$

33. If $A = \begin{bmatrix} 1 & 2 \\ -3 & 0 \end{bmatrix}$ and $B = \begin{bmatrix} -1 & 0 \\ 2 & 3 \end{bmatrix}$, then ______.

(a) AB = BA (b) $B^2 = B$
(c) AB ≠ BA (d) $A^2 = A$

34. Which one of the following statements is true?

(a) If |A| ≠ 0, then $|A \text{ adj } A|^{(n-1)}$ where $A = |a_n|_{n \times n}$.
(b) If A′ − A, then A is a square matrix.
(c) Determinant of a non-square matrix is zero.
(d) Non-singular square matrix does not have a unique inverse.

35. If matrix $A = \begin{bmatrix} 1 & -1 \\ 1 & 1 \end{bmatrix}$, then

(a) $A' = \begin{bmatrix} 1 & 1 \\ 1 & -1 \end{bmatrix}$ (b) $A^{-1} = \begin{bmatrix} 1 & 1 \\ -1 & 1 \end{bmatrix}$

(c) $\text{adj } A = \begin{bmatrix} 1 & 1 \\ -1 & 1 \end{bmatrix}$ (d) $\lambda A = \begin{bmatrix} \lambda & -\lambda \\ 1 & 1 \end{bmatrix}$

36. If for AX = B, $B = \begin{bmatrix} 9 \\ 52 \\ 0 \end{bmatrix}$ and

$A^{-1} = \begin{bmatrix} 3 & -\frac{1}{2} & -\frac{1}{2} \\ -4 & \frac{3}{4} & \frac{5}{4} \\ 2 & -\frac{3}{4} & -\frac{3}{4} \end{bmatrix}$ then X is equal to ____.

(a) $\begin{bmatrix} 3 \\ \frac{3}{4} \\ -\frac{3}{4} \end{bmatrix}$ (b) $\begin{bmatrix} -\frac{1}{2} \\ -\frac{1}{2} \\ 2 \end{bmatrix}$

(c) $\begin{bmatrix} -4 \\ 2 \\ 3 \end{bmatrix}$ (d) $\begin{bmatrix} 1 \\ 3 \\ 5 \end{bmatrix}$

37. If $A = \begin{pmatrix} 1 & -2 \\ 4 & 5 \end{pmatrix}$ and $f(t) = t^2 - 3t + 7$, then $f(A) + \begin{pmatrix} 3 & 6 \\ 12 & -9 \end{pmatrix} =$ _____.

(a) $\begin{pmatrix} 1 & 0 \\ 0 & 1 \end{pmatrix}$ (b) $\begin{pmatrix} 0 & 0 \\ 0 & 0 \end{pmatrix}$

(c) $\begin{pmatrix} 0 & 1 \\ 1 & 0 \end{pmatrix}$ (d) $\begin{pmatrix} 1 & 1 \\ 0 & 0 \end{pmatrix}$

38. The inverse of the matrix $\begin{bmatrix} 7 & -3 & -3 \\ -1 & 1 & 0 \\ -1 & 0 & 1 \end{bmatrix}$ is ____.

(a) $\begin{bmatrix} 1 & 1 & 1 \\ 3 & 4 & 3 \\ 3 & 3 & 4 \end{bmatrix}$ (b) $\begin{bmatrix} 1 & 3 & 1 \\ 4 & 3 & 8 \\ 3 & 4 & 1 \end{bmatrix}$

(c) $\begin{bmatrix} 1 & 1 & 1 \\ 3 & 3 & 4 \\ 3 & 4 & 3 \end{bmatrix}$ (d) $\begin{bmatrix} 1 & 3 & 3 \\ 1 & 4 & 3 \\ 1 & 3 & 4 \end{bmatrix}$

39. If $\begin{bmatrix} 1 & 2 & x \\ 4 & -1 & 7 \\ 2 & 4 & -6 \end{bmatrix}$ is a singular matrix, then x is equal to _____.

(a) 0 (b) 1
(c) −3 (d) 2

40. If A is a square matrix such that A (adj A) = $\begin{bmatrix} 4 & 0 & 0 \\ 0 & 4 & 0 \\ 0 & 0 & 4 \end{bmatrix}$, the det (adj A) is equal to _____.

(a) 4 (b) 16
(c) 64 (d) 256

41. $A = \begin{bmatrix} 1 & 2 & 2 \\ 2 & 1 & 2 \\ 2 & 2 & 1 \end{bmatrix}$, then $A^3 - 4A^2 - 6A$ is equal to _____.

(a) 0 (b) A
(c) −A (d) 1

42. $\text{adj} \begin{bmatrix} 1 & 0 & 2 \\ -1 & 1 & -2 \\ 0 & 2 & 1 \end{bmatrix} = \begin{bmatrix} 5 & a & -2 \\ 1 & 1 & 0 \\ -2 & -2 & b \end{bmatrix}$, then $[a \quad b]$ is equal to _____.

(a) [−4 1] (b) [−4 −1]
(c) [4 1] (d) [4 −1]

43. If A, B are square matrices of order 3. A is non-singular and AB = 0, then B is a _____.
(a) null matrix
(b) non-singular matrix
(c) singular matrix
(d) unit matrix

44. If $A = \begin{bmatrix} 1 & 0 & 1 \\ 2 & 1 & 0 \\ 3 & 2 & 1 \end{bmatrix}$, that det A is equal to _____.

(a) 2 (b) 5
(c) 3 (d) 4

45. If A is a skew-symmetric matrix, then trace of A is _____.
(a) 0 (b) −1
(c) 1 (d) None of these.

46. If $A = \begin{bmatrix} 0 & -1 & 2 \\ 1 & 0 & 3 \\ -2 & -3 & 0 \end{bmatrix}$. Then $A + 2A^T$ equal to __.

(a) A^T (b) 2^2
(c) A (d) $-A^T$

47. If $A=\begin{bmatrix}1 & -2\\ 3 & 4\end{bmatrix}, B=\begin{bmatrix}2 & 1\\ -4 & 3\end{bmatrix}, C=\begin{bmatrix}2 & 12\\ -7 & 9\end{bmatrix}$ then $5A-3B+C$ is equal to ______.

(a) $\begin{bmatrix}-1 & -1\\ -20 & 20\end{bmatrix}$ (b) $\begin{bmatrix}-1 & 1\\ 20 & 20\end{bmatrix}$

(c) $\begin{bmatrix}1 & 1\\ 20 & 20\end{bmatrix}$ (d) $\begin{bmatrix}1 & -1\\ 20 & 20\end{bmatrix}$

48. Which statement is wrong?

(a) $tr(A)=tr(A^T)$

(b) $tr(A+B)=tr(A)+tr(B)$

(c) $tr(A-B)=tr(A)+tr(B)$

(d) $tr(AB)=tr(A)\cdot tr(B)$

49. If AB = C then A, B, C is equal to

(a) $A_{2\times3}, B_{3\times2}, C_{2\times3}$ (b) $A_{3\times2}, B_{2\times3}, C_{3\times3}$

(c) $A_{3\times3}, B_{2\times3}, C_{3\times3}$ (d) $A_{3\times2}, B_{2\times3}, C_{3\times2}$

50. If $A=\begin{bmatrix}0 & 2\\ 3 & -4\end{bmatrix}$ and $kA=\begin{bmatrix}0 & 3a\\ 2b & 24\end{bmatrix}$, then the values of k, a, b are respectively, _____.

(a) −6, 12, 18 (b) −6, −12, −18

(c) −6, 4, 9 (d) −6, −4, −9

51. If A, B, C is same order non-singular matrices then $(AB^{-1}C)^{-1}$ is equal to _____.

(a) CBA^{-1} (b) $C^{-1}B^{-1}A^{-1}$

(c) $A^{-1}BC^{-1}$ (d) $C^{-1}BA^{-1}$

52. If $A=\begin{bmatrix}2 & 0 & 0\\ 0 & 2 & 0\\ 0 & 0 & 2\end{bmatrix}$ and $B=\begin{bmatrix}1 & 2 & 3\\ 0 & 1 & 3\\ 0 & 0 & 2\end{bmatrix}$, then |AB| is equal to _____.

(a) 32 (b) 16

(c) 8 (d) 4

53. If $A=\begin{bmatrix}1 & 3\\ 3 & 2\\ 2 & 5\end{bmatrix}$ and $B=\begin{bmatrix}-1 & -2\\ 0 & 5\\ 3 & 1\end{bmatrix}$ and A + B − D = 0 (zero-matrix), then D matrix will be:

(a) $\begin{bmatrix}0 & 1\\ 3 & 7\\ 5 & 6\end{bmatrix}$ (b) $\begin{bmatrix}0 & -2\\ -3 & -7\\ -5 & -6\end{bmatrix}$

(c) $\begin{bmatrix}0 & 2\\ 3 & 7\\ 5 & 6\end{bmatrix}$ (d) $\begin{bmatrix}0 & 2\\ 3 & 7\\ 6 & 5\end{bmatrix}$

54. If A is a skew-symmetric matrix and n is odd positive integer, then A^n is

(a) a skew-symmetric matrix.

(b) diagonal matrix.

(c) a symmetric matrix.

(d) None of these.

55. If two matrices A and B are of order $p \times q$ and $r \times s$ respectively, can be subtracted only if _____.

(a) $p=r, q=s$

(b) $p=q, r=s$

(c) $p=q$

(d) None of these.

56. If $A=\begin{bmatrix}3 & 1 & -1\\ 0 & 1 & 2\end{bmatrix}$, then AA′ is a

(a) skew-symmetric matrix.

(b) orthogonal matrix.

(c) symmetric matrix.

(d) None of these.

57. Let A and B be 3 × 3 matrices, then AB = 0 implies

(a) either |A| = 0 or |B| = 0.

(b) A = 0 or B = 0.

(c) A = 0 and B = 0.

(d) |A| = 0 and |B| = 0.

58. For a matrix A of order 2 × 2. If (adj A) = $\begin{bmatrix}10 & 0\\ 0 & 10\end{bmatrix}$ then |A| = _____.

(a) 100 (b) 20

(c) 0 (d) 10

59. If $A=\begin{bmatrix}0 & 2 & 0\\ 0 & 0 & 3\\ -2 & 2 & 0\end{bmatrix}$ and $B=\begin{bmatrix}1 & 2 & 3\\ 3 & 4 & 5\\ 5 & -4 & 0\end{bmatrix}$, then the element of 3rd row and third column in AB will be _____.

(a) −12

(b) 4

(c) −18

(d) None of these.

60. If $A=(a_{ij})_{2\times2}$, where $a_{ij}=i+j$, then A is equal to _____.

(a) $\begin{bmatrix}2 & 3\\ 3 & 4\end{bmatrix}$ (b) $\begin{bmatrix}1 & 2\\ 3 & 4\end{bmatrix}$

(c) $\begin{bmatrix}1 & 1\\ 2 & 2\end{bmatrix}$ (d) $\begin{bmatrix}1 & 2\\ 1 & 2\end{bmatrix}$

Questions from Previous Years

61. If $P=\begin{bmatrix}1 & 0\\ \frac{1}{2} & 1\end{bmatrix}$, $(A,B,C\subseteq R)$ then P^{50} is:

(a) $\begin{bmatrix}1 & 0\\ 25 & 1\end{bmatrix}$ (b) $\begin{bmatrix}1 & 50\\ 0 & 1\end{bmatrix}$

(c) $\begin{bmatrix}1 & 25\\ 0 & 1\end{bmatrix}$ (d) $\begin{bmatrix}1 & 0\\ 50 & 1\end{bmatrix}$

62. Let

$A+2B=\begin{bmatrix}1 & 2 & 5\\ 6 & -3 & 6\\ -5 & 3 & 1\end{bmatrix}$ and $A-2B=\begin{bmatrix}2 & -1 & 5\\ 2 & -1 & 6\\ 0 & 1 & 2\end{bmatrix}$.

If Tr (A) denotes the sum of all diagonal elements of the matrix A, then Tr b (A) – Tr (B) has value equal to ______.

(a) 1 (b) 3
(c) 0 (d) 2

63. Let A be a symmetric matrix of order 2 with integer entries. If the sum of the diagonal elements of A^2 is 1, then the possible number of such matrices is ______.

(a) 6 (b) 1
(c) 4 (d) 12

64. Let A and B be 3×3 real matrices such that A is a symmetric matrix and B is a skew-symmetric matrix. Then the systems of linear equations where X is a 3×1 column matrix of unknown variables and O is a 3×1 null matrix, has

(a) no solution.
(b) exactly two solutions.
(c) infinitely many solutions.
(d) a unique solution.

65. If $A=\begin{bmatrix}\cos\theta & i\sin\theta\\ i\sin\theta & \cos\theta\end{bmatrix}$, $\left(\theta=\frac{\pi}{24}\right)$ and

$A^5=\begin{bmatrix}a & b\\ c & d\end{bmatrix}$, where $i=\sqrt{-1}$, then which of the following is not true?

(a) $0\le a^2+b^2\le 1$ (b) a^2-d^2

(c) $a^2-c^2=1$ (d) $a^2-b^2=\frac{1}{2}$

66. Let α be a root of the question $x^2+x+1=0$ and the matrix $A=\frac{1}{\sqrt{3}}\begin{bmatrix}1 & 1 & 1\\ 1 & \alpha & \alpha^2\\ 1 & \alpha^2 & \alpha^4\end{bmatrix}$, then matrix A^{31} is equal to:

(a) A (b) I_3
(c) A^2 (d) A^3

67. Let $a,b,c\in R$ be all non-zero and satisfy $a^3+b^3+c^3=2$. If the matrix $A=\begin{pmatrix}a & b & c\\ b & c & a\\ c & a & b\end{pmatrix}$ satisfies $A^TA=I$, then a value of *abc* can be:

(a) $-\frac{1}{3}$ (b) $\frac{1}{3}$

(c) 3 (d) $\frac{2}{3}$

68. If $A=\begin{pmatrix}2 & 2\\ 9 & 4\end{pmatrix}$ and $I=\begin{pmatrix}1 & 0\\ 0 & 1\end{pmatrix}$, then $10A^{-1}$ is equal to ______.

(a) $A-4I$ (b) $6I-A$
(c) $A-6I$ (d) $4I-A$

69. Let $A=\begin{bmatrix}\cos\alpha & -\sin\alpha\\ \sin\alpha & \cos\alpha\end{bmatrix}$, $(\alpha\in R)$ such that $A^{32}=\begin{bmatrix}0 & -1\\ 1 & 0\end{bmatrix}$. Then the value of α is ______.

(a) $\frac{\pi}{32}$ (b) 0

(c) $\frac{\pi}{64}$ (d) $\frac{\pi}{16}$

70. Let $P=\begin{bmatrix}1 & 0 & 0\\ 3 & 1 & 0\\ 9 & 3 & 1\end{bmatrix}$ and $Q=\left[q_{ij}\right]$ be two 3×3 matrices such that $Q-P^5=I_3$. Then $\frac{q_{21}+q_{31}}{q_{32}}$ is equal to ______.

(a) 10 (b) 135
(c) 15 (d) 9

71. If A is a symmetric matrix and B is a skew-symmetric matrix such that $A+B=\begin{bmatrix}2 & 3\\ 5 & -1\end{bmatrix}$, then AB is equal to ______.

(a) $\begin{bmatrix}-4 & -1\\ -1 & 4\end{bmatrix}$ (b) $\begin{bmatrix}4 & -2\\ -1 & -4\end{bmatrix}$

(c) $\begin{bmatrix}4 & -2\\ 1 & -4\end{bmatrix}$ (d) $\begin{bmatrix}-4 & 2\\ 1 & 4\end{bmatrix}$

72. The total number of matrices $A = \begin{pmatrix} 0 & 2y & 1 \\ 2x & y & -1 \\ 2x & -y & 1 \end{pmatrix}$, $(x, y \in R, x \neq y)$ for which $A^TA = 3I_3$ is ______.

(a) 2 (b) 3
(c) 6 (d) 4

73. Let $A = \begin{pmatrix} 0 & 2q & r \\ p & q & -r \\ p & -q & r \end{pmatrix}$. If $AA^T = I_3$, then $|p|$ is ______.

(a) $\frac{1}{\sqrt{5}}$ (b) $\frac{1}{\sqrt{3}}$
(c) $\frac{1}{\sqrt{2}}$ (d) $\frac{1}{\sqrt{6}}$

74. Let $A = \begin{bmatrix} 1 & 0 & 0 \\ 1 & 1 & 0 \\ 1 & 1 & 1 \end{bmatrix}$ and $B = A^{20}$. Then the sum of the elements of the first column of B is?

(a) 211 (b) 210
(c) 231 (d) 251

75. For two 3×3 matrices A and B, let $A + B = 2B^T$ and $3A + 2B = I_3$, where B^T is the transpose of B and I_3 is 3×3 identity matrix. Then:

(a) $5A + 10B = 2I_3$ (b) $10A + 5B = 3I_3$
(c) $B + 2A = I_3$ (d) $3A + 6B = 2I_3$

76. If $P = \begin{bmatrix} \frac{\sqrt{3}}{2} & \frac{1}{2} \\ -\frac{1}{2} & \frac{\sqrt{3}}{2} \end{bmatrix}$, $A = \begin{bmatrix} 1 & 1 \\ 0 & 1 \end{bmatrix}$ and $Q = PAP^T$, then $P^T Q^{2015} P$ is

(a) $\begin{bmatrix} 0 & 2015 \\ 0 & 0 \end{bmatrix}$ (b) $\begin{bmatrix} 2015 & 0 \\ 1 & 2015 \end{bmatrix}$

(c) $\begin{bmatrix} 1 & 2015 \\ 0 & 1 \end{bmatrix}$ (d) $\begin{bmatrix} 2015 & 1 \\ 0 & 2015 \end{bmatrix}$

77. If $A = \begin{bmatrix} 0 & -1 \\ 1 & 0 \end{bmatrix}$, then which one of the following statements is not correct?

(a) $A^2 + I = A(A^2 - I)$ (b) $A^4 - I = A^2 + I$

(c) $A^3 + I = A(A^3 - I)$ (d) $A^3 - I = A(A - I)$

78. If $A = \begin{bmatrix} 1 & 2 & 2 \\ 2 & 1 & -2 \\ a & 2 & b \end{bmatrix}$ is a matrix satisfying the equation $AA^T = 9I$, where I is a 3×3 identity matrix, then the ordered pair (*a*, *b*) is equal to:

(a) (2, 1) (b) (−2, −1)
(c) (2, −1) (d) (−2, 1)

79. If $A = \begin{bmatrix} 1 & 2 & x \\ 3 & -1 & 2 \end{bmatrix}$ and $B = \begin{bmatrix} y \\ x \\ 1 \end{bmatrix}$ be such that $AB = \begin{bmatrix} 6 \\ 8 \end{bmatrix}$, then ______.

(a) $y = 2x$ (b) $y = -2x$
(c) $y = x$ (d) $y = -x$

80. If *p*, *q*, *r*, are 3 real numbers satisfying the matrix equation, $[p\ q\ r]\begin{bmatrix} 3 & 4 & 1 \\ 3 & 2 & 3 \\ 2 & 0 & 2 \end{bmatrix} = [3\ 0\ 1]$ then $2p + q - r$ equals ______.

(a) −3 (b) −1
(c) 4 (d) 2

81. If P is a 3×3 matrix such that $P^T = 2P + I$, where P^T I is the transpose of P and I is the 3×3 identity matrix, then there exists a column matrix $X = \begin{bmatrix} x \\ y \\ z \end{bmatrix} \neq \begin{bmatrix} 0 \\ 0 \\ 0 \end{bmatrix}$ such that

(a) $PX = \begin{bmatrix} 0 \\ 0 \\ 0 \end{bmatrix}$ (b) $PX = X$

(c) $PX = 2X$ (d) $PX = -X$

82. Let $\omega \neq 1$ be a cube root of unity and S be the set of all non-singular matrices of the form $\begin{bmatrix} 1 & a & b \\ \omega & 1 & c \\ \omega^2 & \omega & 1 \end{bmatrix}$ where each of *a*, *b* and *c* is either ω or ω^2. Then the number of distinct matrices in the set S is ______.

(a) 2 (b) 6
(c) 4 (d) 8

Integer and Numerical Value Type

1. If $A=\begin{bmatrix} t & t+1 \\ t-1 & t \end{bmatrix}$ is a matrix such that $AA' = I_2$ then trace of the matrix must be ____.

2. If $2s=a+b+c$ and $\begin{bmatrix} a^2 & (s-a)^2 & (s-a)^2 \\ (s-b)^2 & b^2 & (s-b)^2 \\ (s-c)^2 & (s-c)^2 & c^2 \end{bmatrix}$ $= ks^3(s-a)(s-b)(s-c)$ then the numerical quantity k should be ____.

3. Let A be an $n\times m$ square matrix given by $A=\begin{bmatrix} 0 & 1 & 1 & 1 & ....1 \\ 1 & 0 & 1 & 1 & .....1 \\ 1 & 1 & 0 & 1 & .....1 \\ ..... & ..... & ..... & ..... & ..... \\ 1 & 1 & 1 & 1 & .....0 \end{bmatrix}$, then in the inverse of A, each diagonal element is equal to $\frac{k-n}{n-1}$ where k is a numerical quantity, then k must be equal to ____.

4. If $A = \begin{pmatrix} 1 & 0 & 0 \\ 1 & 0 & 1 \\ 0 & 1 & 0 \end{pmatrix}$ then $A^3 - A^2 - A + I = 0$ is true. If k is a numerical quantity and $A^{50} = kA^2 - 24I_3$, then k must be ____.

5. If $\begin{vmatrix} \frac{a^2+b^2}{c} & c & c \\ a & \frac{b^2+a^2}{a} & a \\ b & b & \frac{c^2+a^2}{b} \end{vmatrix} = kabc$ where k is a numerical quantity, then k should be ____.

6. Let A be a 3×3 matrix given by $A = (a_{ij})_{3\times3}$. If for every column vector X X′, AX = 0 and $a_{23} = -2006$ then a_{32} must be equal to ____.

Questions from Previous Years

7. The number of elements in the set

$$\left\{A=\begin{pmatrix} a & b \\ 0 & d \end{pmatrix}: a, b, d \in \{-1, 0, 1\} \text{ and } (I-A)^3 = I - A^3\right\}$$

where I is 2×2 identity matrix is ______.

8. If $A = \begin{bmatrix} 1 & 1 & 1 \\ 0 & 1 & 1 \\ 0 & 0 & 1 \end{bmatrix}$ and $M = A + A^2 + A^3 + ... + A^{20}$, then the sum of all the elements of the matrix M is equal to ______.

9. Let $S = \left\{n \in N \left| \begin{pmatrix} 0 & i \\ 1 & 0 \end{pmatrix}^n \begin{pmatrix} a & b \\ c & d \end{pmatrix} = \begin{pmatrix} a & b \\ c & d \end{pmatrix} \right. \forall a, b, c, d \in R\right\}$ where $i = \sqrt{-1}$. Then the number of 2-digit numbers in the set S is ______.

10. Let $A = \begin{bmatrix} 0 & 1 & 0 \\ 1 & 0 & 0 \\ 0 & 0 & 1 \end{bmatrix}$. Then the numbers of 3×3 matrices B with entries from the set {1, 2, 3, 4, 5} and satisfying AB = BA is ______.

11. Let $A = \begin{bmatrix} a_1 \\ a_2 \end{bmatrix}$ and $B = \begin{bmatrix} b_1 \\ b_2 \end{bmatrix}$ be two 2×1 matrices with real entries such that A = XB, where $X = \frac{1}{\sqrt{3}}\begin{bmatrix} 1 & -1 \\ 1 & k \end{bmatrix}$, and $k \in R$. If $a_1^2 + a_2^2 = \frac{2}{3}(b_1^2 + b_2^2)$ and $(k^2+1)b_2^2 \neq 2b_1b_2$, then the value of k is ______.

12. If the matrix $A = \begin{bmatrix} 1 & 0 & 0 \\ 0 & 2 & 2 \\ 3 & 0 & -1 \end{bmatrix}$ satisfying the equation $A^{20} + \alpha A^{19} + \beta A = \begin{bmatrix} 1 & 0 & 0 \\ 0 & 4 & 0 \\ 0 & 0 & 1 \end{bmatrix}$ for some real numbers α and β, then $\beta - \alpha$ is equal to ______.

13. The total number of 3×3 matrices A, having entries from the set 0, 1, 2, 3} such that the sum of all the diagonal entries of AA^T is 9, is equal to ______.

14. Let M be any 3×3 matrix with entries from the set {0, 1, 2}. The maximum number of such matrices, for which the sum of diagonal elements of $M^T M$ is seven, is ______.

15. Let $A = \begin{bmatrix} x & 1 \\ 1 & 0 \end{bmatrix}$, and $A^4 = [a_{ij}]$. If $a_{11} = 109$, then a_{22} is equal to ______.

16. The number of all 3×3 matrices A, with entries from the set $\{-1, 0, 1\}$ such that the sum of the diagonal elements of AA^T is 3, is _______.

17. Let M be a 3×3 matrix satisfying

$$M\begin{bmatrix}0\\1\\0\end{bmatrix}=\begin{bmatrix}-1\\2\\3\end{bmatrix}, M\begin{bmatrix}1\\-1\\0\end{bmatrix}=\begin{bmatrix}1\\1\\-1\end{bmatrix}, \text{and } M\begin{bmatrix}1\\1\\1\end{bmatrix}=\begin{bmatrix}0\\0\\12\end{bmatrix}.$$

Then the sum of the diagonal entries of M is _______.

Assertion and Reason

Each of these questions contains two statements: Statement 1 (Assertion) and Statement 2 (Reason). Each of these questions has four alternative answers. Only one of them is the correct answer. Select the correct answer:

(a) Statement 1 is true, Statement 2 is true; Statement 2 is the correct explanation for Statement 1.
(b) Statement 1 is true, Statement 2 is true; Statement 2 is NOT a correct explanation for Statement 1.
(c) Statement 1 is true, Statement 2 is false.
(d) Statement 1 is false, Statement 2 is true

1. **Statement 1:** The determinants of a matrix $A=\left[a_{ij}\right]_{5\times5}$ where $a_{ij}+a_{ij}=0$ for i and j is zero.

Statement 2: The determinant of a skew-symmetric matrix of odd order is zero.

2. **Statement 1:** The inverse of the matrix $A=\left[a_{ij}\right]_{n\times n}$ where $a_{ij}=0, i\geq j$ is $B=\left[a_{ij}^{-1}\right]_{n\times n}$.

Statement 2: The inverse of singular matrix does not exist.

3. **Statement 1:** If $f_1(x), f_2(x)\ldots, f_g(x)$ are polynomials whose degree ≥ 1, where $f_1(\alpha)= f_2(\alpha)=f_2(\alpha)\ldots=f_9(\alpha)=0$ and

$$A(x) = \begin{bmatrix}f_1(x) & f_2(x) & f_3(x)\\ f_4(x) & f_5(x) & f_6(x)\\ f_7(x) & f_8(x) & f_9(x)\end{bmatrix} \text{ and } \frac{A(x)}{x-\alpha} \text{ is}$$

also a matrix of , 3×3 whose entries are also polynomials.

Statement 2: $x-\alpha$ is a factor of polynomial $f(x)$ if $f(\alpha)=0$.

4. **Statement 1:** For a singular square matrix A, if $AB = AC \Rightarrow B = C$.

Statement 2: If $|A| = 0$ then A^{-1} does not exist.

5. **Statement 1:** $(a_{11}, a_{22},\ldots a_{nn})$ is a diagonal matrix then $A^{-1}=\text{dia}\left(a_{11}^{-1}, a_{22}^{-1},\ldots a_{nn}^{-1}\right)$.

Statement 2: If A = dia (2, 1, −3) and B = dia (1, 1, 2) then det $(AB^{-1}) = 3$.

JEE ADVANCED

Objective Questions

One Option Correct

1. Let A and B are two non-singular square matrices, A^T and B^T be the transpose matrices of A and B respectively, then which of the following is correct?

(a) B^TAB is symmetric matrix if and only if A is symmetric.
(b) B^TAB is symmetric matrix if and only if B is symmetric.
(c) B^TAB is skew-symmetric matrix for every matrix A.
(d) B^TAB is skew-symmetric matrix if B is skew-symmetric.

2. If A and B are two square matrices of order 3×3 which satisfy AB = A and BA = B then $(A+B)^7$ is _______.

(a) $7(A+B)$ (b) $7.I_{3\times3}$
(c) $64(A+B)$ (d) $128I_{3\times3}$

3. If A is a non-diagonal involuntary matrix, then _______.

(a) $A-I=0$ (b) $A+I=0$
(c) $A-I$ is non-zero singular
(d) None of these.

4. If $A^3=0$, then $I+A+A^2$ equals _______.

(a) $I-A$ (b) $(I-A)^{-1}$
(c) $(I+A)^{-1}$ (d) None of these.

5. If A is a diagonal matrix of order 3×3 is commutative with every square matrix of order 3×3 under multiplication and trace (A) = 12, then _______.

(a) $|A|=64$ (b) $|A|=16$
(c) $|A|=12$ (d) $|A|=0$

6. If A is a square matrix of order 3 such that $|A| = 2$ then $\left|\left(\text{adj } A^{-1}\right)^{-1}\right|$ is ________.

(a) 1 (b) 2
(c) 4 (d) 8

7. If A and B are two matrices such that AB = B and BA = A, then ________.

(a) $\left(A^5 - B^5\right)^3 = A - B$

(b) $\left(A^5 - B^5\right)^3 = A^3 - B^3$

(c) $A - B$ is idempotent

(d) $A - B$ is nilpotent

8. Let $A = \begin{bmatrix} 1 & 2 \\ 3 & 4 \end{bmatrix}$ and $B = \begin{bmatrix} a & b \\ c & d \end{bmatrix}$ are two matrices such that AB = BA and $c \neq 0$, then value of $\frac{a-d}{3b-c}$ is:

(a) 0 (b) 2
(c) –2 (d) –1

9. Let $f(\alpha) = \begin{bmatrix} \cos\alpha & -\sin\alpha & 0 \\ \sin\alpha & \cos\alpha & 0 \\ 0 & 0 & 1 \end{bmatrix}$, then $\left(f(\alpha)\right)^{-1}$ is equal to ________.

(a) $f(\alpha)$ (b) $f(-\alpha)$
(c) $f(\alpha - 1)$ (d) none

10. A and B are square matrices and A is non-singular matrix, $\left(A^{-1}BA\right)^n, n \in I^+$ is equal to ________.

(a) $A^{-n}B^nA^n$ (b) $A^nB^nA^{-n}$
(c) $A^{-1}B^nA$ (d) $A^{-n}BA^n$

11. If $\begin{bmatrix} 2 & 1 \\ 3 & 2 \end{bmatrix} A \begin{bmatrix} -3 & 2 \\ 5 & -3 \end{bmatrix} = \begin{bmatrix} 1 & 0 \\ 0 & 1 \end{bmatrix}$, then A =

(a) $\begin{bmatrix} 1 & 1 \\ 1 & 0 \end{bmatrix}$ (b) $\begin{bmatrix} 1 & 1 \\ 0 & 1 \end{bmatrix}$

(c) $\begin{bmatrix} 1 & 0 \\ 1 & 1 \end{bmatrix}$ (d) $-\begin{bmatrix} 1 & 1 \\ 1 & 0 \end{bmatrix}$

12. S_1: Square matrix A is non-singular and symmetric then $\left(\left(A^{-1}\right)^{-1}\right)^{-1}$ is skew-symmetric.

S_2: Adjoint of a symmetric matrix is a symmetric matrix.

S_3: Adjoint of a diagonal matrix is diagonal matrix.

S_4: Product of two invertible square matrices of same order is also invertible.

(a) FTFT (b) FTTF
(c) FTTT (d) TFFT

Questions from Previous Years

13. How many 3×3 matrices M with entries from {0, 1, 2} are there, for which the sum of the diagonal entries of M^TM is 5?

(a) 126 (b) 198
(c) 162 (d) 135

14. Let $P = \begin{bmatrix} 1 & 0 & 0 \\ 4 & 1 & 0 \\ 16 & 4 & 1 \end{bmatrix}$ and/be the identity matrix of order 3. If $Q = \left[q_{ij}\right]$ is a matrix such that $P^{50} - Q = I$, then $\frac{q_{31} + q_{32}}{q_{21}}$ equals ________.

(a) 52 (b) 103
(c) 201 (d) 205

One or More Than One Option Correct

1. Which of the following statement(s) is/are INCORRECT?

(a) Every skew-symmetric matrix is non-invertible.

(b) If A and B are two 3×3 matrices such that AB = O, then at least one of A and B must be null matrix.

(c) If A is a 2×2 matrix, then $\text{adj}(\text{adj}A) = A$

(d) If A and B are two square matrices of order 3 such that $|A| = 5$ and $|B| = 2$, then $|10AB|$ equals 100.

2. For a given matrix $A = \begin{bmatrix} \cos\theta & -\sin\theta \\ \sin\theta & \cos\theta \end{bmatrix}$ which of the following statement/s does/do not hold good? (where $n \in I$)

(a) $A = A^{-1} \forall, \in R$.

(b) A is symmetric, for $\theta = (2n+1)\frac{\pi}{2}$.

(c) A is skew-symmetric for $\theta = n\pi + \frac{\pi}{2}$.

(d) A is skew-symmetric for $\theta = n\pi$.

3. In a square matrix A of order 3, if each element a_{ij} is equal to the sum of the roots of the equation $x^2 - (a+b)x + ab = 0$, each $a_{i,i+1}$ is

equal to the product of the roots, each $a_{i,i-1}$ is unity and the rest of the elements are all zero, then

(a) $|A| = (a^2 + b^2)(a+b)$

(b) A is a symmetric matrix.

(c) If a, b are prime numbers then A is impossible.

(d) $|\text{adj } A| = ((a+b)ab)^2$

4. Let $A = \begin{bmatrix} 1 & 2 & 2 \\ 2 & 1 & 2 \\ 2 & 2 & 1 \end{bmatrix}$, then

(a) $A^2 - 4A - 5I_3 = 0$ (b) $A^{-1} = \frac{1}{5}(A - 4I_3)$

(c) A^3 is not invertible (d) A^2 is invertible

5. If A and B are invertible square matrices of the same order, then which of the following is correct?

(a) $\text{adj}(AB) = (\text{adj } B)(\text{adj } A)$

(b) $(\text{adj } A)' = (\text{adj } A')$

(c) $|\text{adj } A| = |A|^{n-1}$, where n is the order of matrix A.

(d) $\text{adj}(\text{adj } B) = |B|^{n-2} B$, where n is the order of matrix B.

6. If $A = \begin{bmatrix} 3 & -3 & 4 \\ 2 & -3 & 4 \\ 0 & -1 & 1 \end{bmatrix}$, then

(a) $\text{adj}(\text{adj}A) = A$ (b) $|\text{adj}(\text{adj}A)| = 1$

(c) $|\text{adj}A| = 1$ (d) None of these

7. Which of the following statements is always true?

(a) Adjoint of a symmetric matrix is a symmetric matrix.

(b) Adjoint of a unit matrix is a unit matrix.

(c) A (adj A) = (adj A) A

(d) Adjoint of a diagonal matrix is a diagonal matrix.

8. If $A = \begin{bmatrix} 1 & 1 & 1 \\ 1 & 1 & 1 \\ 1 & 1 & 1 \end{bmatrix}$, then

(a) $A^3 = 9A$ (b) $A^3 = 27A$

(c) $A + A = A^2$ (d) A^{-1} does not exist.

Questions from Previous Years

9. Let X and Y be two arbitrary, 3×3, non-zero, skew-symmetric matrices and Z be an arbitrary 3×3, non-zero, symmetric matrix. Then which of the following matrices is (are) skew-symmetric?

(a) $Y^3Z^4 - Z^4Y^3$ (b) $X^{44} + Y^{44}$

(c) $X^4Z^3 - Z^3X^4$ (d) $X^{23} + Y^{23}$

10. Let M be a 2×2 symmetric matrix with integer entries. Then M is invertible if:

(a) the first column of M is the transpose of the second row of M.

(b) the second row of M is the transpose of the first column of M.

(c) M is a diagonal matrix with non-zero entries in the main diagonal.

(d) The product of entries in the main diagonal of M is not the square of an integer.

11. Let ω be a complex cube root of unity with $\omega \neq 1$ and $P = [p_{ij}]$ be an $n\times n$ matrix with $p_{ij} = \omega^{i+j}$. Then $p^2 \neq 0$, when $n =$ _____.

(a) 57 (b) 55

(c) 58 (d) 56

Matrix Type

1. Match the following:

Column I	Column II
(A) A is a real skew-symmetric matrix such that $A^2 + I = 0$. Then	(P) BA – AN
(B) A is a matrix such that $A^2 = A$. If $(I + A)^n = I = \lambda A$, then λ equals $(n \in N)$	(Q) A is of even order.
(C) If for a matrix A, $A^2 = A$, and $B = I - A$, then $AB + BA + I - (I - A)^2$ equals.	(R) A
(D) A is a matrix with complex entries and A* stands for transpose of complex conjugate of A. If A* = A and B* = B, then (AB – BA)* equals.	(S) $2^n - 1$
	(T) ${}^nC_1 + {}^nC_2 \ldots + {}^nC_n$

Integer and Numerical Value Type

1. If $A=\begin{bmatrix}1 & -1 & 1\\ 0 & 2 & -3\\ 2 & 1 & 0\end{bmatrix}$ and $B=(\text{adj } A)$ and $C=5A$, then find the value of $\dfrac{|\text{adj} B|}{|C|}$.

2. If $A=\begin{pmatrix}1 & 2\\ 1 & 1\end{pmatrix}$ and $\phi(x)=(1+x)(1-x)^{-1}$ and $\phi(A)=-\lambda A$, then find the value of λ.

3. If $A=\begin{pmatrix}a & b & c\\ b & c & a\\ c & a & b\end{pmatrix}$, $abc=1$, $A'A=I$, then find maximum value of $a^3+b^3+c^3$.

4. If $A=\begin{bmatrix}1 & 0\\ -1 & 7\end{bmatrix}$ and $A^2=8A+KI_2$, then find the value of $|k|$.

5. Let $z=\dfrac{-1+\sqrt{3}i}{2}$, where $i=\sqrt{-1}$ and $r,s\in\{1,2,3\}$. Let $P=\begin{bmatrix}(-z)^r & z^{2s}\\ z^{2s} & z^r\end{bmatrix}$ and I be the identity matrix of order 2. Then the total number of orders pairs (r, s) for which $P^2=-I$ is ____.

Comprehension Passage

Comprehension 1

Let A be an $m \times n$ matrix. If there exists a matrix L of type $n\ m$ such that $LA=I_n$, then L is called left inverse of A; Similarly, if there exists a matrix R of type $n\ m$ such that $AR=I_m$, then R is called right inverse of A.

For example, to find right inverse of matrix

$$A=\begin{bmatrix}1 & -1\\ 1 & 1\\ 2 & 3\end{bmatrix} \text{ we take } R=\begin{bmatrix}x & y & z\\ u & v & w\end{bmatrix}$$

and solve $AR=I_3$, i.e.,

$$\begin{bmatrix}1 & -1\\ 1 & 1\\ 2 & 3\end{bmatrix}\begin{bmatrix}x & y & z\\ u & v & w\end{bmatrix}=\begin{bmatrix}1 & 0 & 0\\ 0 & 1 & 0\\ 0 & 0 & 1\end{bmatrix}$$

$$\Rightarrow x-u=1 \quad y-v=0 \quad z-w=0$$
$$x+u=0 \quad y+v=1 \quad z+w=0$$
$$2x+3u=0 \quad 2y+3v=0 \quad 2z+3w=1$$

As this system of equations is inconsistent, we say there is no right inverse for matrix A.

1. Which of the following matrices is NOT left inverse of matrix $\begin{bmatrix}1 & -1\\ 1 & 1\\ 2 & 3\end{bmatrix}$?
   (a) $\begin{bmatrix}\frac{1}{2} & \frac{1}{2} & 0\\ -\frac{1}{2} & \frac{1}{2} & 0\end{bmatrix}$ (b) $\begin{bmatrix}2 & -7 & 3\\ -\frac{1}{2} & \frac{1}{2} & 0\end{bmatrix}$
   (c) $\begin{bmatrix}-\frac{1}{2} & \frac{1}{2} & 0\\ \frac{1}{2} & \frac{1}{2} & 0\end{bmatrix}$ (d) $\begin{bmatrix}0 & 3 & -1\\ -\frac{1}{2} & \frac{1}{2} & 0\end{bmatrix}$

2. The number of right inverses for the matrix $\begin{bmatrix}1 & -1 & 2\\ 2 & -1 & 1\end{bmatrix}$ is:
   (a) 0 (b) 1
   (c) 2 (d) infinite

3. For which of the following matrices, the number of left inverses is greater than the number of right inverses?
   (a) $\begin{bmatrix}1 & 2 & 4\\ -3 & 2 & 1\end{bmatrix}$ (b) $\begin{bmatrix}3 & 2 & 1\\ 3 & 2 & 1\end{bmatrix}$
   (c) $\begin{bmatrix}1 & 4\\ 2 & -3\\ 5 & 4\end{bmatrix}$ (d) $\begin{bmatrix}3 & 3\\ 1 & 1\\ 4 & 4\end{bmatrix}$

Comprehension 2

Let $A=[a_{ij}]_3$ be a square matrix of order 3 whose elements are distinct integers from 1, 2 … 9. The matrix is formed so that the sum of numbers in every row, column and diagonal is a multiple of 9.

1. The number of possible combinations of three distinct numbers from 1 to 9 that have a sum of 9 or 18 is ______.
   (a) 10 (b) 7
   (c) 8 (d) 9

2. The element a_{22} must be a multiple of ______.
   (a) 2 (b) 3
   (c) 4 (d) 9

3. The maximum value of trace of the matrix A is:
   (a) 18 (b) 19
   (c) 12 (d) None of these.

ANSWER KEY

JEE Main

Objective Questions

One Option Correct

1. (b) **2.** (a) **3.** (c) **4.** (b) **5.** (d) **6.** (b) **7.** (a) **8.** (a) **9.** (a)
10. (b) **11.** (d) **12.** (b) **13.** (b) **14.** (a) **15.** (c) **16.** (b) **17.** (b) **18.** (b)
19. (d) **20.** (d) **21.** (b) **22.** (a) **23.** (b) **24.** (d) **25.** (a) **26.** (c) **27.** (b)
28. (c) **29.** (b) **30.** (d) **31.** (a) **32.** (d) **33.** (c) **34.** (b) **35.** (c) **36.** (d)
37. (b) **38.** (d) **39.** (c) **40.** (b) **41.** (c) **42.** (c) **43.** (c) **44.** (a) **45.** (a)
46. (a) **47.** (d) **48.** (d) **49.** (b) **50.** (d) **51.** (d) **52.** (b) **53.** (a) **54.** (a)
55. (a) **56.** (c) **57.** (d) **58.** (d) **59.** (b) **60.** (a) **61.** (a) **62.** (d) **63.** (c)
64. (c) **65.** (d) **66.** (d) **67.** (b) **68.** (c) **69.** (c) **70.** (a) **71.** (b) **72.** (d)
73. (c) **74.** (c) **75.** (b) **76.** (c) **77.** (a) **78.** (b) **79.** (a) **80.** (a) **81.** (d)
82. (a)

Integer and Numerical Value Type

1. (0) **2.** (2) **3.** (2) **4.** (25) **5.** (4) **6.** (2006) **7.** (8) **8.** (2020) **9.** (11)
10. (3125) **11.** (1) **12.** (4) **13.** (766) **14.** (540) **15.** (10) **16.** (672) **17.** (9)

Assertion and Reason

1. (a) **2.** (d) **3.** (a) **4.** (d) **5.** (c)

JEE Advanced

Objective Questions

One Option Correct

1. (a) **2.** (c) **3.** (c) **4.** (b) **5.** (a) **6.** (c) **7.** (d) **8.** (d) **9.** (b)
10. (c) **11.** (a) **12.** (c) **13.** (b) **14.** (b)

One or More Than One Option Correct

1. (a, b, d) **2.** (a, b, d) **3.** (a, c) **4.** (a, b, d) **5.** (a, b, c, d) **6.** (a, b, c)
7. (a, b, c, d) **8.** (a, d) **9.** (c, d) **10.** (c, d) **11.** (b, c, d)

Matrix Type

1. A $\rightarrow$ Q; B $\rightarrow$ S, T; C $\rightarrow$ R; D $\rightarrow$ P

Integer and Numerical Value Type

1. (1) **2.** (–A) **3.** (2 and 4) **4.** (7) **5.** (Only one pair)

Comprehension Passage

Comprehension 1

1. (c) **2.** (d) **3.** (c)

Comprehension 2

1. (a) **2.** (b) **3.** (a)

HINTS AND EXPLANATIONS

Worksheet

1. (d)

A is singular $\Rightarrow |A| = 0$

$$\Rightarrow \begin{vmatrix} 1 & -3 & 2 \\ 2 & k & 5 \\ 4 & 2 & 1 \end{vmatrix} = 0$$

$$\Rightarrow 1(k-10)+3(2-20)+2(4-4k)=0$$

$$\Rightarrow 7k+56=0 \Rightarrow k=-8$$

2. (d)

Here, $AA^T = \begin{pmatrix} 2 & -1 \\ -7 & 4 \end{pmatrix}\begin{pmatrix} 2 & -7 \\ -1 & 4 \end{pmatrix} \neq \begin{pmatrix} 1 & 0 \\ 0 & 1 \end{pmatrix}$

$(BB^T)_{11} = (D)^2 + (A)^2 \neq I$

$(AB)_{11} = 8-7=1, (BA)_{11} = 8-7=1$

$\therefore AB \neq BA$ may not be true.

Now,

$$AB = \begin{pmatrix} 2 & -1 \\ -7 & 4 \end{pmatrix}\begin{pmatrix} 4 & 1 \\ 7 & 2 \end{pmatrix}$$

$$= \begin{pmatrix} 8-7 & 2-2 \\ -28+28 & -7+8 \end{pmatrix} = \begin{pmatrix} 1 & 0 \\ 0 & 1 \end{pmatrix}$$

$$(AB)^T = \begin{pmatrix} 1 & 0 \\ 0 & 1 \end{pmatrix} = I$$

3. (c)

$$|A| = \begin{vmatrix} 4 & 1 \\ 2 & 3 \end{vmatrix} = (4\times 3 - 1\times 2) = 12-2=10$$

$$\left(\because \text{if } A = \begin{vmatrix} a_{11} & a_{12} \\ a_{21} & a_{22} \end{vmatrix}, \text{then } |A| = \begin{bmatrix} a_{11} & a_{12} \\ a_{21} & a_{22} \end{bmatrix} = (a_{11}a_{22} - a_{12}a_{21}) \right)$$

4. (c)

$$\text{Here, } [A_{ij}] = \begin{bmatrix} \begin{vmatrix} 0 & 4 \\ 6 & 7 \end{vmatrix} & -\begin{vmatrix} 5 & 4 \\ 4 & 7 \end{vmatrix} & \begin{vmatrix} 5 & 0 \\ 2 & 6 \end{vmatrix} \\ -\begin{vmatrix} 2 & 3 \\ 6 & 7 \end{vmatrix} & \begin{vmatrix} 1 & 3 \\ 2 & 7 \end{vmatrix} & -\begin{vmatrix} 1 & 2 \\ 2 & 6 \end{vmatrix} \\ \begin{vmatrix} 2 & 3 \\ 0 & 4 \end{vmatrix} & -\begin{vmatrix} 1 & 3 \\ 5 & 4 \end{vmatrix} & \begin{vmatrix} 1 & 2 \\ 5 & 0 \end{vmatrix} \end{bmatrix}.$$

$$= \begin{bmatrix} -24 & -27 & 30 \\ 4 & 1 & -2 \\ 8 & 11 & -10 \end{bmatrix}$$

Hence transposing $[A_{ij}]$ we get

$$\text{adj}\,A = \begin{bmatrix} -24 & 4 & 8 \\ -27 & 1 & 11 \\ 30 & -2 & -10 \end{bmatrix}$$

5. (b)

We know that $\text{adj}(\text{adj}\,A) = |A|^{n-2} A$

Now if $n=3$, then $\text{adj}(\text{adj}\,A) = |A|A$

$$= \begin{bmatrix} 1 & 2 & 3 \\ 2 & 3 & 1 \\ 3 & 1 & 2 \end{bmatrix} A$$

$$= \{1(6-1)-2(4-3)+3(2-9)\}A$$

$$= (5-2-21)A = -18A$$

6. (c)

$$A = \begin{bmatrix} 1 & 0 \\ 1 & 1 \end{bmatrix}$$

$$A^{-1} = \frac{1}{1}\begin{bmatrix} 1 & 0 \\ -1 & 1 \end{bmatrix} = \begin{bmatrix} 1 & 0 \\ -1 & 1 \end{bmatrix}$$

$$A^{-2} = \begin{bmatrix} 1 & 0 \\ -1 & 1 \end{bmatrix}\begin{bmatrix} 1 & 0 \\ -1 & 1 \end{bmatrix} = \begin{bmatrix} 1 & 0 \\ -2 & 1 \end{bmatrix}$$

$$A^{-n} = \begin{bmatrix} 1 & 0 \\ -n & 1 \end{bmatrix}$$

7. (c)

Here, let

$$A = k\begin{bmatrix} -1 & 2 & 2 \\ 2 & -1 & 2 \\ 2 & 2 & -1 \end{bmatrix} \therefore A^T = k\begin{bmatrix} -1 & 2 & 2 \\ 2 & -1 & 2 \\ 2 & 2 & -1 \end{bmatrix}$$

Since A is orthogonal $\therefore AA^T = I$

$$\Rightarrow k^2 \begin{bmatrix} -1 & 2 & 2 \\ 2 & -1 & 2 \\ 2 & 2 & -1 \end{bmatrix}\begin{bmatrix} -1 & 2 & 2 \\ 2 & -1 & 2 \\ 2 & 2 & -1 \end{bmatrix}$$

$$\Rightarrow k^2 \begin{bmatrix} 1+4+4 & -2-2+4 & -2+4-2 \\ -2-2+4 & 4+1+4 & 4-2-2 \\ -2+4-2 & 4-2-2 & 4+4+1 \end{bmatrix}$$

$$= k^2 \begin{bmatrix} 9 & 0 & 0 \\ 0 & 9 & 0 \\ 0 & 0 & 9 \end{bmatrix} = 9k^2 I$$

$$\Rightarrow 9k^2 = 1 \Rightarrow k^2 = \frac{1}{9} \Rightarrow k = \pm\frac{1}{3}$$

8. (c)

$$\left[M(\alpha)M(\beta)\right]^{-1} = M(\beta)^{-1}M(\alpha)^{-1}$$

$$\text{Now } M(\alpha)^{-1} = \begin{bmatrix} \cos\alpha & \sin\alpha & 0 \\ -\sin\alpha & \cos\alpha & 0 \\ 0 & 0 & 1 \end{bmatrix}$$

$$= \begin{bmatrix} \cos(-\alpha) & -\sin(-\alpha) & 0 \\ \sin(-\alpha) & \cos(-\alpha) & 0 \\ 0 & 0 & 1 \end{bmatrix} = M(-\alpha)$$

$$M(-\beta)^{-1} = \begin{bmatrix} \cos\beta & 0 & -\sin\beta \\ 0 & 1 & 0 \\ \sin\beta & 0 & \cos\beta \end{bmatrix}$$

$$= \begin{bmatrix} \cos(-\beta) & 0 & \sin(-\beta) \\ 0 & 1 & 0 \\ -\sin(-\beta) & 0 & \cos(-\beta) \end{bmatrix} = M(-\beta)$$

$$\therefore \left[M(\alpha)M(\beta)\right]^{-1} = M(-\beta)M(-\alpha)$$

9. The first part easily follows from $|AB| = |A|\,|B|$.
To prove the part, we note that
$(AB)^{-1}(B^{-1}A^{-1}) = A(BB^{-1})A^{-1}$
$= AIA^{-1} = (AI)A^{-1} = AA^{-1} = I$
Similarly $(B^{-1}A^{-1})(AB) = I$
Thus $B^{-1}A^{-1}$ is the inverse of AB.
Note: We can also show that
$(ABC)^{-1} = C^{-1}B^{-1}A^{-1}$

10. We have

$$\begin{bmatrix} \cos\theta & -\sin\theta \\ \sin\theta & \cos\theta \end{bmatrix}\begin{bmatrix} 1 & \tan\frac{1}{2}\theta \\ -\tan\frac{1}{2}\theta & 1 \end{bmatrix}$$

$$= \begin{bmatrix} \cos\theta + \sin\theta\tan\frac{1}{2}\theta & \cos\theta\tan\frac{1}{2}\theta - \sin\theta \\ \sin\theta - \cos\theta\tan\frac{1}{2}\theta & \sin\theta\tan\frac{1}{2}\theta + \cos\theta \end{bmatrix}$$

$$\begin{bmatrix} \dfrac{\cos\theta\cos\frac{1}{2}\theta + \sin\theta\sin\frac{1}{2}\theta}{\cos\frac{1}{2}\theta} & \dfrac{\cos\theta\sin\frac{1}{2}\theta - \sin\theta\cos\frac{1}{2}\theta}{\cos\frac{1}{2}\theta} \\ \dfrac{\sin\theta\cos\frac{1}{2}\theta - \cos\theta\sin\frac{1}{2}\theta}{\cos\frac{1}{2}\theta} & \dfrac{\sin\theta\sin\frac{1}{2}\theta + \cos\theta\cos\frac{1}{2}\theta}{\cos\frac{1}{2}\theta} \end{bmatrix}$$

$$= \begin{bmatrix} 1 & -\tan\frac{1}{2}\theta \\ \tan\frac{1}{2}\theta & 1 \end{bmatrix}$$

Let $\begin{bmatrix} 1 & \tan\frac{1}{2}\theta \\ -\tan\frac{1}{2}\theta & 1 \end{bmatrix} = A$, then result can be written as

$$\begin{bmatrix} \cos\theta & -\sin\theta \\ \sin\theta & \cos\theta \end{bmatrix} A = \begin{bmatrix} 1 & -\tan\frac{1}{2}\theta \\ \tan\frac{1}{2}\theta & 1 \end{bmatrix}$$

Post multiplying both sides by A^{-1}, we obtain

$$\begin{bmatrix} \cos\theta & -\sin\theta \\ \sin\theta & \cos\theta \end{bmatrix} AA^{-1} = \begin{bmatrix} 1 & -\tan\frac{1}{2}\theta \\ \tan\frac{1}{2}\theta & 1 \end{bmatrix} A^{-1}$$

or

$$\begin{bmatrix} \cos\theta & -\sin\theta \\ \sin\theta & \cos\theta \end{bmatrix}$$

$$I = \begin{bmatrix} 1 & -\tan\frac{1}{2}\theta \\ \tan\frac{1}{2}\theta & 1 \end{bmatrix} \cdot \begin{bmatrix} 1 & \tan\frac{1}{2}\theta \\ -\tan\frac{1}{2}\theta & 1 \end{bmatrix}^{-1}$$

$$\therefore \begin{bmatrix} \cos\theta & -\sin\theta \\ \sin\theta & \cos\theta \end{bmatrix}$$

$$= \begin{bmatrix} 1 & -\tan\frac{1}{2}\theta \\ \tan\frac{1}{2}\theta & 1 \end{bmatrix}\begin{bmatrix} 1 & \tan\frac{1}{2}\theta \\ -\tan\frac{1}{2}\theta & 1 \end{bmatrix}^{-1}$$

Quiz

1. A matrix is an ordered rectangular array of number and functions. The number of functions is called the elements or the entries of the matrix.

2. An $m \times n$ matrix is said to be a square matrix if $m = n$, it means number of rows = number of columns.

3. $m \times n = 12$
$(m,n) = \{(1,12),(2,6),(3,4),(4,3),(6,2),(12,1)\}$

4. A diagonal whose all diagonal elements are equal.

5. Not necessarily.

6. Yes

7. The sum of the elements of a square matrix lying along a principal diagonal is called trace of the matrix. If the matrix is A then its trace matrix is $\text{tr}(A)$.
Suppose $A = [a_{ij}]_{m\times n}$
Then $\text{tr}(A) = \sum_{i=1}^{n} a_{ij} = a_1a_1 + a_{22} + a_{33} + a_{nn}$

8. The matrix obtained by interchanging rows and column of a matrix is called transpose of the matrix and denoted by A^T or A^t.

9. A square matrix is called upper triangular matrix if the all elements below the principal diagonal are zero.

10. A square matrix A is said to be a singular matrix if det A = 0.

JEE Main

Objective Questions

One Option Correct

1. (b)

$$A^{-1} = \frac{\text{adj}\,A}{|A|} = \pm\text{adj}(A), \text{ as } |A| = \pm 1$$

As A is a square matrix with integer entries, so adj (A) is also a square matrix with integer entries.

2. (a)

From the given matrix, we have

$|A| = 5(5\alpha) = 25\alpha$

It is given that $|A^2| = 25$

We know that $|A^2| = |A^2| = 25$ (From property)

$\therefore |A^2| = 25 = 625\alpha^2$

$\Rightarrow 625\alpha^2 = 25$

$\therefore \alpha = \frac{1}{5}$

3. (c)

$\because A^2 - A + I = 0$

$A(A - I) = -I$

$\Rightarrow -A^{-1} = A = A - I$

$\therefore A^{-1} = I - A$

4. (b)

$$A^2 = \begin{bmatrix} 1 & 0 \\ 1 & 1 \end{bmatrix}\begin{bmatrix} 1 & 0 \\ 1 & 1 \end{bmatrix} = \begin{bmatrix} 1 & 0 \\ 2 & 1 \end{bmatrix}$$

$$A^3 = \begin{bmatrix} 1 & 0 \\ 2 & 1 \end{bmatrix}\begin{bmatrix} 1 & 0 \\ 1 & 1 \end{bmatrix} = \begin{bmatrix} 1 & 0 \\ 3 & 1 \end{bmatrix}$$

$$\therefore A^n = \begin{bmatrix} 1 & 0 \\ n & 1 \end{bmatrix}$$

$$nA = \begin{bmatrix} n & 0 \\ n & n \end{bmatrix}$$

$$(n-1)I = \begin{bmatrix} n-1 & 0 \\ 0 & n-1 \end{bmatrix}$$

$$nA - (n-1)I = \begin{bmatrix} 1 & 0 \\ n & 1 \end{bmatrix} = A^n$$

5. (d)

$$A^2 = \begin{bmatrix} 0 & 0 & -1 \\ 0 & -1 & 0 \\ -1 & 0 & 0 \end{bmatrix}\begin{bmatrix} 0 & 0 & -1 \\ 0 & -1 & 0 \\ -1 & 0 & 0 \end{bmatrix}$$

$$= \begin{bmatrix} 1 & 0 & 0 \\ 0 & 1 & 0 \\ 0 & 0 & 1 \end{bmatrix} = I$$

6. (b)

If $B = A^{-1}$ then $AB = I$

$\Rightarrow A(10B) = 10I$

$$\Rightarrow \begin{bmatrix} 1 & -1 & 1 \\ 2 & 1 & -3 \\ 1 & 1 & 1 \end{bmatrix}\begin{bmatrix} 4 & 2 & 2 \\ -5 & 0 & \alpha \\ 1 & -2 & 3 \end{bmatrix} = \begin{bmatrix} 10 & 0 & 0 \\ 0 & 10 & 0 \\ 0 & 0 & 10 \end{bmatrix}$$

$\Rightarrow 1\cdot 2 + 1\cdot\alpha + 1\cdot 3 = 10$

$\Rightarrow \alpha = 5$

7. (a)

$$A^{-1} = \frac{\text{adj}\,A}{|A|}$$

$|A| = abc$

$$\text{adj}\,A = \begin{bmatrix} bc & 0 & 0 \\ 0 & ac & 0 \\ 0 & 0 & ab \end{bmatrix}$$

$$A^{-1} = \frac{1}{abc}\begin{bmatrix} bc & 0 & 0 \\ 0 & ac & 0 \\ 0 & 0 & ab \end{bmatrix}$$

$$A^{-1} = \begin{bmatrix} 1/a & 0 & 0 \\ 0 & 1/b & 0 \\ 0 & 0 & 1/c \end{bmatrix}$$

8. (a)

$$A = \begin{bmatrix} 1 & 2 \\ -5 & 1 \end{bmatrix}$$

$$A^{-1} = \frac{\text{adj}\,A}{|A|} = \frac{1}{11}\begin{bmatrix} 1 & -2 \\ 5 & 1 \end{bmatrix}$$

$$A^{-1} = \begin{bmatrix} \frac{1}{11} & -\frac{2}{11} \\ \frac{5}{11} & \frac{1}{11} \end{bmatrix}$$

$$xA = \begin{bmatrix} x & 2x \\ -5x & x \end{bmatrix}, yI = \begin{bmatrix} y & 0 \\ 0 & y \end{bmatrix}$$

$$A^{-1} = xA + yI$$

$$\begin{bmatrix} \frac{1}{11} & -\frac{2}{11} \\ \frac{5}{11} & \frac{1}{11} \end{bmatrix} = \begin{bmatrix} x+y & 2x \\ -5x & x+y \end{bmatrix}$$

$$\because x+y = \frac{1}{11}$$

$$2x = -\frac{2}{11}$$

$$\because x = -\frac{1}{11}, y = \frac{2}{11}$$

9. (a)

$$\text{adj A} = \begin{bmatrix} 4y & -x \\ -x^2 & 1 \end{bmatrix}$$

$$\text{adj A} + \text{B} = \begin{bmatrix} 4y & -x \\ -x^2 & 1 \end{bmatrix} + \begin{bmatrix} -3 & 1 \\ 1 & 0 \end{bmatrix}$$

$$= \begin{bmatrix} 4y-3 & 1-x \\ -x^2+1 & 1 \end{bmatrix} \quad ...(1)$$

$$\text{adj A} + \text{B} = \begin{bmatrix} 1 & 0 \\ 0 & 1 \end{bmatrix} \text{(given)} \quad ...(2)$$

Equation (1) = (2)

$$\begin{bmatrix} 4y-3 & 1-x \\ 1-x^2 & 1 \end{bmatrix} = \begin{bmatrix} 1 & 0 \\ 0 & 1 \end{bmatrix}$$

Clearly, $x = 1$, $y = 1$

10. (b)

$$A(\text{adj A}) = |A|^n$$

where n is the order of the matrix.

Given A is singular $\Rightarrow |A| = 0$

$$A(\text{adj A}) = |A|^n = 0$$

$\because A(\text{adj A}) = 0$, i.e., zero matrix

11. (d)

A′B is defined only when number of columns in matrix A′ is equal to number of rows in matrix B.

A′B is defined $\Rightarrow A'_{3\times4}\, B_{4\times3}$

BA′ is defined $\Rightarrow B_{4\times3}\, A'_{3\times4}$

So, order of B is 4×3.

12. (b)

$$A = \begin{bmatrix} 0 & 1 \\ 0 & 0 \end{bmatrix}$$

$$A^2 = \begin{bmatrix} 0 & 1 \\ 0 & 0 \end{bmatrix} + \begin{bmatrix} 0 & 1 \\ 0 & 0 \end{bmatrix} = \begin{bmatrix} 0 & 0 \\ 0 & 0 \end{bmatrix}$$

$= 0$ (zero matrix)

$$(aI+bA)^2 = (aI+bA)\cdot(aI+bA)$$

$$= a^2I^2 + abIA + bAaI + b^2A^2$$

$$= a^2I + 2abA + 0$$

$$= a^2I + 2abA$$

13. (b)

$$\begin{bmatrix} x+y+z \\ x-2y-2z \\ x+3y+x \end{bmatrix} = \begin{bmatrix} 0 \\ 3 \\ 4 \end{bmatrix}$$

Comparing both the sides,

$$x+y+z=0$$

$$x-2y-2z=3$$

$$x+3y+z=4$$

Then $x=1$, $y=2$, $z=-3$

$$\text{Thus} \begin{bmatrix} x \\ y \\ z \end{bmatrix} = \begin{bmatrix} 1 \\ 2 \\ -3 \end{bmatrix}$$

14. (a)

A = B + C

B → Symmetric matrix

C → Skew-symmetric matrix

We know, $A = \frac{1}{2}\left[\left(A+A^T\right)+\left(A-A^T\right)\right]$

$B = \frac{1}{2}\left(A+A^T\right)$ and $C = \frac{1}{2}\left(A-A^T\right)$

$$A = \begin{bmatrix} 6 & 8 & 5 \\ 4 & 2 & 3 \\ 9 & 7 & 1 \end{bmatrix}$$

$$A^T = \begin{bmatrix} 6 & 4 & 9 \\ 8 & 2 & 7 \\ 5 & 3 & 1 \end{bmatrix}$$

$$B = \frac{1}{2}\left(A+A^T\right) = \begin{bmatrix} 6 & 6 & 7 \\ 6 & 2 & 5 \\ 7 & 5 & 1 \end{bmatrix}$$

15. (c)

Option (c) is correct because in 2×2 matrices $AB \neq BA$.

16. (b)

$$A^{-1} = \frac{\text{adj A}}{|A|}$$

$$A^{-1} = \frac{1}{(-1)}\begin{bmatrix} 0 & 0 & -1 \\ 0 & -1 & 0 \\ -1 & 0 & 0 \end{bmatrix} = A$$

17. (b)

Since A, B and C are $n \times n$ matrices and if $A^3 + 2A^2 + 3A + 5I = 0$, then A is invertible.

18. (b)

$$A = \begin{bmatrix} 4 & 2 \\ 3 & 4 \end{bmatrix}$$

$$\text{adj } A = \begin{bmatrix} 4 & -2 \\ -3 & 4 \end{bmatrix}$$

$$|\text{adj } A| = 16 - 6 = 10$$

19. (d)

$$|A| = 0$$

$$\begin{vmatrix} 1 & 0 & -k \\ 2 & 1 & 3 \\ k & 0 & 1 \end{vmatrix} = 0$$

$$-1 + k^2 = 0$$

$$k^2 = 1$$

$$k = \pm 1$$

20. (d)

$$A = \begin{bmatrix} 0 & 3 \\ 2 & 0 \end{bmatrix}$$

$$|A| = 0 - 6 = -6$$

$$\text{adj } A = \begin{bmatrix} 0 & -3 \\ -2 & 0 \end{bmatrix}$$

$$A^{-1} = -\frac{1}{6}\begin{bmatrix} 0 & -3 \\ -2 & 0 \end{bmatrix}$$

$$= \frac{1}{6}\begin{bmatrix} 0 & 3 \\ 2 & 0 \end{bmatrix}$$

$$A^{-1} = \frac{1}{6} A$$

$$\text{So, } \lambda = \frac{1}{6}$$

21. (b)

$$A = \begin{bmatrix} 1 & 1 \\ 1 & 1 \end{bmatrix} = 2^0 A$$

$$A^2 = \begin{bmatrix} 2 & 2 \\ 2 & 2 \end{bmatrix} = 2A = 2^1 A$$

$$A^3 = \begin{bmatrix} 4 & 4 \\ 4 & 4 \end{bmatrix} = 4A = 2^2 A$$

$$A^n = 2^{n-1} A$$

$$A^{100} = 2^{99} A$$

22. (a)

$$A = \begin{bmatrix} -2 & 6 \\ -5 & 7 \end{bmatrix}$$

$$\text{adj } A = \begin{bmatrix} 7 & 5 \\ -6 & -2 \end{bmatrix}^T = \begin{bmatrix} 7 & -6 \\ 5 & -2 \end{bmatrix}$$

23. (b)

$$A = \begin{bmatrix} i & -i \\ -i & i \end{bmatrix} \text{ and } B = \begin{bmatrix} 1 & -1 \\ -1 & 1 \end{bmatrix}$$

$$\text{So, } A \cdot A = A^2 = \begin{bmatrix} 2i & -2i^2 \\ -2i^2 & 2i^2 \end{bmatrix}$$

$$A^2 \cdot A = A^3 = \begin{bmatrix} 4i^3 & -4i^3 \\ -4i^3 & 4i^3 \end{bmatrix}$$

$$\text{So, } A^8 = 2^8 \begin{bmatrix} i^8 & -i^8 \\ -i^8 & i^8 \end{bmatrix} = 128 \begin{bmatrix} 1 & -1 \\ -1 & 1 \end{bmatrix}$$

$$= 128 \text{ B}$$

24. (d)

$$\text{Let } A = \begin{bmatrix} \cos 2\theta & -\sin 2\theta \\ \sin 2\theta & \cos 2\theta \end{bmatrix}$$

$$\text{Here, } |A| = |\cos^2 2\theta + \sin^2 2\theta| = 1$$

$$A^{-1} = \frac{1}{|A|}\begin{bmatrix} \cos 2\theta & -\sin 2\theta \\ \sin 2\theta & \cos 2\theta \end{bmatrix}^T$$

$$= \begin{bmatrix} \cos 2\theta & \sin 2\theta \\ -\sin 2\theta & \cos 2\theta \end{bmatrix}$$

25. (a)

$$2X = \begin{bmatrix} 3 & 8 \\ 7 & 2 \end{bmatrix} - \begin{bmatrix} 1 & 2 \\ 3 & 4 \end{bmatrix}$$

$$2X = 2\begin{bmatrix} 1 & 3 \\ 2 & -1 \end{bmatrix}$$

$$X = \begin{bmatrix} 1 & 3 \\ 2 & -1 \end{bmatrix}$$

26. (c)

C, A and C^T are matrix of order $n \times 1, n \times n$ and $1 \times n$ respectively.

Let $C^T AC = k$

$$\text{Then } \left(C^T AC\right)^T = C^T A^T \left(C^T\right)^T$$

$$= C^T A^T C = C^T(-A)C$$

$$= -C^T AC = -k$$

$$k = -k \Rightarrow k = 0$$

C^T AC is null matrix, which shows that C^T AC is a zero matrix of order 1.

27. (b)
Given A is a square matrix
$AA^T = I = A^TA$
$|AA^T| = |I| = |A^TA|$
$|A||A^T| = I = |A^T|\cdot|A| \left(\because A\cdot A^T = |A|^2\right)$
$|A| = \pm 1$

28. (c)
$x + y + z = 9$
$x + y = 5$
Clearly $z = 4$
But $y + z = 7$
So, clearly $y = 3$
Thus, $x = 2, y = 3, z = 4$

29. (b)
If A is the matrix of order $n\times n$
Then $\text{adj}(\text{adj A}) = |A|^{n-2}\,A$
where $|A|$ is denoted for determined of A.

30. (d)
$$\begin{bmatrix} \cos^2\theta & \sin\theta\cos\theta \\ -\sin\theta\cos\theta & \cos^2\theta \end{bmatrix} + \begin{bmatrix} \sin^2\theta & -\sin\theta\cos\theta \\ \sin\theta\cos\theta & \sin^2\theta \end{bmatrix} = \begin{bmatrix} 1 & 0 \\ 0 & 1 \end{bmatrix}$$

31. (a)
$$X = \begin{bmatrix} 3 & 1 \\ 4 & 1 \end{bmatrix}^{-1}\cdot\begin{bmatrix} 5 & -1 \\ 2 & 3 \end{bmatrix}$$
$$X = \frac{1}{(-1)}\begin{bmatrix} 1 & -1 \\ 4 & 1 \end{bmatrix}\cdot\begin{bmatrix} 5 & -1 \\ 2 & 3 \end{bmatrix}$$
$$X = -\begin{bmatrix} 5-2 & -1-3 \\ -20+6 & 4+9 \end{bmatrix}$$
$$= -\begin{bmatrix} 3 & -4 \\ -14 & 13 \end{bmatrix}$$
$$= \begin{bmatrix} -3 & 4 \\ 14 & -13 \end{bmatrix}$$

32. (d)
$$A = \begin{bmatrix} 1 & 0 & -1 \\ 3 & 4 & 5 \\ 0 & 6 & 7 \end{bmatrix}$$
$|A| = 1(28-30) - 1(18-0)$
$= -2 - 18 = -20$
$$A^{-1} = \frac{\text{adj A}}{|A|}$$
A^{-1} of element a_{23} is $-\frac{1}{20}(-8) = \frac{2}{5}$

33. (c)
$$AB = \begin{bmatrix} 1 & 2 \\ -3 & 0 \end{bmatrix}\cdot\begin{bmatrix} -1 & 0 \\ 2 & 3 \end{bmatrix}$$
$$AB = \begin{bmatrix} -1+4 & 0+6 \\ 3+0 & 0+0 \end{bmatrix}$$
$$AB = \begin{bmatrix} 3 & 6 \\ 3 & 0 \end{bmatrix}$$
$$BA = \begin{bmatrix} -1 & 0 \\ 2 & 3 \end{bmatrix}\cdot\begin{bmatrix} 1 & 2 \\ -3 & 0 \end{bmatrix}$$
$$= \begin{bmatrix} -1+0 & -2+0 \\ 2-9 & 4 \end{bmatrix}$$
$$= \begin{bmatrix} -1 & -2 \\ -7 & 4 \end{bmatrix}$$
$AB \neq BA$

34. (b)
A' = A is possible only
(i) when rank of A and A' are same.
(ii) both have same elements in the diagonals.
This is possible only in a square matrix.

35. (c)
$$A = \begin{bmatrix} 1 & -1 \\ 1 & 1 \end{bmatrix}$$
$$\text{adj A} = \begin{bmatrix} 1 & -1 \\ 1 & 1 \end{bmatrix}^T$$
$$\text{adj A} = \begin{bmatrix} 1 & 1 \\ -1 & 1 \end{bmatrix}$$

36. (d)
From $AX = B$
$X = A^{-1}B$
$$X^{-1} = \begin{bmatrix} 3 & -1/2 & -1/2 \\ -4 & 3/4 & 5/2 \\ 2 & -3/4 & -3/4 \end{bmatrix}\begin{bmatrix} 9 \\ 52 \\ 0 \end{bmatrix} = \begin{bmatrix} 1 \\ 3 \\ 5 \end{bmatrix}$$

37. (b)
If $A = \begin{pmatrix} 1 & -2 \\ 4 & 5 \end{pmatrix}$ and $f(t) = t^2 - 3t + 7$
Then $f(A) = A^2 - 3A + 7$
$$f(A) = \begin{pmatrix} -7 & -12 \\ 24 & 17 \end{pmatrix} - \begin{pmatrix} 3 & -6 \\ 12 & 15 \end{pmatrix} + \begin{pmatrix} 7 & 0 \\ 0 & 7 \end{pmatrix}$$
$$f(A) = \begin{pmatrix} -7-3+7 & -12+6+0 \\ 24-12+0 & 17-15+7 \end{pmatrix}$$
$$f(A) = \begin{pmatrix} -3 & -6 \\ 12 & 9 \end{pmatrix}$$

Now, $f(A)+\begin{pmatrix} 3 & 6 \\ -12 & -9 \end{pmatrix}$

$=\begin{pmatrix} -3 & -6 \\ 12 & 9 \end{pmatrix}+\begin{pmatrix} 3 & 6 \\ -12 & -9 \end{pmatrix}$

$=\begin{pmatrix} 3-3 & -6+6 \\ 12-12 & 9-9 \end{pmatrix}=\begin{pmatrix} 0 & 0 \\ 0 & 0 \end{pmatrix}$

38. (d)

Let $A=\begin{bmatrix} 7 & -3 & -3 \\ -1 & 1 & 0 \\ -1 & 0 & 1 \end{bmatrix}$

$\text{adj } A=\begin{bmatrix} 1 & 1 & 1 \\ 3 & 4 & 3 \\ 3 & 3 & 4 \end{bmatrix}^T$

and $|A|=7(1-0)+3(-1-0)-3(0+1)$

$=7-3-3$

$|A|=1$

So, $A^{-1}=\dfrac{(\text{adj } A)^T}{|A|}$

$A^{-1}=\dfrac{\begin{bmatrix} 1 & 3 & 3 \\ 1 & 4 & 3 \\ 1 & 3 & 4 \end{bmatrix}}{1}$

$A^{-1}=\begin{bmatrix} 1 & 3 & 3 \\ 1 & 4 & 3 \\ 1 & 3 & 4 \end{bmatrix}$

39. (c)

$\begin{bmatrix} 1 & 2 & x \\ 4 & -1 & 7 \\ 2 & 4 & -6 \end{bmatrix}=0$

$1(6-28)-2(-24-14)+x(16+2)=0$

$-22+76+18x=0$

$18x=-54$

$x=-3$

40. (b)

$A\,(\text{adj } A)=\begin{bmatrix} 4 & 0 & 0 \\ 0 & 4 & 0 \\ 0 & 0 & 4 \end{bmatrix}$

$|A(\text{adj } A)|=\begin{vmatrix} 4 & 0 & 0 \\ 0 & 4 & 0 \\ 0 & 0 & 4 \end{vmatrix}$

$|A|\times|\text{adj } A|=4^3$

$|\text{adj } A|=|A|^{n-1}$

Here $n=3$

$|\text{adj } A|=|A|^2$

$|A|\cdot|A|^2=64$

$|A|=4$

$\because |\text{adj } A|=|A|^2=16$

41. (c)

$A^2=\begin{bmatrix} 9 & 8 & 8 \\ 8 & 9 & 8 \\ 8 & 8 & 9 \end{bmatrix}$

$A^3=\begin{bmatrix} 41 & 42 & 42 \\ 42 & 41 & 42 \\ 42 & 42 & 41 \end{bmatrix}$

$\because A^3-4A^2-6A=-\begin{bmatrix} 1 & 2 & 2 \\ 2 & 1 & 2 \\ 2 & 2 & 1 \end{bmatrix}=-A$

42. (c)

Cofactor of $\begin{bmatrix} -1 & 0 & 2 \\ -1 & 1 & -2 \\ 0 & 2 & 1 \end{bmatrix}$

$C_{11}=5,\quad C_{12}=1,\quad C_{13}=-2$

$C_{21}=4,\quad C_{22}=1,\quad C_{23}=-2$

$C_{31}=-2,\quad C_{32}=0\quad C_{33}=1$

On comparing the corresponding elements, we get $a=4$, $b=1$

$[a\ b]=[4\ 1]$

43. (c)

$AB=0\Rightarrow|AB|=0$

$\Rightarrow|A||B|=0$

$\Rightarrow|A|=0$ or $|B|=0$

Let $A=\begin{bmatrix} 1 & 0 \\ 0 & 0 \end{bmatrix}$ and $B=\begin{bmatrix} 0 & 0 \\ 0 & 1 \end{bmatrix}$ then $AB=0$

But $A\neq 0$, $B\neq 0$. So, B is singular matrix.

44. (a)

$|A|=\begin{vmatrix} 1 & 0 & 1 \\ 2 & 1 & 0 \\ 3 & 2 & 1 \end{vmatrix}$

$|A|=1(1-0)-0(2-0)+1(4-3)$

$=1+1=2$

45. (a)

Since diagonal element of a skew-symmetric matrix are all zero,

$\therefore \text{tr}(A)=\sum_{t=1}^{n} a_{ij}=0$

46. (a)

$$A+2A^{T}=\begin{bmatrix}0&1&-2\\-1&0&-3\\2&3&0\end{bmatrix}=A^{T}$$

47. (d)

$$5A-3B+C=\begin{bmatrix}5-6+2&-10-3+12\\15+12-7&20-9+9\end{bmatrix}$$

$$=\begin{bmatrix}1&-1\\20&20\end{bmatrix}$$

48. (d)

$$\sum A_{ij}=\sum\left(A^{T}\right)_{ij}$$

$$\sum(A+B)_{ij}=\sum(A)_{ij}+\sum(B)_{ij}$$

$$\sum(A-B)_{ij}=\sum(A)_{ij}-\sum(B)_{ij}$$

$$\sum(AB)_{ij}\neq\sum(A)_{ij}\cdot\sum(B)_{ij}$$

49. (b)

AB is defined only when number of column in matrix A is equal to number of rows in matrix B

So, $\left(A_{3\times2}\right)\times\left(B_{2\times3}\right)=(AB)_{3\times3}$

C is also in 3×3order.

50. (d)

We have, $A=\begin{bmatrix}0&2\\3&-4\end{bmatrix}$ and $kA=\begin{bmatrix}0&3a\\2b&24\end{bmatrix}$

$\Rightarrow kA=\begin{bmatrix}0&2k\\3k&-4k\end{bmatrix}$ and $kA=\begin{bmatrix}0&3a\\2b&24\end{bmatrix}$

$$\Rightarrow\begin{bmatrix}0&2k\\3k&-4k\end{bmatrix}=\begin{bmatrix}0&3a\\2b&24\end{bmatrix}$$

$\Rightarrow 2k=3a,\quad 3k=2b,\quad -4k=24$

$\Rightarrow k=-6\quad a=-4,\quad b=-9$

51. (d)

$$\left(AB^{-1}C\right)^{-1}=C^{-1}\left(B^{-1}\right)^{-1}A^{-1}$$

$$=C^{-1}BA^{-1}$$

52. (b)

$|A|=2\times2\times2=8$

$|B|=1\times1\times2=2$

$\therefore|AB|=|A||B|=2\times8=16$

53. (a)

Let $D=\begin{bmatrix}a&b\\c&d\\e&f\end{bmatrix}$

$$\therefore A+B-D=\begin{bmatrix}1&3\\3&2\\2&5\end{bmatrix}+\begin{bmatrix}-1&-2\\0&5\\3&1\end{bmatrix}-\begin{bmatrix}a&b\\b&d\\c&f\end{bmatrix}$$

$$\Rightarrow\begin{bmatrix}1-1-a&3-2-b\\3+0-c&2+5-d\\2+3-e&5+1-f\end{bmatrix}=\begin{bmatrix}0&0\\0&0\\0&0\end{bmatrix}$$

$\Rightarrow -a=0\Rightarrow a=0,1-b=0,\Rightarrow b=1$

$\Rightarrow 3-c=0\Rightarrow c=3,7-d=0,\Rightarrow d=7$

$\Rightarrow 5-e=0\Rightarrow e=5,6-f=0,\Rightarrow f=6$

$$\therefore D=\begin{bmatrix}0&1\\3&7\\5&6\end{bmatrix}$$

54. (a)

By definition of skew-symmetric matrix.

55. (a)

For subtraction of two matrix, they should be of the same order, i.e., $p=r$, $q=s$.

56. (c)

We have, $AA'=\begin{bmatrix}3&1&-1\\0&1&2\end{bmatrix}\begin{bmatrix}3&0\\1&1\\-1&2\end{bmatrix}$

$$=\begin{bmatrix}9+1+1&0+1-2\\0+1-2&0+1+4\end{bmatrix}$$

$$=\begin{bmatrix}11&-1\\-1&5\end{bmatrix}$$

Clearly AA′ is symmetric as $(AA')=AA'$.

57. (d)

$AB=0$

$\Rightarrow|AB|=0$

$\Rightarrow|A||B|=0$

$\Rightarrow|A|=0$ and $|B|=0$

58. (d)

$$\because A(\text{adj}.A)=10\begin{bmatrix}1&0\\0&1\end{bmatrix}=10/2$$

$\because|A|=10\ \left(\because\ A(\text{adj}.A)=|A|I_n\right)$

59. (b)

In the product AB, the required element

$C_{33}=(-2)\cdot3+2\cdot5+0\cdot0=-6+10=4$

60. (a)

$$A=\begin{bmatrix}a_{11}&a_{12}\\a_{21}&a_{22}\end{bmatrix}=\begin{bmatrix}1+1&1+2\\2+1&2+2\end{bmatrix}=\begin{bmatrix}2&3\\3&4\end{bmatrix}$$

61. (a)

Since, $P = \begin{bmatrix} 1 & 0 \\ \frac{1}{2} & 1 \end{bmatrix}$

$$\Rightarrow P^2 = \begin{bmatrix} 1 & 0 \\ \frac{1}{2} & 1 \end{bmatrix}\begin{bmatrix} 1 & 0 \\ \frac{1}{2} & 1 \end{bmatrix} = \begin{bmatrix} 1 & 0 \\ 1 & 1 \end{bmatrix}$$

$$\Rightarrow P^3 = \begin{bmatrix} 1 & 0 \\ 1 & 1 \end{bmatrix}\begin{bmatrix} 1 & 0 \\ \frac{1}{2} & 1 \end{bmatrix} = \begin{bmatrix} 1 & 0 \\ \frac{3}{2} & 1 \end{bmatrix}$$

$$\Rightarrow P^4 = \begin{bmatrix} 1 & 0 \\ 1 & 1 \end{bmatrix}\begin{bmatrix} 1 & 0 \\ 1 & 1 \end{bmatrix} = \begin{bmatrix} 1 & 0 \\ 2 & 1 \end{bmatrix}$$

$$\therefore P^{50} = \begin{bmatrix} 1 & 0 \\ 25 & 1 \end{bmatrix}$$

62. (d)

$$\because A = \frac{1}{5}\left((A+2B)+2(2A-B)\right)$$

$$= \frac{1}{5}\left(\begin{bmatrix} 1 & 2 & 0 \\ 6 & -3 & 3 \\ -5 & 3 & 1 \end{bmatrix} + \begin{bmatrix} 4 & -2 & 10 \\ 4 & -2 & 12 \\ 0 & 2 & 4 \end{bmatrix}\right)$$

$$= \frac{1}{5}\left(\begin{bmatrix} 5 & 0 & 10 \\ 10 & -5 & 15 \\ -5 & 5 & 5 \end{bmatrix}\right)$$

In the same way, so tr(A) = 1

$$B = \frac{1}{5}\left(2(A+2B)-2(2A-B)\right)$$

$$= \frac{1}{5}\left(\begin{bmatrix} 2 & 4 & 0 \\ 12 & -6 & 6 \\ -10 & 6 & 2 \end{bmatrix} - \begin{bmatrix} 2 & -1 & 5 \\ 2 & -1 & 6 \\ 0 & 1 & 2 \end{bmatrix}\right)$$

So, tr(B) = – 1

Now, Tr (A) – Tr (B) = 1 – (–1) = 2

63. (c)

Let $A = \begin{bmatrix} x & z \\ z & y \end{bmatrix}$

$$A^2 = \begin{bmatrix} x & z \\ z & y \end{bmatrix}\begin{bmatrix} x & z \\ z & y \end{bmatrix} = \begin{bmatrix} x^2+z^2 & xz+yz \\ xz+yz & z^2+y^2 \end{bmatrix}$$

Given that $x^2+y^2+2z^2=1$

$\Rightarrow x=1, y=0, z=0$

or, $x=-1, y=0, z=0$

or, $x=0, y=1, z=0$

or, $x=0, y=-1, z=0$

Hence, totally four matrices are possible.

64. (c)

Let $A^T = A$ and $B^T = -B$

$$C = A^2B^2 - B^2A^2$$

$$C^T = \left(A^2B^2\right)^T \left(B^2A^2\right)^T \quad \left[\because (AB)^T = B^TA^T\right]$$

$$= \left(B^2\right)^T\left(A^2\right)^T - \left(A^2\right)^T\left(B^2\right)^T$$

$$= B^2A^2 - A^2B^2$$

$C^T = -C$

C is skew-symmetric.

So, det (C) = 0

So, the systems have infinite solutions.

65. (d)

$$\because A = \begin{bmatrix} \cos\theta & i\sin i\theta \\ i\sin\theta & \cos\theta \end{bmatrix}$$

$$\therefore A^n = \begin{bmatrix} \cos n\theta & i\sin n\theta \\ i\sin n\theta & \cos n\theta \end{bmatrix}, n \in N$$

$$\because A^5 = \begin{bmatrix} a & b \\ c & d \end{bmatrix}$$

$$\therefore A^5 = \begin{bmatrix} \cos 5\theta & i\sin 5\theta \\ i\sin 5\theta & \cos 5\theta \end{bmatrix} = \begin{bmatrix} a & b \\ c & d \end{bmatrix}$$

$\therefore a = \cos 5\theta, b = i\sin 5\theta = c, d = \cos 5\theta$

$\therefore a^2 - b^2 = \cos^2 5\theta + \sin^2 5\theta = 1$

$a^2 - c^2 = \cos^2 5\theta + \sin^2 5\theta = 1$

$a^2 - d^2 = \cos^2 5\theta - \cos^2 5\theta = 1$

$a^2 + b^2 = \cos^2 5\theta - \sin^2 5\theta = \cos 10\theta = \cos\dfrac{10\pi}{24}$

and $0 < \cos\dfrac{5\pi}{12} < 1 \Rightarrow 0 \le a^2 + b^2 \le 1$

$\therefore a^2 - b^2 = \dfrac{1}{2}$ is wrong.

66. (d)

Solution of $x^2+x+1=0$ is ω, ω^2, 1.

So, $\alpha = \omega$ and

$\omega^4 = \omega^3 \cdot \omega = 1 \cdot \omega = \omega$

$$A^2 = \frac{1}{3}\begin{bmatrix} 1 & 1 & 1 \\ 1 & \omega & \omega^2 \\ 1 & \omega^2 & \omega \end{bmatrix}\begin{bmatrix} 1 & 1 & 1 \\ 1 & \omega & \omega^2 \\ 1 & \omega^2 & \omega \end{bmatrix} = \begin{bmatrix} 1 & 0 & 0 \\ 0 & 0 & 1 \\ 0 & 1 & 0 \end{bmatrix}$$

$\Rightarrow A^4 = I$

$\Rightarrow A^{30} = A^{28} \times A^3 = A^3$

67. (b)

Given $A^TA = I$

$$\Rightarrow \begin{bmatrix} a & b & c \\ b & c & a \\ c & a & b \end{bmatrix}\begin{bmatrix} a & b & c \\ b & c & a \\ c & a & b \end{bmatrix} = \begin{bmatrix} 1 & 0 & 0 \\ 0 & 1 & 0 \\ 0 & 0 & 1 \end{bmatrix}$$

$$\Rightarrow \begin{bmatrix} \sum a^2 & \sum ab & \sum ab \\ \sum ab & \sum a^2 & \sum ab \\ \sum ab & \sum ab & \sum a^2 \end{bmatrix} = \begin{bmatrix} 1 & 0 & 0 \\ 0 & 1 & 0 \\ 0 & 0 & 1 \end{bmatrix}$$

So, $\sum a^2 = 1$ and $\sum ab = 0$

Now, $a^3 + b^3 + c^3 - 3abc$

$= (a+b+c)(a^2+b^2+c^2-ab-bc-ca)$

$= (a+b+c)(1-0)$

$= \sqrt{(a+b+c)^2} = \sqrt{\sum a^2 + 2\sum ab} = \pm 1$

$\Rightarrow 2 - 3abc = 1 \Rightarrow abc = \frac{1}{3}$

or $2 - 3abc = -1 \Rightarrow abc = 1$

68. (c)

Characteristics equation of matrix 'A' is $|A - \lambda I| = 0$.

$$\begin{vmatrix} 2-\lambda & 2 \\ 9 & 4-\lambda \end{vmatrix} = 0 \Rightarrow \lambda^2 - 6\lambda - 10 = 0$$

$\therefore A^2 - 6A - 10I = 0$

$\Rightarrow A^{-1}(A^2) - 6A^{-1} - 10IA^{-1} = 0$

$\Rightarrow 10A^{-1} = A - 6I$

69. (c)

$$A = \begin{bmatrix} \cos\alpha & -\sin\alpha \\ \sin\alpha & \cos\alpha \end{bmatrix}$$

$$A^2 = \begin{bmatrix} \cos\alpha & -\sin\alpha \\ \sin\alpha & \cos\alpha \end{bmatrix}\begin{bmatrix} \cos\alpha & -\sin\alpha \\ \sin\alpha & \cos\alpha \end{bmatrix}$$

$$= \begin{bmatrix} \cos 2\alpha & -\sin 2\alpha \\ \sin 2\alpha & \cos 2\alpha \end{bmatrix}$$

Similarly, $A^4 = A^2.A^2 = \begin{bmatrix} \cos 4\alpha & -\sin 4\alpha \\ \sin 4\alpha & \cos 4\alpha \end{bmatrix}$

and so on. $A^{32} = \begin{bmatrix} \cos 32\alpha & -\sin 32\alpha \\ \sin 32\alpha & \cos 32\alpha \end{bmatrix} = \begin{bmatrix} 0 & -1 \\ 1 & 0 \end{bmatrix}$

Then $\sin 32\alpha = 1$ and $\cos 32\alpha = 0$

$\Rightarrow 32\alpha = n\pi + (-1)^n \frac{\pi}{2}$ and $32\alpha = 2n\pi + \frac{\pi}{2}$

$\Rightarrow \alpha = \frac{n\pi}{32} + (-1)^n \frac{\pi}{64}$ and

$\alpha = \frac{n\pi}{16} + \frac{\pi}{64}$ where $n \in z$

Put $n = 0$, $\alpha = \frac{\pi}{64}$

Hence, required value of α is $\frac{\pi}{64}$.

70. (a)

$$P^2 = \begin{bmatrix} 1 & 0 & 0 \\ 3 & 1 & 0 \\ 9 & 3 & 1 \end{bmatrix}\begin{bmatrix} 1 & 0 & 0 \\ 3 & 1 & 0 \\ 9 & 3 & 1 \end{bmatrix} = \begin{bmatrix} 1 & 0 & 0 \\ 6 & 1 & 0 \\ 27 & 12 & 1 \end{bmatrix}$$

$$\Rightarrow P^4 = \begin{bmatrix} 1 & 0 & 0 \\ 6 & 1 & 0 \\ 27 & 12 & 1 \end{bmatrix}\begin{bmatrix} 1 & 0 & 0 \\ 6 & 1 & 0 \\ 27 & 12 & 1 \end{bmatrix} = \begin{bmatrix} 1 & 0 & 0 \\ 12 & 1 & 0 \\ 90 & 12 & 1 \end{bmatrix}$$

$$\Rightarrow P^5 = \begin{bmatrix} 1 & 0 & 0 \\ 12 & 1 & 0 \\ 90 & 12 & 1 \end{bmatrix}\begin{bmatrix} 1 & 0 & 0 \\ 3 & 1 & 0 \\ 9 & 3 & 1 \end{bmatrix} = \begin{bmatrix} 1 & 0 & 0 \\ 15 & 1 & 0 \\ 135 & 15 & 1 \end{bmatrix}$$

$\because Q - P^5 = I_3$

$$\therefore Q = I_3 + P^5 = \begin{bmatrix} 2 & 0 & 0 \\ 15 & 2 & 0 \\ 135 & 15 & 2 \end{bmatrix}$$

$$\frac{q_{21} + q_{31}}{q_{32}} = \frac{15 + 135}{15} = 10$$

71. (b)

Let $A = \begin{bmatrix} a & c \\ c & b \end{bmatrix}$ and $B = \begin{bmatrix} 0 & d \\ -d & 0 \end{bmatrix}$

Then, $A + B = \begin{bmatrix} a & c+d \\ c-d & b \end{bmatrix} = \begin{bmatrix} 2 & 3 \\ 5 & -1 \end{bmatrix}$

On comparing each term,

$a = 2$, $b = -1$, $c - d = 5$, $c + d = 3$

$\Rightarrow a = 2$, $b = -1$, $c = 4$, $d = -1$

Now, $AB = \begin{bmatrix} 2 & 4 \\ 4 & -1 \end{bmatrix}\begin{bmatrix} 0 & -1 \\ 1 & 0 \end{bmatrix} = \begin{bmatrix} 4 & -2 \\ -1 & -4 \end{bmatrix}$

72. (d)

Given, $A^T A = 3I$

$$\begin{bmatrix} 0 & 2x & 2x \\ 2y & y & -y \\ 1 & -1 & 1 \end{bmatrix}\begin{bmatrix} 0 & 2y & 1 \\ 2x & y & -1 \\ 2x & -y & 1 \end{bmatrix} = 3I$$

$$\Rightarrow \begin{bmatrix} 8x^2 & 0 & 0 \\ 0 & 6y^2 & 0 \\ 0 & 0 & 3 \end{bmatrix} = \begin{bmatrix} 3 & 0 & 0 \\ 0 & 3 & 0 \\ 0 & 0 & 3 \end{bmatrix}$$

$\Rightarrow 8x^2 = 3$ and $6y^2 = 3 \Rightarrow x = \pm\sqrt{\frac{3}{8}}$ and $y = \pm\sqrt{\frac{1}{2}}$

Number of combinations of $(x, y) = 2 \times 2 = 4$

73. (c)

$$A = \begin{bmatrix} 0 & 2q & r \\ p & q & -r \\ p & -q & r \end{bmatrix}$$

$$\therefore A \cdot A^T = \begin{bmatrix} 0 & 2q & r \\ p & q & -r \\ p & -q & r \end{bmatrix} \times \begin{bmatrix} 0 & p & p \\ 2q & q & -q \\ r & -r & r \end{bmatrix}$$

$$= \begin{bmatrix} 4q^2+r^2 & 2q^2-r^2 & -2q^2+r^2 \\ 2q^2-r^2 & p^2+q^2+r^2 & p^2-q^2-r^2 \\ -2q^2+r^2 & p^2-q^2-r^2 & p^2+q^2+r^2 \end{bmatrix}$$

Given, $AA^T = I$

$\therefore 4q^2+r^2 = p^2+q^2+r^2 = 1$

$\Rightarrow p^2 - 3q^2 = 0 \text{ and } r^2 = 1-4q^2$

and $2q^2 - r^2 = 0 \Rightarrow r^2 = 2q^2$

$\therefore p^2 = \frac{1}{2}, q^2 = \frac{1}{6} \text{ and } r^2 = \frac{1}{3}$

$\therefore |p| = \frac{1}{\sqrt{2}}$

74. (c)

Here, $A = \begin{bmatrix} 1 & 0 & 0 \\ 1 & 1 & 0 \\ 1 & 1 & 1 \end{bmatrix}$

$$\therefore A^2 = A.A = \begin{bmatrix} 1 & 0 & 0 \\ 1 & 1 & 0 \\ 1 & 1 & 1 \end{bmatrix} \times \begin{bmatrix} 1 & 0 & 0 \\ 1 & 1 & 0 \\ 1 & 1 & 1 \end{bmatrix} = \begin{bmatrix} 1 & 0 & 0 \\ 2 & 1 & 0 \\ 3 & 2 & 1 \end{bmatrix}$$

also

$$A^3 = A^2.A = \begin{bmatrix} 1 & 0 & 0 \\ 2 & 1 & 0 \\ 3 & 2 & 1 \end{bmatrix} \times \begin{bmatrix} 1 & 0 & 0 \\ 1 & 1 & 0 \\ 1 & 1 & 1 \end{bmatrix} = \begin{bmatrix} 1 & 0 & 0 \\ 3 & 1 & 0 \\ 6 & 3 & 1 \end{bmatrix}$$

and,

$$A^4 = A^3.A = \begin{bmatrix} 1 & 0 & 0 \\ 3 & 1 & 0 \\ 6 & 3 & 1 \end{bmatrix} \times \begin{bmatrix} 1 & 0 & 0 \\ 1 & 1 & 0 \\ 1 & 1 & 1 \end{bmatrix} = \begin{bmatrix} 1 & 0 & 0 \\ 4 & 1 & 0 \\ 10 & 4 & 1 \end{bmatrix}$$

On observing the pattern, we come to a conclusion that,

$$A = \begin{bmatrix} 1 & 0 & 0 \\ n & 1 & 0 \\ \frac{n(n+1)}{2} & n & 1 \end{bmatrix}$$

$$\therefore A^{20} = \begin{bmatrix} 1 & 0 & 0 \\ 20 & 1 & 0 \\ 210 & 20 & 1 \end{bmatrix}$$

Therefore, sum of the first column of $A^{20} = [1+20+=210] = 231$

75. (b)

$A^T + B^T = 2B$

$\because \left[(A+B)^T = (2B^T)^T \right]$

$\Rightarrow B = \frac{A^T+B^T}{2} = A + \left(\frac{B^T+A^T}{2} \right) = 2B^T$

$\Rightarrow 2A + A^T = 3B^T \Rightarrow A = \frac{3B^T - A^T}{2}$

Also,

$3A + 2B = I_3$ (i)

$\Rightarrow 3\left(\frac{3B^T - A^T}{2} \right) + 2\left(\frac{A^T+B^T}{2} \right) = I_3$

$\Rightarrow 11B^T - A^T = 2I_3$ (ii)

Add (i) and (ii)

$35B = 7I_3$

$\Rightarrow B = \frac{I_3}{5} \Rightarrow 11\frac{I_3}{5} - A = 2I_3$

$\Rightarrow 11\frac{I_3}{5} - 2I_3 = A \Rightarrow A = \frac{I_3}{5}$

$\because 5A = 5B = I_3$

$\Rightarrow 10A + 5B = 3I_3$

76. (c)

$$P = \begin{bmatrix} \frac{\sqrt{3}}{2} & \frac{1}{2} \\ -\frac{1}{2} & \frac{\sqrt{3}}{2} \end{bmatrix} P^T = \begin{bmatrix} \frac{\sqrt{3}}{2} & \frac{-1}{2} \\ \frac{1}{2} & \frac{\sqrt{3}}{2} \end{bmatrix}$$

$PP^T = P^TP = I$

$Q^{2015} = (PAP^T)(PAP^T)\_\_\_\_ (2015 \text{ terms})$

$= PA^{2015}P^T$

$P^TQ^{2015}P = A^{2015}$

$$A^2 = \begin{bmatrix} 1 & 1 \\ 0 & 1 \end{bmatrix}\begin{bmatrix} 1 & 1 \\ 0 & 1 \end{bmatrix} = \begin{bmatrix} 1 & 2 \\ 0 & 1 \end{bmatrix}$$

$$A^3 = \begin{bmatrix} 1 & 2 \\ 0 & 1 \end{bmatrix}\begin{bmatrix} 1 & 1 \\ 0 & 1 \end{bmatrix} = \begin{bmatrix} 1 & 3 \\ 0 & 1 \end{bmatrix}$$

$$\therefore A^{2015} = \begin{bmatrix} 1 & 2015 \\ 0 & 1 \end{bmatrix}$$

77. (a)
Given that

$$A = \begin{bmatrix} 0 & -1 \\ 1 & 0 \end{bmatrix}$$

$$A^2 = \begin{bmatrix} -1 & 0 \\ 0 & -1 \end{bmatrix} \Rightarrow A^2 = -I$$

$$A^3 = \begin{bmatrix} 0 & 1 \\ -1 & 0 \end{bmatrix}$$

$$A^4 = \begin{bmatrix} 1 & 0 \\ 0 & 1 \end{bmatrix} = I$$

$A^2 + I = A^3 - A$

$-I + I = A^3 - A$

$A^3 \neq A$

78. (b)
Given that $AA^T = 9I$

$$\begin{bmatrix} 1 & 2 & 2 \\ 2 & 1 & -2 \\ a & 2 & b \end{bmatrix}\begin{bmatrix} 1 & 2 & a \\ 2 & 1 & 2 \\ 2 & -2 & b \end{bmatrix} = \begin{bmatrix} 9 & 0 & 0 \\ 0 & 9 & 0 \\ 0 & 0 & 9 \end{bmatrix}$$

$$\Rightarrow \begin{bmatrix} 1+4+4 & 2+2-4 & a+4+2b \\ 2+2-4 & 4+1+4 & 2a+2-2b \\ a+4+2b & 2a+2-2b & a^2+4+b^2 \end{bmatrix} = \begin{bmatrix} 9 & 0 & 0 \\ 0 & 9 & 0 \\ 0 & 0 & 9 \end{bmatrix}$$

$\Rightarrow a+4+2b = 0 \Rightarrow a+2b = -4$ (i)

$2a+2-2b = 0 \Rightarrow 2a-2b = -2$

$\Rightarrow a-b = -1$ (ii)

Subtract (ii) from (i)

$a + 2b = -4$

$a - b = -1$

$- \quad + \quad +$

$3b = -3$

$B = -1$

and $a = -2$

$(a, b) = (-2, -1)$

79. (a)

Let $A = \begin{bmatrix} 1 & 2 & x \\ 3 & -1 & 2 \end{bmatrix}$ and $B = \begin{bmatrix} y \\ x \\ 1 \end{bmatrix}$

$$AB = \begin{bmatrix} 1 & 2 & x \\ 3 & -1 & 2 \end{bmatrix}\begin{bmatrix} y \\ x \\ 1 \end{bmatrix}$$

$$\Rightarrow \begin{bmatrix} 6 \\ 8 \end{bmatrix} = \begin{bmatrix} y+2x+x \\ 3y-x+2 \end{bmatrix}$$

$$\Rightarrow \begin{bmatrix} 6 \\ 8 \end{bmatrix} = \begin{bmatrix} y+3x \\ 3y-x+2 \end{bmatrix}$$

$\Rightarrow y+3x = 6$ and $3y-x = 6$

On solving, we get

$x = \frac{6}{5}$ and $y = \frac{12}{5}$

$\Rightarrow y = 2x$

80. (a)
Given

$$[p \quad q \quad r]\begin{bmatrix} 3 & 4 & 1 \\ 3 & 2 & 3 \\ 2 & 0 & 2 \end{bmatrix} = [3 \quad 0 \quad 1]$$

$\Rightarrow [3p+3q+2r \;\; 4p+2q \;\; p+3q+2r] = [3 \quad 0 \quad 1]$

$\Rightarrow 3p+3q+2r = 3$ ...(i)

$4p+2q = 0 \Rightarrow q = -2p$ ...(ii)

$p+3q+2r = 1$ ...(iii)

On solving (i), (ii) and (iii), we get

$p = 1, q = -2, r = 3$

$\therefore 2p+q-r = 2(1)+(-2)-(3) = -3$

81. (d)

$P^T = 2P + I$

$\Rightarrow P = 2P^T + I \Rightarrow P = 2(2P+I)+I$

$\Rightarrow P = 4P + 3I \Rightarrow P + I = 0$

$\Rightarrow PX + X = 0 \Rightarrow PX = -X$

82. (a)
For the given matrix to be non-singular

$$\begin{vmatrix} 1 & a & b \\ \omega & 1 & c \\ \omega^2 & \omega & 1 \end{vmatrix} \neq 0$$

$\Rightarrow 1-(a+c)\omega + ac\omega^2 \neq 0 \Rightarrow (1-a\omega)(1-c\omega) \neq 0$

$\Rightarrow a \neq \omega^2$ and $c \neq \omega$, where ω is complex cube root of unity.

As a, b and c are complex cube roots of unity, $\therefore$ a and c can take only one value, i.e., ω while b can take 2 values, i.e., ω and ω^2.

$\therefore$ Total number of distinct matrices in the set $S = 1 \times 1 \times 2 = 2$

Integer and Numerical Value Type

1.

$$AA' = \begin{bmatrix} t^2+(t+1)^2 & t(t-1)+t(t+1) \\ t(t-1)+t(t+1) & (t-1)^2+t^2 \end{bmatrix}$$

$$\Rightarrow AA^t = I_2 \Rightarrow t^2 - t + t^2 + t = 0$$

$$\Rightarrow 2t^2 = 0 \Rightarrow t = 0$$

$$\Rightarrow \text{Tr}\,A = 0$$

2. Let $\alpha = s-a, \beta = s-b, \gamma = s-c$
Then determinant

$$\begin{vmatrix} (\beta+\gamma)^2 & \alpha^2 & \alpha^2 \\ \beta^2 & (\gamma+\alpha)^2 & \beta^2 \\ \gamma^2 & \gamma^2 & (\alpha+\beta)^2 \end{vmatrix}$$

$$= 2\alpha\beta\gamma(\alpha+\beta+\gamma)^3$$

$$= 2(s-a)(s-b)(s-c)s^3$$

$$\Rightarrow k = 2$$

3. Take I_n as $n \times n$ identity matrix and B as another $n \times n$ matrix whose all elements are equal to 1 then A = B – I, and $B^2 = nB$.
Note that $(B-I)(CB-I) = CB^2(C+I)B+I$

If we choose $C = C = \frac{1}{n-1}$, then

$$A^{-1} = (B-I)^{-1}$$

$$= \frac{1}{n-1}B - I$$

The diagonal elements in the matrix $\frac{1}{n-1}B - I$ will be $\frac{1}{n-1} - 1 = \frac{2-n}{n-1}$

$$\Rightarrow k = 2$$

4. We are given that

$$A(A^2 - I) = A^2 - I$$

or $A^3 - A = A^2 - I$

On multiplying by $A^{47}, A^{46}, A^{45}, \ldots\ldots, A$ in succession

$$A^{50} - A^{48} = A^{49} - A^{47}$$

$$A^{49} - A^{47} = A^{48} - A^{46}$$

$A^{48} - A^{46} = A^{47} - A^{45}$ and so on.

On adding all such relations, we get

$$A^{50} - A^{48} = A^2 - I$$

As earlier, we will get

$$A^{48} - A^{46} = A^2 - I$$

$A^{46} - A^{44} = A^2 - I$ and so on.

On adding all such relations, we get

$$A^{50} - A^2 = 24(A^2 - I)$$

$$\Rightarrow A^{50} = 25A^2 - 24I$$

$$\Rightarrow k = 25$$

5. Since on expanding the determinant, it will become a symmetric homogenous function of degree 3 in *a*, *b*, *c* and RHS is also a numerical quantity (*k* is a numerical quantity), *k* can be obtained by assigning numerical values of *a*, *b*, *c*.
On putting, $a = b = c = 1$, we get

$$\begin{vmatrix} 2 & 1 & 1 \\ 1 & 2 & 1 \\ 1 & 1 & 2 \end{vmatrix} = kabc$$

$$\Rightarrow 4 = kabc$$

$$\Rightarrow k = 4$$

6. Let $X = \begin{pmatrix} x_1 \\ x_2 \\ x_3 \end{pmatrix}$

$$X'AX = 0$$

$$\Rightarrow (x_1\ x_2\ x_3)\begin{pmatrix} a_{11} & a_{12} & a_{13} \\ a_{21} & a_{22} & a_{23} \\ a_{31} & a_{32} & a_{33} \end{pmatrix}\begin{pmatrix} x_1 \\ x_2 \\ x_3 \end{pmatrix} = 0$$

$$\Rightarrow a_{11}x_1^2 + a_{22}x_2^2 + a_{33}x_3^2 + (a_{12}+a_{21})x_1x_2 + (a_{13}+a_{31})x_1x_3 + (a_{23}+a_{32})x_2x_3 = 0$$

If this is true for every x_1, x_2, x_3, then

$$a_{11} = a_{22} = a_{33} = 0$$

and $a_{12} = -a_{21}, a_{13} = -a_{31}, a_{23} = -a_{32}$

Now as $a_{23} = -2006$

$$a_{32} = 2006$$

7. Given that

$$(I-A)^2 = I^3 - A^3 - 3A(I-A) = I - A^3$$

$$\Rightarrow 3A(I-A) = 0 \Rightarrow A^2 = A$$

$$\Rightarrow \begin{bmatrix} a^2 & ab+bd \\ 0 & d^2 \end{bmatrix} = \begin{bmatrix} a & b \\ 0 & d \end{bmatrix}$$

$$\Rightarrow a^2 = a, b(a+d-1) = 0, d^2 = d$$

Case I:

$$b = 0 \Rightarrow (a,d) = (0,1),(0,0),(1,1) \rightarrow 4 \text{ ways}$$

Case II:

$a+d=1 \Rightarrow (1,0),(0,1)$ and $b=\pm 1 \rightarrow 4$ ways

$\Rightarrow$ Total 8 matrices.

8. $A^n = \begin{bmatrix} 1 & n & \frac{n(n+1)}{2} \\ 0 & 1 & n \\ 0 & 0 & 1 \end{bmatrix}$

Since, $\sum 1 = 20, \sum_{n=1}^{20} n = \frac{20\times 21}{2} = 210$

and $\frac{1}{2}\sum_{n=1}^{20} n(n+1) = \frac{1}{2}\times\frac{20\times 21\times 22}{3} = 1540$

$\therefore M = A + A^2 + ... + A^{20}$

$= \begin{bmatrix} \sum 1 & \sum n & \sum\frac{n(n+1)}{2} \\ 0 & \sum 1 & \sum n \\ 0 & 0 & \sum 1 \end{bmatrix} = \begin{bmatrix} 20 & 210 & 1540 \\ 0 & 20 & 210 \\ 0 & 0 & 20 \end{bmatrix}$

$\therefore$ Sum of all the elements of M

$= 20 + 20 + 20 + 210 + 1540 = 2020$

9. Given that

$\begin{bmatrix} 0 & i \\ 1 & 0 \end{bmatrix}^n \begin{bmatrix} a & b \\ c & d \end{bmatrix} = \begin{bmatrix} a & b \\ c & d \end{bmatrix}$

$\begin{bmatrix} 0 & i \\ 1 & 0 \end{bmatrix} \begin{bmatrix} a & b \\ c & d \end{bmatrix} = \begin{bmatrix} a & b \\ c & d \end{bmatrix}$

It is possible when

$\begin{bmatrix} 0 & i \\ 1 & 0 \end{bmatrix}^n = I$

Let $A = \begin{bmatrix} 0 & i \\ 1 & 0 \end{bmatrix}$

$\Rightarrow A^2 = \begin{bmatrix} 0 & i \\ 1 & 0 \end{bmatrix}\begin{bmatrix} 0 & i \\ 1 & 0 \end{bmatrix} = \begin{bmatrix} i & 0 \\ 0 & i \end{bmatrix}$

$\Rightarrow A^2 = iI$

$\Rightarrow (A^2)^4 = i^4 I$

$\Rightarrow A^8 = I$

$\Rightarrow n$ is a multiple of 8

So, the number of 2-digit numbers in the set

S = 16, 24, 32, …, 96

$\therefore n(S) = 11$

10. Let $B = \begin{bmatrix} x_1 & x_2 & x_3 \\ y_1 & y_2 & y_3 \\ z_1 & z_2 & z_3 \end{bmatrix}$

$\because AB = BA$

$\Rightarrow \begin{bmatrix} 0 & 1 & 0 \\ 1 & 0 & 0 \\ 0 & 0 & 1 \end{bmatrix}\begin{bmatrix} x_1 & x_2 & x_3 \\ y_1 & y_2 & y_3 \\ z_1 & z_2 & z_3 \end{bmatrix}$

$= \begin{bmatrix} x_1 & x_2 & x_3 \\ y_1 & y_2 & y_3 \\ z_1 & z_2 & z_3 \end{bmatrix}\begin{bmatrix} 0 & 1 & 0 \\ 1 & 0 & 0 \\ 0 & 0 & 1 \end{bmatrix}$

$\Rightarrow \begin{bmatrix} y_1 & y_2 & y_3 \\ x_1 & x_2 & x_3 \\ z_1 & z_2 & z_3 \end{bmatrix} = \begin{bmatrix} x_2 & x_1 & x_3 \\ y_2 & y_1 & y_3 \\ z_2 & z_1 & z_3 \end{bmatrix}$

$\Rightarrow y_1 = x_2, y_2 = x_1\ y_3 = x_3, z_1 = z_2$

$\therefore B = \begin{bmatrix} x_1 & x_2 & x_3 \\ y_1 & y_2 & y_3 \\ z_1 & z_2 & z_3 \end{bmatrix}$

$\therefore$ Five elements can be written independent in $5^5 = 3125$ ways

$\therefore$Number of matrices = 3125

11. Given that A = XB

$\Rightarrow \begin{bmatrix} a_1 \\ a_2 \end{bmatrix} = \frac{1}{\sqrt{3}}\begin{bmatrix} 1 & -1 \\ 1 & k \end{bmatrix}\begin{bmatrix} b_1 \\ b_2 \end{bmatrix}$

On comparing, we get

$\sqrt{3}a_1 = b_1 - b_2 \Rightarrow 3a_1^2 = b_1^2 + b_2^2 - 2b_1b_2$ ....(i)

$\sqrt{3}a_2 = b_1 + kb_2 \Rightarrow 3a_2^2 = b_1^2 + k^2b_2^2 + 2kb_1b_2$....(ii)

Adding (i) and (ii), we get

$3(a_1^2 + a_2^2) = 2b_1^2 + (k^2+1)b_2^2 + 2(k-1)b_1b_2$

Given that $a_1^2 + a_2^2 = \frac{2}{3}(b_1^2 + b_2^2)$

$\Rightarrow 2(b_1^2 + b_2^2) = 2b_1^2 + (k^2-1)b_2^2 + 2(k-1)b_1b_2$

$\Rightarrow (1-k^2)b_2^2 = 2(k-1)b_1b_2$

$\Rightarrow (k-1)[(k+1)b_2^2 + 2b_1b_2]$

$= 0\left[\because (k^2+1)b_2^2 \neq -2b_1b_2\right]$

$\Rightarrow k = 1$

12.

$A^2 = \begin{vmatrix} 1 & 0 & 0 \\ 0 & 4 & 0 \\ 0 & 0 & 1 \end{vmatrix}, A^3 = \begin{vmatrix} 1 & 0 & 0 \\ 0 & 2^3 & 0 \\ 3 & 0 & -1 \end{vmatrix}$

$$A^4 = \begin{vmatrix} 1 & 0 & 0 \\ 0 & 2^4 & 0 \\ 0 & 0 & 1 \end{vmatrix}$$

So, $A^{20} + \alpha A^{19} + \beta A$

$$= \begin{vmatrix} 1 & 0 & 0 \\ 0 & 2^{20} & 0 \\ 0 & 0 & 1 \end{vmatrix} + \alpha = \begin{vmatrix} 1 & 0 & 0 \\ 0 & 2^{19} & 0 \\ 3 & 0 & -1 \end{vmatrix} + \beta = \begin{vmatrix} 1 & 0 & 0 \\ 0 & 2 & 0 \\ 3 & 0 & -1 \end{vmatrix}$$

$$= \begin{vmatrix} 1+\alpha+\beta & 0 & 0 \\ 0 & 2^{20}+\alpha 2^{19}+2\beta & 0 \\ 3\alpha+3\beta & 0 & 1-\alpha-\beta \end{vmatrix} = \begin{vmatrix} 1 & 0 & 0 \\ 0 & 4 & 0 \\ 0 & 0 & 1 \end{vmatrix}$$

On comparing, we get

$\alpha+\beta=0$ and $2^{20}+\alpha.2^{19}+2\beta=4$

$\alpha=-2$ and $\beta=2$

So, $\beta-\alpha=2+2=4$

13. Let the matrix be $A = \begin{bmatrix} a & b & c \\ d & e & f \\ g & h & i \end{bmatrix}$

Then,

$$AA^T = \begin{bmatrix} a^2+b^2+c^2 & - & - \\ - & d^2+e^2+f^2 & - \\ - & - & g^2+h^2+i^2 \end{bmatrix}$$

Sum of diagonal entries,

$a^2+b^2+c^2+d^2+e^2+f^2+g^2+h^2+i^2=9$

Case I: Eight times 0's and one time 3 = 9 cases

Case II: Two times 2, one time 1, and six times $0 = \frac{9!}{6!2!} = 252$ cases

Case III: One time 2, five times 1, and three times $0 = \frac{9!}{5!3!} = 504$ cases

Case IV: Nine times ones = 1 case

$\therefore$ Total cases = 9 + 252 + 504 + 1 = 766

14.

$$\begin{bmatrix} a & b & c \\ d & e & f \\ g & h & i \end{bmatrix}\begin{bmatrix} a & d & g \\ b & e & h \\ c & f & i \end{bmatrix}$$

$a^2+b^2+c^2+d^2+e^2+^2+g^2+h^2+i^2=7$

Case I: Seven (1's) and two (0's)

${}^9C_2 = 36$

Case II: One (2) and three (1's) and five (0's)

$\frac{9!}{5!3!} = 504$

$\therefore$ Total = 540

15.

$$A^2 = \begin{bmatrix} x & 1 \\ 1 & 0 \end{bmatrix}\begin{bmatrix} x & 1 \\ 1 & 0 \end{bmatrix} = \begin{bmatrix} x^2+1 & x \\ x & 1 \end{bmatrix}$$

$$A^4 = \begin{bmatrix} x^2+1 & x \\ x & 1 \end{bmatrix}\begin{bmatrix} x^2+1 & x \\ x & 1 \end{bmatrix}$$

$$= \begin{bmatrix} (x^2+1)^2+x^2 & x(x^2+1)+x \\ x(x^2+1)+x & x^2+1 \end{bmatrix}$$

Given that $(x^2+1)^2+x^2=109$

$x^4+3x^2-108=0$

$\Rightarrow (x^2+12)(x^2-9)=0$

$\therefore x^2=9$

$a_{22}=x^2+1=9+1=10$

16. Let $A=[a_{ii}]_{3\times3}$

It is given that the sum of diagonal elements of AA^T is 3, i.e.,

$tr(AA^T)=3$

$a_{11}^2+a_{12}^2+a_{13}^2+a_{21}^2+..........+a_{33}^2=3$

Possible cases are

$$\left.\begin{matrix} 0,0,0,0,0,0,1,1,1 & \to 1 \\ 0,0,0,0,0,0,-1,-1,-1 & \to 1 \\ 0,0,0,0,0,0,1,1,-1 & \to 3 \\ 0,0,0,0,0,0,-1,1,-1 & \to 3 \end{matrix}\right\} {}^9C_6 \times 8 = 84\times8 = 672$$

17. Let $M = \begin{bmatrix} a_1 & b_1 & c_1 \\ a_2 & b_2 & c_2 \\ a_3 & b_3 & c_3 \end{bmatrix}$

Then $\begin{bmatrix} a_1 & b_1 & c_1 \\ a_2 & b_2 & c_2 \\ a_3 & b_3 & c_3 \end{bmatrix}\begin{bmatrix} 0 \\ 1 \\ 0 \end{bmatrix} = \begin{bmatrix} -1 \\ 2 \\ 3 \end{bmatrix}$

$\Rightarrow b_1=-1, b_2=2, b_3=3$

$$\begin{bmatrix} a_1 & b_1 & c_1 \\ a_2 & b_2 & c_2 \\ a_3 & c_3 & c_3 \end{bmatrix}\begin{bmatrix} 1 \\ -1 \\ 0 \end{bmatrix} = \begin{bmatrix} 1 \\ 1 \\ 1 \end{bmatrix}$$

$\Rightarrow a_1-b_1=1, a_2-b_2=1, a_3-b_3=-1$

$\Rightarrow a_1=0, a_2=3, a_3=2$

and $\begin{bmatrix} a_1 & b_1 & c_1 \\ a_2 & b_2 & c_2 \\ a_3 & b_3 & c_3 \end{bmatrix}\begin{bmatrix} 1 \\ 1 \\ 1 \end{bmatrix} = \begin{bmatrix} 0 \\ 0 \\ 12 \end{bmatrix}$

$\Rightarrow a_3+b_3+c_3=12 \Rightarrow c_3=7$

$\therefore$ Sum of diagonal elements
$= a_1 + b_2 + c_3 = 0 + 2 + 7 = 9$

Assertion and Reason

1. (a)
$A = -A^T \Rightarrow |A| = -|A^T| = -|A|$
$\Rightarrow 2|A| = 0 \Rightarrow |A| = 0$

2. (d)
Statement 1 is false.
$\because A = [a_{ij}]_{n \times n}$ where $a_{ij} = 0, i \geq j$
$\therefore |A| = 0$. Hence A is singular as inverse of A is not defined.
Statement 2 $|A| = 0$
$\therefore$Inverse of A is not defined.

3. (a)
$$A(\alpha) = \begin{bmatrix} f_1(\alpha) & f_2(\alpha) & f_3(\alpha) \\ f_4(\alpha) & f_5(\alpha) & f_6(\alpha) \\ f_7(\alpha) & f_8(\alpha) & f_9(\alpha) \end{bmatrix} = \begin{bmatrix} 0 & 0 & 0 \\ 0 & 0 & 0 \\ 0 & 0 & 0 \end{bmatrix}$$
$x - \alpha$ is a factor of $f_1(x), f_2(x) \ldots\ldots\ldots f_9(x)$
$f(x) = (x - \alpha)\phi(x)$
$f(\alpha) = 0 \Rightarrow x - \alpha$ is a factor of $f(x)$

4. (d)
A^{-1} exists only for non-singular matrix.
$AB = AC \Rightarrow B = C$ if A^{-1} exists
If A^{-1} exists

5. (c)
$\det(AB^{-1}) = \det A \,.\, \det B^{-1} = \dfrac{\det A}{\det B} = \dfrac{-6}{2} = -3$

JEE Advanced

Objective Questions

One Option Correct

1. (a)
$(B^TAB)^T = B^TA^T(B^T)^T = B^TA^TB$
$= B^TAB$ iff A is symmetric.
$\therefore B^TAB$ is symmetric iff A is symmetric.
Also $(B^TAB)^T = B^TA^TB = (-B)A^TB$
$\therefore B^TAB$ is not skew-symmetric if B is skew-symmetric.

2. (c)
$$AB = A, BA = B \Rightarrow A^2 = A \quad \text{and} \quad B^2 = B$$
$$(A+B)^2 = A^2 + B^2 + AB + BA$$
$$= A + B + A + B = 2(A+B)$$
$$(A+B)^3 = (A+B)^2(A+B)$$
$$= 2(A+B)^2 = 2^2(A+B)$$
$$\therefore (A+B)^7 = 2^6(A+B) = 64(A+B)$$

3. (c)
$A^2 = I \Rightarrow A^2 - I = 0$
$\Rightarrow (A+I)(A-I) = 0$
$\therefore$ either $|A+I| = 0$ or
$|A-I| = 0$
If $|A-I| \neq 0$, then
$(A+I)(A-I) = 0 \Rightarrow A + I = 0$
which is not so $\therefore |A-I| = 0$ and $A - I \neq 0$

4. (b)
$A^3 = 0$
$(I + A + A^2)(I - A) = I - A^3 = I$
$\therefore \quad I + A + A^2 = (I - A)^{-1}$

5. (a)
A diagonal matrix is commutative with every square matrix if it is a scalar matrix. So, every diagonal element is 4.
$\therefore |A| = 64$

6. (c)
$$|\text{adj}A^{-1}| = |A^{-1}|^2 = \frac{1}{|A|^2};$$
$$|(\text{adj}A^{-1})^{-1}| = \frac{1}{|\text{adj}A^{-1}|} = |A|^2 = 2^2 = 4$$

7. (d)
Since AB = B and BA = A
$\therefore$ A and B, both are idempotent.
$(A-B)^2 = A^2 - AB - BA + B^2$
$= A - B - A + B = 0$
$\therefore$ A – B is nilpotent.

8. (d)
$$AB = \begin{bmatrix} 1 & 2 \\ 3 & 4 \end{bmatrix}\begin{bmatrix} a & b \\ c & d \end{bmatrix} = \begin{bmatrix} a+2c & b+2d \\ 3a+4c & 3b+4d \end{bmatrix}$$
$$BA = \begin{bmatrix} a & b \\ c & d \end{bmatrix}\begin{bmatrix} 1 & 2 \\ 3 & 4 \end{bmatrix} = \begin{bmatrix} a+3b & 2a+4b \\ c+3d & 2c+4d \end{bmatrix}$$
If AB = BA, then $a + 2c = a + 3b$
$\Rightarrow \quad 2c = 3b \Rightarrow \quad b \neq 0$
$b + 2d = 2a + 4b$
$\Rightarrow \quad 2a - 2d = -3b$
$$\frac{a-d}{3b-c} = \frac{-\frac{3}{2}b}{3b - \frac{3}{2}b} = -1$$

9. (b)

$$(f(\alpha))^{-1} = \begin{bmatrix} \cos\alpha & \sin\alpha & 0 \\ -\sin\alpha & \cos\alpha & 0 \\ 0 & 0 & 1 \end{bmatrix} = f(-\alpha)$$

10. (c)

For $n = 2 \Rightarrow (A^{-1}BA)(A^{-1}BA) = A^{-1}B^2A$

$$(A^{-1}BA)^3 = (A^{-1}BA)^2(A^{-1}BA)$$

$$= (A^{-1}B^2A)(A^{-1}BA)$$

$= A^{-1}B^3A$ and so on.

Thus $(A^{-1}BA)^n = A^{-1}B^nA$

11. (a)

$$\begin{bmatrix} 2 & 1 \\ 3 & 2 \end{bmatrix} A \begin{bmatrix} -3 & 2 \\ 5 & -3 \end{bmatrix} = \begin{bmatrix} 1 & 0 \\ 0 & 1 \end{bmatrix}$$

$$A = \begin{bmatrix} 2 & -1 \\ -3 & 2 \end{bmatrix} \begin{bmatrix} 1 & 0 \\ 0 & 1 \end{bmatrix} \begin{bmatrix} -3 & -2 \\ -5 & -3 \end{bmatrix} (-1)$$

$$= \begin{bmatrix} 2 & -1 \\ -3 & 2 \end{bmatrix} \begin{bmatrix} 3 & 2 \\ 5 & 3 \end{bmatrix} = \begin{bmatrix} 1 & 1 \\ 1 & 0 \end{bmatrix}$$

12. (c)

$$S_1 : A^T = A$$

$$\left((A^{-1})^{-1}\right)^{-1} = A^{-1}$$

$$(A^{-1})^T = (A^T)^{-1} = A^{-1}$$

So $\left((A^{-1})^{-1}\right)^{-1}$ is also symmetric.

S_2: Obvious

S_3: Obvious

S_4: $|A| \neq 0, |B| \neq 0 \Rightarrow |AB| \neq 0$

So, AB is invertible.

13. (b)

Let $M = \begin{bmatrix} a_1 & a_2 & a_3 \\ a_4 & a_5 & a_6 \\ a_7 & a_8 & a_9 \end{bmatrix}$ where $a_i \in \{0, 1, 2\}$

Then $M^T M = \begin{bmatrix} a_1 & a_4 & a_7 \\ a_2 & a_5 & a_8 \\ a_3 & a_6 & a_9 \end{bmatrix} \begin{bmatrix} a_1 & a_2 & a_3 \\ a_4 & a_5 & a_6 \\ a_7 & a_8 & a_9 \end{bmatrix}$

Sum of the diagonal entries in $M^T M = 5$.

$$\Rightarrow (a_1^2 + a_4^2 + a_7^2) + (a_2^2 + a_5^2 + a_8^2)$$

$$+ (a_3^2 + a_6^2 + a_9^2) = 5,$$

which is possible when

Case I: Five a_i's are 1 and four a_i's are zero, which can be done in ${}^9C_4$ ways

$$= \frac{9 \times 8 \times 7 \times 6}{4 \times 3 \times 2 \times 1} = 126$$

Case II: One a_i is 1, one a_i is 2 and the remaining seven a_i's are zero.

It can be done in ${}^9C_1 \times {}^8C_1 = 9 \times 8 = 72$ ways

$\therefore$ Total no. ways = 126 + 72 = 198.

14. (b)

$$P = \begin{bmatrix} 1 & 0 & 0 \\ 4 & 1 & 0 \\ 16 & 4 & 1 \end{bmatrix} = I + \begin{bmatrix} 0 & 0 & 0 \\ 4 & 0 & 0 \\ 16 & 4 & 0 \end{bmatrix} = I + A$$

where $A = \begin{bmatrix} 1 & 0 & 0 \\ 4 & 0 & 0 \\ 16 & 4 & 0 \end{bmatrix} \Rightarrow A^2 = \begin{bmatrix} 0 & 0 & 0 \\ 0 & 0 & 0 \\ 16 & 0 & 0 \end{bmatrix}$

and $A^3 = \begin{bmatrix} 0 & 0 & 0 \\ 0 & 0 & 0 \\ 0 & 0 & 0 \end{bmatrix}, \therefore A^n = O, \forall n \geq 3$

Now

$$P^{50} = (I + A)^{50}$$

$$= {}^{50}C_0 I^{50} + {}^{50}C_1 I^{49} A + {}^{50}C_2 I^{48} A^2 + O$$

$$= I + 50A + 25 \times 48A^2$$

$$\therefore Q = P^{50} - I = 50A + 25 \times 49A^2$$

$$\Rightarrow q_{21} = 50 \times 4 = 200,$$

$$q_{31} = 50 \times 16 + 25 \times 49 \times 16 = 20400$$

and $q_{32} = 50 \times 4 = 200$

$$\therefore \frac{q_{31} + q_{32}}{q_{21}} = \frac{20600}{200} = 103$$

One or More Than One Option Correct

1. (a, b, d)

$$|10AB| = 10^3 |A||B| = (10^3)(5)(2) = 10^4$$

2. (a, b, d)

For skew-symmetric matrix,

$$a_{ii} = 0 \Rightarrow \theta = (2n+1)\frac{\pi}{2}$$

For symmetric matrix, $\sin\theta = 0 \Rightarrow \theta = n\pi$

$A = A^{-1}$ is possible if $\sin\theta = 0$

3. (a, c)

$$|A| = \begin{vmatrix} a+b & ab & 0 \\ 1 & a+b & ab \\ 0 & 1 & a+b \end{vmatrix}$$

$$= (a^2 + b^2)(a + b)$$

4. (a, b, d)

$$A^2-4A-5I_3=\begin{bmatrix}1&2&2\\2&1&2\\2&2&1\end{bmatrix}\begin{bmatrix}1&2&2\\2&1&2\\2&2&1\end{bmatrix}$$

$$-4\begin{bmatrix}1&2&2\\2&1&2\\2&2&1\end{bmatrix}-5\begin{bmatrix}1&0&0\\0&1&0\\0&0&1\end{bmatrix}$$

$$=\begin{bmatrix}9&8&8\\8&9&8\\8&8&9\end{bmatrix}+\begin{bmatrix}-4&-8&-8\\-8&-4&-8\\-8&-8&-4\end{bmatrix}+\begin{bmatrix}-5&0&0\\0&-5&0\\0&0&-5\end{bmatrix}$$

$$=\begin{bmatrix}0&0&0\\0&0&0\\0&0&0\end{bmatrix}=0$$

Also,

$$|A^2|=\begin{bmatrix}9&8&8\\8&9&8\\8&8&9\end{bmatrix}$$

$$=9(81-64)-8(72-64)+8(64-72)$$

$$=9\times17-8\times8+8\times(-8)$$

$$=133-128=5\neq0$$

$\therefore$ A^2 is invertible.

and $A^3=A.A^2=A.(4A-5I_3)=4A^2-5A$

$$=\begin{bmatrix}36&32&32\\32&36&32\\32&32&36\end{bmatrix}+\begin{bmatrix}-5&-10&-10\\-10&-5&-10\\-10&-10&-5\end{bmatrix}$$

$$=\begin{bmatrix}31&22&22\\22&31&22\\22&22&31\end{bmatrix}$$

$\therefore \quad |A^3|\neq0$

$\therefore$ A^3 is invertible

5. (a, b, c, d)

Here, (a), (b), (c), and (d) are the properties of adjoint.

6. (a, b, c)

$$\because |A|=\begin{vmatrix}3&-3&4\\2&-3&4\\0&-1&1\end{vmatrix}=3(-3+4)-2(-3+4)+0=1$$

$$\because \text{adj}(\text{adj A})=|A|^{3-2}\,A$$

$$=A \text{ and } |\text{adj}(\text{adj A})|=|A|=1$$

Also, $|\text{adjA}|=|A|^{3-1}=|A|^2=1^2=1$

7. (a, b, c, d)

Obvious (using properties)

8. (a, d)

$$A^2=\begin{bmatrix}3&3&3\\3&3&3\\3&3&3\end{bmatrix}=3A$$

$$A^3=A^2A=3A, A=3A^2=3.(3A)$$

$$=9A \text{ and } |A|=0$$

$\therefore$ A^{-1} does not exist.

9. (c, d)

$X'=-X, Y'=-Y, Z'=Z$

$$\left(Y^3Z^4-Z^4Y^3\right)'=\left(Z^4\right)'\left(Y^3\right)'-\left(Y^3\right)'\left(Z^4\right)'$$

$$=\left(Z'\right)^4\left(Y'\right)^3-\left(Y'\right)^3\left(Z'\right)^4$$

$$=-Z^4Y^3+Y^3Z^4=Y^3Z^4-Z^4Y^3$$

Therefore $\left(Y^3Z^4-Z^4Y^3\right)$ is a symmetric matrix.

Similarly, $X^{44}+Y^{44}$ is a symmetric matrix and $X^4Z^3-Z^3X^4$ and $X^{23}+Y^{23}$ are skew-symmetric matrices.

10. (c, d)

Let $M=\begin{bmatrix}a&b\\b&c\end{bmatrix}$ where a, b and c are integers.

M is invertible, if $\begin{vmatrix}a&b\\b&c\end{vmatrix}\neq0\Rightarrow ac\neq b^2$

$$\therefore \begin{bmatrix}a\\b\end{bmatrix}=\begin{bmatrix}b\\c\end{bmatrix}\Rightarrow a=b=c\Rightarrow ac=b^2$$

Hence (a) is not correct.

If $[b \quad c]=[a \quad b]\Rightarrow b=a=c\Rightarrow ac=b^2$

$\therefore$ (b) is not correct.

If $M=\begin{bmatrix}a&0\\0&c\end{bmatrix}$, then $|M|=ac\neq0$

$\therefore$ M is invertible

(c) is correct

Since, $ac\neq(\text{integer})^2\Rightarrow ac\neq b^2$

$\therefore$ (d) is correct.

11. (b, c, d)

For $n=3$, $P=\begin{bmatrix}\omega^2&\omega^3&\omega^4\\\omega^3&\omega^4&\omega^5\\\omega^4&\omega^5&\omega^6\end{bmatrix}$

and $P^2=\begin{bmatrix}0&0&0\\0&0&0\\0&0&0\end{bmatrix}$

It shows $P^2 = 0$ if n is a multiple of 3.
So, for $P^2 \neq 0$, n should not be a multiple of 3, hence n can take values 55, 58 and 56.

Matrix Type

1. $A \to Q$; $B \to S, T$; $C \to R$, $D \to P$

(A) $A^2 = -I$ $\therefore$ A is of even order.

(B) $(I+A)^n = C_0 I^n + C_1 IA + C_2 IA^2 + ... + C_n IA^n$

$= C_0 I + C_1 A + C_2 A + ... + C_n A$

$= I + (2^n - 1)A$

$\therefore \quad \lambda = 2^n - 1$

(C) $A^2 = A$ and $B = I - A$

$AB + BA + I - (I + A^2 - 2A)$

$= AB + BA - A + 2A = AB + BA + A$

$= A(I - A) + (I - A)A + A$

$= A - A + A - A + A = A$

(D) $A^* A$, $B^* = B$

$(AB - BA)^* = B^* A^* - A^* B^* = BA - AB$

Integer and Numerical Value Type

1. (1)

$$\frac{|\text{adj}B|}{|C|} = \frac{|\text{adj}(\text{adj}A)|}{|5A|} = \frac{|A|^{(3-1)^2}}{5^3|A|} = \frac{|A|^3}{125}$$

Now $|A| = 5$ $\therefore$ $\dfrac{|\text{adj } B|}{|C|} = 1$

2. (–A)

$$I + A = \begin{pmatrix} 1 & 0 \\ 0 & 1 \end{pmatrix} + \begin{pmatrix} 1 & 2 \\ 1 & 1 \end{pmatrix} = \begin{pmatrix} 2 & 2 \\ 1 & 2 \end{pmatrix}$$

$$\text{and } I - A = \begin{pmatrix} 1 & 0 \\ 0 & 1 \end{pmatrix} - \begin{pmatrix} 1 & 2 \\ 1 & 1 \end{pmatrix} = \begin{pmatrix} 0 & -2 \\ -1 & 0 \end{pmatrix}$$

$$\text{Now, } |I - A| = \begin{vmatrix} 0 & -2 \\ -1 & 0 \end{vmatrix} = 0 - 2 = -2$$

$$\text{adj}(I - A) = \begin{pmatrix} 0 & 2 \\ 1 & 0 \end{pmatrix}$$

$$(I - A)^{-1} = \begin{pmatrix} 0 & -1 \\ -\frac{1}{2} & 0 \end{pmatrix}$$

$$\therefore \phi(A) = (I + A)(I - A)^{-1}$$

$$= \begin{pmatrix} 2 & 2 \\ 1 & 2 \end{pmatrix} \begin{pmatrix} 0 & -1 \\ -\frac{1}{2} & 0 \end{pmatrix} = \begin{pmatrix} -1 & -2 \\ -1 & -1 \end{pmatrix} = -A$$

3. (2 and 4)

$A'A = I$

$\therefore \quad |A'A| = |I|$

$\Rightarrow \quad |A| = \pm 1$

$$\begin{vmatrix} a & b & c \\ b & c & a \\ c & a & b \end{vmatrix} = \pm 1$$

$\Rightarrow 3abc - a^3 - b^3 - c^3 = \pm 1$

$\Rightarrow \quad a^3 + b^3 + c^3 = 2$ and 4

4. (7)

Here $|A - \lambda I| = 0$

$$\begin{vmatrix} 1-\lambda & 0 \\ -1 & 7-\lambda \end{vmatrix} = 0$$

$\Rightarrow \quad (1-\lambda)(7-\lambda) = 0$

$\Rightarrow \quad \lambda^2 - 8\lambda + 7 = 0$

$\Rightarrow \quad A^2 - 8A + 7I_2 = 0$

$\Rightarrow \quad A^2 = 8A - 7I_2$

$\Rightarrow \quad k = -7$

$\Rightarrow \quad |k| = 7$

5. (Only one pair)

$z = \dfrac{-1 + i\sqrt{3}}{2} \Rightarrow z^3 = 1$ and $1 + z + z^2 = 0$

$$P^2 = \begin{bmatrix} (-z)^r & z^{2s} \\ z^{2s} & z^r \end{bmatrix} \begin{bmatrix} (-z)^r & z^{2s} \\ z^{2s} & z^r \end{bmatrix}$$

$$= \begin{bmatrix} z^{2r} + z^{4s} & z^{2s}\left((-z)^r + z^r\right) \\ z^{2s}\left((-z)^r + z^r\right) & z^{4s} + z^{2r} \end{bmatrix}$$

For $P^2 = I$, we should have

$z^{2r} + z^{4s} = -1$ and $\left((-z)^r + z^r\right) = 0$

$\Rightarrow z^{2r} + z^{4s} + 1 = 0$ and $(-z)^r + z^r = 0$

$\Rightarrow r$ is odd and $s = r$ but not a multiple of 3, which is possible when $s = r = 1$.

$\therefore$ Only one pair is there.

Comprehension Passage

Comprehension I

1. (c)

As second row of all the options is same, we are to look at the elements of the first row.

Let left inverse be $\begin{bmatrix} a & b & c \\ d & e & f \end{bmatrix}$, then

$$\begin{bmatrix} a & b & c \\ d & e & f \end{bmatrix}\begin{bmatrix} 1 & -1 \\ 1 & 1 \\ 2 & 3 \end{bmatrix}=\begin{bmatrix} 1 & 0 \\ 0 & 1 \end{bmatrix}$$

$\therefore a+b+2c=1$

$-a+b+3c=0$

i.e. $b=\dfrac{1-5c}{2}, a=\dfrac{1+c}{2}$

Thus, matrices in the options A, B and D are the inverses and matrix in option C is not the left inverse.

2. (d)

Let right inverse is

$$\begin{bmatrix} a & b \\ c & d \\ e & f \end{bmatrix}$$

Now $a-c+2e=1$

$b-d+2f=0$

$2a-c+e=0$

$2b-d+f=1$

Infinite solution

So, answer is (D).

3. (c)

By observation there can't be any left inverse for (B) and (D). So we will check for (A) and (C) only.

For (A) let left inverse be $\begin{bmatrix} a & b \\ c & d \\ e & f \end{bmatrix}$, then

$$\begin{bmatrix} a & b \\ c & d \\ e & f \end{bmatrix}\begin{bmatrix} 1 & 2 & 4 \\ -3 & 2 & 1 \end{bmatrix}=\begin{bmatrix} 1 & 0 & 0 \\ 0 & 1 & 0 \\ 0 & 0 & 1 \end{bmatrix}$$

Now $a-3b=1$, $2a+2b=0$ and $4a+b=0$, which is not possible.

For (C), $\begin{bmatrix} a & b & c \\ d & e & f \end{bmatrix}=\begin{bmatrix} 1 & 4 \\ 2 & -3 \\ 5 & 4 \end{bmatrix}=\begin{bmatrix} 1 & 0 \\ 0 & 1 \end{bmatrix}$

$\Rightarrow a+2b+5c=1, 4a-3b+4c=0,$

$d+2e+5f=0, 4d-3e+4f=1$

$\therefore$ There are infinite number of left inverses.

$$\begin{bmatrix} 1 & 4 \\ 2 & -3 \\ 5 & 4 \end{bmatrix}\begin{bmatrix} a & b & c \\ d & e & f \end{bmatrix}=\begin{bmatrix} 1 & 0 & 0 \\ 0 & 1 & 0 \\ 0 & 0 & 1 \end{bmatrix}$$

$\Rightarrow a+4d=1, 2a-3d=0$ and $5a+4d=0$

which is not possible.

$\therefore$ There is no right inverse.

Comprehension 2

1. (a)

Possible combinations are (1, 2, 6), (1, 3, 5), (1, 8, 9), (2, 3, 4), (2, 7, 9), (3, 6, 9), (3, 7, 8), (4, 5, 9), (4, 6, 8), (5, 6, 7).

Hence total 10 combinations are possible.

2. (b)

In a matrix $\begin{bmatrix} a_{11} & a_{12} & a_{13} \\ a_{21} & a_{22} & a_{23} \\ a_{31} & a_{32} & a_{33} \end{bmatrix}$ a_{22} must be a multiple of 3 (3, 6 or 9) because from the above possible combinations, only 3, 6 and 9 are repeated four times in a row or column or diagonal.

3. (a)

Clearly maximum value of sum of the diagonal elements is 18, which is called the trace of the matrix A.

CHAPTER 4

Determinants

Objectives

In this chapter on Determinants, we shall study the following topics:

1. Determinant and determinants of 2nd and 3rd order
2. Symmetric and skew-symmetric determinants
3. Multiplication of two determinants
4. Area of a triangle
5. System of linear equations
6. Applications of determinants.

Introduction

In the previous chapter, we have studied about matrices. We have also learnt about a system of algebraic equation that can be expressed in the form of matrices.

Consider a system of linear equations in two variables, x and y:

$$a_1x+b_1y=c_1$$
$$a_2x+b_2y=c_2$$

Solving the equations for x and y, we get

$$(a_1b_2-a_2b_1)x=(c_1b_2-c_2b_1) \text{ and}$$
$$(a_1b_2-a_2b_1)y=(a_1c_1-a_2c_1)$$

The expression $(a_1b_2-a_2b_1)$ can be expressed in the form of an array as

$$\begin{vmatrix} a_1 & a_2 \\ b_1 & b_2 \end{vmatrix} \text{ or } \begin{vmatrix} a_1 & b_1 \\ a_2 & b_2 \end{vmatrix}$$

This square is called determinant. It is represented by symbol Δ (delta). Determinants have a wide range of applications in Science, Engineering, Economics and Social Science, etc.

In the present chapter, we will study determinants up to order three only with real entries. Also, we will study various properties of determinants, minors, cofactors and applications of determinants in finding the area of a triangle, adjoin and inverse of a square matrix, consistency and inconsistency of a system of linear equations and solution of linear equations in two or three variables using inverse of a matrix.

Matrices and determinants are two very important tools in the hands of a mathematician. These have wide application in algebra, trigonometric, co-ordinate geometry and calculus. Using matrices and determinants, problems can be handled in a much-simplified way and many calculations can be cut short.

Determinants of 2nd Order

The symbol $\begin{vmatrix} a_1 & b_1 \\ a_2 & b_2 \end{vmatrix}$ is called a determinants of order 2. It has two rows and columns. $a_1b_2-a_2b_1$ is called expansion of determinant $\begin{vmatrix} a_1 & b_1 \\ a_2 & b_2 \end{vmatrix}$. These ideas can be extended to higher order determinants.

Determinants of 3rd Order

The symbol $\Delta=\begin{vmatrix} a_{11} & a_{12} & a_{13} \\ a_{21} & a_{22} & a_{23} \\ a_{31} & a_{32} & a_{33} \end{vmatrix}=|a_{ij}|$ is called the determinant of 3rd order and its value is equal to the number:

$$(-1)^{1+1}a_{11}\begin{vmatrix} a_{22} & a_{23} \\ a_{32} & a_{33} \end{vmatrix}+(-1)^{1+2}a_{12}\begin{vmatrix} a_{21} & a_{23} \\ a_{31} & a_{33} \end{vmatrix}$$
$$+(-1)^{1+3}a_{13}\begin{vmatrix} a_{21} & a_{22} \\ a_{31} & a_{32} \end{vmatrix}$$
$$=a_{11}(a_{22}a_{33}-a_{23}a_{32})-a_{12}(a_{21}a_{33}-a_{31}a_{23})$$
$$+a_{13}(a_{21}a_{32}-a_{31}a_{22}) \quad \text{(A)}$$

(A) is called expansion of determinant Δ along its first row. In fact, the expansion can be carried along any row or column in a similar way.

ILLUSTRATION 4.1

Find the value of determinant $\begin{vmatrix} a+1 & a-2 \\ a+2 & a-1 \end{vmatrix}$.

Solution

$$\begin{vmatrix} a+1 & a-2 \\ a+2 & a-1 \end{vmatrix}$$

$$=(a+1)(a-1)-(a+2)(a-2)$$

(Multiplying diagonally)

$$=(a^2-1)-(a^2-4)=3$$

ILLUSTRATION 4.2

Find the value of $\begin{vmatrix} 1+\cos\theta & \sin\theta \\ \sin\theta & 1-\cos\theta \end{vmatrix}$.

Solution

$$\begin{vmatrix} 1+\cos\theta & \sin\theta \\ \sin\theta & 1-\cos\theta \end{vmatrix}$$

$$=(1+\cos\theta)(1-\cos\theta)-(\sin\theta)(\sin\theta)$$

$$=1-\cos^2\theta-\sin^2\theta=0$$

ILLUSTRATION 4.3

Find the value of $\begin{vmatrix} 1 & 2 & 3 \\ -4 & 3 & 6 \\ 2 & -7 & 9 \end{vmatrix}$.

Solution

$$\begin{vmatrix} 1 & 2 & 3 \\ -4 & 3 & 6 \\ 2 & -7 & 9 \end{vmatrix}=1\begin{vmatrix} 3 & 6 \\ -7 & 9 \end{vmatrix}-2\begin{vmatrix} -4 & 6 \\ 2 & 9 \end{vmatrix}+3\begin{vmatrix} -4 & 3 \\ 2 & -7 \end{vmatrix}$$

$$=1(3\times 9-6(-7))-2(-4\times 9-2\times 6)$$

$$+3[(-4)(-7)-3\times 2]$$

$$=(27+42)-2(-36-12)+3(28-6)=231$$

Minors and Cofactors of Determinants

In the determinant Δ, the minor of the element a_{ij} (where i denotes number of row and j the number of columns, i.e., element common to ith row and jth column) is obtained by removing the ith row and jth column and it is given by M_{ij}

For example, $M_{12}=\begin{vmatrix} a_{21} & a_{23} \\ a_{31} & a_{23} \end{vmatrix}$ and cofactor of a_{ij} is defined as $C_{ij}=(-1)^{i+j}+M_{ij}$, for example,

$$C_{12}=(-1)^{1+2}\begin{vmatrix} a_{21} & a_{23} \\ a_{31} & a_{33} \end{vmatrix} \text{ or } C_{12}=-\begin{vmatrix} a_{21} & a_{23} \\ a_{31} & a_{33} \end{vmatrix}$$

Notes

(i) The sum of product of the element of any row with their corresponding cofactor is equal to the value of determinant, i.e.,

$\Delta=a_{11}C_{11}+a_{12}C_{12}+a_{13}C_{13}$

(ii) The sum of the product of element of any row with corresponding cofactor of another row is equal to zero, i.e.,

$a_{11}C_{21}+a_{12}C_{22}+a_{13}C_{23}=0$

(iii) If order of a determinant (Δ) is 'n' then the value of the determinant formed by replacing every element by its cofactor is Δ^{n-1}

ILLUSTRATION 4.4

Find the cofactor of element 2 in Determinant $\begin{vmatrix} -1 & 0 \\ -2 & 2 \end{vmatrix}$.

Solution

$$F_{22}=(-1)^{2+2}M_{22}=1(-1)=-1$$

ILLUSTRATION 4.5

Find the cofactor of element 0 in Determinant $\begin{vmatrix} -1 & 2 & 1 \\ -2 & 3 & -3 \\ 4 & 0 & -4 \end{vmatrix}$

Solution

$$F_{32}=(-1)^{3+2}\begin{vmatrix} -1 & 1 \\ -2 & -3 \end{vmatrix}$$

$$=-[(-1)(-3)-(-2)(1)]=-[3+2]=-5$$

Properties of Determinant

1. The value of Determinant remains unchanged, if the rows and the columns are interchanged. This is always denoted by $'$ and is called Transpose.

e.g., $D=\begin{vmatrix} a_1 & b_1 & c_1 \\ a_2 & b_2 & c_2 \\ a_3 & b_3 & c_3 \end{vmatrix}$ and $D'=\begin{vmatrix} a_1 & a_2 & a_3 \\ b_1 & b_2 & b_3 \\ c_1 & c_2 & c_3 \end{vmatrix}$

Then D′ = D, D and D′ are transpose of each other.

Remark: Since the Determinant remains unchanged when rows and columns are interchanged, it is obvious that any theorem which is true for 'Rows' must also be true for 'Columns.'

2. If any two rows (or columns) of a Determinant be interchanged, the Determinant is unaltered in numerical value, but is changed in sign only,

e.g., $D=\begin{vmatrix} a_1 & b_1 & c_1 \\ a_2 & b_2 & c_2 \\ a_3 & b_3 & c_3 \end{vmatrix}$ and $D'=\begin{vmatrix} a_2 & b_2 & c_2 \\ a_1 & b_1 & c_1 \\ a_3 & b_3 & c_3 \end{vmatrix}$

Then $D'=-D$.

3. If a Determinant has two rows (or columns) identical, then its value is zero.

e.g., Let $D=\begin{vmatrix} a_1 & b_1 & c_1 \\ a_1 & b_1 & c_1 \\ a_2 & b_2 & c_2 \end{vmatrix}$

Then $D=0$

4. If all the elements of any row (or column) be multiplied by the same number, then the value of Determinant is multiplied by that number.

e.g. $D=\begin{vmatrix} a_1 & b_1 & c_1 \\ a_2 & b_2 & c_2 \\ a_3 & b_3 & c_3 \end{vmatrix}$ and $D'=\begin{vmatrix} ka_1 & kb_1 & kc_1 \\ a_2 & b_2 & c_2 \\ a_3 & b_3 & c_3 \end{vmatrix}$

Then $D'=kD$

5. If each element of any row (or column) can be expressed as a sum of two terms, then the determinant can be expressed as the sum of the Determinant.

e.g.

$$\begin{vmatrix} a_1+x & b_1+y & c_1+z \\ a_2 & b_2 & c_2 \\ a_3 & b_3 & c_3 \end{vmatrix}=\begin{vmatrix} a_1 & b_1 & c_1 \\ a_2 & b_2 & c_2 \\ a_3 & b_3 & c_3 \end{vmatrix}+\begin{vmatrix} x & y & z \\ a_2 & b_2 & c_2 \\ a_3 & b_3 & c_3 \end{vmatrix}$$

6. The value of a Determinant is not altered by adding to the elements of any row (or column) the same multiples of the corresponding elements of any other row (or column).

e.g., $D=\begin{vmatrix} a_1 & b_1 & c_1 \\ a_2 & b_2 & c_2 \\ a_3 & b_3 & c_3 \end{vmatrix}$

and $D'=\begin{vmatrix} a_1+ma_2 & b_1+mb_2 & c_1+mc_2 \\ a_2 & b_2 & c_2 \\ a_3-na_1 & b_3-nb_1 & c_3-nc_1 \end{vmatrix}$

Then $D'=D$

ILLUSTRATION 4.6

Find the value of the determinant $\begin{vmatrix} 19 & 6 & 7 \\ 21 & 3 & 15 \\ 28 & 11 & 6 \end{vmatrix}$.

Solution

Applying $C_1 \to C_1-(C_2+C_3)$ we get

$$\text{Determinant}=\begin{vmatrix} 6 & 6 & 7 \\ 3 & 3 & 15 \\ 11 & 11 & 6 \end{vmatrix}=0 \quad (\because C_1=C_2)$$

ILLUSTRATION 4.7

$\begin{vmatrix} 1 & 1 & 1 \\ b+c & c+a & a+b \\ b+c-a & c+a-b & a+b-c \end{vmatrix}$ equals

(a) 0 **(b)** $ab+bc+ca$

(c) abc **(d)** $a+b+c$

Solution

Determinant

$$\begin{vmatrix} 1 & 1 & 1 \\ b+c & c+a & a+b \\ -(a+b+c) & -(a+b+c) & -(a+b+c) \end{vmatrix}$$

Applying $R_3 \to [R_3+(-2R_2)]$, We get

$$=-(a+b+c)\begin{vmatrix} 1 & 1 & 1 \\ b+c & c+a & a+b \\ 1 & 1 & 1 \end{vmatrix}=0$$

Answer is (a)

ILLUSTRATION 4.8

If $\Delta=\begin{vmatrix} a & b & c \\ x & y & z \\ p & q & r \end{vmatrix}$, then $\begin{vmatrix} ka & kb & kc \\ kx & ky & kz \\ kp & kq & kr \end{vmatrix}$ equals ______.

(a) $k\Delta$ **(b)** $3k\Delta$ **(c)** $k^3\Delta$ **(d)** $k\Delta^3$

Solution

We know that if any row of a determinant is multiplied by k, then the value of the determinant is also multiplied by k. Here all the three rows are multiplied by k, therefore the value of the new determinant will be $k^3\Delta$.

Answer is (c).

ILLUSTRATION 4.9

The value of the determinant $\begin{vmatrix} 265 & 240 & 219 \\ 240 & 225 & 198 \\ 219 & 198 & 181 \end{vmatrix}$ is ______.

(a) 1000 **(b)** 779 **(c)** 679 **(d)** 0

Solution

Applying $C_1 - C_2$ and $C_2 - C_3$, we get

$$\text{Det.} = \begin{vmatrix} 25 & 21 & 219 \\ 15 & 27 & 198 \\ 21 & 17 & 181 \end{vmatrix} = \begin{vmatrix} 4 & 21 & 9 \\ -12 & 27 & -72 \\ 4 & 17 & 11 \end{vmatrix}$$

$$[\text{by } C_1 - C_2, C_3 - 10C_2]$$

$$= \begin{vmatrix} 4 & 21 & 9 \\ 0 & 90 & -45 \\ 0 & -4 & 2 \end{vmatrix} \quad [\text{By } R_2 + 3R_1, R_3 - R_1]$$

$$= 4(180 - 180) = 0$$

ILLUSTRATION 4.10

If x, y, z are unequal and $\begin{vmatrix} x & x^2 & 1+x^3 \\ y & y^2 & 1+y^3 \\ z & z^2 & 1+z^3 \end{vmatrix} = 0$ then the value of xyz is ______.

(a) 0 **(b)** 1 **(c)** −1 **(d)** None of these.

Solution

Writing the given determinants as the sum of two determinants, we have

$$\begin{vmatrix} x & x^2 & 1 \\ y & y^2 & 1 \\ z & z^2 & 1 \end{vmatrix} + \begin{vmatrix} x & x^2 & x^3 \\ y & y^2 & y^3 \\ z & z^2 & z^3 \end{vmatrix} = 0$$

$$\Rightarrow \begin{vmatrix} x & x^2 & 1 \\ y & y^2 & 1 \\ z & z^2 & 1 \end{vmatrix}(1 + xyz) = 0$$

$$\Rightarrow (x-y)(y-z)(z-x)(1+xyz) = 0$$

$$\Rightarrow 1 + xyz = 0 \quad (\because x \neq y \neq z)$$

$$\Rightarrow xyz = -1$$

ILLUSTRATION 4.11

$\begin{vmatrix} a & b & c \\ a^2 & b^2 & c^2 \\ bc & ca & ab \end{vmatrix}$ is equal to

(a) $abc(a-b)(b-c)(c-a)$

(b) $(a-b)(b-c)(c-a)(ab+bc+ca)$

(c) $(a+b+c)(a-b)(b-c)(c-a)$

(d) None of these.

Solution

Multiplying C_1, C_2, C_3 by a, b, c we get

$$\text{Det.} = \frac{1}{\text{abc}} \begin{vmatrix} a^2 & b^2 & c^2 \\ a^3 & b^3 & c^3 \\ abc & abc & abc \end{vmatrix}$$

$$= \begin{vmatrix} a^2 & b^2 & c^2 \\ a^3 & b^3 & c^3 \\ 1 & 1 & 1 \end{vmatrix}$$

$$= \begin{vmatrix} a^2 - b^2 & b^2 - c^2 & c^2 \\ a^3 - b^3 & b^3 - c^3 & c^3 \\ 0 & 0 & 1 \end{vmatrix}$$

$$[\text{by } C_1 - C_2, C_2 - C_3]$$

$$= (a-b)(b-c)\begin{vmatrix} a+b & b+c \\ a^2 + b^2 + ab & b^2 + c^2 + bc \end{vmatrix}$$

$$= (a-b)(b-c)\begin{vmatrix} a-c & b+c \\ (a-c)(a+b+c) & b^2 + c^2 + bc \end{vmatrix}$$

$$[\text{by } C_1 - C_2]$$

$$= (a-b)(b-c)(a-c)\begin{vmatrix} 1 & b+c \\ a+b+c & b^2 + c^2 + bc \end{vmatrix}$$

$$= (a-b)(b-c)(a-c)(-ab - bc - ca)$$

$$= (a-b)(b-c)(c-a)(ab + bc + ca)$$

Remarks

1. The value of a determinant remains unaltered under an operation of the form $R_i \to R_i + mR_j + nR_k \ (j, k \neq i)$. These operations enable us to convert maximum elements into zeros for making evaluation of the determinant simpler.
2. If a determinant $\Delta(x)$ becomes zero on putting $x = \alpha$, then $x - \alpha$ is a factor of $\Delta(x)$.
3. A determinant which has all elements equal to zero except the diagonal elements is equal to the product of the diagonal elements, i.e.,

$$\begin{vmatrix} a & 0 & 0 \\ 0 & b & 0 \\ 0 & 0 & c \end{vmatrix} = abc$$

Also $\begin{vmatrix} a & \alpha & \beta \\ 0 & b & \gamma \\ 0 & 0 & c \end{vmatrix} = \begin{vmatrix} a & 0 & 0 \\ \alpha & b & 0 \\ \beta & \gamma & c \end{vmatrix} = abc$

4. Determinant $\begin{vmatrix} 0 & a & b \\ -a & 0 & c \\ -b & -c & 0 \end{vmatrix} = 0$ (note that upper triangular is reflected along the diagonal, with opposite sign.)

Symmetric and Skew-symmetric Determinant

Symmetric Determinant

A determinant is called Symmetric Determinant if for its every element,

$$a_{ij} = a_{ji} \forall i, j$$

e.g., $\begin{vmatrix} a & h & g \\ h & b & f \\ g & f & c \end{vmatrix}$

Skew-symmetric Determinant

A determinant is called Skew-symmetric determinant if for its every element

$$a_{ij} = -a_{ji} \forall i, j$$

e.g., $\begin{vmatrix} 0 & 3 & -1 \\ -3 & 0 & 5 \\ 1 & -5 & 0 \end{vmatrix}$

Remarks

(i) Every diagonal element of a skew-symmetric determinant is always zero.

(ii) The value of a skew-symmetric determinant of even order is always a perfect square and that of odd order is always zero.

e.g., (order #3) $\begin{vmatrix} 0 & c & -b \\ -c & 0 & a \\ b & -a & 0 \end{vmatrix}$

$= -c(0 - ab) - b(ac - 0) = abc - abc = 0$

(order #2) $\begin{vmatrix} 0 & a \\ -a & 0 \end{vmatrix} = 0 + a^2 = a^2$ (perfect square)

Multiplication of Two Determinants

Multiplication of two second-order determinants is defined as follows:

$$\begin{vmatrix} a_1 & b_1 \\ a_2 & b_1 \end{vmatrix} \times \begin{vmatrix} \ell_1 & m_1 \\ \ell_2 & m_2 \end{vmatrix} = \begin{vmatrix} a_1\ell_1 + b_1\ell_2 & a_1m_1 + b_1m_2 \\ a_2\ell_1 + b_2\ell_2 & a_2m_1 + b_2m_2 \end{vmatrix}$$

Multiplication of two third-order determinants is defined as follows

$$\begin{vmatrix} a_1 & b_1 & c_1 \\ a_2 & b_2 & c_2 \\ a_3 & b_3 & c_3 \end{vmatrix} \times \begin{vmatrix} \ell_1 & m_1 & n_1 \\ \ell_2 & m_2 & n_2 \\ \ell_3 & m_3 & n_3 \end{vmatrix}$$

$$= \begin{vmatrix} a_1\ell_1 + b_1\ell_2 + c_1\ell_3 & a_1m_1 + b_1m_2 + c_1m_3 & a_1n_1 + b_1n_2 + c_1n_3 \\ a_2\ell_1 + b_2\ell_2 + c_2\ell_3 & a_2m_1 + b_2m_2 + c_2m_3 & a_2n_1 + b_2n_2 + c_2n_3 \\ a_3\ell_1 + b_3\ell_2 + c_3\ell_3 & a_3m_1 + b_3m_2 + c_3m_3 & a_3n_1 + b_3n_2 + c_3n_3 \end{vmatrix}$$

Note: In above case the order of the determinant is same. If the order is different, then for their multiplication, first of all they should be expressed in the same order.

Remarks

1. Since value of determinant remains same when rows and columns are interchanged, in finding product of two determinants, we can also multiply rows by columns or column by columns.
2. $D_1D_2 = D_2D_1$

   Important observation

   If $D_1 = \begin{vmatrix} a_1 & a_2 & a_3 \\ b_1 & b_2 & b_3 \\ c_1 & c_2 & c_3 \end{vmatrix}$ and $D_2 = \begin{vmatrix} A_1 & A_2 & A_3 \\ B_1 & B_2 & B_3 \\ C_1 & C_2 & C_3 \end{vmatrix}$

   where A_i = cofactor of a_i, B_i = cofactor of b_i and C_1 = cofactor of c_i then $D_2 = (D_1)^2$

ILLUSTRATION 4.12

If $t_r = a^r + b^r + c^r$, show that

$$\begin{vmatrix} t_0 & t_1 & t_2 \\ t_1 & t_2 & t_3 \\ t_2 & t_3 & t_4 \end{vmatrix} = (a-b)^2(b-c)^2(c-a)^2$$

Solution

We have

$$\begin{vmatrix} t_0 & t_1 & t_2 \\ t_1 & t_2 & t_3 \\ t_2 & t_3 & t_4 \end{vmatrix} = \begin{vmatrix} 1+1+1 & a+b+c & a^2+b^2+c^2 \\ a+b+c & a^2+b^2+c^2 & a^3+b^3+c^3 \\ a^2+b^2+c^2 & a^3+b^3+c^3 & a^4+b^4+c^4 \end{vmatrix}$$

$$= \begin{vmatrix} 1 & 1 & 1 \\ a & b & c \\ a^2 & b^2 & c^2 \end{vmatrix} \begin{vmatrix} 1 & 1 & 1 \\ a & b & c \\ a^2 & b^2 & c^2 \end{vmatrix} = \begin{vmatrix} 1 & 1 & 1 \\ a & b & c \\ a^2 & b^2 & c^2 \end{vmatrix}^2$$

$$= \begin{vmatrix} 1 & 0 & 0 \\ a & b-a & c-a \\ a^2 & b^2-a^2 & c^2-a^2 \end{vmatrix}^2$$

Applying $(C_2 \to C_2 - C_1, C_3 \to C_3 - C_1)$

$$= (b-a)^2(c-a)^2 \begin{vmatrix} 1 & 0 & 0 \\ a & 1 & 1 \\ a^2 & b+a & c+a \end{vmatrix}^2$$

$$= (a-b)^2(c-a)^2[c+a-b-a]^2$$

$$= (a-b)^2(b-c)^2(c-a)^2$$

Area of a Triangle

If ABC is a triangle with $A(x_1, y_1), B(x_2, y_2)$ and $C(x_3, y_3)$ then the area of $\Delta = \frac{1}{2}\begin{vmatrix} x_1 & y_1 & 1 \\ x_2 & y_2 & 1 \\ x_3 & y_3 & 1 \end{vmatrix}$

The area may come out to be either positive or negative, hence modulus of above would give the magnitude of area.

If $\begin{vmatrix} x_1 & y_1 & 1 \\ x_2 & y_2 & 1 \\ z_3 & y_3 & 1 \end{vmatrix}$ turns out to be zero, it means the points A, B and C are collinear and vice-versa.

System of Linear Equations

System of linear equations:

$$a_{11}x_1 + a_{12}x_2 + \ldots + a_{1n}x_n = b_1$$
$$a_{21}x_1 + a_{22}x_2 + \ldots + a_{2n}x_n = b_2$$
$$\vdots \qquad \vdots$$
$$a_{n1}x_1 + a_{n2}x_2 + \ldots + a_{nn}x_n = b_n$$

is said to be consistent if it has at least one solution. Otherwise, it is said to be inconsistent.

I. System of homogeneous equations (linear) in three variables.

$$a_1x + b_1y + c_1z = 0, a_2x + b_2y + c_2z = 0, a_3x + b_3y + c_3z = 0$$

Note that $x = y = z = 0$ (trivial solution) is always a solution. Therefore, this system is always consistent.

Case I: If $\Delta = \begin{vmatrix} a_1 & b_1 & c_1 \\ a_2 & b_2 & c_2 \\ a_3 & b_3 & c_3 \end{vmatrix} \neq 0$, there is unique solution $x = y = z = 0$.

Case II: If $\Delta = 0$, the system has infinite number of solutions (existence of non-trivial solutions.)

Applications of Determinants

Crammer's Rule

Consider three linear simultaneous equations in 'x', 'y', 'z'

$$a_1x + b_1y + c_1z = d_1 \quad \ldots\text{(i)}$$
$$a_2x + b_2y + c_2z = d_2 \quad \ldots\text{(ii)}$$
$$a_3x + b_3y + c_3z = d_3 \quad \ldots\text{(iii)}$$

and if

$$\Delta = \begin{vmatrix} a_1 & b_1 & c_1 \\ a_2 & b_2 & c_2 \\ a_3 & b_3 & c_3 \end{vmatrix} \; \Delta_x = \begin{vmatrix} d_1 & b_1 & c_1 \\ d_2 & b_2 & c_2 \\ d_3 & b_3 & c_3 \end{vmatrix} \; \Delta_y = \begin{vmatrix} a_1 & d_1 & c_1 \\ a_2 & d_2 & c_2 \\ a_3 & d_3 & c_3 \end{vmatrix}$$

$$\Delta_z = \begin{vmatrix} a_1 & b_1 & d_1 \\ a_2 & b_2 & d_2 \\ a_3 & b_3 & d_3 \end{vmatrix}$$

Then using Crammer's rule of determinant, we get

$$\frac{x}{\Delta_x} = \frac{y}{\Delta_y} = \frac{z}{\Delta_z} = \frac{1}{\Delta}$$

i.e., $x=\frac{\Delta_x}{\Delta}, y=\frac{\Delta_y}{\Delta}, z=\frac{\Delta_z}{\Delta}$

Case I: If $\Delta \neq 0$

Then $x=\frac{\Delta_x}{\Delta}, y=\frac{\Delta_y}{\Delta}, z=\frac{\Delta_z}{\Delta}$

$\therefore$ The system is consistent and has unique solutions.

Case II: If $\Delta = 0$ and

(i) if at least one of $\Delta_x, \Delta_y, \Delta_z$ is not zero, then the system of equations is inconsistent, i.e., it has no solution.

(ii) if $d_1 = d_2 = d_3 = 0$ or $\Delta_x, \Delta_y, \Delta_z$ are all zero, then the system of equations has infinitely many solutions.

The above can be shown diagrammatically as follows:

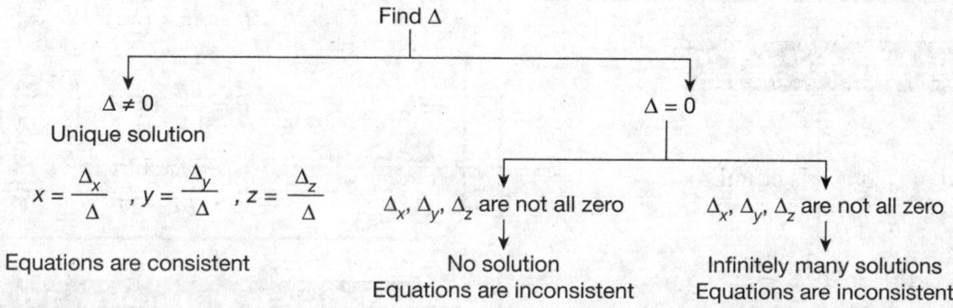

ILLUSTRATION 4.13

If the system of equations $x+4ay+az=0$, $x+3by+bz=0$, $x+2cy+cz=0$ has a non-zero solutions, then a, b, c are in ______.

(a) AP **(b)** GP **(c)** HP **(d)** None of these.

Solution

For a non-zero solution, we must have

$$\Delta = 0 \Rightarrow \begin{vmatrix} 1 & 4a & a \\ 1 & 3b & b \\ 1 & 2c & c \end{vmatrix} = 0$$

$$\Rightarrow \quad bc - 2ac + ab = 0 \Rightarrow b = \frac{2ac}{a+c}$$

$\therefore a,b,c$ are in HP

ILLUSTRATION 4.14

The system of equations $x+y+z=2, 2x+y-z=3, 3x+2y+kz=4$ has unique solution if ______.

(a) $k=0$ **(b)** $k \neq 0$

(c) $-1<k<1$ **(d)** $-2<k<2$

Solution

Given system will have unique solution, if

$$\begin{vmatrix} 1 & 1 & 1 \\ 2 & 1 & -1 \\ 3 & 2 & k \end{vmatrix} \neq 0 \Rightarrow k \neq 0$$

Answer is (b)

ILLUSTRATION 4.15

The value of the determinant $\begin{vmatrix} a+b+2c & a & b \\ c & b+c+2a & b \\ c & a & c+a+2b \end{vmatrix}$ is

(a) $2(a+b+c)$ **(b)** $2(a+b+c)^2$

(c) $2(a+b+c)^3$ **(d)** $(2a+2b+2c)^3$

Solution

Applying $C_1+C_2+C_3$, we get

$$\text{Det.} = 2(a+b+c)\begin{vmatrix} 1 & a & b \\ 1 & b+c+2a & b \\ 1 & a & c+a+2b \end{vmatrix}$$

$$= 2(a+b+c)\begin{vmatrix} 1 & a & b \\ 0 & a+b+c & 0 \\ 0 & 0 & a+b+c \end{vmatrix}$$

$[\text{by } R_2 - R_1, R_3 - R_1]$

$$= 2(a+b+c)^3$$

ILLUSTRATION 4.16

If $\begin{vmatrix} a & b & a\alpha+b \\ b & c & b\alpha+c \\ a\alpha+b & b\alpha+c & 0 \end{vmatrix} = 0$, then a,b,c are in

(a) AP **(b)** GP **(c)** HP **(d)** None of these.

Solution

By the operation $C_3-(\alpha C_1+C_2)$, we get

$$\begin{vmatrix} a & b & 0 \\ b & c & 0 \\ a\alpha+b & b\alpha+c & -(a\alpha^2+2b\alpha+c) \end{vmatrix}=0$$

$\Rightarrow \quad -(a\alpha^2+2b\alpha+c)(ac-b^2)=0$

$\Rightarrow \quad b^2=ac \Rightarrow a, b, c$ are in GP.

ILLUSTRATION 4.17

$$\begin{vmatrix} b^2+c^2 & a^2 & a^2 \\ b^2 & c^2+a^2 & b^2 \\ c^2 & c^2 & a^2+b^2 \end{vmatrix}$$ is equal to

(a) $a^2b^2c^2$ **(b)** $2a^2b^2c^2$

(c) $4a^2b^2c^2$ **(d)** None of these.

Solution

Applying $R_1-(R_2+R_3)$, we get

$$\text{Det.}=\begin{vmatrix} 0 & -2c^2 & -2b^2 \\ b^2 & c^2+a^2 & b^2 \\ c^2 & c^2 & a^2+b^2 \end{vmatrix}$$

$$=2\begin{vmatrix} 0 & -c^2 & -b^2 \\ b^2 & c^2+a^2 & b^2 \\ c^2 & c^2 & a^2+b^2 \end{vmatrix}$$

$$=2\begin{vmatrix} 0 & -c^2 & -b^2 \\ b^2 & a^2 & 0 \\ c^2 & 0 & a^2 \end{vmatrix} \text{(by } R_2+R_1, R_3+R_1)$$

$$=2(a^2b^2c^2+a^2b^2c^2)=4a^2b^2c^2$$

ILLUSTRATION 4.18

If $0<\theta<\pi/2$ and

$$\begin{vmatrix} 1+\sin^2\theta & \cos^2\theta & 4\sin 4\theta \\ \sin^2\theta & 1+\cos^2\theta & 4\sin 4\theta \\ \sin^2\theta & \cos^2\theta & 1+4\sin 4\theta \end{vmatrix}=0,$$

then θ is equal to

(a) $\frac{\pi}{24},\frac{5\pi}{24}$ **(b)** $\frac{5\pi}{24},\frac{7\pi}{24}$

(c) $\frac{7\pi}{24},\frac{11\pi}{24}$ **(d)** None of these.

Solution

Applying R_2-R_1 and R_3-R_1, we get

$$\begin{vmatrix} 1+\sin^2\theta & \cos^2\theta & 4\sin 4\theta \\ -1 & 1 & 0 \\ -1 & 0 & 1 \end{vmatrix}=0$$

$\Rightarrow \quad 2+4\sin 4\theta=0$

$\Rightarrow \quad \sin 4\theta=-\frac{1}{2}$

$\Rightarrow \quad 4\theta=n\pi+(-1)^n\left(-\frac{\pi}{6}\right)$

$\Rightarrow \quad \theta=\frac{n\pi}{4}+(-1)^n\left(-\frac{\pi}{24}\right)$

$\therefore \quad \theta=\frac{7\pi}{24},\frac{11\pi}{24}$

ILLUSTRATION 4.19

$$\begin{vmatrix} {}^xC_1 & {}^xC_2 & {}^xC_3 \\ {}^yC_1 & {}^yC_2 & {}^yC_3 \\ {}^zC_1 & {}^zC_2 & {}^zC_3 \end{vmatrix}$$ is equal to

(a) $xyz(x-y)(y-z)(z-x)$

(b) $\frac{xyz}{6}(x-y)(y-z)(z-x)$

(c) $\frac{xyz}{12}(x-y)(y-z)(z-x)$

(d) None of these.

Solution

$$\text{Det.}=\begin{vmatrix} x & \frac{x(x-1)}{2} & \frac{x(x-1)(x-2)}{6} \\ y & \frac{y(y-1)}{2} & \frac{y(y-1)(y-2)}{6} \\ z & \frac{z(z-1)}{2} & \frac{z(z-1)(z-2)}{6} \end{vmatrix}$$

$$=\frac{xyz}{12}\begin{vmatrix} 1 & x-1 & (x-1)(x-2) \\ 1 & y-1 & (y-1)(y-2) \\ 1 & z-1 & (z-1)(z-2) \end{vmatrix}$$

$$=\frac{xyz}{12}\begin{vmatrix} 1 & x & x^2 \\ 1 & y & y^2 \\ 1 & z & z^2 \end{vmatrix}$$

(by C_2+C_1, C_3+C_1, $3C_2$)

$$=\frac{xyz}{12}(x-y)(y-z)(z-x)$$

ILLUSTRATION 4.20

If $\begin{vmatrix} x^2+x & x+1 & x-2 \\ 2x^2+3x-1 & 3x & 3x-3 \\ x^2+2x+3 & 2x-1 & 2x-1 \end{vmatrix} = Px-12$ then

(a) $P=24$ **(b)** $P=-24$

(c) $P=0$ **(d)** $P=12$

Solution

Applying $R_2 \to R_2-(R_1+R_3)$

$$\text{Determinant} \begin{vmatrix} x^2+x & x+1 & x-2 \\ -4 & 0 & 0 \\ x^2+2x+3 & 2x-1 & 2x-1 \end{vmatrix}$$

Applying $R_1 \to R_1+\frac{1}{4}x^2R_2$ and $R_3 \to R_3+\frac{1}{4}x^2R_2$

$$\text{Det.} = \begin{vmatrix} x & x+1 & x-2 \\ -4 & 0 & 0 \\ 2x+3 & 2x-1 & 2x-1 \end{vmatrix}$$

Applying $= R_3 \to R_3-2R_1$

$$\text{Det.} = \begin{vmatrix} x+0 & x+1 & x-2 \\ -4 & 0 & 0 \\ 3 & -3 & 3 \end{vmatrix}$$

$$= \begin{vmatrix} x & x & x \\ -4 & 0 & 0 \\ 3 & -3 & 3 \end{vmatrix} + \begin{vmatrix} 0 & 1 & -2 \\ -4 & 0 & 0 \\ 3 & -3 & 3 \end{vmatrix}$$

$$= x\begin{vmatrix} 1 & 1 & 1 \\ -4 & 0 & 0 \\ 3 & -3 & 3 \end{vmatrix} + \begin{vmatrix} 0 & 1 & -2 \\ -4 & 0 & 0 \\ 3 & -3 & 3 \end{vmatrix}$$

$$= (24x-12) \Rightarrow P = 24$$

ILLUSTRATION 4.21

If $\Delta_r = \begin{vmatrix} r-1 & n & 6 \\ (r-1)^2 & 2n^2 & 4n-2 \\ (r-1)^3 & 3n^2 & 3n^2-3n \end{vmatrix}$,

then $\sum_{r=1}^{n}\Delta_r$ equals ______.

(a) 1 **(b)** -1 **(c)** 0 **(d)** None of these.

Solution

$$\because \quad \sum_{r=1}^{n}(r-1) = 1+2+\ldots+(n-1) = \frac{n(n-1)}{2}$$

$$\sum_{r=1}^{n}(r-1)^2 = 1^2+2^2+\ldots+(n-1)^2 = \frac{n(n-1)(2n-1)}{6}$$

$$\sum_{r=1}^{n}(r-1)^3 = 1^3+2^3+\ldots+(n-1)^3 = \frac{n^2(n-1)^2}{4}$$

$$\therefore \quad \sum_{r=1}^{n}\Delta_r \begin{vmatrix} \frac{n(n-1)}{2} & n & 6 \\ \frac{1}{6}n(n-1)(2n-1) & 2n^2 & 2(2n-1) \\ \frac{1}{4}n^2(n-1)^2 & 3n^3 & 3n(n-1) \end{vmatrix}$$

$$= \frac{n(n-1)}{12}\begin{vmatrix} 6 & n & 6 \\ 2(2n-1) & 2n^2 & 2(2n-1) \\ 3n(n-1) & 3n^3 & 3n(n-1) \end{vmatrix} = 0$$

ILLUSTRATION 4.22

If a, b, c are the pth, qth and rth terms respectively of a HP, then value of the determinant $\begin{vmatrix} bc & ca & ab \\ p & q & r \\ 1 & 1 & 1 \end{vmatrix}$ is ______.

(a) $p+q+r$ **(b)** 0

(c) $\frac{1}{p}+\frac{1}{q}+\frac{1}{r}$ **(d)** None of these.

Solution

Let A and d be the first term and common difference of the corresponding AP, then

$$\frac{1}{a} = A+(p-1)d, \frac{1}{b} = A+(q-1)d, \frac{1}{c} = A+(r-1)d$$

Now

$$\text{det.} = abc\begin{vmatrix} \frac{1}{a} & \frac{1}{b} & \frac{1}{c} \\ p & q & r \\ 1 & 1 & 1 \end{vmatrix}$$

$$= abc\begin{vmatrix} A+(p-1)d & A+(q-1)d & A+(r-1)d \\ p & q & r \\ 1 & 1 & 1 \end{vmatrix}$$

$$= abc\begin{vmatrix} A & A & A \\ p & q & r \\ 1 & 1 & 1 \end{vmatrix} + abcd\begin{vmatrix} p-1 & q-1 & r-1 \\ p & q & r \\ 1 & 1 & 1 \end{vmatrix} = 0+0=0$$

WORKSHEET

1. Let $f(x)=\begin{vmatrix} x^3 & \sin x & \cos x \\ 6 & -1 & 0 \\ p & p^2 & p^3 \end{vmatrix}$ where p is a constant.

Then show that $\dfrac{d^3}{dx^3}(f(x))$ at $x=0$ is independent of p.

2. If

$$f(x)=\begin{vmatrix} 1 & x & (x+1) \\ 2x & x(x-1) & (x+1)x \\ 3x(x-1) & x(x-1)(x-2) & (x+1)x(x-1) \end{vmatrix}$$

then find the value of $f(100)$.

3. If $a^2+b^2+c^2=-2$ and

$$f(x)=\begin{vmatrix} 1+a^2x & (1+b^2)x & (1+c^2) \\ (1+a^2)x & 1+b^2x & (1+c^2)x \\ (1+a^2)x & (1+b^2)x & 1+c^2x \end{vmatrix},$$

then find out the polynomial of degree of $f(x)$.

4. For the system of equations $a_1x+b_1y+c_1z=0$, $a_2x+b_2y+c_2z=0$ and $a_3x+b_3y+c_3z=0$, if $\begin{vmatrix} a_1 & b_1 & c_1 \\ a_2 & b_2 & c_2 \\ a_3 & b_3 & c_3 \end{vmatrix}=0$, then this system has

(a) infinitely many solutions.
(b) only the trivial solution (0, 0, 0).
(c) no solution.
(d) one trivial and one non-trivial solution.

5. In a third-order determinant, each element of the first column consists of the sum of two terms, each element of the second column consists of the sum of three terms and each element of the third column consists of the sum of four terms. Then, it can be decomposed in n determinants. Find the value of n.

6. Calculate the value of the determinant.

$$\begin{vmatrix} 3 & 2 & 1 & 4 \\ 15 & 29 & 2 & 14 \\ 16 & 19 & 3 & 17 \\ 33 & 39 & 8 & 38 \end{vmatrix}.$$

7. Calculate the value of the determinant.

$$\begin{vmatrix} 0 & x & y & z \\ -x & 0 & c & b \\ -y & -c & 0 & a \\ -z & -b & -a & 0 \end{vmatrix}$$

8. If ω is one of the imaginary cube roots of unity, show that the square of

$$\begin{vmatrix} 1 & \omega & \omega^2 & \omega^3 \\ \omega & \omega^2 & \omega^3 & 1 \\ \omega^2 & \omega^3 & 1 & \omega \\ \omega^3 & 1 & \omega^2 & \omega^2 \end{vmatrix}=\begin{vmatrix} 1 & 1 & -2 & 1 \\ 1 & 1 & 1 & -2 \\ -2 & 1 & 1 & 1 \\ 1 & -2 & 1 & 1 \end{vmatrix}$$

Hence, show that the value of the determinant on the left is $3\sqrt{-3}$.

9. Solve the equations:

$$x+y+z=1$$
$$zx+by+cz=k$$
$$a^2x+b^2y+c^2z=k^2$$

10. Prove that,

$$\begin{vmatrix} a^2 & a^2-(b-c)^2 & bc \\ b^2 & b^2-(c-a)^2 & ca \\ c^2 & c^2-(a-b)^2 & ab \end{vmatrix}$$

$$=(b-c)(c-a)(a-b)(a+b+c)(a^2+b^2+c^2)$$

SOLVED EXAMPLES

Example 1. Show that

$$\begin{vmatrix} 1 & a & bc \\ 1 & b & ac \\ 1 & c & ab \end{vmatrix} = \begin{vmatrix} 1 & a & a^2 \\ 1 & b & b^2 \\ 1 & c & c^2 \end{vmatrix} = (a-b)(b-c)(c-a)$$

Solution

On taking $1/a$ common from 1st row, $1/b$ from 2nd row, and $1/c$ from 3rd row, the determinant on LHS

$$= \frac{1}{a}\cdot\frac{1}{b}\cdot\frac{1}{c}\begin{vmatrix} a & a^2 & abc \\ b & b^2 & abc \\ c & c^2 & abc \end{vmatrix} = \frac{abc}{abc}\begin{vmatrix} a & a^2 & 1 \\ b & b^2 & 1 \\ c & c^2 & 1 \end{vmatrix}$$

$$= -\begin{vmatrix} a & 1 & a^2 \\ b & 1 & b^2 \\ c & 1 & c^2 \end{vmatrix} = (-1)^2\begin{vmatrix} 1 & a & a^2 \\ 1 & b & b^2 \\ 1 & c & c^2 \end{vmatrix}$$

Now applying $R_2 - R_1$, $R_3 - R_1$, the determinant

$$= \begin{vmatrix} 1 & a & a^2 \\ 0 & b-a & b^2-a^2 \\ 0 & c-a & c^2-a^2 \end{vmatrix}$$

$$= \begin{vmatrix} b-a & b^2-a^2 \\ c-a & c^2-a^2 \end{vmatrix} = (a-b)(c-a)\begin{vmatrix} 1 & b+a \\ 1 & c+a \end{vmatrix}$$

$$= (b-a)(c-a)(c-b) = (a-b)(b-c)(c-a)$$

Example 2. If a, b, c are different and

$$\begin{vmatrix} a & a^2 & a^3+1 \\ b & b^2 & b^3+1 \\ c & c^2 & c^3+1 \end{vmatrix} = 0, \text{ prove that } abc = -1.$$

Solution

On expanding the determinant on LHS, we get

$$\begin{vmatrix} a & a^2 & a^3 \\ b & b^2 & b^3 \\ c & c^2 & c^3 \end{vmatrix} + \begin{vmatrix} a & a^2 & 1 \\ b & b^2 & 1 \\ c & c^2 & 1 \end{vmatrix} = 0$$

$$\Rightarrow \quad abc\begin{vmatrix} 1 & a & a^2 \\ 1 & b & b^2 \\ 1 & c & c^2 \end{vmatrix} - \begin{vmatrix} a & 1 & a^2 \\ b & 1 & b^2 \\ c & 1 & c^2 \end{vmatrix} = 0$$

$$\Rightarrow \begin{vmatrix} 1 & a & a^2 \\ 1 & b & b^2 \\ 1 & c & c^2 \end{vmatrix} + \begin{vmatrix} 1 & a & a^2 \\ 1 & b & b^2 \\ 1 & c & c^2 \end{vmatrix} \Rightarrow (1+abc)\begin{vmatrix} 1 & a & a^2 \\ 1 & b & b^2 \\ 1 & c & c^2 \end{vmatrix} = 0$$

$$\Rightarrow \quad (1 + abc)(a - b)(b - c)(c - a) = 0$$

(From Example 1)

$\Rightarrow 1 + abc = 0$ since a, b, c are all different $\Rightarrow abc = -1$.

Example 3. Show that

$$D(a,b,c) = \begin{vmatrix} (b+c)^2 & a^2 & a^2 \\ b^2 & (c+a)^2 & b^2 \\ c^2 & c^2 & (a+b)^2 \end{vmatrix}$$

$$= 2abc(a+b+c)^3.$$

Solution

Applying $C_2 - C_1$, $C_3 - C_1$, we get

$D(a,b,c)$

$$= \begin{vmatrix} (b+c)^2 & (a+b+c)(a-b-c) & (a+b+c)(a-b-c) \\ b^2 & (c+a+b)(c+a-b) & 0 \\ c^2 & 0 & (a+b+c)(a+b-c) \end{vmatrix}$$

$$= (a+b+c)^2\begin{vmatrix} (b+c)^2 & (a-b-c) & (a-b-c) \\ b^2 & (c+a-b) & 0 \\ c^2 & 0 & (a+b-c) \end{vmatrix}$$

On applying $R_1 - (R_2 + R_3)$, we get

$$D(a,b,c) = (a+b+c)^2\begin{vmatrix} 2bc & -2c & -2b \\ b^2 & c+a-b & 0 \\ c^2 & 0 & a+b-c \end{vmatrix}$$

$$= 2(a + b + c)^2\, [bc(c + a - b)(a + b - c) - 0$$

$$+ c(a + b - c)b^2 + bc^2(c + a - b)]$$

(On expanding along 1st row after taking 2 as a common factor.)

$$= 2(a + b + c)^2\, [abc^2 + bca^2 + cab^2]$$

$$= 2abc(a + b + c)^3$$

Example 4. Prove that

$$\begin{vmatrix} (a+b)^2 & ca & bc \\ ca & (b+c)^2 & ab \\ bc & ab & (c+a)^2 \end{vmatrix} = D(a,b,c)$$

where $D(a, b, c)$ is the determinant in Example 3.

Solution

The determinant on LHS

$$= c\cdot a\cdot b\begin{vmatrix} \frac{(a+b)^2}{c} & a & b \\ c & \frac{(b+c)^2}{a} & b \\ c & a & \frac{(c+a)^2}{b} \end{vmatrix}$$

$$= \begin{vmatrix} (a+b)^2 & a^2 & b^2 \\ c^2 & (b+c)^2 & b^2 \\ c^2 & a^2 & (c+a)^2 \end{vmatrix}$$

$$= -\begin{vmatrix} c^2 & (b+c)^2 & b^2 \\ (a+b)^2 & a^2 & b^2 \\ c^2 & a^2 & (c+a)^2 \end{vmatrix}$$

(on interchanging first and second row.)

$$= (-1)^2\begin{vmatrix} (b+c)^2 & c^2 & b^2 \\ a^2 & (a+b)^2 & b^2 \\ a^2 & c^2 & (c+a)^2 \end{vmatrix}$$

on interchanging first and second columns.

$$= (-1)^3\begin{vmatrix} (b+c)^2 & c^2 & b^2 \\ a^2 & c^2 & (c+a)^2 \\ a^2 & (a+b)^2 & b^2 \end{vmatrix}$$

on interchanging 2nd and 3rd rows.

$$= (-1)^4\begin{vmatrix} (b+c)^2 & b^2 & c^2 \\ a^2 & (c+a)^2 & c^2 \\ a^2 & b^2 & (a+b)^2 \end{vmatrix}$$

on interchanging 2nd and 3rd columns.

$= D(a, b, c).$

Example 5. Show that

$$\begin{vmatrix} (b+c)^2 & c^2 & b^2 \\ c^2 & (c+a)^2 & a^2 \\ b^2 & a^2 & (a+b)^2 \end{vmatrix} = a^4b^4c^4D\left(\frac{1}{a},\frac{1}{b},\frac{1}{c}\right).$$

Hence, show that the value of the given determinant is $2(ab + bc + ac)^3$.

Solution

The determinant on LHS

$$= b^2c^2\cdot c^2a^2\cdot b^2a^2\begin{vmatrix} \left(\frac{1}{b}+\frac{1}{c}\right)^2 & \frac{1}{b^2} & \frac{1}{c^2} \\ \frac{1}{a^2} & \left(\frac{1}{c}+\frac{1}{a}\right)^2 & \frac{1}{c^2} \\ \frac{1}{a^2} & \frac{1}{b^2} & \left(\frac{1}{a}+\frac{1}{b}\right)^2 \end{vmatrix}$$

$$= a^4b^4c^4\begin{vmatrix} \left(\frac{1}{b}+\frac{1}{c}\right)^2 & \frac{1}{a^2} & \frac{1}{a^2} \\ \frac{1}{b^2} & \left(\frac{1}{c}+\frac{1}{a}\right)^2 & \frac{1}{b^2} \\ \frac{1}{c^2} & \frac{1}{c^2} & \left(\frac{1}{a}+\frac{1}{b}\right)^2 \end{vmatrix}$$

$$= a^4b^4c^4D\left(\frac{1}{a},\frac{1}{b},\frac{1}{c}\right) = a^4b^4c^4\,2\left(\frac{1}{a}+\frac{1}{b}+\frac{1}{c}\right)^3$$

$$= 2(ab+bc+ca)^3$$

Example 6. If $a+b+c=0$ and $\begin{vmatrix} a-x & c & b \\ c & b-x & a \\ b & a & c-x \end{vmatrix} = 0$

then show that $x=0$ or $x^2 = \frac{3}{2}\left(a^2+b^2+c^2\right)$.

Solution

Applying $C_1 + C_2 + C_3$, we get

$$\begin{vmatrix} a+b+c-x & c & b \\ a+b+c-x & b-x & a \\ a+b+c-x & a & c-x \end{vmatrix} = 0$$

$$\Rightarrow \quad (a+b+c-x)\begin{vmatrix} 1 & c & b \\ 1 & b-x & a \\ 1 & a & c-x \end{vmatrix} = 0$$

$$\Rightarrow \quad (a+b+c-x)\begin{vmatrix} 1 & c & b \\ 0 & b-x-c & a-b \\ 0 & a-c & c-x-b \end{vmatrix} = 0$$

(applying $R_2 - R_1$, $R_3 - R_1$)

$$\Rightarrow (a+b+c-x)[(b-x-c)(c-x-b)-(a-c)(a-b)] = 0$$

$$\Rightarrow \quad (a+b+c-x)[x^2-a^2-c^2+ab+bc+ac] = 0$$

$$\Rightarrow \quad \text{Either } a+b+c = x$$

$$\text{or} \quad x^2 = a^2+b^2+c^2-ab-bc-ac$$

Now, if $x = a+b+c$, then $x = 0$. If $x \neq 0$, then

$$x^2 = a^2+b^2+c^2 - ½[(a+b+c)^2-(a^2+b^2+c^2)]$$

$$= 3/2\,(a^2+b^2+c^2) \qquad [\because a+b+c=0]$$

Thus, either $x = 0$ or $x^2 = 3/2\,(a^2+b^2+c^2)$.

Example 7. If all the entries in a 3×3 determinant are either 1 or –1, show that the value of the determinant is either 0 or ±4.

Solution

We have

$$D = \begin{vmatrix} a_1 & b_1 & c_1 \\ a_2 & b_2 & c_2 \\ a_3 & b_3 & c_3 \end{vmatrix}$$

$$= a_1b_2c_3 + a_2b_3c_1 + a_3b_1c_2 - a_1b_3c_2 - a_3b_2c_1 - a_2b_1c_3 \quad \text{(A)}$$

Since all the nine numbers in the determinant are either 1 or –1, each term in RHS of (A) is either 1 or –1. Therefore, the maximum and minimum values of such a determinant may be 6 and –6 respectively. We will show that these values are not attained.

Indeed, the determinant is equal to 6 if and only if

$$a_1b_2c_3 = 1,\ a_2b_3c_1 = 1,\ a_3b_1c_2 = 1$$

$$\text{and} \quad a_1b_3c_2 = -1,\ a_3b_2c_1 = -1,\ a_2b_1c_3 = -1$$

The above six equations cannot exist simultaneously. To show this, let us multiply all the above equations to arrive at a contradiction $a_1^2b_1^2c_1^2 = a_2^2b_2^2c_2^2 = a_3^2b_3^2c_3^2 = -1$, which is impossible. (Why?)

A similar proof can be given in order to show that the determinant cannot be equal to –6. Now we show that the determinant cannot be an odd number either. From (A),

$$D = x_1 + x_2 + x_3 - x_4 - x_5 - x_6 \ (x_1 = a_1b_2c_3 \text{ etc.})$$

where each x_1 is either 1 or –1

$$= (x_1 - x_4) + (x_2 - x_5) + (x_3 - x_6)$$

$$- y_1 + y_2 + y_3 \text{ where } y_1 - x_1 - x_4 \text{ etc.}$$

Now since each x_i is either 1 or –1, each y_i is either 0 or 2 or –2, consequently $y_1 + y_2 + y_3$ is essentially even. Now, we show that there exists a determinant with entries 1 or –1 whose value is equal to 4 (or –4).

This determinant is $\begin{vmatrix} 1 & -1 & 1 \\ 1 & 1 & -1 \\ -1 & 1 & 1 \end{vmatrix}$ (interchange any two adjacent rows to make the value.

We can also show that such a determinant cannot be equal to 2 or –2. Thus, such a determinant can take only three values namely 0, 4, –4.

Example 8. If all the entries in a 3×3 determinant are either 0 to 1 prove that the value of the determinant cannot be 3.

Solution

$$D = a_1b_2c_3 + a_2b_3c_1 + a_3b_1c_2 - a_3b_2c_1 - a_2b_1c_3$$

It is clear that any term on RHS is either 0 or 1. Since all the entries are either 0 to 1, the maximum possible value of the determinant is 3. The value 3 is attained if and only if

$$a_1b_2c_3 = 1,\ a_2b_3c_1 = 1,\ a_3b_1c_2 = 1$$

$$\text{and} \quad a_1b_3c_2 = 0,\ a_3b_2c_1 = 0,\ a_2b_1c_3 = 0$$

which is impossible, since the last three equations imply that at least three of the nine numbers $a_1, b_1, \ldots\ldots\ldots, c_3$ is zero, whence the first three equations are contradicted (because the first three equations cover the 9 entries of the determinant). Thus, D cannot be equal to 3. We now show that there exists a determinant whose value is 2. This can be easily constructed as $\begin{vmatrix} 0 & 1 & 1 \\ 1 & 0 & 1 \\ 1 & 1 & 0 \end{vmatrix}$.

It can be shown that such a determinant can take values 2, –2 (interchange any two adjacent columns), 1, –1 and 0.

1 is attained in $\begin{vmatrix} 1 & 0 & 0 \\ 0 & 1 & 0 \\ 0 & 0 & 1 \end{vmatrix}$, –1 is attained in $\begin{vmatrix} 0 & 1 & 0 \\ 1 & 0 & 0 \\ 0 & 0 & 1 \end{vmatrix}$

and 0 is attained in $\begin{vmatrix} 1 & 1 & 1 \\ 1 & 1 & 1 \\ 0 & 0 & 0 \end{vmatrix}$

Example 9. Show that the system of equations $ax + 4y + z = 0$, $2y + 3z - 1 = 0$ and $3x - bz + 2 = 0$ has a unique solution if $ab \neq 15$; has infinite solutions if $ab = 15$, $a = 3$, $b = 5$ and has no solution if $ab = 15$, $a \neq 3$, $b \neq 5$.

Solution

The system can be written as

$$ax + 4y + z = 0,\ 0x + 2y + 3z = 1$$

$$\text{and} \quad 3x + 0y - bz = -2$$

$$\Rightarrow \quad D = \begin{vmatrix} a & 4 & 1 \\ 0 & 2 & 3 \\ 3 & 0 & -b \end{vmatrix} = 30 - 2ab,$$

$$D_x = \begin{vmatrix} 0 & 4 & 1 \\ 1 & 2 & 3 \\ -2 & 0 & -b \end{vmatrix} = 4b - 20$$

$$D_y = \begin{vmatrix} a & 0 & 1 \\ 0 & 1 & 3 \\ 3 & -2 & -b \end{vmatrix} = -ab + 6a - 3,$$

$$D_z = \begin{vmatrix} a & 4 & 0 \\ 0 & 2 & 1 \\ 3 & 0 & -2 \end{vmatrix} = -4a + 12$$

Now the system will have unique solution if $D \neq 0 \Rightarrow ab \neq 15$

If $ab = 15$, let us re-write D_x, D_y, D_z, we get

$$D_x = \frac{60}{a} - 20, \quad D_y = 6a - 18, \quad D_z = -4a + 12$$

We note that there is a value of a for which D_x, D_y, D_z vanish simultaneously, which is obviously $a = 3$. Also, for $a \neq 3$ all the determinants become non-zero. Thus, the system will have infinite solutions if $ab = 15$, $a = 3$ and will have no solution if $ab = 15$, $a \neq 3$.

Example 10. Show that the system $ax + y + z = a$, $x + by + z = b$ and $x + y + cz = c$ is inconsistent if $abc - a - b - c + 2 = 0$ and a, b, c are different from unity.

Solution

We have $D = \begin{vmatrix} a & 1 & 1 \\ 1 & b & 1 \\ 1 & 1 & c \end{vmatrix} = abc - a - b - c + 2.$

$$D_x = \begin{vmatrix} a & 1 & 1 \\ b & b & 1 \\ c & 1 & c \end{vmatrix} = abc - a + b + c - 2bc$$

$$D_y = abc - b + a + c - 2ab$$

and $\quad D_z = abc - c + a + b - 2ab$

Now for a system to be inconsistent, it is necessary that $D = 0$.

$$\Rightarrow \quad abc - a - b - c + 2 = 0$$

But for a system to be inconsistent, it is sufficient that $D = 0$ and at least one of D_x, D_y, D_z is non zero. Rewriting D_x as

$$D_x = abc - a - b - c + 2 + 2b + 2c - 2bc - 2$$

$$= 0 + 2b + 2c - 2bc - 2 \quad (\because D = 0) = -2(1 - b)(1 - c)$$

Similarly, $D_y = -2(1 - c)(1 - a)$, $D_z = -2(1 - a)(a - c)$

Now if a, b, c are different from unity, D_x, D_y, D_z cannot vanish, whence the result follows.

RECAPITULATION (FORMULAE, THEOREM, IMPORTANT RESULTS)

1. Determinant of a matrix $A = [a_{11}]_{1\times1}$ is given by $|a_{11}| = a_{11}$

2. Determinant of a matrix $A = \begin{bmatrix} a_{11} & a_{12} \\ a_{21} & a_{22} \end{bmatrix}$ is given by

3. Determinant of a matrix $A = \begin{bmatrix} a_1 & b_1 & c_1 \\ a_2 & b_2 & c_2 \\ a_3 & b_3 & c_3 \end{bmatrix}$ is given by (expanding along R_1)

$$|A| = \begin{vmatrix} a_1 & b_1 & c_1 \\ a_2 & b_2 & c_2 \\ a_3 & b_3 & c_3 \end{vmatrix}$$

$$= a_1 \begin{vmatrix} b_2 & c_2 \\ b_3 & c_3 \end{vmatrix} - b_1 \begin{vmatrix} a_2 & c_2 \\ a_3 & c_3 \end{vmatrix} + c_1 \begin{vmatrix} a_2 & b_2 \\ a_3 & b_3 \end{vmatrix}$$

For any square matrix A, the |A| satisfies the following properties.

4. $|A'| = |A|$, where A' = transpose of A.
5. If we interchange any two rows (or columns), then sign of the determinant changes.
6. If any two rows or any two columns are identical or proportional, then value of the determinant is zero.
7. If we multiply each element of a row or a column of a determinant by constant k, then value of determinant is multiplied by k.
8. Multiplying a determinant by k means multiply elements of only one row (or one column) by k.
9. If $A = [a_{ij}]_{3\times3}$, then $|k \cdot A| = k^3 |A|$.

10. If the elements of a row or a column in a determinant can be expressed as sum of two or more elements, then the given determinant can be expressed as sum of two or more determinants.

11. If, to each element of a row or a column of a determinant, the equimultiples of corresponding elements of other rows or columns are added, then value of the determinant remains same.

12. Area of a triangle with vertices $(x_1, y_1), (x_2, y_2)$ and (x_3, y_3) is given by

$$\Delta = \frac{1}{2}\begin{vmatrix} x_1 & y_1 & 1 \\ x_2 & y_2 & 1 \\ x_3 & y_3 & 1 \end{vmatrix}$$

13. Minor of an element a_{ij} of the determinant of matrix A is the determinant obtained by deleting ith row and jth column and denoted by M_{ij}.

14. Cofactor of a_{ij} of given by $A_{ij} = (-1)^{i+j} M_{ij}$.

15. Value of determinant of a matrix A is obtained by the sum of product of elements of a row (or a column) with corresponding cofactors. For example,

$|A| = a_{11}A_{11} + a_{12}A_{12} + a_{13}A_{13}$

16. If elements of one row (or column) are multiplied with cofactors of elements of any other row (or column), then their sum is zero. For example, $a_{11}A_{21} + a_{12}A_{22} + a_{13}A_{23} = 0$

17. If $A = \begin{bmatrix} a_{11} & a_{12} & a_{13} \\ a_{21} & a_{22} & a_{23} \\ a_{31} & a_{32} & a_{33} \end{bmatrix}$, then

$$adj.\, A = \begin{bmatrix} A_{11} & A_{12} & A_{13} \\ A_{21} & A_{22} & A_{23} \\ A_{31} & A_{32} & A_{33} \end{bmatrix},$$

where A_{ij} is cofactor of a_{ij}.

18. A (*adj* A) = (*adj* A) A = |A| I, where A is square matrix of order n.

19. A square matrix A is said to be singular or non-singular according as $|A| = 0$ or $|A| \neq 0$.

20. If AB = BA = I, where B is square matrix, then B is called inverse of A.

Also, $A^{-1} = B$ or $B^{-1} = A$ and hence $(A^{-1})^{-1} = A$.

21. A square matrix A has inverse if and only if A is non-singular.

22. $A^{-1} = \frac{1}{|A|}(\text{adj} A)$

23. If $a_1x + b_1y + c_1z = d_1$

$a_2x + b_2y + c_2z = d_2$

$a_3x + b_3y + c_3z = d_3$

then these equations can be written as AX = B, where

$$A = \begin{bmatrix} a_1 & b_1 & c_1 \\ a_2 & b_2 & c_2 \\ a_3 & b_3 & c_3 \end{bmatrix}, X = \begin{bmatrix} x \\ y \\ z \end{bmatrix} \text{ and } B = \begin{bmatrix} d_1 \\ d_2 \\ d_3 \end{bmatrix}$$

24. Unique solution of equation AX = B is given by $X = A^{-1}B$, where $|A| \neq 0$.

25. A system of equations is consistent or inconsistent according as whether its solution exists or not.

26. For a square matrix A, in matrix equation AX = B

(i) $|A| \neq 0$, there exists unique solution

(ii) $|A| = 0$ and (adj A) B $\neq 0$, then there exists no solution

(iii) $|A| = 0$ and (adj A) B = 0, then system may or may not be consistent.

QUIZ

Instructions for Questions 1 to 7

Check the correctness of the statements.

1. If two adjacent columns or rows of the determinant are interchanged, the sign of the determinant is changed but its value remains unaltered.

2. If rows and columns of a determinant is interchanged, its value is changed.

3. If two rows or two columns of a determinant are identical, the determinant vanishes.

4. If each constituent in any row, or in any column, is multiplied by the same factor, then the determinant can be expressed as the sum of two other determinants.

5. If each constituent in any row or column consists of two terms, then the sum determinant can be expressed as the sum of two other determinants.
6. If in a determinant we add k terms of a row or a column to another row or a column of the determinant, the value of the determinant becomes k times.
7. If, in a determinant, the elements of a row or column are multiplied by the cofactors of the elements of the other row or column then the sum of the product is zero.
8. Define Minors of a determinant.
9. What do you mean by cofactor of a determinant?
10. What is the minor of the element 3 in the determinant?

$$\Delta = \begin{vmatrix} 1 & 2 & 7 \\ 4 & 5 & 3 \\ 7 & 9 & 6 \end{vmatrix}$$

JEE MAIN

Objective Questions

One Option Correct

1. If $\Delta = \begin{vmatrix} 0 & b-a & c-a \\ (a-b) & 0 & c-b \\ (a-c) & b-c & 0 \end{vmatrix}$ then Δ is equal to ______.

(a) $a+b+c$ (b) $-(a+b+c)$
(c) 0 (d) abc

2. If A and B are square matrices of order 3 such that $|A| = -1, |B| = 3$, then $|3AB| =$ ______.

(a) −27 (b) 81
(c) −9 (d) −81

3. If the value of a third-order determinant is 11, then the value of the determinant formed by its cofactors will be ______.

(a) 14641 (b) 121
(c) 11 (d) 1331

4. Value of $\begin{vmatrix} 3i & 2i & 2i \\ 5 & 4 & -3i \\ i & 2i & 7 \end{vmatrix}$ is ______.

(a) $-2i+12$ (b) $-2i-12$
(c) $2i+12$ (d) $2i-12$

5. $\begin{vmatrix} \sin^2 x & \cos^2 x & 1 \\ \cos^2 x & \sin^2 x & 1 \\ -10 & 12 & 2 \end{vmatrix} =$

(a) $12\cos^2 x - 10\sin^2 x$
(b) $12\sin^2 x - 10\cos^2 x - 2$
(c) $10\sin 2x$
(d) 0

6. If k is a scalar and A is an $n \times n$ square matrix, then $|kA| =$

(a) $k^n |A|^n$ (b) $k^n |A|$
(c) $k|A|$ (d) $k|A|^n$

7. If d is the determinant of a square matrix A of order n, then the determinant of its adjoint is ______.

(a) d^{n-1} (b) d^{m-1}
(c) d (d) d^n

8. The value of the determinant

$\begin{vmatrix} a-b & b-c & c-a \\ x-y & y-z & z-x \\ p-q & q-r & r-p \end{vmatrix}$ is ______.

(a) $(a-x)(y-z)(r-p)$ (b) 0
(c) $abc + pqr + xyz$ (d) None of these.

9. For positive number x, y, z, the numerical value of $\begin{vmatrix} 1 & \log_x y & \log_x z \\ \log_y x & 3 & \log_y z \\ \log_z x & \log_z y & 5 \end{vmatrix}$ is ______.

(a) $\log x \log y \log z$ (b) 1
(c) 0 (d) 8

10. If α, β, γ are roots of $x^3 + ax^2 + b = 0$, then the value of $\begin{vmatrix} \alpha & \beta & \gamma \\ \beta & \gamma & \alpha \\ \gamma & \alpha & \beta \end{vmatrix}$ is ______.

(a) $a^2 - 3b$ (b) $-a^3$
(c) $a^3 - 3b$ (d) a^3

11. If $f(x)=\begin{vmatrix} a & -1 & 0 \\ ax & a & -1 \\ ax^2 & ax & a \end{vmatrix}$, then $f(2x)-f(x)$ equals ______.

(a) $x(2a+3x)$ (b) $ax(2a+3x)$
(c) $a(2a+3x)$ (d) $ax(2x+3a)$

12. The value of the determinant $\Delta=\begin{vmatrix} 1! & 2! & 3! \\ 2! & 3! & 4! \\ 3! & 4! & 5! \end{vmatrix}$ is ______.

(a) 4! (b) 5!
(c) 2! (d) 3!

13. Let a_{ij} denote the element of the ith row and jth column in a 3×3 determinant $(1\le i\le3, 1\le j\le3)$ and let $a_{ij}=-a_{ji}$ for every i and j. Then the determinant has all the principal diagonal elements as ______.

(a) -1 (b) 0
(c) 1 (d) None of these.

14. If [ ] denotes the greatest integer less than or equal to the real number under consideration, and $-1\le x<0; 0\le y<1; 1\le z<2$, then the value of the determinant

$$\begin{vmatrix} [x]+1 & [y] & [z] \\ [x] & [y]+1 & [z] \\ [x] & [y] & [z]+1 \end{vmatrix}$$ is ______.

(a) $[z]$ (b) $[y]$
(c) $[x]$ (d) None of these.

15. If $f(x)=\begin{vmatrix} x^3 & \sin x \\ 1 & 2 \end{vmatrix}$, then $\int_{-a}^{a} f(x)dx$ is ______.

(a) $-\frac{1}{2}$ (b) 3
(c) $\frac{1}{2}$ (d) 0

16. If a, b, c are different, then ______.

$$\begin{vmatrix} 1 & 1 & 1 \\ (x-a)^2 & (x-b)^2 & (x-c)^2 \\ (x-b)(x-c) & (x-c)(x-a) & (x-a)(x-b) \end{vmatrix}$$

(a) $x=\frac{1}{3}(a+b+c)$ (b) $x=\frac{1}{2}(a+b+c)$
(c) $a+b+c=0$ (d) $x=a+b+c$

17. The value of the determinant
$\Delta=\begin{vmatrix} \log x & \log y & \log z \\ \log 2x & \log 2y & \log 2z \\ \log 3x & \log 3y & \log 3z \end{vmatrix}$ is ______.

(a) $6\log(xyz)$ (b) $\log(6xyz)$
(c) $\log(xyz)$ (d) 0

18. If capital letters denote the cofactors of the corresponding small letters in $\Delta=\begin{vmatrix} a_1 & b_1 & c_1 \\ a_2 & b_2 & c_2 \\ a_3 & b_3 & c_3 \end{vmatrix}$, then $\Delta'=\begin{vmatrix} A_1 & B_1 & C_1 \\ A_2 & B_2 & C_2 \\ A_3 & B_3 & C_3 \end{vmatrix}$ is ______.

(a) Δ^2 (b) 2Δ
(c) 0 (d) Δ

19. If a, b and c are non-zero real numbers, then

$\Delta=\begin{vmatrix} b^2c^2 & bc & b+c \\ c^2a^2 & ca & c+a \\ a^2b^2 & ab & a+b \end{vmatrix}$ is equal to ______.

(a) $a^2b^2c^2$ (b) $bc+ca+ab$
(c) abc (d) 0

20. If a, b, c are in AP with common difference d and $\begin{vmatrix} x+1 & x+a & x+b \\ x+a & x+b & x+c \\ x-b+1 & x-1 & x-a+c \end{vmatrix}$ has absolute value 2, then d is ______.

(a) ±1 (b) a
(c) -2 (d) None of these.

21. If a, b, c are roots of the equation $x^3+px+q=0 (p\ne0, q\ne0)$, the value of the determinant $\begin{vmatrix} a & b & c \\ b & c & a \\ c & a & b \end{vmatrix}$ is ______.

(a) p^2-2q (b) p
(c) 0 (d) q

22. If $\frac{1}{a}+\frac{1}{b}+\frac{1}{c}=0$, then $\begin{vmatrix} 1+a & 1 & 1 \\ 1 & 1+b & 1 \\ 1 & 1 & 1+c \end{vmatrix}$ is equal to ______.

(a) $-abc$ (b) 0
(c) abc (d) None of these.

23. The value of the determinant

$$\Delta = \begin{vmatrix} 1 & 1 & 1 \\ {}^{m}C_1 & {}^{m+1}C_1 & {}^{m+2}C_1 \\ {}^{m}C_2 & {}^{m+1}C_2 & {}^{m+2}C_2 \end{vmatrix} \text{ is equal to } \_\_\_\_\_\_.$$

(a) 0 (b) −1
(c) 1 (d) None of these.

24. If x, y, z are different and $\begin{vmatrix} x & x^2 & 1+x^3 \\ y & y^2 & 1+y^3 \\ z & z^2 & 1+z^3 \end{vmatrix} = 0$, then the value of xyz is ______.

(a) −3 (b) −1
(c) −2 (d) None of these.

25. If $\begin{vmatrix} x+a & b & c \\ a & x+b & c \\ a & b & x+c \end{vmatrix} = 0$, then x equals ______

(a) $0, -(a+b+c)$ (b) $0, a+b+c$
(c) $a+b+c$ (d) $-(a+b+c)$

26. If A, B, C are angles of a triangle and

$$\begin{vmatrix} 1 & 1 & 1 \\ 1+\sin A & 1+\sin B & 1+\sin C \\ \sin A+\sin^2 A & \sin B+\sin^2 B & \sin C+\sin^2 C \end{vmatrix} = 0,$$

then triangle ABC is ______.
(a) right-angled isosceles
(b) isosceles
(c) equilateral
(d) None of these.

27. If

$$\begin{vmatrix} x^n & x^{n+2} & x^{n+3} \\ y^n & y^{n+2} & y^{n+3} \\ z^n & z^{n+2} & z^{n+3} \end{vmatrix} = (x-y)(y-z)(z-x)\left(\frac{1}{x}+\frac{1}{y}+\frac{1}{z}\right)$$

then value of n is ______.
(a) 2 (b) −1
(c) −2 (d) 1

28. If a, b, c are sides of a triangle and

$$\begin{vmatrix} a^2 & b^2 & c^2 \\ (a+1)^2 & (b+1)^2 & (c+1)^2 \\ (a-1)^2 & (b-1)^2 & (c-1)^2 \end{vmatrix} = 0, \text{ then}$$

(a) ΔABC is isosceles.
(b) ΔABC is equilateral.
(c) ΔABC is right angled isosceles.
(d) None of these.

29. The value of the determinant $\begin{vmatrix} 1 & ka & k^2+a^2 \\ 1 & kb & k^2+b^2 \\ 1 & kc & k^2+c^2 \end{vmatrix}$ is ______.

(a) $k(a+b-c)(b+c-a)(c+a-b)$
(b) $kabc(a^2+b^2+c^2)$
(c) $k(a+b)(b+c)(c+a)$
(d) $k(a-b)(b-c)(c-a)$

30. If A, B and C are the angles of a triangle then the value of the determinant

$$\begin{vmatrix} -1+\cos B & \cos C+\cos B & \cos B \\ \cos C+\cos A & -1+\cos A & \cos A \\ -1+\cos B & -1+\cos A & -1 \end{vmatrix} \text{ is } \_\_\_\_\_\_.$$

(a) −1 (b) 0
(c) 2 (d) 1

31. If $2^{a_1}, 2^{a_2}, 2^{a_3} \ldots\ldots 2^{a_n}$ are in GP Then

$$\begin{vmatrix} a_1 & a_2 & a_3 \\ a_{n+1} & a_{n+2} & a_{n+3} \\ a_{2n+1} & a_{2n+2} & a_{2n+3} \end{vmatrix} \text{ is equal to } \_\_\_\_\_\_.$$

(a) 0 (b) 2
(c) 2^3 (d) None of these.

32. If p, q, r are non-zero and different real numbers and $\Delta = \begin{vmatrix} 1 & p^2 & p^3 \\ 1 & q^2 & q^3 \\ 1 & r^2 & r^3 \end{vmatrix} = 0$, then ______.

(a) $p^3+q^3+r^3 = 3pqr$ (b) $p+q+r=0$
(c) $\frac{1}{p}+\frac{1}{q}+\frac{1}{r}=0$ (d) None of these.

33. If $\delta r = \begin{vmatrix} 2r & x & n(n+1) \\ 6r^2-1 & y & n^2(2n+3) \\ 4r^3-2nr & z & n^3(n+1) \end{vmatrix}$, then value of $\sum_{r=1}^{n} \delta r$ is independent of ______.

(a) x, y, z, n (b) n only
(c) x only (d) y only

34. If $S_r = \alpha^r + \beta^r + \gamma^r$ then $\Delta = \begin{vmatrix} S_0 & S_1 & S_2 \\ S_1 & S_2 & S_3 \\ S_2 & S_3 & S_4 \end{vmatrix}$ is equal to ______.

(a) $S_6 - S_4$ (b) $S_5 - S_3$
(c) S_6 (d) None of these.

35. If $\Delta(x)=\begin{vmatrix} x & 1+x^2 & x^3 \\ \log(1+x^2) & e^x & \sin x \\ \cos x & \tan x & \sin^2 x \end{vmatrix}$ then ____.

(a) $\Delta(x)=0$

(b) $\Delta'(x)=0$

(c) $\Delta(x)$ is divisible by x

(d) None of these.

36. If ω is a complex cube root of unity, then the value of the determinant $\begin{vmatrix} 1 & \omega & \omega+1 \\ \omega+1 & 1 & \omega \\ \omega & \omega+1 & 1 \end{vmatrix}$ is ____.

(a) ω (b) 4

(c) 2 (d) 0

37. Suppose $D=\begin{vmatrix} a_1 & b_1 & c_1 \\ a_2 & b_2 & c_2 \\ a_3 & b_3 & c_3 \end{vmatrix}$ and

$D'=\begin{vmatrix} a_1+pb_1 & b_1+qc_1 & c_1+ra_1 \\ a_2+pb_2 & b_2+qc_2 & c_2+ra_2 \\ a_3+pb_3 & b_3+qc_3 & c_3+ra_3 \end{vmatrix}$, then ______.

(a) $D'=D(1+p+q+r)$

(b) $D'=D(1+pqr)$

(c) $D'=D$

(d) $D'=D(1-pqr)$

38. If $x^a y^b=e^m$, $x^c y^d=e^n$, $\Delta_1=\begin{vmatrix} m & b \\ n & d \end{vmatrix}$, $\Delta_2=\begin{vmatrix} a & m \\ c & n \end{vmatrix}$ and $\Delta_3=\begin{vmatrix} a & b \\ c & d \end{vmatrix}$, then the value of x and y are, respectively, ______.

(a) Δ_2/Δ_1 and Δ_3/Δ_1

(b) Δ_1/Δ_3 and Δ_2/Δ_3

(c) $e^{\frac{\Delta_1}{\Delta_3}}$ and $e^{\frac{\Delta_2}{\Delta_3}}$

(d) $\log(\Delta_1/\Delta_3)$ and $\log(\Delta_2/\Delta_3)$

39. The values of the determinant

$$\begin{vmatrix} 2 & \tan A\cdot\cot B+\cot A\cdot\tan B & \tan A\cdot\cot C+\cot A\cdot\tan C \\ \tan B\cdot\cot A & 2 & \tan B\cdot\cot C+\cot B\cdot\tan C \\ \tan C\cdot\cot A+\cot C\cdot\tan A & \tan C\cdot\cot B+\cot C\cdot\tan B & 2 \end{vmatrix}$$ is ______.

(a) 2 (b) 0

(c) 1 (d) None of these.

40. The value of a for which the system of equations

$a^3x+(a+1)^3y+(a+2)^3z=0$

$ax+(a+1)y+(a+2)z=0$

$x+y+z=0$

has a non-zero solution is ______.

(a) 1 (b) -1

(c) 0 (d) None of these.

41. $a=1+2+4+\ldots$ upto n terms

$b=1+3+9+\ldots$ upto n terms and

$c=1+5+25+\ldots$ upto n terms

Then $\begin{vmatrix} a & 2b & 4c \\ 2 & 2 & 2 \\ 2^n & 3^n & 5^n \end{vmatrix}=$ ______.

(a) $2^n+3^n+5^n$ (b) $(30)^n$

(c) 0 (d) 10^n

42. If a, b, c are non-zeroes, then the system of equations

$(\alpha+a)x+\alpha y+\alpha z=0$, $\alpha x+(\alpha+b)y+\alpha z=0$, $\alpha x+\alpha y+(\alpha+c)z=0$ has a non-trivial solution if ______.

(a) $\alpha+a+b+c=1$

(b) $\alpha^{-1}=a+b+c$

(c) $\alpha^{-1}=-(a^{-1}+b^{-1}+c^{-1})$

(d) None of these.

43. If $f(x)=\begin{vmatrix} 1 & a & a^2 \\ \sin(n-1)x & \sin nx & \sin(n+1)x \\ \cos(n-1)x & \cos nx & \cos(n+1)x \end{vmatrix}$,

then $\int_0^{\pi/2} f(x)dx$ is equal to

(a) $-\left(1+a+a^2\right)$ (b) $-a+\left(1+a^2\right)$

(c) $a-\left(1+a^2\right)$ (d) $1+a+a^2$

44. If $\Delta_r = \begin{vmatrix} 2^{r-1} & \dfrac{(r+1)!}{1+\frac{1}{r}} & 2r \\ a & b & c \\ 2^n-1 & (n+1)!-1 & n(n+1) \end{vmatrix}$,

then value of $\sum_{r=1}^{n}\Delta_r$ is ______.

(a) $a(n!)+b$ (b) $(n+3)!$

(c) 0 (d) None of these.

45. If $\Delta_1 = \begin{vmatrix} x & b & b \\ a & x & b \\ a & a & x \end{vmatrix}$ and $\Delta_2 = \begin{vmatrix} x & b \\ a & x \end{vmatrix}$, then, ______.

(a) $\dfrac{d}{dx}(\Delta_1) = 3\Delta_2$ (b) $\dfrac{d}{dx}(\Delta_1) = 3\Delta_2^2$

(c) $\Delta_1 = 3\Delta_2^2$ (d) $\Delta_1 = 3(\Delta_2)^{3/2}$

46. If $\begin{vmatrix} x-1 & 5x & 7 \\ x^2-1 & x-1 & 8 \\ 2x & 3x & 0 \end{vmatrix} = ax^3+bx^2+cx+d$,

then the value of c is given by______.

(a) –1 (b) 15

(c) 17 (d) 12

47. Let m be a positive integer and $\Delta_r =$

$\begin{vmatrix} 2r-1 & {}^mC_r & 1 \\ m^2-1 & 2^m & m+1 \\ \sin^2(m^2) & \sin^2(m) & \sin^2(m+1) \end{vmatrix} (0 \le r \le m)$,

then the value of $\sum_{r=0}^{m}\Delta_r$ is given by ______.

(a) m^2-1 (b) 0

(c) 2^m (d) $m^2\left(2^m-1\right)$

48. Two non-zero distinct numbers a, b are used as elements to make determinants of the third order. The number of determinants whose value is zero for all a, b is ______.

(a) 32 (b) 24

(c) $a+b$ (d) 18

49. If α, β, γ are the roots of $x^3+px^2+q=0$

where $q \neq 0$, then, $\Delta = \begin{vmatrix} \frac{1}{\alpha} & \frac{1}{\beta} & \frac{1}{x} \\ \frac{1}{\beta} & \frac{1}{x} & \frac{1}{\alpha} \\ \frac{1}{x} & \frac{1}{\alpha} & \frac{1}{\beta} \end{vmatrix}$ equals to ___.

(a) $-\dfrac{p}{q}$ (b) $\dfrac{1}{q}$

(c) $\dfrac{p^2}{q}$ (d) None of these.

50. If $l_1^2+m_1^2+n_1^2=1$ etc.; and $l_1l_2+m_1m_2+n_1n_2=0$

etc., and $\Delta = \begin{vmatrix} l_1 & m_1 & n_1 \\ l_2 & m_2 & n_2 \\ l_3 & m_3 & n_3 \end{vmatrix}$, then

(a) $|\Delta| = 3$ (b) $|\Delta| = 2$

(c) $|\Delta| = 1$ (d) $\Delta = 0$

51. If x, y, z are integers in AP, lying between 1 and 9, and $x51$, $y41$ and $z31$ are three-digit numbers,

then the value of $\begin{vmatrix} 5 & 4 & 3 \\ x51 & y41 & z31 \\ x & y & z \end{vmatrix}$ is ______.

(a) $x+y+z$ (b) 0

(c) $x-y+z$ (d) $x-y-z$

52. If

$p\lambda^4+q\lambda^3+r\lambda^2+s\lambda+t = \begin{vmatrix} \lambda^2+3\lambda & \lambda-1 & \lambda+3 \\ \lambda^2+1 & 2-\lambda & \lambda-3 \\ \lambda^2-3 & \lambda-4 & 3\lambda \end{vmatrix}$,

then p is equal to ______.

(a) –2 (b) –3

(c) –4 (d) –5

53. If $a+b+c>0$ and $\Delta = \begin{vmatrix} a & b & c \\ b & c & a \\ c & a & b \end{vmatrix}$, then ______.

(a) $\Delta < 0$ (b) $\Delta \le 0$

(c) $\Delta > 0$ (d) $\Delta = 0$

54. If

$f(x) = \begin{vmatrix} 1 & x & x+1 \\ 2x & x(x-1) & (x+1)x \\ 3x(x-1) & x(x-1)(x-2) & (x+1)x(x-1) \end{vmatrix}$,

then $f(500)$ is equal to ______.

(a) 0 (b) 1
(c) 500 (d) –500

55. The sum of two non-integral roots of $\begin{vmatrix} x & 2 & 5 \\ 3 & x & 3 \\ 5 & 4 & x \end{vmatrix} = 0$ is ______.

(a) 5 (b) –5
(c) 18 (d) –18

56. Let $\Delta = \begin{vmatrix} x^2+4x+5 & x+2 & 5 \\ 2x^2+6x+10 & 2x+3 & 10 \\ 4x^2-2x+20 & 4x-1 & 20 \end{vmatrix}$, then

(a) $y = \Delta$ represents a parabola passing through the origin.
(b) $y = \Delta$ represents a straight line through the origin.
(c) $\Delta = 0$ has only one real root.
(d) $\Delta = 0$ has two distinct real roots.

57. If $z = \begin{vmatrix} 2 & 5-i & 7+i \\ 5+i & 2 & 3-i \\ 7-i & 3+i & 7 \end{vmatrix}$ and $\text{Arg}(z) = \alpha$, then α is ______.

(a) 0 (b) $0 < \alpha < \frac{\pi}{2}$
(c) $-\frac{\pi}{2} < \alpha < 0$ (d) π

58. If each element of a 3×3 matrix is multiplied by 3, then the determinant of the newly formed matrix is ______.

(a) $27|A|$ (b) $|A|^3$
(c) $3|A|$ (d) $9|A|$

59. If a, b, c are non-zero real numbers, then $\begin{vmatrix} bc & ca & ab \\ ca & ab & bc \\ ab & bc & ca \end{vmatrix}$ vanishes when ______.

(a) $\frac{1}{b} - \frac{1}{c} - \frac{1}{a} = 0$ (b) $\frac{1}{c} - \frac{1}{b} - \frac{1}{a} = 0$
(c) $\frac{1}{a} + \frac{1}{b} + \frac{1}{c} = 0$ (d) $\frac{1}{a} - \frac{1}{b} - \frac{1}{c} = 0$

60. The value of determinant $\begin{vmatrix} 2a_1b_1 & a_1b_2+a_2b_1 & a_1b_3+a_3b_1 \\ a_1b_2+a_2b_1 & 2a_2a_3 & a_2b_3+a_3b_3 \\ a_1b_3+a_3b_1 & a_3b_2+a_2b_3 & 2a_3b_3 \end{vmatrix}$ is ______.

(a) $a_1a_2a_3b_1b_2b_3$ (b) 0
(c) 1 (d) None of these.

Questions from Previous Years

61. Let A and B be two 3×3 real matrices such that $(A^2 - B^2)$ is an invertible matrix. If $A^5 = B^5$ and $A^3B^2 = A^2B^3$, then the value of the determinant of the matrix $A^3 + B^3$ is equal to ______.

(a) 2 (b) 4
(c) 1 (d) 0

62. The maximum value of
$$f(x) = \begin{vmatrix} \sin^2 x & 1+\cos^2 x & \cos 2x \\ 1+\sin^2 x & \cos^2 x & \cos 2x \\ \sin^2 x & \cos^2 x & \sin 2x \end{vmatrix}, x \in R \text{ is}$$
______.

(a) $\frac{3}{4}$ (b) 5
(c) $\sqrt{7}$ (d) $\sqrt{5}$

63. The value of $\begin{vmatrix} (a+1) & (a+2) & a+21 \\ (a+2) & (a+3) & a+31 \\ (a+3) & (a+4) & a+41 \end{vmatrix}$ is ______.

(a) 0 (b) $(a+2)(a+3)(a+4)$
(c) –2 (d) $(a+1)(a+2)(a+3)$

64. Let $A = \begin{bmatrix} 1 & 2 \\ -1 & 4 \end{bmatrix}$. If $A^{-1} = \alpha I + \beta A, \alpha, \beta \in R, I$ is a 2×2 identity matrix, then $4(\alpha - \beta)$ is equal to ______.

(a) 5 (b) $\frac{8}{3}$
(c) 2 (d) 4

65. Consider the system of linear equations
$-x + y + 2z = 0$
$3x - ay + 5z = 1$
$2x - 2y - az = 7$
Let S_1 be the set of all $a \in R$ for which the system is inconsistent and S_2 be the set all $a \in R$ for which the system has infinitely many solutions. If $n(S_1)$ and $n(S_2)$ denote the number of elements in S_1 and S_2 respectively, then ______.

(a) $n(S_1) = 2, n(S_2) = 2$
(b) $n(S_1) = 1, n(S_2) = 0$
(c) $n(S_1) = 2, n(S_2) = 0$
(d) $n(S_1) = 0, n(S_2) = 2$

66. Let $[\lambda]$ be the greatest integer less than or equal to λ. The set for all values of λ for which the system of linear equations $x + y + z = 4$, $3x + 2y + 5z = 3$, $9x + 4y + (28 + [\lambda])z = [\lambda]$ has a solution is:
(a) R (b) $(-\infty,-9)\cup(-9,\infty)$
(c) $[-9,-8)$ (d) $(-\infty,-9)\cup[-8,\infty)$

67. Let the system of linear equations
$4x + \lambda y + 2z = 0$
$2x - y + z = 0$
$mx + 2y + 3z = 0,\ \lambda, \mu \in R$
have a non-trivial solution. Then which of the following is true?
(a) $\mu = -6, \lambda \in R$ (b) $\lambda = 3, \mu \in R$
(c) $\mu = 6, \lambda \in R$ (d) $\lambda = 2, \mu \in R$

68. Consider the following system of equations:
$x + 2y - 3z = a$
$2x + 6y - 11z = b$
$x - 2y + 7z = c,$
where a, b and c are real constants. Then the system of equations
(a) has infinite number of solutions when $5a = 2b + c$.
(b) has no solution for all a, b and c.
(c) has a unique solution when $5a = 2b + c$.
(d) has a unique solution for all a, b and c.

69. The system of linear equations
$3x - 2y - kz = 10;\ 2x - 4y - 2z = 6;\ x + 2y - z = 5m$
is inconsistent if ______.
(a) $k = 3, m = \frac{4}{5}$ (b) $k \neq 3, m \in R$
(c) $k \neq 3, m \neq \frac{4}{5}$ (d) $k = 3, m \neq \frac{4}{5}$

70. For the system of linear equations
$x - 2y = 1$, $x - y + kz = -2$, $ky + 4z = 6$, $k \in R$,
consider the following statements
(A) The system has unique solution if $k \neq 2, k \neq -2$.
(B) The system has unique solution if $k = -2$.
(C) The system has unique solution if $k = 2$.
(D) The system has no solution if $k = 2$.
(E) The system has infinite number of solutions if $k \neq -2$.
Which of the following statements are correct?
(a) (C) and (D) only
(b) (B) and (E) only
(c) (A) and (E) only
(d) (A) and (D) only

71. The values of λ and μ for which the system of linear equations
$x + y + z = 2$
$x + 2y + 3z = 5$
$x + 3y + \lambda z = \mu$
has infinitely many solutions are, respectively, ______.
(a) 6 and 8 (b) 5 and 7
(c) 5 and 8 (d) 4 and 9

72. Let $\lambda \in R$. The system of linear equations is true for
$2x_1 - 4x_2 + \lambda x_3 = 1$
$x_1 - 6x_2 + x_3 = 2$
$\lambda x_1 - 10x_2 + 4x_3 = 3$
(a) exactly one negative value of λ.
(b) exactly one positive value of λ.
(c) every value of λ.
(d) exactly two values of λ.

73. If the system of linear equations
$x + y + 3z = 0$
$x + 3y + k^2z = 0$
$3x + y + 3z = 0$
has a non-zero solution (x, y, z) for some $k \in R$, then $x + \left(\frac{y}{z}\right)$ is equal to ______.
(a) -3 (b) 9
(c) 3 (d) -9

74. Suppose the vectors x_1, x_2 and x_3 are the solutions of the system of linear equation, $Ax = b$ when the vector b on the right side is equal to b_1, b_2 and b_3 respectively.
If $x_1 = \begin{bmatrix}1\\1\\1\end{bmatrix}, x_2 = \begin{bmatrix}0\\2\\1\end{bmatrix}, x_3 = \begin{bmatrix}0\\0\\1\end{bmatrix}, b_1 = \begin{bmatrix}1\\0\\0\end{bmatrix}, b_2 = \begin{bmatrix}0\\2\\0\end{bmatrix}$ and $b_3 = \begin{bmatrix}0\\0\\2\end{bmatrix}$, then the determinant of A is equal to ______.
(a) 4 (b) 2
(c) $\frac{1}{2}$ (d) $\frac{3}{2}$

75. If the system of equations
$x + y + z = 2$
$2x + 4y - z = 6$
$3x + 2y + \lambda z = \mu$
has infinitely many solutions, then ______.

(a) $\lambda + 2\mu = 14$ (b) $2\lambda - \mu = 5$
(c) $\lambda - 2\mu = -5$ (d) $2\lambda + \mu = 14$

76. Let S be the set of all $\lambda \in R$ for which the system of linear equations
$2x - y + 2z = 2$
$x - 2y + \lambda z = -4$
$x + \lambda y + z = 4$
has no solution. Then the set S
(a) contains more than two elements.
(b) is an empty set.
(c) is a singleton.
(d) contains exactly two elements.

77. Let
$A = \left\{X = (x, y, z)^T : PX = 0 \text{ and } x^2 + y^2 + z^2 = 1\right\}$,

where $P = \begin{bmatrix} 1 & 2 & 1 \\ -2 & 3 & -4 \\ 1 & 9 & -1 \end{bmatrix}$, then the set A
(a) is a singleton.
(b) is an empty set.
(c) contains more than two elements.
(d) contains exactly two elements.

78. The following system of linear equations
$7x + 6y - 2z = 0$
$3x + 4y + 2z = 0$
$x - 2y - 6z = 0$, has
(a) infinitely many solutions, (x, y, z) satisfying $y = 2z$
(b) no solution.
(c) infinitely many solutions, (x, y, z) satisfying $x = 2z$.
(d) only the trivial solution.

79. For which of the following ordered pairs (μ, δ), the system of linear equations
$x + 2y + 3z = 1$
$3x + 4y + 5x = \mu$
$4x + 4y + 4z = \delta$
is inconsistent?
(a) (4, 3) (b) (4, 6)
(c) (1, 0) (d) (3, 4)

80. The system of linear equations
$\lambda x + 2y + 2z = 56$
$2\lambda x + 3y + 5z = 8$
$4x + \lambda y + 6z = 10$
has
(a) no solution when $\lambda = 8$.
(b) a unique solution when $\lambda = -8$.
(c) no solution when $\lambda = 2$.
(d) infinitely many solutions when $\lambda = 2$.

81. If the system of linear equations
$2x + 2ay + az = 0$
$2x + 3by + bz = 0$
$2x + 4cy + cz = 0$,
where $a, b, c \in R$ are non-zero and distinct, has a non-zero solution, then:
(a) $\frac{1}{a}, \frac{1}{b}, \frac{1}{c}$ are in AP (b) a, b, c are in GP
(c) $a + b + c = 0$ (d) a, b, c are in AP

82. Let A be a 3×3 matrix such that $\text{adj } A = \begin{bmatrix} 2 & -1 & 1 \\ -1 & 0 & 2 \\ 1 & -2 & -1 \end{bmatrix}$ and $B = \text{adj}(\text{adj} A)$. If $|A| = \lambda$ and $\left|\left(B^{-1}\right)^T\right| = \mu$, then the ordered pair, is equal to ______.
(a) $\left(3, \frac{1}{81}\right)$ (b) $\left(9, \frac{1}{9}\right)$
(c) (3, 81) (d) $\left(9, \frac{1}{81}\right)$

83. If the matrices $A = \begin{bmatrix} 1 & 1 & 2 \\ 1 & 3 & 4 \\ 1 & -1 & 3 \end{bmatrix}$, $B = \text{adj } A$ and $C = 3A$, then $\frac{|\text{adj} B|}{|C|}$ is equal to ______.
(a) 8 (b) 16
(c) 72 (d) 2

84. If the minimum and the maximum values of the function $f : \left[\frac{\pi}{4}, \frac{\pi}{4}\right] \to R$, defined by
$f(\theta) = \begin{vmatrix} -\sin^2\theta & 1 - \sin^2\theta & 1 \\ -\cos^2\theta & 1 - \cos^2\theta & 1 \\ 12 & 10 & -2 \end{vmatrix}$ are m and M respectively, then the ordered pair (m, M) is equal to ______.
(a) $\left(0, 2\sqrt{2}\right)$ (b) $(-4, 0)$
(c) $(-4, 4)$ (d) $(0, 4)$

85. If $a + x = b + y = c + z + 1$, where a, b, c, x, y, z are non-zero distinct real numbers, then $\begin{vmatrix} x & a + y & x + a \\ y & b + y & y + b \\ z & c + y & z + c \end{vmatrix}$ is equal to ______.

(a) $y(b-a)$ (b) $y(a-b)$
(c) 0 (d) $y(a-c)$

86. Let $A=[a_{ij}]$ and $B=[b_{ij}]$ be two 3×3 real matrices such that $b_{ij}=(3)^{(i+j-2)}a_{ij}$, where $i, j=1,2,3$. If the determinant of B is 81, then the determinant of A is:
(a) 1/3 (b) 3
(c) 1/81 (d) 1/9

87. Let $\theta=\frac{\pi}{5}$ and $A=\begin{bmatrix}\cos\theta & \sin\theta\\ -\sin\theta & \cos\theta\end{bmatrix}$. If $B=A+A^4$, then det (B)
(a) is one. (b) lies in (2, 3).
(c) is zero. (d) lies in (1, 2).

88. If

$$\Delta=\begin{vmatrix}x-2 & 2x-3 & 3x-4\\ 2x-3 & 3x-4 & 4x-5\\ 3x-5 & 5x-8 & 10x-17\end{vmatrix}$$

$=Ax^3+Bx^2+Cx+D$,

then B + C is equal to ______.
(a) –1 (b) 1
(c) –3 (d) 9

89. Let $a-2b+c=1$. If $f(x)=\begin{vmatrix}x+a & x+2 & x+1\\ x+b & x+3 & x+2\\ x+c & x+4 & x+3\end{vmatrix}$, then:
(a) $f(-50)=501$ (b) $f(-50)=-1$
(c) $f(50)=-501$ (d) $f(50)=1$

90. If $\Delta_1=\begin{vmatrix}x & \sin\theta & \cos\theta\\ -\sin\theta & -x & 1\\ \cos\theta & 1 & x\end{vmatrix}$ and

$\Delta_2=\begin{vmatrix}x & \sin2\theta & \cos2\theta\\ -\sin2\theta & -x & 1\\ \cos2\theta & 1 & x\end{vmatrix}$, $x\neq0$, then for all $\theta\in\left(0,\frac{\pi}{2}\right)$:

(a) $\Delta_1-\Delta_2=-2x^3$
(b) $\Delta_1-\Delta_2=x(\cos2\theta-\cos4\theta)$
(c) $\Delta_1\times\Delta_2=2(x^3+x-1)$
(d) $\Delta_1+\Delta_2=-2x^3$

91. Let $A=\begin{vmatrix}2 & b & 1\\ b & b^2+1 & b\\ 1 & b & 2\end{vmatrix}$ where $b>0$. Then the minimum value of $\frac{\det(A)}{b}$ is ______.
(a) $2\sqrt{3}$ (b) $-2\sqrt{3}$
(c) $-\sqrt{3}$ (d) $\sqrt{3}$

92. A value of $\theta\in(0,\pi/3)$, for which

$$\begin{vmatrix}1+\cos^2\theta & \sin^2\theta & 4\cos6\theta\\ \cos^2\theta & 1+\sin^2\theta & 4\cos6\theta\\ \cos^2\theta & \sin^2\theta & 1+4\cos6\theta\end{vmatrix}=0$$ is ______.

(a) $\frac{\pi}{9}$ (b) $\frac{\pi}{18}$
(c) $\frac{7\pi}{24}$ (d) $\frac{7\pi}{36}$

93. Let the numbers 2, b, c be in an AP and $A=\begin{bmatrix}1 & 1 & 1\\ 2 & b & c\\ 4 & b^2 & c^2\end{bmatrix}$. If det(A) = [2, 16], then c lies in the interval:
(a) [2, 3) (b) $(2+2^{3/4},4)$
(c) [4, 6] (d) $[3,2+2^{3/4}]$

94. If $A=\begin{bmatrix}1 & \sin\theta & 1\\ -\sin\theta & 1 & \sin\theta\\ -1 & -\sin\theta & 1\end{bmatrix}$; then for all $\in\left(\frac{3\pi}{4},\frac{5\pi}{4}\right)$, det (A) lies in the interval:
(a) $\left(1,\frac{5}{2}\right]$ (b) $\left[\frac{5}{2},4\right)$
(c) $\left(0,\frac{3}{2}\right]$ (d) $\left(\frac{3}{2},3\right]$

95. Let $d\in R$ and

$$A=\begin{bmatrix}-2 & 4+d & (\sin\theta)^{-2}\\ 1 & (\sin\theta)+2 & d\\ 5 & (2\sin\theta)-d & (-\sin\theta)+2+2d\end{bmatrix},$$

$\theta\in[0,2\pi]$. If the minimum value of det(A) is 8, then a value of d is ______.

(a) –5 (b) –7

(c) $2(\sqrt{2}+1)$ (d) $2(\sqrt{2}+2)$

96. If

$$A=\begin{bmatrix} e^t & e^{-t}\cos t & e^{-t}\sin t \\ e^t & -e^{-t}\cos t-e^{-t}\sin t & -e^{-t}\sin t+e^{-t}\cos t \\ e^t & 2e^{-t}\sin t & -2e^{-t}\cos t \end{bmatrix}$$

then A is

(a) invertible for all $t \in R$.

(b) invertible for only if $t=\pi$.

(c) not invertible for any $t \in R$.

(d) invertible only if $t=\dfrac{\pi}{2}$.

97. If $B=\begin{bmatrix} 5 & 2\alpha & 1 \\ 0 & 2 & 1 \\ \alpha & 3 & -1 \end{bmatrix}$ is the inverse of a 3×3 matrix A, then the sum of all values of α for which det(A) + 1 = 0, is ______.

(a) 0 (b) –1

(c) 1 (d) 2

98. If $\begin{bmatrix} 1 & 1 \\ 0 & 1 \end{bmatrix}\cdot\begin{bmatrix} 1 & 2 \\ 0 & 1 \end{bmatrix}\cdot\begin{bmatrix} 1 & 3 \\ 0 & 1 \end{bmatrix}\ldots\ldots$

$\begin{bmatrix} 1 & n-1 \\ 0 & 1 \end{bmatrix}=\begin{bmatrix} 1 & 78 \\ 0 & 1 \end{bmatrix}$, then the inverse of $\begin{bmatrix} 1 & n \\ 0 & 1 \end{bmatrix}$ is ______.

(a) $\begin{bmatrix} 1 & 0 \\ 0 & 1 \end{bmatrix}$ (b) $\begin{bmatrix} 1 & -13 \\ 0 & 1 \end{bmatrix}$

(c) $\begin{bmatrix} 1 & -12 \\ 0 & 1 \end{bmatrix}$ (d) $\begin{bmatrix} 1 & 0 \\ 13 & 1 \end{bmatrix}$

99. Let A and B be two invertible matrices of order 3×3. If $\det(ABA^T)=8$ and $\det(AB^{-1})=8$, then $\det(BA^{-1}B^T)$ is equal to ______.

(a) $\dfrac{1}{4}$ (b) 1

(c) $\dfrac{1}{16}$ (d) 16

100. If the system of linear equations

$x+y+z=5$

$x+2y+2z=6$

$x+3y+\lambda z=\mu$, $(1, \mu, IR)$, has infinitely many solutions, then the value of $\lambda+\mu$ is ______.

(a) 12 (b) 9

(c) 7 (d) 10

101. If the system of equation $2x+3y-z=0$, $x+ky-2z=0$ and $2x-y+z=0$ has a non-trivial solution (x, y, z), then $\dfrac{x}{y}+\dfrac{y}{z}+\dfrac{z}{x}+k$ is equal to ______.

(a) $\dfrac{3}{4}$ (b) $\dfrac{1}{2}$

(c) $-\dfrac{1}{4}$ (d) –4

102. The set of all values of λ for which the system of linear equations

$x-2y-2z=\lambda x$

$x+2y+z=\lambda y$

$-x-y=\lambda$

has a non-trivial solution, which

(a) is a singleton.

(b) contains exactly two elements.

(c) is an empty set.

(d) contains more than two elements.

103. The number of values of $\theta\in(0,\pi)$ for which the system of linear equations

$x+3y+7z=0$

$-x+4y+7z=0$

$(\sin 3\theta)x+(\cos 2\theta)y+2z=0$

has a non-trivial solution, is ______.

(a) three (b) two

(c) four (d) one

104. If the system of linear equations

$x+ky+3z=0$

$3x+ky-2z=0$

$2x+4y-3z=0$ has a non-zero solution (x, y, z), then $\dfrac{xz}{y^2}$ is equal to ______.

(a) 10 (b) −30

(c) 30 (d) −10

105. If $\begin{vmatrix} x-4 & 2x & 2x \\ 2x & x-4 & 2x \\ 2x & 2x & x-4 \end{vmatrix}=(A+B)(x-A)^2$, then the ordered pair (A, B) is equal to ______.

(a) (–4, 3) (b) (–4, 5)

(c) (4, 5) (d) (–4, –5)

106. Suppose A is 3×3 non-singular matrix and $(A-3I)(A-5I)=0$, where $I=I_3$ and $0=0_3$. If $\alpha A+\beta A^{-1}=4I$, then $\alpha+\beta$ is equal to ______.

(a) 8 (b) 12

(c) 13 (d) 7

107. Let A be any 3×3 invertible matrix. Then which one of the following is not always true?

(a) $\text{adj}(A)=|A|\cdot A^{-1}$

(b) $\text{adj}(\text{adj}(A))=|A|\cdot A$

(c) $\text{adj}(\text{adj}(A))=|A|^2\cdot(\text{adj}(A))^{-1}$

(d) $\text{adj}(\text{adj}(A))=|A|\cdot(\text{adj}(A))^{-1}$

108. If $S=\left\{x\in[0,2\pi]:\begin{vmatrix}0 & \cos x & -\sin x\\ \sin x & 0 & \cos x\\ \cos x & \sin x & 0\end{vmatrix}=0\right\}$,

then $\sum_{x\in S}\tan\left(\frac{\pi}{3}+x\right)$ is equal to ______.

(a) $4+2\sqrt{3}$ (b) $-2+\sqrt{3}$

(c) $-2-\sqrt{3}$ (d) $-4-2\sqrt{3}$

109. If S is the set of distinct values of b for which the following system of linear equations

$x+y+z=1$

$x+ay+z=1$

$ax+by+z=0$ has no solution, then S is

(a) a singleton.

(b) an empty set.

(c) an infinite set.

(d) a finite set containing two or more elements.

110. Let k be an integer such that triangle with vertices $(k-3k)$, $(5, k)$ and $(-k, 2)$ has area 28 sq. units. Then the orthocentre of this triangle is at the point:

(a) $\left(2,\frac{1}{2}\right)$ (b) $\left(2,-\frac{1}{2}\right)$

(c) $\left(1,\frac{3}{4}\right)$ (d) $\left(1,-\frac{3}{4}\right)$

111. If $A=\begin{bmatrix}-4 & -1\\ 3 & 1\end{bmatrix}$, then the determinant of the matrix $\left(A^{2016}-2A^{2015}-A^{2014}\right)$ is ______.

(a) –175 (b) 2014

(c) 2016 (d) –25

112. If $A=\begin{bmatrix}5a & -b\\ 3 & 2\end{bmatrix}$ and $A\,\text{adj}\,A=A\,A^T$, then $5a+b$ is equal to ______.

(a) 4 (b) 13

(c) –1 (d) 5

113. If $\begin{vmatrix}x^2+x & x+1 & x-2\\ 2x^2+3x-1 & 3x & 3x-3\\ x^2+2x+3 & 2x-1 & 2x-1\end{vmatrix}=ax-12$, then a is equal to ______.

(a) 24 (b) –12

(c) –24 (d) 12

114. If $f(\theta)=\begin{vmatrix}1 & \cos\theta & 1\\ -\sin\theta & 1 & -\cos\theta\\ -1 & \sin\theta & 1\end{vmatrix}$ and A and B are respectively the maximum and the minimum values of $f(\theta)$, then (A, B) is equal to ______.

(a) (3, –1) (b) $\left(4,2-\sqrt{2}\right)$

(c) $\left(2+\sqrt{2},2-\sqrt{2}\right)$ (d) $\left(2+\sqrt{2},-1\right)$

115. If A is a 3×3 non-singular matrix such that $AA'=A'A$ and $B=A^{-1}A'$, then BB' equals:

(a) B^{-1} (b) $\left(B^{-1}\right)'$

(c) $I+B$ (d) I

116. If $\Delta_r=\begin{vmatrix}r & 2r-1 & 3r-2\\ \frac{n}{2} & n-1 & a\\ \frac{1}{2}n(n-1) & (n-1)^2 & \frac{1}{2}(n-1)(3n-4)\end{vmatrix}$

then the value of $\sum_{r=1}^{n-1}\Delta_r$

(a) depends only on a.

(b) depends only on n.

(c) depends both on a and n.

(d) is independent of both a and n.

117. If $P=\begin{vmatrix}1 & \alpha & 3\\ 1 & 3 & 3\\ 2 & 4 & 4\end{vmatrix}$ is the adjoint of a 3×3 matrix A and $|A|=4$, then α is equal to ______.

(a) 4 (b) 11

(c) 5 (d) 0

118. The number of values of k, for which the system of equations

$(k+1)\,x+8y=4k$

$kx+(k+3)y=3k-1$ has no solution, is ______.

(a) infinite (b) 1

(c) 2 (d) 3

119. Let $P=\left[a_{ij}\right]$ be a 3×3 matrix and let $Q=\left[b_{ij}\right]$, where $b_{ij}=2^{i+j}a_{ij}$ for $1\le i,\ j\le3$. If the determinant of P is 2, then the determinant of the matrix Q is ______.

(a) 2^{10} (b) 2^{11}
(c) 2^{12} (d) 2^{13}

120. Let $\omega\neq1$ be a cube root of unity and S be the set of all non-singular matrices of the form $\begin{bmatrix}1 & a & b\\ \omega & 1 & c\\ \omega^2 & \omega & 1\end{bmatrix}$ where each of a,b and c is either ω or ω^2. Then the number of distinct matrices in the set S is ______.

(a) 2 (b) 6
(c) 4 (d) 8

121. The number of 3×3 matrices A whose entries are either 0 or 1 and for which the system $A\begin{bmatrix}x\\y\\z\end{bmatrix}=\begin{bmatrix}1\\0\\0\end{bmatrix}$ has exactly two distinct solutions, is ______.

(a) 0 (b) 2^9-1
(c) 168 (d) 2

Integer and Numerical Value Type

1. If $a_1, a_2, a_3, 5, 4, a_6, a_7, a_8, a_9$ are in HP, and $D=\begin{vmatrix}a_1 & a_2 & a_3\\ 5 & 4 & a_6\\ a_7 & a_8 & a_9\end{vmatrix}$, then the value of $[D]$ is (where [.] represents the greatest integer function) ______.

2. If

$$\begin{vmatrix}(\beta+\gamma-\alpha-\delta)^4 & (\beta+\gamma-\alpha-\delta)^2 & 1\\ (\gamma+\alpha-\beta-\delta)^4 & (\gamma+\alpha-\beta-\delta)^2 & 1\\ (\alpha-\beta-\gamma-\delta)^4 & (\alpha-\beta-\gamma-\delta)^2 & 1\end{vmatrix}$$

$=-k(\alpha-\beta)(\alpha-\gamma)(\alpha-\delta)(\beta-\gamma)(\beta-\delta)(\gamma-\delta)$, then the value of k is ______.

3. If

$$f(x)=\begin{vmatrix}1 & x & x+1\\ 2x & x(x+1) & (x+1)x\\ 3x(x-1) & x(x-1)(x-2) & (x+1)x(x-1)\end{vmatrix},$$

then the value of $f(500)$ ______.

4. Let $D_1=\begin{vmatrix}a & b & a+b\\ c & d & c+d\\ a & b & a-b\end{vmatrix}$ and $D_2=\begin{vmatrix}a & c & a+c\\ b & d & b+d\\ a & c & a+b+c\end{vmatrix}$ then the value of $\left|\dfrac{D_1}{D_2}\right|$, where $b\neq0$ and $ad\neq bc$, is ______.

5. If $a_1, a_2, a_3, \ldots, a_{12}$ are in AP and

$$\Delta_1=\begin{vmatrix}a_1a_5 & a_1 & a_2\\ a_2a_6 & a_2 & a_3\\ a_3a_7 & a_3 & a_4\end{vmatrix}\quad \Delta_2=\begin{vmatrix}a_2a_{10} & a_2 & a_3\\ a_3a_{11} & a_3 & a_4\\ a_3a_{12} & a_4 & a_5\end{vmatrix}$$

then $\dfrac{\Delta_2}{\Delta_1}=$ ______.

6. Let $a, b, c\in R$ not all are equal and

$$\Delta_1=\begin{vmatrix}a & b & c\\ b & c & a\\ c & a & b\end{vmatrix},$$

$$\Delta_2=\begin{vmatrix}a+2b & b+3c & c+4a\\ b+2c & c+3a & a+4b\\ c+2a & a+3b & b+4c\end{vmatrix}$$ then $\dfrac{\Delta_2}{\Delta_1}=$ ______.

7. If $\Delta_r=\begin{vmatrix}r & 612 & 915\\ 101r^2 & 2r & 3r\\ r & \dfrac{1}{r} & \dfrac{1}{r^2}\end{vmatrix}$, then the value of $\lim\limits_{n\to\infty}\dfrac{1}{n^3}\sum\limits_{r=1}^{n}\Delta_r$ is ______.

8. If $x=31$, $y=32$ and $z=33$, then the value of

$$\begin{vmatrix}(x^2+1)^2 & (xy+1)^2 & (xz+1)^2\\ (xy+1)^2 & (y^2+1)^2 & (yz+1)^2\\ (xz+1)^2 & (yz+1)^2 & (z^2+1)^2\end{vmatrix}$$ is ______.

9. Let α,β,γ are the real roots of the equation $x^3+ax^2+bx+c=0\,(a, b, c\in R$ and $a\neq0)$. If the system of equations (in u, v and w) given by

$\alpha u+\beta v+\gamma w=0$
$\beta u+\gamma v+\alpha w=0$
$\gamma u+\alpha v+\beta w=0$

has non-trivial solutions, then the value of a^2/b is ______.

10. The value of $|\alpha|$ for which the system of equations

$\alpha x+y+z=\alpha-1$
$x+\alpha y+z=\alpha-1$
$x+y+\alpha z=\alpha-1$
has no solution, is ______.

11. Number of values of θ lying in $[0,100\pi]$ for which the system equations, $(\sin 3\theta)x-y+z=0$; $(\cos 2\theta)x+4y+3z=0$; $2x+7y+7z=0$ has non-trivial solution is ______.

Questions from Previous Years

12. Let

$$M=\left\{A=\begin{pmatrix} a & b \\ c & d\end{pmatrix}: a,b,c,d\in\{\pm 3,\pm 2,\pm 1,0\}\right\}.$$

Define $f: M\to Z$, as $f(A)=\det(A)$, for all $A\in M$, where Z is a set of all integers. Then the number of $A\in M$ such that $f(A)=15$ is equal to ______.

13. If $A=\begin{bmatrix} 0 & -\tan\frac{\theta}{2} \\ \tan\frac{\theta}{2} & 0\end{bmatrix}$ and

$(I_2+A)(I_2-A)^{-1}=\begin{bmatrix} a & -b \\ b & a\end{bmatrix}$ then $13(a^2+b^2)$ is equal to ______.

14. Let $P-\begin{bmatrix} 3 & -1 & -2 \\ 2 & 0 & \alpha \\ 3 & -5 & 0\end{bmatrix}$, where $\alpha\in R$. Suppose $Q=[q_{ij}]$ is a matrix satisfying $PQ=kI_3$, for some non-zero $k\in R$. If $q_{23}=-\frac{k}{8}$, then α^2+k^2 is equal to ______.

15. Let A be a 3×3 real matrix. If det (2Adj (2 Adj(A 3×3 matrix dj(2A))) = 2^{41} , then the value of the det(A^2) equals ______.

16. Let $A\{a_{ij}\}$ be a 3×3 matrix, where

$$a_{ij}=\begin{cases} (-1)^{j-i}\, j-i & \text{if } i<j \\ 2 & \text{if } i=j, \text{ then} \\ (-1)^{i+j} & \text{if } i>j\end{cases}$$

$\det\left(3\mathrm{Adj}\left(2A^{-1}\right)\right)$ is equal to ______.

17. If $A=\begin{bmatrix} 2 & 3 \\ 0 & -1\end{bmatrix}$, then the value of $\det\left(A^4\right)+\det\left(a^{10}-(\mathrm{Adj}(2A))^{10}\right)$ is equal to ______.

18. For real numbers α and β, consider the following system of linear equations:
$x+y-z=2$, $x+2y+\alpha z=1$, $2x-y+z=\beta$
If the system has infinite solutions, then $\alpha+\beta$ is equal to ______.

19. If the system of equations
$kx+y+2z=1$
$3x-y-2z=2$
$-2x-2y-4z=3$ has infinitely many solutions, then k is equal to______.

20. The sum of distinct values of λ for which the system of equations
$(\lambda-1)x+(3\lambda+1)y+2\lambda z=0$
$(\lambda-1)x+(4\lambda-2)y+(\lambda+3)z=0$
$2x+(3\lambda+1)y+3(\lambda-1)z=0$
has non-zero solutions, is______.

21. If the system of equation $x-2y+3z=9$, $2x+y+z=b$, $x-7y+az=24$, has infinitely many solutions, then $a-b$ is equal to______.

22. Let S be the set of all integer solutions, (x, y, z), of the system of equations
$x-2y+5z=0$
$-2x+4y+z=0$
$-7x+14y+9z=0$ such that
$15\le x^2+y^2+z^2\le 150$. Then, the number of elements in the set S is equal to______.

23. If the system of linear equations,
$x+y+z=6$
$x+2y+3z=10$
$3x+2y+\lambda z=\mu$ has more than two solutions, then $\mu-\lambda^2$ is equal to ______.

24. Let k be a positive real number and

let $A=\begin{bmatrix} 2k-1 & 2\sqrt{k} & 2\sqrt{k} \\ 2\sqrt{k} & 1 & -2\sqrt{k} \\ -2\sqrt{k} & 2k & -1\end{bmatrix}$ and

$B=\begin{bmatrix} 0 & 2k-1 & \sqrt{k} \\ 1-2k & 0 & 2\sqrt{k} \\ -\sqrt{k} & -2\sqrt{k} & 0\end{bmatrix}$.

If $\det(\mathrm{adj}\,A)+\det(\mathrm{adj}\,B)=10^6$, then $[k]$ is equal to ______.

25. Let ω be the complex number $\cos\frac{2\pi}{3}+i\sin\frac{2\pi}{3}$. Then the number of distinct complex numbers

z satisfying $\begin{vmatrix} z+1 & \omega & \omega^2 \\ \omega & z+\omega^2 & 1 \\ \omega^2 & 1 & z+\omega \end{vmatrix} = 0$ is equal to ______.

Assertion and Reason

Select the correct answer.

(a) Both Statement 1 and Statement 2 are true and Statement 2 is correct explanation of Statement 1.

(b) Both Statement 1 and Statement 2 are true and Statement 2 is not correct explanation of Statement 1.

(c) Statement 1 is true, Statement 2 is false.

(d) Statement 1 is false, Statement 2 is true.

1. **Statement 1:** The value of determinant $A = \begin{vmatrix} 0 & 2 & 1 \\ -2 & 0 & 3 \\ -1 & -3 & 0 \end{vmatrix}$ is equal to 0.

   Statement 2: If A is skew-symmetric of odd order, then $|A|$ is a perfect square.

2. **Statement 1:** The value of $\begin{vmatrix} 1 & \omega^n & \omega^{2n} \\ \omega^{2n} & 1 & \omega^n \\ \omega^n & \omega^{2n} & 1 \end{vmatrix}$ is zero. ($\because$ where $n \in N$).

   Statement 2: $1+\omega+\omega^2 = 0$

3. **Statement 1:** If a, b, c are distinct and x, y, z are not all zero, given that $ax+by+cz=0$, $bx+cy+cy+az=0, cx+ay+bz=0$, then $a+b+c \neq 0$.

   Statement 2: $a^2+b^2+c^2 > ab+bc+ca$ if a, b, c are distinct.

4. **Statement 1:** If $f(x) = \begin{vmatrix} f_1(x) & f_2(x) & f_3(x) \\ a & b & c \\ d & e & f \end{vmatrix}$ where f_1, f_2, f_3 are differentiable functions and a, b, c, d, e, f are constants, then

   $$\int f(x)dx = \begin{vmatrix} \int f_1(x)dx & \int f_2(x)dx & \int f_2(x)dx \\ a & b & c \\ d & e & f \end{vmatrix} + k$$

   Statement 2: Integration of the sum of several functions is equal to the sum of integration of individual function.

5. **Statement 1:** If $\Delta = \begin{vmatrix} a & b & c \\ x & y & z \\ p & q & r \end{vmatrix}$, then $\begin{vmatrix} ka & kb & kc \\ kx & ky & kz \\ kp & kq & kr \end{vmatrix}$ is equal to $k^3\Delta$.

   Statement 2: If any row of a determinant is multiplied by r, then the value of the determinant is also multiplied by r.

6. Consider the system of equations

   $x-2y+3z=-1$
   $-x+y-2z=k$
   $x-3y+4z=1$

   Statement 1: The system of equations has no solution for $k \neq 3$.

   Statement 2: The determinant $\begin{vmatrix} 1 & 3 & -1 \\ -1 & -2 & k \\ 1 & 4 & 1 \end{vmatrix} \neq 0$, for $k \neq 3$.

JEE ADVANCED

Objective Questions

One Option Correct

1. The system of equations $x+ky+3z=0$, $3x+ky-2z=0$, $2x+3y-4z=0$ possess a non-trivial solution over the set of rationals, then $2k$ is an integral element of the interval ______.

   (a) [10, 20] (b) (20, 30)
   (c) [30, 40] (d) (40, 50)

2. If $p+q+r=a+b+c$, then the value of the determinant $\begin{vmatrix} pa & qb & rc \\ qc & ra & pb \\ rb & pc & qa \end{vmatrix}$ is ______.

   (a) 0 (b) $pa+qb+rc$
   (c) 1 (d) None of these.

3. If a determinant of order 3×3 is formed by using the numbers 1 or –1, then minimum value of the determinant is ______.

(a) –2 (b) –4
(c) 0 (d) –8

4. S_1: The value of the determinant

$$D=\begin{vmatrix} \ell nx & \ell ny & \ell nz \\ \ell n2x & \ell n2y & \ell n2z \\ \ell n3x & \ell n3y & \ell n3z \end{vmatrix} \text{ is } \ell n216xyz.$$

S_2: The roots of $\begin{vmatrix} x & a & b & 1 \\ \lambda & x & b & 1 \\ \lambda & \mu & x & 1 \\ \lambda & \mu & v & 1 \end{vmatrix}=0$ are independent of λ,μ,v,a,b.

S_3: If a, b, c are sides of a scalene triangle, then value of $\begin{vmatrix} a & b & c \\ b & c & a \\ c & a & b \end{vmatrix}$ is negative.

S_4: Let $f(x)=\begin{vmatrix} \frac{1}{x} & \ell n x & x^n \\ 1 & -\frac{1}{n} & (-1)^n \\ 1 & a & a^2 \end{vmatrix}$, if $f^n(x)$ is the nth derivative of $f(x)$ then $f^n(1)$ is independent of a.

(a) FFTT
(b) FTTT
(c) FFFT
(d) TTTT

Question from Previous Year

5. Let $M=\begin{bmatrix} \sin^4\theta & -1-\sin^2\theta \\ 1+\cos^2\theta & \cos^4\theta \end{bmatrix}=\alpha I+\beta M^{-1}$, where $\alpha=\alpha(\theta)$ and $\beta=\beta(\theta)$ are real numbers, and I is the 2×2 identity matrix. If $a*$ is the minimum of the set $\{\alpha(\theta)\text{: }\theta\in[0, 2\pi)\}$ and $\beta*$ is the minimum of the set $\{\beta(\theta)\text{: }\theta\in[0, 2\pi)\}$. Then the value of $a^* + b^*$ is ______.

(a) $-\frac{31}{16}$ (b) $-\frac{17}{16}$

(c) $-\frac{37}{16}$ (d) $-\frac{29}{16}$

One or More Than One Option Correct

1. The system of equation

$x+3y+2z=6$
$x+\lambda y+2z=7$
$x+3y+2z=\mu$ has

(a) a unique solution if $\lambda=2, \mu\neq6$.
(b) infinitely many solutions if $\lambda=4, \mu=6$.
(c) no solution if $\lambda=5, \mu=7$.
(d) no solution if $\lambda=3, \mu=5$.

2. Let M and N be two 3×3 non-singular skew-symmetric matrices such that MN = NM. If P^T denotes the transpose of P, then $M^2N^2(M^TN^{-1})(MN^{-1})$ is equal to ______.

(a) M^2 (b) $-N^2$
(c) $-M^2$ (d) MN

Questions from Previous Years

3. For any 3×3 matrix M, let $|M|$ denote the determinant of M. $E=\begin{bmatrix} 1 & 2 & 3 \\ 2 & 3 & 4 \\ 8 & 13 & 18 \end{bmatrix}$, $P=\begin{bmatrix} 1 & 0 & 0 \\ 0 & 0 & 1 \\ 0 & 1 & 0 \end{bmatrix}$ and $F=\begin{bmatrix} 1 & 3 & 2 \\ 8 & 18 & 13 \\ 2 & 4 & 3 \end{bmatrix}$. If Q is a nonsingular matrix of order 3×3, then which of the following statements is (are) true?

(a) F = PEP and $P^2\begin{bmatrix} 1 & 0 & 0 \\ 0 & 1 & 0 \\ 0 & 0 & 1 \end{bmatrix}$
(b) $|EQ+PFQ^{-1}|=|EQ|+|PFQ^{-1}|$
(c) $|(EF)^3|>|EF|^2$
(d) Sum of the diagonal entries of $P^{-1}EP+F$ is equal to the sum of diagonal entries of $E+P^{-1}FP$

4. For any 3×3 matrix M, let $|M|$ denote the determinant of M. Let I be the 3×3 identity matrix. Let E and F be two 3×3 matrices such that $(I-EF)$ is invertible. If $G=(I-EF)^{-1}$, then which of the following statements is (are) true?

(a) $|FE|=|I-FE||FGE|$
(b) $(I-FE)(I+FGE)=I$
(c) $EFG=GEF$
(d) $(I-FE)(I-FGE)=I$

5. Let $P_1 = I\begin{bmatrix}1&0&0\\0&1&0\\0&0&1\end{bmatrix}, P_2 = \begin{bmatrix}1&0&0\\0&0&1\\0&1&1\end{bmatrix},$

$P_3 = \begin{bmatrix}0&1&0\\1&0&0\\0&0&1\end{bmatrix}, P_4 = \begin{bmatrix}0&1&0\\0&0&1\\1&0&0\end{bmatrix}, P_5 = \begin{bmatrix}0&0&1\\1&0&0\\0&1&0\end{bmatrix},$

$P_6 = \begin{bmatrix}0&0&1\\0&1&0\\1&0&0\end{bmatrix}$ and $x = \sum_{k=1}^{6} P_k \begin{bmatrix}2&1&3\\1&0&2\\3&2&1\end{bmatrix} P_k^T,$

where P_k^T denotes the transpose of the matrix P_k. Then which of the following options is/are correct?

(a) X is a symmetric matrix.
(b) The sum of diagonal entries of X is 18.
(c) $X - 30I$ is an invertible matrix.
(d) If $X = \begin{bmatrix}1\\1\\1\end{bmatrix} = \alpha\begin{bmatrix}1\\1\\1\end{bmatrix}$, then $a = 30$.

6. Let M be a 3×3 invertible matrix with real entries and let I denote the 3×3 identity matrix. If $M^{-1} = \text{adj}(\text{adj}M)$, then which of the following statements is/are always true?

(a) $M = I$ (b) $\det M = I$
(c) $M^2 = I$ (d) $(adjM)^2 = I$

7. Let $M = \begin{bmatrix}0&1&a\\1&2&3\\3&b&1\end{bmatrix}$ and $(\text{adj } M) = \begin{bmatrix}-1&1&-1\\8&-6&2\\-5&3&-1\end{bmatrix}$

where a and b are real numbers. Which of the following options is/are correct?

(a) $a + b = 3$
(b) $\det(\text{adj } M^2) = 81$
(c) $(\text{adj } M)^{-1} + \text{adj}M^{-1} = {}^-M$
(d) If $M\begin{bmatrix}\alpha\\\beta\\\gamma\end{bmatrix} = \begin{bmatrix}1\\2\\3\end{bmatrix}$, then $\alpha - \beta + \gamma = 3$

8. Let $x \in R$ and let $P = \begin{vmatrix}1&1&1\\0&2&2\\0&0&3\end{vmatrix}, Q = \begin{vmatrix}2&x&x\\0&4&0\\x&x&6\end{vmatrix}$

and $R = PQP^{-1}$. Then which of the following options is/are correct?

(a) $\det R = \det\begin{bmatrix}2&x&x\\0&4&0\\x&x&5\end{bmatrix} + 8$

(b) For $x = 1$, there exists a unit vector $\alpha\hat{i} + \beta\hat{j} + \gamma\hat{k}$ for which $R\begin{bmatrix}\alpha\\\beta\\\gamma\end{bmatrix} = \begin{bmatrix}0\\0\\0\end{bmatrix}$

(c) There exists a real number x such that PQ = QP.

(d) For $x = 0$, if $R = \begin{bmatrix}1\\a\\b\end{bmatrix} = 6\begin{bmatrix}1\\a\\b\end{bmatrix}$, then $a + b = 5$

9. Let S be the set of all column matrices $\begin{bmatrix}b_1\\b_2\\b_3\end{bmatrix}$ such that $b_1, b_2, b_3, \in R$ and the system of equations (in real variables)

$-x + 2y + 5z = b_1$

$2x - 4y + 3z = b_2$

$x - 2y + 2z = b_3$

has at least one solution. Then, which of the following system(s) (in real variables) has (have) at least one solution for each $\begin{bmatrix}b_1\\b_2\\b_3\end{bmatrix} \in S$?

(a) $x + 2y + 3z = b_1, 4y + 5z = b_2$ and $x + 2y + 6z = b_3$
(b) $x + y + 3z = b_1, 5x + 2y + 6z = b_2$ and $-2x - y - 3z = b_3$
(c) $-x + 2y - 5z = b_1, 2x - 4y + 10z = b_2$ and $x - 2y + 5z = b_3$
(d) $sx + 2y + 5z = b_1, 2x + 3z = b_2$ and $x + 4y - 5z = b_3$

10. Let $a, \lambda, \mu, \in \mathbb{R}$. Consider the system of linear equations:

$ax + 2y = \lambda$

$3x - 2y = \mu$

Which of the following statement(s) is (are) correct?

(a) If $a = -3$, then the system has infinitely many solutions for all values of λ and μ.
(b) If $a \neq -3$, then the system has a unique solution for all values of λ and μ
(c) If $\lambda + \mu = 0$, then the system has infinitely many solutions for $a = -3$.
(d) If $\lambda + \mu \neq 0$, then the system has no solution for $a = -3$.

11. Which of the following values of α satisfy the equation

$$\begin{vmatrix} (1+\alpha)^2 & (1+2\alpha)^2 & (1+3\alpha)^2 \\ (2+\alpha)^2 & (2+2\alpha)^2 & (2+3\alpha)^2 \\ (3+\alpha)^2 & (3+2\alpha)^2 & (3+3\alpha)^2 \end{vmatrix} = -64\alpha ?$$

(a) –4 (b) 9
(c) –9 (d) 4

12. Let M and N be two 3×3 matrices such that $MN = NM$. Further, if $M \neq N^2$ and $M^2 = N^4$, then,

(a) determinant of $\left(M^2 + MN^2\right)$ is 0.

(b) There is a 3×3 non-matrix U such that $\left(M^2 + MN^2\right)U$ is the zero matrix.

(c) Determinant of $\left(M^2 + MN^2\right) \geq 1$

(d) For 3×3 matrix U, if $\left(M^2 + MN^2\right)U$ equals the zero matrix then U is the zero matrix.

13. For 3×3 matrices M and N, which of the following statement(s) is (are) NOT correct?

(a) $N^T MN$ is symmetric or skew-symmetric, according as M is symmetric or skew-symmetric.

(b) MN – NM is skew-symmetric for all symmetric matrices M and N.

(c) MN is symmetric for all symmetric matrices M and N.

(d) (adj M) (adj N) = adj (MN) for all invertible matrices M and N.

Matrix Type

1. Match the following:

Column I	Column II
(A) Let $\lvert A\rvert = \lvert a_{ij}\rvert_{3\times3} \neq 0$. Each element a_{ij} is multiplied by k^{i-j}. Let |B| the resulting determinant, where $k_1\lvert A\rvert + k_2\lvert B\rvert = 0$. Then $k_1 + k_2 =$	(P) 0
(B) The maximum value of a third-order determinant when each of its entries are ±1 equals	(Q) 4
(C) $\begin{vmatrix} 1 & \cos\alpha & \cos\beta \\ \cos\alpha & 1 & \cos\gamma \\ \cos\beta & \cos\gamma & 1 \end{vmatrix} = \begin{vmatrix} 0 & \cos\alpha & \cos\beta \\ \cos\alpha & 0 & \cos\gamma \\ \cos\beta & \cos\gamma & 0 \end{vmatrix}$ if $\cos^2\alpha + \cos^2\beta + \cos^2\gamma =$	(R) 1
(D) $\begin{vmatrix} x^2+x & x+1 & x-2 \\ 2x^2+3x-1 & 3x & 3x-3 \\ x^2+2x+3 & 2x-1 & 2x-1 \end{vmatrix} = Ax + B$ where A and B are determinants of order 3. Then $A + 2B =$	(S) 2
	(T) $\begin{vmatrix} 1 & 2 \\ 2 & 4 \end{vmatrix}$

Integer and Numerical Value Type

1. Find the coefficient of x in the determinant

$$\begin{vmatrix} (1+x)^{a_1b_1} & (1+x)^{a_1b_2} & (1+x)^{a_1b_3} \\ (1+x)^{a_2b_1} & (1+x)^{a_2b_2} & (1+x)^{a_2b_3} \\ (1+x)^{a_3b_1} & (1+x)^{a_3b_2} & (1+x)^{a_3b_3} \end{vmatrix},$$

where $a_i, b_j \in N$

2. If $f(x) = \begin{vmatrix} \cos(x+\alpha) & \cos(x+\beta) & \cos(x+\gamma) \\ \sin(x+\alpha) & \sin(x+\beta) & \sin(x+\gamma) \\ \sin(\beta-\gamma) & \sin(\gamma-\alpha) & \sin(\alpha-\beta) \end{vmatrix}$

and $f(2) = 6$, then find $\frac{1}{5}\sum_{r=1}^{25} f(r)$,

3. Let $f(x)=\begin{vmatrix} x & 1 & 1 \\ \sin 2\pi x & 2x^2 & 1 \\ x^3 & 3x^4 & 1 \end{vmatrix}$. If $f(x)$ be an odd function and its odd values are equal to $g(x)$, and then find the value of λ. If $f(1)\, g(1) = -4\lambda$

4. If $f(x)$ satisfies the equation
$\begin{vmatrix} f(x+1) & f(x+8) & f(x+1) \\ 1 & 2 & -5 \\ 2 & 3 & \lambda \end{vmatrix} = 0$ for all real x
and f is periodic with period 7, then find the value of $|\lambda|$.

Questions from Previous Years

5. The trace of a square matrix is defined to be the sum of its diagonal entries. If A is a 2×2 matrix such that the trace of A is 3 and the trace of A^3 is -18, then the value of the determinant of A is ______.

6. Let P be a matrix of order 3×3 such that all the entries in P are from the set $\{-1, 0, 1\}$. Then the maximum possible value of the determinant of P is ______.

7. For a real number α, if the system $\begin{bmatrix} 1 & \alpha & \alpha^2 \\ \alpha & 1 & \alpha \\ \alpha^2 & \alpha & 1 \end{bmatrix}\begin{bmatrix} x \\ y \\ z \end{bmatrix} = \begin{bmatrix} 1 \\ -1 \\ 1 \end{bmatrix}$ of linear equations, has infinitely many solutions, then $1+\alpha+a^2 =$ ______.

8. Let $z = \dfrac{-1+\sqrt{3}i}{2}$, where $i=\sqrt{-1}$, and $r, s \in \{1,2,3\}$.
Let $P = \begin{bmatrix} (-z)^r & z^{2s} \\ z^{2s} & z^r \end{bmatrix}$ and I be identity matrix of order 2. Then the total number of ordered pairs (r, s) for which $P^2 = -I$ is ______.

Comprehension Passage

Comprehension 1

Consider the determinant

$$\Delta = \begin{vmatrix} a_1 & a_2 & a_3 \\ b_1 & b_2 & b_3 \\ d_1 & d_2 & d_3 \end{vmatrix}$$

M_1 = Minor of the element of ith row and jth column

C_1 = Cofactor of the element of ith row and jth column

1. Value of $b_1\cdot C_{31} + b_2\cdot C_{32} + b_3\cdot C_{33}$ is
(a) 0 (b) Δ
(c) 2Δ (d) Δ^2

2. If all the elements of the determinant are multiplied by 2, then the value of new determinant is ______.
(a) 0 (b) 8Δ
(c) 2Δ (d) $2^9\cdot\Delta$

3. $a_3 M_{13} - b_3\cdot M_{23} + d_3\cdot M_{33}$ is equal to ______.
(a) 0 (b) 4Δ
(c) 2Δ (d) Δ

Comprehension 2

Let p be an prime number and T_p be the following set of 2×2 matrices:

$$T_p = \left\{ A = \begin{bmatrix} a & b \\ c & a \end{bmatrix} : a,b,c \in \{0,1,2,\ldots,p-1\} \right\}$$

1. The number of A in T_p such that the trace of A is not divisible by p but det (A) is divisible by p is ______.
(a) $(p-1)^2$ (b) $2(p-1)$
(c) $(p-1)^2+1$ (d) $2p-1$

2. The number of A in T_p such that det (A) is not divisible by p is______.
(a) $2p^2$ (b) p^3-5p
(c) p^3-3p (d) p^3-p^2

Comprehension 3

Let A be the set of all 3×3 symmetric matrices all of whose entries are either 0 or 1. Five of these entries are 1 and four of them are 0.

1. The number of matrices in A is______.
(a) 12 (b) 6
(c) 9 (d) 3

2. The number of matrices A in A for which the system of linear equations
$A\begin{bmatrix} x \\ y \\ z \end{bmatrix} = \begin{bmatrix} 1 \\ 0 \\ 0 \end{bmatrix}$ has a unique solution, is
(a) less than 4.
(b) at least 4 but less than 7.
(c) at least 7 but less than 10.
(d) at least 10.

3. The number of matrices A in A for which the system of liner equations
$A\begin{bmatrix} x \\ y \\ z \end{bmatrix} = \begin{bmatrix} 1 \\ 0 \\ 0 \end{bmatrix}$ is inconsistent, is ______.

(a) 0 (b) more than 2
(c) 2 (d) 1

Question from Previous Year

Comprehension 4

Let α, β and γ be real numbers such that the system of linear equations

$x + 2y + 3z = \alpha$

$4x + 5y + 6z = \beta$

$7x + 8y + 9z = \gamma - 1$

is consistent. Let |M| represent the determinant of the matrix

$$M = \begin{bmatrix} \alpha & 2 & \gamma \\ \beta & 1 & 0 \\ -1 & 0 & 1 \end{bmatrix}$$

Let P be the plane containing all those (α, β, γ) for which the above system of linear equations is consistent, and D be the square of the distance of the point (0, 1, 0) from the plane P.

1. The value of |M| is_______.
2. The value of D is_______.

ANSWER KEY

JEE Main

Objective Questions

One Option Correct

1. (c) **2.** (d) **3.** (b) **4.** (d) **5.** (d) **6.** (b) **7.** (a) **8.** (b) **9.** (d)
10. (d) **11.** (b) **12.** (a) **13.** (b) **14.** (a) **15.** (d) **16.** (a) **17.** (d) **18.** (a)
19. (d) **20.** (a) **21.** (c) **22.** (c) **23.** (c) **24.** (b) **25.** (a) **26.** (b) **27.** (b)
28. (a) **29.** (d) **30.** (b) **31.** (a) **32.** (c) **33.** (a) **34.** (d) **35.** (c) **36.** (b)
37. (b) **38.** (c) **39.** (b) **40.** (b) **41.** (c) **42.** (c) **43.** (c) **44.** (c) **45.** (a)
46. (c) **47.** (b) **48.** (a) **49.** (d) **50.** (c) **51.** (b) **52.** (c) **53.** (b) **54.** (a)
55. (b) **56.** (b) **57.** (a) **58.** (a) **59.** (c) **60.** (b) **61.** (d) **62.** (d) **63.** (c)
64. (d) **65.** (c) **66.** (a) **67.** (c) **68.** (a) **69.** (d) **70.** (d) **71.** (c) **72.** (a)
73. (a) **74.** (b) **75.** (d) **76.** (d) **77.** (d) **78.** (c) **79.** (a) **80.** (c) **81.** (a)
82. (a) **83.** (a) **84.** (b) **85.** (b) **86.** (d) **87.** (d) **88.** (c) **89.** (d) **90.** (d)
91. (a) **92.** (a) **93.** (c) **94.** (d) **95.** (a) **96.** (a) **97.** (c) **98.** (b) **99.** (c)
100. (d) **101.** (b) **102.** (a) **103.** (b) **104.** (a) **105.** (b) **106.** (a) **107.** (b) **108.** (c)
109. (a) **110.** (a) **111.** (d) **112.** (d) **113.** (a) **114.** (c) **115.** (d) **116.** (d) **117.** (b)
118. (b) **119.** (d) **120.** (a) **121.** (a)

Integer and Numerical Value Type

1. (2) **2.** (−64) **3.** (0) **4.** (−2) **5.** ($-2d^4$) **6.** (25) **7.** (2)
8. $\{2(x-y)^2(y-z)^2(z-x)^2\}$ **9.** (3) **10.** (−2) **11.** (201) **12.** (16) **13.** (13) **14.** (−1)
15. (4) **16.** (108) **17.** (16) **18.** (5) **19.** (21) **20.** (0 and 3) **21.** (5) **22.** (8)
23. (13) **24.** (4) **25.** (0)

Assertion and Reason

1. (c) **2.** (a) **3.** (d) **4.** (a) **5.** (a) **6.** (a)

JEE Advanced

Objective Questions

One Option Correct

1. (c) **2.** (a) **3.** (b) **4.** (a) **5.** (d)

One or More Than One Option Correct

1. (b, c, d) **2.** (c) **3.** (a, b, d) **4.** (a, b, c) **5.** (a, b, d) **6.** (b, c, d)
7. (a, c, d) **8.** (a, d) **9.** (a, d) **10.** (b, c, d) **11.** (b, c) **12.** (a, b) **13.** (c, d)

Matrix Type

1. A → P, T; B → Q; C → R; D → P, T

Integer and Numerical Value

1. (0) **2.** (150) **3.** (–1) **4.** (4) **5.** (5) **6.** (4) **7.** (1) **8.** (Only one pair)

Comprehension Passage

Comprehension 1

1. (a) **2.** (b) **3.** (d)

Comprehension 2

1. (d) **2.** (d)

Comprehension 3

1. (a) **2.** (b) **3.** (b)

Comprehension 4

1. 1.00 **2.** 1.50

HINTS AND EXPLANATIONS

Worksheet

1. $f(x)=\begin{vmatrix} x^3 & \sin x & \cos x \\ 6 & -1 & 0 \\ p & p^2 & p^3 \end{vmatrix}$

$f'''(x)=\begin{vmatrix} 6 & -\cos x & \sin x \\ 6 & -1 & 0 \\ 6 & p^2 & p^3 \end{vmatrix}$

$f'''(0)=\begin{vmatrix} 6 & -1 & 0 \\ 6 & -1 & 0 \\ p & p^2 & p^3 \end{vmatrix}=0$

2. $f(x)$

$=\begin{vmatrix} 1 & x & x+1 \\ 2x & x(x-1) & x(x+1) \\ 3x(x-1) & x(x-1)(x-2) & (x+1)x(x-1) \end{vmatrix}$

Applying $C_3 \to C_3 - C_2$, we get $f(x)=0$
$\Rightarrow f(100)=0$

3. On applying $C_1+C_2+C_3$

$f(x)=\begin{vmatrix} 1+2x+\lambda x & (1+b^2)x & (1+c^2)x \\ 1+2x+\lambda x & 1+b^2x & (1+c^2)x \\ 1+2x+\lambda x & (1+b^2)x & 1+c^2x \end{vmatrix}$

Where $\lambda=a^2+b^2+c^2$
But $a^2+b^2+c^2=-2$

$\Rightarrow f(x)=\begin{vmatrix} 1 & (1+b^2)x & (1+c^2)x \\ 1 & 1+b^2x & (1+c^2)x \\ 1 & (1+b^2)x & 1+c^2x \end{vmatrix}=(1-x)^2$

On applying R_2-R_1, R_3-R_1

4. This is the case of non-trivial solution and system of equations has infinitely many solutions.

5. Number of determinants corresponding to 1st column = 2
Number of determinants corresponding to 2nd column = 3
Number of determinants corresponding to 3rd column = 4
Total no. = 2×3×4 = 24
from multiplication theorem.

6. Let $\Delta=\begin{vmatrix} 3 & 2 & 1 & 4 \\ 15 & 29 & 2 & 14 \\ 16 & 19 & 3 & 17 \\ 33 & 39 & 8 & 38 \end{vmatrix}$

Applying $C_1-3C_2, C_2-2C_3, C_4-4C_3$

$$\Delta=\begin{vmatrix}0&0&1&0\\9&25&2&6\\7&13&3&5\\9&23&8&6\end{vmatrix}$$

Expanding the above determinant by C_3, we get,

$$\Delta=\begin{vmatrix}9&25&6\\7&13&5\\9&23&6\end{vmatrix}$$

Again applying C_1-C_3, C_2-4C_3, we have,

$$\Delta=\begin{vmatrix}3&1&6\\2&-7&5\\3&-1&6\end{vmatrix}$$

Expanding $\Delta=-2(12-15)=6$

7. Let $\Delta=\begin{vmatrix}0&x&y&z\\-x&0&c&b\\-y&-c&0&a\\-z&-b&-a&0\end{vmatrix}$

Multiplying the second column by a and dividing the determinant by a.

$$\Delta=\frac{1}{a}\begin{vmatrix}0&ax&y&z\\-x&0&c&b\\-y&-ac&0&a\\-z&-ab&-a&0\end{vmatrix}$$

Applying $C_2+cC_4+bC_3$, we get,

$$\Delta=\frac{1}{a}\begin{vmatrix}0&ax+by+cz&y&z\\-x&0&c&b\\-y&0&0&a\\-z&0&-a&0\end{vmatrix}$$

Expanding the above determinant in terms of 2nd column, we have

$$\Delta=\frac{ax-by+cz}{a}\begin{vmatrix}-x&b&a\\-y&0&a\\-z&-a&0\end{vmatrix}$$

Taking –1 as common factor from 1st column

$$\Delta=\frac{ax-by+cz}{a^2}\begin{vmatrix}x&c&b\\y&0&a\\z&-a&0\end{vmatrix}$$

$$\Delta=\frac{ax-by+cz}{a^2}\begin{vmatrix}ax&ac&ab\\y&0&a\\z&-a&0\end{vmatrix}$$

Multiplying R_1 by a and dividing the determinant by a

Applying $R_1+cR_3-bR_2$, we have,

$$\Delta=\frac{ax-by+cz}{a^2}\begin{vmatrix}ax-by+cz&0&0\\y&0&a\\z&-a&0\end{vmatrix}$$

Expanding $\Delta=\dfrac{ax-by+cz}{a^2}\left[(ax-by+cz).a^2\right]$

$$=(ax-by+cz)^2$$

8. Denote the LHS determinant by Δ.

Then $\Delta=\begin{vmatrix}1&\omega&\omega^2&\omega^3\\\omega&\omega^2&\omega^3&1\\\omega^2&\omega^3&1&\omega\\\omega^3&1&\omega^2&\omega^2\end{vmatrix}$

$\because \omega$ is a cube root of 1.

$\therefore$ Replacing ω^3 by 1,

Then

$$\Delta^2=\begin{vmatrix}1&\omega&\omega^2&1\\\omega&\omega^2&1&1\\\omega^2&1&1&\omega\\1&1&\omega&\omega^2\end{vmatrix}\times\begin{vmatrix}1&\omega&\omega^2&1\\\omega&\omega^2&1&1\\\omega^2&1&1&\omega\\1&1&\omega&\omega^2\end{vmatrix}$$

By actual multiplication, we get

$$\Delta^2=\begin{vmatrix}1+\omega^2+\omega^4+1&\omega+\omega^3+\omega^2+1&\omega^2+\omega+\omega^2+\omega&1+\omega+\omega^2+\omega^2\\\omega+\omega^3+\omega^2+1&\omega^2+\omega^4+1+1&\omega^3+\omega^2+1+\omega&\omega+\omega^2+\omega+\omega^2\\\omega^2+\omega+\omega^2+\omega&\omega^3+\omega^2+1+\omega&\omega^4+1+1+\omega^2&\omega^2+1+\omega+\omega^2\\1+\omega+\omega^3+\omega^2&\omega+\omega^2+\omega+\omega^2&\omega^2+1+\omega+\omega^3&1+1+\omega^2+\omega^4\end{vmatrix}$$

$(\because 1+\omega+\omega^2=0, \omega^4=\omega, \omega+\omega^2=-1)$

Then $\Delta^2=\begin{vmatrix}1&1&-2&1\\1&1&1&-2\\-2&1&1&1\\1&-2&1&1\end{vmatrix}$

9. The given equations are

$x+y+z-1=0$

$zx+by+cz-k=0$

$a^2x+b^2y+c^2z-k^2=0$

We have,

$$\frac{x}{\begin{vmatrix}1 & 1 & -1\\ b & c & -k\\ b^2 & c^2 & -k^2\end{vmatrix}}=\frac{-y}{\begin{vmatrix}1 & 1 & -1\\ a & c & -k\\ a^2 & c^2 & -k^2\end{vmatrix}}$$

$$=\frac{z}{\begin{vmatrix}1 & 1 & -1\\ a & b & -k\\ a^2 & b^2 & -k^2\end{vmatrix}}=\frac{-1}{\begin{vmatrix}1 & 1 & 1\\ a & b & c\\ a^2 & b^2 & c^2\end{vmatrix}}$$

$$\therefore x\begin{vmatrix}1 & 1 & 1\\ a & b & c\\ a^2 & b^2 & c^2\end{vmatrix}=\begin{vmatrix}1 & 1 & 1\\ k & b & c\\ k^2 & b^2 & c^2\end{vmatrix}$$

or $x(a-b)(b-c)(c-a)=(k-b)(b-c)(c-k)$

$$\therefore x=\frac{(k-b)(c-k)}{(a-b)(c-a)}$$

$$\text{Similarly, } y\begin{vmatrix}1 & 1 & 1\\ 1 & b & c\\ a^2 & b^2 & c^2\end{vmatrix}=-\begin{vmatrix}1 & 1 & 1\\ k & a & c\\ k^2 & a^2 & c^2\end{vmatrix}$$

$y(a-b)(b-c)(c-a)=-(k-a)(a-c)(c-k)$

$$\therefore y=\frac{(k-b)(c-k)}{(a-b)(b-c)}$$

Similarly for z.

10. Let $\Delta=\begin{vmatrix}a^2 & a^2-(b-c)^2 & bc\\ b^2 & b^2-(c-a)^2 & ca\\ c^2 & c^2-(a-b)^2 & ab\end{vmatrix}$

Applying $C_2-2C_3-2C_1$ we get

$$\Delta=\begin{vmatrix}a^2 & -a^2-b^2-c^2 & bc\\ b^2 & -b^2-c^2-a^2 & ca\\ c^2 & -c^2-a^2-b^2 & ab\end{vmatrix}$$

$$\Delta=-\left(a^2+b^2+c^2\right)\begin{vmatrix}a^2 & 1 & bc\\ b^2 & 1 & ca\\ c^2 & 1 & ab\end{vmatrix}$$

$$=\left(a^2+b^2+c^2\right)\begin{vmatrix}1 & a^2 & bc\\ 1 & b^2 & ca\\ 1 & c^2 & ab\end{vmatrix} \text{ by } (C_1 \leftrightarrow C_2)$$

Now clearly the factors of the above determinant are

$(b-c)(c-a)(a-b)(a+b+c)$

Thus

$\Delta=K\left(a^2+b^2+c^2\right)(b-c)(c-a)(a-b)(a+b+c)$

The value of k when found shall be 1.

$\therefore \Delta=\left(a^2+b^2+c^2\right)(b-c)(c-a)(a-b)(a+b+c)$

Quiz

1. True
2. False
3. True
4. True
5. True
6. False
7. True
8. Minor of an element a_{ij} of a determinant is the determinant obtained by deleting its ith row and jth column in which element a_{ij} lies. Minor of an element a_{ij} is denoted by M_{ij}.
9. Cofactor of an element a_{ij}, denoted by A_{ij} is defined by

   $A_{ij}=(+1)^{i+j}M_{ij}$, where M_{ij} is minor of a_{ij}.
10. Since 3 lies in the second row and third column its minor M_{23} is

    $M_{23}=\begin{vmatrix}1 & 7\\ 7 & 9\end{vmatrix}=9-49=-40$

    (Obtained by deleting R_2 and C_3 in Δ)

JEE Main

Objective Questions

One Option Correct

1. (c) Δ is a skew-symmetric determinant because $a_{ij}=-a_{ji}$ and the principal diagonal elements are equal to zero.

   So, odd order of skew-symmetric determinant is zero.

2. (d) $|3AB| = 3\times3\times3|A||B| = 27(-1)(3) = -81$

3. (b) We know that $\Delta^C = \Delta^{3-1} = \Delta^2$
$= (11)^2 = 121$

4. (d) Expand from R_1

5. (d) By operating $C_1 = C_1 + C_2$, we get

$$\begin{vmatrix} 1 & \cos^2 x & 1 \\ 1 & \sin^2 x & 1 \\ 2 & 12 & 2 \end{vmatrix} = 0$$

6. (b) As k is common from each row of $|kA|$ and there are n rows in A.

7. (a) We have, $|\text{adj } A| = |A|^{n-1} = d^{n-1}$

8. (b) Operate $C_1 = C_1 + C_2 + C_3$, then the value of the determinant = 0.

9. (d) Using $\log_b a = \dfrac{\log a}{\log b}$

$$\Delta = \frac{1}{\log x \log y \log z}\begin{vmatrix} \log x & \log y & \log z \\ \log x & 3\log y & \log z \\ \log x & \log y & 5\log z \end{vmatrix}$$

$$= \begin{vmatrix} 1 & 1 & 1 \\ 1 & 3 & 1 \\ 1 & 1 & 5 \end{vmatrix} = 8$$

10. (d) Since α, β, γ are roots of $x^3 + ax^2 + b = 0$
$\therefore \alpha+\beta+\gamma = -a, \alpha\beta+\beta\gamma+\gamma\alpha = 0, \ \alpha\beta\gamma = -b$

$$\text{Now,} \begin{vmatrix} \alpha & \beta & \gamma \\ \beta & \gamma & \alpha \\ \gamma & \alpha & \beta \end{vmatrix}$$

$$= -(\alpha+\beta+\gamma)(\alpha^2+\beta^2+\gamma^2-\alpha\beta-\beta\gamma-\gamma\alpha)$$

$$= -(\alpha+\beta+\gamma)\left[(\alpha+\beta+\gamma)^2 - 3(\alpha\beta+\beta\gamma+\gamma\alpha)\right]$$

$$= -(-a)(a^2-0) = a^3$$

11. (b) Applying $R_2 = R_2 - xR_1, R_3 = R_3 - xR_2$, we get

$$f(x) = \begin{vmatrix} a & -1 & 0 \\ 0 & a+x & -1 \\ 0 & 0 & a-x \end{vmatrix} = a(a+x)^2$$

$$\therefore f(2x) - f(x) = a(a+2x)^2 - a(a+x)^2$$
$$= ax(2a+3x)$$

12. (a) $\Delta = \begin{vmatrix} 1 & 2 & 6 \\ 2 & 6 & 24 \\ 6 & 24 & 120 \end{vmatrix}$

Operate $R_3 = R_3 - 6R_1, R_2 = R_2 - 2R_1$ and we get

$$= \begin{vmatrix} 1 & 2 & 6 \\ 0 & 2 & 12 \\ 0 & 12 & 84 \end{vmatrix} = 168 - 144 = 24 = 4!$$

13. (b) For skew-symmetric matrix, all diagonal elements = 0

$[\because a_{11} = -a_{11} \Rightarrow a_{11} = 0 \text{ and so on}]$

14. (a) Since $-1 \le x < 0$
$\therefore [x] = -1$
Also, $0 \le y < 1 \therefore [y] = 0$
and $1 \le z < 2 \therefore [z] = 1$

$$\text{So,} \begin{vmatrix} 0 & 0 & 1 \\ -1 & 1 & 1 \\ -1 & 0 & 2 \end{vmatrix} = 1 = [z]$$

15. (d) $\displaystyle\int_{-a}^{a} f(x)d(x) = \begin{vmatrix} \int_{-a}^{a} x^3 dx & \int_{-a}^{a} \sin x dx \\ 1 & 2 \end{vmatrix}$

$$= \begin{vmatrix} 0 & 0 \\ 1 & 2 \end{vmatrix} = 0$$

16. (a) Put $x-a = \alpha, x-b = \beta, x-c = \gamma$

$$\Delta = \begin{vmatrix} 1 & 1 & 1 \\ \alpha^2 & \beta^2 & \gamma^2 \\ \beta\gamma & \gamma\alpha & \alpha\beta \end{vmatrix} = \frac{1}{\alpha\beta\gamma}\begin{vmatrix} \alpha & \beta & \gamma \\ \alpha^3 & \beta^3 & \gamma^3 \\ \alpha\beta\gamma & \alpha\beta\gamma & \alpha\beta\gamma \end{vmatrix}$$

$$= \begin{vmatrix} 1 & 1 & 1 \\ \alpha & \beta & \gamma \\ \alpha^3 & \beta^3 & \gamma^3 \end{vmatrix} = (\alpha-\beta)(\beta-\gamma)(\alpha+\beta+\gamma)$$

$$= (a-b)(b-c)(c-a)(3x-a-b-c) = 0$$

$$\Rightarrow x = \frac{1}{3}(a+b+c) \qquad [\because a \ne b \ne c]$$

17. (d)

$$\Delta = \begin{vmatrix} \log x & \log y & \log z \\ \log 2 + \log x & \log 2 + \log y & \log 2 + \log z \\ \log 3 + \log x & \log 3 + \log y & \log 3 + \log z \end{vmatrix}$$

Operate $R_2 - R_1$ and $R_3 - R_1$ and get

$$\Delta = \begin{vmatrix} \log x & \log y & \log z \\ \log 2 & \log 2 & \log 2 \\ \log 3 & \log 3 & \log 3 \end{vmatrix} = 0$$

18. (a) $\Delta.\Delta' = \begin{vmatrix} \Delta & 0 & 0 \\ 0 & \Delta & 0 \\ 0 & 0 & \Delta \end{vmatrix} = \Delta^3$

$\Rightarrow \Delta' = \Delta^2$

19. (d) $\Delta = \frac{1}{abc}\begin{vmatrix} ab^2c^2 & abc & ab+ac \\ a^2bc^2 & abc & bc+ab \\ a^2b^2c & abc & ac+bc \end{vmatrix}$

$= \frac{a^2b^2c^2}{abc}\begin{vmatrix} bc & 1 & ab+ac \\ ac & 1 & bc+ab \\ ab & 1 & ac+bc \end{vmatrix}$

Operate $C_1 = C_1 + C_3$

$= abc\begin{vmatrix} ab+bc+ca & 1 & ab+ac \\ ab+bc+ca & 1 & bc+ab \\ ab+bc+ca & 1 & ac+bc \end{vmatrix}$

$= abc(ab+bc+ca)\begin{vmatrix} 1 & 1 & ab+ac \\ 1 & 1 & bc+ab \\ 1 & 1 & ac+bc \end{vmatrix}$

$= 0$

20. (a) Let $a = 1+d, b = 1+2d, c = 1+3d$

$\therefore \Delta = \begin{vmatrix} x+1 & x+1+d & x+1+2d \\ x+1+d & x+1+2d & x+1+3d \\ x-2d & x-1 & x+2d \end{vmatrix}$

$(C_2 - C_1, C_3 - C_2)$

$= \begin{vmatrix} x+1 & d & d \\ x+1+d & d & d \\ x-2d & 2d-1 & 2d+1 \end{vmatrix}$

$R_2 = R_2 - R_1$ and $R_3 = R_3 - R_1$

$= \begin{vmatrix} x+1 & d & d \\ d & 0 & 0 \\ -2d-1 & d-1 & d+1 \end{vmatrix}$

$= -d^2\left[d+1-(d-1)\right] = -2d^2$

$\therefore |\Delta| = 2d^2 = 2$ (given)

$\Rightarrow d = \pm 1$

21. (c) Since a, b, c are the roots of $x^3 + px + q = 0$

$\therefore a+b+c = 0$

$\therefore \begin{vmatrix} a & b & c \\ b & c & a \\ c & a & b \end{vmatrix}$

Operating $R_1 + R_2 + R_3$

$= \begin{vmatrix} a+b+c & a+b+c & a+b+c \\ b & c & a \\ c & a & b \end{vmatrix} = 0$

22. (c) $\begin{vmatrix} 1+a & 1 & 1 \\ 1 & 1+b & 1 \\ 1 & 1 & 1+c \end{vmatrix}$

By expanding the determinant, we get

$abc\left(1+\frac{1}{a}+\frac{1}{b}+\frac{1}{c}\right) = abc(1+0) = abc$

23. (c) $\begin{vmatrix} 1 & 1 & 1 \\ m & m+1 & m+2 \\ \frac{m(m-1)}{2} & \frac{(m+1)m}{2} & \frac{(m+2)(m+1)}{2} \end{vmatrix}$

Operate $C_3 - C_2, C_2 - C_1$ and get

$= \begin{vmatrix} 1 & 0 & 0 \\ m & 1 & 1 \\ \frac{m(m-1)}{2} & m & m+1 \end{vmatrix} = 1$

24. (b) Given equation $\begin{vmatrix} x & x^2 & 1 \\ y & y^2 & 1 \\ z & z^2 & 1 \end{vmatrix} + \begin{vmatrix} x & x^2 & x^3 \\ y & y^2 & y^3 \\ z & z^2 & z^3 \end{vmatrix} = 0$

$\Rightarrow \begin{vmatrix} x & x^2 & 1 \\ y & y^2 & 1 \\ z & z^2 & 1 \end{vmatrix} + xyz\begin{vmatrix} 1 & x & x^2 \\ 1 & y & y^2 \\ 1 & z & z^2 \end{vmatrix} = 0$

$\Rightarrow \begin{vmatrix} x & x^2 & 1 \\ y & y^2 & 1 \\ z & z^2 & 1 \end{vmatrix}(1+xyz) = 0$

$\Rightarrow (x-y)(y-z)(z-x)(1+xyz) = 0$

$\Rightarrow xyz + 1 = 0 \qquad [\because x \neq y \neq z]$

$\Rightarrow xyz = -1$

25. (a) Operate $C_1 + C_2 + C_3$ and get

$\begin{vmatrix} x+a+b+c & b & c \\ x+a+b+c & x+b & c \\ x+a+b+c & b & x+c \end{vmatrix} = 0$

$\Rightarrow (x+a+b+c)\begin{vmatrix} 1 & b & c \\ 1 & x+b & c \\ 1 & b & x+c \end{vmatrix} = 0$

Operate $R_2 - R_1, R_3 - R_1$

$$\Rightarrow (x+a+b+c)\begin{vmatrix} 1 & b & c \\ 0 & x & 0 \\ 0 & 0 & x \end{vmatrix} = 0$$

$$\Rightarrow x^2(x+a+b+c) = 0$$

$$x = 0, -(a+b+c)$$

26. (b) Operating $C_2 - C_1$ and $C_3 - C_1$ on the given determinant, we have

$$\begin{vmatrix} 1 & 0 & 0 \\ 1+\sin A & \sin B - \sin A & \sin C - \sin A \\ \sin A + \sin^2 A & \sin B - \sin A + \sin^2 B - \sin^2 A & \sin C - \sin A + \sin^2 C - \sin^2 A \end{vmatrix} = 0$$

$$\Rightarrow (\sin B - \sin A)(\sin C - \sin A) \times \begin{vmatrix} 1 & 1 \\ 1+\sin B + \sin A & 1 + \sin C + \sin A \end{vmatrix} = 0$$

$$\Rightarrow (\sin B - \sin A)(\sin C - \sin A)(\sin C - \sin B) = 0$$

$\Rightarrow \sin A = \sin B$ or $\sin C = \sin A$ or $\sin C = \sin B$

$\Rightarrow A = B$ or $B = C$ or $C = A$

$\Rightarrow \Delta$ABC is isosceles

27. (b) On expansion, we can see that degree of the determinant is $n+(n+2)+(n+3)$, i.e., $3n+5$ and the degree of that expression on RHS is 2.

$3n+5 = 2 \Rightarrow n = -1$

28. (a) Operating $R_2 \to R_2 - R_3$ and taking 4 common.

$$\Delta = 4\begin{vmatrix} a^2 & b^2 & c^2 \\ a & b & c \\ (a-1)^2 & (b-1)^2 & (c-1)^2 \end{vmatrix}$$

(Operating $R_3 \to R_3 + 2R_2 - R_1$)

$$= 4\begin{vmatrix} a^2 & b^2 & c^2 \\ a & b & c \\ 1 & 1 & 1 \end{vmatrix} = 4(a-b)(b-c)(c-a) = 0$$

$\Rightarrow$ at least one of the factors is zero.

$\Rightarrow$ The triangle is isosceles.

29. (d) $$\begin{vmatrix} 1 & ka & k^2+a^2 \\ 1 & kb & k^2+b^2 \\ 1 & kc & k^2+c^2 \end{vmatrix} = \begin{vmatrix} 1 & ka & k^2 \\ 1 & kb & k^2 \\ 1 & kc & k^2 \end{vmatrix} + \begin{vmatrix} 1 & ka & a^2 \\ 1 & kb & b^2 \\ 1 & kc & c^2 \end{vmatrix}$$

$$= 0 + k\begin{vmatrix} 1 & a & a^2 \\ 1 & b & b^2 \\ 1 & c & c^2 \end{vmatrix}$$

$$= k(a-b)(b-c)(c-a)$$

30. (b) We have

$$\begin{vmatrix} -1+\cos B & \cos C + \cos B & \cos B \\ \cos C + \cos A & -1 + \cos A & \cos A \\ -1 + \cos B & -1 + \cos A & -1 \end{vmatrix}$$

$$= \begin{vmatrix} -1 & \cos C & \cos B \\ \cos C & -1 & \cos A \\ \cos B & \cos A & -1 \end{vmatrix}$$

[Applying $C_1 \to C_1 - C_3$ and $C_2 \to C_2 - C_3$]

$$= \frac{1}{a}\begin{vmatrix} -a & \cos C & \cos B \\ a\cos C & -1 & \cos A \\ a\cos B & \cos A & -1 \end{vmatrix} = \frac{1}{a}\begin{vmatrix} 0 & \cos C & \cos B \\ 0 & -1 & \cos A \\ 0 & \cos A & -1 \end{vmatrix}$$

[Applying $C_1 \to C_1 + bC_2 + cC_3$] $= 0$

31. (a) $2^{a_1}, 2^{a_2}, \ldots, 2^{a_n}$ are in GP.

Then $a_1, a_2, \ldots a_n$ are in AP.

Let $a_1 = \alpha, a_2 = \alpha + \beta$

$$= \begin{vmatrix} \alpha & \beta & \alpha + 2\beta \\ \alpha + n\beta & \beta & \alpha + (n+2)\beta \\ \alpha + 2n\beta & \beta & \alpha + (2n+2)\beta \end{vmatrix} \left(\begin{array}{c} \because C_3 \to C_3 - C_1 \\ C_2 \to C_2 - C_1 \end{array} \right)$$

$$= \begin{vmatrix} \alpha & \beta & 2\beta \\ \alpha + n\beta & \beta & 2\beta \\ \alpha + 2n\beta & \beta & 2\beta \end{vmatrix} = 0$$

32. (c) Operating $R_2 \to R_2 - R_1, R_3 \to R_3 - R_1$, we get

$$\Delta = \begin{vmatrix} 1 & p^2 & p^3 \\ 0 & q^2 - p^2 & q^3 - p^3 \\ 0 & r^2 - p^2 & r^3 - p^3 \end{vmatrix}$$

$$= (q-p)(r-p)(r-q)(qr + qr + pq)$$

$$= pqr(q-p)(r-p)(r-q)\left(\frac{1}{p} + \frac{1}{q} + \frac{1}{r}\right)$$

$$\therefore \Delta = 0 \Rightarrow \frac{1}{p} + \frac{1}{q} + \frac{1}{r} = 0$$

33. (a) $$\sum_{r=1}^{n} S_r = \begin{vmatrix} \sum_{r=1}^{n} 2r & x & n(n+1) \\ \sum_{r=1}^{n}(6r^2-1) & y & n^2(2n+3) \\ \sum_{r=1}^{n}(4r^3-2nr) & z & n^3(n+1) \end{vmatrix}$$

$$= \begin{vmatrix} n(n+1) & x & n(n+1) \\ n(n+1)(2n+1)-n & y & n^2(2n+3) \\ n^2(n+1)^2-n^2(n+1) & z & n^3(n+1) \end{vmatrix}$$

$$= \begin{vmatrix} n(n+1) & x & n(n+1) \\ n^2(2n+3) & y & n^2(2n+3) \\ n^3(n+1) & z & n^3(n+1) \end{vmatrix} = 0$$

and hence independent of x, y, z, n.

34. (d) Δ can be written as

$$\Delta = \begin{vmatrix} 1 & 1 & 1 \\ \alpha & \beta & \gamma \\ \alpha^2 & \beta^2 & \gamma^2 \end{vmatrix} \times \begin{vmatrix} 1 & 1 & 1 \\ \alpha & \beta & \gamma \\ \alpha^2 & \beta^2 & \gamma^2 \end{vmatrix}$$

(Row by row multiplication)

$= (\alpha-\beta)^2(\beta-\gamma)^2(\gamma-\alpha)^2$

So, (d) is the answer.

35. (c) Let $\Delta(x) = A + Bx + Cx^2 + Dx^3 + \ldots$

$$\Delta(0) = \begin{vmatrix} 0 & 1 & 0 \\ 0 & 1 & 0 \\ 1 & 0 & 0 \end{vmatrix} = 0 \Rightarrow A = 0$$

$$\Delta'(0) = \begin{vmatrix} 1 & 1 & 0 \\ 0 & 1 & 0 \\ 1 & 0 & 0 \end{vmatrix} + \begin{vmatrix} 0 & 1 & 0 \\ 0 & 1 & 1 \\ 1 & 0 & 0 \end{vmatrix} + \begin{vmatrix} 0 & 1 & 0 \\ 0 & 1 & 0 \\ 0 & 1 & 0 \end{vmatrix}$$

$\Delta'(0) = 1$

$\Delta(x) = x + Cx^2 + Dx^3 + \ldots$

$\Delta(x)$ is divisible by x.

36. (b) The given determinant

$= (1-\omega^2-\omega) - \omega(\omega+1-\omega^2) + (\omega+1)\left[(\omega+1)^2 - \omega\right]$

$= 1-\omega^2-\omega-\omega^2-\omega+\omega^3+(\omega+1)(\omega^2+2\omega+1-\omega)$

$= 2-2\omega^2-2\omega+(\omega+1)(1+\omega+\omega^2)$

$= 2(1-\omega-\omega^2)+(\omega+1)(0)$

$= 2(1+1) = 4$

37. (b) Operate $C_1 + pC_2, C_2 + qC_3$ on D and get

$$D = \begin{vmatrix} a_1+pb_1 & b_1+qc_1 & c_1 \\ a_2+pb_2 & b_2+qc_2 & c_2 \\ a_3+pb_3 & b_3+qc_3 & c_3 \end{vmatrix}$$

$$\text{Now } D' = \begin{vmatrix} a_1+pb_1 & b_1+qc_1 & c_1 \\ a_2+pb_2 & b_2+qc_2 & c_2 \\ a_3+pb_3 & b_3+qc_3 & c_3 \end{vmatrix} + r\begin{vmatrix} a_1+pb_1 & b_1+qc_1 & a_1 \\ a_2+pb_2 & b_2+qc_2 & a_2 \\ a_3+pb_3 & b_3+qc_3 & a_3 \end{vmatrix}$$

$$= D + r\begin{vmatrix} a_1+pb_1 & b_1+qc_1 & a_1 \\ a_2+pb_2 & b_2+qc_2 & a_2 \\ a_3+pb_3 & b_3+qc_3 & a_3 \end{vmatrix}$$

$$= D + pr\begin{vmatrix} b_1 & b_1+qc_1 & a_1 \\ b_2 & b_2+qc_2 & a_2 \\ b_3 & b_3+qc_3 & a_3 \end{vmatrix}$$

[$C_1 = C_1 - C_3$ and taking out p]

$$= D + pqr\begin{vmatrix} b_1 & c_1 & a_1 \\ b_2 & c_2 & a_2 \\ b_3 & c_3 & a_3 \end{vmatrix}$$

[$C_2 = C_2 - C_1$ and taking out q]

$= (1+pqr)D$

38. (c) Given, $x^a y^b = e^m, x^c y^d = e^n$

$\Rightarrow a\log x + b\log y = m$ and $c\log x + d\log y = n$

By Cramer's rule,

$\log x = \dfrac{\Delta_1}{\Delta_2}$ and $\log y = \dfrac{\Delta_2}{\Delta_3}$

$\Rightarrow\ x = e^{\Delta_1/\Delta_3}$ and $y = e^{\Delta_2/\Delta_3}$

39. (b) The given determinant can be expressed as the product of two determinants

$$\begin{vmatrix} \tan A & \cot A & 0 \\ \tan B & \cot B & 0 \\ \tan C & \cot C & 0 \end{vmatrix} \times \begin{vmatrix} \cot A & \tan A & 0 \\ \cot B & \tan B & 0 \\ \cot C & \tan C & 0 \end{vmatrix} = 0$$

40. (b) The system of equations has a non-zero solution

$$\therefore \begin{vmatrix} a^3 & (a+1)^3 & (a+2)^3 \\ a & a+1 & (a+2) \\ 1 & 1 & 1 \end{vmatrix} = 0$$

Operating $C_2 - C_1$ and $C_3 - C_2$ we get

$$\Rightarrow \begin{vmatrix} a^3 & 3a^2+3a+1 & 3(a+1)^2+3(a+1)+1 \\ a & 1 & 1 \\ 1 & 0 & 0 \end{vmatrix} = 0$$

$$\Rightarrow 3a^2+3a+1-\left[3(a+1)^2+3(a+1)+1\right]=0$$

$$\Rightarrow -6a-6=0$$
$$\Rightarrow a=-1$$

41. (c)

Here, $a=\dfrac{1(2^n-1)}{2-1}, b=\dfrac{1(3^n-1)}{3-1}$ and $c=\dfrac{1(5^n-1)}{5-1}$

$$\therefore \begin{vmatrix} a & 2b & 4c \\ 2 & 2 & 2 \\ 2^n & 3^n & 5^n \end{vmatrix} = \begin{vmatrix} 2^n-1 & 3^n-1 & 5^n-1 \\ 2 & 2 & 2 \\ 2^n & 3^n & 5^n \end{vmatrix}$$

Operate $R_3-R_1, \frac{1}{2}R_2$

$$=2\begin{vmatrix} 2^n-1 & 3^n-1 & 5^n-1 \\ 1 & 1 & 1 \\ 1 & 1 & 1 \end{vmatrix}=0$$

42. (c) The system has a non-trivial solution of

$$\begin{vmatrix} \alpha+a & \alpha & \alpha \\ \alpha & \alpha+b & \alpha \\ \alpha & \alpha & \alpha+c \end{vmatrix}=0$$

i.e., if $\begin{vmatrix} \alpha+a & \alpha & \alpha \\ -a & b & 0 \\ -a & 0 & c \end{vmatrix}=0$

$$[R_2=R_2-R_1 \text{ and } R_3=R_3-R_1]$$

if $\alpha\begin{vmatrix} -a & b \\ -a & 0 \end{vmatrix}+c\begin{vmatrix} \alpha+a & \alpha \\ -a & b \end{vmatrix}=0$

i.e., if $\alpha ab+c(\alpha b+ab+a\alpha)=0$

i.e., if $\alpha(ab+bc+ca)=-abc$

i.e., if $\dfrac{1}{\alpha}=-\left(\dfrac{1}{a}+\dfrac{1}{b}+\dfrac{1}{c}\right)$

43. (c) Applying $C_1+C_3-2(\cos x)C_2$.

$$f(x)=\begin{vmatrix} 1+a^2-2a\cos x & a & a^2 \\ \sin(n-1)x+\sin(n+1)x-2\cos x.\cos nx & \sin nx & \sin(n+1)x \\ \cos(n-1)x+\cos(n+1)x-2\cos x.\cos nx & \cos nx & \cos(n+1)x \end{vmatrix}$$

$$=\begin{vmatrix} (1+a^2-2a\cos x) & a & a^2 \\ 0 & \sin nx & \sin(n+1)x \\ 0 & \cos nx & \cos(n+1)x \end{vmatrix}$$

$$=(1+a^2-2a\cos x)\left[\sin nx\cos(n+1)x-\cos nx\sin(n+1)x\right]$$

$$=-(1+a^2-2a\cos x)\sin x$$

$$=\int_0^{\pi/2} f(x)dx=-(1+a^2)\int_0^{\pi/2}\sin x dx$$

$$=+a\int_0^{\pi/2}\sin 2x dx=a-(1+a^2)$$

44. (c) $\sum_{r=1}^{n}2^{r-1}=2^n-1$

$$2\sum_{r=1}^{n}r=2\left\{\frac{n(n+1)}{2}\right\}=n(n+1)$$

$$\sum_{r=1}^{n}\frac{(r+1)!}{1+\frac{1}{r}}=\sum_{r=1}^{n}r(r!)=\sum_{r=1}^{n}(r+1-1)(r!)$$

$$=\sum_{r=1}^{n}\{(r+1)!-r!\}=(n+1)!-1$$

$$\therefore \sum_{r=1}^{n}\Delta_r=\begin{vmatrix} \sum_{r=1}^{n}2^{r-1} & \sum_{r=1}^{n}\frac{(r+1)!}{\left(1+\frac{1}{r}\right)} & 2\sum_{r=1}^{n}r \\ a & b & c \\ 2^n-1 & (n+1)!-1 & n(n+1) \end{vmatrix}$$

$$=\begin{vmatrix} 2^n-1 & (n+1)!-1 & n(n+1) \\ a & b & c \\ 2^n-1 & (n+1)!-1 & n(n+1) \end{vmatrix}=0$$

[$\because R_1$ and R_3 are identical.]

45. (a)

$$\frac{d}{dx}(\Delta_1)=\begin{vmatrix} 1 & 0 & 0 \\ a & x & b \\ a & a & x \end{vmatrix}+\begin{vmatrix} x & b & b \\ 0 & 1 & 0 \\ a & a & x \end{vmatrix}+\begin{vmatrix} x & b & b \\ a & x & b \\ 0 & 0 & 1 \end{vmatrix}$$

$$=\begin{vmatrix} x & b \\ a & x \end{vmatrix}+\begin{vmatrix} x & b \\ a & x \end{vmatrix}+\begin{vmatrix} x & b \\ a & x \end{vmatrix}$$

$$=3\Delta_2$$

46. (c) Differentiating both the sides, we get

$$\begin{vmatrix} 1 & 5 & 0 \\ x^2-1 & x-1 & 8 \\ 2x & 3x & 0 \end{vmatrix} + \begin{vmatrix} x-1 & 5x & 7 \\ 2x & 1 & 0 \\ 2x & 3x & 0 \end{vmatrix}$$

$$= \begin{vmatrix} x-1 & 5x & 7 \\ x^2-1 & x-1 & 8 \\ 2 & 3 & 0 \end{vmatrix} = 3ax^2+2bx+c$$

Putting $x=0$, we get

$$c = \begin{vmatrix} 1 & 5 & 0 \\ -1 & -1 & 8 \\ 0 & 0 & 0 \end{vmatrix} + \begin{vmatrix} -1 & 0 & 7 \\ 0 & 1 & 0 \\ 0 & 0 & 0 \end{vmatrix} + \begin{vmatrix} -1 & 0 & 7 \\ -1 & -1 & 8 \\ 2 & 3 & 0 \end{vmatrix}$$

$$c = 0+0+17 = 17$$

47. (b)

$$\sum_{r=0}^{m}\Delta_r = \begin{vmatrix} \sum_{r=0}^{m}(2r-1) & \sum_{r=0}^{m}{}^mC_r & \sum_{r=0}^{m}1 \\ m^2-1 & 2^m & m+1 \\ \sin^2(m^2) & \sin^2(m) & \sin^2(m+1) \end{vmatrix}$$

But $\sum_{r=0}^{m}(2r-1) = \frac{1}{2}(m+1)(2m-1-1) = m^2-1$

$$\sum_{r=0}^{m}{}^mC_r = 2^m, \sum_{r=0}^{m}1 = m+1$$

$$\sum_{r=0}^{m}\Delta_r = \begin{vmatrix} m^2-1 & 2^m & m+1 \\ m^2-1 & 2^m & m+1 \\ \sin^2(m^2) & \sin^2(m) & \sin^2(m+1) \end{vmatrix} = 0$$

48. (a) As the determinant has to vanish for all a, b, we have at least two rows or two columns identical.
After filling the first column in $2\times2\times2$ ways and filling the next column likewise, the remaining column can be filled in $2\times2\times2$ ways. So, the number of ways will be $2^3\times2^3 = 2^6$
But each of the two ways give the same determinant.
∴ The required number of determinants are 32.

49. (d) We have $\beta\gamma+\gamma\alpha+\alpha\beta = 0$
We can write Δ as

$$\Delta = \frac{1}{\alpha^3\beta^3\gamma^3}\begin{vmatrix} \beta\gamma & \gamma\alpha & \alpha\beta \\ \gamma\alpha & \alpha\beta & \beta\gamma \\ \alpha\beta & \beta\gamma & \gamma\alpha \end{vmatrix}$$

$$= \frac{1}{\alpha^3\beta^3\gamma^3}\begin{vmatrix} \alpha\beta+\gamma\alpha+\beta\gamma & \gamma\alpha & \alpha\beta \\ \alpha\beta+\gamma\alpha+\beta\gamma & \alpha\beta & \beta\gamma \\ \alpha\beta+\gamma\alpha+\beta\gamma & \beta\gamma & \gamma\alpha \end{vmatrix}$$

[∵ using $C_1 \to C_1+C_2+C_3$]

$$= \frac{1}{\alpha^3\beta^3\gamma^3}\begin{vmatrix} 0 & \gamma\alpha & \alpha\beta \\ 0 & \alpha\beta & \beta\gamma \\ 0 & \beta\gamma & \gamma\alpha \end{vmatrix} = 0$$

50. (c) We have,

$$\Delta^2 = \Delta.\Delta = \begin{vmatrix} l_1 & m_1 & n_1 \\ l_2 & m_2 & n_2 \\ l_3 & m_3 & n_3 \end{vmatrix} . \begin{vmatrix} l_1 & m_1 & n_1 \\ l_2 & m_2 & n_2 \\ l_3 & m_3 & n_3 \end{vmatrix}$$

$$= \begin{vmatrix} l_1^2+m_1^2+n_1^2 & l_1l_2+m_1m_2+n_1n_2 & l_1l_3+m_1m_3+n_1n_3 \\ l_1l_2+m_1m_2+n_1n_2 & l_2^2+m_2^2+n_2^2 & l_2l_3+m_2m_3+n_2n_3 \\ l_1l_3+m_1m_3+n_1n_3 & l_2l_3+m_2m_3+n_2n_3 & l_3^2+m_3^2+n_3^2 \end{vmatrix}$$

$$= \begin{vmatrix} 1 & 0 & 0 \\ 0 & 1 & 0 \\ 0 & 0 & 1 \end{vmatrix} = 0$$

$\Delta = \pm1$
$|\Delta| = 1$

51. (b)

$$\Delta = \begin{vmatrix} 5 & 4 & 3 \\ 100x+50+1 & 100y+40+1 & 100z+30+1 \\ x & y & z \end{vmatrix}$$

$$\Delta = \begin{vmatrix} 5 & 4 & 3 \\ 1 & 1 & 1 \\ x & y & z \end{vmatrix} \quad [R_2 \to R_2 - 100R_3 - 10R_1]$$

$= x-2y+z$
$= 0$ [∵ x, y, z are in AP]

52. (c) We divide LHS by λ^4 and C_1 by λ^2, C_2 by λ and C_3 by λ on the RHS to obtain

$$p+q\left(\frac{1}{\lambda}\right)+r\left(\frac{1}{\lambda}\right)^2+s\left(\frac{1}{\lambda}\right)^3+t\left(\frac{1}{\lambda}\right)^4$$

$$= \begin{vmatrix} 1+\frac{3}{\lambda} & 1-\frac{1}{\lambda} & 1+\frac{3}{\lambda} \\ 1+\frac{1}{\lambda^2} & \frac{2}{\lambda}-1 & 1-\frac{3}{\lambda} \\ 1-\frac{3}{\lambda^2} & 1-\frac{4}{\lambda} & 3 \end{vmatrix}$$

Taking limit as $\lambda \to \infty$, we get

$$p=\begin{vmatrix}1&1&1\\1&-1&1\\1&1&3\end{vmatrix}=\begin{vmatrix}1&1&1\\0&-2&0\\0&0&2\end{vmatrix}=-4$$

$[\because$ Using $R_2 \to R_2 - R_1, R_3 \to R_3 - R_1]$

53. (b) Applying $C_1 \to C_1 + C_2 + C_3$, we get

$$\Delta=\begin{vmatrix}a+b+c&b&c\\a+b+c&c&a\\a+b+c&a&b\end{vmatrix}$$

$$\Delta=(a+b+c)\begin{vmatrix}1&b&c\\1&c&a\\1&a&b\end{vmatrix}$$

Operating $R_2 \to R_2 - R_1$ and $R_3 \to R_3 - R_1$, we get

$$\Delta=(a+b+c)\begin{vmatrix}1&b&c\\0&c-b&a-c\\0&a-b&b-c\end{vmatrix}$$

Expanding along C_1, we get

$$\Delta=-\frac{1}{2}(a+b+c)\left[(b-c)^2+(c-a)^2+(a-b)^2\right]\le 0$$

$[\because a+b+c>0]$

54. (a) Taking x common from R_2 and $x(x-1)$ common from R_3, we get

$$f(x)=x^2(x-1)\begin{vmatrix}1&x&x+1\\2&(x-1)&x+1\\3&(x-2)&x+1\end{vmatrix}$$

Applying $C_3 \to C_3 - C_2$, we get

$$f(x)=x^2(x-1)\begin{vmatrix}1&x&1\\2&(x-1)&2\\3&(x-2)&3\end{vmatrix}$$

$f(x)=0$

Thus $f(500)=0$

55. (b)

Here, applying $C_1 = C_1 - C_3$ $\begin{vmatrix}x-5&2&5\\0&x&3\\5-x&4&x\end{vmatrix}=0$

or, $(x-5)\begin{vmatrix}1&2&5\\0&x&3\\-1&4&x\end{vmatrix}=0$

$(x-5).\left[18-x(5+x)\right]=0$

The non-integral roots are the roots of $x^2+5x-18=0$

Sum of non-integral roots $=-5$

56. (b)

$$y=\Delta=\begin{vmatrix}2x&2&5\\3x&3&10\\-x&-1&20\end{vmatrix} C_1 \to (C_1 - xC_2 - C_3)$$

$$y=\Delta=x\begin{vmatrix}2&2&5\\3&3&10\\-1&-1&20\end{vmatrix} C_2 \to C_2 - \frac{x}{5}C_3$$

57. (a) $z=-94$ [Expanding the determinant]

$\text{Arg}(z)=0$

58. (a)

$\because |kA| = k^n|A|$ for a $n\times n$ matrix A

If matrix is of 3×3 order then

$n=3$

i.e., $|kA|=k^3|A|$

59. (c)

$$\begin{vmatrix}bc&ca&ab\\ca&ab&bc\\ab&bc&ca\end{vmatrix}=0$$

$\Rightarrow 3a^2b^2c^2-\left[(ab)^3+(bc)^3+(ca)^3\right]=0$

$\Rightarrow (ab+bc+ca)$

$\left(a^2b^2+b^2c^2+c^2a^2-ab^2c-bc^2a-ca^2b\right)=0$

$\Rightarrow ab+bc+ca=0$

$\Rightarrow \frac{1}{a}+\frac{1}{b}+\frac{1}{c}=0$

60. (b) The given determinant can be written as a product of two zero determinants

$$\begin{vmatrix}a_1&b_1&0\\a_2&b_2&0\\a_3&b_3&0\end{vmatrix}\times\begin{vmatrix}b_1&a_1&0\\b_2&a_2&0\\b_3&a_3&0\end{vmatrix}=0$$

61. (d)

Let $C=A^2-B^2; |C|\ne 0$

It is given that

$A^5=B^5$ and $A^3B^2=A^2B^2$

Now, $A^5-A^3B^2=B^5-A^2B^3$

$\Rightarrow A^3\left(A^2-B^2\right)+B^3\left(A^2-B^2\right)=0$

$\Rightarrow \left(A^3+B^3\right)\left(A^2-B^2\right)=0$

Post multiplying inverse of C,

$\left(A^3+B^3\right).CC^{-1}=0.C^{-1}$

$\Rightarrow (A^3+B^3).I = 0 \Rightarrow A^3+B^3 = 0 \qquad (\because I \neq 0)$

Hence, $|A^3+B^3| = 0$

62. (d) Apply $C_1 \to C_1 + C_2$

$$f(x) = \begin{vmatrix} 2 & 1+\cos^2 x & \cos 2x \\ 2 & \cos^2 x & \cos 2x \\ 1 & \cos^2 x & \sin 2x \end{vmatrix}$$

Apply $R_1 \to R_1 - 2R_3$ and $R_2 \to R_2 - 2R_3$

$$= \begin{vmatrix} 0 & \sin^2 x & \cos 2x - 2\sin 2x \\ 0 & -\cos^2 x & \cos 2x - 2\sin 2x \\ 1 & \cos^2 x & \sin 2x \end{vmatrix}$$

$= (\cos 2x - 2\sin 2x)(\sin^2 x + \cos^2 x)$

$= \cos 2x - 2\sin 2x = \sqrt{5}\left(\frac{1}{\sqrt{5}}\cos 2x - \frac{2}{\sqrt{5}}\sin 2x\right)$

$= \sqrt{5}\sin(\alpha - 2x) \qquad \left(\text{where} \sin\alpha = \frac{1}{\sqrt{5}}\right)$

$\therefore$ Maximum value of $\sin(\alpha - 2x)$ is 1.

$\therefore$ Maximum value of $f(x) = \sqrt{5}$

63. (c)

$$\begin{vmatrix} (a+1)(a+2) & (a+2) & 1 \\ (a+2)(a+3) & (a+3) & 1 \\ (a+3)(a+4) & (a+4) & 1 \end{vmatrix}$$

$R_3{'}! R_3 - R_2{'}! R_2 - R_1$

$$= \begin{vmatrix} (a+1)(a+2) & (a+2) & 1 \\ 2(a+2) & 1 & 0 \\ 2(a+3) & 1 & 0 \end{vmatrix} = -2$$

64. (d) Given that:

$A = \begin{bmatrix} 1 & 2 \\ -1 & 4 \end{bmatrix} \Rightarrow |A| = 6$

$\because A^{-1} = \frac{\text{adj}.A}{|A|} = \frac{1}{6}\begin{bmatrix} 4 & -2 \\ 1 & 1 \end{bmatrix} = \begin{bmatrix} \frac{2}{3} & -\frac{1}{3} \\ \frac{1}{6} & \frac{1}{6} \end{bmatrix}$

$$\begin{bmatrix} \frac{2}{3} & -\frac{1}{3} \\ \frac{1}{6} & \frac{1}{6} \end{bmatrix} = \begin{bmatrix} \alpha & 0 \\ 0 & \alpha \end{bmatrix} + \begin{bmatrix} \beta & 2\beta \\ -\beta & 4\beta \end{bmatrix}$$

$$= \begin{bmatrix} \alpha+\beta & 2\beta \\ -\beta & \alpha+4\beta \end{bmatrix}$$

$\Rightarrow \beta = -\frac{1}{6}$ and $\alpha + \beta = \frac{2}{3}$

$\Rightarrow \alpha = \frac{5}{6}$

Hence, $4(\alpha - \beta) = 4(1) = 4$

65. (c) $\Delta = \begin{vmatrix} -1 & 1 & 2 \\ 3 & -a & 5 \\ 2 & -2 & -a \end{vmatrix}$

$= -1(a^2+10) - 1(-3a-10) + 2(-6+2a)$

$= -a^2 - 10 + 3a + 10 - 12 + 4a$

$\Delta = -a^2 + 7a - 12$

$\Delta = -[a^2 - 7a + 12]$

$\Delta = -[(a-3)(a-4)] = 0 \Rightarrow a = 3, a = 4 \qquad \ldots(i)$

$\Delta_1 = \begin{vmatrix} 0 & 1 & 2 \\ 1 & -a & 5 \\ 7 & -2 & -a \end{vmatrix}$

$= 0 - 1(-a-35) + 2(-2+7a)$

$= a + 35 - 4 + 14a$

$= 15a + 131 = 0 \Rightarrow a = -5 \qquad \ldots(ii)$

Now $\Delta_1 = 15a + 31$

From (i) and (ii), we conclude that for $a = 3$ and $4\ \Delta_1 \neq 0$

The system of equations is inconsistent.

$\Rightarrow n(S_1) = 2$

For infinite solution: $\Delta = 0$

and $\Delta_1 = \Delta_2 = \Delta_3 = 0$

Not possible.

$\therefore n(S_2) = 0$

66. (a)

$D = \begin{vmatrix} 1 & 1 & 1 \\ 3 & 2 & 5 \\ 9 & 4 & 28+[\lambda] \end{vmatrix} = 24 - [\lambda] + 15 = -[\lambda] - 9$

If $[\lambda] + 9 \neq 0$ then there is a unique solution.

If $[\lambda] + 9 = 0$ then $D_1 = D_2 = D_3 = 0$

$\Rightarrow$infinite solutions

Therefore λ can be any real number.

67. (c) For non-trivial solution,

$$\begin{vmatrix} 4 & \lambda & 2 \\ 2 & -1 & 1 \\ \mu & 2 & 3 \end{vmatrix} = 0$$

$\Rightarrow \mu\lambda + 2\mu - 6\lambda - 12 = 0$

$\Rightarrow (\lambda+2)(\mu-6)=0$
$\Rightarrow \lambda=-2$ or $\mu=6$

68. (a) $\Delta=\begin{vmatrix}1 & 2 & -3\\2 & 6 & -11\\1 & -2 & 7\end{vmatrix}=(20)-2(25)-3(-10)=0$

$\Delta_1=\begin{vmatrix}a & 2 & -3\\b & 6 & -11\\c & -2 & 7\end{vmatrix}=4(5a-2b-c)$

$\Delta_2=\begin{vmatrix}1 & a & -3\\2 & b & -11\\1 & c & 7\end{vmatrix}=-5(5a-2b-c)$

$\Delta_3=\begin{vmatrix}1 & 2 & a\\2 & 6 & b\\1 & -2 & c\end{vmatrix}=-2(5a-2b-c)$

For infinite solution $\Delta=\Delta_1=\Delta_2=\Delta_3=0$
$\Rightarrow 5a=2b+c$

69. (d) $\Delta=\begin{vmatrix}3 & -2 & -k\\2 & -4 & -2\\1 & 2 & -1\end{vmatrix}=0$

$\Rightarrow 24-2(0)-k(8)=0 \Rightarrow k=3$

$\Delta_x=\begin{vmatrix}10 & -2 & -3\\6 & -4 & -2\\5m & 2 & -1\end{vmatrix}$

$=10(8)-2(-10m+6)-3(12+20m)$
$=8(4-5m)$

$\Delta_y=\begin{vmatrix}3 & 10 & -3\\2 & 6 & -2\\1 & 5m & -1\end{vmatrix}$

$=3(-6+10m)+10(0)-3(10m-6)=0$

$\Delta_z=\begin{vmatrix}3 & -2 & 10\\2 & -4 & 6\\1 & 2 & 5m\end{vmatrix}$

$=3(-20m-12)-2(6-10m)+10(8)$
$=40m-32=8(5m-4)$

For the equations to be inconsistent,

$k=3 \text{ and } m\neq\frac{4}{5}$

70. (d) $D=\begin{vmatrix}1 & -2 & 0\\1 & -1 & k\\0 & k & 4\end{vmatrix}=4-k^2$

So, A is correct and B,C,E are incorrect.
If $k=2$

$D_1=\begin{vmatrix}1 & -2 & 0\\-2 & -1 & 2\\6 & 2 & 4\end{vmatrix}=-48\neq 0$

So, there is no solution.
D is correct.

71. (c)

$D=\begin{vmatrix}1 & 1 & 1\\1 & 2 & 3\\1 & 3 & \lambda\end{vmatrix}=0 \Rightarrow \lambda=5$

$D_x=\begin{vmatrix}2 & 1 & 1\\5 & 2 & 3\\\mu & 3 & 5\end{vmatrix}=0 \Rightarrow \mu=8$

72. (a)

$\because \begin{vmatrix}2 & -4 & \lambda\\1 & -6 & 1\\\lambda & -10 & 4\end{vmatrix}=0 \Rightarrow 3\lambda^2-7\lambda-12=0$

$\Rightarrow \lambda=3$ or $-\frac{2}{3}$

$D_1=\begin{vmatrix}1 & -4 & \lambda\\2 & -6 & 1\\3 & -10 & 4\end{vmatrix}=2(3-\lambda)$

$\therefore$ When $\lambda=-\frac{2}{3}$, $D_1\neq 0$

Hence, equations will be inconsistent when $\lambda=-\frac{2}{3}$.

73. (a) Since the system of linear equations has non-zero solution,

$\therefore \Delta=0$

$\Rightarrow \begin{vmatrix}1 & 2 & 3\\1 & 3 & k^2\\3 & 1 & 3\end{vmatrix}=0$

$\Rightarrow 1(9-k^2)-1(3-3k^2)+3(1-9)=0$
$\Rightarrow 9-k^2-3+3k^2-24=0$
$\Rightarrow 2k^2=18 \Rightarrow k^2=9, k=\pm 3$

So, the equations are

$x+y+3z=0$ ...(i)
$x+3y+9z=0$ ...(ii)
$3x+y+3z=0$ ...(iii)

Now, subtracting equation (ii) from (i),

$-2y-6z=0 \Rightarrow y=-3z \Rightarrow \frac{y}{z}=-3$ ...(iv)

Subtracting equation (iii) from (i),
$-2x=0 \Rightarrow x=0$

So, $x+\frac{y}{z}=0-3=-3$

74. (b) Given that $Ax=b$ has solutions x_1,x_2,x_3 and b is equal to b_1,b_2 and b_3

$\therefore x_1+y_1+z_1=1$

$\Rightarrow 2y_1+z_1=2 \Rightarrow z_1=2$

Determinant of coefficient matrix

$$|A|=\begin{vmatrix}1 & 1 & 1\\ 0 & 2 & 1\\ 0 & 0 & 1\end{vmatrix}=2$$

75. (d) $\begin{vmatrix}1 & 1 & 1\\ 2 & 4 & -1\\ 3 & 2 & \lambda\end{vmatrix}=0$

[The equation has many solutions.]

$\Rightarrow -15+6+2\lambda=0 \Rightarrow \lambda=\frac{9}{2}$

$$\therefore D_Z=\begin{vmatrix}1 & 1 & 2\\ 2 & 4 & 6\\ 3 & 2 & 2\mu\end{vmatrix}=0 \Rightarrow \mu=5$$

$\therefore 2\lambda+\mu=14$

76. (d) $\Delta=\begin{vmatrix}2 & -1 & 2\\ 1 & -2 & \lambda\\ 1 & \lambda & 1\end{vmatrix}=-(\lambda-1)(2\lambda+1)$

$$\Delta_1=\begin{vmatrix}2 & -1 & 2\\ -4 & -2 & \lambda\\ 4 & \lambda & 1\end{vmatrix}=-2(\lambda^2+6\lambda-4)$$

For no $\Delta=0$ and at least one of Δ_1,Δ_2 and Δ_3 is non-zero.

$\therefore \Delta=0 \Rightarrow \lambda=1,-\frac{1}{2}$ and $\Delta_1 \neq 0$

Hence, $S=\left\{1,-\frac{1}{2}\right\}$

77. (d) $\because |P|=1(-3+36)-2(2+4)+1(-18-3)=0$

Given that PX = 0

$\therefore$ System of equations

$x+2y+z=0; 2x-3y+4z=0$

and $x+9y-z=0$ has infinitely many solutions.

Let $z=k \in R$ and solve the above equations. We get

$x=-\frac{11k}{7}, y=\frac{2k}{7}, z=k$

But given that $x^2+y^2+z^2=1$

$\therefore k=\pm\frac{7}{\sqrt{174}}$

$\therefore$ Two solutions only.

78. (c) The given system of linear equations:

$7x+6y-2z=0$ ...(i)

$3x+4y+2z=0$ ...(ii)

$x-2y-6z=0$ ...(iii)

Now, determinant of coefficient matrix

$$\Delta=\begin{vmatrix}7 & 6 & -2\\ 3 & 4 & 2\\ 1 & -2 & -6\end{vmatrix}$$

$=7(-20)-6(-20)-2(-10)$

$=-140+120+20=0$

So, there are infinite non-trivial solutions.

From equation (i) $+3\times$ equation (iii); we get

$10x-20z=0 \Rightarrow x=2z$

Hence, there are infinitely many solutions (x,y,z) satisfying $x=2z$.

79. (a) From the given linear equation, we get

$$D=\begin{vmatrix}1 & 2 & 3\\ 3 & 4 & 5\\ 4 & 4 & 4\end{vmatrix}(R_3 \to R_3-2R_2+3R_3)$$

$$=\begin{vmatrix}1 & 2 & 3\\ 3 & 4 & 5\\ 0 & 0 & 0\end{vmatrix}=0$$

Now, let $P_3=4x+4y+4z-\delta=0$. If the system has solutions, it will have infinite solution.

So, $P_3 \equiv \alpha P_1+\beta P_2$

Hence, $3\alpha+\beta=4$ and $4\alpha+2\beta=4$

$\Rightarrow \alpha=2$ and $\beta=-2$

So, for infinite solution $2\mu-2=\delta$

$\Rightarrow$ For $2\mu \neq \delta+2$, the system is inconsistent.

80. (c) $D=\begin{vmatrix}\lambda & 2 & 2\\ 2\lambda & 3 & 5\\ 4 & \lambda & 6\end{vmatrix}$

$D=\lambda^2+6\lambda-16$

$D=(\lambda+8)(2-\lambda)$

For no solutions, $D=0$

$\Rightarrow \lambda=-8, 2$

When $\lambda=2$

$$D_1=\begin{vmatrix}5 & 2 & 2\\ 8 & 3 & 5\\ 10 & 2 & 6\end{vmatrix}$$

$= 5[18-10]-2[48-50]+2[16-30]$
$= 40+4-28 \neq 0$
There exist no solutions for $\lambda = 2$.

81. (a) For non-zero solution,

$$\begin{vmatrix} 2 & 2a & a \\ 2 & 3b & b \\ 2 & 4c & c \end{vmatrix} = 0 \Rightarrow \begin{vmatrix} 1 & 2a & a \\ 1 & 3b & b \\ 1 & 4c & c \end{vmatrix} = 0$$

$\Rightarrow (3bc-4bc)-(2ac-4ac)+(2ab-3ab)=0$
$\Rightarrow -bc+2ac-ab=0$
$\Rightarrow ab+bc+2ac$
$\Rightarrow \frac{2}{b}=\frac{1}{a}+\frac{1}{c}$
$\Rightarrow \frac{1}{a},\frac{1}{b},\frac{1}{c}$ are in AP.

82. (a) $|\text{adj}.A| = |A|^2 = 9$ $\quad \left[\because |\text{adj}.A| = |A|^{n-1}\right]$

$\Rightarrow |A| = \pm 3 = \lambda \Rightarrow |\lambda| = 3$
$\Rightarrow |B| = |\text{adj}.A|^2 = 81$
$\mu = \left|\left(B^{-1}\right)^T\right| = \left|B^{-1}\right| = |B|^{-1} = \frac{1}{|B|} = \frac{1}{81}$

83. (a)

$$|A| = \begin{vmatrix} 1 & 1 & 2 \\ 1 & 3 & 4 \\ 1 & -1 & 3 \end{vmatrix} = ((9+4)-1(3-4)+2(-1-3))$$

$= 13+1-8=6$
$|\text{adj}.B| = |\text{adj}.(\text{adj}.A)| = |A|^{(n-1)^2} = |A|^4 = (36)^2$
$|C| = |3A| = 3^3 \times 6$
Hence, $\frac{|\text{adj}.B|}{|C|} = \frac{36\times 36}{3^3\times 6} = 8$

84. (b) Applying $C_2 \to C_2 - C_1$

$$f(\theta) = \begin{vmatrix} -\sin^2\theta & -1 & 1 \\ -\cos^2\theta & -1 & 1 \\ 12 & -2 & -2 \end{vmatrix}$$

$= 4\left(\cos^2\theta - \sin^2\theta\right)$
$= 4\cos 2\theta, \theta \in \left(\frac{\pi}{4},\frac{\pi}{2}\right)$

Max. $f(\theta) = M = 0$
Min. $f(\theta) = \text{m} = -4$
So, $(m, M) = (-4, 0)$

85. (b) Use properties of determinant,

$$\begin{vmatrix} x & a+y & x+a \\ y & b+y & y+b \\ z & c+y & z+c \end{vmatrix} = \begin{vmatrix} x & a & x+a \\ y & b & y+b \\ z & c & z+c \end{vmatrix} + y\begin{vmatrix} x & 1 & x+a \\ y & 1 & y+b \\ z & 1 & z+c \end{vmatrix}$$

$$= 0 + y\begin{vmatrix} x & 1 & x+a \\ y-x & 0 & 0 \\ z-x & 0 & -1 \end{vmatrix} \quad \begin{bmatrix} R_2 \to R_2 - R_1 \\ R_3 \to R_3 - R_1 \end{bmatrix}$$

$= -y(x-y) = -y(b-a) = y(a-b)$

86. (d) It is given that $|B| = 81$

$$\therefore |B| = \begin{vmatrix} b_{11} & b_{12} & b_{13} \\ b_{21} & b_{22} & b_{23} \\ b_{31} & b_{32} & b_{33} \end{vmatrix} = \begin{vmatrix} 3^0 b_{11} & 3^1 b_{12} & 3^2 b_{13} \\ 3^1 b_{21} & 3^2 b_{22} & 3^3 b_{23} \\ 3^2 b_{31} & 3^3 b_{32} & 3^4 b_{33} \end{vmatrix}$$

$\Rightarrow 81 = 3^2.3^2.3^1|A|$
$\Rightarrow 3^4 = 3^6|A| \Rightarrow |A| = \frac{1}{9}$

87. (d) $\because A = \begin{bmatrix} \cos\theta & \sin\theta \\ -\sin\theta & \cos\theta \end{bmatrix}$

$\therefore A^n = \begin{bmatrix} \cos n\theta & \sin n\theta \\ -\sin n\theta & \cos n\theta \end{bmatrix}, n \in N$

$\therefore B = A + A^4$

$= \begin{bmatrix} \cos\theta & \sin\theta \\ -\sin\theta & \cos\theta \end{bmatrix} + \begin{bmatrix} \cos 4\theta & \sin 4\theta \\ -\sin 4\theta & \cos 4\theta \end{bmatrix}$

$$\therefore B = \begin{bmatrix} \cos\frac{\pi}{5}+\cos\frac{4\pi}{5} & \sin\frac{\pi}{5}+\sin\frac{4\pi}{5} \\ -\sin\frac{\pi}{5}-\sin\frac{4\pi}{5} & \cos\frac{\pi}{5}+\cos\frac{4\pi}{5} \end{bmatrix}$$

Then, $\det(B) = 2\sin\left(\frac{\pi}{5}\right)\begin{vmatrix} 0 & 1 \\ -1 & 0 \end{vmatrix}$

$= \frac{\sqrt{10-2\sqrt{5}}}{2} \approx \frac{2.35}{2} \approx 1.175 \therefore \det B \in (1,2)$

88. (c) $\Delta = \begin{vmatrix} x-2 & 2x-3 & 3x-4 \\ 2x-3 & 3x-4 & 4x-5 \\ 3x-5 & 5x-8 & 10x-17 \end{vmatrix}$

$$\Rightarrow \Delta = \begin{vmatrix} x-2 & x-1 & x-1 \\ 2x-3 & x-1 & x-1 \\ 3x-5 & 2x-3 & 5x-9 \end{vmatrix} \quad \begin{bmatrix} C_3 \to C_3 - C_2 \\ C_2 \to C_2 - C_1 \end{bmatrix}$$

$$\Rightarrow \Delta = \begin{vmatrix} x-2 & x-1 & x-1 \\ x-1 & 0 & 0 \\ 3x-5 & 2x-3 & 5x-9 \end{vmatrix} \quad [R_2 \to R_2 - R_1]$$

$\Rightarrow \Delta = -(x-1)\left[(x-1)(5x-9)-(x-1)(2x-3)\right]$

$\Rightarrow \Delta = -(x-1)\left[\left(5x^2-14x+9\right)-\left(2x^2-5x+3\right)\right]$

$= -3x^3+12x^2-15x+6$ So, $B+C=-3$

89. (d) If $f(x)=\begin{vmatrix} x+a & x+2 & x+1 \\ x+b & x+3 & x+2 \\ x+c & x+4 & x+3 \end{vmatrix}$

$R_1 = R_1 + R_3 - 2R_2$

$\Rightarrow f(x)=\begin{vmatrix} 1 & 0 & 0 \\ x+b & x+3 & x+2 \\ x+c & x+4 & x+3 \end{vmatrix}$

$\Rightarrow f(x)=1 \Rightarrow f(50)=1$

90. (d) $\Delta_1=\begin{vmatrix} x & \sin\theta & \cos\theta \\ -\sin\theta & -x & 1 \\ \cos\theta & 1 & x \end{vmatrix}$

$=\left(x-x^2-1\right)-\sin\theta\left(x\sin\theta-\cos\theta\right)+\cos\theta\left(-\sin\theta+x\cos\theta\right)$

$=-x^3-x+x\sin^2\theta+\sin\theta\cos\theta-\cos\theta\sin\theta+x\cos^2\theta$

$=-x^3-x+x=-x^3$

Similarly, $\Delta_2=-x^3$. Then, $\Delta_1+\Delta_2=-2x^3$

91. (a) $|A|=\begin{vmatrix} 2 & b & 1 \\ b & b^2+1 & b \\ 1 & b & 2 \end{vmatrix}$

$=2\left(2b^2+2-b^2\right)-b(2b-b)+1\left(b^2-b^2-1\right)$

$=2b^2+4-b^2-1=b^2+3$

$\frac{|A|}{b}=b+\frac{3}{b}$

$\because \frac{b+\frac{3}{b}}{2}\geq\left(b\frac{3}{b}\right)^{\frac{1}{2}} \Rightarrow b+\frac{3}{b}\geq 2\sqrt{3}$

$\therefore \frac{|A|}{b}\geq 2\sqrt{3}$

Minimum value of $\frac{|A|}{b}$ is $2\sqrt{3}$.

92. (a) $C_1 \to C_1 + C_2$

$\begin{vmatrix} 2 & \sin^2\theta & 4\cos 6\theta \\ 2 & 1+\sin^2\theta & 4\cos 6\theta \\ 2 & \sin^2\theta & 1+4\cos 6\theta \end{vmatrix}=0$

$R_1 \to R_1 - R_2, R_2 \to R_2 - R_3$

$\begin{vmatrix} 0 & -1 & 0 \\ 1 & 1 & -1 \\ 1 & \sin^2\theta & (1+4\cos 6\theta) \end{vmatrix}=0$

On expanding, we get $2+4\cos 6\theta=0$

$\cos 6\theta=-\frac{1}{2} \because \theta\in\left(0,\frac{\pi}{3}\right)\Rightarrow 6\theta\,(0,2\pi)$

Therefore, $6\theta=\frac{2\pi}{3}$ or $\frac{4\pi}{3}\Rightarrow\theta=\frac{\pi}{9}$ or $\frac{2\pi}{9}$

93. (c) Consider, $|A|=\begin{vmatrix} 1 & 1 & 1 \\ 2 & b & c \\ 4 & b^2 & c^2 \end{vmatrix}$

$C_2 \to C_2 - C_1, C_3 \to C_3 - C_1$

$|A|=\begin{vmatrix} 1 & 0 & 0 \\ 2 & b-2 & c-2 \\ 4 & (b-2)(b+2) & (c-2)(c+2) \end{vmatrix}$

$=(b-2)(c-2)(c-b)$

$\therefore 2,b,c$ are in AP.

$\therefore (b-2)=(c-b)=d$ and $c-2=2d$

$\Rightarrow |A|=d.2d.d=2d^3$

$\because |A|\in[2,16]\Rightarrow 1\leq d^3\leq 8\Rightarrow 1\leq d\leq 2$

$4\leq 2d+2\leq 6\Rightarrow 4\leq c\leq 6$

94. (d) $|A|=\begin{vmatrix} 1 & \sin\theta & 1 \\ -\sin\theta & 1 & \sin\theta \\ -1 & -\sin\theta & 1 \end{vmatrix}$

$=\begin{vmatrix} 0 & 0 & 2 \\ -\sin\theta & 1 & \sin\theta \\ -1 & -\sin\theta & 1 \end{vmatrix} R_1 \to R_1 + R_3$

$=2\left(\sin^2\theta+1\right)$

Since, $\theta\in\left(\frac{3\pi}{4},\frac{5\pi}{4}\right)\Rightarrow \sin^2\theta\in\left(0,\frac{1}{2}\right)$

$\therefore \det(A)\in[2,3) \quad [2,3)\subset\left(\frac{3}{2},3\right]$

95. (a) $\det(A)=\begin{vmatrix} -2 & 4+d & \sin\theta-2 \\ 1 & \sin\theta+2 & d \\ 5 & 2\sin\theta-d & -\sin\theta+2+2d \end{vmatrix}$

Applying $R_3 \to R_3 - 2R_2 + R_1$ we get

$\det(A)=\begin{vmatrix} -2 & 4+d & \sin\theta-2 \\ 1 & \sin\theta+2 & d \\ 1 & 0 & 0 \end{vmatrix}$

$= d(4+d)-(\sin^2\theta-4)$

$\Rightarrow \det(A)=d^2+4d+4-\sin^2\theta=(d+2)^2-\sin^2\theta$

Minimum value of $\det(A)$ is attained when $\sin^2\theta=1$

$\therefore (d+2)^2-1=8 \Rightarrow (d+2)^2=9 \Rightarrow d+2=\pm 3$

$\Rightarrow d=-5$ or 1

96. (a) $\det(A)=|A|$

$$=\begin{vmatrix} e^t & e^{-t}\cos t & e^{-t}\sin t \\ e^t & -e^{-t}\cos t-e^{-t}\sin t & -e^{-t}\sin t+e^{-t}\cos t \\ e^t & 2e^{-t}\sin t & -2e^{-t}\cos t \end{vmatrix}$$

$$=e^t.e^{-t}.e^{-t}\begin{vmatrix} 1 & \cos t & \sin t \\ 1 & -\cos t-\sin t & -\sin t+\cos t \\ 1 & 2\sin t & -2\cos t \end{vmatrix}$$

$$=e^{-t}\begin{vmatrix} 0 & 2\cos t+\sin t & 2\sin t-\cos t \\ 0 & -\cos t-3\sin t & -\sin t+3\cos t \\ 1 & 2\sin t & -2\cos t \end{vmatrix}$$

$$R_1 \to R_1 - R_2$$
$$R_2 \to R_2 + R_3$$

$$=e^{-t}\left((-5\sin t)(-\sin t+3\cos t)-5\cos t(-\cos t-3\sin t)\right)$$

$=5e^{-t}\neq 0, \forall t\in R$

$\therefore A$ is invertible.

97. (c) $\because B=A^{-1}\Rightarrow |B|=\dfrac{1}{|A|}$

Now, $|B|=\begin{vmatrix} 5 & 2\alpha & 1 \\ 0 & 2 & 1 \\ \alpha & 3 & -1 \end{vmatrix}=2\alpha^2-2\alpha-25$

Given, $\det.(A)+1=0$

$\Rightarrow \dfrac{1}{2\alpha^2-2\alpha-25}+1=0$

$\Rightarrow \dfrac{2\alpha^2-2\alpha-24}{2\alpha^2-2\alpha-25}=0$

$\Rightarrow \alpha=4,-3 \Rightarrow$ Sum of values $=1$

98. (b)

$$\begin{bmatrix} 1 & 1 \\ 0 & 1 \end{bmatrix}\begin{bmatrix} 1 & 2 \\ 0 & 1 \end{bmatrix}\begin{bmatrix} 1 & 3 \\ 0 & 1 \end{bmatrix}\begin{bmatrix} 1 & 4 \\ 0 & 1 \end{bmatrix}\cdots\begin{bmatrix} 1 & n-1 \\ 0 & 1 \end{bmatrix}=\begin{bmatrix} 1 & 78 \\ 0 & 1 \end{bmatrix}$$

$$\Rightarrow \begin{bmatrix} 1 & 1+2+3+\ldots+(n-1) \\ 0 & 1 \end{bmatrix}=\begin{bmatrix} 1 & 78 \\ 0 & 1 \end{bmatrix}$$

$\Rightarrow \dfrac{(n-1)n}{2}=78 \Rightarrow n^2-n-15=0$

$\Rightarrow n=13$

Now, the matrix $\begin{bmatrix} 1 & n \\ 0 & 1 \end{bmatrix}=\begin{bmatrix} 1 & 13 \\ 0 & 1 \end{bmatrix}$

Then, the required inverse of $\begin{bmatrix} 1 & 13 \\ 0 & 1 \end{bmatrix}=\begin{bmatrix} 1 & -13 \\ 0 & 1 \end{bmatrix}$

99. (c) Let $|A|=a, |B|=b$

$\Rightarrow |A^T|=a|A^{-1}|=\dfrac{1}{a}, |B^T|=b, |B^{-1}|=\dfrac{1}{b}$

$\because |ABA^T|=8 \Rightarrow |A||B||A^T|=8$ ...(i)

$\Rightarrow a.b.a=8 \Rightarrow a^2b=8$

$\because |AB^{-1}|=8 \Rightarrow |A||B^{-1}|=8 \Rightarrow a.\dfrac{1}{b}=8$ ...(ii)

From (i) and (ii),

$a=4, b=\dfrac{1}{2}$

Then, $|BA^{-1}B^T|=|B||A^{-1}||B^T|=b\cdot\dfrac{1}{a}\cdot b=\dfrac{b^2}{a}=\dfrac{1}{16}$

100. (d) Given system of linear equations: $x+y+z=5$; $x+2y+2z=6$ and $x+3y+\lambda z=\mu$ have infinite solution.

$\therefore \Delta=0, \Delta x=\Delta y=\Delta z=0$

Now, $\Delta=\begin{vmatrix} 1 & 1 & 1 \\ 1 & 2 & 2 \\ 1 & 3 & \lambda \end{vmatrix}=0$

$\Rightarrow 1(2\lambda-6)-1(\lambda-2)+1(3-2)=0$

$\Rightarrow 2\lambda-6-\lambda+2+1=0 \Rightarrow \lambda=3$

$\Delta y=\begin{vmatrix} 1 & 5 & 1 \\ 1 & 6 & 2 \\ 1 & \mu & 3 \end{vmatrix}=0 \Rightarrow \begin{vmatrix} 1 & 5 & 1 \\ 0 & 1 & 1 \\ 0 & \mu-5 & 2 \end{vmatrix}=0$

$\Rightarrow 1(2-\mu+5)=0 \Rightarrow \mu=7$

$\therefore \lambda+\mu=10$

101. (b) The given system of equations has a non-trivial solution.

$\Rightarrow \Delta=0 \Rightarrow \begin{vmatrix} 2 & 3 & -1 \\ 1 & k & -2 \\ 2 & -1 & 1 \end{vmatrix}=0 \Rightarrow k=\dfrac{9}{2}$

$\therefore$ The equations are $2x+3y-z=0$ ...(i)

$2x-y+z=0$ ...(ii)

$2x+9y-4z=0$ ...(iii)

Subtracting (ii) from (i), $2y=z$

$\therefore z=-4x$ and $2x+y=0$

$\therefore \frac{x}{y}+\frac{y}{z}+\frac{z}{x}+k=\frac{-1}{2}+\frac{1}{2}-4+\frac{9}{2}=\frac{1}{2}$

102. (a) Consider the given system of linear equations

$x(1-\lambda)-2y-2z=0$

$x+(2-\lambda)y+z=0$

$-x-y-\lambda z=0$

Now, for a non-trivial solution, the determinant of coefficient matrix is zero.

$$\begin{vmatrix} 1-\lambda & -2 & -2 \\ 1 & 2-\lambda & 1 \\ -1 & -1 & -\lambda \end{vmatrix}=0$$

$\Rightarrow (1-\lambda)^3=0$

$\lambda=1$

103. (b) Since the system of linear equations has non-trivial solution, the determinant of coefficient matrix = 0

i.e., $\begin{vmatrix} \sin 3\theta & \cos 2\theta & 2 \\ 1 & 3 & 7 \\ -1 & 4 & 7 \end{vmatrix}=0$

$\sin 3\theta(21-28)-\cos 2\theta(7+7)+2(4+3)=0$

$\sin 3\theta+2\cos 2\theta-2=0$

$3\sin\theta-4\sin^3\theta+2-4\sin^2\theta-2=0$

$4\sin^3\theta+4\sin^2\theta-3\sin\theta=0$

$\sin\theta\left(4\sin^2\theta+4\sin\theta-3\right)=0$

$\sin\theta\left(4\sin^2\theta+6\sin\theta-2\sin\theta-3\right)=0$

$\sin\theta\left[2\sin\theta(2\sin\theta-1)+3(2\sin\theta-1)\right]=0$

$\sin\theta(2\sin\theta-1)(2\sin\theta+3)=0$

$\sin\theta=0, \sin\theta=\frac{1}{2} \quad \left(\because \sin\theta\neq-\frac{3}{2}\right)$

$\theta=\frac{\pi}{6},\frac{5\pi}{6}$

Hence, for two values of θ, the system of equations has non-trivial solution

104. (a) For non-zero solution of the system of linear equations;

$$\begin{vmatrix} 1 & k & 3 \\ 3 & k & -2 \\ 2 & 4 & -3 \end{vmatrix}=0$$

$\Rightarrow k=11$

Now the equations become

$x+11y+3z=0$ ...(i)

$3x+11y-2z=0$ ...(ii)

$2x+4y-3z=0$ ...(iii)

Adding equations (i) and (iii), we get

$3x+15y=0$

$\Rightarrow x=-5y$

Now put $x=-5y$ in equation (i), we get

$-5y+11y+3z=0$

$\Rightarrow z=-2y$

$\therefore \frac{xz}{y^2}=\frac{(-5y)(-2y)}{y^2}=10$

105. (b)

Here, $\begin{vmatrix} x-4 & 2x & 2x \\ 2x & x-4 & 2x \\ 2x & 2x & x-4 \end{vmatrix}=(A+Bx)(x-A)^2$

Put $x=0 \Rightarrow \begin{vmatrix} -4 & 0 & 0 \\ 0 & -4 & 0 \\ 0 & 0 & -4 \end{vmatrix}=A^3 \Rightarrow A^3=(-4)^3$

$\Rightarrow A=-4$

$\Rightarrow \begin{vmatrix} x-4 & 2x & 2x \\ 2x & x-4 & 2x \\ 2x & 2x & x-4 \end{vmatrix}=(Bx-4)(x+4)^2$

Now take x common from both the sides.

$\therefore \begin{vmatrix} 1-\frac{4}{x} & 2x & 2x \\ 2x & 1-\frac{4}{x} & 2x \\ 2x & 2x & 1-\frac{4}{x} \end{vmatrix}=\left(B-\frac{4}{x}\right)\left(1+\frac{4}{x}\right)^2$

Now take $x\to\infty$, then $\frac{1}{x}\to 0$

$\Rightarrow \begin{vmatrix} 1 & 2 & 2 \\ 2 & 1 & 2 \\ 2 & 2 & 1 \end{vmatrix}=B \Rightarrow B=5$

$\therefore$ ordered pair (A, B) is (–4, 5)

106. (a) We have $(A-3I)(A-5I)=0$

$\Rightarrow A^2-8A+15I=0$

Multiplying both sides by A^{-1}, we get;

$A^{-1}A.A-8A^{-1}A+15A^{-1}I=A^{-1}0$

$\Rightarrow A-8I+15A^{-1}=0$

$A+15A^{-1}=8I$

$\frac{A}{2}+\frac{15A^{-1}}{2}=4I$

$\therefore \alpha+\beta=\frac{1}{2}+\frac{15}{2}=\frac{16}{2}=8$

107. (b)

108. (c) Since the given determinant is equal to zero.

$$\Rightarrow 0(0-\cos x\sin x)-\cos x(0-\cos^2 x)-\sin x(\sin^2 x-0)=0$$

$\Rightarrow \cos^3 x-\sin^3 x=0$

$\Rightarrow \tan^3=1 \Rightarrow \tan x=1$

$$\therefore \sum_{x\in s}\tan\left(\frac{\pi}{3}+x\right)=\sum_{x\in s}\frac{\tan\pi/3+\tan x}{1-\tan\frac{\pi}{3}.\tan x}$$

$$\sum_{x\in S}\frac{\sqrt{3}+1}{1-\sqrt{3}}$$

$$\sum_{x\in S}\frac{\sqrt{3}+1}{1-\sqrt{3}}\times\frac{1+\sqrt{3}}{1+\sqrt{3}}\Rightarrow\sum_{x\in s}\frac{1+3+2\sqrt{3}}{-2}$$

$=-2-\sqrt{3}$

109. (a) $D=\begin{vmatrix}1&1&1\\1&a&1\\a&b&1\end{vmatrix}=0$

$\Rightarrow 1[a-b]-1[1-a]+1[b-a^2]=0$

$\Rightarrow (a-1)^2=0$

$\Rightarrow a=1$

For $a=1$, the first two equations are identical

i.e., $x+y+z=1$

To have no solution with $x+by+z=0$

b should be equal to 1.

So $b=\{1\}\Rightarrow$ It is singleton set.

110. (a) Let $A(k,-3k)$, $B(5,k)$ and $C(-k+2)$,

We have

$$\frac{1}{2}\left\|\begin{matrix}k&-3k&1\\5&k&1\\-k&2&1\end{matrix}\right\|=28$$

$\Rightarrow 5k^2+13k-46=0$

or $5k^2+13k+66=0$

Now, $5k^2+13k-46=0$

$$\Rightarrow k=\frac{-13\pm\sqrt{1089}}{10}\therefore k=\frac{-23}{5};k=2$$

Since k is an integer, $\therefore k=2$

Also $5k^2+13k+66=0$

$$\Rightarrow k=\frac{-13\pm\sqrt{-1151}}{10}$$

So no real solution exists.

$A(2,-6)$, $B(5,2)$ and $C(-2,2)$

For orthocentre $H(\alpha,\beta)$

$BH\perp AC$

$$\therefore\left(\frac{\beta-2}{\alpha-5}\right)\left(\frac{8}{-4}\right)=-1$$

$\Rightarrow \alpha-2\beta=1$ ...(i)

Also $CH\perp AB$

$$\therefore\left(\frac{\beta-2}{\alpha+2}\right)\left(\frac{8}{3}\right)=-1$$

$\Rightarrow 3\alpha+8\beta=1$ ...(ii)

Solving (i) and (ii), we get

$\alpha=2,\ \beta=\frac{1}{2}$

Orthocentre is $\left(2,\frac{1}{2}\right)$

111. (d) $A=\begin{bmatrix}-4&-1\\3&1\end{bmatrix}\Rightarrow A^2=\begin{bmatrix}-4&-1\\3&1\end{bmatrix}\begin{bmatrix}-4&-1\\3&1\end{bmatrix}$

$=\begin{bmatrix}13&3\\-9&-2\end{bmatrix}$ and $|A|=1$

Now, $A^{2016}-2A^{2015}-A^{2014}=A^{2014}(A^2-2A-I)$

$\Rightarrow|A^{2016}-2A^{2015}-A^{2014}|=|A^{2014}||A^2-2A-I|$

$=|A|^{2014}\begin{vmatrix}20&5\\-15&-5\end{vmatrix}=-25$

112. (d) Given that $A(\text{adj}.A)=AA^T$

Pre-multiply by A^{-1} both sides, we get

$\Rightarrow A^{-1}A(\text{adj }A)=A^{-1}AA^T$

$\text{adj}.A=A^T$

$\Rightarrow\begin{bmatrix}2&b\\-3&5a\end{bmatrix}=\begin{bmatrix}4a&3\\-b&2\end{bmatrix}$

$\Rightarrow a=\frac{2}{5}$ and $b=3$

$\Rightarrow 5a+b=5$

113. (a) Let $\begin{vmatrix}x^2+x&x+1&x-2\\2x^2+3x-1&3x&3x-3\\x^2+2x+3&2x-1&2x-1\end{vmatrix}=ax-12$

Put $x=-1$, we get

$\begin{vmatrix}0&0&-3\\-2&-3&0\\2&-3&-3\end{vmatrix}=-a-12$

$\Rightarrow -3(6+6)=-a-12\Rightarrow -36+12=a$

$\Rightarrow a=24$

114. (c) Let $f(\theta)=\begin{vmatrix} 1 & \cos\theta & 1 \\ -\sin\theta & 1 & -\cos\theta \\ -1 & \sin\theta & 1 \end{vmatrix}$

$=(1+\sin\theta\cos\theta)-\cos\theta(-\sin\theta-\cos\theta)+1\left(-\sin^2\theta-1\right)$

$=1+\sin\theta\cos\theta+\sin\theta\cos\theta+\cos^2\theta-\sin^2\theta+1$

$=2+2\sin\theta\cos\theta+\cos 2\theta$

$=2+\sin 2\theta+\cos 2\theta$ ...(i)

Now, maximum value of (i) is

$2+\sqrt{1^2+1^2}=2+\sqrt{2}$

and minimum value of (i) is

$2-\sqrt{1^2+1^2}=2-\sqrt{2}$

115. (d) $BB'=B\left(A^{-1}A'\right)'=B(A')'\left(A^{-1}\right)'$

$=BA\left(A^{-1}\right)'=\left(A^{-1}A'\right)\left(A\left(A^{-1}\right)'\right)$

$=A^{-1}A.A'.\left(A^{-1}\right)'$ {as $AA'=A'A$}

$=I\left(A^{-1}A\right)'=I.I=I^2=I$

116. (d) $\sum_{r=1}^{n-1} r=1+2+3+\ldots+(n-1)=\frac{n(n-1)}{2}$

$\sum_{r=1}^{n-1}(2r-1)=1+3+5+\ldots+\left[2(n-1)-2\right]=(n-1)^2$

$\sum_{r=1}^{n-1}(3r-2)=1+4+7+\ldots+(3n-3-2)$

$=\frac{(n-1)(3n-4)}{2}$

$\therefore \sum_{r=1}^{n-1}\Delta_r=\begin{vmatrix} \sum r & \sum(2r-1) & \sum(3r-2) \\ \frac{n}{2} & n-1 & a \\ \frac{n(n-1)}{2} & (n-1)^2 & \frac{(n-1)(3n-4)}{2} \end{vmatrix}$

$\sum_{r=1}^{n-1}\Delta_r$ consists of $(n-1)$ determinants in LHS and in RHS every constituent of first row consists of $(n-1)$ elements and hence it can be splitted into sum of $(n-1)$ determinants.

$\therefore \sum_{r=1}^{n-1}\Delta_r=\begin{vmatrix} \frac{n(n-1)}{2} & (n-1)^2 & \frac{(n-1)(3n-4)}{2} \\ \frac{n}{2} & n-1 & a \\ \frac{n(n-1)}{2} & (n-1)^2 & \frac{(n-1)(3n-4)}{2} \end{vmatrix}=0$

($\because R_1$ and R_3 are identical)

Hence, value of $\sum_{r=1}^{n-1}\Delta_r$ is independent of both 'a' and 'b'.

117. (b) $|P|=1(12-12)-\alpha(4-6)+3(4-6)=2\alpha-6$

Now, adj. $A=P\Rightarrow|\text{adj}.A|=|P|$

$\Rightarrow|A|^2=|P|$

$\Rightarrow|P|=16$

$\Rightarrow 2\alpha-6=16$

$\Rightarrow\alpha=11$

118. (b) Since the system of equations have no solution

$\therefore \frac{k+1}{k}=\frac{8}{k+3}\neq\frac{4k}{3k-1}$

($\because$ the system has no solution)

$\Rightarrow k^2+4k+3=8k\Rightarrow k^2-4k+3=0\Rightarrow k=1,3$

If $k=1$ then $\frac{8}{1+3}\neq\frac{4.1}{2}$, which is false and if $k=3$ then $\frac{8}{6}\neq\frac{4.3}{9-1}$ which is true, therefore $k=3$.

Hence for only one value of k, the system has no solution.

119. (d) $|Q|=\begin{vmatrix} 2^2a_{11} & 2^3a_{12} & 2^4a_{13} \\ 2^3a_{21} & 2^4a_{22} & 2^5a_{23} \\ 2^4a_{31} & 2^5a_{32} & 2^6a_{33} \end{vmatrix}$

$=2^2.2^3.2^4\begin{vmatrix} a_{11} & a_{12} & a_{13} \\ 2a_{21} & 2a_{22} & 2a_{23} \\ 2^2a_{31} & 2^2a_{32} & 2^2a_{33} \end{vmatrix}$

$=2^9.2.2^2\begin{vmatrix} a_{11} & a_{12} & a_{13} \\ a_{21} & a_{22} & a_{23} \\ a_{31} & a_{32} & a_{33} \end{vmatrix}$

$=2^{12}\times|P|=2^{12}\times 2=2^{13}$

120. (a) For the given matrix to be non-singular

$\begin{vmatrix} 1 & a & b \\ \omega & 1 & c \\ \omega^2 & \omega & 1 \end{vmatrix}\neq 0$

$\Rightarrow 1-(a+c)\omega+ac\omega^2\neq 0\Rightarrow(1-a\omega)(1-c\omega)\neq 0$

$\Rightarrow a\neq\omega^2$ and $c\neq\omega^2$, where ω is complex cube roots of unity.

As a, b and c are complex cube roots of unity.

$\therefore a$ and c can take only one value, i.e., ω while b can take 2 values, i.e., ω and ω^2.

$\therefore$ Total number of distinct matrices in the set S.

$=1\times1\times2=2$

121. (a) Since, $A\begin{bmatrix}x\\y\\z\end{bmatrix}=\begin{bmatrix}1\\0\\0\end{bmatrix}$ is linear equation in three variables and that could have only unique, no solution or infinitely many solution.

$\therefore$ It is not possible to have two solutions.

Hence, number of matrices A is zero.

Integer and Numerical Value Type

1. We have $D=\begin{vmatrix}a_1 & a_2 & a_3\\5 & 4 & a_6\\a_7 & a_8 & a_9\end{vmatrix}$

Since $a_n=\frac{20}{n};d=\frac{1}{20}$

Hence, $D=\begin{vmatrix}20 & \frac{20}{2} & \frac{20}{3}\\\frac{20}{4} & \frac{20}{5} & \frac{20}{6}\\\frac{20}{7} & \frac{20}{8} & \frac{20}{9}\end{vmatrix}=\frac{(20)^3}{4\times 7}=\begin{vmatrix}1 & \frac{1}{2} & \frac{1}{3}\\1 & \frac{4}{5} & \frac{2}{3}\\1 & \frac{7}{8} & \frac{7}{9}\end{vmatrix}$

$R_1\to R_1-R_2$ and $R_2\to R_2-R_3$

$$=\frac{(20)^3}{4\times 7}\begin{vmatrix}0 & \frac{-3}{10} & \frac{-1}{3}\\0 & \frac{-3}{40} & \frac{-1}{9}\\1 & \frac{7}{8} & \frac{7}{9}\end{vmatrix}=\frac{50}{21}$$

$\Rightarrow [D]=2$

2.

Let $D=\begin{vmatrix}(\beta+\gamma-\alpha-\delta)^4 & (\beta+\gamma-\alpha-\delta)^2 & 1\\(\gamma+\alpha-\beta-\delta)^4 & (\gamma+\alpha-\beta-\delta)^2 & 1\\(\alpha-\beta-\gamma-\delta)^4 & (\alpha-\beta-\gamma-\delta)^2 & 1\end{vmatrix}$

Applying $R_1\to R_1-R_3, R_2\to R_2-R_3$

$$=\begin{vmatrix}(\beta+\gamma-\alpha-\delta)^4-(\alpha+\beta-\gamma-\delta)^4 & (\beta+\gamma-\alpha-\delta)^2-(\alpha+\beta-\gamma-\delta)^2 & 0\\(\gamma+\delta-\beta-\delta)^4-(\alpha+\beta-\gamma-\delta)^4 & (\gamma+\alpha-\beta-\delta)^2-(\alpha+\beta-\gamma-\delta)^2 & 0\\(\alpha+\beta-\gamma-\delta)^4 & (\alpha+\beta-\gamma-\delta)^2 & 1\end{vmatrix}$$

$$=4(\beta-\delta)(\gamma-\alpha).4(\alpha-\delta)(\gamma-\beta)\times\begin{vmatrix}(\beta+\gamma-\alpha-\delta)^2+(\alpha+\beta-\gamma-\delta)^2 & 1 & 0\\(\gamma+\alpha-\beta-\delta)^2+(\alpha+\beta-\gamma-\delta)^2 & 1 & 0\\(\alpha+\beta-\gamma-\delta)^4 & (\alpha+\beta-\gamma-\delta)^2 & 1\end{vmatrix}$$

Apply $R_1\to R_1-R_2$

$$=16(\beta-\delta)(\gamma-\alpha)(\alpha-\delta).4(\gamma-\delta)(\beta-\alpha)\times\begin{vmatrix}1 & 0 & 0\\(\gamma+\alpha-\beta-\delta)^2+(\alpha+\beta-\gamma-\delta)^2 & 1 & 0\\(\alpha+\beta-\gamma-\delta)^4 & (\alpha+\beta-\gamma-\delta)^2 & 1\end{vmatrix}$$

$$=-64(\alpha-\beta)(\alpha-\gamma)(\alpha-\delta)(\beta-\gamma)(\beta-\delta)(\gamma-\delta)$$

3. Taking x common from R_2 and $x(x-1)$ common from R_3, we get

$$f(x)=x^2(x-1)\begin{vmatrix}1 & x & x+1\\2 & x-1 & x+1\\3 & x-2 & x+1\end{vmatrix}$$

Applying $C_3\to C_3-C_2$, we get

$$f(x)=x^2(x-1)\begin{vmatrix}1 & x & 1\\2 & x-1 & 2\\3 & x-2 & 3\end{vmatrix}=0$$

Thus, $f(500)=0$

4. Using $C_3\to C_3-(C_1+C_2)$ in D_1 and D_2, we have

$$\frac{D_1}{D_2}=\frac{-2b(ad-bc)}{b(ad-bc)}=-2$$

5. $\Delta_1=\begin{vmatrix}a_1^2+4a_1d & a_1 & d\\a_2^2+4a_2d & a_2 & d\\a_3^2+4a_3d & a_3 & d\end{vmatrix},[C_3\to C_3-C_2]$

where d is the common difference of AP.

$$=d\begin{vmatrix}a_1^2 & a_1 & 1\\a_2^2 & a_2 & 1\\a_3^2 & a_3 & 1\end{vmatrix}+4d\begin{vmatrix}a_1 & a_1 & d\\a_2 & a_2 & d\\a_3 & a_3 & d\end{vmatrix}$$

$=d(a_1-a_2)(a_2-a_3)(a_3-a_1)=-2d^4$

Similarly, $\Delta_2=-2d^4$

6. $\Delta_2=\underbrace{\begin{vmatrix}a & b+3c & c+4a\\b & c+3a & a+4b\\c & a+3b & b+4c\end{vmatrix}}_{\Delta'}+2\underbrace{\begin{vmatrix}b & b+3c & c+4a\\c & c+3a & a+4b\\a & a+3b & b+4c\end{vmatrix}}_{\Delta''}$

Applying $C_3 \to C_3 - 4C_1$ in Δ' and $C_2 \to C_2 - C_1$ in Δ'', we get

$$\Delta_2 = \begin{vmatrix} a & b+3c & c \\ b & c+3a & a \\ c & a+3b & b \end{vmatrix} + 2\times 3\begin{vmatrix} b & c & c+4a \\ c & a & a+4b \\ a & b & b+4c \end{vmatrix}$$

Applying $C_2 \to C_2 - 3C_3$ in Δ' and $C_3 \to C_3 - C_2$ in Δ'', we get

$$= \begin{vmatrix} a & b & c \\ b & c & a \\ c & a & b \end{vmatrix} + 6\begin{vmatrix} b & c & 4a \\ c & a & 4b \\ a & b & 4c \end{vmatrix}$$

$$= \begin{vmatrix} a & b & c \\ b & c & a \\ c & a & b \end{vmatrix} + 24\begin{vmatrix} a & b & c \\ b & c & a \\ c & a & b \end{vmatrix}$$

$$= 25\begin{vmatrix} a & b & c \\ b & c & a \\ c & a & b \end{vmatrix} = 25\Delta_1$$

$$\Rightarrow \frac{\Delta_2}{\Delta_1} = 25$$

7. $$\Delta_r = \begin{vmatrix} r & 612 & 915 \\ 101r^2 & 2r & 3r \\ r & \frac{1}{r} & \frac{1}{r^2} \end{vmatrix}$$

$$= r^2\begin{vmatrix} 1 & 612 & 915 \\ 101 & 2r & 3r \\ 1 & \frac{1}{r} & \frac{1}{r^2} \end{vmatrix}$$

(Taking r common from C_1 and R_2)

$$= \begin{vmatrix} 1 & 612 & 915 \\ 101 & 2 & 3 \\ r^2 & r & 1 \end{vmatrix}$$

$$\therefore \lim_{n\to\infty} \frac{1}{n^3}\sum_{r=1}^{n}\begin{vmatrix} 1 & 612 & 915 \\ 101 & 2 & 3 \\ r^2 & r & 1 \end{vmatrix}$$

$$= \lim_{n\to\infty} \frac{1}{n^3}\begin{vmatrix} 1 & 612 & 915 \\ 101 & 2 & 3 \\ \frac{n(n+1)(2n+1)}{6} & \frac{n(n+1)}{2} & n \end{vmatrix}$$

$$= \lim_{n\to\infty}\begin{vmatrix} 1 & 612 & 915 \\ 101 & 2 & 3 \\ \frac{n(n+1)(2n+1)}{6n^2} & \frac{n+1}{2n^2} & \frac{1}{n^2} \end{vmatrix}$$

$$= \begin{vmatrix} 1 & 612 & 915 \\ 101 & 2 & 3 \\ \frac{1}{3} & 0 & 0 \end{vmatrix}$$

$$= \frac{1}{3}(3\times 612 - 2\times 915)$$

$$= 2$$

8. $\{2(x-y)^2(y-z)^2(z-x)^2\}$

$$\begin{vmatrix} (x^2+1)^2 & (xy+1)^2 & (xz+1)^2 \\ (xy+1)^2 & (y^2+1)^2 & (yz+1)^2 \\ (xz+1)^2 & (yz+1)^2 & (z^2+1)^2 \end{vmatrix}$$

$$= \begin{vmatrix} 1 & x & x^2 \\ 1 & y & y^2 \\ 1 & z & z^2 \end{vmatrix}\begin{vmatrix} 1 & 1 & 1 \\ 2x & 2y & 2z \\ x^2 & y^2 & z^2 \end{vmatrix}$$

$$= 2(x-y)^2(y-z)^2(z-x)^2$$

9. Equation $x^3 + ax^2 + bx + c = 0$ has roots α, β, γ. Therefore,

$\alpha + \beta + \gamma = -a$

$\alpha\beta + \beta\gamma + \gamma\alpha = b$

Since the given system of equations has non-trivial solutions, we have

$$\begin{vmatrix} \alpha & \beta & \gamma \\ \beta & \gamma & \alpha \\ \gamma & \alpha & \beta \end{vmatrix} = 0$$

or $\alpha^3 + \beta^3 + \gamma^3 - 3\alpha\beta\gamma = 0$

or $(\alpha+\beta+\gamma)\left[\alpha^2+\beta^2+\gamma^2-\alpha\beta-\beta\gamma-\gamma\alpha\right] = 0$

or $(\alpha+\beta+\gamma)\left[(\alpha+\beta+\gamma)^2 - 3(\alpha\beta+\beta\gamma+\gamma\alpha)\right] = 0$

$\Rightarrow -a\left[a^2 - 3b\right] = 0$

or $a^2 / b = 3$

10. System of equations:

$\alpha x + y + z = \alpha - 1$ ...(i)

$x + \alpha y + z = \alpha - 1$ ...(ii)

$x + y + \alpha z = \alpha - 1$ ...(iii)

Since the system has no solution, therefore, (i) $\Delta = 0$ and (ii) $\alpha - 1 \neq 0$

$$\begin{vmatrix} \alpha & 1 & 1 \\ 1 & \alpha & 1 \\ 1 & 1 & \alpha \end{vmatrix} = 0, \alpha \neq 1$$

$R_1 \to R_1 - R_3, R_2 \to R_2 - R_3$

$$\begin{vmatrix} \alpha-1 & 0 & 1-\alpha \\ 0 & \alpha-1 & 1-\alpha \\ 1 & 1 & \alpha \end{vmatrix} = 0$$

$(\alpha-1)[\alpha(\alpha-1)-(1-\alpha)]+(1-\alpha)[-\alpha-1]=0$

$(\alpha-1)[\alpha(\alpha-1)+(\alpha-1)]+(\alpha-1)^2=0$

$(\alpha-1)^2[(\alpha+1)+1]=0$

$\alpha = 1, 1, -2$ or $\alpha = 1, -2$

Since system has no solution, $\alpha \neq 1$.

$\therefore \alpha = -2$

11. The system has a non-trivial solution if

$$\Delta = \begin{vmatrix} \sin 3\theta & -1 & 1 \\ \cos 2\theta & 4 & 3 \\ 2 & 7 & 7 \end{vmatrix} = 0$$

$\Rightarrow (28-21)\sin 3\theta - (-7-7)\cos 2\theta + 2(-3-4) = 0$

$\Rightarrow \sin 3\theta + 2\cos 2\theta - 2 = 0$

$\Rightarrow 3\sin\theta - 4\sin^3\theta + 2 - 4\sin^2\theta - 2 = 0$

$\Rightarrow 4\sin^3\theta + 4\sin^2\theta - 3\sin\theta = 0$

$\Rightarrow \sin\theta(4\sin^2\theta + 4\sin\theta - 3) = 0$

$\Rightarrow \sin\theta(2\sin\theta+3)(2\sin\theta-1) = 0$

So, $\sin\theta = 0; 1/2$, since $\sin\theta \neq -3/2$

Now, $\sin\theta = 0$ is for $0, \pi, 2\pi, \ldots, 100\pi$, total 101 values

$\sin\theta = \frac{1}{2}$ for $\theta = \frac{\pi}{3}$ and $\frac{2\pi}{3}$.

So, there will be 100 values lying in $[0, 100\pi]$

Therefore, total number of values of θ is 201.

12. $f(A) = |A| = ad - bc = 15$

$\Rightarrow (ad, bc) = (9, -6)$ or $(6, -9)$

Case I: If $ad = 9$ and $bc = -6$ then

All possible ways of $ad = (3,3), (-3,-3)$

and all possible ways of

$bc = (2,-3), (-2,3), (3,-2), (-3,2)$

$\therefore$ Total ways $= 2 \times 4 = 8$

Case II: If $ad = 6$ and $bc = -9$, then

All possible ways of

$ad = (3,2), (-3,-2), (2,3)(-2,-3)$

and all possible ways of $bc = (3,-3), (-3,3)$

Total ways = 8

Total such matrices are = 16

13. $I_2 + A = \begin{bmatrix} 1 & -\tan\frac{\theta}{2} \\ \tan\frac{\theta}{2} & 1 \end{bmatrix}$ ...(i)

$I_2 - A = \begin{bmatrix} 1 & \tan\frac{\theta}{2} \\ -\tan\frac{\theta}{2} & 1 \end{bmatrix}$

$\Rightarrow (I_2 - A)^{-1} = \frac{1}{\sec^2\frac{\theta}{2}} \begin{bmatrix} 1 & -\tan\frac{\theta}{2} \\ \tan\frac{\theta}{2} & 1 \end{bmatrix}$ ...(ii)

Now, $(I_2 + A)(I_2 - A)^{-1}$

$= \frac{1}{1+\tan^2\frac{\theta}{2}} \begin{bmatrix} 1-\tan^2\frac{\theta}{2} & -2\tan\frac{\theta}{2} \\ 2\tan\frac{\theta}{2} & 1-\tan^2\frac{\theta}{2} \end{bmatrix}$

$= \begin{bmatrix} \cos\theta & -\sin\theta \\ \sin\theta & \cos\theta \end{bmatrix} = \begin{bmatrix} a & -b \\ b & a \end{bmatrix}$

On comparing,

$a = \cos\theta$ and $b = \cos\theta$, then $13(a^2 + b^2) = 13$

14. Given that $PQ = kI$

$|P|.|Q| = k^3$

$\Rightarrow |P| = 2k \neq 0 \Rightarrow$ P is an invertible matrix.

$\because PQ = kI$

$\therefore Q = kP^{-1}I$ $\quad [\because P^{-1}P = I]$

$Q = \frac{\text{adj.}P}{2}$

$q_{23} = -\frac{k}{8}$

$\therefore \frac{(3\alpha+4)}{2} = -\frac{k}{8} \Rightarrow k = 4$

$\therefore |P| = 2k \Rightarrow k = 10 + 6\alpha$ ...(i)

Put value of k in (i), we get $\alpha = -1$

15. $\because |KA| = K^n|A|$, if A is an $n \times n$ matrix.

$|\text{adj}.A| = |A|^{n-1}$, if A is an $n \times n$ matrix.

$|2\,\text{adj}.(2\,\text{adj}.(\text{adj}.(2A)))| = 2^{41}$

$\Rightarrow 2^3|2\,\text{adj}.(2\,\text{adj}.(\text{adj}.(2A)))| = 2^{38}$

$\Rightarrow |2\,\text{adj}.(2\,\text{adj}.(\text{adj}.(2A)))| = 2^{38}$

$\Rightarrow (|2\,\text{adj}.(\text{adj}.(2A))|)^2 = 2^{38}$

$\Rightarrow |2\,\text{adj}.(\text{adj}.(2A))| = 2^{19}$

$\Rightarrow 2^3|\text{adj}.(\text{adj}.(2A))| = 2^{19}$

$\Rightarrow |\text{adj}.(\text{adj}.(2A))| = 2^{16}$

$\Rightarrow \left(|\text{adj.}(2A)|\right)^2 = 2^{16}$

$\Rightarrow |\text{adj.}(2A)| = 2^8$

$\Rightarrow |(2A)|^2 = 2^8$

$\Rightarrow |(2A)| = 2^4$

$\Rightarrow 2^3|A| = 2^4$

$\Rightarrow |A| = 2$

So, $|A^2| = 4$

16. Since, the matrix A is

$$A = \begin{bmatrix} 2 & -1 & 1 \\ -1 & 2 & -1 \\ 1 & -1 & 2 \end{bmatrix}$$

$\Rightarrow |A| = 4$

$\because \det\left[3\,\text{adj.}\left(2A^{-1}\right)\right] = \left|3.2^2\,\text{adj.}\left(A^{-1}\right)\right|$

$= 12^3\left|\text{adj.}\left(A^{-1}\right)\right| = 12^3\left|A^{-1}\right|^2 = \dfrac{12^3}{|A|^2} = \dfrac{12^3}{16} = 108$

17. $\because A = \begin{bmatrix} 2 & 3 \\ 0 & -1 \end{bmatrix}, A^2 = \begin{bmatrix} 4 & 3 \\ 0 & 1 \end{bmatrix}, \begin{bmatrix} 8 & 9 \\ 0 & -1 \end{bmatrix}, \ldots$

Now by mathematical induction we can conclude that

$$A^n = \begin{bmatrix} 2^n & 2^n - (-1)^n \\ 0 & (-1)^n \end{bmatrix}$$

Also, $2A.(\text{adj.}(2A)) = |2A|$

$\Rightarrow A.\text{adj.}(2A) = -4I$ $\qquad [\because |A| = -2]$

Now, $\left|A^{10} - (\text{adj.}2A)^{10}\right| = \dfrac{\left|A^{20} - A^{10}\left(\text{adj}(2A)\right)^{10}\right|}{|A|^{10}}$

$= \dfrac{\left|A^{20} - 2^{20}I\right|}{|A|^{10}}$ $\qquad$ ...(i)

$\because A^{20} - A^{20}.I = \begin{bmatrix} 2^{20} & 2^{20}-1 \\ 0 & 1 \end{bmatrix} - 2^{20}\begin{bmatrix} 1 & 0 \\ 0 & 1 \end{bmatrix}$

$= \begin{bmatrix} 0 & 2^{20}-1 \\ 0 & 1-2^{20} \end{bmatrix}$

$\Rightarrow \left|A^{20} - 2^{20}I\right| = 0$

From (i),

$\left|A^{10} - (\text{adj.}(2\text{A}))^{10}\right| = 0$

So, $\det\left(A^4\right) + \det\left(A^{10} - \left(\text{adj.}(2A)\right)^{10}\right)$

$= |A|^4 + 0 = (-2)^4 = 16$

18. For infinite many solutions,

$\Delta = \Delta_1 = \Delta_2 = \Delta_3 = 0$

$$\Delta = \begin{vmatrix} 1 & 1 & -1 \\ 1 & 2 & \alpha \\ 2 & -1 & 1 \end{vmatrix} = 0$$

$\Rightarrow 1(2+\alpha) - 1(1-2\alpha) - 1(-1-4) = 0$

$\Rightarrow 3\alpha + 6 = 0 \Rightarrow \alpha = -2$

$$\Delta_3 = \begin{vmatrix} 1 & 1 & 2 \\ 1 & 2 & 1 \\ 2 & -1 & \beta \end{vmatrix} = 0$$

$\Rightarrow 1(2\beta+1) - 1(\beta-2) + 2(-1-4) = 0$

$\Rightarrow \beta - 7 = 0 \Rightarrow \beta = 7$

So, $\alpha + \beta = 5$

19. $\because \text{D} = 0$

$$\therefore \text{D}_y = \begin{vmatrix} k & 1 & 2 \\ 3 & 2 & -2 \\ -2 & 3 & -4 \end{vmatrix} = 0$$

$\Rightarrow -2k + 16 + 26 = 0$

$\Rightarrow k = 21$

20. For non-zero solution, $\Delta = 0$

$$\Rightarrow \begin{vmatrix} \lambda-1 & 3\lambda+1 & 2\lambda \\ \lambda-1 & 4\lambda-2 & \lambda+3 \\ 2 & 3\lambda+1 & 3(\lambda-1) \end{vmatrix} = 0$$

$\Rightarrow 6\lambda^3 - 36\lambda^2 + 54\lambda = 0 \Rightarrow 6\lambda\left[\lambda^2 - 6\lambda + 9\right] = 0$

$\Rightarrow \lambda = 0, \lambda = 3$ $\qquad$ [Distinct values]

21. For infinitely many solutions,

$\Delta = \Delta_1 = \Delta_2 = \Delta_3 = 0$

$$\Delta = \begin{vmatrix} 1 & -2 & 3 \\ 2 & 1 & 1 \\ 1 & -7 & a \end{vmatrix} = 0$$

$\Rightarrow (a+7) - 2(1-2a) + 3(-15) = 0 \Rightarrow a = 8$

$$\Delta_3 = \begin{vmatrix} 1 & -2 & 9 \\ 2 & 1 & b \\ 1 & -7 & 24 \end{vmatrix} = 0$$

$\Rightarrow (24+7b) - 2(b-48) + 9(-15) = 0$

$\Rightarrow b = 3$ $\qquad \therefore a - b = 5$

22. The given system of equations,

$x - 2y + 5z = 0$ $\qquad$ ...(i)

$-2x + 4y + z = 0$ $\qquad$ ...(ii)

$-7x+14y+9z=0 \quad \ldots\text{(iii)}$
From equation, $2\times$ (i) + (ii) $\Rightarrow z=0$
Put $z=0$ in equation (i), we get $x=2y$
$\because 15\le x^2+y^2+z^2\le 150 \Rightarrow 15\le 4y^2+y^2 \Rightarrow 150$
$[\because x=2y, z=0] \Rightarrow 3\le y^2\le 30$
$\Rightarrow y=\pm 2, \pm 3, \pm 4, \pm 5 \Rightarrow$ 8 solutions.

23. $x+y+z=6 \quad \ldots\text{(i)}$
$x+2y+3z=10 \quad \ldots\text{(ii)}$
$3x+2y+\lambda z=\mu \quad \ldots\text{(iii)}$
From (i) and (ii),
If $z=0 \Rightarrow x+y+6$ and $x+2y=10$
$\Rightarrow y=4, x=2$
$(2,4,0)$
If $y=0 \Rightarrow x+z=6$ and $x+3y=10$
$\Rightarrow z=2$ and $x=4(4,0,2)$
So, $3x+2y+\lambda z=\mu$, must pass through (2, 4, 0) and (4, 0, 2)
So, $6+8=\mu \Rightarrow \mu=14$
and $12+2\lambda=\mu$
$12+2\lambda=14 \Rightarrow \lambda=1$
So, $\mu-\lambda^2=14-1=13$

24. $|A|=\begin{vmatrix} 2k-1 & 2\sqrt{k} & 2\sqrt{k} \\ 2\sqrt{k} & 1 & -2k \\ -2\sqrt{k} & 2k & -1 \end{vmatrix}$

$$=\begin{vmatrix} 2k-1 & 0 & 2\sqrt{k} \\ 2\sqrt{k} & 1+2k & -2k \\ -2\sqrt{k} & 1+2k & -1 \end{vmatrix} [C_2 \to C_2 - C_3]$$

$$=\begin{vmatrix} 2k-1 & 0 & 2\sqrt{k} \\ 4\sqrt{k} & 0 & 1-2k \\ -2\sqrt{k} & 1+2k & -1 \end{vmatrix} [R_2 \to R_2 - R_3]$$

$$=(1+2k)(8k-4k+4k^2+1)=(2k+1)^3$$

Since B is skew-symmetric of odd order,
$\therefore |B|=0$
Hence, $|\text{adj.A}|+|\text{adj.B}|=|A|^2+|B|^2=10^6$
$\Rightarrow (2k+1)^6=10^6 \Rightarrow k=4.5, \therefore [k]=4$

25. Given: $\omega=\cos\frac{2\pi}{3}+i\sin\frac{2\pi}{3}=\frac{-1+i\sqrt{3}}{2}$
$\therefore 1+\omega+\omega^2=0$ and $\omega^3=1$

Then $\begin{vmatrix} z+1 & \omega & \omega^2 \\ \omega & z+\omega^2 & 1 \\ \omega^2 & 1 & z+\omega \end{vmatrix}=0$

$$\Rightarrow \begin{vmatrix} z+1+\omega+\omega^2 & \omega & \omega^2 \\ z+1+\omega+\omega^2 & z+\omega^2 & 1 \\ z+1+\omega+\omega^2 & 1 & z+\omega \end{vmatrix}=0$$

$[C_1 \to C_1 + C_2 + C_3]$

$$\Rightarrow z\begin{vmatrix} 1 & \omega & \omega^2 \\ 1 & z+\omega^2 & 1 \\ 1 & 1 & z+\omega \end{vmatrix}=0$$

$$\Rightarrow z\left[1\left(z^2+z\omega+z\omega^2+\omega^3-1\right)-\omega\left(z+\omega-1\right)+\omega^2\left(1-z-\omega^2\right)=0\right]$$

$$\Rightarrow z\left[z^2\right]=0 \Rightarrow z^3=0 \Rightarrow z=0$$

Assertion and Reason

1. (c) $A=\begin{vmatrix} 0 & 2 & 1 \\ -2 & 0 & 3 \\ -1 & -3 & 0 \end{vmatrix}$

$A=0(0+9)-2(0+3)+1(6-0)$
$A=0$
Odd order skew-symmetric determinant is always equal to zero.
Even order skew-symmetric determinant is always equal to a perfect square.
So, Assertion is true but Reason is not true.

2. (a) Apply $C_1=C_1+C_2+C_3$ and taking common from C_1.

$$=1+\omega^n+\omega^{2n}\begin{vmatrix} 1 & \omega^n & \omega^{2n} \\ 1 & 1 & \omega^n \\ 1 & \omega^{2n} & 1 \end{vmatrix}$$

$1+\omega^n+\omega^{2n}=0 \quad [\because \text{where } n\in N]$
Assertion and Reason both are true and Assertion follows from Reason.

3. (d) $D=\begin{vmatrix} a & b & c \\ b & c & a \\ c & a & b \end{vmatrix}=-\left(a^3+b^3+c^3-3abc\right)$

$$=-(a+b+c).\left(a^2+b^2+c^2-ab-bc-ca\right)$$

Now the other bracket cannot vanish in the light of the correct statement 2.
The system can have non-zero solutions (x, y, z are not all zero), if $D=0$
So, $a+b+c=0$
Assertion is false.

4. (a) $D(x)=\lambda_1 f_1(x)-\lambda_2 f_2(x)+\lambda_3 f(x)$ where
$\lambda_1=(bf-ce);\lambda_2=(af-cd);\lambda_3=(ae-bd)$
Then,
$\int Ddx=\int\lambda_1 f_1(x)dx+\int\lambda_2 f_2(x)dx+\int\lambda_3 f_3(x)dx+k$

$$=\begin{vmatrix}\int f_1(x)dx & \int f_2(x)dx & \int f_3(x)dx\\ a & b & c\\ d & e & f\end{vmatrix}$$

Assertion is true and follows from Statement 2.

5. (a)

If $\Delta=\begin{vmatrix}a & b & c\\ x & y & z\\ p & q & r\end{vmatrix}$ Then $\begin{vmatrix}ka & kb & kc\\ kx & ky & kz\\ kp & kq & kr\end{vmatrix}$

Taking common k from each of three rows or columns.
So, this determinant is equal to $k^3\Delta$.
Thus, assertion and reason, both are true and assertion follows from reason.

6. (a) The given system of equations are
$x-2y+3z=-1$
$-x+y-2z=k$
$x-3y+4z=1$

Here $D=\begin{vmatrix}1 & -2 & 3\\ -1 & 1 & -2\\ 1 & -3 & 4\end{vmatrix}=0$

and $D_2=\begin{vmatrix}1 & -1 & 3\\ -1 & k & -2\\ 1 & 1 & 4\end{vmatrix}=k-3\neq 0$ if $k\neq 3$

$\therefore$ If $k\neq 3$, the system has no solutions.
Therefore, Statement 1 is true and Statement 2 is a correct explanation for statement 1.

JEE Advanced

Objective Questions

One Option Correct

1. (c) For the given system to have a non-trivial solution, we must have

$$\begin{vmatrix}1 & k & 3\\ 3 & k & -2\\ 2 & 3 & -4\end{vmatrix}=0\Rightarrow\begin{vmatrix}1 & k & 3\\ 0 & -2k & -11\\ 0 & 3-2k & -10\end{vmatrix}=0$$

[Applying $R_2\to R_2-3R_1, R_3\to R_3-2R_1$]

$\Rightarrow 20k+11(3-2k)=0\quad\Rightarrow\quad k=\dfrac{33}{2}$

2. (a) $\begin{vmatrix}pa & qb & rc\\ qc & ra & pb\\ rb & pc & qa\end{vmatrix}$

$=pqr\left(a^3+b^3+c^3\right)-abc\left(p^3+q^3+r^3\right)$
$=pqr(3abc)-abc(3pqr)=0$

3. (b) Let $\begin{vmatrix}a_{11} & a_{12} & a_{13}\\ a_{21} & a_{22} & a_{23}\\ a_{31} & a_{32} & a_{33}\end{vmatrix}$

$C_2\to C_2-\dfrac{a_{12}}{a_{11}}C_1\quad C_3\to C_3-\dfrac{a_{13}}{a_{11}}C_1$

$$\begin{vmatrix}a_{11} & 0 & 0\\ a_{21} & \left(a_{22}-\dfrac{a_{12}}{a_{11}}\times a_{21}\right) & \left(a_{23}-\dfrac{a_{13}}{a_{11}}\times a_{21}\right)\\ a_{31} & \left(a_{32}-\dfrac{a_{12}}{a_{11}}\times a_{31}\right) & \left(a_{32}-\dfrac{a_{13}}{a_{11}}\times a_{31}\right)\end{vmatrix}$$

so minimum value = – 4

4. (a) $S_1: D=\begin{vmatrix}\ln x & \ln y & \ln z\\ \ln 2x & \ln 2y & \ln 2y\\ \ln 3x & \ln 3y & \ln 3z\end{vmatrix}$

$$=\begin{vmatrix}\ln x & \ln y & \ln z\\ \ln 2+\ln x & \ln 2+\ln y & \ln 2+\ln z\\ \ln 3+\ln x & \ln 3+\ln y & \ln 3+\ln z\end{vmatrix}$$

$$=\begin{vmatrix}\ln x & \ln y & \ln z\\ \ln 2 & \ln 2 & \ln 2\\ \ln 3 & \ln 3 & \ln 3\end{vmatrix}=0$$

$$S_2: 0=\begin{vmatrix}x & a & b & 1\\ \lambda & x & b & 1\\ \lambda & \mu & x & 1\\ \lambda & \mu & v & 1\end{vmatrix}$$

$$=\begin{vmatrix}x & a & b & 1\\ \lambda-x & x-a & 0 & 0\\ \lambda-x & \mu-a & x-b & 0\\ \lambda-x & \mu-a & v-b & 0\end{vmatrix}\quad\begin{bmatrix}R_2\to R_2-R_1\\ R_3\to R_3-R_1\\ R_4\to R_4-R_1\end{bmatrix}$$

$$=(\lambda-x)\begin{vmatrix}1 & x-a & 0\\ 1 & \mu-a & x-b\\ 1 & \mu-a & v-b\end{vmatrix}\quad\begin{cases}R_2\to R_2-R_1\\ R_3\to R_3-R_1\end{cases}$$

$$=(\lambda-x)\begin{vmatrix}1 & x-a & 0\\ 0 & \mu-x & x-b\\ 0 & \mu-x & v-b\end{vmatrix}=(\lambda-x)(\mu-x)(v-x)$$

$$S_3 : \begin{vmatrix} a & b & c \\ b & c & a \\ c & a & b \end{vmatrix}$$

$$= -\frac{a+b+c}{2}\left((a-b)^2+(b-c)^2+(c-a)^2\right) < 0$$

$$S_4 : f(x) = \begin{vmatrix} \frac{1}{x} & \ln x & x^n \\ 1 & -\frac{1}{n} & (-1)^n \\ 1 & a & a^2 \end{vmatrix}$$

$$= f'(x) = \begin{vmatrix} -\frac{1}{x^2} & \frac{1}{x} & nx^{n-1} \\ 1 & -\frac{1}{n} & (-1)^n \\ 1 & a & a^2 \end{vmatrix}$$

$$f^{(n)}(x) = \begin{vmatrix} (-1)^n \frac{n!}{x^{n+1}} & (-1)^{n-1}\frac{(n-1)!}{x^n} & n! \\ 1 & -\frac{1}{n} & (-1)^n \\ 1 & a & a^2 \end{vmatrix}$$

$$f^{(n)} = \begin{vmatrix} (-1)^n n! & (-1)^{n-1}(n-1)! & n! \\ 1 & -\frac{1}{n} & (-1)^n \\ 1 & a & a^2 \end{vmatrix}$$

$$= (-1)^n n! \begin{vmatrix} 1 & -\frac{1}{n} & (-1)^n \\ 1 & -\frac{1}{n} & (-1)^n \\ 1 & a & a^2 \end{vmatrix} = 0$$

5. (d) $M = \begin{bmatrix} \sin^4\theta & -1-\sin^2\theta \\ 1+\cos^2\theta & \cos^4\theta \end{bmatrix}$

$|M| = \sin^4\theta\cos^4\theta + \sin^4\theta\cos^4\theta$

$$M^{-1} = \frac{1}{|M|}\begin{bmatrix} \cos^4\theta & 1+\sin^2\theta \\ -1-\cos^2\theta & \sin^4\theta \end{bmatrix}$$

Given that $M = \alpha I + \beta M^{-1}$

$$\Rightarrow \begin{bmatrix} \sin^4\theta & -1-\sin^2\theta \\ 1+\cos^2\theta & \cos^4\theta \end{bmatrix}$$

$$= \begin{bmatrix} \alpha & 0 \\ 0 & a \end{bmatrix} + \frac{\beta}{|M|}\begin{bmatrix} \cos^4\theta & 1+\sin^2\theta \\ -1-\cos^2\theta & \sin^4\theta \end{bmatrix}$$

$\Rightarrow \frac{\beta}{|M|} = -1$ and $\alpha + \frac{\beta}{|M|}\cos^4\theta = \sin^4\theta$

$\Rightarrow \alpha = \sin^4\theta + \cos^4\theta$

and $\beta = -\left[2+\sin^2\theta\cos^2\theta + \sin^4\theta\cos^4\theta\right]$

Now, $\alpha = \left(\sin^2\theta+\cos^2\theta\right)^2 - 2\sin^2\theta\cos^2\theta$

$= 1 - 2\sin^2\theta\cos^2\theta = 1 - \frac{1}{2}\sin^2 2\theta$

For α to be minimum, $\sin^2 2\theta$ is maximum, i.e., 1.

$\therefore \alpha^* = 1 - \frac{1}{2} = \frac{1}{2}$

Also, $\beta = -\left[2 + \frac{1}{4}\sin^2 2\theta + \frac{1}{16}\sin^2 2\theta\right]$

For β to be minimum, $\sin^2 2\theta$ is maximum, i.e., 1.

$\therefore \beta^* = -\left[2+\frac{1}{4}+\frac{1}{16}\right] = -\frac{32+4+1}{16} = \frac{-37}{16}$

$\therefore \alpha^* + \beta^* = \frac{1}{2} - \frac{37}{16} = \frac{-29}{16}$

One or More Than One Option Correct

1. (b, c, d) $x+3y+2z=6$ ...(i)

$x+\lambda y+2z=7$ ...(ii)

$x+3y+2z=\mu$ ...(iii)

(a) If $\lambda = 2$, then D = 0, therefore unique solution is not possible.

(b) If $\lambda = 4, \mu = 6$

$x+3y = 6-2z$

$x+4y = 7-2z$

$\therefore y = 1$ and $x = 3-2z$

Substituting in equation (iii),

$3-2z+3+2z = 6$ is satisfied

Infinite solutions.

(c) $\lambda = 5, \mu = 7$

Consider equations (ii) and (iii),

$x+5y = 7-2z$

$x+3y = 7-2z$

$y = 0, x = 7-2z$ are solutions

Substitute in (ii),

$7-2z+2z = 6$

$\Rightarrow$ The equation is not satisfied.

No solution.

(d) if $\lambda = 3, \mu = 5$

Then equations (i) and (ii) have no solutions.

No solution.

2. (c) [Since a skew-symmetric matrix of order 3 cannot be non-singular, the data given in the question is inconsistent.]

Now $M^2N^2\left(M^TN\right)^{-1}\left(MN^{-1}\right)^T$

$= M^2N^2N^{-1}\left(M^T\right)^{-1}\left(N^{-1}\right)^T M^T$

$= M^2N\left(M^T\right)^{-1}\left(N^{-1}\right)^T M^T = -M^2NM^{-1}N^{-1}M$

$\left[\because M^T = -M,\ N^T = -N \text{ and } \left(N^{-1}\right)^T = \left(N^T\right)^{-1}\right]$

$= -M(NM)(NM)^{-1}M \qquad [\because MN = NM]$

$= -MM = -M^2$

3. (a, b, d)

(a) We have

$$PEP = \begin{pmatrix}1&0&0\\0&0&1\\0&1&0\end{pmatrix}\begin{pmatrix}1&2&3\\2&3&4\\8&13&18\end{pmatrix}\begin{pmatrix}1&0&0\\0&0&1\\0&1&0\end{pmatrix}$$

$$\begin{pmatrix}1&2&3\\8&13&18\\2&3&4\end{pmatrix}\begin{pmatrix}1&0&0\\0&0&1\\0&1&0\end{pmatrix} = \begin{pmatrix}1&3&2\\8&18&13\\2&4&3\end{pmatrix} = F$$

$$P^2 = \begin{pmatrix}1&0&0\\0&0&1\\0&1&0\end{pmatrix}\begin{pmatrix}1&0&0\\0&0&1\\0&1&0\end{pmatrix} = \begin{pmatrix}1&0&0\\0&1&0\\0&0&1\end{pmatrix}$$

(b) $\left|EQ + PFQ^{-1}\right| = |EQ| + \left|PFQ^{-1}\right|$

$|E| = 0$ and $|F| = 0$ and $|Q| \neq 0$

$|EQ| = |E||Q| = 0, \left|PFQ^{-1}\right| = \dfrac{|P||F|}{|Q|} = 0$

$\left[\because \left|Q^{-1}\right| = \dfrac{1}{|Q|}\right]$

$T = EQ + PFQ^{-1}$

$TQ = EQ^2 + PF = EQ^2 + P^2EP$

$= EQ^2 + EP = E\left(Q^2 + P\right) \qquad$ [from (a)]

$|TQ| = \left|E\left(Q^2+P\right)\right| \Rightarrow |T||Q| = |E|\left|Q^2+P\right| = 0$

$\Rightarrow |T| = 0 \qquad (\text{as } |Q| \neq 0)$

(c) $\left|(EF)^3\right| > |EF|^2$

Here $0 > 0$ (false)

(d) as $P^2 = I \Rightarrow P^{-1} = P$

So, $Tr\left(P^{-1}EP + F\right) = Tr(PEP + F) = Tr(2F)$

Also, $Tr\left(E + P^{-1}FP\right) = Tr(E + PFP) = Tr(2E)$

Given that

$Tr(E) = Tr(F)$

$\therefore Tr(2E) = Tr(2F)$

$Tr(2PEP) = 2Tr(PEP) = 2Tr(EPP) = 2Tr(E)$

4. (a, b, c)

Since $(I - EF)$ is invertible therefore,

$|I - EF| \neq 0; G = (I - EF)^{-1} \Rightarrow G^{-1} = I - EF$

We have $G.G^{-1} = I = G^{-1}G$

$\Rightarrow G(I - EF) = I = (I - EF)G$

$\Rightarrow G - GEF = I = G - EFG \qquad \ldots(i)$

$\Rightarrow GEF = EFG$ (option (c) is correct.)

$(I - FE)(I + FGE) = I + FGE - FE - FEFGE$

$= I + FGE - FE - F(G - I)E$ [from (i)]

$= I + FGE - FE - FGE + FE$

$= I$ [Option (b) is correct.]

$(I - FE)(I + FGE) = I \qquad \ldots(ii)$

(So, option 'd' is incorrect)

Now $FE(I + FGE)$

$= FE + FEFGE$

$= FE + F(G - I)E$ [from (i)]

$= FE + FGE - FE = FGE$

$\Rightarrow |FE||I + FGE| = |FGE|$

$\Rightarrow |FE| \times \dfrac{1}{|I - FE|} = |FGE| \quad (\text{from (ii)})$

$\Rightarrow |FE| = |I - FE||FGE|$

(Option (a) is correct.)

5. (a, b, d)

We observe that

$P_1' = P_1, P_2' = P_2, P_3' = P_3, P_4' = P_5, P_5' = P_4, P_6' = P_6$

Also, $P_kP_k' = I$ for $k = 1$ to 6.

Now $X = \sum_{k=1}^{6} P_k A P_k'$,

where $A = \begin{bmatrix}2&1&3\\1&0&2\\3&2&1\end{bmatrix}$ and $A' = A$

$$X' = \left(\sum_{k=1}^{6} P_k A P_k'\right)' = \sum_{k=1}^{6}\left(P_k A P_k'\right)'$$

$$= \sum_{k=1}^{6}\left(P_k'\right)' A' P_X' = \sum_{k=1}^{6} P_k A P_k' = X$$

$\therefore$ X is a symmetric matrix.

$\therefore$ (a) is correct.

Sum of diagonals entries of X = Trace X

$$= \text{Trace}\sum_{k=1}^{6}\left(P_k A P_k'\right)$$

$$= \sum_{k=1}^{6}\text{Trace}\left(P_k A P_k'\right)$$

$$= \sum_{k=1}^{6}\text{Trace}\left(P_k A P_k'\right)$$

[using Trace AB = Trace BA]

$$= \sum_{k=1}^{6}\text{Trace}\,(AI) = \sum_{k=1}^{6}\text{Trace A} = 6\times\text{Trace A}$$

$$= 6\times(2+0+1) = 18$$

$\therefore$ (b) is correct.

$$X\begin{bmatrix}1\\1\\1\end{bmatrix} = \sum_{k=1}^{6}P_k A\left(P_k'\begin{bmatrix}1\\1\\1\end{bmatrix}\right) = \sum_{k=1}^{6}P_k A\begin{bmatrix}1\\1\\1\end{bmatrix} = \sum_{k=1}^{6}P_k\begin{bmatrix}6\\3\\6\end{bmatrix}$$

$$= \begin{bmatrix}2&2&2\\2&2&2\\2&2&2\end{bmatrix}\begin{bmatrix}6\\3\\6\end{bmatrix} = \begin{bmatrix}30\\30\\30\end{bmatrix} = 30\begin{bmatrix}1\\1\\1\end{bmatrix} = \alpha\begin{bmatrix}1\\1\\1\end{bmatrix}$$

$\therefore \alpha = 30$

$\therefore$ (d) is correct.

Also, $X\begin{bmatrix}1\\1\\1\end{bmatrix} = 30\begin{bmatrix}1\\1\\1\end{bmatrix} \Rightarrow (X-30\,I)\begin{bmatrix}1\\1\\1\end{bmatrix} = 0$

$\Rightarrow X - 30\,I = 0 \Rightarrow |X - 30\,I| = 0$

$\Rightarrow X - 30\,I$ is not invertible.

$\therefore$ (c) is incorrect.

6. (b, c, d)

Since matrix M is invertible matrix,

$\det(M) \neq 0$

$M^{-1} = \text{adj.}(\text{adj.}M)$

$M^{-1} = \det(M).M$

$$\left[\because \text{adj.}(\text{adj.}M) = (\det M)^{3-2}.M\right]$$

$M^{-1}M = \det(M).M^2$

$I = \det(M).M^2 \qquad \ldots(i)$

$$\left[\because \det(\det M) = (\det M)^3 \text{ and } \det\left(M^2\right) = (\det M)^2\right]$$

$I = \det(M) \qquad \ldots(ii)$

From (i), $I = M^2$

$(\text{adj.}M)^2 = \text{adj}\left(M^2\right) = \text{adj.}I = I.$

7. (a, c, d)

$$M = \begin{bmatrix}0&1&a\\1&2&3\\3&b&1\end{bmatrix};\text{adj.}M = \begin{bmatrix}-1&1&-1\\8&-6&2\\-5&3&-1\end{bmatrix}$$

Cofactor of a_{11} in $M = 2-3b = -1 = a_{11}$ in adj. $M \Rightarrow b = 1$

Cofactor of a_{12} in $M = 3-2a = -1 = a_{13}$ in adj. $M \Rightarrow a = 2$

$\therefore a+b = 2+1 = 3$

$\therefore$ (a) is correct

$$|M| = \begin{vmatrix}0&1&2\\1&2&3\\3&1&1\end{vmatrix} = -(1-9)+2(1-6) = 8-10 = -2$$

$$\left|\text{adj.}M^2\right| = \left|M^2\right|^2 = |M|^4 = (-2)^4 = 16$$

$\therefore$ (b) is incorrect.

Also, $(\text{adj.}M)^{-1} = \text{adj.}M^{-1}$

$\therefore (\text{adj.}M)^{-1} + \text{adj.}M^{-1} = 2\text{adj.}\left(M^{-1}\right)$

$$= \frac{2\left|M^{-1}\right|}{\left|M^{-1}\right|}\text{adj.}\left(M^{-1}\right) = 2\left|M^{-1}\right|.\left(M^{-1}\right)^{-1}$$

$$= 2\times\frac{1}{2}\times M = -M$$

$\therefore$ (c) is correct

$$\text{Now, } M\begin{bmatrix}\alpha\\\beta\\\gamma\end{bmatrix} = \begin{bmatrix}1\\2\\3\end{bmatrix} \Rightarrow \begin{bmatrix}\alpha\\\beta\\\gamma\end{bmatrix} = M^{-1}\begin{bmatrix}1\\2\\3\end{bmatrix}$$

$$= \frac{1}{-2}\begin{bmatrix}-1&1&-1\\8&-6&2\\-5&3&-1\end{bmatrix}\begin{bmatrix}1\\2\\3\end{bmatrix}$$

$$\Rightarrow \begin{bmatrix}\alpha\\\beta\\\gamma\end{bmatrix} = \frac{1}{-2}\begin{bmatrix}-2\\2\\-2\end{bmatrix} = \begin{bmatrix}1\\-1\\1\end{bmatrix}$$

$\therefore \alpha - \beta + \gamma = 1-(-1)+1 = 3$

$\therefore$ (d) is correct.

8. (a, d)

$$P = \begin{vmatrix}1&1&1\\0&2&2\\0&0&3\end{vmatrix};Q = \begin{vmatrix}2&x&x\\0&4&0\\x&x&6\end{vmatrix}$$

$R = PQP^{-1}$

$\Rightarrow \det R = \det\left(PQP^{-1}\right) = |P||Q|\left|P^{-1}\right|$

$= |Q| = 4\left(12-x^2\right)$

Also, $\det\begin{bmatrix}2 & x & x\\ 0 & 4 & 0\\ x & x & 5\end{bmatrix}+8=4(12-x^2)$

$\therefore \det R=\det\begin{bmatrix}2 & x & x\\ 0 & 4 & 0\\ x & x & 5\end{bmatrix}+8\ \forall x\in R$

(a) is correct.

For $x=1$, $\det R=4(12-1)\neq 0$

$\therefore R\begin{bmatrix}\alpha\\ \beta\\ \gamma\end{bmatrix}=\begin{bmatrix}0\\0\\0\end{bmatrix}$ will have only trivial solution

i.e., $\alpha=0, \beta=0$ and $\gamma=0$

(b) is incorrect.

For $PQ=QP, a_{11}$ in $PQ=a_{11}$ in QP

$\Rightarrow 2+x=2\Rightarrow x=0$

Now a_{12} in $PQ=2x+4$ and a_{12} in $QP=2+2x$

$\Rightarrow 2x+4=2+2x\Rightarrow 4=2$, which is not possible

$\therefore PQ=QP$, is not possible for any real x.

(c) is incorrect.

For

$x=0$,

$$R=\begin{bmatrix}1 & 1 & 1\\ 0 & 2 & 2\\ 0 & 0 & 3\end{bmatrix}\begin{bmatrix}2 & 0 & 0\\ 0 & 4 & 0\\ 0 & 0 & 6\end{bmatrix}\begin{bmatrix}1 & -1/2 & 0\\ 0 & 1/2 & -1/3\\ 0 & 0 & 1/3\end{bmatrix}$$

$$=\begin{bmatrix}2 & 4 & 6\\ 0 & 8 & 12\\ 0 & 0 & 18\end{bmatrix}\begin{bmatrix}1 & -1/2 & 0\\ 0 & 1/2 & -1/3\\ 0 & 0 & 1/3\end{bmatrix}=\begin{bmatrix}2 & 1 & 2/3\\ 0 & 4 & 4/3\\ 0 & 0 & 6\end{bmatrix}$$

Now $R\begin{bmatrix}1\\a\\b\end{bmatrix}=6\begin{bmatrix}1\\a\\b\end{bmatrix}\Rightarrow\begin{bmatrix}2 & 1 & 2/3\\ 0 & 4 & 4/3\\ 0 & 0 & 6\end{bmatrix}\begin{bmatrix}1\\a\\b\end{bmatrix}=\begin{bmatrix}6\\6a\\6b\end{bmatrix}$

$\Rightarrow 2+a+\frac{2}{3}b=6\Rightarrow 3a+2b=12$ ...(i)

and $4a+\frac{4}{3}b=6a\Rightarrow 3a-2b=0$ ...(ii)

On solving (i) and (ii),

we get $a=2, b=3\Rightarrow a+b=5$

(d) is correct.

9. (a, d) $-x+2y+5z=b_1$

$2x-4y+3z=b_2$

$x-2y+2z=b_3$

has at least one solution.

$\therefore D=\begin{vmatrix}-1 & 2 & 5\\ 2 & -4 & 3\\ 1 & -2 & 3\end{vmatrix}$ and $D_1=D_2=D_3=0$

$\Rightarrow D_1=\begin{vmatrix}b_1 & 2 & 5\\ b_2 & -4 & 3\\ b_3 & -2 & 2\end{vmatrix}$

$=-2b_1-14b_2+26b_3=0\Rightarrow b_1+7b_2=13b_3$

(a) $D=\begin{vmatrix}1 & 2 & 3\\ 0 & 4 & 5\\ 1 & 2 & 6\end{vmatrix}=1(24-10)+1(10-12)$

$=14-2=12\neq 0$

Here, $D\neq 0\Rightarrow$ unique solution for any b_1, b_2, b_3.

(b) $D=\begin{vmatrix}1 & 1 & 3\\ 5 & 2 & 6\\ -2 & -1 & -3\end{vmatrix}$

$=1(-6+6)-1(-15+12)+3(-5+4)=0$

For at least one solution,

$D_1=D_2=D_3=0$

Now, $D_1=\begin{vmatrix}b_1 & 1 & 3\\ b_2 & 2 & 6\\ b_3 & -1 & -3\end{vmatrix}$

$=b_1(-6+6)-b_2(-3+3)+b_3(6-6)=0$

$D_2=\begin{vmatrix}1 & b_1 & 3\\ 5 & b_2 & 6\\ -2 & b_3 & -3\end{vmatrix}$

$=-b_1(-15+12)+b_2(-3+6)-b_3(6-15)$

$=3b_1+3b_2+9b_3=0\Rightarrow b_1+b_2+b_3=0$

It does not satisfy equation (i).

It has no solution.

(c) $D=\begin{vmatrix}-1 & 2 & -5\\ 2 & -4 & 10\\ 1 & -2 & 5\end{vmatrix}$

$=-1(-20+20)-2(10-10)-5(-4+4)=0$

Here, $b_2=-2b_1$ and $b_3=-b_1$ satisfies equation (i)

The planes are parallel.

(d) $D=\begin{vmatrix}1 & 2 & 5\\ 2 & 0 & 3\\ 1 & 4 & -5\end{vmatrix}$

$=1(0-12)-2(-10-3)+5(8-0)=54$

$D\neq 0$

It has unique solution for any b_1, b_2, b_3.

10. (b, c, d) $ax+2y=\lambda$

$3x-2y=\mu$

For unique solution, $\frac{a}{3}\neq\frac{2}{-2}\Rightarrow a\neq -3$

∴ (b) is correct.

For infinite solutions and $a=-3$

$$\frac{-3}{3}=\frac{2}{-2}=\frac{\lambda}{\mu}=-1 \text{ or } \lambda+\mu=0$$

∴ (c) is correct.

Also if $\lambda+\mu\neq 0$, then $\frac{-3}{3}=\frac{2}{-2}\neq\frac{\lambda}{\mu}$

⇒The system has no solution.

∴ (d) is correct.

11. (b, c) $\begin{vmatrix}(1+\alpha)^2 & (1+2\alpha)^2 & (1+3\alpha)^2\\(2+\alpha)^2 & (2+2\alpha)^2 & (2+3\alpha)^2\\(3+\alpha)^2 & (3+2\alpha)^2 & (3+3\alpha)^2\end{vmatrix}$

$$\begin{vmatrix}(1+\alpha)^2 & (1+2\alpha)^2 & (1+3\alpha)^2\\2\alpha+3 & 4\alpha+3 & 6\alpha+3\\2\alpha+5 & 4\alpha+5 & 6\alpha+5\end{vmatrix}=-648\alpha$$

$[R_2\to R_2-R_1, R_3\to R_3-R_2]$

$$2\begin{vmatrix}(1+\alpha)^2 & (1+2\alpha)^2 & (1+3\alpha)^2\\2\alpha+3 & 4\alpha+3 & 6\alpha+3\\1 & 1 & 1\end{vmatrix}=-648\alpha$$

$[R_3\to R_3-R_2]$

$$\begin{vmatrix}(1+\alpha)^2 & \alpha(3\alpha+2) & \alpha(5\alpha+2)\\2\alpha+3 & 2\alpha & 2\alpha\\1 & 1 & 1\end{vmatrix}=-324\alpha$$

$[C_2-C_1, C_3-C_2]$

$\therefore 2\alpha^2(-2\alpha)=-324\alpha$

$\Rightarrow \alpha^3-81\alpha=0$

$\Rightarrow \alpha=0,9,-9$

12. (a, b) Given: $MN=NM, M\neq N^2$ and $M^2=N^4$

Then $M^2=N^4\Rightarrow (M+N^2)(M-N^2)=0$

⇒(i) $M+N^2=0$ and $M-N^2\neq 0$

(ii) $|M+N^2|=0$ and $|M-N^2|\neq 0$

In each case $|M+N^2|=0$

$\therefore |M^2+MN^2|=|M||M+N^2|=0$

∴ (a) is correct and (c) is not correct.

Also, we know if $|A|=0$, then there can be many matrices U, such that $AU=0$

$\therefore (M^2+MN^2)U=0$ will be true for many values of U.

∴ (b) is correct.

Again if $AX=0$ and $|A|=0$, then X can be non-zero.

∴ (d) is not correct.

13. (c, d)

(a) $(N'MN)'=(MN)'N=N'M'N=N'MN$

or $-N'MN$,

According as M is symmetric matrix, hence (a) is correct.

(b) $(MN-NM)'=(MN)'-(NM)'$

$=N'M'-M'N'=NM-MN$

$=-(MN-NM)$

Hence $(MN-NM)$ is skew-symmetric matrix.

Therefore (b) is correct.

(c) $(MN)'=N'M'=NM\neq MN$

Hence (c) is incorrect.

(d) $(\text{adj.}M)(\text{adj.}N)=\text{adj.}(MN)$ is incorrect.

Matrix Type

1. A → P, T; B → Q; C → R; D → P, T

(A) $|A|=\begin{vmatrix}a_{11} & a_{12} & a_{13}\\a_{21} & a_{22} & a_{23}\\a_{31} & a_{32} & a_{33}\end{vmatrix}$

$$|B|=\begin{vmatrix}a_{11} & k^{-1}a_{12} & k^{-2}a_{13}\\a_{21} & a_{22} & k^{-1}a_{23}\\k^2a_{31} & ka_{32} & a_{33}\end{vmatrix}$$

$$=\frac{1}{k^3}\begin{vmatrix}k^2a_{11} & ka_{12} & a_{13}\\k^2a_{21} & ka_{22} & a_{23}\\k^2a_{31} & ka_{32} & a_{33}\end{vmatrix}$$

$=|A|$

$k_1|A|+k_2|B|=0$

$k_1+k_2=0$

(B) $\begin{vmatrix}1 & -1 & 1\\1 & 1 & -1\\1 & 1 & 1\end{vmatrix}=4$

(C) $\begin{vmatrix}1 & \cos\alpha & \cos\beta\\\cos\alpha & 1 & \cos\gamma\\\cos\beta & \cos\gamma & 1\end{vmatrix}$

$$=\begin{vmatrix}0 & \cos\alpha & \cos\beta\\\cos\alpha & 0 & \cos\gamma\\\cos\beta & \cos\gamma & 0\end{vmatrix}$$

$$\Rightarrow \sin^2\gamma - \cos\alpha(\cos\alpha - \cos\beta\cos\gamma) + \cos\beta(\cos\alpha\cos\gamma - \cos\beta)$$
$$= -\cos\alpha(-\cos\beta\cos\gamma) + \cos\beta(\cos\alpha\cos\gamma)$$
$$\Rightarrow \sin^2\gamma - \cos^2\alpha + 2\cos\alpha\cos\beta\cos\gamma - \cos^2\beta = 2\cos\alpha\cos\beta\cos\gamma$$
$$\Rightarrow \sin^2\gamma = \cos^2\alpha + \cos^2\beta$$
$$\Rightarrow \cos^2\alpha + \cos^2\beta + \cos^2\gamma = 1$$

(D) $\begin{vmatrix} x^2+x & x+1 & x-2 \\ 2x^2+3x-1 & 3x & 3x-3 \\ x^2+2x+3 & 2x-1 & 2x-1 \end{vmatrix}$

$R_2 \to R_2 - (R_1 + R_3)$

$$= \begin{vmatrix} x^2+x & x+1 & x-2 \\ -4 & 0 & 0 \\ x^2+2x+3 & 2x-1 & 2x-1 \end{vmatrix}$$
$$= 4\begin{vmatrix} x+1 & x-2 \\ 2x-1 & 2x-1 \end{vmatrix}$$
$$= 4\begin{vmatrix} x+1 & -3 \\ 2x-1 & 0 \end{vmatrix} = (24x-12)$$

$\therefore$ A = 24, B = −12

$\therefore$ A + 2B = 0

Integer and Numerical Value Type

1. (0) Let $\begin{vmatrix} (1+x)^{a_1b_1} & (1+x)^{a_1b_2} & (1+x)^{a_1b_3} \\ (1+x)^{a_2b_1} & (1+x)^{a_2b_2} & (1+x)^{a_2b_3} \\ (1+x)^{a_3b_1} & (1+x)^{a_3b_2} & (1+x)^{a_3b_3} \end{vmatrix}$

$$= \lambda_0 + \lambda_1 x + \lambda_2 x^2 + \lambda_3 x^3 + \ldots$$

For λ_1 differentiate w.r.t. x and put $x = 0$

So $\lambda_1 = 0$

2. (150) Clearly $f'(x) = 0$

$\therefore f(x) = c = 6$

$$\therefore \sum_{r=1}^{25} f(r) = \sum_{r=1}^{25} 6 = 150$$

3. (–1) $f(-x) = -f(x) = g(x)$

$\therefore f(x).g(x) = -(f(x))^2$ or

$$f(1)g(1) = -(f(1))^2 = \begin{vmatrix} 1 & 1 & 1 \\ 0 & 2 & 1 \\ 1 & 3 & 1 \end{vmatrix}^2 = -4$$

$$\Rightarrow \lambda f(1)g(1) = 4$$
$$\Rightarrow \lambda(-4) = 4$$
$$\Rightarrow \lambda = -1$$

4. (4) On solving, we get

$$(2\lambda+15)f(x+1) - (\lambda+10)f(x+8) - f(x+1) = 0$$
$$(2\lambda+14)f(x+1) = (\lambda+10)f(x+8)$$

Since f is periodic with period 7

$$\therefore f(x+1) = f(x+8)$$
$$\Rightarrow 2\lambda + 14 = \lambda + 10$$
$$\Rightarrow |\lambda| = 4$$

5. (5) The trace of A is 3.

Let $A = \begin{bmatrix} x & y \\ z & 3-x \end{bmatrix}$

Now $A^2 = \begin{bmatrix} x & y \\ z & 3-x \end{bmatrix}\begin{bmatrix} x & y \\ z & 3-x \end{bmatrix}$

$$= \begin{bmatrix} x^2+yz & xy+3y-xz \\ xz+3z-xz & yz+9+x^2-6x \end{bmatrix}$$
$$\Rightarrow A^3 = \begin{bmatrix} x^2+yz & xy+3y-xz \\ xz+3z-xz & yz+9+x^2-6x \end{bmatrix}\begin{bmatrix} x & y \\ z & 3-x \end{bmatrix}$$
$$= \begin{bmatrix} x^2+xyz+3yz & x^2y+y^2z+9y-3xy \\ 3xz+yz^2+9z+x^2z-6xz & 6yz+27+3x^2-18x-xyz-9x-x^3+6x^2 \end{bmatrix}$$

Given that trace of A^3 is −18

$$\therefore x^3 + xyz + 3yz + 3yz + 27 + 3x^2 - 18x - xyz - 9x - x^3 + 6x^2 = -18$$
$$\Rightarrow 9yz + 9x^2 - 27x + 27 = -18$$
$$\Rightarrow yz + x^2 - 3x + 3 = -2$$
$$\Rightarrow 3x - x^2 - yz = 5 \quad \ldots\text{(i)}$$

Now, $|A| = 3x - x^2 - yz$

$|A| = 5$ from (i)

6. (4) Let $\det(P) = \begin{vmatrix} a_1 & b_1 & c_1 \\ a_2 & b_2 & c_2 \\ a_3 & b_3 & c_3 \end{vmatrix}$

$$= a_1(b_2c_3 - b_3c_2) - a_2(b_1c_3 - b_3c_1) + a_3(b_1c_2 - b_2c_1)$$

Now, maximum value of det(P) = 6

If $a_1 = 1, a_2 = -1, a_3 = 1, b_2c_3 = b_1c_3 = b_1c_2 = 1$ and $b_3c_2 = b_3c_1 = b_3c_1 = b_2c_1 = -1$

But it is not possible as

$(b_2c_3)(b_3c_1)(b_1c_2) - -1$ and $(b_1c_3)(b_3c_2)(b_2c_1) = 1$

i.e., $b_1b_2b_3c_1c_2c_3 = 1$ and −1.

A similar contradiction occurs when

$a_1 = 1, a_2 = 1, a_3 = 1, b_2c_1 = b_3c_1 = b_1c_2 = 1$ and $b_3c_2 = b_1c_3 = b_1c_2 = -1$

Now, for the value to be 5, one of the terms

must be zero but that will make 2 terms zero, which means answer cannot be 5.

Now $\begin{vmatrix} 1 & 1 & 1 \\ -1 & 1 & 1 \\ 1 & -1 & 1 \end{vmatrix} = 4$

Therefore, maximum value is 4.

7. (1) For infinite solutions,

$$\begin{vmatrix} 1 & \alpha & \alpha^2 \\ \alpha & \alpha & \alpha \\ \alpha^2 & \alpha & 1 \end{vmatrix} = 0 \Rightarrow \left(1-\alpha^2\right)^2 = 0 \Rightarrow \alpha = \pm 1$$

For $\alpha = 1$, the system will have no solution and for $\alpha = -1$, all three equations reduce to $x - y + z = 1$ giving infinite dependent solutions.

$\therefore 1 + \alpha + \alpha^2 = 1 - 1 + 1 = 1$

8. (Only one pair)

$$z = \frac{-1+i\sqrt{3}}{2} \Rightarrow z^3 = 1 \text{ and } 1 + z + z^2 = 0$$

$$P^2 = \begin{bmatrix} (-z)^r & z^{2s} \\ z^{2s} & z^r \end{bmatrix}\begin{bmatrix} (-z)^r & z^{2s} \\ z^{2s} & z^r \end{bmatrix}$$

$$= \begin{bmatrix} z^{2r} + z^{4s} & z^{2s}\left((-z)^r + z^r\right) \\ z^{2s}\left((-z)^r + z^r\right) & z^{4s} + z^{2r} \end{bmatrix}$$

For $P^2 = -I$, we should have

$z^{2r} + z^{4s} = -1$ and $z^{2s}\left((-z)^r + z^r\right) = 0$

$\Rightarrow z^{2r} + z^{4s} + 1 = 0$ and $(-z)^r + z^r = 0$

$\Rightarrow r$ is odd and $s = r$ but not a multiple of 3, which is possible when $s = r = 1$

$\therefore$ only one pair is there.

Comprehension Passage

Comprehension 1

1. (a) $b_1.C_{31} + b_2.C_{32} + b_3.C_{33}$

$$= b_1\begin{vmatrix} a_2 & a_3 \\ b_2 & b_3 \end{vmatrix} - b_2\begin{vmatrix} a_1 & a_3 \\ b_1 & b_3 \end{vmatrix} + b_3\begin{vmatrix} a_1 & a_2 \\ b_1 & b_2 \end{vmatrix} = 0$$

2. (b) Value of determinant $= 2^3\Delta = 8\Delta$

3. (d) $a_3M_{13} - b_3.M_{23} + d_3.M_{33}$

$= a_3C_{13} + b_3.C_{23} + d_3.C_{33} = \Delta$ by definition

Comprehension 2

1. (d) Given, $A = \begin{bmatrix} a & b \\ c & a \end{bmatrix}, a, b, c \in \{0, 1, 2, \ldots, p-1\}$

If A is skew-symmetric matrix, then $a = 0, b = -c$

$\therefore |A| = -b^2$

Thus, P divides $|A|$, only when $b = 0$. ...(i)

Again, if A is symmetric matrix, then $b = c$ and $|A| = a^2 - b^2$

Thus, p divides $|A|$, if either p divides $(a-b)$ or p divides $(a+b)$.

p divides $(a-b)$, only when $a = b$,

i.e., $a = b \in \{0, 1, 2, \ldots, (p-1)\}$

i.e., p choices ...(ii)

p divides $(a+b)$

$\Rightarrow p$ choices, including $a = b = 0$ included in equation (i)

$\therefore$ Total number of choices are $(p + p - 1) = 2p - 1$

2. (d) The number of matrices for which p does not divide $\text{Tr}(A) = (p-1)p^2$. Of these, $(p-1)^2$ are such that p divides $|A|$. The number of matrices for which p divides $\text{Tr}(A)$ and p does not divide $|A|$ are $(p-1)^2$.

$\therefore$ Required number

$= (p-1)p^2 - (p-1)^2 + (p-1)^2$

$= p^3 - p^2$

Comprehension 3

1. (a) Each element of set A is 3×3 symmetric matrix with five of its entries as 1 and four of its entries as 0. In the diagonal, we can keep either two 0's and one 1 or no zero and three 1's so that the left over 0's and 1's are even in number. Therefore, taking two 0's and one 1 in diagonal, the possible cases are $\frac{3!}{2!}\times\frac{3!}{2!} = 9$ and taking 3 ones in diagonal, the possible cases are $1\times\frac{3!}{2!} = 3$.

$\therefore$ Total number of elements of $A = 9 + 3 = 12$

2. (b)

Given system of linear equations: $A\begin{bmatrix} x \\ y \\ z \end{bmatrix} = \begin{bmatrix} 1 \\ 0 \\ 0 \end{bmatrix}$

This system will have unique solution if $|A| \neq 0$.

$\therefore$ The possible matrix A are

$$\begin{bmatrix} 1 & 0 & 1 \\ 0 & 0 & 1 \\ 1 & 1 & 0 \end{bmatrix}, \begin{bmatrix} 1 & 1 & 0 \\ 1 & 0 & 1 \\ 0 & 1 & 0 \end{bmatrix}, \begin{bmatrix} 0 & 0 & 1 \\ 0 & 1 & 1 \\ 1 & 1 & 0 \end{bmatrix},$$

$$\begin{bmatrix} 0 & 1 & 1 \\ 1 & 1 & 0 \\ 1 & 0 & 0 \end{bmatrix}, \begin{bmatrix} 0 & 1 & 1 \\ 1 & 0 & 0 \\ 1 & 0 & 1 \end{bmatrix} \text{ and } \begin{bmatrix} 0 & 1 & 0 \\ 1 & 0 & 1 \\ 0 & 1 & 1 \end{bmatrix}$$

i.e., 6 possible matrices

3. (b)

For the given system to be inconsistent $|A| = 0$.

$\therefore$ The possible matrix A are

$$\begin{bmatrix} 1 & 0 & 0 \\ 0 & 1 & 1 \\ 0 & 1 & 1 \end{bmatrix}, \begin{bmatrix} 1 & 0 & 1 \\ 0 & 1 & 0 \\ 1 & 0 & 1 \end{bmatrix}, \begin{bmatrix} 1 & 1 & 0 \\ 1 & 1 & 0 \\ 0 & 0 & 1 \end{bmatrix} \text{ and}$$

(i) (ii) (iii)

$$\begin{bmatrix} 1 & 1 & 1 \\ 1 & 0 & 0 \\ 1 & 0 & 0 \end{bmatrix}, \begin{bmatrix} 0 & 1 & 0 \\ 1 & 1 & 1 \\ 0 & 1 & 0 \end{bmatrix}, \begin{bmatrix} 0 & 0 & 1 \\ 0 & 0 & 1 \\ 1 & 1 & 1 \end{bmatrix}$$

(iv) (v) (vi)

On solving $A\begin{bmatrix} x \\ y \\ z \end{bmatrix} = \begin{bmatrix} 1 \\ 0 \\ 0 \end{bmatrix}$

We find for $A =$ (i),

by Cramer's rule $D_1 = 0 = D_2 = D_3$

$\therefore$ There are infinite solutions.

For $A =$ (ii),

by Cramer's rule $D_1 \neq 0$

$\Rightarrow$ No solution, i.e., inconsistent.

Similarly, we find the system as inconsistent in cases (iii), (v) and (vi).

Hence for four cases, the system is inconsistent.

Comprehension 4

$$\Delta = \begin{vmatrix} 1 & 2 & 3 \\ 4 & 5 & 6 \\ 7 & 8 & 9 \end{vmatrix} = 0$$

Given that the system of equations is consistent,

$$\therefore \Delta_x = \begin{vmatrix} \alpha & 2 & 3 \\ \beta & 5 & 6 \\ \gamma-1 & 8 & 9 \end{vmatrix} = 0$$

$$\Rightarrow \alpha - 2\beta + \gamma = 1$$

$$|M| = \begin{vmatrix} \alpha & 2 & \gamma \\ \beta & 1 & 0 \\ -1 & 0 & 1 \end{vmatrix} = \alpha - 2\beta + \gamma = 1$$

The plane containing all those (α, β, γ):

Perpendicular distance from (0, 1, 0) $= \left|\frac{3}{\sqrt{6}}\right| = x$

$$\Rightarrow D = x^2 = \frac{9}{6} = 1.5$$

1. 1.00

2. 1.50

CHAPTER 5

Continuity and Differentiability

Objectives

In this chapter on Continuity and Differentiability, we shall study the following topics:

1. Continuity and its understanding in terms of existence of limits
2. Discontinuity and its type
3. Arithmetic operations on continuous functions
4. Properties of a function continuous on a closed interval
5. Differentiability at a point
6. Differentiability of $f(x)$ over closed interval $[a, b]$
7. Differentiation from first principle.

Continuity

The word "Continuity" has similar connotation to the term "Continuous" in daily life. In mathematics, Continuity means the continuity in the graph of a given function.

For example, consider a simple function, given as $y = f(x) = x, x \in [0,6]$ or $y = x$.

From your earlier classes, you can plot the graph as,

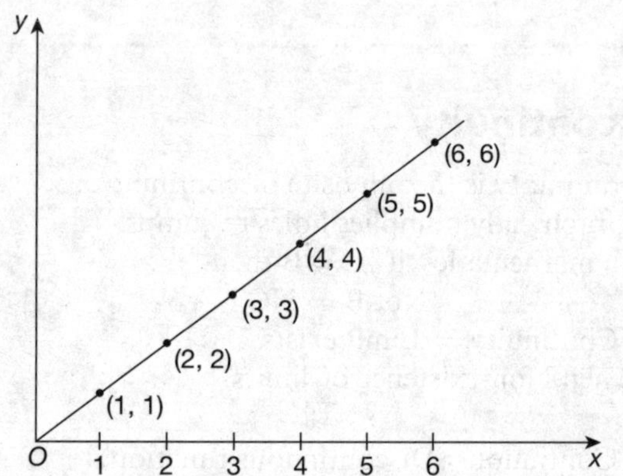

Consider the thick line, which is clearly a continuous line. This is what we can graphically understand when we consider the function as being continuous.

Now, if we slightly change the above function, in following way:

$$y = f(x)$$

$$= x, x \in [0,1) \cup [1,2) \cup [2,3) \cup [3,4) \cup [4,5) \cup [5,6)$$

and $\quad y = f(x) = 0$ for $x = 1, 2, 3, 4, 5, 6$

If by using graphical technique, you plot the graph, it will come out like this

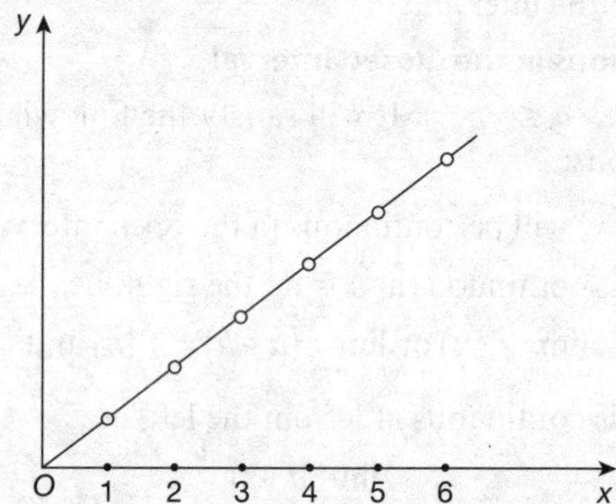

Now, you can see the same graph being discontinuous at $x = 1, 2, 3, 4, 5, 6$, and we will get the same value. So, the graph becomes discontinuous at $x=1$. Similarly, for $x = 2, 3, 4, 5$ and 6.

In simple terms, the existence of any holes or jumps in graphs indicate discontinuity.

Understanding Continuity in Terms of Existence of Limits

You have already studied about the concept of limits in earlier classes.

There is a very popular understanding in Mathematics, which is, Continuity implies that Limit exists.

i.e., if a function is continuous at $x = c$

then $\quad \lim_{x \to c} f(x)$ exists.

and, $\quad \lim_{x \to c^+} f(x) = \lim_{x \to c^-} f(x) = f(c)$

i.e., whether we approach the function from right side of c (c^+) or from left side of $c(c^-)$, we will get the same result.

Limiting Value

Right-hand limit (RHL) is also alternatively written as

$$x = c + h$$

where 'h' is a very small term.

Thus, $x \to c$

$$\Rightarrow \qquad h \to 0$$

or, $\lim_{h\to 0} f(c+h)$ is the right-hand limit.

For left-hand limit (LHL) (i.e., approaching 'c' from left side of 'c'),

$$x \to c$$

$$\Rightarrow \qquad h \to 0$$

$$\text{or,} \qquad \lim_{h\to 0} (c-h)$$

For a function to be continuous at $x = c$

$$\lim_{h\to 0} f(c+h) = \lim_{h\to 0} f(c-h) = f(c)$$

Continuity in the Open Interval

(a, b) or $]\,a, b\,[$ i.e. $a < x < b$

$f(x)$ will be continuous if it is continuous at every point of the interval.

Continuous in the Closed Interval

$[a, b]$, i.e., $a \le x \le b$. It will satisfy the following three conditions:

(i) $f(x)$ will be continuous in the open interval (a, b)

(ii) It is continuous at a from the right, i.e.,

$$\lim_{x\to a+0} f(x) \text{ or } \lim_{h\to 0} f(a+h) = f(a) \text{ and}$$

(iii) It is continuous at b from the left, *i.e.*,

$$\lim_{x\to b-0} f(x)$$

$$\text{or } \lim_{h\to 0} f(b-h) = f(b)$$

ILLUSTRATION 5.1

Let $f(x) = x^2$, then for any real number c.

$$\lim_{x\to c} f(x) = \lim_{x\to c} x^2 = c^2 = f(c)$$

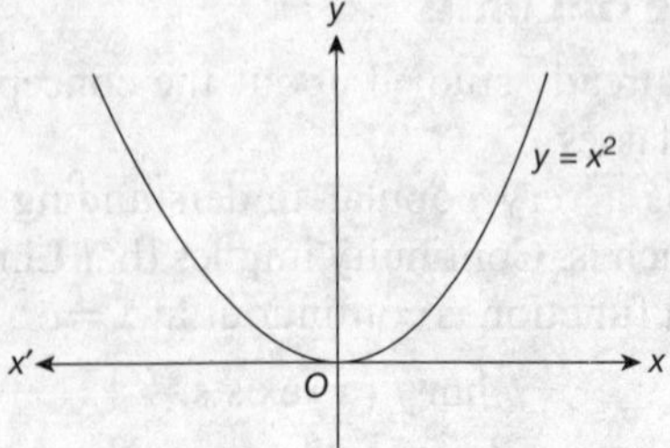

Hence, the function $f(x) = x^2$ is continuous for all values of x.

We have earlier seen that $\lim_{x\to c} f(x)$ may exist without the function being defined at $x = c$. However, for the continuity of a function at $x = c$, the function must be defined at $x = c$.

ILLUSTRATION 5.2

The function $f(x) = |\,x\,|$ is continuous at every value of x. For $x > 0$, we have $f(x) = x$, a polynomial. For $x < 0$, we have $f(x) = -x$, another polynomial. So f is continuous.

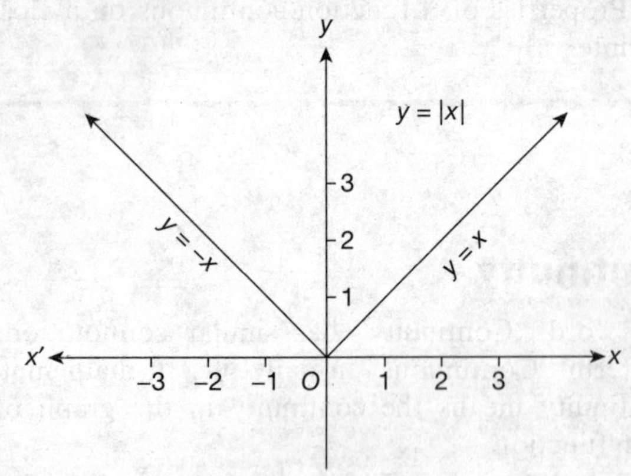

Finally at the origin $\lim_{x\to 0} |x| = 0 = f(0)$, i.e., $\lim_{x\to 0} f(x) = f(0)$, so f is continuous at 0 also.

Discontinuity

Discontinuity is the opposite of continuity.

Graphically it implies holes or jumps.

In mathematics, if A $\Rightarrow$ B then

Not B $\Rightarrow$ NOT A

$\therefore$ If Continuity $\Rightarrow$ Limit exists

Then Non-existence of limits

$\Downarrow$

Not Continuous (Discontinuous function)

$f(x)$ will be discontinuous at $x = a$ in any of the following cases:

(i) $\lim_{x\to a^-} f(x)$ and $\lim_{x\to a^+} f(x)$ exist but are not equal.

(ii) $\lim_{x\to a^-} f(x)$ and $\lim_{x\to a^+} f(x)$ exist and are equal but not equal to $f(a)$.

(iii) $f(a)$ is not defined.

(iv) At least one of the limits does not exist.

Types of Discontinuities

(1) Discontinuity of First Kind: The point $x = a$ will be a point of discontinuity of first type if both right-hand and left-hand limits at $x = a$ exist but are not equal, i.e.,

$$\lim_{h\to 0} f(a-h) \neq \lim_{h\to 0} f(a+h).$$

(2) Discontinuity of Second Kind: The point $x = a$ will be a point of discontinuity of second type if either or both the right-hand and left-hand limits do not exist or if either or both the limits

$$\lim_{h\to 0} f(a-0) \text{ or } \lim_{h\to 0} f(a+0) \text{ are infinite.}$$

(3) Removable Discontinuity: If

$$\lim_{h\to 0} f(a+0) = \lim_{h\to 0} f(a-0) \text{ i.e.,}$$

limit exists but is not equal to $f(a)$, i.e., value, then the function is said to have a removable discontinuity. We can re-define the function at $x = a$ so that limit = value and thus the discontinuity has been removed and the function has become continuous now.

ILLUSTRATION 5.3

If $f(x) = \begin{cases} \dfrac{x^3-1}{x-1}, & x \neq 1 \\ 4 & x = 1 \end{cases}$ check whether the function is continuous of or not.

$$\lim_{x\to 1} f(x) = \lim_{x\to 1} \frac{(x-1)(x^2+x+1)}{x-1}$$

$$\lim_{x\to 1} f(x) = \lim_{x\to 1} (x^2+x+1) = 3$$

$$\lim_{x\to 1} f(x) \neq f(1)$$

$\Rightarrow$ The function is not continuous at $x = 1$.

ILLUSTRATION 5.4

A function $f(x)$ is defined as follows:

$$f(x) = \frac{\sin x}{x}, \text{ when } x \neq 0$$

$$= 4, \quad x = 0$$

Is (x) continuous at $x = 0$? If not, redefine it so that it becomes continuous at $x = 0$.

$$\lim_{x\to 0} f(x) = \lim_{x\to 0} \frac{\sin x}{x}$$

$$\lim_{x\to 0} f(x) = 1 \text{ while } f(0) = 4$$

$\therefore \lim_{x\to 0} f(x) \neq f(0)$. Hence $f(x)$ is discontinuous at $x = 0$.

In order to make $f(x)$ continuous at $x = 0$, function is redefined as,

$$f(x) = \begin{cases} \dfrac{\sin x}{x}, & x \neq 0 \\ 1, & x = 0 \end{cases}$$

ILLUSTRATION 5.5

$f(x) = \dfrac{x^2-a^2}{x-a}$ is discontinuous at $x = a$ as $f(a)$ does not exist. Similarly, if $y = 1/x$, it is also discontinuous at $x = 0$ as it is not defined.

ILLUSTRATION 5.6

$f(x) = \begin{cases} (1+2x)^{1/x} & x \neq 0 \\ k, & x = 0 \end{cases}$ For what value of k, $f(x)$ is continuous at $x = 0$?

$$\lim_{x\to 0} f(x) = \lim_{x\to 0} (1+2x)^{1/x}$$

$$= \lim_{x\to 0} (1+2x)^{\frac{1}{2x}\cdot 2} = e^2$$

$f(x)$ will be continuous at $x = 0$ only when $\lim_{x\to 0} f(x) = f(0)$

$$\Rightarrow \qquad k = e^2.$$

ILLUSTRATION 5.7

Find the points of discontinuity of $y = \dfrac{1}{u^2+u-2}$ where $u = \dfrac{1}{x-1}$.

The function $u = f(x)$ and $f(x) = \dfrac{1}{x-1}$ is discontinuous at $x = 1$.

The function

$$y = g(x) = \frac{1}{u^2+u-2} = \frac{1}{(u+2)(u-1)}$$

is discontinuous at $u = -2$ and $u = 1$

$$\text{when } u = -2, \frac{1}{x-1} = -2 \qquad x = \frac{1}{2}$$

$$\text{when } u = 1, \frac{1}{x-1} = 1 \qquad x = 2.$$

Hence the composite function $y = g(f(x))$ is discontinuous at three points $x = \dfrac{1}{2}, 1, 2$.

ILLUSTRATION 5.8

ABC is an isosceles triangle inscribed in a circle of radius r. If AB = AC and h is the attitude from A to BC, then the ΔABC has perimeter $P = 2(\sqrt{2hr - h^2} + \sqrt{2hr})$ and area $A = \ldots$. Also, $\lim\limits_{h \to 0} \frac{A}{P^3} = \ldots$

Solution

Given, $P = 2(\sqrt{2hr - h^2} + \sqrt{2hr})$

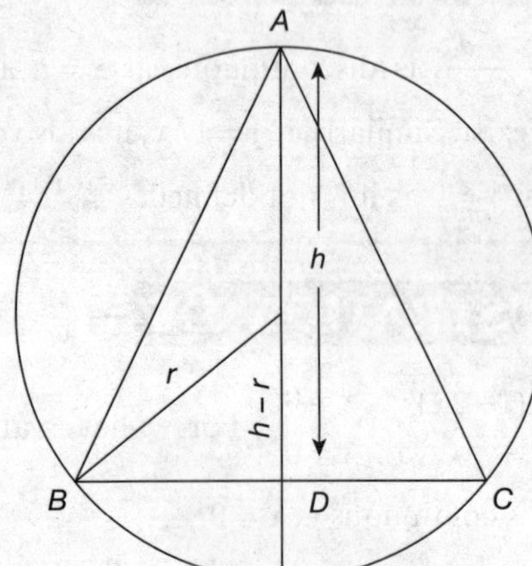

Here, $BD = \sqrt{r^2 - (h-r)^2} = \sqrt{2hr - h^2}$

$$\therefore \quad A = \frac{1}{2}.2BD.h = \left(\sqrt{2hr - h^2}\right)h$$

$$\therefore \quad \lim_{h \to 0} \frac{A}{P^3} = \lim_{h \to 0} \frac{h\sqrt{2hr - h^2}}{8(\sqrt{2hr - h^2} + \sqrt{2hr})^3}$$

$$= \lim_{h \to 0} \frac{h^{3/2}\left(\sqrt{2r - h}\right)}{8\,h^{3/2}\,(\sqrt{2r - h} + \sqrt{2hr})^3}$$

$$= \frac{1}{8}.\frac{\sqrt{2r}}{(\sqrt{2r} + \sqrt{2r})^3} = \frac{1}{128\,r}.$$

Arithmetical Operations on Continuous Functions

Continuity of a Composite Function

If the functions $f(x)$ and $g(x)$ are continuous at the point $x = x_0$, then the functions

(1) $f(x) \pm g(x)$; (2) $f(x) \cdot g(x)$; (3) $\frac{f(x)}{g(x)} (g(x_0)) \neq 0)$

are also continuous at this point.

If the function $u = \phi(x)$ is continuous at the point $x = x_0$ and the function $y = f(u)$ is continuous at the point $u_0 = \phi(x_0)$, then the composite function $y = f[\phi(x)]$ is continuous at the point $x = x_0$.

ILLUSTRATION 5.9

Given the function $f(x) = 1/(1-x)$. Find the points of discontinuity of the composite function.

$$y = f\{f[f(x)]\}$$

Solution

The point $x = 1$ is a discontinuity of the function

$$v = f(x) = \frac{1}{1-x}.$$

If $x \neq 1$, then

$$u = f[f(x)] = \frac{1}{1 - \frac{1}{1-x}} = \frac{x-1}{x}.$$

Hence, the point $x = 0$ is a discontinuity of the function $u = f[f(x)]$.

If $x \neq 0, x \neq 1$, then

$$y = f\{f[f(x)]\} = \frac{1}{1-(x-1)/x} = x$$

is continuous everywhere.

Thus, the points of discontinuity of this composite function are $x = 0, x = 1$, both of them being removable.

Properties of a Function Continuous on a Closed Interval

I. The function $f(x)$, continuous on the interval $[a, b]$, possesses the following properties:

(1) $f(x)$ is bounded on $[a, b]$

(2) $f(x)$ has the minimum and maximum values on $[a, b]$

(3) If $m = \min\limits_{a \le x \le b} f(x), M = \max\limits_{a \le x \le b} f(x)$, then for any A satisfy which $f(x_0) = A$.

In particular, if $f(a).f(b) < 0$, then we can find a point $c(a < c < b)$ such that $f(c) = 0$.

II. Continuity of an Inverse Function

If the function $y = f(x)$ is defined, continuous and strictly monotonic on the interval X, then there exists a single-valued inverse function $x = \varphi(y)$ defined, continuous and also strictly monotonic in the range of the function $y = f(x)$.

ILLUSTRATION 5.10

Does the equation $\sin x - x + 1 = 0$ have a root?

Solution

The function

$$f(x) = \sin x - x + 1$$

is continuous over the entire number scale. Besides, this function changes sign, since $f(0) = 1$, and $f(3\pi/2) = -3\pi/2$. Hence, by property (3) within the interval $[0, 3\pi/2]$ there is at least one root of the given equation.

ILLUSTRATION 5.11

Prove that any algebraic equation of an odd power with real coefficients

$$a_0 x^{2n+1} + a_1 x^{2n} + \ldots + a_{2n} x + a_{2n+1} = 0$$

has at least one real root.

Solution

Consider the function

$$f(x) = a_0 x^{2n+1} + a_1 x^{2n} + \ldots + a_{2n} x + a_{2n+1},$$

which is continuous throughout the number scale.

Let, for the sake of determinacy, $a_0 > 0$. Then

$$\lim_{x \to +\infty} f(x) = +\infty, \text{ and } \lim_{x \to -\infty} f(x) = -\infty.$$

Hence, we can find numbers a, b, $a < b$ such that $f(a) < 0$; $f(b) > 0$. By property (3), between a and b there exists a number c such that $f(c) = 0$, which proves that the equation has at least one real root.

ILLUSTRATION 5.12

Let the function $f(x)$ be continuous on $[a, b]$ and let the equation $f(x) = 0$ have a finite number of roots on the interval $[a, b]$. Arrange them in the ascending order:

$$a < x_1 < x_2 < x_3 < \ldots ,(x_n, b)$$

Prove that in each of the intervals

$$(a, x_1), (x_1, x_2), (x_2, x_3), \ldots ,(x_n, b)$$

the function $f(x)$ retains the same sign.

Solution

If the function changed its sign at a certain interval, then we could find one more root of the function, which contradicts the condition. To determine the sign of the function at any of the indicated intervals, it is sufficient to compute the value of the function at an arbitrary point of the appropriate interval.

ILLUSTRATION 5.13

Does the function

$$f(x) = x^3/4 - \sin \pi x + 3$$

take on the value $2\frac{1}{3}$ within the interval $[-2, 2]$?

Solution

The function $f(x) = \frac{x^3}{4} - \sin \pi x + 3$ is continuous within the interval $[-2, 2]$. Furthermore, at the end-points of this interval, it attains the values

$$f(-2) = 1; f(2) = 5.$$

Since $1 < 2\frac{1}{3} < 5$, then, by property (3), within the interval $[-2, 2]$ there exists at least one point x such that $f(x) = 2\frac{1}{3}$.

ILLUSTRATION 5.14

Prove that for any function of the form

$$y = a_0 x^{2n+1} + a_1 x^{2n-1} + a_2 x^{2n-3} + \ldots + a_n x + a_{n+1}$$

where $a_0, a_1, a_2, \ldots, a_n, a_{n+1}$ are positive numbers, there exists an inverse function increasing and continuous throughout the number scale.

Solution

As is known, the functions $x, x^3, x^5, \ldots, x^{2n+1}$ increase throughout the entire number scale. Then, since the coefficients a_i ($i = 0, 1, \ldots, n + 1$) are positive, the function $f(x) = a_0 x^{2n+1} + a_1 x^{2n-1} + a_2 x^{2n-3} + \ldots + a_n x + a_{n+1}$ also increases. Furthermore, it is continuous. Therefore, for a function of the form, there exists an inverse function increasing and continuous over the entire number scale.

Misconception 1

This example establishes only the existence of an inverse function $x = g(y)$, but gives no analytic expression for it. It is not always possible to express it in radicals. The problems of the existence of an inverse function and the problems of expressing an inverse function analytically should not be confused with one another.

Differentiability at a Point

A function $f(x)$ is said to be differentiable at $x = a$ if the following limits are satisfied:

$$\underbrace{\lim_{h\to 0}\frac{f(a+b)-f(a)}{h}}_{\text{Right-hand limit (RHL)}} = \underbrace{\lim_{h\to 0}\frac{f(a-b)-f(a)}{-h}}_{\text{Left-hand limit (LHL)}}$$

In case of RHL, we approach the point from right side of $x = a$ and in case of LHL, we approach the point from left side of $x = a$.

Graphically, we can understand it as follows:

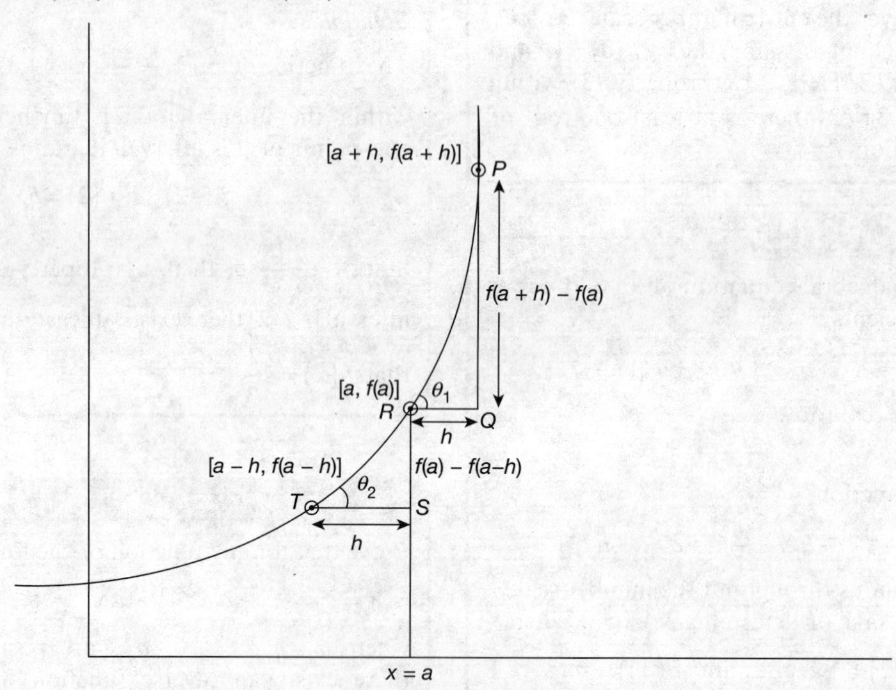

$$P[(a+h), f(a+h)]$$

$$R[a, f(a)]$$

$$T[(a-h), f(a-h)]$$

Look at triangles PQR and RST.

In the above graphical representation, if we approach the point $[a, f(a)]$ from the right, i.e., from $a + h(h \to 0)$

we get, $\dfrac{PQ}{QR} = \dfrac{f(a+h)-f(a)}{h} = \tan\theta_1$

Similarly, if we approach the point $[a, f(a)]$ from left, i.e., from $a - h(h \to 0)$

we get, $\dfrac{RS}{ST}(\text{in } \Delta RST) = \dfrac{f(a)-f(a-h)}{h} = \tan\theta_2$

In the limiting case, the tangent at the point $R[a, f(a)]$ has to be one single tangent.

In other words, $\tan\theta_1$ (approaching from Right)

$= \tan\theta_2$ (approaching from Left)

$\Rightarrow \qquad \theta_1 = \theta_2 = \theta$

So, differentiability implies that a single tangent exists at point $x = a$ for $f(x)$ to be differentiable at $x = a$.

Consider a case of a function which may be continuous but not differentiable at point $x = 0$.

Consider $f(a) = |X|$ at $x = 0$.

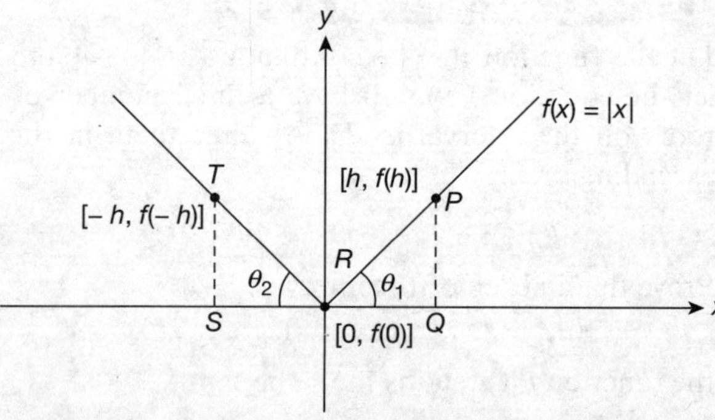

Consider the continuity of $|X|$ at $X = 0$.

Graphically, there is no jump at $X = 0$.

Also, there is no hole at $X = 0$ (remember our discussion during continuity)

So, graphically, we can state that $f(x) = |X|$ is continuous at $X = 0$.

What about differentiability?

From our mathematical expression

$$\text{RHL} = \lim_{h\to 0}\frac{f(0+h)-f(0)}{h}$$

$$= \lim_{h\to 0}\frac{f(h)-f(0)}{h}$$

$$f(x)=|X| = x \text{ for } x>0 : \begin{matrix} f(h)=h \\ f(0)=0 \end{matrix}$$

$$\therefore \quad \text{RHL} = \lim_{h\to 0}\frac{h-0}{h}=1$$

Similarly,

$$\text{LHL} = \lim_{h\to 0}\frac{f(0-h)-f(0)}{-h} = \lim_{h\to 0}\frac{f(-h)-f(0)}{-h}$$

$$f(x)=|X|=-x \text{ for } x<0$$

$$\therefore \quad f(-h)=+h$$

$$f(0)=0$$

$$\therefore \quad \text{LHL} = \frac{+h-0}{-h}=-1$$

$$\therefore \quad \text{RHL} \neq \text{LHL}$$

$\therefore$ The function is not differentiable at $x=0$ even though it continuous.

Thus, a sharp change (lack of smoothness of curve) in the curve at any point indicates non-differentiability even though there may be continuity.

Graphically also, we can see that

$$\tan\theta_1 = -\tan\theta_2$$

i.e., the slopes while approaching point $R[0, f(0)]$ from left and right sides are unequal.

At what is differentiable at a point $x=a$ must also be continuous at that point. On the other hand, if a function is continuous at a point $x=a$, it is not necessarily differentiable at $x=a$.

(A) Differentiability implies continuity, always.

Since the function is differentiable,

$$\therefore \quad L' = R' = \text{finite}$$

$$\lim_{h\to 0}\frac{f(a+h)-f(a)}{h} \text{ exists and is finite} \quad \ldots(1)$$

or $$\lim_{h\to 0} f(a+h) = \lim_{h\to 0}\left[f(a+h)-f(a)\right]+f(a)$$

$$= \lim_{h\to 0} h\left[\frac{f(a+h)-f(a)}{h}\right]+f(a)$$

$$= \lim h.\lim\frac{f(a+h)-f(a)}{h}+f(a)$$

$$= 0+f(a)=f(a)$$

Similarly, $\lim_{h\to 0} f(a-h)=f(a)$

Since $\lim f(a+h)=\lim f(a-h)=f(a)$, the function is continuous.

(B) Continuity does not necessarily imply differentiability.

We will illustrate the above by the following example:

$$\text{Consider } f(x)=\begin{cases} x\sin\frac{1}{x}, & x\neq 0 \\ 0, & x=0 \end{cases}$$

Consider its differentiability at $x=0$

$$R'(0) = \lim_{h\to 0}\frac{f(0+h)-f(0)}{h}$$

$$= \lim_{h\to 0}\frac{h\sin(1/h)}{h} = \lim_{h\to 0}\sin\frac{1}{h}$$

The above limit does not exist as $\sin\frac{1}{h}$ is an oscillating function. Similarly, $L'(0)$ does not exist. Hence the function is not differentiable at $x=0$.

Continuity $R=L=V$

$$R = \lim_{h\to 0} h\sin\frac{1}{h}=0$$

$$L = \lim_{h\to 0}(-h)\sin\left(-\frac{1}{h}\right)=0$$

$$V=0$$

Since $R=L=V$ the function is continuous but as shown above, it is not differentiable. But a function which is differentiable at a point must be continuous at that point.

Note: Continuity and differentiability are established at change points.

As we have learnt earlier $A \Rightarrow B$

then not $B \Rightarrow$ not A

Extending this logic, Differentiability $\Rightarrow$ Continuity

then not continuous $\Rightarrow$ not differentiable

ILLUSTRATION 5.15

(a) A function f is defined as

$$f(x)=\frac{x^2-4x+3}{x^2-1} \text{ for } x\neq 1.$$

$= 2$ for $x=1$

Test the continuity of the function at $x=1$.

$$\lim_{x\to 1}\frac{(x-1)(x-3)}{(x-1)(x+1)} = \lim_{x\to 1}\frac{x-3}{x+1} = \frac{1-3}{1+1} = \frac{-2}{2} = -1$$

but $f(1)=2$

$\therefore$ Not continuous at $x=1$

(b) A function f is defined as

$$f(x)=\frac{x^2-4x+3}{x^2+2x-3} \text{ for } x\neq 1$$

$$=-1/2 \text{ for } x=1.$$

$$\lim_{x\to 1}\frac{(x-1)(x-3)}{(x-1)(x+)}=\lim_{x\to 1}\frac{x-3}{x+3}=\frac{1-3}{1+3}$$

$$=\frac{-2}{4}=-\frac{1}{2}=f(1)$$

$\therefore$ The function is continuous.

ILLUSTRATION 5.16

$f(x)=|x-a|$ is not differentiable at $x=a$

$$f'(a+)=\lim_{h\to 0+}\frac{f(a+h)-f(a)}{h}$$

$$=\lim_{h\to 0+}\frac{|a+h-a|-|a-a|}{h}$$

$$=\lim_{h\to 0+}\frac{|h|}{h}=1$$

Similarly, $f'(a-)=\lim_{h\to 0-}\dfrac{f(a+h)-f(a)}{h}$

$$=\lim_{h\to 0-}\frac{|h|}{h}=-1$$

$\Rightarrow \quad f'(a+)\neq f'(a-)$

Therefore, f is not differentiable at $x=a$

Every differentiable function is continuous but not vice-versa.

ILLUSTRATION 5.17

Examine the differentiability of the following functions at $x=0$:

(i) $|x^2|$

Solution

(i) $\text{LHD}=\lim_{x\to 0-}\dfrac{f(x)-f(0)}{x-0}$

$$=\lim_{x\to 0-}\frac{|x^2|-0}{x}=\lim_{h\to 0}\frac{|(0-h)^2|-0}{0-h}$$

[Putting $x=0-h, h>0$]

$$=\lim_{h\to 0}\frac{h^2}{-h}=\lim_{h\to 0}(-h)=0$$

$$\text{RHD}=\lim_{x\to 0+}\frac{f(x)-f(0)}{x-0}$$

$$=\lim_{x\to 0+}\frac{|x^2|-0}{x}=\lim_{h\to 0}\frac{h^2}{h}=\lim_{h\to 0}(h)=0$$

[Putting $x=0+h, h>0$]

Since LHD = RHD

$\therefore |x^2|$ is differentiable at $x=0$.

Differentiability of $f(x)$ Over Closed Interval $[a, b]$

This can be understood as

$$(a,b)\cup\{a\}\cup\{b\}$$

i.e., we have to satisfy that $f(x)$ is differentiable in open interval (a, b).

PLUS, $f(X)$ should be differentiable at end point $\{a\}$ and $\{b\}$.

At $\{a\}$, we have to check only RHL.

At $\{b\}$, we have to check only LHL.

We can summarize it as: A function $f(x)$, defined on $[a, b]$ is said to be differentiable on $[a, b]$ if it is differentiable at every (a, b) and both exists.

$$\lim_{h\to 0+}\frac{f(a+h)-f(a)}{h} \text{ and } \lim_{h\to 0-}\frac{f(b+h)-f(b)}{h}$$

Differentiation from First Principles

Consider a simple function $f(x)=x^2$

If LHL and RHL exist and are equal then the expression for LHL or RHL is also the derivative of the function $f(x)$.

Consider any general point $x=c$

$$\text{RHL}=\lim_{h\to 0}\frac{f(c+h)-f(c)}{h}$$

$$=\lim_{h\to 0}\frac{(c+h)^2-c^2}{h}$$

$$=\lim_{h\to 0}\frac{c^2+2ch+h^2-c^2}{h}$$

$$=\lim_{h\to 0}\frac{2ch+h^2}{h}=\lim_{h\to 0}2c+h=2c$$

Similarly,

$$\text{LHL}=\lim_{h\to 0}\frac{f(c-h)-f(c)}{-h}$$

$$=\lim_{h\to 0}\frac{(c-h)^2-c^2}{-h}$$

$$=\lim_{h\to 0}\frac{c^2-2ch+h^2-c^2}{-h}=2c.$$

Generalizing, as c represents any x,

$$\text{RHL} = \text{LHL} = 2c$$

which is what we already know.

$$\frac{d}{dx}\left(x^2\right) = 2x$$

Note: We can use first principle to find

$$\frac{d}{dx}\left(x^n\right) = nx^{n-1}$$

$$\frac{d}{dx}(\sin x) = \cos x$$

$$\frac{d}{dx}\left(a^x\right) = a^x \log e^a$$

$$\frac{d}{dx}\left(\log e^x\right) = \frac{1}{x}$$

and $$\frac{d}{dx}\left(\sin^{-1}x\right) = \frac{1}{\sqrt{1-x^2}}$$

Thus, through mathematical rigor, we can find $\frac{d}{dx}$ (Derivative) of practically all the known functions, which are listed below:

List of Derivatives of Important Functions

(i) $\frac{d}{dx}\left(x^n\right) = nx^{n-1}$

(ii) $\frac{d}{dx}(x) = 1$

(iii) $\frac{d}{dx}\left(\frac{1}{x}\right) = -\frac{1}{x^2}$

(iv) $\frac{d}{dx}\left(\frac{1}{x^n}\right) = \frac{n}{x^n+1},\ x \neq 0$

(v) $\frac{d}{dx}(\sin x) = \cos x$

(vi) $\frac{d}{dx}(\cos x) = -\sin x$

(vii) $\frac{d}{dx}(\tan x) = \sec^2 x$

(viii) $\frac{d}{dx}(\cot x) = -\text{cosec}^2 x$

(ix) $\frac{d}{dx}(\text{cosec}\, x) = -\text{cosec}\, x \cot x$

(x) $\frac{d}{dx}\left(a^x\right) = a^x \log_e a \left[\text{and also } \frac{d}{dx}\left(e^x\right) = e^x\right]$

(xi) $\frac{d}{dx}(\log_e x) = \frac{1}{x}$

(xii) $\frac{d}{dx}\left(\sin^{-1}x\right) = \frac{1}{\sqrt{1-x^2}}$

(xiii) $\frac{d}{dx}\left(\cos^{-1}x\right) = \frac{-1}{\sqrt{1-x^2}}$

(xiv) $\frac{d}{dx}\left(\tan^{-1}x\right) = \frac{1}{1+x^2}$

(xv) $\frac{d}{dx}\left(\cot^{-1}x\right) = \frac{-1}{1+x^2}$

(xvii) $\frac{d}{dx}\left(\text{cosec}^{-1}x\right) = \frac{-1}{|x|\sqrt{x^2-1}}$

(xviii) $\frac{d}{dx}\left(\sec^{-1}x\right) = \frac{-1}{|x|\sqrt{x^2-1}}$

While you may memorize the above list, you are advised to find $\frac{d}{dx}$ of at least a few functions from first principles as we did for $f(x) = x^2$.

WORKSHEET – 1

1. Evaluate $\lim\limits_{x\to 1}\left(\frac{x-1}{2x^2-7x+5}\right)$.

2. Evaluate $\lim\limits_{x\to 0}\sqrt{\frac{x-\sin x}{x+\cos^2 x}}$.

3. $\lim\limits_{x\to -\infty}\left[\frac{\left(x^4\sin\left(\frac{1}{x}\right)+x^2\right)}{\left(1+|x|^3\right)}\right] = $ ____.

4. If $f(x) = \begin{cases} \sin x, x \neq n\pi, n = 0, \pm 1, \pm 2, \ldots \\ 2, \quad \text{otherwise} \end{cases}$

and $g(x) = \begin{cases} x^2+1, & x \neq 0, 2 \\ 4, & x = 0 \\ 5, & x = 2 \end{cases}$, then $\lim\limits_{x\to 0} g\left[f(x)\right]$ is ____.

(5–8) Discuss the continuity of the following functions at the indicated points:

5. $f(x)=\dfrac{e^{1/x}-1}{e^{1/x}+1}, x \neq 0, f(0)=0$

6. $f(x)=\dfrac{|x-3|}{x-3}, x \neq 0, f(3)=0$

7. $f(x)=10^{\frac{-1}{(x-3)^2}}, x \neq 3, f(3)=0$

8. $f(x)=(-1)^{[x]}$

(9–11) Test the following functions for continuity

9. $f(x)=\dfrac{2x^5-8x^2+11}{x^4+4x^3+8x^2+8x+4}$;

10. $f(x)=\dfrac{3\sin^3 x+\cos^2 x+1}{4\cos x-2}$;

11. $f(x)=\dfrac{x^3\cos x+x^2\sin x}{\cos(1/\sin x)}$

12. Given a function on the interval $[-2,+2]$:

$$f(x)=\begin{cases} x^2+2 & \text{if } -2\le x\le 0, \\ -(x^2+2) & \text{if } 0\le x\le 2. \end{cases}$$

Is there a point on this closed interval at which $f(x)=0$?

Derivative of Composite Functions and Chain Rule

Composite Function

A composite function is a function of some other function.

For example, $y=(x^3+1)^2$

Here y is a function of (x^3+1) and (x^3+1) is a function of x.

Let $u=(x^3+1)$, then $y=u^3$.

y is a function of u and u is a function of x. Therefore, y is a function of x.

$$\frac{dy}{dx}=\frac{dy}{du}\times\frac{du}{dx}$$

Chain rule is used to find derivatives of such functions.

Chain Rule

If $y=f(u)$ is a differential function of u and $u=g(x)$ is a differentiable function of x, then $y=f(g(x))$ is a differentiable function of x and $\dfrac{dy}{dx}=\dfrac{dy}{du}\times\dfrac{du}{dx}$ or, equivalently

$$\frac{d}{dx}[f(g(x))]=f'(g(x))g'(x).$$

ILLUSTRATION 5.18

Differentiate the functions w.r.t x.

(i) $(x^2+2x+5)^3$

(ii) $\cos\sqrt{x}$

(iii) $\cos(\sin x)$

Solution

(i) Let $u=(x^2+2x+5)^3$

then $y=u^3$

$$\frac{dy}{du}=3u^2,\quad \frac{du}{dx}=2x+2$$

$$\frac{dy}{dx}=\frac{dy}{du}\times\frac{du}{dx}$$

$$=3(x^2+2x+5)^2.(2x+2)$$

$$=6(x^2+2x+5)^2.(x+1)$$

(ii) Let, $y=\cos\sqrt{x}$

Putting $\sqrt{x}=u$, we get

$y=\cos u$ and $u=\sqrt{x}=x^{\frac{1}{2}}$

Now $\dfrac{dy}{du}=-\sin u$ and

$$\frac{du}{dx}=\frac{1}{2}x^{\frac{1}{2}-1}=\frac{1}{2}x^{-\frac{1}{2}}=\frac{1}{2\sqrt{x}}$$

$$\frac{dy}{dx}=\frac{dy}{du}\times\frac{du}{dx}\ \text{[Chain rule]}$$

$$\frac{dy}{dx}=-\sin u\times\frac{1}{2\sqrt{x}}=-\frac{\sin\sqrt{x}}{2\sqrt{x}}$$

(iii) We have $y=\cos(\sin x)$

Let $\sin x=t$, then $y=\cos t$ and $t=\sin x$

$$\therefore \frac{dy}{dt}=-\sin t \text{ and } \frac{dt}{dx}=\cos x$$

Now $\dfrac{dy}{dx}=\dfrac{dy}{dt}\times\dfrac{dt}{dx}$ [Chain rule]

$$=-\sin t\times\cos x$$

$$=-\cos x\sin(\sin x)$$

ILLUSTRATION 5.19

If $y=\sqrt{\left(\frac{1-x}{1+x}\right)}$, prove that $\left(1-x^2\right)\frac{dy}{dx}+y=0.$

Solution

$$y=\sqrt{\left(\frac{1-x}{1+x}\right)}=\left(\frac{1-x}{1+x}\right)^{\frac{1}{2}}$$

Then $\qquad \frac{dy}{dx}=\frac{1}{2}\left(\frac{1-x}{1+x}\right)^{\frac{1}{2}}\cdot\frac{d}{dx}\left(\frac{1-x}{1+x}\right)$

$$=\frac{1}{2}\left(\frac{1-x}{1+x}\right)^{\frac{1}{2}}\cdot\left[\frac{(1+x)\frac{d}{dx}(1-x)-(1-x)\frac{d}{dx}(1+x)}{(1+x)^2}\right]$$

$$=\frac{1}{2}\left(\frac{1-x}{1+x}\right)^{\frac{1}{2}}\cdot\left[\frac{(1+x)(-1)-(1-x)(1)}{(1+x)^2}\right]$$

$$=\frac{1}{2}\left(\frac{1-x}{1+x}\right)^{\frac{1}{2}}\cdot\left[\frac{-1-x-1+x}{(1+x)^2}\right]$$

$$=\frac{1}{2}\left(\frac{1-x}{1+x}\right)^{\frac{1}{2}}\times\frac{-2}{(1+x)^2}$$

$$\Rightarrow \quad \frac{dy}{dx}=-\left(\frac{1-x}{1+x}\right)^{\frac{1}{2}}\cdot\frac{1}{(1+x)^2}$$

$$\therefore \quad \left(1-x^2\right)\frac{dy}{dx}=-\left(\frac{1+x}{1-x}\right)^{\frac{1}{2}}\times\frac{1-x^2}{(1+x)^2}$$

$$=-\left(\frac{1-x}{1+x}\right)^{\frac{1}{2}}=-y$$

$$\Rightarrow \quad \left(1-x^2\right)\frac{dy}{dx}+y=0.$$

ILLUSTRATION 5.20

If $x\sqrt{1+y}+y\sqrt{1+x}=0$ $(x\neq y)$, then prove that

$$\frac{dy}{dx}=-\frac{1}{(1+x)^2}$$

Solution

We have $x\sqrt{1+y}+y\sqrt{1+x}=0$

$$\Rightarrow \quad x\sqrt{1+y}=-y\sqrt{1+x}$$

Squaring both sides, we get

$$x^2(1+y)=y^2(1+x)$$

$$\Rightarrow \quad x^2+x^2y=y^2+y^2x\Rightarrow x^2-y^2=x^2-y^2$$
$$=-x^2y+y^2x$$

$$\Rightarrow \quad (x-y)(x+y)=-xy(x-y)$$

Since $x-y\neq 0$, dividing both sides by $x-y$, we get

$$x+y=-xy \Rightarrow y+xy=-x$$

$$\Rightarrow \quad y(1+x)=-x \text{ or } y=\frac{-x}{1+x}$$

Now differentiating both sides, we get

$$\frac{dy}{dx}=\frac{(1+x)(-1)-(-x)(1)}{(1+x)^2}$$

$$\Rightarrow \quad \frac{dy}{dx}=\frac{-1-x+x}{(1+x)^2}=\frac{-1}{(1+x)^2}$$

hence, $\qquad \frac{dy}{dx}=-\frac{1}{(1+x)^2}$

Misconception 2

$\frac{dx}{dy}$ is the reciprocal of $\frac{dy}{dx}$ but $\frac{d^2x}{dy^2}$ is not the reciprocal of $\frac{d^2y}{dx^2}$.

Indeed $\frac{d^2x}{dy^2}=-\left(\frac{dy}{dx}\right)^{-2}\frac{d^2y}{dx^2}$

Misconception 3

$\frac{dy}{dx}=\frac{dy/dt}{dx/dt},\frac{d^2y}{dx^2}\neq\frac{d^2y/dt^2}{d^2x/dt^2}$. Indeed

$$\frac{d^2y}{dx^2}=\frac{\left(\frac{dx}{dt}\right)\left(\frac{d^2y}{dt^2}\right)-\frac{d^2x}{dt^2}\cdot\frac{dt}{dt}}{\left(\frac{dx}{dt}\right)^2}$$

Misconception 4

The derivative of a function may be coming mechanically but it may not have any meaning at all. For example, let $f(x) = \log(\log \sin x)$

then $$f'(x) = \frac{1}{\log \sin x}.\cos x$$

$$\Rightarrow \quad f'\left(\frac{\pi}{6}\right) = \frac{1}{\log\frac{1}{2}}.\sqrt{3} = -\frac{\sqrt{3}}{\log 2}$$

But we note that $\log \sin x \le 0$ for all permissible x. Therefore $\log(\sin x)$ is not defined for any x.

WORKSHEET – 2

1. Find $\sin(\tan^{-1} e^{-x})$.
2. Find the derivative of $\sin^{-1}\left(\frac{2x}{1+x^2}\right)$ with respect to $\cos^{-1}\left(\frac{1-x^2}{1+x^2}\right)$ in the domain of $f(x) = \frac{1}{\sqrt{(x-1)}}$.
3. Find the derivative of $\tan^{-1}\left(\frac{\sqrt{(1+x^2)}-1}{x}\right)$ with respect to $\tan^{-1} x$.
4. If $y = x - x^2$, then find the derivative of y^2 w.r.t. x^2.
5. Find the derivative of $f(\tan x)$ with respect to $g(\sec x)$ at $x = \pi/4$, if $f'(1) = 2$, $g'(\sqrt{2}) = 4$.

Derivative of Inverse Trigonometric Functions

We are already familiar with the values of trigonometric functions. For example, $\sin\frac{\pi}{6} = \frac{1}{2}$, $\tan\frac{\pi}{3} = \sqrt{3}$, etc. There are situations where we have to find the angle when the values of trigonometric functions are given. In the example $\sin\frac{\pi}{6} = \frac{1}{2}$

Here $\frac{\pi}{6}$ is the angle for which the value of sin is $\frac{1}{2}$.

This information we write as $\frac{\pi}{6} = \sin^{-1}\frac{1}{2}$.

In general, $\theta = \sin^{-1} x$ is often read as sine inverse of x or arc sin of x

Let $y = \sin^{-1} x \Rightarrow x = \sin y$

The domain of $\sin^{-1} x$ is $[-1, 1]$ and the range $\left[-\frac{\pi}{2}, \frac{\pi}{2}\right]$

The following table gives the range and domain of inverse trigonometric functions.

Function	Domain	Range
$y = \sin^{-1} x$	$[-1, 1]$	$\left[-\frac{\pi}{2}, \frac{\pi}{2}\right]$
$y = \cos^{-1} x$	$[-1, 1]$	$[0, \pi]$
$y = \tan^{-1} x$	$(-\infty, \infty)$	$\left(-\frac{\pi}{2}, \frac{\pi}{2}\right)$
$y = \cot^{-1} x$	$(-\infty, \infty)$	$(0, \pi)$
$y = \sec^{-1} x$	$(-\infty, -1] \cup [1, \infty]$	$[0, \pi], y \ne \frac{\pi}{2}$
$y = \text{cosec}^{-1} x$	$(-\infty, -1] \cup [1, \infty]$	$\left[-\frac{\pi}{2}, \frac{\pi}{2}\right], y \ne 0$

The following table gives the derivatives of the inverse trigonometric functions:

Inverse trigonometric functions	Derivatives of inverse trigonometric functions
$\sin^{-1} x$	$\frac{d}{dx}(\sin^{-1} x) = \frac{1}{\sqrt{1-x^2}}, -1 < x < 1$

Inverse trigonometric functions	Derivatives of inverse trigonometric functions
$\cos^{-1} x$	$\frac{d}{dx}\left(\cos^{-1} x\right) = -\frac{1}{\sqrt{1-x^2}}, -1 < x < 1$
$\tan^{-1} x$	$\frac{d}{dx}\left(\tan^{-1} x\right) = \frac{1}{1+x^2}, x \in R$
$\cot^{-1} x$	$\frac{d}{dx}\left(\cot^{-1} x\right) = -\frac{1}{1+x^2}, x \in R$
$\sec^{-1} x$	$\frac{d}{dx}\left(\sec^{-1} x\right) = \frac{1}{\lvert x\rvert\sqrt{x^2-1}}$, $x > 1$ or $x < -1$
$\operatorname{cosec}^{-1} x$	$\frac{d}{dx}\left(\operatorname{cosec}^{-1} x\right) = -\frac{1}{\lvert x\rvert\sqrt{x^2-1}}$, $x > 1$ or $x < -1$

ILLUSTRATION 5.21

Differentiate the $\cos^{-1}(\sin x)$ w.r.t x.

Solution

Let $y = \cos^{-1}(\sin x)$

$$\frac{dy}{dx} = \frac{-1}{\sqrt{1-\sin^2 x}} \times \frac{d}{dx}(\sin x)$$

$$= \frac{-1}{\cos x} \times \cos x = -1.$$

ILLUSTRATION 5.22

Differentiate $\sin^{-1}(\cos x)$ for $0 < x < \frac{\pi}{2}$.

Solution

Let $y = \sin^{-1}(\cos x)$, then

$$\frac{dy}{dx} = \frac{d}{dx}\left[\sin^{-1}(\cos x)\right] = \frac{1}{\sqrt{1-\cos^2 x}} \times \frac{d}{dx}(\cos x)$$

$$= \frac{1}{\sqrt{\sin^2 x}}.(-\sin x) = -\frac{\sin x}{\sin x} = -1.$$

ILLUSTRATION 5.23

Differentiate $\tan^{-1}\frac{x}{2} + \tan^{-1}\frac{2}{x}$

Solution

Let $y = \tan^{-1}\frac{x}{2} + \tan^{-1}\frac{2}{x}$

$$\therefore \quad \frac{dy}{dx} = \frac{d}{dx}\left(\tan^{-1}\frac{x}{2}\right) + \frac{d}{dx}\left(\tan^{-1}\frac{2}{x}\right)$$

$$= \frac{1}{1+\left(\frac{x}{2}\right)^2}.\frac{d}{dx}\left(\frac{x}{2}\right) + \frac{1}{1+\left(\frac{2}{x}\right)^2}.\frac{d}{dx}\left(\frac{2}{x}\right)$$

$$= \frac{1}{1+\frac{x^2}{4}}.\frac{1}{2} + \frac{1}{1+\frac{4}{x^2}}\left(\frac{-2}{x^2}\right) = \frac{2}{4+x^2} - \frac{2}{x^2+4} = 0$$

ILLUSTRATION 5.24

Prove that

$$\frac{d}{dx}\left[2x\tan^{-1}x - \log\left(1+x^2\right)\right] = 2\tan^{-1}x.$$

Solution

$$\frac{d}{dx}\left[2x\tan^{-1}x - \log\left(1+x^2\right)\right]$$

$$= \frac{d}{dx}\left(2x\tan^{-1}x\right) - \frac{d}{dx}\left[\log\left(1+x^2\right)\right]$$

$$= 2x\left(\frac{1}{1+x^2}\right) + \left(\tan^{-1}x\right).(2) - \frac{1}{\left(1+x^2\right)}.\frac{d}{dx}\left(1+x^2\right)$$

$$= \frac{2x}{1+x^2} + 2\tan^{-1}x - \frac{2x}{\left(1+x^2\right)} = 2\tan^{-1}x.$$

Misconception 5

We may get different answers by different methods due to lack of understanding of mathematical definitions/results and conditions.

For example, let $f(x) = \sin^{-1}\left(2x\sqrt{1-x^2}\right)$.

If we put $x - \sin\theta$

we get $f(x) = \sin^{-1}(\sin 2\theta) = 2\theta = 2\sin^{-1}x$

$$\Rightarrow \quad f'(x) = \frac{2}{\sqrt{1-x^2}}$$

But if we put $x = \cos .$, we get $f(x) = \sin^{-1}(\sin 2\theta)$

$$= 2\theta - 2\cos^{-1}x \Rightarrow f'(x) = \frac{-2}{\sqrt{1-x^2}}$$

Where does the mistake lie? When we carefully examine our steps, we note that we have used $\sin^{-1}(\sin 2\theta) = 2\theta$, *which is true only when*

$$-\frac{\pi}{2} \le 2\theta \le \frac{\pi}{2} \text{ or } -\frac{\pi}{4} \le \theta \ge \frac{\pi}{4}$$

or $$-\frac{1}{\sqrt{2}} \le x \le \frac{1}{\sqrt{2}}$$

Thus, the derivative of $\sin^{-1}x$ is $\frac{1}{\sqrt{1-x^2}}$ when

$$x \in \left[-\frac{1}{\sqrt{2}}, \frac{1}{\sqrt{2}}\right]$$

Now, if $x \in \left[\frac{1}{\sqrt{2}}, 1\right]$ and $x = \sin\theta$, then $\theta \in \left(\frac{\pi}{4}, \frac{\pi}{2}\right)$

$$\Rightarrow \quad 2\theta \in \left[\frac{\pi}{2}, \pi\right]$$

Misconception 6

Using derivatives, we can show that a given function is a constant. At times, we may prove some identities within a given domain. Suppose that we want to show that $\sin^{-1}x + \cos^{-1}x = \frac{\pi}{2}$.

Let $\quad f(x) = \sin^{-1}x + \cos^{-1}x \ (|x| \le 1)$.

Then $\quad f'(x) = \frac{1}{\sqrt{1-x^2}} - \frac{1}{\sqrt{1-x^2}} = 0$

$\Rightarrow f(x)$ is constant.

$\Rightarrow f(x) = f$ (any permissible value)

$$= f(0) \text{ in particular} = \frac{\pi}{2}.$$

Derivative of Implicit Functions

If $f(x, y) = 0$, then differentiate each term w.r.t. x regarding y as a function of x and then collect the terms of $\frac{dy}{dx}$ together on left-hand side and remaining terms on right-hand side and then find $\frac{dy}{dx}$.

Note in case of implicit functions, both x and y are present in $\frac{dy}{dx}$.

Alternative Method

If $f(x, y) = 0$,

$$\text{then } \frac{dy}{dx} = \frac{\left(\frac{df}{dx}\right)}{\left(\frac{df}{dy}\right)}$$

$$= -\frac{\text{diff. of '}f\text{' w.r.t. '}x\text{' treating } y \text{ as constant}}{\text{diff. of '}f\text{' w.r.t. '}y\text{' treating } x \text{ as constant}}$$

In general,

If $f(x_1, x_2, x_3, \ldots\ldots, x_n) = 0$ and $x_2, x_3, \ldots\ldots, x_n$ are the functions of x_1, then

$$\frac{df}{dx_1} = \frac{\partial f}{\partial x_1} + \frac{\partial f}{\partial x_2}\cdot\frac{dx_2}{dx_1} + \frac{\partial f}{\partial x_3}\cdot\frac{dx_3}{dx_1} + \ldots. + \frac{\partial f}{\partial x_n}\cdot\frac{dx_n}{dx_1}$$

ILLUSTRATION 5.25

If $ax^2 + 2hxy + by^2 + 2gx + 2fy + c = 0$, find $\frac{dy}{dx}$.

Solution

$$ax^2 + 2hxy + by^2 + 2gx + 2fy + c = 0$$

Differentiating both sides w.r.t. x, we get,

$$2ax + 2h\left(x\frac{dy}{dx} + y\cdot 1\right) + 2by\frac{dy}{dx} + 2g + 2f\frac{dy}{dx} + 0 = 0$$

$$\Rightarrow (2ax + 2hy + 2g) + (2hx + 2by + 2f)\frac{dy}{dx} = 0$$

$$\therefore \frac{dy}{dx} = -\frac{(ax + hy + g)}{(hx + by + f)}$$

Alternative Method

Let $f(x, y) \equiv ax^2 + 2hxy + by^2 + 2gx + 2fy + c = 0$

$$\therefore \quad \frac{\partial f}{\partial x} = 2ax + 2hy + 0 + 2g + 0 + 0$$

$$= (2ax + 2hy + 2g)\left(\begin{array}{c}\text{Differentiating w.r.t. '}x\text{'}\\ \text{treating } y \text{ as constant}\end{array}\right)$$

and $$\frac{\partial f}{\partial x} = 0 + 2hx + 2by + 0 + 2f + 0$$

$$= 2hx + 2by + 2f\left(\begin{array}{c}\text{Differentiating w.r.t. '}y\text{'}\\ \text{treating } x \text{ as constant}\end{array}\right)$$

Hence

$$\frac{dy}{dx} = -\frac{\frac{\partial f}{\partial x}}{\frac{\partial f}{\partial y}} = -\frac{(2ax + 2hy + 2g)}{(2hx + 2by + 2f)} = -\frac{(ax + hy + g)}{(hx + by + f)}$$

ILLUSTRATION 5.26

If $\sin y = x\sin(a+y)$, prove that $\dfrac{dy}{dx} = \dfrac{\sin^2(a+y)}{\sin a}$

Solution

We have $\sin y = x\sin(a+y)$

$$\Rightarrow \quad x = \frac{\sin y}{\sin(a+y)}$$

Differentiating both sides w.r.t. 'y', we get

$$\frac{dx}{dy} = \frac{\sin(a+y)\cdot\cos y - \sin y\cdot\cos(a+y)}{\sin^2(a+y)}$$

(Using quotient rule)

$$= \frac{\sin(a+y-y)}{\sin^2(a+y)} = \frac{\sin a}{\sin^2(a+y)}$$

Hence $\dfrac{dy}{dx} = \dfrac{\sin^2(a+y)}{\sin a}$

ILLUSTRATION 5.27

If $\sqrt{(1-x^{2n})}+\sqrt{(1-y^{2n})} = a(x^n - y^n)$, then prove that $\dfrac{dy}{dx} = \dfrac{x^{n-1}}{y^{n-1}}\sqrt{\left(\dfrac{1-y^{2n}}{1-x^{2n}}\right)}$.

Solution

We have $\sqrt{(1-x^{2n})}+\sqrt{(1-y^{2n})} = a(x^n - y^n)$ ...(i)

Putting $\quad x^n = \sin\theta \Rightarrow \theta = \sin^{-1}x^n$

and $\quad y^n = \sin\phi \Rightarrow \phi = \sin^{-1}y^n$ ...(ii)

Then (i), becomes $\cos\theta + \cos\phi = a(\sin\theta - \sin\phi)$

$$\Rightarrow 2\cos\left(\frac{\theta+\phi}{2}\right)\cos\left(\frac{\theta-\phi}{2}\right)$$

$$= a\cdot 2\cos\left(\frac{\theta+\phi}{2}\right)\sin\left(\frac{\theta-\phi}{2}\right)$$

$$\Rightarrow \cot\left(\frac{\theta-\phi}{2}\right) = a \Rightarrow \theta-\phi = 2\cot^{-1}a$$

$$\Rightarrow \quad \sin^{-1}x^n - \sin^{-1}y^n = 2\cot^{-1}a \quad \text{(From (ii))}$$

Differentiating both sides w.r.t. 'x', we get

$$\frac{1}{\sqrt{(1-x^{2n})}}\cdot nx^{n-1} - \frac{1}{\sqrt{(1-y^{2n})}}\cdot ny^{n-1}\frac{dy}{dx} = 0$$

$$\Rightarrow \quad \frac{dy}{dx} = \frac{x^{n-1}}{y^{n-1}}\sqrt{\left(\frac{1-y^{2n}}{1-x^{2n}}\right)} \quad \text{(Remember)}$$

Corollary

(i) For $n = 1$

$$\sqrt{(1-x^2)}+\sqrt{(1-y^2)} = a(x-y),$$

then $\dfrac{dy}{dx} = \sqrt{\left(\dfrac{1-y^2}{1-x^2}\right)}$

(ii) For $n = 2$

$$\sqrt{(1-x^4)}+\sqrt{(1-y^4)} = a(x^2-y^2),$$

then $\dfrac{dy}{dx} = \dfrac{x}{y}\sqrt{\left(\dfrac{1-y^4}{1-x^4}\right)}$

(iii) For $n = 3$

$$\sqrt{(1-x^6)}+\sqrt{(1-y^6)} = a(x^3-y^3),$$

then $\dfrac{dy}{dx} = \dfrac{x^2}{y^2}\sqrt{\left(\dfrac{1-y^6}{1-x^6}\right)}$

Exponential Functions

An exponential function is a function of the form $f(x) = b^x$ where b is a positive real number and $b \neq 1$. The domain of f is the set of all real numbers.

Some salient features of the exponential functions are given below:

1. Domain of the exponential function is R. (the set of all real numbers)
2. Range of the exponential function is the set of all positive real numbers.
3. y-intercept is 1. The point (0, 1) is always on the graph of the exponential function.
4. For large negative values of x, the exponential function is very close to 0.

The base e:

The sum of the series $1+\dfrac{1}{2!}+\dfrac{1}{3!}+\ldots$ is a number between 2 and 3 and is denoted by e.

The exponential function $f(x) = e^x$ whose base is the number e is called natural exponential function.

The graph of $y = e^x$ lies between the graph of $y = 2^x$ and 3^x $(2 < e < 3)$.

ILLUSTRATION 5.28

Differentiate the following functions w.r.t x:

(i) $\log(\cos e^x)$

(ii) $e^{\sin^{-1}x}, x > 0$

Solution

(i) Let $y = \log\left(\cos e^x\right)$

Putting $y = \log u$, $u = \cos t$ and $t = e^x$

$$\frac{dy}{du} = \frac{1}{u}, \frac{du}{dt} = -\sin t, \frac{dt}{dx} = e^x$$

Now, $\frac{dy}{dx} = \frac{dy}{du} \times \frac{du}{dt} \times \frac{dt}{dx} = \frac{1}{u} \times (-\sin t) \times e^x$

$$= -\frac{e^x \sin e^x}{\cos e^x} = -e^x \tan e^x,$$

where $e^x \neq (2n-1)\frac{\pi}{2}, n \in \mathrm{N}$.

(ii) Let $y = e^{\sin^{-1}x}$

Putting $\sin^{-1}x = t$, we get

$y = e^t$ and $t = \sin^{-1}x$

$$\Rightarrow \frac{dy}{dt} = e^t \text{ and } \frac{dt}{dx} = \frac{1}{\sqrt{1-x^2}}$$

$$\Rightarrow \frac{dy}{dx} = \frac{dy}{dt} \times \frac{dt}{dx} = e^t \times \frac{1}{\sqrt{1-x^2}}$$

$$= e^{\sin^{-1}x} \times \frac{1}{\sqrt{1-x^2}}$$

$$\Rightarrow \frac{dy}{dx} = \frac{e^{\sin^{-1}x}}{\sqrt{1-x^2}}$$

ILLUSTRATION 5.29

Differentiate $e^{2x}\sin 3x$ w.r.t x:

Solution

Let $y = e^{2x}\sin 3x$

$$\frac{dy}{dx} = e^{2x}.\frac{d}{dx}(\sin 3x) + \sin 3x.\frac{d}{dx}\left(e^{2x}\right)$$

$$= e^{2x}(3\cos 3x) + \sin 3x\left(2e^{2x}\right) = (3\cos 3x + 2\sin 3x)e^{2x}.$$

Logarithmic Function to the base *a* – where $a > 0$ and $a \neq 1$ is defined by $y = \log_a x = a^y$

Note: If the base of the logarithmic function is e then we call it natural logarithmic function and it is represented by a special symbol lnl (derived from Latin word logarithmic naturalist)

Therefore $y = \ln x$ if and only if $x = e^y$.

Logarithmic Differentiation

The process of taking logarithms before differentiation is called Logarithmic Differentiation. When the function consists of a single term which is a variable raised to a variable power, we first take logarithms and then differentiate.

Let us differentiate u^v when u and v are both functions of x.

Let $y = u^v$

Taking logarithms of both sides,

$$\log y = \log u^v = v \log u$$

Differentiating w.r.t. x, we get

$$\frac{1}{y}.\frac{dy}{dx} = v.\frac{1}{u}.\frac{du}{dx} + \log u.\frac{dv}{dx}$$

$$\Rightarrow \frac{dy}{dx} = y\left[\frac{v}{u}.\frac{du}{dx} + \log u.\frac{dv}{dx}\right]$$

$$= u^v\left[\frac{v}{u}.\frac{du}{dx} + \log u.\frac{dv}{dx}\right]$$

ILLUSTRATION 5.30

Differentiate w.r.t. $x : \log(\log x), x > 0$

Solution

Let $y = \log(\log x)$

Putting $y = \log t$, $t = \log x$

$$\Rightarrow \frac{dy}{dt} = \frac{1}{t}, \frac{dt}{dx} = \frac{1}{x}$$

$$\frac{dy}{dx} = \frac{dy}{dt} \times \frac{dt}{dx}$$

$$= \frac{1}{t} \times \frac{1}{x} = \frac{1}{\log x}.\frac{1}{x} = \frac{1}{x \log x}$$

ILLUSTRATION 5.31

Differentiation w.r.t. $x : \cos\left(\log x + e^x\right), x > 0$

Solution

Let $y = \cos\left(\log x + e^x\right)$

Putting $y = \cos t$ and $t = \log x + e^x$, we get

$$\therefore \frac{dy}{dt} = -\sin t \text{ and } \frac{dt}{dx} = \frac{1}{x} + e^x$$

Now $\frac{dy}{dx} = \frac{dy}{dt} \times \frac{dt}{dx} = -\sin t \times \left(\frac{1}{x} + e^x\right)$

$$= -\sin\left(\log x + e^x\right) \times \left(\frac{1}{x} + e^x\right)$$

$$= -\left(\frac{1}{x} + e^x\right)\sin\left(\log x + e^x\right), x > 0$$

ILLUSTRATION 5.32

Differentiate the following functions w.r.t. x:

(i) $\log\left(\sqrt{x}+\frac{1}{\sqrt{x}}\right)$

(ii) $\log\left(e^{mx}+e^{-mx}\right)$

Solution

(i) Let $y=\log\left(\sqrt{x}+\frac{1}{\sqrt{x}}\right)$

then $\frac{dy}{dx}=\frac{1}{\left(\sqrt{x}+\frac{1}{\sqrt{x}}\right)}\cdot\frac{d}{dx}\left(\sqrt{x}+\frac{1}{\sqrt{x}}\right)$

$$=\frac{1}{\left(\sqrt{x}+\frac{1}{\sqrt{x}}\right)}\cdot\frac{d}{dx}\left(x^{\frac{1}{2}}+x^{-\frac{1}{2}}\right)$$

$$=\left(\frac{\sqrt{x}}{x+1}\right)\cdot\left(\frac{1}{2\sqrt{x}}-\frac{1}{2}x^{-\frac{3}{2}}\right)$$

$$=\left(\frac{\sqrt{x}}{x+1}\right)\cdot\left(\frac{1}{2\sqrt{x}}-\frac{1}{2x\sqrt{x}}\right)=\left(\frac{\sqrt{x}}{x+1}\right)\left(\frac{x-1}{2x\sqrt{x}}\right)$$

$$=\frac{x-1}{2x(x+1)}.$$

(ii) Let $y=\log\left(e^{mx}+e^{-mx}\right)$

$$\frac{dy}{dx}=\frac{1}{e^{mx}+e^{-mx}}\cdot\frac{d}{dx}\left(e^{mx}+e^{-mx}\right)$$

$$=\frac{1}{e^{mx}+e^{-mx}}\left(me^{mx}-me^{-mx}\right)=\frac{m\left(e^{mx}-e^{-mx}\right)}{e^{mx}+e^{-mx}}.$$

ILLUSTRATION 5.33

If $y=x^x$, then find $\frac{dy}{dx}$

Solution

We have

$$y=x^x=e^{x\ln x} \quad \text{....(i)}$$

$$\therefore \quad \frac{dy}{dx}=e^{x\ln x}\left\{x\cdot\frac{1}{x}+\ln x\cdot 1\right\}$$

$$=x^x(1+\ln x) \text{ (from (i))}$$

I Alternative Method: (Short-cut Method)

$$\because \quad y=x^x$$

$$\therefore \quad \frac{dy}{dx}=\frac{d}{dx}\underset{\substack{\downarrow\\ \text{Variable}}}{(x)}^{x\to\text{Constant}}+\frac{d}{dx}\underset{\substack{\downarrow\\ \text{Constant}}}{(x)}^{x\to\text{Variable}}$$

$$=x\cdot(x)^{x-1}\cdot(1)+x^x\cdot\ln x(1)$$

$$\left(\text{using } \frac{d}{dx}x^n=nx^{n-1} \text{ and } \frac{d}{dx}a^x=a^x\ln a\right)$$

$$=x^x(1+\ln x)$$

II Alternative Method

$$\because \quad y=x^x \text{ (Hence Base = Power = } x\text{)}$$

$$\therefore \quad \frac{dy}{dx}=x^x\left\{\frac{x}{x}\cdot(1)+\ln(x)\cdot 1\right\}$$

$$=x^x(1+\ln x)$$

ILLUSTRATION 5.34

If $f(x)=(x+1)(x+2)(x+3)....(x+n)$, then prove that $f'(o)=n!\sum_{r=1}^{n}\frac{1}{r}$

Solution

$$f(x)=(x+1)(x+2)(x+3)....(x+n)$$

$$=\prod_{r=1}^{n}(x+r)$$

$$\therefore \quad \ln a\, f(x)=\sum_{r=1}^{n}\ln(x+r)$$

Differentiating both sides w.r.t. x, we have,

$$\frac{1}{f(x)}\cdot f'(x)=\sum_{r=1}^{n}\frac{1}{(x+r)}$$

Put $x=0$, then

$$f'(0)=f(0)\sum_{r=1}^{n}\frac{1}{r}=(1.2.3.......n)\sum_{r=1}^{n}\frac{1}{r}$$

$$=n!\sum_{r=1}^{n}\frac{1}{r}$$

ILLUSTRATION 5.35

If $x^y+y^x=a^b$ (where, a, b are constants), then find $\frac{dy}{dx}$.

Solution

We have $x^y+y^x=a^b$

$$\Rightarrow \quad e^{\ln x^y}+e^{\ln y^x}=a^b \quad \Rightarrow e^{y\ln x}+e^{x\ln y}=a^b$$

Differentiating both sides w.r.t. x, we get

$$\frac{d}{dx}\left(e^{y\ln x}\right)+\frac{d}{dx}\left(e^{x\ln y}\right)=\frac{d}{dx}(a^b)$$

$$\Rightarrow e^{y\ln x}\frac{d}{dx}(y\ln x)+e^{x\ln y}\frac{d}{dx}(x\ln y)=0$$

$$\Rightarrow x^y\left\{y\cdot\frac{1}{x}+\ln x\cdot\frac{dy}{dx}\right\}+y^x\left\{x\cdot\frac{1}{y}\frac{dy}{dx}+\ln y\cdot 1\right\}=0$$

$$\Rightarrow \left(x^y\ln x+x\cdot y^{x-1}\right)\frac{dy}{dx}+\left(y\cdot x^{y-1}+y^x\ln y\right)=0$$

Hence $\dfrac{dy}{dx}=-\left\{\dfrac{yx^{y-1}+y^x\cdot\ln y}{x^y\ln x+x\cdot y^{x-1}}\right\}$

Derivative of Parametric Functions

Let x and y be two functions of variable t (parameter) such that $x=f(t)$ and $y=g(t)$,

then $$\frac{dy}{dx}=\frac{\left(\frac{dy}{dt}\right)}{\left(\frac{dx}{dt}\right)}=\frac{g'(t)}{f'(t)}$$

ILLUSTRATION 5.36

If $x=a(t+\sin t)$, $y=a(1-\cos t)$, then find $\dfrac{dy}{dx}$.

Solution

Since $x = a(t + \sin t)$ and $y = a(1 - \cos t)$

$\therefore \quad \dfrac{dx}{dt}=a(1+\cos t)$ and $\dfrac{dy}{dt}=a\sin t$

Now $$\frac{dy}{dx}=\frac{\frac{dy}{dt}}{\frac{dx}{dt}}=\frac{a\sin t}{a(1+\cos t)}=\frac{\sin t}{1+\cos t}$$

$$=\frac{2\sin(t/2)\cos(t/2)}{2\cos^2(t/2)}=\tan(t/2)$$

ILLUSTRATION 5.37

If $x=a\cos^3\theta$, $y=a\sin^3\theta$, then find $\dfrac{dy}{dx}$.

Solution

Since $x=a\cos^3\theta$

$$\therefore \quad \frac{dx}{d\theta}=a\frac{d\left(\cos^3\theta\right)}{d(\cos\theta)}\cdot\frac{d(\cos\theta)}{d\theta} \text{ (Using chain rule)}$$

$$=3a\cos^2\theta(-\sin\theta)=-3a\cos^2\theta\sin\theta$$

and $\quad y=a\sin^3\theta$

$$\therefore \quad \frac{dy}{d\theta}=a\frac{d\left(\sin^3\theta\right)}{d(\sin\theta)}\cdot\frac{d(\sin\theta)}{d\theta}=3a\sin^2\theta\cdot\cos\theta$$

Now $$\frac{dy}{dx}=\frac{\left(\frac{dy}{d\theta}\right)}{\left(\frac{dx}{d\theta}\right)}=\frac{3a\sin^2\theta\cos\theta}{-3a\cos^2\theta\sin\theta}=-\tan\theta$$

Higher-order Derivatives

If $y=f(x)$ be a differentiable function of x such that its second, third, …, nth derivatives exists.

The first, second, third, …, nth derivatives of $y = f(x)$ are denoted respectively by

$$\frac{dy}{dx},\frac{d^2y}{dx^2},\frac{d^3y}{dx^3},\ldots\ldots,\frac{d^ny}{dx^n}$$

and also denoted by $y',y'',y''',\ldots y^{(n)}$

$$y_1,y_2,y_3,\ldots\ldots,y_n$$

$$f',f'',f''',\ldots.f^{(n)}$$

$$Dy,D^2y,D^3y,\ldots\ldots,D^ny$$

The symbols $y^{(n)}(a),y_n(a),f^{(n)}(a),D^ny(a),\left.\dfrac{d^ny}{dx^n}\right|_{x=a}$ denote the value of the nth derivative at $x=a$.

Some Standard Results

(i) $\dfrac{d^n}{dx^n}(ax+b)^m=\dfrac{m!}{(m-n)!}\cdot a^n\cdot(ax+b)^{m-n}$, $m\in W$, $x\in N$ and $m>n$

Note: $\dfrac{d^n}{dx^n}x^n=n!, x\in N$

(ii) $\dfrac{d^n}{dx^n}\left(\dfrac{1}{ax+b}\right)=\dfrac{(-1)^n\,n!a^n}{(ax+b)^{n+1}}, x\in N$

(iii) $\dfrac{d^n}{dx^n}\left(\ln(ax+b)\right)=\dfrac{(-1)^{n-1}\cdot(n-1)!}{(ax+b)^n}a^n, n\in N$

(iv) $\dfrac{d^n}{dx^n}(e^{mx})=m^n\cdot e^{mx}, m\in R, n\in N$

(v) $\dfrac{d^n}{dx^n}\left(\sin(ax+b)\right)=a^n\sin\left(ax+b+\dfrac{n\pi}{2}\right), n\in N$

(vi) $\dfrac{d^n}{dx^n}\left(\cos(ax+b)\right)=a^n\cos\left(ax+b+\dfrac{n\pi}{2}\right), n\in N$

(vii) $\dfrac{d^n}{dx^n}\left(e^{ax}\sin(bx+c)\right)=r^n\cdot e^{ax}\sin(bx+c+n\phi)$, $n\in\mathrm{N}$

where $r=\sqrt{(a^2+b^2)},\phi=\tan^{-1}(b/a)$

(viii) $\dfrac{d^n}{dx^n}\left(e^{ax}\cdot\cos(bx+c)\right)=r^n\cdot e^{ax}\cos(bx+c+n\phi)$, $n\in\mathrm{N}$

where $r=\sqrt{(a^2+b^2)},\phi=\tan^{-1}(b/a)$

ILLUSTRATION 5.38

If $y=\dfrac{\ln x}{x}$, then prove that $\dfrac{d^2y}{dx^2}=\dfrac{2\ln x-3}{x^3}$.

Solution

We have $y=\dfrac{\ln x}{x}$

Differentiating both sides w.r.t. x, we have,

$$\therefore\quad \frac{dy}{dx}=\frac{x\cdot\frac{1}{x}-\ln x\cdot 1}{x^2}\ \text{(by Quotient rule)}$$

Again differentiating both sides w.r.t. x, we get

$$\frac{d^2y}{dx^2}=\frac{x^2\cdot\left(0-\frac{1}{x}\right)-(1-\ln x)\cdot 2x}{x^4}=\frac{2\ln x-3}{x^3}$$

ILLUSTRATION 5.39

If $y=\left(\dfrac{1}{x}\right)^x$, then show that $y''(1)=0$

Solution

We have $y=\left(\dfrac{1}{x}\right)^x=x^{-x}=e^{-x\ln x}$

$$\therefore y'(x)=\frac{dy}{dx}=e^{-x\ln x}\left((-x)\cdot\frac{1}{x}+\ln x\cdot(-1)\right)$$

$$\therefore y'(x)=-e^{-x\ln x}(1+\ln x)\qquad\text{(i)}$$

Again differentiating both sides w.r.t. x, we get

$$y''(x)=-e^{-x\ln x}\cdot\frac{d}{dx}(1+\ln x)$$

$$-(1+\ln x)\frac{d}{dx}\left(e^{-x\ln x}\right)$$

$$=-e^{-x\ln x}\cdot\left(\frac{1}{x}\right)-(1+\ln x)\cdot -e^{-x\ln x}$$

$$\cdot(1+\ln x)\qquad\text{[from (i)]}$$

$$\therefore y''(1)=-e^{-o}\left(\frac{1}{1}\right)-(1+0)^2\times -e^{-o}=-1+1=0$$

Rolle's Theorem

Rolle's Theorem, named after the French mathematician Michel Rolle (1652–1719), gives conditions that guarantee the existence of an extreme value in the interior of a closed interval.

Statement: If a function $f(x)$ is such that

(i) it is continuous in the closed interval $[a,b]$,

(ii) it is differentiable in the open interval (a,b),

(iii) $f(a)=f(b)$,

then there exists at least one value 'c' of x in the open interval (a, b) such that $f'(c)=0$.

ILLUSTRATION 5.40

Verify Rolle's Theorem for the function $f(x)=x^2-5x+4$ on $[1,4]$.

Solution

$f(x)$, being a polynomial function, is continuous in the closed interval [1, 4].

$f'(x)=2x-5$, which exists in the open interval (1, 4)

$f(x)$ is derivable in the open interval (1, 4)

$$f(1)=(1)^2-5(1)+4=1-5+4=0$$

$$f(4)=(4)^2+4=16-20+4=0$$

$$\therefore\quad f(1)=f(4)$$

Thus $f(x)$ satisfies all the three conditions of Rolle's Theorem

$\therefore$ There must be exist at least one number c such that

$$1<c<4 \text{ and } f'(c)=0$$

Now, $f'(c)=0\Rightarrow 2c-5=0\ \Rightarrow\ c=\dfrac{5}{2}=2.5\in(1,4)$

Hence, Rolle's Theorem is verified.

ILLUSTRATION 5.41

Verify Rolle's Theorem for the following function:

$$f(x)=\sqrt{4-x^2} \text{ in } [-2, 2]$$

Solution

$f(x)=\sqrt{4-x^2}$ has a unique and definite value for each x in $[-2, 2]$

$\therefore f(x)$ is continuous in $[-2, 2]$

$$f'(x)=\frac{1}{2}(4-x^2)^{-\frac{1}{2}}.(-2x)=\frac{-x}{\sqrt{4-x^2}}$$

which exists for all x in $(-2, 2)$.

For $-2<x<2$, $f(x)$ is derivable in $(-2, 2)$.

Also $f(-2)=0=f(2)$.

$\therefore f(x)$ satisfies all the three conditions of Rolle's Theorem.

$\therefore$ There exists at least one number 'c' in $(-2, 2)$ such that $f'(c)=0$.

$$\Rightarrow \quad \frac{-c}{\sqrt{4-c^2}}=0 \text{ or } -c=0$$

i.e., $\quad c=0\in(-2, 2)$

Hence, Rolle's Theorem is verified.

ILLUSTRATION 5.42

If $2a+3b+6c=0$, then at least one root of the equation $ax^2+bx+c=0$ lies in the interval (0, 1).

Solution

We have $2a+3b+6c=0$

Let $f'(x)=ax^2+bx+c$, then

$$f(x)=\int(ax^2+bx+c)dx$$

$$\Rightarrow \quad f(x)=\frac{ax^3}{3}+\frac{bx^2}{2}+cx+k$$

$f(x)$, being a polynomial function, is continuous and differentiable everywhere.

Now, $\quad f(x)=\dfrac{2ax^3+3bx^2+6cx+6k}{6}$

$$f(1)=\frac{2a+3b+6c+6k}{6}=\frac{0+6k}{6}=k$$

$$[\because 2a+3b+6c=0]$$

and $\quad f(0)=\dfrac{6k}{6}=k$

$\therefore \quad f(0)=f(1)$

Thus, $f(x)$ satisfies all the three conditions of Rolle's Theorem.

$$f(x)=0$$

$\Rightarrow$ The equation $ax^2+bx+c=0$ has at least one root in the interval (0, 1).

Lagrange's Mean Value Theorem

Rolle's theorem can be used to prove another theorem – the Mean Value Theorem.

Statement: If a function $f(x)$ is such that

(i) it is continuous in the closed interval $[a, b]$,

(ii) it is derivable in the open interval (a, b), then there exists at least one value 'c' of 'x' in the open interval $]a, b[$, such that $\dfrac{f(b)-fa)}{b-a}=f'(c)$.

Geometrical Interpretation of Lagrange's Mean Value Theorem

Let $A[a, f(a)]$ and $B[b, f(b)]$ be two points on the graph of the function f.

The number $\dfrac{f(b)-f(a)}{b-a}$ gives the slope of the chord AB joining the two points, and also gives the slope of the tangent to the graph, at the point $[c, f(c)]$.

Then geometrically Lagrange's Mean Value Theorem asserts that:

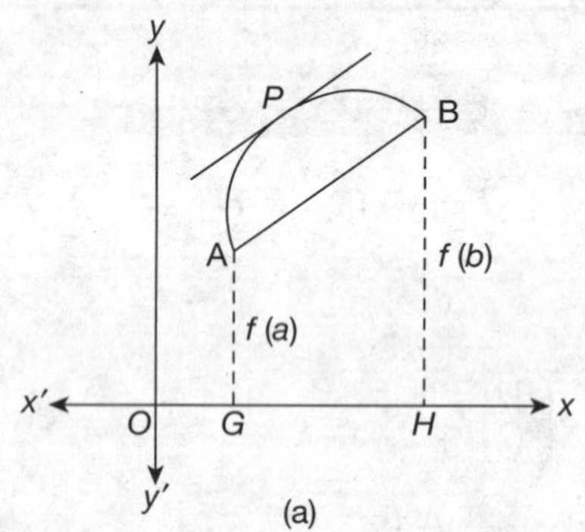

(a)

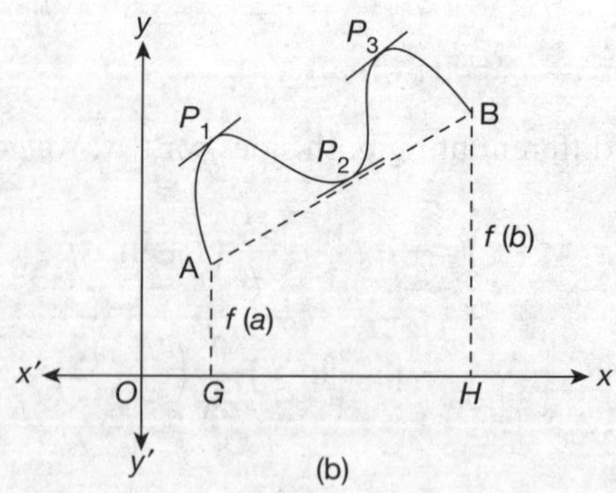

(b)

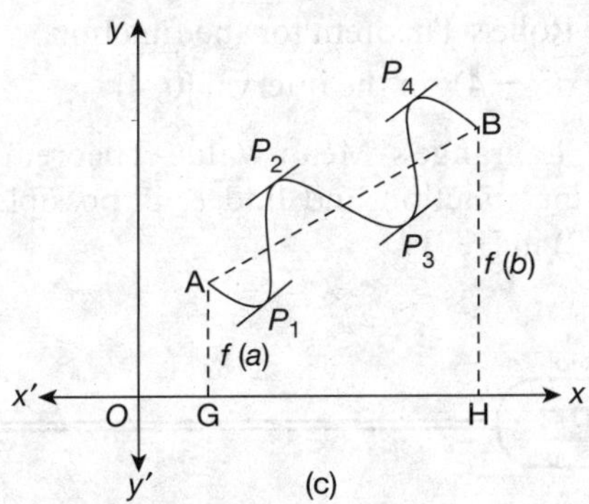

(c)

If a curve is continuous between two points A and B, and possesses a unique tangent at each point of the curve between A and B, then there is at least one point on the curve lying between A and B where the tangent is parallel to the chord AB.

ILLUSTRATION 5.43

Verify the conditions of Mean Value Theorem and in each case find a point c in the interval as stated by the Mean value Theorem, for the following functions, given by

(i) $f(x)=|x| \text{ on} [-1,1]$

(ii) $f(x)=3x^2-2$ on $[2, 3]$

Solution

(i) $f(x)=|x|$

$$\therefore \quad f(x)=\begin{cases} -x, & \text{for} \quad x<0 \\ 0, & \text{for} \quad x=0 \\ x, & \text{for} \quad x>0 \end{cases}$$

The function is continuous in the closed interval $[-1, 1]$ but $f'(x)$ does not exist at $x=0$. As such it is not derivable in the open interval $(-1,1)$.

$\therefore$ the Lagrange's Mean Value Theorem does not apply.

(ii) $f(x)=3x^2-2$ ...(1)

$f(x)$ being a polynomial function is continuous and derivable for all real values of x.

$f(x)$ is continuous in the closed interval and $f(x)$ is derivable in the open interval $(2, 3)$.

Thus $f(x)$ satisfies both the conditions of LMV theorem.

$\therefore$ There must exist at least one number c such that $2<c<3$ and

$$\frac{f(b)-f(a)}{b-a}=f'(c)$$

Here $a=2, b=3$

$$\frac{f(3)-f(2)}{3-2} f'(c) \qquad ...(2)$$

Now from (1), $f(2)=3(2)^2-2=10$

and $f(3)=3(3)^2-2=25$

$\therefore$ From (2), we get $\frac{25-10}{3-2}=f'(c)$

and $f'(x)=6x \Rightarrow f'(c)=6c$

so $\frac{15}{1}=6c \Rightarrow c=\frac{15}{6}=\frac{5}{2}\in(2, 3)$.

Hence, LMV theorem is verified.

WORKSHEET – 3

1. If $y=x\sin y$, prove that $x\frac{dy}{dx}=\frac{y}{(1-x\cos y)}$

2. If $y=\frac{1}{x}$, then show that $\frac{dy}{\sqrt{(1+y^4)}}+\frac{dx}{\sqrt{(1+x^4)}}=0$

3. If $x=a(\cos\theta+\theta\sin\theta), y=a(\sin\theta-\theta\cos\theta)$, then find $\frac{dy}{dx}$.

4. If $x=a\left(\cos t+\ln\tan\frac{t}{2}\right)$ and $y=a\sin t$, then find $\frac{dy}{dx}$.

5. If $x=e^t\left(t+\frac{1}{t}\right)$ and $y=e^{-t}\left(t-\frac{1}{t}\right)$, then find $\frac{dy}{dx}$.

6. If $x^y=e^{x-y}$, then prove that $\frac{dy}{dx}=\frac{\ln x}{(1+\ln x)^2}$

7. If $y=\dfrac{(x+1)^2\sqrt{(x-2)}}{(x+4)^3 e^x}$, then find $\dfrac{dy}{dx}$.

8. If $y=\left(1+\dfrac{1}{x}\right)^x+x^{(1+1/x)}+(\sin x)^{\tan x}$, then find $\dfrac{dy}{dx}$.

9. Verify Rolle's Theorem for the function $f(x) = x(x-4)^2$ in the interval [0, 4]

10. Verify Lagrange's Mean Value Theorem for the following function and find c, if possible: $f(x) = x(x-2)$ in [1, 3]

SOLVED EXAMPLES

Example 1. Determine the values of a and b such that the function

$$f(x)=\begin{cases} -2\sin x & x\le \dfrac{-\pi}{2} \\ a\sin x+b & \dfrac{-\pi}{2}\le x\le \dfrac{\pi}{2} \\ \cos x & x\ge \dfrac{\pi}{2}\end{cases}$$

is continuous everywhere.

Solution

Since the functions given in the representation of $f(x)$ are sin x, cos x and constants, they are continuous everywhere except possibly at transition points which are $-\dfrac{\pi}{2}$ and $\dfrac{\pi}{2}$.

Thus, in order to be continuous at $-\dfrac{\pi}{2}$, we must have

$$-2\sin\left(-\frac{\pi}{2}=a\sin\left(-\frac{\pi}{2}\right)\right)+b$$

This is equivalent to saying

$$\lim_{x\to\left(-\frac{\pi}{2}\right)^{-1}} f(x)=-2\sin\left(-\frac{\pi}{2}\right)$$

and $$\lim_{x\to\left(-\frac{\pi}{2}\right)^{+1}} f(x)=a\sin\left(-\frac{\pi}{2}\right)+b)$$

$$\Rightarrow \quad 2=-a+b \quad \text{...(A)}$$

Now last point of second graph $=\left(\dfrac{\pi}{2},a+b\right)$ and first point of third graph $=\left(\dfrac{\pi}{2},0\right)$

Thus, for continuity at $\dfrac{\pi}{2}$, we must have $a+b=0$

On solving (A) and (B), we get $a=-1$, $b=1$.

Example 2. Let $f(x)=\begin{cases}\dfrac{1-\cos 4x}{x^2}, & x<0\\ a, & x=0\\ \dfrac{\sqrt{x}}{\left(16+\sqrt{x}\right)^{1/2}-4}, & x>0\end{cases}$

Determine the value of a for which $f(x)$ becomes continuous at 0.

Solution

$$\lim_{x\to 0^-} f(x)=\lim_{h\to 0}\frac{1-\cos 4h}{h^2}=\lim_{h\to 0}\frac{2\sin^2 2h}{h^2}=8$$

$$\lim_{x\to 0^+} f(x)=\lim_{h\to 0}\frac{\sqrt{h}}{\left(16+\sqrt{h}\right)^{1/2}-4}$$

$$=\lim_{h\to 0}\frac{\sqrt{h}\left(16+\sqrt{h}\right)^{1/2}+4}{\sqrt{h}}=8$$

Thus for continuity at $f(0)$ 'a' must be equal to 8 $\Rightarrow a=8$.

Example 3. Determine the values of a, b for which $f(x)=\begin{cases}ax^2-bx+5, & x\le 3\\ bx^2-2ax-1, & x>3\end{cases}$ is differentiable at $x=3$.

Solution

First of all, $f(x)$ should be continuous at $x=3$.

$$\Rightarrow \quad 9a-3b+5=9b-6a-1 \quad \text{...(A)}$$

$f'(3)$ exists if $Lf'(3)=Rf'(3)$

$$\Rightarrow \quad 6a-b=6b-2a \quad \text{...(B)}$$

On solving (A) and (B), we get

$$a=-\frac{42}{9},\ b=-\frac{48}{9}$$

Example 4. If $f(a) = 2$, $f'(a) = 1$ and $g(a) = -1$, $g'(a) = 2$, then show that

$$\lim_{x\to a}\frac{g(x)f(a)-g(a)f(x)}{x-a}=5.$$

Solution

$$\text{LHS}=\lim_{x\to a}\frac{g(x)f(a)-g(a)f(a)+f(a)g(a)-g(a)f(x)}{x-a}$$

$$=\lim_{x\to a}\frac{f(a)[g(x)-g(a)]-g(a)[f(x)-f(a)]}{x-a}$$

$$=\lim_{x\to a}f(a)\frac{g(x)-g(a)}{x-a}-\lim_{x\to a}g(a)\frac{f(x)-f(a)}{x-a}$$

$$=f(a)g'(a)-g(a)f'(a)=2(2)-(-1)1=5$$

Example 5. Let $f(x)=\begin{cases}\dfrac{x^2}{2}, & 0\le x<1\\ 2x^2-3x+\dfrac{3}{2}, & 1\le x\le 2\end{cases}$, prove that

(i) f is continuous at 1

(ii) f' is continuous at 1

(iii) f' is discontinuous at 1.

Solution

(i) We easily note that

$$\lim_{x\to 1^-}f(x)=\lim_{x\to 1^+}f(x)=\frac{1}{2}$$

(ii) $Lf'(1)=(x)_{x=1}=1$ and $Rf'(1)=(4x-3)_{x=1}=1$

$\Rightarrow f'(1)$ exists.

Let us represent $f'(x)$ in the form of $f(x)$.

Indeed $f'(x)=\begin{cases}x, & 0<x<1\\ 1, & x=1\\ 4x-3, & 1\le x\le 2\end{cases}$

from which we note that $f'(x)$ is continuous at 1.

(iii) Now $Lf''(1) = 1$ and $Rf''(1) = 4$

$\Rightarrow f''(x)$ is discontinuous at 1.

Example 6. Let $f(x)=\begin{cases}xe^{-\left(\frac{1}{|x|}+\frac{1}{x}\right)}, & x\ne 0\\ 0, & x=0\end{cases}$ Show that

(i) f is continuous at 0

(ii) f is not differentiable at 0

Solution

If $x < 0$, $|x| = -x$ and if $x > 0$, $|x| = x$, we have

$$f(x)=\begin{cases}x, & \text{if } x<0\\ 0, & \text{if } x=0\\ xe^{-\frac{1}{2x}}, & \text{if } x>0\end{cases}$$

from which it follows that f is continuous at 0.

Since $\lim_{x\to 0^+}f(x)=\lim_{h\to 0}he^{-\frac{1}{2h}}=0\times 0=0$

$\lim_{x\to 0^-}f(x)$ is obviously zero. Again $Lf'=1$ (since $\frac{d}{dx}x=1$)

$$Rf'(1)=\lim_{h\to 0}\frac{f(0+h)-f(0)}{h}=\lim_{h\to 0}\frac{he^{-\frac{1}{2h}}-0}{h}$$

$=\lim_{h\to 0}he^{-\frac{1}{2h}}=0\Rightarrow f$ is not differentiable at zero.

Example 7. If $f(0) = g(0) = 0$ and $f'(0)$, $g(0)$ exists ($g'(0)\ne 0$), prove that

$\lim_{x\to 0}\frac{f(x)}{g(x)}=\frac{f'(x)}{g'(x)}$. Hence deduce LH rule.

Solution

$$\text{LHS}=\lim_{x\to 0}\frac{\frac{f(x)-f(0)}{x-0}}{\frac{g(x)-g(0)}{x-0}}\ (\text{since } f(0)=g(0)=0)$$

$$\Rightarrow\quad \frac{f'(0)}{g'(0)}\ (\text{since } g'(0)\ne 0)$$

To deduce LH rule consider

$$\lim_{x\to 0}\frac{f(x)}{g(x)}=\lim_{h\to 0}\frac{f(a+h)}{g(a+h)}$$

$$=\lim_{h\to 0}\frac{f(a+h)-f(a)}{g(a+h)-g(a)}\ (\text{since } f(a)=g(a)=0)$$

$$=\lim_{h\to 0}\frac{\frac{f(a+h)-f(a)}{h}}{\frac{g(a+h)-g(a)}{h}}=\frac{f'(a)}{g'(a)}$$

provided $f'(a)\,.\,g'(a)$ exists and $g'(a)\ne 0$.

Example 8. Let $f(x)$ be a function satisfying $f(-x) = f(x)$ for all x. If $f'(0)$ exists, find its value.

Solution

$f'(0)$ exists $\Rightarrow Lf'(0) = Rf'(0) = f'(0)$

$$\Rightarrow \quad \lim_{h\to 0}\frac{f(0-h)-f(0)}{-h} = \lim_{h\to 0}\frac{f(0+h)-f(0)}{h}$$

$$\Rightarrow \quad \lim_{h\to 0}\frac{f(-h)-f(0)}{-h} = \lim_{h\to 0}\frac{f(h)-f(0)}{h}$$

$$\Rightarrow \quad -\lim_{h\to 0}\frac{f(h)-f(0)}{h} = \lim_{h\to 0}\frac{f(h)-f(0)}{h}$$

$$\Rightarrow \quad -f'(0) = f'(0) \text{ (since } f'(0) = \lim_{h\to 0}\frac{f(h)-f(0)}{h}\text{)}$$

$$\Rightarrow \quad f'(0) = 0$$

Note: If we assume the existence of $f'(x)$ in the neighborhood of zero, then on differentiating the equation $f(-x) = f(x)$, we get

$$f'(-x)(-1) = f'(x)$$

On putting $x = 0$, we get

$$-f'(0) = f'(0) \Rightarrow f'(0) = 0.$$

Example 9. Show that $f(x) = [x]\sin^2\pi x$ is continuous for all x but $[x]\cos^2\pi x$ is discontinuous for integral values of x (where $[x]$ denotes greatest integer $\le x$).

Solution

Since $[x]$ is continuous for all x except integers and $\sin^2\pi x$ is universally continuous, $[x]\sin^2\pi x$ is continuous for all x except possibly at integers.

Now, if x is an integer, say, n, then

$\lim_{x\to n^-}[x]\sin^2\pi x = (n-1)\sin^2\pi x = 0$ and

$\lim_{x\to n^+}[x]\sin^2\pi x = n\sin^2\pi x = n\sin^2\pi x = 0$ and $f(x) = 0$.

Thus $f(x)$ is continuous for all x. (For any integer x, $\sin\pi x = 0$). Now, let $g(x) = [x]\cos^2\pi x$, then $f(x) + g(x) = [x]$ which is discontinuous at integer points.

$\Rightarrow$ $g(x)$ is discontinuous at integer points. (The other part can be done independently also.)

Example 10. Test the following composite functions for continuity:

(a) $y = \cos x^n$, where n is a natural number

(b) $y = \cos\log x$

(c) $y = \sqrt{1/2 - \cos^2 x}$.

Solution

(a) We have a composite function $y = \cos u$, where $u = x^n$. The function $y = \cos u$ is continuous at any point u, and the function $u = x^n$ is continuous at any value of x. Therefore, the function $y = \cos x^n$ is continuous throughout the entire number scale.

(b) $y = \cos\log(x)$

Since $\cos(x)$ is continuous everywhere and $\log_e x$ is also continuous everywhere in the domain $x \in (0, \infty)$, hence $\cos x(\log x)$ is also continuous in $x \in (0)$.

(c) Here $y = \sqrt{1/2 - u^2}$, where $u = \cos x$. The function $\sqrt{1/2 - u^2}$ is defined and continuous on the interval $\left|-\sqrt{2}/2, \sqrt{2}/2\right|$, the function $u = \cos x$ is continuous throughout the entire number scale. Therefore, the function $y = \sqrt{1/2 - \cos^2 x}$ is continuous at all values of x for which

$$|\cos x| \le \sqrt{2}/2, \text{ i.e. } \begin{cases} \pi/4 + 2\pi n \le x \le 3\pi/4 + 2\pi n, \\ 5\pi/4 + 2\pi n \le x \le 7\pi/4 + 2\pi n. \end{cases}$$

Example 11. Show that the function

$$f(x) = \begin{cases} 2^x + 1 \text{ for } -1 \le x < 0 \\ 2^x \quad \text{ for } x = 0, \\ 2^x - 1 \text{ for } 0 < x \le 1, \end{cases}$$

defined and bounded in the interval $[-1, 1]$, has neither maximum, 1 nor minimum values.

Solution

In the interval $[-1, 0)$ the function increases from 3/2 to 2, and in $(0, 1]$ it increases from 0 to 1; it does not attain either the value 2 or 0. Therefore, the function is bounded but never reaches its upper and lower bounds. This is because there is a discontinuity at the point $x=0$.

Example 12. Show that on any interval $[a, b]$ of length greater than unity, the function $f(x) = x - [x]$ attains its minimum value but never reaches its maximum.

Solution

In any interval $[n, n + 1)$, where n is an integer, the given function $f(x)$ increases from 0 to 1, never attaining the maximum. Hence, $0 \le f(x) \le 1$ for any x. Since on the interval $[a, b]$ we can find at least one internal integral, point n, $f(n) = 0$ and $\lim_{x\to n-0} f(x) = 1$, but $f(x) \ne 1$ for any x. It means that the function reaches its minimum value but never reaches its maximum. This is because there is a discontinuity at the point $x = n$ (see figure).

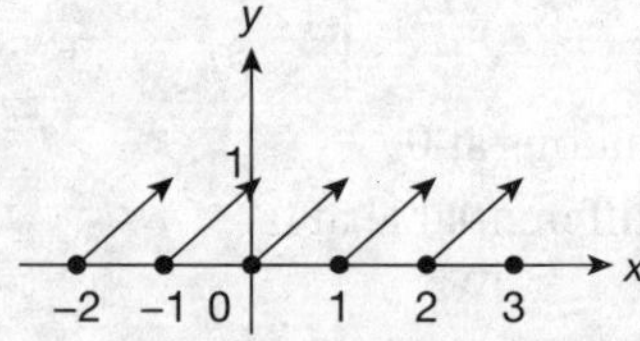

Example 13. Prove that the function $y = \sqrt[2n+1]{x}$ (a natural number) is continuous throughout the number scale, considering it as a function inverse to $y = x^{2n+1}$.

Solution

The function $y = x^{2n+1}$ is continuous and increases from $-\infty$ to ∞ over the entire number scale. Hence, the inverse function $x = \sqrt[2n+1]{y}$ is defined for all y, continuous and increasing. Denoting the independent variable again as x, we find that the function $y = \sqrt[2n+1]{x}$ possesses the required properties.

Example 14. Prove that there exists only one continuous function $x = x(y)(-\infty < y < \infty)$ which satisfies the Kepler equation:

$$x - \varepsilon \sin x = y \; (0 < \varepsilon < 1)$$

Solution

Let us show that $y(x)$ is an increasing function. Let $x_1 < x_2$ be arbitrary points on the number scale. Then

$$y(x_2) - y(x_1) = (x_2 - \varepsilon \sin x_2) - (x_1 - \varepsilon \sin x_1)$$
$$= (x_2 - x_1) - \varepsilon(\sin x_2 - \sin x_1).$$

Estimate the absolute value of the difference $|\sin x_2 - \sin x_1|$:

$$|\sin x_2 - \sin x_1| = 2\left|\sin\frac{x_2 - x_1}{2}\right|\left|\cos\frac{x_2 + x_1}{2}\right| \le$$

$$\le 2\left|\sin\frac{x_2 - x_1}{2}\right| \le 2\frac{|x_2 - x_1|}{2} = |x_2 - x_1| = (x_2 - x_1).$$

Since $0 < \mu < 1$,

$$\varepsilon|\sin x_2 - \sin x_1| < (x_2 - x_1),$$

whence

$$(x_2 - x_1) - \varepsilon(\sin x_2 - \sin x_1) = y(x_2) - y(x_1) > 0$$

Since $y(x)$ is a continuous function in the interval $(-\infty, \infty)$, the inverse function x is a single-valued and continuous function of y.

RECAPITULATION (FORMULAE, THEOREM, IMPORTANT RESULTS)

1. A continuous function in the closed interval $a \le x \le b$ has the following properties:

(a) If $f(a)$ and $f(b)$ are of opposite signs, then there exits at least one, and in general, odd solution of the equation $f(x) = 0$ for any x in the open interval (a, b).

(b) If λ is any real number between $f(a)$ and $f(b)$, then there exists at least one solution of the equation $f(x) = \lambda$ in the open interval (a, b).

2. Geometrical meaning of continuity

(i) The function 'f' will be continuous at $x = a$ if there is no break in the graph of the function $y = f(x)$ at the point $(a, f(a))$.

(ii) The function $f(x)$ will be continuous in the closed interval $[a, b]$ if the graph of $y = f(x)$ is an unbroken line (curved or straight) from the point $(a, f(a))$ to $(b, f(b))$.

3. A polynomial is a continuous function at each point

$$f(x) = a_0x^n + a_1x^{n-1} + \ldots + a_n$$

where $a_0, a_1, \ldots, a_n$ are fixed real constants and $n, n-1, n-2, \ldots$ are positive integers.

4. Every rational function is continuous in its domain. Every rational function is of the form $\frac{p(x)}{q(x)}$ where $p(x)$ and $q(x)$ are polynomial functions. Therefore, it is continuous at all those points where $q(x)$ is not zero. In other words, they are continuous at all points of its domain.

5. The domain of f' consists of those points where f is differentiable. f is differentiable on an open interval (a, b) if it is differentiable at each point in (a, b) and we say that f is a differentiable function if it is differentiable on $(-\infty, \infty)$. At points where fis not differentiable, we say that derivative of f does not exist.

6. If R $f'(a)$ or L $f'(a)$ does not exist or if both R $f'(a)$ and L $f'(a)$ exist but are not equal, then f is not differentiable at a.

QUIZ

1. The limit of a function exists at a point. Is it necessary that the function is continuous at that point?

2. A function $f(x)$ is continuous in an open interval (a, b) if it is continuous at every point in (a, b). Is the statement true or false?

3. Which of the following functions are continuous everywhere?
(i) $\sin x$ (ii) e^x (iii) a^x

4. Are the functions $\sin^{-1}x$ and $\cos^{-1}x$ continuous everywhere? If not, then in what interval is it continuous?

5. Is a constant function continuous everywhere?

6. Is the function $\cos(x^3+1)$ is continuous or discontinuous? If continuous, give the argument.

7. Define exponential function.

8. What is natural logarithmic function?

9. What is an implicit function? Explain with an example.

10. What is the domain and range of $\tan^{-1}x$?

11. The derivative of an odd function is always an even function. True or false?

12. What is the inverse of $\log_e x$ function?

13. If $f(x)$ and $g(x)$ have no derivative at a point then $f(x)g(x)$ has no derivative at that point. True or false?

14. What is the inverse of the function $f(x)=\left[1-(x-5)^3\right]^{\frac{1}{5}}$?

JEE MAIN

Objective Questions

One Option Correct

1. The function $f(x)=[x]^2-[x^2]$ (where, $[x]$ is the greatest integer less than or equal to (x), is discontinuous at
(a) all integers.
(b) all integers except 0 and 1.
(c) all integers except 0.
(d) all integers except 1.

2. Let [.] denotes the greatest integer function and $f(x)=[\tan^2 x]$, then
(a) $\lim_{x\to 0} f(x)$ does not exist.
(b) $f(x)$ is continuous at $x=0$.
(c) $f(x)$ is not differentiable at $x=0$.
(d) $f'(0)=1$.

3. The function $f(x)=[x]\cos\left(\frac{2x-1}{2}\right)\pi$, [.] denotes the greatest integer function, which is discontinuous at
(a) all x.
(b) all integer points.
(c) no x.
(d) x which is not an integer.

4. If $f(x)=x\left(\sqrt{x}+\sqrt{x+1}\right)$, then
(a) $f(x)$ is continuous but not differentiable at $x=0$.
(b) $f(x)$ is differentiable at $x=0$.
(c) $f(x)$ is not differentiable at $x=0$.
(d) None of the above.

5. The function $f(x)=\frac{\log(1+ax)-\log(1-bx)}{x}$ is not defined at $x=0$. The value which should be assigned to f at $x=0$, so that it is continuous at $x=0$, is ______.
(a) $a-b$
(b) $a+b$
(c) $\log a+\log b$
(d) None of the above.

6. $\lim_{x\to\frac{\pi}{4}}\frac{\int_2^{\sec^2 x} f(t)\,dt}{x^2-\frac{\pi^2}{16}}$ equals ______.
(a) $\frac{8}{\pi}f(2)$ (b) $\frac{2}{\pi}f(2)$
(c) $\frac{2}{\pi}f\left(\frac{1}{2}\right)$ (d) $4f(2)$

7. $\lim_{n\to\infty}\frac{1}{n}\sum_{r=1}^{2n}\frac{r}{\sqrt{n^2+r^2}}$ equals ______.
(a) $1+\sqrt{5}$ (b) $\sqrt{5}-1$
(c) $-1+\sqrt{2}$ (d) $1+\sqrt{2}$

8. If $f:R\to R$ is defined by
$$f(x)=\begin{cases}\frac{\cos 3x-\cos x}{x^2} & for\ x\neq 0\\ \lambda & for\ x=0\end{cases}$$
and if f is continuous at $x=0$, then $\lambda=$______.

(a) –2 (b) –4
(c) –6 (d) –8

9. If $f(2)=4$ and $f'(2)=1$ then $\lim_{x\to 2}\frac{xf(2)-2f(x)}{x-2}=$ ______.

(a) –2 (b) 1
(c) 2 (d) 3

10. If $x=a\left\{\cos\theta+\log\tan\left(\frac{\theta}{2}\right)\right\}$ and $y=a\sin\theta$ then $\frac{dy}{dx}=$ ______.

(a) $\cot\theta$ (b) $\tan\theta$
(c) $\sin\theta$ (d) $\cos\theta$

11. If $y=\sin(\log_e x)$ then $x^2\frac{d^2y}{dx^2}+x\frac{dy}{dx}=$ ______.

(a) $\sin(\log_e x)$ (b) $\cos(\log_e x)$
(c) y^2 (d) $-y$

12. If $f(x)=\begin{cases}x-5, & \text{for } x\le 1\\ 4x^2, & \text{for } 1<x<2\\ 3x+4 & \text{for } x\ge 2\end{cases}$ then $f'(2^+)$ is equal to ______.

(a) 0 (b) 2
(c) 3 (d) 4

13. If $2x^2-3xy+y^2+x+2y-8=0$, then $\frac{dy}{dx}$ is equal to ______.

(a) $\frac{3y-4x-1}{2y-3x+2}$ (b) $\frac{3y+4x+1}{2y+3x+2}$
(c) $\frac{3y-4x+1}{2y-3x-2}$ (d) $\frac{3y-4x+1}{2y+3x+2}$

14. $x=\cos\theta,\ y=\sin 5\theta \Rightarrow (1-x^2)\frac{d^2y}{dx^2}-x\frac{dy}{dx}$ is equal to ______.

(a) $-5y$ (b) $5y$
(c) $25y$ (d) $-25y$

15. If $f(x)=\begin{cases}\frac{1-\sqrt{2}\sin x}{\pi-4x} & \text{if } x\ne\frac{\pi}{4}\\ a & \text{if } x=\frac{\pi}{4}\end{cases}$ is continuous at $\frac{\pi}{4}$, then a is equal to ______.

(a) 4 (b) 2
(c) 1 (d) $\frac{1}{4}$

16. Observe the following statements:

I. $f(x)=ax^{41}+bx^{-40}\Rightarrow\frac{f''(x)}{f(x)}=1640x^{-2}$

II. $\frac{d}{dx}\tan^{-1}\left(\frac{2x}{1-x^2}\right)=\frac{1}{1+x^2}$

Which of the following is correct?
(a) I is true, but II is false.
(b) Both I and II are true.
(c) Neither I nor II is true.
(d) I is false, but II is true.

17. If $f(x)=\frac{x}{1+|x|}$, for $x\in$ R, then $f'(0)$ is equal to ______.

(a) 0 (b) 1
(c) 2 (d) 3

18. It is given that $f'(a)$ exists, then $\lim_{x\to a}\frac{xf(a)-f(x)}{x-a}$ is equal to ______.

(a) $f(a)-af'(a)$ (b) $f'(a)$
(c) $-f'(a)$ (d) $f(a)+af'(a)$

19. Let $y=t^{10}+1$ and $x=t^8+1$, then $\frac{d^2y}{dx^2}$ is equal to ______.

(a) $\frac{5}{2}t$ (b) $20t^8$
(c) $\frac{5}{16t^6}$ (d) None of these.

20. If $y^{x+e^{x+e^{x..\infty}}}$, $\frac{dy}{dx}=$ ______.

(a) $\frac{x}{1-x}$ (b) $\frac{y}{1+y}$
(c) $\frac{y}{1-y}$ (d) $\frac{1-y}{y}$

21. If $y=\cos^{-1}(\cos x)$, then $y'(x)$ is equal to ______.

(a) 1 for all x.
(b) –1 for all x.
(c) 1 in 2nd and 3rd quadrants.
(d) –1 in 3rd and 4th quadrants.

22. If $x=$ A$\cos 4t$, then $\frac{d^2x}{dt^2}$ is equal to ______.

(a) $-16x$ (b) $16x$
(c) x (d) $-x$

23. The differential coefficient of $f(\sin x)$ with respect to x where $f(x)=\log x$ is ______.

(a) $\tan x$ (b) $\cot x$

(c) $f(\cos x)$ (d) $\frac{1}{x}$

24. If $f(x)=\begin{cases}\frac{1-\cos x}{x}, & x\neq 0\\ k, & x=0\end{cases}$ is continuous at $x=0$, then the value of k is ______.

(a) 0 (b) $\frac{1}{2}$

(c) $\frac{1}{4}$ (d) $-\frac{1}{2}$

25. If $y=\tan^{-1}(\sec x-\tan x)$, then $\frac{dy}{dx}$ is equal to ______.

(a) 2 (b) –2

(c) $\frac{1}{2}$ (d) $-\frac{1}{2}$

26. The value of k for which the function $f(x)=\begin{cases}\frac{1-\cos 4x}{8x^2}, x\neq 0\\ k \quad ,x=0\end{cases}$ is continuous $x=0$, is ______.

(a) $k=0$ (b) $k=1$

(c) $k=-1$ (d) None of these.

27. If $f(x)=\cos x\ \cos 2x\ \cos 4x\ \cos 8x\ \cos 16x$, then $f'\left(\frac{\pi}{4}\right)$ is ______.

(a) $\sqrt{2}$ (b) $\frac{1}{\sqrt{2}}$

(c) 0 (d) $\frac{\sqrt{3}}{2}$

28. If $f(x)$ is differentiable function and $f''(0)=a$, the $\lim\limits_{x\to 0}\frac{2f(x)-3f(2x)+f(4x)}{x^2}$ is equal to ______.

(a) $3a$ (b) $2a$

(c) $5a$ (d) $4a$

29. If $y=\cos^{-1}\left(\frac{1-\ln x}{1+\ln x}\right)$, then $\frac{dy}{dx}$ at $x=e$ is ______.

(a) $-\frac{1}{e}$ (b) $-\frac{1}{2e}$

(c) $\frac{1}{2e}$ (d) $\frac{1}{e}$

30. If $f(x)=2x^6+3x^4+4x^2$, then $f'(x)$ is

(a) an even function.

(b) an odd function.

(c) neither even nor odd.

(d) None of these.

31. If $y=\tan^{-1}\left(\frac{\sqrt{1+x^2}-1}{x}\right)$, then $y'(0)$ is ______.

(a) $\frac{1}{2}$ (b) 0

(c) 1 (d) It does not exist.

32. The value of $f(0)$ so that function $f(x)=\frac{1-\cos(1-\cos x)}{x^4}$ is continuous everywhere is ______.

(a) $\frac{1}{8}$ (b) $\frac{1}{2}$

(c) $\frac{1}{4}$ (d) $\frac{1}{3}$

33. Let $f(x)=g(x).\frac{e^{1/x}-e^{-1/x}}{e^{1/x}+e^{-1/x}}$, where g is a continuous function. Then $\lim\limits_{x\to 0}f(x)$ does not exist if

(a) $g(x)$ is any constant function.

(b) $g(x)=x$.

(c) $g(x)=x^2$.

(d) $g(x)=xh(x)$, where $h(x)$ is a polynomial.

34. Let $[x]$ denote the greatest integer less than or equal to x. If $f(x)=[x \sin \pi x]$, then $f(x)$ is

(I) continuous at $x=0$.

(II) continuous in $(-1, 0)$.

(III) differentiable at $x=1$.

(IV) differentiable $(-1, 1)$.

(a) I, II and IV (b) II, III and IV

(c) I, III and IV (d) I, II and IV

35. If $x^y=e^{x-y}$ then $\frac{dy}{dx}$ at $x=1$ is equal to ______.

(a) 0 (b) –2

(c) 1 (d) None of these.

36. A function f is defined by $f(x)=2+(x-2)^{2/3}$ in [0, 2]. Which of the following is not correct?

(a) f is not derivable in (0, 2).

(b) f is continuous in [0, 2].

(c) $f(0)=f(2)$.

(d) Rolle's theorem is true in [0, 2].

37. $\frac{d}{dx}\left(\tan^{-1}\sqrt{\frac{1+\cos\frac{x}{2}}{1-\cos\frac{x}{2}}}\right)$ is equal to ______.

(a) $-\frac{1}{4}$ (b) $\frac{1}{4}$

(c) $-\frac{1}{2}$ (d) $\frac{1}{2}$

38. Let $f(x)=\frac{\cos^2 x-\sin^2 x-1}{\sqrt{x^2+4}-2}(x\neq 0)$

$= a \qquad (x=0)$

Then the value of a in order that $f(x)$ may be continuous at $x=0$ is ______.

(a) –8 (b) 8

(c) –4 (d) 4

39. Let $f(x)=\sin x, g(x)=x^2$ and $h(x)=\log_e x$. If $F(x)=(h\,o\,g\,o\,f)(x)$, then $F''(x)$ is equal to ______.

(a) $a\operatorname{cosec}^3 x$

(b) $2\cot x^2-4x^2\operatorname{cosec}^2 x^2$

(c) $2x\cot x^2$

(d) $-2\operatorname{cosec}^2 x$

40. If $y=\cos\left(\sin x^2\right)$ then at $x=\sqrt{\frac{\pi}{2}}, \frac{dy}{dx}$ is equal to ______.

(a) –2 (b) 2

(c) $-2\sqrt{\frac{\pi}{2}}$ (d) 0

41. The first derivative of the function $\left[\cos^{-1}\left(\sin\sqrt{\frac{1+x}{2}}\right)+x^x\right]$ with respect to x at $x=1$ is ______.

(a) $\frac{3}{4}$ (b) 0

(c) $\frac{1}{2}$ (d) $-\frac{1}{2}$

42. The function $f(x)=\frac{\log(1+ax)-\log(1-bx)}{x}$ is not defined at $x=0$. The value of which should be assigned to f at $x=0$, so that it is continuous at $x=0$, is ______.

(a) $a-b$ (b) $a+b$

(c) $\log a+\log b$ (d) $\log a-\log b$

43. The function $f(x)=[x]\cos\left[\frac{2x-1}{2}\right]\pi$, where [ ] denotes the greatest integer function, is discontinuous at

(a) all x.

(b) no x.

(c) all integer points.

(d) x which is not an integer.

44. Differential coefficient of $\sqrt{\sec\sqrt{x}}$ is ______.

(a) $\frac{1}{4\sqrt{x}}\sec\sqrt{x}\sin\sqrt{x}$

(b) $\frac{1}{4\sqrt{x}}\left(\sec\sqrt{x}\right)^{3/2}\sin\sqrt{x}$

(c) $\frac{1}{2}\sqrt{x}.\sec\sqrt{x}\sin\sqrt{x}$

(d) $\frac{1}{2}\sqrt{x}\left(\sec\sqrt{x}\right)^{3/2}\sin\sqrt{x}$

45. If $y=e^{1+\log_e x}$ then the value $\frac{dy}{dx}$ is equal to ______.

(a) e (b) 1

(c) 0 (d) $\log_e xe$

46. The function which is continuous for all real values of $x$and differentiable at $x=0$ is ______.

(a) $x^{1/2}$ (b) $|x|$

(c) $\log x$ (d) $\sin x$

47. Function $f(x)=\begin{Bmatrix} x-1 & x<2 \\ 2x-c & x\geq 2\end{Bmatrix}$ is a continuous function

(a) for $x=2$ only.

(b) for all real value of x such that $x\neq 2$.

(c) for all integer value of x only.

(d) None of these.

48. $\lim_{h\to 0}\frac{f\left(2h+2+h^2\right)-f(2)}{f\left(h-h^2+1\right)-f(1)}$, given that $f'(2)=6$ and $f'(1)=4$

(a) does not exist.

(b) is equal to $-3/2$.

(c) is equal to 3/2.

(d) is equal to 3.

49. If $\lim_{x\to 0}\frac{\{(a-n)nx-\tan x\}\sin nx}{x^2}=0$, where n is non-zero real number, then a is equal to ______.

(a) 0 (b) $\frac{n+1}{n}$

(c) n (d) $n+\frac{1}{n}$

50. The integer n for which $\lim_{x\to 0}\frac{(\cos x-1)(\cos x-e^x)}{x^n}$ is a finite non-zero number, is ______.

(a) 1 (b) 2
(c) 3 (d) 4

51. $\lim_{x\to 0}\frac{x\tan 2x-2x\tan x}{(1-\cos 2x)^2}$ is ______.

(a) 2 (b) −2
(c) $\frac{1}{2}$ (d) $-\frac{1}{2}$

52. $\lim_{x\to 1}\frac{\sqrt{1-\cos 2(x-1)}}{x-1}$

(a) exists and it equals $\sqrt{2}$.
(b) exists and it equals $-\sqrt{2}$.
(c) does not exist because $x-1\to 0$.
(d) does not exist because left-hand limit is not equal to right-hand limit.

53. The value of $\lim_{x\to 0}\frac{\sqrt{\frac{1}{2}(1-\cos^2 x)}}{x}$ is ______.

(a) 1 (b) −1
(c) 0 (d) None of these.

54. If $f(x)=\begin{cases}\frac{\sin[x]}{[x]}, & [x]\neq 0\\ 0, & [x]\neq 0\end{cases}$

where, $[x]$ denotes the greatest integer less than or equal to x, then $\lim_{x\to 0} f(x)$ equals ______.

(a) 1 (b) 0
(c) −1 (d) None of these.

55. $\lim_{n\to\infty}\left(\frac{1}{1-n^2}+\frac{2}{1-n^2}+\ldots+\frac{n}{1-n^2}\right)$ is equal to ______.

(a) 0 (b) $-\frac{1}{2}$

(c) $\frac{1}{2}$ (d) None of these.

56. If $f(a)=2, f'(a)=1, g(a)=-1, g'(a)=2$, then the value of $\lim_{x\to a}\frac{g(x)f(a)-g(a)f(x)}{x-a}$ is ______.

(a) −5 (b) $\frac{1}{5}$

(c) 5 (d) None of these.

57. If $G(x)=-\sqrt{25-x^2}$, then $\lim_{x\to 1}\frac{G(x)-G(1)}{x-1}$ has the value ______.

(a) $\frac{1}{\sqrt{24}}$ (b) $\frac{1}{5}$

(c) $-\sqrt{24}$ (d) None of these.

58. For $x>0$, $\lim_{x\to 0}\left[(\sin x)^{1/x}+\left(\frac{1}{x}\right)^{\sin x}\right]$ is ______.

(a) 0 (b) −1
(c) 1 (d) 2

59. Let $f:\mathrm{R}\to\mathrm{R}$ be such that $f(1)=3$ and $f'(1)=6$. Then, $\lim_{x\to 0}\left[\frac{f(1+x)}{f(1)}\right]^{1/x}$ equals ______.

(a) 1 (b) $e^{\frac{1}{2}}$

(c) e^2 (d) e^3

60. For $x\in\mathrm{R}$, $\lim_{x\to\infty}\left(\frac{x-3}{x+2}\right)^x$ is equal to ______.

(a) e (b) e^{-1}
(c) e^{-5} (d) e^5

61. Let $f(x)=\begin{cases}(x-1)\sin\frac{1}{x-1} & \text{if } x\neq 1\\ 0 & \text{if } x=1\end{cases}$ then which one of the following is true?

(a) f is differentiable at $x = 0$ and at $x = 1$.
(b) f is differentiable at $x = 0$ but not at $x = 1$.
(c) f is differentiable at $x = 1$ but not at $x = 0$.
(d) f is neither differentiable at $x = 0$ nor at $x = 1$.

62. If f is a real-valued differentiable function satisfying $|f(x) - f(y)| \leq (x-y)^2, x, y\in R$ and $f(0) = 0$, then $f(1)$ is equal to ______.

(a) 2 (b) 1
(c) −1 (d) 0

63. Suppose $f(x)$ is differentiable at $x = 1$ and $\lim_{h\to 0}\frac{1}{h}f(1+h)=5$, then $f'(1)$ is equal to ______.

(a) 5 (b) 6
(c) 3 (d) 4

64. If $x=e^{y+e^{y+\ldots\infty}}, x>0$, then $\frac{dy}{dx}$ is ______.

(a) $\frac{x}{1+x}$ (b) $\frac{1}{x}$

(c) $\frac{1-x}{x}$ (d) $\frac{1+x}{x}$

65. If $x^2+y^2=t-\frac{1}{t}=x^4+y^4=t^2+\frac{1}{t^2}$, then the value of $x^3y\frac{dy}{dx}=$ ______.

(a) 2 (b) 1
(c) 0 (d) None

66. If $f(x)=x(\sqrt{x}+\sqrt{x+1})$, then
(a) $f(x)$ is continuous but not differentiable at $x=0$.
(b) $f(x)$ is differentiable at $x=0$.
(c) $f(x)$ is not differentiable at $x=0$.
(d) None of these.

67. The number of points at which the function $f(x)=(|x-1|x-2|+\cos x)$ where $x\in[0,4]$, is not continuous, is ______.

(a) 1 (b) 2
(c) 3 (d) 0

68. When $y=\sin^{-1}\left(\frac{2t}{1+t^2}\right), x=\tan^{-1}\left(\frac{2t}{1-t^2}\right)$ where $-1<t<1$, then $\frac{dy}{dx}$ is equal to ______.

(a) 0 (b) 1
(c) –1 (d) ∞

69. $f(x)=\begin{cases}\frac{\sin 3x}{\sin x}, & x\neq 0\\ k, & x=0\end{cases}$ is continuous, if k is ______.

(a) 3 (b) 0
(c) –3 (d) –1

70. The function $f(x)=\frac{e^{\tan x}-1}{e^{\tan x}+1}$ is discontinuous at x equal to ______.

(a) $(n\pi+\pi)$ (b) $\left(n\pi+\frac{\pi}{2}\right)$

(c) $\left(n\pi+\frac{\pi}{4}\right)$ (d) $\left(n\pi+\frac{\pi}{8}\right)$

71. If $y=\int_x^{x^2}\sqrt{5-t^2}\,dt$, then value of $\frac{dy}{dx}$ at $x=\sqrt{2}$ is ______.

(a) $1-\sqrt{3}$ (b) $\sqrt{3}\left(2\sqrt{6}-1\right)$

(c) $2\sqrt{2}-\sqrt{3}$ (d) None

72. For the function $f(x)=\frac{e^{1/x}-1}{e^{1/x}+1}, x=0$, which of the following is correct?
(a) $\lim_{x\to 0} f(x)$ does not exist.
(b) $\lim_{x\to 0} f(x)=1$.
(c) $\lim_{x\to 0} f(x)$ exist but $f(x)$ is not continuous at $x=0$.
(d) $f(x)$ is continuous at $x=0$.

73. Let $y=x^{x^{x^{\cdots\cdots}}}$, then $\frac{dy}{dx}$ is equal to ______.

(a) yx^{y-1} (b) $\frac{y^2}{x(1-y\log x)}$

(c) $\frac{y}{x(1+y\log x)}$ (d) None.

74. A function f on R into itself is continuous at a point a in R, iff for each $\in>0$, there exists, $\delta>0$ such that ______.
(a) $|f(x)-f(a)|<\in\Rightarrow|x-a|<\delta$
(b) $|f(x)-f(a)|>\in\Rightarrow|x-a|>\delta$
(c) $|x-a|>\delta\Rightarrow|f(x)-f(a)|>\in$
(d) $|x-a|<\delta\Rightarrow|f(x)-f(a)|<\in$

75. The set of all points of discontinuity of the function $f(x)=\frac{1+\cos 5x}{1-\cos 4x}$ is ______.

(a) $\{0,\pi/2,\pi/4\}$ (b) $\{0,\pi/2,\pi/6\}$
(c) $\{0,\pi/2,\pi\}$ (d) $\{0,\pi/2\}$

76. If $y=\sqrt{\log x+\sqrt{\log x+\sqrt{\log x+\sqrt{\log x+\ldots\infty}}}}$ then $\frac{dy}{dx}$ is equal to ______.

(a) $\frac{x}{2y-1}$ (b) $\frac{x}{2y+1}$

(c) $\frac{1}{x(2y-1)}$ (d) $\frac{1}{x(1-2y)}$

77. If $x^y=y^x$, then $x(x-y\log x)\frac{dy}{dx}$ is equal to ______.

(a) $y(y-x\log y)$ (b) $y(y+x\log y)$
(c) $x(x+y\log x)$ (d) $x(y-x\log y)$

78. If $f:\mathrm{R}\to\mathrm{R}$ is an even function which is twice differentiable on R and $f''(\pi)=1$, then $f''(-\pi)$ is equal to ______.

(a) –1 (b) 0
(c) 1 (d) 2

79. If $x\sqrt{1+y}+y\sqrt{1+x}=0$, then $\frac{dy}{dx}$ is equal to ______.

(a) $\frac{1}{(1+x)^2}$ (b) $-\frac{1}{(1+x)^2}$

(c) $\frac{1}{1+x^2}$ (d) $\frac{1}{1-x^2}$

80. If $f:\mathrm{R}\to\mathrm{R}$ is an even function having derivatives of orders, then an odd function among the following is ______.

(a) f'' (b) f'''

(c) $f'+f''$ (d) $f''+f'''$

81. If $f(x)=\begin{cases}\frac{x-1}{2x^2-7x+5}, & \text{for } x\neq 1\\ -\frac{1}{3}, & \text{for } x=1\end{cases}$ then $f'(1)$ is equal to ______.

(a) $-\frac{1}{9}$ (b) $-\frac{2}{9}$

(c) $-\frac{1}{3}$ (d) $\frac{1}{3}$

82. If $f:\mathrm{R}\to\mathrm{R}$ defined by

$f(x)=\begin{cases}a^2\cos^2 x+b^2\sin^2 x & ,x\le 0\\ e^{ax+b} & ,x>0\end{cases}$ is a continuous function, then ______.

(a) $b=2\log|a|$ (b) $2b=\log|a|$

(c) $b=\log|2a|$ (d) $b^2=\log|a|$

83. If $y=\frac{a^{\cos^{-1}x}}{1+a^{\cos^{-1}x}}, z=a^{\cos^{-1}x}$, then $\frac{dy}{dx}$ is equal to ______.

(a) $\frac{1}{1+a^{\cos^{-1}x}}$ (b) $\frac{1}{(1+a^{\cos^{-1}x})^2}$

(c) $\frac{-1}{1+a^{\cos^{-1}x}}$ (d) None

84. If $y=\sqrt{\sin\sqrt{x}}$, then $\frac{dy}{dx}$ is equal to ______.

(a) $\frac{\cos\sqrt{x}}{4\sqrt{x}\sqrt{\sin\sqrt{x}}}$ (b) $\frac{1}{2\sqrt{\cos x}\sqrt{x}}$

(c) $\frac{1}{2\sqrt{\sin x}}$ (d) $\frac{\sqrt{\cos x}\sqrt{x}}{2x}$

85. If $y=\tan^{-1}\left[\frac{\sin x+\cos x}{\cos x-\sin x}\right]$, then $\frac{dy}{dx}$ is ______.

(a) $\frac{1}{2}$ (b) $\frac{\pi}{4}$

(c) 0 (d) 1

86. The set of points where $f(x)=\frac{x}{1+|x|}$ is differentiable is ______.

(a) $(-\infty,-1)\cup(-1,\infty)$ (b) $(-\infty,\infty)$

(c) $(0,\infty)$ (d) $(-\infty,0)\cup(0,\infty)$

87. If $f(x)=\begin{cases}xe^{-\left(\frac{1}{|x|}+\frac{1}{x}\right)}, & x\neq 0\\ 0 & x=0\end{cases}$ then $f(x)$ is

(a) continuous as well as differentiable for all x.

(b) continuous for all x but not differentiable at $x=0$.

(c) neither differentiable nor continuous at $x=0$.

(d) discontinuous everywhere.

88. If $\sqrt{(1-x^2)}+\sqrt{(x-y^2)}=a(x-y)$, then $\frac{dy}{dx}=$ ______.

(a) $\sqrt{\frac{(1-x^2)}{(1-y^2)}}$ (b) $\sqrt{\frac{(1-y^2)}{(1-x^2)}}$

(c) $\left(\frac{1-x^2}{1-y^2}\right)$ (d) $\left(\frac{1-y^2}{1-x^2}\right)$

89. If function $f(x)=\begin{cases}x, & \text{if } x \text{ is rational}\\ 1-x & \text{if } x \text{ is irrational}\end{cases}$ then the number of points at which $f(x)$ is continuous, is ______.

(a) ∞ (b) 1

(c) 0 (d) None of these.

90. Let $f(x)=\begin{cases}\frac{\tan x-\cot x}{x-\frac{\pi}{4}}, & x\neq\frac{\pi}{4}\\ a, & x=-\frac{\pi}{4}\end{cases}$ the value of a so that $f(x)$ is continuous at $x=\frac{\pi}{4}$ is ______.

(a) 2 (b) 4

(c) 3 (d) 1

91. If $y=\sin^{-1}\frac{x}{2}+\cos^{-1}\frac{x}{2}$, then the value of $\frac{dy}{dx}$ is ______.

(a) 1 (b) –1

(c) 0 (d) 2

92. If y is a function of x and $\log(x+y)=2xy$, then the value of $y'(0)$ is equal to ______.

(a) 1 (b) –1
(c) 2 (d) 0

93. Let $f(x)=\begin{cases} x^p \sin\frac{1}{x}, & x\neq 0, \\ 0, & x=0, \end{cases}$ then $f(x)$ is continuous but not differentiable at $x = 0$, if ______.

(a) $0<p\le 1$ (b) $1\le p<\infty$
(c) yx^{y-1} (d) $p=0$

94. $y=x^{x^{x^{\cdots}}}$, then $\frac{dy}{dx}$ is equal to ______.

(a) yx^{y-1} (b) $\frac{y^2}{x(1-y\log x)}$
(c) $\frac{y}{x(1+y\log x)}$ (d) None of these.

95. If $f(x)=x\tan^{-1}x$, then $f'(1)$ is equal to ______.

(a) $\frac{1}{2}+\frac{\pi}{4}$ (b) $-\frac{1}{2}+\frac{\pi}{4}$
(c) $-\frac{1}{2}-\frac{\pi}{4}$ (d) $\frac{1}{2}-\frac{\pi}{4}$

96. If $y=\sec^{-1}\frac{x+1}{x-1}+\sin^{-1}\frac{x-1}{x+1}$, then $\frac{dy}{dx}$ is equal to ______.

(a) 1 (b) 0
(c) $\frac{x-1}{x+1}$ (d) $\frac{x+1}{x-1}$

97. $f(x) = \sin |x|$. Then $f(x)$ is not differentiable at
(a) $x=0$ only. (b) all x.
(c) multiples of π. (d) multiples of $\frac{\pi}{2}$.

98. If $y=e^{1+\log_e x}$, then value of $\frac{dy}{dx}$ is equal to ______.

(a) e (b) 1
(c) 0 (d) $\log_e xe^{\log_e ex}$

99. $\frac{d}{dx}\left(\tan^{-1}\frac{\sqrt{1+x^2}-1}{x}\right)$ is equal to ______.

(a) $\frac{1}{1+x^2}$

(b) $\frac{x^2}{2\sqrt{1+x^2}\,(\sqrt{1-x^2}-1)}$

(c) $\frac{2}{1+x^2}$

(d) $\frac{1}{2(1+x^2)}$

100. Define f on R into itself by

$$f(x)=\begin{cases} x\sin\frac{1}{x}, & \text{when } x\neq 0 \\ 0, & \text{when } x=0 \end{cases}$$

Then
(a) f is continuous at 0 but not differentiable at 0.
(b) f is both continuous and differentiable at 0.
(c) f is differentiable but not continuous at 0.
(d) None of the above.

101. If $x=a\sin\theta$ and $y=b\cos\theta$, then $\frac{d^2y}{dx^2}$ is ______.

(a) $\frac{a}{b^2}\sec^2\theta$ (b) $\frac{-b}{a}\sec^2\theta$
(c) $\frac{-b}{a^2}\sec^3\theta$ (d) $\frac{b}{a^2}\sec^3\theta$

102. If $f(x)g(x)$ be twice differentiable functions on [0, 2] satisfying $f''(x) = g''(x)$, $f'(1) = 2g'(1) = 4$ and $f(2) = 3g(2) = 9$, then $f(x) - g(x)$ at = 4 equals ______.
(a) 0 (b) 10
(c) 8 (d) 2

103. If $y = x\log\left(\frac{x}{a+bx}\right)$ then $x^3\frac{d^2y}{dx^2}$ is equal to ______.

(a) $x\frac{dy}{dx}-y$ (b) $\left(x\frac{dy}{dx}-y\right)^2$
(c) $y\frac{dy}{dx}-x$ (d) $\left(y\frac{dy}{dx}-x\right)^2$

104. $y=\log\left\{\left(\frac{1+x}{1-x}\right)^{1/4}\right\}-\frac{1}{2}\tan^{-1}(x)$, then $\frac{dy}{dx}$ is equal to ______.

(a) $\frac{x}{1-x^2}$ (b) $\frac{x^2}{1-x^4}$
(c) $\frac{x}{1+x^4}$ (d) $\frac{x}{1-x^4}$

105. $f(x)=e^x\sin x$, then $f^{(6)}(x)$ is equal to ______.
(a) $e^{6x}\sin 6x$ (b) $-8e^x\cos x$
(c) $8e^x\sin x$ (d) $8e^x\cos x$

106. If $f:\mathrm{R}\to$ is defined by

$$f(x)=\begin{cases} \frac{x+2}{x^2+3x+2} & \text{if } x\in R-(-1,-2) \\ -1 & \text{if } x=-2 \\ 0 & \text{if } x=-1 \end{cases}$$ then f

is continuous on the set ______.

(a) R (b) R – {–2}
(c) R – {–1} (d) R – {–1, –2)

107. If $f(x) = 10\cos x + (13 + 2x)\sin x$ then $f''(x) + f(x)$ is ______.
(a) $\cot x$ (b) $4\cos x$
(c) $\sin x$ (d) $4\sin x$

108. If $x > 0, x^y = e^{x-y}$, then $\frac{dy}{dx}$ is equal to ______.
(a) $\frac{1}{(1+\log x)^2}$ (b) $\frac{\log x}{(1+\log x)^2}$
(c) $\left(\frac{\log x}{1+\log x}\right)^2$ (d) $\frac{(\log)^2}{1+\log x}$

109. Let $f(x) = e^x, g(x) = \sin^{-1} x$ and $h(x) = f(g(x))$, then $\frac{h'(x)}{h(x)}$ is equal to ______.
(a) $\sin^{-1} x$ (b) $\frac{1}{\sqrt{1-x^2}}$
(c) $-\frac{1}{\sqrt{1-x^2}}$ (d) $e^{\sin^{-1} x}$

110. $\frac{d}{dx}\left(\tan^{-1}\sqrt{\frac{1+\cos\frac{x}{2}}{1-\cos\frac{x}{2}}}\right)$ is equal to ______.
(a) $-\frac{1}{4}$ (b) $\frac{1}{4}$
(c) $-\frac{1}{2}$ (d) $\frac{1}{2}$

111. If $ax^2 + 2hxy + by^2 = 1$, then $\frac{d^2y}{dx^2}$ is equal to ______.
(a) $\frac{h^2+ab}{(hx+by)^2}$ (b) $\frac{h^2-ab}{(hx+by)^2}$
(c) $\frac{h^2+ab}{(hx+by)^3}$ (d) $\frac{h^2-ab}{(hx+by)^3}$

112. If the function

$$f(x) = \begin{cases} x + a\sqrt{2}\sin & , \quad 0 \le x \le \frac{\pi}{4} \\ 2x\cot x + b & , \quad \frac{\pi}{4} \le x \le \frac{\pi}{2} \\ a\cos 2x - b\sin x & , \quad \frac{\pi}{2} < x \le \pi \end{cases}$$

is continuous in the interval $[0, \pi]$, then ______.
(a) $a = \frac{\pi}{6}, b = \frac{\pi}{12}$ (b) $a = -\frac{\pi}{6}, b = \frac{\pi}{12}$
(c) $a = -\frac{\pi}{6}, b = \frac{-\pi}{12}$ (d) $a = \frac{\pi}{6}, b = -\frac{\pi}{12}$

113. Let a function f: R → R satisfy the equation $f(x + y) = f(x) + f(y)$ for all x, y, If the function $f(x)$ is continuous at $x = 0$, then
(a) $f(x) = 0$ is continuous for all x.
(b) $f(x)$ is continuous for all positive real x.
(c) $f(x)$ is continuous for all x.
(d) None of these.

114. If $f(x) = \begin{cases} \frac{x\log x}{\log(1+x^2)}, & x \ne 0 \\ 0 & , x = 0 \end{cases}$ then $f(x)$ is
(a) continuous as well as differentiable at $x = 0$.
(b) continuous but not differentiable at $x = 0$.
(c) differentiable but not continuous at $x = 0$.
(d) neither continuous nor differentiable at $x = 0$.

115. At the point $x = 0$, for the function $f(x) = \frac{1}{2-|x|}$, which of the following are true?
(I) $f(x)$ is continuous.
(II) $f(x)$ is discontinuous.
(III) $f(x)$ is differentiable.
(IV) $f(x)$ is not differentiable.
(a) I and II (b) II and IV
(c) I and III (d) I and IV

116. A function is defined as $f(x) = \begin{cases} e^x & , \quad x \le 0 \\ |x-1|, & x > 0 \end{cases}$ then $f(x)$ is
(I) continuous at $x = 0$.
(II) continuous at $x = 1$.
(III) differentiable at $x = 0$.
(IV) differentiable at $x = 1$.
(a) I and II (b) II and IV
(c) I and III (d) I and IV

117. The function $f(x) = 1 + |\sin x|$ is
(a) continuous everywhere.
(b) differentiable nowhere.
(c) not differentiable at $x = 0$.
(d) not differentiable at infinite number of points.
(a) I, II and III (b) II, III and IV
(c) I, III and IV (d) I, II and IV

118. Over (0, 2) the function $[x^2 - x + 1]$ (where $[x]$ denotes greatest integer $\le x$) is discontinuous at ______.
(I) $\frac{1+\sqrt{2}}{2}$ (II) $\frac{1+\sqrt{3}}{2}$ (III) $\frac{1+\sqrt{5}}{2}$ (IV) 1.

(a) I and II (b) III and IV
(c) I and III (d) II and IV

119. For the function $f(x) = (\pi - x)\dfrac{\cos x}{|\sin x|}; x \neq \pi$, $f(\pi) = 1$, which of the following statements are true?

(I) $f(\pi - 0) = -1$
(II) $f(\pi + 0) = 1$
(III) $f(x)$ is continuous at $x = \pi$
(IV) $f(x)$ is differentiable at $x = \pi$.

(a) I and II (b) II and IV
(c) I and III (d) I and IV

Questions from Previous Years

120. Let $f : R \to R$ be defined as

$$f(x)\begin{cases} \dfrac{»\left|x^2-5x+6\right|}{\mu\left(5x-x^2-6\right)} & , \ x<2 \\ e^{\frac{\tan(x-2)}{x-[x]}} & , \ x>2 \\ \mu & , \ x=2 \end{cases}$$

where $[x]$ is the greatest integer less than or equal to x. If is continuous at $x = 2$, then $\lambda + \mu$ is equal to _______.

(a) $e(-e+1)$ (b) $e(e-2)$
(c) 1 (d) $2e-1$

121. Let a function $f : \mathrm{R} \to \mathrm{R}$ be defined as

$$f(x) = \begin{cases} \sin x - e^x & \text{if} \quad x \le 0 \\ a + [-x] & \text{if} \quad 0 < x < 1 \\ 2x - b & \text{if} \quad x \ge 1 \end{cases}$$

where $[x]$ is the greatest integer less than or equal to x. If f is continuous on R, then $(a+b)$ is equal to _______.

(a) 4 (b) 3
(c) 2 (d) 5

122. Let $\alpha \in \mathrm{R}$ be such that the function

$$f(x) = \begin{cases} \dfrac{\cos^{-1}(1-\{x\}^2)\sin^{-1}(1-\{x\})}{\{x\}-\{x\}^3}, & x \neq 0 \\ \alpha, & x = 0 \end{cases}$$

is continuous at $x = 0$, where $\{x\} = x - [x]$, $[x]$ is the greatest integer less than or equal to x.

(a) No such α exists. (b) $\alpha = \dfrac{\pi}{\sqrt{2}}$

(c) $\alpha = 0$ (d) $\alpha = \dfrac{\pi}{4}$

123. If $f : \mathrm{R} \to \mathrm{R}$ is a function defined by $f(x) = [x-1]\cos\left(\dfrac{2x-1}{2}\right)\pi$, where [.] denotes greatest integer function, then f is:

(a) discontinuous at all integral values of x except at $x = 1$.
(b) continuous only at $x = 1$.
(c) continuous for every real x.
(d) discontinuous only at $x = 1$.

124. If the function f defined on $\left(\dfrac{\pi}{6}, \dfrac{\pi}{3}\right)$ by

$$f(x) = \begin{cases} \dfrac{\sqrt{2}\cos x - 1}{\cot x - 1}, & x \neq \dfrac{\pi}{4} \\ k, & x = \dfrac{\pi}{4} \end{cases}$$ is continuous,

then k is equal to _______.

(a) $\dfrac{1}{2}$ (b) 2

(c) 1 (d) $\dfrac{1}{\sqrt{2}}$

125. The function $f(x) = \left|x^2 - 2x - 3\right| . e^{\left|9x^2 - 12x + 4\right|}$ is not differentiable at exactly:

(a) four points. (b) three points.
(c) two points. (d) one point.

126. If $\begin{cases} \dfrac{1}{|x|} & ; \ |x| \ge 1 \\ ax^2 + b & ; \ |x| < 1 \end{cases}$ is differentiable at every point of the domain, then the values of a and b are respectively _______.

(a) $-\dfrac{1}{2}, \dfrac{3}{2}$ (b) $\dfrac{5}{2}, -\dfrac{3}{2}$

(c) $\dfrac{1}{2}, -\dfrac{3}{2}$ (d) $\dfrac{1}{2}, \dfrac{1}{2}$

127. Let the function $f : R \to R$ and $g : R \to R$ be defined as:

$$f(x) = \begin{cases} x+2, & x<0 \\ x^2, & x \ge 0 \end{cases} \text{ and } g(x) = \begin{cases} x^3, & x<1 \\ 3x-2 & x \ge 1 \end{cases}$$

Then, the number of points in R where $(fog)(x)$ is NOT differentiable is equal to _______.

(a) 3 (b) 0
(c) 2 (d) 1

128. Let $f(x)$ be a differentiable function at $x = a$ $x = a$ with $f'(a) = 2$ and $f(a) = 4$. Then $\lim_{h\to 0} \frac{xf(a) - af(x)}{x-a}$ equals _______.

(a) $4-2a$ (b) $a+4$
(c) $2a-4$ (d) $2a+4$

129. $\lim_{x\to 0} \frac{x+2\sin x}{\sqrt{x^2+2\sin x+1} - \sqrt{\sin^2 x - x + 1}}$ is _______.

(a) 6 (b) 2
(c) 3 (d) 1

130. If $\lim_{x\to 1} \frac{x^2 - ax + b}{x-1} = 5$, then $a+b$ is equal to _______.

(a) –4 (b) 1
(c) –7 (d) 5

131. If $\lim_{x\to 1} \frac{x^4-1}{x-1} = \lim_{x\to k} \frac{x^3-k^3}{x^2-k^2}$, then k is _______.

(a) $\frac{4}{3}$ (b) $\frac{3}{8}$
(c) $\frac{3}{2}$ (d) $\frac{8}{3}$

132. $\lim_{x\to 0} \frac{\sin^2 x}{\sqrt{2} - \sqrt{1+\cos x}}$ equals _______.

(a) $4\sqrt{2}$ (b) $\sqrt{2}$
(c) $2\sqrt{2}$ (d) 4

133. $\lim_{x\to \frac{\pi}{4}} \frac{\cot^3 x - \tan x}{\cos\left(x + \frac{\pi}{4}\right)}$ is _______.

(a) $4\sqrt{2}$ (b) 4
(c) 8 (d) 2

134. $\lim_{x\to 0} \frac{x\cot(4x)}{\sin^2 x \cot^2(2x)}$ is equal to _______.

(a) 0 (b) 1
(c) 4 (d) 2

135. $\lim_{y\to 0} \frac{\sqrt{1+\sqrt{1+y^4}} - \sqrt{2}}{y^4}$

(a) exists and equals $\frac{1}{4\sqrt{2}}$.

(b) does not exist.

(c) exists and equals $\frac{1}{2\sqrt{2}}$.

(d) exists and equals $= \frac{1}{2\sqrt{2}\left(\sqrt{2}+1\right)}$.

136. $\lim_{x\to \pi/2} \frac{\cot x - \cos x}{(\pi - 2x)^3}$ equals _______.

(a) $\frac{1}{24}$ (b) $\frac{1}{16}$
(c) $\frac{1}{8}$ (d) $\frac{1}{4}$

137. $\lim_{x\to 0} \frac{\sin\left(\pi \cos^2 x\right)}{x^2}$ is equal to _______.

(a) $\frac{\pi}{2}$ (b) 1
(c) $-\pi$ (d) π

138. $\lim_{x\to 0} \frac{(1-\cos 2x)(3+\cos x)}{x\tan 4x}$ is equal to _______.

(a) 4 (b) 3
(c) 2 (d) $\frac{1}{2}$

139. If $\lim_{x\to\infty} \left(\frac{x^2+x+1}{x+1} - ax - b\right) = 4$, then _______.

(a) $a = 1, b = 4$ (b) $a = 1, b = -4$
(c) $a = 2, b = -3$ (d) $a = 2, b = 3$

140. Let $f: R \to R$ be a differentiable function, satisfying $f'(3) + f'(2) = 0$. Then $\lim_{x\to 0} \left(\frac{1+f(3+x)-f(3)}{1+f(2-x)-f(2)}\right)^{\frac{1}{x}}$ is equal to _______.

(a) e (b) e^{-1}
(c) e^2 (d) 1

141. $\lim_{x\to 1^-} \frac{\sqrt{\pi} - \sqrt{2\sin^{-1} x}}{\sqrt{1-x}}$ is equal to _______.

(a) $\sqrt{\frac{\pi}{2}}$ (b) $\sqrt{\frac{2}{\pi}}$
(c) $\sqrt{\pi}$ (d) $\frac{1}{\sqrt{2\pi}}$

142. Let $[x]$ denote the greatest integer less than or equal to x. Then $\lim_{x\to 0} \frac{\tan\left(\pi \sin^2 x\right) + \left(|x| - \sin x[x]\right)^2}{x^2}$

(a) equals π. (b) does not exist.
(c) equals –1. (d) equals 1.

143. For each $t \in R$, let $[t]$ be the greatest integer less than or equal to t. Then,

$$\lim_{x\to 1^+} \frac{\left(1-|x|+\sin|1-x|\right)\sin\left(\frac{\pi}{2}[1-x]\right)}{|1-x|[1-x]}$$

(a) equals 0. (b) does not exist.
(c) equals –1. (d) equals 1.

144. For each $x \in \mathbb{R}$, let $[x]$ be the greatest integer less than or equal to x. Then,

$\lim_{x\to 0^-} \frac{x([x]+|x|)\sin[x]}{|x|}$ is equal to ______.

(a) 0 (b) sin 1
(c) – sin 1 (d)

145. If α and β are the roots of the equation

$375x^2 - 25x - 2 = 0$, then $\lim_{n\to\infty}\sum_{r=1}^{n}\alpha^r + \lim_{n\to\infty}\sum_{r=1}^{n}\beta^r$

is equal to ______.

(a) $\frac{21}{346}$ (b) $\frac{29}{358}$
(c) $\frac{1}{12}$ (d) $\frac{7}{116}$

146. For each $t \in \mathbb{R}$, let $[t]$ be the integer less than or equal to t. Then,

$$\lim_{x\to 0^+} x\left(\left[\frac{1}{x}\right]+\left[\frac{2}{x}\right]+\ldots+\left[\frac{15}{x}\right]\right)$$

(a) is equal to 0.
(b) is equal to 15.
(c) is equal to 120.
(d) does not exist (in R).

147. Let $f(x) = \frac{1-x(1+|1-x|)}{|1-x|}\cos\left(\frac{1}{1-x}\right)$ for $x \neq 1$.

Then

(a) $\lim_{x\to 1^+} f(x) = 0$.

(b) $\lim_{x\to 1^-} f(x)$ does not exist.

(c) $\lim_{x\to 1^-} f(x) = 0$.

(d) $\lim_{x\to 1^+} f(x)$ does not exist.

148. Let $p = \lim_{x\to 0^+}(1+\tan^2\sqrt{x})^{1/2x}$, then $\log p$l is equal to ______.

(a) 2 (b) 1
(c) $\frac{1}{2}$ (d) $\frac{1}{4}$

149. Let $\alpha(a)$ and $\beta(a)$ be the roots of the equation $\left(\sqrt[3]{1+a}-1\right)x^2 - \left(\sqrt{1+a}-1\right)x + \left(\sqrt[6]{1+a}-1\right) = 0$, where $a > -1$. Them $\lim_{a\to 0^+} \pm(a)$ and $\lim_{a\to 0^+}\beta(a)$ are ______.

(a) $-\frac{5}{2}$ and 1 (b) $-\frac{1}{2}$ and -1
(c) $-\frac{7}{2}$ and 2 (d) $-\frac{9}{2}$ and 3

150. If $\lim_{x\to 0}\left[1+x\log(1+b^2)\right]^{\frac{1}{x}} = 2b\sin^2\theta, b>0$ and $\theta \in (-\pi, \pi]$, then the value of θ is ______.

(a) $\pm\frac{\pi}{4}$ (b) $\pm\frac{\pi}{3}$
(c) $\pm\frac{\pi}{6}$ (d) $\pm\frac{\pi}{2}$

Integer and Numerical Value Type

1. $\lim_{h\to 0}\frac{\log(1+2h)-2\log(1+h)}{h^2} = \ldots.$

2. If $f(x) = \begin{cases} \sin x, & x \neq n\pi, n = 0, \pm1, \pm2, \ldots \\ 2, & \text{otherwise} \end{cases}$

and $g(x) = \begin{cases} x^2+1, & x \neq 0, 2 \\ 4, & x = 0 \\ 5, & x = 2 \end{cases}$, then

$\lim_{x\to 0} g[f(x)]$ is ______.

3. ABC is an isosceles triangle inscribed in a circle of radius r. If AB = AC and h is the altitude from A to BC, then the ΔABC has perimeter $P = 2\left(\sqrt{2hr-h^2}+\sqrt{2hr}\right)$ and area $A =$ ______.

Also, $\lim_{h\to 0}\frac{A}{P^3} =$ ______.

4. $\lim_{x\to -\infty}\left[\frac{\left(x^4\sin\left(\frac{1}{x}\right)+x^2\right)}{\left(1+|x|^3\right)}\right] =$ ______.

5. Let

$(x) = \begin{cases} (x^3+x^2-16x+20)/(x-2)^2, & \text{if } x \neq 2 \\ k, & \text{if } x = 2 \end{cases}.$

6. $\lim_{x\to 1}(1-x)\tan\frac{\pi x}{2} =$ ______.

7. $\lim_{x\to 0}\frac{\int_0^{x^2}\cos^2 t\, dt}{x\sin x} =$ ______.

8. If $\lim_{x \to a}[f(x)g(x)]$ exists, then both $\lim_{x \to a} f(x)$ and $\lim_{x \to a} g(x)$ exist.

9. Use the formula $\lim_{x \to 0} \frac{a^x - 1}{x} = \log_e a$, to find $\lim_{x \to 0} \frac{2^x - 1}{(1+x)^{1/2} - 1}$.

10. Evaluate $\lim_{h \to 0} \frac{(a+h)^2 \sin(a+h) - a^2 \sin a}{h}$.

11. Evaluate $\lim_{x \to 0} \sqrt{\frac{x - \sin x}{x + \cos^2 x}}$.

12. Evaluate $\lim_{x \to 1}\left(\frac{x-1}{2x^2 - 7x + 5}\right)$.

13. $\lim_{x \to 0}\left(\frac{1+5x^2}{1+3x^2}\right)^{1/x^2} =$ ______.

14. $\lim_{x \to \infty}\left(\frac{x+6}{x+1}\right)^{x+4} =$ ______.

15. Find $\lim_{x \to 0}\left\{\tan\left(\frac{\pi}{4}+x\right)\right\}^{1/x}$.

16. A discontinuous function $y = f(x)$ satisfying $x^2 + y^2 = 4$ is given by $(x) =$ ______.

Questions from Previous Years

17. Let $[t]$ denote the greatest integer $\le t$. The number of points where the function

$$f(x) = [x]\left|x^2 - 1\right| + \sin\left(\frac{\pi}{[x]+3}\right) - [x+1],\ x \in (-2, 2)$$

is not continuous is ______.

18. Let $a, b \in R, b \neq 0$, Define a function

$$f(x) = \begin{cases} a \sin\frac{\pi}{2}(x-1), & \text{for } x \le 0 \\ \frac{\tan 2x - \sin 2x}{bx^3}, & \text{for } x > 0 \end{cases}$$

If f is continuous at $x = 0$, then $10 - ab$ is equal to ______.

19. Let $f : [0, 3] \to \mathrm{R}$ be defined by $f(x) = \min\{x - [x], 1 + [x] - x\}$ where $[x]$ is the greatest integer less than or equal to x. Let P denote the set containing all $x \in [0, 3]$ where f is discontinuous, and Q denote the set containing all $x \in (0, 3)$ where f is not differentiable. Then the sum of number of elements in P and Q is equal to ______.

20. Let $f : \mathrm{R} \to \mathrm{R}$ be a function defined as

$$f(x) = \begin{cases} 3\left(1 - \frac{|x|}{2}\right) & \text{if } |x| \le 2 \\ 0 & \text{if } |x| > 2 \end{cases}$$

Let $g : \mathrm{R} \to \mathrm{R}$ be given by $g(x) = f(x+2) - f(x-2)$.
If n and m denote the number of points in R where g is not continuous and not differentiable, respectively, then $n + m$ is equal to ______.

21. The number of points at which the function $f(x) = |2x+1| - 3|x+2| + \left|x^2 + x - 2\right|, x \in \mathrm{R}$ is not differentiable is ______.

22. If $y = y(x)$ is an implicit function of x such that $\log_e(x+y) = 4xy$, then $\frac{d^2y}{dx^2}$ at $x = 0$ is equal to ______.

Assertion and Reason

(a) Both Statement 1 and Statement 2 are true and Statement 2 is the correct explanation for Statement 1.

(b) Both Statement 1 and Statement 2 are True but Statement 2 is NOT a correct explanation for Statement 1.

(c) Statement 1 is true, Statement 2 is false.

(d) Statement 1 is false, Statement 2 is true.

1. **Assertion:** $\left|x^3\right|$ is differentiable at $x = 0$.
Reason: If $f(x)$ is differentiable at $x = a$ then $|f(x)|$ is also differentiable at $x = a$.

2. **Assertion:** $f(x) = \sin x + [x]$ is discontinuous at $x = 0$.
Reason: If $g(x)$ is continuous and $h(x)$ is discontinuous at $x = a$, then $g(x) + h(x)$ will necessarily be discontinuous at $x = a$.

3. **Assertion:** $f(x) = |x| \sin x$ is differentiable at $x = 0$.
Reason: If $f(x)$ is not differentiable and $g(x)$ is differentiable at $x = a$, then $f(x).\ g(x)$ can still be differentiable at $x = a$.

4. **Assertion:** $f(x) = |[x]\ x|$ in $x \in (-1, 2]$, where [.] represents greatest integer function, is non-differentiable at $x = 2$.
Reason: Discontinuous function is always non-differentiable.

5. **Assertion:** Sum of left-hand derivative and right-hand derivative of $f(x)=\left|x^2-5x+6\right|$ at $x=2$ is equal to zero.
Reason: Sum of left-hand derivative and right-hand derivative of $f(x) = |(x-a)(x-b)|$ at $x = a$ $(a < b)$ is equal to zero, (where $a, b \in R$)

6. **Assertion:** If $f: R \rightarrow R$ is a continuous function such that $f(x) = f(3x)$ $\forall x \in R$, then f is constant function.
Reason: If f is continuous at $x = \lim_{x\to a} g(x)$, then $\lim_{x\to a} f(g(x)) = f\left(\lim_{x\to a} g(x)\right)$

7. **Assertion:** If f is continuous and differentiable in $(a-\delta, a+\delta)$, where $a, \delta \in R$ and $\delta > 0$, then $f'(x)$ is continuous at $x = a$.
Reason: Every differentiable function at $x = a$ is continuous at $x = a$.

JEE ADVANCED

Objective Questions

One Option Correct

1. Number of points where the function $f(x)$ = max $(|\tan x|, \cos|x|)$ is non-differentiable in the interval $(-\pi, \pi)$ is ________.
(a) 4 (b) 6
(c) 3 (d) 2

2. The function $f(x) = \text{maximum}\left\{\sqrt{x(2-x)}, 2-x\right\}$ is non-differentiable at x equal to ________.
(a) 1 (b) 0, 2
(c) 0, 1 (d) 1, 2

3. A function $f: R \rightarrow R$ satisfies $\sin x.\cos y.(f(2x+2y) - f(2x-2y)) = \cos x \sin y\,(f(2x+2y) + f(2x-2y))$
If $f'(0) = \frac{1}{2}$, then ________.
(a) $f''(x) = f(x) = 0$ (b) $4f''(x) + f(x) = 0$
(c) $f''(x) + f(x) = 0$ (d) $4f''(x) - f(x) = 0$

4. Let $f: R \rightarrow R$ be any function and $g(x) = \frac{1}{f(x)}$. Then g is
(a) onto if f is onto.
(b) one–one if f is one–one.
(c) continuous if f is continuous.
(d) differentiable if f is differentiable.

5. If $f(x) = [x] (\sin kx)^p$ is continuous for real x, then ________.
(a) $k \in \{n\pi, n \in I\}, p > 0$
(b) $k \in \{2n\pi, n \in I\}, p > 0$
(c) $k \in \{2n\pi, n \in I\}, p \in R - \{0\}$
(d) $k \in \{2n\pi, n \in I, n \neq 0\}, p \in R - \{0\}$

6. If $f(x) = \begin{cases} x+2 & x<0 \\ -x^2-x & 0 \le x < 1 \\ x & x \ge 1 \end{cases}$
then the number of points of discontinuity of $|f(x)|$ is ________.
(a) 1 (b) 2
(c) 3 (d) None of these.

7. $f(x) = \begin{cases} \dfrac{e^{e/x} - e^{-e/x}}{e^{1/x} + e^{-1/x}}, & x \neq 0 \\ k, & x = 0 \end{cases}$
(a) f is continuous at $x = 0$, when $k = 0$.
(b) f is not continuous at $x = 0$ for any real k.
(c) $\lim_{x\to 0} f(x)$ exists infinitely.
(d) None of these.

8. The correct statement for the function $f(x) = \begin{cases} x, & x \in Q \\ -x, & x \in R-Q \end{cases}$ is
(a) continuous everywhere.
(b) $f(x)$ is a periodic function.
(c) discontinuous everywhere except at $x = 0$.
(d) $f(x)$ is an even function.

9. If $f(x) = \text{sgn}(x)$ and $g(x) = x(1 - x^2)$, then the number of points of discontinuity of function $f(g(x))$ is
(a) exactly two.
(b) exactly three.
(c) finite and more than 3.
(d) infinitely many.

10. The value of Arg z + Arg $\bar{z}$, $z = x + iy, \forall x, y \in R$ is (Arg z stands for principal argument of z):
(a) 0
(b) Non-zero real number

(c) Any real number
(d) Can't say

11. If $f(x) = \text{maximum}\left(\cos, \frac{1}{2}, \{\sin x\}\right), 0 \le x \le 2\pi,$ where { . } represents fractional part function, then number of points at which $f(x)$ is continuous but not differentiable, is ______.

(a) 1 (b) 2
(c) 3 (d) 4

12. Function $f(x) = \begin{cases} 2x\tan x - \dfrac{\pi}{\cos x}, & x \ne \dfrac{\pi}{2} \\ k, & x = \dfrac{\pi}{2} \end{cases}$ is continuous at $x = x = \frac{\pi}{2}$ if $k =$ ______.

(a) −2
(b) 2
(c) $\frac{1}{2}$
(d) No such values of k exist.

13. If $f(x) = \begin{cases} x^2\{e^{1/x}\} & x \ne 0 \\ k & x = 0 \end{cases}$ is continuous at $x = 0$, then ({} denotes fractional part function)

(a) it is differentiable at $x = 0$.
(b) $k = 1$.
(c) continuous but not differentiable at $x = 0$.
(d) continuous everywhere in its domain.

14. S_1: If f is continuous and g is discontinuous at $x = a$, then $f(x)$. $g(x)$ is discontinuous at $x = a$.

S_2: $f(x) = \sqrt{2-x} + \sqrt{x-2}$ is not continuous at $x = 2$.

S_3: $e^{-|x|}$ is differentiable at $x = 0$.

S_4: If $f(x)$ is differentiable everywhere, then $|f|^2$ is differentiable everywhere.

(a) TTFF (b) TIFT
(c) FIFT (d) FFFT

15. S_1: If $\lim_{x\to a} f(x) = \lim_{x\to a}[f(x)]$ (where [ ] denotes greatest integer function) and $f(x)$ is non-constant continuous function, then $f(a)$ is an integer.

S_2: $\cos|x| + |x|$ is differentiable at $x = 0$.

S_3: If a function has a tangent at $x = a$ then it must be differentiable at $x = a$.

S_4: If $f(x)$ and $g(x)$ both are discontinuous at any point, then their composition may be differentiable at that point.

(a) FTFT (b) TFFT
(c) TFFF (d) FFFT

16. Consider the following statements:

S_1 : If $f'(a^+)$ and $f'(a^-)$ exist finitely at a point, then f is continuous at $x = a$.

S_2 : The function $f(x) = 3\tan 5x - 7$ is differentiable at all points in its domain.

S_3 : The existence of $\lim_{x\to c}(f(x) + g(x))$ does not imply the existence of $\lim_{x\to c} f(x)$ and $\lim_{x\to c} g(x)$.

S_4 : If $f(x) < g(x)$ then $f'(x) < g'(x)$.

State, in order, whether S_1, S_2, S_3, S_4, are true or false.

(a) TTTF (b) FFTF
(c) TFTF (d) TFTT

17. Let $f(x) = \begin{cases} x^2\left|\cos\dfrac{\pi}{x}\right|, & x \ne 0, x \in R, \\ 0, & x = 0 \end{cases}$

then f is

(a) differentiable both at $x = 0$ and at $x = 2$
(b) differentiable at $x = 0$ but not differentiable at $x = 2$
(c) not differentiable at $x = 0$ but differentiable at $x = 2$
(d) differentiable neither at $x = 0$ nor at $x = 2$.

One or More Than One Option Correct

1. The function $f(x) = \begin{cases} [x^2] + \text{sgn}(x-1), & x \le 2 \\ \dfrac{\sin\{2x-3\}}{x-2}, & x > 2 \end{cases}$, where [.] and {.} denote GIF and FPF respectively is

(a) discontinuous at $x = 1$.
(b) differentiable at $x = 2$.
(c) continuous and differentiable at $x = 2$.
(d) discontinuous at $x = 2$.

2. $f(x) = \sin^{-1}(3x - 4x^3)$ is

(a) always differentiable
(b) non-differentiable at 2 points
(c) not continuous at 2 points
(d) not differentiable at 3 points

3. Let 'f' be a function whose graph is obtained by summing the ordinate values of the graph of functions 'g' and 'h' where 'g' is obtained by shifting graph of $\phi(x) = \begin{cases} 1-|x|, & |x| \le 1 \\ 0, & |x| > 1 \end{cases}$ by one unit leftwards and 'h' is obtained by shifting the graph of $\phi(x)$ by one unit rightwards. Then $f(x)$ is

(a) continuous everywhere.
(b) differentiable everywhere.
(c) non-differentiable at 3 points.
(d) non-differentiable at 5 points.

4. Which of the following function(s) has/have removable discontinuity at $x = 1$?
(a) $f(x)=\dfrac{1}{\ell n|x|}$
(b) $f(x)=\dfrac{x^2-1}{x^3-1}$
(c) $f(x)=2^{-2^{\frac{1}{1-x}}}$
(d) $f(x)=\dfrac{\sqrt{x+1}-\sqrt{2x}}{x^2-x}$

5. A function $f(x)$ satisfies the relation $f(x + y) = f(x) + f(y) + xy(x + y)\ \forall\ x, y \in R$. If $f'(0) = 1$, then
(a) $f(x)$ is a polynomial function.
(b) $f(x)$ is an exponential function.
(c) $f(x)$ is twice differentiable for all $x \in R$.
(d) $f''(3) = 8$

6. Let $f(x)=\int_{-2}^{x}|t+1|\,dt$, then
(a) $f(x)$ is continuous in $[-1, 1]$.
(b) $f(x)$ is differentiable in $[-1, 1]$.
(c) $f'(x)$ is continuous in $[-1, 1]$.
(d) $f'(x)$ is differentiable in $[-1, 1]$.

7. If $f(x)=\dfrac{[x]+1}{\{x\}+1}$ for $f:\left[0,\dfrac{5}{2}\right)\to\left(\dfrac{1}{2},3\right]$, where $[\,.\,]$ represents greatest integer function and $\{\,.\,\}$ represents fractional part of x, then which of the following is true?
(a) $f(x)$ is injective discontinuous function.
(b) $f(x)$ is subjective non-differentiable function.
(c) $\min\left(\lim\limits_{x\to1^-}f(x),\ \lim\limits_{x\to1^+}f(x)\right)=f(1)$.
(d) max (x values of point of discontinuity) = $f(1)$.

8. If $f(x) = 0$ for $x < 0$ and $f(x)$ is differentiable at $x = 0$, then for $x > 0$, $f(x)$ may be
(a) x^2 (b) x
(c) $\sin x$ (d) $-x^{3/2}$

Questions from Previous Years

9. Let the function $f:R\to R$ be defined by $f(x)=x^3-x^2+(x-1)\sin x$ and let $g: R\to R$ be an arbitrary function. Let $fg:R\to R$ be the product function defined by $(fg)(x)=f(x)g(x)$. Then which of the following statements is/are true?
(a) If g is continuous at $x=1$, then fg is differentiable at $x=1$.
(b) If fg is differentiable at $x=1$, then g is continuous at $x=1$.
(c) If g is differentiable at $x=1$, then fg is differentiable at $x=1$.
(d) If fg is differentiable at $x=1$, then g is differentiable at $x=1$.

10. Let $f:R\to R$ given by

$$f(x)\begin{cases} x^5+5x^4+10x^3+10x^2+3x+1, & x<0; \\ x^2-x+1, & 0\le x<1; \\ \dfrac{2}{3}x^3-4x^2+7x-\dfrac{8}{3}, & 1\le x<3; \\ (x-2)\log_e(x-2), & x\ge 3 \end{cases}$$

Then which of the following options is/are correct?
(a) f' has a local maximum at $x = 1$.
(b) f is increasing on $(-\infty,0)$.
(c) f' is NOT differentiable at $x = 1$.
(d) f is onto.

11. Let $f:R\to R$ and $g:R\to R$ be two non-constant differentiable functions. If

$f'(x)=\left(e^{(f(x)-g(x))}\right)g'(x)$ for all $x\in R$,

and $f(1)=g(2)=1$, then which of the following statement (s) is (are) TRUE?
(a) $f(2)<1-\log_e 2$ (b) $f(2)>1-\log_e 2$
(c) $g(1)>1-\log_e 2$ (d) $g(1)<1-\log_e 2$

12. Let $f;\left[-\dfrac{1}{2},2\right]\to R$ and $g:\left[-\dfrac{1}{2},2\right]\to R$ be functions defined by $f(x)=\left[x^2-3\right]$ and $g(x)=|x|f(x)+|4x-7|f(x)$, where $[y]$ denotes the greatest integer less than or equal to y for $y\in R$. Then
(a) f is discontinuous exactly at three points in $\left[-\dfrac{1}{2},2\right]$.
(b) f is discontinuous exactly at four points in $\left[-\dfrac{1}{2},2\right]$.

(c) g is NOT differentiable exactly at four points in $\left(-\frac{1}{2}, 2\right)$.

(d) g is NOT differentiable exactly at five points in $\left(-\frac{1}{2}, 2\right)$.

13. Let $a, b \in R$ and $f : R \to R$ be defined by $f(x) = a\cos\left(\left|x^3 - x\right|\right) + b|x|\sin\left(\left|x^3 + x\right|\right)$

Then f is

(a) differentiable at $x = 0$ if $a = 0$ and $b = 1$.

(b) differentiable at $x = 1$ if $a = 1$ and $b = 0$.

(c) NOT differentiable at $x = 0$ if $a = 1$ and $b = 0$.

(d) NOT differentiable at $x = 1$ if $a = 1$ and $b = 1$.

14. Let $g : R \to R$ be a differentiable function with $g(0) = 0$, $g'(0) = 0$ and $g'(1) \neq 0$. Let $f(x) = \begin{cases} \frac{x}{|x|} g(x), & x \neq 0 \\ 0, & x = 0 \end{cases}$ and $h(x) = e^{|x|}$ for all $x \in R$. Let $(foh)(x)$ denote $f(h(x))$ and $(hof)(x)$ denote $h(f(x))$. Then which of the following is (are) true?

(a) f is differentiable at $x = 0$.

(b) h is differentiable at $x = 0$.

(c) foh is differentiable at $x = 0$.

(d) hof is differentiable at $x = 0$.

15. For any positive integer n, define $f_n : (0, \infty) \to R$ as $f_n(x) = \sum_{j=1}^{n} \tan^{-1}\left(\frac{1}{1 + (x + j)(x + j - 1)}\right)$ for all $x \in (0, \infty)$. (Here, the inverse trigonometric function $\tan^{-1}x$ assumes values in $\left(-\frac{\pi}{2}, \frac{\pi}{2}\right)$.) Then, which of the following statement(s) is (are) TRUE?

(a) $\sum_{j=1}^{5} \tan(f_j(0) = 55$

(b) $\sum_{j=1}^{10} (1 + f_j'(0)) \sec(f_j(0)) = 10$

(c) For any fixed positive integer n, $\lim_{x \to \infty} \tan(f_n(x)) = \frac{1}{n}$.

(d) For any fixed positive integer n, $\lim_{x \to \infty} \sec^2(f_n(x)) = 1$.

16. Let $[x]$ be the greatest integer less than or equal to x. Then, at which of the following point(s) the function $f(x) = x\cos(\pi(x + [x]))$ is discontinuous?

(a) $x = -1$ (b) $x = 1$

(c) $x = 0$ (d) $x = 2$

17. For every pair of continuous function $f, g : [0, 1] \to R$ such that max $\{f(x) : x \in [0, 1]\}$ = max $\{g(x) : x \in [0, 1]\}$ the correct statement(s) is (are)

(a) $[f(c)]^2 + 3f(c) = [g(c)^2 + 3g(c)$ for some $c \in [0, 1]$

(b) $[f(c)]^2 + f(c) = [g(c)^2 + 3g(c)$ for some $c \in [0, 1]$

(c) $[f(c)]^2 + 3f(c) = [g(c)^2 + g(c)$ for some $c \in [0, 1]$

(d) $[f(c)]^2 = [g(c)^2$ for some $c \in [0, 1]$

18 Let

$$f(x) = \lim_{n \to \infty} \left[\frac{n^n (x + n)\left(x + \frac{n}{2}\right)\cdots\left(x + \frac{n}{n}\right)}{n!\left(x^2 + n^2\right)\left(x^2 + \frac{n^2}{4}\right)\cdots\left(x^2 + \frac{n^2}{n^2}\right)}\right]^{\frac{x}{n}},$$

for all $x = 0$. Then

(a) $f\left(\frac{1}{2}\right) \geq f(1)$ (b) $f\left(\frac{1}{3}\right) \leq f\left(\frac{2}{3}\right)$

(c) $f'(2) \leq 0$ (d) $\frac{f'(3)}{f(3)} \geq \frac{f'(2)}{f(2)}$

Matrix Type

1. Match Column I with Column II.

Column I	Column II
(A) Number of points of discontinuity of $f(x) = \tan^2 x - \sec^2 x$ in $(0, 2\pi)$ is	(P) 4
(B) Number of points at which $f(x) = \sin^{-1} x + \tan^{-1} x + \cot^{-1} x$ is non-differentiable in $(-1, 1)$ is	(Q) 3
(C) Number of points of discontinuity of $y = [\sin x]$, $x \in [0, 2\pi]$ where [.] represents greatest integer function	(R) 2
(D) Number of points where $y = \left|(x-1)^3\right| + \left|(x-2)^5\right| + \lvert x-3 \rvert$ is non-differentiable	(S) 1
	(T) 0

2. Match Column I with Column II.

Column I | **Column II**

(A) Number of points where the function

$$f(x)=\begin{cases} 1+\left[\cos\dfrac{\pi x}{2}\right] & 1<x\le 2 \\ 1-\{x\} & 0\le x<1 \\ |\sin \pi x| & -1\le x<0 \end{cases}$$

and $f(1)=0$ is continuous but non-differentiable

(B) $f(x)=\begin{cases} x^2 e^{1/x} & , \quad x\neq 0 \\ 0 & , \quad x=0 \end{cases}$, then $f'(0^-)=$

(C) The number of points at which $g(x)=\dfrac{1}{1+\dfrac{2}{f(x)}}$ is not differentiable, where $f(x)=\dfrac{1}{1+\dfrac{1}{x}}$

(D) Number of points where tangent does not exist for the curve $y=\text{sgn}(x^2-1)$

(P) 0

(Q) 1

(R) 2

(S) 3

(T) 4

3. Match Column I with Column II.

Column I | **Column II**

(A) $f(x) = |x^3|$ is

(B) $f(x) = \sqrt{|x|}$ is

(C) $f(x) = |\sin^{-1} x|$ is

(D) $f(x) = \cos^{-1} |x|$ is

(P) continuous in $(-1, 1)$

(Q) discontinuous in $(-1, 1)$

(R) differentiable in $(0, 1)$

(S) not differentiable at least at one point in $(-1, 1)$

(T) differentiable in $(-1, 1)$

Integer and Numerical Value Type

1. If $f(x)=\begin{cases} \dfrac{\tan\left[x^2\right]\pi}{ax^2}+ax^3+b & , \quad 0\le x \\ 2\cos \pi x+\tan^{-1} x & , \quad 1<x\le 2 \end{cases}$

differentiable in [0, 2], then $a = \dfrac{1}{k_1}$ and $b = \dfrac{\pi}{4}-\dfrac{26}{k_2}$. Find $k_1^2+k_2^2$ {where [.] denotes greatest integer function}.

2. Let $f(x) = \begin{cases} |x|^p \sin\dfrac{1}{x}+x|\tan|^q & , \quad x\neq 0 \\ 0 & , \quad x=0 \end{cases}$

be differentiable at $x = 0$, then find the least possible value of $[p + q]$, (where [.] represents greatest integer function).

3. If $f(x) = \sin^{-1} 2x\sqrt{1-x^2}$ and the values of $f'(1/2)+f'(-1/2)$ is $\dfrac{\lambda}{\sqrt{3}}$ then find λ.

4. Let $f(x)$ is differentiable function and $f(0) = 1$. Also if $f(x)$ satisfies $f(x + y +1) = \left(\sqrt{f(x)}+\sqrt{f(y)}\right)^2$ $\forall x, y \in R$ and $f(x) = a(x + 1)^b$ then find $(a + b)$.

Questions from Previous Years

5. Let $\alpha, \beta \in R$ be such that $\lim\limits_{x\to 0}\dfrac{x^2 \sin(□x)}{\alpha x-\sin x}=1$. Then, $6(\alpha+\beta)$ equals ______.

6. Let m and n be two positive integers greater than 1. If $\lim\limits_{\alpha\to 0}\left(\dfrac{e^{\cos(\pm^n)}-e}{\pm^m}\right)=-\left(\dfrac{e}{2}\right)$, then the value of $\dfrac{m}{n}$ is ______.

7. Let e denote the base of the natural logarithm. The value of the real number a for which the right-hand limit $\lim\limits_{x\to 0^+}\dfrac{(1-x)^{\frac{1}{x}}-e^{-1}}{x^a}$ is equal to a non-zero real number, is ______.

8. The largest value of the non-negative integer a for which $\lim\limits_{x\to 1}\left(\dfrac{-ax+\sin(x+1)+a}{x+\sin(x-1)-1}\right)^{\frac{1-x}{1-\sqrt{x}}}=\dfrac{1}{4}$ is ______.

Comprehension Passage

Comprehension 1

Let a function be defined as

$$f(x)=\begin{cases} [x] & , \quad -2\le x\le -\dfrac{1}{2} \\ 2x^2-1 & , \quad -\dfrac{1}{2}<x\le 2 \end{cases}$$ where [.] denotes greatest integer function.

Answer the following question by using the above information.

1. The number of points of discontinuity of $f(x)$ is ______.
   (a) 1 (b) 2
   (c) 3 (d) None of these.

2. The function $f(x - 1)$ is discontinuous at the points ______.
   (a) $-1, -\frac{1}{2}$ (b) $-\frac{1}{2}, 1$
   (c) $0, \frac{1}{2}$ (d) 0, 1

3. Number of points where $|f(x)|$ is not differentiable is ______.
   (a) 1 (b) 2
   (c) 3 (d) 4

Comprehension 2

Consider two functions $y = f(x)$ and $y = g(x)$ defined as $f(x) =$ ______. $\begin{cases} ax^2 + b & , \quad 0 \le x \le 1 \\ 2bx + 2b & , \quad 1 < x \le 3 \\ (a-1)x + 2a - 3 & , \quad 3 < x \le 4 \end{cases}$ and

$$g(x) = \begin{cases} cx^2 + d & , \quad 0 \le x \le 2 \\ dx + 3 - c & , \quad 2 < x < 3 \\ x^2 + b + 1 & , \quad 3 \le x \le 4 \end{cases}$$

1. $f(x)$ is continuous at $x = 1$ but not differentiable at $x = 1$, if
   (a) $a = 1, b = 0$ (b) $a = 1, b = 2$
   (c) $a = 3, b = 1$ (d) a and b are integers

2. $g(x)$ is continuous at $x = 2$, if ______.
   (a) $c = 1, d = 2$ (b) $c = 2, d = 3$
   (c) $c = 1, d = -1$ (d) $c = 1, d = 4$

3. If $f(x)$ is continuous and differentiable at x = 3, then
   (a) $a = -\frac{1}{3}, b = \frac{2}{3}$ (b) $a = \frac{2}{3}, b = -\frac{1}{3}$
   (c) $a = \frac{1}{3}, b = -\frac{2}{3}$ (d) $a = 2, b = \frac{1}{2}$

Comprehension 3

Left-hand derivative and right-hand derivative of a function $f(x)$ at a point $x = a$ are defined as

$$f'(a^-) = \lim_{h \to 0^+} \frac{f(a) - f(a-h)}{h} = \lim_{h \to 0^-} \frac{f(a+h) - f(a)}{h}$$

and

$$f'(a^+) = \lim_{h \to 0^+} \frac{f(a+h) - f(a)}{h} = \lim_{h \to 0^-} \frac{f(a) - f(a-h)}{h}$$

$$= \lim_{x \to a^+} \frac{f(a) - f(x)}{a - x}$$

respectively.

Let f be a twice differentiable function.

1. If f is odd, which of the following is left-hand derivative of f at $x = -a$?
   (a) $\lim_{h \to 0^-} \frac{f(a-h) - f(a)}{-h}$
   (b) $\lim_{h \to 0^-} \frac{f(h-a) - f(a)}{h}$
   (c) $\lim_{h \to 0^+} \frac{f(a) + f(a-h)}{-h}$
   (d) $\lim_{h \to 0^-} \frac{f(-a) - f(-a-h)}{-h}$

2. If f is even, which of the following is right-hand derivative of f' at $x = a$?
   (a) $\lim_{h \to 0^-} \frac{f'(a) + f'(-a+h)}{h}$
   (b) $\lim_{h \to 0^+} \frac{f'(a) + f'(-a-h)}{h}$
   (c) $\lim_{h \to 0^-} \frac{-f'(-a) + f'(-a-h)}{-h}$
   (d) $\lim_{h \to 0^+} \frac{f'(a) + f'(-a+h)}{-h}$

3. The statement
   $$\lim_{h \to 0} \frac{f(-x) - f(-x-h)}{h} = \lim_{h \to 0} \frac{f(x) - f(x-h)}{-h}$$
   implies that
   (a) f is odd.
   (b) f is even.
   (c) f is neither odd nor even.
   (d) Nothing can be concluded.

Comprehension 4

There are two systems S_1 and S_2 of definitions of limit and continuity. In system S_1 the definitions are as usual. In system S_2 the definition of limit is as usual but the continuity is defined as follows:

A function $f(x)$ is defined to be continuous at $x = a$ if

(i) $\left| \lim_{x \to a^-} f(x) - \lim_{x \to a^+} f(x) \right| \le 1$ and

(ii) $f(a)$ lies between the values of $\lim_{x \to a^-} f(x)$ and $\lim_{x \to a^+} f(x)$ if $\lim_{x \to a^-} f(x) \neq \lim_{x \to a^+} f(x)$, else $f(a) = \lim_{x \to a^-} f(x) = \lim_{x \to a^+} f(x)$

1. If $f(x) = \begin{cases} x + 2.7 & , \quad x < 0 \\ 2.9 & , \quad x = 0 \\ 2x + 3 & , \quad x > 0 \end{cases}$ and

$g(x) = \begin{cases} 3x + 3 & , \quad x < 0 \\ 2.8 & , \quad x = 0 \\ -x^2 + 2.7 & , \quad x > 0 \end{cases}$ then consider the following statements:

(i) $f(x)$.is discontinuous under the system S_1.
(ii) $f(x)$ is continuous under the system S_2.
(iii) $g(x)$ is continuous under the system S_2.
Which of the following options is correct?
(a) Only (i) is true.
(b) Only (i) and (ii) are true.
(c) Only (ii) and (iii) are true.
(d) (i), (ii), (iii) are true.

2. If each of $f(x)$ and $g(x)$ is continuous at $x = a$ in S_2, then in S_2 which of the following is continuous?
(a) $f + g$
(b) $f - g$
(c) $f \cdot g$
(d) None of these.

3. Which of the following is incorrect?
(a) A continuous function under the definition in S_1 must also be continuous under the definition in S_2.
(b) A continuous function under the definition in S_2 must also be continuous under the definition in S_1.
(c) A discontinuous function under the definition in S_1 must also be discontinuous under the definition in S_2.
(d) A discontinuous function under the definition in S_1 must be continuous under the definition in S_2.

ANSWER KEY

JEE Main

Objective Questions

One Option Correct

1. (b) **2.** (b) **3.** (c) **4.** (c) **5.** (b) **6.** (a) **7.** (b) **8.** (b) **9.** (c)
10. (b) **11.** (d) **12.** (c) **13.** (c) **14.** (d) **15.** (d) **16.** (b) **17.** (a) **18.** (a)
19. (c) **20.** (c) **21.** (d) **22.** (a) **23.** (b) **24.** (a) **25.** (d) **26.** (b) **27.** (a)
28. (a) **29.** (c) **30.** (b) **31.** (a) **32.** (a) **33.** (a) **34.** (d) **35.** (a) **36.** (c)
37. (c) **38.** (a) **39.** (d) **40.** (d) **41.** (a) **42.** (b) **43.** (a) **44.** (b) **45.** (a)
46. (d) **47.** (c) **48.** (d) **49.** (d) **50.** (c) **51.** (c) **52.** (c) **53.** (d) **54.** (d)
55. (b) **56.** (c) **57.** (a) **58.** (c) **59.** (c) **60.** (c) **61.** (b) **62.** (d) **63.** (a)
64. (c) **65.** (b) **66.** (b) **67.** (d) **68.** (b) **69.** (a) **70.** (b) **71.** (c) **72.** (a)
73. (b) **74.** (a) **75.** (c) **76.** (c) **77.** (a) **78.** (c) **79.** (b) **80.** (b) **81.** (b)
82. (a) **83.** (d) **84.** (a) **85.** (d) **86.** (b) **87.** (b) **88.** (b) **89.** (c) **90.** (b)
91. (c) **92.** (a) **93.** (a) **94.** (b) **95.** (a) **96.** (b) **97.** (a) **98.** (a) **99.** (d)
100. (a) **101.** (c) **102.** (b) **103.** (b) **104.** (b) **105.** (b) **106.** (c) **107.** (b) **108.** (b)
109. (b) **110.** (a) **111.** (d) **112.** (d) **113.** (c) **114.** (a) **115.** (d) **116.** (a) **117.** (c)
118. (b) **119.** (a) **120.** (a) **121.** (b) **122.** (a) **123.** (c) **124.** (a) **125.** (c) **126.** (a)
127. (d) **128.** (a) **129.** (b) **130.** (c) **131.** (d) **132.** (a) **133.** (c) **134.** (b) **135.** (a)
136. (b) **137.** (d) **138.** (c) **139.** (b) **140.** (d) **141.** (b) **142.** (b) **143.** (a) **144.** (c)
145. (c) **146.** (c) **147.** (d) **148.** (c) **149.** (b) **150.** (d)

Integer and Numerical Value Type

1. (–1) **2.** (1) **3.** $\left(\frac{1}{128r}\right)$ **4.** (–1) **5.** (7) **6.** $\left(\frac{2}{\pi}\right)$ **7.** (1) **8.** (False) **9.** ($\log_e 4$)
10. $a^2 \cos a + 2a \sin a$ **11.** (0) **12.** $\left(-\frac{1}{3}\right)$ **13.** (e^2) **14.** (e^5) **15.** (e^2) **16.** $\left(\sqrt{4 - x^2}\right)$
17. (discontinous at $x = 0, 1$) **18.** (14) **19.** (5) **20.** (4) **21.** $\left(1 \text{ and } -\frac{1}{2}\right)$ **22.** (40)

Assertion and Reason

1. (c) **2.** (a) **3.** (a) **4.** (a) **5.** (a) **6.** (a) **7.** (d)

JEE Advanced

Objective Questions

One Option Correct

1. (a) **2.** (d) **3.** (b) **4.** (b) **5.** (a) **6.** (a) **7.** (b) **8.** (c) **9.** (b)
10. (d) **11.** (d) **12.** (a) **13.** (a) **14.** (c) **15.** (b) **16.** (a) **17.** (b)

One or More Than One Option Correct

1. (a, d) **2.** (a, b, c, d) **3.** (a, d) **4.** (b, d)
5. (a, c, d) **6.** (a, b, c, d) **7.** (a, b, d) **8.** (a, d)
9. (a, c) **10.** (a, c, d) **11.** (b, c) **12.** (b, c)
13. (a, b) **14.** (a, d) **15.** (d) **16.** (a, b, d)
17. (a, d) **18.** (b, c)

Matrix Type

1. A → R; B → T; C → R; D → Q **2.** A → Q; B → P; C → S; D → P
3. A → P, T, R; B → P, R, S; C → P, R, S; D → P, R, S

Integer and Numerical Value Type

1. (180) **2.** (1) **3.** $\left(\frac{4}{\sqrt{3}}\right)$ **4.** $((x+1)^2)$ **5.** (7) **6.** (2) **7.** $\left(-\frac{1}{2e}\right)$ **8.** (2)

Comprehension Passage

Comprehension 1

1. (b) **2.** (c) **3.** (c)

Comprehension 2

1. (c) **2.** (a) **3.** (d)

Comprehension 3

1. (a) **2.** (a) **3.** (b)

Comprehension 4

1. (d) **2.** (d) **3.** (b)

HINTS AND EXPLANATIONS

Worksheet – 1

1. $$\lim_{x\to 1}\left\{\frac{x-1}{(x-1)(2x-5)}\right\} = \lim_{x\to 1}\frac{1}{(2x-5)} = -\frac{1}{3}$$

2. $$\lim_{x\to 0}\sqrt{\frac{x-\sin x}{x+\cos^2 x}} = \frac{\lim\limits_{x\to 0}(x-\sin x)^{\frac{1}{2}}}{\lim\limits_{x\to 0}(x+\cos^2 x)^{\frac{1}{2}}}$$

$$= \frac{\lim\limits_{x\to 0}\left[x\left(1-\frac{\sin x}{x}\right)\right]^{1/2}}{\lim\limits_{x\to 0}(0+1)^{1/2}} = \frac{0.0}{1} = 0$$

3. $$\lim_{x\to -\infty}\left(\frac{x^4\sin\frac{1}{x}+x^2}{1+|x|^3}\right) = \lim_{x\to -\infty}\frac{x^4\sin\frac{1}{x}+x^2}{1-x^3}$$

On dividing by x^3, we get

$$\lim_{x\to -\infty}\frac{\frac{\sin(1/x)}{\frac{1}{x}}+\frac{1}{x}}{\frac{1}{x^3}-1} = \frac{1+0}{0-1} = -1.$$

4. Given $f(x)=\begin{cases}\sin x, & x\neq n\pi, \quad n=0,\pm1,\pm2 \ldots\\ 2, & \text{otherwise}\end{cases}$

$$g[f(x)]=\begin{cases}\{f(x)\}^2+1, & f(x)\neq 0,2\\ 4, & f(x)=2\\ 5, & f(x)=2\end{cases}$$

$$\therefore g[f(x)]=\begin{cases}(\sin^2 x)+1, & x\neq, n\pi=0,\pm1,\ldots\\ 5, & x=n\pi\end{cases}$$

Now, $\lim\limits_{x\to 0} g[f(x)]=\lim\limits_{x\to 0}(\sin^2 x)+1=1.$

5. $\lim\limits_{x\to 0^+} f(x)=\lim\limits_{h\to 0} f(0+h)=\lim\limits_{h\to 0} f(h)$

$=\lim\limits_{h\to 0}\dfrac{e^{1/h}-1}{e^{1/h}+1}=\lim\limits_{h\to 0}\dfrac{1-e^{-1/h}}{1+e^{-1/h}}=\dfrac{1-0}{1+0}$ (since $h>0$)

$\lim\limits_{x\to 0^-} f(x)=\lim\limits_{h\to 0} f(0-h)=\lim\limits_{h\to 0} f(-h)$

$=\lim\limits_{h\to 0}\dfrac{e^{-1/h}-1}{e^{-1/h}+1}=\dfrac{0-1}{0+1}=-1.$

Since RHL $\neq$ LHL, $f(x)$ cannot be continuous at 0, whatever be the value of $f(0)$.

6. $\lim\limits_{x\to 3^+} f(x)=\lim\limits_{h\to 0} f(h+3)=\lim\limits_{h\to 0}\dfrac{|h|}{h}=\lim\limits_{h\to 0}\dfrac{h}{h}=1$

$\lim\limits_{x\to 3^-} f(x)=\lim\limits_{h\to 0} f(3-h)=\lim\limits_{h\to 0}\dfrac{h}{-h}=-1.$

Thus $f(x)$ is discontinuous at $x=3$.

7. $\lim\limits_{x\to 3^+} f(x)=\lim\limits_{h\to 0} f(h+3)=\lim\limits_{h\to 0} 10^{-1/h^2}=0$

$\lim\limits_{x\to 3^-} f(x)=\lim\limits_{h\to 0} f(3-h)=\lim\limits_{h\to 0} 10^{-1/h^2}=0$

Thus RHL = LHL and $f(0)=0$ (given) $\Rightarrow f(x)$ is continuous at $x=3$.

8. If n is an odd integer,

$\lim\limits_{x\to n^-}(-1)^{[x]}=-1,\ \lim\limits_{x\to n^-}(-1)^{[x]}=+1$

and if n is an even integer, then

$\lim\limits_{x\to n^-}(-1)^{[x]}=+1,\ \lim\limits_{x\to n^-}(-1)^{[x]}=-1$

Thus $f(x)$ is discontinuous at all integer points.

(9 – 11) Test the following functions for continuity:

9. A function f representing a ratio of two continuous functions (polynomials in this case) is discontinuous only at points for which the denominator becomes zero. But in our case,

$x^4+4x^3+8x^2+8x+4=(x^2+2x+2)^2,$

and since $x^2+2x+2=(x+1)^a+1>0$ at any x, the denominator never becomes zero. Hence, the function $f(x)$ is continuous throughout the entire number scale.

10. The function $f(x)$ suffers discontinuities only at points for which the denominator equals zero, i.e., at points which are the roots of the equation

$4\cos x-2=0$ or $\cos x=1/2$.

Hence,

$x=x_n=\pm\pi/3+2\pi n\ \ (n=0,\pm1,\pm2,\ldots).$

Thus, the function $f(x)$ is continuous everywhere, except at the point x_n.

11. Just as in the preceding question, the numerator is continuous throughout the entire number scale. As far as the denominator is concerned, according to the theorem on continuity of a composite, it is continuous since the function $\cos u$ is continuous everywhere. Hence, the denominator is continuous everywhere, except at the points $x=k\pi$ (k an integer). Besides, we must exclude the points at which $\cos(1/\sin x)=0$, i.e, the points at which $1/\sin x=(2p+1)\pi/2$ (p is an integer), or $\sin x=2/[(2p+1)\pi]$. Thus, the function $f(x)$ is continuous everywhere except at the points $x=k\pi$ and $x=(-1)^n$ arc $\sin\dfrac{2}{(2p+1)\pi}+n\pi(k,p,n=0,\pm1,\pm2,\ldots).$

12. At the end-points of the interval $[-2,+2]$, the given function has different signs:

$f(-2)=+6;\ f(+2)=-6.$

But it is easy to notice that it does not become zero at any point of the interval $[-2,+2]$. Indeed, $x^2+2>0$ and $-(x^2+2)<0$ at any x; this is due to the fact that $f(x)$ has a discontinuity at the point $x=0$.

Worksheet – 2

1. Let $y=\sin(\tan^{-1}e^{-x})$

Putting $y=\sin u$, $u=\tan^{-1}t$, and $t=e^{-x}$, we get

$\dfrac{dy}{du}=\cos u,\ \dfrac{du}{dt}=\dfrac{1}{1+t^2}$ and $\dfrac{dt}{dx}=-e^{-x}$

Now, $\dfrac{dy}{dx}=\dfrac{dy}{du}\times\dfrac{du}{dt}\times\dfrac{dt}{dx}$

$$= \cos u \times \frac{1}{1+t^2} \times \left(-e^{-x}\right)$$

$$= \cos\left(\tan^{-1} t\right) \times \frac{1}{1+e^{-2x}}\left(-e^{-x}\right)$$

$$= \cos\left(\tan^{-1} e^{-x}\right) \times \frac{1}{1+e^{-2x}}\left(-e^{-x}\right)$$

$$= \frac{-e^{-x}\cos\left(\tan^{-1} e^{-x}\right)}{1+e^{-2x}}.$$

2. Domain of $\frac{1}{\sqrt{(x-1)}}$ is $(1, \infty)$ and when $x > 1$,

$$\sin^{-1}\left(\frac{2x}{1+x^2}\right) = \pi - \cos^{-1}\left(\frac{1-x^2}{1+x^2}\right) \quad \ldots\text{(i)}$$

Let $u = \sin^{-1}\left(\frac{2x}{1+x^2}\right)$ and $v = \cos^{-1}\left(\frac{1-x^2}{1+x^2}\right)$

From (i), $u = \pi - v \Rightarrow \frac{du}{dv} = -1.$

3. Let $\quad u = \tan^{-1} x\left(\frac{\sqrt{(1+x^2)}-1}{x}\right) \quad \ldots\text{(i)}$

and $v = \tan^{-1} x \Rightarrow x = \tan v \quad \ldots\text{(ii)}$

From (i) and (ii), we have

$$u = \tan^{-1}\left(\frac{\sqrt{(1+\tan^2 v)}-1}{\tan v}\right)$$

$$= \tan^{-1}\left(\frac{\sec v - 1}{\tan v}\right) = \tan^{-1}\left(\frac{1-\cos v}{\sin v}\right)$$

$$= \tan^{-1}\left(\frac{2\sin^2 v/2}{2\sin v/2 \cos v/2}\right)$$

$$= \tan^{-1}\left(\tan(v/2)\right)$$

$$= \frac{v}{2}$$

$$\Rightarrow u = \frac{v}{2}$$

$$\therefore \frac{du}{dv} = \frac{1}{2}$$

4. Let $u = y^2 = \left(x - x^2\right)^2 \quad \ldots\text{(i)}$

And $v = x^2 \Rightarrow x = \sqrt{v} \quad \ldots\text{(ii)}$

From (i) and (ii), we have

$$u = \left(\sqrt{v} - v\right)^2$$

$$= v + v^2 - 2v^{3/2}$$

$$\therefore \quad \frac{du}{dv} = 1 + 2v - 3\sqrt{v}$$

$$= 1 + 2x^2 - 3x \qquad \left(\because v = x^2\right)$$

Hence, $\frac{du}{dv} = 1 - 3x + 2x^2.$

5. Let $u = f(\tan x)$

$$\therefore \frac{du}{dx} = f'(\tan x)\cdot \sec^2 x \quad \ldots\text{(i)}$$

and let $v = g(\sec x)$

$$\frac{dv}{dx} = g'(\sec x)\cdot \sec x \tan x \quad \ldots\text{(ii)}$$

From (i) and (ii), we have

$$\frac{du}{dv} = \frac{\left(\frac{du}{dx}\right)}{\left(\frac{dv}{dx}\right)} = \frac{f'(\tan x)\cdot \sec^2 x}{g'(\sec x)\cdot \sec x \tan x}$$

$$= \frac{f'(\tan x)}{g'(\sec x)}\cdot\frac{1}{\sin x}$$

$$\therefore \left.\frac{du}{dv}\right|_{x=\pi/4} = \frac{f'(1)}{g'(\sqrt{2})}\cdot\frac{1}{(1/\sqrt{2})} = \frac{2\cdot\sqrt{2}}{4} = \frac{1}{\sqrt{2}}$$

Note: In general, if $y = f(u)$, then

$$\frac{dy}{dx} = f'(u)\cdot\frac{du}{dx}.$$

Worksheet – 3

1. We have $y = x \sin y$

$$\Rightarrow x = \frac{y}{\sin y} \quad \ldots\text{(i)}$$

Differentiating both sides w.r.t. y, we have

$$\frac{dx}{dy} = \frac{\sin y \cdot 1 - y\cdot \cos y}{\sin^2 y} \text{ (By quotient rule)}$$

$$\Rightarrow \frac{dy}{dx} = \frac{\sin^2 y}{(\sin y - y\cos y)} = \frac{\sin^2 y}{y\left(\frac{\sin y}{y} - \cos y\right)}$$

$$= \frac{\left(\frac{\sin y}{y}\right)^2 \cdot y}{\left(\frac{\sin y}{y}\right) - \cos y}$$

$$= \frac{\frac{1}{x^2}\cdot y}{\left(\frac{1}{x} - \cos y\right)} \quad \text{(From (i))}$$

$$=\frac{y}{x(1-x\cos y)}$$

Hence $x\frac{dy}{dx}=\frac{y}{(1-x\cos y)}$.

2. $\sqrt{(1+y^4)}=\sqrt{\left(1+\frac{1}{x^4}\right)}\qquad \left(\because y=\frac{1}{x}\right)$

$$\Rightarrow \frac{\sqrt{1+y^4}}{\sqrt{1+x^4}}=\frac{1}{x^2} \quad \ldots(i)$$

But $y=\frac{1}{x}\quad \therefore \frac{dy}{dx1}=-\frac{1}{x^2}$ ...(ii)

From (i) and (ii), $\frac{\sqrt{(1+y^4)}}{\sqrt{(1+x^4)}}=-\frac{dy}{dx}$

$$\Rightarrow \frac{dy}{\sqrt{(1+y^4)}}=-\frac{dx}{\sqrt{(1+x^4)}}$$

$$\Rightarrow \frac{dy}{\sqrt{(1+y^4)}}+\frac{dx}{\sqrt{(1+x^4)}}=0.$$

3. Since $x=a(\cos\theta+\theta\sin\theta)$

$$\therefore \frac{dx}{d\theta}=a(-\sin\theta+\theta\cos\theta+\sin\theta)=a\theta\cos\theta$$

And $y=a(\sin\theta-\theta cos\theta)$

$$\therefore \frac{dy}{d\theta}=a(\cos\theta-\theta(-\sin\theta)-\cos\theta\cdot 1)=a\theta\sin\theta$$

Hence $\frac{dy}{dx}=\frac{\left(\frac{dy}{d\theta}\right)}{\left(\frac{dx}{d\theta}\right)}=\frac{a\theta\sin\theta}{a\theta\cos\theta}=\tan\theta.$

4. Since $x=a\left(\cos t+\ln\tan\frac{t}{2}\right)$

$$\therefore \frac{dx}{dt}=a(-\sin t+\text{cosec}\, t)\ \left\{\because \int \text{cosec}\, dt=\ln\tan\frac{t}{2}\right\}$$

$$=a\left(\frac{1-\sin^2 t}{\sin t}\right)=a\frac{\cos^2 t}{\sin t}$$

and $y=a\sin t\quad \therefore \frac{dy}{dt}=a\cos t$

Hence $\frac{dy}{dx}=\frac{\left(\frac{dy}{dt}\right)}{\left(\frac{dx}{dt}\right)}=\frac{a\cos t}{\left(a\frac{\cos^2 t}{\sin t}\right)}=\tan t.$

5. Since $x=e^t\left(t+\frac{1}{t}\right)$

$$\therefore \frac{dx}{dt}=e^t\left(1-\frac{1}{t^2}\right)+\left(t+\frac{1}{t}\right)e^t$$

$$=e^t\left(1-\frac{1}{t^2}+t+\frac{1}{t}\right)=e^t\left(\frac{t^2+t^3+t-1}{t^2}\right)$$

and $y=e^{-t}\left(t-\frac{1}{t}\right)$

$$\therefore \frac{dy}{dt}=e^{-t}\left(1+\frac{1}{t^2}\right)+\left(t-\frac{1}{t}\right)\left(-e^{-t}\right)$$

$$=e^{-t}\left(1+\frac{1}{t^2}-t+\frac{1}{t}\right)=e^{-t}\left(\frac{t^2+1-t^3+t}{t^2}\right)$$

Now $\frac{dy}{dx}=\frac{\left(\frac{dy}{dt}\right)}{\left(\frac{dx}{dt}\right)}=e^{-2t}\left(\frac{1+t+t^2-t^3}{-1+t+t^2-t^3}\right).$

6. Given $x^y=e^{x-y}$.

Taking logarithm of both sides, we get

$y\ln x=x-y$

or $y(1+\ln x)=x\Rightarrow y=\frac{x}{(1+\ln x)}$

$$\therefore \frac{dy}{dx}=\frac{(1+\ln x)\cdot 1-x\cdot\left(\frac{1}{x}\right)}{(1+\ln x)^2}\ \text{(From quotient rule)}$$

$$=\frac{\ln x}{(1+\ln x)^2}$$

7. We have $y=\frac{(x+1)^2\sqrt{(x-2)}}{(x+4)^3 e^x}$ ...(i)

Here, y has product and quotient of many functions. Now taking logarithm on both sides, we get

$$\ln y=2\ln(x+1)+\frac{1}{2}\ln(x-2)-3\ln(x+4)-x$$

Differentiating both sides w.r.t. x, we get

$$\frac{1}{y}\frac{dy}{dx}=\frac{2}{(x+1)}+\frac{1}{2(x-2)}-\frac{3}{(x+4)}-1$$

$$\therefore \frac{dy}{dx}=y\left\{\frac{2}{x+1}+\frac{1}{2(x-2)}-\frac{3}{(x+4)}-1\right\}$$

Hence,

$$\frac{dy}{dx}=\frac{(x+1)^2\sqrt{(x-2)}}{(x+4)^3 e^x}$$

$$\left\{\frac{2}{x+1}+\frac{1}{2(x-2)}-\frac{3}{(x+4)}-1\right\} \text{[From (i)]}$$

8. We have $y=\left(1+\frac{1}{x}\right)^x+x^{(1+1/x)}+(\sin x)^{\tan x}$

$$=e^{\ln(1+1/x)^x}+e^{\ln x^{(1+1/x)}}+e^{\ln(\sin x)^{\tan x}}$$

$$=e^{x\ln(1+1/x)}+e^{(1+1/x)\ln x}+e^{\tan x \ln \sin x}$$

$$\therefore \frac{dy}{dx}=e^{x\ln(1+1/x)}\frac{d}{dx}\left\{x\cdot\ln\left(1+\frac{1}{x}\right)\right\}+e^{(1+1/x)\ln x}.$$

$$\frac{d}{dx}\left\{\left(1+\frac{1}{x}\right)\ln x\right\}$$

$$+e^{\tan x \ln \sin x}\frac{d}{dx}\{\tan x\cdot\ln\sin x\}$$

$$=\left(1+\frac{1}{x}\right)^x\left\{\frac{x}{\left(1+\frac{1}{x}\right)}+\ln\left(1+\frac{1}{x}\right)\cdot 1\right\}$$

$$+x^{\left(1+\frac{1}{x}\right)}\cdot\left\{\left(1+\frac{1}{x}\right)\cdot\frac{1}{x}+\ln x\cdot\left(-\frac{1}{x^2}\right)\right\}$$

$$+(\sin x)^{\tan x}\{\tan x\cdot\cot x+\ln\sin x\cdot\sec^2 x\}.$$

Hence, $\frac{dy}{dx}=\left(1+\frac{1}{x}\right)^x\left\{\frac{x^2}{(x+1)}+\ln\left(1+\frac{1}{x}\right)\right\}$

$$+x^{(1+1/x)}\left\{\frac{x+1-\ln x}{x^2}\right\}+(\sin x)^{\tan x}$$

$$\{1+\sec^2 x\cdot\ln\sin x\}$$

9. $f(x) = x(x-4)^2 = x^3 - 8x^2 + 16x$, which is a polynomial in x.
As polynomial function is continuous and differentiable at each real number, so $f(x)$ is continuous in [0, 4] and differentiable in (0, 4)
Also $f(0) = (0)(0-4)^2 = 0$ and $f(4) = 4(4-4)^2 = 0$
$\therefore f(0) = f(4)$.
Now, $f(x)$ satisfies all the three conditions of Rolle's Theorem in [0, 4], there must exist some $c\in(0,4)$ such that $f'(c) = 0$.
Now $f'(x)=3x^2-16x+16$

$$f'(x)=0\Rightarrow 3x^2-16x+16=0$$

$$\Rightarrow(x-4)(3x-4)=0$$

$$\Rightarrow x=4 \text{ or } \frac{4}{3}$$

so there exists $c=\frac{4}{3}\in(0,4)$ such that $f'(c)=0$.

Hence, Rolle's theorem is verified.

10. $f(x) = x(x-2) = x^2 - 2x$
$f(x)$ being a polynomial function, is continuous in the closed interval [1, 3].
$f'(x) = 2x - 2$
This exists in the open interval (1, 3).
$\therefore f(x)$ is derivable in the open interval (1, 3).
Thus, $f(x)$ satisfies both the conditions of LMV theorem.
$\therefore$ There must exist at least one number $c\in(1,3)$ such that

$$f'(c)=\frac{f(3)-f(1)}{3-1}\Rightarrow 2c-2=\frac{3-(-1)}{3-1}$$

$$\left[\because f(3)=(3)^2-2(3)=3, f(1)=(1)^2-2(1)=-1\right]$$

$$\Rightarrow 2c-2=2 \qquad \Rightarrow c=2$$

$$=2\in(-1,3)$$

Hence Lagrange's Mean Value theorem is verified.

Quiz

1. No
2. False
3. All
4. If is continuous in interval $[-1,1]$.
5. Yes
6. Continuous
   Since (x^3+1) is a polynomial function, it is continuous.
   Assuming $x^3+1=u$, we have $\sin u$, which is a continuous function.
7. An exponential function is a function of the form $f(x)=b^x$ where b is a positive real number and $b\neq 1$
   The domain of f is the set of all real numbers.
8. If the base of logarithmic function is e, then it is called natural logarithmic function.
9. A function where it is not possible to express the function in form $y=f(x)$, is known as implicit function. Example $x^2+y^2=2xy$

Here it is difficult to express y as a function of x. In general, way say that x and y are given implicity. Neither of one is expressed as an explicit function of the other.

10. Domain $(-\infty, \infty)$

Range $\left(-\frac{\pi}{2}, \frac{\pi}{2}\right)$

11. True

12. e^x

13. False

14. Let $y = \left[1-(x-5)^3\right]^{\frac{1}{5}}$

$y^5 = 1-(x-5)^3$

$(x-5)^3 = \left(1-y^5\right)$

$x = 5+(1-y^5)^{\frac{1}{3}}.$

JEE Main

Objective Questions

One Option Correct

1. (b)

Case I When $x \in I$

$f(x) = [x]^2 - \left[x^2\right] = x^2 - x^2 = 0$

Case II When $x \notin I$

If $0 < x < 1$, then $[x] = 0$

and $0 < x^2 < 1$, then $\left[x^2\right] = 0$

Next, if $1 \le x^2 < 2 \Rightarrow 1 \le x \le \sqrt{2}$

$\Rightarrow \quad [x] = 1$ and $\left[x^2\right] = 1.$

Therefore, $f(x) = [x]^2 - \left[x^2\right] = 0$, if $1 \le x < \sqrt{2}$.

Therefore, $f(x) = 0$, if $0 \le x < \sqrt{2}$.

This shows that $f(x)$ is discontinuous at $x = 1$.

Therefore, $f(x)$ is discontinuous in $(-\infty, 0) \cup \left[\sqrt{2}, \infty\right)$ on many other points,

Therefore, (b) is the answer.

2. (b) Given $f(x) = \left[\tan^2 x\right]$

Now, $-45° < x < 45°$

$\Rightarrow \tan(-45°) < \tan x < \tan(45°)$

$\Rightarrow -\tan 45° < \tan x < \tan(45°)$

$\Rightarrow -1 < \tan x < 1$

$\Rightarrow 0 < \tan^2 x < 1 \Rightarrow \left[\tan^2 x\right] = 0$

i.e., $f(x)$ is zero for all values of x from $x = -45°$ to $45°$

Thus, $f(x)$ exists when $x \to 0$ and also it is continuous at $x = 0$. Also $f(x)$ is differentiable at $x = 0$ and has a value of zero.

Therefore, (b) is the answer.

3. (c) Here, $f(x) = [x]\cos\left(\frac{2x-1}{2}\right)\pi$

$$\therefore f(x) = \begin{cases} -\cos\left(\frac{2x-1}{2}\right)\pi & , \quad -1 \le x < 0 \\ 0 & , \quad 0 \le x < 1 \\ \cos\left(\frac{2x-1}{2}\right)\pi & , \quad -1 \le x < 2 \\ 2\cos\left(\frac{2x-1}{2}\right)\pi & , \quad 2 \le x < 3 \end{cases}$$

which shows RHL = LHL at $x = n \in$ Integer as, if $x = 1$

$\Rightarrow \lim_{x \to 1^+} \cos\left(\frac{2x-1}{2}\right)\pi = 0$ and $\lim_{x \to 1^-} 0 = 0$

Also, $f(1) = 0$

$\therefore$ Continuous at $x = 1$

Similarly, when $x = 2$,

$\lim_{x \to 2^+} f(x) = \lim_{x \to 2^-} f(x) = 0$

Thus, function is discontinuous at no x.

Hence, option (c) is the correct answer.

4. (c) Given, $f(x) = x\left(\sqrt{x} + \sqrt{x+1}\right)$

$\Rightarrow f(x)$ would exist when $x \ge 0$ and $x+1 \ge 0$.

$\Rightarrow f(x)$ would exist when $x \ge 0$.

$\therefore f(x)$ is not continuous at $x = 0$ because LHL does not exist.

Hence, option (c) is correct.

5. (b) For $f(x)$ to be continuous, we must have

$f(0) = \lim_{x \to 0} f(x)$

$= \lim_{x \to 0} \frac{\log(1+ax) - \log(1-bx)}{x}$

$= \lim_{x \to 0} \frac{a\log(1+ax)}{ax} + \frac{b\log(1-bx)}{-bx}$

$= a.1 + b.1\left[\text{using } \lim_{x \to 0} \frac{\log(1+x)}{x} = 1\right]$

$= a + b$

$\therefore f(0) = (a+b).$

6. (a) $\lim_{x\to\frac{\pi}{4}} \frac{\int_2^{\sec^2 x} f(t)dt}{x^2-\frac{\pi^2}{16}}$

$= \lim_{x\to\frac{\pi}{4}} \frac{f(\sec^2 x)2\sec x \sec x \tan x}{2x}$

[using L' Hospital's rule]

$= \frac{2f(2)}{\pi/4} = \frac{8}{\pi}f(2).$

7. (b) Let

$I = \lim_{n\to\infty} \frac{1}{n}\sum_{r=1}^{2n} \frac{r}{\sqrt{n^2+r^2}} = \lim_{n\to\infty} \frac{1}{n}\sum_{r=1}^{2n} \frac{r}{n\sqrt{1+(r/n)^2}}$

$= \lim_{n\to\infty} \frac{1}{n}\sum_{r=1}^{2n} \frac{r/n}{\sqrt{1+(r/n)^2}}$

$= \int_0^2 \frac{x}{\sqrt{1+x^2}}dx = \left[\sqrt{1+x^2}\right]_0^2 = \sqrt{5}-1.$

8. (b) $\lambda = \lim_{x\to 0} \frac{\cos 3x - \cos x}{x^2}$

$= \lim_{x\to 0} \frac{-3\sin 3x + \sin x}{2x}$

$= -\frac{3}{2}\times 3 + \frac{1}{2} = -4$

$\Rightarrow$ (b) is correct.

9. (c) $\lim_{x\to 2} \frac{xf(2)-2f(x)}{x-2}$

$= \lim_{x\to 2} \frac{f(2)-2f'(x)}{1}$

$= 4-2\times 1 = 2$

$\Rightarrow$ (c) is correct.

10. (b) $\frac{dx}{d\theta} = a\left(-\sin\theta + \frac{1}{\tan\frac{\theta}{2}}.\sec^2\frac{\theta}{2}.\frac{1}{2}\right)$

$= a\left(\frac{1}{\sin\theta} - \sin\theta\right) = \frac{a\cos^2\theta}{\sin\theta}$

$\frac{dy}{d\theta} = a\cos\theta$

$\Rightarrow \frac{dy}{dx} = \tan\theta$

$\Rightarrow$ (b) is correct.

11. (d) $y = \sin(\log_e x)$

$\frac{dy}{dx} = \cos(\log x)\frac{1}{x}$

$x\frac{dy}{dx} = \cos(\log x)$

$\Rightarrow x\frac{d^2y}{dx^2} + \frac{dy}{dx} = -\sin(\log x)\frac{1}{x}$

$\Rightarrow x\frac{d^2y}{dx^2} + \frac{dy}{dx} = -\frac{y}{x}$

$\Rightarrow x\frac{d^2y}{dx^2} + x\frac{dy}{dx} = -y$

$\Rightarrow$ (d) is correct.

12. (c) Right hand derivative at 2

$= \left|\frac{d}{dx}(3x+4)\right|_{x=2}$

$= 3$

(c) is correct.

13. (c) $4x - 3xy' - 3y + 2yy' + 1 - 2y' = 0$

$\Rightarrow (4x-3y+1) = (3x-2y+1+2)y'$

$\Rightarrow$ (c) is correct.

14. (d) Given $x = \cos\theta$, $y = \sin 5\theta$

$\Rightarrow \frac{dx}{d\theta} = -\sin\theta,\ \frac{dy}{d\theta} = 5\cos 5\theta$

$\Rightarrow \frac{dy}{d\theta} = \frac{\frac{dy}{d\theta}}{\frac{dx}{d\theta}} = -\frac{5\cos 5\theta}{\sin\theta}$

Differentiating again w.r.t x, we get

$\frac{d^2y}{dx^2} = \frac{d}{d\theta}\left(\frac{dy}{dx}\right).\frac{d\theta}{dx}$

$= \frac{d}{d\theta}\left(\frac{-5\cos 5\theta}{\sin\theta}\right)\frac{1}{-\sin\theta}$

$= \left(\frac{\sin\theta\sin 5\theta\cdot 25 + 5\cos 5\theta\cos\theta}{\sin^2\theta}\right)\frac{1}{-\sin\theta}$

$= -\frac{25\sin 5\theta}{\sin^2\theta} - \frac{5\cos\theta\cos\theta}{\sin^3\theta}$

$\therefore (1-x^2)\frac{d^2y}{dx^2} - x\frac{dy}{dx}$

$= (1-\cos^2\theta)\left(\frac{-25\sin 5\theta}{\sin^2\theta} - \frac{5\cos 5\theta\cos\theta}{\sin^3\theta}\right) - \cos\theta\left(\frac{-5\cos 5\theta}{\sin\theta}\right)$

$$= \sin^2\theta\left(\frac{-25\sin 5\theta}{\sin^2\theta} - \frac{5\cos 5\theta\cos\theta}{\sin^3\theta}\right) + \frac{\frac{5\cos\theta\cos 5\theta}{\sin\theta}}{}$$

$$= -25\sin 5\theta - \frac{5\cos\theta\cos 5\theta}{\sin\theta} + \frac{5\cos\theta\cos 5\theta}{\sin\theta} = -25y$$

$\Rightarrow$ (d) is correct.

15. (d) From definition of continuity, we get

$$a = \lim_{x\to\pi/4}\frac{1-\sqrt{2}\sin x}{\pi - 4x}, \frac{0}{0} \text{ form.}$$

$$= \lim_{x\to\pi/4}\frac{-\sqrt{2}\cos x}{-4} = \frac{1}{4}.$$

16. (b) $f(x) = ax^{41} + bx^{-40}$

$f'(x) = 41ax^{40} - 40bx^{-41}$

$f''(x) = 41\times 40ax^{39} + 40\times 41bx^{-42}$

$= \dfrac{41\times 40}{x^2}\left[ax^{41} + bx^{-40}\right]$

$\Rightarrow$ Statement 1 is true.

Now $\dfrac{d}{dx}\tan^{-1}\left(\dfrac{2x}{1-x}\right) = \dfrac{1}{1+x^2}$ is also true putting $x = \tan\theta$ one can verify

Hence (b) is correct.

17. (a) $f(x) = \begin{cases} \dfrac{x}{1+x} & \forall x > 0 \\ 0 & for\, x = 0 \\ \dfrac{x}{1-x} & \forall x < 0 \end{cases} \Rightarrow f'(x) = \begin{cases} \dfrac{1}{(1+x)^2} \\ 0 \\ \dfrac{1}{(1-x)^2} \end{cases}$

Here $f'(0) = 0$

LHL $f'(0) = 0 = \text{RHL}\, f'(0) \Rightarrow f(x)$ is differentiable at $0. f(x)$ is differentiable at all other pts.

18. (a) Given $f'(a)$ exist then, $\lim\limits_{x\to a}\dfrac{xf(a) - f(x)}{x-a}$

$$= \lim_{x\to a}\frac{(x-a)f(a) - f(x) - f(x) + af(a)}{x-a}$$

$$= f(a) - \lim_{x\to a}\frac{f(x) - f(a) + af(a) + f(a)}{x-a}.$$

19. (c) $y = t^{10} + 1, x = t^8 + 1$

$\dfrac{dy}{dt} = 10t^9, \dfrac{dx}{dt} = 8t^7$

$\Rightarrow \dfrac{dy}{dx} = \dfrac{10}{8} = t^2.$

$$\frac{d^2y}{dx^2} = \frac{d}{dx}\left(\frac{dy}{dx}\right) = \frac{d}{dt}\left(\frac{dy}{dx}\right).\frac{dt}{dx}$$

$$= \frac{\frac{d}{dt}\left(\frac{10}{8}t^2\right)}{\frac{dx}{dt}} = \frac{\frac{20}{8}t}{8t^7} = \frac{5}{16t^6}.$$

20. (c) Taking logarithms,

$\log y = x + y$

Differentiation w.r.t. x both sides,

$\dfrac{1}{y}\dfrac{dy}{dx} = \dfrac{dy}{dx} + 1$

$\Rightarrow \left(\dfrac{1}{y} - 1\right)\dfrac{dy}{dx} = 1$

$\dfrac{dy}{dx} = \dfrac{y}{1-y}$

$\Rightarrow$ (c) is correct.

21. (d) $\cos^{-1}(\cos x) = x$ if $0 \le x \le \pi$

$\cos^{-1}(\cos x) = 2\pi - x\ \pi \le x \le 2\pi$

$\Rightarrow$ (d) is correct.

22. (a) $x = A\cos 4t$

$\dfrac{dx}{dt} = -4A\sin 4t$

$\Rightarrow \dfrac{d^2x}{dt^2} = -16A\cos 4t = -16x$

$= -16x$

$\Rightarrow$ (a) is correct.

23. (b) $f(x) = \log x$

$f(\sin x) = \log\sin x$

$\therefore \dfrac{d}{dx}(\log\sin x) = \dfrac{1}{\sin x}(\cos x) = \cot x$

$\Rightarrow$ (b) is correct.

24. (a) $f(x) = \begin{cases} \dfrac{1-\cos x}{x}, & \text{when } x \ne 0 \\ k, & \text{when } k = 0 \end{cases}$

$= \begin{cases} \dfrac{2\sin^2 x/2}{x}, & \text{when } x \ne 0 \\ k, & \text{when } k = 0 \end{cases}$

$= \begin{cases} \dfrac{2\sin^2\frac{x}{2}}{\left(\frac{x}{2}\right)^2}\times\dfrac{x}{4}, & \text{when } x \ne 0 \\ k, & \text{when } k = 0 \end{cases}$

$\Rightarrow k = 0.$

25. (d) $\tan y = \sec x - \tan x$

$$\frac{d}{dx}(\tan y) = \sec x \tan x - \sec^2 x$$

$$\Rightarrow \sec^2 y \frac{dy}{dx} = \sec x(\tan x - \sec x)$$

$$\Rightarrow \frac{dy}{dx} = \frac{\sec x(\tan x - \sec x)}{\sec^2 y}$$

$$= \frac{\sec x(\tan x - \sec x)}{1+(\sec x - \tan x)^2}$$

$$= \frac{\sec x(\tan x - \sec x)}{2\sec x(\sec x - \tan x)} = -\frac{1}{2}$$

$\Rightarrow$ (d) is correct.

26. (b) $f(x) = \begin{cases} \frac{1-\cos 4x}{8x^2}, & x \neq 0 \\ k, & x = 0 \end{cases} = \begin{cases} \left(\frac{\sin 2x}{2x}\right)^2 & x \neq 0 \\ k & x = 0 \end{cases}$

$\Rightarrow k = 1$

27. (a) $f(x) = \cos x \cos 2x \cos 4x \cos 8x \cos 16x$

$$= \frac{1}{2\sin x}(2\sin x \cos x)\cos 2x \cos 4x \cos 8x \cos 16x$$

$$= \frac{1}{2^2 \sin x}\sin 4x \cos 4x \cos 8x \cos 16x$$

$$= \frac{1}{2^5 \sin x}\sin 32x$$

Now $f'(x) = \frac{1}{2^5}\frac{32\sin x \cos 32x - \sin 32x \cos x}{(\sin x)^2}$

$$f'(\pi/4) = \frac{1}{2^5}\frac{32(\sin\pi/4)\cos(32\pi/4) - \sin(32\pi/4)\cos(\pi/4)}{(\sin\pi/4)^2}$$

$$= \frac{1}{2^5}\frac{32\frac{1}{\sqrt{2}} - 0}{\frac{1}{2}} = \sqrt{2}$$

$\Rightarrow$ (a) is correct.

28. (a) Let $f(x) = a\frac{x^2}{2}$ such that $f'(x) = ax$

and $f''(x) = a$, i.e., $f''(0) = a$

$$\therefore \lim_{x\to 0}\left[\frac{2f(x)}{x^2} - \frac{3f(2x)}{x^2} + \frac{f(4x)}{x^2}\right]$$

$$= \lim_{x\to 0}\left[\frac{2a}{x^2}\frac{x^2}{2} - \frac{3}{x^2}\frac{a4x^2}{2} + \frac{a}{2}\frac{a(4x)^2}{x^2}\right]$$

$$= \lim_{x\to 0}[a - 6a + 8a] = 3a.$$

29. (c) $\cos y = \frac{1-\log y}{1+\log x}$

$$\Rightarrow -\sin y\frac{dy}{dx} = \frac{(1+\tan x)\left(-\frac{1}{x}\right) - \frac{1}{x}(1-\tan x)}{(1+\tan x)^2}$$

$$\frac{dy}{dx} = \frac{-2/x}{(1+\log x)^2}\left(-\frac{1}{\sin y}\right)$$

$$\therefore \left.\frac{dy}{dx}\right|_{x=e} = \frac{-2/e}{(1+1)^2}\left(-\frac{1}{\sqrt{1-\left(\frac{1-\tan e}{1+\tan e}\right)^2}}\right)$$

$$= \frac{2}{e\times 4} = \frac{1}{2e}$$

$\Rightarrow$ (c) is correct.

30. (b) Here $f(x) = 2x^6 + 3x^4 + 4x^2$,

$f(-x) = f(x) \Rightarrow$ even function.

$\therefore f'(x) = 12x^5 + 12x^3 + 8x$

$f'(x) = -f'(x)\,\forall x \Rightarrow f'(x)$ is an odd function.

$\Rightarrow$ (b) is correct.

31. (a) $y = \tan^{-1}\left(\frac{\sqrt{1+x^2}-1}{x}\right)$

$$= \tan^{-1}\frac{1-\cos\theta}{\sin\theta}, \text{ when } x = \tan\theta$$

$$= \frac{\theta}{2} = \frac{1}{2}\tan^{-1}x$$

$$\therefore \left.\frac{dy}{dx}\right|_{x=0} = \left.\frac{1}{2(1+x^2)}\right|_{x=0} = \frac{1}{2}$$

$\Rightarrow$ (a) is correct.

32. (a) $\lim_{x\to 0} x\frac{1-\cos(1-\cos x)}{x^2}, \left(\frac{0}{0}\text{ form}\right)$

$$= \lim_{x\to 0}\frac{\sin(1-\cos x)}{4x^2}\frac{\sin x}{x}$$

$$= \lim_{x\to 0}\frac{\sin(2\sin^2 x/2)}{2\sin^2 x/2}\times\lim_{x\to 0}\frac{\sin\frac{x}{2}}{\left(\frac{x}{2}\right)^2}.\frac{2}{4}.\frac{1}{4}.$$

$$= \frac{1}{8}$$

$\Rightarrow$ (b) is correct.

33. (a) $\lim_{x\to 0} g(x)\frac{e^{1/x} - e^{-1/x}}{e^{1/x} + e^{-1/x}} = 0$

$$\left(\because \lim_{x\to 0^+}\frac{e^{1/x} - e^{-1/x}}{e^{1/x} + e^{-1/x}} = 1 \;\; \lim_{x\to 0^-}\frac{e^{1/x} - e^{-1/x}}{e^{1/x} + e^{-1/x}} = -1\right)$$

34. (d) For $-1<x<0$, $-1\le \sin(\pi x)\le 0$

$\Rightarrow 0\le x\sin\pi x<1$

$\Rightarrow [x\sin\pi x]=0$ for $-1<x<0$

For $0<x<1$, $0<\sin\pi x<1$

$\Rightarrow 0<x\sin\pi x<1$

$\Rightarrow [x\sin\pi x]=0$

Thus $f(x)=0$ for $-1<x\le 1$

when $1<x<2$, $-1<\sin\pi x\le 0$,

$\Rightarrow -1<x\sin\pi x<0$

$\Rightarrow [x\sin\pi x]=-1$

Since $f(x)=0$ for $-1<x\le 1$

f is differentiable in $(-1,1)$ and hence continuous in $(-1,1)$, and in particular when $x=0$. Note that f is not differentiable at $x=1$.

35. (a) Here $y=\dfrac{x}{1+\log x}$

$\Rightarrow y'=\dfrac{(1+\log x)-1}{(1+\log x)^2}$

$\Rightarrow y'|_{x=1}=0$

$\Rightarrow$ (a) is correct.

36. (c) $f(x)=2+(x-2)^{2/3}$

$\Rightarrow$ (c) is correct.

37. (c) Let $y=\tan^{-1}\sqrt{\dfrac{1+\cos x}{1-\cos x}}=\tan^{-1}\sqrt{\dfrac{2\cos^2 x/2}{2\sin^2 x/2}}$

$=\tan^{-1}\cot\dfrac{x}{2}=\tan^{-1}\tan\left(\dfrac{\pi}{2}-\dfrac{\pi}{2}\right)=\dfrac{\pi}{2}-\dfrac{\pi}{2}$

$\dfrac{dy}{dx}=-\dfrac{1}{2}$

$\Rightarrow$ (c) is correct.

38. (a) Here $\lim\limits_{x\to 0}\dfrac{\cos^2 x-\sin^2 x-1}{\sqrt{x^2+4}-2}=a$

$\Rightarrow \lim\limits_{x\to 0}\dfrac{-2\sin^2 x}{x^2}\times\left(\sqrt{x^2+4}+2\right)=a$

$\Rightarrow -8=a$

$\Rightarrow$ (a) is correct.

39. (d) $f(x)=\sin x$, $gof(x)=(\sin x)^2$, $g(x)=x^2$

$h(x)=\log_e x$, $F(x)=hogof(x)=\log_e(\sin x)^2$

$\therefore F'(x)=2\dfrac{1}{\sin x}\cos x=2\cot x$

$F''(x)=-2\operatorname{cosec}^2 x$

$\Rightarrow$ (d) is correct.

40. (d) $y=\cos(\sin x^2)$

$\left.\dfrac{dy}{dx}\right|_{x=\sqrt{\frac{\pi}{2}}}=\sin(\sin x^2)\times\cos^2\times 2x|_{x=\sqrt{\frac{\pi}{2}}}$

$=-2\sqrt{\dfrac{\pi}{2}}\times\cos\dfrac{\pi}{2}.\sin\left(\sin\dfrac{\pi}{2}\right)=0$

$\Rightarrow$ (d) is correct.

41. (a) Let $u=\cos^{-1}\cos\left(\dfrac{\pi}{2}-\sqrt{\dfrac{1+x}{2}}\right)=\dfrac{\pi}{2}-\sqrt{\dfrac{1+x}{2}}$

$\Rightarrow \left.\dfrac{du}{dx}\right|_{x=1}=-\dfrac{1}{4}$

and $v=x^x \Rightarrow \left.\dfrac{du}{dx}\right|_{x=1}=1$

$\Rightarrow$ (a) is correct.

42. (b) $f(x)=\dfrac{a\log(1+ax)}{ax}-(-b)\log\dfrac{1-bx}{-bx}$

$\lim\limits_{x\to 0}f(x)=a\lim\limits_{x\to 0}\dfrac{\log(1+ax)}{ax}-(-b)\lim\limits_{x\to 0}\log\dfrac{1-bx}{-bx}$

$=a+b$

From the definition of continuity, we get (b) is correct.

43. (a) $f(x)=p\cos\dfrac{2x-1}{2}\pi$ for $p\le x<p+1$ when $p\in Z$

$\lim\limits_{x\to p^-}f(x)=\lim\limits_{h\to p^-}(p-1)\cos\left[\dfrac{(2p-h)-1}{2}\right]$

$\pi.p-1\le x<p$

$=(p-1)\cos\left(\dfrac{2p-1}{2}\right)\pi=0$

Similarly, $\lim\limits_{x\to p^+}f(x)=0$ for $p\le x<p+1$

$f(p)=p\cos\left(\dfrac{2p-1}{2}\right)\pi=0\Rightarrow f(x)$ is continuous at all integer points.

$f(x)$ is a cosine function between two integer points.

Hence, for all x, $f(x)$ continuous, (a) is correct.

44. (b) Let $y=\sqrt{\sec\sqrt{x}}$

Then $y'=\dfrac{\sqrt{\sec\sqrt{x}}\tan\sqrt{x}}{2\sqrt{\sec\sqrt{x}}}\dfrac{1}{2\sqrt{x}}$

$=\dfrac{1}{4\sqrt{x}}\left(\sec\sqrt{x}\right)^{3/2}\sin\sqrt{x}$

$\Rightarrow$ (b) is correct.

45. (a) Let $y = e^{1+\log_e x} = e.x$

$\therefore y' = e.(a)$ is correct.

46. (d) We know all trigonometric functions are continuous and differentiable at its domain.

Hence (d) is correct.

47. (c) $\lim_{x\to 2^-} f(x) = \lim_{x\to 2^-}(x-1) = 1$

and $\lim_{x\to 2^+} f(x) = \lim_{x\to 2^+}(2x-3) = 1$

$\lim_{x\to 2^+} f(2) = 2.2 - 3 = 1$

Hence, $f(x)$ is continuous at $x = 2$.

$\Rightarrow$ (c) is correct.

48. (d) Here, $\lim_{h\to 0}\dfrac{f(2h+2+h^2)-f(2)}{f(h-h^2+1)-f(1)}$

$[\because f'(2) = 6 \text{ and } f'(1) = 4, \text{given}]$

Applying L′ Hospital rule,

$$= \lim_{h\to 0}\frac{\{f'(2h+2+h^2)\}\cdot(2+2h)-0}{\{f'(h-h^2+1)\}\cdot(1-2h)-0} = \frac{f'(2)\cdot 2}{f'(1)\cdot 1}$$

$$= \frac{6.2}{4.1} = 3\,[\text{using } f'(2) = 6 \text{ and } f'(1) = 4]$$

49. (d) Given, $\lim_{x\to 0}\dfrac{\{(a-n)nx-\tan x\}\sin nx}{x^2} = 0$

$$\Rightarrow \lim_{x\to 0}\left\{(a-n)n - \frac{\tan x}{x}\right\}\frac{\sin nx}{nx}\times n = 0$$

$$\Rightarrow \{(a-n)n-1\}n = 0 \Rightarrow (a-n)n = 1 \Rightarrow a = n + \frac{1}{n}.$$

50. (c) $\lim_{x\to 0}\dfrac{(\cos x-1)(\cos x - e^x)}{x^n}$

$$= \lim_{x\to 0}\frac{\left(-2\sin^2\frac{x}{2}\right)\left\{\left(1-\frac{x^2}{2!}+\frac{x^4}{4!}-\ldots\right)-\left(1+x+\frac{x^2}{2!}+\frac{x^3}{3!}+\ldots\right)\right\}}{x^n}$$

$$= \lim_{x\to 0}\frac{\left(-2\sin^2\frac{x}{2}\right)\left(-x-\frac{2x^2}{2!}-\frac{x^3}{3!}-\ldots\right)}{4\left(\frac{x}{2}\right)^2 x^{n-2}}$$

$$= \lim_{x\to 0}\frac{\sin^2\frac{x}{2}\left(1+x+\frac{x^2}{3!}+\ldots\right)}{2\left(\frac{x^2}{2}\right)x^{n-3}}$$

The above limit is finite, if $n-3=0$, i.e., $n=3$.

51. (c) $\lim_{x\to 0}\dfrac{x\tan 2x - 2x\tan x}{(1-\cos 2x)^2}$

Note: Try to make all trigonometric functions in the same angle. It is called 3rd golden rule of trigonometry.

$$= \lim_{x\to 0}\frac{x\dfrac{2\tan x}{1-\tan^2 x} - 2x\tan x}{(2\sin^2 x)^2}$$

$$= \lim_{x\to 0}\frac{2x\tan x\left[\dfrac{1}{1-\tan^2 x}-1\right]}{4\sin^4 x}$$

$$= \lim_{x\to 0}\frac{2x\tan x\left[\dfrac{1-1+\tan^2 x}{1-\tan^2 x}\right]}{4\sin^4 x}$$

$$= \lim_{x\to 0}\frac{2x\tan^3 x}{2\sin^4 x(1-\tan^2 x)} = \lim_{x\to 0}\frac{1}{2}\frac{x\left(\frac{\tan x}{x}\right)^3}{\sin^4 x(1-\tan^2 x)}$$

$$= \frac{1}{2}\lim_{x\to 0}\frac{\left(\frac{\tan x}{x}\right)^3}{\left(\frac{\sin x}{x}\right)^4(1-\tan^2 x)} = \frac{1.(1)^3}{2(1)^4(1-0)} = \frac{1}{2}.$$

52. (c) LHL $= \lim_{x\to 1^-}\dfrac{\sqrt{1-\cos 2(x-1)}}{x-1}$

$$= \lim_{x\to 1^-}\frac{\sqrt{2\sin^2(x-1)}}{x-1} = \sqrt{2}\lim_{x\to 1^-}\frac{|\sin(x-1)|}{x-1}$$

Put $x = 1-h, h>0$, for $x\to 1^-, h\to 0$

$$= \sqrt{2}\lim_{h\to 0}\frac{|\sin(-h)|}{-h} = \sqrt{2}\lim_{h\to 0}\frac{\sin h}{-h} = -\sqrt{2}$$

Again, RHL $= \lim_{x\to 1^+}\dfrac{\sqrt{1-\cos 2(x-1)}}{x-1}$

$$= \lim_{x\to 1^+}\sqrt{2}\frac{|\sin(x-1)|}{x-1}$$

Put $x = 1+h, h>0$

For $x\to 1^+, h\to 0$

$$= \lim_{h\to 0}\sqrt{2}\frac{|\sin h|}{h} = \lim_{h\to 0}\sqrt{2}\frac{\sin h}{h} = \sqrt{2}$$

$\therefore$ LHL $\neq$ RHL.

Hence, $\lim_{x\to 1} f(x)$ does not exist.

53. (d) $\lim_{x\to 0}\dfrac{\sqrt{\frac{1}{2}(1-\cos^2 x)}}{x} = \lim_{x\to 0}\dfrac{1}{\sqrt{2}}\cdot\dfrac{|\sin x|}{x}$

At $x = 0$

$$\text{RHL} = \lim_{h\to 0}\frac{1}{\sqrt{2}}.\frac{\sin h}{h} = \frac{1}{\sqrt{2}}$$

$$\text{and LHL} = \lim_{h\to 0}\frac{1}{\sqrt{2}}.\frac{\sin h}{-h} = -\frac{1}{\sqrt{2}}$$

Here, RHL ≠ LHL

∴ Limit does not exist.

54. (d) Since, $f(x)=\begin{cases}\dfrac{\sin[x]}{[x]}, & [x]\neq 0\\ 0, & [x]\neq 0\end{cases}$

$$\Rightarrow f(x)=\begin{cases}\dfrac{\sin[x]}{[x]}, & x\in R-[0,1)\\ 0, & 0\le x<1\end{cases}$$

At $x=0$

RHL $= \lim_{x\to 0^+} 0 = 0$

$$\text{and LHL} = \lim_{x\to 0^-}\frac{\sin[x]}{[x]} = \lim_{h\to 0}\frac{\sin[0-h]}{[0-h]}$$

$$= \lim_{h\to 0}\frac{\sin(-1)}{-1} = \sin 1$$

∴ Limit does not exist.

55. (b) $\lim_{n\to\infty}\left(\frac{1}{1-n^2}+\frac{2}{1-n^2}+\ldots+\frac{n}{1-n^2}\right)$

$$= \lim_{n\to\infty}\frac{1+2+3+\ldots+n}{(1-n^2)} = \lim_{n\to\infty}\frac{n(n+1)}{2(1-n)(1+n)}$$

$$\Rightarrow \lim_{n\to\infty}\frac{n}{2(1-n)} = -\frac{1}{2}.$$

56. (c) Given,

$f(a)=2,\ f'(a)=1,\ g(a)=-1,\ g'(a)=2$

$$\therefore \lim_{x\to a}\frac{g(x)f(a)-g(a)f(x)}{x-a}$$

$$= \lim_{x\to a}\frac{g'(x)f(a)-g(a)f'(x)}{1-0}$$

[using L′ Hospital's rule]

$= g'(a)f(a)-g(a)f'(a) = 2(2)-(-1)(1) = 5$

57. (a) Given, $G(x) = -\sqrt{25-x^2}$

$$\therefore \lim_{x\to 1}\frac{G(x)-G(1)}{x-1} = \lim_{x\to 1}\frac{G'(x)-0}{1-0}$$

[using L′ Hospital's rule]

$$= G'(1) = \frac{1}{\sqrt{24}}$$

$$\left[\because\ G(x) = -\sqrt{25-x^2} \Rightarrow G'(x) = \frac{2x}{2\sqrt{25-x^2}}\right]$$

58. (c) Here, $\lim_{x\to 0}(\sin x)^{1/x} + \lim_{x\to 0}\left(\frac{1}{x}\right)^{\sin x}$

$$= 0 + \lim_{x\to 0} e^{\log\left(\frac{1}{x}\right)^{\sin x}}$$

$$= e^{\lim_{x\to 0}\frac{\log(1/x)}{\operatorname{cosec} x}}\quad \left[\begin{array}{l}\lim_{x\to 0}(\sin x)^{1/x}\to 0\\ \text{as, (decimal)}^{\infty}\to 0\end{array}\right]$$

Applying L′ Hospital's rule, we get

$$e^{\lim_{x\to 0}\frac{x\left(-\frac{1}{x^2}\right)}{-\operatorname{cosec} x\cot x}} = e^{\lim_{x\to 0}\frac{\sin x}{x}\tan x} = e^0 = 1.$$

59. (c) Let $y = \left[\frac{f(1+x)}{f(1)}\right]^{\frac{1}{x}}$

$$\Rightarrow \log y = \frac{1}{x}\left[\log f(1+x) - \log f(1)\right]$$

$$\Rightarrow \lim_{x\to 0}\log y = \lim_{x\to 0}\left[\frac{1}{f(1+x)}.f'(1+x)\right]$$

$$= \frac{f(1)}{f(1)} = \frac{6}{3}$$

$$\Rightarrow \log\left(\lim_{x\to 0} y\right) = 2$$

$$\Rightarrow \lim_{x\to 0} y = e^2.$$

60. (c) For $x\in R$, $\lim_{x\to\infty}\left(\frac{x-3}{x+2}\right)^x$

$$= \lim_{x\to\infty}\frac{(1-3/x)^x}{(1+2/x)^x} = \frac{e^{-3}}{e^2} = e^{-5}.$$

61. (b) Let $t = x-1$

Since $\lim_{t\to 0} t\sin\frac{1}{t} = 0$ and

$$f'(t)\Big|_{t=0} = \lim_{h\to 0}\frac{f(0+h)-f(0)}{h}$$

$= \lim_{h\to 0}\sin\frac{1}{h}$, which does not exist.

Hence given $f(x)$ is not differentiable at $x=1$.

$f'(O^+) = \cos 1 - \sin 1 = f(O^-)$

Given $f(x)$ is differentiable at $x=0$.

62. (d) $\lim_{x\to y}\frac{|f(x)-f(y)|}{|x-y|} \le \lim_{x\to y}|x-y|$

$\Rightarrow |f'(y)| \le 0$

$f(y) = f(0) = 0$ (given)

$f(1) = 0.$

63. (a) $\lim_{x\to y}\frac{1}{h}f(1+h)=5\text{ (given)}$

$$f'(1)=\lim_{h\to 0}\frac{f(1+h)-f(1)}{h}$$

$$=5-\lim_{h\to 0}\frac{f(1)}{h}$$

as $f(x)$ is differentiable at $x=1$.

$\lim_{h\to 0}\frac{f(1)}{h}$ must be finite, i.e., only if $f(1)=0$.

64. (c) Taking logarithm on both sides,
$\log x=y+x$ when $x=e^{y+x}$
Differentiate w.r.t. x,

$$\frac{1}{x}=\frac{dy}{dx}+1\Rightarrow\frac{dy}{dx}=\frac{1-x}{x}.$$

65. (b) $x^2+y^2=t-\frac{1}{t}$

and $x^4+y^4=t^2+\frac{1}{t^2}=\left(t-\frac{1}{t}\right)^2+2$

$$=\left(x^2+y^2\right)^2+2$$

$$\Rightarrow x^2y^2=-1.$$

$$\Rightarrow y^2=-\frac{1}{x^2}$$

Differentiating w.r.t. x, we get

$$x^3y\frac{dy}{dx}=1$$

66. (b) $f'(x)=\left(\sqrt{x}-\sqrt{x+1}\right)+\frac{2\left(\sqrt{x+1}-\sqrt{x}\right)}{2\sqrt{x}\sqrt{x+1}}$

For $f'\left(0^+\right)=-1=f''\left(0^-\right)$

and $f'(x)=-1\,\forall\,x\neq 0$.

So, $f(x)$ is differentiable and hence continuous at $x=0$.

67. (d) We have

$$f(x)=|x-1|+|x-2|+\cos x,\ x\in[0,4]$$

$=-2x+3+\cos x$, when $0\le x\le 1$

$=1+\cos x\ 1\le x\le 2$

$=2x-3+\cos x\ 2\le x\le 4$

$$\lim_{x\to 1^-}f(x)=1+\cos 1=\lim_{x+1^+}f(x)=f(1)$$

$$\lim_{x\to 2^-}f(x)=1+\cos 2=\lim_{x+2^+}f(x)=f(2)$$

68. (b) Put
$t=\tan\theta$. then $y=2\tan^{-1}$, and $x=2\tan^{-1}t$

$$\therefore\frac{y}{x}=1\Rightarrow y=x\Rightarrow\frac{dy}{dx}=1$$

69. (a)

$\lim_{x\to 0}\frac{\sin 3x}{\sin x}=3=k$ by definition of continuity.

70. (b) Let $x=n\pi+\frac{\pi}{2},n\in Z$

$\Rightarrow f(x)=\infty$

$\therefore f(x)$ will be discontinuous at

$x=n\pi+\frac{\pi}{2},n\in Z.$

71. (c) $\frac{dy}{dx}=\sqrt{5-x^4}\cdot 2x-\sqrt{5-x^2}\cdot 1$

$$\Rightarrow\left(\frac{dy}{dx}\right)_{x=\sqrt{2}}=2\sqrt{2}-\sqrt{3}.$$

72. (a) $\because$ RHL $=1$, LHL $=-1$

$\Rightarrow\lim_{x\to 0}\frac{e^{1/x}-1}{e^{1/x}+1}$ does not exist.

73. (b) $y=x^y$

$\log y=y\log x$

Differentiating w.r.t. x both sides, we get

$$\frac{1}{y}\frac{dy}{dx}=\frac{y}{x}+\log x\left(\frac{dy}{dx}\right)$$

$$\Rightarrow\frac{dy}{dx}=\frac{y^2}{x}+y\log x\frac{dy}{dx}$$

$$\Rightarrow(1-y\log x)\frac{dy}{dx}=\frac{y^2}{x}$$

$$\Rightarrow\frac{dy}{dx}=\frac{y^2/x}{1-y\log x}.$$

74. (a) From the definition of continuity.

75. (c) $1-\cos 4x=0\Rightarrow\cos 4x=1$

$\Rightarrow 4x=0,2\pi,4\pi,\ldots$

$\Rightarrow x=0,\pi/2,\pi,\ldots$

76. (c) $y=\sqrt{\log x+y}$

$\Rightarrow y^2=\log x+y\Rightarrow y^2-y=\log x$

Differentiate w.r.t. x both sides.

$$\therefore(2y-1)\frac{dy}{dx}=\frac{1}{x}.$$

77. (a) Taking logarithm,

$y\log x=x\log y$

$\therefore$ Differentiating w.r.t. x

$$\frac{y}{x}+\log x\frac{dy}{dx}=\frac{x}{y}\frac{dy}{dx}+\log y$$

$$\Rightarrow\quad x(x-y\log x)\frac{dy}{dx}=y(y-x\log y).$$

78. (c) As $f(x)$ is an even function and twice differentiable (given),
$f'(x)$ is an odd function and $f''(x)$ is an even function.
Hence $f''(\pi)=1=f''(-\pi)$.

79. (b) $x\sqrt{1+y}=-y\sqrt{1+x}$
$\Rightarrow x^2(1+y)=y^2(1+x)$
$\Rightarrow x^2-y^2=-xy(x-y)$
$\Rightarrow x+y=-xy$
$\Rightarrow y=-\dfrac{x}{1+x}$
$\therefore \dfrac{dy}{dx}=-\dfrac{1}{(1+x)^2}$.

80. (b) $(-x)=f(x)$
$\Rightarrow f'(-x)=-f'(x)$
$\Rightarrow -f''(-x)=-f''(x)$
$\therefore f'''(-x)=-f'''(x)$.
$\Rightarrow f'''(x)$ is even.

81. (b) $f(x)=\dfrac{x-1}{2x^2-7x+5}$, for $x\neq 1$
$=\dfrac{1}{2x-5}$, for $x\neq 1$

$$f'(x)\big|_{x=1}=\left.\frac{-2}{(2x-5)^2}\right|_{x=1}=-\frac{2}{9}.$$

82. (a) $\lim\limits_{x\to 0}\left(a^2\cos^2 x+b^2\sin^2 x\right)=f(0)$.
$\Rightarrow a^2=e^b$
$\Rightarrow 2\log|a|=b$

83. (d) $y=\dfrac{a^{\cos^{-1}x}}{1+a^{\cos^{-1}x}}==\dfrac{z}{1+z}$
$\Rightarrow \dfrac{dy}{dz}=\dfrac{1}{(1+z)^2}$
and $\dfrac{dz}{dx}=a^{\cos^{-1}x}\log a.\dfrac{-1}{\sqrt{1-x^2}}$

$$\frac{dy}{dx}=\frac{dy/dz}{dz/dx}=\frac{\dfrac{1}{\left(1+a^{\cos^{-1}x}\right)^2}}{a^{\cos^{-1}x}\log a\left(-\dfrac{1}{\sqrt{1-x^2}}\right)}.$$

84. (a) $\dfrac{dy}{dx}=\dfrac{1}{2\sqrt{\sin\sqrt{x}}}\times\cos\sqrt{x}\times\dfrac{1}{2\sqrt{x}}$
$=\dfrac{\cos\sqrt{x}}{4\sqrt{x}\sqrt{\sin\sqrt{x}}}$

85. (d) $y=\tan^{-1}\left[\dfrac{\tan x+1}{1-\tan x}\right]$
$=\tan^{-1}\left[\tan\left(x-\dfrac{\pi}{4}\right)\right]=x-\dfrac{\pi}{4}$
$\dfrac{dy}{dx}=1$

86. (b) $f(x)=\dfrac{p(x)}{q(x)}$ where $p(x)$ and $q(x)$ are differential on R and R $-[0]$ respectively.

Now, $\lim\limits_{x\to 0}\dfrac{f(x)-f(0)}{x-0}=\lim\limits_{x\to 0}\dfrac{\frac{x}{1+|x|}}{x}$
$=\lim\limits_{x\to 0}\dfrac{x}{1+|x|}=1.$

Power $f(x)$ is always differentiable on R.

87. (b) Here $f(0-0)=f(0)=f(0+0)=0$
$\Rightarrow f(x)$ is continuous at $x=0$.
$f'(0)\lim\limits_{h\to 0}e^0=1$ and $Rf'(0)=\lim\limits_{h\to 0}e^{-2./x}=0$
as $Lf'(0)\neq Rf'(0)$
$\Rightarrow (x)$ is not differentiable at $x=0$.

88. (b) Let $x=\sin\theta$, $y=\sin\alpha$.
Thus given equation
$\Rightarrow \sin^{-1}x-\sin^{-1}y=2\cot^{-1}(a)$.
Now differentiating w.r.t. x, we get,

$$\frac{1}{\sqrt{1-x^2}}-\frac{y'}{\sqrt{1-y^2}}=0\Rightarrow\frac{dy}{dx}=\frac{\sqrt{1-y^2}}{\sqrt{1-x^2}}.$$

89. (c) $f(x)=\begin{cases}x & \text{if } x \text{ is rational}\\ 1-x & \text{if } x \text{ is irrational}\end{cases}$

Real number = Rational number + Irrational number.
$=x+(1-x)=1$
at point $f(x)$ is not continuous.

90. (b) $\lim\limits_{x\to\pi/4}\dfrac{\tan x-\tan\left(\dfrac{\pi}{2}-x\right)}{x-\pi/4}$
$=\lim\limits_{x\to\pi/4}\dfrac{\tan(x-\pi/2+x)}{(2x-\pi/2)}[1+\tan x\cot x]$
$=2[1+1]=4.$

91. (c) $y=\sin^{-1}\dfrac{x}{2}+\cos^{-1}\dfrac{x}{2}=\pi/2$
$\therefore \dfrac{dy}{dx}=0$

92. (a) $\log(x+y)=2xy$

Differentiating both side w.r.t. x, we get

$$\frac{1}{x+y}\left(1+\frac{dy}{dx}\right)=2\left(x\frac{dy}{dx}+y\right)$$

$$\Rightarrow 2y^2-1=\frac{dy}{dx}$$

For $x=0,\ y=1\Rightarrow\frac{dy}{dx}=1.$

93. (a) As $f(x)$ is continuous at $x=0\Rightarrow\lim_{x\to0}f(x)=f(0)=0$

$$\lim_{x\to0}x^p\sin\left(\frac{1}{x}\right)=0\Rightarrow p>0$$

$f(x)$ will be differentiable at $x=0$,

if $\lim_{x\to0}\frac{f(x)-f(0)}{x-0}$ exist.

i.e., $\lim_{x\to0}\frac{x^p\sin\frac{1}{x}-0}{x}$ exist.

i.e., $\lim_{x\to0}x^{p-1}$, $\lim_{x\to0}\sin\left(\frac{1}{x}\right)$ exist.

$\Rightarrow p-1>0\Rightarrow p>0.$

When $p\le1$, $\lim_{x\to0}x^p\sin\left(\frac{1}{x}\right)$ does not exist.

$\Rightarrow 0<p\le1.$

94. (b) $y=x^y$

$\Rightarrow \log y=y\log x$

$$\Rightarrow\frac{1}{y}\frac{dy}{dx}=(\log x)\frac{dy}{dx}+\frac{y}{x}$$

$$\Rightarrow\left(\frac{1}{y}-\log x\right)\frac{dy}{dx}=\frac{y}{x}$$

$$\Rightarrow\frac{dy}{dx}=\frac{y^2}{x(1-y\log x)}.$$

95. (a) $f(x)=x\tan^{-1}x$

$$f'(x)=\frac{x}{1+x^2}+\tan^{-1}x$$

$$f'(1)=\frac{1}{2}+\frac{\pi}{4}.$$

96. (b) $y=\sec^{-1}\frac{x+1}{x-1}+\sin^{-1}\frac{x-1}{x+1}$

$$=\cos^{-1}\frac{x+1}{x-1}+\sin^{-1}\frac{x-1}{x+1}=\pi/2$$

$$\therefore\frac{dy}{dx}=0.$$

97. (a) $|x|$ is not differentiable at $x=0$.

$\therefore$ $\sin|x|$ is also not differentiable at $x=0$.

98. (a) $y=e^{1+\log x}$

$=e.e^{\log x}=e^x$

$$\frac{dy}{dx}=e.$$

99. (d) $y'=\frac{d}{dx}\left(\tan^{-1}\frac{\sqrt{1+x^2}-1}{x}\right)$

$$=\frac{d}{dx}\frac{1}{2}\tan^{-1}x$$

Put $x=\cos\theta=\frac{\sqrt{1+x^2}-1}{x}$

$$=\frac{\sec\theta-1}{\tan\theta}$$

$$=\frac{1-\cos\theta}{\sin\theta}=\tan\frac{\theta}{2}$$

$$=\frac{1}{2(1+x^2)}$$

100. (a) $\lim_{x\to0}x\sin\frac{1}{x}=0,\ f'(0)=\lim_{n\to0}\frac{f(0+n)-f(0)}{n}$

$=\lim_{n\to0}\sin\frac{1}{n}$, which does not exist.

101. (c) Here, $x=a\sin\theta$ and $y=b\cos\theta$

$\frac{dx}{d}=a\cos\theta$ and $\frac{dy}{d}=-b\sin\theta$

$$\Rightarrow\quad\frac{dy}{dx}=-\frac{b}{a}\tan\theta$$

Now, $\frac{d^2y}{dx^2}=\frac{d}{dx}\left(\frac{dy}{dx}\right)$

$$=\frac{d}{d\theta}\left(-\frac{b}{a}\tan\theta\right)/\frac{dx}{d\theta}$$

$$=-\frac{b}{a}\sec^2\theta/a\cos\theta$$

$$=-\frac{b}{a^2}\sec^3\theta.$$

102. (b) $f''(x)=g''(x)$

$\Rightarrow f'(x)=g'(x)+A$

Put $x=1$, then

$f'(1)=g(1)+A$

$\Rightarrow 4=2+A\Rightarrow A=2$

$\Rightarrow f'(x)=g'(x)+2$

$\Rightarrow f(x)=g(x)+2x+B$

Put $x = 2$

$f(2) = g'(2) + 4 + B$

$\Rightarrow g = 3 + 4 + B \Rightarrow B = 2$

$\Rightarrow f(x) = g(x) = 2x + 2$

$\Rightarrow f(4) - g(4) = 10.$

103. (b) $y = x\log\left(\dfrac{x}{a+bx}\right)$, then $\dfrac{y}{x} = \log\dfrac{x}{a+bx}$

Differentiating $xy' - y = \dfrac{ax}{a+bx}$

Again differentiating, we get

$$xy'' = \frac{a^2}{(a+bx)^2}$$

$$\Rightarrow x^3y'' = \left(\frac{ax}{a+bx}\right)^2 = (xy' - y)^2.$$

104. (b) $y = \dfrac{1}{4}\log\dfrac{1+x}{1-x} - \dfrac{1}{2}\tan^{-1}x$

$$\frac{dy}{dx} = \frac{1}{4}\cdot\frac{1-x}{1+x}\cdot\frac{2}{(1-x)^2} - \frac{1}{2(1+x^2)}$$

$$= \frac{1}{2}\cdot\frac{1}{1-x^2} - \frac{1}{2(1+x^2)} = \frac{x^2}{1-x^4}$$

105. (b) $f(x) = e^x \sin x$

$f'(x) = e^x\cos x + e^x \sin x$

$f^2(x) = -e^x\sin x + e^x\cos x + e^x\cos + e^x\sin x$

$= 2e^x\cos x.$

$f^3(x) = 2e^x(-\sin x + \cos x)$

$f^4(x) = 2[e^x(-\cos x - \sin x) + e^x(-\sin x + \cos x)]$

$= -2.2\, e^x \sin x.$

$f^5(x) = -4\left[e^x\ \cos x + e^x \sin x\right]$

$f^6(x) = -4[e^x(-\cos x + \cos x)]$

$= -8e^x\ \cos x.$

106. (c) $\lim\limits_{x\to -2}\dfrac{x+2}{x^2+3x+2}, \left(\dfrac{0}{0}\text{form}\right)$

$= \lim\limits_{x\to -2}\dfrac{1}{2x+3}$

$= -1 = f(-2) \Rightarrow f(x)$ is continuous at $x = -2$.

and $\lim\limits_{x\to -1}\dfrac{x+2}{x^2+3x+2}, \left(\dfrac{0}{0}\text{form}\right)$

$= \lim\limits_{x\to -1}\dfrac{1}{2x+3}$

$= +1 \neq f(-1) \Rightarrow f(x)$ is discontinuous at $x = -1$.

107. (b) $f(x) = 10\cos x + (13 + 2x)\sin x$

$f'(x) = -8\sin x + (13+2x)\cos x$

$f''(x) = -6\cos x - (13+2x)\sin x$

$\therefore f''(x) + f(x) = 4\cos x.$

108. (b) $x^y = e^{x-y}$

$\Rightarrow e^{y\log x} = e^{x-y}$

$\Rightarrow y\log x = x - y$

$\Rightarrow y = \dfrac{x}{1+\log x}$

$\therefore \dfrac{dy}{dx} = \dfrac{\log x}{(1+\log x)^2}.$

109. (b) $h(x) = e^{\sin^{-1}x}$

$h'(x) = e^{\sin^{-1}x}\dfrac{1}{\sqrt{1-x^2}}$

$\dfrac{h'(x)}{h(x)} = \dfrac{1}{\sqrt{1-x^2}}.$

110. (a) $\dfrac{d}{dx}\left[\tan^{-1}\sqrt{\dfrac{2\cos^2\frac{x}{4}}{2\sin^2\frac{x}{4}}}\right] = \dfrac{d}{dx}\tan^{-1}\left(\cot\dfrac{x}{4}\right)$

$$= \frac{d}{dx}\tan^{-1}\tan\left(\frac{\pi}{2} - \frac{x}{4}\right)$$

$$= \frac{d}{dx}\left(\frac{\pi}{2} - \frac{x}{4}\right) = -\frac{1}{4}.$$

111. (d) On differentiating, we get

$$2ax + 2h\left(1\cdot y + x\frac{dy}{dx}\right) + 2by\frac{dy}{dx} = 0$$

$$\Rightarrow \frac{dy}{dx} = -\frac{ax+hy}{hx+by}$$

$$\Rightarrow \frac{d^2y}{dx^2} = -\frac{\left(a + h\dfrac{dy}{dx}\right)(hx+by) - (ax+hy)\left(h + b\dfrac{dy}{dx}\right)}{(hx+by)^2}$$

$$= -\frac{(ab-h^2)y + x\dfrac{dy}{dx}(h^2-ab)}{(hx+by)^2}$$

$$= \frac{(h^2-ab)\left(y + x\dfrac{ax+hy}{hx+by}\right)}{(hx+by)^2}$$

$$= \frac{h^2 - ab}{(hx+by)^3} \left(\because ax^2 + 2hxy + by^2 = 1\right)$$

112. (d) To examine the function at $x = \frac{\pi}{4}, \frac{\pi}{2}$. At all other points, this function is continuous.

$$\lim_{x \to \frac{\pi}{4}^+} f(x) = \lim_{x \to \frac{\pi}{4}^+} 2x \cot x + b = \frac{\pi}{2} + b$$

$$\lim_{x \to \frac{\pi}{4}^-} f(x) = \lim_{x \to \frac{\pi}{4}^-} f(x)\left(x + a\sqrt{2}\sin x\right) = \frac{\pi}{4} + a$$

$$\Rightarrow a - b = \frac{\pi}{4}$$

Similarly, at $x \to \frac{\pi^+}{2}, \frac{\pi^-}{2}$

$$b = -\frac{a}{2}$$

$$\Rightarrow a = \frac{\pi}{6}, b = -\frac{\pi}{12}.$$

113. (c)

114. (a)

115. (d) $f(x) = \begin{cases} \frac{1}{2-x}, & x \ge 0 \\ \frac{1}{2+x}, & x < 0 \end{cases}$

at $x = 0$, $f(x) = \frac{1}{2}$

$$\lim_{x \to 0^+} f(x) = \lim_{x \to 0^+} \frac{1}{2-x} = \frac{1}{2}$$

$$\lim_{x \to 0^-} f(x) = \lim_{x \to 0^-} \frac{1}{2+x} = \frac{1}{2}$$

It can be shown that $f(x)$ is not differentiable at $x = 0$.

116. (a) $\lim_{x \to 0^+} f(x) = \lim_{x \to 0^+} |x-1| = 1$

$\lim_{x \to 0^-} f(x) = \lim_{x \to 0^-} e^x = 1\ f(0) = 1$

$\Rightarrow f(x)$ is continuous at $x = 0$

$\lim_{x \to 1^+} f(x) = \lim_{x \to 1^+} x - 1 = 0$

$\lim_{x \to 1^-} f(x) = \lim_{x \to 1^-} 1 - x = 0\ f(1) = 0$

$\Rightarrow f(x)$ is continuous at $x = 1$

At $x = 0$

$$\lim_{h \to 0^+} \frac{f(0+h) - f(0)}{h} = \lim_{h \to 0^+} \frac{1-h-1}{h} = -1$$

$$\lim_{h \to 0^-} \frac{f(0+h) - f(0)}{h} = \lim_{h \to 0^-} \frac{e^h - 1}{h} = 1$$

117. (c)

118. (b)

119. (a) $\lim_{x \to \pi^-} f(x) = \lim_{x \to \pi^-} (\pi - x) \cdot \frac{\cos x}{\sin x}$

Let $\pi - x = h$

$$\lim_{h \to 0^-} h \cdot \frac{\cos(\pi - h)}{\sin(\pi - h)} = \lim_{h \to 0^-} \frac{h \cdot (-\cos h)}{\sin h} = -1$$

$$\lim_{h \to 0^+} \frac{-h \cdot \cos(h)}{-\sin h} = \pm 1$$

120. (a) RHL $= \lim_{x \to 2^+} f(x) = \lim_{x \to 2^+} e^{\frac{\tan(x-2)}{x-2}} = e^1$

$$\left[\because \lim_{x \to 2^+} [x] = 2\right]$$

LHL $= \lim_{x \to 2^-} f(x)$

$$= \lim_{x \to 2^-} \frac{-\lambda(x-2)(x-3)}{\mu(x-2)(x-3)} = -\frac{\lambda}{\mu}$$

Given that $f(x)$ is continuous at $n = 2$

$$\therefore \mu = e = -\frac{\lambda}{\mu} \Rightarrow \mu = e, \lambda = -e^2$$

$$\lambda + \mu = e(-e+1)$$

121. (b) Since function $f(x)$ is continuous at $x = 0$

$\Rightarrow \lim_{x \to 0^+} f(x) = \lim_{x \to 0^-} f(x)$

$\Rightarrow a = 0$

Since function $f(x)$ is continuous at $x = 1$

$\therefore \lim_{x \to 1^+} f(x) = \lim_{x \to 1^-} f(x)$

$\Rightarrow 2(1) - b = a + (-1)$

$\Rightarrow b = 2 - a + 1 \Rightarrow b = 3$

So, $a + b = 3$

122. (a) We know that when $x \to 0^-$, then $\{x\} = 1 - h$ where $h \to 0$

$$\text{LHL} = \lim_{h \to 0} \frac{\cos^{-1}\left(1-(1-h)^2\right)\sin^{-1}(h)}{(1-h)(1-(1-h)^2)}$$

$$= \lim_{h \to 0} \frac{\cos^{-1}\left(1-(1-h)^2\right)\sin^{-1}(h)}{(1-h)h(2-h)} = \frac{\frac{\pi}{2} \times 1}{1 \times 2} = \frac{\pi}{4}$$

and when $x \to 0^+$ then $\{x\} = h$ where $h \to 0$

$$\text{RHL} = \lim_{h\to 0}\frac{\cos^{-1}\left(1-h^2\right)\sin^{-1}(1-h)}{h-h^3}$$

$$= \lim_{h\to 0}\frac{\sin^{-1}\sqrt{1-(1-h^2)^2}\sin^{-1}(1-h)}{h\left(1-h^2\right)}$$

$$= \lim_{h\to 0}\frac{\sin^{-1}\left(h\sqrt{2-h^2}\right)\sin^{-1}(1-h)}{h\left(1-h^2\right)}$$

$$= \lim_{h\to 0}\frac{\sin^{-1}\left(h\sqrt{2-h^2}\right)}{h\sqrt{2-h^2}}\times\frac{h\sqrt{2-h^2}}{h\left(1-h^2\right)}\times\sin^{-1}(1-h)$$

$$= 1\times\frac{\sqrt{2}}{1}\times\frac{\pi}{2}=\frac{\pi}{\sqrt{2}}$$

LHL $\neq$ RHL

So, $f(x)$ is discontinuous at $x = 0$.

123. (c) For $x = n, n \in Z$

$$\text{LHL} = \lim_{x\to n^-} f(x) = \lim_{x\to n^-}[x-1]\cos\left(\frac{2x-1}{2}\right)\pi = 0$$

$$\text{RHL} = \lim_{x\to n^+} f(x) = \lim_{x\to n^+}[x-1]\cos\left(\frac{2x-1}{2}\right)\pi = 0$$

$f(n) = 0$

$\Rightarrow$ LHL = RHL = $f(n)$

$\Rightarrow f(x)$ is continuous for every real x.

124. (a) The given function is

$$f(x)=\begin{cases}\dfrac{\sqrt{2}\cos x-1}{\cot x-1}, & x\neq\dfrac{\pi}{4}\\ k, & x=\dfrac{\pi}{4}\end{cases}$$

$\because$ Function $f(x)$ is continuous, so it is continuous at $x = \pi/4$.

$$\therefore f\left(\frac{\pi}{4}\right)=\lim_{x\to\frac{\pi}{4}} f(x)$$

$$\Rightarrow k = \lim_{x\to\frac{\pi}{4}}\frac{\sqrt{2}\cos x-1}{\cot x-1}$$

Put $x = \dfrac{\pi}{4}+h$, when $x\to\dfrac{\pi}{4}$, then $h\to 0$

$$k = \lim_{h\to 0}\frac{\sqrt{2}\cos\left(\frac{\pi}{4}+h\right)-1}{\cot\left(\frac{\pi}{4}+h\right)-1}$$

$$= \lim_{h\to 0}\frac{\sqrt{2}\left[\frac{1}{\sqrt{2}}\cos h-\frac{1}{\sqrt{2}}\sin h\right]-1}{\frac{\cot h-1}{\cot h+1}-1}$$

$$\left[\begin{array}{l}\because \cos(x+y)=\cos x\cos y-\sin x\sin y\\ \text{and } \cot(x+y)=\dfrac{\cot x\cot y-1}{\cot y+\cot x}\end{array}\right]$$

$$= \lim_{h\to 0}\frac{\cos h-\sin h-1}{\frac{-2}{1+\cot h}}$$

$$= \lim_{h\to 0}\left[\frac{(1-\cos h)+\sin h}{2\sin h}(\sin h+\cos h)\right]$$

$$= \lim_{h\to 0}\left[\frac{2\sin^2\frac{h}{2}+2\sin\frac{h}{2}\cos\frac{h}{2}}{4\sin\frac{h}{2}\cos\frac{h}{2}}(\sin h+\cos h)\right]$$

$$= \lim_{h\to 0}\left[\frac{\sin\frac{h}{2}+\cos\frac{h}{2}}{2\cos\frac{h}{2}}\times(\sin h+\cos h)\right]$$

$$\Rightarrow k = \frac{1}{2}.$$

125. (c) Given $f(x)=\left|x^2-2x-3\right|.e^{\left|9x^2-12x+4\right|}$

$$f(x)=\left|(x-3)(x+1)\right|.e^{(3x-2)^2}$$

$$f(x)=\begin{cases}(x-3)(x+1).e^{(3x-2)^2} & ; \quad x\in(3,\infty)\\ -(x-3)(x+1).e^{(3x-2)^2} & ; \quad x\in[-1,3]\\ (x-3).(x+1).e^{(3x-2)^2} & ; \quad x\in(-\infty,-1)\end{cases}$$

Hence at $x=-1, 3$ $f(x)=0$

Clearly, non-differentiable at $= -1$ and $x = 3$.

126. (a) $f(x)$ must be continuous at $x = \pm 1$

$\Rightarrow 1 = a + b$ ...(i)

For the function $f(x)$ to be differentiable at $x = 1$

$$f(x)=\begin{cases}1+x & x\geq 1\\ ax^2+b & x<1\end{cases}$$

$$\text{LHD} = \lim_{h\to 0}\frac{a(1-h)^2+b-1}{-h}=\frac{-2ah+h^2}{-h}=2a$$

$$\text{RHD} = \lim_{h\to 0}\frac{\frac{1}{1+h}-1}{h}=-1$$

$\because f(x)$ is differentiable at every point of the domain.

$\Rightarrow$ LHD = RHD

$\Rightarrow 2a = -1$

$\Rightarrow a = \frac{-1}{2}, b = \frac{3}{2}$.

127. (d) Since, LHL $= \lim_{x\to 0^-} f(x) = 2$

RHL $= \lim_{x\to 0^+} f(x) = 0$

$\therefore f(x)$ is not continuous at $x = 0$ only. So, $f(x)$ is not differentiable at $x = 0$ only.

But LHL $= \lim_{x\to 1^-} g(x) = 1$ and

RHL $= \lim_{x\to 1^+} g(x) = 1$

So, $g(x)$ is continuous on 1

$$\text{LHD} = \lim_{h\to 0} \frac{f(1-h) - f(1)}{-h} = 3$$

$$\text{and RHD} = \lim_{h\to 0} \frac{f(1+h) - f(1)}{h} = 3$$

$\therefore g(x)$ is differentiable also on 1.

Hence, $f(g(x))$ is non-differentiable at $x = 0$ only.

128. (a) $\lim_{x\to a} \frac{xf(a) - af(x)}{x-a} \left[\frac{0}{0} \text{ form}\right]$

On applying L′ Hospital rule, we get

$$= \lim_{x\to a} \frac{f(a) - af'(x)}{1}$$

$f(a) - af'(a) = 4 - 2a$

$[\because\ f'(a) = 2$ and $f(a) = 4]$.

129. (b) Let

$$P = \lim_{x\to 0} \frac{x + 2\sin x}{\sqrt{x^2 + 2\sin x + 1} - \sqrt{\sin^2 x - x + 1}} \left[\frac{0}{0} \text{ form}\right]$$

On rationalization, we get

$$P = \lim_{x\to 0} \frac{(x + 2\sin x)}{x^2 + 2\sin x + 1 - \sin^2 x + x - 1}$$

$$\times\left(\sqrt{x^2 + 2\sin x + 1} + \sqrt{\sin^2 x - x + 1}\right)$$

$$= \lim_{x\to 0}\left(\sqrt{x^2 + 2\sin x + 1} + \sqrt{\sin^2 x - x + 1}\right)$$

$$\times \lim_{x\to 0} \frac{x + 2\sin x}{x^2 - \sin^2 x + 2\sin x + x}$$

$$= 2\times \lim_{x\to 0} \frac{x + 2\sin x}{x^2 - \sin^2 x + 2\sin x + x} \left[\frac{0}{0} \text{ form}\right]$$

Now applying the L′ Hospital's rule, we get

$$P = 2\times \lim_{x\to 0} \frac{1 + 2\cos x}{2x - \sin 2x + 2\cos x + 1}$$

$$= 2\frac{(1+2)}{0 - 0 + 2 + 1} \text{[on applying limit]}$$

$$= 2\times\frac{3}{2} = 2$$

$$\Rightarrow \lim_{x\to 0} \frac{x + 2\sin x}{\sqrt{x^2 + 2\sin x + 1} - \sqrt{\sin^2 x - x + 1}} = 2.$$

130. (c) It is given that $\lim_{x\to 1} \frac{x^2 - ax + b}{x-1} = 5$ ...(i)

Since, limit exist and equal to 5 and denominator zero at $x = 1$, so numerator $x^2 - ax + b$ should be zero at $x = 1$,

So $1 - a + b = 0 \Rightarrow a = 1 + b$ ...(ii)

On putting the value of 'a' from Eq. (ii) in Eq. (i), we get

$$\lim_{x\to 1} \frac{x^2 - (1+b)x + b}{x-1} = 5$$

$$\Rightarrow \lim_{x\to 1} \frac{(x^2 - x) - b(x-1)}{x-1} = 5$$

$$\Rightarrow \lim_{x\to 1} \frac{(x-1)(x-b)}{x-1} = 5 \Rightarrow \lim_{x\to 1}(x-b) = 5$$

$\Rightarrow 1 - b = 5 \Rightarrow b = -4$ ...(iii)

On putting value of 'b' from Eq. (iii) in Eq. (ii), we get

$a = -3 \Rightarrow a + b = -7$.

131. (d) Given,

$$\lim_{x\to 1} \frac{x^4 - 1}{x-1} = \lim_{x\to k} \frac{x^3 - k^3}{x^2 - k^2}$$

$$= \lim_{x\to 1} \frac{(x-1)(x+1)(x^2+1)}{x-1}$$

$$= \lim_{x\to k} \frac{(x-k)(x^2 + k^2 + xk)}{(x-k)(x+k)} \Rightarrow 2\times 2 = \frac{3k^2}{2k}$$

$$\Rightarrow k = \frac{8}{3}.$$

132. (a) Given limit is

$$\lim_{x\to 0} \frac{\sin^2 x}{\sqrt{2} - \sqrt{1+\cos x}} \left[\frac{0}{0}, \text{Form}\right]$$

$$= \lim_{x\to 0} \frac{\sin^2 x}{\sqrt{2} - \sqrt{1+\cos x}} \left[\because 1 + \cos x = 2\cos^2\frac{x}{2}\right]$$

$$= \lim_{x\to 0} \frac{\sin^2 x}{\sqrt{2}\left(1 - \cos\frac{x}{2}\right)}$$

$$= \lim_{x\to 0} \frac{\sin^2 x}{\sqrt{2}\times 2\sin^2\left(\frac{x}{4}\right)} \left[\because 1-\cos\frac{x}{2} = 2\sin^2\frac{x}{4}\right]$$

$$= \lim_{x\to 0} \frac{x^2}{2\sqrt{2}\left(\frac{x}{4}\right)^2}$$

$$= \frac{16}{2\sqrt{2}} = 4\sqrt{2}.\left[\lim_{x\to 0}\sin x = \lim_{x\to 0} x\right]$$

133. (c) Given, limit $= \lim_{x\to\frac{\pi}{4}} \frac{\cot^3 x - \tan x}{\cos\left(x+\frac{\pi}{4}\right)}$

$$= \lim_{x\to\pi/4} \frac{1-\tan^4 x}{\frac{1}{\sqrt{2}}(\cos x - \sin x)} \times \frac{1}{\tan^3 x}$$

$$\left[\because \cot x = \frac{1}{\tan x}\right]$$

$$= \lim_{x\to\pi/4} \frac{\left(1-\tan^2 x\right)}{\cos x - \sin x} \times \frac{\sqrt{2}\left(1+\tan^2 x\right)}{\tan^3 x}$$

$$= \lim_{x\to\pi/4} \frac{\cos^2 x - \sin^2 x}{\cos x - \sin x} \times \frac{\sqrt{2}\left(\sec^2 x\right)}{\cos^2 x\tan^3 x}$$

$$\left[\because 1+\tan^2 x = \sec^2 x\right]$$

$$= \lim_{x\to\pi/4} \frac{(\cos x-\sin x)(\cos x+\sin x)}{(\cos x - \sin x)} \times \frac{\sqrt{2}\sec^4 x}{\tan^3 x}$$

$$\left[\because \left(a^2-b^2\right) = (a-b)(a+b)\right]$$

$$= \lim_{x\to\pi/4} \frac{\sqrt{2}\sec^4 x}{\tan^3 x}(\cos x + \sin x)$$

$$= \frac{\sqrt{2}(\sqrt{2})^4}{(1)^3}\left(\frac{1}{\sqrt{2}}+\frac{1}{\sqrt{2}}\right) \text{ [on applying limit]}$$

$$= 4\sqrt{2}\left(\frac{2}{\sqrt{2}}\right) = 8.$$

134. (b) $\lim_{x\to 0} \frac{x\cot 4x}{\sin^2 x\cot^2 2x} = \lim_{x\to 0} \frac{x}{\tan 4x}.\frac{1}{\sin^2 x}\frac{\tan^2 2x}{1}$

$$= \lim_{x\to 0} \frac{1}{4}\frac{4x}{(\tan 4x)}\frac{x^2}{\sin^2 x}.\frac{\tan^2 2x}{x^2}$$

$$= \lim_{x\to 0} \frac{1}{4}\frac{4x}{(\tan x)}\left(\frac{x}{\sin x}\right)^2.\left(\frac{\tan 2x}{2x}\right)^2.\frac{4}{1}$$

$$= \frac{1}{4}.1.1.1.\frac{4}{1} = 1.\left[\because \lim_{x\to 0}\frac{x}{\sin x} = 1 = \lim_{x\to 0}\frac{\tan x}{x}\right]$$

135. (a) Clearly, $\lim_{y\to 0} \frac{\sqrt{1+\sqrt{1+y^4}} - \sqrt{2}}{y^4}$

$$= \lim_{y\to 0} \frac{\sqrt{1+\sqrt{1+y^4}} - \sqrt{2}}{y^4} \times \frac{\sqrt{1+\sqrt{1+y^4}} + \sqrt{2}}{\sqrt{1+\sqrt{1+y^4}} + \sqrt{2}}$$

[rationalizing the numerator]

$$= \lim_{y\to 0} \frac{\left(1+1+y^4\right) - \sqrt{2}}{y^4\left(\sqrt{1+\sqrt{1+y^4}} + \sqrt{2}\right)}$$

$$\left[\because (a+b)(a-b) = a^2 - b^2\right]$$

$$= \lim_{y\to 0} \frac{\sqrt{1+y^4} - 1}{y^4\left(\sqrt{1+\sqrt{1+y^4}} + \sqrt{2}\right)} \times \frac{\sqrt{1+y^4}+1}{\sqrt{1+y^4}+1}$$

[again, rationalizing the numerator]

$$= \lim_{y\to 0} \frac{y^4}{y^4\left(\sqrt{1+\sqrt{1+y^4}} + \sqrt{2}\right)\left(\sqrt{1+y^4}+1\right)}$$

$$= \frac{1}{2\sqrt{2}\times 2}$$

(by cancelling y^4 and then by direct substitution)

$$= \frac{1}{4\sqrt{2}}.$$

136. (b) $\lim_{x\to\pi/2} \frac{\cot x - \cos x}{(\pi - 2x)^3} = \lim_{x\to\pi/2} \frac{1}{8}.\frac{\cos x(1-\sin x)}{\sin x\left(\frac{\pi}{2}-x\right)^3}$

$$= \lim_{h\to 0} \frac{1}{8}.\frac{\cos\left(\frac{\pi}{2}-h\right)\left[1-\sin\left(\frac{\pi}{2}-h\right)\right]}{\sin\left(\frac{\pi}{2}-h\right)\left(\frac{\pi}{2}-\frac{\pi}{2}+h\right)^3}$$

$$= \frac{1}{8}\lim_{h\to 0} \frac{\sin h(1-\cos h)}{\cos h.h^3}$$

$$= \frac{1}{8}\lim_{h\to 0} \frac{\sin h\left(2\sin^2\frac{h}{2}\right)}{\cos h.h^3} = \frac{1}{4}\lim_{h\to 0} \frac{\sin h.\sin^2\left(\frac{h}{2}\right)}{h^3\cos h}$$

$$= \frac{1}{4}\lim_{h\to 0}\left(\frac{\sin h}{h}\right)\left(\frac{\sin\frac{h}{2}}{\frac{h}{2}}\right)^2.\frac{1}{\cos h}.\frac{1}{4} = \frac{1}{4}\times\frac{1}{4} = \frac{1}{16}$$

137. (d) $\lim_{x\to 0} \frac{\sin\left(\pi\cos^2 x\right)}{x^2} = \lim_{x\to 0} \frac{\sin\pi(1-\sin^2 x)}{x^2}$

$$= \lim_{x\to0}\frac{\sin\left(\pi-\pi\sin^2x\right)}{x^2}$$

$$= \lim_{x\to0}\frac{\sin\left(\pi\sin^2x\right)}{x^2}\left[\because \sin(\pi-\theta)=\sin\theta\right]$$

$$= \lim_{x\to0}\frac{\sin\pi\sin^2x}{\pi\sin^2x}\times(\pi)\left(\frac{\sin^2x}{x^2}\right)$$

$$= \pi\left[\because \lim_{\theta\to0}\frac{\sin\theta}{\theta}=1\right]$$

138. (c) We have,

$$\lim_{x\to0}\frac{(1-\cos2x)(3+\cos x)}{x\tan4x}=\lim_{x\to0}\frac{2\sin^2x(3+\cos x)}{x\times\frac{\tan4x}{4x}\times4x}$$

$$= \lim_{x\to0}\frac{2\sin^2x}{x^2}\times\lim_{x\to0}\frac{(3+\cos x)}{4}\times\frac{1}{\lim\limits_{x\to0}\frac{\tan4x}{4x}}$$

$$= 2\times\frac{4}{4}\times1\left[\because \lim_{\theta\to0}\frac{\sin\theta}{\theta}=1 \text{ and } \lim_{\theta\to0}\frac{\tan\theta}{\theta}=1\right]=2$$

139. (b) Plan $\left(\frac{\infty}{\infty}\right)$ form

$$\lim_{x\to\infty}\frac{a_0x^n+a_1x^{n-1}+\ldots+a_n}{b_0x^m+b_1x^{m-1}+\ldots+b_m}$$

$$=\begin{cases}0, & \text{if } n<m\\ \frac{a_0}{b_0}, & \text{if } n=m\\ +\infty, & \text{if } n>m \text{ and } a_0b_0>0\\ -\infty, & \text{if } n>m \text{ and } a_0b_0<0\end{cases}$$

Description of situation: As to make degree of numerator equal to degree of denominator.

$$\therefore \lim_{x\to\infty}\left(\frac{x^2+x+1}{x+1}-ax-b\right)=4$$

$$\Rightarrow \lim_{x\to\infty}\frac{x^2+x+1-ax^2-ax-bx-b}{x+1}=4$$

$$\Rightarrow \lim_{x\to\infty}\frac{x^2(1-a)+x(1-a-b)+(1-b)}{x+1}=4$$

Here, we make degree of numerator = degree of denominator.

$\therefore 1-a=0\Rightarrow a=1$

and $\lim\limits_{x\to\infty}\frac{x(1-a-b)+(1-b)}{x+1}=4$

$\Rightarrow 1-a-b=4$

$\Rightarrow b=-4 \quad [\because (1-a)=0]$

140. (d) Let $l=\lim\limits_{x\to0}\left(\frac{1+f(3+x)-f(3)}{1+f(2-x)-f(2)}\right)^{\frac{1}{x}}$ [1^∞ form]

$$\Rightarrow \quad l=e^{\lim\limits_{x\to0}\frac{1}{x}\left(1-\frac{1+f(3+x)-f(3)}{1+f(2-x)-f(2)}\right)}$$

$$= e^{\lim\limits_{x\to0}\left[\frac{1+f(2-x)-f(2)-1-f(3+x)+f(3)}{x(1+f(2-x)-f(2)}\right]}$$

$$= e^{\lim\limits_{x\to0}\left[\frac{f(2-x)-f(3+x)+f(3)-f(2)}{x(1+f(2-x)-f(2)}\right]}$$

On applying L′ Hospital rule, we get

$$l=e^{\lim\limits_{x\to0}\left[\frac{-f'(2-x)-f'(3+x)}{1-xf'(2-x)+f(2-x)-f(2)}\right]}$$

On applying limit, we get

$$l=e^{\left(\frac{-f'(2)-f'(3)}{1-0+f(2)-f(2)}\right)}=e^0=1$$

So, $\lim\limits_{x\to0}\left(\frac{1+f(3+x)-f(3)}{1+f(2-x)-f(2)}\right)^{\frac{1}{x}}=1.$

141. (b) Let $L=\lim\limits_{x\to1^-}\frac{\sqrt{\pi}-\sqrt{2\sin^{-1}x}}{\sqrt{1-x}}$, then

$$L=\lim_{x\to1^-}\frac{\sqrt{\pi}-\sqrt{2\sin^{-1}x}}{\sqrt{1-x}}\times\frac{\sqrt{\pi}+\sqrt{2\sin^{-1}x}}{\sqrt{\pi}+\sqrt{2\sin^{-1}x}}$$

$$=\lim_{x\to1^-}\frac{\pi-2\sin^{-1}x}{\sqrt{1-x}}\times\frac{1}{\sqrt{\pi}+\sqrt{2\sin^{-1}x}}$$

$$=\lim_{x\to1^-}\frac{\pi-2\left(\frac{\pi}{2}-\cos^{-1}x\right)}{\sqrt{1-x}}\times\frac{1}{\sqrt{\pi}+\sqrt{2\sin^{-1}x}}$$

$$=\lim_{x\to1^-}\frac{2\cos^{-1}x}{\sqrt{1-x}}\times\lim_{x\to1^-}\frac{1}{\sqrt{\pi}+\sqrt{2\sin^{-1}x}}$$

$$=\frac{1}{2\sqrt{\pi}}\lim_{x\to1^-}\frac{2\cos^{-1}x}{\sqrt{1-x}}\qquad\left[\because \lim_{x\to1^-}\sin^{-1}x=\frac{\pi}{2}\right]$$

Put $x=\cos\theta$, then as $x\to1^-,\theta\to0^+$

Now, $L=\frac{1}{2\sqrt{\pi}}\lim\limits_{\theta\to0^+}\frac{2\theta}{\sqrt{1-\cos\theta}}$

$$=\frac{1}{2\sqrt{\pi}}\lim_{\theta\to0^+}\frac{2\theta}{\sqrt{2}\sin\left(\frac{\theta}{2}\right)}$$

$$=\frac{1}{2\sqrt{\pi}}.\sqrt{2}\lim_{\theta\to0^+}\frac{2.\left(\frac{\theta}{2}\right)}{\sin\left(\frac{\theta}{2}\right)}$$

$$= \frac{1}{2\sqrt{\pi}} 2\sqrt{2} = \sqrt{\frac{2}{\pi}}.$$

142. (b) At $x = 0$,

$$\text{RHL} = \lim_{x\to 0^+} \frac{\tan(\pi\sin^2 x) + (|x| - \sin x[x])^2}{x^2}$$

$$= \lim_{x\to 0^+} \frac{\tan(\pi\sin^2 x) + (x - \sin(x.0))^2}{x^2}$$

$$\left[\begin{array}{c}\because |x| = x \text{ for } x > 0 \\ \text{and } [x] = 0 \text{ for } 0 < x < 1\end{array}\right]$$

$$= \lim_{x\to 0^+} \frac{\tan(\pi\sin^2 x) + x^x}{x^2}$$

$$= \lim_{x\to 0^+} \left(\frac{\tan(\pi\sin^2 x)}{\pi\sin^2 x} . \frac{\pi\sin^2 x}{x^2} + 1\right)$$

$$= \pi \lim_{x\to 0^+} \frac{\tan(\pi\sin^2 x)}{\pi\sin^2 x} . \lim_{x\to 0^+} \frac{\sin^2 x}{x^2} + 1$$

$$= \pi + 1 \left[\begin{array}{c}\because \lim_{x\to 0} \frac{\tan x}{x} = 1 \\ \text{and } \lim_{x\to 0} \frac{\sin x}{x} = 1\end{array}\right]$$

and LHL

$$= \lim_{x\to 0^-} \frac{\tan(\pi\sin^2 x) + (|x| - \sin x[x])^2}{x^2}$$

$$= \lim_{x\to 0^-} \frac{\tan(\pi\sin^2 x) + (-x - \sin(x(-1))^2}{x^2}$$

$$\left[\begin{array}{c}\because |x| = -x \text{ for } x < 0 \\ \text{and } [x] = -1 \text{ for } -1 < x < 0\end{array}\right]$$

$$= \lim_{x\to 0^-} \frac{\tan(\pi\sin^2 x) + (x + \sin(-x))^2}{x^2}$$

$$= \lim_{x\to 0^-} \frac{\tan(\pi\sin^2 x) + (x - \sin x)^2}{x^2}$$

$$[\because \sin(-\theta) = -\sin\theta]$$

$$= \lim_{x\to 0^-} \left(\frac{\tan(\pi\sin^2 x) + x^2 + \sin^2 x - 2x\sin x}{x^2}\right)$$

$$= \lim_{x\to 0^-} \left(\frac{\tan(\pi\sin^2 x)}{x^2} + 1 + \frac{\sin^2 x}{x^2} - \frac{2x\sin x}{x^2}\right)$$

$$= \lim_{x\to 0^-} \left(\frac{\tan(\pi\sin^2 x)}{\pi\sin^2 x} . \frac{\pi\sin^2 x}{x^2} + 1 + \frac{\sin^2 x}{x^2} - \frac{\sin x}{x}\right)$$

$$= \lim_{x\to 0^-} \frac{\tan(\pi\sin^2 x)}{\pi\sin^2 x} . \lim_{x\to 0^-} \frac{\pi\sin^2 x}{x^2} + 1 +$$

$$\lim_{x\to 0^-} \frac{\sin^2 x}{x^2} - 2 \lim_{x\to 0^-} \frac{\sin x}{x}$$

$$= \pi + 1 + 1 - 2 = \pi$$

$\because$ RHL $\neq$ LHL

$\therefore$ Limit does not exist.

143. (a) $\lim_{x\to 1^+} \frac{(1 - |x| + \sin|1 - x|)\sin\left(\frac{\pi}{2}[1 - x]\right)}{|1 - x|[1 - x]}$

Put $x = 1 + h$, then $x \to 1^+ \Rightarrow h \to 0^+$

$$\therefore \lim_{x\to 1^+} \frac{(1 - |x| + \sin|1 - x|)\sin\left(\frac{\pi}{2}[1 - x]\right)}{|1 - x|[1 - x]}$$

$$= \lim_{h\to 0^+} \frac{(1 - |h + 1| + \sin|-h|\sin\left(\frac{\pi}{2}[-h]\right)}{|-h|[-h]}$$

$$= \lim_{h\to 0^+} \frac{(1 - (h + 1) + \sin h)\sin\left(\frac{\pi}{2}[-h]\right)}{h[-h]}$$

$(\because |-h| = h \text{ and } |h + 1| = h + 1 \text{ as } h > 0)$

$$= \lim_{h\to 0^+} \frac{(-h + \sin h)\sin\left(\frac{\pi}{2}[-h]\right)}{h(-1)}$$

$(\because [x] = -1$ for $-1 < x < 0$ and

$h \to 0^+ \Rightarrow -h \to 0^-)$

$$= \lim_{h\to 0^+} \frac{(-h + \sin h)}{-h} \sin\left(\frac{-\pi}{2}\right)$$

$$= \lim_{h\to 0^+} \left(\frac{\sin h}{h}\right) - \lim_{h\to 0^+} \left(\frac{h}{h}\right)$$

$$= \lim_{h\to 0^+} \left(\frac{\sin h}{h}\right) - \lim_{h\to 0^+} \left(\frac{h}{h}\right) = 1 - 1 = 0.$$

$$\left[\because \lim_{h\to 0^+} \frac{\sin h}{h} = 1\right]$$

144. (c)

$$\lim_{x\to 0^-} \frac{x([x] + |x|)\sin[x]}{|x|} = \lim_{x\to 0^-} \frac{x([x] - x)\sin[x]}{-x}$$

$(\because |x| = -x, \text{ is } x < 0)$

$$= \lim_{x\to 0^-} \frac{x(-1-x)\sin(-1)}{-x} \quad \left(\because \lim_{x\to 0^-}[x]=-1\right)$$

$$= \lim_{x\to 0^-} \frac{-x(x+1)\sin(-1)}{-x} = \lim_{x\to 0^-}(x+1)\sin(-1)$$

$= (0+1)\sin(-1)$ (by direct substitution)

$= -\sin 1 \qquad (\because \sin(-\theta) = -\sin\theta)$

145. (c) Given α and β are the roots of the quadratic equation $375x^2 - 25x - 2 = 0$.

$$\therefore \alpha+\beta = \frac{25}{375} = \frac{1}{15} \quad \text{...(i)}$$

$$\text{and } \alpha\beta = -\frac{2}{375} \quad \text{...(ii)}$$

Now, $\lim_{n\to\infty}\sum_{r=1}^{n}\alpha^r + \lim_{n\to\infty}\sum_{r=1}^{n}\beta^r$

$= (\alpha+\alpha^2+\alpha^3+\ldots+$ upto infinite terms) $+(\beta+\beta^2+\beta^3+\ldots+$ upto infinite terms)

$$= \frac{\alpha}{1-\alpha} + \frac{\beta}{1-\beta}$$

$$\left(S_\infty = \frac{a}{1-r} \text{ for geometrical progression}\right)$$

$$= \frac{\alpha(1-\beta)+\beta(1-\alpha)}{(1-\alpha)(1-\beta)}$$

$$= \frac{(\alpha+\beta)-2\alpha\beta}{1-(\alpha+\beta)+\alpha\beta}$$

The value $\alpha+\beta = \frac{1}{15}$ and $\alpha\beta = \frac{-2}{375}$ from Eqs. (i) and (ii) respectively we get

$$= \frac{\frac{1}{15}+\frac{4}{375}}{1-\frac{1}{15}-\frac{2}{375}} = \frac{29}{175-25-2} = \frac{29}{348} = \frac{1}{12}$$

146. (c) $\lim_{x\to 0^+} x\left(\left[\frac{1}{x}\right]+\left[\frac{2}{x}\right]+\ldots+\left[\frac{15}{x}\right]\right)$

We know, $[x] = x - \{x\}$

$$\therefore \left[\frac{1}{x}\right] = \frac{1}{x} - \left\{\frac{1}{x}\right\}$$

Similarly, $\left[\frac{n}{x}\right] = \frac{n}{x} - \left\{\frac{n}{x}\right\}$

$\therefore$ Given limit

$$= \lim_{x\to 0^+} x\left(\frac{1}{x}-\left\{\frac{1}{x}\right\}+\frac{2}{x}-\left\{\frac{2}{x}\right\}+\ldots\frac{15}{x}-\left\{\frac{15}{x}\right\}\right)$$

$$= \lim_{x\to 0^+}(1+2+3+\ldots+15) - x\left(\left\{\frac{1}{x}\right\}+\left\{\frac{2}{x}\right\}+\ldots+\left\{\frac{15}{x}\right\}\right)$$

$$= 120 - 0 = 120 \left[\begin{array}{l} \because 0 \le \left\{\frac{n}{x}\right\} < 1, \text{ therefore} \\ 0 \le x\left\{\frac{n}{x}\right\} < x \Rightarrow \lim_{x\to 0^+} x\left\{\frac{n}{x}\right\} = 0 \end{array}\right]$$

147. (d) $f(x) = \dfrac{1-x(1+|1-x|)}{|1-x|}\cos\left(\dfrac{1}{1-x}\right)$

Now, $\lim_{x\to 1^-} f(x) = \lim_{x\to 1^-} \dfrac{1-x(1+1-x)}{1-x}\cos\left(\dfrac{1}{1-x}\right)$

$$= \lim_{x\to 1^-}(1-x)\cos\left(\frac{1}{1-x}\right) = 0$$

and $\lim_{x\to 1^+} f(x) = \lim_{x\to 1^+} \dfrac{1-x(1-1+x)}{x-1}\cos\left(\dfrac{1}{1-x}\right)$

$= \lim_{x\to 1^-} -(x+1).\cos\left(\dfrac{1}{x+1}\right)$, which does not exist.

148. (c) Given, $p = \lim_{x\to 0^+}(1+\tan^2\sqrt{x})^{\frac{1}{2x}}$

$$= e^{\lim_{x\to 0^+}\frac{\tan^2\sqrt{x}}{2x}} = e^{\frac{1}{2}\lim_{x\to 0^+}\left(\frac{\tan\sqrt{x}}{\sqrt{x}}\right)^2} = e^{\frac{1}{2}}$$

$$\therefore \log p = \log e^{\frac{1}{2}} = \frac{1}{2}$$

149. (b) **Description of Situation:** As for given equation when $a \to 0$ the equation reduces to identity in x,

i.e., $ax^2+bx+c=0, \forall x \in \mathrm{R}$ or $a=b=c\to 0$

Thus, first we should make the above equation independent of coefficients, as 0.

Let $a+1=t^6$. Thus, when $a\to 0, t\to 1$

$$\therefore (t^2-1)x^2+(t^3-1)x+(t-1)=0$$

$$\Rightarrow (t-1)\{(t+1)x^2+(t^2+t+1)x+1\}=0, \text{ as } t\to 1$$

$$2x^2+3x+1=0$$

$$\Rightarrow 2x^2+2x+x+1=0$$

$$\Rightarrow (2x+1)(x+1)=0$$

Thus, $x=-1, -1/2$

or $\lim_{a\to 0^+}\alpha(a) = -1/2$

and $\lim_{a\to 0^+}\beta(a) = -1$.

150. (d) Here, $\lim_{x\to0}\left[1+x\log(1+b^2)\right]^{\frac{1}{x}}$

$= e^{\lim_{x\to0}\{x\log(1+b^2)\}.\frac{1}{x}}$

$= e^{\log(1+b^2)} = (1+b^2)$ ...(i)

Given, $\lim_{x\to0}\{1+x\log(1+b^2)\}^{1/x} = 2b\sin^2\theta$

$\Rightarrow (1+b^2) = 2b\sin^2\theta$

$\therefore \sin^2\theta = \frac{1+b^2}{2b}$ ...(ii)

By AM $\geq$ GM, $\frac{b+\frac{1}{b}}{2} \geq \left(b.\frac{1}{b}\right)^{1/2}$

$\Rightarrow \frac{b^2+1}{2b} \geq 1$ ...(iii)

From Eqs (ii) and (iii),

$\sin^2\theta = 1$

$\Rightarrow \theta = \pm\frac{\pi}{2}$, as $\theta \in (-\pi, \pi]$

Integer and Numerical Value Type

1. $\lim_{h\to0}\frac{\log(1+2h)-2\log(1+h)}{h^2}$

Applying L′ Hospital's rule, we get

$= \lim_{h\to0}\frac{\frac{2}{1+2h}-\frac{2}{1+h}}{2h}$

$= \lim_{h\to0}\frac{2+2h-2-4h}{2h(1+2h)(1+h)}$

$= \lim_{h\to0}\frac{-1}{(1+2h)(1+h)} = -1.$

2. Given,

$$f(x) = \begin{cases}\sin x, & x \neq n\pi, \quad n = 0,\pm1,\pm2,\ldots\\ 2, & \text{otherwise}\end{cases}$$

$$g[f(x)] = \begin{cases}\{f(x)\}^2+1, & f(x)\neq 0,2\\ 4, & f(x)=0\\ 5, & f(x)=2\end{cases}$$

$$\therefore \quad g[f(x)] = \begin{cases}(\sin^2 x)+1, & x \neq, n\pi = 0,\pm1\\ 5, & x = n\pi\end{cases}$$

Now, $\lim_{x\to0} g[f(x)] = \lim_{x\to0}(\sin^2 x)+1 = 1.$

3. Given, $P = 2\left(\sqrt{2hr-h^2}+\sqrt{2hr}\right)$

Here, BC $= \sqrt{r^2-(h-r)^2} = \sqrt{2hr-h^2}$

$\therefore\ A = \frac{1}{2}.2\text{BC}.h = \left(\sqrt{2hr-h^2}\right)h$

$\therefore \lim_{h\to0}\frac{A}{P^3} = \lim_{h\to0}\frac{h\sqrt{2hr-h^2}}{8(\sqrt{2hr-h^2}+\sqrt{2hr})^3}$

$= \lim_{h\to0}\frac{h^{3/2}\left(\sqrt{2r-h}\right)}{8\,h^{3/2}(\sqrt{2r-h}+\sqrt{2r})^3}$

$= \frac{1}{8}.\frac{\sqrt{2r}}{\left(\sqrt{2r}+\sqrt{2r}\right)^3} = \frac{1}{128\,r}.$

4. $\lim_{x\to-\infty}\left[\frac{\left(x^4\sin\left(\frac{1}{x}\right)+x^2\right)}{\left(1+|x|^3\right)}\right] = \lim_{x\to-\infty}\frac{x^4\sin\frac{1}{x}+x^2}{1-x^3}$

On dividing by x^3, we get

$\lim_{x\to-\infty}\frac{\frac{\sin(1/x)}{\frac{1}{x}}+\frac{1}{x}}{\frac{1}{x^3}-1} = \frac{1+0}{0-1} = -1.$

5. $f(x) = \begin{cases}(x^3+x^2-16x+20)/(x-2)^2, & \text{if } x\neq 2\\ k, & \text{if } x = 2\end{cases}$

Since, continuous at $x = 2$

$\Rightarrow f(2) = \lim_{x\to2}\frac{x^3+x^2-16x+20}{(x-2)^2}$,

[using L′ Hospital's rule]

$= \lim_{x\to2}\frac{3x^2+2x-16}{2(x-2)} = \lim_{x\to2}\frac{6x+2}{2} = 7$

$\therefore\ k = 7.$

6. $\lim_{x\to1}(1-x)\tan\frac{\pi x}{2}$

Put $x-1 = y$

$\therefore -\lim_{y\to0} y\tan\frac{\pi}{2}(y+1) = -\lim_{y\to0} y\left[-\cot\left(\frac{\pi}{2}y\right)\right]$

$= \lim_{y\to0}\left(\frac{y\frac{\pi}{2}}{\tan\frac{\pi}{2}y}\right).\frac{2}{\pi} = \frac{2}{\pi}.$

7. $\lim_{x\to 0}\dfrac{\int_0^{x^2}\cos^2 t\,dt}{x\sin x}$

Applying L′ Hospital's rule, we get

$$=\lim_{x\to 0}\frac{\cos^2(x^2).2x-0}{x\cos x+\sin x}$$

$$=\lim_{x\to 0}\frac{2.\cos^2(x^2)}{\cos x+\dfrac{\sin x}{x}}=\frac{2}{1+1}=1$$

8. If $\lim_{x\to a}[f(x)g(x)]$ exists, then both $\lim_{x\to a}f(x)$ and $\lim_{x\to a}g(x)$ may or may not exist. Hence, it is a false statement.

9. $\lim_{x\to 0}\dfrac{2^x-1}{\sqrt{1+x}-1}\times\dfrac{\sqrt{1+x}+1}{\sqrt{1+x}+1}$

$$=\lim_{x\to 0}\frac{(2^x-1)(\sqrt{1+x}+1)}{x}$$

$$=\log_e(2).(2)=2\log_e(2)=\log_e 4.$$

10. Here, $\lim_{h\to 0}\dfrac{(a+h)^2\sin(a+h)-a^2\sin a}{h}$

$$=\lim_{h\to 0}\frac{a^2[\sin(a+h)-\sin a]}{h}+\frac{h[2a\sin(a+h)+h\sin(a+h)]}{h}$$

$$=\lim_{h\to 0}\frac{a^2.2\cos\left(a+\dfrac{h}{2}\right).\sin\dfrac{h}{2}}{2.\dfrac{h}{2}}+(2a+h)\sin(a+h)$$

$$=a^2\cos a+2a\sin a.$$

11. $\lim_{x\to 0}\sqrt{\dfrac{x-\sin x}{x+\cos^2 x}}$

$$=\frac{\lim_{x\to 0}(x-\sin x)^{1/2}}{\lim_{x\to 0}(x-\cos^2 x)^{1/2}}=\frac{\lim_{x\to 0}\left[x\left(1-\dfrac{\sin x}{x}\right)\right]^{1/2}}{\lim_{x\to 0}(0+1)^{1/2}}$$

$$=\frac{0.0}{1}=0.$$

12. Evaluate $\lim_{x\to 1}\left(\dfrac{x-1}{2x^2-7x+5}\right)$.

$$\lim_{x\to 1}\left\{\frac{x-1}{(x-1)(2x-5)}\right\}=\lim_{x\to 1}\frac{1}{(2x-5)}=-\frac{1}{3}.$$

13. $\lim_{x\to 0}\left(\dfrac{1+5x^2}{1+3x^2}\right)^{1/x^2}$

$$=\frac{\lim_{x\to 0}[(1+5x^2)^{1/5x^2}]^5}{\lim_{x\to 0}[(1+3x^2)^{1/3x^2}]^3}=\frac{e^5}{e^3}=e^2.$$

14. $\lim_{x\to\infty}\left(\dfrac{x+6}{x+1}\right)^{x+4}$

$$=\lim_{x\to\infty}\left(1+\frac{5}{x+1}\right)^{x+4}=e^{\lim\limits_{x\to\infty}\frac{5(x+4)}{x+1}}=e^5.$$

15. $\lim_{x\to 0}\left\{\tan\left(\dfrac{\pi}{4}+x\right)\right\}^{1/x}$

$$=\lim_{x\to 0}\left\{\frac{\tan\dfrac{\pi}{4}+\tan x}{1-\tan\dfrac{\pi}{4}\tan x}\right\}^{1/x}=\lim_{x\to 0}\left\{\frac{1+\tan x}{1-\tan x}\right\}^{1/x}$$

$$=\lim_{x\to 0}\frac{[(1+\tan x)^{1/\tan x}]^{\tan x/x}}{[(1-\tan x)^{-1/\tan x}]^{-\tan x/x}}=\frac{e^1}{e^{-1}}=e^2.$$

16. Given, $x^2+y^2=4\ \Rightarrow\ y=\sqrt{4-x^2}$

or $f(x)=\sqrt{4-x^2}$.

17. Given that

$$(x)=[x]|x^2-1|+\sin\left(\frac{\pi}{[x]+3}\right)-[x+1]$$

$$f(x)=\begin{cases}3-2x^2 & , \ -2<x<-1\\ x^2 & , \ -1\le x<0\\ \dfrac{\sqrt{3}}{2}+1 & , \ 0\le x<1\\ x^2+1+\dfrac{1}{\sqrt{2}} & , \ 1\le x<2\end{cases}$$

We conclude from above $f(x)$ is discontinuous at $x=0,1$.

18. $f(x)=\begin{cases}a\sin\dfrac{\pi}{2}(x-1), & \text{for } x\le 0\\ \dfrac{\tan 2x-\sin 2x}{bx^3}, & \text{for } x>0\end{cases}$

For continuity at '0',

LHL = RHL = $f(0)$

RHL = $\lim_{x\to 0^+}f(x)=f(0)$

$$\Rightarrow \lim_{x\to 0^+} \frac{\tan 2x - \sin 2x}{bx^3} = -a$$

$$\Rightarrow \lim_{x\to 0^+} \frac{\sin 2x(1-\cos 2x)}{\cos 2x . bx^3} = -a$$

$$\Rightarrow \lim_{x\to 0^+} \frac{\frac{\sin 2x}{2x} \times 2x.2\frac{\sin^2 x}{x^2} \times x^2}{\cos 2x . bx^3} = -a$$

$$\Rightarrow \frac{4}{b} = -a \Rightarrow ab = -4.$$

Now, $10 - ab = 10 + 4 = 14$.

19. We have $x - [x] = \{x\}$

and $1 + [x] - x = 1 - \{x\}$

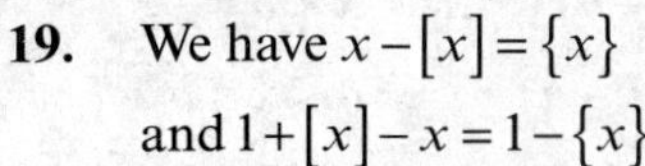

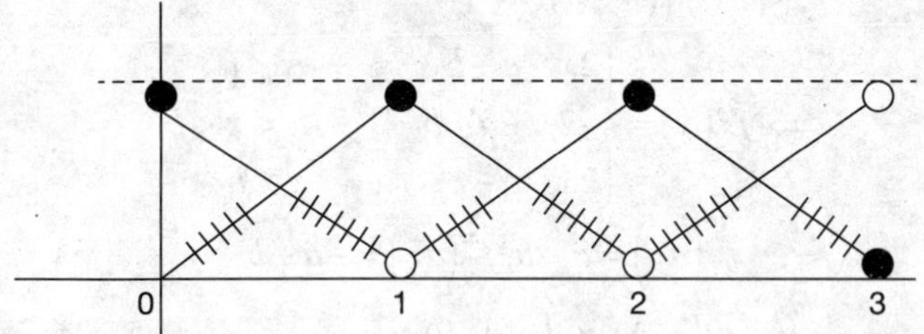

$1 - \{x\} = 1 - x; 0 \le x \le 1$ and $\{x\} = x$

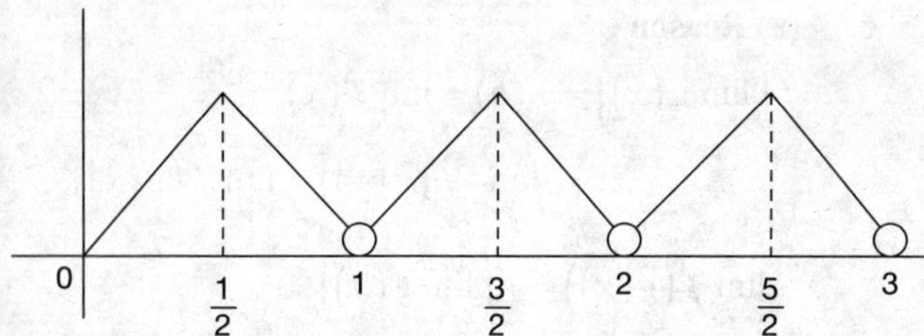

Non-differentiable at

$x = \frac{1}{2}, 1, \frac{3}{2}, 2, \frac{5}{2}$

20. $f(x) = \begin{cases} 3\left(1 - \frac{|x|}{2}\right) & \text{if} \quad -2 \le x \le 2 \\ 0 & \text{if} \quad x \in (-\infty, -2) \cup (2, \infty) \end{cases}$

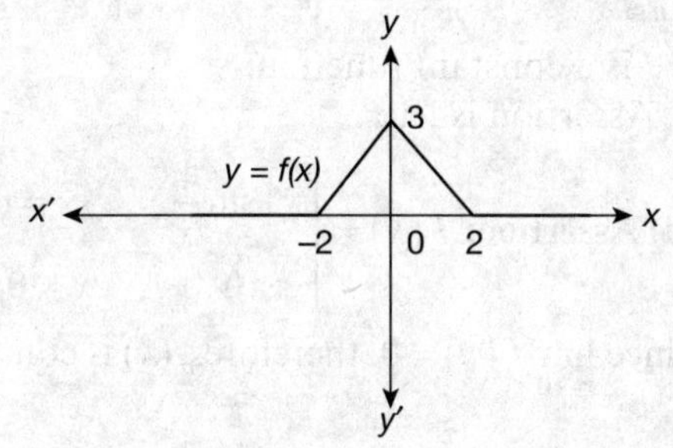

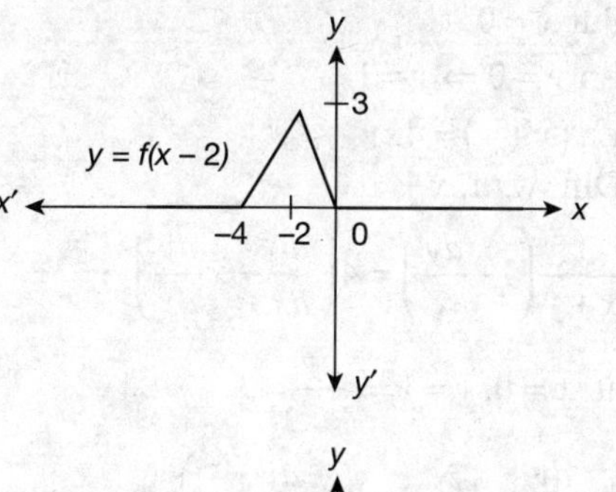

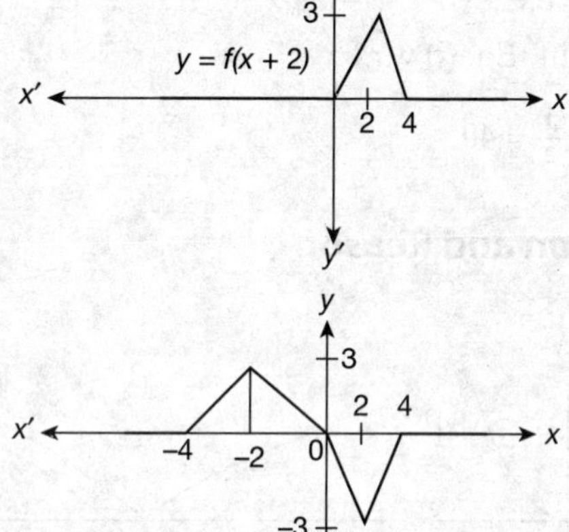

$y = g(x) = f(x+2) - f(x-2)$

$\because g(x)$ is continuous in all real number.

Therefore, $n = 0$. But $g(x)$ is not differentiable at 4 points $(x = -4, -2, 2, 4)$. Hence, $m = 4$

Now $m + n = 4$.

21. $f(x) = |2x+1| - 3|x+2| + |x^2 + x - 2|$

$= |2x+1| |-3| |x+2| + |(x+2)(x-1)|$

$$f(x) = \begin{cases} x^2 + 2x + 3, & x < -2 \\ -x^2 - 6x - 5, & -2 \le x < -\frac{1}{2} \\ -x^2 - 2x - 3, & -\frac{1}{2} \le x < 1 \\ x^2 - 7, & x \ge 1 \end{cases}$$

$$f(x) = \begin{cases} 2x + 2, & x < -2 \\ -2x - 6, & -2 \le x < -\frac{1}{2} \\ -2x - 2, & -\frac{1}{2} \le x < 1 \\ 2x, & x \ge 1 \end{cases}$$

Check at $-2, -\frac{1}{2}, 1$.

Non-differentiable at $x = 1$ and $-\frac{1}{2}$.

22. Put $x = 0$

$\text{Ln } y = 0 \Rightarrow y = 1$

$\text{Ln}(x+y) = 4xy$

Diff. w.r.t. x

$$\frac{1}{x+y}\left(1+\frac{dy}{x}\right) = 4\left(\frac{dy}{dx}+x\frac{dy}{dx}\right)$$

at $x = 0,\ y = 1 \Rightarrow \frac{dy}{dx} = 3$

$$1+\frac{dy}{dx} = 4(x+y)\left(x\frac{dy}{dx}+y\right) \quad \text{(i)}$$

Diff. Eq. (i) w.r.t x

$$\frac{d^2y}{dx^2} = 40.$$

Assertion and Reason

1. (c)

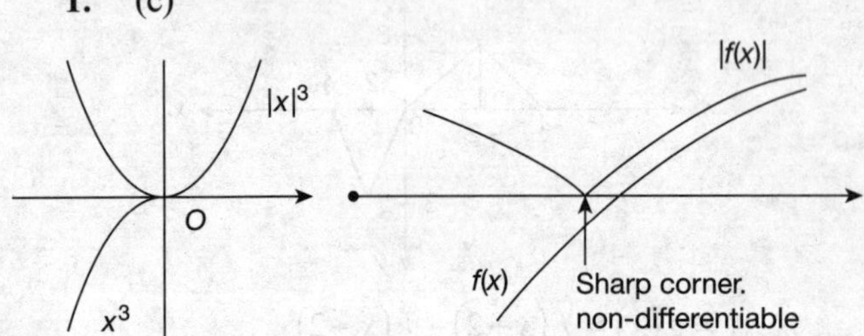

2. (a) $\lim\limits_{x\to 0^+}(\sin x + [x]) = 0$

$\lim\limits_{x\to 0^-}(\sin x + [x]) = -1$

Limit does not exist.

$\lim\limits_{x\to a}(f(x)+h(x)) = \lim\limits_{x\to a} f(x) + \lim\limits_{x\to a} h(x)$

$\neq f(a) + h(a) \therefore f(x) + h(x)$ is a discontinuous function.

3. (a) $f(x) = |x| \sin x$

$$\text{LHD} = \lim_{h\to 0}\frac{|0-h|\sin(0-h)-0}{h}$$

$$= \lim_{h\to 0}\frac{-h\sin h}{h} = 0$$

$$\text{RHD} = \lim_{h\to 0}\frac{|0-h|\sin(0+h)-0}{h};$$

$f(x)$ is differentiable at $x = 0$.

4. (a) $f(2) = 4$

$f(2^-) = \text{Lim}_{h\to 2^-} |[x]x| = 2$

Discontinuous $\Rightarrow$ Non-differentiable

5. (a)

Assertion

$$f(x) = \begin{cases} x^2-5x+6 & , \quad x \le 2 \\ -x^2+5x-6 & , \quad 2 \le x \le 3 \\ x^2-5x+6 & , \quad x \ge 3 \end{cases}$$

$$\Rightarrow f'(x) = \begin{cases} 2x-5 & , \quad x < 2 \\ -2x+5 & , \quad 2 < x < 3 \\ 2x-5 & , \quad x > 3 \end{cases}$$

$f'(2^-) + f'(2') = -1 + 1 = 0$

Reason

$$f(x) = \begin{cases} (x-a)(x-b) & , \quad x < a \\ -(x-a)(x-a) & , \quad a \le x \le b \\ (x-a)(x-b) & , \quad x > b \end{cases}$$

$$\Rightarrow f'(x) = \begin{cases} 2x-a-b & , \quad x < a \\ -2x+a+b & , \quad a < x < b \\ 2x-a-b & , \quad x > b \end{cases}$$

$\therefore\ f'(a^-) = a-b,\ f'(a^+) = -a+b$

$\therefore\ f'(a^-) + f'(a^+) = 0$

Reason explains Assertion.

6. (a) **Reason**

$$f\left(\lim_{x\to a} g(x)\right) = f(b) = \lim_{x\to b} f(x)$$

$$= \lim_{g(x)\to b} f(g(x)) = \lim_{x\to a} f(g(x))$$

$$\therefore \lim_{x\to a} f(g(x)) = f\left(\lim_{x\to a} g(x)\right)$$

$\therefore$ Reason is true.

Assertion

Since f is continuous on R

and $f(x) = f\left(\frac{x}{3}\right) = f\left(\frac{x}{3^2}\right) \ldots\ldots\ldots\ldots = f\left(\frac{x}{3^n}\right)$

and $\lim\limits_{n\to\infty}\frac{x}{3^n} = 0$

$$\therefore \lim_{n\to\infty} f(x) = \lim_{n\to\infty} f\left(\frac{x}{3^n}\right) = f\left(\lim_{x\to 0}\frac{x}{3^n}\right) = f(0)$$

$\therefore f$ is a constant function.

$\therefore$ Assertion is true.

7. (d) **Assertion:** $f(x) = \begin{cases} x^2 \sin\frac{1}{x} & , \quad x \neq 0 \\ 0 & , \quad x = 0 \end{cases}$

Since $\lim\limits_{x\to 0} f(x) = 0$, therefore, $f(x)$ is continuous.

$$f'(0)=\lim_{x\to 0}\frac{x^2\sin\frac{1}{x}-0}{x}=0$$

$$f'(x)=\begin{cases}2x\sin\frac{1}{2}-\cos\frac{1}{x} & , \quad x\neq 0\\ 0 & , \quad x=0\end{cases},$$

which is clearly not continuous at $x=0$.

$\therefore$ Assertion is false.

Reason: is true (standard result)

JEE Advanced

Objective Questions

One Option Correct

1. (a)

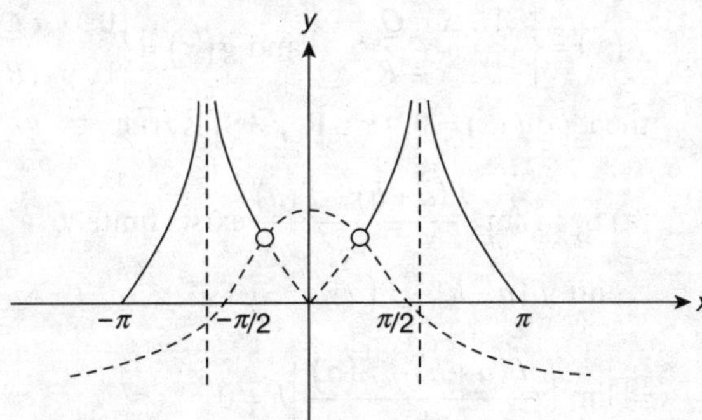

The functions is not differentiable at two points between $x=-\pi/2$ and $x=\pi/2$. Also, the function is not continuous at $x=\frac{\pi}{2}$ and $x=-\frac{\pi}{2}$.

Hence, the function is not differentiable at four points.

2. (d)

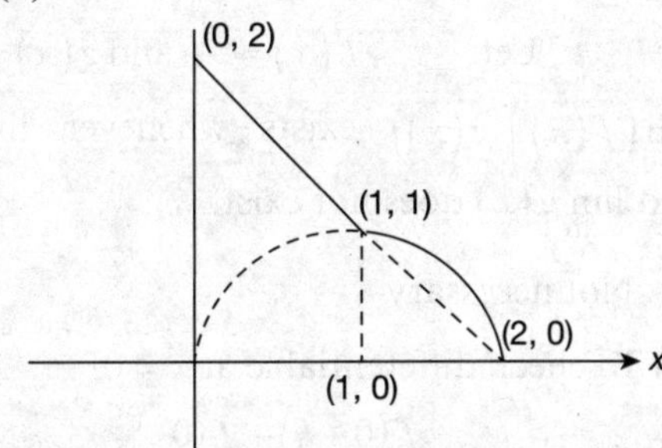

3. (b) $\dfrac{f(2x+2y)}{f(2x-2y)}=\dfrac{\sin(x+y)}{\sin(x-y)}$

$$\Rightarrow \frac{f(\alpha)}{\sin\frac{\alpha}{2}}=\frac{f(\beta)}{\sin\frac{\beta}{2}}=k$$

$$f(x)=k\sin\frac{x}{2}$$

$$f'(x)=\frac{k}{2}\cos\frac{x}{2},\ f''(x)=\frac{-k}{4}\sin\frac{x}{2}$$

$$4f''(x)+f(x)=0$$

4. (b) $f:\mathrm{R}\to\mathrm{R},\qquad g(x)=\dfrac{1}{f(x)}$

$g'(x)=-\dfrac{1}{f(x)^2}f'(x)\Rightarrow g$ is one–one if f is one–one.

5. (a) $f(x)=[x](\sin kx)^p$

$(\sin kx)^p$ is continuous and differentiable function $\forall x\in\mathrm{R}, k\in\mathrm{R}$ and $p>0$.

[X] is discontinuous at $x\in 1$

For $k=n\pi,\ n\in\mathrm{I}$

$f(x)=[x](\sin(n\pi x))^p$

$\lim_{x\to a}f(x)=0,\ a\in\mathrm{I}$

and $f(a)=0$

So. $f(x)$ becomes continuous for all $x\in\mathrm{R}$.

6. (a) $f(x)=\begin{cases}x+2 & x<0\\ -x^2-2 & 0\le x<1\\ x & x\ge 1\end{cases}$

$$\therefore |f(x)|=\begin{cases}-x-2 & x<-2\\ x+2 & -2\le x<0\\ x^2+2 & 0\le x<1\\ x & x\ge 1\end{cases}$$

Discontinuous at $x=1$ $\therefore$ Number of points of discount = 1

7. (b) $\lim_{x\to 0^+}\dfrac{e^{e/x}-e^{-e/x}}{e^{1/x}+e^{-1/x}}=\lim_{x\to 0^+}\dfrac{e^{\frac{e-1}{ex}}\left(1-e^{-2e/x}\right)}{\left(1+e^{-2/x}\right)}=+\infty$

$$\lim_{x\to 0^-}\frac{e^{e/x}-e^{-e/x}}{e^{1/x}+e^{-1/x}}=\lim_{x\to 0^-}\frac{e^{-e/x}\left(e^{2e/x}-1\right)}{e^{-e/x}\left(e^{+2/x}+1\right)}$$

$$=\lim_{x\to 0^-}e^{-\left(\frac{e-1}{x}\right)}\left(\frac{e^{2e/x}-1}{e^{2/x}+1}\right)=-\infty$$

Limit does not exist, So $f(x)$ is discontinuous.

8. (c) $\lim_{x\to a}f(x)=\lim_{x\to a}x=a, x\in Q$

$\lim_{x\to a}f(x)=\lim_{x\to a}(-x)=-a, x\in R\sim Q$

Limit exists $\Leftrightarrow a=0$

9. (b) $f\left(g(x)\right)=\begin{cases}1 & , & x<-1\\ 0 & , & x=-1\\ -1 & , & -1<x<0\\ 0 & , & x=0\\ 1 & , & 0<x<1\\ 0 & , & x=1\\ -1 & , & x>1\end{cases}$

$\therefore$ Points of discontinuity are $x=-1, 0, 1$

10. (d) Let $z=-2+0i$, then $\bar{z}=-2-0i$ $\therefore$ Arg (z) + Arg $(\bar{z})=2\pi\neq 0$
If $z=2+3i$

Arg $(2+3i)$ is $\tan^{-1}\frac{3}{2}$.

Arg $(2-3i)$ is $\tan^{-1}\left(-\frac{3}{2}\right)$.

Arg $(2+3i)$ + Arg $(2-3i)=0$.

11. (d) From figure,

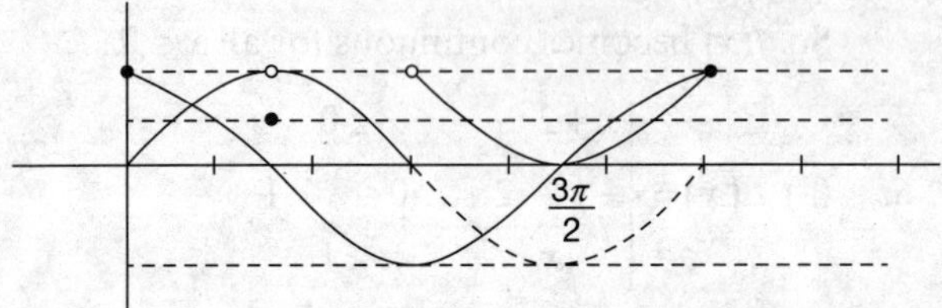

there are 4 points.

12. (a) $\lim_{x\to\frac{\pi}{2}}\left(2x\tan x-\frac{\pi}{\cos x}\right)$

$=\lim_{x\to\frac{\pi}{2}}\left(\frac{2x\sin x-\pi}{\cos x}\right)\lim_{h\to 0}\left(\frac{2\left(\frac{\pi}{2}+h\right)\cos h-\pi}{-\sin h}\right)$

$=\lim_{x\to 0}-\frac{2h\cos h}{\sin h}=-2$ $\therefore k=-2$

13. (a) $\lim_{x\to 0}(fx)=0$

$\{\because \lim_{x\to 0}x^2=0$ and $\left\{e^{1/x}\right\}$ is a bounded function$\}$

$\because k=0$

$\lim_{x\to 0}\frac{f(0+x)-f(0)}{x}=\lim_{x\to 0}x\left\{e^{1/x}\right\}=0$

$\therefore f'(0)=0$

Not continuous at $x=\log_2 e, \log_3 e,\ldots$ etc.

14. (c) S_1: False (take $f(x)=0$, $x\in$ R.
S_2: Domain of $f(x)$ is $\{2\}$
$\therefore f(x)$ is not continuous at $x=2$.

S_3: $e^{-|x|}$ is not differentiable at $x=0$.

S_4: Derivative of $|f|^2$ is 0 wherever $f(x)$ is 0 and the derivative of $|f|^2$ is $2|f(x)|.\frac{f(x)}{|f(x)|}=2f(x)$ wherever $f(x)\neq 0$

15. (b) S_1: $\lim_{x\to a}f(x)=f(a)$

$\therefore \lim_{x\to a}\left[f(x)\right]$ is an integer and

$\lim_{x\to a}\left[f(x)\right]\lim_{x\to a}f(x)=f(a)$

$\therefore S_1$ is true.
S_2: Derivative of $\cos|x|$ at $x=0$ is 0 but derivative of $|x|$ does not exist at $x=0$.
$\therefore S_2$ is false
S_3: False: Consider the function $x^{1/3}$.
S_4: Let

$f(x)=\begin{cases}1, & x\in Q\\ -1, & x\in R\sim Q\end{cases}$ and $g(x)=\begin{cases}0, & x\in Q\\ 1, & x\in R\sim Q\end{cases}$

then $g\ o\ t\ (x)=0, \forall x\in$ R $\therefore$ S_4 is true.

16. (a) S_1 : $\lim_{h\to 0^+}\frac{f(a+h)-f(a)}{h}$ exists finitely.

$\therefore \lim_{h\to 0^+}f(a+h)-f(a)$

$=\lim_{x\to 0^+}\left(\frac{f(a+h)-f(a)}{h}\right)h=0$

$\Rightarrow \lim_{h\to 0^+}f(a+h)-f(a)$

Similarly $\lim_{h\to 0^-}f(a+h)-f(a)$

$\therefore f$ is continuous at $x=a$

S_2 : The function is not differentiable at $5x=(2n+1)\frac{\pi}{2}$ only, which is not in its domain.

S_3: Let $f(x)=\frac{1}{x^2}$ and $g(x)=-\frac{1}{x^2}$, $\lim_{x\to 0}\left(f(x)+g(x)\right)$ exists whenever $\lim_{x\to 0}f(x)$ and $\lim_{x\to 0}g(x)$ does not exist.

S_4 : Not necessary.

17. (b) To check differentiable at $x=0$

$R\left\{f'(0)\right\}=\lim_{h\to 0}\frac{f(0+h)-f(0)}{h}$

$=\lim_{h\to 0}\frac{h^2\left|\cos\frac{\pi}{h}\right|-0}{h}$

$=\lim_{h\to 0}h.\left|\cos\frac{\pi}{h}\right|=0$

$$L\{f'(0)\} = \lim_{h\to 0}\frac{f(0-h)-f(0)}{-h}$$

$$= \lim_{h\to 0}\frac{h^2\left|\cos\left(-\frac{\pi}{h}\right)\right|-0}{-h}$$

$$= 0$$

So, $f(x)$ is differentiable at $x = 0$

Now check the differentiability at $x = 2$, you will find that it is not differentiable at $x = 2$.

One or More Than One Option Correct

1. (a, d) $\lim_{x\to 2^-} f(x) = 3+1 = 4$

$\lim_{x\to 2^-} f(x) = 3+1 = 4$

$\Rightarrow \lim_{x\to 2^+} f(x) = \lim_{x\to 2^+}\frac{\sin(2x-4)}{x-2} = 2$

2. (a, b, c, d)

$f(x)$ is non-derivable at $x = \pm\frac{1}{2}, \pm 1$

3. (a, d)

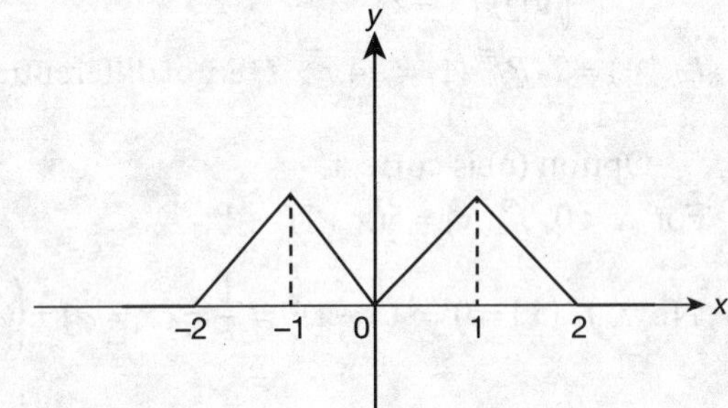

4. (b, d)

(a) $\lim_{x\to 1} f(x)$ does not exist.

(b) $\lim_{x\to 1} f(x) = \frac{2}{3}$

$\therefore f(x)$ has removable discontinuity at $x = 1$

(c) $\lim_{x\to 1} f(x)$ does not exist.

(d) $\lim_{x\to 1} f(x) = \frac{-1}{2\sqrt{2}}$

$\therefore f(x)$ has removable discontinuity at $x = 1$.

5. (a, c, d)

$f(x + y) = f(x) + f(y) + xy\,(x + y)$

$f(0) = 0 \therefore \lim_{h\to 0}\frac{f(h)}{h} = -1$

$$\therefore \lim_{h\to 0}\frac{f(x+h)-f(x)}{h}$$

$$= \lim_{h\to 0}\frac{f(x)+f(h)+xh(x+h)-f(x)}{h}$$

$$= \lim_{h\to 0}\frac{f(h)}{h} + \lim_{h\to 0} x(x+h) = -1+x^2$$

$$\therefore f'(x) = -1+x^2$$

$$\therefore f(x) = \frac{x^3}{3} - x + c$$

$\therefore f(x)$ is a polynomial function, $f(x)$ is twice differentiable for all $x \in R$ and $f'(3) = 3^2 - 1 = 8$.

6. (a, b, c, d)

$$f(x) = \int_{-2}^{x}|t+1|\,dt \quad = -\int_{-2}^{-1}(t+1)\,dt + \int_{-1}^{x}(t+1)\,dt$$

$$= \frac{1}{2} + \left(\frac{t^2}{2}+t\right)_{-1}^{x} \quad = \frac{x^2}{2} + x + 1 \text{ for } x \geq -1$$

$f(x)$ is a quadratic polynomial.

$\therefore f(x)$ is continuous as well as differentiable in $[-1, 1]$

also $f'(x)$ is continuous as well as differentiable in $[-1, 1]$

7. (a, b, d)

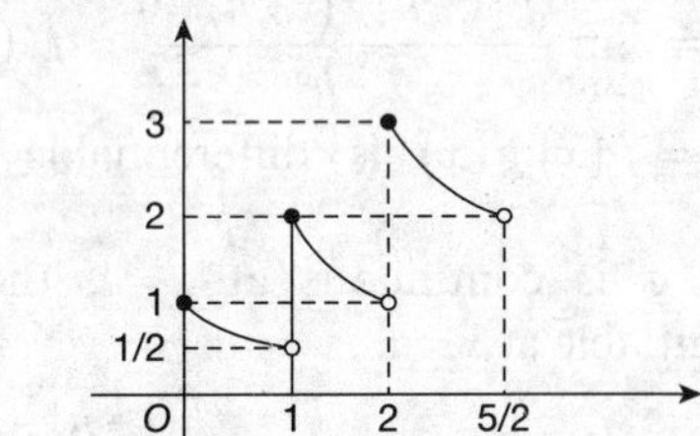

$$f(x) = \begin{cases} \dfrac{1}{x+1}, & 0 \leq x < 1 \\ \dfrac{2}{x}, & 1 \leq x < 2 \\ \dfrac{3}{x-1}, & 2 \leq x < \dfrac{5}{2} \end{cases}$$

Clearly, $f(x)$ is discontinuous and objective function.

$\lim_{x\to 1^-} f(x) = \frac{1}{2}$

$\lim_{x\to 1^+} f(x) = 2$

$$\min\left(\lim_{x\to 1^-} f(x), \lim_{x\to 1^+} f(x)\right) = \frac{1}{2} \neq f(1)$$

$\max(1, 2) = 2 = f(1)$.

8. (a, d)

Both $x^2, -x^{3/2}$ have their RHL = 0 and RHD = 0.

9. (a, c)

Let $f: \mathrm{R} \to \mathrm{R}$ is defined by
$f(x) = (x^2 + \sin x)(x-1)$.

Then, $f(1^+) = f(1^-) = f(1) = 0$.

Let $(fg): \mathrm{R} \to \mathrm{R}$ is defined by
$(fg)(x) = f(x).g(x)$.

Let $fg(x) = h(x) = f(x).g(x)$ then h: $\mathrm{R} \to \mathrm{R}$.

$h'(x) = f'(x)g(x) + f(x)g'(x)$

If g is differentiable at $x = 1$

$h'(1) = f'(1)g(1) + 0$, $[\because f(1) = 0]$

$\Rightarrow$ if $g(x)$ is differentiable then $h(x)$ is also differentiable (true).

$\Rightarrow$ if $g(x)$ is differentiable at $x = 1$, then fg is also differentiable at $x = 1$.

If $g(x)$ is continuous at $x = 1$, then $g(1^+) = g(1^-) = g(1)$.

$$h'(1^+) = \lim_{h\to 0^+} \frac{h(1+h) - h(1)}{h}$$

$$h'(1^+) = \lim_{h\to 0^+} \frac{f(1+h)g(1+h) - 0}{h} = f'(1)g(1)$$

$$h'(1^-) = \lim_{h\to 0^+} \frac{f(1-h)g(1-h) - 0}{-h} = f'(1)g(1)$$

$\Rightarrow h(x = f(x).g(x)$ is differentiable at $x = 1$ (True)

So, if g is continuous at $x = 1$, then fg is differentiable at $x = 1$.

$$\text{Options (b),(d)}: h'(1^+) \lim_{h\to 0^+} \frac{h(1+h) - h(1)}{-h}$$

$$h'(1^+) \lim_{h\to 0^+} \frac{f(1+h)g(1+h)}{h} = f'(1)g(1^+)$$

$$h'(1^-) \lim_{h\to 0^+} \frac{f(1+h)g(1-h)}{-h} = f'(1)g(1^-)$$

$\Rightarrow g(1^+) = g(1^-)$

So, it does not mean that if fg is differentiable at $x = 1$, then g is continuous or differentiable at $x = 1$.

10. (a, c, d)

$$f(x) = \begin{cases} x^5 + 5x^4 + 10x^3 + 10x^2 + 3x + 1, & x < 0; \\ x^2 - x + 1, & 0 \le x < 1; \\ \frac{2}{3}x^3 - 4x^2 + 7x - \frac{8}{3}, & 1 \le x < 3; \\ (x-2)\log_e(x-2), & x \ge 3 \end{cases}$$

$$= \begin{cases} (x+1)^5, & x < 0; \\ \left(x - \frac{1}{2}\right)^2 + \frac{3}{4}, & 0 \le x < 1; \\ \frac{2}{3}x^3 - 4x^2 + 7x - \frac{8}{3}, & 1 \le x < 3; \\ (x-2)\log_e(x-2), & x \ge 3 \end{cases}$$

For $x = 0$, $f(x) = 1$.

For $x < 0$, $f(x) = (x+1)^5 - 2x$

It decreases to $-\infty$.

$\therefore f(x) \in (-\infty, 1]$ for $x \le 0$

For $x = 3$, $f(x) = \frac{1}{3}$.

For $x \ge 3$, $f(x)$ increases to ∞.

$\therefore f(x) \in \left[\frac{1}{3}, \infty\right)$ for $x \ge 3$

On combining the two $f(x) \in R \Rightarrow f$ is onto.

$\therefore$ Option (d) is correct.

$$f'(x) = \begin{cases} 5(x+1)^4 - 2, & x < 0 \\ 2x - 1, & 0 \le x < 1 \\ 2(x-2)^2 - 1, & 1 \le x < 3 \\ \log_e(x-2), & x \ge 3 \end{cases}$$

$Lf'')1) = 2$, $Rf''(1) = -4, \Rightarrow f'$ is not differentiable at $x = 1$

$\therefore$ Option (c) is correct.

For $x < 0$, $f'(x) = 5(x+1)^4 - 1$

Now, $f'(x) = 0 \Rightarrow (x+1)^4 = \frac{1}{5} \Rightarrow x = -1 \pm \left(\frac{1}{5}\right)^{1/4}$

$\Rightarrow f$ is changing its nature at $x = -1 - \left(\frac{1}{5}\right)^{\frac{1}{4}}$

$\therefore f$ is not increasing on $(-\infty, 0)$

$\therefore$ Option (b) is incorrect.

$$f'(x) = \begin{cases} 2x - 1, & 0 \le x < 1 \\ 2(x-2)^2 - 1, & 1 \le x < 3 \end{cases}$$

From its graph $f(x)$ has local maxima at $x = 1$.

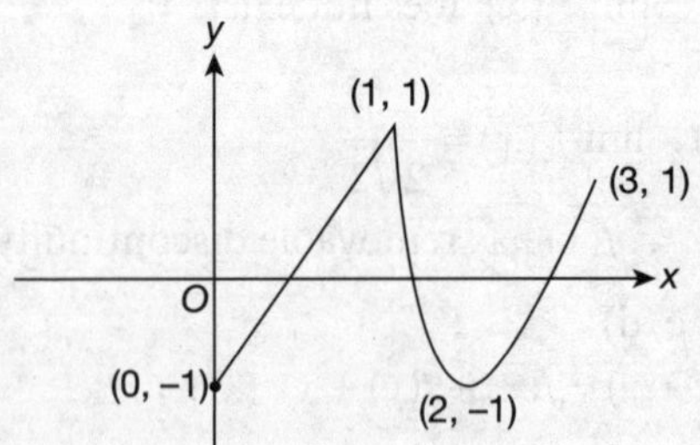

$\therefore$ Option (a) is correct.

11. (b, c)

Given,

$f'(x) = \left(e^{(f(x)-g(x))}\right) g'(x) \forall x \in R,$

$\Rightarrow e^{-f(x)} f'(x) = e^{-g(x)} g'(x)$

Integrating both sides, we get

$-e^{-f(x)} = -e^{-g(x)} + c \Rightarrow -e^{-f(x)} + e^{-g(x)} = c$

$\Rightarrow -e^{-f(1)} + e^{-g(1)} = -e^{-f(2)} + e^{-g(2)}$

$\therefore -e^{-1} + e^{-g(1)} = -e^{-f(2)} + e^{-1}\ [\because f(1) = g(2) = 1]$

$\Rightarrow e^{-f(2)} + e^{-g(1)} = \frac{2}{e}$

$\Rightarrow e^{-f(2)} < \frac{2}{e}$ and $e^{-g(1)} < \frac{2}{e}$

$\Rightarrow -f(2) < \text{In}2 - 1$ and $-g(1) < \text{In}2 - 1$

$\Rightarrow$ $f(2) > 1 - \text{In}2$ and $g(1) > 1 - \text{In}2$

12. (b, c)

$$f(x) = [x^2 - 3] = [x^2] - 3 = \begin{cases} -3, & -1/2 \le x < 1 \\ -2, & 1 \le x < \sqrt{2} \\ -1, & \sqrt{2} \le x < \sqrt{3} \\ 0, & \sqrt{3} \le x < 2 \\ 1, & x = 2 \end{cases}$$

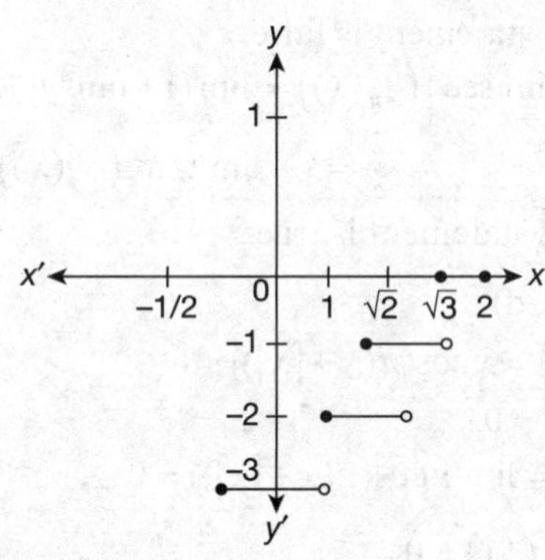

Clearly, $f(x)$ is discontinuous at 4 points. Option (b) is correct.

and $g(x) = |x| f(x) + |4 \pm x - 7| f(x)$

$= (|x| + |4x - 7|) f(x)$

$= (|x| + |4x - 7|)[x^2 - 3]$

$$= \begin{cases} (-x - 4x - 7)(-3), & -1/2 \le x < 0 \\ (x - 4 + 7)(-3), & 0 \le x < 1 \\ (x - 4x + 7)(-2), & 1 \le x < \sqrt{2} \\ (x - 4x + 7)(-1), & \sqrt{2} \le x < \sqrt{3} \\ (x - 4x + 7)(0), & \sqrt{3} \le x < 7/4 \\ (x + 4x - 7)(0), & 7/4 \le x < 2 \\ (x + 4x - 7)(1), & x = 2 \end{cases}$$

$$\therefore g(x) = \begin{cases} 15x + 21, & -1/2 \le x < 0 \\ 9x - 21, & 0 \le x < 1 \\ 6x - 14, & 1 \le x < \sqrt{2} \\ 3x - 7, & \sqrt{2} \le x < \sqrt{3} \\ 0, & \sqrt{3} \le x < 2 \\ 5x - 7, & x = 2 \end{cases}$$

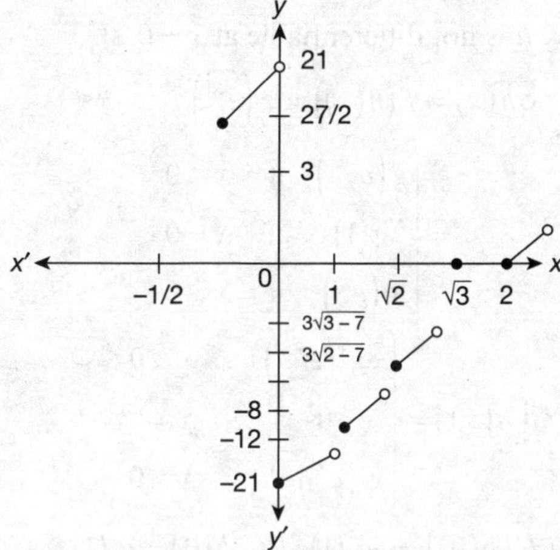

Clearly, $g(x)$ is not differentiable at 4 points, when $x \in (-1/2, 2)$.

$\therefore$ Option (c) is correct.

13. (a, b)

$f(x) = a\cos(|x^3 - x|) + b|x|\sin(|x^3 + x|)$

(a) If $a = 0, b = 1$

$\Rightarrow f(x) = |x|\sin|x^3 + x| = x\sin(x^3 + x)$

which is differentiable everywhere.

Options (b), (c): If $a = 1, b = 0$

$\Rightarrow f(x) = \cos^3(|x^3 - x)| = \cos^3(x^3 - x)$

Which is differentiable everywhere.

(d) When

$a = 1, b = 1, f(x) = \cos(x^3 - x) + x\sin(x^3 + x)$

Hence only options (a) and (b) are the correct options.

14. (a, d)

$$f(x) = \begin{cases} \frac{x}{|x|} g(x), & x \ne 0 \\ 0, & x = 0 \end{cases} = \begin{cases} -g(x), & x < 0 \\ 0, & x = 0 \\ g(x), & x > 0 \end{cases}$$

$$f'(x) = \begin{cases} -g'(x), & x < 0 \\ 0, & x = 0 \\ g'(x), & x > 0 \end{cases}$$

$\therefore Lf'(0) = -g'(0) = 0$ and $Rf'(0) = g'(0) = 0$

$\therefore f$ is differentiable at $x=0$

$$h(x)=\begin{cases} e^{|x|}, & x<0 \\ e^{x}, & x\ge 0\end{cases}$$

$$h'(x)=\begin{cases} -e^{-x}, & x<0 \\ e^{x}, & x\ge 0\end{cases}$$

$\therefore Lh'(0)=-1, Rh'(0)=1$

$\therefore h$ is not differentiable at $x=0$

$$f o h(x)=f(h(x))=f\left(e^{|x|}\right)$$

$$=\begin{cases} g\left(e^{-x}\right) & \text{if } x<0 \\ g(1) & \text{if } x=0 \\ g\left(e^{x}\right) & \text{if } x>0\end{cases}$$

$$f'[h(x)]=\begin{cases} -g'\left(e^{-x}\right), & x<0 \\ 0, & x=0 \\ g'\left(e^{x}\right)e^{x}, & x>0\end{cases}$$

$Lf'(h(0))=-g'(1), Rf')h(0)=g'(1)$

$g'(1)\ne 0, Lf'(h(0))\ne Rf'(h(0))$

foh is not differentiable at $x=0$.

$$h o f(x)=\begin{cases} e^{|f(x)|}, & x\ne 0 \\ 1, & x=0\end{cases}$$

$$Lh'(f(0))=\lim_{k\to 0}\frac{h(f(0))-h(f(0-k))}{k}$$

$$=\lim_{k\to 0}\frac{1-e^{|g(-k)|}}{k}=\lim_{k\to 0}\frac{1-e^{|g(-k)|}}{|g(-k)|}\times\frac{|g(-k)|}{k}$$

$$=1\times 0=0$$

$$\left(\because g'(0)=0\Rightarrow \lim_{k\to 0}\frac{g(-k)}{k}=\lim_{k\to 0}\frac{g(k)}{k}=0\right)$$

$$Rh'(f(0))=\lim_{k\to 0}\frac{h(f(1+k))-h(f(0))}{k}$$

$$=\lim_{k\to 0}\frac{e^{|g(k)|}-1}{k}=\lim_{k\to 0}\frac{e^{|g(k)|}-1}{|g(k)|}\times\frac{|g(k)|}{k}=0$$

$\because Lh'(f(0))=Rh'(f(0)=0$

$\therefore h o f$ is differentiable at $x=0$.

15. (d)

We have,

$$f_n(x)=\sum_{j=1}^{n}\tan^{-1}\left(\frac{1}{1+(x+j)(x+j-1)}\right)$$

for all $x\in(0,\infty)$.

$$\Rightarrow f_n(x)=\sum_{j=1}^{n}\tan^{-1}\left(\frac{(x+j)-(x+j-1)}{1+(x+j)(x+j-1)}\right)$$

$$\Rightarrow f_n(x)=\sum_{j=1}^{n}\tan^{-1}\left[\tan^{-1}(x+j)-\tan^{-1}(x+j-1)\right]$$

$$\Rightarrow f_n(x)=\left(\tan^{-1}(x+j)-\tan^{-1}x\right)+$$
$$\left(\tan^{-1}(x+2)-\tan^{-1}(x+1)\right)+$$
$$\left(\tan^{-1}(x+3)-\tan^{-1}(x+3)\right)+\ldots+$$
$$\left(\tan^{-1}(x+n)-\tan^{-1}(x+n-1)\right)$$

$$\Rightarrow f_n(x)=\tan^{-1}(x+n)-\tan^{-1}x.$$

This statement is false as $x\ne 0$, i.e., $x\in(0,\infty)$.

(b) This statement is also false as $0\in(0,\infty)$.

(c) $f_n(x)=\tan^{-1}(x+n)-\tan^{-1}x$

$$\lim_{x\to\infty}\tan f_n(x)=\lim_{x\to\infty}\tan(\tan^{-1}(x+n)-\tan^{-1}x)$$

$$\Rightarrow \lim_{x\to\infty}\tan(f_n(x))=\lim_{x\to\infty}\tan\left(\tan^{-1}\frac{n}{1+nx+x^2}\right)$$

$$=\lim_{x\to\infty}\frac{n}{1+nx+x^2}=0$$

$\therefore$ (c) statement is false.

(d) $\lim\limits_{x\to\infty}\sec^2(f_n(x))=\lim\limits_{x\to\infty}(1+\tan^2 f_n(x))$

$$=1+\lim_{x\to\infty}\tan^2(f_n(x))=1+0=1$$

$\therefore$ (d) statement is true.

16. (a, b, d)

$f(x)=x\cos(\pi(x+[x]))$

At $x=0$,

$\lim\limits_{x\to 0}=\lim\limits_{x\to 0}x\cos(\pi(x+[x]))=0$

and $f(x)=0$.

$\therefore$ It is continuous at $x=0$ and clearly discontinuous at other in integer points.

17. (a, d)

We take two cases.

Case I: Let f and g attain their common maximum value at p.

$\Rightarrow f(p)=g(p)$

where $p\in[0,1]$.

Case II: Let f and g attain their common maximum value at different points.

$\Rightarrow f(a)=M$ and $g(c)=M$

$\Rightarrow f(a)-g(a)>0$ and $f(b)-g(b)<0$

$\Rightarrow f(c)-g(c)=0$ for some $c\in[0,1]$ for all cases …(i)

Option(a)$\Rightarrow f^2(c)-g^2(c)+3[f(c)-g(c)]=0$ which is true from Eq. (i).

Option (d) $\Rightarrow f^2(c)-g^2(c)=0$ which is true from Eq. (i)

Now, if we take $f(x)=1$ and $g(x)=1, \forall x\in[0,1]$ Options (b) and (c) do not hold. Hence, option (a) and (d) are correct.

18. (b, c)

Here,

$$f(x)=\lim_{n\to\infty}\left[\frac{n^n(x+n)\left(x+\frac{n}{2}\right)\ldots\left(x+\frac{n}{n}\right)}{n!\left(x^2+n^2\right)\left(x^2+\frac{n^2}{4}\right)\ldots\left(x^2+\frac{n^2}{n^2}\right)}\right]^{\frac{x}{n}}, x>0$$

Taking log on both sides, we get

$$\log\{f(x)\}=\lim_{n\to\infty}\left[\frac{n^n(x+n)\left(x+\frac{n}{2}\right)\ldots\left(x+\frac{n}{n}\right)}{n!\left(x^2+n^2\right)\left(x^2+\frac{n^2}{4}\right)\ldots\left(x^2+\frac{n^2}{n^2}\right)}\right]^{\frac{x}{n}}$$

$$=\lim_{n\to\infty}\frac{x}{n}.\log\left[\frac{\prod_{r=1}^{n}\left(x+\frac{1}{r/n}\right)}{\prod_{r=1}^{n}\left(x^2+\frac{1}{(r/n)^2}\right)\prod_{r=1}^{n}\left(\frac{r}{n}\right)}\right]$$

$$=x\lim_{n\to\infty}\frac{1}{n}\sum_{r=1}^{n}\log\left[\frac{x+\frac{n}{r}}{\left(x^2+\frac{n^2}{r^2}\right)\frac{r}{n}}\right]$$

$$=x\lim_{n\to\infty}\frac{1}{n}\sum_{r=1}^{n}\log\left[\frac{\frac{r}{n}.x+1}{\frac{r^2}{n^2}.x^2+1}\right]$$

Converting summation into definite integration, we get

$$\log_e\{f(x)\}=x\int_0^1\log\left(\frac{xt+1}{x^2t^2+1}\right)dt$$

Put, $tx=z$

$\Rightarrow \quad xdt=dz$

$$\therefore \log_e\{f(x)\}=x\int_0^x\log\left(\frac{1+z}{1+z^2}\right)\frac{dz}{x}$$

$$\Rightarrow \quad \log_e\{f(x)\}=\int_0^x\log\left(\frac{1+z}{1+z^2}\right)dz$$

Using Newton–Leibnitz formula, we get

$$\frac{1}{f(x)}.f'(x)=\log\left(\frac{1+x}{1+x^2}\right) \quad \ldots\text{(i)}$$

Here, at $x=1$

$$\frac{f'(1)}{f(1)}=\log(1)=0$$

$\therefore f'(1)=0$

Now, sign scheme of $f'(x)$ is shown below:

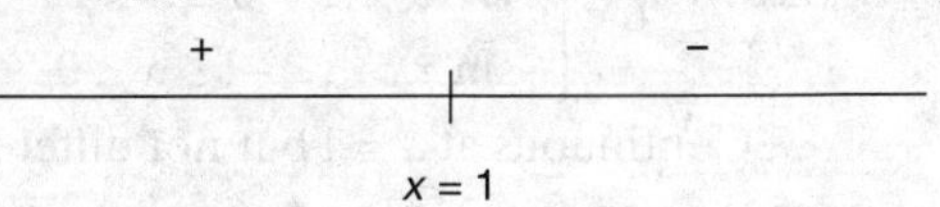

$\therefore$ At $x=1$, function attains maximum.

Since, $f(x)$ increases on $(0,1)$.

$\therefore f(1)>f(1/2)$

$\therefore$ Option (a) is incorrect.

$\therefore f(1/3)>f(2/3)$

$\therefore$ Option (b) is correct.

Also, $f'(x)<0$, when $x>1\Rightarrow f'(2)<0$

$\therefore$ Option (c) is correct.

Also, $\frac{f'(x)}{f(x)}=\log\left(\frac{1+x}{1+x^2}\right)$

$$\therefore \frac{f'(3)}{f(3)}-\frac{f'(2)}{f(2)}=\log\left(\frac{4}{10}\right)-\log\left(\frac{3}{5}\right)$$

$$=\log(2/3)<0\Rightarrow\frac{f'(3)}{f(3)}<\frac{f'(2)}{f(2)}$$

$\therefore$ Option (d) is incorrect.

Matrix Type

1. $A\to R; B\to T; C\to R; D\to Q$

(A) $\tan^2 x$ is discontinuous at $x=\frac{\pi}{2},\frac{3\pi}{2}$

$\sec^2 x$ is discontinuous at $x=\frac{\pi}{2},\frac{3\pi}{2}$

$\Rightarrow$ Number of discontinuities $=2$

(B) Since

$$f(x)=\sin^{-1}x+\tan^{-1}x+\cot^{-1}x=\sin^{-1}x+\frac{\pi}{2}$$

$\therefore$ $f(x)$ is differentiable in $(-1,1)\Rightarrow$ No. of points non-differentiable $=0$

$$\text{(C) } y=[\sin x]=\begin{cases}0, & 0\le x\le\frac{\pi}{2}\\ 1, & x=\frac{\pi}{2}\\ 0, & \frac{\pi}{2}<x\le\pi\\ -1, & \pi<x<2\pi\\ 0, & x=2\pi\end{cases}$$

$\therefore$ Points of discontinuity are $\frac{\pi}{2}, \pi$.

(D) $y = |(x-1)^3| + |(x-2)^5| + |x-3|$ is not differentiable at $x = 3$ only.

2. A → Q; B → P; C → S; D → P

(A) $f(x) = \begin{cases} 0 & , \quad 1 < x \le 2 \\ 1-x & , \quad 0 \le x < 1 \\ -\sin \pi x & , \quad -1 \le x < 0 \end{cases}$

Continuous at $x = 1$ but not differentiable.

(B) $f'(0^-) = \lim_{h \to 0^-} \frac{h^2 e^{-1/h} - 0}{-h} = \lim_{h \to 0^-} \left(-he^{-1/h}\right) I$

(C) $g(x) = \frac{1}{1 + \frac{1}{x}(2+2x)} = \frac{x}{3x+2}$.

Thus the points where $g(x)$ is not differentiable are $x = 0, -1, -\frac{2}{3}$.

(D) Vertical tangents exist at $x = 1$ and $x = -1$ else where horizontal tangents exist.

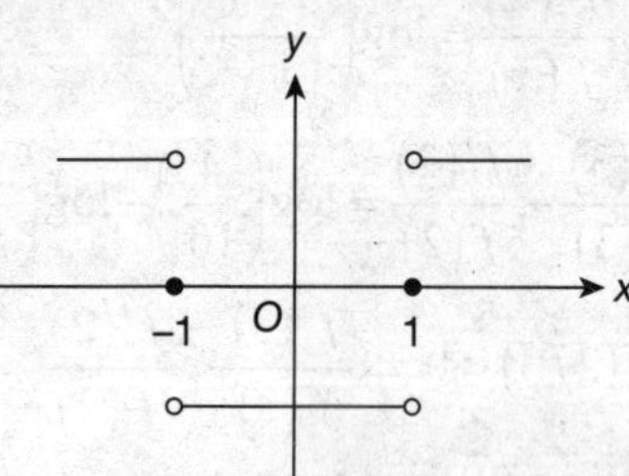

$\therefore$ Number of points where tangents do not exist is 0.

3. A → P, T, R; B → P, R, S; C → P, R, S; D → P, R, S

(A) $f(x) = |x^3|$ is continuous and differentiable.

(B) $f(x) = \sqrt{|x|}$ is continuous.

$f'(x) = \frac{1}{2\sqrt{|x|}} \cdot \frac{x}{|x|}$ {does not exist at $x = 0$}.

(C) $f(x) = |\sin^{-1} x|$ is continuous.

$f'(x) = \frac{\sin^{-1} x}{|\sin^{-1} x|} \cdot \frac{1}{\sqrt{1-x^2}}$ {does not exist at $x = 0$}

(D) $f(x) = \cos^{-1} x$ is continuous.

$f'(x) = \frac{-1}{\sqrt{1-x^2}} \cdot \frac{x}{|x|}$ {does not exist at $x = 0$}.

Integer and Numerical Value Type

1. $f(x) = \begin{cases} ax^3 + b & , \quad 0 \le x \le 1 \\ 2\cos \pi X + \tan^{-1} x & , \quad 1 < x \le 2 \end{cases}$

as $\tan\left[x^2\right]\pi = \tan n\pi, n \in \text{I}$

$f'(x) = \begin{cases} 3ax^2 & , \quad 0 < x < 1 \\ -2\pi \sin \pi x + \frac{1}{1+x^2} & , \quad 1 < x < 2 \end{cases}$

As the function is differentiable in [0, 2]

$\Rightarrow$ Function is differentiable at $x = 1$.

$\therefore f'(1^-) = f'(1^+) \Rightarrow 3a = \frac{1}{2} \Rightarrow a = \frac{1}{6}$.

Function will also be continuous at $x = 1$.

$\therefore \lim_{x \to 1^-} f(x) = \lim_{x \to 1^-} f(x) \Rightarrow a + b = -2 + \frac{\pi}{4}$

$\therefore b = -2 - \frac{1}{6} + \frac{\pi}{4} = \frac{\pi}{4} - \frac{13}{6} \Rightarrow k_1 = 6$ and $k_2 = 12$

$\Rightarrow k_1^2 + k_2^2 = 180$

2. $\lim_{x \to 0^+} \frac{|x|^p \sin\frac{1}{x} + x|\tan x|^q - 0}{x}$

$= \lim_{x \to 0^+} \left(x^{p-1} \sin\frac{1}{x} + x|\tan x|^q\right) = 0$

for $p - 1 > 0$ and $q > 0$ ...(i)

$\lim_{x \to 0^-} \frac{|x|^p \sin\frac{1}{x} + x|\tan x|^q - 0}{x}$

$= \lim_{x \to 0^-} (-1)^p x^{p-1} \sin\frac{1}{x} + |\tan x|^q = 0$

for $p - 1 > 0$ and $q > 0$ ...(ii)

$\therefore f(x)$ is differentiable if $p > 1$ and $q > 0$, i.e., $p + q > 1$.

$\therefore$ Least possible value of $[p+q] = 1$.

3. (i) $f(x) = \sin^2\left(2x\sqrt{1-x^2}\right)$

$= \begin{cases} -\pi - 2\sin^{-1} x & , \quad -1 \le x < -\frac{1}{\sqrt{2}} \\ 2\sin^{-1} x & , \quad -\frac{1}{\sqrt{2}} \le x \le \frac{1}{\sqrt{2}} \\ \pi - 2\sin^{-1} x & , \quad \frac{1}{\sqrt{2}} < x \le 1 \end{cases}$

$$f'(x)=\begin{cases}-\dfrac{2}{\sqrt{1-x^2}} & -1\le x<-\dfrac{1}{\sqrt{2}}\\ \dfrac{2}{\sqrt{1-x^2}} & -\dfrac{1}{\sqrt{2}}\le x\le\dfrac{1}{\sqrt{2}}\\ \dfrac{-2}{\sqrt{1-x^2}} & \dfrac{1}{\sqrt{2}}<x\le 1\end{cases}$$

$$f'(1/2)=\frac{4}{\sqrt{3}},\ f'(-1/2)=\frac{4}{\sqrt{3}}$$

4. $f(x+y+1)=\left(\sqrt{f(x)}+\sqrt{f(y)}\right)^2$

$f(1)=4$

$$\frac{\partial}{\partial x}f'(x+y+1)=2\left(\sqrt{f(x)}+\sqrt{f(y)}\right).\frac{f'(x)}{2\sqrt{f(x)}}$$

$$\frac{\partial}{\partial y}f'(x+y+1)=\left(\sqrt{f(x)}+\sqrt{f(y)}\right).\frac{f'(x)}{\sqrt{f(x)}}$$

$$\Rightarrow \frac{f'(x)}{\sqrt{f(x)}}=\frac{f'(y)}{\sqrt{f(y)}}=k$$

$$\therefore \int\frac{f'(x)}{\sqrt{f(x)}}dx=\int k\,dx\Rightarrow 2\sqrt{f(x)}=kx+c$$

put $x=0$

$\Rightarrow \quad 4=k+2 \qquad \Rightarrow k=2$

$\therefore \quad 2\sqrt{f(x)}=2(x+1) \quad \Rightarrow f(x)=(x+1)^2$

5. Here, $\lim\limits_{x\to0}\dfrac{x^2\sin(\beta x)}{ax-\sin x}=1$

$$\Rightarrow \lim_{x\to0}\frac{x^2\left(\beta x-\dfrac{(\beta x)^3}{3!}+\dfrac{(\beta x)^5}{5!}-\ldots\right)}{ax-\left(x-\dfrac{x^3}{3!}+\dfrac{x^5}{5!}-\ldots\right)}=1$$

$$\Rightarrow \lim_{x\to0}\frac{x^3\left(\beta-\dfrac{\beta^3x^2}{3!}+\dfrac{\beta^5x^4}{5!}-\ldots\right)}{(\alpha-1)x+\dfrac{x^2}{3!}+\dfrac{x^5}{5!}-\ldots}=1$$

Limit exists only, when $\alpha-1=0$

$\Rightarrow \alpha=1$ …(i)

$$\therefore \lim_{x\to0}\frac{x^3\left(\beta-\dfrac{\beta^3x^2}{3!}+\dfrac{\beta^5x^4}{5!}\ \ldots\right)}{x^3\left(\dfrac{1}{3!}-\dfrac{x^2}{5!}-\ldots\right)}=1$$

$\Rightarrow\ 6\beta=1$ …(ii)

From Eqs (i) and (ii), we get

$6(\alpha+\beta)=6\alpha+6\beta=6+1=7$

6. Given, $\lim\limits_{\alpha\to0}\dfrac{e^{\cos(\alpha^n)}-e}{\alpha^m}=-\dfrac{e}{2}$

$$\Rightarrow \lim_{\alpha\to0}\frac{e\left\{e^{\cos(\alpha^n)-1}-1\right\}}{\cos(\alpha^n)-1}.\frac{\cos(\alpha^n)-1}{\alpha^m}=\frac{-e}{12}$$

$$\Rightarrow \lim_{\alpha\to0}e\left\{\frac{e^{\cos(\alpha^n)-1}-1}{\cos(\alpha^n)-1}\right\}.\lim_{\alpha\to0}\frac{-2\sin^2\dfrac{\alpha^n}{2}}{\alpha^m}=-e/2$$

$$\Rightarrow e\times1\times(-2)\lim_{\alpha\to0}\frac{\sin^2\left(\dfrac{\alpha^n}{2}\right)}{\dfrac{\alpha^n}{4}}.\frac{\alpha^{2n}}{4\alpha^m}=\frac{-e}{2}$$

$$\Rightarrow e\times1\times-2\times1\times\lim_{\alpha\to0}\frac{\alpha^{2n}-m}{4}=\frac{-e}{2}$$

For this to be exist, $2n-m=0$

$$\Rightarrow \frac{m}{n}=2$$

7. The right-hand limit

$$\lim_{x\to0^+}\frac{(1-x)^{\frac{1}{x}}-e^{-1}}{x^a}$$

$$=\lim_{x\to0^+}\frac{e^{\left\{\frac{1}{x}\log_e(1-x)\right\}}-e^{-1}}{x^a}$$

$$=\lim_{x\to0^+}\frac{e^{\frac{1}{x}\left(-x-\frac{x^2}{2}-\frac{x^3}{3}-\ldots\right)}-e^{-1}}{x^a}$$

$$=\lim_{x\to0^+}\frac{e^{-1}.e^{\left(-\frac{x}{2}-\frac{x^2}{3}-\frac{x^3}{4}-\ldots\right)}-e^{-1}}{x^a}$$

$$=e^{-1}\lim_{x\to0^+}\frac{e^{-\left(\frac{x}{2}+\frac{x^2}{3}+\frac{x^3}{4}+\ldots\right)}-1}{x^a}$$

The above limit will be non-zero, if $a=1$. And at $a=1$, the value of the limit is ,

$$=e^{-1}\left(-\frac{1}{2}\right)=-\frac{1}{2e}$$

8. Plan $\lim\limits_{x\to0}\dfrac{\sin x}{x}=1$

Given, $\lim_{x\to 1}\left\{\frac{\sin(x-1)+a(1-x)}{(x-1)+\sin(x-1)}\right\}^{\frac{(1+\sqrt{x})(1-\sqrt{x})}{1-\sqrt{x}}}=\frac{1}{4}$

$$\lim_{x\to 1}\left\{\frac{\frac{\sin(x-1)}{(x-1)}-a}{1+\frac{\sin(x-1)}{(x-1)}}\right\}^{1+\sqrt{x}}=1/4$$

$\Rightarrow\left(\frac{1-a}{2}\right)^2=\frac{1}{4}$

$\Rightarrow (a-1)^2=1$

$\Rightarrow a=2$ or 0

Hence, the maximum value of a is 2.

Comprehension Passage

Comprehension 1

1. (b)

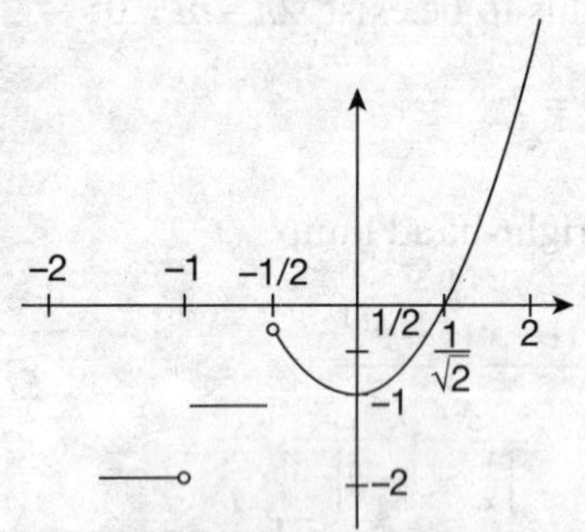

1, −1/2 are two points of discontinuity

2. (c)

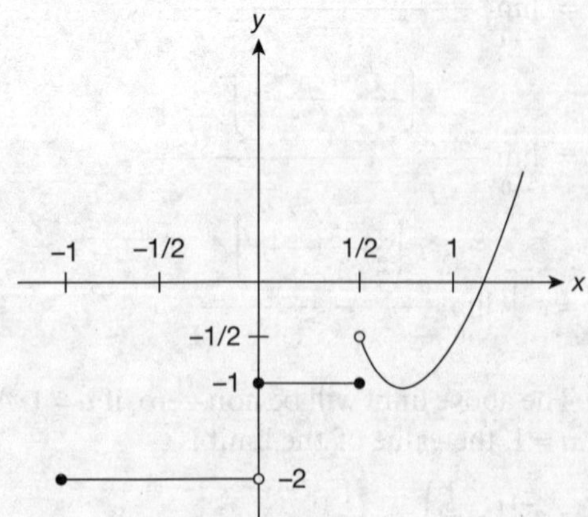

Discontinues at 0, 1/2

3. (c)

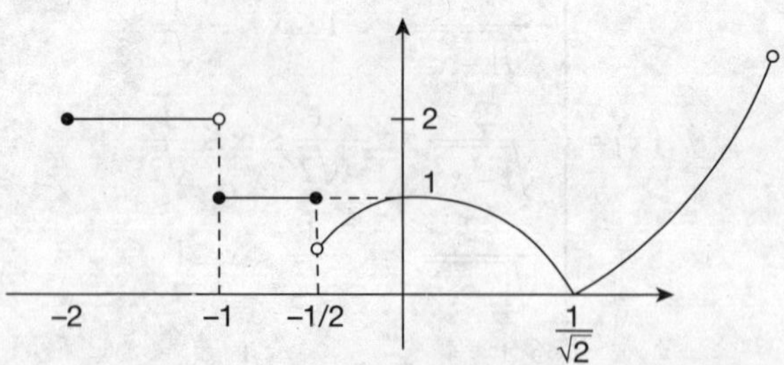

At −1, −1/2, $1/\sqrt{2}$ the function is not differentiable

Comprehension 2

1. (c) $\lim_{x\to 1^-} f(x)=a+b$

$\lim_{x\to 1^+} f(x)=4b$

For continuity, $a+b=4b$ I., $a=3b$ ...(i)

$$f'(1^+)=\lim_{h\to 0^+}\frac{f(1+h)-f(1)}{h}$$

$$=\lim_{h\to 0^+}\frac{2b(1+h)+2b-(a+b)}{h}$$

$$=\lim_{h\to 0^+}\frac{3b-a+2bh}{h}=2b$$

$$f'(1^-)=\lim_{h\to 0^+}\frac{f(1-h)-f(1)}{-h}$$

$$=\lim_{h\to 0^+}\frac{a(1-h)^2+b-a-b}{-h}$$

$$=\lim_{h\to 0^+}\frac{a(-2h+h^2)}{-h}=2a$$

$2a\neq 2b,\ a\neq b$

2. (a) $\lim_{x\to 2^-} g(x)=\lim_{x\to 2^-} cx^2+d=4c+d$

$\lim_{x\to 2^+} g(x)=\lim_{x\to 2^+}(dx+3-c)=2d+3-c$

$g(2)=4c+d$

$\therefore\ 4c+d=2d+3-c$

$\therefore\ d=5c-3$

3. (d) $\lim_{x\to 3^-} f(x)=8b,\ \lim_{x\to 3^-} f(x)=3(a-1)+2a-3$

$=5a-6$

Since $f(x)$ is continuous at $x=3$

$\therefore I\,8b=5a-6$ ...(i)

$$f'(3^-)=\lim_{h\to 0^+}\frac{f(3-h)-f(3)}{-h}$$

$$=\lim_{h\to 0^+}\frac{2b(3-h)+2b-8b}{-h}$$

$$f'(3^+) = \lim_{h\to 0^+} \frac{f(3+h)-f(3)}{h}$$

$$= \lim_{h\to 0^+} \frac{(a-1)(3+h)+2a-3-8b}{h}$$

Since f is differentiable at $x = 3$

$\therefore \lim_{h\to 0^+} (a-1)(3+h)+2a-3-8b = 0$

i.e. $5a-8b-6=0$

$\therefore f'(3^+) = a-1.$

Thus $aI = 2b$ ...(ii)

From (i) and (ii), we get $a = 2, b = \frac{1}{2}$.

Comprehension 3

1. (a) L.H.D $= \lim_{h\to 0^-} \frac{f-}{h} = \lim_{h\to 0^-} \frac{-f(a-h)+f(a)}{h}$

$= \lim_{h\to 0^-} \frac{f(a-h)-f(a)}{-h}$.

2. (a) If f is even, then $f'(-x) = -f'(x)$

$$\therefore f'(a^+) = \lim_{h\to 0^-} \frac{f'(a-h)-f'(a)}{-h}$$

$$= \lim_{h\to 0^-} \frac{f'(a)-f'(a-h)}{h}$$

$$= \lim_{h\to 0^-} \frac{f'(a)+f'(h-a)}{h}.$$

3. (b) $\lim_{h\to 0} \frac{f(-x)-f(-x-h)}{h} = f'(-x)$

and $\lim_{h\to 0^-} \frac{f(x)-f(x-h)}{-h} = -f'(x)$

$\Rightarrow f'(-x) = -f'(x) \therefore f'(x)$ is an odd function

$\therefore f$ is an even function.

Comprehension 4

1. (d) If $f(x) = \begin{cases} x+2.7, & x<0 \\ 2.9, & x=0 \\ 2x+3, & x>0 \end{cases}$

and $g(x) = \begin{cases} 3x+3, & x<0 \\ 2.8, & x=0 \\ -x^2+2.7, & x>0 \end{cases}$

Then $\lim_{x\to 0^-} f(x) = 2.7$, $\lim_{h\to 0^+} f(x) = 3$

$\therefore |3-2.7| = 0.3 < 1$ and $f(0) = 2.9$ lies in $(2, 7, 3)$

$\therefore f(x)$ is continuous under the system S_2

$g(x)$ is also continuous under the system S_2 under system S_1, since $\lim_{x\to 0} f(x)$ does not exist.

$\therefore f(x)$ is not continuous.

$\therefore$ (i), (ii) and (iii) are all true.

2. (d) Let $f(x) = \begin{bmatrix} x+2.7, & x<0 \\ 2.9, & x=0 \\ 2x+3, & x>0 \end{bmatrix}$

and $g(x) = \begin{bmatrix} 3x+3, & x<0 \\ 2.9, & x=0 \\ -x^2+2.75, & x>0 \end{bmatrix}$

$\therefore (f+g)(x) = \begin{bmatrix} 4x+5.7, & x<0 \\ 5.8, & x=0 \\ 2x-x^2+5.75, & x>0 \end{bmatrix}$

$\therefore \lim_{x\to 0^-} (f+g)(x) = 5.7$

and $\lim_{x\to 0^+} (f+g)(x) = 5.75$

$\therefore \left| \lim_{x\to 0^-} (f+g) - \lim_{x\to 0^+} (f+g) \right| = .05 < 1$ is satisfied

$\therefore (f+g)(0) = 5.8$ which do not lie in $(5.7, 5.75)$

$\therefore f+g$ is not continuous.

Similarly, we can show that $f-g$ and $f.g$ are not continuous under S_2.

3. (b) A function continuous under system S_2 may not be continuous under system S_1.

CHAPTER 6

Application of Derivatives

Objectives

In this chapter on Application of Derivatives, we shall study the following topics:

1. Rate of change of quantities
2. Monotonicity, increasing and decreasing functions
3. Tangents and Normal
4. Use of derivative in approximation
5. Maxima and Minima
6. The greatest and least value of a function.

Introduction

We have learnt to find the derivatives of various functions in the previous chapter. In the present chapter we will study the application of derivatives in various fields such as Science, Engineering, Economics, Commerce, etc.

The subject matter of study of this chapter can be categorized as follows:

(i) Rate of changes of quantities
(ii) Increasing and decreasing functions
(iii) Tangents and normals
(iv) Use of derivative in approximation
(v) Maxima and minima

Rate of Change of Quantities

The origin of calculus was basically to solve the problem of instantaneous rate of change of velocity by Newton. It could be applied to two quantities where one varies with respect to the other. The independent quantity is called Independent Variable.

Let us take an example. In straight-line motion the distance 's' covered by a body depends upon time 't'.

The change in s from a time t to $(t+\Delta t)$ is Δs and $\Delta s = f(t+\Delta t) - f(t)$

The average rate of change of s with respect to t is

$$\frac{\Delta s}{\Delta t} = \frac{f(t+\Delta t)-f(t)}{\Delta t}$$

If $\Delta t \to 0$

$$\lim_{\Delta t\to 0}\frac{\Delta s}{\Delta t} = \lim_{\Delta t\to 0}\frac{f(t+\Delta t)-f(t)}{\Delta t}$$

$$\frac{ds}{dt} = f'(t)$$

($\frac{ds}{dt}$ = instantaneous rate of change of s with respect to t)

In general, if a quantity 'y' varies with respect to a second quantity 'x' under a certain rule (function), $y = f(x)$ then $\frac{dy}{dx}$ or $f'(x)$ denotes the rate of change of y with respect x.

Monotonocity

Increasing Function

A function $f(x)$ is an increasing function at the point $x = a$ if $f'(a) \ge 0$, i.e., for every $x_1, x_2 \in D_1, x_1 > x_2 \Rightarrow f(x_1) \ge f(x_2)$.

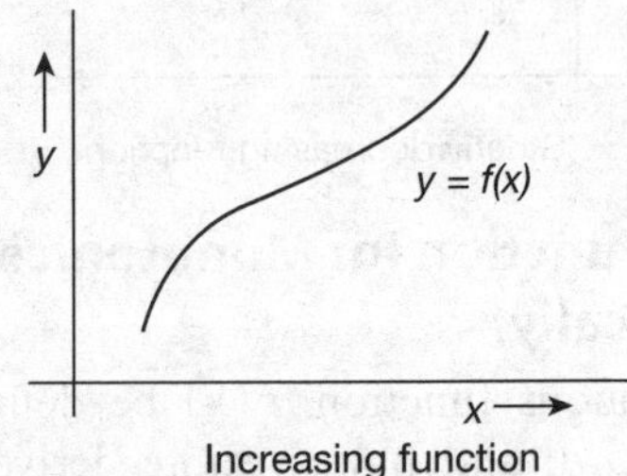

Increasing function

Decreasing Function

A function $f(x)$ is a decreasing function at the point if $f'(a) \le 0$, i.e., for every $x_1, x_2 \in D$, $x_1 > x_2 \Rightarrow f(x_1) \le f(x_2)$.

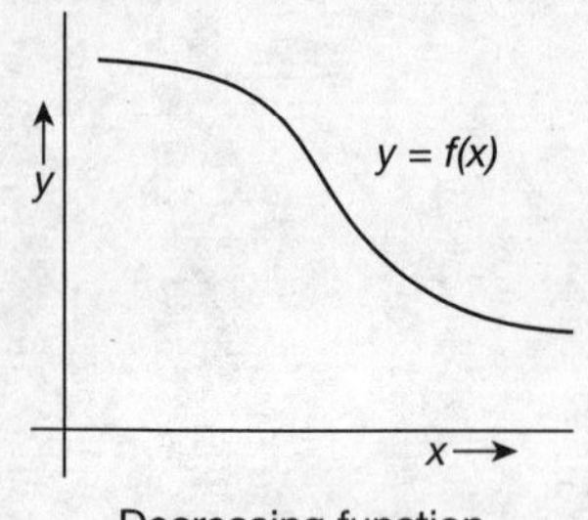

Decreasing function

Strictly Increasing Function

A function $f(x)$ is said to be strictly increasing in D if for every $x_1, x_2 \in D, x_1 > x_2 \Rightarrow f(x_1) > f(x_2)$. It means that the value of $f(x)$ will keep on increasing with an increase in the value of x ($f_1(x), f_2(x)$ and $f_3(x)$ are all strictly increasing functions).

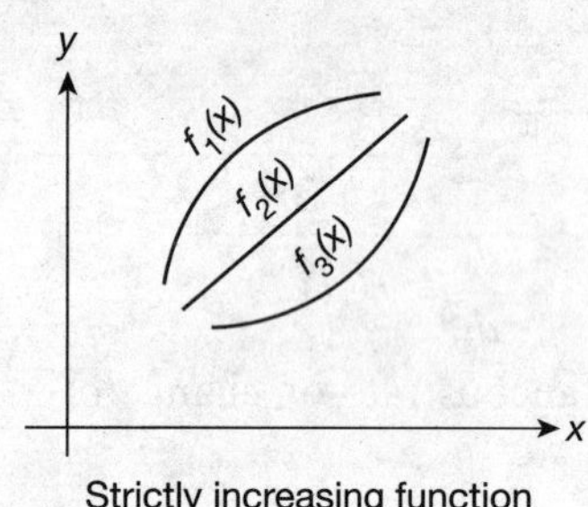

Strictly increasing function

Strictly Decreasing Function

A function $f(x)$ is said to be strictly decreasing in D if for every $x_1, x_2 \in D, x_1 > x_2 \Rightarrow f(x_1) < f(x_2)$. It means that the value of $f(x)$ would decrease with an increase in the value of x ($f_1(x), f_2(x)$ and $f_3(x)$ are all strictly decreasing functions).

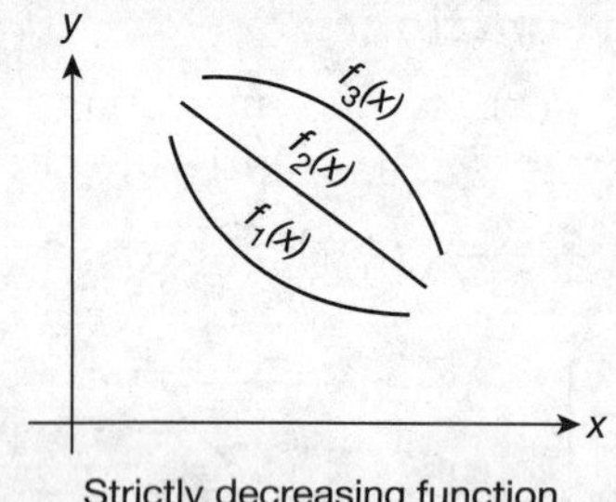

Strictly decreasing function

Testing a Function for Monotonicity (Geometrically)

Let a continuous function $f(x)$ be defined on the interval $I \equiv [a,b]$ and have a finite derivative in the interval. Then,

(A) $f(x)$ is non-decreasing (monotonically increasing) on $[a,b]$ if and only if $f'(x) \ge 0, \forall x \in (a,b)$. It is increasing (strictly increasing) if and only if $f'(x) > 0, \forall x \in (a,b)$

Geometrically, for monotonically increasing of angles θ_1, θ_2 are +ve.

$$f'(x) = \tan\theta \ge 0$$

i.e.,
$$0° \le \theta \le 90°$$

(θ in first quadrant)

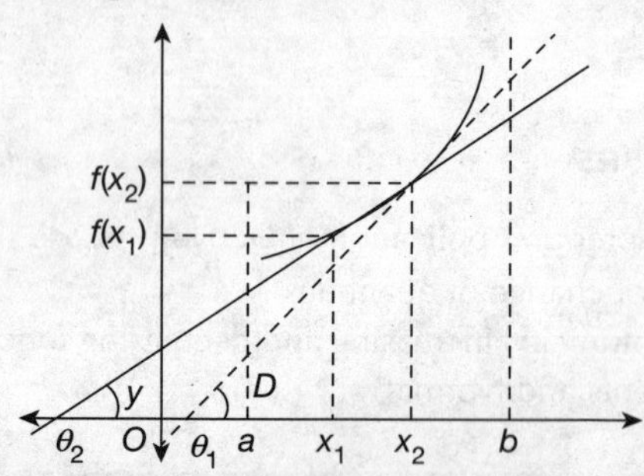

(B) $f(x)$ is non-increasing (monotonically decreasing) on $[a,b]$ if and only if $f'(x) \le 0, \forall x \in (a,b)$. It is decreasing (strictly decreasing) if and only if $f'(x) < 0, \forall x \in (a,b)$.

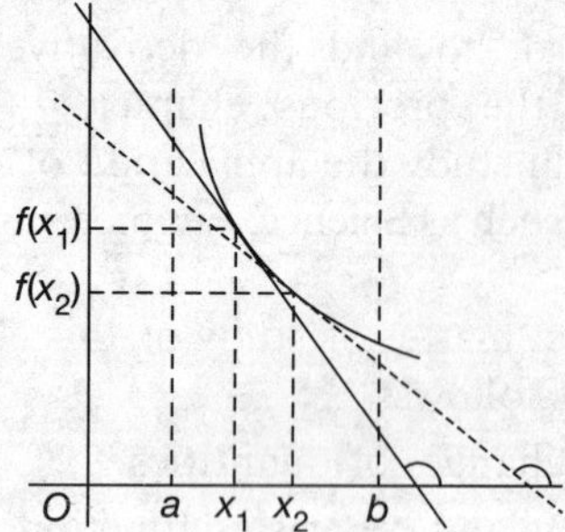

(i) A function $f(x)$ is said to be monotonic in the interval $[a,b]$ if either $f(x)$ is monotonically increasing or monotonically decreasing in $[a,b]$.

(ii) If f be monotonically increasing on (a,b), then $f(x^+)$ and $f(x^-)$ exist at every point x of (a,b). If $a < x < y < b$ then $f(x^+) \le f(y^-)$.

(iii) A monotonic function is one-one but the converse is not true.

(iv) Let $f(x)$ be monotonic in $[a,b]$, then $f(x)$ has at most one zero in $[a,b]$.

ILLUSTRATION 6.1

Find the intervals where $y = \frac{3}{2}x^4 - 3x^2 + 1$ is increasing or decreasing.

Solution

$$\frac{dy}{dx} = 6x^3 - 6x \Rightarrow \frac{dy}{dx} = 6x(x^2 - 1)$$

$$\frac{dy}{dx} = 6x(x-1)(x+1)$$

−	+	−	+
	−1	0	+1

The sign of $\frac{dy}{dx}$ is positive in the interval $(-1,0)\cup(1,\infty)$ and negative in the interval $(-\infty, -1)\cup(0, 1)$.

Hence the function is increasing in $[-1,0]\cup[1,\infty)$ and decreasing $(-\infty,-1]\cup[0,1]$.

ILLUSTRATION 6.2

Find the interval where $y = \tan x - x$ is increasing.

Solution

$$\frac{dy}{dx} = \sec^2 x - 1$$

Since $\sec^2 x \geq 1 \Rightarrow \frac{dy}{dx} \geq 0$

Hence, y will always be increasing.

ILLUSTRATION 6.3

Find the intervals of monotonicity of the function $f(x) = \frac{|x-1|}{x^2}$.

Solution

The given function $f(x)$ can be written as

$$f(x) = \frac{|x-1|}{x^2} = \begin{cases} \frac{1-x}{x^2}; & x < 1, x \neq 0 \\ \frac{x-1}{x^2}; & x \geq 1 \end{cases}$$

Consider $x < 1$, $f'(x) = -\frac{2}{x^3} + \frac{1}{x^2} = \frac{x-2}{x^3}$

for increasing, $f'(x) > 0 \Rightarrow \frac{x-2}{x^3} > 0.$

$\Rightarrow \quad x(x-2) > 0 \Rightarrow x \in (-\infty, 0)\cup(2,\infty).$

Combining with $x < 1$, we get $f(x)$ is increasing in $x < 0$ and decreasing in $x \in (0,1)$

Consider $x \geq 1$.

$$f'(x) = -\frac{1}{x^2} + \frac{2}{x^3} = \frac{2-x}{x^3}$$

For increasing $f'(x) > 0$

$(2 - x) > 0 \quad (\because x^3 \text{ is positive})$

$(x - 2) < 0 \Rightarrow x < 2.$

Combining with $x > 1$, $f(x)$ is increasing in $x \in (1, 2)$ and decreasing in $x \in (2, \infty)$

Combining (1) and (2), we get,

$f(x)$ is strictly increasing on $x \in (-\infty, 0)\cup(1, 2)$ and strictly decreasing on

$$x \in (0,1)\cup(2,\infty).$$

ILLUSTRATION 6.4

Show that $x \geq \log(1+x)$ for all $x \in (-1, \infty)$.

Solution

Let $f(x) = x - \log(1 + x)$

Differentiating both sides

$$f'(x) = 1 - \frac{1}{1+x}$$

$$f'(x) = \frac{x}{1+x}$$

$$f'(x) = \frac{1}{1+x}$$

Case I: $f'(x) < 0$

$$\frac{x}{(1+x)} < 0 \text{ when } x \in (-1, 0)$$

Therefore $-1 < x < 0 \Rightarrow f(x) \geq f(0)$;

But $f(0) = 0 \Rightarrow f(x) \geq 0$

$x - \log(1+x) \geq 0$ for all $x \in (-1, 0)$ ...(i)

Case II: $x \in [0, \infty)$

In this interval, $f'(x) > 0$

$\Rightarrow f(x)$ is an increasing function.

Therefore, $0 \leq x < \infty \Rightarrow f(x) \geq f(0)$

$$x - \log(1+x) \geq 0$$

$x \geq \log(1+x)$ for all $x \in [0, \infty)$ ...(ii)

Combining (i) and (ii), $x \geq \log(1+x)$ for all $x \in (-1, \infty)$.

ILLUSTRATION 6.5

Determine the intervals of monotonicity for the following functions:

(i) $f(x) = 4x^3 - 12x^2 + 18x + 20$

(ii) $f(x) = \cos(\pi/x)$

Solution

(i) $f(x)$ is defined everywhere. Now differentiate w.r.t. x, we get $f'(x) = 12x^2 - 24x + 18$.

$f'(x) = 0 \Rightarrow 12x^2 - 24x + 18 = 0$

$\Rightarrow 2x^2 - 4x + 3 = 0 \Rightarrow D < 0$

$\Rightarrow f'(x) > 0 \, \forall x \in R$

$\Rightarrow f(x)$ is increasing for all x

$\Rightarrow$ interval of increasing $(-\infty, \infty)$

(ii) The function $f(x) = \cos(\pi/x)$ is continuous and differentiable throughout, except at the point $x = 0$; now, $f'(x) = \frac{\pi}{x^2}\sin\left(\frac{\pi}{x}\right)$

Sign of $f'(x)$ is same as of $\sin\left(\frac{\pi}{x}\right)$. Then

Case I: $\sin\left(\frac{\pi}{x}\right) > 0$ if $2k\pi < \frac{\pi}{x} < (2k+1)$ $\pi (k = 0, \pm 1, \pm 2, \ldots)$

Case II: $\sin\left(\frac{\pi}{x}\right) < 0$ is $(2k+1)\pi < \frac{\pi}{x} < (2k+2)\pi$

Hence the function increases in $\left(\frac{1}{2k+1}, \frac{1}{2k}\right)$ and decreases in $\left(\frac{1}{2k+2}, \frac{1}{2k+1}\right)(k \in I)$

ILLUSTRATION 6.6

In the interval (1,2), the function $f(x) = 2|x-1| + 3|x-2|$ is:

(a) Monotonically increasing

(b) Monotonically decreasing

(c) Not monotonic

(d) Constant

Solution (b)

$x \in (1,2) \Rightarrow f(x) = 2(x-1) - 3(x-2) = -x + 4$

$\Rightarrow f'(x) = -1 < 0 \, \forall x$

$\therefore f(x)$ is monotonic decreasing in (1, 2).

ILLUSTRATION 6.7

If a function $f(x) = \cos|x| - 2ax + b$ is an increasing function on the whole number line, then the value of a is _____.

(a) b (b) $\frac{b}{2}$ (c) $a \le \frac{-1}{2}$ (d) $a > \frac{-3}{2}$

Solution

$$\therefore \quad \frac{d}{dx}\cos|x| = -\sin x, \text{for } x \in R.$$

$$\therefore \quad f'(x) = -\sin x - 2a$$

Now $f(x)$ is an increasing function; therefore,

$$f'(x) > 0 \Rightarrow -\sin x - 2a > 0 \Rightarrow a < \frac{-1}{2}\sin x \Rightarrow a \le \frac{-1}{2}$$

ILLUSTRATION 6.8

Find the interval in which $f(x) = 2x^2 - \log|x| (x \ne 0)$ is monotonic increasing.

Solution

$$f(x) = 2x^2 - \log|x|$$

$\Rightarrow$ for $x > 0$ [Case I]

$$f(x) = 2x^2 - \log x$$

$$f'(x) = 4x - \frac{1}{x} = \frac{4x^2 - 1}{x}.$$

For $x < 0$ [Case II]

$$f(x) = 2x^2 - \log(-x)$$

$$f'(x) = 4x + \frac{1}{x} = \frac{4x^2 + 1}{x}$$

Case I: $\{x > 0\}$

$$f'(x) > 0 \Rightarrow \frac{4x^2 - 1}{x} > 0$$

$$\Rightarrow \quad \frac{(2x-1)(2x+1)}{x} > 0$$

$$\left(-\frac{1}{2}, 0\right) \cup \left(\frac{1}{2}, \infty\right)$$

Case II: $\{x < 0\}$

$$f'(x) > 0 \Rightarrow \frac{4x^2 + 1}{x} > 0$$

$4x^2 + 1 > 0$ (always)

$x < 0 \Rightarrow$ denominator is always –ve.

$\therefore f'(x)$ will never be > 0.

When $x < 0$

Combining only $\left(\frac{1}{2}, \infty\right)$ is the interval.

$$f'(x) = 4x - 1/x|$$

$f(x)$ is monotonic increasing when

$$f'(x) > 0 \Rightarrow 4x - \frac{1}{x} > 0 \Rightarrow \frac{4x^2 - 1}{x} > 0.$$

$$4x^2 - 1 > 0 \text{ when } x > 0$$

$$4x^2 - 1 < 0 \text{ when } x < 0$$

But $x > 0, 4x^2 - 1 > 0 \Rightarrow x^2 > 1/4$

$\Rightarrow \quad |x| > 1/2 \Rightarrow x \in (1/2, \infty)$

But $x < 0, 4x^2 - 1 < 0 \Rightarrow x^2 < 1/4$

$\Rightarrow \quad |x| < 1/2 \Rightarrow x \in (-1/2, \infty)$

$\therefore \quad x \in (-1/2, 0) \cup (1/2, \infty)$.

Tangents and Normals

The derivative of a function $y = f(x)$ represents the slope of the tangent to the curve at the general point (x, y).

Equation of Tangent and Normal

Let (x_1, y_1) be any point on $y = f(x)$ curve. Slope of the tangent at (x_1, y_1) will be $\left(\frac{dy}{dx}\right)_{(x_1, y_1)}$

Thus, the equation of the tangent at (x_1, y_1) would be

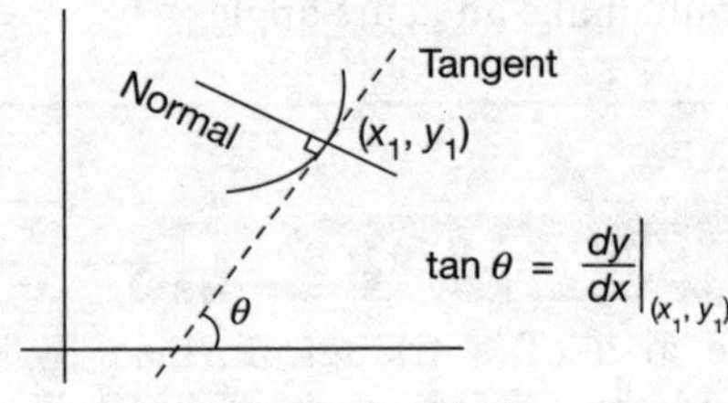

$$(y - y_1) = \left(\frac{dy}{dx}\right)_{(x_1, y_1)} (x - x_1)$$

or $$\left(y - f(x_1)\right) = \left(\frac{dy}{dx}\right)_{(x_1, y_1)} (x - x_1)$$

Similarly, the equation of the normal at (x_1, y_1) would be

$$\tan\theta = \left.\frac{dy}{dx}\right|_{(x_1, y_1)}$$

$$(y - y_1) = \frac{-1}{\left(\frac{dy}{dx}\right)_{(x_1, y_1)}} (x - x_1)$$

provided that $\left(\frac{dy}{dx}\right)_{(x_1, y_1)} \neq 0$

Note:

(i) If $\left(\frac{dy}{dx}\right)_{(x_1, y_1)} = 0$, then the equation of the normal would be $x = x_1$

(ii) If the tangent is parallel to y-axis then the slope of the curve

$$\frac{dy}{dx} = \infty \text{ or } \frac{dx}{dy} = 0.$$

Then equation of tangent would be $x = x_1$ and equation of normal would be $y = y_1$.

Length of Tangent, Normal, Sub-tangent and Sub-normal

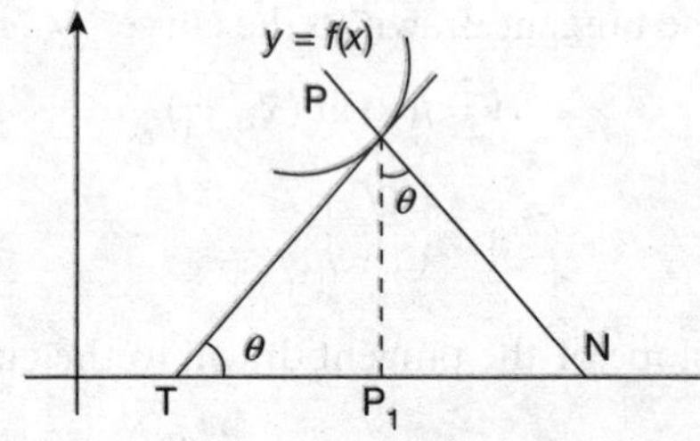

$$PT = PP_1 \operatorname{cosec}\theta$$

$$PT = y\sqrt{1 + \cot^2\theta}$$

$$\textbf{PT (length of tangent)} = y\sqrt{1 + \left(\frac{dx}{dy}\right)^2}$$

$$PN = PP_1 \sec\theta$$

$$PN = y\sqrt{1 + \left(\frac{dy}{dx}\right)^2} \quad \textbf{(length of normal)}$$

$$TP_1 = y\cot\theta$$

$$TP_1 = \left|y\frac{dx}{dy}\right| \text{ (sub-tangent)}$$

$$NP_1 = y\tan\theta$$

$$NP_1 = \left|y\frac{dy}{dx}\right| \text{ (sub-normal)}$$

Angle of Intersection of Two Curves

Let $y = f(x)$ and $y = g(x)$ be two given intersecting curves. The angle of intersection of these curves is defined as the acute angle between the tangents that can be drawn to the given curves at the point of intersection.

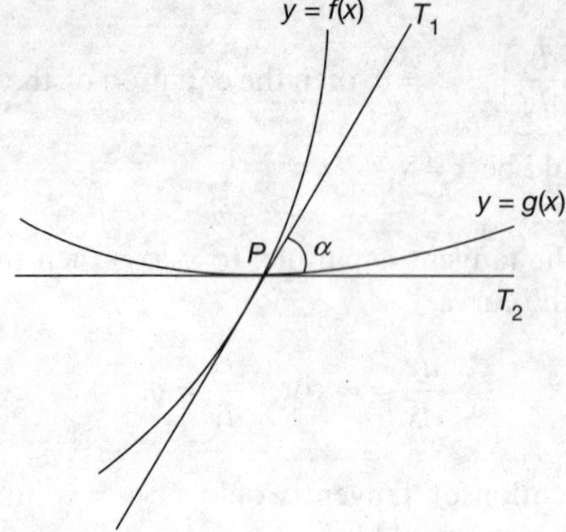

Let (x_1, y_1) be the point of intersection.

$$y = f(x_1) = g(x_1)$$

Slope of the tangent drawn to the curve

$$y = f(x) \text{ at } (x_1, y_1)$$

$$m_1 = \left(\frac{df}{dx}\right)_{(x_1,y_1)}$$

Similarly slope of the tangent drawn to the curve

$$y = g(x) \text{ at } (x_1, y_1)\ m_2 = \left(\frac{dg}{dx}\right)_{(x_1,y_1)}$$

If α' be the angle of intersection (acute), then

$$\tan\alpha = \left|\frac{m_1 - m_2}{1 + m_1 m_2}\right|.$$

If $\alpha = 0$, then $m_1 = m_2$. Thus, the given curves would touch each other at (x_1, y_1) (and NOT intersect).

If $\alpha = \pi/2$, then $m_1 m_2 = -1$. Thus, the given curves would meet at right angle at (x_1, y_1). In other words, the curves cut orthogonally at (x_1, y_1).

ILLUSTRATION 6.9

Find the equation of the tangent and normal at (1, 3), on $y = x^3 + x^2 + 1$ curve.

Solution

$$\frac{dy}{dx} = 3x^2 + 2x + 1$$

$$\left.\frac{dy}{dx}\right|_{(1,1)} = 5 + 1 \Rightarrow \left.\frac{dy}{dx}\right|_{(1,1)} = 6$$

Equation of tangent having 6 as its slope and passing through (1, 3) is

$$(y - 3) = 6(x - 1)$$

$$y = 6x - 3$$

Equation of normal having (–1/6) as its slope and passing through (1, 3) is

$$(y-3) = -\frac{1}{6}(x-1)$$

$$6(y - 3) = -(x - 1) \Rightarrow 6y - 18 = -x + 1$$

$$6y + x - 19 = 0$$

ILLUSTRATION 6.10

Prove that a tangent drawn at any point to the curve $f(x) = x^5 + 3x^3 + 4x + 8$ would make an acute angle with the x-axis.

Solution

$$f(x) = x^5 + 3x^3 + 4x + 8$$

$$f'(x) = 5x^4 + 9x^2 + 4$$

Clearly, $f'(x) > 0\ \forall\ x \in R$. Thus, the tangent drawn at any point would have positive slope and hence would make an acute angle.

ILLUSTRATION 6.11

Find the area of a triangle formed by positive x-axis and the tangent and the normal to circle $x^2 + y^2 = 9$ at $(2, \sqrt{5})$.

Solution

$$x^2 + y^2 = 9$$

Differentiating w.r.t. x we get

$$2x+2y\frac{dy}{dx}=0 \Rightarrow \frac{dy}{dx}=\frac{-x}{y}$$

$$2x+2y\frac{dy}{dx}=0 \Rightarrow \frac{dy}{dx}=\frac{-x}{y} \Rightarrow \left(\frac{dy}{dx}\right)_{(2,\sqrt{5})}=\frac{-2}{\sqrt{5}}$$

Hence the equation of tangent at $(2,\sqrt{5})$ is

$$\left(y-\sqrt{5}\right)=-\frac{2}{\sqrt{5}}(x-2).$$

Equation of normal at $\left(2,\sqrt{5}\right)$

$$\left(y-\sqrt{5}\right)=\frac{\sqrt{5}}{2}(x-2).$$

Graphically draw and find the co-ordinates of the vertices of the triangle. Then find the areas.

Use of Derivatives in Approximation

1. Take a suitable function according to the given quantity.
2. Then, find $f'(x)$ and put in $f(a+h)=f(a)+hf'(a)$
3. Put a number in place of a (in case of decimal value, a is taken as the nearest integer) and h such that $f(a+h)$ becomes equals to the quantity whose approximate value is to be calculated.
4. If approximate change in the value of function $y=f(x)$ is to be found, use $\delta y=\frac{dy}{dx}\delta x$.

ILLUSTRATION 6.12

Find the approximate volume of the metal in a hollow spherical shell whose internal and external radii are 3 cms and 3.0005 cms respectively.

Solution

Let x be the radius and y be the volume of the solid sphere

$$y=\frac{4}{3}\pi x^3 \qquad \text{...(i)}$$

$$\therefore \qquad \frac{dy}{dx}=4\pi x^2 \qquad \text{...(i)}$$

Now volume of hollow shell = Change in volume of the solid sphere when its radius increases from 3 cm to 3.0005 cm.

Let $x=3, x+\delta x=3.0005$ and δy, change in volume.

Then $\delta y=\frac{dy}{dx}\delta x=4\pi x^2(0.0005)$

$=4\pi(9)(0.0005)=0.018\pi$ c.c

Maxima and Minima

If a function $y=f(x)$ is defined in the interval I, then an interior point x_0 of this interval is called the point of maximum of the function $f(x)$ if there exists a neighborhood U of point x_0 such that inequality $f(x)\le f(x_0)$ holds true within it or equivalently for small h (positive) $f(x)\le(x_0)\forall x\in(x_0-h,x_0+h)$. Similarly, x_0 is called the point of minimum if $f(x)\ge f(x_0)\forall x\in(x_0-h,x_0+h)$.

The Necessary Condition for Existence of an Extremum

Let $f(x)$ be defined in the domain D. Then $x_0\in D$ is point of extremum if

(i) $f'(x_0)=0$

(ii) $f'(x_0)$ does not exist but is continuous at x_0.

Critical Point: The points x_0 at which $f'(x_0)=0$ or does not exist are called critical points. So maximum and minimum can exist only at critical points.

Sufficient Conditions for Existence of an Extremum

First Derivative Condition

Let the function $f(x)$ be continuous at some neighborhood of x_1 and x_2 (critical points).

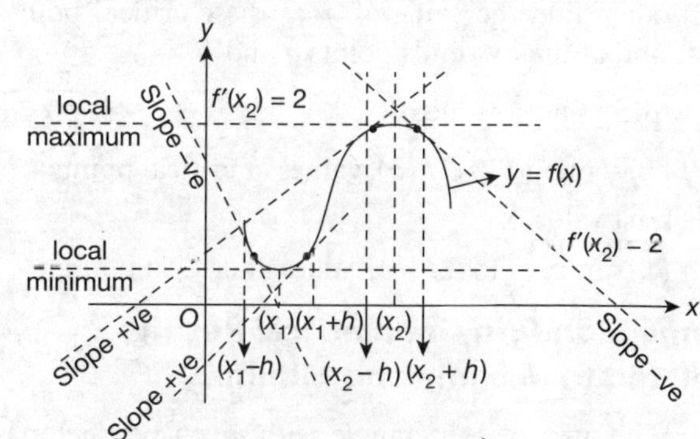

Case I: If $f'(x)<0$ for $x<x_1$ and $f'(x)>0$ for $x>x_1\left(x\in(x_1-h,x+h)\right)$| (If moving from left to right through the point x_1, the derivative changes sign from –ve to +ve), then at the point x_1 the function attains a minimum.

Case II: If $f'(x)>0$ for $x<x_2$ and $f'(x)<0$ for $x>x_2$ (If moving from left to right through the point x_2 the derivative changes from +ve to –ve), then at the point x_2 the function attains a maximum.

Case III: If the derivative does not change sign in moving through the critical point, then there is no extremum.

Second Derivative Condition

Let the function be twice differentiable at the critical point, $x_0, f'(x_0)=0$: If $f''(x_0)<0$ then at x_0 the function has a maximum, if $f''(x_0)>0$ then at x_0, the function has a minimum. But if $f''(x_0)=0$, then existence of an extremum at this point remains open (i.e., cannot be decided by this condition).

Higher Order Derivatives Condition

Let $n\ge 2$ and

$f'(x_0)=f''(x_0)=0=f'''(x_0)=\ldots=f^{(n-1)}(x_0)=0,$

but $f^{(n)}(x_0)\neq 0$ (where $f^{(n)}(x)$ denotes n th derivative).

If n is even

$$\Rightarrow\begin{cases} f^{(n)}(x_0)<0, & x_0 \text{ point of maximum}\\ f^{(n)}(x_0)>0, & x_0 \text{ is point of minimum}\end{cases}$$

If n is odd, then there is no extremum at the point x_0.

Greatest and Least Values of a Function (Absolute Maximum and Absolute Minimum)

The greatest (or least) values of a continuous function on $[a,b]$ is attained either at the critical point or at either of the end-points a and b.

To find absolute maximum or absolute minimum:

(i) First locate the critical point of f in (a,b)

(ii) Then find the value of f at these critical points and at the two ends points a and b.

Now the greatest value of

$$f(x)\equiv\max\{f(a),f(b),\text{values at critical points}\}$$

The least value of

$$f(x)\equiv\min\{f(a),f(b),\text{values at critical points}\}$$

Some Standard Geometrical Results Related to Maxima and Minima

(i) The area of a rectangle with given perimeter is greatest when it is a square.

(ii) The perimeter of a rectangle with given area is least when it is a square.

(iii) The greatest rectangle inscribed in a given circle is a square.

(iv) The greatest triangle inscribed in a given circle is equilateral.

(v) The semi-vertical angle of a cone with given slant height and maximum volume is $\tan^{-1}\sqrt{2}$.

(vi) The height of a cylinder of maximum volume inscribed in a sphere of radius a is $\dfrac{2a}{\sqrt{3}}$.

ILLUSTRATION 6.13

Find the points of maxima and minima for the function $f(x)=x^3-9x^2+15x-11$.

Solution

Let $f(x)=x^3-9x^2+15x-11$.

Then $f'(x)=3x^2-18x+15=3(x^2-6x+5)$.

For maxima and minima,

$$f'(x)=0\Rightarrow x^2-6x+5=0$$

$$\Rightarrow\quad (x-1)(x-5)=0\Rightarrow x=1,5$$

Again, $f''(1)=-12<0\Rightarrow x=1$ is a point of maxima.

And $f''(5)=12>0\Rightarrow x=5$ is a point of minima.

ILLUSTRATION 6.14

Find the greatest value of x^3-12x^2+45x in the interval [0, 7].

Solution

Let $f(x)=x^3-12x^2+45$. Then

$$f'(x)=3x^2-24x+45=3(x-3)(x-5) \text{ and}$$

$$f''(x)=6x-24$$

Now, for maximum and minimum values,

$$f'(x)=0\Rightarrow 3(x-3)(x-5)=0\Rightarrow x=3,5$$

Again, $f''(3)=-6<0$

$\Rightarrow$ The function is maximum at $x=3$ and $f''(5)=6>0$.

$\Rightarrow$ The function is minimum at $x=5$.

Now $f(0)=0, f(3)=54, f(5)=50, f(7)=70$

$\Rightarrow$ The greatest value in [0, 7]

$$=\max\{0,54,50,70\}=70.$$

ILLUSTRATION 6.15

The minimum value of the function $x^x (x > 0)$ is at ______.

(a) $x = 1$ (b) $x = e$
(c) $x = e^{-1}$ (d) None of these.

Solution

Let $y = x^x \Rightarrow \log y = x \log x \Rightarrow \frac{d}{dx}(\log y)$

$= 1 + \log x$ and $\frac{d^2}{dx^2}(\log y)$

$= \frac{1}{x} = x^{-1}$

Now, for minimum value of y or $\log y$

$\frac{d}{dx}(\log y) = 0 \Rightarrow 1 + \log x = 0 \Rightarrow x$

$= e^{-1}$. Again for $x = e^{-1}$

$\frac{d^2}{dx^2}(\log y) = e > 0 \Rightarrow y$ is minimum at $x = e^{-1}$

Correct answer is (c).

ILLUSTRATION 6.16

If $x = -1$ and $x = 2$ are extreme points of the function $y = a \log x + bx^2 + x$, then ______.

(a) $a = 2, b = \frac{1}{2}$ (b) $a = 2, b = -\frac{1}{2}$

(c) $a = -2, b = \frac{1}{2}$ (d) $a = -2, b = -\frac{1}{2}$

Solution

$$\frac{dy}{dx} = \frac{a}{x} + 2bx + 1$$

Since $x = -1$ and $x = 2$ are extreme points, $\frac{dy}{dx}$ at these points must be zero.

So $-a - 2b + 1 = 0$ and

$\frac{a}{2} + 4b + 1 = 0 \Rightarrow a + 2b - 1 = 0$ and $a + 8b + 2 = 0$

$\Rightarrow$ $a = 2, b = -\frac{1}{2}$.

Correct option is (b)

SOLVED EXAMPLES

Example 1. Find the maximum and minimum values of the following function:

$$y = 2x^3 - 15x^2 + 36x + 11$$

Solution

$$y = 2x^3 - 15x^2 + 36x + 11 \quad \ldots\text{(i)}$$

$$\frac{dy}{dx} = 6x^2 - 30x + 36 \quad \ldots\text{(ii)}$$

For maximum or minimum value of y, $\frac{dy}{dx} = 0$

$$6x^2 - 30x + 36 = 0$$

$$x^2 - 5x + 6 = 0$$

$\therefore$ $x = 2, 3$

From (ii), $\frac{d^2y}{dx^2} = 12x - 30 \quad \ldots\text{(iii)}$

At $x = 2$, $\frac{d^2y}{dx^2} = 12 \times 2 - 30 = -6 < 0$

At $x = 2$, y has maximum value and this maximum value of y

$$= 2 \times 2^3 - 15 \times 2^2 + 36 \times 2 + 11 = 39.$$

at $x = 3, \frac{d^2y}{dx^2} = 12 \times 3 - 30 = 6 > 0$

Then y has minimum value at $x = 3$ and this minimum value of y

$$= 2 \times 3^2 - 15 \times 3^2 + 36 \times 3 + 1 = 38$$

Hence maximum value of y is 39 and minimum value is 38.

II Method

y has maximum at $x = 2$ and from (i) maximum value of $y = 39$.

y has minimum value at $x = 3$ and the minimum value of $y = 38$

Example 2. Find the maximum and minimum value of

(i) $x(1-x)^2 \quad 0 \le x \le 2$

(ii) $\dfrac{40}{3x^4 - 8x^3 - 18x^2 + 60}$

Solution

(i) $y = x(1-x)^2 \quad 0 \le x \le 2$

For closed interval we use monotonic function. Here y is M.I. in $\left[0, \frac{1}{3}\right] \cup [1, 2]$ and $f(2) > f\left(\frac{1}{3}\right)$

So, $f(2)$ is maximum.

Also M.D. in $\left[\frac{1}{3}, 1\right]$ so $f(1)$ is minimum. So max. value = 2; min. value = 0

(ii) $y = \dfrac{40}{z}$

where $z = 3x^4 + 8x^3 - 18x^2 + 60$

$\therefore \quad \dfrac{dz}{dx} = 12x^3 + 24x^2 - 36x$

and $\dfrac{d^2z}{dx^2} = 36x^2 + 48x - 36$

Now, $\dfrac{dy}{dx} = \dfrac{-40}{z^2} \cdot \dfrac{dz}{dx}$, y is max. or min.

If $\dfrac{dy}{dx} = 0 \Rightarrow \dfrac{dz}{dx} = 0$

$\Rightarrow \ 12x^3 + 24x^2 - 36x = 0 \Rightarrow x^3 + 2x^2 - 3x = 0$

$\Rightarrow \ x(x-1)(x+3) \ \Rightarrow x = 0, 1, -3$

$$\frac{d^2y}{dx^2} = \frac{80}{z^3}\left[\frac{dz}{dx}\right]^2 - \frac{40}{z^2}\frac{d^2z}{dx^2}$$

at $x = 1$, $\dfrac{d^2y}{dx^2} < 0$

$\therefore y$ is max. when $x = 1$

at $x = 0$, $\dfrac{d^2y}{dx^2} > 0$.

$\therefore y$ is min. when $x = 0$.

So $y_{\min} = \dfrac{40}{60} = \dfrac{2}{3}$

Similarly, at $x = -3$, y is max.

Example 3. Show that the following function has neither a maximum nor a minimum:

$$x^3 - 3x^2 + 6x + 4$$

Solution

Let $\quad y = x^3 - 3x^2 + 6x + 4 \quad$ ...(i)

$$\frac{dy}{dx} = 3x^2 - 6x + 6 \quad \text{...(ii)}$$

For max. or min. values y, $\dfrac{dy}{dx} = 0$.

$\therefore \quad 3x^2 - 6x + 6 = 0$

$\therefore \quad x = \dfrac{6 \pm \sqrt{36-72}}{6} = \dfrac{6+6i}{6} = 1 + i.$

$\therefore \dfrac{dy}{dx} \ne 0$ for any real values of x and $\dfrac{dy}{dx}$ exists at all x.

Therefore, y has neither max. nor min. value.

II Method

$$\frac{dy}{dx} = 3(x^3 - 2x + 2)$$

When $3(x^3 - 2x + 2) = 0$, x is imaginary.

Sign scheme for $3(x^3 - 2x + 2)$ is as follows.

y is an increasing function.

$\therefore y$ has neither max. nor min. value.

Example 4. Investigate for maximum and minimum values the function given by

$$y = \sin x + \cos 2x$$

Solution

We have, $y = \sin x + \cos 2x$

$$\frac{dy}{dx} = \cos x - 2\sin 2x = \cos x - 4\sin x\cos x$$

$\dfrac{dy}{dx} = 0$ when $\cos x = 0$ or $\sin x = \dfrac{1}{4}$.

We consider values of x between 0 and 2π only for the given function if it is periodic with period 2π. Now,

$$\cos x = 0 \Rightarrow x = \pi/2 \text{ and } \frac{3\pi}{2}, \text{ and } \sin x = \frac{1}{4}$$

$\Rightarrow \quad x = \sin^{-1}\dfrac{1}{4}$ and $\pi - \sin^{-1}\dfrac{1}{4}$

$\sin^{-1}\dfrac{1}{4}$ lying between 0 and $\dfrac{\pi}{2}$.

Now, $\dfrac{d^2y}{dx^2} = -\sin x - 4\cos 2x$

$$x = \frac{\pi}{2} \Rightarrow \frac{d^2y}{dx^2} = 3 > 0$$

$$x = \frac{3\pi}{2} \Rightarrow \frac{d^2y}{dx^2} = 6 > 0$$

$$x = \sin^{-1}\frac{1}{4} \Rightarrow \frac{d^2y}{dx^2} = -\sin x - 4(1 - 2\sin^2 x) = \frac{-15}{4} < 0$$

$$x = \pi - \sin^{-1}\left(\frac{1}{4}\right) \Rightarrow \frac{d^2y}{dx^2} = \frac{-15}{4} < 0$$

Therefore, y is max. for $x = \sin^{-1}\frac{1}{4}, \pi - \sin^{-1}\left(\frac{1}{4}\right)$ and is max. for $x = \frac{\pi}{2}, \frac{3\pi}{2}$

Putting the values of x in $\sin x + \cos 2x$ we see that $\frac{9}{8}, \frac{9}{8}$ are the two maximum values and 0, –2 are the two minimum values.

Example 5. Show that maximum value of $\left(\frac{1}{x}\right)^x$ is $(e)^{1/e}$

Solution

Let $y = \left(\frac{1}{x}\right)^x$

$$\Rightarrow \quad \log y = -x \log x$$

$$\frac{1}{y}\frac{dy}{dx} = -(1 + \log x)$$

$$\frac{dy}{dx} = -(1 + \log x)\left(\frac{1}{x}\right)^x$$

$$\frac{dy}{dx} = 0 \Rightarrow \log x + 1 = 0 \Rightarrow x = e^{-1}$$

Again, $\frac{d^2y}{dx^2} = -\frac{1}{x}\left(\frac{1}{x}\right)^x + (1 + \log x)^2\left(\frac{1}{x}\right)^x$

at $x = e^{-1}$ $\frac{d^2y}{dx^2} = -e\left(e^{1/e}\right) < 0.$

$\Rightarrow$ y has max. for $x = e^{-1}$ and the maximum value is $(e)^{1/e}$

Example 6. $A(p^2, -p)$; $B(q^2, q)$; $C(r^2, -r)$ be the vertices of a triangle ABC. A parallelogram AFDE is drawn with vertices D, E, F on the line segments BC, CA and AB respectively. Show that maximum area of such a parallelogram is

$$\frac{1}{4}(p+q)(q+r)(p-r).$$

Solution

Since AFDE is a parallelogram,

$$\angle BFD = \angle DEC = \angle A$$

$$\angle EDC = \angle B, \angle FDB = \angle C$$

Let $BD = x$. Then $FL = \frac{x}{\cot B + \cot C}, EM = \frac{a - x}{\cot B + \cot C}$

in any triangle ABC, altitude $AL = \left(\frac{a}{\cot B + \cot C}\right)$

$\Rightarrow$ Area of parallelogram AFDE (say, s)

= Area of ΔABC – Area ΔFBD – Area ΔEDC

$$= \Delta - \frac{1}{2(\cot B + \cot C)}\left[x^2 + (a - x)^2\right]$$

For maxima, minima, $\frac{ds}{dx} = 0$

$\Rightarrow 2x - 2(a - x) = 0 \Rightarrow x = a/2$ whence $BD = a/2$.

$\Rightarrow$ F and E are mid-points of AB, AE respectively.

$\Rightarrow$ Area of parallelogram of maximum area

Half the area of the given triangle.

$$\frac{1}{2}\cdot\frac{1}{2}\begin{vmatrix} p^2 & p & 1 \\ q^2 & q & 1 \\ r^2 & r & 1 \end{vmatrix} = \frac{1}{4}(p+q)(q+r)(p-r).$$

Example 7. A sector with central angle α is cut out of a circle to make a cone. Show that the volume of such a cone will be maximum if $\alpha = 2\pi\left(\frac{2}{3}\right)^{1/2}$

Solution

Let the radius of the circle = R

Then arc AB = αR, which will become the base of the cone.

Thus, if r be the base radius of the cone, then

$$2\pi r = \alpha R \Rightarrow r = \frac{\alpha R}{2\pi}.$$

Also, R of the original circle will become slant height of the cone.

$\Rightarrow$ Height of the cone $= \sqrt{R^2 - r^2}$

$\Rightarrow$ Volume V of the cone is given by $V = \pi r^2 h$

$$= \pi r^2\sqrt{R^2 - r^2}$$

$$\Rightarrow \frac{dV}{dr} = 2\pi r\sqrt{R^2 - r^2} + \pi r^2 \cdot \frac{1}{2}\left(R^2 - r^2\right)^{-\frac{1}{2}}(-2r)$$

$$= 2\pi r\sqrt{R^2 - r^2} - \frac{\pi r^2}{\sqrt{R^2 - r^2}} = \pi\left(\frac{2rR^2 - 3r^3}{\sqrt{R^2 - r^2}}\right)$$

From maxima, minima $\frac{dV}{dr} = 0$

$$\Rightarrow 2rR^2 - 3r^3 = 0 \Rightarrow r^2 = \frac{2R^2}{3}$$

$$\Rightarrow \frac{\alpha^2 R^2}{4\pi^2} = \frac{2}{3}R^2 \Rightarrow \alpha^2 - \frac{8\pi^2}{3}$$

$$\Rightarrow \alpha = 2\pi\left(\frac{2}{3}\right)^{1/2}.$$

We can easily verify that $\frac{d^2V}{dr^2} < 0.$

RECAPITULATION (FORMULAE, THEOREM, IMPORTANT RESULTS)

1. A function f is said to be
   (a) increasing on an interval (a,b) if
   $x_1 < x_2$ in $(a,b) \Rightarrow f(x_1) \le f(x_2)$ for all
   $x_1, x_2 \in (a,b)$.
   Alternatively, if $f'(x) \ge 0$, for each x in (a,b).
   (b) decreasing on (a,b) if
   $x_1 < x_2$ in $(a,b) \Rightarrow f(x_1) \ge f(x_2)$ for all
   $x_1, x_2 \in (a,b)$
   Alternatively, if $f'(x) \le 0$ for each x in (a,b).
2. The equation of the tangent at (x_0, y_0) to the curve $y = f(x)$ is given by
$$y - y_0 = \left.\frac{dy}{dx}\right]_{(x_0,y_0)} (x - x_0).$$
3. If $\frac{dy}{dx}$ does not exist at the point (x_0, y_0), then the tangent at this point is parallel to the y-axis and its equation is $x = x_0$.
4. If tangent to a curve $y = f(x)$ at $x = x_0$ is parallel to x-axis, then $\left.\frac{dy}{dx}\right]_{(x_0,y_0)} = 0$.
5. Equation of the normal to the curve $y = f(x)$ at a point (x_0, y_0) is given by
$$y - y_0 = \frac{-1}{\left.\frac{dy}{dx}\right]_{(x_0,y_0)}}(x - x_0).$$
6. If $\frac{dy}{dx}$ at the point (x_0, y_0) is zero, then equation of the normal is $x = x_0$.
7. If $\frac{dy}{dx}$ at the point (x_0, y_0) does not exist, then the normal is parallel to x-axis and its equation is $y = y_0$.
8. Let $y = f(x)$, Δx be a small increment in x and Δy be the increment in y corresponding to the increment in x, i.e., $\Delta y = f(x + \Delta x) - f(x)$. Then dy given by
$$dy = f'(x)dx \text{ or } dy = \left(\frac{dy}{dx}\right)\Delta x$$
   is a good approximation of Δy when $dx = \Delta x$ is relatively small and we denote it by $dy \approx \Delta y$.
9. A point c in the domain of a function f at which either $f'(c) = 0$ or f is not differentiable is called a *critical point* of f.
10. **First Derivative Test:** Let f be a function defined on an open interval I. Let f be continuous at a critical point c in I. Then,
    (i) If $f'(x)$ changes sign from positive to negative as x increases through c, i.e., if $f'(x) > 0$ at every pint sufficiently close to and to the left of c, and $f'(x) < 0$ at every point sufficiently close to and to the right of c, then c is a point of *local maxima*.
    (ii) If $f'(x)$ changes sign from negative to positive as x increases through c, i.e., if $f'(x) < 0$ at every point sufficiently close to and to the left of c, and $f'(x) > 0$ at every point sufficiently close to and to the right of c, then c is a point of *local minima*.
    (iii) If $f'(x)$ does not change sign as x increase through c, then c is neither a point of local maxima nor a point of local minima. In fact, such a point is called *point of inflexion*.
11. **Second Derivative Test:** Let f be a function defined on an interval I and $c \in I$. Let f be twice differentiable at c. Then
    (i) $x = c$ is a point of local maxima if $f'(c) = 0$ and $f''(c) < 0$
    The value of $f(c)$ is local maximum value of f.
    (ii) $x = c$ is a point of local minima if $f'(c) = 0$ and $f''(c) > 0$
    In this case, $f(c)$ is local minimum value of f.
    (iii) The test fails if $f'(c) = 0$ and $f''(c) = 0$.
    In this case, we go back to the first derivative test and find whether c is a point of maxima, minima or a point of inflexion.
12. Working rule for finding absolute maxima and/or absolute minima.
    Step 1: Find all critical points of f in the interval, i.e., find points x where either $f'(x) = 0$ or f is not differentiable.
    Step 2: Take the end points of the interval.
    Step 3: At all these points (listed in Step 1 and 2), calculate the values of f.
    Step 4: Identify the maximum and minimum values of f out of the values calculated in Step 3. This maximum value will be the absolute maximum value of f and the minimum value will be the absolute minimum value of f.

QUIZ

1. What is the rate of change of area of a circle w.r.t. to its radius?
2. The radius of a circle is increasing uniformly at the ratio of 2 cm/s. What is the rate of change of area of the circle when its radius is 10 cm.
3. Is the statement correct? "If I be an open interval contained in the domain of a real-valued function, then f is said to be increasing on I if $x_1 < x_2 \Rightarrow f(x_1) \le f(x_2)$ for all $x_1, x_2 \in I$".
4. Write down the condition for strictly decreasing function in I.
5. $f(x) = ax + b$ where a and b are constants and $a > 0$. What is nature of the function: Increasing or decreasing?
6. Is $\cos x$ is an increasing or decreasing function in the interval $\left(0, \frac{\pi}{2}\right)$?
7. What is the slope of the tangent to the curve $y = x^3 - 3x + 2$ at the point whose x co-ordinate is 3?
8. Using differential calculus find the approximate value of $\sqrt{25.3}$
9. Find the local maxima and local minima, if any, for the function $\cos x, 0 < x < \pi$
10. Find the approximate change in volume of a cube of side x metre due to an increase in side by 2%.

JEE MAIN

Objective Questions

One Option Correct

1. The value of c for which the conclusion of Mean Value Theorem holds for the function $f(x) = \log_e x$ on the interval [1, 3] is ________.

(a) $2\log_3 e$ (b) $\frac{1}{2}\log_e 3$

(c) $\log_3 e$ (d) $\log_e 3$

2. The function $f(x) = \tan^{-1}(\sin x + \cos x)$ is an increasing function in ________.

(a) $\left(\frac{\pi}{4}, \frac{\pi}{2}\right)$ (b) $\left(-\frac{\pi}{2}, \frac{\pi}{4}\right)$

(c) $\left(0, \frac{\pi}{2}\right)$ (d) $\left(-\frac{\pi}{2}, \frac{\pi}{2}\right)$

3. The function $f(x) = \frac{x}{2} + \frac{2}{x}$ has a local minimum at ________.

(a) $x = -2$ (b) $x = 0$

(c) $x = 1$ (d) $x = 2$

4. If $x^m y^n = (x - y)^{m+n}$; then $\frac{dy}{dx}$ is ________.

(a) $\frac{x+y}{xy}$ (b) xy

(c) $\frac{x}{y}$ (d) $\frac{y}{x}$

5. Let f be differentiable for all x. If $f(1) = -2$ and $f'(x) \ge 2$ for $x \in [1, 6]$, then

(a) $f(6) < 5$ (b) $f(6) = 5$

(c) $f(6) \ge 8$ (d) $f(6) < 8$

6. The real number x, when added to its inverse, gives the minimum value of sum at x equal to ________.

(a) 2 (b) 1

(c) –1 (d) –2

7. The value of θ, $\theta \in [0, \pi/2]$ for which the sum of intercepts on co-ordinate axes by tangent at point $\left(3\sqrt{3}\cos\theta, \sin\theta\right)$ of ellipse $\frac{x^2}{27} + y^2 = 1$ is minimum, at ________.

(a) $\frac{\pi}{6}$ (b) $\frac{\pi}{4}$

(c) $\frac{\pi}{3}$ (d) $\frac{\pi}{2}$

8. The length of longest interval in which the function 3 sin x – 4 $\sin^3 x$ keeps increasing is ________.

(a) $\frac{\pi}{3}$ (b) $\frac{\pi}{2}$

(c) $\frac{3\pi}{2}$ (d) π

9. The equation of tangent to the curve $\frac{x^2}{3}-\frac{y^2}{2}=1$ which is parallel to $y = x$, is ________.

(a) $y=x\pm1$ (b) $y=x-\frac{1}{2}$

(c) $y=x+\frac{1}{2}$ (d) $y=1-x$

10. If $f(x)=\int_{x^2}^{x^2+1} e^{-t^2}\,dt$, then $f(x)$ increases in ________.

(a) (–2, 2) (b) no value of x

(c) $(0, \infty)$ (d) $(-\infty, 0)$

11. The angle between the tangent drawn from the point (1, 4) to the parabola $y^2 = 4x$ is ________.

(a) $\frac{\pi}{6}$ (b) $\frac{\pi}{4}$

(c) $\frac{\pi}{3}$ (d) $\frac{\pi}{2}$

12. If $y=a\log x+bx^2+x$, has its extremum value at $x=-1$ and $x=2$, then ________.

(a) $a=2, b=-1$ (b) $a=2, b=-\frac{1}{2}$

(c) $a=-\frac{1}{2}, b=\frac{1}{2}$ (d) None of these.

13. The number of values of x where $f(x) = \cos x + \cos\sqrt{2}x$ attains its maximum is ________.

(a) 1 (b) 0

(c) 2 (d) infinite

14. The slope of the tangent at the point (h, h) to the circle $x^2 + y^2 = a^2$ is ________.

(a) 0 (b) 1

(c) –1 (d) will depend on h

15. Maximum value of $\frac{\log x}{x}$ in interval $[2, \infty]$ is ________.

(a) $\frac{\log c}{2}$ (b) 0

(c) $\frac{1}{e}$ (d) 1

16. Let $f(x)=x^3-6x^2+12x-3$. Then at $x=2$, $f(x)$ has ________.

(a) a maxima

(b) a minima

(c) Both (a) and (b).

(d) Neither maxima nor minima.

17. $2x^3-6x+5$ is an increasing function, if ________.

(a) $0<x<1$ (b) $-1<x<1$

(c) $x<-1$ or $x>1$ (d) $-1<x<-1/2$

18. The greatest value of the function

$$f(x)=\frac{\sin 2x}{\sin\left(x+\frac{\pi}{4}\right)}$$

on the interval $(0, \pi/2]$ is ________.

(a) $\frac{1}{\sqrt{2}}$ (b) $\sqrt{2}$

(c) 1 (d) $-\sqrt{2}$

19. For the curve $y^n = a^{n-1}x$, if the subnormal at any point is a constant, then n is equal to ________.

(a) 1 (b) 2

(c) –2 (d) –1

20. If tangent to the curve $x = at^2$, $y = 2at$ is perpendicular to x-axis, then its point of contact is ________.

(a) (a, a) (b) $(0, a)$

(c) $(0, 0)$ (d) $(a, 0)$

21. For the function $f(x)=x^2-6x+8, 2\le x\le 4$, the value of x for which $f'(x)$ vanishes has ________.

(a) $\frac{9}{4}$ (b) $\frac{5}{2}$ (c) 3 (d) $\frac{7}{2}$

22. From mean value theorem $f(b)-f(a)=(b-a)f'(x_1)$ $a<x_1<b$ is $f(x)=\frac{1}{x}$, then x equal to ________.

(a) $\frac{2ab}{a+b}$ (b) $\frac{b-a}{b+a}$

(c) $\sqrt{ab}$ (d) $\frac{a+b}{2}$

23. The condition for $f(x)=x^3+px^2+qx+r$, $(x\in R)$ to have no extreme value, is ________.

(a) $p^2<3q$ (b) $2p^2<q$

(c) $p^2<\frac{1}{4}q$ (d) $p^2>3q$

24. A particle moves along the curve $y = x^2 + 2x$. Then the point on the curve such that x- and y-co-ordinates of the particle change with the same rate is ________.

(a) (1, 3) (b) $\left(\frac{1}{2}, \frac{5}{2}\right)$

(c) $\left(-\frac{1}{2}, -\frac{3}{4}\right)$ (d) (–1, –1)

25. The largest interval lying in $\left(-\frac{\pi}{2}, \frac{\pi}{2}\right)$ for which the function

$\left[f(x) = 4^{-x^2} + \cos^{-1}\left(\frac{x}{2} - 1\right) + \log(\cos x)\right]$ is

defined, is ________.

(a) $[0, \pi]$ (b) $\left(-\frac{\pi}{2}, \frac{\pi}{2}\right)$

(c) $\left(-\frac{\pi}{4}, \frac{\pi}{2}\right)$ (d) $\left[0, \frac{\pi}{2}\right)$

26. If x is real, the maximum value of $\frac{3x^2+9x+17}{3x^2+9x+7}$ is ________.

(a) 41 (b) 1

(c) $\frac{17}{7}$ (d) $\frac{1}{4}$

27. A triangular park is enclosed on two sides by a fence and on the third side by a straight river bank. The two sides having fence are of same length x. the maximum area enclosed by the park is ________.

(a) $\sqrt{\frac{x^3}{8}}$ (b) $\frac{1}{2}x^2$

(c) πx (d) $\frac{3}{2}x^2$

28. A function $y = f(x)$ has a second-order derivative $f''(x) = 6(x - 1)$. If its graph passes through the point (2, 1) and at that point the tangent to the graph in $y = 3x - 5$, then the function is ________.

(a) $(x - 1)^2$ (b) $(x - 1)^3$

(c) $(x + 1)^3$ (d) $(x + 1)^2$

29. The normal to the curve $x = a(1 + \cos\theta)$, $y = a\sin\theta$ at θ always passes through the fixed point ________.

(a) $(a, 0)$ (b) $(0, a)$

(c) $(0, 0)$ (d) (a, a)

30. If the function $f(x) = 2x^3 - 9ax^2 + 12a^2x + 1$, where $a > 0$ attains its maximum and minimum at p and q respectively such that $p^2 = q$, then a equals ________.

(a) 3 (b) 1

(c) 2 (d) 1/2

31. In the interval (0, 1), the mean value theorem is not true for ________.

(a) $f(x) = \begin{cases} \frac{1}{2} - x, & x < \frac{1}{2} \\ \left(\frac{1}{2} - x\right)^2, & x \geq \frac{1}{2} \end{cases}$

(b) $f(x) = \begin{cases} \frac{\sin x}{x}, & x \neq 0 \\ 1, & x = 0 \end{cases}$

(c) $f(x) = x|x|$

(d) $f(x) = |x|$

32. For what value of a, $f(x) = -x^3 + 4ax^2 + 2x - 5$ is decreasing $\forall\, x$?

(a) (1, 2) (b) (3, 4)

(c) R (d) No value of a.

33. The length of the subtangent to the curve $x^2 + xy + y^2 = 7$ at (1, –3) is ________.

(a) 3 (b) 5

(c) 15 (d) $\frac{3}{5}$

34. The maximum value $x^3 - 3x$ in the interval [0, 2] is ________.

(a) –2 (b) 0

(c) 2 (d) 1

35. If $f(x) = \frac{a\sin x + b\cos x}{c\sin x + d\cos x}$ is decreasing for x, then ________.

(a) $ad + bc < 0$ (b) $ad - bc < 0$

(c) $ab - cd > 0$ (d) $ab - cd < 0$

36. The minimum value of $\left(1 + \frac{1}{\sin^n \alpha}\right)\left(1 + \frac{1}{\cos^n \alpha}\right)$ is ________.

(a) 1 (b) 2

(c) $(1 + 2^{n/2})^2$ (d) None of these.

37. If $f(x) = 2x + \cot^{-1} x - \log(\sqrt{1+x^2} - x)$, then $f(x)$

(a) decreases in $(-\infty, \infty)$.

(b) decreases in $[0, \infty)$.

(c) neither increases nor decreases in $(0, \infty)$.

(d) increases in $(-\infty, \infty)$.

38. The maximum value of $x^{1/x}$ is ________.

(a) $1/e^e$ (b) e
(c) $e^{1/e}$ (d) $1/e$

39. The function f defined by $f(x) = 4x^4 - 2x + 1$ is increasing for ________.

(a) $x < 1$ (b) $x > 0$
(c) $x < \frac{1}{2}$ (d) $x > \frac{1}{2}$

40. The minimum value of $x^2 + \frac{1}{1+x^2}$ is at ________.

(a) $x = 0$ (b) $x = 1$
(c) $x = 4$ (d) $x = 3$

41. If a tangent to the curve $y = 6x - x^2$ is parallel to the line $4x - 2y - 1 = 0$, then the point of tangency on the curve is ________.

(a) (2, 8) (b) (8, 2)
(c) (6, 1) (d) (4, 2)

42. The interval in which the function $f(x) = \frac{4x^2+1}{x}$ is decreasing, is ________.

(a) $\left(-\frac{1}{2}, \frac{1}{2}\right)$ (b) $\left[-\frac{1}{2}, \frac{1}{2}\right]$
(c) (–1, 1) (d) [–1, 1]

43. A right-circular cylinder which is open at the top and has a given surface area, will have the greatest volume if its height h and radius r are related by ________.

(a) $2h = r$ (b) $h = 4r$
(c) $h = 2r$ (d) $h = r$

44. On the interval [0, 1], the function $x^{25}(1-x)^{75}$ takes its maximum value at the point ________.

(a) 9 (b) $\frac{1}{2}$
(c) $\frac{1}{3}$ (d) $\frac{1}{4}$

45. Function $f(x) = 2x^3 - 9x^2 + 12x + 99$ is monotonically decreasing, when

(a) $x < 2$ (b) $x > 2$
(c) $x > 1$ (d) $1 < x < 2$

46. If $f(x) = \frac{x}{\sin x}$ and $g(x) = \frac{x}{\tan x}$, where $0 < x \leq 1$, then in this interval

(a) both $f(x)$ and $g(x)$ are increasing functions.
(b) both $f(x)$ and $g(x)$ are decreasing functions.
(c) $f(x)$ is an increasing function.
(d) $g(x)$ is an increasing function.

47. If the lengths of tangent, subtangent, normal and subnormal for the curve $y = x^2 + x - 1$ at (1, 1) are A, B, C and D respectively, then their increasing order is ________.

(a) B, D, A, C (b) B, A, C, D
(c) A, B, C, D (d) B, A, D, C

48. If the function $y = \sin^{-1} x$, then $(1 - x^2)\frac{d^2y}{dx^2}$ is equal to ________.

(a) $-x\frac{dy}{dx}$ (b) 0
(c) $x\frac{dy}{dx}$ (d) $x\left(\frac{dy}{dx}\right)^2$

49. The minimum value of $\frac{\log x}{x}$ in the interval $[2, \infty]$ ________.

(a) is zero (b) is $\frac{1}{e}$
(c) is $\frac{\log e}{2}$ (d) does not exist

50. If $f(x) = \frac{x^2-1}{x^2+1}$, for every real number x, then the minimum value of f

(a) does not exist because f is unbounded.
(b) is not attained even though f is bounded.
(c) is equal to 1.
(d) is equal to –1.

51. Two towns A and B are 60 km apart. A school is to be built to serve 150 students in town A and 50 students in town B. If the total distance to be traveled by all 200 students is to be as small as possible, then the school should be built at

(a) town B.
(b) 45 km from town A.
(c) town A.
(d) 45 km from town B.

52. If $f(x) = \begin{cases} 3x^2 + 12x - 1, & -1 \leq x \leq 2 \\ 37 - x, & 2 < x \leq 3 \end{cases}$ then

(a) $f(x)$ is increasing on [–1, 2].
(b) $f(x)$ is discontinuous on [–1, 3].
(c) $f(x)$ has the maximum value at $x = 2$
(d) None of these.

53. The curve $y = \frac{2x}{1+x^2}$ has

(a) exactly 3 points of inflexion separated by a point of maximum and a point of minimum.
(b) exactly 2 points of inflexion with a point of maximum lying between them.

(c) exactly 2 points of inflexion with a point of minimum lying between them.
(d) exactly 3 points of inflexion separated by two points of maximum.

54. If $a+b+c=0$, then the quadratic equation $3ax^2+2bx+c=0$ has
(a) at least one root in [0, 1].
(b) one root in [2, 3] and the other in [−2, −1].
(c) imaginary roots.
(d) None of these.

55. The triangle formed by the tangent to the curve $f(x) = x^2 + bx - b$ at the point (1, 1) and the co-ordinate axes, lies in the first quadrant. If its area is 2, then the value of b is ________.
(a) −1 (b) 3
(c) −3 (d) 1

56. Let (h, k) be a fixed point, were $h > 0$, $k > 0$. A straight line passing through this point cuts the positive direction of co-ordinate axes at the points P and Q. Then the minimum area of the triangle OPQ, (O being the origin) must be ________.
(a) $\frac{1}{2}hk$ (b) $2hk$
(c) hk (d) None

57. If $f(x) = x^2 + 2bx + c^2$ and $g(x) = -x^2 - 2cx + b^2$ such that min $f(x)$ > max $g(x)$, then the relation between b and c, is ________.
(a) no real value of b and c
(b) $0 < c < b\sqrt{2}$
(c) $|c| < |b|\sqrt{2}$
(d) $|c| > |b|\sqrt{2}$

58. If the distance 'S' metre travelled by a particle in t seconds is given by $S = t^3 - 3t^2$, then the velocity of the particle when the acceleration is zero, (in m/s) is ________.
(a) 3 (b) −2
(c) −3 (d) 2

59. A point is moving on $y = 4 - 2x^2$. The x-co-ordinate of the point is decreasing at the rate of 5 units per second. Then the rate at which the y-co-ordinate of the point is changing when the point is at (1, 2) is ________.
(a) 5 unit/s (b) 10 unit/s
(c) 15 unit/s (d) 20 unit/s

60. Match the points on the curve $2y^2 = x + 1$ with the slopes of normal at those points and choose the correct answer,

	Point	Slope of the normal
A.	(7, 2)	(1) $-4\sqrt{2}$
B.	$\left(0, \frac{1}{\sqrt{2}}\right)$	(2) −8
C.	(1, −1)	(3) 4
D.	$(3, \sqrt{2})$	(4) 0
		(5) $-2\sqrt{2}$

	A	B	C	D
(a)	2	4	3	1
(b)	2	5	3	1
(c)	2	3	5	1
(d)	2	5	1	3

61. Let the function $g:(-\infty,\infty)\to\left(-\frac{\pi}{2},\frac{\pi}{2}\right)$ be given by $g(u) = 2\tan^{-1}(e^u) - \frac{\pi}{2}$. Then g is
(a) even and strictly increasing in $(0,\infty)$
(b) odd and strictly decreasing in $(-\infty,\infty)$
(c) odd and strictly increasing in $(-\infty,\infty)$
(d) neither even and nor odd, but is strictly increasing in $(-\infty,\infty)$

62. If $f(x) = xe^{x(1-x)}$, then $f(x)$ is
(a) increasing on $[-1/2, 1]$.
(b) decreasing on R
(c) Increasing on R.
(d) Decreasing on $[-1/2, 1]$.

63. For all $x \in (0,1)$,
(a) $e^x < 1+x$ (b) $\log_e(1+x) < x$
(c) $\sin x > x$ (d) $\log_e x > x$

64. If the normal to the curve $y = f(x)$ at the point (3, 4) makes an angle $\frac{3\pi}{4}$ with the positive x-axis, then $f'(3) =$ ________.
(a) −1 (b) $-\frac{3}{4}$
(c) $\frac{4}{3}$ (d) 1

65. Let $f(x) = \int e^x(x-1)(x-2)dx$. Then f decreases in the interval ________.
(a) $(-\infty, -2)$ (b) $(-2, -1)$
(c) (1, 2) (d) $(2, +\infty)$

66. Consider the following statements in S and R

S: Both $\sin x$ and $\cos x$ are decreasing functions in the interval $\left(\frac{\pi}{2}, \pi\right)$.

R: If a differentiable function decreases in an interval (a, b), then its derivative also decreases in (a, b).

Which of the following is true?

(a) Both S and R are wrong.

(b) Both S and R are correct, but R is not the correct explanation of S.

(c) S is correct and R is the correct explanation of S.

(d) S is correct and R is wrong.

67. The function $f(x) = \sin^4 x + \cos^4 x$ increases if ________.

(a) $0 < x < \pi/8$ (b) $\pi/4 < x < 3\pi/8$

(c) $3\pi/8 < x < 5\pi/8$ (d) $5\pi/8 < x < 3\pi/4$

68. The slope of the tangent to a curve $y = f(x)$ at $[x, f(x)]$ is $2x+1$. If the curve passes through the point (1, 2), then the area bounded by the curve, the x- axis and the line $x = 1$ is ________.

(a) $\frac{5}{6}$ (b) $\frac{6}{5}$

(c) $\frac{1}{6}$ (d) 6

69. The function $f(x) = \frac{\ln(\pi + x)}{\ln(e + x)}$ is

(a) increasing on $(0, \infty)$.

(b) decreasing on $(0, \infty)$.

(c) increasing on $(0, \pi/e)$, decreasing on $(\pi/e, \infty)$.

(d) decreasing on $(0, \pi/e)$, increasing on $(\pi/e, \infty)$.

70. The function defined by $f(x) = (x+2)e^{-x}$

(a) decreases for all x.

(b) decreases in $(-\infty, -1)$ and increases in $(-1, \infty)$.

(c) Increases for all x.

(d) Decreases in $(-1, \infty)$ and increases in $(-\infty, -1)$.

71. If $a+b+c = 0$, then the quadratic equation $3ax^2 + 2bx + c = 0$ has

(a) at least one root in [0, 1].

(b) one root in [2, 3].

(c) imaginary roots.

(d) None of these.

Questions from Previous Years

72. A wire of length 20 m is to be cut into two pieces. One of the pieces is to be made into a square and the other into a regular hexagon. Then the length of the side (in metres) of the hexagon, so that the combined area of square and the hexagon is minimum, is ________.

(a) $\frac{5}{2+\sqrt{3}}$ (b) $\frac{10}{2+3\sqrt{3}}$

(c) $\frac{5}{3+\sqrt{3}}$ (d) $\frac{10}{3+2\sqrt{3}}$

73. Let $A = [a_{ij}]$ be a 3×3 matrix, where

$$a_{ij} = \begin{cases} 1 & , \text{ if } i = j \\ -x & , \text{ if } |i-j| = 1 \\ 2x+1 & , \text{ otherwise} \end{cases}$$

Let a function $f: \mathrm{R} \to \mathrm{R}$ be defined as $f(x) = \det(\mathrm{A})$. Then the sum of maximum and minimum values of f on R is equal to ________.

(a) $-\frac{20}{27}$ (b) $\frac{88}{27}$

(c) $\frac{20}{27}$ (d) $-\frac{88}{27}$

74. The sum of all the local minimum values of the twice differentiable function $f: \mathrm{R} \to \mathrm{R}$ defined by $f(x) = x^3 - 3x^2 - \frac{3f''(2)}{2}x + f''(1)$ is ________.

(a) −22 (b) 5

(c) −27 (d) 0

75. The range of $a \in \mathrm{R}$ for which the function

$$f(x) = (4a-3)(x + \log_e 5) + 2(a-7)\cot\left(\frac{x}{2}\right)$$

$\sin^2\left(\frac{x}{2}\right), x \neq 2n\pi, n \in N$ has critical points, is ________.

(a) $(-3, 1)$ (b) $[1, \infty]$

(c) $(-\infty, -1)$ (d) $\left[-\frac{4}{3}, 2\right]$

76. Let $f(x) = 3\sin^4 x + 10\sin^3 x + 6\sin^2 x - 3$, $X \in \left[-\frac{\pi}{6}, \frac{\pi}{2}\right]$

Then f is:

(a) Increasing in $\left(-\frac{\pi}{6}, \frac{\pi}{2}\right)$.

(b) Decreasing in $\left(0, \frac{\pi}{2}\right)$.

(c) Increasing in $\left(-\frac{\pi}{6}, 0\right)$.

(d) Decreasing in $\left(-\frac{\pi}{6}, 0\right)$.

77. Let f be a real-valued function, defined on $R-\{-1,1\}$ and given by

$$f(x) = 3\log_e\left|\frac{x-1}{x+1}\right| - \frac{2}{x-1}$$

Then in which of the following intervals, function $f(x)$ is increasing?

(a) $(-\infty, \frac{1}{2}] - \{1,1\}$

(b) $(-1, \frac{1}{2}]$

(c) $(-\infty, \infty) - \{-1,1\}$

(d) $(-\infty, -1) \cup \left([\frac{1}{2}, \infty] - \{1\}\right)$

78. Let m and M be respectively the minimum and maximum value of

$$\begin{vmatrix} \cos^2 x & 1+\sin^2 x & \sin 2x \\ 1+\cos^2 x & \sin^2 x & \sin 2x \\ \cos^2 x & \sin^2 x & 1+\sin 2x \end{vmatrix}$$

Then the ordered pair (m, M) is equal to ________.

(a) $(-3, 3)$ (b) $(-3, -1)$

(c) $(-4, -1)$ (d) $(1, 3)$

79. The set of all real values of λ for which the function $f(x) = (1-\cos^2 x).(\lambda + \sin x), x \in \left(-\frac{\pi}{2}, \frac{\pi}{2}\right)$, has exactly one maxima and exactly minima, is ________.

(a) $\left(-\frac{1}{2}, \frac{1}{2}\right) - \{0\}$ (b) $\left(-\frac{3}{2}, \frac{3}{2}\right)$

(c) $\left(-\frac{1}{2}, \frac{1}{2}\right)$ (d) $\left(-\frac{3}{2}, \frac{3}{2}\right) - \{0\}$

80. The function, $f(x) = (3x-7)x^{2/3}$, $x \in R$, is increasing for all x lying in ________.

(a) $(-\infty, 0) \cup \left(\frac{14}{15}, \infty\right)$ (b) $(-\infty, 0) \cup \left(\frac{3}{7}, \infty\right)$

(c) $\left(-\infty, \frac{14}{15}\right)$ (d) $\left(-\infty, \frac{14}{15}\right) \cup (0, \infty)$

81. Let f be any function continuous on $[a,b]$ and twice differentiable on (a,b). If for all $x \in (a,b), f2(x) > 0$ and $f''(x) < 0$, then for any $c \in (a,b)$, $\frac{f(c)-f(a)}{f(b)-f(c)}$ is greater than ________.

(a) $\frac{b+a}{b-a}$ (b) 1

(c) $\frac{b-c}{c-a}$ (d) $\frac{c-a}{b-c}$

82. The position of a moving car at time t is given by $f(t) = at^2 + bt + c, t > 0$, where a, b and c are real numbers greater than 1. Then the average speed of the car over the time interval $[t_1, t_2]$ is attained at the point ________.

(a) $(t_2 - t_1)/2$ (b) $a(t_2 - t_1) + b$

(c) $(t_1 + t_2)/2$ (d) $2a(t_1 - t_2) + b$

83. If the surface area of a cube is increasing at a rate of 3.6 cm^2 / sec, retaining its shape; then the rate of change of its volume (in cm^2 / sec), when the length of a side of the cube is 10 cm, is ________.

(a) 18 (b) 10

(c) 20 (d) 9

84. If a function $f(x)$ defined by

$$f(x) = \begin{cases} ae^x + be^{-x}; & -1 \le x < 1 \\ cx^2 & ; \ 1 \le x < 3 \\ ax^2 + 2cx \ ; & 3 \le x < 4 \end{cases}$$

be continuous for some $a, b, c \in R$ and $f'(0) + f'(2) = e$, then the value of a is ________.

(a) $\frac{1}{e^2 - 3e + 13}$ (b) $\frac{e}{e^2 - 3e - 13}$

(c) $\frac{e}{e^2 + 3e + 13}$ (d) $\frac{e}{e^2 - 3e + 13}$

85. A spherical iron ball of 10 cm radius is coated with a layer of ice of uniform thickness that melts at a rate of $50\text{cm}^3/\text{min}$. When the thickness of ice is 5 cm, then the rate (in cm/min) at which of the thickness of ice decreases is ______.

(a) $\frac{5}{6\pi}$ (b) $\frac{1}{54\pi}$

(c) $\frac{1}{36\pi}$ (d) $\frac{1}{18\pi}$

86. If the tangent to the curve, $y = f(x) = x\log_e x, (x > 0)$ at a point $(c, f(c))$ is parallel to the line segment joining the points (1, 0)) and (*e*, *e*) then C is equal to ______.

(a) $\frac{e-1}{e}$ (b) $e^{\left(\frac{1}{e-1}\right)}$

(c) $e^{\left(\frac{1}{1-e}\right)}$ (d) $\frac{e}{e-1}$

87. Which of the following points lies on the tangent to the curve $x^4e^y + 2\sqrt{y+1} = 3$ at the point (1, 0)?

(a) (2, 2) (b) (2, 6)

(c) (–2, 6) (d) (–2, 4)

88. The length of the perpendicular from the origin, on the normal to the curve, $x^2 + 2xy - 3y^2 = 0$, at the point (2, 2) is ______.

(a) $\sqrt{2}$ (b) $4\sqrt{2}$

(c) 2 (d) $2\sqrt{2}$

89. Let $f(x) = x^2 + \frac{1}{x^2}$ and $g(x) = x - \frac{1}{x}$, $x \in R - \{-1, 0, 1\}$.

If $h(x) = \frac{f(x)}{g(x)}$, then the local minimum value of $h(x)$ is ______.

(a) –3 (b) $-2\sqrt{2}$

(c) $2\sqrt{2}$ (d) 3

90. Twenty metres of wire is available for fencing off a flower bed in the form of a circular sector. Then the maximum area (in sq. m) of the flower-bed is ______.

(a) 30 (b) 12.5

(c) 10 (d) 25

91. Let P be a point on the parabola, $x^2 = 4y$. If the distance of P from the centre of the circle, $x^2 + y^2 + 6x + 8 = 0$ is minimum, then the equation of the tangent to the parabola at P, is ______.

(a) $x + 4y - 2 = 0$ (b) $x + 2y = 0$

(c) $x + y + 1 = 0$ (d) $x - y + 3 = 0$

92. The normal to the curve $y(x-2)(x-3) = x + 6$ at the point where the curve intersects the *y*-axis passes through the point ______.

(a) $\left(\frac{1}{2}, \frac{1}{3}\right)$ (b) $\left(-\frac{1}{2}, -\frac{1}{2}\right)$

(c) $\left(\frac{1}{2}, \frac{1}{2}\right)$ (d) $\left(\frac{1}{2}, -\frac{1}{3}\right)$

93. The eccentricity of an ellipse whose centre is at the origin is $\frac{1}{2}$. If one of its directives is $x = -4$, then the equation of the normal to it at $\left(1, \frac{3}{2}\right)$ is ______.

(a) $x + 2y = 4$ (b) $2y - x = 2$

(c) $4x - 2y = 1$ (d) $4x + 2y = 7$

94. The function f defined by $f(x) = x^3 - 3x^2 + 5x + 7$ is

(a) increasing in R.

(b) decreasing in R.

(c) decreasing in $(0, \infty)$ and increasing in $(-\infty, 0)$.

(d) increasing in $(0, \infty)$ and decreasing in $(-\infty, 0)$.

95. A wire of length 2 units is cut into two parts which are bent respectively to form a square of side = *x* units and a circle of radius = *r* units. If the sum of the areas of the square and the circle so formed is minimum, then ______. (2016)

(a) $x = 2r$ (b) $2x = r$

(c) $2x = (\pi + 4)r$ (d) $(4 - \pi)x = \pi r$

96. The minimum distance of a point on the curve $y = x^2 - 4$ from the origin is ______.

(a) $\frac{\sqrt{15}}{2}$ (b) $\sqrt{\frac{19}{2}}$

(c) $\sqrt{\frac{15}{2}}$ (d) $\frac{\sqrt{19}}{2}$

97. Consider

$f(x)=\tan^{-1}\left(\sqrt{\frac{1+\sin x}{1-\sin x}}\right), x\in\left(0,\frac{\pi}{2}\right)$

A normal to $y = f(x)$ at $x = \frac{\pi}{6}$ also passes through the point:

(a) $\left(\frac{\pi}{6},0\right)$ (b) $\left(\frac{\pi}{4},0\right)$

(c) (0, 0) (d) $\left(0,\frac{2\pi}{3}\right)$

98. If the tangent at a point P, with parameter t, on the curve $X = 4t^2+3,\ Y = 8t^3-1,\ t\in R$, meets the curve again at a point Q, then the co-ordinates of Q are ________.

(a) $\left(16t^2+3,-64t^3-1\right)$ (b) $\left(4t^2+3,-8t^3-2\right)$

(c) $\left(t^2+3,t^3-1\right)$ (d) $\left(t^2+3,-t^3-1\right)$

99. The normal to the curve, $x^2 + 2xy - 3y^2 = 0$ at (1, 1)

(a) meets the curve again in the third quadrant.
(b) meets the curve again in the fourth quadrant.
(c) does not meets the curve again.
(d) meets the curve again in the second quadrant.

100. The equation of a normal to the curve $\sin y = x\sin\left(\frac{\pi}{3}+y\right)$ at $x = 0$ is ________.

(a) $2x-\sqrt{3}y=0$ (b) $2x+\sqrt{3}y=0$

(c) $2y-\sqrt{3}x=0$ (d) $2y+\sqrt{3}y=0$

101. Let k and K be the minimum and the maximum values of the function $f(x)=\frac{(1+x)^{0.6}}{1+x^{0.6}}$ in [0, 1] respectively. Then the ordered pair (k,K) is equal to:

(a) $\left(2^{-0.4},1\right)$ (b) $\left(2^{-0.4},2^{0.6}\right)$

(c) $\left(2^{-0.6},1\right)$ (d) $\left(1,2^{0.6}\right)$

102. Two ships A and B are sailing straight away from a fixed point O along routes such that $\angle AOB$ is always $120°$. At a certain instance, OA = 8 km, OB = 6 km and the ship A is sailing t the rate of 20 km/hr. Then the distance between A and B changes at the rate (in km/hr) of ______.

(a) $\frac{260}{\sqrt{37}}$ (b) $\frac{260}{37}$

(c) $\frac{80}{\sqrt{37}}$ (d) $\frac{80}{37}$

103. If $x = -1$ and $x = 2$ are extreme points of $f(x)=\alpha\log|x|+\beta x^2+x$, then

(a) $\alpha=2,\ \beta=-\frac{1}{2}$ (b) $\alpha=2,\ \beta=\frac{1}{2}$

(c) $\alpha=-6,\ \beta=\frac{1}{2}$ (d) $\alpha=-6,\ \beta=-\frac{1}{2}$

104. For the curve $y = 3\sin\theta\cos\theta$, $x = e^{\theta}\sin\theta$, $0\le\theta\le\pi$, the tangent is parallel to x-axis when θ is ______.

(a) $\frac{3\pi}{4}$ (b) $\frac{\pi}{2}$

(c) $\frac{\pi}{4}$ (d) $\frac{\pi}{6}$

105. If an equation of a tangent to the curve is, $y-\cos(x+y), -1\le x\le 1+\pi, x+2y=k$ is equal to:

(a) 1 (b) 2

(c) $\frac{\pi}{4}$ (d) $\frac{\pi}{2}$

106. The cost of running a bus from A to B, is $\Delta\left(av+\frac{b}{v}\right)$, where v km/h is the average speed of the bus. When the bus travels at 30 km/h, the cost comes out be ₹75 while at 40 km/h, it is ₹65. Then the most economical speed (in km/h) of the bus is:

(a) 45 (b) 50
(c) 60 (d) 40

107. A spherical balloon is being inflated at the rate of 35 cc/min. The rate of increase in the surface area (in cm²/min) of the balloon when its diameter is 14 cm, is:

(a) 10 (b) $\sqrt{10}$
(c) 100 (d) $10\sqrt{10}$

108. The real number k for which the equation, $2x^3+3x+k=0$ has two distinct real roots in [0, 1]

(a) lies between 1 and 2.
(b) lies between 2 and 3.
(c) lies between 1 and 0.
(d) does not exist.

Integer and Numerical Value Type

1. If the segment of the tangent to the curve
$$y=\frac{a}{2}\log\frac{a+\sqrt{a^2-x^2}}{a-\sqrt{a^2-x^2}}-\sqrt{a^2-x^2}$$
contained between y-axis and the point of tangency is λa, then the numerical quantity λ should be equal to ______.

2. If the angle of intersection of two curves $x^2=4ay$ and $2y^2=ax$ is $90°$ and $\tan^{-1}\frac{3}{k}$, then the numerical non-zero quantity k must be ______.

3. The segment of the tangent to the curve $x^{2/3}+y^{2/3}=16$, contained between x- and y-axis, has length ______.

4. All the values of m for which the function $f(x)=\frac{x^3}{3}-(m-3)\frac{x^3}{2}+mx$ is increasing in $[0,\infty)$ lie in the interval $[0,k]$ then the numerical quantity k must be equal to ______.

5. If $4x^2+y^2=1$, then the maximum value of $12x^2-3y^2+16xy$ must be ______.

6. If A + B + C = π and maximum value of $\cos A+\cos B+k\cos C$ is $\frac{1}{\lambda k}+k$, then λ must be equal to ______.

7. A 12 cm long wire is bent to form a triangle with one of the angles as $60°$. The length of the side opposite to the angle $60°$, when the area of the triangle is maximum, is ______.

8. The absolute minimum value of the function x^2+4x+5 is ______.

9. Assuming that the petrol burn in driving a motor boat varies as the cube of its velocity, the most economical speed when going against current of 8 km/hour, is ______.

Questions from Previous Years

10. Let x be an integer such that all real roots of the polynomial $2x^5+5x^4+10x^3+10x^2+10x+10$ lie in the interval $(a, a+1)$. Then, $|a|$ is equal to ______.

11. If 'R' is the least value of 'a' such that the function $f(x)=x^2+ax+1$ is increasing on [1, 2] and 'S' is the greatest value of 'a' such that function $f(x)=x^2+ax+1$ is decreasing on [1, 2], then the value of $|R-S|$ is ______.

12. A wire of length 36 m is cut into two pieces, one of the pieces is bent to form a square and the other is bent to form a circle. If the sum of the areas of the two figures is minimum, and the circumference of the circle is k (metre), then $\left(\frac{4}{\pi}+1\right)k$ is equal to ______.

13. Let
$$f(x)=\begin{vmatrix}\sin^2 x & -2+\cos^2 x & \cos 2x\\ 2+\sin^2 x & \cos^2 x & \cos 2x\\ \sin^2 x & \cos^2 x & 1+\cos 2x\end{vmatrix}, x\in[0,\pi]$$
Then the maximum value of $f(x)$ is equal to ______.

14. Let $f:[-1,1]\to R$ be defined as $f(x)=ax^2+bx+c$ for all $x\in[-1,1]$, where $a,b,c\in R$ such that $f(-1)=2, f'(-1)=1$ and for $x\in(-1,1)$ the maximum value of $f''(x)$ is $\frac{1}{2}$. If $f(x)\le\alpha$, $x\in\in[-1,1]$, then the least value of α is equal to ______.

15. Let AD and BC be two vertical poles at A and B respectively on a horizontal ground. If AD = 8 m, BC = 11 m and AB = 10 m, then the distance (in metres) of a point M on AB from the point A such that MD^2+MC^2 is minimum is ______.

16. Let $f(x)$ be a polynomial of degree 3 such that $f(-1)=10$, $f(1)=-6$, $f(x)$ has a critical point at $x=-1$ and $f'(x)$ has a critical point at $x=1$. Then $f(x)$ has a local minima at $x=$ ______.

17. Let $p(x)$ be a real polynomial of least degree, which has a local maximum at $x=1$ and a local minimum at $x=3$. If $p(1)=6$ and $p(3)=2$, then $p'(0)$ is ______.

18. Let $f: IR\to IR$ be defined as $f(x)=|x|+|x^2-1|$. The total number of points at which $f(x)$ attains either a local maximum or a local minimum is ______.

19. Let f be a function defined on R (the set of all real numbers) such that

$f'(x) = 2010(x-2009)(x-2010)^2(x-2011)^3$ $(x-2012)^4$ for all $x \in R$.
If g is a function defined on R with values in the interval $(0, \infty)$ such that $f(x) = \ln(g(x))$, for all $x \in R$ then the number of points in R at which g has a local maximum is ______.

20. Let f be a real-valued differentiable function on R (the set of all real numbers) such that $f(1) = 1$. If the y-intercept of the tangent at any point $P(x, y)$ on the curve $y = f(x)$ is equal to the cube of the abscissa of P, then find the value of $f(-3)$.

21. Let $p(x)$ be a polynomial of degree 4 having extremum at $x = 1, 2$ and $\lim_{x\to 0}\left(1+\frac{p(x)}{x^2}\right) = 2$.
Then the value of $p(2)$ is _____.

22. The maximum value of the function $f(x) = 2x^3 - 15x^2 + 36x - 48$ on the set $A = \{x \mid x^2 + 20 \le 9x\}$ is ______.

Assertion and Reason

Which of the following is correct?

(a) Both Statement 1 and Statement 2 are true and Statement 2 is the correct reason for Statement 1.

(b) Both Statement 1 and Statement 2 are true but Statement 2 is not the correct reason for Statement 1.

(c) Statement 1 is true, Statement 2 is false.

(d) Statement 1 is false, Statement 2 is true.

1. **Statement 1:** Both $\sin x$ and $\cos x$ are decreasing functions in $\left(\frac{\pi}{2}, \pi\right)$.

Statement 2: If a differentiable function decreases in an interval (a, b), then its derivative also decreases in (a, b).

2. **Statement 1:** If a function is continuous and differentiable in $[a, b]$ and $f(a) \neq f(b)$, then there will not exist a point $c \in (a, b)$ such that $f'(c) = 0$.
Statement 2: Rolle's theorem

3. **Statement 1:** If $f(0) = 0, f'(x) = \log\left(x+\sqrt{1+x^2}\right)$, then $f(x)$ is positive for all x.

Statement 2: $f(x)$ is increasing for $x > 0$ and decreasing for $x < 0$.

4. **Statement 1:** The equation $x^2 = x\sin x + \cos x$ has only one solution.
Statement 2: The derivative of the function $x^2 - x\sin x - \cos x$l is $x(2 - \cos x)$.

5. **Statement 1:** Two conics can have, at the most, four common tangents.
Statement 2: The general equation of second degree $ax^2 + 2hxy + hy^2 + 2gx + 2fy + c = 0$ represents a conic, if

$$abc + 2fgh - af^2 - bg^2 - ch^2 = 0.$$

6. **Statement 1:** $f(x) = xe^{-x}$ has the maximum at $x = 1$ because
Statement 2: $f'(1) = 0$ and $f''(1) < 0$

7. **Statement 1:** $f(x) = 2x^3 - 9x^2 + 12x - 3$ is increasing outside the interval (1, 2)
Statement 2: $f'(x) < 0$ for $x \in (1, 2)$.

8. **Statement 1:** The curves $y^2 = 16x$ and $x^2 = 16y$ are orthogonal curves.
Statement 2: Two curves are said to be orthogonal, if they intersect each other at right angle at each point of intersection.

9. **Statement 1:** Shortest distance between $|x| + |y| = 2$ and $x^2 + y^2 = 16$ is $4 - \sqrt{2}$
Statement 2: Shortest distance between the two differentiable curves lies along the common normal.

10. **Statement 1:** If $f(x)$ is an increasing function with concavity upwards, then concavity of $f'(x)$ is also upwards.

Statement 2: If $f(x)$ is a decreasing function with concavity upwards, then concavity of $f'(x)$ is also upwards.

JEE ADVANCED

Objective Questions

One Option Correct

1. Tangent of acute angle between the curves $y=|x^2-1|$ and $y=|x^3-3|$ at their points of intersection is ______.

(a) 0 (b) $\frac{4\sqrt{2}}{3}$

(c) $\frac{4\sqrt{2}}{7}$ (d) $2\sqrt{2}$

2. The curve $y-e^{xy}+x=0$ has a vertical tangent at ______.

(a) (1, 1) (b) (0, 1)

(c) (1, 0) (d) no point

3. A particle moving on a curve has the position at time t given by $x=f'(t)\sin t+f''(t)\cos t, y=f'(t)\cos t-f'(t)\sin t$, where f is a thrice differentiable function. Then the velocity of the particle at time t is ______.

(a) $f'(t)+f''(t)$ (b) $f'(t)-f'''(t)$

(c) $f'(t)+f'''(t)$ (d) $f'(t)-f''(t)$

4. $f(x)$ and $g(x)$ are differentiable in [0, 1] such that $f(0) = 2$, $g(0) = 0$, $f(1) = 6$, $g(1) = 2$, then Rolle's theorem is applicable for which of the following in [0, 1]?

(a) $f(x) - g(x)$ (b) $f(x) - 2g(x)$

(c) $f(x) + 3g(x)$ (d) None of these.

5. If $f(x) = \frac{a\sin x+b\cos x}{c\sin x+d\cos x}$ is monotonically increasing, then

(a) $ad \geq bc$ (b) $ad < bc$

(c) $ad \leq bc$ (d) $ad > bc$

6. Let $f(x) = x^3+ax^2+bx+5\sin^2 x$ be an increasing function in the set of real numbers R. Then a and b satisfy the condition

(a) $a^2-3b-15\geq 0$ (b) $a^2-3b+15\leq 0$

(c) $a^2-3b-15\leq 0$ (d) $a>0$ and $b>0$

7. If $\frac{a_0}{n+1}+\frac{a_1}{n}+\frac{a_2}{n-1}+........+\frac{a_{n-1}}{2}+a_n=0$, then the equation $a_0x^n+a_1x^{n-1}+a_2x^{n-2}+......+a_{n-1}x+a_n=0$ has

(a) exactly one root in (0, 1).

(b) at least one root in (0, 1).

(c) no root in (0, 1).

(d) at the most one root in (0, 1).

8. $f:[0,4]\to R$ is a differentiable function. Then for some $a, b \in (0, 4)$, $f^2(4) - f^2(0) =$

(a) $8f'(a) f(b)$ (b) $4f'(b) f(a)$

(c) $2f(b) f(a)$ (d) $f'(b) f(a)$

9. The set of values of the parameter 'a' for which the function; $f(x) = 8ax - a \sin 6x - 7x - \sin 5x$ increases and has no critical points for all $x \in$ R, is:

(a) [–1, 1] (b) (–∞, –6)

(c) (6, +∞) (d) [6, +∞)

10. The function $f(x)=\frac{\ell n(\pi+x)}{\ell n(e+x)}$ is:

(a) Increasing on $(0,\infty)$.

(b) Decreasing on $(0,\infty)$.

(c) Increasing on $\left(0,\frac{\pi}{e}\right)$, decreasing on $\left(\frac{\pi}{e},\infty\right)$.

(d) Decreasing on $\left(0,\frac{\pi}{e}\right)$, increasing on $\left(\frac{\pi}{e},\infty\right)$.

11. Let $f(x) = \{x\}$, where {.} denotes the fractional part. For $f(x)$, $x = 5$ is

(a) a point of local maxima.

(b) a point of local minima.

(c) neither a point of local minima nor maxima.

(d) None of these.

12. The equation $x^3-3x+[a]=0$, where [.] denotes the greatest integer function, will have three real and distinct roots if ______.

(a) $a\in(-\infty,2)$ (b) $a\in(0,2)$

(c) $a\in(\infty,-2)\cup(0,\infty)$ (d) $a\in[-1,2]$

13. A function is defined as $f(x)=ax^2-b|x|$, where a and b are constants. $x=0$ will be a point of maxima of $f(x)$ if ______.

(a) $a>0, b>0$ (b) $a>0, b<0$

(c) $a<0, b<0$ (d) $a<0, b>0$

14. If $f(x)=2x^3-3(a+1)x^2+6ax-12$ has local maximum at x_1 and local minimum at x_2 and if $2x_1=x_2$ then value of a is ______.

(a) 1 (b) $\frac{1}{2}$

(c) –1 (d) 2

15. Let $f(x)$ be defined as

$$f(x)=\begin{cases}\tan^{-1}\alpha-5x^2 & 0<x<1\\ -6x & x\geq 1\end{cases}$$

$f(x)$ can have a local maximum at $x = 1$ if value of α is ________.

(a) 0 (b) –1
(c) –tan 1 (d) – 2

16. A truck is to be driven 300 km on a highway at a constant speed of x kmph. Speed rules of the highway required that $30 \leq x \leq 60$. The fuel costs ₹10 per litre and is consumed at the rate of $2+\frac{x^2}{600}$ litres per hour. The wages of the driver are ₹200 per hour. The most economical speed to drive the truck, in kmph, is ________.

(a) 30 (b) 60
(c) $30\sqrt{3.3}$ (d) $20\sqrt{3.3}$

17. Slope of tangent to the curve $y = 2e^x \sin\left(\frac{\pi}{4}-\frac{x}{2}\right)\cos\left(\frac{\pi}{4}-\frac{x}{2}\right)$, where $0 \leq x \leq 2\pi$ is minimum at $x =$ ________.

(a) 0 (b) π
(c) 2π (d) None of these.

18. The lateral edge of a regular hexagonal pyramid is 1 cm. If the volume is maximum, then its height must be equal to ________.

(a) $\frac{1}{3}$ (b) $\frac{2}{3}$
(c) $\frac{1}{\sqrt{3}}$ (d) 1

Questions from Previous Years

19. Consider all rectangles lying in the region

$$\left\{(x,y)\in \mathrm{R}\times\mathrm{R}: 0\leq x\leq\frac{\pi}{2} \text{ and } 0\leq y\leq 2\sin(2x)\right\}$$

and having one side on the x-axis. The area of the rectangle which has the maximum perimeter among all such rectangles, is ________.

(a) $\frac{3\pi}{2}$ (b) π
(c) $\frac{\pi}{2\sqrt{3}}$ (d) $\frac{\pi\sqrt{3}}{2}$

20. If $f: \mathrm{R} \to \mathrm{R}$ is a twice differentiable function such that $f''(x) > 0$ for all $x \in \mathrm{R}$, and $f\left(\frac{1}{2}\right)=\frac{1}{2}$, $f(1) = 1$, then ________.

(a) $f'(1) \leq 0$ (b) $0 < f'(1) \leq \frac{1}{2}$
(c) $\frac{1}{2} < f'(1) \leq 1$ (d) $f'(1) > 1$

21. The least value of $a \in R$ for which $4\alpha x^2+\frac{1}{x}\geq 1$, for all $x > 0$, is ________.

(a) $\frac{1}{64}$ (b) $\frac{1}{32}$
(c) $\frac{1}{27}$ (d) $\frac{1}{25}$

One or More Than One Option Correct

1. The real values of λ for which equation $x^3 + x^2 - x + \lambda = 0$ has three distinct real roots, lies in interval ________.

(a) [– 2, 2] (b) [– 1, 0]
(c) (– 1, 0) (d) $\left[-1,\frac{1}{2}\right]$

2. If α, β, γ are roots of $x^3+\left(a^4+4a^2+1\right)x=x^2+a^2$ then minimum value of $\sum\left\{\frac{\alpha}{\beta}+\left(\frac{\alpha}{\beta}\right)^{-1}\right\}$ is ________.

(a) $\log_2 8$ (b) $\log_2 16$
(c) $\log_3 27$ (d) $\log_4 256$

3. If $f(x) = x^3 - 3x^2 + 2x$ and $f(x) = \lambda$ has exactly two solutions which are of opposite signs, then λ equals ________.

(a) $\frac{-2\sqrt{3}}{9}$ (b) $\frac{-2}{3\sqrt{3}}$
(c) $\frac{2}{3\sqrt{3}}$ (d) $\frac{2\sqrt{3}}{9}$

4. The abscissa of a point on the curve $xy=(a+x)^2$, at which the normal cuts off numerically equal intercepts from the co-ordinate axes is ________.

(a) $-\frac{a}{\sqrt{2}}$ (b) $\sqrt{2}a$
(c) $\frac{\sqrt{2}a}{2}$ (d) $-\sqrt{2}a$

5. For function $f(x)=\frac{\ln x}{x}$, which of the following statements are true?

(a) $f(x)$ has a horizontal tangent at $x = e$.
(b) $f(x)$ cuts the x-axis only at one point.
(c) $f(x)$ is many-one function.
(d) $f(x)$ has one vertical tangent.

6. The equation of tangent drawn to the curve $y=(x+1)^3$, from origin is ______.

(a) $y=3x$ (b) $y=-3x$

(c) $4y=27x$ (d) $y=0$

7. For the curve represented parametrically by the equations, $x=2\,\ell n\cot t+1$ and $y=\tan t+\cot t$

(a) tangent at $t=\pi/4$ is parallel to x-axis.

(b) normal at $t=\pi/4$ is parallel to y-axis.

(c) tangent at $t=\pi/4$ is parallel to the line $y=x$.

(d) tangent and normal intersect at the point (2, 1).

8. Let $g'(x)>0$ and $f'(x)<0, \forall x\in R$, then

(a) $g(f(x+1))>g(f(x-1))$.

(b) $f(g(x-1))>f(g(x+1))$.

(c) $g(f(x+1))<g(f(x-1))$.

(d) $g(g(x+1))<g(g(x-1))$

9. If $f(x)=x^3-x^2+100x+1001$, then ______.

(a) $f(2000)>f(2001)$

(b) $f\left(\frac{1}{1999}\right)>f\left(\frac{1}{2000}\right)$

(c) $f(x+1)>f(x-1)$

(d) $f(3x-5)>f(3x)$

10. If the derivative of an odd cubic polynomial vanishes at two different values of 'x', then

(a) coefficient of x^3 and x in the polynomial must be same in sign.

(b) coefficient of x^3 and x in the polynomial must be different in sign.

(c) the values of x where derivative vanishes are closer to origin as compared to the respective roots on either side of origin.

(d) the values of x where derivative vanishes are far from origin as compared to the respective roots on either side of origin.

11. Let $f(x)=(x-1)^4(x-2)^n, n\in N$. $f(x)$ has

(a) local minimum at $x=2$ if n is even.

(b) local minimum at $x=1$ if n is odd.

(c) local maximum at $x=1$ if n is odd.

(d) local minimum at $x=1$ if n is even.

12. For the function, $f(x)=x\cos\frac{1}{x}, x\geq 1$

(a) for at least one x in the interval $[1,\infty)$, $f(x+2)-f(x)<2$.

(b) $\lim_{x\to\infty} f'(x)=1$.

(c) for all x in the interval $[1,\infty)$, $f(x+2)-f(x)>2$.

(d) $f'(x)$ is strictly decreasing in the interval $[1,\infty)$.

13. Let $h(x)=f(x)-(f(x))^2+(f(x))^3$ for every real number x. Then

(a) h is increasing whenever f is increasing.

(b) h is increasing whenever f is decreasing.

(c) h is decreasing whenever f is decreasing.

(d) Nothing can be said in general.

14. If $f(x)=\begin{cases}3x^2+12x-1, & -1\leq x\leq 2\\ 37-x & 2<x\leq 3\end{cases}$ then:

(a) $f(x)$ is increasing on $[-1,2]$.

(b) $f(x)$ is continues on $[-1,3]$.

(c) $f(2)$ does not exist.

(d) $f(x)$ has the maximum value at $x=2$.

15. Let f and g be increasing and decreasing functions, respectively from $[0,\infty)$. Let $h(x)=f(g(x))$. If $h(0)=0$, then $h(x)-h(1)$ is

(a) always zero.

(b) always negative.

(c) always positive.

(d) None of these.

Questions from Previous Years

16. Let $f:\mathbb{R}\to\mathbb{R}$ be defined by $f(x)=\frac{x^2-3x-6}{x^2+2x+4}$.

Then which of the following statements is (are) TRUE?

(a) f is decreasing in the interval $(-2,-1)$.

(b) f is increasing in the interval (1, 2).

(c) f is onto.

(d) Range of f is $\left[-\frac{3}{2},2\right]$.

17. Let $f(x)=\frac{\sin\pi x}{x^2}, x>0$

Let $x_1<x_2<x_3<\ldots<x_n<\ldots$ be all the points of local maximum of f and $y_1<y_2<y_3<$

$\ldots < y_n < \ldots$ be all the points of local minimum of f. Then which of the following options is/are correct?

(a) $x_{n+1} - x_n > 2$.

(b) $x_n \in \left(2n, 2n + \frac{1}{2}\right)$ for every n.

(c) $|x_n - y_n| > 1$ for every n.

(d) $x_1 < y_1$

18. Let $f : R \to R$ be given by $f(x) = (x-1)(x-2)(x-5)$.

Define $F(x) = \int_0^x f(t)\,dt,\ x > 0$.

Then which of the following options is/are correct?

(a) F has a local maximum at $x = 2$.
(b) F has a local minimum at $x = 1$.
(c) F has two local maxima and one local minimum in $(0, \infty)$.
(d) $F(x) < 0$ for all $x \in (0, 5)$.

19. Define the collections $\{E_1, E_2, E_3, \ldots\}$ of ellipse and $\{R_1, R_2, R_3, \ldots\}$ of rectangles as follows:

$E_1 : \frac{x^2}{9} + \frac{y^2}{4} = 1$;

R_1 : rectangle of largest area, with sides parallel to the axes, inscribed in E_1;

E_n : ellipse $\frac{x^2}{a_n^2} + \frac{y^2}{b_n^2} = 1$ of largest area inscribed in $R_{n-1}, n > 1$;

R_n : rectangle of largest area, with sides parallel to the axes, inscribed in $E_n, n > 1$.

Then which of the following options is/are correct?

(a) The eccentricities of E_{18} and E_{19} are NOT equal.
(b) The length of latus rectum of E_9 is $\frac{1}{6}$.
(c) $\sum_{n=1}^{N} (\text{area of } R_n) < 24,$ for each positive integer N.
(d) The distance of a focus from the centre in E_9 is $\frac{\sqrt{5}}{32}$.

20. If $f(x) = \begin{vmatrix} \cos(2x) & \cos(2x) & \sin(2x) \\ -\cos x & \cos x & -\sin x \\ \sin x & \sin x & \cos x \end{vmatrix}$, then

(a) $f'(x) = 0$ at exactly three points in $(-\pi, \pi)$.
(b) $f'(x) = 0$ at more than three points in $(-\pi, \pi)$.
(c) $f(x)$ attains its maximum at $x = 0$.
(d) $f(x)$ attains its minimum at $x = 0$.

21. If $f : R \to R$ is a differentiable function such that $f'(x) > 2f(x)$ for all $x \in R$, and $f(0) = 1$, then

(a) $f(x)$ is increasing in $(0, \infty)$.
(b) $f(x)$ is decreasing in $(0, \infty)$.
(c) $f(x) > e^{2x}$ in $(0, \infty)$.
(d) $f(x) < e^{2x}$ in $(0, \infty)$.

22. Let $f : R \to (0, \infty)$ and $g : R \to R$ be twice differentiable functions such that f'' and g'' are continuous functions on R. Suppose $f'(2) = g(2) = 0, f''(2) \neq 0$ and $g'(2) \neq 0$. If $\lim_{x \to 2} \frac{f(x)g(x)}{f'(x)g'(x)} = 1$, then

(a) f has a local minimum at $x = 2$.
(b) f has a local maximum at $x = 2$.
(c) $f''(2) > f(2)$
(d) $f(x) - f''(x) = 0$ for at least one $x \in R$.

23. Let $f : (0, \infty) \to R$ be given by $f(x) = \int_{\frac{1}{x}}^{x} e^{-\left(t + \frac{1}{t}\right)} \frac{dt}{t}$.

Then

(a) $f(x)$ is monotonically increasing on $[1, \infty)$.
(b) $f(x)$ is monotonically decreasing on (0, 1).
(c) $f(x) + f\left(\frac{1}{x}\right) = 0$, for all $x \in (0, \infty)$.
(d) $f(2^x)$ is an odd function of x on R.

24. A rectangular sheet of fixed perimeter with sides having their lengths in the ratio 8 : 15 is converted into an open rectangular box by folding, after removing squares of equal area from all four corners. If the total area of removed squares is 100, the resulting box has maximum volume. Then the lengths of the sides of the rectangular sheet are ________.

(a) 24
(b) 32
(c) 45
(d) 60

Matrix Type

1. Match the following functions with the intervals in which they are increasing:

Column I	Column II
(A) $x-e^x$	(P) $(-\infty, \infty)$
(B) $(x-2)^5(2x+1)^4$	(Q) $\left(\frac{1}{2}, \infty\right)$
(C) $2x^2-\log x$	(R) $\left(-\infty, -\frac{1}{2}\right)\cup\left(\frac{11}{18}, \infty\right)$
(D) $\log\left(x+\sqrt{1+x^2}\right)$	(S) $(-\infty, 0)$

2. Let $y=f(x)$ be a differentiable curve. Let the tangent at P meet the axes at Q and R while normal at P meets the x-axis at G. Let M be the foot of the perpendicular from O origin on normal at P. Then match the following curves with the distances which will be constant for all positions of P on the curve:

Column I	Column II
(A) $\sqrt{x}+\sqrt{y}=\sqrt{a}$	(P) $PR-OR$
(B) $x^2+y^2=a^2$	(Q) QR
(C) $4x^2+4y^2=x$	(R) $OQ+OR$
(D) $x=a\cos^3 t, y=a\sin^3 t$	(S) OM

3. Match the following pairs of curves with their angles of intersections:

Column I	Column II
(A) $\sin x, \cos x$	(P) $90°$
(B) $x^2=4ay, y=\frac{8a^3}{x^2+4a^2}$	(Q) $\tan^{-1}2\sqrt{2}$
(C) $x^2-y^2=5, \frac{x^2}{18}+\frac{y^2}{8}=1$	(R) $\tan^{-1}3$

4. Match the following with their 'c' of Rolle's theorem:

Column I	Column II
(A) $f(x)=e^x\sin x, [0, \pi]$	(P) $\frac{\pi}{3}$
(B) $f(x)=e^x(\sin x-\cos x), \left[\frac{\pi}{4}, \frac{5\pi}{4}\right]$	(Q) $\frac{3\pi}{4}$
(C) $f(x)=x+\sin 2x, [0, \pi]$	(R) π
	(S) $\frac{2\pi}{3}$

5. Match the following inequalities with the intervals in which they are true:

Column I	Column II
(A) $\frac{x}{1+x}<\log(1+x)$	(P) $(0, \infty)$
(B) $x-\frac{x^2}{2}<\log(1+x)$	
(C) $\log(1+x)<x$	(Q) $(-1, 0)\cup(0, \infty)$
(D) $\frac{1}{\log(1+x)}-\frac{1}{x}<1$	

6. Match Column I with Column II.

Column I	Column II
(A) If a portion of the tangent at any point on the curve $x=at^3, y=at^4$ between the axes is divided by the point of contact in the ratio $m:n$ externally, then $\lvert n+m\rvert$ is equal to (m and n are co-prime)	(P) 1
(B) The area of triangle formed by normal at the point (1, 0) on the curve $x=e^{\sin y}$ with axes is	(Q) $\frac{1}{2}$
(C) If the angle between curves $x^2y=1$ and $y=e^{2(1-x)}$ at the point (1, 1) is θ then $\tan\theta$ is equal to	(R) 7
(D) The length of sub-tangent at any point on the curve $y=be^{x/3}$ is equal to	(S) 3
	(T) 0

7. Match Column I with Column II.

Column I	Column II
(A) Number of values of x lying in $\left(0,\frac{\pi}{2}\right)$ at which $f(x)=\ell n(\sin x)$ is not monotonic, is	(P) 0
(B) If the greatest interval in which the function $f(x)=x^3-3x+2$ is decreasing is $[a, b]$, then $a + b =$	(Q) 2
(C) If $f(x)=\frac{x^2+2}{[x]},1\le x\le 3$ (where [.] is the greatest integer function), then least value of $f(x)$ is	(R) –3
(D) If the set of all possible values of 'a' such that $f(x)=e^{2x}-(a+1)e^x+2x$ is monotonically increasing for all $x \in R$ is $(-\infty,k]$, then $k=$	(S) 3
	(T) –2

8. Match Column I with Column II.

Column I	Column II
(A) Number of points which are local extreme of $f(x)=\begin{cases}(2+x)^3 & ; -3\le x\le -1\\ x^{2/3} & ; -1<x<2\end{cases}$	(P) 1
(B) If $a + b = 1$; $a > 0$, $b > 0$, then the minimum value of $\sqrt{\left(1+\frac{1}{a}\right)\left(1+\frac{1}{b}\right)}$ is	(Q) 2
(C) The maximum value attained by $y = 10 - \lvert x - 10\rvert, -1 \le x \le 3$ is	(R) 3
(D) If P $(t^2, 2t)$, $t \in [0, 2]$ is an arbitrary point on parabola $y^2 = 4x$ and Q is at the foot of perpendicular from focus S on the tangent at P, then maximum area of triangle PQS is	(S) 4
	(T) 5

(Questions 9–11): By appropriately matching the information given in the three columns of the following table

Let $f(X)=X+\log_e X-X\log_e X, X\in(0,\infty)$

Column 1 contains information about zero of $f(X)$, $f'(X)$ and $f''(X)$.

Column 2 contains information about the limiting behaviour of $f(X)$, $f'(X)$ and $f''(X)$ at infinity.

Column 3 contains information about increasing/decreasing nature of $f(X)$ and $f'(X)$.

Column 1	Column 2	Column 3
(I) $f(X)=0$ for some $X\in(1,e^2)$	(i) $\lim_{X\to\infty} f(X)=0$	(P) f is increasing in (0, 1)
(II) $f'(x)=0$ for some $X\in(1,e)$	(ii) $\lim_{X\to\infty} f(X)=-\infty$	(Q) f is increasing in (e,e^2)
(III) $f'(x)=0$ for some $X\in(0,1)$	(iii) $\lim_{X\to\infty} f'(X)=-\infty$	(R) f' is increasing in (0, 1)
(IV) $f''(x)=0$ for some $X\in(1,e)$	(iv) $\lim_{X\to\infty} f''(X)=-\infty$	(S) f' is decreasing in (e,e^2)

9. Which of the following options is the only correct combination?
(a) (I) (i) (P).
(b) (II) (ii) (Q)
(c) (III) (iii) (R).
(d) (IV) (iv) (S)

10. Which of the following options is the only correct combination?
(a) (I) (ii) (R)
(b) (II) (iii) (S)
(c) (III) (iv) (P)
(d) (IV) (i) (S)

Question from Previous Year

11. Which of the following options is the only incorrect combination?
(a) (I) (iii) (P)
(b) (II) (iv) (Q)
(c) (III) (i) (R)
(d) (II) (iii) (P)

Integer and Numerical Value Type

1. If all the values of parameter m for which the function

$$\frac{x^3}{3}+(m-1)x^2+(m+5)x+7$$

is increasing for all $x>0$ lie in the interval $[-k,\infty)$, then the numerical quantity k should be equal to ______.

2. If over any interval a chord of the curve

$$y=\frac{x^4}{4}-\frac{3x^2}{2}+\lambda x-3$$

is parallel to three tangents, then λ must lie between $-k$ and k. k must be ______.

3. If $p(x)=51x^{101}-2323x^{100}-45x+1035$, then, using Rolle's theorem, at least one root lies between $45^{1/100}$ and k. The best possible value of k must be ______.

4. If $x=\sec\theta-\cos\theta$
$y=\sec''\theta-\cos''\theta$, then
If the slope of the tangent to the above curve is $n\sqrt{\frac{y^2+k}{x^2+k}}$, the numerical quantity k should be equal to ______.

5. If the function given by

$$y=\frac{1}{3}\log\frac{x+1}{\sqrt{x^2-x+1}}+\frac{1}{\sqrt{3}}\tan^{-1}\frac{2x+1}{\sqrt{3}}$$

is increasing for all $x>-k$, then the numerical quantity k should be equal to ______.

6. The least positive integer k, such that $f(x)=(2x+1)^{50}(3x-4)^{60}$ is increasing in $[k,\infty)$ is ______.

7. If the angle of intersection of two curves $x^2=4ay$ and $y=\frac{8a^3}{x^2+4a^2}$, $(a>0)$ is $\tan^{-1}k$, then numerical quantity k, should be equal to ______.

8. The function $\frac{2\sin x+5\cos x}{8\sin x-k\cos x}$ (k constant) is increasing in its domain, if k is a less than a negative integer l. Then $-l$ must be ______.

9. If the line $\frac{x}{3}+\frac{y}{4}=1$ touches the curve $y=4e^{-\frac{x}{3}}$ at $(0,l)$, then l must be equal to ______.

10. If the normals to the curve $x^3+y^3=3axy$ which are drawn at a point where it is intersected by the parabola $y^2=ax$, have slopes k, then the integer k must be equal to ______.

11. All the values of parameter m for which the function $(m-2)x^3-12\times 2+3(m+4)x-11$ is some monotonic for all x, lie in the interval $(-\infty,-k)\cup[2,\infty)$. Then the numerical quantity k, should be ______.

12. Let α be the angle in radians between $\frac{x^2}{36}+\frac{y^2}{4}=1$ and the circle $x^2+y^2=12$ at their points of intersection.
If $\alpha=\tan^{-1}\frac{k}{2\sqrt{3}}$, then find the value of k^2.

13. The chord of parabola $y=-p^2x^2+5px-4$ touches the curve $y=\frac{1}{(1-x)}$ at the point $x=2$ and is bisected that point. Find the number of values of p.

Questions from Previous Years

14. Let the function $f:(0,\pi)\to R$ be defined by $f(\theta)=(\sin\theta+\cos\theta)^2+(\sin\theta-\cos\theta)^4$.
Suppose the function f has a local minimum at θ precisely when $\theta\in\{\lambda_1\pi,\ldots\lambda_r\pi\pi\}$, where $0<\lambda_1<\ldots<\lambda_r<1$. Then the value of $\lambda_1+\ldots+\lambda_r$ is ________.

15. A cylindrical container is to be made from certain solid material with the following constraints: It has a fixed inner volume of V mm^3, has a 2 mm thick solid wall and is open at the top. The bottom of the container is a solid circular disc of thickness 2 mm and is of radius equal to the outer radius of the container.
If the volume of the material used to make the container is minimum when the inner radius of the container is 10 mm, then the value of $\frac{V}{250\pi}$ is ________.

16. A vertical line is passing through the point $(h, 0)$ intersects the ellipse $\frac{X^2}{4}+\frac{Y^2}{3}=1$ at the points P and Q. Let the tangents to the ellipse at P and Q meet at the point R. If $\Delta(h)=$ area of the triangle $\Delta PQR, \Delta_1=\max_{1/2\le h\le 1}\Delta(h)$ and $\Delta(h)$ and $\Delta_2=\min_{1/2\le h\le 1}\Delta(h)$, then $\frac{8}{\sqrt{5}}\Delta_1-8\Delta_2=$ ______.

Comprehension Passage

Comprehension 1

In second degree curves, a line which once touches the curve cannot meet the curve again but in cubic and other non-algebraic curves, the tangent can meet the curve again. If we solve the equation of tangent and a cubic curve, we will, in general, get three roots, two of which will be equal, since they will correspond to the point where the tangent was initially drawn.

1. If P is a point (β, β^3) different from (0, 0) on the curve $y=x^3$. The tangent at P meets the curve again at Q and the tangent at Q meets the curve again at R. Then abscissa of the point R must be ______.

(a) 8β (b) -4β
(c) 4β (d) -2β

2. The tangent at (t, t^2-t^3) on the curve $y=x^2-x^3$ meets the curve again at Q, then abscissa of Q must be ______.

(a) $1+2t$ (b) $1-2t$
(c) $-1-2t$ (d) $2t-1$

3. If the tangent at 't' of the curve $y=8t^3-1$, $x=4t^2+3$ meets the curve at t' and is normal to the curve at that point, then the value of t must be ______.

(a) $\pm\frac{1}{\sqrt{3}}$ (b) $\pm\frac{1}{\sqrt{2}}$
(c) $\pm\frac{\sqrt{2}}{3}$ (d) None of these.

Comprehension 2

Let P be a point on the curve $y=f(x)$. The tangent and normal at P meets x-axis at T and N. The perpendicular from P on x-axis is PM. We define sub-tangent

$$=TM=y\cot\psi=\frac{y}{\frac{dy}{dx}}.$$

$$\text{Sub-normal}=MN=y\tan\psi=y\frac{dy}{dx}.$$

At times PT and PN are respectively called length of the tangent and length of the normal.

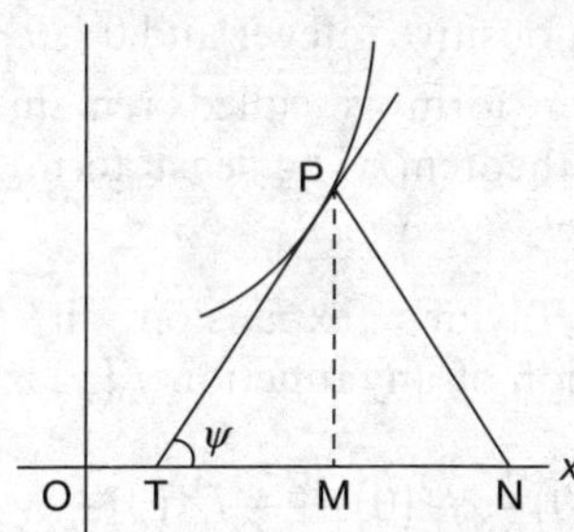

1. The of the sub-normal at any point 't' of the curve $x=a(t+\sin t)$, $y=a(1-\cos t)$ must be ______.

(a) $4a\sin^2\frac{t}{2}$ (b) $2a\tan\frac{t}{2}$
(c) $2a\sin^3\frac{t}{2}$ (d) $2a\sin^3\frac{t}{2}\sec\frac{t}{2}$

2. The value of n for which the length of the sub-normal at any point of the curve $y=a^{1-n}x^n$ must be constant, is ______.

(a) -1 (b) $-\frac{1}{2}$ (c) $\frac{1}{2}$ (d) 1

3. Let OM be the perpendicular from origin upon tangent at any point P on the curve $\frac{x^2}{a^2}+\frac{y^2}{b^2}=1$. Then length of the normal at P varies ______.

(a) as OM^2 (b) as OM^3
(c) as OM (d) as $\frac{1}{OM}$

Comprehension 3

We are familiar with expansion of functions like e^x, $\log_e(1+x)(-1<x\le 1)(1+x)^n$ (n is not a positive integer, $|x|\le 1$) in form of infinite series. In general, functions under certain conditions can be expanded by using Taylor's theorem, which states that

$$f(x+h)=f(x)+hf'(x)+\frac{h^2}{2!}f''(x)$$
$$=\frac{h^3}{3!}f'''(x)+\ldots\infty$$

which is sometimes written as

$$f(x+h)=f(x)+hf'(x)+\frac{h^2}{2!}f''(x)+\ldots+$$
$$\frac{h^{n-1}}{(n-1)!}f^{(n-1)}(x)+\frac{h^n(1-\theta)^{n-p}}{(n-1)!\,p}f^{(n)}(a+\theta h)$$

where (i) $f^{(n-1)}(x)$ and $f^{(n)}(x)$ denote $(n-1)$th term and nth derivative of $f(x)$, which are continuous.

(ii) p is a positive integer and $0 < \theta < 1$

The latter form is called remainder form of Taylor's theorem. The least term is called the remainder.

1. From Taylor's expansion, it follows that expansion of any function $f(x)$ is ______.

(a) $f(0) - xf'(0) + \frac{x^2}{2!} f''(0) + \ldots$

(b) $f(0) + xf'(0) + \frac{x^2}{2!} f''(0) + \frac{x^3}{3!} f'''(0) + \ldots$

(c) $f(0) + xf' + x^2 f''(0) + x^3 f'''(0) + \ldots$

(d) None of these.

2. By Taylor's theorem, the coefficient of x^2 in the expansion of $\log(1 + \sin x)$ must be ______.

(a) $-\frac{1}{3}$ (b) $\frac{1}{2}$

(c) $-\frac{1}{2}$ (d) 1

3. The coefficient of x^2 in the expansion of $e^{m\sin^{-1}x}$ must be ______.

(a) m (b) $\frac{m}{2}$

(c) $\frac{m^2}{2}$ (d) $-\frac{m^2}{2}$

Comprehension 4

Let the tangent to the cubic curve $x^3 + y^3 = a^3$ at $P(x_1, y_1)$ meet curve again at $Q(h, k)$.

$$\text{Put } A = \frac{h}{x_1}, B = \frac{k}{y_1}. A \neq B, \text{then}$$

1. $x_1^{\,3}$ must be equal to ______.

(a) $\frac{a^3(1-B)}{A-B}$ (b) $\frac{a^3(1-A)}{A-B}$

(c) $\frac{a^3(1-B)}{A+B}$ (d) None of these.

2. Which of the following is true:

(a) $(A+B)^2 - AB(A+B) + AB + 1 = 0$

(b) $(A+B)^2 + AB(A+B) + AB + 1 = 0$

(c) $(A+B)^2 - AB(A+B) - AB - 1 = 0$

(d) None of these.

3. The value of $\frac{h}{x_1} + \frac{k}{y_1}$ must be ______.

(a) 1 (b) −1 (c) 0 (d) 3

Comprehension 5

Let P and Q be any two points on the curve $ay^2 = x^3$ such that PQ always passes through a fixed point $R(at^2, at^3)$.

1. If $P(at_1^{\,2}, at_1^{\,3})$ and $Q(at_2^{\,2}, at_2^{\,3})$ are arbitrary points on the curve then ______.

(a) $t = \frac{t_1 t_2}{t_1 + t_2}$ (b) $t = \frac{t_1 - t_2}{t_1 + t_2}$

(c) $t = -\frac{t_1 t_2}{t_1 + t_2}$ (d) None of these.

2. If tangent at P and Q intersect at (x, y) then $t_1^{\,2} + t_2^{\,2} + t_1 t_2$ must be equal to ______.

(a) $\frac{2x}{a}$ (b) $\frac{3x}{a}$

(c) $-\frac{3x}{a}$ (d) None of these.

3. The locus of point of intersection of tangent at P and Q is the curve ______.

(a) $(3tx + 2y)^2 = 2a^3t^3y$

(b) $(3tx + 2y)^2 = -2a^3t^3y$

(c) $(3tx + 2y)^2 = 3a^3t^3y$

(d) $(3tx + 2y)^2 = -3a^3t^3y$

Comprehension 6

Let $a(t)$ is a function of t such that $\frac{da}{dt}$ =2 for all values of t and $a = 0$ when $t = 0$. Further $y = m(t)\, x + c(t)$ is tangent to the curve $y = x^2 - 2ax + a^2 + a$ at the point whose abscissa is 0. Then

1. If the rate of change of distance of vertex of $y = x^2 - 2ax + a^2 + a$ from the origin with respect to t is k, then $k =$ ______.

(a) 2 (b) $2\sqrt{2}$

(c) $\sqrt{2}$ (d) $4\sqrt{2}$

2. If the rate of change of $c(t)$ with respect to t, when $t = k$, is ℓ, then ______.

(a) $16\sqrt{2} - 2$ (b) $8\sqrt{2} + 2$

(c) $10\sqrt{2} + 2$ (d) $16\sqrt{2} + 2$

3. The rate of change of $m(t)$, with respect to t, at $t = \ell$ is ______.

(a) −2 (b) 2 (c) −4 (d) 4

Comprehension 7

Let $y = a\sqrt{x} + bx$ be curve, $(2x - y) + \lambda(2x + y - 4) = 0$ be family of lines.

1. If curve has slope $-\frac{1}{2}$ at (9, 0) then a tangent belonging to the family of lines is
(a) $x + 2y - 5 = 0$ (b) $x - 2y + 3 = 0$
(c) $3x - y - 1 = 0$ (d) $3x + y - 5 = 0$

2. A line of the family cutting positive intercepts on axes and forming triangle with co-ordinate axes, then minimum length of the line segment between axes is ______.
(a) $\left(2^{2/3} - 1\right)^{3/2}$ (b) $\left(2^{2/3} + 1\right)^{3/2}$
(c) $7^{3/2}$ (d) 27

3. Two perpendicular focal chords of curve form diagonals of a quadrilateral. Minimum area of quadrilateral is ______.
(a) 16 (b) 32
(c) 64 (d) 50

ANSWER KEY

JEE Main

Objective Questions

One Option Correct

1. (a) **2.** (b) **3.** (d) **4.** (d) **5.** (c) **6.** (a) **7.** (a) **8.** (a) **9.** (a)
10. (d) **11.** (c) **12.** (b) **13.** (a) **14.** (c) **15.** (c) **16.** (d) **17.** (c) **18.** (c)
19. (b) **20.** (c) **21.** (c) **22.** (c) **23.** (a) **24.** (c) **25.** (d) **26.** (a) **27.** (b)
28. (b) **29.** (a) **30.** (c) **31.** (a) **32.** (d) **33.** (c) **34.** (c) **35.** (b) **36.** (c)
37. (d) **38.** (b) **39.** (d) **40.** (a) **41.** (a) **42.** (a) **43.** (d) **44.** (d) **45.** (d)
46. (c) **47.** (d) **48.** (c) **49.** (d) **50.** (d) **51.** (c) **52.** (a) **53.** (a) **54.** (a)
55. (c) **56.** (b) **57.** (d) **58.** (c) **59.** (d) **60.** (b) **61.** (c) **62.** (a) **63.** (b)
64. (d) **65.** (c) **66.** (d) **67.** (b) **68.** (a) **69.** (b) **70.** (d) **71.** (a) **72.** (d)
73. (d) **74.** (c) **75.** (d) **76.** (d) **77.** (d) **78.** (b) **79.** (d) **80.** (a) **81.** (d)
82. (c) **83.** (d) **84.** (d) **85.** (d) **86.** (b) **87.** (c) **88.** (d) **89.** (c) **90.** (d)
91. (c) **92.** (c) **93.** (c) **94.** (a) **95.** (a) **96.** (a) **97.** (d) **98.** (d) **99.** (b)
100. (b) **101.** (a) **102.** (a) **103.** (a) **104.** (c) **105.** (d) **106.** (c) **107.** (a) **108.** (d)

Integer and Numerical Value Type

1. (1) **2.** (5) **3.** (64) **4.** (9) **5.** (5) **6.** (2) **7.** (4) **8.** (2) **9.** (12)
10. (2) **11.** (2) **12.** (36) **13.** (6) **14.** ($\leq$ 5) **15.** (5 m) **16.** (3) **17.** (9) **18.** (5)
19. (1) **20.** (9) **21.** (0) **22.** (7)

Assertion and Reason

1. (c) **2.** (d) **3.** (a) **4.** (d) **5.** (b) **6.** (a) **7.** (b) **8.** (d) **9.** (d) **10.** (d)

JEE Advanced

Objective Questions

One Option Correct

1. (c) **2.** (c) **3.** (c) **4.** (b) **5.** (a) **6.** (b) **7.** (b) **8.** (a) **9.** (d)
10. (b) **11.** (b) **12.** (d) **13.** (a) **14.** (b) **15.** (d) **16.** (b) **17.** (b) **18.** (c)
19. (c) **20.** (d) **21.** (c)

One or More Than One Option Correct

1. (a, d) **2.** (a, c, d) **3.** (a, b) **4.** (a, c) **5.** (a, b, c) **6.** (c, d) **7.** (a, b) **8.** (b, c) **9.** (b, c)
10. (b, c) **11.** (a, c, d) **12.** (b, c, d) **13.** (a, c) **14.** (a, b, c) **15.** (a) **16.** (a, b) **17.** (a, b, c)
18. (a, b, d) **19.** (b, c) **20.** (b, c) **21.** (a, c) **22.** (a, d) **23.** (a, c, d) **24.** (a, c)

Matrix Type Questions

1. A → S; B → R; C → Q; D → Q
2. A → R; B → S; C → P; D → Q
3. A → Q; B → R; C → P
4. A → Q; B → R; C → P, S
5. A → P, Q; B → P, Q; C → P, Q; D → P, Q
6. A → R; B → Q; C → T; D → S
7. A → P; B → P; C → S; D → S
8. A → Q; B → R; C → R; D → T

9. (b) **10.** (b) **11.** (c)

Integer and Numerical Value Type

1. (1) **2.** (2) **3.** (46) **4.** (4) **5.** (1) **6.** (2) **7.** (3) **8.** (20) **9.** (4)

10. (0) **11.** (4) **12.** (16) **13.** (2) **14.** $\left(\frac{1}{2}\right)$ **15.** (4) **16.** (9)

Comprehension Passage

Comprehension 1

1. (c) **2.** (b) **3.** (c)

Comprehension 2

1. (d) **2.** (c) **3.** (d)

Comprehension 3

1. (b) **2.** (c) **3.** (c)

Comprehension 4

1. (a) **2.** (c) **3.** (b)

Comprehension 5

1. (c) **2.** (b) **3.** (b)

Comprehension 6

1. (b) **2.** (d) **3.** (c)

Comprehension 7

1. (b) **2.** (b) **3.** (b)

HINTS AND EXPLANATIONS

Quiz

1. $A = \pi r^2 \; \frac{dA}{dr} = 2\pi r$

2. $\frac{dr}{dt} = 2\,\text{cm}$

Area of circle $= \pi r^2 \; \frac{dA}{dt} = 2\pi \frac{dr}{dr} = 2\pi \times 2$

$= 4\pi$ cm²/s.

3. Yes, correct

4. If $x_1 < x_2$ then $f(x_1) > f(x_2)$ for all $x_1, x_2 \in I$.

5. Let $x_1, x_2 \in R$ and $x_1 < x_2 \Rightarrow ax_1 < ax_2$

$ax_1 + b < ax_2 + b$

$\Rightarrow f(x_1) < f(x_2)$

Therefore, f is strictly increasing.

6. Let $f(x) = \cos x \; f'(x) = -\sin x$

For $0 < x < \frac{\pi}{2}$, $\sin x > 0$

$-\sin x < 0 \Rightarrow f'(x) < 0$ for the interval $\left(0, \frac{\pi}{2}\right)$.

Therefore f is strictly increasing.

7. $y = x^3 - 3x + 2$ y at $x = 3$ is 20.

The point on the curve is $(3, 20)$.

$\frac{dy}{dx} = 3x^2 - 3$

$\frac{dy}{dx}$ at $(3,20)$ is $3.(3)^2 - 3 = 24$.

8. $y = \sqrt{x}$

Let $x = 25$ and $\Delta x = 0.3$.

$\Delta y = \sqrt{x + \Delta x} - \sqrt{x}$

$\frac{dy}{dx} = \frac{1}{2\sqrt{x}} \; dy = \frac{dx}{2\sqrt{x}} = \frac{0.3}{2\sqrt{25}} = 0.03$

Therefore, approximate value is $5 + .03 = 5.03$

9. $f'(x) = -\sin x$

For maxima or minima,

$f'(x) = 0 \Rightarrow \sin x = 0$

$\sin x$ is not zero for the given interval.

Therefore, $\cos x$ has neither minima nor maxima for the interval $0 < x < \pi$.

10. 2% of $x = 0.02x$

$V = x^3 \Rightarrow \frac{dV}{dx} = 3x^2 \; \Delta v = \frac{dV}{dx}\Delta x$

$\Delta v = 3x^2 \times .02x = 0.6x^3$.

JEE Main

Objective Questions

One Option Correct

1. (a) $f'(x)=\frac{1}{x}$

Using mean value theorem

$$f'(c)=\frac{f(3)-f(1)}{3-1}$$

$$\Rightarrow \frac{1}{c}=\frac{\log 3-\log 1}{2}\Rightarrow c=2\log_3 e$$

2. (b) $f'(x)=\frac{1}{1+(\sin x+\cos x)^2}(\cos x-\sin x)$

$$=\frac{\sqrt{2}\cos\left(x+\frac{\pi}{4}\right)}{1+(\sin x+\cos x)^2}$$

$f'(x)>0$ if $-\frac{\pi}{2}<x+\frac{\pi}{4}<\frac{\pi}{2}$, i.e., $-\frac{3\pi}{4}<x<\frac{\pi}{4}$

Hence $\left(-\frac{\pi}{2},\frac{\pi}{4}\right)$ is correct.

3. (d) $f(x)=\frac{x}{2}+\frac{2}{x}=\frac{x^2+4}{2x}$

$$f'(x)=\frac{(2x)(2x)-(x^2+4)(2)}{(2x)^2}$$

$$=\frac{2(x^2-4)}{4x^2}=\frac{x^2-4}{2x^2}$$

For maxima and minima:

$f'(x)=0\Rightarrow x=\pm 2$

$$f''(x)=\frac{x^2(2x)-(x^2-4)2x}{2x^4}=\frac{8x}{2x^4}=\frac{4}{x^3}$$

$f''(2)>0\Rightarrow$ at $x=2$ local minima exist.

4. (d) $x^m y^n=(x+y)^{m+n}$

Taking log on both sides, we get

$m\log x+n\log y=(m+n)\log(x+y)$

On differentiating, w.r.t. x, we get

$$\frac{m}{x}+\frac{n}{y}\frac{dy}{dx}=\frac{(m+n)}{(x+y)}\left(1+\frac{dy}{dx}\right)$$

$$\Rightarrow\left(\frac{m+n}{x+y}-\frac{n}{y}\right)\frac{dy}{dx}=\frac{m}{x}-\frac{m+n}{x+y}$$

$$\Rightarrow\frac{dy}{dx}=\frac{mx+my-mx-nx}{x(x+y)}\times\frac{y(x+y)}{my+ny-nx-ny}=\frac{y}{x}$$

5. (c) Here $f(1)=-2$ and $f'(x)\ge 2$ [given]

$$\therefore \frac{dy}{dx}\ge 2$$

$$\int_{f(1)}^{f(6)} dy\ge 2\int_1^6 dx$$

i.e., $f(6)-f(1)\ge 2\times 5$

$\therefore f(6)\ge 10+f(10$

$f(6)\ge 10+(-2)$

$\therefore f(6)\ge 8$

6. (a) Here $f(x)=x+\frac{1}{x}\Rightarrow f'(x)=1-\frac{1}{x^2}$

When $f'(x)=0$ then $x=\pm 1$, $f''(x)=+\frac{2}{x^3}$

$f''(1)>0\Rightarrow$ at $x=1$, $f(x)$ has a minimum.

$\therefore$ Min. $f(x)=1+\frac{1}{1}=2$

7. (a)

$\frac{x^2}{27}+y^2=1$ on differentiating both sides, we get

$$\therefore \frac{2x}{27}+2y\frac{dy}{dx}=0$$

$$\therefore \frac{dy}{dx}=-\frac{1}{27}\frac{x}{y}$$

$$\therefore \left.\frac{dy}{dx}\right|_{\substack{x=3\sqrt{3}\cos\theta\\ y=\sin\theta}}=-\frac{1}{3\sqrt{3}}\frac{\cos\theta}{\sin\theta}.$$

$\therefore$ Equation of tangent at $(3\sqrt{3}\cos\theta,\sin\theta)$ is

$$y-\sin\theta=-\frac{\cos\theta}{\sin\theta}\frac{1}{3\sqrt{3}}(x-3\sqrt{3}\cos\theta)$$

When $y=0$, we get x intercept $=\frac{3\sqrt{3}}{\cos\theta}$ and

When $x=0$, we get y intercept $=\frac{1}{\sin\theta}$

Hence $f(\theta)=\frac{3\sqrt{3}}{\cos\theta}+\frac{1}{\sin\theta}$

$$\therefore f'(\theta)=3\sqrt{3}\frac{\sin\theta}{\cos^2\theta}-\frac{\cos\theta}{\sin^2\theta}$$

when $f'(\theta)=0$, we get $\tan^3\theta=\left(\frac{1}{\sqrt{3}}\right)^3$

$\therefore \theta=\pi/6$

Now $f''(\theta) = 3\sqrt{3}\dfrac{\cos^3\theta + 2\sin^2\theta}{\cos^4\theta} + \dfrac{2\cos\theta}{\sin^3\theta}$.

$f''\left(\dfrac{\pi}{6}\right) > 0$

$\Rightarrow$ at $x = \pi/6$, $f(\theta)$ attains minima.

8. (a) $y = 3\sin x - 4\sin^3 x = \sin 3x$

$\dfrac{dy}{dx} = 3\cos 3x$

y is increasing if

$\cos 3x \geq 0$

$\Rightarrow -\dfrac{\pi}{2} \leq 3x \leq \dfrac{\pi}{2} \Rightarrow -\dfrac{\pi}{6} \leq x \leq \dfrac{\pi}{6}$

$\Rightarrow$ Length of the interval $= \dfrac{\pi}{6} - \left(-\dfrac{\pi}{6}\right) = \dfrac{\pi}{3}$.

9. (a) $\dfrac{x^2}{3} - \dfrac{y^2}{2} = 1$ (i)

$\therefore \dfrac{2}{3}x - y\dfrac{dy}{dx} = 0 \Rightarrow \dfrac{dy}{dx} = \dfrac{2}{3}\dfrac{x}{y}$ (ii)

Slope of the given line $y = x$ is 1.

$\dfrac{2}{3}\dfrac{x}{y} = 1 \Rightarrow x = \dfrac{3y}{2}$

Putting the value of $x = \dfrac{3y}{2}$ in equation (i), we get

$y^2 = 4 \Rightarrow y = \pm 2$ and $x = \pm 3$.

$\therefore$ Equation of tangents

$y - 2 = x - 3$ and $y + 2 = x + 3$

i.e., $y = x - 1$ and $y = x + 1$

$\therefore y = x \pm 1$

10. (d) $f'(x) = 2xe^{-(x^2+1)^2} - 2xe^{-x^4}$

$= 2x\left[e^{-(x^2+1)^2} - e^{-x^4}\right] > 0 \,\forall x \left\langle 0 \text{ as } e^{-x^4} \right\rangle e^{-(x^2+1)^2}$

11. (c) $y - 4 = m(x-1)$ is the tangent from (1, 4) to the parabola $y^2 = 4x$

$\therefore y = mx + (4 - m)$

$\therefore 4 - m = \dfrac{1}{m}$

Comparing condition of tangency, i.e., $e = \dfrac{a}{m}$

$\Rightarrow m = 2 \pm \sqrt{3}$

$\therefore m_1 = 2 + \sqrt{3}$ and $m_2 = 2 - \sqrt{3}$

$\therefore \tan\theta = \left|\dfrac{2\sqrt{3}}{1+1}\right| = \sqrt{3}$

$\therefore \theta = \pi/3$

12. (b) $y = a\log x + bx^2 + x$

$y' = \dfrac{a}{x} + 2bx + 1$

When $y' = 0$ at $x = -1$ and $x = 2$.

$\therefore -a - 2b + 1 = 0$ and $\dfrac{a}{2} + 4b + 1 = 0$

Solving, we get

$a = 2, b = -\dfrac{1}{2}$

13. (a) $f(x) = \cos x + \cos\sqrt{2}x$

$\therefore f'(x) = -\sin x - \sqrt{2}\sin\sqrt{2}x$

When $f'(x) = 0$, we get $x = 0$ as the only solution.

Now $f''(x) = -\left(\cos x + 2\cos\sqrt{2}x\right)$

$f''(x) < 0$ when $x = 0$

Hence at $x = 0$ when $f(x)$ attains maxima.

$\therefore$ There exists only one maxima.

14. (c) $x^2 + y^2 = a^2$

$\therefore 2x + 2y\dfrac{dy}{dx} = 0$

$\therefore \dfrac{dy}{dx} = -\dfrac{x}{y}$

$\therefore$ Slope of tangent at $(h, h) = -1$.

15. (c) $y = \dfrac{\log x}{x}$

$\therefore y' = \dfrac{1 - \log x}{x^2}$

$y'' = \dfrac{-x - 2x(1 - \log x)}{x^4} = \dfrac{-3x + 2x\log x}{x^4}$

When $y' = 0 \Rightarrow \log x = 1 \Rightarrow x = e$

$y''|_e < 0$

$\therefore y_{\max} = \dfrac{\log e}{e} = \dfrac{1}{e}$.

16. (d) $f(x) = x^3 - 6x^2 + 12x - 3$

$f'(x) = 3x^2 - 6x + 12$

$= 3(x-2)^2 > 0 \,\forall x$

Hence $f(x)$ has neither maximum nor minimum at $x = 2$.

17. (c) $f(x)=2x^3-6x+5$

$f'(x)=6x^2-6$

$=6(x^2-1)>0$ for $x\in(-\infty,-1)\cup(1,\infty)$

18. (c) $f(x)=\dfrac{\sin 2x}{\sin\left(x+\dfrac{\pi}{4}\right)}=\sqrt{2}\dfrac{(\sin x+\cos x)^2-1}{(\sin x+\cos x)}$

$=\sqrt{2}\dfrac{t^2-1}{t}$ where $\sin x+\cos x=t, \dfrac{dt}{dx}=\cos x-\sin x$

For maximum, $\dfrac{dt}{dx}=0\Rightarrow \tan x=1\Rightarrow x=\pi/4$

$\dfrac{d^2t}{dx^2}=-\sin x-\cos x=-\dfrac{2}{\sqrt{2}}$ for $x=\pi/4$

$\dfrac{d^2t}{dx^2}<0$

$\therefore$ at $x=\pi/4$ $f(x)$ attains maximum.

$\therefore$ max. $f(x)=f\left(\dfrac{\pi}{4}\right)=1.$

19. (b) We know length of the subnormal of $y=f(x)$ is $y\dfrac{dy}{dx}$

Here, $y^n=a^{n-1}x$

$\therefore n\log y=(n-1)\log a+\log x$

$\therefore \dfrac{n}{y}\dfrac{dy}{dx}=\dfrac{1}{x}$

$\therefore y\dfrac{dy}{dx}=\dfrac{y^2}{n}.\dfrac{1}{x}$

$=\dfrac{a^{\left(\frac{n-1}{n}\right)2}x^{\frac{2}{n}}}{n.x}=\dfrac{1}{n}\cdot a^{\left(\frac{n-1}{n}\right)2}\cdot x^{\frac{2}{n}-1}$

$\therefore\ y\dfrac{dy}{dx}=$ constant $\Rightarrow \dfrac{2}{n}-1=0\Rightarrow n=2$

20. (c) $x=at^2\Rightarrow \dfrac{dx}{dt}=2at$

$y=2at\Rightarrow \dfrac{dy}{dt}=2a$

$\therefore \dfrac{dy}{dx}=\dfrac{2a}{2at}=\dfrac{1}{t}$

When tangent with slope $\dfrac{1}{t}$ is $\perp$ to x-axis, then $t=0$.

Then point of contact is $(0,0)$.

21. (c) $f'(x)=2x-6=0\Rightarrow x=3.$

22. (c) $f(x)=\dfrac{1}{x}\Rightarrow f'(x)=-\dfrac{1}{x^2}$

$\therefore \dfrac{f(b)-f(a)}{b-a}=f'(x_1)$

$\Rightarrow \dfrac{\frac{1}{b}-\frac{1}{a}}{b-a}=f'(x_1)$

$\Rightarrow -\dfrac{1}{ab}=-\dfrac{1}{x_1^2}$

$\Rightarrow x_1=\sqrt{ab}$

So $x=\sqrt{ab}$

23. (a) $f(x)=x^3+px^2+qx+r$

$f'(x)=3x^2+2px+q$

$D=$ Discriminant $=4p^2-12q$

D should be less than zero for no real roots. Hence, when $p^2<3q$, no extreme values exist.

24. (c) $y=x^2+2x$ ...(i)

Here $\dfrac{dy}{dt}=\dfrac{dx}{dt}\Rightarrow\dfrac{dy}{dt}=1$

From (i), we get $\dfrac{dy}{dx}=2x+2$

$\therefore 1=2x+2$

$\Rightarrow x=-\dfrac{1}{2}\ \therefore\ y=-\dfrac{3}{4}$

The required point is $p\left(-\dfrac{1}{2},-\dfrac{3}{4}\right)$.

25. (d) Here $\cos^{-1}\left(\dfrac{x}{2}-1\right)$ is defined only when

$-1\le\dfrac{x}{2}-1\le1$

i.e., $0\le x\le4$

$\log(\cos x)$ is defined only when $\cos x>0$.

i.e., $-\dfrac{\pi}{2}<x<\dfrac{\pi}{2}$

4^{-x^2} is also defined in $\left(-\dfrac{\pi}{2},\dfrac{\pi}{2}\right)$.

$\therefore f(x)$ is defined within $\left[0,\dfrac{\pi}{2}\right)$.

26. (a) $y=\dfrac{3x^2+9x+17}{3x^2+9x+17}=1+\dfrac{10}{3\left(x^2+3x+\dfrac{7}{3}\right)}$

$=1+\dfrac{10}{3\left[\left(x+\dfrac{3}{2}\right)^2+\dfrac{1}{12}\right]}$

y will be maximum when $x=-\dfrac{3}{2}$ and max. value of $y=41$.

27. (b) As the figure is isosceles, we know area will be maximum when it is a right-angle triangle.

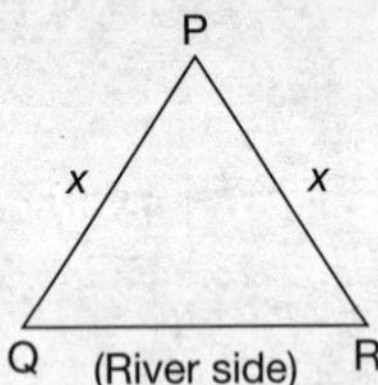

Hence maximum area of PQR $=\dfrac{1}{2}x^2$.

28. (b) $f''(x)=6(x-1)$ and slope of tangent = 3
Integrating both sides, we get

$f'(x)=3(x-1)^2+e$

$\therefore f'(2)=3(2-1)^2+e\Rightarrow 3=3+e\Rightarrow e=0$

$\therefore f'(x)=3(x-1)^2$

$\Rightarrow f(x)=(x-1)^3+e_1$ which pass through $(2,1)$.

$\Rightarrow e_1=0$

$\therefore f(x)=(x-1)^3$

29. (a) $\dfrac{dx}{d\theta}=-a\sin\theta$ and $\dfrac{dy}{d\theta}=a\cos\theta$

$\therefore \dfrac{dy}{dx}=-\dfrac{\cos\theta}{\sin\theta}$ = slope of the tangent.

$\therefore$ Equation of the normal through $P[a(1+\cos\theta), a\sin\theta]$

$y-a\sin\theta=\dfrac{\sin\theta}{\cos\theta}[x-a(1+\cos\theta)]$

Now proceeding from given option, one can get $(a,0)$ is the fixed point.

30. (c) $f(x)=2x^3-9ax^2+12a^2x+1$

$f'(x)=6x^2-18ax+12a^2$

When $f'(x)=0\Rightarrow x^2-3ax+4a^2=0$

$\Rightarrow x=a$ or $2a$.

Now, $f''(x)=12x-18a$

$f''(a)\langle 0\forall a\rangle 0$ at $x=1, f(x)$ attains maximum $p=a$ and $f''(2a)>0\forall a>0$.

Hence at $x=2a$, $f(x)$ attains minima.

$\therefore q=2a$

As $p^2=q\Rightarrow a^2=2a$

$\therefore a=2$ as $a>0$.

31. (a) The function given in choice (a) is not differentiable at $x=\dfrac{1}{2}$. (The other functions are either differentiable or not differentiable at boundary.)

32. (d) $f(x)=-x^3+4ax^2+2x-5$

$f'(x)=-3x^2+8ax+2$

$f'(x)<0$ no value a.

33. (c) $x^2+xy+y^2=7$

$\therefore$ Differentiating w.r.t. x, we get

$2x+x\dfrac{dy}{dx}+y+2y\dfrac{dy}{dx}=0$

$\therefore \dfrac{dy}{dx}=-\dfrac{(2x+y)}{(x+2y)}$

$\therefore$ Length of subtangent $=\dfrac{y}{dy/dx}=-\dfrac{y(x+2y)}{2x+y}$

At $(1,-3)$, required length $=\dfrac{-(-3)(1-6)}{2-3}=15$.

34. (c) $y=x^3-x$

$y'=3x^2-3$

$y''=6x>0\forall x\in[0,2]$

When $y'=0$ then $x=\pm1$

$y''|_{x=1}>0$

$\therefore$ Max.$(y)_{x=1}=(1)^3-3(1)=-2$

Max.$(y)_{x=2}=(2)^3-3(2)=2$

35. (b) $f'(x)=\dfrac{(ad-bc)}{(c\sin x+d\cos x)^2}$

$\therefore f'(x)<0\Rightarrow ad-bc<0$.

36. (c)

$f'(\alpha)=\left(1+\dfrac{1}{\sin^n\alpha}\right)\left(\dfrac{n\sin\alpha}{\cos^{n+1}\alpha}\right)+\left(1+\dfrac{1}{\cos^n\alpha}\right)\left(-\dfrac{n\cos\alpha}{\sin^{n+1}\alpha}\right)$

$= \frac{-n}{\sin^{n+1}\alpha\cos^{n+1}\alpha}.\cos 2\alpha$

$= -2^{n+1}n\frac{\cos 2\alpha}{\sin^{n+1}2\alpha}$

when $f'(\alpha) = 0 \Rightarrow \cos 2\alpha = 0 \Rightarrow \frac{\pi}{4}$

$f''(\alpha)\Big|_{\alpha=\frac{\pi}{4}} > 0$

$\therefore f(\alpha)\Big|_{\alpha=\frac{\pi}{4}} = \left[1+\left(\sqrt{2}\right)^n\right]\left[1+\left(\sqrt{2}\right)^n\right] = \left[1+2^{\frac{n}{2}}\right]^2.$

37. (d) $f(x) = 2x + \cot^{-1}x - \log(\sqrt{1+x^2} - x)$

$f'(x) = 2 - \frac{1}{1+x^2} - \frac{1}{\sqrt{1+x^2}}\left[\frac{2x}{2\sqrt{1+x^2}} - 1\right]$

$= \frac{1+2x^2}{1+x^2} + \frac{1}{\sqrt{1+x^2}} > 0$ always $-\infty < x < \infty$

$\Rightarrow f(x)$ increases in $(-\infty, \infty)$.

38. (b) $y = x^{1/x}$

$\log y = \frac{1}{x}\log x$

Differentiating both sides, w.r.t. x, we get

$\frac{1}{y}\frac{dy}{dx} = \frac{1}{x}\frac{d}{dx}(\log x) + (\log x)\frac{d}{dx}\left(\frac{1}{x}\right)$

$\frac{dy}{dx} = \frac{y}{x}\frac{1}{x} + y\log x\left(-\frac{1}{x^2}\right)$

$= \frac{y}{x^2} - \frac{y\log x}{x^2} = \frac{y}{x^2}(1-\log x)$

$\frac{d^2y}{dx^2} = \frac{x^2y' - 2xy}{(x^2)^2} - \frac{y'\log x + \frac{y}{x} - 2xy\log x}{(x^2)^2}$

Now $\frac{y}{x^2}(1-\log x) = 0 \Rightarrow \log x = 1 \Rightarrow x = e$

$\frac{d^2y}{dx^2} = \frac{-2e.e^{\frac{1}{e}} - e^{\frac{1}{e}-1}}{e} < 0.$

Hence, for $x = e, y$ attains maximum.

39. (d)

Here, $f'(x) = 16x^3 - 2 \Rightarrow f'(x) > 0$ when $x^3 > \frac{1}{8}$.

i.e., $\left(x - \frac{1}{2}\right)\left(x^2 + x + 1\right) > 0$, i.e., $x > \frac{1}{2}$

40. (a) $f(x) = x^2 + \frac{1}{1+x^2}$

which is always positive for any value of proceed options.

For $x = 0$, $f(0) = 1$

$x = 1$, $f(1) = \frac{3}{2} > f(0)$

$x = 4$, $f(4) = \frac{273}{17} > f(0)$

$x = 3$, $f(3) = \frac{91}{10} > f(0)$

Hence minimum value of $f(x)$ is at $x = 0$.

41. (a) Slope of the line $4x - 2y = 1$ is $m_1 = 2$

$\frac{dy}{dx} = 6 - 2x \therefore 6 - 2x = 2 \Rightarrow x = 2$

$\therefore$ For $x = 2, y = 6(2) - (2)^2 = 8$

$\therefore$ Point of tangency $(2, 8)$.

42. (a) $f(x) = \frac{4x^2+1}{x} = 4x + \frac{1}{x}, x \neq 0$

$f'(x) = 4 - \frac{1}{x^2} = \left(2 + \frac{1}{x}\right)\left(2 - \frac{1}{x}\right)$

$f'(x) < 0$ when $-\frac{1}{2} < x < \frac{1}{2}, x \neq 0.$

43. (d) Let r be the radius and h be the height of a cylinder, of given surface area S.

Then $S = \pi r^2 + 2\pi rh \Rightarrow h = \frac{S - \pi r^2}{2\pi r}$ (i)

$\therefore v = \pi r^2 h = \pi r^2\left(\frac{S - \pi r^2}{2\pi r}\right) = \frac{Sr - \pi r^3}{2}$

$\therefore \frac{dv}{dr} = \frac{S - 3\pi r^2}{2}$

For $\frac{dv}{dr} = 0 \Rightarrow \frac{S - 3\pi r^2}{2} = 0$

$\Rightarrow S = 3\pi r^2 \Rightarrow \pi r^2 + 2\pi rh = 3\pi r^2 \Rightarrow r = h$

$\frac{d^2v}{dr^2} = -3\pi r < 0$

Hence v is maximum, then $r = h$.

44. (d) $f(x) = x^{25}(1-x)^{75}$

$\therefore f'(x) = 25x^{24}(1-x)^{74}[1-4x]$

$\therefore f'(x)=0 \Rightarrow x=0,1,\frac{1}{4}$ in the interval $\left(0,\frac{1}{4}\right)$.

So at $x=\frac{1}{4}$, $f(x)$ attains maximum $\left(\frac{1}{4},1\right)$

$f'(x)<0.$

45. (d) $f(x)=2x^3-9x^2+12x+99$

$$\therefore f'(x)=6x^2-18x+12 = 6(x^2-3x+2) = 6(x-1)(x-2)$$

$\therefore f'(x)$, when $x\in(-\infty,1)\cup(2,\infty)$

$f'(x)<0$, when $1<x<2$.

46. (c) $f(x)=\frac{x}{\sin x}$ and $g(x)=\frac{x}{\tan x}$

$$f'(x)=\frac{\sin x - x\cos x}{\sin^2 x}$$

and $g'(x)=\frac{\tan x - x\sec^2 x}{\tan^2 x}$

Let $h(x)=\sin x - x\cos x$

$$h'(x)=\cos x-\cos x+x\sin x = x\sin x>0\ \forall\ 0<x\le 1$$

Let $p(x)=\tan x - x\sec^2 x$

$p'(x)=-2x\sec^2 x\tan x<0\ \forall\ 0<x\le 1$

$\therefore f'(x)>0$ and $g'(x)<0\ \forall\ 0<x\le 1.$

47. (d) $y=x^2+x-1$

$\therefore \frac{dy}{dx}=2x+1$ Hence $\left.\frac{dy}{dx}\right|_{x=1}=3$

$$\text{Length of tangent} = A = \frac{y\sqrt{1+\left(\frac{dy}{dx}\right)^2}}{\frac{dy}{dx}} = \frac{1\sqrt{1+3^2}}{3}=\frac{\sqrt{10}}{3}$$

$$\text{Length of subtangent} = B = \frac{y}{dy/dx}=\frac{1}{3}$$

$$\text{Length of normal} = C = y\sqrt{1+\left(\frac{dy}{dx}\right)^2} = 1.\sqrt{1+3^2}=\sqrt{10}$$

$$\text{Length of subnormal} = D = y\frac{dy}{dx}=1\times 3=3$$

Arrange in increasing order, we get
$B<A<D<C$.

48. (c) $y=\sin^{-1}x$

$$\therefore \frac{dy}{dx}=\frac{1}{\sqrt{1-x^2}}$$

$$\therefore \frac{d^2y}{dx^2}=\frac{-(-2x)}{2(1-x^2)^{3/2}}=\frac{x}{(1-x^2)\sqrt{1-x^2}}$$

$$\Rightarrow (1-x^2)\frac{d^2y}{dx^2}=x.\frac{1}{\sqrt{1-x^2}}=x\frac{dy}{dx}$$

49. (d) $f'(x)=\frac{1-\log x}{x^2}$

$\Rightarrow x=e$

Now $f(2)=\frac{\log 2}{2}>0$

$f(e)=\frac{\log e}{e}=\frac{1}{e}$

and $f(\infty)=0$

$\Rightarrow$ Minimum value over $[2,\infty)$ approaches zero, i.e., it does not exist.

50. (d) $y=\frac{x^2-1}{x^2+1}$

$\therefore y'=\frac{4x}{(x^2+1)^2}$ $\therefore y'=0\Rightarrow x=0$

$$y''=\frac{(x^2+1)^2 4-16x^2(x^2+1)}{(x^2+1)^4} = \frac{-4(x^2-1)}{(x^2+1)^3}>0$$

For $x=0\Rightarrow$ minimum exist at $x=0$.

Hence $\left.y\right|_{x=0}=-1$

51. (c)

150 ← 60 → 50
A C B
← x → ← (60 – x) →

Suppose the school is built at a distance x from town A.

Then $f(x)=150\times x+(60-x)50$

is to be minimised.

$f(x)=100x+3000$

$f'(x)=100>0$

$\Rightarrow f(x)$ is increasing function.

Minimum value of f is at $x=0$.

School is to be built at A.

52. (a) Between $[-1, 2]$, $f'(x)=6x+12>0$ for the given interval. Therefore, $f(x)$ is increasing on $[-1, 2]$.

53. (a) $y=\dfrac{2x}{1+x^2}$

$$\frac{dy}{dx}=\frac{2(1+x^2)-4x^2}{(1+x^2)^2}=\frac{2-2x^2}{(1+x^2)^2}$$

$\Rightarrow$ Maximum and minimum occur at $x=1$ and -1

(–) –1 (+) 1 (–) (for $f'(x)$)

$$\text{Again, }\frac{d^2y}{dx^2}=\frac{-4x(1+x^2)^2-(2-2x^2)2(1+x^2)2x}{(1+x^2)^4}$$

$$=\frac{-4x[1+x^2+2-2x^2]}{(1+x^2)^3}$$

$$=\frac{12x(3-x^2)}{(1+x^2)^3}$$

$\Rightarrow$ Inflexion points occur $0, -\sqrt{3}$ and $\sqrt{3}$

(–) $-\sqrt{3}$ (+) 0 (–) $\sqrt{3}$ (+) (for $f''(x)$)

54. (a) Define $f(x)=ax^3+bx^2+cx$, which is continuous and differentiable in [0, 1] with $f(0)=f(1)=0$.

By Rolle's theorem $3ax^3+2bx^2+x=0$ for at least one x in [0, 1].

55. (c) $\dfrac{dy}{dx}=2x+b\big|_{(1,1)}\, b+2$

Equation of tangent $y-1=(b+2)(x-1)$.

The co-ordinates are $(0,-b,-1),\left(\dfrac{b+1}{b+2},0\right)$.

Area of triangle formed $=-\dfrac{1}{2}\dfrac{(b+1)(b+1)}{b+2}=+2$

$$\Rightarrow \frac{(b+1)^2}{b+2}=-4$$

$$\Rightarrow (b+3)^2=0$$

$$\Rightarrow b=-3.$$

56. (b)

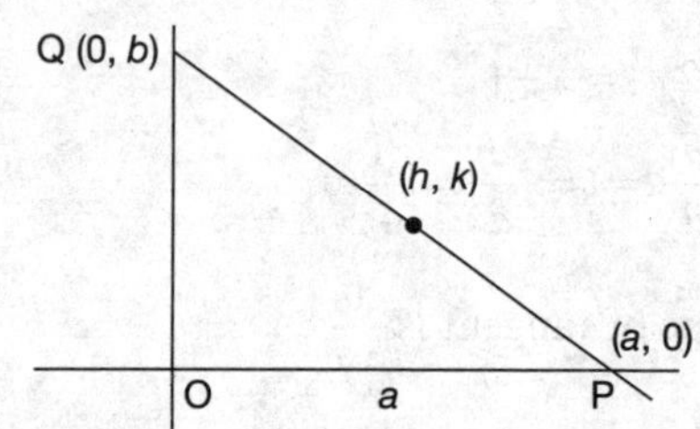

The area S is given by

$S=\dfrac{1}{2}ab$ where $\dfrac{h}{a}+\dfrac{k}{b}=1$

Now $\dfrac{h}{a}+\dfrac{k}{b}=1$

$\Rightarrow \left(\dfrac{h}{a}\right)\left(\dfrac{k}{b}\right)$ will be maximum

When $\dfrac{h}{a}=\dfrac{k}{b}=\dfrac{1}{2}$

or $a=-2h, b=2k$

or $\dfrac{ab}{hk}$ will be minimum when $a=2h, b=2k$

$\Rightarrow S$ will be minimum when $a=2h, b=2k$.

57. (d) $\min f(x)>\max g(x)$

$$\Rightarrow \frac{4\times1\times2c^2-4b^2}{4}>\frac{4(-1)(b^2)-(-2c)^2}{4(-1)}$$

$$\Rightarrow c^2>2b^2$$

$$\Rightarrow |c|^2>2|b|^2$$

$$\Rightarrow |c|>|b|\sqrt{2}.$$

58. (c) $S=t^3-3t^2$

$$v=\frac{dS}{dt}=3t^2-6t$$

$$\text{Acceleration}=\frac{d^2S}{dt^2}=6t-6$$

When $\dfrac{d^2S}{dt^2}=0\Rightarrow t=1\text{sec}.$

$$\therefore \left.\frac{dS}{dt}\right|_{t=1}=3(1)^2-6(1)=-3 \text{ m/s}.$$

59. (d) Here $\frac{dx}{dt} = -5$ [Given]

and $y = 4 - 2x^2$

$\therefore \frac{dy}{dt} = -4x\frac{dx}{dt}$

$= 4(1).(-5) = 20$ unit/sec.

60. (b) $2y^2 = x + 1$

$4y\frac{dy}{dx} = 1 \Rightarrow \frac{dy}{dx} = \frac{1}{4y}$

Slope of tangent is $\frac{1}{4y}$

$\therefore$ Slope of normal $m = -4y$

A. For point (7, 2), $m = -8 \Rightarrow (2)$ is matched.

B. For point $\left(0, \frac{1}{\sqrt{2}}\right), m = -2\sqrt{2} \Rightarrow (5)$ is matched.

C. For point $(1, -1), m = 4 \Rightarrow (3)$ matched.

D. For point $(3, \sqrt{2}), m = -4\sqrt{2} \Rightarrow (1)$ matched.

61. (c) Given $g(u) = 2\tan^{-1}(e^u) - \frac{\pi}{2}$.

$\therefore g(-u) = 2\tan^{-1}(e^{-u}) - \frac{\pi}{2} = 2\tan^{-1}\left(\frac{1}{e^u}\right) - \frac{\pi}{2}$

$= 2\cot^{-1}(e^u) - \frac{\pi}{2} = 2\left[\frac{\pi}{2} - \tan^{-1}(e^u)\right] - \frac{\pi}{2}$

$= \frac{\pi}{2} - 2\tan^{-1}(e^u) = -g(u)$

$\therefore g$ is an odd function.

Also $g'(u) = \frac{2e^u}{1 + e^{2u}} > 0, \forall u \in (-\infty, \infty)$

$\therefore g$ is strictly increasing on $(-\infty, \infty)$.

62. (a) $f(x) = xe^{x(1-x)}$

$\Rightarrow f'(x) = e^{x(1-x)} + (1 - 2x)xe^{x(1-x)}$

$= -e^{x(1-x)}(2x + 1)(x - 1)$

Critical points are $x = -\frac{1}{2}$ and 1

– + –

−1/2 1

Hence, $f(x)$ is increasing on $[-1/2, 1]$.

63. (b) Let $f(x) = e^x - 1 - x$

$\Rightarrow f'(x) = e^x - 1 > 0$ for $x \in (0,1)$

$\therefore f(x)$ is an increasing function.

$\therefore f(x) > f(0), \forall x \in (0,1)$

$\Rightarrow e^x - 1 - x > 0 \Rightarrow e^x > 1 + x$

$\therefore$ (a) is not correct.

(b)Let $g_e(x) = \log_e(1 + x) - x$

$\Rightarrow g'(x) = \frac{1}{1+x} - 1 = -\frac{x}{1+x} < 0, \forall x \in (0,1)$

Hence, $g(x)$ is decreasing on (0, 1).

$\Rightarrow g(x) < g(0)$

$\Rightarrow \log_e(1 + x) - x < 0 \Rightarrow \log_e(1 + x) < x$

$\therefore$ (b) is correct.

Similarly, it can be shown that (c) and (d) are not correct

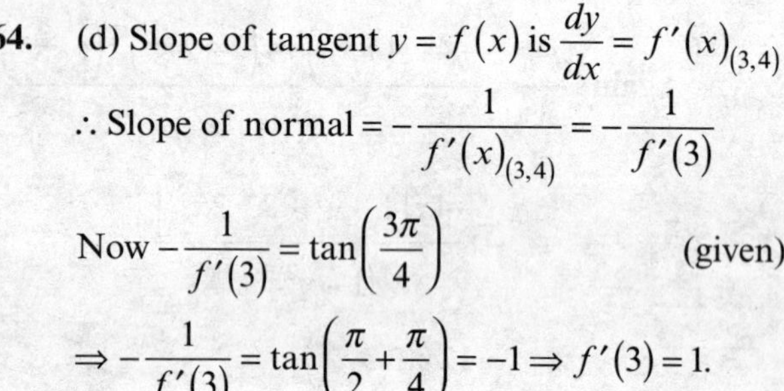

64. (d) Slope of tangent $y = f(x)$ is $\frac{dy}{dx} = f'(x)_{(3,4)}$

$\therefore$ Slope of normal $= -\frac{1}{f'(x)_{(3,4)}} = -\frac{1}{f'(3)}$

Now $-\frac{1}{f'(3)} = \tan\left(\frac{3\pi}{4}\right)$ (given)

$\Rightarrow -\frac{1}{f'(3)} = \tan\left(\frac{\pi}{2} + \frac{\pi}{4}\right) = -1 \Rightarrow f'(3) = 1.$

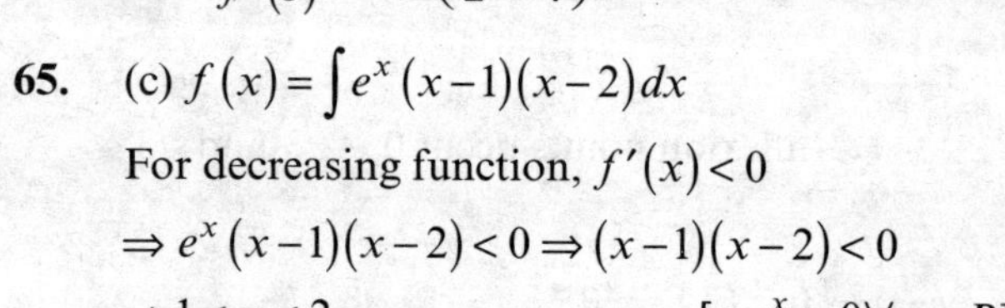

65. (c) $f(x) = \int e^x(x - 1)(x - 2)dx$

For decreasing function, $f'(x) < 0$

$\Rightarrow e^x(x - 1)(x - 2) < 0 \Rightarrow (x - 1)(x - 2) < 0$

$\Rightarrow 1 < x < 2$ $[\because e^x > 0 \forall x \in R]$

66. (d) From the graph, it is clear that both $\sin x$ and $\cos x$ in the interval $(\pi/2, \pi)$ are decreasing functions.

$\therefore S$ is correct

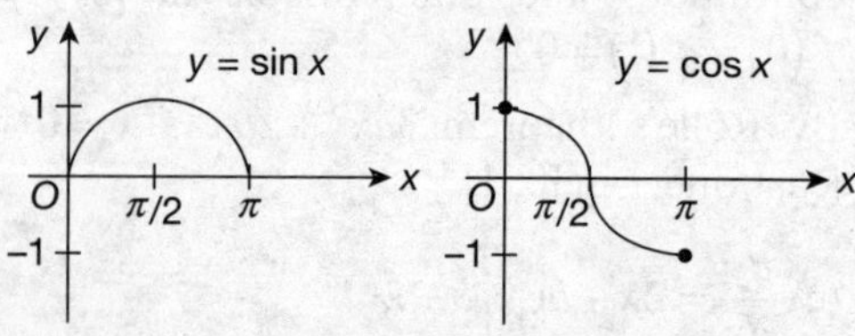

Consider $f(x) = \sin x$ on $\left(0, \frac{\pi}{2}\right) \Rightarrow f'(x) = \cos x$

From the graph, it is clear that $f(x)$ is increasing on $(0, \pi/2)$ but

$f'(x)$ is decreasing on $(0, \pi/2)$

$\therefore R$ is wrong.

67. (b) Given: $f(x)=\sin^4 x+\cos^4 x$

$\Rightarrow f'(x)=4\sin^3 x\cos x-4\cos^3 x\sin x$

$=-2.\sin 2x\cos 2x=-\sin 4x$

Now for $f(x)$ to be increasing function

$f'(x)>0\Rightarrow -\sin 4x>0\Rightarrow \sin 4x<0$

$\Rightarrow \pi<4x<2\pi\Rightarrow \frac{\pi}{4}<x<\frac{\pi}{2}$

Hence, $f(x)$ increasing on $(\pi/4,\pi/2)$

Now, $\frac{\pi}{2}=\frac{4\pi}{8}>\frac{3\pi}{8}$

$\therefore f(x)$ will be increasing on $(\pi/4,3\pi/8)$.

68. (a) Given that slope of tangent at $(x,f(x))$ is $2x+1$.

$\Rightarrow f'(x)=2x+1\Rightarrow f(x)=x^2+x+c$

Since the curve passes through (1, 2), $\therefore f(1)=2$

$\Rightarrow 2=1+1+c\Rightarrow c=0 \therefore f(x)=x^2+x$

The graph of $f(x)$ is an upward parabola, which touches the x-axis at the origin.

$\therefore$ Required function is $=\int_0^1 (x^2+x)dx$

$=\left[\frac{x^3}{3}+\frac{x^2}{2}\right]_0^1=\frac{1}{3}+\frac{1}{2}=\frac{5}{6}$

69. (b) Given: $f(x)=\frac{\ln(\pi+x)}{\ln(e+x)}$

$\Rightarrow f'(x)=\frac{\left(\frac{1}{\pi+x}\right)\ln(e+x)-\frac{1}{(e+x)}\ln(\pi+x)}{[\ln(e+x)]^2}$

$=\frac{(e+x)\ln(e+x)-(\pi+x)\ln(\pi+x)}{(e+x)(\pi+x)(\ln(e+x))^2}<0$

On $(0,\infty)$ since $1<e<\pi$

$\therefore f(x)$ decreases on $(0,\infty)$.

70. (d) Given: $f(x)=(x+2)e^{-x}$

$\therefore f'(x)=-(x+2)e^{-x}+e^{-x}=-(x+1)e^{-x}$

Put $f'(x)=0\Rightarrow x=-1$

For $x\in(-1,\infty)$, $f'(x)>0$

and for $x\in(-1,\infty)$, $f'(x)<0$

$\therefore f(x)$ is increasing on $(-\infty,-1)$ and decreasing on $(-1,-\infty)$

71. (a) Consider the function $f(x)=ax^3+bx^2+cx$ on [0, 1].

Since $f(x)$ is a polynomial,

$\therefore f(x)$ is continuous on [0, 1].

And hence, it is differentiable on (0, 1)

Now $f(0)=f(1)=0$ $[\because a+b+c=0]$

By Rolle's theorem, there exists $x\in(0,1)$ such that

$f'(x)=0\Rightarrow 3ax^2+2bx+c=0$

Hence, the equation $3ax^2+2bx+c=0$ has at least one root in [0, 1].

72. (d) Let the wire be cut into two pieces of length, x and $20-x$.

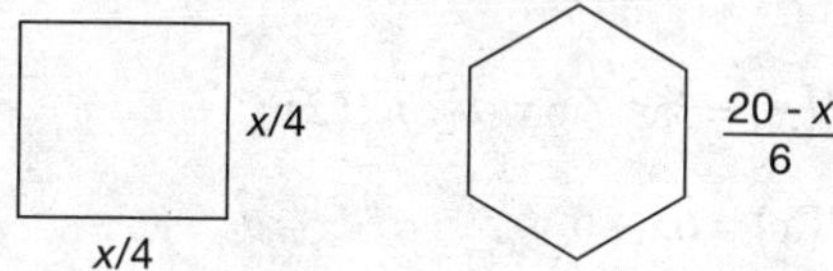

Area of square $=\left(\frac{x}{4}\right)^2$

Area of regular hexagon $=6\times\frac{\sqrt{3}}{4}\left(\frac{20-x}{6}\right)^2$

Total area of square and hexagon

$=A(x)=\frac{x^2}{16}+\frac{3\sqrt{3}}{2}\frac{(20-x)^2}{36}$

$=A'(x)=\frac{2x}{16}+\frac{3\sqrt{3}\times 2}{2\times 36}(20-x)(-1)$

Now for critical point

$A'(x)=0\Rightarrow x=\frac{40\sqrt{3}}{3+2\sqrt{3}}$

$\therefore$ Required length of side of regular hexagon

$=\frac{1}{6}(20-x)$

$=\frac{1}{6}\left(20-\frac{40\sqrt{3}}{3+2\sqrt{3}}\right)=\frac{10}{3+2\sqrt{3}}$

73. (d) $A=\begin{bmatrix}1 & -x & 2x+1\\ -x & 1 & -x\\ 2x+1 & -x & 1\end{bmatrix}$

$|A|=4x^3-4x^2-4x$

$\because f(x)=\det(A)$

$\therefore f'(x)=4(3x^2-2x-1)$

$\therefore f(x) = 4x^3 - 4x^2 - 4x$

For critical points $f'(x) = 0$

$\Rightarrow 3x^3 - 2x - 1 = 0$

$\Rightarrow x = 1; x = \frac{-1}{3}$

$\therefore \min f(x) = f(1) = -4,$

$\max \text{ of } f(x) + = f\left(-\frac{1}{3}\right) = \frac{20}{27}$

$\therefore$ Sum of maximum and minimum

$= \frac{20}{27} - 4 = -\frac{88}{27}.$

74. (c) $f(x) = x^3 - 3x^2 - \frac{3f''(2)}{2}x + f''(1)$

$f'(x) = 3x^2 - 6x - \frac{3}{2}f''(2)$

$f'(x) = 6x - 6$

Then, $f''(2) = 12 - 6 = 6$ and $f''(1) = 0$

$\because f(x) = x^3 - 3x^2 - 9x$

$f'(x) = 3x^2 - 6x - 9 = 3(x+1)(x-3)$

For critical points, $f'(x) = 0$

$\Rightarrow 3x^2 - 6x - 9 = 0$

$\Rightarrow x = -1, 3$

Now, $f'(x) = 6x - 6$

$f''(-1) = -12 < 0 \Rightarrow x = -1$ is the point of maxima.

$f'(3) = 12 > 0 \Rightarrow x = 3$ is the point of minima.

So, local minimum value is,

$f(2) = 27 - 27 - 9 \times 3 = -27.$

75. (d)

$f(x) = (4a-3)(x + \log 5) + 2(a-7)\sin\frac{x}{2}\cos\frac{x}{2}$

$f(x) = (4a-3)(x + \log 5) + (a-7)\sin x$

For critical points:

$f'(x) = 0$

$(4a-3) + (a-7)\cos x = 0$

$\Rightarrow \cos x = \frac{3-4a}{a-7}$

$\because -1 \le \frac{3-4a}{a-7} \le 1$ $\quad [\because -1 \le \cos x \le 1]$

$\Rightarrow \frac{4a-3}{a-7} - 1 \le 0$

$\Rightarrow a \in \left[-\frac{4}{3}, 7\right)$ ...(ii)

$\frac{4a-3}{a-7} + 1 \ge 0$ and

$\Rightarrow a \in [-\infty, 2) \cup (7, \infty)$ ...(ii)

From (i) and (ii)

$a \in \left[-\frac{4}{3}, 2\right]$

76. (d)

(Sign scheme: $-$ on $\left(-\frac{\pi}{6}, 0\right)$, $+$ on $\left(0, \frac{\pi}{2}\right)$)

Given that

$f(x) = 3\sin^4 x + 10\sin^3 x + 6\sin^2 x - 3,$

$x \in \left[-\frac{\pi}{6}, \frac{\pi}{2}\right]$

$f'(x) = 12\sin^3 x\cos x + 30\sin^2 x\cos x + 12\sin x\cos x$

$= 6\sin x\cos x\left(2\sin^2 x + 5\sin x + 2\right)$

$= 6\sin x\cos x(2\sin x + 1)(\sin x + 2) = 0$

$\Rightarrow x = -\frac{\pi}{6}, 0, \frac{\pi}{2}$

Decreasing in $\left(-\frac{\pi}{6}, 0\right)$.

77. (d)

$f'(x) = 3\left(\frac{1}{x-1} - \frac{1}{x+1}\right) + \frac{2}{(x-1)^2}$

$= \frac{6}{(x^2-1)} + \frac{2}{(x-1)^2}$

$= \frac{6(x-1) + 2(x+1)}{(x-1)^2(x+1)} = \frac{8x-4}{(x-1)^2(x+1)} > 0$

(Sign scheme: $+$ on $(-\infty, -1)$, $-$ on $\left(-1, \frac{1}{2}\right)$, $+$ on $\left(\frac{1}{2}, 1\right)$, $+$ on $(1, \infty)$)

$\Rightarrow x \in (-\infty, -1) \cup \left[\frac{1}{2}, \infty\right) - \{1\}$

78. (b) $C_1 \to C_1 + C_2$

$$\text{Let } f(x) = \begin{vmatrix} 2 & 1+\sin^2 x & \sin 2x \\ 2 & \sin^2 x & \sin 2x \\ 1 & \sin^2 x & 1+\sin 2x \end{vmatrix}$$

$R_1 \to R_1 - 2R_3; R_2 \to R_2 - 2R_3$

$$= \begin{vmatrix} 0 & \cos^2\theta & -(2+\sin 2x) \\ 0 & -\sin^2 x & -(2+\sin 2x) \\ 1 & \sin^2 x & 1+\sin 2x \end{vmatrix} = -2 - 2\sin 2x$$

$f'(x) = -2\cos 2x = 0$

$\Rightarrow \cos 2x = 0 \Rightarrow x = \dfrac{\pi}{4}, \dfrac{3\pi}{4}$

$f''(x) = 4\sin 2x$

So, $f''\left(\dfrac{\pi}{4}\right) = 4 > 0$ (minima)

$m = f\left(\dfrac{\pi}{4}\right) = -2 - 1 = -3$

$f''\left(\dfrac{3\pi}{4}\right) = -4 < 0$ (maxima)

$M = f\left(\dfrac{3\pi}{4}\right) = -2 + 1 = -1$

So, $(m, M) = (-3, -1)$

79. (d)

$f(x) = (1-\cos^2 x)(\lambda + \sin x) = \sin^2 x(\lambda + \sin x)$

$\Rightarrow f(x) = \lambda \sin^2 x + \sin^3 x$ ...(i)

$\Rightarrow f'(x) = \sin x \cos x[2\lambda + 3\sin x] = 0$

$\Rightarrow \sin x = 0$ and $\sin x = -\dfrac{2\lambda}{3} \Rightarrow x = \alpha$ (let)

So, $f(x)$ will change its sign at $x = 0, \alpha$ because there is exactly one maxima and one minima in $\left(-\dfrac{\pi}{2}, \dfrac{\pi}{2}\right)$.

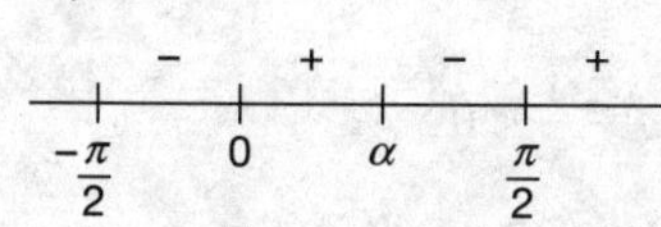

OR

– + – +

$-\frac{\pi}{2}$ α 0 $\frac{\pi}{2}$

Now, $\sin x = -\dfrac{2\lambda}{3}$

$\Rightarrow -1 \le -\dfrac{2\lambda}{3} \le 1 \Rightarrow -\dfrac{3}{2} \le \lambda \le \dfrac{3}{2} - \{0\}$

$\because$ If $\lambda = 0 \Rightarrow f(x) = \sin^3 x$ (from (i)) which is monotonic, then no maxima/minima.

So, $\lambda \in \left(-\dfrac{3}{2}, \dfrac{3}{2}\right) - \{0\}$

80. (a)

$f(x) = (3x-7)x^{2/3}$

$f'(x) = 3x^{2/3} + (3x-7).\dfrac{2}{3}x^{-1/3} = \dfrac{15x-14}{3x^{1/3}}$

+ – +

$-\infty$ 0 $\frac{14}{15}$ ∞

For increasing function,

$f'(x) > 0$ then $x \in (-\infty, 0) \cup \left(\dfrac{14}{15}, \infty\right)$

81. (d) Since, function $f(x)$ is twice differentiable and continuous in $x \in [a,b]$. Then, by Lagrange's Mean Value Theorem for $x \in [a,c]$

$\dfrac{f(c)-f(a)}{c-a} = f'(\alpha), \alpha \in (a,c)$

Again, by Lagrange's Mean Value Theorem for $x \in [c,b]$

$\dfrac{f(b)-f(c)}{b-c} = f'(\beta), \beta \in (c,b)$

$f''(x) < 0 \Rightarrow f'(x)$ is decreasing.

$f'(\alpha) > f'(\beta) \Rightarrow \dfrac{f(c)-f(a)}{c-a} > \dfrac{f(b)-f(c)}{b-c}$

$\Rightarrow \dfrac{f(c)-f(a)}{f(b)-f(c)} > \dfrac{c-a}{b-c}$ ($f(x)$ is increasing)

82. (c) Average speed $= f'(t) = \dfrac{f(t_2)-f(t_1)}{t_2 - t_1}$

$2at + b = a(t_1 + t_2) + b \Rightarrow t = \dfrac{t_1 + t_2}{2}$

83. (d) Let the side of cube be a

$S = 6a^2 \Rightarrow \dfrac{dS}{dt} = 12a.\dfrac{da}{dt} \Rightarrow 3.6 = 12a.\dfrac{da}{dt}$

$\Rightarrow 12(10)\frac{da}{dt} = 3.6 \Rightarrow \frac{da}{dt} = 0.03$

$V = a^3 \Rightarrow \frac{dV}{dt} = 3a^2.\frac{da}{dt} = 3(10)^2.\left(\frac{3}{100}\right) = 9$

84. (d) Since function $f(x)$ is continuous at $x = 1, 3$

$\therefore f(1) = f(1^+) \Rightarrow ae + be^{-1} = c$ (i)

$f(3) = f(3^+)$ (ii)

$\Rightarrow 9c = 9a + c \Rightarrow c = 3a$

From (i) and (ii)

$b = ae(3 - e)$

$$f(x) = \begin{cases} ae^x - be^{-x} & -1 < x < 1 \\ 2cx & 1 < x < 3 \\ 2ax + 2c & 3 < x < 4 \end{cases}$$

$f'(0) = a - b,\ f'(2) = 4c$

Given, $f'(0) + f'(2) = e$

$a - b + 4c = e$ (iv)

From eqs (i), (ii), (iii) and (iv)

$a - 3ae + ae^2 + 12a = e$

$\Rightarrow 13a - 3ae + ae^2 = e \Rightarrow a = \frac{e}{e^2 - 3e + 13}$

85. (d) Let the thickness of ice layer be $= x$ cm

Total volume $V = \frac{4}{3}\pi(10 + x)^3$

$\frac{dV}{dt} = 4\pi(10 + x)^2 \frac{dx}{dt}$ (i)

Since it is given that

$\frac{dV}{dt} = 50\,\text{cm}^3/\text{min}$ (ii)

From (i) and (ii), $50 = 4\pi(10 + x)$

$\Rightarrow 50 = 4\pi(10 + 5)^2 \frac{dx}{dt}$ [$\because$ thickness of ice $x = 5$]

$\Rightarrow \frac{dx}{dt} = \frac{1}{18\pi}$ cm/min.

86. (b) The given tangent to the curve is,

$y = \log_e x$

$\Rightarrow \frac{dy}{dx} = 1 + \log_e x \Rightarrow \left.\frac{dy}{dx}\right] = 1 + \log_e c$ (Slope)

$\because$ The tangent is parallel to line joining (1, 0), (e, e)

$\therefore 1 + \log_e c = \frac{e - 0}{e - 1}$

$\Rightarrow \log_e c = \frac{e}{e - 1} - 1 \Rightarrow \log_e c = \frac{1}{e - 1} \Rightarrow c = e^{\frac{1}{e-1}}$

87. (c) The given curve is $x^4 \cdot e^y + 2\sqrt{y + 1} = 3$.

Differentiating w.r.t. x we get

$(4x^3 + x^4.y')e^y + \frac{y'}{\sqrt{1 + y}} = 0$

$\Rightarrow \left(\frac{dy}{dx}\right) = \frac{-4x^3e^y}{\left(\frac{1}{\sqrt{y + 1}} + e^y x^4\right)} \Rightarrow \left(\frac{dy}{dx}\right)_{(1,0)} = -2$

$\therefore$ Equation of tangent

$y - 0 = -2(x - 1) \Rightarrow 2x + y = 2$

Only point (–2, 6) lies on the tangent.

88. (d) Given equation of curve is

$x^2 + 2xy - 3y^2 = 0$

$\Rightarrow 2x + 2y + 2xy' - 6yy' = 0$

$\Rightarrow x + y + xy' - 3yy' = 0$

$\Rightarrow y'(x - 3y) = -(x + y) \Rightarrow \frac{dy}{dx} = \frac{x + y}{3y - x}$

Slope of normal $= \frac{-dx}{dy} = \frac{x - 3y}{x - 3y}$

Normal at point (2, 2) $= \frac{2 - 6}{2 + 2} = -1$

Equation of normal to curve $= y - 2 = -1(x - 2)$

$\Rightarrow x + y = 4$

$\therefore$ Perpendicular distance from origin

$= \left|\frac{0 + 0 - 4}{\sqrt{2}}\right| = 2\sqrt{2}.$

89. (c)

Here,

$h(x) = \frac{x^2 + \frac{1}{x^2}}{x - \frac{1}{x}} = \left(x - \frac{1}{x}\right) + \frac{2}{x - \frac{1}{x}}.$

When $x - \frac{1}{x} < 0$

$\therefore x - \frac{1}{x} + \frac{2}{x - \frac{1}{x}} \le -2\sqrt{2}$

Hence, $-2\sqrt{2}$ will be local maximum value of $h(x)$.

When $x-\frac{1}{x}>0$

$\therefore x-\frac{1}{x}+\frac{2}{x-\frac{1}{x}}\geq 2\sqrt{2}$

Hence, $2\sqrt{2}$ will be local minimum value of $h(x)$.

90. (d) We have

Total length $= r+r+r\theta = 20$

$\Rightarrow 2r+r\theta = 20 \Rightarrow \theta = \frac{20-2r}{r}$ ...(i)

$A = \text{Area} = \frac{\theta}{2\pi}\times\pi r^2\theta = \frac{1}{2}r^2\left(\frac{20-2r}{r}\right)$

$A = 10r - r^2$

For A to be maximum

$\frac{dA}{dr} = 0 \Rightarrow 10-2r = 0$

$\Rightarrow r = 5$

$\frac{d^2A}{dr^2} = -2 < 0$

$\therefore$ For $r = 5, A$ is maximum.

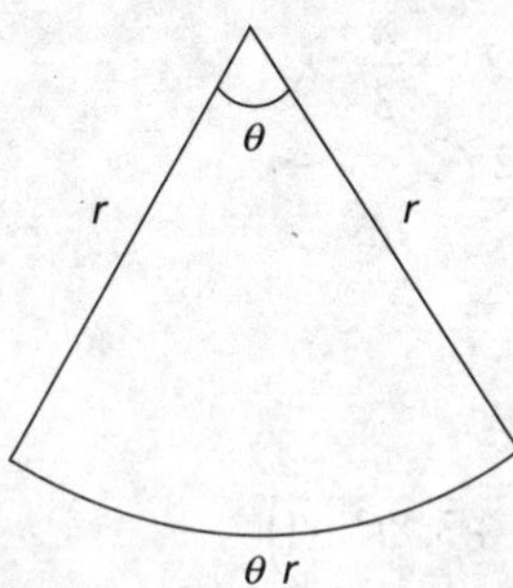

From (i)

$\theta = \frac{20-2(5)}{5} = \frac{10}{5} = 2$

$A = \frac{2}{2\pi}\times\pi(5)^2 = 25$ sq. m

91. (c)
Let $P(2t,t^2)$ be at any point on the parabola.
Centre of the given circle $C = (-g,-f) = (-3,0)$
For PC to be minimum, it must be the normal to the parabola at P.

Slope of line PC $= \frac{y_2-y_1}{x_2-x_1} = \frac{t^2-0}{2t+3}$

Also, slope of tangent to parabola at P, $\frac{dy}{dx} = \frac{x}{2} = t$

$\therefore$ Slope of normal $= \frac{-1}{t}$

$\therefore \frac{t^2-0}{2t+3} = \frac{-1}{t}$

$\Rightarrow t^3+2t+3 = 0 \Rightarrow (t+1)(t^2-t+3) = 0.$

$\therefore$ Real root of the above equation is $t = -1$

Co-ordinate of $P = (2t,t^2) = (-2,1)$

Slope of tangent to parabola at P $= t = -1$
Therefore, equation of tangent is:
$(y-1) = (-1)(x+2)$
$\Rightarrow x+y+1 = 0$

92. (c)

We have $y = \frac{x+6}{(x-2)(x-3)}$]

At y-axis, $x = 0 \Rightarrow y = 1$
On differentiating, we get

$\frac{dy}{dx} = \frac{(x^2-5x+6)(1)-(x+6)(2x-5)}{(x^2-5x+6)^2}$

$\frac{dy}{dx} = 1$ at point (0, 1)
$\therefore$ Slope of normal $= -1$
Now equation of normal is $y-1 = -1(x-0)$
$\Rightarrow y-1 = -x$
$x+y = 1$
$\therefore \left(\frac{1}{2},\frac{1}{2}\right)$ satisfy it.

93. (c)

Eccentricity of an ellipse $= \frac{1}{2}$

Now, $-\frac{a}{c} = -4 \Rightarrow a = 4\times\frac{1}{2} = 2 \Rightarrow a = 2$

We have $b^2 = a^2(1-e^2) = a^2\left(1-\frac{1}{4}\right) = 4\times\frac{3}{4} = 3$

$\therefore$ Equation of ellipse is $\frac{x^2}{4}+\frac{y^2}{3} = 1$

Now differentiating, we get

$\Rightarrow \frac{x}{2}+\frac{2y}{3}\times y' = 0 \Rightarrow y' = -\frac{3x}{4y}$

$y'\left|\left(1,\frac{3}{2}\right)\right| = -\frac{3}{4}\times\frac{2}{3} = -\frac{1}{2}$

Slope of normal = 2

$\therefore$ Equation of normal at $\left(1, \frac{3}{2}\right)$ is

$$y - \frac{3}{2} = 2(x-1) \Rightarrow 2y - 3 = 4x - 4$$

$$\therefore 4x - 2y = 1$$

94. (a) $f(x) = x^3 - 3x^2 + 5x + 7$

For increasing

$$f'(x) = 3x^2 - 6x + 5 > 0$$

$$\Rightarrow x \in R$$

For decreasing

$$f'(x) = 3x^2 - 6x + 5 < 0$$

95. (a) $4x + 2\pi r = 2 \Rightarrow 2x + \pi r = 1$

$$S = x^2 + \pi r^2$$

$$S = \left(\frac{1-\pi r}{2}\right)^2 + \pi r^2$$

$$\frac{dS}{dr} = 2\left(\frac{1-\pi r}{2}\right)\left(\frac{-\pi}{2}\right) + 2\pi r$$

$$\Rightarrow \frac{-\pi}{2} + \frac{\pi^2 r}{2} + 2\pi r = 0 \Rightarrow r = \frac{1}{\pi + 4}$$

$$\Rightarrow x = \frac{2}{x+4} \Rightarrow x = 2r$$

96. (a) $D = \sqrt{\alpha^2 + (\alpha^2 - 4)^2}$

$$D^2 = \alpha^2 + \alpha^4 + 16 - 8\alpha^2 = \alpha^4 - 7\alpha^2 + 16$$

$$\frac{dD^2}{d\alpha} = 4\alpha^3 - 14\alpha = 0$$

$$2\alpha(2\alpha^2 - 7) = 0$$

$$\alpha^2 = \frac{7}{2}$$

$$D^2 = \frac{49}{2} - \frac{49}{2} + 16 = -\frac{49}{4} + 16 = \frac{15}{4}$$

$$D = \frac{\sqrt{15}}{2}$$

O

$(\alpha, \alpha^2 - 4)$

97. (d) $f(x) = \tan^{-1}\left(\sqrt{\frac{1+\sin x}{1-\sin x}}\right)$

$$= \tan^{-1}\left(\sqrt{\frac{\left(\sin\frac{x}{2} + \cos\frac{x}{2}\right)^2}{\left(\sin\frac{x}{x} - \cos\frac{x}{2}\right)^2}}\right) = \tan^{-1}\left(\frac{1+\tan\frac{x}{2}}{1-\tan\frac{x}{2}}\right)$$

$$= \tan^{-1}\left(\tan\left(\frac{x}{4} + \frac{x}{2}\right)\right)$$

$$\Rightarrow y = \frac{\pi}{4} + \frac{x}{2} \Rightarrow \frac{dy}{dx} = \frac{1}{2}$$

Slope of normal $= \frac{-1}{\left(\frac{dy}{dx}\right)} = -2$

Equation of normal at $\left(\frac{\pi}{6}, \frac{\pi}{4} + \frac{\pi}{12}\right)$

$$y - \left(\frac{\pi}{4} + \frac{\pi}{12}\right) = -2\left(x - \frac{\pi}{6}\right)$$

$$y - \frac{4\pi}{12} = -2x + \frac{2\pi}{6}$$

$$y - \frac{\pi}{3} = -2x + \frac{\pi}{3}$$

$$y = 2x + \frac{2\pi}{3}$$

$$\left(0, \frac{2\pi}{3}\right)$$

98. (d) $P(4t^2 + 3, 8t^3 - 1)$

$\frac{dy/dt}{dt/dt} = \frac{dy}{dx} = 3t$ (Slope of tangent at P)

Let $Q = (4\lambda^2 + 3, 8\lambda^3 - 1)$

Slope of PQ $= 3t$

$$\frac{8t^3 - 8\lambda^3}{4t^2 - 4\lambda^2} = 3t$$

$$\Rightarrow t^3 - 3\lambda^2 t + 2\lambda^3 = 0$$

$$(t-\lambda)\cdot(t^2 + t\lambda - 2\lambda^2) = 0$$

$$(t-\lambda)^2 \cdot (t + 2\lambda) = 0$$

$$t = \lambda \text{ or } \lambda = \frac{-t}{2}$$

$$\therefore Q\left[t^2 + 3, -t^3 - 1\right]$$

99. (b) Given curve is

$x^2+2xy-3y^2=0$ (i)

Differentiate w.r.t. x

$$2x+2x\frac{dy}{dx}+2y-6y\frac{dy}{dx}=0$$

$$\left(\frac{dy}{dx}\right)_{(1,1)}=1$$

Equation of normal at (1, 1) is

$y=2-x$ (ii)

Solving equations (i) and (ii), we get

$x=1, 3$

Points of intersection (1, 1), (3, −1)

Normal cuts the curve again in 4th quadrant

100. (b) Given curve is $\sin y = x\sin\left(\frac{\pi}{3}+y\right)$

Differentiate with respect to x, we get

$$\cos y\frac{dY}{dX}=\sin\left(\frac{\pi}{3}+y\right)+x\cos\left(\frac{\pi}{3}+y\right)\frac{dy}{dx}$$

$$\Rightarrow \frac{dy}{dx}=\frac{\sin\left(\frac{\pi}{3}+y\right)}{\cos y-x\cos\left(\frac{\pi}{3}+y\right)}$$

$$\frac{dy}{dx}\text{ at }(0, 0)=\frac{\sqrt{3}}{2}$$

$\Rightarrow$ Equation of normal is $y=0=-\frac{2}{\sqrt{3}}(x-0)$

$\Rightarrow 2x+\sqrt{3}y=0$

101. (a) Let

$$f(x)=\frac{(1+x)^{\frac{3}{5}}}{1+x^{\frac{3}{5}}}\text{ and } x\in[0,1]$$

$$\Rightarrow f'(x)=\frac{\left(1+x^{\frac{3}{5}}\right)\frac{3}{5}(1+x)^{-\frac{2}{5}}-\frac{3}{5}(1+x)^{\frac{3}{5}}\left(x^{\frac{-2}{5}}\right)}{(1+x^{\frac{3}{5}})^2}$$

$$=\frac{3}{5}\left[\left(1+x^{\frac{3}{5}}\right)(1+x)^{-\frac{2}{5}}-(1+x)^{\frac{3}{5}}x^{\frac{-2}{5}}\right]$$

$$=\frac{3}{5}\left[\frac{1+x^{\frac{3}{5}}}{(1+x)^{\frac{2}{5}}}-\frac{(1+x)^{\frac{3}{5}}}{x^{\frac{2}{5}}}\right]$$

$$=\frac{x^{\frac{2}{5}}+x-1-x}{x^{\frac{2}{5}}(1+x)^{\frac{2}{5}}}=\frac{x^{\frac{2}{5}}-1}{x^{\frac{2}{5}}(1+x)^{\frac{2}{5}}}<0$$

Also, $f(0)=1\Rightarrow f(x)\in\left[2^{-0.4},1\right]$

$f(a)=2^{-0.4}$

102. (a)

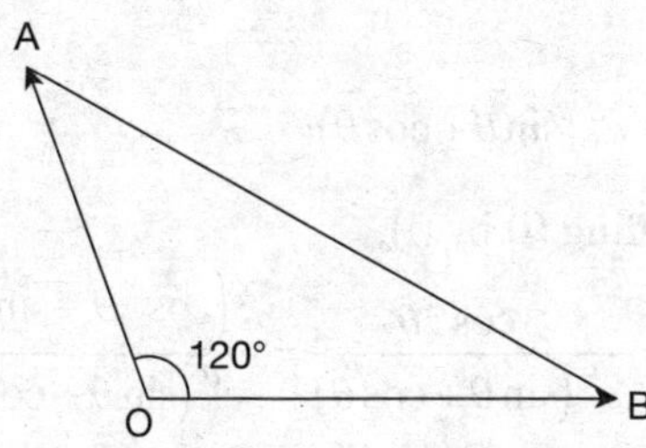

Let OA = x km, OB = y km, AB = R

$$(\text{AB})^2=(\text{OA})^2+(\text{OB})^2-2(\text{OA})(\text{OB})\cos 120^\circ$$

$$R^2=x^2+y^2-2xy\left(-\frac{1}{2}\right)=x^2+y^2+xy \quad \text{(i)}$$

R at x = 6 km, and y = 8 km

$$R=\sqrt{6^2+8^2+6\times 8}=2\sqrt{37}$$

Differentiating equation (i) with respect to t

$$2R\frac{dR}{dt}=2x\frac{dx}{dt}+2y\frac{dy}{dt}+\left(x\frac{dy}{dt}+y\frac{dx}{dt}\right)$$

$$=\frac{1}{2R}\left[2\times 8\times 20+2\times 6\times 30+(8\times 30+6\times 20)\right]$$

$$\frac{dR}{dt}=\frac{1}{2\times 2\sqrt{37}}[1040]=\frac{260}{\sqrt{37}}$$

103. (a) Let $f(x)=\alpha\log|x|+\beta x^2+x$.

Differentiate both sides,

$$f'(x)=\frac{\alpha}{x}+2\beta x+1$$

Since, $x=-1$ and $x=2$ are extreme points.

Therefore $f'(x)=0$ at these points.

Put $x=-1$ and $x=2$ in $f'(x)$, we get

$-\alpha-2\beta+1=0\Rightarrow\alpha+2\beta=1$ ...(i)

$\frac{\alpha}{2}+4\beta+1=0\Rightarrow\alpha+8\beta=-2$ ...(ii)

One solving (i) and (ii), we get

$$6\beta=-3\Rightarrow\beta=-\frac{1}{2}$$

$\therefore \alpha=2$

104. (c) Given $y=3\sin\theta.\cos\theta$

$$\frac{dy}{d\theta}=3\left[\sin\theta(-\sin\theta)+\cos\theta(\cos\theta)\right]$$

$\frac{dy}{d\theta}=3\left[\cos^2\theta-\sin^2\theta\right]=3\cos 2\theta$ ...(i)

and $x=e^{\theta}\sin\theta$

$\frac{dy}{d\theta}=e^{\theta}\cos\theta+\sin\theta e^{\theta}$

$\frac{dx}{d\theta}=e^{\theta}(\sin\theta+\cos\theta)$ ...(ii)

Dividing (i) by (i),

$\frac{dy}{dx}=\frac{3\cos 2\theta}{e^{\theta}(\sin\theta+\cos\theta)}=\frac{3(\cos^2\theta-\sin^2\theta)}{e^{\theta}(\sin\theta+\cos\theta)}$

$\frac{dy}{dx}=\frac{3(\cos\theta-\sin\theta)}{e^{\theta}}$

Given tangent is parallel to x-axis, then $\frac{dy}{dx}=0.$

$0=\frac{3(\cos\theta-\sin\theta)}{e^{\theta}}$

or $\cos\theta-\sin\theta=0\Rightarrow\cos\theta=\sin\theta$

$\Rightarrow\tan\theta=1\Rightarrow\tan\theta=\frac{\tan\pi}{4}\Rightarrow\theta=\frac{\pi}{4}$

105. (d) Let $y=\cos(x+y)$

$\Rightarrow\frac{dy}{dx}=-\sin(x+y)\left(1+\frac{dy}{dx}\right)$ ...(i)

Now, given equation of tangent is $x+2y=k$

$\Rightarrow$ Slope $=\frac{-1}{2}$

So, $\frac{dy}{dx}=\frac{-1}{2}$ put this value in (i), we get

$\frac{-1}{2}=-\sin(x+y)\left(1-\frac{1}{2}\right)$

$\Rightarrow\sin(x+y)=1$

$\Rightarrow x+y=\frac{\pi}{2}\Rightarrow y=\frac{\pi}{2}-x$

Now, $\frac{\pi}{2}-x=\cos(x+y)$

$\Rightarrow x=\frac{\pi}{2}$ and $y=0$

Thus, $x+2y=k\Rightarrow\frac{\pi}{2}=k$

106. (c) Let cost $C=av+\frac{b}{v}$

According to given question,

$30a+\frac{b}{30}=75$ ...(i)

$40a+\frac{b}{40}=65$ ...(ii)

On solving (i) and (ii), we get

$a=\frac{1}{2}$ and $b=1800$

Now, $C=av+\frac{b}{v}$

$\Rightarrow\frac{dC}{dv}=a-\frac{b}{v^2}$

$\frac{dC}{dv}=0\Rightarrow a-\frac{b}{v^2}=0$

$\Rightarrow v=\sqrt{\frac{b}{a}}=\sqrt{3600}\Rightarrow v=60$ kmph.

107. (a) Volume of sphere $V=\frac{4}{3}\pi r^3$

$\frac{dV}{dt}=\frac{4}{3}.\pi.3r^2.\frac{dr}{dt}$

$35=4\pi r^2.\frac{dr}{dt}$ or $\frac{dr}{dt}=\frac{35}{4\pi r^2}$ ...(i)

Surface area of sphere $=S=4\pi r^2$

$\frac{dS}{dt}=4\pi\times 2r\times\frac{dr}{dt}=8\pi r.\frac{dr}{dt}$

$\frac{dS}{dt}=\frac{70}{r}$ (By using (i))

Now, diameter = 14 cm, $r=7$

$\therefore\frac{dS}{dt}=10$

108. (d) $f(x)=2x^3+3x+k$

$f'(x)=6x^2+3>0\ \forall x\in R$ $(\because x^2>0)$

$\Rightarrow f(x)$ is strictly increasing function.

$\Rightarrow f(x)=0$ has only one real root, so two roots are not possible.

Integer and Numerical Value Type

1. Equation of tangent at (x_1, y_1) is

$y-y_1=m(x-x_1)$

where $m=\left(\frac{dy}{dx}\right)_{(x_1,y_1)}$

This meets y-axis at $(0, y_1-mx_1)$.

Now, the required segment length

$=\sqrt{(0-x_1)^2+(y_1-mx_1-y_1)^2}$

$= x_1\sqrt{1+m^2}$

It can be easily shown that $m = \left(\dfrac{dy}{dx}\right)_{(x_1,y_1)}$

$= -\dfrac{\sqrt{a^2-x_1^2}}{x_1}$,

whence $x_1\sqrt{1+m^2} = a \Rightarrow \lambda = 1.$

2. $\left(\dfrac{dy}{dx}\right)_{\text{first}} = \dfrac{x}{2a}$

$\left(\dfrac{dy}{dx}\right)_{\text{first}} = \dfrac{a}{4\sqrt{ax}}$

The two curves intersect at (0, 0) and $(2a, 1)$

$\Rightarrow \dfrac{dy}{dx} = 1, \dfrac{1}{4}$

$\Rightarrow \tan\theta = \dfrac{1-\dfrac{1}{4}}{1+\dfrac{1}{4}} = \dfrac{3}{5}.$

3. Take $x = 64\cos^3 t, y = 64\sin t$ as any point on $x^{2/3} + y^{2/3} = 16.$

Then $\dfrac{dx}{dt} = -192\cos^2 t(-\sin t)$

$\dfrac{dx}{dt} = 192\sin^2 t\cos t$

$\Rightarrow \dfrac{dy}{dx} = -\dfrac{\sin t}{\cos t}$

$\Rightarrow$ Equation of tangent at t^{-1}.

$y - 64\sin^{-3} t = -\dfrac{\sin t}{\cos t}\left(x - 64\cos^3 t\right)$

$\dfrac{y}{\sin t} - 64\sin^2 t = -\dfrac{x}{\cos x} + 64\cos^2 t$

or $\dfrac{x}{64\cos t} + \dfrac{y}{64\sin t} = 1$

$\Rightarrow$ Segment of tangent contained between a and b.

$= \sqrt{64\cos^2 t + 64\sin^2 t} = 64$

$\Rightarrow k = 64.$

4. $f'(x) = x^2 - (m-3)x + m$

If $f(x)$ is MI over $[0,\infty]$, then

$f'(x)$ must be ≥ 0 for all x is $[0,\infty]$

$\Rightarrow f'(x)$ is either ≥ 0 for all x or both roots of $f'(x) = 0$ are negative.

Now, $f'(x) \geq 0$ for all x.

If $(m-3)2 - 4m \leq 0$

$\Rightarrow m \in [1,9]]$

Again, both roots of $f'(x) = 0$ are negative if $D > 0, m - 3\langle 0, m\rangle 0$

On taking intersection, we get $m \in [0,1]$.

Thus, $f(x)$ is MI for all $x \geq 0$ if $m \in [0,9]$

$\Rightarrow k = 9.$

5. Put $x = \dfrac{1}{2}\cos\theta, y = \sin\theta$ $\quad (\because 4x^2 + y^2 = 1)$

Now $12x^2 - 3y^2 + 16xy$

$= 3\cos^2\theta - 3\sin^2\theta + 8\sin\theta\cos\theta$

$= 3\cos 2\theta + 4\sin 2\theta$

$\Rightarrow$ max value $= 5$

6. Let $\cos A + \cos B + k\cos C = y$ then

$2\cos\dfrac{A+B}{2}\cos\dfrac{A-B}{2} + K\left(1 - 2\sin^2\dfrac{C}{2}\right) = y$

On arranging as a quadratic in $\sin\dfrac{C}{2}$

$\left(\because \cos\dfrac{A+B}{2} = \sin\dfrac{C}{2}\right)$, we get

$2k\sin^2\dfrac{C}{2} - 2\cos\dfrac{A-B}{2}\sin\dfrac{C}{2} + y - k = 0$

Now $D \geq 0$

$\Rightarrow k(y-k) \leq \dfrac{1}{2}\cos^2\dfrac{A-B}{2} \leq \dfrac{1}{2}$

$\Rightarrow y \leq \dfrac{1}{2k} + k$

$\Rightarrow \lambda = 2$

7. Let the sides be x, y and $12 - x - y$.
The area A is given by

$A = \dfrac{1}{2}xy\sin 60° = \sqrt{3} \times y$

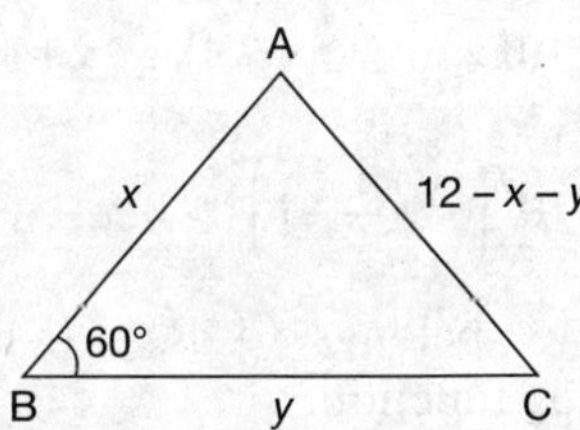

Also, $\cos 60° = \dfrac{x^2 + y^2 - (12-x-y)^2}{2xy}$

$\Rightarrow y = \dfrac{8x-48}{x-8}$

Now area $A = 8\sqrt{3}\cdot\dfrac{x^2-6x}{x-8}$

$\dfrac{dA}{dx} = 8\sqrt{3}\dfrac{x^2-16x+48}{(x-8)^2}$

$\Rightarrow \dfrac{dA}{dx} = 0 \Rightarrow x = 4, 12$

$x = 12$ is not possible (∵ perimeter = 12)

$\Rightarrow x = 4$ whence $y = 4$

and $12 - x - y$ is also 4.

8. Let $f(x) = x^4 + 4x + 5$.

Then, $f'(x) = 4x^3 + 4 = 4(x+1)(x^2-x+1)$

$\Rightarrow$ f(x) is MD over $(-\infty, -1]$ and MI over $[-1, \infty)$

$\Rightarrow$ Absolute minimum occurs at $x = -1$

$\Rightarrow$ Absolute minimum value $= f(-1) = 2$.

9. Let P be the petrol burnt per hour, then $P = kv^3$, where k is a constant.

In covering a distance d, the time taken $= \dfrac{d}{v-8}$

The petrol burnt Q in covering distance d is

$Q = \dfrac{dkv^3}{v-8}$

$\dfrac{dQ}{dv} = dk\left[\dfrac{3v^2(x-8)-v^3}{(v-8)^2}\right]$

$\dfrac{dQ}{dv} = 0 \Rightarrow v = 12$

For this value of v, $\dfrac{d^2Q}{dv^2} > 0$

Thus, Q is minimum when $v = 12$.

10. Let $f(x) = 2x^5 + 5x^4 + 10(x^3 + x^2 + x + 1)$

$\because f(-1) = 3$

and $f(-2) = -34$

$f'(x) = 10(x^4 + 2x^3 + 3x^2 + 2x + 1)$

$= 10^2\left(x + \dfrac{1}{x} + 1\right)^2 > 0.$

For all x belongs to R, $f(x)$ is a strictly increasing function.

Hence, roots of $f(x)$ lies in $(-2, -1)$

Clearly, $|a| = 2$.

11. Given $f(x) = x^2 + ax + 1$

$f'(x) = 2x + a$.

When $f(x)$ is increasing on [1, 2],

$2x + a \geq 0 \forall x \in [1,2]$

$a \geq -2x \forall x \in [1,2]$

$R = -4$

When $f(x)$ is decreasing on [1, 2]

$2x + a \leq 0 \forall x \in [1,2]$

$a \leq -2 \forall x \in [1,2]$

$S = -2$

$|R - S| = |-4 + 2| = 2$

12. Let then $x + y = 36$

x is perimeter of square and y is perimeter of circle side of square $= \dfrac{x}{4}$.

Sum Areas $= \left(\dfrac{x}{4}\right)^2 + \pi\left(\dfrac{y}{2\pi}\right)^2 \{\because (a)^2 + \pi(R)^2\}$

$\because y = 36 - x$

$= \dfrac{x^2}{16} + \dfrac{(36-x^2)}{4\pi}$

For min area:

$x = \dfrac{144}{\pi + 4}$

$\Rightarrow$ Radius $= y = 36 - \dfrac{144}{\pi+4}$.

$\Rightarrow k = \dfrac{36\pi}{\pi+4}$(Circumference of circle)

$= \left(\dfrac{4}{\pi} + 1\right)k = 36$

13. Applying $R_1 \to R_1 - R_2$ and $R_2 \to R_2 - R_3$, we get

$f(x) = \begin{vmatrix} -2 & -2 & 0 \\ 2 & 0 & -1 \\ \sin^2 x & \cos^2 x & 1+\cos 2x \end{vmatrix}$

Applying $C_2 \to C_2 - C_1$, we get

$= \begin{vmatrix} -2 & 0 & 0 \\ 2 & -2 & -1 \\ \sin^2 x & \cos^2 x & 1+\cos 2x \end{vmatrix}$

$= -2[-2 - 2\cos 2x + \cos 2x]$

$= -2[-2-\cos 2x] = 2[2+\cos 2x]$

$\therefore$ Maximum value of $\cos 2x = 1$

$\therefore$ Maximum value of $f(x) = 2(2+1) = 6.$

14. $f(x) = ax^2 + bx + c$

$\Rightarrow f(-1) = a - b + c = 2$ ...(i)

$f'(x) = 2ax + b$

$\Rightarrow f'(-1) = -2a + b = 1$ ...(ii)

$f''(x) = 2a = \frac{1}{2}$

$\Rightarrow a = \frac{1}{4}.$ ...(iii)

Solving (i), (ii) and (iii), we get $b = \frac{3}{2}$ and $c = \frac{13}{4}$

$\because f'(x) > 0, \forall x \in [-1,1]$

So, $f(x) \le f(1)$

$\Rightarrow f(x) \le 5$

15.

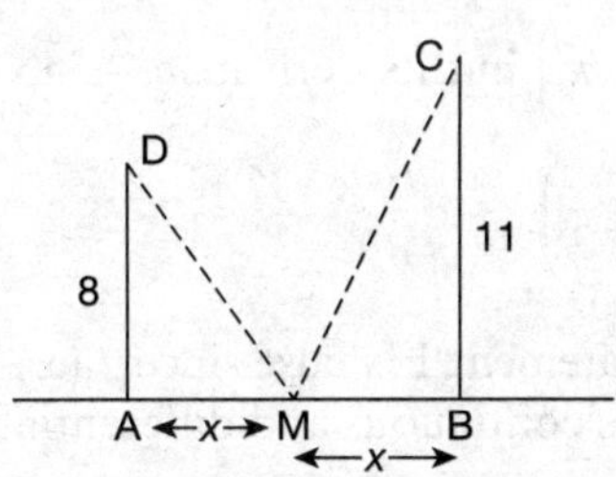

Let AM $= x\ m$

$\therefore (\text{MD})^2 + (\text{MC})^2 = 64 + x^2 + 121 + (10-x)^2 = f(x)$ (say)

$f'(x) = 2x - 2(10-x) = 0$

$\Rightarrow 4x = 20 \Rightarrow x = 5$

$f''(x) = 2 - 2(-1) > 0$

$\therefore f(x)$ is minimum at $x = 5$ m.

16. Let $f(x) = ax^3 + bx^2 + cx + d$

$f(-1) = 10$ and $f(1) = -6$

$-a + b - c + d = 10$ ...(i)

$a + b + c + d = -6$ ...(ii)

Solving equations (i) and (ii), we get

$a = \frac{1}{4}, d = \frac{35}{4}$

$b = \frac{-3}{4}, c = -\frac{9}{4}$

$\Rightarrow f(x) = a(x^3 - 3x^2 - 9x) + d$

$f'(x) = \frac{3}{4}(x^2 - 2x - 3) = 0$

$\Rightarrow x = 3, -1$

Local minima exists at $x = 3$

17. $p(x)$ has a local maximum at $x = 1$ and a local minimum at $x = 3$ and $p(x)$ is a real polynomial of least degree.

Hence, let

$p'(x) = k(x-1)(x-3) = k(x^2 - 4x + 3)$

$\Rightarrow p(x) = k\left(\frac{x^3}{3} - 2x^2 + 3x\right) + c.$

Now, $p(1) = 6$ and $p(3) = 2$

$\Rightarrow \frac{4}{3}k + C = 6 \text{ and } 0 + C = 2 \Rightarrow k = 3$

$\therefore p'(x) = 3(x-1)(x-3) \Rightarrow p'(0) = 9.$

18.

$$f(x) = |x| + |x^2 - 1| = \begin{cases} -x + x^2 - 1, & x < -1 \\ -x - x^2 + 1, & -1 \le x \le 0 \\ x - x^2 + 1, & 0 < x < 1 \\ x^2 + x - 1, & x \ge 1 \end{cases}$$

$$\therefore f'(x) = \begin{bmatrix} 2x - 1 & , & x < -1 \\ -2x - 1 & , & -1 \le x \le 0 \\ -2x + 1 & , & 0 < x < 1 \\ 2x + 1 & , & x > 1 \end{bmatrix}$$

Critical points are $\frac{1}{2}, \frac{-1}{2}, -1, 0$ and 1.

–ve	+ve	–ve	+ve	–ve	+ve
	–1	–1/2	0	1/2	1

We observe at five points $f'(x)$ changes its sign.

$\therefore$ There are 5 points at which $f(x)$ attains either local maximum or local minimum.

19. Let $g(x) = e^{f(x)}, \forall x \in \text{R}$

$\Rightarrow g'(x) = e^{f(x)} \cdot f'(x)$

$\Rightarrow f'(x)$ changes its sign from positive to negative in the neighbourhood of $x = 2009$

$\Rightarrow f(x)$ has local maxima at $x = 2009$

So, the number of local maximum is one.

20. The equation of tangent to the curve $y = f(x)$ at the point $P(x, y)$ is

$$(X-x)\frac{dy}{dx}-(Y-y)=0 \Rightarrow X\frac{dy}{dx}-Y=x\frac{dy}{dx}-y$$

Its $y-\text{intercept}=y-x\frac{dy}{dx}=x^3 \Rightarrow \frac{dy}{dx}-\frac{y}{x}=-x^2$

$$\text{I.F.}=e^{\int -\frac{1}{x}dx}=\frac{1}{x}$$

$$\therefore y.\frac{1}{x}=\int(-x^2)\frac{1}{x}dx=\frac{-x^2}{2}+c, y=\frac{-x^3}{2}+cx$$

Since, $f(1)=1 \Rightarrow$ At $x=1, y=1$

$$\therefore 1=\frac{-1}{2}+c \Rightarrow c=3/2. \therefore y=-\frac{x^3}{2}+\frac{3x}{2}$$

At $x=-3$, $y=\frac{27}{2}-\frac{9}{2}=9, \therefore f(-3)=9$.

21. Let $p(x)=ax^4+bx^3+cx^2+dx+e$

Now $\lim_{x\to 0}\left[1+\frac{p(x)}{x^2}\right]=2 \Rightarrow \lim_{x\to 0}\frac{p(x)}{x^2}=1$ ...(i)

$\Rightarrow p(0)=0 \Rightarrow e=0$

On applying L' Hospital rule to Eq. (i), we get

$$\lim_{x\to 0}\frac{p'(x)}{2x}=1 \Rightarrow p'(0)=0 \Rightarrow d=0$$

Again on applying L' Hospital rule, we get

$$\lim_{x\to 0}\frac{p''(x)}{2}=1 \Rightarrow p''(0)=2 \Rightarrow c=1$$

$\therefore p(x)=ax^4+bx^3+x^2$

$\Rightarrow p'(x)=4ax^3+3bx^2+2x$

As $p(x)$ has extremum at $x=1$ and 2.

$\therefore p'(1)=0$ and $p'(2)=0$

$\Rightarrow 4a+3b+2=0$ ...(i)

and $8a+3b+1=0$ ...(ii)

On solving (i) and (ii), we get $a=\frac{1}{4}$ and $b=-1$

$$\therefore p(x)=\frac{1}{4}x^4-x^3+x^2, \therefore p(2)=0$$

22. Given, $f(x)=2x^3-15x^2+36x-48$

and $A=\{x \mid x^2+20\le 9x\}$

$\Rightarrow A=\{x \mid x^2-9x+20\le 0\}$

$\Rightarrow A=\{x|(x-4)(x-5)\le 0\} \Rightarrow A=[4,5]$

Also,

$$f'(x)=6x^2-30x+36=6(x^2-5x+6)$$
$$=6(x-2)(x-3)$$

Clearly $\forall x \in A$. $f'(x)>0$

$\therefore f$ is strictly increasing function on A.

$\therefore$ Maximum value of f on $A=f(5)$

$=2\times 5^3-15\times 5^2+36\times 5-48=7$.

Assertion and Reason

1. (c) Statement 1 is true since $\frac{d}{dx}(\sin x)=\cos x$, $\frac{d}{dx}(\cos x)=-\sin x$. But derivatives are negative in $\left(\frac{\pi}{2},\pi\right)$.

The Statement 2 is false since $\cos x$ is decreasing in $\left(\frac{\pi}{2},\pi\right)$ but its derivative $-\sin x$ is increasing in $\left(\frac{\pi}{2},\pi\right)$.

2. (d) Statement 1 is false since $f(x)=x^2-5x+6$, [0, 3] is continuous and differentiable

$f(0)\ne f(3)$ but $f'\left(\frac{5}{2}\right)=0, \frac{5}{2}\in(0,3)$.

The Statement 2 is true since it is Rolle's theorem.

3. (a) $f'(x)=\log\left(x+\sqrt{1+x^2}\right)$

$$=-\log\left(\sqrt{1+x^2}-x\right)$$

$\Rightarrow f'(x)>0$ when $x>0$

and $f'(x)<0$ when $x<0$

$\Rightarrow f(x)$ is increasing when $x>0$

$\Rightarrow f(x)>f(0) \Rightarrow f(x)>0$.

Again, $f(x)$ is decreasing in $(-\infty,0)$.

$\Rightarrow f(x)>f(0)$

$\Rightarrow f(x)>0$

$\Rightarrow f(x)$ is positive for all x.

Thus, Statement 1 is true and follows from Statement 2.

4. (d) $f(x)=x^2-x\sin x-\cos x$

$f'(x)=2x-\sin x-x\cos x+\sin x=x(2-\cos x)$

$\Rightarrow$ Statement 2 is true.

But $f(-\pi)=\pi^2+1>0, f(0)\langle 0, f(\pi)\rangle 0.$

$\Rightarrow f(x)=0$ for atleast one x in $(-\pi,0)$ and for at least one x is $(0, \pi)$.

$\Rightarrow$ Statement 1 is false.

5. (b) Statement 1 is true since conics are of second degree. Therefore, in general $y=mx+c$ will meet every conic in two points.
Statement 2 is true but it is not the correct explanation for A.

6. (a) $f(x)=xe^{-x}$

$f'(x)=e^{-x}-xe^{-x}=e^{-x}(1-x)$

For maximum or minimum,

$f'(x)=0\Rightarrow x=1$

$f''(x)=-e^{-x}-\left(e^{-x}-xe^{-x}\right)$

$=-2e^{-x}+xe^{-x}$

$=e^{-x}(x-2)<0$ for $x=1$

$\therefore$ at $x=1, f(x)$ attains maxima.

7. (b) $f(x)=6x^2-18x+12$

$=6\left(x^2-3x+12\right)=6(x-1)(x-2)$

$\Rightarrow f'(x)$ is non-negative for x lying outside the interval [1, 2]

8. (d) Point of intersection (0, 0), (16, 16)
At (0, 0) both curves are orthogonal.
At (16, 16), for $y^2 = 16x$.

$$\left(\frac{dy}{dx}\right)_{(16,16)}=\frac{1}{2}$$

for $x^2=16y$

$$\left(\frac{dy}{dx}\right)_{(16,16)}=2$$

$m_1.m_2\neq -1$. Hence, they are not orthogonal at (16, 16).

9. (d)

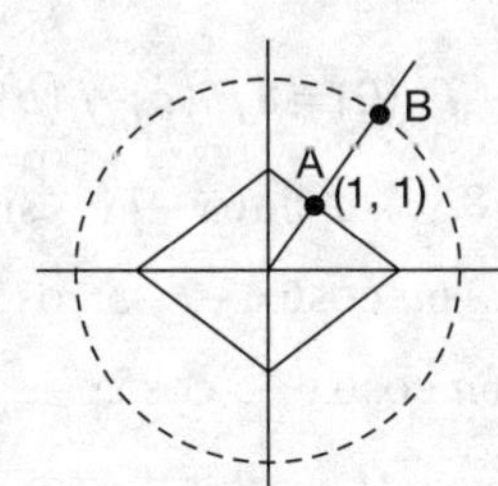

Common normal is $y=x$
Solving $x+y=2\Rightarrow(1, 1)$
and $x^2+y^2=16\Rightarrow\left(2\sqrt{2},2\sqrt{2}\right)$

The distance $AB=4-\sqrt{2}$.
But as curves are not smooth, checks at slope points. The co-ordinates in 1st quadrant are (2, 0) and (4, 0). Here, distance = 2.
$\therefore 4-\sqrt{2}$ is not shortest.

10. (d) Let $g(x)$ is the inverse function of $f(x)$. Then $f(g(x)) = x$.

$\therefore f'^{(g(x))}.g'^{(x)}=1$ i.e., $g'(x)=\dfrac{1}{f'(g(x))}$

$\therefore g''(x)=-\dfrac{1}{\left(f'(g(x))\right)^2}.f''(g(x)).g'(x)$

In statement 1, $f''(g(x))>0$ and $g'(x)>0$
$\Rightarrow g''(x)<0$
$\Rightarrow$ concavity of $f^{-1}(x)$ is downwards
$\therefore$ Statement is false
In statement 2, $f''(g(x))>0$ and $g'(x)<0$
$\Rightarrow g''(x)>0$
$\Rightarrow$ concavity of $f^{-1}(x)$ is upwards
$\therefore$ The statement is true.

JEE Advanced

Objective Questions

One Option Correct

1. (c) Points of intersection of curves $y=|x^2-1|$ and $y=|x^2-3|$ are $\left(\pm\sqrt{2},1\right)$.

$y=|x^2-1|$ and $y=|x^2-3|$

$\frac{dy}{dx}=2x$ and $\frac{dy}{dx}=-2x$

$m_1=\left.\frac{dy}{dx}\right|_{(\sqrt{2},1)}=2\sqrt{2}$ and $m_2=\left.\frac{dy}{dx}\right|_{(\sqrt{2},1)}=-2\sqrt{2}$

$\tan\theta=\left|\frac{4\sqrt{2}}{7}\right|$

2. (c) $y-e^{xy}+x=0$

$\therefore \frac{dy}{dx}-e^{xy}\left(y+x\frac{dy}{dx}\right)+1=0$

i.e., $\frac{dy}{dx}-y(x+y)-x(x+y)\frac{dy}{dx}+1=0$

i.e., $\left[1-x(x+y)\right]\frac{dy}{dx}=y(x+y)-1.$

For the vertical tangents,

$1-x(x+y)=0$ i.e., $y=\dfrac{1-x^2}{x}$

i.e., $\dfrac{1}{x}-x-e^{1-x^2}+x=0$

i.e., $e^{1-x^2}=\dfrac{1}{x}$ $\therefore x=1$ and $y=0$

3. (c) Given, $x=f'(t)\sin t+f''(t)\cos t$

$y=f'(t)\cos t-f''(t)\sin t$

From given equation,

$$\frac{dx}{dt}=\{f'(t)+f'''(t)\}\cos t$$

and $\dfrac{dy}{dx}=-\{f'(t)+f'''(t)\}\sin t$

Velocity

$$\sqrt{\left(\frac{dx}{dt}\right)^2+\left(\frac{dy}{dt}\right)^2}$$

$$=\sqrt{\{f'(t)+f'''(t)\}^2\cos^2 t+\{f'(t)+f'''(t)\}^2\sin^2 t}$$

$$=f'(t)+f'''(t)$$

4. (b) Since $f(x)$ and $g(x)$ are differentiable in [0, 1], then $f(x)$ and $g(x)$ are continuous.

Let $h(x)=f(x)-2g(x)$

$h(0)=2$, $h(1)=2$

$\therefore$ Rolle's Theorem is applicable

5. (a) Here, $f(x)=\dfrac{a\sin x+b\cos x}{c\sin x+d\cos x}$ is monotonically increasing.

Then $f'(x)\geq 0$

$$\text{Now } f'(x)=\frac{(c\sin x+d\cos x)(a\cos x-b\sin x)-(a\sin x+b\cos x)(\cos x-d\sin x)}{(c\sin x+d\cos x)^2}$$

$$=\frac{ac\sin x\cos x-bc\sin^2 x+ad\cos^2 x-bd\sin x\cos x-ac\sin x\cos x+ad\sin^2 x-bc\cos^2 x+bd\sin x\cos x}{(c\sin x+d\cos x)^2}$$

$$=\frac{ad\left(\sin^2 x+\cos^2 x\right)-bc\left(\sin^2 x+\cos^2 x\right)}{(c\sin x+d\cos x)^2}=\frac{ad-bc}{(c\sin x+d\cos x)^2}$$

Here $f'(x)\geq 0\ \forall x\in R\Rightarrow ad-bc\geq 0$

$ad\geq bc$

6. (b) $f(x)=x^3+ax^2+bx+5\sin^2 x$

$f'(x)=3x^2+2ax+b+10\sin x\cos x\geq 0\ \forall x\in R$

ie., $3x^2+2ax+b+5\sin 2x\geq 0, \forall x\in R$

If $3x^2+2ax+b-5\geq 0\ \forall x\in R$

$\therefore 4a^2-12(b-5)\leq 0$

$a^2-3b+15\leq 0$

7. (b) Consider the function

$$f(x)=\frac{a_0x^{n+1}}{n+1}+\frac{a_1x^n}{n}+\frac{a_2x^{n-1}}{n-1}+\ldots+\frac{a_{n-1}x^2}{2}+a_nx$$

Then $f(0)=0$ and $f(1)=0$.

Hence $f'(x)=0$ has at least one solution in (0, 1)

8. (a) Here, f is a differentiable function. Then, f is a continuous function.

So by L.M.V. theorem, for any $a\in(0,4)$

$$f'(a)=\frac{f(4)-f(0)}{4-0} \quad \ldots(1)$$

Again from mean value theorem, for any $b\in(0,4)$

$$f(b)=\frac{f(4)+f(0)}{2} \quad \ldots(2)$$

Now multiplying (1) and (2), we get

$$\frac{f^2(4)-f^2(0)}{8}=f'(a)\cdot f(b)$$

$$\Rightarrow f^2(4)-f^2(0)=8f'(a)\cdot f(b)$$

9. (d) $f(x)=8ax-a\sin 6x-7x-\sin 5x$

$f'(x)=8a-6a\cos 6x-7-5\cos 5x$

$=8a-7-6a\cos 6x-5\cos 5x$

$f(x)$ is an increasing function.

$f'(x) \geq 0 \therefore 8a - 7 \geq 6a + 5$

$\Rightarrow 2a \geq 12$

$a \geq 6$

$a \in [6, \infty)$

10. (b) $f'(x) = \dfrac{(e+x)\ell n(e+x) - (\pi+x)\ell n(\pi+x)}{(e+x)(\pi+x)\ell n^2(e+x)}$

Let $g(t) = t\,\ell n t$

$g'(t) = 1 + \ell n t$

$\Rightarrow g(t)$ is increasing for $t \in \left(\dfrac{1}{e}, \infty\right)$.

For $x > 0$,

$\dfrac{1}{e} < e + x < \pi + x$

$g(e+x) < g(\pi+x)$

$\Rightarrow (e+x)\ell n(e+x) - (\pi+x)\ell n(\pi+x) < 0$

$\Rightarrow f'(x) < 0 \; \forall x \in (0, \infty)$

11. (b)

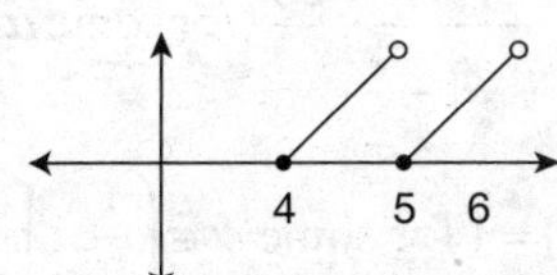

$f(5^-) > f(5) < f(5^+)$

$\therefore x = 5$ is local minima.

12. (d)

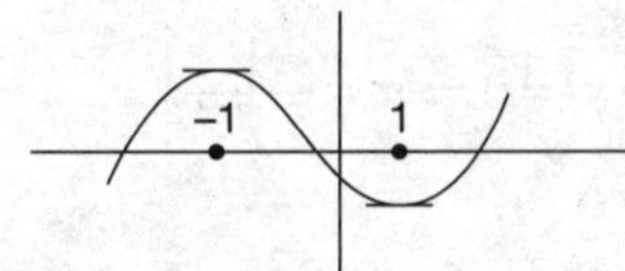

$f(x) = x^3 - 3x + k, k = [a]$

$f'(x) = 3(x-1)(x+1)$

-1 is maxima and 1 is minima for three roots

$f(-1)\,f(1) < 0$

$\Rightarrow (k+2)(k-2) < 0$

$k \in (-2, 2) \Rightarrow -2 < [a] < 2$

$\Rightarrow -1 \leq a < 2$

13. (a) When $a > 0$, $b > 0$, the graph will be

i.e., maxima at $x = 0$.

14. (b) $f(x) = 2x^3 - 3(a+1)x^2 + 6ax - 12$

$f'(x) = 6\{x^2 - (a+1)x + a\} = 0$

$6(x-1)(x-a) = 0 \Rightarrow x = 1, a$

$f''(x) = 12x - 6(a+1)$

$f''(1) = 6 - 6a > 0$

$f''(a) = 12a - 6a - 6 = 6a - 6$

$= 6(a-1)$

$\therefore$ If $a < 1$ then $x = a$ is a local max

$\therefore 2a = 1 \therefore a = 1/2$

If $a > 1$, then $x = a$ is a local minimum.

$\therefore a = 2$

15. (d) $f(1) = -6$

For maximum at $x = 1$

$\lim\limits_{x \to 1^{-1}} f(x) = \tan^{-1}\alpha - 5 < -6$

$\Leftrightarrow \tan^{-1}\alpha < -1 \Leftrightarrow \alpha < -\tan 1$

16. (b) Time taken by the truck $= \dfrac{300}{x}$ hours

$\therefore$ Fuel consumed $= \left(2 + \dfrac{x^2}{600}\right)\dfrac{300}{x}$ liter

$\therefore$ Expenses on travelling

$E = 200 \times \dfrac{300}{x} + \left(2 + \dfrac{x^2}{600}\right)\dfrac{3000}{x} = \dfrac{60000}{x} + \dfrac{6000}{x} + 5x$

$= \dfrac{66000}{x} + 5x \;\therefore\; \dfrac{dE}{dx} = -\dfrac{66000}{x^2} + 5 < 0$

for all $x \in [30, 60]$

$\therefore$ most economical speed is 60 kmph.

17. (b) The slope of the tangent to the curve $y = 2e^x \sin\left(\dfrac{\pi}{4} - \dfrac{x}{2}\right)\cos\left(\dfrac{\pi}{4} - \dfrac{x}{2}\right) = e^x \cos x$ is

$S = \dfrac{dy}{dx} = e^x(-\sin x + \cos x)$

Now, $\dfrac{dS}{dx} = e^x(-\sin x + \cos x - \cos x - \sin x)$

$= -2e^x \sin x$

$\dfrac{dS}{dx} = 0 \Rightarrow -2e^x \sin x = 0$

$\Rightarrow x = 0, \pi, 2\pi \; (\because 0 \leq x \leq 2\pi)$

Value of S at $x = 0$ is 1, value of S at $x = \pi$ is $-e^{\pi}$, value of S at $x = 2\pi$ is $e^{2\pi}$

$\therefore S$ is minimum at $x = \pi$.

18. (c)

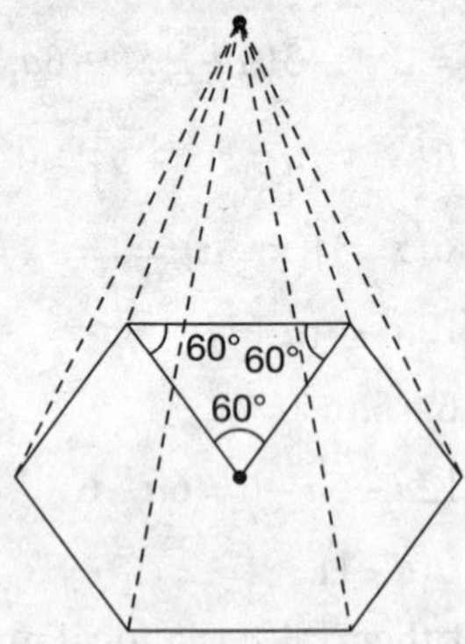

Volume $= \frac{1}{3}$ (Area of base × Height)

$$= \frac{1}{3}\left(\frac{\sqrt{3}}{4}a^2\right) \times 6 \times h$$

$$\Rightarrow a = \sqrt{1-h}^{\,2}$$

$$\Rightarrow V = \frac{\sqrt{3}}{2}(h)\left(1-h^2\right)$$

$$V' = \frac{\sqrt{3}}{2}\left[1-h^2+h(-2h)\right] = 0$$

$$h = \frac{1}{\sqrt{3}}$$

$$V'' < 0 \text{ at } h = \frac{1}{\sqrt{3}}$$

$\Rightarrow h = \frac{1}{\sqrt{3}}$ is local maxima.

(What is the global maxima, think?)

19. (c)

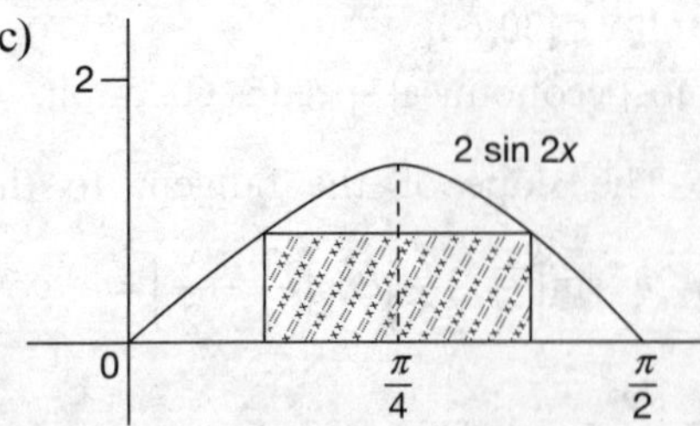

Let perimeter of rectangle be $P(x)$ then

$$P(x) = 2\left(\frac{\pi}{2} - 2x\right) + 4\sin 2x = \pi - 4x + 4\sin 2x$$

$$\text{Now, } \frac{dP}{dx} = 0 - 4 + 8\cos 2x$$

$$\frac{dP}{dx} = 0 \Rightarrow x = \frac{\pi}{6}$$

$$\frac{d^2P}{dx^2} = -16\sin 2x$$

$$\left(\frac{d^2P}{dx^2}\right)_{\left(x=\frac{\pi}{6}\right)} = -16\sin\frac{\pi}{3} < 0$$

$\therefore x = \frac{\pi}{6}$ is point of local maxima

$\therefore$ Maximum length of one side of rectangle

$$= 2\sin 2\frac{\pi}{3} = \sqrt{3}$$

Length of the other side $= \frac{\pi}{6}$

Hence, required area of rectangle

$$= \frac{\pi}{6} \times \sqrt{3} = \frac{\pi}{2\sqrt{3}}.$$

20. (d) $f''(x) > 0 \,\forall\, x \in R, f\left(\frac{1}{2}\right) = \frac{1}{2}, f(1) = 1$

$\therefore f'(x)$ is an increasing function on R.

By Lagrange's Mean Value theorem,

$$f'(\alpha) = \frac{f(1) - f\left(\frac{1}{2}\right)}{1 - \frac{1}{2}}, \text{for some } \alpha \in \left(\frac{1}{2}, 1\right)$$

$$\Rightarrow f'(\alpha) = 1 \text{ for some } \alpha \in \left(\frac{1}{2}, 1\right)$$

$$\therefore f'(1) > 1$$

21. (c) Let $f(x) = 4\alpha x^2 + \frac{1}{x} \Rightarrow f'(x) = 8\alpha x - \frac{1}{x^2}$

Put $f'(x) = 0 \Rightarrow x = \frac{1}{2\alpha^{1/3}}$

For $x > 0$

$$f_{\min} = 1 \Rightarrow 4\alpha\left(\frac{1}{2\alpha^{1/3}}\right)^2 + 2\alpha^{1/3} = 1$$

$$\Rightarrow 3\alpha^{\frac{1}{3}} = 1 \text{ or } \alpha = \frac{1}{27}.$$

One or More Than One Option Correct

1. (a, d)

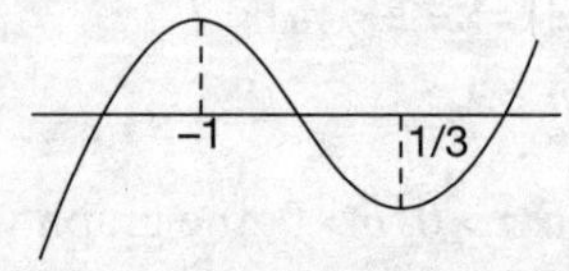

$$f'(x) = 3x^2 + 2x - 1 = (x+1)(3x-1)$$

$$\Rightarrow f(-1) > 0 \text{ and } f\left(\frac{1}{3}\right) < 0$$

$$\Rightarrow -1 < \lambda < \frac{5}{27}$$

2. (a, c, d)

$$\sum\left(\frac{\alpha}{\beta}+\frac{\beta}{\alpha}\right)=\frac{\alpha}{\beta}+\frac{\beta}{\alpha}+\frac{\beta}{\gamma}+\frac{\gamma}{\beta}+\frac{\gamma}{\alpha}+\frac{\alpha}{\gamma}=\frac{1}{\alpha}+\frac{1}{\beta}+\frac{1}{\gamma}-3$$

$$=\frac{\alpha\beta+\beta\gamma+\gamma\alpha}{\alpha\beta\gamma}-3=\frac{a^4+4a^2+1}{a^2}-3$$

$$=a^2+\frac{1}{a^2}+1 \Rightarrow \text{minimum value is 3}$$

3. (a, b)

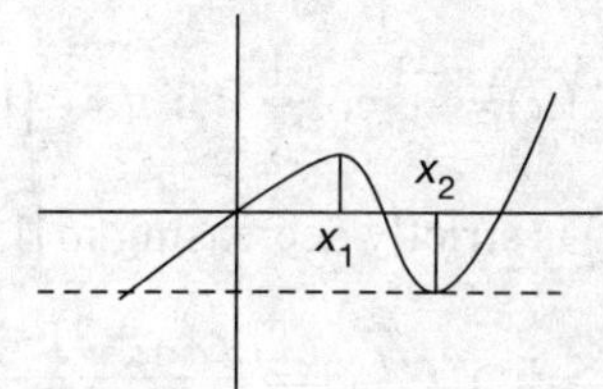

$f'(x)=0$ for $x_1=1-\frac{1}{\sqrt{3}}$ and $x_2=1+\frac{1}{\sqrt{3}} \Rightarrow \lambda$ should be $f\left(1+\frac{1}{\sqrt{3}}\right)$

4. (a, c)

$$xy=(a+x)^2$$

$$y+xy'=2(a+x)$$

$$y'=\pm 1$$

$$\therefore y\pm x=2(a+x)$$

$$\frac{(a+x)^2}{x}\pm x=2(a+x)$$

$$\Rightarrow \pm x=2(a+x)-\frac{(a+x)^2}{x}\pm x^2=(2+x)[x-a]$$

$$\Rightarrow \pm x^2=x^2-a^2\ 2x^2=a^2$$

$$\Rightarrow x=\pm\frac{a}{\sqrt{2}}$$

5. (a, b, c)

$$f(x)=\frac{\ln x}{x} \quad \text{...(i)}$$

$\because$ Domain is R^+

$$\because f'(x)=\frac{x.\frac{1}{x}-\ln x}{x^2}=\frac{1-\ln x}{x^2}$$

(A) for horizontal tangent $f'(x)=0$

$\Rightarrow \ln x=1 \Rightarrow x=e$ True

(B) If (i) cuts the x-axis

$\Rightarrow \frac{\ln x}{x}=0 \Rightarrow x=1$ True

(C) $f'(x)$ is +ve if $x\in(0, e)$ and $f'(x)$ is −ve if $x\in(e,\infty)$

$\therefore f(x)$ is not monotonic.

$\therefore f(x)$ is many-one. True

(D) For vertical tangent $f'(x)=\infty$

$$\Rightarrow \frac{1-\ln x}{x^2}=\infty \Rightarrow \frac{x^2}{1-\ln x}=0$$

$\Rightarrow x=0$, which is not in the domain of $f(x)$ False

6. (c, d)

Note: The curve is not passing through origin.

Let (α,β) be the point of contact.

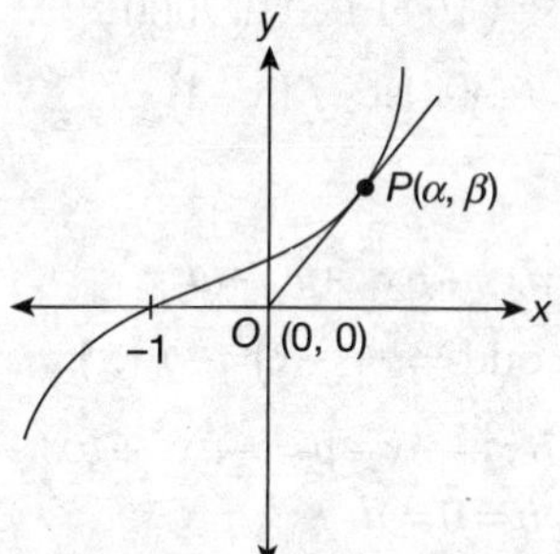

$$\left(\frac{dy}{dx}\right)_P=\frac{\beta-0}{\alpha-0} \quad \text{...(i)}$$

$$\Rightarrow 3(\alpha+1)^2=\frac{\beta}{\alpha}$$

Also (α,β) lies on the curve.

$$\therefore 3(\alpha+1)^2=\frac{(\alpha+1)^3}{\alpha} \Rightarrow (\alpha+1)^2\left\{3-\frac{\alpha+1}{\alpha}\right\}=0$$

$$\alpha=-1 \text{ or } \alpha=\frac{1}{2}$$

$\therefore$ Equation of tangent is

$$y=0 \text{ or } y=\frac{27}{4}x$$

7. (a, b)

$$\frac{dx}{dt}=\frac{2}{\cot t}\left(-\operatorname{cosec}^2 t\right)=\frac{-2}{\sin t\cos t}$$

$$\frac{dy}{dt}=\sec^2 t-\operatorname{cosec}^2 t=\frac{\sin^2 t-\cos^2 t}{\sin^2 t\cos^2 t}$$

$$\frac{dx}{dt} at\left(t=\frac{\pi}{4}\right)=\frac{-2}{1/2}=-4$$

$$\frac{dy}{dt} at\left(t=\frac{\pi}{4}\right)=2-2=0$$

Here, $\frac{dy}{dx}=0 \Rightarrow -\frac{dx}{dy}=\infty$

8. (b, c)

$\because g(x)$ is increasing and $f(x)$ is decreasing

$\Rightarrow g(x+1)>g(x-1)$ and $f(x+1)<f(x-1)$

$\Rightarrow f\{g(x+1)\}<f\{g(x-1)\}$ and

$g\{f(x+1)\}<g\{f(x-1)\}$

9. (b, c)

$f(x)=x^3-x^2+100x+1001$

$f'(x)=3x^2-2x+100>0 \quad \forall x\in R$

$\therefore f(x)$ is increasing (strictly)

$\therefore \quad f\left(\frac{1}{1999}\right)>f\left(\frac{1}{2000}\right)$

$f(x+1)>f(x-1)$

10. (b, c)

$f(x)=ax^3+bx^2+cx+d$

$f(x)$ is odd $\Rightarrow f(-x)=-f(x)$

$-ax^3-bx^2-cx-d=-ax^3+bx^2-cx+d.$

It gives $b=0=d$

$f(x)=ax^3+cx=x(ax^2+c)$

$f'(x)=3ax^2+c=0$ only when $x^2=-\frac{c}{3a}$ is positive.

$\Rightarrow c$ and a are of different sign

Let $-\frac{c}{a}=k$, so non-zero root of $f(x)$ is $\pm\sqrt{k}$.

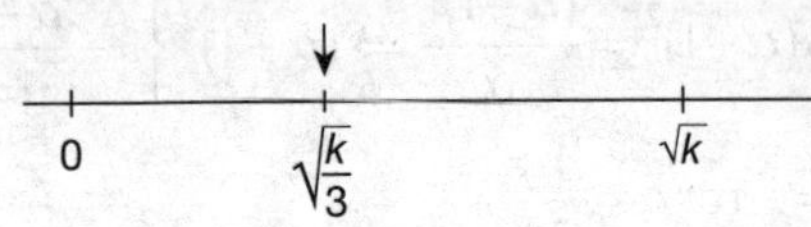

Also $\pm\sqrt{\frac{k}{3}}$ is closer to origin than "$\pm\sqrt{k}$"

11. (a, c, d)

$f(x)=(x-1)^4(x-2)^n, n\in N \quad ...(i)$

$\therefore f'(x)=4(x-1)^3(x-2)^n+(x-1)^4 n(x-2)^{n-1}$

$=(x-1)^3(x-2)^{n-1}(4x-8+nx-n)$

$=(x-1)^3(x-2)^{n-1}[(n+4)x-(n+8)]$

If n is odd, then $f'(x)>0$ if $x<1$ and sufficiently close to 1
and $f'(x)<0$ if $x>1$ and sufficiently close to 1.
$\therefore x=1$ is point of local maximum.
Similarly, if n is even, then $x=1$ is a point of local minimum.
Further if n is even, then $f'(x)<0$ for $x<2$ and sufficiently close 2 and $f'(x)>0$ for $x>2$ and sufficiently close to 2.
$\therefore x=2$ is a point of local minimum.

12. (b, c, d)

Given: $f(x)=x\cos\frac{1}{x}, x\ge 1$

$\therefore f'(x)=\cos\frac{1}{x}+\frac{1}{x}\sin\frac{1}{x}$

$\lim\limits_{x\to\infty} f'(x)=1$

Also, $f''(x)=\frac{-1}{x^3}\cos\frac{1}{x}<0, \forall x\in[1,\infty)$

$\Rightarrow f'(x)$ is strictly decreasing in $[1,\infty)$.

$\therefore f'(x)>\lim\limits_{x\to\infty} f'(x)\Rightarrow \frac{f(x+2)-f(x)}{(x+2)-x}>1$

$\Rightarrow f(x+2)-f(x)>2.$

13. (a, c)

$h(x)=f(x)-(f(x))^2+(f(x))^3 \ \forall x\in \mathrm{R}$

$h'(x)=f'(x)\left[1-2f(x)+3(f(x))^2\right]$

$=3f'(x)\left[(f(x))^2-\frac{2}{3}f(x)+\frac{1}{3}\right]$

$=3f'(x)\left[\{f(x)-1/3\}^2+2/9\right].$

Here, $h'(x)<0$ whenever $f'(x)<0$ and $h'(x)>0$ whenever $f'(x)>0$.

Hence, $h(x)$ increases (decreases) whenever $f(x)$ increases (decreases).

14. (a, b, c)

$f(x)=\begin{cases}3x^2+12x-1, & -1\le x\le 2\\ 37-x & 2<x\le 3\end{cases}$

Hence, on $[-1,2]$, $f'(x)=6x+12$

For $-1\le x\le 2, -6\le 6x\le 12$

$\Rightarrow 6\le 6x+12\le 24\Rightarrow f'(x)>0, \forall x\in[-1,2]$

$\therefore f$ is increasing on $[-1,2]$.

Also, $f(x)$ being polynomial for $x\in[-1,2)\cup(2,3]$

$f(x)$ is continuous on, $[-1,3]$ except possibly at $x=2$.

At $x=2$

L.H.L

$=\lim\limits_{h\to 0} f(2-h)=\lim\limits_{h\to 0} 3(2-h)^2+12(2-h)-1=35$

R.H.L $= \lim_{h\to 0} f(2+h) = \lim_{h\to 0} 37-(2+h) = 35$

and $f(2) = 3.2^2 + 12.2 - 1 = 35.$

Thus, L.H.L = R.H.L $= f(2) \Rightarrow f(x)$ is continuous at $x = 2$

Hence, $f(x)$ is continuous on $[-1,3]$

Again at $x = 2$

$$\text{RD} = \lim_{h\to 0} \frac{f(2+h)-f(2)}{h}$$

$$= \lim_{h\to 0} \frac{37-(2+h)-35}{h} = 1$$

$$\text{LD} = \lim_{h\to 0} \frac{f(2)-f(2-h)}{h}$$

$$= \lim_{h\to 0} \frac{35-3(2-h)^2-12(2-h)+1}{h} = 24$$

Thus LD $\neq$ RD $\Rightarrow f'(2)$ does not exist.

Hence, $f(x)$ cannot have maximum value at $x = 2$.

15. (a) Since g is decreasing in $[0,\infty)$

$\therefore$ For $x \geq y$, $g(x) \leq g(y)$

Also, $g(x)$, $g(y) \in [0,\infty)$ and f is increasing from $[0,\infty)$ to $[0,\infty)$.

Now $g(x), g(y) \in [0,\infty)$ and $g(x) \leq g(y)$

$\Rightarrow f(g(x)) \leq f(g(y))$, where $x \geq y$

$\therefore h(x) \leq h(y)$

$\Rightarrow h$ is decreasing function from $[0,\infty)$ to $[0,\infty)$

$\therefore h(x) \leq h(0)$, $\forall\ x \geq 0$ but $h(0) = 0$ (given)

$\therefore h(x) \leq 0 \forall x \geq 0$ ...(i)

Also $h(x) \geq 0 \forall x \geq$ [as $h(x) \in [0,\infty)$] ...(ii)

From (i) and (ii) we get, $h(x) = 0, \forall x \geq 0$

Hence, $h(x) - h(1) = 0 - 0 = 0 \forall x \geq 0$.

16. (a, b) Given that $f(x) = \dfrac{x^2-3x-6}{x^2+2x+4}$

$$f'(x) = \frac{(x^2+2x+4)(2x-3)-(x^2-3x-6)(2x+2)}{(x^2+2x+4)^2}$$

$$f'(x) = \frac{5x(x+4)}{(x^2+2x+4)^2}$$

$f'(x)$: + on $(-\infty, -4)$, − on $(-4, 0)$, + on $(0, \infty)$

$\therefore$ Options (a) and (b) are true.

$f(-4) = \dfrac{11}{6}, f(0) = -\dfrac{3}{2}$,

Also, $f(x)$ is maximum at -4 and minimum at 0.

Range: $\left[-\dfrac{3}{2}, \dfrac{11}{6}\right]$, clearly $f(x)$ is into.

17. (a, b, c)

$f(x) = \dfrac{\sin \pi x}{x^2}, x > 0$

For points of local max/min, put $f'(x) = 0$

$$\Rightarrow \frac{\pi x^2 \cos \pi x - 2x \sin \pi x}{x^4} = 0$$

$$\Rightarrow \frac{\cos \pi x(\pi x - 2\tan \pi x)}{x^3} = 0$$

$$\Rightarrow \cos \pi x = 0 \Rightarrow x = \frac{1}{2}, \frac{3}{2}, \frac{5}{2}, \frac{7}{2}, \ldots$$

and $\pi x - 2\tan \pi x = 0$, which can be solved by drawing the graphs of $y = \pi x$ and $y = 2\tan \pi x$, as follows:

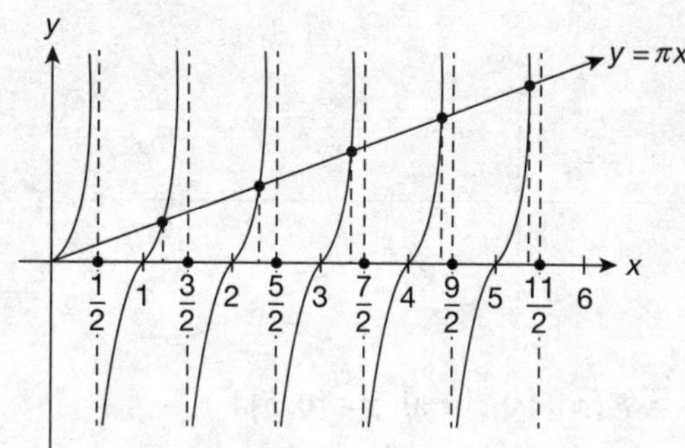

Plotting the stationary points on number line and finishing the sign of $f'(x)$ in different intervals, we observe

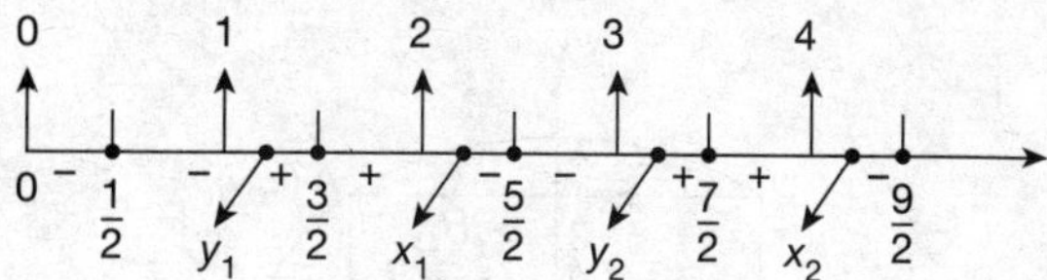

i.e., $x_{n+1} - x_n > 2$ for every n

$x_n \in \left(2x, 2n + \dfrac{1}{2}\right)$ for every n

$|x_n - y_n| > 1$ for every n

$x_1 > y_1$

Hence, (a, b, c) are correct, but (d) is incorrect.

18. (a, b, d)

$F(x) = (x-1)(x-2)(x-5)$.

$F(x)=\int_0^x f(t)dt,\ x>0$

$\Rightarrow F'(x)=f(x)=(x-1)(x-2)(x-5)$

Put $F'(x)=0 \Rightarrow x=1,2,5$

$F\ \ (x)=(x-2)(x-5)+(x-1)(x-5)+(x-1)(x-2)$

$F(1)=+\text{ve},\ F''(2)=-\text{ve},\ F(5)=+\text{ve}$

$\therefore F(x)$ has local minima at $x=1$ and $x=5$ and local maxima at $x=2$.

Also $F(x)=\int_0^x (t-1)(t-2)(t-5)dt$

$$\int_0^x (t^3-8t^2+17t-10)dt=\frac{x^4}{4}-\frac{8x^3}{3}+\frac{17x^2}{2}-10x$$

$$=\frac{x}{12}(3x^3-32x^2+102x-120)$$

$F(1), F(2), F(5)<0$

Hence, approximate graph of $F(x)$ for $x\in(0,5)$ is

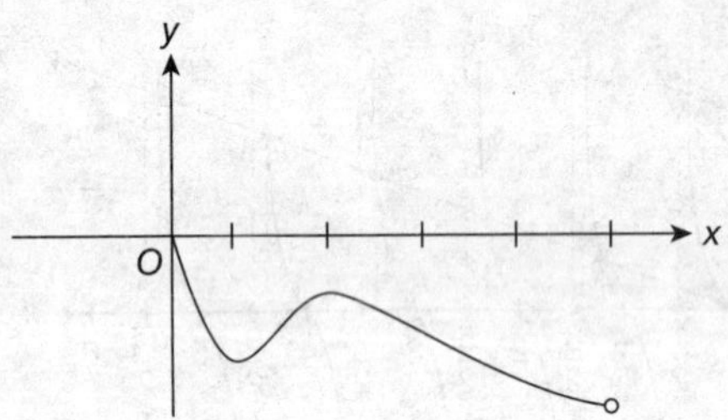

$\therefore F(x)<0$, for all $x\in(0,5)$.

19. (b, c)

$E_1:\dfrac{x^2}{9}+\dfrac{y^2}{4}=1$

R_1 : Rectangle ABCD with largest area.

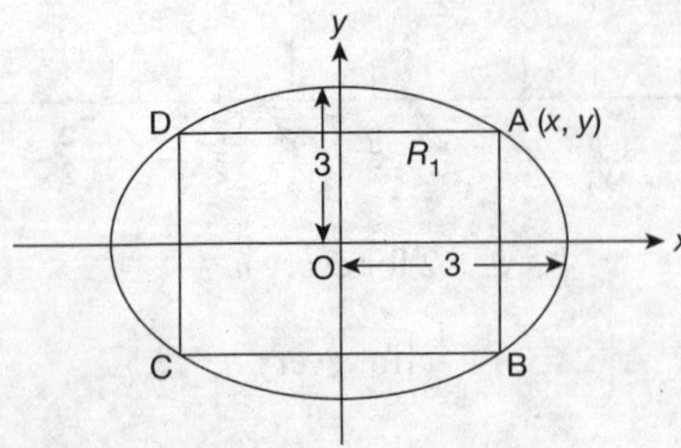

Area of $R_1=A=2x\times 2y$

$$\Rightarrow A=4x\times\frac{2}{3}\sqrt{9-x^2}=\frac{8}{3}x\sqrt{9-x^2}$$

$$\therefore \frac{dA}{dx}=\frac{8}{3}\left[\sqrt{9-x^2}-\frac{x^2}{\sqrt{9-x^2}}\right]$$

$$\Rightarrow \frac{8}{3}\left[\frac{9-2x^2}{\sqrt{9-x^2}}\right]=0\Rightarrow x=\frac{3}{\sqrt{2}},\ y=\frac{2}{3}\sqrt{9-\frac{9}{2}}=\frac{2}{\sqrt{2}}$$

$\Rightarrow$ For $E_2: a=\dfrac{3}{\sqrt{2}}, b=\dfrac{2}{\sqrt{2}}$

Similarly, for $E_3: a=\dfrac{3}{(\sqrt{2})^2}, b=\dfrac{2}{(\sqrt{2})^2}$ and so on.

Now eccentricity depends on $\dfrac{b}{a}$, which is same for all E_n. Therefore, eccentricity for all the E_n's will remain

$$\sqrt{1-\frac{4}{9}}=\frac{\sqrt{5}}{3}$$

(a) is correct

For $E_9; a=\dfrac{3}{(\sqrt{2})^8}, b=\dfrac{2}{(\sqrt{2})^8}$

$$\Rightarrow \text{Length of latus rectum}=\frac{2b^2}{a}=\frac{2\times\frac{4}{256}}{\frac{3}{16}}=\frac{1}{6}$$

Hence (b) is correct.

$$\text{Area of } R_1=4\times\frac{3}{\sqrt{2}}\times\frac{2}{\sqrt{2}}=\frac{24}{2}$$

$$\text{Area of } R_2=4\times\frac{3}{(\sqrt{2})^2}\times\frac{2}{(\sqrt{2})^2}=\frac{24}{2^2}$$

$$\text{Area of } R_3=4\times\frac{3}{(\sqrt{2})^3}\times\frac{2}{(\sqrt{2})^3}=\frac{2}{(2)^3} \text{ and so on}$$

$$\sum_{n=1}^{\infty}\text{area of } R_n=\frac{24}{2}+\frac{24}{2^2}+\frac{24}{2^3}+\ldots=\frac{12}{1-\frac{1}{2}}=24$$

$$\Rightarrow \sum_{n=1}^{N}(\text{area of } R_n)<24 \text{ for each positive integer N.}$$

Hence,
(c) is correct

For $E_9: a=\dfrac{3}{(\sqrt{2})^8}, b=\dfrac{2}{(\sqrt{2})^8}, e=\dfrac{\sqrt{5}}{3}$

Thus, focus $=(ae,0)=\left(\dfrac{\sqrt{5}}{16},0\right)$

Hence, distance of focus from centre $=\dfrac{\sqrt{5}}{16}$

(d) is incorrect.

20. (b, c)

$$f(x)=\begin{vmatrix}\cos 2x & \cos 2x & \sin 2x\\ -\cos x & \cos x & -\sin x\\ \sin x & \sin x & \cos x\end{vmatrix}$$

$$\Rightarrow f(x)=\begin{vmatrix}0 & \cos 2x & \sin 2x\\ -2\cos x & \cos x & -\sin x\\ 0 & \sin x & \cos x\end{vmatrix}$$

$[C_1 \to C_1 - C_2]$

$\Rightarrow f(x)=2\cos 3x\cos x \Rightarrow f(x)=\cos 4x+\cos 2x$

$\therefore f_{\max}=2$ at $x=0$

$f'(x)=-4\sin 4x-2\sin 2x=-2\sin 2x[4\cos 2x+1]$

$f'(x)=0 \Rightarrow \sin 2x=0$ or $\cos 2x=\dfrac{-1}{4}$

$\Rightarrow x=-\dfrac{\pi}{2},0,\dfrac{\pi}{2}$, which is true for some $x\in(-\pi,\pi)$

$\therefore f'(x)=0$ at more than three points in $(-\pi,\pi)$.

21. (a, c)

$f'(x)-2f(x)>0$

$\Rightarrow e^{-2x}\dfrac{df(x)}{dx}-2e^{-2x}f(x)>0$

$\Rightarrow \dfrac{d}{dx}\left(e^{-2x}f(x)\right)>0$

$\Rightarrow e^{-2x}f(x)$ is an increasing function

$\therefore$ For $x>0$, $f(x)>f(0)$

$\Rightarrow e^{-2x}f(x)>1 \Rightarrow f(x)>e^{2x}$ in $(0,\infty)$

Also, $f'(x)>2e^{2x}>0$

$\therefore f$ is an increasing function in $(0,\infty)$

22. (a, d)

$\lim\limits_{x\to 2}\dfrac{f(x)g(x)}{f'(x)g'(x)}=1\left[\dfrac{0}{0}\text{form}\right]$

$\Rightarrow \lim\limits_{x\to 2}\dfrac{f'(x)g(x)+f(x)g'(x)}{f''(x)g'(x)+f'(x)g''(x)}=1[\text{LH Rule}]$

$\Rightarrow \dfrac{g'(2)f(2)}{f''(2)g'(2)}=1 \Rightarrow f(2)=f''(2)$

$\therefore f(x)-f''(x)=0$ for atleast one $x\in R$

$\therefore$ Range of $f(x)$ is $(0,\infty)$

$\because f(x)>0, \forall x\in R$

$\Rightarrow f(2)>0 \Rightarrow f''(2)>0$

$\Rightarrow f$ has a local minimum at $x=2$.

23. (a, c, d)

$$f(x)=\int_{1/x}^{x} e^{-\left(t+\frac{1}{t}\right)}\frac{dt}{t}$$

$$\therefore f'(x)=\frac{e^{-\left(x+\frac{1}{x}\right)}}{x}+\frac{x}{x^2}e^{-\left(\frac{1}{x}+x\right)}=\frac{2}{x}e^{-\left(x+\frac{1}{x}\right)}$$

For $x\in[1,\infty), f'(x)>0$

$\therefore f$ is monotonically increasing on $[1,\infty)$, (a) is correct.

For $x\in(0,1), f'(x)>0$

Hence, f is monotonically increasing on (0, 1).

$\therefore$ (b) is not correct.

$$f(x)+f\left(\frac{1}{x}\right)=\int_{1/x}^{x} e^{-\left(t+\frac{1}{t}\right)}\frac{dt}{t}+\int_{x}^{1/x} e^{-\left(t+\frac{1}{t}\right)}\frac{dt}{t}=0$$

$\therefore$ (c) is correct

Replacing x by 2^x in $f(x)+f\left(\dfrac{1}{x}\right)=0$,

we get $f(2^x)+f(2^{-x})=0$ or $f(2^x)=-f(2^{-x})$

Hence, $f(2^x)$ is an odd function.

$\therefore$ (d) is correct.

24. (a, c)

Let length and breadth of rectangular sheet be $8x$, and $15x$ respectively. Also let y be the length of square cut off from each corner.

$\therefore$ Volume of box $=(8x-2y)(15x-2y)y$

$\Rightarrow V=120x^2y-46xy^2+4y^3$

$\dfrac{dV}{dy}=120x^2-92xy+12y^2$

Now, $4y^2=100 \Rightarrow y=5$

Put $\dfrac{dV}{dy}=0$ for maximum value of V.

$\Rightarrow 30x^2-115x+75=0$

$\Rightarrow x=3,\dfrac{5}{6}$

For $x=3$, sides are 45 and 24.

Matrix Type

1. A → S; B → R; C → Q; D → Q

All matchings easily follow by solving $f'(x)>0$ in each case.

(In case (iii), $x>0$ should also be considered.)

2. A → R; B → S; C → P; D → Q

(B → S) It is very evident that every normal to $x^2+y^2=a^2$ passes through (0, 0)

$\Rightarrow$ OM = 0 for all normal.

(A → R) Any tangent to the curve in (A)

$$\frac{x}{a\cos^2 t}+\frac{y}{a\sin^2 t}=1$$

$\Rightarrow$ Sum of intercepts is constant.

(D → Q) Any tangent to curve in (D)

$$\frac{x}{a\cos t}+\frac{y}{a\sin t}=1$$

$\Rightarrow \sqrt{a^2\cos^2 t+a^2\sin^2 t}=1$

$\Rightarrow$ QR is constant.

We must have (C → P).

3. A → Q; B → R; C → P

(A → Q) $y=\sin x$ and $y=\cos x$ intersect at $\left(\frac{\pi}{4},\frac{1}{\sqrt{2}}\right)$ and have slopes $+\frac{1}{\sqrt{2}}$ and $-\frac{1}{\sqrt{2}}$ for their tangents at this point of intersection.

$$\Rightarrow \tan\theta=\frac{m_1-m_2}{1+m_1m_2}=2\sqrt{2}.$$

B → R) The two curves intersect at $(2a,a)$.

We easily get

$$\left(\frac{dy}{dx}\right)_1=\frac{2x}{4a}\text{and}\left(\frac{dy}{dx}\right)_2=\frac{-16a^3}{(x^2 4a^2)^2}\text{and }\tan\theta\text{ as 3.}$$

(C) The left curve must match with (P) (Or else we can proceed as above.)

4. A → Q; B → R; C → P, S

(A → Q) $f'(x)=0\Rightarrow \sin x+\cos x=0$

$$\Rightarrow \tan x=-1\Rightarrow x=\frac{3\pi}{4}$$

(B → R) $f'(x)=0\Rightarrow \sin x=0$

$$\Rightarrow x=\pi\ (\because e^x\neq 0)$$

(C → P, S) $f'(x)=0\Rightarrow \cos 2x=-\frac{1}{2}$

$$\Rightarrow 2x=\frac{2\pi}{3},\frac{4\pi}{3}\Rightarrow x=\frac{\pi}{3},\frac{2\pi}{3}.$$

5. A → P, Q; B → P, Q; C → P, Q; D → P, Q

(A → P, Q) $f(x)=\log(1+x)-\frac{x}{1+x}$

$$f(0)=0,\ f'(x)=\frac{x}{(1+x)^2}>0\text{ for }x>0$$

$\Rightarrow f(x)$ is MI over $(0,\infty)$

$\Rightarrow f(x)>f(0)$

$$\Rightarrow \log(1+x)-\frac{x}{1+x}>0$$

Again, $f'(x)<0$ for $x\in(-1,0)$

$\Rightarrow f'(x)$ is MD over $(-1,0)$

$\Rightarrow f(x)>f(0)$ on $(-1,0)$

$\Rightarrow f(x)>0$ for $x\in(-1,0)$

The identical proofs are therefore (B) and (C).

$\Rightarrow$ (A → P, Q)

(A → P, Q)

For (D), we consider RHS = LHS

$$=\frac{x\log(1+x)+\log(1+x)-x}{x\log(1+x)}$$

If $x>0$, denominator is positive.

If $x\in(-1,0)$ then also denominator is positive.

Thus, it is sufficient to consider

$f(x)=x\log(1+x)+\log(1+x)-x$

$f(0)=0$

$f'(x)=\log(1+x)$

$\Rightarrow f'(x)>0$ for $(0,\infty)$ and <0 for $(-1,0)$

$\Rightarrow f(x)$ is MI in $(0,\infty)$ and MD in $(-1,0)$

$\Rightarrow f(x)>f(0)$ for $x\in(-1,\infty)$

Thus, even in the last case (a) and (b) both are matching.

6. A → R; B → Q; C → T; D → S

(A) $\frac{dy}{dx}=\frac{4t}{3}$

Tangent is $y-at^4=\frac{4t}{3}(x-at^3)$

x-intercept $=\frac{at^3}{4}$ y-intercept $=-\frac{at^4}{3}$

If P divides AB in ratio $\lambda:1$

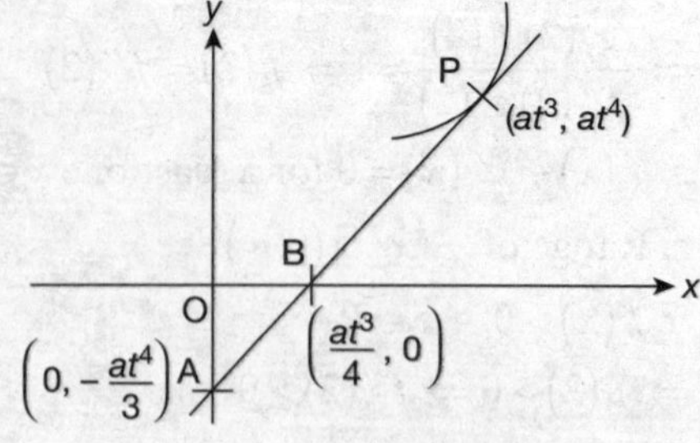

$$\Rightarrow at^3 = \frac{\lambda.0 + \frac{at^3}{4}}{\lambda + 1}$$

$$\Rightarrow \lambda = \frac{-3}{4}$$

$$\therefore \frac{m}{n} = \frac{3}{4}$$

$\therefore m = 3, n = 4$

$\therefore m + n = 7$

(B) $\frac{dy}{dx} = e^{\sin y}\cos y$: slope of normal $= -1$

Equation of normal is $x + y = 1$

$\Rightarrow$ Area $= \frac{1}{2}$

(C) $y = \frac{1}{x^2} : \frac{dy}{dx} = -\frac{2}{x^3}$: Slope of tangent $= -2$

$y = e^{2-2x} : \frac{dy}{dx} = e^{2-2x}.(-2)$:

Slope of tangent $= -2 \therefore \tan\theta = 0$

(D) Length of sub-tangent $= \left|\frac{y}{y'}\right| = \left|\frac{be^{x/3}}{b\frac{1}{3}e^{x/3}}\right| = 3$

7. A → P; B → P; C → S; D → S

(A) $f(x) = \ell n(\sin x)$

$$f'(x) = \frac{\cos x}{\sin x} > 0$$

$\therefore$ Required number of values of x is 0

(B) $f'(x) = 3x^2 - 3 \le 0$ if $-1 \le x \le 1$

$\therefore a = -1, b = 1$

$\therefore a + b = 0$

(C) $f(x) = \begin{cases} x^2 + 2 & , \; 1 \le x < 2 \\ \frac{x^2+2}{2} & , \; 2 \le x < 3 \\ \frac{x^2+2}{3} & , \; x = 3 \end{cases}$

$f'(x) = \begin{cases} 2x & , \; 1 < x < 2 \\ x & , \; 2 < x < 3 \end{cases}$

$\therefore$ Least value of $f(x)$ in [1, 2) is 3.

Least value of $f(x)$ in [2, 3) is 3.

$$f(3) = \frac{11}{3}$$

$\therefore$ Least value of $f(x)$ is 3.

(D) $f(x) = e^{2x} - (a+1)e^x + 2x$

$f'(x) = 2e^{2x} - (a+1)e^x + 2$

Now, $2e^{2x} - (a+1)e^x + 2 \ge 0$ for all $x \in R$

i.e., $2\left(e^x + \frac{1}{e^x}\right) - (a+1) \ge 0$ for all $x \in R$

i.e., $4 - (a+1) \ge 0$

i.e., $a \le 3 \therefore \quad a = 3$

8. A → Q; B → R; C → R; D → T

(A) By graph it is clear that at $x = -1$ is local max. and $x = 0$ is local min.

(B) $a + b = 1$

$$\sqrt{\left(1+\frac{1}{a}\right)\left(1+\frac{1}{b}\right)} = \sqrt{1+\frac{1}{a}+\frac{1}{b}+\frac{1}{ab}} = \sqrt{1+\frac{2}{ab}}$$

$$\sqrt{ab} < \frac{a+b}{2} = \frac{1}{2}$$

$$\therefore ab < \frac{1}{4} \Rightarrow \frac{1}{ab} > 4$$

$$\therefore \sqrt{\left(1+\frac{1}{a}\right)\left(1+\frac{1}{b}\right)} \ge \sqrt{1+8} = 3$$

(C) $\therefore y = 10 - (10 - x) = x$

$\therefore$ Maximum value of $y = 3$.

(D) Equation of tangent at P is $ty = x + t^2$.

It intersects the line $x = 0$ at Q

$\therefore$ Co-ordinates of Q are $(0, t)$

$$\therefore \text{area of } \Delta PQS = \frac{1}{2}\begin{vmatrix} 0 & t & 1 \\ 1 & 0 & 1 \\ t^2 & 2t & 1 \end{vmatrix}$$

$$= \frac{1}{2}[-t(1-t^2) + 2t] = \frac{1}{2}(t + t^3)$$

$$\frac{dA}{dt} = \frac{1}{2}(3t^2 + 1) > 0 \; \forall t \in [0, 2]$$

$\therefore$ Area is maximum for $t = 2$

Maximum area $= \frac{1}{2}[2 + 8] = 5$.

9. (b) The only correct combination is (II) (ii) (Q).

10. (b) The only correct combination is (II) (iii) (S).

11. (c) The only incorrect combination is (III) (i) (R).

Integer and Numerical Value Type

1. $f'(x)=x^2+2(m-1)x+(m+5)$

To be increasing for $x>0$ we must have $f'(x)>0$ for all $x>0$.

$\Rightarrow$ Either $f'(x)>0$ for all x, or both roots of $f'(x)=0$ must be negative.

Now $f'(x)>0$ for all x

$4(m-1)^2-4(m+5)<0$

$\Rightarrow m\in(-1,4)$

Again, both roots are negative if $m\in[4,\infty)$

Thus $m\in[-1,\infty)\Rightarrow k=1$.

2. $\dfrac{dy}{dx}=x^3-3x+\lambda$

We must have $\dfrac{dy}{dx}=0$ for three values.

$\Rightarrow$ Equation $x^3-3x+\lambda=0$ has three real roots.

Let $g(x)=x^3-3x+\lambda$

Then $g'(x)=0$ for $x=1,-1$.

For three roots of $g(x)>0$

$g(1)g(-1)<0$

$\Rightarrow -2<\lambda<2\Rightarrow k=2$.

3. Define $g(x)=\int p(x)dx+c$

$=\dfrac{x^{102}}{2}-23x^{101}-\dfrac{45}{2}x^2+1035x+c$

Since $g\left(45^{1/100}\right)=c$

We must guess a value of x for which $g(x)=c$ (For the applicability of Rolle's Theorem)

We note that $g(46)=c$

$\Rightarrow g'(x)=p(x)=0$ will have atleast one root in $\left(45^{1/100},46\right)$

$\Rightarrow k=46$.

4. $\dfrac{dx}{d\theta}=\sec\theta\tan\theta+\sin\theta$

$=\tan\theta(\sec\theta+\cos\theta)$

$\dfrac{dy}{d\theta}=n\sec^{n-1}\theta\sec\theta\tan\theta-n\cos^{n-1}\theta(-\sin\theta)$

$\Rightarrow\dfrac{dy}{dx}=\dfrac{n\left(\sec^n\theta+\cos^n\theta\right)}{\sec\theta+\cos\theta}$

$=n\sqrt{\dfrac{(\sec^n\theta-\cos^n\theta)^2+4}{(\sec\theta-\cos\theta)^2+4}}$

$=n\sqrt{\dfrac{y^2+4}{x^2+4}}\Rightarrow k=4$.

5. The domain of the given function is $x>-1$ ($\because x^2-x+1>0$ for all x and $\tan^{-1}x$ is defined for all x.)

We can easily find $\dfrac{dy}{dx}$ which is given by

$\dfrac{dy}{dx}=\dfrac{1}{x^3+1}=\dfrac{1}{(x+1)\left(x^2-x+1\right)}$

$\Rightarrow\dfrac{dy}{dx}>0$ in its domain

$\Rightarrow k=1$

6. We can get

$f'(x)=220(2x+1)^{4x}(3x-4)^{50}(2x+1)(3x-4)(3x$

$\Rightarrow f'(x)\geq 0$ in $\left(-\infty,-\dfrac{1}{2}\right)\cup\left(\dfrac{4}{3},x\right)$

$\Rightarrow$ Least positive integer such that $f'(x)\geq 0$ in $[k,\infty]$ is 2.

$\Rightarrow k=2$.

7. The two curves intersect only at one point $(2a,a)$.

$\left(\dfrac{dy}{dx}\right)_{(2a,a)}=1$ (For first curve)

$\left(\dfrac{dy}{dx}\right)_{(2a,a)}=-\dfrac{1}{2}$ (For second curve)

$\Rightarrow\tan\theta=\left|\dfrac{m_1-m_2}{1+m_1m_2}\right|=3$

$\Rightarrow k=3$

8. $f'(x)=\dfrac{-2k-40}{(8\sin x-\cos x)^2}$

$\Rightarrow$ In its domain $f(x)$ is increasing if

$-2k-40>0$

$\Rightarrow k<-20$

$\Rightarrow -l=20$.

9. For the second curve $\dfrac{dy}{dx}=-\dfrac{4}{3}e^{-x/3}$

$\Rightarrow$ Equation of tangent at (x_1,y_1) is

$y-y_1=-\dfrac{4}{3}e^{-x_1/3}(x-x_1)$

$\left(\because y_1-4e^{-x_1/3}\right)$

$\Rightarrow 3y - 3y_1 = y_1(x - x_1)$

$\Rightarrow xy_1 + 3y = 3y_1 + x_1y_1$

This should be the same as

$\frac{x}{3} + \frac{y}{4} = 1$ or $4x + 3y = 12.$

On comparing we get $\frac{y_1}{4} = \frac{3}{3} = \frac{3y_1 + x_1y_1}{12}$

We easily get $y_1 = 4$, $x_1 = 0$

$\Rightarrow$ Line touches curve at (0, 4)

$\Rightarrow l = 4$

10. On differentiating the first curve,

$$3x^2 + 3y^2\frac{dy}{dx} = 3a\left[1y + x\frac{dy}{dx}\right]$$

$\Rightarrow$ Slope of the normal $= -\dfrac{1}{\frac{dy}{dx}}$

$$= -\frac{y^2 - ax}{ay - x^2}.$$

On solving the two curves, we get (from second curve $x = y^2/a$)

$$\frac{y^6}{a^3} + y^3 = 3ay.\frac{y^2}{a}.$$

If $y = 0$ then $x = 0$ but $\frac{dy}{dx}$ is not defined at $(0,0)$ therefore $y \neq 0$.

If $y \neq 0$ then on cancelling y^3 we get $y^3 = 2a^3$

$\Rightarrow y = a\sqrt[3]{2}$ whence $x = a.2^{2/3}$.

Slope of normal $= \frac{0}{\text{non zero}} = 0$

$\Rightarrow k = 0.$

11. If the function is monotonic then either $f'(x) > 0$ for all x or $f'(x) < 0$ for all x.

$f'(x) = 3\left[(m-2)x^2 - 8x + (m+4)\right]$

Now $f'(x) > 0$ for all x if

$m - 2 > 0$ and $64 - 4(m-2)(m+4) < 0$

$\Rightarrow m > 2$ and $m^2 + 2m - 8 > 0$

$\Rightarrow m > 2$ and $(m+4)(m-2) > 0$

$\Rightarrow m > 2$ and $m \in (-\infty, -4) \cup (2, \infty)$

$\Rightarrow m \in (2, \infty).$

Again $f(x) < 0$ for all x, if

$m - 2 < 0$ and $64 - 4(m-2)(m+4) < 0$

We will easily get $m \in (-\infty, -4)$.

Combining the two sets of values, we get $m \in (-\infty, -4) \cup (2, \infty)$

$\Rightarrow k = 4$

12. For the points of intersection, we have

$$\frac{12 - y^2}{36} + \frac{y^2}{4} = 1$$

$\Rightarrow y = \pm\sqrt{3}$ and $x = \pm 3$

Consider the point $P(3, \sqrt{3})$

Equation of the tangent at P to the circle is $3x + \sqrt{3}y = 12$

$\therefore$ Slope of this tangent is $-\sqrt{3}$

Equation of the tangent at P to the ellipse is

$$\frac{x}{12} + \frac{\sqrt{3}}{4}y = 1$$

$\therefore$ Slope of this tangent is $-\frac{1}{3\sqrt{3}}$.

If α is the angle between these tangents, then

$$\tan\alpha = \frac{2}{\sqrt{3}}$$

$\therefore \alpha = \tan^{-1}\frac{2}{\sqrt{3}}$ $\therefore$ $k = 4$ and hence $k^2 = 16$.

13. Given $y = -p^2x^2 + 5px - 4$ ...(1)

$$y = \frac{1}{(1-x)} \quad ...(2)$$

$\because$ Chord touches curve (2) at $x = 2$, which provides $y = -1$.

Let (x_1y_1) and (x_2y_2) be the ends of the chord, touching point is middle point

$$\left(\frac{x_1 + x_2}{2}, \frac{y_1 + y_2}{2}\right) \Rightarrow \frac{x_1 + x_2}{2} = 2; x_1 + x_2 = 4$$

and $y_1 + y_2 = -2$

$\because (x_1, y_1)$ and (x_2, y_2) satisfy the curve.

$\therefore y_1 = -p^2x^2_1 + 5px_1 - 4$ ...(3)

$y_2 = -p^2x^2_2 + 5px_2 - 4$ ...(4)

Subtracting (4) from (3), we get

$y_1 - y_2 = -p^2x^2_1 + 5px_1 - 4 + p^2x^2_2 + 5px_2 - 4$

$y_1 - y_2 = -p^2\left(x^2_1 - x^2_2\right) + 5p(x_1 - x_2)$

$$\frac{y_1 - y_2}{x_1 - x_2} = -p^2(x_1 + x_2) + 5p = -4p^2 + 5p$$

again $\left(\frac{dy}{dx}\right)_{\text{at } x=2} = \frac{1}{(1-x)^2} = \frac{1}{1} = -4p^2 + 5p$

$4p^2 - 5p + 1 = 0 \Rightarrow p = 1, \frac{1}{4}.$

14. $f(\theta) = (\sin\theta + \cos\theta)^2 + (\sin\theta - \cos\theta)^4.$

$= (1 + \sin 2\theta) + (1 - \sin 2\theta)^2$

$f(\theta) = \sin^2 2\theta - \sin 2\theta + 2$

$f'(\theta) = 2(\sin 2\theta).(2\cos 2\theta) - 2\cos 2\theta$

$= 2\cos 2\theta(2\sin 2\theta - 1) = 0$

$\cos 2\theta = 0 \Rightarrow \theta = \frac{\pi}{4}, \frac{3\pi}{4}$

$\sin 2\theta = \frac{1}{2} \Rightarrow \theta = \frac{\pi}{12}, \frac{5\pi}{12}.$

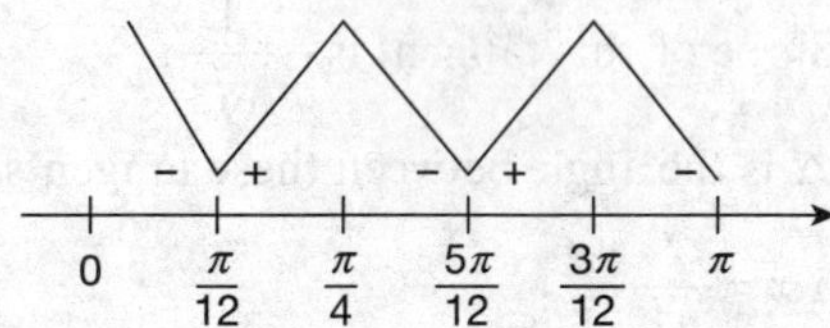

So, $f(\theta)$ is minimum at $\theta = \frac{\pi}{12}, \frac{5\pi}{12}.$

$\therefore \lambda_1 + \lambda_2 = \frac{1}{12} + \frac{5}{12} = \frac{6}{12} = \frac{1}{2}$

15. Let r be the internal radius, R be the external radius and h be the internal height of the cylinder.

Now, $V = \pi r^2 h \Rightarrow h = \frac{V}{\pi r^2}$

Also, volume of material

$= M = \pi\left[(r+2)^2 - r^2\right]h + \pi(r+2)^2 \times 2$

$= 4\pi(r+1).\frac{V}{\pi r^2} + 2\pi(r+2)^2$

$\Rightarrow M = 4V\left[\frac{1}{r} + \frac{1}{r^2}\right] + 2\pi(r+2)^2$

$\Rightarrow \frac{dM}{dr} = 4V\left[\frac{-1}{r^2} - \frac{2}{r^3}\right] + 4\pi(r+2)$

For minimum value of M, put $\frac{dM}{dr} = 0.$

$\Rightarrow \frac{-4V}{r^3}(r+2) + 4\pi(r+2) = 0.$

$\Rightarrow \frac{4V}{r^3} = 4\pi \text{ or } r^3 = \frac{V}{\pi} = 1000 \Rightarrow \frac{V}{250\pi} = 4$

16. Vertical line $x = h$, meets the ellipse $\frac{x^2}{4} + \frac{y^2}{3} = 1$ at

$P\left(h, \frac{\sqrt{3}}{2}\sqrt{4-h^2}\right)$ and $Q\left(h, \frac{-\sqrt{3}}{2}\sqrt{4-h^2}\right)$

By symmetry, tangents at P and Q will meet each other at x-axis.

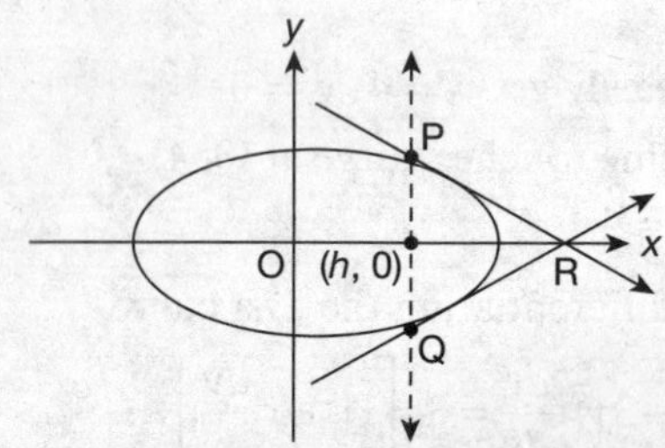

Tangent at P is $\frac{xh}{4} + \frac{y\sqrt{3}}{6}\sqrt{4-h^2} = 1$

which meets $x-$ axis at $R\left(\frac{4}{h}, 0\right)$

Area of $\Delta PQR = \frac{1}{2} \times \sqrt{3}\sqrt{4-h^2} \times \left(\frac{4}{h} - h\right)$

i.e., $\Delta(h) = \frac{\sqrt{3}}{2}\frac{\left(4-h^2\right)^{3/2}}{h}$

$\frac{d\Delta}{dh} = -\sqrt{3}\left[\frac{\sqrt{4-h^2}\left(h^2+2\right)}{h^2}\right] < 0$

$\Rightarrow \Delta(h)$ is a decreasing function.

Hence, $\frac{1}{2} \le h \le 1 \Rightarrow \Delta_{\max} = \Delta\left(\frac{1}{2}\right)$ and

$\Delta_{\min} = \Delta(1)$

$\therefore \Delta_1 = \frac{\sqrt{3}}{2}\frac{\left(4-\frac{1}{4}\right)^{3/2}}{\frac{1}{2}} = \frac{45}{8}\sqrt{5}$

and $\Delta_2 = \frac{\sqrt{3}}{2}\frac{3\sqrt{3}}{1} = \frac{9}{2}.$

Hence, $\frac{8}{\sqrt{5}}\Delta_1 - 8\Delta_2 = 45 - 36 = 9.$

Comprehension Passage

Comprehension 1

1. (c) The equation of tangent at $P\left(\beta, \beta^3\right)$ is

$y - \beta^3 = 3\beta^2(x - \beta).$

On solving it with the given curve, we get

$x^3 - \beta^3 = 3\beta^2(x - \beta)$

$\Rightarrow (x-\beta)\left[x^2+\beta^2+x\beta-3\beta^2\right]=0$

$\Rightarrow x=\beta,\ \beta,\ -2\beta$

Thus, abscissa of Q must be -2β and that of $R=-2[-2\beta]=4\beta$.

2. (b) As earlier, the equation of tangent at P is

$y-\left(t^2-t^3\right)=\left(2t-3t^2\right)(x-t)$

On solving with the curve,

$x^2-x^3-\left(t^2-t^3\right)=\left(2t-3t^2\right)(x-t)$

$\Rightarrow\left(x^2-t^2\right)-\left(x^3-t^3\right)=\left(2t-3t^2\right)(x-t)$

$\Rightarrow(x-t)\left[(x+t)-\left(x^2+t^2+xt\right)-\left(2t-3t^2\right)\right]=0$

We easily get $(x-t)^2(x-(1-2t))=0$.

$x=t, t,$ and $1-2t$.

3. (c) The equation of tangent at 't'

$y-\left(8t^3-1\right)=3t\left[x-\left(4t^2+3\right)\right]$.

If this meets the curve again at 't' then

$\left(8t'^3-1\right)-\left(8t^3-1\right)=3t\left[4t'^2+3-\left(4t^2+3\right)\right]$.

We will get $t'=-\dfrac{t}{2}$ ($t'=t$ will be rejected)

But this line is normal at t' and slope of the normal at t' is $-\dfrac{1}{3t'}=+\dfrac{2}{3t}$.

But slope of the tangent at $t'=3t$.

Since they are same lines $3t=\dfrac{2}{3t}$.

$\Rightarrow t=\pm\dfrac{\sqrt{2}}{3}$

Comprehension 2

1. (d) $\dfrac{dx}{dt}=a(1+\cos t), \dfrac{dy}{dt}=a\sin t$

$\Rightarrow\dfrac{dy}{dt}=\dfrac{a\sin t}{a(1+\cos t)}$

$=\dfrac{2\sin\frac{t}{2}\cos\frac{t}{2}}{2\cos^2\frac{t}{2}}=\tan\dfrac{t}{2}$

$\Rightarrow$ Sub-normal $=y\dfrac{dy}{dx}$

$=a(1-\cos t)\tan\dfrac{t}{2}$

$=2a\sin^3\dfrac{t}{2}\sec\dfrac{t}{2}$

2. (c) $\dfrac{dy}{dx}=na^{1-n}x^{n-1}$

ω

$\Rightarrow$ Sub-normal $=y\dfrac{dy}{dx}$

$=a^{1-n}.x^n.na^{1-n}x^{n-1}$

$=na^{2-2n}x^{2n-1}$

If this is independent of $x, 2n-1=0$.

$\Rightarrow n=\dfrac{1}{2}$.

3. (d) $\dfrac{x^2}{a^2}+\dfrac{y^2}{b^2}=1$

$\Rightarrow\dfrac{dy}{dx}=-\dfrac{b^2x}{a^2y}$

Now, length of the normal l at $P(x, y)$

$=y\left[1+\left(\dfrac{dy}{dx}\right)^2\right]^{1/2}=\dfrac{1}{a^2}\sqrt{a^4y^2+b^4x^2}$

Now equation of tangent at any point (x, y) of ellipse is $\dfrac{Xx}{a^2}+\dfrac{Yy}{b^2}=1$

$\Rightarrow$ Length of perpendicular from origin.

$$=\left|\frac{-1}{\sqrt{\frac{x^2}{a^4}+\frac{y^2}{b^4}}}\right|=\frac{a^2b^2}{\sqrt{a^4y^2+b^4x}}=\frac{b^2}{l}$$

Thus $l=\dfrac{b^2}{\text{OM}}\Rightarrow l\propto\dfrac{1}{\text{OM}}$.

Comprehension 3

1. (b) On putting $x=0$ and $h=x$ in Taylor's series, we get

$f(x)=f(0)+xf'(0)+\dfrac{x^2}{2!}f''(0)+\ldots$

(This is called Maclaurin's expansion.)

2. (c) Coefficient of $x^2=(f\ \ (0))/2!$

Since $f(x)=\log(1+\sin x)$

$f'(x)=\dfrac{\cos x}{1+\sin x}$

$f\ \ =\dfrac{-\sin x(1+\sin x)-\cos^2 x}{(1+\sin x)^2}$

$=\dfrac{-\sin x-1}{(1+\sin x)^2}\Rightarrow f\ \ (0)=-1.$

3. (c) $y = e^{m\sin^{-1}x}$

$y_1 = e^{m\sin^{-1}x}.\dfrac{m}{\sqrt{1-x^2}} = \dfrac{my}{\sqrt{1-x^2}}$

$\left(1-x^2\right)y_1^2 = m^2y^2$

$\Rightarrow \left(1-x^2\right)2y_1y_2 - 2x\,y_1^2 = m^2.2yy_1$

or $\left(1-x^2\right)y_2 - xy_1 = m^2y$

On putting $x = 0$, we get $(y_2) = m^2(y)_0$.
But $(y)_0 = 1$

$\Rightarrow y_2(0) = m^2$

$\Rightarrow$ Coefficient of x^2 in the expansion

$= \dfrac{1}{2!}m^2 = \dfrac{m^2}{2}.$

Comprehension 4

1. (a) The equation of the tangent at (x_1, y_1) of $x^3 + y^3 = a^3$ is $xx_1^{\ 2} + yy_1^{\ 2} = a^3$. If this tangent passes through (h,k), we have

$hx_1^{\ 2} + ky_1^{\ 2} = a^3$

Also, $x_1^{\ 3} + y_1^{\ 3} = a^3$

$h^3 + k^3 = a^3$

Now, let $A = \dfrac{h}{x_1}$, $B = \dfrac{k}{y_1}$, then the three relations become

$Ax_1^{\ 3} + By_1^{\ 3} = a^3, x_1^{\ 3} + y_1^{\ 3} = a^3,$

$A^3x_1^{\ 3} + B^3y_1^{\ 3} = a^3.$

Solving first two for $x_1^{\ 3}$ and $y_1^{\ 3}$, we get

$x_1^{\ 3} = \dfrac{a^3(1-B)}{A-B}, y_1^{\ 3} = -\dfrac{b^3(1-A)}{A-B}.$

2. (c) On putting in the least relation and cancelling a^3, we get

$A^3(1-B) - B^3(1-A) = A - B$

$\Rightarrow \left(A^3 - B^3\right) - AB\left(A^2 - B^2\right) = A - B$

$\Rightarrow (A-B)\left[A^2 + B^2 + AB - AB(A+B) - 1\right] = 0$

$\Rightarrow (A-B)[(A+B)^2 - AB(A+B) - AB - 1] = 0$

$\Rightarrow (A+B)^2 - AB(A+B) - AB - 1 = 0$

$(\because A \neq B)$

3. (b) Solving the quadratic for $A+B$, we get

$A + B = \dfrac{AB \pm (AB+2)}{2}$

On taking – sign, we get $A + B = -1$

$\Rightarrow \dfrac{h}{x_1} + \dfrac{k}{y_1} = -1 \Rightarrow (b)$ is true

We have to rule out $A = B$ and $A + B = AB + 1$

If $A = B$, i.e., $\dfrac{h}{x_1} = \dfrac{k}{y_1} = \alpha$ (say), then the points (h,k) and (x_1, y_1) coincide.

If $A + B = AB + 1$, then

$\Rightarrow (1-A)(1-B) = 0$

$\Rightarrow A = 1, B = 1$

$\Rightarrow (x_1, y_1)$ and (h,k) coincide.

Comprehension 5

1. (c) Since Slope PR = Slope QR

$\dfrac{at^3 - at_1^{\ 3}}{at^2 - at_1^{\ 3}} = \dfrac{at^3 - t_2^{\ 2}}{at^2 - at_2^{\ 2}}$

from which we get

$t = -\dfrac{t_1t_2}{t_1 + t_2}$ (A)

2. (b) Now equation of tangent at P and Q are

$2y - 3t_1x = -at_1^{\ 3}$

$2y - 3t_2x = -at_2^{\ 3}$ (B)

These tangents intersect at (x, y), where

$x = \dfrac{a}{3}\left(t_1^{\ 2} + t_2^{\ 2} + t_1t_2\right), y = \dfrac{a}{2}t_1t_2(t_1 + t_2)$

3. (b) On eliminating t_1, t_2 in (A) and (B), we get the locus of point of intersection of tangents gives as $(3tx + 2y)^2 = -2a^3t^3y$.

Comprehension 6

1. (b) $\dfrac{da}{dt} = 2 \Rightarrow a = 2t + c$

$\because c = 0$ $\{\because a = 0$, when $t = 0\}$

$\therefore a = 2t$

$\therefore$ The curve $y = x^2 - 2ax + a^2 + a$ becomes

$y = x^2 - 4tx + 4t^2 + 2t$

If $x = 0$, then $y = 4t^2 + 2t$

$\dfrac{dy}{dx} = 2x - 4t \therefore \left.\dfrac{dy}{dx}\right|_{at\,x=0} = -4t$

$\therefore$ Equation of the tangent is,

$$y-(4t^2+2t)=-4t(x-0)$$

i.e., $y=-4tx+4t^2+2t$

Vertex of $y=x^2-4tx+4t^2+2t$ is $(2t, 2t)$

$\therefore$ Distance of vertex from the origin $=2\sqrt{2}\ t$.

2. (d) $\therefore$ rate of change of distance of vertex from origin with respect to $t=2\sqrt{2}$

i.e., $k=2\sqrt{2}$

$c(t)=4t^2+2t$

$$\therefore \frac{dc}{dt}=8t+2$$

$$\therefore \left.\frac{dc}{dt}\right|_{at\ t=2\sqrt{2}}=16\sqrt{2}+2$$

$\therefore \ell=16\sqrt{2}+2$

3. (c) $m(t)=-4t$

$$\therefore \frac{dm}{dt}=-4$$

$$\therefore \left.\frac{dm}{dt}\right|_{\text{at } t=\ell}=-4$$

Comprehension 7

1. (b) Putting $x=9$, $y=0$ in the given equation of the curve,

$$0=3a+9b-\frac{1}{2}=\frac{a}{2\times 3}+b$$

$\Rightarrow a=-3b$ (i)

$$\frac{dy}{dx}=\frac{a}{2\sqrt{x}}+b$$

$$\left.\frac{dy}{dx}\right|_{(9,0)}=\frac{a}{6}+b=-\frac{1}{2} \quad \text{(ii)}$$

using (i) and (ii), we get $b=-1$ and $a=3$

$\therefore y=3\sqrt{x}-x$.

Point (1, 2) lies on the curve as well as it is point of intersection of the family of lines.

$$\frac{dy}{dx}=\frac{3}{2\sqrt{x}}-1$$

$$\frac{dy}{dx}\text{ at }(1,2)\text{ is }\frac{1}{2}$$

$$y-2=\frac{1}{2}(x-1)$$

$$x-2y+3=0$$

2. (b)

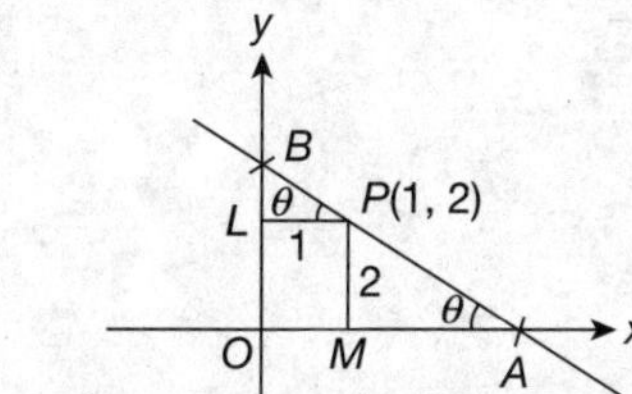

$$\sin\theta=\frac{2}{PA}$$

$$PA=2\operatorname{cosec}\theta$$

$$\cos\theta=\frac{1}{BP}$$

$$BP=\sec\theta$$

$$AB=AP+BP=2\operatorname{cosec}\theta+\sec\theta$$

Minimum value of AB $=\left(2^{2/3}+1\right)^{3/2}$

3. (b) Let $y-2=m(x-1)$ be a focal chord.

$y=mx+2-m$

Length of diagonal $=\sqrt{(x_2-x_1)^2+m^2(x_2-x_1)^2}$

$$=\frac{4}{m^2}(m^2+1).$$

Length of diagonal of perpendicular chord

$$=4m^2\left(\frac{1}{m^2}+1\right)=4(1+m^2)$$

Area

$$=\frac{1}{2}\frac{4}{m^2}(1+m^2)4(1+m^2)=8\left(m^2+2+\frac{1}{m^2}\right)$$

$\Rightarrow$ Minimum area $=16+8\times 2=32$.

Integrals

Objectives

In this chapter on Integrals, we shall study the following topics:

1. Integral of a function
2. Comparison of differentiation and integration
3. Properties of indefinite integral
4. Integration by transformation
5. Integration by substitution
6. Integration by parts
7. Integration of rational function.

Introduction

We have learnt differential calculus in earlier classes as well as in the present class also.

The concept of Integral is linked with differential calculus. Suppose a function f is differential in an interval I, i.e., its derivative f' exists at each point I. A natural question arises: If we know the derivative f', can we find the function f? yes we could find by a mathematical tool known as integration.

Basically, integration is the inverse process of differentiation.

Integral of a Function

To explain the meaning of the word 'integral', we first consider an example.

We know, $\frac{d}{dx}(\sin x) = \cos x$

The function $\cos x$ is the derivative of the function $\sin x$. On the other hand, we say that the function $\sin x$ is the integral of $\cos x$.

More generally, if $\frac{d}{dx}(f(x)) = g(x)$

We say that $f(x)$ is the integral of $g(x)$.

In other words, $f(x)$ is the integral of $g(x)$ when and only when $g(x)$ is the derivative of $f(x)$. For this reason, integral is sometimes described as **'anti-derivative'**. It is also called 'indefinite integral' as opposed to the 'definite integral' which you will study later in the chapter.

Notation

If $f(x)$ is an integral of $g(x)$, we write

$$\int g(x)\,dx = f(x)$$

and read the symbol $\int$ as 'integral'. In this notation dx separately does not have a meaning. This symbol dx conveys that x is the variable of integration. It means that the function $g(x)$, as a function of x, is taken, and we are finding a function $f(x)$ whose derivative (with respect to x) is $g(x)$.

ILLUSTRATION 7.1

Write the integration formula corresponding to

$$\frac{d}{dx}(\cot x) = -\text{cosec}^2 x$$

Solution

The corresponding formula is

$$\int(-\text{cosec}^2 x)\,dx = \cot x$$

Remark

We have to be careful while considering functions whose domain is not the whole real line. For instance, when we say $\frac{d}{dx}(\log x) = \frac{1}{x}$, we know that $\log x$ is defined only for positive x, and therefore, its derivative is also taken only for positive x. But, if we write $\int \frac{1}{x}\,dx = \log x$, the given function $\frac{1}{x}$ is defined for all non-zero values of x, and therefore, its integral must also be defined there, but $\log x$ is not defined for negative values of x. This is overcome in the manner given below.

Formula

$$\int \frac{1}{x} dx = \log|x|$$

Note: The above formula is to be applied for x lying in an interval (δ,∞) or an interval $(-\infty,-\delta)$ for a fixed $\delta > 0$. It does not apply for any $x \in R$ or any x in an interval containing 0.

Constant of Integration

We know that the differential coefficient of a constant is zero. Thus if 'c' is an arbitrary constant independent of x, then

$$\frac{d}{dx}\left[f(x)+c\right] = F(x)$$

where $$\frac{d}{dx}(f(x)) = F(x)$$

$\therefore$ $$\int F(x)dx = f(x)+c \quad \text{...(i)}$$

In order to find the general form of the integral of $f(x)$, an arbitrary constant c is added to the integral of $f(x)$.

Equation (i) is called the definite integral with respect to x.

The indefinite integral of a function is also called as the antiderivative of that function.

Comparison of Differentiation and Integration

(i) Both are operation on functions. Both give functions as answers.

(ii) Both are linear. This means:

(a) In order to differentiate the sum of two functions, it is enough to differentiate the two functions separately, and then add the two derivatives obtained. Similarly, in order to integrate the sum of two functions, it is enough to integrate them separately and then add the two integrals obtained.

(b) The constant can be taken outside the differential or integral sign as shown below:

$$\frac{d}{dx}(cf(x)) = c\frac{d}{dx}f(x)$$

and $$\int cf(x)dx = c\int f(x)dx$$

(iii) You have already seen that not all functions are differentiable. Similarly, not all functions are integrable. You will learn about this in higher classes.

(iv) The derivative of a function, when it exists, is a unique function. The integral of a function is not so. It always varies up to a constant.

(v) When a polynomial function $p(x)$ is differentiated, the result is a polynomial function whose degree is 1 less than that of $p(x)$. When a polynomial function $p(x)$ is integrated, the result is a polynomial function whose degree is 1 more than that of $p(x)$.

(vi) Let $f(x)$ be a function obtainable in a finite number of steps, by taking sums, products, quotients, number-multiples and composites, from polynomial, exponential, logarithmic and trigonometric functions. For instance,

$$f(x) = \sin\left(\log x^2\right) + 2\log\operatorname{cosec} x^2$$

We have already studied how to differentiate all such functions. However, it is not the case with integration. We do not have methods to integrate all such functions. We can integrate only some of them.

(vii) We can speak of the derivative of a function at a point. We never speak of the integral of a function at a point.

(viii) The derivative has a geometric meaning, namely, the slopes of the tangents to a curve. Similarly, the integral also has a geometric meaning, namely, the area of some regions.

(ix) The derivative is used for finding some physical quantities like the velocity of a moving particle, Similarly, the integral is also used in calculating some physical quantities like center of mass, momentum, etc.

(x) Differentiation is a process involving limits. So is integration, as will be seen in the next chapter.

(xi) Differentiation and integration are inverses of each other.

$$\frac{d}{dx}\left(\int f(x)dx\right) = f(x) \text{ and } \int\left(\frac{d}{dx}f(x)\right)dx = f(x)+C$$

Whenever these exist. In other words, if for a function $f(x)$ both these processes are applied one after the other, we get back the function $f(x)$ up to constant.

Properties of Indefinite Integration

(a) $\int k f(x)dx = k\int f(x)dx$

(b) $\int\left[f_1(x) \pm f_2(x) \pm f_3(x) \pm .....f_n(x)\right]dx$

$= \int f_1(x)dx \pm \int f_2(x)dx \pm \int f_3(x)dx \pm .... \pm \int f_n(x)dx$

(c) $\int \frac{f'(x)}{f(x)}dx = \log_e|f(x)| + c$

(d) $\int f'(x)\left[f(x)\right]^n dx = \frac{\left[f(x)\right]^{n+1}}{n+1} + c$

(e) If $\int f(x)dx = g(x)$, then

$$\int f(ax+b)dx = \frac{1}{a}g(ax+b)+c$$

Some Standard Formulae

Integrals of Basic Functions

(a) $\int x^n dx = \frac{x^{n+1}}{n+1}+c \ (n \neq -1)$

(b) For $n=-1$, $\int \frac{1}{x}dx = \log|x|+c \ (x \neq 0)$

(c) $\int \sin x\, dx = -\cos x + c$

(d) $\int \cos x\, dx = \sin x + c$

(e) $\int \sec^2 x\, dx = \tan x + c$

(f) $\int \operatorname{cosec}^2 x\, dx = -\cot x + c$

(g) $\int \sec x \tan x\, dx = \sec x + c$

(h) $\int \operatorname{cosec} x \cot x\, dx = -\operatorname{cosec} x + c$

(i) $\int \frac{dx}{\sqrt{1-x^2}} = \sin^{-1} x + c; |x| < 1$

(j) $\int \frac{dx}{1+x^2} = \tan^{-1} x + c$

(k) $\int \frac{dx}{x\sqrt{x^2-1}} = \sec^{-1}|x| + c; |x| > 1$

(l) $\int a^x\, dx = \frac{a^x}{\log a} + c\ ; a \neq 1, a > 0$

(m) $\int e^x\, dx = e^x + c$

Integrals of tan *x*, cot *x*, sec *x*, cosec *x*

These integrals can be evaluated by using

$$\int \frac{f'(x)}{f(x)}dx = \log|f(x)| + c$$

(i) $\int \tan x\, dx = \log|\sec x| + c$

(ii) $\int \cot x\, dx = \log|\sin x| + c$

(iii) $\int \sec x\, dx = \int \frac{\sec x(\sec x + \tan x)}{(\sec x + \tan x)}dx$

$$\int \sec x\, dx = \log|\sec x + \tan x| + c$$

$$\int \sec x\, dx = \log\left|\tan\left(\frac{x}{2}+\frac{\pi}{4}\right)\right| + c$$

(iv) $\int \operatorname{cosec} x\, dx = \log|\operatorname{cosec} x - \cot x| + c$

$$= \log|\tan x/2| + c$$

The following methods of integration aim at reducing a given integrand into one of the standard integrals:

1. Method of transformation
2. Method of substitution
3. Integration by parts

Integration by Transformation

This method of integration transforms the integrand into standard integrand or into an algebraic sum of standard integrand which can be integrated easily.

ILLUSTRATION 7.2

$$\int \frac{(x-2)}{x^2-3x+2}dx, x \neq 2$$

$$\int \frac{(x-2)}{(x-2)(x-1)}dx = \int \frac{dx}{x-1} = \log(x-1)+c$$

ILLUSTRATION 7.3

$$\int \frac{e^{2x}-e^{4x}}{e^x-e^{-x}}dx$$

$$\Rightarrow \int \frac{e^{2x}(1-e^{2x})}{e^{-x}(e^{2x}-1)}dx \Rightarrow \int -e^{3x}\, dx \Rightarrow -\frac{e^{3x}}{3}+c$$

ILLUSTRATION 7.4

$$\int \frac{(x-3)^2}{\sqrt{x}}dx$$

$$\int \frac{x^2-6x+9}{\sqrt{x}}dx = \int \left(x^{3/2}-6x^{1/2}+9x^{-1/2}\right)dx$$

$$= \frac{2}{5}x^{5/2} - 4x^{3/2} + 18x^{1/2} + c$$

ILLUSTRATION 7.5

$$\int \frac{x+2}{x+4}dx$$

$$\int \frac{x+4-2}{x+4}dx = \int\left(1-\frac{2}{x+4}\right)dx = x - 2\log(x+4)+c$$

Integration of Trigonometric Function by Transformation

ILLUSTRATION 7.6

$$\int \frac{\cos^3 x}{\sin^2 2x} dx$$

$$\int \frac{\cos^3 x}{4\sin^2 x\cos^2 x} dx = \int \frac{1}{4}\cot x \operatorname{cosec} x\, dx$$

$$= -\frac{1}{4}\operatorname{cosec} x + c$$

ILLUSTRATION 7.7

$$\int \sin^2 x\, dx$$

$$\Rightarrow \quad \frac{1}{2}\int (1-\cos 2x)\, dx \Rightarrow \frac{x}{2} - \frac{1}{2}\left(\frac{\sin 2x}{2}\right) + c$$

$$\Rightarrow \quad \frac{x}{2} - \frac{\sin 2x}{4} + c$$

ILLUSTRATION 7.8

$$\int \frac{dx}{1+\sin x}$$

$$\int \frac{1-\sin x}{\cos^2 x} dx$$

$$\Rightarrow \int \left[\left(\sec^2 x - \tan x \sec x\right)\right] dx = \tan x - \sec x + c$$

ILLUSTRATION 7.9

$$\int \sin 2x \sin 3x\, dx$$

$$\Rightarrow \quad \frac{1}{2}\int 2\sin 2x \sin 3x\, dx$$

$$\Rightarrow \quad \frac{1}{2}\int (\cos x - \cos 5x)\, dx = \frac{1}{2}\sin x - \frac{\sin 5x}{10} + c$$

ILLUSTRATION 7.10

$$\int \sin^2 x \cos^2 x\, dx$$

$$\Rightarrow \quad \frac{1}{4}\int \sin^2 2x\, dx \Rightarrow \frac{1}{4}\int \frac{(1-\cos 4x)}{2} dx$$

$$\Rightarrow \quad \frac{1}{8}\left(x - \frac{\sin 4x}{4}\right) + c$$

Integration by Substitution

If the integral is in the form of $\int f\left[g(x)\right]g'(x)dx$,

We substitute $g(x) = t$ and $g'(x)dx = dt$. Then the above integral reduces to $\int f(t)dt$.

This substitution process can reduce the integrand to the form where we can evaluate the integral by using the standard result.

ILLUSTRATION 7.11

$$\int \sqrt{2x-3}\, dx$$

Let $\quad 2x - 3 = z$

$$2dx = dz$$

$$\Rightarrow \quad \int \frac{\sqrt{z}\, dz}{2} \Rightarrow \frac{1}{2}\frac{(z)^{3/2}}{(3/2)} \Rightarrow \frac{1}{3}(2x-3)^{3/2} + c$$

ILLUSTRATION 7.12

$$\int (2x+3)\sqrt{x^2+3x}\, dx$$

Put $\quad x^2 + 3x = t$

$$(2x + 3)dx = dt$$

$$\int \sqrt{t}\, dt \Rightarrow \frac{(t)^{3/2}}{3/2} + c \Rightarrow \frac{2}{3}\left(x^2+3x\right)^{3/2} + c$$

Some Important Substitutions

Expression	Type of substitution
$a^2 + x^2$	$x = a\tan\theta$ or $x = a\cot\theta$
$a^2 - x^2$	$x = a\sin\theta$ or $x = a\cos\theta$
$x^2 - a^2$	$x = a\sec\theta$ or $x = a\operatorname{cosec}\theta$
$\sqrt{\frac{a-x}{a+x}}$ or $\sqrt{\frac{a+x}{a-x}}$	$x = a\cos 2\theta$
$\sqrt{\frac{x-\alpha}{\beta-x}}$ or $\sqrt{(x-\alpha)(\beta-x)}$	$x = \alpha\cos^2\theta + \beta\sin^2\theta$

TYPE I

Some standard formulae derived using substitution

1. $\int \frac{dx}{a^2+x^2} = \frac{1}{a}\tan^{-1}\left(\frac{x}{a}\right) + c$

2. $\int \frac{dx}{x^2-a^2} = \frac{1}{2a}\log\left|\frac{x-a}{x+a}\right| + c$

3. $\int \frac{dx}{a^2-x^2} = \frac{1}{2a}\log\left|\frac{a+x}{a-x}\right| + c$

4. $\int \frac{dx}{\sqrt{a^2+x^2}} = \log\left|x+\sqrt{x^2+a^2}\right| + c$

5. $\int \frac{dx}{\sqrt{a^2-x^2}} = \log\left|x+\sqrt{x^2-a^2}\right| + c$

6. $\int \frac{dx}{\sqrt{a^2-x^2}} = \sin^{-1}\left(\frac{x}{a}\right) + c$

7. $\int \frac{dx}{x\sqrt{x^2-a^2}} = \frac{1}{a}\sec^{-1}\left(\frac{x}{a}\right) + c$

8. $\int \sqrt{a^2+x^2}\,dx = \frac{x}{2}\sqrt{a^2+x^2} + \frac{a^2}{2}\log\left|x+\sqrt{a^2+x^2}\right| + c$

9. $\int \sqrt{x^2-a^2}\,dx = \frac{x}{2}\sqrt{x^2-a^2} - \frac{a^2}{2}\log\left|x+\sqrt{x^2-a^2}\right| + c$

10. $\sqrt{a^2-x^2}\,dx = \frac{x}{2}\sqrt{a^2-x^2} + \frac{a^2}{2}\sin^{-1}\left(\frac{x}{a}\right) + c$

TYPE II

Integrals of the type

$$\int \frac{dx}{ax^2+bx+c}, \int \frac{dx}{\sqrt{ax^2+bx+c}}, \int \sqrt{ax^2+bx+c}\,dx$$

To find integrals in these forms, first express $ax^2 + bx + c$ in the form of a perfect square and then apply formulae given in Type I

ILLUSTRATION 7.13

$$\int \frac{dx}{x^2+x+1} = \int \frac{dx}{(x+1/2)^2 + \left(\sqrt{3}/2\right)^2}$$

$$= \frac{1}{\sqrt{3}/2}\tan^{-1}\left(\frac{x+1/2}{\sqrt{3}/2}\right)$$

$$\Rightarrow \quad \frac{2}{\sqrt{3}}\tan^{-1}\left(\frac{2x+1}{\sqrt{3}}\right) + c$$

ILLUSTRATION 7.14

$$\int \frac{dx}{\sqrt{x^2+x+1}} \cdot \int \frac{dx}{\sqrt{\left(x+\frac{1}{2}\right)^2 + \left(\frac{\sqrt{3}}{2}\right)^2}}$$

$$= \log\left|x+\frac{1}{2} + \left(x+\frac{1}{2}\right)^2 + \left(\frac{\sqrt{3}}{2}\right)^2\right|$$

$$= \log\left|2x+1+2\sqrt{x^2+x+1}\right| + c$$

ILLUSTRATION 7.15

$$\int \sqrt{x^2+x+1}\,dx$$

$$\Rightarrow \quad \int \sqrt{\left(x+\frac{1}{2}\right)^2 + \left(\frac{\sqrt{3}}{2}\right)^2}\,dx$$

$$\Rightarrow \frac{\left(x+\frac{1}{2}\right)}{2}\sqrt{x^2+x+1} + \frac{3}{4\times 2}\log\left|x+\frac{1}{2}+\sqrt{x^2+x+1}\right|$$

$$\Rightarrow \frac{(2x+1)}{4}\sqrt{x^2+x+1} + \frac{3}{8}\log\left|2x+1+2\sqrt{x^2+x+1}\right| + c$$

TYPE III

Integrals of the type

(i) $\int \frac{px+q}{ax^2+bx+c}dx$

(ii) $\int \frac{(px+q)dx}{\sqrt{ax^2+bx+c}}$

put $px+q = A(2ax+b)+B$ and then solve by usual methods

(iii) $\int \frac{p(x)dx}{ax^2+bx+c}$, where $p(x)$ is a polynomial of degree greater than or equal to 2 divide $p(x)$ by $ax^2 + bx + c$ and express integrand as $\int \left[Q(x) + \frac{R(x)}{ax^2+bx+c}\right]dx$

ILLUSTRATION 7.15

$$\int \sqrt{x^2+x+1}\,dx$$

$$\Rightarrow \quad \int \sqrt{\left(x+\frac{1}{2}\right)^2 + \left(\frac{\sqrt{3}}{2}\right)^2}\,dx$$

$$\Rightarrow \frac{\left(x+\frac{1}{2}\right)}{2}\sqrt{x^2+x+1} + \frac{3}{4\times 2}\log\left|x+\frac{1}{2}+\sqrt{x^2+x+1}\right|$$

$$\Rightarrow \frac{(2x+1)}{4}\sqrt{x^2+x+1} + \frac{3}{8}\log\left|2x+1+2\sqrt{x^2+x+1}\right| + c$$

ILLUSTRATION 7.16

$$I = \int \frac{(3x+1)dx}{\sqrt{x^2+4x+1}}$$

Using the above techniques, we can write

$$3x + 1 = A(2x + 4) + B$$

$$A = 3/2 \quad B = -5$$

$$I=\int \frac{\left[\frac{3}{2}(2x+4)-5\right]dx}{\sqrt{x^2+4x+1}}$$

$$I=\frac{3}{2}\int \frac{(2x+4)\,dx}{\sqrt{x^2+4x+1}}-5\int \frac{dx}{\sqrt{x^2+4x+1}}$$

$$I_1=\frac{3}{2}\int \frac{(2x+4)\,dx}{\sqrt{x^2+4x+1}}\quad t=x^2+4x+1$$

$$I_1=\frac{3}{2}\int \frac{dt}{\sqrt{t}} \Rightarrow I_1=3\sqrt{t}+c$$

$$\Rightarrow I_1=3\sqrt{x^2+4x+1}+c$$

$$\Rightarrow I_2=5\int \frac{dx}{\sqrt{x^2+4x+1}}$$

$$I_2=5\int \frac{dx}{\sqrt{(x+2)^2-3}}$$

$$I_2=5\log\left|x+2+\sqrt{x^2+4x+1}\right|+c$$

$$I=3\sqrt{x^2+4x+1}-5\log\left|x+2+\sqrt{x^2+4x+1}\right|+c$$

ILLUSTRATION 7.17

$$\int \frac{(x^2+1)\,dx}{x^2+x+1}$$

$$I=\int \frac{\left[x^2+x+1-\frac{1}{2}(2x+1)+\frac{1}{2}\right]dx}{x^2+x+1}$$

$$I=\int \frac{(x^2+x+1)}{x^2+x+1}dx-\frac{1}{2}\int \frac{(2x+1)}{x^2+x+1}dx+\frac{1}{2}\int \frac{dx}{x^2+x+1}$$

$$I=x-\frac{1}{2}\log_e\left|x^2+x+1\right|+\frac{1}{2}\int \frac{dx}{\left(x+\frac{1}{2}\right)^2+\left(\frac{\sqrt{3}}{2}\right)^2}$$

$$I=x-\frac{1}{2}\log_e\left|x^2+x+1\right|+\frac{1}{2}\frac{(2\times 1)}{\sqrt{3}}\tan^{-1}\left(\frac{2x+1}{\sqrt{3}}\right)$$

$$I=x-\frac{1}{2}\log_e\left|x^2+x+1\right|+\frac{1}{\sqrt{3}}\tan^{-1}\left(\frac{2x+1}{\sqrt{3}}\right)$$

TYPE IV

Integrals of the type

$$\int \frac{dx}{a\sin^2 x+b\cos^2 x},\ \int \frac{dx}{a+b\sin^2 x},\ \int \frac{dx}{a+b\cos^2 x}$$

$$\int \frac{dx}{(a\sin x+b\cos x)^2},\ \int \frac{f(\tan x)\,dx}{a+b\sin^2 x+\cos^2 x+d\sin x\cos x}$$

In all the above integrals, divide numerator and denominator by $\cos^2 x$, then put $\tan x = t$ and $\sec^2 x\, dx = dt$. After employing these substitutions, integral will transform into standard integral.

ILLUSTRATION 7.18

$$\int \frac{dx}{\sin^2 x+3\cos^2 x}\ \text{(dividing Nr and Dr by } \cos^2 x)$$

$$I=\int \frac{\sec^2 x\,dx}{\tan^2 x+3}\quad \tan x=t,\ \frac{dt}{dx}=\sec^2 x$$

$$I=\int \frac{dt}{t^2+(\sqrt{3})^2}=\frac{1}{\sqrt{3}}\tan^{-1}\left(\frac{t}{\sqrt{3}}\right)$$

$$=\frac{1}{\sqrt{3}}\tan^{-1}\left(\frac{\tan x}{\sqrt{3}}\right)+c$$

TYPE V

Integrals of the type

$$\int \frac{dx}{a\sin x+b\cos x},\ \int \frac{dx}{a+b\sin x},$$

$$\int \frac{dx}{a+b\cos x},\ \int \frac{f(\tan x/2)\,dx}{a\sin x+b\cos x+c}$$

where $f(\tan x/2)$ is a polynomial.

Replace $\sin x=\frac{2\tan x/2}{1+\tan^2 x/2}$ and $\cos x=\frac{1-\tan^2 x/2}{1+\tan^2 x/2}$

and put $\tan x/2 = t$ and $\frac{1}{2}\sec^2(x/2)\,dx = dt$

ILLUSTRATION 7.19

$$\int \frac{dx}{4+5\sin x}$$

$$I=\int \frac{dx}{4+\frac{5(2\tan x/2)}{1+\tan^2 x/2}}=\int \frac{\sec^2 x/2\,dx}{4\tan^2 x/2+10\tan x/2+4}$$

Put $\quad t=\tan x/2\ \ dt=\frac{1}{2}\sec^2 x/2\,dx$

$$I=\int \frac{dt}{2\left(t^2+\frac{5t}{2}+1\right)}=\frac{1}{2}\int \frac{dt}{\left(t+\frac{5}{4}\right)^2-\frac{9}{16}}=\frac{1}{3}\log\left|\frac{4t+2}{4t+8}\right|$$

$$I=\frac{1}{3}\log\left|\frac{2\tan x/2+1}{2\tan x/2+4}\right|+c$$

Integration by Parts

Integrals of the Type

$$\int \underset{\text{I}}{f(x)}\,\underset{\text{II}}{g(x)}\,dx = f(x)\int g(x)\,dx - \int\left(f'(x)\int g(x)\,dx\right)dx$$

In the above integration, $f(x)$ is taken as first function and $g(x)$ as second function. Depending upon the ease of integration, we will select functions as first and second. However, the following order is useful in deciding the first part. The function on the left is taken as the first part and the function on the right is taken as second part.

Inverse, Logarithmic, Algebraic, Trigonometric, Exponents

The above rule to decide between first and second part is known as ILATE rule.

ILLUSTRATION 7.20

$$\int \log x\,dx$$

Taking $\log(x)$ as first function and dx as second function

$$\int \log x\,dx = \log x\int dx - \int\left(\frac{1}{x}\right)x\,dx = x\log x - x$$

$$\int \log x\,dx = x(\log x - 1) + c$$

ILLUSTRATION 7.21

$\int x\tan^{-1}x$ is equal to

(a) $\frac{1}{2}(x^2+1)\tan^{-1}x - x + c$

(b) $\frac{1}{2}(x^2+1)\tan^{-1}x + x + c$

(c) $\frac{1}{2}(x^2+1)\tan^{-1}x - \frac{1}{2}x + c$

(d) $\frac{1}{2}(x^2-1)\tan^{-1}x - \frac{1}{2}x + c$

Solution

Integrating by parts, taking x as second part

$$I = \frac{x^2}{2}\tan^{-1}x - \int\frac{1}{1+x^2}.\frac{x^2}{2}dx$$

$$= \frac{1}{2}x^2\tan^{-1}x - \frac{1}{2}\int\left(1 - \frac{1}{1-x^2}\right)dx$$

$$= \frac{1}{2}x^2\tan^{-1}x - \frac{1}{2}x + \frac{1}{2}\tan^{-1}x + c$$

$$= \frac{1}{2}(x^2+1)\tan^{-1}x - \frac{1}{2}x + c$$

ILLUSTRATION 7.22

Evaluate $\int \sec^3 x\,dx$

Solution

$$I = \int \sec x.\sec^2 x\,dx = \sec x\tan x - \int \sec x\tan x.\tan x\,dx$$

$$= \sec x\tan x - \int \sec x(\sec^2 x - 1)dx$$

$$= \sec x\tan x - \int \sec^3 x\,dx + \int \sec x\,dx$$

$$= \sec x\tan x - I + \log|\sec x + \tan x|$$

$$I = \frac{1}{2}\{\sec x\tan x + \log|\sec x + \tan|\} + c$$

ILLUSTRATION 7.23

$$\int\frac{x^2\,dx}{(x\sin x + \cos x)^2}$$

Solution

$$\text{Let } I = \int\frac{x^2dx}{(x\sin x + \cos x)^2}$$

$$= \int x\sec x\frac{x\cos x}{(x\sin x + \cos x)^2}dx$$

$$\text{Now} \int\frac{x\cos x\,dx}{(x\sin x + \cos x)^2} = \int\frac{dt}{t^2}$$

(put $x\sin x + \cos x = t \Rightarrow x\cos x\,dx = dt$)

$$= -\frac{1}{t} = -\frac{1}{x\sin x + \cos x}$$

Therefore,

$$I = x\sec x\left(-\frac{1}{x\sin x + \cos x}\right) + \int\frac{\sec x + x\sec x\tan x}{x\sin x + \cos x}dx$$

$$= \frac{x\sec x}{x\sin x + \cos x} + \int\frac{\sec x(x\sin x + \cos x)}{\cos x(x\sin x + \cos x)}dx$$

$$= -\frac{x\sec x}{x\sin x + \cos x} + \int \sec^2 x\,dx$$

$$= -\frac{x\sec x}{x\sin x + \cos x} + \tan x + c$$

Integrals of the Type

$$\int e^x[f(x) + f'(x)]dx$$

$$\Rightarrow \quad \int e^x f(x)dx = e^x f(x) - \int f'(x)e^x\,dx$$

$$\int e^x f(x)dx + \int f'(x)e^x dx = e^x f(x),$$

Hence

$$\int e^x\left[f(x)+f'(x)\right]dx = e^x f(x)$$

ILLUSTRATION 7.24

Evaluate $\int \frac{(\log x)}{(1+\log x)^2}dx.$

Solution

Substitute

$$\log x = t \Rightarrow \frac{1}{x}dx = dt \Rightarrow \frac{1}{x}dx = dt \Rightarrow dx = e^t dt$$

$$\text{Now } I = \int \frac{e^t . t\, dt}{(1+t)^2} = \int e^t \left\{\frac{1+t-1}{(1+t)^2}\right\}dt$$

$$= \int e^t \left\{\frac{1}{1+t} - \frac{1}{(1+t)^2}\right\}dt = e^t \frac{1}{(1+t)} + c$$

$$= x\frac{1}{(1+\log x)} + c$$

ILLUSTRATION 7.25

$\int e^x \frac{x-1}{(x+1)^3}dx$ equals ______.

(a) $-\frac{e^x}{x+1}+c$ **(b)** $\frac{e^x}{x+1}+c$

(c) $\frac{e^x}{(x+1)^2}+c$ **(d)** $-\frac{e^x}{(x+1)^2}+c$

Solution

$$I = \int e^x\left[\frac{x+1-2}{(x+1)^3}\right]dx = \int e^x\left(\frac{1}{(x+1)^2} - \frac{2}{(x+1)^3}\right)dx$$

Thus the given integral is of the form

$$\int e^x\{f(x)+f'(x)\}dx$$

$$I = e^x f(x) = \frac{e^x}{(x+1)^2} + c$$

ILLUSTRATION 7.26

$\int \frac{e^{2\tan^{-1}x}(1+x)^2}{(1+x^2)}dx$ is equal to ______.

(a) $xe^{\tan-1x}+c$ **(b)** $xe^{2\tan-1x}+c$

(c) $2xe^{2\tan-1x}+c$ **(d)** None of these

Solution

Putting the value of $2\tan^{-1}x = t$

$$I = \frac{1}{2}\int e^t\{1+\tan(t/2)\}^2\, dt$$

$$= \frac{1}{2}\int e^t\left[\sec^2\frac{t}{2} + 2\tan\frac{t}{2}\right]dt = \frac{1}{2}e^t\,(2\tan t/2)$$

$$= e^t \tan\frac{t}{2} = xe^{2\tan^{-1}x} + c$$

Integration of Rational Function

Method of Partial Fractions for Rational Functions

Integrals of the type $\int \frac{P(x)}{Q(x)}dx$ can be integrated by resolving the integrand into partial fractions. We proceed as follows:

(i) Check degree of $P(x)$ and $Q(x)$.

(ii) If degree of $P(x)$ > degree of $Q(x)$, then divide $P(x)$ by $Q(x)$ till its degree is less, i.e., put in the form

$$\frac{P(x)}{Q(x)} = R(x) + \frac{F(x)}{Q(x)}$$

where degree of $F(x) < $ degree of $Q(x)$

Case 1: When the denominator contains non-repeated linear factors. That is,

$$Q(x) = (x-\alpha_1)(x-\alpha_2)....(x-\alpha_n)$$

In such a case write $F(x)$ and $Q(x)$ as:

$$\frac{F(x)}{Q(x)} = \frac{A_1}{x-\alpha_1} + \frac{A_2}{x-\alpha_2} + ... + \frac{A_n}{x-\alpha_n}$$

where $A_1, A_2, ... A_n$ are constants to be determined by comparing the coefficient of various powers of x on both sides after taking LCM.

ILLUSTRATION 7.27

Evaluate the integral $\int \frac{x}{(x-2)(x-3)}dx.$

Solution

Let $\frac{x}{(x-2)(x-3)} = \frac{A}{x-2} + \frac{B}{x-3}$

$\therefore$ $x = A(x-3) + B(x-2)$

Put $x = 2$ and then $x = 3$

$2 = -A;\ 3 = B$

$$\therefore \int \frac{x}{(x-2)(x-3)}dx = \int \frac{-2}{(x-2)}dx + \int \frac{3}{(x-3)}dx$$

$$= -2\ln|x-2| + 3\ln|x-3| + C$$

ILLUSTRATION 7.28

Evaluate $\int \tan^3 x.\sec^2 x\,dx$.

Solution

Let $\tan x = t \Rightarrow \sec^2 x\,dx = dt$

$$\therefore \quad \int \tan^3 x.\sec^2 x\,dx = \int t^3 dt = \frac{t^4}{4} + c = \frac{\tan^4 x}{4} + c$$

ILLUSTRATION 7.29

Evaluate $\int \frac{x}{x^2 - x - 2} dx$.

Solution

Here, $I = \int \frac{x}{(x-2)(x+1)} dx$

$$= \frac{1}{3}\int\left(\frac{2}{x-2} + \frac{1}{x+1}\right)dx$$

$$= \frac{1}{3}\left[2\log(x-2) + \log(x+1)\right] + c$$

$$= \frac{1}{3}\log\left[(x-2)^2(x+1)\right] + c$$

ILLUSTRATION 7.30

Evaluate $\int \frac{2x}{x^2 + 3x + 2} dx$.

Solution

$$I = \int \frac{2x}{(x+1)(x+2)} dx = 2\int \frac{x}{(x+1)(x+2)} dx$$

$$= \frac{2}{(2-1)}\int \frac{(2-1)x}{(x+1)(x+2)} dx$$

$$= 2\int\left[\frac{2(x+1)-(x+2)}{(x+1)(x+2)}\right]dx$$

$$= 2\int\left[\frac{2}{x+2} - \frac{1}{x+1}\right]dx$$

$$= \int \frac{4}{x+2} dx - \int \frac{2}{x+1} dx$$

$$= 4\log(x+2) - 2\log(x+1) + c = 2\log\frac{(x+2)^2}{(x+1)} + c$$

ILLUSTRATION 7.31

$\int \frac{x^2+1}{(x-1)(x-2)} dx$ equals ______.

(a) $\log\left[\frac{(x-2)^5}{(x-1)^2}\right] + c$ **(b)** $x + \log\left[\frac{(x-2)^5}{(x-1)^2}\right] + c$

(c) $x + \log\left[\frac{(x-1)^5}{(x-2)^5}\right] + c$ **(d)** None of these.

Solution

Here, since the highest powers of x in Nr and Dr are equal and the coefficients of x^2 are also equal, therefore

$$\frac{x^2+1}{(x-1)(x-2)} \equiv 1 + \frac{A}{x-1} + \frac{B}{x-2}$$

On solving, we get $A = -2$, $B = 5$

Thus $$\frac{x^2+1}{(x-1)(x-2)} \equiv 1 - \frac{2}{x-1} + \frac{5}{x-2}$$

The above method is used to obtain the value of constant corresponding to non-repeated linear factor in the Dr.

Now $$I = \int\left(1 - \frac{2}{x-1} + \frac{5}{x-2}\right)dx$$

$$= x - 2\log(x-1) + 5\log(x-2) + c$$

$$= x + \log\left[\frac{(x-2)^5}{(x-1)^2}\right] + c$$

ILLUSTRATION 7.32

$\int \frac{e^x}{e^{2x} + 5e^x + 6} dx$ equals ______.

(a) $\log\left(\frac{e^x+3}{e^x+2}\right) + c$ **(b)** $\log\left(\frac{e^x+2}{e^x+3}\right) + c$

(c) $\frac{1}{2}\log\left(\frac{e^x+2}{e^x+3}\right) + c$ **(d)** None of these.

Solution

Put $e^x = t \Rightarrow e^x dx = dt$

$$\therefore I = \int \frac{dt}{t^2 + 5t + 6} = \int \frac{dt}{(t+2)(t+3)} = \int\left(\frac{1}{t+2} - \frac{1}{t+3}\right)dt$$

$$= \log\left(\frac{t+2}{t+3}\right) + c = \log\left(\frac{e^x+2}{e^x+3}\right) + c$$

Case 2: When the denominator contains repeated as well as non-repeated linear factors. This is,

$$Q(x)=(x-\alpha_1)^2(x-\alpha_3)...(x-\alpha_n)$$

In such a case, express $f(x)$ and $g(x)$ as:

$$\frac{F(x)}{Q(x)}=\frac{A_1}{x-\alpha_1}+\frac{A_2}{(x-\alpha_1)^2}+\frac{A_3}{x-\alpha_3}+....+\frac{A_n}{(x-\alpha_n)}$$

where $A_1, A_2,...A_n$ are constants to be determined by comparing the coefficient of various power of x on both sides, after taking LCM.

ILLUSTRATION 7.33

Evaluate the integral $\int \frac{3x-2}{(x+1)^2(x+3)}dx$.

Solution

Let $\frac{3x-2}{(x+1)^2(x+3)}=\frac{A}{x+1}+\frac{B}{(x+1)^2}+\frac{C}{x+3}$

Multiplying both sides by $(x+1)^2(x+3)$, we get

$$3x-2=A(x+1)(x+3)+B(x+3)+C(x+1)^2$$

Put $x=-1$, $-5=B(2) \Rightarrow B=-\frac{5}{2}$

Put $x=-3$, $-11=C(4) \Rightarrow C=\frac{-11}{4}$

Equating coefficient of $x^2, 0=A+C \Rightarrow A=-C=\frac{11}{4}$

$$\therefore \int \frac{3x-2}{(x+1)^2(x+3)}dx$$

$$=\int\left[\frac{11}{4(x+1)}-\frac{5}{2(x+1)^2}-\frac{11}{4(x+3)}\right]dx$$

$$=\frac{11}{4}\log|x+1|-\frac{5}{2}.\frac{(x+1)^{-1}}{-1}-\frac{11}{4}\log|x+3|+C$$

$$=\frac{11}{4}\log\left|\frac{x+1}{x+3}\right|+\frac{5}{2(x+1)}+C$$

Case 3: When the denominator contains a non-repeated quadratic factor which cannot be factorized further,

$$Q(x)=(ax^2+bx+c)(x-\alpha_3)(x-\alpha_4)...(x-\alpha_n)$$

In such a case, express $F(x)$ and $Q(x)$ as:

$$\frac{F(x)}{Q(x)}=\frac{A_1x+A_2}{ax^2+bx+c}+\frac{A_3}{x-\alpha_3}+...+\frac{A_n}{x-\alpha_n}$$

where $A_1, A_2,...A_n$ are constants to be determined by comparing the coefficients of various powers of x on both sides after taking LCM.

Case 4: When the denominator contains a repeated quadratic factor which cannot be factorized further. That is,

$$Q(x)=(ax^2+bx+c)^2(x-\alpha_5)(x-\alpha_6)....(x-\alpha_n)$$

In such a case, write $F(x)$ and $Q(x)$ as:

$$\frac{F(x)}{Q(x)}=\frac{A_1x+A_2}{ax^2+bx+c}+\frac{A_3x+A_4}{(ax^2+bx+c)^2}+\frac{A_5}{x-\alpha_5}+...+\frac{A_n}{(x-\alpha_n)}$$

where $A_1, A_2,...A_n$ are constants to be determined by comparing the coefficients of various powers of x on both sides after taking LCM.

ILLUSTRATION 7.34

Evaluate $\int \frac{2x^2}{(x^2+4)(x^2+1)}dx$.

Solution

Let $x^2=Z$

$$\frac{2x^2}{(x^2+4)(x^2+1)}=\frac{2Z}{(Z+4)(Z+1)}=\frac{A}{(Z+4)}+\frac{B}{(Z+1)}$$

$\therefore 2Z=A(Z+1)+B(Z+4)$

Put $Z=-1$ and then $Z=-4$, we get

$$-2=3B \Rightarrow B=-2/3$$

$$-8=-3A \Rightarrow A=8/3$$

$$\therefore \int \frac{2x^2}{(x^2+4)(x^2+1)}=\int \frac{8}{3(x^2+4)}-\frac{2}{3(x^2+1)}dx$$

$$=\frac{8}{3}\int\frac{1}{x^2+4}dx-\frac{2}{3}\int\frac{1}{x^2+1}dx$$

$$=\frac{8}{3}\times\frac{1}{2}\tan^{-1}\frac{x}{2}-\frac{2}{3}\tan^{-1}\frac{x}{1}+C$$

$$=\frac{4}{3}\tan^{-1}\frac{x}{2}-\frac{2}{3}\tan^{-1}x+C$$

Case 5: When the denominator cannot be factorized:

(i) $\int \frac{dx}{ax^2+bx+c}$ In this case,

$$ax^2+bx+c=a^2\left[\left(x+\frac{b}{2a}\right)^2-\frac{b^2-4ac}{4a^2}\right]$$

(ii) $\int \frac{px+q}{ax^2+bx+c}dx$ In this case, put

$px+q=A$ [differentiation of ax^2+bx+c] $+B$

$\Rightarrow px+q=A(2ax+b)+B$

WORKSHEET – 1

Evaluate the following:

1. $\int \frac{\sqrt{9-x^2}}{x^4} dx$
2. $\int \frac{dx}{9\sin^2 x + 16\cos^2 x}$
3. $\int \sqrt[3]{1+3\sin x} \cos x \, dx$
4. $\int \frac{dx}{(\sin^{-1} x)^5 \sqrt{1-x^2}}$
5. $\int \frac{\sin 2x}{1+\cos^2 x} dx$
6. $\int \frac{dx}{x^2+x+1}$
7. $\int \frac{dx}{\sqrt[4]{(x-1)^3 (x+2)^5}}$
8. $\int \frac{dt}{(t+2)\sqrt{t^2+2t-3}}$
9. $\int \frac{x^2+1}{x^4+1} dx$
10. $\int \sqrt{\tan x} \, dx$

Definite Integration

In geometrical and other applications of integral calculus, it becomes necessary to find the difference in the values of an integral of a function $f(x)$ for two assigned values of the independent variable x, say a, b. This difference is called definite integration of $f(x)$ over the interval [a, b] and is denoted by $\int_a^b f(x) dx$

Thus, $\int_a^b f(x) dx = f(b) - f(a)$, where $f(x)$ is an integral of $f(x)$.

The difference $|f(b) - f(a)|$ is sometimes denoted as

$$|F(x)|_a^b = F(b) - F(a)$$

Where a is called the lower limit and the upper limit is b.

Properties of Definite Integral

1. $\int_a^b f(x) dx = \int_a^b f(t) dt$

   The integral remains unaffected with the change of variable.

2. $\int_a^b f(x) dx = -\int_b^a f(x) dx$

   Interchanging of limits, changes the sign of the integral also.

3. $\int_a^b f(x) dx = \int_a^c f(x) dx + \int_c^b f(x) dx$

4. $\int_0^a f(x) dx = \int_0^a f(a-x) dx$

   Putting $x = a - t$ so that $dx = -dt$

   $\int_0^a f(x) dx = -\int_a^0 f(a-t) dt$

   $\int_0^a f(x) dx = -\int_a^0 f(a-x) dx$

5. $\int_{-a}^a f(x) dx = 2\int_0^a f(x) dx$, when $f(x)$ is even

   $= 0$, when $f(x)$ is odd

6. $\int_0^{2a} f(x) dx = 2\int_0^a f(x) dx$ if $f(2a - x) = f(x)$

   $= 0$, if $f(2a - x) = -f(x)$

7. $\int_a^b f(x) dx = \int_a^b f(a+b-x) dx$

8. $\int_0^{nT} f(x) dx = n\int_0^T f(x) dx$

   where $f(x)$ is a periodic function with period T and n is an integer.

9. $\int_0^\infty f(x) dx = \lim_{n\to\infty} \int_0^n f(x) dx$

10. $\left|\int_a^b f(x) dx\right| \le \int_a^b |f(x)| dx$

11. If the function $f(x)$ and $g(x)$ are defined on $[a, b]$ and differentiable at all points $x \in [f(a), g(b)]$, then

$$\frac{d}{dx}\left[\int_{f(x)}^{g(x)} h(t) dt\right] = h(g(x))g'(x) - h(f(x))f'(x)$$

12. If $f(x)$ is a function defined in the interval $[a, b]$, then

$$m(b-a) \le \int_a^b f(x) dx \le M(b-a)$$

   Where m is the least and M is the greatest value of the function $f(x)$ in the interval $[a, b]$

13. **Mean Value Theorem**

   If the function $f(x)$ is continuous in the interval $[a, b]$, then $\int_a^b f(x) dx = f(c)(b-a)$ where $a < c < b$

ILLUSTRATION 7.35

$$\int_0^{\pi}\frac{e^{\cos x}}{e^{\cos x}+e^{-cox}}dx$$

Let $$I=\int_0^{\pi}\frac{e^{\cos x}\,dx}{e^{\cos x}+e^{-\cos x}}$$

$$I=\int_0^{\pi}\frac{e^{\cos(\pi-x)}}{e^{\cos(\pi-x)}+e^{-\cos(\pi-x)}}$$

$$2I=\int_0^{\pi}\frac{e^{-\cos x}\,dx}{e^{-\cos x}+e^{\cos x}}$$

$$2I=\int_0^{\pi}\frac{e^{\cos x}+e^{-\cos x}\,dx}{e^{\cos x}+e^{-\cos x}}$$

$$2I=\int_0^{\pi}dx\Rightarrow I=\frac{\pi}{2}$$

ILLUSTRATION 7.36

$$\int_0^{\pi/2}\log\sin x\,dx$$

$$I=\int_0^{\pi/2}\log\sin\left(\frac{\pi}{2}-x\right)dx \text{ (using Property 4)}$$

$$I=\int_0^{\pi/2}\log\cos x\,dx$$

$$2I=\int_0^{\pi/2}\log\sin x\cos x\,dx$$

$$2I=\int_0^{\pi/2}\log\sin 2x\,dx-\int_0^{\pi/2}\log 2\,dx$$

Take $$I_1=\int_0^{\pi/2}\log\sin 2x\,dx\Rightarrow I_1=\int_0^{\pi}\frac{(\log\sin t)}{2}dt$$

$$t=2x\ dt=2dx$$

$$I_1=\int_0^{\pi/2}\log\sin t\,dt \quad \text{(using Property 6)}$$

$$I_1=\int_0^{\pi/2}\log\sin x\,dx$$

$$2I=\int_0^{\pi/2}\log\sin x\,dt\int_0^{\pi/2}\log 2dx$$

$$2I=I-\frac{\pi}{2}\log 2\ \Rightarrow I=\frac{\pi}{2}\log\left(\frac{1}{2}\right)$$

$$\int_0^{\pi/2}\log\sin x dx=\frac{\pi}{2}\log\left(\frac{1}{2}\right),$$

memorise the result for further use.

ILLUSTRATION 7.37

$$\int_{-7}^{7}\frac{x^4\sin x}{x^6+2}dx$$

Let $$I=\int_{-7}^{7}\frac{x^4\sin x}{x^6+2}dx$$

$$f(x)=\frac{x^4\sin x}{x^6+2}$$

$$f(-x)=-f(x).$$

Hence $f(x)$ is an odd function.

By Property 5, $$I=\int_{-7}^{7}f(x)dx$$

$$I=0$$

ILLUSTRATION 7.38

$$\int_{\pi/6}^{\pi/3}\frac{dx}{1+\sqrt{\tan x}}$$

$$I=\int_{\pi/6}^{\pi/3}\frac{dx}{1+\sqrt{\tan x}}$$

$$I=\int_{+\pi/6}^{\pi/3}\frac{\sqrt{\cos x}}{\sqrt{\cos x}+\sqrt{\sin x}}$$

$$I=\int_{\pi/6}^{\pi/3}\frac{\sqrt{\cos\left(\frac{\pi}{3}+\frac{\pi}{6}-x\right)}}{\sqrt{\cos\left(\frac{\pi}{3}+\frac{\pi}{6}-x\right)}+\sqrt{\sin\left(\frac{\pi}{3}+\frac{\pi}{6}-x\right)}}$$

$$I=\int_{\pi/6}^{\pi/3}\frac{\sqrt{\sin x}\,dx}{\sqrt{\sin x}+\sqrt{\cos x}}$$

$$2I=\int_{\pi/6}^{\pi/3}\frac{\sqrt{\sin x}+\sqrt{\cos x}\,dx}{\sqrt{\sin x}+\sqrt{\cos x}}$$

$$2I=\int_{\pi/6}^{\pi/3}dx\Rightarrow 2I=\frac{\pi}{3}-\frac{\pi}{6}\Rightarrow I=\frac{\pi}{12}$$

ILLUSTRATION 7.39

Show that $1 \le \int_0^1 e^{x^2}\, dx \le e$

Using formulae (Property 12), we can get

$$m(1-0) \le \int_0^1 e^{x^2}\, dx \le M(1-0)$$

Let $\quad f(x) = e^{x^2}$

$$f'(x) = 2x \times e^{x^2} = 0 \Rightarrow x = 0$$

Min. value of $f(x)$ is $m = 1$ and max. value $M = e$

$$1 \le \int_0^1 e^{x^2}\, dx \le e(1-0)$$

$$1 \le \int_0^1 e^{x^2}\, dx \le e$$

ILLUSTRATION 7.40

Show that $4 \le \int_1^3 \sqrt{3+x^3}\, dx \le 2\sqrt{30}$

Using Property 12, we can get,

$$m(3-1) \le \int_1^3 \sqrt{3+x^3}\, dx \le M(3-1)$$

$$f(x) = \sqrt{3+x^2}$$

$$f'(x) = \frac{1(x)}{\sqrt{3+x^2}}, f'(x)$$

is increasing function in [1, 3].

Hence, $\quad m = \sqrt{3+1} \quad m = 2$

$$M = \sqrt{3+27} \Rightarrow M = \sqrt{30}$$

$$4 \le \int_1^3 \sqrt{3+x^3}\, dx \le 2\sqrt{30}$$

WORKSHEET – 2

1. Evaluate $\int_0^4 \left(x - 2\sqrt{x} + x^2\right) dx$.

2. Evaluate $\int_0^{n^2} \left[\sqrt{x}\right] dx$, where [x] denotes greatest integer function.

3. If $f(x) = \begin{cases} x^2, & 0 < x < 2 \\ 3x-4, & 2 \le x < 3 \end{cases}$, find $\int_0^3 f(x)\, dx$.

4. If $f(x) = \begin{cases} x^2, & 0 \le x \le 1 \\ \sqrt{x}, & 1 \le x \le 2 \end{cases}$, find $\int_0^2 f(x)\, dx$.

5. $\int_0^{\pi/2} (2\log\sin x - \log\sin 2x)\, dx$ equals ______.

6. $\int_{-1}^{1} \frac{\sin x - x^2}{3 - |x|} dx$ is equal to ______.

7. If $I_n = \int_0^{\pi/2} x^n \sin x\, dx, n > 1,$ show that $I_n + n(n-1)I_{n-2} = n(\pi/2)^{n-1}$.

8. $\int_2^4 \frac{\sqrt{x^2-4}}{x} dx$ equals ______.

9. $\int_{-4}^{-5} e^{(x+5)^2} dx + 3\int_{1/3}^{2/3} e^{9(x-2/3)^2} dx$ is equal to ______.

10. Prove that $\int_0^{2\pi} \frac{dx}{10 + 3\cos x}$ lies between $\frac{2\pi}{13}$ and $\frac{2\pi}{7}$.

SOLVED EXAMPLES

Example 1: Evaluate $\int \frac{dx}{x\sqrt{x^4 - 1}}$

Solution

Substitute $x^2 = \sec\theta \Rightarrow 2x\, dx = \sec\theta \tan\theta\, d\theta$

$$I = \frac{1}{2}\int \frac{\sec\theta \tan\theta\, d\theta}{\sec\theta \tan\theta} = \frac{1}{2}\theta + c$$

$$= \frac{1}{2}\sec^{-1}\left(x^2\right) + c$$

Example 2: Evaluate: $\int x\sqrt{1+x^2}\,dx$

Solution

Let $\quad 1+x^2=t$

then $\quad 2x\,dx=dt$

or $\quad x\,dx=1/2\,dt.$

$$\therefore \quad \int x\sqrt{1+x^2}\,dx=\int\frac{1}{2}\sqrt{t}\,dt=\frac{t^{3/2}}{3}+c$$

$$=\frac{1}{3}\left(1+x^2\right)^{3/2}+c$$

Example 3: Evaluate $\int\frac{\left(a+b\tan^{-1}x\right)^3}{1+x^2}dx.$

Solution

Let $\quad a+b\tan^{-1}x=t$

$$\therefore \quad b\left(\frac{1}{1+x^2}\right)dx=dt \text{ or } \left(\frac{1}{1+x^2}\right)dx=\frac{dt}{b}$$

$$\therefore \quad \int\frac{\left(a+b\tan^{-1}x\right)^3}{1+x^2}dx=\frac{1}{b}\int t^3dt=\frac{1}{b}\frac{t^4}{4}+c$$

$$\Rightarrow \quad \frac{1}{4b}\left(a+b\tan^{-1}x\right)^4+c$$

Example 4: $\int\frac{1}{x\log x}dx$ is equal to ______.

(a) $\log(x\log x)+c$ **(b)** $\log(\log x+x)+c$

(c) $\log x+c$ **(d)** $\log(\log x)+c$

Solution

$$\int\frac{1}{x\log x}dx=\int\frac{1}{x}\cdot\frac{1}{\log x}dx$$

Put $\quad \log x=t,\ \frac{1}{x}dx=dt$

$$\therefore \quad \int\frac{1}{x}\cdot\frac{1}{\log x}dx=\int\frac{1}{t}dt$$

$$\therefore \quad \int\frac{1}{t}dt=\log t+c=\log(\log x)+c$$

(putting the value of $t=\log x$)

Example 5: $\int\frac{dx}{\sqrt{3-5x-x^2}}$ equals ______.

(a) $\sin^{-1}\left(\frac{2x+5}{\sqrt{37}}\right)+c$ **(b)** $\cos^{-1}\left(\frac{2x+5}{\sqrt{37}}\right)+c$

(c) $\sin^{-1}(2x+5)+c$ **(d)** None of these.

Solution

$$I=\int\frac{dx}{\sqrt{\frac{37}{4}-\left(x+\frac{5}{2}\right)^2}}$$

$$=\sin^{-1}\left(\frac{x+5/2}{\sqrt{37}/2}\right)+c=\sin^{-1}\left(\frac{2x+5}{\sqrt{37}}\right)+c$$

Example 6: Evaluate $\int\frac{dx}{\sqrt[3]{\sin^{11}x\cos x}}.$

Solution

We have $\quad I=\int\sin^{-11/3}x\cos^{-1/3}x\,dx$

Substitute $\quad \sin x=t\Rightarrow\cos x\,dx=dt$

$$\therefore \quad I=\int t^{-11/3}\left(1-t^2\right)^{-2/3}dt$$

Here $\quad \frac{m+1}{n}+p=\frac{-\frac{11}{3}+1}{2}-\frac{2}{3}=-\frac{4}{3}-\frac{2}{3}=-2$

(an integer)

Substitute

$$1-t^2=t^2\alpha^3\Rightarrow t=\frac{1}{\sqrt{1+\alpha^2}}$$

$$\Rightarrow \quad dt=\frac{-3\alpha^2}{2\left(1+\alpha^3\right)^{3/2}}d\alpha$$

$$I=\frac{3}{2}\int\frac{1}{\left(1+\alpha^2\right)^{-11/6}}\left(\frac{\alpha^3}{1+\alpha^3}\right)^{-2/3}\frac{\alpha^2}{\left(1+a^3\right)^{3/2}}d\alpha$$

$$=-\frac{3}{2}\int\left(1+\alpha^3\right)d\alpha$$

$$=-\frac{3}{2}\left\{\alpha+\frac{\alpha^4}{4}\right\}+c=-\frac{3}{2}\left(\cot^2x\right)^{1/3}-\frac{3}{8}\left(\cot^2x\right)^{4/3}+c$$

$$=-\frac{3}{2}\left\{4\left(\cot^2x\right)^{1/3}+\left(\cot^2x\right)^{4/3}\right\}+c$$

Example 7: Evaluate: $\int\frac{x^2-1}{x^4+4x^2+1}dx.$

Solution

Dividing numerator and denominator by x^2, we get

$$I=\int\frac{1-\left(1/x^2\right)}{x^2+4+1/x^2}dx=\int\left(\frac{1-\left(1/x^2\right)}{\left[\left\{x+1/x\right\}^2+2\right]}\right)dx$$

Now taking $x+1/x=t\Rightarrow\left[1-\frac{1}{x^2}\right]dx=dt$, we get

$$I=\int\frac{dt}{t^2+2}=\frac{1}{\sqrt{2}}\tan^{-1}\left(\frac{t}{\sqrt{2}}\right)+c$$

$$=\frac{1}{\sqrt{2}}\tan^{-1}\left[\frac{x+1/x}{\sqrt{2}}\right]+c=\frac{1}{\sqrt{2}}\tan^{-1}\left(\frac{x^2+1}{\sqrt{2}x}\right)+c$$

Example 8: $\int\frac{4e^2+6e^{-x}}{9e^x-4e^{-x}}dx$ is equal to ______.

(a) $\frac{19}{36}x+\frac{35}{36}\log\left(9e^x-4e^{-x}\right)+c$

(b) $-\frac{19}{36}x+\frac{35}{36}\log\left(9e^x-4e^{-x}\right)+c$

(c) $\frac{1}{36}x+\frac{1}{36}\log\left(9e^x-4e^{-x}\right)+c$

(d) None of these.

Solution

Suppose $4e^x+6e^{-x}=A\left(9e^x-4e^{-x}\right)+B\left(9e^x+4e^{-x}\right)$

By comparing $4=9A+9B,$

$6=-4A+4B$

or $\text{A}+\text{B}=\frac{4}{9},-\text{A}+\text{B}=\frac{3}{2}$

After solving, $A=-\frac{19}{36},B=\frac{35}{36}$

$$\therefore\quad I=\int\left[-\frac{19}{36}+\frac{35}{36}\left(\frac{9e^x+4e^{-x}}{9e^x-4e^{-x}}\right)\right]dx$$

$$=-\frac{19}{36}x+\frac{35}{36}\log\left(9e^x-4e^{-x}\right)+c$$

Example 9. Evaluate: $\int_1^2 x^4dx.$

Solution

$$\int_1^2 x^4dx=\left[\frac{x^5}{5}\right]_1^2=\frac{32}{5}-\frac{1}{5}=\frac{31}{5}$$

Example 10. Evaluate: $\int_1^2\frac{1}{\sqrt{4-x^2}}dx$

Solution

$$\int_1^2\frac{1}{\sqrt{4-x^2}}dx=\left[\sin^{-1}\left(\frac{x}{2}\right)\right]_1^2$$

$$=\sin^{-1}(1)-\sin^{-1}\left(\frac{1}{2}\right)$$

$$=\frac{\pi}{2}-\frac{\pi}{6}=\frac{\pi}{3}$$

Example 11. Evaluate $\int_0^{\pi/2}\sin^2 x\,dx.$

Solution

$$\int_0^{\pi/2}\sin^2 x\,dx=\int_0^{\pi/2}\left(\frac{1-\cos 2x}{2}\right)dx$$

$$=\frac{1}{2}\left[x-\frac{\sin 2x}{2}\right]_0^{\pi/2}=\frac{1}{2}\left[\frac{\pi}{2}-0\right]=\frac{\pi}{4}$$

Example 12. $\int_1^2\log x\,dx$ equals ______.

(a) 2log 2 **(b)** $\log\left(\frac{2}{e}\right)$

(c) $\log\left(\frac{4}{e}\right)$ **(d)** None of these.

Solution

$$I=\int_1^2 1\cdot\log x\,dx$$

(Integrating by parts, by taking 1 as a second function)

$$=\{x\cdot\log x\}_1^2-\int_1^2\left(\frac{1}{x}.x\right)dx=(2\log 20\log 1)-[x]_1^2$$

$$=(2\log 2-0)-(2-1)$$

$$=\log 4-\log e=\log\left(\frac{4}{e}\right)$$

Example 13. Evaluate $\int_0^2|1-x|dx.$

Solution

$$\because |1-x|=\begin{cases}1-x, & \text{when } 0<x<1\\ x-1, & \text{when } 1\le x<2\end{cases}$$

$$\therefore\quad I=\int_0^1(1-x)dx+\int_1^2(x-1)\,dx$$

$$=\left[x-\frac{x^2}{2}\right]_0^1+\left[\frac{x^2}{2}-x\right]_1^2=\left(\frac{1}{2}-0\right)+\left(0+\frac{1}{2}\right)=1$$

Example 14. If $f(x)=\begin{cases}2x+1, & \text{when } 0<x<1\\ x^2+2, & \text{when } 1\le x<2\end{cases}$, then the value of $\int_0^2 f(x)dx$ is ______.

(a) $\frac{-19}{3}$ **(b)** $\frac{19}{3}$

(c) $\frac{3}{19}$ **(d)** None of these.

Solution

$$\int_0^2 f(x)dx = \int_0^1 f(x)dx + \int_1^2 f(x)dx$$

$$= \int_0^1 (2x+1)dx + \int_1^2 (x^2+2)dx$$

$$= \left[x^2+x\right]_0^1 + \left[\frac{x^3}{3}+2x\right]_1^2$$

$$= 2-0+\left(\frac{20}{3}-\frac{7}{3}\right) = \frac{19}{3}$$

Example 15. Show that $\int_0^\pi x f(\sin x)dx = \frac{\pi}{2}\int_0^\pi f(\sin x)dx$.

Solution

Let $I = \int_0^\pi x f(\sin x)dx$

$$= \int_0^\pi (\pi - x) f(\sin(\pi - x))dx$$

$$= \int_0^\pi (\pi - x) f(\sin x)dx$$

$$= \pi\int_0^\pi f(\sin x)dx - \int_0^\pi x f(\sin x)dx$$

$$= \pi\int_0^\pi f(\sin x)dx - I$$

$$\Rightarrow \quad 2I = \pi\int_0^\pi f(\sin x)dx$$

$$\Rightarrow \quad I = \frac{\pi}{2}\int_0^\pi f(\sin x)dx$$

Example 16. Evaluate $\int_0^{\pi/2} \frac{\sqrt{\sin^7 x}}{\sqrt{\sin^7 x}+\sqrt{\cos^7 x}}dx$

Solution

$$I = \int_0^{\pi/2} \frac{\sqrt{\sin^7 x}}{\sqrt{\sin^7 x}+\sqrt{\cos^7 x}}dx$$

$$\Rightarrow \quad I = \int_0^{\pi/2} \frac{\sqrt{\cos^7 x}}{\sqrt{\sin^7 x}+\sqrt{\cos^7 x}}dx$$

$$\therefore \quad 2I = \int_0^{\pi/2} dx = (x)_0^{\pi/2}$$

$$\therefore \quad 2I = \frac{\pi}{2} \Rightarrow I = \frac{\pi}{4}$$

Example 17. Evaluate $\int_0^1 \log\left(\frac{1}{x}-1\right)dx$.

Solution

$$I = \int_0^1 \log\left(\frac{1}{x}-1\right)dx \quad \ldots(1)$$

$$\Rightarrow \quad I = \int_0^1 \log\left[\frac{1-(1-x)}{1-x}\right]dx$$

$$= \int_0^1 \log\left(\frac{x}{x-1}\right)dx = -\int_0^1 \log\left(\frac{1-x}{x}\right)dx = -I$$

$$\Rightarrow \quad 2I \Rightarrow I = 0$$

Example 18. Evaluate $\int_{\pi/6}^{\pi/3} \frac{dx}{1+\sqrt{\tan x}}$.

Solution

$$I = \int_{\pi/6}^{\pi/3} \frac{\sqrt{\cos x}}{\sqrt{\sin x}+\sqrt{\cos x}}dx \quad \ldots(1)$$

$$\therefore \quad I = \int_{\pi/6}^{\pi/3} \frac{\sqrt{\cos\left(\frac{\pi}{6}+\frac{\pi}{3}-x\right)}}{\sqrt{\sin\left(\frac{\pi}{6}+\frac{\pi}{3}-x\right)}+\sqrt{\cos\left(\frac{\pi}{6}+\frac{\pi}{3}-x\right)}}dx$$

$$\Rightarrow \quad I = \int_{\pi/6}^{\pi/3} \frac{\sqrt{\sin x}}{\sqrt{\cos x}+\sqrt{\sin x}}dx \quad \ldots(2)$$

Adding (1) and (2), we get

$$\therefore \quad 2I = \int_{\pi/6}^{\pi/3} dx = (x)_{\pi/6}^{\pi/3} = \frac{\pi}{6} \Rightarrow I = \frac{\pi}{12}$$

Example 19. Evaluate $\int_1^4 \frac{\sqrt{x}}{\sqrt{5-x}+\sqrt{x}}dx$

Solution

Here $\quad I = \int_1^4 \frac{\sqrt{x}}{\sqrt{5-x}+\sqrt{x}}dx = \frac{4-1}{2} = \frac{3}{2}$

Example 20. $\int_0^{\pi/2} \frac{\sqrt{\sin x}}{\sqrt{\sin x}+\sqrt{\cos x}}dx$ equals ______.

(a) $\frac{\pi}{2}$ **(b)** $\frac{\pi}{4}$

(c) π **(d)** 2π

Solution

Using Property 4, we have

$$I = \int_0^{\pi/2} \frac{\sqrt{\cos x}}{\sqrt{\cos x}+\sqrt{\sin x}}dx$$

Adding it to the given integral we have,

$$2I = \int_0^{\pi/2} dx = [x]_0^{\pi/2} = \frac{\pi}{2} \Rightarrow I = \frac{\pi}{4}$$

Example 21. $\int_0^{\pi/2} \frac{\cos x - \sin x}{1+\sin x \cos x} dx$ equals ______.

(a) 0 **(b)** 1

(c) $\frac{\pi}{2}$ **(d)** $\frac{\pi}{4}$

Solution

$$I = \int_0^{\pi/2} \frac{\cos\left(\frac{\pi}{2}-x\right) - \sin\left(\frac{\pi}{2}-x\right)}{1+\sin\left(\frac{\pi}{2}-x\right)\cos\left(\frac{\pi}{2}-x\right)} dx$$

$$= \int_0^{\pi/2} \frac{\sin x - \cos x}{1+\sin x \cos x} dx = -I$$

$\Rightarrow \quad 2I = 0 \Rightarrow I = 0$

Example 22: Let f be a positive function. If $I_1 = \int_{1-k}^{k} x f\{x(1-x)\} dx, I_2 = \int_{1-k}^{k} f[x(1-x)]dx$, where $2k - 1 > 0$, then the value of I_1/I_2 is equal to ______.

(a) 2 **(b)** k

(c) 1/2 **(d)** 1

Solution

$$I_1 = \int_{1-k}^{k} (k+1-k-x) f[(k+1-k-x)\times(1-k-1+k+x)]dx$$

$$= \int_{1-k}^{k} (1-x) f[(1-x)(x)]dx$$

$$= \int_{1-k}^{k} f[x(1-x)]dx - \int_{1-k}^{k} xf[x(1-x)]dx = I_2 - I_1$$

$\Rightarrow \quad 2I_1 = I_2$

$\therefore \quad \frac{I_1}{I_2} = \frac{1}{2}$

20.

Example 23: Evaluate $\int_0^{\pi/2} \log \sin x \, dx$

Solution

Let $I = \int_0^{\pi/2} \log \sin x \, dx$ ...(1)

Then $I = \int_0^{\pi/2} \log \sin\left(\frac{\pi}{2}-x\right)dx = \int_0^{\pi/2} \log(\cos x)dx$ ...(2)

Adding (1) and (2) we get

$$2I = \int_0^{\pi/2} \log(\sin x \cos x)dx = \int_0^{\pi/2} \log\left(\frac{\sin 2x}{2}\right)dx$$

$$= \int_0^{\pi/2} \log(\sin 2x)dx - \int_0^{\pi/2} (\log 2)dx$$

$$= \frac{1}{2}\int_0^{\pi} \log(\sin t)dt - (\log 2)\frac{\pi}{2}$$

$$\left(\text{putting } 2x = t \Rightarrow dx = \frac{dt}{2}\right)$$

$$= \frac{1}{2} 2 \int_0^{\pi/2} \log(\sin t)dt - \frac{2\pi}{2}\log 2 (\text{using, } \int_0^{2a} f(x)dx$$

$$= 2\int_0^{a} f(x)dx, (f(2a-x) = f(x))$$

$$= I - \frac{\pi}{2}\log 2 \Rightarrow I = \frac{-\pi}{2}\log 2$$

Example 24: $\int_0^1 \frac{6x^2+1}{4x^3+2x+3}$ is equal to ______.

(a) $-\frac{1}{2}\log 3$ **(b)** $\frac{1}{2}\log 3$

(c) 2 log 3 **(d)** None of these.

Solution

Let $4x^3 + 2x + 3 = t$

$\therefore \quad 2(6x^2+1)dx = dt$

$x = 0;\ t = 3$, at $x = 1;\ t = 9$

$$\therefore\ I = \int_3^9 \frac{1}{2}\frac{dt}{t} = \frac{1}{2}[\log t]_3^9 = \frac{1}{2}[\log 9 - \log 3] = \frac{1}{2}\log 3$$

Example 25: $\int_0^\infty \frac{x \log x}{(1+x^2)} dx$ equals ______

(a) 0 **(b)** log 7

(c) 5 log 13 **(d)** None of these.

Solution

Here,

$$\int_0^\infty \frac{x\log x}{(1+x^2)^2}dx = \int_0^1 \frac{x\log x}{(1+x^2)^2}dx + \int_1^\infty \frac{x\log x}{(1+x^2)^2}dx$$

$$I = I_1 + I_2$$

Putting $x = \frac{1}{t}$ in second integrand

$$dx = -\frac{1}{t^2}dt$$

$$\therefore\quad I_2=\int_1^0\frac{\frac{1}{t}\log\left(\frac{1}{t}\right)}{\left(1+\frac{1}{t^2}\right)^2}\left(-\frac{1}{t^2}\right)dt=-\int_0^1\frac{t\log t}{\left(1+t^2\right)^2}dt=-I_1$$

$$I=I_2+I_1=-I_1+I_1=0$$

Example 26: Evaluate $\int_0^\infty \frac{x}{(1+x)\left(1+x^2\right)}dx$

Solution

Substitute $x=\tan\theta, dx=\sec^2\theta d\theta$ (limits of integration are changed to 0 and $\pi/2$)

$$\therefore\quad I=\int_0^\infty\frac{x}{(1+x)\left(1+x^2\right)}dx$$

$$=\int_0^{\pi/2}\frac{\tan\theta}{(1+\tan)\sec^2\theta}\sec^2\theta\,d\theta$$

$$=\int_0^{\pi/2}\frac{\tan\theta}{1+\tan\theta}d\theta$$

$$\Rightarrow\quad I=\int_0^{\pi/2}\frac{\sin\theta}{\sin\theta+\cos\theta}d\theta \quad \ldots(1)$$

$$\Rightarrow\quad I=\int_0^{\pi/2}\frac{\sin\left(\frac{\pi}{2}-\theta\right)}{\sin\left(\frac{\pi}{2}-\theta\right)+\cos\theta\left(\frac{\pi}{2}-\theta\right)}d\theta$$

$$\Rightarrow\quad I=\int_0^{\pi/2}\frac{\cos\theta}{\cos\theta+\sin\theta}d\theta \quad \ldots(2)$$

Adding (1) and (2), we get

$$2I=\int_0^{\pi/2}d\theta=\frac{\pi}{2}\Rightarrow I=\frac{\pi}{4}$$

Example 27: Compute the following limits (using definite integral):

(1) $\lim_{n\to\infty}\frac{1}{n}\left(\sqrt{1+\frac{1}{n}}+\sqrt{1+\frac{2}{n}}+\ldots.+\sqrt{1+\frac{n}{n}}\right)$

Solution $\quad=\lim_{n\to\infty}\frac{1}{n}\sum_{r=1}^{n}\sqrt{1+\frac{r}{n}}$

$$=\int_0^1\sqrt{1+x}\,dx=\left[\frac{(1+x)^{3/2}}{\frac{3}{2}}\right]_0^1$$

$$=\frac{2^{3/2}}{\frac{3}{2}}-\frac{1}{\frac{3}{2}}=\frac{2}{3}\{2\sqrt{2}-1\}$$

(2) $\lim_{n\to\infty}\frac{\pi}{2n}\left[1+\cos\frac{\pi}{2n}+\cos\frac{2\pi}{2n}+\ldots..+\cos\frac{(n-1)\pi}{2n}\right]$

Solution $\quad=\lim_{n\to\infty}\frac{(\pi/2)}{n}\sum_{r=0}^{n-1}\cos\left(r\left(\frac{\pi/2}{n}\right)\right)$

[**Note:** Here, interval $[0, \pi/2]$ is divided into n equal parts]

$$=\int_0^{\pi/2}\cos x\,dx=[\sin x]_0^{\pi/2}=1$$

RECAPITULATION (FORMULAE, THEOREM, IMPORTANT RESULTS)

1. Integration is the inverse process of differentiation. In differential calculus, we are given a function and we have to find the derivative or differential of this function, but in the integral calculus, we are to find a function whose differential is given. Thus, integration is a process which is the inverse of differentiation.

   Let $\frac{d}{dx}F(x)=f(x)$. Then we write

   $\int f(x)dx=F(x)+C$. These integrals are called indefinite integrals or general integrals; C is called constant of integration. All these integrals differ by a constant.

2. From the geometric point of view, an indefinite integral is collection of family of curves, each of which is obtained by translating one of the curves parallel to itself upwards or downwards along the y-axis.

3. Some properties of indefinite integrals are as follows:

First Fundamental Theorem of Integral Calculus

Let the area function be defined by $A(x)=\int_a^x f(x)dx$, for all $x\geq a$, where the function f is assumed to be continuous on $[a, b]$. Then $A'(x)=f(x)$ for all $x\in[a,b]$.

Second Fundamental Theorem of Integral Calculus

Let f be a continuous function of x defined on the closed interval $[a, b]$ and let F be another function such that $\frac{d}{dx}F(x) = f(x)$ for all x in the domain of f. Then

$\int_a^b f(x)dx = [F(x)+C]_a^b = F(b) - F(a)$.

This is called the definite integral of f over the range $[a, b]$, where a and b are called the limits of integration, a being the lower limit and b the upper limit.

Integration by Partial Fractions

Recall that a rational function is the ratio of two polynomials of the form $\frac{P(x)}{Q(x)}$, where $P(x)$ and $Q(x)$ are polynomials in x and $Q(x) \neq 0$. If degree of the polynomial $P(x)$ is greater than the degree of the polynomial $Q(x)$, then we may divide $P(x)$ by $Q(x)$ so that $\frac{P(x)}{Q(x)} = T(x) + \frac{P_1(x)}{Q(x)}$, where $T(x)$ is a polynomial in x and degree of $P_1(x)$ is less than the degree of $Q(x)$. $T(x)$, being a polynomial, can be easily integrated. $\frac{P_1(x)}{Q(x)}$ can be integrated expressing $\frac{P_1(x)}{Q(x)}$ as the sum of partial fractions of the following type:

1. $\frac{px+q}{(x-a)(x-b)} = \frac{A}{x-a} + \frac{B}{x-b}, a \neq b$

2. $\frac{px+q}{(x-a)^2} = \frac{A}{x-a} + \frac{B}{(x-a)^2}$

3. $\frac{px^2+qx+r}{(x-a)(x-b)(x-c)} = \frac{A}{x-a} + \frac{B}{x-b} + \frac{C}{x-c}$

4. $\frac{px^2+qx+r}{(x-a)^2(x-b)} = \frac{A}{x-a} + \frac{B}{(x-a)^2} + \frac{C}{x-b}$

5. $\frac{px^2+qx+r}{(x-a)(x^2+bx+c)} = \frac{A}{x-a} + \frac{Bx+C}{x^2+bx+c}$

Where $x^2 + bx + c$ cannot be factorized further.

Integration by Substitution

A change in the variable of integration often reduces an integral to one of the fundamental integrals. Thc method in which we change the variable to some other variable is called the method of substitution. When the integrand involves some trigonometric functions, we use some well-known identities to find the integrals. Using substitution technique, we obtain the following standard integrals.

(i) $\int \tan x\, dx = \log|\sec x| + C$

(ii) $\int \cot x\, dx = \log|\sin x| + C$

(iii) $\int \sec x\, dx = \log|\sec x + \tan| + C$

(iv) $\int \cos ecx\, dx = \log|\cos ecx - \cot x| + C$

Integrals of Some Special Functions

(i) $\int \frac{dx}{x^2 - a^2} = \frac{1}{2a}\log\left|\frac{x-a}{x+a}\right| + C$

(ii) $\int \frac{dx}{a^2 - x^2} = \frac{1}{2a}\log\left|\frac{a+x}{a-x}\right| + C$

(iii) $\int \frac{dx}{x^2 + a^2} = \frac{1}{a}\tan^{-1}\frac{x}{a} + C$

(iv) $\int \frac{dx}{\sqrt{x^2 - a^2}} = \log\left|x + \sqrt{x^2 - a^2}\right| + C$

(v) $\int \frac{dx}{\sqrt{a^2 - x^2}} = \sin^{-1}\frac{x}{a} + C$

(vi) $\int \frac{dx}{\sqrt{x^2 + a^2}} = \log\left|x + \sqrt{x^2 + a^2}\right| + C$

Integration by Parts

For given functions f_1 and f_2, we have

$\int f_1(x).f_2(x)dx$

$= f_1(x)\int f_2(x)dx - \int\left[\frac{d}{dx}f_1(x).\int f_2(x)dx\right]dx$,

i.e., the integral of the product of two functions = first function × integral of the second function – integral of {differential coefficient of the first function × integral of the second function}. Care must be taken in choosing the first function and the second function. Obviously, we must take that function as the second function, whose integral is well known to us.

$$\int e^x[f(x) + f'(x)]dx = \int e^x f(x)dx + C$$

Some Special Types of Integrals

(i) $\int \sqrt{x^2 - a^2}\, dx$

$= \frac{x}{2}\sqrt{x^2 - a^2} - \frac{a^2}{2}\log\left|x + \sqrt{x^2 - a^2}\right| + C$

(ii) $\int\sqrt{x^2+a^2}\,dx$

$$=\frac{x}{2}\sqrt{x^2+a^2}+\frac{a^2}{2}\log\left|x+\sqrt{x^2+a^2}\right|+C$$

(iii) $\int\sqrt{a^2-x^2}\,dx=\frac{x}{2}\sqrt{a^2-x^2}+\frac{a^2}{2}\sin^{-1}\frac{x}{a}+C$

(iv) Integrals of the types $\int\frac{dx}{ax^2+bx+c}$ or $\int\frac{dx}{\sqrt{ax^2+bx+c}}$ can be transformed into standard form by expressing

$$ax^2+bx+c=a\left[x^2+\frac{b}{a}x+\frac{c}{a}\right]$$

$$=a\left[\left(x+\frac{b}{2a}\right)^2+\left(\frac{c}{a}-\frac{b^2}{4a^2}\right)\right]$$

(v) Integrals of the types $\int\frac{px+q\,dx}{ax^2+bx+c}$ or $\int\frac{px+q\,dx}{\sqrt{ax^2+bx+c}}$ can be transformed into standard form by expressing

$$px+q=A\frac{d}{dx}\left(ax^2+bx+c\right)+B=A(2ax+b)+B,$$

where A and B are determined by comparing coefficients on both sides.

We have defined $\int_a^b f(x)dx$ as the area of the region bounded by the curve $y=f(x), a\le x\le b$, the x-axis and the ordinates $x=a$ and $x=b$. Let x be a given point in $[a, b]$. Then $\int_a^x f(x)dx$ represents the Area function $A(x)$. This concept of area function leads to the Fundamental Theorems of Integral Calculus.

QUIZ

1. What do you mean by anti-derivative?
2. What is basic difference between the derivative of a polynomial and integration of a polynomial?
3. For the integral of $\int\sin^3 x\cos^2 x\,dx$ suggest a substitution.
4. What trigonometric identities will be used to integrate $\int\cos^2 x\,dx$?
5. The value of integral $\int\frac{dx}{x^2-a^2}$ is ______.
6. Integrate $\int\frac{dx}{x^2-16}$.
7. If $F(x)$ be the anti-derivative of $f(x)$ then the value of $\int_a^b f(x)$ is ______.
8. Evaluate the integral $\int_1^2 x^2dx$.
9. What is the value of integral $\int\sqrt{x^2+a^2}\,dx$?
10. What is the substitution to integral $\int\cos 6x\sqrt{1+\sin 6x}\,dx$.

JEE MAIN

Objective Questions

One Option Correct

1. The value of $\sqrt{2}\int\frac{\sin dx}{\sin(x-\pi/4)}$ is ______.

(a) $x-\log\left|\sin\left(x-\frac{\pi}{4}\right)\right|+c$

(b) $x-\frac{\pi}{4}+\log\left|\sin\left(x-\frac{\pi}{4}\right)\right|+c$

(c) $x-\log\left|\cos\left(x-\frac{\pi}{4}\right)\right|+c$

(d) $x+\log\left|\cos\left(x-\frac{\pi}{4}\right)\right|+c$

2. The value of the integral, $\int_3^6\frac{\sqrt{x}}{\sqrt{9-x}+\sqrt{x}}dx$ is ______.

(a) $\frac{3}{2}$ (b) 2

(c) 1 (d) $\frac{1}{2}$

3. $\int_{-3\pi/2}^{-\pi/2} [(x+\pi)^3 + \cos(x+3\pi)]dx$ is equal to ______.

(a) $\left(\frac{\pi^4}{32}\right)+\left(\frac{\pi}{2}\right)$ (b) $\frac{\pi}{2}$

(c) $\frac{\pi}{4}-1$ (d) $\frac{\pi^4}{32}$

4. The value of $\int_1^a [x] f'(x)dx; a>1$ where $[x]$ denotes the greatest integer not exceeding x is ______.

(a) $[a]f(a)-\{f(1)+f(2)+\ldots+f([a])\}$

(b) $[a]f([a])-\{f(1)+f(2)+\ldots+f(a)\}$

(c) $a f([a])-\{f(1)+f(2)+\ldots+f(a)\}$

(d) $a f(a)-\{f(1)+f(2)+\ldots+f([a])\}$

5. $\int\left(\frac{(\log x-1)}{1+(\log x)}\right)^2 dx$ is equal to ______.

(a) $\frac{xe^2}{1+x^2}+c$ (b) $\frac{x}{(\log x)^2+1}+c$

(c) $\frac{\log x}{(\log x)^2+1}+c$ (d) $\frac{x}{x^2+1}+c$

6. If $I_1=\int_0^1 2^{x^2}dx, I_2=\int_0^1 2^{x^3}dx, I_3=\int_0^2 2^{x^2}dx$ and $I_4=\int_0^2 2^{x^2}dx$, then ______.

(a) $I_3=I_4$ (b) $I_3>I_4$

(c) $I_2>I_1$ (d) $I_1>I_2$

7. If $\int\frac{\sin x}{\sin(x-\alpha)}dx = Ax+B\log\sin(x-\alpha),+C,$ then value of (A,B) is ______.

(a) $(\sin\alpha,\cos\alpha)$ (b) $(\cos\alpha,\sin\alpha)$

(c) $(-\sin\alpha,\cos\alpha)$ (d) $(-\cos\alpha,\sin\alpha)$

8. $\int\frac{dx}{\cos x-\sin x}$ is equal to ______.

(a) $\frac{1}{\sqrt{2}}\log\left|\tan\left(\frac{x}{2}-\frac{\pi}{8}\right)\right|+c$

(b) $\frac{1}{\sqrt{2}}\log\left|\cot\left(\frac{x}{2}\right)\right|+c$

(c) $\frac{1}{\sqrt{2}}\log\left|\tan\left(\frac{x}{2}-\frac{3\pi}{8}\right)\right|+c$

(d) $\frac{1}{\sqrt{2}}\log\left|\tan\left(\frac{x}{2}+\frac{3\pi}{8}\right)\right|+c$

9. If $f(x),=\frac{e^x}{1+e^x}, I_1=\int_{f(-a)}^{f(a)} xg\{x(1-x)\}dx$ and $I_2=\int_{-f(-a)}^{f(a)} g\{x(1-x)\}dx$, then the value of $\frac{I_2}{I_1}$ is ______.

(a) 2 (b) –3

(c) –1 (d) 1

10. If $f(y)=e^y, g(y)=y; y>0$ and $F(t)=\int_0^t f(t-y)g(y),dy$, then

(a) $F(t)=1-e^{-1}(1+t)$ (b) $F(t)=e^t-(1+t)$

(c) $F(t)=te^t$ (d) $F(t)=te^{-t}$

11. Let $f(x)$ be a function satisfying $f'(x)=f(x)$, with $f(0)=1$. and $g(x)$ be a function that satisfies $f(x)+g(x)=x^2$. Then the value of integral $\int_0^1 f(x)g(x)dx$ is ______.

(a) $e-\frac{e^2}{2}-\frac{5}{2}$ (b) $e+\frac{e^2}{2}-\frac{3}{2}$

(c) $e-\frac{e^2}{2}-\frac{3}{2}$ (d) $e+\frac{e^2}{2}+\frac{3}{2}$

12. $\int_0^{10}|x(x-1)(x-2)|dx$ is equal to ______.

(a) 160.05 (b) 1600.5

(c) 16.005 (d) None of these.

13. $\int_0^3|x^3+x^2+3x|dx$ is equal to ______.

(a) $\frac{171}{2}$ (b) $\frac{171}{4}$

(c) $\frac{170}{4}$ (d) $\frac{170}{3}$

14. The value of $\int\frac{1}{\left[(x-1)^3(x+2)^5\right]^{1/4}}dx$ is ______.

(a) $\frac{4}{3}\left(\frac{x-1}{x+2}\right)^{1/4}+c$ (b) $\frac{4}{3}\left(\frac{x+1}{x+2}\right)^{1/4}+c$

(c) $\frac{4}{3}\left(\frac{x+1}{x-2}\right)^{1/4}+c$ (d) $\frac{4}{3}\left(\frac{x-1}{x-2}\right)^{1/4}+c$

15. $\int \frac{dx}{x^2+4x+13}$ is equal to ______.

(a) $\log(x^2+4x+13)+c$ (b) $\frac{1}{3}\tan^{-1}\left(\frac{x+2}{3}\right)+c$

(c) $\log(2x+4)+c$ (d) $\frac{2x+4}{(x^2+4x+13)^2}+c$

16. The value of the $\int \frac{\sin x+\cos x}{3+\sin 2x}dx$ is ______.

(a) $\frac{1}{4}\ln\left(\frac{2-\sin x+\cos x}{2+\sin x+\cos x}\right)+c$

(b) $\frac{1}{2}\ln\left(\frac{2+\sin x}{2-\sin x}\right)+c$

(c) $\frac{1}{4}\ln\left(\frac{1+\sin x}{1-\sin x}\right)+c$

(d) None of the above.

17. The value of $\int_{-\pi/2}^{\pi/2} \frac{\sin^2 x-\cos^2 x}{\sin^3 x+\cos^3 x}dx$ is ______.

(a) 0 (b) 1
(c) 2 (d) 3

18. The value of $\int \frac{dx}{x+\sqrt{x-1}}$ is ______.

(a) $\log\left|x+\sqrt{x-1}\right|+\sin^{-1}\left(\sqrt{\frac{x-1}{x}}\right)+c$

(b) $\log\left|x+\sqrt{x-1}\right|+c$

(c) $\log\left|x+\sqrt{x-1}\right|-\frac{2}{\sqrt{3}}\tan^{-1}\left(\frac{2\sqrt{x-1}+1}{\sqrt{3}}\right)+c$

(d) None of the above.

19. If $\int \frac{\cos 4x+1}{\cot x-\tan x}dx = k\cos 4x+c$, then

(a) $k=-\frac{1}{2}$ (b) $k=-\frac{1}{8}$

(c) $k=-\frac{1}{4}$ (d) None of these.

20. If $\int \frac{1}{x+x^5}dx = f(x)+c$, then the value of $\int \frac{x^4}{x+x^5}dx$ equals ______.

(a) $\log x - f(x)+c$ (b) $f(x)+\log x+c$

(c) $f(x)-\log x+c$ (d) None of the above.

21. Let $I=\int_0^1 \frac{\sin x}{\sqrt{x}}$ and $J=\int_0^1 \frac{\cos x}{\sqrt{x}}dx$. Then which one of following is true?

(a) $I<\frac{2}{3}$ and $J<2$

(b) $I<\frac{2}{3}$ and $J>2$

(c) $I>\frac{2}{3}$ and $J<2$

(d) $I>\frac{2}{3}$ and $J>2$

22. The value of $\int_{-\pi}^{\pi} \frac{\cos^2 x}{1+a^x}dx, a<0$, is ______.

(a) $\frac{\pi}{a}$ (b) 2π

(c) $a\pi$ (d) $\frac{\pi}{2}$

23. If $\int_0^{\pi} xf(\sin x)dx = A\int_0^{\pi/2} f(\sin x)dx = A...$, then A is ______.

(a) 0 (b) π

(c) $\frac{\pi}{4}$ (d) 2π

24. If $f(a+b-x)=f(x)$, then $\int_a^b xf(x)dx$ is equal to ______.

(a) $\frac{a+b}{2}\int_a^b f(b-x)dx$

(b) $\frac{a+b}{2}\int_a^b f(x)dx$

(c) $\frac{b-a}{2}\int_a^b f(x)dx$

(d) $4\frac{a+b}{2}\int_a^b f(a+b-c)dx$

25. The value of the integral $I=\int_0^1 x(1-x)^n dx$ is ______.

(a) $\frac{1}{n+1}$ (b) $\frac{1}{n+2}$

(c) $\frac{1}{n+1}-\frac{1}{n+2}$ (d) $\frac{1}{n+1}+\frac{1}{n+2}$

26. If $f(t)=\int_{-t}^{t} \frac{e^{-|x|}}{2}dx$, then $\lim_{t\to\infty} f(t)$ is equal to ______.

(a) 1 (b) $\frac{1}{2}$

(c) 0 (d) –1

27. If $\int \sqrt{\frac{x}{a^3-x^3}}\,dx = g(x)+c$, then $g(x)$ is equal to ______.

(a) $\frac{2}{3}\cos^{-1} x$ (b) $\frac{2}{3}\sin^{-1}\left(\frac{x^3}{a^3}\right)$

(c) $\frac{2}{3}\sin^{-1}\left(\sqrt{\frac{x^3}{a^3}}\right)$ (d) $\frac{2}{3}\cos^{-1}\left(\frac{x}{a}\right)$

28. If $\int \frac{dx}{x^2+2x+2} = f(x)+c$, then $f(x)$ is equal to ______.

(a) $\tan^{-1}(x+1)$ (b) $2\tan^{-1}(x+1)$

(c) $-\tan^{-1}(x+1)$ (d) $3\tan^{-1}(x+1)$

29. If $\int \frac{\sin x}{\cos x(1+\cos x)}dx = f(x)+c$ then $f(x)$ is equal to ______.

(a) $\log\left|\frac{1+\cos x}{\cos x}\right|$ (b) $\log\left|\frac{\cos x}{1+\cos x}\right|$

(c) $\log\left|\frac{\sin x}{1+\sin x}\right|$ (d) $\log\left|\frac{1+\sin x}{\sin x}\right|$

30. $\int \frac{x^{49}\tan^{-1}(x^{50})}{(1+x^{100})}dx = k(\tan^{-1}(x^{50}))^2+c$, then k is equal to ______.

(a) $\frac{1}{50}$ (b) $-\frac{1}{50}$

(c) $\frac{1}{100}$ (d) $-\frac{1}{100}$

31. $\int \frac{dx}{(x+100)\sqrt{x+99}} = f(x)+c \Rightarrow f(x) =$ ______.

(a) $2(x+100)^{1/2}$ (b) $3(x+100)^{1/2}$

(c) $2\tan^{-1}(\sqrt{x+99})$ (d) $2\tan^{-1}(\sqrt{x+100})$

32. $\int \frac{3-x^2}{1-2x+x^2}e^x\,dx = e^x f(x)+c \Rightarrow f(x) =$ ____.

(a) $\frac{1+x}{1-x}$ (b) $\frac{1-x}{1+x}$

(c) $\frac{1+x}{x-1}$ (d) $\frac{x-1}{1+x}$

33. $\int \frac{\sqrt{\cot x}}{\sin x\cos x}dx = -f(x)+c \Rightarrow f(x) =$ ______.

(a) $2\sqrt{\tan x}$ (b) $-2\sqrt{\tan x}$

(c) $-2\sqrt{\cot x}$ (d) $2\sqrt{\cot x}$

34. $\int \frac{1+x+\sqrt{x+x^2}}{\sqrt{x}+\sqrt{1+x}}dx$ is equal to ______.

(a) $\frac{1}{2}\sqrt{1+x}+c$ (b) $\frac{2}{3}(1+x)^{3/2}+c$

(c) $\sqrt{1+x}+c$ (d) $2(1+x)^{3/2}+c$

35. $\int \frac{dx}{1-\cos x-\sin x}$ is equal to ______.

(a) $\log\left|1+\cot\frac{x}{2}\right|+c$ (b) $\log\left|1-\tan\frac{x}{2}\right|+c$

(c) $\log\left|1-\cot\frac{x}{2}\right|+c$ (d) $\log\left|1+\tan\frac{x}{2}\right|+c$

36. $\int \frac{dx}{7+5\cos x}$ is equal to ______.

(a) $\frac{1}{\sqrt{3}}\tan^{-1}\left(\frac{1}{\sqrt{3}}\tan\frac{x}{2}\right)+c$

(b) $\frac{1}{\sqrt{6}}\tan^{-1}\left(\frac{1}{\sqrt{6}}\tan\frac{x}{2}\right)+c$

(c) $\frac{1}{7}\tan^{-1}\left(\tan\frac{x}{2}\right)+c$

(d) $\frac{1}{4}\tan^{-1}\left(\tan\frac{x}{2}\right)+c$

37. $\int \frac{3^x\,dx}{\sqrt{9^x-1}}$ is equal to ______.

(a) $\frac{1}{\log 3}\log|3^x+\sqrt{9^x-1}|+c$

(b) $\frac{1}{\log 3}\log|3^x-\sqrt{9^x-1}|+c$

(c) $\frac{1}{\log 9}\log|3^x+\sqrt{9^x-1}|+c$

(d) $\frac{1}{\log 3}\log|9^x+\sqrt{9^x-1}|+c$

38. $\int \frac{dx}{\sqrt{x}(x+9)}$ is equal to ______.

(a) $\frac{2}{3}\tan^{-1}\sqrt{x}+c$ (b) $\frac{2}{3}\tan^{-1}\left(\frac{\sqrt{x}}{3}\right)+c$

(c) $\tan^{-1}(\sqrt{x})+c$ (d) $\tan^{-1}\left(\frac{\sqrt{x}}{3}\right)+c$

39. $\int (x+1)^2 e^x dx$ is equal to ______.

(a) xe^x+c (b) x^2x^x+c

(c) $(x+1)e^x+c$ (d) $(x^2+1)e^x+c$

40. $\int \frac{\sin^6 x}{\cos^8 x}dx$ is equal to ______.

(a) $\tan 7x+c$ (b) $\frac{\tan^7 x}{7}+c$

(c) $\frac{\tan 7x}{7}+c$ (d) $\sec^7 x+c$

41. The value of α which satisfies $\int_{\pi/2}^{\alpha} \sin x dx = \sin 2\alpha$, $[\alpha \in (0, 2\pi)]$ is equal to ______.

(a) $\frac{\pi}{2}$ (b) $\frac{3\pi}{2}$

(c) $\frac{7\pi}{6}$ (d) All the above.

42. The value of $I = \int_0^{\pi/2} \frac{(\sin x + \cos x)^2}{\sqrt{1+\sin 2x}} dx$ is ______.

(a) 0 (b) 1

(c) 2 (d) 3

43. The value of $\lim_{x\to 0} \frac{\int_0^{x^2} \sec^2 t dt}{x \sin x}$ is ______.

(a) 3 (b) 2

(c) 1 (d) 0

44. $\lim_{n\to\infty} \frac{1+2^4+3^4+\ldots+n^4}{n^5} - \lim_{n\to\infty} \frac{1+2^3+3^3\ldots+n^3}{n^5}$ is ______.

(a) $\frac{1}{30}$ (b) zero

(c) $\frac{1}{4}$ (d) $\frac{1}{5}$

45. $\int_0^{\pi/2} \sin^8 x \cos^2 x dx$ is equal to ______.

(a) $\frac{\pi}{512}$ (b) $\frac{3\pi}{512}$

(c) $\frac{5\pi}{512}$ (d) $\frac{7\pi}{512}$

46. If $\int \frac{e^x - 1}{e^x + 1} dx = f(x) + c$, then $f(x)$ is equal to ______.

(a) $2 \log (e^x + 1)$ (b) $\log (e^{2x} - 1)$

(c) $2\log (e^x + 1) - x$ (d) $\log (e^{2x} + 1)$

47. $\int \tan^{-1}\left(\sqrt{\frac{1-x}{1+x}}\right) dx$ is equal to ______.

(a) $\frac{1}{2}(x\cos^{-1} x - \sqrt{1-x^2}) + c$

(b) $\frac{1}{2}(x\cos^{-1} x + \sqrt{1-x^2}) + c$

(c) $\frac{1}{2}(x\sin^{-1} x - \sqrt{1-x^2}) + c$

(d) $\frac{1}{2}(x\sin^{-1} x + \sqrt{1-x^2}) + c$

48. $\int \frac{\sin x + 8\cos x}{4\sin x + 6\cos x} dx$ is equal to ______.

(a) $x + \frac{1}{2} \log (4 \sin x + 6 \cos x) + c$

(b) $2x + \log (2 \sin x + 3 \cos x) + c$

(c) $x + 2\log(2 \sin x + 3 \cos x) + c$

(d) $\frac{1}{2} \log (4 \sin x + 6 \cos x) + c$

49. Observe the following statements:

A: $\int \left(\frac{x^2-1}{x^2}\right) e^{\frac{x^2-1}{x}} dx = e^{\frac{x^2+1}{x}} + c$

R: $\int f'(x) e^{f(x)} dx = f(x) + c$

Then which of the following is true?

(a) Both A and R are true but R is not the correct reason for A.

(b) Both A and R are true and R is the correct reason for A.

(c) A is true, R is false.

(d) A is false, R is true.

50. $\int_0^{\pi/2} \frac{dx}{1+\tan^3 x}$ is equal to ______.

(a) π (b) $\frac{\pi}{2}$

(c) $\frac{\pi}{4}$ (d) $\frac{3\pi}{2}$

51. $\int_0^{\pi/2} \frac{200 \sin x + 100 \cos x}{\sin x + \cos x} dx$ is equal to ______.

(a) 50π (b) 25π

(c) 75π (d) 150π

52. $\int_0^{\pi} \frac{\theta \sin\theta}{1+\cos^2\theta} d\theta$ is equal to ______.

(a) $\frac{\pi^2}{2}$ (b) $\frac{\pi^3}{3}$

(c) π^2 (d) $\frac{\pi^2}{4}$

53. If $\int \sin^{-1}\left(\frac{2x}{1+x^2}\right) dx = f(x) - \log(1+x^2) + c$, then $f(x)$ is equal to ______.

(a) $2x\tan^{-1} x$ (b) $-2x\tan^{-1} x$

(c) $x\tan^{-1} x$ (d) $-x\tan^{-1} x$

54. $\int_{-\pi/2}^{\pi/2} \log\left(\frac{2-\sin\theta}{2+\sin\theta}\right) d\theta =$ ______.

(a) 0 (b) 1
(c) 2 (d) – 1

55. $\int_0^2 \frac{2x-2}{2x-x^2}dx =$ ______.

(a) 0 (b) 2
(c) 3 (d) 4

56. $\int \frac{dx}{a^2\sin^2 x + b^2\cos^2 x}$ is equal to ______.

(a) $\frac{1}{ab}\tan^{-1}\left(\frac{a\tan x}{b}\right)+c$

(b) $\tan^{-1}\left(\frac{a\tan x}{b}\right)+c$

(c) $\frac{1}{ab}\tan^{-1}\left(\frac{b\tan x}{a}\right)+c$

(d) $\tan^{-1}\left(\frac{b\tan x}{a}\right)+c$

57. $\int_0^1 \sin\left(2\tan^{-1}\sqrt{\frac{1+x}{1-x}}\right)dx$ is equal to ______.

(a) $\frac{\pi}{6}$ (b) $\frac{\pi}{4}$

(c) $\frac{\pi}{2}$ (d) π

58. $\int_0^3 \frac{3x+1}{x^2+9}dx$ is equal to ______.

(a) $\log(2\sqrt{2})+\frac{\pi}{12}$ (b) $\log(2\sqrt{2})+\frac{\pi}{2}$

(c) $\log(2\sqrt{2})+\frac{\pi}{6}$ (d) $\log(2\sqrt{2})+\frac{\pi}{3}$

59. $\int_2^3 \frac{dx}{x^2-x}$ is equal to ______.

(a) $\log\frac{2}{3}$ (b) $\log\frac{4}{3}$

(c) $\log\frac{8}{3}$ (d) $\log\frac{1}{3}$

60. $\int_{-\pi/2}^{\pi/2} \sin^4 x\cos^6 x\,dx$ is equal to ______.

(a) $\frac{3\pi}{128}$ (b) $\frac{3\pi}{256}$

(c) $\frac{3\pi}{572}$ (d) $\frac{3\pi}{64}$

Questions from Previous Years

61. The value of the integral

$$\int \frac{\sin\theta.\sin 2\theta\left(\sin^6\theta+\sin^4\theta+\sin^2\theta\right)\sqrt{2\sin^4\theta+3\sin^2\theta+6}}{1-\cos 2\theta}d\theta$$ is ______.

(where c is a constant of integration)

(a) $\frac{1}{18}\left[9-2\cos^6\theta-3\cos^4\theta-6\cos^2\theta\right]^{\frac{3}{2}}+c$

(b) $\frac{1}{18}\left[11-18\cos^2\theta+9\cos^4\theta-2\cos^2\theta\right]^{\frac{3}{2}}+c$

(c) $\frac{1}{18}\left[9-2\sin^6\theta-3\sin^4\theta-6\sin^2\theta\right]^{\frac{3}{2}}+c$

(d) $\frac{1}{18}\left[11-18\sin^2\theta+9\sin^4\theta-2\sin^2\theta\right]^{\frac{3}{2}}+c$

62. The function $f(x)$ that satisfies the condition $f(x)=x+\int_0^{\pi/2}\sin x.\cos y f(y)dy$, is ______.

(a) $x+\frac{2}{3}(\pi-2)\sin x$ (b) $x+(\pi+2)\sin x$

(c) $x+\frac{\pi}{2}\sin x$ (d) $x+(\pi-2)\sin x$

63. If $[x]$ is the greatest integer $\le x$, then $\pi^2\int_0^2\left(\sin\frac{\pi x}{2}\right)(x-[x])^{[x]}dx$ is equal to ______.

(a) $2(\pi-1)$ (b) $4(\pi-1)$
(c) $4(\pi+1)$ (d) $2(\pi+1)$

64. The value of $\int_{-1/\sqrt{2}}^{1/\sqrt{2}}\left(\left(\frac{x+1}{x-1}\right)^2+\left(\frac{x-1}{x+1}\right)^2\right)dx$ is ______.

(a) $\log_e 4$ (b) $\log_e 16$
(c) $2\log_e 16$ (d) $4\log_e\left(3+2\sqrt{2}\right)$

65. If the value of the integral $\int_0^5 \frac{x+[x]}{e^x-[x]}dx = \alpha e^{-1}+\beta$, where $\alpha, \beta\in R$, $5\alpha+6\beta=0$, and $[x]$ denotes the greatest integer less than or equal to x; then the value of $(\alpha+\beta)^2$ is equal to ______.

(a) 100 (b) 25
(c) 16 (d) 36

66. Let a be a positive real number such that $\int_0^a e^{x-[x]}dx = 10e-9$ where $[x]$ is the greatest integer less than or equal to x. Then a is equal to ______.

(a) $10-\log_e(1+e)$ (b) $10+\log_e 2$

(c) $10+\log_e 3$ (d) $10+\log_e(1+e)$

67. Let $g(t)=\int_{-\pi/2}^{\pi/2}\cos\left(\frac{\pi}{4}t+f(x)\right)dx$, where $f(x)=\log_2\left(x+\sqrt{x^2+1}\right)$, $x\in R$. Then which one of the following is correct?

(a) $g(1)=g(0)$ (b) $\sqrt{2}g(1)=g(0)$

(c) $g(1)=\sqrt{2}g(0)$ (d) $g(1)+g(0)=0$

68. If the integral $\int_0^{10}\frac{[\sin 2\pi x]}{e^{x-[x]}}dx = ae^{-1}+\beta e^{-\frac{1}{2}}+\gamma$, where α, β, Υ are integers and $[x]$ denotes the greatest integer less than or equal to x, then the value of $\alpha+\beta+\Upsilon$ is equal to:

(a) 25 (b) 10

(c) 0 (d) 20

69. Consider the integral $I=\int_0^{10}\frac{[x]e^{[x]}}{e^{x-1}}dx$, where $[x]$ denotes the greatest integer less than or equal to x. Then the value of I is equal to:

(a) $9(e-1)$ (b) $45(e-1)$

(c) $9(e+1)$ (d) $45(e+1)$

70. If $I_n=\int_{\frac{\pi}{4}}^{\frac{\pi}{2}}\text{con}^n\,dx$, then:

(a) $\frac{1}{I_2+I_4},\frac{1}{I_3+I_5},\frac{1}{I_4+I_6}$ are in GP.

(b) $\frac{1}{I_2+I_2},\frac{1}{I_3+I_5},\frac{1}{I_4+I_6}$ are in AP.

(c) $I_2+I_4,(I_3+I_5)^2,I_4+I_6$ are in GP.

(d) $\frac{1}{I_2+I_4},\frac{1}{I_3+I_5},\frac{1}{I_4+I_6}$ are in AP.

71. The value of the integral, $\int_1^3\left[x^2-2x-2\right]dx$, where $[x]$ denotes the greatest integer less than or equal to x, is ______.

(a) $-\sqrt{2}-\sqrt{3}+1$ (b) $-\sqrt{2}-\sqrt{3}-1$

(c) –5 (d) –4

72. $\lim_{x\to 1}\left(\frac{\int_0^{(x-1)^2}t\cos(t^2)dt}{(x-1)\sin(x-1)}\right)$

(a) is equal to $\frac{1}{2}$ (b) is equal to 1

(c) is equal to $-\frac{1}{2}$ (d) It does not exist.

73. If $\int\left(e^{2x}+2e^x-e^{-x}-1\right)$

$\int\left(+2e^x-e^{-x}-1\right)e^{\left(e^x+e^{-x}\right)}dx = g(x)e^{\left(e^x+e^{-x}\right)}+c$, where c is a constant of integration, then $g(0)$ is equal to ______.

(a) e (b) e^2

(c) 1 (d) 2

74. If $\int\frac{\cos\theta}{5+7\sin\theta-2\cos^2\theta}d\theta = A\log_2|B(\theta)|+C$, where C is a constant of integration, then $\frac{B(\theta)}{A}$ can be ______.

(a) $\frac{2\sin\theta+1}{\sin\theta+3}$ (b) $\frac{2\sin\theta+1}{5(\sin\theta+3)}$

(c) $\frac{5(\sin\theta+3)}{2\sin\theta+1}$ (d) $\frac{5(2\sin\theta+1)}{\sin\theta+3}$

75. The integral $\int\left(\frac{x}{x\sin x+\cos x}\right)^2 dx$ is equal to ______ (where C is a constant of integration.)

(a) $\tan x-\frac{x\sec x}{x\sin x+\cos x}+C$

(b) $\sec x+\frac{x\tan x}{x\sin x+\cos x}+C$

(c) $\sec x-\frac{x\tan x}{x\sin x+\cos x}+C$

(d) $\tan x+\frac{x\sec x}{x\sin x+\cos x}+C$

76. Let $f(x)=\int\frac{\sqrt{x}}{(1+x)^2}dx\,(x\ge 0)$.
Then $f(3)-f(1)$ is equal to ______.

(a) $-\frac{\pi}{12}+\frac{1}{2}+\frac{\sqrt{3}}{4}$ (b) $\frac{\pi}{6}+\frac{1}{2}-\frac{\sqrt{3}}{4}$

(c) $-\frac{\pi}{6}+\frac{1}{2}+\frac{\sqrt{3}}{4}$ (d) $\frac{\pi}{12}+\frac{1}{2}-\frac{\sqrt{3}}{4}$

77. If

$$\int\sin^{-1}\left(\sqrt{\frac{x}{1+x}}\right)dx=A(x)\tan^{-1}\left(\sqrt{x}\right)+B(x)+C,$$

where C is a constant of integration, then the ordered pair $(A(x)B(x))$ is ______.

(a) $\left(x+1,-\sqrt{x}\right)$ (b) $\left(x+1,\sqrt{x}\right)$

(c) $\left(x-1,-\sqrt{x}\right)$ (d) $\left(x-1,\sqrt{x}\right)$

78. The integral $\int\frac{dx}{(x+4)^{8/7}(x-3)^{6/7}}$ is equal to ______ (where C is a constant of integration.)

(a) $\left(\frac{x-3}{x+4}\right)^{1/7}+C$

(b) $-\left(\frac{x-3}{x+4}\right)^{-1/7}+C$

(c) $\frac{1}{2}\left(\frac{x-3}{x+4}\right)^{-1/7}+C$

(d) $-\frac{1}{13}\left(\frac{x-3}{x+4}\right)^{-13/7}+C$

79. If $\int\frac{d\theta}{\cos^2\theta(\tan 2\theta+\sec 2\theta)}=\lambda\tan\theta+2|f(\theta)|+C$ where C is a constant of integration, then the ordered pair $(\lambda, f(\theta))$ is ______.

(a) $(1,1-\tan\theta)$ (b) $(-1,1-\tan\theta)$

(c) $(-1,1+\tan\theta)$ (d) $(1,1+\tan\theta)$

80. The integral $\int_1^2 e^x.x^2(2+\log_e x)dx$ equals ______.

(a) $e(4e+1)$ (b) $4e^2-1$

(c) $e(4e-1)$ (d) $e(2e-1)$

81. If $I_1=\int_0^1\left(1-x^{50}\right)^{100}dx$ and $I_2=\int_0^1\left(1-x^{50}\right)^{101}dx$ such that $I_2=\alpha I_1$ then α equals to ______.

(a) $\frac{5049}{5050}$ (b) $\frac{5050}{5049}$

(c) $\frac{5050}{5051}$ (d) $\frac{5051}{5050}$

82. The value of $\int_{-\pi/2}^{\pi/2}\frac{1}{1+e^{\sin x}}dx$ is ______.

(a) $\frac{\pi}{4}$ (b) π

(c) $\frac{\pi}{2}$ (d) $\frac{3\pi}{2}$

83. Let $f(x)=|x-2|$ and $g(x)=f(f(x))$, $x\in[0,4]$. Then $\int_0^3(g(x)-f(x))dx$ is equals to

(a) 1 (b) 0

(c) $\frac{1}{2}$ (d) $\frac{3}{2}$

84. The integral

$$\int_{\pi/6}^{\pi/3}\tan^3 x.\sin^2 3x\left(2\sec^2 x.\sin^2 3x+3\tan x.\sin 6x\right)dx$$

is equal to ______.

(a) $\frac{7}{18}$ (b) $-\frac{1}{9}$

(c) $-\frac{1}{18}$ (d) $\frac{9}{2}$

85. The value of $\int_0^{2\pi}\frac{x\sin^8 x}{\sin^8 x+\cos^8 x}dx$ is equal to ______.

(a) 2π (b) $2\pi^2$

(c) π^2 (d) 4π

86. If $I=\int_1^2\frac{dx}{\sqrt{2x^3-9x^2+12x+4}}$, then ______.

(a) $\frac{1}{8}<I^2<\frac{1}{4}$ (b) $\frac{1}{9}<I^2<\frac{1}{8}$

(c) $\frac{1}{16}<I^2<\frac{1}{9}$ (d) $\frac{1}{6}<I^2<\frac{1}{2}$

87. If $f(a+b+1-x)=f(x)$, for all x, where a and b are fixed positive real numbers, then $\frac{1}{a+b}\int_a^b x(f(x)+f(x+1))dx$ is equal to ______.

(a) $\int_{a+1}^{b+1} f(x)dx$ (b) $\int_{a-1}^{b-1} f(x)dx$

(c) $\int_{a-1}^{b-1} f(x+1)dx$ (d) $\int_{a+1}^{b+1} f(x+1)dx$

88. The value of α for which $4\alpha\int_{-1}^{2} e^{-\alpha|x|}dx = 5$, is ______.

(a) $\log_e 2$ (b) $\log_e\left(\frac{3}{2}\right)$

(c) $\log_{e\sqrt{2}}$ (d) $\log_e\left(\frac{4}{3}\right)$

89. If θ_1 and θ_2 be respectively the smallest and the largest values of θ in $(0, 2\pi) - \{\pi\}$ which satisfy the equation, $2\cot^2\theta - \frac{5}{\sin\theta} + 4 = 0$, then $\int_{\theta_1}^{\theta_2} \cos^2 3\theta\, d\theta$ is equal to ______.

(a) $\frac{\pi}{3}$ (b) $\frac{2\pi}{3}$

(c) $\frac{\pi}{3} + \frac{1}{6}$ (d) $\frac{\pi}{9}$

90. Let a function $f:[0,5] \to R$ be continuous, $f(1) = 3$ and F be defined as: $F(x) = \int_1^x t^2 g(t)\,dt$, where $g(t) = \int_1^t f(u)du$. Then for the function F, the point $x = 1$ is ______.

(a) a point of local minima
(b) not a critical point
(c) a point of local maxima
(d) a point of inflection

91. Let $f: R \to$ R be a continuously differentiable function such that $f(2) = 6$ and $f'(2) = \frac{1}{48}$. If $\int_6^{\frac{x}{2}} 4t^3 dt = (x-2)g(x)$, then $\lim_{x\to 2} g(x)$ is equal to ______.

(a) 18 (b) 24
(c) 12 (d) 36

92. If $\int_0^{\frac{x}{2}} \frac{\cot x}{\cot x + \operatorname{cosec} x} dx = m(\pi + n)$, then $m.n$ is equal to ______.

(a) $-\frac{1}{2}$ (b) 1

(c) $\frac{1}{2}$ (d) -1

93. The value of $\int_0^{\pi} |\cos x|^3\, dx$ is ______.

(a) 0 (b) $\frac{4}{3}$

(c) $\frac{2}{3}$ (d) $-\frac{4}{3}$

94. The integral

$$\int \frac{\sin^2 x \cos^2 x}{\left(\sin^5 x + \cos^3 x \sin^2 x + \sin^3 x \cos^2 x + \cos^5 x\right)^2} dx$$

is equal to ______.

(a) $\frac{-1}{3(1+\tan^3 x)} + C$ (b) $\frac{1}{1+\cot^3 x} + C$

(c) $\frac{-1}{1+\cot^3 x} + C$ (d) $\frac{1}{3(1+\tan^3 x)} + C$

95. If $I_1 = \int_0^1 e^{-x}\cos^2 x\,dx$; $I_2 = \int_0^1 e^{-x^2}\cos^2 x\,dx$ and $I_3 = \int_0^1 e^{-x^3}\,dx$; then ______.

(a) $I_2 > I_3 > I_1$ (b) $I_3 > I_1 > I_2$

(c) $I_2 > I_1 > I_3$ (d) $I_3 > I_2 > I_1$

96. The integral $\int_{\frac{\pi}{4}}^{\frac{3\pi}{4}} \frac{dx}{1+\cos x}$ is equal to ______.

(a) -1 (b) -2
(c) 2 (d) 4

97. Let $I_n = \int \tan^n x\,dx, (n > 1)$ $.I_4 + I_6 = a\tan^5 x + bx^5 + C$, where C is the constant of integration. Then the ordered pair (a, b) is equal to ______.

(a) $\left(-\frac{1}{5}, 0\right)$ (b) $\left(-\frac{1}{5}, 1\right)$

(c) $\left(\frac{1}{5}, 0\right)$ (d) $\left(\frac{1}{5}, -1\right)$

98. The integral $\int \frac{dx}{\left(1+\sqrt{x}\right)\sqrt{x-x^2}}$ is equal to ______ (where C is a constant of integration.)

(a) $-2\sqrt{\frac{1+\sqrt{x}}{1-\sqrt{x}}}+C$ (b) $-1\sqrt{\frac{1-\sqrt{x}}{1+\sqrt{x}}}+C$

(c) $-2\sqrt{\frac{1-\sqrt{x}}{1+\sqrt{x}}}+C$ (d) $2\sqrt{\frac{1+\sqrt{x}}{1-\sqrt{x}}}+C$

99. For $x \in R$, $x \neq 0$, if $y(x)$ is a differentiable function such that $x\int_1^x y(t)dt=(x+1)\int_1^x ty(t)dt$, then $y(x)$ equals:

(a) $Cx^3e^{\frac{1}{x}}$ (b) $\frac{C}{x^2}e^{-\frac{1}{x}}$

(c) $\frac{C}{x}e^{-\frac{1}{x}}$ (d) $\frac{C}{x^3}e^{-\frac{1}{x}}$

100. The integral $\int \frac{dx}{x^2\left(x^4+1\right)^{3/4}}$ equals ______.

(a) $-\left(x^4+1\right)^{\frac{1}{4}}+c$ (b) $-\left(\frac{x^4+1}{x^4}\right)^{\frac{1}{4}}+c$

(c) $\frac{\left(x^4+1\right)^{\frac{1}{4}}}{x^4}+c$ (d) $\left(x^4+1\right)^{\frac{1}{4}}+c$

101. The integral $\int_2^4 \frac{\log x^2}{\log x^2+\log\left(36-12x+x^2\right)}dx$ is equal to

(a) 1 (b) 6
(c) 2 (d) 4

102. If m is a non-zero number and $\int \frac{x^{5m-1}+2x^{4m-1}}{\left(x^{2m}+x^m+1\right)}dx = f(x)+c$, then $f(x)$ is ______.

(a) $\frac{x^{5m}}{2m\left(x^{2m}+x^m+1\right)^2}$

(b) $\frac{x^{4m}}{2m\left(x^{2m}+x^m+1\right)^2}$

(c) $\frac{2m\left(x^{5m}+x^{4m}\right)}{\left(x^{2m}+x^m+1\right)^2}$

(d) $\frac{\left(x^{5m}-x^{4m}\right)}{2m\left(x^{2m}+x^m+1\right)^2}$

103. The integral $\int_0^{\pi}\sqrt{1+4\sin^2\frac{x}{2}-4\sin\frac{x}{2}}\,dx$ is equal to ______.

(a) $4\sqrt{3}-4$ (b) $4\sqrt{3}-4-\frac{\pi}{3}$

(c) $\pi-4$ (d) $\frac{2\pi}{3}-4-4\sqrt{3}$

104. The integral $\int \frac{xdx}{2-x^2+\sqrt{2-x^2}}$ equals ______.

(a) $\log\left|1+\sqrt{2+x^2}\right|+c$

(b) $-\log\left|1+\sqrt{2-x^2}\right|+c$

(c) $-x\log\left|1-\sqrt{2-x^2}\right|+c$

(d) $x\log\left|1-\sqrt{2+x^2}\right|+c$

105. For $0 \le x \le \frac{\pi}{2}$, the value of $\int_0^{\sin^2 x}\sin^{-1}\left(\sqrt{t}\right)dt+\int_0^{\cos^2 x}\cos^{-1}\left(\sqrt{t}\right)dt$ equals ______.

(a) $\frac{\pi}{4}$ (b) 0

(c) 1 (d) $-\frac{\pi}{4}$

106. The integral $\int \frac{\sec^2 x}{(\sec x+\tan x)^{\frac{9}{2}}}dx$ equals ______ (for some arbitrary constant K.)

(a) $-\frac{1}{(\sec x+\tan x)^{\frac{11}{2}}}\left\{\frac{1}{11}-\frac{1}{7}(\sec x+\tan x)^2\right\}+K$

(b) $\frac{1}{(\sec x+\tan x)^{\frac{11}{2}}}\left\{\frac{1}{11}-\frac{1}{7}(\sec x+\tan x)^2\right\}+K$

(c) $-\frac{1}{(\sec x+\tan x)^{\frac{11}{2}}}\left\{\frac{1}{11}+\frac{1}{7}(\sec x+\tan x)^2\right\}+K$

(d) $\frac{1}{(\sec x+\tan x)^{\frac{11}{2}}}\left\{\frac{1}{11}+\frac{1}{7}(\sec x+\tan x)^2\right\}+K$

107. The value of the integral $\int_{-\pi/2}^{\pi/2}\left(x^2+\ln\frac{\pi+x}{\pi-x}\right)\cos x\,dx$ is ______.

(a) 0 (b) $\frac{\pi^2}{2}-4$

(c) $\frac{\pi^2}{2}+4$ (d) $\frac{\pi^2}{2}$

108. The value of $\int_{\sqrt{\ln 2}}^{\sqrt{\ln 3}}\frac{x\sin x^2}{\sin x^2+\sin\left(\ln 6-x^2\right)}dx$ is ______.

(a) $\frac{1}{4}\ln\frac{3}{2}$ (b) $\frac{1}{2}\ln\frac{3}{2}$

(c) $\ln\frac{3}{2}$ (d) $\frac{1}{6}\ln\frac{3}{2}$

109. Let f be a real-valued function defined on the interval $(-1,1)$ such that $e^{-x}f(x)=2+\int_0^x\sqrt{t^4+1}\,dt$, for all $x\in(-1,1)$, and let f^{-1} be the inverse function of f. Then $\left(f^{-1}\right)'(2)$ is equal to ______.

(a) 1 (b) $\frac{1}{3}$

(c) $\frac{1}{2}$ (d) $\frac{1}{e}$

110. The value of $\lim_{x\to 0}\frac{1}{3}\int_0^x\frac{t\ln(1+t)}{t^4+4}=dt$ is ______.

(a) 0 (b) $\frac{1}{12}$

(c) $\frac{1}{24}$ (d) $\frac{1}{64}$

111. Let f be a non-negative function defined on the interval [0, 1]. If $\int_0^x\sqrt{1-(f'(t))^2}\,dt=\int_0^x f(t)dt$, $0\le x\le 1$, and $f(0)=0$, then

(a) $f\left(\frac{1}{2}\right)<\frac{1}{2}$ and $f\left(\frac{1}{3}\right)>\frac{1}{3}$

(b) $f\left(\frac{1}{2}\right)>\frac{1}{2}$ and $f\left(\frac{1}{3}\right)>\frac{1}{3}$

(c) $f\left(\frac{1}{2}\right)<\frac{1}{2}$ and $f\left(\frac{1}{3}\right)<\frac{1}{3}$

(d) $f\left(\frac{1}{2}\right)>\frac{1}{2}$ and $f\left(\frac{1}{3}\right)<\frac{1}{3}$

112. Let $I=\int\frac{e^x}{e^{4x}+e^{2x}+1}dx$, $J=\int\frac{e^{-x}}{e^{-4x}+e^{-2x}+1}dx$. Then, for an arbitrary constant C, the value of $J-I$ equals ______.

(a) $\frac{1}{2}\log\left(\frac{e^{4x}-e^{2x}+1}{e^{4x}+e^{2x}+1}\right)+C$

(b) $\frac{1}{2}\log\left(\frac{e^{2x}+e^x+1}{e^{2x}-e^x+1}\right)+C$

(c) $\frac{1}{2}\log\left(\frac{e^{2x}-e^x+1}{e^{2x}+e^x+1}\right)+C$

(d) $\frac{1}{2}\log\left(\frac{e^{4x}+e^{2x}+1}{e^{4x}-e^{2x}+1}\right)+C$

Integer and Numerical Value Type

1. If $a>0$ and $A=\int_0^a\cos^{-1}x\,dx$, then $\int_{-a}^a\left(\cos^{-1}x-\sin^{-1}\sqrt{1-x^2}\right)dx=\pi a-\lambda A$ then λ is ______.

2. Let f be a function from R → R, defined by $f(x)=\sin x+x$. If $\int_0^\pi f^{-1}(x)dx=\frac{\pi^2}{2}-k$, then the value of k must be ______.

3. If the value of the integral $\int_0^\pi x^2\operatorname{cosec}^2 x\,dx$ is $\pi\log k$, then the numerical quantity k should be ______.

4. If $\int_0^\infty\frac{x\,dx}{(1+x)(1+x^2)}=\frac{\pi}{k}$, then the numerical quantity k should be equal to ______.

5. If $\int\frac{\cos 6x+\cos 9x}{1-2\cos 5x}dx=-\frac{\sin 4x}{k}-\sin x+c$, then the numerical quantity k should be ______.

6. If λ is a root of trigonometric equation tan x = x, the value of integral $\int_0^1 2\sin^2\lambda x\,dx$ is

$\dfrac{\lambda^2}{k+\lambda^2}$, where k is a numerical quantity, then k must be equal to ______.

7. If $f(x)=\sin x+\int_{-\pi/2}^{\pi/2}(\sin x+t\cos x)f(t)dt$, then $f(x)$ may be equal to $-\dfrac{1}{k}\sin x-\dfrac{2}{k}\cos x$, where k is a numerical quantity, which equals ______.

Questions from Previous Years

8. If $\int\dfrac{\sin x}{\sin^3 x+\cos^3 x}dx=\alpha\log_e|1+\tan x|+\beta\log_e|1-\tan x+\tan^2 x|+\gamma\tan^{-1}\left(\dfrac{2\tan x-1}{\sqrt{3}}\right)+C$, where C is a constant of integration, then the value of $18(\alpha+\beta+\gamma^2)$ is ______.

9. If $(x)=\int\dfrac{5x^8+7x^6}{(x^2+1+2x^7)^2}dx$, $(x\geq 0)$, $f(0)=0$ and $f(1)=\dfrac{1}{K}$, then the value of K is ______.

10. For real numbers α, β, Υ and δ, if

$$\int\frac{(x^2-1)+\tan^{-1}\left(\frac{x^2+1}{x}\right)}{(x^4+3x^2+1)\tan^{-1}\left(\frac{x^2+1}{x}\right)}dx$$

$$=\alpha\log_2\left(\tan^{-1}\left(\frac{x^2+1}{x}\right)\right)+\beta\tan^{-1}\left(\frac{\gamma(x^2-1)}{x}\right)+\delta\tan^{-1}\left(\frac{x^2+1}{x}\right)+C$$

where C is an arbitrary constant, then the value of $10(\alpha+\beta\gamma+\delta)$ is equal to ______.

11. If $x\phi(x)=\int_5^x(3t^2-2\phi'(t))dt$, $x>-2$, and $\phi(0)=4$, then $\phi(2)$ is ______.

12. Let the domain of the function $f(x)=\log_4(\log_5(\log_3(18x-x^2)))$ be (a, b). Then the valuc of the integral $\int_a^b\dfrac{\sin^3 x}{(\sin^3 x+\sin^3(a+b-x))}dx$ is equal to ______.

13. If $\int_0^{\pi}(\sin^3 x)e^{-\sin^2 x}dx=\alpha-\dfrac{\beta}{e}\int_0^1\sqrt{t}\,e^t\,dt$, then $\alpha+\beta$ is equal to ______.

14. Let $I_n=\int_1^e x^{19}(\log|x|)^n\,dx$, where $n\in N$. If $(20)I_{10}-\alpha I_9+\beta I_8$, for natural numbers α and β, then $\alpha-\beta$ equal to ______.

15. If $[.]$ represents the greatest integer function, then the value of $\left|\int_0^{\sqrt{\frac{\pi}{2}}}\left[[x^2]-\cos x\right]dx\right|$ is ______.

16. If $I_{m,n}=\int_0^1 x^{m-1}(1-x)^{n-1}dx$, for $m,n\geq 1$ and $\int_0^1\dfrac{x^{m-1}+x^{n-1}}{(1+x)^{m+n}}dx=\alpha I_{m,n}$, $\alpha\in R$, then α equals ______.

17. If $\int_{-a}^{a}(|x|+|x-2|)dx=22$, $(a>2)$ and $[x]$ denotes the greatest integer $\leq x$, then $\int_a^a(x+[x])dx$ is equal to ______.

18. Let $f:(0,2)\rightarrow R$ be defined as $f(x)=\log_2\left(1+\tan\left(\dfrac{\pi x}{4}\right)\right)$. Then $\lim\limits_{n\to\infty}\dfrac{2}{n}\left(f\left(\dfrac{1}{n}\right)+f\left(\dfrac{2}{n}\right)+...+f(1)\right)$ is equal to ______.

19. Let $f(x)$ and $g(x)$ be two functions satisfying $f(x^2)+g(4-x)=4x^3$ and $g(4-x)+g(x)=0$, then the value $\int_{-4}^{4}f(x^2)dx$ is ______.

20. If [.] represents the greatest integer function, then the value of $\left|\int_0^{\sqrt{\frac{\pi}{2}}}\left[[x^2]-\cos x\right]dx\right|$ is ______.

21. Let $\{x\}$ and $[x]$ denote the fractional part of x and the greatest integer $\leq x$ respectively of a real number x. If $\int_0^n \{x\}dx, \int_0^n [x]dx$ and $10(n^2 - n)$, $(n \in N, n > 1)$ are three consecutive terms of a GP, then n is equal to ______.

22. The integral $\int_0^2 ||x-1|-x|dx$ is equal to ______.

23. Let $[t]$ denote the greatest integer less than or equal to t. Then the value of $\int_1^2 |2x-[3x]|dx$ is ______.

Assertion and Reason

1. **Statement 1**: If x > 0, $x \neq 1$ then $\int\left(\log_x e - (\log_x e)^2\right)dx = x\log_x e + C$

   Statement 2: $\int e^x(f(x)+f'(x))dx = e^x f(x) + C$ and $e^t = x$ if $t = \ell n x$

   (a) Statement 1 is True, Statement 2 is True; Statement 2 is a correct explanation for Statement 1.
   (b) Statement 1 is True, Statement 2 is True; Statement 2 is NOT a correct explanation for Statement 1.
   (c) Statement 1 is True, Statement 2 is false
   (d) Statement 1 is False, Statement 2 is True

2. **Statement 1**: If $y = \sin^{-1} x$, then $\int \sin^{-1} x\, dx = \int y\cos y\, dy + c$

   Statement 2: If $y = f^{-1}(x)$, then $\int f^{-1}(x)\, dx = \int yf'(y)dy + c$

   (a) Statement 1 is True, Statement 2 is True; Statement 2 is a correct explanation for Statement 1
   (b) Statement 1 is True, Statement 2 is True; Statement 2 is NOT a correct explanation for Statement 1
   (c) Statement 1 is True, Statement 2 is False
   (d) Statement 1 is False, Statement 2 is True

3. **Statement 1**: $\int 2^{\tan^{-1} x} d\left(\cot^{-1} x\right) = \frac{2^{\tan^{-1} x}}{\ell n 2} + c$

   Statement 2: $\frac{d}{dx}\left(a^x + c\right) = a^x \ell n\, a$

   (a) Statement 1 is True, Statement 2 is True; Statement 2 is a correct explanation for Statement 1.
   (b) Statement 1 is True, Statement 2 is True; Statement 2 is NOT a correct explanation for Statement 1.
   (c) Statement 1 is True, Statement 2 is False.
   (d) Statement 1 is False, Statement 2 is True.

4. **Statement 1**:

   $\int e^{\sin^{-1} x}\left(1 - \frac{x}{\sqrt{1-x^2}}\right)dx = e^{\sin^{-1} x}.\sqrt{1-x^2} + c$

   Statement 2:

   $\int e^{g(x)}\left(g'(x)f(x) + f'(x)\right)dx = e^{g(x)}f(x) + c$

   (a) Statement 1 is True, Statement 2 is True; Statement 2 is a correct explanation for Statement 1.
   (b) Statement 1 is True, Statement 2 is True; Statement 2 is NOT a correct explanation for Statement 1.
   (c) Statement 1 is True, Statement 2 is False.
   (d) Statement 1 is False, Statement 2 is True.

5. **Statement 1**: $\int_0^{\pi/4} \sec x \sqrt{\frac{1-\sin x}{1+\sin x}}dx = 2 - \sqrt{2}$

   Statement 2:

   $\int_0^{\pi/4} \frac{\sec x}{1+2\sin^2 x}dx = \frac{1}{2}\log\left(\sqrt{2}-1\right) + \frac{\pi}{6\sqrt{2}}$

   (a) Statement 1is True, Statement 2 is True; Statement 2 is a correct explanation for Statement 1.
   (b) Statement 1 is True, Statement 2 is True; Statement 2 is NOT a correct explanation for Statement 1.
   (c) Statement 1 is True, Statement 2 is False.
   (d) Statement 1 is False, Statement 2 is True.

6. **Statement1**: $\int_0^1 2^x x^2 dx = \frac{2}{\ell n 2} - \frac{4}{(\ell n 2)^2} + \frac{2}{(\ell n 2)^3}$

   Statement 2: $\int_0^1 e^x (x-1)^n dx = 16 - 6e$, then value of n is 3.

   (a) Statement 1is True, Statement 2 is True; Statement 2 is a correct explanation for Statement 1.
   (b) Statement 1 is True, Statement 2 is True; Statement 2 is NOT a correct explanation for Statement 1.
   (c) Statement 1 is True, Statement 2 is False.
   (d) Statement 1 is False, Statement 2 is True.

7. **Statement 1**: If $\frac{d}{dx}F(x)=\frac{e^{\sin x}}{x}, x>0$ and $\int_1^4 \frac{3e^{\sin x^3}}{x}dx = F(k)-F(1)$, then one possible value of K is 64.

Statement 2: If *f*(*x*) is a function satisfying $f\left(\frac{1}{x}\right)+x^2 f(x)=0 \forall x \in R_0$, then $\int_{\sin\theta}^{\operatorname{cosec}\theta} f(x)dx = \sin\theta - \operatorname{cosec}\theta$.

(a) Statement 1is True, Statement 2 is True; Statement 2 is a correct explanation for Statement 1.
(b) Statement 1 is True, Statement 2 is True; Statement 2 is NOT a correct explanation for Statement 1.
(c) Statement 1 is True, Statement 2 is False.
(d) Statement 1 is False, Statement 2 is True.

8. **Statement 1**: $\int_{-\pi}^{\pi} \frac{2x\left(1+\sin^2 x\right)}{1+\cos^2 x}dx = 0$

Statement 2:

$$\int_0^{\pi/2} \sin^2 x\cos^2 x(\sin x+\cos x)dx = \frac{4}{15}$$

(a) Statement 1is True, Statement 2 is True; Statement 2 is a correct explanation for Statement 1.
(b) Statement 1 is True, Statement 2 is True; Statement 2 is NOT a correct explanation for Statement 1.
(c) Statement 1 is True, Statement 2 is False.
(d) Statement 1 is False, Statement 2 is True.

9. **Statement 1**: $\int_0^n \{x\}dx = \frac{n}{2}$, where {.} represents fractional part function and $n \in N$.

Statement 2: $\int_0^n [x]dx = \frac{n(n-1)}{2}$, where [.] represents greatest integer function and $n \in N$.

(a) Statement 1is True, Statement 2 is True; Statement 2 is a correct explanation for Statement 1.
(b) Statement 1 is True, Statement 2 is True; Statement 2 is NOT a correct explanation for Statement 1.
(c) Statement 1 is True, Statement 2 is False.
(d) Statement 1 is False, Statement 2 is True.

10. **Statement 1**:

$$\int_0^{\pi} x\sin x\cos^2 x dx = \frac{\pi}{2}\int_0^{\pi} \sin x\cos^2 x dx$$

Statement 2: $\int_a^b x f(x)dx = \frac{a+b}{2}\int_a^b f(x)dx$

(a) Statement 1is True, Statement 2 is True; Statement 2 is a correct explanation for Statement 1.
(b) Statement 1 is True, Statement 2 is True; Statement 2 is NOT a correct explanation for Statement 1.
(c) Statement 1 is True, Statement 2 is False.
(d) Statement 1 is False, Statement 2 is True.

11. **Statement 1**: $\int_0^{\frac{\pi}{3}} \frac{dx}{8\cos^2 x+1} = \frac{\pi}{18}$

Statement 2: $\int_0^{\frac{2\pi}{3}} \frac{dx}{8\cos^2 x+1} = -\frac{\pi}{18}$

(a) Statement 1is True, Statement 2 is True; Statement 2 is a correct explanation for Statement 1.
(b) Statement 1 is True, Statement 2 is True; Statement 2 is NOT a correct explanation for Statement 1.
(c) Statement 1 is True, Statement 2 is False.
(d) Statement 1 is False, Statement 2 is True.

12. **Statement 1**: Let f be real valued function such that f(2)=2 and f'(2)=1, then $\lim_{x\to 2}\int_2^{f(x)} \frac{4t^3}{x-2}dt=12$.

Statement 2: Let $f(x)=\int_{u(x)}^{v(x)} g(t)dt$, then $f'(x)=g(v(x))v'(x)-g(u(x))u'(x)$.

(a) Statement 1is True, Statement 2 is True; Statement 2 is a correct explanation for Statement 1.
(b) Statement 1 is True, Statement 2 is True; Statement 2 is NOT a correct explanation for Statement 1.
(c) Statement 1 is True, Statement 2 is False.
(d) Statement 1 is False, Statement 2 is True.

13. **Statement 1**: $1 \le \int_0^{\pi/2} \frac{\sin x}{x} \le \frac{\pi}{2}$.

Statement 2: If $f(x)$ is continuous in $[a, b]$ and m and ℓ are greatest and least value of $f(x)$ in $[a, b]$, then $\ell(b-a) \le \int_a^b f(x)dx \le m(b-a)$

(a) Statement 1is True, Statement 2 is True; Statement 2 is a correct explanation for Statement 1.

(b) Statement 1 is True, Statement 2 is True; Statement 2 is NOT a correct explanation for Statement 1.

(c) Statement 1 is True, Statement 2 is False.

(d) Statement 1 is False, Statement 2 is True.

JEE ADVANCED

Objective Questions

One Option Correct

1. If $\int \frac{2\cos x - \sin x + \lambda}{\cos x + \sin x - 2} dx = A\ell n|\cos x + \sin x - 2| + Bx + C$, then the ordered triplet A, B, λ is ______.

(a) $\left(\frac{1}{2}, \frac{3}{2}, -1\right)$ (b) $\left(\frac{3}{2}, \frac{1}{2}, -1\right)$

(c) $\left(\frac{1}{2}, -1, -\frac{3}{2}\right)$ (d) $\left(\frac{3}{2}, -1, \frac{1}{2}\right)$

2. If $y^2 = x^2 - x + 1$ and $I_n = \int \frac{x^n}{y} dx$ and $AI_3 + BI_2 + CI_1 = x^2 y$, then ordered triplet A, B, C is ______.

(a) $\left(\frac{1}{2}, -\frac{1}{2}, 1\right)$ (b) $(3,1,0)$

(c) $(1,-1,2)$ (d) $\left(3, -\frac{5}{2}, 2\right)$

3. $\int 4\cos\left(x + \frac{\pi}{6}\right)\cos 2x.\cos\left(\frac{5\pi}{6} + x\right)dx$ = ______.

(a) $-\left(x + \frac{\sin 4x}{4} + \frac{\sin 2x}{2}\right) + c$

(b) $-\left(x + \frac{\sin 4x}{4} - \frac{\sin 2x}{2}\right) + c$

(c) $-\left(x - \frac{\sin 4x}{4} + \frac{\sin 2x}{2}\right) + c$

(d) $-\left(x - \frac{\sin 4x}{4} + \frac{\cos 2x}{2}\right) + c$

4. If

$$\int \frac{e^{x-1}}{(x^2 - 5x + 4)} 2x dx = AF(x-1) + BF(x-4) + C$$

and $F(x) = \int \frac{e^x}{x} dx$, then ordered set of A and B ordered set is ______.

(a) $\left(-\frac{2}{3}, \frac{8}{3}\right)$ (b) $\left(-\frac{2}{3}, \frac{8e^3}{3}\right)$

(c) $\left(\frac{8}{3}, \frac{2}{3}\right)$ (d) $\left(-\frac{2}{3}, -\frac{8e^3}{3}\right)$

5. $\int x^{2/3}\left(1 + x^{1/2}\right)^{-13/3} dx =$ ______.

(a) $\frac{3}{5}\left(1 + x^{-1/2}\right)^{-10/3} + C$

(b) $\frac{3}{5}\left(1 + x^{1/2}\right)^{-10/3} + C$

(c) $-\frac{\left(1 + x^{1/2}\right)}{13} + C$

(d) None of these.

6. Evaluate $\int x^2 \log\left(1 - x^2\right).dx$ and hence find the value of:

$$\frac{1}{1.5} + \frac{1}{2.7} + \frac{1}{3.9} + ...$$

(a) $\frac{1}{4}\log 2$ (b) $\frac{2}{7} - \frac{2}{3}\log 2$

(c) $\frac{8}{9} - \frac{2}{3}\log 2$ (d) $\frac{2}{3}\log 2 - \frac{8}{9}$

7. If

$$\int \frac{\cos^4 x\,dx}{\sin^3 x\left(\sin^5 x+\cos^5 x\right)^{3/5}} = -\frac{1}{2}\left(1+\cos^A x\right)^B + C,$$

then find AB.

(a) 5 (b) $\frac{2}{5}$

(c) 2 (d) 1

8. $I_n = \int (\log x)^n dx$, then $I_n + nI_{n-1} =$ ______.

(a) $n(\log x)^n$ (b) $(n\log x)^{n-1}$

(c) $(\log x)^{n-1}$ (d) $(n\log x)^n$

9. If A is a square matrix and e^A is defined as $e^A = I + A + \frac{A^2}{2!} + \frac{A^3}{3!}\ldots\ldots = \frac{1}{2}\begin{bmatrix} f(x) & g(x) \\ g(x) & f(x) \end{bmatrix}$, where $A = \begin{bmatrix} x & x \\ x & x \end{bmatrix}$ and $0 < x < 1$, I is an identify matrix, then $\int \frac{g(x)}{f(x)} dx$ is equal to ______.

(a) $\log\left(e^x + e^{-x}\right) + C$ (b) $\log\left(e^x - e^{-x}\right) + C$

(c) $\log\left(e^{2x} - 1\right) + C$ (d) None of these.

10. The value of

$$\int e^{(x\sin x+\cos x)}\left(\frac{x^4\cos^3 x - x\sin x + \cos x}{x^2\cos^2 x}\right)dx$$ is

equal to ______.

(a) $e^{x\sin x+\cos x}.\left(x + \frac{1}{\cos x}\right) + C$

(b) $e^{x\sin x+\cos x}.\left(x\cos x + \frac{1}{x}\right) + C$

(c) $e^{x\sin x+\cos x}.\left(x - \frac{1}{x\cos x}\right) + C$

(d) None of these.

11. If $f(x) = \sqrt{x}, g(x) = e^x - 1$, and $\int fog(x)dx = Afog(x) + B\tan^{-1}(fog(x)) + C$, then $A + B$ is equal to ______.

(a) 1 (b) 2

(c) 3 (d) None of these.

12. $\int \frac{x + \sqrt[3]{x^2} + \sqrt[6]{x}}{x\left(1+\sqrt[3]{x}\right)}dx$ is ______.

(a) $\frac{3}{2}x^{2/3} + \tan^{-1} x^6 + C$

(b) $\frac{3}{2}x^{2/3} + 6\tan^{-1} x^6 + C$

(c) $\frac{3}{2}x^{2/3} + 6\tan^{-1} x^{1/6} + C$

(d) None of these.

13. If $\int \frac{dx}{a+\cos x} = f\left(\tan\frac{x}{2}\right) + C$, then:

S_1: f is a log function for $a = 0$.

S_2: f is an inverse trigonometric function for $|a| < 1$.

S_3: f is a polynomial function for $a = 1$.

S_4: f is rational function but not polynomial for $a = 1$.

(a) TTTF (b) TFTF

(c) FTTF (d) TFFT

14. S_1: $\int e^x\left(\ln x + \frac{1}{x^2}\right)dx = e^x\left(\ln x - \frac{1}{x^2}\right) + C$

S_2: $\int \frac{dx}{\sin\frac{x}{2}\sqrt{\cos\frac{x}{2}}} = -\ln\left|\frac{1+\sqrt{\cos\frac{x}{2}}}{1-\sqrt{\cos\frac{x}{2}}}\right| - 2\tan^{-1}\sqrt{\cos\frac{x}{2}} + C$

S_3: $\int \frac{dx}{\sin x\cos^3 x} = \frac{1}{2}\tan^2 x + \ln(\tan x) + C$

S_4: $\int \frac{\sin x + \cos x}{\sin 2x + 3}dx = \frac{1}{2}\tan^{-1}\left(\frac{\sin x - \cos x}{2}\right) + C$

(a) TTTF (b) FFTF

(c) FTTF (d) TFFT

15. $\int_0^\pi \frac{\sin x}{1+\cos^2 x}dx = \pi\frac{\cos\alpha}{1+\sin^2\alpha}$

(a) for no value of α

(b) for exactly two values of α in $(0, \pi)$

(c) for at least one α in $\left(\frac{\pi}{2}, \pi\right)$

(d) for exactly one α in $\left(0, \frac{\pi}{2}\right)$

16. Let $\lambda = \int_0^1 \frac{dx}{1+x^3}, p = \lim_{n\to\infty}\left[\frac{\prod_{r=1}^{n}\left(n^3+r^3\right)}{n^{3n}}\right]^{1/n}$,

then ln p is equal to ______.

(a) $\ln 2 - 1 + \lambda$ (b) $\ln 2 - 3 + 3\lambda$
(c) $2\ln 2 - \lambda$ (d) $\ln 4 - 3 + 3\lambda$

17. $\int_0^{2\pi} \frac{e^{|\sin x|}\cos x}{1+e^{\tan x}} dx =$ ______.

(a) e^{π} (b) 1
(c) $e^{\pi} - 1$ (d) 0

18. Consider the integrals

$I_1 = \int_0^1 e^{-x}\cos^2 x\, dx, I_2 = \int_0^1 e^{-x^2}\cos^2 x\, dx, I_3 = \int_0^1 e^{-\frac{x^2}{2}}\cos^2 x\, dx, I_4 = \int_0^1 e^{-\frac{x^2}{2}} dx.$

Then,

(a) $I_2 > I_4 > I_1 > I_3$ (b) $I_2 < I_4 < I_1 < I_3$
(c) $I_1 < I_2 < I_3 < I_4$ (d) $I_1 > I_2 > I_3 > I_4$

19. $\int_0^2 \sqrt{x+\sqrt{x+\sqrt{x+....\infty}}}\, dx$ is equal to ______ $(x > 0)$.

(a) $\frac{19}{6}$
(b) $\frac{17}{6}$
(c) $\frac{13}{6}$
(d) It cannot be determined.

20. If a function $f(x)$ which satisfies the relation $f(x) = e^x + \int_0^1 e^x f(t)dt$, then $f(x)$ is ______.

(a) $\frac{e^x}{2-e}$ (b) $(e-2)e^x$
(c) $2e^x$ (d) $\frac{e^x}{2}$

21. If $m = \int_{-2}^{0} \frac{|\sin x|}{\left[\frac{x}{\pi}\right]+\frac{1}{2}} dx$ and $n = \int_0^2 \frac{|\sin x|}{\left[\frac{x}{\pi}\right]+\frac{1}{2}} dx$ where [.] represents the greatest integer function, then ______.

(a) $m = n$ (b) $m = -n$
(c) $m = 2n$ (d) $m = -2n$

22. If $I_m = \int_1^e (\ell nx)^m dx, m \in N$, then $I_{10} + 10I_9$ is equal to ______.

(a) e^{10}
(b) $\frac{e^{10}}{10}$
(c) e
(d) $e - 1$

23. If $S_n = \frac{1}{2n} + \frac{1}{\sqrt{4n^2-1}} + \frac{1}{\sqrt{4n^2-4}} + ... + \frac{1}{\sqrt{3n^2+2n-1}}, n \in N$, then $\lim_{n\to\infty} S_n$ is equal to ______.

(a) $\frac{\pi}{2}$ (b) 2
(c) 1 (d) $\frac{\pi}{6}$

24. Let $A = \int_0^1 \frac{e^t}{t+1} dt$, then the value of $\int_0^1 \frac{te^{t^2}}{t^2+1} dt$ is ______.

(a) A^2 (b) $\frac{1}{2}A$
(c) $2A$ (d) $\frac{1}{2}A^2$

25. The function $f(x) = \int_1^x \frac{1}{t} dt$ satisfies ______.

(a) $f(x+y) = f(x) + f(y)$
(b) $f\left(\frac{x}{y}\right) = f(x) + f(y)$
(c) $f(xy) = f(x) + f(y)$
(d) None of these.

26. If f is a function with period P, then $\int_a^{a+p} f(x)dx$ is

(a) equal to $f(a)$. (b) equal to $f(P)$.
(c) independent of a. (d) None of these.

27. $\int_{-\pi}^{\pi} (\cos px - \sin qx)^2 dx$ is equal to ______.

(a) 0 (b) $\frac{\pi}{2}$
(c) π (d) 2π

28. $\int_0^1 \tan^{-1}\left(\frac{2x-1}{1+x-x^2}\right)dx$ is equal to ______.

(a) 0 (b) 1
(c) −1 (d) None of these.

29. The value of $\int_a^b (x-a)^3(b-x)^4\,dx$ is $\frac{(b-a)^m}{n}$.
Then (m, n) is ______.

(a) (6, 260) (b) (8, 280)
(c) (4, 240) (d) None of these.

30. The absolute value of $\int_{10}^{19} \frac{\sin x}{1+x^8}dx$, is

(a) less than 10^{-7} (b) more than 10^{-7}
(c) less than 10^{-6} (d) more than 10^{-6}

31. Evaluate $\int_0^a \ln(\cot a+\tan x)dx; a\in\left(0,\frac{\pi}{2}\right)$.

(a) $a\ln(\sin a)$ (b) $-a\ln(\sin a)$
(c) $\ln(\cos a)$ (d) $-\ln(\cos a)$

32. S_1: In $\int_{-1}^{1} f(\cot^{-1}x)dx$ putting $\cot^{-1}x = t$ may change the limits to $\int_{3\pi/4}^{\pi/4}$ ......

S_2: If $f(x)$ has removable discontinuities at finite number of points in (a, b) then if $\int f(x)dx = F(x), \int_a^b f(x)dx = F(b)-F(a)$.

S_3: If $f(x)$ has an infinite discontinuity in (a, b), then we can always write $\int_a^b f(x)dx = F(b)-F(a)$ where $\int f(x)dx = F(x)$.

S_4: If $f(x):[0,1]\to R$ has single-point continuity in (0, 1) then $\int_0^1 f(x)dx$ can be evaluated.

(a) FTTF (b) TFFT
(c) FFFF (d) TTFF

33. S_1: $\int_{-1}^{\infty} \frac{d}{dx}\left(x\sin\frac{1}{x}\right)dx = 2\sin^2\left(\frac{\pi}{4}-\frac{1}{2}\right)$

S_2: $\int_0^{10\pi} \sin^2 x\cos^4 x\,dx = \frac{5\pi}{8}$

S_3: If f(x) is even then $\int_0^x f(x)dx$ will always be odd.

S_4: If f(x) is even and $\int_0^x f(x)dx$ is odd where $f(x)$ is a continuous function then $f(x) = 0$ must have exactly three roots in (−1, 1).

(a) FTTF (b) TFFT
(c) TTTF (d) FFFT

34. S_1: $\int_0^{\infty} \frac{\log(1+x^2)}{1+x^2}dx = \pi\log 2$

S_2: $\int_0^{\infty} e^{\sin^2 x}\cos^3(2n+1)x\,dx = 0$

S_3: $\int_0^4 |x-1|dx = 5$

S_4: $\int_{-2}^{2} (1-x^2)dx = \frac{4}{3}$

(a) FTTF (b) TFFT
(c) TTTF (d) FFFT

Questions from Previous Years

35. The following integral: $\int_{\pi/4}^{\pi/2} (2\text{cosec}, x)^{17}dx$ is equal to ______.

(a) $\int_0^{\log(1+\sqrt{2})} 2(e^u+e^{-u})^{16}du$

(b) $\int_0^{\log(1+\sqrt{2})} 2(e^u+e^{-u})^{17}du$

(c) $\int_0^{\log(1+\sqrt{2})} 2(e^u-e^{-u})^{17}du$

(d) $\int_0^{\log(1+\sqrt{2})} 2(e^u-e^{-u})^{16}du$

36. Let $f:\left[\frac{1}{2},1\right]\to R$ (the set of all real number) be a positive, non-constant and differentiable function such that $f'(x)<2f(x)$ and $f\left(\frac{1}{2}\right)=1$. Then the value of $\int_{1/2}^{1} f(x)dx$ is ______.

(a) $(2e-1,2e)$ (b) $(e-1,2e-1)$

(c) $\left(\frac{e-1}{2},e-1\right)$ (d) $\left(0,\frac{e-1}{2}\right)$

One or More Than One Option Correct

1. Let f be a real-valued function defined on the interval $(0, \infty)$ by $f(x)=\ln x+\int_0^x \sqrt{1+\sin t}\,dt$. Then which of the following statement(s) is (are) true?

(a) $f''(x)$ exists for all $x\in(0, \infty)$.

(b) $f'(x)$ exists for all $x\in(0, \infty)$ and f' is continuous on $(0, \infty)$, but not differentiable on $(0, \infty)$.

(c) There exists $\alpha>1$ such that $|f'(x)|<|f(x)|$ for all $x\in(0, \infty)$.

(d) There exists $\beta > 0$ such that $|f(x)|+|f'(x)|\le\beta$ for all $x\in(0, \infty)$.

2. The value(s) of $\int_0^1 \frac{x^4(1-x)^4}{1+x^2}dx$ is (are)______.

(a) $\frac{22}{7}-\pi$ (b) $\frac{2}{105}$

(c) 0 (d) $\frac{71}{15}-\frac{3\pi}{2}$

3. If $I_n=\int_{-\pi}^{\pi}\frac{\sin nx}{(1+\pi^x)\sin x}dx \quad n=0,1,2...,$ then ______.

(a) $I_n=I_{n+2}$ (b) $\sum_{m=1}^{10} I_{2m+1}=10\pi$

(c) $\sum_{m=1}^{10} I_{2m}=0$ (d) $I_n=I_{n+1}$

4. Let $f(x)$ be a non-constant twice-differentiable function defined on $(-\infty,\infty)$ such that $f(x)=f(1-x)$ and $f'\left(\frac{1}{4}\right)=0$. Then,

(a) $f''(x)$ vanishes at least twice on [0, 1].

(b) $f'\left(\frac{1}{2}\right)=0$.

(c) $\int_{-1/2}^{1/2} f\left(x+\frac{1}{2}\right)\sin x\,dx=0$.

(d) $\int_0^{1/2} f(t)e^{\sin\pi t}dt=\int_{1/2}^{1} f(1-t)e^{\sin\pi t}dt$.

5. $\int_0^{\pi/2}\frac{\cos\theta-\sin\theta}{(1+\cos\theta)(1+\sin\theta)}d\theta$ equals ______.

(a) $\cos^{-1}(1)$ (b) $\cos^{-1}\left(\frac{1}{\sqrt{2}}\right)$

(c) $\sin^{-1}(0)$ (d) $\cos^{-1}(-1)$

6. If the value of the definite integral $\int_0^1\frac{\sin^{-1}\sqrt{x}}{x^2-x+1}dx$ is $\frac{\pi^2}{\sqrt{n}}$ where $n\in N$, then the value of n is divisible by prime number ______.

(a) 2 (b) 3

(c) 5 (d) 9

7. Let

$J=\int_{-5}^{-4} f(x)dx$ and $f(x)=(3-x^2)\tan(3-x^2)$

and $K=\int_{-2}^{-1}(6-6x+x^2)\tan(6x-x^2-6)dx$,

then

(a) $J-K=0$.

(b) $f(x)$ is an even function.

(c) $J+K=0$.

(d) $f(x)$ is an odd function.

8. If $f(x)$ is integral over [1, 2], then $\int_1^2 f(x)dx$ is equal to______.

(a) $\lim_{n\to\infty}\frac{1}{n}\sum_{r=1}^{n} f\left(\frac{r}{n}\right)$ (b) $\lim_{n\to\infty}\frac{1}{n}\sum_{r=n+1}^{2n} f\left(\frac{r}{n}\right)$

(c) $\lim_{n\to\infty}\frac{1}{n}\sum_{r=1}^{n} f\left(\frac{r+n}{n}\right)$ (d) $\lim_{n\to\infty}\frac{1}{n}\sum_{r=1}^{2n} f\left(\frac{r}{n}\right)$

9. If $I_n = \int_0^1 \frac{dx}{(1+x^2)^n}, n \in N$, then which of the following statements hold good?

(a) $2nI_{n+1} = 2^{-n} + (2n-1)I_n$

(b) $I_2 = \frac{\pi}{8} + \frac{1}{4}$

(c) $I_2 = \frac{\pi}{8} - \frac{1}{4}$

(d) $I_3 = \frac{\pi}{16} - \frac{5}{48}$

10. If $f(2-x) = f(2+x)$ and $f(4-x) = f(4+x)$ and $f(x)$ is a function for which $\int_0^2 f(x)dx = 5$, then $\int_0^{50} f(x)dx$ is equal to ______.

(a) 125 (b) $\int_{-4}^{46} f(x)dx$

(c) $\int_1^{51} f(x)dx$ (d) $\int_2^{52} f(x)dx$

11. If $F(x) = \frac{1}{x^2}\int_4^x \left(4t^2 - 2F'(t)\right)dt$, then $F'(4)$ equals ______.

(a) $\frac{32}{9}$ (b) $\frac{64}{9}$

(c) $\frac{F(8)}{28}$ (d) $\frac{11F(8)}{28}$

12. If $I = \int_0^{\pi/2} e^{-\alpha \sin x} dx$, where $\alpha \in (0, \infty)$, then ______.

(a) $I < \frac{\pi}{2}$ (b) $I > \frac{\pi}{2}\left(e^{-\alpha} + 1\right)$

(c) $I > \frac{\pi}{2}e^{-\alpha}$ (d) $I > 0$

Questions from Previous Years

13. Let $f : \left[-\frac{\pi}{2}, \frac{\pi}{2}\right] \to \mathbb{R}$ be continuous function such that $f(0) = 1$ and $\int_0^{\frac{\pi}{3}} f(t)dt = 0$. Then which of the following statements is (are) true?

(a) The equation $f(x) - 3\cos 3x = 0$ has at least one solution in $\left(0, \frac{\pi}{3}\right)$

(b) The equation $f(x) - 3\sin 3x = -\frac{6}{\pi}$ has at least one solution in $\left(0, \frac{\pi}{3}\right)$

(c) $\lim\limits_{x\to\infty} \frac{x\int_0^x f(t)dt}{1 - e^{x^2}} = -1$

(d) $\lim\limits_{x\to\infty} \frac{\sin x \int_0^x f(t)dt}{x^2} = -1$

14. Let $\psi_1 : [0, \infty) \to R$, $\psi_2 : [0, \infty) \to R$, $f: (0, \infty) \to R$ and $g: (0, \infty) \to R$ be functions such that $f(0) = g(0)$,

$\psi_1(x) = e^{-x} + x, x \geq 0,$

$\psi_2(x) = x^2 - 2x, -2e^{-x} + 2, x \geq 0,$

$f(x) = \int_{-x}^{x} (|t| - t^2)e^{-t^2} dt, x > 0$ and

$g(x) = \int_0^{x^2} \sqrt{t}e^{-t} dt, x > 0.$

Which of the following statements is true?

(a) $f\left(\sqrt{\ln 3}\right) + g\left(\sqrt{\ln 3}\right) = \frac{1}{3}$

(b) For every $x > 1$, there exists an $\alpha \in (1, x)$ such that $\psi_1(x) = 1 + az$

(c) For every $x > 0$, there exists a $\beta \in (0, x)$ such that $\psi_2(x) = 2x(\psi_1(\beta) - 1)$

(d) f is an increasing function on the interval $\left[0, \frac{3}{2}\right]$

15. Which of the following statements is true?

(a) $\psi_1(x) \leq 1$, for all $x > 0$

(b) $\psi_2(x) \leq 0$, for all $x > 0$

(c) $f(x) \geq 1 - e^{-x^2} - \frac{2}{3}x^3 + \frac{2}{5}x^5$ for all $x \in \left(0, \frac{1}{2}\right)$

(d) $g(x) \leq \frac{2}{3}x^3 + \frac{2}{5}x^5 + \frac{1}{7}x^7$ for all $x \in \left(0, \frac{1}{2}\right)$

16. Let b be a non-zero real number Suppose $f : \mathbb{R} \to \mathbb{R}$ is a differentiable function such that $f(0) = 1$. If the derivative f' of f satisfies the equation $f'(x) = \frac{f(x)}{b^2 + x^2}$

For all $x \in \mathbb{R}$, then which of the following statements is/are true?

(a) If $b > 0$, then f is an increasing function.
(b) If $b < 0$, then f is an decreasing function.
(c) $f(x)f(-x) = 1$ For all $x \in \mathbb{R}$.
(d) $f(x) - f(-x) = 0$ For all $x \in \mathbb{R}$.

17. Let $f : \mathbb{R} \to \mathbb{R}$ and $f(x+y) = f(x) + f(y) + f(x)f(y)$ and $f(x) = xg(x)$ for all $x \in \mathbb{R}$. If $\lim_{x\to 0} g(x) = 1$, then which of the following statements is/are true?
(a) f is differentiable at every $x \in \mathbb{R}$.
(b) If $g(0) = 1$, then g is differentiable at every $x \in \mathbb{R}$.
(c) The derivative $f'(1)$ is equal to 1.
(d) The derivative $f'(0)$ is equal to 1.

18. Which of the following inequalities is/ are true?
(a) $\int_0^1 x\cos x\,dx \ge \frac{3}{8}$ (b) $\int_0^1 x\sin x\,dx \ge \frac{3}{10}$
(c) $\int_0^1 x^2\cos x\,dx \ge \frac{1}{2}$ (d) $\int_0^1 x^2\sin x\,dx \ge \frac{2}{9}$

19. For $a \in R\,|a| > 1$, let

$$\lim_{x\to\infty}\left(\frac{1+\sqrt[3]{2}+\ldots\sqrt[3]{n}}{n^{7/3}\left(\frac{1}{(an+1)^2}+\frac{1}{(an+2)^2}+\ldots+\frac{1}{(an+n)^2}\right)}\right)=54$$

Then the possible value(s) of a is/are ______.
(a) –9 (b) 7
(c) –6 (d) 8

20. If $I = \sum_{k=1}^{98}\int_k^{k+1}\frac{k+1}{x(x+1)}dx$, then ______.
(a) $I > \log_e 99$ (b) $I < \log_e 99$
(c) $I < \frac{49}{50}$ (d) $I > \frac{49}{50}$

21. If $g(x) = \int_{\sin x}^{\sin(2x)} \sin^{-1}(t)\,dt$, then ______.
(a) $g'\left(\frac{\pi}{2}\right) = -2\pi$ (b) $g'\left(-\frac{\pi}{2}\right) = 2\pi$
(c) $g'\left(\frac{\pi}{2}\right) = 2\pi$ (d) $g'\left(-\frac{\pi}{2}\right) = -2\pi$

22. Let $f : R \to (0,1)$ be a continuous function. Then, which of the following function(s) has (have) the value zero at some point in the interval (0, 1)?
(a) $x^9 - f(x)$
(b) $x - \int_0^{\frac{\pi}{2}-x} f(t)\cos t\,dt$
(c) $e^x - \int_0^x f(t)\sin t\,dt$
(d) $f(x) + \int_0^{\frac{\pi}{2}} f(t)\sin t\,dt$

23. Let

$$f(x) = \lim_{n\to\infty}\left(\frac{n^n(x+n)\left(x+\frac{n}{2}\right)\ldots\left(x+\frac{n}{n}\right)}{n!\left(x^2+n^2\right)\left(x^2+\frac{n^2}{2}\right)\ldots\left(x^2+\frac{n^2}{n^2}\right)}\right)$$

for all $x > 0$. Then, ______.
(a) $f\left(\frac{1}{2}\right) \ge f(1)$ (b) $f\left(\frac{1}{3}\right) \le f\left(\frac{2}{3}\right)$
(c) $f'(2) \le 0$ (d) $\frac{f'(3)}{f(3)} \ge \frac{f'(2)}{f(3)}$

24. Let $f'(x) = \frac{192x^3}{2+\sin^4 \pi x}$ for all $x \in \mathbb{R}$ with $f\left(\frac{1}{2}\right) = 0$. If $m \le \int_{1/2}^{1} f(x)dx \le M$, then the possible values of m and M are ______.
(a) $m = 13, M = 24$ (b) $m = \frac{1}{4}, M = \frac{1}{2}$
(c) $m = -11, M = 0$ (d) $m = 1, M = 12$

25. Let $f(x) = 7\tan^8 x + 7\tan^6 x - 3\tan^4 x - 3\tan^2 x$ for all $x \in \left(-\frac{\pi}{2}, \frac{\pi}{2}\right)$. Then the correct expression(s) is (are) ______.
(a) $\int_0^{\pi/4} xf(x)dx = \frac{1}{12}$ (b) $\int_0^{\pi/4} f(x)dx = 0$
(c) $\int_0^{\pi/4} xf(x)dx = \frac{1}{6}$ (d) $\int_0^{\pi/4} f(x)dx = 1$

26. The option(s) with the values of a and L that satisfy the following equation is (are) ______.

$$\frac{\int_0^{4\pi} e^t\left(\sin^6 at + \cos^4 at\right)dt}{\int_0^{\pi} e^t\left(\sin^6 at + \cos^4 at\right)dt} = L\,?$$

(a) $a=2, L=\dfrac{e^{4\pi}-1}{e^{\pi}-1}$ (b) $a=2, L=\dfrac{e^{4\pi}+1}{e^{\pi}+1}$

(c) $a=4, L=\dfrac{e^{4\pi}-1}{e^{\pi}-1}$ (d) $a=4, L=\dfrac{e^{4\pi}+1}{e^{\pi}+1}$

Matrix Type

1.

Column I	Column II
(A) If $I=\int \frac{\sin x-\cos x}{\lvert\sin x-\cos x\rvert}dx$, where $\frac{\pi}{4}<x<\frac{3\pi}{8}$, then I equal to	(P) $\sin x$
(B) If $\int \frac{x^2}{(x^3+1)(x^3+2)}dx = \frac{1}{3}f\left(\frac{x^3+1}{x^3+2}\right)+C$ then $f(x)$ is equal to	(Q) $x+c$
(C) If $\int \sin^{-1}x.\cos^{-1}x\,dx = f^{-1}(x)\left[\frac{\pi}{2}x-xf^{-1}(x)-2\sqrt{1-x}\right]+\frac{\pi}{2}\sqrt{1-x^2}+2x+C$ then $f(x)$ is equal to	(R) $\ln\lvert x\rvert$
(D) If $\int \frac{dx}{xf(x)}=f(f(x))+C$, then $f(x)$ is equal to	(S) $\sin^{-1}x$
	(T) $-x+c$

2.

Column I	Column II
(A) If $F(x)=\int \frac{x+\sin x}{1+\cos}dx$ and $F(0)=0$, then $F(\pi/2)=$	(P) $-\frac{\pi}{2}$
(B) Let $F(x)=\int e^{\sin^{-1}x}\left(1-\frac{x}{\sqrt{1-x^2}}\right)dx$ and $F(0)=1$. If $F(1/2)=\frac{k\sqrt{3}e^{\pi/6}}{\pi}$, then $k=$	(Q) $\frac{\pi}{3}$
(C) Let $F(x)=\int \frac{dx}{(x^2+1)(x^2+9)}$ and $F(0)=0$, if $F(\sqrt{3})=\frac{5}{36}k$, then $k=$	(R) $\frac{\pi}{4}$
(D) Let $F(x)=\int \frac{\sqrt{\tan x}}{\sin x\cos x}dx$ and $F(0)=0$ if $F(\pi/4)=\frac{2k}{\pi}$, then $k=$	(S) π
	(T) $\frac{\pi}{2}$

3.

Column I	Column II
(A) If $\int_0^1 \frac{\cos x}{1+x}dx=k$ and $\int_{6\pi-3}^{6\pi} \frac{\cos(x/3)}{6\pi+3-x}dx=mk$, then m is	(P) 4
(B) If $\int_{-1}^{1}\left(\sin^{-1}\left[x+\frac{3}{4}\right]\right)dx=\frac{\pi}{k}$, then k is ______{where [.] denotes greatest integer function.}	(Q) 1
(C) If $f(x)=\max(x-\lvert x\rvert, x+[x])$ and $\int_{-3}^{3} f(x)dx=-\frac{3}{k}$, then k is ______{where [.] denotes greatest integer function.}	(R) 0
(D) If $\int_0^{20}\sqrt{1-\cos\pi x}\,dx=\frac{10k\sqrt{2}}{\pi}$, then k is	(S) 2
	(T) 3

4.

Column I	Column II
(A) $\int_4^{10}\frac{[x^2]dx}{[x^2-28x+196]+[x^2]}$ = ______{where [.] denotes greatest integer function}	(P) $\frac{1}{101}$

(Continued)

Column I	Column II
(B) $\int_{-1}^{2}\frac{\lvert x\rvert}{x}dx =$ ______	(Q) 3
(C) $\lim_{n\to\infty}\frac{1^{99}+2^{99}+.....+n^{99}}{n^{100}}=$	(R) $\frac{1}{3}$
(D) $5050\int_{-1}^{1}\sqrt{x^{200}}\,dx=\frac{1}{\alpha}$, then $\alpha=$	(S) 1
	(T) $\frac{1}{100}$

5.

Column I	Column II
(A) $\int_{-1}^{1}\frac{3x^2}{1+4^{\tan x}}dx=$	(P) 7
(B) $\int_{6}^{8}\frac{\sin x^2 dx}{\sin x^2+\sin(x-14)^2}=$	(Q) $\frac{1}{2}$
(C) $\frac{1}{156}\int_{1}^{13}[x]dx=$	(R) 1
(D) $\frac{1}{\pi\ell n2}\int_{\pi/2}^{0}\ell n\sin 2x\,dx=$	(S) 2
	(T) 0

6.

Column I	Column II
(A) $Lim_{n\to\infty}\sum_{r=1}^{n-1}\frac{\sqrt{n^2-r^2}}{n^2}=$	(P) $\frac{\pi}{4}$
(B) $\int_{0}^{\frac{\pi}{4}}\left(\sqrt{\tan x}+\sqrt{\cot x}\right)dx=$	(Q) $\frac{\pi}{\sqrt{2}}$
(C) $\int_{-1}^{1}\sin^3 x\cos^2 x\,dx=$	(R) $-\frac{\pi}{4}$
(D) $\int_{0}^{\frac{\pi}{2}}\frac{\sqrt{\sin^3 x dx}}{\sqrt{\sin^3 x}+\sqrt{\cos^3 x}}=$	(S) $\frac{\pi}{2}$
	(T) 0

Question from Previous Year

7.

List I	List II
(A) The number of polynomials $f(x)$ with non-negative integer coefficients of degree ≤ 2, satisfying $f(0)=0$ and $\int_0^1 f(x)dx=1$, is	(P) 8
(B) The number of points in the interval $\left[-\sqrt{13},\sqrt{13}\right]$ at which $f(x)=\sin\left(x^2\right)+\cos\left(x^2\right)$ attains its maximum value, is	(Q) 2
(C) $\int_{-2}^{2}\frac{3x^2}{\left(1+e^x\right)}dx$ equals	(R) 4
(D) $\frac{\left(\int_{-\frac{1}{2}}^{\frac{1}{2}}\cos 2x\log\left(\frac{1+x}{1-x}\right)dx\right)}{\left(\int_{0}^{\frac{1}{2}}\cos 2x\log\left(\frac{1+x}{1-x}\right)dx\right)}$	(S) 0

	A	B	C	D
(a)	R	Q	S	P
(b)	R	Q	P	S
(c)	Q	R	S	P
(d)	Q	R	P	S

Integer and Numerical Value Type

1. If $\int\frac{x\cos\alpha+1}{\left(x^2+2x\cos\alpha+1\right)^{3/2}}dx$

$=\frac{\lambda x}{\sqrt{x^2+2x\cos\alpha+1}}+c$, then find λ.

2. If $\int\left(\frac{x-1}{x+1}\right)\frac{dx}{\sqrt{x^3+x^2+x}}$

$=\lambda\tan^{-1}\sqrt{\left(x+\frac{1}{x}+1\right)}+c$, then find λ.

3. If Limit

$\frac{1}{n}\left[\cos^{2p}\frac{\pi}{2n}+\cos^{2p}\frac{2\pi}{2n}+\cos^{2p}\frac{3\pi}{2n}+...+\cos^{2p}\frac{\pi}{2}\right]$

$=\frac{\lambda}{\pi}\left[\frac{2p-1}{2p}\cdot\frac{2p-3}{2p-2}\cdots\frac{1}{2},\frac{\pi}{2}\right], p\in N,$

then find λ.

4. If $\int\limits_0^{2\pi} \frac{dx}{a+b\cos x+c\sin x} = \frac{\lambda\pi}{\sqrt{a^2-b^2-c^2}}$ where $a>\sqrt{b^2+c^2}>0$, then find λ.

5. A function $f:R\to R$ satisfies the equation $f(x+y)=f(x)+f(y)\forall x,y\in R$ and is continuous throughout the domain. If $I_1+I_2+I_3+I_4+I_5=450$ where

$I_n = n\int\limits_0^n f(x)dx$ and $f(x)=\lambda x$, then find λ.

6. If $\int\limits_o^1 \frac{1}{(5+2x-2x^2)(1+e^{(2-4x)})}dx$

$=\frac{1}{\sqrt{\lambda}}\log\frac{\sqrt{11}+1}{\sqrt{10}}$,then find λ.

Questions from Previous Years

7. Let $f:\mathbb{R}\to\mathbb{R}$ be a differentiable function such that its derivative f' is continuous and $f(\pi)=-6$.
If $F:[0,\pi]\to\mathbb{R}$ is defined by $F(x)=\int_0^x f(t)dt$
and if $\int\limits_0^{\pi}(f'(x)+F(x))\cos x\,dx=2$, then the value of $f(0)$ is ______.

8. If $I=\frac{2}{\pi}\int\limits_{-\pi/4}^{\pi/4}\frac{dx}{(1+e^{\sin x})(2-\cos 2x)}$, then $27I^2$ equals ______.

9. The value of integral $\int\limits_0^{\pi/2}\frac{3\sqrt{\cos\theta}}{\left(\sqrt{\cos\theta}+\sqrt{\sin\theta}\right)^5}d\theta$ equals ______.

10. The value of the integral $\int\limits_0^{\frac{1}{2}}\frac{1+\sqrt{3}}{\left((x+1)^2(1-x)^6\right)^{\frac{1}{4}}}dx$ is ______

11. Let $f:R\to R$ be a differentiable function such that $f(0)=0,\ f\left(\frac{\pi}{2}\right)=3$ and $f'(0)=1$

If $g(x)=\int\limits_x^{\frac{\pi}{2}}[f'(t)\operatorname{cosec}t-\operatorname{cosec}t\,f(t)]dt$ for $x\in\left(0,\frac{\pi}{2}\right]$, then $\lim\limits_{x\to 0}g(x)=$

12. For each positive integer n, let
$y_n=\frac{1}{n}(n+1)(n+2)....(n+n)^{\frac{1}{n}}$
For $x\in\mathbb{R}$, let $[x]$ be the greatest integer less than or equal to x. If $\lim\limits_{n\to\infty}y_n=L$, then the value of $[L]$ is ______

13. The total number of distinct $x\in[0,1]$ for which $\int\limits_0^x\frac{t^2}{1+t^4}dt=2x-1$ is ______.

14. If $\alpha=\int\limits_0^1\left(e^{9x+3\tan^{-1}x}\right)\left(\frac{12+9x^2}{1+x^2}\right)dx$ where $\tan^{-1}x$ takes only principal values, then the value of $\left(\log_e|1+\alpha|-\frac{3\pi}{4}\right)$ is ______.

15. Let $f:\mathbb{R}\to\mathbb{R}$ be a continuous odd function, which vanishes exactly at one point and $f(1)=\frac{1}{2}$. Suppose that $F(x)=\int\limits_{-1}^x f(t)dt$ for all $x\in[-1,2]$ and $G(x)=\int\limits_{-1}^x t|f(f(t))|dt$ for $x\in[-1,2]$. If $\lim\limits_{x\to1}\frac{F(x)}{G(x)}=\frac{1}{14}$, then the value of $f\left(\frac{1}{2}\right)$ is ______.

16. Let $f:R\to R$ be a function defined by $f(x)=\begin{cases}[x]x\le 2\\ 0,\ x\le 2\end{cases}$ where $[x]$ is the greatest integer less than or equal to x.
If $I=\int\limits_{-1}^2\frac{xf(x^2)}{2+f(x+1)}dx$, then the value of $4I$ is ______.

17. The value of $\int\limits_0^1 4x^3\left\{\frac{d^2}{dx^2}(1-x^2)^5\right\}dx$ is ______.

Comprehension Passage

Comprehension 1

If A is a square matrix and e^A is defined as

$e^A=I+A+\frac{A^2}{2!}+\frac{A^3}{3!}.......\infty=\frac{1}{2}\begin{bmatrix}f(x)&g(x)\\g(x)&f(x)\end{bmatrix}$, where $A=\begin{bmatrix}x&x\\x&x\end{bmatrix}$, and I being the identify matrix, then

1. $\int\frac{g(x)}{f(x)}dx=$ ______.

(a) $\log\left(e^x+e^{-x}\right)+C$ (b) $\log\left|e^x-e^{-x}\right|+C$

(c) $\log\left|e^{2x}-1\right|+C$ (d) $\log\left(e^{2x}+1\right)-x+c$

2. $\int \frac{f(x)}{\sqrt{g(x)}}dx=$ ______.

(a) $\frac{1}{2\sqrt{e^x-1}}-\operatorname{cosec}^{-1}\left(e^x\right)+C$

(b) $\frac{2}{\sqrt{e^x-e^{-x}}}-\sec^{-1}\left(e^x\right)+C$

(c) $\sqrt{e^{2x}-1}+\sec^{-1}\left(e^x\right)+C$

(d) $\sqrt{e^{2x}-1}+\tan^{-1}\left(\sqrt{e^{2x}-1}\right)+c$

Comprehension 2

Integrals of the form $\int \frac{p_m(x)}{\sqrt{ax^2+bx+c}}dx$, where $p_m(x)$ is a polynomial of degree *m*, are calculated by the reduction formula.

$$\int \frac{p_m(x)}{\sqrt{ax^2+bx+c}}dx$$

$$=p_{m-1}(x)\sqrt{ax^2+bx+c}+\lambda\int\frac{dx}{\sqrt{ax^2+bx+c}}$$

where $p_{m-1}(x)$ is a polynomial of degree $(m-1)$ and λ is some constant number.

e.g., If $I=\int\frac{x^3-x-1}{\sqrt{x^2+2x+2}}dx$, then applying the above formula, we can write

$$\int\frac{x^3-x-1}{\sqrt{x^2+2x+2}}dx$$

$$=\left(Ax^2+Bx+C\right)\sqrt{x^2+2x+2}+\lambda\int\frac{dx}{\sqrt{x^2+2x+2}}$$

Differentiating both sides, we get

$$\int\frac{x^3-x-1}{\sqrt{x^2+2x+2}}=\left(Ax^2+Bx+C\right).\frac{2(x+1)}{2\sqrt{x^2+2x+2}}+$$

$$(2Ax+B)\sqrt{x^2+2x+2}+\frac{\lambda}{\sqrt{x^2+2x+2}}$$

$$x^3-x-1=\left(Ax^2+Bx+C\right)(x+1)+(2Ax+B)$$

$$\left(x^2+2x+2\right)+\lambda$$

On comparing coefficients of like powers of *x*, we obtain the values of *A*, *B*, *C* and λ.

1. If $\int\frac{x^3-6x^2+11x-6}{\sqrt{x^2+4x+3}}dx$

$$=\left(Ax^2+Bx+C\right)\sqrt{x^2+4x+3}+\lambda\int\frac{dx}{\sqrt{x^2+4x+3}},$$

then value of 'A' is ______.

(a) $\frac{1}{3}$ (b) 1

(c) 3 (d) −1/3

2. In Q. 1, the value of *C* is ______.

(a) −37 (b) $-\frac{14}{3}$

(c) $\frac{14}{3}$ (d) 37

3. In Q. 1 value of λ is ______.

(a) 66 (b) −66

(c) $\frac{37}{3}$ (d) $-\frac{37}{3}$

Comprehension 3

Let $I_{n,m}=\int\sin^n x\cos^m x.dx$. Then we can relate $I_{n,m}$ with each of the following:

(i) $I_{n-2,m}$ (ii) $I_{n+2,m}$ (iii) $I_{n,m-2}$

(iv) $I_{n,m+2}$ (v) $I_{n-2,m+2}$ (vi) $I_{n+2,m-2}$

Suppose we want to establish a relation between $I_{n.m}$ and $I_{n,m-2}$, then we set

$$P(x)=\sin^{n+1}x\cos^{m-1}x \quad ...(1)$$

In $I_{n,m}$ and $I_{n,m-2}$ the exponent of cos *x* is *m* and $m-2$ respectively. The minimum of the two is $m-2$. Adding 1 to the minimum we get $m-2+1=m-1$. Now choose the exponent $m-1$of cos *x* in *P*(*x*). Similarly, choose the exponent of sin*x* for *P*(*x*).

Now differentiating both sides of (1), we get

$$P'(x)=(n+1)\sin^n x\cos^m x-(m-1)\sin^{n+2}x\cos^{m-2}x$$

$$=(n+1)\sin^n x\cos^m x-(m-1)\sin^n x$$

$$\left(1-\cos^2 x\right)\cos^{m-2}x$$

$$=(n+1)\sin^n x\cos^m x-(m-1)\sin^n x\cos^{m-2}x+$$

$$(m-1)\sin^n x\cos^m x$$

$$=(n+m)\sin^n x\cos^m x-(m-1)\sin^n x\cos^{m-2}x$$

Now integrating both sides, we get

$$\sin^{n+1}x\cos^{m-1}x=(n+m)I_{n,m}-(m-1)I_{n,m-2}$$

Similarly, we can establish the other relations.

1. The relation between $I_{4,2}$ and $I_{2,2}$ is ______.

(a) $I_{4,2}=\frac{1}{6}\left(-\sin^3 x\cos^3 x+3I_{2,2}\right)$

(b) $I_{4,2}=\frac{1}{6}\left(\sin^3 x\cos^3 x+3I_{2,2}\right)$

(c) $I_{4,2}=\frac{1}{6}\left(\sin^3 x\cos^3 x-3I_{2,2}\right)$

(d) $I_{4,2}=\frac{1}{4}\left(-\sin^3 x\cos^3 x+2I_{2,2}\right)$

2. The relation between $I_{4,2}$ and $I_{6,2}$ is ______.

(a) $I_{4,2}=\frac{1}{5}\left(\sin^5 x\cos^3 x+8I_{6,2}\right)$

(b) $I_{4,2}=\frac{1}{5}\left(-\sin^5 x\cos^3 x+8I_{6,2}\right)$

(c) $I_{4,2}=\frac{1}{5}\left(\sin^5 x\cos^3 x-8I_{6,2}\right)$

(d) $I_{4,2}=\frac{1}{6}\left(\sin^5 x\cos^3 x+8I_{6,2}\right)$

3. The relation between $I_{4,2}$ and $I_{4,4}$ is ______.

(a) $I_{4,2}=\frac{1}{3}\left(\sin^5 x\cos^3 x+8I_{4,4}\right)$

(b) $I_{4,2}=\frac{1}{3}\left(-\sin^5 x\cos^3 x+8I_{4,4}\right)$

(c) $I_{4,2}=\frac{1}{3}\left(\sin^5 x\cos^3 x-8I_{4,4}\right)$

(d) $I_{4,2}=\frac{1}{3}\left(\sin^5 x\cos^3 x+6I_{4,4}\right)$

Comprehension 4

Consider the polynomial $f(x)=1+2x+3x^2+4x^3$. Let s be the sum of all distinct real roots of $f(x)$ and let $t=|s|$

1. The real numbers lies in the interval ______.

(a) $\left(-\frac{1}{4},0\right)$ (b) $\left(-11,-\frac{3}{4}\right)$

(c) $\left(-\frac{3}{4},-\frac{1}{2}\right)$ (d) $\left(0,\frac{1}{4}\right)$

2. The area bounded by the curve $y=f(x)$ and the lines $x=0$, $y=0$ and $x=t$, lies in the interval ______.

(a) $\left(\frac{3}{4},3\right)$ (b) $\left(\frac{21}{11},\frac{11}{16}\right)$

(c) (9, 10) (d) $\left(0,\frac{21}{11}\right)$

3. The function $f'(x)$ is

(a) increasing in $\left(-t,-\frac{1}{4}\right)$ and decreasing in $\left(-\frac{1}{4},t\right)$.

(b) decreasing in $\left(-t,-\frac{1}{4}\right)$ and increasing in $\left(-\frac{1}{4},t\right)$.

(c) increasing in $(-t,t)$.

(d) decreasing in $(-t,t)$.

Comprehension 5

Consider the function $f:(-\infty,\infty)\to(-\infty,\infty)$ defined by $f(x)=\frac{x^2-ax+1}{x^2+ax+1}, 0<a<2$

1. Which of the following is true?

(a) $(2+a)^2 f''(1)+(2-a)^2 f''(-1)=0$

(b) $(2-a)^2 f''(1)-(2+a)^2 f''(-1)=0$

(c) $f'(1)f'(-1)=(2-a)^2$

(d) $f'(1)f'(-1)=-(2+a)^2$

2. Which of the following is true?

(a) $f(x)$ is decreasing on $(-1,1)$ and has a local minimum at $x=1$.

(b) $f(x)$ is increasing on $(-1,1)$ and has a local minimum at $x=1$.

(c) $f(x)$ is increasing on $(-1,1)$ but has neither a local maximum nor a local minimum at $x=1$.

(d) $f(x)$ is decreasing on $(-1,1)$ but has neither a local maximum nor a local minimum at $x=1$.

3. Let $g(x)=\int_0^{e^x}\frac{f'(t)}{1+t^2}dt$. Which of the following is true?

(a) $g'(x)$ is positive on $(-\infty,0)$ and negative on $(0,\infty)$.

(b) $g'(x)$ is negative on $(-\infty,0)$ and positive on $(0,\infty)$.

(c) $g'(x)$ changes sign on both $(-\infty,0)$ and $(0,\infty)$.

(d) $g'(x)$ does not change sign on $(-\infty,\infty)$.

Comprehension 6

The definite integral of any discontinuous or non-differentiable function is normally solved by the property

$$\int_a^b f(x)dx=\int_a^c f(x)dx+\int_c^b f(x)dx,$$

where $c\in(a,b)$ is the point of discontinuity or non-differentiability.

1. The value of $A=\int_1^{\infty}\left[\operatorname{cosec}^{-1}x\right]dx$, {where [.] denotes greatest integer function}, is equal to ______.
(a) cosec 1 – 1 (b) 1
(c) 1 – sin 1 (d) None of these.

2. The value of $B=\int_1^{100}\left[\sec^{-1}x\right]dx$, {where [.] denotes greatest integer function}, is equal to ______.
(a) sec 1 (b) 100 – sec 1
(c) 99 – sec 1 (d) None of these.

3. The value of integral $\int_A^B\left[\tan^{-1}x\right]dx$, {where [.] denotes greatest integer function}, is equal to ______.
(a) tan 1 (b) 100 – tan 1 – sec 1
(c) 99 – sec 1 (d) none of these

Comprehension 7

Using integral

$$\int_0^{\pi/2}\ln(\sin x)dx=-\int_0^{\pi/2}\ln(\sec x)dx=-\frac{\pi}{2}\ln 2,$$

$$\int_0^{\pi/2}\ell n(\tan x)dx=0$$

and $\int_0^{\pi/4}\ell n(1+\tan x)dx=\frac{\pi}{8}\ell n2,$

1. $\int_{-\pi/4}^{\pi/4}\ell n\left(\frac{\sin x+\cos x}{\cos x-\sin x}\right)dx=$ ______.
(a) $\pi\ell n2$ (b) $\frac{\pi\ell n2}{2}$
(c) 0 (d) $-\pi\ell n2$

2. $\int_{-\pi/4}^{\pi/4}\ell n(\sin x+\cos x)dx=$ ______.
(a) $\frac{\pi\ell n2}{2}$ (b) $\frac{-\pi\ell n2}{4}$
(c) $\pi\ell n2$ (d) 0

3. $\int_0^{\pi/4}\ell n(\sin 2x)dx=$ ______.
(a) $\frac{-\pi\ell n2}{2}$ (b) $\pi\ell n2$
(c) $\frac{\pi\ell n2}{4}$ (d) $-\frac{\pi\ell n2}{4}$

Questions from Previous Years

Comprehension 8

The integral $\int_a^b f(x)dx$ can be represented as a limit of a sum of infinite series

$$\int_a^b f(x)dx=\lim_{n\to\infty}\sum_{r=na+c}^{nb+c}\frac{1}{n}f\left(\frac{r}{n}\right)$$

where $na+c\le r\le nb+c, r, n\in N, c\in R$ and any limit of sum of series of the same form can be changed to definite integral by replacing ______.

(1) $\lim_{n\to\infty}\sum\to\int$ (2) $\frac{1}{n}\to dx$ (3) $\frac{r}{n}\to x$

(4) Lower limit $=\lim_{n\to\infty}\left(\frac{r}{n}\right)_{\min}=\lim_{n\to\infty}\left(\frac{na+c}{n}\right)=a$

(5) Upper limit $=\lim_{n\to\infty}\left(\frac{r}{n}\right)_{\max}=\lim_{n\to\infty}\left(\frac{nb+c}{n}\right)=b$

1. Find the value of

$$\lim_{n\to\infty}\left(\frac{n}{(n+1)\sqrt{2n+1}}+\frac{n}{(n+2)\sqrt{2(2n+2)}}+\frac{n}{(n+3)\sqrt{3(2n+3)}}+....+\frac{1}{2n\sqrt{3}}\right).$$

(a) $\frac{\pi}{3}$ (b) $\frac{\pi}{2}$
(c) $\frac{\pi}{4}$ (d) None of these.

2. The nth term of the corresponding series of $\int_0^1\tan^{-1}x\,dx$ is ______.
(a) $\frac{\pi}{4n}$ (b) $\frac{1}{n}\tan^{-1}(n-1)$
(c) $\frac{\pi}{2n}$ (d) $\tan^{-1}n$

3. $\lim_{n\to\infty}\sum_{r=0}^{2n-1}\frac{1}{n}\sec^2\left(\frac{r}{n}\right)$ is ______.
(a) sec 2 (b) tan 2
(c) sec² (d) not defined

Comprehension 9

Let $g_i:\left[\frac{\pi}{8},\frac{3\pi}{8}\right]\to\mathbb{R}$, $i=1, 2$ and $f:\left[\frac{\pi}{8},\frac{3\pi}{8}\right]\to\mathbb{R}$ be functions such that $g_1(x)=|4x-\pi|$ and $f(x)=\sin^2 x$, for all $x\in\left[\frac{\pi}{8},\frac{3\pi}{8}\right]$.

Define $S_i = \int_{\frac{\pi}{8}}^{\frac{3\pi}{8}} f(x).g_i(x)dx, i=1,2$

1. The value of $\frac{16S_1}{\pi}$ is ______.

2. The value of $\frac{48S_2}{\pi^2}$ is ______.

Comprehension 10

Let $f_1:(0,\infty)\to\mathbb{R}$ and $f_2:(0,\infty)\to\mathbb{R}$ be defined by

$$f_1(x)=\int_0^x \prod_{j=1}^{21}(t-j)^j\,dt,\ x>0$$

and $f_2(x)=98(x-1)^{50}-600(x-1)^{49}+2450,\quad x>0$ where for any positive integer n and real numbers a_1, a_2, ..., a_n, $\prod_{i=1}^{n} a_i$ denotes the product of a_1, a_2,..., a_n. Let m_1 and n_1 respectively, denote the number of points of local minima and the number of points of local maxima of function f_i, $i=1,2$ in the interval $(0,\infty)$.

1. The value of $2m_1+3n_1+m_1n_1$ is ______.

2. The value of $6m_2+4n_2+8m_2n_2$ is ______.

Comprehension 11

Let $F:\mathbb{R}\to\mathbb{R}$ be a thrice differentiable function. Suppose that $F(1)=0, F(3)=-4$ and $F(x)<0$ for all $x\in\left(\frac{1}{2},3\right)$. Let $f(x)=xF(x)$ for all $x\in\mathbb{R}$

1. The correct statement(s) is (are) ______.

(a) $f'(1)<0$

(b) $f'(2)<0$

(c) $f'(x)\neq 0$ for any $x\in(1,3)$

(d) $f'(x)=0$ for some $x\in(1,3)$

2. If $\int_1^3 x^2F'(x)dx=-12$ and $\int_1^3 x^2F''(x)dx=40$, then the correct expression(s) is (are)

(a) $9f'(3)+f'(1)-32=0$

(b) $\int_1^3 f(x)dx=12$

(c) $9f'(3)-f'(1)+32=0$

(d) $\int_1^3 f(x)dx=-12$

Comprehension 12

Given that for each $a\in(0,1)$, $\lim_{h\to 0^+}\int_h^{1-h} t^{-a}(1-t)^{a-1}dt$ exists.

Let this limit be $g(a)$. In addition, it is given that the function $g(a)$ is differentiable on (0, 1).

1. The value of $g\left(\frac{1}{2}\right)$ is ______.

(a) π (b) 2π

(c) $\frac{\pi}{2}$ (d) $\frac{\pi}{4}$

2. The value of $g'\left(\frac{1}{2}\right)$ is ______.

(a) $\frac{\pi}{2}$ (b) π

(c) $-\frac{\pi}{2}$ (d) 0

ANSWER KEY

JEE Main

Objective Questions

One Option Correct

1. (b)	**2.** (a)	**3.** (b)	**4.** (a)	**5.** (b)	**6.** (d)	**7.** (b)	**8.** (d)	**9.** (a)
10. (b)	**11.** (c)	**12.** (b)	**13.** (b)	**14.** (a)	**15.** (b)	**16.** (d)	**17.** (a)	**18.** (c)
19. (b)	**20.** (a)	**21.** (a)	**22.** (d)	**23.** (b)	**24.** (b)	**25.** (c)	**26.** (a)	**27.** (c)
28. (a)	**29.** (a)	**30.** (a)	**31.** (c)	**32.** (a)	**33.** (a)	**34.** (b)	**35.** (c)	**36.** (b)
37. (a)	**38.** (b)	**39.** (d)	**40.** (b)	**41.** (d)	**42.** (c)	**43.** (c)	**44.** (d)	**45.** (d)
46. (c)	**47.** (a)	**48.** (a)	**49.** (a)	**50.** (c)	**51.** (c)	**52.** (d)	**53.** (a)	**54.** (a)
55. (a)	**56.** (a)	**57.** (b)	**58.** (a)	**59.** (b)	**60.** (b)	**61.** (b)	**62.** (d)	**63.** (b)
64. (b)	**65.** (b)	**66.** (b)	**67.** (b)	**68.** (c)	**69.** (b)	**70.** (d)	**71.** (b)	**72.** (None)
73. (d)	**74.** (d)	**75.** (a)	**76.** (d)	**77.** (a)	**78.** (a)	**79.** (c)	**80.** (c)	**81.** (c)
82. (c)	**83.** (a)	**84.** (c)	**85.** (c)	**86.** (b)	**87.** (c)	**88.** (a)	**89.** (a)	**90.** (a)
91. (a)	**92.** (d)	**93.** (b)	**94.** (a)	**95.** (d)	**96.** (c)	**97.** (c)	**98.** (c)	**99.** (d)

100. (b) **101.** (a) **102.** (b) **103.** (b) **104.** (b) **105.** (a) **106.** (c) **107.** (b) **108.** (a)
109. (b) **110.** (b) **111.** (c) **112.** (c)

Integer and Numerical Value Type

1. (2) **2.** (2) **3.** (2) **4.** (4) **5.** (4) **6.** (1) **7.** (3) **8.** (3) **9.** (4)
10. (6) **11.** (4) **12.** (1) **13.** (5) **14.** (1) **15.** (1) **16.** (1) **17.** (3) **18.** (1)
19. (512) **20.** (1) **21.** (21) **22.** (1.5) **23.** (1)

Assertion and Reason

1. (a) **2.** (a) **3.** (d) **4.** (a) **5.** (c) **6.** (b) **7.** (c) **8.** (b) **9.** (b)
10. (c) **12.** (d) **13.** (a)

JEE Advanced

Objective Questions

One Option Correct

1. (b) **2.** (d) **3.** (a) **4.** (b) **5.** (a) **6.** (d) **7.** (c) **8.** (a) **9.** (a)
10. (c) **11.** (d) **12.** (c) **13.** (b) **14.** (c) **15.** (d) **16.** (b) **17.** (d) **18.** (c)
19. (a) **20.** (a) **21.** (b) **22.** (c) **23.** (d) **24.** (b) **25.** (c) **26.** (c) **27.** (d)
28. (a) **29.** (b) **30.** (a) **31.** (b) **32.** (d) **33.** (c) **34.** (c) **35.** (a) **36.** (d)

One or More Than One Option Correct

1. (b, c) **2.** (a) **3.** (a, b, c) **4.** (a, b, c, d) **5.** (a, c) **6.** (a, b) **7.** (b, c) **8.** (b, c)
9. (a, b) **10.** (a, b, d) **11.** (a, d) **12.** (a, c, d) **13.** (a, b, c) **14.** (c) **15.** (d)
16. (a, c) **17.** (a, b, d) **18.** (a, b, d) **19.** (a, d) **20.** (b, d) **21.** (None)
22. (a, b) **23.** (b, c) **24.** (d) **25.** (a, b) **26.** (a, c)

Matrix Type

1. A → Q; B → R; C → P; D → R
2. A → T; B → T; C → R; D → S
3. A → Q; B → P; C → S; D → S
4. A → Q; B → S; C → T; D → T
5. A → R; B → R; C → Q; D → Q
6. A → P; B → Q; C → T; D → P
7. A → Q; B → R; C → P; D → S

Integer and Numerical Value Type

1. (1) **2.** (2) **3.** (2) **4.** (2) **5.** (4) **6.** (11) **7.** (4) **8.** (4) **9.** (0.50)
10. (2) **11.** (2) **12.** (1) **13.** (1) **14.** (9) **15.** (7) **16.** (0) **17.** (2)

Comprehension Passage

Comprehension 1

1. (a) **2.** (c)

Comprehension 2

1. (a) **2.** (d) **3.** (b)

Comprehension 3

1. (a) **2.** (a) **3.** (b)

Comprehension 4

1. (c) **2.** (a) **3.** (b)

Comprehension 5

1. (a) **2.** (a) **3.** (b)

Comprehension 6

1. (a) **2.** (b) **3.** (b)

Comprehension 7

1. (c) **2.** (b) **3.** (d)

Comprehension 8

1. (a) **2.** (a) **3.** (b)

Comprehension 9

1. (2.00) **2.** (1.50)

Comprehension 10

1. (57.00) **2.** (6.00)

Comprehension 11

1. (a, b, c) **2.** (c, d)

Comprehension 12

1. (a) **2.** (d)

HINTS AND EXPLANATIONS

Worksheet – 1

1. $\int\sqrt{\dfrac{9-x^2}{x^4}}dx$ Substitute $x=\dfrac{1}{t}$, thus $dx=-\dfrac{1}{t^2}$

$\Rightarrow \int\dfrac{\sqrt{9t^2-1}}{\dfrac{1}{t^3}}\left(-\dfrac{1}{t^2}\right)dt \Rightarrow -\int t\sqrt{9t^2-1}dt$

Now substitute $\sqrt{9t^2-1}=z$

$\therefore 9t^2-1=z^2$

$\therefore 18t\,dt=2z\,dz$

$\therefore tdt=\dfrac{zdz}{9}$

$\Rightarrow -\dfrac{1}{9}\int z.zdz=-\dfrac{1}{9}\int z^2dz.$

This is an integrable from.

2. $\int\dfrac{dx}{9\sin^2x+16\cos^2x}=\dfrac{1}{16}\int\dfrac{dx}{\left(\dfrac{3}{4}\right)^2\sin^2x+\cos^2x}$

$=\dfrac{1}{16}\int\dfrac{1}{\left(\dfrac{3}{4}\right)^2\tan^2x+1}.\sec^2xdx$

Now substitute $\dfrac{3}{4}\tan x=t.$|

Thus $\dfrac{3}{4}\sec^2xdx=dt$ or $\sec^2xdx=\dfrac{4}{3}dt$

$\Rightarrow \dfrac{1}{16}\int\dfrac{1}{1+t^2}\left(\dfrac{4}{3}\right)dt \Rightarrow \dfrac{1}{12}\int\dfrac{1}{1+t^2}dt$

This is an integrable form.

3. $\int\sqrt[3]{1+3\sin x}\cos xdx$

Substitute $\sqrt[3]{1+3\sin x}=t$ or

$1+3\sin x=t^3$

Thus $3\cos xdx=3t^2dt$ or $\cos xdx=t^2dt$

$\Rightarrow \int t.t^2dt=\int t^3dt$ (integrable form)

4. $\int\dfrac{dx}{\left(\sin^{-1}x\right)^5\sqrt{1-x^2}}$

Clearly, substitute $\sin^{-1}x=t$

so that $\dfrac{1}{\sqrt{1-x^2}}dx=dt \Rightarrow \int\dfrac{1}{t^5}dt$

5. $\int\dfrac{\sin 2x}{1+\cos^2x}dx$

Put $1+\cos^2x=t$ $\therefore 2\cos x\sin xdx=-dt$

$\sin 2xdx=-dt \Rightarrow -\int\dfrac{1}{t}dt$

6. Here, $I=\int\dfrac{dx}{x^2+x+1}=\int\dfrac{dx}{(x+1/2)^2+(3/4)}$

$=\dfrac{2}{\sqrt{3}}\tan^{-1}\left(\dfrac{x+\dfrac{1}{2}}{\dfrac{\sqrt{3}}{2}}\right)+c$

$=\dfrac{2}{\sqrt{3}}\tan^{-1}\left(\dfrac{2x+1}{\sqrt{3}}\right)+c$

7. $I=\int\dfrac{dx}{\sqrt[4]{(x-1)^3(x+2)^5}}=\int\dfrac{dx}{\left(\dfrac{x-1}{x+2}\right)^{3/4}(x+2)^2}$

Substitute $\dfrac{x-1}{x+2}=t^4$

$\Rightarrow x=\dfrac{2t^4+1}{1-t^4}, x+2=\dfrac{3}{1-t^4}$

Now, $dx=\dfrac{\left(1-t^4\right)8t^3-\left(2t^4+1\right)\left(-4t^3\right)}{\left(1-t^4\right)^2}$

$=\dfrac{8t^3-8t^7+8t^7+4t^3}{\left(1-t^4\right)^2}dt$

$=\dfrac{12t^3}{\left(1-t^4\right)^2}dt$

Then $I=\int\dfrac{12t^3}{\left(1-t^4\right)^2}.\dfrac{1}{t^3\dfrac{9}{\left(1-t^4\right)^2}}dt$

$=\dfrac{4}{3}\int dt=\dfrac{4}{3}t+c=\dfrac{4}{3}\left(\dfrac{x-1}{x+2}\right)^{\frac{1}{4}}+c$

8. Substituting $z=1/(t+2)$ or $t+2=1/z$

$\Rightarrow dt=-1/z^2dz$

now the integrable reduces to,

$$I=-\int\frac{\left(-\frac{1}{z^2}\right)dz}{\left(\frac{1}{z}\right)\sqrt{\left(\frac{1}{z}-2\right)^2+2\left(\frac{1}{z}-2\right)-3}}$$

$$=-\int\frac{\left(\frac{1}{z}\right)dz}{\sqrt{\frac{1}{z^2}-\frac{2}{z}-3}}$$

$$=-\int\frac{dz}{\sqrt{1-2z-3z^2}}=-\int\frac{dz}{\sqrt{1-3\left(z^2+\frac{2}{3}z+\frac{1}{9}\right)+\frac{1}{3}}}$$

$$=-\int\frac{dz}{\sqrt{\frac{4}{3}-3\left(z+\frac{1}{3}\right)^2}}$$

$$=-\frac{1}{\sqrt{3}}\int\frac{dz}{\sqrt{\frac{4}{9}-\left(z+\frac{1}{3}\right)^2}}$$

$$=-\frac{1}{\sqrt{3}}\int\frac{dz}{\sqrt{\left(\frac{2}{3}\right)^2-\left(z+\frac{1}{3}\right)^2}}$$

$$=-\frac{1}{\sqrt{3}}\sin^{-1}\frac{z+\frac{1}{3}}{\frac{2}{3}}+c=-\frac{1}{\sqrt{3}}\sin^{-1}\frac{t+5}{2(t+2)}+c$$

9. Dividing numerator and denominator by x^2, we get,

$$I=\int\frac{1+\left(\frac{1}{x^2}\right)}{x^2+\left(\frac{1}{x^2}\right)}dx=\int\frac{1+\left(\frac{1}{x^2}\right)}{\left[\left\{x-\left(\frac{1}{x}\right)\right\}^2+2\right]}dx$$

Now taking $x-\frac{1}{x}=t\Rightarrow\left[1+\frac{1}{x^2}\right]dx=dt$, we get

$$I=\int\frac{dt}{t^2+2}=\frac{1}{\sqrt{2}}\tan^{-1}\left(\frac{t}{\sqrt{2}}\right)+c$$

$$=\frac{1}{\sqrt{2}}\tan^{-1}\left(\frac{x^2-1}{\sqrt{2}x}\right)+c$$

10. $\int\sqrt{\tan x}\,dx=\frac{1}{2}\int\left(\sqrt{\tan x}+\sqrt{\cot x}\right)dx+$

$$\frac{1}{2}\left(\sqrt{\tan x}-\sqrt{\cot x}\right)dx$$

$$=\frac{1}{\sqrt{2}}\int\frac{\sin x+\cos x}{\sqrt{2\sin x.\cos x}}dx+\frac{1}{\sqrt{2}}\int\frac{\sin x-\cos x}{\sqrt{2\sin x.\cos x}}dx$$

$$=\frac{1}{\sqrt{2}}\int\frac{\cos x+\sin x}{\sqrt{1-(\sin x-\cos x)^2}}dx-$$

$$\frac{1}{\sqrt{2}}\int\frac{\cos x-\sin x}{\sqrt{(\sin x+\cos x)^2-1}}dx$$

Worksheet – 2

1. $\frac{56}{3}$

$$\int_0^4\left(x-2\sqrt{x}+x^2\right)dx=\left[\frac{x^2}{2}-\frac{4x^{\frac{3}{2}}}{3}+\frac{x^3}{3}\right]_0^4$$

$$=\left(8-\frac{32}{3}+\frac{64}{3}\right)-(0)=\frac{56}{3}$$

2. We have $0\le\sqrt{x}\le n$

Therefore,

$$\int_0^{n^2}\left[\sqrt{x}\right]dx=\int_0^1\left[\sqrt{x}\right]dx+\int_1^4\left[\sqrt{x}\right]dx+$$

$$\int_4^9\left[\sqrt{x}\right]dx+\ldots+\int_{(n-1)^2}^{n^2}\left[\sqrt{x}\right]dx$$

$$=\int_0^1 0dx+\int_1^4 dx+\int_4^9 2dx+\ldots+\int_{(n-1)^2}^{n^2}(n-1)dx$$

$$=(4-1)+2(9-4)+3(16-9)+\ldots+(n-1)\left(n^2-(n-1)^2\right)$$

$$=\left(2^2-1^2\right)+2\left(3^2-2^2\right)+3\left(4^2-3^2\right)+\ldots+(n-1)\left(n^2-(n-1)^2\right)$$

$$=-\left\{1^2+2^2+3^2+\ldots+(n-1)^2+n^2\right\}+n^3$$

$$=\frac{-1}{6}n(n+1)(2n+1)+n^3$$

$$=\frac{n}{6}\left\{6n^2-(n+1)(2n+1)\right\}$$

$$=\frac{n}{6}\left\{6n^2-2n^2-3n-1\right\}$$

$$=\frac{n}{6}\left\{4n^2-3n-1\right\}=\frac{n}{6}(n-1)(4n+1)$$

3. $\dfrac{37}{6}$

$$\int_0^3 f(x)\,dx = \int_0^2 f(x)\,dx + \int_2^3 f(x)\,dx$$

$$= \int_0^2 x^2 dx + \int_2^3 (3x-4)\,dx$$

$$= \left(\frac{x^3}{3}\right)_0^2 + \left(\frac{3x^2}{2} - 4x\right)_2^3$$

$$= \frac{8}{3} + \frac{27}{12} - 12 - 6 + 8 = \frac{37}{6}$$

4. We have,

$$\int_0^2 f(x)\,dx = \int_0^1 f(x)\,dx + \int_1^2 f(x)\,dx$$

$$= \int_0^1 x^2 dx + \int_1^2 \sqrt{x}\,dx = \left[\frac{x^3}{3}\right]_0^1 + \left[\frac{x^{\frac{3}{2}}}{3/2}\right]_1^2$$

$$= \frac{1}{2} + \frac{2^{3/2}}{3/2} - \frac{1}{3/2} = \frac{2}{3}\left(2\sqrt{2}\right) - \frac{1}{3}$$

$$= \frac{1}{3}\left[4\sqrt{2} - 1\right]$$

5. $I = \int_0^{\pi/2} (2\log\sin x - \log 2\sin x\cos x)\,dx$

$$= \int_0^{\pi/2} (2\log\sin x - \log 2 - \log\sin x - \log\cos x)\,dx$$

$$= \int_0^{\pi/2} \log\sin x\,dx - \int_0^{\pi/2} \log 2\,dx - \int_0^{\pi/2} \log\cos x\,dx$$

$$= -\left(\frac{\pi}{2}\right)\log 2$$

6. $I = \int_{-1}^{1} \frac{\sin x - x^2}{3-|x|}\,dx = \int_{-1}^{1} \frac{\sin x}{3-|x|}\,dx - \int_{-1}^{1} \frac{x^2}{3-|x|}\,dx$

$$= 0 - 2\int_{-1}^{1} \frac{x^2}{3-|x|}\,dx$$

$$\left[\because \frac{\sin x}{3-|x|} \text{ is an odd and } \frac{x^2}{3-|x|} \text{ is an even function}\right]$$

$$= -2\int_0^1 \frac{x^2}{3-|x|}$$

7. $I_n = \int_0^{\pi/2} x^n \sin x\,dx$ ...(i)

Now $I_n = \left[x^n(-\cos x)\right]_0^{\pi/2} - \int_0^{\pi/2} n x^{n-1}(-\cos x)\,dx$

$$= n\int_0^{\pi/2} x^{n-1}\cos x\,dx$$

$$= n\left\{\left[x^{n-1}\sin x\right]_0^{\pi/2} - \int_0^{\pi/2} (n-1)x^{n-2}\sin x\,dx\right\}$$

$$= n\left(\frac{\pi}{2}\right)^{n-1} - n(n-1)I_{n-2}$$

$$\Rightarrow I_n + n(n-1)I_{n-2} = n\left(\frac{\pi}{2}\right)^{n-1}$$

8. Put $x = 2\sec t$, then

$$I = \int_0^{\pi/3} \frac{2\tan t}{2\sec t}\cdot 2\sec t\cdot\tan t\cdot dt = 2\int_0^{\pi/3} \tan^2 t\,dt$$

$$= 2\int_0^{\pi/3} \left(\sec^2 t - 1\right)dt = 2\left[\tan t - t\right]_0^{\pi/3}$$

$$= 2\left[\sqrt{3} - \pi/3\right] = \frac{2}{3}\left(3\sqrt{3} - \pi\right)$$

9. Putting $x = -t - 4$ in the first integral and $x = \frac{t}{3} + \frac{1}{3}$ in the second integral.

$$I_1 = \int_{-4}^{-5} e^{(x+5)^2}\,dx = -\int_0^1 e^{(-t+1)^2}\,dt = -\int_0^1 e^{(t-1)^2}\,dt$$

$$I_2 = 3\int_{1/3}^{2/3} e^{9(x-2/3)^2}\,dx = 3\int_0^1 e^{9(t/3-1/3)^2}\,dt = \int_0^1 e^{(t-1)^2}\,dt$$

$$\therefore I = I_1 + I_2 = 0$$

10. Let $f(x) = \frac{1}{10+3\cos x}$. Then $f(x)$ is continuous on $[0, 2\pi]$ with maximum value $\frac{1}{7}$ and minimum value $\frac{1}{13}$.

Therefore $\frac{1}{13} \le f(x) \le \frac{1}{7}$

$$\Rightarrow \frac{1}{13}\int_0^{2\pi} dx \le \int_0^{2\pi} f(x)\,dx \le \frac{1}{7}\int_0^{2\pi} dx$$

$$\Rightarrow \frac{2\pi}{13} \le \int_0^{2\pi} \frac{dx}{10+3\cos x} \le \frac{2\pi}{7}$$

Quiz

1. Anti-derivative is the inverse process of differentiation. This is known as Integration. We know $\frac{d}{dx}(\sin x)=\cos x$. The anti-derivative of $\cos x$ is $\sin x$. Or we can say that integration of cos x gives sin x.

2. When a polynomial function P is differentiated, the result is a polynomial whose degree is 1 less than the degree of the function whereas when a polynomial function is integrated, the result is a polynomial and degree is 1 more than the function P.

3. Put $\cos x = t$

4. $\cos 2x = 2\cos^2 x - 1$

5. $\frac{1}{2a}\log\left|\frac{x-a}{x+a}\right|+c$

6. $\int\frac{dx}{(x)^2-(4)^2}=\frac{1}{8}\log\left|\frac{x-4}{x+4}\right|+c$
 (using the formula of Q.5)

7. $F(b)-F(a)$

8. $\int x^2 dx = \frac{x^3}{3}+c$
 Now putting higher and lower limits, we have
 $\frac{(2)^3}{3}-\frac{(1)^3}{3}=\frac{8}{3}-\frac{1}{3}=\frac{7}{3}$

9. $\frac{1}{2}x\sqrt{x^2+a^2}+\frac{a^2}{2}\log\left|x+\sqrt{x^2+a^2}\right|+c$

10. $t = 1+\sin 6x$

JEE Main

Objective Questions

One Option Correct

1. (b) Put $x-\frac{\pi}{4}=t \Rightarrow dx = dt$

$$\therefore\ I=\sqrt{2}\int\frac{\sin\left(t+\frac{\pi}{4}\right)}{\sin t}dt=\int\frac{\sin t+\cos t}{\sin t}dt$$

$$=\int dt+\int\cot t dt$$

$$=t+\log\sin t dt$$

$$=x-\frac{\pi}{4}+\log\sin\left|x-\frac{\pi}{4}\right|+c$$

2. (a) $I=\int_3^6\frac{\sqrt{x}}{\sqrt{9-x}+\sqrt{x}}dx$

$$=\int_3^6\frac{\sqrt{9-x}}{\sqrt{9-(9-x)}+\sqrt{9-x}}dx$$

applying $\int_3^b f(x)dx=\int_a^b f(a+b-x)$

$$=\int_3^6\frac{\sqrt{9-x}}{\sqrt{9-x}+\sqrt{x}}dx$$

On adding, we get

$$\Rightarrow 2I=\int_3^6 dx \Rightarrow I=\frac{3}{2}$$

3. (b) Put $x+\pi=t$

Then $I=\int_{-\pi/2}^{\pi/2}\left[t^3+\cos^2(2\pi+t)\right]dt$

$$=0+2\int_0^{\pi/2}\cos^2 t dt=\frac{\pi}{2}$$

4. (a) $\int_1^a [x]f'(x)dx$

$$=\int_1^2 f'[x]dx+\int_2^3 2f'(x)dx+\ldots+\int_{[a]}^a [a]f'(x)dx$$

$$=[a]f(a)-f(1)-f(2)\ldots-f[a]$$

5. (b) We can check, taking the options one by one, as follows:

Let $\frac{x}{1+(\log x)^2}=t$

$$\frac{dt}{dx}=\frac{1+(\log x)^2-x\left(2\log x.\frac{1}{x}\right)}{\left[1+(\log x)^2\right]^2}$$

$$=\frac{1+(\log x)^2-2\log x}{\left[1+(\log x)^2\right]^2}$$

$$=\frac{(\log x-1)^2}{\left[1+(\log x)^2\right]^2}$$

6. (d) $I_1=\int_0^1 2^{x^2}dx \quad I_2=\int_0^1 2^{x^3}dx$

$$I_3=\int_0^2 2^{x^2}dx \quad I_4=\int_0^2 2^{x^3}dx$$

We know for $0 < x\langle 1, x^2\rangle x^3 \Rightarrow 2^{x^3} < 2^{x^2}$

Then $I_1 > I_2 \Rightarrow$ (d) correct.

and for $x > 1, 2^{x^3} > 2^{x^2}$

$I_1 < I_2$,

and for $1 < x < 2, x^3 > x^2$,

$\Rightarrow I_4 > I_3$

7. (b) $\int \frac{\sin x}{\sin(x-\alpha)}dx$

Put $x-\alpha = t, x = t+\alpha, dx = dt$

$= \int \frac{\sin(t+\alpha)}{\sin t}dt = \int \cos\alpha dt + \sin\alpha \int \frac{\cos t}{\sin t}dt$

$= t\cos\alpha + \sin\alpha \log\sin t + c$

$= (x-\alpha)\cos\alpha + \sin\alpha\log\sin(x-\alpha) + c$

8. (d) $\int \frac{1}{\cos x - \sin x}dx$

$= \frac{1}{\sqrt{2}}\int \frac{1}{\frac{1}{\sqrt{2}}\cos x - \frac{1}{\sqrt{2}}\sin x}dx$

$= \frac{1}{\sqrt{2}}\int \frac{1}{\cos\frac{\pi}{4}\cos x - \sin\frac{\pi}{4}\sin x}dx$

$= \frac{1}{\sqrt{2}}\int \frac{1}{\cos\left(x+\frac{\pi}{4}\right)}dx = \frac{1}{\sqrt{2}}\int \sec\left(x+\frac{\pi}{4}\right)dx$

$= \frac{1}{\sqrt{2}}\log\left|\tan\frac{\pi}{4} + \left(\frac{x}{2}+\frac{\pi}{8}\right)\right| + c$

$= \frac{1}{\sqrt{2}}\log\left|\tan\left(\frac{x}{2}+\frac{3\pi}{8}\right)\right| + c$

9. (a) $f(a) = \frac{e^a}{1+e^a}$ and $f(-a) = \frac{1}{1+e^a}$

$\therefore f(a) + f(-a) = 1$

$f(a) = 1 - f(-a)$

$= 1 - y$ when $f(-a) = y$.

Now $I_1 = \int_y^{1-y} x\, g\{x(1-x)\}dx$ ...(i)

$= \int_y^{1-y} (1-x)\, g\{x(1-x)\}dx$ ...(ii)

$I_1 + I_1 = \int_y^{1-y} g\{x(1-x)\}dx$

$= \int_{f(-a)}^{f(a)} g\{x(1-x)\}dx$

$2I_1 = I_2$

$\therefore \frac{I_2}{I_1} = 2$

10. (b) $F(t) = \int_0^t f(t-y)g(y)dy$

$= \int_0^t e^{t-y}.ydy = e^t\int_0^t e^{-y}ydy$

$= e^t\left[-ye^{-y}\Big|_0^t - \int_0^t 1\times(-e^{-y})dy\right]$

$= e^t\left[-te^{-t} - (e^{-t} - e^0)\right]$

$= e^t\left[-t\times e^{-t} - e^{-t} + 1\right] = e^t - (t+1)$

11. (c) Here, $f'(x) = f(x)$ and $f(0) = 1$.

Then $f(x) = e^x \Rightarrow g(x) = x^2 - e^x$

Now, $\int_0^1 f(x)g(x)dx = \int e^x(x^2 - e^x)dx$

$= \int x^2e^xdx - \int e^{2x}dx$

$= \left[x^2e^x - 2(xe^x - e^x)\right]_0^1 - \frac{1}{2}(e^2 - e^0)$

$= (e-2) - \frac{1}{2}e^2 + \frac{1}{2} = e - \frac{1}{2}e^2 - \frac{3}{2}$

12. (b) $\int_0^{10}|x(x-1)(x-2)|dx = \int_0^{10}|x^3 - 3x^2 + 2x|dx$

$= \int_0^1(x^3 - 3x^2 + 2x)dx + \int_1^2(-x^3 + 3x^2 - 2x)dx + \int_2^{10}(x^3 - 3x^2 - 2x)dx$

$= \left(\frac{x^4}{4} - x^3 + x^2\right)\Bigg|_0^1 + \left(-\frac{x^4}{4} + x^3 - x^2\right)\Bigg|_1^2 + \left(\frac{x^4}{4} - x^3 + x^2\right)\Bigg|_2^{10}$

$= 1600.5$

13. (b) $\int_0^3 |x^3 + x^2 + 3x|dx$

$= \int_0^3 (x^3 + x^2 + 3x)dx = \frac{x^4}{4} + \frac{x^3}{3} + \frac{3x^2}{2}\Bigg|_0^3$

14. (a) Put $\left(\frac{x-1}{x+2}\right)=t \Rightarrow \frac{1}{(x+2)^2}\,dx=\frac{1}{3}\,dt$

$\therefore I=\frac{1}{3}\int t^{-3/4}dt=\frac{4}{3}\left(\frac{x-1}{x+2}\right)^{\frac{1}{4}}+c$

15. (b) $\int\frac{dx}{x^2+4x+13}=\int\frac{dx}{(x+2)^2+(3)^2}$

$=\frac{1}{3}\tan^{-1}\frac{x+2}{3}+c$

16. (d) Let $\sin x-\cos x=t$.

$\therefore(\cos x+\sin x)dx=dt$

$3+\sin 2x=4-t^2$

$\therefore I=\int\frac{dt}{4-t^2}$

$=\frac{1}{4}\ln\left|\frac{2-\sin x+\cos x}{2+\sin x-\cos x}\right|+c$

17. (a) Let $I=\int\limits_{-\pi/2}^{\pi/2}\frac{\sin^2 x-\cos^2 x}{\sin^3 x+\cos^3 x}\,dx.$

Let $f(x)=\frac{\sin^2 x-\cos^2 x}{\sin^3 x+\cos^3 x}$.

$f(-x)=-f(x)$

$\Rightarrow I=0$ (Odd function property)

18. (c) $\int\frac{1}{x+\sqrt{x-1}}\,dx=\int\frac{2t}{t^2+t+1}\,dt$

when $x-1=t^2$, $dx=2t\,dt$

$=\int\frac{(2t+1)-1}{t^2+t+1}\,dt=\int\frac{(2t+1)-1}{\left(t+\frac{1}{2}\right)^2+\left(\frac{\sqrt{3}}{2}\right)^2}\,dt$

$=\int\frac{2t+1}{t^2+t+1}\,dt-\int\frac{1}{t^2+t+1}\,dt$

$=\log\left|t^2+t+1\right|-\frac{2}{\sqrt{3}}\tan^{-1}\frac{2\left(t+\frac{1}{2}\right)}{\sqrt{3}}+c$

$=\log\left|x+\sqrt{x-1}\right|-\frac{2}{\sqrt{3}}.\tan^{-1}\frac{2\sqrt{x-1}+1}{\sqrt{3}}+c$

19. (b) $\int\frac{\cos 4x+1}{\cot x-\tan x}\,dx$

$=\frac{1}{2}\int\frac{\cos 4x+1}{\cos^2 x-\sin^2 x}\times 2\sin x\cos x\,dx$

$=\frac{1}{2}\int\frac{2\cos^2 2x}{\cos 2x}\sin 2x\,dx$

$=\frac{1}{2}.\frac{1}{4}\int\sin 4x d(4x)=-\frac{1}{8}\cos 4x+c$

20. (a) $\int\frac{1}{x+x^5}\,dx=\int\frac{x^3}{x^4\left(1+x^4\right)}\,dx$

$=\frac{1}{4}\int\frac{1}{x^4\left(1+x^4\right)}\,d\left(x^4\right)$

$=\frac{1}{4}\int\left[\frac{1}{x^4}-\frac{1}{x^4+1}\right]d\left(x^4\right)$

$=\frac{1}{4}\log\left|\frac{x^4}{1+x^4}\right|+c$

$\therefore f(x)=\frac{1}{4}\log\left|\frac{x^4}{1+x^4}\right|+c=\log x-\frac{1}{4}\left|\frac{1}{1+x^4}\right|+c$

$\int\frac{x^4}{x\left(1+x^4\right)}\,dx=\frac{1}{4}\log\left|1+x^4\right|+c$

$=\log|x|-f(x)+c$

21. (a) Since $I=\int\limits_0^1\frac{\sin x}{\sqrt{x}}\,dx<\int\limits_0^1\frac{x}{\sqrt{x}}\,dx$

(because in $x\in(0,1), x>\sin x$.)

$\therefore I<\int\limits_0^1\sqrt{x}dx=\frac{2}{3}x\Big|_0^1$

$I<\frac{2}{3}$.

Similarly, $J=\int\limits_0^1\frac{\cos x}{\sqrt{x}}\,dx.<\int\limits_0^1 x^{-\frac{1}{2}}\,dx=2=2x^{\frac{1}{2}}\Big|_0^1$

$\Rightarrow J<2$

22. (d) $I=\int\limits_{-\pi}^{\pi}\frac{\cos^2 x}{1+a^x}\,dx$ ...(i)

Put $x=-y$ in I

$dx=-dy$

$=\int\limits_{\pi}^{-\pi}\frac{(\cos y)^2}{1-a^{-y}}(-dy)=\int\limits_{-\pi}^{\pi}\frac{a^x\cos^2 x}{1+a^x}\,dx$ ...(ii)

Adding (i) and (ii)

$\therefore 2I=\int\limits_{-\pi}^{\pi}\cos^2 x\,dx=2\int\limits_0^{\pi}\frac{1+\cos 2x}{2}\,dx$

$2I=\left[x+\frac{1}{2}\sin 2x\right]_0^{\pi}=\pi$

$\therefore I=\frac{\pi}{2}$

23. (b) $I=\int\limits_0^{\pi}x\,f(\sin x)dx$

$=\int\limits_0^{\pi}(\pi-x)\,f[\sin(\pi-x)]dx$

$$= \int_0^{\pi} (\pi - x) f(\sin x)\, dx$$

$$I + I = \pi \int_0^{\pi} f(\sin x)\, dx$$

$$\Rightarrow I = \frac{\pi}{2} \int_0^{\pi} f(\sin x)\, dx = 2 \cdot \frac{\pi}{2} \int_0^{\pi/2} f(\sin x)\, dx$$

24. (b) $I = \int_a^b x f(x) dx$

$$\therefore I = \int_a^b (a+b-x) f(a+b-x) dx$$

$$= (a+b) \int_a^b f(x) dx - \int_a^b x f(x) dx$$

$$\Rightarrow 2I = (a+b) \int_a^b f(x) dx$$

$$I = \frac{a+b}{2} \int_a^b f(x) dx$$

25. (c) Put $1 - x = t \Rightarrow x = 1 - t$

Given integral $= \int_1^0 t^n (1-t)(-dt)$

$$\int_0^1 \left(t^n - t^{n+1}\right) dt = \frac{1}{n+1} - \frac{1}{n+2}$$

26. (a) $f(t) = 2\int_0^t \frac{e^{-|x|}}{2} dx$ [$\because$ Integrand is even.]

$$= \int_0^t e^{-|x|} dx = \int_0^t e^{-x} dx$$

$$= \left|-e^{-x}\right|_0^t = -e^{-t} + e^{-0}$$

$$= 1 - e^{-t}$$

$$\Rightarrow \lim_{t \to \infty} f(t) = 1$$

27. (c) $I = \int \left(\frac{x}{a^3 - x^3}\right)^{\frac{1}{2}} dx = \int \frac{x^{\frac{1}{2}}}{\sqrt{\left(a^{3/2}\right)^2 - \left(x^{3/2}\right)^2}} dx$

Put $x^{3/2} = t \Rightarrow x^{\frac{1}{2}}\, dx = \frac{2}{3}\, dt$

$$I = \frac{2}{3} \int \frac{dt}{\sqrt{\left(a^{3/2}\right)^2 - t^2}} = \frac{2}{3} \sin^{-1}\left(\frac{x}{a}\right)^{3/2} + c$$

28. (a) $\int \frac{1}{x^2 + 2x + 2} dx$

$$= \int \frac{1}{(x+1)^2 + 1} dx = \tan^{-1}(x+1) + c$$

29. (a) $\int \frac{\sin x}{\cos x(1 + \cos x)} dx = -\int \frac{dt}{t(1+t)}$

[Put $\cos x = t$]

$$= -\int \left(\frac{1}{t} - \frac{1}{t+1}\right) dt$$

$$= -\log\left|\frac{t}{t+1}\right| + c = \log\left|\frac{t+1}{t}\right| + c$$

$$= \log\left|\frac{1 + \cos x}{\cos x}\right| + c$$

30. (a) $\int \frac{x^{49} \tan^{-1}\left(x^{50}\right)}{1 + x^{100}} dx$

$$= \frac{1}{50} \int \frac{\tan^{-1} t}{1 + t^2} dt \qquad \left[\text{Put } t = x^{50}\right]$$

$$= \frac{1}{50}\left[\tan^{-1}\left(x^{50}\right)\right]^2 + c$$

$$k = \frac{1}{50}$$

31. (c)

32. (a) $\int \frac{3 - x^2}{1 - 2x + x^2} e^x\, dx$

$$= \int \left[\frac{2}{\left(1 - x^2\right)} + \frac{\left(1 - x^2\right)}{(1-x)^2}\right] e^x\, dx$$

$$= \int \left[\frac{2}{(1-x)^2} + \left(\frac{1+x}{1-x}\right)\right] e^x dx = \frac{1+x}{1-x} e^x + c$$

as $f(x) = \frac{1+x}{1-x}$

$$\therefore f'(x) = \frac{2}{(1-x)^2}$$

Using $\int \{f(x) + f'(x)] e^x dx = f(x) e^x + c$

33. (a) $\int \frac{\sqrt{\cot x}}{\sin x \cos x} dx = \int \frac{\sec^2 x}{\sqrt{\tan^3 x}} dx = 2\sqrt{\tan x} + c.$

34. (b) $\int \frac{1 + x + \sqrt{x + x^2}}{\sqrt{x} + \sqrt{1+x}} dx$

$$= \int \frac{\left(\sqrt{1+x}\right)^2 \sqrt{x}\sqrt{1+x}}{\sqrt{x} + \sqrt{1+x}} dx$$

$$= \int \sqrt{1+x}\frac{[\sqrt{1+x}+\sqrt{x}]}{\sqrt{x}+\sqrt{1+x}}dx$$

$$= \int \sqrt{1+x}dx = \frac{2}{3}(1+x)^{3/2} + c$$

35. (c) $I = \int \frac{1}{1-\cos x - \sin x}dx$

$$= \int \frac{1}{1 - \frac{1-\tan^2\frac{x}{2}}{1+\tan^2\frac{x}{2}} - \frac{2\tan\frac{x}{2}}{1+\tan^2\frac{x}{2}}}dx$$

$$= \int \frac{\left(1+\tan^2\frac{x}{2}\right)dx}{2\tan^2\frac{x}{2} - 2\tan\frac{x}{2}}$$

Let $\tan\frac{x}{2} = t$

$$\Rightarrow dt = \frac{1}{2}\sec^2\frac{x}{2}dx = \frac{1}{2}\left(1+\tan^2\frac{x}{2}\right)dx.$$

$$\therefore I = \int \frac{2}{2t(t-1)}dt = \int\left(\frac{1}{t-1} - \frac{1}{t}\right)dt$$

$$= \log\left|\frac{t-1}{t}\right| + c = \log\left|1 - \frac{1}{t} + c\right|$$

$$= \log\left|1 - \cot\frac{x}{2}\right| + c$$

36. (b) $I = \int \frac{1}{7+5\cos x}dx = \int \frac{\left(1+\tan^2\frac{x}{2}\right)}{12+2\tan^2\frac{x}{2}}$

$$= \int \frac{2\,dt}{2\left[\left(\sqrt{6}\right)^2 + t^2\right]} \qquad \left[\text{Put } t = \tan\frac{x}{2}\right]$$

$$= \frac{1}{\sqrt{6}}\tan^{-1}\left(\frac{\tan\frac{x}{2}}{\sqrt{6}}\right) + c$$

37. (a) $I = \int \frac{3^x\,dx}{\sqrt{\left(3^2\right)^x - 1}} = \int \frac{3^x}{\sqrt{\left(3^x\right)^2 - 1}}dx$

Let $3^x = t \Rightarrow 3^x \log 2\, dx = dt$

$$I = \frac{1}{\log 3}\log\left|3^x + \sqrt{9^x - 1}\right| + c$$

38. (b)

39. (d)

40. (b)

41. (d) $\int_{\pi/2}^{\alpha} \sin x dx = \sin 2\alpha$

$$\Rightarrow -\cos x\big|_{\pi/2}^{\alpha} = \sin 2\alpha$$

$$\Rightarrow \sin 2\alpha + \cos\alpha = 0$$

$$\Rightarrow \cos\alpha(2\sin\alpha + 1) = 0$$

$$\therefore \cos\alpha = 0 \Rightarrow \alpha = \frac{\pi}{2}, \frac{3\pi}{2}\bigg\}\ 0 < \alpha < 2\pi$$

$$\sin\alpha = \frac{-1}{2} \Rightarrow \alpha = \pi + \frac{\pi}{6} = \frac{7\pi}{6}\bigg\}\ 0 < \alpha < 2\pi$$

42. (c) $\int_0^{\pi/2} \frac{(\sin x + \cos x)^2}{\sqrt{1+\sin 2x}}dx$

$$= \int_0^{\pi/2} \frac{1+\sin 2x}{\sqrt{1+\sin 2x}}dx = \int_0^{\pi/2}\sqrt{1+\sin 2x}dx$$

$$= \int_0^{\pi/2}\sqrt{(\sin x + \cos x)^2}dx = \int_0^{\pi/2}(\sin x + \cos x)dx$$

$$= \int_0^{\pi/2}\sin x dx + \int_0^{\pi/2}\cos x dx$$

$$= \int_0^{\pi/2}\sin\left(\frac{\pi}{2} - x\right)dx + \int_0^{\pi/2}\cos x dx$$

$$= 2\int_0^{\pi/2}\cos x dx = 2\left[\sin x\big|_0^{\pi/2}\right] = 2$$

43. (c) $\lim_{x\to 0}\frac{\int_0^{x^2}\sec^2 t dt}{x\sin x}; \left(\frac{0}{0}\text{form}\right)$

$$= \lim_{x\to 0}\frac{(\sec^2 x^2)(2x)}{x\cos x + \sin x}; \left(\frac{0}{0}\text{form}\right)$$

$$= \lim_{x\to 0}\frac{2x.2\sec^2 x^2.\tan x^2.2x + 2\sec^2 x^2}{-x\sin x + \cos x + \cos x}$$

$$= \frac{2\sec^2 0}{2\cos 0} = 1$$

44. (d)

$$\lim_{x\to\infty}\frac{1+2^4+3^4+\ldots+n^4}{n^5} - \lim_{x\to\infty}\frac{1+2^3+3^3+\ldots+n^3}{n^5}$$

$$= \lim_{x\to\infty}\frac{1}{n}\sum_{r=1}^{n}\left(\frac{r}{n}\right)^4 - \lim_{x\to\infty}\frac{1}{n}\sum_{r=1}^{n}\left(\frac{r}{n}\right)^3$$

$$= \int_0^1 x^4 dx - \lim_{x\to\infty}\frac{1}{n}\times\int_0^1 x^3 dx = \frac{x^5}{5} - 0 = \frac{1}{5}$$

45. (d)

46. (c) $\int \frac{e^x-1}{e^x+1}dx = f(x)+c$

$\Rightarrow \int \frac{e^x}{e^x+1}dx - \int \frac{e^x}{(e^x+1)}dxf(x)+c$

$\Rightarrow \log\left|e^x+1\right| - \log e^x + \log\left|e^x+1\right| + c = f(x)+c$

$\Rightarrow 2\log\left|e^x+1\right| - x + c = f(x)+c$

47. (a) $\int \tan^{-1}\sqrt{\frac{1-x}{1+x}}dx$

Put $x=\cos 2\theta,\ dx=-2\sin 2\theta\ d\theta$

$= \int \theta(-2\sin 2\theta)d\theta$

$= -2\left[\theta\int \sin 2\theta d\theta - \int 1.\left(-\frac{\cos 2\theta}{2}\right)d\theta\right]$

$= -2\left[\theta\left(-\frac{\cos 2\theta}{2}\right)+\frac{\sin 2\theta}{4}\right]+c$

$= 2\left[\frac{x}{2}.\frac{1}{2}\cos^{-1}x - \frac{\sqrt{1-x^2}}{4}\right]+c$

$= \frac{1}{2}\left[x\cos^{-1}x - \sqrt{1-x^2}\right]+c$

48. (a) $I = \int \frac{\sin x + 8\cos x}{4\sin x + 6\cos x}dx,$

Now $\frac{d}{dx}(4\sin x + 6\cos x) = (4\cos x - 6\sin x)$

Let $\sin x + 8\cos x \equiv A(4\cos x - 6\sin x) + B(4\sin x + 6\cos x)$

$\Rightarrow 4A + 6B = 8$ and $-6A + 4B = 1$

$\Rightarrow A = \frac{1}{2}, B = 1$

49. (a) $\int \frac{x^2-1}{x^2}e^{\frac{x^2-1}{x}}dx$

Statement 2 is wrong

as $\int f'(x)e^{f(x)}dx = e^{f(x)}$.

Statement 1 is correct.

50. (c) $I = \int_0^{\pi/2} \frac{1}{1+\tan^3 x} = \int_0^{\pi/2} \frac{\cos^3 x}{\sin^3 x + \cos^4 x}dx$

$= \int_0^{\pi/2} \frac{\sin^3 x}{\cos^3 x + \sin^3 x}dx,$

as $f(x) + f\left(\frac{\pi}{2}-x\right) = 1$

$2I = \int_0^{\pi/2} dx \Rightarrow I = \pi/4$

51. (c) $I = \int_0^{\pi/2} \frac{200\sin x + 100\cos x}{\sin x + \cos x}dx$

$= \int_0^{\pi/2} \frac{200\sin\left(\frac{\pi}{2}-x\right) + 100\cos\left(\frac{\pi}{2}-x\right)}{\sin\left(\frac{\pi}{2}-x\right) + \cos\left(\frac{\pi}{2}-x\right)}dx$

$= \int_0^{\pi/2} \frac{200\cos x + 100\sin x}{\cos x + \sin x}dx$

$2I = 300\int_0^{\pi/2} dx$

$I = 300 \times \frac{\pi}{4} = 75\pi$

52. (d) $I = \int_0^{\pi} \frac{\theta\sin\theta}{1+\cos^2\theta} \quad \left\{\int_0^a f(\theta)d\theta = \int_0^a f(a-\theta)d\theta\right\}$

$= \int_0^{\pi} \frac{(\pi-0)\sin(\pi-\theta)}{1+\cos^2(\pi-\theta)}d\theta$

$= \pi\int_0^{\pi} \frac{\sin\theta}{1+\cos^2\theta}d\theta - I$

$\cos\theta = t \Rightarrow -\sin\theta\ d\theta = dt$

$\Rightarrow 2I = \pi\left\{-\tan^{-1}(\cos\theta)\Big|_0^{\pi}\right\}$

$I = \frac{\pi}{2}.\frac{\pi}{2} = \frac{\pi^2}{4}$

53. (a) $\int \sin^{-1}\left(\frac{2x}{1+x^2}\right)dx$

Put $x = \tan\theta \Rightarrow dx = \sec^2\theta d\theta$

$\frac{2x}{1+x^2} = \sin 2\theta$

$= \int 2\theta.\sec^2\theta\ d\theta$

$= 2\left[\theta\int \sec^2\theta\ d\theta - \int 1\left(\int \sec^2\theta\ d\theta\right)d\theta\right]$

$= 2\left[\theta\int \tan\theta - \int \tan\theta\ d\theta\right]$

$= 2[\theta\tan\theta - \log\sec\theta] + c$

$= 2x\tan^{-1}x - \log\left(1+x^2\right) + c$

54. (a) Here integrand is an odd function. Hence, integral is zero.

55. (a) $\int_0^2 \frac{2x-2}{2x-x^2}dx = -\int \frac{2-2x}{2x-x^2}dx$

$= \log\left|\frac{1}{2x-x^2}\right|_0^2 = 0$

56. (a)

57. (b) $I=\int_0^1 \sin\left(2\tan^{-1}\sqrt{\frac{1+x}{1-x}}\right)dx$

Put $x=\cos\theta \Rightarrow dx=-\sin\theta\, d\theta$

$$\therefore \frac{1+x}{1-x}=\frac{1+\cos\theta}{1-\cos\theta}=\frac{2\cos^2\frac{\theta}{2}}{2\sin^2\frac{\theta}{2}}$$

$$=\cot^2\frac{\theta}{2}=\tan^2\left(\frac{\pi}{2}-\frac{\theta}{2}\right)$$

$x=0 \Rightarrow \theta=\pi/2$ and $x=1 \Rightarrow \theta=0$

$$I=\int_{\pi/2}^{0}\sin 2\left(\frac{\pi}{2}-\frac{\theta}{2}\right)(-\sin\theta)\,d\theta$$

$$I=-\int_{\pi/2}^{0}\sin^2\theta d\theta=\int_0^{\pi/2}\left(\frac{1-\cos 2\theta}{2}\right)d\theta$$

$$=\frac{1}{2}\theta\Big|_0^{\pi/2}-\frac{\sin 2\theta}{4}\Big|_0^{\pi/2}=\frac{\pi}{4}$$

58. (a) $\int_0^3\frac{3x+1}{x^2+9}dx=\frac{3}{2}\int_0^3\frac{2x}{x^2+9}dx+\int_0^3\frac{1}{x^2+(3)^2}dx$

$$=\frac{3}{2}\log\left|x^2+9\right|\Big|_0^3+\frac{1}{3}\tan^{-1}\frac{x}{3}\Big|_0^3$$

$$=\frac{3}{2}\left|\log 18-\log 9\right|+\frac{1}{3}\left[\tan^{-1}(1)-\tan^{-1}(0)\right]$$

$$=\frac{3}{2}\log 2+\frac{\pi}{12}=\log 2\sqrt{2}+\frac{\pi}{12}$$

59. (b) $\int_2^3\frac{1}{x(x-1)}dx=\int_2^3\left[\frac{1}{x-1}-\frac{1}{x}\right]dx$

$$=\log\frac{x-1}{x}\Big|_2^3=\log\frac{2}{3}-\log\frac{1}{2}=\log\frac{4}{3}$$

60. (b) $I=\int_{-\pi/2}^{\pi/2}\sin^4 x\cos^6 x\,dx$

$$=2\int_0^{\pi/2}\sin^4 x\cos^6 x\,dx, \text{ as } f(-x)=f(x)$$

$$=2\cdot\frac{4-1}{4+6}\cdot\frac{4-3}{4+6-2}\cdot\frac{6-1}{6}\frac{6-3}{6-2}\frac{1}{2}\frac{\pi}{2}=\frac{3\pi}{256}$$

$$I_{p,q}=\int_0^{\pi/2}\sin^b x\cos^q x dx$$

$$=\frac{(p-1)}{p+q}\frac{p-3}{p+q-2}\cdots\cdots\frac{(q-1)}{q}\cdot\frac{q-3}{q-2}\cdots\frac{1}{2}\frac{\pi}{2}$$

when p and q are even.

61. (b) $\sin\theta(2\sin\theta)\cos\theta\left(\sin^6\theta+\sin^4\theta+\sin^2\theta\right)$

$$\int\frac{\sqrt{2\sin^4\theta+3\sin^2\theta+6}}{2\sin^2\theta}d\theta$$

$$=\int\cos\theta\sin\theta\left(\sin^5\theta+\sin^3\theta+\sin\theta\right)\sqrt{2\sin^4\theta+3\sin^2\theta d\theta}$$

$$=\int\cos\theta\left(\sin^5\theta+\sin^3\theta+\sin\theta\right)\sqrt{2\sin^6\theta+3\sin^4\theta+6\sin^2\theta}\,d\theta$$

$\sin\theta=t\Rightarrow\cos\theta\,d\theta=dt$

$$\int\left(t^5+t^3+t\right)\sqrt{2t^6+3t^4+6t^2}\,dt$$

$$2t^6+3t^4+6t^2=x$$

$$=12\left(t^5+t^3+t\right)dt=dx$$

$$=\frac{1}{12}\int\sqrt{x}\,dx=\frac{x^{3/2}}{18}$$

$$=\frac{1}{18}\left(2\sin^6\theta+3\sin^4\theta+6\sin^2\theta\right)^{\frac{3}{2}}+C$$

$$=\frac{1}{18}\left(11-18\cos^2\theta+9\cos^4\theta-2\sin^6\theta\right)^{\frac{3}{2}}+C$$

62. (d) Given condition $f(x)=x+\int_0^{\pi/2}\sin x\cos y\,f(y)dy$

$$f(x)=x+\sin x\int_0^{\pi/2}\cos y\,f(y)\,dy$$

Let $K=\int_0^{\pi/2}\cos y\,f(y)dy$

$\Rightarrow f(x)=x+K\sin x$

$\Rightarrow f(y)=y+K\sin y$

Now $K=\int_0^{\pi/2}\cos y(y+K\sin y)dy$

$$K=\int_0^{\pi/2}y\cos dy+K\int_0^{\pi/2}\cos y\sin y dy$$

[Let $\sin y=t$.]

$$K=(y\sin y)_0^{\pi/2}-\int_0^{\pi/2}\sin dy+K\int_0^1 t\,dt$$

$$\Rightarrow K=\frac{\pi}{2}-1+K\left(\frac{1}{2}\right)$$

$$\Rightarrow K=\pi-2$$

So $f(x)=x+(\pi-2)\sin x$

63. (b) $\pi^2\left[\int_0^1 \sin\frac{\pi x}{2}dx+\int_1^2 \sin\frac{\pi x}{2}(x-1)dx\right]$

$=\pi^2\left[\left(-\frac{2}{\pi}\left(\cos\frac{\pi x}{2}\right)\right)_0^1+\right.$

$\left.\left((x-1)\left(-\frac{2}{\pi}\cos\frac{\pi x}{2}\right)\right)_1^2-\int_1^2-\frac{2}{\pi}\cos\frac{\pi x}{2}\,dx\right]$

$=\pi^2\left[0+\frac{2}{\pi}+\frac{2}{\pi}+\frac{2}{\pi}.\frac{2}{\pi}\left(\sin\frac{\pi x}{2}\right)_1^2\right]$

$=4\pi-4=4(\pi-1)$

64. (b) $I=\int_{-1/\sqrt{2}}^{1/\sqrt{2}}\left[\left(\frac{x+1}{x-1}\right)^2+\left(\frac{x-1}{x+1}\right)^2-2\right]^{1/2}dx$

$I=\int_{-1/\sqrt{2}}^{1/\sqrt{2}}\left(\left(\frac{x+1}{x-1}-\frac{x-1}{x+1}\right)^2\right)^{\frac{1}{2}}dx$

$I=\int_{-1/2}^{1/2}\left|\left(\frac{x+1}{x-1}\right)-\left(\frac{x-1}{x+1}\right)\right|dx$

$I=\int_{-1/\sqrt{2}}^{1/\sqrt{2}}\left|\frac{4x}{x^2-1}\right|dx\Rightarrow I=2\cdot 4\int_0^{1/\sqrt{2}}\left|\frac{x}{x^2-1}\right|dx$

$I=-4\int_0^{1/\sqrt{2}}\frac{2x}{x^2-1}dx\Rightarrow I=-4\ln\left|x^2-1\right|_0^{1/\sqrt{2}}$

$\Rightarrow I=4\ln 2\Rightarrow I=\ln 16$

65. (b) Given $I=\int_0^5\frac{x+[x]}{e^{x-[x]}}dx=\alpha e^{-1}+\beta$

$\int_0^1\frac{x}{e^x}dx+\underbrace{\int_1^2\frac{x+1}{e^{x-1}}dx}_{x=t+1}+\underbrace{\int_2^3\frac{x+2}{e^{x-2}}dx}_{x=z+2}+\ldots\underbrace{\int_4^5\frac{x+4}{e^{x-4}}dx}_{x=y+4}$

$\int_0^1\frac{t+2}{e^t}dt+\int_0^1\frac{z+4}{e^x}dz+\ldots+\int_0^1\frac{y+8}{e^x}dx$

$\Rightarrow\int_0^1\frac{5x+20}{e^x}dt=5\int_0^1\frac{x+4}{e^x}dx$

$\Rightarrow 5\int_0^1(x+4)e^{-x}dx$

$\Rightarrow 5\left[(x+4)\frac{e^{-x}}{-1}\right]_0^1-\int_0^1\frac{e^{-x}}{-1}dx$

$=-30e^{-1}+25$

$\alpha=-30$

$\beta=25\Rightarrow 5\alpha+6\beta=0$

$(\alpha+\beta)^2=5^2=25$

66. (b) Let $[a]$ be greatest integer function, $\{a\}$ be the fractional part and $a>0$ then

$n\le a<n+1, n\in W$

$\therefore a=[a]+\{a\}\Rightarrow[a]=n$

Now, $\int_0^a e^{x-[x]}dx=10e-9$

$\Rightarrow\int_0^n e^{\{x\}}dx+\int_0^a e^{x-[x]}dx=10e-9$

$\therefore n\int_0^1 e^x dx+\int_n^a e^{x-n}dx=10e-9$

$\Rightarrow n(e-1)+\left(e^{a-n}-1\right)=10e-9$

$\therefore n=0$ and $\{a\}=\log_e 2$

So, $a=[a]+\{a\}=(10+\log_e 2)$

67. (b) It is given that

$g(t)=\int_{-\pi/2}^{\pi/2}\left(\cos\frac{\pi}{4}t+f(x)\right)dx$

$\Rightarrow g(t)=\pi\cos\frac{\pi}{4}t+\int_{-\pi/2}^{\pi/2}f(x)dx$

$g(t)=\pi\cos\frac{\pi}{4}t\qquad[\because f(x)\text{ is odd function}]$

Then, $g(1)=\frac{\pi}{\sqrt{2}}, g(0)=\pi$. So, $\sqrt{2}g(1)=g(0)$

68. (c) $\int_0^{10}\frac{[\sin 2\pi x]}{e^{\{x\}}}dx=10\int_0^1\frac{[\sin 2\pi x]}{e^x}dx$

$[\because x-[x]=\{x\}]$

$=10\left[\int_0^{1/2}0dx+\int_{1/2}^1-e^{-x}dx\right]=10\left(e^{-1}-e^{-1/2}\right)$

$\Rightarrow\alpha+\beta+\gamma=0$

69. (b) $I=\int_0^{10}[x]e^{[x]-x+1}dx=\int_0^{10}[x]e^{1-\{x\}}dx$

$=(1+2+3+\ldots+9)\int_0^1 e^{1-x}dx$

$[\because\{x\}$ is periodic function with period 1]

$=45\left[-e^{1-x}\right]_0^1=45(e-1)$

70. (d) $I_n=\int_{\frac{\pi}{4}}^{\frac{\pi}{2}}\cot^n x\,dx=\int_{\frac{\pi}{4}}^{\frac{\pi}{2}}\cot^{n-2}x\left(\cot^2 x\right)dx$

$=\int_{\frac{\pi}{4}}^{\frac{\pi}{2}}\cot^{n-2}x\left(\operatorname{cosec}^2 x-1\right)dx$

$$= \int_{\frac{\pi}{4}}^{\frac{\pi}{2}} \cot^{n-2} x \operatorname{cosec}^2 x dx - \int_{\frac{\pi}{4}}^{\frac{\pi}{2}} \cot^{n-2} x dx$$

Let $\cot x = t \Rightarrow -\operatorname{cosec}^2 x\, dx = dt.$

When $x = \frac{\pi}{4} \Rightarrow t = 1$ and when $x = \frac{\pi}{2} \Rightarrow t = 0$

$$\therefore I_n = \int_0^1 t^{n-2} dt - I_{n-2} = \frac{1}{n-1} - I_{n-2}$$

$$\therefore I_n + I_{n-2} = \frac{1}{n-1} \Rightarrow \frac{1}{I_{n-2} + I_n} = n - 1$$

$\therefore \frac{1}{I_2 + I_4} = 3, \frac{1}{I_3 + I_5} = 4$ and $\frac{1}{I_4 + I_6} = 5$ are in AP

71. (b) We have,

$$\int_1^3 \left(\left[(x-1)^2\right] - 3\right) dx$$

$$= \int_1^2 \left[x^2\right] - 3\int_1^3 dx$$

$$= \int_1^3 0\, dx + \int_1^{\sqrt{2}} 2 \cdot dx + \int_{\sqrt{3}}^{2} 3 \cdot dx - 6$$

$$= \sqrt{2} - 1 + 2\left(\sqrt{3} - \sqrt{2}\right) + 3\left(2 - \sqrt{3}\right) - 6$$

$$= -\sqrt{2} - \sqrt{3} - 1$$

72. (None) $\lim_{x \to 1} \frac{\frac{1}{2}\sin(x-1)^4}{(x-1)\sin(x-1)}$

Let $x - 1 = h$ when $x \to 1$ then $h \to 0$

$$\lim_{h \to 0} \frac{\sinh^4}{h^4} \times \frac{h}{\sinh} \times h^2 = 1 \times 1 \times 0 = 0$$

(None of the options are correct.)

73. (d) $\int \left(e^{2x} + 2e^x - e^{-x} - 1\right).e^{\left(e^x + e^{-x}\right)} dx$

$$I = \int \left(e^{2x} + e^x - 1\right) \cdot e^{\left(e^x + e^{-x}\right)} dx + \int \left(e^x - e^{-x}\right) e^{\left(e^x + e^{-x}\right)} dx$$

$$= \int e^x \left(e^x + 1 - e^{-x}\right).e^{\left(e^x + e^{-x}\right)} dx + e^{\left(e^x + e^{-x}\right)}$$

$$= \int \left(e^x - e^{-x} + 1\right) e^{\left(e^x + e^{-x} + x\right)} dx + e^{\left(e^x + e^{-x}\right)}$$

Let $e^x + e^{-x} + x = t \Rightarrow \left(e^x + e^{-x} + 1\right) dx = dt$

$$= \int e^t dt + e^{\left(e^x - e^{-x}\right)} = e^t + e^{\left(e^x + e^{-x}\right)} + C$$

$$= e^{\left(e^x + e^{-x} + x\right)} + e^{\left(e^x + e^{-x}\right)} + C = \left(e^x + 1\right).e^{\left(e^x + e^{-x}\right)} + C$$

So, $g(x) = 1 + e^x$ and $g(0) = 2$

74. (d) Let $\sin\theta = t \Rightarrow \cos\theta d\theta = dt$

$$\int \frac{\cos\theta}{5 + 7\sin\theta - 2\cos^2\theta} d\theta = \frac{dt}{5 + 7t - 2 + 2t^2}$$

$$\Rightarrow \frac{1}{2} \int \frac{dt}{\left(t + \frac{7}{4}\right)^2 - \left(\frac{5}{4}\right)^2} = \frac{1}{5} \ln \left| \frac{t + \frac{1}{2}}{t + 3} \right| + C$$

$$= \frac{1}{5} \ln \left| \frac{2t + 1}{t + 3} \right| + C = \frac{1}{5} \ln \left| \frac{2\sin\theta + 1}{\sin\theta + 3} \right| + C$$

$$\therefore B(\theta) = \frac{2\sin\theta + 1}{2(\sin\theta + 3)} \text{ and } A = \frac{1}{5}.$$

$$\Rightarrow \frac{B(\theta)}{A} = \frac{5(2\sin\theta + 1)}{(\sin\theta + 3)}$$

75. (a) $\int \frac{x^2}{(x\sin x + \cos x)} dx$

$$\because \frac{d}{dx}(x\sin x + \cos x) = x\cos x$$

$$= \int \underset{\text{II}}{\frac{x\cos x}{(x\sin x + \cos x)^2}} \underset{\text{I}}{\left(\frac{x}{\cos x}\right)} dx$$

$$= \frac{x}{\cos x}\left[\frac{-1}{x\sin x + \cos x}\right] - \int \frac{x\sin x + \cos x}{\cos^2 x} \left[\frac{-1}{x\sin x + \cos x}\right] dx$$

$$= \frac{x}{\cos x}\left[\frac{-1}{x\sin x + \cos x}\right] + \int \sec^2 x dx$$

$$= \frac{-x\sec x}{x\sin x + \cos x} + \tan x + C$$

76. (d) $\int \frac{\sqrt{x}}{(1+x)^2} dx \, (x > 0)$

Put $x = \tan^2\theta \Rightarrow 2x\, dx = 2\tan\theta \sec^2\theta\, d\theta$

$$I = \int \frac{2\tan^2\theta \cdot \sec^2\theta}{\sec^4\theta} d\theta = \int 2\sin^2\theta\, d\theta$$

$$= \theta - \frac{\sin 2\theta}{2} + C$$

$$\Rightarrow f(x) = \theta - \frac{1}{2} \times \frac{2\tan\theta}{1 + \tan^2\theta} + C$$

$$f(x) = \theta - \frac{\tan\theta}{1 + \tan^2\theta} + C = \tan^{-1}\sqrt{x} - \frac{\sqrt{x}}{1+x} + C$$

Now $f(3) - f(1) = \tan^{-1}\left(\sqrt{3}\right) - \frac{\sqrt{3}}{1+3} - \tan^{-1}(1) + \frac{1}{2}$

$$= \frac{\pi}{12} + \frac{1}{2} - \frac{\sqrt{3}}{4}$$

77. (a) $I = \int \sin^{-1}\left(\frac{\sqrt{x}}{\sqrt{1+x}}\right) dx = \int \underset{\text{I}}{\tan^{-1}\sqrt{x}} \cdot \underset{\text{II}}{1}\, dx$

$$= x\tan^{-1}\sqrt{x} - \int \frac{1}{1+x} \cdot \frac{1}{2\sqrt{x}} \cdot x\, dx + C$$

$= x\tan^{-1}\sqrt{x}-\frac{1}{2}\int\frac{t\cdot 2t\,dt}{1+t^2}+C\left(\text{Put } x=t^2 \Rightarrow dx=2t\,dt\right)$

$= x\tan^{-1}\sqrt{x}-\int\frac{t^2}{1+t^2}dt+C$

$= x\tan^{-1}\sqrt{x}-t+\tan^{-1}t+C$

$= x\tan^{-1}\sqrt{x}-\sqrt{x}+\tan^{-1}\sqrt{x}+C$

$=(x+1)\tan^{-1}\sqrt{x}-\sqrt{x}+C$

$\Rightarrow A(x)=x+1 \Rightarrow B(x)=-\sqrt{x}$

78. (a) $I=\int\frac{d\theta}{(x+4)^{8/7}(x-3)^{6/7}}$

$=\int\left(\frac{x-3}{x+4}\right)^{\frac{-6}{7}}\frac{1}{(x+4)^2}dx$

Let $\frac{x-3}{x+4}=t^7$

Differentiating on both sides, we get

$\frac{7}{(x+4)^2}dx=7t^6dt$

Hence, $I=\int t^{-6}t^6\,dt=t+C=\left(\frac{x-3}{x+4}\right)^{\frac{1}{7}}+C$

79. (c) $I=\int\frac{d\theta}{\cos^2\theta(\tan 2\theta+\sec 2\theta)}$

$=\int\frac{\sec^2\theta}{\frac{1+\tan^2\theta}{1-\tan^2\theta}+\frac{2\tan\theta}{1-\tan^2\theta}}d\theta$

$=\int\frac{\sec^2\theta\left(1-\tan^2\theta\right)}{\left(1+\tan^2\theta\right)^2}d\theta$

$=\int\frac{\sec^2\theta(1-\tan\theta)}{1+\tan\theta}d\theta$

Let $\tan\theta=t \Rightarrow \sec^2\theta d\theta=dt$, then

$I=\int\left(\frac{1-t}{1+t}\right)dt=\int\left(-1+\frac{2}{1+t}\right)dt$

$=-t+2\log(1+t)+C$

$=-\tan\theta+2\log(1+\tan\theta)+C$

Hence, by comparison $\lambda=-1$ and $f(x)=1+\tan\theta$

80. (c) $I=\int_1^2 e^x x^x(2+\log_e x)dx$

$I=\int_1^2 e^x x^x\left[1+(1+\log_e x)\right]dx$

$=\int_1^2 e^x\left[x^x+x^x(1+\log_e x)\right]dx$

$\because \int e^x\left(f(x)+f'(x)\right)dx=e^x f(x)+c$

$\therefore I=\left[e^x x^x\right]_1^2$

$=e^2\times 4-e\times 1=4e^2-e=e(4e-1)$

81. (c) $I_2=\int_0^1\left(1-x^{50}\right)^{101}dx=\int_0^1\left(1-x^{50}\right)\left(1-x^{50}\right)^{100}dx$

$I_2=\int_0^1\left(1-x^{50}\right)^{100}dx-\int_0^1\underset{\text{I}}{x}\,\underbrace{x^{49}\left(1-x^{50}\right)^{100}}_{\text{II}}dx$

$I_2=I_1+\left[\frac{x}{5050}\left(1-x^{50}\right)^{101}\right]_0^1-\int_0^1\frac{\left(1-x^{50}\right)^{101}}{5050}dx$

$I_2=I_1+0-\frac{I_2}{5050}$

$\Rightarrow\frac{5051}{5050}I_2=I_1 \Rightarrow I_2=\frac{5050}{5051}I_1$

$\Rightarrow\alpha=\frac{5050}{5051}$

82. (c) $=\int_{-\pi/2}^{\pi/2}\frac{1}{1+e^{\sin x}}dx$

$=\int_{-\pi/2}^{0}\frac{1}{1+e^{\sin x}}dx+\int_0^{\pi/2}\frac{1}{1+e^{\sin x}}dx$

$=\int_0^{\pi/2}\left(\frac{1}{1+e^{\sin x}}+\frac{1}{1+e^{-\sin x}}\right)dx$

$=\int_0^{\pi/2}\frac{1+e^{\sin x}}{1+e^{\sin x}}dx=\frac{\pi}{2}$

83. (a) $f(x)=|x-2|=\begin{cases}2-x, & x<2\\ x-2, & x\geq 2\end{cases}$

$g(x)=f(f(x))=\begin{cases}2-f(x), & f(x)<2\\ f(x)-2, & f(x)\geq 2\end{cases}$

$=\begin{cases}2-(2-x), & 2-x<2, \quad x<2\\ (2-x)-2, & 2-x\geq 2, \quad x<2\\ 2-(x-2), & x-2<2, \quad x\geq 2\\ (x-2)-2, & x-2\geq 2, \quad x\geq 2\end{cases}$

$=\begin{cases}-x & 0<x\leq 0\\ x & 0<x<2\\ 4-x & 2\leq x<4\\ x-4 & x\geq 4\end{cases}$

$\therefore \int_0^3\left[g(x)-f(x)\right]dx$

$=\int_0^2 x\,dx+\int_2^3(4-x)dx-\int_0^3|x-2|dx=1$

84. (c) $\int_{\pi/6}^{\pi/3}\left[\frac{1}{2}\frac{d(\tan^4 x)}{dx}\cdot\sin^4 3x+\frac{1}{2}\tan^4 x\cdot\frac{d(\sin^4 3x)}{dx}\right]dx$

$=\frac{1}{2}\int_{\pi/6}^{\pi/3} d\left(\tan^4 x.\sin^4 3x\right)dx$

$=\left[\frac{\tan^4 x\sin^4 3x}{2}\right]_{\pi/6}^{\pi/3}=\frac{9.0}{2}-\frac{\frac{1}{9}.1}{2}=\frac{-1}{18}$

85. (c) $\int_0^{2\pi}\frac{x\sin^8 x}{\sin^8 x+\cos^8 x}dx$

$=\int_0^{\pi}\left[\frac{x\sin^8 x}{\sin^8 x+\cos^8 x}+\frac{(2\pi-x)\sin^8 x}{\sin^8 x+\cos^8 x}\right]dx$

$\left[\because \int_0^{2a} f(x)dx=\int_0^a f(x)dx+\int_0^a f(2a-x)dx\right]$

$=\int_0^{\pi}\frac{2\pi\sin^8 x}{\sin^8 x+\cos^8 x}dx$

$=2\pi\int_0^{\pi/2}\left[\frac{\sin^8 x}{\sin^8 x+\cos^8 x}+\frac{\cos^8 x}{\sin^8 x+\cos^8 x}\right]dx$

$=2\pi\int_0^{\pi/2} 1dx=2\pi\times\frac{\pi}{2}=\pi^2$

86. (b) $f(x)=\frac{1}{\sqrt{2x^3-9x^2+12x+4}}$

$f'(x)=\frac{-1}{2}\left(\frac{\left(6x^2-18x+12\right)}{\left(2x^3-9x^2-12x+4\right)^{3/2}}\right)$

$=\frac{-6(x-1)(x-2)}{2\left(2x^3-9x^2+12x+4\right)^{3/2}}$

$f(1)=\frac{1}{3}$ and $f(2)=\frac{1}{\sqrt{8}}$

It is an increasing function.

$\frac{1}{3}<I<\frac{1}{\sqrt{8}}$

$\frac{1}{9}<I^2<\frac{1}{8}$

87. (c) $I=\frac{1}{(a+b)}\int_a^b x\left[f(x)+f(x+1)\right]dx$

$x\to a+b-x$ ...(i)

$I=\frac{1}{(a+b)}\int_a^b (a+b-x)$

$\left[f(a+b-x)+f(a+b+1-x)\right]dx$

$I=\frac{1}{(a+b)}\int_a^b (a+b-x)\left[f(x+1)+f(x)\right]dx$ ...(ii)

$\left[\because \text{put } x\to x+1 \text{ in } f(a+b+1-x)=f(x)\right]$

Add (i) and (ii).

$2I=\int_a^b\left[f(x+1)+f(x)\right]dx$

$2I=\int_a^b f(x+1)dx+\int_a^b f(x)dx$

$=\int_a^b f(a+b+1-x)dx+\int_a^b f(x)dx$

$2I=2\int_a^b f(x)dx$

$\therefore \int_{a-1}^{b-1} f(x+1)\,dx$ $\left[\because \text{Put } x\to x+1\right]$

88. (a) $4\alpha\left\{\int_{-1}^{0} e^{\alpha x}dx+\int_0^2 e^{-\alpha x}dx\right\}=5$

$\Rightarrow 4\alpha\left\{\left.\frac{e^{\alpha x}}{\alpha}\right|_{-1}^{0}+\left.\frac{e^{-\alpha x}}{-\alpha}\right|_0^2\right\}=5$

$\Rightarrow 4\alpha\left\{\left(\frac{1-e^{-\alpha}}{\alpha}\right)-\left(\frac{e^{-2\alpha}-1}{\alpha}\right)\right\}=5$

$\Rightarrow 4\left(2-e^{-\alpha}-e^{-2\alpha}\right)=5$

Put $e^{-\alpha}=t$

$\Rightarrow 4t^2+4t-3=0$

$\Rightarrow (2t+3)(2t-1)=0$

$\Rightarrow e^{-\alpha}=\frac{1}{2}\Rightarrow \alpha=\log_e 2$

89. (a) $2\cot^2\theta-\frac{5}{\sin\theta}+4=0$

$\frac{2\cos^2\theta}{\sin^2\theta}-\frac{5}{\sin\theta}+4=0$

$\Rightarrow 2\cos^2\theta-5\sin\theta+4\sin^2\theta=0, \sin\theta\neq 0$

$\Rightarrow 2\sin^2\theta-5\sin\theta+2=0$

$\Rightarrow (2\sin\theta-1)(\sin\theta-2)=0$

$\therefore \sin\theta=\frac{1}{2}\Rightarrow \theta=\frac{\pi}{6},\frac{5\pi}{6}$

$\therefore \int_{\pi/6}^{5\pi/6}\cos^2 3\theta d\theta=\int_{\pi/6}^{5\pi/6}\frac{1+\cos 6\theta}{2}d\theta$

$= \frac{1}{2}\left[\theta + \frac{\sin 6\theta}{6}\right]_{\pi/6}^{5\pi/6} = \frac{1}{2}\left[\frac{5\pi}{6} - \frac{\pi}{6} + \frac{1}{6}(0-0)\right]$

$= \frac{1}{2}.\frac{4\pi}{6} = \frac{\pi}{3}$

90. (a) $F(x) = \int_1^x t^2 g(t)dt$

Differentiating by using Leibnitz's rule, we get

$F'(x) = x^2 g(x) = x^2 \int_1^x f(u)du$ ...(i)

At $x = 1$,

$F'(1) = 1\int_1^1 f(u)du = 0$

Now, differentiate Eq. (i).

$F''(x) = x^2 f(x) - 2x\int_1^x f(u)du$

At $x = 1$,

$F''(1) = 1.f(1) - 2\times 1.\int_1^1 f(u)du$

$= f(1) - 2\times 0 = f(1)$

$F''(1) = 3$

Then, for $F'(1) = 0, F''(1) = 3 > 0$

Hence, $x = 1$ is a point of local minima.

91. (a) Given, $\int_6^{f(x)} 4t^3 dt = (x-2)g(x)$

Differentiating both sides,

$4(f(x))^3.f'(x) = g'(x)(x-2) + g(x)$

Putting $x = 2, \frac{4(6)^3\cdot 1}{48} = g(2) \Rightarrow \lim_{x\to 2} g(x) = 18$

92. (d) $\int_0^{\pi/2} \frac{\cot x dx}{\cot x + \operatorname{cosec} x}$

$= \int_0^{\pi/2} \frac{\cot x dx}{1+\cos x} = \int_0^{\pi/2}\left(1 - \frac{1}{1+\cos x}\right)dx$

$= [x]_0^{\pi/2} - \int_0^{\pi/2} \frac{1}{2\cos^2\frac{x}{2}}dx = \frac{\pi}{2} - \frac{1}{2}\int_0^{\pi/2} \sec^2\frac{x}{2}dx$

$= \frac{\pi}{2} - \left(\tan\frac{x}{2}\right)_0^{\frac{\pi}{2}} = \frac{\pi}{2} - [1] = \left(\frac{\pi}{2} - 1\right) = m\pi + mn$

$\therefore m =, n = -2$, Hence, $mn = -1$

93. (b) $I = \int_0^{\pi} |\cos x|^3 dx$

$= 2\int_0^{\pi/2} \cos^3 x\, dx$

$= \frac{2}{4}\int_0^{\pi/2} (3\cos x + \cos 3x)dx$

$[\because \cos 3\theta = 4\cos^3\theta - 3\cos\theta]$

$= \frac{1}{2}\left[3\sin x + \frac{\sin x}{3}\right]_0^{\pi/2}$

$= \frac{1}{2}\left(3 - \frac{1}{3}\right) = \frac{4}{3}$

94. (a) Let I

$= \int \frac{\sin^2 x\cos^2 x}{\left(\sin^5 x + \cos^3 x\sin^2 x + \sin^3 x\cos^2 x + \cos^5 x\right)^2}dx$

$= \int \frac{\sin^2 x\cos^2 x}{\left[\left(\sin^2 x + \cos^2 x\right)\left(\sin^3 x + \cos^3 x\right)\right]^2}dx$

$= \int \frac{\sin^2 x\cos^2 x}{\left(\sin^3 x + \cos^3 x\right)^2}dx = \int \frac{\tan^2 x.\sec^2 x}{\left(1+\tan^3 x\right)^2}dx$

Now, put $\left(1+\tan^3 x\right) = t$

$\Rightarrow 3\tan^2 x\sec^2 x dx = dt$

$\therefore I = \frac{1}{3}\int \frac{dt}{t^2} = -\frac{1}{3t} + C = \frac{-1}{3\left(1+\tan^3 x\right)} + C$

95. (d) Given:

$I_1 = \int_0^1 e^{-x}\cos^2 x dx;$

$I_2 = \int_0^1 e^{-x^2}\cos^2 x dx$ and

$I_3 = \int_0^1 e^{-x^3} dx$

For $x \in (0,1)$

$\Rightarrow x > x^2$ or $-x < -x^2$

and $x^2 > x^3$ or $-x^2 < -x^3$

$\therefore e^{-x^2} < e^{-x^3}$ and $e^{-x} < e^{-x^2}$

$\Rightarrow e^{-x} < e^{-x^2} < e^{-x^3}$

$\Rightarrow e^{-x^3} > e^{-x^2} > e^{-x}$

$\Rightarrow I_3 > I_2 > I_1$

96. (c) $I = \int_{\frac{\pi}{4}}^{\frac{3\pi}{4}} \frac{dx}{1+\cos x}$ ...(i)

$I = \int_{\frac{\pi}{4}}^{\frac{3\pi}{4}} \frac{dx}{1-\cos x}$ ...(ii)

Using $\int_a^b f(x)\,dx = \int_a^b f(a+b-x)\,dx$

Adding (i) and (ii),

$2I = \int_{\frac{\pi}{4}}^{\frac{3\pi}{4}} \frac{2}{\sin^2 x} dx;\ I = \int_{\frac{\pi}{4}}^{\frac{3\pi}{4}} \operatorname{cosec}^2 x dx$

$I = -(\cot x)_{\frac{\pi}{4}}^{\frac{3\pi}{4}} = -\left[\cot\frac{3\pi}{4} - \cot\frac{\pi}{4}\right] = 2$

97. (c) $I_n = \int \tan^n x dx, n > 1$

Let $I = I_4 + I_6$

$= \int\left(\tan^4 x + \tan^6 x\right)dx = \int \tan^4 x \sec x dx$

Let $\tan x = t$

$\Rightarrow \sec^2 x\, dx = dt$

$\therefore I = \int t^4 dt = \frac{t^5}{5} + C$

$= \frac{1}{5}\tan^5 x + C \Rightarrow$ On comparing, we have,

$a = \frac{1}{5}, b = 0$

98. (c) $I = \int \frac{dx}{\left(1+\sqrt{x}\right).\sqrt{x}\sqrt{1-x}}$

Put $1+\sqrt{x} = t \Rightarrow \frac{1}{2\sqrt{x}}dx = dt \Rightarrow 1 = \int \frac{2dt}{t\sqrt{2t-t^2}}$

Again put $t = \frac{1}{z} \Rightarrow dt = \frac{-1}{2^2} dz$

$\Rightarrow I = 2\int \frac{\frac{-1}{z^2}dz}{\frac{1}{z}\sqrt{\frac{2}{z} - \frac{1}{z^2}}} = 2\int \frac{-dz}{\sqrt{2z-1}}$

$= -2\sqrt{2z-1} + c = -2\sqrt{\frac{2}{t} - 1} + c$

$= -2\sqrt{\frac{2-t}{t}} + c = -2\sqrt{\frac{1-\sqrt{x}}{1+\sqrt{x}}} + c$

99. (d) $x\int_1^x y(t)\,dt = x\int_1^x ty(t)\,dt + \int_1^x ty(t)\,dt$

Differentiate w.r.t. to x.

$\int_1^x y(t)dt + x\left[y(x) - y(1)\right]$

$\int_1^x y(t)dt = \int_1^x ty(t)dt + x^2 y(x) - y(1)$

Differentiate again w.r.t. to x.

$y(x) - y(a) = xy(x) - y(a) + 2x\,y(x) + x^2 y^1(x)$

$(1-3x)\,y(x) = x^2 y^1(x)$

$\frac{y^1(x)}{y(x)} = \frac{1-3x}{x^2}$

$\frac{1dy}{ydx} = \frac{1-3x}{x^2} \Rightarrow \ln y = -\frac{1}{x} - 3\ln x$

$\ln\left(y\,x^3\right) = -\frac{1}{x}$

$yx^3 = -e^{-\frac{1}{x}}$

$y = \frac{e^{-\frac{1}{x}}}{x^3}$ or $y = \frac{ce^{-\frac{1}{x}}}{x^3}$

100. (b) $I = \int \frac{dx}{x^2\left(x^4+1\right)^{3/4}} = \int \frac{dx}{x^3\left(1+x^{-4}\right)^{3/4}}$

Let $x^{-4} = y$.

$\Rightarrow -4x^{-3}dx = dy \Rightarrow dx = \frac{-1}{4}x^3 dy$

$\therefore I = \frac{-1}{4}\int \frac{x^3 dy}{x^3(1+y)^{3/4}} = \frac{-1}{4}\int \frac{dy}{(1+y)^{3/4}}$

$= \frac{-1}{4} \times 4(1+y)^{1/4} = -\left(1+x^{-4}\right)^{\frac{1}{4}} + c = -\left(\frac{x^4+1}{x^4}\right)^{\frac{1}{4}} + c$

101. (a) $I = \int_2^4 \frac{\log x^2}{\log x^2 + \log\left(36-12x+x^2\right)} dx$

$I = \int_2^4 \frac{\log x^2}{\log x^2 + \log(6-x)^2} dx$ ...(i)

$I = \int_2^4 \frac{\log(6-x)^2}{\log(6-x)^2 + \log x^2} dx$ ...(ii)

Adding (i) and (ii),

$2I = \int_2^4 dx = [x]_2^4 = 2,\ I = 1$

102. (b) $\int \frac{x^{5m-1}+2x^{4m-1}}{\left(x^{2m}+x^{m}+1\right)^{3}}dx$

$= \int \frac{x^{5m-1}+2x^{4m-1}}{x^{6m}\left(1+x^{-m}+x^{-2m}\right)^{3}}dx$

$= \int \frac{x^{-m-1}+2x^{-2m-1}}{\left(1+x^{-m}+x^{-2m}\right)^{3}}dx$

Put $t = 1+x^{-m}+x^{-2m}$

$\therefore \frac{dt}{dx} = -mx^{-m-1}-2mx^{-2m-1}$

$\Rightarrow \frac{dt}{-m} = \left(x^{-m-1}+2x^{-2m-1}\right)dx$

$\therefore \int \frac{x^{5m-1}+2x^{4m-1}}{\left(x^{2m}+x^{m}+1\right)^{3}}dx = \frac{1}{-m}\int t^{-3}dt = \frac{1}{2mt^{2}}+C$

$= \frac{1}{2m\left(1+x^{-m}+x^{-2m}\right)^{2}}+C$

$= \frac{x^{4m}}{2m\left(x^{2m}+x^{m}+1\right)^{2}}+C$

$\therefore f(x) = \frac{x^{4m}}{2m\left(x^{2m}+x^{m}+1\right)^{2}}$

103. (b) Let $I = \int_{0}^{\pi}\sqrt{1+4\sin^{2}\frac{x}{2}-4\sin\frac{x}{2}}\,dx$

$= \int_{0}^{\pi}\left|2\sin\frac{x}{2}-1\right|dx$

$= \int_{0}^{\pi-3}\left(1-2\sin\frac{x}{2}\right)dx+\int_{\frac{\pi}{3}}^{\pi}\left(2\sin\frac{x}{2}-1\right)dx$

$\left[\because \sin\frac{x}{2}=\frac{1}{2}\Rightarrow\frac{x}{2}=\frac{\pi}{6}\Rightarrow x=\frac{\pi}{3},\frac{x}{2}=\frac{5\pi}{6} \Rightarrow x=\frac{5\pi}{3}>\pi\right]$

$= \left[x+4\cos\frac{x}{2}\right]_{0}^{\frac{\pi}{3}}+\left[-4\cos\frac{x}{2}-x\right]_{\frac{\pi}{3}}^{\pi}$

$= \frac{\pi}{3}+4\frac{\sqrt{3}}{2}-4+\left(0-\pi+4\frac{\sqrt{3}}{2}+\frac{\pi}{3}\right)$

$= 4\sqrt{3}-4-\frac{\pi}{3}$

104. (b) $I = \int \frac{x\,dx}{2-x^{2}+\sqrt{2-x^{2}}}$

Put $t = \sqrt{2-x^{2}}, \frac{dt}{dx} = \frac{1}{2\sqrt{2-x^{2}}}.(-2x)$

$\Rightarrow -t\,dt = x\,dx$

$\therefore I = \int \frac{(-t)\,dt}{t^{2}+t} = -\int \frac{1}{t+1}dt = -\log|t+1|$

$= -\log\left|\sqrt{2-x^{2}}+1\right|+c$

105. (a) Consider

$\int_{0}^{\sin^{2}x}\sin^{-1}\left(\sqrt{t}\right)dt+\int_{0}^{\cos^{2}x}\cos^{-1}\left(\sqrt{t}\right)dt$

Let $I = f(x)$ after integrating and putting the limits.

$f'(x) = \sin^{-1}\sqrt{\sin^{2}x}\,(2\sin x\cos x)-0+\cos^{-1}\sqrt{\cos^{2}x}\,(-2\cos x\sin x)-0$

$\therefore f'(x) = 0 \Rightarrow f(x) = C$ (constant)

Now, we find $f(x)$ at $x = \frac{\pi}{4}$.

$\therefore I = \int_{0}^{1/2}\sin^{-1}\sqrt{t}dt+\int_{0}^{1/2}\cos^{-1}\sqrt{t}dt$

$= \int_{0}^{1/2}\left(\sin^{-1}\sqrt{t}+\cos^{-1}\sqrt{t}\right)dt = \int_{0}^{1/2}\frac{\pi}{2}dt = \frac{\pi}{4} = C$

$\therefore f(x) = \frac{\pi}{4}$

$\therefore$ Required integration $= \frac{\pi}{4}$.

106. (c) $I = \int \frac{\sec^{2}x}{(\sec x+\tan x)^{\frac{9}{2}}}dx$

Let $\sec x+\tan x = t \Rightarrow \sec x-\tan x = \frac{1}{t}$

$\Rightarrow \sec x = \frac{1}{2}\left(t+\frac{1}{t}\right)$ and $\sec x(\sec x+\tan x)dx = dt$

$\Rightarrow \sec x\,dx = \frac{dt}{t}$

$\therefore I = \frac{1}{2}\int \frac{\left(t+\frac{1}{t}\right)dt}{t^{9/2}.t} = \frac{1}{2}\int\left(t^{-9/2}+t^{-13/2}\right)dt$

$= \frac{-1}{7}t^{-\frac{7}{2}}-\frac{1}{11}t^{-\frac{11}{2}}+K$

$$= -\frac{1}{7t^{-\frac{7}{2}}} - \frac{1}{11t^{-\frac{11}{2}}} + K = -\frac{1}{t^{-\frac{11}{2}}}\left(\frac{1}{11}+\frac{t^2}{7}\right)+K$$

$$= \frac{-1}{(\sec x+\tan x)^{11/2}}\left\{\frac{1}{11}+\frac{1}{7}(\sec x+\tan x)^2\right\}+K$$

107. (b) $\int_{-\frac{\pi}{2}}^{\frac{\pi}{2}}\left[x^2+\ln\left(\frac{\pi+x}{\pi-x}\right)\right]\cos x\,dx$

$$= \int_{-\frac{\pi}{2}}^{\frac{\pi}{2}} x^2\cos x dx + \int_{-\frac{\pi}{2}}^{\frac{\pi}{2}} \ln\left(\frac{\pi+x}{\pi-x}\right)\cos x dx$$

$$= 2\int_0^{\frac{\pi}{2}} x^2\cos x dx + 0$$

[$\because$ $x^2\cos x$ is an even function and $\ln\left(\frac{\pi+x}{\pi-x}\right)\cos x$ is an odd function.]

$$= 2\left[x^2\sin x + 2x\cos x - 2\sin x\right]_0^{\pi/2} = \frac{\pi^2}{2}-4$$

108. (a) $I = \frac{1}{2}\int_{\sqrt{\ell n2}}^{\sqrt{\ell n3}} \frac{2x\sin x^2}{\sin x^2+\sin\left(\ell n6-x^2\right)}dx$

Let $x^2 = t \Rightarrow 2x\,dx = dt$

Also, when $x=\sqrt{\ln 2}$, then $t=\ln 2$

and when $x=\sqrt{\ln 3}$, then $t=\ln 3$

$$\therefore I = \frac{1}{2}\int_{\ln 2}^{\ln 3}\frac{\sin t dt}{\sin t+\sin(\ln 6-t)} \quad \text{...(i)}$$

$$\Rightarrow I = \frac{1}{2}\int_{\ln 2}^{\ln 3}\frac{\sin(\ln 6-t)}{\sin t+\sin(\ln 6-t)}dt \quad \text{...(ii)}$$

$$\left[\because \int_a^b f(x)dx = \int_a^b f(a+b-x)dx\right]$$

On adding Eqs (i) and (ii), we get

$$2I = \frac{1}{2}\int_{\ln 2}^{\ln 3} 1dt = \frac{1}{2}(\ln 3-\ln 2) = \frac{1}{2}\ln\frac{3}{2} \Rightarrow I = \frac{1}{4}\ln\frac{3}{2}$$

109. (b) $e^{-x}f(x) = 2+\int_0^x\sqrt{1+t^4}\,dt\,\forall x\in(-1,1)$

On differentiating, we get

$$-e^{-x}f(x)+e^{-x}f'(x) = 0+\sqrt{1+x^4}$$

$$\Rightarrow -f(0)+f'(0)=1 \Rightarrow f'(0)=1+f(0)=1+2=3$$

Now $f^{-1}(f(x)) = x$

$$\Rightarrow \left[\left(f^{-1}\right)'(f(x))\right]f'(x)=1$$

$$\Rightarrow \left(f^{-1}\right)'(f(0))f'(0)=1 \Rightarrow \left(f^{-1}\right)'(2)=\frac{1}{3}$$

110. (b) $\lim_{x\to 0}\frac{1}{x^3}\int_0^x\frac{t\ln(1+t)}{t^4+4}dt \qquad \left[\frac{0}{0}\text{ form}\right]$

Using L' Hospital's rule, we get

$$\lim_{x\to 0}\frac{\frac{x\ln(1+x)}{x^4+4}}{3x^2} = \lim_{x\to 0}\frac{\ln(1+x)}{x}\cdot\frac{1}{3\left(x^4+4\right)}$$

$$= 1\cdot\frac{1}{12} = \frac{1}{12}$$

111. (c) Given that f is a non-negative function defined on [0, 1] and

$$\int_0^x\sqrt{1-(f'(t))^2}\,dt = \int_0^x f(t)dt,\ 0\le x\le 1$$

Differentiating both sides w.r.t to x, we get

$$\sqrt{1-[f'(x)]^2} = f(x)$$

$$\Rightarrow 1-[f'(x)]^2 = [f(x)]^2$$

$$\Rightarrow [f'(x)]^2 = 1-[f(x)]^2$$

$$\Rightarrow \frac{d}{dx}f(x) = \pm\sqrt{1-[f(x)]^2}$$

$$\Rightarrow \pm\frac{d\,f(x)}{\sqrt{1-[f(x)]^2}} = dx$$

Integrating both sides w.r.t to x, we get

$$\pm\int\frac{d\,f(x)}{\sqrt{1-[f(x)]^2}} = \int dx \Rightarrow \pm\sin^{-1}f(x) = x+C$$

$\because$ Given that $f(0)=0 \Rightarrow C=0$

Hence $f(x) = \pm\sin x$

But as $f(x)$ is a non-negative function on [0, 1],

$\therefore f(x) = \sin x$

Now, $\sin x\langle x, \forall x\rangle 0$

$$\therefore f\left(\frac{1}{2}\right)<\frac{1}{2} \text{ and } f\left(\frac{1}{3}\right)<\frac{1}{3}$$

112. (c) Given $I = \int\frac{e^x}{e^{4x}+e^{2x}+1}dx,$

$$J = \int\frac{e^{-x}}{e^{-4x}+e^{-2x}+1}dx = \int\frac{e^{3x}}{e^{4x}+e^{2x}+1}dx$$

$$\therefore J-I=\int \frac{e^x\left(e^{2x}-1\right)}{e^{4x}+e^{2x}+1}dx$$

Let $e^x=t \Rightarrow e^x dx = dt$

$$\therefore J-I=\int \frac{t^2-1}{t^4+t^2+1}dt=\int \frac{1-\frac{1}{t^2}}{t^2+1+\frac{1}{t^2}}dt$$

Let $t-\frac{1}{t}=u \Rightarrow \left(1-\frac{1}{t^2}\right)dt = du.$

$$\therefore J-I=\int \frac{du}{u^2-1}=\frac{1}{2}\log\left|\frac{u-1}{u+1}\right|+c$$

$$=\frac{1}{2}\log\left|\frac{\frac{t^2+1}{t}-1}{\frac{t^2+1}{t}+1}\right|+c=\frac{1}{2}\log\left|\frac{e^{2x}-e^x+1}{e^{2x}+e^x+1}\right|+c$$

Integer and Numerical Value Type

1. $i=\int_{-a}^{a}\left(\cos^{-1}x-\sin^{-1}\sqrt{1-x^2}\right)dx$

$$=\int_{-a}^{0}\cos^{-1}x dx + A - 2\int_0^a \sin^{-1}\sqrt{1-x^2}\,dx$$

$$=\int_0^a(\pi-\cos^{-1}x)dx + A - 2A$$

$$=a\pi-2A \Rightarrow \lambda=2$$

2. $\int_0^x f^{-1}(x)dx=\int_{f^{-1}}^{f^{-1}} tf'(t)dt$

$[f^{-1}(x)=t \text{ or } x=f(t)]$

$$=\left[tf(t)-\int f(t)\right]_{f^{-1}(0)}^{f^{-1}(x)}$$

$$\Rightarrow \int_0^\pi f^{-1}(x)dx=\pi^2-\int_0^\pi (t+\sin t)\,dt$$

$$=\frac{\pi^2}{2}-2 \Rightarrow k=2$$

3. On integrating by parts,

$$\int_0^{\pi/2} x^2 \operatorname{cosec}^2 x\,dx$$

$$=\left|x^2(-\cot x)\right|_0^{\pi/2}-\int_0^{\pi/2}2x(-\cot x)\,dx$$

Since $\lim\limits_{x\to 0} x^2\cot x=\lim\limits_{x\to 0} x.\frac{x}{\tan x}=0$

$$\int_0^{\pi/2} x^2\operatorname{cosec}^2 x\,dx=2\int_0^{\pi/2} x\cot x\,dx$$

Again, on integrating by parts,

$$\text{LHS}=2\left[\left|x(\log\sin x)\right|_0^{\pi/2}-\int_0^{\pi/2}\log\sin x\,dx\right]$$

$$=-2\int_0^{\pi/2}\log\sin x\,dx \qquad \left(\because \lim_{x\to 0} x\log\sin x=0\right)$$

$$=-2\left(-\frac{\pi}{2}\log 2\right)=\pi\log 2$$

$$\Rightarrow k=2$$

4. Let $\frac{x}{(1+x)(1+x^2)}=\frac{A}{1+x}+\frac{Bx+C}{1+x^2}$

Then $x \equiv A(1+x^2)+(Bx+C)(1+x).$

On comparing coefficients

$A+B=0,\ B+C=1,\ A+C=0$

$\Rightarrow B=\frac{1}{2}, C=\frac{1}{2}, A=-\frac{1}{2}$

$\Rightarrow$ Indefinite integral

$$=-\frac{1}{2}\log(1+x)+\frac{1}{4}\log(1+x^2)+\frac{1}{2}\tan^{-1}x$$

$$=\log\frac{(1+x^2)^{1/4}}{(1+x^2)^{1/2}}+\frac{1}{2}\tan^{-1}x$$

$$\lim_{x\to\infty}\frac{(1+x^2)^{1/4}}{(1+x^2)^{1/2}}=\lim_{x\to\infty}\frac{\left(1+\frac{1}{x^2}\right)^{1/4}}{\left(1+\frac{1}{x}\right)^{1/2}}=1$$

$$\Rightarrow \lim_{x\to\infty}\log\frac{(1+x^2)^{1/4}}{(1+x^2)^{1/2}}=0$$

Also, $\lim\limits_{x\to\infty}\frac{1}{2}\tan^{-1}x=\frac{\pi}{4}$

Thus $\int_0^\infty \frac{x\,dx}{(1+x)(1+x^2)}=\frac{\pi}{4}$

$$\Rightarrow k=4$$

5. $\frac{\cos 6x+\cos 9x}{1-2\cos 5x}=\frac{2\cos\frac{15x}{2}\cos\frac{3x}{2}}{1-2\left(2\cos^2\frac{5x}{2}-1\right)}$

$$=\frac{2\left(4\cos^3\frac{5x}{2}-3\cos\frac{5x}{2}\right)\cos\frac{3x}{2}}{3-4\cos^2\frac{5x}{2}}$$

$$=-2\cos\frac{5x}{2}\cos\frac{3x}{2}=-(\cos 4x+\cos x)$$

$$\Rightarrow \text{Integral}=-\frac{\sin 4x}{4}-\sin x+C$$

$$\Rightarrow k=4$$

6. $I = \int_0^1 (1 - \cos 2\lambda x)\, dx$

$$= \left[x - \frac{\sin 2\lambda x}{2\lambda}\right]_0^1 = 1 - \frac{\sin 2\lambda}{2\lambda}$$

$$= 1 - \frac{2\tan\lambda}{2\lambda(1+\tan^2\lambda)}$$

But $\tan\lambda = \lambda$ (as is given)

$$\Rightarrow I = 1 - \frac{2\lambda}{2\lambda(1+\lambda^2)} = \frac{\lambda^2}{1+\lambda^2}$$

$$\Rightarrow k = 1$$

7. RHS $\left(\text{by putting } f(t) = -\frac{1}{k}\sin t - \frac{2}{k}\cos t\right)$

$$= \sin x + \int_{-\pi/2}^{\pi/2} (\sin x + t\cos x)\left(-\frac{1}{k}\sin t - \frac{2}{k}\cos t\right) dt$$

$$= \sin x + I_1 + I_2 + I_3 + I_4$$

$$I_1 = -\frac{\sin x}{k}\int_{-\pi/2}^{\pi/2} \sin t\, dt = 0 \quad (\because \sin t \text{ is odd})$$

$$I_2 = -\frac{2\sin x}{k}\int_{-\pi/2}^{\pi/2} \cos t\, dt = -\frac{4\sin x}{k}$$

$$I_3 = -\frac{\cos x}{k}\int_{-\pi/2}^{\pi/2} \sin t\, dt = -\frac{2\cos x}{k}$$

$$I_4 = -\frac{2\cos x}{k}\int_{-\pi/2}^{\pi/2} \cos t\, dt = 0 \quad (\because t\cos t \text{ is odd})$$

$$\Rightarrow -\frac{1}{k}\sin x - \frac{2}{k}\cos x = \sin x - \frac{4\sin x}{k} - \frac{2\cos x}{k}$$

$$\Rightarrow -\frac{1}{k} = 1 - \frac{4}{k}$$

$$\Rightarrow k = 3$$

8. $I = \int \dfrac{\frac{\sin x}{\cos^3 x}}{1+\tan^3 x} dx$

$$= \int \frac{\tan x.\sec^2 x}{(\tan x + 1)(1 + \tan^2 x - \tan x)} dx$$

Let $\tan x = t \Rightarrow \sec^3 x.dx = dt$

$$= \frac{t}{(t+1)(t^2 - t + 1)} dt$$

$$Let \frac{t}{(t+1)(t^2-t+1)} = \frac{A}{t+1} + \frac{B(2t-1)}{t^2-t+1} + \frac{C}{t^2-t+1}$$

$$\Rightarrow A(t^2 - t + 1) + B(2t-1)(t+1) + C(t+1) = t$$

$$\Rightarrow t^2(A + 2B) + t(-A + B + C) + A - B + C = t$$

$\therefore A + 2B = 0$ ...(i)

$-A + B + C = 1$ ...(ii)

$A - B + C = 0$ ...(iii)

$\Rightarrow C = \frac{1}{2} \Rightarrow A - B = -\frac{1}{2}$ ...(iv)

Solving equations (i) and (iv), we get

$$B = \frac{1}{6},\ A = -\frac{1}{3}$$

$$\therefore I = -\frac{1}{3}\int \frac{dt}{1+t} + \frac{1}{6}\int \frac{2t-1}{t^2-t+1} dt + \frac{1}{2}\int \frac{dt}{t^2-t+1}$$

$$= -\frac{1}{3}\log_e|(1+\tan x)| + \frac{1}{6}\log_e|\tan^2 x - \tan x + 1| + \frac{1}{2}.\frac{2}{\sqrt{3}}\tan^{-1}\left(\frac{\left(\tan x - \frac{1}{2}\right)}{\frac{\sqrt{3}}{2}}\right)$$

$$= -\frac{1}{3}\log_e|(1+\tan x)| + \frac{1}{6}\log_e|\tan^2 x - \tan x + 1| + \frac{1}{\sqrt{3}}\tan^{-1}\left(\frac{2\tan x - 1}{\sqrt{3}}\right) + C$$

$$\alpha = -\frac{1}{3}, \beta = \frac{1}{6}, \gamma = \frac{1}{\sqrt{3}}$$

$$\text{So, } 18(\alpha + \beta + \gamma^2) = 18\left(-\frac{1}{3} + \frac{1}{6} + \frac{1}{3}\right) = 3$$

9. $f(x) = \int \dfrac{5x^8 + 7^6}{x^{14}\left(\frac{1}{x^5} + \frac{1}{x^7} + 2\right)} dx$

$$= \int \frac{5x^{-6} + 7x^{-8}}{\left(2 + \frac{1}{x^5} + \frac{1}{x^7}\right)} dx$$

Put $2 + \frac{1}{x^5} + \frac{1}{x^7} = t$

$$(-5x^{-6} - 7x^{-8})dx = dt$$

$$\Rightarrow f(t) = \int \frac{-dt}{t^2} = \frac{1}{t} + c$$

$$\therefore f(x) = \frac{x^7}{2x^7 + x^2 + 1} + c \quad \left[\because t = 2 + \frac{1}{x^5} + \frac{1}{x^7}\right]$$

$$\because f(0) = 0 \Rightarrow c = 0$$

$$\therefore f(x) = \frac{x^7}{2x^7 + x^2 + 1}$$

$$\Rightarrow f(1) = \frac{1}{4} \Rightarrow k = 4$$

10. $I=\int \frac{x^2-1}{\left(x^4+3x^2+1\right)\tan^{-1}\left(\frac{x^2+1}{x}\right)}dx+\int \frac{1}{x^4+3x^2+1}dx$

$I=I_1+I_2$ ...(i)

For I_1, let $\tan^{-1}\frac{x^2+1}{x}=t$, and $\frac{x^2-1}{x^4+3x^2+1}dx=dt$

$I_1=\int \frac{1}{t}dt$

$=\log t=\log\left|\tan^{-1}\left(\frac{x^2+1}{x}\right)\right|+C_1$ ...(ii)

$I_2=\frac{1}{2}\int \frac{\left(x^2+1\right)-\left(x^2-1\right)}{x^4+3x^2+1}$

$=\frac{1}{2}\int \frac{x^2+1}{x^4+3x^2+1}-\frac{1}{2}\int \frac{x^2-1}{x^4+3x^2+1}$

Divide Nr and Dr by x^2, we get

$=\frac{1}{2}\int \frac{1+\frac{1}{x^2}}{\left(x-\frac{1}{x}\right)^2+5}-\frac{1}{2}\int \frac{1-\frac{1}{x^2}}{\left(x+\frac{1}{x}\right)^2+1}$

$=\frac{1}{2\sqrt{5}}\tan^{-1}\left(\frac{x^2-1}{\sqrt{5}x}\right)-\frac{1}{2}\tan^{-1}\left(\frac{x^2+1}{x}\right)$ ...(iii)

$\alpha=1, \beta=\frac{1}{2\sqrt{5}}, \gamma=\frac{1}{\sqrt{5}}, \delta=\frac{-1}{2}$

So, $10(\alpha+\beta\gamma+\delta)=10\left(1+\frac{1}{10}-\frac{1}{2}\right)=6$

11. $x\phi(x)=\int_5^x 3t^2-2\phi'(t)dt$

$\phi(0)=4, x>-2$

$x\phi(x)=x^3-125-2\left[\phi(x)-\phi(5)\right]$

$x\phi(x)=x^3-125-2\phi(x)-2\phi(5)$

$\phi(0)=4\Rightarrow\phi(5)=-\frac{133}{2}$

$\phi(x)=\frac{x^3+8}{x+2}$

$\phi(2)=\frac{2^3+8}{2+2}=\frac{8+8}{4}=4$

12. For domain, $\log_5\left(\log_3\left(18x-x^2-77\right)\right)>0$

$\Rightarrow\log_3\left(18x-x^2-77\right)>1$

$\Rightarrow 18x-x^2-77>3$

$\Rightarrow x^2-18x+80<0$

$\Rightarrow\left(x-8\right)\left(x-10\right)<0$

+	−	+

$-\infty$ 8 10 ∞

$\therefore x\in(8,10)$

$\Rightarrow a=8$ and $b=10$

$\therefore I=\int_8^{10}\frac{\sin^3 x}{\sin^3 x+\sin^3(18-x)}dx$ ...(i)

$I=\int_8^{10}\frac{\sin^3(18-x)}{\sin^3(18-x)+\sin^3 x}dx$ ...(ii)

Adding (i) and (ii), we get

$2I=\int_8^{10}1\cdot du=10-8=2$

$\therefore I=1$

13. $\int_0^{\pi}\left(\sin^3 x\right).e^{-\sin^2 x}dx=\frac{1}{e}\int_0^{\pi}\sin^2 x.e^{\cos^2 x}.\sin x dx$

Let $\cos x=t$, then $\sin x\, dx=-dt$

$\therefore\int_0^{\pi}\left(\sin^3 x\right)e^{-\sin^2 x}dx$

$=\frac{1}{e}\int_1^{-1}\left(t^2-1\right)e^{t^2}dt=\frac{2}{e}\int_0^1\left(1-t^2\right)e^{t^2}dt$

Let $t^2=z, dt=\frac{dz}{2\sqrt{z}}$.

$=\frac{1}{e}\int_0^1\left(\frac{1}{\sqrt{z}}-\sqrt{z}\right)e^z dz$

$=\frac{1}{e}\left[e^z\cdot 2\sqrt{z}\Big|_0^1-\int_0^1 2e^z\cdot\sqrt{z}dz-\int_0^1\sqrt{z}e^z dz\right]$

$=\frac{1}{e}\left[2e-3\int_0^1 e^t\cdot\sqrt{t}\,dt\right]=2-\frac{3}{e}\int_0^1\sqrt{t}e^t\,dt$

$\Rightarrow\alpha=2$ and $\beta=3$

So, $\alpha+\beta=5$

14. $I_n=\int_1^e \underset{\text{II}}{x^{19}}.\underset{\text{I}}{(\ln x)^n}\,dx$

[using integration by parts]

$$\Rightarrow I_n = \left.\frac{(\ln x)^n .x^{20}}{20}\right|_1^e - \int_1^e n(\ln x)^{n-1}\frac{x^{19}}{20}dx$$

$$\Rightarrow 20I_n = e^{20} - nI_{n-1}$$

So, $20I_{10} = e^{20} - 10I_9$ ...(i)

and $20I_9 = e^{20} - 9I_8$ ...(ii)

$\underline{-\quad -\quad +}$

$20I_{10} = 10I_9 + 9I_8$

$\therefore \alpha = 10$ and $\beta = 9$

So, $\alpha - \beta = 1$

15. $$\left|\int_0^{\sqrt{\frac{\pi}{2}}}\left[\left[x^2\right]-\cos x\right]dx\right|$$

$$= \left|\int_0^1(-1)dx + \int_0^{\sqrt{\frac{\pi}{2}}}0.dx\right|$$

16. Let $x = \frac{1}{y+1} \Rightarrow dx = \frac{-1}{(y+1)^2}dy$

$$I_{m,n} = -\int_{\infty}^{0}\frac{1}{(y+1)^{m-1}}\cdot\frac{y^{n-1}}{(y+1)^{n-1}}\cdot\frac{dy}{(y+1)^2}$$

$$= y\int_0^{\infty}\frac{y^{n-1}}{(1+y)^{m+n}}dy \quad \text{...(i)}$$

Now, $I_{m,n} = \int_0^1 x^{m-1}(1-x)^{n-1}\,d\text{ɩ}$

$$= \int_0^1 (1-x)^{m-1}x^{n-1}dx$$

Similarly,

$$I_{m,n} = \int_0^{\infty}\frac{y^{m-1}}{(1+y)^{m+n}}dy \quad \text{...(ii)}$$

Adding Eqs (i) and (ii), we get

$$2I_{m,n} = \int_0^{\infty}\frac{y^{m-1}+y^{n-1}}{(1+y)^{m+n}}dy$$

$$= \int_0^1\frac{y^{m-1}+y^{n-1}}{(1+y)^{m+n}}dy + \int_0^{\infty}\frac{y^{m-1}+y^{n-1}}{(1+y)^{m+n}}dy$$

$$= I_1 + I_2$$

Put $y = \frac{1}{t}$ and dy $= \frac{-1}{t^2}dt$ in I_2, we get

$$2I_{m,n} = \int_0^1\frac{y^{m-1}+y^{n-1}}{(1+y)^{m+n}}dy - \int_1^0\frac{t^{m-1}+t^{n-1}}{(1+t)^{m+n}}dy$$

$$2I_{m,n} = 2\int_0^1\frac{x^{m-1}+x^{n-1}}{(1+x)^{m+n}}dx$$

$\therefore \alpha = 1$

17. We have,

$$\int_{-a}^{0}(-2x+2)dx + \int_0^2(x+2-x)dx + \int_2^a(2x-2)dx = 22$$

$$\Rightarrow x^2 - 2x\Big|_0^{-a} + 2x\Big|_0^2 + x^2 - 2x\Big|_0^a = 22$$

$$\Rightarrow a^2 + 2a + 4 + a^2 - 2a - (4-4) = 22$$

$$\Rightarrow 2a^2 = 18 \Rightarrow a = 3$$

$$\therefore \int_3^{-3}(x+[x])dx = -(-3-2-1+1+2) = 3$$

18. $$\lim_{n\to\infty}\frac{2}{n}\left(f\left(\frac{1}{n}\right)+f\left(\frac{2}{n}\right)+\ldots+f\left(\frac{n}{n}\right)\right)$$

$$\Rightarrow \lim_{n\to\infty}\frac{2}{n}\sum_{r=1}^{n}\log_2\left(1+\tan\frac{\pi r}{4n}\right)$$

Using integration by limit as a sum formula here,

$nh = 1$, i.e., $h = \frac{1}{n}$

$$\Rightarrow 1 = 2\int_0^1\log_2\left(1+\tan\frac{\pi x}{4}\right)dx \quad \text{...(i)}$$

Using which, we get, $\int_0^a f(x)dx = \int_0^a f(a-x)dx$

$$I = 2\int_0^1\log_2\left(1+\tan\frac{\pi}{4}(1-x)\right)dx$$

$$\Rightarrow I = 2\int_0^1\log_2\left(\frac{2}{1+\tan\frac{\pi x}{4}}\right)dx \quad \text{...(ii)}$$

Add Eqs (i) and (ii), we get

$$\Rightarrow 2I = 2\int_0^1\log_2 2dx \Rightarrow I = \int_0^1 dx = 1$$

19. $$I = \int_{-4}^{4}f(x)^2\,dx = 2.\int_0^4 f(x)^2\,dx$$

$$\because f(x^2) = 4x^3 - g(4-x)$$

$$I = 2\int_0^4\left(4x^3 - g(4-x)\right)dx$$

$$= 2.4\left[\frac{x^4}{4}\right]_0^4 - 2\int_0^4 g(4x)\,dx$$

$I = 512 - 2I_1$, where $I_1 = \int_0^4 g(4-x)\,dx$

Now, $I_1 = \int_0^4 g(4-x)\,dx$

$$= \int_0^4 g(4-(0+4-x))\,dx = \int_0^4 g(x)\,dx$$

$$I_1 = -\int_0^4 g(4-x)\,dx \Rightarrow I_1 = 0$$

So, $I = 512 - 0 = 512$

Hence, $\int_{-4}^{4} f(x)^2\,dx = 512$

20. $I = \int_0^{\sqrt{\pi/2}} ([x^2]+[-\cos x])\,dx$

$$= \int_0^1 0\,dx + \int_1^{\sqrt{\pi/2}} dx + \int_0^{\sqrt{\pi/2}} (-1)\,dx$$

$$= \sqrt{\frac{\pi}{2}} - 1 - \sqrt{\frac{\pi}{2}} = -1$$

$\Rightarrow |I| = 1$

21. $\int_0^n \{x\}\,dx = n\int_0^1 x.dx = \frac{n}{2}$

$$\Rightarrow \int_0^n [x]\,dx = \int_0^n (x - \{x\})\,dx = \frac{n^2}{2} - \frac{n}{2}$$

According to the question,

$\frac{n}{2}, \frac{n^2-n}{2}, 10(n^2-n)$ are in GP

$$\therefore \left(\frac{n^2-n}{2}\right)^2 = \frac{n}{2}\times 10(n^2-n)$$

$\Rightarrow n^2 = 21n \Rightarrow n = 21$

22. $\int_0^2 \big||x-1|-x\big|\,dx$

$$= \int_0^1 |1-x-x|\,dx + \int_1^2 |x-1-x|\,dx$$

$$= \int_0^{1/2} (1-2x)\,dx + \int_{1/2}^{1} (2x-1)\,dx + \int_1^2 dx$$

$$= \left[x - x^2\right]_0^{\frac{1}{2}} + \left[x^2 - x\right]_{\frac{1}{2}}^{1} + [x]_1^2$$

$$= \frac{1}{2} - \frac{1}{4} + (1-1) - \left(\frac{1}{4} - \frac{1}{2}\right) + 2 - 1 = \frac{1}{4} + \frac{1}{4} + 1 = \frac{3}{2}$$

23. $\int_1^2 \big|2x - [3x]\big|\,dx$

$$= \int_1^2 \big|3x - [3x] - x\big|\,dx$$

$$= \int_1^2 \big|\{3x\} - x\big|\,dx = \int_1^2 (x - \{3x\})\,dx$$

$$= \int_1^2 x\,dx - \int_1^2 \{3x\}\,dx = \left[\frac{x^2}{2}\right]_1^2 - 3\int_0^{1/3} 3x\,dx$$

$$= \frac{(4-1)}{2} - 9\left[\frac{x^2}{2}\right]_0^{\frac{1}{3}} = \frac{3}{2} - \frac{1}{2} = 1$$

Assertion and Reason

1. (a) $\int (\log_x e) - (\log_x e)^2\,dx$

$$= \int \left(\frac{1}{\ln x} - \frac{1}{(\ln x)^2}\right) dx$$

$$= \int \left(\frac{1}{t} - \frac{1}{t^2}\right) e^t dt \qquad (\text{where } t = \ln x)$$

$$= \frac{e^t}{t} + C = \frac{x}{\ell n x} + C = x\log_x e + C$$

2. (a) Since $y = f^{-1}(x)$, therefore, $x = f(y)$

$\therefore dx = f'(y)dy$. Thus $\int f^{-1}(x)dx = \int y f'(y)dy + c$

$\therefore$ Statement 2 is true

Statement 1 is true for $f^{-1}(x) = \sin^{-1} x$ in Statement 2.

3. (d) Since $\cot^{-1} x = \frac{\pi}{2} - \tan^{-1} x$,

$\therefore d(\cot^{-1} x) = -d(\tan^{-1} x)$

Thus $\int 2^{\tan^{-1} x} d(\cot^{-1} x) = -\int 2^{\tan^{-1} x} d(\tan^{-1} x)$

$$= -\frac{2^{\tan^{-1} x}}{\ln 2} + c$$

$\therefore$ Statement 1 is false.

Statement 2 is true.

4. (a) **Statement 2**: $\int e^{g(x)} f(x)g'(x)+f'(x)\,dx$

$=\int f(x).e^{g(x)}.g'(x)dx+\int e^{g(x)}.f'(x)dx$

$=f(x).e^{g(x)}-\int f'(x)e^{g(x)}dx+\int e^{g(x)}.f'(x)dx$

$=e^{g(x)}f(x)+c$

Statement 1: Since $g(x)=\sin^{-1}x, f(x)=\sqrt{1-x^2}$

$\therefore \int e^{\sin^{-1}x}\left(1-\frac{x}{\sqrt{1-x^2}}\right)dx$

$=\int e^{\sin^{-1}x}\left(\frac{1}{\sqrt{1-x^2}}.\sqrt{1-x^2}+\frac{-x}{\sqrt{1-x^2}}\right)$

$=e^{\sin^{-1}x}.\sqrt{1-x^2}+c$ by Statement 2.

5. (c) **Statement 1**

$$I=\int_0^{\pi/4}\frac{1}{\cos x}\sqrt{\frac{1-\sin x}{1+\sin x}}dx=\int_0^{\pi/4}\frac{1-\sin x}{\cos x.\cos x}dx$$

$$=\int_0^{\pi/4}\left(\sec^2 x-\sec x\tan x\right)dx$$

$$=\left[\tan x-\sec x\right]_0^{\pi/4}=2-\sqrt{2}$$

Statement 2

$$I=\int_0^{\pi/4}\frac{\sec x dx}{1+2\sin^2 x}=\int_0^{\pi/4}\frac{\cos x}{\left(1-\sin^2 x\right)\left(1+2\sin^2 x\right)}dx$$

$$=\int_0^{\frac{1}{\sqrt{2}}}\frac{dt}{\left(1-t^2\right)\left(1+2t^2\right)}=\frac{1}{3}\int_0^{\frac{1}{\sqrt{2}}}\left(\frac{1}{1-t^2}+\frac{2}{1+2t^2}\right)dt$$

$$=\left(\frac{1}{3}\frac{1}{2}\ln\frac{1+t}{1-t}+\frac{2}{3\sqrt{2}}\tan^{-1}\sqrt{2}t\right)_0^{\frac{1}{\sqrt{2}}}$$

$$=\frac{1}{6}\ln\frac{\sqrt{2}+1}{\sqrt{2}-1}+\frac{\sqrt{2}}{3}\left(\frac{\pi}{4}\right)=\frac{1}{6}\ln\left(3+2\sqrt{2}\right)+\frac{\pi}{6\sqrt{2}}$$

6. (b) **Statement 1**

$$I=\int_0^1 2^x\cdot x^2\,dx=\left[x^2\cdot\frac{2^x}{\ln 2}\right]_0^1-\int_0^1\frac{2x\cdot 2^x}{\ln 2}dx$$

$$=\frac{2}{\ln 2}-\frac{2}{\ln 2}\left[\left(x\cdot\frac{2^x}{\ln 2}\right)_0^1-\int_0^1\frac{2^x}{\ln 2}dx\right]$$

$$=\frac{2}{\ln 2}-\frac{2}{\ln 2}\left[\frac{2}{\ln 2}-\frac{1}{(\ln 2)^2}\left[2^x\right]_0^1\right]$$

$$=\frac{2}{\ln 2}-\frac{4}{(\ln 2)^2}+\frac{2}{(\ln 2)^3}(2-1)$$

$$=\frac{2}{\ln 2}-\frac{4}{(\ln 2)^2}+\frac{2}{(\ln 2)^3}$$

Statement 2

$$I_n=\int_0^1 e^x(x-1)^n\,dx$$

$$=\left[e^x(x-1)^n\right]_0^1-n\int_0^1(x-1)^{n-1}e^x dx$$

$$=-(-1)^n-n\,I_{n-1}$$

$\therefore I_n+n\,I_{n-1}=-(-1)^n$

$I_1+(e-1)=1 \qquad I_1=2-e$

$I_2+2(2-e)=-1 \qquad \therefore I_2=2e-5$

$I_3+3(2e-5)=1 \qquad \therefore I_3=16-6e$

$\therefore n=3$ (true)

7. (c) **Statement 1**

$$\frac{d}{dx}F(x)=\frac{e^{\sin x}}{x},x>0$$

$$\Rightarrow F(x)=\int\frac{e^{\sin x}dx}{x}\quad I=\int_1^4\frac{3e^{\sin x^3}}{x}dx$$

$$=\int_1^4\frac{3x^2e^{\sin x^3}}{x^3}dx=\int_1^{64}\frac{e^{\sin t}}{t}dt$$

$$=\left[F(x)\right]_1^{64}=F(64)-F(1)$$

$\therefore F(64)-F(1)=F(k)-F(1)$

$\therefore k=64$

Statement 2

$$I=\int_{\sin\theta}^{\operatorname{cosec}\theta}f(x)dx=\int_{\operatorname{cosec}\theta}^{\sin\theta}f\left(\frac{1}{t}\right)\left(-\frac{1}{t^2}\right)dt$$

$$=\int_{\operatorname{cosec}\theta}^{\sin\theta}f(t)dt=-I$$

$\therefore I=0$

8. (b) **Statement 1**

$$I=\int_{\pi}^{\pi}\frac{2x\left(1+\sin^2 x\right)}{1+\cos^2 x}dx$$

$I=0 \qquad (\because f(-x)=-f(x))$

Statement 2

$\sin x-\cos x=t\Rightarrow 1-2\sin x\cos x=t^2\Rightarrow\frac{1-t^2}{2}=\sin x\cos x$

$(\sin x+\cos x)dx=dt$

$$I=\int_{-1}^{1}\left(\frac{1-t^2}{2}\right)^2dt=\frac{1}{2}\int_0^1\left(1-2t^2+t^4\right)dt$$

$$=\frac{1}{2}\left(t-\frac{2t^3}{3}+\frac{t^5}{5}\right)_0^1=\frac{4}{15}$$

9. (b) **Statement 1**

$$I=\int_0^n \{x\}dx = n\int_0^1 xdx = n\left[\frac{x^2}{2}\right]_0^1 = \frac{n}{2}$$

$$\left[\because \int_0^{nT} f(x)dx = n\cdot\int_0^T f(x)dx;\text{ where T is the period of } f(x)\right]$$

Statement 2

$$\int_0^n [x]dx = \int_0^n (x-\{x\})dx$$

$$=\int_0^n xdx - \int_0^n \{x\}dx = \left[\frac{x^2}{2}\right]_0^n - n\int_0^1 x.dx$$

$$=\frac{n^2}{2} - n.\left[\frac{x^2}{2}\right]_0^1 = \frac{n}{2}(n-1)$$

10. (c) $\int_a^b xf(x)dx = \int_a^b (a+b-x)f(a+b-x)dx$

$$=(a+b)\int_a^b f(a+b-x)dx - \int_a^b xf(a+b-x)dx$$

$\therefore$ Statement 2 is true only when $f(a+b-x)=f(x)$, which holds in Statement 1.

$\therefore$ Statement 2 is false and Statement 1 is true.

11. (c) $\int_0^{\frac{\pi}{3}} \frac{dx}{8\cos^2 x+1} = \int_0^{\frac{\pi}{3}} \frac{\sec^2 x\, dx}{9+\tan^2 x}$

$$=\frac{1}{3}\tan^{-1}\left(\frac{\tan x}{3}\right)\Bigg|_0^{\frac{\pi}{3}} = \frac{1}{3}\tan^{-1}\left(\frac{1}{\sqrt{3}}\right) = \frac{\pi}{18}$$

$$\int_0^{\frac{2\pi}{3}} \frac{dx}{8\cos^2 x+1} = \int_0^{\frac{2\pi}{3}} \frac{\sec^2 x\, dx}{9+\tan^2 x}$$

$$=\int_0^{\frac{\pi}{3}} \frac{\sec^2 x\, dx}{9+\tan^2 x} + \int_{\frac{\pi}{3}}^{\frac{2\pi}{3}} \frac{\sec^2 x}{9+\tan^2 x}dx$$

[$\because$ the integral consists '$\tan x$' which is not defined at $x=\frac{\pi}{2}$]

$$=\frac{1}{3}\tan^{-1}\left(\frac{\tan x}{3}\right)\Bigg|_0^{\frac{\pi}{3}} + \frac{1}{3}\left|\tan^{-1}\left(\frac{\tan x}{3}\right)\right|_{\frac{\pi}{2}}^{\frac{2\pi}{3}}$$

$$=\frac{\pi}{6}+\frac{1}{3}\left(-\frac{\pi}{6}+\frac{\pi}{2}\right) = \frac{\pi}{6}+\frac{1}{3}\left(\frac{\pi}{3}\right) = \frac{5\pi}{18}$$

Thus Statement 1 is true and Statement 2 is false {because $\tan x$ is not continuous in $(0, 2\pi/3)$}.

12. (d) $\lim\limits_{x\to 2} \frac{4\int_0^{f(x)} t^3 dt}{x-2} = \lim\limits_{x\to 2} \frac{4.f'(x)f(x)^3}{1} = 32$

$\therefore$ Statement 1 is false.

Statement 2 is true, which is a standard result.

13. (a) In $0<x<\frac{\pi}{2}$

$\because f(x)=\frac{\sin x}{x}$ is decreasing function

$$\lim_{x\to 0} f(x) > f(x) > \lim_{x\to\frac{\pi}{2}} f(x)$$

$$\Rightarrow f\left(\frac{\pi}{2}\right) < \frac{\sin x}{x} < \lim_{x\to 0} f(x)$$

$$\Rightarrow \frac{2}{\pi} < \frac{\sin x}{x} \le 1$$

$$\Rightarrow \int_0^{\frac{\pi}{2}} \frac{2}{\pi}dx < \int_0^{\frac{\pi}{2}} \frac{\sin x}{x}dx \le \int_0^{\frac{\pi}{2}} 1dx$$

$$\Rightarrow \frac{2}{\pi}\times\frac{\pi}{2} < \int_0^{\frac{\pi}{2}} \frac{\sin x}{x}dx \le \frac{\pi}{2}$$

$\Rightarrow$ Statement 1 is correct

Statement 2 is also correct and explains Statement 1.

JEE Advanced

Objective Questions

One Option Correct

1. (b) $\frac{d}{dx}\left(A\ln|\cos x+\sin x-2|+Bx+C\right)$

$$=A\frac{\cos x-\sin x}{\cos x+\sin x-2}+B$$

$$=\frac{A\cos x - A\sin x + B\cos x + B\sin x - 2B}{\cos x+\sin x-2}$$

$$\therefore 2=A+B, -1=-A+B, \lambda=-2B$$

$$\therefore A=\frac{3}{2}, B=\frac{1}{2}, \lambda=-1$$

2. (d) $A\int \frac{x^3}{y}dx + B\int \frac{x^2}{y}dx + C\int \frac{x}{y}dx = x^2 y$

Differentiate both sides,

$$\frac{Ax^3+Bx^2+Cx}{y} = 2xy + x^2\frac{dy}{dx}$$

$$Ax^3+Bx^2+Cx=2xy^2+x^2y\frac{dy}{dx}$$

$$Ax^3+Bx^2+Cx=2x\left(x^2-x+1\right)+x^2\left(\frac{2x-1}{2}\right)$$

$$=2x^3-2x^2+2x+x^3-\frac{x^2}{2}$$

$A=3,\ B=-5/2,\ C=2$

3. (a) $\int 4\cos\left(x+\frac{\pi}{6}\right)\cos 2x\cos\left(\frac{5\pi}{6}+x\right)dx$

$$=2\int\left(\cos(2x+\pi)+\cos\frac{2\pi}{3}\right)\cos 2x\,dx$$

$$=2\int\left(-\cos 2x-\frac{1}{2}\right)\cos 2\,xdx$$

$$=\int\left(-2\cos^2 2x-\cos 2x\right)dx$$

$$=-\int(1+\cos 4x+\cos 2x)dx$$

$$=-x-\frac{\sin 4x}{4}-\frac{\sin 2x}{2}+c$$

$$=-\left(x+\frac{\sin 4x}{4}+\frac{\sin 2x}{2}\right)+c$$

4. (b) $\frac{2x}{(x-1)(x-4)}=\frac{D}{x-1}+\frac{E}{x-4}$

$$2x=D(x-4)+E(x-1)$$

$\therefore D=-2/3,\ E=8/3$

$$\therefore\int\frac{e^{x-1}}{(x-1)(x-4)}2x\,dx=\int e^{x-1}\left(\frac{-2/3}{x-1}+\frac{8/3}{x-4}\right)dx$$

$$=-\frac{2}{3}F(x-1)+\frac{8}{3}e^3F(x-4)+C$$

$\therefore\ A=-2/3,\ B=8/3\,e^3$

5. (a) $\int x^{\frac{2}{3}}.x^{-\frac{13}{6}}\left(x^{-1/2}+1\right)^{-13/3}dx$

$$=\int x^{-\frac{3}{2}}\left(x^{-1/2}+1\right)^{-13/3}dx$$

Put $\left(x^{-1/2}+1\right)=t,-\frac{1}{2}x^{-\frac{3}{2}}dx=dt$

$$=-2\int t^{-\frac{13}{3}}dt=-2\frac{t^{-\frac{10}{3}}}{\left(-\frac{10}{3}\right)}+C=\frac{3}{5}\left(x^{-\frac{1}{2}}+1\right)^{-\frac{10}{3}}+C$$

6. (d) We know

$$\log(1-x)=-\left\{x+\frac{x^2}{2}+\frac{x^3}{3}+\ldots\infty\right\}$$

Put x^2 instead of x.

$$\log\left(1-x^2\right)=-\left\{x^2+\frac{x^4}{2}+\frac{x^6}{3}+\ldots\infty\right\}$$

$$\Rightarrow x^2\log\left(1-x^2\right)=-\left\{x^4+\frac{x^6}{2}+\frac{x^8}{3}+\frac{x^{10}}{4}+\ldots\infty\right\}$$

Integrating both sides, we get

$$\int x^2\log\left(1-x^2\right).dx=-\left\{\frac{x^5}{1.5}+\frac{x^7}{2.7}+\frac{x^9}{3.9}+\ldots\infty\right\}$$

Now to find constant, put $x=0$

$0=0+c\Rightarrow c=0$

Applying integration by parts and taking limit 0 to 1,

$$\Rightarrow\left[\frac{x^3}{3}.\log\left(1-x^2\right)\right]_0^1-\int_0^1\frac{x^3}{3}.\frac{(-2x)}{1-x^2}.dx$$

$$\Rightarrow\left[\frac{x^3}{3}.\log\left(1-x^2\right)\right]_0^1+\frac{2}{3}\left[\frac{-x^3}{3}+\frac{1}{2}\log\left|\frac{1+x}{1-x}\right|\right]_0^1$$

Taking $\log\left(1-x^2\right)=\log(1+x)+\log(1-x)$

and $\log\left(\frac{1+x}{1-x}\right)=\log(1+x)-\log(1-x)$

$$\Rightarrow\frac{1}{3}\log 2+\frac{1}{3}\log 2-\frac{2}{3}-\frac{2}{9}+\lim_{x\to 1}\left(\frac{x^3}{3}-\frac{1}{3}\right)\log(1-x)$$

$$\Rightarrow\frac{2}{3}\log 2-\frac{8}{9}=\text{RHS.}$$

$$=\left\{\frac{1}{1.5}+\frac{1}{2.7}+\frac{1}{3.9}+\ldots\right\}=\frac{2}{3}\log(2)-\frac{8}{9}$$

7. (c) $I=\int\frac{\cos^4 xdx}{\sin^3 x\left(\sin^5 x+\cos^5 x\right)^{3/5}}$

$$=\int\frac{\cos^4 xdx}{\sin^6 x\left(1+\cot^5 x\right)^{3/5}}$$

$$=\int\frac{\cot^4 x\,\text{cosec}^2\,x\,dx}{\left(1+\cot^5 x\right)^{3/5}}\text{ put }1+\cot^5 x=t$$

$$\Rightarrow 5\cot^5 x\,\text{cosec}^2\,x\,dx=-dt$$

$$=-\frac{1}{5}\int\frac{dt}{t^{\frac{3}{5}}}=-\frac{1}{5}\frac{t^{-\frac{3}{5}+1}}{-\frac{3}{5}+1}+C=-\frac{1}{2}\left(1+\cot^5 x\right)^{\frac{2}{5}}+C$$

$\therefore A=5, B=\frac{2}{5}$ and $AB=2.$

8. (a) $I_n = \int (\log x)^n dx$

$\therefore I_{n-1} = \int (\log x)^{n-1} dx$

Now $I_n = \int (\log x)^n 1 \cdot dx$

$= (\log x)^n \cdot x - n\int (\log x)^{n-1} \cdot \frac{1}{x} \cdot x\, dx$

$I_n = n(\log x)^n - n\, I_{n-1}$

$I_n + n I_{n-1} = n(\log x)^n$

9. (a) $A = \begin{bmatrix} x & x \\ x & x \end{bmatrix} \Rightarrow A^2 = \begin{bmatrix} 2x^2 & 2x^2 \\ 2x^2 & 2x^2 \end{bmatrix}$,

$A^3 = \begin{bmatrix} 2^2 x^2 & 2^2 x^2 \\ 2^2 x^2 & 2^2 x^2 \end{bmatrix}$... so on

$e^A = I + A + \frac{A^2}{2!} + \frac{A^3}{3!} + \ldots$

$$e^A = \begin{bmatrix} 1 + x + \frac{2x^2}{2!} + \frac{2^2 x^3}{3!} + \ldots & x + \frac{2x^2}{2!} + \frac{2^2 x^3}{3!} + \ldots \\ x + \frac{2x^2}{2!} + \frac{2^2 x^3}{3!} + \ldots & 1 + x + \frac{2x^2}{2!} + \frac{2^2 x^3}{3!} + \ldots \end{bmatrix}$$

$= \frac{1}{2}\begin{bmatrix} e^{2x}+1 & e^{2x}-1 \\ e^{2x}-1 & e^{2x}+1 \end{bmatrix}$

$\Rightarrow f(x) = e^{2x} + 1$ and $g(x) = e^{2x} - 1$

Now, solve all the options.

10. (c) $\int e^{(x\sin x + \cos x)}\left(\frac{x^4 \cos^3 x - x\sin x + \cos x}{x^2 \cos^2 x}\right).dx$

$= \int \left(e^{(x\sin x + \cos x)} x^2 \cos x\right) dx -$

$\int \underset{\text{I}}{e^{x\sin x + \cos x}} \underset{\text{II}}{\left(\frac{x\sin x - \cos x}{(x\cos x)^2}\right)}.dx$

Applying integer by parts,

$= \left\{x.e^{x\sin x + \cos x} - \int e^{x\sin x + \cos x}.dx\right\} -$

$\left\{e^{x\sin x + \cos x}.\frac{1}{x\cos x} - \int e^{x\sin x + \cos x}.dx\right\}$

$= e^{x\sin x + \cos x}\left\{x - \frac{1}{x\cos x}\right\} + C$

11. (d) $fog(x) = \sqrt{e^x - 1}$

$\therefore I = \int \sqrt{e^x - 1}dx = \int \frac{2t^2}{t^2+1}dt \left\{\text{where} \sqrt{e^x - 1} = t\right\}$

$= 2t - 2\tan^{-1} t + C$

$= 2\sqrt{e^x - 1} - 2\tan^{-1}\left(\sqrt{e^x - 1}\right) + C$

$= 2\, fog\,(x) - 2\tan^{-1}\left(fog(x)\right) + C$

$\therefore A + B = 2 + (-2) = 0$

12. (c) Here, LCM of 3 and 6 is 6.

$\therefore$ we put $x = t^6$

$\therefore dx = 6t^5 dt$

Hence, $I = 6\int \frac{(t^6 + t^4 + t)}{t^6(1+t^2)} t^5 dt$

$= 6\int \frac{t^5 + t^3 + 1}{1+t^2} dt = 6\int t^3 dt + 6\int \frac{dt}{1+t^2}$

$= \frac{3}{2}t^4 + 6\tan^{-1} t + C = \frac{3}{2}x^{2/3} + 6\tan^{-1} x^{1/6} + C$

13. (b) $\mathbf{S_1}$: $a = 0$

$\Rightarrow \int \sec x dx = \log\left(\tan\frac{x}{2}\right) + C$

$\Rightarrow f$ is a log function for $a = 0$.

$\mathbf{S_2}: |a| < 1 \Rightarrow \frac{2}{1-a}\int \frac{\frac{1}{2}\sec^2 \frac{x}{2} dx}{\left(\frac{a+1}{1-a}\right) - \tan^2 \frac{x}{2}}$

$= \frac{2}{1-a}\int \frac{dt}{\left(\frac{a+1}{1-a}\right) - t^2}$

$\Rightarrow \frac{1}{\sqrt{1-a^2}} \ln\left(\frac{\sqrt{a+1} + \sqrt{1-a} \times \tan\frac{x}{2}}{\sqrt{a+1} - \sqrt{1-a} \times \tan\frac{x}{2}}\right) + c,$

f is a log function for $|a| < 1$.

$\mathbf{S_3}: a = 1 \Rightarrow \int \frac{dx}{1+\cos x} = \int \frac{1}{2}\sec^2 \frac{x}{2} dx$

$= \tan\frac{x}{2} + C$

$\Rightarrow f$ is a polynomial for $a = 1$

$\mathbf{S_4}$: False

14. (c) $S_1 : I = \int e^x\left(\ln x+\frac{1}{x^2}\right)dx$

$$= \int e^x\left(\ln x+\frac{1}{x}-\frac{1}{x}+\frac{1}{x^2}\right)dx$$

$$= e^x\left(\ln x-\frac{1}{x}\right)+C$$

$$S_2 : I = \int\frac{\sin\frac{x}{2}dx}{\left(1-\cos^2\frac{x}{2}\right)\sqrt{\cos\frac{x}{2}}}$$

Put $\cos\frac{x}{2}=t^2$, then $\sin\frac{x}{2}dx = -4t\,dt$

$$I = \int\frac{-4t\,dt}{\left(1-t^4\right).t} = 4\int\frac{dt}{t^4-1}$$

$$= 2\int\left(\frac{1}{t^2-1}-\frac{1}{t^2+1}\right)dt$$

$$= 2.\left[\frac{1}{2.1}\ln\left|\frac{t-1}{t+1}\right|-\tan^{-1}t\right]+C$$

$$= \ln\left|\frac{\sqrt{\cos\frac{x}{2}}-1}{\sqrt{\cos\frac{x}{2}}+1}\right|-2\tan^{-1}\left(\sqrt{\cos\frac{x}{2}}\right)+C$$

$$S_3 : I = \int\frac{dx}{\sin x\cos^3 x} = \int\frac{\sec^4 x}{\tan x}dx$$

Put $\tan x = t$, then $\sec^2 x dx = dt$

$$I = \int\frac{\left(1+t^2\right).dt}{t} = \int\frac{1}{t}dt+\int t dt = \ln t+\frac{t^2}{2}+C$$

$$= \ln\tan x+\frac{(\tan x)^2}{2}+C$$

$$S_4 : I = \int\frac{\sin x+\cos x}{\sin 2x+3}dx.$$

Put $\sin x-\cos x = t$, then $(\sin x+\cos x)dx = dt$

$1-\sin 2x = t^2 \Rightarrow \sin 2x = 1-t^2$

$$\therefore I = \int\frac{dt}{4-t^2} = \frac{1}{2.2}\ln\left|\frac{2+t}{2-t}\right|+C$$

$$= \frac{1}{4}\ln\left|\frac{2+\sin x-\cos x}{2-\sin x+\cos x}\right|+C$$

15. (d) $\int_0^{\pi}\frac{\sin x}{1+\cos^2 x}dx = 2\int_0^{\pi/2}\frac{\sin x}{1+\cos^2 x}dx$

$$= -2\int_0^{\pi/2}\frac{(-\sin x)dx}{1+\cos^2 x} = -2\left[\tan^{-1}(\cos x)\right]_0^{\pi/2}$$

$$= -2\left(\tan^{-1}0-\tan^{-1}1\right) = \pi/2$$

$$\therefore \frac{\cos\alpha}{1+\sin^2\alpha} = \frac{1}{2} \Rightarrow 2\cos\alpha = 2-\cos^2\alpha$$

$$\Rightarrow \cos^2\alpha+2\cos\alpha-2 = 0$$

$$\cos\alpha = \frac{-2\pm\sqrt{4+8}}{2} = -1\pm\sqrt{3}$$

$$\Rightarrow \cos\alpha = \sqrt{3}-1$$

$\therefore \alpha$ takes exactly one value in $(0, \pi/2)$.

16. (b) $p = \lim\limits_{n\to\infty}\left[\frac{\prod_{r=1}^{n}\left(n^3+r^3\right)}{n^{3n}}\right]^{1/n}$

$$\Rightarrow \ln p = \lim_{n\to\infty}\frac{1}{n}\sum_{r=1}^{n}\ln\left(1+\left(\frac{r}{n}\right)^3\right)$$

$$= \int_0^1\ln\left(1+x^3\right)dx = \ln 2-3+3\lambda$$

17. (d) $I = \int_0^{2\pi}\frac{e^{|\sin x|}\cos x}{1+e^{\tan x}}dx$ ...(i)

$$I = \int_0^{2\pi}\frac{e^{|\sin x|}\cos x}{1+e^{-\tan x}}dx \quad \text{...(ii)}$$

$$\left(\text{using } \int_0^a f(x)dx = \int_0^a f(a-x)dx\right)$$

Adding (i) and (ii), we get

$$2I = \int_0^{2\pi}e^{|\sin x|}\cos x dx = 2\int_0^{\pi}e^{\sin x}\cos x dx$$

$$\{\because f(2a-x) = f(x)\}$$

$$\therefore I = 0\ \left(f(a-x) = -f(x)\right)$$

18. (c) Since $0<x<1 \Rightarrow x>x^2>\frac{x^2}{2}$

$$\Rightarrow -x<-x^2<-\frac{x^2}{2} \Rightarrow e^{-x}<e^{-x^2}<e^{-\frac{x^2}{2}}$$

$$\Rightarrow e^{-x}\cos^2 x<e^{-x^2}\cos^2 x<e^{-\frac{x^2}{2}}\cos^2 x<e^{-\frac{x^2}{2}}$$

$$\therefore \int_0^1 e^{-x}\cos^2 x\,dx<\int_0^1 e^{-x^2}\cos^2 x\,dx<$$

$$\int_0^1 e^{-\frac{x^2}{2}}\cos^2 x\,dx<\int_0^1 e^{-\frac{x^2}{2}}dx$$

$$\therefore I_1<I_2<I_3<I_4$$

19. (a) Let $y=\sqrt{x+y}$

$y^2-y-x=0$

$y=\dfrac{1\pm\sqrt{1+4x}}{2.1}\quad y>1\Rightarrow y=\dfrac{1+\sqrt{1+4x}}{2}$

$$\int_0^2 \frac{1+\sqrt{1+4x}}{2}dx=\left[\frac{x}{2}+\frac{(1+4x)^{3/2}}{\frac{3}{2}.2.4}\right]_0^2$$

$$=\left[\left(1+\frac{27}{12}\right)-\left(0+\frac{1}{12}\right)\right]$$

$$=1+\frac{26}{12}=\frac{19}{6}$$

20. (a) $f(x)=e^x+\int_0^1 e^x f(t)\,dt=e^x+ke^x$

where $k=\int_0^1 f(t)\,dt$

$\therefore k=\int_0^1\left(e^t+ke^t\right)dt=e+ke-1-k$

$\therefore k=\dfrac{e-1}{2-e};$

Thus $f(x)=e^x\left(1+\dfrac{e-1}{2-e}\right)=\dfrac{e^x}{2-e}$

21. (b) Putting $x=-t$ in m,

$$m=\int_2^0\frac{|\sin(-t)|}{\left[\frac{-t}{\pi}\right]+\frac{1}{2}}(-dt)=\int_0^2\frac{|\sin t|}{\left[\frac{-t}{\pi}\right]+\frac{1}{2}}dt$$

$$=\int_2^0\frac{|\sin t|}{-\left[\frac{t}{\pi}\right]-1+\frac{1}{2}}dt\ \{\text{If } x\notin I;\text{then } [-x]=-[x]-1\}$$

$$=-\int_0^2\frac{|\sin t|}{\left[\frac{t}{\pi}\right]+\frac{1}{2}}dt=-n$$

22. (c) $I_{10}=\int_1^e 1\cdot(\ln x)^{10}\,dx$

$$=\left[(\ln x)^{10}x\right]_1^e-\int_1^e 10(\ln x)^9\cdot\frac{1}{x}\cdot x\,dx$$

$$=e-0-10\int_1^e(\ln x)^9\,dx=e-10\,I_9$$

$I_{10}+10I_9=e$

23. (d) $\displaystyle\lim_{n\to\infty}\frac{1}{\sqrt{4n^2}}+\frac{1}{\sqrt{4n^2-1^2}}+\frac{1}{\sqrt{4n^2-2^2}}+\ldots+\frac{1}{\sqrt{4n^2-(n-1)^2}}$

$$=\lim_{n\to\infty}\sum_{r=0}^{n-1}\frac{1}{\sqrt{4n^2-r^2}}=\lim_{n\to\infty}\sum_{r=0}^{n-1}\frac{1}{n}\cdot\frac{1}{\sqrt{4-\left(\frac{r}{n}\right)^2}}$$

$$=\int_0^1\frac{dx}{\sqrt{4-x^2}}=\left(\sin^{-1}\frac{x}{2}\right)_0^1=\frac{\pi}{6}$$

24. (b) $I_1=\int_0^1\dfrac{te^{t^2}}{t^2+1}dt$

Put $t^2=u$, then $t=0,u=0;t=1,u=1$ and $2t\,dt=du$

$\therefore I=\dfrac{1}{2}\int_0^1\dfrac{e^u\,du}{1+u}=\dfrac{1}{2}A$

25. (c) $f(x)=\int_1^x\dfrac{1}{t}dt$

$f'(x)=\dfrac{1}{x}$

$\therefore f(x)=\ln x+c$

at $x=1$, $f(1)=\ln 1+c\ \therefore\ c=0$ [here $f(1)=0$]

$\therefore f(x)=\ln x$

$\therefore f(xy)=\ln(xy)=\ln x+\ln y=f(x)+f(y)$

26. (c) $\therefore I=\int_a^{a+p}f(x)dx$

$$=\int_a^0 f(x)dx+\int_0^p f(x)dx+\int_p^{a+p}f(x)\,dx$$

$$=\int_a^0 f(x)dx+\int_0^p f(x)dx+\int_0^a f(x)dx$$

$$=\int_0^p f(x)dx$$

$\therefore I$ is independent of a.

27. (d) $\int_{-\pi}^{\pi}(\cos px-\sin qx)^2\,dx$

$$=\int_{-\pi}^{\pi}\left(\cos^2 px+\sin^2 qx-2\cos px\sin qx\right)dx,$$

where $\cos px.\sin px$ is an odd function, $f(x)=\cos px.\sin px$

$$= 2\int_0^{\pi}\left(\frac{1+\cos 2px}{2}+\frac{1-\cos 2qx}{2}\right)dx$$

$$= \int_0^{\pi}(2+\cos 2px-\cos 2qx)$$

$$= 2\pi + \left(\frac{\sin 2px}{2p}-\frac{\sin 2qx}{2q}\right)_0^{\pi} = 2\pi$$

28. (a) $I = \int_0^1 \tan^{-1}\left(\frac{2x-1}{1+x-x^2}\right)dx$

$$= \int_0^1 \tan^{-1}\frac{x-(1-x)}{1+x(1-x)}\,dx$$

$$= \int_0^1 \left(\tan^{-1}x - \tan^{-1}(1-x)\right)dx$$

$$= \int_0^1 \tan^{-1} x dx - \int_0^1 \tan^{-1}(1-1+x)dx = 0$$

29. (b) Put $x = a\cos^2\theta + b\sin^2\theta$

$\Rightarrow dx = 2(b-a)\sin\theta\cos\theta d\theta$

$$\therefore I = \int_a^b (x-a)^3(b-x)^4\,dx$$

$$= 2(b-a)^8\int_0^{\pi/2}\sin^7\theta\cos^2\theta d\theta$$

$$= 2(b-a)^8\int_0^{\pi/2}\sin^7\theta\left(1-\sin^2\theta\right)^4\cos\theta d\theta$$

Let $\sin\theta = t \Rightarrow \cos\theta d\theta = dt$

$$\Rightarrow I = 2(b-a)^8\int_0^1 t^7\left(1-t^2\right)^4 dt$$

$$= 2(b-a)^8\int_0^1 t^7\left(1-4t^2+6t^4-4t^6+t^8\right)dt$$

$$= \frac{(b-a)^8}{280}$$

30. (a) $\left|\int_{10}^{19}\frac{\sin x}{1+x^8}dx\right| \le \int_{10}^{19}\frac{|\sin x|}{1+x^8}dx$

$\because \left|\int f(x)dx\right| \le \int |f(x)|dx.$

Now, $\sin x \le 1$ and $1+x^8 > x^8$

$\therefore \frac{1}{1+x^8} < \frac{1}{x^8}$

$\therefore \frac{|\sin x|}{1+x^8} < \frac{1}{x^8}.$

So $\left|\int_{10}^{19}\frac{\sin x}{1+x^8}dx\right| \le \int_{10}^{19}\frac{|\sin x|}{1+x^8}dx < \int_{10}^{19}\frac{1}{x^8}dx < \int_{10}^{19}\frac{1}{x^8}dx$

$= 9\times10^{-8} < 10\times10^{-8} < 10^{-7}$

31. (b) $I = \int_0^a \ln\left(\frac{\cos a}{\sin a}+\frac{\sin x}{\cos x}\right)dx$

$$= \int_0^a \ln\left(\frac{\cos a\cdot\cos x+\sin a\cdot\sin x}{\sin a\cdot\cos x}\right)dx$$

$$= \int_0^a \ln\left(\frac{\cos(a-x)}{\sin a\cos x}\right)dx$$

$$= \int_0^a \ln\{\cos(a-x)\}dx - \int_0^a \ln\sin a\,dx - \int_0^a \ln\cos xdx$$

$$= \int_0^a \ln\cos\{a-(a-x)\}dx - \int_0^a \ln\sin adx - \int_0^a \ln\cos xdx$$

$$= \int_0^a \ln\cos xdx - \int_0^a \ln\sin adx - \int_0^a \ln\cos xdx$$

$= -a\ln(\sin a)$

32. (d) $\mathbf{S_1}$: True

$\cot^{-1}x = t$

When $x=-1$ $\quad t = \frac{3\pi}{4}.$

When $x = 1$ $\quad t = \frac{\pi}{4}$. True.

$\mathbf{S_2}$: True

$\mathbf{S_3}$: False (The discontinuity may not be removable.)

$\mathbf{S_4}$: False

33. (c) $\mathbf{S_1}$: True

$$\int_{-1}^{\infty}\frac{d}{dx}\left(x\sin\frac{1}{x}\right)dx = \int_{-1}^{0}\frac{d}{dx}\left(x\sin\frac{1}{x}\right)dx + \int_0^{\infty}\frac{d}{dx}\left(x\sin\frac{1}{x}\right)dx$$

$$= x\sin\frac{1}{x}\bigg|_{-1}^{0} + \left(x\sin\frac{1}{x}\right)\bigg|_0^{\infty}$$

$$= 0-(-\sin(-1)) + \lim_{x\to\infty} x\sin\frac{1}{x} - \lim_{x\to 0} x\sin\frac{1}{x}$$

$= -\sin 1 + 1 - 0 = 1 - \sin 1$

$$= 1-\cos\left(\frac{\pi}{2}-1\right) = 2\sin^2\left(\frac{\pi}{4}-\frac{1}{2}\right)$$

True

S$_2$: True

$$\int_0^{10\pi} \sin^2 x\cos^4 x dx = 10\int_0^{\pi} \sin^2 x\cos^4 x dx$$

$$= 20\int_0^{\pi/2} \sin^2 x\cos^4 x dx$$

$$= 20\times\frac{1}{6}\times\frac{3}{4}\times\frac{1}{2}.\frac{\pi}{2} = \frac{5\pi}{8}$$

True

S$_3$: True

$$f(x) = \int_0^x f(t)dt$$

$$f(-x) = \int_0^{-x} f(t)dt$$

Put $t = -u = \int_0^x f(-u)(-du) = -\int_0^x f(u)\,du$

$$= -\int_0^x f(t)dt$$

$\therefore f(-x) = f(x)$ so $f(x)$ is an odd function.

S$_4$: False

Consider $f(x) = x^2$,

then $f(x)$ is even and $\int_0^x f(t)dt$

$= \int_0^x t^2 dt = \left.\frac{t^3}{3}\right|_0^x = \frac{x^3}{3}$ is an odd function.

$\therefore f(x) = 0$ i.e., $x^2 = 0$ has only two real roots.

34. (c) $S_1 : \int_0^{\infty} \frac{\log(1+x^2)}{1+x^2} dx = \int_0^{\pi/2} \log\sec^2\theta d\theta$

$$= -2\int_0^{\pi/2} \log\cos\theta d\theta$$

$$S_2 : I = \int_0^{\pi} e^{\sin^2 x}\cos^3(2n+1)xdx$$

$$= \int_0^{\pi} e^{\sin^2 x}\cos^3\{(2n+1)(\pi-x)\}dx$$

$$= -\int_0^{\pi} e^{\sin^2 x}\cos^3(2n+1)xdx = -I$$

$\therefore I = 0$

$$S_3 : \int_0^4 |x-1|dx = \int_0^1 (1-x)dx + \int_1^4 (x-1)dx = 5$$

$$S_4 : \int_{-2}^{2}\left(1-x^2\right)dx = 2\int_0^2\left(1-x^2\right)dx = -\frac{4}{3}$$

35. (a) Let $I = \int_{\frac{\pi}{4}}^{\frac{\pi}{2}} (2\operatorname{cosec} x)^{17}\, dx$

$$= \int_{\frac{\pi}{4}}^{\frac{\pi}{2}} (\operatorname{cosec} x + \cot x + \operatorname{cosec} x - \cot x)^{16}$$

$2\operatorname{cosec} x dx$

$$I = 2\int_{\frac{\pi}{4}}^{\frac{\pi}{2}}\left(\operatorname{cosec} x + \cot x + \frac{1}{\operatorname{cosec} x + \cot x}\right)^{16}.$$

$\operatorname{cosec} x dx$

Let $\operatorname{cosec} x + \cot x = e^u$

$\Rightarrow \left(-\operatorname{cosec} x\cot x - \operatorname{cosec}^2 x\right)dx = e^u du$

$\Rightarrow -\operatorname{cosec} x\, dx = du$

Also at $x = \frac{\pi}{4}, u = \ln\left(\sqrt{2}+1\right)$

and at $x = \frac{\pi}{2}, u = \ln 1 = 0$

$$\therefore I = -2\int_{\ln(\sqrt{2}+1)}^{0} \left(e^u + e^{-u}\right)^{16} du$$

$$= 2\int_0^{\ln(\sqrt{2}+1)} \left(e^u + e^{-u}\right)^{16} du$$

36. (d) We have $f'(x) - 2f(x) < 0$

$\Rightarrow e^{-2x}f'(x) - 2e^{-2x}f(x) < 0$

$\Rightarrow \frac{d}{dx}\left(e^{-2x}f(x)\right) < 0$

$\Rightarrow e^{-2x}f(x)$ is strictly decreasing function on $\left[\frac{1}{2},1\right]$

$\therefore e^{-2x}f(x) < e^{-1}f\left(\frac{1}{2}\right)$ or $f(x) < e^{2x-1}$

Also, given that $f(x)$ is positive function so $f(x) > 0$

$\therefore 0 < f(x) < e^{2x-1}$

$\Rightarrow 0 < \int_{1/2}^{1} f(x)dx < \int_{1/2}^{1} e^{2x-1}dx$

$\Rightarrow 0 < \int_{1/2}^{1} f(x)dx < \left[\frac{e^{2x-1}}{2}\right]_{1/2}^{1}$

$\Rightarrow \int_{1/2}^{1} f(x)dx \in \left(0, \frac{e-1}{2}\right)$

One or More Than One Option Correct

1. (b, c) We have,

$f(x) = \ln x + \int_0^x \sqrt{1+\sin t}\,dt$

$\Rightarrow f'(x) = \frac{1}{x} + \sqrt{1+\sin x}$

which exists $\forall\, x \in (0, \infty)$

and $f'(x)$ has finite value $\forall\, x \in (0, \infty)$, so $f'(x)$ is continuous.

Also, $f''(x) = -\frac{1}{x^2} + \frac{\cos x}{2\sqrt{1+\sin x}}$

which does not exist at the points where

$\sin x = -1$ like $x = \frac{3\pi}{2}, \frac{7\pi}{2}, \ldots$

$\therefore f'(x)$ is not differentiable.

$\therefore$ (a) is false but (b) is true.

Now $\sqrt{1+\sin t} \ge 0 \Rightarrow \int_0^x \sqrt{1+\sin t}\,dt \ge 0\ \forall\, x \in (0, \infty)$

and $\ln x > 0 \forall\, x \in (1, \infty) \Rightarrow f(x) > 0 \forall x \in (1, \infty)$.

For $x \ge e^3$

$f(x) = \ln x + \int_0^x \sqrt{1+\sin t}\,dt \ge 3$

$f'(x) = \frac{1}{x} + \sqrt{1+\sin x} \le \frac{1}{x} + \sqrt{2}, \forall x > 0$

Now for $x \ge e^3$

$\Rightarrow 0 < f'(x) \le \frac{1}{x} + \sqrt{2} < \frac{1}{e^3} + \sqrt{2} < 3\ \forall\, x \in (e^3, \infty)$

$\Rightarrow |f'(x)| < |f(x)|$

$\therefore$ (c) is true.

Also $\lim_{x\to\infty} f(x) = \infty$

$\therefore |f(x)| + |f'(x)|$ is not bounded.

$\therefore$ (d) is wrong.

2. (a) $\int_0^1 \frac{x^4(1-x)^4}{1+x^2}dx$

$= \int_0^1 \left(x^6 - 4x^5 + 5x^4 - 4x^2 + 4 - \frac{4}{1+x^2}\right)dx$

$= \left[\frac{x^7}{7} - \frac{2x^6}{3} + x^5 - \frac{4x^3}{3} + 4x - 4\tan^{-1} x\right]_0^1$

$= \frac{22}{7} - \pi$

3. (a, b, c) We have

$$I_n = \int_{-\pi}^{\pi} \frac{\sin nx}{(1+\pi^x)\sin x}dx \qquad \text{...(i)}$$

$\Rightarrow I_n = \int_{-\pi}^{\pi} \frac{\sin n(-x)}{(1+\pi^{-x})\sin(-x)}dx$

$\left[\because \int_a^b f(x)dx = \int_a^b f(a+b-x)dx\right]$

$$\Rightarrow I_n = \int_{-\pi}^{\pi} \frac{\pi^x \sin nx}{(1+\pi^x)\sin x}dx \qquad \text{...(ii)}$$

Adding equations (i) and (ii), we get

$2I_n = \int_{-\pi}^{\pi} \frac{\sin nx}{\sin x}dx = 2\int_{-\pi}^{\pi} \frac{\sin nx}{\sin x}dx$

[since the integrand is an even function]

$\Rightarrow I_n = \int_0^{\pi} \frac{\sin nx}{\sin x}dx$

Now $I_{n+2} - I_n = \int_0^{\pi} \frac{\sin(n+2)x - \sin nx}{\sin x}dx$

$= 2\int_0^{\pi} \cos(n+1)x\,dx = 2\left[\frac{\sin(n+1)x}{n+1}\right]_0^{\pi} = 0$

$\therefore I_{n+2} = I_n$

Also, $I_1 = \int_0^{\pi} 1dx = \pi$ and $I_0 = 0$

Hence $\sum_{m=1}^{10} I_{2m+1} = I_3 + I_5 + I_7 + \ldots + I_{21}$

$= 10I\,(\because I_{n+2} = I_n) = 10\pi$

and $\sum_{m=1}^{10} I_{2m} = I_2 + I_4 + I_6 + \ldots + I_{20}$

$= 20 \times I_0 \qquad (\because I_{n+2} = I_n)$

$= 20 \times 0 = 0$

4. (a, b, c, d) $\therefore f(x)$ is a non-constant twice differentiable function such that

$f(x)=f(1-x) \Rightarrow f'(x)=-f'(1-x)$ ...(1)

For $x=\frac{1}{2}$, we get $f'\left(\frac{1}{2}\right)=-f'\left(1-\frac{1}{2}\right)$

$\Rightarrow f'\left(\frac{1}{2}\right)+f'\left(\frac{1}{2}\right)=0 \quad [\because f(x)=f(1-x)]$

$\Rightarrow f'\left(\frac{1}{2}\right)=0$

$\Rightarrow$ (b) is correct.

For $x=\frac{1}{4}$, we get $f'\left(\frac{1}{4}\right)=-f'\left(\frac{3}{4}\right)$

$[\because f(x)=f(1-x)]$.

But given that $f'\left(\frac{1}{4}\right)=0$

$\therefore f'\left(\frac{3}{4}\right)=f'\left(\frac{1}{4}\right)=0$

Hence, $f'(x)$ satisfies all conditions of Rolle 's Theorem for

$f''(a_1)=0$ and $f''(b_2)=0$

$\therefore f''(x)$ variables at least twice on [0, 1] $\Rightarrow$ (a) is correct.

Also, using $f(x)=f(1-x)$

$\Rightarrow f\left(x+\frac{1}{2}\right)=f\left(1-x-\frac{1}{2}\right)=f\left(-x+\frac{1}{2}\right)$

$\Rightarrow f\left(x+\frac{1}{2}\right)$ is an even function.

$\Rightarrow \sin x . f\left(x+\frac{1}{2}\right)$ is an odd function.

$\Rightarrow \int_{-1/2}^{1/2} f\left(x+\frac{1}{2}\right)\sin x dx=0,|$

(c) is correct.

5. (a, c) $I=\int_0^{\pi/2}\frac{\cos\theta-\sin\theta}{(1+\cos\theta)(1+\sin\theta)}d\theta$

$=\int_0^{\pi/2}\frac{\sin\theta-\cos\theta}{(1+\sin\theta)(1+\cos\theta)}d\theta$

$\left[\text{using }\int_a^b f(x)dx=\int_a^b f(a+b-x)dx\right]$

$2I=0 \Rightarrow I=0$

6. (a, b) $(i) I=\int_0^1\frac{\sin^{-1}\sqrt{x}}{x^2-x+1}$ ...(1)

$I=\int_0^1\frac{\sin^{-1}\sqrt{1-x}}{x^2-x+1}dx=\int_0^1\frac{\cos^{-1}\sqrt{x}}{x^2-x+1}dx$ ...(2)

$\left(\text{Applying }\int_a^b f(x)dx=\int_a^b f(a+b-x)dx\right)$

$\therefore$ On adding (1) and (2), we get

$$2I=\int_0^1\frac{\sin^{-1}\sqrt{x}+\cos^{-1}\sqrt{x}}{x^2-x+1}dx$$

$$=\frac{\pi}{2}\int_0^1\frac{dx}{x^2-x+1}dx$$

$$=\frac{\pi}{2}\int_0^1\frac{dx}{\left(x-\frac{1}{2}\right)^2+\left(\frac{\sqrt{3}}{2}\right)^2}dx$$

$$2I=\frac{\pi}{2}\frac{1}{\left(\frac{\sqrt{3}}{2}\right)}\left[\tan^{-1}\left(\frac{2x-1}{\sqrt{3}}\right)\right]_0^1=\frac{\pi^2}{3\sqrt{3}}$$

Hence $I=\frac{\pi^2}{6\sqrt{3}}=\frac{\pi^2}{\sqrt{108}}\equiv\frac{\pi^2}{\sqrt{n}}\Rightarrow n=108$

7. (b, c) We have $J=\int_{-5}^{-4}(3-x^2)\tan(3-x^2)\,dx$

$=\int_{-2}^{-1}\left(3-(x-3)^2\right)\tan\left(3-(x-3)^2\right)dx$

$=-K$

8. (b, c)

(A) $\lim_{n\to\infty}\frac{1}{n}\sum_{r=1}^{n}f\left(\frac{r}{n}\right)=\int_0^1 f(x)dx$

(B) $\lim_{n\to\infty}\frac{1}{n}\sum_{r=n+1}^{2n}f\left(\frac{r}{n}\right)=\int_1^2 f(x)dx$

(C) $\lim_{n\to\infty}\frac{1}{n}\sum_{r=1}^{n}f\left(\frac{r+n}{n}\right)=\int_0^1 f(1+x)dx$

$=\int_1^2 f(t)dt=\int_1^2 f(x)dx$

(D) $\lim_{n\to\infty}\frac{1}{n}\sum_{r=1}^{2n}f\left(\frac{r}{n}\right)=\int_1^2 f(x)dx$

9. (a, b) $I_n = \int_0^1 \frac{dx}{(1+x^2)^n} = \int_0^1 (1+x^2)^{-n} dx$

$$= \left.\frac{x}{(1+x^2)^n}\right|_0^1 - \int_0^1 (-n)(1+x^2)^{-n-1}.2x.xdx$$

$$= \frac{1}{2^n} + 2n\int_0^1 \frac{x^2}{(1+x^2)^{n+1}}$$

$$= \frac{1}{2^n} + 2n\int_0^1 \frac{1+x^2-1}{(1+x^2)^{n+1}} dx$$

$$= \frac{1}{2^n} + 2nI_n - 2nI_{n+1}$$

$$\therefore 2nI_{n+1} = 2^{-n} + (2n-1)I_n$$

$$\therefore 2I_2 = \frac{1}{2} + I_1 + \frac{1}{2} + \tan^{-1} x\Big|_0^1$$

$$I_2 = \frac{1}{4} + \frac{\pi}{8}$$

10. (a,b,d) $f(2-x) = f(2+x), f(4-x) = f(4+x)$

$$f(4+x) = f(4-x) = f(2+2-x)$$

$$= f(2-(2-x)) = f(x)$$

$\therefore$ 4 is a period of $f(x)$.

$$\int_0^{50} f(x)dx = \int_0^{48} f(x)dx + \int_0^2 f(x)dx$$

$$= 12\int_0^4 f(x)dx + \int_0^2 f(x)dx$$

$$= 12\left(\int_0^2 f(x)dx + \int_0^2 f(4-x)dx\right)$$

$$= 12\left(\int_0^2 f(x)dx + \int_0^2 f(4+x)dx\right) + 5$$

$$= 24\int_0^2 f(x)dx + 5 = 125$$

Also, $\int_0^{50} f(x)dx = \int_{-4}^{48} f(x)dx = \int_2^{52} f(x)dx = 125$

11. (a, d) $F'(x) = -\frac{2}{x^3}\int_4^x \{4t^2 - 2F'(t)\}dt + \frac{1}{x^2}(4x^2 - 2F'(x))$.

Put $x = 4$

$$F'(4) = 0 + \frac{1}{16}[4(16) - 2F'(4)] = 4 - \frac{F'(4)}{8}$$

$$\Rightarrow F'(4) = \frac{32}{9}$$

$\therefore$ Option (a) is correct

For option (d),

$$F(x) = \frac{1}{x^2}\int_4^k [4t^2 - 2F'(t)]dt$$

$$= \frac{1}{x^2}\left[\frac{4t^3}{3} - 2F(t)\right]_4^x$$

$$\therefore F(8) = \frac{1}{8^2}\left[\frac{4.8^3}{3} - 2F(8) - \frac{4.4^3}{3} + 2F(4)\right]$$

$$= \frac{32}{3} - \frac{F(8)}{32} - \frac{4}{3} + 2(0)$$

$$\therefore F(8) = \frac{32}{33}\left\{\frac{28}{3}\right\} \qquad \therefore F'(4) = \frac{11F(8)}{28}$$

12. (a, c, d) $x \in \left(0, \frac{\pi}{2}\right) \Rightarrow 0 < \sin x < 1$

$$\Rightarrow -\alpha < -\alpha \sin x < 0$$

$$\Rightarrow e^{-\alpha} < e^{-\alpha \sin x} < e^0 = 1$$

$$\Rightarrow \int_0^{\pi/2} e^{-\alpha} dx < \int_0^{\pi/2} e^{-\alpha \sin x} dx < \int_0^{\pi/2} 1\, dx$$

$$\Rightarrow \frac{\pi}{2} e^{-\alpha} < I < \frac{\pi}{2}$$

13. (a, b, c)

(a) Let $g(x) = \int\int_0^{\pi/3} (x)dx - 3\int_0^{\pi/3} \cos 3xdx = 0$

$g(x)$ is a continuous and differentiable function and $g(0) = 0, g\left(\frac{\pi}{3}\right) = 0$

By Rolle's Theorem,

$g'(x) = f(x) - 3\cos x = 0$ has atleast one solution in $\left(0, \frac{\pi}{3}\right)$

(b) Let $h(x) = f(x) - 3\sin 3x + \frac{6}{x}$

Let $h(x) = \int_0^{\pi/3} f(x)dx - 3\int_0^{\pi/3} \sin 3xdx + \int_0^{\pi/3} \frac{6}{\pi}dx$

$$= 0 - 2 + 2 = 0$$

$h(x)$ is continuous and differentiable function and $h(0)=0$ and $h\left(\frac{\pi}{3}\right)=0$

By Rolle's Theorem,

$h'(x)=f(x)-3\sin 3x+\frac{6}{n}=0$

has least one solution in $\left(0,\frac{\pi}{3}\right)$

(C) $\lim\limits_{x\to 0}\frac{x\int_0^x f(t)\,dt}{1-e^{x^2}}=\lim\limits_{x\to 0}\left(\frac{x^2}{1-e^{x^2}}\right)\frac{\int_0^x f(t)\,dt}{x}$

By L'Hospital's Rule,

$=-1\lim\limits_{x\to 0}\frac{f(x)}{1}=-1 \qquad [\because f(0)=1]$

(D) $\lim\limits_{x\to 0}\frac{(\sin x)\int_0^x f(t)\,dt}{x^2}$

$=\lim\limits_{x\to 0}\left(\frac{\sin x}{x}\right)\frac{\int_0^x f(t)\,dt}{x}$

By L'Hospital's Rule,

$=1\lim\limits_{x\to 0}\frac{f(x)}{1}=1 \qquad [\because f(0)=0]$

14. (c) $\because f(x)=\int\limits_{-x}^{x}\left(|t|-t^2\right)e^{-t^2}dt, x>0$

$\therefore f(x)=2\int\limits_0^x\left(t-t^2\right)e^{-t^2}dt$

$f'(x)=2\left(x-x^2\right)e^{-x^2}$ ...(i)

Hence option (d) is wrong.

$g'(x)=xe^{-x^2}2x$ ...(ii)

Add (i) and (ii),

$f'(x)+g'(x)=2xe^{-x^2}$

$f(x)+g(x)=-e^{-x^2}+c$

$\because f(0)=g(0)=0$

$f(x)+g(x)=-e^{-x^2}+1$

$f\left(\sqrt{\ell n3}\right)+g\left(\sqrt{\ln 3}\right)=1-\frac{1}{3}=\frac{2}{3}$

So, option (a) is wrong.

$H(x)=\psi_1(x)-1-\alpha x$

$=e^{-x}+-1-\alpha x, x\ge 1$ and $\alpha\in(1,x)$

$H(1)=e^{-1}+1-1-\alpha<0$

$H'(x)=-e^{-x}+1-\alpha<0\Rightarrow H(x)$ is decreasing

So, option (b) is wrong.

(c) $\psi_2(x)=2(\psi_1(\beta)-1)$

Applying L.M.V.T to $\psi_2(x)$ in $[0,x]$

$\psi_2'(\beta)=\frac{\psi_2(x)-\psi_2(0)}{x}$

$2\beta-2+2e^{-\beta}=\frac{\psi_2(x)-0}{x}$

$\Rightarrow \psi_2(x)=2x(\psi_1(\beta-1))$ has one solution.

So, option (c) is correct.

15. (d) (a) $\psi_1(x)=e^{-x}+x, x\ge 0$

$\psi_1'(x)=1-e^{-x}>0$

$\Rightarrow\psi_1(x)$ is increasing.

Option (a) is incorrect.

(b) $\psi_2(x)=x^2-2x+2-2e^{-x} \quad x\ge 0$

$\psi_2'(x)=2x-2+2e^{-x}=2\psi_1(x)-2\ge 0, \forall x\ge 0$

$\Rightarrow\psi_2(x)$ is increasing

$\Rightarrow\psi_2(x)\ge\psi_2(0) \;\Rightarrow\psi_2(x)\ge 0$

So, option (b) is incorrect.

(C) $f(x)=2\int\limits_0^x\left(t-t^2\right)e^{-t^2}dt$ and $x\in\left(0,\frac{1}{2}\right)$

$=\int\limits_0^x 2te^{-t^2}dt-\int\limits_0^x 2t^2e^{-t^2}dt$

$=-e^{-x^2}\Big|_0^x-\int\limits_0^x 2t^2e^{-t^2}dt$

Let $H(x)=f(x)-1+e^{-x^2}+\frac{2}{3}x^3-\frac{2}{5}x^5,$ $x\in\left(0,\frac{1}{2}\right)$

$H(0)=0$

$H'(x)=2\left(x-x^2\right)e^{-x^2}-2xe^{-x^2}+2x^2-2x^4$

$=-2x^2e^{-x^2}+2x^2-2x^4$

$=2x^2\left(1-x^2-e^{-x^2}\right)$

$\because e^{-x}\ge 1-x\,\forall x\ge 0$

$\Rightarrow H'(x)\le 0$

$\Rightarrow H(x)$ is decreasing.

$\Rightarrow 1-1(x)<0\,\forall\, x\in\left(0,\frac{1}{2}\right)$

Let $P(x)=g(x)-\frac{2}{3}x^3+\frac{2}{5}x^5-\frac{1}{7}x^7\; x\in\left(0,\frac{1}{2}\right)$

$P'(x)=2x^2e^{-x^2}-2x^2+2x^4-x^6$

$=2x^2\left(1-\frac{x^2}{1}+\frac{x^4}{2}-\frac{x^6}{3}+\ldots\right)-2x^2+2x^4-x^6$

$=-\frac{x^8}{3}+\frac{x^{10}}{12}\ldots$

$\Rightarrow P'(x)\le 0$

$\Rightarrow P(x)$ is decreasing.

$\Rightarrow P(x)\le 0$

So, option (d) is correct.

16. (a, c) Given that $f'(x)=\frac{f(x)}{b^2+x^2}$

$\int\frac{f'(x)}{f(x)}dx=\int\frac{dx}{x^2+b^2}$

$\Rightarrow \ln|f(x)|=\frac{1}{b}\tan^{-1}\left(\frac{x}{b}\right)+c$

Now $f(0)=1$.

$\therefore c=0$

$\therefore |f(x)|=e^{\frac{1}{b}\tan^{-1}\left(\frac{x}{b}\right)}$

$\Rightarrow f(x)=\pm e^{\frac{1}{b}\tan^{-1}\left(\frac{x}{b}\right)}$

$\because f(0)=1\;\therefore f(x)=e^{\frac{1}{b}\tan^{-1}\left(\frac{x}{b}\right)}$

$\because f'(x)=e^{\frac{1}{b}\tan^{-1}\frac{x}{b}}\times\frac{1}{b^2+x^2}>0$ for all values of b

$\therefore f(x)$ is increasing function for all values of b.

So, option (a) is true.

$f(-x)=e^{-\frac{1}{b}\tan^{-1}\left(\frac{x}{b}\right)}$

$\therefore f(x).f(-x)=e^0=1$

So, option (c) is true.

17. (a, b, d) $\lim\limits_{x\to 0}f(x)=\lim\limits_{x\to 0}x.\lim\limits_{x\to 0}g(x)=0$

$\because f'(x)=\lim\limits_{h\to 0}\frac{f(x+h)-f(x)}{h}$

$\Rightarrow f'(x)=\lim\limits_{h\to 0}\frac{f(x)+f(h)+f(x).f(h)-f(x)}{h}$

$\Rightarrow f'(x)=\lim\limits_{h\to 0}\frac{f(h)}{h}(1+f(x))$

$\left[\because \lim\limits_{h\to 0}\frac{f(h)}{h}=\lim\limits_{h\to 0}\frac{hg(h)}{h}=0\right]$

$\Rightarrow f'(x)=1+f(x)$

$\Rightarrow \frac{f'(x)}{1+f(x)}=1\Rightarrow\int\frac{f'(x)}{1+f(x)}dx=\int dx$

$\Rightarrow \ln(1+f(x))=x+c$

$\Rightarrow 1+f(x)=e^{x+c}\,(\because f(0)=0\Rightarrow c=0)$

$\Rightarrow f(x)=e^x-1\Rightarrow f'(x)=e^x$

So, $f(x)$ is differentiable at every $x\in R$.

So, option (a) is true.

$f'(0)=e^0=1$

So, option (d) is true.

$g(x)=\frac{f(x)}{x}=\begin{cases}\frac{e^x-1}{x} & ;\; x\ne 0\\ 1 & ;\; x=0\end{cases}$

$\therefore g(x)$ is differentiable at every $x\in R$

So, option (b) is true.

18. (a, b, d) (a) We know that,

$\cos x=1-\frac{x^2}{2!}+\frac{x^4}{4!}-\ldots$

and, $\sin x=x-\frac{x^3}{3!}+\frac{x^5}{5!}-\ldots$

$\because \cos x\ge 1-\frac{x^2}{2}$

$\therefore \int_0^1 x\cos x\ge\int_0^1 x\left(1-\frac{x^2}{2}\right)=\frac{1}{2}-\frac{1}{8}$

$\Rightarrow \int_0^1 x\cos x\ge\frac{3}{8}$

(b) $\because \sin x\ge x-\frac{x^3}{6}$

$\therefore \int_0^1 x\sin x\ge\int_0^1 x\left(x-\frac{x^3}{6}\right)dx$

$\Rightarrow \int_0^1 x\sin x\ge\frac{1}{3}-\frac{1}{30}\Rightarrow\int_0^1 x\sin x\,dx\ge\frac{3}{8}$

(d) $\int_0^1 x^2\sin x\,dx\ge\int_0^1 x^2\left(x-\frac{x^3}{6}\right)dx$

$$\Rightarrow \int_0^1 x^2 \sin x dx \geq \frac{1}{4}-\frac{1}{36} \Rightarrow \int_0^1 x^2 \sin x dx \geq \frac{2}{9}$$

(c) $\because \cos x < 1 \Rightarrow x^2 \cos x < x^2$

$$\therefore \int_0^1 x^2 \cos x dx < \int_0^1 x^2 dx \Rightarrow \int_0^1 x^2 \cos x dx < \frac{1}{3}$$

19. (a, d)

$$\lim_{n\to\infty} \frac{\sqrt[3]{1}+\sqrt[3]{2}+\ldots+\sqrt[3]{n}}{n^{7/3}\left(\frac{1}{(an+1)^2}+\frac{1}{(an+2)^2}+\ldots+\frac{1}{(an+n)^2}\right)}=54$$

$$\Rightarrow \frac{\lim_{n\to\infty} \frac{1}{n}\sum_{r=1}^{n}\left(\frac{r}{n}\right)^{1/3}}{\lim_{n\to\infty} \frac{1}{n}\sum_{r=1}^{n}\frac{1}{\left(a+\frac{r}{n}\right)^2}}=54$$

$$\Rightarrow \frac{\int_0^1 x^{1/3}dx}{\int_0^1 \frac{1}{(a+x)^2}dx}=54 \Rightarrow \frac{\frac{3}{4}}{\frac{1}{a}-\frac{1}{a+1}}=54$$

$$\Rightarrow a^2+a-72=0 \Rightarrow (a+9)(a-8)=0$$

$$\Rightarrow a=8 \text{ or } -9$$

$\therefore$ Options (a) and (d) are correct.

20. (b, d)

$$I=\sum_{k=1}^{98}\int_k^{k+1}\frac{(k+1)}{x(x+1)}dx \; ; \; I=\sum_{k=1}^{98}\int_k^{k+1}\frac{(k+1)}{x(x+1)^2}dx$$

$$\Rightarrow I=\sum_{k=1}^{98}(-(k+1))\left[\frac{1}{k+2}-\frac{1}{k+1}\right]$$

$$\Rightarrow I>\sum_{k=1}^{98}\frac{1}{k+2}$$

$$\Rightarrow I>\frac{1}{3}+\ldots+\frac{1}{100}>\frac{98}{100} \Rightarrow I>\frac{49}{50}$$

Also, $I>\sum_{k=1}^{98}\int_k^{k+1}\frac{(k+1)}{x(x+1)}dx$

$$=\sum_{k=1}^{98}\left[\log_e(k+1)-\log_e k\right]$$

$I<\log_e 99$

21. (None) $g(x)=\int_{\sin x}^{\sin 2x}\sin^{-1}(t)dt$

$$\Rightarrow g'(x)=\sin^{-1}(\sin 2x).2\cos 2x-\sin^{-1}(\sin x).\cos x$$

$$g'\left(\frac{\pi}{2}\right)=\sin^{-1}(\sin\pi).2\cos\pi-\sin^{-1}\left(\sin\frac{\pi}{2}\right)\cos\frac{\pi}{2}=0$$

$$g'\left(-\frac{\pi}{2}\right)=\sin^{-1}(\sin(-\pi)).2\cos(-\pi)-\sin^{-1}\left(\sin\left(\frac{\pi}{2}\right)\right).\cos\left(\frac{-\pi}{2}\right)=0$$

22. (a, b) Let us check the given options one by one.

(a) Let $g(x)=x^9-f(x)$

$\Rightarrow g(0)=-f(0)<0 \qquad [\because f(x)\in(0,1)]$

and $g(1)=1-f(1)>0$

$\therefore x^9-f(x)=0$ for some $x\in(0,1)$.

(b) Let $h(x)=x-\int_0^{\frac{\pi}{2}-x} f(t)\cos t dt$

$$h(0)=-\int_0^{\frac{\pi}{2}} f(t)\cos t dt<0$$

and $h(1)=1-\int_0^{\frac{\pi}{2}-1} f(t)\cos t dt>0$

$\therefore h(0)<0$ and $h(1)>1$

$\Rightarrow h(x)=0$ at some $x\in(0,1)$

$\therefore h(x)=x-\int_0^{\frac{\pi}{2}-x} f(t)\cos t dt=0$ at some $x\in(0,1)$.

(c) $e^x-\int_0^x f(t)\sin t dt$

$\because x\in(0,1)\Rightarrow e^x\in(1,e)$

and $0<f(t)<1$ and $0<\sin t<1, \forall x\in(0,1)$

$$\therefore 0<\int_0^x f(t)\sin t dt<1$$

$\therefore e^x \int_0^x f(t)\sin t\, dt \neq 0$ for any $x\in(0,1)$.

(d) $f(x)+\int_0^{\frac{\pi}{2}} f(t)\sin t dt$ is always positive $\forall x\in(0,1)$

23. (b, c)

$$f(x)=\lim_{n\to\infty}\left[\frac{n^n(x+n)\left(x+\frac{n}{2}\right)\ldots\left(x+\frac{n}{n}\right)}{n!\left(x^2+n^2\right)\left(x^2+\frac{n^2}{4}\right)\ldots\left(x^2+\frac{n^2}{n^2}\right)}\right]^{\frac{x}{n}}$$

$$=\lim_{n\to\infty}\left[\frac{\left(\frac{x}{n}+1\right)\left(\frac{x}{n}+\frac{1}{2}\right)\ldots\left(\frac{x}{n}+\frac{1}{n}\right)}{\left(1.\frac{x^2}{n^2}+1\right)\left(2.\frac{x^2}{n^2}+\frac{1}{2}\right)\ldots\left(n.\frac{x^2}{n^2}+\frac{1}{n}\right)}\right]^{\frac{x}{n}}$$

$\Rightarrow \ln f(x)$

$$=\lim_{n\to\infty}\frac{x}{n}\left[\sum_{r=1}^{n}\ln\left(\frac{x}{n}+\frac{1}{r}\right)-\sum_{r=1}^{n}\ln\left(\frac{rx^2}{n^2}+\frac{1}{r}\right)\right]$$

$$=\lim_{n\to\infty}\frac{x}{n}\left[\sum_{r=1}^{n}\ln\left(1+\frac{rx}{n}\right)-\ln\left(1+\frac{r^2x^2}{n^2}\right)\right]$$

$$=x\int_0^1\ln(1+xy)dy-x\int_0^1\ln\left(1+x^2y^2\right)dy$$

Let $xy=t\Rightarrow x\,dy=dt$

$$\therefore \ln f(x)=\int_0^x\ln(1+t)dt-\int_0^x\ln\left(1+t^2\right)dt$$

$$\ln f(x)=\int_0^x\ln\left(\frac{1+t}{1+t^2}\right)dt\Rightarrow\frac{f'(x)}{f(x)}=\ln\left(\frac{1+x}{1+x^2}\right)$$

$$\Rightarrow\frac{f'(2)}{f(2)}=\ln\left(\frac{3}{5}\right)<0$$

$\Rightarrow f'(2)<0\quad\therefore$ (c) is correct

and $\dfrac{f'(3)}{f(3)}=\ln\left(\dfrac{2}{5}\right)<\dfrac{f'(2)}{f(2)}$

$\therefore$ (d) is not correct.

$\because f'(x)=f(x)\ln\left(\dfrac{1+x}{1+x^2}\right)>0,\forall x\in(0,1)$

$\Rightarrow f'(x)>0\,\forall\,x\in(0,1)$

$\therefore f$ is an increasing function.

$\therefore \dfrac{1}{2}<1\Rightarrow f\left(\dfrac{1}{2}\right)\le f(1)$

$\therefore$ (a) is not correct

and $\dfrac{1}{3}<\dfrac{2}{3}\Rightarrow f\left(\dfrac{1}{3}\right)\le f\left(\dfrac{2}{3}\right)$

$\therefore$ (b) is correct.

24. (d) $f'(x)=\dfrac{192x^3}{2+\sin^4\pi x}$

$$\Rightarrow\frac{192x^3}{3}\le f'(x)\le\frac{192x^3}{2}\Rightarrow 64x^3\le f'(x)\le 96x^3$$

$$\Rightarrow\int_{1/2}^{x}64x^3dx\le\int_{1/2}^{x}f'(x)dx\le\int_{1/2}^{x}96x^3dx$$

$$\Rightarrow\frac{64x^4}{4}-\frac{64}{4}\times\frac{1}{16}\le\int_{\frac{1}{2}}^{x}f'(x)dx\le\frac{96x^4}{4}-\frac{96}{4\times16}$$

$$\Rightarrow 16x^4-1\le\int_{1/2}^{x}f'(x)\le 24x^4-\frac{3}{2}$$

$$\Rightarrow 16x^4-1\le f(x)\le 24x^4-\frac{3}{2}$$

$$\Rightarrow\int_{1/2}^{1}\left(16x^4-1\right)dx\le\int_{1/2}^{1}f(x)dx\le\int_{1/2}^{1}\left(24x^4-\frac{3}{2}\right)dx$$

$$\Rightarrow\left(\frac{16x^5}{5}-x\right)_{\frac{1}{2}}^{1}\le\int_{1/2}^{1}f(x)dx\le\left[\frac{24x^5}{5}-\frac{3}{2}x\right]_{\frac{1}{2}}^{1}$$

$$\Rightarrow 2.6\le\int_{1/2}^{1}f(x)dx\le 3.9$$

$\therefore$ Only (d) is the correct option.

25. (a, b) $f(x)=7\tan^8x+7\tan^6x-3\tan^4x-3\tan^2x$

$$=\left(7\tan^4x-3\right)\left(\tan^4x+\tan^2x\right)$$

$$=\left(7\tan^6x-3\tan^2x\right)\sec^2x$$

$$\int_0^{\pi/4}f(x)dx=\left[\tan^7x-\tan^3x\right]_0^{\pi/4}=1-1=0$$

$$\therefore\int_0^{\pi/4}xf(x)dx=\left[x\left(\tan^7x-\tan^3x\right)\right]_0^{\pi/4}-\int_0^{\pi/4}\left(\tan^7x-\tan^3x\right)dx$$

$$\int_0^{\pi/4}\tan^3x\left(1-\tan^2x\right)\sec^2x\,dx$$

$$=\left[\frac{\tan^4x}{4}-\frac{\tan^6x}{6}\right]_0^{\pi/4}=\frac{1}{12}$$

26. (a, c) Let $f(t)=e^t\left(\sin^6at+\cos^6at\right)$

$\therefore f(k\pi+t)$

$$=e^{k\pi+1}\left(\sin^6a(k\pi+t)+\cos^6a(k\pi+t)\right)=e^{k\pi}f(t)$$

$$\therefore \frac{\int_0^{4\pi} e^t\left(\sin^6 at+\cos^4 at\right)dt}{\int_0^{\pi} e^t\left(\sin^6 at+\cos^4 at\right)dt}$$

$$=\frac{\left(1+e^{\pi}+e^{2x}+e^{3\pi}\right)\int_0^{\pi} e^t\left(\sin^6 at+\cos^4 at\right)dt}{\int_0^{\pi} e^t\left(\sin^6 at+\cos^4 at\right)dt}$$

$$=1+e^{\pi}+e^{2\pi}+e^{3\pi}=\frac{e^{4\pi}-1}{e^{\pi}-1}$$

Matrix Type

1. A → Q; B → R; C → P; D → R

(A) If $\frac{\pi}{4}<x<\frac{3\pi}{8}$, then $\sin x>\cos x$

$$\therefore \int \frac{\sin x-\cos x}{|\sin x-\cos x|}dx=\int 1.dx=x+c$$

(B) $\int \frac{x^2 dx}{\left(x^3+1\right)\left(x^3+2\right)}$

$$=\frac{1}{3}\int 3x^2\left(\frac{1}{x^3+1}-\frac{1}{x^3+2}\right)dx$$

$$=\frac{1}{3}\ell n|x|$$

(C) $\int \sin^{-1} x\cos^{-1} x dx$

$$=\int\left[\frac{\pi}{2}\sin^{-1} x-\left(\sin^{-1} x\right)^2\right]dx$$

$$\Rightarrow \frac{\pi}{2}\left(x\sin^{-1} x+\sqrt{1-x^2}\right)-$$

$$\left(x\left(\sin^{-1} x\right)^2+\sin^{-1} x\sqrt{1-x^2}-x\right)+c$$

By parts

$$\Rightarrow \sin^{-1} x\left[\frac{\pi}{2}x-x\sin^{-1} x-2\sqrt{1-x^2}\right]+$$

$$\frac{\pi}{2}\sqrt{1-x^2}+2x+c$$

$$\therefore f^{-1}(x)=\sin^{-1} x,\ f(x)=\sin x$$

(D) $\int \frac{dx}{x\ln|x|}=\ln|\ln|x||+c$

$$\therefore f(x)=\ln|x|$$

2. A → T; B → T; C → R; D → S

(A) $F(x)=\int \frac{x+\sin x}{1+\cos x}dx$

$$=\int\left(x\frac{1}{2}\sec^2\frac{x}{2}+\tan\frac{x}{2}\right)dx$$

$$=x\tan\frac{x}{2}+c$$

Since $0=F(0)$

$\therefore c=0$ and $F(\pi/2)=\pi/2$

(B) $F(x)=\int e^{\sin^{-1} x}\left(1-\frac{x}{\sqrt{1-x^2}}\right)dx$

$$=\int e^{\sin^{-1} x}\left(\frac{1}{\sqrt{1-x^2}}\sqrt{1-x^2}-\frac{x}{\sqrt{1-x^2}}\right)dx$$

$$F(x)=e^{\sin^{-1} x}\sqrt{1-x^2}+c$$

$F(0)=1+c\Rightarrow c=0$ $\quad (\because F(0)=1)$

$$F\left(\frac{1}{2}\right)=e^{\frac{\pi}{6}}.\frac{\sqrt{3}}{2}=\frac{k\sqrt{3}}{\pi}e^{\pi/6}$$

$$\therefore k=\frac{\pi}{2}$$

(C) $F(x)=\int \frac{dx}{\left(x^2+1\right)\left(x^2+9\right)}$

$$=\frac{1}{8}\int\left(\frac{1}{x^2+1}-\frac{1}{x^2+9}\right)dx$$

$$=\frac{1}{8}\left[\tan^{-1} x-\frac{1}{3}\tan^{-1}\frac{x}{3}\right]+c$$

$$F(0)=c\Rightarrow c=0$$

$$\therefore f\left(\sqrt{3}\right)=\frac{1}{8}\left(\frac{\pi}{3}-\frac{1}{3}.\frac{\pi}{6}\right)=\frac{5\pi}{144}=\frac{5k}{36}$$

$$\therefore k=\frac{\pi}{4}$$

(D) $F(x)=\int \frac{\sqrt{\tan x}}{\sin x\cos x}dx$

$$=\int (\tan x)^{\frac{-1}{2}}\sec^2 x dx$$

$$=2\sqrt{\tan x}+c$$

$$F(0)=c\ \therefore c=0$$

$$\therefore F\left(\frac{\pi}{4}\right)=2=\frac{2k}{\pi} \qquad \therefore k=\pi$$

3. A → Q; B → P; C → S; D → S

(A) $\int_{6\pi-3}^{6\pi} \frac{\cos\left(\frac{x}{3}\right)}{6\pi+3-x}dx$

Let $6\pi - 3t = x$

$$\int_0^1 \frac{3\cos(2\pi - t)}{3(1+t)}dt = \int_0^1 \frac{\cos t}{1+t}dt = \int_0^1 \frac{\cos x}{1+x}dx = k$$

(B) $\int_{-1}^{1}\left(\sin^{-1}\left[x+\frac{3}{4}\right]\right)dx$

$$= \int_{-1}^{-3/4}\left(-\frac{\pi}{2}\right)dx + \int_{-3/4}^{1/4} 0dx + \int_{1/4}^{1}\frac{\pi}{2}dx$$

$$= -\frac{\pi}{8} + \frac{\pi}{2}\times\frac{3}{4} = \frac{2\pi}{8} = \frac{\pi}{4}$$

(C) $\because -|x| \le [x]$ for all +ve x

$-|x| \ge [x]$ for all –ve x

$$\Rightarrow \int_{-3}^{3} f(x)dx = \int_{-3}^{0}(x-|x|)dx + \int_{0}^{3}(x+|x|)dx$$

$$= \int_{-3}^{0} 2xdx + \int_0^3 xdx + \int_0^3 [x]dx = -\frac{3}{2}$$

(D) $\int_0^{20}\sqrt{1-\cos\pi x}\,dx = \int_0^{20}\sqrt{2}\left|\sin\frac{\pi x}{2}\right|dx$

$$= 10\sqrt{2}\int_0^2\left|\sin\frac{\pi x}{2}\right|dx$$

$\left\{\because \left|\sin\frac{\pi x}{2}\right| \text{ is periodic with period 2}\right\}$

$$= 10\sqrt{2}\int_0^2 \sin\left(\frac{\pi x}{2}\right)dx$$

$$= 10\sqrt{2}\left[-\frac{\cos\left(\frac{\pi x}{2}\right)}{\frac{\pi}{2}}\right]_0^2 = \frac{20\sqrt{2}}{\pi}$$

4. A → Q; B → S; C → T; D → T

(A) Let $I = \int_4^{10}\frac{[x^2]dx}{[x^2-28x+196]+[x^2]}$ ...(i)

$$= \int_4^{10}\frac{[x^2]dx}{[(x-14)^2]+[x^2]}$$

Also, $I = \int_4^{10}\frac{[(14-x)^2]dx}{[x^2]+[(14-x)^2]}$ ...(ii)

using $\int_a^b f(x)dx = \int_a^b f(a+b-x)dx$

Adding Eqs (i) and (ii),

$$2I = \int_4^{10} dx = [x]_4^{10} = 6 \qquad \therefore\ I = \frac{6}{2} = 3$$

(B) $\int_{-1}^{2}\frac{|x|}{x}dx = \int_{-1}^{0} -1dx + \int_0^2 1dx$

$$= -[x]_{-1}^{0} + [x]_0^2$$

$$= -[0-(-1)] + [2-0] = 1$$

(C) $\lim_{n\to\infty}\frac{1^{99}+2^{99}+\ldots+n^{99}}{n^{100}}$

$$= \lim_{n\to\infty}\frac{1}{n}\cdot\left[\left(\frac{1}{n}\right)^{99}+\left(\frac{2}{n}\right)^{99}+\ldots+\left(\frac{n}{n}\right)^{99}\right]$$

$$= \lim_{n\to\infty}\frac{1}{n}\cdot\sum_{r=1}^{n}\left(\frac{r}{n}\right)^{99}$$

$$= \lim_{n\to\infty}\sum_{r=1}^{n}\frac{1}{n}\left(\frac{r}{n}\right)^{99} = \int_0^1 x^{99}dx = \left[\frac{x^{100}}{100}\right]_0^1 = \frac{1}{100}$$

(D) $\frac{1}{\alpha} = 5050\int_{-1}^{1}|x|^{100}dx$

$$= 5050\times 2\int_0^1 x^{100}dx = 2\times 5050\left[\frac{x^{101}}{101}\right]_0^1$$

$$= 2\times 5050\left[\frac{1}{101}-0\right]$$

$$\frac{1}{\alpha} = 100 \Rightarrow \alpha = \frac{1}{100}$$

5. A → R; B → R; C → Q; D → Q

(A) $I = \int_{-1}^{1}\frac{3x^2}{1+4^{\tan x}}dx$ ...(i)

$$I = \int_{-1}^{1}\frac{3x^2}{1+4^{-\tan x}}dx$$

$\left(\text{using } \int_a^b f(x)dx = \int_a^b f(a+b-x)dx\right)$

$I = \int_{-1}^{1}\frac{3x^2\,4^{\tan x}}{1+4^{\tan x}}dx$ ...(ii)

Adding (i) and (ii), we get

$$2I = \int_{-1}^{1}3x^2dx,\ \textit{i.e.},\ \ 2I = x^3\Big|_{-1}^{1}\ \ \textit{i.e.},\ I = 1$$

(B) $I = \int_6^8 \frac{\sin x^2 dx}{\sin x^2 + \sin(x-14)^2}$ ...(i)

$$I=\int_6^8 \frac{\sin(14-x)^2\,dx}{\sin x^2+\sin(x-14)^2} \quad \text{...(ii)}$$

$$\left(\text{using} \int_a^b f(x)\,dx=\int_a^b f(a+b-x)\,dx\right)$$

Adding (i) and (ii), we get

$$2I=\int_6^8 dx, \text{ i.e., } I=1$$

(C) $I=\frac{1}{156}\left[\int_1^2 1dx+\int_2^3 2dx+\int_3^4 3dx+\ldots+\int_{12}^{13} 12dx\right]$

$$I=\frac{1}{156}$$

$$[(2-1)+2(3-2)+3(4-3)+\ldots+12(13-12)]$$

$$I=\frac{1}{156}[1+2+3+\ldots+12]=\frac{1}{2}$$

(D) $I=\int_0^{\pi/2} \ln\sin 2x dx=-\frac{\pi}{2}\ln 2$

$$\therefore \frac{1}{\pi\ln 2}\int_{\frac{\pi}{2}}^0 \ln\sin 2x dx=\frac{1}{2}$$

6. A → P; B → Q; C → T; D → P

(A) $L=\lim_{n\to\infty}\sum_{r=1}^{n-1}\frac{\sqrt{n^2-r^2}}{n^2}$

$$=\lim_{n\to\infty}\sum_{r=1}^{n-1}\frac{1}{n}\sqrt{1-\left(\frac{r}{n}\right)^2}$$

$$=\int_0^1\sqrt{1-x^2}\,dx=\frac{\pi}{4}$$

(B) $\int_0^{\frac{\pi}{4}}\left(\sqrt{\tan x}+\sqrt{\cot x}\right)dx$

$$=2\int_0^{\frac{\pi}{4}}\frac{1}{2}\frac{\sqrt{\tan x}+\sqrt{\cot x}}{\tan x+\cot x}(\tan x+\cot x)\,dx$$

$$=2\int_0^{\frac{\pi}{4}}\frac{d\left(\sqrt{\tan x}-\sqrt{\cot x}\right)}{\left(\sqrt{\tan x}-\sqrt{\cot x}\right)^2+2}dx$$

$$=2.\frac{1}{\sqrt{2}}\tan^{-1}\left(\frac{\sqrt{\tan x}+\sqrt{\cot x}}{\sqrt{2}}\right)\Bigg|_0^{\frac{\pi}{4}}=\frac{\pi}{\sqrt{2}}$$

(C) $\int_{-1}^1 \sin^3 x\cos^2 x dx=0$

{∵ the intergrand is odd.}

(D) $I=\int_0^{\frac{\pi}{2}}\frac{\sqrt{\sin^3 x dx}}{\sqrt{\sin^3 x}+\sqrt{\cos^3 x}}$

$$=\int_0^{\frac{\pi}{2}}\frac{\sqrt{\cos^3 x dx}}{\sqrt{\sin^3 x}+\sqrt{\cos^3 x}}$$

$$\therefore\ 2I=\int_0^{\frac{\pi}{2}}1dx=\frac{\pi}{2} \quad \therefore I=\frac{\pi}{4}$$

7. A → Q; B → R; C → P; D → S

$P(2)$ Let $f(x)=ax^2+bx+c$

where $a,b,c\geq 0$ and a,b,c are integers.

$\because f(0)=0\Rightarrow c=0$

$\therefore f(x)=ax^2+bx$

Also $\int_0^1 f(x)\,dx=1$

$$\Rightarrow\left[\frac{ax^3}{3}+\frac{bx^2}{2}\right]_0^1=1\Rightarrow\frac{a}{3}+\frac{b}{2}=1\Rightarrow 2a+3b=6$$

$\because a$ and b are integers

$a=0$ and $b=2$

or $a=3$ and $b=0$

∴ There are only two solutions.

$Q(3)\ f(x)=\sin x^2+\cos x^2$

$f(x)$ is max. $\sqrt{2}$ at $x^2=\frac{\pi}{4}$ or $\frac{9\pi}{4}$

$$\Rightarrow x=\pm\frac{\sqrt{\pi}}{2} \text{ or } \pm\frac{3\sqrt{\pi}}{2}\in\left[-\sqrt{13},\sqrt{13}\right]$$

∴ There are four points.

$$R(1)\,I=\int_{-2}^2\frac{3x^2}{1+e^x}dx=\int_{-2}^2\frac{3x^2}{1+e^{-x}}dx$$

$$\left[\text{Using}\int_a^b f(x)\,dx=\int_a^b f(a+b-x)\,dx\right]$$

$$=\int_{-2}^2\frac{3x^2e^x}{1+e^x}dx$$

$$2I=\int_{-2}^2\frac{3x^2\left(1+e^x\right)}{1+e^x}dx=\int_{-2}^2 3x^2dx$$

$$2I=\left(x^3\right)_{-2}^{2}=8-(-8)=16\Rightarrow I=8$$

$$S(4)=\frac{\int_{-1/2}^{1/2}\cos 2x\log\left(\frac{1+x}{1-x}\right)dx}{\int_{0}^{1/2}\cos 2x\log\left(\frac{1+x}{1-x}\right)dx}=0$$

$\because$ Numerator = 0, the function being odd.
Hence option (d) is correct sequence.

Integer and Numerical Value Type

1. $$\int\frac{x\cos\alpha+1}{\left(x^2+2x\cos\alpha+1\right)^{3/2}}dx$$

$$=\int\frac{x\cos\alpha+1}{\left((x+\cos\alpha)^2+\sin^2\alpha\right)^{3/2}}dx$$

Put $x+\cos\alpha=\sin\alpha\tan\theta$

$$=\int\frac{\cos\alpha(\sin\alpha\tan\theta-\cos\alpha)+1}{\left(\sin^2\alpha\tan^2\theta+\sin^2\alpha\right)^{3/2}}\sin\alpha\cdot\sec^2\theta\,d\theta$$

$$=\int\frac{\cos\alpha\sin\alpha\tan\theta+\sin^2\alpha}{\sin^3\alpha\sec^3\theta}\sin\alpha\sec^2\theta d\theta$$

$$=\int\frac{\cos\alpha\tan\theta+\sin\alpha}{\sin\alpha}\cos\theta d\theta$$

$$=\int(\cot\alpha.\sin\theta+\cos\theta)d\theta=-\cot\alpha\cos\theta+\sin\theta+c$$

$$=-\cot\alpha\frac{\sin\alpha}{\sqrt{x^2+2x\cos\alpha+1}}+\frac{x+\cos\alpha}{\sqrt{x^2+2x\cos\alpha+1}}+c$$

$$=\frac{x}{\sqrt{x^2+2x\cos\alpha+1}}+c$$

2. $$I=\int\left(\frac{x-1}{x+1}\right)\frac{dx}{x\sqrt{x+1+\frac{1}{x}}};$$

$$\text{so, } I=\int\frac{(x-1)dx}{(x+1)x\sqrt{x+1+\frac{1}{x}}}$$

$$=\int\frac{\left(1-\frac{1}{x}\right)\left(1+\frac{1}{x}\right)dx}{(x+1)\left(1+\frac{1}{x}\right)\sqrt{x+1+\frac{1}{x}}}$$

$$=\int\frac{\left(1-\frac{1}{x^2}\right)dx}{\left(x+\frac{1}{x}+2\right)\sqrt{x+\frac{1}{x}+1}}$$

Put $x+1+\frac{1}{x}=t^2$

$$\left(1-\frac{1}{x^2}\right)dx=2t\,dt=\int\frac{2t\,dt}{\left(t^2+1\right)t}$$

$$=2\tan^{-1}t+c=2\tan^{-1}\left(\sqrt{x+\frac{1}{x}+1}\right)+c$$

3. $$\lim_{n\to\infty}\frac{1}{n}\left[\cos^{2p}\frac{\pi}{2n}+\cos^{2p}\frac{2\pi}{2n}+\cos^{2p}\frac{3\pi}{2n}+\ldots+\cos^{2p}\frac{n\pi}{2^n}\right]$$

$$=\int_0^1\cos^{2p}\frac{\pi x}{2}dx$$

$$\text{put }\frac{\pi x}{2}=t=\int_0^{\pi/2}\cos^{2p}t.\frac{2}{\pi}dt=\frac{2}{\pi}\int_0^{\pi/2}\cos^{2p}t\,dt$$

$$=\frac{2}{\pi}\left[\frac{2p-1}{2p}.\frac{2p-3}{2p-2}\ldots\frac{1}{2},\frac{\pi}{2}\right]$$

4. $$\int_0^{2\pi}\frac{dx}{a+b\cos x+c\sin x}=\int_0^{\pi}\frac{dx}{a+b\cos x+c\sin x}+\int_0^{\pi}\frac{dx}{a+b\cos x-c\sin x}$$

$$=2\int_0^{\infty}\frac{dt}{(a-b)t^2+2ct+a+b}+2\int_0^{\infty}\frac{dt}{(a-b)t^2-2ct+a+b}$$

$$=\frac{2}{a-b}\int_0^{\infty}\frac{dt}{t^2+\frac{2c}{a-b}t+\frac{a+b}{a-b}}+\frac{2}{a-b}\int_0^{\infty}\frac{dt}{t^2-\frac{2c}{a-b}t+\frac{a+b}{a-b}}$$

$$=\frac{2}{a-b}\int_0^{\infty}\frac{dt}{\left(t+\frac{c}{a-b}\right)^2+\frac{a^2-b^2-c^2}{(a-b)^2}}+\frac{2}{a-b}\int_0^{\infty}\frac{dt}{\left(t-\frac{c}{a-b}\right)^2+\frac{a^2-b^2-c^2}{(a-b)^2}}$$

$$=\frac{2}{a-b}\frac{a-b}{\sqrt{a^2-b^2-c^2}}\tan^{-1}\frac{(a-b)t+c}{\sqrt{a^2-b^2-c^2}}\Bigg|_0^{\infty}+$$

$$\frac{2}{a-b}\frac{a-b}{\sqrt{a^2-b^2-c^2}}\tan^{-1}\frac{(a-b)t-c}{\sqrt{a^2-b^2-c^2}}\Bigg|_0^{\infty}$$

$$=\frac{2}{\sqrt{a^2-b^2-c^2}}\left(\frac{\pi}{2}+\tan^{-1}\frac{c}{\sqrt{a^2-b^2-c^2}}\right)+$$

$$\frac{2}{\sqrt{a^2-b^2-c^2}}\left(\frac{\pi}{2}+\tan^{-1}\frac{-c}{\sqrt{a^2-b^2-c^2}}\right)$$

$$=\frac{2\pi}{\sqrt{a^2-b^2-c^2}}$$

5. Given that $f(x+y)=f(x)+f(y)$ ...(i)

Putting $x=0$ and $y=0$ in (i), we get

$f(0+0)=f(0)+f(0)\Rightarrow\ f(0)=0$ ...(ii)

Now $f'(x)=\lim\limits_{h\to 0}\dfrac{f(x+h)-f(x)}{h}$

$=\lim\limits_{h\to 0}\dfrac{f(x)+f(h)-f(x)}{h}$

$=\lim\limits_{h\to 0}\dfrac{f(h)}{h}=\lim\limits_{h\to 0}f'(h)=f'(0)$

$\Rightarrow f(x)=\int f'(0)\,dx=xf'(0)+c$ ...(iii)

Putting $x=0$ in Eq. (ii), we get

$f(0)=0+c$

$\Rightarrow\ c=0$ $[\because f(0)=0 \text{ from (ii)}]$

$\Rightarrow f(x)=xf'(0)$

Thus $I_n=n\int_0^n f(x)\,dx=n\int_0^n x\,f'(0)\,dx$

$\Rightarrow I_n=\dfrac{n^3\cdot f'(0)}{2}.$

Therefore, $I_1+I_2+I_3+I_4+I_5$

$=\dfrac{f'(0)}{2}\left(1^3+2^3+3^3+4^3+5^3\right)$

$\Rightarrow 450=\dfrac{f'(0)}{2}.\left\{\dfrac{5.(5+1)}{2}\right\}^2$

$\Rightarrow f'(0)=4$

$\therefore f(x)=4x$ (from Eq. (iii))

6. $I=\displaystyle\int_0^1\frac{1}{\left(5+2x-2x^2\right)\left(1+e^{(2-4x)}\right)}dx$ ...(i)

$\Rightarrow I=\displaystyle\int_0^1\frac{1}{\left(5+2(1-x)-2(1-x)^2\right)\left(1+e^{2-4(1-x)}\right)}dx$

$\left(\text{Applying }\int_a^b f(x)\,dx=\int_a^b f(b+a-x)\,dx\right)$

$\Rightarrow I=\displaystyle\int_0^1\frac{1}{\left(5+2x-2x^2\right)\left(1+e^{(4-2x)}\right)}dx$ ...(ii)

$2I=\displaystyle\int_0^1\frac{1}{\left(5+2x-2x^2\right)}\left(\frac{1}{1+e^{2-4x}}+\frac{1}{1+e^{4x-2}}\right)$ ...(iii)

Now,

$\left(\dfrac{1}{1+e^{2-4x}}+\dfrac{1}{1+e^{4x-2}}\right)=\left(\dfrac{1}{1+e^{y}}+\dfrac{1}{1+e^{-y}}\right)$

$=\dfrac{1+e^{-y}+1+e^{y}}{\left(1+e^{y}\right)\left(1+e^{-y}\right)}=\dfrac{2+e^{y}+e^{-y}}{1+e^{y}+e^{-y}+1}=1$

Hence Eq. (iii) reduces to

$2I=\displaystyle\int_0^1\left(\frac{dx}{5+2x-2x^2}\right)dx$

$I=\dfrac{1}{2}\displaystyle\int_0^1\frac{dx}{\left[\left(\frac{\sqrt{11}}{2}\right)^2-\left(x-\frac{1}{2}\right)^2\right]}$

$=\dfrac{1}{4}.\dfrac{1}{2.\frac{\sqrt{11}}{2}}\left[\ln\dfrac{\frac{\sqrt{11}}{2}+x-\frac{1}{2}}{\frac{\sqrt{11}}{2}-\left(x-\frac{1}{2}\right)}\right]_x^1$

$=\dfrac{1}{4\sqrt{11}}\left[\log\dfrac{\sqrt{11}+1}{\sqrt{11}-1}-\log\dfrac{\sqrt{11}-1}{\sqrt{11}+1}\right]$

$=\dfrac{1}{4\sqrt{11}}\log\left(\dfrac{\sqrt{11}+1}{\sqrt{11}-1}\right)^2$

$=2\cdot\dfrac{1}{4\sqrt{11}}\log\left(\dfrac{\sqrt{11}+1}{\sqrt{11}-1}\right)$

$=2\cdot\dfrac{1}{4\sqrt{11}}\log\left(\dfrac{(\sqrt{11}+1)\ (\sqrt{11}+1)}{(\sqrt{11}-1)\ (\sqrt{11}+1)}\right)$

$$= 2\cdot\frac{1}{4\sqrt{11}}\log\frac{\left(\sqrt{11}+1\right)^2}{\left(\sqrt{10}\right)^2}$$

$$= \frac{1}{\sqrt{11}}\log\frac{\sqrt{11}+1}{\sqrt{10}}.$$

7. $F(x) = \int_0^x f(t).dt$

$$I = \int_0^\pi f'(x).\cos x dx + \int_0^\pi F(x)\cos x dx = 2 \quad ...(i)$$

$$I = \int_0^\pi f'(x).\cos x dx + \int_0^\pi F(x)\cos x dx = 2$$

Using integration by parts,

$$\Rightarrow I = \left(\cos x.f(x)\right)_0^\pi + \int_0^\pi \sin x.f(x)dx + \int_0^\pi F(x)\cos x dx = 2$$

$$\Rightarrow I = 6 - f(0) + \int_0^\pi \sin x.F'(x)dx + \int_0^\pi F(x)\cos x dx = 2$$

$$I = 6 - f(0) + \int_0^\pi F'(x)\sin x dx + \int_0^\pi F(x)\cos x dx = 2$$

$$\Rightarrow I = 6 - f(0) + \left[\sin x F(x)\right]_0^\pi - \int_0^\pi F(x)\cos x dx + \int_0^\pi F(x)\cos x dx = 2$$

$$\Rightarrow I = 6 - f(0) + 0 = 2$$

$$\Rightarrow f(x) = 4$$

8. $I = \frac{2}{\pi}\int_{-\pi/4}^{\pi/4}\frac{dx}{\left(1+e^{\sin x}\right)(2-\cos 2x)} \quad ...(i)$

$$I = \frac{2}{\pi}\int_{-\pi/4}^{\pi/4}\frac{dx}{\left(1+e^{-\sin x}\right)(2-\cos 2x)}$$

$$\left[\text{Using } \int_a^b f(x)dx = \int_a^b f(a+b-x)dx\right]$$

$$I = \frac{2}{\pi}\int_{-\pi/4}^{\pi/4}\frac{e^{\sin x}}{\left(e^{\sin x}+1\right)(2-\cos 2x)} \quad ...(ii)$$

Adding (i) and (ii):

$$2I = \frac{2}{\pi}\int_{-\pi/4}^{\pi/4}\frac{1+e^{\sin x}}{\left(e^{\sin x}+1\right)(2-\cos 2x)}dx$$

$$= \frac{2}{\pi}\int_{-\pi/4}^{\pi/4}\frac{1}{2-\cos 2x}dx$$

$$I = \frac{2}{\pi}\int_0^{\pi/4}\frac{dx}{1+2\sin^2 x}$$

$$= \frac{2}{\pi}\int_0^{\pi/4}\frac{\sec^2 x}{\sec^2 x + 2\tan^2 x}dx = \frac{2}{\pi}\int_0^{\pi/4}\frac{\sec^2 x}{1+3\tan^2 x}dx$$

Put $\tan x = t \Rightarrow \sec^2 x dx = dt$

$At\, x = 0, t = 0; At\, x = \frac{\pi}{4}, t = 1$

$$\therefore I = \frac{2}{\pi}\int_0^1\frac{1}{1+3t^2}dt = \frac{2}{\pi}\left[\frac{1}{\sqrt{3}}\tan^{-1}\left(\sqrt{3}t\right)\right]_0^1$$

$$= \frac{2}{\pi}\left[\frac{1}{\sqrt{3}}\times\frac{\pi}{3}\right] = \frac{2}{3\sqrt{3}}$$

$$\therefore 27I^2 = 4$$

9. $I = \int_0^{\pi/2}\frac{3\sqrt{\cos\theta}}{\left(\sqrt{\cos\theta}+\sqrt{\sin\theta}\right)^5}d\theta \quad ...(i)$

$$I = \int_0^{\pi/2}\frac{3\sqrt{\sin\theta}}{\left(\sqrt{\sin\theta}+\sqrt{\cos\theta}\right)^5}d\theta \quad ...(ii)$$

Adding two values of I, we get:

$$\frac{2}{3}I = \int_0^{\pi/2}\frac{1}{\left(\sqrt{\cos\theta}+\sqrt{\sin\theta}\right)^4}d\theta$$

$$\frac{2}{3}I = \int_0^{\pi/2}\frac{\sec^2\theta}{\left(1+\sqrt{\tan\theta}\right)^4}d\theta$$

$$\left(\text{put } \tan\theta = t^2 \Rightarrow \sec^2\theta d\theta = 2t\,dt\right)$$

$$\Rightarrow \frac{2I}{3} = \int_0^\infty\frac{2+dt}{(1+t)^4}$$

$$\Rightarrow I = 3\int_0^\infty\frac{1+t-1}{(t+1)^4}dt = 3\int_0^\infty\left(\frac{1}{(t+1)^3} - \frac{1}{(t+1)^4}\right)dt$$

$$= 3\left[-\frac{1}{2(t+1)^2} + \frac{1}{3(t+1)^3}\right]_0^\infty = \frac{1}{6}$$

$$\Rightarrow I = \frac{1}{2} = 0.50$$

10. Let $I=\int_0^{\frac{1}{2}}\frac{(1+\sqrt{3})dx}{\left[(1+x)^2(1-x)^6\right]^{1/4}}$

$$=\int_0^{\frac{1}{2}}\frac{(1+\sqrt{3})dx}{(1+x)^2\left[\frac{(1-x)^6}{(1+x)^6}\right]^{1/4}}$$

Now put $\frac{1-x}{1+x}=t\Rightarrow\frac{-2dx}{(1+x)^2}=dt$

$$\therefore I=\int_1^{1/3}\frac{(1+\sqrt{3})dt}{-2t^{6/4}}=\frac{-(1+\sqrt{3})}{2}\times\left|\frac{-2}{\sqrt{t}}\right|_1^{1/3}$$

$$=(1+\sqrt{3})(\sqrt{3}-1)=2$$

11. Given $f(0)=0, f\left(\frac{\pi}{2}\right)=3, f'(0)=1$

$$g(x)=\int_x^{\frac{\pi}{2}}\left[f'(t)\operatorname{cosec} t-\cot t\operatorname{cosec} t\, f(t)\right]dt$$

$$g(x)=\lim_{x\to 0}\int_x^{\frac{\pi}{2}}\frac{d}{dt}\left(f(t)\operatorname{cosec} t\right)dt$$

$$=f\left(\frac{\pi}{2}\right)\operatorname{cosec}\frac{\pi}{2}-f(x)\operatorname{cosec} x$$

$$=3-f(x)\operatorname{cosec}x=3-\frac{f(x)}{\sin x}$$

$$\lim_{x\to 0}g(x)=\lim_{x\to 0}\left(3-\frac{f(x)}{\sin x}\right)=3-\lim_{x\to 0}\frac{f(x)}{\sin x}$$

$$=3-\lim_{x\to 0}\frac{f'(x)}{\cos x}=3-f'(0)=3-1=2$$

12. $y_n=\left(\frac{n+1}{n}\cdot\frac{n+2}{n}\cdot\frac{n+3}{n}\cdots\frac{n+n}{n}\right)^{1/n}$

$$\Rightarrow\log y_n=\frac{1}{n}\sum_{r=0}^{n}\log\left(1+\frac{r}{n}\right)$$

$$\Rightarrow\left(\lim_{n\to\infty}y_n\right)=\lim_{n\to\infty}\frac{1}{n}\sum_{r=0}^{n}\log\left(1+\frac{r}{n}\right)$$

$$\Rightarrow\log L=\int_0^1\log(1+x)dx$$

$$=\left[x\log(1+x)\right]_0^1-\int_0^1\frac{x}{1+x}dx$$

$$=\log 2-\left[x-\log|1+x|\right]_0^1$$

$$=\log 2-1+\log 2=2\log 2-1$$

$$=\log 4-\log e=\log\left(\frac{4}{e}\right)$$

$$\Rightarrow L=\frac{4}{e}\Rightarrow[L]=\left[\frac{4}{e}\right]=1$$

13. Let $f(x)=\int_0^x\frac{t^2}{1+t^4}dt-2x+1$

$$\Rightarrow f'(x)=\frac{x^2}{1+x^4}-2<0\,\forall\, x\in[0,1]$$

$\therefore f$ is decreasing on [0, 1]

Also $f(0)=1$

and $f(1)=\int_0^1\frac{t^2}{1+t^4}dt-1$

For $0\le t\le 1\Rightarrow 0\le\frac{t^2}{1+t^4}<\frac{1}{2}$

$$\therefore\int_0^1\frac{t^2}{1+t^4}dt<\frac{1}{2}\Rightarrow f(1)<0$$

$\therefore f(x)$ crosses x-axis exactly once in [0, 1]

$\therefore f(x)=0$ has exactly one root in [0, 1]

14. $\alpha=\int_0^1 e^{(9x+3\tan^{-1}x)}\left(\frac{12+9x^2}{1+x^2}\right)dx$

Let $9x+3\tan^{-1}x=t\Rightarrow\frac{12+9x^2}{1+x^2}dx=dt$

$$\therefore\log_e\left|1+e^{9+\frac{3\pi}{4}}-1\right|-\frac{3\pi}{4}=9$$

15. $\lim_{x\to 1}\frac{F(x)}{G(x)}=\frac{1}{14}\Rightarrow\lim_{x\to 1}\frac{\int_{-1}^{x}f(t)dt}{\int_{-1}^{x}t|f(f(t))|dt}$

$$\because\int_{-1}^{1}f(t)dt=0\text{ and }\int_{-1}^{1}t|f(f(t))|dt=0$$

$f(t)$ being an odd function

$\therefore$ Using L Hospital's rule, we get

$$\lim_{x\to 1}\frac{f(x)}{x|f(f(x))|}=\frac{1}{14}$$

$$\Rightarrow \frac{f(1)}{|f(f(1))|} = \frac{1}{14} \Rightarrow \frac{1/2}{\left|f\left(\frac{1}{2}\right)\right|} = \frac{1}{14}$$

$$\Rightarrow \left|f\left(\frac{1}{2}\right)\right| = 7 \Rightarrow f\left(\frac{1}{2}\right) = 7$$

16. $I = \int_{-1}^{2} \frac{xf(x^2)}{2+f(x+1)} dx$

$-1 < x < 2 \Rightarrow 0 < x^2 < 4$

Also, $0 < x^2 < 1 \Rightarrow f(x^2) = [x^2] = 0$

$1 \le x^2 < 2 \Rightarrow f(x^2) = [x^2] = 1$

$2 \le x^2 < 3 \Rightarrow f(x^2) = 0$(using definition of f)

$3 \le x^2 < 4 \Rightarrow f(x^2) = 0$(using definition of f)

Also, $1 \le x^2 < 2 \Rightarrow 1 \le x < \sqrt{2}$

$\Rightarrow 2 \le x+1 < \sqrt{2}+1 \Rightarrow f(x+1) = 0$

$$\therefore I = \int_{1}^{\sqrt{2}} \frac{x \times 1}{2+0} dx = \left[\frac{x^2}{4}\right]_1^{\sqrt{2}} = \frac{1}{4}$$

$\Rightarrow 4I = 1$ or $4I - 1 = 0$

17. $\int_0^1 4x^3 \left[\frac{d^2}{dx^2}(1-x^2)^5\right] dx$

$$= 4x^3 \left[\frac{d}{dx}(1-x^2)^5\right]_0^1 - \int_0^1 \left[\frac{d}{dx}(1-x^2)^5\right].12x^2 dx$$

$$= -12x^2(1-x^2)^5\Big|_0^1 + \int_0^1 (1-x^2)^5.24x dx$$

$$= -12\int_0^1 (1-x^2)^5.(-2x)dx = -12\left(\frac{(1-x^2)^6}{6}\right)_0^1 = 2$$

Comprehension Passage

Comprehension 1

$$e^A = I + A + \frac{A^2}{2!} + \ldots$$

$$= \begin{bmatrix} 1+x+\frac{2x^2}{2!}+\frac{2^2x^3}{3!}+\ldots & x+\frac{2x^2}{2!}+\frac{2^2x^3}{3!}+\ldots \\ x+\frac{2x^2}{2!}+\frac{2^2x^3}{3!} & 1+x+\frac{2x^2}{2!}+\frac{2^2x^3}{3!}+\ldots \end{bmatrix}$$

$$= \frac{1}{2}\begin{bmatrix} e^{2x}+1 & e^{2x}-1 \\ e^{2x}-1 & e^{2x}+1 \end{bmatrix}$$

$\therefore f(x) = e^{2x}+1$ and $g(x) = e^{2x}-1$

1. (a) $\int \frac{e^{2x}-1}{e^{2x}+1} dx = \int \frac{e^x - e^{-x}}{e^x + e^{-x}} dx$

2. (c) $\int \frac{e^{2x}+1}{\sqrt{e^{2x}-1}} dx$

$$= \int \frac{e^{2x}}{\sqrt{e^{2x}-1}} dx + \int \frac{e^x}{e^x\sqrt{e^{2x}-1}} dx$$

$$= \sqrt{e^{2x}-1} + \sec^{-1}(e^x) + C$$

Comprehension 2

Differentiating both sides,

$$\frac{x^3-6x^2+11x-6}{\sqrt{x^2+4x+3}} = (Ax^2+Bx+C)\frac{(x+2)}{\sqrt{x^2+4x+3}} + (2Ax+B)\sqrt{x^2+4x+3} + \frac{\lambda}{\sqrt{x^2+4x+3}}$$

$$x^3-6x^2+11x-6 = (Ax^2+Bx+C)(x+2) + (2Ax+B)(x^2+4x+3)+\lambda$$

Comparing coefficients of like powers of x

$x^3: 1 = A+2A \Rightarrow A = 1/3$

$x^2: -6 = 2A+B+8A+B$

$2B = -6-10.\frac{1}{3} \Rightarrow B = -\frac{14}{3}$

$x: 11 = 2B+C+6A+4B$

$C = 11-6\left(-\frac{14}{3}\right)-6.\frac{1}{3} = 11+28-2 = 37$

Constant terms: $-6 = 2C+3B+\lambda$

$\lambda = -6-2.37-3.\left(-\frac{14}{3}\right) = -6-74+14 = -66$

1. (a)

2. (d)

3. (b)

Comprehension 3

1. (a) Let $P = \sin^3 x \cos^3 x$

$$\frac{dP}{dx} = 3\sin^2 x\cos^4 x - 3\sin^4 x\cos^2 x$$

$$= 3\sin^2 x(1-\sin^2 x)\cos^2 x - 3\sin^4 x\cos^2 x$$

$$= 3\sin^2 x\cos^2 x - 6\sin^4 x\cos^2 x$$

$\therefore P = 3I_{2,2} - 6I_{4,2} \quad \therefore I_{4,2} = \frac{1}{6}(-P+3I_{2,2})$

2. (a) Let $P=\sin^5 x\cos^3 x$

$$\therefore \frac{dP}{dx}=5\sin^4 x\cos^4 x-3\sin^6 x\cos^2 x$$

$$=5\sin^4 x\left(1-\sin^2 x\right)\cos^2 x-3\sin^6 x\cos^2 x$$

$$=5\sin^4 x\cos^2 x-8\sin^6 x\cos^2 x$$

3. (b) Let $P=\sin^5 x\cos^3 x$

$$\therefore \frac{dP}{dx}=5\sin^4 x\cos^4 x-3\sin^6 x\cos^2 x$$

$$=5\sin^4 x\cos^4 x-3\sin^4 x\left(1-\cos^2 x\right)\cos^2 x$$

$$=8\sin^4 x\cos^4 x-3\sin^4 x\cos^2 x$$

$$\therefore P=8I_{4,4}-3I_{4,2}$$

$$\therefore I_{4,2}=\frac{1}{3}\left(-P+8I_{4,4}\right)$$

Comprehension 4

1. (c) $f(x)=4x^3+3x^2+2x+1$

$\because f(x)$ is a cubic polynomial, it has at least one real root.

Also,

$$f'(x)=12x^2+6x+2$$

$$=2\left(6x^2+3x+1\right)>0\,\forall x\in R$$

$\therefore f(x)$ is a strictly increasing function.

$\Rightarrow$ There is only one real root of $f(x)=0$.

Also, $f\left(-\frac{1}{2}\right)=1-1+\frac{3}{4}-\frac{1}{2}>0$

and $f\left(-\frac{3}{4}\right)=1-\frac{3}{2}+\frac{27}{16}-\frac{27}{16}<0$

$\therefore$ Real roots lie between $-\frac{3}{4}$ and $-\frac{1}{2}$.

Hence, $s\in\left(-\frac{3}{4},-\frac{1}{2}\right)$

2. (a) $y=f(x), x=0, y=0$ and $x=t$ bounds the area as shown in the figure:

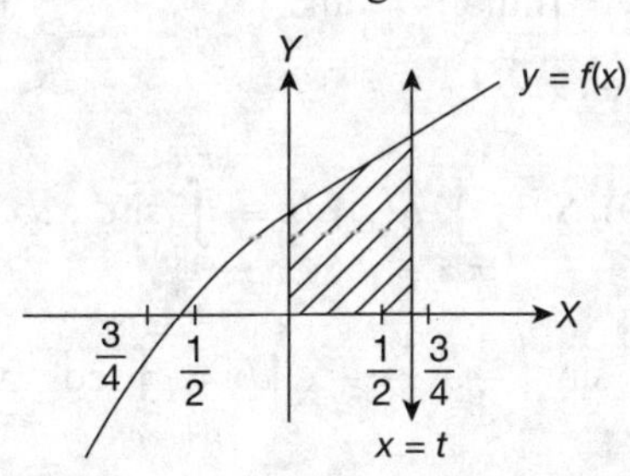

$\therefore$ The required area is given by

$$A=\int_0^t dx=\int_0^t\left(4x^3+3x^2+2x+1\right)dx$$

$$=t^4+t^3+t^2+t=t(t+1)\left(t^2+1\right)$$

Now $\frac{1}{2}<t<\frac{3}{4};\frac{3}{4}<t+1<\frac{7}{4};\frac{5}{4}<t^2+1<\frac{25}{16}$

$$\therefore \frac{1}{2}\times\frac{3}{2}\times\frac{5}{4}<A<\frac{3}{4}\times\frac{7}{4}\times\frac{25}{16}$$

or $A\in\left(\frac{15}{16},\frac{525}{256}\right)\subset\left(\frac{3}{4},3\right)$

3. (b) $f'(x)=2\left(6x^2+3x+1\right)$

$f''(x)=6(4x+1)\Rightarrow$ Critical point $x=-1/4$

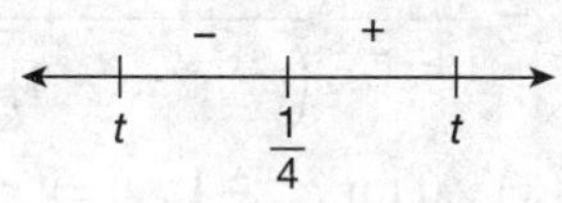

$\therefore$ The function is decreasing in $\left(-t,-\frac{1}{4}\right)$ and increasing in $\left(-\frac{1}{4},t\right)$.

Comprehension 5

1. (a) We have $f(x)=\frac{x^2-ax+1}{x^2+ax+1};0<a<2$

$$\Rightarrow f'(x)=\frac{2a\left(x^2-1\right)}{\left(x^2+ax+1\right)^2}$$

$$\Rightarrow\left(x^2+ax+1\right)^2 f'(x)=2a\left(x^2-1\right)$$

$$\Rightarrow\left(x^2+ax+1\right)^2 f''(x)+2\left(x^2+ax+1\right)$$

$(2x+a)f'(x)=4ax$ ...(i)

Putting $x=-1$ in equation (i), we get

$\left(2-a^2\right)f''(-1)=-4a$ ...(ii)

Putting $x=1$ in equation (i), we get

$\left(2+a^2\right)f''(1)=4a$ ...(iii)

Adding equations (ii) and (iii), we get

$$\left(2+a^2\right)f''(1)+\left(2-a^2\right)f''(-1)=0$$

2. (a) We have $f'(x)=\frac{2a\left(x^2-1\right)}{\left(x^2+ax+1\right)^2}$

$f'(x)=0\Rightarrow x=-1,1$ are the critical points.

$\therefore f(x)$ is decreasing on $(-1,1)$

Also using Eq. (i), $f''(-1)$

$$= \frac{-4a}{(2-a)^2} < 0 \text{ and } f''(1) = \frac{4a}{(2+a)^2} < 0$$

$\therefore x = -1$ is a point of local maximum and $x = 1$ is a point of local minimum.

3. (b) $g(x) = \int_0^{e^x} \frac{f'(t)}{1+t^2}\,dt \Rightarrow g'(x) = \frac{f'(e^x)}{1+e^{2x}}$

$$= \frac{2a(e^{2x}-1)e^x}{(e^{2x}+ae^x+1)^2(1+e^{2x})}$$

$$= \frac{2ae^x}{1+e^{2x}} \cdot \frac{e^{2x}-1}{(e^{2x}+ae^x+1)^2}$$

Now $g'(x) > 0$ for $e^{2x} - 1 > 0 \Rightarrow x > 0$

and $g'(x) < 0$ for $e^{2x} - 1 < 0 \Rightarrow x < 0$

$\therefore g'(x)$ is negative on $(-\infty, 0)$ and positive on $(0, \infty)$.

Comprehension 6

1. (a) $A = \int_1^{\infty} [\operatorname{cosec}^{-1} x]\,dx$

$$= \int_1^{\operatorname{cosec}1} 1dx + \int_{\operatorname{cosec}1}^{\infty} 0dx = \operatorname{cosec} 1 - 1$$

2. (b)

$$= \int_1^{100} [\sec^{-1} x]\,dx = \int_1^{\sec 1} 0dx + \int_{\sec 1}^{100} 1dx = 100 - \sec 1$$

3. (b) $\int_{\operatorname{cosec}1-1}^{100-\sec 1} [\tan^{-1} x]\,dx = 100 - \tan 1 - \sec 1$

Comprehension 7

1. (c) $I = \int_{-\pi/4}^{\pi/4} \ln\left(\frac{\sin x + \cos x}{\cos x - \sin x}\right)dx$

$I = 0 \qquad (\because f(x) = -f(-x))$

2. (b) $I = \int_{-\pi/4}^{\pi/4} \ln(\sin x + \cos x)\,dx$

$$= \int_{-\pi/4}^{\pi/4} \ln(\cos x - \sin x)\,dx$$

$$\therefore 2I = \int_{-\pi/4}^{\pi/4} \ln(\cos^2 x - \sin^2 x)\,dx$$

$$= \int_{-\pi/4}^{\pi/4} \ln(\cos 2x)\,dx = 2\int_0^{\pi/4} \ln(\cos 2x)\,dx$$

$$\therefore I = \int_0^{\pi/4} \ln(\cos 2x)\,dx$$

$$= \frac{1}{2}\int_0^{\pi/2} \ln(\cos t)\,dt = \frac{1}{2}\left(-\frac{\pi}{2}\ln 2\right) = -\frac{\pi}{4}\ln 2$$

3. (d) $I = \int_0^{\frac{\pi}{4}} \ln(\sin 2x)\,dx$

Let $2x = t$, $2\,dx = dt$

$$I = \frac{1}{2}\int_0^{\frac{\pi}{2}} \ln \sin t\,dt = \frac{1}{2}\left(-\frac{\pi}{2}\ln 2\right) = -\frac{\pi \ln 2}{4}$$

Comprehension 8

1. (a) $S = \lim_{n\to\infty} \sum_{r=1}^{n} \frac{n}{(n+r)\sqrt{r(2n+r)}}$

$$= \lim_{n\to\infty} \sum_{r=1}^{n} \frac{1}{n\left(1+\frac{r}{n}\right)\sqrt{\frac{r}{n}\left(2+\frac{r}{n}\right)}}$$

$$S = \int_0^1 \frac{dx}{(1+x)\sqrt{2x+x^2}} = \int_0^1 \frac{dx}{(1+x)\sqrt{(1+x)^2-1}}$$

$$S = \left[\sec^{-1}(1+x)\right]_0^1 = \sec^2 2 - \sec^{-1} 1 = \frac{\pi}{3}$$

2. (a) $f(x) = \tan^{-1} x$

$$T_n = \frac{1}{n} f\left(\frac{n}{n}\right) = \frac{1}{n} f(1) = \frac{\pi}{4n}$$

3. (b) $S = \lim_{n\to\infty} \sum_{r=0}^{2n-1} \frac{1}{n}\sec^2\left(\frac{r}{n}\right) = \int_0^2 \sec^2 x\,dx$

$$= [\tan x]_0^2 = \tan 2$$

Comprehension 9

1. (2.00) $S_i = \int_{\pi/8}^{3\pi/8} f(x)\,dx \equiv \int_{\pi/8}^{3\pi/8} \sin^2 x dx \qquad \text{...(i)}$

$$= \int_{\pi/8}^{3\pi/8} \sin^2\left(\frac{\pi}{8} + \frac{3\pi}{8} - x\right)dx = \int_{\pi/8}^{3\pi/8} \cos^2 x dx \qquad \text{...(ii)}$$

Adding (i) and (ii), we get

$$2S_1 - \int_{\pi/8}^{3\pi/8}\left(\sin^2 x+\cos^2 x\right)dx = \frac{3\pi}{8}-\frac{\pi}{8}=\frac{\pi}{4}$$

$$\Rightarrow \frac{16S_1}{\pi}=2$$

2. (1.50)

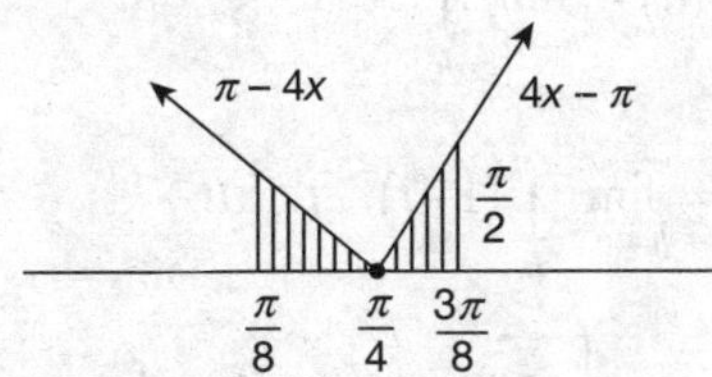

$$S_2 = \int_{\pi/8}^{3\pi/8} f(x)g_2(x)dx$$

$$= \int_{\pi/8}^{3\pi/8} \sin^2 x|4x-\pi|dx \quad \text{...(i)}$$

$$= \int_{\pi/8}^{3\pi/8} \sin^2\left(\frac{\pi}{8}+\frac{3\pi}{8}-x\right)\left|4\left(\frac{\pi}{8}+\frac{3\pi}{8}-x\right)-\pi\right|dx$$

$$= \int_{\pi/8}^{3\pi/8} \left(\cos^2 x\right)|\pi-4\pi|dx$$

$$= \int_{\pi/8}^{3\pi/8} \cos^2 x|4\pi-\pi|dx \quad \text{...(ii)}$$

Adding equations (i) and (ii) we get

$$2S_2 = \int_{\pi/8}^{3\pi/8} |4\pi-\pi|\left(\sin^2 x+\cos^2 x\right)dx$$

$$= \int_{\pi/8}^{3\pi/8} |4\pi-\pi|dx$$

$$=2\times\frac{1}{2}\times\frac{\pi}{8}\times\frac{\pi}{2}=\frac{\pi^2}{16}$$

$$\Rightarrow \frac{48S_2}{\pi^2}=\frac{3}{2}=1.5$$

Comprehension 10

1. (57.00) $f_1(x)=\int_0^x\prod_{j=1}^{21}(t-j)^j\,dt$

$$f_1'(x)=\prod_{j=1}^{21}(x-j)^j$$

$$=(x-1)(x-2)^2(x-3)^3\ldots(x-21)^{21}$$

Checking the sign scheme of $f_1'(x)$ at $x=1,2,3,\ldots 21$, we get

$f_1(x)$ has local minima at $x=1$, 5, 9, 13, 17, 21 and local maxima at 3, 7, 11, 15, 19.

$\Rightarrow m_1=6, n_1=5$

So, $2m_1+3n_1+m_1n_1$

$=2\times6+3\times5+6\times5=57$

2. (6.00) $f_2(x)=98(x-1)^{50}-600(x-1)^{49}+2450$

$\Rightarrow f_2'(x)=2\times49\times50(x-1)^{49}-50\times12\times49(x-1)^{48}$

$=50\times49\times2(x-1)^{48}(x-1-6)$

$=50\times49\times2(x-1)^{48}(x-7)$

$f_2(x)$ has local minimum at $x=7$ and no local maxima.

$\Rightarrow m_2=1, n_2=0$

$6m_2+4n_2+8m_2n_2$

$=6\times1+4\times0+8\times1=6$

Comprehension 11

1. (a,b,c) $f(x)=xF(x)\Rightarrow f'(x)=F(x)+xF'(x)$

$\therefore f'(1)=F(1)+F'(1)=F'(1)<0$

$$\left(\because F'(x)<0, x\in\left(\frac{1}{2},3\right)\right),$$

$f(2)=2F(2)<0$,

$$\left(\because F'(x)<0\Rightarrow F \text{ is decreasing on }\left(\frac{1}{2},3\right) \text{ and } F'(1)=0, F(3)=-4\right)$$

$f'(x)=F(x)+xF'(x)$

For the same reason given above and given $F'(x)<0$,

$\therefore f'(x)\neq 0, x\in(1,3)$

2. (c, d) $\int_1^3 x^2F'(x)dx=-12$

$\Rightarrow\left[x^2F(x)\right]_1^3-\int_1^3 2xF(x)dx=-12$

$\Rightarrow 9F(3)-F(1)-2\int_1^3 xF(x)dx=-12$

$\Rightarrow \int_1^3 xF(x)dx=-12\Rightarrow\int_1^3 f(x)dx=-12 \quad$ (i)

Also, $\int_1^3 x^3F''(x)dx=40$

$$\Rightarrow \left[x^3F'(x)\right]_1^3 - 3\int_1^3 x^2F'(x)\,dx = 40$$

$$\Rightarrow \left[x^2\left(f'(x)-F(x)\right)\right]_1^3 - 3\times(-12) = 40$$

$$\left[\begin{array}{l}\text{Using } xF'(x) = f'(x) - F(x) \text{ and} \\ \qquad\qquad \int_1^3 x^2F'(x)\,dx = -12\end{array}\right.$$

$$\Rightarrow 9\left(f'(3)-F(3)\right) - \left(f'(1)-F(1)\right) = 4$$

$$\Rightarrow 9f'(3) - 9\times(-4) - f'(1) + 0 = 4$$

$$\Rightarrow 9f'(3) - f'(1) + 32 = 0$$

Comprehension 12

1. (a) $g(a) = \lim\limits_{h\to 0^+} \int_h^{1-h} t^{-a}(1-t)^{a-1}\,dt$

$$\therefore g\left(\frac{1}{2}\right) = \lim_{h\to 0^+} \int_h^{1-h} t^{-1/2}(1-t)^{-1/2}\,dt$$

$$= \lim_{h\to 0^+} \int_h^{1-h} \frac{1}{\sqrt{t(1-t)}}\,dt$$

$$= \lim_{h\to 0^+} \int_h^{1-h} \frac{1}{\sqrt{\left(\frac{1}{2}\right)^2 - \left(t-\frac{1}{2}\right)^2}}\,dt$$

$$= \lim_{h\to 0^+} \left[\sin^{-1}\left(\frac{t-\frac{1}{2}}{\frac{1}{2}}\right)\right]_h^{1-h} = \lim_{h\to 0^+} \sin^{-1}(2t-1)\Big]_h^{1-h}$$

$$= \lim_{h\to 0^+} \left[\sin^{-1}(1-2h) - \sin^{-1}(2h-1)\right]$$

$$= \frac{\pi}{2} - \left(-\frac{\pi}{2}\right) = \pi$$

2. (d) $g(a) = \lim\limits_{h\to 0^+} \int_h^{1-h} t^{-a}(1-t)^{a-1}\,dt$

$$g(a) = \lim_{h\to 0^+} \int_h^{1-h} (1-t)^{-a}\,t^{a-1}dt$$

$$\left(\text{Using } \int_a^b f(x)\,dx = \int_a^b f(a+b-x)\,dx\right)$$

Also, $g(1-a) = \lim\limits_{h\to 0^+} \int_h^{1-h} t^{a-1}(1-t)^{-a}\,dt$

Thus, $g(a) = g(1-a)$

$$\Rightarrow g'(a) = -g'(1-a) \Rightarrow g'(a) + g'(1-a) = 0$$

Putting $a = \frac{1}{2}$ we get $g'\left(\frac{1}{2}\right) + g'\left(\frac{1}{2}\right)$

$$= 0 \text{ or } g'\left(\frac{1}{2}\right) = 0.$$

CHAPTER 8

Application of Integrals

Objectives

In this chapter on Application of Integrals, we shall study the following topics:

1. Area under simple curves
2. Area of the region bounded by a curve and a line
3. Area between two curves.

Introduction

In our earlier classes, we have learnt in Mensuration, the formula to calculate the areas of geometrical figures like triangles, rectangles, trapeziums, circles etc. These formulae allow us to calculate the areas of geometrical figures. However, they are inadequate to calculate the areas enclosed by curved surfaces. In the previous chapter we have studied the area bounded by a curve $y = f(x)$, i.e., the area

$$A = \int_a^b f(x)\,dx$$

In the present chapter, we will study the application of integrals to find the areas under simple curves especially lines, circles, parabolas, ellipses.

Computing Areas Bounded by Simple Curves

1. If $y = f(x)$ be a continuous function in $[a,b]$, then area of plane figure bounded by $y = f(x)$, $x = a, x = b$ and x-axis $(b > a)$ is given by

$$A = \int_a^b y\,dx = \int_a^b f(x)\,dx$$

Here $f(x)$ is above the x-axis, therefore $\int_a^b f(x)\,dx$ is

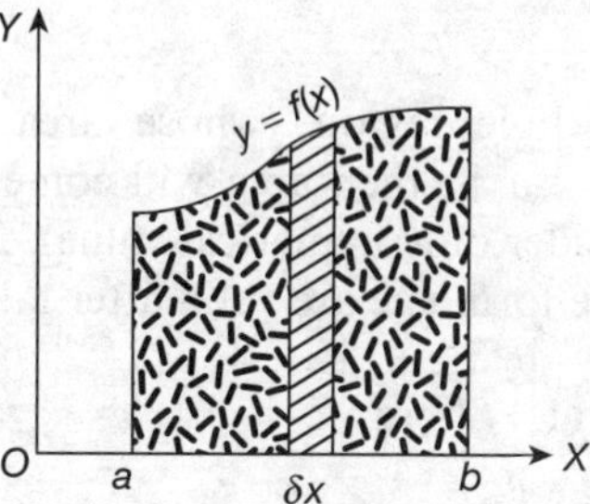

2. Area bounded by the curve $x = f(y)$, y-axis, $y = a, y = b (b > a)$ is given by

$$A = \int_a^b x\,dy = \int_a^b f(y)\,dy$$

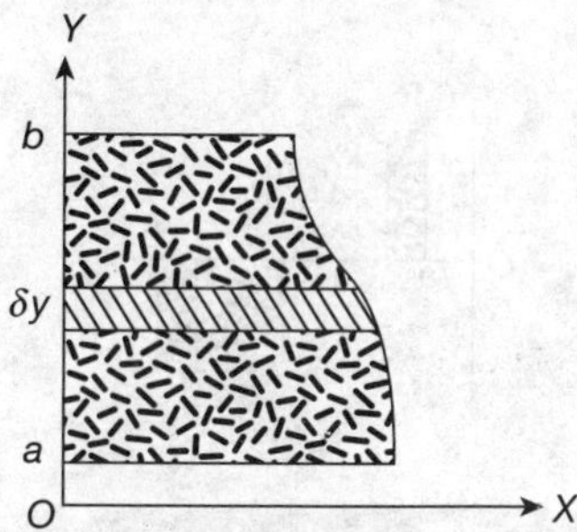

3. Let $y = f(x)$ and $y = g(x)$ be two curves such that the graph of $f(x)$ lies above the graph of $g(x)$ and both are above x-axis; then, area bounded by $y = f(x)$, $y = g(x)$, $x = a, x = b (b > a)$ is given by

$$A=\int_a^b f(x)dx-\int_a^b g(x)dx=\int_a^b [f(x)-g(x)]dx$$

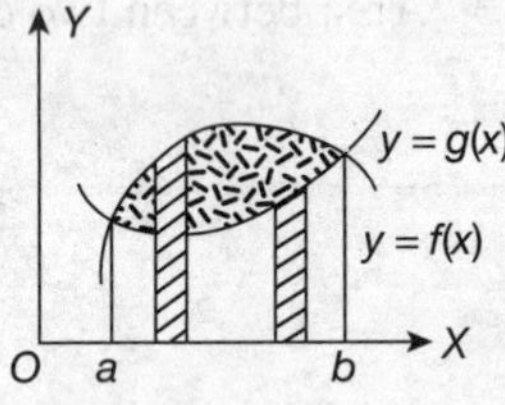

4. If $x=a$ and $x=b$ are abscissas of points of intersection of the given curves,

$$A=\int_a^b (f(x)-g(x))dx$$

Remarks

1. If the whole region (whose area is required) is below x-axis, then area will come out negative (we consider only numerical value). In such cases, the above formulae are valid after taking modulus of integral $\left|\int_a^b f(x)dx\right|$.
2. If some part of the bounded region is below x-axis (see figure), then the above formulae may not work. First, we divide the regions above x-axis and below x-axis and compute area separately.

$$A=\left|\int_a^c f(x)dx\right|+\left|\int_c^b f(x)dx\right|$$

ILLUSTRATION 8.1

Find the area bounded by the curve $y=x^3$, x-axis and ordinates $x=1$ and $x=2$.

Solution

Required area $=\int_1^2 ydx=\int_1^2 x^3dx=\left[\frac{x^4}{4}\right]_1^2=\frac{15}{4}$

ILLUSTRATION 8.2

Find the area bounded by the curve $y=mx$, x-axis and ordinates $x=1$ and $x=2$.

Solution

Required area $=\int_1^2 ydx=\int_1^2 mxdx=\left[\frac{mx^2}{2}\right]_1^2$

$=\frac{m}{2}(4-1)=\left(\frac{3}{2}\right)m$

ILLUSTRATION 8.3

Find the area bounded by the curves $x=a\cos t, y=b\sin t$ in the first quadrant.

Solution

Clearly, the given equations are the parametric equation of ellipse $\frac{x^2}{a^2}+\frac{y^2}{b^2}=1$.

The curve meets the x-axis in the first quadrant at $(a,0)$

$\therefore$ Required area $=\int_0^a ydx=\int_{\pi/2}^0 (b\sin t)(-a\sin t)dt$

$(\because at\, x=0, t=\pi/2 \text{ and } x=a, t=0)$

$$=ab\int_0^{\pi/2}\sin^2 tdt=\left(\frac{\pi ab}{4}\right)$$

ILLUSTRATION 8.4

Find the area bounded by the curve $xy^2=a^2(a-x)$ and x-axis.

Solution

$xy^2=a^2(a-x)\Rightarrow y^2=a^2(a-x)/x$ or

$$y=a\sqrt{\frac{a-x}{x}} \quad \text{...(i)}$$

Here, the power of y is even; so the curve is symmetric w.r.t. x-axis. Putting $x=a$ in equation (i), we get $y=0$ and at $x=0$, we get $y=\infty$.

$\therefore$ Required area

$$A=A_1+A_2=2A_1$$

$$A=2\int_{x=0}^{x=a} ydx=2\int_0^a a\sqrt{\frac{a-x}{x}}dx$$

(From equation (i))

Putting $x = a\sin^2\theta$

$$dx = 2a\sin\theta\cos\theta d\theta$$

When $x = 0 \Rightarrow \theta = 0 \Rightarrow x = a \Rightarrow \theta = \frac{\pi}{2}$

$$\therefore \quad \text{Area} = 2\int_0^{\pi/2} a\frac{\cos\theta}{\sin\theta}.2a\cos\theta\sin\theta d\theta$$

$$= 4a^2\int_0^{\pi/2}\cos^2\theta d\theta$$

$$= 4a^2.\frac{1}{2}.\frac{\pi}{2} = \pi a^2$$

ILLUSTRATION 8.5

Find the area between the curve $y = x(x-1)(x-2)$ and the x-axis.

Solution

The required area

$$= \int_0^1 x(x-1)(x-2)dx + \left|\int_1^2 x(x-1)(x-2)dx\right|$$

$$= \frac{1}{4}\left(x^4 - 4x^3 + 4x^2\right)_0^1 + \frac{1}{4}\left|\left(x^4 - 4x^3 + 4x^2\right)_1^2\right|$$

$$= \frac{1}{4}\left[1 + \left|(16 - 32 + 16) - (1 - 4 + 4)\right|\right] = \frac{1}{4}$$

ILLUSTRATION 8.6

The area between the curve $y = 4 + 3x - x^2$ and x-axis is ______.

(a) $125/6$ **(b)** $125/3$

(c) $125/2$ **(d)** None of these.

Solution

Putting $y = 0$, we get

$x^2 = 3x - 4 = 0 \Rightarrow (x-4)(x+1) = 0$

$\Rightarrow x = -1$ or $x = 4$

$$\therefore \text{Required area} = \int_{-1}^{4}\left(4 + 3x - x^2\right)dx$$

$$= \left(4x + \frac{3x^2}{2} - \frac{x^3}{3}\right)_{-1}^{4} = \frac{125}{6}$$

ILLUSTRATION 8.7

If area bounded by the curve $y = 8x^2 - x^5$ and ordinates $x = 1, x = k$ is $\frac{16}{3}$, then $k =$ ______.

(a) 2 **(b)** $\left[8 - \sqrt{17}\right]^{1/3}$

(c) $\left[\sqrt{17} - 8\right]^{1/3}$ **(d)** -1

Solution

$$\int_1^k \left(8x^2 - x^5\right)dx = \frac{16}{3} \Rightarrow \left[\frac{8x^3}{3} - \frac{x^6}{6}\right]_1^k = \frac{16}{3}$$

$$\Rightarrow \quad \frac{8}{3}\left(k^3 - 1\right) - \left(\frac{k^6 - 1}{6}\right) = \frac{16}{3}$$

$$\Rightarrow \quad 16k^3 - k^6 - 15 = 32$$

$$\Rightarrow \quad k^6 - 16k^3 + 47 = 0 \Rightarrow k^3 = 8 \pm \sqrt{17}$$

$$\Rightarrow \quad k = \left(8 \pm \sqrt{17}\right)^{1/3}$$

ILLUSTRATION 8.8

Find the area between two curves $y = x^3$ and $y = 4x$ in the first quadrant.

Solution

Solving the given equations, we get $x = 0$ and $x = \pm 2$, but we have to obtain area in the first quadrant only. Therefore, required area

$$= \int_0^2 \left(4x - x^3\right)dx = \left(2x^2 - \frac{x^4}{4}\right)_0^2 = 4 \text{ units}$$

ILLUSTRATION 8.9

Find the area between two curves $y^2 = 4ax$ and $x^2 = 4ay$.

Solution

By solving the equation of two given curves $y^2 = 4ax$ and $x^2 = 4ay$, we get $x = 4a$

So required area

$$= \int_0^{4a}\left(\sqrt{4ax} - \frac{x^2}{4a}\right)dx = \left[2\sqrt{a}.\frac{x^{3/2}}{3/2} - \frac{x^3}{12a}\right]_0^{4a}$$

$$= \frac{4\sqrt{a}}{3}(4a)^{3/2} - \frac{64a^3}{12a} = \frac{16}{3}a^2$$

ILLUSTRATION 8.10

Find the area bounded between curve $y = x^3$ and line $y = x$.

Solution

By solving $y = x^3$ and $y = x$, we get $x = 0, -1, 1$

It is clear from the given figure that the required area is symmetrical about the origin.

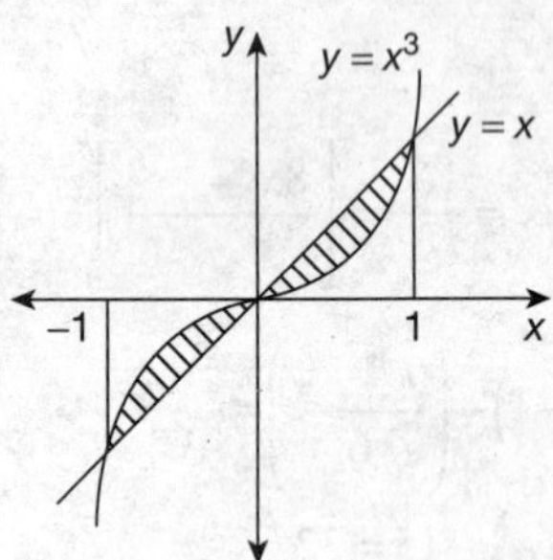

$\therefore$ Required area $= 2\int_0^1 \left(x - x^3\right)dx$

$$= 2\left[\frac{x^2}{2} - \frac{x^4}{4}\right] = \frac{1}{2}$$

ILLUSTRATION 8.11

The area bounded by the circle $x^2 + y^2 = 1$ and the curve $|x| + |y| = 1$ is ______.

(a) $\pi - 2$ **(b)** $\pi - 2\sqrt{2}$

(c) $2\left(\pi - 2\sqrt{2}\right)$ **(d)** None of these.

Solution

By changing x as $-x$ and y as $-y$, both the given equations remain unchanged. So required area will be symmetric w.r.t both the axes, which is shown in figure. So required area is

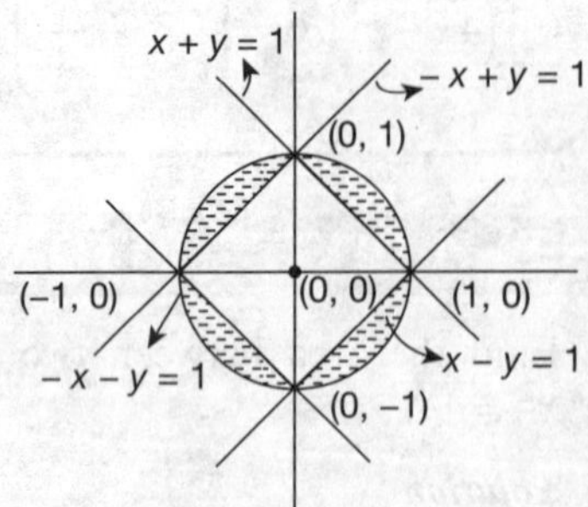

$$= 4\int_0^1 \left[\sqrt{1 - x^2} - (1 - x)\right]dx$$

$$= 4\left[\frac{x}{2}\sqrt{1 - x^2} + \frac{1}{2}\sin^{-1} x - x + \frac{x^2}{2}\right]_0^1$$

$$= 4\left[0 + \frac{1}{2}.\frac{\pi}{2} - 1 + \frac{1}{2}\right] = \pi - 2$$

ILLUSTRATION 8.12

The area between the parabola $x^2 = 4y$ and line $x = 4y - 2$ is ______.

(a) $9/4$ **(b)** $9/8$

(c) $9/2$ **(d)** 9

Solution

Solving the equation of the given curves for x, we get

$x^2 = x + 2$

$\Rightarrow (x - 2)(x + 1) = 0$

$\Rightarrow x = -1, 2$

So, required area

$$= \int_{-1}^{2}\left[\frac{x + 2}{4} - \frac{x^2}{4}\right]dx = \frac{1}{4}\left[\frac{x^2}{2} + 2x - \frac{x^3}{3}\right]_{-1}^{2}$$

$$= \frac{1}{4}\left[(2 + 4 - 8/3) - (1/2 - 2 + 1/3)\right] = 9/8$$

ILLUSTRATION 8.13

The area between the curves $y = \cos x$ and the line $y = x + 1$ in the second quadrant is ______.

(a) 1 **(b)** 2

(c) $3/2$ **(d)** $1/2$

Solution

Let the line $y = x + 1$, meet the x-axis at the point A(0, 1). Also suppose that the curve $y = \cos x$ meets x-axis and y-axis respectively at the points C and A. From the adjoint figure, it is obvious that

Required area = Area of ABC

= Area of OAC – Area of OAB

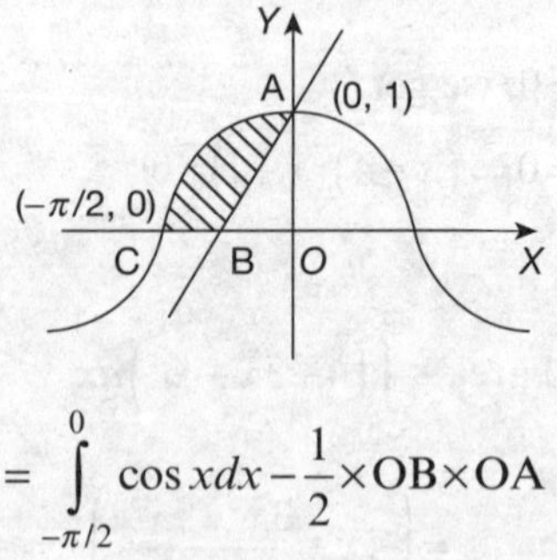

$$= \int_{-\pi/2}^{0} \cos x dx - \frac{1}{2} \times \text{OB} \times \text{OA}$$

$$= 1 - (1/2) = (1/2)$$

ILLUSTRATION 8.14

The area bounded by curves $3x^2+5y=32$ and $y=|x-2|$ is ______.

(a) 25 **(b)** $17/2$

(c) $33/2$ **(d)** 33

Solution

Here, the first curve can be written in the form $x^2=-\frac{5}{3}\left(y-\frac{32}{5}\right)$, which is a parabola whose vertex lies on the y-axis.

Again, the second curve is given by $y=\begin{cases} x-2, x\geq 2 \\ -(x-2), x<2 \end{cases}$ which consists of two perpendicular lines AB and AC as shown in figure.

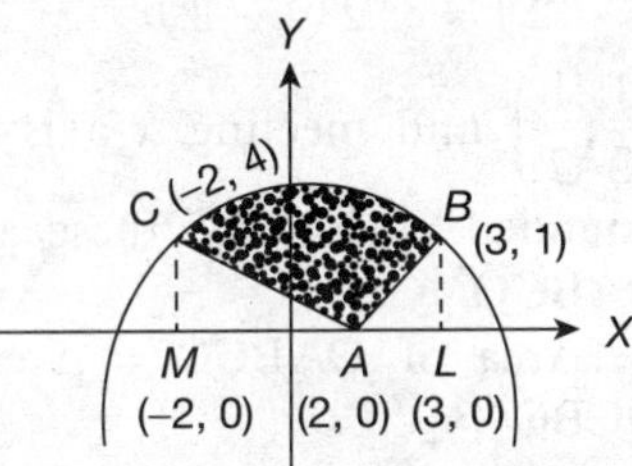

These lines meet the parabola at B (3, 1) and C (–2, 4).

Hence, the required area A is given by

$$A=\int_{-2}^{3} y\,dx-\Delta ABL-\Delta ACM$$

$$=\int_{-2}^{3}\frac{1}{5}\left(32-3x^2\right)dx-\frac{1}{2}.1.1-\frac{1}{2}(4+4)$$

$$=\frac{1}{5}\left[32x-x^3\right]_{-2}^{3}-\frac{17}{2}=\frac{1}{5}[69+56]-\frac{17}{2}=\frac{33}{2}$$

ILLUSTRATION 8.15

Compute the area of the figure bounded by the straight lines $x=0, x=2$ and the curves $y=2^x, y=2x-x^2$.

Solution

$\text{Max}(y)=1$ at $x=1$ and $x=1$ and $2^x\geq 1$. The required region is OPABCO.

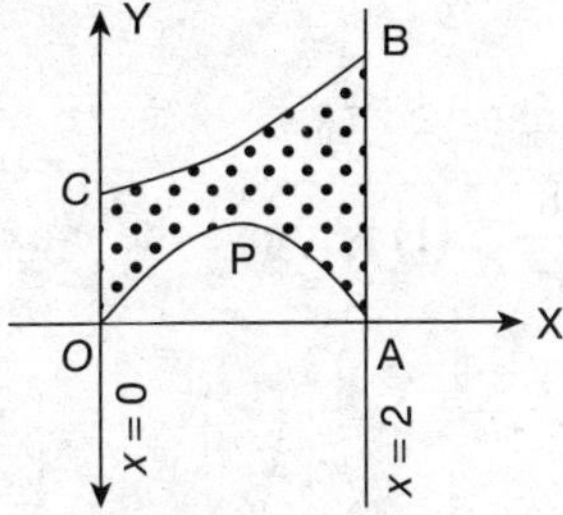

Area = Area of OABCO – Area of OPAO

$$=\int_0^2 2^x\,dx-\int_0^2\left(2x-x^2\right)dx$$

$$=\left[2^x\log_2 e\right]_0^2-\left[x^2-\frac{x^3}{3}\right]_0^2$$

$$=\frac{3}{\ln 2}-\frac{4}{3}$$

WORKSHEET

1. Find the area bounded by the curves $x^2+y^2=25, 4y=\left|4-x^2\right|$ and $x=0$ above the x-axis.
2. Sketch the region bounded by the curves $y=\sqrt{5-x^2}$ and $y=|x-1|$ and find its area.
3. Find the area of the region bounded by the x-axis and the curves defined by $y=\tan x, -\frac{\pi}{3}\leq x\leq\frac{\pi}{3}$ and $y=\cot x, \frac{\pi}{6}\leq x\leq\frac{\pi}{3}$.
4. Find the area bounded by the x-axis, part of the curve $y=\left(1+\frac{8}{x^2}\right)$ and the ordinates at $x=2$ and $x=4$. If the ordinate at $x=a$ divides the area into two equal parts, then find a.

SOLVED EXAMPLES

Example 1. Find the area bounded by the x-axis, part of the curve $y = 1 + \frac{8}{x^2}$ and the ordinate at $x = 2$ and $x = 4$. If the ordinate at $x = a$ divides the area into two equal parts, find a.

Solution

It can be seen that $y \geq 1 \forall x$

Curve, $y = 1 + \frac{8}{x^2}$ ...(1) is above the x-axis

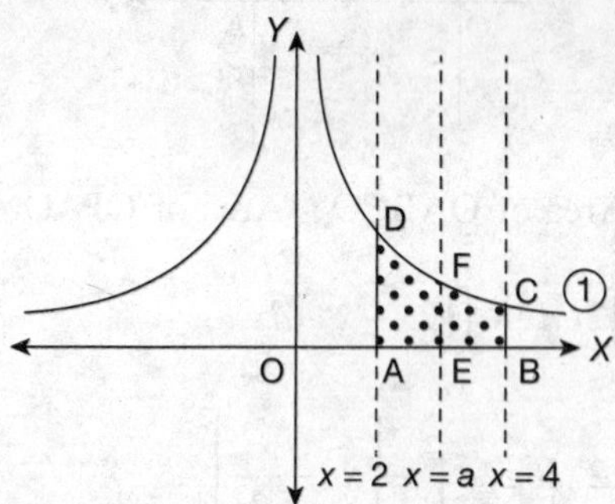

Now, the required region is ABCDA.

$$\text{Area} = \int_2^4 \left(1 + \frac{8}{x^2}\right) dx$$

$$= \left[x - \frac{8}{x}\right]_2^4 = (4-2) - (2-4) = 2 + 2 = 4 \text{ sq. units}$$

Now,

$$\int_2^a \left(1 + \frac{8}{x^2}\right) dx = 2 \Rightarrow \left(a - \frac{8}{a}\right) + 2 = 2 \Rightarrow a - \frac{8}{a} = 0$$

$\Rightarrow a^2 = 8 \Rightarrow 1 = \pm 2\sqrt{2}$

But $2 < a < 4$, therefore $a = 2\sqrt{2}$

Example 2. Indicate the region bounded by the curves $y = x \log x$ and $y = 2x - 2x^2$ and obtain the area enclosed by them.

Solution

Curve $y = x \log x$ ...(1) is defined for $x > 0$

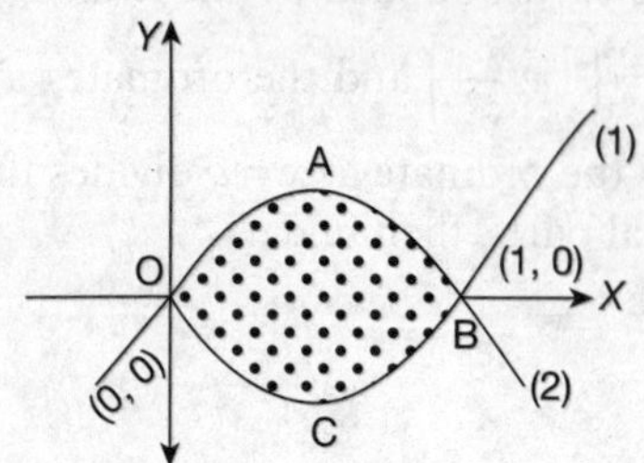

About curve (1):

(1) for $0 < x < 1, y < 0$

(2) $\lim\limits_{x \to 0} x \log x = \lim\limits_{x \to 0^+} \frac{\log x}{(1/x)} = \lim\limits_{x \to 0^+} \frac{1/x}{(-1/x^2)} = 0$

(3) The curve meets x axis at $x = 1$.

For $x > 1 \Rightarrow y > 0$ and $x \to \infty$

$\Rightarrow y \to \infty$.

Curve $y = 2x - 2x^2$ ...(2) can be expressed as $\left(x - \frac{1}{2}\right)^2 = -\frac{1}{2}\left(y - \frac{1}{2}\right)$. It is a parabola with vertex at $\left(\frac{1}{2}, \frac{1}{2}\right)$ and meeting x-axis at $x = 0$ and $x = 1$. The bounded region by these two curves is given by OABCO.

Area of OABCO = Area of OABO + Area of OCBO

$$= \int_0^1 (2x - 2x^2)\, dx + \left|\int_0^1 x \log x\, dx\right|$$

$$= \left[x^2 - \frac{2x^3}{3}\right]_0^1 + \left|\left[\frac{x^2}{2}\log x - \frac{x^2}{4}\right]_0^1\right|$$

$$= \left(1 - \frac{2}{3}\right) + \left|-\frac{1}{4}\right| \qquad \left[\text{Note: } \lim_{x \to 0} \frac{x^2}{2}\log x = 0\right]$$

$$= 1 - \frac{2}{3} + \frac{1}{4} = \frac{12 - 8 + 3}{12} = \frac{7}{12}$$

Example 3. Find the area given by $x + y \leq 6$, $x^2 + y^2 \leq 6y$ and $y^2 \leq 8x$.

Solution

Consider equality sign, then equations of curves are:

$x + y = 6$ ...(1), $x^2 + y^2 - 6y = 0$ ...(2), $y^2 = 8x$ ...(3)

Solving (1) and (2) $\Rightarrow$ A (3, 3)

Solving (1) and (3) $\Rightarrow$ B (2, 4)

Now $x + y \leq 6$ consists of the region below (and on) the line $x + y = 6$ (containing origin); $x^2 + y^2 - 6y \leq 0$ is consists of the interior region and boundary of the circle $x^2 + (y-3)^2 = 3^2$; and $y^2 \leq 8x$ consists of the interior region (and boundary) of the parabola $y^2 = 8x$ (see figure). The required region is OBAO.

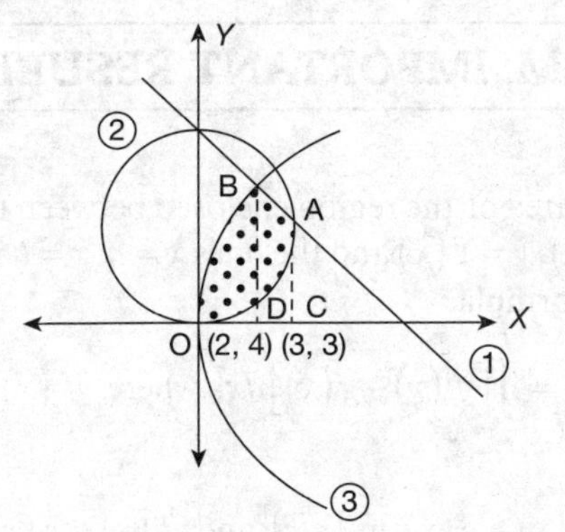

Area of the region = Area of OBDO + Area of BACDB – Area of OACDO.

$$= \int_0^2 2\sqrt{2}\sqrt{x}dx + \int_2^3 (6-x)dx - \int_0^3 \left\{3+\sqrt{9-x^2}\right\}dx$$

$$= 2\sqrt{2}\left[\frac{x^{3/2}}{3/2}\right]_0^2 + \left[6x - \frac{x^2}{2}\right]_2^3 - [3x]_0^3 + \left[\frac{9}{2}\sin^{-1}\frac{x}{3} + \frac{x\sqrt{9-x^2}}{2}\right]_0^3$$

$$= \frac{16}{3} + \frac{7}{2} - 9 + \frac{9\pi}{4} = \frac{1}{12}(27\pi - 2)$$

Example 4. Find the area of the region R given by $R = \{(x, y) : 4 \le x^2 + y^2 \le 2(|x|+|y|)\}$

Solution

The given curve $4 \le x^2 + y^2 \le 2(|x|+|y|)$ is symmetrical about x-axis as well as about y-axis, therefore required area is 4 times the area of the region in 1st quadrant. Now in 1st quadrant, the given inequalities consist of two inequalities:

$$x^2 + y^2 \ge 4 \quad \text{...(i)}$$

and $$x^2 + y^2 - 2x - 2y \le 0 \quad \text{...(ii)}$$

Equation (i) represents exterior (and boundary) of circle $x^2 + y^2 = 4$ and

Equation (ii) represents interior (and boundary) of circle $(x-1)^2 + (y-1)^2 = 2$

Therefore, the required region in 1st quadrant is ABCDA.

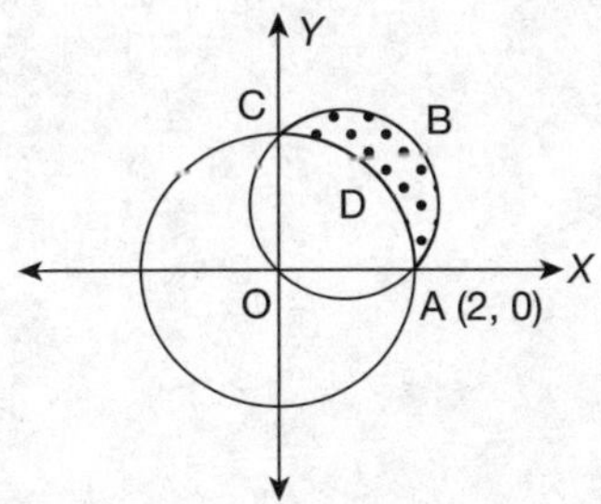

Area = Area of OABCO – Area of OADCO

$$= \int_0^2 \left\{1+\sqrt{2-(x-1)^2}\right\}dx - \int_0^2 \sqrt{4-x^2}dx$$

$$= [x]_0^2 + \left[\sin^{-1}\left(\frac{x-1}{\sqrt{2}}\right) + \frac{(x-1)}{2}\sqrt{2-(x-1)^2}\right]_0^2 - \left[-2\sin^{-1}\frac{x}{2} + \frac{x}{2}\sqrt{4-x^2}\right]_0^2$$

$$= 2 + \sin^{-1}\frac{1}{\sqrt{2}} + \frac{1}{2} - \sin^{-1}\left(-\frac{1}{\sqrt{2}}\right) + \frac{1}{2} - 2\sin^{-1}(1)$$

$$= 2 + 2.\frac{\pi}{4} + 1 - 2\frac{\pi}{2} = \left(3 - \frac{\pi}{2}\right)$$

Therefore, required area $= 4\left(3 - \frac{\pi}{2}\right)$

Example 5. Find the area bounded by parabola $y = x^2 - 2x + 2$, the line tangent to it at the point M (3, 5) and the axis of ordinates.

Solution

Given parabola is $y = x^2 - 2x + 2 \Rightarrow y = (x-1)2 + 1$

$$\Rightarrow \quad (x-1)2 = (y-1) \quad \text{...(1)}$$

Its vertex is (1, 1) and it is above x-axis.

Equation of tangent to (1) at M (3, 5) is given by $y = 4x - 7$ ...(2)

Equation (2) meets x-axis at $B\left(\frac{7}{4}, 0\right)$ and y-axis at $A(0, -7)$

Therefore, required region is ABMNOA.

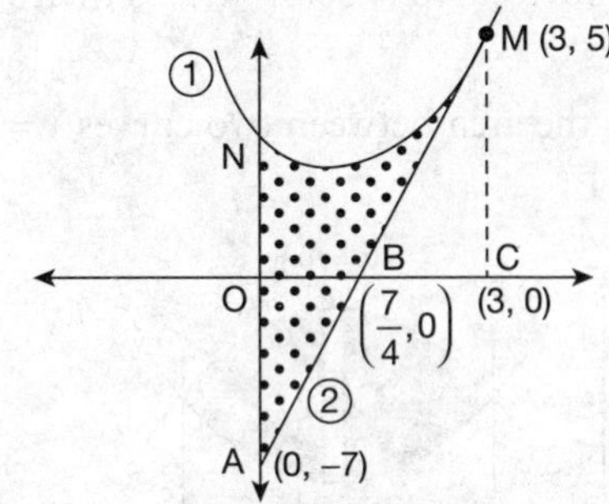

Required area = Area OABO + Area ONMCO – Area BCMB

$$= \frac{1}{2} \times \frac{7}{4} \times 7 + \int_0^3 (x^2 - 2x + 2)dx - \frac{1}{2} \times \left(3 - \frac{7}{4}\right) \times 5$$

$$= \frac{49}{8} + \left[\frac{x^3}{3} - x^2 + 2x\right]_0^3 - \frac{25}{8} = 3 + 6 = 9 \text{ sq.units.}$$

RECAPITULATION (FORMULAE, THEOREM, IMPORTANT RESULTS)

1. The area of the region bounded by the curve $y = f(x)$, x-axis and the lines $x = a$ and $x = b(b > a)$ is given by the formula: Area $= \int_a^b y dx = \int_a^b f(x) dx$.
2. The area of the region bounded by the curve $x = \phi(y)$, y-axis and the lines $y = c$, $y = d$ is given by the formula; Area $= \int_c^d x dy = \int_c^d \phi(y) dy$.
3. The area of the region enclosed between two curves $yf(x), y = g(x)$ and the lines $x = a, x = b$ is given by the formula.

   Area $= \int_a^b [f(x) - g(x)] dx$, where, $f(x) \geq g(x)$ in $[a, b]$
4. If $f(x) \geq g(x)$ in $[a, c]$ and $f(x) \leq g(x)$ in $[c, b]$, $a < c < b$, then

   Area $= \int_a^c [f(x) - g(x)] dx + \int_c^b [g(x) - f(x)] dx$.

QUIZ

1. For the given curve, find the area of the curve within limits $x = a$ to $x = b$.

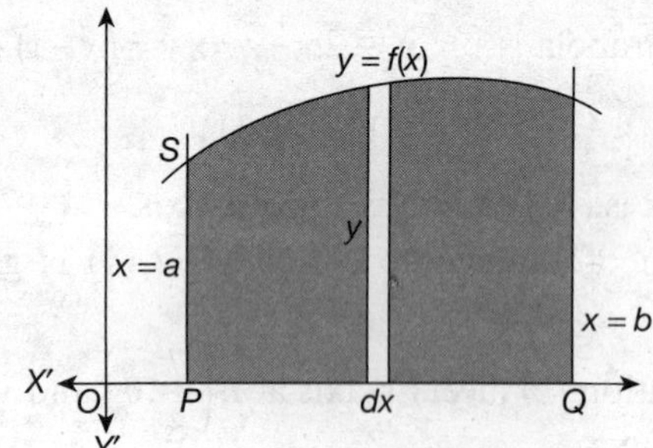

2. If the above curve is below x-axis then what is the area?
3. What is the area between two curves $y = f(x)$ and $y = g(x)$

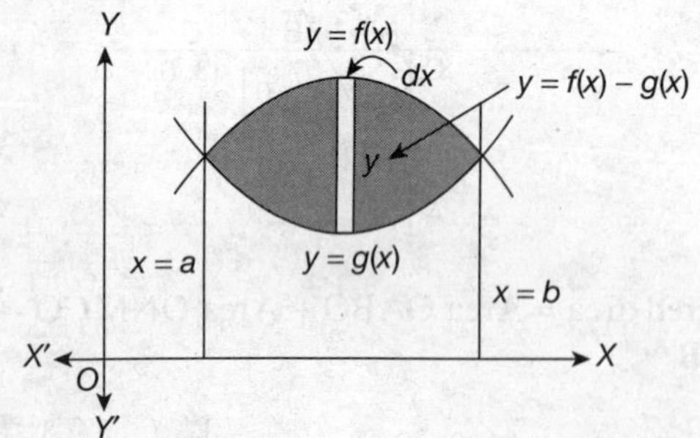

4. Find the area of the region bounded by the two parabolas $y = x^2$ and $y^2 = x$.

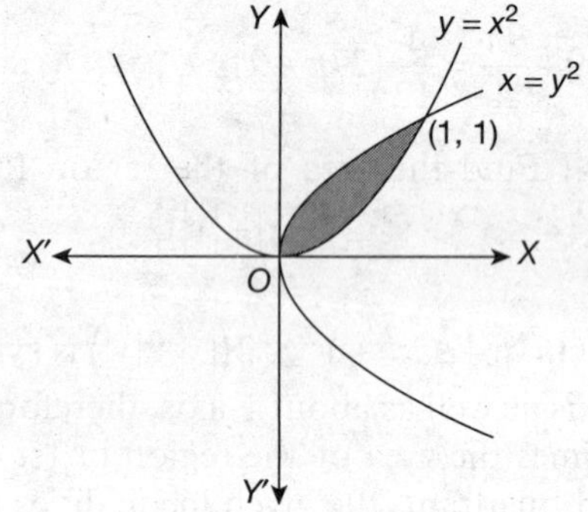

5. If $f(x) \geq g(x)$ in $[a, c]$ and $f(x) \leq g(x)$ in $[c, b]$, $a < c < b$, then what is the area?

JEE MAIN

Objective Questions

One Option Correct

1. The area of the plane region bounded by the curves $x + 2y^2 = 0$ and $x + 3y^2 = 1$ is equal to ______.
(a) $\frac{1}{3}$ (b) $\frac{2}{3}$
(c) $\frac{4}{3}$ (d) $\frac{5}{3}$

2. The area enclosed between the curve $y = \log_e(x+e)$ and the co-ordinate axis is ______.
(a) 3 (b) 4
(c) 1 (d) 2

3. The area of the region bounded by the curves $y = |x-1|$ and $y = 3-|x|$ is ______.
(a) 2 sq. units (b) 3 sq. units
(c) 4 sq. units (d) 6 sq. units

4. The area of the curve $x^2 + y^2 = 2ax$ is ______.
(a) $4\pi a^2$ sq. unit
(b) πa^2 sq. unit
(c) $\frac{1}{2}\pi a^2$ sq. unit
(d) $2\pi a^2$ sq. unit

5. The area bounded by the curve $y^2 = 16x$ and line $y = mx$ is $\frac{2}{3}$ then m is equal to ______.
(a) 3 (b) 4
(c) 1 (d) 2

6. The area of the region bounded by the straight lines $x = 0$ and $x = 2$, and the curves $y = 2^x$ and $y = 2x - x^2$ is equal to ______.
(a) $\frac{2}{\log 2} - \frac{4}{3}$ (b) $\frac{3}{\log 2} - \frac{4}{3}$
(c) $\frac{1}{\log 2} - \frac{4}{3}$ (d) $\frac{4}{\log 2} - \frac{3}{2}$

7. The area bounded by the curves $y = \frac{|4-x^2|}{4}$ and $y = 7-|x|$ is ______.
(a) 16 (b) 28
(c) 32 (d) 40

8. The area of the enclosed region between the two circles $x^2 + y^2 = 4$ and $(x-2)^2 + y^2 = 4$ is ______.
(a) $\frac{8\pi}{3} + 2\sqrt{3}$ (b) $\frac{4\pi}{3} + 2\sqrt{3}$
(c) $\frac{8\pi}{3} - 2\sqrt{3}$ (d) $\frac{4\pi}{3} - 2\sqrt{3}$

9. The area bounded by the curve $y = 1 + |\sin x|$ and $y = 1$ from $x = 0$ to $x = 50\pi$ will be given by ______.
(a) 50π (b) $50\pi - 100$
(c) $50\pi + 100$ (d) 100

10. If a curve $y = a\sqrt{x} + bx$ passes through the point (1, 2) and the area bounded by the curve, line $x = 4$ and x-axis is 8 sq. unit, then the values of a and b are ______.
(a) 2, 3 (b) 3, 2
(c) 2, –1 (d) 3, –1

11. The area included between the curves $x^2 = 4by$ and $y^2 = 4ax$ is
(a) 16 ab (b) $\frac{16\ ab}{3}$
(c) 4 ab (d) $16\pi\ ab$

12. The area lying between the parabola $y^2 = 4ax$ and its latus rectum is ______.
(a) $\frac{1}{3}a$ (b) $\frac{1}{3}a^2$
(c) $\frac{8}{3}a$ (d) $\frac{8}{3}a^2$

13. The area bounded by the curves $a^2 y = x^2(x+a)$ and x-axis is ______.
(a) $\frac{a^2}{4}$ (b) $\frac{a^2}{8}$
(c) $\frac{a^2}{12}$ (d) $\frac{a^2}{16}$

14. The area bounded by the curves $y = |x| - 1$ and $y = -|x| + 1$ is
(a) 1 (b) 2
(c) $2\sqrt{2}$ (d) 4

15. The area bounded by the curves $y = x^2$ and $y = \frac{2}{(1+x^2)}$ is ______.

(a) $2\pi - \frac{1}{3}$ (b) $2\pi + \frac{1}{3}$

(c) $\pi - \frac{2}{3}$ (d) $2\pi + \frac{2}{3}$

16. The area bounded by the curves $y = x\log x$ and $y = 2x - 2x^2$ is ______.

(a) $\frac{1}{12}$ (b) $\frac{5}{12}$

(c) $\frac{7}{12}$ (d) $\frac{11}{12}$

17. If $\int f(x)dx = g(x) + c$, then $\int f^{-1}(x)dx$ is equal to______.

(a) $xf^{-1}(x) + c$

(b) $xf\left(g^{-1}(x)\right) + c$

(c) $xf^{-1}(x) - g\left(f^{-1}(x)\right) + c$

(d) $g^{-1}(x) + c$

18. $\int_0^2 |x-1|\, dx =$ ______.

(a) 0 (b) 2

(c) $\frac{1}{2}$ (d) 1

19. The value of $I = \int_0^1 x\left|x - \frac{1}{2}\right| dx$ is ______.

(a) $\frac{1}{2}$ (b) $\frac{1}{4}$

(c) $\frac{1}{8}$ (d) None of these.

20. If $F(x) = \frac{1}{x^2}\int_4^x (4t^2 - 2F'(t))dt$, the $F'(4)$ equals ______.

(a) $\frac{32}{9}$ (b) $\frac{64}{3}$

(c) $\frac{64}{9}$ (d) None of these.

21. Let $f(x)$, be a non-negative continuous function such that the area bounded by the curve $y = f(x)$, x–axis and the co-ordinates $x = \frac{\pi}{4}$ and $x = \beta > \frac{\pi}{4}$ is $\left(\beta\sin\beta + \frac{\pi}{4}\cos\beta + \sqrt{2}\beta\right)$. Then $f\left(\frac{\pi}{2}\right)$ is ______.

(a) $\left(1 - \frac{\pi}{4} - \sqrt{2}\right)$ (b) $\left(1 - \frac{\pi}{4} + \sqrt{2}\right)$

(c) $\left(\frac{\pi}{4} + \sqrt{2} - 1\right)$ (d) $\left(\frac{\pi}{4} - \sqrt{2} + 1\right)$

22. The parabolas $y^2 = 4x$ and $x^2 = 4y$ divide the square region bounded by the lines $x = 4$, $y = 4$ and the co-ordinate axis. If S_1, S_2, S_3 are respectively the areas of these parts numbered from top to bottom, then $S_1 : S_2 : S_3$ is ______.

(a) 2 : 1 : 2 (b) 1 : 1 : 1

(c) 1 : 2 : 1 (d) 1 : 2 : 3

23. The area between curves $y = x^2, x = y^2$ is ______.

(a) $\frac{1}{9}$ sq. units (b) $\frac{1}{3}$ sq. units

(c) $\frac{1}{\sqrt{3}}$ sq. units (d) $\frac{2}{3}$ sq. units

24. The area bounded by $y = \log x$, x-axis and ordinates $x = 1$, $x = 2$ is ______.

(a) $\frac{1}{2}\log 2$ (b) $\log\frac{2}{e}$

(c) $\log\frac{4}{e}$ (d) $\log 4$

25. The area enclosed between the curves $y = xe^x$, and the line $x = 1$ is ______.

(a) $\frac{1}{e}$ (b) $1 + \frac{2}{3}$

(c) $\frac{2}{e}$ (d) $1 + \frac{1}{e}$

26. The area enclosed by the circle $x^2 + y^2 = a^2$ is equal to ______.

(a) $2\pi a^2$ (b) πa^2

(c) $2\pi a$ (d) πa

27. The area of the region bounded by the curves $y = 2^x, y = 2x - x^2$ and $x = 2$ is

(a) $\frac{1}{\log 2} - \frac{1}{2}$ (b) $\frac{3}{\log 2} - \frac{4}{3}$

(c) $\frac{5}{\log 2} - \frac{5}{4}$ (d) $\frac{6}{\log 2} - \frac{3}{2}$

28. Area of the curve $y = \frac{x}{1 + x^2}$ in the first quadrant between $x = 0$ and at the point where it attains maxima is ______.

(a) $\frac{1}{2}\log 5$ (b) $\frac{1}{2}\log 2$

(c) $\log 5$ (d) $\frac{1}{3}\log 3$

29. Area of the region bounded by $y = |x|$, $x \le 5$ in the first quadrant is ______.

(a) $\frac{11}{2}$ (b) $\frac{17}{2}$

(c) $\frac{25}{2}$ (d) $\frac{27}{2}$

30. The value of k for which the area bounded by $y = x^2 - 3$ and $y = kx + 2$ is the least is ______.

(a) $k = 0$ (b) $k = 1$

(c) $k = -\frac{1}{2}$ (d) $k = -1$

31. In the interval $\left(0, \frac{\pi}{2}\right)$, the area lying between the curves $y = \tan x, y = \cot x$ and x-axis is ______.

(a) 3 log 2 (b) 2 log 2

(c) log 2 (d) $\frac{1}{2}$log 2

32. Area of the region bounded by the curves $y = |x-1|$ and $y = 3 - |x|$ is ______.

(a) 2 (b) 4

(c) 6 (d) 8

33. If the ordinate $x = a$ divides the area bounded by the curve $y = \left(1 + \frac{8}{x^2}\right)$, x-axis and the ordinates $x = 2$, $x = 4$ into two equal parts, then a is equal to ______.

(a) 8 (b) $2\sqrt{2}$

(c) 2 (d) $\sqrt{2}$

34. For $0 \le x \le \pi$, the area bounded by $y = x$ and $y = x + \sin x$ is ______.

(a) 2 (b) 4

(c) 2π (d) 4π

35. Area of the region bounded by the curves $y = \log x$, $y = \sin^4 \pi x$ and $x = 0$ is ______.

(a) $1 + \frac{1}{2}\left(\frac{5}{2} + \frac{2}{\pi}\right)$ (b) $1 - \frac{1}{2}\left(\frac{5}{4} + \frac{4}{\pi}\right)$

(c) $1 + \frac{1}{4}\left(\frac{5}{4} + \frac{4}{\pi}\right)$ (d) $1 - \frac{1}{4}\left(\frac{5}{4} + \frac{4}{\pi}\right)$

36. The value of $\int_{-2}^{3} |1 - x^2| dx$ is ______.

(a) $\frac{28}{3}$ (b) $\frac{14}{-3}$

(c) $\frac{7}{3}$ (d) $\frac{1}{3}$

37. Let $\frac{d}{dx} f(x) = \left(\frac{e^{\sin x}}{x}\right), x > 0.$ If $\int \left(\frac{3}{x}\right) e \sin x^3 dx$ $= f(k) - f(1)$ (where $x \to 1$ to 4), then one of the possible values of k, is ______.

(a) 15 (b) 16

(c) 63 (d) 64

38. $\int_0^3 \left|x^3 + x^2 + 3x\right| dx$ is equal to ______.

(a) $\frac{171}{2}$ (b) $\frac{171}{4}$

(c) $\frac{170}{4}$ (d) $\frac{170}{3}$

39. The value of the integral $\int_0^{\frac{\pi}{2}} \log\left|\tan x\right| dx$ is ______.

(a) $\pi \log 2$ (b) 0

(c) $-\pi \log^2$ (d) None of these.

40. Let $\int \frac{x^{1/2}}{\sqrt{1 - x^3}} dx = \frac{2}{3} gof(x) + c$. Then:

(a) $f(x) = \sqrt{x}$ and $g(x) = x^{2/3}$

(b) $f(x) = x^{3/2}$ and $g(x) = \sin^{-1} x$

(c) $f(x) = x^{2/3}$ and $g(x) = \sin^{-1} x$

(d) $f(x) = \sin^{-1} x$ and $g(x) = \sqrt{x}$

41. The area enclosed within the curve $|x| + |y| = 1$ is ______.

(a) 1 (b) $2\sqrt{2}$

(c) $\sqrt{2}$ (d) 2

42. Area bounded by the curve $y = \log_e x$, $x = 0$, $y \le 0$ and x-axis is ______.

(a) 1 (b) $\frac{1}{2}$

(c) 2 (d) None of these.

43. The area between the parabola $y = x^2$ and the line $y = x$ is ______.

(a) $\frac{1}{6}$ sq. units (b) $\frac{1}{3}$ sq. units

(c) $\frac{1}{2}$ sq. units (d) None of these.

44. The area enclosed between the curves $y = xe^x$, and the line $x = 1$, is ______.

(a) $\frac{5}{3}$ (b) $\frac{5}{4}$

(c) $\frac{5}{12}$ (d) $\frac{12}{5}$

45. The equation of the three sides of a triangle are $x = 2$, $y + 1 = 0$ and $x + 2y = 4$. The co-ordinates of the circumcentre of the triangle are ______.

(a) (4, 0) (b) $(2, -1)$
(c) (0, 4) (d) $(-1, 2)$

46. The area bounded by the curves $y = xe^x$, $y = xe^{-x}$ and the line $x = 1$, is ______.

(a) $\frac{2}{e}$ (b) $1 - \frac{2}{e}$
(c) $\frac{1}{e}$ (d) $1 - \frac{1}{e}$

47. The area (in square units) of the region bounded by the curves $2x = y^2 - 1$ and $x = 0$ is ______.

(a) $\frac{1}{3}$ (b) $\frac{2}{3}$
(c) 1 (d) 2

48. The area of the region enclosed by $y = x^2$ and $y = \sqrt{x}$ is ______.

(a) 1
(b) $\frac{1}{2}$
(c) $\frac{1}{3}$
(d) It cannot be determined.

49. The area of the region bounded by the curve $y = x^2$ and the line $y = 4$ is ______.

(a) $\frac{11}{3}$ (b) $\frac{32}{3}$
(c) $\frac{43}{3}$ (d) $\frac{47}{3}$

50. The area of the region bounded by $y = xe^{-x^2}$, $y = x + 1$, $x = 2$ and y-axis is ______.

(a) $\frac{1}{2e^4} + \frac{7}{2}$ (b) $\frac{1}{2}e^4 + \frac{7}{2}$
(c) $2e^4 + \frac{5}{2}$ (d) $\frac{2}{e^4} + \frac{5}{2}$

51. The area formed by the region bounded by the curves $y = \sin x$, $y = \cos x$ and $x = 0$ in the first quadrant when these curves intersect the first time is______.

(a) $\sqrt{2} - 1$ (b) $\sqrt{2} + 1$
(c) $\sqrt{2}$ (d) $\frac{1}{\sqrt{2}}$

52. The area of the curve $y = (x-1)(x-2)(x-3)$ bounded by x-axis between $x = 1$ and $x = 3$ is ______.

(a) $\frac{1}{2}$ (b) 1
(c) 2 (d) 3

53. The area of the region bounded by the curve $y = \tan x$, line $x = \frac{\pi}{4}$ and the x-axis is ______.

(a) $\log 2$ (b) $\frac{1}{2}\log 2$
(c) $\frac{1}{3}\log 2$ (d) $5\log 2$

54. The area of a triangle whose two vertices lies on the x-axis where it intersects the curve $y = 3 - |x|$ and another point where the curve intersects y-axis is

(a) $\frac{9}{4}$ (b) $\frac{9}{2}$
(c) 9 (d) None of these.

55. The area of the curve $y = x^4 - 2x^3 + x^2 + 3$ between two local minims in the first quadrant is ______.

(a) $\frac{71}{3}$ (b) $\frac{91}{3}$
(c) $\frac{71}{30}$ (d) $\frac{93}{30}$

56. The area bounded by the curve $y = e^{|x|}$, x-axis and lines $x = -a, x = a$ where $a > 0$ is ______.

(a) $a(e-1)$ (b) $2a^2(e^a - 1)$
(c) $2(e^a - 1)$ (d) $\frac{1}{a}e^{a^2}$

57. The area enclosed between the curves $y = \sin^2 x$ and $y = \cos^2 x$ in the first quadrant where they intersect first time is ______.

(a) $\frac{1}{3}$ (b) $\frac{1}{2}$
(c) 1 (d) 2

58. The area of the smaller region bounded by $x^2 + y^2 = 9$ and the line $x = 1$ is ______.

(a) $2 - 3\sec^{-1}3$ (b) $\sqrt{8} - 3\sec^{-1}3$
(c) $9\sec^{-1}3 - \sqrt{8}$ (d) $\sec^{-1}3 - 3\sqrt{8}$

59. The area bounded by the curve $y^2(2a - x) = x^3$ and line $x = 2a$ is ______.

(a) $\frac{3\pi a^2}{7}$ (b) $\frac{3\pi a}{7}$

(c) $\frac{3\pi a^2}{2}$ (d) $3\pi a^2$

60. Let $f(x)=\begin{cases} x^2 : x<0 \\ x : x\geq 0 \end{cases}$ The area bounded by the curve $y=f(x), x^2=9a^2$ and x-axis is $\frac{9a}{2}$. Then the value of a is ______.

(a) $\frac{1}{4}$ (b) $\frac{1}{3}$

(c) $\frac{1}{2}$ (d) 1

Questions from Previous Years

61. The area of the region bounded by the parabola $(y-2)^2=(x-1)$, the tangent to it at the point whose ordinate is 3 and the x-axis is ______.

(a) 9 (b) 10

(c) 4 (d) 6

62. If the area of the bounded region $R=\left\{(x,y): \max\{0, \log_e x\}\leq y\leq 2^x, \frac{1}{2}\leq x\leq 2\right\}$ is $\alpha(\log_e 2)^{-1}+\beta(\log_e 2)+\gamma$, then the value of $(\alpha+\beta-2\gamma)^2$ is equal to ______.

(a) 8 (b) 2

(c) $\frac{13}{3}$ (d) $\frac{7}{3}$

63. Let A_1 be the area of the region bounded by the curves $y=\sin x$, $y=\cos x$ and y-axis in the first quadrant. Also, let A_2 be the area of the region bounded by the curves $y=\sin x$, $y=\cos x$, x-axis and $x=\frac{\pi}{2}$ in the first quadrant. Then, ______.

(a) $A_1 : A_2 = 1:\sqrt{2}$ and $A_1+A_2=1$

(b) $A_1 : A_2 = 1:2$ and $A_1+A_2=1$

(c) $2A_1 = A_2$ and $A_1+A_2=1+\sqrt{2}$

(d) $A_1 = A_2$ and $A_1+A_2=1=\sqrt{2}$

64. The area of the region $R:\{(x,y): 5x^2\leq y\leq 2x^2+9\}$ is ______ sq. units.

(a) $\frac{1}{3}$ (b) $12\sqrt{3}$

(c) $\frac{1}{6}$ (d) $\frac{5}{6}$

65. The area (in sq. units) of the region enclosed by the curves $y=x^2-1$ and $y=1-x^2$ is equal to ______.

(a) $\frac{4}{3}$ (b) $\frac{8}{3}$

(c) $\frac{7}{2}$ (d) $\frac{16}{3}$

66. Consider a region $R=\{(x,y)\in R^2 : x^2\leq y\leq 2x\}$. If a line $y=\alpha$, divides the area of region R into two equal parts, then which of the following is true?

(a) $\alpha^3-6\alpha^2+16=0$ (b) $3\alpha^2-8\alpha^{3/2}+8=0$

(c) $3\alpha^2-8\alpha+8=0$ (d) $\alpha^3-6\alpha^{3/2}-16=0$

67. For $a>0$, let the curves $C_1: y^2=ax$ and $C_2: x^2=ay$ intersect at origin O and a point P. Let the line $x=b\ (0<b<a)$ intersect the chord OP and the x-axis at points Q and R, respectively. If the line $x=b$ bisects the area bounded by the curves, C_1 and C_2, and the area of ΔOQR $=\frac{1}{2}$, then 'a' satisfies the equation:

(a) $x^6-6x^3+4=0$ (b) $x^6-12x^3+4=0$

(c) $x^6+6x^3-4=0$ (d) $x^6-12x^3-4=0$

68. The area (in sq. units) of the region $\{(x,y)\in R^2 : x^2\leq y\leq |3-2x|$, is ______.

(a) $\frac{32}{3}$ (b) $\frac{34}{3}$

(c) $\frac{29}{3}$ (d) $\frac{31}{3}$

69. The area of the region enclosed by the circle $x^2+y^2=2$, which is not common to the region bounded by the parabola $y^2=x$ and the straight line $y=x$ is ______.

(a) $(24\pi-1)$ (b) $(6\pi-1)$

(c) $(12\pi-1)$ (d) $(12\pi-1)/6$

70. The area (in sq. units) of the region $\{(x,y)\in R^2 | 4x^2\leq y\leq 8x+12\}$ is ______.

(a) $\frac{125}{3}$ (b) $\frac{128}{3}$

(c) $\frac{124}{3}$ (d) $\frac{127}{3}$

71. The area (in sq units) of the region $A=\{(x,y): (x-1)[x]\leq y\leq 2\sqrt{x}, 0\leq x\leq 2\}$, where $[t]$ denotes the greatest integer function, is ______.

(a) $\frac{8}{3}\sqrt{2}-\frac{1}{2}$ (b) $\frac{4}{3}\sqrt{2}+1$

(c) $\frac{8}{3}\sqrt{2}-1$ (d) $\frac{4}{3}\sqrt{2}-\frac{1}{2}$

72. The area (in sq units) of the region $\left\{(x,y): 0 \le y \le x^2+1, 0 \le y \le x+1, \frac{1}{2} \le x \le 2\right\}$ is ______.

(a) $\frac{23}{16}$ (b) $\frac{79}{24}$

(c) $\frac{79}{16}$ (d) $\frac{23}{6}$

73. Given: $f(x)=\begin{cases} x, & 0 \le x \le \frac{1}{2} \\ \frac{1}{2}, & x=\frac{1}{2} \\ 1-x, & \frac{1}{2} \le x \le 1 \end{cases}$

and $g(x)=\left(x-\frac{1}{2}\right)$, $x \in R$. Then the area (in sq. units) of the region bounded by the curves, $y=f(x)$ and $y=g(x)$ between the lines, $2x=\sqrt{3}$ is ______.

(a) $\frac{1}{3}+\frac{\sqrt{3}}{4}$ (b) $\frac{\sqrt{3}}{4}-\frac{1}{3}$

(c) $\frac{1}{2}-\frac{\sqrt{3}}{4}$ (d) $\frac{1}{2}+\frac{\sqrt{3}}{4}$

74. If the area (in sq. units) bounded by the parabola $y^2=4\lambda$ and the line $y=\lambda x$, $\lambda>0$, is $\frac{1}{9}$, then λ is equal to ______.

(a) $2\sqrt{6}$ (b) 48

(c) 24 (d) $4\sqrt{3}$

75. The area (in sq. units) bounded by the curves $y=2^x$ and $y=|x+1|$, in the first quadrant is ______.

(a) $\log_e 2+\frac{3}{2}$ (b) $\frac{3}{2}$

(c) $\frac{1}{2}$ (d) $\frac{3}{2}-\frac{1}{\log_e 2}$

76. The area (in sq. units) of the region $\left\{(x,y): \frac{y^2}{2} \le x \le y+4\right\}$ is ______.

(a) $\frac{53}{3}$ (b) 30

(c) 16 (d) 18

77. Let $S(\alpha)=\{(x,y): y^2 \le x, 0 \le x \le \alpha\}$ and $A(\alpha)$ be the area of the region $S(\alpha)$. If for a λ, $0<\lambda<4$, $A(\lambda): A(\alpha)=2:5$, then λ equals ______.

(a) $2\left(\frac{4}{25}\right)^{\frac{1}{3}}$ (b) $2\left(\frac{2}{5}\right)^{\frac{1}{3}}$

(c) $4\left(\frac{2}{5}\right)^{\frac{1}{3}}$ (d) $4\left(\frac{4}{25}\right)^{\frac{1}{3}}$

78. The area (in sq. units) of the region bounded by the parabola, $y=x^2+2$ and the lines, $y=x+1$, $x=0$ and $x=3$ is ______.

(a) $\frac{15}{4}$ (b) $\frac{21}{2}$

(c) $\frac{17}{4}$ (d) $\frac{15}{2}$

79. The area (in sq. units) of the region bounded by the curve $x^2=4y$ and the straight line $x=4y-2$ is ______.

(a) $\frac{5}{4}$ (b) $\frac{9}{8}$

(c) $\frac{7}{8}$ (d) $\frac{3}{4}$

80. The area (in sq. units) of the region $\{x \in R: x \ge 0, y \ge 0, y \ge x-2 \text{ and } y \le \sqrt{x}\}$ is ______.

(a) $\frac{13}{3}$ (b) $\frac{10}{3}$

(c) $\frac{5}{3}$ (d) $\frac{8}{3}$

81. Let $g(x)=\cos x^2$, $f(x)=\sqrt{x}$, and α, β $(\alpha<\beta)$ be the roots of the quadratic equation $18x^2-9\pi x+\pi^2=0$. Then the area (in sq units) bounded by the curve $y=(gof)(x)$ and the lines $x=\alpha, x=\beta$ and $y=0$ is ______.

(a) $\frac{1}{2}(\sqrt{3}+1)$ (b) $\frac{1}{2}(\sqrt{3}-\sqrt{2})$

(c) $\frac{1}{2}(\sqrt{2}-1)$ (d) $\frac{1}{2}(\sqrt{3}-1)$

82. The area (in sq. units) of the region $\{(x,y): x \ge 0, x+y \le 3, x^2 \le 4y \text{ and } y \le 1+\sqrt{x}\}$ is ______.

(a) $\frac{5}{2}$ (b) $\frac{59}{12}$

(c) $\frac{3}{2}$ (d) $\frac{7}{3}$

83. The area (in sq. units) of the region $\{(x,y): y^2 \geq 2x \text{ and } x^2+y^2 \leq 4x, x \geq 0, y \geq 0\}$ is ______.

(a) $\pi - \frac{4\sqrt{2}}{3}$ (b) $\frac{\pi}{2} - \frac{2\sqrt{2}}{3}$

(c) $\pi - \frac{4}{3}$ (d) $\pi - \frac{8}{3}$

84. The area (in sq. units) of the region described by $A = \{(x,y) | y \geq x^2 - 5x + 4, x + y \geq 1, y \leq 0\}$ is ______.

(a) $\frac{19}{6}$ (b) $\frac{17}{6}$

(c) $\frac{7}{2}$ (d) $\frac{13}{2}$

85. The area (in sq. units) of the region described by $\{(x,y): y^2 \leq 2x \text{ and } y \geq 4x - 1\}$ is ______.

(a) $\frac{15}{64}$ (b) $\frac{9}{32}$

(c) $\frac{7}{32}$ (d) $\frac{5}{64}$

86. The area (in sq. units) of the region bounded by the curves $y + 2x^2 = 0$ and $y + 3x^2 = 1$ is equal to ______.

(a) $\frac{3}{5}$ (b) $\frac{1}{3}$

(c) $\frac{4}{3}$ (d) $\frac{3}{4}$

87. The area of the region described by $A = \{(x,y): x^2 + y^2 \leq 1 \text{ and } y^2 \leq 1 - x\}$ is ______.

(a) $\frac{\pi}{2} - \frac{2}{3}$ (b) $\frac{\pi}{2} + \frac{2}{3}$

(c) $\frac{\pi}{2} + \frac{4}{3}$ (d) $\frac{\pi}{2} - \frac{4}{3}$

88. The area (in sq. units) of the region bounded by the curves $y = \sqrt{x}, 2y - x + 3 = 0$, x-axis and lying in the first quadrant is ______.

(a) 9 (b) 36

(c) 18 (d) $\frac{27}{4}$

89. The area under the curve $y = |\cos x - \sin x|$, $0 \leq x \leq \frac{\pi}{2}$ and above the x-axis is ______.

(a) $2\sqrt{2}$ (b) $2\sqrt{2} - 2$

(c) $2\sqrt{2} + 2$ (d) 0

90. Let $f : [-1,2] \to [0,\infty)$ be a continuous function such that $f(x) = f(1-x)$ for all $x \in [-1,2]$. Let $R_1 = \int_{-1}^{2} xf(x)\,dx$ and R_2 be the area of the region bounded by $y = f(x)$, $x = -1$ and the x-axis. Then, ______.

(a) $R_1 = 2R_2$ (b) $R_1 = 3R_2$

(c) $2R_1 = R_2$ (d) $3R_1 = R_2$

91. Let the straight line $x = b$ divide the area enclosed by $y = (1-x)^2$, $y = 0$ and $x = 0$ into two parts, $R_1\ (0 \leq x \leq b)$ and $R_2\ (b \leq x \leq 1)$, such that $R_1 - R_2 = \frac{1}{4}$. Then b equals ______.

(a) $\frac{3}{4}$ (b) $\frac{1}{2}$

(c) $\frac{1}{3}$ (d) $\frac{1}{4}$

Integer and Numerical Value Type

1. The area between the curve $y = \frac{1}{1+x^2}$ and its asymptote, is $k\pi$. Then the numerical value of k must be ______.

2. The area of the curve $xy^2 = 8 - 4x$ and its asymptote is $k\pi$. Then the numerical value of k must be ______.

3. If the area bounded by the circle $x^2 + y^2 = 4$, the parabola $y = x^2 + x + 1$ and the curve $\left[\sin^2\frac{x}{4} + \cos\frac{x}{4}\right]$, {[x] denotes greatest integer} and the x-axis is $\sqrt{3} + \frac{2\pi}{3} - \frac{1}{k}$, then the numerical quantity k should be ______.

4. If the area of closed curve $x = 5\cos t + 2006, y = 3\sin t - 2007$ is $k\pi$, then the numerical quantity k must be ______.

5. The area bounded by parabola $y^2 = 2x$ and line $y = 4x - 1$ is $\frac{K}{32}$ square units then K must be ______.

6. Using integration, find the area of the region bounded by the triangle whose vertices are (1, 0), (2, 2) and (3, 1).

Questions from Previous Years

7. Let a and b respectively be the points of local maximum and local minimum of the function $(x)-2x^3-3x^2-12x$. If A is the total area of the region bounded by $y=f(x)$, then $4A$ is equal to ______.

8. Let T be the tangent to the ellipse $E: x^2+4y^2=5$ at the point (1, 1). If the area of the region bounded by the tangent T, ellipse E, and lines $x=1$ and $x=\sqrt{5}$ is $\alpha\sqrt{5}+\beta+\gamma\cos^{-1}\left(\frac{1}{\sqrt{5}}\right)$, then $|\alpha+\beta+\gamma|$ is equal to ______.

9. Let $f:[-3,1]\to R$ be given as

$$f(x)=\begin{cases}\min\{(x+6),x^2\}, -3\le x\le 0\\ \max\{\sqrt{x},x^2\}, 0\le x\le 1\end{cases}$$

If the area bounded by $y=f(x)$ and x-axis is A, then value of $6A$ is equal to ______.

10. The graphs of sine and cosine functions intersect each other at a number of points and between two consecutive points of intersection, the two graphs enclose the same area A. Then A^4 is equal to ______.

11. The area of the region $S=\{(x,y): 3x^2\le 4y\le 6x+24\}$ is ______.

12. Let P(x) be a real polynomial of degree 3 which vanishes at $x=-3$. Let P(x) have local minima at $x=1$, local maxima at $x=-1$ and $\int_{-1}^{1}P(x)dx=18$. Then the sum of all the coefficients of the polynomial P(x) is equal to ______.

Assertion and Reason

(a) Statement 1 is True, Statement 2 True; Statement 2 is a correct explanation for Statement 1.

(b) Statement 1 is True, Statement 2 is True; Statement 2 is NOT a correct explanation for Statement 1.

(c) Statement 1is True, Statement 2 is False.

(d) Statement 1 is False, Statement 2 is True.

1. **Statement 1**: The area bounded by the curve $|x|+|y|=a$ $(a>0)$ is $2a^2$ and area bounded $|px+qy|+|qx-py|=a$, where $p^2+q^2=1$, is also $2a^2$.
Statement 2: Since $\alpha x+\beta y=0$ is perpendicular to $\beta x-\alpha y=0$, we can take one as x-axis and another as y-axis and therefore the area bounded by $|\alpha x+\beta y|+|\beta x-\alpha y|=a$ is $2a^2$ for all $\alpha,\beta\in R, \alpha\ne 0, \beta\ne 0$.

2. **Statement 1**: Area bounded by parabola $y=x^2-4x+3$ and $y=0$ is 4/3 sq. units.
Statement 2: Area bounded by curve $y=f(x)\ge 0$ and $y=0$ between ordinates $x=a$ and $x=b$ $(b>a)$ is $\int_a^b f(x)dx$.

JEE ADVANCED

Objective Questions

One Option Correct

1. The area enclosed by the curves $y=\sqrt{4-x^2}, y\ge\sqrt{2}\sin\left(\frac{x\pi}{2\sqrt{2}}\right)$ and x-axis is divided by y-axis in the ratio ______.

(a) $\frac{\pi^2-8}{\pi^2+8}$ (b) $\frac{\pi^2-4}{\pi^2+4}$

(c) $\frac{\pi-4}{\pi-4}$ (d) $\frac{2\pi^2}{2\pi+\pi^2-8}$

2. If $f(x)=\sin x\,\forall x\in\left[0,\frac{\pi}{2}\right], f(x)+f(\pi-x)=2\,\forall\, x\in\left(\frac{\pi}{2},\pi\right]$ and $f(x)=f(2\pi-x)\,\forall x\in(\pi,2\pi]$, then the area enclosed by $y=f(x)$ and x-axis is ______.

(a) π (b) 2π

(c) 2 (d) 4

3. The area enclosed by $y=x^3$, its normal at (1, 1) and x-axis is equal to ______.

(a) $\frac{7}{4}$ (b) $\frac{9}{4}$

(c) $\frac{5}{4}$ (d) 2

4. The value of the parameter a such that the area bounded by $y = a^2x^2 + ax + 1$, co-ordinate axes and the line $x = 1$, attains its least value, is equal to ______.

(a) $-\frac{1}{4}$ (b) $-\frac{1}{2}$

(c) $-\frac{3}{4}$ (d) -1

5. The area of the region bounded by $x = 0$, $y = 0$, $x = 2$, $y = 2$, $y \le e^x$ and $y \ge \ell nx$, is ______.

(a) $6 - 4\ln 2$ (b) $4\ln 2 - 2$

(c) $2\ln 2 - 4$ (d) $6 - 2\ln 2$

6. Area bounded by the curve $y = \ln x + \tan^{-1} x$ and x-axis from $x = 1$ to $x = 2$, is ______.

(a) $\frac{5}{2}\ln 2 - \frac{1}{2}\ln 5 + 2\tan^{-1} 2 - \frac{\pi}{3} - 1$

(b) $\frac{5}{2}\ln 2 - \frac{1}{2}\ln 5 + 2\tan^{-1} 2 - \frac{\pi}{4} + 1$

(c) $\frac{5}{2}\ln 2 - \frac{1}{2}\ln 5 + 2\tan^{-1} 2 + \frac{\pi}{4} - 1$

(d) $\frac{5}{2}\ln 2 - \frac{1}{2}\ln 5 + 2\tan^{-1} 2 - \frac{\pi}{4} - 1$

7. The area of the figure bounded by the parabola $(y-2)^2 = x - 1$, the tangent to it, the point with the ordinate 3 and the x-axis is ______.

(a) 3 (b) 6

(c) 9 (d) None of these.

8. The area of the figure bounded by the curve $y = 2x - x^2$ and the straight line $y = -x$ is ______.

(a) $\frac{9}{2}$ (b) 9

(c) $\frac{7}{2}$ (d) 7

9. If the area bounded by the curve $y = x^2, y = -x^2$ and $y^2 = 4x - 3$ is k, then the value of $6k$, is ______.

(a) 2 (b) 3

(c) 0 (d) 4

10. The area of the region enclosed between the two circles $x^2 + y^2 = 1$ and $(x-1)^2 + y^2 = 1$, is ______.

(a) $\left(\frac{2\pi}{3} - \sqrt{\frac{3}{4}}\right)$ (b) $\left(\frac{3\pi}{2} - \frac{\sqrt{3}}{2}\right)$

(c) $\left(\frac{3\pi}{2} + \frac{\sqrt{3}}{2}\right)$ (d) $\left(\frac{2\pi}{3} + \sqrt{\frac{3}{4}}\right)$

11. Which of the following statements are true/false?

S_1: The area between $x^2 = 4by$ and $y^2 = 4ax$ is $\frac{16ab}{3}$.

S_2: The area enclosed by $|x| + |y| = 1$ is 1.

S_3: The smaller area enclosed by $\frac{x^2}{a^2} + \frac{y^2}{b^2} = 1$ and $\frac{x}{a} + \frac{y}{b} = 1$ is $\frac{\pi ab}{4} - \frac{ab}{2}$.

S_4: The area enclosed by $y = [x]$ and $y = \{x\}$ is 1.

(a) TFTT (b) TTTT

(c) TFTF (c) FFTT

Questions from Previous Years

12. Let the functions $f: \mathbb{R} \to \mathbb{R}$ and $g: \mathbb{R} \to \mathbb{R}$ be defined by $f(x) = e^{x-1} - e^{-|x-1|}$ and $g(x) = \frac{1}{2}\left(e^{x-1} + e^{1-x}\right)$. Then the area of the region in the first quadrant bounded by the curves $y = f(x)$, $y = g(x)$ and $x = 0$ is ______.

(a) $\left(2-\sqrt{3}\right) + \frac{1}{2}\left(e - e^{-1}\right)$

(b) $\left(2+\sqrt{3}\right) + \frac{1}{2}\left(e - e^{-1}\right)$

(c) $\left(2-\sqrt{3}\right) + \frac{1}{2}\left(e + e^{-1}\right)$

(d) $\left(2+\sqrt{3}\right) + \frac{1}{2}\left(e + e^{-1}\right)$

13. The area (in sq. units) of the region $\left\{(x, y : xy \le 8, 1 \le y \le x^2\right\}$ is ______.

(a) $8\log_e 2 - \frac{14}{3}$ (b) $16\log_e 2 - \frac{14}{3}$

(c) $8\log_e 2 - \frac{7}{3}$ (d) $16\log_e 2 - 6$

14. Area of the region $\left\{(x, y) \in \mathbb{R}^2 : y \ge \sqrt{|x+3|}, 5y \le x + 9 \le 15\right\}$ is equal to ______.

(a) $\frac{1}{6}$ (b) $\frac{4}{3}$

(c) $\frac{3}{2}$ (d) $\frac{5}{3}$

15. The area enclosed by the curves $y = \sin x + \cos x$ and $y = |\cos x - \sin x|$ over the interval $\left[0, \frac{\pi}{2}\right]$ is ______.

(a) $4\left(\sqrt{2} - 1\right)$ (b) $2\sqrt{2}\left(\sqrt{2} - 1\right)$

(c) $2\left(\sqrt{2} + 1\right)$ (d) $2\sqrt{2}\left(\sqrt{2} + 1\right)$

One or More Than One Option Correct

1. Let C be a curve passing through M (2, 2) such that the slope of the tangent at any point to the curve is reciprocal of the ordinate of that point. If the area bounded by curve C and line $x = 2$ is expressed as a rational number $\frac{p}{q}$ (where p and q are in their lowest form), then value of $(p + q)$ is greatest than prime numbers ______.
   (a) 5 (b) 11
   (c) 17 (d) 19

2. Let $f(x) = \min.\left(\cos^{-1}(\cos x), \cot^{-1}(\cot x)\right)$ and $x \in (\pi, 2\pi)$, then which is true?
   (a) The area enclosed between $f(x) = \min.\left(\cos^{-1}(\cos x), \cot^{-1}(\cot x)\right)$ and x-axis is $\frac{\pi^2}{4}$.
   (b) The area enclosed between $f(x) = \min.\left(\cos^{-1}(\cos x), \cot^{-1}(\cot x)\right)$ and x-axis is $\frac{\pi}{2}$.
   (c) $f(x)$ is non-differentiable at two points.
   (d) $f(x)$ is non-differentiable at one point.

3. Let the curve $a^2 y = x^2 (x + a)$. Then
   (a) the area between curve and negative x-axis is $\frac{a^2}{3}$ sq. unit.
   (b) the area between curve and negative x-axis is $\frac{a^2}{12}$ sq. unit.
   (c) the area between curve and positive x-axis is zero sq. unit.
   (d) the area between curve and negative x-axis is $\frac{a^2}{12}$ sq. unit.

4. Let $f(x) = \sin^{-1}|\sin x|$ and $g(x) = \left(\sin^{-1}|\sin x|\right)^2$. When $x \in [0, 2\pi]$, then
   (a) the area between the curves $f(x)$ and $g(x)$ is $\frac{1}{6} + \frac{\pi^3}{8}$.
   (b) the area between the curves $f(x)$ and $g(x)$ is $\frac{4}{3} + \pi^2\left(\frac{\pi - 3}{6}\right)$.
   (c) $f(x)$ is non-differentiable at two points.
   (d) $f(x)$ is non-differentiable at three points.

5. Let S be the area of the region enclosed by $y = e^{-x^2}$, $y = 0$, $x = 0$ and $x = 1$; then,
   (a) $S \geq \frac{1}{e}$ (b) $S \geq 1 - \frac{1}{e}$
   (c) $S \leq \frac{1}{4}\left(1 + \frac{1}{\sqrt{e}}\right)$
   (d) $S \leq \frac{1}{\sqrt{2}} + \frac{1}{\sqrt{e}}\left(1 - \frac{1}{\sqrt{2}}\right)$

6. The area of the region bounded by the curve $y = e^x$ and lines $x = 0$ and $y = e$ is ______.
   (a) $e - 1$ (b) $\int_1^e \ln(e + 1 - y)\,dy$
   (c) $e - \int_0^1 e^x\,dx$ (d) $\int_1^e \ln y\,dy$

Question from Previous Year

7. If the line $sx = \alpha$ divides the area of region $R = \left\{(x, y) \in R^2; x^3 \leq y \leq x, 0 \leq x \leq 1\right\}$ into two equal parts, then ______.
   (a) $0 < \alpha \leq \frac{1}{2}$ (b) $\frac{1}{2} < \alpha < 1$
   (c) $2\alpha^4 - 4\alpha^2 + 1 = 0$ (d) $\alpha^4 + 4\alpha^2 - 1 = 0$

Matrix Type

1. Match Column I with Column II.

Column I	Column II
(A) The area bounded by the curve $y = x + \sin x$ and its inverse function between the ordinates $x = 0$ to $x = 2\pi$ is $4s$ Then the value of s is	(P) 0
(B) The area bounded by $y = xe^{\lvert x\rvert}$ and lines $\lvert x\rvert = 1$, $y = 0$ is	(Q) 1
(C) The area bounded by the curves $y^2 = x^3$ and $\lvert y\rvert = 2x$ is	(R) $\frac{16}{5}$
(D) The smaller area included between the curves $\sqrt{x} + \sqrt{\lvert y\rvert} = 1$ and $\lvert x\rvert + \lvert y\rvert = 1$ is	(S) $\frac{1}{3}$
	(T) 2

2. Match the Column I with Column II.

Column I	Column II
(A) The area enclosed by $y = \lvert x\rvert$, $\lvert x\rvert = 1$ and $y = 0$ is	(P) 3
(B) The area enclosed by the curve $y = \sin x$, $x = 0$, $x = \pi$ and $y = 0$ is	(Q) 4

(C) If the area of the region bounded by $x^2 \le y$ and $y \le x+2$ is $\frac{k}{4}$, then $k=$	(R) 27
(D) The area of the quadrilateral formed by tangents at the ends of latus rectum of ellipse $\frac{x^2}{9}+\frac{y^2}{5}=1$ is	(S) 18
	(T) 1

Integer and Numerical Value Type

1. Find the area enclosed by the solution set of $[x]\,[y] = 2$, where $[\cdot]$ represent greatest integer function of x.
2. The area of the loop of the curve, $ay^2 = x^2(a-x)$ is $\frac{\lambda a^2}{15}$, then find λ.
3. The area bounded by lines $x=-1$ and $x=1$, $y=x^3$ and $y=g(x)$, where $g(x)=\begin{Bmatrix} x^5 \text{ if } -1 \le x < 0 \\ x \text{ if } 0 \le x \le 1 \end{Bmatrix}$, is $\frac{1}{k}$, then the numerical quantity k must be ______.
4. The curve $y=4-x^2$ cuts the line $y=x+1$ at $A(\alpha,\alpha+1)$ (where $\alpha<-2$) and $B(\beta,\beta+1)$ (where $0<\beta<2$), then area below the x-axis must be $\frac{\alpha^3}{3}+\frac{\alpha^2}{2}-3\alpha-\frac{k}{6}$ then the numerical quantity k should be ______.
5. A farmer F_1 has a land in the shape of a triangle with vertices at P (0, 0), Q (1, 1) and R (2, 0). From this land, a neighboring farmer F_2 takes away the region which lies between the side PQ and a curve of the form $y=x^n\ (n>1)$. If the area of the region taken away by the farmer F_2 is exactly 30% of area of Δ PQR, then the value of n is ______.
6. Let $F(x)=\int_x^{x^2+\frac{\pi}{6}} 2\cos^2 t(dt)$ for all $x\in R$ and $f:\left[0,\frac{1}{2}\right]\to[0,\infty)$ be a continuous function. For $a\in\left[0,\frac{1}{2}\right]$, if $F'(a)+2$ is the area of the region bounded by $x=0, y=0, y=f(x)$ and $x=a$, then $f(0)$ is ______.

Comprehension Passage

Comprehension 1

Asymptotes are the tangents to the curve at infinity. To find the asymptotes of a curve we can use the following methods.

(a) Asymptote parallel to the x-axis is obtained by equating to zero, the coefficient of the highest power to x.

(b) Asymptote parallel to the y-axis is obtained by equating to zero, the coefficient of the highest power of y.

(c) Oblique asymptote: $y=mx+c$

(i) Find $\varphi_n(m)$ by putting $x=1$ and $y=m$ in the highest degree (n) terms of the equation. Similarly, find $\varphi_{n-1}(m)$.

(ii) Solve $\varphi_n(m)=0$ for m.

(iii) Find c by the formula $c=-\frac{\varphi_{n-1}(m)}{\varphi'_n(m)}$ using the value of m as obtained in (ii).

(iv) Obtain the equation of asymptote by putting the values of m and c in $y=mx+c$.

1. The equation of asymptotes of the curve $yx^2-4x^2+x+2=0$ is ______.
   (a) $y-4=0$ and $x=0$ (b) $y=3$ and $x=2$
   (c) $y-4=0$ and $x=2$ (d) $y=3$ and $x=0$
2. The equation of asymptotes of the curve $x^3+y^3-3xy=0$ is ______.
   (a) $y=x+1$ (b) $y+x+1=0$
   (c) $y+x=2$ (d) $y=2x+1$
3. If the equation of asymptotes of the curve $y^2=\frac{x^3}{(2-x)}$ is $ax+by+c=0$, then the value of $|a+b+c|$ is ______.
   (a) 0 (b) 1
   (c) 2 (d) 4

Comprehension 2

Consider the functions defined implicitly by equation $y^3-3y+x=0$ on various intervals in the real line. If $x\in(-\infty,-2)\cup(2,\infty)$, the equation implicitly defines a unique real-valued differentiable function $y=f(x)$. If $x\in(-2,2)$, the equation implicitly defines a unique real-valued differentiable function $y=g(x)$ satisfying $g(0)=0$.

1. If $f\left(-10\sqrt{2}\right)=2\sqrt{2}$, then $f''\left(-10\sqrt{2}\right)=$______.

(a) $\frac{4\sqrt{2}}{7^3 3^2}$ (b) $-\frac{4\sqrt{2}}{7^3 3^2}$

(c) $\frac{4\sqrt{2}}{7^3 3}$ (d) $-\frac{4\sqrt{2}}{7^3 3}$

2. The area of the region bounded by the curve $y=f(x)$, the x-axis and the lines $x=a$ and $x=b$, where $-\infty<a<b<-2$ is______.

(a) $\int_a^b \frac{x}{3\left(\left(f(x)\right)^2-1\right)}dx+bf(b)-af(a)$

(b) $-\int_a^b \frac{x}{3\left(\left(f(x)\right)^2-1\right)}dx+bf(b)-af(a)$

(c) $\int_a^b \frac{x}{3\left(\left(f(x)\right)^2-1\right)}dx-bf(b)+af(a)$

(d) $-\int_a^b \frac{x}{3\left(\left(f(x)\right)^2-1\right)}dx-bf(b)+af(a)$

Comprehension 3

A soft plastic bottle, filled with water of density 1 gm/cc, carries an inverted glass test tube with some air (ideal gas) trapped as shown in the figure. The test tube has a mass of 5 gm and it is made of a thick glass of density 2.5 gm/cc. Initially the bottle is sealed at atmospheric pressure $P_0=10^5$ Pa so that the volume of the trapped air $v_0=3.3$ cc. When the bottle is squeezed from outside at constant temperature, the pressure inside rises and the volume of the trapped air reduces. It is found that the test tube begins to sink at pressure $p_0+\Delta p$ without changing its orientation. At this pressure the volume of trapped air is $v_0-\Delta v$. Let $\Delta v=x$ cc and $\Delta p=y\times10^3$ Pa.

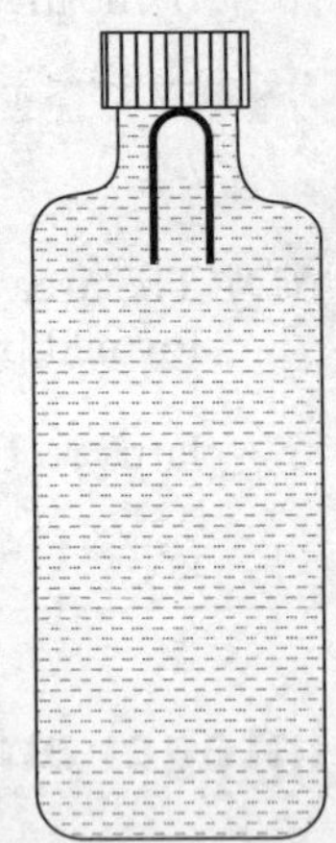

1. The value of x is______.
2. The value of y is______.

ANSWER KEY

JEE Main

Objective Questions

One Option Correct

1. (c)	**2.** (c)	**3.** (c)	**4.** (b)	**5.** (b)	**6.** (b)	**7.** (c)	**8.** (c)	**9.** (c)
10. (d)	**11.** (b)	**12.** (d)	**13.** (c)	**14.** (b)	**15.** (c)	**16.** (c)	**17.** (c)	**18.** (d)
19. (c)	**20.** (a)	**21.** (b)	**22.** (b)	**23.** (b)	**24.** (c)	**25.** (c)	**26.** (b)	**27.** (b)
28. (b)	**29.** (c)	**30.** (a)	**31.** (c)	**32.** (b)	**33.** (b)	**34.** (a)	**35.** (c)	**36.** (a)
37. (d)	**38.** (b)	**39.** (b)	**40.** (b)	**41.** (d)	**42.** (a)	**43.** (a)	**44.** (c)	**45.** (a)
46. (a)	**47.** (b)	**48.** (c)	**49.** (b)	**50.** (a)	**51.** (a)	**52.** (a)	**53.** (b)	**54.** (c)
55. (d)	**56.** (c)	**57.** (c)	**58.** (c)	**59.** (d)	**60.** (c)	**61.** (a)	**62.** (b)	**63.** (a)
64. (b)	**65.** (b)	**66.** (b)	**67.** (b)	**68.** (a)	**69.** (d)	**70.** (b)	**71.** (a)	**72.** (b)
73. (b)	**74.** (c)	**75.** (d)	**76.** (d)	**77.** (d)	**78.** (d)	**79.** (b)	**80.** (b)	**81.** (d)
82. (a)	**83.** (d)	**84.** (a)	**85.** (b)	**86.** (c)	**87.** (c)	**88.** (a)	**89.** (b)	**90.** (c)
91. (b)								

Integer and Numerical Value Type

1. (1)	**2.** (4π)	**3.** (6)	**4.** (15)	**5.** (9)	**6.** (1.5)	**7.** (114)	**8.** (1.25)	**9.** $\frac{41}{6}$
10. (64)	**11.** (27)	**12.** (8)						

Assertion and Reason

1. (c) **2.** (b)

JEE Advanced

Objective Questions

One Option Correct

1. (d) **2.** (b) **3.** (a) **4.** (c) **5.** (a) **6.** (d) **7.** (c) **8.** (a) **9.** (a)
10. (a) **11.** (a) **12.** (a) **13.** (b) **14.** (c) **15.** (b)

One or More Than One Option Correct

1. (a, b, c) **2.** (a, d) **3.** (b, c) **4.** (b, c, d) **5.** (a, b, d) **6.** (b, c, d) **7.** (b, c)

Matrix Type

1. A → T; B → T; C → R; D → S **2.** A → T; B → P; C → S; D → R

Integer and Numerical Value Type

1. (4) **2.** (8) **3.** (3) **4.** (29) **5.** (4) **6.** (3)

Comprehension Passage

Comprehension 1

1. (a) **2.** (b) **3.** (b)

Comprehension 2

1. (b) **2.** (a)

Comprehension 3

1. (3) **2.** (10)

HINTS AND EXPLANATIONS

Worksheet

1. Graphically the area bounded by the curves can be represented as

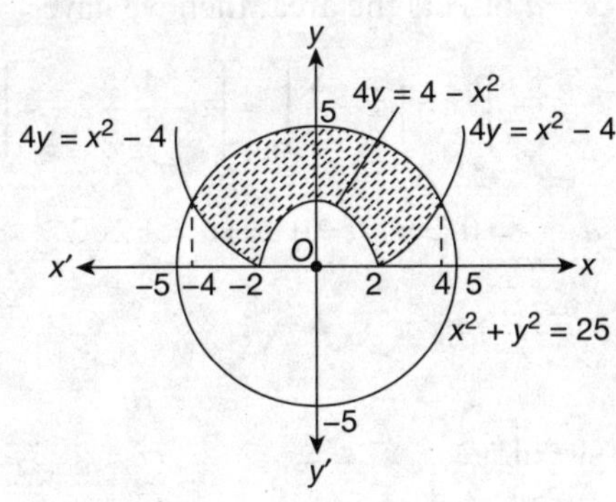

The points of intersection are

$$x^2 + \frac{(4-x^2)^2}{16} = 25$$

$$\Rightarrow (x^2+24)(x^2-16) = 0 \Rightarrow x = \pm 4$$

Required area is

$$= 2\left\{\int_0^4 \sqrt{25-x^2}\,dx - \int_0^2 \frac{4-x^2}{4}\,dx - \int_2^4 \left(\frac{x^2-4}{4}\right)dx\right\}$$

$$= 2\left\{\left(\frac{x}{2}\sqrt{25-x^2} + \frac{25}{2}\sin^{-1}\left(\frac{x}{5}\right)\right)_0^4 - \frac{1}{4}\left(4x - \frac{x^3}{3}\right)_0^2 - \frac{1}{4}\left(\frac{x^3}{3} - 4x\right)_2^4\right\}$$

$$= 2\left[\left[6 + \frac{25}{2}\sin^{-1}\left(\frac{4}{5}\right)\right] - \frac{1}{4}\left[8 - \frac{8}{3}\right] - \frac{1}{4}\left[\left(\frac{64}{3} - 16\right) - \left(\frac{8}{3} - 8\right)\right]\right]$$

$$= 2\left[6 + \frac{25}{2}\sin^{-1}\left(\frac{4}{5}\right) - \frac{4}{3} - \frac{4}{3} - \frac{4}{3}\right]$$

$$= 4 + 25\sin^{-1}\left(\frac{4}{5}\right) \text{ sq.units.}$$

2. The given curves are

$$y = \sqrt{5-x^2} \qquad \text{...(i)}$$

$$y = |x-1| \qquad \text{...(ii)}$$

Squaring both sides of Eq.(i) represents a circle. But as y is a +ve sq. root, ∴ (i) represents upper half of the circle with centre (0, 0) and radius $\sqrt{5}$.

Eq. (ii) represents the curve

$$y=\begin{cases}-x+1 & \text{if } x<1\\ x-1 & \text{if } x\ge 1\end{cases}$$

The graph of these curves is as shown in the figure with point of intersection of $y=\sqrt{5-x^2}$ and $y=-x+1$ as $A(-1,2)$ and of $y=\sqrt{5-x^2}$ and $y=x-1$ as $C(2,1)$

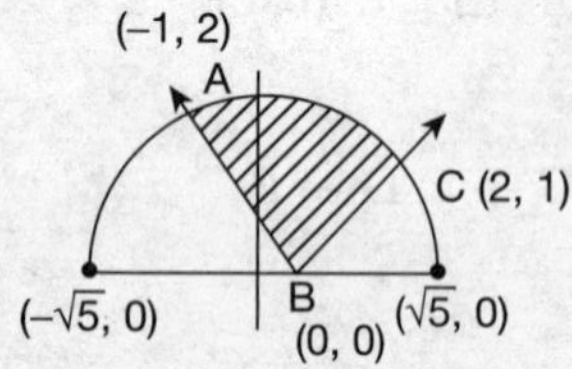

The required is = Shaded area

$$=\int_{-1}^{2}\left(y_{(1)}-y_{(2)}\right)dx=\int_{-1}^{2}\sqrt{5-x^2}\,dx-\int_{-1}^{2}|x-1|\,dx$$

$$=\left[\frac{x}{2}\sqrt{5-x^2}+\frac{5}{2}\sin^{-1}\left(\frac{x}{\sqrt{5}}\right)\right]_{-1}^{2}$$

$$-\int_{-1}^{1}\{-(x-1)dx-\int_{1}^{2}(x-1)dx$$

$$=\left(\frac{2}{2}\sqrt{5-4}+\frac{5}{4}\sin^{-1}\frac{2}{\sqrt{5}}\right)-\left(\frac{-1}{2}\sqrt{5-1}\right.$$

$$\left.+\frac{5}{2}\sin^{-1}\left(\frac{-1}{\sqrt{5}}\right)\right)-\left(\frac{-x^2}{2}+x\right)_{-1}^{1}-\left(\frac{x^2}{2}-x\right)_{1}^{2}$$

$$=2+\frac{5}{2}\left[\sin^{-1}\frac{2}{\sqrt{5}}+\sin^{-1}\frac{1}{\sqrt{5}}\right]-2-\frac{1}{2}$$

$$=\frac{5}{2}\left[\sin^{-1}\frac{2}{\sqrt{5}}+\cos^{-1}\frac{2}{\sqrt{5}}\right]-\frac{1}{2}=\frac{5}{2}\left(\frac{\pi}{2}\right)-\frac{1}{2}$$

$$=\frac{5\pi-2}{4}\text{ square units}$$

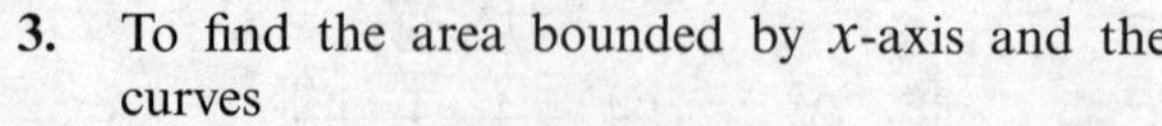

3. To find the area bounded by x-axis and the curves

$y=\tan x, -\pi/3\le x\le\pi/3$ ...(i)

and $y=\cot x, \pi/6\le x\le 3\pi/2$ ...(ii)

The curves intersect at P, where tan x = cot x, which is satisfied at $x=\pi/4$ within the given domain of x.

The required area is shaded area

$$A=\int_{\pi/6}^{\pi/4}\tan x\,dx+\int_{\pi/4}^{\pi/3}\cot x\,dx$$

$$=[\log\sec x]_{\pi/6}^{\pi/4}+[\log\sin x]_{\pi/4}^{\pi/3}$$

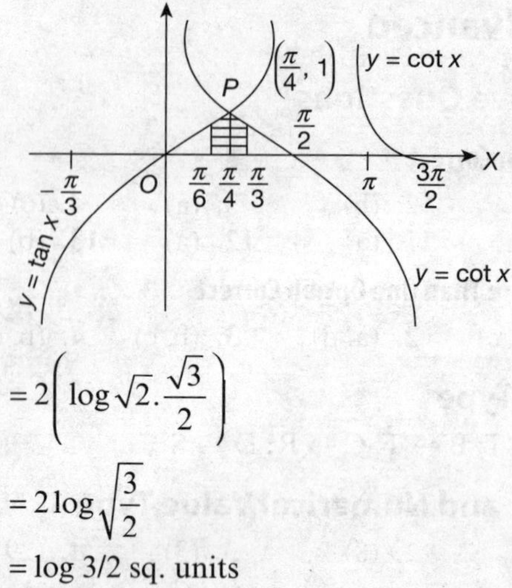

$$=2\left(\log\sqrt{2}.\frac{\sqrt{3}}{2}\right)$$

$$=2\log\sqrt{\frac{3}{2}}$$

$$=\log 3/2\text{ sq. units}$$

4. The equation of the curve is, $y=1+\dfrac{8}{x^2}$

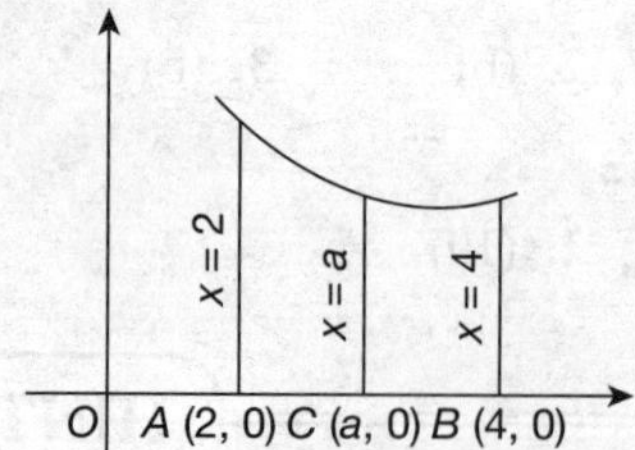

$$\text{Required area}=\int_{2}^{4}y\,dx=\int_{2}^{4}\left(1+\frac{8}{x^2}\right)dx$$

$$=\left[x-\frac{8}{x}\right]_{2}^{4}=4$$

If $x=a$ bisects the area, then we have

$$\int_{2}^{a}\left(1+\frac{8}{x^2}\right)dx=\left[x-\frac{8}{x}\right]_{2}^{a}=\left[a-\frac{8}{a}-2+4\right]=\frac{4}{2}$$

$$\Rightarrow a-\frac{8}{a}=0\ \Rightarrow a^2=0\ \Rightarrow a=\pm 2\sqrt{2}$$

Quiz

1. Area $=\int_{a}^{b}y\,dx$

2. $\left|\int_{a}^{b}f(x)\,dx\right|$

3. $\int_{a}^{b}[f(x)-g(x)]\,dx$

4. The point of intersection of these two parabolas are $O(0,0)$ and $A(1,1)$ as shown in the figure. Here, we can set $y^2=x$ or $y=\sqrt{x}=f(x)$ and

$y = x^2 = g(x)$, where $f(x) \geq g(x)$ in [0, 1].
Therefore, the required area of the shaded region

$$= \int_0^1 [f(x) - g(x)]\,dx$$

$$= \int_0^1 \left[\sqrt{x} - x^2\right] dx = \left[\frac{2}{3}x^{\frac{3}{2}} - \frac{x^3}{3}\right]_0^1 = \frac{2}{3} - \frac{1}{3} = \frac{1}{3}$$

5. Area $= \int_a^c [f(x) - g(x)]\,dx + \int_c^b [g(x) - f(x)]\,dx.$

JEE Main

Objective Questions

One Option Correct

1. (c) Here $x + 3y^2 = 1$ and $x + 2y^2 = 0$ intersect at $x = -2$ and $y = \pm 1$

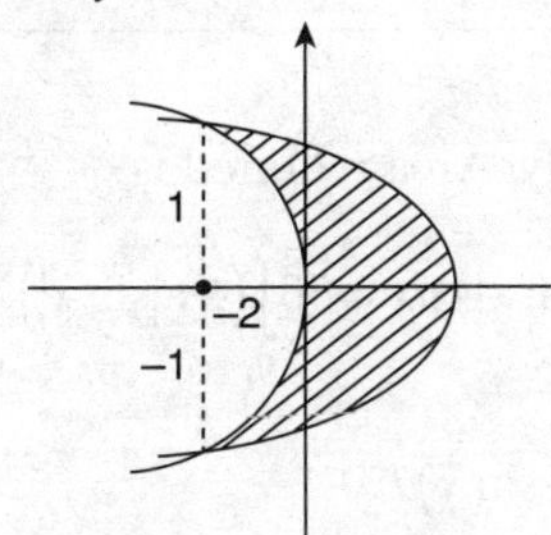

Hence required area $= \int_{-1}^{1} \left(1 - 3y^2 + 2y^2\right) dy$

$$= \int_{-1}^{1} (1 - y^2)\,dy = 2\int_0^1 (1 - y^2) = 2\left[y - \frac{y^3}{3}\right]_0^1 = 2\left(1 - \frac{1}{3}\right)$$

$= \frac{4}{3}$ sq. units.

2. (c) $x + e = 0$ is the asymptote.

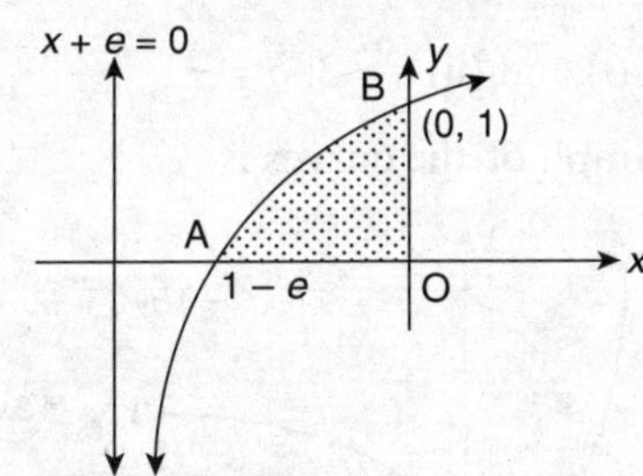

Required area is OBA i.e., $\int_{1-e}^{0} \log_e(x + e)\,dx$

Put $x + e = z \Rightarrow dx = dz$

Area $= \int_1^e \log_e z\,dz = (z \log z - z)\Big|_1^e$

$= (e - e - 0 + 1) = 1$

3. (c) Here $y = |x - 1|$
$= x - 1$ when $x \geq 1$,
$= 1 - x$ when $x < 1$, and
$y = 3 - |x|$
$= 3 - x$ when $x \geq 0$ and $3 - (-x) = 3 + x$ when $x < 0$
Solving, $y = x - 1$ and $y = 3 - x$, we get P (2, 1) and $y = 1 - x$ and $y = 3 + x$ we get Q (−1, 2)
Required area is rectangle APBQ

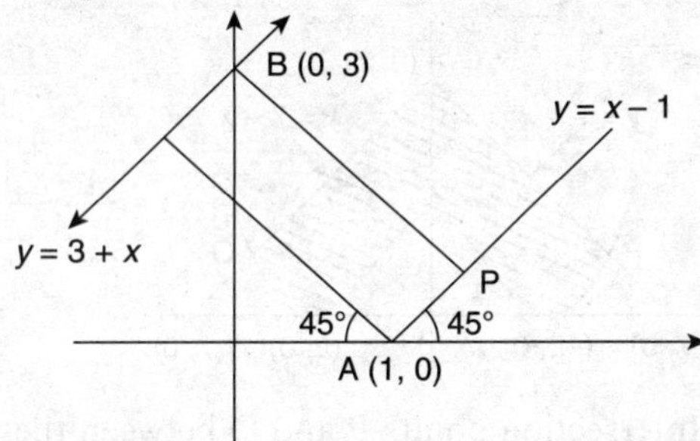

AP × BP

Now $AP = \sqrt{(1-2)^2 + (0-1)^2} = \sqrt{2}$

$BP = \sqrt{(0-2)^2 + (3-1)^2} = 2\sqrt{2}$

Hence AP × BP = 4 sq. units

4. (b) $x^2 + y^2 - 2ax = 0 \Rightarrow (x - a)^2 + y^2 = a^2$, which is the equation of a circle with radius a.
So, area $= \pi a^2$ sq. units.

5. (b) Solving $m^2x^2 = 16x = 0$, we get $x = 0, \frac{16}{m^2}$.

So, required area $= \int_0^{16/m^2} \left(\sqrt{16x} - mx\right) dx$

$\Rightarrow \frac{2}{3} = \frac{8 \times 64}{3 \times m^3} - \frac{2 \times 64}{m^3} \Rightarrow m = 4$

6. (b)

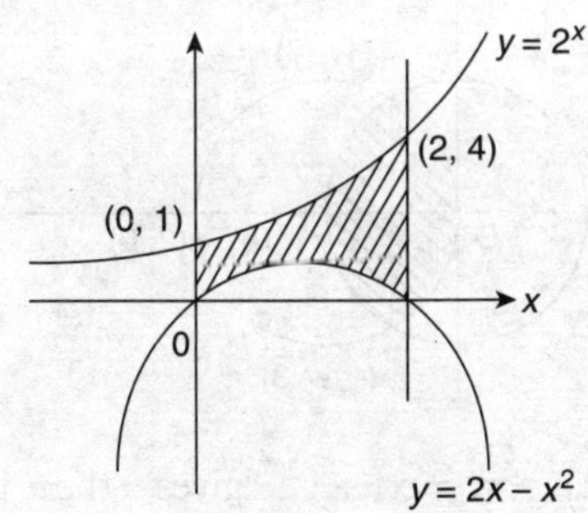

Required area $= \int_0^2 \left(2^x - 2x + x^2\right) dx$

$= \frac{3}{\log 2} - \frac{4}{3}$

7. (c) Equation of the curves can be written as

$$y = \begin{cases} \frac{x^2-4}{4}, x \in (-\infty, -2) \cup (2, \infty) \\ \frac{4-x^2}{4}, x \in (-2, 2) \end{cases}$$

$$y = \begin{cases} 7 - x, x \geq 0 \\ 7 + x, x < 0 \end{cases}$$

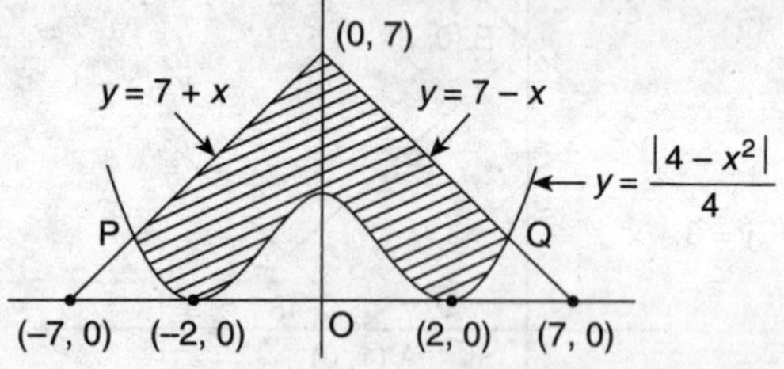

Intersection points P and Q between these two curves are

$$\frac{|4-x^2|}{4} = 7 - |x|$$

P and Q are (−4, 3) and (4, 3) respectively.
Area = 2 × Shaded area in first quadrant

$$= 2 \times \left[\int_0^4 \underset{\text{Line}}{y\, dx} - \int_0^2 \underset{\text{Parabola}}{y\, dx} - \int_2^4 \underset{\text{Parabola}}{y\, dx} \right]$$

$$= 2 \times \left[\int_0^4 (7 - x) dx - \int_0^2 \frac{4-x^2}{4} dx - \int_2^4 \frac{x^2-4}{4} dx \right]$$

$= 2 \times [16] = 32$ sq. units

8. (c) Graphically, required area is represented as shaded area

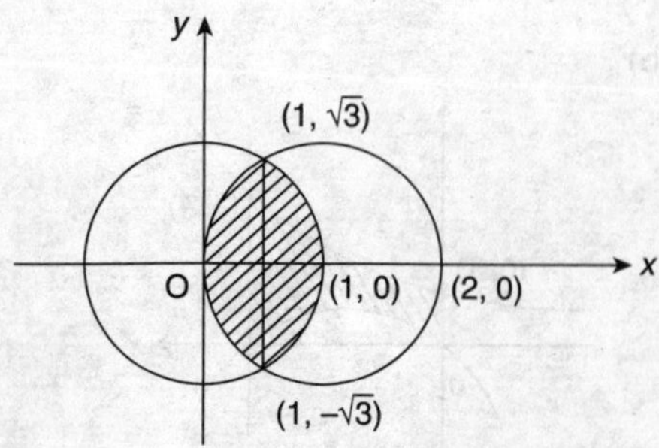

$(x-2)^2 + y^2 = x^2 + y^2$ gives the points of intersection, which are $x = 1$ and $y = \pm\sqrt{3}$

Required area

$$= 2\left[\int_0^1 \sqrt{4-(x-2)^2}\, dx + \int_1^2 \sqrt{4-x^2}\, dx \right]$$

$$= 2\left[\frac{1}{2}(x-2)\sqrt{4-(x-2)^2} + \frac{1}{2} \times 4 \sin^{-1}\left(\frac{x-2}{2}\right) \right]_0^1 +$$

$$2\left[\frac{1}{2} x\sqrt{4-x^2} + \frac{1}{2} \times 4 \sin^{-1} \frac{x}{2} \right]_1^2$$

$$= \left[\left(-\sqrt{3} - 4 \times \frac{\pi}{6}\right) + 4 \times \frac{\pi}{2} \right] + \left[4 \times \frac{\pi}{2} - \sqrt{3} - 4 \times \frac{\pi}{6} \right]$$

$= \frac{8\pi}{3} - 2\sqrt{3}$ sq. units

9. (c) Graph of the curve $y = 1 + |\sin x|$ is

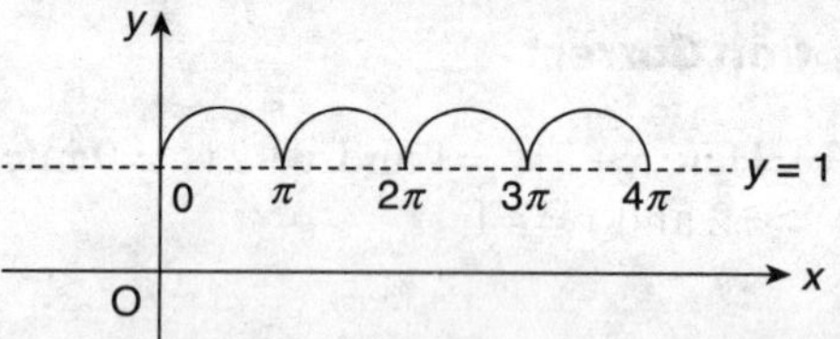

Obviously, Area required is

$$\int_0^{50\pi} (1 + |\sin x|)\, dx = 50 \int_0^{\pi} (1 + |\sin x|)\, dx$$

$$= 50 \int_0^{\pi} (1 + \sin x) dx$$

$= 50[x - \cos x]_0^{\pi} = 50(\pi + 2)$
$= 50\pi + 100$ sq.units

10. (d) Since curve passes through (1, 2)
$\Rightarrow a + b = 2$ ...(i)

Required area $= \int_0^4 \left(a\sqrt{x} + bx\right) dx = 8$

$\Rightarrow \frac{2a}{3} \cdot 8 + 8b = 8 \Rightarrow 2a + 3b = 3$ ...(ii)

From (i) and (ii) $a = 3, b = -1$

11. (b) Graph of the curves is

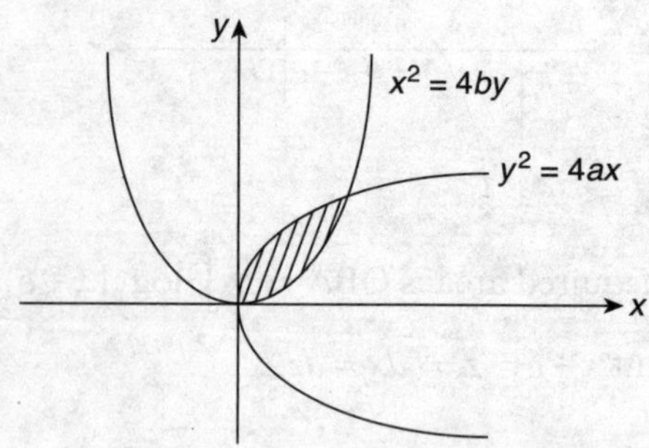

Intersection of these two curves is

$x^2 = 4b \times \sqrt{4ax} \Rightarrow x\left(x^3 - 64b^2a\right) = 0$

$\Rightarrow x = 0,\ x^3 = 64b^2a$

If $x = \alpha$ is the point of intersection, then $\alpha^3 = 64b^2a$

Now, Area is $\int_0^{\alpha}\left(2\sqrt{a}\sqrt{x} - \frac{x^2}{4b}\right)dx$

$= \left[2\sqrt{a} \times x^{\frac{3}{2}} \times \frac{2}{3} - \frac{x^3}{12b}\right]_0^{\alpha} = \frac{4}{3}\sqrt{a}\alpha^{3/2} - \frac{\alpha^3}{12b}$

$= \frac{4}{3}\sqrt{a} \times 8b\sqrt{a} - \frac{64b^2a}{12b}$

$= \frac{32}{3}ab - \frac{64}{12}ab = \frac{16ab}{3}$ sq.units

12. (d) Graph of the curve is

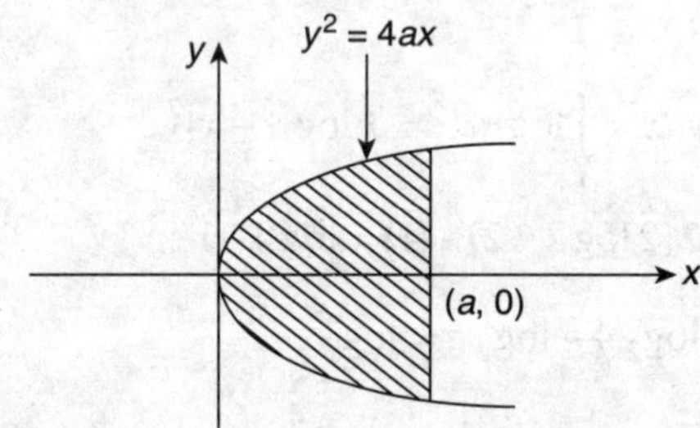

Required area $= 2\int_0^a 2\sqrt{a}\sqrt{x}dx$

$= 2\left[2\sqrt{a}x^{3/2} \times \frac{2}{3}\right]_0^a = \frac{8}{3}a^2$ sq. units

13. (c) Points of intersection of the curve and x-axis are $-a$ and 0.

Therefore,

Required area $= \frac{1}{a^2}\int_a^0 x^2(x+a)dx = \frac{a^2}{12}$ sq.units

14. (b) The graph of these functions is

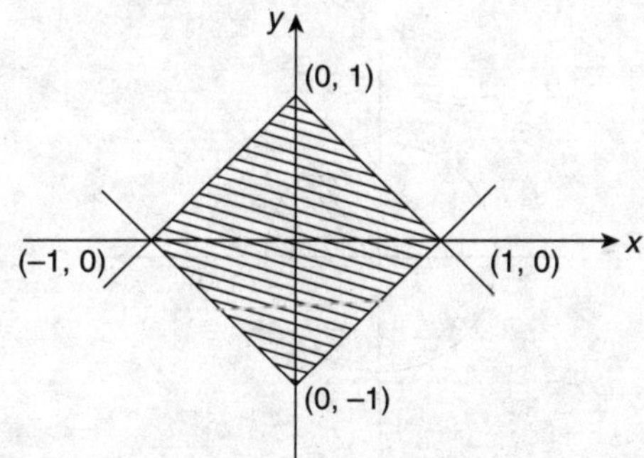

This is obviously a square and area $= \sqrt{2} \times \sqrt{2} = 2$ sq. units

15. (c) Graphically, this can be represented as

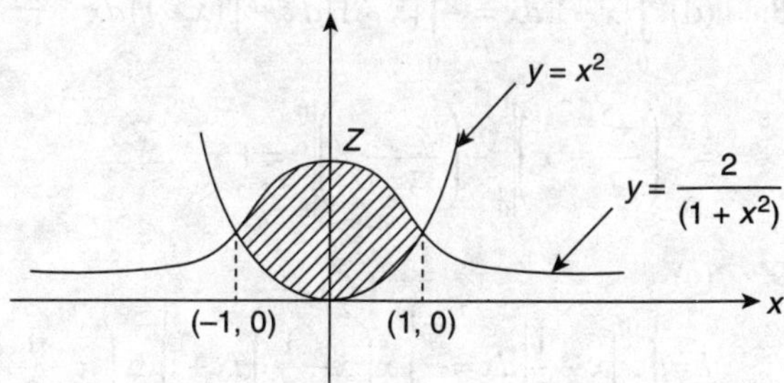

Point of intersection of these two curves is

$y = \frac{2}{y} - 1$

$\Rightarrow y^2 + y - 2 = 0$

$\Rightarrow (y-1)(y+2) = 0$

$\Rightarrow y = -2, 1 \Rightarrow y = 1$ as $y \geq 0$

At $y = 1$, $x = \pm 1$

The co-ordinates are $(1, 1), (-1, 1)$

Required area $= 2\left[\int_0^1 \frac{2}{1+x^2}dx - \int_0^1 x^2dx\right]$

$= 2\left[\left[2\tan^{-1}x\right]_0^1 - \frac{x^3}{3}\Big|_0^1\right]$

$= 2\left[\frac{2\pi}{4} - \frac{1}{3}\right] = \pi - \frac{2}{3}$ sq. units

16. (c) Graphically, the region can be represented as

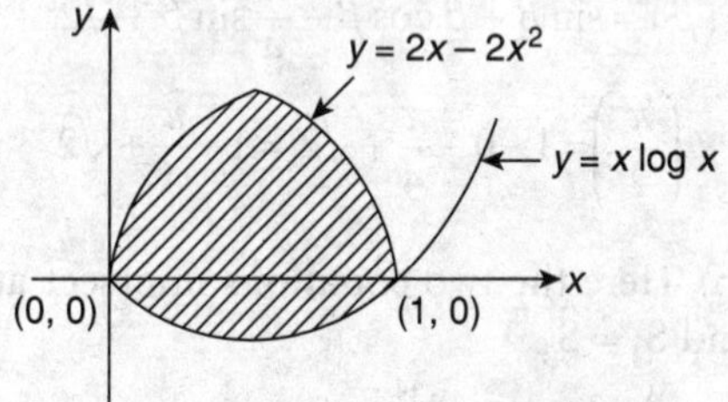

Required area $= \int_0^1\left(2x - 2x^2\right)dx - \int_0^1 x\log x dx$

$= \left[\frac{2x^2}{2} - \frac{2x^3}{3}\right]_0^1 - \left[\frac{x^2}{2}\log x - \frac{x^2}{4}\right]_0^1$

$= \frac{7}{12}$ sq. units

17. (c) $\int f(x)dx = g(x) + c \therefore \int f^{-1}(x)dx$

Put $f'(x) = t,\ x = f(t),\ dx = f'(t)dt$

$\therefore \int f^{-1}(x)dx = \int tf'(t)dt = tf(t) - \int f(t)dt$

$= tf(t) - g(t) + c$

18. (d) $\int_0^2 |x-1|\,dx = -\int_0^1 (x-1)\,dx + \int_1^2 (x-1)\,dx$

$$= -\left(\frac{x^2}{2}-x\right)\Bigg|_0^1 + \left(\frac{x^2}{3}-x\right)\Bigg|_1^2 = 1$$

19. (c)

$$I = \int_0^1 x\left|x-\frac{1}{2}\right|dx = -\int_0^{1/2} x\left(x-\frac{1}{2}\right)dx + \int_{1/2}^1 x\left(x-\frac{1}{2}\right)dx$$

$$= -\left(\frac{x^3}{3}-\frac{x^2}{4}\right)\Bigg|_0^{1/2} + \left(\frac{x^3}{3}-\frac{x^4}{4}\right)\Bigg|_{1/2}^1 = \frac{1}{8}$$

20. (a)

$$F'(x) = -\frac{2}{x^3}\int_4^x \left(4t^2-2F'(t)\right)dt + \frac{1}{x^2}\left(4x^2-2F'(x)\right)$$

$$\Rightarrow F'(4) = -\frac{2}{4^3}\times \text{zero} + \frac{1}{16}(64-2F'(4))$$

$$\Rightarrow F'(4) = \frac{32}{9}$$

21. (b) Here $\int_{\pi/4}^{\beta} f(x)\,dx$ is give area

$$= \beta \sin\beta + \frac{\pi}{4}\cos\beta + \sqrt{2}\,\beta$$

Differentiating w.r.t. x, we get

$$f(\beta) = \sin\beta + \beta\cos\beta - \frac{\pi}{4}\sin\beta + \sqrt{2}$$

$$\therefore f\left(\frac{\pi}{2}\right) = 1 + 0 - \frac{\pi}{4} + \sqrt{2} = 1 - \frac{\pi}{4} + \sqrt{2}$$

22. (b) Here the two equations intersect at A(4, 4) and $S_1 = S_3$

$$S_1 = \int_0^4 \frac{x^2}{4}\,dx = \frac{x^3}{12}\Bigg|_0^4 = \frac{16}{3} \text{ sq. units.}$$

$$\text{Now } S_2 + S_3 = \int_0^4 \sqrt{4x}\,dx$$

$$= 2\left[\frac{x^{3/2}}{2/3}\Bigg|_0^4\right]$$

$$= \frac{32}{3} \text{ sq. units}$$

$$\Rightarrow S_2 = \frac{16}{3} \text{ sq. units}$$

$$\therefore S_1 : S_2 : S_3 = 1:1:1$$

23. (b) Graph of the given curve is

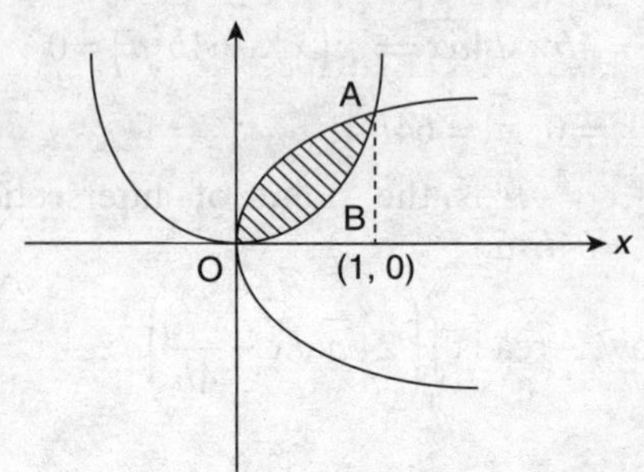

Curves $y = x^2$ and $x = y^2$ intersect at two real points which are obtained from the equation $x = x^4$, i.e., $x = 0$ and 1. The other two are complex roots.

Here, A (1, 1)

$$\text{Required area} = \int_0^1 \left(\sqrt{x} - x^2\right)dx = \frac{1}{3} \text{ sq. units}$$

24. (c)

$$\int_1^2 y\,dx = \int_1^2 \log x\,dx = x\log x - x\Big|_1^2$$

$$= 2(2\log 2 - 2) - (1) = \log 4 - 1$$

$$= \log_2 4 - \log_2 e = \log\frac{4}{e}$$

25. 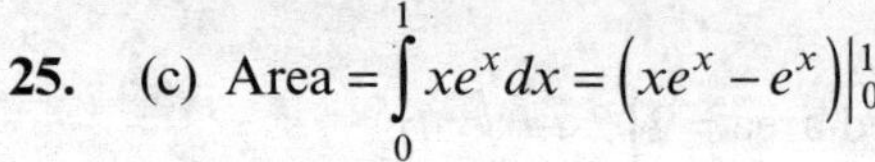

(c) Area $= \int_0^1 xe^x\,dx = \left(xe^x - e^x\right)\Big|_0^1$

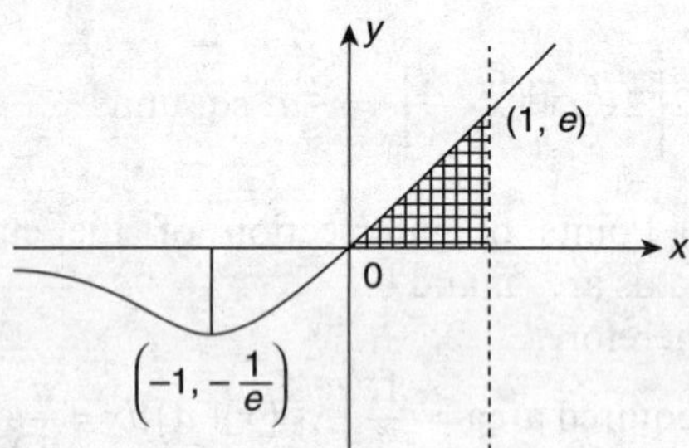

26. (b)

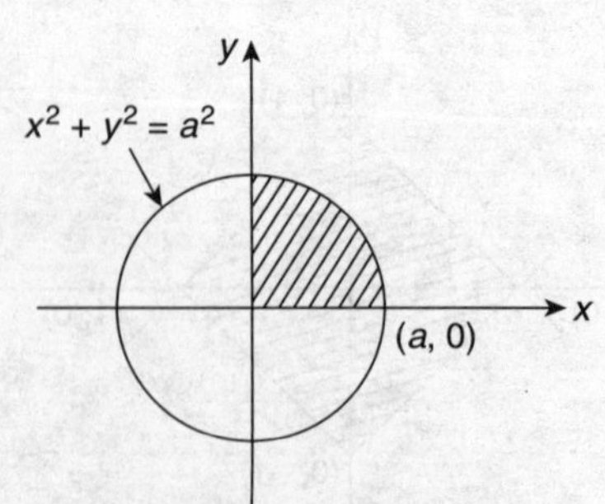

Area = 4 × Area in the first quadrant

$$= 4\times\int_0^a y\,dx = 4\int_0^a \sqrt{a^2-x^2}$$

$$= 4\left[\frac{x}{2}\sqrt{a^2-x^2}+\frac{a^2}{2}\sin^{-1}\frac{x}{a}\right]_0^a$$

$$= 4\left(\frac{a^2}{2}\right)\frac{\pi}{2} = \pi a^2 \text{ sq. units}$$

27. (b) Let us draw the graphs of these curves

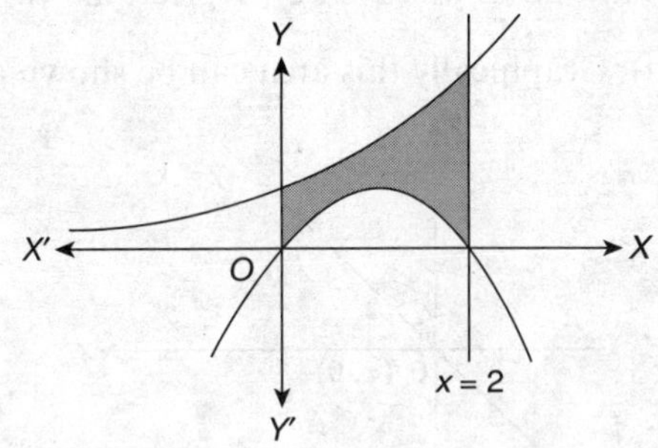

Required area $= \int_0^2 \left(2^x-\left(2x-x^2\right)\right)dx$

$$= \left[\frac{2^x}{\log 2}-x^2+\frac{x^3}{3}\right]_0^2$$

$$= \frac{4}{\log 2}-4+\frac{8}{3}-\frac{1}{\log 2}$$

$$= \frac{3}{\log 2}-\frac{4}{3}$$

28. (b) Graph of the curve $y=\dfrac{x}{1+x^2}$ is

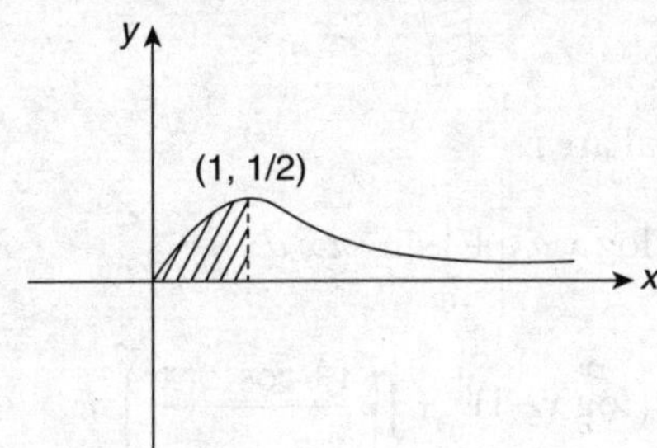

and $\dfrac{dy}{dx}=\dfrac{1-x^2}{\left(1+x^2\right)^2}$

$\Rightarrow x = 1$ is a point of local as well as absolute maxima. $\Rightarrow$ We have to find

Area $= \int_0^1 y\,dx = \int_0^1 \frac{x}{1+x^2}dx$

$$= \frac{1}{2}\left[\log\left(1+x^2\right)\right]_0^1 = \frac{1}{2}(\log 2-\log 1)$$

$$= \frac{1}{2}\log 2 \text{ sq. units}$$

29. (c) Graph of the curve is

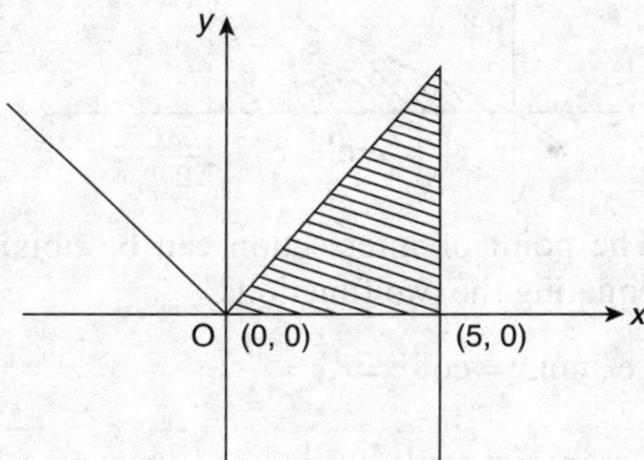

Required area is in the first quadrant

$$= \int_0^5 x\,dx = \left.\frac{x^2}{2}\right|_0^5 = \frac{25}{2}\text{ sq.units}$$

30. (a)

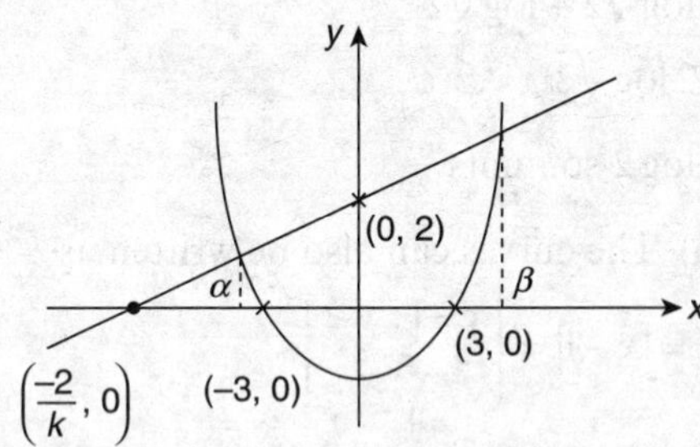

The line $y = kx + 2$ cuts the y-axis at (0, 2) and x-axis at $\left(-\dfrac{2}{k},0\right)$.

Suppose α,β are abscissa of points of intersection of the parabola at line $y = kx + 2$

Then the required area

$$= \int_\alpha^\beta (kx+2)\,dx - \int_\alpha^\beta \left(x^2-3\right)dx$$

$$= \left[\frac{kx^2}{2}+2x\right]_\alpha^\beta - \left[\frac{x^3}{3}-3x\right]_\alpha^\beta$$

Using $\alpha+\beta=-k,\ \alpha\beta=-5$

$|\beta-\alpha| = \sqrt{k^2+20}$

We get, $f(k)=\dfrac{\sqrt{\left(k^2+20\right)}}{6}$

which is minimum at $k = 0$.

31. (c) Graph of the curve is,

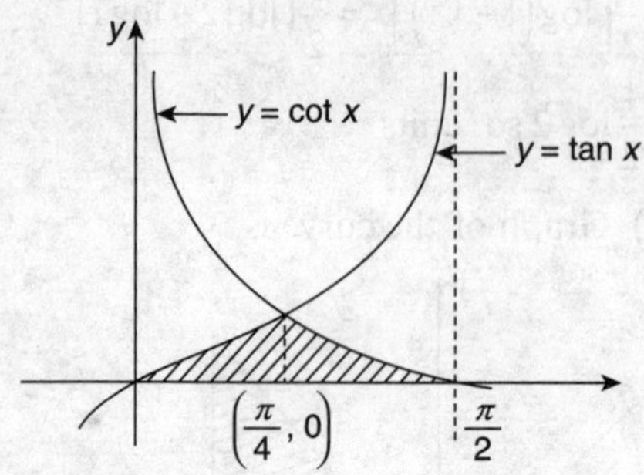

The point of intersection can be obtained by equating the two functions.

i.e., $\tan x = \cot x \Rightarrow x = \dfrac{\pi}{4}$

$$\text{Required area} = \int_0^{\pi/4} \tan x\, dx + \int_{\pi/4}^{\pi/2} \cot x\, dx$$

$$= \left[\log|\sec x|\right]_0^{\pi/4} + \left[\log|\sin x|\right]_{\pi/4}^{\pi/2}$$

$$= \log\sqrt{2} + \log\sqrt{2}$$

$$= 2\log\sqrt{2}$$

$$= \log 2 \text{ sq. units}$$

32. (b) The curves can also be written as

$$y_1 = |x-1| = \begin{cases} x-1 : x \ge 1 \\ 1-x : x \le 1 \end{cases} \quad \text{...(i)}$$

$$y_2 = 3-|x| = \begin{cases} 3-x : x \ge 0 \\ 3+x : x < 0 \end{cases} \quad \text{...(ii)}$$

These two curves meet at $(-1, 2)$ and $(2, 1)$
Now, the graph of these functions is

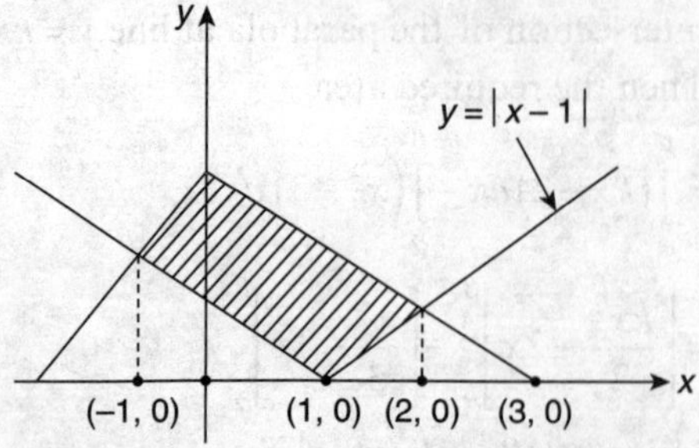

Required area

$$= \int_{-1}^{0} (y_2 - y_1)\,dx + \int_0^1 (y_2 - y_1)\,dx + \int_1^2 (y_2 - y_1)\,dx$$

$$= \int_{-1}^{0} (2+2x)\,dx + \int_0^1 2\,dx + \int_1^2 (4-2x)\,dx$$

$$= 1 + 2 + (4-3) = 4 \text{ sq.units}$$

33. (b) $\displaystyle\int_2^a \left(1+\frac{8}{x^2}\right)dx = \int_a^4 \left(1+\frac{8}{x^2}\right)dx$

$$\Rightarrow \left(x - \frac{8}{x}\right)_2^a = \left(x - \frac{8}{x}\right)_a^4$$

$$\Rightarrow a - \frac{8}{a} + 2 = 2 - a + \frac{8}{a}$$

$$2\left(a^2 - 8\right) = 0$$

$$\Rightarrow a = \pm 2\sqrt{2} \text{ or } a = 2\sqrt{2} \text{ (neglecting –ve sign)}$$

34. (a) Graphically this area can be shown as

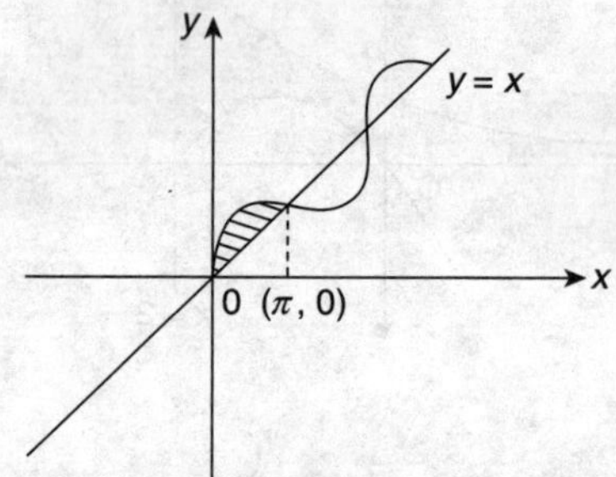

$$\text{Area} = \int_0^{\pi} (x + \sin x - x)\,dx = -\cos x\big|_0^{\pi} = 2 \text{ sq.units}$$

35. (c) Graphically, the region can be represented by

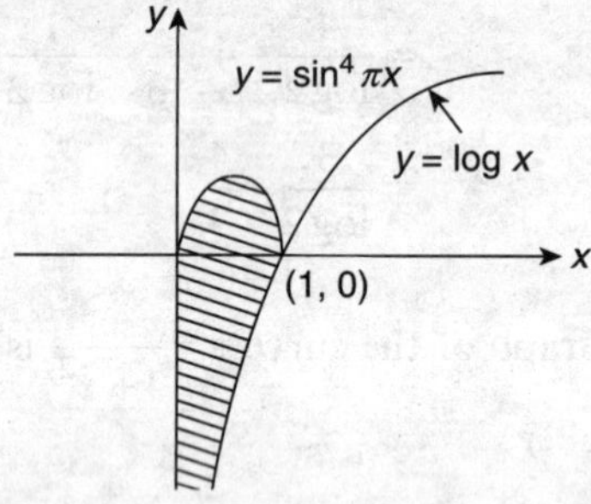

Total area

$$= \left|\int_0^1 \log x\,dx\right| + \int_0^1 \sin^4 \pi x\,dx$$

$$= \left|x(\log x - 1)\right|_0^1 + \int_0^1 \left(\frac{1+\cos 2\pi x}{2}\right)^2 dx$$

$$= |-1-0| + \frac{1}{16} \times 4 \int_0^{\frac{1}{4}} (5 + \cos 4\pi x + 8\cos 2\pi x)\,dx$$

$$= 1 + \frac{1}{4}\left[5x + \frac{\sin 4\pi x}{4\pi} + 8 \cdot \frac{\sin 2\pi x}{2\pi}\right]_0^{\frac{1}{4}}$$

$$= 1 + \frac{1}{4}\left(\frac{5}{4} + \frac{4}{\pi}\right) \text{sq.units}$$

36. (a)

$$\int_{-2}^{3}\left|1-x^2\right|dx=\int_{-2}^{-1}\left(x^2-1\right)dx-\int_{-1}^{1}\left(x^2-1\right)dx+\int_{1}^{3}\left(x^2-1\right)dx$$

$$=\left(\frac{x^3}{3}-x\right)\Bigg|_{-2}^{-1}-\left(+\frac{x^3}{3}-x\right)\Bigg|_{-1}^{1}+\left(\frac{x^3}{3}-x\right)\Bigg|_{1}^{3}$$

$$=\frac{2}{3}+\frac{2}{3}-\left(\frac{1}{3}-1\right)-\left(\frac{-1}{3}+1\right)+(9-3)+\left(1-\frac{1}{3}\right)=\frac{28}{3}$$

37. (d) Integrating, we get

$$f(x)=\int\frac{e^{\sin x}}{x}dx,\ x>0$$

Again $\int_{1}^{4}\frac{3}{x}e^{\sin x^3}dx=\int_{1}^{4}\frac{3x^2}{x^3}e^{\sin x^3}dx=\int_{1}^{64}\frac{e^{\sin t}}{t}dt$

Put $x^3=t, 3x^2dx=dt$ and when $x=1$, $t=1$,

$x=4$, t = 64 = F(64) – F(1)

$\therefore k=64$

38. (b)

$$=\int_{0}^{3}\left|x^3+x^2+3x\right|dx$$

$$=\int_{0}^{3}\left(x^3+x^2+3x\right)dx=\frac{x^4}{4}+\frac{x^3}{3}+\frac{3x^2}{2}\Bigg|_{0}^{3}$$

39. (b) $\int_{0}^{\pi/2}\log|\tan x|\,dx=\int_{0}^{\pi/2}\log\tan x\,dx$

$$=\int_{0}^{\pi/2}\log\tan\left(\frac{\pi}{2}-x\right)dx=\int_{0}^{\pi/2}\log\cot x\,dx$$

$$\Rightarrow\int_{0}^{\pi/2}\log\tan x\,dx+\int_{0}^{\pi/2}\log\cot x\,dx$$

$$=\int_{0}^{\pi/2}\log(\tan x\cdot\cot x)\,dx=\int_{0}^{\pi/2}\log 1\,dx=0$$

40. (b) Given integral $=\int\frac{x^{1/2}}{\sqrt{1-\left(x^{3/2}\right)^2}}$

Put $x^{3/2}=t$

$$\Rightarrow\frac{3}{2}x^{\frac{1}{2}}\,dx=dt$$

$$=\frac{2}{3}\int\frac{dt}{\sqrt{1-t^2}}=\frac{2}{3}\sin^{-1}x^{3/2}+c$$

41. (d) Required area = area ABCD

Here OA = OB = OC = OD = 1

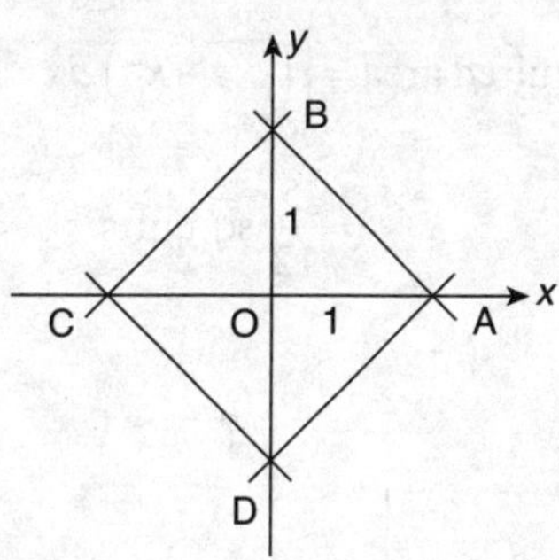

$\Rightarrow$ AB = BC = CD = CD = $\sqrt{2}$

Hence ABCD is a square with side $\sqrt{2}$ units.

Required area $=\left(\sqrt{2}\right)^2=2$ sq. units

42. (a) Area bounded by the curve is

$$\int_{0}^{1}\log x\,dx=\left|x\log x-x\right|_{0}^{1}=1\text{ sq. unit}$$

(on neglecting – sign and using $\lim_{x\to 0^+}x\log x=0$)

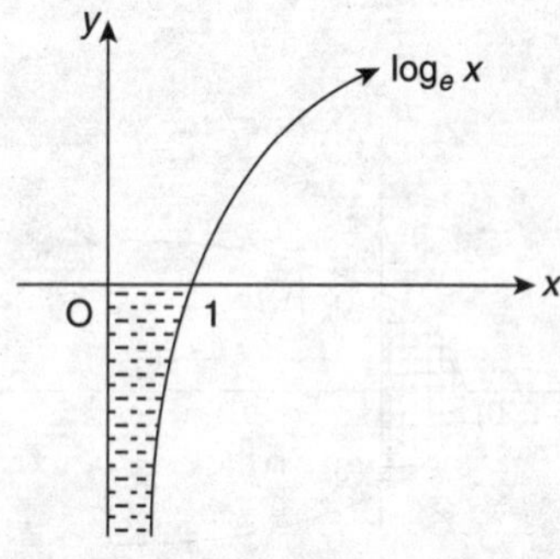

43. (a) Point of intersection is $x^2=x\Rightarrow x=0$ and 1

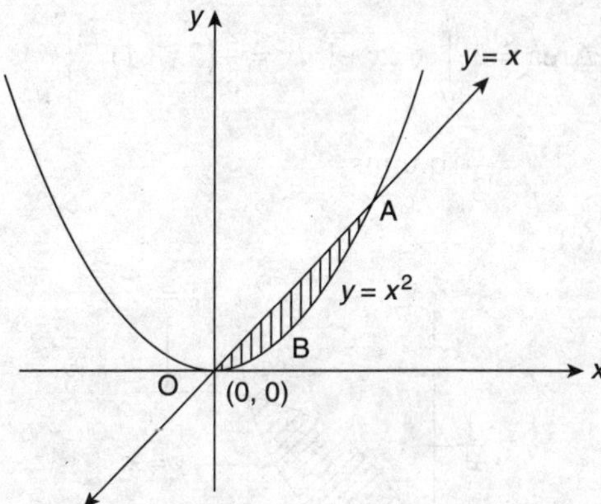

Required area is

$$OAB=\int_{0}^{1}\left(x-x^2\right)dx=\frac{x^2}{2}-\frac{x^3}{3}\Bigg|_{0}^{1}=\frac{1}{6}$$

44. (c) Here $x^3 = \sqrt{x} \Rightarrow x = 0, 1$ and other roots are imaginary.

$\therefore$ Required area $= \int_0^1 \left(\sqrt{x} - x^3\right) dx = \frac{2}{3}x^{3/2} - \frac{x^4}{4}\Big|_0^1$

$= \frac{5}{12}$ sq.units

45. (a)

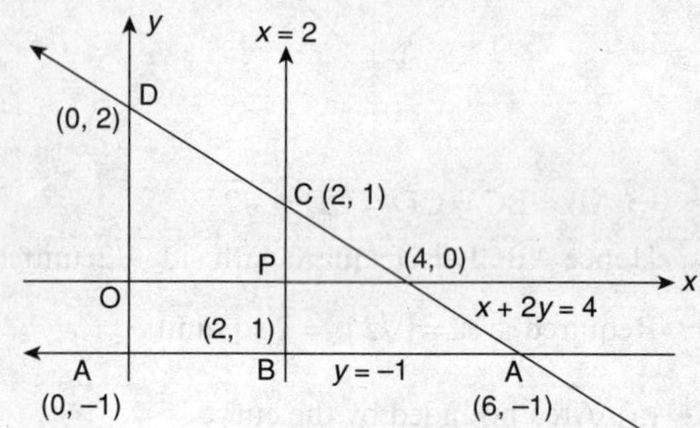

46. (a) Required area $= \int_0^1 \left(xe^x - xe^{-x}\right)dx$

$= \left(xe^x - e^x\right)\Big|_0^1 - \left(-xe^{-x} - e^{-x}\right)\Big|_0^1 = \frac{2}{e}$ sq.units

47. (b)

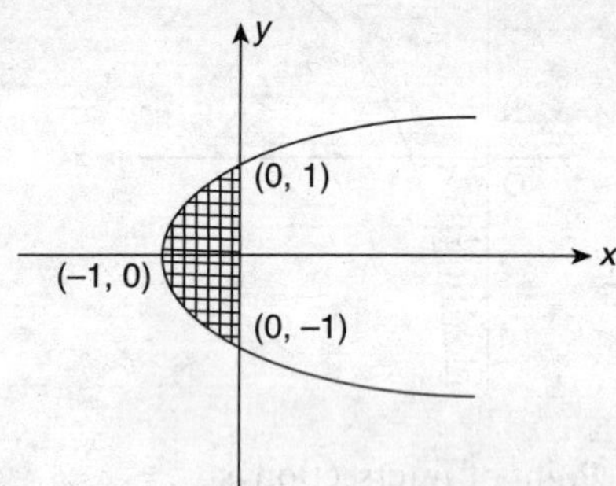

Area $= 2\int_{-1/2}^{0} \sqrt{2x+1}\, dx = \frac{2}{3}(2x+1)^{3/2}\Big|_{-1/2}^{0}$

$= \frac{2}{3}$ sq.units

48. (c)

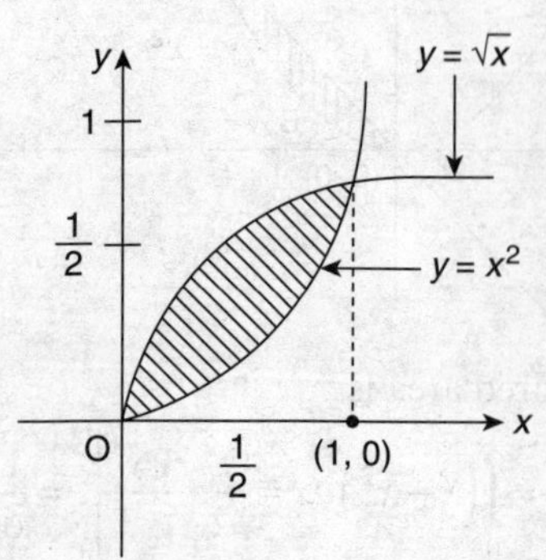

Area $= \int_a^b \left[\text{Upper function}\right] - \left[\text{Lower function}\right] dx$

a and b can be determined by finding the points where these two curves are intersecting,

i.e., $x^2 = \sqrt{x}$

$\Rightarrow x = 0$ or $x = 1$

$\Rightarrow$ Area $= \int_0^1 \left(\sqrt{x} - x^2\right) dx$

$= \left(\frac{2}{3}x^{3/2} - \frac{1}{3}x^3\right)\Big|_0^1 = \frac{1}{3}$ sq.units

49. (b)

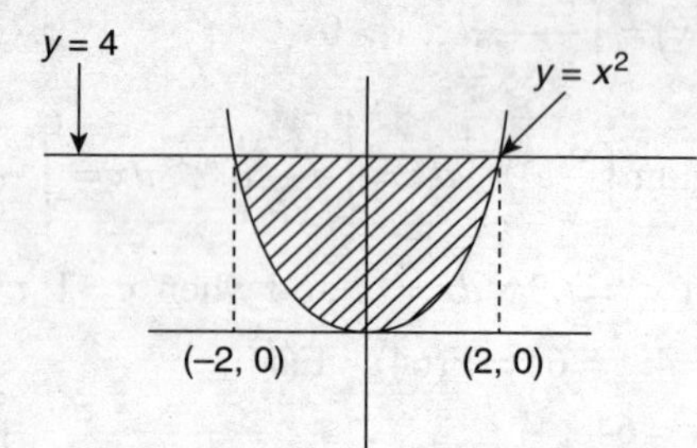

Area of the region will be $\int_{-2}^{2} y\, dx$,

It may be written as,

Area $= 2\int_0^2 \left(4 - x^2\right) dx = 2\left(4x - \frac{x^3}{3}\right)\Big|_0^2$

$= \frac{32}{3}$ sq.units

50. (a)

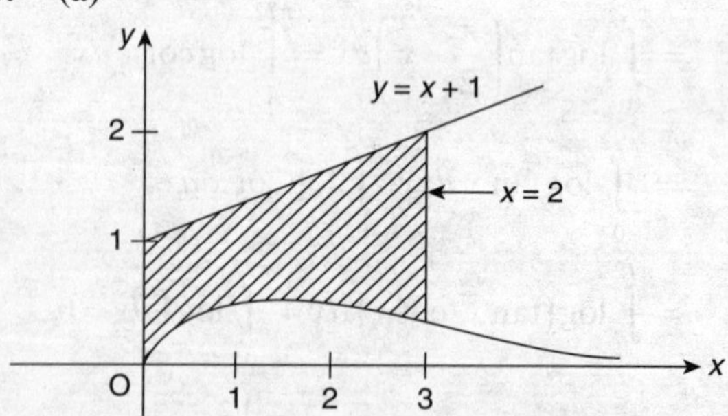

Area $= \int_0^2 \left[(x+1) - xe^{-x^2}\right] dx$

$= \left[\frac{1}{2}x^2 + d + \frac{1}{2}e^{-x^2}\right]_0^2$

$= \frac{1}{2e^4} + \frac{7}{2}$ sq.units

51. (a)

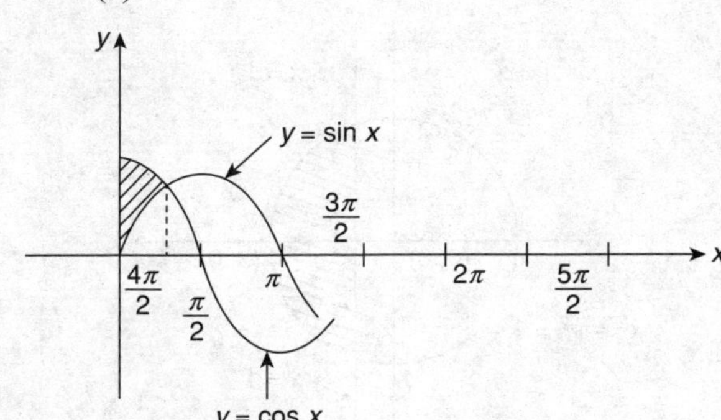

Point of intersection is $\sin x = \cos x \Rightarrow x = \frac{\pi}{4}$

$$\text{Area} = \int_0^{\frac{\pi}{4}} (\cos x - \sin x)\, dx$$

$$= [\sin x + \cos x]_0^{\pi/4}$$

$$= \left[\frac{1}{\sqrt{2}} + \frac{1}{\sqrt{2}} - 1\right]$$

$$= \left(\sqrt{2} - 1\right) \text{ sq. units}$$

52. (a) The curve cuts x-axis at $x = 1, 2$ and 3 and y -axis at –6.

Therefore, graph of the curve is

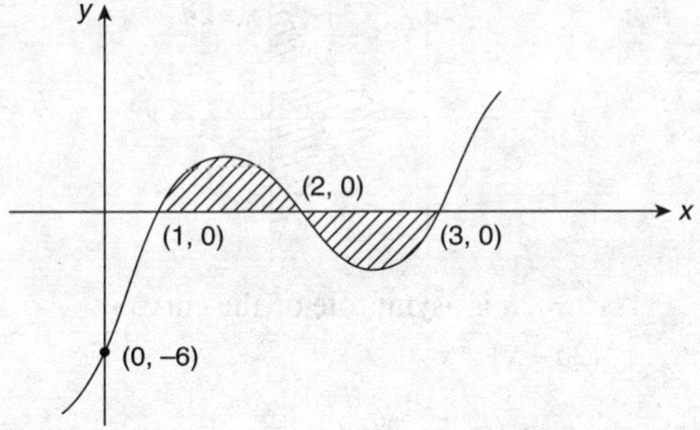

∴ Total area

$$= \int_1^2 \left(x^3 - 6x^2 + 11x - 6\right) dx + \left|\int_2^3 \left(x^3 - 6x^2 + 11x - 6\right) dx\right|$$

$$= \left.\frac{x^4}{4} - 2x^3 + \frac{11x^2}{2} - 6x\right|_1^2 + \left|\frac{x^4}{4} - 2x^3 + \frac{11x^2}{2} - 6x\right|_2^3$$

$$= \frac{1}{4} + \frac{1}{4} = \frac{1}{2} \text{ sq.units}$$

53. (b) Graph of the curve $y = \tan x$ along with the line $x = \frac{\pi}{2}$ and the x-axis is given below

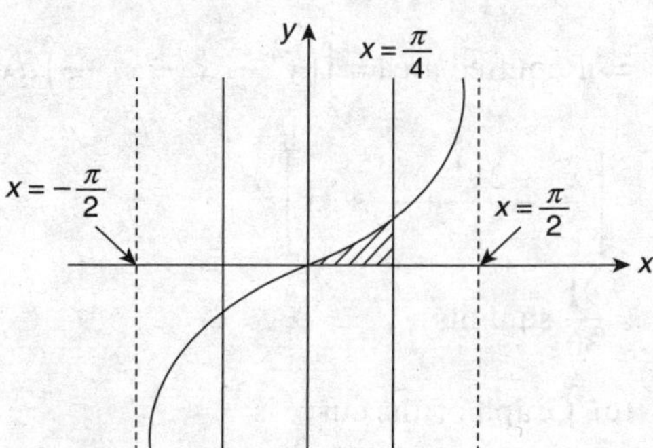

Area of the bounded region

$$= \int_0^{\frac{\pi}{4}} \tan x dx = [\log \sec x]_0^{\pi/4}$$

$$= \log\sqrt{2} - \log 1$$

$$= \frac{1}{2}\log 2 \text{ sq.units}$$

54. (c) Graph of the given curve is

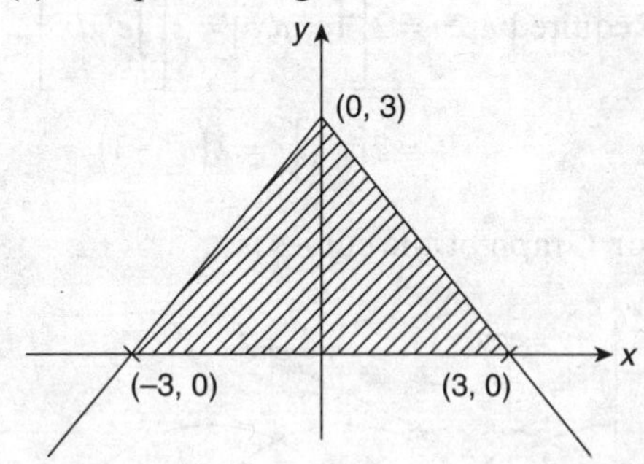

$$\text{Required area} = 2\int_0^3 (3 - x) dx = 2\left[3x - \frac{x^2}{2}\right]_0^3$$

$$= 2\left[9 - \frac{9}{2}\right] = 9 \text{ sq.units}$$

55. (d) $y' = 4x^3 - 6x^2 + 2x$

To get local minima of the function

$y' = 0$

$\Rightarrow 4x^3 - 6x^2 + 2x = 0$

$\Rightarrow 2x(2x - 1)(x - 1) = 0$

$\Rightarrow x = 0, \frac{1}{2}, 1$ are the points of local extrema

Now, $y'' = 12x^2 - 12x + 2$

$y'' > 0$ for $x = 0, 1$

$y'' < 0$ for $x = \frac{1}{2}$

$\Rightarrow$ Local minima at $x = 0, 1$

$$\Rightarrow \text{Required area} = \int_0^1 \left(x^4 - 2x^3 + x^2 + 3\right) dx$$

$$= \left[\frac{x^5}{5} - \frac{2x^4}{4} + \frac{x^3}{3} + 3x\right]_0^1$$

$$= \frac{91}{30} \text{ sq.units}$$

56. (c) Graph of the curve is

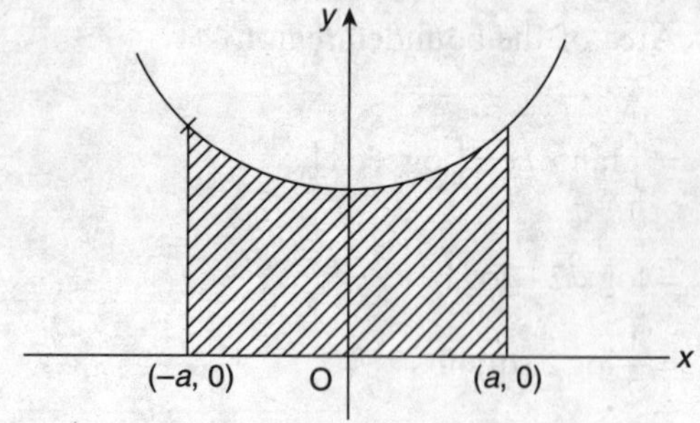

$$\text{Required area} = 2\left[\int_0^a e^{|x|} dx\right] = 2\left[\int_0^a e^x dx\right]$$

$$= 2\left[e^x\right]_0^a = 2\left(e^a - 1\right)$$

57. (c) Graph of the curves is

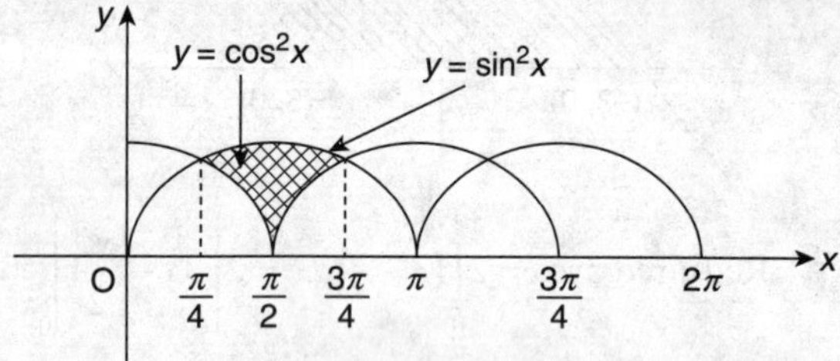

Intersecting points are

$$\cos^2 x = \sin^2 x \Rightarrow x = \frac{\pi}{4}, \frac{3\pi}{4}$$

$$\Rightarrow \text{Required area} = \int_{\frac{\pi}{4}}^{\frac{3\pi}{4}} \left(\sin^2 x - \cos^2 x\right) dx$$

$$= -\int_{\frac{\pi}{4}}^{\frac{3\pi}{4}} \cos 2x dx = -\frac{\sin 2x}{2}\Bigg|_{\frac{\pi}{4}}^{\frac{3\pi}{4}}$$

$$= -\frac{1}{2}\left[\sin\frac{3\pi}{2} - \sin\frac{\pi}{2}\right]$$

$$= -\frac{1}{2}[-1-1] = 1 \text{ sq.units}$$

58. (c) Graph of the curve is

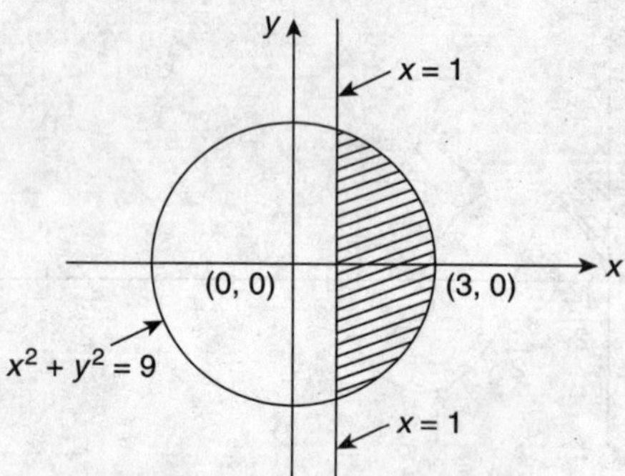

$$\text{Required area} = 2\left[\int_1^3 \sqrt{9 - x^2}\, dx\right]$$

$$= 2\left[\frac{x}{2}\sqrt{9 - x^2} + \frac{9}{2}\sin^{-1}\frac{x}{3}\right]_1^3$$

$$= 2\left[\frac{9}{2}\sin^{-1}\frac{3}{3} - \frac{1}{2}\sqrt{8} - \frac{9}{2}\sin^{-1}\frac{1}{3}\right]$$

$$= -\sqrt{8} + 9\left(\sin^{-1}\frac{3}{3} - \sin^{-1}\frac{1}{3}\right)$$

$$= 9\sec^{-1} 3 - \sqrt{8} \text{ sq. units}$$

59. (d) Graph of the curve

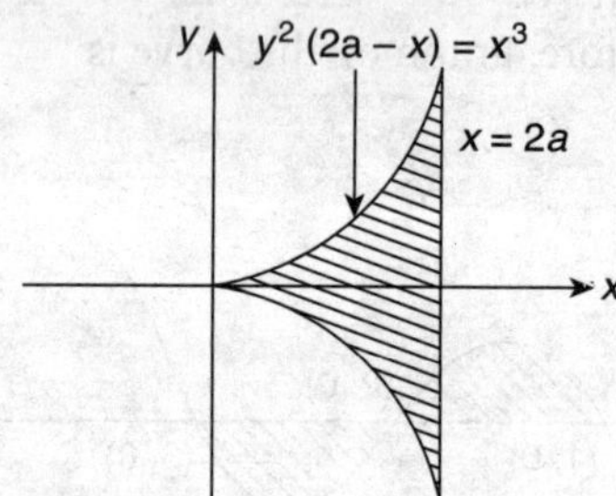

As $x = 2a$ is asymptote of the curve

$$y^2\left(2a - x\right) = x^3$$

$$\text{Area} = 2\int_0^{2a} \frac{x^{3/2}}{\sqrt{2a - x}} dx$$

Substituting $x = 2a\sin^2\theta$

$$\text{Area} = 2\left[\frac{3\pi a^2}{2}\right] = 3\pi a^2 \text{ sq.units}$$

60. (c) The following graph shows the required area in shaded form:

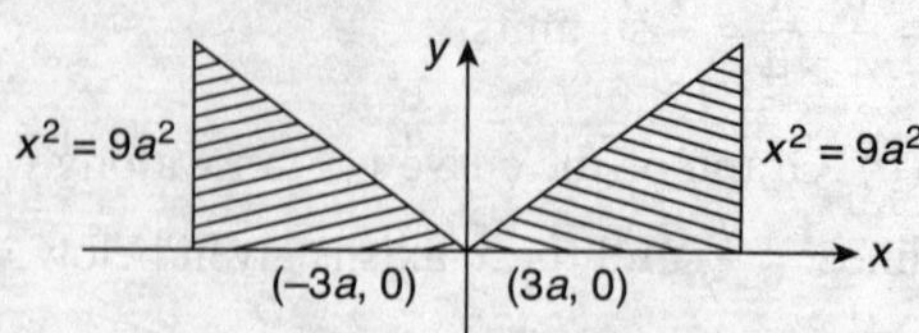

Required area $= \int_{-3a}^{0} x^2\,dx + \int_{0}^{3a} x\,dx$

$$= 9a^3 + \frac{9a^2}{2} = \frac{9a}{2}$$

$\Rightarrow 2a^2 + a - 1 = 0 \Rightarrow a = -1, \frac{1}{2}$

Now, $a > 0$, i.e, $a = \frac{1}{2}$

61. (a) Since ordinate is 3

So, $y = 3 \Rightarrow x = 2$

Point of contact is (2, 3)

Given equation of parabola $(y-2)^2 = (x-1)$

Differentiate w.r.t x,

$2(y-2)y' = 1$

$\Rightarrow y' = \frac{1}{2(y-2)} \Rightarrow y'_{(2,3)} = \frac{1}{2}$

$\Rightarrow \frac{y-3}{x-2} = \frac{1}{2} \Rightarrow x - 2y + 4 = 0$

Equation of tangent to the given parabola is,

$x - 2y + 4 = 0$

Required area of the bounded region

$= \int_0^3 \left((y-2)^2 + 1 - (2y-4)\right)dy = 9$ sq. units

62. (b)

$R = \left\{(x,y) : \max(0, \log_e x) \le y \le 2^x, \frac{1}{2} \le x < 2\right\}$

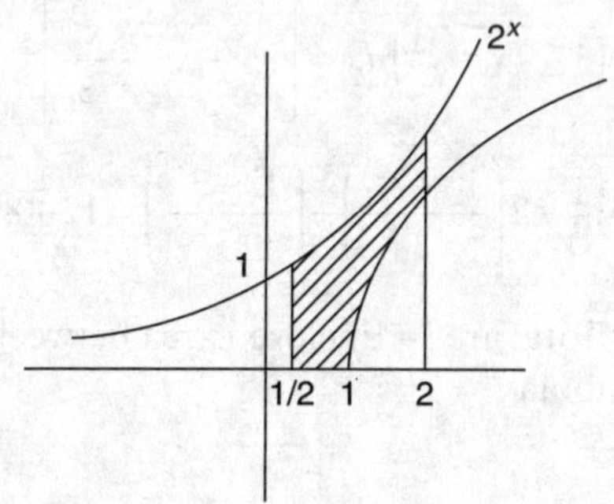

$\int_{\frac{1}{2}}^{2} 2^x\,dx - \int_1^2 \log_e x\,dx \Rightarrow \left[\frac{2^x}{\log_e 2}\right]_{\frac{1}{2}}^{2} - \left[x(\log_e x - 1)\right]$

$\Rightarrow \frac{(2)^2 - 2^{\frac{1}{2}}}{\log_e 2} - (2\log_e 2 - 1)$

$\Rightarrow \frac{\left(2^2 - \sqrt{2}\right)}{\log_e 2} - 2\log_e 2 + 1$

$\therefore \alpha = 2^2 - \sqrt{2}, \beta = -2, \gamma = 1$

So, $(\alpha + \beta - 2\gamma)^2$

$\Rightarrow \left(2^2 - \sqrt{2} - 2 - 2\right)^2 \Rightarrow \left(-\sqrt{2}\right)^2 = 2$

63. (a)

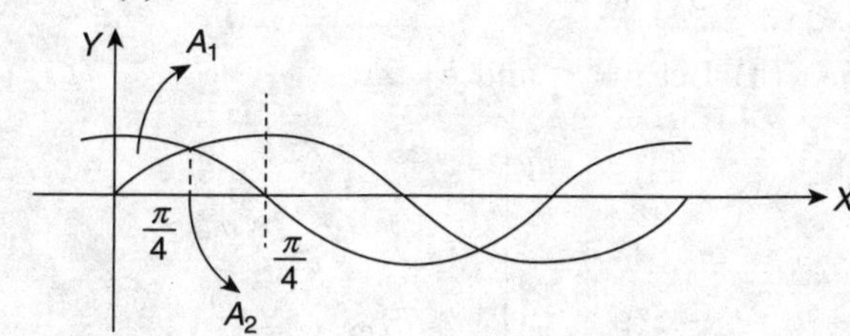

$A_1 = \int_0^{\pi/4} (\cos x \sin x)\,dx = [\sin x + \cos x]_0^{\pi/4} = \sqrt{2} - 1$

$A_2 = \int_0^{\pi/4} \sin x\,dx + A_1 = \int_{\pi/4}^{\pi/2} \cos x\,dx$

$= [-\cos x]_0^{\pi/4} + [\sin x]_{\pi/4}^{\pi/2}$

$= \sqrt{2}\left(\sqrt{2} - 1\right)$

$\therefore A_1 : A_2 = 1 : \sqrt{2}$ and $A_1 + A_2 = 1$

64. (b)

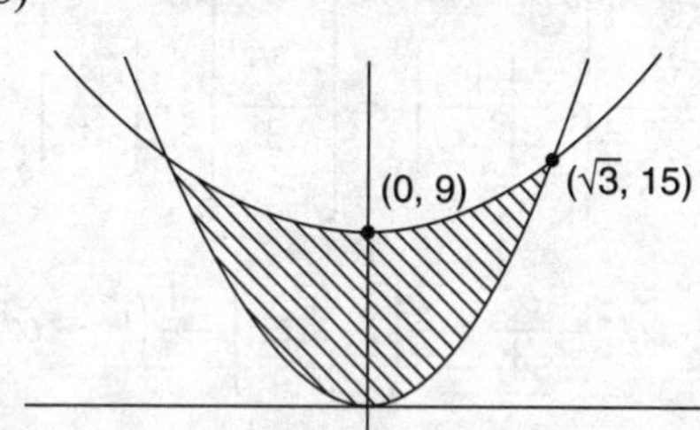

Required area $= 2\int_0^{\sqrt{3}} \left(2x^2 + 9 - 5x^2\right)dx$

$= 2\left[9x - x^3\right]_0^{\sqrt{3}} = 2\left[9\sqrt{3} - 3\sqrt{3}\right] = 12\sqrt{3}$

65. (b) Required area

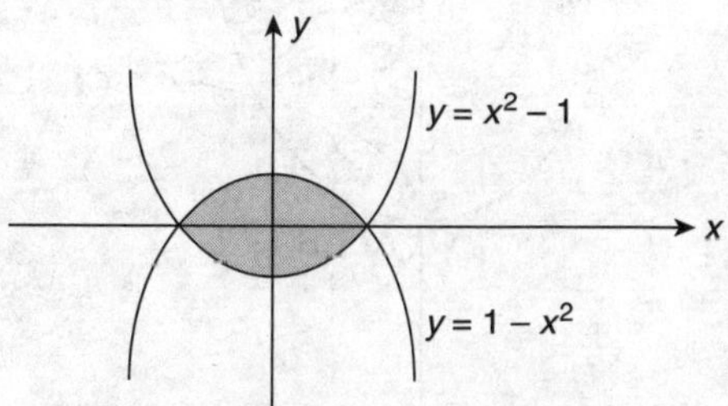

Area $= 2\int_0^1 (1 - x^2) - (x^2 - 1)\,dx = 4\int_0^1 (1 - x^2)\,dx$

$$= 4\left(x - \frac{x^3}{3}\right)\Bigg|_0^1 = 4\left(1 - \frac{1}{3}\right)$$

$$= 4 \cdot \frac{2}{3} = \frac{8}{3} \text{ sq.units}$$

66. (b) Let $y = x^2$ and $y = 2x$

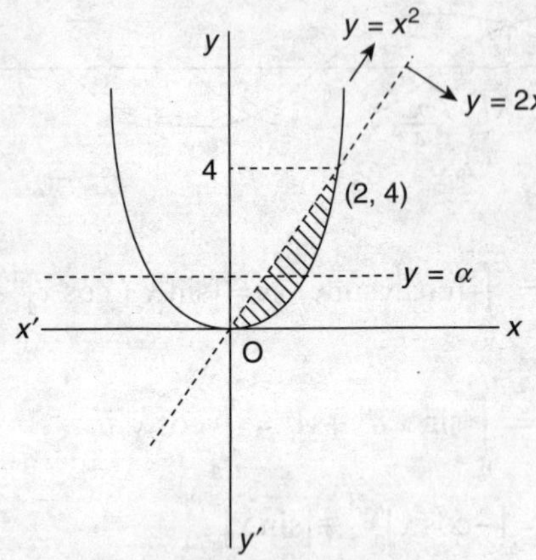

According to the question,

$$\therefore \int_0^{\alpha}\left(\sqrt{y} - \frac{y}{2}\right)dy = \int_{\alpha}^{4}\left(\sqrt{y} - \frac{y}{2}\right)dy$$

$$\Rightarrow \left[\frac{y^{3/2}}{\frac{3}{2}}\right]_0^{\alpha} - \left[\frac{y^2}{4}\right]_0^{\alpha} = \left[\frac{y^{3/2}}{\frac{3}{2}}\right]_{\alpha}^{4} - \left[\frac{y^2}{4}\right]_{\alpha}^{4}$$

$$\Rightarrow \frac{2}{3}\alpha^{3/2} - \frac{\alpha^2}{4} = \frac{2}{3}\left(8 - \alpha^{\frac{3}{2}}\right) - \frac{1}{4}\left(16 - \alpha^2\right)$$

$$\Rightarrow \frac{4}{3}\alpha^{3/2} - \frac{\alpha^2}{2} = \frac{4}{3}$$

$$\Rightarrow 8\alpha^{3/2} - 3\alpha^2 = 8$$

$$\therefore 3\alpha^2 - 8\alpha^{\frac{3}{2}} + 8 = 0$$

67. (b) Given equations are, $x^2 = ay$ and $y^2 = ax$

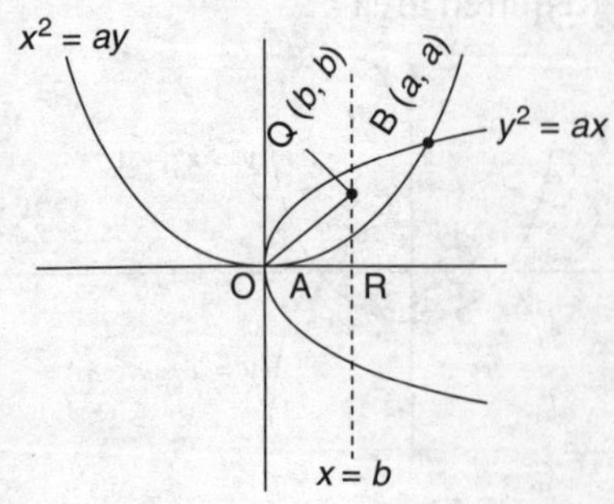

After solving, we get $x = a, y = a$

Now, co-ordinates of B are (a, a) and A is $(0, 0)$

Now, co-ordinates of Q are (b, b)

$$\therefore \frac{1}{2}b^2 = \frac{1}{2} \Rightarrow b = 1$$

Area bounded by the curves and $x = 1$ is

$$\int_0^1\left(\sqrt{a}x^{1/2} - \frac{x^2}{a}\right)dx = \frac{1}{2}\int_0^a\left(\sqrt{a}x^{1/2} - \frac{x^2}{a}\right)dx$$

$$\Rightarrow \frac{2}{3}\sqrt{a} - \frac{1}{3a} = \frac{a^2}{6}$$

$$\Rightarrow 4a\sqrt{a} - 2 = a^3$$

$$\Rightarrow a^6 + 4a^3 + 4 = 16a^3$$

$$\Rightarrow a^6 - 12a^3 + 4 = 0$$

68. (a) Point of intersection of $y = x^2$ and $y = -2x + 3$ is obtained by $x^2 + 2x - 3 = 0$

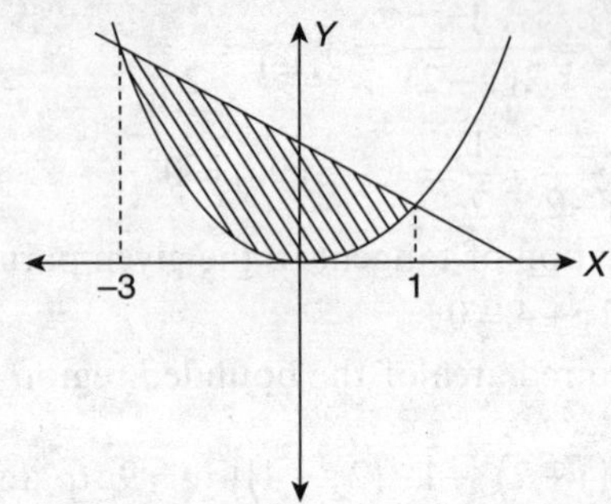

$$\Rightarrow x = -3, 1$$

So, required area $= \int_{-3}^{1}(\text{line} - \text{parabola})\,dz$

$$= \int_{-3}^{1}\left(3 - 2x - x^2\right)dx = \left[3x - x^2 - \frac{x^3}{3}\right]_{-3}^{1}$$

$$= (3)4 - 2\left(\frac{1^2 - 3^2}{2}\right) - \left(\frac{1^3 + 3^3}{3}\right) = 12 + 8 - \frac{28}{3} = \frac{32}{3}$$

69. (d) Total area – Enclosed area between line and parabola

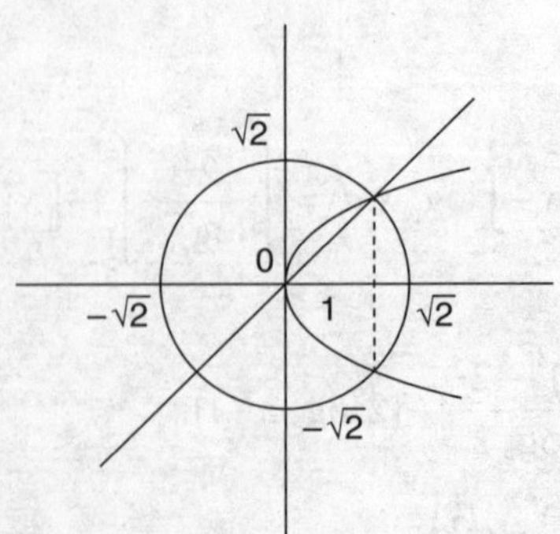

$$= 2\pi - \int_0^1 \sqrt{x} - x\,dx$$

$$= 2\pi - \left(\frac{2x^{3/2}}{3} - \frac{x^2}{2}\right)_0^1$$

$$= 2\pi - \left(\frac{2}{3} - \frac{1}{2}\right) = 2\pi - \left(\frac{1}{6}\right) = \frac{12\pi - 1}{6}$$

70. (b)

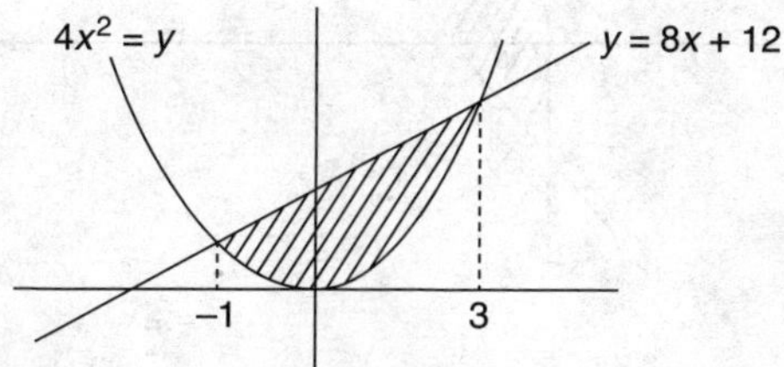

The given curves are

$4x^2 = y$ ...(i)

$y = 8x + 12$ ...(ii)

From equations (i) and (ii),

$4x^2 = 8x + 12$

$\Rightarrow x^2 - x - 3 = 0$

$\Rightarrow x^2 - 2x - 3 = 0$

$\Rightarrow x^2 - 3x + x - 3 = 0$

$\Rightarrow (x+1)(x-3) = 0$

$\Rightarrow x = -1, 3$

Required area bounded by curves is given by

$$A = \int_{-1}^{3} \left(8x + 12 - 4x^2\right) dx$$

$$A = \frac{8x^2}{2} + 12x - \frac{4x^3}{3}\Bigg|_{-1}^{3}$$

$$= \left(4(9) + 36 - 36\right) - \left(4 - 12 + \frac{4}{3}\right)$$

$$= 36 + 8 - \frac{4}{3} = 44 - \frac{4}{3} = \frac{132 - 4}{3} = \frac{128}{3}$$

71. (a) $[x] = 0$ when $x \in [0,1)$ and $[x] = 1$ when $x \in [1,2)$

$$y = \begin{cases} 0 & 0 \le x < 1 \\ x - 1 & 1 \le x < 2 \end{cases}$$

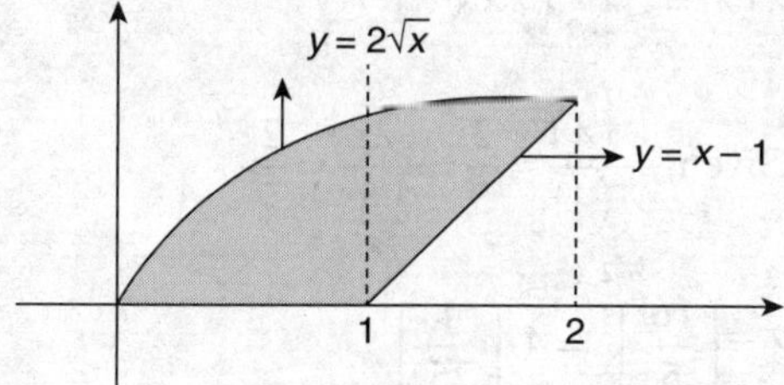

$$\therefore A = \int_0^2 2\sqrt{x}\, dx - \frac{1}{2}(1)(1)$$

$$= \frac{4x^{\frac{3}{2}}}{3}\Bigg|_0^2 - \frac{1}{2} = \frac{8\sqrt{2}}{3} - \frac{1}{2}$$

72. (b)

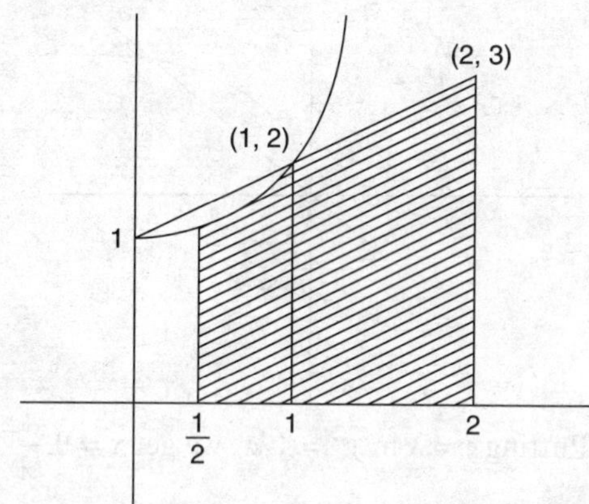

$$\text{Required area} = \int_{1/2}^{1} \left(x^2 + 1\right) dx + \int_{1}^{2} (x+1) dx$$

$$= \left[\frac{x^3}{3} + x\right]_{\frac{1}{2}}^{1} + \left[\frac{x^2}{2} + x\right]_1^2 = \left[\frac{4}{3} - \frac{13}{24}\right] + \frac{5}{2} = \frac{79}{24}$$

73. (b) Co-ordinates of

$P\left(\frac{1}{2}, 0\right), Q\left(\frac{\sqrt{3}}{2}, 0\right), R\left(\frac{\sqrt{3}}{2}, 1 - \frac{\sqrt{3}}{2}\right)$ and $S\left(\frac{1}{2}, \frac{1}{2}\right)$

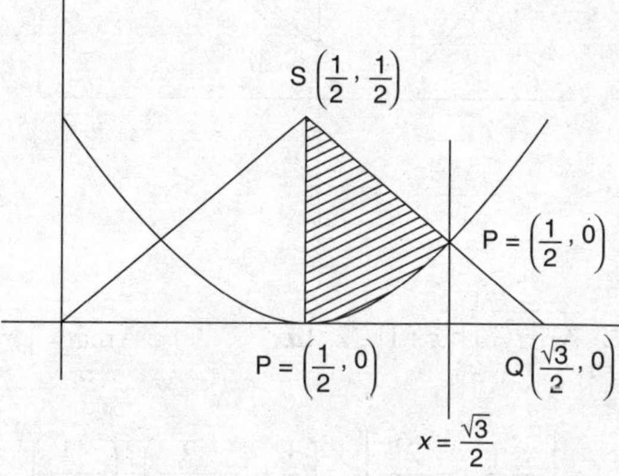

Required area = Area of trapezium PQRS

$$- \int_{1/2}^{\sqrt{3}/2} \left(x - \frac{1}{2}\right)^2 dx$$

$$=\frac{1}{2}\left(\frac{\sqrt{3}-1}{2}\right)\left(\frac{1}{2}+1-\frac{\sqrt{3}}{2}\right)-\frac{1}{3}\left(\left(x-\frac{1}{2}\right)^3\right)_{1/2}^{\sqrt{3}/2}$$

$$=\frac{\sqrt{3}}{4}-\frac{1}{3}$$

74. (c) Given parabola $y^2 = 4\lambda x$ and the line $y = \lambda x$

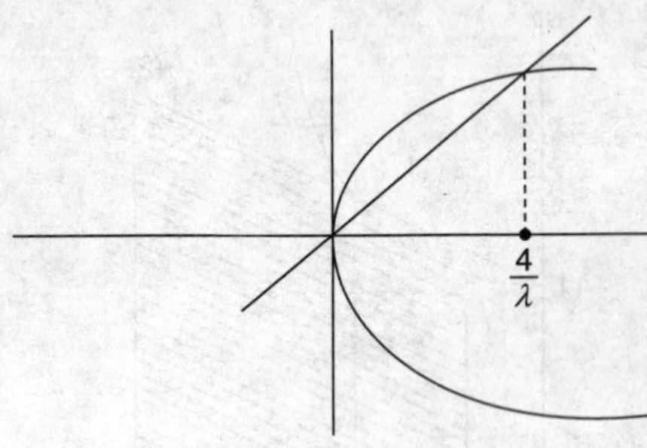

Putting $y = \lambda$ in $y^2 = 4\lambda x$, we get $x = 0, \frac{4}{\lambda}$

$$\therefore \text{ Required area} = \int_0^{\frac{4}{\lambda}} \left(2\sqrt{\lambda x}-\lambda x\right) dx$$

$$=\frac{2\sqrt{\lambda}.x^{3/2}}{3/2}-\frac{\lambda x^2}{2}\bigg|_0^{4/\lambda}=\frac{32}{3\lambda}-\frac{8}{\lambda}=\frac{8}{3\lambda}=\frac{1}{9}$$

$\Rightarrow \lambda = 24$

75. (d)

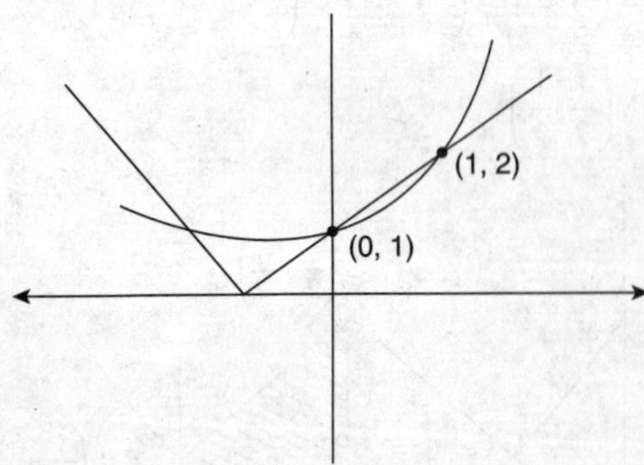

$$\text{Area} = \int_0^1 \left((x+1)-2^x\right)dx \qquad \left(\because \text{Area} = \int y\, dx\right)$$

$$=\left[\frac{x^2}{2}+x-\frac{2^x}{\ln 2}\right]_0^1=\left(\frac{1}{2}+1-\frac{2}{\ln 2}\right)-\left(\frac{-1}{\ln 2}\right)$$

$$=\frac{3}{2}-\frac{1}{\ln 2}$$

76. (d)

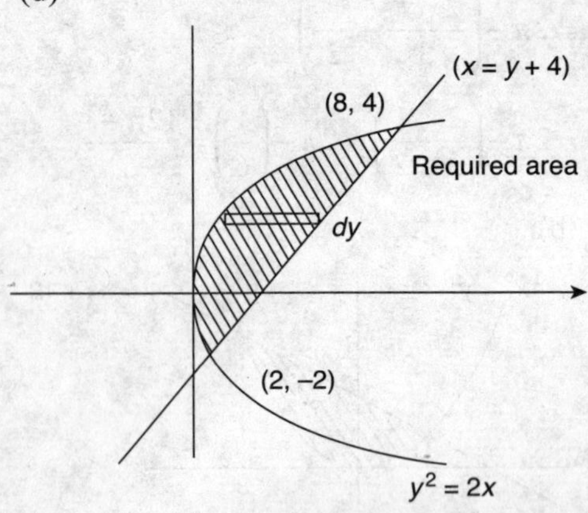

Given region, $A=\left\{(x,y):\frac{y^2}{2}\le x\le y+4\right\}$

$$\text{Hence, area} = \int_{-2}^{4} x dy = \int_{-2}^{4}\left(y+4-\frac{y^2}{2}\right)dy$$

$$=\left[\frac{y^2}{2}+4y-\frac{y^3}{6}\right]_{-2}^{4}=\left(8+16-\frac{64}{6}\right)-\left(2-8+\frac{8}{6}\right)$$

$$=\left(24-\frac{32}{3}\right)-\left(-6+\frac{4}{3}\right)=\frac{40}{3}+\frac{14}{3}=\frac{54}{3}=18$$

77. (d)

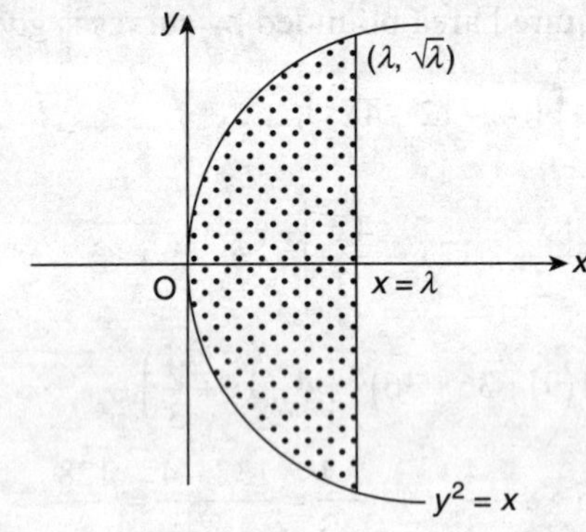

$$\text{Area of the region} = 2\times\int_0^{\lambda} y\, dx = 2\int_0^{\lambda}\sqrt{x}\, dx$$

$$=2\times\frac{2}{3}\pi^{\frac{3}{2}}$$

$$A(\lambda)=2\times\frac{2}{3}\left(\lambda\times\sqrt{\lambda}\right)=\frac{4}{3}\lambda^{\frac{3}{2}}$$

$$\text{Given, } \frac{A(\lambda)}{A(4)}=\frac{2}{5}\Rightarrow\frac{\lambda^{\frac{3}{2}}}{8}=\frac{2}{5}$$

$$\lambda=\left(\frac{16}{5}\right)^{\frac{2}{3}}=4\cdot\left(\frac{4}{25}\right)^{\frac{1}{3}}$$

78. (d)

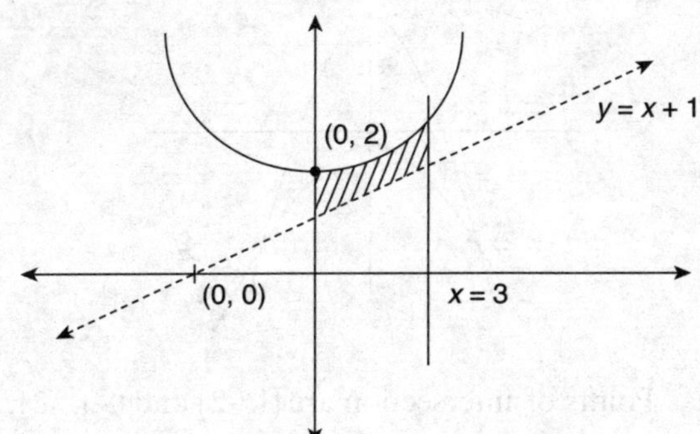

Area of the bounded region

$$\int_0^3 \left[\left(x^2+2\right)-\left(x+1\right)\right]dx$$

$$=\left[\frac{x^3}{3}-\frac{x^2}{2}+x\right]_0^3=9-\frac{9}{2}+3=\frac{15}{2}$$

79. (b)

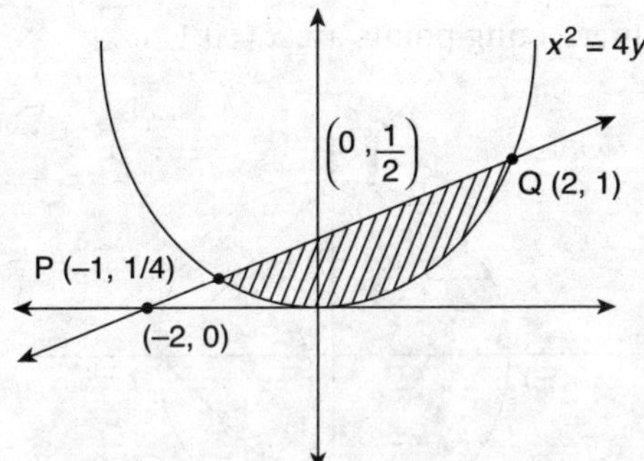

Let points of intersection of the curve and the line be P and Q.

$$x^2=4\left(\frac{x+2}{4}\right)\Rightarrow x^2-x-2=0\Rightarrow x=2,1$$

The points are (2, 1) and $\left(-1,\frac{1}{4}\right)$

$$\text{Area}=\int_{-1}^{2}\left[\left(\frac{x+2}{4}\right)-\left(\frac{x^2}{4}\right)\right]dx$$

$$=\left[\frac{x^2}{8}+\frac{1}{2}x-\frac{x^3}{12}\right]_{-1}^{2}$$

$$=\left(\frac{1}{2}+1-\frac{2}{3}\right)-\left(\frac{1}{8}-\frac{1}{2}+\frac{1}{12}\right)=\frac{9}{8}$$

80. (b) The intersection point of $y=x-2$ and $y=\sqrt{x}$ is (4, 2).

The required area $=\int_0^4 \sqrt{x}\,dx-\frac{1}{2}\times 2\times 2=\frac{16}{3}-2=\frac{10}{3}$

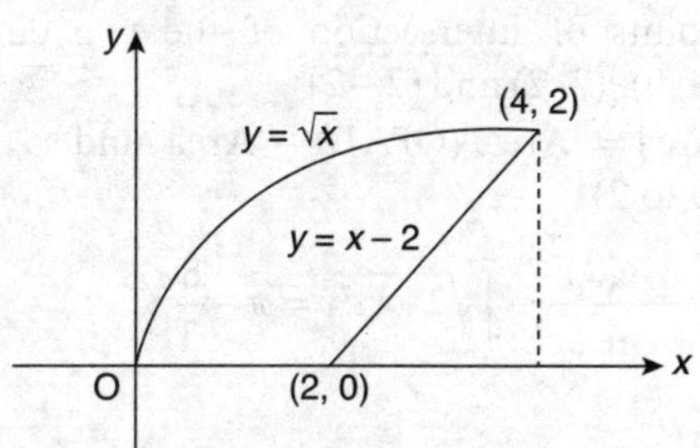

81. (d) Here, $18x^2-9\pi x+\pi^2=0$

$$\Rightarrow(3x-\pi)(6x-\pi)=0\Rightarrow\alpha=\frac{\pi}{6},\beta=\frac{\pi}{3}$$

Also, $gof(x)=\cos x$

$$\therefore \text{ Req. area}=\int_{\pi/6}^{\pi/3}\cos x\,dx=\frac{\sqrt{3}-1}{2}$$

82. (a)

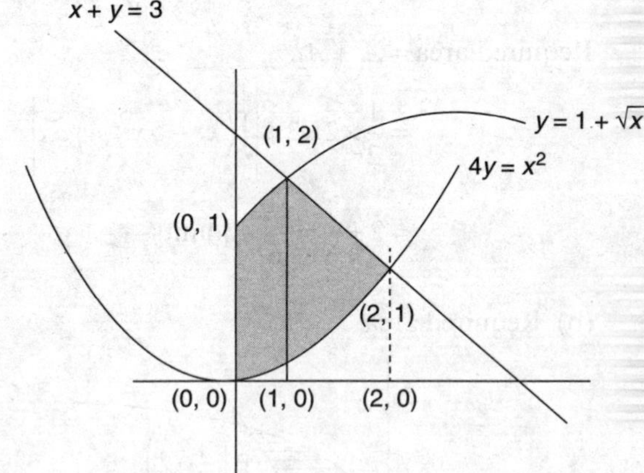

Area of shaded region

$$=\int_0^1\left(1+\sqrt{x}\right)dx+\int_1^2(3-x)\,dx-\int_0^2\frac{x^2}{4}dx$$

$$=[x]_0^1+\left[\frac{x^{\frac{3}{2}}}{\frac{3}{2}}\right]_0^1+[3x]_1^2-\left[\frac{x^2}{2}\right]_1^2-\left[\frac{x^3}{12}\right]_0^2=\frac{5}{2}\text{sq.units}$$

83. (d)

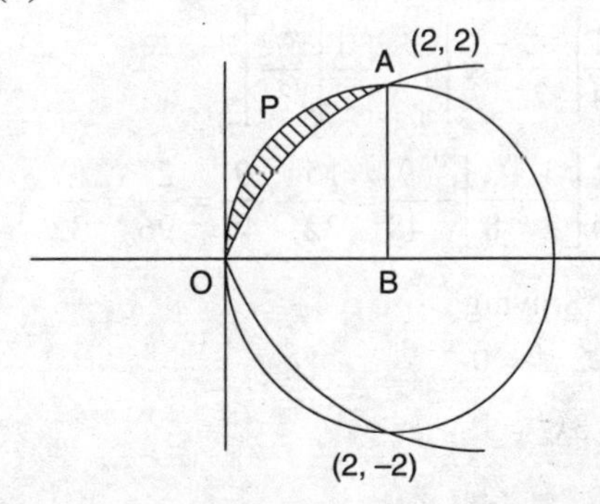

Points of intersection of the two curves are (0, 0), (2, 2) and $(2,-2)$

Area = Area (OPAB) – Area under parabola (0 to 2)

$$=\frac{\pi\times(2)^2}{4}-\int_0^2\sqrt{2}\sqrt{x}dx=\pi-\frac{8}{3}$$

84. (a)

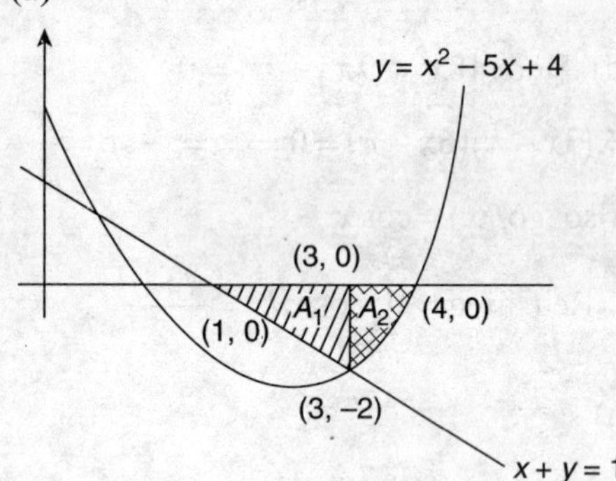

Required area = $A_1 + A_2$

$$=\frac{1}{2}\times2\times2+\left|\int_3^4\left(x^2-5x+4\right)dx\right|$$

$$=2+\frac{7}{6}=\frac{19}{6}\text{ sq.units}$$

85. (b) Required area

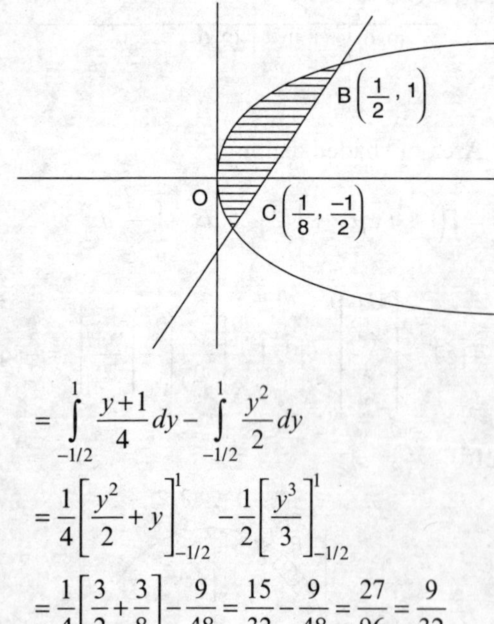

$$=\int_{-1/2}^{1}\frac{y+1}{4}dy-\int_{-1/2}^{1}\frac{y^2}{2}dy$$

$$=\frac{1}{4}\left[\frac{y^2}{2}+y\right]_{-1/2}^{1}-\frac{1}{2}\left[\frac{y^3}{3}\right]_{-1/2}^{1}$$

$$=\frac{1}{4}\left[\frac{3}{2}+\frac{3}{8}\right]-\frac{9}{48}=\frac{15}{32}-\frac{9}{48}=\frac{27}{96}=\frac{9}{32}$$

86. (c) Solving

$y+2x^2=0$

$y+3x^2=1$

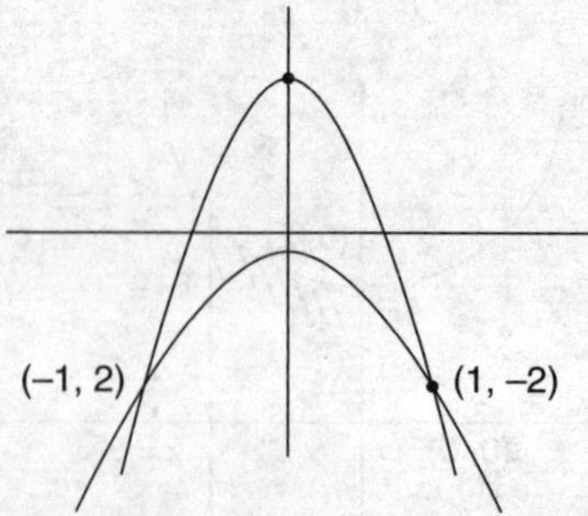

Points of intersection are $(1,-2)$ and $(-1,-2)$

$$\text{Area}=2\int_0^1\left(\left(1-3x^2\right)-\left(-2x^2\right)\right)dx$$

$$2\int_0^1\left(1-x^2\right)dx=2\left(x-\frac{x^3}{3}\right)_0^1=\frac{4}{3}$$

$=15-6=9$ sq units

87. (c) The given curves are $x^2+y^2=1$ and $y^2=1-x$

Intersecting points are $x=0,1$

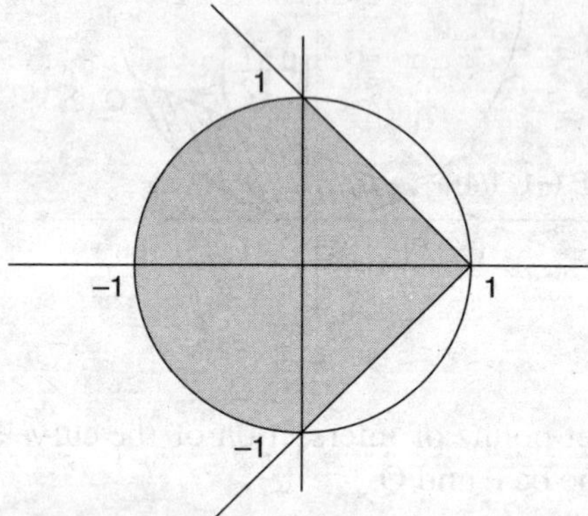

Area of shaded portion is the required area. So,

Required Area = Area of semi-circle + Area bounded by parabola

$$=\frac{\pi r^2}{2}+2\int_0^1\sqrt{1-x}\,dx=\frac{\pi}{2}+2\int_0^1\sqrt{1-x}\,dx$$

($\because$ radius of circle = 1)

$$=\frac{\pi}{2}+2\left[\frac{(1-x)^{3/2}}{-3/2}\right]_0^1=\frac{\pi}{2}-\frac{4}{3}(-1)=\frac{\pi}{2}+\frac{4}{3}\text{ sq.units}$$

88. (a) Given curves are

$y=\sqrt{x}$ ...(i)

and $2y-x+3=0$ ...(ii)

On solving (i) and (ii), we get $y=-1, 3$

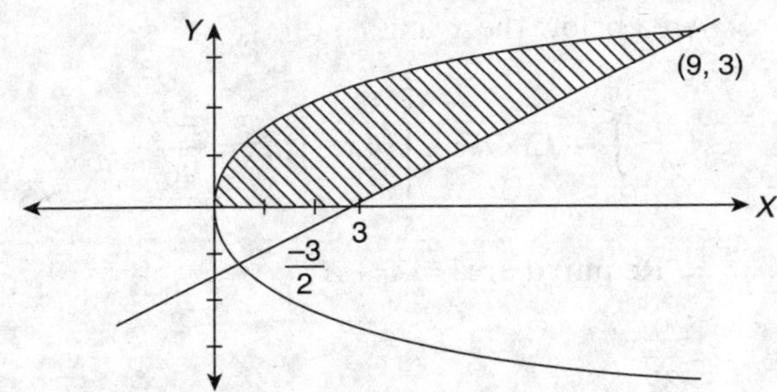

$$\text{Required area} = \int_0^3 \left\{(2y+3) - y^2\right\} dy$$

$$= \left[y^2 + 3y - \frac{y^3}{3}\right]_0^3 = 9$$

89. (b) $y = |\cos x - \sin x|$

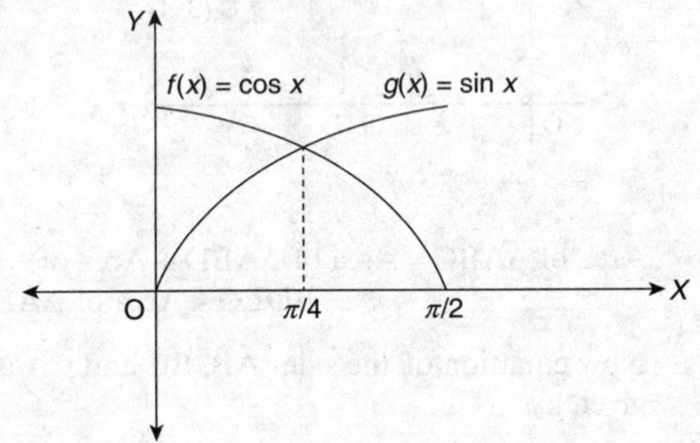

$$\text{Required area} = 2\int_0^{\pi/4} (\cos x - \sin x)\, dx$$

$$= 2[\sin x + \cos x]_0^{\pi/4}$$

$$= 2\left[\frac{2}{\sqrt{2}} - 1\right] = (2\sqrt{2} - 2) \text{ sq.units}$$

90. (c) $R_1 = \int_{-1}^{2} xf(x)\,dx = \int_{-1}^{2} (1-x) f(1-x)\,dx$

$$\left[\because \int_a^b f(x)dx = \int_a^b f(a+b-x)dx\right]$$

$$\Rightarrow R_1 = \int_{-1}^{2} (1-x) f(x)\, dx$$

$$[\because f(x) = f(1-x) \text{ on } (-1,2)]$$

$$\text{Now, } R_1 + R_1 = \int_{-1}^{2} xf(x)\, dx + \int_{-1}^{2} (1-x) f(x)\, dx$$

$$\Rightarrow 2R_1 = \int_{-1}^{2} f(x)dx = R_2$$

91. (b)

$$R_1 = \int_0^b (x-1)^2\, dx = \left[\frac{(x-1)^3}{3}\right]_0^b = \frac{(b-1)^3 + 1}{3}$$

$$R_2 = \int_b^1 (x-1)^2\, dx = \left[\frac{(x-1)^3}{3}\right]_b^1 = -\frac{(b-1)^3}{3}$$

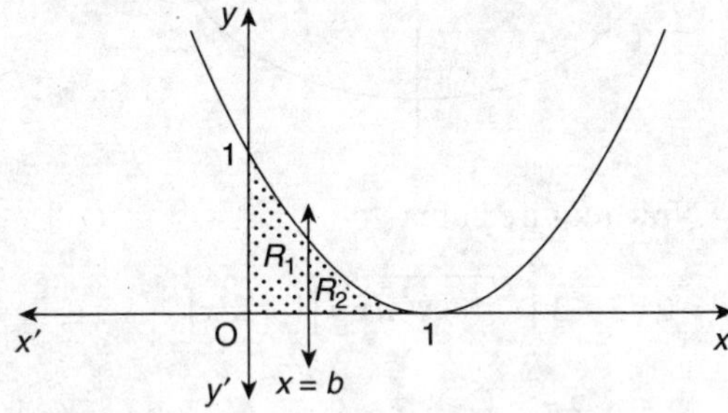

$$\because R_1 - R_2 = \frac{1}{4} \Rightarrow \frac{2(b-1)^3}{3} + \frac{1}{3} = \frac{1}{4}$$

$$\Rightarrow (b-1)^3 = -\frac{1}{8} \Rightarrow b - 1 = \frac{-1}{2} \therefore b = \frac{1}{2}$$

Integer and Numerical Value Type

1. The asymptote is $y = 0$ or x-axis.

$$\Rightarrow \text{Area} = 2\int_0^\infty \frac{dx}{1+x^2} = 2 \cdot \frac{\pi}{2} = \pi \Rightarrow k = 1$$

2. $2\int_0^2 \sqrt{\frac{8-4x}{x}}\, dx = 4\pi$ (Put $x = 2\sin^2\theta$)

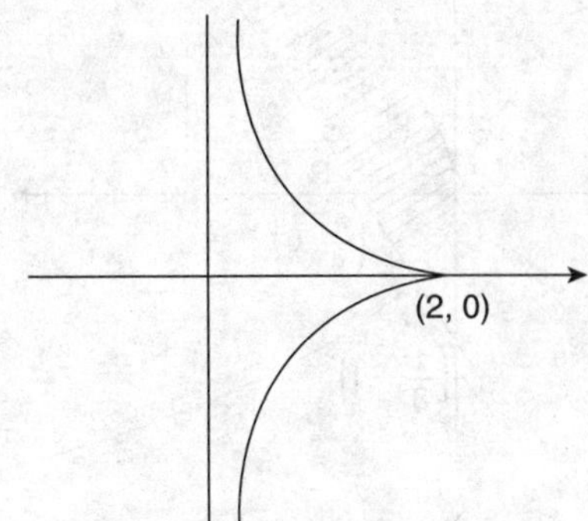

3. Since $\sin^2 \frac{x}{4} + \cos\frac{x}{2}$ lies between 1 and 2 for $x \in [-2, 2]$

$$\Rightarrow \left[\sin^2 \frac{x}{4} + \cos\frac{x}{4}\right] = 1$$

Thus, the problem is reduced to finding the area bounded by $x^2 + y^2 = 4$, $y = x^2 + x + 1$ and line $y = 1$ (See figure)

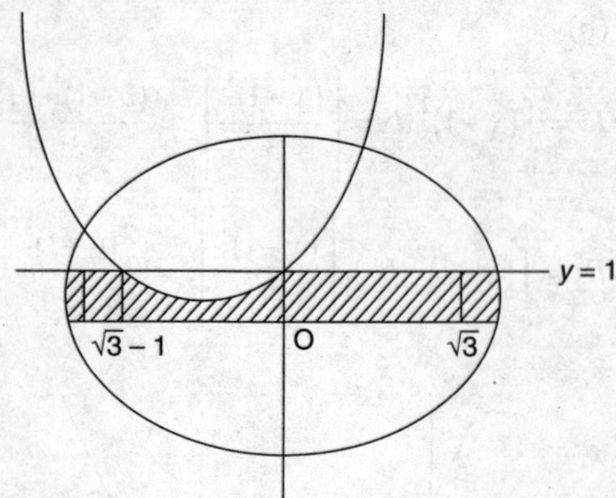

Now required area

$= 2\sqrt{3} + 2\int_{\sqrt{3}}^{2} \sqrt{4 - x^2}\, dx - [1 — dx]$

$= \sqrt{3} + \frac{2\pi}{3} - \frac{1}{6} \Rightarrow k = 6$

4. On eliminating t, we get

$\frac{(x - 2006)^2}{25} + \frac{(y + 2007)^2}{9} = 1$

which is an ellipse whose area will depend on the length of semi-major and minor axes only.

Area $= \pi ab = \pi(5)(3) = 15\pi \Rightarrow k = 15$

5.

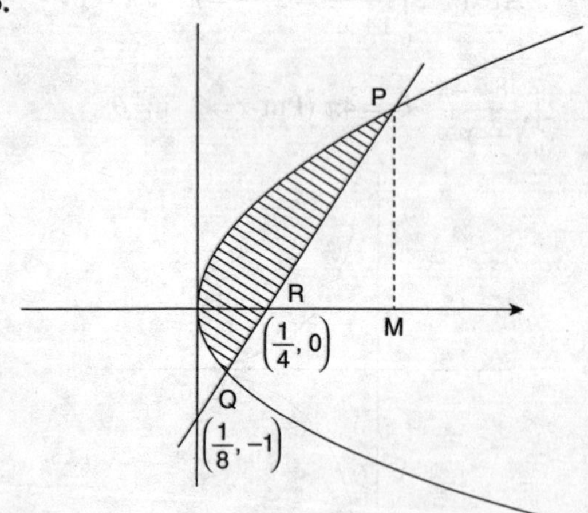

The intersection points are $P\left(\frac{1}{2}, 1\right)$ and $Q\left(\frac{1}{8}, -\frac{1}{2}\right)$

Area above the x-axis

$A_1 = \int_0^{1/2} \sqrt{2\times}\, dx - \int_{\frac{1}{4}}^{\frac{1}{2}} (4x - 1)\, dx = \frac{5}{24}$

Area below the x-axis

$A_2 = \int_0^{1/8} -\sqrt{2\times}\, dx - \int_{\frac{1}{8}}^{\frac{1}{4}} (4x - 1)\, dx = \frac{-7}{96}$

$\Rightarrow$ Required area $= A_1 + A_2$

$= \frac{9}{32}$

$\Rightarrow k = 9$

6. Let A (1, 0), B (2, 2) and C (3, 1) be the vertices of a triangle ABC.

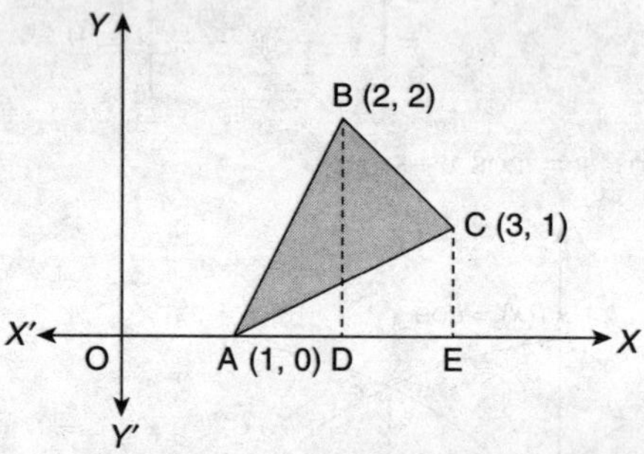

Area of ΔABC = Area of ΔABD + Area of trap BDEC – Area of ΔAEC

Now equation of the sides AB, BC and CA are given by

$y = 2(x - 1), y = 4 - x, y = \frac{1}{2}(x - 1)$, respectively.

Hence, area of ΔABC

$= \int_1^2 2(x - 1)\, dx + \int_2^3 (4 - x)\, dx - \int_1^3 \frac{x - 1}{2}\, dx$

$= 2\left[\frac{x^2}{2} - x\right]_1^2 + \left[4x - \frac{x^2}{2}\right]_2^3 - \frac{1}{2}\left[\frac{x^2}{2} - x\right]_1^3$

$= 2\left[\left(\frac{2^2}{2} - 2\right) - \left(\frac{1}{2} - 1\right)\right] + \left[\left(4\times3 - \frac{3^2}{2}\right) - \left(4\times2 - \frac{2^2}{2}\right) - \frac{1}{2}\left[\left(\frac{3^2}{2} - 3\right) - \left(\frac{1}{2} - 1\right)\right]\right]$

$= \frac{3}{2}$

7. Given, $f(x) = 2x^3 - 3x^2 - 12x$

$f'(x) = 6x^2 - 6x - 12 = 6(x - 2)(x + 1)$

Point = (2, –20) and (–1, 7)

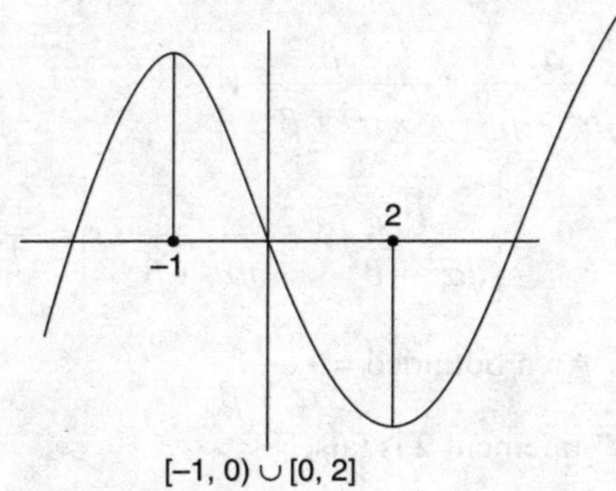

$$A=\int_{-1}^{0}\left(2x^3-3x^2-12x\right)dx+\int_{0}^{2}\left(12x+3x^2-2x^3\right)dx$$

$$A=\left(\frac{x^4}{2}-x^3-6x^2\right)_{-1}^{0}+\left(6x^2+x^3-\frac{x^4}{2}\right)_{0}^{2}$$

$$=-\left(\frac{1}{2}+1-6\right)-\left(8-8-24\right)$$

$$A=\frac{57}{2},4A=4\times\frac{57}{2}=114$$

8. The given ellipse is $x^2+4y^2=5$

So, the tangent at a point P to ellipse is $x+4y=5$

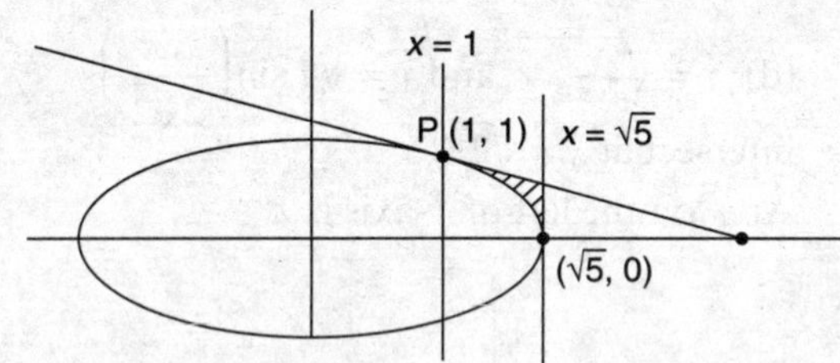

Required bounded area

$$=\int_{1}^{\sqrt{5}}\left(\frac{5-x}{4}-\frac{\sqrt{5-x^2}}{2}\right)dx$$

$$=\left[\frac{5x}{4}-\frac{x^2}{8}-\frac{x}{4}\sqrt{5-x^2}-\frac{5}{2}\sin^{-1}\frac{x}{\sqrt{5}}\right]_{1}^{\sqrt{5}}$$

$$=\frac{5}{4}\sqrt{5}-\frac{5}{4}-\frac{5}{4}\cos^{-1}\left(\frac{1}{\sqrt{5}}\right)$$

Then by comparison

$$\alpha=\frac{5}{4},\beta=-\frac{5}{4},\gamma=-\frac{5}{4}.$$

So, $|\alpha+\beta+\gamma|=1.25$

9. Required area

$$=\int_{-3}^{-2}(x+6)dx+\int_{-2}^{0}x^2dx+\int_{0}^{1}\sqrt{x}\,dx$$

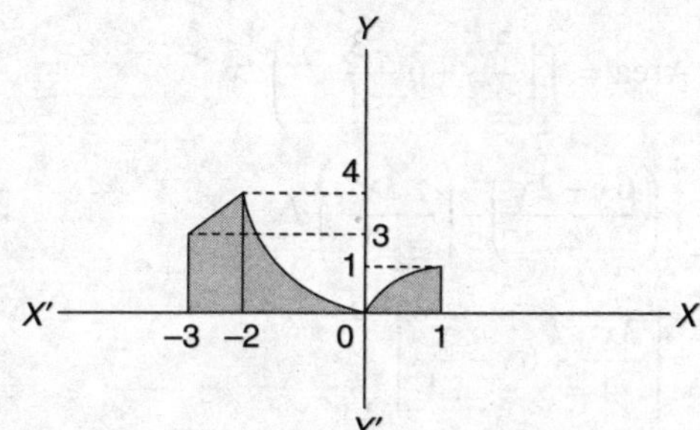

$$A=\frac{1}{2}(3+4)+\frac{1}{3}(8)+\frac{2}{3}=\frac{7}{2}+\frac{8}{3}+\frac{2}{3}=\frac{41}{6}$$

10. (64)

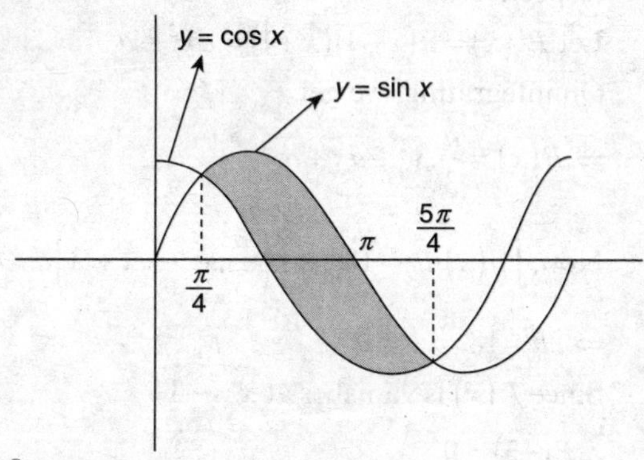

$$A=\int_{\pi/4}^{5\pi/4}(\sin x-\cos x)dx$$

$$=\left[-\cos x-\sin x\right]_{\pi/4}^{5\pi/4}$$

$$=\left[\frac{1}{\sqrt{2}}+\frac{1}{\sqrt{2}}\right]-\left[-\frac{1}{\sqrt{2}}-\frac{1}{\sqrt{2}}\right]$$

$$\Rightarrow A=\sqrt{2}+\sqrt{2}=2\sqrt{2}\Rightarrow A^4=64$$

11. (27)

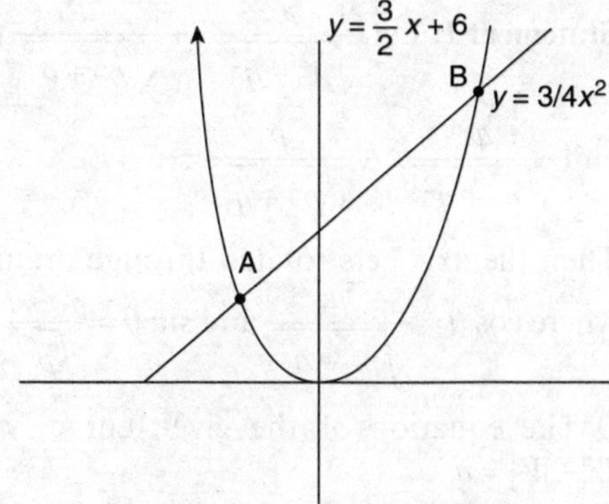

For A and B

$3x^2=6x+24\Rightarrow x^2-2x-8=0$

$\Rightarrow x=-2,4$

$$\text{Area} = \int_{-2}^{4}\left(\frac{3}{2}x+6-\frac{3}{4}x^2\right)dx$$

$$\int_{-2}^{4}\left(\frac{6x+2x}{4}\right)-\left(-\frac{3x^2}{4}\right)dx$$

$$=\left[\frac{3x^2}{4}+6x-\frac{x^3}{4}\right]_{-2}^{4}$$

$20-(-7)\Rightarrow 27$

12. $P'(x)$ is second degree polynomial and $P(x)$ has local minima and local maxima at 1 and -1 respectively.

Let $P'(x)=a(x-1)(x+1)=ax^2-a$

On integrating, we get

$\Rightarrow P(x)=\frac{a}{3}.x^3-ax+b$

Now, $\int_{-1}^{1}p(x)dx=18\Rightarrow\int_{-1}^{1}\left(\frac{a}{3}.x^3-ax+\lambda\right)dx=18$

$\Rightarrow 2b=18\Rightarrow b=9$

Since $P(x)$ is vanishes at $x=-3$

$\therefore P(-3)=0$

$\Rightarrow\frac{a}{3}(-3)^3-a(-3)+9=0\Rightarrow a=\frac{3}{2}$

So, $P(x)=\frac{1}{2}x^3-\frac{3}{2}x+9$

$\Rightarrow$ Sum of all the coefficients

$=\frac{1}{2}-\frac{3}{2}+9=8$

Assertion and Reason

1. (c)

Statement 1: Let $\frac{p}{\sqrt{p^2+q^2}}x+\frac{q}{\sqrt{p^2+q^2}}y=U$

and $\frac{q}{\sqrt{p^2+q^2}}x-\frac{p}{\sqrt{p^2+q^2}}=y=V$

Then the axis gets rotated through an angle θ, where $\cos\theta=\frac{p}{\sqrt{p^2+q^2}}$ and $\sin\theta=\frac{q}{\sqrt{p^2+q^2}}$

$\therefore$ The equation of the given curve becomes $|U|+|V|=a$

$\therefore$ The area bounded $=2a^2$

$\therefore$ Statement 1 is true.

Statement 2: The equation of the curve is $|\alpha x+\beta y|+|\beta x-\alpha y|=a$, which is equivalent to

$$\left|\frac{\alpha}{\sqrt{\alpha^2+\beta^2}}x+\frac{\beta}{\sqrt{\alpha^2+\beta^2}}y\right|+\left|\frac{\beta}{\sqrt{\alpha^2+\beta^2}}x-\frac{\alpha}{\sqrt{\alpha^2+\beta^2}}y\right|=\frac{\alpha}{\sqrt{\alpha^2+\beta^2}}$$

$\therefore$ Area bounded $=\frac{2a^2}{\alpha^2+\beta^2}$

$\therefore$ Statement 2 is false.

2. (b) Area $=\int_{1}^{3}-\left(x^2-4x+3\right)dx$

$$=-\left(\frac{x^3}{3}-\frac{4x^2}{2}+3x\right)\Bigg|_1^3=\frac{4}{3}$$

$\therefore$ Statement 1 is true.

Statement 2 is true but it does not explain Statement 1.

JEE Advanced

Objective Questions

One Option Correct

1. (d) $y=\sqrt{4-x^2}$, and $y=\sqrt{2}\sin\left(\frac{x\pi}{2\sqrt{2}}\right)$ intersect at $x=\sqrt{2}$

Area on the left of y-axis is π

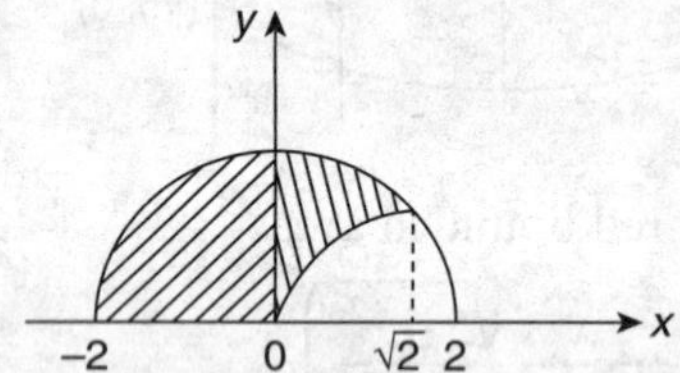

Area to the right of y-axis

$$=\int_{0}^{\sqrt{2}}\left(\sqrt{4-x^2}-\sqrt{2}\sin\frac{x\pi}{2\sqrt{2}}\right)dx$$

$$=\left(\frac{x\sqrt{4-x^2}}{2}+\frac{4}{2}\sin^{-1}\frac{x}{2}\right)\Bigg|_0^{\sqrt{2}}+\frac{4}{\pi}\cos\frac{x\pi}{2\sqrt{2}}\Bigg|_0^{\sqrt{2}}$$

$$=\left(1+2.\frac{\pi}{4}\right)+\frac{4}{\pi}(0-1)$$

$$=1+\frac{\pi}{2}-\frac{4}{\pi}=\frac{2\pi+\pi^2-8}{2\pi}$$

$$\therefore \text{ratio}=\frac{2\pi^2}{2\pi+\pi^2-8}$$

2. (b) $f(x)=\sin x$

$f(x)+f(\pi-x)=2$

$f(x)=2-f(\pi-x)=2-\sin(\pi-x)=2-\sin x$

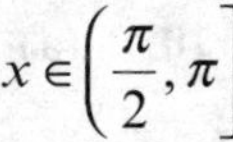

$x\in\left(\frac{\pi}{2},\pi\right]$

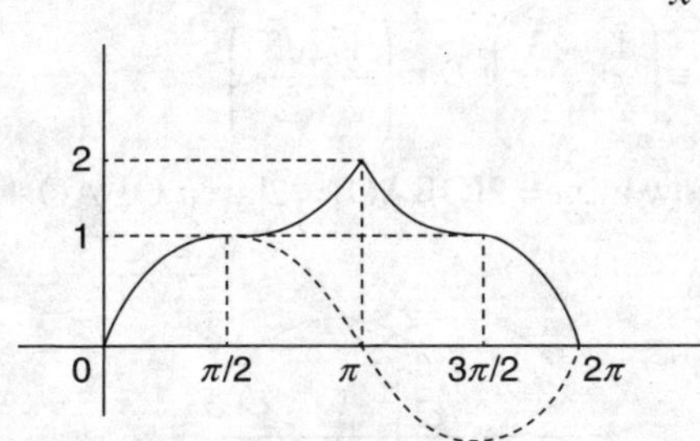

$f(x)=f(2\pi-x)$

$\therefore f(x+\pi)=f(\pi-x)$

So, the curve is symmetric w.r.t. line $x=\pi$ for $(\pi,2\pi]$

$f(x)=f(2\pi-x)=-\sin x$

$$\text{Area}=2\left(\int_0^{\pi/2}\sin x\,dx\int_{\pi/2}^{\pi}(2-\sin x)\,dx\right)$$

$$=2\left(1+2\times\frac{\pi}{2}-1\right)=2\pi$$

3. (a) $y=x^3,\frac{dy}{dx}=3x^2$

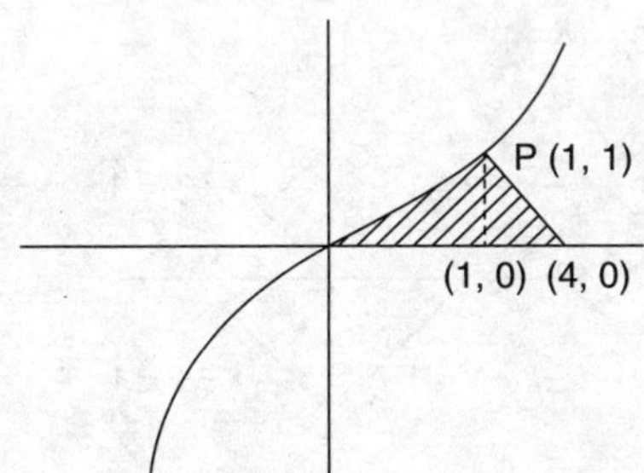

$$\left(\frac{dy}{dx}\right)_{(1,1)}=3$$

Normal at $P(1,1)$ is $y-1=-\frac{1}{3}(x-1)$

$3y+x=4$ ...(i)

So, the intersecting point of normal at x-axis is (4, 0)

$$\text{Area}=\int_0^1 x^3dx+\frac{1}{2}(3\times 1)=\left[\frac{x^4}{4}\right]_0^1+\frac{3}{2}=\frac{7}{4}$$

4. (c) $y=a^2x^2+ax+1$

$$\therefore \text{ Area } A=\int_0^1\left(a^2x^2+ax+1\right)dx$$

$$=\left(a^2\frac{x^3}{3}+\frac{ax^2}{2}+x\right)\Bigg|_0^1$$

$$=\frac{a^2}{3}+\frac{a}{2}+1=\frac{2a^2+3a+6}{6}$$

$$=\frac{1}{3}\left(a^2+\frac{3}{2}a+\frac{9}{16}\right)+1-\frac{9}{48}$$

$=\frac{1}{3}\left(a+\frac{3}{4}\right)^2+\frac{39}{48}$ attains its least value,

when $a=-\frac{3}{4}$

5. (a) $A=\int_1^2 \ell nx\,dx$

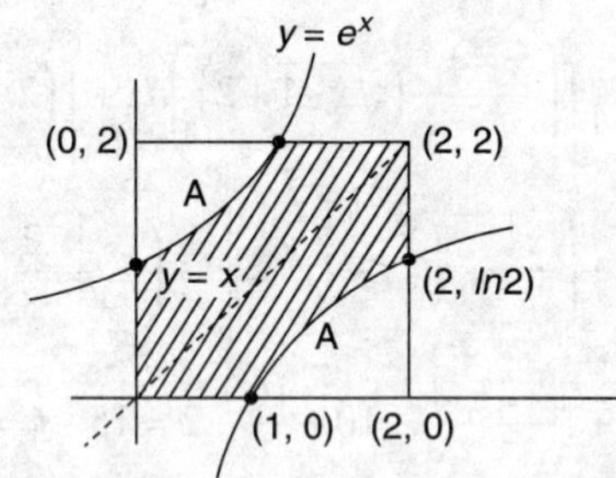

$$=(x\,\ell nx)\Big|_1^2-(x)\Big|_1^2$$

$$=2\ell n2-1$$

$$A'=4-2(2\ell n2-1)=6-4\ell n2$$

6. (d) $\text{Area}=\int_1^2\left(\ell nx+\tan^{-1}x\right)dx$

$$=x\ell n\,x\Big|_1^2-\int_1^2 1\,dx+x\tan^{-1}x\Big|_1^2-\int_1^2\frac{x}{1+x^2}\,dx$$

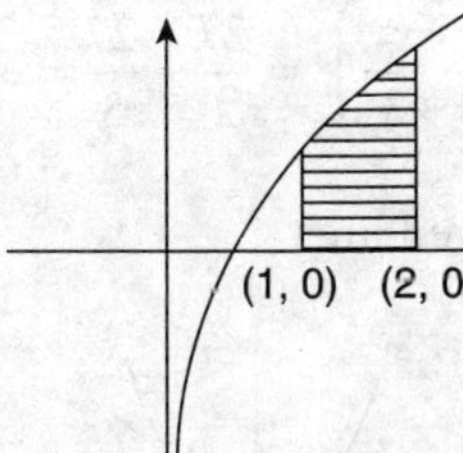

$$=\frac{5}{2}\ell n2-\frac{1}{2}\ell n5+2\tan^{-1}2-\frac{\pi}{4}-1$$

7. (c) The curve is $y^2-4y-x+5=0$

Equation of tangent at P (2, 3) is

$3y-2(y+3)-\frac{1}{2}(x+2)+5=0$

$y-6-\frac{1}{2}x-1+5=0$

$x-2y+4=0$

If the tangent intersects x-axis at Q (–4, 0) and the line $x=1$ at $S\left(1,\frac{5}{2}\right)$

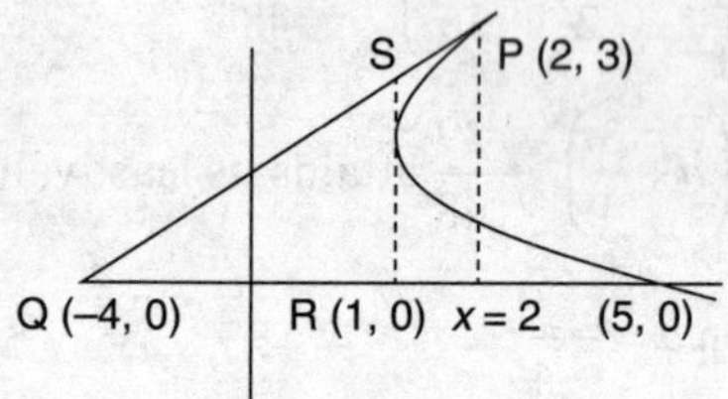

$\therefore$ Area of the ΔQRS $=\frac{1}{2}\times5\times\frac{5}{2}=\frac{25}{4}$

$\therefore$ Area of the bounded region

$=\frac{25}{4}+\int_1^2\left(\frac{x+4}{2}-\left(\sqrt{x-1}+2\right)\right)dx+\int_1^5\left(2-\sqrt{x-1}\right)dx$

$=\frac{25}{4}+\left(\frac{x^2}{4}-\frac{2}{3}(x-1)^{3/2}\right)\Big|_1^2+\left(2x-\frac{2}{3}(x-1)^{3/2}\right)\Big|_1^5$

$=\frac{25}{4}+1-\frac{2}{3}-\frac{1}{4}+10-\frac{16}{3}-2=15-6=9$

8. (a) $A=\int_0^3\left(2x-x^2+x\right)dx$

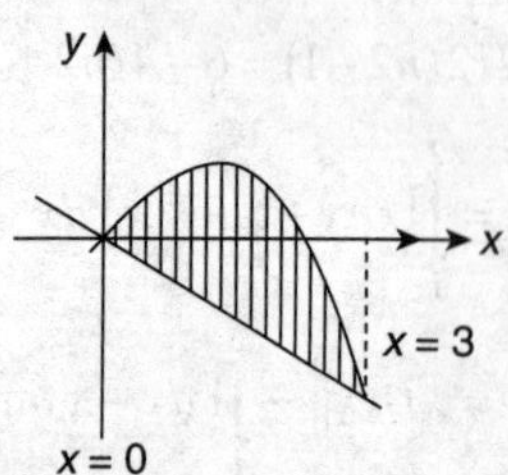

$A=\left(3\frac{x^2}{2}-\frac{x^3}{3}\right)^3=\frac{27}{2}-\frac{27}{3}=\frac{27}{6}=\frac{9}{2}$

9. (a) Required area

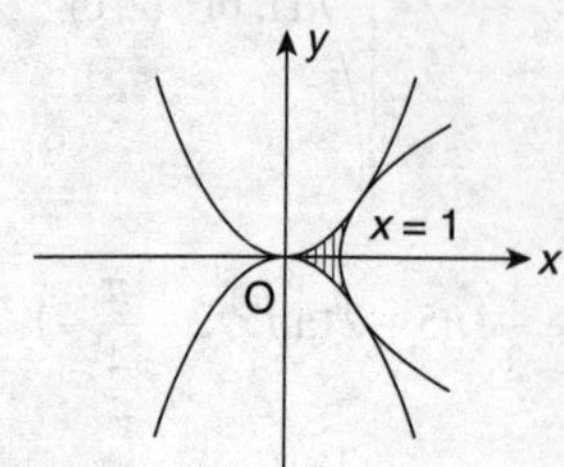

$=2\left[\int_0^1 x^2\,dx-\int_{3/4}^1\sqrt{4x-3}\,dx\right]=\frac{1}{3}=k$

$\Rightarrow 6k=2$

10. (a) Solving the given equation of circle, we get $A\equiv\left(\frac{1}{2},\frac{\sqrt{3}}{2}\right);\ D\equiv\left(\frac{1}{2},\frac{\sqrt{3}}{2}\right)$

Now area = 2[OBAO] = 2[area OEAO + EBAE]

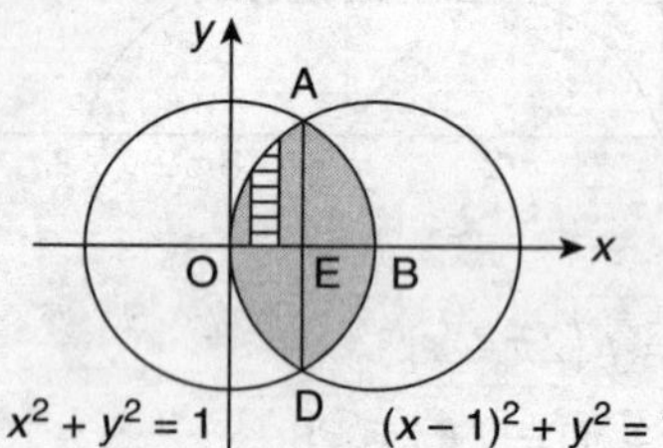

$=2\left[\int_0^{x_E}\sqrt{\left[1-(x-1)^2\right]}dx+\int_{x_E}^{x_B}\sqrt{1-x^2}\,dx\right]$

$=2\left[\int_0^{\frac{1}{2}}\sqrt{1-(x-1)^2}\,dx+\int_{\frac{1}{2}}^1\sqrt{1-x^2}\,dx\right]$

$=\frac{2\pi}{3}-\frac{\sqrt{3}}{2}$ square units

11. (a) S_1: Obvious

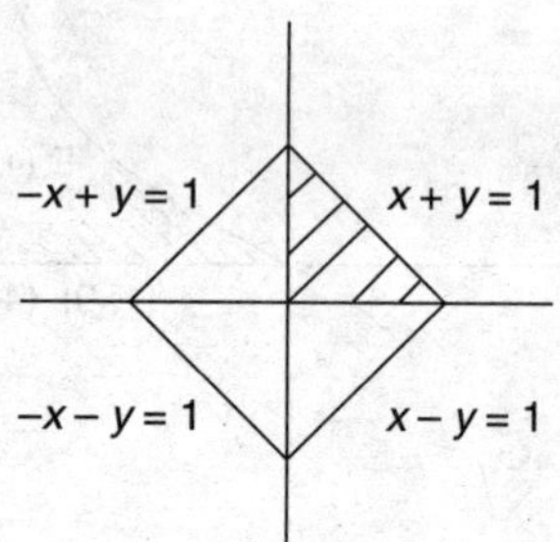

S_2: Area $=4\left(\frac{1}{2}.1.1\right)=2$

S_3: Area $=\frac{1}{4}$ (ellipse area) – ΔOAB

$=\frac{\pi ab}{4}-\frac{ab}{2}$

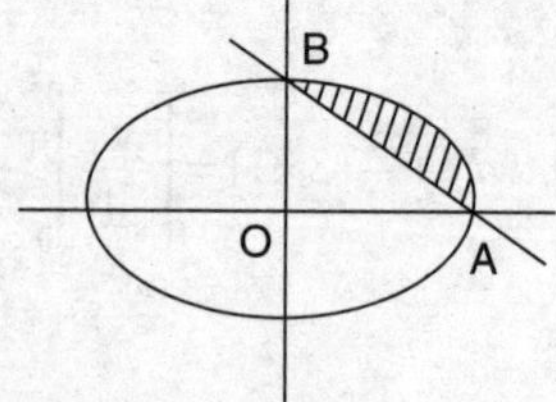

S_4: Area $= 2.\left(\frac{1}{2}1.1\right) = 1$

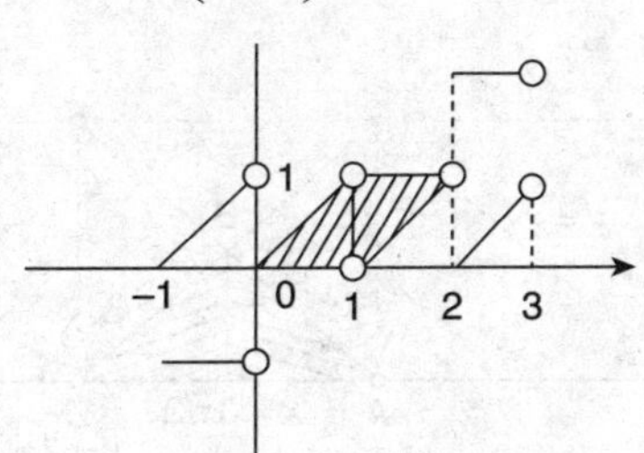

12. (a) $\because f(x) = e^{x-1} - e^{|x-1|}$

$\therefore f(x) = \begin{cases} 0 & x \le 1 \\ e^{x-1} - e^{1-x} & x \ge 1 \end{cases}$

and $g(x) = \frac{1}{2}\left(e^{x-1} + e^{1-x}\right)$

If $f(x) = g(x)$

$\Rightarrow e^{x-1} - e^{-(x-1)} = \frac{e^{x-1} + e^{1-x}}{2}$

$\Rightarrow e^{2(x-1)} = 3$

$\Rightarrow x = \frac{1}{2}\ln 3 + 1$

$\Rightarrow x = 1 + \ln\sqrt{3}$

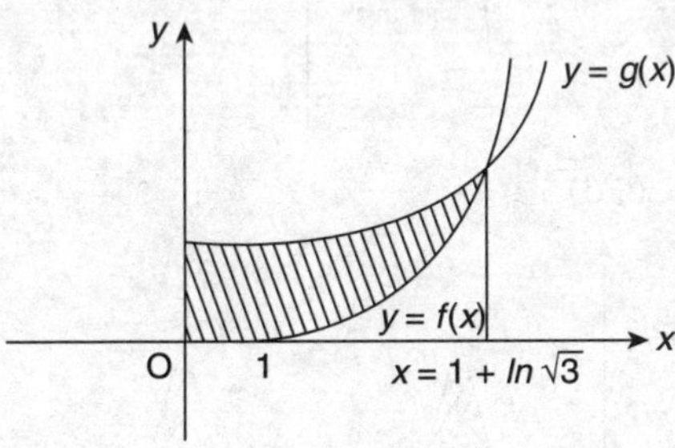

So bounded area $= \int_0^{\frac{1}{2}\ln 3+1} g(x)\,dx - \int_0^{\frac{1}{2}\ln 3+1} f(x)\,dx$

$= \frac{1}{2}\left[e^{x-1} - e^{1-x}\right]_0^{\frac{1}{2}\ln 3+1} - \left[e^{x-1} + e^{1-x}\right]_1^{\frac{1}{2}\ln 3+1}$

$= 2 - \sqrt{3} + \frac{1}{2}\left(e - \frac{1}{e}\right)$

13. (b) $xy \le 8, 1 \le y \le x^2$

Intersection points of $xy = 8$ and $y = 1$ is (8, 1); $xy = 8$ and $y = x^2$ is (2, 4) and $y = x^2$ and $y = 1$ is (1, 1)

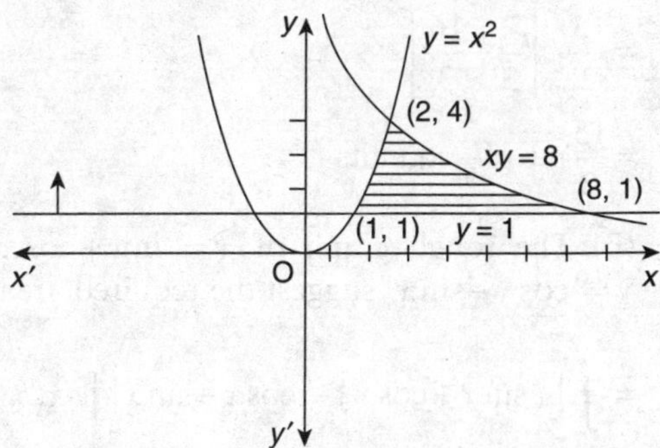

Required area $= \int_1^2 x^2\,dx + \int_2^8 \frac{8}{x}\,dx - \int_1^8 1\,dx$

$= \left[\frac{x^3}{3}\right]_1^2 + \left[8\ln x\right]_2^8 - \left[x\right]_1^8$

$= \frac{8}{3} - \frac{1}{3} + 8\ln 8 - 8\ln 2 - (8-1)$

$= \frac{7}{3} + 24\ln 2 - 8\ln 2 - 7 = 16\ln 2 - \frac{14}{3}$

14. (c) $y \ge \sqrt{|x+3|} \Rightarrow y^2 = |x+3|$

$\Rightarrow y^2 = \begin{cases} -(x+3)\, if\ x < -3 \\ (x+3)\, if\ x > -3 \end{cases}$ ...(i)

Also $y \le \frac{x+9}{5}$ and $x \le 6$ ...(ii)

Solving (i) and (ii), we get intersection points as (1, 2), (6, 3), (–4, 1), (–39, –6)

The graph of given region is as follows:

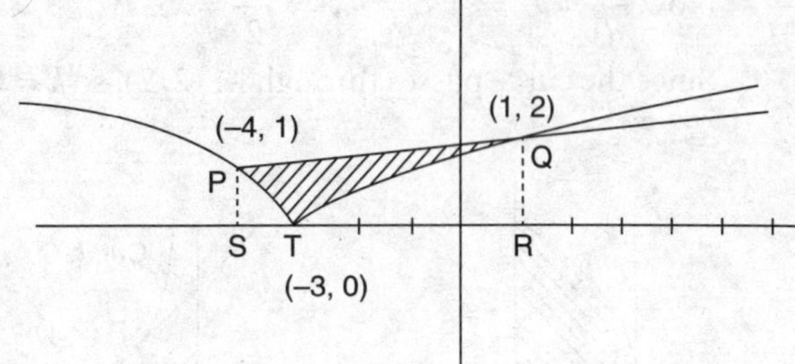

Required area = Area (trap PQRS) – Area (PST + TQR)

$= \frac{1}{2} \times (1+2) \times 5 - \left[\int_{-4}^{-3} \sqrt{-x-3}\,dx + \int_{-3}^{1} \sqrt{x+3}\,dx\right]$

$= \frac{15}{2} - \left[\left(\frac{2(-x-3)^{3/2}}{-3}\right)_{-4}^{-3} + \left(\frac{2(x+3)^{3/2}}{3}\right)_{-3}^{1}\right]$

$$=\frac{15}{2}-\left[\frac{2}{3}+\frac{16}{3}\right]$$

$$=\frac{15}{2}-6=\frac{3}{2}\text{ sq.units}$$

15. (b) The rough graph of $y=\sin x+\cos x$ and $y=|\cos x-\sin x|$ suggest the required area is

$$=\int_0^{\pi/2}\left[(\sin x+\cos x)-|\cos x-\sin x|\right]dx$$

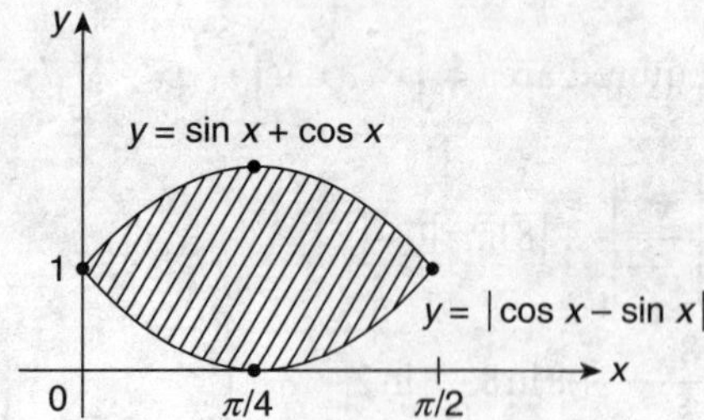

$$=\int_0^{\pi/2}2\sin x\,dx+\int_{\pi/4}^{\pi/2}2\cos x\,dx$$

$$=2\left[(-\cos x)_0^{\pi/4}+(\sin x)_{\pi/4}^{\pi/2}\right]$$

$$=2\sqrt{2}\left(\sqrt{2}-1\right)$$

One or More Than One Option Correct

1. (a, b, c)

Let P (x, y) be any point on the curve C.

Now, $\frac{dy}{dx}=\frac{1}{y}\Rightarrow ydy=dx\Rightarrow\frac{y^2}{2}=x+k$

Since the curve passes through M (2, 2), so $k=0$

$\Rightarrow y^2=2x$

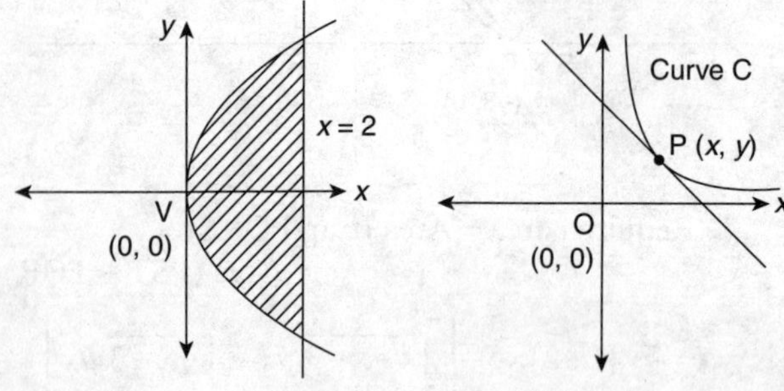

Hence, required area

$$=2\int_0^2\sqrt{2x}dx=2\sqrt{2}\times\frac{2}{3}\left(x^{3/2}\right)_0^2$$

$$=\frac{4}{3}\sqrt{2}\times2\sqrt{2}=\frac{16}{3}\text{(square unit)}$$

$\Rightarrow p+q=19$

2. (a, d)

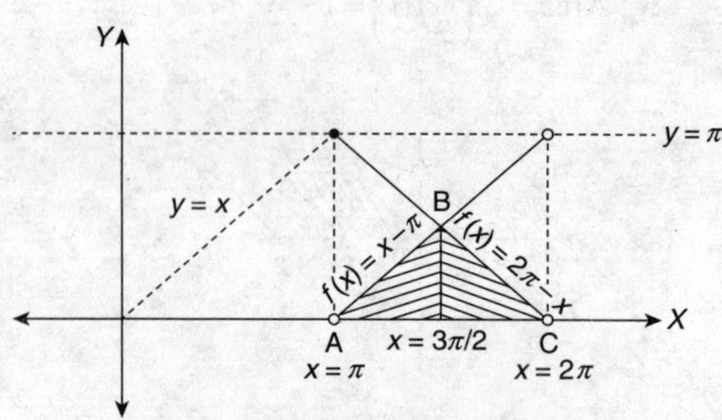

Required area = Area of shaded portion of $\Delta ABC=\frac{1}{2}\times\pi\times\frac{\pi}{2}=\frac{\pi^2}{4}=\frac{\pi^2}{k}$

3. (b, c) The curve $y=\frac{x^2(x+a)}{a^2}$ has $x=0$ as its repeated root and cuts the x-axis at $x=-a$

Hence, required area

$$=\int_{-a}^0 y\,dx=\int_{-a}^0\frac{x^3+ax^2dx}{a^2}=\frac{a^2}{12}\text{ sq.units}$$

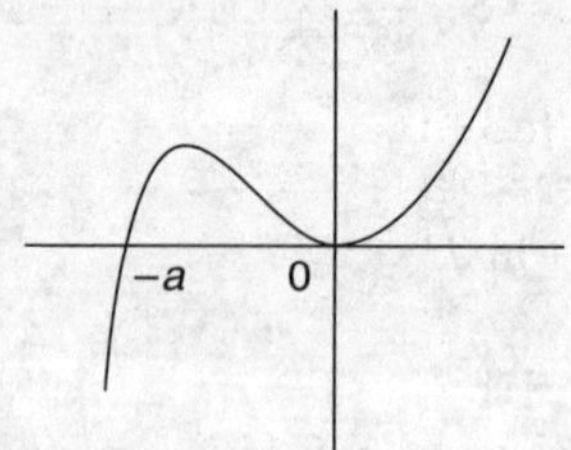

4. (b, c, d)

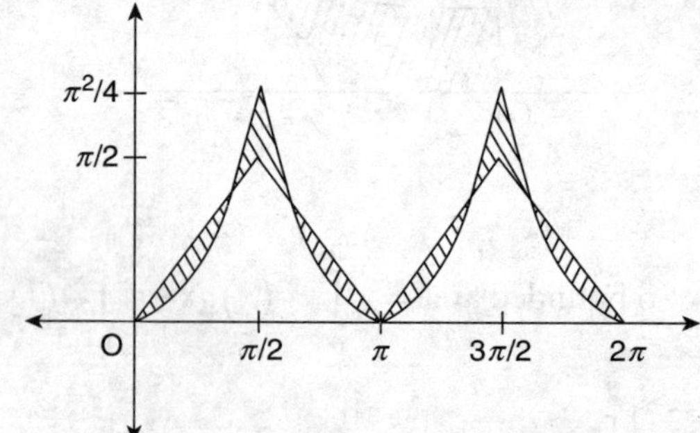

$$\text{Area}=4\int_0^1\left(x-x^2\right)dx+4\int_0^1\left(x-x^2\right)dx$$

$$=\frac{2}{3}+\frac{\pi^3-3\pi^2}{6}+\frac{2}{3}=\frac{4}{3}+\pi^2\left(\frac{\pi-3}{6}\right)$$

5. (a, b, d) The given curve $y=e^{-x^2}$

Draw a rough sketch of the curve starting at $x=0, y=1$ and ending at $x=1, y=1/e$

$\because y = e^{-x^2}$

$\Rightarrow \frac{dy}{dx} = -2xe^{-x^2} < 0 \forall x \in (0,1)$

$\therefore y = e^{-x^2}$ is decreasing on (0, 1)

Hence its graph is as shown in figure given below:

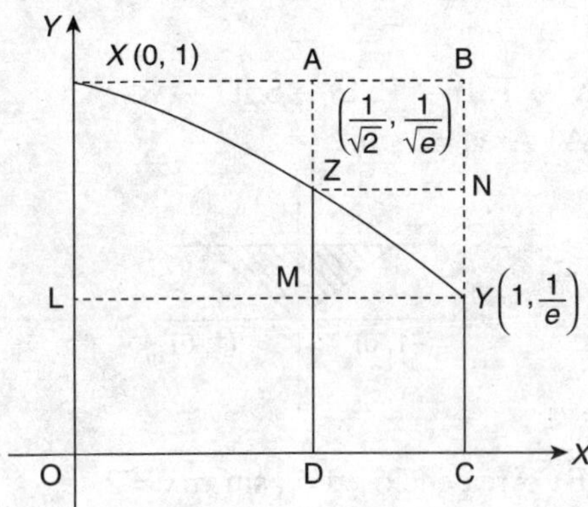

Now, S = area enclosed by curve = XYCO

and area of rectangle OCYL $= \frac{1}{e}$

Clearly $S > \frac{1}{e}$ $\therefore$ (a) is true.

For $x \in [0,1]$

$\Rightarrow x^2 < x$

$\Rightarrow -x^2 > -x \Rightarrow e^{-x^2} \geq e^{-x} \forall x \in [0,1]$

$\Rightarrow \int_0^1 e^{-x^2} dx > \int_0^1 e^{-x} dx = 1 - \frac{1}{e}$

$\Rightarrow S > 1 - \frac{1}{e}$ $\therefore$ (b) is true

Now S < Area of rectangle XADO + Area of rectangle ZDCN

$\Rightarrow S < \frac{1}{\sqrt{2}} \times 1 + \left(1 - \frac{1}{\sqrt{2}}\right)\frac{1}{\sqrt{e}}$

$\therefore S < \frac{1}{\sqrt{2}} + \frac{1}{\sqrt{e}}\left(1 - \frac{1}{\sqrt{2}}\right)$ $\because$ (d) is true

Also as $\frac{1}{4}\left(1 + \frac{1}{\sqrt{e}}\right) < 1 - \frac{1}{e}$ $\therefore$ (c) is incorrect

6. (b, c, d) The area bounded by the curve $y = e^x$ and lines $x = 0$ and $y = e$ is as shown in the graph.

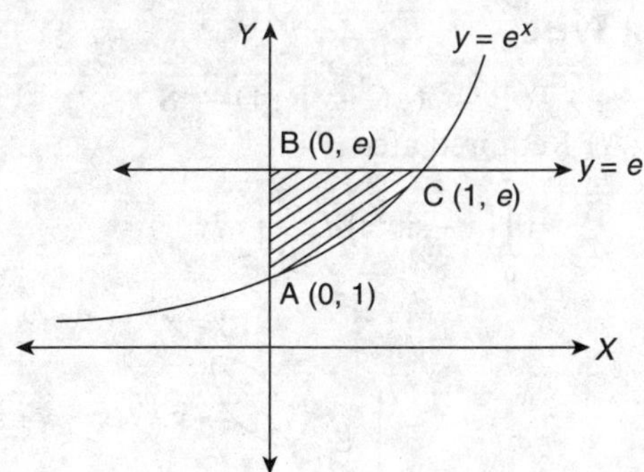

Required area $= \int_0^1 (e - e^x) dx$

$= e - \int_0^1 e^x dx. = 1$

Also, required area

$= \int_0^e x dy = \int_1^e \ln y dy$ (where $e^x = y \Rightarrow x = \ln y$)

$= \int_1^e \ln(e + 1 - y) dy \left[\because \int_a^b f(x) dx = \int_a^b f(a + b - x) dx\right]$

7. (b, c) $\int_0^\alpha (x - x^3) dx = \frac{1}{2}\int_0^1 (x - x^3) dx$

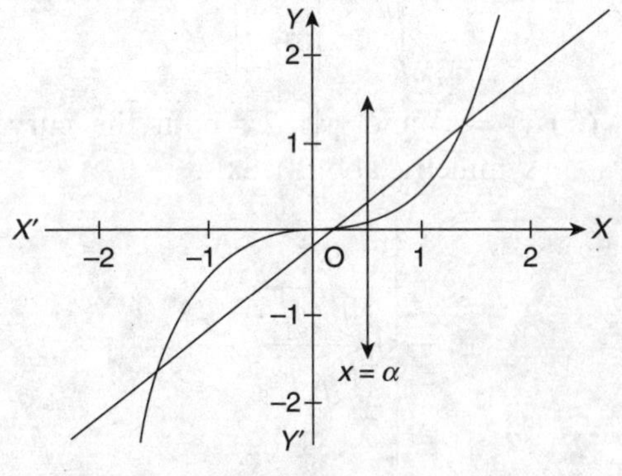

$\Rightarrow \left(\frac{x^2}{2} - \frac{x^4}{4}\right)_0^\alpha = \frac{1}{2}\left(\frac{x^2}{2} - \frac{x^4}{4}\right)_0^1$

$\Rightarrow \frac{\alpha^2}{2} - \frac{\alpha^4}{4} = \frac{1}{2}\left(\frac{1}{2} - \frac{1}{4}\right)$

$4\alpha^2 - 2\alpha^4 = 1 \Rightarrow 2\alpha^4 - 4\alpha^2 + 1 = 0$

$\Rightarrow \alpha^2 = \frac{4 \pm \sqrt{16 - 8}}{4} = 1 \pm \frac{1}{\sqrt{2}}$

$\therefore 0 < \alpha < 1 \Rightarrow \alpha^2 = 1 + \frac{1}{\sqrt{2}}$

$\Rightarrow \sqrt{0.29} > \sqrt{0.25} = \frac{1}{2}$ also $\alpha < 1 \Rightarrow \frac{1}{2} < \alpha < 1$

Matrix Type

1. A → T; B → T; C → R; D → S

(A) Required area = $4s$

$$s=\int_0^{\pi}(x+\sin x)\,dx-\int_0^{\pi}x\,dx$$

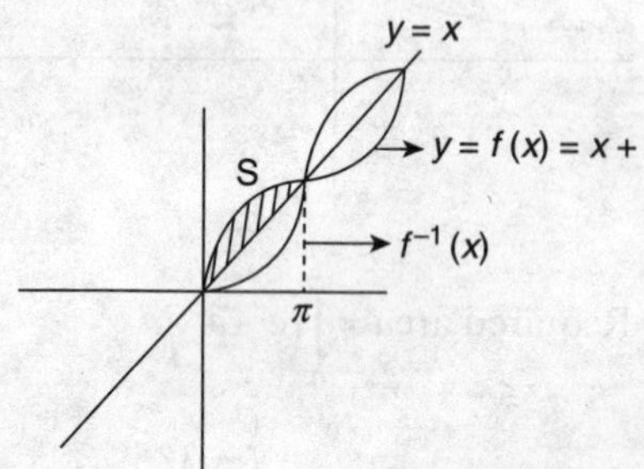

$$=\frac{\pi^2}{2}-\cos\pi+\cos 0-\frac{\pi^2}{2}=2 \text{ sq.units}$$

(B) Required area

$$=2\int_0^1 xe^x\,dx=2\left[xe^x-e^x\right]_0^1=2$$

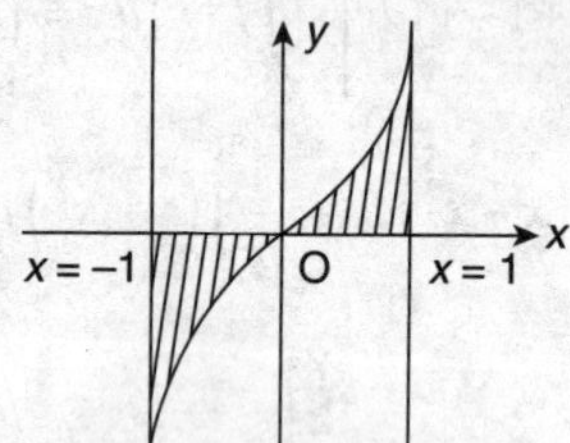

(C) $y^2=x^3$ and $|y|=2x$ both the curves are symmetric about y-axis.

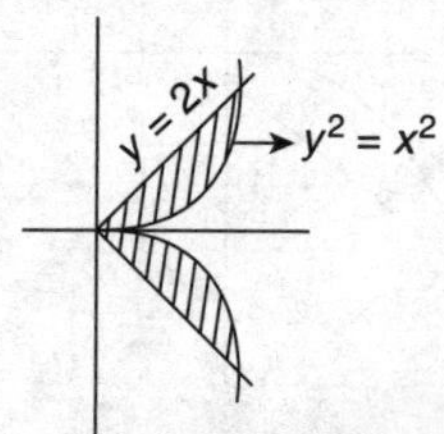

$4x^2=x^3\Rightarrow x=0,4$

Required area $=2\int_0^4\left(2x-x^{3/2}\right)dx=\frac{16}{5}$

(D) $\sqrt{x}+\sqrt{|y|}=1$

The above curve is symmetric about x-axis.

$\sqrt{|y|}=1-\sqrt{x}$ and $\sqrt{x}=1-\sqrt{|y|}$

$\Rightarrow$ for $x>0, y>0\sqrt{y}=1-\sqrt{x}$

$$\frac{1}{2\sqrt{y}}\frac{dy}{dx}=-\frac{1}{2\sqrt{x}}$$

$$\frac{dy}{dx}=-\sqrt{\frac{x}{y}}$$

$\frac{dy}{dx}<0$, function is decreasing

required area

$$1-2\int_0^1\left(1-\sqrt{x}\right)dx=\frac{1}{3}$$

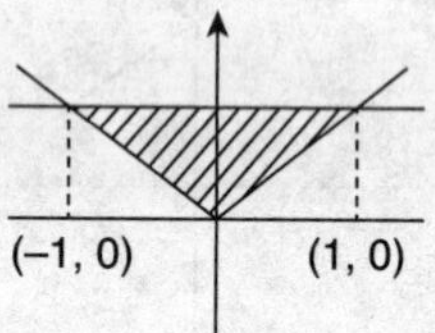

2. A → T; B → P; C → S; D → R

(A) Area = 1 unit

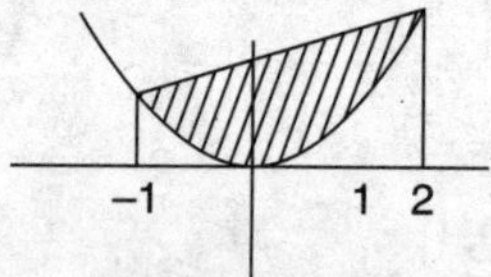

(B) Area enclosed $=\int_0^{\pi}\sin x\,dx=2$

(C) The line $y=x+2$ intersects $y=x^2$ at $x=-1$ and $x=2$

The area of the shaded region

$$\frac{15}{2}-\int_{-1}^{2}x^2\,dx=\frac{9}{2}$$

(D) Here, $a^2=9, b^2=5, b^2=a^2\left(1-e^2\right)$

$$\Rightarrow e^2=\frac{4}{9}\Rightarrow e=\frac{2}{3}$$

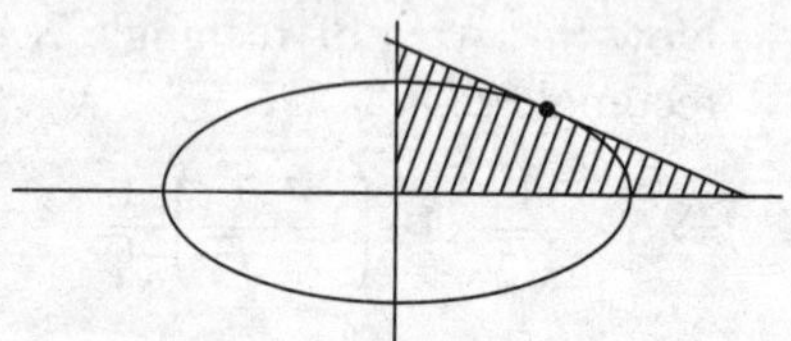

Equation of tangent at $\left(2,\frac{5}{3}\right)$ is

$$\frac{2x}{9}+\frac{y}{3}=1$$

x intercept $=\frac{9}{2}$, y intercept $=3$

Area $=4\times\frac{9}{2}\times3\times\frac{1}{2}=27$ sq.units

Integer and Numerical Value Type

1. (4) $[x][y]=2$

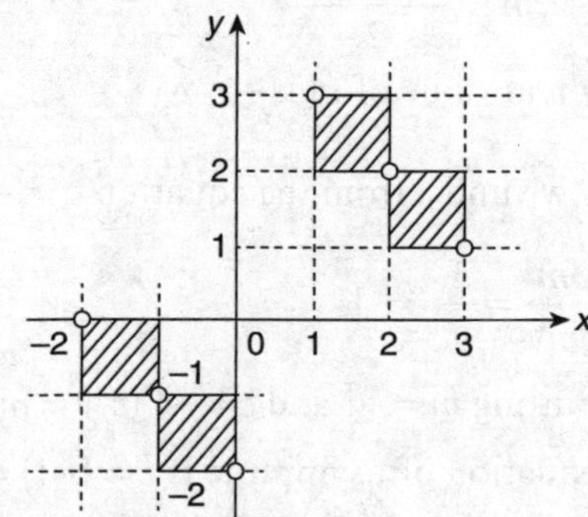

Here four cases arise:

(1) $[x]=2$ and $[y]=1$

$\Rightarrow 2\le x<3$ and $1\le y<2$

(2) $[x]=1$ and $[y]=2 \Rightarrow 1\le x<2$ and $2\le y<3$

(3) $[x]=-2$ and $[y]=-1$

$\Rightarrow -2\le x<-1$ and $-1\le y<0$

(4) $[x]=-1$ and $[y]=-2$

$\Rightarrow -1\le x<0$ and $-2\le y<-1$

Area enclosed by solution set = 4

2. (8) $ay^2 = x^2(a-x)\, y = \pm x\sqrt{\dfrac{a-x}{a}}$

Area $= 2\displaystyle\int_2^a x\sqrt{\frac{a-x}{a}}\,dx$, put $x = a\cos\theta$,

$dx = -a\sin\theta\, d\theta$

$$= 2\int_0^{\pi/2} a\cos\theta\sqrt{2}\sin\frac{\theta}{2}\,a\sin\theta\, d\theta$$

$$= 2\sqrt{2}a^2\int_0^{\pi/2}\left(1-2\sin^2\frac{\theta}{2}\right)2\sin^2\frac{\theta}{2}\cos\frac{\theta}{2}\,d\theta$$

put $\sin\dfrac{\theta}{2} = t, \cos\dfrac{\theta}{2}\,d\theta = 2dt$

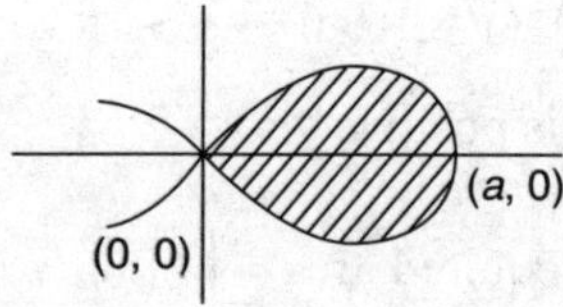

$$= 8\sqrt{2}a^2\int_0^{1/\sqrt{2}}\left(1-2t^2\right)t^2\,dt = 8\sqrt{2}a^2\int_0^{1/\sqrt{2}}\left(t^2-2t^4\right)dt$$

$$= 8\sqrt{2}a^2\left(\frac{t^3}{3}-\frac{2t^5}{5}\right)^{1/\sqrt{2}} = 8\sqrt{2}a^2\left(\frac{1}{6\sqrt{2}}-\frac{2}{20\sqrt{2}}\right)$$

$$= 8a^2\left(\frac{1}{6}-\frac{1}{10}\right) = \frac{8a^2}{15}$$

3.

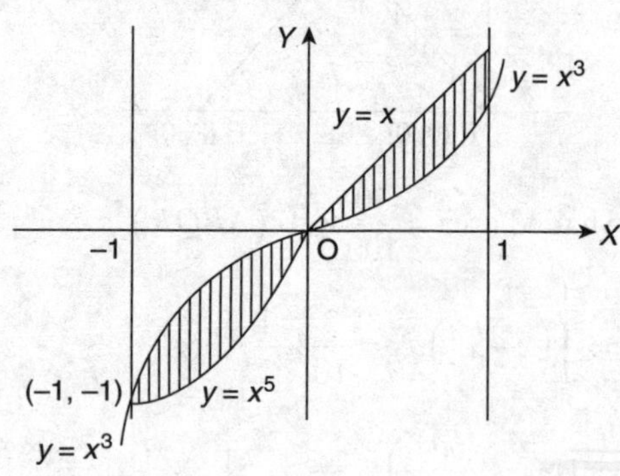

$$\text{Area bounded} = \left|\int_1^0 x^5\,dx - \int_{-1}^0 x^3\,dx\right| + \int_0^1 x\,dx - \int_0^1 x^3\,dx$$

$= \dfrac{1}{3}$ sq.units $\Rightarrow k = 3$

4.

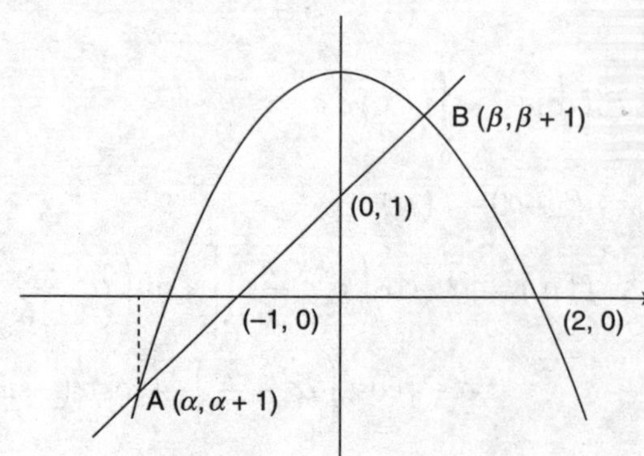

For intersection

$x^2 + x - 3 = 0$

$\Rightarrow \alpha + \beta = -1, \alpha\beta = -3$

Now area below x-axis

$$= \int_\alpha^{-1}(x+1)\,dx - \int_\alpha^{2}\left(4-x^2\right)dx$$

$$= -\frac{\alpha^3}{3} - \frac{\alpha^2}{2} + 3\alpha + \frac{29}{6}$$

But this area is negative

$\Rightarrow$ Actual area $= \dfrac{\alpha^3}{3} + \dfrac{\alpha^3}{2} - 3\alpha - \dfrac{29}{6}$

$\Rightarrow k = 29$

5. (4)

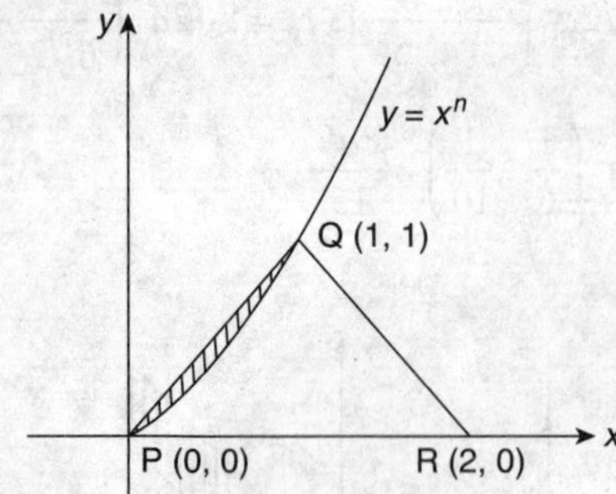

Shaded area $= \frac{30}{100} \times Ar(\Delta PQR)$

$\Rightarrow \int_0^1 (x - x^n)\,dx = \frac{3}{10} \times \frac{1}{2} \times 2 \times 1$

$\Rightarrow \left(\frac{x^2}{2} - \frac{x^{n+1}}{n+1}\right)_0^1 = \frac{3}{10} \Rightarrow \frac{1}{2} - \frac{1}{n+1} = \frac{3}{10} \Rightarrow n = 4$

6. (3) $F(x) = \int_x^{x^2+\pi/6} 2\cos^2 t\,dt$

$F'(\alpha) = 2\cos^2\left(\alpha^2 + \frac{\pi}{6}\right) \cdot 2\alpha - 2\cos^2 \alpha$

$F'(\alpha) + 2 = \int_0^\alpha f(x)\,dx$

$\Rightarrow F''(\alpha) = f(\alpha)$

$\therefore f(\alpha) = 4\alpha.2\cos\left(\alpha^2 + \frac{\pi}{6}\right)\left[-\sin\left(\alpha^2 + \frac{\pi}{6}\right)\right]$

$\cdot 2\alpha + 4\cos^2\left(\alpha^2 + \frac{\pi}{6}\right) - 4\cos\alpha(-\sin\alpha)$

$\therefore f(0) = 4\cos^2\frac{\pi}{6} = 4 \times \frac{3}{4} = 3$

Comprehension Passage

Comprehension 1

1. (a) $(y-4)x^2 + x + 2 = 0$
The coefficient of the highest power of x, i.e., x^2 is $y - 4 = 0$
$y - 4 = 0$ is the asymptote parallel to the x-axis.
The coefficient of the highest power of y is x, so $x = 0$ is also an asymptote.

2. (b) $\varphi_3(m) = 1 + m^3, \varphi_2(m) = -3m$
Putting $\varphi_3(m) = 0$ or $m^3 + 1 = 0$
Then $(m+1)(m^2 - m + 1) = 0$
$m = -1, m = \frac{1 \pm \sqrt{1-4}}{2}$
Only real value of m is -1
Now we find c from the equation $c = -\frac{\varphi_{n-1}(m)}{\varphi'_n(m)}$
$c = \frac{3m}{3m^2} = \frac{1}{m} = -1$
On putting $m = -1$ and $c = -1$ in $y = mx + c$, the equation of asymptote is $y = (-1)x + (-1)$ or $x + y + 1 = 0$

3. (b) The coefficient of the highest power of y is $(2 - x)$, so $x = 2$ is asymptotic.

Comprehension 2

1. (b)
We have $y^3 - 3y + x = 0 \Rightarrow 3y^2\frac{dy}{dx} - 3\frac{dy}{dx} + 1 = 0$
$\Rightarrow \frac{dy}{dx} = \frac{1}{3(1-y^2)}$ or $f'(x) = \frac{1}{3\left[1 - [f(x)]^2\right]}$
Also, $y^2\frac{d^2y}{dx^2} + 6y\left(\frac{dy}{dx}\right)^2 - 3\frac{d^2y}{dx^2} = 0$
$\Rightarrow \frac{d^2y}{dx^2} = \frac{2y}{1-y^2}\left(\frac{dy}{dx}\right)^2$
$\Rightarrow f''(x) = \frac{2f(x)}{9\left[1 - [f(x)]^2\right]^3}$
$\therefore f''(-10\sqrt{2}) = \frac{-4\sqrt{2}}{3^2 \times 7^3}$

2. (a)
For $x < -2$
We have, $3y - y^3 \langle -2 \Rightarrow y^3 - 3y - 2 \rangle 0$
$\Rightarrow (y+1)^2(y-2) > 0 \Rightarrow y > 2 \forall x < -2$
$\Rightarrow f(x)$ is positive $\forall x < -2$
Hence, required area $= \int_a^b f(x)dx = \int_a^b 1.f(x)dx$
$= xf(x)]_a^b - \int_a^b xf'(x)dx$

$$= bf(b) - af(a) - \int_a^b \frac{x.1}{3\left[1-(f(x))^2\right]}dx$$

$$= \int_a^b \frac{x}{3\left[(f(x))^2 - 1\right]} + bf(b) - af(a)$$

Comprehension 3

We are given:

- Test-tube mass $= m = 5$ g
- Glass density $= p_g = 2.5$ g/cc
- Water density $= p_w = 1$ g/cc
- Initial air volume $= v_0 = 3.3$ cc
- Atmospheric pressure $= P_0 = 10^5$ Pa
- Final air volume $= v = v_0 - \Delta v = 3.3 - x$
- Final pressure $= P = P_0 + \Delta p = 10^5 + 1000y$
- Temperature is constant, so use **Boyle's law**: $PV =$ constant

Also, **buoyancy condition** at the point of sinking:
Net force $= 0 \Rightarrow$ **weight = buoyant force**

Step 1: Volume of test tube (glass + air)

Let total volume $= V_t$ and volume of trapped air $= v$

Glass mass = 5 g, density = 2.5 g/cc $\Rightarrow$ glass volume $= \frac{5}{2.5} = 2$ cc

So, total volume of test tube $= V_t = v + 2$

Step 2: Buoyant force = weight

At sinking:

Buoyant force $= \rho_w \times g \times V_t$

Weight $= mg = 5$ g

Set buoyant force = weight:

$1 \times g \times (v+2) = 5g \Rightarrow v + 2 = 5 \Rightarrow v = 3$ cc

So, **when it just sinks**, the trapped air volume is 3 cc $\Rightarrow$

$\Delta v = v_0 - v = 3.3 - 3 = 0.3$ cc $\Rightarrow x = 0.3$

Step 3: Use Boyle's Law to find pressure

$P_0 v_0 = (P_0 + \Delta p)(v_0 - \Delta v)$

Substitute values:

$10^5 \times 3.3 = (10^5 + \Delta p)(3)$

$$\Rightarrow \frac{10^5 \times 3.3}{3} = 10^5 + \Delta p \Rightarrow 1.1 \times 10^5 = 10^5 + \Delta p$$

$\Rightarrow \Delta p = 0.1 \times 10^5 = 10^4$ Pa $= 10000$ Pa

$\Rightarrow y = 10$

1. The value of x is 3.
2. The value of y is 10.

CHAPTER 9

Differential Equations

Objectives

In this chapter on Differential Equations, we shall study the following topics:

1. Basic concepts
2. Formation of differential equation
3. Solution of differential equation
4. Separation of variables
5. Reducing to separation of variable by proper substitution
6. Homogeneous differential equation
7. Linear differential equation
8. Differential equation reducible to linear differential equation
9. Bernoulli's equation.

Introduction

For a given function g, if we find a function $f(x)$ such that $\frac{dy}{dx} = g(x)$ where $y = f(x)$ then the equation $\frac{dy}{dx} = g(x)$ is known as differential equation (DE). These equations have a wide range of applications in many fields such as Physics, Chemistry, Economics, Geology and even in Biology, etc. Hence, in-depth study of differential equations is very essential for modern scientific investigations.

In this chapter, we will study some basic concepts related to differential equations, general and particular solutions of differential equations, formation of differential equations, some methods to solve a first-order–first-degree differential equation and some applications of differential equations in different areas.

An equation involving derivative/derivatives of the dependent variable with respect to independent variable/variables is called a differential equation. It is expressed as shown in the following examples:

Examples

(i) $\frac{dy}{dx} = \sin x$

(ii) $\frac{dy}{dx} + xy = \cot x$

(iii) $\frac{d^2 y}{dx^2} - 5\frac{dy}{dx} + 6y = x^2$

(iv) $\left(\frac{d^4 y}{dx^4}\right)^3 - 4\frac{dy}{dx} + 4y = 5\cos 3x$

(v) $x^2 \frac{d^2 y}{dx^4} + \sqrt{1 + \left(\frac{dy}{dx}\right)^2} = 0$

Some Basic Concepts and Theory

(i) A differential equation involving derivatives of the dependent variable with respect to only one independent variable is called an ordinary differential equation and a differential equation involving derivatives with respect to more than one independent variable is called a partial differential equation.

(ii) **Degree of a differential equation** is defined if it is a polynomial equation in its derivatives. Degree (when defined) of a differential equation is the highest power (positive integer only) of the highest order derivative in it.

(iii) **Order of a differential equation** is the order of the highest-order derivative occurring in the differential equation.

(iv) A relation between involved variables, which satisfy the given differential equation is called its solution. The solution which contains as many arbitrary constants as the order of the differential equation is called the general solution and the solution free from arbitrary constants is called the particular solution.

(v) To form a differential equation from a given function, we differentiate the function successively as many times as the number of arbitrary constants in the given function and then eliminate the arbitrary constants.

(vi) The order of a differential equation representing a family of curves is same as the number of arbitrary constants present in the equation corresponding to the family of curves.

(vii) 'Variable separable method' is used to solve such an equation in which the variables can be separated completely, i.e., terms containing x should remain with dx and terms containing y should remain with dy.

(viii) A function $F(x, y)$ is said to be a homogeneous function of degree n if $F(\lambda x, \lambda y) = \lambda^n F(x, y)$ for some non-zero constant λ.

(ix) A differential equation which can be expressed in the form $\frac{dy}{dx} = F(x, y)$ or $\frac{dx}{dy} = G(x, y)$, where $F(x, y)$ and $G(x, y)$ are homogeneous functions of degree zero, is called a homogeneous differential equation.

(x) To solve a homogeneous differential equation of the type $\frac{dy}{dx} = F(x, y)$, we make the substitution $y = vx$ and to solve a homogenous differential equation of the type $\frac{dx}{dy} = G(x, y)$, we make the substitution $x = vy$.

(xi) A differential equation of the form $\frac{dy}{dx} + Py = Q$, where P and Q are constants or function of x only is known as a first-order linear differential equation. The solution of such a differential equation is given by $y\,(\text{I.F.}) = \int (Q \times \text{I.F.})\, dx + C$, where I.F. is the Integrating Factor $= e^{\int P dx}$.

(xii) Another form of first-order linear differential equation is $\frac{dx}{dy} + P_1 x = Q_1$, where P_1 and Q_1 are constants or functions of y only. The solution of such a differential equation is given by $x\,(\text{I.F.}) = \int (Q_1 \times \text{I.F.})\, dy + C$, where I.F. $= e^{\int P_1 dy}$.

ILLUSTRATION 9.1

Find order and degree of following DEs:

(i) $\frac{d^2y}{dx^2} = \left[1 + \left(\frac{dy}{dx}\right)^4\right]^{5/3}$

(ii) $k^2\left(\frac{d^2y}{dx^2}\right)^2 = \left[1 + \left(\frac{dy}{dx}\right)^2\right]^3$

Solution

(i) After removing fraction in the derivative, we have

$$\left(\frac{d^2y}{dx^2}\right)^3 = \left(1 + \left(\frac{dy}{dx}\right)^4\right)^5$$

Order = 2, Degree = 3

(ii) Order = 2, Degree = 2

Formation of a Differential Equation

(i) Write down the given equation.

(ii) Differentiate it successively with respect to x that many number of times as equal to the arbitrary constants.

(iii) Hence on eliminating the arbitrary constants, it results in a differential equation which involves

$$x, y, \frac{d^2y}{dx^2} \ldots.$$

ILLUSTRATION 9.2

Form the differential equation corresponding to $y^2 = m(a^2 - x^2)$ (a and m are parameters.)

Solution

We are given that $y^2 = m(a^2 - x^2)$ ...(i)

Since the given equation contains two arbitrary constants, so we shall differentiate it two times and we shall get a differential equation of the second order.

Differentiating both sides of (i) w.r.t. x, we get

$$2y\frac{dy}{dx} = m(-2x) \Rightarrow y\frac{dy}{dx} = -mx \quad \text{...(ii)}$$

Differentiating both sides of (ii) w.r. to x, we get

$$y\frac{d^2y}{dx^2} + \left(\frac{dy}{dx}\right)^2 = -m \quad \text{...(iii)}$$

From (ii) and (iii), we get

$$x\left[y\frac{d^2y}{dx^2}+\left(\frac{dy}{dx}\right)^2\right]=y\frac{dy}{dx}$$

This is the required differential equation.

ILLUSTRATION 9.3

The differential equation of all circles which passes through the origin and whose centre lies on y-axis, is ________.

(a) $(x^2-y^2)\frac{dy}{dx}-2xy=0$

(b) $(x^2-y^2)\frac{dy}{dx}+2xy=0$

(c) $(x^2-y^2)\frac{dy}{dx}-xy=0$

(d) $(x^2-y^2)\frac{dy}{dx}+xy=0$

Solution

The system of circles pass through origin and centre lies on y-axis is

$$x^2+y^2-2ay=0\Rightarrow 2x+2y\frac{dy}{dx}-2a\frac{dy}{dx}=0$$

$$\Rightarrow \quad 2a=2y+2x\frac{dx}{dy}$$

Therefore, the required differential equation is

$$x^2+y^2-2y^2-2xy\frac{dy}{dx}=0\Rightarrow (x^2-y^2)\frac{dx}{dy}-2xy=0$$

ILLUSTRATION 9.4

Find the differential equation from $y=ke^{\sin^{-1}x}+3$.

Solution

The given equation is $y=ke^{\sin^{-1}x}+3$ ...(i)

Differentiating (1) w.r.t. x,

$$\frac{dy}{dx}=ke^{\sin-1x}\frac{1}{\sqrt{1-x^2}}+0$$

$$\Rightarrow \quad \frac{dy}{dx}=(y-3)\frac{1}{\sqrt{1-x^2}} \quad \text{[Using (i)]}$$

$\Rightarrow\sqrt{1-x^2}=\frac{dy}{dx}=y-3$, which is the required differential equation.

ILLUSTRATION 9.5

Find the differential equation of the family of curves $y=e^x(A\cos x+B\sin x)$, where A and B are arbitrary constants.

Solution

Differentiating $y=e^x(A\cos x+B\sin x)$ w.r.t. x, we get,

$$\frac{dy}{dx}=e^x(-A\sin x+B\cos x)+e^x(A\cos x+B\sin x)$$

$$\Rightarrow \frac{dy}{dx}=e^x(-A\sin x+B\cos x)+y$$

Differentiating again w.r.t. x we get

$$\frac{d^2y}{dx^2}=e^x(-A\cos x-B\sin x)+$$

$$e^x(-A\sin x+B\cos x)+\frac{dy}{dx}$$

$$\Rightarrow \quad \frac{d^2y}{dx^2}=2\left(\frac{dy}{dx}-y\right)$$

Solution of a Differential Equation

Any relation between dependent and independent variables, which satisfies the given DE is called solution or integral of the DE. A solution of a DE does not involve the derivatives w.r.t. independent variables.

1. A solution of DE, consisting of independent arbitrary constants (in number, equal to the order of the DE) is called general solution or complete primitive
2. The solution of DE for particular values of one or more of the arbitrary constants is called a particular solution of DE.

ILLUSTRATION 9.6

Show that $y=ce^{2x}$ is a solution of $\frac{dy}{dx}=2y$.

Solution

Differentiate $y=ce^{2x}$ w.r.t. x, we get $\frac{dy}{dx}=2ce^{2x}$

Now $\quad$ LHS $=\frac{dy}{dx}=2ce^{2x}=2y=$ RHS

Separation of Variable in Differential Equation

In this type of DE, it is possible to take all the terms involving x and dx to one side, and those involving y and dy to the other side. Thus separating the variables x and y and then integrating, DE can be written as $f\{x)dx = g(y)dy$.

ILLUSTRATION 9.7

Solve: $3e^x \tan y\, dx + (1-e^x)\sec^2 y\, dy = 0$

Solution

Separating the variables: $\dfrac{3e^x}{1-e^x}dx + \dfrac{\sec^2 y}{\tan y}dy = 0$

Integrating,

$\Rightarrow \quad -3\log(1-e^x) + \log(\tan y) = \log C_1$

$\Rightarrow \quad \tan y = C_1(1-e^x)^3$

(In fact, writing C_1 or c makes no difference)

ILLUSTRATION 9.8

The solution of the differential equation $\dfrac{dy}{dx} = \sec x(\sec x + \tan x)$ is:

(a) $y = \sec x + \tan x + c$

(b) $y = \sec x + x + \cot x + c$

(c) $y = \sec x - \tan x + c$

(d) None of these

Solution

$\dfrac{dy}{dx} = \sec x(\sec x + \tan x) \Rightarrow \dfrac{dy}{dx} = \sec^2 x + \sec x \tan x$

Now integrating both sides, we get

$$y = \tan x + \sec x + c$$

Option (a)

ILLUSTRATION 9.9

The solution of the differential equation

$(x^2 - yx^2)\dfrac{dy}{dx} + y^2 + xy^2 = 0$ is:

(a) $\log\left(\dfrac{x}{y}\right) = \dfrac{1}{x} + \dfrac{1}{y} + c$

(b) $\log\left(\dfrac{y}{x}\right) = \dfrac{1}{x} + \dfrac{1}{y} + c$

(c) $\log(xy) = \dfrac{1}{x} + \dfrac{1}{y} + c$

(d) $\log(xy) + \dfrac{1}{x} + \dfrac{1}{y} + c$

Solution

The given equation

$(x^2 - yx^2)\dfrac{dy}{dx} + y^2 + xy^2 = 0 \Rightarrow \dfrac{1-y}{y^2}dy + \dfrac{1-x}{x^2}dx = 0$

On integrating, we get the required solution

$$\log\left(\frac{x}{y}\right) = \frac{1}{x} + \frac{1}{y} + c$$

Option (a)

ILLUSTRATION 9.10

The solution of the differential equation $x\cos y\, dy = (xe^x \log x + e^x)dx$ is:

(a) $\sin y = \dfrac{1}{x}e^x + c$

(b) $\sin y + e^x \log x + c = 0$

(c) $\sin y + ex\log x + c$

(d) None of these

Solution

$$x\cos y\, dy = (xe^x \log x + e^x)dx$$

$$\Rightarrow \quad \cos y\, dy = \left(e^x \log x + \frac{e^x}{x}\right)dx$$

On integration $\sin y = e^x \log x + c$

Option (d)

ILLUSTRATION 9.11

The solution of the equation $\dfrac{dy}{dx} = y(e^x + 1)$ is:

(a) $y + e^{(e^x + x + c)} = 0$

(b) $\log y = e^x + x + c$

(c) $\log y + e^x = x + c$

(d) None of these.

Solution

$$\frac{dy}{dx}=y(e^x+1)\Rightarrow\frac{dy}{y}=(e^x+1)dx$$

Integrating both sides, we get $\log y=e^x+x+c$

Option (b)

Reducing to Separation of Variable by Proper Substitution

Sometimes the variables in equations are not separable but a suitable substitution can reduce them to variable separable form.

ILLUSTRATION 9.12

$$\frac{dy}{dx}=\sec(x+y)$$

Solution

Put $x+y=v\Rightarrow 1+\frac{dy}{dx}=\frac{dv}{dx}$

Now the equation reduces to

$$\frac{dv}{dx}-1=\sec v\Rightarrow\frac{dv}{dx}=\sec v+1\Rightarrow\frac{1}{1+\sec v}dv=dx$$

$$\Rightarrow\qquad\frac{\cos v}{1+\cos v}dv=dx$$

Integrating:

$$\int dv-\frac{1}{2}\int\sec^2\frac{v}{2}dv=\int dx+c\Rightarrow v-\frac{1}{2}2\tan\frac{v}{2}=x+c$$

$$\Rightarrow\quad x+y-\tan\frac{x+y}{2}=x+c\Rightarrow y-\tan\frac{x+y}{2}=c$$

ILLUSTRATION 9.13

Solve the equation

$$\frac{dy}{dx}=(x+y-1)+((x+y)/\log(x+y))$$

Solution

Put

$$x+y=v\Rightarrow 1+\frac{dy}{dx}=\frac{dv}{dx}\Rightarrow\frac{dv}{dx}=\left(v+\frac{v}{\log v}\right)$$

$$\Rightarrow\qquad\frac{\log v}{v(1+\log v)}dv=dx$$

Integrating, $\int\frac{\log v}{v(1+\log v)}dv=\int dx+c$, put

$$\log v=u\Rightarrow\frac{1}{v}dv=du\Rightarrow\int\frac{u}{1+u}du=x+c$$

$$\Rightarrow\qquad u-\log(1+u)=x+c$$

$$\Rightarrow\qquad\log(x+y)-\log\{1+\log(x+y)\}=x+c$$

Homogeneous Differential Equation

A DE of the form $\frac{dy}{dx}=\frac{f(x,y)}{g(x,y)}$ where $f(x, y)$ and $g(x, y)$ are homogeneous expressions in x and y of the same degree is called a homogeneous equation. It can also be put in the form $\frac{dy}{dx}=F(y/x)$

Working Method

Step 1: Put $y=vx$ so that $\frac{dy}{dx}=v+x\frac{dv}{dx}$

Step 2: Substitute in the given equation and we obtain an equation of the form of variables separable in v and x.

Step 3: After integrating, replace v by $\frac{y}{x}$.

ILLUSTRATION 9.14

Solve: $(x^3+y^3)dx-3xy^2dy=0$

Solution

The given DE can be written as $\frac{dy}{dx}=v+x\frac{dv}{dx}$

Put $\qquad y=vx$ and $\frac{dy}{dx}=v+x\frac{dv}{dx}$

$$v+x\frac{dv}{dx}=\frac{1+v^3}{3v^2}\Rightarrow x\frac{dv}{dx}=\frac{1+v^3}{3v^2}-v=\frac{1-2v^3}{3v^2}$$

$$\Rightarrow\qquad\frac{3v^2}{1-2v^3}dv=\frac{dx}{x}$$

Integrating:

$$\int\frac{3v^2}{1-2v^3}dv=\int\frac{dx}{x}+\log c$$

$$\Rightarrow \quad -\frac{1}{2}\log\left(1-2v^3\right)=\log x+\log c$$

$$\Rightarrow \quad \frac{1}{\left(1-2v^3\right)^{1/2}}=cx$$

$$\Rightarrow \quad c^2x^2\left(1-\frac{2y^3}{x^3}\right)=1 \Rightarrow \left(x^3-2y^3\right)=cx$$

(Remember we have replaced $\frac{1}{c^2}$ by c)

ILLUSTRATION 9.15

Solve:

$$\left(x\cos\left(\frac{y}{x}\right)+y\sin\left(\frac{y}{x}\right)\right)y-\left(y\sin\frac{y}{x}-x\cos\frac{y}{x}\right)x\frac{dy}{dx}=0$$

Solution

Put $\frac{y}{x}=v \Rightarrow y=vx \Rightarrow \frac{dy}{dx}=v+x\frac{dv}{dx}$

Then given differential equation reduced to:

$$\Rightarrow \quad \left(vx^2\cos v+v^2x^2\sin v-v^2x^2\sin v+vx^2\cos v\right)-x^2\left(vx\sin v-x\cos v\right)\frac{dv}{dx}=0$$

$$\Rightarrow \quad 2v\cos v-x\left(v\sin v-\cos v\right)\frac{dv}{dx}=0$$

$$\Rightarrow x\left(v\sin v-\cos v\right)\frac{dv}{dx}=2v\cos v \Rightarrow \frac{dx}{dv}=\frac{xv\sin v-x\cos v}{2v\cos v}$$

$$\Rightarrow \quad \frac{2}{x}dx=\left(\tan v-\frac{1}{v}\right)dv$$

Integrating:

$$\int\frac{2}{x}dx+\log c=\int\left(\tan v-\frac{1}{v}\right)dv$$

$$\Rightarrow \quad 2\log x+2\log c=\log(\sec v)-\log v$$

$$\Rightarrow \quad \log\left(cx^2\right)=\log\left(\frac{\sec v}{v}\right)$$

$$\Rightarrow \quad vcx^2=\sec v \Rightarrow \frac{y}{v}cx^2=\sec\left(\frac{y}{x}\right) \Rightarrow \boxed{cxy=\sec\frac{y}{x}}$$

ILLUSTRATION 9.16

The solution of differential equation $\left(1+2e^{x/y}\right)dx+2e^{x/y}(1-x/y)dy=0$ is ________.

(a) $x+2ye^{x/y}=c$ **(b)** $x-2ye^{x/y}=c$

(c) $2x+2ye^{x/y}=c$ **(d)** None of these

Solution

The equation is homogeneous of degree 0. The appearance of x/y throughout the equation suggest the use of the transformation $x=vy, dx=v\,dy+y\,dv$.

Then

$$\left(1+2e^v\right)\left(v\,dy+y\,dv\right)+2e^v(1-v)dy=0$$

$$\Rightarrow \quad \left(v+2e^v\right)dy+y\left(1+2e^v\right)dv=0$$

$$\Rightarrow \quad \frac{dy}{y}+\frac{1+2e^v}{v+2e^v}dv=0$$

Integrating and replacing v by x/y, we get,

$$\ln y+\ln\left(v+2e^v\right)=\ln C \text{ and } x+2ye^{x/y}=C$$

Reducing Homogeneous Form

Differential equation of type:

$\frac{dy}{dx}=\frac{ax+by+c}{a_1x+b_1y+c_1}$ can be reduced to homogeneous form in the following way:

If $\frac{a}{a_1}\neq\frac{b}{b_1}$ then put $x = X + h$ and $y = Y + k$ where h and k are constants to be determined.

Now the equation reduces to

$$\frac{dY}{dX}=\frac{aX+bY+ah+bk+c}{a_1X+b_1Y+a_1h+b_1k+c_1}$$

Now equating $ah+bk+c=a_1h+b_1k+c_1=0$, and we have

$$\frac{dY}{dX}=\frac{aX+bY}{a_1X+b_1Y} \quad \text{(homogeneous)}$$

Note: If $\frac{a}{a_1}=\frac{b}{b_1}$ then put $ax + by = v$

(The equation reduces to variable separable)

ILLUSTRATION 9.17

Solve: $\frac{dy}{dx}=\frac{x+2y-3}{2x+y-3}$

Solution

Put $x=X+h$ and $y=Y+k$,

$$\frac{dY}{dX}=\frac{X+2Y+h+2k-3}{2X+Y+2h+k-3} \quad ...(i)$$

Now equating

$$h+2k-3=2h+k-3=0 \Rightarrow h=1, k=1$$

Now (i) becomes $\frac{dY}{dX}=\frac{X+2Y}{2X+Y}$

Put $Y=vX \Rightarrow \frac{dY}{dX}=v+X\frac{dv}{dX}$

Then, $v+X\frac{dv}{dX}=\frac{1+2v}{2+v} \Rightarrow X\frac{dv}{dX}=\frac{1-v^2}{2+v}$

$$\Rightarrow \frac{2+v}{1-v^2}dv=\frac{1}{X}dX \Rightarrow \frac{2}{1-v^2}dv+\frac{v}{1-v^2}dv=\frac{1}{X}dX$$

$$\Rightarrow \log\left(\frac{1+v}{1-v}\right)-\frac{1}{2}\log\left(1-v^2\right)=\log X+\log c$$

$$\Rightarrow \frac{1}{2}\log(1+v)-\frac{3}{2}\log(1-v)=\log(cX)$$

$$\Rightarrow \log\left(\frac{(1+v)^{1/2}}{(1-v)^{3/2}}\right)=\log(cX) \Rightarrow \frac{(1+v)^{1/2}}{(1-v)^{3/2}}=cX$$

$$\Rightarrow \frac{(Y+X)^{1/2}}{(X-Y)^{3/2}}=c \Rightarrow (x+y-2)^{1/2}=c(x-y)^{3/2}$$

$$\Rightarrow (x+y-2)=c(x-y)^3$$

ILLUSTRATION 9.18

Solve: $\frac{dy}{dx}=\frac{2x+3y+1}{4x+6y+1}$

Solution

Put

$$2x+3y=v \Rightarrow 2+3\frac{dy}{dx}=\frac{dv}{dx} \Rightarrow \frac{dy}{dx}=\frac{1}{3}\left\{\frac{dv}{dx}-2\right\}$$

Then,

$$\frac{1}{3}\left\{\frac{dv}{dx}-2\right\}=\frac{v-1}{2v+1} \Rightarrow \frac{dv}{dx}-2=\frac{3v+3}{2v+1}$$

$$\Rightarrow \frac{dv}{dx}=\frac{7v+5}{2v+1} \Rightarrow \frac{2v+1}{7v+5}dv=dx$$

$$\Rightarrow \frac{\frac{2}{7}(7v+5)-\frac{3}{7}}{7v+5}dv=dx \Rightarrow \frac{2}{7}dv-\frac{3}{7}\left(\frac{1}{7v+5}\right)dv=dx$$

Integrating,

$$\frac{2}{7}v-\frac{3}{49}\log(7v+5)=x+c$$

$$\Rightarrow 14v-3\log(7v+5)=49x+c$$

$$\Rightarrow 28x+42y-3\log(14x+21y+5)=49x+c$$

$$\Rightarrow -3\log(14x+21y+5)=21x-342y+c$$

ILLUSTRATION 9.19

Solution of the differentiation equation $(3y-7x+7)dx+(7y-3x+3)dy=0$ is ________.

(a) $c=(2x+y-1)^5(y-3x+1)^2$

(b) $c=(x+y+1)^5(y-x+1)^2$

(c) $c=(x+2y-1)^5(2y-x+1)^2$

(d) None of these

Solution

The given differential equation is

$$\frac{dy}{dx}=\frac{7x-3y-7}{-3x+7y+3} \quad ...(i)$$

Substituting $x=X+h, y=Y+k$, we obtain

$$\frac{dY}{dX}=\frac{(7X-3Y)+(7h-3k-7)}{(-3X+7Y)+(-3h+7k+3)}$$

Choosing h and k such that $7h-3k-7=0$ and $-3h+7k+3=0$, we get

$$h=1, k=0$$

Under the above transformations, equation (i) can be written as

$$\frac{dY}{dX}=\frac{7X-3Y}{-3X+7Y}$$

Putting $Y=VX$ so that $\frac{dY}{dX}=V+X\frac{dY}{dX}$, we get

$$V+X\frac{dV}{dX}=\frac{-3V+7}{7V-3}\Rightarrow X\frac{dV}{dX}=\frac{7-7V^2}{7V-3}$$

$$\Rightarrow \quad -7\frac{dX}{X}=\frac{7}{2}\cdot\left(\frac{2V}{V^2-1}\right)dV-\left(\frac{3}{V^2-1}\right)dV$$

Integrating, we get

$$-7\log X=\frac{7}{2}\log\left(V^2-1\right)-\frac{3}{2}\log\frac{V-1}{V+1}\log C$$

$$\Rightarrow \quad \log C=\log X^7+\log\left[(V+1)^5(V-1)^2\right]$$

$$\Rightarrow C=(V+1)^5(V-1)^2X^7 \text{ or } C=(Y+X)^5(Y-X)^2$$

$\Rightarrow$ which is the required solution.

Linear Differential Equation

A differential equation of the form $\frac{dy}{dx}+Py=Q$ where P and Q are either constants or functions of x is called Linear Differential Equation (LDE).

Working Method

(i) Write the equation in the above form

(ii) Find integrating factor (I.F.) $=e^{\int P dx}$

(iii) Solution: $ye^{\int Pdx}=\int Qe^{\int Pdx}dx+c$

Note: If DE is represented as $\frac{dx}{dy}=P_1x=Q_1$ where P_1 and Q_1 are constants or functions of y, the solution is given by: $xe^{\int P_1 dy}=\int Q_1e^{\int P_1 dy}dy+c$

ILLUSTRATION 9.20

Solve: $\frac{dy}{dx}+(\cot x)y=\sin x$

Solution

It is an LDE. Here, $P=\cot x$ and $Q=\sin x$. Then

$$\text{I.F.}=e^{\int\cot x dx}e^{\log\sin x}=\sin x$$

The solution is given by:

$$y\sin x=\int\sin^2 x dx+c\Rightarrow y\sin x=\frac{1}{2}\left(x-\frac{1}{2}\sin 2x\right)+c$$

ILLUSTRATION 9.21

Solve: $\left(1+y^2\right)dx=\left(\tan^{-1}y-x\right)dy$

Solution

Given

$$\frac{dy}{dx}=\frac{1+y^2}{\tan^{-1}y-x}\Rightarrow\frac{dx}{dy}=\frac{\tan^{-1}y}{1+y^2}-\frac{x}{1+y^2}$$

$$\Rightarrow \quad \frac{dx}{dy}+\frac{1}{1+y^2}x=\frac{\tan^{-1}y}{1+y^2}$$

Now, I.F. $=e^{\int\frac{1}{1+y^2}dy}=e^{\tan^{-1}}$

Therefore, the solution is given by

$$xe^{\tan^{-1}y}=\int e^{\tan^{-1}y}\frac{\tan^{-1}y}{1+y^2}dy+c=\int te^t dt+c\left(t=\tan^{-1}y\right)$$

$$=te^t-e^t+c=\left(\tan^{-1}y\right)e^{\tan^{-1}y}-e^{\tan^{-1}y}+c$$

$$\Rightarrow \quad xe^{\tan^{-1}y}=e^{\tan^{-1}y}\left(\tan^{-1}y\right)+c$$

ILLUSTRATION 9.22

If $y+\frac{d}{dx}(xy)=x(\sin x+\log x)$, find $y(x)$.

Solution

We have,

$$y+y+x\frac{dy}{dx}=x(\sin x+\log x)$$

$$\Rightarrow \quad \frac{dy}{dx}+\frac{2}{x}y=(\sin x+\log x)$$

(It is linear differential equation of the form $\frac{dy}{dx}+Py=Q$)

Integrating factor I.F. $=e^{\int\frac{2}{x}dx}=x^2$

Then, $yx^2=\int x^2(\sin x+\log x)dx+c$

$$=-x^2\cos x+2x\sin x+2\cos x+\frac{x^3}{3}\log x-\frac{1}{3}\frac{x^3}{3}+c$$

$$\Rightarrow yx^2=-x^2\cos x+2x\sin x+2\cos x+\frac{x^3}{3}\log x-\frac{1}{9}x^3+c$$

ILLUSTRATION 9.23

The solution of $\left(1-x^2\right)\frac{dy}{dx}+2xy=x\sqrt{1-x^2}$ is ________.

(a) $y\frac{1}{\left(1-x^2\right)}=\frac{1}{\sqrt{1-x^2}}+c$

(b) $x\frac{1}{\left(1-x^2\right)}=\frac{1}{\sqrt{1-x^2}}+c$

(c) $xy\frac{1}{\left(1-x^2\right)}=\frac{1}{\sqrt{1-x^2}}+c$

(d) None of these.

Solution

$$\frac{dy}{dx}+\frac{2x}{\left(1-x^2\right)}y=\frac{x\sqrt{1-x^2}}{\left(1-x^2\right)}$$

$$\int Pdx=\int\frac{2x}{1-x^2}dx=-\ln\left(1-x^2\right)$$

$$\text{I.F.}=e^{\int Pdx}=e^{-\ln\left(1-x^2\right)}=\left(\frac{1}{1-x^2}\right)=\frac{-1}{2}\int\frac{-2x}{\left(1-x^2\right)^{\frac{3}{2}}}dx+c$$

Put $1-x^2=t$

$$\Rightarrow\quad -2x\,dx=dt=\frac{-1}{2}\int t^{-3/2}dt+c=t^{-1/2}+c$$

$$y\frac{1}{\left(1-x^2\right)}=\frac{1}{\sqrt{1-x^2}}+c$$

ILLUSTRATION 9.24

Solve: $\frac{dz}{dx}+\frac{z}{x}\log z=\frac{z}{x^2}(\log z)^2$ ...(i)

Solution

$$\frac{1}{z(\log z)^2}\frac{dz}{dx}+\frac{1}{x\log z}=\frac{1}{x^2}$$

Let $\frac{1}{\log z}=y\Rightarrow-\frac{1}{z(\log z)^2}\frac{dz}{dx}=\frac{dy}{dx}$

Then (i) is transformed to:

$$-\frac{dy}{dx}+\frac{1}{x}y=\frac{1}{x^2}\Rightarrow\frac{dy}{dx}-\frac{1}{x}y=-\frac{1}{x^2}$$

$$\text{I.F.}=e^{-\int\frac{1}{x}dx}=\frac{1}{x}$$

Solution is given by

$$\frac{y}{x}=-\int\frac{1}{x^3}dx+c\Rightarrow\frac{y}{x}=\frac{1}{2x^2}+c$$

$$\Rightarrow\quad\frac{y}{x}=\frac{1}{2x^2}+c\Rightarrow\frac{1}{x\log z}=\frac{1}{2}+cx^2$$

ILLUSTRATION 9.25

The solution of differential equation $\frac{dy}{dx}-y\tan x=-y^2\sec x$ is ________.

(a) $y^{-1}\sec x=\cot x+c$

(b) $y^{-1}\cos x=\tan x+c$

(c) $y^{-1}\sec x=\tan x+c$

(d) None of these.

Solution

$$\frac{1}{y^2}\frac{dy}{dx}-\frac{1}{y}\tan x=-\sec x$$

$$\frac{1}{y}=v;\frac{-1}{y^2}\frac{dy}{dx}=\frac{dv}{dx}\Rightarrow\frac{-dv}{dx}-v\tan x=-\sec x$$

$$\frac{dv}{dx}+v\tan x=\sec x.\text{ Here }P=\tan x,Q=\sec x$$

$$\text{I.F.}=e^{\int\tan xdx}=\sec x,v\sec x=\int\sec^2xdx+$$

Hence the solution is $y^{-1}\sec x=\tan x+c$

Differential Equation Reducible to Linear Differential Equation

If the equation is of the form $f'(y)\frac{dy}{dx}+Pf(y)=Q$ where P and Q are constants, or function of x then put $f(y)=v,f'(y)\frac{dy}{dx}=\frac{dv}{dx}$ and equation reduces to $\frac{dv}{dx}+Pv=Q$ which is an LDE in v and x.

Bernoulli's Equations

An equation of the form $\frac{dy}{dx}+Py=Qy^n$ where P and Q are functions of x or constants except 0 and 1. Divide it by y^n we get $y^{-n}\frac{dy}{dx}+Py^{1-n}=Q$, now put $y^{1-n}=v$, it reduces to LDE in v and x.

ILLUSTRATION 9.26

Solve: $y\left(2x^2y+e^x\right)dx-\left(e^x+y^3\right)dy=0$

Solution

$$2y^2x^2dx-y^3dy+ye^xdx-e^xdy=0$$

Dividing by y^2, we get

$$2x^2dx-ydy+\frac{ye^xdx-e^xdy}{y^2}=0$$

$$\Rightarrow \quad 2x^2dx-ydy+d\left(\frac{e^x}{y}\right)=0$$

Integrating, $\frac{2x^3}{3}-\frac{y^2}{2}+\frac{e^x}{y}=c$

$$\Rightarrow \quad 6e^x+4x^3y-3y^3=6cy$$

Solution of Differential Equation of the Type $\frac{d^2y}{dx^2}=f(x)$

These equations are solved by integrating them twice

On integrating $\frac{d^2y}{dx^2}=f(x)$, we get

$$\frac{dy}{dx}=\int f(x)dx+C_1$$

$$\Psi(x)dx+C_1,\text{ where }=\Psi(x)=\int f(x)dx$$

$\therefore\ y=\int\Psi dx+C_1x+C_2$ (Required solution)

ILLUSTRATION 9.27

Solve the differential equation $\frac{d^2y}{dx^2}=\log(x)$

Solution

Upon integration, we get

$$\frac{dy}{dx}=\int\log x dx+C_1=\log x.x-\int\frac{1}{x}.xdx+C_1$$
$$=x\log x-x+C_1$$

On integrating once more, we get

$$y=\int x\log xdx-\int xdx+C_1x+C_2$$

$$=\log x.\frac{x^2}{2}-\int\frac{1}{x}.\frac{x^2}{2}dx-\frac{x^2}{2}C_1x+C_2$$

$$=\frac{x^2}{2}(\log x-1)-\frac{x^2}{4}+C_1x+C_2$$

$$=\frac{x^2}{2}(\log x-1)-\frac{x^2}{4}+C_1x+C_2$$

$$y=\frac{x^2}{4}[2\log x-3]+C_1x+C_2$$

ILLUSTRATION 9.28

Solve: $\frac{d^2y}{dx^2}=x+\sin x$, given that $y=0, \frac{dy}{dx}=-1$ when $x=0$

Solution

On integrating both sides, we get

$$\frac{dy}{dx}=\frac{x^2}{2}-\cos x+C_1$$

Put $x=0$ and $\frac{dy}{dx}=-1$

$$-1=0-1+C_1$$

$$\therefore \quad C_1=0$$

$$\therefore \quad \frac{dy}{dx}=\frac{x^2}{2}-\cos x$$

Again on integration, we get $y=\frac{x^3}{6}-\sin x+C_2$

Put $y=0$ at $x=0$

$$0=0-0+C_2$$

$$\therefore \quad C_2=0$$

Hence the solution is $y=\frac{x^3}{6}-\sin x$.

ILLUSTRATION 9.29

The solution of the differential equation $x\dfrac{d^2y}{dx^2}=1$, given that $y=1, \dfrac{dy}{dx}=0$, when $x=1$, is:

(a) $y=x\log x+x+2$ **(b)** $y=x\log x-x+2$

(c) $y=x\log x+x$ **(d)** $y=x\log x-x$

Solution

$$x\frac{d^2y}{dx^2}=1\Rightarrow\frac{d^2y}{dx^2}=\frac{1}{x}\Rightarrow\frac{dy}{dx}=\log x+c_1$$

$\Rightarrow$ $y=x\log x-x+c_1x+c_2$

(on integrating twice)

Given $y=1$ and $\dfrac{dy}{dx}=0$ at $x=1\Rightarrow c_1=0$ and $c_2=2$

Therefore, the required solution is $y=x\log x-x+2$.

ILLUSTRATION 9.30

The curve passing through the point (0, 1) satisfying the equation $\sin\left(\dfrac{dy}{dx}\right)=a$ is:

(a) $\cos\left(\dfrac{y+1}{x}\right)=a$ **(b)** $\cos\left(\dfrac{x}{y+1}\right)=a$

(c) $\sin\left(\dfrac{y-1}{x}\right)=a$ **(d)** $\sin\left(\dfrac{x}{y-1}\right)=a$

Solution

$$\sin\left(\frac{dy}{dx}\right)=a\Rightarrow\frac{dy}{dx}=\sin^{-1}a\Rightarrow dy=\left(\sin^{-1}a\right)dx$$

On integration, we get $y=x\sin^{-1}a+c$

But it passes through (0, 1), so $1=0+c\Rightarrow c=1$

Hence $y=x\sin^{-1}a+1\Rightarrow\dfrac{y-1}{x}=\sin^{-1}a$

$$\Rightarrow \quad \sin\left(\frac{y-1}{x}\right)=a$$

ILLUSTRATION 9.31

The equation of a curve through point (1, 0) which satisfies the differential equation $\left(1+y^2\right)dx-xydy=0$ is:

(a) $x^2+y^2=1$ **(b)** $x^2-y^2=1$

(c) $2x^2+y^2=2$ **(d)** None of these

Solution

We have $\dfrac{dx}{x}=\dfrac{ydy}{1+y^2}$.

On integrating, we get

$$\log|x|=\frac{1}{2}\log\left(1+y^2\right)+\log c \text{ or } |x|=c\sqrt{\left(1+y^2\right)}$$

But it passes through (1, 0), so we get $c=1$

$\therefore$ Solution is $x^2-y^2=1$.

WORKSHEET

1. Find the differential equation of the family of curves $y=\text{A}e^{2x}+\text{B}.e^{-2x}$.
2. Find the general solution of the differential equation $\dfrac{dy}{dx}=\dfrac{y}{x}$.
3. Solve the differential equation $\dfrac{dy}{dx}+\dfrac{y}{x}=x^2$.
4. Find the differential equation of the family of lines passing through the origin.
5. Find the differential equation of all non-horizontal lines in a plane.
6. The general solution of the differential equation $\dfrac{dy}{dx}=x^5+x^2-\dfrac{2}{x}$ is:

   (a) $y=\dfrac{x^6}{6}+\dfrac{x^3}{3}-2\log|x|+c$

   (b) $y=\dfrac{x^6}{6}-\dfrac{x^3}{3}-2\log|x|+c$

   (c) $y=\dfrac{x^6}{6}-\dfrac{x^3}{3}+2\log|x|+c$

   (d) None of these.

7. The solution of the differential equation $\frac{dy}{dx}+\frac{1+\cos 2y}{1-\cos 2x}=0$ is:

(a) $\tan y + \cot x = c$ (b) $\tan y \cot x = c$

(c) $\tan y - \cot x = c$ (d) None of these.

8. The solution $\frac{dy}{dx}=1+x+y+xy$ is:

(a) $\log(1-y)=x+\frac{x^3}{2}+c$

(b) $\log(1+y)=x+\frac{x^3}{2}+c$

(c) $\log(1+y)=x-\frac{x^2}{2}-c$

(d) None of these.

9. The solution of differential equation $dy/dx+y=e^x$ is ______.

(a) $y=e^x+ce^{-x}$

(b) $y=e^{-x}+ce^x$

(c) $y=(1/2)e^x+ce^{-x}$

(d) $y=(1/2)e^{-x}+ce^x$

10. The solution of differential equation $\left(2x-10y^3\right)\frac{dy}{dx}+y=0$ is:

(a) $xy^2=y^5+c$ (b) $xy^2+2y^5=c$

(c) $xy^2=2y^5+c$ (d) None of these.

SOLVED EXAMPLES

Example 1. Given that $\frac{dy}{dx}=ye^x$ and $x=0$, $y=e$. Find the value of y when $x=1$.

Solution

$$\frac{dy}{dx}=ye^x \Rightarrow \int\frac{dy}{y}=\int e^x dx \Rightarrow \log y=e^x+c$$

Substituting $x=0$ and $y=e$, we get

$$\log e=e^0+c,\text{ i.e., } c=0 \quad (\because \log e=1)$$

Therefore, $\log y=e^x$.

Now, substituting $x=1$ in the above, we get $\log y=e$

$$\Rightarrow \quad y=e^e.$$

Example 2. Find the equation of a curve whose tangent at any point on it, different from origin, has slope $y+\frac{y}{x}$.

Solution

Given $\frac{dy}{dx}=y+\frac{y}{x}=y\left(1+\frac{1}{x}\right)$

$$\Rightarrow \quad \frac{dy}{y}=\left(1+\frac{1}{x}\right)dx$$

Integrating both sides, we get

$$\log y=x+\log x+c \Rightarrow \log\left(\frac{y}{x}\right)=x+c$$

$$\Rightarrow \quad \frac{y}{x}=e^{x+c}=e^x\cdot e^c \Rightarrow \frac{y}{x}=k\cdot e^x$$

$$\Rightarrow \quad y=kx\cdot e^x$$

Example 3. Find the equation of a curve passing through the point (1, 1) if the perpendicular distance of the origin from the normal at any point P (x, y) of the curve is equal to the distance of P from the x-axis.

Solution

Let the equation of normal at P (x, y) be

$\mathrm{Y}-y=\frac{-dx}{dy}(\mathrm{X}-x)$, i.e.,

$$\mathrm{Y}+\mathrm{X}\frac{dx}{dy}-\left(y+x\frac{dx}{dy}\right)=0 \quad \ldots\text{(i)}$$

Therefore, the length of perpendicular from origin to (i) is

$$\frac{y+x\frac{dx}{dy}}{\sqrt{1+\left(\frac{dx}{dy}\right)^2}} \quad \ldots\text{(ii)}$$

Also, the distance between P and x-axis is $|y|$. Thus, we get

$$\frac{y+x\dfrac{dx}{dy}}{\sqrt{1+\left(\dfrac{dx}{dy}\right)^2}}=|y|$$

$$\Rightarrow \quad \left(y+x\frac{dx}{dy}\right)^2=y^2\left[1+\left(\frac{dx}{dy}\right)^2\right]$$

$$\Rightarrow \quad \frac{dx}{dy}\left[\frac{dx}{dy}(x^2-y^2)+2xy\right]=0 \Rightarrow \frac{dx}{dy}=0$$

or
$$\frac{dx}{dy}=\frac{2xy}{y^2-x^2}$$

Case I: $\dfrac{dx}{dy}=0 \Rightarrow dx=0$

Integrating both sides, we get $x = k$, Substituting $x = 1$, we get $k = 1$.

Therefore, $x = 1$ is the equation of the curve (not possible, so rejected).

Case II: $\dfrac{dx}{dy}=\dfrac{2xy}{y^2-x^2} \Rightarrow \dfrac{dy}{dx}=\dfrac{y^2-x^2}{2xy}$.

Substituting $y = vx$, we get

$$v+x\frac{dv}{dx}=\frac{v^2x^2-x^2}{2vx^2} \Rightarrow x\cdot\frac{dv}{dx}=\frac{v^2-1}{2v}-v$$

$$=\frac{-(1+v^2)}{2v} \Rightarrow \frac{2v}{1+v^2}dv=\frac{-dx}{x}$$

Integrating both sides, we get

$$\log(1+v^2)=-\log x+\log c$$

$$\Rightarrow \quad \log(1+v^2)(x)=\log c \Rightarrow (1+v^2)x=c$$

$\Rightarrow x^2+y^2=cx$. Substituting $x = 1$, $y = 1$, we get $c = 2$.

Therefore, $x^2+y^2-2x=0$ is the required equation.

Example 4. Find the equation of a curve passing through $\left(1, \dfrac{\pi}{4}\right)$ if the slope of the tangent to the curve at any point P (x, y) is $\dfrac{y}{x}-\cos^2\dfrac{y}{x}$.

Solution

According to the given condition,

$$\frac{dy}{dx}=\frac{y}{x}-\cos^2\frac{y}{x} \qquad \ldots\text{(i)}$$

This is a homogeneous differential equation. Substituting $y = vx$, we get

$$v+x\frac{dv}{dx}=v-\cos^2 v \Rightarrow x\frac{dv}{dx}=-\cos^2 v$$

$$\Rightarrow \quad \sec^2 v\,dv=-\frac{dx}{x} \Rightarrow \tan v=-\log x+c$$

$$\Rightarrow \quad \tan\frac{y}{x}+\log x=c$$

Substituting $x = 1$, $y=\dfrac{\pi}{4}$, we get, $c = 1$. Thus, we get

$\tan\left(\dfrac{y}{x}\right)+\log x=1$, which is the required equation.

Example 5. Solve $x^2\dfrac{dy}{dx}-xy=1+\cos\left(\dfrac{y}{x}\right)$, $x\neq 0$ and $x=1, y=\dfrac{\pi}{2}$

Solution

The given equation can be written as

$$x^2\frac{dy}{dx}-xy=2\cos^2\left(\frac{y}{2x}\right),\ x\neq 0$$

$$\Rightarrow \quad \frac{x^2\dfrac{dy}{dx}-xy}{2\cos^2\left(\dfrac{y}{2x}\right)}=1 \Rightarrow \frac{\sec^2\left(\dfrac{y}{2x}\right)}{2}\left[x^2\frac{dy}{dx}-xy\right]=1$$

Dividing both sides by x^3, we get

$$\frac{\sec^2\left(\dfrac{y}{2x}\right)}{2}\left[\frac{x\dfrac{dy}{dx}-y}{x^2}\right]=\frac{1}{x^3} \Rightarrow \frac{d}{dx}\left[\tan\left(\frac{y}{2x}\right)\right]=\frac{1}{x^3}$$

Integrating both sides, we get

$$\tan\left(\frac{y}{2x}\right)=\frac{-1}{2x^2}+k$$

Substituting $x = 1$, $y=\dfrac{\pi}{2}$, we get

$k=\dfrac{3}{2}$, therefore, $\tan\left(\dfrac{y}{2x}\right)=-\dfrac{1}{2x^2}+\dfrac{3}{2}$ is the required solution.

Example 6. State the type of the differential equation for the equation $xdy-ydx=\sqrt{x^2+y^2}\,dx$ and solve it.

Solution

The given equation can be written as

$xdy=(\sqrt{x^2+y^2}+y)dx$, i.e.,

$$\frac{dy}{dx}=\frac{\sqrt{x^2+y^2}+y}{x} \quad \ldots\text{(i)}$$

Clearly RHS of (1) is a homogeneous function of degree zero. Therefore, the given equation is a homogeneous differential equation. Substituting $y=vx$, we get from (1),

$$v+x\frac{dv}{dx}=\frac{\sqrt{x^2+v^2x^2}+vx}{x}\text{, i.e., } v+x\frac{dv}{dx}=\sqrt{1+v^2}+v$$

$$x\frac{dv}{dx}=\sqrt{1+v^2} \Rightarrow \frac{dv}{\sqrt{1+v^2}}=\frac{dx}{x} \quad \ldots\text{(ii)}$$

Integrating both sides of (2), we get

$$\log(v+\sqrt{1+v^2})=\log x+\log c \Rightarrow v+\sqrt{1+v^2}=cx$$

$$\Rightarrow \quad \frac{y}{x}+\sqrt{1+\frac{y^2}{x^2}}=cx \Rightarrow y+\sqrt{x^2+y^2}=cx^2$$

Example 7. The solution of differential equation $\left(x^2+xy\right)dy=\left(x^2+y^2\right)dx$ is:

(a) $\log|x|+2\log|x-y|-\frac{y}{x}+c=0$

(b) $\log|x|+2\log|x+y|-\frac{y}{x}+c=0$

(c) $\log|x|-2\log|x-y|-\frac{y}{x}+c=0$

(d) $\log|x|-2\log|x+y|-\frac{y}{x}+c=0$

Solution

The given equation is $\frac{dy}{dx}=\frac{x^2+y^2}{x^2+xy}$ ...(i)

This is homogenous in x and y. Put $y=vx$ ...(ii)

So that $\frac{dy}{dx}=v+x\frac{dv}{dx}$ ...(iii)

From (i), (ii) and (iii)

$$v+x\frac{dv}{dx}=\frac{x^2+v^2x^2}{x^2+vx^2}=\frac{1+v^2}{1+v}$$

$$\Rightarrow \quad x\frac{dv}{dx}=\frac{1+v^2}{1+v}-v=\frac{1+v^2-v-v^2}{1+v}=\frac{1-v}{1+v}$$

$$\Rightarrow \quad \frac{1+v}{1-v}dv=\frac{dx}{x}$$

$$\text{Integrating,} \int\frac{dx}{x}=\int\frac{1+v}{1-v}dv=\int\frac{2-(1-v)}{1-v}dv$$

$$=2\int\frac{1}{1-v}dv-\int 1.dv \Rightarrow \log|x|=-2\log|1-v|-v+c$$

$$\Rightarrow \quad \log|x|=-2\log\left|1-\frac{y}{x}\right|-\frac{y}{x}+c$$

$$\Rightarrow \quad \log|x|=-2\log\left|\frac{x-y}{x}\right|-\frac{y}{x}+c$$

$$\Rightarrow \quad \log|x|=-2\log|x-y|+2\log|x|-\frac{y}{x}+c$$

$$\Rightarrow \log|x|-2\log|x-y|-\frac{y}{x}+c=0,$$

which is required solution.

Example 8. The solution of differential equation $\frac{dy}{dx}=\frac{x+2y+3}{2x+y-3}$ is:

(a) $x=1-\dfrac{\left(1+\frac{y-1}{x-1}\right)^{\frac{1}{2}}}{\left(1-\frac{y-1}{x-1}\right)^{\frac{3}{2}}}-c$

(b) $2-\dfrac{\left(1+\frac{y-1}{x-1}\right)^{\frac{1}{2}}}{\left(1-\frac{y-1}{x-1}\right)^{\frac{3}{2}}}+c$

(c) $x=1+\dfrac{\left(1+\frac{y-1}{x-1}\right)^{\frac{1}{2}}}{\left(1-\frac{y-1}{x-1}\right)^{\frac{3}{2}}}c$

(d) None of these.

Solution

Put $x=X+h, y=Y+k$

$$\frac{dy}{dx}=\frac{X+h+2(Y+k)-3}{2(X+h)+Y+K-3}=\frac{X+2Y+h+2k-3}{2X+y+2h+k-3}$$

Equating $h+2k-3=0$ and $2h+k-3=0$

Solving, we get $3k-3=0, k=1$ and $h=1$

$$\frac{dy}{dx}=\frac{X+2Y}{2X+Y}=\frac{1+2\left(\frac{Y}{X}\right)}{2+\left(\frac{Y}{X}\right)}$$

Put $\frac{Y}{X}=v$ or $Y=vX;\frac{dY}{dX}=v+X\frac{dV}{dX}$ or $X\frac{dv}{dX}+v=\frac{1+2v}{2+v}$

or $$X\frac{dv}{dX}=\frac{1+2v}{2+v}-v=\frac{1-v^2}{2+v}$$

$$\frac{dX}{X}=\frac{2+v}{1-v^2}dv=\frac{2+v}{(1-v)(1+v)}dv$$

$$=\left[\frac{1}{2}\left(\frac{1}{1+v}\right)+\frac{3}{2}\left(\frac{1}{1-v}\right)\right]dv$$

$$\ln X=\frac{1}{2}\ln(1+v)-\frac{3}{2}\ln(1-v)+\ln c$$

or $$\ln X=\ln(1+v)^{\frac{1}{2}}-\ln(1-v)^{\frac{3}{2}}+\ln c$$

$$\ln X=\ln\frac{(1+v)^{\frac{1}{2}}}{(1-v)^{\frac{3}{2}}}c=\ln\frac{\left(1+\frac{Y}{X}\right)^{\frac{1}{2}}}{\left(1-\frac{Y}{X}\right)^{\frac{3}{2}}}c\Rightarrow X=\frac{\left(1+\frac{Y}{X}\right)^{\frac{1}{2}}}{\left(1-\frac{Y}{X}\right)^{\frac{3}{2}}}C$$

Put $X=x-1$ and $Y=y-1\Rightarrow x=1+\frac{\left(1+\frac{y-1}{x-1}\right)^{1/2}}{\left(1-\frac{y-1}{x-1}\right)^{3/2}}c$

RECAPITULATION (FORMULAE, THEOREM, IMPORTANT RESULTS)

1. An equation involving derivatives of the dependent variable with respect to independent variable (variables) is known as a differential equation.
2. Order of a differential equation is the order of the highest order derivative occurring in the differential equation.
3. Degree of a differential equation is defined if it is a polynomial equation in its derivatives.
4. Degree (when defined) of a differential equation is the highest power (positive integer only) of the highest order derivative in it.
5. A function which satisfies the given differential equation is called its solution. The solution which contains as many arbitrary constants as the order of the differential equation is called a general solution and the solution free from arbitrary constants is called a particular solution.
6. To form a differential equation from a given function we differentiate the function successively as many times as the number of arbitrary constants in the given function and then eliminate the arbitrary constants.
7. Variable separable method is used to solve such an equation in which the variables can be separated completely, i.e., terms containing y should remain with dy and terms containing x should remain with dx.
8. A differential equation which can be expressed in the form $\frac{dy}{dx}=f(x,y)$ or $\frac{dy}{dx}=g(x,y)$ where, $f(x, y)$ and $g(x, y)$ are homogenous functions of degree zero is called a homogeneous differential equation.
9. A differential equation of the form $\frac{dy}{dx}+Py=Q$, where P and Q are constants or functions of x only is called a first-order linear differential equation.

QUIZ

1. Define a differential equation.
2. What is an ordinary differential equation? Give an example.
3. What is the order of a differential equation?
4. What is the order and degree of the equation $\left(\frac{d^3y}{dx^3}\right)^2+\left(\frac{d^2y}{dx^2}\right)^3+\left(\frac{dy}{dx}\right)^4+y^5=0$?
5. What is the number of arbitrary constants in the general solution of a differential equation of order 1?
6. What is the number of arbitrary constants in the particular solution of differential equation of order 2?
7. Verify that $y=\sqrt{1+x^2}$ is a solution of differential equation $y'=\frac{xy}{1+x^2}$.
8. What is the particular solution of differential equation $\frac{dy}{dx}=-4xy^2$ given for $x=0$, $y=1$?

JEE MAIN

Objective Questions

One Option Correct

1. The integrating factor of the differential equation $\frac{dy}{dx}+\frac{1}{x}y=3x$ is ______.
   (a) x (b) $\ln x$
   (c) 0 (d) ∞
2. The solution of the differential equation $\sec^2 x\tan y dx+\sec^2 y\tan x dy=0$ is:
   (a) $\tan y\tan x=c$ (b) $\frac{\tan y}{\tan x}=c$
   (c) $\frac{\tan^2 x}{\tan t}=c$ (d) None of these.
3. The differential equation of all straight lines passing through origin is:
   (a) $y=\sqrt{x}\frac{dy}{dx}$ (b) $\frac{dy}{dx}=y+x$
   (c) $\frac{dy}{dx}=y-x$ (d) None of these.
4. To reduce the differential equation $\frac{dy}{dx}+P(x)y=Q(x).y^n$ to the linear form, the substitution is ______.
   (a) $v=\frac{1}{y^n}$ (b) $v=\frac{1}{y^{n-1}}$
   (c) $v=y^n$ (d) $v=y^{n-1}$
5. Integrating factor of the differential equation $\frac{dy}{dx}+P(x)y=Q(x)$ is ______.
   (a) $\int P\,dx$ (b) $\int Q\,dx$
   (c) $e^{\int P\,dx}$ (d) $e^{\int Q\,dx}$
6. $\frac{d^3y}{dx^2}+2\left[1+\frac{d^2y}{dx^2}\right]=1$, has degree and order is ______.
   (a) 1, 3 (b) 2, 3
   (c) 3, 2 (d) 3, 1
7. Consider the differential equation
   $$\left[1+\left(\frac{dy}{dx}\right)^2\right]^{3/2}=k\left(\frac{d^2y}{dx^2}\right).$$
   Find the degree and order of the equation.
   (a) 2, 2 (b) 3, 2
   (c) 2, 3 (d) None of the above.
8. Find the integral factor of expansion
   $$\left(x^2+1\right)\frac{dy}{dx}+2xy=x^2-1.$$
   (a) x^2+1 (b) $\frac{2x}{x^2+1}$
   (c) $\frac{x^2-1}{x^2+1}$ (d) None of the above.

9. The differential equation of $y=\sec\left(\tan^{-1}x\right)$, is:

(a) $\left(1+x^2\right)\dfrac{dy}{dx}=y+x$ (b) $\left(1+x^2\right)\dfrac{dy}{dx}=y-x$

(c) $\left(1+x^2\right)\dfrac{dy}{dx}=xy$ (d) $\left(1+x^2\right)\dfrac{dy}{dx}=\dfrac{x}{y}$

10. The slope of a curve at any point is the reciprocal of twice the ordinate at the point and it passes through the point (4, 3). The equation of the curve is _______.

(a) $x^2=y+5$ (b) $y^2=x-5$

(c) $y^2=x+5$ (d) $x^2=y-5$

11. The general solution of the differential equation $\left(1+y^2\right)dx+\left(1+x^2\right)dy=0$ is:

(a) $x-y=c(1-xy)$ (b) $x-y=c(1+xy)$

(c) $x+y=c(1-xy)$ (d) $x+y=c(1+xy)$

12. The order of the differential equation of all tangents to the parabola $y=x^2$ is _______.

(a) 1 (b) 2
(c) 3 (d) 4

13. The general solution of the differential equation $\dfrac{dy}{dx}+\dfrac{1+\cos 2y}{1-\cos 2x}=0$, is given by:

(a) $\tan y+\cot x=c$ (b) $\tan y-\cot x=c$
(c) $\tan x-\cot y=c$ (d) $\tan x+\cot y=c$

14. The degree of the differential equation

$$\left(1+\left(\frac{dy}{dx}\right)^2\right)^{3/4}=\left(\frac{d^2y}{dx^2}\right)^{1/3}$$, is _______.

(a) $\dfrac{1}{3}$ (b) 4

(c) 9 (d) $\dfrac{3}{4}$

15. The solution of the differential equation $x\dfrac{dy}{dx}+y=x\cos x+\sin x$, given that $y=1$, when $x=\dfrac{\pi}{2}$, is _______.

(a) $y=\sin x-\cos x$ (b) $y=\cos x$

(c) $y=\sin x$ (d) $y=\sin x+\cos x$

16. The solution of $x\,dy-y\,dx+x^2e^2dx=0$ is:

(a) $\dfrac{y}{x}+e^x=c$ (b) $\dfrac{x}{y}+e^x=c$

(c) $x+e^y=c$ (d) $y+e^x=c$

17. The degree and order of the differential equation of all parabolas, whose axis is *x*-axis, are _______.

(a) 2, 1 (b) 1, 2
(c) 3, 2 (d) None of these.

18. The solution of differential equation $(\sin x+\cos x)dy+(\cos x-\sin x)dx=0$, is:

(a) $e^x(\sin x+\cos x)+c=0$

(b) $e^y(\sin x+\cos x)=c$

(c) $e^y(\cos x-\sin x)=c$

(d) $e^x(\sin x-\cos x)=c$

19. Equation of curve through point (1, 0) which satisfies the differential equation $\left(1+y^2\right)dx-xydy=0$ is:

(a) $x^2+y^2=4$ (b) $x^2-y^2=1$

(c) $2x^2+y^2=2$ (d) None of these.

20. Solution of differential equation $\dfrac{dy}{dx}+ay=e^{mx}$, is:

(a) $(a+m)y=e^{mx}+c$

(b) $y=e^{mx}+ce^{-ax}$

(c) $(a+m)=e^{mx}+c$

(d) $(a+m)y=e^{mx}+ce^{-ax}$

21. A solution of the differential equation $\left(\dfrac{dy}{dx}\right)^2-x\left(\dfrac{dy}{dx}\right)+y=0$ is _______.

(a) $y=2$ (b) $y=2x$
(c) $y=2x-4$ (d) $y=2x^2-4$

22. If $y=(x)$ and $\dfrac{2+\sin x}{y+1}\left(\dfrac{dy}{dx}\right)=-\cos x, y(0)=1$, then $y\left(\dfrac{\pi}{2}\right)$ equals _______.

(a) $\dfrac{1}{3}$ (b) $\dfrac{2}{3}$

(c) $-\dfrac{1}{3}$ (d) 1

23. The solution of primitive integral equation $\left(x^2+y^2\right)dy = dydx$ is $y = y(x)$. If $y(1) = 1$ and $Y(x_0) = e$, then x_0 is ______.

(a) $\sqrt{2\left(e^2+1\right)}$ (b) $\sqrt{2\left(e^2-1\right)}$

(c) $\sqrt{3e}$ (d) $\sqrt{\dfrac{e^2+1}{2}}$

24. The equation of the curve which passes through (1, 1) and whose slope is given by $\dfrac{2y}{x}$ is______.

(a) $y = x^2$ (b) $x^2 - y^2 = 0$

(c) $2x^2 + y^2 = 3$ (d) None of these.

25. The solution of the differential equation $\left(\dfrac{dy}{dx}\right)+2xy = y$ is______.

(a) $y = ce^{x-x^2}$ (b) $y = ce^{x^2-x}$

(c) $y = ce^{x}$ (d) $y = ce^{-x^2}$

26. The integrating factor of differential equation $\dfrac{dy}{dx}+y\sec x = \tan x$ must be:

(a) $\sec x+\tan x$ (b) $\log(\sec x+\tan x)$

(c) $e^{\sec x}$ (d) $\sec x$

27. The solution of the differential equation $\dfrac{dy}{dx}=\dfrac{x+y}{x}$ satisfying the condition $y(1)=1$ is ______.

(a) $y = xe^{(x-1)}$ (b) $y = x\log x + x$

(c) $y = \log x + x$ (d) $y = x\log x + x^2$

28. The solution of the differential equation $\left(1+x^2\right)\dfrac{dy}{dx}+1+y^2 = 0$ is:

(a) $\tan^{-1}x-\tan^{-1}y = \tan^{-1}C$

(b) $\tan^{-1}y-\tan^{-1}x = \tan^{-1}C$

(c) $-\tan^{-1}y+\tan^{-1}x = \tan^{-1}C$

(d) $\tan^{-1}y+\tan^{-1}x = \tan^{-1}C$

29. The differential equation $x\dfrac{dy}{dx}-y = x^3$, has the general solution:

(a) $y-x^3 = 2cx$ (b) $2y-x^3 = cx$

(c) $2y+x^2 = 2cx$ (d) $y+x^2 = 2cx$

30. The differential equation of all Simple Harmonic Motion of given period $2\pi/n$ is:

(a) $\dfrac{d^2x}{dt^2}+nx=0$ (b) $\dfrac{d^2x}{dt^2}+n^2x=0$

(c) $\dfrac{d^2x}{dt^2}-n^2x=0$ (d) $\dfrac{d^2x}{dt^2}+\dfrac{1}{n^2}x=0$

31. The solution of $\dfrac{dy}{dx}+\sqrt{\left[\dfrac{1-y^2}{1-x^2}\right]}=0$ is:

(a) $\tan^{-1}x + \cot^{-1}x = c$

(b) $\sin^{-1}x + \sin^{-1}y = c$

(c) $\sec^{-1}x + \operatorname{cosec}^{-1}x = c$

(d) None of these.

32. The differential equation of all circles which pass through the origin and whose centers lie on y-axis is:

(a) $\left(x^2-y^2\right)\dfrac{dy}{dx}-2xy=0$

(b) $\left(x^2-y^2\right)\dfrac{dy}{dx}+2xy=0$

(c) $\left(x^2-y^2\right)\dfrac{dy}{dx}-2xy=0$

(d) $\left(x^2-y^2\right)\dfrac{dy}{dx}+2xy=0$

33. Solution of the differential equation $\dfrac{dy}{dx}+\dfrac{y}{x}=x^2$ is:

(a) $y=\dfrac{x^2}{4}+cx^{-2}$ (b) $y=x^{-1}+cx^{-3}$

(c) $y=\dfrac{x^3}{4}+cx^{-1}$ (d) $xy=x^2+c$

34. The order of the differential equation whose general solution is given by $y=(C_1+C_2)\sin(x+C_3)-C_4e^{x+C_5}$ must be ______.

(a) 4 (b) 5

(c) 2 (d) 3

35. The differential equation of the family of the curve $x^2+y^2-2ax=0$ is:

(a) $x^2-y^2-2xyy''=0$ (b) $y^2-x^2=2xyy'$

(c) $x^2+y^2+2xyy''=0$ (d) None of these.

36. The equation of curve through point (1, 0) which satisfies the differential equation $(1+y^2)dx - xydy = 0$, is:

(a) $x^2 + y^2 = 4$ (b) $x^2 - y^2 = 1$
(c) $2x^2 + y^2 = 2$ (d) None of these.

37. The solution of the equation $\frac{dy}{dx} = \frac{x+y}{x-y}$ is:

(a) $c(x^2+y^2)^{1/2} + e^{\tan^{-1}(y/x)} = 0$
(b) $c(x^2+y^2)^{1/2} = e^{\tan^{-1}(y/x)}$
(c) $c(x^2-y^2)^{1/2} = e^{\tan^{-1}(y/x)}$
(d) None of these.

38. The order and degree of the differential equation $\left(1+3\frac{dy}{dx}\right)^{2/3} = 4\frac{d^3y}{dx^3}$ are:

(a) $\left(1,\frac{2}{3}\right)$ (b) (3,1)
(c) (3,3) (d) (1,2)

39. The solution of the differential equation $ydx + (x + x^2y)dy = 0$ is:

(a) $\log y = cx$ (b) $-\frac{1}{xy} + \log y = c$
(c) $\frac{1}{xy} + \log y = c$ (d) $-\frac{1}{xy} = c$

40. If $x\frac{dy}{dx} = y(\log y - \log x + 1)$, then the solution of the equation is:

(a) $y\log\left(\frac{x}{y}\right) = cx$ (b) $x\log\left(\frac{y}{x}\right) = cy$
(c) $\log\left(\frac{y}{x}\right) = cx$ (d) $\log\left(\frac{x}{y}\right) = cx$

41. If $x^2 + y^2 = 1$, then:

(a) $yy'' - 2(y')^2 + 1 = 0$ (b) $yy'' + (y')^2 + 1 = 0$
(c) $yy'' - 2(y')^2 - 1 = 0$ (d) $yy'' + 2(y')^2 + 1 = 0$

42. The solution of the equation $x\frac{dy}{dx} = y - x\tan\left(\frac{y}{x}\right)$ is:

(a) $x\sin\left(\frac{x}{y}\right) + c = 0$ (b) $x\sin y + c = 0$
(c) $x\sin\left(\frac{y}{x}\right) = c$ (d) None of these.

43. If $y^2 = ae^{-2x} + \frac{2}{5}(\cos x - 2\sin x)$, then $y\frac{dy}{dx} + y^2 + \sin x =$ ________.

(a) −1 (b) 1
(c) 0 (d) None of these.

44. Which of the following is the integrating factor of $x\log x\frac{dy}{dx} + y = 2\log x$?

(a) x (b) e^x
(c) $\log x$ (d) $\log(\log x)$

45. If p and q are the order and degree of differential equation $y\frac{dy}{dx} + x^3\left(\frac{d^2y}{dx^2}\right) + xy = \cos x$, then ________.

(a) $p < q$ (b) $p = q$
(c) $p > q$ (d) None of these.

46. The solution of the differential equation $\frac{dy}{dx} = \frac{x^2+xy+y^2}{x^2}$ is:

(a) $\tan^{-1}\frac{x}{y} = \log y + C$
(b) $\tan^{-1}\frac{y}{x} = \log x + C$
(c) $\tan^{-1}\frac{x}{y} = \log x + C$
(d) $\tan^{-1}\frac{x}{y} = \log x + C$

47. The solution of the differential equation $\frac{dy}{dx} - ky = 0, y(0) = 1$ approaches zero when $x \to \infty$, if ________.

(a) $k = 0$ (b) $k > 0$
(c) $k < 0$ (d) None of these.

48. The differential equation of all parabolas whose axes are parallel to y-axis is:

(a) $\frac{d^3y}{dx^3} = 0$ (b) $\frac{d^2y}{dx^2} = C$
(c) $\frac{d^3y}{dx^3} + \frac{d^2y}{dx^2} = 0$ (d) $\frac{d^2y}{dx^2} + 2\frac{dy}{dx} = C$

49. The order of the differential equation of all circles of radius r, having centre on y-axis and passing through the origin is ________.

(a) 1 (b) 2
(c) 3 (d) 4

50. The solution of $\cos(x+y)dy = dx$ is:

(a) $y = \tan\left(\frac{x+y}{2}\right)+c$ (b) $y+\cos^{-1}\left(\frac{y}{x}\right)=c$

(c) $y = x\sec\left(\frac{y}{x}\right)+c$ (d) None of these.

51. The solution of $\frac{dy}{dx}+2y\tan x = \sin x$, is ________.

(a) $y\cdot\sec^3 x=\sec^2 x+c$ (b) $y\cdot\sec^2 x=\sec x+c$

(c) $y\cdot\sin x=\tan x+c$ (d) None of these.

52. The differential equation of all circle in the first quadrant which touch the co-ordinate axis is of the order ______.

(a) 1 (b) 2
(c) 3 (d) None of these.

53. The solution of differential equation

$\left(1+y^2\right)+\left(x-e^{\tan^{-1}y}\right)\frac{dy}{dx}=0$ is:

(a) $2xe^{\tan^{-1}y}=e^{2\tan^{-1}y}+k$

(b) $xe^{2\tan^{-1}y}=e^{2\tan^{-1}y}+k$

(c) $xe^{\tan^{-1}y}=\tan^{-1}y+k$

(d) $(x-2)=ke^{-\tan^{-1}y}$

54. The solution of the differential equation $\frac{dy}{dx}+y\sec^2 x=\tan x\sec^2 x$ is:

(a) $y=\tan x-1+ce^{-\tan x}$

(b) $y^2=\tan x-1+ce^{\tan x}$

(c) $ye^{\tan x}=\tan x-1+c$

(d) $ye^{\tan x}=\tan x-1+c$

55. The solution of the differential equation $\frac{dy}{dx}-\frac{\tan y}{x}=\frac{\tan y\sin y}{x^2}$ is:

(a) $\frac{x}{\sin y}+\log x=c$ (b) $\frac{y}{\sin y}+\log x=c$

(c) $\log y+x=c$ (d) $\log x+y=c$

56. The solution of the differential equation $\frac{dy}{dx}+\frac{2x}{1+x^2}y=\frac{1}{\left(1+x^2\right)^2}$ is:

(a) $y(1-x^2)=\tan^{-1}x+c$
(b) $y(1+x^2)=\tan^{-1}x+c$
(c) $y(1+x^2)^2=\tan^{-1}x+c$
(d) $y(1-x^2)^2=\tan^{-1}x+c$

57. The degree and order of the differential equation of the family of all parabolas whose axis is x-axis are, respectively,

(a) 2, 3 (b) 2, 1
(c) 1, 2 (d) 1, 1

58. The differential equation representing the family of curves $y^2=2c(x+\sqrt{c})$, where $c>0$ is a parameter. The order and degree of the DE is:

(a) order 1, degree 2
(b) order 1, degree 1
(c) order 1, degree 3
(d) order 2, degree 2

59. The differential equation whose solution is $Ax^2+By^2=1$, where A and B are arbitrary constants is of

(a) first order and first degree.
(b) second order and first degree.
(c) second order and second degree.
(d) first order and second degree.

60. The differential equation of the family of circles with fixed radius 5 units and centre on the line $y=2$ is:

(a) $(y-2)^2(y')^2=25-(y-2)^2$
(b) $(x-2)^2(y')^2=25-(y-2)^2$
(c) $(x-2)(y')^2=25-(y-2)^2$
(d) $(y-2)(y')^2=25-(y-2)^2$

61. A differential equation representing the family of parabolas with axis parallel to y-axis and whose length of latus rectum is the distance of the point (2, –3) from the line $3x+4y=5$, is given by:

(a) $10\frac{d^2y}{dx^2}=11$ (b) $11\frac{d^2x}{dy^2}=10$

(c) $10\frac{d^2x}{dy^2}=11$ (d) $11\frac{d^2y}{dx^2}=10$

62. If $\frac{dy}{dx}=\frac{2^x y+2^y.2^x}{2^x+2^{x+y}\log_e 2}$, $y(0)=0$, then for $y=1$, the value of x lies in the interval:

(a) (1, 2) (b) $\left(\frac{1}{2},1\right]$

(c) (2, 3) (d) $\left(0,\frac{1}{2}\right]$

63. Let $y=y(x)$ satisfy the equation $\frac{dy}{dx}-|A|=0$, for all $x>0$, where

$$A=\begin{pmatrix} y & \sin x & 1 \\ 0 & -1 & 1 \\ 2 & 0 & \frac{1}{x} \end{pmatrix}.$$

If $y(\pi)=\pi+2$, then the value of $y\left(\frac{\pi}{2}\right)$ is:

(a) $\frac{\pi}{2}+\frac{4}{\pi}$ (b) $\frac{\pi}{2}-\frac{1}{\pi}$

(c) $\frac{3\pi}{2}-\frac{1}{\pi}$ (d) $\frac{\pi}{2}-\frac{4}{\pi}$

64. Let $y=y(x)$ be a solution of the differential equation $\cos x(3\sin x+\cos x+3)dy=(1+y\sin x(3\sin x+\cos x+3))dx$, $0\le x\le\frac{\pi}{2}$, $y(0)=0$. Then $y\left(\frac{\pi}{3}\right)$ is equal to

(a) $2\log_e\left(\frac{2\sqrt{3}+10}{11}\right)$ (b) $2\log_e\left(\frac{3\sqrt{3}+8}{4}\right)$

(c) $2\log_e\left(\frac{2\sqrt{3}+9}{6}\right)$ (d) $2\log_e\left(\frac{\sqrt{3}+7}{2}\right)$

65. The differential equation of the family of curves, $x^2=4b(y+b)$, $b\in R$, is:

(a) $x(y')^2=x+2yy'$ (b) $x(y')^2=2yy'-x$

(c) $xy'=y'$ (d) $x(y')^2=x-2yy'$

66. The general solution of the different equation $\sqrt{1+x^2+y^2+x^2y^2}+xy\frac{dy}{dx}=0$ is

(a) $\sqrt{1+y^2}+\sqrt{1+x^2}=\frac{1}{2}\log_e\left(\frac{\sqrt{1+x^2}+1}{\sqrt{1+x^2}-1}\right)+C$

(b) $\sqrt{1+y^2}-\sqrt{1+x^2}=\frac{1}{2}\log_e\left(\frac{\sqrt{1+x^2}+1}{\sqrt{1+x^2}-1}\right)+C$

(c) $\sqrt{1+y^2}+\sqrt{1+x^2}=\frac{1}{2}\log_e\left(\frac{\sqrt{1+x^2}-1}{\sqrt{1+x^2}+1}\right)+C$

(d) $\sqrt{1+y^2}-\sqrt{1+x^2}=\frac{1}{2}\log_e\left(\frac{\sqrt{1+x^2}-1}{\sqrt{1+x^2}+1}\right)+C$

(where C is a constant of integration)

67. If $y=\left(\frac{2}{\pi}x-1\right)\operatorname{cosec} x$ is the solution of the differential equation, $\frac{dy}{dx}+p(x)y=\frac{2}{\pi}\operatorname{cosec} x$, $0<x<\frac{\pi}{2}$, then the function $p(x)$ is equal to:

(a) $\cot x$ (b) $\operatorname{cosec} x$

(c) $\sec x$ (d) $\tan x$

68. If $y=y(x)$ is the solution of the differential equation $\frac{5+e^x}{2+y}.\frac{dy}{dx}+e^x=0$ satisfying $y(0)=1$, then a value of $y(\log_e 13)$ is:

(a) 1 (b) –1

(c) 0 (d) 2

69. The solution of the differential equation $\frac{dy}{dx}-\frac{y+3x}{\log_e(y+3x)}+3=0$ is:

(a) $x-\frac{1}{2}(\log_e(y+3x))^2=C$

(b) $x-\log_e(y+3x)=C$

(c) $y+3x-\frac{1}{2}(\log_e x)^2=C$

(d) $x-2\log_e(y+3x)=C$

(where C is a constant of integration)

70. Let $f:(0,\infty)\to(0,\infty)$ be a differential function such that $f(1)=e$ and

$$\lim_{t\to x}\frac{t^2 f^2(x)-x^2 f^2(t)}{t-x}=0.$$

If $f(x)=1$, then x is equal to:

(a) $\frac{1}{e}$ (b) $2e$

(c) $\frac{1}{2e}$ (d) e

71. The solution curve of the differential equation, $\left(1+e^{-x}\right)\left(1+y^2\right)\frac{dy}{dx}=y^2$, which passes through the point (0, 1), is

(a) $y^2+1=y\left(\log_e\left(\frac{1+e^{-x}}{2}\right)+2\right)$

(b) $y^2+1=y\left(\log_e\left(\frac{1+e^{x}}{2}\right)+2\right)$

(c) $y^2=1+y\log_e\left(\frac{1+e^{x}}{2}\right)$

(d) $y^2=1+y\log_e\left(\frac{1+e^{-x}}{2}\right)$

72. If $x^3dy+xy\,dx=x^2dy+2y\,dx$; $y(2)=e$ and $x>1$, then $y(4)$ is equal to:

(a) $\frac{3}{2}+\sqrt{e}$ (b) $\frac{3}{2}\sqrt{e}$

(c) $\frac{1}{2}+\sqrt{e}$ (d) $\frac{\sqrt{e}}{2}$

73. Let $y=y(x)$ be the solution of the differential equation,

$$\frac{2+\sin x}{y+1}\cdot\frac{dy}{dx}=-\cos x, y>0, y(0)=1.$$

If $y=(\pi)=\alpha$ and $\frac{dy}{dx}$ at $x=\pi$ is b, then the ordered pair (a, b) is equal to:

(a) $\left(2,\frac{3}{2}\right)$ (b) $(1,-1)$

(c) $(1,1)$ (d) $(2,1)$

74. If a curve $y=f(x)$, passing through the point (1, 2), is the solution of the differential equation, $2x^2dy=\left(2xy+y^2\right)dx$, then $f\left(\frac{1}{2}\right)$ is equal to:

(a) $\frac{1}{1+\log_e 2}$ (b) $\frac{1}{1-\log_e 2}$

(c) $1+\log_e 2$ (d) $\frac{-1}{1+\log_e 2}$

75. If $f'(x)=\tan^{-1}(\sec x+\tan x)$, $-\frac{\pi}{2}<x<\frac{\pi}{2}$, and $f(0)=0$, then $f(1)$ is equal to:

(a) $\frac{\pi+1}{4}$ (b) $\frac{1}{4}$

(c) $\frac{\pi-1}{4}$ (d) $\frac{\pi+2}{4}$

76. If $\frac{dy}{dx}=\frac{xy}{x^2+y^2}$; $y(1)=1$, then a value of x satisfying $y(x)=e$ is:

(a) $\frac{1}{2}\sqrt{3}e$ (b) $\frac{e}{\sqrt{2}}$

(c) $\sqrt{2}e$ (d) $\sqrt{3}e$

77. Let $f(x)=\left(\sin\left(\tan^{-1}x\right)+\sin\left(\cot^{-1}x\right)\right)^2-1, |x|>1.$ If $\frac{dy}{dx}=\frac{1}{2}\frac{d}{dx}\left(\sin^{-1}(f(x))\right)$ and $y\left(\sqrt{3}\right)=\frac{\pi}{6}$, then $y\left(-\sqrt{3}\right)$ is equal to:

(a) $\frac{2\pi}{3}$ (b) $-\frac{\pi}{6}$

(c) $\frac{5\pi}{6}$ (d) $\frac{\pi}{3}$

78. Let $y=y(x)$ be a solution of the differential equation, $\sqrt{1-x^2}\frac{dy}{dx}+\sqrt{1-y^2}=0$, $|x|<1$

If $y\left(\frac{1}{2}\right)=\frac{\sqrt{3}}{2}$; then $y\left(\frac{-1}{\sqrt{2}}\right)$ is equal to:

(a) $\frac{\sqrt{3}}{2}$ (b) $-\frac{1}{\sqrt{2}}$

(c) $\frac{1}{\sqrt{2}}$ (d) $-\frac{\sqrt{3}}{2}$

79. If $y = y(x)$ is the solution of the differential equation, $e^y = e^x$ such that $y(0) = 0$, then $y(1)$ is equal to:

(a) $1 + \log_e 2$ (b) $2 + \log_e 2$

(c) $2e$ (d) $\log_e 2$

80. Let $y = y(x)$ be the solution of the differential equation $\cos x \frac{dy}{dx} + 2y \sin 2x$, $x \in \left(0, \frac{\pi}{2}\right)$. If $y(\pi/3) = 0$, then $y(\pi/4)$ is equal to:

(a) $2 - \sqrt{2}$ (b) $2 + \sqrt{2}$

(c) $\sqrt{2} - 2$ (d) $\frac{1}{\sqrt{2}} - 1$

81. Let $y = y(x)$ be the solution of the differential equation, $xy' - y = x^2(x\cos x + \sin x)$, $x > 0$. If $y(\pi) = \pi$, then $y''\left(\frac{\pi}{2}\right) + y\left(\frac{\pi}{2}\right)$ is equal to:

(a) $2 + \frac{\pi}{2}$ (b) $1 + \frac{\pi}{2} + \frac{\pi^2}{4}$

(c) $2 + \frac{\pi}{2} + \frac{\pi^2}{4}$ (d) $1 + \frac{\pi}{2}$

82. Let $y = y(x)$ be the solution curve of the differential equation, $(y^2 - x)\frac{dy}{dx} = 1$, satisfying $y(0) = 1$. This curve intersects the x-axis at a point whose abscissa is:

(a) $2 - e$ (b) $-e$

(c) 2 (d) $2 + e$

83. The general solution of the differential equation $(y^2 - x^2)dx - xydy = 0$ $(x \neq 0)$ is:

(a) $y^2 - 2x^2 + cx^3 = 0$ (b) $y^2 + 2x^3 + cx^2 = 0$

(c) $y^2 + 2x^2 + cx^3 = 0$ (d) $y^2 - 2x^3 + cx^2 = 0$

(where c is a constant of integration)

84. If $\cos x \frac{dy}{dx} - y\sin x = 6x$, $\left(0 < x < \frac{\pi}{2}\right)$ and $y\left(\frac{\pi}{3}\right) = 0$, then $y\left(\frac{\pi}{6}\right)$ is equal to:

(a) $\frac{\pi^2}{2\sqrt{3}}$ (b) $-\frac{\pi^2}{2}$

(c) $-\frac{\pi^2}{2\sqrt{3}}$ (d) $-\frac{\pi^2}{4\sqrt{3}}$

85. The solution of the differential equation, $\frac{dy}{dx} = (x - y)^2$, when $y(1) = 1$ is:

(a) $\log_e \left|\frac{2 - x}{2 - y}\right| = x - y$

(b) $-\log_e \left|\frac{1 - x + y}{1 + x - y}\right| = 2(x - 1)$

(c) $-\log_e \left|\frac{1 + x - y}{1 - x + y}\right| = x + y - 2$

(d) $\log_e \left|\frac{2 - y}{2 - x}\right| = x + (y - 1)$

86. The curve amongst the family of curves represented by the differential equation, $(x^2 - y^2)dx + 2xydy = 0$, which passes through $(1, 1)$, is:

(a) A circle with centre on the x-axis.

(b) An ellipse with major axis along the y-axis.

(c) A circle with centre on the y-axis.

(d) A hyperbola with transverse axis along the x-axis.

87. Let $f : [0,1] \to R$ be such that $f(xy) = f(x).f(y)$, for all $x, y \in [0,1]$ and $f(0) \neq 0$. If $y = y(x)$ satisfies the differential equation, $\frac{dy}{dx} = f(x)$ with $y(0) = 1$, then $y\left(\frac{1}{4}\right) + y\left(\frac{3}{4}\right)$ is equal to:

(a) 3 (b) 4

(c) 2 (d) 5

88. Consider the differential equation,

$$y^2 dx + \left(x - \frac{1}{y}\right)dy = 0.$$

If the value of y is 1 when $x = 1$, then the value of x for which $y = 2$, is:

(a) $\frac{5}{2} + \frac{1}{\sqrt{e}}$ (b) $\frac{3}{2} - \frac{1}{\sqrt{e}}$

(c) $\frac{1}{2} + \frac{1}{\sqrt{e}}$ (d) $\frac{3}{2} - \sqrt{e}$

89. Let $y = y(x)$ be the solution of the differential equation, $\frac{dy}{dx} + y = y\tan x = 2x + x^2 \tan x$, $x \in \left(-\frac{\pi}{2}, \frac{\pi}{2}\right)$, such that $y(0) = 1$. Then:

(a) $y\left(\frac{\pi}{4}\right)+y\left(-\frac{\pi}{4}\right)=\frac{\pi^2}{4}+2$

(b) $y'\left(\frac{\pi}{4}\right)+y'\left(-\frac{\pi}{4}\right)=-\sqrt{2}$

(c) $y\left(\frac{\pi}{4}\right)-y\left(-\frac{\pi}{4}\right)=\sqrt{2}$

(d) $y'\left(\frac{\pi}{4}\right)-y'\left(-\frac{\pi}{4}\right)=\pi-\sqrt{2}$

90. The solution of the differential equation $x\frac{dy}{dx}+2y=x^2$ $(x\neq 0)$ with $y(1)=1$, is:

(a) $y=\frac{4}{5}x^3+\frac{1}{5x^2}$ (b) $y=\frac{x^3}{5}+\frac{1}{5x^2}$

(c) $y=\frac{x^2}{4}+\frac{3}{4x^2}$ (d) $y=\frac{3}{4}x^2+\frac{1}{4x^2}$

91. Let $y=y(x)$ be the solution of the differential equation, $(x^2+1)^2\frac{dy}{dx}+2x(x^2+1)y=1$ such that $y(0)=0$. If $\sqrt{a}\,y(1)=\frac{\pi}{32}$ then the value of a is:

(a) $\frac{1}{4}$ (b) $\frac{1}{2}$

(c) 1 (d) $\frac{1}{16}$

92. Let $y=y(x)$ be the solution of the differential equation $x\frac{dy}{dx}+y=x\ \log_e x$, $(x>1)$. If $2y(2)=\log_e 4-1$, then $y(e)$ is equal to:

(a) $-\frac{e}{2}$ (b) $-\frac{e^2}{2}$

(c) $\frac{e}{4}$ (d) $\frac{e^2}{4}$

Questions from Previous Years

93. If $y(x)$ is the solution of the differential equation $\frac{dy}{dx}+\left(\frac{2x+1}{x}\right)y=e^{-2x}$, $x>0$, where $y(1)=\frac{1}{2}e^{-2}$, then:

(a) $y(\log_e 2)=\log_e 4$

(b) $y(\log_e 2)=\frac{\log_e 2}{4}$

(c) $y(x)$ is decreasing in $\left(\frac{1}{2},1\right)$

(d) $y(x)$ is decreasing in (0, 1)

94. Let $y=y(x)$ be the solution of the differential equation, $x\frac{dy}{dx}+2y=x^2$ satisfying $y(1)=1$, then $y\left(\frac{1}{2}\right)$ is equal to:

(a) $\frac{7}{64}$ (b) $\frac{1}{4}$

(c) $\frac{49}{16}$ (d) $\frac{13}{16}$

95. The differential equation representing the family of ellipse having foci either on the x-axis or on the y-axis with centre at the origin and passing through the point (0, 3) is:

(a) $xyy'+y^2-9=0$

(b) $x+yy''=0$

(c) $xyy''+x(y')^2-yy'=0$

(d) $xyy'-y^2+9=0$

96. The curve satisfying the differential equation. $(x^2-y^2)dx+2xydy=0$ and passing through the point (1, 1) is:

(a) A circle of radius two

(b) A circle of radius one

(c) A hyperbola

(d) An ellipse.

97. Let $y=y(x)$ be the solution of the differential equation, $\sin x\frac{dy}{dx}+y\cos x=4x$, $x\in(0,\pi)$, then $y\left(\frac{\pi}{6}\right)$ is equal to:

(a) $\frac{-8}{9\sqrt{3}}\pi^2$ (b) $-\frac{8}{9}\pi^2$

(c) $-\frac{4}{9}\pi^2$ (d) $\frac{4}{9\sqrt{3}}\pi^2$

98. Let $y=y(x)$ be the solution of the differential equation, $\frac{dy}{dx}+2y=f(x)$, where $f(x)=\begin{cases}1, & x\in[0,1]\\0, & \text{otherwise}\end{cases}$. If $y(0)=0$, then $y\left(\frac{3}{2}\right)$ is ______.

(a) $\frac{e^2-1}{2e^3}$ (b) $\frac{e^2-1}{e^3}$

(c) $\frac{1}{2e}$ (d) $\frac{e^2+1}{2e^4}$

99. If $(2+\sin x)\frac{dy}{dx}+(y+1)\cos x=0$ and $y(0)=1$, then $y\left(\frac{\pi}{2}\right)$ is equal to:

(a) $\frac{4}{3}$ (b) $\frac{1}{3}$

(c) $-\frac{2}{3}$ (d) $-\frac{1}{3}$

100. The curve satisfying the differential equation, $ydx-(x+3y^2)dy=0$ and passing through the point (1, 1), also passes through the point:

(a) $\left(\frac{1}{4},-\frac{1}{2}\right)$ (b) $\left(-\frac{1}{3},\frac{1}{3}\right)$

(c) $\left(\frac{1}{3},-\frac{1}{3}\right)$ (d) $\left(\frac{1}{4},\frac{1}{2}\right)$

101. If $f(x)$ is a differentiable function in the interval $(0, \infty)$ such that $f(a)=1$ and $\lim\limits_{t\to x}\frac{t^2 f(x)-x^2 f(t)}{t-x}=1$, for each $x>0$ then $f\left(\frac{3}{2}\right)$ is equal to:

(a) $\frac{23}{18}$ (b) $\frac{13}{6}$

(c) $\frac{25}{9}$ (d) $\frac{31}{18}$

102. If the differential equation representing the family of all circles touching x-axis at the origin is $(x^2-y^2)\frac{dy}{dx}=g(x)y$, then $g(x)$ equals:

(a) $\frac{1}{2}x$ (b) $2x^2$

(c) $2x$ (d) $\frac{1}{2}x^2$

Integer and Numerical Value Type

1. A spherical rain drop evaporates at a rate proportional to its surface area at any instant t. The differential equation giving the rate of change of radius of the rain drop is ______.

2. If length of tangent at any point the curve $y = f(x)$ intercepted between the point and the x-axis is of length 1. Find the equation of the curve.

3. A right-circular cone with radius R and height H contains a liquid which evaporates at a rate proportional to its surface area in contact with air (proportionality constant = $k > 0$). Find the time after which the cone is empty.

4. A curve passing through the point (1, 1) has the property that the perpendicular distance of the origin from the normal at any point P of the curve is equal to the distance of P from the x-axis. Determine the equation of the curve.

5. A and B are two separated reservoirs of water. The capacity of reservoir A is double the capacity of reservoir B. Both the reservoirs are filled completely with water, their inlets are closed and then the water is released simultaneously from both the reservoirs. The rate of flow of water out of each reservoir at any instant of time is proportional to the quantity of water in the reservoir at the time.
One hour after the water is released, the quantity of water in reservoir A is $1\frac{1}{2}$ times the quantity of water in reservoir B. After how many hours do both the reservoirs have the same quantity of water?

Question from Previous Years

6. Let $y'(x)+y(x)g'(x)=g(x), g'(x), y(0)=0$, $x\in R$, where $f'(x)$ denotes $\frac{df(x)}{dx}$ and $g(x)$ is a given non-constant differential equation on R with $g(0)=g(2)=0$. Then the value of y (2) is ______.

7. The difference between degree and order of a differential equation that represents the family of curves given by $y^2=a\left(x+\frac{\sqrt{a}}{2}\right)$, $a>0$ is ______.

8. If $y^{1/4}+y^{-1/4}=2x$ and $(x^2-1)\frac{d^2y}{dx^2}+\alpha x\frac{dy}{dx}+\beta y=0$, then $|\alpha-\beta|$ is equal to ______.

9. If $y = y(x), y \in \left[0, \frac{\pi}{2}\right)$ is the solution of the differential equation

$\sec y \frac{dy}{dx} - \sin(x+y) - \sin(x-y) = 0$ with $y(0) = 0$,

then $5y'\left(\frac{\pi}{2}\right)$ is equal to ______.

10. Let a curve $y = f(x)$ pass through the point $\left(2, (\log_e 2)^2\right)$ and have slope $\frac{2y}{x \log_e x}$ for all positive real values of x. Then the value of $f(e)$ is equal to ______.

11. Let $y = y(x)$ be the solution of the differential equation

$$\left((x+2)e^{\left(\frac{y+1}{x+2}\right)} + (y+1)\right)dx = (x+2)dy,\ y(1) = 1.$$

If the domain of $y = y(x)$ is an open interval (α, β) then $|\alpha + \beta|$ is equal to ______.

12. Let a curve $y = y(x)$ be given by the solution of the differential equation

$$\cos\left(\frac{1}{2}\cos^{-1}\left(e^{-x}\right)\right)dx = \sqrt{e^{2x} - 1}\,dy.$$

If it intersects the y-axis at $y = -1$, and the intersection point of the curve with x-axis is $(\alpha, 0)$, then e^{α} is equal to ______.

13. Let $y = y(x)$ be the solution of the differential equation $x\,dy - y\,dx = \sqrt{(x^2 - y^2)}\,dx$, $x \ge 1$, with $y(1) = 0$. If the area bounded by the line $x = 1$, $x = e^{\pi}$, $y = 0$ and $y = y(x)$ is $\alpha e^{2\pi} + \beta$, then the value of $10(\alpha + \beta)$ is equal to ______.

14. If the curve, $y = y(x)$ represented by the solution of the differential equation $\left(2xy^2 - y\right)dx + x\,dy = 0$, passes through the intersection of the lines, $2x - 3y = 1$ and $3x + 2y = 8$, then $|y(1)|$ is equal to ______.

15. Let $y = y(x)$ be the solution of the differential equation, $e^{y}\frac{dy}{dx} - 2e^{y}\sin x + \sin x \cos^2 x = 0$, $y\left(\frac{\pi}{2}\right) = 0$. If $y(0)\log_e\left(\alpha + \beta e^{-2}\right)$, then $4(\alpha + \beta)$ is equal to ______.

16. If $y = y(x)$ is the solution of the equation, $e^{\sin} y \cos y \frac{dy}{dx} + e^{\sin y}\cos x = \cos x$, $y(0) = 0$, then $1 + y\left(\frac{\pi}{6}\right) + \frac{\sqrt{3}}{2}y\left(\frac{\pi}{3}\right) + \frac{1}{\sqrt{2}}y\left(\frac{\pi}{4}\right)$ is equal to ______.

Assertion and Reason

Select the correct answer:

(a) Statement 1 is true, Statement 2 is true; Statement 2 is the correct explanation for Statement 1.

(b) Statement 1 is true, Statement 2 is true; Statement 2 is NOT a correct explanation for Statement 1.

(c) Statement 1 is true, Statement 2 is false.

(d) Statement 1 is false, Statement 2 is true

1. **Statement 1:** Solution of $\left(1 + x\sqrt{x^2 + y^2}\right)dx + y\left(-1 + \sqrt{x^2 + y^2}\right)dy = 0$ is $x - \frac{y^2}{2} + \frac{1}{3}(x^2 + y^2)^{\frac{3}{2}} + c = 0.$

   Statement 2: Solution of $(1 + x)y\,dx + (1 - xy)x\,dy = 0$ is $\ell n\frac{x}{y} - \frac{1}{xy} = \text{c}$.

2. **Statement 1:** The equation of the curve passing through (3, 9) satisfies the differential equation $\frac{dy}{dx} = x + \frac{1}{x^2}$ is $6xy = 3x^3 + 29x - 6$

   Statement 2: The solution of the DE $\left(\frac{dy}{dx}\right)^2 - \frac{dy}{dx}\left(e^{x} + e^{-x}\right) + 1 = 0$ is $y = c_1 e^{x} + c_2 e^{-x}$.

3. **Statement 1:** The equation of the curve through the point (1, 0) which satisfies the differential equation $\left(1 + y^2\right)dx - xy\,dy = 0$ is $x^2 - y^2 = 1$.

   Statement 2: The DE $\frac{dy}{dx} = f(x).g(y)$ can be solved by separating variables $\frac{dy}{g(y)} = f(x)dx$.

4. **Statement 1:** The DE of all circles in a plane must be of order 3.

   Statement 2: There is only one circle passing through three non-collinear points.

5. **Statement 1:** The differential equation $y^3dy+\left(x+y^2\right)dx=0$ becomes homogeneous if we put $y^2=t$.

   Statement 2: All differential equations of first order, first degree become homogeneous if we put $y=tx$

6. **Statement 1:** Order of a differential equation represents number of arbitrary constants in the general solution.

   Statement 2: Degree of differential equation represents number of family of curves.

7. Let a solution $y=y(x)$ of the differential equation $x\sqrt{x^2-1}\,dy-y\sqrt{y^2-1}\,dx=0$ satisfy $y(2)=\frac{2}{\sqrt{3}}$.

   Statement 1: $y(x)=\sec\left(\sec^{-1}x-\frac{\pi}{6}\right)$

   Statement 2: $y(x)$ is given by $\frac{1}{y}=\frac{2\sqrt{3}}{x}-\sqrt{1-\frac{1}{x^2}}$

JEE ADVANCED

Objective Questions

One Option Correct

1. Solution of differential equation

$$x^2=1+\left(\frac{x}{y}\right)^{-1}\frac{dy}{dx}+\frac{\left(\frac{x}{y}\right)^{-2}\left(\frac{dy}{dx}\right)^2}{2!}+\frac{\left(\frac{x}{y}\right)^{-3}\left(\frac{dy}{dx}\right)^3}{3!}+\ldots,$$

is:

(a) $y^2=x^2\left(\ell n x^2-1\right)+c$

(b) $y=x^2\left(\ell n x-1\right)+c$

(c) $y^2=x\left(\ell n x-1\right)+c$

(d) $y=x^2e^{x^2}+c$

2. A curve which passes through (1, 2) and whose sub-normal at every point is 2, is:

(a) $2x^2=y$ (b) $y^2=2x+2$

(c) $y^2=x+3$ (d) $y^2=4x$

3. The orthogonal trajectory of system of curve $y=ax^2$ which does not pass through the origin, is _______.

(a) ellipse (b) parabola

(c) circle (d) hyperbola

4. If gradient of a curve at any point $P(x,y)$ is $\frac{x+y+1}{2y+2x+1}$ and it is passes through origin, then curve is:

(a) $2(x+3y)=\ell n\left|\frac{3x+3y+2}{2}\right|$

(b) $x+3y=\ell n\left|\frac{3x+3y+2}{2}\right|$

(c) $3y+x=\ell n(3x+2y+1)$

(d) $6y-3x=\ell n\left|\frac{3x+3y+2}{2}\right|$

5. The solution of the differential equation $y_1y_3=3y_2^{\,2}$ is:

(a) $x=A_1y^2+A_2y+A_3$ (b) $x=A_1y+A_2$

(c) $x=A_1y^2+A_2y$ (d) None of these.

6. The order of the differential equation whose general solution is $y=c_1\cos 2x+c_2\cos^2x+c_3\sin^2x+c_4$, is

(a) 2 (b) 4

(c) 3 (d) None of these.

7. If the solution of the differential equation $\frac{dy}{dx}=\frac{1}{x\cos y+\sin 2y}$ is $x=ce^{\sin y}-k(1+\sin y)$, then $k=$ _______.

(a) 1 (b) 2

(c) 3 (d) 4

8. The differential equation of all parabola having their axis of symmetry coinciding with the axis of x, is:

(a) $y\frac{d^2y}{dx^2}+\left(\frac{dy}{dx}\right)^2=0$ (b) $y\frac{d^2x}{dy^2}+\left(\frac{dx}{dy}\right)^2=0$

(c) $y\frac{d^2y}{dx^2}+\frac{dy}{dx}=0$ (d) None of these.

9. The solution of the differential equation $\sqrt{a+x}\frac{dy}{dx}+xy=0$ is:

(a) $y=Ae^{\frac{2}{3}(2a-x)\sqrt{x+a}}$ (b) $y=Ae^{-\frac{2}{3}(a-x)\sqrt{x+a}}$

(c) $y=Ae^{\frac{2}{3}(2a+x)\sqrt{x+a}}$ (d) $y=Ae^{-\frac{2}{3}(2a-x)\sqrt{x+a}}$

10. The solution of differential equation $\left(x^2-1\right)\frac{dy}{dx}+2xy=\frac{1}{x^2-1}$ is:

(a) $y\left(x^2-1\right)=\frac{1}{2}\log\left|\frac{x-1}{x+1}\right|+C$

(b) $y\left(x^2+1\right)=\frac{1}{2}\log\left|\frac{x-1}{x+1}\right|+C$

(c) $y\left(x^2-1\right)=\frac{5}{2}\log\left|\frac{x-1}{x+1}\right|+C$

(d) None of these.

11. The degree of the differential equation satisfying $\sqrt{1-x^2}+\sqrt{1-y^2}=a(x-y)$ is ______.

(a) 1 (b) 2
(c) 3 (d) None of these.

12. The equation of the curve satisfying the differential equation $y_2\left(x^2+1\right)=2xy_1$, passing through the point (0, 1) and having slope of tangent at $x=0$ as 3, is:

(a) $y=x^2+3x+2$ (b) $y^2=x^2+3x+1$

(c) $y=x^3+3x+1$ (d) None of these.

13. The equation of the curve passing through (3, 4) and satisfying the differential equation $y\left(\frac{dy}{dx}\right)^2+(x-y)\frac{dy}{dx}-x=0$ can be:

$\mathbf{S_1}$: $x-y+1=0$

$\mathbf{S_2}$: $x+y-7=0$

$\mathbf{S_3}$: $x^2+y^2=25$

$\mathbf{S_4}$: $x^2+y^2-5x=10$

(a) TFTF (b) TTFF
(c) TTFT (d) FTFT

14. $\mathbf{S_1}$: The order of differential equation $\sqrt{1+\frac{d^2y}{dx^2}}=x$ is 1.

$\mathbf{S_2}$: Solution of the differential equation $xdy-ydx=\sqrt{x^2+y^2}\,dx$ is $y+\sqrt{x^2+y^2}=cx^2$.

$\mathbf{S_3}$: $\frac{d^2y}{dx^2}=2\left(\frac{dy}{dx}-y\right)$ is a differential equation of family of curves $y=e^x\left(A\cos x+B\sin x\right)$.

$\mathbf{S_4}$: The solution of differential equation $\left(1+y^2\right)+\left(x-2e^{\tan^{-1}y}\right)\frac{dy}{dx}=0$ is $xe^{\tan^{-1}y}=e^{2\tan^{-1}y}+k$

(a) TFTF (b) FTFT
(c) FTTT (d) FFTF

Question from Previous Years

15. If $y=y(x)$ satisfies the differential equation $8\sqrt{x}\left(\sqrt{9+\sqrt{x}}\right)dy=\left(\sqrt{4+\sqrt{9+\sqrt{x}}}\right)^{-1}dx$, $x>0$ and $y(0)=\sqrt{7}$, then $y(256)=$______.

(a) 3 (b) 9
(c) 16 (d) 80

16. The function $y=f(x)$ is the solution of the differential equation $\frac{dy}{dx}+\frac{xy}{x^2-1}=\frac{x^4+2x}{\sqrt{1-x^2}}$ in (−1. 1) satisfying $f(0)=0$. Then $\int_{-\frac{\sqrt{3}}{2}}^{\frac{\sqrt{3}}{2}} f(x)d(x)$ is:

(a) $\frac{\pi}{3}-\frac{\sqrt{3}}{2}$ (b) $\frac{\pi}{3}-\frac{\sqrt{3}}{4}$

(c) $\frac{\pi}{6}-\frac{\sqrt{3}}{4}$ (d) $\frac{\pi}{6}-\frac{\sqrt{3}}{2}$

One or More Than One Option Correct

1. A tangent is drawn at the point (x_1, y_1) to the curve $y = f(x)$, which intersects the x-axis at $(x_{i+1}, 0)$. Now again a tangent is drawn at (x_{i+1}, y_{i+1}) to the curve which intersects the x-axis at $(x_{i+2}, 0)$ and the process is repeated n times. If $x_1, x_2, \ldots,$ forms an AP with common difference $\log_2 e$, and the curve passes through $(0,2)$ and $(-2,k)$, then k is divisible by ______.

(a) 2 (b) 4
(c) 8 (d) 5

2. The solution of the differential equation $\left(e^{x^2} + e^{y^2}\right) y \frac{dy}{dx} + e^{x^2}\left(xy^2 - x\right) = 0$ is:

(a) $e^{x^2}\left(y^2 - 1\right) + e^{y^2} = c$

(b) $e^{y^2}\left(x^2 - 1\right) + e^{x^2} = c$

(c) $e^{y^2}\left(y^2 - 1\right) + e^{x^2} = c$

(d) $e^{-y^2}\left(y^2 - 1\right) + e^{-x^2} - ce^{-x^2} e^{-y^2} = 0$

3. The solution of the differential equation $\left(y + x\sqrt{xy}\,(x+y)\right)dx + \left(y\sqrt{xy}\,(x+y) - x\right)dy = 0$ is:

(a) $\dfrac{x^2 + y^2}{2} + \tan^{-1}\sqrt{\dfrac{y}{x}} = \lambda$

(b) $\dfrac{x^2 + y^2}{2} + 2\tan^{-1}\left(\sqrt{\dfrac{x}{y}}\right) = \lambda$

(c) $\dfrac{x^2 + y^2}{2} + 2\cot^{-1}\sqrt{\dfrac{y}{x}} = \lambda$

(d) $\dfrac{x^2 + y^2}{2} + \cot^{-1}\sqrt{\dfrac{x}{y}} = \lambda$

4. If $y = e^{-x}\cos x$ and $y_n + k_n y = 0$ where $y_n = \dfrac{d^n y}{dx^n}$ and $k_n, n \in N$ are constants, then ______.

(a) $k_4 = 4$ (b) $k_8 = -16$
(c) $k_{12} = 20$ (d) $k_{16} = -24$

5. The differential equation $\dfrac{d^2 y}{dx^2} + y + \cot^2 x = 0$ must be satisfied by:

(a) $y = 2 + c_1\cos x + \sqrt{c_2}\sin x$

(b) $y = \cos x.\,\ell n\left(\tan\dfrac{x}{2}\right) + 2$

(c) $y = 2 + c_1\cos x + c_2\sin x + \cos x \log\left(\tan\dfrac{x}{2}\right)$

(d) All the above.

6. A solution of the differential equation $\left(x^2 y^2 - 1\right)dy + 2xy^3 dx = 0$, is:

(a) $1 + x^2 y^2 = cx$ (b) $1 + x^2 y^2 = cy$

(c) $y = 0$ (d) $y = -\dfrac{1}{x^2}$

7. The solution of $\left(\dfrac{dy}{dx}\right)\left(x^2 y^3 + xy\right) = 1$ is:

(a) $1/x = 2 - y^2 + Ce^{-y^2/2}$

(b) the solution of an equation which is reducible to linear equation.

(c) $2/x = 1 - y^2 + e^{-y/2}$

(d) $\dfrac{1 - 2x}{x} = -y^2 + Ce^{-y^2/2}$

Questions from Previous Years

8. For any real numbers α and β, let $y_{\alpha\beta}(x)$, $x \in \mathbb{R}$, be the solution of the differential equation $\dfrac{dy}{dx} + \alpha y = xe^{\beta x}$, $y(1) = 1$. Let $S\{y_{\alpha\beta}(x) : \alpha, \beta \in \mathbb{R}\}$. Then which of the following function(s) belong(s) to the set S?

(a) $f(x) = \dfrac{x^2}{2}e^{-x} + \left(e - \dfrac{1}{2}\right)e^{-x}$

(b) $f(x) = -\dfrac{x^2}{2}e^{-x} + \left(e + \dfrac{1}{2}\right)e^{-x}$

(c) $f(x) = \dfrac{e^x}{2}\left(x - \dfrac{1}{2}\right) + \left(e - \dfrac{e^2}{2}\right)e^{-x}$

(d) $f(x) = \dfrac{e^x}{2}\left(\dfrac{1}{2} - x\right) + \left(e + \dfrac{e^2}{2}\right)e^{-x}$

9. Let Γ denote a curve $y = y(x)$, which is in the first quadrant and let the point (1, 0) lie on it. Let the tangent to Γ at a point P intersect the y-axis at y_p. If PY_p has length 1 for each point P and Γ, then which of the following options is/are correct?

(a) $y = -\log_e\left(\frac{1+\sqrt{1-x^2}}{x}\right) + \sqrt{1-x^2}$

(b) $xy' - \sqrt{1-x^2} = 0$

(c) $y = \log_e\left(\frac{1+\sqrt{1-x^2}}{x}\right) - \sqrt{1-x^2}$

(d) $xy' + \sqrt{1-x^2} = 0$

10. Let $f:[0,\infty) \to \mathbb{R}$ be a continuous function such that $f(x) = 1-2x+\int_0^x e^{x-t} f(t)\,dt$ for all $x \in [0,\infty)$. Then which of the following statement(s) is (are) true?

(a) The curve $y=f(x)$ passes through the point (1, 2).

(b) The curve $y=f(x)$ passes through the point (2, −1).

(c) The area of the region $\{(x, y) \in [0,1]\times\mathbb{R} : f(x) \le y \le \sqrt{1-x^2}\}$ is $\frac{\pi-2}{4}$.

(d) The area of the region $\{(x, y)\} \in [0,1]\times\mathbb{R} : f(x) \le y \le \sqrt{1-x^2}\}$ is $\frac{\pi-1}{4}$.

11. A solution curve of the differential equation $(x^2+xy+4x+2y+4)\frac{dy}{dx} - y^2 = 0$, passes through the point (1, 3). Then the solution curve

(a) intersects $y = x+2$ exactly at one point.

(b) intersects $y = x+2$ exactly at two points.

(c) intersects $(y = x+2)^2$.

(d) does not intersect $(y = x+3)^2$.

12. Let $f:(0,\infty) \to \mathbb{R}$ be a differential function such that $f'(x) = 2 - \frac{f(x)}{x}$ for all $x \in (0,\infty)$ and $f(1) \ne 1$. Then:

(a) $\lim_{x\to 0+} f'\left(\frac{1}{x}\right) = 1$

(b) $\lim_{x\to 0+} xf\left(\frac{1}{x}\right) = 2$

(c) $\lim_{x\to 0+} x^2 f'(x) = 0$

(d) $|f(x)| \le 2$ for all $x \in (0,2)$

13. Consider the family of all circles whose centres lie on the straight line $y = x$. If this family of circles is represented by the differential equation $Py'' + Qy' + 1 = 0$, where P, Q are functions of x, y and y' $\left(\text{here, } y' = \frac{dy}{dx}, y'' = \frac{d^2y}{dx^2}\right)$, then which of the following statements is (are) true?

(a) $P = y + x$

(b) $P = y - x$

(c) $P + Q = 1 - x + y + y' + (y')^2$

(d) $P - Q = x + y - y' - (y')^2$

14. Let $y(x)$ be a solution of the differential equation $(1+e^x)y' + ye^x = 1$. If $y(0) = 2$, then which of the following statement is (are) true?

(a) $y(-4) = 0$

(b) $y(-2) = 0$

(c) $y(x)$ has a critical point in the interval (−1, 0)

(d) $y(x)$ has no critical point in the interval (−1, 0)

15. A curve passes through the point $\left(1, \frac{\pi}{6}\right)$. Let the slope of the curve at each point (x, y) be $\frac{y}{x} + \sec\left(\frac{y}{x}\right)$, $x > 0$. Then the equation of the curve is:

(a) $\sin\left(\frac{y}{x}\right) = \log x + \frac{1}{2}$

(b) $\operatorname{cosec}\left(\frac{y}{x}\right) = \log x + 2$

(c) $\sec\left(\frac{2y}{x}\right) = \log x + 2$

(d) $\cos\left(\frac{2y}{x}\right) = \log x + \frac{1}{2}$

Matrix Type

1. Match Column I with Column II.

Column I	Column II
(A) Solution of $y-\frac{xdy}{dx}=y^2+\frac{dy}{dx}$ is	(P) $xy^2=2y^5+c$
(B) Solution of $\left(2x-10y^3\right)\frac{dy}{dx}+y=0$ is	(Q) $\sec y=x+1+ce^x$
(C) Solution of $\sec^2 ydy+\tan ydx=dx$ is	(R) $(x+1)(1-y)=cy$
(D) Solution of $\sin y\frac{dy}{dx}=\cos y(1-x\cos y)$ is	(S) $xy^2=2y^5+x+c$
	(T) $\tan y=1+ce^{-x}$

2. Match Column I with Column II.

Column I	Column II
(A) $xdy=y(dx+ydy), y>0$ $y(1)=1$ and $y(x_0)=-3$, then $x_0=$	(P) $\frac{1}{4}$
(B) If $y(t)$ is solution of $(t+1)\frac{dy}{dt}-yt=1$, $y(0)=-1$, then $y(1)=$	(Q) -15
(C) $\left(x^2+y^2\right)dx=xydx$ and $y(1)=1$ and $y(x_0)=e$, then $x_0=$	(R) $-\frac{1}{2}$
(D) $\frac{dy}{dx}+\frac{2y}{x}=0, y(1)=1$, then $y(2)=$	(S) 16
	(T) $\sqrt{3}e$

3. Match Column I with Column II.

Column I	Column II
(A) $(\sin x+\cos x)dy+(\cos x-\sin x)dx=0$	(P) $\sec y+\tan y=c(\operatorname{cosec} x+\cot x)$
(B) $\sin xdy+\cos y\, dx=0$	(Q) $2\left(x^2+y^2\right)+2(x\sin 2x+y\sin 2y)+(\cos 2x+\cos 2y)=c$
(C) $x^{-1}\cos^2 ydy+y^{-1}\cos^2 xdx=0$	(R) $y=c-\log\lvert\sin x+\cos x\rvert$
(D) $\tan x\sec^2 ydy+\tan y\sec^2 xdx=dx$	(S) $\cot y+\sec y=c(\sin x+\tan x)$
	(T) $\tan x\tan y=x+c$

4. Match the following:

Column I	Column II
(A) $\int_0^{\pi/2}(\sin x)^{\cos x}\left\{\cos x\cot x-\log(\sin x)^{\sin x}\right\}dx$	(P) 1
(B) Area bounded by $-4y^2=x$ and $x-1=-5y^2$	(Q) 0
(C) The angle of intersection of the curves $y=3^{x-1}\log x$ and $y=x^x-1$ is	(R) $3e^{y/2}$
(D) If $\frac{dy}{dx}=\frac{2}{x+y}$ passing through (1, 0), then $(x+y+2)$ is	(S) $\frac{4}{3}$

Integer and Numerical Value Type

1. A and B are two separate reservoirs of water. The capacity of reservoir A is double the capacity of reservoir B. The reservoirs are filled completely with water, their inlets are closed and then the water is released simultaneously from both the reservoirs. The rate of flow of water out of each reservoir at any instant of time is proportional to the quantity of water in the reservoir at that time. One hour after the water is released, the quantity of water in reservoir A is 1.5 times the quantity of water in reservoir B. After $(\lambda \log_{4/3} 2)$ hours, both the reservoirs have the same quantity of water, then find the value of λ.

2. Let then curve $y = f(x)$ passing through $(4, -2)$ satisfy the differential equation,
$y(x+y^3)dx = x(y^3 - x)dy$ and
$$y = g(x) = \int_{\frac{1}{8}}^{\sin^2 x} \sin^{-1}\sqrt{t}\,dt + \int_{\frac{1}{8}}^{\cos^2 x} \cos^{-1}\sqrt{t}\,dt, 0 \le x \le \frac{\pi}{2}.$$
The area of the region bounded by curves, $y = f(x), y = g(x)$ and $x = 0$ is $\frac{\lambda}{8}\left(\frac{3\pi}{16}\right)^4$, then find the value of λ.

3. By eliminating the constant in the equation $x^2 - y^2 = c(x^2 + y^2)^2$, the resulting differential equation is $y' = \frac{x(\lambda y^2 - x^2)}{y(\lambda x^2 - y^2)}$, then find the value of λ.

4. The differential equation,
$(x^2 + 4y^2 + 4xy)dy = (2x + 4y + 1)dx$ has solution
$$y = \ell n((x+2y)^2 + \lambda(x+2y) + 2) - \frac{3}{2\sqrt{2}}\ell n\left|\frac{x+2y+2-\sqrt{2}}{x+2y+2+\sqrt{2}}\right| + c.$$
Then find the value of λ.

5. The curve passing through the origin if the middle point of the segment of its normal from any point of the curve to the x-axis lies on the parabola $2y^2 = x$ where $y^2 = \lambda x + 1 - e^{2x}$, then find the value of λ.

Questions from Previous Years

6. Let $f: \mathbb{R} \to \mathbb{R}$ be a differentiable function with $f(0) = 1$ and satisfy the equation
$f(x+y) = f(x)f'(y) + f'(x)f(y)$ for all $x, y \in \mathbb{R}$.
Then, the value of $log_e(f(4))$ is ______.

7. Let $f: \mathbb{R} \to \mathbb{R}$ be a differentiable function with $f(0) = 0$. If $y = f(x)$ satisfies the differential equation $\frac{dy}{dx} = (2+5y)(5y-2)$, then the value of $\lim_{x \to -\infty} f(x)$ is ______.

Comprehension Passage

Comprehension 1

The differential equation $\frac{dy}{dx} = f(x)g(y)$ can be solved by separating variable $\frac{dy}{g(y)} = f(x)dx$

1. The equation of the curve to the point (1, 0) which satisfies the differential equation $(1+y^2)dx - xy\,dy = 0$ is:

   (a) $x^2 + y^2 = 1$ (b) $x^2 - y^2 = 1$

   (c) $x^2 + y^2 = 2$ (d) $x^2 - y^2 = 2$

2. The solution of the differential equation $\frac{dy}{dx} + \frac{1+y^2}{\sqrt{1-x^2}} = 0$ is:

   (a) $\tan^{-1} y + \sin^{-1} x = c$

   (b) $\tan^{-1} x + \sin^{-1} y = c$

   (c) $\tan^{-1} y . \sin^{-1} x = c$

   (d) $\tan^{-1} y - \sin^{-1} x = c$

3. If $\frac{dy}{dx} = 1 + x + y + xy$ and $y(-1) = 0$, then $y =$

   (a) $e^{\frac{(1-x)^2}{2}}$ (b) $e^{\frac{(1+x)^2}{2}} - 1$

   (c) $\ell n(1+x) - 1$ (d) $1 + x$

Comprehension 2

A differential equation of the form $\frac{d^2y}{dx^2}+a_1\frac{dy}{dx}+a_2y=0$ is called linear differential equation where a_1 and a_2 are functions of x only. In case a_1 and a_2 are constants, the solution of the linear differential equation can be easily written from following facts:

(i) $y=0$ is a solution of differential equation.

(ii) If $y=f(x)$ is a solution then $y=cf(x)$ is also a solution.

(iii) If $y=f_1(x)$ and $y=f_2(x)$ are two solutions then $y=f_1(x)+f_2(x)$ will also be a solution.

(iv) If the distinct roots of a quadratic equation $m^2+a_1m+a_2=0$ are m_1 and m_2 (real or imaginary) then solution of differential equation is $y=c_1e^{m_1x}+c_2e^{m_2x}$

In case the roots are complex, the solution can be transformed to the form $e^{ax}(c_1\cos bx+c_2\sin bx)$ by using Euler's theorem.

(v) In case roots of $m^2+a_1m+a_2=0$ are equal (say m_1), the differential equation can be made linear by putting $\frac{dy}{dx}-m_1y=v$

The linear differential equation

$$\frac{d^2y}{dx^2}+a_1\frac{dy}{dx}+a_1y=f(x)$$

can also be satisfied by some other functions which are not of the above type. Such functions are called particular integrals

1. If $y=e^x$ is a solution of differential equation $\frac{d^2y}{dx^2}-y=0$, then which of the following is not a solution?

(a) e^{-x} (b) $e-x$

(c) ae^x+be^{-x} (d) e^x+c

2. Which of the following is solution of the equation $\frac{d^2y}{dx^2}-6\frac{dy}{dx}+9y=0$?

(a) c_1+c_2x (b) $(c_1+c_2x)e^{3x}$

(c) $c_1\cos 3x+c_2\sin 3x$ (d) None of these.

3. A particular integral solution of the equation $\frac{d^2y}{dx^2}+a^2y+\frac{\cos ax}{a}=0$ is:

(a) $c_1\cos(ax)+c_2\sin(ax)$

(b) $\frac{x\sin ax}{2a^2}+\frac{x\cos ax}{2a^2}$

(c) $\frac{x\sin ax}{2a^2}+\frac{\cos ax}{2a^2}$

(d) $\frac{-x\sin ax}{2a^2}-\frac{\cos ax}{2a^2}$

ANSWER KEY

JEE Main

One Option Correct

1. (a)	**2.** (a)	**3.** (d)	**4.** (d)	**5.** (c)	**6.** (a)	**7.** (a)	**8.** (a)	**9.** (c)
10. (c)	**11.** (c)	**12.** (a)	**13.** (b)	**14.** (b)	**15.** (c)	**16.** (a)	**17.** (b)	**18.** (b)
19. (b)	**20.** (d)	**21.** (c)	**22.** (a)	**23.** (c)	**24.** (a)	**25.** (a)	**26.** (a)	**27.** (b)
28. (d)	**29.** (b)	**30.** (b)	**31.** (b)	**32.** (a)	**33.** (c)	**34.** (d)	**35.** (b)	**36.** (b)
37. (b)	**38.** (c)	**39.** (b)	**40.** (c)	**41.** (b)	**42.** (c)	**43.** (c)	**44.** (c)	**45.** (c)
46. (b)	**47.** (c)	**48.** (a)	**49.** (a)	**50.** (a)	**51.** (b)	**52.** (a)	**53.** (a)	**54.** (a)
55. (a)	**56.** (b)	**57.** (d)	**58.** (c)	**59.** (b)	**60.** (a)	**61.** (d)	**62.** (a)	**63.** (a)
64. (a)	**65.** (a)	**66.** (a)	**67.** (a)	**68.** (b)	**69.** (a)	**70.** (a)	**71.** (c)	**72.** (b)
73. (c)	**74.** (a)	**75.** (a)	**76.** (d)	**77.** (b)	**78.** (c)	**79.** (a)	**80.** (c)	**81.** (a)
82. (a)	**83.** (b)	**84.** (c)	**85.** (b)	**86.** (a)	**87.** (a)	**88.** (b)	**89.** (d)	**90.** (c)
91. (d)	**92.** (c)	**93.** (c)	**94.** (c)	**95.** (c)	**96.** (b)	**97.** (b)	**98.** (a)	**99.** (b)
100. (b)	**101.** (d)	**102.** (c)						

Integer and Numerical Value Type

1. $(-\lambda)$ 2. $(\pm x + c)$ 3. $\left(\frac{H}{k}\right)$ 4. $(x^2 + y^2 = 2x)$ 5. $\left(\log_{3/4}\frac{1}{2}\right)$ 6. (0) 7. (2)
8. (17) 9. (2) 10. (1) 11. (4) 12. (2) 13. (4) 14. (1) 15. (4) 16. (1)

Assertion and Reason

1. (b) 2. (c) 3. (a) 4. (b) 5. (c) 6. (b) 7. (c)

JEE Advanced

Objective Questions

One Option Correct

1. (a) 2. (d) 3. (a) 4. (d) 5. (a) 6. (a) 7. (b) 8. (a) 9. (a)
10. (a) 11. (a) 12. (c) 13. (a) 14. (c) 15. (a) 16. (b)

One or More Than One Option Correct

1. (a, b, c) 2. (a, d) 3. (b, c) 4. (a, b) 5. (b, c)
6. (b, c) 7. (a, b, d) 8. (a, c) 9. (c, d) 10. (b, c)
11. (a, d) 12. (a) 13. (b, c) 14. (a, c) 15. (a)

Matrix Type

1. A → R; B → P; C → T; D → Q
2. A → Q; B → R; C → T; D → P
3. A → R; B → P; C → Q; D → T
4. A → P; B → S; C → Q; D → R

Integer and Numerical Value Type

1. (1) 2. (1) 3. (3) 4. (4) 5. (2) 6. (2) 7. (0.4)

Comprehension Passage

Comprehension 1

1. (b) 2. (a) 3. (b)

Comprehension 2

1. (b) 2. (b) 3. (d)

HINTS AND EXPLANATIONS

Worksheet

1. $y = Ae^{2x} + B.e^{-2x}$

$\frac{dy}{dx} = 2Ae^{2x} - 2B.e^{-2x}$ and $\frac{d^2y}{dx^2} = 4Ae^{2x} + 4Be^{-2x}$

Thus, $\frac{d^2y}{dx^2} = 4y$, i.e., $\frac{d^2y}{dx^2} - 4y = 0$.

2. $\frac{dy}{dx} = \frac{y}{x} \Rightarrow \frac{dy}{y} = \frac{dx}{x} \Rightarrow \int\frac{dy}{y} = \int\frac{dx}{x}$

$\Rightarrow \log y = \log x + \log c \Rightarrow y = cx$

3. The equation is of the type $\frac{dy}{dx} + Py = Q$, which is a linear differential equation.

Now I.F. $= \int\frac{1}{x}dx = e^{\log x} = x$.

Therefore, solution of the given differential equation is

$y.x = \int x\,x^2 dx$, i.e., $yx = \frac{x^4}{4} + c$

Hence, $y = \frac{x^3}{4} + \frac{c}{x}$.

4. Let $y = mx$ be the family of lines through origin.

Therefore, $\frac{dy}{dx} = m$

Eliminating m, we get $y = \frac{dy}{dx}\cdot x$ or $x\frac{dy}{dx} - y = 0$.

5. The general equation of all non-horizontal lines in a plane is $ax + by = c$, where $a \neq 0$.

Therefore, $a\dfrac{dx}{dy} + b = 0$.

Again, differentiating both sides w.r.t. y, we get

$a\dfrac{d^2x}{dy^2} = 0 \Rightarrow \dfrac{d^2x}{dy^2} = 0$.

6. We have $\dfrac{dy}{dx} = x^5 + x^2 - \dfrac{2}{x}$

Integrating,

$$y = \int\left(x^5 + x^2 - \frac{2}{x}\right)dx + c$$

$$= \int x^5 dx + \int x^2 dx - 2\int \frac{1}{x}dx + c$$

$$\Rightarrow y = \frac{x^6}{6} + \frac{x^3}{3} - 2\log|x| + c,$$

which is the required general solution.

7. $\dfrac{dy}{dx} = -\dfrac{1+\cos 2y}{1-\cos 2x} = \dfrac{dy}{dx} = -\dfrac{2\cos^2 y}{2\sin^2 x}$

$\Rightarrow \sec^2 y\, dy = -\text{cosec}^2 x\, dx$

On integrating both sides, we get $\tan y = \cot x + c$

$\Rightarrow \tan y - \cot x = c$

8. We are given that $\dfrac{dy}{dx} = 1 + x + y + xy$

$$\Rightarrow \frac{dy}{dx} = (1+x) + y(1+x)$$

$$\Rightarrow \frac{dy}{dx} = (1+x)(1+y)$$

$$\Rightarrow \frac{1}{1+y}dy = (1+x)dx$$

Integrating both sides, we get

$$\int \frac{1}{1+y}dy = \int(1+x)dx$$

$\Rightarrow \log(1+y) = x + \dfrac{x^2}{2} + C$, which is the required solution.

9. $\dfrac{dy}{dx} + y = e^x$

This is linear equation of the form $\dfrac{dy}{dx} + Py = Q$

where P = 1, Q = e^x

So integrating factor (I.F.) $= e^{\int P dx} = e^x$

Thus solution is given by

$$ye^{\int Pdx} = \int Q\, ye^{\int Pdx}dx + c$$

$$\Rightarrow ye^x = \int e^x e^x dx + c$$

$$\Rightarrow ye^x = \frac{e^{2x}}{2} + c$$

$$\Rightarrow y = \frac{e^x}{2} + ce^{-x}$$

10. The given equation can be written as:

$$\frac{dx}{dy} + \frac{2}{y}x = 10y^2 \qquad \text{...(1)}$$

[Linear equation in x]

Here P $= \dfrac{2}{y}$ 'Q' $= 10y^2$

$$\text{I.F.} = e^{\int Pdy} = e^{\int \frac{2}{y}dy} = e^{2\log|y|} = e^{\log y^2} = y^2$$

Multiplying (1) by y^2, we get

$$y^2\frac{dx}{dy} + 2yx = 10y^4 \Rightarrow \frac{d}{dy}\left(x \cdot y^2\right) = 10y^4$$

Integrating $xy^2 = 10\int y^4 dy + c \Rightarrow xy^2 = 2y^5 + c$,

which is required solution.

Quiz

1. In general, an equation involving independent and/or dependent variable (variables) as well as derivative of the dependent variable with respect to independent variable is called a differential equation.

2. A differential equation involving derivatives of the dependent variable with respect to only one independent variable is called ordinary differential equation.

Example: $5\dfrac{d^2y}{dx^2} + \left(\dfrac{dy}{dx}\right)^3 = 0$

3. Order of a differential equation is defined as the order of highest order derivative of the dependent variable with respect to the independent variable involved in the given differential equation. For example, $\dfrac{d^2y}{dx^2} + y = 0$ is a differential equation of Order 2.

4. Order 3; Degree 2

5. 4

6. Zero

7. Differentiating y w.r.t x

We have $y' = \dfrac{1}{2}\dfrac{1 \times 2x}{\sqrt{1+x^2}} = \dfrac{x}{\sqrt{1+x^2}}$

Now, take the equation $y' = \dfrac{xy}{1+x^2}$

$$\text{RHS} = \frac{xy}{1+x^2}$$

Substituting the value of y

$$= \frac{x\sqrt{1+x^2}}{1+x^2} = \frac{x}{\sqrt{1+x^2}}$$

Hence, LHS = RHS verified.

8. The given differential equation can be written as

$$\frac{dy}{y^2} = -4x\,dx \quad (y \neq 0)$$

or $\int \frac{dy}{y^2} = -4\int x\,dx = -\frac{1}{y} = -\frac{4}{2}.x^2 + c$

or $\quad y = \frac{1}{2x^2 - c}$

for $x = 0$ and $y = 1$

JEE Main

Objective Questions

One Option Correct

1. (a) I.F. $= e^{\int \frac{1}{x}dx} = e^{\log x} = x$

2. (a) $-\int \frac{\sec^2 x}{\tan x}dx = \int \frac{\sec^2 y}{\tan y}dy$

$-\log\tan x = \log\tan y + \log k$

$\Rightarrow -\log\tan x\tan y = \log k$

$\Rightarrow \tan x\tan y = \frac{1}{k}$

$\Rightarrow \tan x\tan y = c$

3. (d) $y = mx$

$\frac{y}{x} = m$

Differentiating both sides, w.r.t x we get

$\frac{xy' - y}{x^2} = 0 \Rightarrow \frac{y'}{x} = \frac{y}{x^2}$

4. (d) $\frac{1}{y^n}\frac{dy}{dx} + \frac{1}{y^{n-1}}P(x) = Q(x)$

Put $v = \frac{1}{y^n} \Rightarrow v' = \frac{1}{(1-n)y^{n-1}}.\frac{dy}{dx}$

5. (c) I.F $= e^{\int P\,dx}$

6. (a) The highest-order derivative is 3 and its degree is 1.

7. (a) Rationalizing both sides, we get

$$\left(1+\left(\frac{dy}{dx}\right)^2\right)^3 = k^2\left(\frac{d^2y}{dx^2}\right)^2$$

The order is two, and the degree of $\frac{d^2y}{dx^2}$ is 2.

$\Rightarrow$ Order = 2, Degree = 2

8. (a) $= e^{\int \frac{2x}{x^2+1}dx} = e^{\log(x^2+1)} = (x^2+1)$

9. (c) $y = \sec(\tan^{-1}x)$

$\Rightarrow \sec^{-1}y = \tan^{-1}x$

$\Rightarrow \frac{dy}{dx} = \frac{x\sec(\tan^{-1}x)}{1+x^2}$

$\Rightarrow (1+x^2)\frac{dy}{dx} = xy$

10. (c) Slope $= \frac{dy}{dx} = \frac{1}{2y}$

$\Rightarrow\ 2y\,dy = dx\ y^2 = x + c$

It passes through $(4,3) \Rightarrow 9 = 4 + c\ c = 5$

Therefore, equation of the curve is $y^2 = x + 5$

11. (c) The given differential equation can be written as

$\int \frac{dx}{1+x^2} = -\int \frac{dy}{1+y^2} + c_1$

$\Rightarrow \tan^{-1}x + \tan^{-1}y = \tan^{-1}c$

$\Rightarrow x + y = c(1-xy)$

12. (a) Slope of the tangent is

$\frac{dy}{dx} = 2x$

The order is 1.

13. (b) The differential equation can be written as

$\int \frac{dy}{1+\cos 2y} = -\int \frac{dx}{1-\cos 2x} + c$

$\Rightarrow \frac{1}{2}\int \sec^2 y\,dy = -\frac{1}{2}\int \operatorname{cosec}^2 x + c$

$\Rightarrow \frac{1}{2}\tan y - \frac{1}{2}\cot x = c$

$\Rightarrow \tan y - \cot x = c$

14. (b)

15. (c) $\frac{dy}{dx} = \frac{1}{x}y = \cos x + \frac{\sin x}{x}$

I.F. $= e^{\int P dx} = x$

$xy = \int (x\cos x + \sin x)dx + c$

$= x\sin x - \int \sin x\,dx + \int \sin x\,dx + c$

$= x\sin x + c$

$\frac{\pi}{2}(1)=\frac{\pi}{2}(1)+c \Rightarrow c=0$

$y=\sin x$

16. (a) $x\,dy-y\,dx+x^2e^x dx=0$

$\frac{dy}{dx}-\frac{y}{x}=-xe^x$

$\text{I.F.}=\frac{1}{x}$

$y\frac{1}{x}=-\int e^x dx+c=-e^x+c$

17. (b) $y^2=4A(x-B)$

$2y\frac{dy}{dx}=4A$

$yy'=2A$

$y'y''+(y')^2=0$

Hence, Degree =1, Order = 2

18. (b) $(\sin x+\cos x)dy+(\cos x-\sin x)dx=0$

$\Rightarrow \frac{dy}{dx}=\frac{\sin x-\cos x}{\sin x+\cos x}$

$\Rightarrow \int dy=\frac{\sin x-\cos x}{\sin x+\cos x}dx$

$y=-\log|\sin x+\cos x|+\log c$

$=\log\left|\frac{c}{\sin x+\cos x}\right|$

$\Rightarrow e^y(\sin x+\cos x)=c$

19. (b) $(1+y^2)dx=xydy$

$\int\frac{1}{x}dx=\frac{1}{2}\int\frac{2y}{1+y^2}dy$

$\Rightarrow \log x+c=\frac{1}{2}\log(1+y^2)$

At $x=1, y=0$

$\Rightarrow 0+c=0 \quad c=0$

$x^2-y^2=1$

20. (d) $\frac{dy}{dx}+ay=e^{mx}$

$\text{I.F.}=e^{\int Pdx}=e^{a\int dx}\,e^{ax}$

$ye^{ax}=\int e^{mx}.e^{mx}dx+c$

$=\int e^{mx+ax}.e^{mx}dx+c$

$=\frac{e^{mx+ax}}{m+a}+c$

$(a+m)y=e^{mx}+ce^{-ax}$

21. (c)

22. (a)

23. (c)

24. (a) Slope $=\frac{dy}{dx}=\frac{2y}{x}$

$\Rightarrow \frac{1}{2}\int\frac{dy}{y}=\int\frac{dx}{x}+c$

$\Rightarrow \frac{1}{2}\log y=\log x+\log c$

$\Rightarrow y=cx^2$

It passes through $(1,1)$.

$\Rightarrow c=1$

$\Rightarrow y=x^2$

25. (a) $y=ce^{x-x^2}$

$\Rightarrow y'=ce^{x-x^2}(1-2x)$

Substituting the value of y and y' in the LHS of the differential equation

$\Rightarrow ce^{x-x^2}(1-2x)+2xce^{x-x^2}$

$\Rightarrow ce^{x-x^2}=y=\text{RHS}$

26. (a)

27. (b) $\text{I.F.}=e^{-\int\frac{1}{x}dx}=\frac{1}{x}$

$\therefore y\left(\frac{1}{x}\right)=\int\frac{1}{x}dx+c=\log x+c$

$y(1)=1\Rightarrow c=1$

28. (d) The differential equation can be written as

$\frac{dx}{1+x^2}=-\frac{dy}{1+y^2}$

Integrating both sides, we get

$\int\frac{dx}{1+x^2}=-\int\frac{dy}{1+y^2}+c$

$\Rightarrow \tan^{-1}x=-\tan^{-1}y+\tan^{-1}c$

{as a constant can be written as $\tan^{-1}c$ also}

$\Rightarrow \tan^{-1}x+\tan^{-1}y=\tan^{-1}c$

29. (b) The differential equation can be written as

$\frac{dy}{dx}-\frac{y}{x}=x^2$

which is of the form $\frac{dy}{dx}+Py=Q$

$P=-\frac{1}{x}$ and $Q=x^2$

$\Rightarrow$ I.F. $= e^{\int -\frac{1}{x}dx}$

I.F. $= e^{-\log x} = \frac{1}{x}$

Therefore,

$$\frac{1}{x}\frac{dy}{dx} - \frac{y}{x^2} = x$$

$$\Rightarrow \frac{y}{x} = \int x\,dx + c$$

$$\Rightarrow \frac{y}{x} = \frac{x^2}{2} + c$$

or $y = \frac{x^3}{2} + cx$ or $2y - x^3 = cx$

30. (b) The equation of simple harmonic motion is

$x = a \sin \omega t$

$x =$ Displacement of a particle at time t

$a =$ Amplitude with angular velocity ω.

$$\Rightarrow \frac{dx}{dt} = \omega \times a \cos \omega t$$

$\Rightarrow$ Velocity $= a\omega \cos \omega t$

$\Rightarrow$ Acceleration $= \frac{d^2x}{dt^2} = -a\omega^2 \sin \omega t = -\omega^2 x$

$$\Rightarrow \frac{d^2x}{dt^2} + \omega^2 x = 0$$

Now, $\omega = \frac{2\pi}{T} = \frac{2\pi}{\frac{2\pi}{n}} = n$

$$\Rightarrow \frac{d^2x}{dt^2} + n^2 x = 0$$

31. (b) The differential equation can be written as

$$\frac{dy}{\sqrt{1-y^2}} = -\frac{dx}{\sqrt{1-x^2}}$$

Integrating both sides, we get

$$\int \frac{dy}{\sqrt{1-y^2}} = -\int \frac{dx}{\sqrt{1-x^2}} + c$$

$$\Rightarrow \sin^{-1} y = -\sin^{-1} x + c$$

$$\Rightarrow \sin^{-1} y + \sin^{-1} x = c$$

32. (a)

33. (c) $\frac{dy}{dx} + \frac{1}{x}y = x^2$

I.F. $= e^{\int \frac{1}{x}dx} = e^{\log x} = x$

$$yx = \int x\,x^2 dx + c = \frac{x^4}{4} + c$$

$$y = \frac{x^3}{4} + cx^{-1}$$

34. (d)

35. (b) $x^2 + y^2 - 2ax = 0$

Differentiating w.r.t x, we get

$$2x + 2yy' = 2a$$

$$\Rightarrow x + yy' = a$$

$$\therefore x^2 + y^2 - 2(x + yy')x = 0$$

$$\Rightarrow y^2 - x^2 = 2xyy'$$

36. (b) $(1+y^2)dx = xy\,dy$

$$\Rightarrow \int \frac{1}{x}dx = \frac{1}{2}\int \frac{2y}{1+y^2}dy$$

$$\log x = \frac{1}{2}\log|1+y^2| + \log k$$

$$x = k(1+y^2) \Rightarrow k = 1$$

$$x = 1 + y^2$$

37. (b)

38. (c) Rationalizing the differential equation

$$\left(1+\frac{dy}{dx}\right)^2 = 8\left(\frac{d^3y}{dx^3}\right)^3$$

$\Rightarrow$ Order = 3, Degree = 3.

39. (b) Here $y\,dx + x\,dy = -x^2 y\,dy$

$$\Rightarrow \frac{ydx + xdy}{x^2y^2} = -\frac{1}{y}dy$$

$$\Rightarrow d\left(-\frac{1}{xy}\right) = -\frac{1}{y}dy$$

$\therefore$ Integrating both sides, we get

$$-\frac{1}{xy} + \log y = c$$

40. (c)

$\frac{dy}{dx} = \frac{y}{x}.\left[\log\frac{y}{x} + 1\right]$ is homogeneous equation.

Put $y = vx$, then $v + x\frac{dv}{dx} = \frac{dy}{dx}$

$$\Rightarrow \frac{dv}{v \log v} = \frac{dx}{x}$$

Integrating, we get $\log\left(\frac{y}{x}\right) = cx$

41. (b)

42. (c) The differential equation can be written as

$$\frac{dy}{dx} = \frac{y}{x} - \tan\frac{y}{x}$$

Let $y = vx$

or $\frac{dy}{dx} = v + x\frac{dv}{dx}$

$$\Rightarrow v + x\frac{dv}{dx} = v - \tan v$$

$$\Rightarrow \int \frac{dx}{x} = -\int \frac{dv}{\tan v} + c$$

$$\Rightarrow \log x = -\int \frac{\cos v\, dv}{\sin v} + c$$

$$\Rightarrow \log x = -\log \sin\left(\frac{y}{x}\right) + c$$

$$\Rightarrow \log\left(x \sin\frac{y}{x}\right) = \log c \qquad [\text{Taking } c \text{ as } \log c]$$

$$\Rightarrow x\sin\frac{y}{x} = c$$

43. (c) $y^2 = ae^{-2x} + \frac{2}{5}(\cos x - 2\sin x)$

$$\Rightarrow 2y\frac{dy}{dx} = -2ae^{-2x} + \frac{2}{5}(-\sin x - 2\cos x)$$

$$\Rightarrow y\frac{dy}{dx} = -ae^{-2x} - \frac{1}{5}(\sin x + 2\cos x)$$

Substituting value of y^2 and $y\frac{dy}{dx}$ in the differential equation, we get

$$-ae^{-2x} - \frac{1}{5}(\sin x + 2\cos x) + ae^{-2x} + \frac{2}{5}(\cos x - 2\sin x) + \sin x = 0$$

44. (c) $(x\log x)\left(\frac{dy}{dx}\right) + y = 2\log x$

Dividing both sides by $x\log x$, we get

$$\Rightarrow \frac{dy}{dx} + y\cdot\frac{1}{x\log x} = \frac{2}{x}$$

The above equation is of the form $\frac{dy}{dx} + P_y = Q$

Here, $P = \frac{1}{x\log x}$ and $Q = \frac{2}{x}$.

Integrating factor $= e^{\int f dx}$

Considering $\int Pdx$,

$$\Rightarrow \int Pdx = \int \frac{t}{x\log x}dx$$

Put $\log x = t$

$$\Rightarrow \frac{1}{x}\log x = dt$$

Putting the values, we get

$$\Rightarrow \int Pdx = \int \frac{1}{t}dt = \log t$$

$$\Rightarrow \int Pdx = \log(\log x)$$

$$\therefore e^{\int Pdx} = e^{\log(\log x)}$$

$$\because e^{\log x} = x$$

$$\therefore \text{I.F.} = \log x$$

45. (c)

46. (b) The differential equation can be written as

$$\frac{dy}{dx} = 1 + \frac{y}{x} + \left(\frac{y}{x}\right)^2$$

Let $y = vx$

$$\frac{dy}{dx} = v + x\frac{dv}{dx}$$

$$\Rightarrow v + x\frac{dv}{dx} = 1 + v + v^2$$

$$\Rightarrow x\frac{dv}{dx} = 1 + v^2$$

$$\Rightarrow \int \frac{dv}{1+v^2} = \int \frac{dx}{x} + c$$

$$\Rightarrow \tan^{-1}\left(\frac{y}{x}\right) = \log x + c$$

47. (c) $\frac{dy}{dx} - ky = 0$

$$\Rightarrow \frac{dy}{dx} = ky$$

Integrating both sides, we get

$$\Rightarrow \int \frac{dy}{y} = \int k\, dx + c$$

$$\Rightarrow \log y = kx + c$$

Now $y(0) = 1$

$\Rightarrow c = 0$ as $\log 1 = 0$

Therefore, $y = e^{kx}$

As per the condition $y \to 0$ as $x \to \infty$

This is possible only when $k < 0$.

48. (a) Any line parallel to y- axis is of the form $x = h$

$\Rightarrow$ Let the vertex be (h, k)

$\therefore$ The equation of parabola becomes

$$(x-h)^2 = 4a(y-k)$$

or $y - k = \frac{(x-h)^2}{4a}$

Differentiating,

$$\frac{dy}{dx} = \frac{2(x-h)}{4a}$$

$$\Rightarrow \frac{d^2y}{dx^2} = \frac{2}{4a};\ \frac{d^3y}{dx^3} = 0$$

49. (a)

50. (a)

51. (b) The differential equation $\frac{dy}{dx}+2y\tan x=\sin x$ is of the form $\frac{dy}{dx}+Py=Q$, where $P=2\tan x$ and $Q=\sin x$

Therefore, I.F. $=e^{\int 2\tan x dx}=e^{[\log(\sec x)^2]}=\sec^2 x$

$\Rightarrow \sec^2 x\frac{dy}{dx}+2\tan x\sec^2 x\, y$

$=\sin x\sec^2 x$

Integrating, we get

$y\sec^2 x=\int \sin x\sec^2 x\,dx+c$

$y\sec^2 x=\sec x+c$

52. (a)

53. (a) $(1+y^2)+\left(x-e^{\tan^{-1}y}\right)\frac{dy}{dx}=0$

$\Rightarrow \quad (1+y^2)\frac{dx}{dy}+x=e^{\tan^{-1}y}$

which is a linear differential equation of the form.

$\frac{dx}{dy}+Px=Q$

I.F. $=e^{\int\frac{1}{1+y^2}dy}=e^{\tan^{-1}y}$

$\therefore$ The solution is

$xe^{\tan^{-1}y}=\int e^{\tan^{-1}y}.\frac{e^{\tan^{-1}y}}{1+y^2}dy+c$

$=\frac{1}{2}e^{2\tan^{-1}y}+c$

$\therefore 2xe^{\tan^{-1}y}=e^{2\tan^{-1}y}+c$

54. (a)

55. (a) $\frac{dy}{dx}-\frac{\tan y}{x}=\frac{\tan y\sin y}{x^2}$

$\Rightarrow \cot y\,\text{cosec}\,y\frac{dy}{dx}-\frac{\text{cosec}^2 y}{x}=\frac{1}{x^2}$

which is Bernoulli's equation.

Put $-\text{cosec}\,y=v$;

$\frac{dv}{dx}+\frac{v}{x}=\frac{1}{x^2}$

$\therefore$ I.F. $=e^{\int\frac{1}{x}dx}=x$

$\therefore vx=\int x\frac{1}{x^2}dx+c$

$-x\,\text{cosec}\,y=\ln x+c$

$\frac{x}{\sin y}+\ln x=c$

56. (b) $\frac{dy}{dx}+\frac{2x}{1+x^2}y=\frac{1}{\left(1+x^2\right)^2}$

$\therefore$ I.F. $=e^{\int\frac{2x}{1+x^2}dx}=e^{\log\left(1+x^2\right)}=\left(1+x^2\right)$

$\therefore y\left(1+x^2\right)=\int\left(1+x^2\right)\frac{1}{\left(1+x^2\right)^2}dx$

$=\int\frac{1}{1+x^2}dx=\tan^{-1}x+c$

57. (d)

58. (c) Differentiating, we get

$2yy'=2c\Rightarrow c=yy'$

$\therefore y^2=2yy'\left(x+\sqrt{yy'}\right)$

$\Rightarrow\left(y^2-2yy'x\right)^2=4(yy')^3$

$\Rightarrow$ Degree = 3 and Order = 1

59. (b) Differentiating w.r.t x we get

$2Ax+2By\,y'=0$ ...(i)

Again differentiating, we get

$2Ax+2B\left[yy''+(y')^2\right]=0$ ...(ii)

Eliminating A and B by solving equation (i) and (ii) we get

$yy''+(y')^2=\frac{y}{x}y'$

$\Rightarrow$ Degree is one and Order is 2.

60. (a) $(x-a)^2+(y-2)^2=5^2$ is equation of family of circles.

On differentiating w.r.t x on both sides, we get

$x+(y-2)y'=a$

Hence,

$\left[x-x-(y-2)y'\right]^2+(y-2)^2=5^2$

$\Rightarrow(y-2)^2(y')^2=25-(y-2)^2$

61. (d) Since, the length of latus rectum is the distance of the point (2, −3) from the line $3x+4y=5$.

$4a=\frac{|3(2)+4(-3)-5|}{5}=\frac{11}{5}$

$\therefore(x-h)^2=\frac{11}{5}(y-k)$ ...(i)

Differentiate (i) w.r.t x; we get

$2(x-h)=\frac{11}{5}\frac{dy}{dx}$

Again differentiate w.r.t x; we get

$2=\frac{11}{5}\frac{d^2y}{dx^2}\Rightarrow \frac{11d^2y}{dx^2}=10$

62. (a) $\frac{dy}{dx}=\frac{2^x\left(y+2^y\right)}{2^x\left(1+2^y\log_e 2\right)}$

$\int\frac{\left(1+2^y\log_e^2\right)}{\left(y+2^y\right)}dy=\int dx$

$\Rightarrow \log_e|y+2y|=x+c$

Put $x=0,\ y=0\Rightarrow c=0$

$\Rightarrow x=\log_e 1y+2^y\,|$

$\Rightarrow$ Put $y=1,\ x=\log_e 3$

$\because 3\in\left(e,e^2\right)\Rightarrow x\in(1,2)$

63. (a) $\because |A|=-\frac{y}{x}+2\sin x+2$

Now, $\frac{dy}{dx}=|A|$

$\Rightarrow \frac{dy}{dx}=-\frac{y}{x}+2\sin x+2$

$\frac{dy}{dx}+\frac{y}{x}=2\sin x+2$

Then, I.F. $=e^{\int\frac{1}{x}dx}=x$

$\Rightarrow yx=\int x(2\sin x+2)\,dx$

$xy=x^2-2x\cos x+2\sin x+C$ ...(i)

$\because y(\pi)=\pi+2$

Use in Eq. (i), then

$C=0$.

Now from (i),

$xy=x^2-2x\cos x+2\sin x$

Now, put $x=\frac{\pi}{2}$ in the above equation.

$\frac{\pi}{2}y=\left(\frac{\pi}{2}\right)^2-2.\frac{\pi}{2}\cos\frac{\pi}{2}+2\sin\frac{\pi}{2}$

$\frac{\pi}{2}y=\frac{\pi^2}{4}+2\Rightarrow y=\frac{\pi}{2}+\frac{4}{\pi}$

64. (a) $\frac{dy}{dx}-y\tan x=\frac{1}{\cos x(3\sin x+\cos x+3)}$

I.F. $=e^{-\int\tan x\,dx}=e^{\log\cos x}=\cos x$

$\therefore \cos x\cdot y=\int\frac{dx}{3\sin x+\cos x+3}$

$\Rightarrow y.\cos x=\int\frac{\sec^2\frac{x}{2}}{4+6\tan\frac{x}{2}+2\tan^2\frac{x}{2}}dx$

Let $\tan\frac{x}{2}=t\Rightarrow\frac{1}{2}\sec^2\frac{x}{2}dx=dt$

$\Rightarrow y\cos x=\int\frac{dt}{t^2+3t+2}=\int\left(\frac{1}{t+1}-\frac{1}{t+2}\right)dt$

$\Rightarrow y\cos x=\ln\left|\frac{\tan\frac{x}{2}+1}{\tan\frac{x}{2}+2}\right|+C$

$\because y(0)=0\Rightarrow C=\ln 2$

Then $y\left(\frac{\pi}{3}\right)=2\left[\ln\left(\frac{\sqrt{3}+1}{2\sqrt{3}+1}\right)+\ln 2\right]$

65. (a) Since, $x^2=4b(y+b)$

$x^2=4by+4b^2$

$2x=4by'$

$\Rightarrow b=\frac{x}{2y'}$

So, differential equation is

$x^2=\frac{2x}{y'}.y+\left(\frac{x}{y'}\right)^2$

$x(y')^2=2yy'+x$

$\sqrt{1+x^2}\cdot\sqrt{1+y^2}=-xy\frac{dy}{dx}$

66. (a) $\int\frac{\sqrt{1+x^2}}{x}dx=-\int\frac{y}{\sqrt{1+y^2}}dy$

Let $x=\tan\theta\Rightarrow dx=\sec^2\theta\,d\theta$

$\Rightarrow\int\frac{\sec^3\theta\,d\theta}{\tan\theta}=-\int\frac{2y}{2\sqrt{1+y^2}}dy$

$\Rightarrow\int\frac{\sin^2\theta+\cos^2\theta}{\sin\theta\cdot\cos^2\theta}d\theta=-\sqrt{1+y^2}$

$\Rightarrow\int(\tan\theta\cdot\sec\theta+\text{cosec}\theta)\,d\theta=-\sqrt{1+y^2}$

$\Rightarrow\sec\theta+\log_e|\text{cosec}\theta-\cot\theta|=-\sqrt{1+y^2}+C$

$\therefore\sqrt{1+x^2}+\log_e\left|\frac{\sqrt{1+x^2}-1}{x}\right|=-\sqrt{1+y^2}+C$

$\Rightarrow\sqrt{1+y^2}+\sqrt{1+x^2}=\frac{1}{2}\ln\left(\frac{\sqrt{1+x^2}+1}{\sqrt{1+x^2}-1}\right)+C$

67. (a) $\because y=\left(\frac{2}{\pi}x-1\right)\operatorname{cosec} x$

$$\frac{dy}{dx}=\frac{2}{\pi}\operatorname{cosec} x-\left(\frac{2}{\pi}x-1\right)\operatorname{cosec} x.\cot x$$

$$=\operatorname{cosec} x\left[\frac{2}{\pi}-\left(\frac{2}{\pi}x-1\right)\cot x\right]$$

$$\Rightarrow \frac{dy}{dx}-\frac{2}{\pi}\cos ex=y\cot x \quad \text{...(i)}$$

It is given that,

$$\Rightarrow \frac{dy}{dx}-\frac{2}{\pi}\operatorname{cosec} x=-yp(x) \quad \text{...(i)}$$

By comparison of (i) and (ii), we get

$p(x) = \cot x$

68. (b)

$$\frac{5+e^x}{2+y}\cdot\frac{dy}{dx}=-e^x$$

$$\int\frac{dy}{2+y}=-\int\frac{e^x}{5+e^x}dx$$

$$\Rightarrow \log_e|2+y|\cdot\log_e\left|5+e^x\right|=\log_e C$$

$$\Rightarrow \left|(2+y)\left(5+e^x\right)\right|=C \because y(0)=1$$

$$C=18$$

$$\therefore (2+y)\cdot\left(5+e^x\right)=18$$

When $x=\log_e 13$ then $(2+y).18=18$

$$\Rightarrow 2+y=\pm 1$$

$$\therefore y=-1,-3$$

$$\therefore y(\ln 13)=-1$$

69. (a) Let $y + 3x = t$

$$\Rightarrow \frac{dy}{dx}+3=\frac{dt}{dx}$$

Putting these values in the given differential equation

$$\frac{dt}{dx}=\frac{t}{\log_e t}$$

$$\Rightarrow \int\frac{\log_e t}{t}dt=\int dx$$

$$\Rightarrow \frac{(\log_e t)^2}{2}=x-C$$

$$\Rightarrow x-\frac{1}{2}(\ln(y+3x))^2=C$$

70. (a) $\lim\limits_{t\to x}\dfrac{t^2f^2(x)-x^2f^2(t)}{t-x}=0$

$$\Rightarrow \lim_{t\to x}\frac{2tf^2(x)-2x^2f(t)\cdot f'(t)}{1}=0$$

Using L'Hospital's rule

$$\Rightarrow f(x)=xf'(x)$$

$$\int\frac{f'(x)}{f(x)}dx=\int\frac{1}{x}dx$$

$$\log_e f(x)=\log_e x+\log_e C$$

$$\Rightarrow f(x)=Cx,$$

$$\Rightarrow C=e;\text{so } f(x)=ex$$

When $f(x)=1=ex\Rightarrow x=\frac{1}{e}$

71. (c) $\int\left(\dfrac{y^2+1}{y^2}\right)dy=\int\dfrac{e^x dx}{e^x+1}$

$$\Rightarrow y-\frac{1}{y}=\log_e\left|e^x+1\right|+c$$

$\because$ It passes through (0, 1).

$$\therefore c=-\log_e 2$$

$$\Rightarrow y^2-1=y\log_e\left(\frac{e^x+1}{2}\right)$$

$$\Rightarrow y^2=1+y\log_e\left(\frac{e^x+1}{2}\right)$$

72. (b) $x^3dy+xy\,dyx=2y\,dx+x^2dy$

$$\Rightarrow \left(x^3-x^2\right)dy=(2-x)y\,dx$$

$$\Rightarrow \frac{dy}{y}=\frac{2-x}{x^2(x-1)}dx$$

$$\Rightarrow \int\frac{dy}{y}=\int\frac{2-x}{x^2(x-1)}dx$$

Let $\dfrac{2-x}{x^2(x-1)}=\dfrac{A}{x}+\dfrac{B}{x^2}+\dfrac{C}{x-1}$

$$\Rightarrow 2-x=A(x-1)+B(x-1)+Cx^2$$

Compare the coefficients of x, x^2 and constant term.

C = 1, B = −2 and A = −1

$$\therefore \int\frac{dy}{y}=\int\left\{\frac{-1}{x}-\frac{2}{x^2}+\frac{1}{x-1}\right\}dx$$

$$\Rightarrow \ln y=-\ln x+\frac{2}{x}+\ln|x-1|+C$$

$$\therefore y(2)=e$$

$$\Rightarrow 1=-\ln 2+1+0+C \qquad [\because \log e=1]$$

$$\Rightarrow C=\ln 2$$

$$\Rightarrow \ln y=-\ln x+\frac{2}{x}+\ln|x-1|+\ln 2$$

At $x = 4$,

$\Rightarrow \ln y(4) = -\ln 4 + \frac{1}{2} + \ln 3 + \ln 2$

$\Rightarrow \ln y(4) = \ln\left(\frac{3}{2}\right) + \frac{1}{2} = \ln\left(\frac{3}{2}e^{1/2}\right)$

$[\because \log m + \log n = \log(mn)]$

$\Rightarrow y(4) = \frac{3}{2}e^{1/2}$

73. (c) The given differential equation is

$\frac{2+\sin x}{y+1}\frac{dy}{dx} = -\cos x, y > 0$

$\Rightarrow \frac{dy}{y+1} = -\frac{\cos x}{2+\sin x}dx$

Integrate both sides,

$\int \frac{dy}{y+1} = \int \frac{(-\cos x)dx}{2+\sin x}$

$\ln|y+1| = -\ln|2+\sin x| + \ln C$

$\Rightarrow \ln|y+1| + \ln|2+\sin x| = \ln C$

$\Rightarrow \ln|(y+1)(2+\sin x)| = \ln C$

$\because y(0) = 1 \Rightarrow \ln 4 = \ln C \Rightarrow C = 4$

$\therefore (y+1)(2+\sin x) = 4$

$\Rightarrow y = \frac{4}{2+\sin x} - 1$

$\therefore y = \frac{2-\sin x}{2+\sin x} \Rightarrow y(\pi) = \frac{2-\sin \pi}{2+\sin \pi} = 1$

$\Rightarrow a = 1$

Now, $\frac{dy}{dx} = \frac{(2+\sin x)(-\cos x) - (2-\sin x)\cdot\cos x}{(2+\sin x)^2}$

$\left.\frac{dy}{dx}\right|_{x=\pi} = 1 \Rightarrow b = 1$

Ordered pair (a, b) = (1, 1)

74. (a) $\frac{dy}{dx} = \frac{2xy+y^2}{2x^2}$

It is a homogenous differential equation.

$\therefore$ Put $y = vx$

$\Rightarrow v + x\frac{dv}{dx} = v + \frac{v^2}{2} \Rightarrow \int 2\frac{dv}{v^2} = \int \frac{dx}{x}$

$\Rightarrow \frac{-2}{v} = \log_e x + c \Rightarrow \frac{-2x}{y} = \log_e x + c$

Put $x = 1$, $y = 2$, we get $c = -1$

$\Rightarrow \frac{-2x}{y} = \log_e x - 1$

Hence, put $x = \frac{1}{2} \Rightarrow y = \frac{1}{1+\log_e 2}$

75. (a)

$f'(x) = \tan^{-1}(\sec x + \tan x)$

$= \tan^{-1}\left(\frac{1+\sin x}{\cos x}\right) = \tan^{-1}\left(\frac{1-\cos\left(\frac{\pi}{2}+x\right)}{\sin\left(\frac{\pi}{2}+x\right)}\right)$

$= \tan^{-1}\left(\frac{2\sin^2\left(\frac{\pi}{4}+\frac{x}{2}\right)}{2\sin\left(\frac{\pi}{4}+\frac{x}{2}\right)\cos\left(\frac{\pi}{4}+\frac{x}{2}\right)}\right)$

$= \tan^{-1}\left(\tan\left(\frac{\pi}{4}+\frac{x}{2}\right)\right) = \frac{\pi}{4}+\frac{x}{2}$

Integrate both sides, we get

$\int (f'(x))dx = \int\left(\frac{\pi}{4}+\frac{x}{2}\right)$

$f(x) = \frac{\pi}{4}x + \frac{x^2}{4} + C$

$Q = f(0) = 0$

$C = 0 \Rightarrow f(x) = \frac{\pi}{4} + \frac{x^2}{4}$

So, $f(1) = \frac{\pi+1}{4}$

76. (d) The given differential equation,

$\frac{dy}{dx} = \frac{xy}{x^2+y^2}$

Put $y = vx \Rightarrow \frac{dy}{dx} = v + \frac{dv}{dx}$

Then, $v + x\frac{dv}{dx} = \frac{vx^2}{x^2+v^2x^2} = \frac{v}{1+v^2}$

$\Rightarrow \frac{1+v^2}{v^3}dv = -\frac{1}{x}dx$

$\Rightarrow \int\left(\frac{1}{v^3}+\frac{1}{v}\right)dv = \int\frac{-1}{x}$

$\Rightarrow \frac{-1}{2}\left(\frac{1}{v^2}\right) + \ln v = -\ln x + c$

$\Rightarrow -\frac{x^2}{2y^2} = -\ln y + c \qquad \left[\because v = \frac{y}{x}\right]$

When $x = 1$, $y = 1$, then $-\frac{1}{2} = c$

$\Rightarrow x^2 = y^2(1+2\ln y)$

At $y = e$, $x^2 = e^2(3)$

$\Rightarrow x = \pm\sqrt{3}e$

So, $x = \sqrt{3}e$

77. (b)

$\frac{dy}{dx} = \frac{1}{2}\frac{d}{dx}\left(\sin^{-1} f(x)\right)$

$2y = \sin^{-1} f(x) + C = \sin^{-1}\left(\sin\left(2\tan^{-1x}\right)\right) + C$

$\Rightarrow 2\left(\frac{\pi}{6}\right) = \sin^{-1}\left(\sin\left(\frac{2\pi}{3}\right)\right) + C$

$\frac{\pi}{3} = \frac{\pi}{3} + C \therefore C = 0$

for $x = -\sqrt{3}, 2y = \sin^{-1}\left(\sin\left(\frac{-2\pi}{6}\right)\right) + 0$

$\Rightarrow 2y = \frac{-\pi}{3} \Rightarrow y = \frac{-\pi}{6}$

78. (c) The given differential equation is

$\frac{dy}{\sqrt{1-y^2}} + \frac{dx}{\sqrt{1-x^2}} = 0 \Rightarrow \sin^{-1} y + \sin^{-1} + c$

At $x = \frac{1}{2}, y = \frac{\sqrt{3}}{2} \Rightarrow c = \frac{\pi}{2}$

$\Rightarrow \sin^{-1} y = \cos^{-1} x$

Hence, $y\left(-\frac{1}{\sqrt{2}}\right) = \sin\left(\cos^{-1}\left(-\frac{1}{\sqrt{2}}\right)\right)$

$= \sin\left(\pi - \cos^{-1}\left(\frac{1}{\sqrt{2}}\right)\right) = \frac{1}{\sqrt{2}}$

79. (a) Let $e^y = t$

$e^y \frac{dy}{dx} = \frac{dt}{dx}$

$\therefore \frac{dt}{dx} - t = e^x \qquad \left[\because e^y \frac{dy}{dx} - e^y = e^x\right]$

I.F. $= e^{\int -1.dx} = e^x$

$t\left(e^{-x}\right) = \int e^x.e^x dx \Rightarrow e^{y-x} x + c$

Put $x = 0$, $y = 0$, then we get $c = 1$

$e^{y-x} = x + 1$

$y = x + \log_e(x+1)$

Put $x = 1 \therefore y = 1 + \log_e 2$

80. (c) $\frac{dy}{dx} + 2y\tan x = 2\sin x$

I.F. $= e^{\int 2\tan x dx} = \sec^2 x$

The solution of the differential equation is

$y \times \text{I.F.} = \int \text{I.F.} \times 2\sin x\, dx + C$

$\Rightarrow y \cdot \sec^2 x = \int 2\sin x \cdot \sec^2 x\, dx + C$

$\Rightarrow y\sec^2 x = 2\sec x + C \qquad ...(i)$

When $x = \frac{\pi}{3}, y = 0$; then $C = -4$

$\therefore$ From (i), $y\sec^2 x = 2\sec x - 4$

$\Rightarrow y = \frac{2\sec x - 4}{\sec^2 x} \Rightarrow y\left(\frac{\pi}{4}\right) = \sqrt{2} - 2$

81. (a) $\frac{dy}{dx} - \frac{y}{x} = x(x\cos x + \sin x)$

I.F.$= e^{-\int \frac{1}{x}dx} = \frac{1}{x} \therefore \int d\left(\frac{y}{x}\right) = \int (x\cos x + \sin x)dx$

$\Rightarrow \frac{y}{x} = x\sin x + C \qquad \because y(\pi) = \pi \Rightarrow C = 1$

$y = x^2 \sin x + x \Rightarrow y\left(\frac{\pi}{2}\right) = \frac{\pi^2}{4} + \frac{\pi}{2}$

$y' = 2x\sin x + x^2 \cos x + 1$

$y'' = 2\sin x - x^2 \sin x \Rightarrow y''\left(\frac{\pi}{2}\right) = 2 - \frac{\pi^2}{4}$

$\therefore y''\left(\frac{\pi}{2}\right) + y\left(\frac{\pi}{2}\right) = 2 - \frac{\pi^2}{4} + \frac{\pi^2}{4} + \frac{\pi}{2} = 2 + \frac{\pi}{2}$

82. (a) The given differential equation is $\frac{dx}{dy} + x = y^2$

Comparing with $\frac{dx}{dy} + Px = Q$, where $P = 1$, $Q = y^2$

Now, I.F. $= e^{\int 1.dy} = e^y$

$x.e^y = \int (y^2)e^y.dy = y^2.e^y - \int 2y.e^y.dy$

$= y^2 e^y - 2\left(y.e^y - e^y\right) + C$

$\Rightarrow x.e^y = y^2 e^y - 2ye^y + 2e^y + C$

$\Rightarrow x = y^2 - 2y + 2 + C.e^{-y}$

As $y(0) = 1$, satisfies the given differential eqn.,

$\therefore$ put $x = 0$, $y = 1$ in Eq. (i)

$0 = 1 - 2 + 2 + \frac{C}{e} \qquad C = -e$

$y = 0, x = 0 - 0 + 2 + (-e)\left(e^{-0}\right) \qquad x = 2 - e$

83. (b) The given differential equation can be written as,

$y^2 dx - xydx = x^3 dx$

$\Rightarrow \frac{(ydx - xdy)y}{x^2} = xdx \Rightarrow -yd\left(\frac{y}{x}\right) = xdx$

$\Rightarrow -\frac{y}{x}.d\left(\frac{y}{x}\right) = dx \Rightarrow -\frac{1}{2}\left(\frac{y}{x}\right)^2 = x + c_1$

$\Rightarrow 2x^3 + cx^2 + y^2 = 0$

[Here, $c = 2c_1$]

84. (c)
$\cos xdy-(\sin x)ydx=6xdx$

$\Rightarrow \int d(y\cos x)$

$=\int 6xdx \Rightarrow y\cos x=3x^2+C$ ...(i)

Given, $y\left(\frac{\pi}{3}\right)=0$

Putting $x=\frac{\pi}{3}$ and $y=0$ in equation (i), we get

$(10)\times\left(\frac{1}{2}\right)=\frac{3\pi^2}{9}+C\Rightarrow C\frac{-\pi^2}{3}$

So, from equation (i) $y\cos x=3x^2-\frac{\pi^2}{3}$

Now, put $x=\frac{\pi}{6}$in the above equation,

$y\frac{\sqrt{3}}{2}=\frac{3\pi^2}{36}-\frac{\pi^2}{3}\Rightarrow\frac{\sqrt{3}y}{2}=\frac{-3\pi^2}{12}\Rightarrow y=\frac{-\pi^2}{2\sqrt{3}}$

85. (b) The given differential equation

$\frac{dy}{dx}=(x-y)^2$ (i)

Let $x-y=t\Rightarrow 1-\frac{dy}{dx}=\frac{dt}{dx}$

$\Rightarrow\frac{dy}{dx}=1-\frac{dt}{dx}$

Now, from equation (i)

$\left(1-\frac{dt}{dx}\right)=t^2$

$\Rightarrow 1-t^2=\frac{dt}{dx}\Rightarrow\int dx=\int\frac{dt}{1-t^2}$

$\Rightarrow -x=\frac{1}{2\times 1}\ln\left|\frac{t-1}{t+1}\right|+c$

$\Rightarrow -x=\frac{1}{2}\ln\left|\frac{x-y-1}{x-y+1}\right|+c$

Q: The given condition $y(1)=1$

$-1=\frac{1}{2}\ln\left|\frac{1-1-1}{1-1+1}\right|+c\Rightarrow c=-1$

Hence, $2(x-1)=-\ln\left|\frac{1-x+y}{1-y+x}\right|$

86. (a) $\left(x^2-y^2\right)dx+2xy\,dy=0$

$y^2dx-2xydy=x^2dx$

$2xydy-y^2dx=-x^2dx$

$d\left(xy^2\right)=-x^2dx$

$\frac{xd\left(y^2\right)-y^2d(x)}{x^2}-dx$

$d\left(\frac{y^2}{x}\right)=-dx\int d\left(\frac{y^2}{x}\right)=-\int dx$

$\frac{y^2}{x}=-x+C$ ...(i)

Since the above curve passes through the point (1, 1)

$\frac{1^2}{1}=-1+C\Rightarrow C=2$

Now, the curve (i) becomes

$y^2=-x^2+2x$

$\Rightarrow y^2=-(x-1)^2+1$

$(x-1)^2+y^2=1$

The above equation represents a circle with centre (1, 0) and the centre lies on the x-axis.

87. (a) $f(xy)=f(x).f(y)$ ...(i)

Put $x=y=0$ in (i) to get $f(0)=1$

Put $x=y=1$ in (i) to get $f(1)=0$ or $f(1)=1$

$f(1)=0$ is rejected else $y=1$ in (i) gives $f(x)=0$ imply $f(0)=0$

Hence, $f(0)=1$ and $f(1)=1$

By first principle of derivative formula,

$f'(x)=\lim_{h\to 0}\frac{f(x+h)-f(x)}{h}$

$=\lim_{h\to 0}f(x)\left(\frac{f\left(1+\frac{h}{x}\right)-f(1)}{h}\right)$

$\Rightarrow f'(x)=\frac{f(x)}{x}f'(1)$

$\Rightarrow\frac{f'(x)}{f(x)}=\frac{k}{x}\Rightarrow\ln f(x)=k\ln x+c$

$f(1)=1\Rightarrow\ln 1=k\ln 1+c\Rightarrow c=0$

$\Rightarrow\ln f(x)=k\ln x\Rightarrow f(x)=x^k$ but $f(0)=1$

$\Rightarrow k=0$

$\therefore f(x)=1$

$\frac{dy}{dx}=f(x)=1\Rightarrow y=x+c, y(0)=1\Rightarrow c=1$

$\Rightarrow y=x+1$

$\therefore y\left(\frac{1}{4}\right)+y\left(\frac{3}{4}\right)=\frac{1}{4}+1+\frac{3}{4}+1=3$

88. (b) Consider the differential equation,

$y^2dx+\left(x-\frac{1}{y}\right)dy=0$

$\Rightarrow \frac{dx}{dy} + \left(\frac{1}{y^2}\right)x = \frac{1}{y^3}$

I.F. $= e^{\int \frac{1}{y^2}dy} = e^{-\frac{1}{y}}$

$\therefore x.e^{-\frac{1}{y}} = \int e^{-\frac{1}{y}} \frac{1}{y^3}dy + c$

Put $-\frac{1}{y} = u \Rightarrow \frac{1}{y^2}dy = du$

$\Rightarrow x.e^{-\frac{1}{y}} = -\int ue^u du + c = -ue^u + e^u + c$

$\Rightarrow x.e^{-\frac{1}{y}} = e^{-\frac{1}{y}}\left(\frac{1}{y}+1\right) + c$ At $y = 1, x = 1$

$1 = 2 + ce \Rightarrow c = -\frac{1}{e} \Rightarrow x = \left(1+\frac{1}{y}\right) - \frac{1}{e}e^{\frac{1}{y}}$

On putting $y = 2$, we get $x = \frac{3}{2} - \frac{1}{\sqrt{e}}$

89. (d) The given differential equation is,

$\frac{dy}{dx} + y\tan x = 2x + x^2 \tan x$

Here, $P = \tan x, Q = 2x + x^2 \tan x$

I.F. $= e^{\int \tan x dx} = e^{\ln|\sec x|} = |\sec x|$

$\therefore y(\sec x) = \int (2x + x^2 \tan x)\sec x dx$

$= \int x^2 \tan x \sec x dx + \int 2x \sec x dx = x^2 \sec x + c$

Given $y(0) = 1 \Rightarrow c = 1$

$\therefore y = x^2 + \cos x$ ...(i)

Now put $x = \frac{\pi}{4}$ and $x = \frac{-\pi}{4}$ in eq. (i)

$y\left(\frac{\pi}{4}\right) = \frac{\pi^2}{16} + \frac{1}{\sqrt{2}}$ and $y\left(-\frac{\pi}{4}\right) = \frac{\pi^2}{16} + \frac{1}{\sqrt{2}}$

$\Rightarrow y\left(\frac{\pi}{4}\right) - y\left(-\frac{\pi}{4}\right) = 0$

$\frac{dy}{dx} = 2x - \sin x$

$\therefore y'\left(\frac{\pi}{4}\right) = \frac{\pi}{2} - \frac{1}{\sqrt{2}}$ and $y'\left(-\frac{\pi}{4}\right) = -\frac{\pi}{2} + \frac{1}{\sqrt{2}}$

$\Rightarrow y'\left(\frac{\pi}{4}\right) - y'\left(-\frac{\pi}{4}\right) = \pi - \sqrt{2}$

90. (c) $\frac{dy}{dx} + \frac{2}{x}y = x \quad y(1) = 1$ (given)

Since the above differential equation is the linear differential equation,

I.F. $= e^{\int \frac{2}{x}dx} = x^2$

Now, the solution of the linear differential equation

$y \times x^2 = \int x^3 dx$

$\Rightarrow yx^2 = \frac{x^4}{4} + C \quad \because y(1) = 1$

$\therefore 1 \times 1 = \frac{1}{4} + C \Rightarrow C = \frac{3}{4}$

$\therefore$ The solution becomes, $y = \frac{x^2}{4} + \frac{3}{4x^2}$

91. (d) $(1+x^2)^2 \frac{dy}{dx} + 2x(1+x^2)y = 1$

$\Rightarrow \frac{dy}{dx} + \left(\frac{2x}{1+x^2}\right)y = \frac{1}{(1+x^2)^2}$

Since the above differential equation is a linear differential equation

$\therefore$ I.F. $= e^{\int \frac{2}{1+x^2}dx} = e^{\log(1+x^2)} = 1 + x^2$

Then, the solution of the differential equation.

$\Rightarrow y(1+x^2) = \int \frac{dx}{1+x^2} + c$

$\Rightarrow y(1+x^2) = \tan^{-1}x + c$ ...(i)

If $x = 0$ then $y = 0$ (given)

$\Rightarrow 0 = 0 + c$

$\Rightarrow c = 0$

Then, equation (i) becomes

$\Rightarrow y(1+x^2) = \tan^{-1}x$

Now put $x = 1$ in above equation, then

$2y = \frac{\pi}{4}$

$\Rightarrow 2\left(\frac{\pi}{32\sqrt{a}}\right) = \frac{\pi}{4} \quad \left[\sqrt{a}y(1) = \frac{\pi}{32}\right]$

$\Rightarrow \sqrt{a} = \frac{1}{4} \quad \Rightarrow a = \frac{1}{16}$

92. (c) Consider the differential equation,

$\frac{dy}{dx} + \frac{y}{x} = \log_e x$

I.F. $= e^{\int \frac{1}{x}dx} = x$

$yx = \int x \ln x dx$

$\Rightarrow xy = \ln x.\frac{x^2}{2} - \int \frac{1}{x}.\frac{x^2}{2}dx$

$\Rightarrow xy = \frac{x^2}{2}.\ln x - \frac{x^2}{4} + c$

Given, $2y(2) = \log_e 4 - 1$

$2y = 2\ln 2 - 1 + c$

$\Rightarrow \ln 4 - 1 = \ln 4 - 1 + c$

i.e., $c=0$

$\Rightarrow xy=\frac{x^2}{2}\ln x-\frac{x^2}{4}\Rightarrow y=\frac{x}{2}\ln x-\frac{x}{4}$

$\Rightarrow y(e)=\frac{e}{2}-\frac{e}{4}=\frac{e}{4}$

93. (c) The given differential equation is,

$\frac{dy}{dx}+\left(2+\frac{1}{x}\right)y=e^{-2x},x>0$

$\text{I.F.}=e^{\int\left(2+\frac{1}{x}\right)dx}=e^{2x+\ln x}=xe^{2x}$

The complete solution is given by

$y(x).xe^{2x}=\int xe^{2x}.e^{-2x}dx+c$

$=\int xdx+c$

$y(x).e^{2x}.x=\frac{x^2}{2}+c$

Given, $y(1)=\frac{1}{2}e^{-2}$

$\therefore \frac{1}{2}e^{-2}.e^2.1=\frac{1}{2}+c\Rightarrow c=0$

$\therefore y(x)=\frac{x^2}{2}.\frac{e^{-2x}}{x}\ y(x)=\frac{x}{2}.e^{-2x}$

Differentiating both sides with respect to x,

$y'(x)=\frac{e^{-2x}}{2}(1-2x)<0\,\forall\, x\in\left(\frac{1}{2},1\right)$

Hence, $y(x)$ is decreasing in $\left(\frac{1}{2},1\right)$

94. (c) Since, $x\frac{dy}{dx}+2y=x^2$

$\Rightarrow \frac{dy}{dx}+\frac{2}{x}y=x$

$\text{I.F.}=e^{\int\frac{2}{x}dx}=e^{2\ln x}=e^{\ln x^2}=x^2$

The solution of the differential equation is:

$y.x^2=\int x.x^2dx \quad y.x^2=\frac{x^4}{4}+C \quad$ ...(i)

$\because y(1)=1 \quad \therefore C=\frac{3}{4}$

Then, from equation (i),

$y.x^2=\frac{x^4}{4}+\frac{3}{4}$

$\therefore y=\frac{x^2}{4}+\frac{3}{4x^2} \quad \therefore y\left(\frac{1}{2}\right)=\frac{1}{16}+3=\frac{49}{16}$

95. (c) $\frac{x^2}{a}+\frac{y^2}{b}=1$

Since it passes through (0, 3),

$\therefore \frac{0}{a^2}+\frac{9}{b^2}=1$

$\Rightarrow b^2=9$

$\therefore$ Equation of ellipse becomes

$\frac{x^2}{a^2}+\frac{y^2}{9}=1$

Differential w.r.t x, we get

$\frac{2x}{a^2}+\frac{2y}{9}\frac{dy}{dx}=0$

$\Rightarrow \frac{y}{x}\left(\frac{dy}{dx}\right)=\frac{-9}{a^2}$

Again differentiating w.r.t x, we get,

$\frac{y}{x}\frac{d^2y}{dx^2}+\frac{x\frac{dy}{dx}-y}{x^2}\frac{dy}{dx}=0$

$\Rightarrow xyy''+x(y')^2-yy'=0$

96. (b) $\left(x^2-y^2\right)dx+2xydy=0$

$\Rightarrow \frac{dy}{dx}=\frac{y^2-x^2}{2xy}$

Let $y=vx$

$\frac{dy}{dx}=v+x\frac{dv}{dx}$

$\Rightarrow v+x\frac{dv}{dx}=\frac{v^2x^2-x^2}{2vx^2}\Rightarrow v+x\frac{dv}{dx}=\frac{v^2-1}{2v}$

$x\frac{dv}{dx}=\frac{v^2-1}{2v}-v=-\left(\frac{1+v^2}{2v}\right)$

$\int\frac{2v}{1+v^2}dv=-\int\frac{dx}{x}+\ln c$

$\ln(1+v^2)=-\ln x+\ln c=\ln\frac{c}{x}$

$1+v^2=\frac{c}{x}$

$x^2+y^2=\frac{c}{x}$

which passes through (1, 1). Therefore,

$c=2$

So, $x^2+y^2-2x=0$

Therefore, the curve is a circle.

97. (b) Consider the given differential equation the

$\sin xdy+y\cos xdx=4xdx$

$\Rightarrow d(y.\sin x)=4xdx$

Integrate both sides

$\Rightarrow y.\sin x=2x^2+C \quad$ ...(i)

$\Rightarrow y(x) = \dfrac{2x^2}{\sin x} + c$ ...(ii)

$\because$ equation (ii) passes through $\left(\dfrac{\pi}{2}, 0\right)$

$\Rightarrow 0 = \dfrac{\pi^2}{2} + C \qquad \Rightarrow C = -\dfrac{\pi^2}{2}$

Now, put the value of C in (i)

Then, $y \sin x = 2x^2 - \dfrac{\pi^2}{2}$ is the solution

$\therefore y\left(\dfrac{\pi}{6}\right) = \left(2.\dfrac{\pi^2}{36} - \dfrac{\pi^2}{2}\right)2 = -\dfrac{8\pi^2}{9}$

98. (a)

When $x \in [0,1]$, then $\dfrac{dy}{dx} + 2y = 1 \Rightarrow y = \dfrac{1}{2} + C_1 e^{-2x}$

$\because y(0) = 0 \Rightarrow y(x) = \dfrac{1}{2} - \dfrac{1}{2}e^{-2x}$

Here, $y(1) = \dfrac{1}{2} - \dfrac{1}{2}e^{-2} = \dfrac{e^2 - 1}{2e^2}$

When $x \in [0,1]$, then $\dfrac{dy}{dx} + 2y = 0 \Rightarrow y = c_2 e^{-2x}$

$\because y(1) = \dfrac{e^2-1}{2} \Rightarrow \dfrac{e^2-1}{2} = c^2 e^{-2} \Rightarrow C_2 = \dfrac{e^2-1}{2}$

$\therefore y(x)\left(\dfrac{e^2-1}{2}\right)e^{-2x} \Rightarrow y\left(\dfrac{3}{2}\right) = \dfrac{e^2-1}{2e^3}$

99. (b) We have $(2 + \sin x)\dfrac{dy}{dx} + (y+1)\cos x = 0$

$\Rightarrow \dfrac{d}{dx}(2 + \sin x)(y+1) = 0$

On integrating, we get

$(2 + \sin x)(y+1) = C$

At $x = 0, y = 1$ we have

$(2 + \sin 0)(1+1) = C$

$\Rightarrow C = 4$

$\Rightarrow y + 1 = \dfrac{4}{2 + \sin x}$

$y = \dfrac{4}{2 + \sin x} - 1$

Now $y\left(\dfrac{\pi}{2}\right) = \dfrac{4}{2 + \sin\dfrac{\pi}{2}} - 1 = \dfrac{4}{3} - 1 = \dfrac{1}{3}$

100. (b) $y\,dx - x\,dy - 3y^2 dy = 0$

$\Rightarrow \dfrac{dx}{dy} = \dfrac{x}{y} + 3y \Rightarrow \dfrac{dx}{dy} - \dfrac{x}{y} = 3y$

$\text{I.F.} = e^{-\int \frac{1}{y} dy} = e^{-\ln y} = \dfrac{1}{y}$

$\therefore$ Solution is $\dfrac{x}{y} = \int 3y.\dfrac{1}{y}dy \Rightarrow \dfrac{x}{y} = 3y + C$

which passes through (1, 1)

$\therefore 1 = 3 + C \Rightarrow C = -2$

$\therefore$ Solution becomes $\Rightarrow x = 3y^2 - 2y$

which also passes through $\left(-\dfrac{1}{3}, \dfrac{1}{3}\right)$

101. (d) Let $L = \lim\limits_{t \to x} \dfrac{t^2 f(x) - x^2 f(t)}{t -} = 1$

Applying L.H. rule

$L = \lim\limits_{t \to x} \dfrac{2t\, f(x) - x^2 f'(t)}{1} = 1$

$2xf(x) - x^2 f'(x) = 1$

Solving the above differential equation, we get

$f(x) = \dfrac{2}{3}x^2 + \dfrac{1}{3x}$

Put $x = \dfrac{3}{2}$

$f\left(\dfrac{3}{2}\right) = \dfrac{2}{3} \times \left(\dfrac{3}{2}\right)^2 + \dfrac{1}{3} \times \dfrac{2}{3} = \dfrac{3}{2} + \dfrac{2}{9} = \dfrac{27+4}{18} = \dfrac{31}{18}$

102. (c) Since the family of all circles touches x-axis at the origin

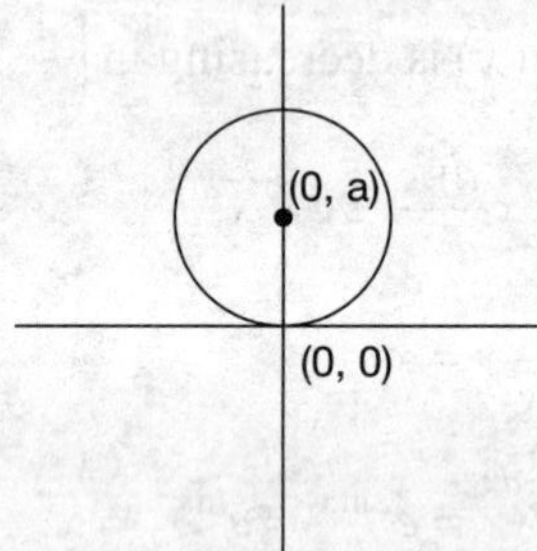

$\therefore$ The equation is $(x)^2 + (y-a)^2 = a^2$

where $(0, a)$ is the centre of the circle.

$\Rightarrow x^2 + y^2 + a^2 - 2ay = a^2$

$\Rightarrow x^2 + y^2 - 2ay = 0$ ...(i)

Differentiate both sides w.r.t x, we get

$2x + 2y\dfrac{dy}{dx} - 2a\dfrac{dy}{dx} = 0$

$\Rightarrow x + y\dfrac{dy}{dx} - a\dfrac{dy}{dx} = 0$

$\Rightarrow \dfrac{x + y\dfrac{dy}{dx}}{\dfrac{dy}{dx}} = a$

Put value of a in equation (i), we get

$$x^2+y^2-2y\left[\frac{y\frac{dy}{dx}+x}{\frac{dy}{dx}}\right]=0$$

$$\Rightarrow \left(x^2+y^2\right)\frac{dy}{dx}-2y^2\frac{dy}{dx}-2xy=0$$

$$\Rightarrow \left(x^2+y^2-2y^2\right)\frac{dy}{dx}=2xy$$

$$\Rightarrow \left(x^2-y^2\right)\frac{dy}{dx}=2xy\equiv g(x)y$$

Hence, $g(x)=2x$

Integer and Numerical Value Type

1. Since, rate of change of volume $\propto$ surface area

$$\Rightarrow \frac{dV}{dt}\propto \text{SA}$$

$$\Rightarrow 4\pi r^2.\frac{dr}{dt}=-\lambda 4\pi r^2$$

$\frac{dr}{dt}=-\lambda$ is the required differential equation.

2. Since the length of tangent $=\left|y\sqrt{1+\left(\frac{dx}{dy}\right)^2}\right|=1$

$$\Rightarrow y^2\left(1+\left(\frac{dx}{dy}\right)^2\right)=1$$

$$\therefore \frac{dy}{dx}=\pm\frac{y}{\sqrt{1-y^2}}$$

$$\Rightarrow \int\frac{\sqrt{1-y^2}}{y}dy=\pm\int xdx$$

$$\Rightarrow \int\frac{\sqrt{1-y^2}}{y}dy=\pm x+C$$

Put $y=\sin\theta \Rightarrow dy=\cos\theta d\theta$

$$\therefore \int\frac{\cos\theta}{\sin\theta}.\cos\theta d\theta=\pm x+C$$

$$\Rightarrow \int\frac{\cos^2\theta}{\sin^2\theta}.\sin\theta d\theta=\pm x+C$$

Again put $\cos\theta=t\Rightarrow -\sin\theta d\theta=dt$

$$\therefore -\int\frac{t^2}{1-t^2}dt=\pm x+C$$

$$\Rightarrow \int\left(1-\frac{1}{1-t^2}\right)dt=\pm x+C$$

$$\Rightarrow t-\log\left|\frac{1+t}{1-t}\right|=\pm x+C$$

$$\Rightarrow \sqrt{1-y^2}-\log\left|\frac{1+\sqrt{1-y^2}}{1-\sqrt{1-y^2}}\right|=\pm x+C$$

3. Given, the liquid evaporates at a rate proportional to its surface area.

$$\Rightarrow \frac{dV}{dt}\propto -S \quad \text{...(i)}$$

We know that, volume of cone $=\frac{1}{3}\pi r^2 h$

and surface area $=\pi r^2$

or $V=\frac{1}{3}\pi r^2 h$ and $S=\pi r^2$ ...(ii)

where, $\tan\theta=\frac{R}{H}$ and $\frac{r}{h}=\tan\theta$ ...(iii)

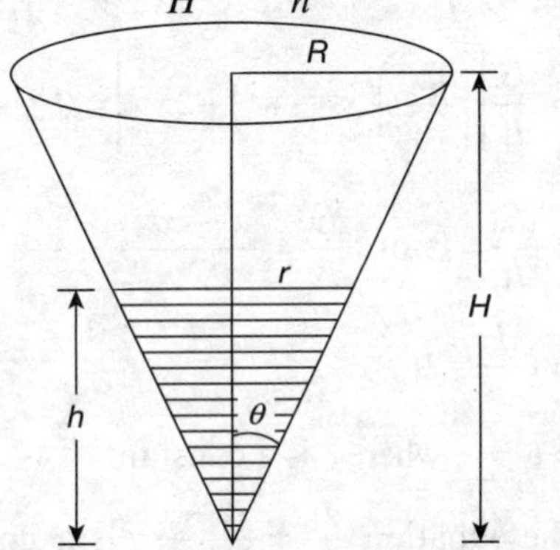

From equations (ii) and (iii), we get

$V=\frac{1}{3}\pi r^3\cot\theta$ and $S=\pi r^2$ ...(iv)

On substituting Eq. (iv) in Eq. (i), we get

$$\frac{1}{3}\cot\theta.3r^2\frac{dr}{dt}=-k\pi r^2$$

$$\Rightarrow \cot\theta\int_R^0 dr=-k\int_0^T dt$$

$$\Rightarrow \cot\theta(0-R)=-k(T-0)$$

$\Rightarrow R\cot\theta=k \Rightarrow H=kT$ [from Eq. (iii)]

$$\Rightarrow T=\frac{H}{k}$$

$\therefore$ Required time after which the cone is empty,

$$T=\frac{H}{k}$$

4. Equation of normal at point (x,y) is

$$Y-y=-\frac{dx}{dy}(X-x) \quad \text{...(i)}$$

Distance of perpendicular from the origin to Eq. (i)

$$=\frac{\left|y+\frac{dx}{dy}.x\right|}{\sqrt{1+\left(\frac{dx}{dy}\right)^2}}$$

Also, distance between P and X-axis is $|y|$.

$$\therefore \frac{\left|y+\frac{dx}{dy}.x\right|}{\sqrt{1+\left(\frac{dx}{dy}\right)^2}}=|y|$$

$$\Rightarrow y^2+\frac{dx}{dy}.x^2+2xy\frac{dx}{dy}=y^2\left[1+\left(\frac{dx}{dy}\right)^2\right]$$

$$\Rightarrow \left(\frac{dx}{dy}\right)^2\left(x^2-y^2\right)+2xy\frac{dx}{dy}=0$$

$$\Rightarrow \frac{dx}{dy}\left[\left(\frac{dx}{dy}\right)\left(x^2-y^2\right)+2xy\right]=0$$

$$\Rightarrow \frac{dx}{dy}=0 \text{ or } \frac{dy}{dx}=\frac{y^2-x^2}{2xy}$$

But $\frac{dx}{dy}=0$

$\Rightarrow x=c$, where c is a constant.

The equation $\frac{dy}{dx}=\frac{y^2-x^2}{2xy}$ is a homogeneous equation.

Put $y=vx\Rightarrow \frac{dy}{dx}=v+x\frac{dv}{dx}$

$$v+x\frac{dy}{dx}=\frac{v^2x^2-x^2}{2x^2v}$$

$$\Rightarrow x\frac{dv}{dx}=\frac{v^2-1}{2v}-v=\frac{v^2-1-2v^2}{2v}=-\frac{v^2+1}{2v}$$

$$\Rightarrow \frac{-2v}{v^2+1}dv=\frac{dx}{x}$$

$$\Rightarrow c_1-\log\left(v^2+1\right)=\log|x|$$

$$\Rightarrow \log|x|\left(v^2+1\right)=c_1\Rightarrow |x|\left(\frac{y^2}{x^2}+1\right)=e^{c_1}$$

$\Rightarrow x^2+y^2=\pm e^{c_1}x$ or $x^2+y^2=\pm e^{c}x$ is passing through (1, 1).

$\therefore 1+1=\pm e^c.1$

$\Rightarrow \pm e^c=2$

Hence, the required curve is $x^2+y^2=2x$

5. $\frac{dV}{dt}\propto V$ for each reservoir.

$\frac{dV}{dx}\propto V_A\Rightarrow \frac{dV_A}{dt}=-K_1V_A$ [K_1 is the proportionality constant]

$$\Rightarrow \int_{V_A}^{V'_A}\frac{dV_A}{V_A}=-K_1\int_0^t dt$$

$$\Rightarrow \log\frac{V'_A}{V_A}=-K_1t\Rightarrow V'_A=V_A.e^{-K_1t} \quad \text{...(i)}$$

Similarly for B, $V'_B=V_B.e^{-K_2t}$ ...(ii)

On dividing Eq. (i) by Eq. (ii), we get

$$\frac{V'_A}{V'_B}=\frac{V_A}{V_B}.e^{-(K_1-K_2)t}$$

It is given that at $t=0, V_A=2V_B$ and at $t=\frac{3}{2}, V'_A=\frac{3}{2}V'_B$

Thus, $\frac{3}{2}=2.e^{-(K_1-K_2)t}\Rightarrow e^{-(K_1-K_2)}=\frac{3}{4}$ ...(iii)

Now, let at $t=t_0$ both the reservoirs have some quantity of water. Then,

$V'_A=V'_B$

From Eq. (iii), $2e^{-(K-K_2)}=1$

$$\Rightarrow 2.\left(\frac{3}{4}\right)^{t_0}=1;\ t_0=\log_{3/4}(1/2)$$

6. The given equation is $\frac{dy}{dx}+g'(x)y=g(x)g'(x)$

I.F. $=e^{\int g'(x)dx}=e^{g(x)}$

Hence solution is $y.e^{g(x)}=\int e^{g(x)}g(x).g'(x)dx$

Put $g(x)=t$ so that $g'(x)dx=dt$

$$y.e^{g(x)}=\int e^t t\,dt=e^t(t-1)+c$$

$$\therefore y.e^{g(x)}=e^{g(x)}\left[g(x)-1\right]+c$$

As $y(0)=0$ and $g(0)=0$, $\therefore c=1$

$$\therefore y.e^{g(x)}=e^{g(x)}\left[g(x)-1\right]+1$$

As $g(2)=0$, putting $x=2$ we get

$$y(2).e^{g(2)}=e^{g(2)}\left[g(2)-1\right]+1\Rightarrow y(2)=0$$

7. Given that

$$y^2=a\left(x+\frac{\sqrt{a}}{2}\right) \quad \text{(i)}$$

Differentiating it w.r.t x

$$2y.\frac{dy}{dx}=a \quad \text{(ii)}$$

From (i) and (ii),

$$y^2=2y\frac{dy}{dx}\left(x+\frac{1}{2}\sqrt{2y\frac{dy}{dx}}\right)$$

$$y-2x\frac{dy}{dx}=\frac{dy}{dx}.\sqrt{2y\frac{dy}{dx}}$$

$$\Rightarrow \left(y-2x\frac{dy}{dx}\right)^2=2y^2\left(\frac{dy}{dx}\right)^3$$

$\Rightarrow$ Order 1 and Degree 3

So, difference = 2

8. It is given that,

$$y^{\frac{1}{4}}+\frac{1}{y^{\frac{1}{4}}}=2x$$

$$\Rightarrow y^{\frac{1}{4}}=x+\sqrt{x^2-1}\text{ or }x-\sqrt{x^2-1}$$

$$\text{So, }\frac{1}{4}\frac{1}{y^{\frac{3}{4}}}\frac{dy}{dx}=1+\frac{x}{\sqrt{x^2-1}}$$

$$\Rightarrow \frac{1}{4}\frac{1}{y^{\frac{3}{4}}}\frac{dy}{dx}=\frac{y^{\frac{1}{4}}}{\sqrt{x^2-1}}$$

$$\Rightarrow \frac{dy}{dx}=\frac{4y}{\sqrt{x^2-1}} \qquad \text{(i)}$$

$$\text{Now, }\frac{d^2y}{dx^2}=4\frac{\left(\sqrt{x^2-1}\right)y'-\frac{yx}{\sqrt{x^2-1}}}{x^2-1}$$

$$\Rightarrow \left(x^2-1\right)y''=4\left(\sqrt{x^2-1}y'-\frac{xy}{\sqrt{x^2-1}}\right)$$

$$\Rightarrow \left(x^2-1\right)y''=4\left(4y-\frac{xy'}{4}\right)\quad\text{(from (i))}$$

$$\Rightarrow \left(x^2-1\right)y''+xy'-16y=0$$

So, $|\alpha-\beta|=17$

9. $\sec y\frac{dy}{dx}=2\sin x\cos y$

$$\left[\because \sin(x+y)+\sin(x-y)=2\sin x.\cos y\right]$$

$$\Rightarrow \int \sec^2 y dy=2\int \sin x dx$$

$$\Rightarrow \tan y=-2\cos x+c$$

$\because y(0)=0$

$\therefore c=2$

$$\Rightarrow \tan y=-2\cos x+2\Rightarrow \text{at } x=\frac{\pi}{2}$$

$$\Rightarrow \tan y=2$$

$$\sec^2 y\frac{dy}{dx}=2\sin x\Rightarrow \left(1+\tan^2 y\right)\frac{dy}{dx}=2\sin x$$

Put $x=\frac{\pi}{2}$ and $\tan y=2$, we get

$$\Rightarrow 5\frac{dy}{dx}=2$$

10. Given, that $\frac{dy}{dx}=\frac{2y}{x\log_e x}$

$$\Rightarrow \int\frac{1}{y}dy=2\int\frac{1}{x\log_e x}dx$$

$$\Rightarrow \log_e y=2\log_e\left(\log_e x\right)+c$$

Put $x=2, y=\left(\log_e 2\right)^2$, we get $c=0$

$$\Rightarrow y=\left(\log_e x^2\right)$$

$$\Rightarrow f(e)=1$$

11. Let $x+2=u\Rightarrow dx=du$

and $y+1=v\Rightarrow dy=v$

$$\Rightarrow \left(ue^{\frac{v}{u}}+v\right)du=u\,dv$$

$$\Rightarrow e^{\frac{v}{u}}+\frac{v}{u}=\frac{dv}{du}$$

Let $\frac{v}{u}=t$, then

$$\frac{dv}{du}=t+\frac{u\,dt}{du}$$

$$\Rightarrow e^t+t=t+\frac{u\,dt}{du}$$

$$\Rightarrow \int\frac{1}{u}du=\int e^{-t}dt$$

$$\log u=-e^{-t}+C$$

$$\log|x+2|=-e^{-\left[\frac{y+1}{x+2}\right]}+C$$

Put $x=1$ and $y=1$

$$\Rightarrow C=\log 3+e^{-\frac{2}{3}}$$

$$\Rightarrow e^{-\left[\frac{y+1}{x+2}\right]}=\log 3+e^{-\frac{2}{3}}-\log|x+2|$$

$$\Rightarrow -\left(\frac{y+1}{x+2}\right)=\log\left|\log 3+e^{-\frac{2}{3}}-\log|x+2|\right|$$

$$\Rightarrow y=-1-(x+2)\log\left|\log 3+e^{-\frac{2}{3}}-\log|x+2|\right|$$

$$=-1-(x+2)\log\left|\log\left|\frac{3}{x+2}\right|+e^{-\frac{2}{3}}\right|$$

Domain:

$$\log\left|\frac{3}{x+2}\right|>-e^{-\frac{2}{3}}$$

$$\Rightarrow \left|\frac{3}{x+2}\right|>e^{-e^{-\frac{2}{3}}}$$

$$\Rightarrow |x+2|<3e^{e^{-\frac{2}{3}}}$$

$$\Rightarrow -3e^{e^{-\frac{2}{3}}}-2<x<3e^{e^{-\frac{2}{3}}}-2$$

So $\alpha+\beta=-4$

$$\Rightarrow |\alpha+\beta|=4$$

12. It is given that,

$$\cos\left(\frac{1}{2}\cos^{-1}\left(e^{-x}\right)\right)dx = \sqrt{e^{2x}-1}dy$$

$$dy = \frac{\cos\left(\frac{1}{2}\cos^{-1}\left(e^{-x}\right)\right)}{\sqrt{e^{2x}-1}}dx$$

Let $\frac{1}{2}\cos^{-1}\left(e^{-x}\right) = \theta$

$\Rightarrow e^{-x} = \cos^2\theta$

$\Rightarrow \ln\left(e^{-x}\right) = \ln(\cos 2\theta)$

$\Rightarrow x = \ln(\sec 2\theta)$

$\Rightarrow dx = 2\tan 2\theta d\theta$

So, $\int dy = \int \frac{\cos\frac{1}{2}\cos^{-1}\left(e^{-x}\right)}{\sqrt{e^{2x}-1}}dx$

$y = \int 2\cos\theta d\theta = 2\sin\theta + C$

$= \sqrt{2}\sqrt{1-\cos 2\theta} + C = \sqrt{2}\sqrt{1-e^{-x}} + C$

$\because y(0) = 1 \Rightarrow C = -1$

$y = \sqrt{2\left(1-e^{-x}\right)} - 1 \because y(\pm) = 0$

$\sqrt{2\left(1-e^{-x}\right)} - 1 = 0$

$\Rightarrow e^{-\alpha} = \frac{1}{2} \Rightarrow e^{\alpha} = 2$

13. $xdy - ydx = \sqrt{x^2-y^2}dx \Rightarrow \frac{dy}{dx} = \frac{\sqrt{x^2-y^2}+y}{x}$

Let $y = vx$

$\Rightarrow \frac{dy}{dx} = v + x\frac{dv}{dx}$

$\Rightarrow v + x\frac{dv}{dx} = \sqrt{1-v^2} + v$

$\Rightarrow \int \frac{1}{\sqrt{1-v^2}}dv = \int \frac{1}{x}dx$

$\Rightarrow \sin^{-1}v = \log x + C$

$\Rightarrow \sin^{-1}\left(\frac{y}{x}\right) = \log x + C$

Put $x = 1$ and $y = 0$ we get $C = 0$

Hence, $y = x.\sin(\log x)$

$\therefore$ Required bounded area

$$A = \int_1^{e^\pi} x.\sin(\log x)dx$$

Let $\log x = t \Rightarrow x = e^t$

$\Rightarrow dx = e^t dt$

When $x = 1, t = 0$ and $x = e^\pi, t = \pi$

$$A = \int_0^\pi e^t.\sin t.e^t dt$$

$$A = \int_0^\pi e^{2t}.\sin t dt$$

$$= \left[\frac{e^{2t}}{1^2+2^2}.(2\sin t - \cos t)\right]_0^\pi$$

$$= \frac{1}{5}\left(e^{2\pi}(0+1) - 1(0-1)\right) = \frac{1}{5}.e^{2\pi} + \frac{1}{5} = \alpha.e^{2\pi} + \beta$$

$\therefore \alpha = \frac{1}{5},\ \beta = \frac{1}{5}$

So, $10(\alpha+\beta) = 4$

14. Given, $2x - 3y = 1$ ...(i)

and $3x + 2y = 8$ ...(ii)

On solving equations (i) and (ii), we get the point of intersection (2, 1)

$\because \left(2xy^2 - y\right)dx + x\,dy = 0$

$\Rightarrow 2x\,dx = \frac{y\,dx - x\,dy}{y^2}$

$\Rightarrow \int 2xdx = \int d\left(\frac{x}{y}\right)$

$\Rightarrow x^2 = \frac{x}{y} + C$

Putting $x = 2$ and $y = 1$, we get $C = 2$

$\therefore y(x) = \frac{x}{x^2-2}$

Now, $y(1) = \frac{1}{1-2} = -1 \Rightarrow |y(1)| = 1$

15. Let $e^y = u \Rightarrow e^y\frac{dy}{dx} = \frac{du}{dx}$

$\Rightarrow \frac{du}{dx} - (2\sin x)u = -\sin x\cos^2 x$

$\text{I.F.} = e^{-2\int \sin x dx = e^{2\cos x}}$

$\Rightarrow u.e^{2\cos x} = \int e^{2\cos x}.\left(-\sin x\cos^2 x\right)dx$

Let $\cos x = t \Rightarrow -\sin x dx = dt$

$\therefore e^y e^{2\cos x} = \int e^{2t}.t^2 dt$

$= \frac{1}{2}t^2.e^{2t} - \int te^{2t}dt$

$= \frac{1}{2}t^2.e^{2t} - \frac{1}{2}t.e^{2t} + \frac{1}{4}e^{2t} + C$

$\Rightarrow e^y.e^{2\cos x}$

$= \frac{1}{2}.\cos^2 x.e^{2\cos x} - \frac{1}{2}\cos x.e^{2\cos x} + \frac{e^{2\cos x}}{4} + C$

Putting $x=\frac{\pi}{2}, y=0 \Rightarrow C=\frac{3}{4}$

$\Rightarrow e^y=\frac{1}{2}\cos^2 x-\frac{1}{2}\cos x+\frac{1}{4}+\frac{3}{4}.e^{-2\cos x}$

$\Rightarrow y=\log\left[\frac{\cos^2 x}{2}-\frac{\cos x}{2}+\frac{1}{4}+\frac{3}{4}e^{-2\cos x}\right]$

Putting $x=0$, we get

$\Rightarrow y=\log\left[\frac{1}{4}+\frac{3}{4}e^{-2}\right]\Rightarrow \alpha=\frac{1}{4},\ \beta=\frac{3}{4}$

$\therefore 4(\alpha+\beta)=4$

16. Let $e^{\sin y}=t$

$\Rightarrow e^{\sin y}.\cos y\frac{dy}{dx}=\frac{dt}{dx}$

$\therefore \frac{dt}{dx}+t\cos x=\cos x$

$\Rightarrow$ I.F. $=e^{\int\cos x dx}=e^{\sin x}$

$\Rightarrow t.e^{\sin x}=\int e^{\sin x}.\cos dx+C$

$\Rightarrow e^{\sin y}.e^{\sin x}=e^{\sin x}+C$

Put $x=0, y=0$, then $C=0$

$\Rightarrow e^{\sin x+\sin y}=e^{\sin x}\Rightarrow \sin y=0$

$\Rightarrow y=0$

$\Rightarrow y(x)=0$

Hence, $1+y\left(\frac{\pi}{6}\right)+\frac{\sqrt{3}}{2}y\left(\frac{\pi}{3}\right)+\frac{1}{\sqrt{2}}y\left(\frac{\pi}{4}\right)=1$

Assertion and Reason

1. (b) **Statement 2:** $\ln\frac{x}{y}-\frac{1}{xy}=c$

i.e., $\ln x-\ln y-\frac{1}{xy}=c$

differentiating w.r.t x

$\frac{1}{x}-\frac{1}{y}\frac{dy}{dx}+\frac{y+x\frac{dy}{dx}}{(xy)^2}=0$

$xy^2dx-x^2ydy+ydx+xdy=0$

$(1+xy)ydx+x(1-xy)dy=0$

$\therefore$ Statement 2 is true.

Statement 1: $x-\frac{y^2}{2}+\frac{1}{3}(x^2+y^2)^{3/2}+c=0$

$1-y\frac{dy}{dx}+\frac{1}{2}\sqrt{x^2+y^2}\left(2x+2y\frac{dy}{dx}\right)=0$

$dx-y\,dy+x\sqrt{x^2+y^2}dx+y\sqrt{x^2+y^2}dy=0$

$\left(1+x\sqrt{x^2+y^2}\right)dx+y\left(-1+\sqrt{x^2+y^2}\right)dy=0$

$\therefore$ Statement 1 is true.

2. (c)

Statement 1:

$\frac{dy}{dx}=x+\frac{1}{x^2}$ $\quad: dy=xdx+\frac{1}{x^2}.dx : y=\frac{x^2}{2}-\frac{1}{x}+c$

$x=3, y=9$ $\quad: 9=\frac{9}{2}-\frac{1}{3}+c$ $\quad: \frac{9}{2}+\frac{1}{3}=c$

$c=\frac{27+2}{6}=\frac{29}{6}: y=\frac{x^2}{2}-\frac{1}{x}+\frac{29}{6}$

Statement 1 is true.

Statement 2:

$\frac{dy}{dx}=t$

$t^2-t(e^x+e^{-x})+1=0; e^xt^2-t(e^{2x}+1)+e^x=0;$

$e^xt(t-e^x)-(t-e^x)=0; (t-e^x)(te^x-1)=0$

$dy=e^xdx$

$y=e^x$ ...(i)

$e^x\frac{dy}{dx}=1$

$y=-e^{-x}$ ...(ii)

$y=c_1e^x+c_2e^{-x}$ cannot be the solution of the given differential equation.

3. (a) $(1+y^2)dx=xy\,dy$

$\frac{dx}{x}=\frac{y\,dy}{1+y^2}$

$\ln x=\frac{1}{2}\ln(1+y^2)+c$

$x=1, y=0\Rightarrow c=0$

$x^2-y^2=1$

Statement 1is true

Statement 2 is true and it explains Statement 1

4. (b) **Statement 1:** Equation of all circles can be given by $x^2+y^2+2gx+2fy+c=0$; it will be of Order 3

Statement 2 is obviously true but it does not explain Statement 1.

5. (c) **Statement 1:** $y^3\frac{dy}{dx}+(x+y^2)=0$

$y^2=t$

$2y\frac{dy}{dx}=\frac{dt}{dx}$

$\frac{1}{2}\frac{dt}{dx}.t+x+t=0$

is a homogenous equation.

Statement 2 is obviously false.

6. (b) Statement 1 is obviously true.
Statement 2 is also obviously true
But Statement 2 does not explain Statement 1.

7. (c) The given differential equation is

$x\sqrt{x^2-1}dy - y\sqrt{y^2-1}dx = 0$

$\Rightarrow \int \frac{dy}{y\sqrt{y^2-1}} = \int \frac{dx}{x\sqrt{x^2-1}} \Rightarrow \sec^{-1}y = \sec^{-1}x + c$

$\Rightarrow y = \sec\left[\sec^{-1}x + c\right], \because y(2) = \frac{2}{\sqrt{3}}$

$\Rightarrow \frac{2}{\sqrt{3}} = \sec\left(\sec^{-1}2 + c\right) \Rightarrow \sec^{-1}\frac{2}{\sqrt{3}} - \sec^{-1}2 = c$

$\Rightarrow c = \frac{\pi}{6} - \frac{\pi}{3} = -\frac{\pi}{6}, \therefore y = \sec\left[\sec^{-1}x - \frac{\pi}{6}\right]$

$\therefore$ Statement 1 is true

Also, $\frac{1}{y} = \cos\left[\cos^{-1}\frac{1}{x} - \frac{\pi}{6}\right]$

$= \cos\left(\cos^{-1}\frac{1}{x}\right)\cos\frac{\pi}{6} + \sin\left(\cos^{-1}\frac{1}{x}\right)\sin\frac{\pi}{6}$

$\Rightarrow \frac{1}{y} = \frac{\sqrt{3}}{2x} + \frac{1}{2}\sqrt{1-\frac{1}{x^2}}$

$\therefore$ Statement 2 is false

JEE Advanced

Objective Questions

One Option Correct

1. (a)

$x^2 = e^{\left(\frac{x}{y}\right)^{-1}\left(\frac{dy}{dx}\right)} \Rightarrow x^2 = e^{\left(\frac{y}{x}\right)\left(\frac{dy}{dx}\right)}$

$\Rightarrow \ln x^2 = \frac{y}{x}\frac{dy}{dx} \Rightarrow \int x\ln x^2 dx = \int y\,dy$

Put $x^2 = t \Rightarrow 2xdx = dt \Rightarrow \frac{1}{2}\int \ln t\,dt = \frac{y^2}{2}$

$\Rightarrow c + t\ln t - t = y^2 \Rightarrow y^2 = x^2\ln x^2 - x^2 + c$

2. (d) Length of subnormal $= y\frac{dy}{dx} = 2$

$ydy = 2dx \Rightarrow \frac{y^2}{2} = 2x + c$

Since (1, 2) lies on it,

$\therefore c = 0 \therefore$ The curve is $y^2 = 4x$

3. (a)

$y = ax^2 \Rightarrow \frac{dy}{dx} = 2ax \Rightarrow a = \frac{1}{2x}\frac{dy}{dx}$

Now equation of curve

$y = \frac{1}{2x}\frac{dy}{dx}.x^2 \Rightarrow \frac{x}{2}\frac{dy}{dx} = y$

Orthogonal trajectory $\frac{dx}{dy} = \frac{2y}{x}$

$\frac{x^2}{2} = -y^2 + c \Rightarrow y^2 + \frac{x^2}{2} = c$. This is the equation of an ellipse

4. (d) $\frac{dy}{dx} = \frac{(x+y)+1}{2(x+y)+1}$

put $x+y=t, 1+\frac{dy}{dx} = \frac{dt}{dx}$

$\Rightarrow \frac{dt}{dx} = \frac{t+1}{2t+1} + 1 = \frac{3t+2}{2t+1}$

$\Rightarrow \int \frac{2t+1}{3t+2}dt = \int dx \Rightarrow \frac{2t}{3} - \frac{1}{9}\ln(3t+2) = x + c$

$\Rightarrow 6(x+y) - \ln(3x+3y+2) = 9x + c$

$\Rightarrow \ln(3x+3y+2) = 6y - 3x + c$

Since it passes through (0, 0) hence equation of curve is $6y - 3x = \ln\left|\frac{3x+3y+2}{2}\right|$

5. (a)

$y_1y_3 = 3y_2^2 \Rightarrow \int \frac{y_3}{y_2} = \int \frac{3y_2}{y_1} \Rightarrow \ln y_2 = 3\ln y_1 + \ln c$

$\Rightarrow y_2 = cy_1^3 \Rightarrow \int \frac{y_2}{y_1} = \int cy_1 \Rightarrow -\frac{1}{y_1} = cy + d$

$\Rightarrow -dx = (cy+d)dy \Rightarrow -x = \frac{cy^2}{2} + dy + e$

6. (a)

$y = c_1\cos 2x + c_2\cos^2 x + c_3\sin^2 x + c_4$

$\Rightarrow y = c_1\cos 2x + \frac{c_2}{2}(1+\cos 2x) + \frac{c_3}{2}(1-\cos 2x) + c_4$

$\Rightarrow y = \left(\frac{c_2}{2} + \frac{c_3}{2} + c_4\right) + \left(c_1 + \frac{c_2}{2} - \frac{c_3}{2}\right)\cos 2x$

$\Rightarrow y = A + B\cos 2x$

That means there are two independent parameters. Thus, the order of the differential equation will be 2.

7. (b)

$\frac{dy}{dx} = \frac{1}{x\cos y + 2\sin y\cos y}$

$\therefore \frac{dy}{dx} = x\cos y + 2\sin y\cos y$

$\Rightarrow \frac{dx}{dy} + (-\cos y)x = 2\sin y\cos y$

$\therefore$ I.F. $= e^{-\int \cos y\,dy} = e^{-\sin y}$

$\therefore$ The solution is

$$x.e^{-\sin y} = 2\int e^{-\sin y}.\sin y\cos y\,dy$$

$$\Rightarrow x.e^{-\sin y} = -2\sin y e^{-\sin y} - 2\int\left(-e^{-\sin y}\right)\cos y\,dy$$

$$\Rightarrow x.e^{-\sin y} = -2\sin y e^{-\sin y}$$

$$+2\int e^{-\sin y}\cos y\,dy = -2\sin y e^{-\sin y} - 2e^{-\sin y} + c$$

$$\Rightarrow x = -2\sin y - 2 + c\,e^{\sin y} = ce^{\sin y} - 2(1+\sin y)$$

$$\therefore k = 2$$

8. (a) Directrix is $\perp$ to x axis; let $x = \alpha$ and focus on x axis be $(\beta, 0)$. Now

$$(x-\beta)^2 + y^2 = (x-\alpha)^2$$

$$\beta^2 - 2\beta x + y^2 = \alpha^2 - 2\alpha x$$

$$y^2 = 2(\beta-\alpha)x + \alpha^2 - \beta^2$$

In general, $y^2 = mx + c$ (two arbitrary constant m and c)

$$2y\frac{dy}{dx} = m;\ 2y\frac{d^2y}{dx^2} + \left(\frac{dy}{dx}\right)^2 = 0$$

9. (a)

$$\int\frac{dy}{y} = -\int\frac{x}{\sqrt{a+x}}dx$$

$$\Rightarrow \ln y = -\left\{\frac{2}{3}(a+x)^{3/2} - 2a(a+x)^{1/2}\right\} + k$$

$$\Rightarrow \ln y = \frac{2}{3}(2a-x)\sqrt{a+x} + \ln A$$

$$\Rightarrow y = Ae^{\frac{2}{3}(2a-x)\sqrt{x+a}}$$

10. (a) The given differential equation is

$$(x^2-1)\frac{dy}{dx} + 2xy = \frac{1}{x^2-1}$$

$$\Rightarrow \frac{dy}{dx} + \frac{2x}{x^2-1}y = \frac{1}{\left(x^2-1\right)^2} \quad \text{(i)}$$

This is a linear differential equation of the form $\frac{dy}{dx} + Py = Q$, where $P = \frac{2x}{x^2-1}$ and $Q = \frac{1}{\left(x^2-1\right)^2}$

$$\therefore \text{I.F.} = e^{\int P\,dx} = e^{\int\frac{2x}{x^2-1}dx} = e^{\log\left(x^2-1\right)} = \left(x^2-1\right)$$

$\therefore$ Solution is y (I.F.) $= \int Q(\text{I.F.})dx$

$$\Rightarrow y\left(x^2-1\right)$$

$$= \int\frac{1}{x^2-1}dx \Rightarrow y\left(x^2-1\right) = \frac{1}{2}\log\left|\frac{x-1}{x+1}\right| + C$$

This is the required solution.

11. (a) Put $x = \sin\theta, y = \sin\phi$, so that the given equation is reduce to

$$\cos\theta + \cos\phi = a(\sin\theta - \sin\phi)$$

$$\Rightarrow 2\cos\frac{\theta+\phi}{2}\cos\frac{\theta-\phi}{2} = 2a\cos\frac{\theta+\phi}{2}\sin\frac{\theta-\phi}{2}$$

$$\Rightarrow \cot\frac{\theta-\phi}{2} = a \Rightarrow \theta - \phi = 2\cot^{-1}a$$

$$\Rightarrow \sin^{-1}x - \sin^{-1}y = 2\cot^{-1}a$$

Differentiating, we get

$$\frac{1}{\sqrt{1-x^2}} - \frac{1}{\sqrt{1-y^2}}\frac{dy}{dx} = 0$$

So the degree is one.

12. (c)

$$\frac{d^2y}{dx^2}\left(x^2+1\right) = 2x\frac{dy}{dx} \Rightarrow \int\frac{\frac{d^2y}{dx^2}}{\frac{dy}{dx}}dx = \int\frac{2x}{x^2+1}dx$$

$$\Rightarrow \ln\left(\frac{dy}{dx}\right) = \ln\left(x^2+1\right) + \ln c$$

$$\Rightarrow \frac{dy}{dx} = c\left(x^2+1\right) \Rightarrow c = 3 \text{ as at}$$

$$x = 0\frac{dy}{dx} = 3 \Rightarrow dy = 3\left(x^2+1\right)dx$$

$$\Rightarrow y = x^3 + 3x + 1$$

13. (a)

$$y\left(\frac{dy}{dx}\right)^2 + x\frac{dy}{dx} - y\frac{dy}{dx} - x = 0;$$

$$y\frac{dy}{dx}\left(\frac{dy}{dx} - 1\right) + x\left(\frac{dy}{dx} - 1\right) = 0$$

$$\left(y\frac{dy}{dx} + x\right)\left(\frac{dy}{dx} - 1\right) = 0$$

$\therefore$ Either $ydy + xdx = 0$ or $y - x = c$

Since the curves pass through the point (3, 4)

$\therefore x^2 + y^2 = 25$ or $x - y + 1 = 0$

14. (c) **S$_1$**: Order of the differential equation is 2

$$\mathbf{S_2}: \frac{xdy - ydx}{\sqrt{x^2+y^2}} = dx \Rightarrow \frac{\frac{xdy - ydx}{x^2}}{\sqrt{\left(1+\frac{y^2}{x^2}\right)}} = \frac{dx}{x}$$

$$\therefore \ln\left|\frac{y}{x} + \sqrt{1+\frac{y^2}{x^2}}\right| = \ln|cx| \therefore \frac{y}{x} + \frac{\sqrt{x^2+y^2}}{x} = cx$$

i.e., $y + \sqrt{x^2+y^2} = cx^2$

S_3: $y = e^x(A\cos x + B\sin x)$

$$\frac{dy}{dx} = e^x(A\cos x + B\sin x) + e^x(-A\sin x + B\cos x)$$

$$= y + e^x(-A\sin x + B\cos x)$$

$$\Rightarrow \frac{d^2y}{dx^2} = \frac{dy}{dx} + e^x(-A\sin x + B\cos x) + e^x(-A\cos x - B\sin x)$$

$$\Rightarrow \frac{d^2y}{dx^2} = \frac{dy}{dx} + e^x(-A\sin x + B\cos x) - y$$

$$\Rightarrow \frac{d^2y}{dx^2} = \frac{dy}{dx} + \frac{dy}{dx} - y - y = 2\left(\frac{dy}{dx} - y\right)$$

S_4: $(1+y^2)\frac{dx}{dy} + x = 2e^{\tan^{-1}y}$

$$\Rightarrow \frac{dy}{dx} + \frac{1}{1+y^2}x = 2\frac{e^{\tan-1y}}{1+y^2}$$

$$\text{I.F.} = e^{\int\frac{1}{1+y^2}dy} = e^{\tan-1}y$$

$$\Rightarrow xe^{\tan-1}y$$

$$= 2\int e^{\tan^{-1}y}.\frac{e^{\tan^{-1}y}}{1+y^2}dy \Rightarrow xe^{\tan^{-1}y} = e^{2\tan^{-1}} + k$$

15. (a) Given differential equation can be written as

$$\int dy = \int \frac{1}{\left(\sqrt{4+\sqrt{9+\sqrt{x}}}\right)(9+\sqrt{x})8\sqrt{x}}dx$$

Putting $\sqrt{4+\sqrt{9+\sqrt{x}}} = t$, we get

$$\frac{1}{2\sqrt{4+\sqrt{9+\sqrt{x}}}\cdot 2\sqrt{9+\sqrt{x}}\cdot 2\sqrt{x}}dx = dt$$

$$\therefore \int dy = \int dt \Rightarrow y = t + c$$

$$\Rightarrow y = \sqrt{4+\sqrt{9+\sqrt{x}}} + c$$

Now, $y(0) = \sqrt{7} \Rightarrow c = 0$

$$\therefore y = \sqrt{4+\sqrt{9+\sqrt{x}}} \Rightarrow y(256) = 3$$

16. (b) $\frac{dy}{dx} - \frac{x}{1-x^2}y = \frac{x^4+2x}{\sqrt{1-x^2}}$

$$\text{I.F. } e^{\int\frac{-x}{1-x^2}dx} = e^{\frac{+1}{2}\log(1-x^2)} = \sqrt{1-x^2}$$

Hence, the solution is given by

$$y\sqrt{1-x^2} = \int\sqrt{1-x^2}.\frac{x^4+2x}{\sqrt{1-x^2}}dx$$

$$y\sqrt{1-x^2} = \frac{x^5}{5} + x^2 + c$$

$$f(0) = 0 \Rightarrow Atx = 0, y = 0 \Rightarrow c = 0$$

$$\therefore y\sqrt{1-x^2} = \frac{x^5}{5} + x^2$$

$$\Rightarrow y = f(x) = \frac{\frac{x^5}{5} + x^2}{\sqrt{1-x^2}}, \qquad \therefore I = \int_{-\frac{\sqrt{3}}{2}}^{\frac{\sqrt{3}}{2}} \frac{\frac{x^5}{5} + x^2}{\sqrt{1-x^2}}dx$$

$$= 2\int_0^{\frac{\sqrt{3}}{2}} \frac{x^2}{\sqrt{1-x^2}}dx \qquad \left(\because \frac{x^5}{\sqrt{1-x^2}} \text{ is odd}\right)$$

Put $x = \sin\theta$ $dx = \cos\theta\, d\theta$

$$\therefore I = 2\int_0^{\frac{\pi}{3}} \sin^2\theta d\theta = \int_0^{\frac{\pi}{3}}(1-\cos 2\theta)d\theta$$

$$= \left(\theta - \frac{\sin 2\theta}{2}\right)_0^{\frac{\pi}{3}} = \frac{\pi}{3} - \frac{\sqrt{3}}{4}$$

One or More Than One Option Correct

1. (a, b, c)

Tangent at P (x_1, y_1) to $y = f(x)$

$$y - y_1 = \frac{dy}{dx}(x - x_1)$$

This cuts the x-axis at $x_2 = x_1 - \frac{y_1}{\left(\frac{dy}{dx}\right)}$

$\because x_1, x_2, \ldots\ldots x_n \to$ AP

$$\Rightarrow x_2 - x_1 = \frac{y_1}{\left(\frac{dy}{dx}\right)} = 1/\ln 2 \Rightarrow y = \lambda e^{-x\ln 2};$$

putting $(0, 2) \Rightarrow \lambda = 2 \Rightarrow y = 2^{1-x}$

2. (a, d)

$$y^2 = t \Rightarrow \left(e^{x^2} + e^t\right)\frac{dt}{dx} + 2e^{x^2}(xt - x) = 0$$

$$e^{x^2} = z \Rightarrow e^{x^2}(2x)\frac{dx}{dt} = \frac{dz}{dt} \Rightarrow \frac{dz}{dt} + \frac{z}{t-1} = -\frac{e^t}{(t-1)}$$

3. (b, c)

$$\frac{\left[\frac{ydx - xdy}{y^2}\right]}{\left(\frac{x}{y}+1\right)\sqrt{\frac{x}{y}}} + d\left(\frac{x^2+y^2}{2}\right) = 0$$

$$\Rightarrow \frac{x^2+y^2}{2} + 2\tan^{-1}\sqrt{\frac{x}{y}} = \lambda$$

4. (a, b)

$y = e^{-x}\cos x$

$y_1 = -e^{-x}\cos x - e^{-x}\sin x = -\sqrt{2}e^{-x}\cos\left(x-\frac{\pi}{4}\right)$

$y_2 = +\left(\sqrt{2}\right)^2 e^{-x}\cos\left(x-\frac{\pi}{2}\right)$

$y_3 = \left(-\sqrt{2}\right)^3 e^{-x}\cos\left(x-\frac{3\pi}{4}\right)$

$y_4 = +\left(\sqrt{2}\right)^4 e^{-x}\cos(x-\pi) = -4e^{-x}\cos x$

$\Rightarrow y_4 + 4y = 0 \quad k_4 = 4$

Differentiating it again, 4 times,

$y_8 + 4y_4 = 0 \Rightarrow y_8 - 16y = 0 \Rightarrow k_8 = -16$

$y_{12} + 4y_8 = 0 \Rightarrow y_{12} + 64y = 0 \Rightarrow k_{12} = 64$

Similarly, $k_{16} = -256$

5. (b, c)

(a) $\frac{dy}{dx} = -c_1\sin x + \sqrt{c_2}\cos x$

$\Rightarrow \frac{d^2y}{dx^2} = -c_1\cos x - \sqrt{c_2}\sin x = 2 - y$

$\frac{d^2y}{dx^2} + y - 2 = 0$

(b) $\frac{dy}{dx} = \cos x\frac{\sec^2 x/2}{2\tan\frac{x}{2}} - \sin x\ln\left(\tan\frac{x}{2}\right)$

$\frac{dy}{dx} = \cot x - \sin x\ln\left(\tan\frac{x}{2}\right)$

$\frac{d^2y}{dx^2} = -\cos ec^2 x - \left(\sin x.\frac{1}{\sin x} + \cos x\ln\left(\tan\frac{x}{2}\right)\right)$

$\frac{d^2y}{dx^2} = -\cot^2 x - 2 - \cos x\ln\left(\tan\frac{x}{2}\right)$

$\frac{d^2y}{dx^2} + y + \cot^2 x = 0$

(c) $\frac{dy}{dx} = -c_1\sin x + c_2\cos x$

$+\frac{d}{dx}\left(\cos x\ln\left(\tan\frac{x}{2}\right)\right)$

$\frac{d^2y}{dx^2} = -(c_1\cos x + c_2\sin x)$

$+\frac{d^2}{dx^2}\left(\cos x\ln\left(\tan\frac{x}{2}\right)\right)$

$\frac{d^2y}{dx^2} = -c_1\cos x - c_2\sin x - \cot^2 x - 2$

$-\cos x\ln\left(\tan\frac{x}{2}\right)$

$\frac{d^2y}{dx^2} + y + \cot^2 x = 0$

6. (b, c)

$\left(x^2y^2 - 1\right)dy + 2xy^3dx = 0$

$x^2y^2dy + 2xy^3dx = dy$

$x^2dy + 2xy\,dx = \frac{dy}{y^2};\int d\left(x^2y\right) = \int\frac{dy}{y^2} + c$

$\Rightarrow x^2y = \frac{y^{-1}}{-1} + c \Rightarrow x^2y^2 = -1 + cy$

$\Rightarrow 1 + x^2y^2 = cy$

7. (a, b, d)

$\frac{dy}{dx}\left(x^2y^3 + xy\right) = 1$

$\Rightarrow \frac{dx}{dy} = x^2y^3 + xy \Rightarrow \frac{dy}{dx} - xy = x^2y^3$

$\Rightarrow -\frac{1}{x^2}\frac{dx}{dy} + \frac{y}{x} = -y^3$

Put, $\frac{1}{x} = t$ and differentiating with respect to

$y - \frac{1}{x^2}\frac{dx}{dy} = \frac{dt}{dy}$

$\Rightarrow \frac{dt}{dy} + ty = -y^3 \Rightarrow e^{\frac{y^2}{2}}t = \int -y^3e^{y^2/2}dy$

$\Rightarrow \frac{1}{x} = 2 - y^2 + ce^{\frac{y^2}{2}}$

8. (a, c) Integrating factor $= e^{\int x\,dx} = e^{\alpha x}$

Solution: $ye^{\alpha x} = \int xe^{(\alpha+\beta)x} = dx$

Case I

If $\alpha + \beta = 0$, then $ye^{\alpha x} = \frac{x^2}{2} + \text{C}$

Put $x = 1$ and $y = 1$

$\Rightarrow \text{C} = e^{\alpha} - \frac{1}{2}$

$\Rightarrow \lim_{x\to\infty}\left|\frac{5f(x)-2}{5f(x)+2}\right| = \lim_{x\to\infty} e^{20x} = e^{-\infty} = 0$

$\Rightarrow 5\lim_{x\to\infty} f(x) - 2 = 0 \Rightarrow \lim_{x\to\infty} f(x) = \frac{2}{5} = 0.4$

9. (c, d)

Tangent to the curve $y = y(x)$ at point $P(x, y)$ is given by

$Y - y = y(x)(X - x)$

It intersects y-axis at Y_p, putting $x = 0$

$Y - y = -xy'(x) \Rightarrow Y = y - xy'(x)$

$\therefore Y_p(0, y - xy'(x))$

Given

$PY_p = 1 \Rightarrow \sqrt{(x-0)^2 + (y - y + xy'(x))^2} = 1$

$\Rightarrow x^2 + x^2(y'(x))^2 = 1 \Rightarrow y'(x) = \pm\frac{\sqrt{1-x^2}}{x}$

Now, $y = y(x)$ lies in the first quadrant and its tangent passes through (1, 0). Therefore, it has to be a decreasing function; so the derivative should be negative.

$\therefore y'(x) = \frac{-\sqrt{1-x^2}}{x} \qquad \left[\text{or } xy'(x) + \sqrt{1-x^2} = 0\right]$

$\Rightarrow y(x) = -\int \frac{\sqrt{1-x^2}}{x} dx$

Put $x = \sin\theta \, dx \cos\theta \, d\theta$

$y = -\int \frac{\cos\theta}{\sin\theta} \cos\theta d\theta = -\int (\text{cosec}\,\theta - \sin\theta) d\theta$

$y = +\log|\text{cosec}\theta + \cot\theta| - \cos\theta + c$

$y = \log\left|\frac{1+\sqrt{1-x^2}}{x}\right| - \sqrt{1-x^2} + c$

For $x = 1$ and $y = 0$, we get $c = 0$

$y = \ln\left(\frac{1+\sqrt{1-x^2}}{x}\right) - \sqrt{1-x^2}$

10. (b, c) $f(x) = 1 - 2x + \int_0^x e^{x-t} f(t) dt$

$\Rightarrow f(x) = 1 - 2x + e^x \int_0^x e^{-t} f(t) dt$

$\Rightarrow f'(x) = -2 + e^x \int_0^x e^{-t} f(t) dt + e^x \left[e^{-x} f(x)\right]$

$\Rightarrow f'(x) = -2 + \left[f(x) - 1 + 2x\right] + f(x)$

$\Rightarrow f'(x) - 2f(x) = 2x - 3$

Its linear differential equation

$\text{I.F.} = e^{\int -2dx} = e^{-2}$

Solution: $f(x) \times e^{-2x} = \int e^{-2x}(2x-3) dx$

$\Rightarrow f(x) \times e^{-2x} = \frac{e^{-2x}}{-2}(2x-3) - \int \frac{e^{-2x}}{-2} \times 2dx$

$\Rightarrow e^{-2x} f(x) = \frac{e^{-2x}}{-2}(2x-3) + \frac{e^{-2x}}{-2} + c$

$\Rightarrow f(x) = -x + \frac{3}{2} + \frac{3}{-2} + ce^{2x}$

$\Rightarrow f(x) = -x + 1 + ce^{2x}$

From definition of the function, $f(0) = 1$

$\therefore 1 = 1 + c \, c = 0, \; f(x) = 1 - x$

Clearly curve $y = 1 - x$, does not pass through (1, 2) but it passes through (2, −1)

$\therefore$ (a) is false and (b) is true

Also the area of the region

$1 - x \le y \le \sqrt{1-x^2}$, is shown in the figure by the shaded region and is given by

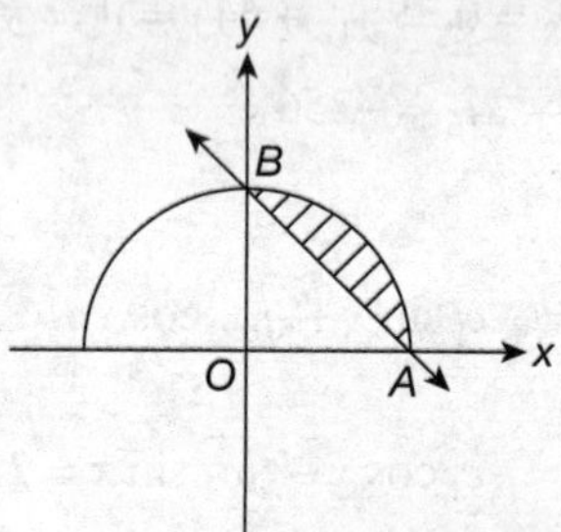

Area of quadrant − Area of ΔOAB

$= \frac{1}{4} \times \pi \times 1^2 - \frac{1}{2} \times 1 \times 1 = \frac{\pi - 2}{4}$

$\therefore$ (c) is true, (d) is false

11. (a, d) $\left[(x+2)^2 + y(x+2)\right]\frac{dy}{dx} = y^2$

$\Rightarrow y^2 \frac{dx}{dy} - (x+2)y = (x+2)^2$

$\Rightarrow \frac{1}{(x+2)^2}\frac{dx}{dy} - \frac{1}{(x+2)y} = \frac{1}{y^2}$

Let $\frac{-1}{x+2} = u, \therefore \frac{1}{(x+2)^2}\frac{dx}{dy} = \frac{du}{y^2}$

$\Rightarrow \frac{du}{dy} + \frac{1}{y}u = \frac{1}{y^2}, \therefore \text{I.F.} = e^{\log y} = y$

Hence, solution is $u \times y = \int y \times \frac{1}{y^2} dy = \log y + c$

$\Rightarrow \frac{-y}{x+2} = \log y + c$

Since it passes through (1, 3), $\therefore c = -1 - \log 3$

$\therefore \frac{-y}{x+2} = \log y - 1 - \log 3$

$\Rightarrow \log\frac{y}{3} = 1 - \frac{y}{x+2} \qquad \text{...(i)}$

Put $y = x + 2$ in equation (i) to find the point of

intersection of (i) and $y = x+2$.

$\log\frac{y}{3} = 0 \Rightarrow y = 3 \Rightarrow x = 1$

$\therefore$ Put $y = (x+2)^2$ in equation (i), to find the point of intersection of (i) and $y = (x+2)^2$

$$\log\frac{(x+2)^2}{3} = 1-(x+2)$$

$$\Rightarrow \log\frac{(x+2)^2}{3} + (x+2) = 1$$

$$\because \log\frac{(x+2)^2}{3} > \frac{4}{3} > 1, \forall x > 0$$

$\therefore$ LHS > 2, $\forall\ x > 0 \Rightarrow$ no solution

Put $y = (x+3)^2$ in equation (i), to find the point of intersection of (i) and $y = (x+3)^2$

$$\log\frac{(x+3)^2}{x+2} = 1 - \frac{(x+3)^2}{x+2} = -\frac{x^2+5x+7}{x+2}$$

$$\Rightarrow \log\frac{(x+3)^2}{x+2} = -\left[x+2+\frac{x+3}{x+2}\right]$$

Now for $x > 0, (x+2) > 2$ and $\frac{x+3}{x+2} > 1$

$$\therefore x+2+\frac{x+3}{x+2} > 3 \Rightarrow -\left[x+2+\frac{x+3}{x+2}\right] < -3$$

But for $x > 0$, $\frac{(x+3)^2}{x+2} > \frac{9}{2} \Rightarrow \log\frac{(x+3)^2}{x+2} > 0$

Thus LHS > 0 and RHS < −3

Hence, there is no solution for equation (ii)

12. (a) $f'(x) = 2 - \frac{f(x)}{x} \Rightarrow f'(x) + \frac{1}{x}f(x) = 2$

I.F. $= e^{\log x} = x \qquad \therefore f(x).x = \int 2x dx = x^2 + c$

$\Rightarrow f(x) = x + \frac{c}{x}, c \neq 0$ as $f(x) \neq 1$

(a) $\lim_{x\to 0^+} f'\left(\frac{1}{x}\right) = \lim_{x\to 0^+}\left(1-cx^2\right) = 1$

(b) $\lim_{x\to 0^+} xf\left(\frac{1}{x}\right) = \lim_{x\to 0^+} x\left(\frac{1}{x}+cx\right) \lim_{x\to 0^+} 1+cx^2 = 1$

(c) $\lim_{x\to 0^+} x^2 f'x\left(\frac{1}{x}\right) = \lim_{x\to 0^+} x^2\left(1-\frac{c}{x^2}\right)$

$= \lim_{x\to 0^+} x^2 - c = -c$

(d) For $c \neq 0$, $f(x)$ is unbounded as $0 < x < 2$

$\Rightarrow \frac{c}{2} < \frac{c}{x} < \infty \Rightarrow \frac{c}{2} < x + \frac{c}{x} < \infty$

13. (b, c) Let the equation of circle be

$x^2 + y^2 + 2gx + 2gy + c = 0$

$\Rightarrow 2x + 2yy' + 2g + 2gy' = 0$

$\Rightarrow x + yy' + g + gy' = 0$

On differentiating again, we get

$$1 + yy'' + (y')^2 + gy'' = 0 \Rightarrow g = -\left[\frac{1+(y')^2+yy'}{yy'}\right]$$

On substituting the value of given equation (i), we get

$$x + yy'' - \left(\frac{1+(y')^2+yy'}{yy'}\right)y' = 0$$

$$\Rightarrow xy'' + yy'y'' - 1 - (y')^2 - yy' - y'$$
$$-(y')^3 - yy'y'' = 0$$

$$\Rightarrow (x-y)y'' - y'\left(1+y'+(y')^2\right) = 1$$

$$\Rightarrow (y-x)y'' + \left[1+y'+(y')^2\right]y' + 1 = 0$$

$\because Py'' + Qy' + 1 = 0$

$\therefore P = y - x, Q = 1 + y' + (y')^2$

Hence, $P + Q = 1 - x + y + y' + (y')^2$

14. (a, c)

$$\frac{dy}{dx} + \frac{e^x}{1+e^x}y = \frac{1}{1+e^x}$$

I.F. $= 1 + e^x$. Hence, the solution is

$y\left(1+e^x\right) = x + c$

Now, $y(0) = 2 \Rightarrow c = 4$,

$\therefore y = \frac{x+4}{e^x+1} \Rightarrow y(-4) = 0$

Also, $\frac{dy}{dx} = \frac{\left(e^x+1\right) - e^x(x+4)}{\left(e^x+1\right)^2}$

From critical point, put $\frac{dy}{dx} = 0$

$\Rightarrow e^x(x+3) = 1 \Rightarrow x + 3 = e^{-x}$

Its solution will be the intersection point of $y = x+3$ and $y = e^{-x}$

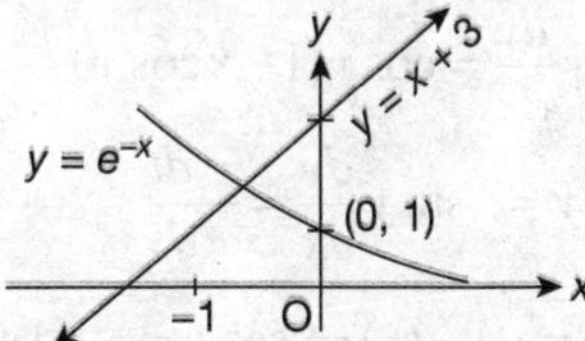

Clearly, there is a critical point in (−1, 0)

15. (a) $\frac{dy}{dx}=\frac{y}{x}+\sec\frac{y}{x}$

Put $y=vx, \therefore \frac{dy}{dx}=v+x\frac{dv}{dx}$

$\Rightarrow x\frac{dv}{dx}=\sec v \Rightarrow \int \cos v dv=\int\frac{dx}{x}$

$\Rightarrow \sin v=\log x+c \qquad (\because x>0)$

$\Rightarrow \sin\frac{y}{x}=\log x+c$

Since it passes through $\left(1,\frac{\pi}{6}\right) \Rightarrow c=\frac{1}{2}$

Hence, $\sin\frac{y}{x}=\log x+\frac{1}{2}$

Matrix Type

1. A → R; B → P; C → T; D → Q

(A) $\frac{ydx-xdy}{y^2}=dx+\frac{dy}{y^2}$

$\Rightarrow d\left(\frac{x}{y}\right)=dx+\frac{dy}{y^2} \Rightarrow \frac{x}{y}=x-\frac{1}{y}+k$

$\Rightarrow x=xy-1+ky \Rightarrow (x+1)(1-y)=cy$

(B) $\left(2x-10y^3\right)\frac{dy}{dx}+y=0 \Rightarrow \frac{dy}{dx}=\frac{y}{10y^3-2x}$

$\Rightarrow \frac{dx}{dy}=\frac{10y^3-2x}{y}$

$\frac{dx}{dy}=10y^2-2\frac{x}{y} \Rightarrow \frac{dx}{dy}+\frac{2}{y}x=10y^2$

$\Rightarrow xy^2=10\frac{y^5}{5}+c$

$\Rightarrow xy^2=2y^5+c$

(C) $\sec^2 y\frac{dy}{dx}+\tan y=1$ put

$\tan y=t, \sec^2 y\frac{dy}{dx}=\frac{dt}{dx}$

$\Rightarrow \frac{dt}{dx}=1-t \Rightarrow \ln(1-t)-cx \Rightarrow 1-t=e^{-cx}$

$\Rightarrow t=1-e^{-cx} \Rightarrow \tan y=1+ce^{-x}$

(D) $\sin y\frac{dy}{dx}=\cos y(1-x\cos y)$

Put $\cos y=t, \sin y\frac{dy}{dx}=-\frac{dt}{dx}$

$\Rightarrow -\frac{dt}{dx}=t(1-tx) \Rightarrow \sec y=x+1+ce^{x}$

2. A → Q; B → R; C → T; D → P

(A) $xdy=ydx+y^2dy \Rightarrow \frac{xdy-ydx}{y^2}=dy$

$\Rightarrow -d\left(\frac{x}{y}\right)=dy$

$-\frac{x}{y}=y+c$ put $x=1, y=1 \Rightarrow c=-2$

$-\frac{x}{y}=y-2$ Now $\frac{x_0}{3}=-5 \Rightarrow x_0=-15$

(B) $\frac{dy}{dt}-\frac{t}{t+1}y=\frac{1}{t+1}$

I.F. $=e^{-\int\frac{t+1-1}{t+1}dt}=e^{-t+\ell n(t+1)}=(t+1)e^{-t}$

The solution is $(t+1)e^{-t}y=-e^{-t}+c$; put $t=0$ and $y=-1 \Rightarrow c=0$

Now, at $t=1$

$\therefore 2e^{-1}y=-e^{-1}$

$y=-\frac{1}{2}$

(C) $\left(x^2+y^2\right)dy=xy\,dx$

$\frac{dy}{dx}=\frac{xy}{x^2+y^2}$; put $y=vx$, $\frac{dy}{dx}=v+x\frac{dv}{dx}$

$\therefore \frac{v}{1+v^2}=v+x\frac{dv}{dx}$

$\Rightarrow \ln v-\frac{1}{2v^2}=-\ln x+c$

Put $x=1$, $y=1$, $v=1$, then $c=-\frac{1}{2}$

$\therefore \ln\frac{y}{x}-\frac{1}{2}\frac{x^2}{y^2}=-\ln x-\frac{1}{2}$ put $y=e$

$\therefore x_0=\sqrt{3}e$

(D) $\frac{dy}{dx}+2\frac{y}{x}=0$

$x^2y=C$ put $x=1$, $y=1$ and we get C = 1

Put $x=2 \Rightarrow y=\frac{1}{4}$

3. A → R; B → P; C → Q; D → T

(A) $(\sin x+\cos x)\,dy=dx\,(\sin x-\cos x)$

$\Rightarrow \int dy=\int\frac{\sin x-\cos x}{\sin x+\cos x}dx+c$

$\Rightarrow y=-\ln|(\sin x+\cos x)|+c$

(B) $\int\frac{dx}{\sin x}=-\int\frac{dy}{\cos y}+\ln c$

$-\ln|\text{cosec}\, x+\cot x| = -\ln = |\sec y+\tan y| + \ln c$

$|\sec y+\tan y| = c|\text{cosec}\, x+\cot x|$

(C) $\frac{1}{x}\cos^2 y\, dy + \frac{1}{y}\cos^2 x\, dx = 0$

$\int x\cos^2 x\, dx = -\int y\cos^2 y\, dy + k$

$\Rightarrow (x^2+y^2)+(x\sin 2x + y\sin 2y)$

$+\frac{\cos 2x}{2}+\frac{\cos 2y}{2} = k$

(D) $d(\tan x\tan y) = dx$

$\Rightarrow \tan x\tan y = x + c$

4. A → P; B → S; C → Q; D → R

(A) $I = \int_0^{\pi/2} (\sin x)^{\cos x}$

$\{\cos x.\cot x - \log(\sin x)^{\sin x}\}dx$

$= \int_0^{\pi/2} \frac{d}{dx}(\sin x)^{\cos x}\, dx = 1$

(B) The point of intersection of $-4y^2 = x$ and $x-1=-5y^2$ is $(-4,-1)$ and $(-4,1)$.

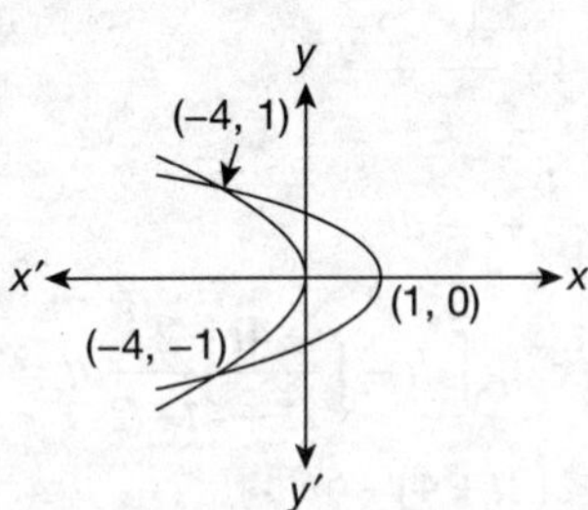

∴ Required area

$= 2\left\{\left[\int_0^1 (1-5y^2)\,dy - \int_0^1 -4y^2 dy\right]\right\}$

$= 2\int_0^1 (1-y^2)\,dy = 2\left[y-\frac{y^3}{3}\right]_0^1 = \frac{4}{3}$ sq units

(C) The point of intersection $y = 3^{x-1}\log x$ and $y = x^x - 1$ is (1, 0).

Hence, $\frac{dy}{dx} = \frac{3^{x-1}}{x} + 3^{x-1}.\log 3.\log x$

$\therefore \left(\frac{dy}{dx}\right)_{(1,0)} = 1$

For $y = x^x - 1$

$\Rightarrow \frac{y}{dx} = x^x(1+\log x)$

$\therefore \left(\frac{dy}{dx}\right)_{(1,0)} = 1$

If θ is angle between the curves, then $\tan\theta = 0$.

$\Rightarrow \theta = 0°$

(D) $\frac{dy}{dx} = \frac{2}{x+y}$

$\Rightarrow \frac{dx}{dy} - \frac{x}{2} = \frac{y}{2} \Rightarrow xe^{-y/2} = \frac{1}{2}.\int y.e^{-y/2}dy$

$\Rightarrow xe^{-y/2} = \frac{1}{2}\left[\frac{ye^{-y/2}}{-1/2} - \frac{e^{-y/2}}{(1/2)^2}\right] + k$

$\Rightarrow x+y+2 = ke^{y/2}$

It passing through (1, 0)

$\Rightarrow k = 3$

$\Rightarrow x+y+2 = 3e^{y/2}$

Integer and Numerical Value Type

1. For reservoir A, $\frac{dV_A}{dt} = -k_1 V_A$

$\Rightarrow \frac{dV_A}{V_A} = -k_1 dt \Rightarrow \ln V_A = -k_1 t + c$

Initially, $t = 0$, $V_A = V_{OA}$

$\therefore \ln V_{OA} = c \Rightarrow \ln V_A = -k_1 t + \ln V_{OA}$

$\Rightarrow \ln\left(\frac{V_A}{V_{OA}}\right) = -k_1 t \Rightarrow V_A = V_{OA}e^{-k_1 t}$ ...(1)

For reservoir B, $\frac{dV_B}{dt} = -k_2 V_B$

Similarly, $V_B = V_{OB}e^{-k_2}$ ...(2)

Given that at $t = 0$, $V_{OA} = 2V_{OB}$

At $t = 1$ hour, $V_A = \frac{3}{2}V_B \Rightarrow V_{OA}e^{-k_1} = \frac{3}{2}V_{OB}e^{-k_2}$

$\Rightarrow 2V_{OB}e^{-k_1} = \frac{3}{2}V_{OB}e^{-k_2}$ $(\because V_{OA} = 2V_{OB})$

$\Rightarrow e^{(k_1-k_2)} = \frac{4}{3} \Rightarrow (k_1-k_2) = \ln\frac{4}{3}$ ...(3)

After time t: $V_A = V_B$

$\Rightarrow V_{OA}e^{-k_1 t} = V_{OB}e^{-k_2 t}$

$\Rightarrow 2V_{OB}e^{-k_1 t} = V_{OB}e^{-k_2 t}$ $(\because V_{OA} = 2V_{OB})$

$\Rightarrow 2 = e^{(k_1-k_2)^t}$

$\Rightarrow \ln 2 = (k_1-k_2)t$ ...(4)

From (3) and (4) we get $t = \frac{\ln 2}{\ln(4/3)} \Rightarrow t = \log_{4/3} 2$

2. Since given differential equation is

$y(x+y^3)dx = x(y^3-x)dy$

$\Rightarrow (xy\,dx + x^2dy) + y^4dx - y^3x\,dy = 0$

$\Rightarrow x(y\,dx + x\,dy) + y^3(y\,dx - x\,dy) = 0$

$\Rightarrow xd(xy) = y^3(x\,dy - y\,dx)$

$\Rightarrow xd(xy) = x^2y^3d.\left(\frac{y}{x}\right)$

$\Rightarrow \frac{d(xy)}{(xy)^2} = \left(\frac{y}{x}\right)d\left(\frac{y}{x}\right)$

On integrating $-\frac{1}{xy} = \frac{1}{2}\left(\frac{y}{x}\right)^2 + c$

at (4, −2)

$\Rightarrow \frac{1}{8} = \frac{1}{2}\left(-\frac{2}{4}\right)^2 + c \Rightarrow c = 0$

$\therefore y^3 = -2x \Rightarrow y = (-2x)^{1/3}$

Since $= y = g(x)$

$= \int_{1/8}^{\sin^2 x} \sin^{-1}\sqrt{t}\,dt + \int_{1/8}^{\cos^2 x} \cos^{-1}\sqrt{t}\,dt$

$\therefore \frac{dy}{dx} = g'(x) = x.\sin 2x + x.(-\sin 2x) = 0$

$\therefore y = c_1$ (constant)

Put $\sin x = \cos x = \frac{1}{\sqrt{2}}$

$\therefore c_1 = \int_{1/8}^{1/2}(\sin^{-1}\sqrt{t} + \cos^{-1}\sqrt{t})dt = \frac{\pi}{2}\left(\frac{1}{2} - \frac{1}{8}\right) = \frac{3\pi}{16}$

$\therefore y = g(x) = \frac{3\pi}{16}$

Hence, area between $y = \frac{3\pi}{16}$ and

$y = (-2x)^{1/3}$

$= \int_0^{3\pi/16} x\,dy = \left|\int_0^{3\pi/16}\left(-\frac{y^3}{2}\right)dy\right| = \frac{1}{8}.\left(\frac{3\pi}{16}\right)^4$ sq.units

3. The given equation contains one constant. Differentiating the equation once, we get

$2x - 2yy' = 2c(x^2+y^2)(2x+2yy')$

But $c = \frac{x^2-y^2}{(x^2+y^2)^2}$

Substituting for c, we get

$(x - yy') = \frac{(x^2+y^2)(x^2-y^2)}{(x^2+y^2)^2}.2(x+yy')$

$\Rightarrow (x^2+y^2)(x-yy') = 2(x^2-y^2)(x+yy')$

$\Rightarrow yy'[(x^2+y^2) + 2(x^2-y^2)]$

$= x(x^2+y^2) - 2x(x^2-y^2)$

$\Rightarrow yy'(3x^2-y^2) = x(3y^2-x^2)$

$\Rightarrow y' = \frac{x(3y^2-x^2)}{y(3x^2-y^2)}$

4. $(x^2+4y^2+4xy)dy = (2x+4y+1)dx$

$\frac{dy}{dx} = \frac{2(x+2y)+1}{(x+2y)^2}$ ...(1)

Put, $x+2y = t$ and differentiate with respect to x.

$1 + \frac{2dy}{dx} = \frac{dt}{dx}$

From equation (i), we get $\left(\frac{dt}{dx} - 1\right)\frac{1}{2} = \frac{2t+1}{t^2}$

$\Rightarrow \frac{dt}{dx} - 1 = \frac{4t+2}{t^2}$

$\Rightarrow \int \frac{t^2}{t^2+4t+2}dt$

$= \int dx \Rightarrow \int dt - \int \frac{4t+2}{t^2+4t+2}dt = x + c_1$

$\Rightarrow t - \int \frac{2(2t+4)-6}{t^2+4t+2}dt = x + c_1$

$\Rightarrow t - 2\ln(t^2+4t+2) + 6\frac{1}{2\sqrt{2}}\ln\left|\frac{t+2-\sqrt{2}}{t+2+\sqrt{2}}\right| = x + c_1$

$\Rightarrow y = \ln((x+2y)^2 + 4(x+2y) + 2)$

$-\frac{3}{2\sqrt{2}}\ln\left|\frac{x+2y+2-\sqrt{2}}{x+2y+2+\sqrt{2}}\right| + c$, where $c = \frac{c_1}{2}$

5. Suppose the point at which normal is drawn is $p \equiv (x, y)$

Equation of the normal is $Y - y = -\frac{dx}{dy}(X - x)$

At X-axis, $Y = 0$

$\therefore$ Coordinate of X $= \frac{y\,dy}{dx} + x$

$\therefore$ The point where it cuts the X-axis is $\left(y\frac{dy}{dx}+x,0\right)$, say A

The mid-point of PA is $\left(x+\frac{y\frac{dy}{dx}}{2},\frac{y}{2}\right)$, which lies on the curve $2y^2 = x$

$\therefore y^2 = 2x + y\frac{dy}{dx} \Rightarrow y\frac{dy}{dx}+2x = y^2$ ...(i)

Putting $y^2 = t$ and differentiating with respect to x, $2y\frac{dy}{dx}=\frac{dt}{dx}$

Now from equation (i), we get $\frac{1}{2}\frac{dt}{dx}+2x=t$

$\Rightarrow \frac{dt}{dx}-2t=-4x;\text{I.F.}=e^{\int -2dx}=e^{-2x}$

$\Rightarrow te^{-2x}=-\int e^{-2x}4x+c$

$\Rightarrow y^2e^{-2x}=-\left\{\frac{4xe^{-2x}}{-2}+\frac{1}{2}\int 4e^{-2x}dx\right\}+c$

$\Rightarrow y^2e^{-2x}=+2xe^{-2x}+e^{-2x}+c$

at $x=0$ $y=0 \Rightarrow c=-1$ $\therefore y^2=2x-e^{2x}+1$

6. (2)

$f(x+y)=f(x)f'(y)+f'(x)f(y)$ ...(i)

On putting $x=y=0$, we get

$f(0)=2f'(0)f(0)\Rightarrow f'(0)=\frac{1}{2}$ $[\because f(0)=1]$

On putting $y=0$ in equation (i), we get

$f(x)=f(x)f'(0)+f'(x)f(0)$

$\Rightarrow f'(x)=\frac{f(x)}{2}\Rightarrow \int\frac{f'(x)}{f(x)}dx=\frac{1}{2}\int dx$

$\left[\because f(0)=1 \text{ and } f'(0)=\frac{1}{2}\right]$

$\Rightarrow \log_e f(x)=\frac{x}{2}+\log_e c$

$\Rightarrow f(x)=ce^{x/2}\Rightarrow f(x)=e^{x/2}$ $[\because f(0)=1]$

$\Rightarrow \log_e(f(x))=\frac{x}{2} \Rightarrow \log_e(f(4))=2$

7. (0.4)

$\frac{dy}{dx}=(5y+2)(5y-2)=25\left(y+\frac{2}{5}\right)\left(y-\frac{2}{5}\right)$

$\Rightarrow \frac{1}{25}\int\frac{dy}{\left(y+\frac{2}{5}\right)\left(y-\frac{2}{5}\right)}=\int dx$

$\Rightarrow \frac{1}{25}\int\frac{5}{4}\left[\frac{1}{y-\frac{2}{5}}-\frac{1}{y+\frac{2}{5}}\right]dy=\int dx$

$\Rightarrow \frac{1}{25}\times\frac{5}{4}\ln\left|\frac{y-\frac{2}{5}}{y+\frac{2}{5}}\right|=x+c$

$\Rightarrow \frac{1}{20}\ln\left|\frac{5y-2}{5y+2}\right|=x+c$

As, $f(0)=0, \therefore 0=0+c\Rightarrow c=0$

Hence, $\left|\frac{5y-2}{5y+2}\right|=e^{20x}$

$\Rightarrow \lim_{x\to-\infty}\left|\frac{5f(x)-2}{5f(x)+2}\right|=\lim_{x\to-\infty}e^{20x}=e^{-\infty}=0$

$\Rightarrow 5\lim_{x\to-\infty}f(x)=-2=0 \Rightarrow \lim_{x\to-\infty}f(x)=\frac{2}{5}=0.4$

Comprehension Passage

Comprehension 1

1. (b) $\frac{dx}{x}=\frac{y\,dy}{1+y^2}\Rightarrow \ln x=\frac{1}{2}.\ln(1+y^2)+c$

From the given condition $c=0$

$\therefore x^2-y^2=1$

2. (a) $\frac{dy}{dx}+\frac{1+y^2}{\sqrt{1-x^2}}=0\Rightarrow\frac{dy}{1+y^2}+\frac{dx}{\sqrt{1-x^2}}=0$

$\Rightarrow \tan^{-1}y+\sin^{-1}x=c$

3. (b) $\frac{dy}{dx}=(1+x).(1+y)$ gives $y=e^{\frac{(1+x)^2}{2}}-1$

Comprehension 2

1. (b) Obviously, option (b) is not a solution.

2. (b) Among the four options, (b) is the solution of the differential equation.

3. (d) Among the four options (d) is satisfying.

CHAPTER

10 Vectors

Objectives

In this chapter on Vectors, we shall study the following topics:

1. Vectors, their types and notations
2. Addition of vectors
3. Resolution of vectors
4. Subtraction of vectors
5. Vectors in terms of position vectors
6. Scalar or dot product of two vectors
7. Vector product or cross product of two vectors
8. Scalar triple product
9. Vector triple product
10. Vector equation of straight line
11. Vector equation of a plane.

Introduction

All physical quantities can be divided into two categories — **scalars** and **vectors**.

Scalars are quantities that have only magnitude but no direction and can be completely specified if its magnitude is given.

Mass, volume, distance, speed, current, temperature are some of the examples of scalar quantities.

Vectors are quantities that have both magnitude and direction and can be completely specified only if both its magnitude and direction are given. In addition, a vector must also obey the rules of vector algebra.

Displacement, force, intensity of gravitational field, intensity of electric and magnetic fields are some of the examples of vector quantities.

A vector quantity is represented by putting an arrow over the symbol representing it. Hence, for example, $\vec{A}$ stands for a vector quantity.

Notation and Representation of Vectors

Vectors are represented by $\vec{a}, \vec{b}, \vec{c}$ and their magnitude (modulus) is represented by a, b, c or $|\vec{a}|, |\vec{b}|, |\vec{c}|$, ... Vectors are represented by directed line segments.

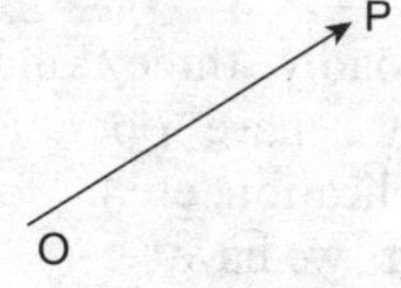

For example, line segment $\overrightarrow{OP}$ represents a vector with magnitude OP (length of line segment), arrow denotes its direction. O is initial point and P is terminal point.

Graphical Representation of a Vector

A vector is denoted by $\overrightarrow{OA}$, i.e., with an arrow over OA. The point O is known as origin while A is the terminus, as shown in the figure below. The magnitude of the vector is given by the length OA.

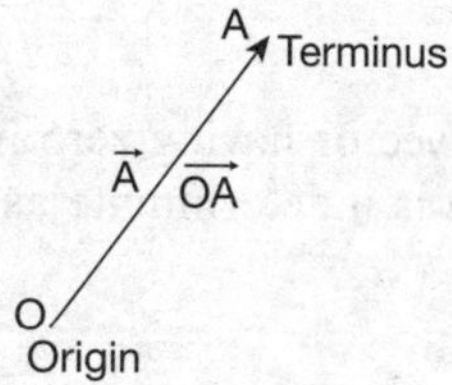

Some Specific Vectors

Like Vectors

Two vectors are said to be like vectors if they have same direction but different magnitude. Figure 10.1 shows two vectors $\vec{A}$ and $\vec{B}$ which have different magnitude but are parallel to each other.

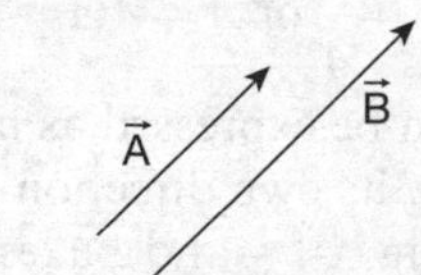

Figure 10.1 Like vectors

Unlike Vectors

The vectors having opposite direction and different magnitude, are called unlike vectors. Figure 10.2 shows two such vectors $\vec{A}$ and $\vec{B}$ which have different magnitudes and are antiparallel to each other.

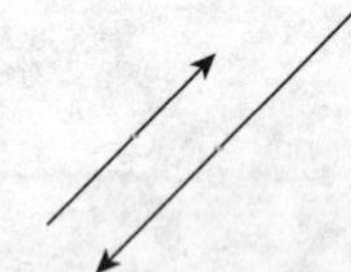

Figure 10.2 Unlike vectors

Equal Vectors

Two vectors are said to be equal, if they have the same magnitude and direction. Figure 10.3 shows two vectors $\vec{A}$ and $\vec{B}$ having the same magnitude and same direction and therefore $\vec{A} = \vec{B}$.

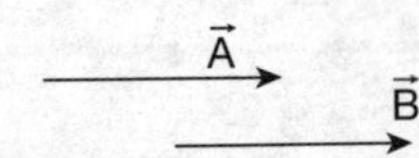

Figure 10.3 Equal vectors

Opposite Vectors

The vectors having same magnitude but opposite direction, are known as opposite vectors. Figure 10.4 shows two such vectors $\vec{A}$ and $\vec{B}$ having the same magnitude and opposite direction and therefore $\vec{A} = -\vec{B}$

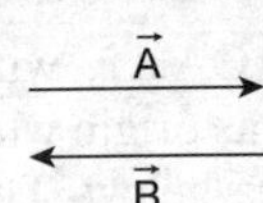

Figure 10.4 Opposite vectors

Null Vector

It is defined as a vector having zero magnitude. It has a direction which is indeterminate as its magnitude is zero.

Unit Vector

A vector divided by its magnitude is called a unit vector along the direction of the vector. Obviously, the unit vector has unit magnitude and its direction is the same as that of the given vector.

A unit vector in the direction of some vector $\vec{A}$ is written as $\hat{A}$ and is read as 'A cap' or 'A hat'. Therefore, by definition

$$\hat{A} = \frac{\vec{A}}{A} \text{ or } \vec{A} = \hat{A}\,A$$

Thus, any vector can be expressed as magnitude times the unit vector along its own direction.

Unit vectors along, x-, y- and z-axes are represented by $\hat{i}, \hat{j}$ and $\hat{k}$ respectively.

Co-initial Vectors

Vectors are said to be co-initial, if they have a common initial point. In Fig. 10.5, $\vec{A}$ and $\vec{B}$ originating from the same point O as their origin are called co-initial vectors.

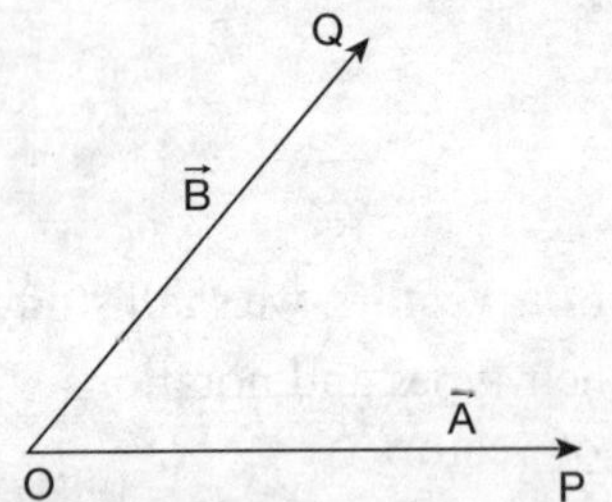

Figure 10.5 Co-initial vectors

Collinear Vectors

Two vectors having equal or unequal magnitudes, which either act along the same line [Fig. 10.6 (a)] or along the parallel lines in the same direction [Fig. 10.6 (b)] or along the parallel lines in opposite direction [Fig. 10.6 (c)], are called collinear vectors. Like, unlike, equal and opposite vectors are collinear.

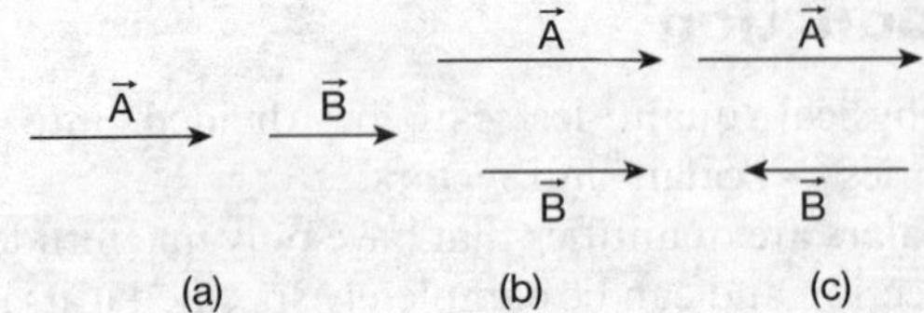

Figure 10.6 Collinear vectors

Coplanar Vectors

Vectors lying in the same plane are called coplanar vectors. Figure 10.7 (a) shows three vectors $\vec{X}, \vec{Y}$ and $\vec{Z}$ along mutually ^ axes x, y and z respectively. These vectors are non-coplanar but the vectors $\vec{x}-\vec{y}$, $\vec{y}-\vec{z}$ and $\vec{z}-\vec{x}$ are coplanar (Fig. 10.7 (b)).

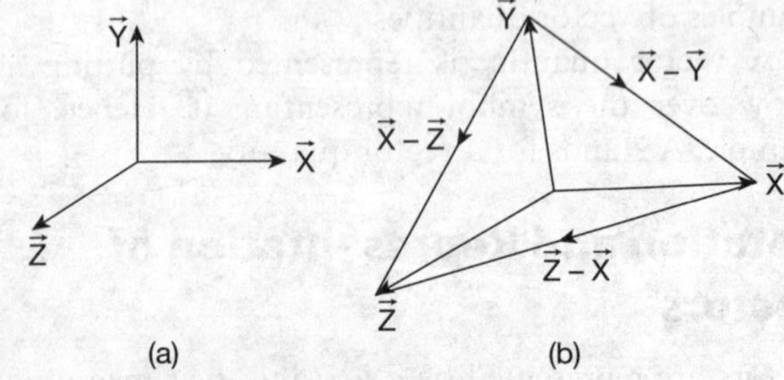

Figure 10.7 Coplanar vectors

Operation on Vectors

There can be broadly three kinds of operations on vectors (like we have operations of addition, subtraction, multiplication, etc.).

Similarly, for vectors we have

(A) Addition of vectors
Say, two vectors $\vec{a} = 2\hat{i} + 3\hat{j}$

$$\vec{b} = 3\hat{i} - 5\hat{j}$$

Then $\vec{a} + \vec{b} = (2\hat{i} + 3\hat{j}) + (3\hat{i} - 5\hat{j}) = 5\hat{i} - 2\hat{j}$

(B) Subtraction of vectors
In the above example,

$$\vec{a} - \vec{b} = (2\hat{i} + 3\hat{j}) - (3\hat{i} - 5\hat{j})$$
$$= (-\hat{i} + 8\hat{j})$$

(C) Multiplication of vectors
This can be of two types:

(i) Scalar Product (or dot product) of vectors

$$\vec{a}.\vec{b}$$

(ii) Vector product (or cross product) of vector

$\vec{a} \times \vec{b}$ (called, $\vec{a}$ cross $\vec{b}$).

Addition of Vectors

Because the vectors have directions also along with their magnitudes, hence they cannot be added according to the simple rules of algebra as applied to scalars. In order to make it clear, let us add two displacement vectors $\vec{A}$ and $\vec{B}$ having different possible directions with respect to each other.

Case I: When two vectors $\vec{A}$ and $\vec{B}$ are in the same direction
If a body moves a distance of 1 cm first and then 2 cm, due east in both the cases and the two displacement vectors are represented by $\vec{A}$ and $\vec{B}$ (figure), then the sum of two displacement vectors, i.e., $\vec{A} + \vec{B}$, will be equal to a displacement of 3 cm along east.

$\xrightarrow{\vec{A}} + \xrightarrow{\vec{B}} = \xrightarrow{\vec{C} = \vec{A} + \vec{B}}$ East

Case II: When two Vectors $\vec{A}$ and $\vec{B}$ are in opposite directions
If the body moves a distance of 1 cm due west and then through 2 cm due east, (figure), then sum $(\vec{A} + \vec{B})$ of two displacement vectors will be equal to a displacement of 1 cm due east.

$\xleftarrow{\vec{A}} + \xrightarrow{\vec{B}} = \xrightarrow{\vec{D} = \vec{A} + \vec{B}}$ East

Case III: When two vectors are inclined to each other
We consider here the simple case of two mutually perpendicular vectors. Suppose the body moves a distance of 1 cm due east and then a distance of 2 cm due north. In figure given below the two displacement vectors are shown by $\overrightarrow{AB}$ and $\overrightarrow{BC}$, according to some scale. It is clear that the magnitude of the displacement vector $\overrightarrow{AC}$ is $\sqrt{5}$ cm and direction of displacement is due north of east measured in terms of θ.

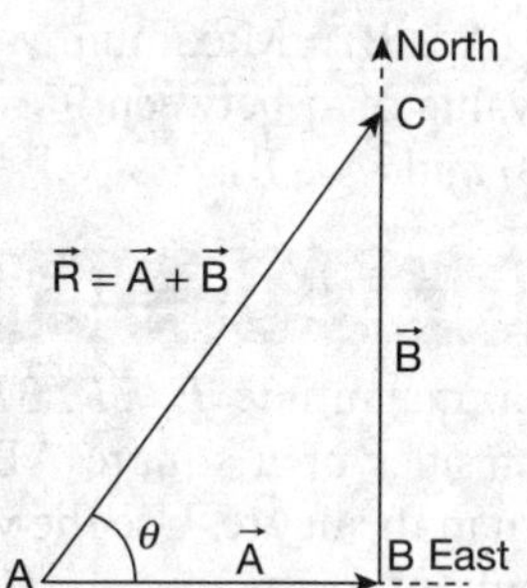

Therefore, if instead of moving a distance of 1 cm along AB and then a distance of 2 cm along BC, the body moves a distance of $\sqrt{5}$ cm along AC at an angle θ with AB, it reaches the same point. It means that the displacement of $\sqrt{5}$ cm along AC produces the same effect as the displacements of 1 cm along AB and 2 cm along BC together. This implies that displacement vector $\overrightarrow{AC}$ is the sum or resultant of displacement vectors $\overrightarrow{AB}$ and $\overrightarrow{BC}$, i.e., $\overrightarrow{AB} + \overrightarrow{BC} = \overrightarrow{AC}$

If vectors $\overrightarrow{AB}$, $\overrightarrow{BC}$ and $\overrightarrow{AC}$ are represented by vectors $\vec{A}$, $\vec{B}$ and $\vec{R}$ respectively, then $\vec{A} + \vec{B} = \vec{R}$ where $\vec{R}$ is called as the resultant of vectors $\vec{A}$ and $\vec{B}$. Thus we conclude that

(i) The resultant of two or more vectors is a single vector which produces the same effect as the individual vectors do together.
(ii) Vectors are not added algebraically but are added geometrically.

The triangle law of vector addition states that if three vectors $\vec{A}$, $\vec{B}$ and $\vec{C}$ can be represented in magnitude and direction by the sides of a triangle taken in order then,

$$\vec{A} + \vec{B} + \vec{C} = 0$$

$$\Rightarrow \quad \vec{A} + \vec{B} = -\vec{C}$$

$\vec{C}$ is the resultant of $\vec{A}$ and $\vec{B}$ and its magnitude is given by

$$C = |\vec{C}| = \sqrt{A^2 + B^2 + 2\ AB\cos\theta}$$

where q is the angle between $\vec{A}$ and $\vec{B}$.
The resultant $\vec{C}$ is inclined with $\vec{A}$ by an angle a, where

$$\tan\alpha = \frac{B\sin\theta}{A + B\cos\theta}.$$

for A = B,

$$\tan\alpha = \frac{\sin\theta}{1+\cos\theta} = \frac{2\sin\theta/2.\cos\theta/2}{1+2\cos^2\frac{\theta}{2}-1} = \tan\theta/2$$

$$\Rightarrow \quad \alpha = \theta/2$$

Hence, the resultant of two equal vectors bisects the angle between them. Further,
if $\theta = 0$, C = (A + B) = Maximum value of C and

if $\theta = 180°$, C = (A − B) = Maximum value of C.
For any other value of q between 0° and 180°, C lies between (A + B) and (A – B).

ILLUSTRATION 10.1

Four forces of magnitude P, $2P$, $3P$ and $4P$ act along the four sides of a square ABCD in cyclic order as shown in the figure. Use the vector method to find the resultant force.

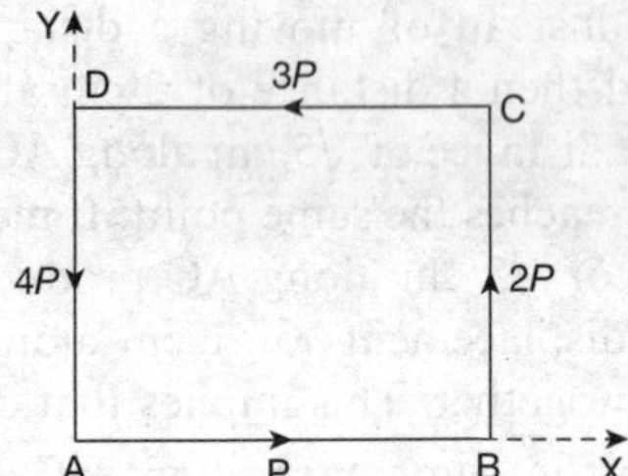

Solution

Let us assume x-axis along the side AB and y-axis along the side AD of the square ABCD.

Since a parallel translation of a vector does not bring any change in it, we can assume that the force $3P$ is acting along the negative direction of x-axis and the force $4P$ is acting along the negative direction of y-axis,

It follows from figure, that effectively a force $2P$ acts along the negative y-direction,

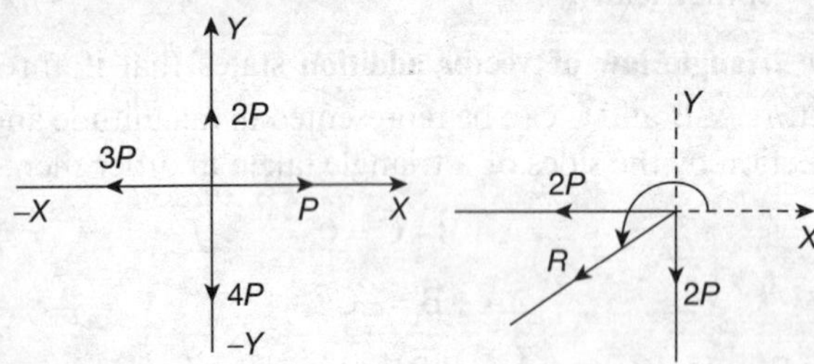

The resultant of these two equal forces $2P$ inclined at an angle of 90° is R, where

$$R = \sqrt{(2P)^2 + (2P)^2} = 2\sqrt{2}P$$

Since the forces are equal, their resultant R bisects the angle between them and hence, is inclined to either force at an angle of 45°.

Hence, the resultant of the four given forces is $2\sqrt{2}P$ and is inclined with the axis of x at an angle of 225° (180° + 45°) measured in the anticlockwise direction.

Properties of Vector Addiction

Vector addition has the following properties.

(i) Binary operation: The sum of two vectors is always a vector

(ii) Commutativity: For any two vectors $\vec{a}$ and $\vec{b}$, $\vec{a}+\vec{b}=\vec{b}+\vec{c}$

(iii) Associativity: For any three vectors $\vec{a},\vec{b}$ and $\vec{c}, \vec{a}+(\vec{b}+\vec{c})=(\vec{a}+\vec{b})+\vec{c}$

(iv) Identity: Zero vector is the identity for addition. For any vector a

$$\vec{0}+\vec{a}=\vec{a}=\vec{a}+\vec{0}$$

(v) Additive inverse: For every vector $\vec{a}$ its negative vector $-\vec{a}$ exists such that

$$\vec{a}+(-\vec{a})=(-\vec{a})+\vec{a}=0$$

i.e., $(-a)$ is the additive inverse of the vector $\vec{a}$.
Also if $\vec{a}=a_1\hat{i}+a_2\hat{j}+a_3\hat{k}$
Then $-\vec{a}=-a_1\hat{i}-a_2\hat{j}-a_3\hat{k}$

ILLUSTRATION 10.2

If position vectors of A and B are $3\hat{i}+4\hat{j}-5\hat{k}$ and $2\hat{i}-3\hat{j}+4\hat{k}$ then AB equals ______.

(a) $-\hat{i}-7\hat{j}+9\hat{k}$ (b) $3\hat{i}+2\hat{j}+\hat{k}$
(c) $\hat{i}+5\hat{j}+\hat{k}$ (d) None of these.

Solution

$$\overrightarrow{AB} = \text{p.v. of B} - \text{p.v. of A}$$
$$= (2\hat{i}-3\hat{j}+4\hat{k})-(3\hat{i}+4\hat{j}-5\hat{k})$$
$$=(2-3)\hat{i}+(-3-4)\hat{j}+(4+5)\hat{k}$$
$$=-\hat{i}-7\hat{j}+9\hat{k}$$

ILLUSTRATION 10.3

If $\vec{a}=2\hat{i}-3\hat{j}+4\hat{k}$ and $\vec{b}=\hat{i}+2\hat{j}-\hat{k}$, then $\vec{a}+\vec{b}$ equals ______.

(a) $\hat{i}+\hat{j}+3\hat{k}$ (b) $3\hat{i}-\hat{j}+3\hat{k}$
(c) $\hat{i}-\hat{j}-3\hat{k}$ (d) $2\hat{i}+\hat{j}+\hat{k}$

Solution

$$\vec{a}+\vec{b}=(2+1)\hat{i}+(-3+2)\hat{j}+(4-1)\hat{k}=3\hat{i}-\hat{j}+3\hat{k}$$

ILLUSTRATION 10.4

Solution

$$\overrightarrow{AB}=(2\hat{i}+3\hat{j}-4\hat{k})-(\hat{i}-\hat{j}+2\hat{k})$$
$$=\hat{i}+4\hat{j}-6\hat{k} \therefore = \sqrt{(1)^2+(4)^2+(6)^2}$$
$$=\sqrt{1+16+36}=\sqrt{53}$$

ILLUSTRATION 10.5

If the modulus of vectors $\vec{a}=\lambda\hat{i}+2\hat{j}-3\hat{k}$ and $\vec{b}=\sqrt{\lambda}\hat{i}+\sqrt{13}\hat{j}$ be equal then the value of λ is ______.

(a) 0 (b) 1 (c) 0 or 1 (d) None of these.

Solution

Since $|\vec{a}|=|\vec{b}| \Rightarrow \sqrt{\lambda^2+(2)^2+(-3)^2}$

$$=\sqrt{(\sqrt{\lambda})^2+(\sqrt{13})^2} \Rightarrow \lambda^2+13=\lambda+13$$

$\therefore$ $\lambda^2-\lambda=0 \Rightarrow \lambda=0 \text{ or } 1.$

Parallelogram Law of Addition

If two vectors are represented by two adjacent sides of a parallelogram, then their sum is represented by the diagonal of the parallelogram whose initial point is the same as the initial point of the given vectors. This is known as parallelogram law of addition of vectors.

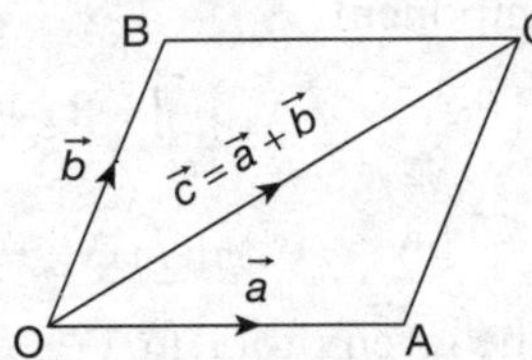

Thus if $\overrightarrow{OA}=\vec{a}, \overrightarrow{OB}=\vec{b}$ and $\overrightarrow{OC}=\vec{c}$

Then $\overrightarrow{OA}+\overrightarrow{OB}=\overrightarrow{OC}$

i.e., $\vec{a}+\vec{b}=\vec{c}$

where $\overrightarrow{OC}$ is a diagonal of the parallelogram OABC.

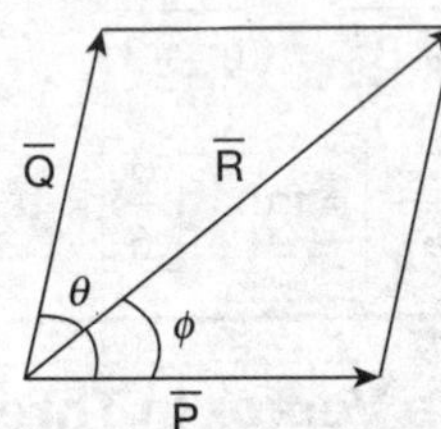

$$R=\sqrt{P^2+Q^2+2PQ\cos\theta} \text{ and } \tan\phi=\frac{Q\sin\theta}{P+Q\cos\theta}$$

If $\theta=0°$ then $R=P+Q$ (maximum)

$\theta=\pi$ then $R=P-Q$ (minimum)

If $\theta=\frac{\pi}{2}$ then $R=\sqrt{P^2+Q^2}$ and $\tan\phi=\frac{Q}{P}$

Law of Polygons

If several vectors, when added, form a closed polygon, their resultant is zero.

ILLUSTRATION 10.6

The length of diagonal AC of a parallelogram ABCD whose two adjacent sides AB and AD are represented respectively by $2\hat{i}+4\hat{j}-5\hat{k}$ and $\hat{i}+2\hat{j}+3\hat{k}$ is ______.

(a) 3 (b) 4 (c) 5 (d) 7

Solution

$$\because \overrightarrow{AC}=\overrightarrow{AB}+\overrightarrow{AD}$$
$$=3\hat{i}+6\hat{j}-2\hat{k}$$

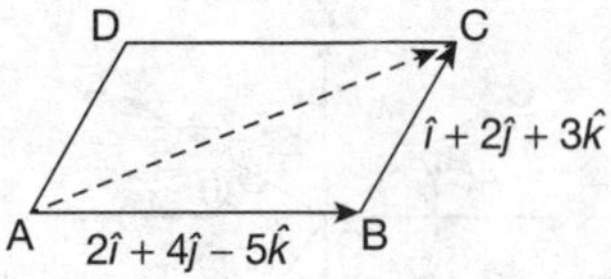

$\therefore$ Length of the diagonal $\overline{AC}=|\overline{AC}|$

$$=\sqrt{3^2+6^2+(-2)^2}=7$$

Vector Resolution of a Vector (Components of a Vector)

A vector $\vec{r}=a\hat{i}+b\hat{j}+c\hat{k}$ has its x, y and z components as a, b and c respectively.

The vector resolution can be orthogonal or non-orthogonal. Orthogonal resolution means the components of the vector are mutually perpendicular, otherwise the components are non-orthogonal. Orthogonal components of vector $\bar{R}$ are as shown.

$$P=R\cos\theta \qquad \bar{R}=\bar{P}+\bar{Q}$$
$$Q=R\sin\theta$$

Non-orthogonal components of $\bar{R}$ are as shown:

$$P=\frac{R\sin\beta}{\sin(\alpha+\beta)}$$

$$Q=\frac{R\sin\alpha}{\sin(\alpha+\beta)}$$

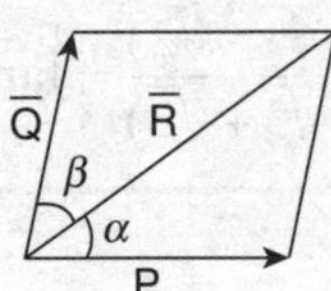

Addition in component form: If the vectors are defined in terms of $\hat{i}, \hat{j}$ and $\hat{k}$, i.e., if $\vec{a} = a_1\hat{i} + a_2\hat{j} + a_3\hat{k}$ and $\vec{b} = b_1\hat{i} + b_2\hat{j} + b_3\hat{k}$ then their sum is defined as

$$\vec{a} + \vec{b} = (a_1 + b_1)\hat{i} + (a_2 + b_2)\hat{j} + (a_3 + b_3)\hat{k}$$

ILLUSTRATION 10.7

The figure shows three vectors $\overrightarrow{OA}, \overrightarrow{OB}$ and $\overrightarrow{OC}$ having same magnitude. Find the direction of $\overrightarrow{OA} + \overrightarrow{OB} - \overrightarrow{OC}$.

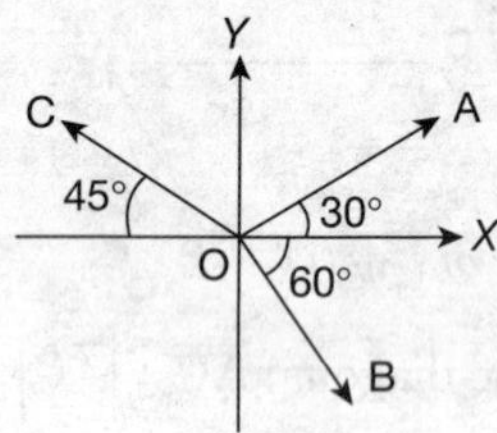

Solution

Let $OA = OB = OC = P$

x-component of $\overrightarrow{OA} = \text{P}\cos 30° = \dfrac{P\sqrt{3}}{2}$

x-component of $\overrightarrow{OB} = \text{P}\cos 60° = \dfrac{\text{P}}{2}$

x-component of $\overrightarrow{OC} = \text{P}\cos 135° = -\dfrac{\text{P}}{\sqrt{2}}$

$\therefore$ x-component of

$$\left(\overrightarrow{OA} + \overrightarrow{OB} - \overrightarrow{OC}\right) = \frac{\text{P}\sqrt{3}}{2} + \frac{\text{P}}{2} + \frac{\text{P}}{\sqrt{2}} = \frac{\text{P}}{2}(\sqrt{3} + 1 + \sqrt{2})$$

Similarly,

y-component of $\overrightarrow{OA} = \text{P}\cos 60° = \dfrac{\text{P}}{2}$

y-component of $\overrightarrow{OB} = \text{P}\cos 150° = -\dfrac{\text{P}\sqrt{3}}{2}$

y-component of $\overrightarrow{OC} = \text{P}\cos 45° = \dfrac{\text{P}}{\sqrt{2}}$

$\therefore$ y-component of

$$\left(\overrightarrow{OA} + \overrightarrow{OB} - \overrightarrow{OC}\right) = \frac{\text{P}}{2} + \left(-\frac{\text{P}\sqrt{3}}{2}\right) - \frac{\text{P}}{\sqrt{2}}$$

$$= \frac{\text{P}}{2}(1 - \sqrt{3} - \sqrt{2})$$

Thus, angle of $\left(\overrightarrow{OA} + \overrightarrow{OB} - \overrightarrow{OC}\right.$ with the x-axis

$$= \tan^{-1}\frac{(\text{P}/2)(1 - \sqrt{3} - \sqrt{2})}{(\text{P}/2)(1 + \sqrt{3} + \sqrt{2})} = \tan^{-1}\frac{(1 - \sqrt{3} - \sqrt{2})}{(1 + \sqrt{3} + \sqrt{2})}$$

ILLUSTRATION 10.8

Can a vector be zero if any of its components is not zero?

Solution

In terms of components in three dimensions, $\vec{A} = \vec{i}\,A_x + \vec{j}\,A_y + \vec{k}\,A_z$ with $A = \sqrt{A_x^2 + A_y^2 + A_z^2}$. So, if any of A_x, $A_y$and A_z is not zero, the vector cannot be zero, i.e., for a vector to be zero (or null) all its component must be zero.

ILLUSTRATION 10.9

The x and y components of vector $\vec{A}$ are 4 m and 6 m respectively. The x and y components of vector $\vec{A} + \vec{B}$are 10 m and 9 m respectively. Calculate for the vector $\vec{B}$, the following:

(a) its x and y components, (b) its length, and (c) the angle it makes with x-axis.

Solution

In terms of components

$$\vec{A} + \vec{B} = \left(\vec{i}\,A_x + \vec{j}\,A_y\right) + \left(\vec{i}\,B_x + \vec{j}\,B_y\right)$$

i.e., $\quad \vec{A} + \vec{B} = \vec{i}\left(A_x + B_x\right) + \vec{j}\left(A_x + B_y\right)$

According to the given problem,

$$A_x + B_x = 10 \text{ and } A_y + B_y = 9$$

and $A_x = 4$ and $A_y = 6$

So, (a) $B_x = 10 - 4 = 6$ m and $B_y = 9 - 6 = 3$m

(b) $B = \sqrt{B_x^2 + B_y^2} = \sqrt{6^2 + 3^2} = \sqrt{45}$ m $= 3\sqrt{5}$m

(c) $\theta = \tan^{-1}\left[\dfrac{B_y}{B_x}\right] = \tan^{-1}\left[\dfrac{3}{6}\right] = \tan^{-1}\left[\dfrac{1}{2}\right] = 26.6°$

Component of a Vector in Three Dimensions

A vector can be, in general, resolved into any three mutually perpendicular components. If a vector $\vec{A}$ makes angles *a*, *b* and *g* with *x*-, *y*- and *z*-axes respectively, then,

$$\vec{A} = A\cos\alpha\,\hat{i} + A\cos\beta\,\hat{j} + A\cos\gamma\,\hat{k}$$

$$\Rightarrow \quad \vec{A} = A_x\hat{i} + A_y\hat{j} + A_z\hat{k}$$

where,

$A_x = A\cos\alpha$, the component of $\vec{A}$ along x-axis.
$A_y = A\cos\beta$, the component of $\vec{A}$ along y-axis.
$A_z = A\cos\gamma$, the component of $\vec{A}$ along z-axis.

Subtraction of Vectors

If $\vec{a}$ and $\vec{b}$ are two vectors, then their subtraction $\vec{a}-\vec{b}$ is defined as $\vec{a}-\vec{b}=\vec{a}+(-\vec{b})$, where $-\vec{b}$ is the negative of $\vec{b}$ having magnitude equal to that of $\vec{b}$ and direction opposite of $\vec{b}$.

If $\vec{a}=a_1\hat{i}+a_2\hat{j}+a_3\hat{k}$

and $\vec{b}=b_1\hat{i}+b_2\hat{j}+b_3\hat{k}$

then $\vec{a}-\vec{b}=\left(\vec{a}_1-\vec{b}_1\right)\hat{i}+\left(\vec{a}_2-\vec{b}_2\right)\hat{j}+\left(\vec{a}_3-\vec{b}_3\right)\hat{k}$

Remark

(i) $\vec{a}-\vec{b}\neq\vec{b}-\vec{a}$

(ii) $(\vec{a}-\vec{b})-\vec{c}\neq\vec{a}-(\vec{b}-\vec{c})$

(iii) Since any one side of a triangle is less than the sum and greater than the difference of the other two sides, so for any two vectors $\vec{a}$ and $\vec{b}$, we have

$$\left|\vec{a}+\vec{b}\right|\leq|\vec{a}|+\left|\vec{b}\right|$$

$$\left|\vec{a}+\vec{b}\right|\geq|\vec{a}|-\left|\vec{b}\right|$$

$$\left|\vec{a}-\vec{b}\right|\leq|\vec{a}|+\left|\vec{b}\right|$$

$$\left|\vec{a}-\vec{b}\right|\geq|\vec{a}|-\left|\vec{b}\right|$$

ILLUSTRATION 10.10

If $\vec{a}=2\hat{i}-5\hat{j}+2\hat{k}$ the find $3\vec{a}$ and $-5\vec{a}$.

Solution

$\because\vec{a}=2\hat{i}-5\hat{j}+2\hat{k}$, then $3\vec{a}=6\hat{i}-15\hat{j}+6\hat{k}$ and $-5\vec{a}=-10\hat{i}+25\hat{j}-10\hat{k}$

ILLUSTRATION 10.11

A particle is moving eastward with a velocity of 5 m/s. in 10 s, the velocity changes to 5 m/s northward. The average acceleration in this time is ______.

(a) zero (b) $\frac{1}{\sqrt{2}}$ ms^{-2} towards north-east

(c) $\frac{1}{\sqrt{2}}$ ms^{-2} towards south-west

(d) $\frac{1}{2}$ms^{-2} towards north-west.

Solution

Let us consider the x-axis eastward and y-axis northward. Then, the initial velocity $\vec{v}_i$ is given by

$$\vec{v}_i=5\hat{i}$$

and the final velocity $\vec{v}_f$, after time $t=10$ s, is given by

$$\vec{v}_f=5\,\hat{j}$$

The change in velocity is, then, $\vec{v}_f-\vec{v}_i$ and the acceleration, defined as the rate of change of velocity, is

$$\vec{a}=\frac{\vec{v}_f-\vec{v}_i}{t}=\frac{5\hat{j}-5\hat{i}}{10}=\frac{\hat{j}}{2}-\frac{\hat{i}}{2}$$

$$\Rightarrow\quad a=|\vec{a}|=\sqrt{\left(\frac{1}{2}\right)^2+\left(\frac{1}{2}\right)^2}=\frac{1}{\sqrt{2}}\text{ ms}^{-2}$$

If this acceleration makes an angle q with the axis of x, then

$$\tan\theta=\frac{1/2}{-1/2}=-1=\tan 135°$$

$$\Rightarrow\quad \theta=135°$$

Thus, the average acceleration of the particle is $\frac{1}{\sqrt{2}}$ ms^{-2} along the north-west direction. Hence, the correct answer is (b).

Vectors in Terms of Position Vectors of End Points

If $\overrightarrow{AB}$ be any given vector and also suppose that the position vectors of initial point A and terminal point B are a and b respectively,

Then $\overrightarrow{AB}=\overrightarrow{OB}-\overrightarrow{OA}=\vec{b}-\vec{a}$

i.e., $\overrightarrow{AB}$ = p.v. of point B – p.v. of point A

= p.v. of terminal point – p.v. of initial point

Distance Between Two Points

Let A and B be two given points whose co-ordinates are, respectively, (x_1,y_1,z_1) and (x_2,y_2,z_2).

If $\vec{a}$ and $\vec{b}$ are position vectors of A and B relative to point O, then

$$\vec{a}=x_1\hat{i}+y_1\hat{j}+z_1\hat{k}$$

$$\vec{b}=x_2\hat{i}+y_2\hat{j}+z_2\hat{k}$$

Now $\overrightarrow{AB}=\overrightarrow{OB}-\overrightarrow{OA}=\vec{b}-\vec{a}$

$$=(x_2-x_1)\hat{i}+(y_2-y_1)\hat{j}+(z_2-z_1)\hat{k}$$

Distance between the points $\vec{A}$ and $\vec{B}$ = magnitude of

$$\overline{AB}=\sqrt{(x_2-x_1)^2+(y_2-y_1)^2+(z_2-z_1)^2}$$

Multiplication of Vectors

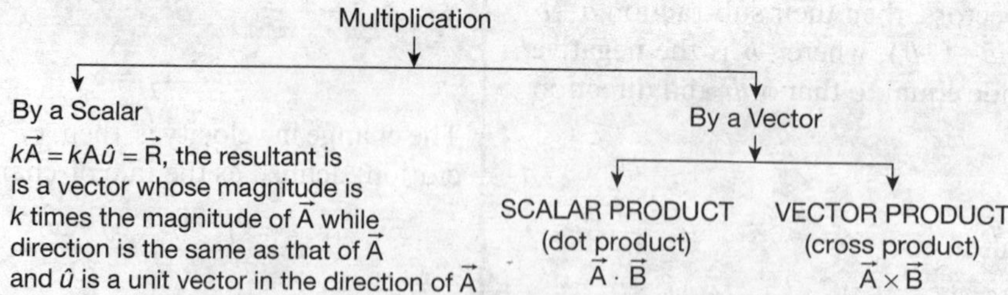

Scalar Product of Two Vectors

The scalar product of two vectors in defined as scalar quantity having magnitude equal to the product of the magnitude of two vectors and the cosine of the smaller angle between them.

Mathematically, if θ is the angle between vectors $\vec{A}$ and $\vec{B}$, then $\vec{A}.\vec{B} = AB\ \cos\theta$

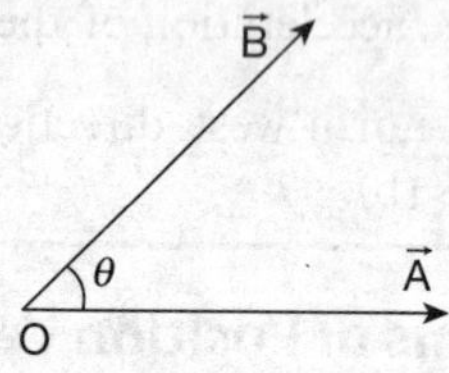

The above result can also be expressed as $\vec{A}.\vec{B} = A(B\cos\theta) = B(A\cos\theta)$ where $B\cos\theta$ is the magnitude of the component of $\vec{B}$ along the direction of vector $\vec{A}$ and $A\cos\theta$ is the magnitude of component of $\vec{A}$ along the direction of vector $\vec{B}$. Therefore, the dot product of two vectors can also be interpreted as the product of the magnitude of one vector and the magnitude of the component of other vector along the direction of first vector.

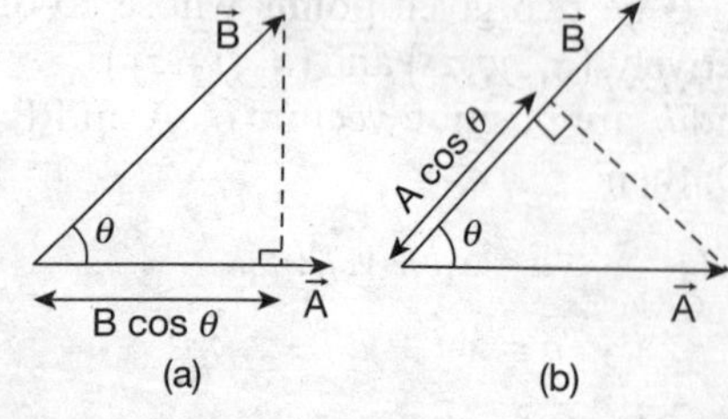

(iii) The dot product of two vectors can be positive or zero or negative depending upon whether θ is less than 90° or equal to 90° or $90° < \theta < 180°$.

(iv) The dot product of two vectors is always commutative, i.e.,

$$\vec{A}.\vec{B} = \vec{B}.\vec{A}$$

$$\vec{A}.\vec{B} = AB\cos\theta$$

(θ is the angle between $\vec{A}$ and $\vec{B}$ measured in anticlockwise direction)

and $\vec{B}.\vec{A} = BA\cos(-\theta) = AB\cos\theta$

Thus $\vec{A}.\vec{B} = \vec{B}.\vec{A}$

(v) The dot product of a vector with itself gives the square of its magnitude, i.e.,

$$\vec{A}.\vec{A} = A\ A\cos 0° = A^2$$

(vi) The dot product of two mutually perpendicular vectors is zero, i,e., if two vectors $\vec{A}$ and $\vec{B}$ are perpendicular then

$$\vec{A}.\vec{B} = AB\ \cos 90° = 0$$

(vii) The dot product obeys the distributive law i.e., $\vec{A}.(\vec{B}+\vec{C}) = \vec{A}.\vec{B} + \vec{A}.\vec{C}$

(viii) Two vectors are collinear if their dot product is numerically equal to product their magnitudes, i.e., when $\theta = 0°$ or $180°$, $|\vec{A}.\vec{B}| = AB$

(ix) In case of unit vectors $\hat{i}, \hat{j}$ and $\hat{k}$, we have following two properties:

(a) $\hat{i}.\hat{i} = \hat{j}.\hat{j} = \hat{k}.\hat{k} = 1$ (as $\theta = 0°$)

(b) $\hat{i}.\hat{j} = \hat{j}.\hat{k} = \hat{k}.\hat{i} = 0$ (as $\theta = 90°$)

(x) Dot product of two vectors in terms of their rectangular components in three dimensions:

$$\vec{A}.\vec{B} = (A_x\hat{i} + A_y\hat{j} + A_z\hat{k}).(B_x\hat{i} + B_y\hat{j} + B_z\hat{k})$$
$$= A_xB_x + A_yB_y + A_zB_z$$

(xi) Examples of some physical quantities which can be expressed as scalar product of two vectors:

(a) Work (W) is defined as the scalar product of force ($\vec{F}$) and the displacement ($\vec{s}$), i.e., $W = \vec{F}.\vec{s}$

(b) Power (P) is defined as the scalar product of force ($\vec{F}$) and the velocity ($\vec{v}$), i.e., $P = \vec{F}.\vec{v}$

(c) Magnetic flux (ϕ) linked with a surface is defined as the dot product of magnetic induction ($\vec{B}$) and area vector (A), i.e., $\phi = \vec{B}.\vec{A}$

(Magnetic flux ϕ is a part of Class XII syllabus)

ILLUSTRATION 10.12

Find the angle between the vectors

$$\vec{A} = \hat{i} + 2\hat{j} + 3\hat{k} \text{ and } \vec{B} = 2\hat{i} + 3\hat{j} + 4\hat{k}$$

Solution

We have, $\vec{A}.\vec{B} = AB\cos\theta$

$$\Rightarrow \quad \cos\theta = \frac{\vec{A}.\vec{B}}{AB} \quad \text{(i)}$$

Now, $\vec{A}.\vec{B} = A_xB_x + A_yB_y + A_zB_z$

$$\Rightarrow \quad \vec{A}.\vec{B} = 1\times2 + 2\times3 + 3\times4$$

$$\Rightarrow \quad \vec{A}.\vec{B} = 20 \quad \text{(ii)}$$

Also, $A = \sqrt{A_x^2 + A_y^2 + A_z^2} = \sqrt{1^2 + 2^2 + 3^2} = \sqrt{14}$ (iii)

and $B = \sqrt{B_x^2 + B_y^2 + B_z^2} = \sqrt{2^2 + 3^2 + 4^2} = \sqrt{29}$ (iv)

Hence, $\cos\theta = \frac{\vec{A}.\vec{B}}{AB} = \frac{.20}{\sqrt{14}\sqrt{29}} = \frac{20}{\sqrt{406}}$

$$\Rightarrow \quad \theta = \cos^{-1}\left(\frac{20}{\sqrt{406}}\right)$$

ILLUSTRATION 10.13

The angle between two vectors $= 2\hat{i} + 3\hat{j} + \hat{k}$ and $\hat{i} + 2\hat{j} + 4\hat{k}$ is

(a) 0° (b) 60° (c) 90° (d) 180°

Solution

Let $\vec{A} = -2\hat{i} + 3\hat{j} + \hat{k}$

and $\vec{B} = \hat{i} + 2\hat{j} - 4\hat{k}$

Then,

$$\vec{A}.\vec{B} = \left(-2\hat{i} + 3\hat{j} + \hat{k}\right).\left(\hat{i} + 2\hat{j} - 4\hat{k}\right)$$

$$\Rightarrow \quad \vec{A}.\vec{B} = -2\times1 + 3\times2 - 1\times4 = 0$$

$$\Rightarrow \quad \vec{A}.\vec{B} = 0$$

Since $\vec{A}.\vec{B}$ is zero, the given vectors are at right angle to each other. Hence, the correct answer is (c).

Scalar Product in Particular Cases

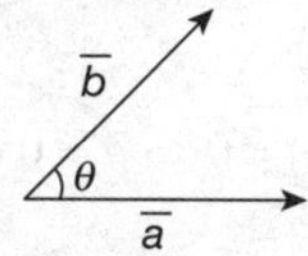

1. $\vec{a}.\vec{b} = \vec{b}.\vec{a}$
2. $\vec{a}.(\vec{b} + \vec{c}) = \vec{a}.\vec{b} + \vec{a}.\vec{c}$
3. $(m\vec{a}).\vec{b} = m(\vec{a}.\vec{b}) = \vec{a}.(m\vec{b})$
4. If $\theta = 0 \Rightarrow \vec{a}.\vec{b} = |\vec{a}||\vec{b}|$ (like vectors)
5. If $\theta = \pi \Rightarrow \vec{a}.\vec{b} = -|\vec{a}||\vec{b}|$ (unlike vectors)
6. If $\hat{a}$ and $\hat{b}$ are unit vectors then $\hat{a}.\hat{b} = \cos\theta$ (where θ is angle between them.)
7. $\vec{a}.\vec{a} = |\vec{a}|^2 \Rightarrow |\vec{a}| = \sqrt{\vec{a}.\vec{a}}$
8. If $\vec{a} \perp \vec{b} \Rightarrow \vec{a}.\vec{b} = 0$ but $\vec{a}.\vec{b} = 0 \Rightarrow \vec{a} = \vec{0}$ or $\vec{b} = 0$ or $\vec{a} \perp \vec{b}$.
9. If $\hat{i}, \hat{j}$ and $\hat{k}$ are unit vectors along the rectangular co-ordinates OX, OY and OZ then $\hat{i}.\hat{i} = \hat{j}.\hat{j} = \hat{k}.\hat{k} = 1$, $\hat{i}.\hat{j} = \hat{j}.\hat{k} = \hat{k}.\hat{i} = 0$.
10. $\vec{a}.\vec{b} = \vec{a}.\vec{c} \nRightarrow \vec{b} = \vec{c}$

    In fact, $\vec{a}.\vec{b} = \vec{a}.\vec{c} \Rightarrow \vec{a}.(\vec{b} - \vec{c}) = 0 \Rightarrow \vec{a} = 0$ or $\vec{b} = \vec{c}$ or $\vec{a} \perp (\vec{b} - \vec{c})$.
11. $(\vec{a}.\vec{b}).\vec{c}$ is meaningless.
12. Scalar product is not binary operation.

Remark

(a) $(\vec{a}.\vec{b}).\vec{b}$ is not defined

(b) $(\vec{a} + \vec{b})^2 = |\vec{a}|^2 + 2\vec{a}.\vec{b} + |\vec{b}|^2$

(c) $(\vec{a} - \vec{b})^2 = |\vec{a}|^2 - 2\vec{a}.\vec{b} + |\vec{b}|^2$

(d) $(\vec{a} + \vec{b}).(\vec{a} - \vec{b}) = |\vec{a}|^2 - |\vec{b}|^2$

(e) $|\vec{a} + \vec{b}| = |\vec{a}| + |\vec{b}| \Rightarrow \vec{a} \parallel \vec{b}$

(f) $|\vec{a} + \vec{b}|^2 = |\vec{a}|^2 + |\vec{b}|^2 \Rightarrow \vec{a} \perp \vec{b}$

(g) $|\vec{a} + \vec{b}| = |\vec{a} - \vec{b}| \Rightarrow \vec{a} \perp \vec{b}$

ILLUSTRATION 10.14

Determine the values of c such that for all x (real), the vectors $cx\hat{i} - 6\hat{j} + 3\hat{k}$ and $x\hat{i} + 2\hat{j} + 2cx\hat{k}$ make an obtuse angle with each other.

Solution

If θ is the angle between the given vectors then

$$\cos\theta = \frac{cx^2 - 12 + 6cx}{\sqrt{c^2x^2 + 45}\sqrt{x^2 + 4c^2x^2 + 4}}$$

If θ is obtuse then

$$\cos\theta < 0 \Rightarrow cx^2 + 6cx - 12 < 0 \forall x \in R,$$

which is possible if $c < 0$ and $36c^2 + 48c < 0 \Rightarrow c < 0$ and $12c(3c + 4) < 0$

$\Rightarrow c < 0$ and $c > -\frac{4}{3} \Rightarrow -\frac{4}{3} < c < 0$ (but for $c = 0$, $cx^2 + 6cx - 12 < 0 \ \forall x$)

Hence, $-\frac{4}{3} < c \le 0$.

ILLUSTRATION 10.15

If $\vec{a}, \vec{b}, \vec{c}$ are unit vectors such that $\vec{a}$ is perpendicular to the plane of $\vec{b}$ and $\vec{c}$, then find $|\vec{a} + \vec{b} + \vec{c}|$ when the angle between $\vec{b}$ and $\vec{c}$ is $\pi/3$.

Solution

We have $|\vec{a}| = |\vec{b}| = |\vec{c}| = 1, \vec{a}.\vec{b} = \vec{a}.\vec{c} = 0$ and

$\vec{b}.\vec{c} = \cos\frac{\pi}{3} = \frac{1}{2}$.

Now

$$|\vec{a}+\vec{b}+\vec{c}|^2 = (\vec{a}+\vec{b}+\vec{c}).(\vec{a}+\vec{b}+\vec{c})$$

$$= |\vec{a}|^2 + |\vec{b}|^2 + |\vec{c}|^2 + 2\vec{b}.\vec{c} \ (\because \vec{a}.\vec{c} = \vec{a}.\vec{b} = 0)$$

$$= 1+1+1+2.\frac{1}{2} = 4 \Rightarrow |\vec{a}+\vec{b}+\vec{c}| = 2$$

Scalar Product in Terms of Components

Let $\vec{a}$ and $\vec{b}$ be two vectors such that $\vec{a} = \vec{a}_1\hat{i} + \vec{a}_2\hat{j} + \vec{a}_3\hat{k}$ and $\vec{b} = \vec{b}_1\hat{i} + \vec{b}_2\hat{j} + \vec{b}_3\hat{k}$.

Then $\vec{a}.\vec{b} = \vec{a}_1\vec{b}_1 + \vec{a}_2\vec{b}_2 + \vec{a}_3\vec{b}_3$

In particular,

$$\vec{a}.\vec{a} = |\vec{a}|^2 = \vec{a}_1^{\,2} + \vec{a}_2^{\,2} + \vec{a}_3^{\,2}$$

For any vector $\vec{a}$,

$$\vec{a} = (\vec{a}.\hat{i})\hat{i} + (\vec{a}.\hat{j})\hat{j} + (\vec{a}.\hat{k})\hat{k}.$$

ILLUSTRATION 10.16

If $\vec{a} = 3\hat{i} + 2\hat{j} + \hat{k}$ and $\vec{b} = \hat{i} - 2\hat{j} + 5\hat{k}$, then find $\vec{a}.\vec{b}$.

Solution

$$\vec{a} = (3)(1) + (2)(-2) + (1)(5) = 3 - 4 + 5 = 4.$$

Angle Between Two Vectors

(i) If $\vec{a}$ and $\vec{b}$ be two vectors and θ be the angle between them, then

$$\cos\theta = \frac{\vec{a}.\vec{b}}{|\vec{a}||\vec{b}|} = \frac{\vec{a}}{a}.\frac{\vec{b}}{b} = \hat{a}.\hat{b}$$

(ii) If $\vec{a} = \vec{a}_1\hat{i} + \vec{a}_2\hat{j} + \vec{a}_3\hat{k}$ and $\vec{b} = \vec{b}_1\hat{i} + \vec{b}_2\hat{j} + \vec{b}_3\hat{k}$, then

$$\cos\theta = \frac{\vec{a}_1\vec{b}_1 + \vec{a}_2\vec{b}_2 + \vec{a}_3\vec{b}_3}{\sqrt{a_1^2 + a_2^2 + a_3^2}\sqrt{b_1^2 + b_2^2 + b_3^2}}$$

Remark: If $\vec{a}$ and $\vec{b}$ are perpendicular to each other, then $a_1b_1 + a_2b_2 + a_3b_3 = 0$

ILLUSTRATION 10.17

Find the angle between the vectors $4\hat{i} + \hat{j} + 2\hat{k}$ and $2\hat{i} + 2\hat{j} - \hat{k}$.

Solution

Let the required angle is θ.

$$\therefore \theta = \cos^{-1}\left(\frac{4.2 + 1.2 + 3(-1)}{\sqrt{16+1+9}\sqrt{4+4+1}}\right) = \cos^{-1}\left(\frac{7}{3\sqrt{26}}\right)$$

Components of *b* Along and Perpendicular to *a*

(i) Component along $a = \overline{\text{OM}}$

$$= \text{OM}\,\hat{a} = (b\cos\theta)\hat{a}$$

$$= \frac{(ab\cos\theta)}{a}\hat{a}$$

$$= \frac{(a.b)}{a^2}.a$$

(ii) Component perpendicular to $\vec{a} = \overline{\text{MB}}$

$$= \overline{\text{MO}} + \overline{\text{OB}}$$

$$= \overline{\text{OB}} - \overline{\text{OM}}$$

$$= \vec{b} - \frac{(\vec{a}.\vec{b})}{a^2}\vec{a}$$

ILLUSTRATION 10.18

Find the work done by a force represented by $4\hat{i} + \hat{j} - 3\hat{k}$, which displaces a particle from the point $A(\hat{i} + 2\hat{j} + 3\hat{k})$ to the point $B(5\hat{i} + 4\hat{j} + \hat{k})$.

Solution

Here $F = 4\hat{i} + \hat{j} - 3\hat{k}$

$d = \overline{\text{AB}} = (5-1)\hat{i} + (4-2)\hat{j} + (1-3)\hat{k} = 4\hat{i} + 2\hat{j} - 2\hat{k}$

Work done by the given force $= F.\vec{d}$

$= (4\hat{i} + \hat{j} - 3\hat{k}).(4\hat{i} + 2\hat{j} - 2\hat{k}) = 16 + 2 + 6 = 24$ units.

ILLUSTRATION 10.19

If for three vectors $\vec{a}, \vec{b}, \vec{c}, \vec{a}.\vec{b} = \vec{a}.\vec{c}$, then which of the following is correct?

(a) $\vec{b} = \vec{c}$ (b) $\vec{a} = 0$

(c) $\vec{a} = 0$ or $\vec{b} = \vec{c}$ or $\vec{a} \perp (\vec{b} - \vec{c})$ (d) $\vec{a} \perp (\vec{b} - \vec{c})$

Solution

$\vec{a}.\vec{b} = \vec{a}.\vec{c} \Rightarrow \vec{a}.(\vec{b} - \vec{c}) = 0 \Rightarrow \vec{a} = 0$ or $\vec{b} - \vec{c} = 0$ or $\vec{a} \perp (\vec{b} - \vec{c})$

$\Rightarrow \vec{a} = 0$ or $\vec{b} = \vec{c}$ or $\vec{a} \perp (\vec{b} - \vec{c})$

ILLUSTRATION 10.20

For any two vectors $\vec{a}$ and $\vec{b}.(\vec{a}\times\vec{b})^2$ equals ________.

(a) $|\vec{a}|^2|\vec{b}|^2-(\vec{a}\vec{b})^2$ (b) $|\vec{a}|^2+|\vec{b}|^2$

(c) $|\vec{a}|^2-|\vec{b}|^2$ (d) None of these.

Solution

$$(\vec{a}\times\vec{b})^2=|\vec{a}\times\vec{b}|^2=(\vec{a}\vec{b}\sin\theta)^2=\vec{a}^2\vec{b}^2\sin^2\theta$$

$$=\vec{a}^2\vec{b}^2(1-\cos^2\theta)=\vec{a}^2\vec{b}^2-(\vec{a}\vec{b}\cos\theta)^2$$

$$=|\vec{a}|^2|\vec{b}|^2-(\vec{a}\vec{b})^2$$

Vector or Cross Product of Two Vectors

If $\vec{a}$ and $\vec{b}$ be two vectors and $\theta(0\le\theta\le\pi)$ be the angle between them, their vector (or cross) product is defined to be a vector whose magnitude is $ab\sin\theta$ and whose direction is perpendicular to the plane of $\vec{a}$ and $\vec{b}$ such that $\vec{a},\vec{b}$ and $\vec{a}\times\vec{b}$ form a right-handed system.

$$\therefore \quad \vec{a}\times\vec{b}=|\vec{a}||\vec{b}|\sin\theta\hat{n}$$
$$=\vec{a}\vec{b}\sin\theta\hat{n}$$

Vector Product in Terms of Components

If $\vec{a}=a_1\hat{i}+a_2\hat{j}+a_3\hat{k}$ and $\vec{b}=b_1\hat{i}+b_2\hat{j}+b_3\hat{k}$ then

$$\vec{a}\times\vec{b}=(a_2b_3-a_3b_2)\hat{i}+(a_3b_1-a_1b_3)\hat{j}+(a_1b_2-a_2b_1)\hat{k}$$
$$=\begin{vmatrix}\hat{i} & \hat{j} & \hat{k}\\ a_1 & a_2 & a_3\\ b_1 & b_2 & b_3\end{vmatrix}$$

Angle between two vectors: If θ is the angle between $\vec{a}$ and $\vec{b}$, then

$$\sin\theta=\frac{|\vec{a}\times\vec{b}|}{|\vec{a}||\vec{b}|}$$

If $\hat{n}$ is the unit vector perpendicular to the plane of $\vec{a}$ and $\vec{b}$, then

$$\hat{n}=\frac{\vec{a}\times\vec{b}}{|\vec{a}\times\vec{b}|}$$

The direction of vector $\vec{A}\times\vec{B}$ is the same as that of unit vector $\hat{n}$. It is decided by any of the following two rules:

(a) **Right-handed screw rule:** Rotate a right-handed screw form vector $\vec{A}$ to $\vec{B}$ through the smaller angle between them; then the direction of motion of screw gives the direction of vector $\vec{A}\times\vec{B}$ [Fig. 10.8 (i).]

(b) **Right-hand thumb rule:** Bend the fingers of the right hand in such a way that they point in the direction of rotation; then the thumb points in the direction of vector $\vec{A}\times\vec{B}$ [Fig. 10.8 (ii).]

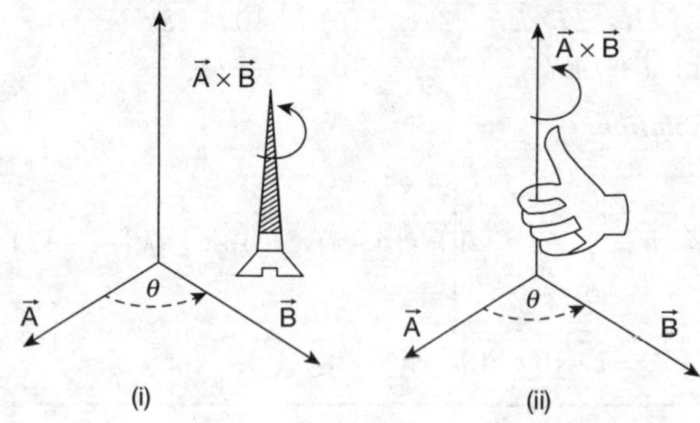

Figure 10.8

In case of vector $\vec{B}\times\vec{A}$, the direction of advancement of screw, on rotating it form vector $\vec{B}\times\vec{A}$, will be just opposite to that observed in the case of vector $\vec{A}\times\vec{B}$. Hence, the vector $\vec{B}\times\vec{A}$ is directed in just the opposite direction to that of vector $\vec{A}\times\vec{B}$, i.e., $\vec{B}\times\vec{A}=-(\vec{A}\times\vec{B})$

(iii) The cross product of the two vectors does not obey commutative law. As discussed above $\vec{A}\times\vec{B}=-(\vec{B}\times\vec{A})$, i.e., $\vec{A}\times\vec{B}\neq(\vec{B}\times\vec{A})$

(iv) The cross product follows the distributive law, i.e.,

$$\vec{A}\times(\vec{B}+\vec{C})=\vec{A}\times\vec{B}+\vec{A}\times\vec{C}$$

(v) The cross product of two vectors with itself is a NULL vector, i.e.,

$$\vec{A}\times\vec{A}=(A)(A)\sin 0°\ \hat{n}=0.$$

(vi) The cross product of two vectors represents the area of the parallelogram formed by them.

Figure 10.9 shows a parallelogram PQRS whose adjacent sides $\overline{PQ}$ and $\overline{PS}$ are represented by vectors $\vec{A}$ and $\vec{B}$ respectively.

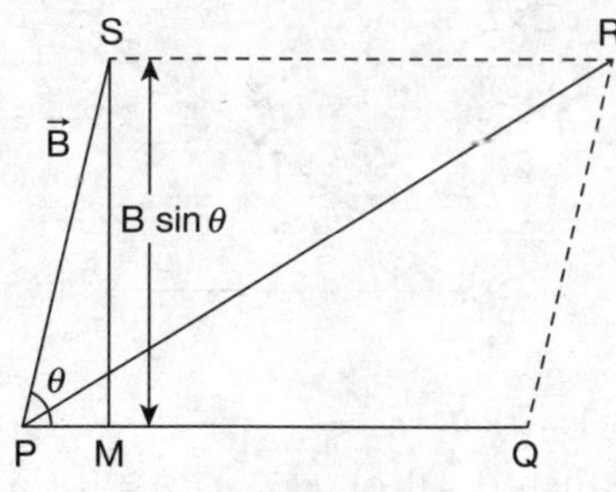

Figure 10.9

Now, area of parallelogram
= PQ × SM = AB sin θ

ILLUSTRATION 10.21

If $\vec{a} = 2\hat{i} + 2\hat{j} - \hat{k}$ and $\vec{b} = 6\hat{i} - 3\hat{j} + 2\hat{k}$, then $\vec{a} \times \vec{b}$ equals

(a) $2\hat{i} - 2\hat{j} - \hat{k}$ (b) $\hat{i} - 10\hat{j} - 18\hat{k}$

(c) $\hat{i} + \hat{j} + \hat{k}$ (d) $6\hat{i} - 3\hat{j} + 2\hat{k}$

Solution

$$\vec{a} \times \vec{b} = \begin{vmatrix} \hat{i} & \hat{j} & \hat{k} \\ 2 & 2 & -1 \\ 6 & -3 & 2 \end{vmatrix} = \hat{i}(4-3) - \hat{j}(4+6) + \hat{k}(-6-12)$$

$$= \hat{i} - 10\hat{j} - 18\hat{k}$$

ILLUSTRATION 10.22

If the angle between $\hat{i} - 2\hat{j} + 3\hat{k}$ and $2\hat{i} + \hat{j} + \hat{k}$ is θ then $\sin\theta$ equals ________.

(a) $5/\sqrt{7}$ (b) $5/21$

(c) $5/2\sqrt{7}$ (d) $3/\sqrt{14}$

Solution

We know that $\sin\theta = \dfrac{|a \times b|}{|a||b|}$.

Now $\vec{a} \times \vec{b} = -5\hat{i} + 5\hat{j} + 5\hat{k}$

$$\therefore |\vec{a} \times \vec{b}| = \sqrt{(5)^2 + (5)^2 + (5)^2} = \sqrt{75} = 5\sqrt{3}$$

$$|\vec{a}| = \sqrt{1+4+9}, |\vec{b}| = \sqrt{4+1+1}$$

$$\therefore \sin\theta = \frac{5\sqrt{3}}{\sqrt{1+4+9}\sqrt{4+1+1}} = \frac{5\sqrt{3}}{\sqrt{14}\sqrt{6}} = \frac{5}{\sqrt{28}} = \frac{5}{2\sqrt{7}}$$

Properties of Vector Product

1. In general, $\vec{a} \times \vec{b} \neq \vec{b} \times \vec{a}$. In fact, $\vec{a} \times \vec{b} = -\vec{b} \times \vec{a}$.
2. For scalar m, $m\vec{a} \times \vec{b} = m(\vec{a} \times \vec{b}) = \vec{a} \times m\vec{b}$.

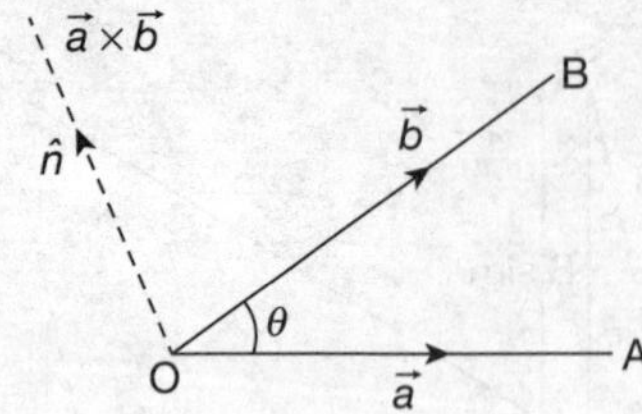

3. $\vec{a} \times (\vec{b} + \vec{c}) = \vec{a} \times \vec{b} + \vec{a} \times \vec{c}$
4. If $\vec{a} \parallel \vec{b}$ then $\theta = 0$ or $\pi \Rightarrow \vec{a} \times \vec{b} = 0$ (but $\vec{a} \times \vec{b} = \vec{0} \Rightarrow \vec{a} = \vec{0}$ or $\vec{b} = \vec{0}$ $\vec{a} \parallel \vec{b}$). In particular, $\vec{a} \times \vec{a} = 0$.
5. If $\vec{a} \perp \vec{b}$ then $\vec{a} \times \vec{b} = |\vec{a}||\vec{b}|\hat{n}$

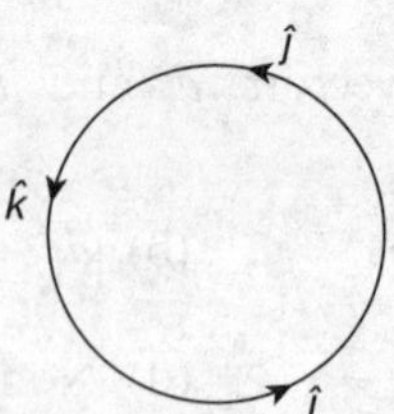

6. $\hat{i} \times \hat{i} = \hat{j} \times \hat{j} = \hat{k} \times \hat{k} = 0$ and $\hat{i} \times \hat{j} = \hat{k}, \hat{j} \times \hat{k} = \hat{i}$ and $\hat{k} \times \hat{i} = \hat{j}$ (use cyclic system.)
7. Unit vector perpendicular to $\vec{a}$ and $\vec{b}$ is given by $\pm\dfrac{\vec{a} \times \vec{b}}{|\vec{a} \times \vec{b}|}$
8. If θ is the angle between $\vec{a}$ and $\vec{b}$, then sin , $= \dfrac{|\vec{a} \times \vec{b}|}{|\vec{a}||\vec{b}|}$
9. If $\vec{a} = a_1\hat{i} + a_2\hat{j} + a_3\hat{k}$ and $\vec{b} = b_1\hat{i} + b_2\hat{j} + b_3\hat{k}$, then
$$\vec{a} \times \vec{b} = \begin{vmatrix} \hat{i} & \hat{j} & \hat{k} \\ a_1 & a_2 & a_3 \\ b_1 & b_2 & b_3 \end{vmatrix}$$

Geometrical Interpretation of Vector Product

The vector product of the vectors $\vec{a}$ and $\vec{b}$ represents a vector whose modulus is equal to the area of the parallelogram whose two adjacent sides are represented by $\vec{a}$ and $\vec{b}$. Therefore $|\vec{a} \times \vec{b}|$ = area of a parallelogram whose adjacent sides are $\vec{a}$ and $\vec{b}$. Further it should be noted that if $\vec{a}, \vec{b}$ represent two diagonals of a parallelogram,

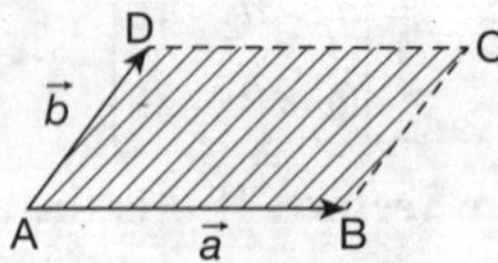

then the area of the parallelogram $= \dfrac{1}{2}|\vec{a} \times \vec{b}|$

Remark: Area of a quadrilateral ABCD $= \dfrac{1}{2}|\overrightarrow{AC} \times \overrightarrow{BD}|$

ILLUSTRATION 10.23

Find the area of a parallelogram whose two adjacent sides are represented by

$$\vec{a} = 3\hat{i} + \hat{j} + 2\hat{k} \text{ and } \vec{b} = 2\hat{i} - 2\hat{j} + 4\hat{k}.$$

Solution

Area of parallelogram $= |\vec{a} \times \vec{b}|$

Now $\vec{a} \times \vec{b} = \begin{vmatrix} \hat{i} & \hat{j} & \hat{k} \\ 3 & 1 & 2 \\ 2 & -2 & 4 \end{vmatrix} = 8\hat{i} - 8\hat{j} - 8\hat{k}$

$\therefore$ Area $= |8\hat{i} - 8\hat{j} - 8\hat{k}| = 8\sqrt{3}$ units

ILLUSTRATION 10.24

If $\ell\hat{i} + m\hat{j} + n\hat{k}$ is a unit vector which is perpendicular to vectors $2\hat{i} - \hat{j} + \hat{k}$ and $3\hat{i} + 4\hat{j} - \hat{k}$ then $|\ell|$ is equal to _________.

(a) $-\dfrac{3}{\sqrt{155}}$ (b) $\sqrt{\dfrac{3}{155}}$

(c) $\dfrac{3}{\sqrt{155}}$ (d) None of these.

Solution

The given vectors are $2\hat{i} - \hat{j} + \hat{k}$ and $3\hat{i} + 4\hat{j} - \hat{k}$.

$$= \frac{(2\hat{i} - \hat{j} + \hat{k}) \times (3\hat{i} + 4\hat{j} - \hat{k})}{|(2\hat{i} - \hat{j} + \hat{k}) \times (3\hat{i} + 4\hat{j} - \hat{k})|}$$

$$= \frac{\hat{i}(1-4) - \hat{j}(-2-3) + \hat{k}(8+3)}{\sqrt{9+25+121}}$$

$$= \frac{-3\hat{i} + 5\hat{j} + 11\hat{k}}{\sqrt{155}}$$

$$\therefore \quad |\ell| = \left|\frac{-3}{\sqrt{155}}\right| = \frac{3}{\sqrt{155}}$$

ILLUSTRATION 10.25

If $\vec{a}, \vec{b}, \vec{c}$ be three non-zero vectors, then $|(\vec{a} \times \vec{b}).\vec{c}| = |\vec{a}||\vec{b}||\vec{c}|$, if _________.

(a) $\vec{a}.\vec{b} = \vec{b}.\vec{c} = 0$ (b) $\vec{b}.\vec{c} = \vec{c}.\vec{a} = 0$

(c) $\vec{c}.\vec{a} = \vec{a}.\vec{b} = 0$ (d) $\vec{a}.\vec{b} = \vec{b}.\vec{c} = \vec{c}.\vec{a} = 0$

Solution

$|(\vec{a} \times \vec{b}).\vec{c}| = |\vec{a}||\vec{b}||\vec{c}| \Leftrightarrow |(\vec{a} \times \vec{b})||\vec{c}|\cos\theta = |\vec{a}||\vec{b}||\vec{c}|$

(where θ is the angle between $\vec{a} \times \vec{b}$ and $\vec{c}$)

$\Leftrightarrow \quad |\vec{a}||\vec{b}||\vec{c}|\sin\phi\cos\theta = |\vec{a}||\vec{b}||\vec{c}|$

(where ϕ is the angle between $\vec{a}$ and $\vec{b}$)

$\Leftrightarrow \quad \sin\phi = 1, \cos\theta = 1 \Leftrightarrow \phi = 90°, \theta = 0°$

$\Leftrightarrow \quad \vec{a}.\vec{b} = 0, \vec{a}.\vec{c} = 0, \vec{b}.\vec{c} = 0$

ILLUSTRATION 10.26

The area of parallelogram whose diagonals are respectively $3\hat{i} + \hat{j} - 2\hat{k}$ and $\hat{i} - 3\hat{j} + 4\hat{k}$ is ________.

(a) $5\sqrt{2}$ (b) $5\sqrt{3}$ (c) $2\sqrt{5}$ (d) $3\sqrt{5}$

Solution

Area of parallelogram $= \dfrac{1}{2}|\vec{a} \times \vec{b}|$

where $\vec{a} = 3\hat{i} + \hat{j} - 2\hat{k}$ and $\vec{b} = \hat{i} - 3\hat{j} + 4\hat{k}$

Now $\vec{a} \times \vec{b} = \begin{vmatrix} \hat{i} & \hat{j} & \hat{k} \\ 3 & 1 & -2 \\ 1 & -3 & 4 \end{vmatrix} = -2\hat{i} - 14\hat{j} - 10\hat{k}.$

$\therefore$ Area of parallelogram

$$= \frac{1}{2}|-2\hat{i} - 14\hat{j} - 10\hat{k}| = \sqrt{1+49+25} = 5\sqrt{3}$$

ILLUSTRATION 10.27

Two vectors A and B lie in the same *xy* plane of a rectangular coordinate system. The vector A has components A_x and A_y along the *x* and *y* axes. The vector B has components B_x and B_y along the *x* and *y* axes. Find the magnitude and direction of A × B.

Solution

Let **i**, **j** and **k** be unit vectors along the *x*, *y* and *z* axes of a rectangular coordinate system. Then

$$\mathbf{A} = A_x\mathbf{i} + A_y\mathbf{j} \text{ and } \mathbf{B} = B_x\mathbf{i} + B_y\mathbf{j}$$

Now $\quad \mathbf{A} \times \mathbf{B} = (A_x\mathbf{i} + A_y\mathbf{j}) \times (B_x\mathbf{i} + B_y\mathbf{j})$

$= A_xB_x\mathbf{i}\times\mathbf{i} + A_xB_y\mathbf{i}\times\mathbf{j} + A_yB_x\mathbf{j}\times\mathbf{i} + A_yB_y\mathbf{j}\times\mathbf{j}$

We know that

$$\mathbf{i}\times\mathbf{i} = \mathbf{j}\times\mathbf{j} = 0,\ \mathbf{i}\times\mathbf{j} = \mathbf{k} \text{ and } \mathbf{j}\times\mathbf{i} = -\mathbf{k}$$

$$\therefore \quad \mathbf{A}\times\mathbf{B} = (A_xB_y - A_yB_x)\,\mathbf{k}$$

Thus, the vector **A × B** is in the direction of *z*-axis. The magnitude of **A × B** is given by $|\mathbf{A}\times\mathbf{B}| = (A_xB_y - A_yB_x)$.

ILLUSTRATION 10.28

If **A × B = A × C**, is it correct to conclude that **B = C**?

Solution

Given that **A × B = A × C**

$\therefore$ AB $\sin\theta_1\hat{n}_1 = AC\sin\theta_2\hat{n}_2$

where $\hat{n}_1$ and $\hat{n}_2$ are unit vectors in the direction of vectors **(A × B)** and **(A × C)** respectively.

Here, we consider the following four cases:

(i) If $\theta_1 = \theta_2$ and $\hat{n}_1 = \hat{n}_2$, then
B = C and **B** = **C**

(ii) If $\theta_1 = \theta_2$ and $\hat{n}_1 \neq \hat{n}_2$, then
B = C but **B** ≠ **C**

(iii) If $\theta_1\,{}^1\,q_2$ and $\hat{n}_1{}^1\,\hat{n}_2$, then
B ≠ C and **B** ≠ **C**

(iv) If $\theta_1\,{}^1\,\theta_2$ and $\hat{n}_1 = \hat{n}_2$, then
B ≠ C and **B** ≠ **C**

Relative Motion in Two Dimensions

$\vec{r}_A$ = position of A with respect to O
$\vec{r}_B$ = position of B with respect to O
$\vec{r}_{AB}$ = position of A with respect to B.

$$\vec{r}_{AB} = \vec{r}_A - \vec{r}_B$$

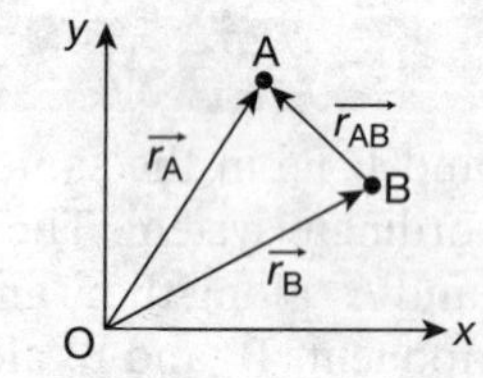

$$\therefore \quad \frac{d(\vec{r}_{AB})}{dt} = \frac{d(\vec{r}_A)}{dt} - \frac{d(\vec{r}_B)}{dt}$$

$$\Rightarrow \quad \vec{V}_{AB} = \vec{V}_A - \vec{V}_B$$

$$\frac{d(\vec{V}_{AB})}{dt} = \frac{d(\vec{V}_A)}{dt} - \frac{d(\vec{V}_B)}{dt}$$

$$\Rightarrow \quad \vec{a}_{AB} = \vec{a}_A - \vec{a}_B$$

ILLUSTRATION 10.29

Consider the situation shown in figure:

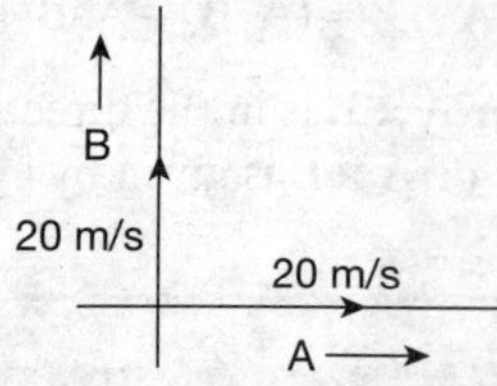

(i) The velocity of B with respect to A

$$\vec{V}_{AB} = \vec{V}_B - \vec{V}_A = 20\hat{j} - 20\hat{i}$$

(ii) The velocity of A with respect to B

$$\vec{V}_{AB} = \vec{V}_A - \vec{V}_B = 20\hat{i} - 20\hat{j}$$

ILLUSTRATION 10.30

Objects A and B have velocities 10 m/s. A is moving along East while B is moving towards North from the same point as shown. Find velocity of A relative to B ($\vec{V}_{AB}$)

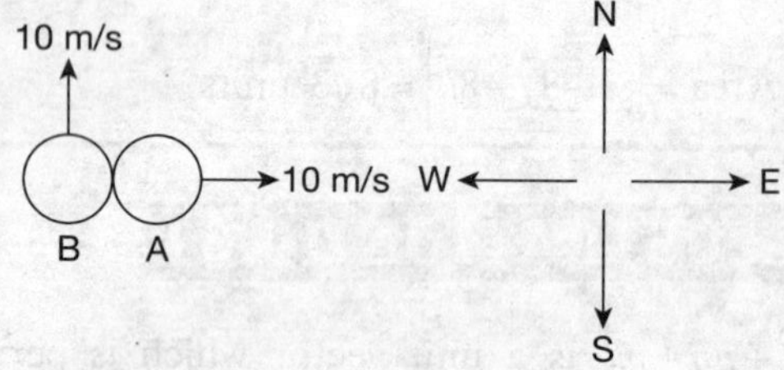

Solution

$$\vec{V}_{AB} = \vec{V}_A - \vec{V}_B$$

$$\therefore \quad |\vec{V}_{AB}| = 10\sqrt{2}$$

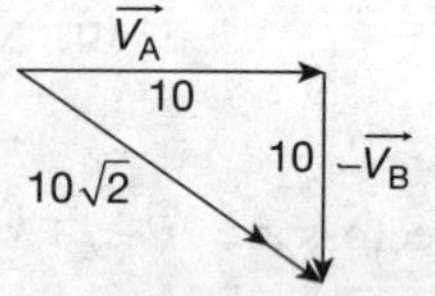

Remark: $|\vec{V}_A - \vec{V}_B| = \sqrt{V_A^2 + V_B^2 V_B \cos\theta}$

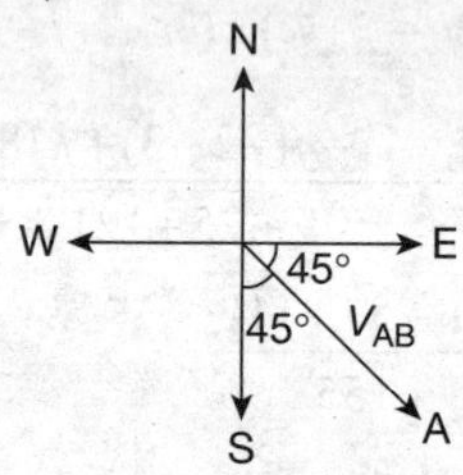

ILLUSTRATION 10.31

Two particles A and B are projected in air. A is thrown horizontally, B is thrown vertically up. What is the separation between them after 1 second?

Solution

$$\vec{a}_{AB} = \vec{a}_A - \vec{a}_B = 0$$

$\vec{V}_A = 10$

$\vec{V}_{AB}$ $\quad -\vec{V}_A = 10$

10 m/s

B A 10 m/s

$$\vec{V}_{AB} = \sqrt{10^2 + 10^2} = 10\sqrt{2}$$

$$\therefore \quad S_{AB} = V_{AB}\, t = (10\sqrt{2})\ \text{m}$$

Area of Triangle

(i) Area of triangle ABC $= \frac{1}{2}\left|\overline{AB} \times \overline{AC}\right|$

(ii) If $\vec{a}, \vec{b}, \vec{c}$ are position vectors of vertices of a ΔABC then its

$$\text{Area} = \frac{1}{2}\left|\left(\vec{a}\times\vec{b}\right)+\left(\vec{b}\times\vec{c}\right)+\left(\vec{c}\times\vec{a}\right)\right|$$

Remark: Three points with position vectors $\vec{a}, \vec{b}, \vec{c}$ are collinear if

$$\left(\vec{a}\times\vec{b}\right)+\left(\vec{b}\times\vec{c}\right)+\left(\vec{c}\times\vec{a}\right)=0$$

ILLUSTRATION 10.32

Find the area of ΔABC if position vectors of its vertices A, B, C are $\hat{i}+\hat{j}, \hat{j}+\hat{k}$ and $\hat{k}+\hat{i}$ respectively.

Solution

$$\overline{AB} = \left(\hat{j}+\hat{k}\right)-\left(\hat{i}+\hat{j}\right)=\hat{k}-\hat{i}$$

$$\overline{AC} = \left(\hat{k}+\hat{i}\right)-\left(\hat{i}+\hat{j}\right)=\hat{k}-\hat{j}$$

$$\therefore \quad \overline{AB}\times\overline{AC} = \begin{vmatrix} \hat{i} & \hat{j} & \hat{k} \\ -1 & 0 & 1 \\ 0 & -1 & 1 \end{vmatrix} = \hat{i}+\hat{j}+\hat{k}$$

$$\therefore \text{Area of } \Delta ABC = \frac{1}{2}\left|\overline{AB}\times\overline{AC}\right| = \frac{1}{2}\sqrt{1+1+1} = \frac{\sqrt{3}}{2}$$

Scalar Triple Product

If $\vec{a}, \vec{b}, \vec{c}$ are three vectors, then their scalar triple product is defined as the dot product of two vectors $\vec{a}$ and $\vec{b}\times\vec{c}$. It is generally denoted by $\vec{a}.(\vec{b}\times\vec{c})$ or $[\vec{a}\vec{b}\vec{c}]$. It is read as box product of $\vec{a}, \vec{b}, \vec{c}$. Similarly other scalar triple products can be defined as $(\vec{b}\times\vec{c}).\vec{a}$, $(\vec{c}\times\vec{a}).\vec{b}$.

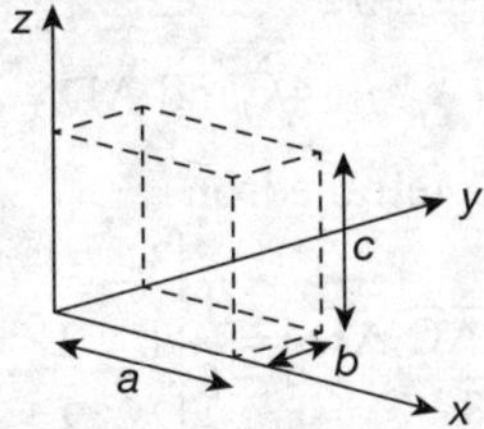

Remark: Scalar triple product always result in a scalar quantity (number).

Geometrical Interpretation

The scalar triple product of three vectors is equal to the volume of the parallelepiped whose three coterminous edges are represented by the given vector.

Therefore $(\vec{a}\times\vec{b}).\vec{c} = [\vec{a}\vec{b}\vec{c}] =$ Volume of the parallelepiped whose coterminous edges are $\vec{a}, \vec{b}$ and $\vec{c}$.

Formula for Scalar Triple Product

(i) If $\vec{a} = a_1\ell + a_2 m + a_3 n, \vec{b} = b_1\ell + b_2 m + b_3 n$ and $\vec{c} = c_1\ell + c_2 m + c_2 n$, then

$$\left[\vec{a}\vec{b}\vec{c}\right] = \begin{vmatrix} a_1 & a_2 & a_3 \\ b_1 & b_2 & b_3 \\ c_1 & c_2 & c_3 \end{vmatrix} = [\ell m n]$$

(ii) $\vec{a} = a_1 i + a_2 j + a_3 k$, $\vec{b} = b_1 i + b_2 j + b_3 k$ and $\vec{c} = c_1 i + c_2 j + c_3 k$, then

$$\left[\vec{a}\vec{b}\vec{c}\right] = \begin{vmatrix} a_1 & a_2 & a_3 \\ b_1 & b_2 & b_3 \\ c_1 & c_2 & c_3 \end{vmatrix}$$

(iii) For any three vectors $\vec{a}, \vec{b}$ and $\vec{c}$

(a) $\left[\vec{a}+\vec{b} \quad \vec{b}+\vec{c} \quad \vec{c}+\vec{a}\right] = 2\left[\vec{a}\vec{b}\vec{c}\right]$

(b) $\left[\vec{a}-\vec{b} \quad \vec{b}-\vec{c} \quad \vec{c}-\vec{a}\right] = 0$

(c) $\left[\vec{a}+\vec{b} \quad \vec{b}+\vec{c} \quad \vec{c}+\vec{a}\right] = \left[\vec{a}\vec{b}\vec{c}\right]^2$

Properties of Scalar Triple Product

(i) The position of (.) and (×) can be interchanged, i.e., $\vec{a}.(\vec{b}\times\vec{c}) = (\vec{a}\times\vec{b}).\vec{c}$

But $(\vec{a}\times\vec{b}).\vec{c} = \vec{c}.(\vec{a}\times\vec{b})$

So $[\vec{a}\vec{b}\vec{c}] = [\vec{b}\vec{c}\vec{a}] = [\vec{c}\vec{a}\vec{b}]$

Therefore, if we don't change the cyclic order of a, b and c, then the value of scalar triple product is not changed by interchanging dot and cross.

(ii) If the cyclic order of vectors is changed, then the sign of the scalar triple product is changes, i.e.,

$$\vec{a}.\left[\vec{b}\times\vec{c}\right] = -\vec{a}.\left(\vec{c}\times\vec{b}\right)$$

or

$$\left[\vec{a}\vec{b}\vec{c}\right] = -\left[\vec{a}\vec{c}\vec{b}\right]$$

From the above points (i) and (ii) we have

$$\left[\vec{a}\vec{b}\vec{c}\right] = \left[\vec{b}\vec{c}\vec{a}\right] = \left[\vec{c}\vec{a}\vec{b}\right]$$
$$= -\left[\vec{a}\vec{c}\vec{b}\right] = -\left[\vec{b}\vec{a}\vec{c}\right] = -\left[\vec{c}\vec{b}\vec{a}\right]$$

(iii) The scalar triple product of three vectors when two of them are equal or parallel, is zero, i.e.,

$$\left[\vec{a}\vec{b}\vec{b}\right] = \left[\vec{a}\vec{b}\vec{a}\right] = 0$$

(iv) The scalar triple product of three mutually perpendicular unit vectors is ±1. Thus, $[\hat{i}\hat{j}\hat{k}] = 1, [\hat{i}\hat{k}\hat{j}] = -1$

(v) If two of the three vectors $\vec{a}, \vec{b}, \vec{c}$ are parallel, then $[\vec{a}\vec{b}\vec{c}] = 0$

(vi) $\vec{a},\vec{b},\vec{c}$ are three coplanar vectors if $[\vec{a}\vec{b}\vec{c}]=0$, i.e., the necessary and sufficient condition for three non-zero collinear vectors to be coplanar is

$$\left[\vec{a}\vec{b}\vec{c}\right]=0$$

(vii) For any vectors $\vec{a},\vec{b},\vec{c},\vec{d}$

$$\left[\vec{a}+\vec{b}\vec{c}\vec{d}\right]=\left[\vec{a}\vec{c}\vec{d}\right]+\left[\vec{b}\vec{c}\vec{d}\right]$$

(viii) For a right-handed system, $[\vec{a}\vec{b}\vec{c}]>0$ and for left-handed system, $[\vec{a}\vec{b}\vec{c}]<0$; where $\vec{a},\vec{b},\vec{c}$ are non-coplanar.

(ix) $[\vec{a}+\vec{b},\vec{b}+\vec{c},\vec{c}+\vec{a}]=2[\vec{a}\vec{b}\vec{c}]$

(x) $[\vec{a}-\vec{b}\ \vec{b}-\vec{c}\ \vec{c}-\vec{a}]$ is always zero.

(xi) $[\vec{l}\vec{m}\vec{n}][\vec{a}\vec{b}\vec{c}]=\begin{vmatrix}\vec{l}.\vec{a} & \vec{l}.\vec{b} & \vec{l}.\vec{c}\\ \vec{m}.\vec{a} & \vec{m}.\vec{b} & \vec{m}.\vec{c}\\ \vec{n}.\vec{a} & \vec{n}.\vec{b} & \vec{n}.\vec{c}\end{vmatrix}$, where $\vec{l},\vec{m},\vec{n}$ and $\vec{a},\vec{b},\vec{c}$ are non-coplanar vectors.

(xii) $[\vec{a}\vec{b}\vec{c}](\vec{p}\times\vec{q})=\begin{vmatrix}\vec{p}.\vec{a} & \vec{q}.\vec{a} & \vec{a}\\ \vec{p}.\vec{b} & \vec{q}.\vec{b} & \vec{b}\\ \vec{p}.\vec{c} & \vec{q}.\vec{c} & \vec{c}\end{vmatrix}$

(xiii) $(\vec{a}\times\vec{b}).(\vec{c}\times\vec{d})=(\vec{a}.\vec{c})(\vec{b}.\vec{d})-(\vec{a}.\vec{d})(\vec{b}.\vec{c})$

$$=\begin{vmatrix}\vec{a}.\vec{c} & \vec{a}.\vec{d}\\ \vec{b}.\vec{c} & \vec{b}.\vec{d}\end{vmatrix}$$

ILLUSTRATION 10.33

If $\vec{a}=2\hat{i}+3\hat{j},\vec{b}=\hat{i}+\hat{j}-\hat{k}$ and $\vec{c}=3\hat{i}-\hat{k}$ represents three coterminous edges of a parallelepiped, then the volume of that parallelepiped is ________.

(a) 2 (b) 4 (c) 6 (d) 10

Solution

Volume $=\left[\vec{a}\vec{b}\vec{c}\right]$

$$=\begin{vmatrix}2 & -3 & 0\\ 1 & 1 & -1\\ 3 & 0 & -1\end{vmatrix}=-2+9-3=4$$

ILLUSTRATION 10.34

For any three vectors $\vec{a},\vec{b},\vec{c}$ $[\vec{a}+\vec{b}\ \vec{b}+\vec{c}\ \vec{c}+\vec{a}]$ equals ________.

(a) $\left[\vec{a}\,\vec{b}\,\vec{c}\right]$ (b) $2\left[\vec{a}\,\vec{b}\,\vec{c}\right]$

(c) $\left[\vec{a}\,\vec{b}\,\vec{c}\right]^2$ (d) 0

Solution

$$\left[\vec{a}+\vec{b}\,\vec{b}+\vec{c}\,\vec{c}+\vec{a}\right]$$

$$=\left(\vec{a}+\vec{b}\right).\left[\left(\vec{b}+\vec{c}\right)\times\left(\vec{c}+\vec{a}\right)\right]$$

$$=\left(\vec{a}+\vec{b}\right).\left(\vec{b}\times\vec{c}+\vec{b}\times\vec{a}+\vec{c}\times\vec{c}+\vec{c}\times\vec{a}\right)$$

$$=\left(\vec{a}+\vec{b}\right).\left(\vec{b}\times\vec{c}+\vec{b}\times\vec{a}+\vec{c}\times\vec{a}\right)\ [\because \vec{c}\times\vec{c}=0]$$

$$=\left[\vec{a}\,\vec{b}\,\vec{c}\right]+\left[\vec{a}\,\vec{b}\,\vec{a}\right]+\left[\vec{a}\,\vec{c}\,\vec{a}\right]+\left[\vec{b}\,\vec{b}\,\vec{c}\right]+\left[\vec{b}\,\vec{b}\,\vec{a}\right]+\left[\vec{b}\,\vec{c}\,\vec{a}\right]$$

$$=\left[\vec{a}\,\vec{b}\,\vec{c}\right]+\left[\vec{b}\,\vec{c}\,\vec{a}\right]=2\left[\vec{a}\,\vec{b}\,\vec{c}\right]$$

Volume of Tetrahedron

(i) If $\vec{a},\vec{b},\vec{c}$ are position vectors of vertices A, B and C with respect to O, then volume of tetrahedron OABC.

$$=\frac{1}{6}\left[\vec{a}\,\vec{b}\,\vec{c}\right]$$

(ii) If $\vec{a},\vec{b},\vec{c},\vec{d}$ are position vectors of vertices A, B, C, D of a tetrahedron ABCD, then

$$\text{its volume}=\begin{cases}\frac{1}{6}\left[\overline{\text{AB}}\ \overline{\text{AC}}\ \overline{\text{AD}}\right]\\ \frac{1}{6}\left[b-a\ c-a\ d-a\right]\end{cases}$$

ILLUSTRATION 10.35

If the vertices of any tetrahedron be $\vec{a}=\hat{j}+2\hat{k},\vec{b}=3\hat{i}+\hat{k},\vec{c}=4\hat{i}+3\hat{j}+6\hat{k}$ and $\vec{d}=2\hat{i}+3\hat{j}+2\hat{k}$, then find its volume.

Solution

Let the position vector of the vertices A, B, C, D with respect to 0 be $\vec{a},\vec{b},\vec{c}$ and $\vec{d}$ respectively. Then $\overline{\text{AB}}=\vec{b}-\vec{a}=3\hat{i}-\hat{j}-\hat{k}$,

$$\overline{\text{AC}}=4\hat{i}+2\hat{j}+4\hat{k}\text{ and }\overline{\text{AD}}=2\hat{i}+2\hat{j}$$

Now, volume of tetrahedron

$$=\frac{1}{6}\left[\overline{\text{AB}}\ \overline{\text{AC}}\ \overline{\text{AD}}\right]=\frac{1}{6}\begin{vmatrix}3 & -1 & -1\\ 4 & 2 & 4\\ 2 & 2 & 0\end{vmatrix}=-6$$

$\therefore$ Required volume = 6 units.

ILLUSTRATION 10.36

If A, B, C, D are any four points, then $\left|\overline{\text{AB}}\times\overline{\text{CD}}+\overline{\text{BC}}\times\overline{\text{AD}}+\overline{\text{CA}}\times\overline{\text{BD}}\right|$ equals ________.

(a) Area of ΔABC (b) 2(Area of Δ ABC)
(c) 3(Area of ΔABC) (d) 4(Area of ΔABC)

Solution

Let a, b, c and d be position vectors of points A, B, C and D respectively, then

$$\overline{AB} \times \overline{CD} = (\vec{b} - \vec{a}) \times (\vec{d} - \vec{c}) = \vec{b} \times \vec{d} - \vec{b} \times \vec{c} - \vec{a} \times \vec{d} + \vec{a} \times \vec{c}$$

Similarly,

$$\overline{BC} \times \overline{AD} = \vec{c} \times \vec{d} - \vec{c} \times \vec{a} - \vec{b} \times \vec{d} + \vec{b} \times \vec{a}$$

$$\overline{CA} \times \overline{BD} = \vec{a} \times \vec{d} - \vec{a} \times \vec{b} - \vec{c} \times \vec{d} + \vec{c} \times \vec{b}$$

Therefore, given expression

$$= \left|2\left(\vec{b} \times \vec{a} - \vec{b} \times \vec{c} + \vec{a} \times \vec{c}\right)\right| = 2\left|\left(\vec{a} \times \vec{b} + \vec{b} \times \vec{c} + \vec{c} \times \vec{a}\right)\right|$$

= 4(Area of ΔABC)

ILLUSTRATION 10.37

a, b, c, d are the position vectors of four coplanar points A, B, C and D respectively. If $(\vec{a} - \vec{d}).(\vec{b} - \vec{c}) = 0 = (\vec{b} - \vec{d}).(\vec{c} - \vec{a})$, then for the Δ ABCD, D is ________.

(a) incentre (b) orthocentre
(c) circumcentre (d) centroid

Solution

$$\left(\vec{b} - \vec{d}\right).\left(\vec{c} - \vec{a}\right) = 0 \Rightarrow \left(\vec{a} - \vec{d}\right) \perp \left(\vec{b} - \vec{c}\right) = \overline{AD} \perp \overline{BC}$$

Similarly $\left(\vec{b} - \vec{d}\right).\left(\vec{c} - \vec{a}\right) = 0 \Rightarrow \overline{BD} \perp \overline{AC}$

∴ D is the orthocentre of Δ ABC.

Relation Between Two Parallel Vectors

(i) If $\vec{a}$ and $\vec{b}$ be two parallel vectors, then there exists a scalar k such that $\vec{a} = k\vec{b}$,

i.e., there exist two non-zero scalar quantities x and y so that $x\vec{a} + y\vec{b} = 0$.

If a and b be two no-zero non-parallel vectors, then $x\vec{a} + y\vec{b} = 0 \Rightarrow x = 0$ and $y = 0$ obviously.

$$x\vec{a} + y\vec{b} = 0 \Rightarrow \begin{cases} \vec{a} = 0, \vec{b} = 0 \\ \text{or} \\ x = 0, y = 0 \\ \text{or} \\ \vec{a} \parallel \vec{b} \end{cases}$$

(ii) If $\vec{a} = a_1\hat{i} + a_2\hat{j} + a_3\hat{k}$ and $\vec{b} = b_1\hat{i} + b_2\hat{j} + b_3\hat{k}$, then from the property of parallel vector we have $\vec{a} \parallel \vec{b}$.

$$\Rightarrow \quad \frac{a_1}{b_1} = \frac{a_2}{b_2} = \frac{a_3}{b_3}$$

ILLUSTRATION 10.38

The value of λ when $\vec{a} = 2\hat{i} - 3\hat{j} + \hat{k}$ and $\vec{b} = 8\hat{i} + \lambda\hat{j} + 4\hat{k}$ are parallel is ________.

(a) 4 (b) –6 (c) –12 (d) 1

Solution

Since $\vec{a} \parallel \vec{b} \Rightarrow \frac{a_1}{b_1} = \frac{a_2}{b_2} = \frac{a_3}{b_3}$

$$\therefore \quad \frac{2}{3} = -\frac{3}{\lambda} = \frac{1}{4} \Rightarrow \lambda = -12$$

Collinearity of Three Points

(i) If $\vec{a}, \vec{b}, \vec{c}$ be position vectors of three points A, B and C respectively and x, y, z be three scalars so that all are not zero, then the necessary and sufficient conditions for three points to be collinear is that

$$x\vec{a} + y\vec{b} + z\vec{c} = 0 \text{ and } x + y + z = 0$$

(ii) Three points A, B and C are collinear, if any two vectors $\overline{AB}, \overline{BC}$ and $\overline{CA}$ are parallel, i.e., one of them is a scalar multiple of any one of the remaining vectors.

ILLUSTRATION 10.39

If $2\vec{a} - 3\vec{b}, \vec{b}$ and $\vec{a} - \vec{b}$ are position vectors of three points A, B and C, then they are ________.

(a) collinear (b) non-collinear
(c) Can't say anything. (d) None of these.

Solution

$\because 1(2\vec{a} - 3\vec{b}) + 1(\vec{b}) - 2(\vec{a} - \vec{b}) = 0$ and $1 + 1 - 2 = 0$

So the given vectors are collinear.

ILLUSTRATION 10.40

If $A \equiv (2\hat{i} + 3\hat{j}), B \equiv (p\hat{i} + 9\hat{j})$ and $C \equiv (\hat{i} - \hat{j})$ are collinear, then the value of p is ________.

(a) $1/2$ (b) $3/2$ (c) $7/2$ (d) $5/2$

Solution

$$\overline{AB} = (p-2)\hat{i} + 6\hat{j}, \overline{AC} = -\hat{i} - 4\hat{j}$$

Now A, B, C are collinear

$$\Leftrightarrow \quad \overline{AB} \parallel \overline{AC} \Leftrightarrow \frac{p-2}{-1} = \frac{6}{-4} \Leftrightarrow p = 7/2$$

Coplanar and Non-coplanar Vector

(i) If $\vec{a},\vec{b},\vec{c}$ be three coplanar vectors, then a vector c can be expressed uniquely as a linear combination of remaining two vectors, i.e.,

$$\vec{c} = \lambda\vec{a} + \mu\vec{b}$$

where λ and μ are suitable scalars.
Again, $\vec{c} = \lambda\vec{a} + \mu\vec{b} \Rightarrow$ vectors $\vec{a},\vec{b}$ and $\vec{c}$ are coplanar.
If $\vec{a},\vec{b},\vec{c}$ be three non-coplanar vectors, then there exist three non-zero scalars x, y, z so that

$$x\vec{a} + y\vec{b} + z\vec{c} = 0$$

(ii) If $\vec{a},\vec{b},\vec{c}$ be three non-coplanar non zero vector then

$$x\vec{a} + y\vec{b} + z\vec{c} = 0 \Rightarrow x = 0, y = 0, z = 0$$

(iii) Any vector r can be expressed uniquely as a linear combination of three non-coplanar and non-zero vectors $\vec{a},\vec{b}$ and $\vec{c}$, i.e., $r = x\vec{a} + y\vec{b} + z\vec{c}$ where x, y and z are scalars.

Linear Combination of Vectors

If $\vec{a}$ and $\vec{b}$ are two non-collinear vectors, then any vector $\vec{r}$ in the plane of $\vec{a}$ and $\vec{b}$ is uniquely expressed as $\vec{r} = x\vec{a} + y\vec{b}$, where x and y are scalars. (Similar linear combination (unique) exists for three non-coplanar vectors, $\vec{r} = x\vec{a} + y\vec{b} + z\vec{c}$).

Linearly Dependent and Independent Vectors

Vector $\vec{a}_1\vec{a}_2,\ldots,\vec{a}_n$ are said to be linearly dependent if for scalars $x_1, x_2,\ldots, x_n$ (not all zero), $x_1\vec{a}_1 + x_2\vec{a}_2 + \ldots + x_n\vec{a}_n$.
If $x_1\vec{a}_1 + \ldots + x_n\vec{a}_n = \vec{0} \Rightarrow x_1 = x_2 = \ldots x_n = 0$, then the vectors are called linearly independent.

Remark

1. If $\vec{a},\vec{b}$ and $\vec{c}$ are three non-coplanar vectors, then the vectors are linearly independent.
2. If $\vec{a} = a_1\hat{i} + a_2\hat{j} + a_3\hat{k}, \vec{b} = b_1\hat{i} + b_2\hat{j} + b_3\hat{k}$ and $\vec{c} = c_1\hat{i} + c_2\hat{j} + c_3\hat{k}$ are linearly dependent, then
$$\begin{vmatrix} a_1 & b_1 & c_1 \\ a_2 & b_2 & c_2 \\ a_3 & b_3 & c_3 \end{vmatrix} = 0$$
3. If $\vec{a},\vec{b},\vec{c}$ are non-coplanar, then $x_1\vec{a} + y_1\vec{b} + z_1\vec{c}$, $x_2\vec{a} + y_2\vec{b} + z_1\vec{c}$ and $x_3\vec{a} + y_3\vec{b} + z_3\vec{c}$ are coplanar iff
$$\begin{vmatrix} x_1 & x_2 & x_3 \\ y_1 & y_2 & y_3 \\ z_1 & z_2 & z_3 \end{vmatrix} = 0$$
4. Two non-zero, non-collinear vectors, or any three non-coplanar vectors are linearly independent.
5. Two collinear or any three coplanar or any four vectors in 3-D are linearly dependent.
6. Three points with position vectors $\vec{a},\vec{b}$ and $\vec{c}$ are collinear iff there exist scalars x, y and z (not all zero) such that $x\vec{a} + y\vec{b} + z\vec{c} = \vec{0}$ where $x + y + z = 0$.
7. Four points with position vectors $\vec{a},\vec{b},\vec{c}$ and $\vec{d}$ are said to be coplanar iff there exist scalars x, y, z and w (not all zero) such that $x\vec{a} + y\vec{b} + z\vec{c} + w\vec{d} = \vec{0}$ where $x + y + z + w = 0$.

ILLUSTRATION 10.41

If the vectors $x\hat{i} + \hat{j} + \hat{k}$, $\hat{i} + y\hat{j} + \hat{k}$ and $\hat{i} + \hat{j} + z\hat{k}$ are coplanar when $x \neq 1$, $y \neq 1$ and $z \neq 1$, then show that $\frac{1}{1-x} + \frac{1}{1-y} + \frac{1}{1-z} = 1$.

Solution

For the given vectors to be coplanar.

$$\begin{vmatrix} x & 1 & 1 \\ 1 & y & 1 \\ 1 & 1 & z \end{vmatrix} = 0 \Rightarrow \begin{vmatrix} x-1 & 1-y & 0 \\ 0 & y-1 & 1-z \\ 1 & 1 & z \end{vmatrix} = 0$$

(Operations $R_1 - R_2$ and $R_2 - R_3$)

$$\Rightarrow \quad (x-1)(1-y)(1-z)\begin{vmatrix} 1 & 1 & 0 \\ 0 & 1 & 1 \\ \frac{1}{x-1} & \frac{1}{1-y} & \frac{z}{1-z} \end{vmatrix} = 0$$

$$(x-1)(1-y)(1-z)\left[\left(\frac{-z}{1-z} - \frac{1}{1-y}\right) + \frac{1}{x-1}\right] = 0$$

$$\Rightarrow \quad \frac{1}{1-x} + \frac{1}{1-y} = \frac{-z}{1-z} (\because x \neq 1, y \neq 1, z \neq 1)$$

$$\Rightarrow \quad \frac{1}{1-x} + \frac{1}{1-y} + \frac{1}{1-z} = \frac{1}{1-z} - \frac{z}{1-z} = \frac{1-z}{1-z} = 1$$

ILLUSTRATION 10.42

If the middle points of sides BC, CA and AB in triangle ABC are respectively D, E, F, then position vector of centre of triangle DEF, when position vector of A, B, C are respectively $\hat{i} + \hat{j}, \hat{j} + \hat{k}, \hat{k} + \hat{i}$ is ________.

(a) $\frac{1}{3}(\hat{i} + \hat{j} + \hat{k})$ (b) $(\hat{i} + \hat{j} + \hat{k})$
(c) $2(\hat{i} + \hat{j} + \hat{k})$ (d) $\frac{2}{3}(\hat{i} + \hat{j} + \hat{k})$

Solution

The position vector of points D, E, F are respectively

$\frac{\hat{i}+\hat{j}}{2}+\hat{k}, \hat{i}+\frac{\hat{k}+\hat{j}}{2}$ and $\frac{\hat{i}+\hat{k}}{2}+\hat{j}$

So, position vector of centre of ΔDEF

$$=\frac{1}{3}\left[\frac{\hat{i}+\hat{j}}{2}+\hat{k}+\hat{i}+\frac{\hat{k}+\hat{j}}{2}+\frac{\hat{i}+\hat{k}}{2}+\hat{j}\right]=\frac{2}{3}\left[\hat{i}+\hat{j}+\hat{k}\right]$$

ILLUSTRATION 10.43

The position vector of the point A is $6\vec{b}-2\vec{a}$ and the point P divides any line AB in the ratio 1 : 2. If the position vector of P is $\vec{a}-\vec{b}$, then position. of B will be ________.

(a) $7\vec{a}+15\vec{b}$ (b) $7\vec{a}+15\vec{b}$

(c) $15\vec{a}-7\vec{b}$ (d) $15\vec{a}+7\vec{b}$

Solution

Let the position vector of B be *r*

$$\therefore \vec{a}-\vec{b}=\frac{\vec{r}+2\left(6\vec{b}-2\vec{a}\right)}{1+2} \Rightarrow 3\vec{a}-3\vec{b}=\vec{r}+12\vec{b}-4\vec{a}$$

$$\Rightarrow \vec{r}=7\vec{a}-15\vec{b}$$

ILLUSTRATION 10.44

If $3\hat{i}-2\hat{j}+5\hat{k}$ and $-2\hat{i}+p\hat{j}-q\hat{k}$ are collinear vectors, then ________.

(a) $p=\frac{4}{3}, q=-\frac{10}{3}$ (b) $p=\frac{10}{3}, q=\frac{4}{3}$

(c) $p=-\frac{4}{3}, q=\frac{10}{3}$ (d) $p=\frac{4}{3}, q=\frac{10}{3}$

Solution

Given vectors are parallel, so we have

$$\frac{3}{-2}=\frac{-2}{p}=\frac{5}{-q}\Rightarrow p=\frac{4}{3}, q=\frac{10}{3}.$$

ILLUSTRATION 10.45

If $\overrightarrow{AO}+\overrightarrow{OB}=\overrightarrow{BO}+\overrightarrow{OC}$, then A, B, C are ________.

(a) coplanar (b) collinear

(c) non-collinear (d) None of these.

Solution

$$\overrightarrow{AO}+\overrightarrow{OB}=\overrightarrow{BO}+\overrightarrow{OC}\Rightarrow \overrightarrow{AB}=\overrightarrow{BC}$$

∴ $\overrightarrow{AB} \parallel \overrightarrow{BC}$ and a point B is common to both vectors $\overrightarrow{AB}$ and $\overrightarrow{BC}$. Hence A, B, C are collinear.

ILLUSTRATION 10.46

Let $A=(x+4y)\vec{a}+(2x+y+1)\vec{b}$ and $B=(y-2x+2)\vec{a}+(2x-3y-1)\vec{b}$ where $\vec{a}$ and $\vec{b}$ are non collinear vectors, if 3A = 2B; then ________.

(a) $x=1, y=2$ (b) $x=2, y=1$

(c) $x=2, y=-1$ (d) $x=-1, y=2$

Solution

$$3A=3(x+4y)\vec{a}+(2x+y+1)\vec{b}$$

$$2B=2(y-2x+2)\vec{a}+(2x-3y-1)\vec{b}$$

$$\therefore \quad 3A=2B\Rightarrow 3(x+4y)=2(y-2x+2),$$

$$3(2x+y+1)=2(2x-3y-1)\Rightarrow 7x+10y=4$$

and $$2x+9y=-5$$

$$\Rightarrow \quad x=2,\ y=-1$$

Vector Triple Product and Application of Vectors to Geometry

The vector triple product of three vectors $\vec{a}, \vec{b}, \vec{c}$ is defined as the vector product of two vectors $\vec{a}$ and $\vec{b}\times\vec{c}$. It is denoted by $\vec{a}\times(\vec{b}\times\vec{c})$.

Properties

(i) Expansion formula for vector triple product is given by

$$\vec{a}\times(\vec{b}\times\vec{c})=(\vec{a}.\vec{c})\vec{b}-(\vec{a}.\vec{b})\vec{c}$$

$$(\vec{b}\times\vec{c})\times\vec{a}=(\vec{b}.\vec{a})\vec{c}-(\vec{c}.\vec{a})\vec{b}$$

(ii) If $\vec{a}=a_1\hat{i}+a_2\hat{j}+a_3\hat{k}$, $\vec{b}=b_1\hat{i}+b_2\hat{j}+b_3\hat{k}$ and $\vec{c}=c_1\hat{i}+c_2\hat{j}+c_3\hat{k}$ then

$$\vec{a}\times(\vec{b}\times\vec{c})=\begin{vmatrix}\hat{i} & \hat{j} & \hat{k}\\ a_1 & a_2 & a_3\\ b_2c_3-b_3c_2 & b_3c_1-b_1c_3 & b_1c_2-b_2c_1\end{vmatrix}$$

(iii) $$[\vec{a}\times\vec{b}\ \vec{b}\times\vec{c}\ \vec{c}\times\vec{a}]=[\vec{a}\,\vec{b}\,\vec{c}]^2=\begin{vmatrix}\vec{a}.\vec{a} & \vec{a}.\vec{b} & \vec{a}.\vec{c}\\ \vec{b}.\vec{a} & \vec{b}.\vec{b} & \vec{b}.\vec{c}\\ \vec{c}.\vec{a} & \vec{c}.\vec{b} & \vec{c}.\vec{c}\end{vmatrix}$$

Note that if $\vec{a},\vec{b},\vec{c}$ are non-coplanar vectors, then $\vec{a}\times\vec{b}$, $\vec{b}\times\vec{c}$ and $c\times\vec{a}$ will also be non-coplanar vectors.

(iv) $(\vec{a}\times\vec{b})\times(\vec{c}\times\vec{d})=[\vec{a}\vec{b}\vec{d}]\vec{c}-[\vec{a}\vec{b}\vec{d}]=[\vec{a}\vec{c}\vec{d}]\vec{b}-[\vec{b}\vec{c}\vec{d}]\vec{a}$

Note that $(\vec{a}\times\vec{b})\times(\vec{c}\times\vec{d})=\vec{0}\Rightarrow$ plane containing the values $\vec{a}$ and $\vec{b}$ and $\vec{c}$ and $\vec{d}$ are parallel.

Similarly $(\vec{a}\times\vec{b}).(\vec{c}\times\vec{d})=0\Rightarrow$ the two planes are perpendicular.

Remark: **(i)** Vector triple product is a vector quantity.

(ii) $\vec{a}\times(\vec{b}\times\vec{c})\neq(\vec{a}\times\vec{b})\times\vec{c}$

(iii) Vector $\vec{a}\times(\vec{b}\times\vec{c})$ is coplanar with $\vec{b}$ and $\vec{c}$.

(iv) The direction of $\vec{a}\times(\vec{b}\times\vec{c})$ is perpendicular to $\vec{a}$ and parallel to $\vec{b}$ and $\vec{c}$.

ILLUSTRATION 10.47

$\hat{i}\times(\hat{j}\times\hat{k})+\hat{j}\times(\hat{k}\times\hat{i})+\hat{k}(\hat{i}\times\hat{j})$ equals ________.

(a) $\hat{i}$ (b) $\hat{j}$ (c) $\hat{k}$ (d) 0

Solution

$$\hat{i}\times(\hat{j}\times\hat{k})+\hat{j}\times(\hat{k}\times\hat{i})+\hat{k}\times(\hat{i}\times\hat{j})$$

$$\Rightarrow \hat{i}\times\hat{i}+\hat{j}\times\hat{j}+\hat{k}\times\hat{k}=0+0+0=0$$

Vector Equation of Straight Line

Case I: Vector equation of a straight line passing through a point $\vec{a}$ and parallel to vector $\vec{b}$.

Let O be the origin and $\overline{OA}=\vec{a}$. Let P be any point on the line with $\overline{OP}=\vec{r}$.

Since $\overline{AP}\parallel\vec{b}$, for some scalar t,

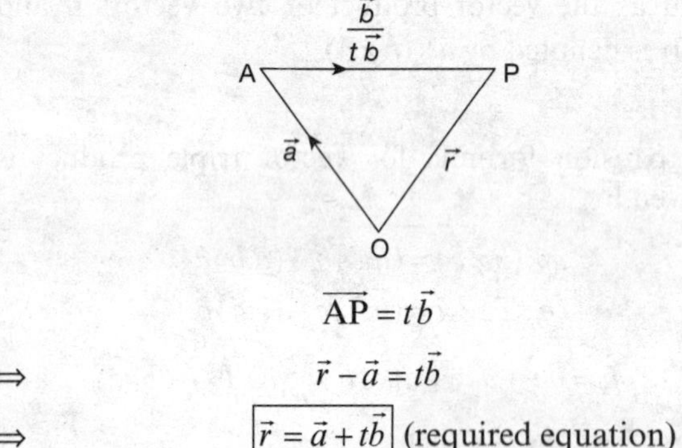

$$\overline{AP}=t\vec{b}$$

$$\Rightarrow \vec{r}-\vec{a}=t\vec{b}$$

$$\Rightarrow \boxed{\vec{r}=\vec{a}+t\vec{b}} \text{ (required equation)}$$

Case II: Vector equation of a straight line passing through two points $\vec{a}$ and $\vec{b}$.

$$\boxed{\vec{r}=\vec{a}+t(\vec{b}-\vec{a})}$$

Vector Equation of a Plane

Case I: Vector equation of a plane passing through the point $\vec{a}$ and parallel to two given vectors $\vec{b}$ and $\vec{c}$.

Let O be the origin, $\overline{OA}=\vec{a}$. Let AB and AC be the lines on the plane $\|$ to vectors $\vec{b}$ and $\vec{c}$. Let $P(\vec{r})$ be any point on the plane, draw PL$\|$AC, (AL$\|$AB).

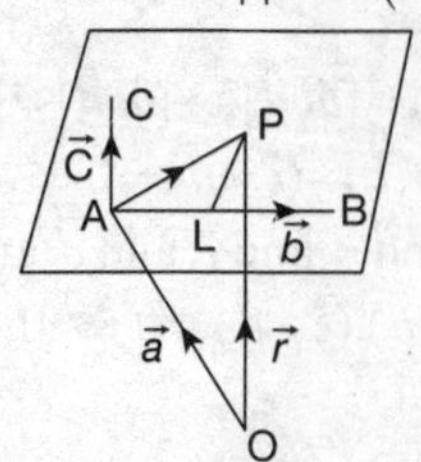

Now $\overline{AL}=s\vec{b}$ and $\overline{LP}=t\vec{c}$ (s, t are scalars)

Now $\overline{AP}=\overline{AL}+\overline{LP}=s\vec{b}+t\vec{c}$

and $\overline{OP}=\overline{OA}+\overline{AP}=\vec{a}+s\vec{b}+t\vec{c}$

Required equation: $\boxed{\vec{r}=\vec{a}+s\vec{b}+t\vec{c}}$

Another form: $\boxed{[\vec{r}\vec{b}\vec{c}]=[\vec{a}\vec{b}\vec{c}]}$

Case II: Vector equation of a plane passing through the point $\vec{a},\vec{b},\vec{c}$ is given by

$$\boxed{\vec{r}=(1-s-t)\vec{a}+s\vec{b}+t\vec{c}} \text{ (where } s,t \text{ are scalars)}$$

Another form: $\boxed{\vec{r}.(\vec{b}\times\vec{c}+\vec{c}\times\vec{a}+\vec{a}\times\vec{b})=[\vec{a}\vec{b}\vec{c}]}$

Case III: Vector equation of plane passing through a point $\vec{a}$ and perpendicular to vector $\vec{n}$:

$$\boxed{(\vec{r}-\vec{a}).\vec{n}=0}$$

Perpendicular Distance of a Point from a Line

Let a straight line pass through $A(\vec{a})$ and be $\|$ to $\vec{b}$. Then perpendicular distance of point $(\vec{r})$ from the line.

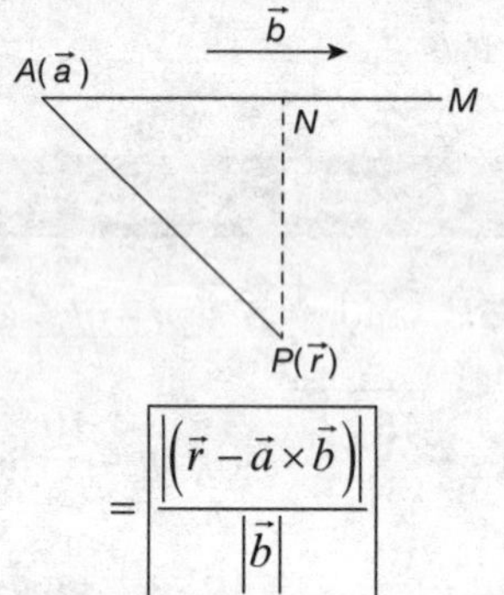

$$=\boxed{\frac{|(\vec{r}-\vec{a}\times\vec{b})|}{|\vec{b}|}}$$

Another form: $\boxed{\left[(\vec{r}-\vec{a})^2-\left\{\frac{(\vec{r}-\vec{a}).\vec{b}}{|\vec{b}|}\right\}^2\right]^{1/2}}$

Perpendicular Distance of a Point from a Plane

Case I: When plane passes through a point $\vec{a}$ and is $\|$ to $\vec{b}$ and $\vec{c}$.

Distance of point $P(\vec{r})$ from the plane $\boxed{\frac{(\vec{r}-\vec{a}).(\vec{b}\times\vec{c})}{|\vec{b}\times\vec{c}|}}$

Case II: When plane passes through the points $\vec{a},\vec{b}$ and $\vec{c}$.

$$\boxed{\frac{(\vec{r}-\vec{a}).(\vec{b}\times\vec{c}+\vec{c}\times\vec{a}+\vec{a}\times\vec{b})}{|\vec{b}\times\vec{c}+\vec{c}\times\vec{a}+\vec{a}\times\vec{b}|}}$$

Shortest Distance Between Two Non-intersecting Lines

Let there be two non-intersecting lines passing through $\vec{a}$ and $\vec{b}$ and are || to $\vec{c}$ and $\vec{d}$. If PQ be the shortest distance between them, then

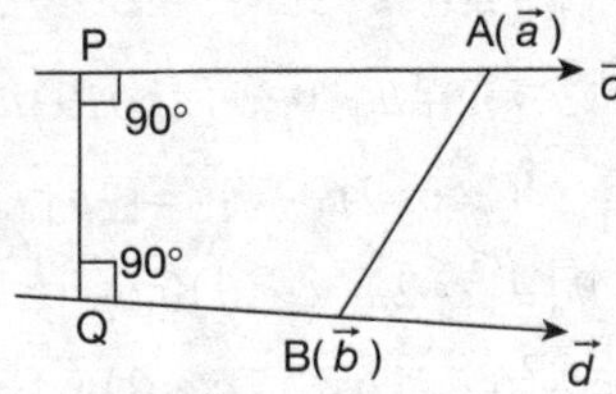

PQ = projection of $\overrightarrow{AB}$ on the vector $\vec{n}$ (where $\vec{n} = \vec{c} \times \vec{d}$)

Then $\boxed{PQ = \dfrac{(\vec{b} - \vec{a}).(\vec{c} \times \vec{d})}{|\vec{c} \times \vec{d}|}}$

Vector Equation of the Bisector of the Angle Between Two Straight Lines

Let two lines AB and AC meet at A whose position vector is $\vec{a}$. Let $\vec{b}$ and $\vec{c}$ be vectors || to AB and AC respectively. Also let P be a point on internal bisector of ∠BAC where $\overrightarrow{OP} = \vec{r}$.

From P draw line || AB which meets AC at M.

Now ∠PAM = ∠PAB = ∠APM

$$\Rightarrow \text{AM} = \text{PM} = t \text{ (say)}$$

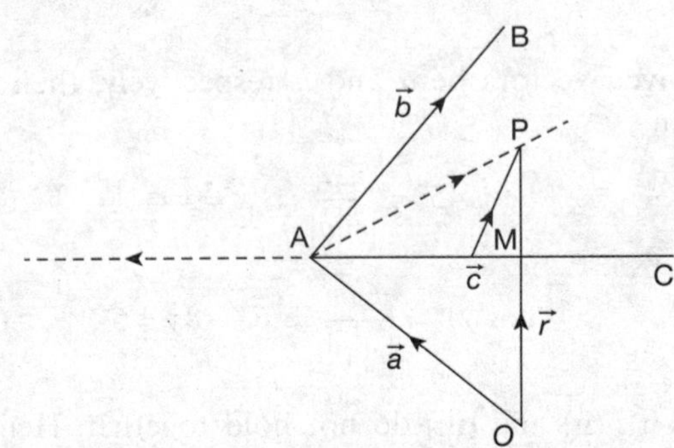

$\overrightarrow{AM}$ is collinear with $\vec{c}$ and $\overrightarrow{MP}$ is collinear with $\vec{b}$.

Then $\overrightarrow{AM} = t\dfrac{\vec{c}}{|\vec{c}|}, \overrightarrow{MP} = t\dfrac{\vec{b}}{|\vec{b}|}$

Now in ΔAPM, $\overrightarrow{AP} = \overrightarrow{AM} + \overrightarrow{MP} = t\left(\dfrac{\vec{c}}{|\vec{c}|} + \dfrac{\vec{b}}{|\vec{b}|}\right)$

In ΔOAP, $\overrightarrow{OP} = \overrightarrow{OA} + \overrightarrow{AP}$

$$\Rightarrow \qquad \vec{r} = \vec{a} + t\left(\frac{\vec{b}}{|\vec{b}|} + \frac{\vec{c}}{|\vec{c}|}\right),$$

required equation of internal bisector.

Equation of external bisector is given by

$$\vec{r} = \vec{a} + t\left(\frac{\vec{b}}{|\vec{b}|} - \frac{\vec{c}}{|\vec{c}|}\right)$$

ILLUSTRATION 10.48

Show that the perpendicular distance of the point A whose position vector is $\vec{a}$ from the line $\vec{r} = \vec{b} + t\vec{c}$ is

$$\left|\vec{b} - \vec{a} + \frac{\{(\vec{a} - \vec{b}).\vec{c}\}\vec{c}}{|\vec{c}|^2}\right|$$

Solution

We have

$\overrightarrow{BC}$ = projection of vector $\overrightarrow{BA}$ along the given line

$$= \frac{(\vec{a} - \vec{b}).\vec{c}}{|\vec{c}|^2}\vec{c}$$

Now in ΔABC we have

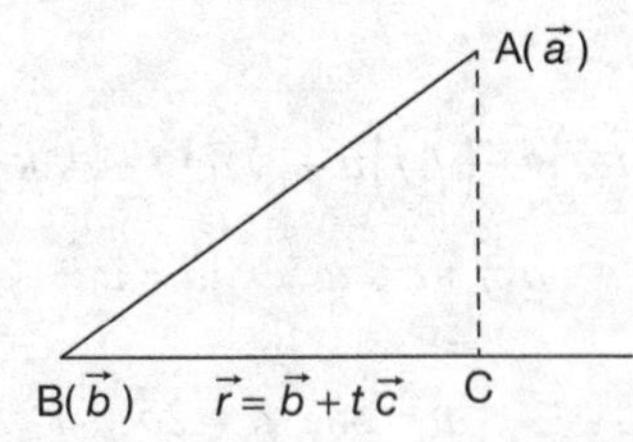

$$\overrightarrow{AC} = \overrightarrow{AB} + \overrightarrow{BC} = (\vec{b} - \vec{a}) + \frac{(\vec{a} - \vec{b}).\vec{c}}{|\vec{c}|^2}\vec{c}$$

Therefore perpendicular distance

$$= |\overrightarrow{AC}| = \left|\vec{b} - \vec{a} + \frac{(\vec{a} - \vec{b}).\vec{c}}{|\vec{c}|^2}\vec{c}\right|$$

ILLUSTRATION 10.49

Find the distance of the point $B(\hat{i} + 2\hat{j} + 3\hat{k})$ from the line which passes through $A(4\hat{i} + 2\hat{j} + 2\hat{k})$ and which is parallel to the vector $C = 2\hat{i} + 3\hat{j} + 6\hat{k}$.

Solution

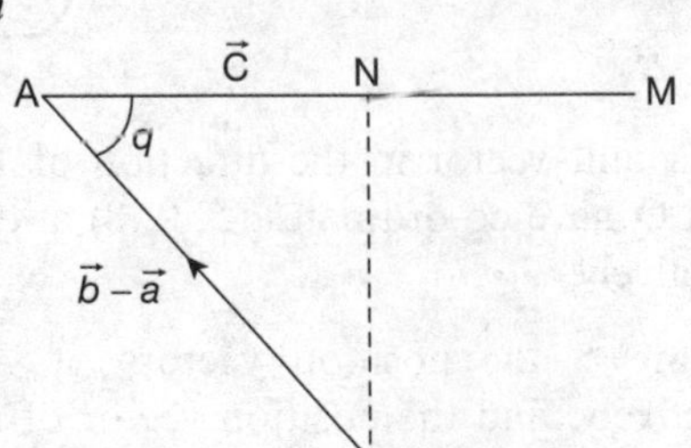

Given,

Position vector of A $= \vec{a} = 4\hat{i} + 2\hat{j} + 2\hat{k}$

Position vector of B $= \vec{b} = \hat{i} + 2\hat{j} + 3\hat{k}$

and position vector of C $= \vec{c} = 2\hat{i} + 3\hat{j} + 6\hat{k}$

Now $\overrightarrow{AB} = \vec{b} - \vec{a} = -3\hat{i} + \hat{k}$

Then $\overrightarrow{AB} \times \vec{c} = -3\hat{i} + 20\hat{j} - 9\hat{k}$

Now $|\overrightarrow{AB} \times \vec{c}| = |\overrightarrow{AB}||\vec{c}|\sin\theta = |\overrightarrow{AB}||\vec{c}|\frac{BN}{|\overrightarrow{AB}|} = |\vec{c}|BN$

$$\Rightarrow \quad BN = \frac{|-3\hat{i} + 20\hat{j} - 9\hat{k}|}{|2\hat{i} + 3\hat{j} + 6\hat{k}|} = \frac{\sqrt{490}}{\sqrt{49}} = \sqrt{10}$$

ILLUSTRATION 10.50

For any vector $\vec{a}, \vec{u} = \hat{i} \times (\vec{a} \times \hat{i}) + \hat{j} \times (\vec{a} \times \hat{j}) + \hat{k} \times (\vec{a} \times \hat{k})$ equals ________.

(a) $2\vec{a}$ (b) $-2\vec{a}$ (c) $\vec{a}$ (d) $-\vec{a}$

Solution

$$\vec{u} = (\hat{i}.\hat{i})\vec{a} - (\hat{i}.\hat{a})\hat{i} + (\hat{j}.\hat{j})\vec{a} - (\hat{j}.\vec{a})\hat{j} + (\hat{k}.\hat{k})\vec{a} - (\hat{k}.\vec{a})\hat{k}$$

$$= \vec{a} - a_1\hat{i} + \vec{a} - a_2\hat{j} + \vec{a} - a_3\hat{k} \quad [\vec{a} = a_1\hat{i} + a_2\hat{j} + a_3\hat{k} \text{ (say)}]$$

$\therefore \quad u = 3\vec{a} - \vec{a} = 2\vec{a}$

ILLUSTRATION 10.51

Let $\vec{a}, \vec{b}, \vec{c}$ such that $|\vec{a}| = 1, |\vec{b}| = 1$ and $|\vec{c}| = 2$ and if $\vec{a} \times (\vec{a} \times \vec{c}) + \vec{b} = 0$ then the acute angle between $\vec{a}$ and $\vec{c}$ is ________.

(a) $\frac{\pi}{3}$ (b) $\frac{\pi}{4}$ (c) $\frac{\pi}{6}$ (d) None of these.

Solution

If the angle between $\vec{a}$ and $\vec{c}$ is θ, then

$$\vec{a}.\vec{c} = |\vec{a}||\vec{c}|\cos\theta = 1.2\cos\theta = 2\cos\theta$$

But $\vec{a} \times (\vec{a} \times \vec{c}) + \vec{b} = 0$

$\Rightarrow \quad (\vec{a}.\vec{c})\vec{a} - (\vec{a}.\vec{a})\vec{c} + \vec{b} = 0 \Rightarrow (2\cos\theta)\vec{a} - 1.\vec{c} = -\vec{b}$

$\Rightarrow \quad [(2\cos\theta)\vec{a} - \vec{c}]^2 = [-\vec{b}]^2$

$\Rightarrow \quad 4\cos^2\theta|\vec{a}|^2 - 2.(2\cos\theta)\vec{a}.\vec{c} + |\vec{c}|^2 = |\vec{b}|^2$

$= 4\cos^2\theta - 4\cos\theta(2\cos\theta) + 4 = 1$

$\Rightarrow \quad 4(1 - \cos^2\theta) = 1 [\because |\vec{a}| = 1, |\vec{b}| = 1]$

$\Rightarrow \quad \sin\theta = 1/2$

$\Rightarrow \quad \theta = \frac{\pi}{6}$

ILLUSTRATION 10.52

A unit vector in xy-plane which makes 45° angle with vector $\hat{i} + \hat{j}$ and 60° angle with vector $3i - 4j$ will be ________.

(a) $\hat{i}$ (b) $(\hat{i} + \hat{j})/\sqrt{2}$

(c) $(\hat{i} - \hat{j})/\sqrt{2}$ (d) None of these.

Solution

Let the required vector be $\vec{r} = x\hat{i} + y\hat{j}$

$\therefore \quad x^2 + y^2 = 1 \quad \ldots$(i)

If given vectors be a and b respectively, then as given

$$\cos 45° = \frac{\vec{r}.\vec{a}}{|\vec{r}||\vec{a}|} \Rightarrow x = y = 1 \quad \ldots\text{(ii)}$$

$$\cos 60° = \frac{\vec{r}.\vec{b}}{|\vec{r}||\vec{b}|} \Rightarrow 6x - 8y = 5 \quad \ldots\text{(ii)}$$

But (i), (ii) and (iii) do not hold together. Hence such a vector is not possible.

WORKSHEET

1. Find a unit vector in the direction of $\overrightarrow{PQ}$ where P and Q have co-ordinates (5, 0, 8) and (3, 3, 2), respectively.
2. If $\vec{a}$ are $\vec{b}$ the position vectors of A and B, respectively, find the position vector of a point C in BA produced such that BC = 1.5 BA.
3. Using vectors, find the value of $\boldsymbol{k}$ such that the points ($\boldsymbol{k}$, – 10, 3), (1, –1, 3) and (3, 5, 3) are collinear.
4. Find a vector of magnitude 6, which is perpendicular to both the vectors $2\hat{i} - \hat{j} + 2\hat{k}$ and $4\hat{i} - \hat{j} + 3\hat{k}$.

5. Find the angle between the vectors $2\hat{i}-\hat{j}+\hat{k}$ and $3\hat{i}+4\hat{j}-\hat{k}$.

6. If $\vec{a}+\vec{b}+\vec{c}=0$, show that $\vec{a}\times\vec{b}=\vec{b}\times\vec{c}=\vec{c}\times\vec{a}$. Interpret the result geometrically.

7. If A, B, C, D are the points with position vectors $\hat{i}+\hat{j}-\hat{k}, 2\hat{i}-\hat{j}+3\hat{k}, 2\hat{i}-3\hat{k}, 3\hat{i}-2\hat{j}+\hat{k}$ respectively, find the projection of $\overline{AB}$ along $\overline{CD}$.

8. Using vectors, find the area of the triangle ABC with vertices A (1, 2, 3), B (2, – 1, 4) and C (4, 5, – 1).

9. Prove that in any triangle ABC, cos A = $\frac{b^2+c^2-a^2}{2bc}$, where a, b, c are the magnitudes of the sides opposite to the vertices A, B, C, respectively.

10. Show that area of the parallelogram whose diagonals are given by $\vec{a}$ and $\vec{b}$ is $\frac{|\vec{a}\times\vec{b}|}{2}$. Also find the area of the parallelogram whose diagonals are $2\hat{i}-\hat{j}+\hat{k}$ and $\hat{i}+3\hat{j}-\hat{k}$.

SOLVED EXAMPLES

Example 1. Show that for a plane triangle, $\frac{A}{\sin\alpha}=\frac{B}{\sin\beta}=\frac{C}{\sin\gamma}$ where a, b and g are the angles opposite to side A, B and C respectively.

Solution

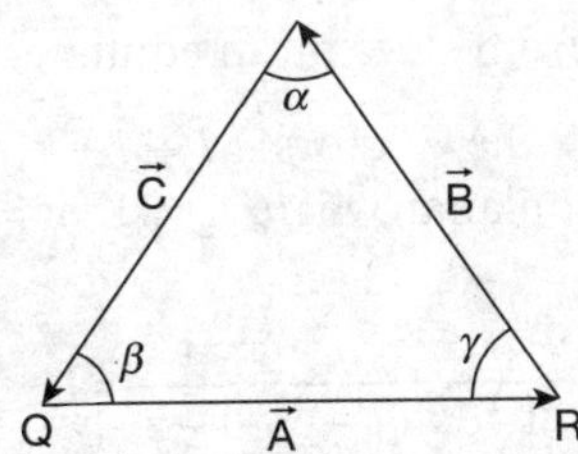

The figure shows the three sides of a triangle PQR, represented by vectors $\vec{A}, \vec{B}$ and $\vec{C}$, taken in order. Let α, β and γ be the three angles of the triangle. From the triangle law of addition of vectors,

$$\vec{A}+\vec{B}+\vec{C}=0 \quad \ldots(1)$$

Taking cross product with vector $\vec{B}$ on both sides of the equation (1), we get,

$$\vec{A}\times\vec{B}+\vec{B}\times\vec{B}+\vec{C}\times\vec{B}=0$$

or $\quad \vec{A}\times\vec{B}+\vec{C}\times\vec{B}=0$ or $\vec{A}\times\vec{B}-\vec{B}\times\vec{C}=0 \quad \ldots(2)$

Taking cross product with $\vec{C}$ on both sides of the equation (1), we get

$$\vec{A}\times\vec{C}+\vec{B}\times\vec{C}+\vec{C}\times\vec{C}=0$$

or $\quad -\vec{C}\times\vec{A}+\vec{B}\times\vec{C}=0$ or $\vec{B}\times\vec{C}=\vec{C}\times\vec{A} \quad \ldots(3)$

From (2) and (3), we get

$$\vec{B}\times\vec{C}=\vec{C}\times\vec{A}=\vec{A}\times\vec{B}$$

or $|\vec{B}\times\vec{C}|=|\vec{C}\times\vec{A}|=|\vec{A}\times\vec{B}|$ or BC sin α = CA sin β = AB sin γ

Dividing the above equation by ABC, we get

$$\frac{\sin\alpha}{A}=\frac{\sin\beta}{B}=\frac{\sin\gamma}{C}$$

$$\Rightarrow \quad \frac{A}{\sin\alpha}=\frac{B}{\sin\beta}=\frac{C}{\sin\gamma}$$

Example 2. If vectors $a\hat{i}+\hat{j}+\hat{k}, \hat{i}+b\hat{j}+\hat{k}$ and $\hat{i}+\hat{j}+c\hat{k}(a\neq b\neq c\neq 1)$ are coplanar, then $\frac{1}{1-a}+\frac{1}{1-b}+\frac{1}{1-c}$ equals _______.

(a) 1 (b) 0 (c) –1 (d) None of these.

Solution

Since the vectors are coplanar,

$$\therefore \begin{vmatrix} a & 1 & 1 \\ 1 & b & 1 \\ 1 & 1 & c \end{vmatrix}=0 \Rightarrow \begin{vmatrix} a & 1 & 1 \\ 1-a & b-1 & 0 \\ 0 & 1-b & c-1 \end{vmatrix}=0$$

[Using R2 – R1, R3 – R2]

$$\Rightarrow \quad a(b-1)(c-1)-(1-a)\{(c-1)-(1-b)]=0$$

$$\Rightarrow \quad a(1-b)(1-c)+(1-a)(1-c)+(1-a)(1-b)=0$$

$$\Rightarrow (a-1+1)(1-b)(1-c)+(1-a)(1-c)+(1-a)(1-b)=0$$

$$\Rightarrow \quad (1-b)(1-c)+(1-a)(1-c)+(1-a)(1-b)=(1-a)(1-b)(1-c)$$

$$\Rightarrow \quad \frac{1}{1-a}+\frac{1}{1-b}+\frac{1}{1-c}=1$$

Example 3. Let $\vec{a}=2\hat{i}+\hat{j}-2\hat{k}$ and $\vec{b}=\hat{i}+\hat{j}$. If vector $\vec{c}$ is such that $\vec{a}.\vec{c}=|\vec{c}|, |\vec{c}-\vec{a}|=2\sqrt{2}$ and angle between $(a\times b)$ and c is 30°, then $|(\vec{a}\times\vec{b})\times\vec{c}|$ is equal to ______.

(a) $\frac{2}{3}$ (b) $\frac{3}{2}$ (c) 2 (d) 3

Solution

$$|\vec{c}-\vec{a}|^2=(\vec{c}-\vec{a}).(\vec{c}-\vec{a})=(2\sqrt{2})^2$$

$\Rightarrow \quad \vec{c}^2+\vec{a}^2-2\vec{c}.\vec{a}=8 \Rightarrow \vec{c}^2+(4+1+4)-2\vec{c}.\vec{a}=8$

$\Rightarrow \quad \vec{c}^2+9-2|\vec{c}|=8\left[\because \vec{a}.\vec{c}=|\vec{c}|\right]$

$\Rightarrow \quad \vec{c}^2-2|\vec{c}|+1=0 \Rightarrow \vec{c}^2-2\vec{c}+1=0$

$\Rightarrow \quad (\vec{c}-1)^2=0 \Rightarrow \vec{c}=1$

$$\left|\left(\vec{a}\times\vec{b}\right)\times\vec{c}\right|=\left|\vec{a}\times\vec{b}\right||\vec{c}|\sin 30^\circ=1\times\frac{1}{2}\left|\vec{a}\times\vec{b}\right|=\frac{1}{2}\left|\vec{a}\times\vec{b}\right|$$

But $\vec{a}\times\vec{b}=\begin{vmatrix}\hat{i} & \hat{j} & \hat{k}\\ 2 & 1 & -2\\ 1 & 1 & 0\end{vmatrix}=2\hat{i}-2\hat{j}+\hat{k}$

$\therefore \quad \left|\vec{a}\times\vec{b}\right|=\sqrt{4+4+1}=3$

$\therefore \quad \left|\left(\vec{a}\times\vec{b}\right)\times\vec{c}\right|=\frac{3}{2}$

Example 4. If vectors $\vec{a}=\hat{i}-\hat{j}+\hat{k}, \vec{b}=-3\hat{i}+7\hat{j}-3\hat{k}$ and $\vec{c}=3\hat{i}+p\hat{j}+5\hat{k}$ are coplanar, then the value of p is ______.

(a) 2 (b) 6 (c) –2 (d) –6

Solution

$\vec{a}, \vec{b}, \vec{c}$ are coplanar $\Rightarrow \left[\vec{a}\vec{b}\vec{c}\right]=0 \Rightarrow \begin{vmatrix}1 & -1 & 1\\ 1 & 2 & -1\\ 3 & p & 5\end{vmatrix}=0$

$\Rightarrow \quad (10+p+3)-(6-5-p)=0$

$\Rightarrow \quad p=-6$

Example 5. If $\vec{a}=-3\hat{i}+8\hat{j}+5\hat{k}, \vec{b}=-3\hat{i}+7\hat{j}-3\hat{k}$ and $\vec{c}=7\hat{i}-5\hat{j}-3\hat{k}$ are the coterminous edges of a parallelepiped, then its volume is ______.

(a) 108 (b) 210 (c) 272 (d) 308

Solution

Required volume $=\left[\vec{a}\vec{b}\vec{c}\right]$

$$=\begin{vmatrix}-3 & 7 & 5\\ -3 & 7 & -3\\ 7 & -5 & -3\end{vmatrix}=|-3(-21-15)-7(9+21)+5(15-49)|$$

$$=|108-210-170|=272$$

Example 6. If $\hat{i}-\hat{j}+2\hat{k}, 2\hat{i}+\hat{j}-\hat{k}$ and $3\hat{i}-\hat{j}+2\hat{k}$ are position vectors of a triangle, then its area is ______.

(a) 26 (b) 13 (c) $2\sqrt{13}$ (d) $\sqrt{13}$

Solution

If A, B, C are the given vertices, then

$$\overline{\text{AB}}=\hat{i}+2\hat{j}-3\hat{k}, \overline{AC}=2\hat{i}$$

$\therefore \quad \overline{\text{AB}}\times\overline{\text{AC}}=\left(\hat{i}+2\hat{j}-3\hat{k}\right)\times 2\hat{i}=-4\hat{k}-6\hat{j}$

$\Rightarrow \quad \left|\overline{\text{AB}}\times\overline{\text{AC}}\right|=\sqrt{16+36}=\sqrt{213}$

$\therefore \quad$ Area of $\Delta\text{ABC}=\frac{1}{2}\left|\overline{AB}\times\overline{AC}\right|=\sqrt{13}$

Example 7. Prove that the vectors $3\hat{i}+5\hat{j}+2\hat{k}, 2\hat{i}-3\hat{j}-5\hat{k}$ and $5\hat{i}+2\hat{j}-3\hat{k}$ form the sides of an equilateral triangle.

Solution

Let $\vec{a}=3\vec{i}+5\vec{j}+2\vec{k}, \vec{b}=2\vec{i}-3\vec{j}-5\vec{k}$ and $\vec{c}=5\vec{i}+2\vec{j}-3\vec{k}$.

Clearly $\vec{a}+\vec{b}=(3+2)\vec{i}+(5-3)\vec{j}+(2-5)\vec{k}$

$$=5\vec{i}+2\vec{j}-3\vec{k}=\vec{c}$$

$\therefore$ The vectors $\vec{a}, \vec{b}, \vec{c}$ are represented by the three sides of a triangle.

Now $\quad |\vec{a}|=|3\vec{i}+5\vec{j}+2\vec{k}|$

$$=\sqrt{3^2+5^2+2^2}=\sqrt{38}$$

$$\left|\vec{b}\right|=|2\vec{i}-3\vec{j}-5\vec{k}|=\sqrt{38}$$

$$|\vec{c}|=|5\vec{i}+2\vec{j}-3\vec{k}|=\sqrt{38}$$

$\therefore \quad |\vec{a}|=\left|\vec{b}\right|=|\vec{c}|$

So the vectors from sides of an equilateral triangle.

Example 8. If the vectors $x\vec{i}+\vec{j}+\vec{k}, \vec{i}+y\vec{j}+\vec{k}$ and $\vec{i}+\vec{j}+z\vec{k}$ are coplanar where $x\neq 1, y\neq 1, z\neq 1$ then prove that

$$\frac{1}{1-x}+\frac{1}{1-y}+\frac{1}{1-z}=1$$

Solution

The vectors are coplanar if we can find two scalars l, m such that

$$x\vec{i}+\vec{j}+\vec{k}=\lambda\left(\vec{i}+y\vec{j}+\vec{k}\right)+\mu\left(\vec{i}+\vec{j}+z\vec{k}\right)$$

$\Rightarrow \quad x=\lambda+\mu, 1=\lambda y+\mu, 1=\lambda+\mu z$

$$1-x=1-\lambda-\mu,\ y=\frac{1-\mu}{\lambda}, z=\frac{1-\lambda}{\mu}$$

$\therefore \quad 1-y=\frac{\lambda-1+\mu}{\lambda}$ and $1-z=\frac{\mu-1+\lambda}{\mu}$

$\therefore \quad \frac{1}{1-x}+\frac{1}{1-y}+\frac{1}{1-z}$

$$=\frac{1}{1-\lambda-\mu}+\frac{\lambda}{\lambda+\mu-1}+\frac{\mu}{\lambda+\mu-1}$$

$$=\frac{1-\lambda-\mu}{1-\lambda-\mu}=1$$

Example 9. Find the vector c, directed along internal bisector of the angle between the vectors $a=7\vec{i}-4\vec{j}-4k$ and $b=-2\hat{i}-\hat{j}+2\hat{k}$ with $|c|=5\sqrt{6}$.

Solution

The required vector c is given by $\lambda\left(\frac{a}{|a|}+\frac{b}{|b|}\right)$

Now $\frac{a}{|a|}=\frac{1}{9}(7i-4j-4k)$ and $\frac{b}{|b|}=\frac{1}{3}(-2j-j+2k)$

$$\Rightarrow \quad c=\lambda\left[\frac{1}{9}i-\frac{7}{9}j+\frac{2}{9}k\right]=\left|c^2\right|=\lambda^2\times\frac{54}{81}$$

Therefore $\lambda^2=225$ or $l=\pm 15$

$$c=\pm\frac{5}{3}[i-7j+2k]$$

Example 10. A vector $\vec{a}=(x, y, z)$ makes an obtuse angle with y-axis equal angles with $\vec{b}=(y, -2z, 3x)$ and $\vec{c}=(2z, 3x, -y)$ and $\vec{a}$ is perpendicular to $\vec{d}=(1, -1, 2)$ if $|\vec{a}|=2\sqrt{3}$ then find vector $\vec{a}$.

Solution

$\vec{a}=(x, y, z)$

$\vec{a}$ makes an obtuse angle with y-axis

$\therefore\ y<0$ and given

$$\frac{\vec{a}.\vec{b}}{|\vec{a}||\vec{b}|}=\frac{\vec{a}.\vec{c}}{|\vec{a}||\vec{c}|}$$

$$\Rightarrow \quad \frac{\vec{a}.\vec{b}}{|\vec{b}|}=\frac{\vec{a}.\vec{c}}{|\vec{c}|}$$

$$\Rightarrow \quad \frac{xy-2yz+3zx}{\sqrt{y^2+4z^2+9x^2}}=\frac{2zx+3xy-yz}{\sqrt{4z^2+9x^2+y^2}}$$

$$\Rightarrow \quad xy-2yz+3zx=2zx+3xy-yz$$

$$\Rightarrow \quad 2xy+yz-xz=0 \qquad \ldots(1)$$

and given $\vec{a}.\vec{d}=0$

$\therefore\ x-y+2z=0$

$$z=\frac{y-x}{2} \qquad \ldots(2)$$

From (1) and (2), we get

$$2xy+\frac{(y-x)^2}{2}=0$$

$$(y+x)^2=0$$

$$y=-x \qquad \ldots(3)$$

$z=-x$ (from 2)

$\therefore\ \vec{a}=(x,-x,-x)$

$$|\vec{a}|=\sqrt{x^2+x^2+x^2}=x\sqrt{3}=2\sqrt{3}$$

$\therefore\ x=2, y=-2, z=-2$

$\therefore\ \vec{a}=(2, -2, -2)$

Example 11. If the vectors a, b, c are coplanar, show that

$$\begin{vmatrix} a & b & c \\ a.b & a.b & a.c \\ b.a & b.b & b.c \end{vmatrix}=0$$

Solution

Let $a=a_1\hat{i}+a_2\hat{j}+3\hat{k},\ b=b_1\hat{i}+b_2\hat{j}+b_3\,k\ c=c_1\hat{i}+c_2\hat{j}+c_3\hat{k}$

$$\Delta=\begin{vmatrix} a & b & c \\ a.a & a.b & a.c \\ b.a & b.b & b.c \end{vmatrix}=\begin{vmatrix} a & a.a & b.a \\ b & a.b & b.b \\ c & a.c & b.c \end{vmatrix}$$

$$=\begin{vmatrix} a_1 & a_2 & a_3 \\ b_1 & b_2 & b_3 \\ c_1 & c_2 & c_3 \end{vmatrix}\begin{vmatrix} \hat{i} & \hat{j} & \hat{k} \\ a_1 & a_2 & a_3 \\ b_1 & b_2 & b_3 \end{vmatrix}=0$$

(because, a, b, c are coplanar.)

RECAPITULATION (FORMULAE, THEOREM, IMPORTANT RESULTS)

1. Two vectors are said to be collinear if they are either parallel or coincident. The necessary and sufficient conditions for two vectors $\vec{a}$ and $\vec{b}$ to be collinear is that there exists a scalar λ such that $\vec{a}=\lambda\vec{b}$. Further, if $\vec{a}$ and $\vec{b}$ are non-collinear and $x\vec{a}+y\vec{b}=0$, then $x=0, y=0$ essentially.
2. Three points A, B, C with position vectors $\vec{a},\vec{b},\vec{c}$ will be collinear, iff there exist scalars $x, y, z, (x^2+y^2+z^2\neq 0)$ such that $x\vec{a}+y\vec{b}+z\vec{c}=0$, $x+y+z=0$.
3. Two vectors are always coplanar. Three vectors $\vec{a},\vec{b},\vec{c}$ will be coplanar iff
   (i) one of them can be expressed as a linear combination of remaining two, i.e., (say) $\vec{c}=x\vec{a}+y\vec{b}$
   or
   (ii) There exists a scalar $x, y, z (x^2+y^2+z^2\neq 0)$ such that $x\vec{a}+y\vec{b}+z\vec{c}=0$
   or
   (iii) $[\vec{a}\vec{b}\vec{c}]=0$ where $[\vec{a}\vec{b}\vec{c}]$ is scalar the triple product $\vec{a}.(\vec{b}\times\vec{c})$.
4. Three vectors $\vec{a},\vec{b},\vec{c}$ are non-coplanar if the relation $x\vec{a}+y\vec{b}+z\vec{c}$ is possible only when $x=0, y=0, z=0$.
   or

Three vectors $\vec{a},\vec{b},\vec{c}$ will be non-coplanar iff $[\vec{a}\vec{b}\vec{c}]\neq 0$. Further, if $\vec{a},\vec{b},\vec{c}$ are non-coplanar, then any other vector $\vec{d}$ in place must be a linear combination of $\vec{a},\vec{b}$ and $\vec{c}$, i.e., $\vec{d}$ can be written as $\vec{d}=x\vec{a}+y\vec{b}+z\vec{c}$.

Note the following carefully

(i) A trio of non-coplanar vectors is called a basis. If the vectors in the basis are mutually perpendicular it is called normal basis. Further, if the magnitude of each vector in the basis is unity, it is called orthonormal basis. For instance $\{i, j, k\}$ forms an orthonormal basis.

(ii) If $\vec{a}$ and $\vec{b}$ are non-collinear then $\vec{a},\vec{b}$ and $\vec{a}\times\vec{b}$ can also form a basis since $\vec{a},\vec{b}$ and $\vec{a}\times\vec{b}$ are non-coplanar.

(iii) Since $[\vec{b}\times\vec{c},\vec{c}\times\vec{a},\vec{a}\times\vec{b}]=[\vec{a}\vec{b}\vec{c}]^2$, we conclude that if $\vec{a},\vec{b},\vec{c}$ are non-coplanar then so are $\vec{b}\times\vec{c},\vec{c}\times\vec{a}$ and $\vec{a}\times\vec{b}$.

Thus, a basis can also be taken as a trio $\vec{b}\times\vec{c},\vec{c}\times\vec{a}$ and $\vec{a}\times\vec{b}$.

5. Results on scalar triple product $[\vec{a}\vec{b}\vec{c}]$ or $\vec{a}.(\vec{b}\times\vec{c})$.

(i) If the cyclic order is maintained the value of $[\vec{a}\vec{b}\vec{c}]$ remains same.
Thus $[\vec{a}\vec{b}\vec{c}]=[\vec{b}\vec{c}\vec{a}]=[\vec{c}\vec{a}\vec{b}]=\beta$ (say)
and $[\vec{a}\vec{c}\vec{b}]=[\vec{c}\vec{b}\vec{a}]=[\vec{b}\vec{a}\vec{c}]=-\beta$.

(ii) If any two vectors in $[\vec{a}\vec{b}\vec{c}]$ are same, $[\vec{a}\vec{b}\vec{c}]=0$.

(iii) If $\vec{a},\vec{b},\vec{c}$ are coplanar $[\vec{a}\vec{b}\vec{c}]=0$

(iv) If λ is a scalar, then
$[\lambda\vec{a},\vec{b},\vec{c}]=\lambda[\vec{a}\vec{b}\vec{c}]$

More generally,
$[\lambda_1\vec{a},\lambda_2\vec{b},\lambda_3\vec{c}]=\lambda_1\lambda_2\lambda_3[\vec{a}\vec{b}\vec{c}]$

(v) $(\vec{a}\times\vec{b})\times(\vec{b}\times\vec{c})=[\vec{a}\vec{b}\vec{c}]\vec{b}$
[**Remark**: $\vec{b}$ is common in both brackets]

(vi) $\left(\vec{a}\times\vec{b}\right).\left(\vec{c}\times\vec{d}\right)=\begin{vmatrix}\vec{a}.\vec{c} & \vec{a}.\vec{d}\\ \vec{b}.\vec{c} & \vec{b}.\vec{d}\end{vmatrix}$
(Scalar product of four vectors)

(vii) The volume of a parallelepiped whose edges are along $\vec{a},\vec{b},\vec{c}$ must be $[\vec{a}\vec{b}\vec{c}]$.

(viii) The volume of a tetrahedron OABC where position vectors of vertices A,B and C with reference to O are $\vec{a},\vec{b},\vec{c}$ is $\frac{1}{6}\left[\vec{a}\vec{b}\vec{c}\right]$.

(ix) $\left[\vec{a}\vec{b}\vec{c}\right]^2=\begin{vmatrix}\vec{a}.\vec{a} & \vec{a}.\vec{b} & \vec{a}.\vec{c}\\ \vec{b}.\vec{a} & \vec{b}.\vec{b} & \vec{b}.\vec{c}\\ \vec{c}.\vec{a} & \vec{c}.\vec{b} & \vec{c}.\vec{c}\end{vmatrix}$

(x) Four points A,B,C and D with position vectors $\vec{a}, \vec{b}, \vec{c}$ and $\vec{d}$ wil be coplanar if
$[\vec{d}\vec{b}\vec{c}]+[\vec{d}\vec{c}\vec{a}]+[\vec{d}\vec{a}\vec{b}]=[\vec{a}\vec{b}\vec{c}]$.
(With several other permutations,) note that the points will be coplanar if
$[\vec{d}-\vec{a},\vec{d}-\vec{b},\vec{d}-\vec{c}]=0$

(xi) $[\vec{b}+\vec{c},\vec{c}+\vec{a},\vec{a}+\vec{b}]=2[\vec{a}\vec{b}\vec{c}]$. More generally,
$[x\vec{b}+y\vec{c}, z\vec{c}+w\vec{a}, t\vec{a}+s\vec{b}]$
$=xzt[\vec{a}\vec{b}\vec{c}]+yws[\vec{c}\,\vec{a}\,\vec{b}]$
(x,y,z,w,t,s) are scalars)

(xii) $[\vec{b}\times\vec{c},\vec{c}\times\vec{a},\vec{a}\times\vec{b}]=[\vec{a}\vec{b}\vec{c}]^2$

QUIZ

1. What are direction cosines?

2. In the figure, identify the following vectors.

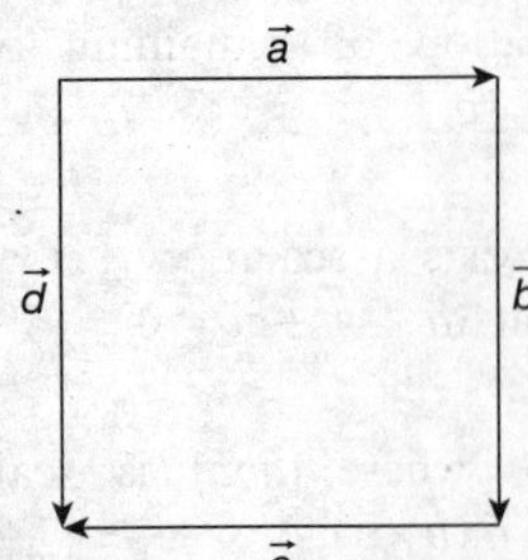

(i) Coinitial (ii) Equal

(ii) Collinear but not equal

3. Let $\vec{a}=\hat{i}+2\hat{j}$ and $\vec{b}=2\hat{i}+\hat{j}$. Is $|\vec{a}|=|\vec{b}|$? Are the vectors $\vec{a}$ and $\vec{b}$ equal?

4. Find a vector in the direction of vector $\vec{a}=\hat{i}-2\hat{j}$ that has magnitude 7 units.

5. Find the angle between two vectors $\vec{a}$ and $\vec{b}$ with magnitude 1 and 2 respectively and when $\vec{a}.\vec{b}=1$.

6. What is the relation among magnitude r direction ratios (a,b,c) and direction cosines (l,m,n) of the vectors?

7. What is the vector sum of the three sides of a triangle, taken in order?

8. If either $\vec{a}=\vec{0}$ or $\vec{b}=\vec{0}$ then $\vec{a}\times\vec{b}=\vec{0}$. Is the converse true? Justify your answer with an example.

9. If $\hat{p}$ is a unit vector along line l, then what is the projection of a vector $\vec{a}$ on the line l?

10. What is the distributive property of scalar product over addition for three vectors $\vec{a}, \vec{b}$ and $\vec{c}$.

JEE MAIN

Objective Questions

One Option Correct

1. $\hat{a}, \hat{b}, \hat{c}$ are three unit vectors, $\hat{a}$ and $\hat{b}$ are perpendicular to each other and vector $\hat{c}$ is equally inclined to both $\hat{a}$ and $\hat{b}$ at an angle θ. If $\hat{c} = \alpha\hat{a} + \beta\hat{b} + \gamma(\hat{a}\times\hat{b})$, where α, β, γ are constants, then
(a) $\alpha = \beta = -\cos\theta$, $\gamma^2 = \cos 2\theta$
(b) $\alpha = \beta = \cos\theta$, $\gamma^2 = \cos 2\theta$
(c) $\alpha = \beta = \cos\theta$, $\gamma^2 = -\cos 2\theta$
(d) $\alpha = \beta = -\cos\theta$, $\gamma^2 = -\cos 2\theta$

2. Let a, b and c be distinct non-negative numbers. If the vectors $a\hat{i} + a\hat{j} + c\hat{k}$, $\hat{i} + \hat{k}$ and $c\hat{i} + c\hat{j} + b\hat{k}$ lie in a plane, then c is
(a) equal to zero.
(b) the harmonic mean of a and b.
(c) the geometric mean of a and b.
(d) the arithmetic mean of a and b.

3. If $\vec{a}, \vec{b}, \vec{c}$ are non-coplanar vectors and l is a real number, then
$\left[\lambda(\vec{a}+\vec{b}) \;\; \lambda^2\vec{b} \;\; \lambda\vec{c}\right] = \left[\vec{a} \;\; \vec{b}+\vec{c} \;\; \vec{b}\right]$ for
(a) exactly three values of l.
(b) exactly two values of l.
(c) exactly one value of l .
(d) no value of l.

4. A particle is acted upon by constant forces $4\hat{i} + \hat{j} - 3\hat{k}$ and $3\hat{i} + \hat{j} - \hat{k}$ which displace it from a point $\hat{i} + 2\hat{j} + 3\hat{k}$ to the point $5\hat{i} + 4\hat{j} + \hat{k}$. The work done in standard units by the forces is given by _______.
(a) 40 (b) 30
(c) 25 (d) 15

5. Let $\vec{u}, \vec{v}, \vec{w}$ be such that $|\vec{u}| = 1$, $|\vec{v}| = 2$, $|\vec{w}| = 3$. If the projection $\vec{v}$ along $\vec{u}$ is equal to that of $\vec{w}$ along $\vec{u}$ and $\vec{v}, \vec{w}$ are perpendicular to each other, then $|\vec{u} - \vec{v} + w|$ equals ________.
(a) 2 (b) $\sqrt{7}$
(c) $\sqrt{14}$ (d) 14

6. The summation of two unit vectors is a third unit vector, then the modulus of the difference of the unit vectors is ________.
(a) $\sqrt{3}$ (b) $1-\sqrt{3}$
(c) $1+\sqrt{3}$ (d) $-\sqrt{3}$

7. If $\vec{A}\times(\vec{B}\times\vec{C}) = \vec{B}\times(\vec{C}\times\vec{A})$ and $[\vec{A}\,\vec{B}\,\vec{C}] \neq 0$ then $\vec{A}\times(\vec{B}\times\vec{C})$ is equal to
(a) $\vec{0}$ (b) $\vec{A}\times\vec{B}$
(c) $\vec{B}\times\vec{C}$ (d) $\vec{C}\times\vec{A}$

8. Let $\vec{a} = \hat{i} - \hat{k}$, $\vec{b} = x\hat{i} + y\hat{j} + (1-x)\hat{k}$, $\vec{c} = y\hat{i} + x\hat{j} + (1+x-y)\hat{k}$, then $[\vec{a}, \vec{b}, \vec{c}]$ depends upon
(a) only x. (b) only y.
(c) neither x or y. (d) both x and y.

9. If $\vec{a}, \vec{b}, \vec{c}$ are non-coplanar unit vectors such that $\vec{a}\times(\vec{b}\times\vec{c}) = \frac{\vec{b}+\vec{c}}{\sqrt{2}}$, then the angle between $\vec{a}$ and $\vec{c}$ is _______.
(a) $\pi/4$ (b) $3\pi/4$
(c) $3\pi/2$ (d) π

10. The non-zero vectors $\vec{a}, \vec{b}$ and $\vec{c}$ are related by $\vec{a} = 8\vec{b}$ and $\vec{c} = -7\vec{b}$. Then the angle between $\vec{a}$ and $\vec{c}$ is _______.
(a) $\frac{\pi}{4}$ (b) $\frac{\pi}{2}$
(c) π (d) 0

11. The vector $\vec{a} = \alpha\hat{i} + 2\hat{j} + \beta\hat{k}$ lies in the plane of the vectors $\vec{b} = \hat{i} + \hat{j}$ and $\vec{c} = \hat{j} + \hat{k}$ and bisects the angle between $\vec{b}$ and $\vec{c}$. Then which one of the following gives possible values of α and β?
(a) $\alpha = 1, \beta = 2$ (b) $\alpha = 2, \beta = 1$
(c) $\alpha = 1, \beta = 1$ (d) $\alpha = 2, \beta = 2$

12. Let $\vec{a} = \hat{i} + \hat{j} + \hat{k}, \vec{b} = \hat{i} - \hat{j} + 2\hat{k}$ and $\vec{c} = x\hat{i} + (x-2)\hat{j} - \hat{k}$. If the vector $\vec{c}$ lies in the plane of $\vec{a}$ and $\vec{b}$, then x equals _______.
(a) 0 (b) 1
(c) –4 (d) 2

13. The values of a, for which the points A, B, C with position vectors $2\hat{i} - \hat{j} + \hat{k}; \hat{i} - 3\hat{j} - 5\hat{k}$ and $a\hat{i} - 3\hat{j} + \hat{k}$ respectively are the vertices of a right-angled triangle with $C = \frac{\pi}{2}$ are
(a) –2 and –1. (b) –2 and 1.
(c) 2 and –1 (d) 2 and 1.

14. If C is the midpoint of AB and P is any point outside AB, then
(a) $\overrightarrow{PA}+\overrightarrow{PB}+2\overrightarrow{PC}=\vec{0}$ (b) $\overrightarrow{PA}+\overrightarrow{PB}+\overrightarrow{PC}=\vec{0}$
(c) $\overrightarrow{PA}+\overrightarrow{PB}=2\overrightarrow{PC}$ (d) $\overrightarrow{PA}+\overrightarrow{PB}=\overrightarrow{PC}$

15. The distance between the line $\vec{r}=2\hat{i}-2\hat{j}+3\hat{k}+\lambda(\hat{i}+\hat{j}+4\hat{k})$ and plane $\vec{r}.(\hat{i}-5\hat{j}+\hat{k})=5$ is ______.
(a) $\frac{3}{10}$ (b) $\frac{10}{3}$
(c) $\frac{10}{9}$ (d) $\frac{10}{3\sqrt{3}}$

16. Let $\vec{a}=\hat{i}-\hat{k},\vec{b}=x\hat{i}+\hat{j}+(1-x)\hat{k}$ and $\vec{c}=y\hat{i}+x\hat{j}+1(1+x-y)\hat{k}$. Then $[\vec{a}\vec{b}\vec{c}]$ depends on
(a) both x and y. (b) neither x nor y.
(c) only y. (d) only x.

17. Let $\vec{a},\vec{b}$ and $\vec{c}$ be three non-zero vectors such that no two of these are collinear. If the vector $\vec{a}+2\vec{b}$ is collinear with $\vec{c}$ and $\vec{b}+3\vec{c}$ is collinear with $\vec{a}$ (λ being some non-zero scalar), then $\vec{a}+2\vec{b}+6\vec{c}$ equals ______.
(a) $\lambda\vec{a}$ (b) $\lambda\vec{b}$
(c) $\lambda\vec{c}$ (d) 0

18. If $\vec{a},\vec{b},\vec{c}$ are non-coplanar vectors and λ is a real number, then the vectors $\vec{a}+2\vec{b}+3\vec{c},\lambda\vec{b}+3\vec{c}$ and $(2\lambda-1)\vec{c}$ are non-coplanar for
(a) all values of λ.
(b) all except one value of λ.
(c) all except two values of λ.
(d) no value of λ.

19. Let $\vec{a},\vec{b}$ and $\vec{c}$ be non-zero vectors such that $(\vec{a}\times\vec{b})\times\vec{c}=\frac{1}{3}|\vec{b}||\vec{c}||\vec{a}|$. If θ is the acute angle between the vectors $\vec{b}$ and $\vec{c}$, then $\sin\theta$ equals ______.
(a) $\frac{1}{3}$ (b) $\frac{\sqrt{2}}{3}$
(c) $\frac{2}{3}$ (d) $\frac{2\sqrt{2}}{3}$

20. If $\vec{a},\vec{b},\vec{c}$ are 3 vectors, such that $\vec{a}+\vec{b}+\vec{c}=0$, $|\vec{a}|=1,|\vec{b}|=2,|\vec{c}|=3$, then $\vec{a}.\vec{b}+\vec{b}.\vec{c}+\vec{c}.\vec{a}$ is equal to ______.
(a) 0 (b) −7
(c) 7 (d) 1

21. If $\vec{u},\vec{v}$ and $\vec{w}$ are three non-coplanar vectors, then $(\vec{u}+\vec{v}-\vec{w}).(\vec{u}-\vec{v})\times(\vec{v}-\vec{w})$ equals ______.
(a) $-\vec{v}.(\vec{u}\times\vec{w})$ (b) $\vec{u}.\vec{v}\times\vec{w}$
(c) $\vec{u}.\vec{w}\times\vec{v}$ (d) $3\vec{u}.\vec{v}\times\vec{w}$

22. Let $\vec{u}=\hat{i}+\hat{j},\vec{v}=\hat{i}-\hat{j}$ and $\vec{w}=\hat{i}+2\hat{j}+\hat{k}$. If $\hat{n}$ is a unit vector such that $\vec{u}.\hat{n}=0$ and $\vec{v}.\hat{n}=0$, then $|\vec{w}.\hat{n}|$ is equal to ______.
(a) 0 (b) 1
(c) 2 (d) 3

23. If three vectors $\vec{a},\vec{b},\vec{c}$ are such that $[\vec{a}\vec{b}\vec{c}]=4$, then $[\vec{a}\times\vec{b}\;\vec{b}\times\vec{c}\;\vec{c}\times\vec{a}]$ has the value ______.
(a) 8 (b) 16
(c) 4 (d) 64

24. The magnitude of the cross product of two vectors is $\sqrt{3}$ times the dot product. The angle between the vectors is ______.
(a) $\frac{\pi}{6}$ (b) $\frac{\pi}{3}$
(c) $\frac{\pi}{2}$ (d) $\frac{\pi}{4}$

25. Let $\vec{a},\vec{b}$ and $\vec{c}$ be three vectors. Then scalar triple product $[\vec{a}\vec{b}\vec{c}]$ is equal to ______.
(a) $[\vec{b}\vec{a}\vec{c}]$ (b) $[\vec{a}\vec{c}\vec{b}]$
(c) $[\vec{c}\vec{b}\vec{a}]$ (d) $[\vec{b}\vec{c}\vec{a}]$

26. If $|\vec{a}-\vec{b}|=|\vec{a}|=|\vec{b}|=1$, then the angle between $\vec{a}$ and $\vec{b}$ is ______.
(a) $\frac{\pi}{3}$ (b) $\frac{3\pi}{4}$
(c) $\frac{\pi}{2}$ (d) $0°$

27. $(\vec{a}.\hat{i})^2+(\vec{a}.\hat{j})^2+(\vec{a}.\hat{k})^2$ is equal to ______.
(a) 1 (b) $\vec{a}$
(c) $-\vec{a}$ (d) $|\vec{a}|^2$

28. If sum of two unit vectors is also an unit vector, then the angle between the two vectors is ______.
(a) $\frac{\pi}{3}$ (b) $\frac{\pi}{2}$
(c) $\frac{\pi}{4}$ (d) $\frac{2\pi}{3}$

29. The value of c so that for all real x, the vectors $x\hat{i}-6\hat{j}-3\hat{k},x\hat{i}+2\hat{j}+2cx\hat{k}$ make an obtuse angle are ______.
(a) $c<0$ (b) $0<c<\frac{4}{3}$
(c) $-\frac{4}{3}<c<0$ (d) $c>0$

30. If $\hat{i},\hat{j},\hat{k}$ are unit normal vectors and $\vec{a}$ is a vector such that $\vec{a}\times\vec{r}=\hat{j}$, then $\vec{a}.\vec{r}$ is ______.
(a) −1 (b) 0
(c) 1 (d) None of these.

31. For any vector $\vec{a}$, the value of $\hat{i}\times(\vec{a}\times\hat{i})+\hat{j}\times(\vec{a}\times\hat{j})+\hat{k}\times(\vec{a}\times\hat{k})$ is ______.

(a) $\vec{a}$ (b) $2\vec{a}$
(c) $3\vec{a}$ (d) $\vec{0}$

32. A vector perpendicular to the plane containing the vectors $\hat{i}-2\hat{j}-\hat{k}$ and $3\hat{i}-2\hat{j}-\hat{k}$ is inclined to the vector $\hat{i}+\hat{j}+\hat{k}$ at an angle ______.

(a) $\tan^{-1}\sqrt{14}$ (b) $\sec^{-1}\sqrt{14}$
(c) $\tan^{-1}\sqrt{15}$ (d) $\sec^{-1}\sqrt{15}$

33. A vector of magnitude 4 which is equally inclined to the vector $\hat{i}+\hat{j}, \hat{j}+\hat{k}, \hat{k}+\hat{i}$ is ______.

(a) $\frac{4}{\sqrt{3}}\left(\hat{i}-\hat{j}-\hat{k}\right)$ (b) $\frac{4}{\sqrt{3}}\left(\hat{i}+\hat{j}-\hat{k}\right)$
(c) $\frac{4}{\sqrt{3}}\left(\hat{i}+\hat{j}+\hat{k}\right)$ (d) None of these.

34. Unit vectors $\vec{a}, \vec{b}$ and $\vec{c}$ are coplanar. A unit vector $\vec{d}$ is perpendicular to them. If $\left(\vec{a}\times\vec{b}\right)\times\left(\vec{c}\times\vec{d}\right)=\frac{1}{6}\hat{i}-\frac{1}{3}\hat{j}+\frac{1}{3}\hat{k}$ and the angle between $\vec{a}$ and $\vec{b}$ is 30°, then $\vec{c}$ is ______.

(a) $\frac{\left(\hat{i}-2\hat{j}+2\hat{k}\right)}{3}$ (b) $\frac{\left(2\hat{i}+\hat{j}-\hat{k}\right)}{3}$
(c) $\frac{\left(-2\hat{i}-2\hat{j}+\hat{k}\right)}{3}$ (d) $\frac{\left(-\hat{i}+2\hat{j}-2\hat{k}\right)}{3}$

35. The unit vector perpendicular to $\hat{i}-\hat{j}$ and coplanar with $\hat{i}+2\hat{j}$ and $2\hat{i}+3\hat{j}$ is ______.

(a) $\frac{2\hat{i}-5\hat{j}}{\sqrt{29}}$ (b) $2\hat{i}+5\hat{j}$
(c) $\frac{1}{\sqrt{2}}\left(\hat{i}+\hat{j}\right)$ (d) $\left(\hat{i}+\hat{j}\right)$

36. If $\vec{a}, \vec{b}, \vec{c}, \vec{d}$ are the position vectors of points A, B, C and D respectively such that $(\vec{a}-\vec{d}).(\vec{b}-\vec{c})=(\vec{b}-\vec{d})(\vec{c}-\vec{a})=0$ then D is the

(a) centroid of ΔABC.
(b) circumcentre of ΔABC.
(c) orthocentre of ΔABC.
(d) None of these.

37. The angle between two vectors $\sqrt{3}(\vec{a}\times\vec{b})$ and $\vec{b}-(\vec{a}.\vec{b})\vec{a}$ is ______.

(a) $\frac{\pi}{2}$ (b) 0
(c) $\frac{\pi}{4}$ (d) $\frac{\pi}{3}$

38. If $\vec{a}, \vec{b}, \vec{c}$ are any three coplanar unit vectors, then ______.

(a) $\vec{a}.(\vec{b}\times\vec{c})=1$ (b) $\vec{a}.(\vec{b}\times\vec{c})=3$
(c) $(\vec{a}\times\vec{b}).\vec{c}=0$ (d) $(\vec{c}\times\vec{a}).\vec{b}=1$

39. Let $\vec{a}, \vec{b}$ and $\vec{c}$ be three vectors. Then scalar triple product $[\vec{a}, \vec{b}, \vec{c}]$ is equal to ______.

(a) $[\vec{b}, \vec{a}, \vec{c}]$ (b) $[\vec{a}, \vec{c}, \vec{b}]$
(c) $[\vec{c}, \vec{b}, \vec{a}]$ (d) $[\vec{b}, \vec{c}, \vec{a}]$

40. The vectors $\overline{AB}=3\hat{i}+5\hat{j}+4\hat{k}$ and $\overline{AC}=5\hat{i}-5\hat{j}+2\hat{k}$ are the sides of a triangle ABC. The length of the median through A is ______.

(a) $\sqrt{13}$ units (b) $2\sqrt{5}$ units
(c) 5 units (d) 10 units

41. If $\vec{a}, \vec{b}, \vec{c}$ are three non-coplanar vectors, then $(\vec{a}+\vec{b}+\vec{c}).[(\vec{a}+\vec{b})\times(\vec{a}\times\vec{c})]$ is ______.

(a) 0 (b) $2[\vec{a}\,\vec{b}\,\vec{c}]$
(c) $-[\vec{a}\,\vec{b}\,\vec{c}]$ (d) $[\vec{a}\,\vec{b}\,\vec{c}]$

42. If the position vectors of the vertices A, B, C of a triangle ABC are $7\hat{i}+10\hat{k}, -\hat{i}+6\hat{j}+6\hat{k}$ and $-4\hat{i}+9\hat{j}+6\hat{k}$ respectively, then triangle is

(a) equilateral.
(b) isosceles.
(c) scalene.
(d) right-angled and isosceles also.

43. If $\left|\vec{a}\times\vec{b}\right|=\left|\vec{a}.\vec{b}\right|$, then the angle between $\vec{a}$ and $\vec{b}$ is ______.

(a) $\frac{\pi}{4}$ (b) $\frac{\pi}{2}$
(c) π (d) $\frac{2\pi}{3}$

44. $\left[\hat{i}\,\hat{k}\,\hat{j}\right]+\left[\hat{k}\,\hat{j}\,\hat{i}\right]+\left[\hat{j}\,\hat{k}\,\hat{i}\right]=$

(a) 1 (b) 3
(c) −3 (d) −1

45. Find the length of the longer diagonal of the parallelogram constructed on $5\vec{a}+2\vec{b}$ and $\vec{a}-3\vec{b}$, if it is given that $|\vec{a}|=2\sqrt{2}, \left|\vec{b}\right|=3$ and angle between $\vec{a}$ and $\vec{b}$ is $\frac{\pi}{4}$.

(a) 15 (b) $\sqrt{113}$
(c) $\sqrt{593}$ (d) $\sqrt{369}$

46. The value of a for which the volume of the parallelepiped formed by the vectors $\vec{i}+a\vec{j}+\vec{k}, \vec{j}+a\vec{k}$ and $a\vec{i}+\vec{k}$ is minimum is ______.

(a) −3 (b) 3
(c) $\frac{1}{\sqrt{3}}$ (d) $-\sqrt{3}$

47. Let $\vec{a}=\hat{i}-\hat{k}, \vec{b}=x\hat{i}+y\hat{j}+(1-x)\hat{k}$ and $\vec{c}=y\hat{i}+x\hat{j}+(1+x-y)\hat{k}$. Then $[\vec{a},\vec{b},\vec{c}]$ depends on

(a) only x. (b) only y.
(c) neither x or y. (d) both x and y.

48. A vector $\vec{a}$ has components $2p$ and 1 with respect to a rectangular Cartesian system. The system is rotated through a certain angle about the origin in the counter-clockwise sense. If $\vec{a}$ has components $p+1$ and 1 with respect to the new system, then _______.

(a) $p=0$ (b) $p=1$ or $p=-\frac{1}{3}$
(c) $p=-1$ or $p=2$ (d) $p=1$ or $p=-1$

49. $(\vec{a}.\hat{i})\hat{i}+(\vec{a}.\hat{j})\hat{j}+(\vec{a}.\hat{k})\hat{k}$ is equal to _______.

(a) $\vec{a}$ (b) $2\vec{a}$
(c) $3\vec{a}$ (d) $\vec{0}$

50. Let $\overrightarrow{OA}=\vec{a}, \overrightarrow{OB}=10\vec{a}+2\vec{b}$, and $\overrightarrow{OC}=\vec{b}$, where O, A and C are non-collinear points. Let p denote the area of the quadrilateral OABC and q denote the area of the parallelogram with OA and OC as adjacent sides. Then $\frac{p}{q}$ is equal to _______.

(a) 4 (b) 6
(c) $\frac{|\vec{a}-\vec{b}|}{2|\vec{a}|}$ (d) $\frac{|\vec{a}+\vec{b}|}{2|\vec{a}|}$

51. If $\vec{a}=\hat{i}+\hat{j}-\hat{k}, \vec{b}=-\hat{i}+\hat{k}, \vec{c}=2\hat{i}+\hat{j}$, the value of λ such that $\vec{a}+\lambda\vec{c}$ is perpendicular to $\vec{b}$ is _______.

(a) 1 (b) –1
(c) 0 (d) None of these.

52. If $a=4\hat{i}+6\hat{j}$ and $b=3\hat{j}+4\hat{k}$, then the component of $\vec{a}$ along $\vec{b}$ is _______.

(a) $\frac{18}{10\sqrt{3}}(3\hat{j}+4\hat{k})$ (b) $\frac{18}{25}(3\hat{j}+4\hat{k})$
(c) $\frac{18}{\sqrt{3}}(3\hat{j}+4\hat{k})$ (d) $3\hat{j}+4\hat{k}$

53. A unit vector a makes an angle $\frac{\pi}{4}$ with z-axis, if $\vec{a}+\hat{i}+\hat{j}$ is a unit vector, then $\vec{a}$ is equal to _______.

(a) $\frac{\hat{i}}{2}+\frac{\hat{j}}{2}+\frac{\hat{k}}{2}$ (b) $\frac{\hat{i}}{2}+\frac{\hat{j}}{2}-\frac{\hat{k}}{\sqrt{2}}$
(c) $-\frac{\hat{i}}{2}-\frac{\hat{j}}{2}+\frac{\hat{k}}{\sqrt{2}}$ (d) $\frac{\hat{i}}{2}-\frac{\hat{j}}{2}-\frac{\hat{k}}{\sqrt{2}}$

54. The position vectors of P and Q are respectively $\vec{a}$ and $\vec{b}$. If R is a point on $\overrightarrow{PQ}$ such that $\overrightarrow{PR}=5\overrightarrow{PQ}$, then the position vector of R is _______.

(a) $5\vec{b}-4\vec{a}$ (b) $5\vec{b}+4\vec{a}$
(c) $4\vec{b}-5\vec{a}$ (d) $4\vec{b}+5\vec{a}$

55. If the points with position vectors $60\vec{i}+9\vec{j}, 40\vec{i}-8\vec{j}$ and $a\vec{i}-52\vec{j}$ are collinear, then $a=$ _______.

(a) –40 (b) –20
(c) 20 (d) 40

56. If the position vectors of A, B and C are respectively $2\vec{i}-\vec{j}+\vec{k}, \vec{i}-3\vec{j}-5\vec{k}$ and $3\vec{i}-4\vec{j}-4\vec{k}$, then $\cos^2 A=$ _______.

(a) 0 (b) $\frac{6}{41}$
(c) $\sqrt{\frac{35}{41}}$ (d) 1

57. Let $\vec{a}$ be the unit vector, $\vec{b}=2\vec{i}+\vec{j}-\vec{k}$ and $\vec{c}=\vec{i}+3\vec{k}$. The maximum value of $[\vec{a}\,\vec{b}\,\vec{c}]$ is _______.

(a) –1 (b) $\sqrt{10}+\sqrt{6}$
(c) $\sqrt{10}-\sqrt{6}$ (d) $\sqrt{59}$

58. If the points whose position vectors are $2\hat{i}+\hat{j}+\hat{k}$, $6\hat{i}-\hat{j}+2\hat{k}$ and $14\hat{i}-5\hat{j}+p\hat{k}$ are collinear, then the value of p is _______.

(a) 2 (b) 4
(c) 6 (d) 8

59. The ratio in which $\hat{i}+2\hat{j}+3\hat{k}$ divides the join of $-2\hat{i}+3\hat{j}+5\hat{k}$ and $7\hat{i}-\hat{k}$ is _______.

(a) 1 : 2 (b) 2 : 3
(c) 3 : 4 (d) 1 : 4

60. If $\vec{a}=\hat{i}-\hat{j}-\hat{k}$ and $\vec{b}=+\lambda\hat{i}-3\hat{j}+\hat{k}$ and the orthogonal projection of $\vec{b}$ on $\vec{a}$ is $\frac{4}{3}(\hat{i}-\hat{j}-\hat{k})$, then λ is equal to _______.

(a) 0 (b) 2
(c) 12 (d) –1

Questions from Previous Years

61. Let a vector $\vec{a}$ be coplanar with vectors $\vec{b}=2\hat{i}+\hat{j}+\hat{k}$ and $\vec{c}=\hat{i}-\hat{j}+\hat{k}$. If $\vec{a}$ is perpendicular to $\vec{d}=3\hat{i}+2\hat{j}+6\hat{k}$ and $|\vec{a}|=\sqrt{10}$. Then a possible value of $[\vec{a}\ \vec{b}\ \vec{c}]+[\vec{a}\ \vec{b}\ \vec{d}]+[\vec{a}\ \vec{c}\ \vec{d}]$ is equal to _______.

(a) –42 (b) –40
(c) –29 (d) –38

62. Let O be the origin. Let $\overrightarrow{OP} = x\hat{i} + y\hat{j} - \hat{k}$ and $\overrightarrow{OQ} = -\hat{i} + 2\hat{j} + 3x\hat{k}, x, y \in R, x > 0$, be such that $|\overrightarrow{PQ}| = \sqrt{20}$ and the vector $\overrightarrow{OP}$ be perpendicular to $\overrightarrow{OQ}$. If $\overrightarrow{OR} = 3\hat{i} + z\hat{j} - 7\hat{k}, z \in R$ is coplanar with $\overrightarrow{OP}$ and $\overrightarrow{OQ}$, then the value of $x^2 + y^2 + z^2$ is equal to ______.

(a) 7 (b) 9
(c) 2 (d) 1

63. Let the position vectors of two points P and Q be $3\hat{i} - \hat{j} + 2\hat{k}$ and $\hat{i} + 2\hat{j} - 4\hat{k}$, respectively. Let R and S be two points such that the direction ratios of lines PQ and QS are $(4, -1, 2)$ and $(-2, 1, -2)$, respectively. Let lines PR and QS intersect at T. If the vector $\overrightarrow{TA}$ is perpendicular to both $\overrightarrow{PR}$ and $\overrightarrow{QS}$ and the length of vector $\overrightarrow{TA}$ is $\sqrt{5}$ units, then the modulus of a position vector of A is ______.

(a) $\sqrt{482}$ (b) $\sqrt{227}$
(c) $\sqrt{5}$ (d) $\sqrt{171}$

64. If vectors $\vec{a}_1 = x\hat{i} - \hat{j} + \hat{k}$ and $\vec{a}_2 = \hat{i} + y\hat{j} + z\hat{k}$ are collinear, then a possible unit vector parallel to the vector $x\hat{i} + y\hat{j} + z\hat{k}$ is ______.

(a) $\frac{1}{\sqrt{3}}(\hat{i} - \hat{j} + \hat{k})$ (b) $\frac{1}{\sqrt{3}}(\hat{i} + \hat{j} - \hat{k})$
(c) $\frac{1}{\sqrt{2}}(\hat{i} - \hat{j})$ (d) $\frac{1}{\sqrt{2}}(-\hat{j} + \hat{k})$

65. Let $\vec{a}, \vec{b}$ and $\vec{c}$ be three vectors mutually perpendicular to each other and have same magnitude. If a vector $\vec{r}$ satisfies $\vec{a} \times \{(\vec{r} - \vec{b}) \times \vec{a}\} + \vec{b} \times \{(\vec{r} - \vec{c}) \times \vec{b}\} + \vec{c} \times \{(\vec{r} - \vec{a}) \times \vec{c}\} = \vec{0}$, then $\vec{r}$ is equal to ______.

(a) $\frac{1}{3}(\vec{a} + \vec{b} + \vec{c})$ (b) $\frac{1}{3}(2\vec{a} + \vec{b} - \vec{c})$
(c) $\frac{1}{2}(\vec{a} + \vec{b} + \vec{c})$ (d) $\frac{1}{2}(\vec{a} + \vec{b} + 2\vec{c})$

66. Let $\vec{a} = \hat{i} + 2\hat{j} - 3\hat{k}$ and $\vec{b} = 2\hat{i} - 3\hat{j} + 5\hat{k}$. If $\vec{r} \times \vec{a} = \vec{b} \times \vec{r}, \vec{r}.(a\hat{i} + 2\hat{j} + \hat{k}) = 3$ and $\vec{r}.(2\hat{i} + 5\hat{j} - \alpha\hat{k}) = 1, \alpha \in R$, then the value of $\alpha + |\vec{r}|^2$ is equal to ______.

(a) 15 (b) 13
(c) 9 (d) 11

67. If $\vec{a}$ and $\vec{b}$ are perpendicular, then $\vec{a} \times (\vec{a} \times (\vec{a} \times (\vec{a} \times \vec{b})))$ is equal to:

(a) $\vec{a} \times \vec{b}$ (b) $\vec{0}$
(c) $\frac{1}{2}|\vec{a}|^4 \vec{b}$ (d) $|\vec{a}|^4 \vec{b}$

68. If the volume of a parallelepiped, whose coterminous edges are given by the vectors $\vec{a} = \hat{i} + \hat{j} + n\hat{k}, \vec{b} = 2\hat{i} + 4\hat{j} - n\hat{k}$ and $\vec{c} = \hat{i} + n\hat{j} + 3\hat{k}\ (n \geq 0)$, is 158 cu. units, then ______.

(a) $\vec{a}.\vec{c} = 17$ (b) $\vec{b}.\vec{c} = 10$
(c) $n = 7$ (d) $n = 9$

69. Let x_0 be the point of local maxima of $f(x) = \vec{a}.(\vec{b} \times \vec{c})$, where $\vec{a} = x\hat{i} - 2\hat{j} + 3\hat{k}$, $\vec{b} = -2\hat{i} + x\hat{j} - \hat{k}$ and $\vec{c} = 7\hat{i} - 2\hat{j} + x\hat{k}$. Then the value of $\vec{a}.\vec{b} + \vec{b}.\vec{c} + \vec{c}.\vec{a}$ at $x = x_0$ is ______.

(a) –4 (b) –30
(c) 14 (d) –22

70. Let the volume of a parallelepiped whose coterminous edges are given by $\vec{u} = \hat{i} + \hat{j} + \lambda\hat{k}, \vec{v} = \hat{i} + \hat{j} + 3\hat{k}$ and $\vec{w} = 2\hat{i} + \hat{j} + \hat{k}$ be 1 cu. unit. If θ be the angle between the edges $\vec{u}$ and $\vec{w}$, then $\cos\theta$ can be ______.

(a) $\frac{7}{6\sqrt{6}}$ (b) $\frac{7}{6\sqrt{3}}$
(c) $\frac{5}{7}$ (d) $\frac{5}{3\sqrt{3}}$

71. Let $\vec{u}$ be a vector coplanar with the vectors $\vec{a} = 2\hat{i} + 3\hat{j} - \hat{k}$ and $\vec{b} = \hat{j} + \hat{k}$. If $\vec{u}$ is perpendicular to $\vec{a}$ and $\vec{u}.\vec{b} - 24$, then $|\vec{u}|^2$ is equal to ______.

(a) 315 (b) 256
(c) 84 (d) 336

72. Let $\vec{a} = 2\hat{i} + \hat{j} - 2\hat{k}$ and $\vec{b} = \hat{i} + \hat{j}$. Let $\vec{c}$ be a vector such that $|\vec{c} - \vec{a}| = 3$, $|(\vec{a} \times \vec{b}) \times \vec{c}| = 3$ and the angle between $\vec{c}$ and $\vec{a} \times \vec{b}$ be 30°. Then $\vec{a}.\vec{c}$ is equal to ______.

(a) $\frac{1}{8}$ (b) $\frac{25}{8}$
(c) 2 (d) 5

73. Let $\vec{a}, \vec{b}$ and $\vec{c}$ be three unit vectors such that $\vec{a} \times (\vec{b} \times \vec{c}) = \frac{\sqrt{3}}{2}(\vec{b} + \vec{c})$. If $\vec{b}$ is not parallel to $\vec{c}$, then the angle between $\vec{a}$ and $\vec{b}$ is ______.

(a) $\frac{2\pi}{3}$ (b) $\frac{5\pi}{6}$
(c) $\frac{3\pi}{4}$ (d) $\frac{\pi}{2}$

74. Let $\vec{a}, \vec{b}$ and $\vec{c}$ be three non-zero vectors such that no two of them are collinear and

$(\vec{a}\times\vec{b})\times\vec{c}=\frac{1}{3}|\vec{b}||\vec{c}||\vec{a}|$. If θ is the angle between vectors $\vec{b}$ and $\vec{c}$, then a value of $\sin\theta$ is ______.

(a) $\frac{2}{3}$ (b) $\frac{-2\sqrt{3}}{3}$

(c) $\frac{2\sqrt{2}}{3}$ (d) $\frac{-\sqrt{2}}{3}$

75. If $[\vec{a}\times\vec{b}\;\vec{b}\times\vec{c}\;\vec{c}\times\vec{a}]=\lambda[\vec{a}\;\vec{b}\;\vec{c}]^2$ then λ is equal to ______.

(a) 0 (b) 1

(c) 2 (d) 3

76. If $|\vec{a}|=2, |\vec{b}|=3$ and $|2\vec{a}-\vec{b}|=5$, then $|2\vec{a}+\vec{b}|$ equals ______.

(a) 17 (b) 7

(c) 5 (d) 1

77. If the vectors $\overline{AB}=3\hat{i}+4\hat{k}$ and $\overline{AC}=5\hat{i}-2\hat{j}+4\hat{k}$ are the sides of a triangle ABC, then the length of the median through A is ______.

(a) $\sqrt{18}$ (b) $\sqrt{72}$

(c) $\sqrt{33}$ (d) $\sqrt{45}$

78. If $\vec{a}$ and $\vec{b}$ are vectors such that $|\vec{a}+\vec{b}|=\sqrt{29}$ and $\vec{a}\times(2\hat{i}+3\hat{j}+4\hat{k})=(2\hat{i}+3\hat{j}+4\hat{k})\times\vec{b}$, then a possible value of $(\vec{a}+\vec{b}).(-7\hat{i}+2\hat{j}+3\hat{k})$ is ______.

(a) 0 (b) 3

(c) 4 (d) 8

79. L$\vec{a}=\hat{i}+\hat{j}+\hat{k}, \vec{b}=\hat{i}-\hat{j}+\hat{k}$ and $\vec{c}=\hat{i}-\hat{j}-\hat{k}$ be three vectors. A vector $\vec{v}$ in the plane of $\vec{a}$ and $\vec{b}$, whose projection on $\vec{c}$ is $\frac{1}{\sqrt{3}}$, is given by ______.

(a) $\hat{i}-3\hat{j}+3\hat{k}$ (b) $-3\hat{i}-3\hat{j}-\hat{k}$

(c) $3\hat{i}-\hat{j}+3\hat{k}$ (d) $\hat{i}+3\hat{j}-3\hat{k}$

80. Let P, Q, R and S be the points on the plane with position vectors $-2\hat{i}-\hat{j}, 4\hat{i}, 3\hat{i}+3\hat{j}$ and $-3\hat{i}+2\hat{j}$ respectively. The quadrilateral PQRS must be a

(a) parallelogram, which is neither a rhombus nor a rectangle.

(b) square.

(c) rectangle, but not a square.

(d) rhombus, but not a square.

81. Two adjacent sides of a parallelogram ABCD are given by

$\overrightarrow{AB}=2\hat{i}+10\hat{j}+11\hat{k}$ and $\overrightarrow{AD}=\hat{i}+2\hat{j}+2\hat{k}$

The sides AD is rotated by an acute angle α in the plane of the parallelogram so that AD becomes AD′. If AD′ makes a right angle with the side AB, then the cosines of the angle α is given by ______.

(a) $\frac{8}{9}$ (b) $\frac{\sqrt{17}}{9}$

(c) $\frac{1}{9}$ (d) $\frac{4\sqrt{5}}{9}$

82. If $\vec{a},\vec{b},\vec{c}$ and $\vec{d}$ are unit vectors such that $(\vec{a}\times\vec{d}).(\vec{c}\times\vec{d})=1$ and $\vec{a}.\vec{c}=\frac{1}{2}$, then

(a) $\vec{a},\vec{b},\vec{c}$ are non-coplanar.

(b) $\vec{b},\vec{c},\vec{d}$ are non-coplanar.

(c) $\vec{b}.\vec{d}$ are non-parallel.

(d) $\vec{a}.\vec{d}$ are parallel and $\vec{b}.\vec{c}$ are parallel

Integer and Numerical Value Type

1. From the following system for unknown vectors $\vec{x}$ and $\vec{y}$ ($\vec{a}$ and $\vec{b}$ are given vectors) $\vec{x}+\vec{y}=\vec{a}, \vec{x}\times\vec{y}=\vec{b}, \vec{x}.\vec{a}=$ ______.

2. If $\vec{a}\times(\vec{b}\times\vec{c})+(\vec{a}.\vec{b})\vec{b}=(4-2\beta-\sin\alpha)\vec{b}+(\beta^2-1)\vec{c}, (\vec{c}.\vec{c})\vec{a}=\vec{c}$ and $\vec{b}\times\vec{c}\neq 0$, then the value of the scalar α, is ______.

3. The position vectors of two points A and C are $9\hat{i}-\hat{j}+7\hat{k}$ and $7\hat{i}-2\hat{j}+7\hat{k}$ respectively. The point of intersection of vectors $AB=4\hat{i}-\hat{j}+3\hat{k}$ and $CD=2\hat{i}-\hat{j}+2\hat{k}$ is P. If vector PQ is perpendicular to AB and CD and PQ = 15 units, then the position vectors of Q is $6\hat{i}-9\hat{j}-\mu\hat{k}$. μ must be equal to ______.

4. Vector $\overline{AB}=3\hat{i}-\hat{j}+\hat{k}$ and $\overline{CD}=-3\hat{i}+2\hat{j}+4\hat{k}$ are not coplanar. The position vectors of points A and C are $6\hat{i}-7\hat{j}+4\hat{k}$ and $-9\hat{j}+2\hat{k}$ respectively. If the position vectors of a point P on the line AB and a point Q on the line CD, such that $\overline{PQ}$ is perpendicular to $\overline{AB}$ and $\overline{CD}$ both, is $3\hat{i}+p\hat{j}+3\hat{k}$, then p must be ______.

Questions from Previous Years

5. Let be two vectors. If $\vec{p}=2\hat{i}+3\hat{j}+\hat{k}$ and $\vec{q}=\hat{i}+2\hat{j}+\hat{k}$, vector $\vec{r}=(\alpha\hat{i}+\beta\hat{j}+\gamma\hat{k})$ is perpendicular to each of the vectors $(\vec{p}+\vec{q})$ and $(\vec{p}-\vec{q})$, and $|\vec{r}|=\sqrt{3}$, then $|\alpha|+|\beta|+|\gamma|$ is equal to ______.

6. For $p>0$, a vector $\vec{v}_2=2\hat{i}+(p+1)\hat{j}$ is obtained by rotating the vector $\vec{v}_1=\sqrt{3}p\hat{i}+\hat{j}$ by an angle θ about the origin in counter-clockwise direction. If $\tan\theta = \frac{(\alpha\sqrt{3}-2)}{(4\sqrt{3}+3)}$, then the value of α is equal to ______.

7. Let $\vec{a} = \hat{i}+\hat{j}+\hat{k}, \vec{b}$ and $\vec{c} = \hat{j}-\hat{k}$ be three vectors such that $\vec{a}\times\vec{b} = \vec{c}$ and $\vec{a}.\vec{b} = 1$. If the length of projection vector of the vector $\vec{b}$ on the vector $\vec{a}\times\vec{c}$ is l, then the value of $3l^2$ is equal to ______.

8. $\vec{a} = \hat{i}+a\hat{j}+3\hat{k}$ and $3\hat{i}-a\hat{j}+\hat{k}$. If the area of the parallelogram whose adjacent sides are represented by the vectors $\vec{a}$ and $\vec{b}$ is $8\sqrt{3}$ square units, then $\vec{a}.\vec{b}$ is equal to ______.

9. A line I passing through the origin is perpendicular to the lines

$$I_1 : \vec{r} = (3+t)\hat{i}+(1+2t)\hat{j}+(4+2t)\hat{k}$$
$$I_2 : \vec{r} = (3+2s)\hat{i}+(3+2s)\hat{j}+(2+s)\hat{k}$$

If the co-ordinates of the point in the first octant on I_2 at a distance of $\sqrt{17}$ from the point of intersection of I and I_1 are (a,b,c) then $18(a+b+c)$ is equal to ______.

10. Let $\vec{a} = \hat{i}+5\hat{j}+\pm\hat{k}, \vec{b} = \hat{i}+3\hat{j}+{}^2\hat{k}$ and $\vec{c} = -\hat{i}+2\hat{j}-3\hat{k}$ be three vectors such that, $|\vec{b}\times\vec{c}| = 5\sqrt{3}$ and $\vec{a}$ be perpendicular to $\vec{b}$. Then the greatest amongst the values of $|\vec{a}|^2$ is ______.

11. If $\vec{a} = 2\hat{i}+\hat{j}+2\hat{k}$, then the value of $\left|\hat{i}\times(\vec{a}\times\hat{i})\right|^2 + \left|\hat{j}\times(\vec{a}\times\hat{j})\right|^2 + \left|\hat{k}\times(\vec{a}\times\hat{k})\right|^2$ is equal to ______.

12. Let $\vec{b}, \vec{b}$ and $\vec{c}$ be three vectors such that $|\vec{a}| = \sqrt{3}, |\vec{b}| = 5, \vec{b}.\vec{c} = 10$ and the angle between $\vec{b}$ and $\vec{c}$ be $\frac{\pi}{3}$. If $\vec{b}$ is perpendicular to the vector $\vec{b}\times\vec{c}$, then $|\vec{a}\times(\vec{b}\times\vec{c})|$ is equal to ______.

13. If $\vec{a}$ and $\vec{b}$ are unit vectors, then the greatest value of $\sqrt{3}|\vec{a}+\vec{b}| + |\vec{a}-\vec{b}|$ is ______.

14. If $\vec{x}$ and $\vec{y}$ be two non-zero vectors such that $|\vec{x}+\vec{y}| = |\vec{x}|$ and $2\vec{x}+\lambda\vec{y}$ is perpendicular to $\vec{y}$, then the value of λ is ______.

15. Let the vectors $\vec{a}, \vec{b}, \vec{c}$ be such that $|\vec{a}| = 2, |\vec{b}| = 4$ and $|\vec{c}| = 4$. If the projection of $\vec{b}$ on $\vec{a}$ is equal to the projection of $\vec{c}$ on $\vec{a}$ and $\vec{b}$ is perpendicular to $\vec{c}$, then the value of $|\vec{a}+\vec{b}-\vec{c}|$ is ______.

16. Let $\vec{a}, \vec{b}$ and $\vec{c}$ be three unit vectors such that $|\vec{a}-\vec{b}|^2 + |\vec{a}-\vec{c}|^2 = 8$. Then $|\vec{a}+2\vec{b}|^2 + |\vec{a}+2\vec{c}|^2$ is equal to ______.

17. The projection of the line segment joining the points $(1,-1,3)$ and $(2,-4,11)$ on the line joining the points $(-1,2,3)$ and $(3,-2,10)$ is ______.

18. Let the position vectors of points A and B be $\hat{i}+\hat{j}+\hat{k}$ and $2\hat{i}+\hat{j}+3\hat{k}$, respectively. A point P divides the line segment AB internally in the ratio $\lambda:1 (\lambda > 0)$. If O is the region and $\overline{\text{OB}}.\overline{\text{OP}} - 3\left|\overline{\text{OA}}\times\overline{\text{OP}}\right|^2 = 6$, then λ is equal to ______.

19. If the vectors $\vec{p} = (a+1)\hat{i}+a\hat{j}+a\hat{k}$, $\vec{q} = a\hat{i}+(a+1)\hat{j}+a\hat{k}$ and $\vec{r} = a\hat{i}+a\hat{j}+(a+1)\hat{k}$ $(a \in \text{R})$ are coplanar and $3(\vec{p}.\vec{q})^2 - \lambda|\vec{r}\times\vec{q}|^2 = 0$, then the value of λ is ______.

20. If $\vec{a}, \vec{b}$ and $\vec{c}$ are unit vectors satisfying $|\vec{a}-\vec{b}|^2 + |\vec{b}-\vec{c}|^2 + |\vec{c}-\vec{a}|^2 = 9$, then $|2\vec{a}+5\vec{b}+5\vec{c}|$ is ______.

21. Let $\vec{a} = -\hat{i}-\hat{k}, \vec{b} = -\hat{i}+\hat{j}$ and $\vec{c} = \hat{i}+2\hat{j}+3\hat{k}$ be three given vectors. If $\vec{r}$ is a vector such that $\vec{r}\times\vec{b} = \vec{c}\times\vec{b}$ and $\vec{r}.\vec{a} = 0$, then the value of $\vec{r}.\vec{b}$ is ______.

22. If $\vec{a}$ and $\vec{b}$ are vectors in space given by $\vec{a} = \frac{\hat{i}-2\hat{j}}{\sqrt{5}}$ and $\vec{b} = \frac{2\hat{i}+\hat{j}+3\hat{k}}{\sqrt{14}}$, then find the value of $(2\vec{a}+\vec{b}).[(\vec{a}\times\vec{b})\times(\vec{a}-2\vec{b})]$.

Assertion and Reason

(a) Statement 1 is True, Statement 2 True; Statement 2 is a correct explanation for Statement 1.

(b) Statement 1 is True, Statement 2 is True; Statement 2 is NOT a correct explanation for Statement 1.

(c) Statement 1is True, Statement 2 is False.

(d) Statement 1 is False, Statement 2 is True.

1. **Statement 1**: If $\vec{a} = 3\hat{i}+\hat{k}, \vec{b} = -\hat{i}+2\hat{j}+\hat{k}$, $\vec{c} = \hat{i}+\hat{j}+\hat{k}$ and $\vec{d} = 2\hat{i}-\hat{j}$, then there exist real numbers α, β, γ such that $\vec{a} = \alpha\vec{b}+\beta\vec{c}+\gamma\vec{d}$.
Statement 2: $\vec{a}, \vec{b}, \vec{c}, \vec{d}$ are four vectors in a 3-dimensional space. If $\vec{b}, \vec{c}, \vec{d}$, are non-coplanar, then there exist real numbers α, β, γ such that $\vec{a} = \alpha\vec{b}+\beta\vec{c}+\gamma\vec{d}$.

2. **Statement 1**: Let $A(\vec{a}), B(\vec{b})$ and $C(\vec{c})$ be three points such that $\vec{a} = 2\hat{i} + \hat{j} + \hat{k}, \vec{b} = 3\hat{i} - \hat{j} + 3\hat{k}$ and $\vec{c} = -\hat{i} + 7\hat{j} - 5\hat{k}$, then OABC is a tetrahedron.
Statement 2: Let $A(\vec{a}), B(\vec{b})$ and $C(\vec{c})$ be three points such that $\vec{a}, \vec{b}$ and $\vec{c}$ are non-coplanar, then OABC is a tetrahedron, where O is the origin.

3. **Statement 1**: Let $\vec{a} = 2\hat{i} + 3\hat{j} - \hat{k}, \vec{b} = 4\hat{i} + 6\hat{j} - 2\hat{k}$, then $\vec{a} \times \vec{b} = \vec{0}$.
Statement 2: If $\vec{a} \neq 0, \vec{b} \neq 0$ and $\vec{a}$ and $\vec{b}$ are non-collinear vectors, then $\vec{a} \times \vec{b} = ab\sin\theta\hat{n}$, where θ is the smaller angle between the vectors $\vec{a}$ and $\vec{b}$ and $\hat{n}$ is a unit vector such that $\vec{a}, \vec{b}, \hat{n}$ taken in this order form a right-handed orientation.

4. **Statement 1**: Let a, b, c and d be position vectors of four points A, B, C and D, and $3a - 2b + 5c - 6d = 0$. Then points A, B, C and D are coplanar.
Statement 2: Three non-zero, linearly dependent co-initial vectors ($\overrightarrow{PQ}, \overrightarrow{PR}$ and $\overrightarrow{PS}$) are coplanar.

5. **Statement 1**: Let $\vec{a} = 3\hat{i} - \hat{j}, \vec{b} = 2\hat{i} + \hat{j} - 3\hat{k}$. If $\vec{b} = \vec{b}_1 + \vec{b}_2$ such that $\vec{b}_1$ is collinear with $\vec{a}$ and $\vec{b}_2$ is perpendicular to $\vec{a}$ is possible, then $\vec{b}_2 = \hat{i} + 3\hat{j} - 3\hat{k}$.
Statement 2: If $\vec{a}$ and $\vec{b}$ are non-zero, non-collinear vectors, then $\vec{b}$ can be expressed as $\vec{b} = \vec{b}_1 + \vec{b}_2$, where $\vec{b}_1$ is collinear with $\vec{a}$ and $\vec{b}_2$ is perpendicular to $\vec{a}$.

JEE ADVANCED

Objective Questions

One Option Correct

1. If $\vec{a}, \vec{b}$ are unit vectors such that $(\vec{a} + \vec{b}).\{(2\vec{a} + 3\vec{b}) \times (3\vec{a} - 2\vec{b})\} = 0$, then angle between $\vec{a}$ and $\vec{b}$ is ______.
(a) 0 (b) $\pi/2$
(c) π (d) indeterminate

2. If $\vec{a}' = \hat{i} + \hat{j}, \vec{b}' = \hat{i} - \hat{j} + 2\hat{k}$ and $\vec{c}' = 2\hat{i} + \hat{j} - \hat{k}$, then altitude of the parallelepiped formed by the vectors $\vec{a}, \vec{b}, \vec{c}$ having base formed by $\vec{b}$ and $\vec{c}$ is ($\vec{a}, \vec{b}, \vec{c}$ and $\vec{a}', \vec{b}', \vec{c}'$ are reciprocal systems of vectors) ______.
(a) 1 (b) $\frac{3\sqrt{2}}{2}$
(c) $\frac{1}{\sqrt{6}}$ (d) $\frac{1}{\sqrt{2}}$

3. A unit vector $\vec{a}$ in the plane of $\vec{b} = 2\hat{i} + \hat{j}$ and $\vec{c} = \hat{i} - \hat{j} + \hat{k}$ is such that $\vec{a}\hat{\ }\vec{b} = \vec{a}\hat{\ }\vec{d}$ where $\vec{d} = \hat{j} + 2\hat{k}$ is ______.
(a) $\frac{\hat{i} + \hat{j} + \hat{k}}{\sqrt{3}}$ (b) $\frac{\hat{i} - \hat{j} + \hat{k}}{\sqrt{3}}$
(c) $\frac{2\hat{i} + \hat{j}}{\sqrt{5}}$ (d) $\frac{2\hat{i} - \hat{j}}{\sqrt{5}}$

4. Let $\vec{r}, \vec{a}, \vec{b}$ and $\vec{c}$ be four non-zero vectors such that $\vec{r}.\vec{a} = 0, |\vec{r} \times \vec{b}| = |\vec{r}||\vec{b}|, |\vec{r} \times \vec{c}| = |\vec{r}||\vec{c}|$, then $[abc] =$ ______.
(a) $|a||b||c|$ (b) $-|a||b||c|$
(c) 0 (d) None of these.

5. If $\vec{a}_1, \vec{a}_2$ and $\vec{a}_3$ are non-coplanar vectors and $(x + y - 3)\vec{a}_1 + (2x - y + 2)\vec{a}_2 + (2x + y + \lambda)\vec{a}_3 = \vec{0}$ holds for some x and y, then λ is ______.
(a) $\frac{7}{3}$ (b) 2
(c) $-\frac{10}{3}$ (d) $\frac{5}{3}$

6. If $\vec{a}, \vec{b}, \vec{c}$ are such that $[\vec{a}, \vec{b}, \vec{c}] = 1, \vec{c} = \lambda\vec{a} \times \vec{b}, \vec{a}\hat{\ }\vec{b} < \frac{2\pi}{3}$ and $|\vec{a}| = \sqrt{2}, |\vec{b}| = \sqrt{3}, |\vec{c}| = \frac{1}{\sqrt{3}}$, then the angle between $\vec{a}$ and $\vec{b}$ is ______.
(a) $\frac{\pi}{6}$ (b) $\frac{\pi}{4}$
(c) $\frac{\pi}{3}$ (d) $\frac{\pi}{2}$

7. If the plane $\vec{r}.(4\hat{i} + 3\hat{j}) + 2 = 0$ is rotated about its line of intersection with the plane $\vec{r}.(\hat{k}) = 0$ through an angle 45°, then the length of perpendicular from the origin on the new position of the plane is ______.
(a) $\frac{1}{5}$ (b) $\frac{2}{5}$
(c) $\frac{\sqrt{2}}{5}$ (d) $\frac{\sqrt{3}}{5}$

8. If $\vec{a} = \hat{i} + \hat{j}, \vec{b} = \hat{j} + \hat{k}, \vec{c} = \hat{k} + \hat{i}$, then in the reciprocal system of the vectors $\vec{a}, \vec{b}, \vec{c}$, the reciprocal $\vec{a}'$ of vector $\vec{a}$ is ______.
   (a) $\frac{\hat{i} + \hat{j} + \hat{k}}{2}$ (b) $\frac{\hat{i} - \hat{j} + \hat{k}}{2}$
   (c) $\frac{-\hat{i} - \hat{j} + \hat{k}}{2}$ (d) $\frac{\hat{i} + \hat{j} - \hat{k}}{2}$

9. If $\vec{a} + 2\vec{b} + 3\vec{c} = \vec{0}$, then $\vec{a} \times \vec{b} + \vec{b} \times \vec{c} + \vec{c} \times \vec{a}$ is equal to ______.
   (a) $6(\vec{b} \times \vec{c})$ (b) $6(\vec{c} \times \vec{a})$
   (c) $6(\vec{a} \times \vec{b})$ (d) None of these.

10. Let $\vec{a}, \vec{b}, \vec{c}$ be three non-coplanar vectors and $\vec{r}$ be any vector in space such that $\vec{r}.\vec{a} = 1, \vec{r}.\vec{b} = 2$ and $\vec{r}.\vec{c} = 3$. If $[\vec{a}\,\vec{b}\,\vec{c}] = 1$. Then $\vec{r}$ is equal to ______.
    (a) $\vec{a} + 2\vec{b} + 3\vec{c}$
    (b) $(\vec{b} \times \vec{c}) + 2(\vec{c} \times \vec{a}) + 3(\vec{a} \times \vec{b})$
    (c) $(\vec{b}.\vec{c})\vec{a} + 2(\vec{c}.\vec{a})\vec{b} + 3(\vec{a}.\vec{b})\vec{c}$
    (d) None of these.

11. In a right-angled ΔABC, the hypotenuse AB $= p$, then $\overrightarrow{AB}.\overrightarrow{AC} + \overrightarrow{BC}.\overrightarrow{BA} + \overrightarrow{CA}.\overrightarrow{CB}$ equals ______.
    (a) $2p^2$ (b) p^2
    (c) $p^2/2$ (d) None of these.

12. If three vectors $\vec{a} = x\vec{i} + y\vec{j} + z\vec{k}, \vec{b} = \vec{j}$ and $\vec{c}$ are such that $\vec{a}, \vec{c}$ and $\vec{b}$ form a right-handed system, then $\vec{c}$ is ______.
    (a) $z\vec{i} - x\vec{k}$ (b) $\vec{0}$
    (c) $y\vec{j}$ (d) $-z\vec{i} + x\vec{k}$

13. S_1: If $\vec{a} \neq 0, \vec{b} \neq 0$ and $\vec{a}, \vec{b}$ are non-collinear, then $\vec{a}, \vec{b}$ and $\vec{a} + \vec{b}$ are coplanar vectors.
    S_2: Vector $3\hat{i} - 4\hat{j} + 7\hat{k}, 4\hat{i} + 7\hat{j} - 2\hat{k}$ and $\hat{i} + 11\hat{j} - 9\hat{k}$ form a triangle.
    S_3: $\left[\vec{a}\ \vec{a} + \vec{b}\ \vec{a} + \vec{b} + \vec{c}\right] = 2\left[\vec{a}\,\vec{b}\,\vec{c}\right]$
    S_4: If $\left(\vec{a} \times \vec{b}\right) \times \vec{c} = \vec{a} \times \left(\vec{b} \times \vec{c}\right) \neq \vec{0}$, then $\vec{c} \times \left(\vec{a} \times \vec{b}\right) = \vec{0}$.
    (a) TTFF (b) FFTT
    (c) FTFT (d) TTTF

14. S_1: If $\lambda \in R, \lambda \neq 0$ and $\vec{a} \neq \vec{0}$, then vectors $\vec{a}$ and $\lambda\vec{a}$ are non-parallel vectors.
    S_2: Minimum value of $\vec{a}.\vec{b}$ is 0.
    S_3: If $\vec{a}$ and $\vec{b}$ are non-collinear vectors, then $\left|\left(\vec{a} + \vec{b}\right) \times \left(\vec{a} - \vec{b}\right)\right|$ is equal to twice the area of the parallelogram formed by $\vec{a}$ and $\vec{b}$.
    S_4: $\left(\vec{a} \times \vec{b}\right)^2 = \begin{vmatrix} \vec{a}.\vec{a} & \vec{b}.\vec{a} \\ \vec{a}.\vec{b} & \vec{b}.\vec{b} \end{vmatrix}$
    (a) TTFF (b) FFTT
    (c) FTFT (d) TTTF

15. S_1: $|\vec{a}| = |\vec{b}| = |\vec{a} - \vec{b}| = 1$, then angle between $\vec{a}$ and $\vec{b}$ is $\frac{\pi}{2}$.
    S_2: $\frac{\vec{a} + \vec{b}}{2}$ is a vector in the direction of angle bisector of vectors $\vec{a}$ and $\vec{b}$.
    S_3: $\left(\vec{a}\,.\,\hat{i}\right)^2 + \left(\vec{a}\,.\hat{j}\right)^2 + \left(\vec{a}\,.\,\hat{k}\right)^2$ is equal to $\vec{a}^2$.
    S_4: The vector perpendicular to $-\hat{i} + \hat{j} + \hat{k}$ and coplanar with $\hat{i} + \hat{j} + \hat{k}$ and $-\hat{i} + \hat{j} + \hat{k}$ is ____.
    (a) TFTF (b) FTFT
    (c) FTTF (d) FFTF

16. S_1: If $\vec{a}, \vec{b}$ and $\vec{c}$ are three non-coplanar vectors and $\vec{d}.\vec{a} = \vec{d}.\vec{b} = \vec{d}.\vec{c} = 0$, then $\vec{d}$ is a vector.
    S_2: If $\vec{a}, \vec{b}$ and $\vec{c}$ are three non-coplanar vectors, then the vector $\vec{a} \times \vec{b} + \vec{b} \times \vec{c} + \vec{c} \times \vec{a}$ and $\vec{a} + \vec{b} + \vec{c}$ are perpendicular to each other.
    S_3: If $\vec{a}$ and $\vec{c}$ are perpendicular to each other, then the vectors $\vec{a} \times (\vec{b} \times \vec{c})$ and $(\vec{a} \times \vec{b}) \times \vec{c}$ are also perpendicular to each other.
    S_4: If $\vec{a}, \vec{b}, \vec{c}$ and $\vec{d}$ are coplanar vectors, then $(\vec{a} \times \vec{b}) \times (\vec{c} \times \vec{d}) = \vec{0}$.
    (a) TTFF (b) FFTT
    (c) FTFT (d) TFTT

One or More Than One Option Correct

1. If vectors $\vec{a}$ and $\vec{b}$ are non-collinear, then $\frac{\vec{a}}{|\vec{a}|} + \frac{\vec{b}}{|\vec{b}|}$ is
   (a) a unit vector.
   (b) in the plane of $\vec{a}$ and $\vec{b}$.
   (c) equally inclined to $\vec{a}$ and $\vec{b}$.
   (d) perpendicular to $\vec{a} \times \vec{b}$.

2. A, B, C and D are four points such that $\overrightarrow{AB} = m(2\hat{i} - 6\hat{j} + 2\hat{k}), \overrightarrow{BC} = m(\hat{i} - 2\hat{j})$ and $\overrightarrow{CD} = n(-6\hat{i} + 15\hat{j} - 3\hat{k})$. If CD intersects AB at some point E, then ______.
   (a) $m \geq 1/2$ (b) $n \geq 1/3$
   (c) $m = n$ (d) $m < n$

3. If in ΔABC, $\overline{AB} = \frac{\vec{u}}{|\vec{u}|} - \frac{\vec{v}}{|\vec{v}|}$ and $\overline{AC} = \frac{2\vec{u}}{|\vec{u}|}$, where $|\vec{u}| \neq |\vec{v}|$, then
   (a) $1 + \cos 2A + \cos 2B + \cos 2C = 0$.
   (b) $\sin A = \cos C$.

(c) projection of AC on BC equal to BC.
(d) projection of AB on BC is equal to AB.

4. If vectors $\vec{b} = \left(\tan\alpha, -1, 2\sqrt{\sin\alpha/2}\right)$ and $\vec{c} = \left(\tan\alpha, \tan\alpha - \frac{3}{\sqrt{\sin\alpha/2}}\right)$ are orthogonal and vector $\vec{a} = (1, 3, \sin 2\alpha)$ makes an obtuse angle with the z-axis, the value of α is ______.
(a) $\alpha = (4n+1)\pi + \tan^{-1}2$
(b) $\alpha = (4n+1)\pi - \tan^{-1}2$
(c) $\alpha = (4n+2)\pi + \tan^{-1}2$
(d) $\alpha = (4n+2)\pi - \tan^{-1}2$

5. Let P_1 denote the equation of a plane to which the vector $(\hat{i}+\hat{j})$ is normal and which contains the line whose equation is $\vec{r} = \hat{i}+\hat{j}+\hat{k}+\lambda(\hat{i}-\hat{j}-\hat{k})$ and P_2 denote the equation of the plane containing the line L and a point with position vector $\hat{j}$. Which of the following holds good?
(a) The equation of P_1 is $x + y = 2$.
(b) The equation of P_2 is $\vec{r}.(\hat{i}-2\hat{j}+\hat{k}) = 2$.
(c) The acute angle between P_1 and P_2 is $\cot^{-1}(\sqrt{3})$.
(d) The angle between the plane P_2 and the line L is $\tan^{-1}\sqrt{3}$.

6. If $\vec{a}$ and $\vec{b}$ unequal unit vectors such that $(\vec{a}-\vec{b})\times[(\vec{b}+\vec{a})\times(2\vec{a}+\vec{b})] = \vec{a}+\vec{b}$, then smaller angle θ between $\vec{a}$ and $\vec{b}$ is ______.
(a) $\frac{\pi}{2}$ (b) 0
(c) π (d) $\frac{\pi}{4}$

7. In a four-dimensional space where unit vectors along axes are $\hat{i}, \hat{j}, \hat{k}$ and $\hat{\ell}$ and $\vec{a}_1, \vec{a}_2, \vec{a}_3, \vec{a}_4$ are four non-zero vectors such that no vector can be expressed as linear combination of others and $(\lambda-1)(\vec{a}_1-\vec{a}_2)+\mu(\vec{a}_2+\vec{a}_3)+\gamma(\vec{a}_3+\vec{a}_4-2\vec{a}_2)+\vec{a}_3+\delta\vec{a}_4 = \vec{0}$, then ______.
(a) $\lambda = 1$ (b) $\mu = -\frac{2}{3}$
(c) $\gamma = \frac{2}{3}$ (d) $\delta = \frac{1}{3}$

8. A vector $(\vec{d})$ is equally inclined to three vectors $\vec{a} = \hat{i}-\hat{j}+\hat{k}, \vec{b} = 2\hat{i}+\hat{j}$ and $\vec{c} = 3\hat{j}-2\hat{k}$. Let $\vec{x}, \vec{y}, \vec{z}$ be three vectors in the plane of $\vec{a}, \vec{b}; \vec{b}, \vec{c} : \vec{c}, \vec{a}$ respectively. Then, ______.
(a) $\vec{x}.\vec{d} = 14$
(b) $\vec{y}.\vec{d} = 3$
(c) $\vec{z}.\vec{d} = 0$
(d) $\vec{r}.\vec{d} = 0$ where $\vec{r} = \lambda\vec{x}+\mu\vec{y}+\delta\vec{z}$

9. Identify the statement(s) which is/are incorrect?
(a) $\vec{a}\times\left[\vec{a}\times\left(\vec{a}\times\vec{b}\right)\right] = \left(\vec{a}\times\vec{b}\right)\left(\vec{a}^2\right)$
(b) If $\vec{a}, \vec{b}, \vec{c}$ are non-zero, non-coplanar vectors, and $\vec{v}.\vec{a} = \vec{v}.\vec{b} = \vec{v}.\vec{c} = 0$, then $\vec{v}$ must be null vector.
(c) If $\vec{a}$ and $\vec{b}$ lie in a plane normal to the plane containing vectors $\vec{a}\times\vec{b}, \vec{c}\times\vec{d}; \vec{a}, \vec{b}, \vec{c}, \vec{d}$ are non-zero vector, then $\left(\vec{a}\times\vec{b}\right)\times\left(\vec{c}\times\vec{d}\right) = \vec{0}$
(d) If $\vec{a}, \vec{b}, \vec{c}$ and $\vec{a}', \vec{b}', \vec{c}'$ are reciprocal system of vectors, then $\vec{a}.\vec{b}' + \vec{b}.\vec{c}' + \vec{c}.\vec{a}' = 3$

10. $\vec{a}, \vec{b}, \vec{c}$ are three co-planar unit vectors such that $\vec{a}+\vec{b}+\vec{c} = 0$. If three vectors $\vec{p}, \vec{q}, \vec{r}$ are parallel to $\vec{a}, \vec{b}, \vec{c}$ respectively and have integral but different magnitudes, then among the following options, $|\vec{p}+\vec{q}+\vec{r}|$ can take a value equal to ______.
(a) 1 (b) 0
(c) $\sqrt{3}$ (d) 2

11. Let a and b be positive real numbers. Suppose $\overrightarrow{PQ} = a\hat{i}+b\hat{j}$ and $\overrightarrow{PS} = a\hat{i}-b\hat{j}$ are adjacent sides of a parallelogram PQRS. Let $\vec{u}$ and $\vec{v}$ be the projection vectors of $\vec{w} = \hat{i}+\hat{j}$ along $\overrightarrow{PQ}$ and $\overrightarrow{PS}$, respectively. If $|\vec{u}|+|\vec{v}| = |\vec{w}|$ and if the area of the parallelogram PQRS is 8, then which of the following statements is/are TRUE?
(a) $a + b = 4$
(b) $a - b = 2$
(c) The length of the diagonal PR of the parallelogram PQRS is 4.
(d) $\vec{w}$ is an angle bisector of the vectors $\overrightarrow{PQ}$ and $\overrightarrow{PS}$

Questions from Previous Years

12. Let O be the origin. Let $\overrightarrow{OA} = 2\hat{i}+2\hat{j}+\hat{k}, \overrightarrow{OB} = \hat{i}-2\hat{j}+2\hat{k}$ and $\overrightarrow{OC} = \frac{1}{2}\left(\overrightarrow{OB}-\lambda\overrightarrow{OA}\right)$ for some $\lambda > 0$. If $\left|\overrightarrow{OB}\times\overrightarrow{OA}\right| = \frac{9}{2}$, then which of the following statements is(are) TRUE?
(a) Projection of $\overrightarrow{OC}$ on $\overrightarrow{OA}$ is $-\frac{3}{2}$.
(b) Area of the triangle OAB is $\frac{9}{2}$.

(c) Area of the triangle ABC is $\frac{9}{2}$.

(d) The acute angle between the diagonals of the parallelogram with adjacent sides $\overrightarrow{OA}$ and $\overrightarrow{OC}$ is $\frac{\pi}{3}$.

13. Let ΔPQR be a triangle. Let $\vec{a} = \overrightarrow{QR}, \vec{b} = \overrightarrow{RP}$ and $\vec{c} = \overrightarrow{PQ}$. If $|\vec{a}| = 12, |\vec{b}| = 4\sqrt{3}, \vec{b}.\vec{c} = 24$, then which of the following is (are) true?

(a) $\frac{|\vec{c}|^2}{2} - |\vec{a}| = 12$ (b) $\frac{|\vec{c}|^2}{2} + |\vec{a}| = 30$

(c) $|\vec{a}\times\vec{b} + \vec{c}\times\vec{a}| = 48\sqrt{3}$ (d) $\vec{a}.\vec{b} = -72$

14. Let $\vec{x}, \vec{y}$ and $\vec{z}$ be three vectors, each of magnitude $\sqrt{2}$ and the angle between each pair of them be $\frac{\pi}{3}$. If $\vec{a}$ is a non-zero vector perpendicular to $\vec{x}$ and $\vec{y}\times\vec{z}$ and $\vec{b}$ is a non-zero vector perpendicular to $\vec{y}$ and $\vec{z}\times\vec{x}$, then ________.

(a) $\vec{b} = (\vec{b}.\vec{z})(\vec{z}-\vec{x})$ (b) $\vec{a} = (\vec{a}.\vec{y})(\vec{y}-\vec{z})$

(c) $\vec{a}.\vec{b} = -(\vec{a}.\vec{y})(\vec{b}.\vec{z})$ (d) $\vec{a} = -(\vec{a}.\vec{y})(\vec{z}-\vec{y})$

15. The vector(s) which is/are coplanar with vectors $\hat{i}+\hat{j}+2\hat{k}$ and $\hat{i}+2\hat{j}+\hat{k}$, and perpendicular to the vector $\hat{i}+\hat{j}+\hat{k}$ is/are ________.

(a) $\hat{j}-\hat{k}$ (b) $-\hat{i}+\hat{j}$

(c) $\hat{i}-\hat{j}$ (d) $-\hat{j}+\hat{k}$

Matrix Type

1. Match Column I with Column II.

Column I	Column II
(A) If $\vec{a}+\vec{b} = \hat{j}$ and $2\vec{a}-3\vec{b} = 3\hat{i}+\frac{\hat{j}}{2}$, then modulus of cosine of the angle between $\vec{a}$ and $\vec{b}$ is	(P) 1
(B) If $\lvert\vec{a}\rvert = \lvert\vec{b}\rvert = \lvert\vec{c}\rvert$, the angle between each pair of vectors is $\frac{\pi}{3}$ and $\lvert\vec{a}+\vec{b}+\vec{c}\rvert = \sqrt{6}$, then $\lvert\vec{a}\rvert =$	(Q) $5\sqrt{3}$
(C) Area of the parallelogram whose diagonals represent the vectors $3\hat{i}+\hat{j}-2\hat{k}$ and $\hat{i}-3\hat{j}+4\hat{k}$ is	(R) 2
(D) If $\vec{a}$ is perpendicular to $\vec{b}+\vec{c}, \vec{b}$ is perpendicular to $\vec{c}+\vec{a}, \vec{c}$ is perpendicular to $\vec{a}+\vec{b}, \lvert\vec{a}\rvert = 2, \lvert\vec{b}\rvert = 3$ and $\lvert\vec{c}\rvert = 6$, then $\lvert\vec{a}+\vec{b}+\vec{c}\rvert =$	(S) 7
	(T) $\frac{3}{5}$

2. Match Column I with Column II.

Column I	Column II
(A) The area of the triangle whose vertices are the points with rectangular Cartesian co-ordinates $(1,2,3),(-2,1,-4),(3,4,-2)$ is	(P) 0
(B) The value of $(\vec{a}\times\vec{b}).(\vec{c}\times\vec{d})+(\vec{b}\times\vec{c}).(\vec{a}\times\vec{d})+(\vec{c}\times\vec{a}).(\vec{b}\times\vec{d})$ is	(Q) 28
(C) A square PQRS of side length P is folded along the diagonal PR so that the point Q reaches at Q' and planes PRQ' and PRS are perpendicular to one another, the shortest distance between PQ' and RS is $\frac{kP}{\sqrt{6}}$, then $k =$	(R) 2
(D) $\vec{a} = 2\hat{i}+3\hat{j}-\hat{k}, \vec{b} = -\hat{i}+2\hat{j}-4\hat{k}, \vec{c} = \hat{i}+\hat{j}+\hat{k}$ and $\vec{d} = 3\hat{i}+2\hat{j}+\hat{k}$, then $(\vec{a}\times\vec{b}).(\vec{c}\times\vec{d}) =$	(S) $\frac{\sqrt{1218}}{2}$
	(T) 21

Questions from Previous Years

3. Match Column I with Column II.

Column I	Column II
(A) In a triangle ΔXYZ, let a, b, and c be the lengths of the sides opposite to the angles X, Y and Z, respectively. If $2(a^2-b^2) = c^2$ and $\lambda = \frac{\sin(X-Y)}{\sin Z}$, then possible values of n for which $\cos(n\pi\lambda) = 0$ is (are)	(P) 1
(B) In a triangle ΔXYZ, let a, b and c be the lengths of the sides opposite to the angles X, Y and Z respectively. If $1+\cos 2X - 2\cos 2Y = 2\sin X \sin Y$ then the possible value(s) of $\frac{a}{b}$ is (are)	(Q) 2

(*Continued*)

Column I	Column II
(C) In R^2, let $\sqrt{3}\hat{i}+\hat{j}, \hat{i}+\sqrt{3}\hat{j}$ and $\beta\hat{i}+(1-\beta)\hat{j}$ be the position vectors of X, Y and Z with respect to the origin O, respectively. If the distance of Z from the bisector of the acute angle of $\overrightarrow{OX}$ with $\overrightarrow{OY}$ is $\frac{3}{\sqrt{2}}$, then the possible value(s) of $\lvert\beta\rvert$ is (are)	(R) 3
(D) Suppose that $F(\alpha)$ denotes the area of the region bounded by $x=0$, $x=2$, $y^2=4x$ and $y=\lvert\alpha x-1\rvert+\lvert\alpha x-2\rvert+\alpha x$, where $\alpha\in\{0,1\}$. Then the value(s) of $F(\alpha)+\frac{8}{3}\sqrt{2}$, when $\alpha=0$ and $\alpha=1$, is (are)	(S) 5
	(T) 6

4. Match Column I with Column II.

Column I	Column II
(A) In R^2, if the magnitude of the projection vector of the vector $\alpha\hat{i}+\beta\hat{j}$ on $\sqrt{3}\hat{i}+\hat{j}$ is $\sqrt{3}$ and if $\alpha=2+\sqrt{3}\beta$, then the possible value of $\lvert\alpha\rvert$ is/are	(P) 1
(B) Let a and b be real numbers such that the function $f(x)=\begin{cases}-3ax^2-2, & x<1\\ bx+a^2, & x\geq 1\end{cases}$ is differentiable for all $x\in$ R. Then possible value of a is (are)	(Q) 2
(C) Let $\omega\neq 1$ be a complex cube root of unity. If $(3-3\omega+2\omega^2)^{4n+3}+(3-3\omega+2\omega^2)^{4n+3}+$ then possible value(s) of n is (are)	(R) 3
(D) Let the harmonic mean of two positive real numbers a and b be 4. If q is a positive real number such that $a, 5, q, b$ is an arithmetic progression, then the value(s) of $\lvert q-a\rvert$ is(are)	(S) 4
	(T) 5

5. Match Column I with Column II.

Column I	Column II
(A) Let $y(x)=\cos(3\cos^{-1}x), x\in[-1,1]$, $x\neq\pm\frac{\sqrt{3}}{2}$. Then $\frac{1}{y(x)}\left\{(x^2-1)\frac{d^2y(x)}{dx^2}+x\frac{dy(x)}{dx}\right\}$ equals	(P) 1
(B) Let $A_1, A_2, \ldots, A_n (n>2)$ be the vertices of a regular polygon of n sides with its centre at the origin. Let $\vec{a}_k$ be the position vector of the point $A_k, k=1,2,\ldots,n$. If $\left\lvert\sum_{k=1}^{n-1}(\vec{a}_k\times\vec{a}_{k+1})\right\rvert=\left\lvert\sum_{k=1}^{n-1}(\vec{a}_k.\vec{a}_{k+1})\right\rvert$, Then the minimum value of n is	(Q) 2
(C) If the normal from the point $P(h,1)$ on the ellipse $\frac{x^2}{6}+\frac{y^2}{3}=1$ is perpendicular to the line $x+y=8$, then the value of h is	(R) 8
(D) The number of positive solutions satisfying the equation $\tan^{-1}\left(\frac{1}{2x+1}\right)+\tan^{-1}\left(\frac{1}{4x+1}\right)=\tan^{-1}\left(\frac{2}{x^2}\right)$ is	(S) 9

6. Match Column I with Column II.

Column I	Column II
(A) The volume of a parallelepiped determined by vectors $\vec{a}, \vec{b}$ and $\vec{c}$ is 2. Then the volume of the parallelepiped determined by vectors $2(\vec{a}\times\vec{b}), 3(\vec{b}\times\vec{c})$ and $2(\vec{c}\times\vec{a})$ is	(P) 100
(B) The volume of a parallelepiped determined by vectors $\vec{a}, \vec{b}$ and $\vec{c}$ is 5. Then the volume of the parallelepiped determined by vectors $3(\vec{a}+\vec{b}), 3(\vec{b}+\vec{c})$ and $2(\vec{c}+\vec{a})$ is	(Q) 30

(Continued)

Column I	Column II
(C) The area of a triangle with adjacent sides determined by vectors $\vec{a}$ and $\vec{b}$ is 20. Then the area of the triangle with adjacent sides determined by vectors $(2\vec{a}+3\vec{b})$ and $(\vec{a}-\vec{b})$ is	(R) 24
(D) The area of a parallelogram with adjacent sides determined by vectors $\vec{a}$ and $\vec{b}$ is 30. Then the area of the parallelogram with adjacent sides determined by vectors $(\vec{a}+\vec{b})$ and $\vec{a}$ is	(S) 60

7. Match the statements given in Column I with the values given in Column II.

Column I	Column II
(A) If $\vec{a}=\hat{j}+\sqrt{3}\hat{k}, \vec{b}=-\hat{j}+\sqrt{3}\hat{k}$ and $\vec{c}=2\sqrt{3}\hat{k}$ form a triangle, then the internal angle of the triangle between $\vec{a}$ and $\vec{b}$ is	(P) $\frac{\pi}{6}$
(B) If $\int_a^b (f(x)-3x)dx = a^2-b^2$, then the value of $f\left(\frac{\pi}{6}\right)$ is	(Q) $\frac{2\pi}{3}$
(C) The value of $\frac{\pi^2}{\ell n 3}\int_{7/6}^{5/6} \sec(\pi x)dx$ is	(R) $\frac{\pi}{3}$
(D) The maximum value of $\left\|\text{Arg}\left(\frac{1}{1-z}\right)\right\|$ for $\|z\|=1, z\neq 1$ is given by	(S) π
	(T) $\frac{\pi}{2}$

Integer and Numerical Value Type

1. If V_1, V_2, V_3 are volumes of parallelepiped, triangular prism and tetrahedral prism and tetrahedron respectively the three coterminous edges of all three figures are the vectors ________.

2. If $\vec{p},\vec{q},\vec{r}$ are mutually perpendicular vectors, where $|\vec{p}|=|\vec{q}|=|\vec{r}|$ and $\vec{p}\times\{(\vec{x}-\vec{q})\times\vec{p}\}+\vec{q}\times\{(\vec{x}-\vec{r})\times\vec{q}\}+\vec{r}\times\{(\vec{x}-\vec{p})\times\vec{r}\}=\vec{0}$ and $\vec{x}=\frac{1}{\lambda}(\vec{p}+\vec{q}+\vec{r})$, then find the value of λ.

3. In a regular tetrahedron, let θ be the angle between any edge and a face not containing the edge. If $\cos^2\theta=\frac{a}{b}$ where $a,b\in I^+$, also a and b are coprime, then find the value of $10a+b$.

Questions from Previous Years

4. Let $\vec{u},\vec{v}$ and $\vec{w}$ be vectors in three-dimensional space, where $\vec{u}$ and $\vec{v}$ are unit vectors which are not perpendicular to each other and $\vec{u}.\vec{w}=1, \vec{v}.\vec{w}=1, \vec{w}.\vec{w}=4$. If the volume of the parallelepiped, whose adjacent sides are represented by the vectors $\vec{u},\vec{v}$ and $\vec{w}$ is $\sqrt{2}$, then the value of $|3\vec{u}+5\vec{v}|$ is ________.

5. Let $\vec{a}=2\hat{i}+\hat{j}-\hat{k}$ and $\vec{b}=\hat{i}+2\hat{j}+\hat{k}$ be two vectors.
Consider a vector $\vec{c}=\alpha\hat{a}+\beta\hat{b}, \alpha,\beta\in \text{R}$. If the projection of $\vec{c}$ on the vector $(\vec{a}+\vec{b})$ is $3\sqrt{2}$, then the minimum value of $(\vec{c}-(\vec{a}\times\vec{b})).\vec{c}$ equals ________.

6. Let $\vec{a}$ and $\vec{b}$ be two unit vectors such that $\vec{a}.\vec{b}=0$. For some $x, y\in\mathbb{R}$, let $\vec{c}=x\vec{a}+y\vec{b}+(\vec{a}\times\vec{b})$. If $|\vec{c}|=2$ and the vector $\vec{c}$ is inclined at the same angle α to both $\vec{a}$ and $\vec{b}$, then the value of $8\cos^2\alpha$ is ________.

7. Suppose that $\vec{p},\vec{q}$ and $\vec{r}$ are three non-coplanar vectors in $\mathbb{R}^3$. Let the components of a vector $\vec{s}$ along $\vec{p},\vec{q}$ and $\vec{r}$ be 4, 3 and 5, respectively. If the components of this vector $\vec{s}$ along $(-\vec{p}+\vec{q}+\vec{r}),(\vec{p}-\vec{q}+\vec{r})$ and $(-\vec{p}-\vec{q}+\vec{r})$ are x, y and z, respectively, then the value of $2x+y+z$ is ________.

8. Let $\vec{a},\vec{b}$ and $\vec{c}$ be three non-coplanar unit vectors such that the angle between every pair of them is $\frac{\pi}{3}$. If $\vec{a}\times\vec{b}+\vec{b}\times\vec{c}=p\vec{a}+q\vec{b}+r\vec{c}$, where p,q and r are scalars, then the value of $\frac{p^2+2q^2+r^2}{q^2}$ is ________.

9. Consider the set of eight vectors $\text{V}=\{a\hat{i}+b\hat{j}+c\hat{k}: a,b,c\in\{-1,1\}\}$. Three non-coplanar vectors can be chosen from V in 2^p ways. Then p is ________.

Comprehension Passage

Comprehension 1

Let $\vec{a} = 2\hat{i} + 3\hat{j} - 6\hat{k}$, $\vec{b} = 2\hat{i} - 3\hat{j} + 6\hat{k}$ and $\vec{c} = -2\hat{i} + 3\hat{j} + 6\hat{k}$. Let $\vec{a}_1$ be the projection of $\vec{a}$ on $\vec{b}$ and $\vec{a}_2$ be the projection of $\vec{a}_1$ on $\vec{c}$, then

1. $\vec{a}_2 =$ ________.

(a) $\frac{943}{49}\left(2\hat{i} - 3\hat{j} - 6\hat{k}\right)$ (b) $\frac{943}{49^2}\left(2\hat{i} - 3\hat{j} - 6\hat{k}\right)$

(c) $\frac{943}{49}\left(-2\hat{i} + 3\hat{j} + 6\hat{k}\right)$ (d) $\frac{943}{49^2}\left(-2\hat{i} + 3\hat{j} + 6\hat{k}\right)$

2. $\vec{a}_1.\vec{b} =$ ________.

(a) −41 (b) $-\frac{41}{7}$
(c) 41 (d) 287

3. Which of the following is true?

(a) $\vec{a}$ and $\vec{a}_2$ are collinear.
(b) $\vec{a}_1$ and $\vec{c}$ are collinear.
(c) $\vec{a}, \vec{a}_1, \vec{b}$ are coplanar.
(d) $\vec{a}, \vec{a}_1, \vec{a}_2$ are coplanar.

Comprehension 2

ABCD is a parallelogram. L is a point on BC which divides BC in the ratio 1 : 2. AL intersects BD at P. M is a point on DC which divides DC in the ratio 1 : 2 and AM intersects BC at Q.

1. Point P divided AL in the ratio ________.

(a) 1 : 2 (b) 1 : 3
(c) 3 : 1 (d) 2 : 1

2. Point Q divides DB in the ratio ________.

(a) 1 : 2 (b) 1 : 3
(c) 3 : 1 (d) 2 : 1

3. PQ : DB = ________.

(a) 2 : 3 (b) 1 : 3
(c) 1 : 2 (d) 3 : 4

Comprehension 3

Three vectors $\hat{a}, \hat{b}$ and $\hat{c}$ are forming a right-handed system, if $\hat{a} \times \hat{b} = \hat{c}, \hat{b} \times \hat{c} = \hat{a}, \hat{c} \times \hat{a} = \hat{b}$, then answer the following questions.

1. If vector $3\hat{a} - 2\hat{b} + 2\hat{c}$ and $-\hat{a} - 2\hat{c}$ are adjacent sides of a parallelogram, then an angle between the diagonals is ________.

(a) $\frac{\pi}{4}$ (b) $\frac{\pi}{3}$
(c) $\frac{\pi}{2}$ (d) $\frac{2\pi}{3}$

2. If $\vec{x} = \hat{a} + \hat{b} - \hat{c}, \vec{y} = -\hat{a} + \hat{b} - 2\hat{c}, \vec{z} = -\hat{a} + 2\hat{b} - \hat{c}$, then a unit vector normal to the vectors $\vec{x} + \vec{y}$ and $\vec{y} + \vec{z}$ is ________.

(a) $\vec{a}$ (b) $\vec{b}$
(c) $\vec{c}$ (d) None of these.

3. Vectors $2\hat{a} - 3\hat{b} + 4\hat{c}, \hat{a} + 2\hat{b} - \hat{c}$ and $x\hat{a} - \hat{b} + 2\hat{c}$ are coplanar, then $x =$ ________.

(a) $\frac{8}{5}$ (b) $\frac{5}{8}$
(c) 0 (d) 1

4. Let $\vec{x} = \vec{a} + \vec{b}, \vec{y} = 2\vec{a} - \vec{b}$, then the point of intersection of straight lines $\vec{r} \times \vec{x} = \vec{y} \times \vec{x}, \vec{r} \times \vec{y} = \vec{x} \times \vec{y}$ is ________.

(a) $2\vec{b}$ (b) $3\vec{b}$
(c) $3\vec{a}$ (d) $2\vec{a}$

5. $\hat{a}.\left(\hat{b} \times \hat{c}\right) + \hat{b}.\left(\hat{c} \times \hat{a}\right) + \hat{c}.\left(\hat{a} \times \hat{b}\right)$ is equal to ________.

(a) 1 (b) 3
(c) 0 (d) −12

ANSWER KEY

JEE Main

Objective Questions

One Option Correct

1. (c)	**2.** (c)	**3.** (d)	**4.** (a)	**5.** (c)	**6.** (a)	**7.** (a)	**8.** (c)	**9.** (b)
10. (c)	**11.** (c)	**12.** (d)	**13.** (d)	**14.** (c)	**15.** (d)	**16.** (b)	**17.** (d)	**18.** (c)
19. (d)	**20.** (b)	**21.** (d)	**22.** (b)	**23.** (b)	**24.** (b)	**25.** (d)	**26.** (a)	**27.** (d)
28. (d)	**29.** (c)	**30.** (d)	**31.** (b)	**32.** (d)	**33.** (c)	**34.** (a)	**35.** (c)	**36.** (c)
37. (a)	**38.** (c)	**39.** (d)	**40.** (c)	**41.** (c)	**42.** (c)	**43.** (a)	**44.** (a)	**45.** (a)

46. (c) **47.** (c) **48.** (b) **49.** (a) **50.** (b) **51.** (b) **52.** (b) **53.** (c) **54.** (a)
55. (a) **56.** (c) **57.** (d) **58.** (b) **59.** (a) **60.** (b) **61.** (a) **62.** (b) **63.** (d)
64. (a) **65.** (c) **66.** (a) **67.** (d) **68.** (b) **69.** (d) **70.** (b) **71.** (d) **72.** (c)
73. (b) **74.** (c) **75.** (b) **76.** (c) **77.** (c) **78.** (c) **79.** (c) **80.** (a) **81.** (b)
82. (c)

Integer and Numerical Value Type

1. (1) **2.** (1) **3.** (9) **4.** (8) **5.** (3) **6.** (6) **7.** (2) **8.** (2) **9.** (44)
10. (90) **11.** (18) **12.** (30) **13.** (4) **14.** (1) **15.** (6) **16.** (2) **17.** (8) **18.** (0.8)
19. (1) **20.** (3) **21.** (9) **22.** (5)

Assertion and Reason

1. (b) **2.** (d) **3.** (b) **4.** (a) **5.** (d)

JEE Advanced

Objective Questions

One Option Correct

1. (d) **2.** (d) **3.** (b) **4.** (c) **5.** (c) **6.** (b) **7.** (c) **8.** (d) **9.** (a)
10. (b) **11.** (b) **12.** (a) **13.** (a) **14.** (b) **15.** (d) **16.** (d)

One or More Than One Option Correct

1. (b, c, d) **2.** (a, b) **3.** (a, b, c) **4.** (b, c) **5.** (a, c) **6.** (a, c) **7.** (a, b, d) **8.** (c, d) **9.** (a, c, d)
10. (c, d) **11.** (a, c) **12.** (a, b, c) **13.** (a, c, d) **14.** (a, b, c) **15.** (a, d)

Matrix Type

1. A $\rightarrow$ T; B $\rightarrow$ P; C $\rightarrow$ Q; D $\rightarrow$ S
2. A $\rightarrow$ R; B $\rightarrow$ P; C $\rightarrow$ Q; D $\rightarrow$ S
3. A $\rightarrow$ P, R, S; B $\rightarrow$ P; C $\rightarrow$ P, Q; D $\rightarrow$ S, T
4. A $\rightarrow$ Q; B $\rightarrow$ P, Q; C $\rightarrow$ P, Q, S, T; D $\rightarrow$ Q, T
5. A $\rightarrow$ S; B $\rightarrow$ R; C $\rightarrow$ Q; D $\rightarrow$ P
6. A $\rightarrow$ R; B $\rightarrow$ S; C $\rightarrow$ P; D $\rightarrow$ Q
7. A $\rightarrow$ Q; B $\rightarrow$ P; C $\rightarrow$ S; D $\rightarrow$ T

Integer and Numerical Value Type

1. (6 : 3 : 1) **2.** (2) **3.** (13) **4.** (7) **5.** $(6((\alpha - 1)^2 + 3))$ **6.** (3) **7.** (9) **8.** (4)
9. (5)

Comprehension Passage

Comprehension 1

1. (b) **2.** (a) **3.** (c)

Comprehension 2

1. (c) **2.** (b) **3.** (c)

Comprehension 3

1. (a) **2.** (d) **3.** (a) **4.** (c) **5.** (b)

HINTS AND EXPLANATIONS

Worksheet

1. P (5, 0, 8) Q (3, 3, 2)

$\overline{PQ} = (3-5)i + (3-0)j + (2-8)k$

$= -2i + 3j - 6k$

$|\overline{PQ}| = \sqrt{(-2)^2 + (3)^2 + (-6)^2} = \sqrt{4+9+36} = 7$

$\therefore$ Unit vector along $\overline{PQ} = \dfrac{\overline{PQ}}{|\overline{PQ}|} = \dfrac{-2i+3j-6k}{7}$

$= -\dfrac{2}{7}i + \dfrac{3}{7}j - \dfrac{6}{7}k$

2. Given $\overline{BC} = 1.5\overline{BA}$

B 2 A 1 C; $\vec{b}$, $\vec{a}$, $\vec{c}$; 0

$\therefore \dfrac{\overline{BC}}{\overline{BA}} = 1.5 = \dfrac{3}{2}$

$\therefore$ Point A divides BC in the ration 2 : 1

Now P.V. of A $= \dfrac{2\text{P.V. of C} + 1 \times \text{P.V. of B}}{2+1}$ (section formula)

$\vec{a} = \dfrac{2\vec{c} + \vec{b}}{3}$

$3\vec{a} = 2\vec{c} + \vec{b}$

$\vec{c} = \dfrac{3\vec{a} - \vec{b}}{2}$

3. Points A (*k*, –10, 3), B (1, –1, 3) and C (3, 5, 3) are collinear if

area of ΔABC = 0

$2.\vec{c} \quad |\overrightarrow{AB} \times \overrightarrow{AC}| = 0$

$\overrightarrow{AB} = (1-k)i + 4j$

$\overrightarrow{AC} = (3-k)i + 15j$

$\therefore \overrightarrow{AB} \times \overrightarrow{AC} = \begin{vmatrix} \hat{i} & \hat{j} & \hat{k} \\ 1-k & 9 & 0 \\ 3-k & 15 & 0 \end{vmatrix} = 0$

$= |\hat{k}(15(1-k) - 9(3-k))| = 0$

$\Rightarrow k = -2$

4. A vector $\perp$r to both $\vec{a} = 2i - j + 2k$ and $\vec{b} = 4i - j + 3k$ is

$\vec{a} \times \vec{b} = \begin{vmatrix} i & j & k \\ 2 & -1 & 2 \\ 4 & -1 & 3 \end{vmatrix}$

$= i(-1) - j(-2) + k(2)$

$= -i + 2j + 2k$

$|\vec{a} \times \vec{b}| = \sqrt{1+4+4}$

$\therefore$ A vector of magnitude 6 and $\perp$r to both $\vec{a}$ and $\vec{b}$.

$= 6\dfrac{\vec{a} \times \vec{b}}{|\vec{a} \times \vec{b}|}$

$= \dfrac{6}{3}(-i + 2j + 2k)$

$= -2i + 2j + 4k$

5. Angle θ between $\vec{a} = 2i - j + k$ and $\vec{b} = 3i + 4j - k$ is given by

$\cos\theta = \dfrac{\vec{a}.\vec{b}}{|\vec{a}||\vec{b}|}$

$= \dfrac{(2i - j + k).(3i + 4j - k)}{\sqrt{4+1+1}\sqrt{9+16+1}}$

$= \dfrac{6-4-1}{\sqrt{6}\sqrt{26}} = \dfrac{1}{\sqrt{156}}$

$\therefore \theta = \cos^{-1}\dfrac{1}{2\sqrt{39}}$

6. $\vec{a} + \vec{b} + \vec{c} = 0$

$\Rightarrow \vec{a} + \vec{b} = -\vec{c}$

Taking cross product of both sides with $\vec{c}$,

$\vec{a} \times \vec{c} + \vec{b} \times \vec{c} = -\vec{c} \times \vec{c}$

$\Rightarrow \vec{b} \times \vec{c} = \vec{c} \times \vec{a}$

Similarly, we can show that

$\vec{c} \times \vec{a} = \vec{a} \times \vec{b}$

$\therefore \vec{a} \times \vec{b} = \vec{b} \times \vec{c}$ ...(i)

Similarly, $\vec{b} \times \vec{c} = \vec{c} \times \vec{a}$ ...(ii)

Using (i) and (ii)

$\vec{a} \times \vec{b} = \vec{b} \times \vec{c} = \vec{c} \times \vec{a}$

7. $\overrightarrow{AB} = i - 2j + k, \overrightarrow{CD} = i - 2j + k$

Projection of $\overrightarrow{AB}$ on CD $= \dfrac{\overrightarrow{AB}.\overrightarrow{CD}}{|CD|}$

$= \dfrac{(i - 2j + 4k).(i - 2j) + 4k}{|CD|}$

$= \dfrac{21}{\sqrt{21}} = \sqrt{21}$

8. Area of triangle ABC $= \frac{1}{2}|\overrightarrow{AB} \times \overrightarrow{AC}|$

$= \frac{1}{2}\begin{vmatrix} i & j & k \\ 1 & -3 & 1 \\ 3 & 3 & -4 \end{vmatrix}$

$= \frac{1}{2}|i(4) - j(-7) + k(12)|$

$= \frac{1}{2}|9i + 7j + 12k|$

$= \frac{1}{2}\sqrt{81+49+144}$

9. In ΔABC

Clearly, $\vec{a} + \vec{b} + \vec{c} = 0$

$\vec{b} + \vec{c} = -\vec{a}$

$\therefore (\vec{b}+\vec{c}).(\vec{b}+\vec{c}) = (-\vec{a}).(\vec{a})$

$|\vec{b}|^2 + |\vec{c}|^2 + 2\vec{b}.\vec{c} = |\vec{a}|^2$

$b^2 + c^2 + 2\overrightarrow{CA}.\overrightarrow{AB} = a^2$

$b^2 + c^2 - 2\overrightarrow{AB}.\overrightarrow{AC} = a^2$

$b^2 + c^2 - 2|\vec{c}||\vec{b}| = \cos A = a^2$

$\therefore a^2 = b^2 + c^2 - 2bc\cos A$

10. Let $\overrightarrow{OA} = \vec{p}, \overrightarrow{OC} = \vec{q}$

Clearly, $\overrightarrow{OA} + \overrightarrow{AC} = \overrightarrow{OC}$

$\vec{p} + \vec{b} = \vec{q}$ ...(i)

and $\overrightarrow{OA} + \overrightarrow{AB} = \overrightarrow{OB}$

$\vec{p} + \vec{q} = \vec{a}$ ...(ii)

Using equations (i) and (ii), we get

$\vec{p} = \frac{\vec{a}-\vec{b}}{2}, \vec{q} = \frac{\vec{a}+\vec{b}}{2}$

Now, area(llgm OABC) $= |\overrightarrow{OA} \times \overrightarrow{OC}|$

$= |\vec{p} \times \vec{q}|$

$= \left|\frac{\vec{a}-\vec{b}}{2} \times \frac{\vec{a}+\vec{b}}{2}\right|$

$= \frac{1}{4}|(\vec{a}-\vec{b}) \times (\vec{a}+\vec{b})|$

$= \frac{1}{4}|\vec{a}\times\vec{a} + \vec{a}\times\vec{b} - \vec{b}\times\vec{a} - \vec{b}\times\vec{b}|$

$= \frac{1}{4} \times 2|\vec{a}\times\vec{b}|$

$= \frac{1}{2}|\vec{a}\times\vec{b}|$

Now area of llgm whose diagonals are along

$\vec{a} = 2i - j + k$ and $\vec{b} = i + 3j - k$

$= \frac{1}{2}|\vec{a}\times\vec{b}|$

$= \frac{1}{2}\begin{vmatrix} i & j & k \\ 2 & -1 & 1 \\ 1 & 3 & -1 \end{vmatrix}$

$= \frac{1}{2}|i(-2) - j(-3) + k(7)|$

$= \frac{1}{2}|-2i + 3j + 7k|$

$= \frac{1}{2}\sqrt{4+9+49} = \frac{1}{2}\sqrt{62}$

Quiz

1. If $\vec{r}$ is the position vector of a point $P(x, y, z)$ then the angles α, β and γ made by the vector $\vec{r}$ with positive direction of x, y, z axes respectively, are called direction cosines.

2. (i) Vectors $\vec{b}$ and $\vec{d}$ are coinitial
   (ii) Vectors $\vec{b}$ and $\vec{d}$ are equal
   (iii) Vectors $\vec{a}$ and $\vec{c}$ collinear but not equal.

3. We have $|\vec{a}| = \sqrt{1^2+2^2} = \sqrt{5}$ and

$|\vec{b}| = \sqrt{2^2+1^2} = \sqrt{5}$

4. The unit vector in the direction of the given vector $\vec{a}$ is

$\hat{a} = \frac{1}{|\vec{a}|}\vec{a} = \frac{1}{\sqrt{5}}(\hat{i} - 2\hat{j}) = \frac{1}{\sqrt{5}}\hat{i} - \frac{2}{\sqrt{5}}\hat{j}$

Therefore, the vector having magnitude equal to 7 and in the direction of $\vec{a}$ is

$7\hat{a} = 7\left(\frac{1}{\sqrt{5}}\hat{i} - \frac{2}{\sqrt{5}}\hat{j}\right) = \frac{7}{\sqrt{5}}\hat{i} - \frac{14}{\sqrt{5}}\hat{j}$

5. Given $\vec{a}.\vec{b} = 1, |\vec{a}| = 1$ and $|\vec{b}| = 2$.

We have $\theta = \cos^{-1}\left(\frac{\vec{a}.\vec{b}}{|\vec{a}||\vec{b}|}\right) = \cos^{-1}\left(\frac{1}{2}\right) = \frac{\pi}{3}$

6. $l = \frac{a}{r}, m = \frac{b}{r}$ and $n = \frac{c}{r}$

7. Zero

8. Let us take two non-zero vectors $\vec{a}$ and $\vec{b}$ which are collinear. Then $|\vec{a}\times\vec{b}| = ab\sin\theta = ab \times 0 = 0$

So, the converse is not true.

9. $\vec{a}.\vec{p}$

10. $\vec{a}.(\vec{b}+\vec{c}) = \vec{a}.\vec{b}+\vec{a}.\vec{c}$

JEE Main

Objective Questions

One Option Correct

1. (c) $|\vec{a}| = |\vec{b}| = |\vec{c}| = 1$ and $\vec{a}\cdot\vec{b} = 0$

We get $\vec{c}\cdot\vec{a} = \alpha = \cos\theta$

$\vec{c}\cdot\vec{b} = \beta = \cos\theta$

$\vec{c}\cdot(\vec{a}\times\vec{b}) = \gamma(\vec{a}\times\vec{b}) = \gamma$

$\therefore \gamma^2 = [\vec{c}\,\vec{a}\,\vec{b}]\cdot[\vec{c}\,\vec{a}\,\vec{b}]$

$$= \begin{vmatrix} \vec{c}\cdot\vec{c} & \vec{c}\cdot\vec{a} & \vec{c}\cdot\vec{b} \\ \vec{a}\cdot\vec{c} & \vec{a}\cdot\vec{a} & \vec{a}\cdot\vec{b} \\ \vec{b}\cdot\vec{c} & \vec{b}\cdot\vec{a} & \vec{b}\cdot\vec{b} \end{vmatrix}$$

$$= \begin{vmatrix} 1 & \cos\theta & \cos\theta \\ \cos\theta & 1 & 0 \\ \cos\theta & 0 & 1 \end{vmatrix} = -\cos 2\theta$$

2. (c) We must have

$$\begin{vmatrix} a & a & c \\ 1 & 0 & 1 \\ c & c & b \end{vmatrix} = 0$$

$\Rightarrow c^2 = ab$

$\Rightarrow c$ is the geometric mean of a and b.

3. (d) $[\lambda(\vec{a}+\vec{b}), \lambda^2\vec{b}, \lambda\vec{c}] = [\vec{a}, \vec{b}+\vec{c}, \vec{b}]$

$\Rightarrow \{\lambda(\vec{a}+\vec{b})\}\cdot(\lambda^2\vec{b}\times\lambda\vec{c}) = \vec{a}\cdot\{(\vec{b}+\vec{c})\times\vec{b}\}\times 2$

If $[\vec{a},\vec{b},\vec{c}] = \beta$ then

$\Rightarrow \lambda^4[\beta+0] = \vec{a}\cdot(\vec{c}\times\vec{b}) = -\beta$

$\Rightarrow \lambda^4 = -1 \quad [\because \beta \neq 0]$

which is not possible.

4. (a) $\vec{F} = (4\hat{i}+\hat{j}-3\hat{k}) + 3\hat{i}+\hat{j}-\hat{k}$

$= 7\hat{i}+2\hat{j}-4\hat{k}$

$W = \vec{F}\cdot\overrightarrow{AB}$

$= (7\hat{i}+2\hat{j}-4\hat{k})\cdot(4\hat{i}+2\hat{j}-2\hat{k})$

$= 40$ units

5. (c) Projection of $\vec{v}$ along $\vec{u} = \dfrac{\vec{v}.\vec{u}}{|\vec{u}|}$

and projection of $\vec{w}$ along $\vec{u} = \dfrac{\vec{w}.\vec{u}}{|\vec{u}|}$

If $\dfrac{\vec{v}\cdot\vec{u}}{|\vec{u}|} = \dfrac{\vec{w}\cdot\vec{u}}{|\vec{u}|}$

Then $\vec{v}\cdot\vec{u} = \vec{w}\cdot\vec{u}$ ...(i)

Also, $\vec{v}\cdot\vec{w} = 0$ ...(ii)

Now

$|\vec{u}-\vec{v}+\vec{w}|^2 = u^2+v^2+w^2-2\vec{u}\cdot\vec{v}-2\vec{v}\cdot\vec{w}-2\vec{w}\cdot\vec{u}$

$= 1+4+9 = 14$

$\Rightarrow |\vec{u}-\vec{v}+\vec{w}| = \sqrt{14}$

6. (a) $|\hat{a}| = 1, |\hat{b}| = 1, |\hat{a}+\hat{b}| = 1$

$\Rightarrow |\hat{a}+\hat{b}|^2 = 1 \Rightarrow 1+1+2\hat{a}\cdot\hat{b} = 1$

$\Rightarrow \hat{a}\cdot\hat{b} = -\dfrac{1}{2}$

Now, $|\hat{a}-\hat{b}|^2 = 1+1-2\hat{a}\cdot\hat{b} = 3$

$\Rightarrow |\hat{a}-\hat{b}| = \sqrt{3}$

7. (a) $\vec{A}\times(\vec{B}\times\vec{C}) = \vec{B}\times(\vec{C}\times\vec{A})$

$\Rightarrow (\vec{A}\cdot\vec{C})\vec{B} - (\vec{A}\cdot\vec{B})\vec{C} = (\vec{B}\cdot\vec{A}) - (\vec{B}\cdot\vec{C})\vec{A}$

$\Rightarrow (\vec{B}\cdot\vec{C})\vec{A} + (\vec{A}\cdot\vec{C})\vec{B} - 2(\vec{A}\cdot\vec{B})\vec{C} = 0$

$\Rightarrow \vec{B}\cdot\vec{C} = 0, \vec{A}\cdot\vec{C} = 0, \vec{A}\cdot\vec{B} = 0$

$[\because \vec{A}, \vec{B}, \vec{C}$ are non-coplanar$]$

$\Rightarrow \vec{A}\times(\vec{B}\times\vec{C}) = (\vec{A}\cdot\vec{C})\vec{B} - (\vec{A}\cdot\vec{B})\vec{C} = 0$

8. (c)

$$[\vec{a}\,\vec{b}\,\vec{c}] = \begin{vmatrix} 1 & 0 & -1 \\ x & 1 & 1-x \\ y & x & 1+x-y \end{vmatrix} = 1$$

9. (b) $(\vec{a}\cdot\vec{c})\vec{b} - (\vec{a}\cdot\vec{b})\vec{c} = \dfrac{1}{\sqrt{2}}\vec{b} + \dfrac{1}{\sqrt{2}}\vec{c}$

$\Rightarrow a.b = -\dfrac{1}{\sqrt{2}} \Rightarrow |\vec{a}||\vec{b}|\cos\theta_1 = \cos\dfrac{3\pi}{4}$

10. (c) Angle θ between $\vec{a}$ and $\vec{c}$ is given by

$\cos\theta = \dfrac{\vec{a}\cdot\vec{c}}{|\vec{a}||\vec{c}|}$

Now, $|\vec{a}| = 8|\vec{b}|, |\vec{c}| = 7|\vec{b}|$

$\Rightarrow \vec{a}.\vec{c} = -56|\vec{b}|^2$

$\Rightarrow \cos\theta = \dfrac{-56|\vec{b}|^2}{8|\vec{b}|7|\vec{b}|} = -1 \Rightarrow \theta = \pi$

11. (c)

$$\vec{a}=\lambda\left(\frac{\hat{i}+\hat{j}}{|\hat{i}+\hat{j}|}+\frac{\hat{j}+\hat{k}}{|\hat{j}+\hat{k}|}\right)$$

$$=\frac{\lambda}{\sqrt{2}}\left(\hat{i}+2\hat{j}+\hat{k}\right)$$

$\vec{b}=\hat{i}+\hat{j}$, $\vec{a}$, $\vec{c}=\hat{j}+\hat{k}$

$\alpha=1,\beta=1$

12. (d) $\begin{vmatrix}1 & 1 & 1\\ 1 & -1 & 2\\ x & x-2 & -1\end{vmatrix}=0$

$\Rightarrow 1-2(x-2)-1(-1-2x)+1(x-2+x)=0$

$\Rightarrow x=-2$

13. (d) If $C=\pi/2$ then

$\overline{\text{AC}}\cdot\overline{\text{BC}}=0$

$\Rightarrow\left[(a-2)\hat{i}-2\hat{j}\right]\cdot\left[(a-1)\hat{i}+4\hat{k}\right]=0$

$\Rightarrow(a-2)(a-1)=0\Rightarrow a=1,a=2$

14. (c) Take P as origin

$\overrightarrow{\text{PC}}=\dfrac{\overrightarrow{\text{PA}}+\overrightarrow{\text{PB}}}{2}$

B C A

$\Rightarrow 2\overrightarrow{\text{PC}}=\overrightarrow{\text{PA}}+\overrightarrow{\text{PB}}$

15. (d) The given line is parallel to the plane since

$\left(\hat{i}+\hat{j}+4\hat{k}\right)\cdot\left(\hat{i}-5\hat{j}+\hat{k}\right)=0$

$\Rightarrow$ perpendicular distance $(2,-2,3)$ from plane $x-5y+z=5$ is the required distance, which is equal to $\dfrac{2+10+3-5}{\sqrt{27}}$.

16. (b)

$$\left[\vec{a}\,\vec{b}\,\vec{c}\right]=\begin{vmatrix}1 & 0 & -1\\ x & 1 & 1-x\\ y & x & 1+x-y\end{vmatrix}=1$$

which does not depend upon x or y.

17. (d) $\vec{a}+2\vec{b}=\alpha\vec{c},\vec{b}+3\vec{c}=\beta\vec{a}$

Eliminating $\vec{b}$ we get

$\vec{a}+2\left(\beta\vec{a}-3\vec{c}\right)=\alpha\vec{c}$

$\Rightarrow(1+2\beta)\vec{a}+(-6-\alpha)\vec{c}=0$

since $\vec{a}$ and $\vec{c}$ are non-collinear

$1+2\beta=0,-6-\alpha=0$

$\Rightarrow\beta=-\dfrac{1}{2},\alpha=-6$

$\Rightarrow\vec{a}+2\vec{b}=-6\vec{c}$

$\vec{a}+2\vec{b}+6\vec{c}=0$

18. (c) The scalar triple product

$=\left(\vec{a}+2\vec{b}+3\vec{c}\right)\cdot\left\{\left(\lambda\vec{b}+3\vec{c}\right)\times(2\lambda-1)\vec{c}\right\}$

$=\left(\vec{a}+2\vec{b}+3\vec{c}\right)\cdot\left\{\lambda(2\lambda-1)\vec{b}\times\vec{c}+0\right\}$

$=\lambda(2\lambda-1)\left[\vec{a}\,\vec{b}\,\vec{c}\right]$

which is non-zero except for two values of λ, namely $\lambda=0,\lambda=\dfrac{1}{2}$

19. (d)

$\left(\vec{a}\times\vec{b}\right)\times\vec{c}=\dfrac{1}{3}|\vec{b}||\vec{c}|\,|\vec{a}$ ...(i)

$\Rightarrow\vec{c}\times\left(\vec{a}\times\vec{b}\right)=\dfrac{1}{3}bc\vec{a}$

$\Rightarrow\left(-\vec{c}\cdot\vec{b}\right)\vec{a}+(\vec{c}\cdot\vec{a})\vec{b}=\dfrac{1}{3}bc\vec{a}$

$\Rightarrow\left(\dfrac{1}{3}bc-\vec{b}\cdot\vec{c}\right)\vec{a}+(\vec{c}\cdot\vec{a})\vec{b}=0$

If $\vec{a}$ and $\vec{b}$ are collinear then given equation (i) cannot be true since $|\vec{a}|,|\vec{b}|,|\vec{c}|\neq 0$.

Thus $\vec{a}$ and $\vec{b}$ are non-collinear.

whence $\dfrac{1}{3}bc=\vec{b}\cdot\vec{c},\vec{c}\cdot\vec{a}=0$

$\Rightarrow\dfrac{1}{3}|\vec{b}||\vec{c}|=|\vec{b}||\vec{c}|\cos\theta$

$\Rightarrow\cos\theta=\dfrac{1}{3}$

$\Rightarrow\sin\theta=\dfrac{2\sqrt{2}}{3}$

20. (b) $\left|\vec{a}+\vec{b}+\vec{c}\right|=a^2+b^2+c^2+2\left(\vec{a}\cdot\vec{b}+\vec{b}\cdot\vec{c}+\vec{c}\cdot\vec{a}\right)$

$\Rightarrow 0=1+4+9+2\left(\vec{a}\cdot\vec{b}+\vec{b}\cdot\vec{c}+\vec{c}\cdot\vec{a}\right)$

$\Rightarrow\vec{a}\cdot\vec{b}+\vec{b}\cdot\vec{c}+\vec{c}\cdot\vec{a}=-7$

21. (d) $(\vec{u}+\vec{v}-\vec{w})\cdot\{(\vec{u}-\vec{v})\times(\vec{v}-\vec{w})\}$

$=(\vec{u}+\vec{v}-\vec{w})\cdot(\vec{u}\times\vec{v}-\vec{u}\times\vec{w}-0+\vec{v}\times\vec{w})$

$=\beta-(-\beta)+\beta$ (where $\beta=[\vec{u}\vec{v}\vec{w}]$)

$=3\beta=3[\vec{u}\vec{v}\vec{w}]$

22. (b) Let $\hat{n}=x\hat{i}+y\hat{j}+z\hat{k}$ then $x^2+y^2+z^2=1$

$x+y=0,x-y=0$

$\Rightarrow x=0,y=0$

$\Rightarrow\hat{n}=\pm\hat{k}$

$\Rightarrow\vec{w}\cdot\hat{n}=\pm 1$

$\Rightarrow|\vec{w}\cdot\hat{n}|=1$

23. (b) Since $\left[\vec{a}\times\vec{b}\ \ \vec{b}\times\vec{c}\ \vec{c}\times\vec{a}\right]=\left[\vec{a}\,\vec{b}\,\vec{c}\right]^2$

24. (b) $|\vec{a}\times\vec{b}| = \sqrt{3}|\vec{a}\cdot\vec{b}|$

$\Rightarrow |\vec{a}||\vec{b}|\sin\theta = \sqrt{3}|\vec{a}||\vec{b}|\cos\theta$

$\Rightarrow \tan\theta = \sqrt{3} \Rightarrow \theta = \dfrac{\pi}{3}$

25. (d) (d) is correct since cyclic order is maintained.

26. (a) $|\vec{a}-\vec{b}| = 1$

$\Rightarrow 1+1-2\vec{a}\cdot\vec{b} = 1$

$\Rightarrow \vec{a}\cdot\vec{b} = \dfrac{1}{2}$

$\Rightarrow \theta = \dfrac{\pi}{3}$

27. (d) If $\vec{a} = x\hat{i}+y\hat{j}+z\hat{k}$

Then $\vec{a}\cdot\hat{i} = x$, etc.

$\Rightarrow (\vec{a}\cdot\hat{i})^2 + (\vec{a}\cdot\hat{j})^2 + (\vec{a}\cdot\hat{k})^2 = x^2+y^2+z^2 = |\vec{a}|^2$

28. (d) $|\hat{a}| = |\hat{b}| = |\hat{a}+\hat{b}| = 1$

$\Rightarrow \hat{a}\cdot\hat{b} = -\dfrac{1}{2}$

$\Rightarrow$ Angle between $\hat{a}$ and $\hat{b}$ is 120°.

29. (c) If θ is obtuse, $\vec{a}\cdot\vec{b} < 0$ for all x.

$\Rightarrow cx^2 - 6cx - 12 < 0$

$\Rightarrow c < 0$ and $36c + 48 < 0$

$\Rightarrow -\dfrac{4}{3} < c < 0$

30. (d) $(\vec{a}\cdot\vec{r}) + |\vec{a}\times\vec{r}|^2 = |\vec{a}||\vec{r}|^2$

$\Rightarrow (\vec{a}\cdot\vec{r})^2 + \hat{j}\cdot\hat{j} = |\vec{a}|^2|\vec{r}|^2 \Rightarrow \vec{a}\cdot\vec{r} = \sqrt{a^2r^2-1}$

31. (b) $\hat{i}\times(\vec{a}\times\hat{i}) = (\hat{i}\cdot\hat{i})\vec{a} - (\hat{i}-\vec{a})\hat{i} = \vec{a} - (\hat{i}\cdot\vec{a})\hat{i}$

$\hat{j}\times(\vec{a}\times\hat{i}) = \vec{a} - (\hat{j}\cdot\vec{a})\hat{j}$

$\hat{k}\times(\vec{a}\times\hat{k}) = \vec{a} - (\hat{k}\cdot\vec{a})\hat{k}$

Sum $= 3\vec{a} - \{(\hat{i}\cdot\vec{a})\hat{i} + (\hat{j}\cdot\vec{a})\hat{j} + (\hat{k}\cdot\vec{a})\hat{k}\}$

$= 3\vec{a} - \vec{a} = 2\vec{a}$

[$\because$ If $\vec{a} = x\hat{i}+y\hat{j}+z\hat{k}$ then on taking dot product with $i\vec{a}\cdot\hat{i} = x$ etc.]

32. (d) A vector perpendicular to $\hat{i}-2\hat{j}-\hat{k}$ and $3\hat{i}-2\hat{j}-\hat{k}$

$$= \begin{vmatrix} \hat{i} & \hat{j} & \hat{k} \\ 1 & -2 & -1 \\ 3 & -2 & -1 \end{vmatrix} = \hat{j}(2) + 4\hat{k}$$

$\Rightarrow$ Angle θ with $\hat{i}+\hat{j}+\hat{k}$ is given by

$\cos\theta = \dfrac{-2+4}{\sqrt{3}\sqrt{20}} = \dfrac{2}{\sqrt{60}} = \dfrac{1}{\sqrt{15}}$

$\Rightarrow \sec\theta = \sqrt{15}$

$\Rightarrow \theta = \sec^{-1}\sqrt{15}$

33. (c) If $x\hat{i}+y\hat{j}+z\hat{k}$ be a vector of magnitude 4, which is equally inclined to the three vectors.

Then $\sqrt{x^2+y^2+z^2} = 4$ and

$\dfrac{x+y}{4\sqrt{2}} = \dfrac{y+z}{4\sqrt{2}} = \dfrac{z+x}{4\sqrt{2}}$

$\Rightarrow x+y = \lambda, y+z = \lambda, z+x = \lambda$

$\Rightarrow x = y = z = \dfrac{\lambda}{2}$

On putting the value of x, y, z in $x^2+y^2+z^2 = 16$

we get, $\dfrac{3\lambda^2}{4} = 16 \Rightarrow \lambda = \dfrac{8}{\sqrt{3}}$

$\Rightarrow x\hat{i}+y\hat{j}+z\hat{k} = \pm\dfrac{4}{\sqrt{3}}(\hat{i}+\hat{j}+\hat{k})$

34. (a) $\vec{a}.\vec{b} = \dfrac{\sqrt{3}}{2}$ ...(i)

$(\vec{a}\times\vec{b})\times(\vec{c}\times\vec{d}) = \{(\vec{a}\times\vec{b})\cdot\vec{d}\}\vec{c} - \{(\vec{a}\times\vec{b})\cdot\vec{c}\}\vec{d}$

$= [\vec{a}\,\vec{b}\,\vec{c}]\vec{c}$ – zero [$\because \vec{a}, \vec{b}, \vec{c}$ are coplanar]

$= \dfrac{1}{6}\hat{i} - \dfrac{1}{3}\hat{j} + \dfrac{1}{3}\hat{k}$ (given)

$= \dfrac{1}{6}(\hat{i}-2\hat{j}+2\hat{k})$

$\Rightarrow \vec{c} = \lambda\left(\dfrac{\hat{i}-2\hat{j}+2\hat{k}}{3}\right)$

35. (c) Any vector $\vec{a}$ coplanar with $\hat{i}+2\hat{j}$ and $2\hat{i}+3\hat{j}$ is

$\vec{a} = x(\hat{i}+2\hat{j}) + y(2\hat{i}+3\hat{j})$

$= (x+2y)\hat{i} + (2x+3y)\hat{j}$

If $\vec{a}$ is perpendicular to $\hat{i}-\hat{j}$ then

$x+2y-(2x+3y) = 0$

$\Rightarrow y = -x$

But $(x+2y)^2 + (2x+3y)^2 = 1$ $[\because |\vec{a}| = 1]$

$\Rightarrow x = \pm\dfrac{1}{\sqrt{2}}$ whence $y = \mp\dfrac{1}{\sqrt{2}}$

$\Rightarrow \vec{a} = \dfrac{\hat{i}+\hat{j}}{\sqrt{2}}$ $\left[\text{if } x = -\dfrac{1}{\sqrt{2}}, y = \dfrac{1}{\sqrt{2}}\right]$

36. (c) We have $\overrightarrow{AD}\cdot\overrightarrow{BC}=\overrightarrow{BD}\cdot\overrightarrow{AC}$

$\Rightarrow$ D is orthocentre of triangle ABC

37. (a) $\sqrt{3}\left(\vec{a}\times\vec{b}\right)\cdot\left\{\vec{b}-\left(\vec{a}\cdot\vec{b}\right)\vec{a}\right\}=0$

$\Rightarrow$ They are perpendicular to each other.

38. (c)

$\left[\vec{a}\,\vec{b}\,\vec{c}\right]=0$ for any three coplanar vectors $\vec{a},\vec{b},\vec{c}$

39. (d) $\left[\vec{a}\,\vec{b}\,\vec{c}\right]=\left[\vec{b}\,\vec{c}\,\vec{a}\right]$

since cyclic order is maintained.

40. (c) $\overrightarrow{AD}=\dfrac{\overrightarrow{AB}+\overrightarrow{AC}}{2}=4\hat{i}+3\hat{k}$

$\Rightarrow\left|\overrightarrow{AD}\right|=5$

A
B D C

41. (c) $\left(\vec{a}+\vec{b}+\vec{c}\right)\cdot\left[\left(\vec{a}+\vec{b}\right)\times\left(\vec{a}+\vec{c}\right)\right]$

$=\left(\vec{a}+\vec{b}+\vec{c}\right)\cdot\left(\vec{a}\times\vec{a}+\vec{a}\times\vec{c}+\vec{b}\times\vec{a}+\vec{b}\times\vec{c}\right)$

$=\left[\vec{a}\,\vec{b}\,\vec{c}\right]+\left[\vec{b}\,\vec{a}\,\vec{c}\right]+\left[\vec{c}\,\vec{b}\,\vec{a}\right]=-\left[\vec{a}\,\vec{b}\,\vec{c}\right]$

42. (c) $\overrightarrow{OA}=7\hat{i}+10\hat{k}$

$\overrightarrow{OB}=-\hat{i}+6\hat{j}+6\hat{k},\ \overrightarrow{OC}=-4\hat{i}+9\hat{j}+6\hat{k}$

$\overrightarrow{AB}=-8\hat{i}+6\hat{j}-4\hat{k},\ \overrightarrow{BC}=-3\hat{i}+3\hat{j}$

$\overrightarrow{CA}=11\hat{i}+4\hat{j}+4\hat{k}$

Now $\left|\overrightarrow{AB}\right|=\sqrt{116},\left|\overrightarrow{BC}\right|=\sqrt{18},\left|\overrightarrow{AC}\right|=\sqrt{153}$

$\Rightarrow$ The triangle is scalene.

43. (a) $|\vec{a}||\vec{b}|\sin\theta=|\vec{a}||\vec{b}|\cos\theta$

$\Rightarrow\tan\theta=1\Rightarrow\theta=\dfrac{\pi}{4}$

44. (a) If $\left[\hat{i}\,\hat{j}\,\hat{k}\right]=\beta$

Then $\left[\hat{i}\,\hat{j}\,\hat{k}\right]=-\beta,\left[\hat{j}\,\hat{k}\,\hat{i}\right]=\beta$

$\Rightarrow$ Sum $=\beta-\beta+\beta=\beta=1$

45. (a) Longer diagonal vector

$=\left(5\vec{a}+2\vec{b}\right)+\left(\vec{a}-3\vec{b}\right)=6\vec{a}-\vec{b}$

Now $\left|6\vec{a}-\vec{b}\right|^2=36|\vec{a}|^2+|\vec{b}|^2-12\vec{a}\cdot\vec{b}$

$=36(8)+9-12\times2\sqrt{2}\times3\times\dfrac{1}{\sqrt{2}}=225$

$\Rightarrow\left|6\vec{a}-\vec{b}\right|=15$

46. (c) Volume $=\begin{vmatrix}1 & a & 1\\ 0 & a & 1\\ a & 0 & 1\end{vmatrix}$

$=a^3-a+1$

which will be minimum if

$\dfrac{d}{da}\left(a^3-a+1\right)=0$

The second derivative is positive

$\Rightarrow a=\dfrac{1}{\sqrt{3}}$

47. (c) $\left[\vec{a}\,\vec{b}\,\vec{c}\right]=\begin{vmatrix}1 & 0 & -1\\ x & 1 & 1-x\\ y & x & 1+x-y\end{vmatrix}=1$

48. (b) We must have

$(2p)^2+1^2=(p+1)^2+1$

$\Rightarrow p=1,p=-\dfrac{1}{3}$

49. (a) Any vector $\vec{a}$ is given by

$\vec{a}=x\hat{i}+y\hat{j}+z\hat{k}$

Taking dot product of both sides with $\hat{i}$, we get

$\vec{a}\cdot\hat{i}=x$

Similarly, $\vec{a}\cdot\hat{j}=y,\vec{a}\cdot\hat{k}=z$

$\Rightarrow\vec{a}=\left(\vec{a}\cdot\hat{i}\right)\hat{i}+\left(\vec{a}\cdot\hat{j}\right)\hat{j}+\left(\vec{a}\cdot\hat{k}\right)\hat{k}$

50. (b) Area of quadrilateral OABC

$=\Delta\overline{OAB}+\Delta\overline{OBC}$

$=\dfrac{1}{2}\left\{\vec{a}\times\left(10\vec{a}+2\vec{b}\right)\right\}+\dfrac{1}{2}\left\{\left(10\vec{a}+2\vec{b}\right)\times\vec{b}\right\}$

$=6\left(\vec{a}\times\vec{b}\right)$

$\Rightarrow p=6\left|\vec{a}\times\vec{b}\right|$

C B
$\vec{b}$ $10\vec{a}+2\vec{b}$
O $\vec{a}$ A

Also, $q=$ Area of parallelogram whose adjacent sides are along $\vec{a}$ and $\vec{b}=\left|\vec{a}\times\vec{b}\right|$

$\Rightarrow\dfrac{p}{q}=6$

51. (b) $\vec{a}+\lambda\vec{c}=\hat{i}+\hat{j}-\hat{k}+\lambda\left(2\hat{i}+\hat{j}\right)$

$=(2\lambda+1)\hat{i}+\left(2+\lambda\hat{j}\right)-\hat{k}$

If $(\vec{a}+\lambda\vec{c})\cdot\vec{b}=0$, then

$-(2\lambda+1)-1=0$

$\Rightarrow-2\lambda-1-1=0\Rightarrow\lambda=-1$

52. (b) $\cos\theta = \dfrac{\vec{a}.\vec{b}}{|\vec{a}||\vec{b}|} = \dfrac{18}{\sqrt{52}\times 5}$

$\Rightarrow$ Component of $\vec{a}$ along $\vec{b}$

$= (|\vec{a}|\cos\theta)$ unit vector along $\vec{b}$

$= \sqrt{52}\times\dfrac{18}{\sqrt{52}\times 5}\cdot\dfrac{3\hat{j}+4\hat{k}}{5}$

$= \dfrac{18}{25}(3\hat{j}+4\hat{k})$

53. (c) Let $\vec{a} = x\hat{i}+y\hat{j}+z\hat{k}$

Then $x^2+y^2+z^2=1$ ...(i)

Also $\hat{a}+\hat{i}+\hat{j}$ is a unit vector

$\Rightarrow (x+1)^2+(y+1)^2+z^2=1$ ...(ii)

Since angle between $\vec{a}$ and $\hat{k}$ is $\pi/4$.

$\cos\dfrac{\pi}{4} = \dfrac{\vec{a}\cdot\hat{k}}{|\vec{a}||\hat{k}|} = z$

$\Rightarrow z = \dfrac{1}{\sqrt{2}}$ ...(iii)

From (ii)

$x^2+y^2+2(x+y)+2+\dfrac{1}{2}=1$

$\Rightarrow 1-\dfrac{1}{2}+2(x+y)+2+\dfrac{1}{2}=1$

$\Rightarrow x+y=-1$ ...(iv)

Also $x^2+y^2=\dfrac{1}{2}$ ...(v)

$\Rightarrow x=-\dfrac{1}{2}, y=-\dfrac{1}{2}$ and $z=\dfrac{1}{\sqrt{2}}$

Thus $\vec{a}$ is $-\dfrac{\hat{i}}{2}-\dfrac{\hat{j}}{2}+\dfrac{\hat{k}}{\sqrt{2}}$

54. (a) $\overline{PR} = 5\overline{PQ}$

$\overline{PQ}+\overline{QR} = 5\overline{PQ} \Rightarrow \overline{PQ} = \dfrac{1}{4}\overline{QR}$

$\vec{b} = \dfrac{1\times(\text{p.v. of }\vec{R})+4\vec{a}}{5}$

P($\vec{a}$) Q($\vec{b}$) R — 1 : 4

$\Rightarrow$ p.v. of $\vec{R} = 5\vec{b}-4\vec{a}$

55. (a) We must have $\overline{AB} = \lambda\overline{BC}$

$= -20\hat{i}-11\hat{j} = \lambda[(a-40)\hat{i}-44\hat{j}]$

$\Rightarrow -11 = -44\lambda = \lambda = \dfrac{1}{4} \Rightarrow -20 = \lambda(a-40)$

$\Rightarrow a = -40$

56. (c)

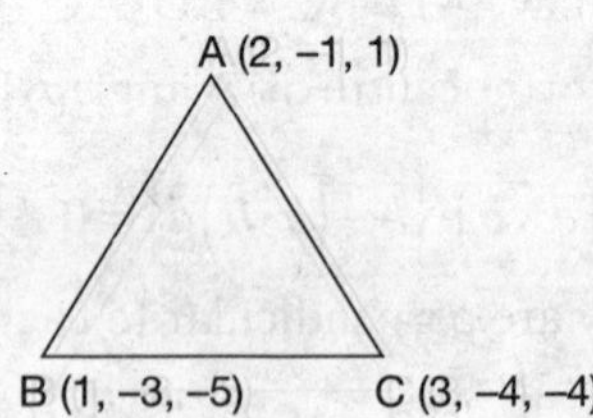

$BC = \sqrt{(3-1)^2+(-4+3)^2+(-4+5)^2} = \sqrt{6}$

$AC = \sqrt{(3-2)^2+(-4+1)^2+(-4-1)^2} = \sqrt{35}$

$AB = \sqrt{(2-1)^2+(-1+3)^2+(1+5)^2} = \sqrt{41}$

$\Rightarrow \cos A = \dfrac{41+35-6}{2\sqrt{35}\sqrt{41}} = \sqrt{\dfrac{35}{41}}$

57. (d) $\vec{b}\times\vec{c} = \begin{vmatrix} i & j & k \\ 2 & 1 & -1 \\ 1 & 0 & 3 \end{vmatrix}$

$= 3\hat{i}-\hat{j}(6+1)+\hat{k}(0-1)$

$= 3\hat{i}-7\hat{j}-\hat{k}$

Now $[\vec{a}\,\vec{b}\,\vec{c}] = \vec{a}\cdot(\vec{b}\times\vec{c})$

$\Rightarrow |\vec{a}||\vec{b}||\vec{c}|\cos\theta$ whence

(θ is the angle between $\vec{a}$ and $\vec{b}\times\vec{c}$)

$= \sqrt{59}\cos\theta$

$\Rightarrow$ Maximum value of $[\vec{a}\,\vec{b}\,\vec{c}] = \sqrt{59}$

58. (b) Let the points be designated by A, B, C respectively, then

$\overline{AB} = (6\hat{i}-\hat{j}+2\hat{k})-(2\hat{i}+\hat{j}+\hat{k})$

$= 4\hat{i}-2\hat{j}+\hat{k}$

$\overline{BC} = 8\hat{i}-4\hat{j}+(p-2)\hat{k}$

If points A, B, C are collinear, then $\overline{AB} = \lambda\overline{BC}$ for same scalar λ

$\Rightarrow \dfrac{4}{8} = \dfrac{-2}{-4} = \dfrac{1}{p-2} \Rightarrow p = 4$

59. (a)

A (–2, 3, 5) —λ— P (1, 2, 3) —1— B (7, 0, –1)

If the ratio is $\lambda:1$

$1 = \dfrac{7\lambda-2}{\lambda+1} \Rightarrow \lambda = \dfrac{1}{2}$

$2 = \dfrac{0+1\times 3}{\lambda+1} \Rightarrow \lambda = \dfrac{1}{2}$

$3 = \frac{-\lambda+5}{\lambda+1} \Rightarrow \lambda = \frac{1}{2}$

Since all λ are same, such a division is possible and the ratio is 1 : 2.

60. (b) Orthogonal projection of $\vec{b}$ on $\vec{a}$ is

$\frac{\vec{b}\cdot\vec{a}}{|\vec{a}|} = \frac{\lambda+3-1}{\sqrt{3}} = \frac{\lambda+2}{\sqrt{3}}$

$\Rightarrow$ Orthogonal projection of vector $\frac{\lambda+2}{\sqrt{3}}\cdot\frac{\vec{b}}{|\vec{a}|}$

$= \frac{(\lambda+2)}{\sqrt{3}}\frac{\hat{i}-\hat{j}-\hat{k}}{\sqrt{3}} = \frac{(\lambda+2)(\hat{i}-\hat{j}-\hat{k})}{3}$

If this is equal to $\frac{4}{3}(\hat{i}-\hat{j}-\hat{k})$ then $\lambda = 2$

61. (a) Let $\vec{a} = x\hat{i} + y\hat{j} + z\hat{k}$

Since $\vec{a}, \vec{b}$ and $\vec{c}$ are coplanar

$\therefore \left[\vec{a}\,\vec{b}\,\vec{c}\right] = \begin{vmatrix} x & y & z \\ 2 & 1 & 1 \\ 1 & -1 & 1 \end{vmatrix} = 0$

$\Rightarrow 2x - y - 3z = 0$ ...(i)

and $\vec{a}\cdot\vec{d} = 0$

$3x + 2y + 6z = 0$ ...(ii)

From (i) and (ii)

$\frac{x}{\begin{vmatrix} -1 & -3 \\ 2 & 6 \end{vmatrix}} = \frac{-y}{\begin{vmatrix} 2 & -3 \\ 3 & 6 \end{vmatrix}} = \frac{z}{\begin{vmatrix} 2 & -1 \\ 3 & 2 \end{vmatrix}} = \lambda$

$\Rightarrow x = 0, y = -21\lambda, z = 7\lambda$

Given $x^2 + y^2 + z^2 = 10$

$441\lambda^2 + 49\lambda^2 = 10$

$\therefore \vec{a} = -3\hat{i} + \hat{k}$ or $\vec{a} = 3\hat{j} - \hat{k}$

Now, $\left[\vec{a}\,\vec{b}\,\vec{c}\right] + \left[\vec{a}\,\vec{b}\,\vec{d}\right] + \left[\vec{a}\,\vec{c}\,\vec{d}\right] = \left[\vec{a}\,\vec{b}+\vec{c}\,\vec{d}\right]$

$= \begin{vmatrix} 0 & -3 & 1 \\ 3 & 0 & 2 \\ 3 & 2 & 6 \end{vmatrix}$ or $\begin{vmatrix} 0 & 3 & -1 \\ 3 & 0 & 2 \\ 3 & 2 & 6 \end{vmatrix}$

= 42 or −42

62. (b) Since, $\overrightarrow{OP} \perp \overrightarrow{OQ}$

$\therefore \overrightarrow{OP}\cdot\overrightarrow{OQ} = 0 \Rightarrow -x+2y-3x=0 \Rightarrow 2x=y$ ...(i)

and $\left|\overrightarrow{OQ}-\overrightarrow{OP}\right|^2 = 20$

$\Rightarrow (x+1)^2 + (y-2)^2 + (3x+1)^2 = 20$

$\Rightarrow 10x^2 + 8x + y^2 - 4y = 14$

$\Rightarrow 14x^2 = 14$ [From (i)]

$\Rightarrow x = 1$ $(\because x > 0)$

$\Rightarrow y = 2$ [From (i)]

Since, $\overrightarrow{OP}, \overrightarrow{OQ}$ and $\overrightarrow{OR}$ are coplanar

$\therefore \left[\overrightarrow{OP}\;\overrightarrow{OQ}\;\overrightarrow{OR}\right] = 0$

$\Rightarrow \begin{vmatrix} 1 & 2 & -1 \\ -1 & 2 & 3 \\ 3 & z & -7 \end{vmatrix} = 0 \Rightarrow z = -2$

So, $x^2 + y^2 + z^2 = 9$

63. (d) Since $P(3,-1,2)$ and d.r. = 4, −1, 2

$\therefore$ Equation of PR:

$\frac{x-3}{4} = \frac{y+1}{-1} = \frac{z-2}{2}$ ...(i)

$Q(1, 2, -4)$ and d.r. = −2, 1, −2

$\therefore$ Equation of QS:

$\frac{x-1}{-2} = \frac{y-2}{1} = \frac{z+4}{-2}$ ...(ii)

On solving (i) and (ii) we get, the point of intersection of PQ and QS is $T(11, -3, 6)$

$\because \overrightarrow{TA} \perp \overrightarrow{PR}$ and $\overrightarrow{TA} \perp \overrightarrow{QS}$

$\therefore \overrightarrow{TA} = k\,\overrightarrow{PR}\times\overrightarrow{QS} = k\begin{vmatrix} \hat{i} & \hat{j} & \hat{k} \\ 4 & -1 & 2 \\ -2 & 1 & -2 \end{vmatrix} = k\left(4\hat{j}+2\hat{k}\right)$

$\because \left|\overrightarrow{TA}\right| = \sqrt{5} \Rightarrow k = \pm\frac{1}{2}$

So, $\overrightarrow{TA} = \pm\left(2\hat{j}+\hat{k}\right)$

Position vector of $A = 11\hat{i} - 3\hat{j} + 6\hat{k} \pm \left(2\hat{j}+\hat{k}\right)$

$\Rightarrow 11\hat{i} - \hat{j} + 7\hat{k}$ or $11\hat{i} - 5\hat{j} + 5\hat{k}$

Hence, $\left|\overrightarrow{OA}\right| = \sqrt{171}$

64. (a) Since, $\overrightarrow{a_1}$ and $\overrightarrow{a_2}$ are collinear,

$\overrightarrow{a_2} = \lambda\overrightarrow{a_1}$

$\Rightarrow \hat{i} + y\hat{j} + z\hat{k} = \lambda\left(x\hat{i} - \hat{j} + \hat{k}\right)$

$1 = \lambda x, y = -\lambda, z = \lambda$

Now, $x\hat{i} + y\hat{j} + z\hat{k} = \frac{1}{\lambda}\hat{i} - \lambda\hat{j} + \lambda\hat{k}$

Unit vector parallel to

$x\hat{i} + y\hat{j} + z\hat{k} = \frac{\frac{1}{\lambda}\hat{i} - \lambda\hat{j} + \lambda\hat{k}}{\sqrt{\frac{1}{\lambda^2} + \lambda^2 + \lambda^2}}$

$$=\frac{\hat{i}+\lambda^2\hat{j}+\lambda^2\hat{k}}{\sqrt{1+2\lambda^4}}$$

Let $\lambda = 1$, possible unit vector $=\dfrac{\hat{i}-\hat{j}+\hat{k}}{\sqrt{3}}$

65. (c) Let $\vec{r} = x\vec{a} + y\vec{b} + 2\vec{c}$

Given that $|\vec{a}| = |\vec{b}| = |\vec{c}|$ and $\vec{a}\cdot\vec{b} = \vec{b}\cdot\vec{c} = \vec{c}\cdot\vec{a} = 0$

$$\vec{a}\times\left[(\vec{r}-\vec{b})\times\vec{a}\right]+\vec{b}\times\left[(\vec{r}-\vec{c})\times\vec{b}\right]$$
$$+\vec{c}\times\left[(\vec{r}-\vec{a})\times\vec{c}\right]=\vec{0}$$
$$\Rightarrow(\vec{a}\cdot\vec{a})(\vec{r}-\vec{b})-\left[\vec{a}\cdot(\vec{r}-\vec{b})\right]\vec{a}+(\vec{b}\cdot\vec{b})(\vec{r}-\vec{c})$$
$$-\left[\vec{b}\cdot(\vec{r}-\vec{c})\right]\cdot\vec{b}+(\vec{c}\cdot\vec{c})(\vec{c}\cdot\vec{a})-\left[\vec{c}\cdot(\vec{r}-\vec{a})\right]\vec{c}=0$$
$$\Rightarrow|\vec{a}|^2(\vec{r}-\vec{b})-(\vec{a}\cdot\vec{r})\vec{a}+|\vec{b}|^2(\vec{r}-\vec{c})-(\vec{b}-\vec{r})\vec{b}$$
$$+|\vec{c}|^2(\vec{c}-\vec{a})-(\vec{c}-\vec{r})\vec{c}=0$$
$$=|\vec{a}|^2\left[3\vec{r}-(\vec{a}+\vec{b}+\vec{c})\right]$$
$$-(\vec{r}\cdot\vec{a})\vec{a}+(\vec{r}\cdot\vec{b})\vec{b}+(\vec{r}\cdot\vec{c})\vec{c}=0$$
$$\left[\because|\vec{a}|^2=|\vec{b}|^2=|\vec{c}|^2\right]$$
$$=|\vec{a}|^2\left[3\vec{r}-(\vec{a}+\vec{b}+\vec{c})-(x\vec{a}+y\vec{b}+z\vec{c})\right]=\vec{0}$$
$$\therefore 3\vec{r}-(\vec{a}+\vec{b}+\vec{c})-\vec{r}=\vec{0}$$
$$\Rightarrow\vec{r}=\frac{\vec{a}+\vec{b}+\vec{c}}{2}$$

66. (a) Given that $\vec{r}\times\vec{a} = -\vec{r}\times\vec{b}$

$\therefore \vec{r}$ is parallel to $\vec{a}+\vec{b}$ then

$\vec{r}\times(\vec{a}+\vec{b}) = 0$

Let $\vec{r} = \lambda(\vec{a}+\vec{b}) = \lambda(3\hat{i}-\hat{j}+2\hat{k})$

$\because \vec{r}\cdot(\alpha\hat{i}+2\hat{j}+\hat{k}) = 3 \Rightarrow \alpha\lambda = 1$ ...(i)

$\vec{r}\cdot(2\hat{i}+5\hat{j}-\alpha\hat{k}) = -1 \Rightarrow \lambda - 2\alpha\lambda = -1$

$\Rightarrow \lambda = 1$ and $\alpha = 1$ [from (i)]

$\because \vec{r} = 3\hat{i}-\hat{j}+2\hat{k}$

$\therefore |\vec{r}| = \sqrt{9+1+4} = \sqrt{14}$

So, $\alpha + |\vec{r}|^2 = 1+14 = 15$

67. (d) Let $\hat{n}$ be a unit vector in the direction of $\vec{a}\times\vec{b}$.

Then $\vec{a}, \vec{b}$ and are mutually perpendicular

$\Rightarrow \vec{a}\times\vec{b} = |\vec{a}||\vec{b}|\hat{n}$

$$\Rightarrow\vec{a}\times(\vec{a}\times\vec{b})=|\vec{a}||\vec{b}|(\vec{a}\times\hat{n})=-|\vec{a}|^2|\vec{b}|\vec{b}$$
$$\left[\because\vec{a}\times\hat{n}=-|\vec{a}|\vec{b}\right]$$
$$\Rightarrow\vec{a}\times(\vec{a}\times(\vec{a}\times\vec{b}))=-|\vec{a}|^3|\vec{b}|\hat{n}$$
$$\Rightarrow\vec{a}\times(\vec{a}\times(\vec{a}\times(\times\vec{b})))=|\vec{a}|^4|\vec{b}||\vec{b}|=|\vec{a}|^4\vec{b}$$

68. (b) We know that the volume of parallelepiped $=[\vec{a}\ \vec{b}\ \vec{c}]$

$$\begin{vmatrix}1 & 1 & n\\ 2 & 4 & -n\\ 1 & n & 3\end{vmatrix}=158$$
$$\Rightarrow(12+n^2)-1(6+n)+n(2n-4)=158$$
$$\Rightarrow 3n^2-5n-152=0$$
$$\Rightarrow 3n^2-24n+19n-152=0$$
$$\Rightarrow 3n(n-8)+19(n-8)=0$$
$$\Rightarrow n=8 \text{ or } n=\frac{-19}{3}$$

$\Rightarrow n = 8$ $(\because n \geq 0)$

$\therefore \bar{a} = \hat{i}+\hat{j}+8\hat{k}, \bar{b} = 2\hat{i}+4\hat{j}-8\hat{k}$ and

$\bar{c} = \hat{i}+8\hat{j}+3\hat{k}$

$\bar{a}\cdot\bar{c} = 1+8+24 = 33$

$\bar{b}\cdot\bar{c} = 2+32-24 = 10$

69. (d) $f(x) = \bar{a}\cdot(\bar{b}\times\bar{c})$

$$=\begin{vmatrix}\hat{i} & \hat{j} & \hat{k}\\ x & -2 & 3\\ -2 & x & -1\\ 7 & -2 & x\end{vmatrix}=x^3-27x+26$$
$$\Rightarrow f(x)=x^3-27x+26$$
$$\Rightarrow f'(x)=3x^2-27$$

For critical point $f'(x) = 0$

$\Rightarrow 3x^2-27=0 \Rightarrow x=-3, 3$

+   −   +
−3 (Max.)   3 (Min.)

The local maxima of $f(x)$ is, $x_0 = -3$.

Then $\bar{a}\cdot\bar{b}+\bar{b}\cdot\bar{c}+\bar{c}\cdot\bar{a}$

$= -2x-2x-3-14-2x-x+7x+4+3x = 3x-13$

So, value at $x = x_0, = \bar{a}\cdot\bar{b}+\bar{b}\cdot\bar{c}+\bar{c}\cdot\bar{a} = 3x-13$

$= 3\times(-3)-13 = -22$

70. (b) It is given that $\vec{u} = \hat{i} + \hat{j} + \lambda\hat{k}$, $\vec{v} = \hat{i} + \hat{j} + 3\hat{k}$ and $\vec{w} = 2\hat{i} + \hat{j} + \hat{k}$

Volume of parallelepiped $= [\vec{u} \cdot \vec{v} \cdot \vec{w}]$

$$\Rightarrow \pm 1 = \begin{vmatrix} 1 & 1 & \lambda \\ 1 & 1 & 3 \\ 2 & 1 & 1 \end{vmatrix} \Rightarrow -\lambda + 3 = \pm 1 \Rightarrow \lambda = 2 \text{ or } \lambda = 4$$

For $\lambda = 2$

$$\cos\theta = \frac{2+1+2}{\sqrt{6}\sqrt{6}} = \frac{5}{6}$$

For $\lambda = 4$

$$\cos\theta = \frac{2+1+4}{\sqrt{6}\sqrt{18}} = \frac{7}{6\sqrt{3}}$$

71. (d) $\because \vec{u}, \vec{a}$ and $\vec{b}$ are coplanar

$$\therefore \vec{u} = \lambda(\vec{a} \times \vec{b}) \times \vec{a} = \lambda\{\vec{a}^2 \cdot \vec{b} - (\vec{a} \cdot \vec{b})\vec{a}\}$$

$$= \lambda\{-4\hat{i} + 8\hat{j} + 16\hat{k}\} = \lambda'\{-\hat{i} + 2\hat{j} + 4\hat{k}\}$$

Also, $\vec{u}.\vec{b} = 24 \quad \Rightarrow \lambda' = 4$

$$\therefore \vec{u} = -4\hat{i} + 8\hat{j} + 16\hat{k} \Rightarrow |\vec{u}|^2 = 336$$

72. (c) Given: $\vec{a} = 2\hat{i} + \hat{j} - 2\hat{k}, \vec{b} = \hat{i} + \hat{j}$

$\Rightarrow |\vec{a}| = 3$

$\therefore \vec{a} \times \vec{b} = 2\hat{i} - 2\hat{j} + \hat{k}$

$|\vec{a} \times \vec{b}| = \sqrt{2^2 + 2^2 + 1^2} = 3$

We have $(\vec{a} \times \vec{b}) \times \vec{c} = |\vec{a} \times \vec{b}||\vec{c}|\sin 30n$

$$\Rightarrow |(\vec{a} \times \vec{b}) \times \vec{c}| = 3|\vec{c}| \cdot \frac{1}{2} \Rightarrow 3 = 3|\vec{c}| \cdot \frac{1}{2}$$

$\therefore |\vec{c}| = 2$

Now $|\vec{c} - \vec{a}| = 3$

On squaring, we get

$$\Rightarrow |\vec{c}|^2 + |\vec{a}|^2 - 2\vec{a} \cdot \vec{c} = 9 \Rightarrow 4 + 9 - 2\vec{a}.\vec{c} = 9$$

$\Rightarrow \vec{a} \cdot \vec{c} = 2 \qquad [\because \vec{c} \cdot \vec{a} = \vec{a} \cdot \vec{c}]$

73. (b) $\vec{a} \times (\vec{b} \times \vec{c}) = \frac{\sqrt{3}}{2}(\vec{b} + \vec{c})$

$$\Rightarrow (\vec{a} \cdot \vec{c})\vec{b} - (\vec{a} \cdot \vec{b})\vec{c} = \frac{\sqrt{3}}{2}\vec{b} + \frac{\sqrt{3}}{2}\vec{c}$$

On comparing both sides,

$$\vec{a} \cdot \vec{b} = -\frac{\sqrt{3}}{2} \Rightarrow \cos\theta = -\frac{\sqrt{3}}{2}$$

[$\because \vec{a}$ and $\vec{b}$ are unit vectors]

where θ is the angle between $\vec{a}$ and $\vec{b}$.

$$\theta = \frac{5\pi}{6}$$

74. (c) $(\vec{a} \times \vec{b}) \times \vec{c} = \frac{1}{3}|\vec{b}||\vec{c}|\vec{a}$

$$\Rightarrow -\vec{c} \times (\vec{a} \times \vec{b}) = \frac{1}{3}|\vec{b}||\vec{c}|\vec{a}$$

$$\Rightarrow -(\vec{c} \cdot \vec{b})\vec{a} + (\vec{c} \cdot \vec{a})\vec{b} = \frac{1}{3}|\vec{b}||\vec{c}|\vec{a}$$

$$\Rightarrow -|\vec{b}||\vec{c}|\cos\theta\, \vec{a} + (\vec{c} \cdot \vec{a})\vec{b} = \frac{1}{3}|\vec{b}||\vec{c}|\vec{a}$$

$\because \vec{a}, \vec{b}, \vec{c}$ are non-collinear, the above equation is possible only when

$$-\cos\theta = \frac{1}{3} \text{ and } \vec{c} \cdot \vec{a} = 0$$

$$\Rightarrow \cos\theta = -\frac{1}{3} \Rightarrow \sin\theta = \frac{2\sqrt{2}}{3}; \theta \in \text{II quad}$$

75. (b) LHS $= (\vec{a} \times \vec{b}) \cdot [(\vec{b} \times \vec{c}) \times (\vec{c} \times \vec{a})]$

$$= (\vec{a} \times \vec{b}) \cdot [(\vec{b} \times \vec{c}.\vec{a})\vec{c} - (\vec{b} \times \vec{c} \cdot \vec{c})\vec{a}]$$

$$= (\vec{a} \times \vec{b}) \cdot [[\vec{b}\ \vec{c}\ \vec{a}]\vec{c}] \qquad [\because \vec{b} \times \vec{c} \cdot \vec{c} = 0]$$

$$= [\vec{a}\,\vec{b}\,\vec{c}] \cdot (\vec{a} \times \vec{b}.\vec{c}) = [\vec{a}\,\vec{b}\,\vec{c}]^2$$

$$= [\vec{a} \times \vec{b}\ \vec{b} \times \vec{c}\ \vec{c} \times \vec{a}] = [\vec{a}\ \vec{b}\ \vec{c}]^2 \qquad \text{So } \lambda = 1$$

76. (c) Given $|2\vec{a} - \vec{b}| = 5$

$$\sqrt{(2|\vec{a}|)^2 + |\vec{b}|^2 - 2 \times |2\vec{a}||\vec{b}|\cos\theta} = 5$$

Putting values of $|\vec{a}|$ and $|\vec{b}|$, we get

$$(2 \times 2)^2 + (3)^2 - 24\cos\theta = 25$$

$$\Rightarrow \cos\theta = 0 \Rightarrow \theta = \frac{\pi}{2}$$

$$|2\vec{a} + \vec{b}| = \sqrt{16 + 9 + 24\cos\theta} = \sqrt{25} = 5$$

77. (c) We have,

$\overline{AB} + \overline{BC} + \overline{CA} = 0 \Rightarrow \overline{BC} = \overline{AC} - \overline{AB}$

Let M be the mid-point of BC.

Now, $\overline{BM} = \dfrac{\overline{AC} - \overline{AB}}{2} \qquad \left(\because \overline{BM} = \dfrac{\overline{BC}}{2}\right)$

Also, we have

$\overline{AB} + \overline{BM} + \overline{MA} = 0$

$$\Rightarrow \overline{AB} + \frac{\overline{AC} - \overline{AB}}{2} = \overline{AM}$$

$$\Rightarrow 1\overline{AM} = \frac{\overline{AB} + \overline{AC}}{2}$$

$= 4\hat{i} - \hat{j} + 4\hat{k}$

$\Rightarrow |\overline{AM}| = \sqrt{33}$

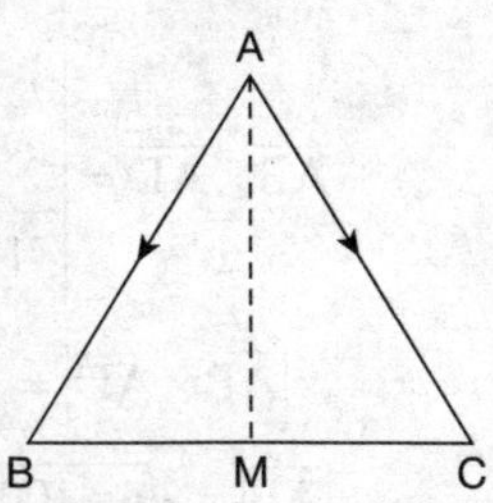

78. (c) Given that

$\vec{a}\times\left(2\hat{i}+3\hat{j}+4\hat{k}\right)=\left(2\hat{i}+3\hat{j}+4\hat{k}\right)\times\vec{b}$

$\Rightarrow\left(\vec{a}+\vec{b}\right)\times\left(2\hat{i}+3\hat{j}+4\hat{k}\right)=\vec{0}$

But $\vec{a}+\vec{b}\neq 0$ and $2\hat{i}+3\hat{j}+4\hat{k}=\vec{0}$

$\therefore\left(\vec{a}+\vec{b}\right)\parallel\left(2\hat{i}+3\hat{j}+4\hat{k}\right)$

Let $\vec{a}+\vec{b}=\lambda\left(2\hat{i}+3\hat{j}+4\hat{k}\right)$

Also given that $\left|\vec{a}+\vec{b}\right|=\sqrt{29}\Rightarrow\lambda=\pm1$

$\therefore\vec{a}+\vec{b}=\pm\left(2\hat{i}+3\hat{j}+4\hat{k}\right)$

So, $\left(\vec{a}+\vec{b}\right)\cdot\left(-7\hat{i}+2\hat{j}+3\hat{k}\right)=\pm4$

79. (c) We know that vector in the plane of $\vec{a}$ and $\vec{b}$ is $\vec{v}=\vec{a}+\lambda\vec{b}$

$\Rightarrow\vec{v}=(1+\lambda)\hat{i}+(1-\lambda)\hat{j}+(1+\lambda)\hat{k}$

$\because$ Projection of $\vec{v}$ on $\vec{c}$ is $\dfrac{1}{\sqrt{3}}$

$\therefore\dfrac{\vec{v}\cdot\vec{c}}{|\vec{c}|}=\dfrac{1}{\sqrt{3}}\Rightarrow\dfrac{(1+\lambda)-(1-\lambda)-(1+\lambda)}{\sqrt{3}}=\dfrac{1}{\sqrt{3}}$

$\Rightarrow 1-\lambda=-1$

$\Rightarrow\lambda=2\Rightarrow\lambda=2$

$\vec{v}=3\hat{i}-\hat{j}+3\hat{k}$

80. (a) Since $\overrightarrow{PQ}=6\hat{i}+\hat{j}, \overrightarrow{QR}=-\hat{i}+3\hat{j}, \overrightarrow{SR}=6\hat{i}+\hat{j}$, $\overrightarrow{SP}=-\hat{i}+3\hat{j}$.

Here $\overrightarrow{PQ}=\overrightarrow{SR}$; $\overrightarrow{QR}=\overrightarrow{PS}$ and $\overrightarrow{PQ}\cdot\overrightarrow{PS}\neq 0$

Also $\left|\overrightarrow{PQ}\right|\neq\left|\overrightarrow{QR}\right|$

$\Rightarrow$ PQRS is a parallelogram but neither a rhombus nor a rectangle.

81. (b)

$$\overrightarrow{AB}\times\overrightarrow{AD}=\begin{vmatrix}\hat{i} & \hat{j} & \hat{k}\\ 2 & 10 & 11\\ -1 & 2 & 2\end{vmatrix}=-2\hat{i}-15\hat{j}+14\hat{k}$$

$\therefore\left|\overrightarrow{AB}\times\overrightarrow{AD}\right|=\sqrt{4+225+196}=\sqrt{425}$

$\left|\overrightarrow{AB}\right|=\sqrt{4+100+121}=\sqrt{225}=15$

$\left|\overrightarrow{AD}\right|=\sqrt{1+4+4}=3$

$\therefore\sin(90-\alpha)=\dfrac{\left|\overrightarrow{AB}\times\overrightarrow{AD}\right|}{\left|\overrightarrow{AB}\right|\left|\overrightarrow{AD}\right|}$

$\therefore\sin(90-\alpha)=\dfrac{\sqrt{425}}{15\times3}=\dfrac{\sqrt{17}}{9}$

$\Rightarrow\cos\alpha=\dfrac{\sqrt{17}}{9}$

82. (c) Given $\vec{a},\vec{b},\vec{c}$ and $\vec{d}$ are unit vectors,

Let $\vec{a}\times\vec{b}=(\sin\alpha)\overrightarrow{n_1}$ and $\vec{c}\times\vec{d}=(\sin\beta)\overrightarrow{n_2}$

where $\overrightarrow{n_1}$ and $\overrightarrow{n_2}$ are unit normal vectors.

Then $\left(\vec{a}\times\vec{b}\right)\cdot\left(\vec{c}\times\vec{d}\right)=1$

$\Rightarrow(\sin\alpha)\overrightarrow{n_1}\cdot(\sin\beta)\overrightarrow{n_2}=1$

$\Rightarrow\sin\alpha\sin\beta\overrightarrow{n_1}\cdot\overrightarrow{n_2}=1$

$\Rightarrow\sin\alpha\sin\beta\cos\gamma=1$

where γ is the angle between $\overrightarrow{n_1}$ and $\overrightarrow{n_2}$.

$\Rightarrow\alpha=\dfrac{\pi}{2},\beta=\dfrac{\pi}{2}$ and $\gamma=0°$

If $\gamma=0°\Rightarrow\vec{a}\times\vec{b}\parallel\vec{c}\times\vec{d}$

Let $\vec{a}\times\vec{b}=\lambda\left(\vec{c}\times\vec{d}\right)$

$\Rightarrow\left(\vec{a}\times\vec{b}\right)\cdot\vec{c}=\lambda\left(\vec{c}\times\vec{d}\right)\cdot\vec{c}=0$ and

$\left(\vec{a}\times\vec{d}\right)\cdot\vec{d}=\lambda\left(\vec{c}\times\vec{d}\right)\cdot\vec{d}=0$

$\left[\vec{a}\ \vec{b}\ \vec{c}\right]=\left[\vec{a}\ \vec{b}\ \vec{d}\right]=0$

$\therefore\vec{a},\vec{b},\vec{c}$ are coplanar and $\vec{a},\vec{b},\vec{d}$ are coplanar.

$\Rightarrow\vec{a},\vec{b},\vec{c},\vec{d}$ are coplanar

and $\alpha=90°\Rightarrow\vec{a}\perp\vec{b}$ and $\beta=90°\Rightarrow\vec{c}\perp\vec{d}$

But angle between $\vec{a}$ and $\vec{c}$ is $\pi/3$ $\left(\because\vec{a}\cdot\vec{c}=\dfrac{1}{2}\right)$

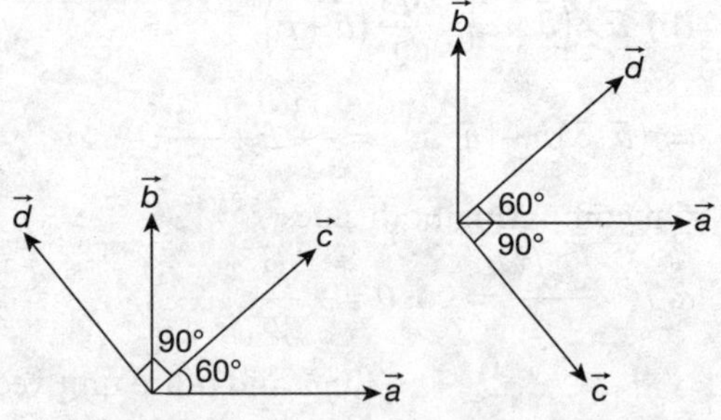

The possible cases are shown in the above figures and in any case $\vec{b}$ and $\vec{d}$ are non-parallel vectors.

Integer and Numerical Value Type

1. From the first relation

$\vec{a}\cdot\vec{x}+\vec{a}\cdot\vec{y}=|\vec{a}|^2$

$\Rightarrow \vec{a}\cdot\vec{y}=|\vec{a}|^2-1$ ...(i)

$(\because \vec{x}\cdot\vec{a}=1)$

Now taking the cross product of both sides of the second equation, we get

$\vec{a}\times(\vec{x}\times\vec{y})=\vec{a}\times\vec{b}$

$\Rightarrow (\vec{a}\cdot\vec{y})\vec{x}-(\vec{a}\cdot\vec{x})\vec{y}=\vec{a}\times\vec{b}$

$\Rightarrow \left(|\vec{a}|^2-1\right)\cdot\vec{x}-\vec{y}=\vec{a}\times\vec{b}$ ...(ii)

Now, the first equation is $\vec{x}\times\vec{y}=\vec{a}$

If we add (ii) and this equation, we get

$|\vec{a}|^2\cdot\vec{x}=\vec{a}+\left(\vec{a}\times\vec{b}\right)$

$$\Rightarrow \vec{x}=\frac{\vec{a}+\left(\vec{a}\times\vec{b}\right)}{|\vec{a}|^2}$$

$\Rightarrow \lambda=1$

2. From the second relation $\vec{a}\cdot\vec{c}=1$. From the first relation

$(\vec{a}\cdot\vec{c})\vec{b}-\left(\vec{a}\cdot\vec{b}\right)\vec{c}+\left(\vec{a}\cdot\vec{b}\right)\vec{b}$

$=(4-2\beta-\sin\alpha)\vec{b}+\left(b^2-1\right)\vec{c}$

Since $\vec{b}$ and $\vec{c}$ are non-collinear, equate their coefficients to zero to get $\alpha=1$.

3. The equations of $\overline{\text{AB}}$ and $\overline{\text{CD}}$ are

$\vec{r}=9\hat{i}-\hat{j}+7\hat{k}+\lambda\left(4\hat{i}-\hat{j}+3\hat{k}\right)$ and

$\vec{r}=7\hat{i}-2\hat{j}+7\hat{k}+\mu\left(2\hat{i}-\hat{j}+2\hat{k}\right)$

whence their point of intersection P is determined as (1, 1, 1).

Now $\overline{\text{PQ}}=\lambda\overline{\text{AB}}\times\overline{\text{CD}}$ use $\left|\overline{\text{PQ}}\right|=15$ to get λ we will get $\bar{\text{D}}=6\hat{i}-9\hat{j}-9\hat{k}$

$\Rightarrow \mu=9$

4. We have $\overrightarrow{\text{OP}}=\overrightarrow{\text{OA}}+\lambda\overrightarrow{\text{AB}}$

$(\because$ P lies on AB)

$=(6+3\lambda)\hat{i}+(7-\lambda)\hat{j}+(4+\lambda)\hat{k}$ ...(i)

$\overrightarrow{\text{OQ}}=\overrightarrow{\text{CD}}+\mu\overrightarrow{\text{CD}}$

$=-9\hat{j}+2\hat{k}+\mu\left(-3\hat{i}+2\hat{j}+4\hat{k}\right)$

$=-3\mu\hat{i}+(2\mu-9)\hat{j}+(4\mu+2)\hat{k}$ ...(ii)

Now

$\overrightarrow{\text{PQ}}=(-3\mu-3\lambda-6)\hat{i}+(2\mu-9-7+\lambda)\hat{j}$
$+(4\mu+2-4-\lambda)\hat{k}$

$=(-3\mu-3\lambda-6)\hat{i}+(2\mu+\lambda-16)\hat{j}+(4\mu-\lambda-2)\hat{k}$

Now as $\overrightarrow{\text{PQ}}$ is perpendicular to $\overrightarrow{\text{AB}}$ and $\overrightarrow{\text{CD}}$, $\overrightarrow{\text{PQ}}\cdot\overrightarrow{\text{AB}}=0$, $\overrightarrow{\text{PQ}}\cdot\overrightarrow{\text{CD}}=0$

$=3(-3\mu-3\lambda-6)-(2\mu+\lambda-16)+(4\mu-\lambda-2)=0$

$3(-3\mu-3\lambda-6)+2(2\mu+\lambda-16)+4(4\mu-\lambda-2)=0$

$\Rightarrow 11\lambda+7\mu+4=0, 7\lambda+29\mu-22=0$

$\Rightarrow \lambda=-1, \mu=1$

Now from (i) and (ii),

$\overrightarrow{\text{OP}}=3\hat{i}+8\hat{j}+3\hat{k}$ (on putting $\lambda=-1$)

and $\overrightarrow{\text{OQ}}=-3\hat{i}-7\hat{j}+6\hat{k}$ (on putting $\mu=1$)

$\Rightarrow p=8$

5. $\vec{p}+\vec{q}=3\hat{i}+5\hat{j}+2\hat{k}$ and $\vec{p}-\vec{q}=\hat{i}+\hat{j}$

$$\text{Now, }(\vec{p}+\vec{q})\times(\vec{p}-\vec{q})=\begin{vmatrix}\hat{i}&\hat{j}&\hat{k}\\3&5&2\\1&1&0\end{vmatrix}$$

$=-2\hat{i}+2\hat{j}-2\hat{k}$

$\therefore \vec{r}=\lambda\left(-2\hat{i}+2\hat{j}-2\hat{k}\right)$

$|\vec{r}|=\pm2\sqrt{3}\lambda=\sqrt{3}\Rightarrow\lambda=\pm\frac{1}{2}$

$\therefore |\alpha|+|\beta|+|\gamma|=1+1+1=3$

6. It is given that,

$|\vec{v}|=|\vec{v}_2|$

$\Rightarrow 3p^2+1=4+(p+1)^2$

$\Rightarrow 2p^2-2p-4=0\Rightarrow p^2-p-2=0$

$p=2,-1$ (Rejected)

$$\cos\theta=\frac{\vec{v}_1\cdot\vec{v}_2}{|\vec{v}_1||\vec{v}_2|}=\frac{2\sqrt{3}p+(p+1)}{\sqrt{(p+1)^2+4}\sqrt{3p^2+1}}$$

$$\Rightarrow \cos\theta=\frac{4\sqrt{3}+3}{\sqrt{13}\sqrt{13}}=\frac{4\sqrt{3}+3}{13}$$

$$\Rightarrow \tan\theta=\frac{\sqrt{112-24\sqrt{3}}}{4\sqrt{3}+3}=\frac{6\sqrt{3}-2}{4\sqrt{3}+3}=\frac{\alpha\sqrt{3}-2}{4\sqrt{3}+3}$$

$\Rightarrow \alpha=6$

7. $\vec{a}\times\vec{b}=\vec{c}$

$\therefore \vec{a}$ and $\vec{c}$ are perpendicular.

$\therefore |\vec{a}\times\vec{c}|=|\vec{a}||\vec{c}|$

Now, $\left(\vec{a}\times\vec{b}\right)\cdot\vec{c}=|\vec{c}|^2=2=\left[\vec{a}\;\vec{b}\;\vec{c}\right]$

Given that projection of $\vec{b}$ on $\vec{a}\times\vec{c}\,\ell$

$$\Rightarrow \frac{\left|\vec{b}\cdot(\vec{a}\times\vec{c})\right|}{|\vec{a}\times\vec{c}|}=\ell \Rightarrow \frac{\left|\vec{b}\;\vec{a}\;\vec{c}\right|}{|\vec{a}||\vec{c}|}=\ell$$

$$\Rightarrow \frac{\left|-\left[\vec{a}\,\vec{b}\,\vec{c}\right]\right|}{\sqrt{3}\cdot\sqrt{2}}=\ell$$

$$\Rightarrow \frac{2}{\sqrt{6}}=\ell \Rightarrow 3\ell^2=2$$

8. $\vec{a}\times\vec{b}=\begin{vmatrix}\hat{i} & \hat{j} & \hat{k}\\ 1 & a & 3\\ 3 & -a & 1\end{vmatrix}$

$=4a\hat{i}+8\hat{j}-4a\hat{k}=4\left(a\hat{i}+2\hat{j}-a\hat{k}\right)$

Area of parallelogram $=\left|\vec{a}\times\vec{b}\right|$

$\Rightarrow 8\sqrt{3}=4\sqrt{2a^2+4}\Rightarrow a=\pm 2$

$\vec{a}\cdot\vec{b}=3-a^2+3=2$

9. $I_1:\vec{r}=\left(3\hat{i}-\hat{j}+4\hat{k}\right)+t\left(\hat{i}+2\hat{j}+2\hat{k}\right)$ and

$I_2:\vec{r}=\left(3\hat{i}+3\hat{j}+2\hat{k}\right)+s\left(2\hat{i}+2\hat{j}+\hat{k}\right)$

$$\therefore \vec{b}_1\times\vec{b}_2=\begin{vmatrix}\hat{i} & \hat{j} & \hat{k}\\ 1 & 2 & 2\\ 2 & 2 & 1\end{vmatrix}=-2\hat{i}+3\hat{j}-2\hat{k}$$

So, $I:\vec{r}=\lambda\left(-2\hat{i}+3\hat{j}-2\hat{k}\right)$

$\therefore$ Point of intersection of I and I_1 is $P(2,-3,2)$

Let a point I_2 be $Q(2\mu+3,2\mu+3,\mu+2)$

$\because PQ=\sqrt{17}$

$\Rightarrow (2\mu+1)^2+(2\mu+6)^2+\mu^2=17$

$\Rightarrow \mu=-\dfrac{10}{9}$ or -2

Given that Q lies in 1st octant, then $Q\left(\dfrac{7}{9},\dfrac{7}{9},\dfrac{8}{9}\right)$

Hence, $18(a+b+c)=44$

10. Since, $\vec{a}$ is perpendicular to $\vec{b}$

$\Rightarrow \vec{a}\cdot\vec{b}=0$

$\Rightarrow 1+15+\alpha\beta=0\Rightarrow \alpha\beta=-16$ ...(i)

Also,

$\left|\vec{b}\times\vec{c}\right|^2=75\Rightarrow\left(10+\beta^2\right)14-(5-3\beta)^2=75$

$\Rightarrow 5\beta^2+30\beta+40=0\Rightarrow\beta=-4,-2$

Then, from (i)

$\alpha=4,8$

$\Rightarrow |a|^2_{\max}=(26+\alpha)_{\max}=90$

11. $\hat{i}\times\left(\vec{a}\times\hat{i}\right)=\left(\hat{i}\cdot\hat{i}\right)\vec{a}-\left(\hat{i}\cdot\vec{a}\right)\hat{i}=\hat{j}+2\hat{k}$

Similarly, $\hat{j}\times\left(\vec{a}\times\hat{j}\right)=2\hat{i}+2\hat{k}$,

$\hat{k}\times\left(\vec{a}\times\hat{k}\right)=2\hat{i}+\hat{j}$

$\therefore \left|\hat{j}+2\hat{k}\right|^2+\left|2\hat{i}+2\hat{k}\right|^2+\left|2\hat{i}+\hat{j}\right|^2=5+8+5=18$

12. $\vec{b}\cdot\vec{c}=10\Rightarrow\left|\vec{b}\right||\vec{c}|\cos\left(\dfrac{\pi}{3}\right)=10$

$\Rightarrow 5\cdot|\vec{c}|\cdot\dfrac{1}{2}=10\Rightarrow|\vec{c}|=4$

Since $\vec{b}$ is perpendicular to the vector $\vec{b}\times\vec{c}$, then $\vec{a}\cdot\left(\vec{b}\times\vec{c}\right)=0$.

Now, $\left|\vec{a}\times\left(\vec{b}\times\vec{c}\right)\right|=|\vec{a}|\left|\vec{b}\times\vec{c}\right|\sin\left(\dfrac{\pi}{2}\right)$

$=\sqrt{3}\times\left|\vec{b}\right||\vec{c}|\sin\dfrac{\pi}{3}\times 1$

Hence, $\left|\vec{a}\times\left(\vec{b}\times\vec{c}\right)\right|=30$.

13. Let the angle between $\vec{a}$ and $\vec{b}$ be θ.

$\left|\vec{a}+\vec{b}\right|=\sqrt{1+1+2\cos\theta}=2\left|\cos\dfrac{\theta}{2}\right|\left[\because|a|=|b|=1\right]$

Similarly, $\left|\vec{a}-\vec{b}\right|=2\left|\sin\dfrac{\theta}{2}\right|$

So, $\sqrt{3}\left|\vec{a}+\vec{b}\right|+\left|\vec{a}-\vec{b}\right|=2\left[\sqrt{3}\left|\cos\dfrac{\theta}{2}\right|+\left|\sin\dfrac{\theta}{2}\right|\right]$

$\because$ Maximum value of $(a\cos\theta+b\sin\theta)=\sqrt{a^2+b^2}$

$\therefore$ Maximum value $=2\sqrt{\left(\sqrt{3}\right)^2+(1)^2}=4$

14. $\because|\vec{x}+\vec{y}|=|\vec{x}|$

Squaring both sides, we get

$|\vec{x}|^2+2\vec{x}\cdot\vec{y}+|\vec{y}|^2=|\vec{x}|^2$

$\Rightarrow 2\vec{x}\cdot\vec{y}+\vec{y}\cdot\vec{y}=0$ ...(i)

Also, $2\vec{x}+\lambda\vec{y}$ and $\vec{y}$ are perpendicular

$\therefore 2\vec{x}\cdot\vec{y}+\lambda\vec{y}\cdot\vec{y}=0$ ...(ii)

Comparing (i) and (ii), $\lambda=1$

15. $\because$ Projection of $\vec{b}$ on $\vec{a}$ = Projection of $\vec{c}$ on $\vec{a}$

$\therefore \vec{a}\cdot\vec{b} = \vec{a}\cdot\vec{c}$

Given, $\vec{b}\cdot\vec{c} = 0$

$\because \left|\vec{a}+\vec{b}-\vec{c}\right|^2 = |\vec{a}|^2 + \left|\vec{b}\right|^2 + |\vec{c}|^2 - 2\vec{a}\cdot\vec{b} - 2\vec{b}\cdot\vec{c} - 2\vec{a}\cdot\vec{c}$

$= 4+16+16 = 36$

$\Rightarrow \left|\vec{a}+\vec{b}-\vec{c}\right| = 6$

16. $|\vec{a}| = \left|\vec{b}\right| = |\vec{c}| = 1$

$\left|\vec{a}-\vec{b}\right|^2 + |\vec{a}-\vec{c}|^2 = 8$

$\Rightarrow \vec{a}\cdot\vec{b} + \vec{a}\cdot\vec{c} = -2$

Now, $\left|\vec{a}+2\vec{b}\right|^2 + |\vec{a}+2\vec{c}|^2$

$= 2|\vec{a}|^2 + 4\left|\vec{b}\right|^2 + 4|\vec{c}|^2 + 4\left(\vec{a}\cdot\vec{b}+\vec{a}\cdot\vec{c}\right) = 2$

17. Let $P(1,-1,3)$, $Q(2,-4,11)$, $R(-1,2,3)$ and $S(3,-2,10)$

Then, $\overrightarrow{PQ} = \hat{i} - 3\hat{j} + 8\hat{k}$

Projection of $\overrightarrow{PQ}$ on $\overrightarrow{RS}$

$= \dfrac{\overrightarrow{PQ}\cdot\overrightarrow{RS}}{\left|\overrightarrow{RS}\right|} = \dfrac{4+12+56}{\sqrt{(4)^2+(4)^2+(7)^2}} = 8$

18. Let position vector of A and B be $\vec{a}$ and $\vec{b}$ respectively.

$\therefore$ Position vector of P is $\overrightarrow{OP} = \dfrac{\lambda\vec{b}+\vec{a}}{\lambda+1}$

λ : 1
A P B
(1, 1, 1) (2, 1, 3)

Given $\overrightarrow{OB}\cdot\overrightarrow{OP} - 3\left|\overrightarrow{OA}\times\overrightarrow{OP}\right|^2 = 6$

$\Rightarrow \vec{b}\cdot\left(\dfrac{\lambda\vec{b}+\vec{a}}{\lambda+1}\right) - 3\left|\vec{a}\times\dfrac{\lambda\vec{b}+\vec{a}}{\lambda+1}\right|^2 = 6$

$\Rightarrow \dfrac{\vec{a}\cdot\vec{b}+\lambda\left|\vec{b}\right|^2}{\lambda+1} - \dfrac{3\lambda^2}{(\lambda+1)^2}\left|\vec{a}\times\vec{b}\right|^2 = 6$

$\left(\because \vec{a}\times\vec{b} = 2\hat{i}-\hat{j}-\hat{k} \text{ and } \vec{a}\cdot\vec{b} = 6\right)$

$\Rightarrow \dfrac{6+14\lambda}{\lambda+1} - \dfrac{18\lambda^2}{(\lambda+1)^2} = 6$

$\Rightarrow 6 + \dfrac{8\lambda}{\lambda+1} - \dfrac{18\lambda^2}{(\lambda+1)^2} = 6$

Let $\dfrac{\lambda}{\lambda+1} = t$

$\Rightarrow 18t^2 - 8t = 0 \Rightarrow 2t(9t-4) = 0$

$\Rightarrow t = 0, \dfrac{4}{9}$

$\therefore \dfrac{\lambda}{\lambda+1} = \dfrac{4}{9} \Rightarrow \lambda = \dfrac{4}{5} = 0.8$

19. $\begin{vmatrix} a+1 & a & a \\ a & a+1 & a \\ a & a & a+1 \end{vmatrix} = 0$

$\Rightarrow 3a+1 = 0$

$\Rightarrow a = -\dfrac{1}{3}$

The given vectors are

$\vec{p} = \dfrac{2}{3}\hat{i} - \dfrac{1}{3}\hat{j} - \dfrac{1}{3}\hat{k} = \dfrac{1}{3}\left(2\hat{i}-\hat{j}-\hat{k}\right)$

$\vec{q} = \dfrac{1}{3}\left(-\hat{i}+2\hat{j}-\hat{k}\right) \vec{r} = \dfrac{1}{3}\left(-\hat{i}-\hat{j}+2\hat{k}\right)$

Now, $\vec{p}\cdot\vec{q} = \dfrac{1}{9}(-2-2+1) = -\dfrac{1}{3}$

$\vec{r}\times\vec{q} = \dfrac{1}{9}\begin{vmatrix} i & j & k \\ -1 & 2 & -1 \\ -1 & -1 & 2 \end{vmatrix}$

$= \dfrac{1}{9}\left(i(4-1) - j(-2-1) + k(1+2)\right)$

$= \dfrac{1}{9}(3i+3j+3k) = \dfrac{i+j+k}{3}$

$|\vec{r}\times\vec{q}| = \dfrac{1}{3}\sqrt{3} \Rightarrow |\vec{r}\times\vec{q}|^2 = \dfrac{1}{3}$

$3(\vec{p}\cdot\vec{q})^2 - \lambda|\vec{r}\times\vec{q}|^2 = 0$

$\Rightarrow 3\cdot\dfrac{1}{9} - \lambda\cdot\dfrac{1}{3} = 0 \Rightarrow \lambda = 1$

20. $\because |\vec{a}| = \left|\vec{b}\right| = |\vec{c}| = 1$

$\left|\vec{a}-\vec{b}\right|^2 + \left|\vec{b}-\vec{c}\right|^2 + |\vec{c}-\vec{a}|^2 = 9$

$\Rightarrow 2\left(|\vec{a}|^2 + \left|\vec{b}\right|^2 + |\vec{c}|^2\right) - 2\left(\vec{a}\cdot\vec{b} + \vec{b}\cdot\vec{c} + \vec{c}\cdot\vec{a}\right) = 9$

$\Rightarrow \vec{a}\cdot\vec{b} + \vec{b}\cdot\vec{c} + \vec{c}\cdot\vec{a} = \dfrac{-3}{2}$

Also,

$\left|\vec{a}+\vec{b}+\vec{c}\right|^2 = |\vec{a}|^2 + \left|\vec{b}\right|^2 + |\vec{c}|^2 + 2\left(\vec{a}\cdot\vec{b} + \vec{b}\cdot\vec{c} + \vec{c}\cdot\vec{a}\right)$

$= 1+1+1+2\times\left(\dfrac{-3}{2}\right) = 0$

$\Rightarrow \vec{a}+\vec{b}+\vec{c} = 0 \Rightarrow \left(\vec{b}+\vec{c}\right) = -\vec{a}$

$\therefore \left|2\vec{a}+5\left(\vec{b}+\vec{c}\right)\right| = |2\vec{a}-5\vec{a}| = |-3\vec{a}| = 3$

21. $\because \vec{r}\times\vec{b}=\vec{c}\times\vec{b}\Rightarrow\vec{r}\times\vec{b}-\vec{c}\times\vec{b}=\vec{0}$

$\Rightarrow(\vec{r}-\vec{c})\times\vec{b}=\vec{0}\Rightarrow\vec{r}-\vec{c}\parallel\vec{b}$

Let $\vec{r}-\vec{c}=\lambda\vec{b}\Rightarrow\vec{r}=\vec{c}+\lambda\vec{b}$

$\Rightarrow\vec{r}=\hat{i}+2\hat{j}+3\hat{k}-\lambda\hat{i}+\lambda\hat{j}=(1-\lambda)\hat{i}+(2+\lambda)\hat{j}+3\hat{k}$

$\therefore \vec{r}\cdot\vec{a}=0\Rightarrow -1+\lambda-3=0\Rightarrow\lambda=4$

$\therefore \vec{r}=-3\hat{i}+6\hat{j}+3\hat{k}$

Now, $\vec{r}\cdot\vec{b}=3+6=9$

22. We have $|\vec{a}|=1|\vec{b}|=1$ and $\vec{a}\cdot\vec{b}=0$

Now, $(2\vec{a}+\vec{b})\cdot\left[(\vec{a}\times\vec{b})\times(\vec{a}-2\vec{b})\right]$

$=(2\vec{a}+\vec{b})\cdot\left[(\vec{a}\cdot\vec{a})\vec{b}-(\vec{a}\cdot\vec{b})\vec{a}-2(\vec{b}\cdot\vec{a})+2(\vec{b}\cdot\vec{b})\vec{a}\right]$

$=(2\vec{a}+\vec{b}).\left[(\vec{a})^2\vec{b}+2(\vec{b})^2\vec{a}\right]$

$=(2\vec{a}+\vec{b})\cdot(2\vec{a}+\vec{b})=4|\vec{a}|^2+|\vec{b}|^2 \qquad (\because \vec{a}\cdot\vec{b}=0)$

$=4+1=5$

Assertion and Reason

1. (b) Both the statements are correct but Statement 2 is not correct explanation of Statement 1 because vectors $\vec{b},\vec{c},\vec{d}$ in Statement 1 are coplanar.

2. (d) Statement 1 is false and Statement 2 is true.

Since $\vec{a}\cdot(\vec{b}\times\vec{c})=0$

$\therefore \vec{a},\vec{b},\vec{c}$ are coplanar.

3. (b) Statement 1 is correct and Statement 2 is correct but Statement 2 is not correct explanation of Statement 1.

4. (a)

$3\vec{a}-2\vec{b}+5\vec{c}-6\vec{d}$

$=(2\vec{a}-2\vec{b})+(-5\vec{a}+5\vec{c})+(6\vec{a}+6\vec{d})$

$=-2\overline{AB}+5\overline{AC}-6\overline{AD}=\vec{0}$

$\therefore \overline{AB},\overline{AC}$ and $\overline{AD}$ are linearly dependent. Hence by statement 2, the statement 1 is true.

5. (d) Statement 1

$$\vec{b}_1=\left(\frac{(2\hat{i}+\hat{j}-3\hat{k}).(3\hat{i}-\hat{j})}{|3\hat{i}-\hat{j}|}\right)\frac{3\hat{i}-\hat{j}}{|3\hat{i}-\hat{j}|}=\frac{3\hat{i}}{2}-\frac{\hat{j}}{2}$$

$\therefore \vec{b}_2=2\hat{i}+\hat{j}-3\hat{k}-\frac{3\hat{i}}{2}+\frac{\hat{j}}{2}=\frac{\hat{i}}{2}+\frac{3\hat{j}}{2}-3\hat{k}$

$\therefore$ Statement 1 is false

Statement 2 is true

JEE Advanced

Objective Questions

One Option Correct

1. (d) $0=(\vec{a}+\vec{b})\cdot(2\vec{a}+3\vec{b})\times(3\vec{a}-2\vec{b})$

$=(\vec{a}+\vec{b})\cdot(-4\vec{a}\times\vec{b}-9\vec{a}\times\vec{b})$

$=-13(\vec{a}+\vec{b})\cdot(\vec{a}\times\vec{b})$

which is true for all values of $\vec{a}$ and $\vec{b}$.

2. (d) Volume of the parallelepiped formed by $\vec{a}',\vec{b}',\vec{c}'$ is 4.

$\therefore$ Volume of the parallelepiped formed by $\vec{a},\vec{b},\vec{c}$ is $\frac{1}{4}$.

$\vec{b}\times\vec{c}=\frac{(\vec{c}'\times\vec{a}')\times\vec{c}}{4}=\frac{1}{4}\vec{a}'$

$\therefore |\vec{b}\times\vec{c}|=\frac{\sqrt{2}}{4}=\frac{1}{2\sqrt{2}}$

$\therefore$ Altitude $=\frac{1}{4}\times2\sqrt{2}=\frac{1}{\sqrt{2}}$

3. (b) Let $\vec{a}=\lambda\vec{b}+\mu\vec{c}$

Then $\frac{\vec{a}\cdot\vec{b}}{ab}=\frac{\vec{a}\cdot\vec{d}}{ad}$

i.e., $\frac{(\lambda\vec{b}+\mu\vec{c})\cdot\vec{b}}{b}=\frac{(\lambda\vec{b}+\mu\vec{c})\cdot\vec{d}}{d}$

i.e., $\frac{\left[\lambda(2\hat{i}+\hat{j})+\mu(\hat{i}-\hat{j}+\hat{k})\right]\cdot(2\hat{i}+\hat{j})}{\sqrt{5}}$

$=\frac{\left[\lambda(2\hat{i}+\hat{j})+\mu(\hat{i}-\hat{j}+\hat{k})\right]\cdot(\hat{j}+2\hat{k})}{\sqrt{5}}$

i.e., $\lambda(4+1)+\mu(2-1)=\lambda(1)+\mu(-1+2)$,

i.e., $4\lambda=0$, i.e., $\lambda=0$

$\therefore \hat{a}=\frac{\hat{i}-\hat{j}+\hat{k}}{\sqrt{3}}$

4. (c) $\vec{r}\cdot\vec{a}=0, |\vec{r}\times\vec{b}|=|\vec{r}||\vec{b}|$ and $|\vec{r}\times\vec{c}|=|\vec{r}||\vec{c}|$

$\Rightarrow \vec{r}\perp\vec{a},\vec{b},\vec{c}$

$\therefore \vec{a},\vec{b},\vec{c}$ are coplanar $\qquad \therefore [\vec{a}\,\vec{b}\,\vec{c}]=0$

5. (c) Since $\vec{a}_1,\vec{a}_2$ and $\vec{a}_3$ are non-coplanar vectors

$\therefore x+y-3=0 \qquad$ ...(i)

$2x-y+2=0 \qquad$ ...(ii)

$2x+y+\lambda=0 \qquad$ ...(iii)

From (i) and (ii), $x = 1/3,\ y = 8/3$

$\therefore$ From (iii), $\lambda = -\dfrac{10}{3}$

6. (b) $\vec{c} = \lambda(\vec{a} \times \vec{b})$

$1 = (\vec{a} \times \vec{b}) \cdot \vec{c} = \dfrac{\vec{c} \cdot \vec{c}}{\lambda} = \dfrac{|\vec{c}|^2}{\lambda} = \dfrac{1}{3\lambda}$

$\lambda = \dfrac{1}{3}$

$\vec{c}^2 = \lambda^2 (\vec{a} \times \vec{b})^2$

$\dfrac{1}{3} = \dfrac{1}{9}(a^2 b^2 \sin^2 \theta) = \dfrac{1}{9} \times 2 \times 3 \sin^2 \theta$

$\sin^2 \theta = \dfrac{1}{2} \qquad \therefore \theta = \dfrac{\pi}{4}$

7. (c) Plane $4x + 3y + 2 = 0$ ...(i)

$z = 0$ ...(ii)

Any plane passing through the intersecting line of the given planes is

$4x + 3y + \lambda z + 2 = 0$ ...(iii)

Thus for suitable value of λ, plane (iii) will be inclined at 45° to the plane (i)

Direction cosine of the normal to the plane (iii) are

$\left(\dfrac{4}{\sqrt{25+\lambda^2}}, \dfrac{3}{\sqrt{25+\lambda^2}}, \dfrac{\lambda}{\sqrt{25+\lambda^2}}\right)$

Direction cosines of the normal to the plane (i) are

$\left(\dfrac{4}{5}, \dfrac{3}{5}, \dfrac{0}{5}\right)$; $\cos 45° = \dfrac{16+9+0}{5\sqrt{25+\lambda^2}} \Rightarrow \lambda = \pm 5$

$\therefore$ the required plane is $4x + 3y + 5z - 2 = 0$.

Length of perpendicular from the origin to the above plane $= \dfrac{2}{\sqrt{50}} = \dfrac{\sqrt{2}}{5}$

8. (d)

$\vec{a}' = \dfrac{\vec{b} \times \vec{c}}{[\vec{a}\,\vec{b}\,\vec{c}]} = \dfrac{\hat{i} + \hat{j} - \hat{k}}{2}$

9. (a) $\vec{a} + 2\vec{b} + 3\vec{c} = \vec{0} \Rightarrow \vec{a} \times \vec{b} + 3\vec{c} \times \vec{b} = \vec{0}$

i.e. $\vec{a} \times \vec{b} = 3\vec{b} \times \vec{c},\ \vec{a} \times \vec{c} + 2\vec{b} \times \vec{c} = \vec{0}$,

i.e. $2\vec{b} \times \vec{c} = \vec{c} \times \vec{a}$

$\therefore \vec{a} \times \vec{b} + \vec{b} \times \vec{c} + \vec{c} \times \vec{a}$

$= 3\vec{b} \times \vec{c} + \vec{b} \times \vec{c} + 2\vec{b} \times \vec{c} = 6\vec{b} \times \vec{c}$

10. (b) Let $\vec{r} = \ell(\vec{b} \times \vec{c}) + m(\vec{c} \times \vec{a}) + n(\vec{a} \times \vec{b})$

$\vec{r} \cdot \vec{a} = \ell[\vec{a}\,\vec{b}\,\vec{c}]$

$\Rightarrow \ell = 1$

Similarly, $m = 2, n = 3$

$\therefore \vec{r} = (\vec{b} \times \vec{c}) + 2(\vec{c} \times \vec{a}) + 3(\vec{a} \times \vec{b})$

11. (b) We have,

$\overrightarrow{AB} \cdot \overrightarrow{AC} = |\overrightarrow{AB}||\overrightarrow{AC}| \cos(90° - \theta) = AB \cdot AC \sin\theta$

$\overrightarrow{BC} \cdot \overrightarrow{BA} = |\overrightarrow{BC}| \cdot |\overrightarrow{BA}| \cos\theta = BC \cdot BA \cos\theta$

$\overrightarrow{CA} \cdot \overrightarrow{CB} = |\overrightarrow{CA}| \cdot |\overrightarrow{CB}| \cos 90° = 0$

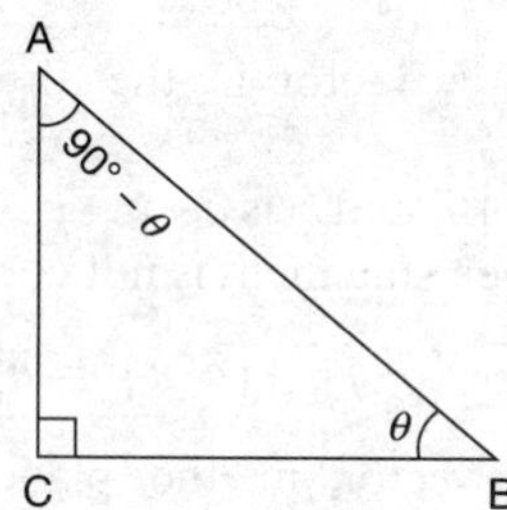

$\therefore \overrightarrow{AB} \cdot \overrightarrow{AC} + \overrightarrow{BC} \cdot \overrightarrow{BA} + \overrightarrow{CA} \cdot \overrightarrow{CB}$

$= AB \cdot AC \sin\theta + BC \cdot BA \cos\theta$

$= AB\left[AC \cdot \dfrac{AC}{AB} + BC \cdot \dfrac{BC}{AB}\right]$

$= AC^2 + BC^2 = AB^2 = p^2$.

12. (a) Since $\vec{a}, \vec{c}, \vec{b}$ form a right-handed system; therefore

$\vec{c} = \vec{b} \times \vec{a} = \begin{vmatrix} \vec{i} & \vec{j} & \vec{k} \\ 0 & 1 & 0 \\ x & y & z \end{vmatrix} = z\vec{i} - x\vec{k}$

13. (a) S_1: Obvious

S_2: $(4\hat{i} + 7\hat{j} - 2\hat{k}) - (3\hat{i} - 4\hat{j} + 7\hat{k}) = \hat{i} + 11\hat{j} - 9\hat{k}$

$\therefore$ The vectors form a triangle.

S_3: $\vec{a} \cdot (\vec{a} + \vec{b}) \times (\vec{a} + \vec{b} + \vec{c}) = \vec{a} \cdot (\vec{a} + \vec{b}) \times \vec{c} = [\vec{a}\,\vec{b}\,\vec{c}]$

S_4: $(\vec{a} \times \vec{b}) \times \vec{c} = \vec{a} \times (\vec{b} \times \vec{c}) \Rightarrow (\vec{b} \cdot \vec{c})\vec{a} = (\vec{a} \cdot \vec{b})\vec{c}$

$\Rightarrow (\vec{b} \cdot \vec{c})\vec{a} - (\vec{a} \cdot \vec{b})\vec{c} = \vec{0}$

$\vec{c} \times (\vec{a} \times \vec{b}) = (\vec{b} \cdot \vec{c})\vec{a} - (\vec{a} \cdot \vec{c})\vec{b}$

$= (\vec{a} \cdot \vec{b})\vec{c} - (\vec{a} \cdot \vec{c})\vec{b} = (\vec{b} \times \vec{c}) \times \vec{a} \neq \vec{0}$

14. (b) S_1: $\vec{a}$ and $\lambda\vec{a}$ are parallel vectors.

S_2: $\vec{a} \cdot \vec{b}$ may take negative values also.

S_3: $|(\vec{a} + \vec{b}) \times (\vec{a} - \vec{b})| = |-(\vec{a} \times \vec{b}) + (\vec{b} \times \vec{a})| = 2|\vec{b} \times \vec{a}|$

$\mathbf{S_4}:(\vec{a}\times\vec{b})^2=(\vec{a}\times\vec{b})\cdot(\vec{a}\times\vec{b})=\vec{a}\cdot(\vec{b}\times(\vec{a}\times\vec{b}))$

$=\vec{a}\cdot((\vec{b}.\vec{b})\vec{a}-(\vec{a}\cdot\vec{b})\vec{b})$

$=(\vec{a}\cdot\vec{a})(\vec{b}\cdot\vec{b})-(\vec{a}\cdot\vec{b})(\vec{b}\cdot\vec{a})$

15. (d) $\mathbf{S_1}$: $|\vec{a}|=|\vec{b}|=|\vec{a}-\vec{b}|=1$

$a^2+b^2-2\vec{a}\cdot\vec{b}=1$

$\therefore$ Angle between $\vec{a}$ and $\vec{b}$ is $\frac{\pi}{3}$.

$\mathbf{S_2}$: $\frac{\vec{a}+\vec{b}}{2}$ A vector in the direction of angle bisector of $\vec{a}$ and $\vec{b}$ is $\vec{a}+\vec{b}$.

$\therefore$ The given statement is not correct in general.

$\mathbf{S_3}$: $(\vec{a}\cdot\hat{i})^2+(\vec{a}\cdot\hat{j})^2+(\vec{a}\cdot\hat{k})^2=|\vec{a}|^2=\vec{a}^2$

$\mathbf{S_4}$: Any vector in the plane $\hat{i}+\hat{j}+\hat{k}$ and $-\hat{i}+\hat{j}+\hat{k}$ is of the form

$\alpha(\hat{i}+\hat{j}+\hat{k})+\beta(-\hat{i}+\hat{j}+\hat{k})$

$=(\alpha-\beta)\hat{i}+(\alpha+\beta)\hat{j}+(\alpha+\beta)\hat{k}.$

This will be perpendicular with $\hat{i}-\hat{j}-\hat{k}$ is $(\alpha-\beta)-(\alpha+\beta)-(\alpha+\beta)=0$

$\Rightarrow\alpha=-3\beta$

Hence, the required vector is of the form $(-4\hat{i}-2\hat{j}-2\hat{k})$

$\therefore$ The statement is false.

16. (d) $\mathbf{S_1}$: Let $\vec{d}=d_1\vec{i}+d_2\vec{j}+d_3\vec{k}$

Then $[\vec{d}\,\vec{j}\,\vec{k}]-d_1$

$$\text{Now}[\vec{d}\,\vec{j}\,\vec{k}][\vec{a}\,\vec{b}\,\vec{c}]=\begin{vmatrix}\vec{d}\cdot\vec{a} & \vec{d}\cdot\vec{b} & \vec{d}\cdot\vec{c}\\ \vec{j}\cdot\vec{a} & \vec{j}\cdot\vec{b} & \vec{j}\cdot\vec{c}\\ \vec{k}\cdot\vec{a} & \vec{k}\cdot\vec{b} & \vec{k}\cdot\vec{c}\end{vmatrix}=0$$

$\because$ The first row is zero.

But since $\vec{a},\vec{b},\vec{c}$ are non-coplanar vectors

$\therefore[\vec{a}\,\vec{b}\,\vec{c}]\neq0 \qquad \therefore d_1=0$

Hence $[\vec{d}\,\vec{j}\,\vec{k}]=0$

It can similarly be shown that $d_2=0, d_3=0$. Hence $\vec{d}$ is a zero vector.

$\mathbf{S_2}$: Since $\vec{a},\vec{b},\vec{c}$ are non-coplanar vectors, therefore $[\vec{a}\,\vec{b}\,\vec{c}]\neq0$.

Now $\{\vec{a}+\vec{b}+\vec{c}\}\cdot\{\vec{a}\times\vec{b}+\vec{b}\times\vec{c}+\vec{c}\times\vec{a}\}$

$=\vec{a}\cdot(\vec{a}\times\vec{b})+\vec{a}\cdot(\vec{b}\times\vec{c})+\vec{a}\cdot(\vec{c}\times\vec{a})+\vec{b}\cdot(\vec{a}\times\vec{b})$

$+\vec{b}\cdot(\vec{b}\times\vec{c})+\vec{b}\cdot(\vec{c}\times\vec{a})+\vec{c}\cdot(\vec{a}\times\vec{b})$

$+\vec{c}\cdot(\vec{b}\times\vec{c})+\vec{c}\cdot(\vec{c}\times\vec{a})$

$=0+[\vec{a}\,\vec{b}\,\vec{c}]+0+0+0+[\vec{b}\,\vec{c}\,\vec{a}]+[\vec{c}\,\vec{a}\,\vec{b}]+0+0$

$=3[\vec{a}\,\vec{b}\,\vec{c}]\neq0.$

Hence, $\vec{a}+\vec{b}+\vec{c}$ is not perpendicular to $\vec{a}\times\vec{b}+\vec{b}\times\vec{c}+\vec{c}\times\vec{a}$.

$\mathbf{S_3}$: Since $\vec{a}\perp\vec{c}$

$\therefore\vec{a}\cdot\vec{c}=0$

Now, $\vec{a}\times(\vec{b}\times\vec{c})=(\vec{a}\cdot\vec{c})\vec{b}-(\vec{a}\cdot\vec{b})\vec{c}$

$=0-(\vec{a}\cdot\vec{b})\vec{c}=-(\vec{a}\cdot\vec{b})\vec{c}$ and

$(\vec{a}\times\vec{b})\times\vec{c}=-\{(\vec{c}\cdot\vec{b})\vec{a}-(\vec{c}\cdot\vec{a})\vec{b}\}$

$=-(\vec{c}\cdot\vec{b})\vec{a}-0=-(\vec{b}\cdot\vec{c})\vec{a}$

Hence,

$\{\vec{a}\times(\vec{b}\times\vec{c})\}\cdot\{(\vec{a}\times\vec{b})\times\vec{c}\}=(\vec{a}\cdot\vec{b})(\vec{b}\cdot\vec{c})\vec{c}\cdot\vec{a}=0$

$\therefore(\vec{a}\times\vec{b})\times\vec{c}$ is perpendicular to $\vec{a}\times(\vec{b}\times\vec{c})$

$\mathbf{S_4}$: Writing $\vec{a}\times\vec{b}=\vec{p}$, we have,

$(\vec{a}\times\vec{b})\times(\vec{c}\times\vec{d})=\vec{p}\times(\vec{c}\times\vec{d})=(\vec{p}.\vec{d})\vec{c}-(\vec{p}.\vec{c})\vec{d}$

$=(\vec{a}\times\vec{b}.\vec{d})\vec{c}-(\vec{a}\times\vec{b}.\vec{c})\vec{d}=0\vec{c}-0\vec{d}=\vec{0};$

Since $\vec{a}\cdot\vec{b}\times\vec{d}=0$ because $\vec{a},\vec{b},\vec{d}$ are coplanar, and $\vec{a}\cdot\vec{b}\times\vec{c}=0$ because $\vec{a},\vec{b},\vec{c}$ are coplanar.

One or More Than One Option Correct

1. (b, c, d) Obviously, $\frac{\vec{a}}{|\vec{a}|}+\frac{\vec{b}}{|\vec{b}|}$ is a vector in the plane of $\vec{a}$ and $\vec{b}$ and hence perpendicular to $\vec{a}\times\vec{b}$. It is also equally inclined to $\vec{a}$ and $\vec{b}$ as it is along the angle bisector.

2. (a, b)

Let $\overline{\text{EB}}=p,\overline{\text{AB}}$

and $\overline{\text{CE}}=q\overline{\text{CD}}$.

Then $0<p$ and $q\leq1$

Since $\overline{\text{EB}}+\overline{\text{BC}}+\overline{\text{CE}}=\vec{0}$

D B E A C

$pm(2\hat{i}-6\hat{j}+2\hat{k})+(\hat{i}-2\hat{j})+qn(-6\hat{i}+15\hat{j}-3\hat{k})=\hat{0}$

or $(2pm+1-6qn)\hat{i}+(-6pm-2+15qn)\hat{j}$

$$+(2pm-6qn)\hat{k}=\vec{0}$$

$\Rightarrow 2pm-6qn+1=\vec{0}, -6pm-2+15qn=\vec{0}$

$2pm-6qn=\vec{0}$

Solving these, we get

$p=\dfrac{1}{(2m)}$ and $q=\dfrac{1}{(3n)}$

$\therefore 0<\dfrac{1}{(2m)}\le 1$ and $0<\dfrac{1}{(3n)}\le 1$

$\Rightarrow m\ge\dfrac{1}{2}$ and $n\ge\dfrac{1}{3}$

3. (a, b, c) $\overrightarrow{AB}+\overrightarrow{BC}=\overrightarrow{AC}$

$\overrightarrow{BC}=\dfrac{2\vec{u}}{|\vec{u}|}-\dfrac{\vec{u}}{|\vec{u}|}+\dfrac{\vec{v}}{|\vec{v}|}=\dfrac{\vec{u}}{|\vec{u}|}+\dfrac{\vec{v}}{|\vec{v}|}$

$\overrightarrow{AB}\cdot\overrightarrow{BC}=\left(\dfrac{\vec{u}}{|\vec{u}|}-\dfrac{\vec{v}}{|\vec{v}|}\right)\left(\dfrac{\vec{u}}{|\vec{u}|}+\dfrac{\vec{v}}{|\vec{v}|}\right)$

$=(\hat{u}-\hat{v})\cdot(\hat{u}+\hat{v})=1-1=0$

$\Rightarrow \angle B=90^\circ$

$\Rightarrow 1+\cos 2A+\cos 2B+\cos 2C=0$

4. (b, c) Since $\vec{a}=(1,3,\sin 2\alpha)$ makes an obtuse angle with z-axis, its z-component is negative. Thus,

$-1\le\sin 2\alpha<0$ ...(i)

But $\vec{b}\cdot\vec{c}=0$

$\tan^2\alpha-\tan\alpha-6=0$

$\therefore (\tan\alpha-3)(\tan\alpha+2)=0\Rightarrow \tan\alpha=3,-2$

(i) Now $\tan\alpha=3$. Therefore,

$\sin 2\alpha=\dfrac{2\tan\alpha}{1+\tan^2\alpha}=\dfrac{6}{1+9}=\dfrac{3}{5}$

(not possible as sin 2α < 0)

Now, if $\tan\alpha=-2$,

(ii) $\Rightarrow \sin 2\alpha=\dfrac{2\tan\alpha}{1+\tan^2\alpha}=\dfrac{-4}{1+4}=\dfrac{-4}{5}$

$\Rightarrow \tan 2\alpha>0$

Hence, 2α is the third quadrant. Also, $\sqrt{\sin\alpha/2}$ is meaningful. If $0<\sin\alpha/2<1$, then $\alpha=(4n+1)\pi-\tan^{-1}2$ and

$\alpha=(4n+2)\pi-\tan^{-1}2.$

5. (a, c) Plane P_1 contains the lines $\vec{r}=\hat{i}+\hat{j}+\hat{k}+\lambda(\hat{i}-\hat{j}-\hat{k})$. Hence, it contains the point $\hat{i}+\hat{j}+\hat{k}$ and is normal to the vector $(\hat{i}+\hat{j})$. Hence, equation of plane is

$(\vec{r}-(\hat{i}+\hat{j}+\hat{k})).(\hat{i}+\hat{j})=0$ or $x+y=2$

Plane P_2 contains the line

$\vec{r}=\hat{i}+\hat{j}+\hat{k}+\lambda(\hat{i}-\hat{j}-\hat{k})$ and point $\hat{j}$

Hence, equation of the plane is

$$\begin{vmatrix} x-0 & y-1 & z-0 \\ 1-0 & 1-1 & 1-0 \\ 1 & -1 & -1 \end{vmatrix}=0 \quad\text{or}\quad x+2y-z=2$$

If θ is the acute angle between P_1 and P_2, then

$\cos\theta=\dfrac{\vec{n}_1\cdot\vec{n}_2}{|\vec{n}_1||\vec{n}_2|}=\left|\dfrac{(\hat{i}+\hat{j})\cdot(\hat{i}+2\hat{j}-\hat{k})}{\sqrt{2}.\sqrt{6}}\right|$

$=\dfrac{3}{\sqrt{2}\cdot\sqrt{6}}=\dfrac{\sqrt{3}}{2}$

$\theta=\cos^{-1}\dfrac{\sqrt{3}}{2}=\dfrac{\pi}{6}$

As L is contained in $P_2\Rightarrow\theta=0$

6. (a, c) $(\vec{a}-\vec{b})\times[(\vec{b}+\vec{a})\times(2\vec{a}+\vec{b})]=\vec{b}+\vec{a}$

$\Rightarrow\{(\vec{a}-\vec{b}).(2\vec{a}+\vec{b})\}(\vec{b}+\vec{a})-$

$$\{(\vec{a}-\vec{b}).(\vec{b}+\vec{a})\}(2\vec{a}+\vec{b})=\vec{b}+\vec{a}$$

$\Rightarrow(2-\vec{a}\cdot\vec{b}-1)(\vec{b}+\vec{a})=\vec{b}+\vec{a}$

$\Rightarrow$ Either $\vec{b}+\vec{a}=\vec{0}$ or $1-\vec{a}\cdot\vec{b}=1$

$\Rightarrow$ Either $\vec{b}=-\vec{a}$ or $\vec{a}\cdot\vec{b}=0$

$\Rightarrow$ Either $\theta=\pi$ or $\theta=\dfrac{\pi}{2}$.

7. (a, b, d)

$(\lambda-1)(\vec{a}_1-\vec{a}_2)+\mu(\vec{a}_2+\vec{a}_3)+\gamma(\vec{a}_3+\vec{a}_4-2\vec{a}_2)$

$$+\vec{a}_3+\delta\vec{a}_4=\vec{0}$$

i.e., $(\lambda-1)\vec{a}_1+(1-\lambda+\mu-2\gamma)\vec{a}_2+(\mu+\gamma+1)\vec{a}_3$

$$+(\gamma+\delta)\vec{a}_4=\vec{0}$$

Since, $\vec{a}_1,\vec{a}_2,\vec{a}_3,\vec{a}_4$ are linearly independent,

$\therefore \lambda-1=0, 1-\lambda+\mu-2\gamma=0, \mu+\gamma+1=0$ and $\gamma+\delta=0$

i.e., $\lambda=1-0, 1-\lambda+\mu-2\gamma=0, \mu+\gamma+1=0$ and $\gamma+\delta=0$

i.e., $\lambda=1, \mu=-\dfrac{2}{3}, \gamma=-\dfrac{1}{3}, \delta=\dfrac{1}{3}$

8. (c, d) Since $\left[\vec{a}\ \vec{b}\ \vec{c}\right]=0$

$\therefore \vec{a}, \vec{b}$ and $\vec{c}$ are coplanar vectors.

Further, since $\vec{d}$ is equally inclined to $\vec{a}, \vec{b}$ and $\vec{c}$

$\therefore \vec{d}\cdot\vec{a}=\vec{d}\cdot\vec{b}=\vec{d}\cdot\vec{c}=0$

$\therefore \vec{d}\cdot\vec{x}=\vec{d}\cdot\vec{y}=\vec{d}\cdot\vec{z}=0$

$\therefore \vec{d}\cdot\vec{r}=0$

9. (a, c, d)

(a) $\vec{a}\times\left[\vec{a}\times\left(\vec{a}\times\vec{b}\right)\right]=\vec{a}\times\left[\left(\vec{a}\cdot\vec{b}\right)\vec{a}-\left(\vec{a}\cdot\vec{a}\right)\vec{b}\right]$

$=-\left(\vec{a}\cdot\vec{a}\right)\left(\vec{a}\times\vec{b}\right)$

$\therefore$ (a) is not correct.

(b) $\vec{v}.\vec{a}=\vec{0}\Rightarrow\vec{v}=\vec{0}$ or $\vec{v}\perp\vec{a}$

$\vec{v}\cdot\vec{b}=\vec{0}\Rightarrow\vec{v}=\vec{0}$ or $\vec{v}\perp\vec{b}$

$\vec{v}\cdot\vec{c}=\vec{0}\Rightarrow\vec{v}=\vec{0}$ or $\vec{v}\perp\vec{c}$

$\therefore \vec{v}=\vec{0}$ or $\vec{v}\perp\vec{a},\vec{b},\vec{c}$

$\therefore \vec{v}=\vec{0}$

(c) $\left(\vec{a}\times\vec{b}\right)\cdot\left(\vec{c}\times\vec{d}\right)=\vec{0}$

$\therefore$ The statement is incorrect.

(d) $\vec{a}\times\vec{b}'+\vec{b}\cdot\vec{c}'+\vec{c}\cdot\vec{a}'=0.$

(Property of reciprocal system)

(d) is incorrect.

10. (c, d) Let $\vec{a},\vec{b},\vec{c}$ lie in the x-y plane.

Let $\vec{a}=\hat{i}, \vec{b}=-\frac{1}{2}\hat{i}+\frac{\sqrt{3}}{2}\hat{j}$ and $\vec{c}=-\frac{1}{2}\hat{i}-\frac{\sqrt{3}}{2}\hat{j}$

$\therefore |\vec{p}+\vec{q}+\vec{r}|=|\lambda\vec{a}+\mu\vec{b}+\nu\vec{c}|$

$=\left|\lambda\hat{i}+\mu\left(-\frac{1}{2}\hat{i}+\frac{\sqrt{3}}{2}\hat{j}\right)+\nu\left(-\frac{1}{2}\hat{i}-\frac{\sqrt{3}}{2}\hat{j}\right)\right|$

$=\left|\left(\lambda-\frac{\mu}{2}-\frac{\nu}{2}\right)\hat{i}+\frac{\sqrt{3}}{2}(\mu-\nu)\hat{j}\right|$

$=\sqrt{\left(\lambda-\frac{\mu}{2}-\frac{\nu}{2}\right)^2+\frac{3}{4}(\mu-\nu)^2}$

$=\sqrt{\lambda^2+\mu^2+\nu^2-\lambda\mu-\lambda\nu-\mu\nu}$

$=\frac{1}{\sqrt{2}}\sqrt{(\lambda-\mu)^2+(\mu-\nu)^2+(\nu-\lambda)^2}\geq\frac{1}{\sqrt{2}}\sqrt{1+1+4}$

$\geq\sqrt{3}$

$\therefore |\vec{p}+\vec{q}+\vec{r}|$ can take a value to $\sqrt{3}$, 2.

11. (a, c) Since $\vec{u}$ be the projection vector of $\vec{w}$ along $\overline{PQ}$.

$\therefore |\vec{u}|=\frac{|a+b|}{\sqrt{a^2+b^2}}$

and $\vec{v}$ be the projection vector of $\vec{w}$ along $\overline{PS}$

$\therefore |\vec{v}|=\frac{(a-b)}{\sqrt{a^2+b^2}}$

Given that $|\vec{v}|+|\vec{u}|=|\vec{w}|\Rightarrow\frac{|a+b|+|a-b|}{\sqrt{a^2+b^2}}=\sqrt{2}$

Case I: If $a\geq b$ then

$\sqrt{2}a=\sqrt{a^2+b^2}\Rightarrow a=b$

Case II: If $a\leq b$ then

$\sqrt{2}b=\sqrt{a^2+b^2}\Rightarrow a=b$

So, from (I) and (II)

$a=b$

Now, area of PQRS $=\left|\overline{PQ}\times\overline{PS}\right|=2ab=8$

So $a=b=2, \therefore a+b=4$

So, option (a) is true.

Length of diagonal is $|2a\hat{i}|=|4\hat{i}|=4$

So, option (c) is true.

12. (a, b, c) $\overrightarrow{OB}\times\overrightarrow{OC}=\frac{1}{2}\overrightarrow{OB}\times\left(\overrightarrow{OB}-\lambda\overrightarrow{OA}\right)$

$=\frac{\lambda}{2}\left(\overrightarrow{OA}\times\overrightarrow{OB}\right)$ ...(i)

Since, $\overrightarrow{OA}\cdot\overrightarrow{OB}=0$ therefore, $\overrightarrow{OA}\perp\overrightarrow{OB}$

$\therefore \left|\overrightarrow{OB}\times\overrightarrow{OC}\right|=\left|\frac{\lambda}{2}\left(\overrightarrow{OA}\times\overrightarrow{OB}\right)\right|$ [from (i)]

$=\frac{|\lambda|}{2}\left|\overrightarrow{OA}\right|\left|\overrightarrow{OB}\right|\sin\frac{\pi}{2}$

$\Rightarrow\frac{9}{2}=\frac{|\lambda|}{2}\times3\times3$

$\Rightarrow\lambda=1$ (given, $\lambda>0$)

So, $\overrightarrow{OC}=\frac{\overrightarrow{OB}-\overrightarrow{OA}}{2}$

$=\frac{1}{2}\left(-\hat{i}-4\hat{j}+\hat{k}\right)$

(a) Projection $\overrightarrow{OC}$ on $\overrightarrow{OA}$

$$=\frac{\overrightarrow{OC}.\overrightarrow{OA}}{\left|\overrightarrow{OA}\right|}=\frac{\frac{1}{2}(-2-8+1)}{3}=-\frac{3}{2}$$

(b) Area of ΔOAB $=\frac{1}{2}\left|\overrightarrow{OA}\times\overrightarrow{OB}\right|=\frac{9}{2}$

(c) Area of ΔABC

$$= \frac{1}{2}\left|\overrightarrow{AB}\times\overrightarrow{AC}\right| = \left\|\begin{matrix} \hat{i} & \hat{j} & \hat{k} \\ -1 & -4 & 1 \\ \frac{-5}{2} & -4 & -\frac{1}{2} \end{matrix}\right\|$$

$$= \frac{1}{2}\left|6\hat{i}-3\hat{j}-6\hat{k}\right| = \frac{9}{2}$$

(d) Let the acute angle between the diagonals be θ, then

$$\cos\theta = \frac{\left(\overrightarrow{OA}+\overrightarrow{OC}\right)\cdot\left(\overrightarrow{OA}-\overrightarrow{OC}\right)}{\left|\overrightarrow{OA}+\overrightarrow{OC}\right|\left|\overrightarrow{OA}-\overrightarrow{OC}\right|} = \frac{18}{3\sqrt{2}\sqrt{90}}$$

$$\therefore \theta \neq \frac{\pi}{3}$$

13. (a, c, d) From the given information,

$\vec{a}+\vec{b}+\vec{c}=\vec{0}$

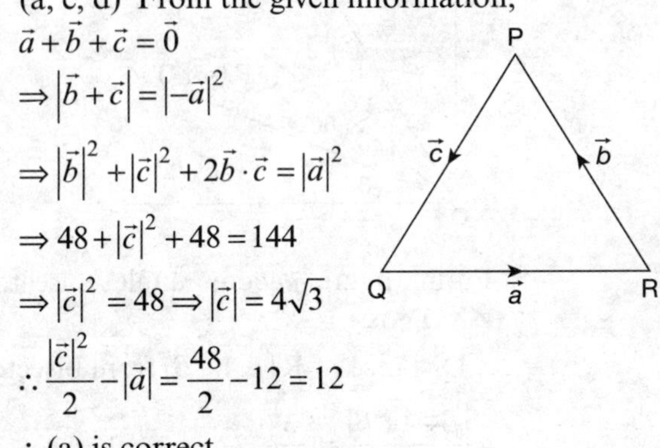

$\Rightarrow \left|\vec{b}+\vec{c}\right| = \left|-\vec{a}\right|^2$

$\Rightarrow \left|\vec{b}\right|^2 + \left|\vec{c}\right|^2 + 2\vec{b}\cdot\vec{c} = \left|\vec{a}\right|^2$

$\Rightarrow 48 + \left|\vec{c}\right|^2 + 48 = 144$

$\Rightarrow \left|\vec{c}\right|^2 = 48 \Rightarrow \left|\vec{c}\right| = 4\sqrt{3}$

$\therefore \frac{\left|\vec{c}\right|^2}{2} - \left|\vec{a}\right| = \frac{48}{2} - 12 = 12$

$\therefore$ (a) is correct.

Also $\left|\vec{b}\right| = \left|\vec{c}\right| \Rightarrow \angle Q = \angle R$

and $\cos(180 - \mathrm{P}) = \frac{\vec{b}.\vec{c}}{\left|\vec{b}\right|\left|\vec{c}\right|} = \frac{1}{2}$

$\Rightarrow \angle\mathrm{P} = 120° \therefore \angle\mathrm{Q} = \angle\mathrm{R} = 30°$

And $\vec{a}\cdot\vec{b} = 12\times4\sqrt{3}\times\cos150 = -72$

$\therefore$ (d) is correct.

Now, $\vec{a}\times\left(\vec{a}+\vec{b}+\vec{c}\right) = \vec{0} \Rightarrow \vec{a}\times\vec{b} = \vec{c}\times\vec{a}$

$\therefore \left|\vec{a}\times\vec{b}+\vec{c}\times\vec{a}\right| = 2\left|\vec{a}\times\vec{b}\right|$

$= 2\times12\times4\sqrt{3}\times\sin150 = 48\sqrt{3}$

$\therefore$ (c) is correct.

14. (a, b, c) Given that $|\vec{x}| = |\vec{y}| = |\vec{z}| = \sqrt{2}$

and angle between each pair is $\frac{\pi}{3}$.

Since $\vec{a}$ is perpendicular to both $\vec{x}$ and $\vec{y}\times\vec{z}$,

$\therefore \vec{a} = \lambda\left[\vec{x}\times(\vec{y}\times\vec{z})\right] = \lambda\left[(\vec{x}.\vec{z})\vec{y} - (\vec{x}.\vec{y})\vec{z}\right]$

$= \lambda\left[\left(\sqrt{2}.\sqrt{2}\cos\frac{\pi}{3}\right)\vec{y} - \left(\sqrt{2}.\sqrt{2}\cos\frac{\pi}{3}\right)\vec{z}\right]$

$= \lambda(\vec{y}-\vec{z})$

Since $\vec{b}$ is perpendicular to both $\vec{y}$ and $\vec{z}\times\vec{x}$

$\vec{b} = \mu\left[\vec{y}\times(\vec{z}\times\vec{x})\right] = \mu\left[(\vec{y}.\vec{x})\vec{z} - (\vec{y}.\vec{z})\vec{x}\right]$

$= \mu\left[\left(\sqrt{2}\cdot\sqrt{2}\cos\frac{\pi}{3}\right)\vec{z} - \left(\sqrt{2}\cdot\sqrt{2}\cos\frac{\pi}{3}\right)\vec{x}\right]$

$= \mu(\vec{z}-\vec{x})$

Now, $\vec{b}\cdot\vec{z} = \mu[\vec{z}\cdot\vec{z} - \vec{x}\cdot\vec{z}] = \mu(2-1) = \mu$

$\therefore \vec{b} = \left(\vec{b}\cdot\vec{z}\right)(\vec{z}-\vec{x})$

$\therefore$ (a) is correct.

Also, $\vec{a}\cdot\vec{y} = \lambda(\vec{y}\cdot\vec{y} - \vec{z}\cdot\vec{y}) = \lambda(2-1) = \lambda$

$\therefore \vec{a} = (\vec{a}\cdot\vec{y})(\vec{y}-\vec{z})$

$\therefore$ (b) is also correct.

$\vec{a}\cdot\vec{b} = \lambda\mu(\vec{y}\cdot\vec{z} - \vec{y}\cdot\vec{x} - \vec{z}\cdot\vec{z} + \vec{z}\cdot\vec{x})$

$= \lambda\mu(1-1-2+1) = -\lambda\mu = -(\vec{a}\cdot\vec{y})\left(\vec{b}\cdot\vec{z}\right)$

$\therefore$ (c) is correct.

$-(\vec{a}\cdot\vec{y})(\vec{z}-\vec{y}) = \lambda(\vec{z}-\vec{y}) = -\vec{a}$

$\therefore$ (d) is not correct.

15. (a, d)

Let $\vec{a} = \hat{i}+\hat{j}+2\hat{k}, \vec{b} = \hat{i}+2\hat{j}+\hat{k}, \vec{c} = \hat{i}+\hat{j}+\hat{k}$

Vector coplanar with $\vec{a}$ and $\vec{b}$ is

$\vec{r} = \vec{a}+\lambda\vec{b}$

$\Rightarrow \vec{r} = (1+\lambda)\hat{i} + (1+2\lambda)\hat{j} + (2+\lambda)\hat{k}$

$\because \vec{r}\perp\vec{c} \Rightarrow \vec{r}\cdot\vec{c} = 0$

$\Rightarrow 1+\lambda+1+2\lambda+2+\lambda = 0 \Rightarrow \lambda+\mu = 0$

$\Rightarrow \lambda = -\mu$

$\Rightarrow 4\lambda+4 = 0$

$\Rightarrow \lambda = -1$

Any scalar multiple of $\vec{r}$ is also a solution.

$\therefore$ (a) and (d) are the correct options.

Matrix Type

1. A $\to$ T; B $\to$ P; C $\to$ Q; D $\to$ S

(A) $\vec{a}+\vec{b} = \hat{j}$ and $2\vec{a}-\vec{b} = 3\hat{i}+\frac{\hat{j}}{2}$

$\therefore \vec{a} = \hat{i}+\frac{\hat{j}}{2}, \vec{b} = -\hat{i}+\frac{\hat{j}}{2}$

$\therefore \cos\theta = -\frac{3}{5}$

(B) $\left|\vec{a}+\vec{b}+\vec{c}\right| = \sqrt{6}$

$\Rightarrow \vec{a}^2+\vec{b}^2+\vec{c}^2+2\vec{a}\cdot\vec{b}+2\vec{b}\cdot\vec{c}+2\vec{c}\cdot\vec{a} = 6$

$\therefore |\vec{a}| = 1$

(C) $\begin{vmatrix} \hat{i} & \hat{j} & \hat{k} \\ 3 & 1 & -2 \\ 1 & -3 & 4 \end{vmatrix} = -2\hat{i} - 14\hat{j} - 10\hat{k}$

$\therefore$ Area $= 5\sqrt{3}$

(D) $\vec{a}$ is perpendicular $\vec{b}+\vec{c} \Rightarrow \vec{a}\cdot\vec{b} + \vec{a}\cdot\vec{c} = 0$ ...(i)

$\vec{b}$ is perpendicular $\vec{a}+\vec{c} \Rightarrow \vec{b}\cdot\vec{c} + \vec{b}\cdot\vec{c} = 0$ ...(ii)

$\vec{c}$ is perpendicular $\vec{a}+\vec{b} \Rightarrow \vec{c}\cdot\vec{a} + \vec{c}\cdot\vec{b} = 0$ ...(iii)

From (i), (ii) and (iii), we get

$\vec{a}\cdot\vec{b} = \vec{b}\cdot\vec{c} = \vec{c}\cdot\vec{a} = 0$

$\therefore \left|\vec{a}+\vec{b}+\vec{c}\right| = 7$

2. A $\to$ R; B $\to$ P; C $\to$ Q; D $\to$ S

(A) $\overline{\text{OA}} = \hat{i} + 2\hat{j} + 3\hat{k}, \overline{\text{OB}} = -2\hat{i} + \hat{j} - 4\hat{k},$

$\overline{\text{OC}} = 3\hat{i} + 4\hat{j} - 2\hat{k}$

Area $= \frac{1}{2}\left|\overline{\text{AB}} \times \overline{\text{AC}}\right| = \frac{\sqrt{1218}}{2}$

(B) $\left(\left(\vec{a}\times\vec{b}\right)\times\vec{c}\right)\cdot\vec{d} + \left(\left(\vec{b}\times\vec{c}\right)\times\vec{a}\right)\cdot\vec{d} + \left(\left(\vec{c}\times\vec{a}\right)\times\vec{b}\right)\cdot\vec{d} = 0$

(C) Taking P as origin, position vector of Q, R and S are $P\hat{i}, P\hat{i}+P\hat{j}, P\hat{j}$

Equation of PQ′ and RS are

$\vec{r} = t\left(\hat{i}+\hat{j}+\sqrt{2}\hat{k}\right), \vec{r} = P\hat{i} + P\hat{j} + \lambda\hat{i}$

$\therefore$ Shortest distance $= \frac{2P}{\sqrt{6}}$

$\therefore k = 2$

(D) $\left(\vec{a}\cdot\vec{c}\right)\left(\vec{b}\cdot\vec{d}\right) - \left(\vec{b}\cdot\vec{c}\right)\left(\vec{a}\cdot\vec{d}\right) = 21$

3. A $\to$ P, R, S; B $\to$ P; C $\to$ P, Q; D $\to$ S, T

(A) Since, $2\left(a^2 - b^2\right) = c^2$

$\Rightarrow 2\left(\sin^2 x - \sin^2 y\right) = \sin^2$

$\Rightarrow 2\sin(x+y)\sin(x-y) = \sin^2 z$

$\Rightarrow 2\sin(x-y) = \sin z \quad (\because \sin(x+y) = \sin z)$

$\Rightarrow \frac{\sin(x-y)}{\sin z} = \frac{1}{2} = \lambda$ (Given)

$\therefore \cos(n\pi) = 0 \Rightarrow \cos\frac{n\pi}{2} = 0 \Rightarrow n = 1, 3, 5$

(B) $1 + \cos 2X - 2\cos 2Y = 2\sin X \sin Y$

$\Rightarrow 2\cos^2 X - 2\left(1 - 2\sin^2 Y\right) = 2\sin X \sin Y$

$\Rightarrow 1 - \sin^2 X - 1 + 2\sin^2 Y = 2\sin X \sin Y$

$\Rightarrow \sin^2 X + \sin X \sin Y - 2\sin^2 Y = 0$

$\Rightarrow \left(\sin X - \sin Y\right)\left(\sin X + 2\sin Y\right) = 0$

$\Rightarrow \frac{\sin X}{\sin Y} = 1 \text{ or } -2$

$\therefore \frac{a}{b} = 1$

(C) $P\left(\sqrt{3}, 1\right), Q\left(1, \sqrt{3}\right), R\left(\beta, 1-\beta\right)$

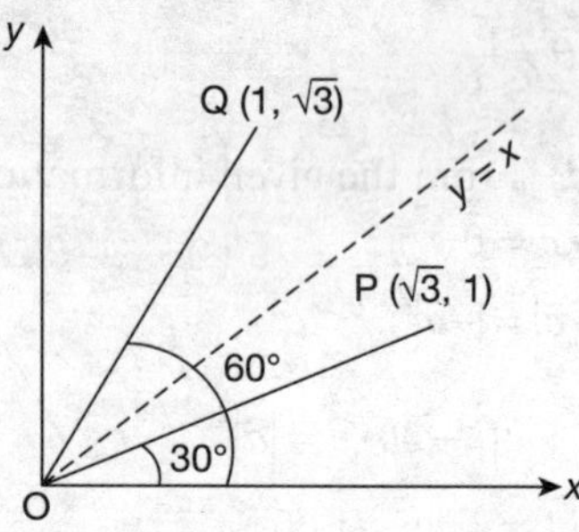

From figure, acute angle bisector of $\angle$XOY is $y = x$.

$\therefore$ Distance of $R\left(\beta, 1-\beta\right)$ from bisector

$= \left|\frac{\beta - 1 + \beta}{\sqrt{2}}\right| = \frac{3}{\sqrt{2}}$

$\Rightarrow 2 - 1 = \pm 3$ or $\beta = 2$ or -1

$\therefore |\beta| = 1, 2$

(D) For $\alpha = 0, y = 3$

For $\alpha = 1, y = |x-1| + |x-2| + x$

$\alpha = 0$

Case I

$F(\alpha)$ is the area bounded by

$x = 0, x = 2, y^2 = 4x$

and $y = 3$

$\therefore F(\alpha) = \int_0^2 \left(3 - 2\sqrt{x}\right)dx$

$= \left|3x - \frac{4x\sqrt{x}}{3}\right|_0^2 = 6 - \frac{8\sqrt{2}}{3}$

$\therefore F(\alpha) + \frac{8}{3}\sqrt{2} = 6$

Case II

$F(\alpha)$ is the area bounded by

$x = 0, x = 2, y^2 = 4x$ and

$y=|x-1|+|x-2|+x$

$$=\begin{cases}3-x, 0\le x<1\\ x+1, 1\le x<2\\ 3x-3, x\ge 2\end{cases}$$

$$\therefore \text{F}(\alpha)=\int_0^1\left(3-x-2\sqrt{x}\right)dx+\int_1^2\left(x+1-2\sqrt{x}\right)dx$$

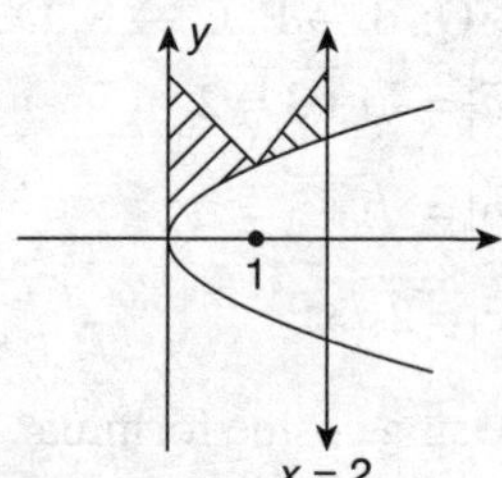

$$=\left(3x-\frac{x^2}{2}-\frac{4x}{3}\sqrt{x}\right)_0^1+\left(\frac{x^2}{2}+x-\frac{4}{3}x\sqrt{x}\right)_1^2$$

$$=3-\frac{1}{2}-\frac{4}{3}+2+2-\frac{8\sqrt{2}}{3}-\frac{1}{2}-1+\frac{4}{3}=5-\frac{8\sqrt{2}}{3}$$

$$\text{F}(\alpha)+\frac{8\sqrt{2}}{3}=5$$

4. $\text{A}\to\text{Q}$; $\text{B}\to\text{P, Q}$; $\text{C}\to\text{P, Q, S, T}$; $\text{D}\to\text{Q, T}$

(A) Projection of $\alpha\hat{i}+\beta\hat{j}$ on

$$\sqrt{3}\hat{i}+\hat{j}\;\frac{\sqrt{3}\alpha+\beta}{2}=\sqrt{3}$$

$$\Rightarrow \alpha=\frac{2\sqrt{3}-\beta}{\sqrt{3}}$$

$$\therefore \frac{2\sqrt{3}-\beta}{\sqrt{3}}=2+\sqrt{3}\beta\Rightarrow\beta=0\Rightarrow\alpha=2$$

(B) LHD $=f'(1)=-6a$ and RHD $=f'(1)=b$

$-6a=b$ ...(i)

Also, f is continuous at $x=1$,

$\therefore -3a-2=b+a^2$

$\Rightarrow a^2-3a+2=0$ (from (i))

$\Rightarrow a=1, 2$

(C) $\left(3-3\omega+2\omega^2\right)^{4n+3}+\left(2+3\omega-3\omega^2\right)^{4n+3}$

$+\left(-3+2\omega+3\omega^2\right)^{4n+3}=0 \quad \left[\because \omega^3=1\right]$

$$\Rightarrow\left(3-3\omega+2\omega^2\right)^{4n+3}+\left(\frac{2\omega^2+3-3\omega}{\omega^2}\right)^{4n+3}+\left(\frac{-3\omega+2\omega^2+3}{\omega}\right)^{4n+3}=0$$

$$\Rightarrow\left(3-3\omega+2\omega^2\right)^{4n+3}\left[1+\omega^{4n+3}+\left(\omega^2\right)^{4n+3}\right]=0$$

$\Rightarrow 4n+3$ should be an integer other than multiple of 3.

$\therefore n=1, 2, 4, 5$

(D) $\because$ HM of a and b must be 4

$$\therefore \frac{2ab}{a+b}=4\Rightarrow ab=2a+2b \qquad \text{...(i)}$$

Also $a+q=10$ or $a=10-q$

and $b+5=2q$ or $b=2q-5$

Putting values of a and b in equation (i), we get

$$q=4 \text{ or } \frac{15}{2}\Rightarrow a=6 \text{ or } \frac{5}{2}$$

$\therefore |q-a|=2$ or 5.

5. $\text{A}\to\text{S}$; $\text{B}\to\text{R}$; $\text{C}\to\text{Q}$; $\text{D}\to\text{P}$

A – S; $y=\cos\left(3\cos^{-1}x\right)$

$y=\cos\left[\cos^{-1}\left(4x^3-3x\right)\right]$

$y=4x^3-3x$

$$\Rightarrow\frac{dy}{dx}=12x^2-3 \text{ and } \frac{d^2y}{dy^2}=24x$$

$$\therefore \frac{1}{4}\left\{\left(x^2-1\right)\frac{d^2y}{dy^2}+x\frac{dy}{dx}\right\}$$

$$=\frac{1}{4x^3-3x}\left\{\left(x^2-1\right)24x+x\left(12x^2-3\right)\right\}$$

$$=\frac{1}{4x^3-3x}\left\{36x^3-27x\right\}$$

$$=\frac{9\left\{4x^3-3x\right\}}{4x^3-3x}=9$$

B – R : Since $\overrightarrow{a_1}, \overrightarrow{a_2}, \ldots \overrightarrow{a_n}$ are position vectors of vertices $\text{A}_1, \text{A}_2, \text{A}_3, \ldots, \text{A}_n$ of a regular polygon of n sides with its centre at origin.

$\therefore \left|\overrightarrow{a_1}\right|=\left|\overrightarrow{a_2}\right|=\ldots=\left|\overrightarrow{a_n}\right|=\lambda$

Now, $\overrightarrow{a_k}\times\overrightarrow{a_{k+1}}=\lambda^2\sin\frac{2\pi}{n}\hat{n}$

and $\overrightarrow{a_k}.\overrightarrow{a_{k+1}}=\lambda^2\cos\frac{2\pi}{n}$

$$\therefore \left|\sum_{k=1}^{n-1}\left(\vec{a}_k\times\vec{a}_{k+1}\right)\right|=\left|\sum_{k=1}^{n-1}\left(\vec{a}_k-\vec{a}_{k+1}\right)\right|$$

$$\Rightarrow (n-1)\lambda^2\sin\frac{2\pi}{n}=(n-1)\lambda^2\cos\frac{2\pi}{n}$$

$\Rightarrow \tan\dfrac{2\pi}{n} = 1 \Rightarrow \dfrac{2\pi}{n} = \dfrac{\pi}{4} \Rightarrow n = 8$

C – Q; Normal from $P(h,1)$ on $\dfrac{x^2}{6}+\dfrac{y^2}{3}=1$ is

$$\frac{x-h}{\frac{h}{6}} = \frac{y-1}{\frac{1}{3}}$$

$\Rightarrow 2(x-h) = h(y-1) \Rightarrow 2x - hy - h = 0$

Slope of normal $= \dfrac{2}{h}$

It is perpendicular to $x + y = 8$

$\therefore \dfrac{2}{h}\times -1 = -1 \Rightarrow h = 2$

D – P; $\tan^{-1}\left(\dfrac{1}{2x+1}\right)+\tan^{-1}\left(\dfrac{1}{4x+1}\right)=\tan^{-1}\left(\dfrac{2}{x^2}\right)$

$$\Rightarrow \tan^{-1}\left(\frac{\frac{1}{2x+1}+\frac{1}{4x+1}}{1-\frac{1}{2x+1}\cdot\frac{1}{4x+1}}\right)=\tan^{-1}\left(\frac{2}{x^2}\right)$$

$$\Rightarrow \tan^{2}\left(\frac{6x+2}{8x^2+6x}\right)=\tan^{-1}\left(\frac{2}{x^2}\right)$$

$$\Rightarrow \frac{3x+1}{4x^2+3x}=\frac{2}{x^2} \Rightarrow 3x^2-7x-6=0$$

$\Rightarrow x = 3$ or $-\dfrac{2}{3}$

Since $x > 0$

$\therefore$ Only one +ve solution is there.

6. A $\to$ R; B $\to$ S; C $\to$ P; D $\to$ Q

(A) Given that $\left[\vec a\,\vec b\,\vec c\right] = 2$

$\therefore \left[2(\vec a\times\vec b)3(\vec b\times\vec c)\vec c\times\vec a\right]$

$= 6\left[\vec a\times\vec b\,\vec b\times\vec c\,\vec c\times\vec a\right]$

$= 6\left[\vec a\,\vec b\,\vec c\right]^2 = 6\times 4 = 24$

$\therefore$ (A) – (R)

(B) Given that $\left[\vec a\,\vec b\,\vec c\right] = 5$

$\therefore \left[3(\vec a+\vec b)\vec b+\vec c\,2(\vec c+\vec a)\right]$

$= 6\left[\vec a+\vec b \quad \vec b+\vec c \quad \vec c+\vec a\right]$

$= 6\times 2\left[\vec a\,\vec b\,\vec c\right] = 6\times 2\times 5 = 60$

$\therefore$ (B) – (S)

(C) Given that $\dfrac{1}{2}\left|\vec a\times\vec b\right| = 20 \Rightarrow \left|\vec a\times\vec b\right| = 40$

$\therefore \dfrac{1}{2}\left|(2\vec a+3\vec b)\times(\vec a-\vec b)\right| = \dfrac{1}{2}\left|-2\vec a+\vec b+3\vec b\times\vec a\right|$

$= \dfrac{1}{2}\times 5\left|\vec a\times\vec b\right| = \dfrac{5}{2}\times 40 = 100$

$\therefore$ (C) – (P)

(D) Given that $\left|\vec a\times\vec b\right| = 30$

$\therefore \left|(\vec a+\vec b)\times\vec a\right| = \left|(\vec b\times\vec a)\right| = 30$

$\therefore$ (D) – (Q)

7. A $\to$ Q; B $\to$ P; C $\to$ S; D $\to$ T

A. $|\vec a| = \sqrt{1+3} = 2$

$|\vec b| = \sqrt{1+3} = 2$

$|\vec c| = 2\sqrt{3}$

Using cosine formula

$$\cos C = \frac{|\vec a|^2+|\vec b|^2-|\vec c|^2}{2|\vec a||\vec b|} = \frac{4+4-12}{2\times2\times2} = \frac{1}{2}$$

$\Rightarrow \cos\theta = -\dfrac{1}{2} \Rightarrow \theta = \dfrac{2\pi}{3}$

$\Rightarrow$ A – Q

B. $\displaystyle\int_a^b (f(x)-3x)\,dx = a^2 - b^2$

$\Rightarrow \displaystyle\int_a^b f(x)\,dx + \frac{3}{2}\left[-b^2+a^2\right] = a^2 - b^2$

$\Rightarrow \displaystyle\int_a^b f(x)\,dx = -\frac{1}{2}(a^2-b^2) = \int_a^b x\,dx$

$\Rightarrow f(x) = x \Rightarrow f\left(\dfrac{\pi}{6}\right) = \dfrac{\pi}{6}$

$\Rightarrow B - P$

$$\frac{\pi^2}{\ln 3}\int_{\frac{7}{6}}^{\frac{5}{6}} \sec(\pi x)\,dx = \frac{\pi^2}{\pi\ln 3}\Big[\ln|\sec\pi x+\tan\pi x|\Big]_{7/6}^{5/6}$$

$$= \frac{\pi}{\ln 3}\left[\ln\left|\sec\frac{5\pi}{6}+\tan\frac{5\pi}{6}\right| - \ln\left|\sec\frac{7\pi}{6}+\tan\frac{7\pi}{6}\right|\right]$$

$$= \frac{\pi}{\ell n 3}\left[\ln\left|-\frac{2}{\sqrt3}-\frac{1}{\sqrt3}\right| - \ln\left|-\frac{2}{\sqrt3}+\frac{1}{\sqrt3}\right|\right]$$

$$= \frac{\pi}{\ln 3}\ln 3 = \pi$$

$\therefore C - S$

For $|z| = 1$ and $z \neq 1$. Let $z = \cos\theta + i\sin\theta$

Then $1-z=1-\cos\theta-i\sin\theta$

$=2\sin^2\dfrac{\theta}{2}-2i\sin\dfrac{\theta}{2}\cos\dfrac{\theta}{2}$

or $1-z=2\sin\dfrac{\theta}{2}\left[\sin\dfrac{\theta}{2}-i\cos\dfrac{\theta}{2}\right]$

$\therefore \dfrac{1}{1-z}=\dfrac{1}{2}\left[1+i\cot\dfrac{\theta}{2}\right]$

Clear that real part of $\dfrac{1}{1-z}$ is always $\dfrac{1}{2}$

$\therefore$ Locus of $\dfrac{1}{1-z}$ is $x=\dfrac{1}{2}$

$\left|\text{Arg}\left(\dfrac{1}{1}-z\right)\right|$ is maximum when value of ϕ approaches to $\dfrac{\pi}{2}$ but will not be attained.

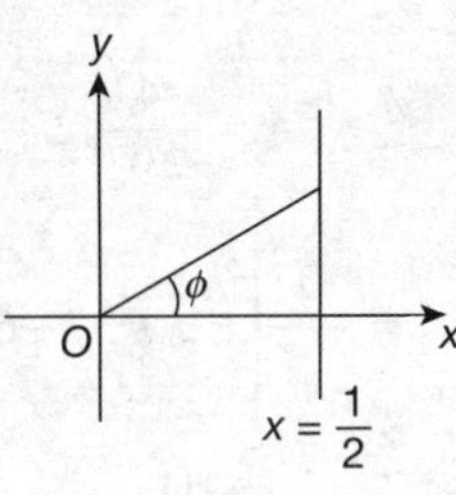

$\therefore D-\text{T}$

Integer and Numerical Value Type

1. $V_1=\left[\vec{a}\,\vec{b}\,\vec{c}\right]$

$V_2=\dfrac{1}{2}\left[\vec{a}\,\vec{b}\,\vec{c}\right]$

$V_3=\dfrac{1}{6}\left[\vec{a}\,\vec{b}\,\vec{c}\right]$

$V_1:V_2:V_3=1:\dfrac{1}{2}:\dfrac{1}{6}$

$=6:3:1$

$V_1=\begin{vmatrix}1 & -1 & -6\\ 1 & -1 & 4\\ 2 & -5 & 3\end{vmatrix}$

$=1(-3+20)+1(3-8)-6(-5+2)=17-5+18=30$

$\therefore V_1+V_2+V_3=30+15+5=50$

2. $\sum\left[\vec{p}\times\{(\vec{x}-\vec{q})\times\vec{p}\}\right]=\vec{0}$

$\Rightarrow \sum\left[\vec{p}\times(\vec{x}\times\vec{p})\right]-\sum\left[\vec{p}\times(\vec{q}\times\vec{p})\right]=\vec{0}$

$\Rightarrow \sum\vec{p}^2\vec{x}-\sum(\vec{p}\cdot\vec{x})\vec{p}-\sum\vec{p}^2\vec{q}+\sum(\vec{p}\cdot\vec{q})\vec{p}=\vec{0}$

$\Rightarrow 3\vec{p}^2\vec{x}-\vec{p}^2\vec{x}-\vec{p}^2\left(\vec{p}+\vec{q}+\vec{r}\right)=\vec{0}$

$\Rightarrow 2\vec{p}^2\vec{x}=\vec{p}^2\left(\vec{p}+\vec{q}+\vec{r}\right)$

$\Rightarrow \vec{x}=\dfrac{1}{2}\left(\vec{p}+\vec{q}+\vec{r}\right)$

3. Let OABC be the tetrahedron. Let G be the centroid of the face OAB, then $\text{GA}=\dfrac{1}{\sqrt{3}}\text{AC}$.

Then $\cos\theta=\dfrac{\text{GA}}{\text{CA}}=\dfrac{1}{\sqrt{3}}$

$\therefore \cos^2\theta=\dfrac{1}{3}$

$\therefore a=1$ and $b=3$

$\therefore 10a+b=13$

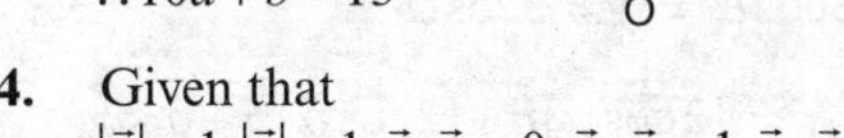

4. Given that $|\vec{u}|=1;|\vec{v}|=1;\vec{u}\cdot\vec{v}\neq 0;\vec{u}\cdot\vec{w}=1;\vec{v}\cdot\vec{w}=1$ and

$\vec{w}\cdot\vec{w}=|\vec{w}|^2=4\Rightarrow|\vec{w}|=2$

Volume of parallelepiped $=\left[\vec{u}\;\vec{v}\;\vec{w}\right]=\sqrt{2}$

$\Rightarrow\left[\vec{u}\;\vec{v}\;\vec{w}\right]^2=\begin{vmatrix}\vec{u}\cdot\vec{u} & \vec{u}\cdot\vec{v} & \vec{u}\cdot\vec{w}\\ \vec{v}\cdot\vec{u} & \vec{v}\cdot\vec{v} & \vec{v}\cdot\vec{w}\\ \vec{w}\cdot\vec{u} & \vec{w}\cdot\vec{v} & \vec{w}\cdot\vec{w}\end{vmatrix}=2$

$\Rightarrow\begin{vmatrix}1 & \vec{u}\cdot\vec{v} & 1\\ \vec{u}\cdot\vec{v} & 1 & 1\\ 1 & 1 & 4\end{vmatrix}=2\Rightarrow\vec{u}\cdot\vec{v}=\dfrac{1}{2}$

Now, $|3\vec{u}+5\vec{v}|^2=(3\vec{u}+5\vec{v})(3\vec{u}+5\vec{v})$

$=9|\vec{u}|^2+25|\vec{v}|^2+30\vec{u}\cdot\vec{v}$

$=9+25+15=49$

$\therefore|3\vec{u}+5\vec{v}|=\sqrt{49}=7$

5. Given that $\vec{a}=2\hat{i}+\hat{j}-\hat{k},\vec{b}=\hat{i}+2\hat{j}+\hat{k}$

$\Rightarrow\vec{a}+\vec{b}=3\hat{i}+3\hat{j}$

Given that projection of $\vec{c}$ on $\vec{a}+\vec{b}=3\sqrt{2}$

$\Rightarrow\dfrac{\vec{c}\cdot\left(\vec{a}+\vec{b}\right)}{\left|\vec{a}+\vec{b}\right|}=3\sqrt{2}$

$\Rightarrow\dfrac{3(2\alpha+\beta)+3(\alpha+2\beta)}{3\sqrt{2}}=3\sqrt{2}$

$\Rightarrow\alpha+\beta=2$ ...(i)

$\vec{c}=\alpha\vec{a}+\beta\vec{b}=(2\alpha+\beta)\hat{i}+(\alpha+2\beta)\hat{j}+(-\alpha+\beta)\hat{k}$

$\vec{c}=(\alpha+2)\hat{i}+(4-\alpha)\hat{j}+(2-2\alpha)\hat{k}$

using equation (i).

Now $\vec{c}$ is in the plane of $\vec{a}$ and $\vec{b}$ $\left(\because\vec{c}=\alpha\vec{a}+\beta\vec{b}\right)$

$\therefore\left(\vec{a}\times\vec{b}\right)\cdot\vec{c}=0$

Hence $\left(\vec{c}-\left(\vec{a}\times\vec{b}\right)\right)\cdot\vec{c}=\vec{c}\cdot\vec{c}$

$=(\alpha+2)^2+(4-\alpha)^2+(2-2\alpha)^2=6\left(\alpha^2-2\alpha+4\right)$

$=6\left((\alpha-1)^2+3\right)$

6. Given that $|\vec{a}| = |\vec{b}| = 1, \vec{a}\cdot\vec{b} = 0$ and $|\vec{c}| = 2$

$\vec{c}$ makes angle α with both $\vec{a}$ and $\vec{b}$.

Also, $\vec{c} = x\vec{a} + y\vec{b} + \vec{a}\times\vec{b}$

$\vec{c}\cdot\vec{a} = |\vec{c}||\vec{a}|\cos 2 = 2\cos\alpha \Rightarrow x = 2\cos\alpha$

$\vec{c}\cdot\vec{b} = 2\cos\alpha \Rightarrow y = 2\cos\alpha$

$|\vec{c}|^2 = \vec{c}\cdot\vec{c} = \left|(2\cos\alpha)\vec{a} + (2\cos\alpha)\vec{b} + \vec{a}\times\vec{b}\right|^2$

$\Rightarrow (2)^2 = 4\cos^2\alpha + 4\cos^2\alpha + \left|\vec{a}\times\vec{b}\right.$

$\Rightarrow 4 = 8\cos^2\alpha + 1 \quad \left(\because \left|\vec{a}\times\vec{b}\right| = 1\times 1\times\sin 90° = 1\right)$

$\Rightarrow 8\cos^2\alpha = 3$

7. $\vec{s} = 4\vec{p} + 3\vec{q} + 5\vec{r}$ ...(i)

$\vec{s} = x(-\vec{p}+\vec{q}+\vec{r}) + y(\vec{p}-\vec{q}+\vec{r}) + z(-\vec{p}-\vec{q}+\vec{r})$ ...(ii)

Comparing (i) and (ii), we get

$\Rightarrow -x+y-z = 4$

$x-y-z = 3$

$x+y+z = 5$

Solving above equations, we get

$x = 4, y = \frac{9}{2}, z = \frac{-7}{2}$

$\therefore 2x+y+z = 9$

8. $\vec{a}\cdot\vec{b} = \vec{b}\cdot\vec{c} = \vec{c}\cdot\vec{a} = \cos\frac{\pi}{3} = \frac{1}{2}$

Given $p\vec{a} + q\vec{b} + r\vec{c} = \vec{a}\times\vec{b} + \vec{b}\times\vec{c}$

Taking its dot product with $\vec{a}, \vec{b}, \vec{c}$, we get

$\left[\vec{a}\ \vec{b}\ \vec{c}\right] = p|\vec{a}|^2 + q\left(\vec{b}\cdot\vec{a}\right) + r\left(\vec{c}-\vec{a}\right)$

$= p + \frac{1}{2}q + \frac{1}{2}r$ ...(i)

Given that $\frac{1}{2}p + q + \frac{1}{2}r = 0$ ...(ii)

and $\frac{1}{2}p + \frac{1}{2}q + r = \left[\vec{a}\ \vec{b}\ \vec{c}\right]$ ...(iii)

From (i) and (iii), $p = r$.

Using (ii) $q = -p$

$\therefore \frac{p^2+2q^2+r^2}{q^2} = \frac{p^2+2p^2+p^2}{p^2} = 4$

9. Given eight vectors are

(1, 1, 1), (–1, –1, –1); (–1, 1, 1), (1, –1, –1); (1, –1, 1), (–1, 1, –1); (1, 1, –1), (–1, –1, 1)

These are the four diagonals of a cube and their opposites.

For three non-coplanar vectors first we select 3 groups of diagonals and its opposite in $^4C_3$ ways. Then one vector from each group can be selected in $2\times2\times2$ ways.

$\therefore$ Total ways $= {}^4C_3\times2\times2\times2 = 32 = 2^5$

$\therefore p = 5$

Comprehension Passage

Comprehension 1

1. (b)

$\vec{a}_1 = \left[\left(2\hat{i}+3\hat{j}-6\hat{k}\right)\frac{\left(2\hat{i}-3\hat{j}+6\hat{k}\right)}{7}\right]\frac{2\hat{i}-3\hat{j}+6\hat{k}}{7}$

$= \frac{-41}{49}\left(2\hat{i}-3\hat{j}+6\hat{k}\right)$

$\vec{a}_2 = \frac{-41}{49}\left(\left(2\hat{i}-3\hat{j}+6\hat{k}\right)\frac{\left(-2\hat{i}+3\hat{j}+6\hat{k}\right)}{7}\right)\frac{\left(-2\hat{i}+3\hat{j}+6\hat{k}\right)}{7}$

$= \frac{-41}{(49)^2}(-4-9+36)\left(-2\hat{i}+3\hat{j}+6\hat{k}\right)$

$= \frac{943}{49^2}\left(2\hat{i}-3\hat{j}-6\hat{k}\right)$

2. (a)

$\vec{a}_1\cdot\vec{b} = \frac{-41}{49}\left(2\hat{i}-3\hat{j}+6\hat{k}\right).\left(2\hat{i}-3\hat{j}+6\hat{k}\right) = -41$

3. (c) $\vec{a}, \vec{a}_1, \vec{b}$ are coplanar, because $\vec{a}_1, \vec{b}$ are collinear.

Comprehension 2

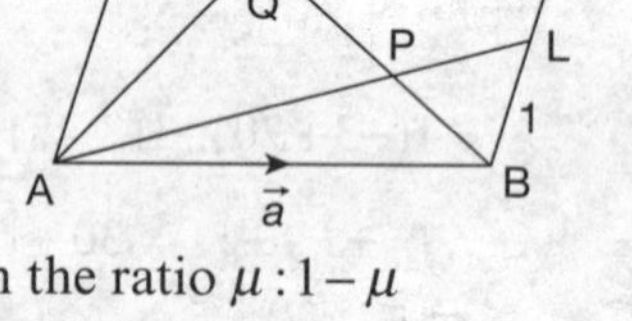

$\overline{BL} = \frac{1}{3}\vec{b}$

$\therefore \overline{AL} = \vec{a} + \frac{1}{3}\vec{b}$

Let $\overline{AP} = \lambda\overline{AL}$

and P divides DB in the ratio $\mu : 1-\mu$

Then $\overline{AP} = \lambda\vec{a} + \frac{\lambda}{3}\vec{b}$ ...(i)

Also $\overline{AP} = \mu\vec{a} + (1-\mu)\vec{b}$ ...(ii)

From (i) and (ii) $\lambda\vec{a} + \frac{\lambda}{3}\vec{b} = \mu\vec{a} + (1-\mu)\vec{b}$

$\therefore \lambda = \mu$ and $\frac{\lambda}{3} = 1-\mu$

$\therefore \lambda = \frac{3}{4}$

$\therefore$ P divides AL in the ratio 3 : 1 and P divides

DB in the ratio 3 : 1. Similarly, Q divides DB in the ratio 1 : 3.

Thus $DQ = \frac{1}{4}DB$ and $PB = \frac{1}{4}DB$

$\therefore PQ = \frac{1}{2}DB$, i.e., $PQ : DB = 1 : 2$

1. (c)
2. (b)
3. (c)

Comprehension 3

1. (a) The diagonals are

$\vec{d}_1 = 3\hat{a} - 2\hat{b} + 2\hat{c} + (-\hat{a} - 2\hat{c}) = 2\hat{a} - 2\hat{b}$

$\vec{d}_2 = 3\hat{a} - 2\hat{b} + 2\hat{c} - (-\hat{a} - 2\hat{c}) = 4\hat{a} - 2\hat{b} + 4\hat{c}$

Angle between them

$$= \cos^{-1}\frac{\vec{d}_1.\vec{d}_2}{|\vec{d}_1|.|\vec{d}_2|} = \cos^{-1}\left(\frac{8+4}{2\sqrt{2}\,(6)}\right) = \cos^{-1}\frac{1}{\sqrt{2}} = \frac{\pi}{4}$$

2. (d)

$\vec{x} + \vec{y} = 2\vec{b} - 3\vec{c}$ and $\vec{y} + \vec{z} = -2\hat{a} + 3\hat{b} - 3\hat{c}$

$$\therefore (\vec{x} + \vec{y}) \times (\vec{y} + \vec{z}) = \begin{vmatrix} \hat{a} & \hat{b} & \hat{c} \\ 0 & 2 & -3 \\ -2 & 3 & -3 \end{vmatrix} = 3\hat{a} + 6\hat{b} + 4\hat{c}$$

$\therefore$ Required unit vector $= \dfrac{3\hat{a} + 6\hat{b} + 4\hat{c}}{\sqrt{61}}$.

3. (a) $\begin{vmatrix} 2 & -3 & 4 \\ 1 & 2 & -1 \\ x & -1 & 2 \end{vmatrix} = 0$

$\Rightarrow 2(4-1) + 3(2+x) + 4(-1-2x) = 0 \Rightarrow x = \frac{8}{5}$

4. (c) $\vec{r} \times \vec{x} = \vec{y} \times \vec{x} \Rightarrow (\vec{r} - \vec{y}) \times \vec{x} = \vec{0} \Rightarrow \vec{r} = \vec{y} + \lambda\vec{x}$

$\vec{r} \times \vec{y} = \vec{x} \times \vec{y} \Rightarrow (\vec{r} - \vec{x}) \times \vec{y} = \vec{0} \Rightarrow \quad \vec{r} = \vec{x} + \mu\vec{y}$

$\vec{y} + \lambda\vec{x} = \vec{x} + \mu\vec{y}$

$(2\vec{a} - \vec{b}) + \lambda(\vec{a} + \vec{b}) = (\vec{a} + \vec{b}) + \mu(2\vec{a} - \vec{b})$

$\Rightarrow 2 + \lambda = 1 + 2\mu,\ -1 + \lambda = 1 - \mu \Rightarrow \mu = 1,\ \lambda = 1$

5. (b) $\hat{a} \times \hat{b} = \hat{c}$

$\Rightarrow \hat{c} \cdot \hat{a} \times \hat{b} = \hat{c} \cdot \hat{c} = 1$

$\Rightarrow \hat{a} \cdot (\hat{b} \times \hat{c}) + \hat{b} \cdot (\hat{c} \times \hat{a}) + \hat{c} \cdot (\hat{a} \times \hat{b}) = 3$

CHAPTER 11

Three-dimensional Geometry

Objectives

In this chapter on Three-dimensional Geometry, we shall study the following topics:

1. Direction cosines and direction ratio of a line joining two points
2. Equations of lines and planes in space under different conditions
3. Angles between two lines, two planes, and a line and a plane
4. Shortest distance between two skew lines
5. The distance of a point from a plane.

Introduction

In Class XI, we studied two-dimensional Analytical Geometry. Also, we were introduced to three-dimensional geometry confined to only Cartesian methods. Now that we have learned vectors in the previous chapter, we will use vector concepts in three-dimensional geometry.

In the present chapter, we shall study about the direction cosines (DC's) and direction ratios (DR's) of a line joining two points and also discuss about the equations of lines and planes in space under different conditions, angle between two lines, two planes, a line and a plane, shortest distance between two skew lines and distance of a point from a plane. Most of the above results are obtained in vector form. Nevertheless, we shall also translate these results in the Cartesian form, which, at times, present a clearer geometric and analytic picture of the situation.

WORKSHEET

1. Find the vector equation of the line which is parallel to the vector $3\hat{i}-2\hat{j}+6\hat{k}$ and which passes through the point (1, –2, 3).

2. Show that the lines $\dfrac{x-1}{2}=\dfrac{y-2}{3}=\dfrac{z-3}{4}$ and $\dfrac{x-4}{5}=\dfrac{y-1}{2}=z$ intersect. Also, find their point of intersection.

3. Prove that the lines $x=py+q, z=ry+s$ and $x=p'y+q', z=r'y+s'$ are perpendicular if $pp'+rr'+1=0$.

4. Find the equation of a plane, which bisects perpendicularly, the line joining the points A (2, 3, 4) and B (4, 5, 8) at right angles.

5. If the line drawn from the point (–2, – 1, – 3) meets a plane at right angle at the point (1, – 3, 3), find the equation of the plane.

6. Find the equation of the plane through the points (2, 1, 0), (3, –2, –2) and (3, 1, 7).

7. Find the equations of the two lines through the origin, which intersect the line $\dfrac{x-3}{2}=\dfrac{y-3}{1}=\dfrac{z}{1}$ at angles of $\pi/3$ each.

8. Find the angle between the lines whose direction cosines are given by the equations $l+m+n=0$, $l^2+m^2-n^2=0$.

9. If a variable line in two adjacent positions has direction cosines l, m, n and $l+\delta l, m+\delta m, n+\delta n$, show that the small angle $\delta\theta$ between the two positions is given by $\delta\theta^2=\delta l^2+\delta m^2+\delta n^2$.

10. O is the origin and A is (a, b, c). Find the direction cosines of the line OA and the equation of plane through A at right angle to OA.

11. Two systems of rectangular axis have the same origin. If a plane cuts them at distances a, b, c and a', b', c', respectively, from the origin, prove that
$\frac{1}{a^2}+\frac{1}{b^2}+\frac{1}{c^2}=\frac{1}{a'^2}+\frac{1}{b'^2}+\frac{1}{c'^2}$

12. Find the distance of a point (2, 4, –1) from the line $\frac{x+5}{1}=\frac{y+3}{4}=\frac{z-6}{-9}$.

13. Find the equation of the plane through the points (2, 1, –1) and (–1, 3, 4) and perpendicular to the plane $x - 2y + 4z = 10$.

14. Find the shortest distance between the lines given by $\vec{r}=(8+3\lambda\hat{i}-(9+16\lambda)\hat{j}+(10+7\lambda)\hat{k}$ and $\vec{r}=15\hat{i}+29\hat{j}+5\hat{k}+\mu(3\hat{i}+8\hat{j}-5\hat{k})$.

15. The plane $ax + by = 0$ is rotated about its line of intersection with the plane $z = 0$ through an angle α. Prove that the equation of the plane in its new position is $ax+by\pm\left(\sqrt{a^2+b^2}\tan\alpha\right)z=0$.

SOLVED EXAMPLES

Example 1. Find the co-ordinates of the point where the line joining A (3, 4, 1) and B (5, 1, 6) crosses the xy-plane.

Solution

Let the line joining A and B crosses the xy-plane at the point M and let M divide AB in the ratio $\lambda:1$ internally. Co-ordinates of M are

$$\left[\frac{5\lambda+3}{\lambda+1},\frac{\lambda+4}{\lambda+1},\frac{6\lambda+1}{\lambda+1}\right]$$

Since the point M lies on the xy-plane, its z-co-ordinate is zero.

$$\therefore\ \frac{6\lambda+1}{\lambda+1}=0\Rightarrow 6\lambda+1=0.\text{ or }\lambda=-\frac{1}{6}$$

The ratio is $\lambda:1=\frac{1}{6}:1$, i.e., 1 : 6 externally.

Hence the co-ordinates of M are

$$\left[\frac{5(-1/6)+3}{(-1/6)+1},\frac{(-1/6)+4}{(-1/6)+1},0\right],\text{ i.e.,}\left(\frac{13}{5},\frac{23}{5},0\right)$$

Example 2. A (l, 8, 4), B (0, –11, 3), C (2, –3, –1) are three points and D is the foot of the perpendicular from A on BC. Find the co-ordinate of D.

Solution

Let D be the foot of the ⊥ from A on BC. If D divides BC in the ratio k : 1, then co-ordinates of D are

A (1, 8, 4)

B (0, –11, 3) D C (2, –3, –1)

$$\left(\frac{2k+0}{k+1},\frac{-3k-11}{k+1},\frac{-k+3}{k+1}\right)$$

Direction ratios of AD are

$$\frac{2k}{k+1}-1,\frac{-3k-11}{k+1}-8,\frac{-k+3}{k+1}-4$$

$$\Rightarrow\ \frac{k-1}{k+1},\frac{-11k-19}{k+1},\frac{-5k-1}{k+1}$$

Direction ratios of BC are 2 – 0, –3 +11, –1 – 3, i.e., 2, 8, –4

Since AD ⊥ BC

$$\therefore\ \frac{k-1}{k+1}\times2+\frac{-11k-19}{k+1}\times8+\frac{-5k-1}{k+1}\times(-4)=0$$

$$\Rightarrow\ \frac{2k-2-88k-152+20k+4}{k+1}=0$$

$$\Rightarrow\ -66k-150=0\ \Rightarrow 66k=-150$$

$$\therefore\ k=-\frac{25}{11}$$

Hence the co-ordinates of D are $\left(\frac{25}{7},\frac{23}{7},-\frac{4}{7}\right)$.

Example 3. Find the projection of the line segment joining the points (–1, 0, 3) and (2, 5, 1) on the line whose direction ratios are 6, 2, 3.

Solution

The direction cosines ℓ,m,n of the line are given by

$$\frac{\ell}{6}=\frac{m}{2}=\frac{n}{2}=\frac{\sqrt{\ell^2+m^2+n^2}}{\sqrt{6^2+2^2+3^2}}=\frac{1}{\sqrt{49}}=\frac{1}{7}$$

$$\therefore\ \ell=\frac{6}{7},m=\frac{2}{7},n=\frac{3}{7}$$

The required projection is given by

$$=\ell(x_2-x_1)+m(y_2-y_1)+n(z_2-z_1)$$

$$=\frac{6}{7}[2-(-1)]+\frac{2}{7}(5-0)+\frac{3}{7}(1-3)$$

$$=\frac{6}{7}\times3+\frac{2}{7}\times5+\frac{3}{7}\times-2=\frac{18}{7}+\frac{10}{7}-\frac{6}{7}=\frac{18+10-6}{7}=\frac{22}{7}$$

Example 4. Find the ratio in which the plane $2x + 3y + 5z = 1$ divides the line joining the points (1, 0, –3) and (1, –5, 7).

Solution

Let the required ratio be $k : 1$.

The co-ordinates of the point which divides the join of (1, 0, –3) and (1, –5, 7) in the ratio $k : 1$ are

$$\left(\frac{k+1}{k+1}, \frac{-5k}{k+1}, \frac{7k-3}{k+1}\right)$$

Since this point lies on $2x + 3y + 5z = 1$,

$$\therefore \quad 2\left(\frac{k+1}{k+1}\right)+3\left(-\frac{5k}{k+1}\right)+5\left(\frac{7k-3}{k+1}\right)=1$$

$$\Rightarrow \quad 2k+2-15k+35k-15=k+1$$

$$\Rightarrow \quad 22k-13=k+1$$

$$\Rightarrow \quad 21k=14; \quad k=\frac{14}{21}=\frac{2}{3}$$

Hence the required ratio =

$$k:1=\frac{2}{3}:1=2:3$$

Example 5. The locus of a point, which moves in such a way that its distance from the origin is thrice the distance from xy-plane is _______.

(A) $x^2-8y^2-8z^2=0$ (B) $x^2 - 8y^2 + z^2 = 0$

(C) $-8x^2+y^2+z^2=0$ (D) $x^2+y^2-8z^2=0$

Solution

Let the point be P (x, y, z), then its distance from origin is

$\text{OP}=\sqrt{x^2+y^2+z^2}$

The distance of P from xy-plane is z.

According to the question $\sqrt{x^2+y^2+z^2}=3z$.

Thus, the required locus is $x^2+y^2-8z^2=0$.

Example 6. The foot of the perpendicular drawn from the point P (1, 0, 3) to the join of points A (4, 7, 1) and B (3, 5, 3) is _______.

(1) (5, 7, 1) (2) $\left(\frac{5}{3},\frac{7}{3},\frac{17}{3}\right)$

(3) $\left(\frac{2}{3},\frac{5}{3},\frac{7}{3}\right)$ (4) $\left(\frac{5}{3},\frac{2}{3},\frac{7}{3}\right)$

Solution

Drawn PL$\perp$AB. Suppose L divides AB in the ratio $k : 1$, then co-ordinates of L are

$$\left(\frac{3k+4}{k+1}, \frac{5k+7}{k+1}, \frac{3k+1}{k+1}\right)$$

Direction ratios of AB are 1, 2, -2 and Direction ratios of, PL are

$$\left(\frac{2k+3}{k+1}, \frac{5k+7}{k+1}, \frac{-2}{k+1}\right)$$

Since PL$\perp$AB, we have

$$\left(\frac{2k+3}{k+1}\right)\times(1)+\left(\frac{5k+7}{k+1}\right)\times(2)+\left(-\frac{2}{k+1}\right)\times(-2)=0$$

$\Rightarrow k=-7/4$

$\therefore L \text{ is} \left(\frac{5}{3},\frac{7}{3},\frac{17}{3}\right)$

Example 7: If a line makes angles $\alpha,\beta,\gamma,\delta$ with four diagonals of a cube, then $\cos^2\alpha+\cos^2\beta+\cos^2\gamma+\cos^2\delta$ equals _______.

(a) 3 (b) 4

(c) 4/3 (d) 3/4

Solution

Let OA, OB, OC be coterminous edges of a cube and OA= OB = OC = a, then co-ordinates of its vertices are O (0, 0, 0), A (a, 0, 0), B (0, a, 0), C (0, 0, c), L (0, a, a), N (a, a, 0) and P (a, a, a)

Direction ratio of diagonal AL, BM, CN and OP are

$$\left(-\frac{1}{\sqrt{3}},\frac{1}{\sqrt{3}},\frac{1}{\sqrt{3}}\right),\left(\frac{1}{\sqrt{3}},-\frac{1}{\sqrt{3}},\frac{1}{\sqrt{3}}\right),\left(\frac{1}{\sqrt{3}},\frac{1}{\sqrt{3}},-\frac{1}{\sqrt{3}}\right),$$
$$\left(\frac{1}{\sqrt{3}},\frac{1}{\sqrt{3}},\frac{1}{\sqrt{3}}\right)$$

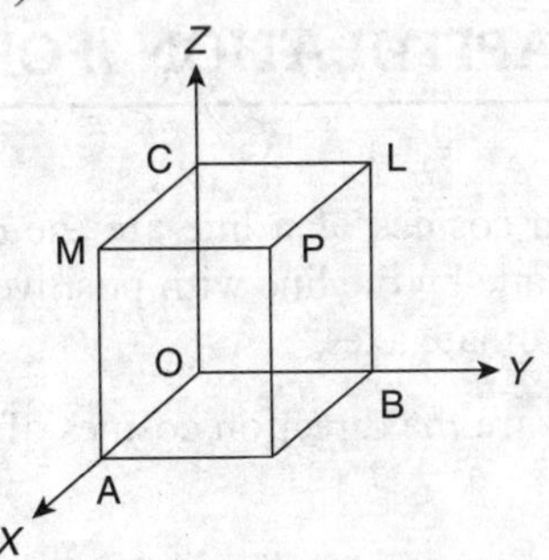

Let ℓ,m,n be the direction cosines of the given line, then

$$\cos\alpha=\ell\left(-\frac{1}{\sqrt{3}}\right)+m\left(\frac{1}{\sqrt{3}}\right)+n\left(\frac{1}{\sqrt{3}}\right)$$
$$=\frac{-\ell+m+n}{\sqrt{3}}$$

Similarly $\cos\beta=\frac{\ell-m+n}{\sqrt{3}}$, $\cos\gamma=\frac{\ell+m+n}{\sqrt{3}}$ and

$\cos\delta=\frac{\ell+m+n}{\sqrt{3}}$

$\therefore \quad \cos^2\alpha+\cos^2\beta+\cos^2\gamma+\cos^2\delta=\frac{4}{3}$

Example 8. If $l_1\, m_1\, n_1$ and $l_2\, m_2\, n_2$ are DC's of the two lines inclined to each other at angle θ, then the DC's of the internal bisector of the angle between these lines are ________.

(a) $\frac{l_1+l_2}{2\sin\theta/2}, \frac{m_1+m_2}{2\sin\theta/2}, \frac{n_1+n_2}{2\sin\theta/2}$

(b) $\frac{l_1+l_2}{2\cos\theta/2}, \frac{m_1+m_2}{2\cos\theta/2}, \frac{n_1+n_2}{2\cos\theta/2}$

(c) $\frac{l_1-l_2}{2\sin\theta/2}, \frac{m_1-m_2}{2\sin\theta/2}, \frac{n_1-n_2}{2\sin\theta/2}$

(d) $\frac{l_1-l_2}{2\cos\theta/2}, \frac{m_1-m_2}{2\cos\theta/2}, \frac{n_1-n_2}{2\cos\theta/2}$

Solution

Let OA and OB be two lines with DC's $l_1\, m_1\, n_1$ and $l_2\, m_2\, n_2$. Let OA = OB = 1. Then the co-ordinates of A and B are $(l_1\, m_1\, n_1)$ and $(l_2\, m_2\, n_2)$ respectively. Let OC be the bisector of $\angle$ AOB. Then C is the mid-point of AB and so its co-ordinates are

$$\left(\frac{l_1+l_2}{2}, \frac{m_1+m_2}{2}, \frac{n_1+n_2}{2}\right)$$

$\therefore$ DR's of OC are $\frac{l_1+l_2}{2}, \frac{m_1+m_2}{2}+\frac{n_1+n_2}{2}$.

We have $OC=\sqrt{\left(\frac{l_1+l_2}{2}\right)^2+\left(\frac{m_1+m_2}{2}\right)^2+\left(\frac{n_1+n_2}{2}\right)^2}$

$$=\frac{1}{2}\sqrt{\left(l_1^2+m_1^2+n_1^2\right)+\left(l_2^2+m_2^2+n_2^2\right)+2\left(l_1l_2+m_1m_2+n_1n_2\right)}$$

$$=\frac{1}{2}\sqrt{2+2\cos\theta} \qquad [\because \cos\theta=l_1l_2+m_1m_2+n_1n_2]$$

$$\frac{1}{2}\sqrt{2(1+\cos\theta)}=\cos\left(\frac{\theta}{2}\right)$$

$\therefore$ DC's of $\overline{OC}$ are $\frac{l_1+l_2}{2(OC)}, \frac{m_1+m_2}{2(OC)}, \frac{n_1+n_2}{2(OC)}$

or $\frac{l_1+l_2}{2\cos\theta/2}, \frac{m_1+m_2}{2\cos\theta/2}, \frac{n_1+n_2}{2\cos\theta/2}$

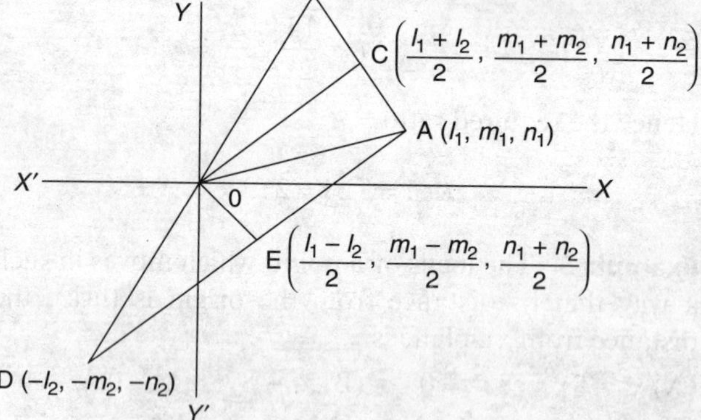

RECAPITULATION (FORMULAE, THEOREM, IMPORTANT RESULTS)

1. Direction cosines of a line are the cosines of the angles made by the line with positive directions of the co-ordinate axes.

2. If l, m, n are the direction cosines of a line, then $l^2+m^2+n^2=1$

3. The direction cosines of a line joining two points P (x_1, y_1, z_1) and Q (x_2, y_2, z_2) are

   $\frac{x_2-x_1}{PQ}, \frac{y_2-y_1}{PQ}, \frac{z_2-z_1}{PQ},$

   where $PQ=\sqrt{(x_2-x_1)^2+(y_2-y_1)^2+(z_2-z_1)^2}$

4. The direction ratios of a line are the numbers which are proportional to the direction cosines of the line.

5. If l, m, n are the direction cosines and a, b, c are direction ratios of a line, then

   Then

   $$l=\frac{\pm a}{\sqrt{a^2+b^2+c^2}}; m=\frac{\pm b}{\sqrt{a^2+b^2+c^2}}; n=\frac{\pm c}{\sqrt{a^2+b^2+c^2}}$$

6. Skew lines are lines in the space which are neither parallel nor intersecting. They lie in different planes.

7. The angle between skew lines is the angle between two intersecting lines drawn from any point (preferably through the origin) parallel to each of the skew lines.

8. If l_1, m_1, n_1 and l_2, m_2, n_2 are the direction cosines of two lines and θ is the acute angle between the two lines, then $\cos\theta=|l_1l_2+m_1m_2+n_1n_2|$

9. If a_1, b_1, c_1 and a_2, b_2, c_2 are the directions ratios of two lines and θ is the acute angle between the two lines, then

$$\cos\theta = \left|\frac{a_1a_2 + b_1b_2 + c_1c_2}{\sqrt{a_1^2 + a_2^2 + a_3^2}.\sqrt{a_1^2 + a_2^2 + a_3^2}}\right|$$

10. The vector equation of a line that passes through a given point whose position vector is $\vec{a}$ and parallel to a given vector $\vec{b}$ .is $\vec{r} = \vec{a} + \lambda\vec{b}$.

11. The equation of a line through a point (x_1, y_1, z_1) and having directions cosines l, m, n (or, direction ratios a, b and c) is

$$\frac{x - x_1}{l} = \frac{y - y_1}{m} = \frac{z - z_1}{n} \text{ or } \left(\frac{x - x_1}{a} = \frac{y - y_1}{b} = \frac{z - z_1}{c}\right).$$

12. The vector equation of a line that passes through two points whose positions vectors are $\vec{a}$ and $\vec{b}$ is $\vec{r} = \vec{a} + \lambda(\vec{b} - \vec{a})$.

13. The cartesian equation of a line that passes through two points (x_1, y_1, z_1) and (x_2, y_2, z_2) is $\frac{x - x_1}{x_2 - x_1} = \frac{y - y_1}{y_2 - y_1} = \frac{z - z_1}{z_2 - z_1}$.

14. If θ is the acute angle between the lines $\vec{r} = \vec{a}_1 + \lambda\vec{b}_1$ and $\vec{r} = \vec{a}_2 + \lambda\vec{b}_2$, then θ is given by $\cos\theta = \frac{|\vec{b}_1.\vec{b}_2|}{|\vec{b}_1||\vec{b}_2|}$ or $\theta = \cos^{-1}\frac{|\vec{b}_1.\vec{b}_2|}{|\vec{b}_1||\vec{b}_2|}$.

15. If $\frac{x - x_1}{l_1} = \frac{y - y_1}{m_1} = \frac{z - z_1}{n_1}$ and $\frac{x - x_2}{l_1} = \frac{y - y_2}{m_2} = \frac{z - z_2}{n_2}$ are equations of two lines, then the acute angle θ between the two lines is given by $\cos\theta = |l_1l_2 + m_1m_2 + n_1m_2|$.

16. The shortest distance between two skew lines is the length of the line segment perpendicular to both the lines.

17. The shortest distance between the lines $\vec{r} = \vec{a}_1 + \lambda\vec{b}_1$ and $\vec{r} = \vec{a}_2 + \lambda\vec{b}_2$ is

$$\left|\frac{(\vec{b}_1 \times \vec{b}_2).(\vec{a}_2 - \vec{a}_1)}{|\vec{b}_1 \times \vec{b}_2|}\right|.$$

18. The shortest distance between the lines: $\frac{x - x_1}{a_1} = \frac{y - y_1}{b_1} = \frac{z - z_1}{c_1}$ and $\frac{x - x_2}{a_2} = \frac{y - y_2}{b_2} = \frac{z - z_2}{c_2}$

is $$\frac{\begin{vmatrix} x_2 - x_1 & y_2 - y_1 & z_2 - z_1 \\ a_1 & b_1 & c_1 \\ a_2 & b_2 & c_2 \end{vmatrix}}{\sqrt{(b_1c_2 - b_2c_1)^2 + (c_1a_2 - c_2a_1)^2 + (a_1b_2 - a_2b_1)^2}}$$

19. The distance between parallel lines $\vec{r} = \vec{a}_1 + \mu\vec{b}$ and $\vec{r} = \vec{a}_2 + \lambda\vec{b}$ is $\left|\frac{\vec{b} \times (\vec{a}_2 - \vec{a})}{|\vec{b}|}\right|$.

20. The vector equation of a plane which is at a distance p from the origin, where $\hat{n}$ is the unit vector normal to the plane, is $\vec{r}.\hat{n} = p$.

21. The equation of a plane which is at a distance p from the origin with direction cosines of the normal to the plane as l, m, n is $lx + my + nz = p$.

22. The equation of a plane through a point whose position vector is $\vec{a}$ and perpendicular to the vector $\vec{n}$ is $(\vec{r} - \vec{a}).\vec{n} = 0$ or $\vec{r}.\vec{n} = d$, where $d = \vec{a}.\vec{n}$.

23. The equation of a plane perpendicular to a given line with direction ratios a, b, c and passing through a given point (x_1, y_1, z_1) is $a(x - x_1) + b(y - y_1) + c(z - z_1) = 0$.

24. The equation of a plane passing through three non-collinear points (x_1, y_1, z_1), (x_2, y_2, z_2) and (x_3, y_3, z_3) is $\begin{vmatrix} x - x_1 & y - y_1 & z - z_1 \\ x_2 - x_1 & y_2 - y_1 & z_2 - z_1 \\ x_3 - x_1 & y_3 - y_1 & z_3 - z_1 \end{vmatrix} = 0$

25. The vector equation of a plane that contains three non-collinear points having position vectors $\vec{a}, \vec{b}, \vec{c}$ is $(\vec{r} - \vec{a}).[(\vec{b} - \vec{a}) \times (\vec{c} - \vec{a})] = 0$

26. The equation of a plane that cuts the co-ordinates axes at $(a, 0, 0)$, $(0, b, 0)$ and $(0, 0, c)$ is $\frac{x}{a} + \frac{y}{b} + \frac{x}{c} = 1$.

27. Vector equation of any plane that passes through the intersection planes $\vec{r}.\vec{n}_1 = d_1$ and $\vec{r}.\vec{n}_2 = d_2$ is $(\vec{r}.\vec{n}_1 - d_1) + \lambda(\vec{r}.\vec{n}_2 - d_2) = 0$, where λ is any non-zero constant.

28. Cartesian equation of any plane that passes through the intersection of two given planes $A_1x + B_1y + C_1z + D_1 = 0$ and $A_2x + B_2y + C_2z + D_2 = 0$ is $(A_1x + B_1y + C_1z + D_1) + \lambda(A_2x + B_2y + C_2z + D_2) = 0$.

29. Two lines $\vec{r} = \vec{a}_1 + \lambda\vec{b}_1$ and $\vec{r} = \vec{a}_2 + \lambda\vec{b}_2$ are coplanar if $(\vec{a}_2 - \vec{a}_1).(\vec{b}_1 \times \vec{b}_2) = 0$.

30. Two lines $\frac{x - x_1}{a_1} = \frac{y - y_1}{b_1} = \frac{z - z_1}{c_1}$ and $\frac{x - x_2}{a_2} = \frac{y - y_2}{b_2} = \frac{z - z_2}{c_2}$ are coplanar if

$$\begin{vmatrix} x_2 - x_1 & y_2 - y_1 & z_2 - z_1 \\ a_1 & b_1 & c_1 \\ a_2 & b_2 & c_2 \end{vmatrix} = 0.$$

31. In vector form, if θ is the acute angle between the two planes, $\vec{r}.\vec{n}_1 = d_1$ and $\vec{r}.\vec{n}_2 = d_2$, then $\theta = \cos^{-1} \dfrac{|\vec{n}_1.\vec{n}_2|}{|\vec{n}_1|.|\vec{n}_2|}$.

32. The acute angle θ between the line $\vec{r} = \vec{a} + \lambda\vec{b}$ and plane $\vec{r}.\vec{n} = d$ is given by $\sin\theta = \dfrac{|\vec{b}.\vec{n}|}{|\vec{b}|.|\vec{n}|}$.

QUIZ

1. If a given line in space does not pass through origin, then how will you find the direction cosine?
2. What is the relation between direction ratios and direction cosines?
3. If a line has direction ratios $2, -1, -2$, what is its direction cosine?
4. If a line makes angle 90°, 60° and 30° with the positive direction of x, y and z-axis respectively, then what is its direction cosine?
5. What are the direction cosines of x, y and z-axis?
6. What is the equation of the line passing through (x_1, y_1, z_1) if the direction cosines are l, m and n?
7. What is the Cartesian form of equation of a line which passes through two points (x_1, y_1, z_1) and (x_2, y_2, z_2)?
8. What is the equation of a plane in normal form? Give the answer in vector as well as Cartesian mode.
9. What is the equation of the plane with intercepts 2, 3 and 4 on the x, y and z-axis respectively?
10. What is the angle between the two planes $3x - 6y + 2z = 7$ and $2x + 2y - 2z = 5$?

JEE MAIN

Objective Questions

One Option Correct

1. The angle between the lines $2x = 3y = -z$ and $6x = -y = -4z$ is ______.

(a) 45° (b) 30°
(c) 0° (d) 90°

2. If the angle θ between the line: $\dfrac{x+1}{1} = \dfrac{y-1}{2} = \dfrac{z-2}{2}$ and the plane $2x - y + \sqrt{\lambda}z + 4 = 0$ is such that $\sin\theta = \dfrac{1}{3}$, then value of λ is ______.

(a) $\dfrac{3}{4}$ (b) $\dfrac{-4}{3}$
(c) $\dfrac{5}{3}$ (d) $\dfrac{-3}{5}$

3. The distance between two parallel planes $2x + y + 2z = 8$ and $4x + 2y + 4z + 5 = 0$ is ______.

(a) $\dfrac{3}{2}$ (b) $\dfrac{5}{2}$
(c) $\dfrac{7}{2}$ (d) $\dfrac{9}{2}$

4. If non-zero numbers a, b, c are in HP, then the straight line $\dfrac{x}{a} + \dfrac{y}{b} + \dfrac{1}{c} = 0$ always passes through a fixed point. That point is ______.

(a) $(1, -2)$ (b) $\left(1, -\dfrac{1}{2}\right)$
(c) $(-1, 2)$ (d) $(-1, -2)$

5. A line makes the same angle θ, with each of the x- and z-axis. If the angle β, which it makes with y-axis, is such that $\sin^2\beta = 3\sin^2\theta$, then $\cos^2\theta$ is equal to ______.

(a) $\dfrac{2}{3}$ (b) $\dfrac{1}{5}$
(c) $\dfrac{3}{5}$ (d) $\dfrac{2}{5}$

6. A line with direction cosines proportional to 2, 1, 2, meets each of the lines $x = y + a = z$ and $x + a = 2y = 2z$. The co-ordinates of each of the points of intersection are given by ______.

(a) $(3a, 3a, 3a), (a, a, a)$
(b) $(3a, 2a, 3a), (a, a, a)$
(c) $(3a, 2a, 3a), (a, a, 2a)$
(d) $(2a, 3a, 3a), (2a, a, a)$

7. If the straight lines $x=1+s, y=-3-\lambda s, z=1+\lambda s$ and $x=\frac{t}{2}, y=1+t, z=2-t$, with parameters s and t respectively, are coplanar. Then λ is equal to ______.

(a) -2 (b) -1

(c) $-\frac{1}{2}$ (d) 0

8. The lines $\frac{x-2}{1}=\frac{y-3}{1}=\frac{z-4}{-k}$ and $\frac{x-1}{k}=\frac{y-4}{2}=\frac{z-5}{1}$ are coplanar if ______.

(a) $k=0$ or -1 (b) $k=1$ or -1

(c) $k=0$ or -3 (d) $k=3$ or -3

9. Two systems of a rectangular axes have the same origin. If a plane separates them at distances a,b,c and a',b',c' from the origin, then ______.

(a) $\frac{1}{a^2}+\frac{1}{b^2}+\frac{1}{c^2}+\frac{1}{a'^2}+\frac{1}{b'^2}+\frac{1}{c'^2}=0$

(b) $\frac{1}{a^2}+\frac{1}{b^2}-\frac{1}{c^2}+\frac{1}{a'^2}+\frac{1}{b'^2}+\frac{1}{c'^2}=0$

(c) $\frac{1}{a^2}-\frac{1}{b^2}-\frac{1}{c^2}+\frac{1}{a'^2}-\frac{1}{b'^2}-\frac{1}{c'^2}=0$

(d) $\frac{1}{a^2}+\frac{1}{b^2}+\frac{1}{c^2}-\frac{1}{a'^2}-\frac{1}{b'^2}-\frac{1}{c'^2}=0$

10. A plane meets axis at A, B, C and the centroid of triangle ABC is at (1, 2, 3). then the equation of plane is ______.

(a) $\frac{x}{1}+\frac{y}{2}+\frac{z}{3}=1$ (b) $\frac{x}{3}+\frac{y}{6}+\frac{z}{9}=1$

(c) $\frac{x}{1}+\frac{y}{2}+\frac{z}{3}=\frac{1}{3}$ (d) None of these.

11. If M denotes the mid-point of the line joining $A(4\hat{i}+5\hat{j}-10\hat{k})$ and $B(-\hat{i}+2\hat{j}+\hat{k})$, then equation of the plane through M and perpendicular to AB is ______.

(a) $\vec{r}.(-5\hat{i}-3\hat{j}+11\hat{k})+\frac{135}{2}=0$

(b) $\vec{r}.\left(\frac{3}{2}\hat{i}+\frac{7}{2}\hat{j}-\frac{9}{2}\hat{k}\right)+\frac{135}{2}=0$

(c) $\vec{r}.(4\hat{i}+5\hat{j}-10\hat{k})+4=0$

(d) $\vec{r}.(-\hat{i}+2\hat{j}+\hat{k})+4=0$

12. A plane $x+y+z-a\sqrt{3}$ touches the sphere $2x^2+2y^2+2z^2-2x+4y+4z+3=0$. Then the value of a is ______.

(a) $+\frac{1}{\sqrt{3}}$ (b) $\frac{1}{2\sqrt{3}}$

(c) $1-\frac{1}{\sqrt{3}}$ (d) $1+\frac{1}{\sqrt{3}}$

13. The equation of the plane passing through the mid-point of the line of join of the points (1, 2, 3) and (3, 4, 5) and perpendicular to it is ______.

(a) $x+y+z=9$ (b) $x+y+z=-9$

(c) $2x+3y+4z=9$ (d) $2x+3y+4z=-9$

14. The equation of a line is $6x-2=3y-1=2z-2$. The direction ratios of the line are ______.

(a) 1, 2, 3 (b) 1, 1, 1

(c) $\frac{1}{3},\frac{1}{3},\frac{1}{3}$ (d) $\frac{1}{3},\frac{-1}{3},\frac{1}{3}$

15. Let O be the origin and P be the point at a distance 3 unit from origin. If direction ratios of OP are $(1,-2,-2)$, then co-ordinates of P is given by ______.

(a) $(1,-2,-2)$ (b) $(3,-6,-6)$

(c) $\left(\frac{1}{3},-\frac{2}{3},-\frac{2}{3}\right)$ (d) $\left(\frac{1}{9},-\frac{2}{9},-\frac{2}{9}\right)$

16. The reflection of the plane $2x-3y+4z-3=0$ in the plane $x-y+z-3=0$ is the plane ______.

(a) $4x-3y+2z-15=0$

(b) $x-3y+2z-15=0$

(c) $4x+3y-2z+15=0$

(d) None of the above.

17. The distance from the point (1, 6, 3) to the line $\vec{r}=(\hat{j}+2\hat{k})+\lambda(\hat{i}+2\hat{j}+3\hat{k})$ is ______.

(a) $\sqrt{13}$ (b) 13

(c) $2\sqrt{3}$ (d) None of these.

18. The plane $\frac{x}{2}+\frac{y}{3}+\frac{z}{1}=1$ meets the co-ordinate axis at the point A, B, C respectively, then the area bounded by the triangle ABC, is ______.

(a) 6 sq. unit (b) $\frac{7}{2}$ sq. unit

(c) $\frac{5}{2}$ sq. unit (d) None of these.

19. If A, B, C are points (1, 0, 4), (0, –1, 5) and (2, –3, 1) respectively, then the co-ordinates of the foot of the perpendicular drawn from A to the line BC are ______.

(a) $\left(\frac{1}{2},-\frac{1}{2},\frac{9}{2}\right)$ (b) (1, – 2, 3)

(c) $\left(\frac{3}{2},-\frac{3}{2},\frac{5}{2}\right)$ (d) $\left(\frac{1}{3},\frac{-4}{3},\frac{13}{3}\right)$

20. A mirror and a source of light are situated at the origin O and at a point on OX respectively. A ray of light from the source strikes the mirror and is reflected. If the direction ratios of the normal to the plane are 1, –1, 1, then direction cosines of the reflected rays are ______.

(a) $\frac{1}{3}, \frac{2}{3}, \frac{2}{3}$ (b) $-\frac{1}{3}, \frac{2}{3}, \frac{2}{3}$

(c) $-\frac{1}{3}, -\frac{2}{3}, -\frac{2}{3}$ (d) $-\frac{1}{3}, -\frac{2}{3}, \frac{2}{3}$

21. If the straight lines $\frac{x-1}{k} = \frac{y-2}{2} = \frac{z-3}{3}$ and $\frac{x-2}{3} = \frac{y-3}{k} = \frac{z-1}{2}$ intersect at a point, then the integer k is equal to ______.

(a) 5 (b) 2
(c) – 2 (d) – 5

22. Let L be the line of intersection of the planes $2x+3y+z=1$ and $x+3y+2z=2$. If L makes an angle α with the positive x-axis, then $\cos\alpha$ is equal to ______.

(a) $\frac{1}{\sqrt{3}}$ (b) $\frac{1}{2}$

(c) 1 (d) $\frac{1}{\sqrt{2}}$

23. The two lines $x = ay + b, z = c + d$ and $x = dy + b'; z = cy + d'$ are perpendicular to each other if ______.

(a) $aa' + cc' = 1$ (b) $\frac{a}{a'} + \frac{c}{c'} = -1$

(c) $\frac{a}{a'} + \frac{c}{c'} = 1$ (d) $aa' + cc' = -1$

24. The image of the point (−1, 3, 4) in the plane $x - 2y = 0$ is ______.

(a) $(15, 11, 4)$ (b) $\left(-\frac{17}{3}, -\frac{19}{3}, 1\right)$

(c) $(8, 4, 4)$ (d) None of these.

25. If the x-co-ordinate of a point P on the join of Q (2, 2, 1) and R (5, 1, −2) is 4, then its z-co-ordinate is ______.

(a) 2 (b) 1
(c) –1 (d) –2

26. The equation of the plane through the points (2, 2, 1) and (9, 3, 6) and perpendicular to the plane $2x + 6y + 6z - 1 = 0$ is ______.

(a) $3x + 4y + 5z = 9$ (b) $3x + 4y - 5z = 9$

(c) $3x + 4y - 5z + 9 = 0$ (d) None of these.

27. The equation of the plane which meets the axes in A, B, C such that the centroid of the triangle ABC is $\left(\frac{1}{3}, \frac{1}{3}, \frac{1}{3}\right)$, is given by ______.

(a) $x + y + z = 1$ (b) $x + y + z = 2$

(c) $\frac{x}{3} + \frac{y}{3} + \frac{z}{3} = 3$ (d) $x + y + z = \frac{1}{3}$

28. If O is the origin and A is the point (a, b, c), then the equation of the plane through A and at right angles to OA is

(a) $a(x-a) - b(y-b) - c(z-c) = 0.$
(b) $a(x+a) + b(y+b) + c(z+c) = 0.$
(c) $a(x-a) + b(y-b) + c(z-c) = 0.$
(d) None of these.

29. The values of a for which the system of equation $ax + y + z = 0, x - ay + z = 0, x + y + z = 0$ possesses none-zero solutions, are given by ______.

(a) 1, 2 (b) 1, −1
(c) 0 (d) None of these.

30. The values of x so that the four points A = (0, 2, 0), B = (1, x, 0), C = (1, 2, 0) and D = (1, 2, 1) are coplanar, is ______.

(a) 0 (b) 1
(c) 2 (d) 3

31. The co-ordinates of a point P are (3, 12, 4) with respect to origin O. Then the direction cosines of OP are ______.

(a) $\frac{3}{\sqrt{13}} \frac{1}{\sqrt{13}} \frac{2}{\sqrt{13}}$ (b) $\frac{3}{13}, \frac{12}{13}, \frac{4}{14}$

(c) $\frac{1}{4}, \frac{1}{3}, \frac{1}{2}$ (d) 2, 12, 4

32. The angle between the lines $x + 2y + 3z = 0 = 3x + 2y + z$ and y-axis is ______.

(a) $2\sec^{-1} 3$ (b) $\sec^{-1}\frac{3}{5}$

(c) $\frac{1}{2}\sec^{-1}\frac{3}{4}$ (d) $\sec^{-1}\frac{9}{4}$

33. A line with direction ratios 4, 1, 1 is drawn intersect the lines $L_1 = 5x - 6 = 4y + 3 = z$ and $L_2 = 2x - 4 = 3y + 5 = z$. It lies in the plane ______.

(a) $14x - 21y - 15z = 6$
(b) $15x - 76y + 16z - 75 = 0$
(c) $14x + 21y + 15z = 6$
(d) $15x + 76y - 16z = 0$

34. The lines $L_1 : y+z=1=z+x+\lambda$ and $L_2 : x-y=0=z-x-1$ are coplanar for ______.

(a) $\lambda=2$ (b) $\lambda=\frac{1}{2}$

(c) $\lambda=-2$ (d) for no value of λ.

35. Under what condition does the equation $x^2+y^2+z^2+2ux+2vy+2wz+d=0$ represent a real sphere ?

(a) $u^2+v^2+w^2=d^2$ (b) $u^2+v^2+w^2>d$

(c) $u^2+v^2+w^2<d$ (d) $u^2+v^2+w^2<d^2$

36. The direction ratios *l, m, n* of two lines are connected by the relation $l+m+n=0$. the angle between them is ______.

(a) $\frac{\pi}{3}$ (b) $\frac{\pi}{4}$

(c) $\frac{\pi}{2}$ (d) 0

37. The equation of the plane through (1, 0, −1) and (3, 2, 2) and parallel to the line $\frac{x-1}{1}=\frac{y-1}{-2}=\frac{z-2}{3}$ is ______.

(a) $4x+y+2z=6$ (b) $4x-y-2z-6=0$

(c) $4x-y+2z=6$ (d) $4x+y+2z=-6$

38. The shortest distance between the lines $\vec{r}=6\hat{i}+2\hat{j}+2\hat{k}+\lambda(\hat{i}-2\hat{j}+2\hat{k})$ and $r=4\hat{i}-\hat{k}+\mu(3\hat{i}-2\hat{j}-2\hat{k})$ is ______.

(a) 0 (b) + 2 units

(c) + 30 units (d) 9 units

39. A line is uniquely determined if

(a) it passes through a given point and has a given direction.

(b) it passes through two given points.

(c) it passes through a point and is parallel to a line.

(d) it passes through two given points and is perpendicular to the plane.

40. A plane passes through a fixed point (*a, b, c*) and cuts the axes at A, B, C. The locus of a point equidistant from origin A, B and C must be______.

(a) $\frac{x}{a}+\frac{y}{b}+\frac{z}{c}=2$ (b) $\frac{a}{x}+\frac{b}{y}+\frac{c}{z}=2$

(c) $\frac{a}{x}+\frac{b}{y}+\frac{c}{z}=1$ (d) $\frac{a}{x}+\frac{b}{y}+\frac{c}{z}=\frac{1}{2}$

41. The line passing through the points $(5,1,a)$ and $(3,b,1)$ crosses the yz-plane at the point $\left(0,\frac{17}{2},\frac{-13}{2}\right)$. Then ______.

(a) $a=4, b=6$ (b) $a=6, b=4$

(c) $a=8, b=2$ (d) $a=2, b=8$

42. If a line makes an angle of $\frac{\pi}{4}$ with the positive direction of each of *x*-axis and *y*-axis, then the angle that the line makes with the positive direction of the *z*-axis is ______.

(a) $\frac{\pi}{6}$ (b) $\frac{\pi}{3}$

(c) $\frac{\pi}{4}$ (d) $\frac{\pi}{2}$

43. The point of intersection of the lines $\frac{x-1}{2}=\frac{y-2}{3}=\frac{z-3}{4}$ and $\frac{x-4}{5}=\frac{y-1}{2}=z$ is ______.

(a) (0, 0, 0) (b) (1, 1, 1)

(c) (−1,−1,−1) (d) (1, 2, 3)

44. The image of the point (5, 4, 6) in the plane $x+y+2z-15=0$ is ______.

(a) (3, 2, 2) (b) (2, 3, 2)

(c) (2, 2, 3) (d) (− 5, −4, −6)

45. The length of perpendicular from (1, 6, 3) to the line $\frac{x}{1}=\frac{y-1}{2}=\frac{z-2}{3}$ is ______.

(a) 3 (b) $\sqrt{11}$

(c) $\sqrt{13}$ (d) 5

46. The plane $2x+3y+4z=1$ meets the co-ordinate axes in A, B, C. The centroid of the triangle ABC is ______.

(a) (2, 3, 4) (b) $\left(\frac{1}{2},\frac{1}{3},\frac{1}{4}\right)$

(c) $\left(\frac{1}{6},\frac{1}{9},\frac{1}{12}\right)$ (d) $\left(\frac{3}{2},\frac{3}{3},\frac{3}{4}\right)$

47. The vector equation of the sphere whose centre is the point (1, 0, 1) and radius is 4, is ______.

(a) $\left|\vec{r}-\left(\hat{i}+\hat{k}\right)\right|=4$ (b) $\left|\vec{r}+\left(\hat{i}+\hat{k}\right)\right|=4^2$

(c) $\vec{r}.\left(\hat{i}+\hat{k}\right)=4$ (d) $\vec{r}.\left(\hat{i}+\hat{k}\right)=4^2$

48. The equation of a plane which cuts equal intercepts of unit lengths on the axes, is ______.

(a) $x+y+z=0$ (b) $x+y+z=1$

(c) $x+y-z=1$ (d) $\frac{x}{a}+\frac{y}{b}+\frac{z}{a}=1$

49. The shortest distance between the line $L_1=2x+y+z-1=0=3x+y+2z-2$ and $L_2=x=y=z$ is ______.

(a) $\sqrt{2}$ (b) $\frac{\sqrt{3}}{2}$

(c) $\frac{1}{\sqrt{3}}$ (d) $\frac{1}{\sqrt{2}}$

50. What is the equation of the plane through z-axis and parallel to the line $\frac{x-1}{\cos\theta}=\frac{y+2}{\sin\theta}=\frac{z-3}{0}$?
(a) $x\cot\theta+y=0$ (b) $x\tan\theta-y=0$
(c) $x+y\cot\theta=0$ (d) $x-y\tan\theta=0$

51. What are the direction ratios of the line determined by the planes $x - y + 2z = 1$ and $x+y-z=3$?
(a) (−1, 3, 2) (b) (−1, −3, 2)
(c) (2, 1, 3) (d) (2, 3, 2)

52. The equation of the plane which makes with co-ordinate axes, a triangle with its centroid (α, β, γ) is ______.
(a) $\alpha x+\beta y+\gamma z=3$ (b) $\alpha x+\beta y+\gamma z=1$
(c) $\frac{x}{\alpha}+\frac{y}{\beta}+\frac{z}{\gamma}=3$ (d) $\frac{x}{\alpha}+\frac{y}{\beta}+\frac{z}{\gamma}=1$

53. If A, B, C are points (1, 0, 4), (0, –1, 5) and (2, –3, 1) respectively, then the co-ordinates of foot of the perpendicular drawn from A to the line BC are ______.
(a) $\left(\frac{1}{2},-\frac{1}{2},\frac{9}{2}\right)$ (b) (1, −2, 3)
(c) $\left(\frac{3}{2},-\frac{3}{2},\frac{5}{2}\right)$ (d) $\left(\frac{1}{3},-\frac{4}{3},\frac{13}{3}\right)$

54. The equation of the plane passes through (2, 3, 4) and parallel to the plane $x+2y+4z=5$ is ______.
(a) $x+2y+4z=10$ (b) $x+2y+4z=3$
(c) $x+2y+4z=24$ (d) $x+2y+4z=2$

55. The distance of the point (2, 3, 4) from the plane $3x-6y+2z+18=0$ is
(a) 2 (b) 9
(c) 10 (d) 1

56. The direction cosines to the normal plane $x+2y-3z=12$ are ______.
(a) $\frac{1}{\sqrt{10}}=,\frac{2}{\sqrt{10}},\frac{-3}{\sqrt{10}}$ (b) $\frac{1}{\sqrt{14}},\frac{2}{\sqrt{14}},\frac{-3}{\sqrt{14}}$
(c) $\frac{1}{\sqrt{10}},\frac{2}{\sqrt{10}},\frac{3}{\sqrt{10}}$ (d) $\frac{1}{\sqrt{14}},\frac{2}{\sqrt{14}},\frac{3}{\sqrt{14}}$

57. The equation of the plane passing through mid-point of line the joining the points P (3, 4, 5) and Q (1, 2, 3) and which is perpendicular to PQ is ______.
(a) $2x+3y+4z=-9$ (b) $x+y+z=9$
(c) $2x+3y+4z=9$ (d) $x+y+z=-9$

58. The lines $2x+3y+7z-k=0=x-y-z+1$ and $2x-y+5z-6=0=x+y+z-3$ will be coplanar for k equal to ______.
(a) 14 (b) 12
(c) 13 (d) 20

59. The planes $2x+5y+3z=0, 7y-5z+4=0$ and $2x+5y+3z=0, 7y-5z+4=0$
(a) are equidistant from the origin.
(b) meet a line.
(c) meet at a point.
(d) are perpendicular to line with DR's 1, 0, 1.

60. The number of integral points on the surface $x^2+y^2+z^2=1$ is ______.
(a) 2^3 (b) 3
(c) 2 (d) 6

61. The distance of centroid from the x-axis of the triangle formed by the points (2, −4, 3), (3, −1, −2) and (−2, 5, 8) is ______.
(a) 1 (b) 0
(c) 3 (d) $\sqrt{10}$

62. If two points are P (7, −5, 11) and Q (−2, 8, 13), then the projection of PQ on a straight line with direction cosines $\frac{1}{3},\frac{2}{3},\frac{2}{3}$ is ______.
(a) 7 (b) 8
(c) 9 (d) None of these.

Questions from Previous Years

63. Let α be the angle between the lines whose direction cosines satisfy the equations $l + m + n = 0$ and $l^2+m^2-n^2=0$. Then the value of $\sin^4 a+\cot^4 a$ is ______.
(a) $\frac{3}{4}$ (b) $\frac{5}{8}$
(c) $\frac{1}{2}$ (d) $\frac{3}{8}$

64. A plane passes through the points A (1, 2, 3), B (2, 3, 1) and C (2, 4, 2). If O is the origin and P is (2, −1, −1), then the projection of $\overline{OP}$ on this plane is of length:
(a) $\sqrt{\frac{2}{5}}$ (b) $\sqrt{\frac{2}{7}}$
(c) $\sqrt{\frac{2}{3}}$ (d) $\sqrt{\frac{2}{11}}$

65. The line $x = ay - 1 = z - 2$ and $x=3y-2=bz-2, (ab\neq 0)$ are coplanar, if ______.

(a) $b=1, a\in R-\{0\}$ (b) $a=1, b\in R-\{0\}$
(c) $a=2, b=2$ (d) $a=2, b=3$

66. Let the foot of the perpendicular from a point P (1, 2, −1) to the straight line $L:\frac{x}{1}=\frac{y}{0}=\frac{z}{-1}$ be N. Let a line be drawn from P parallel to the plane $x+y+2z=0$, which meets L at point Q. If α is the acute angle between the lines PN and PQ, then $\cos\alpha$ is equal to________.

(a) $\frac{1}{\sqrt{5}}$ (b) $\frac{\sqrt{3}}{2}$
(c) $\frac{1}{\sqrt{3}}$ (d) $\frac{1}{2\sqrt{3}}$

67. If the mirror image of the point (1, 3, 5) with respect to the plane $4x-5y+2z=8$ is (α,β,γ), then $5(\alpha+\beta+\gamma)$ equals ______.
(a) 43 (b) 47
(c) 41 (d) 39

68. A plane P meets the co-ordinates axes at A, B and C respectively. The centroid of ΔABC is given to be (1, 1, 2). Then the equation of the line through this centroid and perpendicular to the plane P is _______.
(a) $\frac{x-1}{2}=\frac{y-1}{1}=\frac{z-2}{1}$
(b) $\frac{x-1}{1}=\frac{y-1}{1}=\frac{z-2}{2}$
(c) $\frac{x-1}{2}=\frac{y-1}{2}=\frac{z-2}{1}$
(d) $\frac{x-1}{1}=\frac{y-1}{2}=\frac{z-2}{2}$

69. If (*a*, *b*, *c*) is the image of the point (1, 2, −3) on the line, $\frac{x+1}{2}=\frac{y-3}{-2}=\frac{z}{-1}$, then $a+b+c$ is equal to _______.
(a) 2 (b) −1
(c) 3 (d) 1

70. The lines $\vec{r}=(\hat{i}-\hat{j})+l(2\hat{i}+\hat{k})$ and $\vec{r}=(2\hat{i}-\hat{j})+m(\hat{i}+\hat{j}-\hat{k})$
(a) do not intersect for any values of l and m.
(b) intersect for all values of l and m.
(c) intersect when $l=2$ and $m=\frac{1}{2}$.
(d) intersect when $l=1$ and $m=2$.

71. The shortest distance between the lines $\frac{x-3}{3}=\frac{y-8}{-1}=\frac{z-3}{1}$ and $\frac{x+3}{-3}=\frac{y+7}{2}=\frac{z-6}{4}$ is _______.

(a) $2\sqrt{30}$ (b) $\frac{7}{2}\sqrt{30}$
(c) $3\sqrt{30}$ (d) 3

72. The shortest distance between the lines $\frac{x-1}{0}=\frac{y+1}{-1}=\frac{z}{1}$ and $x+y+z+1=0$, $2x-y+x+3=0$ is _______.
(a) 1 (b) $\frac{1}{\sqrt{3}}$
(c) $\frac{1}{\sqrt{2}}$ (d) $\frac{1}{2}$

73. If for some $\alpha\in R$, the lines $L_1:\frac{x+1}{2}=\frac{y-2}{-1}=\frac{z-1}{1}$ and $L_2:\frac{x+2}{\alpha}=\frac{y+1}{5-\alpha}=\frac{z+1}{1}$ are coplanar, then the line L_2 passes through the point _____.
(a) (10, 2, 2) (b) (2, −10, −2)
(c) (10, −2, −2) (d) (−2, 10, 2)

74. The distance of the point (1, −2, 3) from the plane $x-y+z=5$ measured parallel to the line $\frac{x}{2}=\frac{y}{3}=\frac{z}{-6}$ is_______.
(a) $\frac{7}{5}$ (b) 1
(c) $\frac{1}{7}$ (d) 7

75. The foot of the perpendicular drawn from the point (4, 2, 3) to the line joining the points (1, −2, 3) and (1, 1, 0) lies on the plane _______.
(a) $2x+y-z=1$ (b) $x-y-2z=1$
(c) $x-2y+z=1$ (d) $x+2y-z=1$

76. The plane which bisects the line joining the points (4, −2, 3) and (2, 4, −1) at right angles also passes through the point _______.
(a) (4, 0, 1) (b) (0, −1, 1)
(c) (4, 0, −1) (d) (0, 1, −1)

77 The plane passing through the points (1, 2, 1), (2, 1, 2) and parallel to the line, $2x=3y$, $z=1$ also passes through the point _______.
(a) (0, 6, −2) (b) (−2, 0, 1)
(c) (0, −6, 2) (d) (2, 0, −1)

78. A plane passing through the point (3, 1, 1) contains two lines whose direction ratios are 1, −2, 2 and 2, 3, −1 respectively. If this plane also passes through the point $(\alpha, -3, 5)$, then α is equal to ______.

(a) 5 (b) −10
(c) 10 (d) −5

79. If for some α and β in R, the intersection of the following three planes $x + 4y - 2z = 1$, $x + 7y - 5z = \beta$, $x + 5y + \alpha z = 5$ is a line in R^3, then $\alpha + \beta$ is equal to ______.

(a) 0 (b) 10
(c) 2 (d) −10

80. The mirror image of the point (1, 2, 3) in a plane is $\left(-\frac{7}{3}, -\frac{4}{3}, -\frac{1}{3}\right)$. Which of the following points lie on this plane?

(a) (1, 1, 1) (b) (1, −1, 1)
(c) (−1, −1, 1) (d) (−1, −1, −1)

81. Let P be a plane passing through the points (2, 1, 0), (4,1, 1) and (5, 0, 1) and R be any point (2, 1, 6). Then the image of R in the plane P is ______.

(a) (6, 5, 2) (b) (6, 5, −2)
(c) (4, 3, 2) (d) (3, 4, −2)

82. If the length perpendicular from the point $(\beta, 0, \beta)$ $(\beta \neq 0)$ to the line, $\frac{x}{1} = \frac{y-1}{0} = \frac{z+1}{-1}$ is $\sqrt{\frac{3}{2}}$, then β is equal to ______.

(a) 1 (b) 2
(c) −1 (d) −2

83. The vertices B and C of a ΔABC lie on the line, $\frac{x+2}{3} = \frac{y-1}{0} = \frac{z}{4}$ such that BC = 5 units. Then the area (in sq. units) of this triangle, given that the point A has co-ordinates (1, −1, 2), is ______.

(a) $5\sqrt{17}$ (b) $2\sqrt{34}$
(c) 6 (d) $\sqrt{34}$

84. If a point R (4, y, z) lies on the line segment joining the points P (2, −3, 4) and Q (8, 0, 10), then the distance of R from the origin is ______.

(a) $2\sqrt{14}$ (b) $2\sqrt{21}$
(c) 6 (d) $\sqrt{53}$

85. A perpendicular is drawn from a point on the line $\frac{x-1}{2} = \frac{y+1}{-1} = \frac{z}{1}$ to the plane $x + y + z = 3$ such that the foot of the perpendicular Q also lies on the plane $x - y + z = 3$. Then the co-ordinates of Q are ______.

(a) (1, 0, 2) (b) (2, 0, 1)
(c) (−1, 0, 4) (d) (4, 0, −1)

86. The length of the perpendicular from the point (2, −1, 4) on the straight line, $\frac{x+3}{10} = \frac{y-2}{-7} = \frac{z}{1}$ is

(a) greater than 3 but less than 4.
(b) less than 2.
(c) greater than 2 but less than 3.
(d) greater than 4.

87. Two lines $\frac{x-3}{1} = \frac{y+1}{3} = \frac{z-6}{-1}$ and $\frac{x+5}{7} = \frac{y-2}{-6} = \frac{z-3}{4}$ intersect at the point R. The reflection of R in the xy-plane has co-ordinates ______.

(a) (2, −4, −7) (b) (2, 4, 7)
(c) (2, −4, 7) (d) (−2, 4, 7)

88. If the line $\frac{x-2}{3} = \frac{y+1}{2} = \frac{z-1}{-1}$ intersects the plane $2x + 3y - z + 13 = 0$ at a point P and the plane $3x + y + 4z = 16$ at a point Q, then PQ is equal to ______.

(a) 14 (b) $\sqrt{14}$
(c) $2\sqrt{7}$ (d) $2\sqrt{14}$

89. A plane which bisects the angle between the two given planes $2x - y + 2z - 4 = 0$ and $x + 2y + 2z - 2 = 0$, passes through the point ______.

(a) (1, −4, 1) (b) (1, 4, −1)
(c) (2, 4, 1) (d) (2, −4, 1)

90. A plane passing through the points (0, −1, 0) and (0, 0, 1) and making an angle $\frac{\pi}{4}$ with the plane $y - z + 5 = 0$, also passes through the point ______.

(a) $(-\sqrt{2}, 1, -4)$ (b) $(\sqrt{2}, -1, 4)$
(c) $(-\sqrt{2}, -1, -4)$ (d) $(\sqrt{2}, 1, 4)$

91. The equation of a plane containing the line of intersection of the planes $2x - y - 4 = 0$ and $y + 2z - 4 = 0$ and passing through the point (1, 1, 0) is ______.

(a) $x - 3y - 2z = -2$ (b) $2x - z = 2$
(c) $x - y - z = 0$ (d) $x + 3y + z = 4$

92. The perpendicular distances from the origin to the plane containing the two lines, $\frac{x+2}{3} = \frac{y-2}{5} = \frac{z+5}{7}$ and $\frac{x-1}{1} = \frac{y-4}{4} = \frac{z+4}{7}$, is ______.

(a) $11\sqrt{6}$ (b) $\frac{11}{\sqrt{6}}$

(c) 11 (d) $6\sqrt{11}$

93. If an angle between the line, $\frac{x+1}{2}=\frac{y-2}{1}=\frac{z-3}{-2}$ and the plane, $x-2y-kx=3$ is $\cos^{-1}\left(\frac{2\sqrt{2}}{3}\right)$, then the value of k is ______.

(a) $\sqrt{\frac{5}{3}}$ (b) $\sqrt{\frac{3}{5}}$

(c) $-\frac{3}{5}$ (d) $-\frac{5}{3}$

94. Let S be the set of all real values of λ such that a plane passing through the points $(-\lambda^2, 1, 1)$, $(1, -\lambda^2, 1)$ and $(1, 1, -\lambda^2)$ also passes through the point $(-1, -1, 1)$. Then S is equal to ______.

(a) $\{\sqrt{3}\}$ (b) $\{\sqrt{3}, -\sqrt{3}\}$

(c) $\{1, -1\}$ (d) $\{3, -3\}$

95. The plane containing the line $\frac{x-3}{2}=\frac{y+2}{-1}=\frac{z-1}{3}$ and also containing its projection on the plane $2x + 3y - z = 5$, contains which one of the following points?

(a) (2, 2, 0) (b) (–2, 2, 2)

(c) (0, –2, 2) (d) (2, 0, –2)

96. If the point $(2, \alpha, \beta)$ lies on the plane which passes through the points (3, 4, 2) and (7, 0, 6) and is perpendicular to the plane $2x - 5y = 15$, then $2\alpha - 3\beta$ is equal to ______.

(a) 12 (b) 7

(c) 5 (d) 17

97. The plane which bisects the line segment joining the points (−3, −3, 4) and (3, 7, 6) at right angles, passes through which one of the following points?

(a) (−2, 3, 5) (b) (4, −1, 7)

(c) (2, 1, 3) (d) (4, 1, −2)

98. The length of the projection of the line segment joining the points (5, −1, 4) and (4, −1, 3) on the plane, $x + y + z = 7$ is ______.

(a) $\frac{2}{3}$ (b) $\frac{1}{3}$

(c) $\sqrt{\frac{2}{3}}$ (d) $\frac{2}{\sqrt{3}}$

99. The angle between the lines whose direction cosines are given by the equation, $l+3m+5n=0$ and $5lm-2mn+6nl=0$ is ______.

(a) $\cos^{-1}\left(\frac{1}{8}\right)$ (b) $\cos^{-1}\left(\frac{1}{6}\right)$

(c) $\cos^{-1}\left(\frac{1}{3}\right)$ (d) $\cos^{-1}\left(\frac{1}{4}\right)$

100. If the angle between the lines, $\frac{x}{2}=\frac{y}{2}=\frac{z}{1}$ and $\frac{5-x}{-2}=\frac{7y-14}{P}=\frac{z-3}{4}$ is $\cos^{-1}\left(\frac{2}{3}\right)$, then P is equal to ______.

(a) $-\frac{7}{4}$ (b) $\frac{2}{7}$

(c) $-\frac{4}{7}$ (d) $\frac{7}{2}$

101. A plane bisects the line segment joining the points (1, 2, 3) and (−3, 4, 5) at right angles. Then this plane also passes through the point ______.

(a) (−3,2,1) (b) (3,2,1)

(c) (1,2,−3) (d) (−1,2,3)

102. If the image of the point P(1, – 2, 3) in the plane, $2x+3y-4z+22=0$, measured parallel to line, $\frac{x}{1}=\frac{y}{4}=\frac{z}{5}$ is Q, then PQ is equal to ______.

(a) $6\sqrt{5}$ (b) $3\sqrt{5}$

(c) $2\sqrt{42}$ (d) $\sqrt{42}$

103. The distance of the point (1, – 5, 9) from the plane $x-y+z=5$ measured along the line $x = y = z$ is ______.

(a) $\frac{10}{\sqrt{3}}$ (b) $\frac{20}{3}$

(c) $3\sqrt{10}$ (d) $10\sqrt{3}$

104. The equation of the plane containing the line $2x-5y+z=3$; $x+y+4z=5$, and parallel to the plane, $x+3y+6z=1$, is ______.

(a) $x+3y+6z=7$ (b) $2x+6y+12z=-13$

(c) $2x+6y+12z=13$ (d) $x+3y+6z=-7$

105. A symmetrical form of the line of intersection of the planes $x = ay+b$ and $z = cy+d$ is ______.

(a) $\frac{x-b}{a}=\frac{y-1}{1}=\frac{z-d}{c}$

(b) $\frac{x-b-a}{a}=\frac{y-1}{1}=\frac{z-d-c}{c}$

(c) $\frac{x-a}{b}=\frac{y-0}{1}=\frac{z-c}{d}$

(d) $\frac{x-b-a}{b}=\frac{y-1}{0}=\frac{z-d-c}{d}$

106. The distance between two parallel planes $2x+y+2z=8$ and $4x+2y+4z+5=0$ is ______.

(a) $\frac{3}{2}$ (b) $\frac{5}{2}$

(c) $\frac{7}{2}$ (d) $\frac{9}{2}$

107. The equation of a plane passing through the line of intersection of the planes $x+2y+3z=2$ and $x-y+z=3$ and at a distance $\frac{2}{\sqrt{3}}$ from the point (3, 1, −1) is ______.

(a) $5x-11y+z=17$ (b) $\sqrt{2}x+y=3\sqrt{2}-1$

(c) $x+y+z=\sqrt{3}$ (d) $x-\sqrt{2}y=1-\sqrt{2}$

108. The point P is the intersection of the straight line joining the points $Q(2,3,5)$ and $R(1,-1,4)$ with the plane $5x-4y-z=1$. If S is the foot of the perpendicular drawn from the point $T(2,1,4)$ to QR, then the length of the line segment PS is ______.

(a) $\frac{1}{\sqrt{2}}$ (b) $\sqrt{2}$

(c) 2 (d) $2\sqrt{2}$

109. If the distance of the point $P(1,-2,1)$ from the plane $x+2y-2z=\alpha$, where $\alpha>0$, is 5, then the foot of the perpendicular from P to the plane is ______.

(a) $\left(\frac{8}{3},\frac{4}{3},-\frac{7}{3}\right)$ (b) $\left(\frac{4}{3},-\frac{4}{3},\frac{1}{3}\right)$

(c) $\left(\frac{1}{3},\frac{2}{3},\frac{10}{3}\right)$ (d) $\left(\frac{2}{3},-\frac{1}{3},\frac{5}{2}\right)$

110. The equation of the plane containing the straight line $\frac{x}{2}=\frac{y}{3}=\frac{z}{4}$ and perpendicular to the plane containing the straight lines $\frac{x}{3}=\frac{y}{4}=\frac{z}{2}$ and $\frac{x}{4}=\frac{y}{2}=\frac{z}{3}$ is ______.

(a) $x+2y-2z=0$ (b) $3x+2y-2z=0$

(c) $x-2y+z=0$ (d) $5x+2y-4z=0$

111. A line with positive direction cosines passes through the point $P(2,-1,2)$ and makes equal angles with the co-ordinate axes. The line meets the plane $2x+y+z=9$ at point Q. The length of the line segment PQ equals ______.

(a) 1 (b) $\sqrt{2}$

(c) $\sqrt{3}$ (d) 2

112. Let $P(3,2,6)$ be a point in space and Q be a point on the line $\vec{r}=(\hat{i}-\hat{j}+2\hat{k})+\mu(-3\hat{i}+\hat{j}+5\hat{k})$. Then the value of μ for which the vector $\overrightarrow{PQ}$ is parallel to the plane $x-4y+3z=1$ is ______.

(a) $\frac{1}{4}$ (b) $-\frac{1}{4}$

(c) $\frac{1}{8}$ (d) $-\frac{1}{8}$

Integer and Numerical Value Type

1. A line OP makes, with the x-axis, an angle of measure 120° and with y-axis an angle of measure 60°. Find the angle made by the line with the z-axis.

2. Find the measure of the angle between the lines whose direction ratios are 1, −2, 7 and 3, −2, −1.

3. A non-zero vector $\vec{a}$ is parallel to the line of intersection of the plane determined by the vectors $\hat{i}, \hat{j}+\hat{j}$ and the plane determined by the vectors $\hat{i}-\hat{j}, \hat{i}+\hat{k}$. What is the angle between $\vec{a}$ and $\hat{i}-2\hat{j}+2\hat{k}$?

4. The lines $\frac{x+4}{3}=\frac{y+6}{5}=\frac{z-1}{-2}$ and $3x-2y+z+5=0=2x+3y+4z-k$ are coplanar for $k=$ ______.

5. The lines $x+2+z-4=2x+y+z-4=0$, $3x-y-z-1=0=x+7y+4z-k$ are coplanar for $k=$ ______.

6. In ΔABC, D divides BC in the ratio 3 : 2 and E divides CA in the ratio 1 : 3. The lines AD and BE meet at H and CH meets AB at F. If the ratio $\frac{\text{AF}}{\text{FB}}=\frac{k}{1}$, then the numerical quantity k should be ______.

7. A variable plane is at a constant distance p from the origin and meets the axes in A, B and C. If the locus of centroid of the tetrahedron OABC is $\frac{1}{x^2}+\frac{1}{y^2}+\frac{1}{z^2}=\frac{\lambda}{p^2}$ then the numerical quantity λ should be ______.

Questions from Previous Years

8. Let $\vec{x}$ be a vector in the plane containing vector $\vec{a}=2\hat{i}-\hat{j}+\hat{k}$ and $\vec{b}=\hat{i}+2\hat{j}-\hat{k}$. If the vector $\vec{x}$ is perpendicular to $(3\hat{i}+2\hat{j}-\hat{k})$ and its projection on $\vec{a}\ \dfrac{-17\sqrt{6}}{2}$, then the value of $|\vec{x}|^2$ is equal to________.

9. If the shortest distance between the lines $\vec{r_1}=\alpha\hat{i}+2\hat{j}+2\hat{k}+\lambda(\hat{i}-2\hat{j}-2\hat{k}),\ \lambda\in R, \alpha>0$ and $\vec{r_2}=-4\hat{i}-\hat{k}+\mu(3\hat{i}-2\hat{j}-2\hat{k}),\ \mu\in R$ is 9, then α is equal to________.

10. Let $(\lambda,2,1)$ be a point on the plane which passes through the point (4, −2, 2). If the plane is perpendicular to the line joining the points (−2, −21, 29) and (−1, −16, 23), then $\left(\dfrac{\lambda}{11}\right)^2-\dfrac{4\lambda}{11}-4$ is equal to________.

11. The square of the distance of the point of intersection of the line $\dfrac{x-1}{2}=\dfrac{y-2}{3}=\dfrac{z+1}{6}$ and the plane $2x-y+z=6$ from the point (−1, −1, 2) is________.

12. Let S be the mirror image of the point Q (1, 3, 4) with respect to the plane $2x-y+z+3=0$ and let R (3, 5, γ) be a point on this plane. Then the square of the length of the line segment SR is__________.

13. Let the line L be the projection of the line $\dfrac{x-1}{2}=\dfrac{y-3}{1}=\dfrac{z-4}{2}$ on the plane $x-2y-z=3$. If d is the distance of the point (0, 0, 6) from L, then d^2 is equal to________.

14. Let Q be the foot of the perpendicular from the point P (7, −2, 13) on the plane containing the lines $\dfrac{x+1}{6}=\dfrac{y-1}{7}=\dfrac{z-3}{8}$ and $\dfrac{x-1}{3}=\dfrac{y-2}{5}=\dfrac{z-3}{7}$. Then $(PQ)^2$ is equal to______.

15. Let the plane $ax+by+cz+d=0$ bisect the line joining the points (4, −3, 1) and (2, 3, −5) at the right angles. If a, b, c, d are integers, then the minimum value of $(a^2+b^2+c^2+d^2)$ is ________.

16. The equation of the planes parallel to the plane $x-2y+2z-3=0$ which are at unit distance from the point (1, 2, 3) is $ax+by+cz+d=0$. If $(b-d)=K(c-a)$, then the positive value of K is________.

17. If the equation of the plane passing through the line of intersection of the planes $2x-7y+4z-3=0$, $3x-5y+4z+11=0$ and the point (−2, 1, 3) is $ax+by+cz-7$ is______.

18. Let P be an arbitrary point having sum of the squares of the distances from the planes $x+y+z=0$, $lx-nz=0$ and $x-2y+z=0$, equal to 9. If the locus of the point p is $x^2+y^2+z^2=9$, then the value of $1-n$ is equal to______.

19. If the foot of the perpendicular drawn from the point (1, 0, 3) on a line passing through (α, 7, 1) is $\left(\dfrac{5}{3},\dfrac{7}{3},\dfrac{17}{3}\right)$, then α is equal to________.

20. If the equation of a plane P, passing through the intersection of the planes, $x+4y-z+7=0$ and $3x+y+5z=8$ is $ax+by+6z=15$ for some $a,b\in R$, then the distance of the point (3, 2, −1) from the plane P is________.

21. Let a plane P contain two lines $\vec{r}=\hat{i}+\lambda(\hat{i}+\hat{j})$, $\lambda\in R$ and $\vec{r}=-\hat{j}+\mu(\hat{j}-\hat{k})$, $\mu\in R$. If Q (α,β,γ) is the foot of the perpendicular drawn from the point M (1, 0, 1) to P, then 3 $(\alpha+\beta+\gamma)$ equals __________.

22. If the distance between the plane, $23x-10y-2z+48=0$ and the plane containing the lines $\dfrac{x+1}{2}=\dfrac{y-3}{4}=\dfrac{z-1}{3}$ and $\dfrac{x+3}{2}=\dfrac{y+2}{6}=\dfrac{z-1}{\lambda}(\lambda\in R)$ is equal to $\dfrac{k}{\sqrt{633}}$, then k is equal to______.

Assertion and Reason

(a) Statement 1 is True, Statement 2 True; Statement 2 is a correct explanation for Statement 1.

(b) Statement 1 is True, Statement 2 is True; Statement 2 is NOT a correct explanation for Statement 1.

(c) Statement 1is True, Statement 2 is False.

(d) Statement 1 is False, Statement 2 is True.

1. **Statement 1**: line $\dfrac{x-1}{3}=\dfrac{y-2}{11}=\dfrac{z+1}{11}$ lies in the plane $11x-3z-14=0$.
Statement 2: A straight line lies on a plane if the line is parallel to the plane and a point of the line lies on the plane.

2. **Statement 1**: Let $A(\vec{i}+\vec{j}+\vec{k})$ and $B(\vec{i}-\vec{j}+\vec{k})$ be two points, then point $P(2\vec{i}+3\vec{j}+\vec{k})$ lies exterior to the sphere with AB as one of its diameters.
Statement 2: If A and B are any two points and P is a point in space such that $\overline{PA}.\overline{PB}>0$, then

point P lies exterior to the sphere with AB as one of its diameters.

3. **Statement 1**: If $\vec{r} = x\vec{i} + y\vec{j} + z\vec{k}$, then the equation $\vec{r} \times (2\vec{i} - \vec{j} + 3\vec{k}) = 3\vec{i} + \vec{k}$ represents a straight line.
Statement 2: If $\vec{r} = x\vec{i} + y\vec{j} + z\vec{k}$, then the equation $\vec{r} \times (\vec{i} + 2\vec{j} - 3\vec{k}) = 2\vec{i} + \vec{j}$ represents a straight line.

4. **Statement 1**: Let θ be the angle between the line $\frac{x-2}{2} = \frac{y-1}{-3} = \frac{z+2}{-2}$ and the plane $x + y - z = 5$.
Then $\theta = \sin^{-1}\frac{1}{\sqrt{51}}$
Statement 2: The angle between a straight line and a plane is the complement of angle between the line and normal to the plane.

5. **Statement 1**: A point on the straight line $2x + 3y - 4z = 5$ and $3x - 2y + 4z = 7$ can be determined by taking $x = k$ and then solving the two equations for yand z, where k is any real number.
Statement 2: If $c' \neq kc$, then the straight line $ax + by + cz + d = 0, kax + kby + c'z + d' = 0$, does not intersect the plane $z = \alpha$, where α is any real number.

JEE ADVANCED

Objective Questions

One Option Correct

1. Equation of the straight line in the plane $\vec{r}.\vec{n} = d$ which is parallel to $\vec{r} = \vec{a} + \lambda\vec{b}$ and passes through the foot of perpendicular drawn form the point $P(\vec{a})$ to the plane $\vec{r}.\vec{n} = d$ is ________ (where $\vec{n}.\vec{b} = 0$).
(a) $\vec{r} = \vec{a} + \left(\frac{d - \vec{a}.\vec{n}}{\vec{n}^2}\right)\vec{n} + \lambda\vec{b}$
(b) $\vec{r} = \vec{a} + \left(\frac{d - \vec{a}.\vec{n}}{\vec{n}}\right)\vec{n} + \lambda\vec{b}$
(c) $\vec{r} = \vec{a} + \left(\frac{\vec{a}.\vec{n} - d}{\vec{n}^2}\right)\vec{n} + \lambda\vec{b}$
(d) $\vec{r} = \vec{a} + \left(\frac{\vec{a}.\vec{n} - d}{\vec{n}}\right)\vec{n} + \lambda\vec{b}$

2. If the foot of the perpendicular from the origin to a plane is P(a, b, c), then the equation of the plane is ________.
(a) $\frac{x}{a} + \frac{y}{b} + \frac{z}{c} = 3$
(b) $ax + by + cz = 3$
(c) $ax + by + cz = a^2 + b^2 + c^2$
(d) $ax + by + cz = a + b + c$

3. The equation of a line in the plane $\pi \equiv 2x - y + z - 4 = 0$, which is perpendicular to the line ℓ, whose equation is $\frac{x-2}{1} = \frac{y-2}{-1} = \frac{z-3}{-2}$ and which passes through the point of intersection of ℓ and π is ________.
(a) $\frac{x-2}{3} = \frac{y-1}{5} = \frac{z-1}{-1}$
(b) $\frac{x-1}{3} = \frac{y-3}{5} = \frac{z-5}{-1}$
(c) $\frac{x+2}{2} = \frac{y+1}{-1} = \frac{z-1}{1}$
(d) $\frac{x-2}{2} = \frac{y-1}{-1} = \frac{z-1}{1}$

4. The equation of a plane which passes through the point of intersection of lines $\frac{x-1}{3} = \frac{y-2}{1} = \frac{z-3}{2}$ and $\frac{x-3}{1} = \frac{y-1}{2} = \frac{z-2}{3}$ and at greatest distance from the point (0, 0, 0) is ________.
(a) $4x + 3y + 5z = 25$ (b) $4x + 3y + 5z = 50$
(c) $3x + 4y + 5z = 49$ (d) $x + 7y - 5z = 2$

5. Let A($\vec{a}$) and B($\vec{b}$) be points on two skew lines $\vec{r} = \vec{a} + \lambda\vec{p}$ and $\vec{r} = \vec{b} + u\vec{q}$ and the shortest distance between the skew lines be 1, where $\vec{p}$ and $\vec{q}$ are unit vectors forming adjacent sides of a parallelogram enclosing an area of $\frac{1}{2}$ units. If the angle between AB and the lines of shortest distance is 60°, then AB = ________.
(a) $\frac{1}{2}$ (b) 2
(c) 1 (d) $\lambda = R - \{0\}$

6. If a plane $2x + 3y + 6z + k = 0$ is tangent to the sphere $x^2 + y^2 + z^2 + 2x - 2y + 2z - 6 = 0$, then the value of k is ________.
(a) 26 (b) 16
(c) −26 (d) None of these.

7. If $P_1 : \vec{r}.\vec{n}_1 - d_1 = 0$ $P_2 : \vec{r}.\vec{n}_2 - d_2 = 0$ and $P_3 : \vec{r}.\vec{n}_3 - d_3 = 0$ are three planes and $\vec{n}_1, \vec{n}_2$ and $\vec{n}_3$ are three non-coplanar vectors, then the three lines $P_1 = 0, P_2 = 0; P_2 = 0, P_3 = 0$ and $P_3 = 0, P_1 = 0$ are ________.
(a) parallel lines (b) coplanar lines
(c) coincident lines (d) concurrent lines

8. If $\vec{a}, \vec{b}$ and $\vec{c}$ are three unit-vectors equally inclined to each other at an angle α, then the angle between $\vec{a}$ and plane of $\vec{b}$ and $\vec{c}$ is ________.
(a) $\theta = \cos^{-1}\left(\frac{\cos\alpha}{\cos\frac{\alpha}{2}}\right)$ (b) $\theta = \sin^{-1}\left(\frac{\cos\alpha}{\cos\frac{\alpha}{2}}\right)$
(c) $\theta = \cos^{-1}\left(\frac{\sin\frac{\alpha}{2}}{\sin\alpha}\right)$ (d) $\theta = \sin^{-1}\left(\frac{\sin\frac{\alpha}{2}}{\sin\alpha}\right)$

9. The reflection of the point P (1, 0, 0) on the line $\frac{x-1}{2} = \frac{y+1}{-3} = \frac{z+10}{8}$ is ________.
(a) $(3, -4, -2)$ (b) $(5, -8, -4)$
(c) $(1, -1, -10)$ (d) $(2, -3, 8)$

10. Let $A(1, 1, 1), B(2, 3, 5), C(-1, 0, 2)$ be three points, then the equation of a plane parallel to the plane ABC which is at a distance 2 from origin, is ________.
(a) $2x - 3y + z + 2\sqrt{14} = 0$
(b) $2x - 3y + z - \sqrt{14} = 0$
(c) $2x - 3y + z + 2 = 0$
(d) $2x - 3y + z - 2 = 0$

11. The points $A(2-x, 2, 2)$, $B(2, 2-y, 2)$, $C(2, 2, 2-z)$ and $D(1, 1, 1)$ are coplanar, then the locus of $P(x, y, z)$ is ________.
(a) $\frac{1}{x} + \frac{1}{y} + \frac{1}{z} = 1$
(b) $x + y + z = 1$
(c) $\frac{1}{1-x} + \frac{1}{1-y} + \frac{1}{1-z} = 1$
(d) None of these.

12. The equation of a plane through (1, 2, 3) and at maximum distance from origin is ________.
(a) $x + 2y + 3z = 14$ (b) $x + y + z = 6$
(c) $x + 2y + 3z + 14 = 0$ (d) $3x + 2y + z = 10$

13. S_1: The radius of the great circle of section of a plane with a sphere is equal to the radius of the sphere.
S_2: Points (x_1, y_1, z_1) and (x_2, y_2, z_2) lie on the same side of the plane $ax + by + cz + d = 0$ if $(ax_1, by_1, cz_1 + d)(ax_2, by_2, cz_2 + d) \geq 0$.
S_3: Point (x_1, y_1, z_1) lies inside of the sphere $x^2 + y^2 + z^2 - 2ux - 2vy + 2wz + d = 0$ if $x_1^2 + y_1^2 + z_1^2 < 2ux_1 + 2uy_1 - 2wz_1 - d$.
S_4: The shortest distance between the planes $3x + 6y - 2z - 11 = 0$ and $3x + 6y - 2z + 3 = 0$ is 2 units.
(a) TFTT (b) FFTT
(c) TTFF (d) FTFT

14. S_1: Plane containing the line $\frac{x-1}{2} = \frac{y+2}{5} = \frac{z-3}{7}$ and a straight line parallel to the line whose DR's are 1, 2, 3 contains the points (1, 7, −4).
S_2: The point (4, 13, 5) lies on the line $\frac{x+2}{3} = \frac{y-3}{5} = \frac{1-z}{2}$.
S_3: The point (2, 1, 1) lies on a tangent plane to the sphere $x^2 + y^2 + z^2 - 2x - 4y - 2z + 2 = 0$.
S_4: Direction ratios of the line $x + y + z - 7 = 0$, $4x + y - 2z + 7 = 0$ are (1, −2, 1)
(a) TFTT (b) FFFT
(c) TTFF (d) FTFT

One or More Than One Option Correct

1. The equation of the line $x + y + z - 1 = 0$, $4x + y - 2z + 2 = 0$ written in the symmetrical form is ________.
(a) $\frac{x+1}{1} = \frac{y-2}{-2} = \frac{z-0}{1}$
(b) $\frac{x}{1} = \frac{y}{-2} = \frac{z-1}{1}$
(c) $\frac{x+1/2}{1} = \frac{y-1}{-2} = \frac{z-1/2}{1}$
(d) $\frac{x-1}{2} = \frac{y+2}{-1} = \frac{z-2}{2}$

2. Consider the planes $3x - 6y + 2z + 5 = 0$ and $4x - 12y + 3z = 3$. The plane $67x - 162y + 47z + 44 = 0$ bisects that angle between the given planes which
(a) contains origin. (b) is acute.
(c) is obtuse. (d) None of these.

3. The plane $\ell x + my = 0$ is rotated about its line of intersection with the plane $z = 0$, through an angle α, then equation of plane in its new position may be
(a) $\ell x + my + z\sqrt{\ell^2 + m^2}\tan\alpha = 0$
(b) $\ell x + my - z\sqrt{\ell^2 + m^2}\tan\alpha = 0$

(c) Data is not sufficient.
(d) None of these.

4. If the straight lines $\frac{x-1}{2}=\frac{y+1}{k}=\frac{z}{2}$ and $\frac{x+1}{5}=\frac{y+1}{2}=\frac{z}{k}$ are coplanar, then the plane(s) containing these two lines is (are) ________.

(a) $y+2z=-1$ (b) $y+z=-1$
(c) $y-z=-1$ (d) $y-2z=-1$

Questions from Previous Years

5. Let L_1 and L_2 be the following lines:
$L_1:\frac{x-1}{1}=\frac{y}{-1}=\frac{z-1}{3}$ and $L_2:\frac{x-1}{-3}=\frac{y}{-1}=\frac{z-1}{1}$.
Suppose the straight line

$$L:\frac{x-\alpha}{l}=\frac{y-1}{m}=\frac{z-\gamma}{-2}$$

lies in the plane containing L_1 and L_2, and passes through the point of intersection of L_1 and L_2. If the line L bisects the acute angle between the lines L_1 and L_2, then which of the following statement is/are TRUE?

(a) $\alpha-\gamma=3$ (b) $l+m=2$
(c) $\alpha-\gamma=1$ (d) $l+m=0$

6. Let $\alpha,\beta,\gamma,\delta$ be real numbers such that $\alpha^2+\beta^2+\gamma^2\neq 0$ and $\alpha+\gamma=1$. Suppose the point $(3, 2, -1)$ is the mirror image of the point $(1, 0, -1)$ with respect to the plane $\alpha x+\beta y+\gamma z=\delta$.
Then which of the following statement(s) is (are) TRUE?

(a) $\alpha+\beta=2$ (b) $\delta-\gamma=3$
(c) $\delta+\beta=4$ (d) $\delta+\beta+\gamma=\delta$

7. Three lines $L_1:\vec{r}=\lambda\hat{i},\lambda\in R$ $L_2:\vec{r}=\hat{k}+\mu\hat{j},\ \mu\in R$ and $L_3:\vec{r}=\hat{i}+\hat{j}+v\hat{k}, v\in R$ are given. For which point(s) Q on L_2 can we find a point P on L_1 and a point R on L_3 so that P, Q and R are collinear?

(a) $\hat{k}-\frac{1}{2}\hat{j}$ (b) $\hat{k}$
(c) $\hat{k}+\hat{j}$ (d) $\hat{k}+\frac{1}{2}\hat{j}$

8. Let L_1 and L_2 denote the lines $\vec{r}=\hat{i}+\lambda(-\hat{i}+2\hat{j}+2\hat{k}),\ \lambda\in R$ and $\vec{r}=\mu(2\hat{i}-\hat{j}+2\hat{k}),\ \mu\in R$ respectively. If L_3 is a line which is perpendicular to both L_1 and L_2 and cuts both of them, then which of the following options describe(s) L_3?

(a) $\vec{r}=\frac{2}{9}(4\hat{i}+\hat{j}+\hat{k})+t(2\hat{i}+2\hat{j}-\hat{k}),\ t\in R$
(b) $\vec{r}=\frac{2}{9}(2\hat{i}-\hat{j}+2\hat{k})+t(2\hat{i}+2\hat{j}-\hat{k}),\ t\in R$
(c) $\vec{r}=t(2\hat{i}+2\hat{j}-\hat{k}),\ t\in R$
(d) $\vec{r}=\frac{1}{3}(2\hat{i}+\hat{k})+t(2\hat{i}+2\hat{j}-\hat{k}),\ t\in R$

9. Let $P_1:2x+y-z=3$ and $P_2:x+2y+z=2$ be two planes. Then, which of the following statement(s) is (are) TRUE?

(a) The line of intersection of P_1 and P_2 has direction ratios $1, 2, -1$
(b) The line $\frac{3x-4}{9}=\frac{1-3y}{9}=\frac{z}{3}$ is perpendicular to the line of intersection of P_1 and P_2
(c) The acute angle between P_1 and P_2 is $60°$.
(d) If P_3 is the plane passing through the point $(4, 2, -2)$ and perpendicular to the line of intersection of P_1 and P_2, then the distance of the point $(2, 1, 1)$ from the plane P_3 is $\frac{2}{\sqrt{3}}$.

10. Consider a pyramid OPQRS located in the first octant$(x\geq 0, y\geq 0, z\geq 0)$ with O as origin, and OP and OR along the x-axis and the y-axis, respectively. The base OPQR of the pyramid is a square with OP = 3. The point S is directly above the mid-point, T of diagonal OQ such that TS = 3. Then

(a) the acute angle between OQ and OS is $\frac{\pi}{3}$.
(b) the equation of the plane containing the triangle OQS is $x-y=0$
(c) the length of the perpendicular from P to the plane containing the triangle OQS is $\frac{3}{\sqrt{2}}$.
(d) the perpendicular distance from O to the straight line containing RS is $\sqrt{\frac{15}{2}}$.

11. In R^3, let L be a straight line passing through the origin. Suppose that all the points on L are at a constant distance from the two planes $P_1:x+2y-z+1=0$ and $P_2:2x-y-z-1=0$. Let M be the locus of the feet of the perpendiculars drawn from the points on L to the plane P. Which of the following points lie(s) on M?

(a) $\left(0,-\frac{5}{6},-\frac{2}{3}\right)$ (b) $\left(-\frac{1}{6},-\frac{1}{3},\frac{1}{6}\right)$
(c) $\left(-\frac{5}{6},0,\frac{1}{6}\right)$ (d) $\left(-\frac{1}{3},0,\frac{2}{3}\right)$

12. In R^3, consider the planes $P_1 : y = 0$ and $P_2 : x + z = 1$. Let P_3 be the plane, different from P_1 and P_2, which passes through the intersection of P_1 and P_2. If the distance of the point (0, 1, 0) from P_3 is 1 and the distance of a point (α, β, γ) from P_3 is 2, then which of the following relations is (are) true?

(a) $2\alpha + \beta + 2\gamma + 2 = 0$ (b) $2\alpha - \beta + 2\gamma + 4 = 0$
(c) $2\alpha + \beta - 2\gamma - 10 = 0$ (d) $2\alpha - \beta + 2\gamma - 8 = 0$

13. From a point $P(\lambda, \lambda, \lambda)$, perpendiculars PQ and PR are drawn respectively on the lines $y = x, z = 1$ and $y = -x, z = -1$. If P is such that $\angle QPR$ is a right angle, then the possible values(s) of λ is/(are) ________.

14. Two lines $L_1 : x = 5, \dfrac{y}{3-\alpha} = \dfrac{z}{-2}$ and $L_2 : x = \alpha, \dfrac{y}{-1} = \dfrac{z}{2-\alpha}$ are coplanar. Then α can take value(s) ________.

(a) 1 (b) 2
(c) 3 (d) 4

15. A line l passing through the origin is perpendicular to the lines:

$l_1 : (3+t)\hat{i} + (-1+2t)\hat{j} + (4+2t)\hat{k}, \quad -\infty < t < \infty$

$l_2 : (3+2s)\hat{i} + (3+2s)\hat{j} + (2+s)\hat{k}, \quad -\infty < s < \infty$

Then, the co-ordinate(s) of the points(s) on l_2 at a distance of $\sqrt{17}$ from the point of intersection of l and l_1 is (are) _______.

(a) $\left(\dfrac{7}{3}, \dfrac{7}{3}, \dfrac{5}{3}\right)$ (b) $(-1, -1, 0)$
(c) $(1, 1, 1)$ (d) $\left(\dfrac{7}{9}, \dfrac{7}{9}, \dfrac{8}{9}\right)$

Matrix Type

1. Match Column I with Column II.

Column I	Column II
(A) $\dfrac{x-1}{2} = \dfrac{y-2}{3} = \dfrac{z-3}{4}$ and $\dfrac{x-1}{3} = \dfrac{y-3}{4} = \dfrac{z-5}{5}$ are	(P) coincident.
(B) $\dfrac{x-1}{2} = \dfrac{y-2}{3} = \dfrac{z-3}{4}$ and $\dfrac{x-3}{2} = \dfrac{y-5}{3} = \dfrac{z-7}{4}$ are	(Q) parallel and different.
(C) $\dfrac{x-2}{5} = \dfrac{y+3}{4} = \dfrac{5-z}{2}$ and $\dfrac{x-7}{5} = \dfrac{y-1}{4} = \dfrac{z-2}{-2}$ are	(R) skew.
(D) $\dfrac{x-3}{2} = \dfrac{y+2}{3} = \dfrac{z-4}{5}$ and $\dfrac{x-3}{3} = \dfrac{y-2}{2} = \dfrac{z-7}{5}$ are	(S) intersecting at a point.
	(T) coplanar.

2. Match Column I with Column II.

Column I	Column II
(A) The foot of perpendicular drawn from point (1, 2, 3) to the line $\dfrac{x-2}{2} = \dfrac{y-1}{3} = \dfrac{z-2}{4}$ is	(P) $\left(\dfrac{107}{29}, \dfrac{30}{29}, \dfrac{69}{29}\right)$
(B) The image of line point (1, 2, 3) on the line $\dfrac{x-2}{2} = \dfrac{y-1}{3} = \dfrac{z-2}{4}$ is	(Q) $\left(\dfrac{88}{29}, \dfrac{125}{29}, \dfrac{69}{29}\right)$
(C) The foot of perpendicular from the point (2, 3, 5) to the plane $2x + 3y - 4z + 17 = 0$ is	(R) $\left(\dfrac{107}{29}, \dfrac{125}{29}, \dfrac{185}{29}\right)$
(D) The image of the point (2, 5, 1) on the plane $3x - 2y + 4z - 5 = 0$ is	(S) $\left(\dfrac{68}{29}, \dfrac{44}{29}, \dfrac{78}{29}\right)$
	(T) $\left(\dfrac{38}{29}, \dfrac{57}{29}, \dfrac{185}{29}\right)$

3. Match Column I with Column II.

Column I	Column II
(A) A line from the origin meets the lines $\dfrac{x-2}{1} = \dfrac{y-1}{-2} = \dfrac{z+1}{1}$ and $\dfrac{x-\frac{8}{3}}{2} = \dfrac{y+3}{-1} = \dfrac{z-1}{1}$ at P and Q respectively. If length PQ = d, then d^2 is	(P) –4

(Continued)

Column I	Column II
(B) The values of x satisfying $\tan^{-1}(x+3)-\tan^{-1}(x-3)=\sin^{-1}\left(\frac{3}{5}\right)$ are	(Q) 0
(C) Non-zero vectors $\vec{a},\vec{b}$ and $\vec{c}$ satisfy $\vec{a}.\vec{b}=0$. $(\vec{b}-\vec{a}).(\vec{b}+\vec{c})=0$ and $2\lvert\vec{b}+\vec{c}\rvert=\lvert\vec{b}-\vec{a}\rvert$. If $\vec{a}=\mu\vec{b}+4\vec{c}$, then the possible values of μ are	(R) 4
(D) Let f be the function on $[-\pi,\pi]$ given by $f(0)=9$ and $f(x)=\sin\left(\frac{9x}{2}\right)$ for $x\neq 0$. The value of $\frac{2}{\pi}\int_{-\pi}^{\pi} f(x)dx$ is	(S) 5
	(T) 6

4. Match Column I with Column II.

Column I	Column II
(A) The number of solutions of the equation $xe^{\sin x}-\cos x=0$ in the interval $\left(0,\frac{\pi}{2}\right)$	(P) 1
(B) Value(s) of k for which the planes $kx+4y+z=0$, $4x+ky+2z=0$, and $2x+2y+z=0$ intersect in a straight line	(Q) 2
(C) Value(s) of k for which $\lvert x-1\rvert+\lvert x-2\rvert+\lvert x+1\rvert+\lvert x+2\rvert=4k$ has integer solution(s)	(R) 3
(D) If $y'=y+1$ and $y(0)=1$, then value(s) of $y(\ln 2)$	(S) 4
	(T) 5

Questions from Previous Years

5. Consider the lines $L_1:\frac{x-1}{2}=\frac{y}{-1}=\frac{z+3}{1}=\frac{z+3}{1}$, $L_2\frac{x-4}{1}=\frac{y+3}{1}=\frac{z+3}{2}$ and the planes $P_1:7x+y+2z=3, P_2=3x+5y-6z=4$. Let $ax+by+cz=d$ be the equation of the plane passing through the point of intersection of lines L_1 and L_2, and perpendicular to planes P_1 and P_2.

Match List I with List II and select the correct answer using the code given below the lists:

List I	List II
P. $a=$	1. 13
Q. $b=$	2. −3
R. $c=$	3. 1
S. $d=$	4. −2

	P	Q	R	S
(a)	3	2	4	1
(b)	1	3	4	2
(c)	3	2	1	4
(d)	2	4	1	3

Integer and Numerical Value Type

1. If the equation of the plane through the straight line $\frac{x-1}{2}=\frac{y+2}{-3}=\frac{z}{5}$ and perpendicular to the plane $x-y+z+2=0$ is $ax-by+cz+4=0$, then the find the value of $a+b+c$.

2. The shortest distance between the lines given by $\vec{r}=3\vec{i}+8\vec{j}+3\vec{k}+\lambda(3\vec{i}-\vec{j}+\vec{k})$ and $\vec{r}=-3\vec{i}-7\vec{j}+6\vec{k}+\mu(-3\vec{i}+2\vec{j}+4\vec{k})$ is $\lambda\sqrt{30}$. Then find the value of λ.

Questions from Previous Years

3. Three lines are given by $\vec{r}=\lambda\hat{i},\lambda\in R$; $\vec{r}=\mu(\hat{i}+\hat{j}),\mu\in R$ and $\vec{r}=v(\hat{i}+\hat{j}+\hat{k}),v\in R$. Let the lines cut the plane $x+y+z=1$ at the points A, B and C respectively. If the area of the triangle ABC is Δ then the value of $(6\Delta)^2$ equals______.

4. Let P be a point in the first octant, whose image Q in the plane $x+y=3$ (that is, the line segment PQ is perpendicular to the plane $x+y=3$ and the mid-point of PQ lies in the plane $x+y=3$) lies on the z-axis. Let the distance P from the x-axis be 5. If R is the image of P on the xy-plane, then the length of PR is_______.

Comprehension Passage

Comprehension 1

Consider two lines $L_1:\frac{x-2}{1}=\frac{y-1}{0}=\frac{z+1}{2}$ and $L_2:\frac{x-3}{1}=\frac{y-1}{1}=\frac{z-0}{-1}$. Let P be the plane which contains the line L_1 and is parallel to L_2 and intersecting co-ordinates axes at A, B, C respectively.

1. The shortest distance between the lines L_1 and L_2 is _______.
(a) $\frac{1}{\sqrt{5}}$ (b) $\frac{1}{\sqrt{6}}$
(c) $\frac{1}{\sqrt{8}}$ (d) $\frac{1}{\sqrt{14}}$

2. The image of origin on the plane P, is _______.
(a) $\left(\frac{2}{7},\frac{-3}{7},\frac{-1}{7}\right)$ (b) $\left(\frac{4}{7},\frac{-6}{7},\frac{-2}{7}\right)$
(c) $\left(\frac{-2}{7},\frac{3}{7},\frac{1}{7}\right)$ (d) $\left(\frac{-4}{7},\frac{6}{7},\frac{2}{7}\right)$

3. Volume of the tetrahedron OABC (where 'O' is origin) is _______.
(a) $\frac{2}{3}$ (b) $\frac{4}{9}$
(c) $\frac{2}{9}$ (d) $\frac{4}{3}$

Comprehension 2

Suppose direction cosines of two lines are given by $u\ell+vm+wn=0$ and $a\ell^2+bm^2+cn^2=0$, where u,v,w,a,b,c are arbitrary constants and ℓ,m,n are direction cosines of the lines.

1. For $u=v=w=1$, both lines satisfy the relation:
(a) $(b+c)\left(\frac{n}{\ell}\right)^2+2b\frac{n}{\ell}+(a+b)=0$
(b) $(c+a)\left(\frac{\ell}{m}\right)^2+2c\frac{\ell}{m}+(b+c)=0$
(c) $(a+b)\left(\frac{m}{n}\right)^2+2a\frac{m}{n}+(c+a)=0$
(d) All of the above.

2. For $u=v=w=1$, If $\frac{n_1n_2}{\ell_1\ell_2}=\frac{a+b}{b+c}$, then:
(a) $\frac{m_1m_2}{\ell_1\ell_2}=\frac{b+c}{c+a}$ (b) $\frac{m_1m_2}{\ell_1\ell_2}=\frac{c+a}{b+c}$
(c) $\frac{m_1m_2}{\ell_1\ell_2}=\frac{a+b}{c+a}$ (d) $\frac{m_1m_2}{\ell_1\ell_2}=\frac{c+a}{a+b}$

3. For $u=v=w=1$, if the lines are perpendicular, then:
(a) $a+b+c=0$ (b) $ab+bc+ca=0$
(c) $ab+bc+ca=3abc$ (d) $ab+bc+ca=abc$

Comprehension 3

Consider the lines

$L_1:\frac{x+1}{3}=\frac{y+2}{1}=\frac{z+1}{2}$ $L_2:\frac{x-2}{1}=\frac{y+2}{2}=\frac{z-3}{3}$

1. The unit vector perpendicular to both L_1 and L_2 is _______.
(a) $\frac{-\hat{i}+7\hat{j}+7\hat{k}}{\sqrt{99}}$ (b) $\frac{-\hat{i}-7\hat{j}+5\hat{k}}{5\sqrt{3}}$
(c) $\frac{-\hat{i}+7\hat{j}+5\hat{k}}{5\sqrt{3}}$ (d) $\frac{7\hat{i}-7\hat{j}-\hat{k}}{\sqrt{99}}$

2. The shortest distance between L_1 and L_2 is _______.
(a) 0 (b) $\frac{17}{\sqrt{3}}$
(c) $\frac{41}{5\sqrt{3}}$ (d) $\frac{17}{5\sqrt{3}}$

3. The distance of the point (1, 1, 1) from the plane passing through the point (−1, −2, −1) and whose normal is perpendicular to both the lines L_1 and L_2 is _______.
(a) $\frac{2}{\sqrt{75}}$ (b) $\frac{7}{\sqrt{75}}$
(c) $\frac{13}{\sqrt{75}}$ (d) $\frac{23}{\sqrt{75}}$

ANSWER KEY

JEE Main

Objective Questions

One Option Correct

1. (d)	**2.** (c)	**3.** (c)	**4.** (a)	**5.** (c)	**6.** (b)	**7.** (a)	**8.** (c)	**9.** (a, d)
10. (b)	**11.** (b)	**12.** (a)	**13.** (a)	**14.** (a)	**15.** (a)	**16.** (a)	**17.** (d)	**18.** (b)
19. (d)	**20.** (d)	**21.** (d)	**22.** (a)	**23.** (d)	**24.** (d)	**25.** (c)	**26.** (b)	**27.** (a)
28. (c)	**29.** (b)	**30.** (c)	**31.** (b)	**32.** (d)	**33.** (b)	**34.** (d)	**35.** (b)	**36.** (a)
37. (b)	**38.** (d)	**39.** (d)	**40.** (b)	**41.** (b)	**42.** (d)	**43.** (c)	**44.** (a)	**45.** (c)

46. (c) **47.** (a) **48.** (b) **49.** (d) **50.** (b) **51.** (a) **52.** (c) **53.** (b) **54.** (c)
55. (a) **56.** (b) **57.** (b) **58.** (b) **59.** (b) **60.** (d) **61.** (d) **62.** (a) **63.** (b)
64. (d) **65.** (a) **66.** (c) **67.** (b) **68.** (c) **69.** (a) **70.** (a) **71.** (c) **72.** (b)
73. (b) **74.** (b) **75.** (a) **76.** (c) **77.** (b) **78.** (a) **79.** (b) **80.** (b) **81.** (b)
82. (c) **83.** (d) **84.** (a) **85.** (b) **86.** (a) **87.** (a) **88.** (d) **89.** (d) **90.** (d)
91. (c) **92.** (b) **93.** (a) **94.** (b) **95.** (d) **96.** (b) **97.** (d) **98.** (c) **99.** (b)
100. (d) **101.** (a) **102.** (c) **103.** (d) **104.** (a) **105.** (b) **106.** (c) **107.** (a) **108.** (a)
109. (a) **110.** (c) **111.** (c) **112.** (a)

Integer and Numerical Value Type

1. ($\gamma = 45°$ or $135°$) **2.** (90°) **3.** $\left(\frac{\pi}{4}\right)$ **4.** (4) **5.** (12) **6.** (2) **7.** (16) **8.** (486)
9. (6) **10.** (8) **11.** (61) **12.** (72) **13.** (26) **14.** (96) **15.** (28) **16.** (−2 or 4)
17. (4) **18.** (0) **19.** (4) **20.** (3) **21.** (5) **22.** (3)

Assertion and Reason

1. (a) **2.** (a) **3.** (d) **4.** (a) **5.** (b)

JEE Advanced

Objective Questions

One Option Correct

1. (a) **2.** (c) **3.** (b) **4.** (b) **5.** (b) **6.** (a) **7.** (d) **8.** (a) **9.** (b)
10. (a) **11.** (a) **12.** (a) **13.** (a) **14.** (b)

One or More Than One Option Correct

1. (a, b, c) **2.** (a, b) **3.** (a, b) **4.** (b, c) **5.** (a, b) **6.** (a, b, c) **7.** (a, d) **8.** (a, b, d) **9.** (c, d)
10. (b, c, d) **11.** (a, b) **12.** (b, d) **13.** (c) **14.** (a, d) **15.** (b, d)

Matrix Type

1. A → S, T; B → P, T; C → Q; D → R
2. A → S; B → P; C → T; D → Q
3. A → T; B → P, R; C → Q, S; D → R
4. A → P; B → Q, S; C → Q, R, S, T; D → R
5. (a)

Integer and Numerical Value Type

1. ($a = 2, b = -3, c = 1$) **2.** (3) **3.** (0.75) **4.** (8)

Comprehension Passage

Comprehension 1

1. (d) **2.** (b) **3.** (c)

Comprehension 2

1. (d) **2.** (b) **3.** (a)

Comprehension 3

1. (b) **2.** (d) **3.** (c)

HINTS AND EXPLANATIONS

Worksheet

1. We know that the equation of line is
$\vec{r} = \vec{a} + \vec{b}\lambda$
Here, $\vec{a} = \hat{i} - 2\hat{j} + 3\hat{k}$ and $\vec{b} = 3\hat{i} - 2\hat{j} + 6\hat{k}$
∴ Equation of line is
$\vec{r} = (\hat{i} - 2\hat{j} + 3\hat{k}) + \lambda(3\hat{i} - 2\hat{j} + 6\hat{k})$
$\Rightarrow \left(x\hat{i} + y\hat{j} + z\hat{k}\right) = \left(\hat{i} - 2\hat{j} + 3\hat{k}\right) + \lambda\left(3\hat{i} - 2\hat{j} + 6\hat{k}\right)$
$\Rightarrow \left(x\hat{i} + y\hat{j} + z\hat{k}\right) - \left(\hat{i} - 2\hat{j} + 3\hat{k}\right) = \lambda\left(3\hat{i} - 2\hat{j} + 6\hat{k}\right)$
$\Rightarrow (x-1)\hat{i} + (y+2)\hat{j} + (z-3)\hat{k} = \lambda\left(3\hat{i} - 2\hat{j} + 6\hat{k}\right)$
Thus, the required equation is
$(x-1)\hat{i} + (y+2)\hat{j} + (z-3)\hat{k} = \lambda\left(3\hat{i} - 2\hat{j} + 6\hat{k}\right)$

2. Given equations are,
$\frac{x-1}{2} = \frac{y-2}{3} = \frac{z-3}{4}$ and $\frac{x-4}{5} = \frac{y-1}{2} = z$
Let, $\frac{x-1}{2} = \frac{y-2}{3} = \frac{z-3}{4} = \lambda$
$\Rightarrow x = 2\lambda + 1, y = 3\lambda + 2$ and $z = 4\lambda + 3$
and, $\frac{x-4}{5} = \frac{y-1}{2} = \frac{z}{1} = \mu$
$\Rightarrow x = 5\mu + 4, y = 2\mu + 1$ and $z = \mu$
If the two lines intersect each other at one point,
Then $2\lambda + 1 = 5\mu + 4 \;\Rightarrow\; 2\lambda - 5\mu = 3$ ...(i)
$3\lambda + 2 = 2\mu + 1 \;\Rightarrow\; 3\lambda - 2\mu = -1$ ...(ii)
and $4\lambda + 3 = \mu \;\Rightarrow\; 4\lambda - \mu = -3$ ...(iii)
Solving equations (i) and (ii), we get
$2\lambda - 5\mu = 3$ [multiply by 3]
$3\lambda - 2\mu = -1$ [multiply by 2]
$\Rightarrow 6\lambda - 15\mu = 9$
$6\lambda - 4\mu = -2$
(−) (+) (+)
$-11\mu = 11 \therefore \mu = -1$
Putting the value of μ in equation (i) we get,
$2\lambda - 5(-1) = 3$
$2\lambda + 5 = 3$
$2\lambda = -2 \Rightarrow \lambda = -1$
$4(-1) - (-1) = -3$
$-4 + 1 = -3$
$-3 = -3$ (satisfied)
So, co-ordinates of the point of intersection are
$x = 5(-1) + 4 = -5 + 4 = -1$
$y = 2(-1) + 1 = -2 + 1 = -1$
$z = -1$
Thus, the given lines intersect each other at (−1, −1, −1).

3. Given, $x = py + q \;\Rightarrow\; y = \frac{x-q}{p}$
and $z = xy + s \;\Rightarrow\; y = \frac{z-s}{r}$
So, the equation becomes
$\frac{x-q}{p} = \frac{y}{1} = \frac{z-s}{r}$ in which d' ratios are
$a_1 = p, b_1 = 1, c_1 = r$
Similarly, $x = p'y + q' \Rightarrow y = \frac{x-q'}{p'}$
and $z = r'y + s' \Rightarrow y = \frac{z-q'}{p'}$
Hence, the equation becomes
$\frac{x-q'}{p'} = \frac{y}{1} = \frac{z-s'}{r'}$ in which $a_2 = p', b_2 = 1, c_2 = r'$
If the lines are perpendicular to each other, then
$a_1a_2 + b_1b_2 + c_1c_2 = 0$
$pp' + 1.1 + rr' = 0$
Thus, $pp' + rr' + 1 = 0$ is the required condition.

4. The given co-ordinates are A (2, 3, 4) and B (4, 5, 8).
Now, the co-ordinates of the mid-point C are (2 + 4/2, 3 + 5/2, 4 + 8/2) = (3, 4, 6)
and the direction ratios of the normal to the plane = direction ratios of AB = 4 – 2, 5 – 3, 8 – 4 = (2, 2, 4)
Equation of the plane is
$a(x - x_1) + b(y - y_1) + c(z - z_1) = 0$
$2(x-3) + 2(y-4) + 4(z-6) = 0$
$2x - 6 + 2y - 8 + 4z - 24 = 0$
$2x + 2y + 4z = 38$
$x + y + 2z = 19$
Thus, the required equation of plane is $x + y + 2z = 19$ or $\vec{r}(\hat{i} + \hat{j} + 2\hat{k}) = 19$.

5. Given, points (−2, −1, −3) and (1, −3, 3)
Direction ratios of the normal to the plane are (1 + 2, −3 + 1, 3 + 3) = (3, −2, 6)
Now, the equation of plane passing through one point (x_1, y_1, z_1) is
$a(x - x_1) + b(y - y_1) + c(z - z_1) = 0$
$3(x-1) - 2(y+3) + 6(z-3) = 0$
$3x - 3 - 2y - 6 + 6z - 18 = 0$
$3x - 2y + 6z - 27 = 0 \Rightarrow 3x - 2y + 6z = 27$

Thus, the required equation of plane is $3x - 2y + 6z = 27$.

6. The given points are (2, 1,0), (3, −2, −2) and (3, 1, 7).
As the equation of the plane passing through the points $(x_1, y_1, z_1), (x_2, y_2, z_2)$ and (x_3, y_3, z_3) is

$$(x-2)(-21)-(y-1)(7+2)+z(3)=0$$
$$-21(x-2)-9(y-1)+3z=0$$
$$-21x+42-9y+9+3z=0$$
$$-21x-9y+3z+51=0 \Rightarrow 7x+3y-z-17=0$$

Thus, the required equation of plane is $7x + 3y - z - 17 = 0$.

7. Any point in the given line is
$x-3/2=y-3/1=z/1=\lambda$
$x=2\lambda+3, y=\lambda+3$ and $z=\lambda$
Let these be the co-ordinates of P.
So, the direction ratios of OP are $(2\lambda+3-0), (\lambda+3-0)$ and $(\lambda - 0) \Rightarrow 2\lambda + 3, \lambda+3, \lambda$
But the direction ratios of the line PQ are 2, 1, 1.
Now, we know that

$$\cos\theta = \frac{a_1a_2+b_1b_2+c_1c_2}{\sqrt{a_1^2+b_1^2+c_1^2}.\sqrt{a_2^2+b_2^2+c_2^2}}$$
$$\cos\frac{\pi}{3} = \frac{2(2\lambda+3)+1(\lambda+3)+1.\lambda}{\sqrt{(2)^2+(1)^2+(1)^2}.\sqrt{(2\lambda+3)^2+(\lambda+3)^2+\lambda^2}}$$
$$\frac{1}{2} = \frac{4\lambda+6+\lambda+3+\lambda}{\sqrt{6}.\sqrt{4\lambda^2+9+12\lambda+\lambda^2+9+6\lambda+\lambda^2}}$$
$$\frac{\sqrt{6}}{2} = \frac{6\lambda+9}{\sqrt{6\lambda^2+18\lambda+18}} = \frac{6\lambda+9}{\sqrt{6}\sqrt{\lambda^2+3\lambda+3}}$$
$$\frac{6}{2} = \frac{3(2\lambda+3)}{\sqrt{\lambda^2+3\lambda+3}} \Rightarrow 3 = \frac{3(2\lambda+3)}{\sqrt{\lambda^2+3\lambda+3}}$$
$$1 = \frac{2\lambda+3}{\sqrt{\lambda^2+3\lambda+3}} \Rightarrow \sqrt{\lambda^2+3\lambda+3} = 2\lambda+3$$

$\lambda^2+3\lambda+3=4\lambda^2+9+12\lambda$ (On squaring on both sides)

$$3\lambda^2+9\lambda+6=0$$
$$\lambda^2+3\lambda+2=0$$
$$(\lambda+1)(\lambda+2)=0$$
$$\lambda=-1,-2$$

So, the direction co-ordinates are:
[2(−1) + 3, −1 + 3, −1] = (1, 2, −1) when $\lambda = -1$
and
[2 (−2) + 3, −2 + 3, −2] = (−1, 1, −2) when $\lambda = -2$.

Thus, the required equations of plane are $x/1 = y/2 = z/-1$ and $x/-1 = y/1 = z/-2$

8. The given equations are,
$l+m+n=0$ ...(i)
$l^2+m^2-n^2=0$ ...(ii)
From equation (i), we have $n = -(1 + m)$
Putting the value of n in equation (ii), we get

$$1^2+m^2+[-(1+m)]^2=0$$
$$1^2=m^2-1^2-m^2-21m=0$$
$$-21m=0$$
$$1m=0 \Rightarrow (-m-n)m=0\ [\text{Since}, 1=-m-n]$$
$$(m+n)m=0 \Rightarrow m=0 \text{ or } m=-n$$
$$\Rightarrow 1=0 \text{ or } 1=-n$$

Now, the direction cosines of the two lines are 0, −n, n and −n, 0, n ⇒ 0, −1, 1 and −1, 0, 1

$$\cos\theta = \frac{(0\hat{i}-\hat{j}+\hat{k}).(-\hat{i}+0\hat{j}+\hat{k})}{\sqrt{(-1)^2+(1)^2}\sqrt{(-1)^2+(1)^2}} = \frac{1}{\sqrt{2}.\sqrt{2}} = \frac{1}{2}$$
$$\theta = \frac{\pi}{3}$$

Thus, the required angle is $\pi/3$.

9. Given that 1, m, n and $l+\delta l, m+\delta m, n+\delta n$ are the direction cosines of a variable line in two positions,
$1^2+m^2+n^2=1$ ...(i) and
$(l+\delta l)^2+(m+\delta m)^2+(n+\delta n)^2=1$ ...(ii)

$$l^2+\delta l^2+2l.\delta l+m^2+\delta m^2+2m.\delta m+n^2 +\delta n^2+2n.\delta n=1$$
$$(l^2+m^2+n^2)+(\delta l^2+\delta m^2+\delta n^2) +2(l.\delta l+m.\delta m+n.\delta n)=1$$
$$1+(\delta l^2+\delta m^2+\delta n^2)+2(l.\delta l+m.\delta m+n.\delta n)=1$$
$$\Rightarrow l.\delta l+m.\delta m+n.\delta n = -\frac{1}{2}(\delta l^2+\delta m^2+\delta n^2)$$

Let $\vec{a}$ and $\vec{b}$ be the unit vectors along a line with d' cosines l, m, n and $(l+\delta l), (m+\delta m), (n+\delta n)$.
$\therefore \vec{a} = l\hat{i}+m\hat{j}+n\hat{k}$ and
$\therefore \vec{b} = (l+\delta l)\hat{i}+(m+\delta m)\hat{j}+(n+\delta n)\hat{k}$

$$\cos\delta\theta = \frac{\vec{a}.\vec{b}}{|\vec{a}||\vec{b}|}$$
$$\cos\delta\theta = \frac{(l\hat{i}+m\hat{j}+n\hat{k}).[(l+\delta l)\hat{i}+(m+\delta m)\hat{j}+(n+\delta n)\hat{k}]}{1.1}$$
$$\left[\because |\vec{a}|=|\vec{b}|=1\right]$$

$\cos\delta\theta = l(l+\delta l)+m(m+\delta m)+n(n+\delta n)$

$\cos\delta\theta = l^2 + l.\delta l + m^2 + m.\delta m + n^2 + n.\delta n$

$\cos\delta\theta = (l^2+m^2+n^2)+(l.\delta l + m.\delta m + n.\delta n)$

$\cos\delta\theta = 1-\frac{1}{2}(\delta l^2+\delta m^2+\delta n^2)$

$1-\cos\delta\theta = \frac{1}{2}(\delta l^2+\delta m^2+\delta n^2)$

$2\sin^2\frac{\delta\theta}{2} = \frac{1}{2}(\delta l^2+\delta m^2+\delta n^2)$

$4\sin^2\frac{\delta\theta}{2} = \delta l^2+\delta m^2+\delta n^2$

$4\left(\frac{\delta\theta}{2}\right)^2 = \delta l^2+\delta m^2+\delta n^2$

$(\delta\theta)^2 = \delta l^2+\delta m^2+\delta n^2$

Hence proved.

10. Given, O (0, 0, 0) and A (a, b, c)
So, the direction ratios of OA $= a-0, b-0, c-0$
$= a, b, c$
and, the direction cosines of line OA

$= \frac{a}{\sqrt{a^2+b^2+c^2}}, \frac{b}{\sqrt{a^2+b^2+c^2}}, \frac{c}{\sqrt{a^2+b^2+c^2}}$

Now, the direction ratios of the normal to the plane are (a, b, c).
We know that the equation of the plane passing through the point A (a, b, c) is

$a(x-a)+b(y-b)+c(z-c)=0$

$ax-a^2+by-b^2+cz-c^2=0$

$ax+by+cz=a^2+b^2+c^2$

Thus, the required equation of the plane is $ax+by+cz=a^2+b^2+c^2$.

11. Let us take OX, OY, OZ and ox, oy, oz to be the two rectangular systems.
The equations of the two planes are

$\frac{X}{a}+\frac{Y}{b}+\frac{Z}{c}=1$.....(i) and $\frac{x}{a'}+\frac{y}{b'}+\frac{z}{c'}=1$....(ii)

Length of the perpendicular from origin to the plane (i) is

$= \left|\frac{\frac{0}{a}+\frac{0}{b}+\frac{0}{c}-1}{\sqrt{\frac{1}{a^2}+\frac{1}{b^2}+\frac{1}{c^2}}}\right| = \frac{1}{\sqrt{\frac{1}{a^2}+\frac{1}{b^2}+\frac{1}{c^2}}}$

Length of the perpendicular from origin to the plane (ii)

$= \left|\frac{\frac{0}{a'}+\frac{0}{b'}+\frac{0}{c'}-1}{\sqrt{\frac{1}{a'^2}+\frac{1}{b'^2}+\frac{1}{c'^2}}}\right| = \frac{1}{\sqrt{\frac{1}{a'^2}+\frac{1}{b'^2}+\frac{1}{c'^2}}}$

As per the condition of the question,

$\frac{1}{\sqrt{\frac{1}{a^2}+\frac{1}{b^2}+\frac{1}{c^2}}} = \frac{1}{\sqrt{\frac{1}{a'^2}+\frac{1}{b'^2}+\frac{1}{c'^2}}}$

Thus, $\frac{1}{a^2}+\frac{1}{b^2}+\frac{1}{c^2} = \frac{1}{a'^2}+\frac{1}{b'^2}+\frac{1}{c'^2}$

12. Given equation of line, $\frac{x+5}{1}=\frac{y+3}{4}=\frac{z-6}{-9}=\lambda$
and any point P (2, 4, −1)
Let Q be any point on the given line.
So, co-ordinates of Q are $x=\lambda-5$, $y=4\lambda-3$, $z=-9\lambda+6$
D' ratios of PQ are $\lambda-5-2$, $4\lambda-3-4$, $-9\lambda+6+1$, i.e., $\lambda-7, 4\lambda-7, -9\lambda+7$
and the d ' ratios of the line are 1, 4, −9
If PQ is the ⊥ line then

$1(\lambda-7)+4(4\lambda-7)-9(-9\lambda+7)=0$

$\lambda-7+16\lambda-28+81\lambda-63=0$

$98\lambda-98=0 \Rightarrow \lambda=1$

So, the co-ordinates of Q are $1-5$, $4\times1-3$, $-9\times1+6$, i.e., $-4, 1, -3$

$\therefore PQ=\sqrt{(-4-2)^2+(1-4)^2+(-3+1)^2}$

$=\sqrt{(-6)^2+(-3)^2+(-2)^2}=\sqrt{36+9+4}=\sqrt{49}=7$

Thus, the required distance is 7 units.

13. We know that equation of the plane passing through two points (x_1, y_1, z_1) and (x_2, y_2, z_2) with its normal's direction ratios is

$a(x-x_1)+b(y-y_1)+c(z-z_1)=0$ ...(i)

Now, if the plane is passing through two points (2, 1, −1) and (−1, 3, 4) then

$a(x_2-x_1)+b(y_2-y_1)+c(z_2-z_1)=0$

$a(-1-2)+b(3-1)+c(4+1)=0$

$-3a+2b+5c=0.$ ...(ii)

As the required plane is perpendicular to the given plane $x-2y+4z=10$, then

$a-2, b+4, c=10$ (iii)

On solving (ii) and (iii) we get,

$\frac{a}{8+10}=\frac{-b}{-12-5}=\frac{c}{6-2}=\lambda$

So, $a=18\lambda$, $b=17\lambda$ and $c=4\lambda$

Thus, the required plane is

$18\lambda(x-2)+17\lambda(y-1)+4\lambda(z+1)=0$

$18x-36+17y-17+4z+4=0$

$\Rightarrow 18x+17y+4z-49=0$

14. The given equations of lines are

$\vec{r}=(8+3\lambda)\hat{i}-(9+16\lambda)\hat{j}+(10+7\lambda)\hat{k}$ ...(i)

and $\vec{r}=15\hat{i}+29\hat{j}+5\hat{k}+\mu(3\hat{i}+8\hat{j}-5\hat{k})$ ...(ii)

Equation (i) can be re-written as

$\vec{r}=8\hat{i}-9\hat{j}+10\hat{k}+\lambda(3\hat{i}-16\hat{j}+7\hat{k})$ ...(iii)

Here $\vec{a}_1=8\hat{i}-9\hat{j}+10\hat{k}$ and $\vec{a}_2=15\hat{i}+29\hat{j}+5\hat{k}$

$\vec{b}_1=3\hat{i}-16\hat{j}+7\hat{k}$ and $\vec{b}_2=3\hat{i}+8\hat{j}-5\hat{k}$

$\vec{a}_2-\vec{a}_1=7\hat{i}+38\hat{j}-5\hat{k}$

$$\vec{b}_1\times\vec{b}_2=\begin{vmatrix}\hat{i} & \hat{j} & \hat{k}\\ 3 & -16 & 7\\ 3 & 8 & -5\end{vmatrix}$$

$$=\hat{i}(80-56)-\hat{j}(-15-21)+\hat{k}(24+48)$$

$$=24\hat{i}+36\hat{j}+72\hat{k}$$

Thus,

$$\text{Shortest distance, SD}=\left|\frac{(\vec{a}_2-\vec{a}_1).(\vec{b}_1\times\vec{b}_2)}{|\vec{b}_1\times\vec{b}_2|}\right|$$

$$=\left|\frac{(7\hat{i}+38\hat{j}-5\hat{k}).(24\hat{i}+36\hat{j}+72\hat{k})}{\sqrt{(24)^2+(36)^2+(72)^2}}\right|$$

$$=\left|\frac{168+1368-360}{\sqrt{576+1296+5184}}\right|=\left|\frac{168+1008}{\sqrt{7056}}\right|=\frac{1176}{84}$$

$=14$ units

15. The given plane are: $ax+by=0$ ...(i)

and $z=0$ ...(ii)

Now, the equation of any plane passing through the line of intersection of planes (i) and (ii) is

$(ax+by)+kz=0 \Rightarrow ax+by+kz=0$ (iii)

Dividing both sides by $\sqrt{a^2+b^2+k^2}$, we get

$$\frac{a}{\sqrt{a^2+b^2+k^2}}x+\frac{b}{\sqrt{a^2+b^2+k^2}}y+\frac{k}{\sqrt{a^2+b^2+k^2}}z=0$$

So, direction cosines of the normal to the plane are

$$\frac{a}{\sqrt{a^2+b^2+k^2}},\frac{b}{\sqrt{a^2+b^2+k^2}},\frac{k}{\sqrt{a^2+b^2+k^2}}$$

and direction cosines of the plane (i) are

$$\frac{a}{\sqrt{a^2+b^2}},\frac{b}{\sqrt{a^2+b^2}},0$$

As α is the angle between the planes (i) and (iii), we get

$$\Rightarrow \cos\alpha=\frac{a.a+b.b+k.0}{\sqrt{a^2+b^2+k^2}.\sqrt{a^2+b^2}}$$

$$\cos\alpha=\frac{a^2+b^2}{\sqrt{a^2+b^2+k^2}.\sqrt{a^2+b^2}}$$

$$\cos\alpha=\frac{\sqrt{a^2+b^2}}{\sqrt{a^2+b^2+k^2}}\Rightarrow\cos^2\alpha=\frac{a^2+b^2}{a^2+b^2+k^2}$$

$$(a^2+b^2+k^2)\cos^2\alpha=a^2+b^2$$

$$a^1\cos^2\alpha+b^2\cos^2\alpha+k^2\cos^2\alpha=a^2+b^2$$

$$k^2\cos^2\alpha=a^2-a^2\cos^2\alpha+b^2-b^2\cos^2\alpha$$

$$k^2\cos^2\alpha=\alpha^2(1-\cos^2\alpha)+b^2(1-\cos^2\alpha)$$

$$k^2\cos^2\alpha=a^2\sin^2\alpha+b^2\sin^2\alpha$$

$$k^2\cos^2\alpha=(a^2+b^2)\sin^2\alpha$$

$$\Rightarrow k^2=(a^2+b^2)\frac{\sin^2\alpha}{\cos^2\alpha}\Rightarrow k=\pm\sqrt{a^2+b^2}.\tan\alpha$$

Putting the value of k in equation (iii), we get $ax+by\pm(\sqrt{a^2+b^2}.\tan\alpha)z=0$, which is the required equation of plane.

Quiz

1. We draw a line through the origin and parallel to the given line. Now take one of the directed lines from the origin and find its direction cosines as two parallel lines have same set of direction cosines.

2. $\frac{l}{a}=\frac{m}{b}=\frac{n}{c}=k$, where a,b,c are direction ratios and l,m,n are direction cosines.

3. Direction cosines are

$$\frac{2}{\sqrt{2^2+(-1)^2+(-2)^2}},\frac{-1}{\sqrt{2^2+(-1)^2+(-2)^2}},\frac{-2}{\sqrt{2^2+(-1)^2+(-2)^2}}\text{ or }\frac{2}{3},\frac{-1}{3},\frac{-2}{3}$$

4. Let the direction cosines of the lines be l,m,n. Then $l=\cos 90°=0, m=\cos 60°=\frac{1}{2}$, $n=\cos 30°=\frac{\sqrt{3}}{2}$

5. The x-axis makes angles $0°, 90°$ and $90°$ respectively with x, y and z-axis. Therefore, the direction cosines of x-axis are $\cos 0°, \cos 90°, \cos 90°$, i.e., 1, 0, 0. Similarly, direction cosines of y-axis and z-axis are 0, 1, 0 and 0, 0, 1 respectively.

6. $\dfrac{x-x_1}{l}=\dfrac{y-y_1}{m}=\dfrac{z-z_1}{n}$

7. $\dfrac{x-x_1}{x_2-x_1}=\dfrac{y-y_1}{y_2-y_1}=\dfrac{z-z_1}{z_2-z_1}$

8. $\vec{r}.\hat{n}=d$ (Vector form)
where $\vec{r}$ is the position vector and $\hat{n}$ is the unit normal vector and
d is the perpendicular distance of the plane from origin ($lx+my+nz=d$ in the C artesian form).

9. Let the equation of the plane be
$$\frac{x}{a}+\frac{y}{b}+\frac{z}{c}=1 \qquad \text{...(i)}$$
Here $a=2, b=3, c=4$
Substituting the values of a, b and c in (i), we get the required equation of the plane as
$$\frac{x}{2}+\frac{y}{3}+\frac{z}{4}=1 \text{ or } 6x+4y+3z=12$$

10. Comparing the given equations of the planes with the equations
$A_1x+B_1y+C_1z+D_1=0$ and
$A_2x+B_2y+C_2z+D_2=0$,
we get $A_1=3, B_1=-6, C_1=2$
$A_2=2, B_2=2, C_2=-2$
$$\cos\theta=\left|\frac{3\times 2+(-6)(2)+(2)(-2)}{\sqrt{\left(3^2+(-6)^2+(-2)^2\right)}\sqrt{\left(2^2+2^2+(-2)^2\right)}}\right|$$
$$=\left|\frac{-10}{7\times 2\sqrt{3}}\right|=\frac{5}{7\sqrt{3}}=\frac{5\sqrt{3}}{21}$$
Therefore, $\theta=\cos^{-1}\left(\dfrac{5\sqrt{3}}{21}\right)$

JEE Main

Objective Questions

One Option Correct

1. (d) Here direction ratios of the lines are $\left(\frac{1}{2},\frac{1}{3},-1\right)$ and $\left(\frac{1}{6},-1,-\frac{1}{4}\right)$.
$$\therefore \cos\theta=\frac{\left(\frac{1}{2}\right)\left(\frac{1}{6}\right)+\left(\frac{1}{3}\right)(-1)+(-1)\left(-\frac{1}{4}\right)}{\sqrt{\left(\frac{1}{2}\right)^2+\left(\frac{1}{3}\right)^2+(-1)^2}\sqrt{\left(\frac{1}{6}\right)^2+(-1)^2+\left(\frac{-1}{4}\right)^2}}=0$$

2. (c) Here,
$$\sin\theta=\frac{(1)(2)+(2)(-1)+2\left(\sqrt{\lambda}\right)}{\sqrt{(1)^2+(1)^2+(2)^2}\sqrt{(2)^2+(-1)^2+\lambda}}$$
i.e., $\dfrac{1}{3}=\dfrac{2\sqrt{\lambda}}{\sqrt{6}\sqrt{5+\lambda}}$
$\Rightarrow 6\sqrt{\lambda}=\sqrt{6}\sqrt{5+\lambda}$
$\Rightarrow \lambda=\dfrac{5}{3}$

3. (c) $\dfrac{|d_1-d_2|}{\sqrt{a^2+b^2+c^2}}$
The planes are $2x+y+2z-8=0$
$2x+y+2z+\dfrac{5}{2}=0$
$$\text{Distance}=\left|\frac{-8-5/2}{\sqrt{2^2+1^2+2^2}}\right|=\frac{7}{2}$$

4. (a) a, b, c are in HP.
$\dfrac{1}{a}+\dfrac{1}{c}=\dfrac{2}{b}$
$\Rightarrow \dfrac{1}{a}-\dfrac{2}{b}+\dfrac{1}{c}=0$
$\Rightarrow \dfrac{x}{a}+\dfrac{y}{b}+\dfrac{1}{c}=0$
The lines always pass through the point (1, –2).

5. (c) Let the line satisfy the given condition such that
$\cos^2\theta+\cos^2\beta+\cos^2\theta=1$
$\Rightarrow 2\cos^2\theta=1-\cos^2\beta=\sin^2\beta$
$\Rightarrow 2\cos^2\theta=3\sin^2\theta$
$\therefore\ 2\cos^2\theta=3\left(1-\cos^2\theta\right)$
$\Rightarrow 5\cos^2\theta=3\Rightarrow\cos^2\theta=\dfrac{3}{5}$

6. (b) When one proceeds from the given abscissa, one can observe that choice (a) lies on the first line while the second line is not covered.
Hence, choice (a) is ruled out.
For choice (b), DR's are $3a-a$, $2a-a$, $3a-a$.

i.e., $2a, a, 2$, i.e., 2, 1, 2 which are DR's of the intersecting line.

7. (a) The given lines can be written as,

$$\frac{x-0}{1/2}=\frac{y-1}{1}=\frac{z-2}{-1}=t$$

and $\frac{x-1}{1}=\frac{y+3}{-\lambda}=\frac{z-1}{\lambda}=s$

When coplanar $\begin{vmatrix} 0-1 & 1-(-3) & 2-1 \\ 1/2 & 1 & -1 \\ 1 & -\lambda & \lambda \end{vmatrix}=0$

$\Rightarrow \frac{-5\lambda}{2}=5 \Rightarrow \lambda=-2$

8. (c) Proceed same as above.

$\begin{vmatrix} 2-1 & 3-4 & 4-5 \\ 1 & 1 & -k \\ k & 2 & 1 \end{vmatrix}=0 \Rightarrow k=0 \text{ or } -3$

9. (a, d) In both systems, the perpendicular distances of the plane

$\frac{x}{a}+\frac{y}{b}+\frac{z}{c}=1$ or $\frac{x}{a'}+\frac{y}{b'}+\frac{z}{c'}=1$

from the origin will be same

$$\Rightarrow \left|\frac{-1}{\sqrt{\frac{1}{a^2}+\frac{1}{b^2}+\frac{1}{c^2}}}\right|=\left|\frac{-1}{\sqrt{\left(\frac{1}{a'}\right)^2+\left(\frac{1}{b'}\right)^2+\left(\frac{1}{c'}\right)^2}}\right|$$

10. (b) Let $\frac{x}{a}+\frac{y}{b}+\frac{z}{c}=1$ meet the axes at A $(a, 0, 0)$, B $(0, b, 0)$ and C $(0, 0, c)$

where centroid is $\left(\frac{a+0+0}{3},\frac{0+b+0}{3},\frac{0+0+c}{3}\right)$

$\therefore 1=\frac{a}{3} \Rightarrow a=3$

$2=\frac{b}{3} \Rightarrow b=6$

$3=\frac{c}{3} \Rightarrow c=9$

$\therefore \frac{x}{3}+\frac{y}{6}+\frac{z}{9}=1$ is the required plane.

11. (b) $\vec{n}=\overrightarrow{OM}=\frac{4-1}{2}\hat{i}+\frac{5+2}{2}\hat{j}+\left(\frac{-10+1}{2}\right)\hat{k}$

$=\frac{3}{2}\hat{i}+\frac{7}{2}\hat{j}-\frac{9}{2}\hat{k}$

$\therefore \vec{r}\cdot\vec{n}=\vec{n}.\vec{A}B$

i.e., $\vec{r}.\left(\frac{3}{2}\hat{i}+\frac{7}{2}\hat{j}-\frac{9}{2}\hat{k}\right)=\frac{3}{2}\times(-5)+\frac{7}{2}(-3)-\frac{9}{2}(11)$

$=-\frac{135}{2}$

12. (a) When a plane touches a sphere, then perpendicular distance of the plane from centre of sphere is equal to radius of the sphere.

Hence, from centre $C_1\left(\frac{1}{2},-1,1\right)$ perpendicular distance of $x+y+z=+a\sqrt{3}$ is

$$\left|\frac{\frac{1}{2}-1+1+a\sqrt{3}}{\sqrt{1^2+1^2+1^2}}\right|=\text{Radius}=\frac{\sqrt{3}}{2}\Rightarrow a=\frac{1}{\sqrt{3}}$$

13. (a) The direction ratios of normal to the required plane must be $3-1, 4-2, 5-3$ or 2, 2, 2.
$\Rightarrow$ Equation of the plane $2x+2y+2z=\lambda$ which pass through mid-point (2, 3, 4) implies $x+y+z=9$ is required equation.

14. (a) Given equation of line can be written as

$$\frac{x-1/3}{1/6}=\frac{y-1/3}{1/3}=\frac{z-1}{1/2}$$

i.e., $\frac{x-1/3}{1}=\frac{y-1/3}{2}=\frac{z-1}{3}$

Hence, required DR's 1, 2, 3.

15. (a) As $\sqrt{1^2+(-2)^2+(-2)^2}=3$, one can easily infer the required co-ordinate (1, −2, −2).

16. (a) Let us take 3 points on the plane $2x-3y+4z-3=0$ as

$A\left(\frac{3}{2},0,0\right), B(0,-1,0), C\left(0,0,\frac{3}{4}\right)$

(assigning zero values to two variables)

Let us find image of A, B, C on the plane $x-y+z-3=0$

Equation of AA'

$$\frac{x-3/2}{1}=\frac{y}{-1}=\frac{z}{1}=r$$

$\Rightarrow x=\frac{3}{2}+r, y=-r, z=r$

Putting in the plane $x-y+z-3=0$

$\frac{3}{2}+r+r+r-3=0 \Rightarrow r=\frac{1}{2}$

Foot M is $\left(2,-\frac{1}{2},\frac{1}{2}\right)$

If A' is (α,β,γ)

$2=\frac{\alpha+3/2}{2},-\frac{1}{2}=\frac{\beta+0}{2},\frac{1}{2}=\frac{\gamma+0}{2}$

$\Rightarrow A'is\left(\frac{5}{2},-1,1\right)$

Similarly, we can get the images of B and C as $\left(\frac{4}{3},\frac{-7}{3},\frac{4}{3}\right)$ and $\left(\frac{3}{2},\frac{3}{2},\frac{9}{4}\right)$.

The plane given in choice (a) is satisfied by these three points.

17. (d) **Note:** One can directly find the perpendicular distance PM to the line L, by the given formula: Perpendicular distance

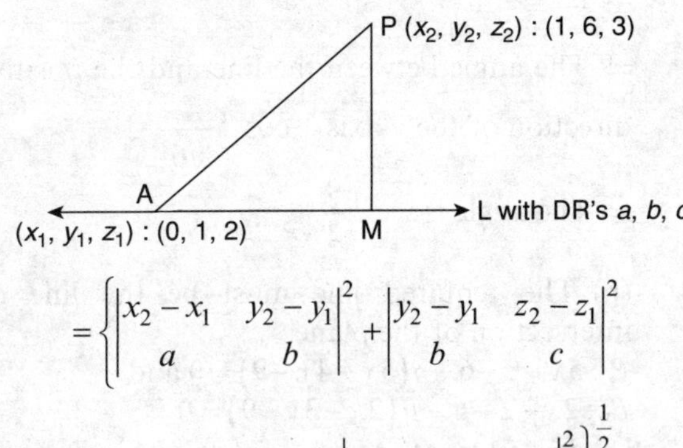

$$=\left\{\begin{vmatrix} x_2-x_1 & y_2-y_1 \\ a & b \end{vmatrix}^2+\begin{vmatrix} y_2-y_1 & z_2-z_1 \\ b & c \end{vmatrix}^2+\begin{vmatrix} z_2-z_1 & x_2-x_1 \\ c & a \end{vmatrix}^2\right\}^{\frac{1}{2}}$$

So PM

$$=\left\{\begin{vmatrix} 1-0 & 6-1 \\ 1 & 2 \end{vmatrix}^2+\begin{vmatrix} 6-1 & 3-2 \\ 2 & 3 \end{vmatrix}^2+\begin{vmatrix} 3-2 & 1-0 \\ 3 & 1 \end{vmatrix}^2\right\}^{\frac{1}{2}}$$

$$=\left\{(-3)^2+(13)^2+(-2)^2\right\}^{\frac{1}{2}}=\sqrt{182}$$

18. (b) A (2, 0, 0), B (0, 3, 0) and C (0, 0, 1) are the intersection points.

$\overline{\text{AB}}=-2\hat{i}+3\hat{j}$

$\overline{\text{AC}}=-2\hat{i}+\hat{k}$

$\therefore$Required area $=\frac{1}{2}\left|\overline{\text{AB}}\times\overline{\text{AC}}\right|$

$=\frac{1}{2}\left|3\hat{i}+2\hat{j}+6\hat{k}\right|$

$=\frac{1}{2}\sqrt{49}=\frac{7}{2}$ sq. units

19. (d) Equation of BC is

$$\frac{x-0}{2-0}=\frac{y-(-1)}{-3-(-1)}=\frac{z-5}{1-5}$$

i.e., $\frac{x}{2}=\frac{y+1}{-2}=\frac{z-5}{-4}$

i.e., $\frac{x}{1}=\frac{y+1}{-1}=\frac{z-5}{-2}(=\lambda)$

Any point on BC, M: $(\lambda,-\lambda-1,5-2\lambda)$

Direction ratios of AM are

$(\lambda-1,-\lambda-1,5-2\lambda-4)$

i.e., $\lambda-1,-\lambda-1,1-2\lambda$.

As AM $\perp$ BC,

$\Rightarrow(\lambda-1).1+(-\lambda-1)(-1)+(1-2\lambda)(-2)=0$

$\Rightarrow 6\lambda=2\Rightarrow\lambda=\frac{1}{3}$

$\therefore M\left(\frac{1}{3},-\frac{1}{3}-1,5-\frac{2}{3}\right)$ i.e., $M\left(\frac{1}{3},-\frac{4}{3},\frac{13}{3}\right)$

20. (d) The equation of plane passing through origin and normal to which has direction ratios 1, −1, 1 is $x-y+z=0$.

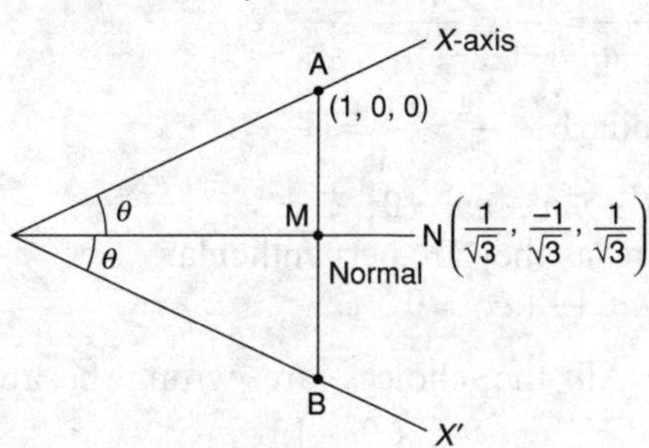

If unit length is cut out from OX, ON and OX′ then

$\overline{\text{OM}}=\frac{\overline{\text{OA}}+\overline{\text{OB}}}{2}\left(\cos\theta=\frac{1}{\sqrt{3}}\right)$

$\overline{\text{OB}}=2\overline{\text{OM}}-\overline{\text{OA}}$

$=2\left(\frac{1}{3}\hat{i}-\frac{1}{3}\hat{j}+\frac{1}{3}\hat{k}\right)-\hat{i}$

$=-\frac{1}{3}\hat{i}-\frac{2}{3}\hat{j}+\frac{2}{3}\hat{k}$

$\Rightarrow\overline{\text{OB}}$ has direction cosines $-\frac{1}{3},-\frac{2}{3},\frac{2}{3}$

21. (d) The given lines intersect at a point

$\Rightarrow\vec{b_1}=k\hat{i}+2\hat{j}+3\hat{k}$,

$\vec{b_2}=3\hat{i}+k\hat{j}+2\hat{k}$

$\vec{a_2}-\vec{a_1}=[-1-(-2)]\hat{i}+[-2-(-3)]\hat{j}+[-3-(-1)]\hat{k}$

are coplanar.

$$\therefore\begin{vmatrix} k & 2 & 3 \\ 3 & k & 2 \\ 1 & 1 & -2 \end{vmatrix}=0\Rightarrow k=\frac{5}{2},-5$$

Hence, Option (d) is correct.

22. (a) Let line l with DR's (l,m,n) intersect the given planes.

$\Rightarrow\ 2l+3m+n=0$

$l+3m+2n=0$

$\Rightarrow\frac{l}{3}=\frac{m}{-3}=\frac{n}{3}$

$\Rightarrow l:m:n=1:-1:1$

$=\frac{1}{\sqrt{3}}:\frac{-1}{\sqrt{3}}:\frac{1}{\sqrt{3}}$

So $\cos\alpha=\frac{1}{\sqrt{3}}$

Hence (a) is correct.

23. (d) Let l, m, n be the direction ratios of the first line, given by the mean y of the two plane equations.
Then $l - am - 0.n = 0$ and
$0.l - cm + n = 0$
$\Rightarrow \frac{1}{-a} = \frac{m}{-1} = \frac{n}{-c}$
Similarly, $\frac{l'}{-a'} = \frac{m'}{-1} = \frac{n'}{-c'}$
$\therefore ll' + mn' + nn' = 0$
Here, as they are perpendicular
$\Rightarrow aa' + 1 + cc' = 0$

24. (d) All the choices are wrong because the correct answer is $\left(\frac{9}{5}, \frac{-13}{5}, 4\right)$.

25. (c) Q (2, 2, 1) —λ— P —l— R (5, 1, –2)

$4 = \frac{5\lambda + 2}{\lambda + 1} \Rightarrow 4\lambda + 4 = 5\lambda + 2$
$\Rightarrow \lambda = 2$
$\Rightarrow$z-co-ordinate $= \frac{2\times(-2) + 1\times(1)}{3} = -1$

26. (b) The plane (b) from the given choices passes through (2, 2, 1) and is perpendicular to the plane $2x + 6y + 6z - 1 = 0$.

27. (a) Here, $A(a,0,0), B(0,b,0), C(0,0,c)$ such that $\frac{a}{3} = \frac{1}{3}, \frac{b}{3} = \frac{1}{3}, \frac{c}{3} = \frac{1}{3}$ in $a = 1, b = 1, c = 1$.
So the required equation is $x + y + z = 1$.

28. (c) Direction ratios of $\overline{OA}$ are perpendicular to the required plane a, b, c.
So, $a(x-a) + b(y-b) + c(z-c) = 0$ is required equation.

29. (b) Here,
$$D = \begin{vmatrix} a & 1 & 1 \\ 1 & -a & 1 \\ 1 & 1 & 1 \end{vmatrix} = a^2 - 1$$
If $D = 0$ then system of equation has infinite non-zero solutions.
$\therefore a^2 - 1 = 0 \Rightarrow a = \pm 1$.

30. (c) Here $\overrightarrow{AB}, \overrightarrow{AC}, \overrightarrow{AD}$ are coplanar, if
$$\begin{vmatrix} 1-0 & x-2 & 0-0 \\ 1-0 & 2-2 & 0-0 \\ 1-0 & 2-2 & 1-0 \end{vmatrix} = 0.$$
i.e., $x - 2 = 0$

31. (b) Here $\sqrt{3^2 + 12^2 + 4^2} = 13$
$\therefore$ DR's are $\frac{3}{13}, \frac{12}{13}, \frac{4}{13}$.

32. (d) $a + 2b + 3c = 0$
$3a + 2b + c = 0$
$\Rightarrow \frac{a}{-1} = \frac{b}{2} = \frac{c}{1}$
$\Rightarrow$ The angle between the line and the positive direction of the y-axis is $\cos^{-1}\frac{2}{\sqrt{6}}$.
$\Rightarrow$ The angle is $\sec^{-1}\frac{9}{4}$.

33. (b) The required line must be the line of intersection of the planes
$P_1 : 5x - z - 6 + »(5x - 4y - 9) = 0$ and
$P_2 : 2x - z - 4 + \mu(2x - 3y - 9) = 0$
i.e., $(5 + 5\lambda)x + (-4)\lambda y - z + (-6 - 9\lambda) = 0$ and
$(2 + 2\mu)x + (-3)\mu y - z + (-4 - 9\mu) = 0$
As the DR's are 4, 1, 1 of the require line which lies in P_1 and P_2.
$\Rightarrow 4(5 + 5\lambda) + (-4\lambda) - 1 = 0$ and
$4(2 + 2\mu) + (-3\mu) - 1 = 0$
$\Rightarrow \lambda = -\frac{19}{16}$ and $\mu = -\frac{7}{5}$
$\therefore P_1 : \left(5 + \frac{-95}{16}\right)x - 4\left(\frac{-19}{16}\right)y - z + \left(-6 + 9\times\frac{19}{16}\right) = 0$
$: -15x + 76y - 16z + 75 = 0$
and $P_2 : 4x - 21y + 5z - 43 = 0$

34. (d) $L_1 = \frac{x}{1} = \frac{y}{1} = \frac{z - »}{1} = r$
$\Rightarrow A : (r, r, r + \lambda)$ is any point on L_2.
A lies on the other line.
$\Rightarrow r + r + \lambda = 1 \Rightarrow 2r + \lambda = 1$ and $2r + 2\lambda = 1$
$\Rightarrow \lambda = 0$

35. (b)

36. (a) $l = -(m + n)$
$\therefore (m + n)m = 0$
$\Rightarrow$ either $m = 0$ or $m + n = 0$
When $m = 0$, $l = -n$
When $m = -n$ $l = 0$
$\therefore (l_1, m_1, n_1) : (0, 1, -1)$
$(l_2, m_2, n_2) : (1, 0, -1)$
$$\cos\theta = \frac{l_1 l_2 + m_1 m_2 + n_1 n_2}{\sqrt{l_1^2 + m_1^2 + n_1^2}\sqrt{l_2^2 + m_2^2 + n_2^2}} = \frac{1}{2}$$

37. (b) Let $A(x-1) + B(y-0) + C(z+1) = 0$ be the required plane
which passes through (3, 2, 2).
$\Rightarrow 2A + 2B + 3C = 0$ ...(i)
Normal vector of the plane and parallel vector of the line are perpendicular to each other.

$\Rightarrow A-2B+3C=0$ ...(ii)

$\therefore \frac{A}{4}=\frac{B}{-1}=\frac{C}{-2}$

$\therefore 4(x-1)-y-2(z+1)=0$

$4z-y-2z=6$

38. (d) $\overrightarrow{a_1}=6\hat{i}+2\hat{j}+2\hat{k}$, $\overrightarrow{b_1}=\hat{i}-2\hat{j}+2\hat{k}$

$\overrightarrow{a_2}=-4\hat{i}-\hat{k}$, $\overrightarrow{b_2}=3\hat{i}-2\hat{j}-2\hat{k}$

Shortes distance $=\left|\frac{(a_2-a_1).(\overrightarrow{b_1}\times\overrightarrow{b_2})}{|\overrightarrow{b_1}\times\overrightarrow{b_2}|}\right|$

Numerator $=\begin{vmatrix}-10 & -2 & -3\\ 1 & -2 & 2\\ 3 & -2 & -2\end{vmatrix}=-108$

Denominator = 12

Shortest distance = 9 units

39. (d)

40. (b) Let $A(x-a)=B(y-b)+C(z-c)=0$

$\Rightarrow Ax+By+Cz+Aa+Bb+Cc$ ...(i)

So A': $\left(\frac{Aa+Bb+Cc}{A},0,0\right)$ is a point on (i)

Similarly B': $\left(0,\frac{Aa+Bb+Cc}{B},0\right)$

and C': $\left(0,0,\frac{Aa+Bb+Cc}{C}\right)$

Let $P(x,y,z)$ be a variable point, then

$OP=OA'=OB'=OC'$

$\Rightarrow x^2+y^2+z^2=\left(x-\frac{k}{A}\right)^2+y^2+z^2$

$=x^2+\left(y-\frac{k}{B}\right)^2+z^2$

$=x^2+y^2+\left(z-\frac{k}{A}\right)^2$

$\Rightarrow x=\frac{k}{2A}, y=\frac{k}{2B}, z=\frac{k}{2C}$

$k=Aa+Bb+Cc$

Now $\frac{a}{x}+\frac{b}{y}+\frac{c}{z}=\frac{2(Aa+Bb+Cc)}{k}=2$

41. (b) Equation of the line passing through the given points is

$\frac{x-3}{5-3}=\frac{y-b}{1-b}=\frac{z-1}{a-1}$ ...(i)

The point $P\left(0,\frac{17}{2},\frac{-13}{2}\right)$ lies on the line represented by equation (i)

$\Rightarrow -\frac{3}{2}=\frac{17/2-b}{1-b}=\frac{-13/2-1}{a-1}$

$\Rightarrow a-1=5$ and $-3(1-b)=2\left(\frac{17}{2}-b\right)$

i.e., $a=6, b=4$

42. (d) We know that $\cos^2\alpha+\cos^2\beta+\cos^2\gamma=1$ for a line in 3D geometry.

Here, $\alpha=\frac{\pi}{4}$ and $\beta=\frac{\pi}{4}$

$\therefore \cos^2\gamma=1-\frac{1}{2}-\frac{1}{2}=0$

$\Rightarrow \gamma=\frac{\pi}{2}$

43. (c) Any point on the first line $M:(1+2\lambda,2+3\lambda,3+4\lambda)$ and any point on the second line $N:(4+5\mu,1+2\mu,\mu)$.

When the lines intersect, M and N coincide at the point of intersection.

Here, $3+4\lambda=\mu$

$2+3\lambda=1+2\mu$

Solving, we get $\lambda=-1$ and $\mu=-1$ which satisfies $1+2» = 4+5\mu$.

So, the point of intersection is $(-1,-1,-1)$

44. (a) The equation of line passing through $P(5,4,6)$ and perpendicular to given plane is

$\frac{x-5}{1}=\frac{y-4}{1}=\frac{z-6}{2}=\lambda$ ...(i)

So the point $M:(5+\lambda,4+\lambda,6+2\lambda)$ which lies on the plane as well as the line (i) is the foot of the perpendicular drawn from P to the plane.

$\therefore (5+\lambda)+(4+\lambda)+2(6+2\lambda)-15=0$

$\Rightarrow \lambda=-1$

$\therefore M:(4,3,4)$

Let the image : $P'(x,y,z)$, then M is the mid-point of PP'.

$\therefore \frac{5+x_1}{2}=4\Rightarrow x_1=3$

$\frac{4+y_1}{2}=3\Rightarrow y_1=2$

$\frac{6+z_1}{2}=4\Rightarrow z_1=2$

$\therefore P'(3,2,2)$.

45. (c) Projection of $\overline{AP}$ along AM line whose direction ratios are $\left(\frac{1}{\sqrt{14}},\frac{2}{\sqrt{14}},\frac{3}{\sqrt{14}}\right)$.

Using the formula for projection we get P.

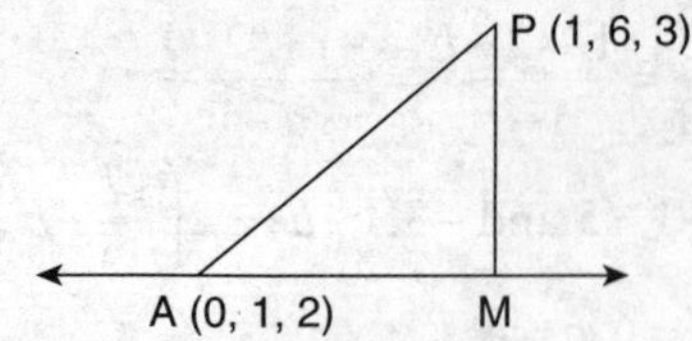

$AM=(x_2-x_1)l+(y_2-y_1)m+(z_2-z_1)n$

$=\frac{(1-0)l}{\sqrt{14}}+\frac{(6-1)2}{\sqrt{14}}+\frac{(3-2)3}{\sqrt{14}}=\sqrt{14}$

$AP=\sqrt{27}$ by distance formula.

$\therefore PM=\sqrt{27-14}=\sqrt{13}$ units.

46. (c) The given plane can be written as

$\frac{x}{1/2}+\frac{y}{1/3}+\frac{z}{1/4}=1$

from which one can infer the points of intersection with axes at

$A\left(\frac{1}{2},0,0\right), B\left(0,\frac{1}{3},0\right)$ and $C\left(0,0,\frac{1}{4}\right)$, so that

Centroid : $\left(\frac{1/2}{3},\frac{1/3}{3},\frac{1/4}{3}\right)$

$:\left(\frac{1}{6},\frac{1}{9},\frac{1}{12}\right)$

47. (a) We know $|r-\vec{a}|=$ radius

$\therefore |r-(i+k)|=4$

is the required vector equation of the sphere.

48. (b) Comparing with $\frac{x}{a}+\frac{y}{b}+\frac{z}{c}=1$

one can easily get, the required equation when $a=b=c=1$.

i.e., $x+y+z=1$

49. (d)

50. (b) The equation of the plane through z-axis is $ax+by=0$

This plane is parallel to the line

$\Rightarrow \cos\theta+b\sin\theta=0 \Rightarrow a=-b\tan\theta$

$\therefore -b\tan\theta x+by=0 \Rightarrow x\tan\theta-y=0$

51. (a) The intersection of the given plane is

$x-y+2z-1+\lambda(x+y-z-3)=0$

$\Rightarrow x(1+\lambda)+y(\lambda-1)+z(2-\lambda)-3\lambda-1=0$

Direction ratio of normal to the above plane is $(1+\lambda,\lambda-1,2-\lambda)$

Since the line formed by intersection of planes and the normal of the planes is perpendicular, by taking option (a)

$-(1+\lambda,\lambda)+3(\lambda-1)+2(2-\lambda)=0$

$\Rightarrow -1-\lambda+3\lambda-3+4-2\lambda=0$

$0=0$

52. (c) Let $\frac{x}{a}+\frac{y}{b}+\frac{z}{c}=1$ be the plane which makes with co-ordinate axes, a triangle at $A(a,0,0), B(0,b,0), C(0,0,c)$.

Hence, centroid $(\alpha,\beta,\gamma):\left(\frac{a}{3},\frac{b}{3},\frac{c}{3}\right)$

$\therefore a=3\alpha, b=3\beta, c=3\gamma$

$\therefore \frac{x}{3\alpha}+\frac{y}{3\beta}+\frac{z}{3\gamma}=1$

53. (b) Any straight line through $A(1,0,4)$ with DR's a_1,b_1,c_1 is

$\frac{x-1}{a_1}=\frac{y}{b_1}=\frac{z-4}{c_1}$

Direction ratios of BC is $(2,-2,-4)$

The equation of BC is $\frac{x}{2}=\frac{y+1}{-2}=\frac{z-5}{-4}=(r)$

$P(1,-2,3)$ satisfies BC.

54. (c) $(x-2)+2(y-3)+4(z-4)=0$ is the required equation.

55. (a) $\left|\frac{3(2)-6(3)+2(4)+18}{\sqrt{3^2+(-6)^2+(2)^2}}\right|=\frac{14}{7}=2$

56. (b) DR's are $(1,2,-3)$ and $\sqrt{1^2+2^2+(-3)^2}=\sqrt{14}$

57. (b) Mid-point of

$PQ\left(\frac{3+1}{2},\frac{4+2}{2},\frac{5+3}{2}\right)$, i.e., $(2,3,4)$

Direction ratio of $PQ:(-2,-2,-2)$

which is the DR of the normal of the required plane.

We know that

$A(x-x_1)+B(y-y_1)+C(z-z_1)=0$

$\Rightarrow -2(x-2)-2(y-3)-2(z-4)=0$

$\Rightarrow x+y+z=9$

58. (b) For coplanarity of two lines, the condition is

$$\begin{vmatrix} x_1-x_2 & y_1-y_2 & z_1-z_2 \\ a_1 & b_1 & c_1 \\ a_2 & b_2 & c_2 \end{vmatrix}=0$$

$\Rightarrow k=12$

59. (b) $Dx=\begin{vmatrix} 0 & -4 & 2 \\ 5 & 7 & -1 \\ 3 & -5 & 4 \end{vmatrix}=0$

Similarly $Dy=0=Dz=D$.

Hence, the system has infinite solution implies they have common line.

60. (d) Since $(\pm1)^2+(0)^2+(0)^2=1$

and $(0)^2+(\pm1)^2+(0)^2=1$

and $(0)^2+(0)^2+(\pm1)^2=1$

$\Rightarrow$ The number of integrals points $2+2+2=6$.

61. (d) Centroid of the triangle is

$\left(\frac{2+3-2}{3},\frac{-4-1+5}{3},\frac{3-2+8}{3}\right)$

or (1, 0, 3)

z
(1, 0, 3) N
P
3 unit
O
y
1 unit
M
x

$OP=\sqrt{(PN)^2+(ON)^2}$

$OP=\sqrt{(OM)^2+(ON)^2}$

$OP=\sqrt{1+9}=\sqrt{10}$ units

62. (a) Projection of PQ on

Given line is $\overrightarrow{PQ}\left(\frac{1}{3}\hat{i}+\frac{2}{3}\hat{j}+\frac{2}{3}\hat{k}\right)$

Q (–2, 8, 13)
P (7, –5, 11)
DC's $\frac{1}{3},\frac{2}{3},\frac{2}{3}$

$\text{Projection}=\left(-9\hat{i}+13\hat{j}+2\hat{k}\right).\left(\frac{1}{3}\hat{i}+\frac{2}{3}\hat{j}+\frac{2}{3}\hat{k}\right)$

$\text{Projection}=-3+\frac{26}{3}+\frac{4}{3}$

Projection = 7

63. (b) Given that,

$l+m+n=0\Rightarrow l=n-m$ ...(i)

and $l^2+m^2-n^2=0$ ...(ii)

Substitute/from (i) into (ii)

$\Rightarrow(n-m)^2+m^2-n^2=0$

$\Rightarrow 2m(m-n)=0$

$\Rightarrow m=0\, or\, m=n$

Case I: If $m=0$,

$l=m$ [from (i)]

We know that,

$l^2+m^2+n^2=1\Rightarrow l^2=\frac{1}{2}$

$\Rightarrow l_1,l_2=\frac{1}{\sqrt{2}},\frac{-1}{\sqrt{2}}$

$\because l=n\Rightarrow n_1,n_2=\frac{1}{\sqrt{2}},\frac{-1}{\sqrt{2}}$

$\therefore\left(\frac{1}{\sqrt{2}},0,\frac{1}{\sqrt{2}}\right)$ or $\left(\frac{-1}{\sqrt{2}},0,\frac{-1}{\sqrt{2}}\right)$ are direction cosines of line l_1.

Case II: If $m=n$

$\Rightarrow l=0$ [from (i)]

$l^2+m^2+n^2=1\Rightarrow m^2=\frac{1}{2}\Rightarrow m_1,m_2=\frac{1}{\sqrt{2}},\frac{-1}{\sqrt{2}}$

$\because m=n\Rightarrow n_1,n_2=\frac{1}{\sqrt{2}},\frac{-1}{\sqrt{2}}$

$\therefore\left(0,\frac{1}{\sqrt{2}},\frac{1}{\sqrt{2}}\right)$ or $\left(0,\frac{-1}{\sqrt{2}},\frac{-1}{\sqrt{2}}\right)$ are direction cosines of line l_2.

$\cos\alpha=l_1l_2+m_1m_2+n_2n_2=0+0\pm\frac{1}{2}=\pm\frac{1}{2}$

$\cos^2\alpha=\frac{1}{4}\Rightarrow\sin^2\alpha=\frac{3}{4}$

So, $\sin^4\alpha+\cos^4\alpha=\frac{5}{8}$

64. (d) We have, $\overline{AB}=\hat{i}+\hat{j}-2\hat{k}$

$\overline{AC}=\hat{i}+2\hat{j}-\hat{k}$

$\therefore$ Normal to plane $\vec{n}=\overline{AB}\times\overline{AC}$

$=\begin{vmatrix}\hat{i}&\hat{j}&\hat{k}\\1&1&-2\\1&2&-1\end{vmatrix}=3\hat{i}-\hat{j}+\hat{k}\ \because\overline{OP}=2\hat{i}-\hat{j}+\hat{k}$

Let $\overline{OP}$ makes an angle θ with the plane.

$\therefore\sin\theta=\frac{\overline{OP}.\vec{n}}{\left|\overline{OP}\right|\times\left|\vec{n}\right|}=\frac{6+1+1}{\sqrt{11}.\sqrt{6}}=\frac{8}{\sqrt{66}}$

$\cos\theta=\sqrt{1-\frac{64}{66}}=\frac{1}{\sqrt{33}}$

$\text{Projection}=\left|\overline{OP}\right|\cos\theta=\sqrt{6}\times\frac{1}{\sqrt{33}}=\sqrt{\frac{2}{11}}$

65. (a) The given lines are

$\frac{x+1}{a}=y=\frac{z-1}{a}$

$\frac{x+2}{3}=y=\frac{z}{3/b}$

Since the lines are co-planar, then

$\begin{vmatrix}a&1&a\\3&1&\frac{3}{b}\\-1&0&-1\end{vmatrix}=0\Rightarrow-\left(\frac{3}{b}-a\right)-1(a-3)=0$

$\Rightarrow a-\frac{3}{b}-a+3=0$

$\Rightarrow b=1,a\in R-\{0\}$

66. (c)

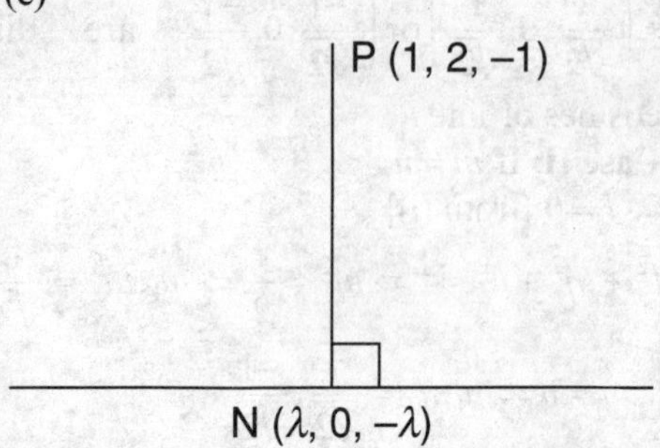

Let $\frac{x}{1}=\frac{y}{0}=\frac{z}{-1}=\lambda$

$\Rightarrow N \equiv (\lambda, 0, -\lambda)$

$\vec{b}=\hat{i}-\hat{k}$

$\because \overrightarrow{PN}.\vec{b}=0$

$\Rightarrow 1(\lambda-1)+(\lambda-1)=0 \Rightarrow \lambda=1$

$\Rightarrow N(1,0,-1)$

Let $Q(\mu,0,-\mu)$

$\because \vec{n}=\vec{i}+\vec{j}+2\vec{k}$

Now,

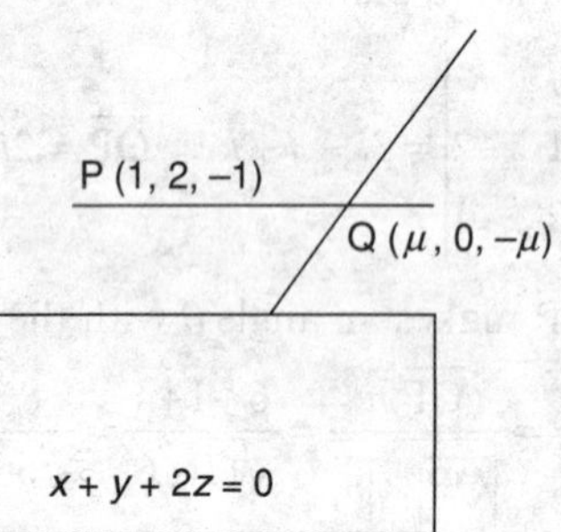

$\because \overrightarrow{PQ}.\vec{n}=0$

$\Rightarrow \mu=-1 \Rightarrow Q(-1,0,1)$

$\overline{PN}=2\hat{j}$ and $\overline{PQ}=2\hat{i}+2\hat{j}-2\hat{k}$

$\Rightarrow \cos\alpha=\frac{1}{\sqrt{3}}$

67. (b) We know that the image of point (x_1, y_1, z_1) with plane $ax+by+cz+d=0$ is

$$\Rightarrow \frac{x-x_1}{2}=\frac{y-y_1}{2}=\frac{z-z_1}{2}=2\frac{(ax_1+by_1+cz_1+d)}{a^2+b^2+c^2}$$

$$\Rightarrow \frac{\alpha-1}{4}=\frac{\beta-3}{-5}=\frac{\gamma-5}{2}=2\frac{(4\times1-5\times3+2\times5-8)}{16+25+4}$$

$$\Rightarrow \frac{\alpha-1}{4}=\frac{\beta-3}{-5}=\frac{\gamma-5}{2}=\frac{2}{5}$$

$$\Rightarrow \alpha=\frac{13}{5}, \beta=1, \gamma=\frac{29}{5}$$

$$\Rightarrow 5|\alpha+\beta+\gamma|=|5\alpha+5\beta+5\gamma|=47$$

68. (c)

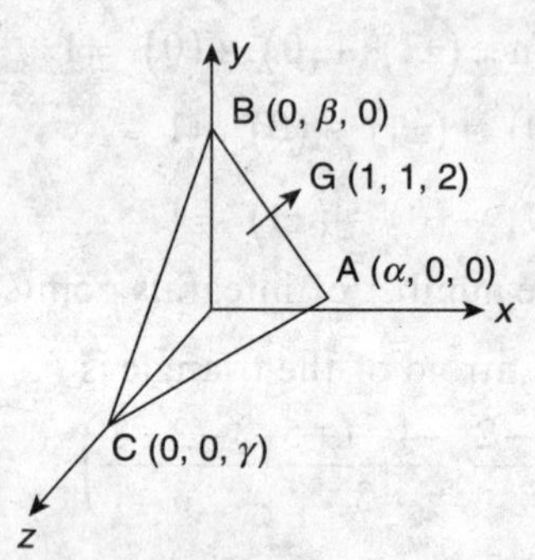

$\therefore \alpha=3, \beta=3$ and $\gamma=6$ as G is centroid.

$\therefore$ The equation of plane is

$$\frac{x}{\alpha}+\frac{y}{\beta}+\frac{z}{\gamma}=1$$

$$\Rightarrow \frac{x}{3}+\frac{y}{3}+\frac{z}{6}=1 \Rightarrow 2x+2y+z=6$$

$\therefore$ The required line is, $\frac{x-1}{2}=\frac{y-1}{2}=\frac{z-2}{1}$

69. (a)

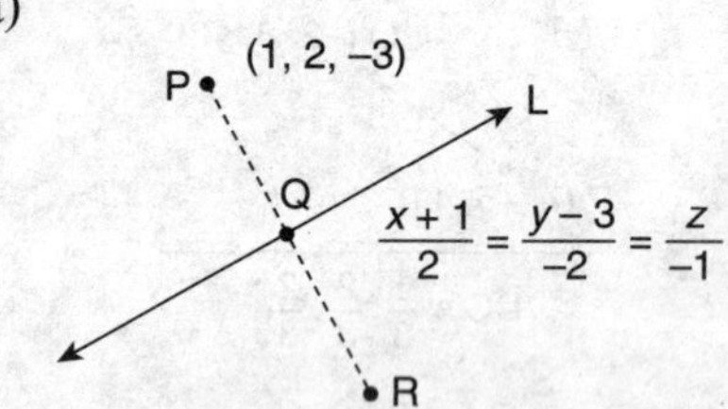

$$\frac{x+1}{2}=\frac{y-3}{-2}=\frac{z}{-1}=\lambda$$

Any point on the line $=Q(2\lambda-1,-2\lambda+3,-\lambda)$

$\therefore$ DR of PQ $=[2\lambda-2,-2\lambda+1,-\lambda+3]$

DR of the given line = [2, −2, −1]

$\because$ PQ is perpendicular to the line L.

$\therefore 2(2\lambda-2)-2(-2\lambda+1)-1(-\lambda+3)=0$

$\Rightarrow 4\lambda-4+4\lambda-2+\lambda-3=0$

$\Rightarrow 9\lambda-9=0 \Rightarrow \lambda=1$

$\because$ Q is the mid-point of PR = Q = (1, 1, −1)

$\therefore$ Co-ordinate of image R = (1, 0, 1) = (a, b, c)

$\therefore a+b+c=2$.

70. (a) $L_1 \equiv \vec{r}=(\hat{i}-\hat{j})+\ell(2\hat{i}+\hat{k})$

$L_2 \equiv \vec{r}=(2\hat{i}-\hat{j})+m(\hat{i}+\hat{j}-\hat{k})$

Equating coeff. of $\hat{i}, \hat{j}$ and $\hat{k}$ of L_1 and L_2

$2\ell+1=m+2$ ...(i)

$-1=-1+m \Rightarrow m=0$ ...(ii)

$\ell=-m$ ...(iii)

$\Rightarrow m=\ell=0$ which does not satisfy equation (i). Hence, the lines do not intersect for any value of ℓ and m.

71. (c) $\overrightarrow{AB} = 6\hat{i} + 15\hat{j} + 3\hat{k}$

$\vec{p} = \hat{i} + 4\hat{j} + 22\hat{k}; \vec{q} = \hat{i} + \hat{j} + 7\hat{k}$

$$\vec{p} \times \vec{q} = \begin{vmatrix} i & j & k \\ 1 & 4 & 22 \\ 1 & 1 & 1 \end{vmatrix} = 6\hat{i} + 15\hat{j} - 3\hat{k}$$

The shortest distance between the lines is

$$= \frac{|\overrightarrow{AB}.(\vec{p} \times \vec{q})|}{|\vec{p} \times \vec{q}|} = \frac{|36 + 225 + 9|}{\sqrt{36 + 225 + 9}} = 3\sqrt{30}$$

72. (b) For line of intersection of planes $x + y + z + 1 = 0$ and $2x - y + z + 3 = 0$:

$$\vec{b} = \begin{vmatrix} \hat{i} & \hat{j} & \hat{k} \\ 1 & 1 & 1 \\ 2 & -1 & 1 \end{vmatrix} = 2\hat{i} + \hat{j} - 3\hat{k}$$

Put $y = 0$, we get $x = -2$ and $z = 1$

$L_2 : \bar{r} = (-2\hat{i} + \hat{k}) + \lambda(2\hat{i} + \hat{j} - 3\hat{k})$ and

$L_1 : \bar{r} = (\hat{i} - \hat{j}) + \mu(-\hat{j} + \hat{k})$(Given)

Now,

$\bar{b}_1 \times \bar{b}_2 = -2[\hat{i} + \hat{j} + \hat{k}]$ and $\bar{a}_2 - \bar{a}_1 = -3\hat{i} + \hat{j} + \hat{k}$

Shortest distance $= \dfrac{1}{\sqrt{3}}$

73. (b) Since, the lines are coplanar

$$\therefore \begin{vmatrix} 1 & 3 & 2 \\ 2 & -1 & 1 \\ \alpha & 5-\alpha & 1 \end{vmatrix} = 0$$

$\Rightarrow 1(-1-5+\alpha) - 3(2-\alpha) + 2(10 - 2\alpha + \alpha) = 0$

$\therefore \alpha = -4$

$\therefore$ Equation of $L_2 : \dfrac{x+2}{-4} = \dfrac{y+1}{9} = \dfrac{z+1}{1}$

$\therefore$ Point $(2, -10, -2)$ lies on line L_2.

74. (b) Equation of line through point P $(1, -2, 3)$ and parallel to the line $\dfrac{x}{2} = \dfrac{y}{3} = \dfrac{z}{-6}$ is

$\dfrac{x-1}{2} = \dfrac{y+2}{3} = \dfrac{z-3}{-6} = \lambda$

So, any point on line $= Q(2\lambda + 1, 3\lambda - 2, -6\lambda + 3)$

Since, this point lies on the plane $x - y + 2 = 5$

$\therefore 2\lambda + 1 - 3\lambda + 2 - 6\lambda + 3 = 5 \Rightarrow \lambda = \dfrac{1}{7}$

$\therefore$ Point of intersection line and plane,

$Q = \left(\dfrac{9}{7}, \dfrac{11}{7}, \dfrac{15}{7}\right)$

$\therefore$ Required distance PQ

$$= \sqrt{\left(\frac{9}{7} - 1\right)^2 + \left(-\frac{11}{7} + 2\right)^2 + \left(\frac{15}{7} - 3\right)^2} = 1$$

75. (a) The equation of the line through points $(1, -2, 3)$ and $(1, 1, 0)$ is

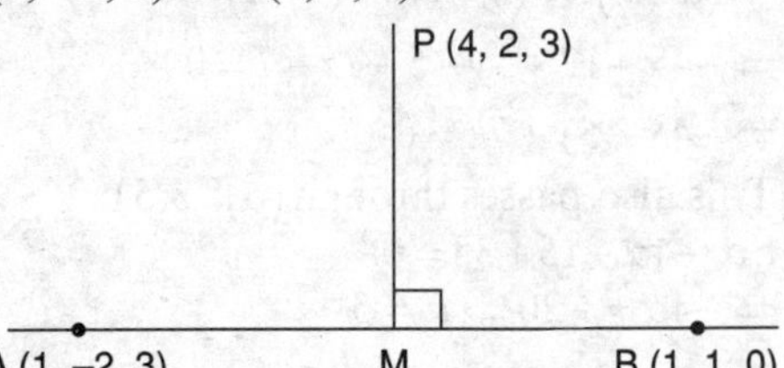

$\dfrac{x-1}{0} = \dfrac{y-1}{-3} = \dfrac{z-0}{3-0} \quad (= \lambda \text{ say})$

$\therefore M(1, -\lambda + 1, \lambda)$

Direction ratios of PM $= [-3, -\lambda - 1, \lambda - 3]$

$\because PM \perp AB$

$\therefore (-3).0 + (-1-\lambda)(-1) + (\lambda - 3).1 = 0$

$\therefore \lambda = 1$

$\therefore$ Foot of perpendicular $= (1, 0, 1)$

This point satisfies the plane $2x + y - z = 1$.

76. (c) Direction ratios of normal to the plane are $<1, -3, 2>$.

The plane passes through $(3, 1, 1)$.

The equation of the plane is,

$1(x-3) - 3(y-1) + 2(z-1) = 0$

$\Rightarrow x - 3y + 2z - 2 = 0$

77. (b) Let the plane passing through $(2, 1, 2)$ be

$a(x-2) + b(y-1) + (z-2) = 0$

It also passes through $(1, 2, 1)$

$\therefore -a + b - c = 0 \Rightarrow a - b + c = 0$

The given line is

$\dfrac{x}{3} = \dfrac{y}{2} = \dfrac{z-1}{0}$, which is parallel to the plane.

$\therefore 3a + 2b + c(0) = 0$

$\Rightarrow \dfrac{a}{0-2} = \dfrac{b}{3-0} = \dfrac{c}{2+3}$

$\Rightarrow \dfrac{a}{2} = \dfrac{b}{-3} = \dfrac{c}{2+3}$

$\Rightarrow \dfrac{a}{2} = \dfrac{b}{-3} = \dfrac{c}{-5}$

$\therefore$ The plane is $2x - 4 - 3y + 3 - 5z + 10 = 0$

$\Rightarrow 2x - 3y - 5z + 9 = 0$

The plane satisfies the point $(-2, 0, 1)$.

78. (a) $\because$ The plane contains two lines

$$\therefore \vec{n} = \begin{vmatrix} \hat{i} & \hat{j} & \hat{k} \\ 1 & -2 & 2 \\ 2 & 3 & -1 \end{vmatrix}$$

$= \hat{i}(2-6) - \hat{j}(-1-4) + \hat{k}(3+4)$

$= -4\hat{i} + 5\hat{j} + 7\hat{k}$

So, equation of the plane is

$-4(x-3)+5(y-1)+7(z-1)=0$

$\Rightarrow -4x+12+5y-5+7z-7=0$

$\Rightarrow -4x+5y+7z=0$

This also passes through $(\alpha,-3,5)$

So, $-4\alpha-15+35=0$

$\Rightarrow -4\alpha=-20 \Rightarrow \alpha=5$

79. (b)

$$\Delta=0 \Rightarrow \begin{vmatrix} 1 & 4 & -2 \\ 1 & 7 & -5 \\ 1 & 5 & \alpha \end{vmatrix}=0$$

$\Rightarrow (7\alpha+25)-(4\alpha+10)+(-20+14)=0$

$\Rightarrow 3\alpha+9=0 \Rightarrow \alpha=-3$

$$\text{Also, } D_z=0 \Rightarrow \begin{vmatrix} 1 & 4 & 1 \\ 1 & 7 & \beta \\ 1 & 5 & 5 \end{vmatrix}=0$$

$\Rightarrow 1(35-5\beta)-(15)+1(4\beta-7)=0 \Rightarrow \beta=13$

Hence, $\alpha+\beta=-3+13=10$

80. (b)

P (1, 2, 3)

$\vec{n}$

m

$Q\left(\frac{-7}{3},\frac{-4}{3},\frac{-1}{2}\right)$

$\vec{n}=\frac{-7}{3}-1,\frac{-4}{3}-2,\frac{-1}{3}-3$

$\vec{n}=\frac{10}{3},\frac{10}{3},\frac{10}{3}$

DR of normal to the plane (1, 1, 1)

Mid-point of P and Q is $\left(\frac{-2}{3},\frac{1}{3},\frac{4}{3}\right)$

$\therefore$Equation of the required plane Q is

$\vec{r}.\vec{n}=\vec{a}.\vec{n}$

$\vec{r}.(\hat{i}+\hat{j}+\hat{k})=\frac{-2}{3}+\frac{1}{3}+\frac{4}{3}$

$\therefore$ Equation of the plane is $x+y+z=1$

81. (b) Equation of plane is $x+y-2z=3$

$\Rightarrow \frac{x-2}{1}=\frac{y-1}{1}=\frac{z-6}{-2}=\frac{-2(2+1-12-3)}{6}$

$\Rightarrow (x,y,z)=(6,5,-2)$

82. (c) Given, $\frac{x}{1}=\frac{y-1}{0}=\frac{z+1}{-1}=p$ (let) and point $P(\beta,0,\beta)$

Any point on line $A=(p,1,-p-1)$

Now, DR of AP $a'' < p-\beta,1-0,-p-1-\beta >$ which is perpendicular to line.

$\therefore (p-\beta)1+0.1-1(-p-1-\beta)=0$

$\Rightarrow p-\beta+p+1+\beta=0 \Rightarrow p=\frac{-1}{2}$

$\therefore$ Point $A\left(\frac{-1}{2},1,-\frac{1}{2}\right)$

Given that distance $AP=\sqrt{\frac{3}{2}} \Rightarrow AP^2=\frac{3}{2}$

$\Rightarrow \left(\beta+\frac{1}{2}\right)^2+1+\left(\beta+\frac{1}{2}\right)^2=\frac{3}{2} \, or \, 2\left(\beta+\frac{1}{2}\right)^2=\frac{1}{2}$

$\Rightarrow \left(\beta+\frac{1}{2}\right)^2=\frac{1}{4} \Rightarrow \beta=0,-1,(\beta\neq 0)$

$\therefore \beta=-1$

83. (d) Let a point D on $BC=(3\lambda-2,1,4\lambda)$

$\overline{AD}=(3\lambda-3)\hat{i}+2\hat{j}+(4\lambda-2)\hat{k}$

$\because \overline{AD}\perp\overline{BC}, \therefore \overline{AD}.\overline{BC}=0$

$\Rightarrow (3\lambda-3)+3+2(0)+(4\lambda-2)=0 \Rightarrow \lambda=\frac{17}{25}$

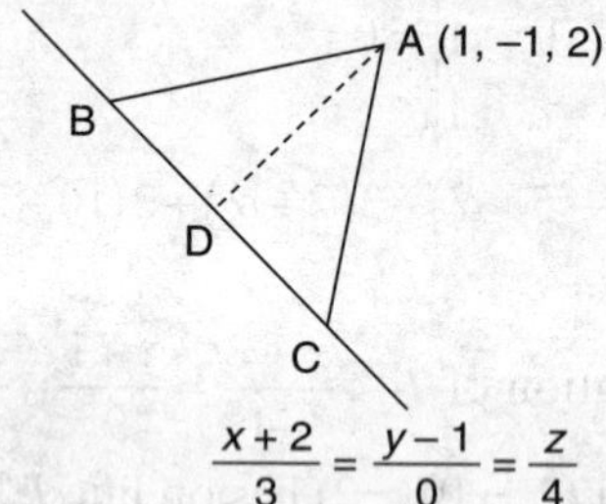

Hence, $D=\left(\frac{1}{25},1,\frac{68}{25}\right)$

$|\overline{AD}|=\sqrt{\left(\frac{1}{25}-1\right)^2+(2)^2+\left(\frac{68}{258}-2\right)^2}$

$=\sqrt{\frac{(24)^2+4(25)^2+(18)^2}{25}}$

$=\sqrt{\frac{3400}{25}}=\frac{2\sqrt{34}}{5}$

Area of triangle $=\frac{1}{2}\times|\overline{BC}|\times|\overline{AD}|$

$=\frac{1}{2}\times 5\times\frac{2\sqrt{34}}{5}=\sqrt{34}\,[\because BC=5]$

84. (a) Here, P, Q, R are collinear.

$\therefore \overrightarrow{PR} = \lambda \overrightarrow{PQ}$

$2\hat{i} + (y+3)\hat{j} + (z-4)\hat{k} = \lambda[6\hat{i} + 3\hat{j} + 6\hat{k}]$

$\Rightarrow 6\lambda = 2, y+3 = 3\lambda, z-4 = 6\lambda$

$\Rightarrow \lambda = \dfrac{1}{3}, y = -2, z = 6$

$\therefore$Point R (4, −2, 6)

Now, $OR = \sqrt{(4)^2 + (-2)^2 + (6)^2} = \sqrt{56} = 2\sqrt{14}$

85. (b) Let co-ordinates of Q be (α, β, γ), then

$\alpha + \beta + \gamma = 3$ ...(i)

$\alpha - \beta + \gamma = 3$ ...(ii)

$\Rightarrow \alpha + \gamma = 3 \text{ and } \beta = 0$

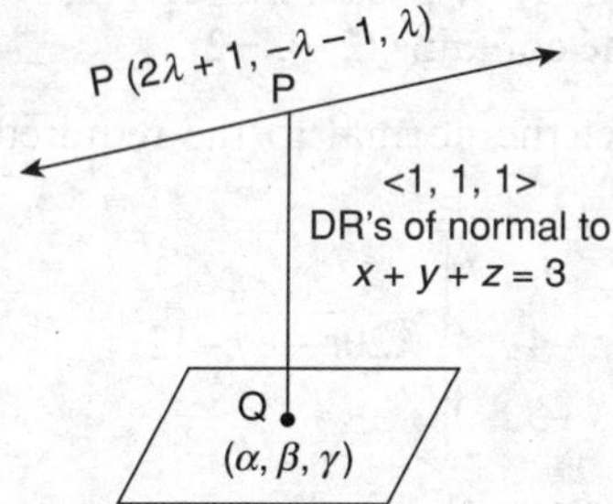

Equating direction ratios of PQ, we get

$\dfrac{\alpha - 2\lambda - 1}{1} + \dfrac{\lambda + 1}{1} = \dfrac{\gamma - \lambda}{1}$

$\Rightarrow \alpha = 3\lambda + 2, \gamma = 2\lambda + 1$

Substituting the values of α and γ in equation (i), we get

$\Rightarrow 5\lambda + 3 = 3 \Rightarrow \lambda = 0$

Hence, point is Q (2, 0, 1)

86. (a) Let P be the foot of perpendicular from point T (2, −1, 4) on the given line. So, P can be assumed as $P(10\lambda - 3, -7\lambda + 2, \lambda)$

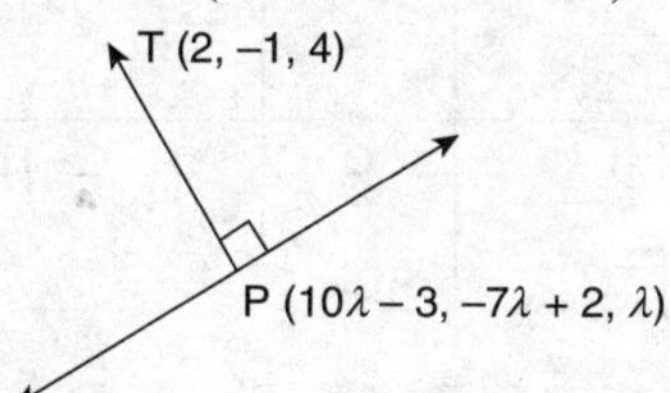

Direction ratios of TP $\propto$ to $10\lambda - 5, -7\lambda + 3, \lambda - 4$

$\therefore$ TP and the given line are perpendicular, so

$10(10\lambda - 5) - 7(-7\lambda + 3) + 1(\lambda - 4) = 0$

$\Rightarrow \lambda = \dfrac{1}{2}$

$\Rightarrow TP = \sqrt{(10\lambda - 5)^2 + (-7\lambda + 3)^2 + (\lambda - 4)^2}$

$= \sqrt{0 + \dfrac{1}{4} + \dfrac{49}{4}} = \sqrt{12.5} = 3.54$

Hence, the length of perpendicular is greater than 3 but less than 4.

87. (a) Let the co-ordinate of R with respect to the line

$\dfrac{x-3}{1} = \dfrac{y+1}{3} = \dfrac{z-6}{-1} = \lambda$

$\dfrac{x+5}{7} = \dfrac{y-2}{-6} = \dfrac{z-3}{4} = \mu$

$L_1 = (\lambda + 3, 3\lambda - 1, -\lambda + 6)$

and co-ordinate of P w.r.t. line

$L_2 = (7\mu - 5, 6\mu + 2, 4\mu + 3)$

$\therefore \lambda - 7\mu = -8, 3\lambda + 6\mu = 3, \lambda + 4\mu = 3$

From the above equation: $\lambda = -1, \mu = 1$

$\therefore$Co-ordinate of point of intersection R = (2, −4, 7).

Image of R w.r.t. *xy* plane = (2, −4, −7).

88. (d) Let the points $P(3\lambda + 2, 2\lambda - 1, -\lambda + 1)$ and $Q(3\mu + 2, 2\mu - 1, -\mu + 1)$

$\because$ P lies on $2x + 3y - z + 13 = 0$

$\therefore 6\lambda + 4 + 6\lambda - 3 + \lambda - 1 + 13 = 0$

$\Rightarrow 13\lambda = -13 \Rightarrow \lambda = -1$

Hence, P (−1, −3, 2)

Similarly, Q lies on $3x + y + 4z = 16$

$\therefore 9\mu + 6 + 2\mu - 1 - 4\mu + 4 = 16$

$\Rightarrow 7\mu = 7 \Rightarrow \mu = 1$

Hence, Q is (5, 1, 0)

Now, $PQ = \sqrt{36 + 16 + 4} = \sqrt{56} = 2\sqrt{14}$

89. (d) The equations of angle bisectors are,

$\dfrac{x + 2y + 2z - 2}{3} = \pm \dfrac{2x - y + 2z - 4}{3}$

$\Rightarrow x - 3y - 2 = 0 \text{ or } 3x + y + 4z - 6 = 0$

(2, −4, 1) lies on the second plane.

90. (d) Let the required plane passing through the points (0, −1, 0) and (0, 0, 1) be $\dfrac{x}{\lambda} + \dfrac{y}{-1} + \dfrac{z}{1} = 1$

and the given plane is $y - z + 5 = 0$

$\therefore \cos\dfrac{\pi}{4} = \dfrac{-1-1}{\sqrt{\left(\dfrac{1}{\lambda^2} + 1 + 1\right)}\sqrt{2}}$

$\Rightarrow \lambda^2 = \dfrac{1}{2} \Rightarrow \dfrac{1}{\lambda} = \pm\sqrt{2}$

Then, the equation of plane is

$\pm\sqrt{2}x - y + z = 1$

Then the point $(\sqrt{2}, 1, 4)$ satisfies the equation of the plane.

91. (c) Let the equation of required plane be
$(2x-y-4)+\lambda(y+2z-4)=0$
$\because$ This plane passes through the point (1, 1, 0)
then $(2-1-4)+\lambda(1+0-4)=0$
$\Rightarrow \lambda=-1$
Then, equation of the required plane is,
$(2x-y-4)-(y+2z-4)=0$
$\Rightarrow 2x-2y-2z=0 \Rightarrow x-y-z=0$

92. (b) $\because$ The plane contains both lines.

$$\therefore \text{DR of plane}=\begin{vmatrix}\hat{i} & \hat{j} & \hat{k}\\ 3 & 5 & 7\\ 1 & 4 & 7\end{vmatrix}=7\hat{i}-14\hat{j}+7\hat{k}$$

Now, equation of the plane is,
$7(x-1)-14(y-4)+7(z+4)=0$
$\Rightarrow x-1-2y+8+z+4=0$
$\Rightarrow x-2y+z+11=0$
Hence, distance from (0, 0, 0) to the plane.
$=\dfrac{11}{\sqrt{1+4+1}}=\dfrac{11}{\sqrt{6}}$

93. (a) Let the angle between line and plane is θ, then

$$\sin\theta=\left|\frac{\vec{b}\cdot\vec{n}}{|\vec{b}|\cdot|\vec{n}|}\right|$$

$$=\left|\frac{(2\hat{i}+\hat{j}-2\hat{k}).(\hat{i}-2\hat{j}-K\hat{k})}{\sqrt{9}\cdot\sqrt{1+4+K^2}}\right|$$

$$=\left|\frac{2-2\_2K}{3\sqrt{5+K^2}}\right|=\frac{2|K|}{3\sqrt{4+K^2}}$$

Since, $\cos\theta=\dfrac{2\sqrt{2}}{3}\Rightarrow \sin\theta=\dfrac{1}{3}$

Then, $\dfrac{2|K|}{3\sqrt{5+K^2}}=\dfrac{1}{3}\Rightarrow 4K^2=5+K^2$

$3K^2=5\Rightarrow K=\pm\sqrt{\dfrac{5}{3}}$

94. (b) Let $A(-\lambda^2,1,1)$, $B(1,-\lambda^2,1)$, $C(1,1,-\lambda^2)$, $D(-1,-1,1)$ lie on same plane, then

$$\begin{vmatrix}1-\lambda^2 & 2 & 0\\ 2 & 1-\lambda^2 & 0\\ 2 & 2 & -\lambda^2-1\end{vmatrix}=0$$

$\Rightarrow(\lambda^2+1)\left((1-\lambda^2)^2-4\right)=0$
$\Rightarrow(3-\lambda^2)(\lambda^2+1)=0\Rightarrow\lambda^2=3$
$\lambda=\pm\sqrt{3}$
Hence, $S=\{-\sqrt{3},\sqrt{3}\}$

95. (d) Let normal to the required plane be $\vec{n}$
$\Rightarrow \vec{n}$ is perpendicular to both vector $2\hat{i}-\hat{j}+3\hat{k}$ and $2\hat{i}+3\hat{j}-3\hat{k}$.

$$\Rightarrow \vec{n}=\begin{vmatrix}\hat{i} & \hat{j} & \hat{k}\\ 2 & -1 & 3\\ 2 & 3 & -1\end{vmatrix}=-8\hat{i}+8\hat{j}+8\hat{k}$$

$\Rightarrow$Equation of the required plane is
$\Rightarrow(x-3)(-8)+(y+2)\times 8+(z-1)\times 8=0$
$\Rightarrow(x-3)(-1)+(y+2)\times 1+(z-1)\times 1=0$
$\Rightarrow x-3-y-2-z+1=0$
$\because x-y-z=4$ passes through (2, 0, −2)
$\therefore$ plane contains (2, 0, −2).

96. (b) Let the normal to the required plane is $\vec{n}$, then

$$\vec{n}=\begin{vmatrix}\hat{i} & \hat{j} & \hat{k}\\ 4 & -4 & 4\\ 2 & -5 & 0\end{vmatrix}=20\hat{i}+8\hat{j}-12$$

$\therefore$ Equation of the plane
$(x-3)\times 20+(y-4)\times 8+(z-2)\times(-12)=0$
$5x-15+2y-8-3z+6=0$
$5x+2y-3z-17=0$ ....(1)
Since, equation of plane (1) passes through $(2,\alpha,\beta)$, then
$10+2\alpha-3\beta-17=0$
$\Rightarrow 2\alpha-3\beta=7$

97. (d)

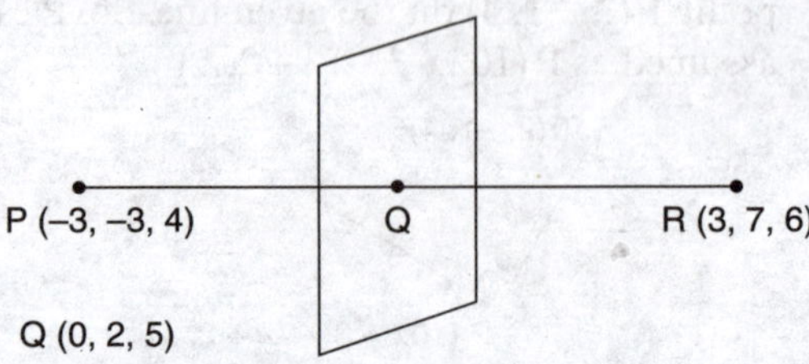

Since direction ratios of normal to the plane is $\vec{n}=6\hat{i}+10\hat{j}+2\hat{k}$
the equation of the plane is
$(x-0)6+(y-2)10+(z-5)2=0$
$3x+5y-10+z-5=0$
$3x+5y+z=15$ ...(i)
Since plane (1) satisfies the point (4, 1, −2)
Hence, required point is (4, 1, −2)

98. (c)

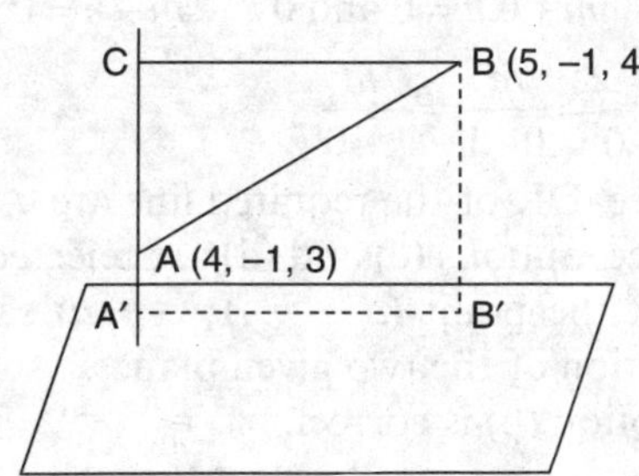

$$AC = \overline{AB}.\widehat{AC} = (\hat{i} + \hat{k}).\frac{(\hat{i} + \hat{j} + \hat{k})}{\sqrt{3}} = \frac{2}{\sqrt{3}}$$

Now, $A'B' = BC = \sqrt{AB^2 - AC^2} = \sqrt{2 - \frac{4}{3}} = \sqrt{\frac{2}{3}}$

$\therefore$ Length of projection $= \sqrt{\frac{2}{3}}$

99. (b) Given

$l + 3m + 5n = 0$ ...(i)

and $5lm - 2mn + 6nl = 0$ ...(ii)

From equation (i) we have

$l = -3m - 5n$

Put the value of l in equation (ii), we get

$5(-3m - 5n)m - 2mn + 6n(-3m - 5n) = 0$

$\Rightarrow 15m^2 + 45mn + 30n^2 = 0$

$\Rightarrow m^2 + 3mn + 2n^2 = 0 \Rightarrow m^2 + 2mn + mn + 2n^2 = 0$

$\Rightarrow (m + n)(m + 2n) = 0$

$\therefore m = -n$ or $m = -2n$

For $m = -n, l = -2n$

and for $m = -2n, l = n$

$\therefore (l, m, n) = (-2n, -n, n)$ or $(l, m, n) = (n, -2n, n)$

$\Rightarrow (l, m, n) = (-2, -1, 1)$ or $(l, m, n) = (1, -2, 1)$

Therefore, angle between the lines is given as:

$$\cos(\theta) = \frac{(-2)(1) + (-1).(-2) + (1)(1)}{\sqrt{6}.\sqrt{6}}$$

$$\Rightarrow \cos(\theta) = \frac{1}{6} \Rightarrow \theta = \cos^{-1}\left(\frac{1}{6}\right)$$

100. (d) Let θ be the angle between the two lines.

Here, direction cosines of $\frac{x}{2} = \frac{y}{2} = \frac{z}{1}$ are 2, 2, 1

Also, the second line can be written as:

$$\frac{x-5}{2} = \frac{y-2}{\frac{P}{7}} = \frac{z-3}{4}$$

$\therefore$ its direction cosines are $2, \frac{P}{7}, 4$

Also, $\cos\theta = \frac{2}{3}$ (given)

$$\because \cos\theta = \left|\frac{a_1a_2 + b_1b_2 + c_1c_2}{\sqrt{a_1^2 + b_1^2 + c_1^2}\sqrt{a_2^2 + b_2^2 + c_2^2}}\right|$$

$$\Rightarrow \frac{2}{3} = \left|\frac{(2\times 2) + \left(2\times\frac{P}{7}\right) + (1\times 4)}{\sqrt{2^2 + 2^2 + 1^2}\sqrt{2^2 + \frac{P^2}{49} + 4^2}}\right|$$

$$= \frac{4 + \frac{2P}{7} + 4}{3\times\sqrt{2^2 + \frac{P^2}{49} + 4^2}}$$

$$\Rightarrow \left(4 + \frac{P}{7}\right)^2 = 20 + \frac{P^2}{49} \Rightarrow 16 + \frac{8P}{7} + \frac{P^2}{49} = 20 + \frac{P^2}{49}$$

$$\Rightarrow \frac{8P}{7} = 4 \Rightarrow P = \frac{7}{2}$$

101. (a) Since the plane bisects the line joining the points (1, 2, 3) and (−3, 4, 5) then the plane passes through the mid-point of the line which is:

$$\left(\frac{1-3}{2}, \frac{2+4}{2}, \frac{5+3}{2}\right) \equiv \left(\frac{-2}{2}, \frac{6}{2}, \frac{8}{2}\right) \equiv (-1, 3, 4)$$

As the plane cuts the line segment at right angle, the direction cosines of the normal of the plane are (−3 − 1, 4 – 2, 5 − 3) = (−4, 2, 2)

So, the equation of the plane is: $-4x + 2y + 2z = \lambda$

As plane passes through (−1, 3, 4), so

$-4(-1) + 2(3) + 2(4) = \lambda \Rightarrow \lambda = 18$

Therefore, equation of plane is: $-4x + 2y + 2z = 18$

Now, only (−3, 2, 1) satisfies the given plane as $-4(-3) + 2(2) + 2(1) = 18$

102. (c) Equation of line PQ is

$$\frac{x-1}{1} = \frac{y+2}{4} = \frac{z-3}{5}$$

Let F be $(\lambda + 1, 4\lambda - 2, 5\lambda + 3)$

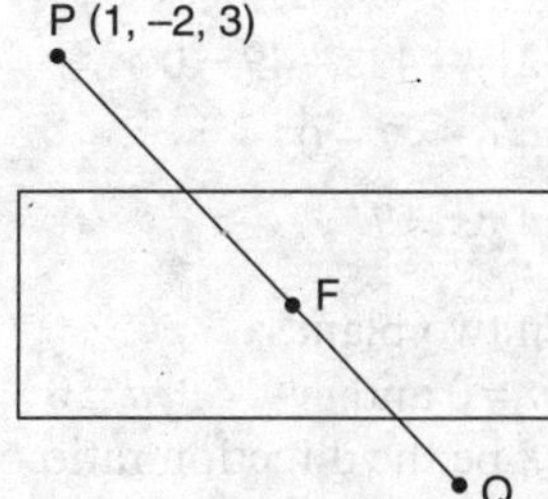

Since F lies on the plane

$\therefore 2(\lambda + 1) + 3(4\lambda - 2) - 4(5\lambda + 3) + 22 = 0$

$2\lambda + 2 + 12\lambda - 6 - 20\lambda - 12 + 22 = 0$

$\Rightarrow -6\lambda+6=0 \Rightarrow \lambda=1$

$\therefore$ F is (2, 2, 8)

$PQ = 2\,PF = 2\sqrt{1^2+4^2+5^2} = 2\sqrt{42}$

103. (d)

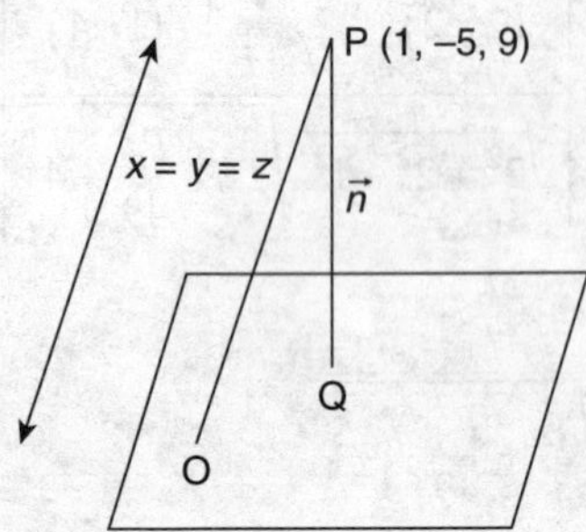

Equation of PO: $\dfrac{x-1}{1}=\dfrac{y+5}{1}=\dfrac{z-9}{1}=\lambda$

$\Rightarrow x=\lambda+1; y=\lambda-5; z=\lambda+9$

Putting these in equation of plane:

$\lambda+1-\lambda+5+\lambda+9=5$

$\Rightarrow \lambda=-10$

$\Rightarrow \text{O is}(-9,-15,-1)$

$\Rightarrow \text{Distance OP}=10\sqrt{3}$

104. (a) Equation of the plane containing the lines $2x-5y+z=3$ and $x+y+4z=5$ is

$2x-5y+z-3+\lambda(x+y+4z-5)=0$

$\Rightarrow (2+\lambda)x+(-5+\lambda)y+(1+4\lambda)z+(-3-5\lambda)$

$=0$ ...(i)

Since the plane (i) is parallel to the given plane $x+3y+6z=1$

$\therefore \dfrac{2+\lambda}{1}=\dfrac{-5+\lambda}{3}=\dfrac{1+4\lambda}{6}\Rightarrow \lambda=-\dfrac{11}{2}$

Hence, equation of the required plane is

$\left(-2-\dfrac{11}{2}\right)x+\left(-5-\dfrac{11}{2}\right)y+\left(1-\dfrac{44}{2}\right)z+$

$\left(-3+\dfrac{55}{2}\right)=0$

$\Rightarrow (4-11)x+(-10-11)y+(2-44)z+(-6+55)=0$

$\Rightarrow -7x-21y-42z+49=0$

$\Rightarrow x+3y+6z-7=0$

$\Rightarrow x+3y+6z=7$

105. (b) Given two planes:

$x-ay-b=0$ and $cy-z+d=0$

Let, l, m, n be the direction ratio of the required line.

Since the required line is perpendicular to normal of both the plane, therefore $l-am+0.n=0$ and $cm-n=0$

$\Rightarrow l-am+0.n=0$ and $0.l+cm-n=0$

$\therefore \dfrac{l}{a-0}=\dfrac{m}{0+1}=\dfrac{n}{c-0}$

Hence, DR of the required line are a, 1, c.

Hence, options (c) and (d) are rejected.

Now, the point ($a+b$, 1, $c+d$) satisfies the equation of the two given planes.

$\therefore$ Option (b) is correct.

106. (c) $2x+y+2z-8=0$ ...(i)

$2x+y+2z+\dfrac{5}{2}=0$ ...(ii)

Distance between (i) and (ii)

$=\left|\dfrac{-8-\dfrac{5}{2}}{\sqrt{2^2+1^2+2^2}}\right|=\left|\dfrac{-21}{6}\right|=\dfrac{7}{2}$

107. (a) Equation of the plane passing through the intersection line of the given planes is

$(x+2y+3z-2)+\lambda(x-y+z-3)=0$

or

$(1+\lambda)x+(2-\lambda)y+(3+\lambda)z+(-2-3\lambda)=0$

$\because$ Its distance from the point (3, 1, −1) is $\dfrac{2}{\sqrt{3}}$.

$\therefore \left|\dfrac{3(1+\lambda)+1(2-\lambda)-1(3+\lambda)+(-2-3\lambda)}{\sqrt{(1+\lambda)^2+(2-\lambda)^2+(3+\lambda)^2}}\right|=\dfrac{2}{\sqrt{3}}$

$\Rightarrow \left|\dfrac{-2\lambda}{\sqrt{3\lambda^2+4\lambda+14}}\right|=\dfrac{2}{\sqrt{3}}$

$\Rightarrow 3\lambda^2+4\lambda+14=3\lambda^2 \Rightarrow \lambda=\dfrac{-7}{2}$

$\therefore$Required equation of the plane is

$(x+2y+3z-2)-\dfrac{7}{2}(x-y+z-3)=0$

or $5x-11y+z=17$

108. (a) Equation of straight line joining Q (2, 3, 5) and R (1, −1, 4) is

$\dfrac{x-2}{-1}=\dfrac{y-3}{-4}=\dfrac{z-5}{1}=\lambda$

Let $P(-\lambda+2,-4\lambda+3,-\lambda+5)$

Since P also lies on $5x-4y-z=1$

$\therefore -5\lambda+10+16\lambda-12+\lambda-5=1$

$\Rightarrow 12\lambda=8\Rightarrow \lambda=\dfrac{2}{3}\therefore P=\left(\dfrac{4}{3},\dfrac{1}{3},\dfrac{13}{3}\right)$

Now let another point S on QR be $(-\mu+2,-4\mu+3,-\mu+5)$

Since S is the foot of perpendicular drawn from

T (2, 1, 4) to QR, where direction ratios of ST are $\mu, 4\mu-2, \mu-1$ and direction ratios of QR are $-1, -4, -1$

$\therefore -\mu-16\mu+8-\mu+1=0 \Rightarrow 8\mu=9 \Rightarrow \mu=\frac{1}{2}$

$\therefore S=\left(\frac{3}{2},1,\frac{9}{2}\right)$

$\therefore$ Distance between P and S

$=\sqrt{\left(\frac{4}{3}-\frac{3}{2}\right)^2+\left(\frac{1}{3}-1\right)^2+\left(\frac{13}{3}-\frac{9}{2}\right)^2}$

$=\sqrt{\frac{1}{36}+\frac{4}{9}+\frac{1}{6}}=\frac{1}{\sqrt{2}}$

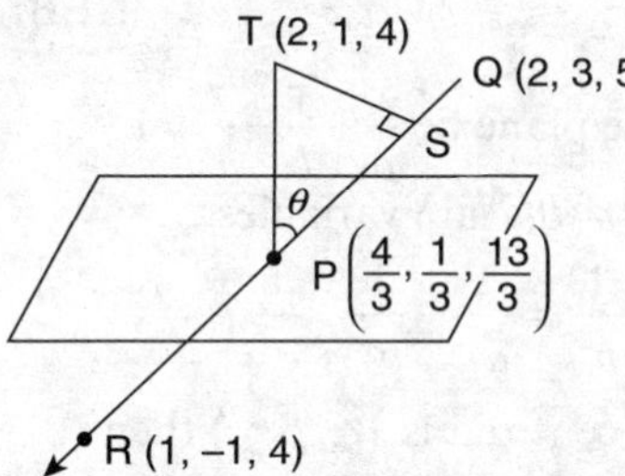

109. (a) Since the perpendicular distance of $x+2y-2z-\alpha=0$ from the point (1, −2, 1) is 5

$\therefore \left|\frac{1-4-2-\alpha}{3}\right|=5$

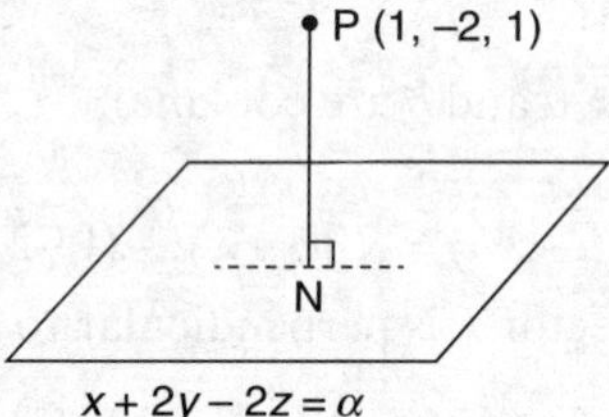

$\Rightarrow \frac{-5-\alpha}{3}=5 \text{ or } -5$

$\Rightarrow \alpha=-20 \text{ or } 10$

But $\alpha>0 \Rightarrow \alpha=10$

$\therefore$ Equation of the plane: $x+2y-2z-10=0$

We know that foot of perpendicular from point (x_1, y_1, z_1) to the plane $ax+by+cz+d=0$ is given by

$$\frac{x-x_1}{a}=\frac{y-y_1}{b}=\frac{z-z_1}{c}=\frac{-(ax_1+by_1+cz_1+d)}{(a^2+b^2+c^2)}$$

$\therefore \frac{x-1}{1}=\frac{y+2}{2}=\frac{z-1}{-2}=\frac{-(1-4-2-10)}{9}=\frac{5}{3}$

$\Rightarrow x=\frac{8}{3}, y=\frac{4}{3}, z=-\frac{7}{3}$

$\therefore$ Foot of perpendicular $\equiv\left(\frac{8}{3},\frac{4}{3},-\frac{7}{3}\right)$

110. (c) Equation of the plane containing two lines $\frac{x}{3}=\frac{y}{4}=\frac{z}{2}$ and $\frac{x}{4}=\frac{y}{2}=\frac{z}{3}$ is given by

$$\begin{vmatrix} x & y & z \\ 3 & 4 & 2 \\ 4 & 2 & 3 \end{vmatrix}=0 \Rightarrow 8x-y-10z=0$$

Now equation of the plane containing the line $\frac{x}{2}=\frac{y}{3}=\frac{z}{4}$ and perpendicular to the plane $8x-y-10z=0$ is given by,

$$\begin{vmatrix} x & y & z \\ 2 & 3 & 4 \\ 8 & -1 & -10 \end{vmatrix}=0$$

$\Rightarrow -26x+52y-26z=0 \text{ or } x-2y+z=0$

111. (c) Since the line makes equal angle with co-ordinate axes and which has positive direction cosines

$\therefore$ DC's $=\frac{1}{\sqrt{3}},\frac{1}{\sqrt{3}},\frac{1}{\sqrt{3}}$

$\Rightarrow$ DR's $=1,1,1$

$\therefore$ Equation of line is

$\frac{x-2}{1}=\frac{y+1}{1}=\frac{z-2}{1}=\lambda$

$\therefore Q(\lambda+2,\lambda-1,\lambda+2)$ be any point on this line where it meets the plane $2x+y+z=9$

$\Rightarrow 2(\lambda+2)+\lambda-1+\lambda+2=9 \Rightarrow \lambda=1$

$\therefore$ Q has co-ordinates (3, 0, 3)

$\therefore PQ=\sqrt{(2-3)^2+(-1-0)^2+(2-3)^2}=\sqrt{3}$

112. (a)

$\because \vec{r}=(\hat{i}-\hat{j}+2\hat{k})+\mu(-3\hat{i}+\hat{j}+5\hat{k})$

$=(1-3\mu)\hat{i}+(-1+\mu)\hat{j}+(2+5\mu)\hat{k}$

Let the co-ordinates of Q be $(-3\mu+1,\mu-1,5\mu+2)$

$\therefore$ DR's of $\overrightarrow{PQ}=-3\mu-2,\mu-3,5\mu-4$

Given that $\overrightarrow{PQ}$ is parallel to the plane $x-4y+3z=1$

$\therefore 1.(-3\mu-2)-4.(\mu-3)+3.(5\mu-4)=0$

$\Rightarrow 8\mu=2 \text{ or } \mu=\frac{1}{4}$

Integer and Numerical Value Type

1. $\alpha=120°$ and $\beta=60°$

$\therefore \cos\alpha=\cos 120°=-\frac{1}{2}$

and $\cos\beta=\cos 60°=\frac{1}{2}$

but $\cos^2\alpha+\cos^2\beta+\cos^2\gamma=1$

$\therefore \left(\frac{-1}{2}\right)^2 + \left(\frac{1}{2}\right)^2 + \cos^2\gamma = 1$

$\cos^2\gamma = 1 - \frac{1}{4} - \frac{1}{4} = \frac{1}{2}$

$\therefore \cos\gamma = \pm\frac{1}{\sqrt{2}}$

$\therefore \gamma = 45°$ or $135°$

2. $\sqrt{1^2+(-2)^2+7^2} = \sqrt{54}$

$\sqrt{3^2+(-2)^2+(-1)^2} = \sqrt{14}$

$\therefore$ The actual direction cosines of the lines are

$\frac{1}{\sqrt{54}}, \frac{-2}{\sqrt{54}}, \frac{7}{\sqrt{54}}$ and $\frac{3}{\sqrt{14}}, \frac{-2}{\sqrt{14}}, \frac{-1}{\sqrt{14}}$

If θ is the angle between the lines, then

$$\cos\theta = \left(\frac{1}{\sqrt{54}}\right)\left(\frac{3}{\sqrt{14}}\right) + \left(\frac{-2}{\sqrt{54}}\right)\left(\frac{7}{\sqrt{14}}\right) + \left(\frac{7}{\sqrt{54}}\right)\left(\frac{-1}{\sqrt{14}}\right)$$

$$= \frac{3+4-7}{\sqrt{54}.\sqrt{14}} = 0 \Rightarrow \theta = 90°$$

3. Clearly, $\vec{a}$ is perpendicular to the normal to the two planes determined by the given pairs of vectors.

$\vec{n}_1$ = Normal vector to the plane determined by $\hat{i}$ and $\hat{i}+\hat{j} = \hat{i}\times(\hat{i}\times\hat{j}) = \hat{k}$

$\vec{n}_2$ = Normal vector to the plane determined by $\hat{i}-\hat{j}$ and $\hat{i}+\hat{k} = (\hat{i}-\hat{j})\times(\hat{i}+\hat{k}) = -\hat{i}-\hat{j}+\hat{k}$

Since $\vec{a}$ is perpendicular to $\vec{n}_1$ and $\vec{n}_2$

$\vec{a} = \lambda(\vec{n}_1+\vec{n}_2) = \lambda(\hat{k}\times(-\hat{i}-\hat{j}+\hat{k}) = \lambda(-\hat{j}+\hat{i})$

Let θ be the angle between $\vec{a}$ and $\hat{i}-2\hat{j}+2\hat{k}$. Then

$$\cos\theta = \frac{\lambda(1+2+0)}{\lambda\sqrt{2}\sqrt{1+4+4}} = \frac{1}{\sqrt{2}} \Rightarrow \quad \theta = \frac{\pi}{4}$$

4. Any point on the first line in symmetrical from is $(3r-4, 5r-6, -2r+1)$. If the lines are coplanar, this point must lie on both the planes which determine the second line.

$\Rightarrow 3(3r-4)-2(5r-6)-2r+1+5 = 0$ and

$\Rightarrow 2(3r-4)+3(5r-6)+4(-2r+1)-k = 0$

We easily get $k = 4$ after eliminating r.

5. Lines will be coplanar if the point of intersection of any three planes lies on the fourth plane. Indeed, the first three planes meet at (1, 1, 1) by inspection. If it lies on fourth plane, we must have $k = 12$.

6. Take C as origin $\overline{CA} = \vec{a}$, $\overline{CB} = \vec{b}$, then D will be $\frac{2}{5}\vec{b}$.

The equations of AD and BE are

$\vec{r} = \vec{a} + \lambda\left(\frac{2}{5}\vec{b} - \vec{a}\right)$ and $\vec{r} = \vec{b} + \mu\left(\frac{1}{4}\vec{a} - \vec{b}\right)$

For intersection, we equate and using non colinearity of $\vec{a}$ and $\vec{b}$ we get $\lambda = \frac{5}{6}, \mu = \frac{2}{3}$

whence H is $\frac{1}{6}\vec{a} + \frac{1}{3}\vec{b}$

Now, we similarly determine the position vector of the point F as $\frac{\vec{a}+2\vec{b}}{3}$ whence $\frac{AF}{FB} = \frac{2}{1} \Rightarrow k = 2$

7. Let the plane be $\frac{x}{a}+\frac{y}{b}+\frac{z}{c} = 1$

where a, b, c are variables.

Then $\frac{1}{p^2} = \frac{1}{a^2}+\frac{1}{b^2}+\frac{1}{c^2}$ ...(1)

If the centroid is, (α, β, γ) then

$\alpha = \frac{a}{4}, \beta = \frac{b}{4}, \gamma = \frac{c}{4}$

On putting a, b, c in (1), we get the locus as

$\frac{1}{x^2}+\frac{1}{y^2}+\frac{1}{z^2} = \frac{16}{p^2}$

$\Rightarrow \lambda = 16$

8. Since, $\bar{x}, \bar{a}$ and $\bar{b}$ are coplanar,

let

$\bar{x} = K(\bar{a}+\lambda\bar{b}) = K[(2+\lambda)\hat{i} - (1-2\lambda)\hat{j} + (1-\lambda)\hat{k}]$

Now vector $\vec{x}$ is perpendicular to $3\hat{i}+2\hat{j}-\hat{k}$.

$\therefore 3(2+\lambda)+(2\lambda-1)2+(1-\lambda)(-1) = 0 \Rightarrow \lambda = \frac{-3}{8}$

Also, $\frac{\vec{x}\cdot\vec{a}}{|\vec{a}|} = \frac{17\sqrt{6}}{2}$

$$\Rightarrow \frac{K\left[2\cdot\frac{13}{8} - 1\cdot\left(-\frac{14}{8}\right) + 1\cdot\frac{11}{8}\right]}{\sqrt{6}} = \frac{17\sqrt{6}}{2}$$

$\Rightarrow K = 8 \Rightarrow \bar{x} = 13\hat{i} - 14\hat{j} + 11\hat{k}$

So, $|\vec{x}|^2 = 486$

9. $\vec{r} = \vec{a}+\lambda\vec{b}$ and $\vec{r} = \vec{c}+\lambda\vec{d}$

$\therefore \vec{a}-\vec{c} = (\alpha+4)\hat{i}+2\hat{j}+3\hat{k}$

$$\frac{\vec{b}\times\vec{d}}{|\vec{b}\times\vec{d}|} = \frac{(2\hat{i}+2\hat{j}+\hat{k})}{3}$$

Then shortest distance between two lines is,

$$\frac{(\vec{a}-\vec{c})\cdot(\vec{b}\times\vec{d})}{|\vec{b}\times\vec{d}|}=9$$

$$\Rightarrow ((\alpha+4)\hat{i}+2\hat{j}+3\hat{k}).\frac{(2\hat{i}+2\hat{j}+\hat{k})}{3}=9$$

$$\Rightarrow 2\alpha+15=27 \Rightarrow \alpha=6$$

10. Since the plane is perpendicular to the line joining points (−2, 21, 29) and (−1, −16, 23)

$\therefore$ Normal vector of plane is

$n=\hat{i}+5\hat{j}-6\hat{k}$

Let A$(\lambda,2,1)$ and (4, −2, 2)

$\because \overline{AB}\perp\overline{n}$

$\Rightarrow (\lambda-4)+5\times4-6(-1)=0$

$\Rightarrow \lambda-4+20+6=0 \Rightarrow \lambda=-22$

Hence, $\left(\frac{\lambda}{11}\right)^2-4\left(\frac{\lambda}{11}\right)-4=8$

11. Let equation of the line

$$\frac{x-1}{2}=\frac{y-2}{3}=\frac{z+1}{6}=\lambda$$

$\Rightarrow x=2\lambda+1, y=3\lambda+2, z=6\lambda-1$

For point of intersection of line and plane

$2(2\lambda+1)-(3\lambda+2)+(6\lambda-1)=6$

$7\lambda=7 \Rightarrow \lambda=1$

Point: (3, 5, 5)

Square of distance from point (−1, −1, 2) is

$(\text{Distance})^2=(3+1)^2+(5+1)^2+(5-2)^2$

$=16+36+9=61$

12. Since R $(3, 5, \gamma)$ lies on the plane $2x-y+z+3=0$.

$\Rightarrow 6-5+\gamma+3=0$

$\Rightarrow \gamma=-4$

So, the point R on the plane is (3, 5, −4)

Now, direction ratios of line QS are 2, −1, 1

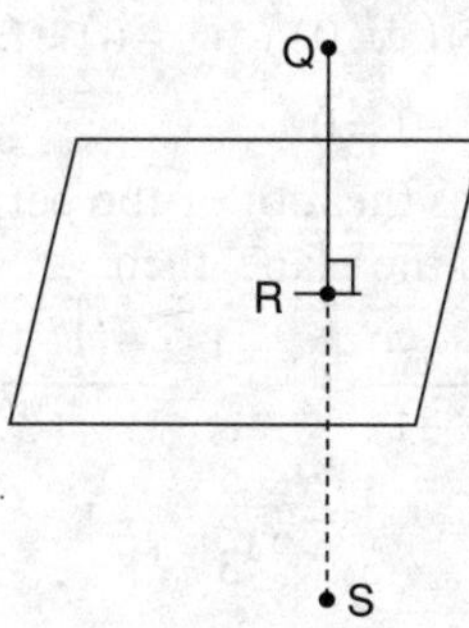

$\therefore$ The equation of line QS is,

$$\frac{x-1}{2}=\frac{y-3}{-1}=\frac{z-4}{1}=\lambda(\text{say})$$

Then, the point on the given plane is,

$F(2\lambda+1,-\lambda+3,\lambda+4)$

$\because$ F lies in the plane

$\therefore 2(2\lambda+1)-(-\lambda+3)+(\lambda+4)+3=0$

$\Rightarrow 6\lambda+6=0 \Rightarrow \lambda=-1$

$\Rightarrow F(-1,4,3)$

Since F is mid-point of the line QS, the co-ordinate of S are (−3, 5, 2).

So, $SR=\sqrt{36+0+36}=\sqrt{72}$

$\Rightarrow SR^2=72.$

13. $L_1:\frac{x-1}{2}=\frac{y-3}{1}=\frac{z-4}{2}$

For foot of perpendicular of (1, 3, 4) on

$x-2y-z-3=0$

$(1+t)-2(3-2t)-(4-t)-3=0$. So foot of $\perp$ is (3, −1, 2)

and point of intersection of L_1 with plane is (−11, −3, −8)

DR's of L is 14, 2, 10

$\cong$ <7, 1, 5>

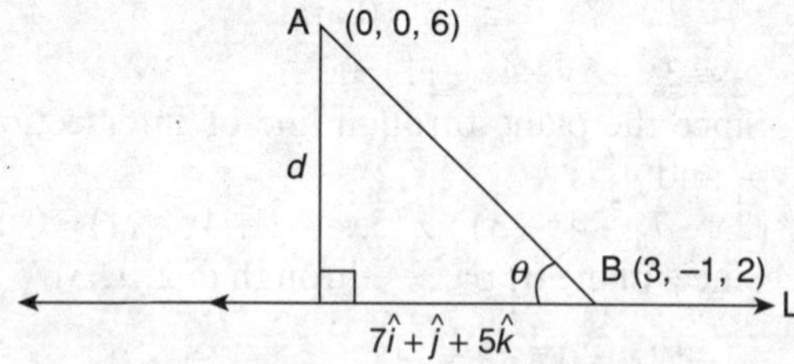

Direction ratio

OAC $=7\lambda+3, \lambda-1, 5\lambda-4$

Here AC is $\perp$ in to projection, so $\lambda=0$

DR $=(3,-1,-4) \Rightarrow d^2=26$

14. Equation of plane containing the line

$$\begin{vmatrix} x+1 & y-1 & z-3 \\ 6 & 7 & 8 \\ 3 & 5 & 7 \end{vmatrix}=0$$

$9(x+1)-18(y-1)+9(z-3)=0$

$x-2y+x=0$

Perpendicular distance from point (7, −2, 13)

$$PQ=\left|\frac{7+4+13}{\sqrt{6}}\right|=4\sqrt{6}$$

$PQ^2=96$

15. Slope of normal to the plane < 2, −6, 6 >

Then, plane is $2x-6y+6z+k=0$ ...(i)

(where $k=\lambda d$)

Since, mid-point of (4, −3, 1) and (2, 3, 5) satisfies the plane (i), then

$\Rightarrow 6-0-12+6=0$

$\Rightarrow k=6$

So, $2x-6y+6z+6=0$

$\Rightarrow x-3y+3z+3=0$

Minimum $\left(a^2+b^2+c^2+d^2\right)$

$=\left(1^2+3^2+3^2+3^2\right)=28$

16. Since, the equation of plane is of the form

$x-2y+2z+d=0$ ...(1)

and it is at unit distance form (1, 2, 3)

$\Rightarrow \left|\frac{1-4+6+d}{3}\right|$

$\Rightarrow |d+3|=3$

$\Rightarrow d=0 \text{ or } -6$

$\therefore$ Plane (1) is $x-2y+2z=0$

or $x-2y+2z-6=0$

$(-2-0)=K(2-1)$

or $(-2+6)=K(2-1)$

$K=-2$ or $K=4$

17. Suppose the plane be

$p_1 \equiv 2x-7y+4z-3=0$ and

$p_2 \equiv 3x-5y+4z+11=0$

Since the plane through line of intersection of p_1 and p_2 is

$(2x-7y+4z-3)+\lambda(3x-5y+4z+11)=0$ ...(i)

Since, plane (i) passes through (−2, 1, 3)

$-2+12\lambda=0 \Rightarrow \lambda=\frac{1}{2}$

So, the required plane is,

$15x-47y+28z-7=0$

Now, compare the above plane with

$ax+by+cz-7=0$

$\therefore a=15, b=-47, c=28$

So, $2a+b+c-7=4$

18. Let P (x, y, z).

A.T.Q

$$\left(\frac{x+y+z}{\sqrt{3}}\right)^2+\left(\frac{lx-nz}{\sqrt{l^2+n^2}}\right)^2+\left(\frac{x-2y+z}{\sqrt{6}}\right)^2=9$$

On simplifying, we get

$$\Rightarrow x^2\left(\frac{1}{2}+\frac{l^2}{l^2+n^2}\right)+y^2+z^2\left(\frac{1}{2}+\frac{n^2}{l^2+n^2}\right)$$

$$+xz\left(1-\frac{ln}{l^2+n^2}\right)=9$$

Comparing with $x^2+y^2+z^2=9$, we get

Clearly $\frac{l^2}{l^2+n^2}=\frac{1}{2} \Rightarrow l=\pm n$ and $\frac{ln}{l^2+n^2}=\frac{1}{2}$

Then $l=n \Rightarrow l-n=0$

19. Since PQ is perpendicular to L,

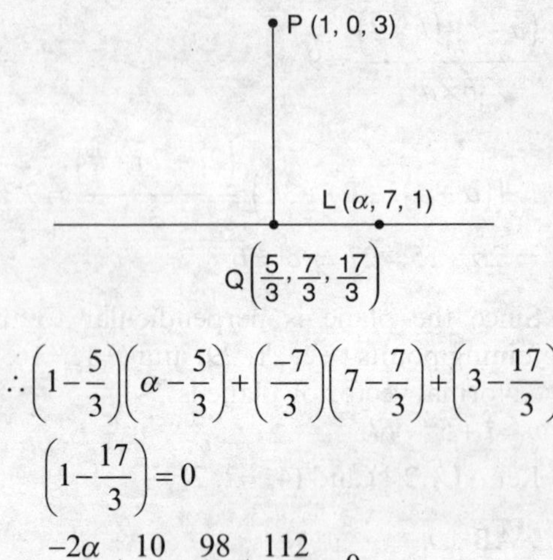

$$\therefore \left(1-\frac{5}{3}\right)\left(\alpha-\frac{5}{3}\right)+\left(\frac{-7}{3}\right)\left(7-\frac{7}{3}\right)+\left(3-\frac{17}{3}\right)$$

$$\left(1-\frac{17}{3}\right)=0$$

$$\Rightarrow \frac{-2\alpha}{3}+\frac{10}{9}-\frac{98}{9}+\frac{112}{9}=0$$

$$\Rightarrow \frac{2\alpha}{3}=\frac{24}{9} \Rightarrow \alpha=4$$

20. Equation of plane P is

$(x+4y-z+7)+\lambda(3x+y+5z-8)=0$

$\Rightarrow x(1+3\lambda)+y(4+\lambda)+z(-1+5\lambda)+(7-8\lambda)=0$

$$\Rightarrow \frac{1+3\lambda}{a}=\frac{4+\lambda}{b}=\frac{5\lambda-1}{6}=\frac{7-8\lambda}{-15}$$

From the last two ratios, $\lambda=-1$

$$\Rightarrow \frac{-2}{a}=\frac{3}{b}=-1$$

$\therefore a=2, b=-3$

$\therefore$The equation of plane is, $2x-3y+6z-15=0$

$$\text{Distance}=\frac{|6-6-6-15|}{7}=\frac{21}{7}=3$$

21. Normal of plane $=\begin{vmatrix}\hat{i} & \hat{j} & \hat{k}\\ 1 & 1 & 0\\ 0 & 0 & -1\end{vmatrix}$

$\vec{n}=-\hat{i}+\hat{j}+\hat{k}$

Direction ratios of normal to the plane = <−1, 1, 1>

Equation of plane

$-1(x-1)+1(y-0)+1(z-0)=0$

$\Rightarrow x-y-z-1=0$

If (x, y, z) is the foot of the perpendicular of M (1, 0, 1) on the plane, then

$$\Rightarrow \frac{x-1}{1}=\frac{y-0}{-1}=\frac{z-1}{-1}=\frac{-(1-0-1-1)}{3}$$

$$\therefore x=\frac{4}{3}, y=-\frac{1}{3}, z=\frac{2}{3}$$

$$\alpha+\beta+\gamma=\frac{4}{3}-\frac{1}{3}+\frac{2}{3}=\frac{5}{3}$$

$$\therefore 3(\alpha+\beta+\gamma)=3\times\frac{5}{3}=5$$

22. Since the line $\frac{x+1}{2}=\frac{y-3}{4}=\frac{z+1}{3}$ contains the point (−1, 3, −1) and line $\frac{x+3}{2}=\frac{y+2}{6}=\frac{z-1}{\lambda}$ contains the point (−3, −2, 1), the distance between the plane $23x - 10y - 2z + 48 = 0$ and the plane containing the lines equals the perpendicular distance of plane.
$23x - 10y - 2z + 48 = 0$ either from (−1, 3, −1) or (−3, −2, 1).

$$=\left|\frac{23(-1)-10(3)-2(-1)}{\sqrt{(23)^2+(10)^2+(-2)^2}}\right|=\frac{3}{\sqrt{633}}$$

It is given that distance between the planes

$$=\frac{k}{\sqrt{633}}\Rightarrow\frac{k}{\sqrt{633}}=\frac{3}{\sqrt{633}}\Rightarrow k=3$$

Assertion and Reason

1. (a) S_1: (1, 2, −1) is a point on the line and 11 +3 − 14 = 0
$\therefore$ The point lies on the plane $11x - 3z - 14 = 0$
Further, $(3 \times 11) + (11(-3)) = 0$
$\therefore$ The line lies in the plane
S_2: Obviously true

2. (a) S_1: $\mathrm{PA}\cdot\mathrm{PB} = 9 > 0$
$\therefore$P is exterior to the sphere.
S_2 is true (standard result)

3. (d) S_2: $\vec{r}\times(\hat{i}+2\hat{j}-3\hat{k})=\begin{vmatrix}\hat{i}&\hat{j}&\hat{k}\\x&y&z\\1&2&-3\end{vmatrix}$

$=\hat{i}(-3y-2z)-\hat{j}(-3x-z)+\hat{k}(2x-y)$
$\therefore -3y - 2z = 2,\ 3x + z = -1,\ 2x - y = 0$
i.e., $-6x - 2z = 2,\ 3x + z = -1$
$\therefore$ Straight line $2x - y = 0,\ 3x + z = -1$

S_1: $\vec{r}\times(2\hat{i}-\hat{j}+3\hat{k})=\begin{vmatrix}\hat{i}&\hat{j}&k\\x&y&z\\2&-1&3\end{vmatrix}$

$=\hat{i}(3y+z)-\hat{j}(3x-2z)+\hat{k}(-x-2y)$
$\therefore 3y + z = 3,\ 3x - 2z = 0,\ -x - 2y = 1$
$3x - 2(3 - 3y) = 0$
$\Rightarrow 3x + 6y = 6 \Rightarrow x + 2y = 2$
Now, $x + 2y = -1$, $x + 2y = 2$ are parallel planes.
$\therefore\ \vec{r}\times(2\hat{i}-j+3\hat{k})=3\hat{i}+\hat{k}$ is not a straight line.

4. (a) $\sin\theta=\left|\frac{2-3+2}{\sqrt{4+9+4\sqrt{3}}}=\frac{1}{\sqrt{51}}\right|$

$\therefore$ Statement 1 is true, Statement 2 is true by definition

5. (b) **Statement 1**
$3y - 4z = 5 - 2k$
$-2y + 4z = i - 3k$
$\therefore x=k, y=12-5k, z=\frac{31-13k}{4}$ is a point on the line for all real values of k.
The statement is true.
Statement 2
Direction ratios of the straight line are $< bc' - kbc, kac - ac', 0 >$
Direction ratios of normal to the plane are $< 0, 0, 1 >$
Now $0\times(bc'-kbc)+0\times(kac-ac')+1\times 0=0$
$\therefore$The straight line is parallel to the plane.
$\therefore$The statement is true but does not explain Statement 1.

JEE Advanced

Objective Questions

One Option Correct

1. (a) Foot of perpendicular from point $A(\vec{a})$ on the plane $\vec{r}.\vec{n}=d$ is $\vec{a}+\frac{(d-\vec{a}.\vec{n})}{|\vec{n}|^2}$

$\therefore$ Equation of line parallel to $\vec{r}=\vec{a}+\lambda\vec{b}$ in the plane $\vec{r}.\vec{n}=d$ is given by

$$\vec{r}=\vec{a}+\frac{(d-\vec{a}.\vec{n})}{|\vec{n}|^2}\vec{n}+\lambda\vec{b}$$

2. (c) Direction ratios of OP are (a, b, c)
$\therefore$ Equation of the plane is
i.e., $xa + yb + zc = a^2 + b^2 + c^2$

3. (b) Let direction ratios of the line be (a, b, c), then
$2a - b + c = 0$
$a - b - 2c = 0$
i.e., $\frac{a}{3}=\frac{b}{5}=\frac{c}{-1}$
$\therefore$ Direction ratios of the line are (3, 5, −1)
Any point on the line is $(2+\lambda, 2-\lambda, 3-2\lambda)$.
It lies on the plane π if
$2(2+\lambda)-(2-\lambda)+(3-2\lambda)=4$
i.e., $4+2\lambda-2+\lambda+3-2\lambda=4$
i.e., $\lambda=-1$
$\therefore$ The point of intersection of the line and the plane is (1, 3, 5)
$\therefore$ Equation of the required line is
$\frac{x-1}{3}=\frac{y-3}{5}=\frac{z-5}{-1}$

4. (b) Let a point $(3\lambda+1, \lambda+2, 2\lambda+3)$ of the first line also lie on the second line.

Then $\frac{3\lambda+1-3}{1} = \frac{\lambda+2-1}{2} = \frac{2\lambda+3-2}{3} \Rightarrow \lambda = 1$

Hence the point of intersection P of the two lines is (4, 3, 5)

The equation of the plane perpendicular to OP where 0 is (0, 0, 0) and passing through P is $4x + 3y + 5z = 50$

5. (b) $1 = \left|(\vec{b}-\vec{a}).\frac{(\vec{p}\times\vec{q})}{|\vec{p}\times\vec{q}|}\right|$

$\Rightarrow |\vec{b}-\vec{a}|\cos 60^\circ = \frac{1}{2}\text{AB} \Rightarrow \text{AB} = 2$

6. (a) Centre and radius of the sphere are $(-1, 1, -1)$ and 3 respectively.

Distance of $(-1, 1, -1)$ from the plane is $\left|\frac{-2+3-6+k}{\sqrt{4+9+36}}\right|$

Since the plane is tangent to the sphere

$\therefore \left|\frac{k-5}{7}\right| = 3$ is $|k-5| = 21$

$\therefore k = -16, 26$

7. (d) $P_1 = P_2 = 0, P_2 = P_3 = 0$ and $P_3 = P_1 = 0$ are lines of intersection of the three planes P_1, P_2 and P_3.

As $\bar{n}_1, \bar{n}_2$ and $\bar{n}_3$ are non-coplanar, planes P_1, P_2 and P_3 will intersect at a unique point. So the given lines will pass through a fixed point.

8. (a) Let θ be the required angle, then θ will be the angle between $\vec{a}$ and $\vec{b}+\vec{c}$ ($\vec{b}+\vec{c}$ lies along the angular bisector $\vec{a}$ and $\vec{b}$.)

$\cos\theta = \frac{\vec{a}.(\vec{b}+\vec{c})}{|\vec{a}||\vec{b}+\vec{c}|}$

$= \frac{2\cos\alpha}{\sqrt{2+2\cos\alpha}} = \frac{\cos\alpha}{\cos\frac{\alpha}{2}}$

$\theta = \cos^{-1}\left(\frac{\cos\alpha}{\cos\alpha/2}\right)$

9. (b) Let reflection of P (1, 0, 0) on the line $\frac{x-1}{2} = \frac{y+1}{-3} = \frac{z+10}{8}$ be (α, β, γ)

then $\left(\frac{\alpha+1}{2}, \frac{\beta}{2}, \frac{\gamma}{2}\right)$ lies on the line.

and $(\alpha-1)\hat{i} + \beta\hat{j} + \gamma\hat{k}$ is perpendicular to $2\hat{i} - 3\hat{j} + 8\hat{k}$.

$\therefore \frac{\frac{\alpha+1}{2}}{2} = \frac{\frac{\beta}{2}+1}{-3} = \frac{\frac{\gamma}{2}+10}{8} = \lambda$

and $2(\alpha-1) - 3(\beta) + \gamma(8) = 0$

$\Rightarrow \alpha = 5, \beta = -8, \gamma = -4$

10. (a) A(1, 1, 1), B(2, 3, 5), C(−1, 0, 2) directions ratios of AB are <1, 2, 4>

Direction ratios of AC are <−2, −1, 1>

∴ Direction ratios of normal to plane ABC are <2, − 3, 1>

∴ Equation of the plane ABC is $2x - 3y + z = 0$

Let the equation of the required plane be $2x - 3y + z = k$, then $\left|\frac{k}{\sqrt{4+9+1}}\right| = 2$

$k = \pm 2\sqrt{14}$

∴Equation of the required plane is $2x - 3y + z + 2\sqrt{14} = 0$

11. (a) Here $\overrightarrow{AB} = x\hat{i} - y\hat{j}$

$\overrightarrow{AC} = x\hat{i} - z\hat{j}$

$\overrightarrow{AD} = (x-1)\hat{i} - \hat{j} - \hat{k}$

As these vectors are coplanar

$\Rightarrow \begin{vmatrix} x & -y & 0 \\ x & 0 & -z \\ x-1 & -1 & -1 \end{vmatrix} = 0 \Rightarrow \frac{1}{x} + \frac{1}{y} + \frac{1}{z} = 1$

12. (a) Direction ratios of normal to the plane is (1, 2, 3)

⇒Equation of plane $(x - 1)1 + (y - 2).2 + (z - 3).3 = 0$

$\Rightarrow x + 2y + 3z = 14$

13. (a) S_1: True by definition.

S_2: False (because by the given condition, at least one point must lie on the plane.)

S3: True (standard result)

S4: True, shortest distance $= \left|\frac{-11-3}{\sqrt{9+36+4}}\right| = 2$

14. (b) S_1: Since

$\begin{vmatrix} 1-1 & 7+2 & -4-3 \\ 2 & 5 & 7 \\ 1 & 2 & 3 \end{vmatrix} = \begin{vmatrix} 0 & 9 & -7 \\ 2 & 5 & 7 \\ 1 & 2 & 3 \end{vmatrix} = 16 \neq 0$

S_2: By the given condition

$\frac{4+2}{3} = \frac{13-3}{5} = \frac{1-5}{2}$ i.e. $2 = 2 = -2$

which is not true

S_3: Let $S \equiv x^2 + y^2 + z^2 - 2x - 4y - 2z + 2 = 0$

Then $S_1 = 4 + 1 + 1 - 4 - 4 - 2 + 2 = -2 < 0$

$\therefore$ The statement is false.

S_4: Let $<a, b, c>$ be direction ratios of the line, then

$a + b + c = 0$

$4a + b - 2c = 0$

i.e., $\frac{a}{-2-1} = \frac{b}{4+2} = \frac{c}{1-4}$

i.e., $\frac{a}{-3} = \frac{b}{6} = \frac{c}{-3}$ i.e. $\frac{a}{1} = \frac{b}{-2} = \frac{c}{1}$

$\therefore$ The statement is true.

One or More Than One Option Correct

1. (a, b, c) $x + y + z - 1 = 0$

$4x + y - 2z + 2 = 0$

$\therefore$ Direction ratios of the line are $<-3, 6, -3>$

i.e., $<1, -2, 1>$

Let $z = k$, then $x = k - 1, y = 2 - 2k$

i.e. $(k - 1, 2 - 2k, k)$ is any point on the line.

$\therefore$ $(-1, 2, 0)$, $(0, 0, 1)$ and $\left(-\frac{1}{2}, 1, \frac{1}{2}\right)$ are points on the line

$\therefore$ (a), (b) and (c) are correct options

2. (a, b) $3x - 6y + 2z + 5 = 0$ ...(i)

$-4x + 12y - 3z + 3 = 0$ ...(ii)

$\frac{3x-6y+2z+5}{\sqrt{9+36+4}} = \frac{-4x+12y-3z+3}{\sqrt{16+144+9}}$

bisects the angle between the planes that contains the origin.

$13(3x - 6y + 2z + 5) = 7(-4x + 12y - 3z + 3)$

$39x - 78y + 26z + 65 = -28x + 84y - 21z + 21$

$67x - 162y + 47z + 44 = 0$ ...(iii)

Further $3 \times (-4) + (-6)(12) + 2 \times (-3) < 0$

$\therefore$ Origin lies in acute angle

3. (a, b) Equation of the required plane is

$\ell x + my + \lambda z = 0$ ...(i)

Angle between (i) and $\ell x + my = 0$ is α.

$\Rightarrow \cos\alpha = \frac{\ell^2 + m^2}{\sqrt{\ell^2 + m^2}\sqrt{\ell^2 + m^2 + \lambda^2}}$

$\Rightarrow \cos^2\alpha = \frac{\ell^2 + m^2}{\ell^2 + m^2 + \lambda^2} \Rightarrow \lambda = \pm\sqrt{\ell^2 + m^2}\tan\alpha$

Hence equation of the plane is

$\ell x + my \pm z\sqrt{\ell^2 + m^2}\tan\alpha = 0$

4. (b, c) Given that the lines are coplanar,

$\therefore \begin{vmatrix} x_2 - x_1 & y_2 - y_1 & z_2 - z_1 \\ a_1 & b_1 & c_1 \\ a_2 & b_2 & c_2 \end{vmatrix} = \begin{vmatrix} 2 & 0 & 0 \\ 2 & k & 2 \\ 5 & 2 & k \end{vmatrix} = 0$

$\Rightarrow k = \pm 2$

For $k = 2$, equation of the plane is given by

$\begin{vmatrix} x-1 & y+1 & z \\ 2 & 2 & 2 \\ 5 & 2 & 2 \end{vmatrix} = 0 \Rightarrow y - z + 1 = 0$

For $k = -2$, the equation of the plane is given by

$\begin{vmatrix} x-1 & y+1 & z \\ 2 & -2 & 2 \\ 5 & 2 & -2 \end{vmatrix} = 0 \Rightarrow y + z + 1 = 0$

5. (a, b) The point of intersection of L_1 and L_2 is $(1, 0, 1)$

$\because$ Line L passes through the point of intersection $(1, 0, 1)$ of L_1 and L_2

$\therefore \frac{1-\alpha}{\ell} = \frac{1}{m} = \frac{1-\gamma}{-2}$ ...(i)

$\because$Line L bisects the acute angle between the lines L_1 and L_2, then

$\vec{r} = \hat{i} + \hat{k} + \lambda\left(\frac{\hat{i} - \hat{j} + 3\hat{k} - 3\hat{i} - \hat{j} + \hat{k}}{\sqrt{11}}\right)$

$\Rightarrow \vec{r} = \hat{i} + \hat{k} + t\left(\hat{i} + \hat{j} - 2\hat{k}\right)$

$\Rightarrow \frac{\ell}{1} = \frac{m}{1} = \frac{-2}{-2} \Rightarrow \ell = m = 1$

From (i), $\frac{1-\alpha}{1} = -1 \Rightarrow \alpha = 2$

and $\frac{1-\gamma}{-2} = -1 \Rightarrow \gamma = -1$

$\therefore \alpha - \gamma = 2 - (1) = 3$ and $l + m = 1 + 1 = 2$

6. (a, b, c)

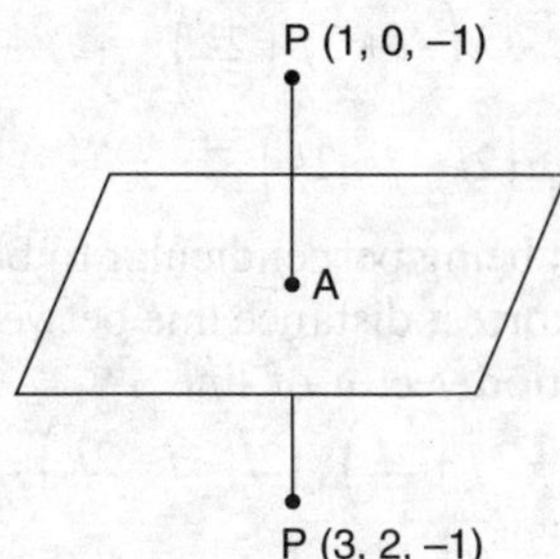

Mid-point of PQ = A (2, 1, −1)

Direction ratios of PQ = 2, 2, 0

Since PQ perpendicular to the plane and the mid-point lies on the plane,

$\therefore$ Equation of the plane:

$2(x-2) + 2(y-1) + 0(z+1) = 0$

$\Rightarrow x - 2 + y - 1 = 0$

$\Rightarrow x + y = 3$

Comparing with $\alpha x + \beta y + \gamma z = \delta$, we get $\alpha = 1, \beta = 1, \gamma = 0$ and $\delta = 3$

$\therefore$ Options (a), (b), (c) are true.

7. (a, d) Let any point $P(\lambda,0,0)$ on $L_1, Q(0,\mu,1)$ on L_2 and R (1, 1, v) on L_3

$\because$ P, Q, R are collinear, $\therefore \overrightarrow{PQ} \parallel \overrightarrow{PR}$

$$\Rightarrow \frac{\lambda}{\lambda-1} = \frac{-\mu}{-1} = \frac{-1}{-v}$$

$$\Rightarrow \mu = \frac{\lambda}{\lambda-1}, v = \frac{\lambda-1}{\lambda}$$

It is clear from the above that $\lambda \neq 0,1$

$$\therefore Q\left(0, \frac{\lambda}{\lambda-1}, 1\right)$$

(a) For $Q = \hat{k} - \frac{1}{2}\hat{j}$

$\frac{\lambda}{\lambda-1} = -\frac{1}{2} \Rightarrow 3\lambda = +1$, which is possible.

(b) For $Q = \hat{k}$

$\frac{\lambda}{\lambda-1} = 0 \Rightarrow \lambda = 0$, not possible

(c) For $Q = \hat{k} + \hat{j}$

$\frac{\lambda}{\lambda-1} = 1 \Rightarrow \lambda = \lambda - 1$, not possible

(d) For $Q = \hat{k} + \frac{1}{2}\hat{j}$

$\frac{\lambda}{\lambda-1} = \frac{1}{2} \Rightarrow 2\lambda = \lambda - 1 \Rightarrow \lambda = -1$, which is possible

Hence options (a) and (d) are correct and options (b) and (c) are incorrect.

8. (a, b, d)

$$L_1 : \vec{r} = \hat{i} + \lambda\left(-i + 2j + 2\hat{k}\right)$$

$$L_2 : \vec{r} = \mu\left(2i - j + 2\hat{k}\right)$$

Since L_3 being perpendicular to both L_1 and L_2, is the shortest distance line between L_1 and L_2

$\therefore$ Direction vector of line

$$L_3 : \left(-\hat{i} + 2\hat{j} + 2\hat{k}\right) \times \left(2\hat{i} - \hat{j} + 2\hat{k}\right)$$

$$= \begin{vmatrix} \hat{i} & \hat{j} & \hat{k} \\ -1 & 2 & 2 \\ 2 & -1 & 2 \end{vmatrix} = 6\hat{i} + 6\hat{j} - 3\hat{k}$$

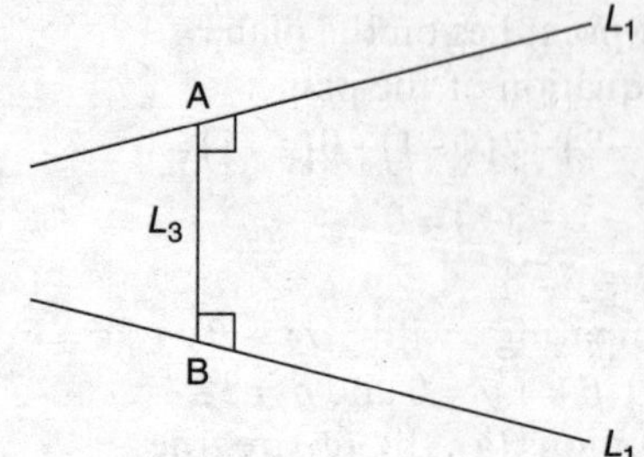

$\because L_1$ and L_2 are skew lines

Let any point on L_1 and L_2 be $A(1-\lambda, 2\lambda, 2\lambda)$ and $(2\mu, -\mu, 2\mu)$

$\therefore$ Direction ratios of AB =

$2\mu + \lambda - 1, -\mu - 2\lambda, 2\mu - 2\lambda$

$\because$AB and L_3 are representing the same line

$$\therefore \frac{2\mu+\lambda-1}{6} = \frac{-\mu-2\lambda}{6} = \frac{2\mu-2\lambda}{-3}$$

$$\Rightarrow 3\lambda + 3\mu = 1 \quad \text{...(i)}$$

$$6\lambda - 3\mu = 0 \quad \text{...(ii)}$$

Solving (i) and (ii) we get: $\lambda = \frac{1}{9}, \mu = \frac{2}{9}$

$$\therefore A\left(\frac{8}{9}, \frac{2}{9}, \frac{2}{9}\right) \text{and } B\left(\frac{4}{9}, \frac{-2}{9}, \frac{4}{9}\right)$$

$\therefore$ Equation of L_3 is given by

$$\vec{r} = \frac{2}{9}\left(4\hat{i} + \hat{j} + \hat{k}\right) + t\left(2i + 2j - \hat{k}\right)$$

$\therefore$ (a) is correct

$$\text{or } \vec{r} = \frac{2}{9}\left(2\hat{i} - \hat{j} + 2\hat{k}\right) + t\left(2i + 2j - \hat{k}\right)$$

$\therefore$ (b) is correct

Also, mid-point of AB is $\left(\frac{2}{3}, 0, \frac{1}{3}\right)$

$\therefore L_3$ can also be written as

$\vec{r} = \frac{1}{3}\left(2\hat{i} + \hat{k}\right) + t\left(2\hat{i} + 2\hat{j} - \hat{k}\right)$, where $t \in R$

$\therefore$ (d) is correct.

Clearly (0, 0, 0) does not lie on

$$\vec{r} = \frac{2}{9}\left(4\hat{i} + \hat{j} + \hat{k}\right) + t\left(2\hat{i} + 2\hat{j} - \hat{k}\right)$$

$\therefore \vec{r} = t(2\hat{i} + 2\hat{j} - \hat{k})$ cannot describe the line L_3.

$\therefore$ (c) is incorrect.

9. (c, d) (a) Direction vector of the line of intersection of two planes is given by $\vec{n_1} \times \vec{n_2}$.

$$\therefore \vec{n_1} \times \vec{n_2} = \begin{vmatrix} \hat{i} & \hat{j} & \hat{k} \\ 2 & 1 & -1 \\ 1 & 2 & 1 \end{vmatrix} = 3\hat{i} - 3\hat{j} + 3\hat{k}$$

$\therefore$ DR's of line of intersection of P_1 and P_2 are 1, −1, 1

$\therefore$ (a) is not correct.

(b) The standard form of the given line is

$$\frac{x - \frac{4}{3}}{3} = \frac{y - \frac{1}{3}}{-3} = \frac{z}{3}$$

$$\because 1 \times 3 + (-1)(-3) + 1(3) = 9 \neq 0$$

$\therefore$ This line is not perpendicular to the line of intersection.

$\therefore$ (b) is not correct.

(c) Let θ be the angle between P_1 and P_2. Then

$$\cos\theta = \left|\frac{2\times1+1\times2+(-1)\times1}{\sqrt{6}\sqrt{6}}\right| = \frac{3}{6} = \frac{1}{2}$$

$\therefore \theta = 60^\circ$

Hence (c) is correct.

(d) Equation of plane P_3:

$1(x-4)-1(y-2)+1(z+2)=0$

$\Rightarrow x-y+z=0$

Distance of (2, 1, 1) from $P_3 = \dfrac{2-1+1}{\sqrt{1+1+1}} = \dfrac{2}{\sqrt{3}}$

$\therefore$ (d) is correct.

10. (b, c, d) According to the question, the co-ordinates of vertices of pyramid OPQRS will be

$O(0,0,0), P(3,0,0), Q(3,3,0), R(0,3,0), S\left(\frac{3}{2},\frac{3}{2},3\right)$

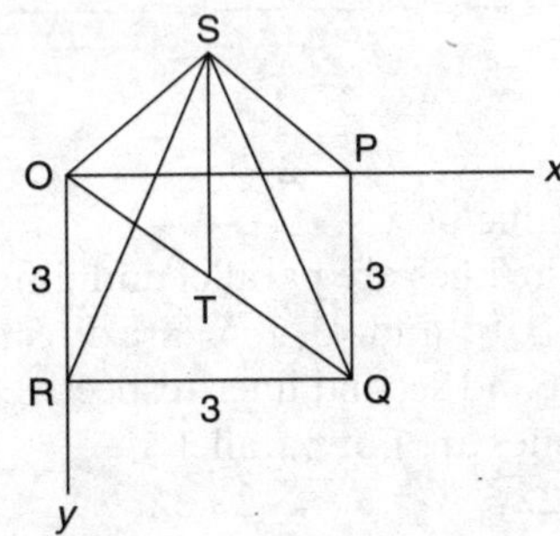

DR's of OQ = 1, 1, 0

DR's of OS = 1, 1, 2

$\therefore$ Acute angle between OQ and OS

$$= \cos^{-1}\left(\frac{2}{\sqrt{2}\times\sqrt{6}}\right) = \cos^{-1}\left(\frac{1}{\sqrt{3}}\right) \neq \frac{\pi}{3}$$

$\therefore$ (a) is not correct.

Equation of plane OQS $= \begin{vmatrix} x & y & z \\ 1 & 1 & 0 \\ 1 & 1 & 2 \end{vmatrix} = 0$

$\Rightarrow 2x-2y=0$ or $x-y=0$

$\therefore$ (b) is correct.

The length of perpendicular from P (3, 0, 0) to the plane $x-y=0$ is $\left|\dfrac{3-0}{\sqrt{2}}\right| = \dfrac{3}{\sqrt{2}}$

$\therefore$ (c) is correct.

Equation of

RS: $\dfrac{x}{\frac{3}{2}} = \dfrac{y-3}{\frac{-3}{2}} = \dfrac{z}{3}$ or $\dfrac{x}{1} = \dfrac{y-3}{-1} = \dfrac{z}{2} = \lambda$

$\therefore$ Any point ON RS is N $(\lambda, -\lambda+3, 2\lambda)$

Since ON is perpendicular to RS,

$\because$ ON $\perp$ RS $\Rightarrow 1\times\lambda - 1(-\lambda+3) + 2\times2\lambda = 0$

$\Rightarrow \lambda = \dfrac{1}{2} \Rightarrow N\left(\dfrac{1}{2},\dfrac{5}{2},1\right)$

$\therefore ON = \sqrt{\dfrac{1}{4}+\dfrac{25}{4}+1} = \sqrt{\dfrac{15}{2}}$

$\therefore$ (d) is correct

11. (a, b) All the points on L are at a constant distance from P_1 and P_2. That means L is parallel to both P_1 and P_2.

$$\vec{b} = \begin{vmatrix} \hat{i} & \hat{j} & \hat{k} \\ 1 & 2 & -1 \\ 2 & -1 & 1 \end{vmatrix} = \hat{i} - 3\hat{j} - 5\hat{k}$$

$\therefore L: \dfrac{x}{1} = \dfrac{y}{-3} = \dfrac{z}{-5} = \lambda\text{(say)}$

$\therefore$ Any point on line L is $(\lambda, -3\lambda, -5\lambda)$

Equation of a line perpendicular to P_1 drawn from any point on L is

$$\frac{x-\lambda}{1} = \frac{y+3\lambda}{2} = \frac{z+5\lambda}{-1} = \mu$$

$\therefore M(\mu+\lambda, 2\mu-3\lambda, -\mu-5\lambda)$

But M lies on P_1, so it satisfies the equation of P_1.

$\therefore \mu+\lambda+4\mu-6\lambda+\mu+5\lambda+1=0 \Rightarrow \mu = \dfrac{-1}{6}$

$\therefore M\left(\lambda-\dfrac{1}{6}, -3\lambda\dfrac{-1}{3}, -5\lambda+\dfrac{1}{6}\right)$

For locus of M;

$x = \lambda - \dfrac{1}{6}, y = -3\lambda - \dfrac{1}{3}, z = 5\lambda + \dfrac{1}{6}$

$\Rightarrow \dfrac{x+1/6}{1} = \dfrac{y+1/3}{-3} = \dfrac{z-1/6}{-5} = \lambda$

On checking the given point, we find $\left(0, \dfrac{-5}{6}, \dfrac{-2}{3}\right)$ and $\left(\dfrac{-1}{6}, \dfrac{-1}{3}, \dfrac{1}{6}\right)$ satisfy the above equation.

12. (b,d) $P_3: (x+z-1)+\lambda y = 0 \Rightarrow x+\lambda y+z-1=0$

Distance of point (0, 1, 0) from P_3:

$$\left|\frac{\lambda-1}{\sqrt{2+\lambda^2}}\right| = 1 \Rightarrow \lambda^2-2\lambda+1 = \lambda^2+2 \Rightarrow \lambda = \frac{-1}{2}$$

Distance of point (α, β, γ) from P_3:

$$\left|\frac{\alpha+\lambda\beta+\gamma-1}{\sqrt{2+\lambda^2}}\right| = 2 \Rightarrow \frac{\alpha-\frac{1}{2}\beta+\gamma-1}{\frac{3}{2}} = \pm2$$

$$\Rightarrow \alpha - \frac{1}{2}\beta + \gamma - 1 = \pm3 \Rightarrow 2\alpha-\beta+2\gamma-2 = \pm6$$

$$\Rightarrow 2\alpha-\beta+2\gamma-8=0 \text{ or } 2\alpha-\beta+2\gamma+4=0$$

13. (c) Given that lines are $x = y, z = 1$.

$\Rightarrow L_1 = \frac{x-0}{1} = \frac{y-0}{1} = \frac{z-1}{0} = \alpha$ ...(i)

$\therefore Q(\alpha, \alpha, 1)$

and $y = -x, z = -1$

$\Rightarrow L_2 = \frac{x-0}{-1} = \frac{y-0}{1} = \frac{z+1}{0} = \beta$ ...(ii)

$\therefore R(-\beta, \beta, -1)$(say)

Direction ratios of PQ are $\lambda - \alpha, \lambda - \alpha, \lambda - 1$ and direction ratios of PR are $\lambda - \beta, \lambda - \beta, \lambda + 1$

$\because$ PQ is perpendicular to L_1

$\therefore \lambda - \alpha = 0 \Rightarrow \lambda = \alpha$ ...(iii)

$\because$ PR is perpendicular to L_2

$\therefore -(\lambda + \beta) + \lambda - \beta = 0 \Rightarrow \beta = 0$

$\therefore$ Direction ratios of PQ are $0, 0, \lambda - 1$

and direction ratios of PR are $\lambda, \lambda, \lambda + 1$

$\because \angle OPR = 90° \Rightarrow (\lambda - 1)(\lambda + 1) = 0 \Rightarrow \lambda = 1$ or -1

But for $\lambda = 1$, we get point Q itself.

$\therefore$We take $\lambda = -1$

14. (a, d) Given that L_1 and L_2 are coplanar, therefore

$$\begin{vmatrix} 5-\alpha & 0 & 0 \\ 0 & 3-\alpha & -2 \\ 0 & -1 & 2-\alpha \end{vmatrix} = 0$$

$\Rightarrow (5-\alpha)\left[6 - 5\alpha + \alpha^2 - 2\right] = 0$

$\Rightarrow (5-\alpha)(\alpha - 1)(\alpha - 4) = 0 \Rightarrow \alpha = 1, 4, 5$

15. (b, d) The given lines are

$\ell_1 : \left(3\hat{i} - \hat{j} + 4\hat{k}\right) + t\left(\hat{i} + 2\hat{j} + 2\hat{k}\right)$

$\ell_2 : \left(3\hat{i} + 3\hat{j} + 2\hat{k}\right) + s\left(2\hat{i} + 2\hat{j} + \hat{k}\right)$

Let the direction vector perpendicular to ℓ_1 and ℓ_2 be

$$\vec{b} = \ell_1 \times \ell_2 = \begin{vmatrix} \hat{i} & \hat{j} & \hat{k} \\ 1 & 2 & 2 \\ 2 & 2 & 1 \end{vmatrix} = -2\hat{i} + 3\hat{j} - 2\hat{k}$$

$\therefore \ell : \frac{x}{2} = \frac{y}{-3} = \frac{z}{2} = \lambda$

Any point on ℓ_1 is $(t+3, 2t-1, 2t+4)$ and any point on ℓ is $(2\lambda, -3\lambda, 2\lambda)$

Let the intersection point of ℓ and ℓ_1 be P.

$t + 3 = 2\lambda, 2t - 1 = -3\lambda, 2t + 4 = 2\lambda$

$\Rightarrow t = -1, \lambda = 1 \therefore \text{P}(2, -3, 2)$

Any point Q on $\ell_2$is $(2s+3, 2s+3, s+2)$

According to the question $PQ = \sqrt{17}$

$\Rightarrow (2s+1)^2 + (2s+6)^2 + s^2 = 17$

$\Rightarrow 9s^2 + 28r + 20 = 0 \Rightarrow s = -2, \frac{-10}{9}$

$\therefore$ Point Q can be $(-1, -1, 0)$ and $\left(\frac{7}{9}, \frac{7}{9}, \frac{8}{9}\right)$

Matrix Type

1. (A → S, T; B → P, T, C → Q; D → R)

(A) Both the lines pass through the point (7, 11, 15)

(B) <2, 3, 4> are direction ratios of both the lines. Also, the point (1, 2, 3) is common to both

$\therefore$ The lines are coincident.

(C) <5, 4 −2> are direction ratios of both the lines

$\therefore$ The lines are parallel.

Also,$x = 2 + 5\lambda, y = -3 + 4\lambda, z = 5 - 2\lambda$

$\therefore \frac{2 + 5\lambda - 7}{5} = \frac{-3 + 4\lambda - 1}{4} = \frac{5 - 2\lambda - 2}{-2}$

i.e. $\lambda - 1 = \lambda - 1 = \frac{3 - 2\lambda}{-2}$

$\therefore$ No value of λ.

Thus, the lines are parallel and different.

(D) <2, 3, 5> and <3, 2, 5> are direction ratios of the first and second lines respectively.

$\therefore$The lines are not parallel.

$x = 3 + 2\lambda, \quad y = -2 + 3\lambda, \quad z = 4 + 5\lambda$

$x = 3 + 3\mu, \quad y = -2 + 2\mu, \quad z = 7 + 5\mu$

are parametric equations of the lines.

Solving $3 + 2\lambda = 3 + 3\mu$ and $-2 + 3\lambda = 2 + 2\mu$,

we get $\lambda = \frac{12}{5}, \mu = \frac{8}{5}$

Now substituting these values in $4 + 5\lambda = 7 + 5\mu$, we get

4 + 12 = 7 + 8, i.e., 16 = 15, which is not true.

$\therefore$ The lines do not intersect.

Hence the lines are skew.

2. (A → S; B → P; C →T; D → Q)

(A) Let the foot of the perpendicular be $Q(2 + 2\lambda, 1 + 3\lambda, 2 + 4\lambda)$

$\therefore 2(2\lambda + 1) + 3(3\lambda - 1) + 4(4\lambda - 1) = 0$

$29\lambda = 5 \Rightarrow \lambda = \frac{5}{29}$

$\therefore$ Foot $= \left(\frac{68}{29}, \frac{44}{29}, \frac{78}{29}\right) \therefore$ (A) → (S)

(B) Let the image be the point (a, b, c), then from previous solution

$\frac{1+a}{2} = \frac{68}{29}, \frac{2+b}{2} = \frac{44}{29}$ and $\frac{3+c}{2} = \frac{78}{29}$

i.e $a=\dfrac{107}{29}, b=\dfrac{30}{29}$ and $c=\dfrac{68}{29}$ $\therefore$ (B) $\to$ (P)

(C) $\dfrac{x-2}{2}=\dfrac{y-3}{3}=\dfrac{z-5}{-4}=-\dfrac{4+9-20+17}{4+9+16}=\dfrac{-10}{29}$

$\therefore a=\dfrac{38}{29}, b=\dfrac{57}{29}$ and $c=\dfrac{185}{29}$

$\therefore$ (C) $\to$ (T)

(D) $\dfrac{x-2}{3}=\dfrac{y-5}{-2}=\dfrac{z-1}{4}=-2\left(\dfrac{6-10+4-5}{29}\right)=\dfrac{10}{29}$

$x=2+\dfrac{30}{29}=\dfrac{88}{29}, y=5-\dfrac{20}{29}=\dfrac{125}{29}, z=1+\dfrac{40}{29}=\dfrac{69}{29}$

$\therefore$ (D) $\to$ (Q)

3. (A $\to$ T; B $\to$ P, R; C $\to$ Q, S; D $\to$ R)

Let the line through origin be $L: \dfrac{x}{a}=\dfrac{y}{b}=\dfrac{z}{c}$ ...(i)

Since line L intersects

$L_1: \dfrac{x-2}{1}=\dfrac{y-1}{-2}=\dfrac{z+1}{1}$ ...(ii)

And $L_2: \dfrac{x-8/3}{2}=\dfrac{y+3}{-1}=\dfrac{z-1}{1}$ ...(iii)

At P and Q

$\therefore$ Lines L and L_1 are coplanar.

$\therefore$ Using $\begin{vmatrix} x_2-x_1 & y_2-y_1 & z_2-z_1 \\ a_1 & b_1 & c_1 \\ a_2 & b_2 & c_2 \end{vmatrix}=0$

We get $\begin{vmatrix} 2 & 1 & -1 \\ a & b & c \\ 1 & -2 & 1 \end{vmatrix}=0 \Rightarrow a+3b+5c=0$ ...(iv)

Also, L and L_2 are coplanar

and $\begin{vmatrix} 8/3 & -3 & 1 \\ a & b & c \\ 2 & -1 & 1 \end{vmatrix}=0 \Rightarrow 3a+b-5c=0$ ...(v)

On solving (iv) and (v), we get

$\dfrac{a}{-15-5}=\dfrac{b}{15+5}=\dfrac{c}{1-9}$ or $\dfrac{a}{5}=\dfrac{b}{-5}=\dfrac{c}{2}$

Hence equation (i) becomes $\dfrac{x}{5}=\dfrac{y}{-5}=\dfrac{z}{2}=\lambda$

Any point on L, $P(5\lambda,-5\lambda,2\lambda)$ which also lies on (ii).

$\therefore \dfrac{5\lambda-8/3}{3}=\dfrac{-5\lambda+3}{-1}=\dfrac{2\lambda-1}{1} \Rightarrow \lambda=2/3$

$\therefore Q\left(\dfrac{10}{3},\dfrac{-10}{3},\dfrac{4}{3}\right)$

Hence $d^2=PQ^2=\left(\dfrac{25}{9}+\dfrac{25}{9}+\dfrac{4}{9}\right)=6$

(B) $\tan^{-1}(x+3)-\tan^{-1}(x-3)=\tan^{-1}\dfrac{3}{4}$

$\left[\because \sin^{-1}\dfrac{3}{5}=\tan^{-1}\dfrac{3}{4}\right]$

$\Rightarrow \tan^{-1}\left(\dfrac{x+3-x+3}{1+x^2-9}\right)=\tan^{-1}\left(\dfrac{3}{4}\right), x^2-9\geq -1$

$\Rightarrow \dfrac{6}{x^2-8}=\dfrac{3}{4} \Rightarrow x^2=16$ or $x=4,-4$

(C) $\because \vec{a}=\mu\vec{b}+4\vec{c} \Rightarrow \vec{c}=\dfrac{\vec{a}-\mu\vec{b}}{4}$

Then $(\vec{b}-\vec{a}).(\vec{b}+\vec{c})=0$

$\Rightarrow (\vec{b}-\vec{a}).\left(\vec{b}+\dfrac{\vec{a}-\mu\vec{b}}{4}\right)=0$

$\Rightarrow (\vec{b}-\vec{a}).\left(\dfrac{4-\mu}{4}\vec{b}+\dfrac{\vec{a}}{4}\right)=0$

$\Rightarrow \dfrac{4-\mu}{4}|\vec{b}|^2-\dfrac{|\vec{a}|^2}{4}=0$

$\Rightarrow (4-\mu)|\vec{b}|^2-|\vec{a}|^2=0$ ...(i)

Also,

$2|\vec{b}+\vec{c}|=|\vec{b}-\vec{a}| \Rightarrow 2^2\left|\dfrac{4-\mu}{4}\vec{b}+\dfrac{\vec{a}}{4}\right|^2=|\vec{b}-\vec{a}|^2$

$\Rightarrow (4-\mu)^2|\vec{b}|^2+|\vec{a}|^2=4|\vec{b}|^2+4|\vec{a}|^2 \left[\because \vec{a}\vec{b}=0\right]$

$\Rightarrow \left[(4-\mu)^2-4\right]|\vec{b}|^2=3|\vec{a}|^2$ ...(ii)

From (i) and (ii), we get

$\dfrac{(4-\mu)^2-4}{4-\mu}=\dfrac{3}{1}$

$\Rightarrow \mu^2-8\mu+12=12-3\mu$

$\Rightarrow \mu^2-5\mu=0$

$\Rightarrow \mu=0$ or 5

(D) $I=\dfrac{2}{\pi}\displaystyle\int_{-\pi}^{\pi}\dfrac{\sin(9x/2)}{\sin(x/2)}dx=\dfrac{2}{\pi}\times 2\int_0^{\pi}\dfrac{\sin 9x/2}{\sin x/2}dx$

$\left[\because f(x) \text{ is an even function}\right]$

Let $\dfrac{x}{2}=\theta \Rightarrow dx=2d\theta$

Also, at $x=0, \theta=0$

and at $x=\pi, \theta=\pi/2$

$\therefore I=\dfrac{8}{\pi}\displaystyle\int_0^{\pi/2}\dfrac{\sin 9\theta}{\sin\theta}d\theta$

$$I=\frac{8}{\pi}\int_0^{\pi/2}\left[\frac{\sin 9\theta-\sin 7\theta}{\sin\theta}+\frac{(\sin 7\theta-\sin 5\theta)}{\sin\theta}+\frac{(\sin 5\theta-\sin 3\theta)}{\sin\theta}+\frac{(\sin 3\theta-\sin\theta)}{\sin\theta}+\frac{\sin\theta}{\sin\theta}\right]d\theta$$

$$=\frac{16}{\pi}\int_0^{\pi/2}(\cos 8\theta+\cos 6\theta+\cos 4\theta+\cos 2\theta)\,d\theta$$

$$+\frac{8}{\pi}\int_0^{\pi/2}d\theta$$

$$=\frac{16}{\pi}\left[\frac{\sin 8\theta}{8}+\frac{\sin 6\theta}{6}+\frac{\sin 4\theta}{4}+\frac{\sin 2\theta}{2}\right]_0^{\pi/2}+\frac{8}{\pi}(\theta)_0^{\pi/2}$$

$$=0+\frac{8}{\pi}\left(\frac{\pi}{2}-0\right)=4$$

4. (A $\to$ P; B $\to$ Q, S; C $\to$ Q, R, S, T; D $\to$ R)

(A) Let us consider two functions

$y=xe^{\sin x}$ and $y=\cos x$

The range of $y=xe^{\sin x}$ is $\left(0,\frac{\pi e}{2}\right)$ and

$\frac{dy}{dx}=e^{\sin x}+xe^{\sin x}\cos x\geq 0$, for $x\leftarrow\left(0,\frac{\pi}{2}\right)$; so,

it is an increasing function on $\left(0,\frac{\pi}{2}\right)$. Their graph are as shown in the figure below:

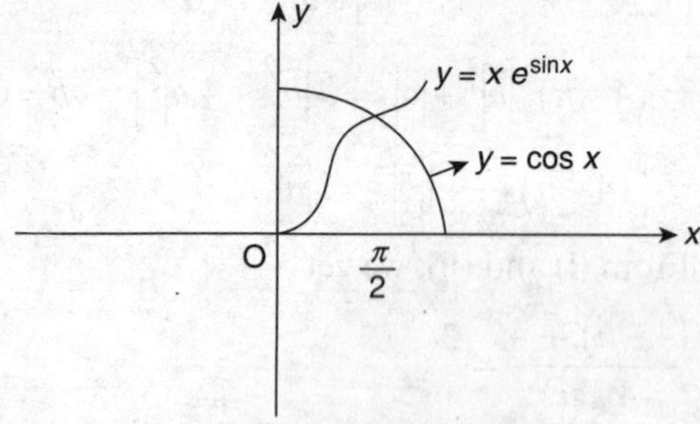

Clearly the two curves meet only at one point. Therefore the given equation has only one solution in $\left(0,\frac{\pi}{2}\right)$.

(B) Since given planes intersect in a staright line

$$\therefore\begin{vmatrix}k & 4 & 1\\ 4 & k & 2\\ 2 & 2 & 1\end{vmatrix}=0$$

$$\Rightarrow k(k-4)-4(4-4)+1(8-2k)=0$$

$$\Rightarrow k^2-6k+8=0\Rightarrow(k-2)(k-4)=0$$

$$\Rightarrow k=2\text{ or }4$$

(C) We have $f(x)=|x-1|+|x-2|+|x+1|+|x+2|$

$$=\begin{cases}-4x, & x\leq -2\\ -2x+4, & -2<x\leq -1\\ 6, & -1<x\leq 1\\ 2x+4, & 1<x\leq 2\\ 4x, & x\geq 2\end{cases}\quad\left[\because |x-1|\begin{cases}x-1, \text{ is } x\geq 1\\ -(x-1) \text{ is } x<1\end{cases}\right]$$

The graph of the above function is as given below:

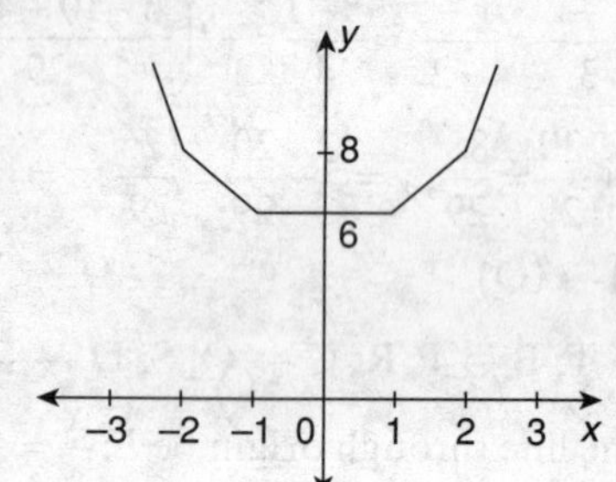

Clearly, from graph, $f(x)\geq 6$

$\Rightarrow 4k\geq 6\Rightarrow k\geq\frac{3}{2}$

$\therefore k=2,3,4,5,6,\ldots\ldots$

(D) $\int\frac{dy}{y+1}=\int dx\Rightarrow \ln|y+1|=x+c$

At $x=0, y=1\Rightarrow c=\ln 2$

$\therefore \ln|y+1|=x+l\,n2\Rightarrow y+1=2e^x\Rightarrow y=2e^x-1$

$\therefore y(\ln 2)=2e^{\ln 2}-1=2\times 2-1=3$

5. (a) Let any point on L_1 be $(2\lambda+1,-\lambda,\lambda-3)$ and that on L_2 be $(\mu+4,\mu-3,2\mu-3)$

For point of intersection of L_1 and L_2,

$2\lambda+1=\mu+4, -\lambda=\mu-3, \lambda-3=2\mu-3$

$\Rightarrow\lambda=,\mu=1$

$\therefore$ Intersection point of L_1 and L_2 is (5, −2, −1)

Equation of the line passing through, (5, −2, −1) and perpendicular to P_1 and P_2 is given by

$$\begin{vmatrix}x-5 & y+2 & z+1\\ 7 & 1 & 2\\ 3 & 5 & -6\end{vmatrix}=0$$

$\Rightarrow x-3y-2z=13$

$\therefore a=1, b=-3, c=-2, d=13$

or $(P)\to(3)(Q)\to(2)(R)\to(4)(S)\to(1)$

Integer and Numerical Value Type

1. Let the equation of a plane containing the line be $\ell(x-1)+m(y+2)+nz=0$.

Then $2\ell-3m+5m=0$ and $\ell-m+n=0$

$\therefore\frac{\ell}{2}=\frac{m}{3}=\frac{n}{1}$

$\therefore$ The plane is $2(x-1)+3(y+2)+z=0$
i.e., $2x+3y+z+4=0$
$\therefore a=2, b=-3, c=1$

2. $\vec{r}=3\vec{i}+8\vec{j}+3\vec{k}+\lambda\left(3\vec{i}-\vec{j}+\vec{k}\right)$ ...(i)

$\vec{r}=3\vec{i}-7\vec{j}+6\vec{k}+\mu\left(-3\vec{i}+2\vec{j}+4\vec{k}\right)$ ...(ii)

Let L and M be points on the lines (i) and (ii) respectively so that LM is perpendicular to both the lines.

Let position vector of L be $3\vec{i}+8\vec{j}+3\vec{k}+\lambda_0(3\vec{i}-\vec{j}+\vec{k})$ and the position vector of M be $-3\vec{i}-7\vec{j}+6\vec{k}+\mu_0(-3\vec{i}+2\vec{j}+4\vec{k})$. Then

$$\overline{LM}=6\vec{i}-15\vec{j}+3\vec{k}-\lambda_0(3\vec{i}-\vec{j}+\vec{k})+\mu_0(-3\vec{i}+2\vec{j}+4\vec{k})$$

since $\overline{LM}$ is perpendicular to both the lines (i) and (ii).

$\therefore \overline{LM}\cdot(3\vec{i}-\vec{j}+\vec{k})=0$ and $\overline{LM}\cdot(-3\vec{i}+2\vec{j}+4\vec{k})=0$

Thus

$-18+15+3-\lambda_0(9+1+1)+\mu_0(-9-2+4)=0$

i.e., $-11\lambda_0-7\mu_0=0$ ...(iii)

and

$18-30+12-\lambda_0(-9-2+4)+\mu_0(9+4+16)=0$

i.e. $7\lambda_0+29\mu_0=0$ ...(iv)

From (iii) and (iv) we get,

$\lambda_0=\mu_0=0$

$\therefore \overline{LM}=-6\vec{i}-15\vec{j}+3\vec{k}$

$\therefore \left|\overline{LM}\right|=\sqrt{36+225+9}=\sqrt{270}=3\sqrt{30}$

Position vector of L is $3\vec{i}+8\vec{j}+3\vec{k}$.

$\therefore$ Equation of the line of shortest distance (LM) is

$\vec{r}=3\vec{i}+8\vec{j}+3\vec{k}+\lambda\left(-6\vec{i}-15\vec{j}+3\vec{k}\right)$

3. Given that lines are $\vec{r}=\lambda\hat{i}$ ...(i)

$\vec{r}=\mu\left(\hat{i}+\hat{j}\right)$ ...(ii)

$\vec{r}=v\left(\hat{i}+\hat{j}+\hat{k}\right)$ ...(iii)

These lines cut the plane $x+y+z=1$ at points $A(\lambda,0,0), B(\mu,\mu,0)$ and $C(v,v,v)$ respectively.

Since A lies on plane $\Rightarrow \lambda=1 \Rightarrow A(1,0,0)$

Since B lies on plane $\Rightarrow \mu+\mu=1\Rightarrow \mu=\dfrac{1}{2}$

$\Rightarrow B\left(\dfrac{1}{2},\dfrac{1}{2},0\right)$

Since, C lies on plane $\Rightarrow v+v+v=1\Rightarrow v=\dfrac{1}{3}$

$\Rightarrow C\left(\dfrac{1}{3},\dfrac{1}{3},\dfrac{1}{3}\right)$

$$\text{Area}(\Delta ABC)=\frac{1}{2}\left|\overline{AB}\times\overline{AC}\right|=\frac{1}{2}\begin{vmatrix}\hat{i} & \hat{j} & \hat{k}\\ \frac{-1}{2} & \frac{1}{2} & 0\\ \frac{-2}{3} & \frac{1}{3} & \frac{1}{3}\end{vmatrix}$$

$$=\frac{1}{2}\left|\frac{1}{6}\hat{i}+\frac{1}{6}\hat{j}+\frac{1}{6}\hat{k}\right|=\frac{1}{2}\times\frac{1}{6}\sqrt{3}=\frac{\sqrt{3}}{12}$$

$$\therefore (6\Delta)^2=36\times\frac{3}{144}=\frac{3}{4}=0.75$$

4. Let the co-ordinates of P be (a, b, c).

So, the co-ordinates of Q are $(0, 0, c)$ and co-ordinates of R are $(a, b, -c)$.

Given that PQ is perpendicular to the plane $x+y=3$,

So, PQ is parallel to the normal of given plane, i.e., $\left(a\hat{i}+b\hat{j}\right)$ is parallel to $\left(\hat{i}+\hat{j}\right)$

$\Rightarrow a=b$

As mid-point of PQ lies on the plane $x+y=3$, so

$\dfrac{a}{2}+\dfrac{b}{2}=3$

$\Rightarrow a+b=6\Rightarrow a=3=b$

Therefore, distance of P from the x-axis

$=\sqrt{b^2+c^2}=5$ (given)

$\Rightarrow b^2+c^2=25$

$\Rightarrow c^2=25-9=16$

$\Rightarrow c=\pm 4$

Hence, PR $=|2c|=8$

Comprehension Passage

Comprehension 1

1. (d) Shortest distance between both lines

$$=\left|\frac{\begin{vmatrix}3-2 & 1-1 & 0+1\\ 1 & 0 & 2\\ 1 & 1 & -1\end{vmatrix}}{\begin{vmatrix}\hat{i} & \hat{j} & \hat{k}\\ 1 & 0 & 2\\ 1 & 1 & -1\end{vmatrix}}\right|$$

$$=\left|\frac{1(0-2)-0(-1-2)+1(1-0)}{-2\hat{i}+3\hat{j}+\hat{k}}\right|$$

$$=\frac{1}{\sqrt{4+9+1}}=\frac{1}{\sqrt{14}}$$

2. (b) Equation of plane P

$$\begin{vmatrix} x-2 & y-1 & z+1 \\ 1 & 0 & 2 \\ 1 & 1 & -1 \end{vmatrix} = 0$$

$$-2(x-2)+(y-1)(3)+(z+1)=0$$

$$-2x+3y+z+2=0$$

$$2x-3y-z-2=0$$

Now image of point 0(0, 0, 0) on plane P

$$\frac{x-0}{2}=\frac{y-0}{-3}=\frac{z-0}{-1}=\frac{-2(-2)}{4+9+1}$$

$$\frac{x}{2}=\frac{y}{-3}=\frac{z}{-1}=\frac{4}{14}$$

Image point is $\left(\frac{4}{7},\frac{-6}{7},\frac{-2}{7}\right)$

3. (c) $0(0,0,0), A(1,0,0), B\left(0,\frac{-2}{3},0\right), C(0,0,-2)$

Volume of tetrahedron OABC = $\frac{1}{6}\left[\overrightarrow{OA}\ \overrightarrow{OB}\ \overrightarrow{OC}\right]$

$$=\frac{1}{6}\begin{vmatrix} 1 & 0 & 0 \\ 0 & -2/3 & 0 \\ 0 & 0 & -2 \end{vmatrix}=\frac{4}{18}=\frac{2}{9} \text{ cu unit}$$

Comprehension 2

1. (d)

2. (b)

3. (a)

(1 to 3)

$$ul+vm+wn=0$$

$$a\ell^2+bm^2+cn^2=0$$

$$a\ell^2+bm^2+c\left\{\frac{(ul+vm)}{w}\right\}^2=0$$

$$\Rightarrow \left(aw^2+cu^2\right)\ell^2+\left(bw^2+cv^2\right)m^2+2cuv\ell m=0$$

$$\Rightarrow \left(aw^2+cu^2\right)\left(\frac{\ell}{m}\right)^2+\left(bw^2+cv^2\right)+2cuv\left(\frac{\ell}{m}\right)$$

$$=0 \quad \text{...(i)}$$

Put $u=v=w=1$ in equation (i), then

$$(a+c)\left(\frac{\ell}{m}\right)^2+2c\left(\frac{\ell}{m}\right)+(b+c)=0$$

Similarly, $(a+b)\left(\frac{m}{n}\right)^2+2a\left(\frac{m}{n}\right)+(c+a)=0$

and $(b+c)\left(\frac{n}{\ell}\right)^2+2b\left(\frac{n}{\ell}\right)+(a+b)=0 \quad$ (ii)

From equation (ii) $\frac{n_1}{\ell_1}.\frac{n_2}{\ell_2}=\frac{a+b}{b+c}$

Similarly, $\frac{\ell_1\ell_2}{b+c}=\frac{m_1m_2}{c+a}=\frac{n_1n_2}{a+b} \quad$ (iii)

$\therefore \frac{m_1m_2}{\ell_1\ell_2}=\frac{c+a}{b+c}$

From equation (iii),

$$\frac{\ell_1\ell_2}{b+c}=\frac{m_1m_2}{c+a}=\frac{n_1n_2}{a+b}=\therefore\frac{\ell_1\ell_2+m_1m_2+n_1n_2}{(b+c)+(b+c)+(a+b)}$$

$\because$ the lines are perpendicular,

$\ell_1\ell_2+m_1m_2+n_1n_2=0$

Then $(b + c) + (c + a) + (a + b)$ must be zero.

$2a + 2b + 2c = 0$

$\Rightarrow a+b+c=0$

Comprehension 3

1. (b) Vector in the direction of $L_1=\vec{b_1}=3\hat{i}+\hat{j}+2\hat{k}$

Vector in the direction of $L_2=\vec{b_2}=\hat{i}+2\hat{j}+3\hat{k}$

$\therefore$ Vector perpendicular to both L_1 and L_2.

$$=\vec{b_1}\times\vec{b_2}=\begin{vmatrix} \hat{i} & \hat{j} & \hat{k} \\ 3 & 1 & 2 \\ 1 & 2 & 3 \end{vmatrix}=-\hat{i}-7\hat{j}+5\hat{k}$$

$\therefore$ Required unit vector

$$=\hat{b}=\frac{-\hat{i}-7\hat{j}+5\hat{k}}{\sqrt{1+49+25}}=\frac{-\hat{i}-7\hat{j}+5\hat{k}}{5\sqrt{3}}$$

2. (d) The shortest distance between L_1 and L_2 is

$$=\frac{\left(\vec{a_2}-\vec{a_1}\right).\vec{b_1}\times\vec{b_2}}{\left|\vec{b_1}\times\vec{b_2}\right|}=\left(\vec{a_2}-\vec{a_1}\right).\hat{b}$$

Since, $a_1=-\hat{i}-2\hat{j}-\hat{k}\ \ a_2=2\hat{i}-2\hat{j}+3\hat{k}$

$\therefore \vec{a_2}-\vec{a_1}=3\hat{i}+4\hat{k}\ \therefore \left(\vec{a_2}-\vec{a_1}\right)\hat{b}$

$$\therefore\left(3\hat{i}+4\hat{k}\right)\left(\frac{-\hat{i}-7\hat{j}+5\hat{k}}{5\sqrt{3}}\right)=\frac{-3+20}{5\sqrt{3}}=\frac{17}{5\sqrt{3}}$$

3. (c) The plane passing through $(-1, -2, -1)$ and having normal along $\vec{b}$ is

$$-1(x+1)-7(y+2)+5(z+1)=0$$

$$\Rightarrow x+7y-5z+10=0$$

$\therefore$ Distance of point (1, 1, 1) from the above plane is

$$=\frac{1+7\times1-5\times1+10}{\sqrt{1+49+25}}=\frac{13}{\sqrt{75}}$$

CHAPTER

12 Probability

Objectives

In this chapter on Probability, we shall study the following topics:

1. Definition – Stochastic experiment, Sample space, Mutually exclusive events, Exhaustive events
2. Set notations and results from set theory
3. Conditional probability
4. Multiplication theorem and total probability
5. Baye's theorem.

Basic Terminology and Definitions

The following terms and definitions are useful to learn probability theory:

Stochastic Experiment

An experiment whose outcome depends upon chance is called a stochastic (or random or statistical) experiment. When a single die is thrown, there are six possible outcomes which depend on chance. Thus, throwing a die is a stochastic experiment. Chance refers to an event whose exact outcome is not known in advance. For example, while throwing a ludo dice, one doesn't know which number will come on the top, among 1, 2, 3, 4, 5 and 6.

Sample Space

The collection of all possible outcomes of a stochastic experiment is called a sample space. The individual members of sample space are called events. When a single die is thrown, the sample space S contains 6 events. Thus, $S = \{1, 2, 3, 5, 6\}$. When two coins are tossed together $S = \{HH, HT, TH, TT\}$. The individual elements of Sample Space are called Sample Points. These Sample Points can be one-dimensional, two-dimensional or n-dimensional. The theory of the Probability is based upon the assumption that all Sample Points are **equally** probable.

Mutually Exclusive Events

Two events are said to be mutually exclusive if occurrence on one excludes the occurrence of the other. Similarly, a set of events $A_1, A_2, \ldots, A_n$ is said to be mutually exclusive if occurrence of one excludes the occurrence of the remaining events. In throwing a single die once if 2 occurs, then all other numbers (1, 3, 4, 5, 6) cannot occur. Thus, all the six events are mutually exclusive.

One card is drawn from a pack of cards. Consider the events

A: Occurrence of King B: Occurrence of Queen

In this example, A and B are mutually exclusive because a single card cannot be King as well as Queen simultaneously. On the contrary, events

A: Occurrence of King B: Occurrence of Club

are not mutually exclusive since A and B can occur together (The King of Clubs).

Equally Likely Events

A set of events is said to be equally likely if we have no reason to accept one event over any other possible events. For example, occurrence of head or tail on throwing a coin.

Exhaustive Events

A set of events is said to be exhaustive if they include all possible cases.

Notations

The events are denoted by capital letters A, B, C, or A_1, A_2 etc. Their probability are denoted by P(A), P(B) etc. The following set of notations will be used:

$P(A \cup B)$ or $P(A + B)$: Probability that either A or B (or both) will occur.

$P(A \cap B)$ or $P(AB)$: Probability that both A and B will occur.

$P(A/B)$: Probability of occurrence of 'A' when 'B' has already occurred. It is called conditional probability of occurrence of A when B has already occurred. It is read as P, A given B.

$P(\overline{A})$: Probability that A will not occur.

A and B are said to be mutually exclusive if $P(A \cap B) = 0$. It is clear that if events A and B are mutually exclusive then $P(A \cup B) = P(A) + P(B)$.

Further, note that if $A_1, A_2, \ldots, A_n$ area n mutually exclusive events then

$P(A_1 \cup A_2 \cup \ldots \cup A_n) = P(A_1) + P(A_2) + \ldots + P(A_n)$

An event A is said to be independent of another event B if $P(A) = P(A/B)$.

Results from Set Theory

For arbitrary Set A, B, C

(a) $\overline{A \cup B} = \overline{A} \cap \overline{B}$

(b) $\overline{A \cap B} = \overline{A} \cup \overline{B}$

(c) $A = (A \cap B) \cup (A \cap \overline{B})$

(Note that $(A \cap B)$ and $(A \cap \overline{B})$ are disjoint).

(d) $A \cup (B \cap C) = (A \cup B) \cap (A \cup C)$

(e) $A \cap (B \cup C) = (A \cap B) \cup (A \cap C)$

(f) For any two events A and B,

$P(A \cap \overline{B}) = P(A) - P(A \cap B)$

Remark: You can mentally confirm these results by drawing appropriate Venn diagrams.

ILLUSTRATION 12.1

A coin, weighted so that $P(H) = \frac{2}{3}$ and $P(T) = \frac{1}{3}$, is tossed. If heads appear, then a number is selected at random from the number 1 through 9; if tails appear, then a number is selected at random from the numbers 1 through 5. Find the probability p that an even number is selected.

Solution

The tree diagram with respective probabilities is

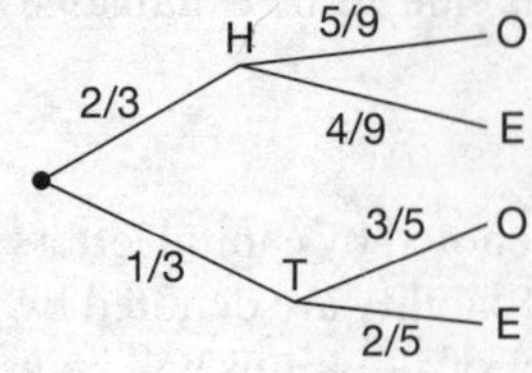

As seen in the diagram, the two paths HE and TE lead to even number.

Thus, $p = P(E) = \frac{2}{3} \cdot \frac{4}{9} + \frac{1}{3} \cdot \frac{2}{5} = \frac{8}{27} + \frac{2}{15} = \frac{58}{135}$

ILLUSTRATION 12.2

Two dice are thrown simultaneously. Find the probabilities of the following events

E_1 : 6 on first dice 5 on second
E_2 : 6 on one dice 5 on the other
E_3 : 6 on at least one dice
E_4 : Same number on both dice
E_5 : A total of 5
E_6 : A total of 7
E_7 : Multiple of 3 on first or total of 10
E_8 : Multiple of 3 on first dice and a total of 10

Solution

When two dice are thrown, the sample space will consist of 36 cases which are listed below:

(1, 1) (1, 2) (1, 3) (1, 4) (1, 5) (1, 6) (2, 1) (2, 2)
(2, 3) (2, 4) (2, 5) (2, 6)

(3, 1) (3, 2) (3, 3) (3, 4) (3, 5) (3, 6) (4, 1) (4, 2)
(4, 3) (4, 4) (4, 5) (4, 6)

(5, 1) (5, 2) (5, 3) (5, 4) (5, 5) (5, 6) (6, 1) (6, 2)
(6, 3) (6, 4) (6, 5) (6, 6)

For E_1, the favourable sample point is (6, 5)

$\Rightarrow \quad P(E_1) = \frac{1}{36}$

For E_2, the favourable sample points are (6, 5) or (5, 6) $\Rightarrow P(E_2) = \frac{2}{36}$

For E_3, the favourable sample points are (6, 1) (6, 2) (6, 3) (6, 4) (6, 5) (6, 6) (1, 6) (2, 6) (3, 6) (4, 6) (5, 6)

For E_4, the favourable sample points are (1, 1) (2, 2) (3, 3) (4, 4) (5, 5) and (6, 6)

$\Rightarrow \quad P(E_4) = \frac{6}{36}$

For E_5, the favourable sample points are (1, 4) (2, 3) (3, 2) and (4, 1) $\Rightarrow P(E_5) = \frac{4}{36}$

Similarly, $P(E_6) = \frac{6}{36}$

For E_7, the favourable sample points are,

(3, 1), (3, 2), (3, 3), (3, 4), (3, 5), (3, 6); 6, 1), (6, 2), (6, 3), (6, 4), (6, 5), (6, 6), (5, 5), (4, 6),

$\Rightarrow \quad P(E_7) = \frac{14}{36}$

From the same, $P(E_8) = \frac{1}{36}$ (since only (6, 4) satisfies the condition).

Note: The student is advised to remember the probabilities of all possible totals when two dice are thrown. The tabulation of these probabilities is called a probability distribution. Let X denote the value of the possible total when two dice are thrown. Then X can take the values 2, 3, 4, …, 12. Let $P(X = i)$ denote the probability that the total is $i(i = 2, 3, \ldots, 12)$. Then $P(i = 2) = \frac{1}{36}$, $P(i = 3) = \frac{2}{36}$ and so on. More systematically we can represent this in the form of a table as follows:

X	$P(X)$
2	1/36
3	2/36
4	3/36
5	4/36
6	5/36
7	6/36
8	5/36
9	4/36
10	3/36
11	2/36
12	1/36

The above table is called probability distribution of the variable (or random variable) X. Note that in any correct table $\sum P(X)$ must be unity.

Conditional Probability

The probability of occurrence of an event A, given that B has already occurred is called the conditional probability of occurrence of A. It is denoted by P(A/B).

$$P(A/B) = \frac{P(A \cap B)}{P(B)} = \frac{n(A \cap B)}{n(B)} \quad (P(B) \neq 0)$$

$$\Rightarrow \quad P(A/B) = \frac{N_{A\cap B}/N}{N_B/N} = \frac{P(A \cap B)}{P(B)}$$

(N = total no. of outcomes)

$$\text{Hence,} \quad P(A \cap B) = \begin{cases} P(B) \cdot P(A/B) & (\text{if } P(B) \neq 0) \\ P(A) \cdot P(B/A) & (\text{if } P(A) \neq 0) \end{cases}$$

Remark

(i) Two events A and B are independent if the occurrence or non-occurrence of A(B) does not affect the probability of occurrence or non-occurrence of B(A), that is, P(B/A) = P(B) provided P(A) ≠ 0.

Here P(B/A) stands for probability of B if A has already happened. Thus, A and B are independent if and only if $P(A \cap B) = P(A) \cdot P(B)$.

(ii) Generally, students find it difficult to distinguish between independent and mutually exclusive events. These events have the following differences:

(a) Independent events are always taken from different experiments, while mutually exclusive events are from only one experiment.

(b) Independent events can happen together but in mutually exclusive events, only one event may happen at one time.

(c) Independent events are represented by the word "and" but mutually exclusive events are represented by the word "or".

(iii) Two non-impossible events cannot be mutually exclusive and independent simultaneously, i.e., if A and B are two non-impossible events (P(A), P(B) ≠ 0) then if events are mutually exclusive ⇒ events are dependent and if events are independent ⇒ events are not exclusive.

(iv) If two events are mutually exclusive as well as independent then at least one of them is an impossible event

(v) If A and B are two independent events then following pairs are also independent:

(a) A′ and B (b) A′ and B′

(c) A and B′

ILLUSTRATION 12.3

For any two events A and B, $P\left(\frac{A}{A \cup B}\right)$ equals

(a) $\frac{P(\bar{A})}{P(A \cup B)}$ **(b)** $\frac{P(\bar{B})}{P(A \cup B)}$

(c) $\frac{P(B)}{P(A \cup B)}$ **(d)** $\frac{P(A \cap B)}{P(A \cup B)}$

Solution

Here

$$P\left(\frac{A}{A \cup B}\right) = \frac{P(A \cap A \cup B)}{P(A \cup B)} = \frac{P(\bar{A})}{P(A \cup B)}$$

You should be able to mentally figure it out through a Venn diagram.

ILLUSTRATION 12.4

Let A and B be two independent events. The probability that both A and B occur together is 1/6 and the probability that neither of them occurs is 1/3. The probability of occurrence of A is:

(a) P(A) = 1/4, P(B) = 1/3

(b) P(A) = 1/2, P(B) = 1/6

(c) P(A) = 1/3, P(B) = 1/2

(d) None of these.

Solution

A and B are independent events

$\therefore$ $P(AB) = 1/6 \Rightarrow P(A)\,P(B) = 1/6$ ...(i)

Further $P(A + B) = 1 - P(\bar{A}\,\bar{B}) = 1 - 1/3 = 2/3$

$\Rightarrow$ $P(A + B) = P(A) + P(B) - P(AB) = 2/3$

$\Rightarrow$ $P(A) + P(B) = 2/3 + 1/6 = 5/6$ ...(ii)

(i), (ii) $\Rightarrow$ $P(A) = 1/3, P(B) = 1/2$

Answer (c)

ILLUSTRATION 12.5

A card is drawn from a pack of playing cards. Find the probability that the drawn card is a court card when it is black.

(a) 3/26 **(b)** 3/13

(c) ½ **(d)** None of these.

Solution

Let A = Event that drawn card is a court card,
i.e., a card of king, queen or jack
B = Event that drawn card is black.

then $P(B) = \frac{26}{52}, P(AB) = \frac{6}{52}$

$$\therefore \quad P\left(\frac{A}{B}\right) = \frac{P(AB)}{P(B)} = \frac{6/52}{26/52} = \frac{6}{26} = \frac{3}{13}$$

Multiplication Theorem on Probability

If $E_1, E_2, \ldots E_{n+1}$ be $(n+1)$ events such that $P(E_1 \cap E_2 \cap \ldots \cap E_n) > 0$ then,

$$P(E_1 \cap E_2 \cap \ldots \cap E_{n+1})$$
$$= P(E_1)\cdot P(E_2/E_1)\cdot P(E_3/E_1 \cap E_2)\ldots P(E_{n+1}/E_1 \cap E_2 \cap \ldots \cap E_n)$$

Special Cases

(i) If $E_1, E_2, \ldots E_n$ are independent events. Then $P(E_1 \cap E_2 \ldots \cap E_n) = P(E_1)P(E_2)\ldots P(E_n)$

(ii) The probability that none of $E_1, \ldots E_n$ happen is $= (1-P(E_1))\cdot(1-P(E_2))\ldots(1-P(E_n))$

(iii) If p is the probability that an independent event happens in one trial, then the probability that it will happen in a succession of k trials $= p^k$

ILLUSTRATION 12.6

A bag contains 2 white and 2 black balls. A ball is drawn at random. If it is white, it is not replaced into the bag. Otherwise, it is replaced along with another ball of the same colour. The process is repeated. Find the probability that the third ball is drawn is black.

Solution

We can get a black ball in the third draw in four cases. They are BBB, BWB, WBB, WWB, where BWB means black ball in first draw, white in the second and black if) the third. Now the required probability

$$= P(BBB) + P(BWB) + P(WBB) + P(WWB)$$
$$= \frac{2}{4}\times\frac{3}{5}\times\frac{4}{6} + \frac{2}{4}\times\frac{2}{5}\times\frac{3}{4} + \frac{2}{4}\times\frac{2}{3}\times\frac{3}{4} + \frac{2}{4}\times\frac{1}{3}\times\frac{2}{2}$$
$$= \frac{1}{5} + \frac{3}{20} + \frac{1}{4} + \frac{1}{6} = \frac{12+9+15+10}{60} = \frac{46}{60} = \frac{23}{30}$$

ILLUSTRATION 12.7

Box-I contains 5 red and 4 white balls while Box-II contains 4 red and 2 white balls. A fair die is thrown. If it turns up a multiple of 3, a ball is drawn from Box-I, else a ball is drawn from Box-II. Find the probability that the ball drawn is white.

Solution

Let A be the event 'a multiple of 3 turns up on the die' and R be the event 'the ball drawn is white' then P(ball drawn is white)

$$= P(A)\cdot P(R/A) + P(\bar{A})P(R/\bar{A})$$
$$= \frac{2}{6}\times\frac{4}{9} + \left(1-\frac{2}{6}\right)\frac{2}{6} = \frac{10}{27}$$

Independence of Three Events

Three events A, B, C are said to be independent if the following four relations are satisfied:

(a) $P(A \cap B) = P(A)\cdot P(B)$

(b) $P(B \cap C) = P(B)\cdot P(C)$

(c) $P(C \cap A) = P(C)\cdot P(A)$

(d) $P(A \cap B \cap C) = P(A)\cdot P(B)\cdot P(C)$

If only (i), (ii), (iii) are satisfied, A, B, C are said to be pair-wise independent, but mutually dependent.

Total Probability

If $\{E_i\}, i = 1, 2, \ldots n$ be n events such that $E_i \cap E_j = 0$ for j and $\neq i \bigcup_{i=1}^{n} E_i = S$ (they are mutually exclusive and exhaustive). Suppose $P(E_i) > 0$ $(i = 1, 2, \ldots n)$. Then for any event A,

$$P(A) = \sum_{i=1}^{n} P(E_i)P(A/E_i)$$
$$= P(E_1)P(A/E_1) + \ldots + P(E_n)P(A/E_n)$$

Baye's Theorem

If $A_1, A_2, \ldots, A_n$ denote a disjoint partitions of the outcome sets whose union is the sample apace S and if $P(A_i) \neq 0$ $(i = 1, 2, \ldots, n)$ then for any arbitrary event B of the sample

$$S, P(A_i/B) = \frac{P(A_i \cap B)}{P(B)} = \frac{P(A_i)P(B/A_i)}{\sum_{i=l}^{n} P(A_i)P(B/A_i)}$$

Baye's Theorem for Future Event

The probability of the materialization of another event C, is given by

$$P(C/B) = \frac{P\sum_{i=l}^{n} P(A_i)P(B/A_i)P(C/A_i)}{\sum_{i=l}^{n} P(A_i)P(B/A_i)}$$

ILLUSTRATION 12.8

In a factory which manufactures bolts, machines A, B and C manufacture respectively 25%, 35% and 40% of the bolts. Of their outputs, 5, 4 and 2 percent are respectively defective bolts. A bolt is drawn at random from the product and is found to be defective. What is the probability that it is manufactured by the machine B?

Solution

Let events B_1, B_2, B_3 be the following:

B_1: the bolt is manufactured by machine A

B_2: the bolt is manufactured by machine B

B_3: the bolt is manufactured by machine C

Clearly, B_1, B_2, B_3 are mutually exclusive and exhaustive events and hence, they represent a partition of the sample space.

Let the event E be 'the bolt is defective'.

The event E occurs with B_1 or with B_2 or with B_3. Given that,

$P(B_1) = 25\% = 0.25$, $P(B_2) = 0.35$ and $P(B_3) = 0.40$

Again $P(E|B_1)$ = Probability that the bolt drawn is defective, given that it is manufactured by machine A = 5% = 0.05

Similarly, $P(E|B_2) = 0.04$, $P(E|B_3) = 0.02$.

Hence, by Bayes' Theorem, we have

$$P(B_2|E) = \frac{P(B_2)P(E|B_2)}{P(B_1)P(E|B_1)P(E|B_2) + P(B_3)P(E|B_3)}$$
$$= \frac{0.35 \times 0.04}{0.25 \times 0.05 + 0.35 \times 0.04 + 0.40 \times 0.02}$$
$$= \frac{0.0140}{0.0345} = \frac{28}{69}$$

ILLUSTRATION 12.9

A man is known to speak the truth 3 out of 4 times. He throws a die and reports that it is a six. Find the probability that it is actually a six.

Solution

Let E be the event that the man reports that six occurs in the throwing of the die and let S_1 be the event that six occurs and S_2 be the event that six does not occur.

Then

$P(S_1)$ = Probability that six occurs $= \frac{1}{6}$

$P(S_2)$ = Probability that six does not occur $= \frac{5}{6}$

$P(E|S_1)$ = Probability that the man reports that six occurs when six has actually occurred on the die

= Probability that the man speaks the truth $= \frac{3}{4}$

$P(E|S_2)$ = Probability that the man reports that six occurs when six has not actually occurred on the die

= Probability that the man does not speak the truth $= 1 - \frac{3}{4} = \frac{1}{4}$

Thus, by Bayes' theorem, we get

$P(S_1|E)$ = Probability that the report of the man that six has occurred is actually a six

$$= \frac{P(S_1)P(E|S_1)}{P(S_1)P(E|S_1) + P(S_2)P(E|S_2)}$$
$$= \frac{\frac{1}{6} \times \frac{3}{4}}{\frac{1}{6} \times \frac{3}{4} + \frac{5}{6} \times \frac{1}{4}} = \frac{1}{8} \times \frac{24}{8} = \frac{3}{8}$$

Hence, the required probability is $\frac{3}{8}$.

ILLUSTRATION 12.10

Pal's gardener is not dependable, the probability that he will forget to water the rose bush is $\frac{2}{3}$. The rose bush is in questionable condition anyhow, and if watered, the probability of its withering is $\frac{1}{2}$; if not watered, the probability of its withering is $\frac{3}{4}$. Pal went out of station and upon returning, he finds that the rose bush has withered, what is the probability that the gardener did not water the bush.

[Here, result is known that the rose bush has withered, therefore. Baye's theorem should be used.]

Solution

Let A = the event that the rose bush has withered

Let A_1 = the event that the gardener did not water.

A_2 = the event that the gardener watered.

By Bayes' theorem, required probability,

$$P(A_1/A)=\frac{P(A_1)\cdot P(A/A_1)}{P(A_1)\cdot P(A/A_1)+P(A_2)\cdot P(A/A_2)} \quad ...(i)$$

Given, $\quad P(A_1)=\frac{2}{3} \quad \therefore P(A_2)=\frac{1}{3}$

$$P\left(\frac{A}{A_1}\right)=\frac{3}{4}, P\left(\frac{A}{A_2}\right)=\frac{1}{2}$$

From (i) $\quad P(A_1/A)=\dfrac{\frac{2}{3}\cdot\frac{3}{4}}{\frac{2}{3}\cdot\frac{3}{4}+\frac{1}{3}\cdot\frac{1}{2}}=\frac{6}{6+2}=\frac{3}{4}$

Binomial Probability Distribution

(i) A probability distribution spells out how a total probability of 1 is distributed over several values of a random variable.

(ii) Mean of any probability distribution of a random variable is given by:

$$\mu=\frac{\sum p_i x_i}{\sum p_i}=\sum p_i x_i \qquad \left(\text{Since} \sum p_i=1\right)$$

(iii) Variance of a random variable is given by,

$$\sigma^2=\sum (x_i-\mu)^2\cdot p_i$$

$$\therefore \quad \sigma^2=\sum p_i x_i^2-\mu^2 \qquad (\text{Note that SD}=+\sqrt{\sigma^2})$$

(iv) The probability distribution for a binomial variate 'X' is given by:

$P(X=r)={}^nC_r p^r q^{n-r}$ where $P(X = r)$ is the probability of r successes.

The recurrence formula $\frac{P(r+1)}{P(r)}=\frac{n-r}{r+1}\cdot\frac{p}{q}$ is very helpful for quickly computing $P(1)\cdot P(2)\cdot P(3)$ etc. if P(0) is known.

(v) Mean of BPD = np; variance of BPD = npq.

(vi) If p represents a person's chance of success in any venture and M the sum of money which he will receive in case of success, then his expectations or probable value = pM.

ILLUSTRATION 12.11

A random variable X has the following probability distribution:

X	0	1	2	3	4	5	6	7
$P(X)$	0	k	$2k$	$2k$	$3k$	k^2	$2k^2$	$7k^2+k$

Determine

(i) k **(ii)** $P(X<3)$

(iii) $P(X>6)$ **(iv)** $P(0<X<3)$

Solution

Use $\sum P(X)=1$ to determine k, $P(X<3)=P(0)+P(1)+P(2)$, $P(X>6)=P(7)$ etc.

ILLUSTRATION 12.12

A pair of dice is thrown 5 times. If getting a doublet is considered as a success, then find the mean and variance of successes.

Solution

In a single throw of a pair of dice, probability of getting a doublet $=\frac{1}{6}$

considering it to be a success, $p=\frac{1}{6}$

$\therefore \quad q=1-\frac{1}{6}=\frac{5}{6}$

Mean $\quad =5\times\frac{1}{6}=\frac{5}{6}$

Variance $\quad =5\times\frac{1}{6}\cdot\frac{5}{6}=\frac{25}{36}$

ILLUSTRATION 12.13

Difference between mean and variance of a binomial variate is 1 and difference between their squares is 11. Find the probability of getting exactly three successes.

Solution

Mean = np and Variance = npq

therefore, $np - npq = 1$ ...(i)

$n^2p^2 - n^2p^2q^2 = 11$ ...(ii)

Also, we know that $p + q = 1$ ...(iii)

Divide equation (ii) by square of (i) and solve, we get, $q = \frac{5}{6}$, $p = \frac{1}{6}$ and $n = 36$.

Hence probability of '3' successes

$$= {}^{36}C_3 \times \left(\frac{1}{6}\right)^3 \times \left(\frac{5}{6}\right)^{33}$$

ILLUSTRATION 12.14

Find the variance of the number obtained on a throw of an unbiased die.

Solution

The sample space of the experiment is S = {1, 2, 3, 4, 5, 6}.

Let X denote the number obtained on the throw. Then X is a random variable which can take values 1, 2, 3, 4, 5, or 6.

Also $P(1) = P(2) = P(3) = P(4) = P(5) = P(6) = \frac{1}{6}$

Therefore, the probability distribution of X is

X	1	2	3	4	5	6
P(X)	$\frac{1}{6}$	$\frac{1}{6}$	$\frac{1}{6}$	$\frac{1}{6}$	$\frac{1}{6}$	$\frac{1}{6}$

Now $$E(X) = \sum_{i=1}^{n} x_i p(x_i)$$

$$= 1\times\frac{1}{6} + 2\times\frac{1}{6} + 3\times\frac{1}{6} + 4\times\frac{1}{6} + 5\times\frac{1}{6} + 6\times\frac{1}{6} = \frac{21}{6}$$

Also

$$E(X^2) = 1^2\times\frac{1}{6} + 2^2\times\frac{1}{6} + 3^2\times\frac{1}{6} + 4^2\times\frac{1}{6} + 5^2\times\frac{1}{6} + 6^2\times\frac{1}{6} = \frac{91}{6}$$

Thus, $$\text{Var}(X) = E(X^2) - (E(X))^2$$

$$= \frac{91}{6} - \left(\frac{21}{6}\right)^2 = \frac{91}{6} - \frac{441}{36} = \frac{35}{12}$$

Binomial Distribution for Repeated Trials

Let an experiment be repeated n times and the probability of happening of any event called success is p and not happening the event called failure is $q = 1 - p$. Then by binomial theorem.

$$(q+p)^n = q^n + {}^nC_1 q^{n-1} p + \ldots + {}^nC_r q^{n-r} p^r + \ldots + p^n$$

Now probability of

(a) Occurrence of the event exactly r times

$$= {}^nC_r q^{n-r} p^r$$

(b) Occurrence of the event at least r times

$$= {}^nC_r q^{n-r} p^r + \ldots + p^n$$

(c) Occurrence of the event at most r times

$$= q^n + {}^nC_1 q^{n-1} p + \ldots + {}^nC_r q^{n-r} p^r$$

ILLUSTRATION 12.15

If m different toys are distributed among n kids, find the probability that a particular kid will get exactly $k\,(< n)$ toys.

Solution

The total number of ways of distributing m different toys among n kids = $\underbrace{n.n.n \ldots n}_{(m \text{ times})} = n^m$.

Now exactly k toys can be given to a particular kid in ${}^mC_k$ ways. The remaining $(m-k)$ toys can be distributed among $(n-1)$ kids in $(n-1)^{m-k}$ ways. Thus, the number of favourable cases is ${}^mC_k (n-1)^{m-k}$. Hence the probability of the required event is $\frac{{}^mC_k (n-1)^{m-k}}{n^m}$.

ILLUSTRATION 12.16

The probability that A hits a target is $\frac{1}{4}$ and the probability that B hits it is $\frac{2}{5}$. What is the probability that the target will be hit if A and B each shoot at the target?

Solution

We are given $P(A) = \frac{1}{4}$ and $P(B) = \frac{2}{5}$. We have to find $P(A \cup B)$. Thus

$$P(A \cup B) = P(A) + P(B) - P(A \cap B)$$

$$= \frac{1}{4} + \frac{2}{5} - \frac{1}{4} \cdot \frac{2}{5} = \frac{1}{4} + \frac{2}{5} - \frac{2}{20} = \frac{5+8-2}{20} = \frac{11}{20}$$

ILLUSTRATION 12.17

A bag contains 17 mobile phones. Out of these, 6 are defective. 4 mobile phones are drawn at random from the bag. What is the probability that at least one of these is defective?

Solution

A : A mobile phone chosen is defective

Probability of all 4 phones drawn are non-defective $= \frac{^{11}C_4}{^{17}C_4}$

Hence probability that at least one is defective

$$= 1 - \frac{^{11}C_4}{^{17}C_4}$$

$$= \frac{^{17}C_4 - {}^{11}C_4}{^{17}C_4} = \frac{\frac{17 \times 16 \times 15 \times 14}{4!} - \frac{11 \times 10 \times 9 \times 8}{4!}}{\frac{17 \times 16 \times 15 \times 14}{4!}}$$

$$= \frac{17 \times 16 \times 15 \times 14 - 11 \times 10 \times 9 \times 8}{17 \times 16 \times 15 \times 14}$$

$$= \frac{16 \times 3 \times 5[17 \times 14 - 11 \times 3]}{16 \times 3 \times 5[17 \times 14]}$$

$$= \frac{238 - 33}{238} = \frac{205}{238}$$

WORKSHEET

1. A natural number is drawn from the set of first 20 natural numbers. Find the probabilities of the following events:

E_1: The number is divisible by 2 (or the number is even).

E_2: The number is divisible by 3.

E_3: The number is divisible by 2 or 3.

E_4: The number is divisible by 2 and 3.

E_5: The number is greater than 13.

E_6: The number is prime or it is divisible by 5.

2. From a bag containing 8 white and 5 black balls, four balls are drawn with replacement. Find the chance that exactly two balls are white.

3. Show that the probability remains same in all the following problems of arrangements in distinct and non-distinct cases.

(i) 5 green balls and 3 red balls (identical or distinct) are arranged in a line. Find the chance that the balls are of the same colour.

(ii) 5 green balls and 3 red balls (identical or distinct) are arranged in a line. Find the chance that all the red balls are together.

(iii) 5 green balls and 3 red balls (identical or distinct) are arranged in a line. Find the chance that no two red balls are together.

4. Four cards are drawn form a pack of cards. What is the chance of getting

(i) exactly two kings.

(ii) different suits.

(iii) the same suit.

(iv) different suits and different denominations.

(v) two pairs.

(vi) exactly one pair.

(vii) two kings and two clubs.

5. Two dice are thrown at a time. Find the probability of the following:

(i) These numbers shown are equal.

(ii) The difference of numbers shown is 1.

6. If three cards are drawn from a pack of 52 cards, what is the chance that all will be Queen?

7. The letters of the word 'SHANU' are written in a row randomly. Then find the probability that vowels occupy the even places.

8. If a card is drawn from a pack of 52 cards, then find the odds in favour of the event that it is a card of a Queen.

9. If A and B are two events such that P(A) = 1/2, P(B) = 1/3 and P(A∪B) = 7/12 then P(A/B) equals ________.

(a) 3/4 (b) 1/4

(c) 1/2 (d) None of these

10. Two dice are tossed four times. Find the probability of getting

(i) equal digits exactly two times.

(ii) equal digits at least two times.

(iii) equal digits at the most two times.

SOLVED EXAMPLES

Example 1. In an examination of 10 multiple choice questions (1 or more can be correct out of 4 options), a student decides to mark the answers at random. Find the probability that he gets exactly two questions correct.

Solution

A student can mark 15 different answers to an MCQ with 4 options, i.e., ${}^4C_1 + {}^4C_2 + {}^4C_3 + {}^4C_4 = 15$

Hence if he marks the answer at random, chance that his answer is correct $\frac{1}{15}$ and being incorrect is $\frac{14}{15}$.

Thus $p = \frac{1}{15}, q = \frac{14}{15}$

$$P(2\,\text{success}) = {}^{10}C_2 \times \left(\frac{1}{15}\right)^2 \times \left(\frac{14}{15}\right)^8$$

Example 2. Let A = {1, 2, 3, 4}. Two numbers are drawn from the set A. What is the probability that one of them is odd and the other is even.

Solution

Whenever more than one item is drawn from a lot, the problem of probability can be solved by unordered and ordered methods. In the unordered method, the order is disregarded. For instance, in the given problem the sample point (1, 2) is same as (2, 1). On the contrary in the ordered method the two sample points are supposed to be different. The sample point (1, 2) means that the number 1 is drawn on the first trial while 2 is drawn on the second trial. On the other hand, the sample point (2, 1) corresponds to the case where 2 is drawn on the first occasion while 1 is drawn on the second. It should be borne in mind that answer of the problem will be same in both the methods.

In the present problem, if we take unordered sample space then

S′ = {(1, 2) (1, 3) (1, 4), (2, 3), (2, 4), (3, 4)}

P(one even, one odd) = $\frac{4}{6}$. Combinatorically, this probability $= \frac{{}^2C_1 \times {}^2C_1}{{}^4C_2} = \frac{4}{6}$

In the ordered method, S′ = {(1, 2), (2, 1), (1, 3), (3, 1), (1, 4), (4, 1), (2, 3), (3, 2), (2, 4), (4, 2)}

$$\text{P (one even one odd)} = \frac{8}{12}$$

Combinatorically, we can argue as follows: When two numbers are drawn from the set A, the sample point is a two-dimensional point (x, y) where x is the number drawn at the first trial while y is the number drawn at the second trial. Since x can take 4 four values and for every value of x, y can take 3 values ⇒ total number of cases = 4 × 3

Now to calculate number of favourable cases, we notice that either x is even, y is odd; or x is odd, y is even

⇒ number of favourable cases = 2 × 2 + 2 × 2 = 8

Thus, the required probability $= \frac{8}{12}$

Thus, the probability remains 2/3, irrespective of the logic or approach being followed. How to handle a problem with bigger data or when 3, 4 or more objects are drawn? Suppose A = {1, 2, 3, 4, …, 20} and let six numbers be drawn from A we want to have 4 even and 2 odd numbers as our outcome. The unordered method easily gives the probability of the required event as

$$\frac{{}^{10}C_4 \times {}^{10}C_2}{{}^{20}C_6} \quad \ldots(*)$$

To solve the problem by ordered method, we note that the sample point is a six-dimensional point as $(x_1, x_2, x_3, x_4, x_5, x_6)$. The first co-ordinate is the first number drawn, the second co-ordinate is the second number drawn and so on.

⇒ Total number of cases = 20 × 19 × 18 × 17 × 16 × 15

Now the favourable sample points are the ones in which four co-ordinates are held by any four of the 10 even numbers 2, 4, 6, …, 20 and two other co-ordinates are held by any two of the 10 odd numbers 1, 3, 5, …, 19

⇒ Total number of favourable cases

$$=\left({}^6C_4\times{}^2C_2\right)(10\times9\times8\times7)\times10\times9$$

Thus, required probability

$$=\frac{(10\times9\times8\times7)\times(10\times9)}{20\times19\times18\times17\times16\times15}\,{}^6C_4 \qquad \ldots(**)$$

It is easy to observe that both answers are same. We also note that the ordered method can be applied to the case where numbers are drawn with replacement, for which case and unordered method cannot be applied. Indeed, when four numbers are drawn with replacement, the total number of cases

$$= 20\times20\times20\times20\times20\times20 = 20^6$$

and the number of favourable cases = $({}^6C_4)$ $({}^2C_2)$ $(10\times10\times10\times10)\times10\times10$

$$= 10^4\times10^2\times{}^6C_4 \qquad \ldots(*)$$

⇒ Probability of getting four even and two odd numbers

$$=\frac{10^6}{20^6}\times{}^6C_4=\left(\frac{10}{20}\right)^6\times{}^6C_4 \qquad \ldots(***)$$

The answers (**) and (***) easily boil down to the answers obtained by most common and popular method of solving a probability problem, which can be stated as follows:

1. Determine the number of general orders in which the event can occur.
2. Find the probability of each case by multiplication rule.
3. Add the probability of each case to get the required probability.

Remark: In case each term is same, this is added as many times as is the number of rough orders.

In the last problem, where four numbers are drawn without replacement, the general orders are EEEEOO (E stands for even O for odd) EOEEOE, OOEEEE etc.

The number of rough orders = The number of ways of arranging six of one kind and two of another kind in a line.

$$=\frac{6!}{4!2!}={}^6C_4$$

Now the probability of any general order (for instance, EEEEOO)

$$=\frac{10}{20}\times\frac{9}{19}\times\frac{8}{18}\times\frac{7}{17}\times\frac{10}{16}\times\frac{9}{15}$$

⇒ Required probability

$$=\frac{10}{20}\times\frac{9}{19}\times\frac{8}{18}\times\frac{7}{17}\times\frac{10}{16}\times\frac{9}{15}\times{}^6C_4$$

We can also solve the problem when the numbers are drawn with replacement. In this case, the probability

$$=\underbrace{\frac{10}{20}\times\frac{10}{20}\times\frac{10}{20}\times\frac{10}{20}}_{\text{even}}\times\underbrace{\frac{10}{20}\times\frac{10}{20}}_{\text{odd}}\times{}^6C_4=\left(\frac{10}{20}\right)^6{}^6C_4$$

Example 3. From a bag containing 8 white and 5 black balls, four balls are drawn one by one without replacement (or drawn simultaneously). Find the chance that exactly two of them are white.

Solution

For your better understanding the problems are solved by multiple methods. Kindly understand all the methods as that will enhance your thinking

We will again discuss all the above methods for solving the problem.

First Method

This method is equivalent to "forming committee of boys and girls", problems of permutations and combinations. We may call it unordered method since the order of the persons in the committee is disregarded. For instance (W_1, W_2, W_3, B_4) is same as (W_1, W_3, B_4, W_2) or (W_3, B_4, W_2, W_1).

Now the total number of sample points counted in this style = ${}^{13}C_4$ ...(*)

and favourable number of sample points counted in this style = ${}^8C_2\times{}^5C_2$ ...(**)

$$\text{Required probability}=\frac{{}^8C_2\times{}^5C_2}{{}^{13}C_4}$$

Second Method

We can use addition and multiplication rules of the probability. We determine the number of "general orders" in which the event can happen. This number is found by arranging some varieties in a line. For instance, in the given problem, we want to draw exactly two white balls. The number of general orders must be equal to number of ways of arranging 2W and 2B in a line, which is equal to $\frac{4!}{2!2!}$ = 6 (The general orders are WWBB, WBWB, BWBW, BWWB, BBWW, WBBW). Now the probability of any of these orders (by multiplication rule, for instance WWBB) = $\frac{8}{13}\times\frac{7}{12}\times\frac{5}{11}\times\frac{4}{10}$.

Therefore, required probability

$$=\frac{8}{13}\times\frac{7}{12}\times\frac{5}{11}\times\frac{4}{10}\times\frac{4!}{2!2!}, \qquad \ldots(**)$$

Which is numerically equal to the last answer. We can call it ordered method of solving probability problems.

Third Method

When four balls are drawn from the bag without replacement, a typical sample point will be a point of four-dimensional space in which all the four co-ordinates are distinct. For instance, (W_1, W_2, W_3, B_4) is a sample point. Another favourable sample point is $(W_1\ W_2, B_3, B_4)$. Note carefully that the sample points are ordered. Therefore, total number of sample points $= 13 \times 12 \times 11 \times 10$

and the total number of favourable sample points $= 8 \times 7 \times 5 \times 4 \times {}^4C_2$

(Out of the four co-ordinates, choose two for white in ${}^4C_2$ ways – then it gives rise to $8 \times 7 \times 5 \times 4$ cases)

Thus, the required probability

$$= \frac{8 \times 7 \times 5 \times 4}{13 \times 12 \times 11 \times 10} \times {}^4C_2 \qquad \ldots(***)$$

Important points

(i) A beginner in probability may think that if the balls are identical, the total number of sample points is just five: WWWW, WWWB, WWBB, WBBB, BBBB, i.e., either all are white or four are white and so on. This reasoning is incorrect since each ball, despite being identical occupies a different co-ordinate or position within the bag and therefore the number of sample points will still be ${}^{13}C_4$ and all these ${}^{13}C_4$ sample points will be equally likely. There is another way to discard the above logic. If the bag contains *m* white and *n* black balls where $m \geq 4$, $n \geq 4$ and we draw four balls, then by the above logic, the probability of the required event will remain $\frac{1}{5}$ for all values of *m*, *n* which is absurd since the probability for a given *n* must approach 0 as $m \to \infty$ (why?).

(ii) The beginner may again think, why, if the balls are distinct, we are writing $\frac{4!}{2!2!}$. It should be noted that the distinctness of the balls has been taken into account in $8 \times 7 \times 5 \times 4$ and the number $\frac{4!}{2!2!}$ is being multiplied because the position of 2 white and 2 black balls can be in $\frac{4!}{2!2!}$ (or ${}^4C_2$) ways.

(iii) The second and third method are equivalent. The third method actually explains why we add and multiply in the second method.

At least/At most:

Note that the probability of at least two white balls

$= P(2 \text{ white}) + P(3 \text{ white}) + P(4 \text{ white})$

$$= \frac{8C_2 \times {}^5C_2}{{}^{13}C_4} + \frac{{}^8C_3 \times {}^5C_1}{{}^{13}C_4} + \frac{{}^8C_4}{{}^{13}C_4}$$

Again, probability of at most one white ball

$$= P(0) + P(1) = \frac{{}^8C_0 \times {}^5C_4}{{}^{13}C_4} + \frac{{}^8C_1 \times {}^5C_3}{{}^{13}C_4}$$

Thus, at most/ at least probability problems are done by adding the probabilities of disjoint events. We can also calculate these probabilities by subtraction. For instance, P(at least 1 white ball) $= P(1) + P(2) + P(3) + P(4) = 1 - P(0)$

Since by fundamental probability law

$$= P(0) + P(1) + P(2) + P(4) = 1,$$

it is much more quicker to calculate $1 - P(0)$ than $P(1) + P(2) + P(3) + P(4)$.

Example 4. 10 coins are tossed simultaneously (or a coin is tossed ten times in succession). Find the chance of getting:

(i) Exactly 8 heads

(ii) At least 8 heads

(iii) At least one head

(iv) Head and tail alternately

(v) Heads, exactly as many times in the first seven throws as in the last three

(vi) Heads, an odd number of times

(vii) A sequence of six of more consecutive heads

(viii) No two consecutive heads

Solution

(i) By binomial formula, the required probability

$$= {}^{10}C_8 \left(\frac{1}{2}\right)^8 \left(1 - \frac{1}{2}\right)^2 = \frac{{}^{10}C_8}{2^{10}}$$

(or there are 2^{10} ten-dimensional sample points, out of which ${}^{10}C_8$ are favourable to the event since positions of eight heads can be chosen in ${}^{10}C_8$ ways in a ten-tuple).

(ii) Answer $= \frac{{}^{10}C_8}{2^{10}} + \frac{{}^{10}C_9}{2^{10}} + \frac{{}^{10}C_{10}}{2^{10}}$

(iii) Answer $= 1 - \frac{{}^{10}C_0}{2^{10}}$

(iv) There are only two major ways in which the event can occur: either the first coin is head or tail

$\Rightarrow$ Number of favourable cases = 2

$\Rightarrow$ Required probability $= \frac{2}{2^{10}}$

(v) Let $P(i, j)$ be the probability that i heads are thrown in the first seven throws while j heads are thrown in the last three.

Then the required probability

$= P(0, 0) + P(1, 1) + P(2, 2) + (3, 3)$

$$= {}^7C_0\left(\frac{1}{2}\right)^0\left(\frac{1}{2}\right)^7 . {}^3C_0\left(\frac{1}{2}\right)^0\left(\frac{1}{2}\right)^3 + {}^7C_1\left(\frac{1}{2}\right)^1\left(\frac{1}{2}\right)^6 . {}^3C_1\left(\frac{1}{2}\right)^1\left(\frac{1}{2}\right)^2$$

(vi) Required probability $= P(1) + P(3) + P(5) + \ldots$

$$= {}^{10}C_1\left(\frac{1}{2}\right)^1\left(\frac{1}{2}\right)^9 + {}^{10}C_3\left(\frac{1}{2}\right)^3\left(\frac{1}{2}\right)^7 + \ldots$$

$$= \frac{{}^{10}C_1 + {}^{10}C_3 + {}^{10}C_5 + \ldots}{2^{10}} = \frac{2^9}{2^{10}} = \frac{1}{2}$$

(vii) Since the number of throws is ten, only one sequence of six or more consecutive heads is possible, which either begins from the first trial or begins from the second trial … or begins from the fifth trial. Observe the following table

Type	Number of ways
H H H H H H H or T	2^4
T H H H H H H H or T (first trial tail)	2^3
[H T H H H H H H H or T T T H H H H H H H or T] (second trial T, H or T before)	2×2^2
[H H T H H H H H H H or T H T T H H H H H H H or T T H T H H H H H H H or T T T T H H H H H H H or T] (third trial T, H or T before)	$2^2 \times 2$
Fourth trial T, H or T before, then six consecutive heads;	$2^3 \times 1$

Thus, required probability $= \dfrac{2^4 + 4 \times 2^3}{2^{10}} = \dfrac{48}{1024}$

(viii) There are two methods of doing this non-elementary problem.

First Method: Recall that m identical red balls. $(m \geq n - 1)$ can be arranged in a line in ${}^{m+n}C_n$ ways if on two red balls are to be together. In the present problem we may take heads as red balls and tails as green balls. Now if we want to have no two red balls together then this may happen with (10T, 0H) or (9T, 1H) or (8T, 2H) … … (5T, 5H) (more than 5H is not possible)

⇒ Required number of favourable ways

$= {}^{11}C_0 + {}^{10}C_1\, {}^9C_2 + {}^8C_3\, {}^7C_4 + {}^6C_5 = 144$

⇒ Required probability $= \dfrac{144}{2^{10}}$

Second Method: We will use method of recursions. Let U_n be the probability that n tosses of coin contains no two consecutive heads, then it is easy to observe that

$$U_1 = 1, U_2 = \frac{3}{4}$$

Now up to n tosses, our condition can be satisfied in two disjoint ways. Either no two consecutive heads occur till $n - 2$ trials and TH occur in the last two trials OR no two consecutive heads occur till $n - 1$ trials and T occurs in the last trial.

The probability of these cases are

$$U_{n-2} \times \frac{1}{4} \text{ and } U_{n-1} \times \frac{1}{2} \text{ respectively.}$$

$$\text{Consequently } U_n = \frac{1}{4}U_{n-2} + \frac{1}{2}U_{n-1}$$

$$\text{Now } U_3 = \frac{1}{4} \times 1 + \frac{1}{2} \times \frac{3}{4} \times \frac{5}{8}, U_4 = \frac{1}{4} \times \frac{3}{4} + \frac{1}{2} \times \frac{5}{8} \text{ etc.}$$

Continuing in this manner we will easily obtain $U_{10} = \dfrac{144}{2^{10}}$

Example 5. Six dice are thrown simultaneously. Find the chance of getting:

(i) Exactly four sixes
(ii) At most one six
(iii) All possible faces
(iv) A total of 20
(v) Six an even number of times
(vi) Exactly three different numbers 1, 2 and 5
(vii) Exactly three different numbers
(viii) Same number on at least four dice

Solution

(i) We have $n = 6$, $p = \frac{1}{6}, q = \frac{5}{6}$

$\Rightarrow$ Required probability $= {}^6C_0\left(\frac{1}{6}\right)^4\left(\frac{5}{6}\right)^2$

(ii) Required probability

$= P(0)+P(1) = {}^6C_0\left(\frac{1}{6}\right)^0\left(\frac{5}{6}\right)^6 + {}^6C_1\left(\frac{1}{6}\right)^1\left(\frac{5}{6}\right)^5$

(iii) Required probability $= \frac{6!}{6^6}\left(\text{or } \frac{6}{6}\times\frac{5}{5}\times\frac{4}{6}\times\ldots\times\frac{1}{6}\right)$

(iv) We must have $x_1 + x_2 + x_3 + x_4 + x_5 + x_6 = 20$ where $0 < x_i \le 6, i = 1,2,\ldots,6$

$\Rightarrow$ Favourable number of ways

$=$ Coeff. of t^{20} in $(t^1 + t^2 + t^3 + \ldots + t^6)^6$

$=$ Coeff. of t^{14} in $(1 + t + t^2 + \ldots + t^5)^6$

$=$ Coeff. of t^{14} in $(1 - t^6)^6 (1 - t)^{-6}$

$=$ Coeff. of t^{14} in $(1 - 6t^6 + 15t^{12} - \ldots)(1 - t)^{-6}$

$= 1 \times$ Coeff. of t^{14} in $(1 - t)^{-6} - 6$ coeff. of $t^8 + 15$ coeff. of t^2

$= 1 \times {}^{19}C_{14} - 6 \cdot {}^{13}C_8 + 15 \cdot {}^7C_2$

(Coeff. of x^r in the expansion of $(1 - x)^{-n}$ is ${}^{n+r-1}C_r$).

(v) Required probability $= P(0) + P(2) + (4) + P(6)$

$= {}^6C_0\left(\frac{1}{6}\right)^0\left(\frac{5}{6}\right)^6 + {}^6C_2\left(\frac{1}{6}\right)^2\left(\frac{5}{6}\right)^{40} + {}^6C_6\left(\frac{1}{6}\right)^6\left(\frac{5}{6}\right)^0$

$= \frac{1}{2}\left[\left(\frac{5}{6}+\frac{1}{6}\right)^6 + \left(\frac{5}{6}-\frac{1}{6}\right)^6\right] = \frac{1}{2}\left[1+\left(\frac{2}{3}\right)^6\right]$

(vi) The favourable number of ways must be same as the number of ways of distributing six distinct objects (six dice) in 3 distinct boxes (the name of the boxes are 1, 2 and 5) such that no box remains. This should be

$3^6 - {}^3C_1\, 2^6 + {}^3C_2\, 1^6 - {}^3C_3\, 0^6 = 540$

$\Rightarrow$ Required probability $= \frac{540}{6^6}$

(vii) Multiply the answer of (vi) by ${}^6C_3$.

(viii) Required probability = P(Four dice show the same number)

+ P(Five dice show the same number)

+ P(Six dice show the same number)

= P(A) + P(B) + P(C)

Let us calculate the number of favourable cases of P(A). If the same number is 1 then the other two numbers can be in 5×5 ways at the two remaining positions. But these non 1's can be in ${}^6C_2$ position. Again, the same number can take six values.

The number of favourable cases for

$$P(A) = 6 \times {}^6C_2 \times 5^2.$$

The number of favourable cases corresponding to P(B) and P(C) must be $6 \times 6C_1 \times 5$ and 6 respectively. The total number of cases in each case must be 6^6.

$\Rightarrow$ Required probability

$$= \frac{6\times{}^6C_2\times5^2+6\times{}^6C_1\times5+6}{6^6} = \frac{2436}{6^6}$$

Problems on Cards

The student must be familiar with the pack of cards. A pack of cards contains 52 cards in which there are four suits. The name of the suits are spade, heart, diamond and clubs. The colour of spade and club is black while that of heart and diamond is red. Each suit has 13 cards and therefore 13 denominations. These are Ace, King, Queen, Jack (face cards) and 10, 9, 8, … 2. Thus, in a pack of cards:

(i) There are 26 black cards and 26 red cards.

(ii) There are four cards of any denomination.

Example 6. Six cards are drawn at random from a pack of cards with replacement. What is the chance that all four suits are represented at least once? What is the chance that exactly six drawings are required to get all the suits?

We will give elementary and advanced solutions for both parts.

Elementary Solution

Let us denote the suits by S, H, D, C. Consider the arrangement of S, H, D, C at six places where each letter should appear at least once. For the sake of simplicity, suppose the first four positions are SHDC. The fifth and sixth places can be in two major ways.

(i) Two letters are alike (possible in 4 ways)

(ii) Two letters are different (possible in ${}^4C_2 = 6$ ways)

But in both cases, various permutations are possible Indeed, in case I the total number of arrangements

$= \frac{6!}{3!}\times 4 = 480$

and in case II the total number of arrangements

$= \frac{6!}{2!2!}\times {}^4C_2 = 1080$

Thus, the total number of arrangements of four letters at six places, such that each of the letter S, H, D, C appear at least once = 480 + 1080 = 1560.

The probability in each case $=\left(\frac{13}{52}\right)^6=\left(\frac{1}{4}\right)^6$.

$\therefore$ Required chance $=\frac{1560}{4^6}$.

To solve the other part, we note that the required event is possible if the first five drawings yield exactly three suits and the sixth drawing gives the suit which does not appear in the first five drawings. Let us suppose that club does not appear in the first five drawings, which is possible if the first five drawings have

(i) Three spades, a heart and a diamond or three hearts, a spade and a diamond or 3 diamonds, a spade and a heart

(ii) SSHHD, HHDDS, SSDDH (with all possible permutations)

Thus, number of arrangements $=\frac{5!}{3!}\times 3\frac{5!}{2!2!}\times 3=150$

Similarly, we can have other suits at the last drawings. Thus, required number of permutations of S, H, D, C such that our requirement is fulfilled = 4 × 150 = 600.

Therefore, required probability $=\frac{600}{4^6}$

Advanced Solution of First Part

Define the events

A_1: Spade does not appear in the six drawings
A_2: Heart does not appear in the six drawings
A_3: Diamond does not appear in the six drawings
A_4: Club does not appear in the six drawings

Required probability $=P(\bar{A}_1\cap\bar{A}_2\cap\bar{A}_3\cap\bar{A}_4)$

$=P(\overline{A_1\cup A_2\cup A_3\cup A_4})$

$=1-P(A_1\cup A_2\cup A_3\cup A_4)$

$=1-\sum P(A_i)+\sum_{i\neq j} P(A_iA_j)-\sum_{i\neq j\neq k} P(A_iA_jA_k)+P(A_1A_2A_3A_4)$

Now, $P(A_i)$ = P(Spade does not appear in six drawings) $(i=1, 2, 3, 4)=\left(\frac{3}{4}\right)^6$

Now, $P(A_1\cap A_2)$ = P(Spade and Heart do not appear in six drawings) $=\left(\frac{2}{4}\right)^6$

The remaining 5 terms in $\sum_{i\neq j} P(A_1\cap A_j)$ will be same.

Finally, $P(A_1\cap A_2\cap A_3)=\left(\frac{1}{4}\right)^6$ and

$P(A_1\cap A_2\cap A_3\cap A_4)=0$ (Why?)

Thus, required probability

$$=1-4\left(\frac{3}{4}\right)^6+6\left(\frac{2}{4}\right)^6-4\left(\frac{1}{4}\right)^6+0$$

$$=\frac{4^6-4.3^6+6.2^6-4}{4^6}=\frac{1560}{4^6}$$

Advanced Solution of Second Part

Consider the distribution of 5 different objects 1, 2, 3, 4, 5 (first drawing, second drawing, … fifth drawing) in three different boxes S, H, D such that no box remains empty (such that exactly three suits S, H, D appear in the first five drawings).

The number of distributions $=3^5-{}^3C_1\,1^5-{}^3C_3\,0^5=150$.

But as there are $4C_3=4$ combinations of four suits which may appear at the first five drawings, the required probability $=\frac{4\times 150}{4^5}\cdot\frac{1}{4}$ (Since the sixth card must be a new suit).

Example 7. A bag contains 3 red, 6 white and 7 blue balls. Two balls are drawn one by one. What is the probability that first ball is white and second ball is blue when the first-drawn ball is not replaced in the bag?

(a) 3/40 **(b)** 7/40
(c) 1/8 **(d)** None of these.

Solution

Let A be the event of drawing first ball white and B be the event of drawing second ball blue. Then,

$$P(A)=6/16,\ P(B/A)=7/15$$

$$\therefore\quad P(AB)=(6/16)\cdot(7/15)=7/40$$

Example 8. A bag contains 4 red and 4 blue balls. Four balls are drawn one by one from the bag. Then the probability that the drawn balls are in alternate colours is ______.

(a) 2/3 **(b)** 3/35
(c) 6/35 **(d)** 3/4

Solution

E_1 = Event that first-drawn ball is red, second is blue and so on.

E_2 = Event that first-drawn ball is blue second is red and so on.

$\therefore \quad P(E_1) = \frac{4}{8} \cdot \frac{4}{7} \cdot \frac{3}{6} \cdot \frac{3}{5}$ and

$P(E_2) = \frac{4}{8} \cdot \frac{4}{7} \cdot \frac{3}{6} \cdot \frac{3}{5}$

$\therefore \quad P(E) = P(E_1) + P(E_2)$

$= \frac{4}{8} \cdot \frac{4}{7} \cdot \frac{3}{6} \cdot \frac{3}{5} + \frac{4}{3} \cdot \frac{4}{7} \cdot \frac{3}{6} \cdot \frac{3}{5} = \frac{6}{35}$

Example 9. If the probability of A to fail in an examination is 0.2 and that of B to fail is 0.3, then the probability that either A or B fails is _____.

(a) 0.5 **(b)** 0.44
(c) 0.56 **(d)** None of these.

Solution
Let A be event for A to fail and B be the event for B to fail, then

$P(A) = 0.2$ and $P(B) = 0.3$

Since A and B are independent events,

$\therefore \quad P(AB) = P(A)\,P(B)$

∴ Required probability

$= P(A + B)$

$= P(A) + P(B) - P(AB) = P(A) + P(B) - P(A)\,P(B)$

$= 0.2 + 0.3 - 0.2 \times 0.3 = 0.5 - 0.06 = 0.44$

Example 10. One person can kill a bird twice in 3 shots, second once in 3 shots and third thrice in 4 shots. If they shot together then what is the probability that the bird will be killed?

(a) 18/19 **(b)** 17/18
(c) 18/17 **(d)** 20/17

Solution
If A, B, C denote events of killing the bird by first, second and third person respectively, then

$P(A) = 2/3,\ P(B) = 1/3,\ P(0) = 3/4$

The bird will be killed if at least one of these three independent events happens. So, required probability

$= 1 - P(\bar{A}_1)P(\bar{A}_2)P(\bar{A}_3)$

$= 1 - (12/3)(11/3)(1 - 3/4) = 1 - 1/3.2/3.1/4 = 17/18$

Example 11. A coin is tossed thrice. If E be the event of showing at least two heads and F the event of showing head in the first throw, then P(E/F) is equal to ________.

(a) 3/4 **(b)** 3/8
(c) 1/2 **(d)** 1/8

Solution
The following are the 8 outcomes from three throws:

HHH, HHT, HTH, HTT, THH, THT, TTH, TTT

Also, $P(E \cap F) = 3/8$ and P(F) 4/8

∴ Required probability = P(E/F)

$= \frac{P(E \cap F)}{P(F)} = \frac{3/8}{4/8} = \frac{3}{4}$

Example 12. Two dice are thrown together 4 times. The probability that both dice will show same numbers twice is ________.

(a) 1/3 **(b)** 25/36
(c) 25/216 **(d)** None of these.

Solution
The probability of showing same number by both dice $p = 6/36 = 116$

In binomial distribution, here $n = 4$, $r = 2$, $p = 1/6$, $q = 5/6$

$\therefore \quad$ Required probability $= {}^nC_r\, q^{n-r}\, p^r$

$= {}^4C_2 \left(\frac{5}{6}\right)^2 \left(\frac{1}{6}\right)^2 = 6\left(\frac{25}{36}\right)\left(\frac{1}{36}\right) = \frac{25}{216}$

Example 13. Two dice are thrown together. The probability of getting the-sum of digits as a multiple of 4 is _______.

(a) 1/9 **(b)** 1/3
(c) 1/4 **(d)** 5/9

Solution
Total exhaustive cases $= 6^2 = 36$

The following 9 pairs are favourable as the sum of their digits are multiples of 4, i.e., 4 or 8 or 12.

(1, 3), (2, 2), (3, 1), (2, 6), (3, S), (4, 4), (5, 3), (6, 2), (6, 6)

$\therefore \quad$ Required probability $= 9/36 = 1/4$

Example 14. A speaks truth in 75% cases and B in 80% cases. What is the probability that they contradict each other in stating the same fact?

(a) 7/20 **(b)** 13/20
(c) 3/20 **(d)** 1/5

Solution
There are two mutually exclusive cases in which they contradict each other, i.e., $\bar{A}B$ and $A\bar{B}$. Hence,

Required probability

$= P(A\bar{B} + \bar{A}B) = P(A\bar{B}) + P(\bar{A}B)$

$= P(A)P(\bar{B}) + P(\bar{A})\,P(B)$

$= \frac{3}{4} \cdot \frac{1}{5} + \frac{1}{4} \cdot \frac{4}{5} = \frac{7}{20}$

Example 15. A target is hit by A, 4 times out of 5 attempts; by B, 3 times out of 4 attempts; and by C, 2 times out of 3 attempts. The probability that the target is hit by two of them is ______.

(a) 25/60 **(b)** 26/60 **(c)** 1/2 **(d)** 5/6

Solution

The following mutually exclusive cases are possible.

(i) $AB\bar{C}$ **(ii)** $A\bar{B}C$ **(iii)** $\bar{A}BC$

Since A, Band C are independent event, therefore $P(AB\bar{C}) = P(A)\,P(B)\,P(\bar{C})$

$$= \frac{4}{5}\cdot\frac{3}{4}\left(1-\frac{2}{3}\right) = \frac{12}{60}$$

Similarly, $P(A\,\bar{B}\,C) = \frac{8}{60}$ and $P(\bar{A}BC) = \frac{6}{60}$

Thus, the required probability

$$= P\left(AB\bar{C}\right)+P\left(A\bar{B}C\right)+P\left(\bar{A}BC\right)$$

$$= \frac{12}{60}+\frac{8}{60}+\frac{6}{60}=\frac{26}{60}$$

Example 16. A box contains 20 cards. The letter I is written on 10 cards and T is written on other 10 cards, 3 cards are chosen randomly and are kept in the same order. The probability of making the word IIT is ______.

(a) 9/80 **(b)** 1/80

(c) 4/27 **(d)** 5/38

Solution

Required probability = P(I on first card) × P(I on second card) × P(T on third card)

$$= \frac{10}{20}\times\frac{9}{19}\times\frac{10}{18}=\frac{5}{38}$$

Example 17. Assuming that for a husband–wife couple the chances of their child being a boy or a girl are the same, the probability of their two children being a boy and a girl is ______.

(a) 1/4 **(b)** 1

(c) 1/2 **(d)** 1/8

Solution

The following four mutually exclusive cases are possible:

(i) Both children are boys

(ii) Both children are girls

(iii) First is a girl and second is a boy

(iv) First is a boy and second is a girl.

Out of these, the last two are favourable cases.

Hence the required probability

$$= \frac{1}{4}+\frac{1}{4}=\frac{1}{2}$$

Example 18. In the throwing of a die, let A be the event 'an odd number turns up', B be the event 'a number divisible by 3 turns up', and C be the event 'a number 4 turns up'. Then find the probability that exactly two of A, B and C occur.

Event A = {1, 3, 5}, Event B = {3, 6} and Event C = {1, 2, 3, 4}

$\therefore$ $A\cap B=\{3\}, B\cap C=\{3\}, A\cap C=\{1,3\}$

and $A\cap B\cap C=\{3\}.$

Thus P(exactly two of A, B and C occur)

$$= P(A\cap B)+P(B\cap C)+P(C\cap A)-3P(A\cap B\cap C)$$

$$= \frac{1}{6}+\frac{1}{6}+\frac{2}{3}-3\times\frac{1}{6}=\frac{1}{6}$$

Example 19. A pair of dice is thrown 5 times. Find the probability of getting a double, twice.

In a single throw of a pair of dice, the probability of getting a double is $\frac{1}{6}$

Considering it to be a success, $p=\frac{1}{6}$

$\therefore$ $q = 1-\frac{1}{6}=\frac{5}{6}$

Number of success $r = 2$

$\therefore$ $P(r=2) = {}^5C_2\,p^2q^3 = 10.\left(\frac{1}{6}\right)^2.\left(\frac{5}{6}\right)^3 = \frac{625}{3888}$

RECAPITULATION (FORMULAE, THEOREM, IMPORTANT RESULTS)

1. The conditional probability of an event E, given the occurrence of the event F is given by $P(E\,|\,F)=\frac{P(E\cap F)}{P(F)}, P(F)\neq 0$
2. $0\le P(E\,|\,F)\le 1, P(E'\,|\,F)=1-P(E\,|\,F)$
   $P((E\cup F)|G)=P(E|G)+P(F|G)-P((E\cap F)|G)$
3. $P(E\cap F)=P(E)P(F\,|\,E), P(E)\neq 0$
   $P(E\cap F)=P(F)P(E\,|\,F), P(F)\neq 0$
4. If E and F are independent, then
   $P(E\cap F)=P(E)P(F)$
   $P(E\,|\,F)=P(E), P(F)\neq 0$
   $P(F\,|\,E)=P(F), P(E)\neq 0$

5. Theorem of total probability

   Let $\{E_1, E_2, ..., E_n)$ be a partition of a sample space and suppose that each of $E_1, E_2, ..., E_n$ his nonzero probability. Let A be any event associated with S, then

   $$P(A) = P(E_1)P(A|E_1) + P(E_2)P(A|E_2) + ... + P(E_n)P(A|E_n)$$

6. Bayes' theorem If $E_1, E_2, ..., E_n$ are events which constitute a partition of sample space S, i.e., $E_1, E_2, ..., E_n$ are pair-wise disjoint and $E_1 \cup E_2 \cup ... \cup E_n = S$ and A be any event with nonzero probability, then

   $$P(E_i | A) = \frac{P(E_i)P(A|E_i)}{\sum_{j=1}^{n} P(E_j)P(A|E_j)}$$

7. A random variable is a real valued function whose domain is the sample space of a random experiment.

8. The probability distribution of a random variable X is the system of numbers

   $X : x_1 \quad x_2 \; ... \; x_n$

   $P(X): p_1 \quad p_2 \; \cdots \; p_n$

   where $p_i > 0, \sum_{i=1}^{n} p_i = 1, i = 1, 2..., n$

9. Let X be a random variable whose possible values $x_1, x_2, x_3, ..., x_n$ occur with probabilities $p_1, p_2, p_3, ..., p_n$ respectively. The mean of X, denoted by μ, is the number $\sum_{i=1}^{n} x_i p_i$.

   The mean of a random variable X is also called the expectation of X, denoted by $E(X)$.

10. Let X be a random variable whose possible values $x_1, x_2, ... x_n$ occur with probabilities $p(x_1), p(x_2), ... p(x_n)$ respectively.

    Let $\mu = E(X)$ be the mean of X. The variance of X, denoted by Var (X) or σ_x^2 is defined as

    $$\sigma_x^2 = \text{Var}(X) = \sum_{i=1}^{n} (x_i - \mu)^2 \, p(x_i)$$

    or equivalently $\sigma_x^2 = E(X - \mu)^2$

    The non-negative number

    $$\sigma_x = \sqrt{\text{Var}(X)} = \sqrt{\sum_{i=1}^{n} (x_i - \mu)^2 \, p(x_i)}$$

    is called the standard deviation of the random variable X.

11. $\text{Var}(X) = E(X^2) - [E(X)]^2$

12. Trials of a random experiment are called Bernoulli trials, if they satisfy the following conditions:

    (i) There should be a finite number of trials.

    (ii) The trials should be independent.

    (iii) Each trial should have exactly two outcomes – success or failure.

    (iv) The probability of success should remain the same in each trial. For binomial distribution $B(n, p), P(X = x) = {}^nC_x q^{n-x}, x = 0, 1, ..., n (q = 1 - p)$

QUIZ

1. If A and B are two independent events, then what is the probability of occurrence of at least one of A and B?
2. An urn contains 4 black and 2 white balls. Two balls are drawn from the urn, one after the other without replacement. What is the probability that both drawn balls are black?
3. Define the notation $P(E/F)$ for two events E and F.
4. What is an independent event?
5. A real-valued function whose domain is the sample space of a random experiment is known as _____
6. What are the conditions of Bernoulli's trial?
7. What is Baye's theorem?
8. A die marked 1, 2, 3 in red and 4, 5, 6 in green is tossed. Let A be the event, "the number is even" and B be the event, "the number is red". Are A and B independent events?
9. A coin tossed 3 times, in which

   A: head on third toss

   B: head on first on two tosses. What is $P(A/B)$?
10. E_1, E_2, E_3 ... are the events and their respective probabilities are p_1, p_2, p_3, ... then what is the probability that the first two events may occur and the rest may fail?

JEE MAIN

Objective Questions

One Option Correct

1. Two events A and B have probabilities 0.25 and 0.50 respectively. The probability that both A and B occur simultaneously is 0.14. Then the probability that neither A nor B occurs is ________.
(a) 0.39 (b) 0.25
(c) 0.11 (d) None

2. If A and B are two events such that $P(A) > 0$, and $P(B) \neq 1$, the $P(\bar{A}/\bar{B})$ is equal to ________.
(a) $1 - P(A/B)$ (b) $1-(\bar{A}/B)$
(c) $\dfrac{1-P(A\cup B)}{P(\bar{B})}$ (d) $\dfrac{P(\bar{A})}{P(\bar{B})}$

3. Three identical dice are rolled. The probability that the same number will appear on each of them is ________.
(a) 1/6 (b) 1/36
(c) 1/18 (d) 3/28

4. $P(A\cup B) = P(A\cap B)$ is possible only when ________.
(a) $P(A) = P(B)$ (b) $P(A) > P(B)$
(c) $P(A) < P(B)$ (d) None

5. A student appears for tests I, II and III. The student is successful if he passes either in tests I and II or test I and III. The probability of the student passing in tests I, II and III are p, q and ½ respectively. If the probability that the student is successful is ½, then ________.
(a) $p = q = 1$ (b) $p = q = ½$
(c) $p = 1, q = 0$ (d) $p = 1, q = ½$

6. A pair of fair dice is rolled together till a sum of either 5 or 7 is obtained. Then the probability that 5 comes before 7 is ________.
(a) 2/5 (b) 3/5
(c) 1/2 (d) 0

7. Let A and B be two events such that $P(A) = 0.3$ and $P(A\cup B) = 0.8$. If A and B are independent events, then P(B) is equal to ________.
(a) 3/7 (b) 4/7
(c) 5/7 (d) 6/7

8. Three faces of a fair die are yellow, two faces are red and one, blue. The die is tossed three times. The probability that the colours, yellow, red and blue appear in the first, second and the third tosses respectively is ________.
(a) 1/36 (b) 5/36
(c) 7/36 (d) 1/2

9. Three of the six vertices of a regular hexagon are chosen at random. The probability that the triangle with these three vertices is equilateral, equals ________.
(a) 1/2 (b) 1/5
(c) 1/10 (d) 1/20

10. There are four machines and it is known that exactly two of them are faulty. They are tested, one by one, in a random order till both the faulty machines are identified. Then the probability that only two tests are needed is ________.
(a) 1/3 (b) 1/6
(c) 1/2 (d) 1/4

11. Two numbers are selected randomly from the set S = {1, 2, 3, 4, 5, 6} without replacement, one by one. The probability that minimum of the two numbers is less than 4 is ________.
(a) 1/15 (b) 14/15
(c) 1/5 (d) 4/5

12. In three throws of two dice, the probability of throwing doublets not more than twice is ________.
(a) 1/6 (b) 5/72
(c) 215/216 (d) None

13. Bag A contains 2 white and 2 red balls and another bag B contains 4 white and 5 red balls. A ball is drawn and is found to be red. The probability that it was drawn from the bag B is ________.
(a) 25/52 (b) 1/2
(c) 10/19 (d) 5/19

14. Let A and B be two events such that $P(A) = 7/20$, $P(B) = 9/20$, $P(A \cup B) = 11/20$, $P(A\cup\bar{A}) = 1$, then the value of $P(\bar{A}\cup\bar{B})$ is equal to ________.
(a) 1/4 (b) 3/4
(c) 1/10 (d) None

15. A bag contains 5 brown and 4 white socks. A man pulls out two socks without replacement. The probability that they are of the same colour is ________.
(a) 5/108 (b) 1/6
(c) 5/18 (d) 4/9

16. In shuffling a pack of playing cards, four cards accidently dropped. The probability that the missing cards should be one from each suit is
(a) $\frac{1}{256}$ (b) $\frac{4}{20825}$
(c) $\frac{2197}{20825}$ (d) None

17. Four persons are selected at random from a group of 3 men, 2 women and 4 children. What is the chance that exactly 2 of them are children?
(a) 9/21 (b) 10/23
(c) 11/24 (d) 10/21

18. Out of 15 tickets marked with the numbers from 1 to 15, three are drawn at random. What is the chance that the numbers on them are in AP?
(a) 7/65 (b) 9/15
(c) 13/261 (d) None

19. Twelve balls are distributed among three boxes. The probability that the first box contains 3 ball is ________.
(a) $\frac{110}{9}\left(\frac{2}{3}\right)^{10}$ (b) $\frac{9}{110}\left(\frac{2}{3}\right)^{10}$
(c) $\frac{{}^{12}C_3}{12^3}.2^9$ (d) $\frac{{}^{12}C_3}{3^{12}}$

20. A fair coin is tossed a fix number of times. If the probability of getting 4 heads equals the probability of getting 7 heads, then the probability of getting 2 heads is ________.
(a) $\frac{25}{2048}$ (b) $\frac{3}{4096}$
(c) $\frac{1}{1024}$ (d) None

21. X is a random variable with probability function $P(X)={}^3C_x\left(\frac{1}{2}\right)^3, x=0,1,2,3$ the value of $E\left(X^2+2\right)$ is ________.
(a) 7 (b) 17/4
(c) 5 (d) 4

22. For a party, 8 guests are invited by a husband and his wife. They sit around a circular table for dinner. The probability that the husband and his wife sit together is ________.
(a) $\frac{2}{7}$ (b) $\frac{2}{9}$
(c) $\frac{1}{9}$ (d) $\frac{4}{9}$

23. The probability that A speaks the truth is $\frac{4}{5}$, while this probability for B is $\frac{3}{4}$. The probability that they contradict each other is ________.
(a) 4/5 (b) 1/5
(c) 7/20 (d) 3/20

24. The mean and the variance of a binomial distribution are 4 and 2 respectively. Then the probability of 2 successes is ________.
(a) $\frac{28}{256}$ (b) $\frac{219}{256}$
(c) $\frac{128}{256}$ (d) $\frac{37}{256}$

25. Three houses are available in a locality. Three persons apply for the houses. Each applies for one house without consulting the others. The probability that all the three apply for the same house is ________.
(a) $\frac{2}{9}$ (b) $\frac{1}{9}$
(c) $\frac{8}{9}$ (d) $\frac{7}{9}$

26. If the letters of word "ASSASSIN" are written down at random in a row, the probability that no two S's occur together, is ________.
(a) 1/14 (b) 1/35
(c) 1/7 (d) None of these.

27. A box contain 100 tickets numbered 1, 2, ..., 100. Two tickets are chosen at random. It is given that the maximum number on the tickets is not more than 10. Then the probability that the minimum on them is 5, is ________.
(a) 1/9 (b) 2/9
(c) 1/3 (d) 2/5

28. If $\frac{1+3p}{3}, \frac{1-p}{4}$ and $\frac{1-2p}{2}$ are the probabilities of three mutually exclusive events, then p lies in the interval $\frac{1+3p}{3}, \frac{1-p}{4}$ and $\frac{1-2p}{2}$, then, ________.
(a) $0 \le p < \frac{1}{2}$ (b) $\frac{1}{3} \le p \le \frac{1}{2}$
(c) $0 \le p \le \frac{1}{3}$ (d) None of these.

29. The probability that at least one of the events A and B occurs is 0.6. If A and B occur simultaneously with the probability 0.2, then $P(\bar{A})+P(\bar{B})$ is ________.

(a) 0.4 (b) 0.8
(c) 1.2 (d) 1.4

30. An unbiased die with faces marked 1, 2, 3, 4, 5 and 6 is rolled four times. Out of four face values obtained, the probability that the minimum face value is not less than 2 and the maximum face value is not greater than 5 is ________.

(a) 16/81 (b) 1/81
(c) 80/81 (d) 65/81

31. A box contains 24 identical balls of which 12 are white and 12 are black. The balls are drawn at random from the box, one at a time, with replacement. The probability that a white ball is drawn for the 4th time on the 7th draw is ________.

(a) 5/64 (b) 27/32
(c) 5/32 (d) 1/2

32. The probability of India winning a test match against West Indies is ½. Assuming independence from match to match, the probability that in a 5-match series, India's second win occurs off the third test is ________.

(a) 1/8 (b) 1/4
(c) 1/2 (d) 2/3

33. Three numbers are chosen at random without replacement from $\{1, 2, ..., 10\}$. The probability that the minimum of the chosen number is 3, or their maximum is 7, is ________.

(a) 3/10 (b) 11/40
(c) 11/50 (d) 1/50

34. If E and F are events with $P(E) \le P(F)$ and $P(E \cap F) > 0$, then

(a) occurrence of E $\Rightarrow$ occurrence of F.
(b) occurrence of F $\Rightarrow$ occurrence of E.
(c) non-occurrence of E $\Rightarrow$ non-occurrence of F.
(d) none of the above implications hold.

35. A fair coin is tossed repeatedly. If tail appears on first four tosses, then the probability of head appearing on the fifth equals ________.

(a) 1/2 (b) 1/32
(c) 31/32 (d) 1/5

36. If three distinct numbers are chosen randomly from the first 100 natural numbers, then the probability that all three of them are divisible by both 2 and 3 is ________.

(a) 4/25 (b) 4/35
(c) 4/33 (d) 4/1155

37. Given a throw of three unbiased dice show different faces, the probability that at least one face shows 6 is ________.

(a) 5/6 (b) 5/18
(c) 1/2 (d) 13/18

38. A six-faced die is so biased that it is twice as likely to show an even number as an odd number, when thrown. If it is thrown twice, the probability that the sum of two numbers thrown is even, is ________.

(a) 4/9 (b) 5/9
(c) 1/9 (d) None of these.

39. The probability that a student is not a swimmer is 1/5. What is the probability that out of 5 students, 4 are swimmers?

(a) ${}^5C_4\left(\frac{4}{5}\right)^4\frac{1}{5}$ (b) $\left(\frac{4}{5}\right)^4\frac{1}{5}$

(c) ${}^3C_1\frac{1}{5}\left(\frac{4}{5}\right)^4 \times {}^5C_4$ (d) None of these.

40. A person writes 4 letters and addresses 4 envelopes. If the letters are placed in the envelopes at random, what is the probability that all letters are not placed in the right envelopes?

(a) 1/24 (b) 11/24
(c) 15/24 (d) 23/24

41. The probability of event A occurring is 0.5 and event B occurring is 0.3. If A and B are mutually exclusive events, then the probability of neither A nor B occurring is ________.

(a) 0.6 (b) 0.5
(c) 0.7 (d) None of these.

42. If A and B are two independent events, the probability that both A and B occur is 1/8 and the probability that neither of them occurs is 3/8. The probability of the occurrence of A is ________.

(a) 1/2 (b) 1/3
(c) 3/4 (d) 1/5

43. If A and B are two events, that the value of the determinant chosen at random from all the determinants of order 2 with entries 0 or 1 only is, positive or negative respectively. Then ________.

(a) $P(A) > P(B)$ (b) $P(A) \le P(B)$
(c) $P(A) = P(B) = 1/2$ (d) None of these.

44. The probability of A = the probability of B = the probability of C $=\frac{1}{4}$, $P(A\cap B\cap C)=0$, $P(B\cap C)=0$, $P(A\cap C)=\frac{1}{8}$, and $P(A\cap B)=0$, the probability that at least one of the events A, B or C will occur, is _______.

(a) $\frac{5}{8}$ (b) $\frac{37}{64}$

(c) $\frac{3}{4}$ (d) 1

45. If $P(A\cap B)=\frac{1}{3}$, $P(A\cup B)=\frac{5}{6}$ and $P(A)=\frac{1}{2}$, then which one of the following is correct?

(a) A and B are independent events
(b) A are mutually exclusive events
(c) $P(A) = P(B)$
(d) $P(A) < P(B)$

46. One hundred identical coins, each with probability p of showing heads are tossed once. If $0<p<1$ and the probability of head showing on 50 coins is equal to that of head showing on 51 coins, the value of p is _______.

(a) $\frac{1}{2}$ (b) $\frac{51}{101}$

(c) $\frac{49}{101}$ (d) None of these.

47. An anti-aircraft gun can take a maximum of four shots at an enemy plane moving away from it. The probabilities of hitting the plane at the first, second, third and fourth shot are 0.4, 0.3, 0.2, and 0.1 respectively. What is the probability that the gun hits the plane?

(a) 0.6977 (b) 0.6967
(c) 0.6976 (d) None of these.

48. Five horses are in a race. Mr A selects two of the horses at random and bets on them. The probability that Mr A selected the winning horse is _______.

(a) 2/5 (b) 4/5
(c) 3/5 (d) 1/5

49. A random variable X has Poisson distribution with mean 2. Then $P(X>1.5)$ equals _______.

(a) $\frac{2}{e^2}$ (b) 0

(c) $1-\frac{3}{e^2}$ (d) $\frac{3}{e^2}$

50. Let A and B be two events such that $P(\overline{A\cup B})=\frac{1}{6}$, $P(A\cap B)=\frac{1}{4}$ and $P(\bar{A})=\frac{1}{4}$, where $\bar{A}$ stands for complement of event A. Then events A and B are

(a) equally likely and mutually exclusive.
(b) equally likely but not independent.
(c) independent but not equally likely.
(d) mutually exclusive and independent.

51. Four fair dice D_1, D_2, D_3 and D_4; each having six faces numbered 1, 2, 3, 4, 5 and 6 are rolled simultaneously. The probability that D_4 shows a number appearing on one of D_1, D_2 and D_3 is _______.

(a) $\frac{91}{216}$ (b) $\frac{108}{216}$

(c) $\frac{125}{216}$ (d) $\frac{127}{216}$

52. A signal which can be green or red with probability $\frac{4}{5}$ and $\frac{1}{5}$ respectively, is received by station A and then transmitted to station B. The probability of each station receiving the signal correctly is $\frac{3}{4}$. If the signal received at station B is green, then the probability that the original signal was green is _______.

(a) $\frac{3}{5}$ (b) $\frac{6}{7}$

(c) $\frac{20}{23}$ (d) $\frac{9}{20}$

53. Let ω be a complex cube root of unity with $\omega\neq 1$. A fair die is thrown three times. If r_1, r_2 and r_3 are the numbers obtained on the die, then the probability that $\omega^{r_1}+\omega^{r_2}+\omega^{r_3}=0$ is _______.

(a) $\frac{1}{18}$ (b) $\frac{1}{9}$

(c) $\frac{2}{9}$ (d) $\frac{1}{36}$

54. An experiment has 10 equally likely outcomes. Let A and B be non-empty events of the experiment. If A consists of 4 outcomes, the number of outcomes that B must have so that A and B are independent, is _______.

(a) 2, 4 or 8 (b) 3, 6 or 9
(c) 4 or 8 (d) 5 or 10

Questions from Previous Years

55. Let $S=\{1,2,3,4,5,6\}$. Then probability that a randomly chosen onto function g from S to S satisfying $g(3)=2g(1)$ is ______.

(a) $\frac{1}{10}$ (b) $\frac{1}{15}$

(c) $\frac{1}{5}$ (d) $\frac{1}{30}$

56. When a certain biased die is rolled, a particular face occurs with probability $\frac{1}{6}-x$ and its opposite face occurs with probability $\frac{1}{6}+x$. All other faces occur with probability $\frac{1}{6}$. Note that opposite faces sum to 7 in any die. If $0<x<\frac{1}{6}$, and the probability of obtaining total sum = 7, when such a die is rolled twice, is $\frac{13}{96}$, then the value of x is ______.

(a) $\frac{1}{16}$ (b) $\frac{1}{8}$

(c) $\frac{1}{9}$ (d) $\frac{1}{12}$

57. The probability that a randomly selected 2-digit number belongs to the set $\{n\in N:(2^n-2)\text{ is a multiple of }3\}$ is equal to ______.

(a) $\frac{1}{6}$ (b) $\frac{2}{3}$

(c) $\frac{1}{2}$ (d) $\frac{1}{3}$

58. Four dice are thrown simultaneously and the numbers shown on these dice are recorded in 2×2 matrices. The probability that such formed matrices have all different entries and are non-singular, is ______.

(a) $\frac{45}{162}$ (b) $\frac{23}{81}$

(c) $\frac{22}{81}$ (d) $\frac{43}{162}$

59. Let A, B and C be three events such that the probability that exactly one of A and B occurs is $(1-k)$, the probability that exactly one of B and C occurs is $(1-2k)$, the probability that exactly one of C and A occurs is $(1-k)$ and the probability of all A, B and C occur simultaneously is k^2, where $0<k<1$. Then the probability that at least one of A, B and C occur is:

(a) greater than $\frac{1}{8}$ but less than $\frac{1}{4}$.

(b) greater than $\frac{1}{2}$.

(c) greater than $\frac{1}{4}$ but less than $\frac{1}{2}$.

(d) exactly equal to $\frac{1}{2}$.

60. Let a computer program generate only the digits 0 and 1 to form a string of binary numbers with probability of occurrence of 0 at even places as $\frac{1}{2}$ and probability of occurrence of 0 at the odd place as $\frac{1}{3}$. Then the probability that '10' is followed by '01' is equal to ______.

(a) $\frac{1}{18}$ (b) $\frac{1}{3}$

(c) $\frac{1}{6}$ (d) $\frac{1}{9}$

61. A fair coin is tossed a fixed number of times. If the probability of getting 7 heads is equal to probability of getting 9 heads, then the probability of getting 2 heads is ______.

(a) $\frac{15}{2^{13}}$ (b) $\frac{15}{2^{12}}$

(c) $\frac{15}{2^{8}}$ (d) $\frac{15}{2^{14}}$

62. A fair die is tossed until six is obtained on it. Let X be the number of required tosses, then the conditional probability $P(X\geq 5>2)$ is ______.

(a) $\frac{125}{216}$ (b) $\frac{11}{36}$

(c) $\frac{5}{6}$ (d) $\frac{25}{36}$

63. Let in a binomial distribution consisting of 5 independent trials, the probabilities of exactly 1 and 2 successes are 0.4096 and 0.2048 respectively. Then the probability of getting exactly 3 successes is equal to ______.

(a) $\frac{80}{243}$ (b) $\frac{32}{625}$

(c) $\frac{40}{243}$ (d) $\frac{128}{625}$

64. A random variable X has the following probability distribution:

X	1	2	3	4	5
P(X)	K^2	$2K$	K	$2K$	$5K^2$

Then, P($X > 2$) is equal to:

(a) $\frac{7}{12}$ (b) $\frac{1}{36}$

(c) $\frac{1}{6}$ (d) $\frac{23}{36}$

65. In a game, two players A and B take turns in throwing a pair of fair dice starting with player A and the total of scores on the two dice, in each throw, is noted. A wins the game if he throws a total of 6 before B throws a total of 7 and B wins the game if he throws a total of 7 before A throws a total of six. The game stops as soon as either of the players wins. The probability of A winning the game is _______.

(a) $\frac{5}{31}$ (b) $\frac{31}{61}$

(c) $\frac{5}{6}$ (d) $\frac{30}{61}$

66. A die is thrown two times and the sum of the scores appearing on the die is observed to be a multiple of 4. Then the conditional probability that the score 4 has appeared at least once is _______.

(a) $\frac{1}{4}$ (b) $\frac{1}{3}$

(c) $\frac{1}{8}$ (d) $\frac{1}{9}$

67. The probability that a randomly chosen 5-digit number is made from exactly two digits is _______.

(a) $\frac{135}{10^4}$ (b) $\frac{121}{10^4}$

(c) $\frac{150}{10^4}$ (d) $\frac{134}{10^4}$

68. Box I contains 30 cards numbered 1 to 30 and Box II contains 20 cards numbered 31 to 50. A box is selected at random and a card is drawn from it. The number on the card is found to be a non-prime number. The probability that the card was drawn from Box I is:

(a) $\frac{2}{3}$ (b) $\frac{8}{17}$

(c) $\frac{4}{17}$ (d) $\frac{2}{5}$

69. Let E^C denote the complement of an event E. Let E_1, E_2 and E_3 be any pairwise independent events with $P(E_1) > 0$ and $P(E_1 \cap E_2 \cap E_3) = 0$. Then $P\left(E_2^C \cap E_3^C / E_1\right)$ is equal to:

(a) $P\left(E_2^C\right) + P(E_3)$ (b) $P\left(E_3^C\right) - P\left(E_2^C\right)$

(c) $P(E_3) - P\left(E_2^C\right)$ (d) $P\left(E_3^C\right) - P(E_2)$

70. In a box, there are 20 cards, out of which 10 are labelled as A and the remaining 10 are labelled as B. Cards are drawn at random, one after the other and with replacement, till a second A-card is obtained. The probability that the second A-card appears before the third B-card is:

(a) $\frac{9}{16}$ (b) $\frac{11}{16}$

(c) $\frac{13}{16}$ (d) $\frac{15}{16}$

71. Let A and B be two independent events such that $P(A) = \frac{1}{3}$ and $P(B) = \frac{1}{6}$. Then, which of the following is TRUE?

(a) $P(A/B) = \frac{2}{3}$ (b) $P(A/B') = \frac{1}{3}$

(c) $P(A'/B') = \frac{1}{3}$ (d) $P(A/(A \cup B)) = \frac{1}{4}$

72. An unbiased coin is tossed 5 times. Suppose that a variable X is assigned the value k when k consecutive heads are obtained for $k = 3, 4, 5$; otherwise X takes the value -1. Then the expected value of X, is _______.

(a) $\frac{3}{16}$ (b) $\frac{1}{8}$

(c) $-\frac{3}{16}$ (d) $-\frac{1}{8}$

73. In a workshop, there are five machines and the probability of any one of them to be out of service on a day is $\frac{1}{4}$. If the probability that at most two machines will be out of service on the same day is $\left(\frac{3}{4}\right)^3 k$, then k is equal to _______.

(a) $\frac{17}{8}$ (b) $\frac{17}{4}$

(c) $\frac{17}{2}$ (d) 4

74. Let a random variable X have a binomial distribution with mean 8 and variance 4. If $(X \le 2) = \frac{k}{2^{16}}$, then k is equal to ________.

(a) 17 (b) 121
(c) 1 (d) 137

75. Minimum number of times a fair coin must be tossed so that the probability of getting at least one head is more than 99 %, is ________.
(a) 5 (b) 6
(c) 8 (d) 7

76. Let A and B be two non-null events such that $A \subset B$. Then, which of the following statements is always correct?

(a) $P(A|B) = P(B) - P(A)$

(b) $P(A|B) \ge P(A)$

(c) $P(A|B) \le P(A)$

(d) $P(A|B) = 1$

77. If the probability of hitting a target by a shooter, in any shot, is $\frac{1}{3}$, then the minimum number of independent shots at the target required by him so that the probability of hitting the target at least once is greater than $\frac{5}{6}$, is ________.
(a) 3 (b) 6
(c) 5 (d) 4

78. An urn contains 5 red and 2 green balls. A ball is drawn at random from the urn. If the drawn ball is green, then a red ball is added to the urn and if the drawn ball is red, then a green ball is added to the urn; the original ball is not returned to the urn. Now, a second ball is drawn at random from it. The probability that the second ball is red is ________.

(a) $\frac{21}{49}$ (b) $\frac{27}{49}$

(c) $\frac{26}{49}$ (d) $\frac{32}{49}$

79. If two different numbers are taken from the set (0, 1, 2, 3, …, 10), then the probability that their sum as well as absolute difference are both multiple of 4, is ________.

(a) $\frac{7}{55}$ (b) $\frac{6}{55}$

(c) $\frac{12}{55}$ (d) $\frac{14}{55}$

80. An unbiased coin is tossed eight times. The probability of obtaining at least one head and at least one tail is ________.

(a) $\frac{255}{256}$ (b) $\frac{127}{128}$

(c) $\frac{63}{64}$ (d) $\frac{1}{2}$

81. A box contains 15 green and 10 yellow balls. If 10 balls are randomly drawn, one by one, with replacement, then the variance of the number of green balls drawn is ________.

(a) $\frac{6}{25}$ (b) $\frac{12}{5}$

(c) 6 (d) 4

82. Let two fair six-faced dice A and B be thrown simultaneously. If E_1 is the event that die A shows up four, E_2 is the event that die B shows up two and E_3 is even that the sum of numbers on both dice is odd, then which of the following statements is NOT true?
(a) E_1 and E_3 are independent.
(b) E_1, E_2 and E_3 are independent.
(c) E_1 and E_2 are independent.
(d) E_2 and E_3 are independent.

83. An experiment succeeds twice as often as it fails. The probability of at least 5 successes in the six trials of this experiment is ________.

(a) $\frac{496}{729}$ (b) $\frac{192}{729}$

(c) $\frac{240}{729}$ (d) $\frac{256}{729}$

84. Let A and B be two events such that $P(\overline{A \cup B}) = \frac{1}{6}$, $P(\overline{A \cap B}) = \frac{1}{4}$ and $P(\bar{A}) = \frac{1}{4}$, where $\bar{A}$ stands for the complement of the event A. Then the events A and B are
(a) independent but not equally likely.
(b) independent and equally likely.
(c) mutually exclusive and independent.
(d) equally likely but not independent.

Integer and Numerical Value Type

1. An urn contains three red balls and n white balls. Mr A draws two balls together from the urn. The probability that they have the same color is 1/ 2. Mr B draws one ball from the urn, notes its color and replaces it. He then draws a second ball from the urn and finds the probability that both balls have the same color is, 5/8. The possible value of n is ________.

2. Suppose A and B are two events with $P(A)=0.5$ and $P(A\cup B)=0.8$. Let $P(B)=p$ if A and B are mutually exclusive and $P(B)=q$ if A and B are independent events, then the value of q/p is ________.

3. Two cards are drawn from a well-shuffled pack of 52 cards. The probability that one is a heart card and the other is a king is p, then the value of $104p$ is ______.

4. An unbiased normal coin is tossed n times. Let
E_1: event that both heads and tails are present in n tosses.
E_2: event that the coin shows up heads at most once.
The value of n for which E_1 and E_2 are independent is ______.

Questions from Previous Years

5. An electric instrument consists of two units. Each unit must function independently for the instrument to operate. The probability that the first unit functions is 0.9 and that of the second unit is 0.8. The instrument is switched on and it fails to operate. If the probability that only the first unit failed and second unit is functioning is p, then $98p$ is equal to ______.

6. Let $B_i\ (i=1,2,3)$ be three independent events in a sample space. The probability that only B_1 occurs is α, only B_2 occurs is β and only B_3 occurs is γ. Let p be the probability that none of the events B_i occurs and these 4 probabilities satisfy the equations $(\alpha\ 2\ \beta)\,p=\alpha\beta$ and $(\beta-3\gamma)\,p=2\beta\gamma$. (All the given probabilities are assumed to lie in the interval (0, 1)). Then $\dfrac{P(B_1)}{P(B_3)}$ is equal to _______.

7. Four fair dice are thrown independently 27 times. Then the expected numbers of times, at least two dice show up a three or a five, is ________.

8. In a bombing attack, there is 50% chance that a bomb will hit the target. At least two independent hits are required to destroy the target completely. Then the minimum number of bombs that must be dropped to ensure that there is at least 99% chance of completely destroying the target, is ________.

9. The probability of a man hitting a target is $\dfrac{1}{10}$. The least number of shots required, so that the probability of his hitting the target at least once is greater than $\dfrac{1}{4}$, is ________.

Assertion and Reason

(a) Statement 1 is True, Statement 2 is True; Statement 2 is a correct explanation for Statement 1.

(b) Statement 1 is True, Statement 2 is True; Statement 2 is NOT a correct explanation for Statement 1

(c) Statement 1 is True, Statement 2 is False

(d) Statement l is False, Statement 2 is True

1. **Statement 1:** $x+y+z=5, xy+yz+zx=3$ and $x+2y\neq 5\ (x,y,z\in R)$, then the probability of x to be only positive is $\dfrac{13}{16}$.
Statement 2: If $x+y+z=5$ and $xy+yz+zx=3$ then maximum and minimum values of x, y and z are same.

2. **Statement 1:** Two non-negative integers are chosen at random. The probability that the sum of their squares is divisible by 5 is 9 / 25.
Statement 2: If the unit place in any number is zero then number is only divisible by 5.

3. **Statement 1:** If A and B are two events such that $0<P(A), P(B)<1$, then
$$P(A/\bar{B})+P(\bar{A}/\bar{B})=\frac{3}{2}.$$
Statement 2: If A and B are two events such that $0<P(A), P(B)<1$, then
$$P(A/B)=\frac{P(A\cap B)}{P(B)} \text{ and}$$
$$P(\bar{B})=P(A\cap\bar{B})+P(\bar{A}\cap\bar{B}).$$

JEE ADVANCED

Objective Questions

One Option Correct

1. Let set A have 7 elements and set B have 5 elements. If one function is selected from all possible defined functions from A to B, then the probability that it is onto is ________.

   (a) $\frac{7!\times 2}{3\times 5^6}$ (b) $\frac{7!}{10\times 5^6}$

   (c) $\frac{7!}{5^6}$ (d) $\frac{7!}{5^7}$

2. 3 integers are chosen at random from the set of first 20 natural numbers. The chance that their product is a multiple of 3, is ________.

   (a) $\frac{194}{285}$ (b) $\frac{1}{57}$

   (c) $\frac{13}{19}$ (d) $\frac{3}{4}$

3. Three critics review a book. Odds in favour of the book are 5 : 2, 4 : 3 and 3 : 4 respectively for the three critics. The probability that majority are in favour of the book is ________.

   (a) $\frac{35}{49}$ (b) $\frac{125}{343}$

   (c) $\frac{164}{343}$ (d) $\frac{209}{343}$

4. Numbers are selected at random, one at a time, from the numbers 00, 01, 02, …, 99 with replacement. An event E occurs if and only if the product of the digits of a selected number is 18. If four numbers are selected, the probability that the event E occurs 3 times is equal to ________.

   (a) $\frac{37}{390625}$ (b) $\frac{34}{390625}$

   (c) $\frac{96}{390625}$ (d) None of these.

5. A ten-digit number is formed without repeating any digit. The probability that the difference of the digits at equal distances from the beginning and the end is always 1, is ________.

   (a) $\frac{17}{1944}$ (b) $\frac{4}{27}$

   (c) $\frac{1}{945}$ (d) $\frac{34}{243}$

6. S_1: 5 letters are selected from the letters of word MISSISIPPI. The probability, that 3 of them are alike and other two are also alike but their kind is different from those of 3 alike, is $\frac{23}{^{10}C_5}$.

   S_2: A submatrix of order 2 × 2 is randomly selected from a matrix of order 5 × 4. The probability that the rows of the selected matrix are adjacent rows of the parent matrix, is $\frac{3}{5}$.

   S_3: A, B and C are three shooters. Their probabilities of hitting a target are $\frac{4}{5}, \frac{3}{4}$ and $\frac{2}{3}$ respectively. If all of them make an attempt on the target and it is found that two of them have missed out, then the probability that A has hit the target is equal to $\frac{2}{5}$.

   (a) TFT (b) FFT

   (c) FTF (d) TTF

7. S_1: Two persons, each make a single throw with a die. The probability they get equal values is P_1. Four persons, each make a single throw and the probability of exactly three being equal is P_2. Then P_1 greater than P_2.

   S_2: Each of A and B throw 2 dice. If A throws 9, then B's probability of throwing a higher number is $\frac{1}{6}$.

   S_3: If $P(A_1 \cup A_2) = 1 - P(A_1^c).P(A_2^c)$ then A_1 and A_2 are independent.

   S_4: If the events A, B, C are independent, then A, B, C are independent.

   (a) TTTT (b) TTFT

   (c) TFTF (d) FTTF

Questions from Previous Years

8. Consider three sets $E_1 = \{1,2,3\}, F_1 = \{1,3,4\}$ and $G_1 = \{2,3,4,5\}$. Two elements are chosen at random, without replacement, from the set E_1 and let S_1 denote the set of these chosen elements.

   Let $E_2 = E_1 - S_1$ and $F_2 = F_1 \cup S_1$. Now two elements are chosen at random, without replacement, from the set F_2 and let S_2 denote the set of these chosen elements.

   Let $G_2 = G_1 \cup S_2$. Finally, two elements are chosen at random, without replacement, from the set G_2 and let S_3 denote the set of these chosen elements.

Let $E_3 = E_2 \cup S_3$. Given that $E_1 = E_3$, let p be the conditional probability of the event $S_1 = \{1,2\}$. Then the value of p is ______.

(a) $\frac{1}{5}$ (b) $\frac{3}{5}$

(c) $\frac{1}{2}$ (d) $\frac{2}{5}$

9. Let C_1 and C_2 be two biased coins such that the probabilities of getting a head in a single toss are $\frac{2}{3}$ and $\frac{1}{3}$, respectively. Suppose α is the number of heads that appear when C_1 is tossed twice, independently, and suppose β is the number of heads that appear when C_2 is tossed twice, independently. Then the probability that the roots of the quadratic polynomial $x^2 - \alpha x + \beta$ are real and equal, is ______.

(a) $\frac{40}{81}$ (b) $\frac{20}{81}$

(c) $\frac{1}{2}$ (d) $\frac{1}{4}$

10. Three randomly chosen non-negative integers x, y and z are found to satisfy the equation $x+y+z=10$. Then the probability that z is even, is ______.

(a) $\frac{36}{55}$ (b) $\frac{6}{11}$

(c) $\frac{1}{2}$ (d) $\frac{5}{11}$

11. A computer producing factory has only two plants T_1 and T_2. Plant T_1 produces 20% and plant T_2 produces 80% of the total computers produced. 7% of computers produced in the factory turn out to be defective: It is known that P(computer turns out to be defective given that it is produced in plant T_1) = 10P(computer turns out to be defective given that it is produced in plant T_2), where P(E) denotes the probability of an event E. A computer produced in the factory is randomly selected and it does not turn out to be defective. Then the probability that it is produced in plant T_2 is ______.

(a) $\frac{36}{73}$ (b) $\frac{47}{79}$

(c) $\frac{78}{93}$ (d) $\frac{75}{83}$

12. Three boys and two girls stand in a queue. The probability that the number of boys ahead of every girl is at least one more than the number of girls ahead of her, is ______.

(a) $\frac{1}{2}$ (b) $\frac{1}{3}$

(c) $\frac{2}{3}$ (d) $\frac{3}{4}$

One or More Than One Option Correct

1. A bag initially contains one red and two blue balls. An experiment consists of selecting a ball at random, noting its colour and replacing it together with an additional ball of the same colour. If three such trials are made, then:

(a) probability that at least one blue ball is drawn is 0.9.

(b) probability that exactly one blue ball is drawn is 0.2.

(c) probability that all the drawn balls are red, given that all the drawn balls are of same colour is 0.2.

(d) probability that at least one red ball is drawn is 0.6.

2. Two real numbers, x and y are selected at random. Given that $0 \le x \le 1; 0 \le y \le 1$. Let A be the event that $y^2 \le x$; B be the event that $x^2 \le y$ then:

(a) $P(A \cap B) = \frac{1}{3}$

(b) A and B are exhaustive events

(c) A and B are mutually exclusive

(d) A and B are independent events.

3. Two persons A and B have $n+1$ and n coins respectively, which they toss simultaneously. Then the probability that A will have more heads than B is ______.

(a) $\frac{1}{2}$ (b) $> \frac{1}{2}$

(c) $< \frac{1}{2}$ (d) $> \frac{1}{3}$

4. A sum of money is rounded off to the nearest rupee. The probability that the round-off error is at most 10 paise, is ______.

(a) 63/300

(b) 11/100

(c) 3/25

(d) 21/100

5. Consider the Cartesian plane R^2 and let X denote the subset of points for which both co-ordinates are integers. A coin of diameter $\frac{1}{2}$ is tossed randomly into the plane. The probability p that the coin covers a point of X satisfies:

(a) $p=\frac{\pi}{16}$ (b) $p<\frac{\pi}{3}$

(c) $p>\frac{\pi}{30}$ (d) $p=\frac{1}{4}$

6. Let X and Y be two events such that $P(X|Y)=\frac{1}{2}$, $P(Y|X)=\frac{1}{3}$ and $P(X\cap Y)=\frac{1}{6}$. Which of the following is (are) correct?

(a) $P(X\cup Y)=\frac{2}{3}$

(b) X and Y are independent

(c) X and Y are not independent

(d) $P(X^c\cap Y)=\frac{1}{3}$

7. A ship is fitted with three engines E_1, E_2 and E_3. The engines function independently of each other with respective probabilities $\frac{1}{2},\frac{1}{4}$ and $\frac{1}{4}$. For the ship to be operational, at least two of its engines must function. Let X denote the event that the ship is operational and let X_1, X_2 and X_3 denote, respectively, the events that the engines E_1, E_2 and E_3 are functioning. Which of the following is(are) true?

(a) $P[X_1^c|X]=\frac{3}{16}$

(b) P [Exactly two engines of the ship are functioning $|X]=\frac{7}{8}$

(c) $P[X|X_2]=\frac{5}{16}$

(d) $P[X|X_1]=\frac{7}{16}$

8. Let E and F be two independent events. The probability that exactly one of them occurs is $\frac{11}{25}$ and the probability of none of them occurring is $\frac{2}{25}$. If $P(T)$ denotes the probability of occurrence of the event T, then

(a) $P(E)=\frac{4}{5}, P(F)=\frac{3}{5}$

(b) $P(E)=\frac{1}{5}, P(F)=\frac{2}{5}$

(c) $P(E)=\frac{2}{5}, P(F)=\frac{1}{5}$

(d) $P(E)=\frac{3}{5}, P(F)=\frac{4}{5}$

9. A and B are two events, defined as follows:
A: It rains today with P(A) = 40%
B: It rains tomorrow with P(B) = 50%
Also, P(It rains today and tomorrow) = 30%
Also, $E_1 : P((A\cap B)/(A\cup B))$ and
$E_2 : P(\{(A\cap\bar{B}) \text{ or } (B\cup\bar{A})\}/(A\cup B))$.
Then which of the following is/are true?

(a) A and B are independent

(b) P(A/B) < P(B/A)

(c) E_1 and E_2 are equiprobable

(d) $P(A/(A\cup B)) = P(B/(A\cup B))$

10. Two numbers are randomly selected and multiplied. Consider two events E_1 and E_2 defined as

E_1: Their product is divisible by 5.
E_2: Unit's places in their product is 5.
Which of the following statement is/are correct?

(a) E_1 is twice as likely to occur as E_2

(b) E_1 and E_2 are disjoint

(c) $P(E_2/E_1)=1/4$

(d) $P(E_1/E_2)=1$

11. Which of the following statement(s) is/are correct?

(a) Three coins are tossed once. At least two of them must land the same way. No matter whether they land heads or tails, the third coin is equally likely to land either the same way or oppositely. So, the chance that all the three coins land the same way is 1/2.

(b) Let 0 < P(B) < 1 and $P(A/B) = P(A/B^C)$. Then A and B are independent.

(c) Suppose an urn contains w white and b black balls and a ball is drawn from it and is replaced along with d additional balls of the same color. Now a second ball is drawn from it. The probability that the second drawn ball is white is independent of the value of d.

(d) A, B, C simultaneously satisfy

$P(ABC) = P(A)\,P(B)\,P(C)$

$P(AB\bar{C}) = P(A)P(B)P(\bar{C})$

$P(A\bar{B}C) = P(A)P(\bar{B})P(C)$

$P(A-BC) = P(\bar{A})P(B)P(C)$

Then A, B, C are independent.

12. $P(A) = 3/8$; $P(B) = 1/2$; $P(A \cup B) = 5/8$, which of the following do/does hold good?
(a) $P(A^C/B) = 2P(A/B^C)$
(b) $P(B) = P(A/B)$
(c) $15P(A^C/B^C) = 8P(B/A^C)$
(d) $P(A/B^C) = (A \cap B)$

Questions from Previous Years

13. There are three bags B_1, B_2, and B_3. The bag B_1 contains 5 red and 5 green balls, B_2 contains 3 red and 5 green balls, and B_3 contains 5 red and 3 green balls. Bags B_1, B_2 and B_3 have probabilities $\frac{3}{10}, \frac{3}{10}$ and $\frac{4}{10}$ respectively of being chosen. A bag is selected at random and a ball is chosen at random from the bag. Then which of the following options is/are correct?
(a) The probability that the selected bag is B_3 and the chosen ball is green equals $\frac{3}{10}$.
(b) The probability that the chosen ball is green, given that the selected bag is B_3, equals $\frac{3}{8}$
(c) The probability that the selected bag is B_3, given that the chosen ball is green, equals $\frac{5}{13}$
(d) The probability that the chosen ball is green equals $\frac{39}{80}$.

14. Let X and Y be two events such that $P(X) = \frac{1}{3}, P(X|Y) = \frac{1}{2}$ and $P(Y|X) = \frac{2}{5}$. Then ________.
(a) $P(Y) = \frac{4}{15}$
(b) $P(X'|Y) = \frac{1}{2}$
(c) $P(X \cap Y) = \frac{1}{5}$
(d) $P(X \cup Y) = \frac{2}{5}$

15. Four persons independently solve a certain problem correctly with probabilities $\frac{1}{2}, \frac{3}{4}, \frac{1}{4}, \frac{1}{8}$. Then the probability that the problem is solved correctly by at least one of them is ________.
(a) $\frac{235}{256}$
(b) $\frac{21}{256}$
(c) $\frac{3}{256}$
(d) $\frac{253}{256}$

Matrix Type

1. Match the following:

Column I	Column II
(A) One ball is drawn from a bag containing 4 balls and is found to be white. The events that the bag contains "1 white", "2 white", "3 white" and "4 white" are equally likely. If the probability that all the balls are white is $\frac{p}{15}$, then the value of p is	(P) 9
(B) From a set of 12 persons, if the number of different selections of a committee, its chair-person and its secretary (possibly same as chair-person) is $13.2^{10}m$, then the value of m is	(Q) 3
(C) If $x, y, z > 0$ and $x + y + z = 1$, then the least value of $\frac{5x}{2-x} + \frac{5y}{2-y} + \frac{5z}{2-z}$ is	(R) 8
(D) If $\sum_{k=1}^{12} 12k.{}^{12}C_k.{}^{11}C_{k-1}$ is equal to $\frac{12 \times 21 \times 19 \times 17 \times \ldots \times 3}{11!} \times 2^{12} \times p$ is	(S) 6

2. A bag contains some white and some black balls, all combinations being equally likely. The total number of balls in the bag is 12. Four balls are drawn at random from the bag without replacement. Now match the entries from the following two columns:

Column I	Column II
(A) Probability that all the four balls are black is equal to	(P) $\frac{14}{33}$
(B) If the bag contains 10 black and 2 white balls then the probability that all four balls are black is equal to	(Q) $\frac{1}{5}$
(C) If all the four balls are black, then the probability that the selected bag contains 10 black balls, is equal to	(R) $\frac{70}{429}$
(D) Probability that two balls are black and two are white is	(S) $\frac{13}{110}$
	(T) $\frac{13}{165}$

3. An urn contains four black and eight white balls. Three balls are drawn from the urn without replacement. Three events are defined on this experiment.
A: Exactly one black ball is drawn.
B: All balls drawn are of the same color.
C: The third-drawn ball is black.
Match the entries of List I with none, one or more entries of List II.

List I	List II
(A) The events A and B are	(P) mutually exclusive
(B) The events B and C are	(Q) independent
(C) The events C and A are	(R) neither independent nor mutually exclusive
(D) The events A, B and C are	(S) exhaustive

4. An urn contains r red balls and b black balls. Now, match the following lists:

List I	List II
(A) If the probability of getting two red balls in the first two draws (without replacement) is ½, then value of r can be	(P) 10
(B) If the probability of getting two red balls in the first two draws (without replacement) is 1/2 and b is an even number, then r can be	(Q) 3
(C) If the probability of getting two red balls in first four draws (with replacement) from the urn is 3/8 and $b = 10$, then r can be	(R) 8
(D) If the probability of getting exactly n red balls in $2n$ draws (with replacement) is equal to probability of getting exactly n black balls in $2n$ draws (with replacement), then the ratio r/b can be	(S) 2

5. Let A and B are two independent events such that P(A) = 1/3 and P(B) = 1/4. Then match the following lists:

List I	List II
(A) $P(A\cup B)$ is equal to	(P) 1/12
(B) $P(A/A\cup B)$ is equal to	(Q) 1/2
(C) $P(B/A'\cap B')$ is equal to	(R) 2/3
(D) $P(A'/B')$ is equal to	(S) 0

Integer and Numerical Value Type

1. An urn contains 3 white balls, 5 black balls and 2 red balls. Two persons draw balls in turn, without replacement. The person who draws a white ball wins the game. If a red ball is drawn, the game is a tie. Suppose A_1 = {the player who begins the game is the winner}, A_2 = {the second participant is the winner} and B = {the game is a tie}. If λ P(B) = 2, then find the value of λ.

2. There are $N+1$ identical boxes, each containing N wall clocks. The rth box contains $(r-1)$ defective and $(N-r+1)$ effective clocks for $1\le r\le N+1$. A box is selected at random and from this box, a clock is chosen at random and is found to be effective, the probability that it is from the kth box is $\dfrac{2N-2k+\lambda}{N^2+N}$. Then find λ.

3. 5 girls and 10 boys sit at random in a row having 15 chairs numbered as 1 to 15. If the probability that the end seats are occupied by the girls and between any two girls, an odd number of boys take seat is $\dfrac{n}{3003}$, then find the value of n.

4. A is a set containing n elements. A subset P of A is chosen at random. The set A is reconstructed by replacing the elements of the subset P. A subset Q of A is again chosen at random. If the probability such that $P\cap Q$ contains 2 elements is $\dfrac{{}^nC_a.3^{n-b}}{4^n}$, then find $a+b$.

Questions from Previous Years

5. The probability that a missile hits a target successfully is 0.75. In order to destroy the target completely, at least three successful hits are required. Then the minimum number of missiles that have to be fired so that the probability of completely destroying the target is NOT less than 0.95 is _________.

6. Two fair dice, each with faces numbered 1, 2, 3, 4, 5 and 6, are rolled together and the sum of the numbers on the faces is observed. This process is repeated till the sum is either a prime number or a perfect square. Suppose the sum turns out to be a perfect square before it turns out to be a prime number. If p is the probability that this perfect square is an odd number, then the value of $14p$ is _______.

7. Let S be the sample space of all 3 × 3 matrices with entries from the set {0, 1}. Let the events E_1 and E_2 be given by $E_1=\{A\in S:\det A=0\}$ and $=\{A\in S:\text{sum of entries of A is 7}\}$.
If a matrix is chosen at random from S, then the conditional probability $P(E_1/E_2)$ equals_____.

8. The minimum number of times a fair coin needs to be tossed, so that the probability of getting at least two heads is at least 0.96 is _______.

9. Of the three independent events E_1, E_2 and E_3, the probability that only E_1 occurs is α, only E_2 occurs is β and only E_3 occurs is γ. Let the probability p that none of events E_1, E_2 or E_3 occurs satisfy the equations $(\alpha\ 2\beta)p=\alpha\beta$ and $(\beta-3\gamma)p=2\beta\gamma$. All the given probabilities are assumed to lie in the interval (0, 1).
Then $\dfrac{\text{Probability of occurrence of } E_1}{\text{Probability of occurrence of } E_3}$
= __________.

Comprehension Passage

Comprehension 1

Sania Mirza is to play with Sharapova in a three-set match. For a particular set the probability of Sania winning the set is y and if she wins, the probability of her winning the next set becomes $\sqrt{y}$; else the probability that she wins the next one becomes y^2. There is no possibility that a set is to be abandoned. R is the probability that Sania wins the first set.

1. If $R=\frac{1}{2}$ then the probability that match will end in the first two sets is nearly equal to ______.
   (a) 0.73 (b) 0.95
   (c) 0.51 (d) 0.36

2. If $R=\frac{1}{2}$ and Sania wins the second set, the probability that she has won first set as well is nearly equal to ______.
   (a) 0.74 (b) 0.46
   (c) 0.26 (d) 0.54

3. If Sania loses the first set, then the values of R such that her probability of winning the match is still larger than that of her loosing are given by ______.
   (a) $R\in\left(\frac{1}{2},1\right)$ (b) $R\in\left(\left(\frac{1}{2}\right)^{\frac{1}{3}},1\right)$
   (c) $R\in\left(\left(\frac{1}{2}\right)^{\frac{3}{2}},1\right)$ (d) No values of R

Comprehension 2

Let $n=10k+r$ when $k,r\in N, 0\le r\le 9$. A number a is chosen at random from the set $\{1,2,\ldots n\}$ and let p_n denote the probability that a^2-1 is divisible by 10.

1. If $r=0$, p_n equals ______.
   (a) $\frac{2k}{n}$ (b) $\frac{(k+1)}{n}$
   (c) $\frac{(2k+1)}{n}$ (d) $\frac{k}{n}$

2. If $r=9$, p_n equals ______.
   (a) $\frac{2k}{n}$ (b) $\frac{2(k+1)}{n}$
   (c) $\frac{(2k+1)}{n}$ (d) $\frac{k}{n}$

3. If $1\le r\le 8$, p_n equals ______.
   (a) $\frac{(2k-1)}{n}$ (b) $\left(\frac{2k}{n}\right)$
   (c) $\frac{(2k+1)}{n}$ (d) $\frac{k}{n}$

Comprehension 3

A player A plays a game against a machine. At each round, he deposits one rupee in a slot and then flips a coin which has a probability p of showing a head. If the flipped coin shows head, he gets back the rupee he deposited and one more rupee from the machine, else he loses his rupee. Let A start with 10 one-rupee coins and $q=1-p$ (the probability of showing a tail). Then:

1. The probability that he will be drained out of all of his one-rupee coins exactly at the eleventh round is ______.
   (a) q^{11} (b) $1-q^{11}$
   (c) $pq^{10}+q^{11}$ (d) 0

2. The probability that all his money will be finished exactly at the twelfth round is ______.
   (a) q^{12} (b) $1-q^{12}$
   (c) ${}^{10}C_1pq^{11}$ (d) ${}^{12}C_2p^2q^{10}$

3. The probability that he is left with no money by the 14th round or earlier is ______.
   (a) $q^{10}\left(1+10pq+65p^2q^2\right)$
   (b) $q^{14}\left(p^2q+36pq+7\right)$
   (c) $q^{12}+3pq^{13}+3p^{13}q+p^{12}$
   (d) $1-{}^{10}C_1pq^{11}-{}^{10}C_2p^2q^{12}$

Comprehension 4

A fair die tossed repeatedly until a six is obtained. Let X denote the number of tosses required.

1. The probability that $X=3$ equals ______.
   (a) $\frac{25}{216}$ (b) $\frac{25}{36}$
   (c) $\frac{5}{36}$ (d) $\frac{125}{216}$

2. The probability that $X\ge 3$ equals ______.
   (a) $\frac{125}{216}$ (b) $\frac{25}{36}$
   (c) $\frac{5}{36}$ (d) $\frac{25}{216}$

3. The conditional probability that $X\ge 6$ given $X>3$ equals ______.
   (a) $\frac{125}{216}$ (b) $\frac{25}{216}$
   (c) $\frac{5}{36}$ (d) $\frac{25}{36}$

Comprehension 5

Let U_1 and U_2 be two urns such that U_1 contains 3 white and 2 red balls, and U_2 contains only 1 white ball. A fair coin is tossed. If head appears, then 1 ball is drawn at random from U_1 and put into U_2. However, if tail appears then 2 balls are drawn at random from U_1 and put into U_2. Now 1 ball is drawn at random from U_2.

1. The probability of the drawn ball from U_2 being white is ______.

(a) $\frac{13}{30}$ (b) $\frac{23}{30}$

(c) $\frac{19}{30}$ (d) $\frac{11}{30}$

2. Given that the drawn ball from U_2 is white, the probability that head appeared on the coin is ______.

(a) $\frac{17}{23}$ (b) $\frac{11}{23}$

(c) $\frac{15}{23}$ (d) $\frac{12}{23}$

Questions from Previous Years

Comprehension 6

Three numbers are chosen at random, one after another with replacement, from the set S = {1, 2, 3, ..., 100}. Let p_1 be the probability that the maximum of chosen numbers is at least 81 and p_2 be the probability that the minimum of chosen numbers is at most 40.

1. The value of $\frac{625}{4}p_1$ is ______.

2. The value of $\frac{125}{4}p_2$ is ______.

Comprehension 7

There are five students S_1, S_2, S_3, S_4 and S_5 in a music class and for them there are five seats R_1, R_2, R_3, R_4 and R_5 arranged in a row, where initially the seat R_i is allotted to the student $S_i, i = 1, 2, 3, 4, 5$. But, on the examination day, the five students are randomly allotted the five seats.

1. The probability that, on examination day, the student S_1 gets the previously allotted seat R_1, and NONE of the remaining students gets the seat previously allotted to him/her is ______.

(a) $\frac{3}{40}$ (b) $\frac{1}{8}$

(c) $\frac{7}{40}$ (d) $\frac{1}{5}$

2. For $i = 1, 2, 3, 4$, let T_i denote the event that the students S_i and S_{i+1} do NOT sit adjacent to each other on the day of the examination. Then, the probability of the event $T_1 \cap T_2 \cap T_3 \cap T_4$ is ______.

(a) $\frac{1}{15}$ (b) $\frac{1}{10}$

(c) $\frac{7}{60}$ (d) $\frac{1}{5}$

Comprehension 8

Let n_1, and n_2 be the number of red and black balls, respectively, in box I. Let n_3 and n_4 be the number of red and black balls respectively, in box II.

1. One of the two boxes, box I and box II, was selected at random and a ball was drawn randomly out of this box. The ball was found to be red. If the probability that this red ball was drawn from box II is $\frac{1}{3}$, then the correct option(s) with the possible values of n_1, n_2, n_3 and n_4 is (are):

(a) $n_1 = 3, n_2 = 3, n_3 = 5, n_4 = 15$

(b) $n_1 = 3, n_2 = 6, n_3 = 10, n_4 = 50$

(c) $n_1 = 8, n_2 = 6, n_3 = 5, n_4 = 20$

(d) $n_1 = 6, n_2 = 12, n_3 = 5, n_4 = 20$

2. A ball is drawn at random from box I and transferred to box II. If the probability of drawing a red ball from box I, after this transfer, is $\frac{1}{3}$, then the correct option(s) with the possible values of n_1 and n_2 is (are):

(a) $n_1 = 4$ and $n_2 = 6$

(b) $n_1 = 2$ and $n_2 = 3$

(c) $n_1 = 10$ and $n_2 = 20$

(d) $n_1 = 3$ and $n_2 = 6$

Comprehension 9

Box 1 contains three cards bearing numbers 1, 2, 3; box 2 contains five cards bearing numbers 1, 2, 3, 4, 5; and box 3 contains seven cards bearing numbers 1, 2, 3, 4, 5, 6, 7. A card is drawn from each of the boxes. Let x_i be number on the card drawn from the ith box, $i = 1, 2, 3$.

1. The probability that $x_1 + x_2 + x_3$ is odd, is ______.

(a) $\frac{29}{105}$ (b) $\frac{53}{105}$

(c) $\frac{57}{105}$ (d) $\frac{1}{2}$

2. The probability that x_1, x_2, x_3 are in an arithmetic progression, is ______.

(a) $\frac{9}{105}$ (b) $\frac{10}{105}$

(c) $\frac{11}{105}$ (d) $\frac{7}{105}$

Comprehension 10

A box B_1 contains 1 white ball, 3 red balls and 2 black balls. Another box B_2 contains 2 white balls, 3 red balls and 4 black balls. A third box B_3 contains 3 white balls, 4 red balls and 5 black balls.

1. If 1 ball is drawn from each of the boxes B_1, B_2 and B_3, the probability that all 3 drawn balls are of the same colour is ______.

(a) $\frac{82}{648}$ (b) $\frac{90}{648}$

(c) $\frac{558}{648}$ (d) $\frac{566}{648}$

2. If 2 balls are drawn (without replacement) from a randomly selected box and one of the balls is white and the other ball is red, the probability that these 2 balls are drawn from box B_2 is ______.

(a) $\frac{116}{181}$ (b) $\frac{126}{181}$

(c) $\frac{65}{181}$ (d) $\frac{55}{181}$

ANSWER KEY

JEE Main

Objective Questions

One Option Correct

1. (a) 2. (c) 3. (b) 4. (a) 5. (c) 6. (a) 7. (c) 8. (a) 9. (c)
10. (a) 11. (d) 12. (c) 13. (c) 14. (b) 15. (d) 16. (c) 17. (d) 18. (a)
19. (a) 20. (a) 21. (c) 22. (b) 23. (c) 24. (a) 25. (b) 26. (a) 27. (a)
28. (b) 29. (c) 30. (a) 31. (c) 32. (b) 33. (b) 34. (d) 35. (a) 36. (d)
37. (c) 38. (b) 39. (a) 40. (d) 41. (d) 42. (a) 43. (b) 44. (a) 45. (a)
46. (b) 47. (c) 48. (a) 49. (c) 50. (c) 51. (a) 52. (c) 53. (c) 54. (d)
55. (a) 56. (b) 57. (c) 58. (d) 59. (b) 60. (d) 61. (a) 62. (d) 63. (b)
64. (d) 65. (d) 66. (d) 67. (a) 68. (b) 69. (d) 70. (b) 71. (b) 72. (b)
73. (a) 74. (d) 75. (d) 76. (b) 77. (c) 78. (d) 79. (b) 80. (b) 81. (b)
82. (b) 83. (d) 84. (a)

Integer and Numerical Value Type

1. (1) 2. (2) 3. (4) 4. (3) 5. (28) 6. (6) 7. (11) 8. ($n \geq 11$)
9. (3)

Assertion and Reason

1. (b) 2. (c) 3. (d)

JEE Advanced

Objective Questions

One Option Correct

1. (a) 2. (a) 3. (d) 4. (c) 5. (c) 6. (a) 7. (a) 8. (a) 9. (b)
10. (b) 11. (c) 12. (a)

One or More Than One Option Correct

1. (a, b, c, d) 2. (a, b) 3. (a, d) 4. (a, d) 5. (a, b, c)
6. (a, b) 7. (b, d) 8. (a, d) 9. (b, c) 10. (c, d)
11. (b, c, d) 12. (a, b, c, d) 13. (b, d) 14. (a, b) 15. (a)

Matrix Type

1. A→S; B→T; C→Q; D→S
2. A→Q; B→P; C→R; D→Q
3. A→P;B→R;C→Q;D→P
4. A→Q, R; B→R; C→P; D→A, B, C, D
5. A→Q; B→R; C→S; D→R

Integer and Numerical Value Type

1. $\left(\frac{2}{5}\right)$ **2.** $-\left(\frac{2N-2K+2}{N^2+N}\right)$ **3.** $\left(\frac{20}{3003}\right)$ **4.** (nC$_2$, $3^{(n-2)}$) **5.** (6) **6.** (8) **7.** (0.5) **8.** (–8) **9.** (6)

Comprehension Passage

Comprehension 1

1. (a) **2.** (a) **3.** (b)

Comprehension 2

1. (a) **2.** (b) **3.** (c)

Comprehension 3

1. (d) **2.** (c) **3.** (a)

Comprehension 4

1. (a) **2.** (b) **3.** (d)

Comprehension 5

1. (b) **2.** (d)

Comprehension 6

1. (76.25) **2.** (24.50)

Comprehension 7

1. (a) **2.** (c)

Comprehension 8

1. (a, b) **2.** (c, d)

Comprehension 9

1. (b) **2.** (c)

Comprehension 10

1. (a) **2.** (d)

HINTS AND EXPLANATIONS

Worksheet

1. $P(E_1) = \frac{10}{20}$ $(\because m = 10, n = 20)$

$P(E_2) = \frac{6}{20}$; $P(E_3) = \frac{10+6-3}{20} = \frac{13}{20}$

(since there are 10 numbers which are divisible by 2, 6 numbers which are divisibly by 3 but three numbers namely, 6, 12, 18 have been counted twice)

$P(E_4) = \frac{3}{20}$; $P(E_5) = \frac{7}{20}$

For E_6, let us list down all favourable cases. Indeed 2, 3, 5, 7, 10, 11, 13, 15, 17, 19, 20, (11 cases are either prime or are divisible by 5.

Thus $P(E_6) = \frac{11}{20}$

2. If four balls are drawn with replacement, we can have sample points of the type (W_1, W_1, W_1, B_1) (i.e., the same ball W_1 is there on the first three co-ordinates)

⇒ Total sample points = 13^4

Now select two co-ordinates from four in $^4C_2$ ways.

Each selection will rise to 8 × 8 × 5 × 5 cases

Thus, required probability = $\frac{^4C_2 \times 8 \times 8 \times 5 \times 5}{13^4}$

Remark

(i) $\frac{8}{13} \times \frac{8}{13} \times \frac{5}{13} \times \frac{5}{13} \times {}^4C_2$

(ii) Generally, all "with replacement" problems can be done by binomial formula. Here $n = 4$ (number of independent trials).

$p = \frac{8}{13}$ (probability of drawing a white ball in one trial).

$q = \frac{5}{13}$, $q = 1 - q$ ⇒ probability of two successes = $^4C_2\, p^2q^2 = {}^4C_2 \left(\frac{8}{13}\right)^2 \left(\frac{5}{13}\right)^2$

3. (i) If all the green balls are distinct and all the red balls are distinct

T = total number of cases = 8!

F = number of favourable cases = 5! × 3! × 2!

$\Rightarrow$ required probability $= \dfrac{F}{T} = \dfrac{5!\times 3!\times 2!}{8!}$

In the case when balls of the same colour are identical,

$T = \dfrac{8!}{5!3!}$, $F = 2$ whence $\dfrac{F}{T}$ remains same.

(ii) We have $F = 6! \times 3! \Rightarrow$ Probability $= \dfrac{6!\times 3!}{8!}$

If the balls of the same colour are identical then there are six positions for keeping the bunch of three identical balls. These positions are – G – G – G – G – G –

$\Rightarrow F = 6$ required probability $= \dfrac{6}{\frac{8!}{5!3!}} = \dfrac{6!\times 3!}{8!}$

which is same as the above answer.

(iii) If the balls are distinct $F = {}^6C_3 \times 3! \times 5!$ If they are identical $F = {}^6C_3$ whence $\dfrac{F}{T}$ remains same in both cases.

4. (i) Required probability $= \dfrac{{}^4C_2 \times {}^{48}C_2}{{}^{52}C_4}$

(ii) Required probability $= \dfrac{52}{52}\times\dfrac{39}{51}\times\dfrac{26}{50}\times\dfrac{13}{49}$

$\left(\text{OR}\ \dfrac{\left({}^{13}C_1\right)^4}{{}^{52}C_4}\right)$

(iii) Required probability

$= \left(\dfrac{13}{52}\times\dfrac{12}{51}\times\dfrac{11}{50}\times\dfrac{10}{49}\right)\times 4$

(iv) Required probability $= \dfrac{52}{52}\times\dfrac{36}{51}\times\dfrac{22}{50}\times\dfrac{10}{49}$

(At first drawing, we have all four suits and all the 13 denominations favourable, giving $13 \times 4 = 52$ favourable cases. At the second stage, we are left with only three favourable suits and each of the three suits contain only 12 favourable denominations since one suit and one denomination have been exhausted. similarly at the third and fourth trials.)

(v) Choose two denominations in ${}^{13}C_2$ ways. Each selection shall give rise to ${}^4C_2 \cdot {}^4C_2$ favourable cases. Therefore, required probability $= \dfrac{\left({}^4C_2 \times {}^4C_2\right)\times {}^{13}C_2}{{}^{52}C_4}$.

(vi) This is difficult! If we have a pair in a set of four cards then the remaining two cards must not bear the denomination of the cards forming the pair. Besides, the remaining two cards must not be of the same denomination, for otherwise there will be two pairs.

Choose a particular pair (say Ace pair) which has ${}^4C_2$ possibilities. Choose two distinct denominations from remaining 12 denominations (${}^{12}C_2$). Each selection will give rise to 4×4 favourable ways. Thus, the desired number of favourable ways $= 13 \times {}^4C_2 \times {}^{12}C_2 \times 4 \times 4$ and required probability

$= \dfrac{13\times {}^4C_2 \times {}^{12}C_2 \times 4 \times 4}{{}^{52}C_4}$.

(vii) This part is all the more difficult. We proceed systematically. There are two major ways in which the event can happen. Let us denote the kings by K_s, K_h, K_d, K_c (where K_C is king of clubs) and club cards by $C_1, C_2, C_3, \ldots, C_{13}$ where C_{13} is king of clubs. Note that $K_C = C_{13}$. There are 36 cards which are neither kings nor clubs.

Now the major ways are

(a) The set of four card does not contain K_C.

(b) The set of four cards contains K_C.

The number of favourable cases corresponding to (a) must be ${}^3C_2 \cdot {}^{12}C_2$.

Now if the set of four cards contains K_C then only one more king is required, which can be chosen in ${}^3C_1$ ways and as 1 club has already been chosen, the remaining 1 club can be chosen in ${}^{12}C_1$ ways but 1 more card has to be there, which can be chosen in ${}^{36}C_1$ ways.

Thus, the number of favourable cases corresponding to (b) are ${}^3C_1 \times {}^{12}C_1 \times {}^{36}C_1$.

Therefore, required probability

$= \dfrac{{}^3C_2 \times {}^{12}C_2 + {}^3C_1 \times {}^{12}C_1 \times {}^{36}C_1}{{}^{52}C_4}$

5. The sample space in a throw of two dice

$S = \{1, 2, 3, 4, 5, 6\} \times \{1, 2, 3, 4, 5, 6\}$

$\therefore$ total no. of cases $n(S) = 6 \times 6 = 36$.

(i) Here E_1 = The event of showing equal number on both dice

$= \{(1, 1)(2, 2)(3, 3)(4, 4)(5, 5)(6, 6)\}$

$\therefore n(E_1) = 6 \quad \therefore P(E_1) = \dfrac{n(E_1)}{n(s)} = \dfrac{6}{36} = \dfrac{1}{6}$

(ii) Here E_2 the event of showing numbers whose difference is 1.

$= \{(1, 2)\ (2, 1)\ (2, 3)\ (3, 2)\ (3, 4)\ (4, 3)\ (4, 5)$
$(5, 4)\ (5, 6)\ (6, 5)\}$

$\therefore n(E_2) = 10 \quad \therefore P(E_2) = \frac{n(E_2)}{n(s)} = \frac{10}{36} = \frac{5}{18}$

6. If the sample space be S then $n(S)$ = the total number of ways of drawing 3 out of 52 cards $= {}^{52}C_3$
Now, if A = the event of drawing three queens then $n(A) = {}^4C_3$

$$\therefore P(E) = \frac{n(A)}{n(s)} = \frac{{}^4C_3}{{}^{52}C_3} = \frac{4}{\frac{52\times 51\times 50}{3\times 2}} = \frac{1}{5525}$$

7. The total no. of ways to arrange 5 letters at 5 places $n(s) = 5!$
In the five-letter word, two place are even (second and fourth) and there are two vowels A and U in the given word. So, we have to arrange 2 vowels at 2 even places and 3 consonants at remaining 3 places.
$\therefore n(A) = 2!.\ 3!$

$$\therefore P(A) = \frac{n(A)}{n(s)} = \frac{2!.3!}{5!} = \frac{1}{10}$$

8. Here $n = 52$, $m = 4$

$\therefore$ Odds in favour of the event $= \frac{4}{52-4} = \frac{4}{48}$

$= 1:12$

Notations

(i) $P(A + B)$ or $P(A \cup B)$
= Probability of happening of A or B
= Probability of happening of the events A or B or both
= Probability of occurrence of at least one event A or B

(ii) $P(AB)$ or $P(A \cap B)$ = Probability of happening of events A and B together.

(iii) $P(A/B)$ = Conditional probability of A when B has happened.

9. $\therefore P(A \cup B) = P(A) + P(B) - P(A \cap B)$

$\therefore \frac{7}{12} = \frac{1}{2} + \frac{1}{3} - P(A \cap B)$

$\Rightarrow P(A \cap B) = 1/4$

Now, $P(A/B) = \frac{P(A\cap B)}{P(B)} = \frac{1/4}{1/3} = \frac{3}{4}$

10. Let A be the event of getting equal digits on the dice. Since number of exhaustive cases is 36 and favourable cases is 6, $P(A) = P = \frac{6}{36} = \frac{1}{6}$,

$P(\bar{A}) = q = 1 - \frac{1}{6} = \frac{5}{6}$

Hence by binomial theorem, we have
$(5/6 + 1/6)^4 = (5/6)^4 + {}^4C_1\,(5/6)^3\,(1/6) + {}^4C_2\,(5/6)^2\,(1/6)^2 + {}^4C_3\,(5/6)(1/6)^3 + (1/6)^4$
Thus, from the above result, we have

(i) Probability of getting equal digits exactly two times
$= {}^4C_2\,(5/6)^2\,(1/6)^2 = \frac{25}{216}$

(ii) Probability of getting equal digits at least two times.
$= {}^4C_2\,(5/6)^2\,(1/6)^2 + {}^4C_3\,(5/6)(1/6)^3 + (1/6)^4$
$= \frac{25}{216} + \frac{20}{1296} + \frac{1}{1296} = \frac{171}{1296}$

(iii) Probability of getting equal digits at the most two times
$= (5/6)^4 + {}^4C_1\,(5/6)^3\,(1/6) + {}^4C_2\,(5/6)^2\,(1/6)^2$
$= \frac{625}{1296} + \frac{500}{1296} + \frac{150}{1296} = \frac{1275}{1296}$

Quiz

1. $1 - P(A')P(B')$

2. Let E and F be respectively the events that first and second ball drawn are black.
$P(E \cap F) = P(E)\,P(F/E)$

$P(E) = \frac{4}{6} = \frac{2}{3}$

$P(F/E) = \frac{3}{5}$

According to multiplication rule of probability

$P(E\cap F) = P(E)P(F/E)$

$= \frac{2}{3} \times \frac{3}{5} = \frac{2}{5}$

3. The probability of the event E is called conditional probability E is given that F has already occurred.

4. If there are two events such that probability of occurrence of one of them is not affected by the occurrence of other, they are called independent events.

5. Random variable

6. The conditions are the following:
(i) There should be a finite number of trials.
(ii) The trials should be independent.
(iii) Each trial has exactly two outcomes: success or failure.
(iv) The probability of success remains the same in each trial.

7. If $E_1, E_2, \ldots E_n$ are n non-empty events which constitute a partition of sample space S, i.e., $E_1, E_2, \ldots E_n$ are pairwise disjoint and $E_1 \cup E_2 \cup \ldots \cup E_n = S$ and A is any event of nonzero probability, then

$$P(E_i|A) = \frac{P(E_i)P(A|E_i)}{\sum_{j=1}^{n} P(E_j)P(A|E_j)}$$

for any $i = 1, 2, 3, \ldots, n$

8. $P(A) = \frac{3}{6} = \frac{1}{2}$

$P(B) = \frac{3}{6} = \frac{1}{2}$

$A \cap B = \{2\}$

$P(A \cap B) = \frac{1}{6}$

$P(A) \cdot P(B) = \frac{1}{2} \times \frac{1}{2} = \frac{1}{4}$

$\frac{1}{4} \neq \frac{1}{6}$

$P(A) \cdot P(B) \neq P(A \cap B)$

Therefore, A and B are not independent event.

9. $n(S) = 8$

$n(B) = 2$

$P(A \cap B) = 1$

$P(B) = \frac{2}{8} = \frac{1}{4}$

$n(A \cap B) = 1$

$$P(A/B) = \frac{P(\cap B)}{P(B)} = \frac{\frac{1}{8}}{\frac{1}{4}} = \frac{1}{2}$$

10. $p_1 p_2 (1 - p_3)(1 - p_4) \ldots$

JEE Main

Objective Questions

One Option Correct

1. (a) $P(A) = .25, P(B) = .50, P(A \cap B) = .14$

Now $P(\bar{A} \cap \bar{B}) = P(\overline{A \cup B}) = 1 - P(A \cup B)$

$= 1 - P(A) - P(B) + P(A \cap B) = 0.39$

2. (c) $P(\bar{A}/\bar{B}) = \frac{P(\bar{A} \cap \bar{B})}{P(\bar{B})} = \frac{P(\overline{A \cup B})}{P(\bar{B})}$

$= \frac{1 - P(A \cup B)}{P(\bar{B})}$

3. (b)

4. (a) $P(A \cup B) = P(A \cap B)$

$\Rightarrow P(A) + P(B) - P(A \cap B) = P(A \cap B)$

$\Rightarrow (P(A) - P(A \cap B)) + (P(B) - P(A \cap B)) = 0$

Since both brackets are non-negative, the above equality is possible only when

$P(A) - P(A \cap B) = 0$, $P(B) - P(A \cap B) = 0$

whence $P(A) = P(B)$

5. (c) $P((A \cap B) \cup (A \cap C)) = ½$

$\Rightarrow P(A \cap B) + P(A \cap C) - P(A \cap B \cap C) = ½$

$\Rightarrow pq + ½p - ½pq = ½ \Rightarrow pq + p = 1$

Out of the given choices (c) is correct

6. (a) P(getting a total 5 with a pair of dice) = 4/36 = 1/9

P(getting a total of 7 with a pair of dice) = 6/36 = 1/6

P(getting neither 5 nor 7) = 26/36 = 13/18

Required probability

$$= \frac{1}{9} + \frac{13}{18} \times \frac{1}{9} + \left(\frac{13}{18}\right)^2 \times \frac{1}{9} + \left(\frac{13}{18}\right)^3 \times \frac{1}{9} + \ldots\infty$$

$$= \frac{a}{1-r} = \frac{1/9}{1 - \frac{13}{18}} = \frac{2}{5}$$

7. (c) Since A and B are independent $P(A \cap B)$

$= P(A) \cdot P(B) \Rightarrow P(A \cup B)$

$= P(A) + P(B) - P(A) \cdot P(B)$

$\Rightarrow 0.8 = .3 + P(B) - .3 \times P(B) \Rightarrow P(B) = 5/7$

8. (a) Required probability $= \frac{3}{6} \times \frac{2}{6} \times \frac{1}{6} = \frac{1}{36}$

9. (c) The favourable cases are two only. Total number of cases $= {}^6C_3$

10. (a) The two faulty machines will be detected in two tests if the first two draws yield either both faulty or both non-faulty

$\Rightarrow$ Required probability $= \frac{2}{4} \times \frac{1}{3} + \frac{2}{4} \times \frac{1}{3} = \frac{1}{3}$

$\Rightarrow$ (a) is correct

11. (d) The favourable cases are

$(1,2), (1,3), (1,4), (1,5), (1,6), (2,3), (2,4), (2,5), (2,6), (3,4), (3,5), (3,6), = 12$

$\Rightarrow$ Required probability $\frac{12}{{}^6C_2} = \frac{4}{5}$.

12. (c) P(getting a doublet with a pair of dice)

$= \frac{6}{36} = \frac{1}{6}$, P (Not getting a doublet) $= \frac{5}{6}$

P(getting a doublet not more than twice) $= 1 - P$ (getting doublet all the three times)

$= 1 - \left(\frac{1}{6}\right)^3 = 1 - \frac{1}{216} = \frac{215}{216}$.

13. (c) $P(\text{red})=\frac{1}{2}\times\frac{2}{4}+\frac{1}{2}\times\frac{5}{9}$

$\Rightarrow P(B/\text{red})=\dfrac{\frac{1}{2}\times\frac{5}{9}}{\frac{1}{2}\times\frac{2}{4}+\frac{1}{2}\times\frac{5}{9}}=\frac{10}{19}.$

14. (b)

15. (d)

16. (c) $\frac{52}{52}\times\frac{39}{51}\times\frac{26}{50}\times\frac{13}{49}$

17. (d) $\Rightarrow$ Required probability $=\dfrac{{}^4C_2\times{}^5C_2}{{}^9C_4}=\dfrac{10}{21}$

18. (a) Make cases with $d=1, d=2, ..., d=7$

19. (a) $\dfrac{{}^{12}C_3.2^9}{3^{12}}=\dfrac{110}{9}\left(\dfrac{2}{3}\right)^{10}$

20. (a) Let the coin be tossed n number of times. then ${}^nC_4\left(\frac{1}{2}\right)^4\left(\frac{1}{2}\right)^{n-4}={}^nC_7\left(\frac{1}{2}\right)^7\left(\frac{1}{2}\right)^{n-7}$

$\Rightarrow {}^nC_4={}^nC_7\Rightarrow n=11\Rightarrow$ P (getting two heads)

$={}^{11}C_2\left(\frac{1}{2}\right)^2\left(\frac{1}{2}\right)^9=\frac{25}{2048}$

21. (c) $E(x^2+2)=(0^2+2)P(0)+(1^2+2)P(1)+(2^2+2)P(2)+(3^2+2)P(3)$

$=2\,{}^3C_0\cdot\left(\frac{1}{2}\right)^3\cdot+3\times{}^3C_1\left(\frac{1}{2}\right)^3\cdot+6\times{}^3C_2\left(\frac{1}{3}\right)^3+$

$11\times{}^3C_3\left(\frac{1}{2}\right)^3=5$

22. (b) Required probability $=\dfrac{2!\times8!}{9!}=\dfrac{2}{9}.$

23. (c) Required probability

$=\frac{4}{5}\times\left(1-\frac{3}{4}\right)+\frac{3}{4}\left(1-\frac{4}{5}\right)=\frac{7}{20}.$

24. (a) $np=4, npq=2\Rightarrow q=\frac{1}{2}, p=\frac{1}{2}, n=8$

$\Rightarrow P(X=2)={}^8C_2\left(\frac{1}{2}\right)^2\left(\frac{1}{2}\right)^6=\frac{28}{256}.$

25. (b) Total cases $=3^3$

($\because$ each person can apply in 3 ways)

Favourable cases = 3 (when all 3 apply for the same house)

$\Rightarrow$ Required probability $=\dfrac{3}{3^3}=\dfrac{1}{9}$

26. (a) Total cases $=\dfrac{8!}{4!2!}$, Now, AA IN can be arranged in $\dfrac{4!}{2!}$ ways and in the five empty spaces of these four letters, the four S's can be arranged in ${}^5C_4=5$ ways

$\Rightarrow$ Total favourable cases $=5\times\dfrac{4!}{2!}$

$\Rightarrow$ required probability $=\dfrac{5\times\frac{4!}{2!}}{\frac{8!}{4!2!}}=\dfrac{1}{14}$

27. (a) Let A: maximum number is not more than 10 and B: minimum on them is 5

Required probability $=P(B/A)=\dfrac{P(B\cap A)}{P(A)}$

Now $P(A)=\dfrac{{}^{10}C_2}{{}^{100}C_2}$ and favourable cases for $P(B\cap A)$ are (5, 6), (5, 7), (5, 8), (5, 9) and (5, 10)

$\Rightarrow P(B/A)=\dfrac{5/{}^{100}C_2}{{}^{10}C_2/{}^{100}C_2}=\dfrac{1}{9}$

28. (b) We must have

$0\le\frac{1+3p}{3}\le1,\ 0\le\frac{1-p}{4}\le1,\ 0\le\frac{1-2p}{2}\le1$

We must also have $0\le\frac{1+3p}{3}+\frac{1-p}{4}+\frac{1-2p}{2}\le1$

($\because$ Events are mutually exclusive)

On solving the inequalities, we easily get

$\frac{1}{3}\le p\le\frac{1}{2}$

29. (c) $P(A\cup B)=.6$, $P(A\cap B)=.2$

Now

$P(\bar{A})+P(\bar{B})=2-[P(A\cup B)+P(A\cap B)]=1.2$

30. (a) The numbers appearing must be from 2, 3, 4, 5.

$\Rightarrow$ Number of favourable cases $=4^4$

Total cases $=6^4$

$\Rightarrow$ Required probability $4^4/6^4=16/81$

31. (c) A white ball is drawn for the 4th time on 7th draw if, in the first six draws, exactly 3 while balls appear followed by a white ball at the 7th draw.

$\Rightarrow$ Required probability

$={}^6C_3\left(\frac{12}{24}\right)^3\left(\frac{12}{24}\right)^3\times\frac{12}{24}=\frac{5}{32}$

32. (b) If W denotes India's win, then the favourable patterns are $W\bar{W}W$, $\bar{W}WW$

$\Rightarrow$ Required probability

$=\frac{1}{2}\times\left(1-\frac{1}{2}\right)\times\frac{1}{2}+\left(1-\frac{1}{2}\right)\times\frac{1}{2}\times\frac{1}{2}=\frac{1}{4}$

33. (b) Define the events, A: minimum of chosen number is 3, B: maximum of chosen number is 7

$P(A \cap B) = P(A) + P(B) - P(A \cap B)$

$P(A) = \frac{^7C_2}{^{10}C_3}$ ($\because$ 3 should be there and the remaining two numbers must be from 7 numbers)

$P(B) = \frac{^6C_2}{^{10}C_3}$

$P(A \cap B) = \frac{3}{^{10}C_3}$ (3 and 7 should be essentially there and one number must be from 4, 5, 6)

$\Rightarrow P(A \cup B) = \frac{11}{40}$

34. (d) Take an example to disprove choices (a), (b), (c). Suppose a dice is thrown twice.
Let E be the event that the first throw is 2 and F be the event that the second throw is 3
Then (a), (b), (c) are disproved.

35. (a) Since the fifth toss is independent of other tosses. the required probability should be ½.
Note: By Laplace law of succession, the chance can also be $\frac{31}{32}$ by applying Baye's Theorem for future event.

36. (d) Numbers which are divisible by both 2 and 3 are multiples of 6 which are $\left[\frac{100}{6}\right] = 16$

$\Rightarrow$ Required probability $= \frac{^{16}C_3}{^{100}C_3} = \frac{4}{1155}$.

37. (c) Define the events as E: 3 dice show different faces; F: At least one face shows 6.
Require probability

$$P(F/E) = \frac{P(F \cap E)}{P(E)} = \frac{\frac{1}{6} \times \frac{5}{6} \times \frac{4}{6} \times 3}{\frac{6 \times 5 \times 4}{6 \times 6 \times 6}} = \frac{1}{2}$$

(Note that $P(F \cap E) = P$ (six on one die and different non-6 numbers on the remaining two dice)

38. (b) Let β be the probability of getting an odd number, then the probability of getting an even number $= 2\beta$
We must have $\beta + 2\beta = 1 \Rightarrow \beta = 1/3$.
Now the required event will occur if either both outcomes are even or both are odd

$\Rightarrow$ Required probability $= \frac{1}{3} \times \frac{1}{3} + \frac{2}{3} \times \frac{2}{3} = \frac{5}{9}$

39. (a)

40. (d) $1 - \frac{1}{4!} = \frac{23}{24}$

41. (d) $P(A) = .5, P(B) = .3, P(A \cap B) = 0$,

$P(\bar{A} \cap \bar{B}) = P(\overline{A \cup B}) = 1 - P(A \cup B)$

$= 1 - [P(A) + P(B) - P(A \cap B)] = 1 - [.8] = .2$

$\Rightarrow$ None of choices (a), (b), (c) are true

$\Rightarrow$ (d) is true .

42. (a) $\left.\begin{array}{l} P(A \cap B) = \frac{1}{8} \Rightarrow P(A)P(B) = \frac{1}{8} \\ P(\bar{A} \cap \bar{B}) = \frac{3}{8} \Rightarrow P(\bar{A})P(\bar{B}) = \frac{3}{8} \end{array}\right]$

$\because$ A and B are independent

which easily yields $P(A) = \frac{1}{2}, P(B) = \frac{1}{4}$

43. (b) $P(A) = \frac{3}{16}$

P(B) should also be equal to $\frac{3}{16}$ since cases are similar

$\Rightarrow$ Choices (a), (c) are wrong. (b) can be called correct $\because P(A) \le P(B)$ is true.

44. (a) $P(A \cup B \cup C) = \frac{1}{4} + \frac{1}{4} + \frac{1}{4} - 0 - \frac{1}{8} - 0 + 0 = \frac{5}{8}$

45. (a) $P(A \cup B) = P(A) + P(B) - P(A \cap B)$

$\Rightarrow \frac{5}{6} = \frac{1}{2} + P(B) - \frac{1}{3}$

$\Rightarrow P(B) = \frac{4}{6} = \frac{2}{3}$

$\Rightarrow P(A \cap B) = P(A) \cdot P(B)$

$\Rightarrow$ A and B are independent.

46. (b) According to the condition $^{100}C_{50}p^{50}(1-p)^{50} = {}^{100}C_{51}p^{51}(1-p)^{49}$. Solve for p.

47. (c) Required probability
$= 0.4 + (1-0.4)(0.3) + (1-0.4)(1-0.3) \times 0.2 + (1-0.4)(1-0.3)(1-0.2) \times 0.1 = 0.6976$.

48. (a) P(on both occasions Mr A selected losing horse) $= \frac{4}{5} \times \frac{3}{4} = \frac{3}{5}$.

Required probability $= 1 - \frac{3}{5} = \frac{2}{5}$

49. (c) $P(X > 1.5) = 1 - P(X \le 1.5) = 1 - P(X \le 1)$

$= 1 - [P(0) + P(1)] = 1 - \frac{e^{-m}m^0}{0!} - \frac{e^{-m}m^1}{1!}$

$= 1 - e^{-m}(1+m)$

But $m = 2$ is given $\Rightarrow P(X > 1.5) = 1 - 3/e^2$.

50. (c) $P(A\cup B)=\frac{5}{6}, P(A\cap B)=\frac{1}{4}, P(A)=\frac{3}{4}$

$\Rightarrow \frac{5}{6}=\frac{3}{4}+P(B)-\frac{1}{4}\Rightarrow P(B)=\frac{1}{3}$

$\Rightarrow P(A\cup B)=P(A)P(B)$ is satisfied.

$\Rightarrow$ A and B are independent but are not equally likely.

51. (a) D_4 can show a number appearing on one of D_1, D_2, and D_3 in the following cases.

Case 1: D_4 shows a number which is shown by exactly one of D_1, D_2 and D_3.

D_4 shows number in $^6C_1$ ways.

One out of D_1, D_2, and D_3 can be selected in $^3C_1$ ways.

[The selected die shows the same number as on D_4 in one way and remaining two dice show a different number in 5 ways each].

$\therefore$ Number of ways

$^6C_1\times{}^3C_1\times1\times5\times5=450$

Case 2: D_4 shows a number which is shown by exactly two of D_1, D_2 and D_3.

Number of ways

$={}^6C_1\times{}^3C_{22}\times1\times1\times5=90$

Case 3: D_4 shows a number which is shown by all three dice D_1, D_2, and D_3.

Number of ways

$^6C_1\times{}^3C_3\times1\times1\times1=6$

$\therefore$ Total number of favourable ways

$= 450 + 90 + 6 = 546$

Total ways $=6\times6\times6\times6$

$\therefore$ Required probability $=\frac{546}{6\times6\times6\times6}=\frac{91}{216}$

52. (c) From the tree diagram

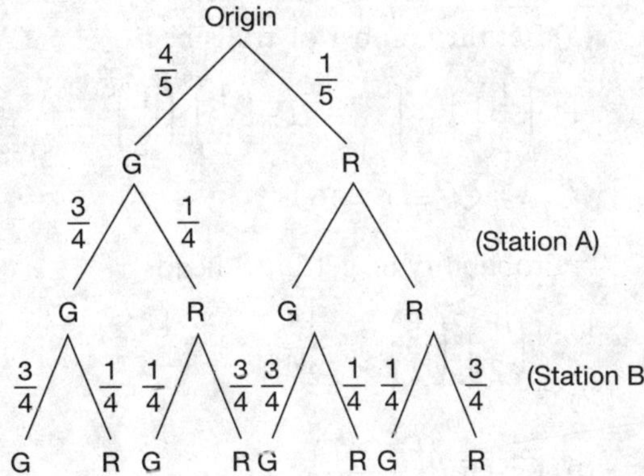

P(original single is green/single received at B is green)

$$=\frac{p(GGG)+P(GRG)}{P(GGG)+P(GRG)+P(RGG)+P(RRG)}$$

$$=\frac{\frac{4}{5}\times\frac{3}{4}\times\frac{3}{4}+\frac{4}{5}\times\frac{1}{4}\times\frac{1}{4}}{\frac{4}{5}\times\frac{3}{4}\times\frac{3}{4}+\frac{4}{5}\times\frac{1}{4}\times\frac{1}{4}+\frac{1}{5}\times\frac{1}{4}\times\frac{3}{4}+\frac{1}{5}\times\frac{3}{4}\times\frac{1}{4}}$$

$$=\frac{\frac{4}{5}\times\frac{10}{16}}{\frac{4}{5}\times\frac{10}{16}+\frac{1}{5}\times\frac{6}{16}}=\frac{40}{40+6}=\frac{40}{46}=\frac{20}{23}$$

53. (c) Given $\omega^{r_1}+\omega^{r_2}+\omega^{r_3}=0$. If ω is a complex cube root of unity then,

Sum of consecutive power ω is zero

$\omega^{3m}+\omega^{3m+1}+\omega^{3m+2}=0$

where m is an integer.

r_1, r_2, r_3 are the numbers obtained on die, these can take any value from 1 to 6.

$\therefore$ m can take values 1 or 2 for r_1 values 0 or 1 for r_2 and values 0 or 1 for r_3.

$\therefore$ Number of ways of selecting r_1, r_2, r_3

$={}^2C_1\times{}^2C_1\times{}^2C_1\times3!$

Also, the total number of ways of getting r_1, r_2, r_3 on die $=6\times6\times6$

$\therefore$ Required probability

$$=\frac{^2C_1\times{}^2C_1\times{}^2C_1\times3!}{6\times6\times6}=\frac{2}{9}$$

54. (d) Given that A and B to be independent events

$\therefore P(A\cap B)=P(A)P(B)$

$\Rightarrow \frac{n(A\cap B)}{n(S)}=\frac{n(A)}{n(S)}\times\frac{n(B)}{n(S)}$

$\Rightarrow \frac{n(A\cap B)}{10}=\frac{4}{10}\times\frac{a}{10}$

$\Rightarrow n(A\cap B)=\frac{5}{2}y$

$\Rightarrow n(A\cap B)$ has to be integer,

we have $b = 5$ or 10

$\therefore n(B)=5 \text{ or } 10$

55. (a) Total number of onto functions = 6!

Since $g(3) = 2g$, (1) can be defined in 3 ways (3, 2), (3, 4), (3, 6)

$\therefore$ Number of onto functions in this condition $=3\times4!$

Required probability $=\frac{3\times4!}{6!}=\frac{1}{10}$

56. (b) Probability of obtaining total sum 7
= Probability of getting opposite faces

$$= 2\left[\left(\frac{1}{6}-x\right)\left(\frac{1}{6}+x\right)+\frac{1}{6}\times\frac{1}{6}+\frac{1}{6}\times\frac{1}{6}\right]$$

$$\Rightarrow 2\left[\left(\frac{1}{6}-x\right)\left(\frac{1}{6}+x\right)+\frac{1}{6}\times\frac{1}{6}+\frac{1}{6}\times\frac{1}{6}\right]=\frac{13}{96}$$

(given)

$$x=\frac{1}{8}$$

57. (c) Total number of 2-digit numbers ${}^{90}C_1 = 90$

Now, $2^n - 2 = (3-1)^n - 2$

$${}^nC_0 3^n - {}^nC_1 3^{n-1} + \ldots + (-1)^{n-1} \times {}^nC_{n-1} 3 + (-1)^n \times {}^nC_n - 2$$

$$3\left(3^{n-1} - n3^{n-2} + \ldots + (-1)^{n-1}.n\right) + (-1)^n - 2$$

$(2^n - 2)$ is a multiple of 3 only when n is odd.

So, required probability $= \frac{45}{90} = \frac{1}{2}$

58. (d) Number of sample spaces = 64w

Let $A = \begin{bmatrix} a & b \\ c & d \end{bmatrix} \Rightarrow |A| = ad - bc$

For singular matrix $|A| = 0$

$\Rightarrow ad = bc$

and a, b, c, d are all different numbers in the set {1, 2, 3, 4, 5, 6}

Now, for $ad = bc$

(i) $6 \times 1 = 2 \times 3$

Total possible ways = 8

(ii) $6 \times 2 = 3 \times 4$

$\therefore$ Total singular matrices = 8 + 8 = 16

Total non-singular matrices
$= 6 \times 5 \times 4 \times 3 - 16 = 344$

Probability $= \frac{344}{64} = \frac{43}{162}$

59. (b) Probability that exactly one of A and B occurs is (1 – k)

$\therefore P(\bar{A} \cap B) + P(A \cap \bar{B}) = 1 - k$

Probability that exactly one of A and C occurs is (1 – k)

$\therefore P(\bar{A} \cap C) + P(A \cap \bar{C}) = 1 - k$

Probability that exactly one of B and C occurs is (1 – k)

$\therefore P(\bar{B} \cap C) + P(B \cap \bar{C}) = 1 - 2k$

The probability that all A, B, and C occur simultaneously is k^2

$\therefore P(A \cap B \cap C) = k^2$

Then, $P(A) + P(B) - 2P(A \cap B) = 1 - k$ ...(i)

$P(C) + P(A) - 2P(A \cap C) = 1 - k$ ...(ii)

$P(B) + P(C) - 2P(B \cap C) = 1 - 2k$ ...(iii)

From (i) + (ii) + (iii)

$$P(A) + P(B) + P(C) - P(A \cap B) - P(C \cap A) = \frac{-4k+3}{2}$$

So, $P(A \cup B \cup C) = \frac{3-4k}{2} + k^2$

$$= \frac{2k^2 - 4k + 3}{2}$$

$$= \frac{2(k-1)^2 + 1}{2} = (k-1)^2 + \frac{1}{2} > \frac{1}{2}$$

$$\Rightarrow P(A \cup B \cup C) > \frac{1}{2}$$

60. (d) Given that

P(0 at even place) $= \frac{1}{2}$

and P(0 at odd place) $= \frac{1}{3}$

$\therefore$ P(1 at even place) $= \frac{1}{2}$

and P(0 at odd place) $= \frac{2}{3}$

('10' is followed by '01' can be if)

...1 0 0 1 (odd position, even position) or ...1 0 0 1 (even position, odd position)

$$P = \left(\frac{2}{3}.\frac{1}{2}.\frac{1}{3}.\frac{1}{2}\right) + \left(\frac{1}{2}.\frac{1}{3}.\frac{1}{2}.\frac{2}{3}\right) = \frac{1}{18} + \frac{1}{18} = \frac{1}{9}$$

61. (a) Let the number of trials be n

$${}^nC_7\left(\frac{1}{2}\right)^7\left(\frac{1}{2}\right)^{n-7} = {}^nC_9\left(\frac{1}{2}\right)^9\left(\frac{1}{2}\right)^{n-9}$$

$${}^nC_7 = {}^nC_9 \Rightarrow n = 16$$

$\therefore$ Probability of getting 2 heads

$${}^{16}C_2\left(\frac{1}{2}\right)^2\left(\frac{1}{2}\right)^{14} = \frac{15}{2^{13}}$$

62. (d) Given $P\left(\frac{x \ge 5}{x > 2}\right)$

$$= \frac{P(x \ge 5 \cap x > 2)}{P(x > 2)}$$

$$= \frac{P(x=5)+P(x=6)+...}{P(x=3)+P(x=4)+...}$$

$$= \frac{\left(\frac{5}{6}\right)^4 \cdot \frac{1}{6} + \left(\frac{5}{6}\right)^5 \cdot \frac{1}{6} + ...}{\left(\frac{5}{6}\right)^2 \cdot \frac{1}{6} + \left(\frac{5}{6}\right)^3 \cdot \frac{1}{6} + ...}$$

$$= \frac{\dfrac{\left(\frac{5}{6}\right)^4 \cdot \frac{1}{6}}{1-\frac{5}{6}}}{\dfrac{\left(\frac{5}{6}\right)^2 \cdot \frac{1}{6}}{1-\frac{5}{6}}} = \left(\frac{5}{6}\right)^2 = \frac{25}{36}$$

63. (b) Here, $n = 5$

Let p and q are probability of success and failure respectively.

$p(x=1) = 5C_1 pq^4 = 0.4096$ and

$p(x=2) = 5C_2 p^2 q^3 = 0.2048$

$\frac{5q}{10p} = \frac{0.4096}{0.2048} \Rightarrow q = 4p$

But $p + q = 1$

$\therefore p = \frac{1}{5}$ and $q = \frac{4}{5}$

$p(\text{exactly3 successes}) = 5c_3 . p^3 q^2$

$= 10 \frac{1}{5^3} . \frac{16}{5^2} = \frac{32}{625}$

64. (d) $\sum P(K) = 1 \Rightarrow 6K^2 + 5k = 1$

$6K^2 + 5K - 1 = 0$

$6K^2 + 6K - K - 1 = 0$

$\Rightarrow K = \frac{1}{6}$ $\quad (K = -1 \text{ rejected})$

$\Rightarrow P(X > 2) = K + 2K + 5K^2$

$= \frac{1}{6} + \frac{2}{6} + \frac{5}{36} = \frac{6+12+5}{36} = \frac{23}{36}$

65. (d) Probability of A getting a total of 6,

$P(A) = \frac{5}{36}$

Probability of B getting a total of 7,

$P(B) = \frac{6}{36} = \frac{1}{6}$

P(A wins)

$= P(A) + P(\overline{A})P(\overline{B})P(A) +$

$P(\overline{A}).P(\overline{B})P(\overline{A})P(\overline{B})P(A) + ...$

$\Rightarrow \frac{5}{36} + \left(\frac{31}{36}\right)\left(\frac{30}{36}\right)\left(\frac{5}{36}\right) + ...\infty$

$\Rightarrow \frac{5}{36}\left(1 + \frac{155}{216} + \left(\frac{155}{216}\right)^2 + ...\infty\right)$

$\Rightarrow \dfrac{\frac{5}{36}}{\frac{61}{216}} = \frac{30}{61} \qquad \left(\because S_x = \frac{a}{1-r}\right)$

66. (d) E_1[the event for getting score a multiple of 4] = (1, 3), (3, 1), (2, 2), (2, 6), (6, 2), (3, 5), (5, 3), (4, 4) and (6, 6)

E_2[4 has appeared at least once] = (1, 4), (2, 4), (3, 4), (4, 4), (5, 4), (6, 4), (4, 1), (4, 2), (4, 3), (4, 5) and (4, 6)

$E_1 \cap E_2 = (4, 4)$

$P\left(\frac{E_2}{E_1}\right) = \frac{1}{9}$

67. (a) Total outcomes $= 9(10^4)$

Favorable outcomes

$= {}^9C_2(2^5 - 2) + {}^9C_1(2^4 - 1) = 36(30) + 9(15)$

$\text{Probability} = \frac{36 \times 30 + 9 \times 15}{9 \times 10^4} = \frac{4 \times 30 + 15}{10^4} = \frac{135}{10^4}$

68. (b) Let B_1 and B_2 be the boxes and N be the number of non-prime numbers.

$\because P(B_1) = P(B_2) = \frac{1}{2}$ and P(non-prime number)

$= P(B_1) \times P\left(\frac{N}{B_1}\right) + P(B_2) \times P\left(\frac{N}{B_2}\right)$

$= \frac{1}{2} \times \frac{20}{30} + \frac{1}{2} \times \frac{15}{20}$

So,

$$P\left(\frac{B_1}{N}\right) = \frac{P(B_1) \times P\left(\frac{N}{B_1}\right)}{P(B_1) \times P\left(\frac{N}{B_1}\right) + P(B_2)P\left(\frac{N}{B_2}\right)}$$

$$= \frac{\frac{1}{2} \times \frac{20}{30}}{\frac{1}{2} \times \frac{20}{30} + \frac{1}{2} \times \frac{15}{20}} = \frac{\frac{1}{3}}{\frac{1}{3} + \frac{15}{40}} = \frac{8}{17}$$

69. (d) $P\left(\frac{E_2^C \cap E_3^c}{E_1}\right) = \frac{P\left[E_1 \cap (E_2^C \cap E_3^C)\right]}{P(E_1)}$

$= \frac{P(E_1) - P[E_1 \cap (E_2 \cup E_1)]}{P(E_1)}$

$\left[\because P(A \cap B^C) = P(A) - P(A \cap B)\right]$

$= \frac{P(E_1) - P[(E_1 \cap E_2) \cup (E_1 \cap E_3)]}{P(E_1)}$

$= \frac{P(E_1) - [P(E_1 \cap E_2) + P(E_1 \cap E_3) - P(E_1 \cap E_2 \cap E_3)]}{P(E_1)}$

$= \frac{P(E_1) - P(E_1 \cap E_2) - P(E_1 \cap E_3) + 0}{P(E_1)}$

$= 1 - (E_2) - P(E_3) \quad [\because P(A \cap B) = P(A) \cdot P(B)]$

$= P(E_2^C) - P(E_3)$ or $P(E_3^C) - P(E_2)$

70. (b) P(second A-card appears before the third B-card)

$= P(AA) + P(ABA) + P(BAA) + P(ABBA) + P(BBAA) + P(BABA)$

$= \frac{1}{4} + \frac{1}{8} + \frac{1}{8} + \frac{1}{16} + \frac{1}{16} + \frac{1}{16} = \frac{11}{16}$

71. (b) A and B are independent events

So, $P\left(\frac{A}{B'}\right) = \frac{P(A \cap B')}{P(B')} = \frac{\frac{1}{3} - \frac{1}{3} \cdot \frac{1}{6}}{\frac{1}{6}} = \frac{1}{3}$

72. (b)

X	0	1	2	3	4	5
P(*k*)	$\frac{1}{32}$	$\frac{12}{32}$	$\frac{11}{32}$	$\frac{5}{32}$	$\frac{2}{32}$	$\frac{1}{32}$

k = No. of times head occur consecutively

Now exception

$\sum xP(k) = (-1) \times \frac{1}{32} + (-1) \times \frac{12}{32} + (-1) \times \frac{11}{32} + 3 \times \frac{5}{32} + 4 \times \frac{2}{32} + 5 \times \frac{1}{32} = \frac{1}{8}$

73. (a) Required probability = when no machine has fault + when only one machine has fault + when only two machines have fault

$= {}^5C_0\left(\frac{3}{4}\right)^5 + {}^5C_1\left(\frac{1}{4}\right)\left(\frac{3}{4}\right)^4 + {}^5C_2\left(\frac{1}{4}\right)^2\left(\frac{3}{4}\right)^3$

$= \frac{243}{1024} + \frac{405}{1024} + \frac{270}{1024} = \frac{918}{1024} = \frac{459}{512} = \frac{27 \times 17}{64 \times 8}$

$= \left(\frac{3}{4}\right)^3 \times k = \left(\frac{3}{4}\right)^3 \times \frac{17}{8}$

$\therefore k = \frac{17}{8}$

74. (d) Given mean $\mu = 8$ and variance $\sigma^2 = 4$

$\Rightarrow \mu = np = 8$ and $\sigma^2 = npq = 4$

$p + q = 1$

$\because p(X \le 2) = \frac{k}{2^{16}}$

$\therefore {}^{16}C_0\left(\frac{1}{2}\right)^{16} + {}^{16}C_1\left(\frac{1}{2}\right)^{16} + {}^{16}C_2\left(\frac{1}{2}\right)^{16} = \frac{k}{2^{16}}$

$k = (1 + 16 + 120) = 137$

75. (d) According to the question,

$1 - \left(\frac{1}{2}\right)^n > \frac{99}{100} \Rightarrow \left(\frac{1}{2}\right)^n < \frac{1}{100} \Rightarrow n \ge 7$

Hence, minimum value is 7.

76. (b) $\because A \subset B$; so $A \cap B = A$

Now, $P\left(\frac{A}{B}\right) = \frac{P(A \cap B)}{P(B)}$

$\Rightarrow P\left(\frac{A}{B}\right) = \frac{P(A)}{P(B)} \qquad \because P(B) \le 1$

$\Rightarrow P\left(\frac{A}{B}\right) \ge P(A)$

77. (c) Let the number of independent shots required to hit the target at least once be n, then

$1 - \left(\frac{2}{3}\right)^n > \frac{5}{6}\left(\frac{2}{3}\right)^n < \frac{1}{6}$

Hence, the above inequality holds when the least value of n is 5.

78. (d) Let G represent drawing a green ball and R represent drawing a red ball.

So, the probability that second drawn ball is red

$= PG \cdot P\left(\frac{R}{G}\right) + P(R)P\left(\frac{R}{R}\right)$

$= \frac{2}{7} \times \frac{6}{7} + \frac{5}{7} \times \frac{4}{7} = \frac{12 + 20}{49} = \frac{32}{49}$

79. (b) Let $A \equiv \{0, 1, 2, 3, 4, ..., 10\}$

$n(S) = {}^{11}C_2 = 55$ where 'S' denotes sample space

Let E be the given event

$\therefore E \equiv \{(0,4), (0,8), (2,6), (2,10), (4,8), (6,10)\}$

$\Rightarrow n(E) = 6$

$\therefore P(E) = \frac{n(E)}{n(S)} = \frac{6}{55}$

80. (b) Required probability

= 1 – {P (All Head) + P(All Tail)}

$=1-\left\{\frac{1}{2^8}+\frac{1}{2^8}\right\} \quad =1-\left\{\frac{1}{2^7}\right\}$

$=1-\left\{\frac{1}{128}\right\}=\frac{127}{128}$

81. (b) We can apply binomial probability distribution

We have $n = 10$

$p=$ Probability of drawing a green ball $=\frac{15}{25}=\frac{3}{5}$

Also $q=1-\frac{3}{5}=\frac{2}{5}$

Variance $= npq = 10\times\frac{3}{5}\times\frac{2}{5}=\frac{12}{5}$

82. (b) $P(E_1)=\frac{1}{6}, P(E_2)=\frac{1}{6}; P(E_3)=\frac{1}{2}$

$P(E_1\cap E_2)=\frac{1}{36}, P(E_2\cap E_3)$

$=\frac{1}{12}, P(E_1\cap E_3)=\frac{1}{12}$

And $P(E_1\cap E_2\cap E_3)=0\neq P(E_1)\cdot P(E_2)\cdot P(E_3)$

$\Rightarrow E_1, E_2, E_3$ are not independent

83. (d) Let $p(F)=p\Rightarrow p(S)=2p$

$\because p+2p=1\Rightarrow p=\frac{1}{3}$

$p(x\geq 5)=p(x=5)+p(x=6)$

$={}^6C_5\left(\frac{2}{3}\right)^5\left(\frac{1}{3}\right)^1+{}^6C_5\left(\frac{2}{3}\right)^6\left(\frac{1}{3}\right)^0$

$=\left(\frac{2}{3}\right)^5\left(6\times\frac{1}{3}+\frac{2}{3}\right)=\frac{256}{729}$

84. (a) Given, $P(\overline{A\cup B})=\frac{1}{6}\Rightarrow P(A\cup B)=1-\frac{1}{6}=\frac{5}{6}$

$P(\overline{A})=\frac{1}{4}\Rightarrow P(A)=1-\frac{1}{4}=\frac{3}{4}$

We know,

$P(A\cup B)=P(A)+P(B)-P(A\cap B)$

$\Rightarrow\frac{5}{6}=\frac{3}{4}+P(B)-\frac{1}{4}$

$\left(\because P(A\cap B)=\frac{1}{4}\right)$

$\Rightarrow P(B)=\frac{1}{3}$

$\because P(A)\neq P(B)$, so they are not equally likely.

Also $P(A)\times P(B)=\frac{3}{4}\times\frac{1}{3}=\frac{1}{4}=P(A\cap B)$

So A and B are independent.

Integer and Numerical Value Type

1. There are n white balls in the urn.

$\Rightarrow$ Probability of Mr A to draw two balls of same colour is

$\frac{{}^3C_2+{}^nC_2}{{}^{n+3}C_2}=\frac{1}{2}$ (given)

or $\frac{6+n(n-1)}{(n+3)(n+2)}=\frac{1}{2}$

or $n^2-7n+6=0$

$\Rightarrow n=1$ or 6 ...(i)

Also required probability for Mr B according to the question is

$\frac{3}{n+3}\frac{3}{n+3}+\frac{n}{n+3}\frac{n}{n+3}=\frac{5}{8}$ (given)

Solving, we get $n^2-10n+9=0$; $n=1$ or 9 ...(ii)

From (i) and (ii), $n=1$

2. When A and B are mutually exclusive,

$P(A\cap B)=0$

$\therefore P(A\cup B)=P(A)+P(B)$ ...(i)

$\Rightarrow 0.8=0.5+p$

or $p=0.3$ ...(ii)

$P(A\cup B)=P(A)+P(B)$

$=P(A)+P(B)-P(A\cup B)$

$=P(A)+P(B)-P(A)P(B)$

$0.3=0.+q-(0.5)q$

$0.3=\frac{q}{2}$

$q=0.6$ ...(iii)

$\Rightarrow\frac{q}{p}=2$

3. Let event A: Card is of Heart but not King (12 cards)

Event B: King but not Heart (3 cards)

Event C: Heart and King (1 card)

Required probability

$p=P(E)=\frac{{}^{12}C_1.{}^3C_1+{}^3C_1.{}^1C_1+{}^1C_1.{}^{12}C_1}{{}^{52}C_2}=\frac{2}{52}$

$\therefore 104\,p=4$

4. $P(E_1)=1-$ [P(all heads) + P(all tails)]

$=1-\left[\frac{1}{2^n}+\frac{1}{2^n}\right]$

$=1-\frac{1}{2^{n-1}}$

$P(E_2)=$ P(no head) + P(exactly one head)

$$= \frac{1}{2^n} + {}^nC_1 \times \frac{1}{2^n} = \frac{n+1}{2^n}$$

$P(E_1 \cap E_2) =$ (exactly on head and $(n-1)$ tail)

$$= {}^nC_1 \times \frac{1}{2} \times \frac{1}{2^{n-1}} = \frac{n}{2^n}$$

If E_1 and E_2 are independent, then

$$\frac{n}{2^n} = \left(1 - \frac{1}{2^{n-1}}\right)\left(\frac{n+1}{2^n}\right)$$

or $n = \left(1 - \frac{1}{2^{n-1}}\right)(n+1)$

or $n = n + 1 - \frac{n+1}{2^{n-1}}$

$n + 1 = 2^{n-1}$

$n = 3$

5. E_1 = first unit is functioning
E_2 = second unit is functioning

$P(\bar{E}_1) = 0.1 P(\bar{E}_2) = 0.2$

$$P = \frac{0.8 \times 0.1}{0.1 \times 0.2 + 0.9 \times 0.2 + 0.1 \times 0.8} = \frac{8}{28}$$

$98P = \frac{8}{28} \times 98 = 28$

6. Let $P(B_1) = p_1, P(B_2) = p_2\, (B_3) = p_3$ given that

$p_1(1-p_2)(1-p_3) = \alpha$ ...(i)

$p_2(1-p_1)(1-p_3) = \beta$ ...(ii)

$p_3(1-p_1)(1-p_2) = \gamma$ ...(iii)

and $(1-p_1)(1-p_2)(1-p_3) = p$ ...(iv)

$$\Rightarrow \frac{p_1}{1-p_1} = \frac{\alpha}{p}, \frac{p_2}{1-p_2} = \frac{\beta}{p} \text{ and } \frac{p_3}{1-p_3} = \frac{\gamma}{p}$$

Also $\beta = \frac{\alpha p}{\alpha + 2p} = \frac{3\gamma p}{p - 2\gamma}$

$\Rightarrow \alpha p - 2\alpha\gamma = 3\alpha\gamma + 6p\gamma$

$\Rightarrow \alpha p - 6p\gamma - 5\alpha\gamma$

$$\Rightarrow \frac{p_1}{1-p_1} - \frac{6p_3}{1-p_3} = \frac{5p_1p_3}{(1-p_1)(1-p_3)}$$

$\Rightarrow p_1 - 6p_3 = 0$

$\Rightarrow \frac{p_1}{p_3} = 6$

7. Probability of getting at least two 3's or 5's in one trial

$$= {}^4C_2\left(\frac{2}{6}\right)^2\left(\frac{4}{6}\right)^2 + {}^4C_3\left(\frac{2}{6}\right)^3\left(\frac{4}{6}\right) + {}^4C_4\left(\frac{2}{6}\right)^4$$

$$= \frac{33}{3^4} = \frac{11}{27}$$

$E(x) = np = 27\left(\frac{11}{27}\right) = 11$

8. Let n bombs be required, then

$$1 - {}^nC_1\left(\frac{1}{2}\right)^1\left(\frac{1}{2}\right)^{n-1} - {}^nC_0\left(\frac{1}{2}\right)^0\left(\frac{1}{2}\right)^n \geq \frac{99}{100}$$

$$\Rightarrow \frac{1}{100} \geq \frac{n+1}{2^n} \Rightarrow 100(n+1) \Rightarrow n \geq 11$$

9. $p = \frac{1}{10} . q = \frac{9}{10}$

P(not hitting target in n trials $= \left(\frac{9}{10}\right)^n$

P(at least one hit) $= 1 - \left(\frac{9}{10}\right)^n$

$\because 1 - \left(\frac{9}{10}\right)^n > \frac{1}{4} \Rightarrow (0.9)^n < 0.75$

$\therefore$ Minimum = 3

Assertion and Reason

1. (b) $Z = 5 - x - y \quad \therefore \quad xy + yz + zx = 3$

$\Rightarrow y^2 + y(x-5) + (3 + x^2 - 5x) = 0$

$\because \; y \in R \quad \therefore D \geq 0$

$\Rightarrow \; (3x - 13)(x+1) \leq 0$

$\Rightarrow x \in \left[-1, \frac{13}{3}\right]$

So maximum value of $x = \frac{13}{3}$

and minimum value of $x = -1$

Then required probability $= \frac{\frac{13}{3} - 0}{\frac{13}{3} + 1} = \frac{13}{16}$

2. (c) Let the two non-negative integers be x and y

Then $x = 5a + \alpha$ and $y = 5b + \beta$, where

$0 \leq \alpha \leq 4, 0 \leq \beta \leq 4$

Now $x^2 + y^2$

$= (5a + \alpha)^2 + (5b + \beta)^2$

$= 25(a^2 + b^2) + 10(\alpha a + b\beta) + \alpha^2 + \beta^2$

$\therefore x^2 + y^2$ is divisible by 5 if and only if 5 divides $\alpha^2 + \beta^2$

The total number of ways of choosing α and β $= 5 \times 5 = 25$

Further, $\alpha^2 + \beta^2$ will be divisible by 5 if

$(\alpha_1, \beta) \in \{(0, 0), (1, 2), (1, 3), (2, 1), (2, 4), (3, 1), (3, 4), (4, 2), (4, 3)\}$

$\therefore$ Favourable number of ways of choosing α

and $\beta = 5\times5 = 25$

Required probability $= \dfrac{9}{25}$.

3. (d) Statement-2 $P\left(\dfrac{A}{B}\right) = \dfrac{P(A\cap B)}{P(B)}$

(by definition)

$P(\bar{B}) = P((A\cup\bar{A})\cap\bar{B}) = P((A\cap\bar{B})\cup(\bar{A}\cap\bar{B}))$

$= P(A\cap\bar{B}) + P(\bar{A}\cap\bar{B})$

Statement 2 is true

Statement 1

$$P(A/\bar{B}) + P(\bar{A}/\bar{B}) = \frac{P(A\cap\bar{B})}{P(\bar{B})} + \frac{P(\bar{A}\cap\bar{B})}{P(\bar{B})}$$

$$= \frac{P(A\cap\bar{B}) + P(\bar{A}\cap\bar{B})}{P(\bar{B})} = \frac{P(\bar{B})}{P(\bar{B})} = 1$$

Statement 1 is false.

JEE Advanced

Objective Questions

One Option Correct

1. (a) Total number of functions from A to $B = n(S) = 5^7$

total number of onto functions from A to B is

$$n(E) = \frac{7!}{3!4!}\times5! + \frac{7!}{3!2!}\times\frac{1}{2!2!}\times5! = \frac{7!\times20}{6}$$

$$\therefore\ P(E) = \frac{n(E)}{n(S)} = \frac{7!\times2}{3\times5^6}$$

2. (a) Total number of ways of selecting 3 integers from 20 natural numbers $= {}^{20}C_3 = 1140$

Their product is a multiple of 3 means, at least one number is divisible by 3. The numbers which are divisible by 3 are 3, 6, 9, 12, 15, 18 and the number of ways of selecting at least one of them is

${}^6C_1\times{}^{14}C_2 + {}^6C_2\times{}^{14}C_1 + {}^6C_3 = 776$

Probability $= \dfrac{776}{1140} = \dfrac{194}{285}$

3. (d) Probability that the first critic favours the book, $P(E_1) = \dfrac{5}{5+2} = \dfrac{5}{7}$

Probability that the second critic favours the book, $P(E_2) = \dfrac{4}{4+3} = \dfrac{4}{7}$

Probability that the third critic favours the book, $P(E_3) = \dfrac{3}{3+4} = \dfrac{3}{7}$

Majority will be in favour if at least two critics favour the book

$$= P(E_1\cap E_2\cap\bar{E}_3) + P(E_1\cap\bar{E}_2\cap E_3) + P(\bar{E}_1\cap E_2\cap E_3) + P(E_1\cap E_2\cap E_3)$$

$$= P(E_1)P(E_2)P(\bar{E}_3) + P(E_1)P(\bar{E}_2)P(E_3) + P(\bar{E}_1\cap E_2\cap E_3) + P(E_1)P(E_2)P(E_3)$$

$$= \frac{5}{7}\times\frac{4}{7}\times\left(1-\frac{3}{7}\right) + \frac{5}{7}\times\left(1-\frac{4}{7}\right)\times\frac{3}{7} + \left(1-\frac{5}{7}\right)\times\frac{4}{7}\times\frac{3}{7} + \frac{5}{7}\times\frac{4}{7}\times\frac{3}{7} = \frac{209}{343}$$

4. (c) Favourable cases are 29, 92, 36, 63

$$\therefore\ \text{Required probability} = {}^4C_3\times\left(\frac{4}{100}\right)^3\frac{96}{100} = \frac{96}{390625}$$

5. (c) Split the digits into pairs, viz.: (0, 1), (1, 2), … (8, 9)

Disjoint pairs out of these are (0, 1), (2, 3), (4, 5), (6, 7), (8, 9)

(there is no other set of 5 disjoints pairs)

Now the two cases are

(a) When the. pair (0, 1) is not used for first and the last place.

The number of ways $= 4\times4!\times2^5$

(b) When the pair (0, 1) is used for first and last place.

The number of ways $= 1\times4!\times2^4$

$\therefore$ Total number of favourable cases $= 9\times4!\times2^4$

$$\therefore\ \text{Probability} = \frac{9\times4!\times2^4}{9\times9!} = \frac{1}{945}$$

6. (a) S_1:

$$P = \frac{{}^4C_3\times{}^3C_2 + {}^4C_3\times{}^2C_2 + {}^3C_3\times{}^4C_3 + {}^3C_3\times{}^2C_2}{{}^{10}C_5}$$

$$= \frac{4\times3+4\times1+1\times6+1}{{}^{10}C_5} = \frac{23}{{}^{10}C_5}$$

S_2: Two adjacent row can be selected out of 5 rows in 4 ways. Total ways of selecting 2 rows is ${}^5C_2$ 4, hence $P = \dfrac{4}{{}^5C_2} = \dfrac{2}{5}$

S_3: Required probability

$$= \frac{P(A\bar{B}\bar{C})}{P(\bar{A}\bar{B}\bar{C}) + P(A\bar{B}\bar{C}) + P(\bar{A}B\bar{C}) + P(\bar{A}\bar{B}C)}$$

$$=\frac{\frac{4}{5}\cdot\frac{1}{4}\cdot\frac{1}{3}}{\frac{1}{5}\cdot\frac{1}{4}\cdot\frac{1}{3}+\frac{4}{5}\cdot\frac{1}{4}\cdot\frac{1}{3}+\frac{1}{5}\cdot\frac{3}{4}\cdot\frac{1}{3}+\frac{1}{5}\cdot\frac{1}{4}\cdot\frac{2}{3}}=\frac{4}{10}=\frac{2}{5}$$

7. (a) $S_1: P_1=\frac{1}{6}$

$P_2={}^4C_3\times{}^6C_1\times\frac{1}{6}\times\frac{1}{6}\times\frac{1}{6}\times\frac{5}{6}=\frac{5}{54}$ $\quad\therefore P_1>P_2$

$\therefore S_1$ is true

S_2 : B throws 10, 11 or 12. The cases are

$6+4, 5+5, 4+6$

$6+5, 5+6, 6+6$

$\therefore$ Probability $=\frac{6}{36}=\frac{1}{6}$

$\therefore S_2$ is true

$S_3: P(A_1\cup A_2)=1-P(A_1\cup A_2)'$

$=1-P(A_1'\cap A_2')$

Also $P(A_1\cup A_2)=1-P(A_1')\,P(A_2')$

$\therefore P(A_1'\cap A_2')=1-P(A_1')\,P(A_2')$

$\therefore A_1'$ and A_2' are independent

$\therefore A_1$ and A_2 are independent

$S_4: P(A\cap B\cap C)=P(A)\,P(B)\,P(C)$

$P(A\cap B\cap \bar{C})=P(A\cap B)-C)$

$=P((A\cap B)-(A\cap B\cap C))$

$=P(A)\,P(B)-P(A)\cdot P(B)\,P(C)$

$=P(A)\,P(B)\,(1-P(C))=P(A)\,P(B)\,P\left(\bar{C}\right)$

$\therefore A, B, \bar{C}$ are independent.

8. (a)

$E_1=\{1, 2, 3\}$

$S_1=\{1, 2\}$ | $S_1=\{2, 3\}$ | $S_1=\{1, 3\}$

$E_2=E_1-S_1=\{3\}$ | $E_2=E_1-S_1=\{1\}$ | $E_2=E_1-S_1=\{1\}$

$F_2=\{1, 2, 3, 4\}$ | $F_2=\{1, 2, 3, 4\}$ | $F_2=\{1, 3, 4\}$

$S_2=\{1, 2\}\ \{1, 3\}\ \{1, 4\}\ \{2, 3\}, \{2, 4\}\ \{3, 4\}$ | $S_2=\{1, 2\}\ \{1, 3\}\ \{1, 4\}\ \{2, 3\}, \{2, 4\}\ \{3, 4\}$ | $S_2=\{1, 3\}\ \{1, 4\}\ \{3, 4\}$

$G_2=\{1, 2, 3, 4, 5\}$ $\{2, 3, 4, 5\}$ | $\{1, 2, 3, 4, 5\}$ $\{2, 3, 4, 5\}$ | $\{1, 2, 3, 4, 5\}$ $\{2, 3, 4, 5\}$

$\{1, 2\}$ $\{2, 3\}$ $\{2, 3\}$ $\{1, 3\}$

$E_3=E_2\cup S_3=E_1=\{1, 2, 3\}$

$$P=\frac{P\left(S_1\cap(E_1=E_3)\right)}{P(E_1=E_3)}=\frac{P\left(A_{1,2}\right)}{P(A)}$$

$P(A)=P\left(A_{1,2}\right)+P\left(A_{1,3}\right)+P\left(A_{2,3}\right)$

(If 1,2 chosen at start; If 1,3 chosen at start; If 2,3 chosen at start)

$P\left(A_{1,2}\right)=\frac{1}{3}\times\underbrace{\frac{1\times{}^3C_1}{{}^4C_2}}_{\text{1 is definitely chosen from }F_2}\times\underbrace{\frac{1}{{}^5C_2}}_{\text{1, 2 chosen from }G_2}$

$=\frac{1}{3}\times\frac{1}{2}\times\frac{1}{10}=\frac{1}{60}$

$P\left(A_{1,3}\right)=\frac{1}{3}\times\underbrace{\frac{1\times{}^2C_1}{{}^3C_2}}_{\text{1 is definitely chosen from }F_2}\times\underbrace{\frac{1}{{}^5C_2}}_{\text{1, 2 chosen from }G_2}$

$=\frac{1}{3}\times\frac{2}{3}\times\frac{1}{10}=\frac{1}{45}$

$P\left(A_{2,3}\right)=\frac{1}{3}\times\left[\underbrace{\frac{{}^3C_2\times 1}{{}^4C_2}\times\frac{1}{{}^4C_2}}_{\text{If 1 is not chosen from }F_2}\times\underbrace{\frac{1\times{}^3C_1}{{}^4C_2}\times\frac{1}{{}^5C_2}}_{\text{If 1 is chosen from }F_2}\right]$

$=\frac{1}{3}\left(\frac{1}{12}+\frac{1}{20}\right)=\frac{2}{45}$

$P(A)=\frac{1}{60}+\frac{1}{45}+\frac{2}{45}=\frac{1}{12}\ \frac{P\left(A_{1,2}\right)}{P(A)}=\frac{1}{5}$

9. (b) Probability of getting head on coin

$C_1 = P(H) = \frac{2}{3}$

Probability of getting head on coin

$C_2 = P(H) = \frac{1}{3}$

For the coin C_1

No. of heads (α)	1	2	3
Probability	$\frac{1}{9}$	$\frac{4}{9}$	$\frac{4}{9}$

For the coin C_2

No. of heads (β)	1	2	3
Probability	$\frac{4}{9}$	$\frac{4}{9}$	$\frac{1}{9}$

For real and equal roots of the given polynomial,

$\alpha^2 - 4\beta = 0 \Rightarrow \alpha^2 = 4\beta$

$\therefore (\alpha, \beta) = (0,0), (2,1)$

So, probability $= \frac{1}{9} \times \frac{4}{9} + \frac{4}{9} \times \frac{4}{9} = \frac{20}{81}$

10. (b) We know the total number of ways of dividing n identical things among r person $= {}^{n+r+1}C_{r-1}$

Total number of non-negative solutions of $x + y + z = 10$ are $= {}^{12}C_2 = 66$

If z is even, then the following cases can arise:

$z = 0 \Rightarrow$ No. of ways of solving $x + y = 10 \Rightarrow {}^{11}C_1$

$z = 2 \Rightarrow$ No. of ways of solving $x + y = 8 \Rightarrow {}^{9}C_1$

$z = 4 \Rightarrow$ No. of ways of solving $x + y = 6 \Rightarrow {}^{7}C_1$

$z = 6 \Rightarrow$ No. of ways of solving $x + y = 4 \Rightarrow {}^{5}C_1$

$z = 8 \Rightarrow$ No. of ways of solving $x + y = 2 \Rightarrow {}^{3}C_1$

$z = 10 \Rightarrow$ No. of ways of solving $x + y = 0 \Rightarrow 1$

$\therefore$ Required probability $= \frac{36}{66} = \frac{6}{11}$

11. (c) Given $P(T_1) = \frac{20}{100}, P(T_2) = \frac{80}{100}, P(D) = \frac{7}{100}$

Let $P\left(\frac{D}{T_2}\right) = P$, then $P\left(\frac{D}{T_1}\right) = 10P$

By total probability,

$$P(D) = P(T_1)P\left(\frac{D}{T_1}\right) + P(T_2)P\left(\frac{D}{T_2}\right)$$

$$\Rightarrow \frac{7}{100} = \frac{20}{100} \times 10P + \frac{80}{100} \times P$$

$$\Rightarrow \frac{7}{280} = P \Rightarrow P = \frac{1}{40}$$

$$\therefore P\left(\frac{D}{T_1}\right) = \frac{10}{40} \text{ and } P\left(\frac{D}{T_2}\right) = \frac{1}{40}$$

$$\Rightarrow P\left(\frac{\bar{D}}{T_1}\right) = 1 - \frac{10}{40} = \frac{30}{40} \text{ and } P\left(\frac{\bar{D}}{T_2}\right) = 1 - \frac{1}{40} = \frac{39}{40}$$

$$P\left(\frac{T_2}{\bar{D}}\right) = \frac{P\left(\frac{\bar{D}}{T_2}\right)P(T_2)}{P\left(\frac{\bar{D}}{T_1}\right)P(T_1) + P\left(\frac{\bar{D}}{T_2}\right)P(T_2)}$$

$$= \frac{\frac{80}{100} \times \frac{39}{40}}{\frac{20}{100} \times \frac{30}{40} + \frac{80}{100} \times \frac{39}{40}} = \frac{156}{186} = \frac{78}{93}$$

12. (a) According to given condition, following cases may arise

GGBBB

GBGBB

GBBGB

BGGBB

BGBGB

Thus, the favorable cases are $= 5 \times 2!3! = 60$

Total ways in which 5 persons can be seated $= 5! = 120$

$\therefore$ Required probability $= \frac{60}{120} = \frac{1}{2}$

One or More Than One Option Correct

1. (a, b, c, d)

(i) $P(E_1) = 1 - P(R\,R\,R)$

$$= 1 - \left[\frac{1}{3}, \frac{2}{4}, \frac{3}{5}\right] = 0.9$$

(ii) $P(E_2) = 3\,P(B\,R\,R) = 3.\frac{2}{3}.\frac{1}{4}.\frac{2}{5} = 0.2$

(iii) $P(E_3) = P(R\,R\,R / R\,R\,R \cup B\,B\,B)$

$$= \frac{P(RRR)}{P(RRR) + P(BBB)}$$

But

$$P(BBB) = \frac{2}{3}.\frac{3}{4}.\frac{4}{5} = \frac{8}{20}$$

$$\Rightarrow P(E_3) = \frac{0.1}{0.1 + 0.4} = 0.2$$

(iv) $P(E_4) = 1 - P(B\,B\,B) = 1 - \frac{2}{5} = 0.6$

2. (a, b) Area of the shaded region

$$= \int_0^1 \left(\sqrt{x} - x^2\right) dx$$

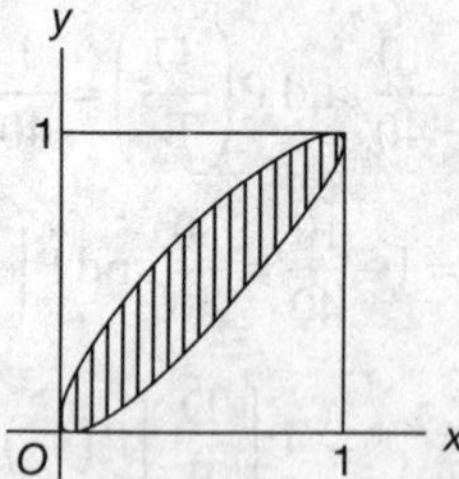

$$= \left(\frac{2}{3}x^{3/2} - \frac{x^3}{3}\right)\Bigg|_0^1 = \frac{2}{3} - \frac{1}{3} = \frac{1}{3}$$

Area of the square = 1

$\therefore$ Probability = 1/3

$A \cup B$ = whole of the region enclosed in $0 \le x \le 1, 0 \le y \le 1$

$\therefore$ A and B are exhaustive events.

$$P(A) = \int_0^1 \sqrt{x}dx = \frac{2}{3}x^{\frac{3}{2}}\Bigg|_0^1 = \frac{2}{3}, P(B) = \frac{1}{3}$$

$$\therefore P(A)P(B) = \frac{2}{9} \ne P(A \cap B)$$

$\therefore$ A and B are not independent

3. (a, d)

Let α and β be the number of heads and tails thrown by A respectively, then $\alpha + \beta = n + 1$ ...(i)

Let γ and δ be the number of heads and tails thrown by B respectively, then $\gamma + \delta = n$ ...(ii)

Required probability is $P(\alpha > \gamma)$ = P(say)

Due to symmetry P(A will have more heads than B) = P(A will have more tails than B) ...(iii)

i.e., $P(\alpha > \gamma) = P(\beta > \delta) = p$

Now $P(\alpha > \gamma) = 1 - P(\alpha \le \gamma)$

Also $\alpha \le \gamma \Leftrightarrow n + 1 - \beta \le n - \delta \Leftrightarrow 1 + \delta \le \beta$

$\Leftrightarrow \delta < \beta$ [from (i) and (ii)

$\therefore P(\alpha \le \gamma) = P(\delta < \beta)$

$$\Rightarrow p = 1 - P(\delta < \beta) = 1 - p \therefore p = \frac{1}{2}$$

4. (a, d)

Sample space is $\{-0.50, -0.49, \ldots -0.01, 0, 0.01 \ldots, 0.49\}$ and the event is $\{-0.10, -0.09, \ldots, 0.10\}$

Hence probability is 21/100

5. (a, b, c)

Let S denote the set of points inside a square with corners $(x, y), (x, y+1), (x+1, y), (x+1, y+1), x$ and y are integers. Clearly each of the four points belong to the set X.

Let P denote the set of points in S with distance less than $\frac{1}{4}$ from any corner point. P consists of four quarter circles, each of radius $\frac{1}{4}$.

A coin, whose centre falls in S, will cover a point of X if and only if its centre falls in P.

Hence, the required probability,

$$p = \frac{\text{Area of P}}{\text{Area of S}} = \frac{\pi\left(\frac{1}{4}\right)^2}{1 \times 1} = \frac{\pi}{16}$$

6. (a, b)

$$\because P(X/Y) = \frac{P(X \cap Y)}{P(\gamma)} \Rightarrow \frac{1}{2} = \frac{1/6}{P(\gamma)} \Rightarrow P(\gamma) = \frac{1}{3}$$

$$\text{Similarly, } P(Y/X) = \frac{P(X \cap \gamma)}{P(X)}$$

$$\Rightarrow \frac{1}{3} = \frac{1/6}{P(X)} \Rightarrow P(X) = \frac{1}{2}$$

(a) $P(X \cup Y) = P(X) + P(Y) - P(X \cap Y)$

$$= \frac{1}{2} + \frac{1}{3} - \frac{1}{6} = \frac{2}{3}$$

$\therefore$ (a) is true.

(b) $\because P(X \cap Y) = P(X)P(Y)$

$\Rightarrow$ X and Y are independent events.

$\therefore$ (b) is true.

But (c) is not true.

(d) $P(X^C \cap Y) = P(X^C) \times P(Y) = \frac{1}{2} \times \frac{1}{3} = \frac{1}{6}$

$\therefore$ (d) is not true.

7. (b, d)

Given that $P(X_1) = \frac{1}{2}, P(X_2) = \frac{1}{4}, P(X_3) = \frac{1}{4}$

P (X) = P(at least 2 engines are functioning)

$$P(X_1 \cap X_2 \cap X_3^C) + P(X_1 \cap X_2^C \cap X_3) + P(X_1^C \cap X_2 \cap X_3) + P(X_1 \cap X_2 \cap X_3)$$

$$= \frac{1}{2} \times \frac{1}{4} \times \frac{3}{4} + \frac{1}{2} \times \frac{3}{4} \times \frac{1}{4} + \frac{1}{2} \times \frac{1}{4} \times \frac{1}{4} + \frac{1}{2} \times \frac{1}{4} \times \frac{1}{4}$$

$$= \frac{1}{4}$$

(a) $P\left(X_1^C/X\right)=\dfrac{P\left(X_1^C\cap X\right)}{P(X)}=\dfrac{P\left(X_1^C\cap X_2\cap X_3\right)}{P(X)}$

$=\dfrac{\frac{1}{2}\times\frac{1}{4}\times\frac{1}{4}}{\frac{1}{4}}=\dfrac{1}{8}$

$\therefore$ (a) is not true.

(b) P [Exactly two engines are functioning/X]

$\dfrac{P[\text{Exactly two engines are functioning})\cap X]}{P(X)}$

$=\dfrac{P\left(X_1^C\cap X_2\cap X_3\right)+P\left(X_1\cap X_2^C\cap X_3\right)+P\left(X_1\cap X_2\cap X_3^C\right)}{P(X)}$

$=\dfrac{\frac{1}{2}\times\frac{1}{4}\times\frac{1}{4}+\frac{1}{2}\times\frac{3}{4}\times\frac{1}{4}+\frac{1}{2}\times\frac{1}{4}\times\frac{3}{4}}{\frac{1}{4}}=\dfrac{7}{8}$

$\therefore$ (b) is true.

(c) $P\left(X/X_2\right)=\dfrac{P\left(X\cap X_2\right)}{P\left(X_2\right)}$

$=\dfrac{P\left(X_1\cap X_2\cap X_3\right)+P\left(X_1^C\cap X_2\cap X_3\right)+P\left(X_1\cap X_2\cap X_3^C\right)}{P\left(X_2\right)}$

$=\dfrac{\frac{1}{2}\times\frac{1}{4}\times\frac{1}{4}+\frac{1}{2}\times\frac{1}{4}\times\frac{1}{4}+\frac{1}{2}\times\frac{1}{4}\times\frac{3}{4}}{\frac{1}{4}}=\dfrac{5}{8}$

$\therefore$ (c) is not true.

(d) $P\left(X/X_1\right)=\dfrac{P\left(X\cap X_1\right)}{P\left(X_1\right)}$

$=\dfrac{P\left(X_1\cap X_2\cap X_3\right)+P\left(X_1\cap X_2^C\cap X_3\right)+P\left(X_1\cap X_2\cap X_3^C\right)}{P\left(X_1\right)}$

$=\dfrac{\frac{1}{2}\times\frac{1}{4}\times\frac{1}{4}+\frac{1}{2}\times\frac{3}{4}\times\frac{1}{4}+\frac{1}{2}\times\frac{1}{4}\times\frac{3}{4}}{\frac{1}{2}}=\dfrac{7}{8}$

$\therefore$ (d) is true.

8. (a, d)

Given that E and F are independent events,

$\therefore P(E\cap F)=P(E)\cdot P(F)$ ...(i)

P (exactly one) $=P\left(E\cap\overline{F}\right)+P\left(\overline{E}\cap F\right)=\dfrac{11}{25}$

$\Rightarrow P(E)P\left(\overline{F}\right)+P\left(\overline{E}\right)P(F)=\dfrac{11}{25}$

$\Rightarrow P(E)(1-P(F))+1(1-P(E))P(F)=\dfrac{11}{25}$

$\Rightarrow P(E)-P(E)P(F)+P(F)-P(E)P(F)=\dfrac{11}{25}$

$\Rightarrow P(E)+P(F)-2P(E).P(F)=\dfrac{11}{25}$ ...(ii)

P (none of them)

$=P\left(\overline{E}\cap\overline{F}\right)=\dfrac{2}{25}\Rightarrow P\left(\overline{E}\right)P\left(\overline{F}\right)=\dfrac{2}{25}$

$\Rightarrow [1-P(E)][1-P(F)]=\dfrac{2}{25}$

$\Rightarrow 1-P(E)-P(F)+P(E)P(F)=\dfrac{2}{25}$ ...(iii)

Adding equations (ii) and (iii), we get

$1-P(E)P(F)=\dfrac{13}{25}$ or

$P(E)P(F)=\dfrac{12}{25}$ ...(iv)

Using the result in equation (ii), we get

$P(E)+P(F)=\dfrac{35}{25}$ ...(v)

Solving (iv) and (v) we get

$P(E)=\dfrac{3}{5}$ and $P(F)=\dfrac{4}{5}$ or

$P(E)=\dfrac{4}{5}$ and $P(F)=\dfrac{3}{5}$

$\therefore$ (a) and (d) are the correct options.

9. (b, c)

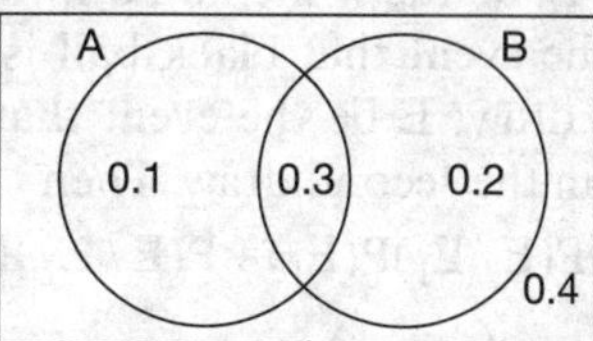

$P(A)=0.4, P(B)=0.5, P(A\cap B)=0.3$

$\Rightarrow P(A\cup B)=P(A)+P(B)-P(A\cap B)=0.6$

Now,

$P(E_1)=\dfrac{P(A\cap B)}{P(A\cup B)}=\dfrac{0.3}{0.6}=\dfrac{1}{2}$

$P(E_2)=\dfrac{P\left(A\cap\overline{B}\right)+P\left(\overline{A}\cap B\right)}{P(A\cup B)}=\dfrac{0.1+0.2}{0.6}=\dfrac{1}{2}$

$P(A/B)=\dfrac{P(A\cap B)}{P(B)}=\dfrac{0.3}{0.5}=\dfrac{3}{5}=0.60;$

$P(B/A)=\frac{0.3}{0.4}=\frac{3}{4}=0.75$

$P\left(A/(A\cup B)\right)=\frac{P(A)}{P(A\cup B)}=\frac{0.4}{0.6}=\frac{2}{3}$

$\frac{P(B)}{P(A\cup B)}=\frac{0.5}{0.6}=\frac{5}{6}$

10. (c, d)

$P(E_1)=1-P$ (unit's place in both is 1, 2, 3, 4, 6,7, 8, 9)

$P(E_1=0\text{ or }5)=1-\left(\frac{4}{5}\right)^2=\frac{9}{25}$

$P(E_2:5)=P(13579)-P(1379)$

$=\frac{1}{4}-\frac{4}{25}=\frac{25-16}{100}=\frac{9}{100}$

$\frac{P(E_2)}{P(E_1)}=\frac{9}{100}\times\frac{25}{9}=\frac{1}{4}$

$P(E_1)=4P(E_2)$

$P(E_2/E_1)=\frac{P(E_2\cap E_1)}{P(E_1)}=\frac{P(E_2)}{P(E_1)}=\frac{1}{4}$

$P(E_1/E_2)=\frac{P(E_1\cap E_2)}{P(E_2)}=\frac{P(E_2)}{P(E_2)}=1$

11. (b, c, d)

Option (a) is false

$P(TTT\text{ or }HHH)=\frac{1}{8}+\frac{1}{8}=\frac{1}{4}$

(b) $\frac{P(A\cap B)}{P(B)}=\frac{P(A\cap B^c)}{1-P(B)}=\frac{P(A)-P(A\cap B)}{1-P(B)}$

$P(A\cap B)[1-P(B)]=P(B)P(A)-P(B)P(A\cap B)$

$P(A\cap B)=P(A)P(B)$

Hence, the given statement is true.

(c) Let E_1 be the white ball drawn in first draw; E_2 be the event that black ball is drawn in the second draw; E be the event that white ball is drawn in the second draw. Then

$P(E)=P(E/E_1)P(E_1)+P(E/E_2)P(E_2)$

$=\frac{d+w}{w+b+d}\left(\frac{w}{w+b}\right)+\frac{w}{w+b+d}\left(\frac{b}{w+b}\right)$

$=\left(\frac{w}{w+b}\right)\left(\frac{d+w}{w+b+d}+\frac{b}{w+b+d}\right)$

$=\left(\frac{w}{w+b}\right)$

12. (a, b, c, d)

$P(A\cup B)=P(A)+P(B)-P(A\cap B)$

or $\frac{5}{8}=\frac{3}{8}+\frac{4}{8}-P(A\cap B)$

or $P(A\cap B)=\frac{2}{8}=\frac{1}{4}$

Now, $P(A^c/B)=\frac{P(A^c\cap B)}{P(B)}$

$=\frac{P(B)-P(A\cap B)}{P(B)}$

$=1-2\left(\frac{1}{4}\right)=\frac{1}{2}$

$2P(A/B^c)=\frac{2P(A\cap B^C)}{P(B^C)}$

$=\frac{P(A)-P(A\cap B)}{1-P(B)}$

$=4\left(\frac{3}{8}-\frac{2}{8}\right)=\frac{1}{2}$

Hence, option (a) is correct

$P(A/B)=\frac{P(A\cap B)}{P(B)}=\frac{1}{4}\times\frac{2}{1}=\frac{1}{2}=P(B)$

Hence, option (b) is correct

$P(A^c/B^c)=\frac{P\left(A^c\cap B^c\right)}{P(B^c)}$

$=\frac{1-P(A\cup B)}{1-P(B)}$

$=2\left(1-\frac{5}{8}\right)=\frac{3}{4}$

$P(B/A^c)=\frac{P\left(B\cap A^C\right)}{1-P(A)}$

$=\frac{P(B)-P(A\cap B)}{5/8}$

$=\frac{1/2-1/4}{5/8}$

$=\frac{1}{4}\times\frac{8}{5}=\frac{2}{5}$

13. (b, d)

R – 5	R – 3	R – 5
G – 5	G – 5	G – 3
B_1	B_2	B_3

$\because P(B_1)=\frac{3}{10},P(B_2)=\frac{3}{10},P(B_3)=\frac{4}{10}$

$P(G/B_1)=\frac{5}{10},P(G/B_2)=\frac{5}{8},P(G/B_3)=\frac{3}{8}$

(a) $P(B_3\cap G)=P(B_3)P(G/B_3)=\frac{4}{10}\times\frac{3}{8}=\frac{3}{20}$

$\therefore$ (a) is not true.

(b) $P(G/B_3)=\frac{3}{8}$

$\therefore$ (b) is true

(c) $\because P(B_3/G)$

$$=\frac{P(G/B_3)P(B_3)}{P(G/B_1)P(B_1)+P(G/B_2)P(B_2)+P(G/B_3)P(B_3)}$$

$$=\frac{\frac{3}{8}\times\frac{4}{10}}{\frac{5}{10}\times\frac{3}{10}+\frac{5}{8}\times\frac{3}{10}+\frac{3}{8}\times\frac{4}{10}}=\frac{\frac{12}{80}}{\frac{15}{100}+\frac{15}{80}+\frac{12}{80}}$$

$$=\frac{12}{80}\times\frac{400}{60+75+60}=\frac{60}{195}=\frac{4}{13}$$

$\therefore$ (c) is not true.

(d) $P(G)=P(G/B_1)P(B_1)+P(G/B_2)P(B_2)+P(G/B_3)P(B_3)$

$$=\frac{5}{10}\times\frac{3}{10}+\frac{5}{8}\times\frac{3}{10}+\frac{3}{8}\times\frac{4}{10}$$

$$=\frac{60+75+60}{400}=\frac{195}{400}=\frac{39}{80}$$

$\therefore$ (d) is true.

14. (a, b)

Given that $P(X)=\frac{1}{3}, P(X/Y)=\frac{1}{2}, P(Y/X)=\frac{2}{5}$

We have $P(X\cap Y)=P(Y/X)P(X)=\frac{2}{5}\times\frac{1}{3}=\frac{2}{15}$

$\therefore$ (c) is not true.

and $P(Y)=\frac{P(X\cap Y)}{P(X/Y)}=\frac{\frac{2}{15}}{\frac{1}{2}}=\frac{4}{15}$

$\therefore$ (a) is true.

$$P(X'/Y)=\frac{P(X'\cap Y)}{P(Y)}=\frac{P(Y)-P(X\cap Y)}{P(Y)}$$

$$=1-P(X/Y)=\frac{1}{2}$$

$\therefore$ (b) is true.

$$P(X\cup Y)=\frac{1}{3}+\frac{4}{15}-\frac{2}{15}=\frac{7}{15}$$

$\therefore$ (d) in not true.

15. (a) P(at least one of them solves the problem) = 1 – P(none of them solves it)

$$=1-\left(1-\frac{1}{2}\right)\left(1-\frac{3}{4}\right)\left(1-\frac{1}{4}\right)\left(1-\frac{1}{8}\right)$$

$$=1-\frac{1}{2}\times\frac{1}{4}\times\frac{3}{4}\times\frac{7}{8}=1-\frac{21}{256}=\frac{235}{256}$$

Matrix Type

1. $A\rightarrow S; B\rightarrow T; C\rightarrow Q; D\rightarrow S$

(A) Let E_1, E_2, E_3, E_4 be the events that the bag contains 1 white, 2 white, 3 white, 4 white ball respectively.

Let $P(E_1)=P(E_2)=P(E_3)=P(E_4)=\frac{1}{4}$

Let W be the event that the ball drawn is white.

Then

$$P(W)=\sum P(E_1)P(W/E_1)$$

$$=\frac{1}{4}\left(\frac{1}{4}+\frac{2}{4}+\frac{3}{4}+\frac{4}{4}\right)=\frac{5}{8}$$

Now $P(E_4/W)=\frac{P(E_4)P(W/E_4)}{P(W)}=\frac{1/4}{5/8}=\frac{2}{5}$

$$\therefore\frac{2}{5}=\frac{p}{15}\Rightarrow p=6$$

(B)

$$^{12}C_1+{}^{12}C_2\left({}^2C_1+2.{}^2C_2\right)+{}^{12}C_3\left({}^3C_1+2{}^3C_2\right)+\ldots+{}^{12}C_{12}\left({}^{12}C_1+2.{}^{12}C_2\right)$$

$$=\left({}^{12}C_1+2\,{}^{12}C_2+3\,{}^{12}C_3+\ldots+12\,{}^{12}C_2\right)+2\left({}^{12}C_2.{}^2C_2+{}^{12}C_3.{}^3C_2+\ldots+{}^{12}C_{12}.{}^{12}C_2\right)$$

$$=\sum_{r=1}^{12}r,\,{}^{12}C_r+12\times11\times\sum_{r=2}^{12}{}^{10}C_{r-2}$$

$$=12\times2^{11}+12\times11\times2^{10}$$

$$=12\times2^{10}(2+11)=13\times2^{10}\times12$$

$$\therefore 13\times2^{10}\times12=13\times2^{10}\times m$$

$$\therefore m=12$$

(C) $\frac{5x}{2-x}+\frac{5y}{2-y}+\frac{5z}{2-z}=5\left[\frac{x}{2-x}+\frac{y}{2-y}+\frac{z}{2-z}\right]$

$$=5\left[\frac{x-2+2}{2-x}+\frac{y-2+2}{2-y}+\frac{z-2+2}{2-z}\right]$$

$$=5\left[-3+2\left[\frac{1}{2-x}+\frac{1}{2-y}+\frac{1}{2-z}\right]\right]$$

Now $2-x+2-y+2-z=5$

$$\therefore\frac{5}{3}\geq\frac{3}{\frac{1}{2-x}+\frac{1}{2-y}+\frac{1}{2-z}}$$

i.e., $\frac{1}{2-x}+\frac{1}{2-y}+\frac{1}{2-z}\geq\frac{9}{5}$

Hence $\frac{5x}{2-x}+\frac{5y}{2-y}+\frac{5z}{2-z}\geq 5\left[-3+2.\frac{9}{5}\right]=3$

$\therefore$ Least value is 3.

(D)

$$\sum_{k=1}^{12}12\cdot k\cdot {}^{12}C_k\cdot {}^{11}C_{k-1}=12^2\sum_{k=1}^{12}\left({}^{11}C_{k-1}\right)^2$$

$$=12^2\cdot\frac{22!}{11!11!}$$

$$=12\cdot\frac{21.19\ldots3}{11!}\cdot 2^{12}\cdot 6$$

$\therefore\ p=6$

2. $A\to Q;\ B\to P;\ C\to R;\ D\to Q$

Let E_i denote the event that the bag contains i black and $(12-i)$ white balls $(i=0, 1, 2, \ldots, 12)$ and A denotes the event that the four balls drawn are all black. Then

$P(E_i)=\frac{1}{13}(i=0,1,2,\ldots 12);P\left(\frac{A}{E_i}\right)=0$ for

$i=0,1,2,3:P\left(\frac{A}{E_i}\right)=\frac{{}^iC_4}{{}^{12}C_4}$ for $i\geq 4$

$$(A)\,P(A)=\sum_{i=0}^{12}P(E_i)P\left(\frac{A}{E_i}\right)$$

$$=\frac{1}{13}\times\frac{1}{{}^{12}C_4}\left[{}^4C_4+{}^5C_4+\ldots+{}^{12}C_4\right]$$

$$=\frac{{}^{13}C_4}{13\times{}^{12}C_4}=\frac{1}{5}$$

(B) Clearly, $P\left(\frac{A}{E_{10}}\right)=\frac{{}^{10}C_4}{{}^{12}C_4}=\frac{14}{33}$

(C) As per Baye's theorem

$$P\left(\frac{E_{10}}{A}\right)=\frac{P(E_{10})P\left(\frac{A}{E_{10}}\right)}{P(A)}=\frac{\frac{1}{13}\times\frac{14}{33}}{\frac{1}{5}}=\frac{70}{429}$$

(D) Let B denotes the probability of drawing 2 white and 2 black balls. Then

$P\left(\frac{B}{E_i}\right)=0$ if $i=0,1$ or $11,12$

$P\left(\frac{B}{E_i}\right)=\frac{{}^iC_2\times{}^{12-i}C_2}{{}^{12}C_4}$ for $i=2,3,\ldots,10$

$$\therefore P(B)=\sum_{i=0}^{12}P(E_i)P\left(\frac{B}{E_i}\right)$$

$$=\frac{1}{13}\times\frac{1}{{}^{12}C_4}\left[{}^2C_2\times{}^{10}C_2+{}^3C_2\times{}^9C_2+{}^4C_2\times{}^8C_2+\ldots+{}^{10}C_2\times{}^2C_2\right]$$

$$=\frac{1}{13}\times\frac{1}{{}^{12}C_4}\left[2\left\{{}^2C_2\times{}^{10}C_2+{}^3C_2\times{}^9C_2+{}^4C_2\times{}^8C_2+{}^5C_2\times{}^7C_2\right\}+{}^6C_2\times{}^6C_2\right]$$

$$=\frac{1}{13}\times\frac{1}{495}(1287)=\frac{1}{5}$$

3. $A\to P;B\to R;C\to Q;D\to P$

$P(A)=\frac{{}^4C_1\,{}^8C_2}{{}^{12}C_3}=\frac{428}{220}=\frac{112}{220}=\frac{28}{55}$

$P(B)=\frac{{}^4C_3\,{}^8C_3}{{}^{12}C_3}=\frac{4+56}{220}=\frac{60}{220}=\frac{3}{11}$

$P(C)=P(\text{WBB or BWB or WWB or BBB})$

$$=\frac{8}{12}\times\frac{4}{11}\times\frac{3}{10}+\frac{4}{12}\times\frac{8}{11}\times\frac{3}{10}+\frac{8}{12}\times\frac{7}{11}\times\frac{4}{10}+\frac{4}{12}\times\frac{3}{11}\times\frac{2}{10}$$

$$=\frac{96+96+224+24}{12\times110}=\frac{440}{12\times110}=\frac{1}{3}$$

$A\cap B=\phi\Rightarrow$ A and B are mutually exclusive.

$P(B\cap C)=P(BBB)=\frac{4\times3\times2}{12\times110}=\frac{1}{55}$

Also,

$P(B)P(C)=\frac{3}{11}\times\frac{1}{3}=\frac{1}{11}$

Hence B and C are neither independent nor mutually exclusive

$P(C\cap A)=P(WWB)\frac{8\times7\times4}{12\times11\times10}=\frac{28}{3\times55}$

$P(C)P(A)=\frac{1}{2}\times\frac{112}{220}=\frac{28}{3\times55}$

Thus, C and A are independent.

Also, A, B, C are mutually exclusive as A and B are mutually exclusive.

4. $A\to Q,R;B\to R;C\to P;D\to A,B,C,D$

(A) $\frac{{}^rC_2}{{}^{r+b}C_2}=\frac{1}{2}2r(r-1)=(r-b)(r+b-1)$

$$=2r(r-1)=(r+b)(r+b-1)$$

or $2r^2-2r=r^2+(2b-1)r+b^2-1$

or $r^2-(1+2b)r+1-b^2=0$

or $b^2+2br+r-r^2-1=0$

or $b=\dfrac{-2r\pm\sqrt{4r^2-4\left(r-r^2-1\right)}}{2}$

$=-r\pm\sqrt{2r^2-r+1}$

Since, b is an integer, possible values of r are 3 and 8

(B) ${}^4C_2\left(\dfrac{r}{r+10}\right)^2\left(\dfrac{10}{r+10}\right)^2=\dfrac{3}{8}$

(C) $\left(\dfrac{r}{r+10}\right)^2\left(\dfrac{10}{r+10}\right)^2=\dfrac{1}{16}$

$\Rightarrow r=10$

D. Probability of getting exactly n red balls in $2n$ draws is always equal to the probability of getting exactly n black balls in $2n$ draws for any value of r and b. Hence the ratio r/b can be 10, 3, 8, 2.

5. A → Q; B → R; C → S; D → R

We have

$P(A\cap B)=P(A)P(B)=\dfrac{1}{12}$

(A) $P(A\cup B)=P(A)+P(B)-P(A\cap B)$

$=\dfrac{1}{3}+\dfrac{1}{4}-\dfrac{1}{12}=\dfrac{1}{2}$

(B) $P\left(\dfrac{A}{A\cup B}\right)=\dfrac{P(A)}{P(A\cup B)}=\dfrac{2}{3}$

(C) $P\left(\dfrac{B}{A'\cap B'}\right)=\dfrac{P\left(B\cap(A'\cap B')\right)}{P(A'\cap B')}$

$=\dfrac{P(\phi)}{P(A'\cap B')}=0$

(D) $P\left(\dfrac{A'}{B}\right)=\dfrac{P(A'\cap B)}{P(B)}=P(A')=\dfrac{2}{3}$

Integer and Numerical Value Type

1. $\Rightarrow P(A_1)=P(\omega)+P(BB\omega)+P(BBBB\omega)$

$=\dfrac{3}{10}+\dfrac{5}{10}\cdot\dfrac{4}{9}\cdot\dfrac{3}{8}+\dfrac{5}{10}\cdot\dfrac{4}{9}\cdot\dfrac{3}{8}\cdot\dfrac{2}{7}\cdot\dfrac{3}{6}$

$=\dfrac{3}{10}+\dfrac{1}{12}+\dfrac{1}{12\times7}=\dfrac{332}{840}=\dfrac{83}{210}$

$\Rightarrow P(A_2)=(B\omega)+(BBB\omega)+(BBBBB\omega)$

$=\dfrac{5}{10}\cdot\dfrac{3}{9}+\dfrac{5}{10}\cdot\dfrac{4}{9}\cdot\dfrac{3}{8}\cdot\dfrac{3}{7}+\dfrac{5}{10}\cdot\dfrac{4}{9}\cdot\dfrac{3}{8}\cdot\dfrac{2}{7}\cdot\dfrac{1}{6}\cdot\dfrac{3}{5}$

$=\dfrac{1}{6}+\dfrac{1}{28}+\dfrac{1}{420}=\dfrac{86}{420}=\dfrac{43}{210}$

$P(B)=1-P(A_1)-P(A_2)=\dfrac{2}{5}$

2. Probability of selection of any box is $=\dfrac{1}{N+1}$

Let E be the event that the wall clock selected is effective. Then

$P(E)=P(B_1)\,P(E/B_1)+P(B_2)\,P(E/B_2)+\ldots+P(B_{N+1})\,P(E/B_{N+1})$

$=\dfrac{1}{(N+1)}\left[1+\dfrac{N-1}{N}+\dfrac{N-2}{N}+\ldots+\dfrac{1}{N}+0\right]$

$=\dfrac{1+2+\ldots+N}{N(N+1)}=\dfrac{1}{2}$

$P(B_k/E)=\dfrac{P(B_k)\cdot P\left(\dfrac{E}{B_k}\right)}{P(E)}=\dfrac{\dfrac{1}{N+1}\dfrac{(N-k+1)}{N}}{\dfrac{1}{2}}$

$=\dfrac{2N-2k+2}{N^2+N}$

3. Total no. of arrangements $=15!$

Extreme chairs are occupied by girls, thus there are four gaps among 5 girls where boys can be seated. Let the number of boys in these four gaps be $2x+1, 2y+1, 2z+1$ and $2t+1$, then

$2x+1+2y+1+2z+1+2t\ \ldots 1=10$

$\Rightarrow x+y+z+t=3$

where x, y, z, t are integers and $0\le x\le3, 0\le y\le3, 0\le z\le3, 0\le t\le3$

∴ The number of ways of selecting positions for boys

= Coefficient of x^3 in $\left(1+x+x^2+x^3\right)^4$

= Coefficient of x^3 in $\left(\dfrac{1-x^4}{1-x}\right)^4$

= Coefficient of x^3 in $\left(1-x^4\right)^4(1-x)^{-4}={}^6C_3=20$

∴ Number of arrangements of boys and girls with given condition $=20\times10!\times5!$

∴ Required probability $=\dfrac{20\times10!\times5!}{15!}=\dfrac{20}{3003}$

4. Let $A=\{a_1, a_2, \ldots a_n\}$. For each $a_i\in A1\sim i\sim n)$, we have the following four cases;

(i) $a_i\in P$ and $a_i\in Q$ (ii) $a_i\notin P$ and $a_i\in Q$

(iii) $a_i\in P$ and $a_i\notin Q$ (iv) $a_i\notin P$ and $a_i\notin Q$

Thus, the total number of ways of choosing P and Q is 4^n

$P\cap Q$ contains exactly two elements in $({}^nC_2)(3^{n-2})$.

∴ Probability of $P\cap Q$ contains two elements

$=\dfrac{{}^nC_2\cdot3^{n-2}}{4^n}$

5. Given that P = Probability of hitting the target

$= 0.75 = \frac{3}{4}$

Z = Probability of missing the target $= 1 - \frac{3}{4} = \frac{1}{4}$

$\therefore P(X = r)$ = Probability of r success

$= {}^nC_r\left(\frac{3}{4}\right)^r\left(\frac{3}{4}\right)^{n-r}$

$P(X \ge 3) = 1 - (P(0) + P(1) + P(2)) \ge 0.95$

$\Rightarrow 1 - {}^nC_0\left(\frac{1}{4}\right)^n - {}^nC_1\left(\frac{3}{4}\right)\left(\frac{1}{4}\right)^{n-1} - {}^nC_1\left(\frac{3}{4}\right)^2\left(\frac{1}{4}\right)^{n-2} \ge 0.95$

$\Rightarrow 1 - \left(\frac{1 + 3n + \frac{9n(n-1)}{2}}{4^n}\right) \ge 0.95$

$\Rightarrow \frac{2 + 6n + 9n^2 - 9n}{2.4^n} \ge 1 - 0.95$

$\Rightarrow 9n^2 - 3n + 2 \le 0.05 \times 4^n \times 2 \le \frac{4^n}{10}$

For $n = 5, 212 \le 102.4$ (Not true)

For $n = 6, 308 \le 409.6$ (True)

Hence, least value of $n = 6$

6. Prime (2, 3, 5, 7, 11) = {(1, 1), (1, 2), (2, 1), (1, 4), (2, 3), (3, 2), (4, 1), (1, 6), (2, 5), (3, 4), (4, 3), (5, 2), (6, 1), (5, 6), (6, 5)}

n (odd prime) = 14

$\therefore$ P(odd prime) $= \frac{14}{36}$

Perfect square = (4, 9) = {(1, 3), (2, 2), (3, 4), (3, 6), (4, 5), (5, 4), (6, 3)}

n (Perfect square) = 7

$\therefore$ P(Perfect square) $= \frac{7}{36}$

Required probability

$$= \frac{\frac{4}{36} + \frac{14}{36} \times \frac{4}{36} + \left(\frac{14}{36}\right)^2 \frac{4}{36} + ..}{\frac{7}{36} + \frac{14}{36} \times \frac{7}{36} + \left(\frac{14}{36}\right)^2 \frac{7}{36} + ..} = P = \frac{4}{7}$$

$\therefore 14P = 14.\frac{4}{7} = 8$

7. Total number of 3×3 matrices with 0 or 1 = 2^9 = 512

E_2 contains those matrices in which sum of entries is 7.

It will contain 7 ones and 2 zeroes.

$\therefore n(E_2) = {}^9C_2 = 36$

$E_1 \cap E_2$ contains those matrices in which there are 7 ones, 2 zeroes and whose det is zero.

Det (A) = 0. This can occur when two rows/columns are identical. e.g.

$$\begin{vmatrix} 0 & 0 & 1 \\ 1 & 1 & 1 \\ 1 & 1 & 1 \end{vmatrix} \text{ or } \begin{vmatrix} 1 & 0 & 0 \\ 1 & 1 & 1 \\ 1 & 1 & 1 \end{vmatrix} \text{ or } \begin{vmatrix} 0 & 1 & 0 \\ 1 & 1 & 1 \\ 1 & 1 & 1 \end{vmatrix}$$

$\therefore n(E_1 \cap E_2) = {}^3C_1 \times {}^3C_1 \times 2 = 18$

$\therefore n(E_1 / E_2) = \frac{P(E_1 \cap E_2)}{P(E_2)} = \frac{18/512}{36/512} = \frac{1}{2} = 0.50$

8. Given that $P(X \ge 2) \ge 0.96$

$\Rightarrow 1 - P(X = 0) - P(X = 1) \ge 0.96$

$\Rightarrow P(X = 0) + P(X = 1) \le 0.04$

$\Rightarrow \left(\frac{1}{2}\right)^n + n\left(\frac{1}{2}\right)^n \le 0.04$

$\Rightarrow \frac{n+1}{2^n} \le \frac{1}{25} \Rightarrow \frac{2^n}{n+1} \ge 25$

$\Rightarrow$ Minimum value of n is 8

9. Let $P(E_1) = x; P(E_2) = y, P(E_3) = z$

P(only E_1)

$= P(E_1 \cap \bar{E}_2 \cap \bar{E}_3) = x(1-y)(1-z) = \alpha$

P(only E_2)

$= P(\bar{E}_1 \cap E_2 \cap \bar{E}_3) = (1-x)y(1-z) = \beta$

P(only E_3)

$= P(\bar{E}_1 \cap \bar{E}_2 \cap E_3) = (1-x)(1-y)z = \gamma$

P(none)

$= P(\bar{E}_1 \cap \bar{E}_2 \cap \bar{E}_3) = (1-x)(1-y)(1-z) = p$

Now given $(\alpha - 2\beta)p = \alpha\beta \Rightarrow x = 2y$

And $(\beta - 3r)p = 2\beta r \Rightarrow y = 3z \quad \therefore x = 6z$

Hence $\frac{P(E_1)}{P(E_3)} = \frac{x}{z} = 6$

Comprehension Passage

Comprehension 1

1. (a) Required probability

$= \frac{1}{2} \times \frac{1}{\sqrt{2}} + \frac{1}{2}\left(1 - \frac{1}{4}\right) = 0.728$

2. (a) $P(I/II)=\frac{P(I).P(II/I)}{P(II)}=\frac{\frac{1}{2}\cdot\frac{1}{\sqrt{2}}}{\frac{1}{2}\cdot\frac{1}{\sqrt{2}}+\frac{1}{8}}=0.74$

3. (b) Probability Sania wins the match = $(1-R)R^2 R = R^3(1-R)$
Probability Sania loses the match
$=(1-R)R^2(1-R)+(1-R)(1-R^2)$
$R^3(1-R)>(1-R)R^2(1-R)+(1-R)(1-R^2)$
$R^3>R^2(1-R)+(1-R^2)=1-R^3 \quad \therefore 2R^3>1$
$\therefore\ R\in\left(\left(\frac{1}{2}\right)^{\frac{1}{3}},1\right)$

Comprehension 2

$n=10k+r,\ k,r\in N,\ 0\le r\le 9$
Unit place of a^2 will contain 0, 1, 4, 5, 6, 9 only.
$\therefore a^2-1$ is divisible by 10 only if unit place of a^2 contain 1.
If unit place of a^2 is 1.
Then unit place of a will be 1 or 9.

1. (a) $n=10k+r$
$r=0$
$n=10k$ for a no. whose unit place is 1 or 9
$\therefore k=1,\ n=10$ for a no. whose unit place is = 2
$\therefore k=2,\ n=20$ for a no. whose unit place is = 4
$\therefore k=k,\ n=10k$ for a no. whose unit place is $=2k$
$\therefore p_n=\frac{2k}{n}$

2. (b) $n=10k+9$
No. with unit place as 1 or 9 = $2(k+1)$
$\therefore p_n=\frac{2(k+1)}{n}$

3. (c) No. with unit place as 1 or 9 = $2k+1$
$p_n=\frac{2k+1}{n}$

Comprehension 3

1. (d) A can be drained out of all his money only in an even numbered round. Therefore, A will not be drained out at the 11th round.

2. (c) To finish at the 12th round, he must have exactly 1 head in the first 10 rounds and a tail in the 11th and the 12th round. The probability of this is ${}^{10}C_1\,pq^{11}$.

3. (a) To drain out at the 14th round, two cases arise
(i) He gets exactly 2 heads in the first 10 rounds
$\therefore$ Probability in this case is ${}^{10}C_2\,p^2q^8\cdot q^4$
$=45p^2q^{12}$
(ii) He gets exactly 1 head in the first 10 rounds and then exactly one head in the next two rounds
$\therefore$ Probability in this case is ${}^{10}C_1\,pq^9\cdot{}^2C_1\,pq\cdot q^2$
$=20p^2q^{12}$
Therefore, required probability = $65\,p^2q^{12}+q^{10}+10\,pq^{11}$

Comprehension 4

1. (a) Let E: getting 6
$P(X=3)$
$=P(\bar{E}\cap\bar{E}\cap\bar{E})=P(\bar{E}).P(\bar{E}).P(E)$
$=\frac{5}{6}\times\frac{5}{6}\times\frac{1}{6}=\frac{25}{216}$

2. (b) $P(X\ge 3)=1-[P(X=1)+P(X=2)]$
$=1-\left[\frac{1}{6}+\frac{5}{6}\times\frac{1}{6}\right]=1-\frac{11}{36}=\frac{25}{36}$

3. (d)
$P(E_1)=P(X\ge 6)=\left(\frac{5}{6}\right)^5\times\frac{1}{6}+\left(\frac{5}{6}\right)^6\times\frac{1}{6}+\ldots\infty$
$=\left(\frac{5}{6}\right)^5\times\frac{1}{6}\left[1+\frac{5}{6}+\left(\frac{5}{6}\right)^2+\ldots\infty\right]$
$=\left(\frac{5}{6}\right)^5\times\frac{1}{6}\times\frac{1}{1-\frac{5}{6}}=\left(\frac{5}{6}\right)^5$
and $P(E_2)P(X>3)=\left(\frac{5}{6}\right)^3\times\frac{1}{6}+\left(\frac{5}{6}\right)^4\times\frac{1}{6}+\ldots\infty$
$=\left(\frac{5}{6}\right)^3$
$\because E_1\cap E_2=X\ge 6=E_1$
$\therefore P(E_1/E_2)=\frac{P(E_1\cap E_2)}{P(E_2)}=\frac{P(E_1)}{P(E_2)}=\frac{\left(\frac{5}{6}\right)^5}{\left(\frac{5}{6}\right)^3}=\frac{25}{36}$

Comprehension 5

1. (b) $P(w)=P(H\cap W)+P(T\cap W)$
$=P(H)P(W/H)+P(T)P(W/T)$
$=\frac{1}{2}\left\{\frac{3}{5}\times 1+\frac{2}{5}\times\frac{1}{2}\right\}+\frac{1}{2}\times$
$\left\{\frac{{}^3C_2}{{}^5C_2}\times 1+\frac{{}^2C_2}{{}^5C_2}\times\frac{1}{3}+\frac{{}^3C_1\cdot{}^2C_1}{{}^5C_2}\times\frac{2}{3}\right\}$
$=\frac{1}{2}\times\frac{8}{10}+\frac{1}{2}\times\left(\frac{3}{10}+\frac{1}{30}+\frac{12}{30}\right)=\frac{23}{30}$

2. (d)

$$P(H/W)=\frac{P(H\cap W)}{P(W)}=\frac{\frac{1}{2}\left[\frac{3}{5}\times1+\frac{2}{5}\times\frac{1}{2}\right]}{\frac{23}{30}}=\frac{12}{33}$$

Comprehension 6

1. Since p_1 = probability that maximum of chosen number is at least 81
$p_1 = 1 -$ probability that maximum of chosen number is less than or equal to 80

$$p_1=1-\frac{80\times80\times80}{100\times100\times100}=1-\frac{64}{125}$$

$$p_1=\frac{61}{125}$$

$$\therefore \frac{625p_1}{4}=\frac{625}{4}\times\frac{61}{125}=\frac{305}{4}=76.25$$

2. Since p_2 = probability that minimum of chosen numbers is greater than or equal to 401

$$=1-\frac{60\times60\times60}{100\times100\times100}=1-\frac{27}{125}=\frac{98}{125}$$

$$\therefore \frac{125}{4}p_2=\frac{125}{4}\times\frac{98}{125}=24.50$$

Comprehension 7

1. (a) Total number of sitting arrangements of 5 students = 5! = 120
No. of derangements of S_2, S_3, S_4 and S_5 seats

$$=4!\left(1-\frac{1}{1!}+\frac{1}{2!}-\frac{1}{3!}+\frac{1}{4!}\right)$$

$$=12-4+1=9$$

$$\therefore \text{Probability}=\frac{9}{120}=\frac{3}{40}$$

2. (c) Total number of seating arrangements of 5 students = 5! = 120
$\therefore$ Favourable cases = 14

$$\therefore \text{Required probability}=\frac{14}{120}=\frac{7}{60}$$

Comprehension 8

1. (a, b) Let $E_1 \equiv$ box I is selected
$E_2 \equiv$ box II is selected
$F \equiv$ ball drawn is red

$$P(E_2/F)=\frac{P(F/E_2)\cdot P(E_2)}{P(F/E_1)\cdot P(E_1)+P(F/E_2)\cdot P(E_2)}$$

$$=\frac{\frac{n_3}{n_3+n_4}\times\frac{1}{2}}{\frac{n_1}{n_1+n_2}\times\frac{1}{2}+\frac{n_3}{n_3+n_4}\times\frac{1}{2}}=\frac{1}{3}$$

On putting the values of the options, we observe that (a) and (b) are the correct options.

2. (c, d) $E_1 \equiv$ Red ball is selected from box I
$E_2 \equiv$ Black ball is selected from box I
$F \equiv$ Second red ball is drawn from box I

$$\therefore P(F)=P(E_1)P(F/E_1)+P(F/E_2)$$

$$=\frac{n_1}{n_1+n_2}\times\frac{n_1-1}{n_1+n_2-1}+\frac{n_2}{n_1+n_2}\times\frac{n_1}{n_1+n_2-1}$$

On putting the value of the options, we observe that (c) and (d) have the correct find options.

Comprehension 9

1. (b) $x_1+x_2+x_3$ will be odd. If two of them are even and one is odd, or if all three are odd.
E_i and O_i denotes the even and odd number respectively, from ith box.
$\therefore$ Required probability

$$=P(E_1E_2O_3)+P(E_1O_2E_3)+P(O_1E_2E_3)+P(O_1O_2O_3)$$

$$=\frac{1}{3}\times\frac{2}{5}\times\frac{4}{7}+\frac{1}{3}\times\frac{3}{5}\times\frac{3}{7}+\frac{2}{3}\times\frac{2}{5}\times\frac{3}{7}+\frac{2}{3}\times\frac{3}{5}\times\frac{4}{7}=\frac{53}{105}$$

2. (c) Let x_1, x_2, x_3 be in AP $\Rightarrow 2x_2 = x_1 + x_3$
$\because$ LHS is even, that means x_1 and x_3 can be both even or both odd.

$$\text{Required probability}=P(E_1E_3)+P(O_1,O_3)$$

$$=\frac{{}^1C_1\times{}^3C_1+{}^2C_1\times{}^4C_1}{{}^3C_1\times{}^5C_1\times{}^7C_1}=\frac{3+8}{3\times5\times7}=\frac{11}{105}$$

Comprehension 10

	B_1	B_2	B_3
	1W	2W	3W
	3R	3R	4R
	2B	4B	5B

1. (a) Probability that all three balls are of same colour
= P(RRR) + P(WWW) + P(BBB)

$$=\frac{3}{6}\times\frac{3}{9}\times\frac{4}{12}+\frac{1}{6}\times\frac{2}{9}\times\frac{3}{12}+\frac{2}{6}\times\frac{4}{9}\times\frac{5}{12}=\frac{82}{648}$$

2. (d) Let E_1, E_2, E_3 be the events to select boxes B_1, B_2 and B_3 respectively.
Let F be the event of getting one white and one red ball.
Then by Baye's theorem,

$$P(E_2/F)=\frac{P(F/E_2)P(E_2)}{P(F/E_1)(E_1)+P(F/E_2)P(E_2)+P(F/E_3)P(E_3)}$$

$$=\frac{\frac{1}{3}\cdot\frac{2\times3}{{}^9C_2}}{\frac{1}{3}\cdot\left(\frac{1\times3}{{}^6C_2}+\frac{2\times3}{{}^9C_2}+\frac{3\times4}{{}^{12}C_2}\right)}=\frac{55}{181}$$